AF735973

SI Base Units

length	meter	m	electric current	ampere	A
mass	kilogram	kg	thermodynamic temperature	Kelvin	K
time	second	s	amount of substance	mole	mol

SI Derived or Related Units

frequency	hertz	Hz	$= s^{-1}$
force	newton	N	$= m\ kg\ s^{-2}$
	kilogram-force	kgf	= 9.80665 N
	dyne	dyne	$= 10^{-5}$ N
energy	joule	J	$= N\ m = kg\ m^2\ s^{-2}$
	erg	erg	$= 10^{-7}$ J
	electron volt	eV	$= 1.60218 \times 10^{-19}$ J
	calorie	cal	= 4.184 J
	bar L	bar L	= 100 J
	atomic unit	au	$= 4.35974 \times 10^{-18}$ J
power	watt	W	$= J\ s^{-1} = V\ A$
volume	liter	L	$= 10^{-3}\ m^3$
pressure	pascal	Pa	$= N\ m^{-2}$
	bar	bar	$= 10^5$ Pa
	atmosphere	atm	= 1.01325 bar
	torr	Torr	= 1/760 atm
	millimeter of mercury	mm Hg	= Torr
electric charge	coulomb	C	= A s
electric potential, electromotive force	volt	V	$= J\ C^{-1}$
electric resistance	ohm	Ω	$= V\ A^{-1}$
electric conductance	siemens	S	$= \Omega^{-1}$
electric capacitance	farad	F	$= C\ V^{-1}$
magnetic flux density	tesla	T	$= V\ s\ m^{-2}$
	gauss	G	$= 10^{-4}$ T
inductance	henry	H	$= V\ A^{-1}\ s$
electric dipole moment	coulomb meter		= C m
	Debye	D	$= 3.3356 \times 10^{-30}$ C m
magnetic moment		$A\ m^2$	$= J\ T^{-1}$

Prefixes

z	a	f	p	n	μ	m	c	d
zepto	atto	femto	pico	nano	micro	milli	centi	deci
10^{-21}	10^{-18}	10^{-15}	10^{-12}	10^{-9}	10^{-6}	10^{-3}	10^{-2}	10^{-1}

da	h	k	M	G	T	P	E	Z
deka	hecto	kilo	mega	giga	tera	peta	exa	zetta
10^{1}	10^{2}	10^{3}	10^{6}	10^{9}	10^{12}	10^{15}	10^{18}	10^{21}

Principles of Physical Chemistry

Principles of Physical Chemistry

Third Edition

Edited by Hans Kuhn, David H. Waldeck, and Horst-Dieter Försterling

WILEY

Published by John Wiley & Sons, Inc., Hoboken, New Jersey.
Published simultaneously in Canada.

Library of Congress Cataloging-in-Publication Data

Names: Kuhn, Hans, 1919-2012 author. | Waldeck, David H. (David Hennessey), 1956-author | Försterling, Horst-Dieter, 1934 author. | John Wiley & Sons, publisher.
Title: Principles of physical chemistry / by Hans Kuhn, David H. Waldeck, Horst-Dieter Försterling.
Description: Third edition. | Hoboken, New Jersey : Wiley, [2025] | Includes index.
Identifiers: LCCN 2023045918 (print) | LCCN 2023045919 (ebook) | ISBN 9781119852667 (hardback) | ISBN 9781119852674 (adobe pdf) | ISBN 9781119852681 (epub)
Subjects: LCSH: Chemistry, Physical and theoretical.
Classification: LCC QD453.3 .K84 2025 (print) | LCC QD453.3 (ebook) | DDC 541--dc23/eng/20231023
LC record available at https://lccn.loc.gov/2023045918
LC ebook record available at https://lccn.loc.gov/2023045919

Cover Design: Wiley
Cover Image: Courtesy of Horst-Dieter Forsterling and David H. Waldeck

Set in 9.5/12.5pt STIXTwoText by Straive, Chennai, India
Printed and bound by CPI Group (UK) Ltd, Croydon, CR0 4YY

C9781119852667_180725

The manufacturer's authorized representative according to the EU General Product Safety Regulation is Wiley-VCH GmbH, Boschstr. 12, 69469 Weinheim, Germany, e-mail: Product_Safety@wiley.com.

Contents

Preface Third Edition

Chemistry comprises the structure, interaction, and transformation of atoms and their assemblies into molecules, solids, and supramolecular structures. With the notable exception of nuclear chemistry, chemists regard matter as a collective interaction of atoms and molecules, and consider the macroscopic properties of matter to arise from a few fundamental principles. In this book on physical chemistry, we emphasize this view by first exploring the nature of atoms and chemical bonds, and then proceeding to a discussion of complex molecules and, subsequently, to more and more complex forms of matter. This sequence of topics is chosen to aid the student in developing a systematic and coherent understanding of the subject. Because it proceeds from the fundamental principles which govern the arrangement of electrons and nuclei into atoms and molecules, this textbook in physical chemistry should be accessible to biologists, chemical engineers, materials scientists, and physicists, as well as chemists.

As a discipline physical chemistry attempts to understand chemistry, in all its complexity, as a consequence of a few fundamental physical laws. This fascinating enterprise includes chemical processes in physics, biology, and engineering, as well. On a deeper level, physical chemistry aims to understand and quantify the tremendous variety that matter takes in our world, from individual atoms to complex life forms. The unifying nature of physical chemistry is emphasized by the book's organization—beginning with atoms and molecules, proceeding to molecular assemblies of increasing complexity, and ending with the emergence of matter that carries information, i.e., the origin of life, a physicochemical process of unique importance. The aim is to show the broad scope and coherence of physical chemistry.

The book begins with the most basic and central question in physical chemistry, "Why do atoms and molecules form from a collection of electrons and nuclei?" The answer to this question requires a fundamentally different way of thinking about the world than we are used to from our common human experiences; and we need the fascinating and strange predictions of quantum mechanics to guide us. In Chapter 1 we consider some astonishing experimental facts that require this new way of thinking, and we show how these facts can be considered to arise from a few fundamental postulates. On this basis Chapter 2 explains the existence of the simplest atom (hydrogen atom H) and the simplest molecule (hydrogen molecule ion H_2^+) in terms of a simple model for an electron confined to a region of space (particle-in-a-box model). Next the book explores the quantum mechanical basis of chemical bonding (Chapters 3–8 and 13) and how we use light to probe molecular structure, a field of study called spectroscopy (Chapters 9–12). We emphasize π electron systems in Chapters 8–10, because they illustrate particularly well how one can use simple theoretical models to understand molecular properties. Throughout these chapters the particle-in-a-box model is used as a touchstone to reinforce and develop the understanding of chemical bonding.

Although some modern experimental methods can probe the properties of individual atoms and molecules, most experiments are performed with collections of atoms and/or molecules. Chapter 13 provides an introduction and overview of some essential aspects of how to think about such collections. This part considers a second fundamental question in physical chemistry, "What is the connection between molecular (or atomic) properties and the macroscopic properties of matter?" Chapters 14–18 show how our atomistic view of matter can be used to predict the properties of a large collection of molecules quantitatively in terms of macroscopic system variables, such as temperature and pressure. In order to present this description most clearly we use simplified models that ignore intermolecular forces, e.g., the ideal gas law. With this fundamental understanding in hand, Chapters 19–21 show how these ideas can be extended to macroscopic chemical systems in which the intermolecular forces are included. Chapters 22 and 23 apply these developments to chemical equilibria in aqueous solutions, biosystems, and electrochemistry.

"How do molecules and atoms (reactants) transform into different molecules and atoms (products)?" For many practicing chemists, this question is the most exciting part of chemistry. We consider chemical reaction rates and reaction dynamics, in Chapters 24 and 25. While much progress has been made in providing a quantitatively accurate picture of reaction dynamics, this area remains a frontier of physical chemistry.

The final chapters of the book address the question, "Why do molecules aggregate into organized assemblies and how does the organization of molecules give rise to new properties?" The treatment of organized systems of molecules, including supramolecular machines, is a fast developing subject which lies at the nexus of biology, chemistry, and physics. Methods and ideas from physical chemistry are a powerful approach to exploring this current frontier, and we develop some of the primary physicochemical considerations for understanding organized molecular assemblies in Chapters 26–29. Chapters 26 (macromolecules) and 27 (interfaces) illustrate different ways of organizing molecules, or molecular subunits, in space. Chapter 28 describes principles for organizing functional molecules to create new behaviors that are characteristic of the supramolecular assembly, distinct from the individual molecule behaviors, and introduces this fascinating and dramatically growing field by discussing selected examples. This part of the book culminates in the final chapter which considers the basic processes in the origin of life and explores some of the underlying physicochemical principles that are important for chemical and biological evolution.

The book aims to develop an intuition for how research is performed by presenting many exciting original ideas and providing instructive examples to inspire thinking. Studying typical examples is an excellent opportunity for learning how to design simple experiments and theoretical models. For these reasons, original research findings and current research findings are included in the examples or as part of problems and exercises to be worked. Rather than always aiming for rigor, we use simplified models and approximations that promote intuition and guide one's thinking at the boundaries of knowledge. Inventing idealized theoretical descriptions of real situations is crucial for understanding new research and developing new knowledge. Lastly, we recommend that you consider the text as a tool through which you can explore and discover the rich field of physical chemistry.

Although the presentation of subject matter is logical and hierarchical, the chapters in the book can be used in almost any order because they are relatively self-contained. The book starts with a simplified introduction to quantum mechanics because it is so important in all parts of physical chemistry and is essential for understanding the nature of the chemical bond. Nevertheless, students in chemistry and biochemistry are trained to think in terms of chemical bonds and processes at the atomic level in their first year classes, so many of the more advanced topics can be explored without the full rigor of the early chapters. Where possible, the logical sequence in a given chapter is constructed to evolve from the more basic and simple picture to a more sophisticated treatment for the more experienced students. If such a section is too difficult in a first reading to study with excitement, consider skipping it and returning at a later stage. Although the sequence of chapters in the book need not determine the sequence of topics in a physical chemistry lecture course, the organization of the chapters in the book and its broad scope aims to emphasize the logical progression and unity of physical chemistry. Whatever order of topics is chosen, it is important to appreciate the unified nature of physical chemistry.

In order to save space in the printed book, part of the material is only provided electronically. These are most of the foundations (containing more basic derivations) and justifications (containing detailed calculations), and the computational exercises (Fortran, MathCad, and Mathematica).

July 2023 *David H. Waldeck and Horst-Dieter Försterling*

Acknowledgement

Here we acknowledge those who have been important to the preparation of this third edition of Principles of Physical Chemistry. We are greatly indebted to Michael Leventhal and his colleagues at John Wiley & Sons, Ltd for their encouragement and support during all stages of this third edition. A number of colleagues have provided important assistance during parts of this work, notably Sean Garrett-Roe and Kenneth Jordan; however, we are especially indebted to Christoph Kuhn, who made a great number of important suggestions. He contributed the step model calculations in Chapters 8 and 9 and the simulations in the origin of life chapter. Also, we are thankful for his comments during proof reading. David Waldeck thanks his wife, Janet, and children, Aaron and Anna, for their support and understanding during this endeavor. Horst-Dieter Försterling is thankful for the great support by his family during the time of developing this book and its precursors, first of all to his wife Inge (who died in 2019) for their encouragement.

We dedicate this book to Hans Kuhn (1919–2012). We have made every effort to present physical chemistry with Hans' spirit of curiosity, enthusiasm, and incisive explanation. As a chemistry student, H.D.F. attended the inspiring lectures on physical chemistry held by Hans since 1953, guided by the idea that quantum theory is the key to understand the whole field of physical chemistry. Moreover, using simplifications and instructive models rather than complicated formalisms, his lectures became a highlight among the chemistry students. Later, these lectures gave rise to a textbook on experimental courses[1] and an introduction into the basic ideas of physical chemistry[2]. Finally, after Hans' retirement and a visit of H.D.F to his home in Switzerland, on a walk through a wonderful small valley, the idea of a more comprehensive textbook on physical chemistry was born, leading to the first edition of this book. During all the time of this cooperation, H.D.F. enjoyed the hospitality and patience of Hans' family, first of all of his wife Elsi, when discussing the topics of the book in his home, mostly working together from early in the morning until late in the evening.

1 H.D. Försterling, H. Kuhn, Physikaliche Chemie in Experimenten, Ein Praktikum, Verlag Chemie, Weinheim 1971.
2 H.D. Försterling, H. Kuhn, Moleküle und Molekülanhäufungen, Springer Verlag Berlin, Heidelberg, New York 1983.

List of Symbols

$P^{\ominus}$	standard pressure = 1 bar
$c^{\ominus}$	standard concentration = 1 mol L^{-1}
$\hat{P}$	$= P/P^{\ominus}$
$\hat{c}$	$= c/c^{\ominus}$
$\boldsymbol{n}$	amount of substance
$U^{\ominus}$, $H^{\ominus}$	U, H under standard conditions
$S^{\ominus}$	S under standard conditions
$G^{\ominus}$	G under standard conditions
$C_V^{\ominus}, C_P^{\ominus}$	C_V, C_P under standard conditions
$\Delta_f H$	molar enthalpy of formation
$\Delta_f G$	molar Gibbs energy of formation
$\Delta_r H$	molar change of H in reaction
$\Delta_r S$	molar change of S in reaction
$\Delta_r G$	molar change of G in reaction
$\Delta_{\text{vap}} H$	molar enthalpy of evaporation
$\Delta_{\text{fus}} H$	molar enthalpy of melting
$C_{V,m}$	molar heat capacity at constant volume
$C_{P,m}$	molar heat capacity at constant pressure
$\hat{n}$	refractive index
$\hat{\varepsilon}$	relative permittivity

Note that the symbols $\hat{P}$ and $\hat{c}$ denote dimensionless quantities: $\hat{P} = P/P^{\ominus}$, $\hat{c} = c/c^{\ominus}$. The symbols $\hat{n}$ and $\hat{\varepsilon}$ are used for the refractive index and the relative permittivity.

a	coefficient, distance
a_0	Bohr radius
b	coefficient
b	contour length per monomer unit
c	coefficient
c_0	speed of light in vacuum
d, d_0	distance, bond length
$\mathrm{d}\tau$	differential volume element
e	elementary charge
e	Euler's number
f	force
g	number of quantum states
h	Planck's constant
$\hbar$	$h/(2\pi)$
h	distance between chain ends
i, j	indices
i	basis of imaginary numbers, $i = \sqrt{-1}$
k	Boltzmann constant
k	magnetic quantum number
k_f	force constant
k_r	rate constant
l	azimuthal quantum number
l	length
l_{segment}	length of segment
m	mass
m_s	spin quantum number
n	quantum number
p	momentum
q	heat
r, r_{coil}	distance, radius of coil
l	length of statistical chain element
m_e	rest mass of electron
m_p	rest mass of proton
s	coordinate
t	time
u, v	speed
x, y, z	coordinates
z_{coll}	collision number
A	area, Helmholtz energy
B	magnetic flux density
C	capacity
C_V, C_P	heat capacity
D	diffusion coefficient
E	energy, cell potential, electromotive force
E_a	energy of activation
$E_{a,m}$	molar energy of activation

F	electric field strength
$\boldsymbol{F}$	Faraday constant
G	Gibbs energy
G_m	molar Gibbs energy
H	enthalpy
H_m	molar enthalpy
H	Hamilton function
$\mathcal{H}$	Hamiltonian operator
I	electric current
I	intensity of light
$\vec{J}$	total angular momentum
K	equilibrium constant
L	length
L	contour length of chain
$\vec{L}$	orbital angular momentum
$\boldsymbol{M}$	molar mass
$\boldsymbol{M}_{\text{monomer}}$	molar mass of monomer
N	number of particles
$\boldsymbol{N}_A$	Avogadro constant
P	pressure
$^{\text{osm}}P$	osmotic pressure
P	bond order
$\mathcal{P}$	probability
Q	electric charge
R	maximum extension of coil
$\boldsymbol{R}$	gas constant
S	entropy
S_m	molar entropy
$\vec{S}$	spin angular momentum
T	absolute temperature
T	kinetic energy
U	internal energy
U_m	molar internal energy
V	volume
V_m	molar volume
V	electric voltage
V	potential energy
Z	atomic number
Z	number of monomer units
Z	partition function
Z_{coll}	number of collisions

α	steric factor, angle
α	polarizability
α	stoichiometric factor
α	Lagrangian multiplicator
β	Lagrangian multiplicator
β	angle
γ, $^{c}\gamma$	activity coefficient
γ	stoichiometric factor
δ	change in length
δ	small change of quantity
δ	chemical shift
ε	energy
ε	molar decadic absorption coefficient
$\hat{\varepsilon}$	relative permittivity
ε_0	permittivity of the vacuum
ζ, ξ	coordinate
η	viscosity
θ	angle
ϑ	angle
κ	electric conductivity
λ	wavelength
λ	free path
μ	dipole moment
μ	chemical potential
μ	reduced mass
μ_0	permeability of the vacuum
μ_B	Bohr magneton
ν	frequency
$\tilde{\nu}$	wavenumber
ξ	coordinate
ρ	electron density
ρ	mass density
σ	cross section
σ	screening constant
τ	lifetime
ϕ	wavefunction, voltage
ϕ	angle
ϕ	fugacity coefficient
φ	quantum efficiency
χ	electric susceptibility
ψ	wavefunction
ω	angular frequency
ω	number of representations
Γ	surface excess
Δ	change of quantity
Δ	Laplace operator
Λ	de Broglie wavelength
Θ	angle
Π	surface pressure
Φ	wavefunction
Ψ	wavefunction
Ω	number of representations
∂	differential operator
∇	Nabla operator

About the Companion Website

This book is accompanied by a companion website:

www.wiley.com/go/Kuhn/physicalchemistry3e

The website includes Foundations & Justifications files.

Introduction

As a discipline physical chemistry attempts to understand chemistry, in all its complexity, as a consequence of a few fundamental physical laws. This fascinating enterprise includes chemical processes in physics, biology, and engineering, as well. On a deeper level, physical chemistry aims to understand and quantify the tremendous variety that matter takes in our world, from individual atoms to complex life forms. The unifying nature of physical chemistry is emphasized in the book by its organization – beginning with atoms and molecules, and proceeding to molecular assemblies of increasing complexity, ending with the emergence of matter that carries information, i.e., the origin of life, a physicochemical process of unique importance. The aim is to show the broad scope and coherence of physical chemistry.

The book begins with the most basic and central question in physical chemistry, "Why do atoms and molecules form from a collection of electrons and nuclei?" The answer to this question requires a fundamentally different way of thinking about the world than we are used to from our daily macroscopic experiences; it requires that the fascinating and strange predictions of quantum mechanics guide us. Thus, Chapter 1 exposes you to some astonishing experimental facts that require this new way of thinking. We show you that the given facts can be seen as consequences of some fundamental postulates. On this basis Chapter 2 explains the existence of the simplest atom (hydrogen atom H) and the simplest molecule (hydrogen molecule ion H_2^+) in terms of a simple model of an electron confined to a region of space (particle-in-a-box). Next the book explores the quantum mechanical basis of chemical bonding (Chapters 3–7 and 13) and how we use light to probe molecular structure, a field of study called spectroscopy (Chapters 8–12). Throughout these chapters the particle-in-a-box model is used as a touchstone to reinforce and develop your understanding of chemical bonding.

Although some modern experimental methods can probe the properties of individual atoms and molecules, most experiments are performed with collections of atoms and/or molecules. Chapter 13 provides an introduction and overview of some essential aspects of how to think about such collections. This chapter begins the book's consideration of a second fundamental question in physical chemistry, "What is the connection between molecular properties and the macroscopic properties of matter?" Chapters 14 through 18 develop our atomistic view of matter and its utility for quantitatively predicting the properties of a large collection of molecules as a function of what is called temperature in the macroscopic description of matter. In order to present this description most clearly we use simplified models, e.g., the ideal gas law, that ignore intermolecular forces. With this fundamental understanding in hand, the book proceeds to develop the macroscopic description of chemical systems with intermolecular forces (Chapters 19–21). Chapters 22 and 23 apply these developments to chemical equilibria in aqueous solutions, biosystems, and electrochemistry.

"How do molecules and atoms (reactants) transform into different molecules and atoms (products)?" For many practicing chemists, this question is the essence of chemistry. We consider chemical reaction rates and reaction dynamics, in Chapters 24 and 25. While much progress has been made in providing a quantitatively accurate picture of reaction dynamics, this area remains a current frontier of physical chemistry.

The final chapters of the book address the question, "Why do molecules aggregate into organized assemblies and how does the organization of molecules give rise to new properties?" First we develop some of the primary physicochemical considerations for understanding organized molecular assemblies; in particular, we explore the properties of macromolecules (Chapter 26) and of interfaces (Chapter 27), because they can be used to organize molecules, or molecular subunits, in space. Chapter 28 explores how the organization of functional molecules can be used to create new behaviors that are characteristic of the supramolecular assembly, distinct from the individual molecule behaviors. The understanding of supramolecular machines and organized assemblies is a new frontier in physical chemistry research. The final chapter explores some of the underlying physicochemical principles that are important for evolution and the origin of life.

Although the sequence of chapters in the book need not determine the sequence of topics in a physical chemistry lecture course, the organization of the chapters in the book and its broad scope aims to emphasize the logical progression and unity of physical chemistry. Whatever order of topics is chosen, it is crucial to present the unifying nature of physical chemistry to the student. Where possible, the logical sequence in a given chapter is constructed to evolve from the more basic and simple picture to a more sophisticated treatment for the more experienced students. If such a section is too difficult in a first reading, it is often still possible to skip it and proceed onward to the next chapter without loss of the general picture. Hence, if a section is too difficult to study with excitement, consider skipping it and returning at a later stage.

The book aims to develop one's intuition on how research works by presenting many exciting original ideas and providing instructive examples that inspire thinking. Rather than always aiming for rigor, the discussion uses simplified models and approximations that can develop our intuition and guide our thinking at the boundaries of knowledge. For these reasons, original research findings and current research findings are included in the examples or as part of problems and exercises to be worked. Lastly, the text is only a tool through which the student must explore and discover the rich field of physical chemistry for her or him self.

1

Wave–Particle Duality

CHAPTER MENU

1.1 Overview of Quantum Mechanics

Matter is composed of atoms that are bound together to form molecules, which can then assemble into larger structures. What are the basic principles that govern the formation of chemical bonds? An understanding of this fascinating phenomenon requires that we abandon familiar ways of thinking. We must learn to think in terms of quantum mechanics, which constitutes the fundamental principles upon which our understanding of chemical bonding is based.

The development of quantum mechanics in the first half of the twentieth century is one of the great paradigm upheavals in science. It required that our basic concepts for describing the material world, as manifest in the postulates of classical mechanics, be replaced by the postulates of quantum mechanics. Although it would be logical to see what follows from these postulates and compare the results with reality, this task is quite complex for even the most simple cases. How then to develop our understanding of quantum mechanics and its implications for chemistry? In this chapter we sketch some of the key historical features of quantum mechanics development and illustrate its fundamental idea, wave–particle duality. Then, in Chapter 2 we apply these ideas to simplified models of an H-atom and H_2^+ molecule before embarking on a more formal development of quantum mechanics to molecular bonding and structure in the later chapters.

1.1.1 Historical Highlights

An important root for growing quantum mechanics was the electromagnetic theory developed by James Clark Maxwell (1831–1879). In 1865 Maxwell predicted the formation of electromagnetic waves, and his theory unified optics with electricity and magnetism. In 1888 Heinrich Hertz (1857–1894) was able to generate and detect electromagnetic waves. Thus the wave nature of light was established and the basis of modern communication techniques was set. Physics based on Newton's mechanics (Isaac Newton 1643–1727) and Maxwell's electrodynamics is now called classical physics.

An important observation that remained unexplained for a long time was the radiation emitted when heating a "black body" such as the burner of an electric stove. A heated body first becomes dull red, then white, and finally blue. The light emitted from the heated body has a maximum intensity at a particular frequency, and this frequency shifts to a higher frequency when increasing the temperature. In complete disagreement with this observation, Maxwell's theory predicted that the radiation intensity should continuously increase with frequency (ultraviolet catastrophe). What is the reason for this discrepancy? Max Planck (1858–1947) modeled the molecules of the black body by oscillators, as had earlier investigators. In classical physics, oscillators can be excited to assume any energy. Max Planck, however, had the revolutionary idea that an oscillator can have only discrete energies $E = nh\nu$ where n is an integer, h is a proportionality constant, and ν is the oscillator's frequency. Using this assumption, he obtained agreement with experiment by setting $h = 6.626 \times 10^{-34}$ J s (Planck radiation law, 1900).

Principles of Physical Chemistry, Third Edition. Hans Kuhn, David H. Waldeck, and Horst-Dieter Försterling.

Companion website: www.wiley.com/go/Kuhn/physicalchemistry3e

Albert Einstein (1879–1955) extended Planck's hypothesis by proposing that the radiation itself existed in packages of energy $E = h\nu$ (photons). On that basis he explained the photoelectric effect discovered by Heinrich Hertz: a metal, illuminated by light, emits electrons with a speed that can be measured (see Section 1.2.1). Maxwell's theory predicted that the speed of the emitted electrons increases proportionally with the intensity of the incident light. In contrast, experiment showed that the number of emitted electrons increases with intensity, but the speed depends on the frequency of the incident light and is independent of intensity. Einstein explained this strong discrepancy in 1905 by his description of light as a beam of photons. All these observations suggest that light has the properties both of waves and of particles (wave–particle duality).

Niels Bohr (1885–1962) saw a correspondence between the actual behavior of electrons and what follows from classical physics. On that basis he proposed a model of the hydrogen atom that explained the emission spectrum of hydrogen (1913).

Louis de Broglie (1892–1987) speculated that matter, like light, may show wave–particle duality. He postulated that the wavelength of an electron moving with speed v is $\Lambda = h/(m_e v)$ where m_e is the electron's mass (1924). de Broglie waves were first observed experimentally in 1926 by C. Davisson, L. H. Germer, G. P. Thomson, and A. Reid. This revolutionary view confirmed the idea behind Bohr's model.

In 1926, Erwin Schrödinger (1887–1961) generalized this way of thinking by postulating an equation, now called the Schrödinger equation, that allowed one to calculate the distribution of electrons over a molecule. Werner Heisenberg (1901–1976) proposed his uncertainty principle in 1927. Together with the view that the Schrödinger equation gives the probability of finding an electron in a given volume element (Max Born, 1882–1970) and the fact that only distinct solutions of the Schrödinger equation are meaningful (exclusion principle by Wolfgang Pauli, 1900–1958, formulated in 1925), the Schrödinger equation is the basis of quantum chemistry.

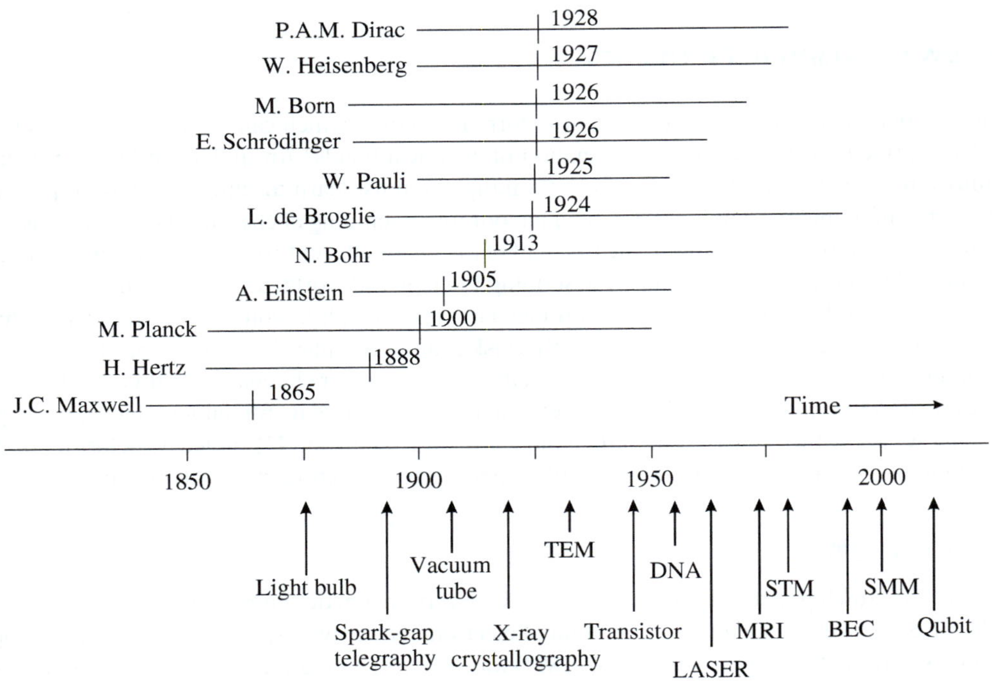

Historical overview: development of ideas (see text) and consequentional realization of devices. Light bulb: Joseph Wilson Swan 1878, Thomas Edison 1880; spark-gap telegraphy: Guglielmo Marconi 1896; vacuum tube (triode): Lee de Forester 1907; X-ray crystallography: William Henry Bragg (Nobel prize 1915), Max von Laue (Nobel prize 1914); TEM: Transmission Electron Microscopy, Ernst Ruska 1933 (Nobel prize 1986); transistor: William Bradford Shockley and Walter Houser Brattain 1945 (Nobel prize 1956); DNA double helix: James Watson, Francis Crick 1953 (Nobel prize 1962); BCS: Bardeen Cooper Schrieffer-theory of superconducting metals 1957 (Nobel prize 1972); LASER: Light amplification by Stimulated Emission of Radiation, Ruby laser Theodore Maiman 1960; Dye laser, Fritz Peter Schäfer, Peter Sorokin 1966; MRI: magnetic resonance imaging, Paul Lauterbur 1973 (Nobel prize 2003); STM: Scanning Tunneling Microscopy, Gerd Binning and Heinrich Rohrer 1981 (Nobel prize 1986); BEC Bose–Einstein-condensate 1995: Eric A. Cornell, Carl E. Wiemann, Wolfgang Ketterle (Nobel prize 2001); SMM: Single Molecule Microscopy, Eric Betzig, Stefan Hell, William Moerner (Nobel prize 2014); Qubit and quantum computing.

This situation was expressed in 1929 by Paul Dirac (1902–1984) in his famous sentence: "The underlying physical laws necessary for the mathematical theory of a large part of physics and the whole of chemistry are thus completely known, and the difficulty is only that the exact application of these laws leads to equations much too complicated to be solved." This is where quantum chemistry—the search for approximation methods—begins.

1.1.2 An Approach to Quantum Mechanics

To develop our quantum mechanical description of atoms and molecules, we start simply and add complexity as it becomes necessary by a comparison with experimental facts.

In a first approach our basic postulate is wave–particle duality. It constitutes a form of the general postulates of quantum mechanics valid only in the special case of free electrons. In order to appreciate better this postulate we compare its predictions with experiments on waves and particles. Our next postulate is the variational principle. In Chapter 2 we use this postulate to gain a basic understanding of the stability, size, and shape of the hydrogen atom, and of the H_2^+ molecular ion. In this way we illustrate the importance of quantum mechanics in understanding the occurrence of chemical bonds.

How best to achieve these goals? First it is necessary to become familiar with a basic process in science: to consider distinct experiments and to identify the logical underlying pattern, in other words to draw conclusions just from these experiments. Keeping this in mind we select key experiments that lead to the postulate of wave–particle duality as directly as possible. In this way it is possible to obtain, right at the beginning, a first idea of the fascinating revolutionary change in thinking that is necessary to understand the fundamental principles of chemistry.

In a second approach we replace the postulates of wave–particle duality by a more general postulate, the time-independent Schrödinger equation (Chapter 3), which holds for a wide range of chemical systems. We will show that the variational principle, considered as a postulate in Chapter 2, is now a theorem, which follows from the time-independent Schrödinger equation (Chapter 5).

In a third step of generalization we replace the time-independent Schrödinger equation (so far considered as a postulate) by postulating operators that lead to an equation called the *time-dependent Schrödinger equation* (Chapter 3, Section 3.5). This equation includes the time-independent Schrödinger equation discussed in Chapter 3 (Sections 3.1–3.4) as a special case. In a fourth step the antisymmetry condition (Pauli exclusion principle) is introduced as a last additional postulate (Chapter 5, Section 5.4.3).

1.2 Light

In classical physics light is described as a wave phenomenon. However, light can also be described as a stream of particles. Some phenomena, such as the photoelectric effect or photochemical reactions, reflect the particle nature of light (see Section 1.2.1). Other phenomena, such as interference in soap films or the diffraction of light from a slit, demonstrate the wave nature of light (see Section 1.2.2). Section 1.2.3 presents a theory of light which combines these two properties into a whole.

1.2.1 Particle Nature of Light: Photoelectric Effect

Electrons can be ejected from the surface of a metal by light (Fig. 1.1). This phenomenon is called the *photoelectric effect*. The speed v of the electrons leaving the surface depends on the color of the light; it is higher for violet light and lower for red light. Increasing the intensity of the light causes more electrons to leave the surface, but does not increase their speed.

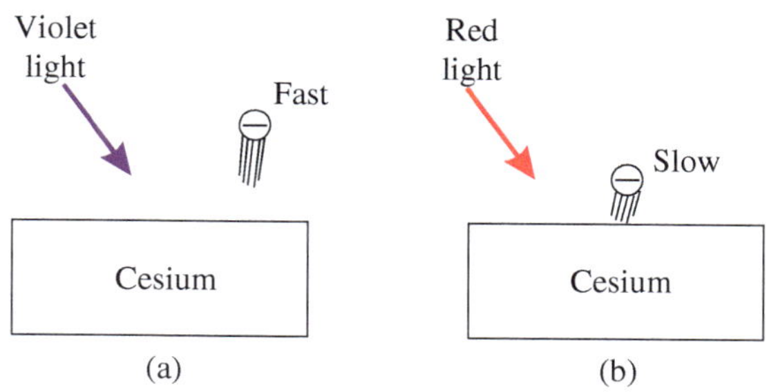

Figure 1.1 Ejection of electrons from a metal by light. (a) Violet light, fast electron. (b) Red light, slow electron.

The sketch in Fig. 1.2a shows an experimental setup for studying this phenomenon. Two metal electrodes are placed inside a vacuum chamber (e.g., an evacuated glass flask), and the electrodes are connected to an external electrical circuit that controls the voltage between them and measures the current flowing in the circuit. Let us choose one of the electrode materials to be composed of Cs, a metal that emits electrons easily, and choose the other electrode to be composed of an inert metal, such as Au or Pt. Now we place a voltage V across the electrodes, with the polarity such that electrons are decelerated in traveling from the Cs electrode (the

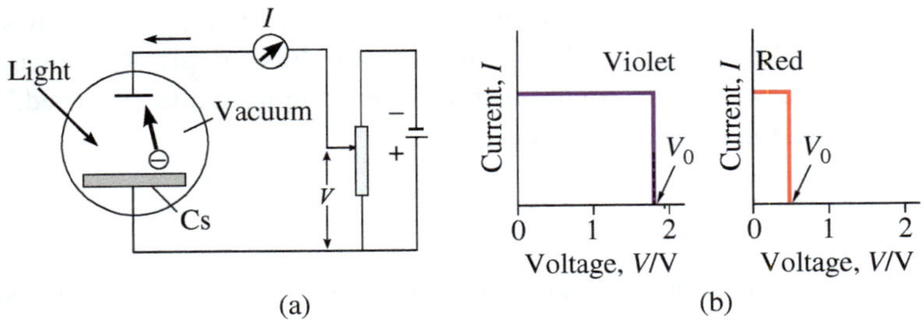

Figure 1.2 Experimental arrangement for the photoelectric effect. (a) Construction of the photocell with decelerating voltage. (b) Current I measured by the ammeter as a function of the decelerating voltage V which decelerates the electron ejected from the Cs metal surface. V_0 is the stopping voltage. Note that the diagram in Fig. 1.2b is oversimplified. The temperature of the metal leads to a broadening of the electron energies so that the fall-off at V_0 is not quite so abrupt. Note that by convention the direction of the current I is defined as opposite to that of the electron flow.

cathode) to the opposite electrode (the anode). When we shine monochromatic visible light on the Cs cathode, we observe current flowing in the circuit as long as V is low enough. If we keep the light intensity constant and increase the size of V, the current remains constant until a certain value, V_0, is reached where the current drops abruptly to zero (Fig. 1.2b). V_0 is called the "stopping voltage." If the experiment is performed with different colors of light the behavior is similar, but the value of V_0 changes. If we perform this experiment at different light intensities, we find that the value of the current increases as the light intensity increases, but the stopping voltage V_0 remains the same. To summarize, the current depends on the intensity of the light, and the stopping voltage V_0 depends on the color of the light.

To analyze the observations we use energy conservation and recall that an electron traveling through a voltage V changes its energy by the amount eV, where $-e$ is the electron's charge. When an electron (mass m_e) leaves the cathode with speed υ, it has kinetic energy $\frac{1}{2}m_e\upsilon^2$, and as it moves through the retarding potential V it converts its kinetic energy into potential energy. The electron will reach the anode if its initial kinetic energy $\frac{1}{2}m_e\upsilon^2$ is greater than or equal to the potential energy change that is required to reach its final potential energy at the anode, $E_{\text{pot}} = eV$. At the stopping voltage V_0, the potential energy is equal to the initial kinetic energy so that

$$E_{\text{kin}} = \frac{1}{2}m_e\upsilon^2 = eV_0. \tag{1.1}$$

For a source of violet light (such as produced by a tunable laser) and Cs metal, the observed stopping voltage is 1.801 V. Given the elementary charge of $e = 1.602 \times 10^{-19}$ C, we obtain

$$E_{\text{kin}} = 1.602 \times 10^{-19}\ \text{C} \times 1.801\ \text{V} = 2.885 \times 10^{-19}\ \text{J}.$$

It will be shown at the end of Section 1.2.2 that it takes 2.080×10^{-19} J to pull an electron out of the metallic bonding of the Cs metal (*work function, E_{work}*). So the violet light supplies each ejected electron with

$$E = E_{\text{kin}} + E_{\text{work}} = (2.885 + 2.080) \times 10^{-19}\ \text{J} = 4.965 \times 10^{-19}\ \text{J}$$

of energy. By repeating the photoelectric effect experiment with light of other colors, we find that yellow light transfers 3.310×10^{-19} J of energy to each electron, and red light transfers 2.837×10^{-19} J of energy to each electron.

We conclude from these experiments that light interacts with electrons in discrete amounts, or energy packets (energy packets of 4.965×10^{-19} J for violet light and those of 2.837×10^{-19} J for red light). These energy packets are called *light quanta* or *photons*. The energy of each photon depends on the color of the light.

1.2.2 Wave Nature of Light: Diffraction

1.2.2.1 Water Surface Waves

The diffraction and interference properties of waves are easily seen in water surface waves. Figure 1.3a illustrates how a wave passing through a small hole in a barrier acts like a source, sending a new wave into the region beyond the barrier. The hole is considered small if it is about 1/10 of the wave's characteristic wavelength. This phenomenon is called *diffraction*.

If two holes are opened in the barrier, then a wave originates from each hole by diffraction (Fig. 1.3b). These waves can overlap to reinforce each other or to cancel each other. This phenomenon is called *interference*. Figure 1.3b illustrates how

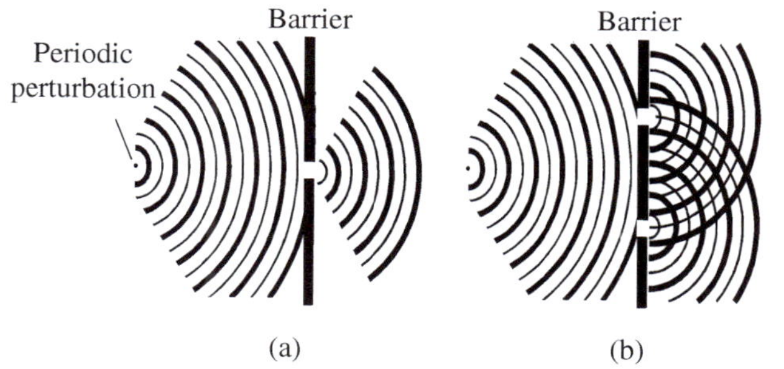

Figure 1.3 (a) Formation of a secondary wave from a small hole (for water waves; the surface of the water is forced to a wave motion by periodically dipping a stick into the water). Thick lines represent maxima of the waves. Thin lines represent minima of the waves. Reflection of the primary waves and disturbance of the secondary waves by the barrier are neglected. The distance between two subsequent thick lines corresponds to the wavelength λ. (b) Interference of two water waves. Amplitude maxima (thick crossing lines); amplitude minima (thin lines crossing); amplitude zero (thin and thick lines crossing).

the waves interfere both constructively (reinforcing each other) and destructively (canceling each other). A wave has an amplitude ψ. For water waves ψ is the distance between the average height of the water's surface and its highest height (peak), or alternatively the distance between the average height and the lowest point (trough). More generally, the amplitude of a wave is the maximum displacement, which for water waves is the distance of the water surface above or below the average height. To describe the interference of the waves we sum the waves' amplitudes at each point in space and time to obtain the net wave displacement. This process is called *superposition*.

1.2.2.2 Light Waves

If light travels through a single slit or a double slit, a pattern of light (a change in brightness as a function of position) can be observed on the screen behind the slits, as illustrated in Fig. 1.4. We can interpret this experimental fact as evidence that light has the properties of a wave. The brightness at a given point P on the screen measures the intensity I of the light at point P:

$$I = \frac{\text{power of light on area } A \text{ of the screen at point P}}{\text{area } A} \tag{1.2}$$

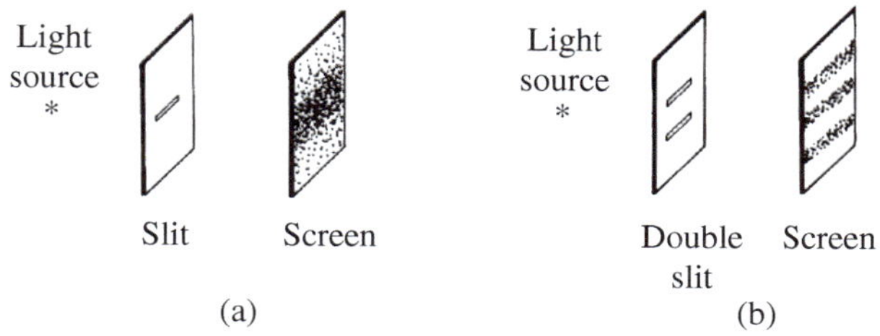

Figure 1.4 (a) Diffraction of light by a single slit. (b) Interference of light from a double slit.

(power = energy per time).[1] Furthermore, we can determine the wavelength λ of the light by measuring the distance between successive minima (positions where the intensity is zero) or maxima (positions where the light intensity is highest) on the screen in Fig. 1.4b. Figure 1.5a shows the intensity pattern of light on a screen when it diffracts from a sharp edge. The light's brightness as a function of position can be quantified by

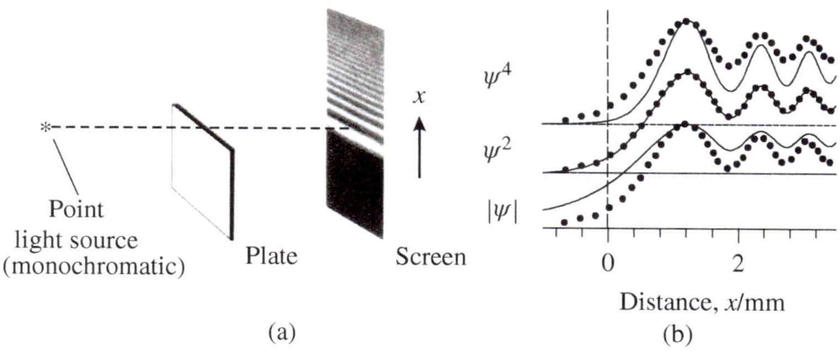

Figure 1.5 Diffraction of light from an edge. (a) Experimental arrangement and interference bands on the screen. (b) Dots: Measured intensity pattern on the screen as a function of distance x from the geometrically constructed shadow boundary. The intensity was measured by replacing the screen by a photographic plate and measuring the density of the photographic image at each point on the screen. Calculated dependences of ψ, ψ^2, and ψ^4on the distance x are shown as solid lines. In each case, the experimental points are adjusted to the first maximum. Only the curve for ψ^2 is in accordance with the experimental intensity pattern.

1 We assume that the area A about point P is sufficiently small to be uniform in brightness. The area equals A if the screen is perpendicular to the direction of the light beam; otherwise, it is $A \cdot \cos\alpha$, where α is the angle between the direction of the light beam and the direction of the normal on the screen.

using a digital camera or some other spatially sensitive photodetector (Fig. 1.5b). Comparison of the light wave's calculated amplitude ψ and the measured intensity shows that the intensity is proportional to ψ^2. Figure 1.5b shows three theoretical curves, for $|\psi|$, ψ^2, and ψ^4, and it is evident that the experimental curve of intensity versus distance is proportional to the calculated curve of ψ^2 versus distance x.

1.2.2.3 Wavelength of Light from Interference Pattern

Because superposition and interference of waves is such an important feature of quantum mechanics, it is worthwhile to show how the interference pattern can be used to determine the wavelength of the light source. Consider two slits with a separation d and a screen at distance a from the slits (Fig. 1.6). At point P_1, which is midway between the slit positions, the distances between each of the slits and the screen are the same. In this case the light waves stay in phase and add constructively when they arrive at the point P_1, and an intensity maximum is therefore observed at P_1. The next intensity maximum occurs at the point P_2. At this position the distances are r_1 (from the upper slit) and $r_1 + \lambda$ (from the lower slit), so that the phase shift is exactly one wavelength, and the two waves add constructively again. In Exercise 1.9 you calculate the wavelength λ from the distance x between the points P_1 and P_2 in Fig. 1.6.

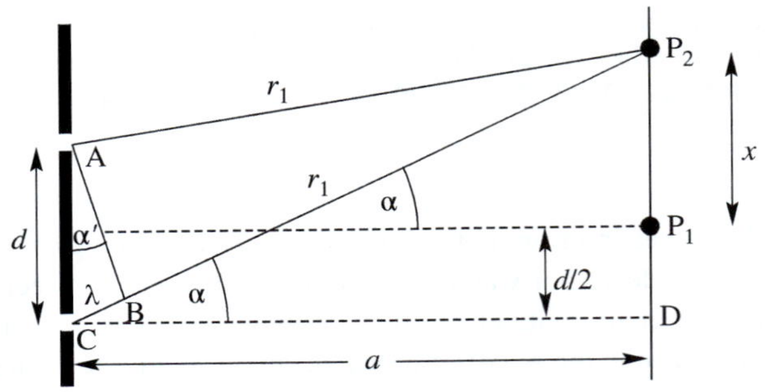

Figure 1.6 Diffraction for a double slit (slits separated by d). The intensity of the light arriving on a screen at distance a is measured at points P_1 and P_2. Usually, the distance x of points P_1 and P_2 is much larger than the distance d of the slits. In this limit the triangle ABC can be treated as a right triangle, leading to the relation $\alpha' \approx \alpha$.

If we perform the double-slit experiment on the same monochromatic light sources used in the photoelectric effect measurements (Section 1.2.1), we find $\lambda = 400$ nm for violet light, $\lambda = 600$ nm for yellow light, and $\lambda = 700$ nm for red light.

1.2.2.4 Relation Between Photon Energy E and Wavelength λ

For a given color of light, the energy E per photon can be measured by a photoelectric effect experiment, and the wavelength λ of the light can be measured by a diffraction experiment. Table 1.1 lists the corresponding energies (per photon) and wavelengths for several colors of light.

Table 1.1 Energies (per Photon) and Wavelengths for Various Colors of Light, as Measured by the Photoelectric Effect and by Diffraction Experiments.

Color	$\frac{E}{10^{-19}\,\text{J}}$	$\frac{\lambda}{\text{nm}}$	$\frac{E\lambda}{10^{-26}\,\text{J m}}$
Violet	4.965	400	19.86
Yellow	3.310	600	19.86
Red	2.837	700	19.86

The fourth column of Table 1.1 gives the product of E and λ for each case. It can be seen that this product is the same for each entry. We can hypothesize that a law of nature makes the product of E and λ the same for each color of light. When the experiments are repeated carefully for different colors of light, the same relationship is always found:

$$E\lambda = 19.86 \times 10^{-26}\ \text{J m}. \tag{1.3}$$

From this relationship we can assign a wavelength λ to a photon whose energy is determined in a photoelectric effect experiment, or alternatively we can determine the wavelength of the light and assign an energy to the photons in the light wave.

We can write Equation (1.3) in a different form by relating the light's wavelength λ to its frequency ν. The product of a wave's frequency ν and its wavelength λ is its speed υ,

$$\nu\lambda = \upsilon \tag{1.4}$$

(see Fig. 1.7). Equation (1.4) is sometimes called a *dispersion relation*. In the photoelectric effect experiment the light is moving through a vacuum, so its speed is $\upsilon = c_0 = 2.998 \times 10^8\ \text{ms}^{-1}$. For 400 nm violet light, we can compute a frequency

$$\nu = \frac{c_0}{\lambda} = \frac{2.998 \times 10^8\ \text{m s}^{-1}}{400 \times 10^{-9}\ \text{m}} = 0.749 \times 10^{15}\ \text{s}^{-1}.$$

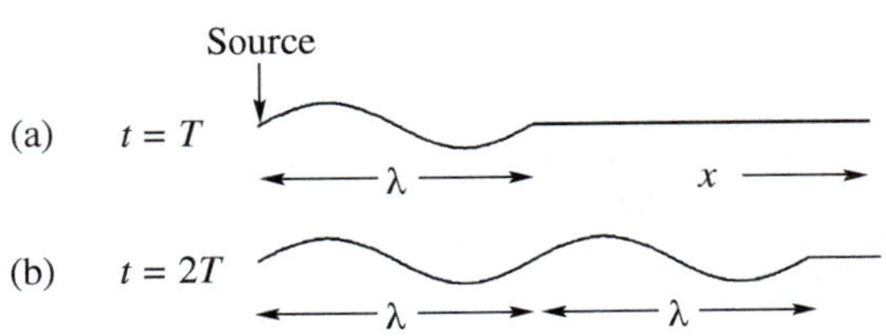

Figure 1.7 A traveling wave. (a) at time $t = T$, (b) at time $t = 2T$. A string is shaken periodically with a frequency ν; the disturbance propagates to the right. After time T (T is the period) the wavefront has traveled the distance $\lambda = \upsilon T$. Since $T = 1/\nu$ (ν = frequency) we obtain $\lambda = \upsilon/\nu$ or $\lambda\nu = \upsilon$.

From Equations (1.3) and (1.4) it follows, with $v = c_0$, that

$$E = \frac{19.86 \times 10^{-26}\ \text{J m}}{\lambda} = \frac{19.86 \times 10^{-26}\ \text{J m}}{c_0} \times \nu = (6.62 \times 10^{-34}\ \text{J s}) \times \nu = h\nu.$$

The constant h is *Planck's constant*. Very precise and exact measurements have established the value $6.6260702 \times 10^{-34}$ J s for h. Thus our law of nature is written most simply as

$$\boxed{E = h\nu} \tag{1.5}$$

This fundamental relationship between frequency and the energy of a light quantum is well established. It is important to note that the value of h obtained by Einstein from the photoelectric effect agrees with the value independently obtained by Planck from black body radiation (see Section 1.1.1).

1.2.2.5 Evaluation of Work Function E_{work}

For a more detailed evaluation, the measured values of the kinetic energy E_{kin} of the emitted photoelectrons are plotted versus the wavenumber, $1/\lambda$, of the incident light in Fig. 1.8. A linear relationship is observed with slope $a = 1.986 \times 10^{-25}$ J m and intercept with the abscissa at wavenumber $1/\lambda_0 = 10{,}470\ \text{cm}^{-1}$:

$$E_{\text{kin}} = a\left(\frac{1}{\lambda} - \frac{1}{\lambda_0}\right) = a\frac{1}{\lambda} - a\frac{1}{\lambda_0} = E - E_{\text{work}}.$$

Then for the energy E of the incident photons and for the work function E_{work} we obtain

$$E = a\frac{1}{\lambda} \quad \text{and} \quad E_{\text{work}} = a\frac{1}{\lambda_0} = 2.080 \times 10^{-19}\ \text{J}.$$

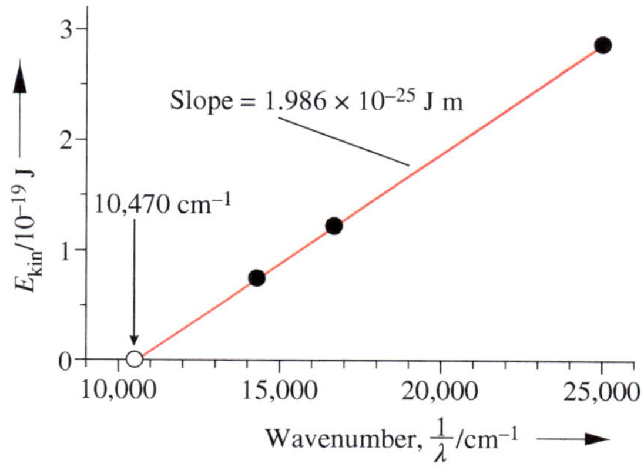

Figure 1.8 Photoelectric effect: evaluation of work function. The kinetic energy E_{kin} measured for the emitted photoelectrons in the experiment depicted in Fig. 1.2 is shown versus the wavenumber, $1/\lambda$, of the incident light (filled circles). At $1/\lambda_0 = 10{,}470\ \text{cm}^{-1}$ the kinetic energy of the photoelectrons becomes zero (open circle); this limiting value of $10{,}470\ \text{cm}^{-1}$ is a material constant for cesium.

1.2.3 Interpretation of the Experiments

We can describe the observed phenomena on the basis of the following postulates:

- *Postulate. Light behaves as a wave with wavelength λ traveling from a light source.*
- *Postulate. The probability that a light quantum is detected in a small area at a position x, y, z is proportional to this area and proportional to ψ^2, where ψ is the amplitude of the light wave at the position x, y, z. Quanta of light of wavelength λ are packages of energy $E = h\nu$, where $\nu = c_0/\lambda$.*

The first postulate does not say that light is a wave phenomenon (which would be in contradiction to the results of the photoelectric effect experiments). It only makes the weaker assertion that light does exactly what we expect a wave to do in certain experiments. The assumption of a wavelength λ also does not imply that a wave really exists, only that the diffraction experiments can be quantitatively analyzed in terms of this quantity.

The second postulate connects the wave and particle descriptions of light. We saw in Section 1.2.1 that the number of photons interacting with matter is proportional to the intensity of light falling on the object. In Section 1.2.2, we found that the intensity I of light is proportional to the square of its amplitude ψ in the wave picture. [2] Thus the number of photons in a given small area on the screen is proportional to ψ^2 and to the area; hence, the probability of detecting one photon by its interaction with matter is proportional to ψ^2 and this area. To summarize, the probability that a light quantum is detected in a small area is proportional to this area and to ψ^2.

It is therefore clear that the properties of light cannot be described by the wave picture alone or by the particle picture alone. We require both features together to describe the observed phenomena, known as *wave–particle duality*.

2 This experimental finding is not surprising; as $I \sim \psi^2$ follows from Maxwell's electrodynamic theory.

1.3 Electrons

Wave–particle duality is also required to describe the behavior of electrons, not just light. It is common to think of atoms, electrons, and the like as particles that follow Newton's laws; however, the wave characteristics of electrons can also be important. Because of their small mass, the wave properties of electrons are easy to observe and are essential for describing the structure of atoms and molecules. The particle nature of electrons is discussed in Section 1.3.1, the wave nature of matter is described in Section 1.3.2, and a unified perspective for electrons is provided in Section 1.3.3.

1.3.1 Particle Nature of Electrons

The particle nature of electrons is evident in many situations, such as the Wilson cloud chamber, Geiger counters, and scintillation on a fluorescent screen. Important particle properties of the electron include its discrete mass $m_e = 9.109383 \times 10^{-31}$ kg and charge $-e = -1.602176 \times 10^{-19}$ C (e = elementary charge). Below we describe the original measurements by Millikan which demonstrated that electrical charge occurs in discrete amounts.

Millikan's oil drop experiment is straightforward in its physical design (see Fig. 1.9), and it demonstrates the discrete nature of electrical charge. Liquid droplets, generated by an aerosol nozzle, are injected into a chamber between two metal plates. By irradiation with X rays the air in the chamber is ionized, and thus a negative electric charge Q is transferred to the droplets. By applying an electric field $F = V/d$ between the two metal plates at distance d, the charged droplets can be trapped and levitated. Any uncharged droplets fall down and are not trapped. With the particles trapped Millikan was able to determine the charge on a droplet by comparing the gravitational force (which depends on the mass of the droplet m_{drop} and the gravitational constant g) acting on it to the electrostatic force (QF). He found that the total charge on the droplets occurred in "clumps." By performing experiments on a large number of droplets and over a wide range of conditions, he found that all of his total charge measurements could be described by an integer multiple of a single value of the charge, that is, $Q = -ne$ where n is the number of elementary charges. He determined the fundamental charge e to be 1.591×10^{-19} C, which is quite close to the more precise modern value of $e = 1.602176 \times 10^{-19}$ C.

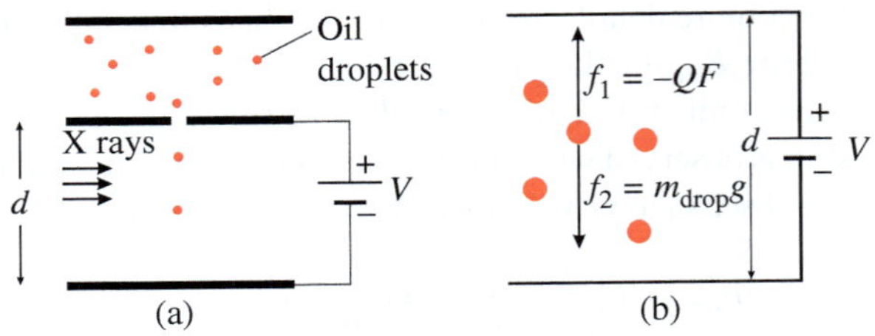

Figure 1.9 Millikan's oil drop experiment. (a) Oil droplets fall from the upper chamber through a hole into the measuring chamber. By irradiation with X rays the air around the droplets is ionized, and thus negative charges are transferred to the droplets. A positive voltage V is applied between the two plates in distance d and the movement of a droplet is followed by using a microscope. (b) Electrostatic force f_1 driving the droplet upward ($F = V/d$) and gravitational force f_2 driving the droplet downward. In the case $f_1 = f_2$ the droplet is trapped.

1.3.2 Wave Nature of Electrons

1.3.2.1 Diffraction Experiments

The wave nature of electrons can be demonstrated by diffraction experiments analogous to those for light. Instead of monochromatic light, electron diffraction experiments use a beam of electrons of a definite speed v. Such beams can be produced from a heated filament in an evacuated tube with an accelerating voltage (Fig. 1.10). Electrons leaving the heated filament acquire kinetic energy eV_a in traveling through the accelerating voltage V_a. Such a device is called an "electron gun." Because the accelerating voltage determines the kinetic energy, we can write the kinetic energy as

$$E_{\text{kin}} = \frac{1}{2} m_e v^2 = eV_a$$

and calculate the average speed as

$$v = \sqrt{\frac{2eV_a}{m_e}}. \tag{1.6}$$

If $V_a = 100$ V, for example, then

$$v = \sqrt{\frac{2 \times 1.602 \times 10^{-19}\ \text{C} \times 100\ \text{V}}{9.109 \times 10^{-31}\ \text{kg}}} = 5.93 \times 10^6\ \text{m s}^{-1}.$$

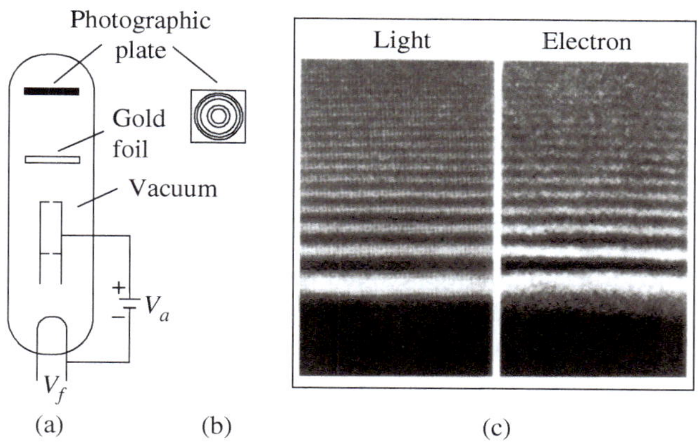

Figure 1.10 Diffraction of electrons. (a) Production of an electron beam (V_f: voltage for heating filament, V_a: accelerating voltage) and diffraction from a gold foil (Davisson–Germer experiment). Diffraction rings on a photographic plate (schematic). (c) Interference bands from diffraction by an edge in the case of monochromatic light ($\lambda = 633$ nm, left, see also Fig. 1.5a) compared to the case of electrons of distinct speed ($v = 1 \times 10^6$ m s^{-1}, right). The enlargement of the photographs is chosen so that the distance between the first and second maxima is the same in each case. This distance is related to the wavelength λ and the distance between edge and screen, see Fig. 1.5.

If an electron beam passes through a double slit, then a diffraction pattern, an image of bright and dark bands (as in Fig. 1.4b), can be observed on a fluorescent screen. The wavelength of the electrons can be calculated from the distance between bright bands, as in the case for light. We call the electron wavelength Λ to distinguish it from the wavelength λ of light.

The double-slit method for electrons requires special techniques, because the wavelength of electrons in practical experiments is very small. The slits must be placed extremely close together to produce an interference pattern. Experimentally, it is easier to use a thin piece of gold foil, in which the crystal lattice of the gold foil acts as a diffraction grating for the electrons, or to use an edge of a solid. The wavelength Λ can be calculated from the spacing between the peaks in the diffraction pattern. Figure 1.10c shows an intensity pattern for electron waves diffracting from the edge of a gold foil and compares it to the case of edge diffraction for light.

1.3.2.2 Relation Between the Electron's Speed v and Wavelength Λ

Table 1.2 lists the wavelengths Λ found for electrons at different accelerating voltages V_a. The last column gives the product of the electron's wavelength Λ and its speed v.

We see that essentially the same value

$$v\Lambda = 7.2 \times 10^{-4}\ \text{m}^2\ \text{s}^{-1}$$

is obtained for both cases. If we try the experiment with still more speeds, we find the same relationship every time. This finding is reminiscent of that for light where the product of the energy and the wavelength was a constant, suggesting that we might try to express the constant 7.2×10^{-4} m^2 s^{-1} in terms of the constant h. For the case of light the product had units of J s, or kg m^2 s^{-1}. The units suggest that we include a mass term, so we could try to divide h by the mass of the electron. We find that

Table 1.2 Speeds and Wavelengths Λ for Electrons Accelerated with Different Voltages V_a.

$\frac{V_a}{\text{V}}$	$\frac{v}{\text{m s}^{-1}}$	$\frac{\Lambda}{\text{m}}$	$\frac{v\Lambda}{\text{m}^2\ \text{s}^{-1}}$
35	3.51×10^6	2.05×10^{-10}	7.20×10^{-4}
100	5.93×10^6	1.22×10^{-10}	7.24×10^{-4}

$$v\Lambda = \frac{h}{m_e} = \frac{6.63 \times 10^{-34}\ \text{J s}}{9.109 \times 10^{-31}\,\text{kg}} = 7.27 \times 10^{-4}\ \text{m}^2\ \text{s}^{-1}.$$

So, within the bounds of experimental error, the new constant $v\Lambda$ really can be written as a combination of the fundamental constants h and m_e! We can hypothesize a new fundamental relationship for electrons as

$$\boxed{\Lambda = \frac{h}{m_e v}} \qquad (1.7)$$

This is called the *de Broglie relationship,* which is the fundamental expression relating wave and particle properties of the electron. As the speed v of the electron increases, its wavelength Λ decreases; in other words, v and Λ are inversely proportional.

1.3.3 Interpretation of the Experiments

The postulates that we developed for light can be recast for electrons:

- *Postulate: An electron traveling from an electron source with speed v behaves as a wave with wavelength* $\Lambda = h/(m_e v)$.

- *Postulate: The probability that an electron is detected in a small area at a position x, y, z is proportional to this area and proportional to ψ^2, where ψ is the amplitude of the electron wave at the position x, y, z (wavefunction).*

This second postulate is called *Born's probability postulate*. Note the essence of our procedure: considering distinct experiments and drawing distinct conclusions (the two postulates).

1.3.4 Formal Similarity Between Electron and Photon

The formal similarity between an electron and a photon can be carried somewhat further by using Einstein's relation $E = mc_0^2$ from relativity theory, which relates the mass of a particle directly to its energy. As we have a way for determining the energy E of a photon we can assign a mass to it, so that

$$m_{\text{photon}} = E/c_0^2. \tag{1.8}$$

Combining Equations (1.5) and (1.8), we obtain

$$E = m_{\text{photon}} c_0^2 = h\nu = \frac{hc_0}{\lambda}$$

or

$$\lambda = \frac{h}{m_{\text{photon}} c_0}, \tag{1.9}$$

which looks very similar to the de Broglie relationship.

Nevertheless, photons and electrons differ in fundamental ways. For a photon, the speed c_0 is a constant of nature. We can only speak of a mass for the photon in the sense of its motion at the speed of light; its rest mass is zero. When a photon interacts with matter, its entire quantum of energy can be transferred to the particle, in which case the photon ceases to exist. For electrons, the mass m_e is a constant of nature, [3] and the wavelength Λ depends on the speed v.

1.4 Questions Arising About Wave–Particle Duality

1.4.1 Single Event: Probability Statement; Collective Behavior: Definite Statement

Wave–particle duality is a basic tenet of our modern theoretical description of the world and its behavior. We have seen that light and electrons behave on the one hand like particles, and on the other hand like waves of wavelength

$$\lambda = \frac{h}{m_{\text{photon}} c_0} \text{(light)} \quad \text{or} \quad \Lambda = \frac{h}{m_e v} \text{(electrons)}.$$

The probability that a photon or an electron appears in a given small range at the position x, y, z is proportional to $\psi^2(x, y, z)$ where ψ is the wavefunction. Experiments with very many photons or electrons lead to interference patterns like those in Fig. 1.11. If we carry out the corresponding experiment with only a few particles, however, then the patterns can look very different. Figure 1.12 shows the results of computer simulations with 26, 182, and 1900 particles. In the first case (Fig. 1.12a) no pattern is seen, in the second case (Fig. 1.12b) the pattern starts to become apparent, and in the third case (Fig. 1.12c) it is clearly evident. With increasing particle number it becomes less and less likely that the observed interference pattern deviates from the pattern that would be produced by an infinite number of particles.

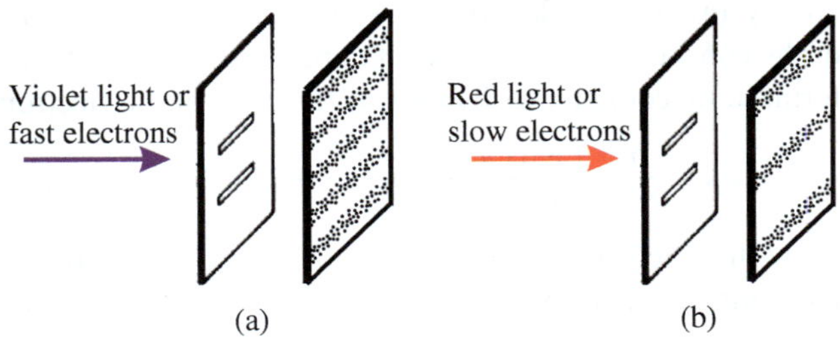

Figure 1.11 Diffraction occurs for both light and electrons. (a) Violet light or fast electrons, (b) red light or slow electrons.

---

3 We have assumed that the mass of the electron is independent of its speed v. This is not completely correct. The energy, and therefore the mass given by relativity theory, increases with increasing speed. The kinetic energy of a particle of rest mass m moving at speed v is

$$E_{\text{kin}} = \frac{1}{2} m v^2 \frac{1}{\sqrt{1 - (v/c_0)^2}}$$

The effect is not significant, however, until the electron's speed v approaches the speed of light c_0. We neglect relativistic effects here.

For simplicity, let us assume that a particle appears in the area of each of the three interference bands of Fig. 1.12c with probability

$$P = \frac{1}{3}.$$

Then, for N particles, the probability that all of them appear in the top band, for example, is

$$P = \left(\frac{1}{3}\right)^N.$$

That is, $\frac{1}{9}$ for $N = 2$, $\frac{1}{81}$ for $N = 4$, and 2×10^{-48} for $N = 100$. A detailed statistical analysis shows that the most likely distribution of particles is $1/3$ in each band and that the accuracy of this prediction improves very rapidly as N increases.

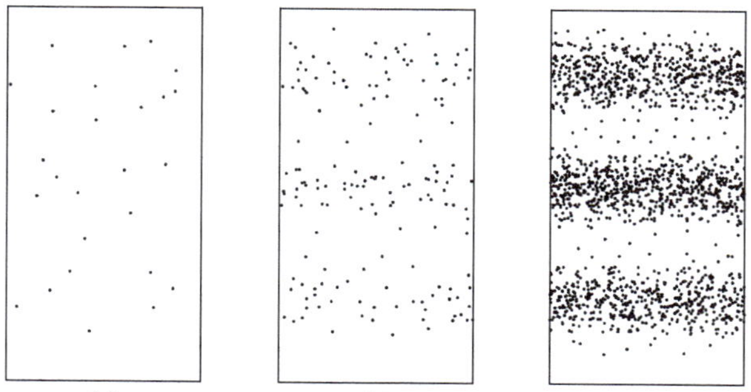

(a) (b) (c)

Figure 1.12 Interference pictures: computer simulation with (a) 26, (b) 182, and (c) 1900 particles.

Only probabilistic statements can be made about single events ("where will the next particle appear on the screen?"). Definite statements can be given for collective behavior: the interference pattern is reproducible with very large numbers of particles. In other words, a causal relationship is possible only for the collective behavior of large numbers of particles but not for single events.

We now consider what happens as the wavelength of the wave–particle becomes small. As with the water waves of Fig. 1.3a, the diffraction becomes less and less noticeable. The diffraction patterns of both light waves and electron waves are transformed gradually into shadow patterns. For light, wave optics goes over to ray optics, and for electrons wave mechanics goes over to classical mechanics as the wavelength becomes small. For the case of very energetic photons or very fast electrons, we can predict with near certainty that the particle will appear on the screen within the geometrically constructed shadow boundary; in this limiting case definite statements about single events are possible.

1.4.2 Wave–Particle Duality and the Need to Abandon Familiar Ways of Thinking

The experiments considered here force us to abandon old ways of thinking. It is therefore of fundamental significance to reflect on the way in which we obtain new knowledge, how conclusions are drawn from experiments, and how this leads to proposing new experiments.

1.4.2.1 You Cannot Trace the Path of a Photon or an Electron

We can detect single photons or electrons as particles by their effect on a detector, such as a fluorescent screen, a photodiode, or a light sensitive microchip in a digital camera. However, it is not possible to trace a path for the particle, to say that the particle has traveled through either one slit or the other. If one slit or the other is closed, something completely new is observed on the screen: photons or electrons appear at places where they were never observed before (at the positions where the waves added through interference in the double-slit case, but not in the single-slit case). Thus it is not useful to ask about a particular trajectory for the electron.

1.4.2.2 Intuitive Aids Assist You to Grasp the Mathematical Formalism Describing Experiments with Photons or Electrons

Water surface waves were introduced in Section 1.2.2 as an intuitive aid, an artifice to express the mathematical formulas describing the phenomena (Figs. 1.4 and 1.11). Particles and waves are also mental pictures that we use to describe physical phenomena concretely. Once the underlying mathematical formulation is discovered, the aids used in our imagination to construct the mathematical expressions (waves in a given medium, particles as really existing entities) should then be thrown away, and the mathematical formalism alone should be used to describe the phenomena (Fig. 1.13).

1.4.2.3 Reflect on the Development of our Everyday View to Understand the Strangeness of the Behavior of Photons or Electrons

The goal of physical sciences is to reduce the multitude of data received by sensory perception to a simple order. Throughout the history of physical sciences this order has been developed and refined. Numerous designs are being utilized to classify our experience as simply and completely as possible, in an attempt to present a logical picture from the totality of data.

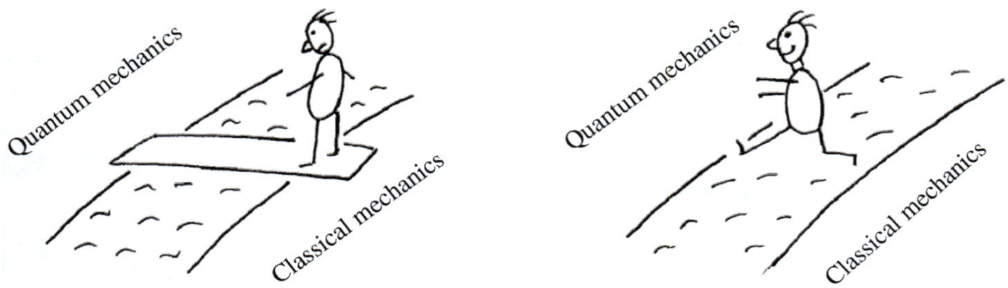

Figure 1.13 Intuitive aids, useful in the beginning, are thrown away.

This process is not different from our own experience of growing up. Infants perceive a multitude of events, and they start to make connections between them. An important step in an infant's development is reached when sensory perceptions are received and correspond to objects in space, changing their location, and form. With this interpretation of visual impressions the classical picture of the world is created in the human mind (Fig. 1.14). With time these pictures become a model of reality that increasingly broadens. More and more new models (or hypotheses) are established and predictions are made. According to the outcome of these predictions of facts the hypotheses are accepted, modified, or rejected. This classical picture of reality succeeds in describing everyday experiences and is therefore deeply rooted in human beings; however, this picture is falsified by the experiments that lead to the postulates of quantum mechanics. Thus, we are forced to change our hypothesis of reality to which we are accustomed.

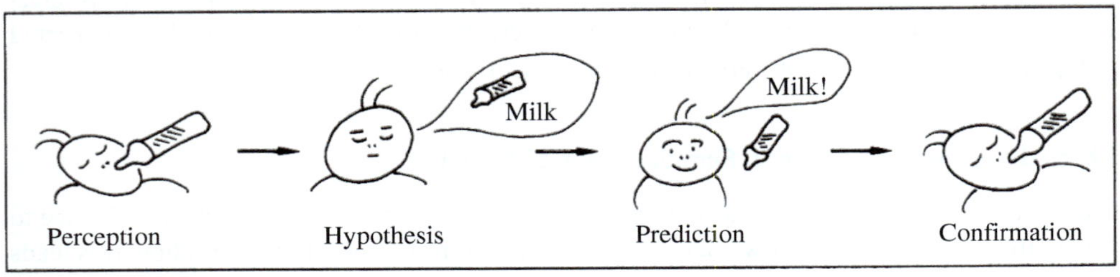

Figure 1.14 Recognition of facts. Hypothesis confirmed (this means it is not falsified).

The unusual model of reality that is taught to us by the experiments in this first chapter must be carefully considered to avoid future pitfalls. Reflect on how you arrive at the given postulates starting from experimental facts and reflect on how you can avoid including more in your picture of reality than the logical ordering scheme of sensory perception requires.

1.4.2.4 Evolutionary and Revolutionary Approach to Quantum Mechanics

We explained experiments by using the classical terms "waves" and "particles," and our path forward is to develop the postulates of quantum mechanics by gradually generalizing (*evolutionary approach*). It would be more logical to delve right into the general postulates of quantum mechanics, attempting to understand the world of chemistry on that mathematically abstract basis without referring to classical concepts (*revolutionary approach*). However, we prefer using the evolutionary approach because humans are more used to thinking in a classical way. Thinking in terms like "as if an electron would be a wave" and "as if the electron would be a particle" is enlightening for a chemistry student's intuitive understanding, whereas immediately being exposed to the general postulates of quantum mechanics might cause more difficulties.

1.5 Conclusion

Light falling on a detector (e.g., a charge-coupled device camera or photographic plate) behaves like a beam of particles, each generating a latent image at the point where the particle hits the screen. The distribution of these bright points is as if the particle were carried by a wave. This effect is most astonishing if the beam of particles passes a double slit. The same behavior is observed in similar experiments with electrons. The wave–particle duality of electrons is the basis for understanding the existence of atoms and molecules and their properties.

1.6 Exercises

E1.1 Consider a microwave source that is generating 2.0 GHz electromagnetic radiation. Compute the wavelength of the microwaves. If this microwave source was used in an oven of width 30 cm, how many wavelengths of the microwave can be included across the oven's width?

Compute the energy per photon for the 2.0 GHz frequency. If a cup containing 250 mL of water is irradiated by this source, how many photons must be absorbed to raise the temperature of the water from 25 °C to 80 °C (a nice temperature for a cup of tea)? For simplicity, assume that the water density is 1.0 g/mL, that the heat capacity is 4.184 J/(g degree), and that they do not change over the temperature range.

E1.2 Consider an ultraviolet light source that generates 300 nm electromagnetic radiation. Compute the frequency of the ultraviolet light. If one photon of this light is absorbed by an organic molecule, how much energy does the molecule gain? Is this energy enough to break a carbon-carbon bond in the molecule? Use a "typical" carbon-carbon bond energy of 5.8×10^{-19} J for your comparison. Perform the same calculations for a photon of wavelength 600 nm and a photon of wavelength 1200 nm. Perform your comparisons using the energy units of J and of eV.

E1.3 Consider an electron with a kinetic energy of 1.0 eV (i.e., it has been accelerated across a 1 V potential difference).

Compute the momentum of this electron. Compare this momentum to that of a "typical" N_2 gas molecule at room temperature (consider the gas molecule to have a speed of 500 m/s).

Compute this electron's speed. At what fraction of light speed (3.00×10^8 m/s) is the electron moving?

Compute this electron's wavelength. Compare this wavelength to the diameter of a hydrogen atom (*ca.* 128 pm).

Perform this same calculation for a 10 eV electron and a 100 eV electron. Comment on the trends in your values. How many electron wavelengths can fit into a hydrogen atom at these different energies?

E1.4 A typical value for a particle's kinetic energy at 25 °C is 6.21×10^{-21} J. Use this value of the kinetic energy to estimate the speed of spheres with different masses; i.e.,

a) ping pong ball (2.60 g)
b) a 10.0 μ diameter polystyrene bead (0.300 g/c m^3)
c) a 50.0 nm radius colloidal particle of Ag (10.5 g/c m^3)
d) Buckminster fullerene (C_{60}) (0.720 kg/mol)
e) He (4.0 amu)

Use these speeds and masses to estimate the de Broglie wavelength of these spheres. Comment on the trend in your wavelengths. For which, if any, of these particles would you expect their wave properties to be important. If the kinetic energy was decreased by 100 times, how would your wavelengths change? Do you think that wave properties would be important under these circumstances?

E1.5 Describe the photoelectric effect experiment.

a) Provide a sketch of the apparatus.
b) State the implications of the experiment.
c) Describe what is observed in the experiment and how it relates to the experiment's implications.

E1.6 Consider the diffraction of photons, electrons, and neutrons from an aperture with diameter *d*. Consider the case where *d* is 1 cm and the case where it is 10^{-7} cm.

a) If you direct a light beam onto the aperture, how large must the wavelength be so that diffraction can be observed? What is the frequency of the light you found?
b) If you direct an electron beam onto the aperture, how large must the speed of the electrons be so that diffraction can be observed?
c) We assume that the de Broglie relationship holds not only for electrons, but also for any particle. How large must the speed of the neutrons be for the aperture to diffract a neutron beam?

Do not be disturbed if the answers to these exercises are not experimentally feasible. The goal is to clarify the content of Equations (1.7) and (1.9).

E1.7 If photons are particles they have momentum. Compute the momentum of a 590 nm photon. Compare this momentum to that of a Na atom moving at a speed of 900 m/s, which is a typical value at 1200 °C. Assume that 590 nm photons collide head on with the sodium atom so that the momentum exchange is twice the photon momentum, how many photons are needed to "stop" the sodium atom?

E1.8 Imagine you build an experimental apparatus in which you can use a photon to eject an electron from an H-atom; i.e., absorb a photon and generate a proton and a free electron. This is called a photoelectron experiment. Imagine that you perform this experiment on single hydrogen atoms, in a chamber where the atoms are surrounded by a thousand electron detectors (numbered *ed1* through *ed1000*), so that all sides (all 4π steradians) are sensed and the detected electron's position can be reported. In more technical language, imagine measuring the full angular distribution of photoejected electrons. In addition, assume that the hydrogen atoms are in their ground electronic state, and that the photons irradiate the sample isotropically with an energy much higher than the ionization energy of the hydrogen atom.

a) If the experiment is performed on a single hydrogen atom, what is the probability that *ed375* detects the photoejected electron? If the experiment is performed on one-hundred hydrogen atoms in succession, what is the probability that *ed375* detects the first photoejected electron? What is the probability that *ed375* detects any photoelectron?
b) Imagine a related experiment in which hydrogen atoms that are initially excited are injected into a chamber and they emit light. Perform experiments of the same type as in part a) but detecting photons instead of electrons. Does your analysis change? Explain!

E1.9 Using Fig. 1.6, calculate the distance x between the intensity maxima, P_1 and P_2, in terms of the wavelength of the incident light and the parameters d and a. If your apparatus has $d = 1.7\ \mu\text{m}$ and $a = 1$ m, and you illuminate the slit with monochromatic light of $\lambda = 640$ nm (red light of a laser pointer), what is the distance x in the figure?

1.7 Problems

P1.1 Consider the following data, taken from O. W. Richardson and K. T. Compton, Phil. Mag. 24 (1913) 575, for the photoemission of electrons from a metal substrate. E_{kin} is the kinetic energy of the photoelectrons and λ is the wavelength of the light. Analyze these data using a least squares analysis. Find the workfunction E_{work} of the metal and determine a value for Planck's constant. The workfunction is the minimum energy that is needed to remove an electron from the metal and place it at the detector some macroscopic distance away.

Sodium		Copper	
E_{kin}/eV	λ/cm	E_{kin}/eV	λ/cm
0.60	43.6×10^{-6}	0.35	26.0×10^{-6}
1.00	36.6×10^{-6}	0.48	25.4×10^{-6}
1.50	31.3×10^{-6}	0.73	23.0×10^{-6}
2.30	25.4×10^{-6}	1.02	21.0×10^{-6}
3.00	21.0×10^{-6}	1.25	20.0×10^{-6}

P1.2 Find a formula for the amplitude distribution on the screen in the double-slit experiment in Fig. 1.6. For arbitrary distances r_1 and r_2 between the slits and the point P, the resulting wave displacement φ on the screen can be written as a superposition of the waves φ_1 and φ_2 originating from each slit, so that

$$\varphi = \varphi_1 + \varphi_2 = A\ \sin\left(\frac{2\pi\left(vt - r_1\right)}{\lambda}\right) + A\ \sin\left(\frac{2\pi\left(vt - r_2\right)}{\lambda}\right)$$

Show that the resulting waveform can be written as

$$\varphi = 2A\ \sin\left(\frac{\pi\left(2vt - r_1 - r_2\right)}{\lambda}\right)\ \cos\left(\frac{\pi\left(r_2 - r_1\right)}{\lambda}\right) = \psi(x)\ \sin\left(\frac{\pi\left(2vt - r_1 - r_2\right)}{\lambda}\right)$$

where the wave's amplitude is $\psi(x) = 2A\cos\left(\pi\left(r_2 - r_1\right)/\lambda\right)$. Simplify this expression for $\psi(x)$ further by realizing that $r_2 - r_1 = dx/a$ for $x \ll a$. Identify where the wave amplitude has extrema: minima and maxima.

P1.3 The essential ideas of the photoelectric effect can be applied to atoms and molecules, as well. Consider a sample of Ne gas that is exposed to high energy electromagnetic radiation; i.e., light. If a 58.5 eV source of UV light shines on a sample of Ne gas one observes two distinct distributions of photoelectron kinetic energies at 17.5 eV and at 44.5 eV. In addition, if X-ray photons of 1.28 keV shine on a sample of Ne three distinct distributions of photoelectron energies can be seen at 417 eV, 1.24 keV, and 1.27 keV. What information about the electronic structure of a neon atoms can you obtain from these data? Make a hypothesis about your observations. How might you test your hypothesis? Describe a set of experiments that could be performed.

P1.4 This problem explores the shape of a diffraction pattern made by a wave passing through a single slit. The intensity profile $I(\theta)$ of a wave passing through a slit of width w is given by

$$I(\theta) = I_{\max}\left(\frac{\sin\left(\frac{\pi\, w}{\lambda}\ \sin(\theta)\right)}{\frac{\pi\, w}{\lambda}\ \sin(\theta)}\right)^2$$

where $I_{\max}$ is the maximum intensity, λ is the wavelength, and θ is the angular distribution of the intensity pattern.

Using a spread sheet program (e.g., Excel) plot this function over the range of −45° to +45° for five different values of the ratio w/λ. Comment on how the diffraction pattern changes as this parameter is changed. In Exercise E1.4 you calculated de Broglie wavelengths for different particles. Based on those calculations and the results of your diffraction calculations, comment on the likelihood of being able to observe diffraction for the different particles in Exercise E1.4.

P1.5 The intensity distribution behind a slit of width w is a special case of the distribution behind a grating with N grooves spaced at distance d. For a grating the intensity is

$$I(\theta) = const \cdot \left(\frac{\sin\left(\frac{\pi N d}{\lambda}\ \sin(\theta)\right)}{\sin\left[\frac{\pi d}{\lambda}\ \sin(\theta)\right]}\right)^2$$

Calculate the intensity distribution for gratings with different numbers of grooves. Choose $d = 1.7\ \mu m$, $\lambda = 500$ nm and perform calculations for $N = 2$, 10, 100, 500, and 1000. Based on your results comment on how the groove number affects the resolution of a wavelength determination.

P1.6 In the early part of the twentieth century, many of the experiments that laid the foundation for quantum mechanics and wave–particle duality were performed. Because of their importance, many of these findings were recognized by awarding the Nobel Prize. Select a Nobel Prize lecture for one of these "classic" studies, explain, and summarize the findings. See https://www.nobelprize.org.

P1.7 In order to gain some elementary experience with probabilities, consider the following thought experiment. Imagine a thin, rectangular, metal film (ca. 100 nm thick). Now place electron detectors on each side (left and right) of the metal film to face each surface of the metal. Imagine shining photons of energy greater than the metal's work function on the film and collecting the electrons. Assume that we can make the film thick enough so that it behaves like the bulk metal, but thin enough that the light source goes completely through the film. In this case, the electrons could escape out of either face of the metal film with equal probability. Can you predict which face (left or right) emits an electron first? How does the distribution (left versus right) of collected electrons evolve as the experiment proceeds, for the first 100 electrons that are detected?

You can answer these questions by making an analogy to the case of a two-sided coin and denote one side "heads" and the other "tails."

a) If the coin is flipped once, will it give heads? How many times will a coin flip yield heads if it is flipped one-hundred times? Comment on your ability to accurately predict the result of these two sets of experiments.
b) Using your coin, perform the experiment by flipping it one-hundred times and recording the results. Before each time you flip the coin make a prediction of the outcome, record the outcome. What percentage of the time was your result correct?
c) Using your data from b), calculate the fraction of total times heads was obtained, after successive flipping of the coin. Make a plot of your fraction versus the number of times you flip the coin. How rapidly do your data converge to the limiting value? Compare your results to your predictions in part a).
d) Apply your understanding to the electron experiment.

P1.8 In Section 1.4 we discussed how the outcome of many events can be predicted with certainty, but that of individual events cannot, since quantum mechanics only predicts probabilities. Let us use this feature of probability to make a certain prediction. You will need to obtain or borrow a globe of the planet earth, hopefully fairly precise and not too fragile. Can you predict the land-to-water ratio of the planet's surface? Can you measure the land-to-water ratio of the planet's surface? Rather than use a deterministic method (e.g., a ruler), we will use a probabilistic method to determine the land to water ratios. Perform the following experiment a number of different times. Toss the globe into the air (or to a friend) and catch it. Record whether your right index finger landed on water or land. Perform this experiment over (and over) and calculate the fraction of times your right index finger ended up on water. Plot your fraction versus the number of trials and determine the value at which it converges. How does this value relate to the known value? Comment on the agreement between your result and the known value. How can your experiment be improved?

2

Essential Aspects of Structure and Bonding

CHAPTER MENU

2.1 Introduction

Electrons and nuclei bind together to form atoms and molecules. Why does this happen? How can we understand atomic structure and bonding? In this chapter we consider the simplest atom, the H atom, and the simplest example for the formation of a covalent bond, the H_2^+ molecular ion, which is formed from a hydrogen atom and a proton. We proceed by extremely simplifying reality, but preserve essential features of chemical bonding. Our aim is to demonstrate how quantum mechanics provides the key to understanding the structure of atoms and the nature of chemical bonds, and to inspire a desire for a deeper understanding of quantum mechanics and its application to electronic structure, which is developed in Chapters 3–8.

Through the simplified approximate approach presented here, three essential features of molecular electronic structure are revealed.

- The size and shape of atoms and molecules are determined by two counteracting effects: lowering of the electrons' potential energy by attractive Coulomb forces with the nuclei and a counteracting increase in the electrons' kinetic energy because of their confinement near the nuclei. For H_2^+, the Coulomb forces (attractive between the protons and the electron and repulsive between protons) result in a net attraction which localizes the electron near the protons; that is, the potential energy E_{pot} relative to an isolated electron and protons is negative. However, the localization of the electron near the protons causes its de Broglie wavelength Λ to decrease, which increases its speed (wave–particle duality) and hence its kinetic energy.
- The lowest energy state of an atom or molecule (ground state) is determined by varying its geometry to minimize its energy, $E = E_{pot} + E_{kin}$ (Variational Principle). When applying this principle to H and H_2^+ with simple approximate wavefunctions, many of the observed properties of H and H_2^+ are well described.
- Electrons can exist only at discrete energies (or states) because of wave–particle duality. This result is in contrast to what follows for systems described by classical mechanics where the excitation energy can assume arbitrary values.

Principles of Physical Chemistry, Third Edition. Hans Kuhn, David H. Waldeck, and Horst-Dieter Försterling.

Companion website: www.wiley.com/go/Kuhn/physicalchemistry3e

2.2 Distinct Energy States

In this section we describe two experimental studies of sodium (Na) atoms that reveal the internal energy structure of the atom. One of the experiments probes the atom by observing its interaction with photons (light quanta), and the other experiment probes the atom by observing its interaction with electrons. Many modern experimental tools apply this same approach, in that they use electrons and photons to probe the internal structure of atoms, molecules, and materials.

2.2.1 Atomic Spectra

If an atom absorbs energy from its environment, for example through the absorption of light (typically visible or UV light) or by collisions in a gas discharge, then the electron(s) can exist in excited states of the atom. The atom's electron(s) can return to the ground state by the emission of light (Fig. 2.1). The dependence of the light's emission (or absorption) probability on the wavelength of the light is called the *spectrum*. The study of spectra and their relation to atomic and molecular structure is called *spectroscopy*. In the early development of spectroscopy such spectra were recorded on photographic film (see Fig. 2.1c). Because of energy conservation the energy of an emitted photon must be the same as the difference ΔE between the higher energy state and the lower energy state of the atom. Hence we have

$$\Delta E = h\,\nu = \frac{h\,c_0}{\lambda} \tag{2.1}$$

or the emitted light has the wavelength

$$\lambda = \frac{h\,c_0}{\Delta E} \tag{2.2}$$

Thus, the structure in the spectrum (i.e., variation of the emission or absorption probability with wavelength) reports on the internal energy structure of the atom.

By way of example consider an atomic vapor of Na, created by heating Na metal to 1200 K in a vacuum sealed chamber. Once in the gas phase it is possible to measure the absorption of light by individual Na atoms (the excitation process in Fig. 2.1b). Alternatively, we can excite the atoms in some way (e.g., an electrical discharge is used in conventional Na lamps for outdoor lighting) and observe their emission of light (the emission process in Fig. 2.1b) as a function of wavelength. The fact that only photons of discrete energies are observed implies that the electrons in the atom only exist in particular energy states. We can use the observable pieces of the line spectrum to identify the energy differences between the atom's internal energy states and construct an energy level profile of an atom or molecule. As a puzzle to piece together, this can be quite challenging. Nevertheless, such puzzles have been solved for most atoms and many molecules, and these spectra are in excellent agreement with quantum mechanical predictions. This field of endeavor is one of our most powerful tools for studying the internal energy structure of atoms and molecules.

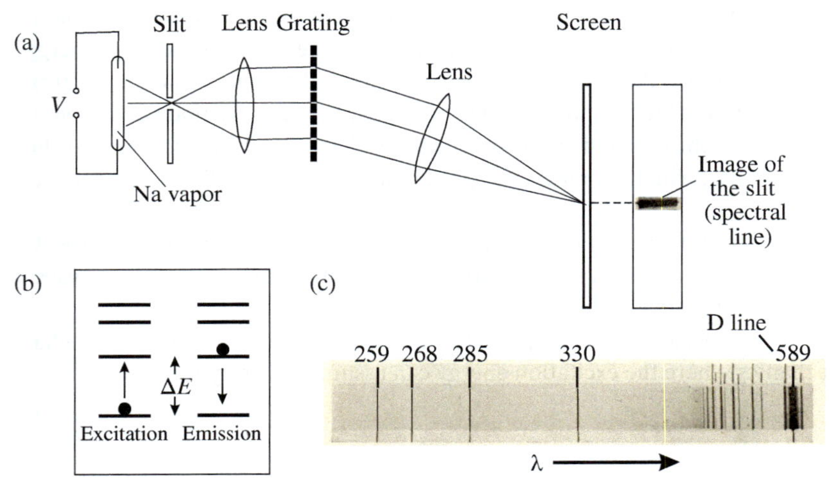

Figure 2.1 (a) Excitation of Na vapor in an evacuated tube by application of a voltage V (glow discharge) and measurement of the wavelength of emitted light using a spectrometer. (b) Energy absorption (excitation from the glow discharge) and energy emission (emission of light with wavelength $\lambda = hc_0/\Delta E$). (c) Emission spectrum of Na vapor. The strong line (Na D line at $\lambda = 589$ nm) and the lines at $\lambda = 259$, 268, 285, and 330 nm correspond to electron transitions into the ground state. The other emission lines in the spectrum arise from electron transitions into excited electronic states (see Chapter 4). ([1], Herzberg, 1937/Dover Publications.)

2.2.2 Franck–Hertz Experiment

Consider the setup in Fig. 2.2a (Franck–Hertz experiment). In a tube filled with Na vapor, electrons are generated by an electron gun (i.e., emitted from the heated cathode and accelerated from the cathode to the grid by the applied voltage V_{grid}) with a kinetic energy

$$E_{\text{kin}} = eV_{\text{grid}} \tag{2.3}$$

The electrons reach the anode if V_{grid} is greater than the opposing voltage V_0 (a typical value being 0.5 V) between grid and anode. On the way to the anode the electrons can collide with Na atoms. If E_{kin} is too small, then the electron cannot excite the sodium atom's electrons and no energy is transferred. In this case the collisions are elastic; that is, the total kinetic energy of the collision is conserved. Moreover, the mass of a sodium atom is much larger than the mass of an electron, so that the loss in the electron's kinetic energy in the collision is negligibly small. Hence, electrons reach the anode with the energy $e(V_{\text{grid}} - V_0)$. The observed current in the measurement rises as V_{grid} increases because more electrons leave the negative charge cloud around the heated cathode. At low grid potentials, V_{grid} small, we observe that the current rises monotonically until it reaches a value where it "levels off" and then decreases. As the grid potential increases further, the current again rises until another drop off is observed, and this behavior continues (see Fig. 2.2b). Moreover, the voltages where the dips occur depend on the nature of the atoms or molecules placed in the apparatus.

We can understand this spectrum by realizing that the internal energy states of atoms and molecules are "quantized"; that is, they have discrete energies. If an electron's kinetic energy E_{kin} is greater than or equal to the energy ΔE needed to excite an electron in a Na atom, then it can transfer some of its energy to the Na atom in the collision; i.e., an inelastic collision can occur. When such a collision has E_{kin} similar to ΔE, the electron's kinetic energy will drop low enough that the electron cannot reach the anode because of the opposing voltage V_0. The observable effect is that the current suddenly decreases (Fig. 2.2b) upon reaching the voltage

$$V_{\text{grid}} = \frac{\Delta E}{e} \tag{2.4}$$

With further increase of V_{grid}, the current rises, because an increasing fraction of the electrons have enough kinetic energy remaining after a collision to reach the anode. This increase lasts until

$$V_{\text{grid}} = \frac{2\Delta E}{e} \tag{2.5}$$

when an electron can transfer the energy ΔE to each of two Na atoms by consecutive collisions. Through continuation of this multiple collision process we obtain a current versus voltage curve with distinct minima and the voltage separation between the minima corresponds to the excitation energy ΔE of the Na atoms ($\Delta E = 3.37 \times 10^{-19}$ J = 2.10 eV).

In addition to the energy loss process observed for the electrons, the Na atoms gain energy that corresponds to electronic excitation. In fact, we can think of the Franck–Hertz apparatus as an electrical discharge that creates excited electronic states of atoms. These electronically excited states relax to the ground state and release energy by the emission of light. In the case of Na, the light emission is primarily yellow with a wavelength $\lambda = 589$ nm, corresponding to $h\nu = 3.37 \times 10^{-19}$ J = 2.10 eV.

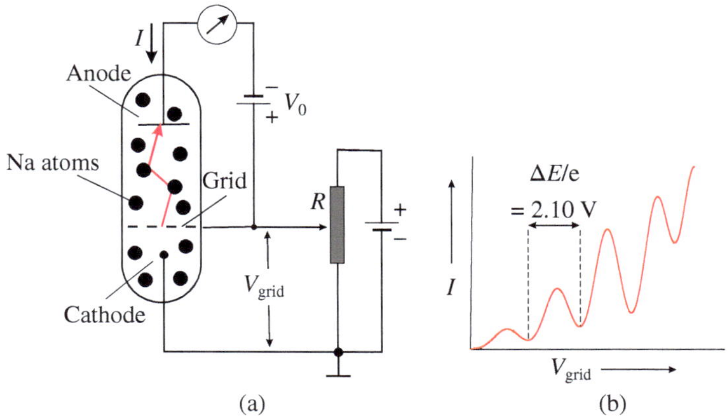

Figure 2.2 Franck–Hertz experiment. (a) Experimental arrangement. R is a variable resistor tuning the grid voltage V_{grid}. The opposing voltage V_0 is 0.5 V. The red line symbolizes the electron path from the cathode to the anode having passed the grid. (b) Anode current I as a function of grid voltage V_{grid}; $\Delta E = 2.10$ eV is the excitation energy of Na.

2.3 Standing Waves

The presence of discrete, or quantized, electronic energy levels can be seen to arise from the wave character of the electron and a wave's resonance behavior. In particular, we find that resonant electron wavelengths, or standing waves, can be associated with the electronic energy states of the atom.

2.3.1 Particle Between Parallel Walls

Electrons in atoms and molecules are confined to a small region around the nuclei by Coulomb forces, that is, the potential well of the electron–nuclear attraction. We consider the simplest case, an H-atom, where there is an electron in the Coulomb field of a proton. In a first attempt to understand the essence of what happens to the electron we carry out a thought experiment, in which we replace reality by an extremely simplifying assumption, but retain the essential feature of an energy well. Rather than realistically describe the Coulomb potential of the proton which traps the electron in an energy well, we consider the electron to be trapped inside a cubic box of corresponding size.

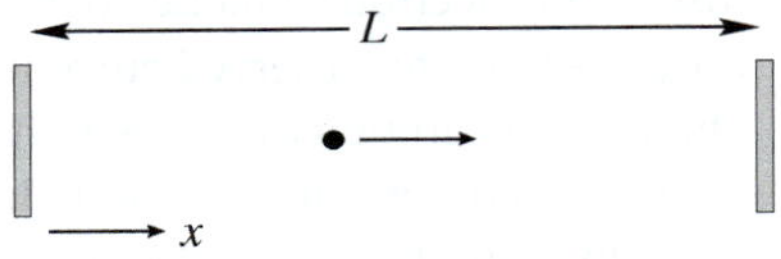

Figure 2.3 Classical view (equal probability to find an electron anywhere between the walls): the electron reflected back and forth between two parallel walls a distance L apart.

Before considering a three-dimensional box, let us simplify the considerations further by restricting our discussion to a one-dimensional box. We ask the following question: What happens to an electron that finds itself between two parallel walls, and is moving with the velocity v perpendicular to the walls (Fig. 2.3)? This simplification allows us to focus on essential features of an electron in a box. Later, in Sections 2.3.3, 2.4, and 2.5.2, we will use a three-dimensional box.

According to Equation (1.7) we can describe the electron as a de Broglie wave of wavelength

$$\boxed{\Lambda = \frac{h}{m_e v}} \tag{2.6}$$

This wave is reflected back and forth between the walls, and the incident wave interferes with the reflected wave. Under most conditions, the superposition of incident and many-fold reflected waves results in a vibration with small, irregularly changing amplitude, because the waves almost cancel each other. For certain values of Λ the incident and reflected waves add constructively. These particular wavelengths are the resonances of the electron wave in the box, and these resonant wavelengths determine the quantized energy levels of the box.

2.3.1.1 Vibrating String

We can find these resonances from the analogous case of a vibrating string that is fixed at both ends (Fig. 2.4). A standing wave appears (i.e., the zeroes and amplitude maxima of the waves remain at the same x positions) if the condition

$$\frac{\Lambda}{2} = L \tag{2.7}$$

is fulfilled, where L is the length of the string. The displacement $\varphi(x, t)$ of a standing wave is

$$\varphi(x, t) = A \cdot \cos \omega t \cdot \sin\left(\frac{\pi x}{L}\right) \tag{2.8}$$

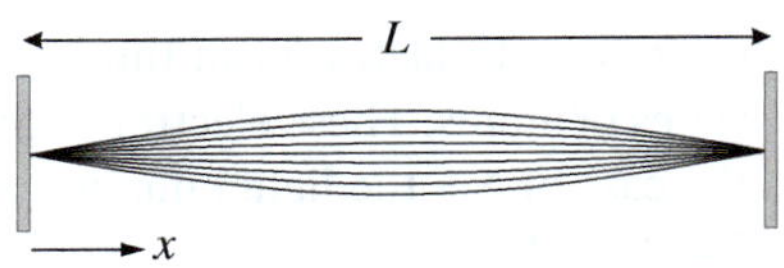

Figure 2.4 Displacement $\varphi(x, t)$ and amplitude $\psi(x)$ of a vibrating string. Each curve displays the displacement at a particular time (fundamental vibration: $\Lambda = 2L$).

where A is a constant and $\omega = 2\pi\nu$. This functional form describes a wave that oscillates in time with a frequency ν and has a sinusoidal space dependence. The amplitude $\psi(x)$ is obtained from Equation (2.8), if we set

$$\cos \omega t = \pm 1 \tag{2.9}$$

(this is the case for $\omega t = 0,\ \pi, 2\pi, \ldots$):

$$\psi(x) = A\ \sin\left(\frac{\pi x}{L}\right) \tag{2.10}$$

We call $\psi(x)$ the spatial part of the wavefunction, or just the wavefunction. *Note that the standing wave along the string is a physical example that we developed to illustrate the mathematical formalism; however, it is no more than an aid that helps us to grasp the features of the basic postulate, which is expressed by the mathematical formalism (Fig. 1.13).*

2.3.1.2 Wavefunction $\psi_n(x)$ and Kinetic Energy T_n (Potential Energy $V = 0$)

For an electron between two parallel walls, we can calculate the speed and hence its kinetic energy by way of the de Broglie relation. From the standing wave's characteristics we see that $\Lambda = 2L$ is the largest possible de Broglie wavelength that also meets the resonance condition of the box. From Equations (2.6) and (2.7), we can determine the speed, and hence the kinetic energy for an electron with this de Broglie wavelength

$$\frac{1}{2}m_e v^2 = \frac{1}{2}m_e\left(\frac{h}{m_e\Lambda}\right)^2 = \frac{1}{2}m_e\left(\frac{h}{m_e 2L}\right)^2 = \frac{h^2}{8m_eL^2} \tag{2.11}$$

We can generalize these results if we consider other possible resonant wavelengths of the vibrating string (Fig. 2.5) and transfer the results to standing waves of electrons. The resonant condition for any standing wave between the walls is

$$\frac{\Lambda}{2} = \frac{L}{n}, \quad n = 1,2,3,\ldots \tag{2.12}$$

This relationship ensures that the wavefunction has a zero at each of the walls and has $n - 1$ nodal points (x values where the wave function $\psi(x)$ changes sign). Sometimes these resonant wavelengths are called modes and the integer n is the mode number. We can generalize the expression (2.10) for the wavefunction

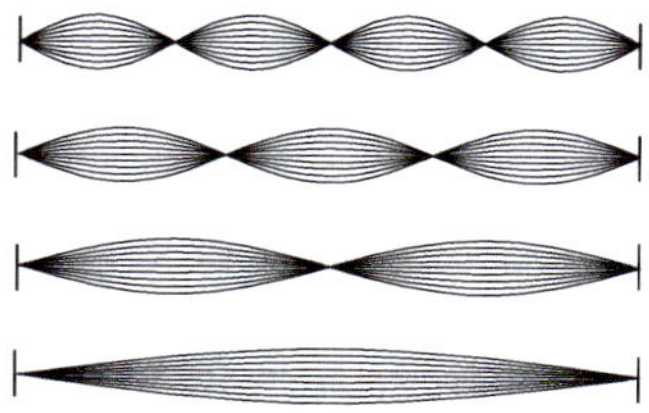

Figure 2.5 Various vibrational states of a string. The $n = 1$ case corresponds to Fig. 2.4.

$$\boxed{\psi_n(x) = A\,\sin\left(\frac{n\pi x}{L}\right)} \tag{2.13}$$

and the expression (2.11) for the kinetic energy

$$\boxed{T_n = \frac{h^2}{8m_eL^2}n^2} \tag{2.14}$$

in terms of the mode number and the length L.

The total energy E of the electron is the sum of the kinetic and potential energies

$$E = T + V \tag{2.15}$$

The potential energy V of an electron in our case does not depend on its position between the walls. If we arbitrarily set $V = 0$, this fixes the zero of energy to the bottom of the box and we obtain the energy diagram in Fig. 2.6 for an electron between two parallel walls. Although the energy of the electron does not depend on the position between the walls, the

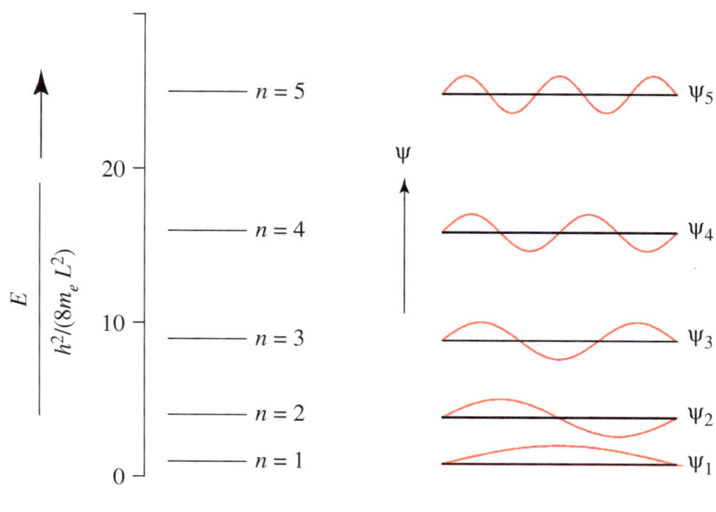

Figure 2.6 Energies E and wavefunctions ψ for an electron between to parallel walls. The energy diagram on the left side of the figure plots the energies for the first five energy levels of the particle in a box, represented as horizontal lines and labeled by the quantum number, n. The diagram on the right plots the wavefunctions, ψ_n, associated with each of these levels in red. The black horizontal line in these plots corresponds to the value zero for the amplitude.

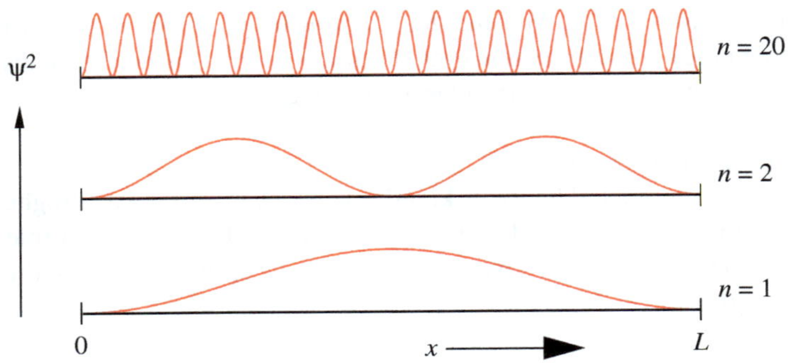

Figure 2.7 ψ^2 (proportional to the probability of finding an electron) as a function of x for an electron between two parallel walls a distance L apart. Comparison of low energy states ($n = 1$, $n = 2$) to a high energy state ($n = 20$).

probability of finding the electron does depend on position. This dependence is apparent in Fig. 2.7, where ψ^2 for an electron between two parallel walls is plotted for $n = 1$, $n = 2$, and $n = 20$. Remember from Section 1.3 that the probability of finding an electron in a region of space is proportional to ψ^2.

2.3.1.3 Electron Cannot Be at Rest: Possible Energies and Probability Distributions

A number of useful observations can be made about this result. First, we see that the electron has kinetic energy even in the lowest energy state ($n = 1$); that is, it cannot be at rest. The case $n = 0$ is not possible because the wave amplitude would be zero everywhere, hence zero probability of finding an electron (i.e., the box is empty). The residual ground-state energy is called the *zero point energy*.

Second, the electron's energy depends on the box size because the box determines the wavelength that the electron can have, by the resonance condition, Equation (2.12). For a box of molecular size, this energy amounts to $E_1 = 6.0 \times 10^{-18}$ J for $L = 100$ pm, whereas it is extremely small for a box of macroscopic size, $E_1 = 6.0 \times 10^{-34}$ J for $L = 1.0$ cm.

Finally, although it is expected classically that the particle should be found with equal probability anywhere between the walls (because the speed is the same everywhere), we see obvious maxima and minima in the wave mechanical treatment, shown in Fig. 2.7. For the ground state it is more probable to find the electron in the middle of the box than on the edges, whereas the electron is equally probable in the middle and near the edges at high energies.

Example 2.1 *Model the outer electron of a Na atom as being confined within two walls with distance $L = 500$ pm. Calculate the energy difference between the ground state ($n = 1$) and the first excited state ($n = 2$) and compare it with the experimental excitation energy of 3.4×10^{-19} J.*

According to Equations (2.14) and (2.15) the energies of the electron are

$$E_n = \frac{h^2}{8 m_e L^2} n^2$$

With $L = 500$ pm we obtain

$$E_n = \frac{(6.62 \times 10^{-34})^2 \mathrm{J^2 s^2}}{8 \cdot 9.11 \times 10^{-31}\ \mathrm{kg} \cdot (500\ \mathrm{pm})^2} n^2 = 2.41 \times 10^{-19}\ \mathrm{J} \cdot n^2$$

and

$$\Delta E = E_2 - E_1 = 2.41 \times 10^{-19}\ \mathrm{J} \cdot (2^2 - 1^2) = 7.2 \times 10^{-19}\ \mathrm{J}$$

This value is on the order of the experimental excitation energy.

2.3.2 Ill-Posed Questions

For the $n = 2$ wavefunction, the question often arises as to how the electron travels from one area of large probability to the other, as there is a location with $\psi^2 = 0$ between them. The postulates of quantum theory allow us to describe the electron in a certain sense as a wave and in another sense as a particle. However, one cannot say that the electron is a particle, and therefore the question of how a particle travels from one area to another is nonsensical in the context of the theory.

The postulates of quantum theory are a basis from which we describe phenomena in nature; questions that extend beyond the postulates' boundaries are not answerable within the theory and make no sense.

2.3.3 Electron in a Cubic Box

Our considerations of the electron between two walls at distance L can be extended to the three-dimensional case (this will be shown in detail in Chapter 3). For now we restrict our discussion to the lowest quantum state of an electron in a cubic box.

For an electron between two walls the kinetic energy in the lowest energy state ($n = 1$) is given by Equation (2.11). For an electron in a cubic box (edge length L) the kinetic energy is the sum of the contributions in the x-, y-, and z-directions:

$$T_1 = \frac{h^2}{8m_e L^2} \times 3 \tag{2.16}$$

The corresponding wavefunction can be written as the product

$$\psi(x, y, z) = A \sin\left(\frac{\pi x}{L}\right) \sin\left(\frac{\pi y}{L}\right) \sin\left(\frac{\pi z}{L}\right) \tag{2.17}$$

where

$$A = \frac{8}{L^3}$$

is a constant (for details see Section 3.3.2) and the side lengths are $L_x = L_y = L_z = L$ for a cubic box. The wavefunction equals zero at the walls (i.e., for $x = 0$ and $x = L$, for $y = 0$ and $y = L$, and for $z = 0$ and $z = L$), and it has a maximum at the center of the cube ($x = y = z = L/2$). According to Chapter 1, the probability of finding the electron at a given point is proportional to $\psi^2(x, y, z)$. Therefore, this probability has a maximum at the center of the cube and it decreases toward the walls. The probability distribution (electron cloud) is plotted in Fig. 2.8a.

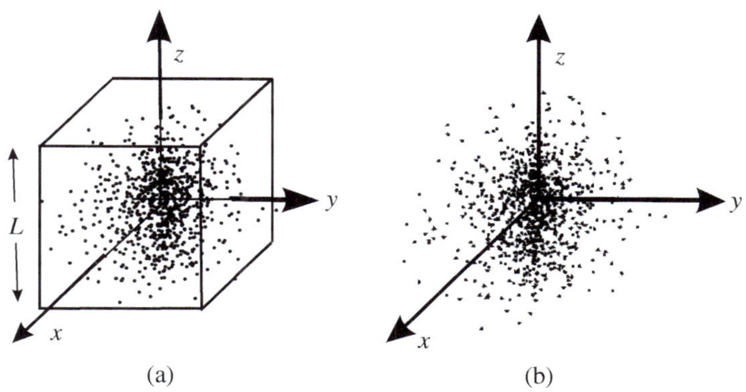

Figure 2.8 Ground-state H atom. Panels (a) and (b) show electron cloud density plots. (a) Variational calculation with box wavefunctions and (b) exact calculation (1s function), see Equation (2.22).

2.4 Ground-State H Atom

The hydrogen atom consists of a proton (mass = 1.6726×10^{-27} kg, charge = $+e$) and an electron. The force to pull the electron apart from the proton is, according to Coulomb's law,

$$f(r) = \frac{1}{4\pi\varepsilon_0}\frac{e^2}{r^2} \tag{2.18}$$

Here, r is the distance from the electron to the nucleus, e is the electron charge, and ε_0 is a constant called the *permittivity of free space*; $\varepsilon_0 = 8.8542 \times 10^{-12}\ \mathrm{C^2\ J^{-1}\ m^{-1}}$. To find the potential energy of the electron in the field of the proton, we integrate over the force and find

$$V(r) = \int_{\infty}^{r} f(r)\,\mathrm{d}r = -\int_{r}^{\infty} f(r)\,\mathrm{d}r = -\frac{1}{4\pi\varepsilon_0}\frac{e^2}{r} \tag{2.19}$$

where we define $V = 0$ at $r = \infty$. Figure 2.9 plots this potential energy curve.

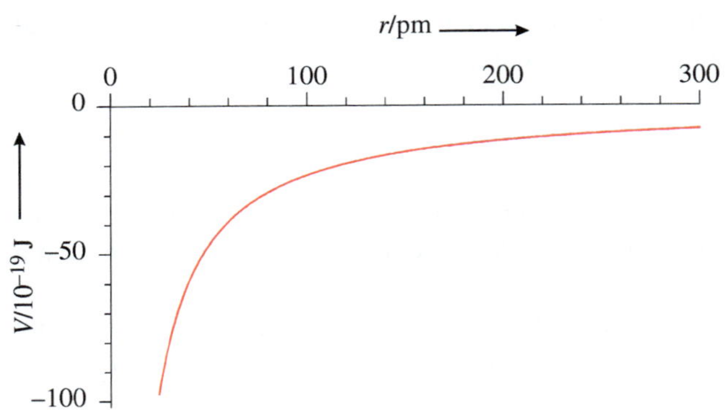

Figure 2.9 Potential energy V of an electron in distance r from a proton (Coulomb field).

2.4.1 Result of Rigorous Treatment (See Chapter 4)

The H atom is the simplest of all atoms, and therefore its behavior is the easiest to understand theoretically. Here we consider the result of the rigorous quantum mechanical treatment of the H atom to compare between reality and the simple box model. You will see again and again (in particular in Chapters 7, 8, 9, and 13) how powerful the box model is in describing molecular features.

The rigorous treatment of the H atom (see Chapter 4) provides a ground-state electronic energy that is given by

$$E = -\frac{1}{4\pi\varepsilon_0}\frac{e^2}{2\,a_0} = -21.8 \times 10^{-19}\ \text{J} = -13.6\ \text{eV} \tag{2.20}$$

where

$$a_0 = 4\pi\varepsilon_0 \frac{h^2}{4\pi^2\, m_e\, e^2} = 52.9\ \text{pm} \tag{2.21}$$

is called the *Bohr radius*. The probability density $\rho(r)$ (see Fig. 2.8b) is found to be

$$\rho(r) = \psi^2(r) = \frac{1}{\pi\, a_0^3}\ \exp\left(-2r/a_0\right) \tag{2.22}$$

The distance r of the electron from the nucleus, measured on many identical systems, has a mean value $\overline{r}$, where the bar symbolizes an average,

$$\overline{r} = \int_{\text{all space}} r \cdot \rho(r)\ \mathrm{d}\tau \tag{2.23}$$

$\rho(r)\ \mathrm{d}\tau$ is the probability of finding the electron in the interval $\mathrm{d}\tau$ ($\mathrm{d}\tau = 4\pi r^2\ \mathrm{d}r$), that is, at a distance between r and $r + \mathrm{d}r$. In Problem 2.8, you will use these results to show that

$$\overline{r} = \frac{3}{2}a_0 = 79.4\ \text{pm} \tag{2.24}$$

The energy E divides into contributions of potential energy V and kinetic energy T. On average $E = \overline{V} + \overline{T}$. In order to obtain $\overline{V}$ we consider the electric charge that is in volume element $\mathrm{d}\tau$, which is $-e\rho\ \mathrm{d}\tau$. According to Equation (2.19) the contribution of this charge to the potential energy is

$$\mathrm{d}V = -\frac{e}{4\pi\varepsilon_0}\frac{e\rho\ \mathrm{d}\tau}{r}$$

so that

$$\overline{V} = -\frac{e^2}{4\pi\varepsilon_0}\int_{\text{all space}} \frac{\rho}{r}\mathrm{d}\tau \tag{2.25}$$

In Problem 2.5, you will use Equation (2.22) to evaluate this integral and show that

$$\overline{V} = -\frac{e^2}{4\pi\varepsilon_0 a_0}$$

Comparison of this result with the result for the total energy in Equation (2.20) gives the following relationships.

$$\overline{V} = 2E = -\frac{e^2}{4\pi\varepsilon_0 a_0} \quad \text{and} \quad \overline{T} = E - \overline{V} = \frac{e^2}{8\pi\varepsilon_0 a_0}$$

Thus, the binding of the electron to the nucleus has a strong negative contribution from the potential energy and a positive contribution from the kinetic energy. In fact, half of the energy stabilization which arises from the decrease in potential energy is counteracted by an increase in the kinetic energy.

As we will see later, the results obtained for the H atom can be transferred in a straightforward manner to more complicated electronic systems. For this reason it is important to have a firm grasp of the structure of the H atom, and it is useful to begin in a simplified manner that will pave the way for an understanding of more complex systems. The probability distribution of Equation (2.22) is similar to the distribution of the electron in a cubic box with side length $L = 7.1\ a_0$, or 376 pm; see Fig. 2.8. The Coulomb field of the nucleus keeps the electron in its proximity, similar to the walls of the cubic box. Can we understand the similarity of the electron clouds in these two cases?

2.4.2 Box Model for H Atom and the Variational Principle

Imagine an electron cloud, or wave, of any size and shape and its corresponding wavefunction ϕ. This wavefunction can be assigned to an energy ε. The variational principle states that

> *the true ground-state wavefunction ψ provides the minimum energy E, all other imagined functions ϕ must have an energy ε that is greater than the energy E. Only in the case of $\phi = \psi$ is the energy ε equal to E.*

The variational method considered here is of crucial importance in quantum chemistry when using approximation methods to describe the electronic structure of molecules. For this reason we apply it to the H atom where a comparison with the exact treatment can be given. In Chapter 5, we will give a rigorous proof of this fundamental principle. For the time being we consider the variational principle as a postulate.

Let us choose an arbitrary function ϕ as an initial guess; we call it a trial function. In the case of ε considerably larger than E, the trial function ϕ will be a bad approximation of the true wavefunction ψ. By adjusting the size and shape of ϕ, we can lower its energy ε, and we can be sure that the resulting energy ε will not be lower than E. We expect that our optimized trial function ϕ will be a reasonable facsimile of the true ground-state wavefunction ψ. However, this is not guaranteed, because the true energy minimum E is usually not known.

Because the electron in the Coulomb field of a proton has a wavefunction qualitatively like an electron in a cubic box in its lowest energy state (Fig. 2.8), we use Equation (2.17) as the trial function. The side length L is a parameter that will be determined by minimizing the energy $\varepsilon = \overline{V} + \overline{T}$ (variational method).

2.4.2.1 Average Potential Energy

In order to obtain $\overline{V}$ we restrict our model to have the proton placed in the center of a cubic box, and we view the electron cloud in Fig. 2.8a as an electron in the Coulomb field of the proton. In this view L is just a parameter that controls the spatial extent of the wavefunction. To find the average potential energy $\overline{V}$ of the electron we must calculate the potential energy of the electron cloud in the Coulomb field of the proton. Before proceeding rigorously, let us first make an estimate. As $L/2$ is about the maximum distance of the electron from the center of the box in Fig. 2.8a, it seems reasonable to approximate the average distance $\overline{r}$ by

$$\overline{r} = \tfrac{1}{4}L$$

By inserting this distance in Equation (2.19) we obtain

$$\overline{V} = -\frac{1}{4\pi\varepsilon_0} \cdot \frac{4e^2}{L}$$

In the rigorous evaluation of Equation (2.25), but using the wavefunction (2.17) of an electron in a cubic box, $\tfrac{1}{4}L$ has to be replaced by $\frac{1}{4.18}L$ (see Justification 2.1, on the book's website):

$$\overline{V} = -\frac{1}{4\pi\varepsilon_0} \cdot \frac{4.18\,e^2}{L} \tag{2.26}$$

2.4.2.2 Average Kinetic Energy

In general, the kinetic energy T assigned to a test function ϕ is dependent on the position of the electron (see Chapter 4), and $\overline{T}$ is obtained by integrating over the space. The box wavefunctions have the very special property that T is independent of position, and $\overline{T}$ equals T. Then, from Equation (2.16) we obtain

$$\overline{T} = \frac{3h^2}{8m_e L^2}$$

For this reason the variational principle can be correctly applied without going into details of quantum-mechanical formalism.

2.4.2.3 Minimization of Total Energy

The total energy ε is a sum of average kinetic and potential energy and may be written as

$$\varepsilon = \overline{V} + \overline{T} = -\frac{1}{4\pi\varepsilon_0}\frac{4.18\, e^2}{L} + \frac{3h^2}{8m_e L^2} \tag{2.27}$$

For sufficiently large L values, corresponding to the proton and electron at infinite distance from each other, the total energy ε becomes zero. For sufficiently small L values, the total energy becomes positive (the term L^2 for the kinetic energy overcomes the term L in the expression for the potential energy) and, for intermediate L values, the total energy becomes negative (Fig. 2.10).

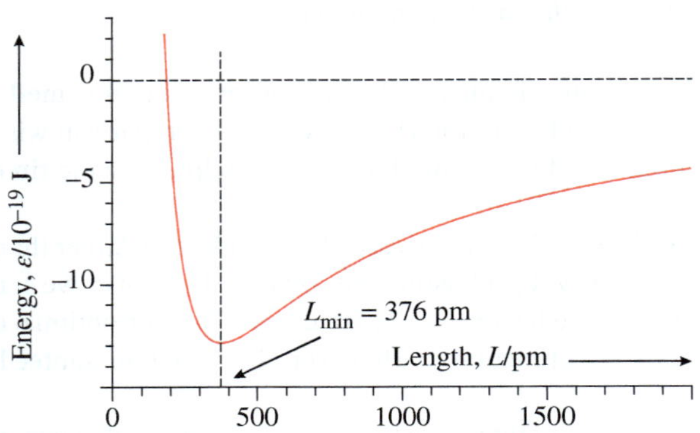

Figure 2.10 Energy found for the H-atom with particle-in-a-box trial wavefunctions. The energy ε is plotted versus L (which is now a parameter that controls the spatial extent of the trial wavefunction) according to Equation (2.27). The minimum energy occurs at $\varepsilon_{min} = -12.9 \times 10^{-19}$ J and corresponds to $L_{min} = 376$ pm, compared to $\varepsilon_{min} = -21.8 \times 10^{-19}$ J for the exact energy of the H-atom.

Decreasing the size of the electron cloud results in a decrease of the potential energy $\overline{V}$ (the electron is more strongly attracted by the nucleus); however, at the same time the kinetic energy $\overline{T}$ increases (the electron is confined to a smaller space). The actual size of the electron cloud follows from a compromise between nuclear attraction and kinetic energy, such that the total energy reaches a minimum. According to Fig. 2.10 and as outlined in Box 2.1, the minimum of the energy, ε_{min}, is reached for

$$\varepsilon_{min} = -\overline{T}_{min} = \frac{1}{2}\overline{V}_{min} = -\frac{1}{4\pi\varepsilon_0}\frac{4.18\, e^2}{2\, L_{min}} = -12.9 \times 10^{-19}\ \text{J} \tag{2.28}$$

with

$$L_{min} = 4\pi\varepsilon_0 \frac{3h^2}{16.7 m_e e^2} = 376\ \text{pm} = 7.1 \cdot a_0 \tag{2.29}$$

The result shows that, on average, half of the Coulomb energy gained in bonding to the proton is converted into kinetic energy. Hence, the net binding energy is $-\varepsilon_{min} = -\overline{V}_{min}/2$. It should be emphasized that this agrees completely with the result obtained in the exact quantum-mechanical calculation ($-E = -\overline{V}/2$). The value obtained for the energy in Equation (2.28) differs substantially from the exact result, namely, by a factor of nearly two. On the other hand, following Equation (2.29), the characteristic average distance of the electron from the nucleus ($\overline{r} = \frac{1}{4.18}L_{min} = \frac{1}{4.18} \cdot 7.1 \cdot a_0 = 1.70 \cdot a_0 = 90$ pm) is near to the average distance calculated for the exact solution ($\overline{r} = \frac{3}{2}a_0 = 79.4$ pm).

Although the box wavefunction, Equation (2.17), mimics the correct probability density qualitatively (compare Figs. 2.8a,b), a more quantitative comparison shows strong deviations from the exact value. At the position of the proton, the density is, according to Equation (2.22),

$$\rho(r = 0) = \psi^2(r)|_{r=0} = \frac{1}{\pi\, a_0^3}$$

for the exact solution, while it is, according to Equations (2.17) and (2.29),

$$\psi^2\left(x, y, z = \frac{L_{\min}}{2}\right) = A^2 = \frac{8}{L_{\min}^3} = \frac{8}{7.1^3 \cdot a_0^3} = \frac{1}{44.7 \cdot a_0^3}$$

for the box model. The exact solution shows much higher (14 times) values for the probability density near the proton than does the box wavefunction, and it shows a steeper decay with distance than does the box wavefunction.

Although introduced in an ad hoc manner, it is instructive to make a few observations about the rigorous result. According to Equation (2.20) the binding energy of the electron to the proton is

$$-E = 21.8 \times 10^{-19}\ \mathrm{J} = 13.6\ \mathrm{eV}$$

This energy value agrees with the experimentally determined *ionization energy* of the H atom (i.e., the minimum energy necessary to remove the electron to an infinite distance from the proton). If ground-state H atoms are being bombarded by photons, the photons must have an energy of at least 13.6 eV in order to release the electron that is bound to the nucleus. Also, the kinetic energy $\overline{T}$ of the electron in the H atom is

$$\overline{T} = \frac{m_e}{2}\overline{v^2} = -E = 21.80 \times 10^{-19}\ \mathrm{J} = 13.6\ \mathrm{eV} \tag{2.30}$$

This energy corresponds to an average speed of 2×10^6 m s $^{-1}$. Note that we can only speak of a speed in the sense that the electron behaves as if it were a moving particle.

2.4.2.4 Electron Speed and Electron Described as a Cloud

The electron's speed, 2×10^6 m s^{-1}, is large compared to the speed with which atoms and molecules move ($\approx 10^3$ m s^{-1} at room temperature, see Chapter 14). Often, we will find that it is reasonable to consider the motion of nuclei to be frozen on the timescale of the electron motion. Imagine the case of two H atoms approaching each other to form an H_2 molecule. The time t required for two H atoms to approach each other by a distance $s = 50$ pm at their average thermal velocity is $t = s/v_{\mathrm{H}} = 5 \times 10^{-14}$ s; during this time the electrons behave as if each electron would traverse the region of its nucleus several thousand times: $t = s/v_e = 50\,\mathrm{pm}/(2.2 \times 10^6\ \mathrm{m\ s^{-1}}) = 2 \times 10^{-17}$ s. Thus the electron in each H atom can be described as a smeared-out cloud of charge $-e$ (electron shell).

In conclusion, the actual electron cloud of an atom or molecule, among all conceivable electron clouds, is the one attributed to the lowest energy. The best approximation among a proposed set of electron clouds is obtained by minimizing the energy assigned to each electron cloud. The energy minimum results from a compromise between the average potential energy $\overline{V}$ ($\overline{V}$ decreases with decreasing size of the electron clouds) and the average kinetic energy $\overline{T}$ ($\overline{T}$ increases with decreasing size). The search for this compromise is a key to understanding the occurrence and properties of atoms and molecules. In the following section this is illustrated for the hydrogen molecular ion, H_2^+. Our procedure is equivalent to what is done in Chapter 6 when applying the Schr ödinger equation. The simple considerations given here are restricted to box wave functions as trial functions where the kinetic energy $\overline{T}$ is given by Equation (2.16).

2.5 Ground-State H_2^+

Now we extend the ideas developed for the H atom and consider the simplest molecule, the H_2^+ molecular ion. H_2^+ is composed of an H atom and a proton (Fig. 2.11). H_2^+ can be produced by the bombardment of H_2 molecules with electrons:

$$H_2 + 1e^- \rightarrow H_2^+ + 2e^- \tag{2.31}$$

or by the photoionization of H_2 with light

$$H_2 + h\nu \rightarrow H_2^+ + e^- \tag{2.32}$$

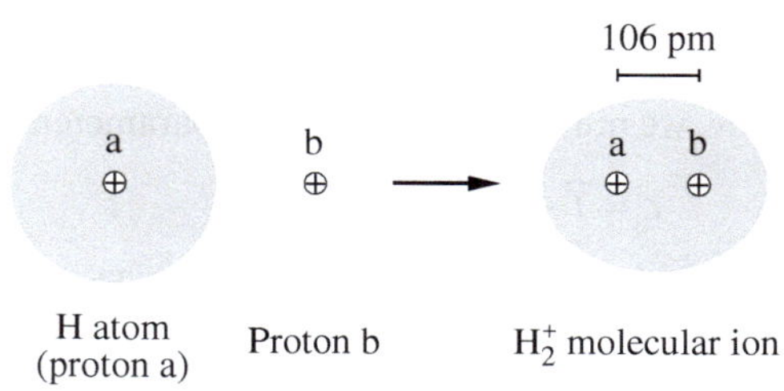

Figure 2.11 Formation of an H_2^+ molecular ion from an H atom (proton a) and a proton b at a distance $d = 106$ pm. Within the gray zone, the charge density is greater than 1% of its value at the nucleus of the H atom.

2.5.1 Forming H_2^+ from an H Atom and a Proton

For H_2^+, the minimum electronic energy is reached at a certain distance between the two protons ($d = 106$ pm) and the establishment of an electron cloud around both protons. In forming the molecular ion, we imagine that the two protons are pulled together by the attractive electrostatic forces between the electron cloud and the protons until they are balanced by the repulsive electrostatic force between the two protons.

The potential energy V depends on the mutual repulsion of the positive charges on the nuclei at the distance d (given by $e^2/(4\pi\varepsilon_0 d)$, positive contribution) and the attraction of the electron charge to the positive charges of nuclei a and b at distances r_a and r_b (negative contributions):

$$V = \frac{1}{4\pi\varepsilon_0}\frac{e^2}{d} - \frac{e^2}{4\pi\varepsilon_0}\left(\frac{1}{r_\mathrm{a}} + \frac{1}{r_\mathrm{b}}\right) \tag{2.33}$$

The fact that the chemical bond is based on Coulomb attraction of the electron cloud and the protons should be emphasized. Relative to the electron cloud in the H atom, the H_2^+ electron cloud is longitudinally extended (the electron cloud is stretched along the bond axis) and laterally compressed (the electron cloud is contracted in the direction perpendicular to the bond axis) by the electrostatic field of the two protons. The net effect is to increase the average kinetic energy of the electron because the changes from lateral compression are larger than those from longitudinal extension. As we found in forming an H atom from a proton and an electron, exactly one-half of the Coulomb stabilization energy that is gained in forming the bond is lost to the electron's kinetic energy.

2.5.2 Box Trial Functions for H_2^+

In analogy to the H atom we imagine that the H_2^+ molecular ion can be considered as an electron confined to a rectangular box (short axes $L_y = L_z = L$, long axis $L_x = bL$) which encloses the two protons at a distance d from each other and is centered on the box's center (Fig. 2.12). This viewpoint is an aid that we use to motivate our choice of the particle in box wavefunctions (2.17) as trial wavefunctions in our variational calculation. As with the analysis for the hydrogen atom, the box lengths are actually parameters that describe the shape and size of the test wavefunction. The average kinetic energy of the electron in its ground state ($n_x = n_y = n_z = 1$) is the sum of the contributions from the electron's speed along the x, y, and z directions

$$\begin{aligned}\overline{T} &= \frac{h^2}{8m_eL_x^2} + \frac{h^2}{8m_eL_y^2} + \frac{h^2}{8m_eL_z^2} = \frac{h^2}{8m_e}\left(\frac{1}{L_x^2} + \frac{1}{L_y^2} + \frac{1}{L_z^2}\right) \\ &= \frac{h^2}{8m_eL^2}\left(\frac{1}{b^2} + 2\right)\end{aligned} \tag{2.34}$$

The attractive electrostatic energy in a volume element $\mathrm{d}\tau$ (= dx dy dz) is given by

$$\mathrm{d}V_\text{attraction} = \frac{-e^2}{4\pi\varepsilon_0}\left(\frac{1}{r_a} + \frac{1}{r_b}\right)\rho\,\mathrm{d}\tau$$

where $\rho\,\mathrm{d}\tau$ is the probability of finding the electron in volume element $\mathrm{d}\tau$. Using Equation (2.33), the average potential energy is given by

$$\overline{V} = \frac{1}{4\pi\varepsilon_0}\frac{e^2}{d} - \frac{e^2}{4\pi\varepsilon_0}\int_\text{all space}\left(\frac{1}{r_a} + \frac{1}{r_b}\right)\rho\,\mathrm{d}\tau \tag{2.35}$$

Now we must optimize the three parameters L, b, and d to obtain the minimum of the total energy

$$\varepsilon = \overline{T} + \overline{V}$$

by applying the Variational Principle. The result (for details see Box 2.2) is

$$L_\text{min} = L_{y,\text{min}} = L_{z,\text{min}} = 301\ \text{pm}, b_\text{min} = 1.36, L_{x,\text{min}} = 409\ \text{pm}$$

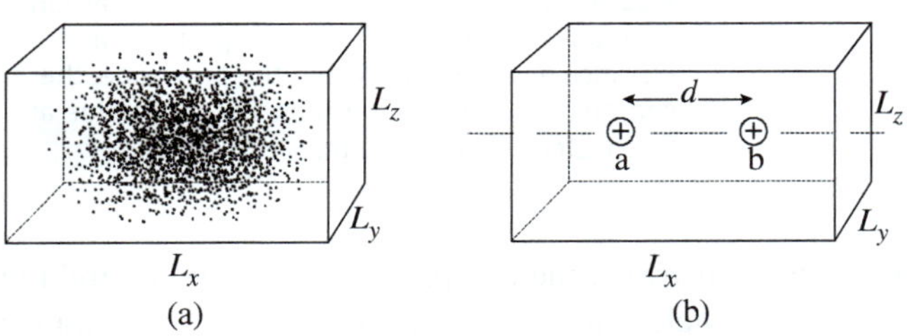

Figure 2.12 Box model of H_2^+. (a) Electron confined to a rectangular box with side lengths L_x and $L_y = L_z$. (b) Protons a and b placed at distance d as indicated; the electron cloud considered unchanged. Electron confined to a rectangular box with side lengths L_x, L_y, and L_z.

(L_x longer, L_y, L_z shorter than the box length for the H atom) and $d_0 = 152$ pm for the equilibrium distance. The minimum energy is

$$\varepsilon_{\text{min}} = -\overline{T}_{\text{min}} = \frac{1}{2}\overline{V}_{\text{min}} = -16.9 \times 10^{-19} \text{ J}$$

Thus, we find that the system (two protons and an electron) is trapped in the energy well formed by the mutual electrostatic interactions.

Comparison of the electron binding energy for the H atom to that found for H_2^+ by using box model wavefunctions shows that H_2^+ has the stronger electron binding energy. In fact, the electronic bond energy can be calculated by finding the difference in the electronic binding energies for each of the species in the model chemical reaction

$$\text{H} + \text{p}^+ \rightarrow \text{H}_2^+$$

The electronic bond energy ε_B is given by

$$\varepsilon_B = -\left(\varepsilon_{\text{H}_2^+} - \varepsilon_{\text{H}}\right) = (16.9 - 12.9) \times 10^{-19} \text{ J} = 4.0 \times 10^{-19} \text{ J} = 2.5 \text{ eV}$$

where the proton contributes zero because it does not have an electron associated with it. The result of the exact treatment (see Chapter 6) is

$$\varepsilon_B = -\left(E_{\text{H}_2^+} - E_{\text{H}}\right) = (26.2 - 21.8) \times 10^{-19} \text{ J} = 4.4 \times 10^{-19} \text{ J} = 2.7 \text{ eV}$$

and in agreement with experiment.[1] The bond energy of H_2^+ is a typical value, about one-third of the binding energy of N_2, which is one of the stronger chemical bonds.

As is evident from the energy values obtained in the calculation, the electronic bond energy from the simple box model wavefunctions (2.5 eV) is in much better agreement with the exact value (2.7 eV) than are the individual electronic binding energies for the H atom (8 eV compared to 13.6 eV) and H_2^+ (10.5 eV compared to 16.4 eV). What is the reason for this result? On the one hand the electronic wavefunctions of H and H_2^+ are badly represented by the box trial functions. They do not possess the proper shape near the protons and they have an abrupt end at the walls of the imagined box, rather than an exponential decay. On the other hand, the physics of bonding is described well:

(1) the increase of the negative charge density between the protons is well approximated,
(2) the compression of the electron cloud with the distance between the two protons is described reasonably well, and
(3) the balance between mean potential energy $\overline{V}$ and mean kinetic energy $\overline{T}$ ($\overline{T}$ is equal to minus half of the decrease in $\overline{V}$) is obtained.

In spite of the poor representation of the electron distribution by using the box model trial functions, it describes these essential features of the bonding.

2.6 Conclusion

We have considered experiments showing that atoms and molecules exist in distinct energy states, and we have seen this energy quantization by discussing a simple theoretical model, an electron enclosed in a box. We used this model, based on the postulates given in Chapter 1 (wave–particle duality), to describe approximately the wavefunctions of the electron of the H atom and of the H_2^+ molecular ion. In order to assess the quality of these functions we introduced the variational principle as a postulate and used it to adjust the wavefunction and to find an energy for the atom (or molecular ion) that is closest to reality. This principle opens the way to a deeper understanding of quantum mechanics and the structure of atoms and molecules as described in the following chapters.

2.7 Box 2.1. H Atom: Energy for Particle in a Box Trial Function

The energy ε in Equation (2.27) depends on the parameter L

$$\varepsilon = \frac{3h^2}{8m_e L^2} - \frac{1}{4\pi\varepsilon_0}\frac{4.18\, e^2}{L}$$

---

1 In a rigorous treatment it must be taken into account that the energy of an H_2^+ ion, as a result of the vibrational motion of the nuclei, lies about 0.2×10^{-19} J higher (see Section 11.3); the energy that is necessary to pull apart the bond must be smaller by this amount. Indeed, the experimental value is $\varepsilon_B = 4.2 \times 10^{-19}$ J.

As we recall from calculus, we can identify the extrema of a function with respect to a variable by taking the derivative and setting it equal to zero. To find the value of the parameter L that lies at the minimum of the energy function we do likewise, and find

$$\left(\frac{d\varepsilon}{dL}\right)_{L=L_{\min}} = \left(-\frac{6h^2}{8m_e L^3} + \frac{1}{4\pi\varepsilon_0}\frac{4.18\,e^2}{L^2}\right)_{L=L_{\min}} = 0$$

If we solve this equation for $L_{\min}$, we find

$$L_{\min} = 4\pi\varepsilon_0 \frac{h^2}{m_e e^2}\frac{3}{4\times 4.18} = 376 \text{ pm}$$

Formally we would need to take a second derivative and analyze it also, in order to prove that this value of L is a minimum. The potential energy and the kinetic energy are given by

$$\overline{V}_{\min} = -\frac{1}{4\pi\varepsilon_0}\frac{4.18\,e^2}{L_{\min}} \quad \text{and} \quad \overline{T}_{\min} = \frac{3h^2}{8m_e L_{\min}^2} = \frac{3h^2}{8m_e L_{\min}}\frac{1}{L_{\min}}$$

at the minimum. If we substitute our expression for $L_{\min}$ into the second term of the kinetic energy expression, we find that

$$\overline{T}_{\min} = \frac{3h^2}{8m_e L_{\min}}\frac{1}{4\pi\varepsilon_0}\frac{m_e e^2}{h^2}\frac{4\times 4.18}{3} = \frac{1}{4\pi\varepsilon_0}\frac{4.18\,e^2}{2L_{\min}} = -\frac{1}{2}\overline{V}_{\min}$$

The total energy is

$$\varepsilon_{\min} = \overline{V}_{\min} + \overline{T}_{\min} = -\frac{1}{4\pi\varepsilon_0}\frac{4.18\,e^2}{2L_{\min}}$$

2.8 Box 2.2. H_2^+ Ground-State Energy from Particle in a Box Trial Functions

Consider the lowest kinetic energy ($n_x = n_y = n_z = 1$) box wavefunctions, Equation (2.17), but place the origin of the coordinate system in the center of the box. Then with $L_x = bL$ and $L_y = L_z = L$ we can write the wavefunction ϕ as

$$\phi = \sqrt{\frac{1}{b}\left(\frac{2}{L}\right)^3}\cdot\cos\left(\frac{\pi x}{bL}\right)\cdot\cos\left(\frac{\pi y}{L}\right)\cdot\cos\left(\frac{\pi z}{L}\right)$$

where x extends from $-bL/2$ to $+bL/2$ and y and z extend from $-L/2$ to $+L/2$. The average kinetic energy of the electron is given by Equation (2.34),

$$\overline{T} = \frac{h^2}{8m_e L^2}\left(\frac{1}{b^2} + 2\right)$$

The potential energy for describing H_2^+ consists of the repulsive interaction of the two atomic nuclei and the attractive interaction between the electron cloud and the nuclei. The lengths L and bL are parameters that control the spatial extent of the electron cloud. According to Equation (2.35) with $\rho = \phi^2$, the average potential energy is

$$\overline{V} = \frac{1}{4\pi\varepsilon_0}\frac{e^2}{d} - \frac{1}{4\pi\varepsilon_0}\int\left[\frac{e^2}{r_a} + \frac{e^2}{r_b}\right]\phi^2\cdot dx\,dy\,dz$$

where the distances r_a and r_b are given by

$$r_a = \sqrt{\left(x+\frac{d}{2}\right)^2 + y^2 + z^2} \quad \text{and} \quad r_b = \sqrt{\left(x-\frac{d}{2}\right)^2 + y^2 + z^2}$$

in Cartesian coordinates. The total energy is the sum of the average kinetic energy and the average potential energy

$$\varepsilon = \overline{T} + \overline{V}$$

and it depends on the internuclear distance d, and the two parameters L and b.

Implementing the variational principle is more involved than the H atom case in Section 2.4.2, and we perform it numerically[2]. We proceed by varying L and b for a series of nuclear distances d until the lowest energy state is found. That is, we set

2 The integral for the potential energy was calculated numerically in a similar way as demonstrated in Justification 2.1, on the book's website, for the H atom: discretization of the box into $60^3 = 216{,}000$ points. The energy minimum is reached for $d = 150$ pm, $L = 300$ pm, and $b = 1.3$. Further refinement of the method results in $d = 152$ pm, $L = 301$ pm, and $b = 1.36$.

the interproton distance at d; and then we compute the energy for different values of b and L, until we find the combination of b and L which gives the lowest electronic energy at that d value. By proceeding in this way for different d values, we calculate how the electronic energy changes with d. Table 2B.1 reproduces some of the values found by this procedure. In Fig. 2B.1 the energies $\overline{T}$, $\overline{V}$, and ε are plotted as a function of the internuclear distance d. From Table 2B.1 and Fig. 2B.1 we discover that the lowest energy state is reached at an interproton distance $d = d_0 = 152$ pm and with the wavefunction parameters $L = 301$ pm and $b = 1.36$.

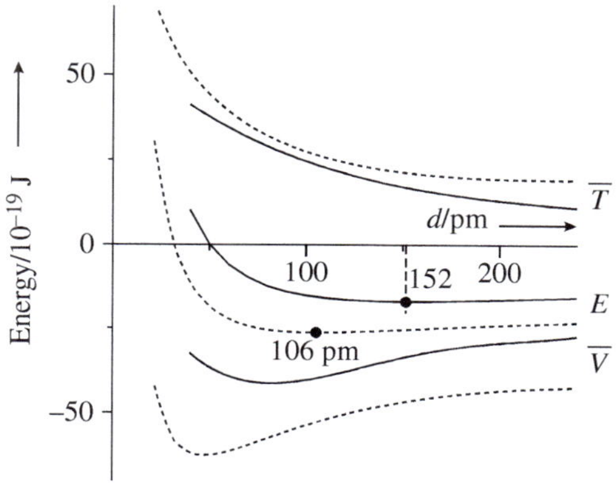

Figure 2B.1 Energy ε, average kinetic energy $\overline{T}$, and average potential energy $\overline{V}$ of the H_2^+ ion as a function of interproton distance d. Solid lines: according to the variational principle with the box trial function. The filled circle indicates the minimum of the energy ε ($\varepsilon_{min} = -16.9 \times 10^{-19}$ J at $d_0 = 152$ pm). The dashed lines indicate the exact solution ($\varepsilon_{min} = -26.8 \times 10^{-19}$ J at $d_0 = 106$ pm).

Table 2B.1 Energy ε of H_2^+, Calculated According to the Variational Principle (Box Wavefunction, $L = L_y = L_z$) for Various Interproton Distances d.

d = 140 pm			d = 150 pm			d = 160 pm		
L/pm	b	$\varepsilon/(10^{-19}$ J)	L/pm	b	$\varepsilon/(10^{-19}$ J)	L/pm	b	$\varepsilon/(10^{-19}$ J)
280	1.3	−16.795	290	1.3	−16.828	290	1.3	−16.627
	1.4	−16.809		1.4	−16.886		1.4	−16.809
	1.5	−16.622		1.5	−16.768		1.5	−16.783
290	1.2	−16.681	300	1.2	−16.696	300	1.3	−16.798
	1.3	**−16.858**		1.3	**−16.914**		1.4	**−16.900**
	1.4	−16.801		1.4	−16.888		1.5	−16.806
300	1.2	−16.763	310	1.2	−16.768	310	1.3	−16.866
	1.3	−16.842		1.3	−16.839		1.4	−16.881
	1.4	−16.696		1.4	−16.810		1.5	−16.739

The minimum electron energies are given by box wavefunctions with the length parameter $L_x = 409$ pm and $L_y = L_z = 301$ pm, and $d = 152$ pm. The electron cloud in the H_2^+ ion has a greater long axis and a smaller minor axis than the electron cloud in the H atom ($L = 376$ pm). From this wavefunction we obtain $\varepsilon_{min} = -16.9 \times 10^{-19}$ J $= -10.5$ eV. The bold typeface highlights the value of energy ε at which the energy has a minimum for a particular choice of d.

2.9 Justifications

2.9.1 Justification 2.1: Particle in a Box Trial Function for the H Atom

See the book's website, www.wiley.com/go/Kuhn/physicalchemistry3e.

2.10 Exercises

E2.1 The potential energy $V(r)$ of interaction between a proton (charge of $+e$) and an electron (charge of $-e$) is given by the Coulomb potential

$$V(r) = -\frac{e^2}{4\pi\varepsilon_0 r}$$

where $e = 1.602 \times 10^{-19}$ C, $\varepsilon_0 = 8.854 \times 10^{-12}$ C^2 m^{-1} J^{-1}, and r is the distance between the particles. Use V to find the expression for the force f_{elec}. Compute the electrostatic force between these two particles at a distance of 300 pm. Express your answer in Newtons. Compute the gravitational force between these two particles at a distance of 300 pm. Note that $m_e = 9.109 \times 10^{-31}$ kg, $m_p = 1.67 \times 10^{-27}$ kg, and $G = 6.67 \times 10^{-11}$ N m^2/kg^2. Compare $f_{gravity}$ to f_{elec}. Is it reasonable to ignore gravitational effects when we describe the interaction of an electron and a proton?

E2.2 Consider two charged particles, an electron and a proton. Compute the electrostatic energy between a proton and an electron at 100 pm, 1.0 nm, and 10.0 nm. Express your answer in J and in electron volts (1 eV = 1.602×10^{-19} J). Evaluate the distance at which the potential energy is 13.6 eV below the dissociation limit (i.e., zero). Compare this distance to the average distance $\bar{r}$ that we found in Section 2.4.2 for the H atom. Extension: Use a spreadsheet program (e.g., Excel) to plot the electrostatic potential energy as a function of the distance.

E2.3 Use the fact that the size of a He atom is about 220 pm in diameter to estimate the kinetic energy of the electrons in this atom. In particular, use the one-dimensional box model to estimate the electron's de Broglie wavelength and then its average kinetic energy. Report your answer in J and in eV. Also estimate the average speed of the electron. At what fraction of the speed of light is the electron moving?

E2.4 Approximate an H_2 molecule as a one-dimensional box of length 216 pm (found from the bond length plus twice the atomic radius) which confines its valence electrons. Determine the kinetic energy of an electron that has a wavelength of twice the box length. Use this kinetic energy with the known ionization potential (or electron binding energy) of H_2, which is 15.43 eV, to estimate the potential energy of the electron.

E2.5 Consider the three one-dimensional wavefunctions sketched in the diagram below. Each wavefunction (A, B, and C) is plotted versus the x coordinate whose scale is set by tic marks at zero and one. Order these wavefunctions by their kinetic energy.

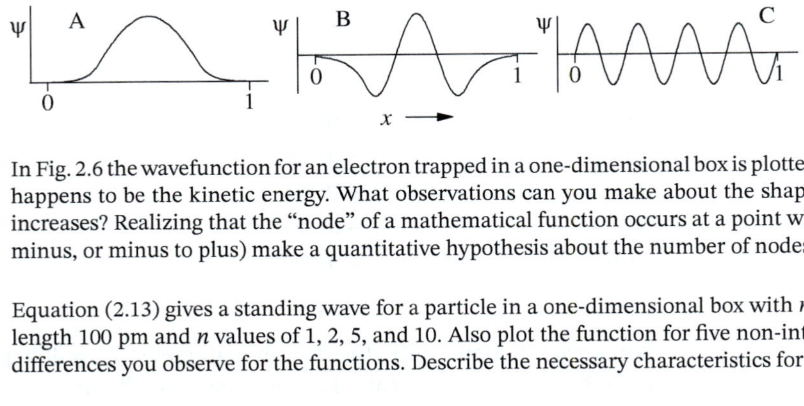

E2.6 In Fig. 2.6 the wavefunction for an electron trapped in a one-dimensional box is plotted for various values of the total energy, which happens to be the kinetic energy. What observations can you make about the shape of the electron wave as the kinetic energy increases? Realizing that the "node" of a mathematical function occurs at a point where the function changes sign (from plus to minus, or minus to plus) make a quantitative hypothesis about the number of nodes and the kinetic energy of an electron wave.

E2.7 Equation (2.13) gives a standing wave for a particle in a one-dimensional box with n as an integer. Plot this function for a box of length 100 pm and n values of 1, 2, 5, and 10. Also plot the function for five non-integer n values between 1 and 2. Describe the differences you observe for the functions. Describe the necessary characteristics for a standing wave of the box.

E2.8 Use the resonance condition of Equation (2.12) to derive the kinetic energy expression of Equation (2.14).

E2.9 In our box model of the hydrogen atom, we found that the ground-state electron wave corresponded to a box length of 376 pm. However, the proton is also contained in this volume of space. Are the wave characteristics of the proton important to consider in this case? Explain.

E2.10 Reproduce Fig. 2.10 using a spreadsheet program, such as Excel, and find the value of L that minimizes the energy. Also determine the minimum value of L analytically, i.e., set the first derivative of the energy with respect to L equal to zero and solve for L. Compare your numerical and analytical results.

E2.11 Use Equation (2.20) to determine the average potential energy of an electron in the H atom, $\overline{V}$. Use this average potential energy to determine the average value of $\overline{1/r}$. Compare your value for $\overline{1/r}$ to the value of $1/a_0$. Comment on your result.

E2.12 To explore the properties of the particle in the box wavefunctions further, perform the MathCAD exercise, titled PinBwaves.xmcd, and answer the questions:

a) For each of the above wavefunctions ($n = 1$ and $n = 2$), find the number of nodes.
b) Plot out the eigenfunctions corresponding to quantum numbers 2 and 3.
c) From the graph, read off the position at which the nodes can be found for the cases of $n = 1,2,3,$ and 4.
d) Plot the probability density, or the square of the wavefunctions for $n = 3$ and $n = 4$.

E2.13 Using Equation (2.1) find the energy that corresponds to the 589 nm Na emission line (see Fig. 2.1). Compare your result to the energy spacing shown in the electron energy loss spectrum of Fig. 2.2.

E2.14 When an electron is bound to a proton it makes a hydrogen atom, for which the accepted atomic diameter is 128 pm.

a) Estimate the largest wavelength that an electron could have if it were trapped inside of a box of this size; assume that the wave has to go to zero at the boundaries of the box. From this wavelength calculate the average kinetic energy for the electron.
b) Perform an estimation similar to that of part a) but perform it for an electron that is confined in i) a benzene ring, ii) a C_{60} molecule, or iii) a 10 nm diameter Au colloid. Estimate the diameters of benzene and C_{60} by using their density and the molecular weights.
c) Based on your calculations, discuss how you expect the kinetic energy of an electron to change as you change its ability to delocalize throughout a molecule or other nanometer-sized particle.

2.11 Problems

P2.1 Figure P2.1 shows an atomic emission spectrum from excited electronic states of the hydrogen atom to its ground electronic state. The first few lines correspond to: $H_\alpha = 656.28$ nm, $H_\beta = 486.13$ nm, $H_\gamma = 434.05$ nm, $H_\delta = 410.17$ nm. Use these wavelengths to determine the energy level spacing between the ground state and each of these excited states. Make an energy level diagram which shows the ground electronic state and these first few excited states. In your diagram, set the zero of energy to correspond to the separated electron and proton and use the fact that the ground electronic state lies 13.6 eV below this value. What do you think the H_∞ label indicates?

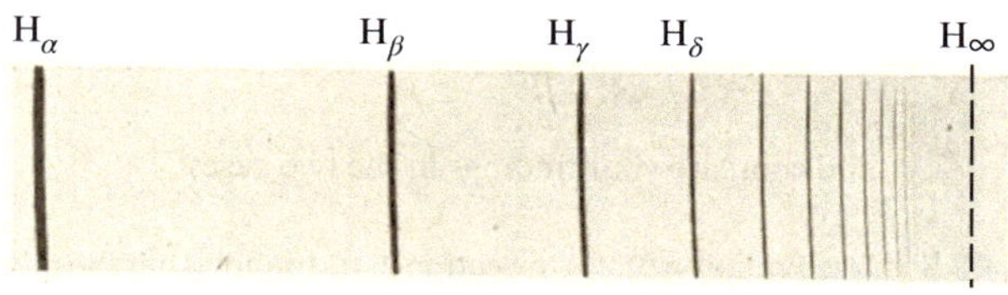

Figure P2.1 This image is taken from G. Herzberg. Atomic Spectra and Atomic Structure, Prentice Hall, NY, 1937.

Compare and contrast the energy level structure you found here for the H-atom to that calculated from Equation (2.14). Do you think that we will need to develop a more detailed model for the structure of the hydrogen atom, than the one given here? Please explain your answer in detail.

P2.2 The figure shows data that were measured using a Franck–Hertz apparatus with Ne gas rather than Na gas. It is reproduced from [2]. Use these data to determine the resonance energy for excitation of Ne atoms. The data are collected for two values of the opposing voltage. You may use either data set.

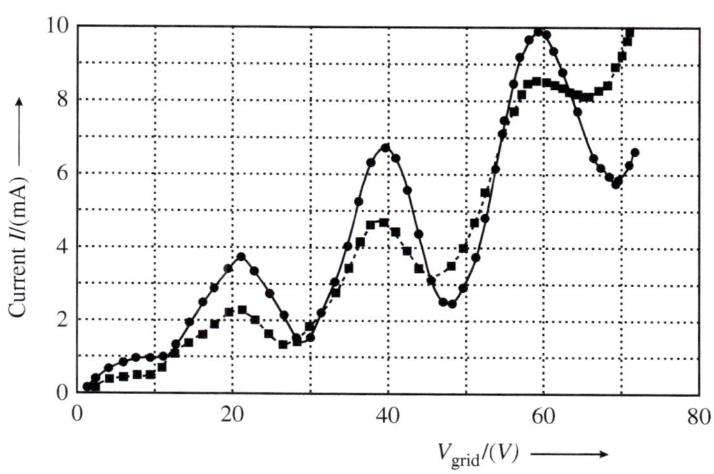

Figure P2.2 This image shows a plot of the measured electron current versus the grid voltage (data are for retarding potentials of 8 V, circles with solid line, and 6V, squares with dashed line) for a Franck-Hertz apparatus containing neon gas.

The authors describe how the gas emits a red glow when the atoms get excited. Using the energy difference that you found from the current–voltage curve, calculate the wavelength of a photon that an excited Ne atom can emit. Could such photons explain the red glow?

P2.3 Show that a standing wave with nodes at the positions $x = 0$ and $x = L$ is formed from the overlap of two traveling waves of the same wavelength moving in opposite directions. Note that a wave traveling from left to right with speed v can be described by

$$\varphi_+ = A \sin\left(\frac{2\pi(vt - x)}{\Lambda}\right)$$

and when this wave is reflected at the right boundary the back traveling wave is described by

$$\varphi_- = -A \sin\left(\frac{2\pi(vt + x)}{\Lambda}\right)$$

P2.4 Since ψ^2 is a probability function, i.e., the probability of finding an electron in a region of space, we can use it to compute the statistical outcome of measurements. Explore this application, for a simple situation in which you know the answer, by computing the average value of an electron's position in the ground electronic state of a box.
Use a spreadsheet program, to numerically determine the average position of an electron for the ground-state wavefunction of the particle in a box of length 100 pm. Take the wavefunction to be

$$\psi(x) = \sqrt{\frac{2}{L}} \sin\left(\frac{\pi x}{L}\right)$$

Remember that the average position $\bar{x}$ for a discrete set of values is given by

$$\bar{x} = \sum_{i=1}^{N} x_i \mathcal{P}_i, \quad \mathcal{P}_i = \rho \Delta x_i$$

where x_i is the value of the displacement in the ith element and $\mathcal{P}_i$ is the probability in the ith element. In our case $\mathcal{P}_i$ is the electron probability density ρ times Δx_i.

Perform the same calculation analytically, for which the average value is given by

$$\bar{x} = \int_0^L x\,\psi^2\,\mathrm{d}x$$

and compare your findings in the two cases.

P2.5 Use Equation (2.25) in Section 2.4.1 to show that the mean potential energy $\overline{V}$ for an electron in the ground state of the H-atom (ρ is given by Equation (2.22)) is $\overline{V} = -e^2/(4\pi\varepsilon_0 a_0)$. Hint: Proceed by converting the integral over all space into spherical coordinates: i.e.,

$$\overline{V} = -\frac{e^2}{4\pi\varepsilon_0} \int_{\text{all space}} \frac{1}{r}\rho\,\mathrm{d}\tau = -\frac{e^2}{4\pi\varepsilon_0} \int_0^{2\pi} \mathrm{d}\phi \int_0^{\pi} \sin\theta\,\mathrm{d}\theta \int_0^{\infty} \frac{1}{r}\rho\, r^2\,\mathrm{d}r$$

P2.6 As we have discussed, the ionization energy of a hydrogen atom is the energy required to remove its electron. The table below provides the measured ionization energies for the hydrogen atom and six atomic ions, each of which contains only one electron. From Coulomb's law, we expect the potential energy of the electron to depend on the nuclear charge, or atomic number. Analyze the data and empirically determine this dependence. Given these ionization energies, estimate the kinetic energy of the electron and the electron's average speed. Do we need to account for relativistic effects?

Atomic Number (Z)	Atom/Ion	Ionization Energy (IE)
1	H	13.60 eV
2	He^+	54.40 eV
3	Li^{2+}	122.43 eV
4	Be^{3+}	217.67 eV
5	B^{4+}	340.11 eV
6	C^{5+}	489.81 eV

P2.7 Given the expression for the energy of an electron trapped in a one-dimensional box, $E_{kin} = [h^2/8m_eL^2)] \cdot n^2$, see Equation (2.14), describe how the energy changes with the size of the box. Make a plot/graph which shows this dependence for $n = 1$, $n = 2$, and $n = 3$. Based on your findings describe how you would expect the electronic energy levels of a chain of hydrogen atoms to change as the number of atoms in the chain is increased. Remember that as you increase the number of atoms in the chain, you are increasing the box length and the number of electrons.

P2.8 Using the probability density given by Equation (2.22), compute the average distance of the electron from the proton in a hydrogen atom.

P2.9 The table shows the variational energy $\varepsilon/(10^{-19}$ J) as a function of the parameters L and b for the box model wavefunctions of H_2^+. The internuclear distance is set at the experimental value of 106 pm. Plot these data and show that they generate a well-defined minimum. Determine the optimized values for L and b. Comment on the results.

L / pm	$b = 1.20$	$b = 1.25$	$b = 1.30$	$b = 1.35$	$b = 1.40$	$b = 1.45$	$b = 1.50$
200	−11.197	−11.779	−12.223	−12.544	−12.753	−12.885	−12.937
220	−13.910	−14.236	−14.451	−14.5637	−14.601	−14.5742	−14.475
240	−15.218	−15.358	−15.405	−15.387	−15.297	−15.157	−14.973
260	−15.644	−15.656	−15.593	−15.466	−15.300	−15.079	−14.839
280	−15.530	−15.441	−15.300	−15.103	−14.878	−14.606	−14.323
300	−15.073	−14.914	−14.715	−14.468	−14.200	−13.898	−13.581
320	−14.403	−14.196	−13.950	−13.674	−13.372	−13.056	−12.714
340	−13.608	−13.368	−13.087	−12.793	−12.468	−12.136	−11.792

References

1 G. Herzberg, *Atomic Spectra and Atomic Structure* (Prentice-Hall, New York, 1937).

2 M.M. Kash and G.C. Shields, Using the Franck-Hertz experiment to illustrate quantization: energy states of the neon atom by electron impact, *J. Chem. Ed.*, 1994, **71**, 466.

3

Schrödinger Equation

CHAPTER MENU

3.1 Introduction

In Chapters 1 and 2 a simplified quantum-mechanical description was developed to illustrate some essential aspects of chemical bonding. For a deeper understanding of atoms and molecules, we must develop a general prescription for calculating the wavefunctions and the energies of electronic states. In this chapter quantum theory is introduced in more formal detail. We introduce the time-independent Schrödinger Equation as a postulate. In the last part of the chapter we generalize our consideration by discussing the time-dependent Schrödinger equation and introducing a distinct set of mathematical operators as the basic postulates of quantum mechanics, replacing the time-independent Schrödinger equation.

It is helpful at this point to make clear how we progress in science. We try to fit a large number of phenomena to a theory, that is, to predict them from a small number of basic assumptions. We gain confidence in a theory when we succeed in explaining a broader and larger range of facts with it. If we find contradictions and if the facts are more successfully interpreted by a new theory, then the new theory will replace the older one. In this way more and more facts are described by increasingly general theories. The basic assumptions are established by the success with which they interpret the widest possible range of observed phenomena and predict new phenomena. In the present case, it will be fascinating to see that quantum theory allows us to understand an immense array of experimental facts.

Beginners who aim to obtain a less rigorous understanding of quantum mechanics can skip Sections 3.5 and 3.6.

3.2 Wave Equation and Schrödinger Equation

3.2.1 Wave Equation

In the preceding chapters we discussed some of the fundamental experiments which led to postulates of quantum mechanics for a simple special case, and we showed how the wave properties of electrons are essential for understanding them.

Principles of Physical Chemistry, Third Edition. Hans Kuhn, David H. Waldeck, and Horst-Dieter Försterling.

Companion website: www.wiley.com/go/Kuhn/physicalchemistry3e

If our particles are going to possess wave properties then it seems reasonable to assume that a wave equation might be found to predict their behavior. It is useful to consider the wave equation of a stretched string as a way to assist our thinking as we begin; later we will discard this physical model and adopt the mathematical equation as our postulate. The wave equation is a differential equation that describes the behavior of a wave in time and space. Differential equations govern many different phenomena: guitar string vibration, heat flow in solids, evolution of populations, planetary motion, diffusion, and electrons.

We recommend that you read Appendix C if you need to review differentiation or differential equations.

3.2.1.1 Vibrating String

We start out with a vibrating string that is fixed at two end points (Fig. 3.1). If the string is struck anywhere, the disturbance spreads with a speed v. If we create an arbitrary displacement φ at time $t = 0$ and afterwards leave the string to itself, then its displacement at any position x will be a function of time t, $\varphi(x, t)$. From classical physics we know that the displacement $\varphi(x, t)$ can be described by the wave equation (see Appendix C.2)

$$\boxed{\frac{\partial^2 \varphi(x,t)}{\partial t^2} = v^2 \frac{\partial^2 \varphi(x,t)}{\partial x^2}} \tag{3.1}$$

As we discussed in Chapter 2, initial displacements which lead to a standing wave are the most interesting case for understanding bound electrons. In this special case $\varphi(x, t)$ is a product of one function that depends only on x and another that depends only on t. We propose a solution

$$\varphi(x,t) = \psi(x)\ \cos(\omega t) \tag{3.2}$$

We can verify that this is a solution to the wave equation by explicitly taking the derivatives, substituting them into Equation (3.1), and solving it. For the derivatives we find

$$\frac{\partial \varphi}{\partial t} = -\psi(x)\,\omega \sin \omega t, \quad \frac{\partial^2 \varphi}{\partial t^2} = -\psi(x)\,\omega^2 \cos \omega t, \quad \frac{\partial^2 \varphi}{\partial x^2} = \cos \omega t\ \frac{d^2 \psi}{dx^2}$$

The displacement φ remains a function of x and t, even though we do not indicate it explicitly. Substituting into Equation (3.1) we have

$$-\psi(x)\,\omega^2\ \cos(\omega t) = v^2\ \cos(\omega t)\ \frac{d^2 \psi(x)}{dx^2} \tag{3.3}$$

Now we can solve this equation to find that

$$\frac{d^2 \psi(x)}{dx^2} = -\left(\frac{\omega}{v}\right)^2 \psi(x) \tag{3.4}$$

where the functional dependence of the wavefunction ψ on x is shown explicitly. By realizing that $\omega = 2\pi\nu$ and $\Lambda\nu = v$ (Λ = wavelength), we can write

$$\frac{\omega}{v} = \frac{2\pi\nu}{\Lambda\nu} = \frac{2\pi}{\Lambda} \tag{3.5}$$

and

$$\boxed{\frac{d^2 \psi(x)}{dx^2} = -\frac{4\pi^2}{\Lambda^2}\ \psi(x)} \tag{3.6}$$

This time-independent equation governs the space dependence of the standing wave.

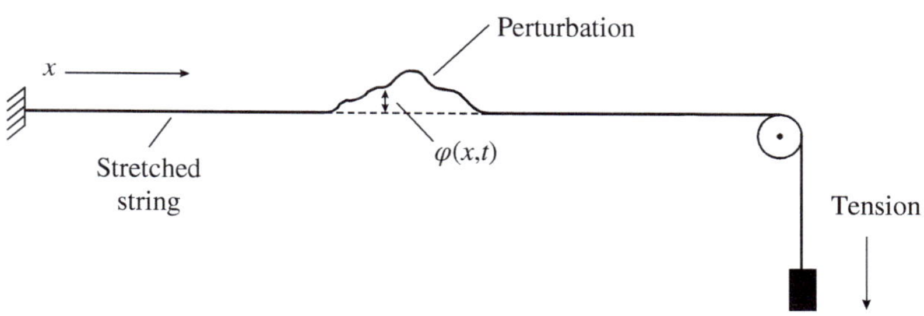

Figure 3.1 Spreading of a disturbance along a stretched string.

Example 3.1 *Verify that Equation (3.6) is consistent with the results found in Section 2.3.1 for standing waves of a string.*

According to Chapter 2, the maximum wave displacement $\psi(x)$ is

$$\psi(x) = A \sin\left(\frac{n\pi x}{L}\right) \qquad \text{with } n = 1, 2, .. \tag{3.7}$$

Is this expression a solution of Equation (3.6)? To show that function (3.7) is a solution of the wave equation we substitute it into Equation (3.6) and verify that the equality holds. To proceed we differentiate $\psi(x)$ twice with respect to x

$$\frac{d^2}{dx^2}\psi(x) = -A\left(\frac{n\pi}{L}\right)^2 \sin\left(\frac{n\pi x}{L}\right) = -\left(\frac{n\pi}{L}\right)^2 \psi(x) \tag{3.8}$$

Now, we substitute for the second derivative of the wavefunction into Equation (3.6) and obtain

$$-\left(\frac{n\pi}{L}\right)^2 \psi(x) = -\frac{4\pi^2}{\Lambda^2}\psi(x) \tag{3.9}$$

By inspection, we see that the equality in (3.9) holds if

$$\frac{n}{L} = \frac{2}{\Lambda} \qquad \text{or} \qquad \Lambda = \frac{2L}{n} \tag{3.10}$$

This result is the resonance condition that was discussed in Chapter 2. In addition, we can see that Equation (3.7) describes a standing wave that has its two zeroes at both ends. More formally, we say that Equation (3.7) obeys the boundary conditions

$$\psi(0) = 0 \quad \text{and} \quad \psi(L) = 0 \tag{3.11}$$

For $n = 1$, corresponding to the largest wavelength which is twice the length of the string, the vibrating string has an antinode in the middle at $x = L/2$. For $n > 1$ the vibrating string has $n - 1$ equally spaced nodes.

3.2.1.2 Electron Between Two Walls

What happens if we try to use the wave equation (Equation (3.6)) to describe the behavior of an electron? If we make use of the de Broglie relation (see Section 1.3.2)

$$\Lambda = \frac{h}{m_e v} \tag{3.12}$$

we can relate Λ to the energy E by way of the kinetic energy T

$$E = T + V = \frac{1}{2}m_e v^2 + V = \frac{1}{2}m_e\left(\frac{h}{m_e\Lambda}\right)^2 + V \tag{3.13}$$

where the potential energy V is a constant. By algebraic rearrangement, we find that

$$\Lambda^2 = \frac{h^2}{2m_e(E - V)} \tag{3.14}$$

If we substitute this result into Equation (3.6) for Λ, we obtain

$$-\frac{h^2}{8\pi^2 m_e}\frac{d^2\psi}{dx^2} + V\psi = E\psi \tag{3.15}$$

This result represents the de Broglie relation in a way that is useful for understanding how the Schrödinger equation, a fundamental postulate of quantum mechanics, was first introduced. Although Equation (3.15) has the same form as the Schrödinger Equation (3.16), this rationale cannot be considered a proof or derivation. We include this logical argument to illustrate how science evolves, by not only professing bold new ideas but also by adapting existing structures and ideas in new ways.

3.2.2 One-Dimensional Schrödinger Equation

Schrödinger assumed, on a trial basis, that Equation (3.15) is also true for the case where $V = V(x)$ is a function of x:

$$\boxed{\textbf{Postulate:} \quad -\frac{h^2}{8\pi^2 m_e}\frac{d^2\psi}{dx^2} + V(x)\psi = E\,\psi} \tag{3.16}$$

with the boundary condition

$$\boxed{\lim_{x\to\pm\infty} \psi(x) = 0} \tag{3.17}$$

which ensures that the particle is confined to a region in space. Solutions of Equation (3.16) that fulfill condition (3.17) are called *eigenfunctions*, and the corresponding energies are called *eigenvalues*. The generalization (3.16) is a postulate. It is freely invented and cannot be derived from anything else. Whether or not this postulate is reasonable can be checked by comparing its implications with experiment. We will see that the Schrödinger Equation (3.16) is very successful and provides the basis for understanding much of chemistry.

First consider the case of an electron trapped in a potential energy well. Take the energy well to be bounded by a wall at $x = 0$ and a wall at $x = L$ with a potential energy $V(x)$ that is infinitely large outside the walls (i.e., $V(x) = \infty$ when $x < 0$ and $x > L$). The wavefunction $\psi(x)$ must be zero everywhere outside the walls; it is the only solution of the Schrödinger Equation (3.16) in these regions. Inside the walls ($0 < x < L$) the Schrödinger Equation (3.16) has the non-trivial solutions given by Equation (3.7), the electron waves are restricted to be inside the walls for all eigenvalues that satisfy the resonance condition $n = 1,2, \ldots$ Combining the resonance condition (Equation (3.10)) with Equation (3.13), we find that

$$E = T + V = \frac{1}{2}m_e\left(\frac{hn}{m_e\,2L}\right)^2 + V = \frac{h^2\,n^2}{8\,m_e\,L^2} + V$$

3.2.2.1 Potential Energy Well with Walls of Finite Height

Although the case of infinitely high walls is a good way to begin, a more realistic model would have walls of finite height. Let us consider a case where the potential energy is $V(x) = 0$ in the region $0 < x < L$ and $V(x) = V_0$ for $x < 0$ and $x > L$. This potential energy diagram is sketched in Fig. 3.2a. In this case, the wavefunction $\psi(x)$ need not vanish at the walls but is required to vanish as the displacement x goes to infinity. Figure 3.2b shows the wavefunctions $\psi(x)$ with the

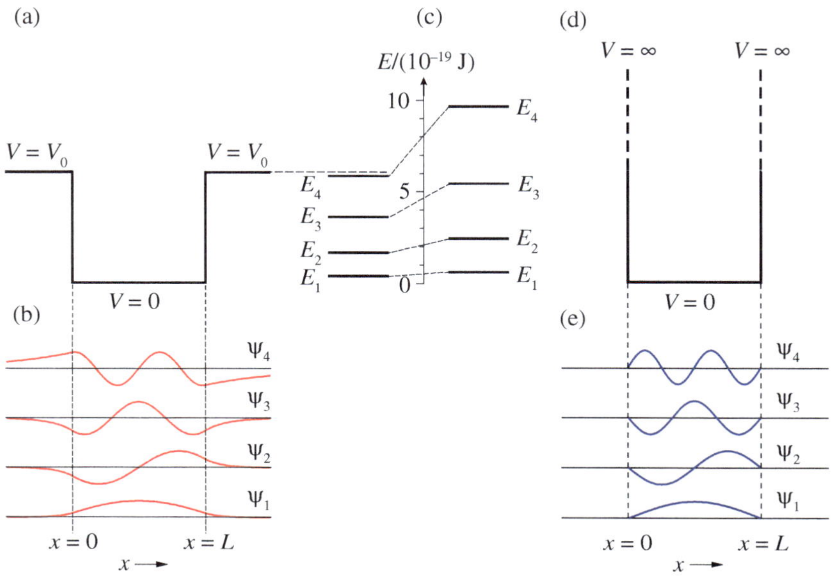

Figure 3.2 Electron in a potential energy well. (a) Potential energy V (potential energy well with finitely high walls) and (b) the corresponding wavefunctions of the lowest four quantum states; the dashed lines mark the positions of the walls. (c) Energies of the four quantum states (left) compared to the energies of the four lowest energy states of a particle in a potential well with infinitely high walls (right). The distance between the walls is $L = 1000$ pm, and the height of the walls is $V_0 = 10\,h^2/(8m_eL^2) = 6.03 \times 10^{-19}$ J. (d) Potential energy V (potential energy well with infinitely high walls) and (e) the corresponding wavefunctions.

energies $E < V_0$ that are obtained by solving Equation (3.16) with the boundary condition (3.17). The basic consideration is outlined in Box 3.1, and a detailed derivation is provided in Foundation 3.1 (available on Wiley's website). As the plots show, the wavefunctions are sinusoidal in the region inside the box, but their amplitude decays monotonically with increasing displacement outside the box. For a given energy level, the sinusoidal part of the wavefunction has the same number of nodes as the corresponding energy states for the particle in a box with infinitely high walls. Now, however, the wavefunction can exist in the region outside the box. This feature allows the wavefunction to spread more and lowers the average kinetic energy, as compared to those for a box with infinitely high walls. Figure 3.2c shows the energy levels of the states with $n = 1, 2, 3, 4$ for the case of infinitely high walls (right) and the case of $V_0 = 6.03 \times 10^{-19}$ J = 3.76 eV (left). For comparison, the potential energy and the corresponding wavefunctions are shown in Fig. 3.2d,e. In the case of finitely high walls with $V_0 = 6.03 \times 10^{-19}$ J = 3.76 eV, only four discrete energy levels are found (bound states).

3.2.2.2 Curvature of ψ

This change in the character of $\psi(x)$ from oscillatory ($E > V(x)$, inside the box) to monotonic ($E < V(x)$, outside the box) is general and can be understood from the mathematical structure of the Schrödinger Equation (3.16). We write Equation (3.16) in the form

$$\frac{d^2\psi}{dx^2} = -\frac{8\pi^2 m_e}{h^2}(E - V(x))\,\psi \tag{3.18}$$

In the region where $E > V(x)$, the wavefunction $\psi(x)$ has the property that its second derivative is opposite in sign to the wavefunction itself. In the region where $E < V(x)$, the wavefunction has the property that $\psi(x)$ and its second derivative have the same sign. Remember that the second derivative is the curvature of the wavefunction, and it tells us how the wavefunction's slope is changing.

Case		ψ	Curvature $\frac{d^2\psi}{dx^2}$
1a	$E > V(x)$	Positive	Negative
1b	$E > V(x)$	Negative	Positive
2a	$E < V(x)$	Positive	Positive
2b	$E < V(x)$	Negative	Negative

As an example, let us consider wavefunction ψ_1 in Fig. 3.2b. Inside the box ($E > V_0$) the wavefunction is positive (case 1a) and the slope decreases with increasing x. At small x the slope is positive and decreases monotonically; it reaches zero in the center of the box and then becomes negative. At $x = L$ we switch to case 2a ($E < V_0$), and the slope increases monotonically with increasing x. Because the wavefunction remains positive and it fulfills the boundary condition (3.17), ψ_1 monotonically approaches zero at sufficiently high x. The situation is a little bit more complicated for wavefunction ψ_4 in Fig. 3.2b. At the left edge of the well we have again case 1a (the slope is decreasing with x), until the wavefunction intersects the abscissa the first time. Beyond the intersection point, the slope increases with x (case 1b), until the second intersection point is reached. Finally, at $x = L$ we have case 2b and the wavefunction monotonically approaches zero.

3.2.2.3 Tunneling

The solution to the Schrödinger equation for the region outside the box where $E < V_0$ is very surprising. In classical physics, the electron would be confined to the region inside the potential energy well because the total energy E is less than the potential energy V_0. This limitation is relaxed in quantum mechanics because the electron's wave character enables it to penetrate part of the way through the wall, as long as it is decaying toward zero. According to the solution of the Schrödinger Equation (see Box 3.1), the probability of observing the electron in the region outside the box decreases exponentially with the distance from the box's edge where $E < V_0$. Furthermore, it decreases more strongly as $E - V_0$ becomes more negative (compare wavefunctions ψ_1 and ψ_4 in Fig. 3.2b). This phenomenon is called *tunneling* and is analogous to an evanescent wave of light, which is well known in optics.

3.2.2.4 Tunneling Probability

In Box 3.1 we show that the solution to the Schrödinger Equation (3.16) in the case of the potential energy well (Fig. 3.2a) in the region where $E < V_0$ is an exponentially decaying function. Let us assume a general exponential form for the wavefunction

$$\psi(x) = A\ \mathrm{e}^{-bx} \tag{3.19}$$

We determine the arbitrary constants b and A by solving the Schrödinger equation. The pre-exponential constant A is an overall proportionality factor for the wavefunction in this region and its value is chosen so that this part of the wavefunction matches smoothly with the sinusoidal part of the wavefunction inside the well (see Box 3.1). Of interest here is the constant b which controls how rapidly the wave decays to zero. First we find the second derivative of the exponential in Equation (3.19)

$$\frac{\mathrm{d}^2\psi}{\mathrm{d}x^2} = Ab^2\ \mathrm{e}^{-bx} = b^2\ \psi \tag{3.20}$$

Then we insert this result into Equation (3.18) and solve for b to find

$$b = \frac{2\pi}{h}\sqrt{2m_e\left(V_0 - E\right)} \tag{3.21}$$

and

$$\psi(x) = A\ \exp\left(-\frac{2\pi}{h}\sqrt{2m_e\left(V_0 - E\right)}\cdot x\right) \tag{3.22}$$

Hence we see that the exponential decays more sharply as $\sqrt{\left(2m_e\left(V_0 - E\right)\right)}$ becomes larger.

Example 3.2 *Using the wavefunction in Equation (3.19), determine how rapidly the probability of finding the electron outside of the well decays, for the case of wavefunctions ψ_1 and ψ_4 in Fig. 3.2b.*

The probability $\mathcal{P}$ of finding the electron at a position $x = L + \Delta x$ (see Fig. 3.2) is given by the electron's probability density, namely

$$\mathcal{P} = \psi^2 = A^2\cdot \mathrm{e}^{-2bx}$$

To find the probability for tunneling beyond the box wall's edge at $x = L$, we find the ratio of probability densities $P_{L+\Delta x}$ at position $x = L + \Delta x$ and P_L at position $x = L$ to be

$$\frac{\mathcal{P}(L+\Delta x)}{\mathcal{P}(L)} = \frac{\mathrm{e}^{-2b(L+\Delta x)}}{\mathrm{e}^{-2bL}} = \mathrm{e}^{-2b\Delta x}$$

where according to Equation (3.21)

$$b = \frac{2\pi}{h}\sqrt{2m_e\left(V_0 - E\right)}$$

Thus we obtain

$$\frac{\mathcal{P}(L+\Delta x)}{\mathcal{P}(L)} = \exp\left(-\frac{4\pi}{h}\sqrt{2m_e\left(V_0 - E\right)}\cdot \Delta x\right) \tag{3.23}$$

The data in Fig. 3.2 are based on $V_0 = 6.03\times 10^{-19}$ J and $L = 1000$ pm. For the case of ψ_1 with $E_1 = 0.42\times 10^{-19}$ J (see table in Box 3.1) we obtain $b = 9.6\ \mathrm{nm}^{-1}$. For a value of $\Delta x = 1$ nm we find

$$\frac{\mathcal{P}(L+1\ \mathrm{nm})}{\mathcal{P}(L)} = \mathrm{e}^{-2\times 9.6\times 1} = 4.59\times 10^{-9}$$

Thus the probability of finding the electron decays by more than 10^8 times per nanometer of distance. Nevertheless electron tunneling is important in many electron transfer reactions and is even used in some electronic devices.

Correspondingly, in the case of ψ_4 the result for E_{barrier} and for b is (using $E_4 = 5.84\times 10^{-19}$ J) $E_{\mathrm{barrier}} = (6.03 - 5.84)\times 10^{-19}$ J $= 0.19\times 10^{-19}$ J and $b = 1.8\ \mathrm{nm}^{-1}$. Again for a value of $\Delta x = 1$ nm we obtain

$$\frac{\mathcal{P}(L+\Delta x)}{\mathcal{P}(L)} = \mathrm{e}^{-2b\Delta x} = \mathrm{e}^{-3.6} = 2.73\times 10^{-2}$$

Comparing the results in the two cases shows the dramatic dependence of the tunneling probability on the energy barrier V_0 relative to the energy E of the quantum state. See Chapter 28 for more discussion of electron tunneling.

Up to here we discussed tunneling of an electron of mass m_e out of a box with walls of finite height, but infinite thickness. Tunneling through a barrier with a finite height and a finite thickness is considered in Foundation 3.6, which is available electronically. The probability of tunneling for a particle with mass m through a barrier of height $(V_0 - E)$ and thickness Δx is

$$\mathcal{P} = \frac{4k^2q^2}{\left(k^2+q^2\right)^2 \cdot \sinh^2(q\Delta x) + 4k^2q^2}$$

where

$$k = \sqrt{\frac{2m}{\hbar^2}E} = \frac{2\pi}{h}\sqrt{2mE} \qquad q = \frac{2\pi}{h}\sqrt{2m(V_0 - E)}$$

In Foundation 3.6 we show that this expression can be simplified for $q\Delta x \gg 1$ as

$$\mathcal{P} = const \cdot \exp\left(-2\Delta x \cdot \sqrt{\frac{2m}{\hbar^2}(V_0 - E)}\right) = const \cdot \exp\left(-\frac{4\pi}{h}\sqrt{2m(V_0 - E)} \cdot \Delta x\right) \tag{3.24}$$

in analogy with Equation (3.23) in Example 3.2.

3.2.2.5 Scanning Tunneling Microscope (STM)

An important application of the tunneling effect is the *Scanning Tunneling Microscope (STM)*. With an STM instrument, a metallic tip of atomic dimensions (e.g., from tungsten) is held very close (distance $d < 1$ nm) to a conducting surface. By applying a small voltage (of about 100 mV) between the surface and tip, a tunneling current results, typically a current of about 2 nA corresponding to 2×10^{-9} A/1.6×10^{-19} C $\approx 10^{10}$ electrons per second (Fig, 3.3). The distance between the surface and tip is controlled by electrically monitoring the tunneling current and using this signal to move a translation stage that controls the tip position.

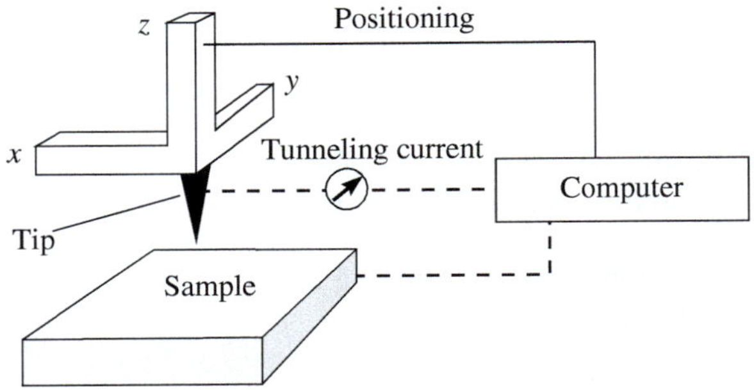

Figure 3.3 Scheme of a scanning tunneling microscope (STM).

From Equation (3.19) we see that the wavefunction ψ and thus the probability density $\rho = \psi^2$ depend exponentially on the width x of the tunneling barrier. This sensitivity on x allows the structure of the surface to be imaged by scanning it with the tiny tungsten tip. In this application, the tip is moved vertically during scanning such that a constant current is measured. The vertical displacement depends on the surface structure of the sample. Figure 3.4 shows an STM image of a conducting oligomer on a graphite surface. The oligomer consists of an octathiophene with hydrocarbon chains acting as spacer layers. The oligomer chains are stacked to form a well-ordered supramolecular array [1].

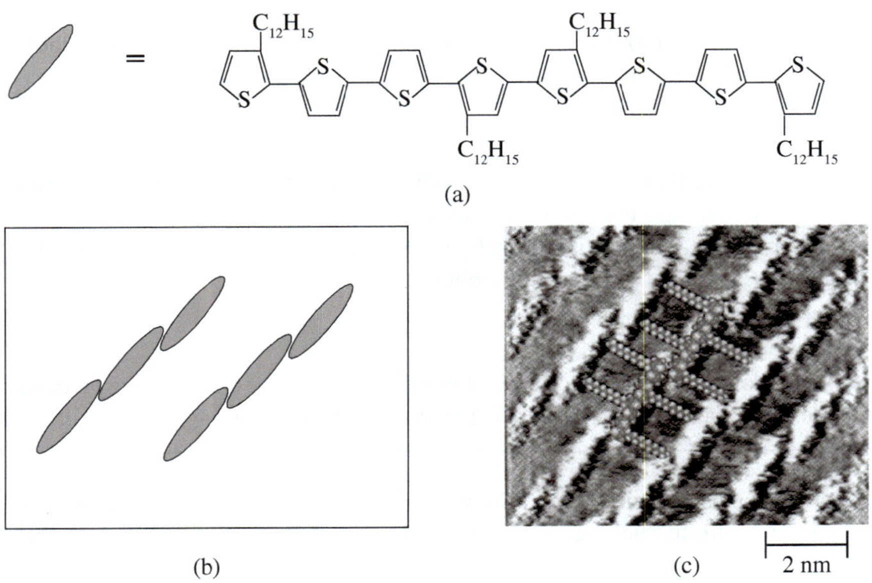

Figure 3.4 Arrangement of chains of octathiophene molecules (monolayers adsorbed on a graphite surface). (a) Structural formula of octathiophene, (b) scheme showing the architecture of the surface assembly, (c) STM image: two molecules in the center of the picture are covered by space-filling models to indicate the four hydrocarbon substituents of each molecule; the space between the octathiophene chains is closely packed by hydrocarbon chains (From [2], Bäuerle et al./JOHN WILEY & SONS, INC.)

As a second example, we consider a surface with copper atoms (Cu(1,1,1)-plane on a single crystal). By shifting iron atoms along the surface, using an atomic force microscope (see Chapter 7) it is possible to arrange them along a circle. The result can be observed in a scanning tunneling microscope (Fig. 3.5). Each of the 48 iron atoms in the circle can be imaged. The ring of Fe atoms creates a barrier, and standing waves of the surface electrons are formed. The density maxima of these waves correspond to regions where the tunneling current is high, and the nodes correspond to regions where the tunneling current is low. The wavelength is 300 pm; this is the wavelength of the electrons injected by the tip.

The standing waves are like the waves of an electron in a box discussed in Chapter 2; only in the present case the box is circular, instead of rectangular. It is fascinating to see that standing waves are not fully defined until the circle of iron atoms has been completed.

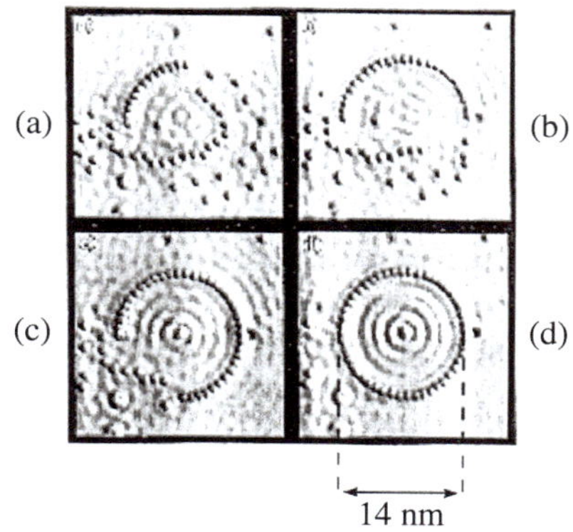

Figure 3.5 Manipulating single atoms on a surface by using an atomic force microscope. Iron atoms are arranged in a circle of 14 nm diameter on a Cu(111) surface (sequence a to d). After completion of the circle standing electron waves inside the circular potential trough are observed (From [3], Lieber et al./JOHN WILEY & SONS, INC.)

3.2.3 Three-Dimensional Schrödinger Equation

Equation (3.16) is true for a one-dimensional wave, where ψ depends only on x; that is, $\psi(x)$. In a molecule, however, an electron's ψ depends on all three spatial coordinates x, y, and z; that is, $\psi(x, y, z)$. To go from a one-dimensional description to a three-dimensional description, the wave equation for an electron must be generalized as

$$\boxed{\textbf{Postulate:}\ -\frac{\hbar^2}{2m_e}\left(\frac{\partial^2}{\partial x^2}+\frac{\partial^2}{\partial y^2}+\frac{\partial^2}{\partial z^2}\right)\psi + V\psi = E\psi} \tag{3.25}$$

with the boundary condition

$$\boxed{\lim_{x,y,z\to\pm\infty}\psi(x,y,z) = 0} \tag{3.26}$$

where we have introduced the notation $\hbar = h/2\pi$ instead of h. Note that the potential energy V is also a function of x, y, z; that is, $V(x, y, z)$. Often, this form of an electron's *Schrödinger Equation* is abbreviated as

$$\left(-\frac{\hbar^2}{2m_e}\nabla^2 + V\right)\psi = E\psi \tag{3.27}$$

where ∇^2 is called the Laplacian operator and is given by

$$\nabla^2 = \frac{\partial^2}{\partial x^2}+\frac{\partial^2}{\partial y^2}+\frac{\partial^2}{\partial z^2} \tag{3.28}$$

An operator is a mathematical entity that prescribes a sequence of mathematical operations that are to be performed. In this case the Laplacian operator tells you to take the second derivative with respect to x, add that to the second derivative with respect to y, and add that to the second derivative with respect to z. In each case the operation is to be performed on a function, in our case the wavefunction. Often an operator may be a combination of different simpler operators. Two of the most common simple operators, which are found in quantum mechanics, are "take the derivative" and "multiply by." Appendix C.4 provides more detail on operators. A further useful abbreviation is given by defining the quantity

$$\mathcal{H} = -\frac{\hbar^2}{2m_e}\nabla^2 + V \tag{3.29}$$

where $\mathcal{H}$ is called the *Hamiltonian operator*.[1] Thus Equation (3.27) can be written in the short form

$$\boxed{\mathcal{H}\psi = E\psi} \tag{3.30}$$

Lastly, we note that the wavefunction ψ does not have a time dependence yet. This will be discussed in Section 3.5.3.

1 The Hamiltonian operator $\mathcal{H}$ is related to the Hamilton function H in classical physics. H is the sum of kinetic and potential energy expressed as a function of the coordinates and momenta.

3.2.3.1 Particle in a Three-Dimensional Box

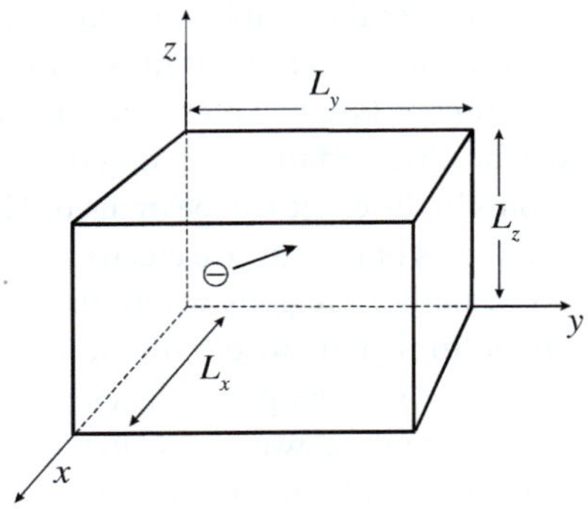

Figure 3.6 Electron in a box with side lengths L_x, L_y, and L_z.

As a simple example for the application of the Schrödinger Equation, we consider a particle in a box with impenetrable walls (outside of the box $V = \infty$, see Fig. 3.6). In addition, we assume that the particle can move freely within the box; this means that $V(x, y, z)$ is constant. Since the zero for the energy can be chosen arbitrarily, we choose the potential energy inside the box to be zero. In this case the Schrödinger Equation (3.25) can be written as

$$\boxed{-\frac{\hbar^2}{2m_e}\left(\frac{\partial^2}{\partial x^2}+\frac{\partial^2}{\partial y^2}+\frac{\partial^2}{\partial z^2}\right)\psi(x,y,z) = E\,\psi(x,y,z)} \tag{3.31}$$

As an initial guess let us choose a wavefunction $\psi(x, y, z)$ that is a product of functions $\psi_x(x)$, $\psi_y(y)$, and $\psi_z(z)$,

$$\psi(x,y,z) = \psi_x(x)\cdot\psi_y(y)\cdot\psi_z(z) \tag{3.32}$$

This method is called the *separation of variables*. We introduce Equation (3.32) in the partial differential Equation (3.31) and obtain

$$-\frac{\hbar^2}{2m_e}\left(\frac{\partial^2\psi_x}{\partial x^2}\psi_y\psi_z+\frac{\partial^2\psi_y}{\partial y^2}\psi_x\psi_z+\frac{\partial^2\psi_z}{\partial z^2}\psi_x\psi_y\right) = E\cdot\psi_x\psi_y\psi_z \tag{3.33}$$

By dividing both sides by $\psi_x\psi_y\psi_z$, we can write

$$\left[-\frac{\hbar^2}{2m_e}\frac{\frac{\partial^2\psi_x}{\partial x^2}}{\psi_x}\right]+\left[-\frac{\hbar^2}{2m_e}\frac{\frac{\partial^2\psi_y}{\partial y^2}}{\psi_y}\right]+\left[-\frac{\hbar^2}{2m_e}\frac{\frac{\partial^2\psi_z}{\partial z^2}}{\psi_z}\right] = E \tag{3.34}$$

Inspection of this differential equation reveals that each of the expressions in brackets depends only on one variable, x, y, or z, respectively. Moreover, the right-hand side of the equation, E, is a constant. If we keep y and z constant and change x arbitrarily, then the first expression in brackets must be a constant, E_x:

$$\left[-\frac{\hbar^2}{2m_e}\frac{\frac{\partial^2\psi_x}{\partial x^2}}{\psi_x}\right] = E_x \tag{3.35}$$

Accordingly, if we keep x and z or x and y constant, we find that the second expression in brackets equals a constant E_y and the third equals a constant E_z, where $E_x + E_y + E_z = E$. In this way we have transformed the partial differential Equation (3.31) into a set of three ordinary differential equations:

$$-\frac{\hbar^2}{2m_e}\frac{\partial^2\psi_x}{\partial x^2} = E_x\psi_x\,, \quad -\frac{\hbar^2}{2m_e}\frac{\partial^2\psi_y}{\partial y^2} = E_y\psi_y\,, \quad -\frac{\hbar^2}{2m_e}\frac{\partial^2\psi_z}{\partial z^2} = E_z\psi_z\,, \tag{3.36}$$

These are differential equations for a particle in a one-dimensional box. The solution for the equation with x is, using Equation (3.7),

$$\psi_x = A_x\sin\left(\frac{n_x\pi x}{L_x}\right), \qquad E_x = \frac{\pi^2\hbar^2}{2m_eL_x^2} = \frac{h^2}{8m_eL_x^2} \tag{3.37}$$

(recall that $\hbar = h/(2\pi)$). The solutions for y and z are obtained correspondingly. Then for $\psi(x, y, z)$ we obtain

$$\psi(x,y,z) = A\,\sin\left(\frac{n_x\pi x}{L_x}\right)\,\sin\left(\frac{n_y\pi y}{L_y}\right)\,\sin\left(\frac{n_z\pi z}{L_z}\right) \tag{3.38}$$

where $A = A_xA_yA_z$ is a constant, $0 < x < L_x$, $0 < y < L_y$, and $0 < z < L_z$. This function is zero at the walls of the box for integer values of n_x, n_y, and n_z, thus fulfilling the boundary condition that the wavefunction must be zero outside the box. In Exercise 3.1 you are asked to confirm that Equation (3.38) is a solution of Equation (3.31).

For the energy E we obtain

$$E = E_x + E_y + E_z = \frac{h^2}{8m_e}\left(\frac{n_x^2}{L_x^2}+\frac{n_y^2}{L_y^2}+\frac{n_z^2}{L_z^2}\right) \tag{3.39}$$

A number of items are worth commenting on here. The energy levels are discrete, or quantized, and the energy level is specified by the values of the integers (n_x, n_y, n_z). We call these integers the *quantum numbers* because they allow us to clearly identify a particular energy and its wavefunction (Equations (3.39) and (3.38), respectively). In Fig. 3.7b we display the energies of a particle in a cubic box ($L_x = L_y = L_z = L$). For comparison, in Fig. 3.7a the energies of a particle between two parallel walls are shown. In addition, we note that a particular energy might have multiple sets of quantum numbers. For example, the set of quantum numbers (2,1,1) and the set of quantum numbers (1,2,1) in Fig. 3.7b correspond to the same energy but different wavefunctions. This property is called *degeneracy*; *i.e.*, a particular energy level may correspond to more than one wavefunction. Pairs of degenerate wavefunctions can be transformed into each other by symmetry operations; see Section 3.4.

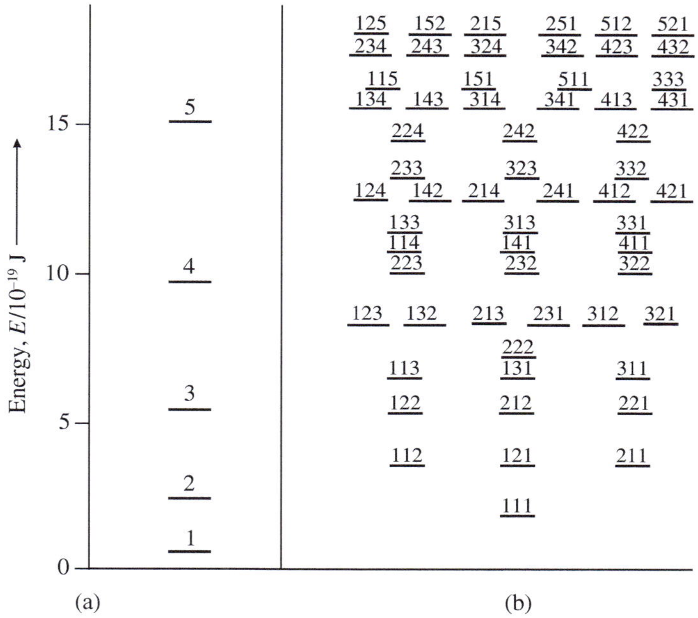

Figure 3.7 Energy diagrams for an electron in a box. (a) Electron between two parallel walls with a distance L ($L = 1000$ pm): $E = \frac{h^2}{8m_eL^2} \cdot n^2$. (b) Electron in a cubic box with side length L ($L = 1000$ pm): $E = \frac{h^2}{8m_eL^2} \cdot (n_x^2 + n_y^2 + n_z^2)$. The numbers above the energy levels correspond to the quantum numbers n_x, n_y, and n_z.

As an abbreviation in discussions about wavefunctions, it is common to identify the quantum numbers rather than the entire wavefunction.

3.3 Normalized Wavefunction: $\rho = \psi^2$

According to the postulates in Chapter 1, the probability $d\mathcal{P}$ that a particle appears in a volume element $d\tau = dx\,dy\,dz$ at a location x, y, z in space is proportional to $\psi^2(x, y, z)$ and $d\tau$:

$$d\mathcal{P} = C\,\psi^2(x,y,z)\,d\tau \tag{3.40}$$

where C is a constant. Moreover, the wavefunction ψ contains an arbitrary proportionality constant (e.g., the constant A in Equation (3.38)). We can choose this proportionality constant such that we can simplify Equation (3.40) by setting $C = 1$:

$$d\mathcal{P} = \psi^2(x,y,z)\,d\tau \tag{3.41}$$

Wavefunctions with the appropriate proportionality constant are called *normalized*.

How can we normalize a wavefunction? The probability $\mathcal{P}$ is understood as the ratio

$$\mathcal{P} = \frac{\text{number of trials in which a given event happens}}{\text{total number of trials}} \tag{3.42}$$

for an infinitely large number of trials. Given that we require the electron to be somewhere, the probability $\mathcal{P}_{\text{total}}$ of finding the electron anywhere in space is

$$\mathcal{P}_{\text{total}} = 1 \tag{3.43}$$

Thus we obtain the following as a normalization condition

$$\mathcal{P}_{\text{total}} = \int_{\text{all space}} d\mathcal{P} = \int_{\text{all space}} \psi^2(x,y,z)\, d\tau = 1 \tag{3.44}$$

3.3.1 One-Dimensional Box Function

Let us illustrate the normalization procedure by considering the wavefunction for an electron that is confined inside two infinitely high potential walls. In this case, the probability $d\mathcal{P}$ of finding an electron in the region dx is given by

$$d\mathcal{P} = \psi^2(x)\, dx = A^2 \sin^2\left(\frac{n\pi x}{L}\right) dx$$

where L is the box length and n is the quantum number. The probability of finding the electron in some region of space dx is determined by the area under the curve of ψ^2 in that region of space. To normalize the wavefunction; that is, scale the function so that the total area under its square is equal to unity, we must determine the constant A. Following the prescription of Equation (3.44) we require that

$$\int_0^L \psi^2(x)\, dx = A^2 \int_0^L \sin^2\left(\frac{n\pi x}{L}\right) dx = 1 \tag{3.45}$$

In order to evaluate the integral of the sine squared, we make the substitution $u = n\pi x/L$ and use the integral table in Appendix B to write

$$\int_0^L \sin^2\left(\frac{n\pi x}{L}\right) dx = \frac{L}{n\pi}\int_0^{n\pi} \sin^2 u\, du = \frac{L}{n\pi}\frac{n\pi}{2} = \frac{L}{2} \tag{3.46}$$

Hence Equation (3.45), becomes

$$A^2\frac{L}{2} = 1 \quad \text{so that} \quad A = \pm\sqrt{\frac{2}{L}} \tag{3.47}$$

Figure 3.8 shows plots of this wavefunction and its square for the case where $n = 1$. A few other points about the wavefunction should be made here. First ψ^2 must be single-valued. This property ensures that it corresponds to a unique probability for observing an electron at a particular location. Second, by using normalized wavefunctions, ψ is determined except for its sign, which can be chosen freely. For an electron between two walls in quantum state $n = 1$, ψ can be set to either

$$\psi = \sqrt{\frac{2}{L}} \cdot \sin\left(\frac{\pi x}{L}\right) \quad \text{or} \quad \psi = -\sqrt{\frac{2}{L}} \cdot \sin\left(\frac{\pi x}{L}\right)$$

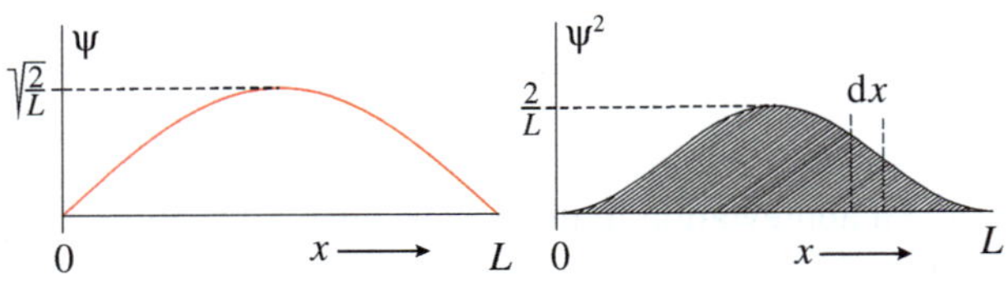

Figure 3.8 Normalized wavefunction $\psi(x) = A\sin(\pi x/L)$ and its square $\psi^2(x)$ with $A = \sqrt{2}/L$ for an electron between two walls a distance L apart and $n = 1$. The dashed area corresponds to the integral $\int \psi^2 \cdot dx$.

Both functions fulfill the Schrödinger Equation. For either choice, the probability density $\rho = \psi^2$ is the same, namely

$$\psi^2 = \frac{2}{L}\sin^2\left(\frac{\pi x}{L}\right)$$

Because the wavefunction ψ is not directly observable, its sign can be arbitrary. Third, the probability density ψ^2 must be normalizable; that is, the integral in Equation (3.44) should exist. The integral can only obtain a finite value if ψ for a bound electron is zero at infinity. So in accordance with (3.26) a normalizable wavefunction must have the property

$$\lim_{x,y,z\to\pm\infty} \psi(x,y,z) = 0$$

It is customary to write the normalization condition as

$$\boxed{\int \psi^2\, d\tau = 1} \tag{3.48}$$

and the integration is considered to proceed over all space.

3.3.2 Three-Dimensional Box Function

The probability density for a three-dimensional wavefunction, $\rho(x, y, z) = \psi^2(x, y, z)$, can be displayed as a cloud (Fig. 3.9). In such plots, the spatial locations of high probability are drawn with a high density of points, and the locations of low probability are drawn with a low density of points. Figure 3.9 shows a cloud for the wavefunction of an electron that is confined to a rectangular box. The nodal planes are clearly evident in these plots. Because the de Broglie wavelength decreases as the number of nodes increases, inspection of the plots shows that the energy associated with each of these wavefunctions increases from left to right.

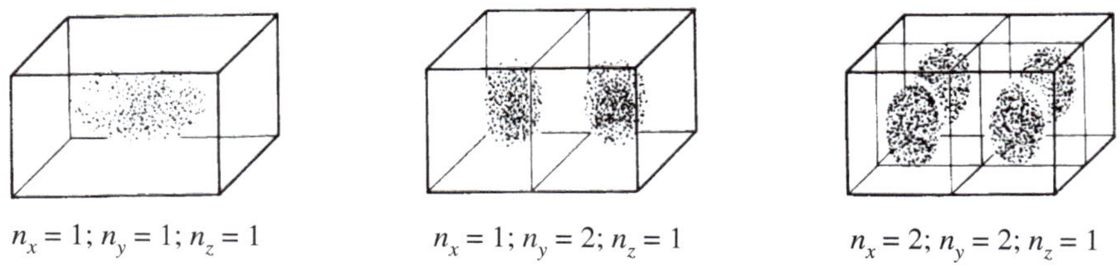

Figure 3.9 Electron in a box. Cloud pictures of the probability density ρ for three different eigenfunctions.

Example 3.3 *In order to properly evaluate the probability of the electron's wavefunction, it should be normalized. Normalize the wavefunction of an electron in a rectangular box.*

In this case, the probability is given by

$$d\mathcal{P} = \psi^2(x, y, z)\,dx\,dy\,dz \tag{3.49}$$

and we will have to perform a three-dimensional integration

$$\mathcal{P}_{total} = 1 = \int_{x=-\infty}^{+\infty} \int_{y=-\infty}^{+\infty} \int_{z=-\infty}^{+\infty} \psi^2(x, y, z)\,dx\,dy\,dz$$

If we substitute the wavefunction from Equation (3.38), we see that each integral can be performed independently, namely

$$\mathcal{P}_{total} = 1 = A^2 \int_0^{L_x} \sin^2 \frac{n_x \pi x}{L_x} dx \cdot \int_0^{L_y} \sin^2 \frac{n_y \pi y}{L_y} dy \cdot \int_0^{L_z} \sin^2 \frac{n_z \pi z}{L_z} dz$$

Each of these integrals may be evaluated like that shown in Equation (3.46). Hence we find

$$A^2 \cdot \frac{L_x}{2} \cdot \frac{L_y}{2} \cdot \frac{L_z}{2} = 1 \tag{3.50}$$

which can be rearranged to yield

$$A = \sqrt{\frac{8}{L_x L_y L_z}} \tag{3.51}$$

3.4 Orthogonality of the Wavefunctions

3.4.1 Non-Degenerate Wavefunctions

The electron wavefunctions have a useful property called *orthogonality*. By way of example, consider the two lowest energy wavefunctions for an electron confined to a one-dimensional box

$$\psi_1 = \sqrt{\frac{2}{L}} \sin\left(\frac{\pi x}{L}\right) \quad \text{and} \quad \psi_2 = \sqrt{\frac{2}{L}} \sin\left(\frac{2\pi x}{L}\right) \tag{3.52}$$

In Fig. 3.10a these functions are plotted along with their product, $\psi_1\psi_2$. From the graphs it is evident that ψ_1 is symmetric about the center point $x = L/2$; e.g., $\psi(\frac{1}{4}L) = \psi(\frac{3}{4}L)$. Also we see that ψ_2 and the product $\psi_1\psi_2$ are antisymmetric; that is,

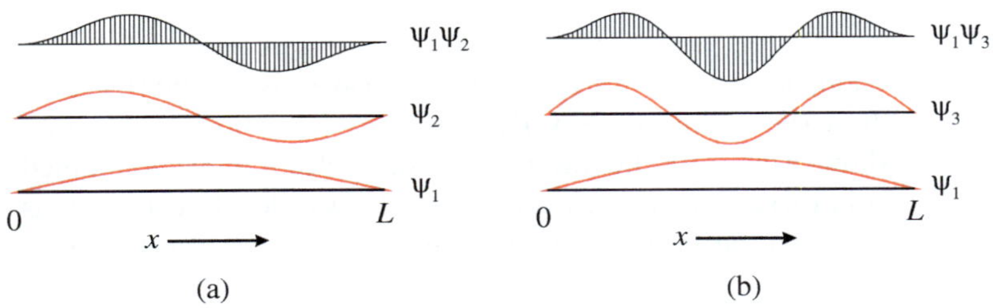

Figure 3.10 Orthogonality of wavefunctions. (a) Plots are shown of two wavefunctions, ψ_1 and ψ_2 for a particle between two parallel walls, and their product. The striped areas above the x-axis (positive sign) and those below the x-axis (negative sign) are of equal magnitude, so the net area is zero. (b) An analogous combination of plots is shown for wavefunctions ψ_1 and ψ_3.

$\psi(\frac{1}{4}L) = -\psi(\frac{3}{4}L)$, and each has a node at $x = L/2$. The net area of the product function, the two striped regions, is zero because the two halves are equal in area and opposite in sign. From these symmetry considerations we can write

$$\int_0^L \psi_1\psi_2 \, \mathrm{d}x = 0 \tag{3.53}$$

Example 3.4 *Show that the ψ_1 wavefunction (Equation (3.52)) and the wavefunction $\psi_3 = \sqrt{2/L}\,\sin(3\pi x/L)$ are orthogonal to each other.*

From the graph in Fig. 3.10b, we might expect that the functions are orthogonal; i.e.,

$$\int_0^L \psi_1 \, \psi_3 \, \mathrm{d}x = 0 \tag{3.54}$$

however let us evaluate the integral analytically to make sure. We first make the substitution $u = \pi x/L$ and obtain

$$\frac{2}{L}\int_0^L \sin\left(\frac{\pi x}{L}\right) \sin\left(\frac{3\pi x}{L}\right) \mathrm{d}x = \frac{2}{\pi}\int_0^\pi \sin u \, \sin 3u \, \mathrm{d}u$$

Now we use a trigonometric relationship (see Appendix B4)

$$\sin u \, \sin 3u = \frac{1}{2}(\cos 2u - \cos 4u)$$

which allows us to substitute for the product of the two sine functions,

$$\frac{2}{\pi}\int_0^L \sin\left(\frac{\pi x}{L}\right) \sin\left(\frac{3\pi x}{L}\right) \mathrm{d}x = \frac{1}{\pi}\int_0^\pi \cos 2u \, \mathrm{d}u - \frac{1}{\pi}\int_0^\pi \cos 4u \, \mathrm{d}u$$

Integrating each of the cosine functions is straightforward, and we find

$$\frac{1}{\pi}\int_0^\pi \cos 2u \, \mathrm{d}u - \frac{1}{\pi}\int_0^\pi \cos 4u \, \mathrm{d}u = \frac{1}{\pi}\left|\frac{\sin 2u}{2}\right|_0^\pi - \frac{1}{\pi}\left|\frac{\sin 4u}{4}\right|_0^\pi = 0$$

Since each of the integrals is zero, their difference is zero, and the wavefunctions are orthogonal.

Because the property of orthogonality is so useful, we will want to ensure that our wavefunctions are orthogonal. Although more difficult to see, with three-dimensional wavefunctions orthogonality also holds. Consider the wavefunctions ψ_{111}, ψ_{112}, and the product function $\psi_{111} \cdot \psi_{112}$ in Fig. 3.11a. Each volume element under the striped plane (Fig. 3.11b) is the mirror image of an element over the plane. In the integral

$$\int_{x=-\infty}^{+\infty}\int_{y=-\infty}^{+\infty}\int_{z=-\infty}^{+\infty} \psi_{111} \, \psi_{112} \, \mathrm{d}x \, \mathrm{d}y \, \mathrm{d}z$$

the two elements make the same contribution with opposite signs. Thus the integral has the value zero.

It can be shown that the orthogonality of wavefunctions of nondegenerate states is valid for solutions of any Schrödinger Equation (see Foundation 3.2). This property is important for understanding the electronic structure of atoms and molecules with many electrons.

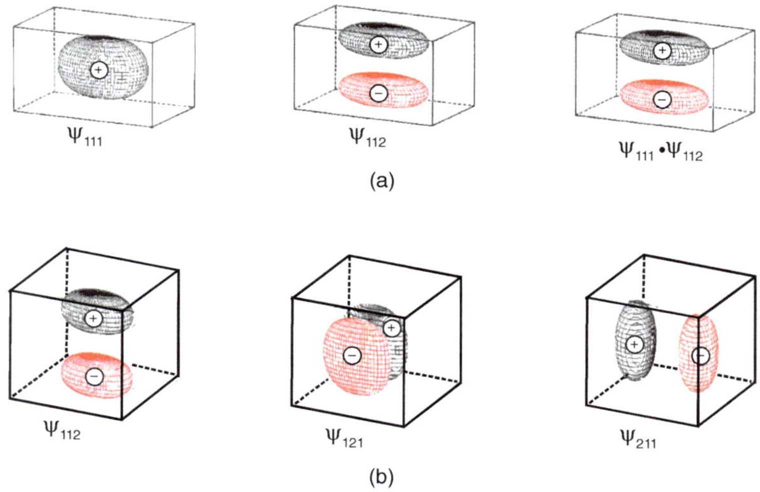

Figure 3.11 (a) Plot of box wavefunctions ψ_{111}, ψ_{112}, and their product function $\psi_{111}\psi_{112}$. (b) Degenerate box wavefunctions ψ_{112}, ψ_{121}, and ψ_{211} in the case of a cubic box.

3.4.2 Degenerate Wavefunctions

A special situation occurs for wavefunctions that correspond to the same energy level. For example, consider the energy level in Fig. 3.7b for which the energy of the wavefunctions 112, 121, and 211 are all the same (Fig. 3.11b). In this case any superposition of wavefunctions from this orthogonal set is a solution of the Schrödinger equation (e.g., $\psi = a\psi_{112} + b\psi_{121} + c\psi_{211}$, where a, b, and c are constants). In general, these functions may not necessarily be orthogonal to each other, so we will need to ensure that we choose such wavefunctions to be orthogonal. A set of n orthogonal wavefunctions corresponding to the same energy level are called an *n-fold degenerate state*.

Example 3.5 *Consider an electron that is confined inside four walls that are arranged in a square. Consider a quantum state with the wavefunction $\psi = a\psi_{12} + b\psi_{21}$, where the subscripts specify the quantum number for the electron-in-a-box wavefunctions along x and y; i.e., ψ_{n_x,n_y}. Find a second quantum state ψ' that has the same energy but is orthogonal to ψ.*

To begin, let us use the normalization requirement to find a relationship between the coefficients a and b. From the normalization conditions for ψ, ψ_{12}, and ψ_{21} we can write that

$$\int \psi^2\,\mathrm{d}\tau = \int \left(a\psi_{12} + b\psi_{21}\right)^2 \mathrm{d}\tau = a^2 \int \psi_{12}^2\,\mathrm{d}\tau + 2\cdot a\cdot b \int \psi_{12}\quad \psi_{21}\,\mathrm{d}\tau + b^2 \int \psi_{21}^2\,\mathrm{d}\tau$$
$$= a^2 + 0 + b^2 = 1, \qquad \text{i.e., } b = \sqrt{1-a^2}$$

where we have used the fact that ψ_{21} and ψ_{12} are orthogonal to each other. Therefore we can write

$$\psi = a\,\psi_{12} + \sqrt{1-a^2}\,\psi_{21}$$

The wavefunction ψ', which corresponds to the same energy as ψ, can be constructed from a linear combination of the same two degenerate wavefunctions ψ_{21} and ψ_{12}. If we begin by using the arbitrary combination $\psi' = c\psi_{12} + d\psi_{21}$and proceeding through a normalization procedure like that used for ψ, we find that $\psi' = c\psi_{12} + \sqrt{1-c^2}\psi_{21}$.

In order to ensure that ψ and ψ' are orthogonal, we require that $\int \psi\psi'\cdot\mathrm{d}\tau = 0$. Substituting our expressions for the wavefunctions we write

$$\int \psi\psi'\cdot\mathrm{d}\tau = a\,c \int \psi_{12}^2\cdot\mathrm{d}\tau + a\sqrt{1-c^2}\int \psi_{12}\psi_{21}\cdot\mathrm{d}\tau + \sqrt{1-a^2}\,c\int \psi_{21}\psi_{12}\cdot\mathrm{d}\tau + \sqrt{1-a^2}\sqrt{1-c^2}\int \psi_{21}^2\cdot\mathrm{d}\tau$$
$$= a\,c\int \psi_{12}^2\cdot\mathrm{d}\tau + 0 + 0 + \sqrt{1-a^2}\sqrt{1-c^2}\int \psi_{21}^2\cdot\mathrm{d}\tau = a\,c\cdot 1 + \sqrt{1-a^2}\sqrt{1-c^2}\cdot 1 = 0$$

Solving this algebraic equation for c, we find that $c = \pm\sqrt{1-a^2}$, so that

$$\psi' = \sqrt{1-a^2}\psi_{12} - a\psi_{21}$$

Thus, we were able to use the normalization condition for each wavefunction and the orthogonality condition (three equations) to find three of the four arbitrary constants (a, b, c, and d), but we are left with one coefficient that is not specified. This means, that any coefficient $0 \leqslant a \leqslant 1$ is acceptable; i.e., satisfies our requirement that the wavefunctions have the same energy and are orthogonal. Figure 3.12a plots the wavefunctions ψ and ψ' for three situations: $a = 0, 0.5$, and $1/\sqrt{2}$.

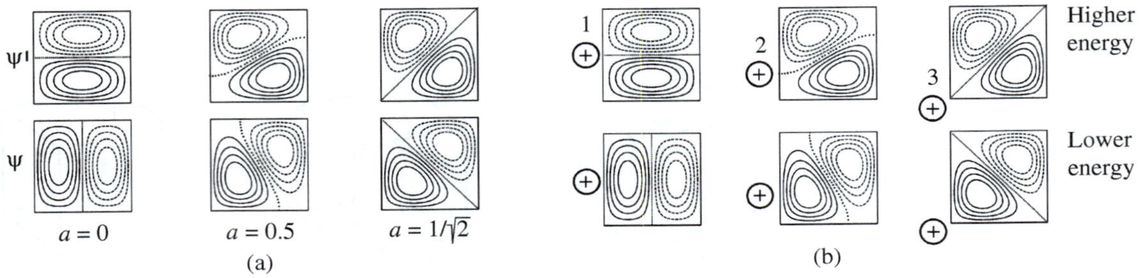

Figure 3.12 (a) Three pairs of degenerate wavefunctions ψ and ψ' are shown: $\psi = a\psi_{12} + \sqrt{1-a^2}\psi_{21}$, $\psi' = \sqrt{1-a^2}\psi_{12} + a\psi_{21}$ Each pair represents a different possible orthogonal set of wavefunctions that satisfy the Schrödinger equation. (b) Perturbation of the energy levels by applying a small charge at positions 1, 2, and 3 (leading to a splitting of the corresponding two energy levels).

3.4.3 Degeneracy Removed in an Electric Field

The ambiguity in describing a degenerate state by a set of orthogonal wavefunctions is removed by a small perturbation, such as a small charge. The charge adds an extra potential energy term to the Schrödinger equation and changes the energy of the different wavefunctions. If this extra term is small enough the energy changes slightly (a few percent or less), and the wavefunction corresponding to the lowest energy state and its orthogonal partner can describe the system reasonably well (Fig. 3.12b). Although the slight difference in energy is relevant, the small change in the wavefunctions, arising from the small perturbation, is often negligible.

This example illustrates the principle underlying the splitting of a spectral line that occurs in the spectroscopy of atoms and molecules by applying an electric field. The splitting and the frequency shift of the spectral lines occurs because the electric field shifts and breaks the degeneracy of the energy level(s) involved in the spectral transtion. This effect was first observed by J. Stark (1874–1957) in 1913 and is called the *Stark effect* [4].

3.5 Bohr Correspondence Principle and Generalized Form of Quantum Mechanics

In the preceding part we restricted ourselves, for simplicity, to systems that can be described by real wavefunctions. Now we include complex wavefunctions.[2] In particular, using real wavefunctions, we have postulated that the probability density is $\rho = \psi^2$. Wavefunctions can be complex, i.e., they can consist of a real part and an imaginary part:

$$\psi = a + \mathrm{i} \cdot b, \qquad \text{where } \mathrm{i} = \sqrt{-1}$$

Since ρ must be a real quantity, our postulate $\rho = \psi^2$ has to be replaced by

$$\rho = \psi\psi^* = |\psi|^2 \tag{3.55}$$

where $\psi^* = a - \mathrm{i} \cdot b$ is the complex conjugate of ψ. Thus the expression

$$\psi\psi^* = (a + \mathrm{i} \cdot b)(a - \mathrm{i} \cdot b) = a^2 - \mathrm{i}^2 \cdot b^2 = a^2 + b^2$$

becomes real. The postulate represented by Equation (3.55) is called the Born interpretation. Using complex wavefunctions we can write the postulates of quantum mechanics in a more elegant and general form.

2 If you are a beginner you are strongly encouraged to study the material on operator algebra in the Mathematical Appendix C.

3.5.1 Bohr Correspondence Principle

In this section we present an alternative approach to forming the postulates of quantum mechanics. In particular, we draw a correspondence between the classical and quantum mechanical pictures which allows the two models to be joined into a comprehensive and self-consistent whole. As we construct an alternative formalism for the quantum theory we want to ensure that it transforms into classical mechanics under the proper set of circumstances; this is called the *Correspondence Principle.*

3.5.1.1 Hamilton's Function

To build this correspondence into our description, we replace our postulate of the Schrödinger Equation by another postulate from which the Schrödinger Equation can be derived. From classical mechanics, we know that the total energy E of a particle of mass m_e is a sum of its potential energy V and its kinetic energy $T = m_e v^2/2 = p^2/(2m_e)$ where the momentum p is the product $m_e v$. For a "classical" particle of mass m_e in three-dimensional space, we can write that

$$H = \frac{1}{2m_e}\left(p_x^2 + p_y^2 + p_z^2\right) + V(x, y, z) = E \tag{3.56}$$

where p_x, p_y, and p_z are the components of the momentum in the three spatial directions and the potential energy is some function of space, but not momentum. H is called *Hamilton's function.* Our postulate will provide a recipe that converts equations of classical mechanics into quantum-mechanical expressions. For example it will allow us to convert Hamilton's function into the Hamiltonian operator in the case that $E = const$.

3.5.1.2 Operators p_{op} and x_{op}

We postulate that the classical momentum becomes the momentum operator p_{op}:

$$\boxed{p_{op} = -\mathrm{i}\frac{h}{2\pi}\frac{\partial}{\partial x} \quad \text{replaces the momentum } p_x} \tag{3.57}$$

and the classical displacement becomes the displacement operator x_{op}:

$$\boxed{x_{op} = x\cdot \quad \text{replaces the displacement } x} \tag{3.58}$$

Using these definitions we can build other operators, even the Hamiltonian operator $\mathcal{H}$. Note that Hamilton's function (Equation (3.56)) has quadratic terms, such as p_x^2. Because the square of p_x means that we multiply twice in succession (i.e., $p_x^2 = p_x \cdot p_x$), for the operator form it means we operate twice in succession. In this case we find that

$$p_x^2 = -\mathrm{i}\frac{h}{2\pi}\frac{\partial}{\partial x}\left(-\mathrm{i}\frac{h}{2\pi}\frac{\partial}{\partial x}\right) = \mathrm{i}^2\left(\frac{h}{2\pi}\right)^2\frac{\partial^2}{\partial x^2} = -\frac{h^2}{4\pi^2}\frac{\partial^2}{\partial x^2} \tag{3.59}$$

3.5.1.3 Hamiltonian Operator and Schrödinger Equation

The corresponding process can be performed for p_y and p_z. In this way, we can use our postulates to transform Hamilton's function into the Hamiltonian operator

$$\mathcal{H} = -\frac{h^2}{8\pi^2 m_e}\left(\frac{\partial^2}{\partial x^2} + \frac{\partial^2}{\partial y^2} + \frac{\partial^2}{\partial z^2}\right) + V(x, y, z) \tag{3.60}$$

This approach gives the time-independent Schrödinger equation.

$$\mathcal{H}\,\psi = E\psi \tag{3.61}$$

If we apply the Hamiltonian operator on the wavefunction ψ then, by correspondence with Hamilton's function (Equation (3.56)), it provides the energy E times the wavefunction ψ.

3.5.2 Electron Moving on a Circle

To illustrate this approach to quantum mechanics further, consider the case of an electron that is confined to the circumference of a circle. This simplified model system is useful for illustrating aspects of electron orbits in atoms and molecules.

3.5.2.1 Representation by Real Wavefunctions

For an electron moving on a circle, the boundary condition for the wavefunction ψ is

$$\psi(\varphi + 2\pi) = \psi(\varphi) \tag{3.62}$$

where φ is the angle of rotation. This condition ensures that $\rho = |\psi|^2$ is single valued; i.e., the probability density is well defined. In the simplest case, this condition is fulfilled for a constant value of the wavefunction along the ring

$$\psi_0 = \frac{1}{\sqrt{2\pi}} \tag{3.63}$$

Also, this condition is fulfilled if the de Broglie wavelength of the electron is $\Lambda = L_c$ (L_c = circumference of the circle), for which the electron can be described by two degenerate standing waves (Fig. 3.13a)

$$\psi_a = \frac{1}{\sqrt{\pi}} \cos\varphi, \quad \psi_b = \frac{1}{\sqrt{\pi}} \sin\varphi \tag{3.64}$$

that have the energy

$$E = \frac{1}{2} m_e v^2 = \frac{1}{2} m_e \left(\frac{h}{m_e \Lambda} \right)^2 = \frac{h^2}{2 m_e L_c^2}$$

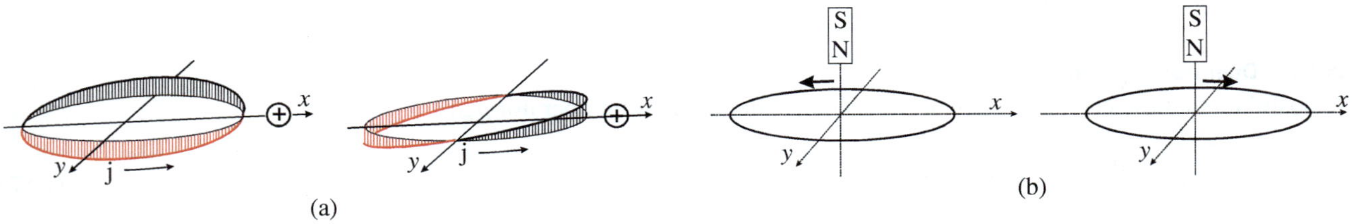

Figure 3.13 Electron on a ring with degeneracy removed. (a) In proximity of a positive charge: standing wave (wavefunctions $A \sin\varphi$ and $A \cos\varphi$, A = normalization factor) with node at charge (left, high energy), and with an antinode at charge (right, low energy). (b) In proximity of a magnet: wavefunctions $Ae^{i\varphi}$ and $Ae^{-i\varphi}$, electron circulating anticlockwise (left, high energy) and clockwise (right, low energy).

3.5.2.2 Representation by Complex Wavefunction: Clockwise and Counterclockwise Rotation

Remembering the ambiguity in our choice of an orthogonal set of degenerate orbitals, we may use complex functions instead of the standing waves:

$$\psi_1 = \frac{1}{\sqrt{2}} (\psi_a + i\psi_b), \quad \psi_{-1} = \frac{1}{\sqrt{2}} (\psi_a - i\psi_b) \tag{3.65}$$

or

$$\psi_1 = \frac{1}{\sqrt{2\pi}} (\cos\varphi + i\sin\varphi), \quad \psi_{-1} = \frac{1}{\sqrt{2\pi}} (\cos\varphi - i\sin\varphi)$$

where $i = \sqrt{-1}$. Using the Euler relations (see Mathematical Appendix C)

$$e^{i\varphi} = \cos\varphi + i\sin\varphi, \qquad e^{-i\varphi} = \cos\varphi - i\sin\varphi$$

we obtain

$$\psi_1 = \frac{1}{\sqrt{2\pi}} e^{i\varphi}, \quad \psi_{-1} = \frac{1}{\sqrt{2\pi}} e^{-i\varphi} \tag{3.66}$$

What is the benefit of using functions with imaginary parts? The traveling waves ψ_1 and ψ_{-1} correspond to an electron with a quantized velocity v in the positive and negative φ-direction (clockwise and counterclockwise), in addition to their energy degeneracy. We have seen in Section 3.4.3 that a degenerate state splits into two non-degenerate states approximated by wavefunction (3.64), when applying an electric field in the x direction, see Figs. 3.12 and 3.13a. The wavefunctions in Equation (3.66) correspond to distinct angular momenta, and their energies split when exposed to a magnetic field.

Degeneracy Removed by Magnetic Field The movement of the electron's negative charge in its orbit creates a ring current which generates a magnetic field perpendicular to the plane of the circle, in the z direction. If we place a magnet above the plane as shown in Fig. 3.13b an attractive force results for the clockwise rotation and a repulsive force for the counter-clockwise rotation. In this way the two energy states become non-degenerate. The splitting of spectral lines into two energy states by applying a magnetic field was first observed by Pieter Zeeman (1865–1943) in 1896 and is called the *Zeeman effect*.

3.5.2.3 General Solution and Connection to Angular Momentum

In classical physics, the motion along a circle in the x, y plane corresponds to a body of mass m orbiting at a distance r, where r is given by $x^2 + y^2 = r^2$, about its center. The angular kinetic energy of the body is

$$T_{\text{ang}} = \frac{1}{2}mv^2 = \frac{1}{2}mr^2\left(\frac{\mathrm{d}\varphi}{\mathrm{d}t}\right)^2 = \frac{1}{2}I\left(\frac{\mathrm{d}\varphi}{\mathrm{d}t}\right)^2$$

where $I = mr^2$ is the moment of inertia and $\mathrm{d}\varphi/\mathrm{d}t$ is the angular velocity. The rotating body possesses an angular momentum along the z-direction, which is given by

$$\vec{L} = mrv = mr^2\frac{\mathrm{d}\varphi}{\mathrm{d}t} = I\frac{\mathrm{d}\varphi}{\mathrm{d}t}$$

By elimination of the angular velocity we obtain

$$T_{\text{ang}} = \frac{|L|^2}{2I}$$

where $|L|^2$ is the dot product of the angular momentum vector with itself, $|L|^2 = \vec{L} \bullet \vec{L}$.

The general definition of the angular momentum is

$$\vec{L} = \vec{r} \times \vec{p} \tag{3.67}$$

For the specific case of a body rotating in the xy-plane we can describe the angular momentum vector in the z-direction by its components xp_y and yp_x as

$$\vec{L_z} = x\,p_y - y\,p_x \tag{3.68}$$

where x and y are cartesian coordinates and $p_y = mv_y$ and $p_x = mv_x$ are the momenta in the x and y directions (see paragraph on vector algebra in the Mathematical Appendix C).

To describe the orbital motion quantum-mechanically, we use our definition of the quantum-mechanical operators and find

$$\mathcal{L}_{\text{op},z} = x_{\text{op}}\ p_{\text{op},y} - y_{\text{op}}\ p_{\text{op},x} = -\frac{ih}{2\pi}\left(x\frac{\partial}{\partial y} - y\frac{\partial}{\partial x}\right) \tag{3.69}$$

To simplify the mathematics, we change from cartesian coordinates to polar coordinates. You show in Problem 3.11 that we can write Equation (3.69) as

$$\mathcal{L}_{z,\text{op}} = -\frac{ih}{2\pi}\frac{\partial}{\partial\varphi} \tag{3.70}$$

which makes the one-dimensional nature of the motion more apparent. By using our operator expression for the angular momentum, we can write the angular kinetic energy as

$$\mathcal{H}\psi = \frac{\mathcal{L}^2_{\text{op},z}}{2I}\psi = -\frac{h^2}{8\pi^2 I}\frac{\partial^2}{\partial\varphi^2}\psi = E\psi \tag{3.71}$$

This equation has the solution $\psi = A\mathrm{e}^{im\varphi}$ (see Appendix C) where A and m are constants which need to be determined. Applying our boundary condition, Equation (3.62), requires that the product $2\pi m$ be either zero ($m = 0$) or a multiple of 2π ($m = \pm1, \pm2, \ldots$). You are asked to confirm these facts in Exercise 3.13. The coefficient A is determined by applying the normalization condition, and we find

$$\psi_m = \frac{1}{\sqrt{2\pi}}\mathrm{e}^{im\varphi}, \qquad m = 0, \pm1, \pm2, \ldots \tag{3.72}$$

Inserting this wavefunction into the Equation (3.71),

$$\mathcal{H}\psi = \frac{\mathcal{L}^2_{\text{op},z}}{2I} = \frac{h^2}{8\pi^2}\frac{\partial^2}{\partial\varphi^2}\left(\frac{1}{\sqrt{2\pi}}\mathrm{e}^{im\varphi}\right) = -\frac{h^2}{8\pi^2 I}\left(-m^2\right)\left(\frac{1}{\sqrt{2\pi}}\mathrm{e}^{im\varphi}\right) = \frac{m^2h^2}{8\pi^2 I}\psi = E\psi$$

To summarize, we find

$$\frac{\mathcal{L}^2_{op,z}}{2I}\psi = E\psi \qquad \text{and} \qquad E = \frac{m^2h^2}{8\pi^2 I} \qquad \text{with } m = 0, \pm 1, \pm 2, \ldots \tag{3.73}$$

Equation (3.73) is an eigenvalue equation, in which $\frac{1}{\sqrt{2\pi}}e^{im\varphi}$ is an eigenfunction of the angular kinetic energy operator and $\frac{m^2h^2}{8\pi^2 I}$ is its eigenvalue. In contrast to the standing wave solutions described above, the traveling wave solutions $\frac{1}{\sqrt{2\pi}}e^{im\varphi}$ are also an eigenfunction of the angular momentum itself.

Example 3.6 *Show that the wavefunction in Equation (3.72) is an eigenfunction for the angular momentum operator* $\mathcal{L}_{op,z}$ *and determine its eigenvalue.*

An eigenvalue equation has the form

$$\boxed{A_{op}\psi = a_n\,\psi}$$

where A_{op} is the mathematical operator (in our case, $\mathcal{L}_{op,z} = -\frac{ih}{2\pi}\frac{\partial}{\partial\varphi}$), ψ is the eigenfunction (in our case, $\frac{1}{\sqrt{2\pi}}e^{im\varphi}$), and a_n is the eigenvalue, which we will determine. To prove that the wavefunction (Equation (3.72)) is an eigenfunction of $\mathcal{L}_{op,z}$, we perform the operation on the wavefunction and show that it produces a constant (the eigenvalue) times the wavefunction; i.e.,

$$\mathcal{L}_{op,z}\psi = -\frac{ih}{2\pi}\frac{\partial}{\partial\varphi}\left(\frac{1}{\sqrt{2\pi}}e^{im\varphi}\right) = \frac{mh}{2\pi}\frac{1}{\sqrt{2\pi}}e^{im\varphi} = L_z\psi \tag{3.74}$$

where L_z is the eigenvalue. Inspection of the result shows that

$$L_z = \frac{mh}{2\pi} = m\hbar \text{ with } m = 0, \ \pm 1, \ \pm 2, \ ..$$

3.5.3 Operator $i\hbar\partial/\partial t$ and the Time-Dependent Schrödinger Equation

In the following we are interested in cases where the probability density ρ depends on space and time

$$\rho(x, y, z, t) = \Psi\ \Psi^* \tag{3.75}$$

where $\Psi(x, y, z, t)$ is the time-dependent Schrödinger wavefunction and Ψ^* is its conjugate. In Section 3.5.1 we obtained the time-independent Schrödinger equation by first considering Hamiltonian's function for the energy E being constant. In the general case E depends on time. We postulate that in this case the operator

$$i\hbar\frac{\partial}{\partial t}$$

replaces the energy. This postulate converts the classical equation

$$H = E$$

into the time-dependent Schrödinger equation

$$\mathcal{H}\Psi = i\hbar\frac{\partial\Psi}{\partial t} \tag{3.76}$$

In the section below, we show that the solution of the time-independent Schrödinger Equation (3.25) is a special case of the time-dependent Schrödinger equation.

Equation (3.76) describes the time evolution of a quantum-mechanical system, via the time evolution of the wavefunction. Because $\partial\Psi/\partial t$ is the first time derivative of the wavefunction, it is possible to calculate Ψ at any time, if we know $\Psi(t = 0)$

and $\mathcal{H}$. Consider that we can solve the time-independent equation

$$\mathcal{H}\psi_n = E_n\psi_n \tag{3.77}$$

where the subscript n acts as a quantum number to label the different standing wave solutions, or "stationary states." Solution of the time-independent equation provides the energy values and the wavefunctions for all possible energy states of the atom. In the following we show that these stationary-state solutions can be used to construct the time-dependent solution for Equation (3.76), so that

$$\Psi = \sum_n c_n\psi_n \exp\left(-\mathrm{i}\frac{2\pi E_n}{h}t\right) \tag{3.78}$$

The time-independent form of the Schrödinger Equation is valid for the special case in which $\rho = \psi\psi^*$ does not depend on time, $\rho = \rho(x, y, z)$, and is sufficient for almost all problems discussed in this book. Schrödinger generalized his postulate for the case that ρ depends on time, and this can be important for understanding aspects of chemical reactions and time-dependent spectroscopy. We have replaced one postulate (namely, the time-independent Schrödinger Equation) by a more general one.

3.5.3.1 Relation Between Time-Independent and Time-Dependent Schrödinger Equation

It should be clarified why the operator $\mathrm{i}\hbar\partial/\partial t$ replaces the energy E for a system in motion. This will be shown in the following by considering the relation between the time-independent and the time-dependent Schrödinger equation.

Consider an electron in a quantum state specified by the quantum number n. This electron has the energy E_n and the probability density distribution $\rho_n = \psi_n^2$, which are determined by

$$\mathcal{H}\psi_n = E_n\psi_n$$

We shall show that this time-independent Schrödinger Equation describes a special case of the time-dependent Schrödinger Equation

$$\mathcal{H}\Psi = \mathrm{i}\hbar\frac{\partial\Psi}{\partial t}$$

where Ψ is a function of position and time. We consider $\mathcal{H}$ as time-independent – that is, the electron moves in a time-independent potential field.

We proceed by assuming that the function

$$\Psi = \psi_n \cdot \mathrm{e}^{\mathrm{i}at}$$

describes the same stationary state as ψ_n, if the constant a is chosen appropriately. If this assumption is true, then the product $\Psi\Psi^*$ must be identical to ψ_n^2. Indeed, this is the case:

$$\Psi\Psi^* = \psi_n \cdot \mathrm{e}^{\mathrm{i}at} \cdot \psi_n \cdot \mathrm{e}^{-\mathrm{i}at} = \psi_n^2$$

Also, Ψ must be a solution of the time-dependent Schrödinger Equation. Substituting our form for the time-dependent wavefunction into the time-dependent Schrödinger equation, we find

$$\mathcal{H}\psi_n\mathrm{e}^{\mathrm{i}at} = \mathrm{i}\hbar\frac{\partial\psi_n\mathrm{e}^{\mathrm{i}at}}{\partial t}$$

which leads to

$$\mathrm{e}^{\mathrm{i}at}\mathcal{H}\psi_n = \mathrm{i}\hbar\psi_n\frac{\mathrm{d}\mathrm{e}^{\mathrm{i}at}}{\mathrm{d}t} = \mathrm{i}\hbar\psi_n \cdot \mathrm{i}a \cdot \mathrm{e}^{\mathrm{i}at}$$

Since $\mathrm{i}^2 = -1$,

$$\mathcal{H}\psi_n = -\hbar a \cdot \psi_n$$

Comparison of this result with the time-independent Schrödinger Equation shows that they are the same if

$$\hbar a = -E_n$$

Thus the time-dependent wavefunction is a traveling wave with the frequency $E_n/(2\pi\hbar) = E_n/h$.

3.5.4 Average Value of an Observable (also Called Expectation Value)

In Chapter 2 the average distance $\overline{r}$ between the proton and an electron was written as

$$\overline{r} = \int_{\text{all space}} \rho \cdot r \, d\tau = \int_{\text{all space}} \psi \cdot r \cdot \psi \, d\tau$$

and the average potential energy $\overline{V}$ for an electron in the 1s state of the hydrogen atom was written as

$$\overline{V} = \int_{\text{all space}} \psi \cdot V(r) \cdot \psi \, d\tau$$

where $\psi(x, y, z)$ is the normalized time-independent wavefunction.

Now we consider the general case of a system represented by the normalized wavefunction $\psi(x, y, z)$, and we seek the average value of an observable $A(x, y, z, p_x, p_y, p_z)$. Generalizing our earlier results in Section 3.5.1.2, we postulate that the average $\overline{A}$ is given by

$$\overline{A} = \int_{\text{all space}} \psi^* \cdot A_{\text{op}} \cdot \psi \, d\tau \tag{3.79}$$

Note that in Equation (3.57) we used $A_{\text{op}} = p_{op,x} = -i\hbar\partial/\partial x$ for the operator of the momentum in the x-direction by replacing p_x by $-i\hbar \cdot \partial/\partial x$. Some examples for the application of Equation (3.79) are shown in Table 3.1.

Table 3.1 Examples for Calculating the Averages of Different Quantities.

Quantity		Operator	Average
Position	$\overline{x}$	$x_{\text{op}} = x$	$\overline{x} = \int \psi^* x_{\text{op}} \psi \, d\tau = \int \psi^* x \psi \, d\tau$
(Position) ^{2}	$\overline{x^2}$	$x_{\text{op}}\left(x_{\text{op}}\right)^2 = x^2$	$\overline{x^2} = \int \psi^* \left(x_{\text{op}}\right)^2 \psi \, d\tau = \int \psi^* x^2 \psi \, d\tau$
Kinetic energy	$\overline{T}$	$T_{\text{op}} = -\frac{\hbar^2}{2m} \cdot d^2/dx^2$	$\overline{T} = \int \psi^* T_{\text{op}} \psi \, d\tau = -\frac{\hbar^2}{2m} \cdot \int \psi^* \frac{d^2}{dx^2} \psi \, d\tau$
Momentum	$\overline{p_x}$	$p_{\text{op,x}} = -i\hbar \cdot d/dx$	$\overline{p_x} = \int \psi^* p_{\text{op,x}} \psi \, d\tau = -i\hbar \cdot \int \psi^* \frac{d}{dx} \psi \, d\tau$
(Momentum) ^{2}	$\overline{(p_x)^2}$	$(p_{\text{op,x}})^2 = -\hbar \cdot d^2/dx^2$	$\overline{(p_x)^2} = \int \psi^* (p_{\text{op,x}})^2 \psi \, d\tau = -\hbar^2 \cdot \int \psi^* \frac{d^2}{dx^2} \psi \, d\tau$

Relation (3.79) will be used in deriving the uncertainty principle (see Foundation 3.3, which is available electronically).

3.5.5 Heisenberg Uncertainty Principle

In quantum mechanics, it is not possible to know simultaneously the position and momentum of a particle exactly. This feature is a manifestation of the *Uncertainty Principle*. The Uncertainty Principle is important because it restricts what is possible to know and measure about a physical system. If we define an uncertainty in the position of the electron Δx and the uncertainty in its momentum Δp in the x-direction in terms of its standard deviation from the average value, then the Uncertainty Principle requires that

$$\boxed{\Delta x \cdot \Delta p \geq \frac{\hbar}{2}} \tag{3.80}$$

Foundation 3.3 (available electronically) provides a rigorous derivation of this equation. Corresponding products can be written for certain other pairs of observables, such as $\Delta E \cdot \Delta t \geq \hbar/2$, see Foundation 3.4 (available electronically).

By way of illustration let us estimate the uncertainty product for a particle that is confined to a one-dimensional box with walls, located at $x = -L/2$ and $x = L/2$. The particle's ground state wavefunction ($n = 1$) is given by

$$\psi(x) = \cos\left(\frac{\pi}{L}x\right) \tag{3.81}$$

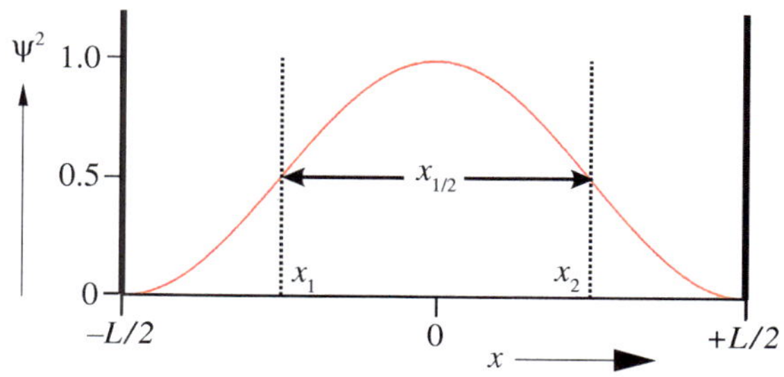

Figure 3.14 Particle confined to a one-dimensional box with walls at $x = -L/2$ and $x = -L/2$. $\psi^2 = \cos^2\left(\frac{\pi}{L}x\right)$ is plotted versus the x-coordinate. The dashed lines indicate the full-width-at-half-maximum $x_{1/2} = x_2 - x_1$.

where the zero point of the coordinate system is at the center of the box. Figure 3.14 shows a plot of the function $\psi^2(x) = \cos^2\left(\frac{\pi}{L}x\right)$, which is proportional to the probability density ρ. We find that it is more probable that the particle can be found near the center of the box than near its walls. To account for this fact, we assume that the uncertainty Δx of the particle's position will be proportional to the full-width-at-half-maximum $x_{1/2}$ of the curve for ψ^2 in Fig. 3.14. To calculate $x_{1/2}$ we set

$$\cos^2\left(\frac{\pi}{L}x_1\right) = \cos^2\left(\frac{\pi}{L}x_2\right) = \frac{1}{2}$$

to obtain

$$x_1 = -\frac{1}{4}L \qquad x_2 = \frac{3}{4}L \qquad x_{1/2} = x_2 - x_1 = \frac{1}{2}L$$

As the probability density is centered around the average value $x = 0$, the deviation from the average is about half of $x_{1/2}$; we approximate the uncertainty in the particle's position by

$$\Delta x = \frac{1}{2} \cdot x_{1/2} = \frac{1}{4}L \tag{3.82}$$

The particle's speed v is, according to the de Broglie relation,

$$v = \frac{h}{m\Lambda} = \frac{h}{2mL}$$

Thus the magnitude of the particle's momentum p is

$$p = mv = \frac{h}{\Lambda} = \frac{h}{2L}$$

Because the particle can move in the positive as well as the negative direction, the average momentum will be zero. Thus, the deviation of the momentum from the average is either p or $-p$

$$\Delta p = \frac{1}{2}(p - (-p)) = p = \frac{h}{2L} \tag{3.83}$$

In this way we find

$$\Delta x \cdot \Delta p = \frac{1}{4}L \cdot \frac{1}{2}\frac{h}{L} = h \cdot 0.125 = \hbar \cdot 0.787 \geq \frac{\hbar}{2} \tag{3.84}$$

This value is in accordance with the Uncertainty Principle, Equation (3.80).

3.5.5.1 Superposition and the Uncertainty Principle

We can extend the consideration above by superimposing waves of different wavelengths, hence different momenta. The waves will interfere and the spatial dependence of the total wavefunction shows structure. This change in the waveform leads to a more precise knowledge of the particle's position and a less precise knowledge of its momentum. We illustrate this effect by considering the superposition

$$\psi = \frac{2}{(n_{max}+1)} \sum_{n=1}^{n_{max}} \cos\left(\frac{2\pi}{\Lambda_0}nx\right) \qquad n = 1, 3, 5, \ldots, n_{max} \tag{3.85}$$

where $\Lambda_0 = 2L$ and we are using cosine functions with the zero point of the x-axis at the center of the box as we did in Fig. 3.14. The restriction of the values of n to uneven integers ($n = 1, 3, 5, \ldots, n_{max}$) is made in order to obtain standing waves with a maximum at the center of the box. The factor $2/(n_{max} + 1)$ ensures that the function has a constant value $\psi = 1.0$ at $x = 0$. Thus, this wavefunction is not normalized.

The ψ obtained from Equation (3.85) is composed of de Broglie wavelengths ranging from $2L$ to $2L/n_{\text{max}}$. Let us use $2L/n_{\text{max}}$ as an estimate of the magnitude for the momentum

$$p = mv = \frac{h}{\Lambda} = \frac{h}{2L} n_{\text{max}}$$

Following the same arguments as in the first example, for the uncertainty of the momentum Δp we obtain

$$\Delta p = \frac{1}{2}(p - (-p)) = p = \frac{h}{2L} n_{\text{max}}$$

In the following we consider the cases $n_{\text{max}} = 5$ and $n_{\text{max}} = 39$ to calculate

$$\Delta p = \frac{h}{L} \cdot 2.5 \text{ for } n_{\text{max}} = 5 \text{ and } \Delta p = \frac{h}{L} \cdot 19.5 \text{ for } n_{\text{max}} = 39$$

From Fig. 3.15 with $\Lambda_0 = 120$ nm and $L = 60$ nm we obtain $\Delta x = \frac{1}{2} x_{1/2} = \frac{1}{2} \cdot 8.96 \text{ nm} = 4.48$ nm. In the calculation of the uncertainty product it will be helpful to express Δx as a fraction of the box length; we find $\Delta x = L \cdot \frac{4.48}{60} = L \cdot 0.0747$ for $n_{\text{max}} = 5$. For $n_{\text{max}} = 39$, we find $\Delta x = \frac{1}{2} \cdot 1.32 \text{ nm} = 0.660$ nm, so that $\Delta x = L \cdot \frac{0.660}{60} = L \cdot 0.0110$. Thus for the uncertainty product we obtain

$$\Delta p \cdot \Delta x = \frac{h}{L} \cdot 2.5 \cdot L \cdot 0.0747 = h \cdot 0.187 \text{ for } n_{\text{max}} = 5$$

and

$$\Delta p \cdot \Delta x = \frac{h}{L} \cdot 19.5 \cdot L \cdot 0.0110 = h \cdot 0.215 \text{ for } n_{\text{max}} = 39$$

Both values are greater than $\hbar/2 = h \cdot 0.0795$, as expected from the Uncertainty Principle. The uncertainty product obtained by this approximate approach is essentially in agreement with that obtained from a rigorous calculation, which is provided in Foundation 3.5 (available electronically)

$$\Delta p \cdot \Delta x = h \cdot 0.130 \text{ for } n_{\text{max}} = 5 \text{ and } \Delta p \cdot \Delta x = h \cdot 0.130 \text{ for } n_{\text{max}} = 39$$

In the approximate method Δp is overestimated, because it is based on the contribution of the wave with the shortest wavelength. Also it appears that the assumption that Δx is just one-half of the full-width-at-half-maximum $x_{1/2}$ overestimates Δx.

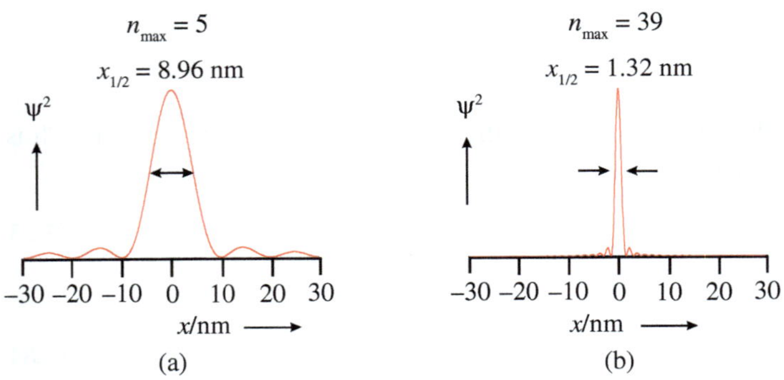

Figure 3.15 Plot of ψ^2 for the wavefunction ψ in Equation (3.85), assuming a value of 120 nm for Λ_0 and 60 nm for L (that is, x is in the range from −30 nm to +30 nm). Comparison of the full-width-at-half-maximum, $x_{1/2}$. (a) Calculation for $n_{\text{max}} = 5$, (b) calculation for $n_{\text{max}} = 39$.

3.6 Summary of Postulates of Quantum Mechanics

1) The probability density ρ of a state of a quantum mechanical system is given by a function Ψ (Schrödinger wavefunction)

$$\boxed{\rho = \Psi\Psi^*}$$

where Ψ^* is the complex conjugate of Ψ. This postulate corresponds with those made in Chapter 1 (pages 7 and 10); namely 'The probability that a particle is detected in a small area at a position x, y, z is proportional to this area and proportional to the probability density ρ'.

2) To every observable in classical mechanics there exists a corresponding operator in quantum mechanics.

$$\boxed{p_{\text{op},x} = -i\hbar \frac{\partial}{\partial x} \quad \text{replaces momentum } p_x}$$

$$\boxed{x_{op} = x \cdot \quad \text{replaces the displacement } x}$$

and

$$\boxed{\mathcal{L}_{op,z} = -i\hbar\left(x\frac{\partial}{\partial y} - y\frac{\partial}{\partial x}\right) \quad \text{replaces the angular momentum } \vec{L}_z}$$

are some examples. The operators have the mathematical property of being *Hermitian*, which ensures that the observable property we compute is a real valued number; see postulate 5 below.

3) In any measurement of an observable associated with the operator A_{op} the only values that will ever be observed are the eigenvalues which satisfy the eigenvalue equation

$$\boxed{A_{op}\psi = a_n\psi}$$

We considered the following eigenvalue equations:

$$\boxed{\mathcal{H}\psi_n = E_n\psi_n} \qquad \text{and} \qquad \boxed{\mathcal{L}_{op,z}\psi = L_z\psi}$$

4) Time-dependent Schrödinger equation

$$\boxed{\mathcal{H}\Psi = i\hbar\frac{\partial\Psi}{\partial t}}$$

5) Average value of observable A (expectation value) is given by

$$\overline{A} = \int_{\text{all space}} \psi^* \cdot A_{op} \cdot \psi \, d\tau$$

and $\overline{A}$ is a real number, This latter requirement means that

$$\int_{-\infty}^{\infty} \psi_i^* A\psi_j \, d\tau = \int_{-\infty}^{\infty} \psi_j A^* \psi_i^* \, d\tau \tag{3.86}$$

where ψ_1 and ψ_2 are eigenfunctions, and we say that our operator A_{op} is *Hermitian*; see Appendix C4. In Foundation 3.2, we show that the Hamiltonian operator has this property.

3.7 Conclusion

This chapter developed a number of important fundamental features of quantum mechanics. Most important for our development of physical chemistry is the introduction of the time-independent Schrödinger equation as a way to calculate the energy levels and find the wavefunctions of a system. This equation was introduced as a postulate. First, it was applied to the model problem of an electron confined to a potential energy, well and the results were shown to be consistent with our more qualitative considerations in Chapter 2. Next, we introduced a different postulatory approach to the construction of quantum-mechanical quantities based on Bohr's correspondence principle.

3.8 Box 3.1. Electron in a Potential Energy Well of Finite Depth

Consider an electron that is confined in a one-dimensional energy well (or box) with walls of finite height (see Fig. 3.2). In this case, the potential energy is given by

$$V(x) = 0 \qquad \text{for } -\frac{L}{2} \le x \le +\frac{L}{2}$$

$$V(x) = V_0 \quad \text{for } x < -\frac{L}{2} \quad \text{and} \quad x > +\frac{L}{2}$$

This potential energy diagram is said to be "piecewise" continuous which means that it is a continuous function in each of the three regions but not at the points where they meet. The way to proceed with such a situation is to solve the Schrödinger equation in each region alone and then require that the wavefunctions in the two regions meet up at the boundary in a smooth way (more on that below). For convenience, we choose the origin of the coordinate system in the center between the walls and consider the case where V_0 is larger than the energy of the electron. If the electron's energy were larger than V_0 then it would not be confined to the well.

We will have two different differential equations. When the electron is localized between the walls the Schrödinger equation is

$$-\frac{h^2}{8\pi^2 m_e}\frac{d^2\psi_{well}}{dx^2} = E\psi_{well} \quad \text{for} \ -\frac{L}{2} \leq x \leq +\frac{L}{2} \tag{3.87}$$

and, when it is in the region outside the walls it is given by

$$-\frac{h^2}{8\pi^2 m_e}\frac{d^2\psi}{dx^2} + V_0\psi = E\psi \quad \text{for } x < -\frac{L}{2} \text{ and } x > +\frac{L}{2} \tag{3.88}$$

3.8.1 Symmetric Solutions

Let us consider the symmetric solutions (that is $\psi(x) = \psi(-x)$) first. We show in Foundation 3.1 (available electronically) that the wavefunction in this case is

$$\psi(x) = \begin{cases} B \cdot e^{bx} & (\text{for } x < -L/2) \\ A \cdot \cos(wx/L) & (\text{for } -L/2 < x < L/2) \\ B \cdot e^{-bx} & \text{for } (x > L/2) \end{cases} \tag{3.89}$$

where b and B are given by

$$b = \frac{w}{L} \cdot \tan\frac{w}{2} \qquad B = A\,\frac{w}{bL}\,\sin\left(\frac{w}{2}\right) \cdot e^{bL/2}$$

In addition, we find that the parameter w is obtained as a solution of the equation

$$w^2 + w^2 \cdot \tan^2\frac{w}{2} - \pi^2 k = 0 \tag{3.90}$$

where $k = V_0 8 m_e L^2/h^2$, and the energy E is given by

$$E = \frac{h^2}{8\pi^2 m_e}\left(\frac{w}{L}\right)^2 = \frac{h^2}{8 m_e L^2}\left(\frac{w}{\pi}\right)^2 \tag{3.91}$$

Equation (3.90) is known as a transcendental equation and must be solved numerically or by graphing it.

3.8.2 Antisymmetric Solutions

The corresponding antisymmetric solutions (that is $\psi(x) = -\psi(-x)$) follow from $\psi = A \cdot \sin(ax)$. In this case the wavefunctions are given by

$$\psi(x) = \begin{cases} -B \cdot e^{bx} & (\text{for } x < -L/2) \\ A \cdot \sin(wx/L) & (\text{for } -L/2 < x < L/2) \\ B \cdot e^{-bx} & \text{for } (x > L/2) \end{cases} \tag{3.92}$$

where b and B are

$$b = -\frac{w}{L \cdot \tan\frac{w}{2}} \qquad B = -A\left(\frac{w}{bL}\right)\cos\left(\frac{w}{2}\right) e^{bL/2}$$

The parameter w is obtained as a solution of the transcendental equation

$$w^2 + w^2/\left(\tan^2\frac{w}{2}\right) - \pi^2 k = 0 \tag{3.93}$$

and the energy E is given by

$$E = \frac{h^2}{8\pi^2 m_e}\left(\frac{w}{L}\right)^2 = \frac{h^2}{8 m_e L^2}\left(\frac{w}{\pi}\right)^2 \tag{3.94}$$

3.8.3 Numerical Evaluation, Fig. 3.2

As an example, we choose $k = 10$ (i.e., $V_0 = 6.025 \times 10^{-19}$ J) which corresponds to the solutions in Fig. 3.2. The two solutions found in the symmetric case are $w = 2.61$ and $w = 7.66$; see Foundation 3.1 (available electronically) for details. These symmetric wavefunctions (ψ_1 and ψ_3 in Fig. 3.2b) have a maximum at $x = 0$ and correspond to the first and third energy states, E_1 and E_3, in Fig. 3.2c. The two solutions found in the antisymmetric case are $w = 5.18$ and $w = 9.78$. These antisymmetric wavefunctions (ψ_2 and ψ_4 in Fig. 3.2b) have a node at $x = 0$ and correspond to the second and fourth energy states, E_2 and E_4, in Fig. 3.2c. The final results for $L = 1000$ pm are as follows[3]:

State	w	$\frac{w}{L} / \text{pm}^{-1}$	bL	$\frac{b}{\text{pm}^{-1}}$	$\frac{B}{A}$	$E \cdot \frac{8m_eL^2}{h^2}$	$\frac{E}{10^{-19}\,\text{J}}$
1 (sym)	2.61	0.00261	9.58	0.00958	31.7	0.69	0.42
2 (anti)	5.18	0.00518	8.47	0.00847	36.1	2.72	1.64
3 (sym)	7.66	0.00766	6.32	0.00632	−18.2	5.95	3.58
4 (anti)	9.78	0.00978	1.75	0.00175	−2.4	9.69	5.84

From these data the wavefunctions and energies displayed in Fig. 3.2(b,c) are calculated. For the unnormalized wavefunctions in Fig. 3.2 we arbitrarily choose $A = 1\ \text{pm}^{-1/2}$.

3.9 Foundations

3.9.1 Foundation 3.1: Electron in a Potential Energy Well of Finite Depth

See the book's website, www.wiley.com/go/Kuhn/physicalchemistry3e.

3.9.2 Foundation 3.2: Hamiltonian Operator and Orthogonality

3.9.2.1 Hamiltonian Operator

First, we prove that the Hamiltonian operator is a *Hermitian operator*, that is, an operator with the property:

$$\int \psi_1 \mathcal{H} \psi_2 \, d\tau = \int \psi_2 \mathcal{H} \psi_1 \, d\tau \tag{3.95}$$

where ψ_1 and ψ_2 are eigenfunctions[4].

We will demonstrate the equality by showing that $\int \psi_1 \mathcal{H} \psi_2 \, d\tau - \int \psi_2 \mathcal{H} \psi_1 \, d\tau = 0$. To simplify, we will prove relation (3.95) for one-dimensional wavefunctions that depend on the variable x. Afterwards the result can be generalized to all three cartesian coordinates: x, y, and z.

Consider the difference $f = \psi_1 \mathcal{H} \psi_2 - \psi_2 \mathcal{H} \psi_1$. Substituting the kinetic energy operator and the potential energy operator in for $\mathcal{H}$ we find

$$\begin{aligned} f &= -\frac{\hbar^2}{2m}\psi_1 \nabla^2 \psi_2 + \frac{\hbar^2}{2m}\psi_2 \nabla^2 \psi_1 + \psi_1 V \psi_2 - \psi_2 V \psi_1 = -\frac{\hbar^2}{2m}\psi_1 \nabla^2 \psi_2 + \frac{\hbar^2}{2m}\psi_2 \nabla^2 \psi_1 \\ &= -\frac{\hbar^2}{2m}\left(\psi_1 \frac{d^2\psi_2}{dx^2} - \psi_2 \frac{d^2\psi_1}{dx^2}\right) = -\frac{\hbar^2}{2m}\frac{d}{dx}\left(\psi_1 \frac{d\psi_2}{dx} - \psi_2 \frac{d\psi_1}{dx}\right) \end{aligned} \tag{3.96}$$

Using this simplified expression for f, we recast Equation (3.95) as

$$\begin{aligned} \int_{-\infty}^{\infty} f dx &= -\frac{\hbar^2}{2m}\int_{-\infty}^{\infty} \frac{d}{dx}\left(\psi_1 \frac{d\psi_2}{dx} - \psi_2 \frac{d\psi_1}{dx}\right) dx \\ &= -\frac{\hbar^2}{2m}\int_{-\infty}^{\infty} d\left(\psi_1 \frac{d\psi_2}{dx} - \psi_2 \frac{d\psi_1}{dx}\right) = -\frac{\hbar^2}{2m}\left|\psi_1 \frac{d\psi_2}{dx}\right|_{-\infty}^{+\infty} - \left|\psi_2 \frac{d\psi_1}{dx}\right|_{-\infty}^{+\infty} \end{aligned} \tag{3.97}$$

3 When solving the transcendental Equations (3.90) and (3.93) two more solutions are obtained for $w = 3.96$ and $w = 8.24$; these solutions, however, result in negative values of b, thus not satisfying the boundary conditions for the wavefunction.

4 Note that this is a special case of our definition of an Hermitian operator in Equation (3.86) (operator A and wavefunction ψ are real quantities).

The solutions of the Schrödinger equation must vanish for $x \to \pm\infty$

$$\lim_{x \to \pm\infty} \psi = 0 \tag{3.98}$$

If the functions become zero, the derivatives must become zero also. Thus

$$\int f \mathrm{d}x = \int \left(\psi_1 \mathcal{H} \psi_2 - \psi_2 \mathcal{H} \psi_1\right) \mathrm{d}x = 0$$

The extension of this derivation to three dimensions is straightforward. In the treatment shown above, the potential energy term can be handled in the same way. The Laplacian operator, $\nabla^2 = \frac{\partial^2}{\partial x^2} + \frac{\partial^2}{\partial y^2} + \frac{\partial^2}{\partial z^2}$, can be analyzed along each of the three directions independently in a manner like that shown above for x. Thus, we find that

$$\int \left(\psi_1 \mathcal{H} \psi_2 - \psi_2 \mathcal{H} \psi_1\right) \mathrm{d}\tau = 0 \tag{3.99}$$

and we have proved that the Hamiltonian operator is a Hermitian operator, see Equation (3.95).

3.9.2.2 Orthogonality

Now we insert the solutions ψ_1 and ψ_2 into the Schrödinger equation (energies E_1 and E_2).

$$\mathcal{H}\psi_1 = E_1 \psi_1 \,, \qquad \mathcal{H}\psi_2 = E_2 \psi_2 \tag{3.100}$$

We multiply the equation with ψ_1 by ψ_2 and integrate over $\mathrm{d}\tau$ and we multiply the equation with ψ_2 by ψ_1 and integrate over $\mathrm{d}\tau$, to find

$$\int \psi_2 \left(\mathcal{H}\psi_1\right) \mathrm{d}\tau = E_1 \int \psi_2 \psi_1 \, \mathrm{d}\tau \quad \text{and} \quad \int \psi_1 \left(\mathcal{H}\psi_2\right) \mathrm{d}\tau = E_2 \int \psi_1 \psi_2 \, \mathrm{d}\tau$$

If we subtract these two equations and interchange ψ_1 and ψ_2 in the right-hand terms, we obtain

$$E_1 \int \psi_1 \psi_2 \, \mathrm{d}\tau - E_2 \int \psi_1 \psi_2 \, \mathrm{d}\tau = \int \psi_2 \mathcal{H} \psi_1 \, \mathrm{d}\tau - \int \psi_1 \mathcal{H} \psi_2 \, \mathrm{d}\tau$$

In Equation (3.99) we proved that the right-hand side equals zero, thus

$$\left(E_1 - E_2\right) \int \psi_1 \psi_2 \, \mathrm{d}\tau = 0 \tag{3.101}$$

Provided that $E_1 - E_2 \neq 0$, we obtain

$$\int \psi_1 \psi_2 \, \mathrm{d}\tau = 0 \tag{3.102}$$

Thus we have proved the orthogonality relation for non-degenerate wavefunctions. No such condition exists for degenerate wavefunctions ($E_1 - E_2 = 0$).

3.9.3 Foundation 3.3: Uncertainty Principle for $\Delta x \cdot \Delta p_x$

See the book's website, www.wiley.com/go/Kuhn/physicalchemistry3e.

3.9.4 Foundation 3.4: Uncertainty Principle for Two Conjugate Variables

See the book's website, www.wiley.com/go/Kuhn/physicalchemistry3e.

3.9.5 Foundation 3.5: Superposition of Wavefunctions: Calculation of Uncertainty Product

See the book's website, www.wiley.com/go/Kuhn/physicalchemistry3e.

3.9.6 Foundation 3.6: Tunneling Through a Barrier

See the book's website, www.wiley.com/go/Kuhn/physicalchemistry3e.

3.10 Exercises

E3.1 Verify that Equation (3.38) is a solution of the Schrödinger Equation (3.31) for a particle in a three-dimensional box.

E3.2 In the spirit of the discussion in Section 3.2.3 consider an electron that is confined to a potential energy well with $V = 0$ in two dimensions and infinitely high walls $V = \infty$ outside of the box. Write the Schrödinger equation for this case; show that the Schrödinger equation is separable; write an expression for the total wavefunction; and normalize the wavefunction.

E3.3 Perform the integration in Equation (3.53) analytically and show that the integral is equal to zero.

E3.4 By inspection, energy order the wavefunctions in Fig. 3.9. By inspection energy order the wavefunctions in Fig. 3.10.

E3.5 Write an expression for the Schrödinger equation which needs to be solved for the electron-in-a-box problem (i.e., consider the box to be one-dimensional and have infinitely high potential walls at x values of 0 and L). Be sure to write out the explicit form of the Hamiltonian operator $\mathcal{H}$. Show that $\sqrt{2/L}\sin(n\pi x/L)$ is a solution of this equation, where n is the quantum number.

E3.6 We can define the uncertainty in a particle's position as $\Delta x = \sigma_x = \sqrt{\left(\overline{x^2} - \overline{x}^2\right)}$. Find the uncertainty in the position of an electron localized in a one-dimensional box of length L and occupying the ground state of the box (i.e., $n = 1$). Evaluate $\overline{x}$ and $\overline{x^2}$ and then calculate Δx.

E3.7 Consider a particle confined to a box whose walls are infinitely high and located at $x = 0$ and $x = L = 10.0$ nm. Find the probability that the particle has a spatial position between 3.00 nm and 7.00 nm for each of the energy levels $n = 1$, $n = 3$, and $n = 10$. In words, explain how you expect the probability to behave as n becomes very large.

E3.8 Consider a particle confined to a box whose walls are infinitely high and located at $x = 0$ and $x = L = 10.0$ nm. Explicitly show how you can use the particle in a box wavefunction and the Schrödinger equation to find a general expression for the particle's energy and evaluate it for the first five energy levels.

E3.9 Perform the integration for $\int_0^{2\pi} \mathrm{d}\varphi \int_0^{\pi} \sin(\theta)\,\mathrm{d}\theta$ and show that it has the value 4π.

E3.10 Consider an electron that is confined to a potential energy well (or box) of length L which is enclosed on each side by a potential energy barrier of height V_0 (see Fig. 3.2). a) Sketch a wavefunction which could be a possible solution for this potential energy profile when $E < V_0$. (Note: Just focus on the qualitative behavior, rather than worrying about quantitative accuracy.) How does the wavefunction change as you decrease the mass of the particle? How does it change as you increase the mass of the particle? b) Sketch a wavefunction which could be a possible solution for this potential energy profile when $E > V_0$, but not much greater than V_0. How does the waveform change with the mass of the particle. Hint: The particle will not be confined to the well anymore, but its relative amounts of kinetic and potential energy will change in different regions of x.

E3.11 Consider the situation where a particle of energy E and mass m is incident on a potential barrier of height V_0. Describe how the tunneling probability of a particle changes as you change a) the mass of the particle, b) the height of the barrier, as compared to the particle's energy (E/V_0), and c) the length of the barrier, L.

E3.12 Using the fact that the displacement along the circumference of a circle is $\mathrm{d}\xi = r\,\mathrm{d}\varphi$, transform the definition of the momentum operator to polar coordinates and show that the wavefunction $\psi = \frac{1}{\sqrt{2\pi}}\exp(+i\varphi)$ is one of its eigenfunctions with a momentum of h/Λ.

E3.13 Show that the wavefunction $\psi(\varphi) = Ae^{ik\varphi}$ is a solution of the Schrödinger Equation for an electron on a ring, namely

$$\mathcal{H}\psi = -\frac{\hbar^2}{2m_e r^2}\frac{\mathrm{d}^2}{\mathrm{d}\varphi^2}\psi = E\psi$$

Apply the boundary condition and show that the energy is quantized.

E3.14 Normalize the wavefunction $\psi(\varphi) = A \cdot \sin(k\varphi)$, which is an energy eigenfunction of the Schrödinger Equation.

E3.15 Classically, the angular momentum $\vec{L}$ of a mass m moving on a circle of radius r with speed v has a magnitude of $\left|\vec{L}\right| = m \cdot v \cdot r$. Consider the case where the mass is the mass of an electron m_e and the speed is given by the de Broglie relation $v = h/(m_e\Lambda)$. By using the boundary condition that $\psi(\varphi + 2\pi) = \psi(\varphi)$, show that the angular momentum is quantized.

E3.16 In Foundation 3.3, we use the orthogonality condition to claim that the integral for a product of cosines is zero when the odd integers n and j are not equal to each other; that is

$$\int_{-L/2}^{L/2} \cos\left(\frac{\pi}{L}jx\right) \cdot \cos\left(\frac{\pi}{L}nx\right) dx = 0$$

for $n \neq j$. Show that this integral is zero by explicitly evaluating it.

E3.17 For the potential well problem in Foundation 3.1 derive the result for the left-hand side of the well.

E3.18 Verify that the wavefunction $\psi = \sqrt{\frac{2}{(n_{\max}+1)}}\sqrt{\frac{2}{L}}\sum_{n=1}^{n_{\max}} \cos\left(\frac{\pi n x}{L}\right)$ is normalized for the case that $n_{\max} = 2$.

E3.19 In this exercise, you are asked to combine basic wave–particle duality ideas from Chapters 1 and 2 with our results for the electron-in-a-box model to estimate the size of a sodium atom. While you are expected to use qualitative reasoning, you also need to perform mathematical calculations and compare values with each other. 1) The excited Na atom has a strong yellow emission wavelength of 589 nm. Assuming that this emission corresponds to relaxation of an electron from the lowest excited state to the ground state, estimate the energy gap between the lowest excited state and the ground state. 2) Using this energy gap and the one-dimensional (1D) electron-in-a-box model, estimate the size of a Na atom. Be sure to account for the fact that Na has 11 electrons and use the Aufbau principle to fill the box model's electronic energy levels. 3) Perform the estimates/calculations in part 2) for a three-dimensional (3D) electron-in-a-box model, also. Compare your results with each other and with literature estimates of the Na atom's size.

3.11 Problems

P3.1 Show that the wavefunction $\psi(x) = A\sin(kx)$ satisfies the Schrödinger equation for an electron in a one-dimensional box. Consider A and k to be constants that must be determined. Determine the constant k by requiring the wavefunction to have the value zero at $\psi(L) = 0$, and verify that the wavefunction has the value zero at $x = 0$. Determine the constant A by applying the normalization condition, Equation (3.44).

P3.2 Consider the electron-in-a-box system like that shown in Fig. 3.2. Consider the length of the box to be 250 pm and the potential energy V_0 to be 14 eV = 2.2×10^{-18} J. These numbers are chosen to mimic the case of a hydrogen atom. Consider the electron in this system to be in the ground-state energy level and imagine performing a photoemission experiment in which you measure the kinetic energy of the electron as a function of the excitation light source's frequency. a) Make a sketch that illustrates your observations; i.e., measured data. b) Draw a sketch of an experimental apparatus that shows the essential components needed in making such a measurement. c) Explain in words how this experiment can demonstrate the particle nature of light. d) Describe how your results would change if the electron was located in the $n = 2$ energy level, rather than the $n = 1$ energy level.

P3.3 Use a data sheet program (e.g., Excel) for this problem. a) Plot the ground electronic state wavefunction for an electron confined to a box of length 100 pm. Plot the square of this wavefunction. b) Since the square of the wavefunction in a region of space represents the probability of detecting an electron in that region of space and our wavefunction is confined to the box, the probability of finding the electron somewhere in the box should be unity, or 100%. Determine what numerical factor you need to multiply your wavefunction by so that the total probability is unity. This numerical factor is called the normalization factor. c) In your data sheet program, change the box length and explore whether the normalization factor changes as you change the size of the box. d) You can also perform this procedure analytically by taking the integral of the square of the electron wavefunction and requiring it to be equal to one; namely

$$\int_0^L \psi^2(x)\, dx = 1$$

Perform this procedure for the wavefunction (the Mathematical Appendix may be useful) and determine the normalization factor. e) Show that your analytical result agrees with your calculated result.

P3.4 A scanning tunneling microscope (STM) uses the quantum-mechanical tunneling of electrons between a sample and a metallic tip to image the surface of a sample. With this device it is possible to achieve a spatial resolution which is of atomic dimension; *ca.* 100 pm.

a) If 100 pm is the uncertainty in an electron's position, what is the minimum uncertainty in its momentum?
b) Suppose you wish to energy resolve the electrons using your STM, in order to measure a spectrum and a spatial image simultaneously. How well can you determine the kinetic energy of 1 eV electrons before the spatial resolution is degraded to 500 pm.

First, perform your estimates assuming a one-dimensional model and then generalize them to account for three dimensions.

P3.5 In the 1980s chemical physicists performed experiments in which Na atoms are trapped in a laser field. In this experiment, the Na atoms are "stopped" with photons so that their kinetic energy is very small. A typical value for the residual uncertainty in the kinetic energy is $\Delta E = 2.0 \times 10^{-28}$ J. Use this uncertainty in the kinetic energy to estimate the uncertainty in an atom's momentum Δp. Subsequently use this Δp to estimate the uncertainty in the atom's position. Express your answer in Bohr radii.

Given that the density of solid Na is 0.97 g/cm^3, use this number to estimate the size of a Na atom. Express your answer in Bohr radii.

Compare the uncertainty in the Na atom's position in the laser trapping experiment to its characteristic size in the solid.

P3.6 This problem requires a spreadsheet program. Use particle in the box wavefunctions with $L = 500$ pm. First, create plots of each wavefunction for the five lowest energy levels over the range of 0 to L. Second, create a superposition (i.e., sum them) of the two lowest wavefunctions and plot it. Third, create a superposition of the three lowest wavefunctions and plot it. Proceed in this manner until you have a superposition state that includes the first five wavefunctions. In addition to these wavefunction plots, plot the probability densities in each case. Describe how the probability density changes with the number of stationary state wavefunctions that you include in the superposition.

P3.7 Problem 2.3 shows that the superposition of two waves can give rise to a standing wave. It is common to describe a complicated electronic state as a superposition of the system's standing wave states. Use the MathCAD exercise, titled PinBsuperposition.mcdx, to explore how the superposition of a particle in the box standing waves can give rise to a diverse array of waveforms. Answer the following questions posed in the worksheet

a) How does the spread of the wavefunction in space relate to the number of contributing eigenstates in the four superposition states you are asked to investigate? Howis the spread of the momentum influenced by the number of states?

b) Comment on how the spread of the wavefunction in position and momentum space changes for the superposition states. Is the Heisenberg Uncertainty principle ever violated? What is the uncertainty product for an eigenstate? You can check this by creating eigenstates rather than superposition states. Does it depend on the eigenstate's quantum number?

c) Comment on the appearance of the superposition states. Are the probability densities symmetric? How does the symmetry of the superposition state depend on the coefficients?

d) Create a superposition state from only odd quantum numbers. Create a superposition state from only even quantum numbers. Plot the probability densities of each state. Plot the wavefunctions of each state. Verbally describe and explain the differences between these states.

P3.8 The MathCAD exercise tunneling.mcdx investigates features of quantum mechanical tunneling through an energy barrier. This exercise describes the tunneling electron as originating from a superposition state. Answer the questions posed in the worksheet; namely

a) Change the mass of the incoming particle. Take for example the mass of the reference particle to be 5 and compare it to the case where the mass is 50. Keep the incident energy E of the particles the same. Use at least four different masses and describe how the particles behavior changes with its mass.

b) Perform a similar series of studies, but keep the mass fixed and change the length (width) of the barrier. How does the barrier width change the tunneling probability?

c) Perform a similar series of studies, but change the ratio of the incoming particle energy to the barrier height, E/V_0, and see how the wavefunctions change. Make sure that your incident energy E is still less than V_0 though, since we are concentrating on tunneling through a barrier.

P3.9 a) Show that the wavefunction $\psi = \sin(kx)$ solves the Schrödinger Equation (3.16) with $V(x) = 0$, but it does not fulfill the condition that $p_{op}\psi = p\psi$.

b) Show that the wavefunction $\psi = Ce^{-ikx}$ solves the Schrödinger Equation (3.16) with $V(x) = 0$, and it fulfills the condition that $p_{op}\psi = p\psi$.

c) Compare and contrast the wavefunctions in parts a) and b). Describe how they describe different aspects of a physical system. Under what conditions might you choose one over the other?

P3.10 Create a superposition of cosine waves and calculate the uncertainty product of a particle's momentum and position, $\Delta p \cdot \Delta x$, for a Gaussian distribution of momenta. Use wavefunctions

$$\cos(ax) \text{ where } a = \pm\frac{2\pi}{\Lambda} \text{ and } p = \pm\frac{h}{\Lambda} = \frac{h}{2\pi}a$$

in which the distribution of coefficients a is given by a Gaussian distribution; namely

$$\frac{d\mathcal{P}}{da} = \frac{b}{\sqrt{\pi}} \cdot e^{-b^2 \cdot a^2}$$

The coefficients a are proportional to the momenta as shown above, and b is a parameter determining the full width at half maximum $a_{1/2}$ of the Gaussian distribution. The standard deviation σ_a for the spread in a values is given by $a_{1/2} = 2\sqrt{2 \cdot \ln 2} \cdot \sigma_a$ (see Appendix C).

Part 1: Use the expression for the Gaussian distribution of a to determine the distribution of a in the probability density. Then find the standard deviation of a in terms of the full-width at half-maximum of the distribution $a_{1/2}$ and use it to show that the uncertainty in the momenta is given by $\Delta p = \frac{h}{2\pi} \cdot \frac{1}{2b}$.

Part 2: Find the uncertainty in x, construct a superposition wavefunction of the form

$$\psi(x) = C \cdot \int_{-\infty}^{\infty} \frac{d\mathcal{P}}{da} \cdot \cos(ax) \cdot da$$

where C is a normalization constant. Then use the probability density $\rho = \psi(x)^2$ to show that $\Delta x = b$.

Part 3: Combine your results to find the uncertainty product. In each case, use the standard deviation for the observable (i.e., x or p) to be its uncertainty.

P3.11 Show that the angular momentum operator can be transformed from cartesian coordinates (Equation (3.69)) to spherical coordinates (Equation (3.70)). The angular momentum operator about the z-axis in cartesian coordinates x and y is written as

$$\mathcal{L}_{op,z} = -\frac{ih}{2\pi}\left(x\frac{\partial}{\partial y} - y\frac{\partial}{\partial x}\right)$$

Hint: start with an equation where this operator acts on an operator $F(\varphi)$ depending on the spherical coordinate

P3.12 Derive the equations for the antisymmetric solution of the potential energy well problem in Foundation 3.1.

P3.13 Following the equation for $\overline{(\Delta A)^2} \cdot \overline{(\Delta B)^2} \leq$ in Foundation 3.4 prove the validity of the uncertainty relations

$$\Delta y \cdot \Delta p_y \geq \hbar/2 \qquad \Delta\varphi \cdot \Delta\mathcal{L}_z \geq \hbar/2$$

P3.14 Consider an electron confined between two walls at a distance L (electron in a box). Imagine that a barrier of height V and width Δx is placed at the center of the box. Discuss qualitatively what will happen with the energies of the electrons as the height of the barrier is increased from zero to infinity.

References

1 P. Bäuerle and J.P. Rabe, Oligothiophene –immer länger? Synthese, Charakterisierung und rastertunnelmikroskopische Abbildung von homologen, isomerenreinen Oligo(alkylthiophenen), *Angew. Chem.*, 1995, **107**, 335.

2 P. Bäuerle, T. Fischer, B. Bidlingmeier, A. Stabel, and J. Rabe, Oligothiophenes—yet longer? Synthesis, characterization, and scanning tunneling microscopy images of homologous, isomerically pure oligo(alkylthiophene)s, *Angew. Chem. Int. Ed. Engl.*, 1995, **34**, 303.

3 C.M. Lieber, J. Liu, and P.E. Sheehan, Understanding and manipulating inorganic materials with scanning probe microscopes, *Angew. Chem. Int. Ed. Engl.*, 1996, **35**, 687; M.F. Crommie, C.P. Lutz, and D.M. Eigler, Confinement of electrons to quantum corrals on a metal surface, *Science*, 1993, **262**, 218.

4 J. Stark and P.S. Epstein, *Der Stark-Effekt* (Battenberg-Verlag, 1965).

5 R. Eisberg and R. Resnick, *Quantum Physics of Atoms, Molecules, Solids, Nuclei and Particles* (Wiley, New York, 1974).

6 L.D. Landau and E.M. Lifshitz, *Quantum Mechanics* (Pergamon, New York, 1977).

4

Hydrogen Atom

CHAPTER MENU

4.1 Introduction

The H atom is of fundamental importance in quantum chemistry because H-atom like wavefunctions are used in constructing approximations for wavefunctions of atoms and molecules. The H-atom wavefunctions are the "orbitals" which most chemists use in rationalizing chemical properties and chemical reactivity. In this chapter we show that the time-independent Schrödinger equation provides a highly accurate description of the H atom, giving rise to discrete energy states referred to as stationary states.

4.2 H Atom in the Ground State

Let us find the stationary states (standing waves) for an electron that is bound to a proton, the H atom. From Coulomb's law, the potential energy $V(r)$ of the electron in the field of the proton is

$$V(r) = -\frac{e^2}{4\pi\varepsilon_0 r} \tag{4.1}$$

where $r = \sqrt{x^2 + y^2 + z^2}$ is the distance between the electron and the nucleus. Hence the Schrödinger Equation for the H atom is:

$$-\frac{\hbar^2}{2m_e}\left(\frac{\partial^2}{\partial x^2} + \frac{\partial^2}{\partial y^2} + \frac{\partial^2}{\partial z^2}\right)\psi(x,y,z) - \frac{1}{4\pi\varepsilon_0}\frac{e^2}{r}\psi(x,y,z) = E\,\psi(x,y,z) \tag{4.2}$$

where $\hbar = h/(2\pi)$; $\hbar$ was introduced by Heisenberg and is frequently used instead of h. Note that a more precise description uses the reduced mass rather than the electron's mass. This Schrödinger equation is more difficult to solve than our electron in a box model. In part this arises because Cartesian coordinates are not the best choice for the H atom (note that a square root appears in the potential energy term, if we express r by x, y, and z). The H atom has spherical symmetry and choosing spherical coordinates, r, ϑ, φ, rather than Cartesian coordinates x, y, z, simplifies the Schrödinger equation; it makes the equation separable into three one-dimensional equaitons – see Foundations 4.1 and 4.2 for details.

Principles of Physical Chemistry, Third Edition. Hans Kuhn, David H. Waldeck, and Horst-Dieter Försterling.

Companion website: www.wiley.com/go/Kuhn/physicalchemistry3e

4.2.1 Wavefunction

As shown in Foundations 4.1 and 4.2, the ground-state wavefunction that solves this Schrödinger Equation depends only on the distance r:

$$\boxed{\psi(r) = \frac{1}{\sqrt{\pi a_0^3}}\, e^{-r/a_0}} \tag{4.3}$$

with

$$\boxed{a_0 = 4\pi\varepsilon_0 \frac{\hbar^2}{m_e e^2} = 52.918 \text{ pm}} \tag{4.4}$$

See Example 4.1. For historical reasons a_0 is called the *Bohr radius*; see Problem 4.1. $\psi(r)$ is plotted in Fig. 4.1a, and 4.1b plots the probability density

$$\boxed{\rho = \psi^2 = \frac{1}{\pi a_0^3}\, e^{-2r/a_0}} \tag{4.5}$$

as a function of r. Figure 4.1c shows a "cloud" representation of ρ. If we evaluate ρ at the location $r = a_0/2 = 26.5$ pm, we find

$$\rho\left(\frac{a_0}{2}\right) = \frac{1}{\pi a_0^3}\, e^{-1} = \frac{1}{\pi a_0^3}\, 0.368$$

So at $r = 26.5$ pm, ρ is less than half as great as it is at the location $r = 0$. At a distance $r = 122$ pm from the nucleus, ρ has already fallen to 1% of its value at $r = 0$.

Such considerations on the size and shape of the electron cloud will be important for understanding the interactions between atoms and molecules in later chapters.

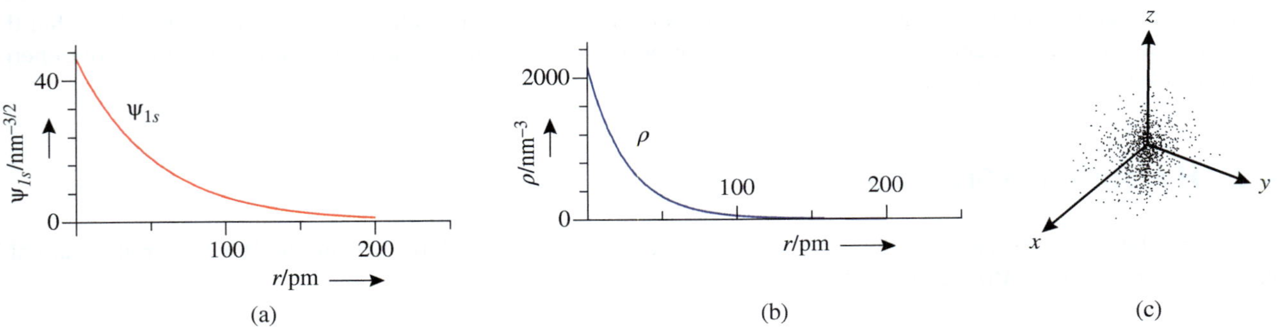

Figure 4.1 Plots of the wavefunction $\psi(r)$ (a) and the probability density $\rho(r)$ (b) versus distance r for the lowest energy state of the electron in a H atom. (c) A cloud picture of $\rho(r)$.

4.2.2 Energy

We can find the energy E for the H atom ground state by substituting $\psi(r)$ in Equation (4.3) into the Schrödinger Equation (4.2) and evaluating it.

Example 4.1 *Show that the wavefunction for the ground state of the H atom, given in Equation (4.3), is a solution to the Schrödinger equation, Equation (4.2), and corresponds to the energy* $E = -\frac{1}{4\pi\varepsilon_0} \cdot \frac{e^2}{2a_0}$.

If we write the distance as $r = \sqrt{x^2 + y^2 + z^2}$ then we can evaluate the derivatives with respect to each of the Cartesian coordinates x, y, and z. For the displacement x we find

$$\frac{\partial \psi}{\partial x} = -\frac{1}{\sqrt{\pi a_0^3}} \cdot \frac{1}{a_0} x \cdot \frac{1}{\sqrt{x^2 + y^2 + z^2}} \cdot e^{-\sqrt{x^2+y^2+z^2}/a_0}$$

and

$$\frac{\partial^2 \psi}{\partial x^2} = -\frac{1}{\sqrt{\pi a_0^3}} \cdot \frac{1}{a_0} \cdot \frac{r - x^2/r - x^2/a_0}{x^2 + y^2 + z^2} \cdot \mathrm{e}^{-\sqrt{x^2+y^2+z^2}/a_0}$$

Similar expressions are obtained for the derivatives according to y and z (x^2 in the numerator has to be replaced by y^2 or z^2, respectively). You are asked to perform this work in Exercise 4.16. By combining these three results we find that

$$\nabla^2 \psi = \frac{\partial^2 \psi}{\partial x^2} + \frac{\partial^2 \psi}{\partial y^2} + \frac{\partial^2 \psi}{\partial z^2} = -\frac{1}{\sqrt{\pi a_0^3}} \cdot \frac{1}{a_0} \cdot \left(\frac{2}{r} - \frac{1}{a_0}\right) \cdot \mathrm{e}^{-r/a_0}$$

We can substitute this expression for the first term on the left-hand side of the Schrödinger Equation (4.2) and write that

$$\frac{\hbar^2}{2m_e} \cdot \frac{1}{\sqrt{\pi a_0^3}} \cdot \frac{1}{a_0} \cdot \left(\frac{2}{r} - \frac{1}{a_0}\right) \cdot \mathrm{e}^{-r/a_0} - \frac{1}{4\pi\varepsilon_0} \cdot \frac{e^2}{r} \cdot \frac{1}{\sqrt{\pi a_0^3}} \cdot \mathrm{e}^{-r/a_0} = E \cdot \frac{1}{\sqrt{\pi a_0^3}} \cdot \mathrm{e}^{-r/a_0}$$

or

$$E = \frac{\hbar^2}{2m_e} \cdot \frac{1}{a_0} \left(\frac{2}{r} - \frac{1}{a_0}\right) - \frac{1}{4\pi\varepsilon_0} \cdot \frac{e^2}{r}$$

Now if we use our definition of the Bohr radius $a_0 = 4\pi\varepsilon_0\hbar^2/\left(m_e e^2\right)$, we can simplify this equation to show that

$$E = \frac{1}{r}\left(\frac{\hbar^2}{m_e a_0} - \frac{e^2}{4\pi\varepsilon_0}\right) - \frac{\hbar^2}{2m_e a_0^2} = \frac{1}{r}\left(\frac{e^2}{4\pi\varepsilon_0} - \frac{e^2}{4\pi\varepsilon_0}\right) - \frac{1}{4\pi\varepsilon_0}\frac{e^2}{2a_0} = -\frac{1}{4\pi\varepsilon_0} \cdot \frac{e^2}{2a_0}$$

Hence we find that the energy of the H atom in the ground state is given by

$$\boxed{E = -\frac{1}{4\pi\varepsilon_0}\frac{e^2}{2a_0} = -21.80 \times 10^{-19}\ \mathrm{J}} \tag{4.6}$$

In addition, we find that Equation (4.6) compares well with the experimental value,[1] -21.787×10^{-19} J. The prediction from the postulates of quantum theory is thus in excellent agreement with experiment.

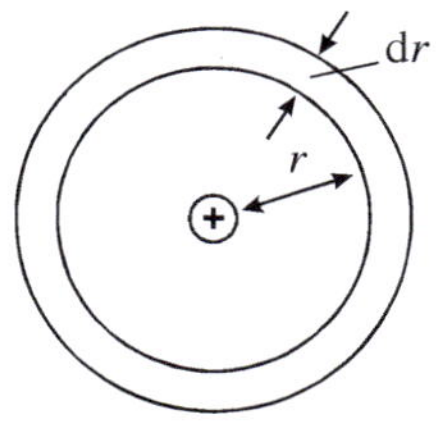

Figure 4.2 Section through a sphere of radius r with a spherical shell of thickness dr (spherical surface area: $4\pi r^2$ and volume element $\mathrm{d}\tau = 4\pi r^2 \cdot \mathrm{d}r$).

4.2.3 Radial Probability Distribution of Electron

It is important to assess how the probability density of the electron changes in space, especially later when we consider chemical bonding and reactivity. For the ground state of the H atom, the electronic wavefunction does not depend on angle (i.e., it is spherical), but it has a strong dependence on the distance from the proton, that is, a radial dependence. The radial probability density is the probability of finding the electron in a spherical shell of distance r, and thickness dr, from the proton (see Fig. 4.2).

For the ground state, the electron's radial probability density is

$$\rho 4\pi r^2\ \mathrm{d}r = \frac{4}{a_0^3} r^2 \mathrm{e}^{-2r/a_0}\ \mathrm{d}r \tag{4.7}$$

The radial probability distribution $\rho 4\pi r^2$ is plotted in Fig. 4.3b. Although the probability density ρ is greatest at the nucleus and then falls off exponentially with distance from the proton (Fig. 4.3a), the radial probability density does not show such a dependence. This occurs because the radial probability density is a product of the probability density ρ and the volume of a spherical shell of thickness dr, namely, $4\pi\ r^2$ dr. Hence the probability of finding an electron in a sphere directly at the

1 This value follows from spectroscopic data as the limit of the Lyman series, Equation (4.23). The small difference between experiment and calculation arises from the neglect of the nucleus' comovement.

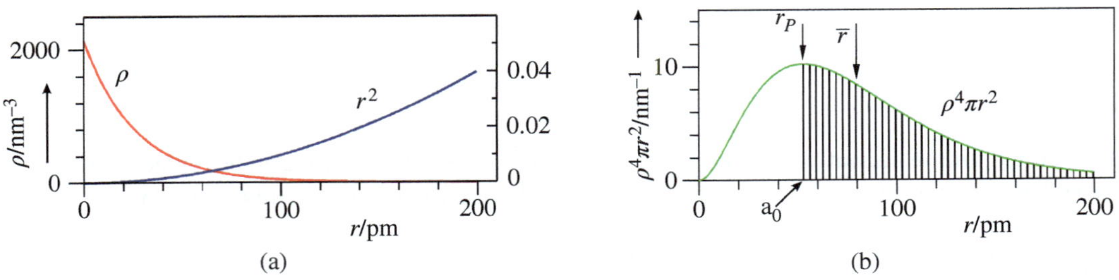

Figure 4.3 Radial probability distribution function for the electron in the H atom ground state. (a) Plot of $\rho = \psi^2$ versus the radius r (left-hand scale) and the quadratic form r^2 (right-hand scale) for the volume element. (b) Plot of $\rho \cdot 4\pi r^2$ versus radius r. The most probable distance, $r_P = a_0$, and the average distance, $\bar{r} = \frac{3}{2}a_0$, are indicated. The shaded region shows the probability for the electron to be at distances $r > r_P$, which is larger than the unshaded region which shows the probability for $r < r_P$.

nucleus ($r = 0$) approaches zero, because the volume element $4\pi r^2\,\mathrm{d}r$ approaches zero. At large distances, the exponential decline of ρ approaches zero more rapidly than the quadratic term becomes large, and the radial probability density approaches zero at large r values. At an intermediate distance the radial probability distribution goes through a maximum value (Fig. 4.3b).

The probability of finding an electron within a spherical shell of radius r and thickness dr should be distinguished from the probability of finding an electron in a volume element dτ. In the latter case, the size of the volume element is held constant, while in the former case only dr is held constant so that the volume element under consideration grows as r^2.

The radial distribution function is useful for quantifying the size of atoms. For the electronic ground state of the H atom, the probability of finding the electron inside a sphere with radius R is

$$\mathcal{P}(r < R) = \int_0^R \rho\, 4\pi r^2\,\mathrm{d}r = 1 - \mathrm{e}^{-2R/a_0}\left[2\left(\frac{R}{a_0}\right)^2 + 2\frac{R}{a_0} + 1\right] \tag{4.8}$$

This probability is 32% for $R = a_0$, 76% for $R = 2a_0$, and 85% for $R = 2.36\ a_0 = 125$ pm. The probability of finding the electron outside a sphere of $r = 125$ pm is 15%, although ρ at $r = 125$ pm is only 1% of the value at $r = 0$.

4.2.4 Most Probable Distance and Average Distance

The *most probable distance* r_P of the electron from the nucleus is the distance at which the radial distribution function $\rho\,4\pi r^2$ has a maximum. We can find the maximum graphically (see Fig. 4.3b) or analytically. To find the maximum analytically, we take the derivative of the radial distribution function with respect to r and set it equal to zero (see Appendix C). For the ground state of the H atom, this procedure gives

$$\frac{\mathrm{d}}{\mathrm{d}r}\left(\rho\, 4\pi r^2\right)_{r=r_P} = \frac{4}{a_0^3}\left(-\frac{2}{a_0}\,r_P^2\,\mathrm{e}^{-2r_P/a_0} + 2r_P\,\mathrm{e}^{-2r_P/a_0}\right) = \frac{4}{a_0^3} r_P\,\mathrm{e}^{-2r_P/a_0}\cdot\left(-\frac{1}{a_0}\,r_P + 1\right) = 0 \tag{4.9}$$

This expression has zeroes at $r_P = 0$ and at $r_P = \infty$, which correspond to the two minimum values, and at

$$r_P = a_0 \tag{4.10}$$

which corresponds to the maximum, see Fig. 4.3b.

The average distance of the electron from the nucleus $\bar{r}$ is given by the following integral

$$\bar{r} = \int_{\text{all space}} r\,\rho\,\mathrm{d}\tau = \int_{r=0}^{\infty} r\,\rho\,4\pi r^2\,\mathrm{d}r \tag{4.11}$$

The integrand is the distance r times the probability $\rho\,\mathrm{d}\tau$ of finding the electron in the volume element $\mathrm{d}\tau = 4\pi r^2\,\mathrm{d}r$ at distance r. By integrating over all possible distances r, the average distance is found. In Example 4.2, we show that the average distance of the electron from the proton is $\bar{r} = \frac{3}{2}a_0$. $\bar{r}$ is greater than r_P because the electron is more likely found at distances $r > r_P$ (shaded area in Fig. 4.3b) than at distances $r < r_P$ (the radial distribution function is not symmetric about the peak). Thus more values of r larger than r_P contribute to the average than do values smaller than r_P.

Example 4.2 *Calculate the average distance of the electron from the proton in the H atom's ground state.*

Following Equation (4.11), the average distance of the electron from the nucleus is given by

$$\bar{r} = \int_0^{\infty} r\,\rho(r)\,4\pi r^2\,\mathrm{d}r$$

By substituting the explicit form for ρ in Equation (4.5) we find that

$$\bar{r} = \int_0^{\infty} r\,\frac{1}{\pi a_0^3}\mathrm{e}^{-2r/a_0}\,4\pi r^2\,\mathrm{d}r = \frac{4}{a_0^3}\int_0^{\infty} r^3\,\mathrm{e}^{-2r/a_0}\,\mathrm{d}r$$

If we make the substitution $\xi = 2r/a_0$ to simplify the argument of the exponential function, the integral becomes

$$\bar{r} = \frac{4}{a_0^3}\left(\frac{a_0}{2}\right)^3\frac{a_0}{2}\int_0^{\infty}\xi^3\,\mathrm{e}^{-\xi}\,\mathrm{d}\xi = \frac{1}{4}a_0\cdot 3! = \frac{3}{2}a_0$$

The integral is given in Appendix B and has the value $3! = 6$, from which we obtain the final result of $\bar{r} = 3a_0/2$.

4.2.5 Average Potential Energy and Virial Theorem

By using the wavefunction (and its probability density), we can calculate many different average properties. The average potential energy $\overline{V}$ is of particular interest. In the following example we show that

$$\overline{V} = \int_{\text{all space}} V\,\rho\,\mathrm{d}\tau = -\frac{1}{4\pi\varepsilon_0}\frac{e^2}{a_0} \tag{4.12}$$

Example 4.3 *Calculate the average potential energy $\overline{V}$ for the ground state of the H atom.*

Since $V = \frac{-e^2}{4\pi\varepsilon_0 r}$ and $\rho = \psi^2 = \frac{1}{\pi a_0^3}\mathrm{e}^{-2r/a_0}$ are only functions of the distance r, we can use Equation (4.12) to write

$$\overline{V} = \int_0^{\infty} V\,\rho(r)\,4\pi r^2\,\mathrm{d}r = \frac{-e^2}{4\pi\varepsilon_0}\frac{1}{\pi a_0^3}4\pi\int_0^{\infty} r\,\mathrm{e}^{-2r/a_0}\,\mathrm{d}r$$

The integral can be solved using the substitution $\xi = 2r/a_0$ and proceeding in a manner analogous to that for calculating the average distance $\bar{r}$, so that

$$\overline{V} = \frac{-e^2}{4\pi\varepsilon_0}\frac{1}{\pi a_0^3}4\pi\frac{a_0^2}{4} = -\frac{1}{4\pi\varepsilon_0}\frac{e^2}{a_0}$$

We know the average potential energy $\overline{V}$ and the total energy E, so we can calculate the average kinetic energy $\overline{T}$. Using conservation of energy, we find that

$$\overline{T} = E - \overline{V} = -\frac{1}{4\pi\varepsilon_0}\frac{e^2}{2a_0} + \frac{1}{4\pi\varepsilon_0}\frac{e^2}{a_0} = \frac{1}{4\pi\varepsilon_0}\frac{e^2}{2a_0} \tag{4.13}$$

The magnitudes of $\overline{T}$ and E are equal, but their signs are opposite:

$$\boxed{E = -\overline{T} = \tfrac{1}{2}\overline{V}} \tag{4.14}$$

Equation (4.14) is a specific example of a general result called the *Virial Theorem* (see Fig. 4.4) and is important for understanding the stability of atoms and molecules. It is valid for all systems consisting of nuclei and electrons at the energy minimum. It tells us that half of the potential energy of stabilization gain by Coulomb forces, when forming the electron cloud from separated electrons and nuclei, is used to speed up the electrons, on average. A derivation of the Virial Theorem is provided in Foundation 4.3 (available on Wiley's website).

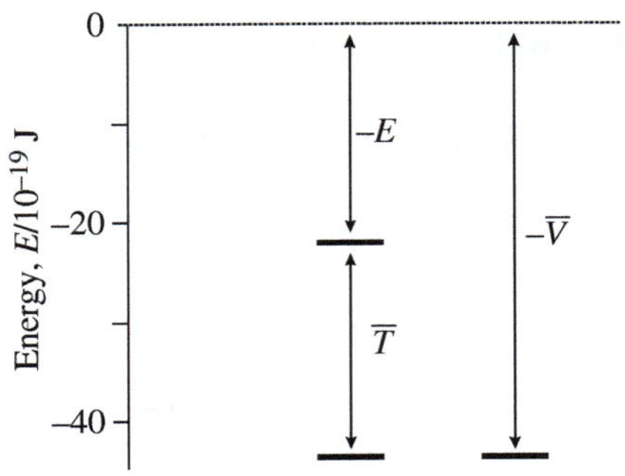

Figure 4.4 The energy plot illustrates the Virial Theorem for the H atom: $E = -\overline{T}$, $\overline{V} = -2\overline{T}$.

4.3 H Atom in Excited States

We now want to calculate the energies and wavefunctions for excited states of the H atom in order to see whether its electronic spectrum can be interpreted using the Schrödinger Equation. Beyond this immediate goal, we will find that features of the H atom wavefunctions are used extensively in chemistry to describe the wavefunctions of more complex systems, so that understanding them in detail is beneficial.

4.3.1 Energies

A detailed solution of the Schrödinger Equation for the H atom is provided in Foundations 4.1 and 4.2 (on Wiley's website). Here we focus on describing the results of the solution.

The energy levels are discrete and given by the energy level formula

$$E_n = -\frac{1}{4\pi\varepsilon_0}\frac{e^2}{2a_0}\frac{1}{n^2} \qquad n = 1, 2, 3, \ldots \tag{4.15}$$

where n is called the *principal quantum number*. The case of $n = 1$ corresponds to the ground electronic state, which we have already discussed. For the quantum number $n = 2$, four wavefunctions are found to have the same energy E_2. One of these is spherically symmetric like the ground state, but the other three are not. For the $n = 3$ case nine solutions are found with the same energy. In general, n^2 wavefunctions are found for each energy level E_n. Figure 4.5a shows the energy levels for the H atom. Superimposed over the energy level diagram is the Coulomb potential for an electron and a proton to emphasize that the electron becomes more weakly bound as the quantum number n increases (Fig. 4.5b). The energy levels become closer together as the quantum number increases and converge at large n on the ionization limit. Because of the weaker binding, the electron cloud is bigger for higher values of n.

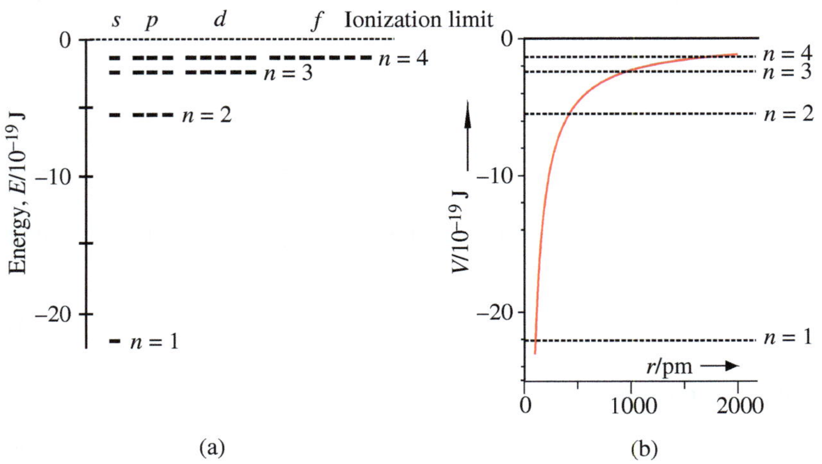

Figure 4.5 (a) Energy diagram of the electron in the H atom with quantum numbers 1, 2, 3, and 4. (b) The Coulomb potential $V = -e^2/(4\pi\varepsilon_0 r)$ for an electron and a proton is superimposed on this energy level diagram.

4.3.2 Wavefunctions

The wavefunction for the electron can be described by the product of an angular function and a radial function, such that

$$\psi(r, \vartheta, \varphi) = Y(\vartheta, \varphi)\ R(r) \tag{4.16}$$

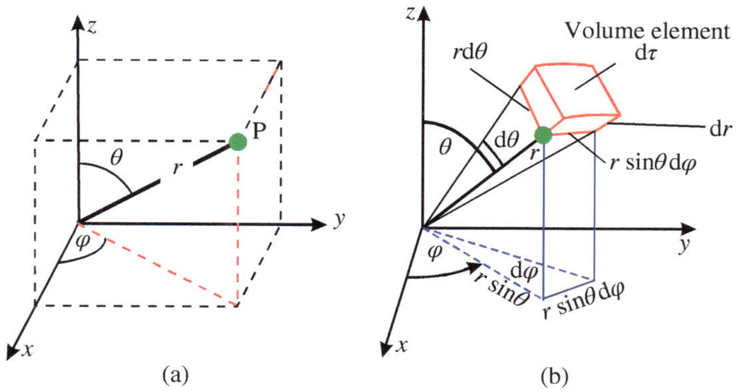

Figure 4.6 (a) Spherical coordinates r, ϑ, and φ of point P. From geometry we see that $z = r\cos\vartheta$; $x = r\sin\vartheta\cos\varphi$; $y = r\sin\vartheta\sin\varphi$. (b) Volume element $\mathrm{d}\tau = (r\sin\vartheta\,\mathrm{d}\varphi)\,(r\mathrm{d}\vartheta)\,(\mathrm{d}r) = r^2\sin\vartheta\,\mathrm{d}r\,\mathrm{d}\varphi\,\mathrm{d}\vartheta$. Note that this relation is valid only for sufficiently small values of $\mathrm{d}r$, $\mathrm{d}\varphi$, and $\mathrm{d}\vartheta$.

(method of separation of variables). The spherical coordinates, r, ϑ, and φ specify the electron position with respect to the proton. Figure 4.6 shows the relation between spherical coordinates and Cartesian coordinates. The distance from the proton is r, the angle ϑ specifies the tilt from the z-axis, and the azimuthal angle φ specifies the angular position about the z-axis, and in the $x-y$ plane. The radial wavefunctions $R(r)$ depend on the principal quantum number n and the angular momentum quantum number l. The angular wavefunctions $Y(\vartheta, \varphi)$ are called spherical harmonics. They depend on the angular momentum quantum number l ($l = 0, 1, 2, \ldots n-1$) and a third quantum number, the magnetic quantum number, which specifies the electron's angular momentum along the z-axis. The k quantum number will be discussed in Section 4.3.5. The spherical harmonics are combinations of sine and cosine functions and correspond to the s, p, d, and f orbitals that you should be familiar with from your introductory chemistry courses. In particular, we obtain the correspondence:

$$\begin{array}{lccccc} l & = & 0, & 1, & 2, & 3, \ldots \\ \text{orbital} & = & s, & p, & d, & f, .. \end{array} \tag{4.17}$$

Because each value of l can have multiple values of the magnetic quantum number, we find that one function belongs to $l = 0$ (s orbitals), three to $l = 1$ (p orbitals), five to $l = 2$ (d orbitals), and seven to $l = 3$ (f orbitals).

The resulting wavefunctions $\psi(r, \vartheta, \varphi)$, along with the functions $Y(\vartheta, \varphi)$ and $R(r)$ are provided in Foundations 4.1 and 4.2 (available on Wiley's website). For describing chemical bonds it is often more reasonable to write the H atom wavefunctions in Cartesian coordinates rather than in polar coordinates. The conversion is explained in Foundation 4.1 (available on Wiley's website) and the wavefunctions in Table 4.1 are obtained.

4.3.2.1 Nodal Surfaces

The plots of the total wavefunctions in Fig. 4.7 show the orbital shape for different n and l values. The s-orbitals ($l = 0$) do not have any nodal planes and are a spherical shape. A nodal plane is clearly evident for the p-orbitals ($l = 1$). In each case the nodal plane is perpendicular to the direction in which the wavefunction is oriented. The plot shown indicates the change in sign of the wavefunction by the color code. The d-orbital ($l = 2$) plots each contain two nodal surfaces that are illustrated in the diagram. Once again the nodal surfaces are planes, except for the d_{z^2} orbital which has two conical surfaces.[2] In general, we find that the angular part of the wavefunction has l nodes (called *angular nodes*). At first, you might expect that orbitals with the same value of n, but different values of l have different energies because of the different numbers of nodes; however, the radial part of the wavefunction also has nodes and the number of nodes for the total wavefunction depends only on n.

The total wavefunctions in Fig. 4.7 include the radial part of the wavefunction, which is a product of an exponentially decaying function and a polynomial. The polynomial part of the function produces an oscillating amplitude for the wave in the region where the total energy is greater than the potential energy (case $E > V$). The exponentially decaying part of the wavefunction dominates at long distances and ensures that the wave is damped to zero in regions of space where the total energy is less than the potential energy (case $E < V$).

As Fig. 4.7 shows, the s–orbital functions ($l = 0$) have a finite amplitude at $r = 0$, and the other radial functions ($l \neq 0$) are zero at $r = 0$. You should also observe that the radial functions have nodes at particular r values, called *radial nodes*, and that the number of radial nodes is given by $n - l - 1$. In the total wavefunction these radial nodes give rise to spherical

2 In this case the apex of the cones is located at the origin and the cones are coaxial with the z-axis. One of the cones is oriented in the positive z direction and the second cone is oriented in the negative z direction.

Table 4.1 Some H Atom Wavefunctions (Principal Quantum Numbers $n = 1, 2,$ and 3) in Cartesian Coordinates ($r = \sqrt{x^2 + y^2 + z^2}$).

Quantum Numbers		
n	l	Wavefunction
$n = 1$	0	$\psi_{1s} = \frac{1}{\sqrt{\pi a_0^3}} e^{-\frac{r}{a_0}}$
$n = 2$	0	$\psi_{2s} = \frac{1}{4\sqrt{2\pi a_0^3}} \left(\frac{r}{a_0} - 2\right) e^{-\frac{r}{2a_0}}$
2	1	$\psi_{2p_x} = \frac{1}{4\sqrt{2\pi a_0^3}} \frac{x}{a_0} e^{-\frac{r}{2a_0}}$
2	1	$\psi_{2p_y} = \frac{1}{4\sqrt{2\pi a_0^3}} \frac{y}{a_0} e^{-\frac{r}{2a_0}}$
2	1	$\psi_{2p_z} = \frac{1}{4\sqrt{2\pi a_0^3}} \frac{z}{a_0} e^{-\frac{r}{2a_0}}$
$n = 3$	0	$\psi_{3s} = \frac{1}{81\sqrt{3\pi a_0^3}} \left(27 - 18\frac{r}{a_0} + 2\frac{r^2}{a_0^2}\right) e^{-\frac{r}{3a_0}}$
3	1	$\psi_{3p_x} = \frac{\sqrt{2}}{81\sqrt{\pi a_0^3}} \left(6 - \frac{r}{a_0}\right) \frac{x}{a_0} e^{-\frac{r}{3a_0}}$
3	1	$\psi_{3p_y} = \frac{\sqrt{2}}{81\sqrt{\pi a_0^3}} \left(6 - \frac{r}{a_0}\right) \frac{y}{a_0} e^{-\frac{r}{3a_0}}$
3	1	$\psi_{3p_z} = \frac{\sqrt{2}}{81\sqrt{\pi a_0^3}} \left(6 - \frac{r}{a_0}\right) \frac{z}{a_0} e^{-\frac{r}{3a_0}}$
3	2	$\psi_{3d_{xy}} = \frac{\sqrt{2}}{81\sqrt{\pi a_0^3}} \frac{x}{a_0} \frac{y}{a_0} e^{-\frac{r}{3a_0}}$
3	2	$\psi_{3d_{yz}} = \frac{\sqrt{2}}{81\sqrt{\pi a_0^3}} \frac{y}{a_0} \frac{z}{a_0} \cdot e^{-\frac{r}{3a_0}}$
3	2	$\psi_{3d_{xz}} = \frac{\sqrt{2}}{81\sqrt{\pi a_0^3}} \frac{x}{a_0} \frac{z}{a_0} \cdot e^{-\frac{r}{3a_0}}$
3	2	$\psi_{3d_{x^2-y^2}} = \frac{1}{81\sqrt{2\pi a_0^3}} \left[\left(\frac{x}{a_0}\right)^2 - \left(\frac{y}{a_0}\right)^2\right] e^{-\frac{r}{3a_0}}$
3	2	$\psi_{d_{z^2}} = \frac{1}{81\sqrt{6\pi a_0^3}} \left[3\left(\frac{z}{a_0}\right)^2 - \left(\frac{r}{a_0}\right)^2\right] e^{-\frac{r}{3a_0}}$

surfaces at which the wavefunction changes sign. The number of nodes in the total wavefunction is the sum of those in the radial and angular parts; hence, the total number of nodes is $(n - l - 1) + l = n - 1$ and does not depend on l.

For a given value of n (e.g., $n = 3$), the energy of the electron is fixed. The curvature of the radial wavefunction is larger for lower values of l than for larger values of l, however. This increased curvature is evident in the higher number of radial nodes. The total kinetic energy of the electron does not change however, because the angular and radial parts of the wavefunction

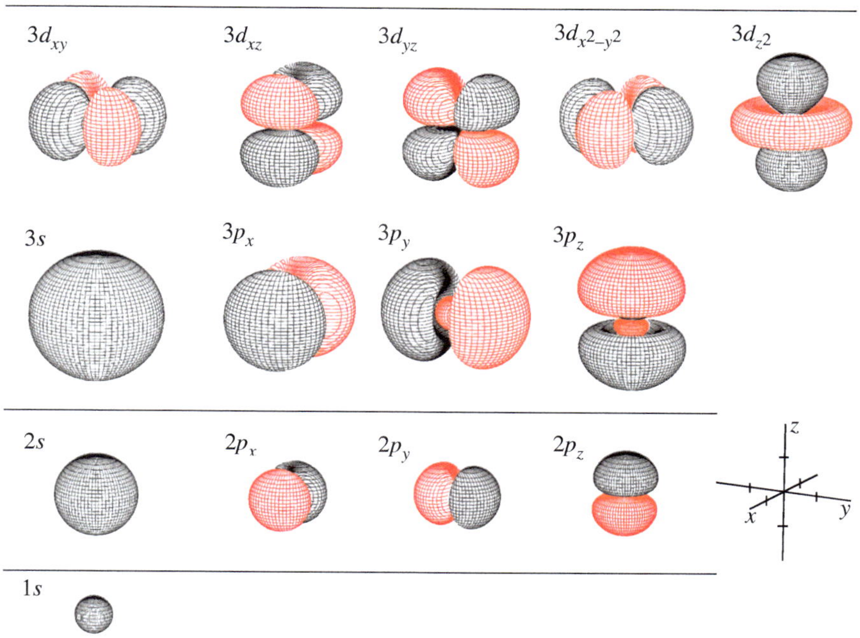

Figure 4.7 Three-dimensional diagrams of the H atom wavefunctions with the quantum numbers $n = 1$ (1s), $n = 2$ (2s, 2p), and $n = 3$ (3s, 3p, 3d). The individual picture elements correspond to the positions in space in which the values of the normalized wavefunctions have an amplitude of $\pm 1 \times 10^{-5}\,\text{pm}^{-3/2}$. The black and red coloring of the regions indicates that the wavefunction in those regions has a different sign. In the coordinate system the distances from the origin to the corresponding tic are 500 pm.

compensate each other for a given value of n; that is, the sum of the radial and angular contributions to the kinetic energy does not depend on l, only on n. The density ρ of $l = 0$ states is spherically symmetric. This is also the case for the sum of the densities of $l = 1, 2, \ldots, n-1$ states as can be seen from Table 4.1, for example the 2p states (note that $r^2 = x^2 + y^2 + z^2$).

2s and 2p Functions Figure 4.8a plots the 2s wavefunction and Fig. 4.8b provides a cloud representation of the probability density in which a planar slice through the origin has been performed. It is evident from this plot that the wavefunction has electron density around the origin, then it decreases as r increases until it strikes zero, and then it increases again before decaying away at long distances. The pattern of zero density appears as a circle on this plot but is actually a spherical shell in the full three-dimensional wavefunction. Figure 4.9a shows the plot for a 2p_y orbital, and Fig. 4.9b provides the cloud representations for 2p_x, 2p_y, and 2p_z. The 2p wavefunctions each have a nodal plane in their angular coordinate. In summary, these plots demonstrate that each $n = 2$ wavefunction has one nodal surface.

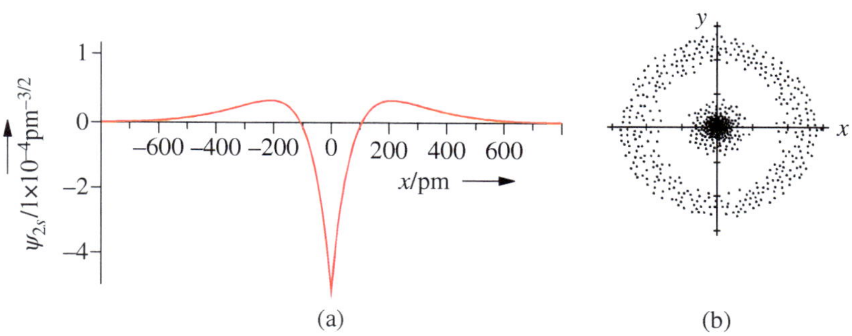

Figure 4.8 2s function for the H atom ($n = 2$). (a) Plot of ψ_{2s} along the abscissa. (b) Cloud display of $\rho = \psi^2$.

4.3.3 Radial Probability Distribution and Average Distance

In Section 4.2.3 we calculated the probability $\mathcal{P}(r < R)$ of finding the electron inside a sphere with radius R for the ground state of the H atom, see Equation (4.8). We apply the same procedure for the H atom in the 2s state. By inserting the corresponding wavefunction from Table 4.1, we find

$$\mathcal{P}(r < R) = \int_0^R \rho\, \mathrm{d}\tau = \int_0^R \rho\, 4\pi r^2\, \mathrm{d}r = \frac{4\pi}{32\pi a_0^3} \int_0^R \left(\frac{r}{a_0} - 2\right)^2 \mathrm{e}^{-r/a_0}\, r^2\, \mathrm{d}r \tag{4.18}$$

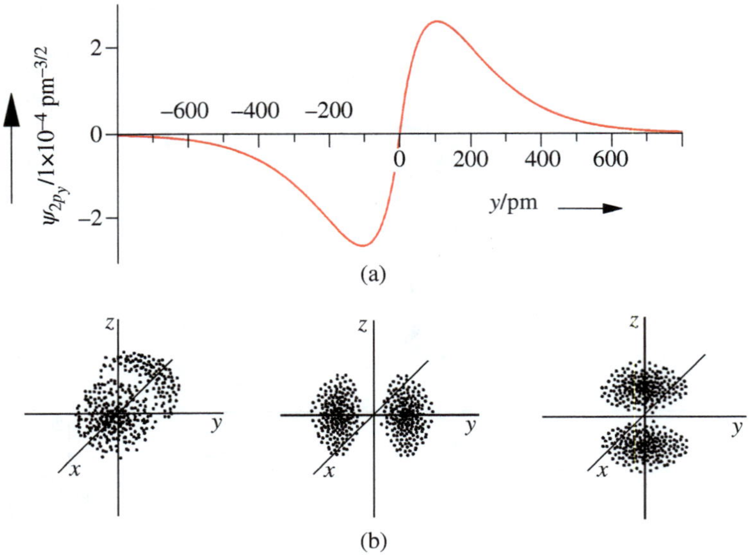

Figure 4.9 2*p* function for the H atom ($n = 2$). (a) Plot of ψ_{2p_y} along the abscissa. (b) Cloud representation of $\rho = \psi^2_{2p_x}$ (left), $\rho = \psi^2_{2p_y}$ (center), and $\rho = \psi^2_{2p_z}$ (right)

with the volume element $d\tau = 4\pi r^2\, dr$. In Problem 4.10, you are asked to show that the result of the integration is

$$\mathcal{P}(r < R) = 1 - e^{-r/a_0} \cdot \left[\frac{1}{8}\left(\frac{R}{a_0}\right)^4 + \frac{1}{2}\left(\frac{R}{a_0}\right)^2 + \frac{R}{a_0} + 1\right] \tag{4.19}$$

The probability is 85% for $R = 8.5a_0 = 450$ pm. This distance is much larger than that for the H atom in the ground state ($R = 2.36a_0 = 125$ pm).

In Section 4.2.4 we calculated the average distance $\bar{r}$ of the electron from the nucleus in the ground state. For the 2*s* state we proceed accordingly.

$$\bar{r}_{2s} = \int_0^\infty \rho \cdot r \cdot 4\pi r^2\, dr = \frac{4\pi}{32\pi a_0^3}\int_0^\infty \left(\frac{r}{a_0} - 2\right)^2 e^{-r/a_0}\, r^3\, dr \tag{4.20}$$

With the substitution $\xi = r/a_0$ we obtain

$$\bar{r}_{2s} = \frac{4\pi}{32\pi a_0^3} \cdot a_0^4 \int_0^\infty (\xi - 2)^2\, e^{-\xi} \xi^3\, d\xi = \frac{a_0}{8} \cdot (4 \cdot 3! - 4 \cdot 4! + 5!) = 6a_0 = 317\text{ pm}$$

Again, this is much more than $r = \frac{3}{2}a_0 = 79$ pm which was calculated in Example 4.2 for the 1s wave function.

For the nonspherical wavefunctions we proceed accordingly, but in this case we must use the volume element (see Appendix B)

$$d\tau = r^2 \sin\vartheta \cdot d\vartheta \cdot d\varphi \cdot dr$$

and integrate over ϑ, φ, and r. As an example, consider the wavefunction

$$\psi_{2p_z} = \frac{1}{4\sqrt{2\pi a_0^3}}\,\frac{z}{a_0}\, e^{-\frac{r}{2a_0}} = \frac{1}{4\sqrt{2\pi a_0^3}}\,\frac{r}{a_0}\cos\vartheta\, e^{-\frac{r}{2a_0}}$$

where we replaced z by $r\cos\vartheta$, according to Fig. 4.6. In Exercise 4.18 you are asked to show that

$$\bar{r}_{2p_z} = \int_{r=0}^\infty \int_{\vartheta=0}^\pi \int_{\varphi=0}^{2\pi} \rho \cdot r\, d\tau = 5a_0 = 264\text{ pm}$$

Why is the average distance for the electron different in 2s and 2p orbitals? See Exercise E4.13.

4.3.4 Emission Spectra

In a glow discharge, the electron in an H atom can be excited into a high-energy state. Once there it can emit light and transform to a lower energy state. Because energy must be conserved, the change in energy of the atom must correspond to the energy of the light that is emitted. By analyzing the light with a spectrophotometer, we can measure the atom's emission spectrum (i.e., intensity of the emitted light versus the wavelength of the light) and find that it contains contributions only from certain wavelengths. A different series of spectral lines is obtained from the set of emissions corresponding to transitions from higher energy electronic states down to each lower energy state. These processes are represented in Fig. 4.10.

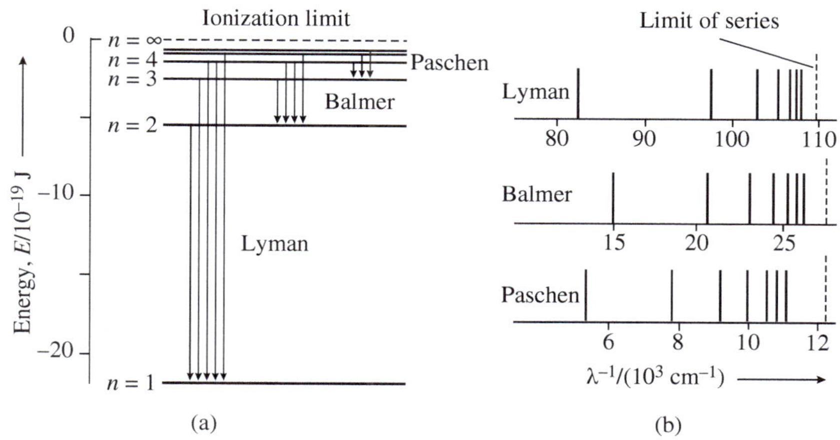

Figure 4.10 Series of spectral lines for the H atom. (a) Energy diagram with transitions. (b) Display of the corresponding spectral lines: Lyman Series (in the ultraviolet), Balmer Series (visible), and Paschen Series (in the infrared).

We can calculate the wavelength of each spectral line from Equation (4.15) using the relation

$$\lambda = \frac{h c_0}{\Delta E} \tag{4.21}$$

(see Chapter 2). For example, the transition from a higher energy level to the level with $n = 2$ corresponds to an energy difference of

$$\Delta E = E_n - E_2 = \frac{1}{4\pi\varepsilon_0} \cdot \frac{e^2}{2a_0} \cdot \left(\frac{1}{4} - \frac{1}{n^2}\right), \quad n = 3, 4, 5, \ldots \tag{4.22}$$

(*Balmer Series*). In Table 4.2, the calculated and measured wavelengths for the Balmer series are compared; we establish that (after application of two minor corrections) there is excellent agreement. This finding provides a further indication that the Schrödinger Equation is a useful postulate.

Table 4.2 Wavelengths λ (in nm) for Transitions in the H Atom from $n = 3, 4, 5$, and 6 to $n = 2$ (Balmer Series). Experiment and Theory.[3]

		Theory	
n	Experiment	Uncorrected	Corrected
3	656.466	656.112	656.466
4	486.271	486.009	486.272
5	434.170	433.937	434.171
6	410.290	410.070	410.291

Another sequence of wavelengths can be identified in the atom's emission spectrum. It is called the *Lyman Series* and the lower energy state in this case is the level with $n = 1$. From this series, the ionization energy E_{ion} of the ground-state H atom can be determined. The ionization energy is given by the energy difference between the ground state and the limit where $n = \infty$. For the Lyman series the transitions are given by

$$\Delta E = \frac{1}{4\pi\varepsilon_0} \cdot \frac{e^2}{2a_0} \cdot \left(1 - \frac{1}{n^2}\right), \quad n = 2, 3, 4, \ldots \tag{4.23}$$

and we find for $n \to \infty$ that

$$E_{\mathrm{ion}} = 21.80 \times 10^{-19}\ \mathrm{J}$$

4.3.5 Degeneracy of H-atom Orbitals Can Be Removed by a Magnetic Field

4.3.5.1 Representation by Real and Complex Wavefunctions

Let us consider an excited H-atom with an electron in a 2*p*-orbital. For simplicity consider the degenerate set of p_x and p_y orbitals

$$\psi_{2p_y} = A \sin\varphi, \qquad \psi_{2p_x} = A \cos\varphi \tag{4.24}$$

3 The experimental values are for vacuum conditions. The theoretical values in column 3 (uncorrected) are calculated from Equations (4.22) and (4.23). Two minor corrections must be applied: The values must be corrected for the comovement of the nucleus (proton mass m_p) in the H atom (leading to an increase of wavelength) and for the effect of relativity: The electron mass increases with velocity, this leads to a decrease of wavelength.

where

$$A = \frac{\cos \vartheta}{4\sqrt{2\pi a_0^3}} \cdot \frac{r}{a_0} \cdot e^{-r/2a_0}$$

As in the case of the electron rotating on a ring, we may instead use complex functions:

$$\psi_{211} = \frac{1}{\sqrt{2}}\left(\psi_{2p_y} + i \cdot \psi_{2p_x}\right), \quad \psi_{21-1} = \frac{1}{\sqrt{2}}\left(\psi_{2p_y} - i \cdot \psi_{2p_x}\right)$$

Then we obtain

$$\psi_{211} = \frac{A}{\sqrt{2}} e^{i\varphi}; \quad \psi_{21-1} = \frac{A}{\sqrt{2}} e^{-i\varphi} \tag{4.25}$$

The first two subscripts denote the quantum numbers $n = 2$ and $l = 1$. The third subscript corresponds to the *magnetic quantum number k* which arises in the rigorous solution of the Schrödinger equation for the H atom, see Foundations 4.1 and 4.2 (available on Wiley's website).

4.3.5.2 Angular Momentum and Energy Splitting in a Magnetic Field

The angular motion of a p electron about the nucleus at a specified distance r proceeds on the surface of a spherical shell. In Foundations 4.1 and 4.2 (available on Wiley's website), we find that the angular kinetic energy of the electron is given by

$$T_{\text{angular}} = \frac{\hbar^2 \, l(l+1)}{2m_e r^2} \tag{4.26}$$

The l quantum number specifies the total orbital angular momentum. To illustrate this result consider our classical notion of angular kinetic energy in terms of a particle's angular momentum $\mathcal{L} = m_e v r$ (where v is the speed and r is the distance from the center of rotation):

$$T_{\text{angular}} = \frac{1}{2} m_e v^2, \quad \mathcal{L}^2 = m_e^2 v^2 r^2, \quad \text{thus } T_{\text{angular}} = \frac{1}{2} \frac{\mathcal{L}^2}{m_e r^2} \tag{4.27}$$

Equating Equations (4.26) and (4.27) we find

$$|\mathcal{L}| = \hbar \sqrt{l(l+1)} \tag{4.28}$$

and it is quantized. For $l = 1$ the angular momentum has a magnitude of $\hbar\sqrt{2}$ and the sign specifies its direction (oriented along positive z or oriented along negative z), depending on the direction of the electron's orbital motion (counter-clockwise or clockwise). Correspondingly, the speed is quantized

$$v = \frac{|\mathcal{L}|}{m_e r} = \frac{\hbar}{m_e r} \sqrt{l(l+1)} \tag{4.29}$$

For the general case of three $2p$ orbitals of an H atom, the quantum number $l = 1$ specifies the total angular momentum with a magnitude of $\hbar\sqrt{l(l+1)} = \hbar\sqrt{2}$. The magnetic quantum number k specifies the component of the angular momentum vector along the z-direction. For $k = 1$, the z-component of the angular momentum is $\hbar$; for $k = -1$, the z-component of the angular momentum is $-\hbar$; and for $k = 0$ the z-component is zero. We can illustrate this result by a vector of length $\hbar\sqrt{2}$ precessing about the z axis such that the z component of the angular momentum is $\hbar$, 0, or $-\hbar$ (Fig. 4.11). In a magnetic field in the z direction the degeneracy in the present case of the $2p$ states (quantum number $l = 1$, see Chapter 3) is removed. A splitting into states $k = 1, 0, -1$ of differing energy takes place (wavefunctions ψ_{211}, $\psi_{2p_z} = \psi_{210}$, and ψ_{21-1} in Foundation 4.1).

The case of three $3p$ orbitals is treated correspondingly (wavefunctions ψ_{311}, ψ_{3p_z}, and ψ_{31-1} in Foundation 4.1). For the case of five $3d$ orbitals we have $l = 2$; thus, these states split into states with $k = 2$, $k = 1$, $k = 0$, $k = -1$, and $k = -2$ (wavefunctions ψ_{322}, ψ_{321}, $\psi_{3dz^2} = \psi_{320}$, ψ_{32-1}, and ψ_{32-2} in Foundation 4.1). In general, an angular momentum state with quantum number l splits into $2l + 1$ energy levels under an applied magnetic field (Zeeman effect).

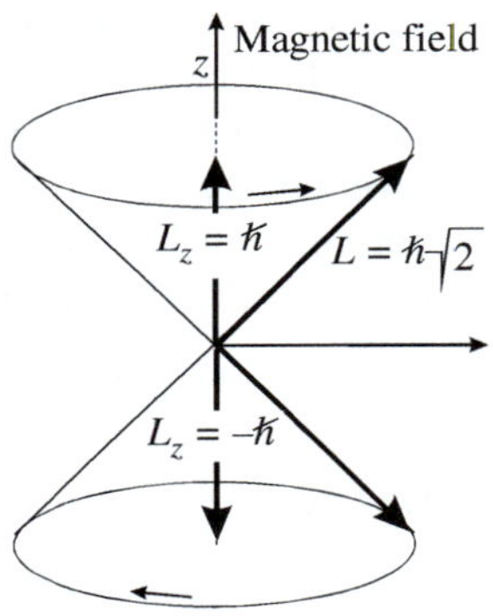

Figure 4.11 H-atom in a magnetic field. The total angular momentum has a magnitude of $|\mathcal{L}| = L = \hbar\sqrt{2}$, and it can be oriented at one of three angles with respect to the z-axis. The z-component of angular momentum is $|\mathcal{L}_z| = L_z = \hbar$ for $\vartheta = 45°$, $|\mathcal{L}_z| = L_z = 0$ for $\vartheta = 90°$, and $|\mathcal{L}_z| = L_z = -\hbar$ for $\vartheta = 135°$.

4.4 Conclusion

The time-independent Schrödinger equation was used to describe the H atom. Comparisons with experiment (see Table 4.2) show excellent agreement between the quantum mechanically determined energy level structure and the energy gaps that are observed through emission spectroscopy. The solutions of the Schrödinger equation were explored to illustrate the spatial and energy features of the atom's electronic charge distribution. In the next chapters we will see other aspects of the H atom wavefunctions' importance.

4.5 Foundations

4.5.1 Foundation 4.1: An Overview of Solution to the Schrödinger Equation for the H-atom

The potential energy $V(r)$ for the electron interacting with the nucleus in the hydrogen atom depends only on the distance r between the charges. In general, the Schrödinger equation has the form

$$-\frac{\hbar^2}{2m_p}\nabla_p^2\psi - \frac{\hbar^2}{2m_e}\nabla_e^2\psi + [V(r) - E]\,\psi = 0$$

where the first term is the kinetic energy of the proton, the second term is the kinetic energy of the electron, and E is the total energy. The solution we develop here ignores the kinetic energy of the proton. A more exact treatment, performed in the center of mass coordinates, gives a similar result to that we develop here, but the mass of the electron m_e is replaced by the reduced mass.

In this approximation, we can write the Schrödinger Equation for the H atom in spherical coordinates r, ϑ, and φ (see Fig. 4.6) as

$$\frac{-\hbar^2}{2m_e r^2}\left[\frac{\partial}{\partial r}\left(r^2\frac{\partial}{\partial r}\right) + \frac{1}{\sin\vartheta}\frac{\partial}{\partial\vartheta}\left(\sin\vartheta\frac{\partial}{\partial\vartheta}\right) + \frac{1}{\sin^2\vartheta}\frac{\partial^2}{\partial\varphi^2}\right]\psi + [V(r) - E]\,\psi = 0 \tag{4.30}$$

where

$$V(r) = \frac{-Ze^2}{4\pi\varepsilon_0 r} \quad \text{with } Z = 1 \text{ for the hydrogen atom} \tag{4.31}$$

The wavefunction ψ is a function of r, ϑ, and φ; $\psi(r, \vartheta, \varphi)$. The transformation from Cartesian coordinates to spherical coordinates for the Laplacian may be found elsewhere.[4] If we multiply both sides of this equation from the left by $(-2m_e r^2/\hbar^2)$, we can separate the radial and angular terms, so that

$$\frac{\partial}{\partial r}\left(r^2\frac{\partial}{\partial r}\right)\psi + \frac{2m_e r^2}{\hbar^2}[E - V(r)]\,\psi = -\left[\frac{1}{\sin\vartheta}\frac{\partial}{\partial\vartheta}\left(\sin\vartheta\frac{\partial}{\partial\vartheta}\right) + \frac{1}{\sin^2\vartheta}\frac{\partial^2}{\partial\varphi^2}\right]\psi$$

This rearrangement allows us to inspect the equation and realize that it is separable. In particular, the left-hand side depends only on the distance r and the right-hand side depends only on the angles ϑ and φ. This observation suggests that we try a solution which is a product of two functions, namely

$$\psi(r, \vartheta, \varphi) = R(r) \cdot Y(\vartheta, \varphi)$$

Substituting this form into the Schrödinger equation and then multiplying by $1/(R \cdot Y)$, we find

$$\frac{1}{R}\frac{\mathrm{d}}{\mathrm{d}r}\left(r^2\frac{\mathrm{d}}{\mathrm{d}r}\right)R + \frac{1}{R}\frac{2m_e r^2}{\hbar^2}(E - V)R = \frac{-1}{Y}\left[\frac{1}{\sin\vartheta}\frac{\partial}{\partial\vartheta}\left(\sin\vartheta\frac{\partial}{\partial\vartheta}\right) + \frac{1}{\sin^2\vartheta}\frac{\partial^2}{\partial\varphi^2}\right]Y$$

This form makes it clear that the two sides of the equation are independent of each other, but they must be equal. Hence, they are each equal to the same constant. If we choose to call the constant β, then we obtain an angular equation

$$\frac{-1}{Y(\vartheta, \varphi)}\left[\frac{1}{\sin\vartheta}\frac{\partial}{\partial\vartheta}\left(\sin\vartheta\frac{\partial}{\partial\vartheta}\right) + \frac{1}{\sin^2\vartheta}\frac{\partial^2}{\partial\varphi^2}\right]Y(\vartheta, \varphi) = \beta \tag{4.32}$$

4 For example, see Chapter 6 of F.B. Hildebrand [1].

and a radial equation

$$\frac{1}{R(r)}\frac{\mathrm{d}}{\mathrm{d}r}\left(r^2\frac{\mathrm{d}}{\mathrm{d}r}\right)R(r)+\frac{1}{R(r)}\frac{2m_e r^2}{\hbar^2}(E-V(r))R(r)=\beta \tag{4.33}$$

In Foundation 4.2 we derive the solutions of the angular and the radial equations. The constant β turns out to be

$$\beta = l(l+1)$$

The resulting energies E depend on the principal quantum number $n = 1, 2, 3, \ldots$

$$E = -\frac{1}{4\pi\varepsilon_0}\cdot\frac{e^2}{2a_0}\cdot\frac{1}{n^2},\quad n = 1, 2, 3, \ldots \tag{4.34}$$

The kinetic energy of the angular motion is

$$T_{\text{angular}} = \frac{\hbar^2}{2m_e r^2}l(l+1) \tag{4.35}$$

The wavefunctions depend on the principal quantum n, on the angular momentum quantum number $l = 0, 1, ..n-1$, and on the magnetic quantum number $k = -l, (-l+1), \ldots -1, 0, 1, \ldots, (l-1), l$ (see Foundation 4.2 for details). The 14 different wavefunctions for the quantum numbers $n = 1$ to $n = 3$ are shown in Table 4F.1a, where $\rho = r/a_0$. Table 4F.1b compares the quantum numbers in Table 4F.1a to the notation in Table 4.1 in the text.

Table 4F.1a Some Total Wavefunctions for a One-Electron Atom with a Nuclear Charge Ze, where $\rho = Zr/a_0$.

Orbital ψ_{nlk}	n	l	k	NormalizationFactor	Functional Form
$\psi_{1s} = \psi_{100}$	1	0	0	$\frac{1}{\sqrt{\pi}}\left(\frac{Z}{a_0}\right)^{3/2}$	$e^{-\rho}$
$\psi_{2s} = \psi_{200}$	2	0	0	$\frac{1}{4\sqrt{2\pi}}\left(\frac{Z}{a_0}\right)^{3/2}$	$(\rho-2)\cdot e^{-\rho/2}$
$\psi_{2p_z} = \psi_{210}$	2	1	0	$\frac{1}{4\sqrt{2\pi}}\left(\frac{Z}{a_0}\right)^{3/2}$	$\rho\cdot\cos\vartheta\cdot e^{-\rho/2}$
ψ_{211}	2	1	1	$\frac{1}{8\sqrt{\pi}}\left(\frac{Z}{a_0}\right)^{3/2}$	$\rho\cdot\sin\vartheta\cdot e^{i\varphi}\cdot e^{-\rho/2}$
ψ_{21-1}	2	1	−1	$\frac{1}{8\sqrt{\pi}}\left(\frac{Z}{a_0}\right)^{3/2}$	$\rho\cdot\sin\vartheta\cdot e^{-i\varphi}\cdot e^{-\rho/2}$
$\psi_{3s} = \psi_{300}$	3	0	0	$\frac{1}{81}\frac{1}{\sqrt{3\pi}}\left(\frac{Z}{a_0}\right)^{3/2}$	$\left[27-18\rho+2\rho^2\right]\cdot e^{-\rho/3}$
$\psi_{3p_z} = \psi_{310}$	3	1	0	$\frac{1}{81}\sqrt{\frac{2}{\pi}}\left(\frac{Z}{a_0}\right)^{3/2}$	$\rho(6-\rho)\cdot\cos\vartheta\cdot e^{-\rho/3}$
ψ_{311}	3	1	1	$\frac{1}{81\sqrt{\pi}}\left(\frac{Z}{a_0}\right)^{3/2}$	$\rho(6-\rho)\cdot\sin\vartheta\cdot e^{i\varphi}\cdot e^{-\rho/3}$
ψ_{31-1}	3	1	−1	$\frac{1}{81\sqrt{\pi}}\left(\frac{Z}{a_0}\right)^{3/2}$	$\rho(6-\rho)\cdot\sin\vartheta\cdot e^{-i\varphi}\cdot e^{-\rho/3}$
$\psi_{3d_{z^2}} = \psi_{320}$	3	2	0	$\frac{1}{81}\sqrt{\frac{1}{6\pi}}\left(\frac{Z}{a_0}\right)^{3/2}$	$\left(3\cos^2\vartheta-1\right)\rho^2\cdot e^{-\rho/3}$
ψ_{321}	3	2	1	$\frac{1}{81}\sqrt{\frac{1}{\pi}}\left(\frac{Z}{a_0}\right)^{3/2}$	$\sin\vartheta\cos\vartheta\cdot\rho^2\cdot e^{i\varphi}\cdot e^{-\rho/3}$
ψ_{32-1}	3	2	−1	$\frac{1}{81}\sqrt{\frac{1}{\pi}}\left(\frac{Z}{a_0}\right)^{3/2}$	$\sin\vartheta\cos\vartheta\cdot\rho^2\cdot e^{-i\varphi}\cdot e^{-\rho/3}$
ψ_{322}	3	2	2	$\frac{1}{81\sqrt{4\pi}}\left(\frac{Z}{a_0}\right)^{3/2}$	$\sin^2\vartheta\cdot\rho^2\cdot e^{i2\varphi}\cdot e^{-\rho/3}$
ψ_{32-2}	3	2	−2	$\frac{1}{81\sqrt{4\pi}}\left(\frac{Z}{a_0}\right)^{3/2}$	$\sin^2\vartheta\cdot\rho^2\cdot e^{-i2\varphi}\cdot e^{-\rho/3}$

The functions presented here are for an arbitrary value of the nuclear charge, Z. If you use $Z = 1$, you can generate the wavefunctions for the hydrogen atom. For $Z = 2$ you will generate the wavefunctions for He^+, and so forth. In addition, the radial distance is expressed in the form of $\rho = Zr/a_0$. Note that the energy expression for a hydrogen-like atom with atomic number Z is

$$E = -\frac{1}{4\pi\varepsilon_0}\cdot\frac{Z^2e^2}{2a_0}\cdot\frac{1}{n^2},\quad n = 1, 2, 3, \ldots \tag{4.36}$$

In Section 4.3.2 the hydrogen wavefunctions ($Z = 1$) from Table 4F.1a are written in Cartesian coordinates. For this purpose, instead of using the pairs of degenerate complex angular functions $\exp(i\varphi)$ and $\exp(-i\varphi)$, we use the equivalent sets of linear combinations

$$\psi_a = A(\exp(i\varphi) + \exp(-i\varphi)), \quad \psi_b = B(\exp(i\varphi) - \exp(-i\varphi)) \tag{4.37}$$

Using the Euler relations (see Appendix C)

$$e^{i\varphi} = \cos\varphi + i\sin\varphi, \quad e^{-i\varphi} = \cos\varphi - i\sin\varphi \tag{4.38}$$

and normalizing, we obtain

$$\psi_a = \frac{1}{\sqrt{\pi}}\cos\varphi, \quad \psi_b = \frac{1}{\sqrt{\pi}}\sin\varphi \tag{4.39}$$

Additionally, we use the relations in Fig. 4.6:

$$x = r\sin\vartheta\cos\varphi, \quad y = r\sin\vartheta\sin\varphi, \quad z = r\cos\vartheta \tag{4.40}$$

The wavefunctions with spherical symmetry (ψ_{1s}, ψ_{2s}, ψ_{3s}) remain unchanged, and the wavefunction ψ_{2pz} becomes (for simplicity, the normalization factor is omitted):

$$\psi_{2pz} = \frac{r}{a_0}\cdot\cos\vartheta\cdot e^{-r/2a_0} = \frac{z}{a_0}\cdot e^{-r/2a_0}$$

The wavefunctions ψ_{2px} and ψ_{2py} can be derived from ψ_{211} and ψ_{21-1}

$$\psi_{2px} = \frac{r}{a_0}\cdot\sin\vartheta\cdot\cos\varphi\cdot e^{-r/2a_0} = \frac{x}{a_0}\cdot e^{-r/2a_0}$$

$$\psi_{2py} = \frac{r}{a_0}\cdot\sin\vartheta\cdot\sin\varphi\cdot e^{-r/2a_0} = \frac{y}{a_0}\cdot e^{-r/2a_0}$$

Most of the remaining wavefunctions are converted correspondingly. Wavefunctions ψ_{3d,x^2-y^2} and $\psi_{3d,xy}$ can be derived from ψ_{322} and ψ_{32-2} but require some trigonomeric manipulations

Table 4F.1b Comparison of wavefunctions $\psi_{n,l,k}$ in Table 4F.1a and notation of the orbitals in Table 4.1.

ψ_{nlk}	**Orbitals in Table 4.1**
ψ_{100}	$1s$
ψ_{200}	$2s$
ψ_{210}	$2p_z$
$\frac{1}{\sqrt{2}}(\psi_{211} + \psi_{21-1})$	$2p_x$
$\frac{1}{\sqrt{2}}(\psi_{211} - \psi_{21-1})$	$2p_y$
ψ_{300}	$3s$
ψ_{310}	$3p_z$
$\frac{1}{\sqrt{2}}(\psi_{311} + \psi_{31-1})$	$3p_x$
$\frac{1}{\sqrt{2}}(\psi_{311} - \psi_{31-1})$	$3p_y$
ψ_{320}	$3d_{z^2}$
$\frac{1}{\sqrt{2}}(\psi_{321} + \psi_{32-1})$	$3d_{xz}$
$\frac{1}{\sqrt{2}}(\psi_{321} - \psi_{32-1})$	$3d_{yz}$
$\frac{1}{\sqrt{2}}(\psi_{322} + \psi_{32-2})$	$3d_{xy}$
$\frac{1}{\sqrt{2}}(\psi_{322} - \psi_{32-2})$	$3d_{x^2-y^2}$

$$\begin{aligned}
\psi_{3d,x^2-y^2} &= \sin^2\vartheta\cdot\left(\frac{r}{a_0}\right)^2\cdot\cos(2\varphi)\cdot e^{-r/3a_0}\\
&= \sin^2\vartheta\cdot\left(\frac{r}{a_0}\right)^2\cdot\left[\cos^2\varphi - \sin^2\varphi\right]\cdot e^{-r/3a_0}\\
&= \left[\sin^2\vartheta\cos^2\varphi\cdot\left(\frac{r}{a_0}\right)^2 - \sin^2\vartheta\sin^2\varphi\cdot\left(\frac{r}{a_0}\right)^2\right]\cdot e^{-r/3a_0}\\
&= \left[\left(\frac{x}{a_0}\right)^2 - \left(\frac{y}{a_0}\right)^2\right]\cdot e^{-r/3a_0}
\end{aligned}$$

Here we have used the relation $\cos(2\varphi) = \cos^2\varphi - \sin^2\varphi$. Wavefunction $\psi_{3d,xy}$ is obtained correspondingly, using $\sin(2\varphi) = 2\cdot\sin\varphi\cos\varphi$.

4.5.2 Foundation 4.2: Solutions to the Angular and Radial Parts of the H-atom Schrödinger Equation

See the book's website, www.wiley.com/go/Kuhn/physicalchemistry3e.

4.5.3 Foundation 4.3: Virial Theorem

See the book's website, www.wiley.com/go/Kuhn/physicalchemistry3e.

4.6 Exercises

E4.1 Sketch the radial and angular parts of a 3s orbital's wavefunction. Identify the number of radial and angular nodes.

E4.2 Sketch the radial and angular parts of a d_{xy} orbital's wavefunction. Identify the number of radial and angular nodes.

E4.3 Sketch the radial and angular parts of a 3d_{z^2} orbital's wavefunction. Identify the number of radial and angular nodes.

E4.4 Sketch the radial and angular parts of a 4p_z orbital's wavefunction. Identify the number of radial and angular nodes.

E4.5 Give the number of nodes for the radial and angular parts of the orbitals in E4.1 through E4.4. Explain the trend observed.

E4.6 Sketch the radial distribution function for a 3s orbital.

E4.7 Consider the 2s orbital of the hydrogen atom. Find the value(s) of r at which the electron probability density, ρ, is zero. Find the value(s) of r at which the electron probability density, ρ, has a maximum.

E4.8 Write the energy level formula for the hydrogen atom's electronic energy and identify the different terms in the equation.

E4.9 a) Consider the case where the electron can be described by the 2p_z wavefunction (or orbital). Sketch both the radial and angular parts of the wavefunction, and state the number of nodes in each part. b) Determine the distance from the nucleus at which it is most probable to find an electron; i.e., the peak of the radial distribution function.

E4.10 Calculate the distance r at which the probability density ρ for the electron in the ground-state hydrogen atom has decreased to 1/1000 of its maximum value.

E4.11 Show that the ψ_{1s} and the ψ_{2p_z} hydrogen atom wavefunctions are orthogonal to each other.

E4.12 Plot the radial distribution functions for the first three s-orbitals of the hydrogen atom, and compare the most probable radius for the electron in each case. Comment on the trend you observe.

E4.13 For $n = 3$, show that the size of the atom (as measured by the radial distribution function) decreases as the l quantum number increases. Provide a physical explanation for this trend.

E4.14 Calculate the normalization constant for the wavefunction of the hydrogen atom in its ground state, $\psi = Ae^{-r/a_0}$. Remember that you must perform a three-dimensional integral. It is best to perform this integral in spherical coordinates; for which we find that

$$\int_0^{2\pi} \mathrm{d}\varphi \int_0^{\pi} \sin\theta \cdot \mathrm{d}\theta \int_0^{\infty} \psi(r)^2\, r^2\, \mathrm{d}r = 4\pi \int_0^{\infty} \psi(r)^2\, r^2\, \mathrm{d}r = 1$$

because ψ is a function of r, but has no explicit dependence on θ or φ. Compare your answer to Equation (4.3).

E4.15 Write the Schrödinger equation for He^+. The Schrödinger equation here is like that in Equation (4.2), but the nuclear charge is $2e$ rather than e. Given that the ground-state wavefunction of He^+ is $\psi = Ae^{-2r/a_0}$ show that

$$E = -\frac{1}{4\pi\varepsilon_0}\frac{2e^2}{a_0}$$

E4.16 In Example 4.1, we find the derivatives of ψ with respect to x. In an analogous way find the derivatives with respect to the variables y and z; and then show that

$$\nabla^2\psi = \frac{\partial^2\psi}{\partial x^2} + \frac{\partial^2\psi}{\partial y^2} + \frac{\partial^2\psi}{\partial z^2} = -\frac{1}{\sqrt{\pi a_0^3}} \cdot \frac{1}{a_0} \cdot \left(\frac{2}{r} - \frac{1}{a_0}\right) \cdot e^{-r/a_0}$$

E4.17 Show that the equality in Equation (4.12) holds. In particular work out the details for how the three-dimensional integral reduces to that used in Example 4.3.

E4.18 Calculate the average distance $\bar{r}$ for an electron in the 2p_z state of the hydrogen atom.

E4.19 It is common for quantum chemists to work in a set of units called "atomic units," which can be obtained by using the Bohr radius a_0 as the unit of distance. By transforming the coordinates (x, y, z) to $(a_0 \cdot x', a_0 \cdot y', a_0 \cdot z')$, show that the Schrödinger equation for the H-atom (Equation (4.2)) can be written as

$$-\frac{1}{2}\nabla^2\psi - \frac{1}{r'}\psi = E_{a.u} \cdot \psi \quad \text{where} \quad E_{a.u} = \frac{1}{2} \cdot \frac{1}{n^2}$$

where $E_{a.u.}$ is the energy in atomic units.

E4.20 If a source of 58.5 nm wavelength photons (typical of that used in photoelectron spectroscopy) irradiates a sample of neutral hydrogen atoms, it is possible to eject electrons from the atoms and generate protons. The most stable hydrogen atoms (i.e., ground-state hydrogen atoms) bind the electron with about 13.6 eV of energy. Convert the 13.6 eV binding energy into Joules. Determine the energy of the 58.5 nm photon. Use energy conservation to deduce the kinetic energy of the photo-ejected electron. If you were trying to measure the electron's kinetic energy in an apparatus analogous to that illustrated in Fig. 1.2, what stopping voltage would you need to apply? Comment on why electron Volts (eV) might be a convenient energy unit for scientists doing photoelectron spectroscopy.

E4.21 Consider the kinetic energy calculations in E1.3. Perform a similar calculation but use a photon wavelength of 23 nm. Perform a similar calculation but use a photon wavelength of 100 nm. Compare your three values of the kinetic energy and comment on them.

Assuming that the electron is bound by 13.6 eV, determine the largest wavelength that a photon may have if it is to eject an electron from a hydrogen atom.

4.7 Problems

P4.1 An early model of the atom, called the Bohr atom model, is of historical significance in the development of quantum mechanics. Before the introduction of wave mechanics, Bohr created an atomic model (Fig. 4P.1).

Figure 4P.1 Bohr atom model; the electron circles at a distance r with the speed v around the nucleus.

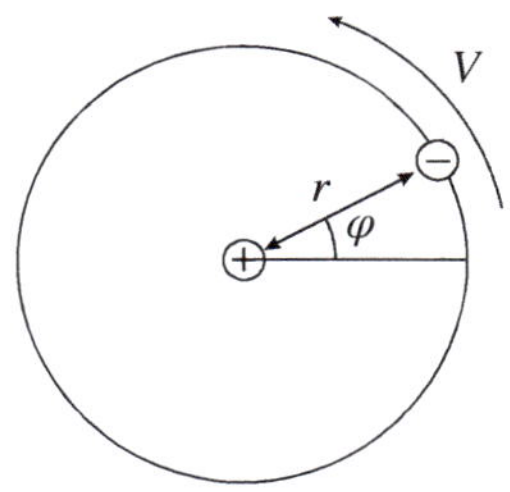

This model views the electron as a particle that orbits about the proton at a distance r and has a velocity v. In this model, Bohr considered that the electrostatic attraction supplies the centripetal force $(m_e v^2/r)$ which holds the electron in a circle. In addition, he postulated that the orbital angular momentum was quantized and given by the formula $m_e vr = h/(2\pi)$ in the ground electronic state. By using the balance of forces and the postulate for the angular momentum, show that the electron's distance from the nucleus is given by

$$r = 4\pi\varepsilon_0 \frac{h^2}{4\pi^2\, m_e e^2}$$

and the electron's speed is given by

$$v = \frac{4\pi^2}{4\pi\varepsilon_0}\frac{e^2}{h}$$

Note that the orbit's radius is the same as that we found for the most probable distance a_0; a_0 is called the *Bohr radius*. Proceed to calculate the kinetic and potential energy from v and r.

P4.2 The energy of the H-atom was calculated with the assumption that the nucleus is at rest; actually, it is the center of mass which is stationary. Account for the movement of the nucleus (see Figure 4P.2) and calculate the energy. To calculate the kinetic and potential energy, you need to replace m_e by the reduced mass

$$m_{\text{red}} = \frac{m_e m_p}{m_e + m_p}$$

Determine the new energy expression by using the Bohr model of the atom. How much does the energy shift from our more approximate solution? Is this shift significant?

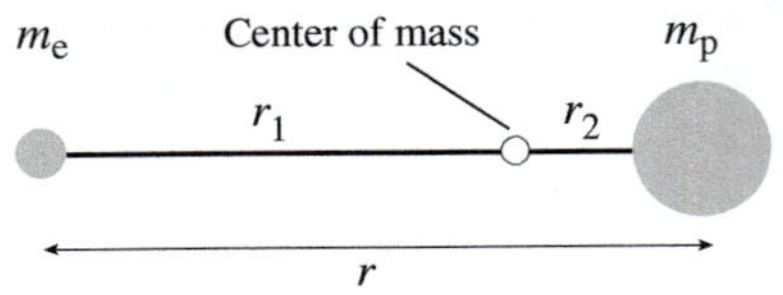

Figure 4P.2 The diagram shows the center of mass for an H-atom. Note that the displacement from the nucleus is exaggerated.

P4.3 The MathCAD exercises included in the module "Hatom.mcdx" ask you to plot and analyze features of the hydrogen atom orbitals.

P4.4 In Section 4.3.3 we calculated the probability $\mathcal{P}(r < R)$ of finding a 2s electron inside a sphere with radius R. Show that the integration in Equation (4.18) leads to the result in Equation (4.19).

P4.5 In Section 4.3.3 we calculated the probability $\mathcal{P}(r < R)$ of finding a $2s$ electron inside a sphere with radius R. Perform a corresponding calculation and obtain an expression for $\mathcal{P}(r < R)$ for a $2p$ electron. Evaluate these expressions for R values of $2a_0$, $4a_0$, $6a_0$, $8a_0$, and $10a_0$.

P4.6 In Section 4.3.3 we calculated the average distance $\bar{r}$ of a $2p_z$ electron from the nucleus. Perform a corresponding calculation for a $2p_x$ and $2p_y$ electron.

P4.7 Consider the $2s$ orbital of the hydrogen atom. Find the value(s) of r at which the electron probability distribution $4\pi r^2 |R(r)|^2$ is a maximum.

P4.8 Plot the radial distribution functions for the first three s-orbitals of the hydrogen atom, and compare the most probable radius for the electron in each case. Analytically determine the value of the most probable radius for the $1s$ orbital and the $2s$ orbital.

P4.9 For a $3s$ orbital, evaluate the average value of the electron's distance from the proton and compare it to its most probable value of $r = 6.96$ Å (see solution to P4.8). Which value is larger and why?

P4.10 Begin with the general expression for the radial probability distribution function $\mathcal{P}(r < R) = \int_0^R \rho\, \mathrm{d}\tau$ and derive a formula for $\mathcal{P}_{2s}$; namely, show that

$$\mathcal{P}_{2s}(r < R) = 1 - \mathrm{e}^{-R/a_0} \cdot \left[\frac{R^4}{8a_0^4} + \frac{R^2}{2a_0^2} + \frac{R}{a_0} + 1\right]$$

P4.11 Prove the identity

$$x\left(\frac{\mathrm{d}^3\psi}{\mathrm{d}x^3}\psi - \frac{\mathrm{d}^2\psi}{\mathrm{d}x^2}\cdot\frac{\mathrm{d}\psi}{\mathrm{d}x}\psi\right) = -2\psi\frac{\mathrm{d}^2\psi}{\mathrm{d}x^2} + \frac{\mathrm{d}}{\mathrm{d}x}\left[\psi^2\frac{\mathrm{d}}{\mathrm{d}x}\left(\frac{x\cdot\mathrm{d}\psi/\mathrm{d}x}{\psi}\right)\right]$$

by performing the differentiation on the right-hand side. This is one step in deriving the Virial Theorem in Foundation 4.3.

P4.12 In Foundation 4.2 we derived the Schrödinger Equation for the radial part R of the wavefunction for the H atom as

$$\frac{\hbar^2}{2m_e r^2}\cdot\frac{\mathrm{d}}{\mathrm{d}r}\left(r^2\frac{\mathrm{d}}{\mathrm{d}r}\right)R(r) + \left[E + \frac{e^2}{4\pi\varepsilon_0}\frac{1}{r} - \frac{\hbar^2}{2m_e r^2}l(l+1)\right]R(r) = 0$$

Here we apply this equation to the problem of standing waves of an electron in a spherical box with diameter D. In this case the potential energy is zero inside the box and the total energy equals the kinetic energy E_{kin}. Furthermore, we restrict our consideration to the case of spherical solutions ($l = 0$) and obtain:

$$\frac{\hbar^2}{2m_e r^2}\cdot\frac{\mathrm{d}}{\mathrm{d}r}\left(r^2\frac{\mathrm{d}}{\mathrm{d}r}\right)R(r) = -E_{kin}R(r)$$

Solve this equation using the separation ansatz

$$R(r) = \frac{1}{r}U(r), \qquad \frac{\mathrm{d}R(r)}{\mathrm{d}r} = \frac{1}{r}\frac{\mathrm{d}U(r)}{\mathrm{d}r} - \frac{1}{r^2}U(r)$$

Find the conditions for standing waves, determine the wavefunction $R(r)$, and find the energy E_{kin} for the lowest quantum state.

P4.13 In this problem you should use your knowledge of differential operators to show that in Foundation 4.3 (rigorous derivation of the Virial theorem) we obtain

$$\mathcal{T}(\mathcal{A}\psi) - \mathcal{A}(\mathcal{T}\psi) = (-i\hbar) \cdot 2 \cdot \mathcal{T}\psi$$

where $T = -\hbar^2/2m$ is the operator of the kinetic energy and the operator A is $A = -i\hbar\left(\frac{\partial^2}{\partial x^2} + \frac{\partial^2}{\partial y^2} + \frac{\partial^2}{\partial z^2}\right)$ This is one step in deriving the Virial Theorem in Foundation 4.3.

References

1 F.B. Hildebrand, *Advanced Calculus for Applications* (Prentice-Hall, Englewood Cliffs, NJ, 1976).
2 L. Pauling and E.B. Wilson, *Introduction to Quantum Mechanics* (McGraw Hill, New York, 1935).
3 J.C. Slater, The virial and molecular structure, *J. Chem. Phys.*, 1933, **1**, 687.

5

Atoms and Variational Principle

CHAPTER MENU

5.1 Introduction

Chapter 4 presented exact quantum-mechanical results for the electron in the hydrogen atom, modeled as an electron interacting with a proton through a Coulomb force. Now we proceed to analyze the electronic structure of multi-electron atoms. As we will discover most of chemistry is concerned with the rearrangement of electrons around the nuclei. Thus, understanding the properties of the electron wavefunctions is critical for understanding bonding and reactivity.

In this chapter, we describe the electronic structure of atoms and the architecture of the periodic table of the elements. For many-electron atoms we are not able to solve the Schrödinger equation exactly, so we use an approximate method which is based on the variational principle. In addition, we must introduce two additional quantum mechanical postulates: Uhlenbeck and Goudsmit's postulate that electrons have an intrinsic angular momentum (called the electron spin) and the Pauli exclusion principle which states that no two electrons can have the same wavefunction.

5.2 Variational Principle

5.2.1 Introducing the Variational Principle

Consider a He atom that has two protons and two neutrons in its nucleus and two electrons in a cloud surrounding it. The case of a two-electron atom is the simplest atomic problem beyond the one-electron atom, and its detailed consideration

Principles of Physical Chemistry, Third Edition. Hans Kuhn, David H. Waldeck, and Horst-Dieter Försterling.

Companion website: www.wiley.com/go/Kuhn/physicalchemistry3e

illustrates the difficulties one encounters in trying to describe many-electron atoms generally. The Schrödinger equation for the He atom is

$$\left[-\frac{\hbar^2}{2m_e}\nabla_1^2 - \frac{\hbar^2}{2m_e}\nabla_2^2 - \frac{2e^2}{4\pi\varepsilon_0 r_1} - \frac{2e^2}{4\pi\varepsilon_0 r_2} + \frac{e^2}{4\pi\varepsilon_0 r_{12}}\right]\psi = E\psi \tag{5.1}$$

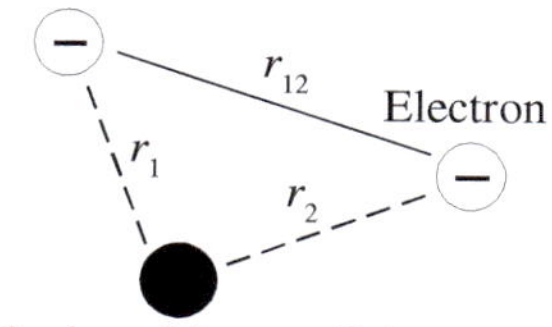

Figure 5.1 The Coulomb interactions in the He atom.

where the origin of the axis system is on the nucleus and we consider only the internal electronic coordinates. Remember that the symbol ∇^2 is a shorthand notation for the second derivatives with respect to x, y, and z, see Equation (3.28). The subscripts 1 and 2 identify the electrons, and the distance variables are defined in Fig. 5.1. The first two terms in the Hamiltonian, $-\left(\hbar^2/(2m_e)\right)\nabla^2$, correspond to kinetic energy, one term for each electron. The next three terms correspond to potential energy. Two of these terms are the Coulomb attraction of each electron to the nucleus, and the third term is the Coulomb repulsion between the electrons. It is the electron–electron repulsion term that causes great difficulties in treating multi-electron atoms accurately. If the electron–electron repulsion term could be ignored then the Hamiltonian would be a sum of two H atom Hamiltonians, with a nuclear charge of $2e$, and the wavefunction would be a product of two H atom wavefunctions, one for each electron. Because the electron–electron repulsion is not zero, the Schrödinger equation is not separable and cannot be solved exactly.

We must rely on an approximate solution to the Schrödinger equation for the electronic structure of He. The variational principle, introduced in Chapter 2, is the basis for a very powerful approach to finding approximate solutions of the Schrödinger equation for many electron atoms. In this section we introduce the variational principle more formally and apply it to the case of the H atom, in order to illustrate its use and compare it with the exact solution. In Section 5.3 we apply the variational method to the Schrödinger equation for the helium atom.

5.2.2 Justification of the Variational Principle

In general we need to calculate the energy E, which belongs to a given wavefunction, by using the Schrödinger equation

$$\mathcal{H}\psi = E\psi \tag{5.2}$$

If we multiply both sides by ψ and integrate over all space $\mathrm{d}\tau$, we can write

$$\int \psi(\mathcal{H}\psi)\cdot \mathrm{d}\tau = E\cdot\int \psi^2\cdot \mathrm{d}\tau \tag{5.3}$$

In Equation (5.3), E can be pulled out of the integral because it is a constant. If we use the normalization condition

$$\int \psi^2\cdot \mathrm{d}\tau = 1$$

we see that

$$E = \int \psi\mathcal{H}\psi\cdot \mathrm{d}\tau \tag{5.4}$$

Equation (5.4) is general and gives a prescription for calculating the energy if we have the exact wavefunction and know the form of the Hamiltonian. In Problem 5.1 you are asked to evaluate Equation (5.4) for the ground electronic state of the hydrogen atom ($\psi_1 = \mathrm{e}^{-r/a_0}/\sqrt{\pi a_0^3}$) and show that it gives our earlier result, $E_1 = -e^2/\left(8\pi\varepsilon_0 a_0\right)$.

5.2.2.1 Trial Function for the Ground State of an Atom

Using diagrams like that in Fig. 5.1, we can deduce the Hamiltonian $\mathcal{H}$ for any multi-electron atom. In particular, we need to include a kinetic energy term for each electron, an electrostatic attraction between each electron and the nucleus, and electron–electron repulsion between each pair of electrons in the atom. Finding the true wavefunction solutions to the Schrödinger equation formed by such Hamiltonians remains an unsolved problem; however, we can obtain accurate values for the electronic energy of atoms by using the variational method.

Consider the trial function ϕ_1 which satisfies the normalization condition ($\int \phi_1^2\cdot \mathrm{d}\tau = 1$), but may not be the exact wavefunction. We can determine the energy ε_1, as defined by:

$$\varepsilon_1 = \int \phi_1\mathcal{H}\phi_1\,\mathrm{d}\tau \tag{5.5}$$

In Foundation 5.1 we prove that ε_1 for an arbitrary function ϕ_1 must be greater than or equal to the true energy E_1 for the lowest quantum state, that is

$$\varepsilon_1 = \int \phi_1 \mathcal{H} \phi_1 \, \mathrm{d}\tau \geq E_1 \tag{5.6}$$

The limiting value $\varepsilon_1 = E_1$ will be reached if ϕ_1 is the exact wavefunction of the lowest quantum state. We can improve the energy we calculate by adjusting the trial function in order to minimize ε_1.

In practice, one proceeds by starting off with a set of functions ϕ_1, varying among these functions until the minimum value for ε_1 is reached, and then perhaps modifying that function further to find an even lower value of ε_1 (see Fig. 5.2). Hence the procedure is called the Variational Method, and we call ε_1 the variational energy.

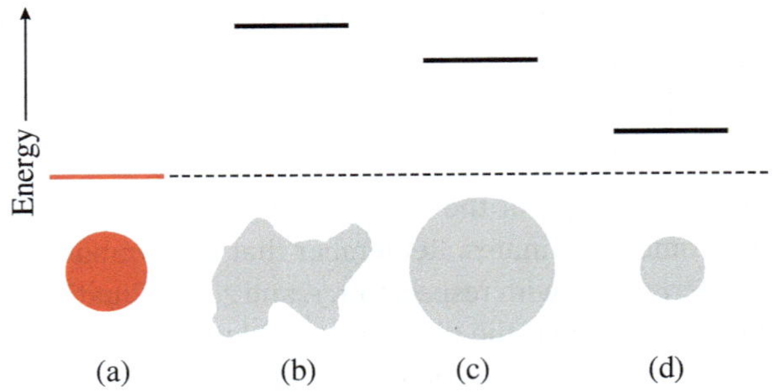

Figure 5.2 Variational method illustrated for the case of an H atom. (a) Exact energy E_1 by solution of the Schrödinger equation (exact wavefunction ψ_1). (b–d) Energies ε_1 for various trial functions ϕ_1 (displayed by schematic electron clouds); note that $\varepsilon_1 \geq E_1$.

Example 5.1 *Minimize the variational energy for the hydrogen atom ground state by optimizing the trial function $\phi_1 = e^{-r/C}/\sqrt{\pi C^3}$ and show that it corresponds to the exact value found in Chapter 4.*

If C is large, then the exponential function falls off gently; if C is small, then the decrease is steep. For different values of C we obtain different diameters of the electron cloud corresponding to the trial function (Fig. 5.3). C is a parameter that we optimize to minimize the energy. Using Equation (5.5) we can write the energy as

$$\varepsilon_1 = \int \phi_1 \mathcal{H} \phi_1 \cdot \mathrm{d}\tau = \frac{1}{\pi C^3} \int \mathrm{e}^{-r/C} \mathcal{H} \mathrm{e}^{-r/C} \cdot \mathrm{d}\tau$$

From Equation (4.2) we know the explicit form for the hydrogen atom's Hamiltonian, and we can write

$$\varepsilon_1 = -\frac{\hbar^2}{2m_e} \frac{1}{\pi C^3} \cdot \int \mathrm{e}^{-r/C} \left(\frac{\partial^2}{\partial x^2} + \frac{\partial^2}{\partial y^2} + \frac{\partial^2}{\partial z^2} \right) \mathrm{e}^{-r/C} \cdot \mathrm{d}\tau - \frac{e^2}{4\pi\varepsilon_0} \frac{1}{\pi C^3} \cdot \int \mathrm{e}^{-r/C} \frac{1}{r} \mathrm{e}^{-r/C} \cdot \mathrm{d}\tau$$

The second term corresponds to the average potential energy $\overline{V_1}$, and it can be found by following the procedure in Example 4.3

$$\overline{V_1} = -\frac{1}{4\pi\varepsilon_0} \frac{e^2}{C}$$

The first term corresponds to the average kinetic energy $\overline{T_1}$. Following the procedure used in Section 4.2.2, the kinetic energy operator acts on the wavefunction to give

$$\left(\frac{\partial^2}{\partial x^2} + \frac{\partial^2}{\partial y^2} + \frac{\partial^2}{\partial z^2} \right) \mathrm{e}^{-r/C} = -\frac{1}{C} \left(\frac{2}{r} - \frac{1}{C} \right) \cdot \mathrm{e}^{-r/C}$$

Because the wavefunction does not depend on the angles ϑ and φ, the volume element is $\mathrm{d}\tau = 4\pi r^2 \cdot \mathrm{d}r$, and the average kinetic energy is

$$\overline{T_1} = \frac{\hbar^2}{2m_e} \frac{1}{\pi C^3} \frac{4\pi}{C} \int_0^\infty \left(\frac{2}{r} - \frac{1}{C} \right) r^2 \cdot \mathrm{e}^{-2r/C} \cdot \mathrm{d}r$$
$$= \frac{\hbar^2}{2m_e} \frac{1}{C^2}$$

Combining the expressions for the average kinetic energy and the average potential energy, we find that

$$\varepsilon_1 = \overline{T} + \overline{V} = \frac{\hbar^2}{2m_e} \frac{1}{C^2} - \frac{1}{4\pi\varepsilon_0} \frac{e^2}{C} \tag{5.7}$$

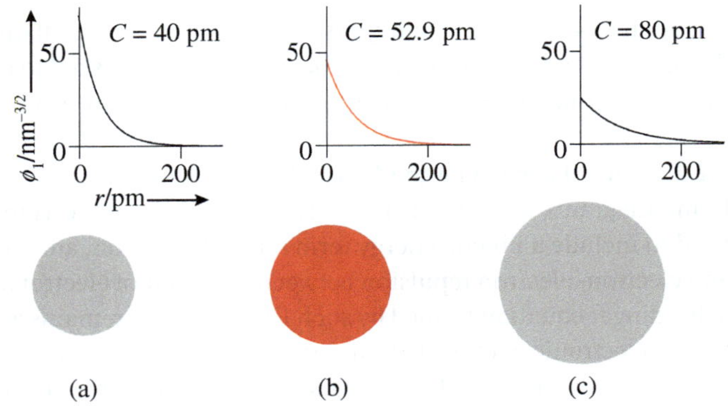

Figure 5.3 H atom trial function. Graphs of the functions and electron cloud pictures (schematic) for the ground state $\phi_1 = \mathrm{e}^{-r/C}$ with various values of C. (a) C = 40 pm, (b) C = 52.9 pm, and (c) C = 80 pm. At the edge of the shaded zone, ϕ_1^2 is 1% of its maximum value.

We must now solve Equation (5.7) to minimize the variational energy. One method is to plot ε_1 as a function of C, see Fig. 5.4. From the curve we can identify a minimum at $C_{\text{min}} = a_0 = 52.92$ pm and $\varepsilon_{1,\text{min}} = -21.8 \times 10^{-19}$ J. Alternatively, we can find the minimum analytically by taking the derivative of ε_1 with respect to C and setting it equal to zero. In this case we find

$$\left(\frac{d\varepsilon_1}{dC}\right)_{C=C_{\text{min}}} = -\frac{\hbar^2}{m_e}\frac{1}{C_{\text{min}}^3} + \frac{1}{4\pi\varepsilon_0}\frac{e^2}{C_{\text{min}}^2} = 0 \tag{5.8}$$

which can be solved to show that

$$C_{\text{min}} = 4\pi\varepsilon_0\frac{\hbar^2}{m_e e^2} = a_0$$

Thus, the minimum occurs when C is equal to the Bohr radius a_0. The smaller the electron cloud (the smaller C), the lower the potential energy and the greater the kinetic energy. The minimum of the energy ε_1 corresponds to a compromise between these two opposing tendencies.

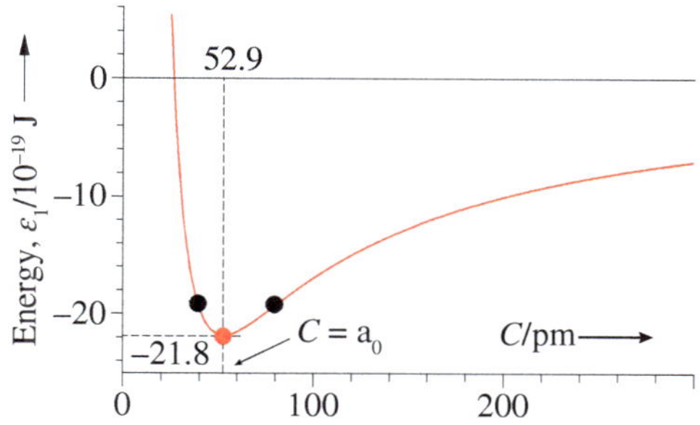

Figure 5.4 Energy ε_1 for the trial function $\phi_1 = e^{-r/C}$ for various values of C with a minimum at $C = a_0 = 52.9$ pm (see Fig. 3b). The full red circle marks the minimum of the curve; the full black circles mark the energies for $C = 40$ pm and $C = 80$ pm, respectively, cases a) and c) in Fig. 5.3.

Note that in this example the trial function is of the same mathematical form as the exact ground-state wavefunction, and therefore the minimum variational energy matches the exact energy.

5.2.2.2 Trial Functions for Excited States

For the true wavefunction ψ_2 of the next higher state, the energy E_2 is given by

$$E_2 = \int \psi_2 \mathcal{H} \psi_2 \cdot d\tau \tag{5.9}$$

in accordance with (5.4). As discussed in Section 5.3.3 and Foundation 5.1 the first excited state wavefunction is orthogonal to the ground-state wavefunction, namely

$$\int \psi_1 \ \psi_2 \cdot d\tau = 0 \tag{5.10}$$

In order to describe the next higher energy state, we consider a trial function ϕ_2, which is normalized and orthogonal to ϕ_1. The orthogonality requirement ensures that the wavefunctions are independent of each other, and the ground-state wavefunction does not contribute in some way to the excited state energy calculation. Mathematically we require that

$$\int \phi_2^2 \cdot d\tau = 1 \quad \text{and} \quad \int \phi_1 \phi_2 \cdot d\tau = 0 \tag{5.11}$$

An approximate value for E_2 can be found by variation of ϕ_2, just like that used for the ground state, and the variational energy ε_2 is

$$\varepsilon_2 = \int \phi_2 \mathcal{H} \phi_2 \cdot d\tau \geq E_2 \tag{5.12}$$

See Foundation 5.1 for more discussion of this point.

Similarly, an approximation for ψ_3 and E_3 can be obtained. When optimizing the energy ε_3 the trial function ϕ_3 is required to be normalized and orthogonal to the best trial functions for the lower energy states ϕ_1 and ϕ_2. We can proceed in an analogous manner to obtain approximate values for consecutive higher excited state energies E_n and wavefunctions ψ_n, having obtained best trial functions of all lower states.

In practically all quantum-chemical problems we must rely on the variational method to select the best of a large number of possible wavefunctions and regard this as an approximation for the true wavefunction. Clearly, the method will be limited by our ability to choose trial wavefunctions that are good approximations to the true wavefunction. The best ideas for trial wavefunctions will come when we have an improved understanding and intuition about atomic and molecular electronic states.

5.3 He-Atom

5.3.1 Ground State of He

5.3.1.1 Product of Atomic Wavefunctions Used as Trial Functions

Although we have to include the repulsion of the two electrons in calculating the ground-state energy of He, let us neglect this repulsion when we make our choice of a trial wavefunction. According to Equation (5.1) and Fig. 5.1 the wavefunction of He depends on the positions of two electrons, x_1, y_1, z_1 and x_2, y_2, z_2. We choose the He trial function $\phi_1(x_1, y_1, z_1, x_2, y_2, z_2)$ for the ground state (denoted by the index 1) as a product of two 1sHe^+ wavefunctions ψ_{He^+}, depending on the coordinates of electron 1 and of electron 2, respectively. Because each of the 1sHe^+ wavefunctions depends only on the electron's distance from the nucleus, ϕ_1 becomes

$$\phi_1(x_1, y_1, z_1, x_2, y_2, z_2) = \psi_{He^+}(r_1) \cdot \psi_{He^+}(r_2) \tag{5.13}$$

with

$$\psi_{He^+}(r_1) = \sqrt{\frac{8}{\pi a_0^3}}\, e^{-2r_1/a_0} \quad \text{and} \quad \psi_{He^+}(r_2) = \sqrt{\frac{8}{\pi a_0^3}}\, e^{-2r_2/a_0} \tag{5.14}$$

Keep in mind that the trial function (5.13) is based on the neglect of the electron repulsion operator, e^2/r_{12}. In Problem 5.2, you are asked to show that the ground-state energy of a He^+ ion is four times the ground-state energy of an H atom.

$$E_{He^+} = -4 \times \frac{1}{4\pi\varepsilon_0}\frac{e^2}{2a_0} = -87.2 \times 10^{-19}\ \text{J} \tag{5.15}$$

In order to obtain a first approximation for the ground-state energy of helium, we apply perturbation theory (see Foundation 5.2). Box 5.1 shows the perturbation theory calculation for the ground-state energy ε_1 of the He atom that is obtained with the trial function (5.13) and finds

$$\begin{aligned}\varepsilon_1 &= -\frac{1}{4\pi\varepsilon_0} \cdot \frac{4e^2}{a_0} + \frac{1}{4\pi\varepsilon_0} \cdot \frac{5e^2}{4a_0} \\ &= -\frac{1}{4\pi\varepsilon_0} \cdot \frac{11}{4} \cdot \frac{e^2}{a_0} = -119.9 \times 10^{-19}\ \text{J} = -74.8\ \text{eV}\end{aligned} \tag{5.16}$$

The first term corresponds to the attraction of both electrons by the nucleus, and for each electron it is the binding energy of an electron in the He^+ ion, see Equation (5.15). The second term corresponds to the Coulomb repulsion of the two electrons described by wavefunction (5.13), which destabilizes the atom.

To compare our calculated energy with experiment, we compute the energy required to ionize the He atom, producing He^+ and a free electron. To obtain the ionization energy E_{ion} of He, we subtract the energy ε_1 from the energy of He^+,

$$\begin{aligned}E_{ion} = E_{He^+} - \varepsilon_1 &= \frac{1}{4\pi\varepsilon_0} \cdot \left(-2\frac{e^2}{a_0} + \frac{11}{4} \cdot \frac{e^2}{a_0}\right) \\ &= \frac{1}{4\pi\varepsilon_0} \cdot \frac{3}{4} \cdot \frac{e^2}{a_0} = 32.7 \times 10^{-19}\ \text{J} = 20.4\ \text{eV}\end{aligned} \tag{5.17}$$

This value is smaller than the experimental value ($E_{ion} = 39.39 \times 10^{-19}$ J = 24.6 eV) by more than 4 eV. This poor agreement results for two reasons: (1) The electronic wavefunction (5.13) does not account for changes in the extent of the electron cloud when proceeding from He^+ to He. (2) The two electrons avoid each other at small distances because of their large Coulomb repulsion (correlation energy). You can imagine that when the two electrons are close together in space, their electron–electron repulsion will be high, but when they are farther apart, for example on opposite sides of the nucleus, their electron–electron repulsion energy will be small.

In order to improve the calculation we can use the trial functions

$$\psi_{He^+}(r_1) = \sqrt{\frac{C^3}{\pi a_0^3}} e^{-Cr_1/a_0} \quad \text{and} \quad \psi_{He^+}(r_2) = \sqrt{\frac{C^3}{\pi a_0^3}} e^{-Cr_2/a_0} \tag{5.18}$$

where C is a variational parameter. In Problem 5.11 we show that the energy minimum is reached for $C = 27/16 = 1.688$, resulting in an energy of

$$\varepsilon_1 = -124.2 \times 10^{-19}\ \text{J} = -77.5\ \text{eV} \tag{5.19}$$

(Fig. 5.5, ground state).[1]

By further improving the trial function (SCF method, see Section 5.3.1.2), an energy of $\varepsilon_1 = -124.8 \times 10^{-19}$ J is reached. If the correlation energy is taken into account, we find excellent agreement with experiment:

$$\varepsilon_1 = -126.6 \times 10^{-19}\ \text{J} = -79.0\ \text{eV} \quad \text{and}$$
$$E_{\text{ion}} = 39.39 \times 10^{-19}\ \text{J} = 24.6\ \text{eV}$$

In an exact description, the electrons in the atom are indistinguishable; and it is not correct to attribute separate, distinguishable wavefunctions to individual electrons. Rather we should treat the system of both electrons 1 and 2 as a whole:

$$\psi(1,2) = \psi(x_1, y_1, z_1, x_2, y_2, z_2) \tag{5.20}$$

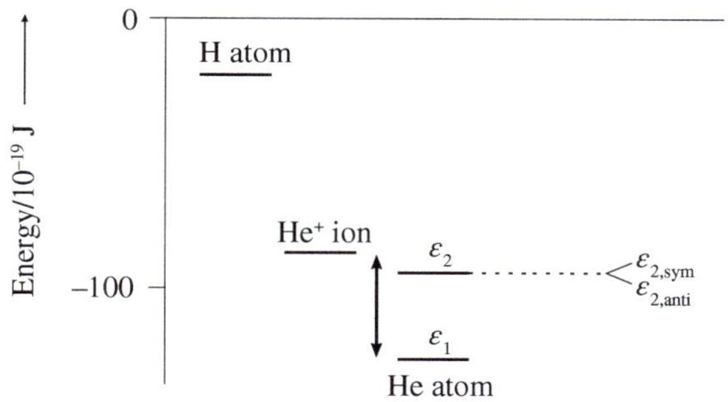

Figure 5.5 Energies of the H atom, the He^+ ion, and the He atom (calculated ground-state energy $\varepsilon_1 = -124.2 \times 10^{-19}$ J, excited state energies $\varepsilon_{2,\text{anti}} = -92.4 \times 10^{-19}$ J and $\varepsilon_{2,\text{sym}} = -89.2 \times 10^{-19}$ J) and calculated ionization energy $E_{\text{ion}} = 37.0 \times 10^{-19}$ J of He (double arrow). For the excited state energies see Section 5.3.3. $\varepsilon_2 = -90.8 \times 10^{-19}$ J is the average of $\varepsilon_{2,\text{anti}}$ and $\varepsilon_{2,\text{sym}}$.

The symbols 1 and 2 are used as a shorthand for the coordinates x_1, y_1, z_1 and x_2, y_2, z_2, respectively. In the present approach we have approximated the exact wavefunction ψ_1 for the ground state of He by a trial function $\phi_1(r_1, r_2)$ that is the product of two 1sHe^+ wavefunctions ϕ_{1s} which each depend on one-electron coordinates. Such one-electron wavefunctions are called orbitals, in this case the orbital for a 1s electron ϕ_{1s}. We rewrite Equation (5.13) for the two-electron wavefunction $\phi_1(1,2)$ as a product of the one-electron orbitals

$$\phi_1(1,2) = \phi_{1s}(1) \cdot \phi_{1s}(2) \tag{5.21}$$

5.3.1.2 Self-Consistent Field Method

An improvement of the approximation in Equation (5.21) is obtained by using

$$\phi_1(1,2) = \phi(1) \cdot \phi(2)$$

as a trial function and varying properties of the orbitals ϕ until a minimum of the corresponding energy ε_1 is reached. This is a very important approximation method introduced by Douglas Hartree (1897–1958); it is called the *self consistent field (SCF) method* (Foundation 5.4). In this method it is assumed that the electrons interact through an average, effective Coulomb potential. By using this method, the result found for the ground-state energy of He is $\varepsilon_1 = -124.8 \times 10^{-19}$ J [2]. The method does not account for the mutual avoidance of the two electrons at small distances because the two electrons are still represented by a product of two (but now best possible) one-electron wavefunctions (orbitals). The difference $\varepsilon_{1,\text{experiment}} - \varepsilon_{1,\text{Hartree}} = (-126.6 + 124.8) \times 10^{-19}\ \text{J} = -1.8 \times 10^{-19}$ J is called the *correlation energy*. Better trial functions must be pair-wavefunctions, which cannot be written as a product of one-electron wavefunctions.

5.3.2 Indistinguishability of Electrons

The wavefunction (5.20) must be restricted by another postulate: electrons must be indistinguishable. This postulate means that the function ψ^2 (which is proportional to the probability ρ) does not change if the coordinates of electrons 1 and 2 are exchanged. Stated more mathematically, the wavefunction (5.20) must fulfill the condition

$$\psi(1,2) = \psi(2,1) \quad \text{or} \quad \psi(1,2) = -\psi(2,1)$$

Note that the trial function ϕ_1 (Equation (5.21)) meets this condition.

1 This is the same result as found by Eckart [1].

5.3.3 Excited States of He Described by the Product of One-Electron Trial Functions

We have considered the indistinguishability of electrons in a very simple case of two electrons occupying the same orbital (more precisely, the same spatial orbital). How can we account for the fact that electrons are indistinguishable in cases where two or more orbitals are involved? This question is of general importance for understanding multi-electron systems. To study this question, we consider the excited state of He. In this discussion, we will attribute a wavefunction to each electron (orbital approximation).

Similar to an H atom, a He atom can be excited in a gas discharge. An electron can be promoted from the ground state to the next higher energy orbital, which we will approximate by a 2*s*-wavefunction. Hence, we might use a product of a 1*s* orbital and a 2*s* orbital as a trial wavefunction,

$$\phi_2(1,2) = \phi_{1s}(1) \cdot \phi_{2s}(2) \tag{5.22}$$

where the coordinates 1 and 2 denote x_1, y_1, and z_1 of electron 1 and x_2, y_2, and z_2 of electron 2, respectively. This trial function fulfills the Schrödinger equation if the repulsion term is neglected; however, it does not fulfill the condition that the two electrons are indistinguishable. If we exchange coordinates 1 and 2 we obtain a completely new function:

$$\phi_2(2,1) = \phi_{1s}(2) \cdot \phi_{2s}(1) \tag{5.23}$$

To build a trial wavefunction that fulfills the indistinguishability condition we use linear combinations of the two functions in Equations (5.22) and (5.23). We have two possibilities. We can create the sum

$$\phi_{2,\text{sym}} = A \cdot \left[\phi_{1s}(1) \cdot \phi_{2s}(2) + \phi_{1s}(2) \cdot \phi_{2s}(1)\right] \tag{5.24}$$

or the difference

$$\phi_{2,\text{anti}} = A \cdot \left[\phi_{1s}(1) \cdot \phi_{2s}(2) - \phi_{1s}(2) \cdot \phi_{2s}(1)\right] \tag{5.25}$$

where A is a constant. These linear combinations are still solutions of the Schrödinger equation, if the repulsion term is neglected. In the first case the function remains unchanged when the two electrons are exchanged (*symmetric function*), and in the second case the sign of the function changes (*antisymmetric function*). In both cases the probability density $\rho = \phi^2$ remains unchanged when the electron coordinates are interchanged. By applying the normalization condition, we find

$$A^2 = \frac{\int \phi_2^2 \cdot d\tau}{\int \left[\phi_{1s}(1) \cdot \phi_{2s}(2) \pm \phi_{1s}(2) \cdot \phi_{2s}(1)\right]^2 \cdot d\tau_1 \cdot d\tau_2} = \frac{1}{1+0+1} = \frac{1}{2}$$

and for the constant we obtain $A = \pm 1/\sqrt{2}$.

Will these two different trial wavefunctions, $\phi_{2,\text{sym}}$ and $\phi_{2,\text{anti}}$, have different energies? Consider trial functions of the form

$$\phi_{1s}(1) = \sqrt{\frac{C^3}{\pi a_0^3}} \cdot e^{-Cr/a_0} \quad \text{and} \quad \phi_{2s}(2) = \sqrt{\frac{C^3}{32\pi a_0^3}} \cdot \left(2 - \frac{Cr}{a_0}\right) e^{-Cr/(2a_0)}$$

where C is the variational parameter, which was calculated to be $C = 27/16$ in the case of the ground state of He. In Problem 5.12, you use these trial functions and find the energies of the symmetric and antisymmetric excited states to be (see Fig. 5.5):

$$\varepsilon_{2,\text{sym}} = -89.2 \times 10^{-19}\ \text{J} \quad \text{and} \quad \varepsilon_{2,\text{anti}} = -92.4 \times 10^{-19}\ \text{J}$$

Also see Box 5.1 and Justification 5.2. Using the ground-state energy from Equation (5.19), $\varepsilon_1 = -124.2 \times 10^{-19}$ J, we calculate the energy gap between these two excited states and the ground state to be

$$\varepsilon_{2,\text{sym}} - \varepsilon_1 = 35.0 \times 10^{-19}\ \text{J}$$

$$\varepsilon_{2,\text{anti}} - \varepsilon_1 = 31.8 \times 10^{-19}\ \text{J}$$

For the energy difference $\Delta E_{\text{theoretical}}$ between both states we obtain

$$\Delta E_{\text{theoretical}} = 35.0 \times 10^{-19}\ \text{J} - 31.8 \times 10^{-19}\ \text{J} = 3.2 \times 10^{-19}\ \text{J}$$

Indeed, different energies for the two states are found experimentally:

$$\varepsilon_{2,\text{sym}} - E_1 = 32.92 \times 10^{-19}\ \text{J} = 20.55\ \text{eV}$$
$$\varepsilon_{2,\text{anti}} - E_1 = 31.67 \times 10^{-19}\ \text{J} = 19.77\ \text{eV}$$

and the energy difference between the two states is

$$\Delta E_{\text{experiment}} = 32.92 \times 10^{-19}\ \text{J} - 31.67 \times 10^{-19}\ \text{J} = 1.25 \times 10^{-19}\ \text{J}$$

The difference between the experimental value and our calculated value arises because our approach uses products of one-electron wavefunctions instead of using two-electron wavefunctions.

The difference in energy between the symmetric and antisymmetric trial functions arises from the difference in how their electron probability amplitude is distributed. The wavefunction (5.24) is symmetric. This means that for all points in space where $r_1 = r_2$ the wavefunction is nonzero, i.e., the probability that both electrons are found at the same place (or nearby) is nonzero. This feature gives rise to significant electron–electron repulsion energy, which destabilizes the corresponding state. In contrast, the wavefunction (5.25) is antisymmetric. In this case for all points in space where $r_1 = r_2$ the wavefunction is zero, i.e., the probability that both electrons are found at the same place is zero. Thus the repulsion energy between the two electrons should be smaller for $\phi_{2,\text{anti}}$ than for $\phi_{2,\text{sym}}$. For more details see Box 5.3 and Justification 5.2.

Although the conclusion that $\phi_{2,\text{anti}}$ is more stable than $\phi_{2,\text{sym}}$ because of electrostatic energy differences is correct, its origin is somewhat more subtle than that indicated by the above analysis. The orbitals used for the wavefunctions 5.24 and 5.25 constrain the two electrons in the $\phi_{2,\text{anti}}$ and $\phi_{2,\text{sym}}$ excited state configurations to have the same radial dependence, but this does not hold true for the "best" variational wavefunctions. The best variational energy, which reproduces the experimental value for ΔE, is found for wavefunction parameters in which the electron–electron repulsion energy in $\phi_{2,\text{anti}}$ is greater than in the $\phi_{2,\text{sym}}$ state [3]. This occurs because the $\phi_{2,\text{anti}}$ state electron cloud is more compact than the $\phi_{2,\text{sym}}$ state's best variational wavefunction. This contraction of the electron cloud decreases the average electron–electron distance for the $\phi_{2,\text{anti}}$ as compared to the $\phi_{2,\text{sym}}$ at the optimized energy—even though the antisymmetric orbital $\phi_{2,\text{anti}}$ inhibits the electrons from getting very close to each other. Overall the contraction of the electron cloud by the favorable electron–nuclear attraction counteracts the increased electron–electron repulsion and ultimately leads to an overall lower energy for the $\phi_{2,\text{anti}}$ than for the $\phi_{2,\text{sym}}$.

The rigorous theoretical calculation for the energy agrees with spectroscopic and electron impact experiments. In the latter case (using a tube filled with He instead of Na vapor, see Fig. 2.2) minima of the current are observed at spacings of 19.8 and 20.6 eV (a 1.28×10^{-19} J difference), corresponding to the excitation of He from its ground state (energy E_1) to the antisymmetric (5.25) and symmetric (5.24) states, respectively. In spectroscopic experiments (light emission from a gas discharge of He) many spectral lines appear corresponding to transitions from higher excited states into the states described by the symmetric (Equation (5.24)) and antisymmetric (Equation (5.25)) wavefunctions.

To summarize, as we construct trial wavefunctions from electronic orbitals to describe an atom's electronic structure, we must ensure that they account for electron indistinguishability.

5.4 Electron Spin and Antisymmetry Postulate

5.4.1 Energy Splitting in a Magnetic Field, Electron Paramagnetic Resonance

If a hydrogen atom is placed in an external magnetic field, a splitting of its ground-state energy level, as displayed in Fig. 5.6, is found. This effect cannot arise from the orbital motion of the electron (see Sections 3.5.2 and 4.3.5) because the electron is in an s-orbital with $l = 0$, and therefore its orbital angular momentum is zero. We cannot understand this phenomenon on the basis of our postulates so far. Although our postulates are sufficient for a fundamental understanding of many aspects of atoms and molecules, they are not yet complete.

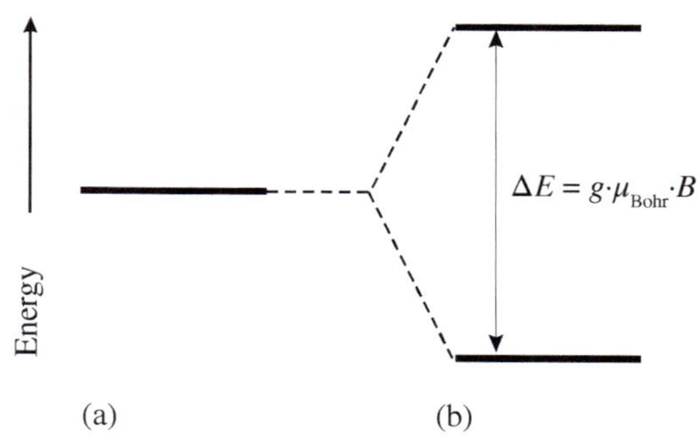

Figure 5.6 Splitting of the energy level of the H atom in an external magnetic field. (a) Without external field and (b) with external field.

Let us begin by describing the magnetic properties through a classical model. We posit that the electron behaves as if it were a charged spinning top; accordingly, it produces a magnetic field and behaves like a small magnet itself. In an external magnetic field, the electron's intrinsic magnetic moment (Bohr magneton μ_{Bohr})

$$\boxed{\mu_{\text{Bohr}} = \frac{e^2\hbar}{2m_e} = 9.274 \times 10^{-24}\ \text{J T}^{-1}} \tag{5.26}$$

is directed parallel or antiparallel to the magnetic field.[2] Thus, the energy of the electron depends on whether its charge distribution is spinning clockwise or anticlockwise and the energy difference in a magnetic field of flux density B is

$$\boxed{\Delta E = g \cdot \mu_{\text{Bohr}} \cdot B} \tag{5.27}$$

In Box 5.2 we use a simple model to derive Equations (5.26) and (5.27) with $g = 2$. Experiment shows that this splitting indeed occurs, but with $g = 2.0023$ for a free electron.[3]

In a magnetic field with $B = 1$ T the energy splitting is $\Delta E = 2.0023 \times 9.274 \times 10^{-24}\ \text{J T}^{-1} \times 1\,\text{T} = 18.54 \times 10^{-24}$ J. This means that the electron can be excited by microwave radiation ("light") with a frequency

$$\nu = \frac{\Delta E}{h} = 2.8 \times 10^{10} \text{s}^{-1} = 28\ \text{GHz}$$

This effect is the basis of *Electron Paramagnetic Resonance* (EPR) spectroscopy and is useful in chemistry for studying molecular species with unpaired electrons, such as organic radicals, transition metal complexes, and triplet states (see Chapter 10).

For more than one electron in an atom or molecule, the magnetic moment from each electron must be added together to obtain the total magnetic moment. Because the magnetic moment of each electron is a vector, they must be added together as vectors. If two electrons (e.g., the two electrons in He) have opposite directions for their magnetic moments then the net magnetic moment is zero, because the electrons necessarily have the same magnitude of their moment and the vector orientations cancel. This case corresponds to spin paired electrons and is a very common situation for ground-state atoms and molecules. Alternatively the electrons' magnetic moments may be oriented in the same direction and add so that the net magnetic moment is twice the value of that for one electron. For the case of unpaired electron spins, the intrinsic magnetic moment of an electron interacts with the effective magnetic field generated by its orbital motion (called spin–orbit coupling) and causes energy splittings among electronic states; see Section 5.6.[4]

5.4.2 Spin Variables

Although the physical model of the electron as a "spinning top" is useful, we cannot derive this result from our previous postulates. Hence we introduce a new postulate

- *An electron has an intrinsic angular momentum which we call spin. The component of the spin angular momentum in the direction of a magnetic field has two possible orientations with the values $\hbar/2$ or $-\hbar/2$. We define the quantum number m_s which specifies the spin of the electron ($m_s = 1/2$ corresponds to spin function α and $m_s = -1/2$ corresponds to spin function β).*

As with the other postulates we have discussed, this postulate is motivated by experimental observations, see Box 5.3.

In quantum mechanics the behavior of the electron spin is described symbolically by introducing a spin variable ζ. The two orientations of the spin of a single unpaired electron in a magnetic field are described by the spin wavefunctions $\alpha(\zeta)$ and $\beta(\zeta)$. Usually we refer to α as the "up" spin with angular momentum $\hbar/2$ and β as the "down" spin with angular momentum $-\hbar/2$. For an H atom, we have two possibilities for the electron's ground-state wavefunction

$$\phi_{1s}(x,y,z) \cdot \alpha(\zeta) \quad \text{or} \quad \phi_{1s}(x,y,z) \cdot \beta(\zeta) \tag{5.28}$$

so that the total electronic wavefunction is a product of a spin wavefunction and a spatial wavefunction. In the absence of an external magnetic field these two wavefunctions would have the same energy, i.e., the ground state is doubly degenerate.[5]

---

2 $1\,\text{T} = 1$ Tesla is the unit of the magnetic flux density. For comparison, the magnetic flux density of the magnetic field of the earth is $B = 4 \times 10^{-5}$ T; whereas, $B = 20$ T can be reached with the strongest magnets.

3 More advanced theories yield a value closer to or equal to $g = 2.0023$ rather than $g = 2$, see "CODATA values of the fundamental constants." NIST.

4 Although the magnetic forces among electrons (and the nucleus) affect an atom's energy, these effects are much smaller (~10,000 times) than those arising from Coulomb forces and are often neglected.

5 Here we have ignored the nuclear spin of the proton. When accounting for the nuclear spin the number of spin substates increases to 4, of which one is more stable than the other three. See Crampton et al. [4].

For a ground-state He atom, which has two electrons, four different spin combinations can be constructed. Using our approximation for the two-electron wavefunction as a product of orbitals (one-electron wavefunctions), we can write the trial wavefunction as

$$\phi_{1s}(1)\,\phi_{1s}(2)\cdot\alpha(1)\alpha(2) \tag{5.29}$$

corresponding to both spins up, and

$$\phi_{1s}(1)\,\phi_{1s}(2)\cdot\beta(1)\beta(2) \tag{5.30}$$

corresponding to both spins down. Here we use the common abbreviation which specifies a number (or index) for the electron (1 or 2) rather than the variables (ζ_1 or ζ_2) of the spin function, as we did in Equation (5.22) for the spatial functions. Each of these two trial wavefunctions satisfies the condition that electrons are indistinguishable; that is, if the electron coordinates are swapped the functions are still the same. The other two cases correspond to antiparallel spins, so that the trial wavefunctions are

$$\phi_{1s}(1)\,\phi_{1s}(2)\cdot\alpha(1)\beta(2)$$

and

$$\phi_{1s}(1)\,\phi_{1s}(2)\cdot\alpha(2)\beta(1)$$

These expressions do not satisfy the condition of indistinguishability for the two electrons. Swapping the electron coordinates interchanges the two functions. On the other hand, the linear combinations

$$\phi_{1s}(1)\,\phi_{1s}(2)\cdot\frac{1}{\sqrt{2}}\,[\alpha(1)\beta(2)+\alpha(2)\beta(1)] \tag{5.31}$$

and

$$\phi_{1s}(1)\,\phi_{1s}(2)\cdot\frac{1}{\sqrt{2}}\,[\alpha(1)\beta(2)-\alpha(2)\beta(1)] \tag{5.32}$$

fulfill the indistinguishability condition. Although these four trial wavefunctions (5.29)–(5.32) satisfy indistinguishability, we find that they do not all manifest in Nature.

5.4.3 Antisymmetry Postulate and Pauli Exclusion Principle

5.4.3.1 Electrons are Described by Antisymmetric Wavefunctions

As shown by Equation (5.27) the effect of electron spin on the energy is relatively small (less than 1 part in 10,000). The more important effect of the spin is its contribution to the total wavefunction's symmetry and results because of the *antisymmetry postulate*. This postulate may be stated as

- *Antisymmetry Postulate*: *The total electronic wavefunction must be antisymmetric with respect to the exchange of two electrons.*

The antisymmetry postulate is certainly a very strange requirement in the context of our daily experience. However, experimental facts force us to accept this strangeness in order to understand the structure and dynamics of atoms and molecules.

The Pauli Exclusion Principle is stated as:

- *In an atom or molecule, at most two electrons can occupy the same spatial orbital.*

It is a consequence of the antisymmetry postulate and was proposed by Wolfgang Pauli in 1925 in order to rationalize the periodic table of the elements in terms of atomic electronic structure. Later, he generalized the Exclusion Principle into the antisymmetry postulate given here and into an even more general form given in Chapter 12 in the section on fermions and bosons. If two electrons occupy the same spatial orbital, as in wavefunction (5.32), they must be described by an antisymmetric spin function, that is, they must have different spins (α or β) in an antisymmetric combination.

5.4.3.2 Antisymmetric Wavefunction for the Ground State of Helium

For the ground state of He we found that the four wavefunctions (5.29)–(5.32) fulfill the indistinguishability condition. However, only the wavefunction (5.32)

$$\phi_{1s}(1)\phi_{1s}(2)\cdot\frac{1}{\sqrt{2}}\,[\alpha(1)\beta(2)-\alpha(2)\beta(1)]$$

causes a change in the sign of the wavefunction and fulfills the antisymmetry postulate. That is, the part of the wavefunction describing the spin changes sign, but the part describing the spatial orbitals does not. Overall, the wavefunction (5.32) is antisymmetric.[6]

5.4.3.3 Antisymmetric Wavefunctions for the Excited States of Helium

Singlet and Triplet States Let us apply these considerations of electron spin and antisymmetry to the discussion of a helium atom's excited state, which was begun in Section 5.3. If we consider the symmetric spatial wavefunction, Equation (5.24), and multiply it by the antisymmetric spin wavefunction $[\alpha(1)\beta(2) - \alpha(2)\beta(1)]/\sqrt{2}$, we find

$$\frac{1}{2}\left[\phi_{1s}(1)\,\phi_{2s}(2) + \phi_{1s}(2)\,\phi_{2s}(1)\right] \cdot [\alpha(1)\,\beta(2) - \alpha(2)\,\beta(1)] \tag{5.33}$$

which is a total wavefunction that satisfies the antisymmetry postulate. See Box 5.3 and Justification 5.2 for more details. In this case we see that the electrons always appear in opposite spin orientations so that their net spin angular momentum is zero. Electron wavefunctions with a net spin angular momentum of zero are called "*singlet states.*"

Correspondingly, if we multiply the antisymmetric spatial wavefunction, Equation (5.25), by a symmetric spin wavefunction we obtain a total wavefunction that also satisfies the antisymmetry postulate. Because there are three possible symmetric spin wavefunctions, Equations (5.29)–(5.31), we find the three total wavefunctions given here

$$\frac{1}{\sqrt{2}}\left[\phi_{1s}(1)\,\phi_{2s}(2) - \phi_{1s}(2)\,\phi_{2s}(1)\right] \cdot [\alpha(1)\,\alpha(2)] \qquad \text{a)}$$

$$\frac{1}{\sqrt{2}}\left[\phi_{1s}(1)\,\phi_{2s}(2) - \phi_{1s}(2)\,\phi_{2s}(1)\right] \cdot [\beta(1)\,\beta(2)] \qquad \text{b)}$$

$$\frac{1}{2}\left[\phi_{1s}(1)\,\phi_{2s}(2) - \phi_{1s}(2)\,\phi_{2s}(1)\right] \cdot [\alpha(1)\,\beta(2) + \alpha(2)\,\beta(1)] \quad \text{c)} \tag{5.34}$$

Each of the wavefunctions in Equation (5.34) has the same energy in the absence of an external magnetic field (see Fig. 5.7a) and correspond to a three-fold degenerate energy level (referred to as a "*triplet*" state). Note that the energy of the excited singlet state ^1S in Fig. 5.7a is higher than that of the triplet state ^3S; as discussed in Section 5.3.3 this energy difference arises from electrostatic effects that manifest from the different symmetry of the spatial part of the wavefunction and are not magnetic in origin.

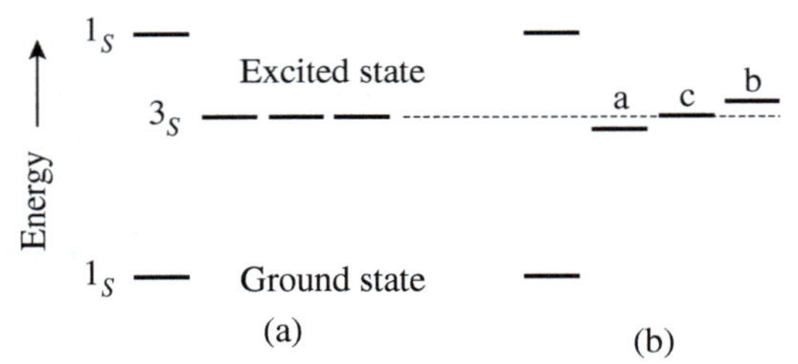

Figure 5.7 He atom. Singlet ground state ^1S, singlet excited state ^1S, and triplet excited state ^3S. (a) Energy in the absence of an external magnetic field (degenerate triplets) and (b) Energy in an external magnetic field (splitting of the ^3S state). The labeling a, b, and c of the three orbitals refers to Equations (5.34a)–(5.34c), respectively. Note that the labeling "S" means that the electronic states have no orbital angular momentum. In this case, they are composed of *s*-orbital trial functions, see Equation (5.13) for the ground state and Equation (5.22) for the excited state. The superscripts "1" and "3" denote a singlet and a triplet state, respectively.

Splitting of Triplet State in an External Magnetic Field In an external magnetic field the three triplet wavefunctions have different energies (Fig. 5.7b) because they have different magnetic moments associated with their spin orientations. In particular, when both electrons are spin up the total spin angular momentum vector has a component of $\hbar$ (half from each electron) along the magnetic field direction and the magnetic moment is oriented against the field (case a). For the wavefunction with both electrons spin down, the total spin angular momentum vector has a component of $-\hbar$, and the magnetic moment is aligned along the applied field's direction (case b). The third case corresponds to the total spin angular momentum vector and the magnetic moment oriented perpendicular to the magnetic field axis, so that their components along the magnetic field direction are zero (case c). Hence these three electronic states would have different energies in an externally applied magnetic field.

6 Applying the Antisymmetry Postulate to the case of a symmetric spin function and an antisymmetric spatial function of the same 1*s* orbital we obtain $\psi = \frac{1}{2}\left[\varphi_{1s}(1)\varphi_{1s}(2) - \varphi_{1s}(2)\varphi_{1s}(1)\right] \cdot [\alpha(1)\beta(2) + \alpha(2)\beta(1)] = 0$. This function is zero everywhere, that is, such a state cannot exist.

Summary To summarize, the first combination (5.33), which is a symmetric spatial wavefunction and an antisymmetric spin wavefunction, corresponds to a singlet excited state and does not split in an applied magnetic field. The second combination (5.34), which is an antisymmetric spatial wavefunction and a symmetric spin wavefunction, corresponds to three degenerate states, called a triplet. These three electronic states have different energies under an applied magnetic field.

5.4.4 Single-Triplet Splitting—Neglecting Magnetic Forces

As outlined above, the energy difference between the singlet and triplet excited states (Fig. 5.7) is caused predominantly by differences in the Coulomb interaction. Additionally, they differ very slightly by magnetic interactions between the two electrons (antiparallel spins in the first case, parallel spins in the second case).

Let us investigate this difference by comparing the singlet-triplet energy splitting to the magnetic energy of interaction for two electrons at a distance of one Bohr radius. We approximate the electrons by two magnets of magnetic moment μ_{Bohr} at a distance of a_0 (the Bohr radius). If the two magnets are antiparallel then the system is stabilized, and if the magnets are parallel the energy is destabilized. The energy difference ΔE_{magn} between these two arrangements is $2\mu_{\text{Bohr}} \cdot B$ according to Equation (5.27), where B is the magnetic flux density from one magnet acting on the other, given by

$$B = \frac{2}{a_0^3} \cdot \frac{\mu_{\text{Bohr}}}{4\pi\varepsilon_0 c_0^2}$$

at a distance a_0 (point dipole approximation). Hence, the energy difference is

$$\Delta E_{\text{magn}} = \frac{4}{a_0^3} \cdot \frac{\mu_{\text{Bohr}}^2}{4\pi\varepsilon_0 c_0^2}$$

Using the fact that $\varepsilon_0 c_0^2 = 1/\mu_0$, where μ_0 is the permeability of the vacuum, we find

$$\Delta E_{\text{magn}} = \frac{4\mu_{\text{Bohr}}^2 \mu_0}{4\pi a_0^3} = 1.1 \times 10^{-22}\ \text{J}$$

The energy of 10^{-22} J is 1/1000 of the energy difference between the singlet and triplet excited states (e.g., 1.25×10^{-19} J in the case of He).

In general, magnetic forces play a small role in comparison to Coulomb forces in the formation of chemical bonds, and we will usually neglect them. More precise calculations of energy levels and particular types of spectroscopies (see Section 5.6 and Chapter 12) will require inclusion of these interactions, however.

5.5 Many-Electron Atoms

We now have the concepts to understand the periodic table and the shell structure of atoms. This remains one of the most important accomplishments of quantum mechanics. Three essential concepts are needed.

- We approximate the electronic structure of atoms with wavefunctions that are similar to those for the hydrogen atom. We call the one-electron hydrogen atom wavefunctions, "orbitals."
- The Pauli Exclusion Principle, in its simple form, restricts the number of electrons that can have the same spatial wavefunction to two. Typically, we say that only two electrons can occupy the same spatial orbital, one with spin up α and one with spin down β.
- The Variational Principle states that, among all imaginable wavefunctions, the ground-state wavefunction corresponds to the minimum of the total energy. In multielectron atoms this minimum is determined predominately by a balance between the stabilization arising from the attractive electron–nuclear Coulomb forces and the destabilization arising from repulsive electron–electron Coulomb forces and increasing electron kinetic energy from the quantum confinement.

5.5.1 Atomic Number Z and Ionization Energy

The elements consist of atoms with a definite atomic number Z (the number of protons in the nucleus) and a matching number of electrons so that the overall charge of the atom is neutral. The atomic numbers for the first 10 elements are shown in Table 5.1.

Table 5.1 Atomic Numbers Z and Ionization Energies (in eV). Taken from the Handbook of Chemistry and Physics (CRC Press, 1992).

Element	Z	$E_{ion,1}$	$E_{ion,2}$	$E_{ion,3}$	$E_{ion,4}$	$E_{ion,5}$	$E_{ion,6}$	$E_{ion,7}$
H	1	13.60						
He	2	24.59	54.42					
Li	3	5.39	75.64	122.45				
Be	4	9.32	18.21	153.90	217.72			
B	5	8.30	25.15	37.93	259.38	340.23		
C	6	11.26	24.38	47.89	64.49	392.09	489.99	
N	7	14.53	29.60	47.45	77.47	97.89	552.07	667.05
O	8	13.62	35.12	54.94	77.41	113.90	138.12	739.29
F	9	17.42	34.97	62.71	87.14	114.24	157.16	185.19
Ne	10	21.56	40.96	63.45	97.12	126.21	157.93	207.28

The electronic energy structure of many-electron atoms is different from that of the H atom. If we could neglect the electrostatic repulsion between electrons, we would find an energy scheme like that of the H atom. The repulsion forces between electrons cause significant changes in the energy of the atom so that it does not follow this simple scheme. Although sophisticated mathematical approaches are needed for quantitative analysis, we can understand the change in the electronic structure qualitatively.

Let us consider the electronic energy levels of the first few elements. Table 5.1 gives the experimental ionization energies for the neutral and ion forms of the first 10 elements. Remember that the ionization energy is the minimum amount of energy required to remove an electron from an atom, given in Table 5.1 as $E_{ion,1}$, or an ion, given as $E_{ion,2}$ through $E_{ion,7}$.

5.5.2 Electronic Structure of the First Three Elements

5.5.2.1 Hydrogen Atom

For ground-state hydrogen, the ionization reaction is

$$H \rightarrow H^{+} + e^{-} \tag{5.35}$$

and the ionization energy is

$$E_{ion} = \frac{1}{4\pi\varepsilon_0} \cdot \frac{e^2}{2a_0} = 21.799 \times 10^{-19}\ \text{J} = 13.606\ \text{eV} \tag{5.36}$$

which is in agreement with the experimental value.

5.5.2.2 Helium Atom and He-Like Systems: H^-, Li^+, Be^{2+}

For the He atom we have to distinguish between the ionization of the first electron (ionization energy $E_{ion,1}$) and of the second one (ionization energy $E_{ion,2}$).

$$He \xrightarrow{E_{ion,1}} He^{+} + e^{-} \xrightarrow{E_{ion,2}} He^{2+} + 2e^{-} \tag{5.37}$$

As discussed in Section 5.3.1, the first ionization energy for He is

$$E_{ion,1} = 39.39 \times 10^{-19}\ \text{J} = 24.59\ \text{eV} \tag{5.38}$$

Now consider the second ionization. In contrast to He, the He^+ ion has only one electron and is a hydrogen-like atom with a nuclear charge of $2e$, its electron should be bound four times more strongly than that of the hydrogen atom; that is

$$E_{ion,2}(He) = \frac{1}{4\pi\varepsilon_0}\frac{e^2}{2a_0}Z^2 = (2)^2 \times E_{ion,1}(H) = 54.42\ \text{eV} \tag{5.39}$$

Both values are in good agreement with experiment (see Table 5.1). Even though both electrons occupy the same 1*s* orbital, the energy needed to remove the first electron is much less than for the second one. This difference arises because the electron–electron repulsion leads to an overall decrease in the binding energy of the electrons in He than for that in He^+. This difference also affects the relative size of the electron cloud for He and He^+ (see Fig. 5.8).

In analogy to He we can understand the dimensions of electron clouds for ions that have only two electrons, like He. Some of these are: H^-, Li^+, Be^{2+} (Fig. 5.8). The electron cloud is less compact in H^- than that of He, and it is more compact in Li^+ (nuclear charge $+3e$) and Be^{2+} (nuclear charge $+4e$) than in He. The change in size of the electron cloud results from the different nuclear charges in the ions. Note that the expansion of the electron cloud for H^- is extremely large—a van der Waals radius (see Section 7.5) of 154 pm as compared to a van der Waals radius of 120 pm for the H atom.

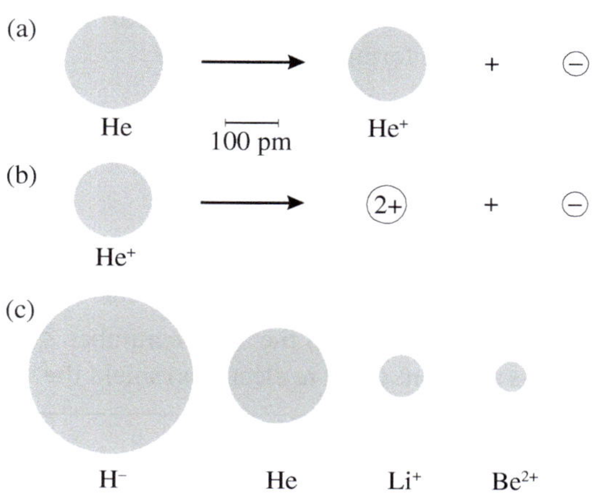

Figure 5.8 and second (b) electrons from the He atom. Because of the repulsion of the two electrons, the electron cloud of He is larger than that of He^+. (c) Relative electron cloud sizes for He-similar atoms with two electrons. At the edge of the gray zone, ϕ_1^2 is 1% of its maximum value.

5.5.2.3 Lithium Atom

Now consider the Li atom, which has three protons in its nucleus and three electrons in its electron cloud. Two of the electrons can share a 1*s* orbital, but the third electron has to occupy another orbital, as required by the Pauli Exclusion principle. From our analysis of the hydrogen atom we expect that it could occupy either a 2*s* or a 2*p* orbital. We expect three different ionization energies for Li.

$$\text{Li} \xrightarrow{E_{\text{ion},1}} \text{Li}^+ + \text{e}^- \xrightarrow{E_{\text{ion},2}} \text{Li}^{2+} + 2\text{e}^- \xrightarrow{E_{\text{ion},3}} \text{Li}^{3+} + 3\text{e}^- \tag{5.40}$$

and they are reported in Table 5.1. The third ionization energy $E_{\text{ion},3}$ corresponds to a one-electron atom with a charge of $3e$ on the nucleus; as you show in Problem 5.2, $E_{\text{ion},3} = Z^2 \cdot E_{\text{ion}}(\text{H atom}) = 9 \times 13.61\ \text{eV} = 122.45\ \text{eV}$. The ionization energy $E_{\text{ion},2}$ is smaller than $E_{\text{ion},3}$ for the same reason found in the analogous He atom case; the electrostatic repulsion between electrons reduces the binding energy.

Shielding Effect The value of $E_{\text{ion},1} = 5.39$ eV for Li is much smaller than $E_{\text{ion},1}$ of He and H, even though the nuclear charge of lithium is $3e$. In part, the weaker binding of this electron arises because it is in the $n = 2$ shell and the average distance from the nucleus of lithium's $n = 2$ electron is larger than for an electron in a 1*s* orbital. For a hydrogen-like atom with an electron in the $n = 2$ shell and $Z = 3$, we expect a binding energy that is

$$E = \frac{1}{4\pi\varepsilon_0}\frac{e^2}{2a_0}\frac{Z^2}{n^2} = \frac{(3)^2}{(2)^2}E_{\text{ion},1}(\text{H}) = 49.05 \times 10^{-19}\ \text{J} = 30.62\ \text{eV}$$

which is still much larger than the ionization energy we observe for $E_{\text{ion},1}(\text{Li})$. A second contribution to the weaker binding of the $n = 2$ electron in Li arises because the 1*s* electrons "shield" the $n = 2$ electron from experiencing the full nuclear charge. Shielding refers to the fact that outer electrons of an atom do not experience the full nuclear charge because of the electrostatic repulsion of the other electrons in the atom which act to counter the attractive interaction with the nucleus. Because of the reduced, or effective, nuclear charge, the outer shell electron is bound more weakly. If the 2*s* electron would be completely shielded by the 1*s* electrons (Z replaced by Z_{eff}), then with $Z_{\text{eff}} = 1$ and $n = 2$ we obtain

$$E = E_{\text{ion},1}(\text{H})\frac{Z_{\text{eff}}^2}{n^2} = E_{\text{ion},1}(\text{H})\frac{1}{4} = 3.4\ \text{eV}$$

which is smaller than the observed ionization energy of 5.39 eV.

Example 5.2 *Consider the Li atom, which has a single valence electron (i.e., the 2s electron). Treat the valence electron as if it were in a hydrogenic orbital, but the charge on the nucleus is a parameter Z_{eff}. Find the effective charge needed for the Li atom that reproduces its experimental ionization potential.*

The experimental ionization potential of Li is 5.39 eV and the valence electron is in the $n = 2$ shell. Using the one-electron atom's energy formula we can write

$$E = 5.39 \text{ eV} = \frac{1}{4\pi\varepsilon_0}\frac{e^2}{2a_0}\frac{Z_{\text{eff}}^2}{n^2} = 13.60 \text{ eV} \times \frac{Z_{\text{eff}}^2}{n^2}$$

Rearranging to solve for Z_{eff}, we find

$$Z_{\text{eff}} = \sqrt{\frac{n^2 \times 5.39 \text{ eV}}{13.60 \text{ eV}}} = 2\sqrt{\frac{5.39}{13.60}} = 1.26$$

This value lies between the atomic number $Z = 3$ for Li and the atomic number $Z = 1$ for the H atom, suggesting that the two 1s electrons (or core electrons) shield the valence electron, but not completely.

An approximate description of the radial part of the atom's electronic wavefunctions, called Slater type orbitals, accounts for this shielding effect (see Box 5.4).

Is the Valence Electron of Li in a 2s or 2p Orbital? We still have not determined whether the third electron of the Li atom occupies a 2s or a 2p orbital, or whether both possibilities are equally likely. The electronic states for these two orbitals have important differences. In particular, a 2s electron's probability density is nonzero close to the nucleus, whereas that for the 2p orbital is zero at the nucleus (Figs. 4.8 and 4.9). Of course if a 2s electron has a finite probability of being at, or near, the nucleus it will not be fully shielded by the core electrons. We say that the 2s electron can "penetrate" to the nucleus. By comparison, the 2p electron will be better shielded by the remaining 1s electrons than is the 2s electron. As a result, the energy level of a 2p electron is higher than for a 2s electron. Thus, we conclude that the ground state for the third electron of the Li atom, according to the variational principle, must occupy the 2s orbital (Fig. 5.9).

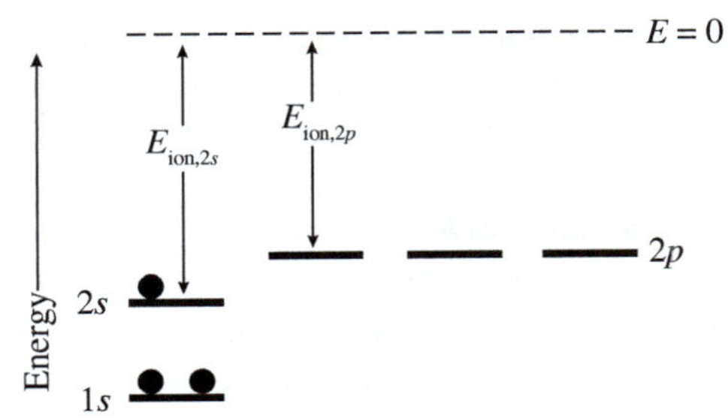

Figure 5.9 Population scheme for the Li atom. The ionization energy from the 2s state ($E_{\text{ion},2s}$) is larger than the corresponding energy from the 2p state ($E_{\text{ion},2p}$), because a 2p electron is more strongly shielded by the 1s electrons than a 2s electron. Therefore the third electron of the Li atom occupies a 2s orbital in its ground state and not a 2p orbital.

5.5.3 Aufbau Principle and Periodic Table

5.5.3.1 Orbital Energies

Repulsion of All Electrons Considered by Search for Self-Consistent Orbitals We will now extend the discussion to all atoms. We will consider that each electron in an atom occupies an orbital in an effective electrostatic field (nuclear field shielded by all other electrons). In order to obtain the relative energies of the electrons in each of the orbitals, we must perform a calculation in a manner that gives a self-consistent description of the probability density for the electron cloud. In Foundation 5.4 we explain the Hartree–Fock self-consistent field (HF-SCF) method for closed shell systems; the case in which 2N electrons occupy N spatial orbitals—called the restricted Hartree–Fock method. Here we describe the basic features of the HF-SCF electronic energy expression.

In the HF-SCF approximation, we assign each electron in the atom to an orbital trial function, which is a product of a spatial part, ϕ_i,and a spin part, α or β. In constructing a total wavefunction, the HF-SCF procedure ensures that electron indistinguishability, and the Pauli exclusion principle are followed by the total wavefunction. As outlined in Foundation 5.4, this method generates a series of one-electron eigenvalue equations that can be solved to find the energy ε_i of each orbital i; namely

$$\varepsilon_i = \varepsilon_i^{(0)} + \sum_j \left(2 \cdot J_{ij} - K_{ij}\right) \tag{5.41}$$

where the sum over j is taken over the other occupied orbitals in the atom. On the right side of this equation, $\varepsilon_i^{(0)}$ is the binding energy for an electron in orbital i in the limit where electron–electron repulsion is neglected. The term J_{ij} is a destabilizing contribution to the energy from the electron–electron repulsion of electrons in the other orbitals of the atom and is given by

$$J_{ij} = \frac{e^2}{4\pi\varepsilon_0} \int \rho_i \frac{1}{r_{12}} \rho_j \cdot d\vec{r}_1 \cdot d\vec{r}_2$$

with

$$\rho_i = \phi_i^2(r_1) \qquad \rho_j = \phi_j^2(r_2)$$

J_{ij} is called the Coulomb integral. The term with K_{ij} is a stabilizing contribution to the energy that is given by

$$K_{ij} = \frac{e^2}{4\pi\varepsilon_0} \int \rho_{ij} \frac{1}{r_{12}} \rho_{ji} \cdot d\vec{r}_1 \cdot d\vec{r}_2$$

with

$$\rho_{ij} = \phi_i(r_1) \cdot \phi_j(r_1) \qquad \rho_{ji} = \phi_j(r_2) \cdot \phi_i(r_2)$$

Note that the integrand for K_{ij} has the electron coordinates swapped (or exchanged) among the different orbitals i and j in the orbital products, and this term appears because of our construction of the total wavefunction with electron indistinguishability. K_{ij} is called the exchange integral.

To proceed with optimizing the electronic energy of the atom by the variational method, we optimize the total electronic energy, varying each orbital ϕ_i, and iterate through the process until a minimum energy is found. Beginning with the initial "guess" for the orbital functions, we vary the parameter(s) of each orbital ϕ_i under the approximation that the other $N-1$ orbitals ϕ_j have the form given by the initial guess. In this process, the variational calculation modifies the trial orbitals of each electron, which changes the average electron densities that should be used to find the average electron–electron repulsion (J_{ij} and K_{ij} terms). So we proceed by performing the calculation over with the improved orbitals. Of course, the orbitals may change again. Hence we must perform the calculation in an iterative manner until the orbitals stop changing their shape and the energy is minimized.

Example 5.3 *Show that Equation (5.41) gives the same energy for the ground state He that we found in Section 5.3.1 using the trial functions*

$$\phi_1(r_1) = \sqrt{\frac{C^3}{\pi a_0^3}} e^{-Cr_1/a_0} \quad \text{and} \quad \phi_2(r_2) = \sqrt{\frac{C^3}{\pi a_0^3}} e^{-Cr_2/a_0}$$

where C is a variational parameter.

Here we deal with a total of two electrons (closed shell system). Using these trial functions, we can write the Coulomb integral J_{12} as

$$J_{12} = \frac{e^2}{4\pi\varepsilon_0} \int \rho_1 \frac{1}{r_{12}} \rho_2 \cdot d\vec{r}_1 \cdot d\vec{r}_2 = \frac{e^2}{4\pi\varepsilon_0} \int \frac{\phi_1^2(r_1) \cdot \phi_2^2(r_2)}{r_{12}} \cdot d\vec{r}_1 \cdot d\vec{r}_2$$

and we can write the exchange integral K_{12} as

$$\begin{aligned} K_{12} &= \frac{e^2}{4\pi\varepsilon_0} \int \rho_{12} \frac{1}{r_{12}} \rho_{21} \cdot d\vec{r}_1 \cdot d\vec{r}_2 \\ &= \frac{e^2}{4\pi\varepsilon_0} \int \phi_1(r_1) \cdot \phi_2(r_1) \frac{1}{r_{12}} \phi_2(r_2) \cdot \phi_1(r_2) \cdot d\vec{r}_1 \cdot d\vec{r}_2 \end{aligned}$$

Because our trial functions are identical, we have that $\phi_1(r_2) = \phi_2(r_2)$ and $\phi_1(r_1) = \phi_2(r_1)$ so that

$$\begin{aligned} K_{12} &= \frac{e^2}{4\pi\varepsilon_0} \int \phi_1(r_1) \cdot \phi_1(r_1) \frac{1}{r_{12}} \phi_2(r_2) \cdot \phi_1(r_2) \cdot d\vec{r}_1 \cdot d\vec{r}_2 \\ &= \frac{e^2}{4\pi\varepsilon_0} \int \frac{\phi_1^2(r_1) \cdot \phi_2^2(r_2)}{r_{12}} \cdot d\vec{r}_1 \cdot d\vec{r}_2 \end{aligned}$$

Thus, in this special case, we arrive at the result

$$K_{12} = J_{12}$$

Using Equation (5.41) we obtain

$$\varepsilon_1 = \varepsilon_1^{(0)} + (2 \cdot J_{12} - J_{12}) = \varepsilon_1^{(0)} + J_{12}$$

which matches the result we obtained in Section 5.3.1 and in Box 5.1.

In certain cases, the restricted Hartree–Fock method can be extended to treat excited states and atoms with an odd number of electrons [6].

5.5.3.2 Self-Consistent Orbital's Energy

Figure 5.10 plots the electronic orbital energies, calculated by the SCF method, versus the atomic number of the atom. You should note that the overall trend is for the orbital energy to decrease as the nuclear charge increases. This trend reflects the favorable Coulomb attraction between the nucleus and the electrons. You should notice that orbitals with different l values have different energies for the multi-electron atom, unlike the one-electron atom. The higher the value of l, the higher is the value of the energy (less stable orbital) for a given n. This trend results from the large differences in the radial parts of the wavefunction for these orbitals. At a given value of n, lower l value wavefunctions penetrate more closely to the nucleus, which results in a favorable electrostatic interaction that stabilizes the orbital.

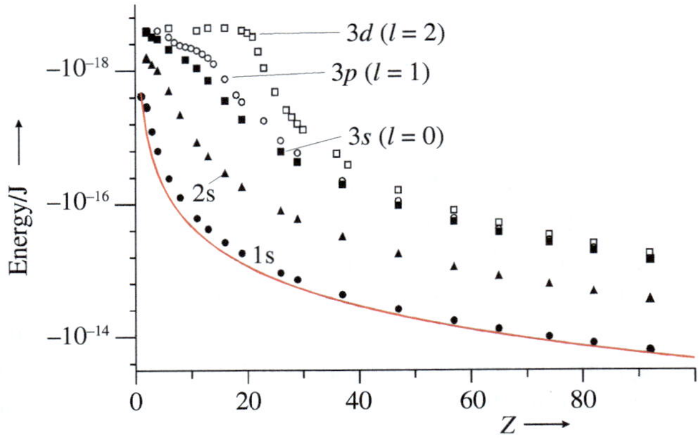

Figure 5.10 SCF orbital energies [7] for the 1s (full circles), 2s (full triangles), 3s (full squares), 3p (open circles), and 3d (open squares) states as a function of atomic number Z. Solid line: energy of the 1s state, calculated according to Equation (5.39): $E = -\left[1/(4\pi\varepsilon_0)\right] \cdot \left[e^2Z^2/(2a_0)\right]$. The orbital energies for the 2p states are not shown, because they differ only slightly from the 2s energies.

In concert with this penetration effect, we also have the shielding effect by the core electrons,[7] so that the outermost shell of electrons (the valence electrons) experiences an effective nuclear charge that is reduced compared with the full nuclear charge. As an example of these effects, consider the relative energy of the 4s and 3d orbital for K and Ca. Because the 3d orbitals are better shielded from the nuclear charge than the 4s orbitals, the 4s orbitals are more stable for K and Ca and fill with electrons before the 3d. For $n \geq 3$ the valence orbitals lie close enough in energy that small differences in the electrostatic forces can affect the most stable configuration. Nevertheless, the orbital picture captures the essence of the electronic structure and provides a useful framework for discussing periodic properties of the elements and electronic structure.

5.5.3.3 Aufbau Principle

Given this energy scheme we can understand the ground-state electronic structure of the atoms throughout the periodic table. This procedure is called the Aufbau ("Building Up") Principle and is outlined below as a set of three rules.

- The atomic orbitals exist with a particular energy sequence (see Fig. 5.10): 1s < 2s < 2p < 3s < 3p < (4s, 3d) .... These orbitals and the wavefunctions are specified by the quantum numbers found for the hydrogen atom wavefunctions, namely $n = 1, 2, 3, \ldots; l = 0, 1, 2, \ldots, n-1; k = 0, \pm 1, \ldots, \pm l$; and $m_s = \pm\frac{1}{2}$.
- We place the appropriate number of electrons into the lowest energy orbital which is available, subject to the constraints of the Pauli Exclusion principle, i.e., no more than two electrons can occupy any single spatial orbital.
- *Hund's rule.* If a set of degenerate orbitals is available, the electrons occupy different members of the set before doubly occupying any one, that is, they adopt a configuration with the greatest number of unpaired electrons. For example, two electrons that occupy a pair of degenerate orbitals (e.g., 2p_x and 2p_y) have their spins parallel in the lowest energy state; hence, the triplet state is lower in energy than the singlet state (see Fig. 5.7). The rationale for this procedure is that placement of the electrons into different spatial orbitals reduces the electron–electron repulsion as compared to the electron–nuclear attraction; e.g., the 2p_x and 2p_y have the same radial probability density (strongly affecting the electron–nuclear attraction) but different angular probability density (strongly affecting the electron–electron repulsion but not the electron–nuclear attraction).

7 The core electrons refer to filled electron shells in the atom. A shell corresponds to all the orbitals associated with a single principal quantum number n.

5.5.3.4 Periodic Table of the Elements

Using these rules, we can predict the ground-state electron configuration, population of electrons in particular atomic orbitals, of the elements and understand the organization of the periodic table. Table 5.2 gives the electronic configurations for the first 10 elements. As discussed above hydrogen has a single electron in a $1s$ orbital, and we represent this by the electronic configuration $1s$. Note that this notation does not specify whether the electron spin is oriented up, α, or down, β. Helium has two electrons with opposing spin orientations in a $1s$ orbital, and we represent this by $1s^2$, in which the superscript 2 indicates that the orbital is doubly occupied. For the Li atom we found that two electrons are in the $1s$ orbital and one is in the $2s$ orbital. We represent its electronic configuration by $1s^2 2s$. Since the principal quantum number of the atom has changed it is placed in the second row of the periodic table. If we proceed to the next element, Be with $Z = 4$, we see that we must have four electrons, and they will occupy the lowest energy orbitals with the electronic configuration of $1s^2 2s^2$.

Table 5.2 Electron Configurations of Elements.

Element	Configuration	Element	Configuration
H	$1s$	C	$1s^2 2s^2 2p^2$
He	$1s^2$	N	$1s^2 2s^2 2p^3$
Li	$1s^2 2s$	O	$1s^2 2s^2 2p^4$
Be	$1s^2 2s^2$	F	$1s^2 2s^2 2p^5$
B	$1s^2 2s^2 2p$	Ne	$1s^2 2s^2 2p^6$

Because of the degeneracy of the three spatial p orbitals, our specification of the electronic configuration for the next few elements is not unique. As we proceed to boron ($Z = 5$), the fifth electron can occupy any of the three $2p$ orbitals ($2p_x$, $2p_y$, or $2p_z$) because they are equivalent in energy. With carbon ($Z = 6$), a new issue arises because the sixth electron could occupy the same spatial orbital and pair up with the other p electron or it could be located in a separate spatial orbital. In the first case, the electrons would be closer together, on average, than those in the second case and generate a larger electron–electron repulsion energy. Thus an electron configuration in which the two electrons occupy different spatial orbitals is preferred, in accordance with Hund's rule, given above. As we proceed across the second row of the periodic table we fill up the p orbitals until we reach the case of neon. For Ne the p orbitals are filled and the charge distribution around the nucleus is spherical. The $1s$ and $2s$ electrons are spherically symmetric; and the $2p$ electrons form a spherical cloud, which can be seen by considering the sum of the $2p$ electron probability densities (see Table 4.1),

$$\begin{aligned}\psi_{2p_x}^2 + \psi_{2p_y}^2 + \psi_{2pz}^2 &= \frac{1}{32\pi a_0^5} \cdot e^{-r/a_0} \cdot (x^2 + y^2 + z^2) \\ &= \frac{1}{32\pi a_0^5} \cdot e^{-r/a_0} \cdot r^2\end{aligned} \tag{5.42}$$

Figure 5.11 shows the *periodic table of the elements*. Each row in the periodic table corresponds to a different principal quantum number n, and as we proceed across a row the orbitals in this shell of the atom are filled with electrons according

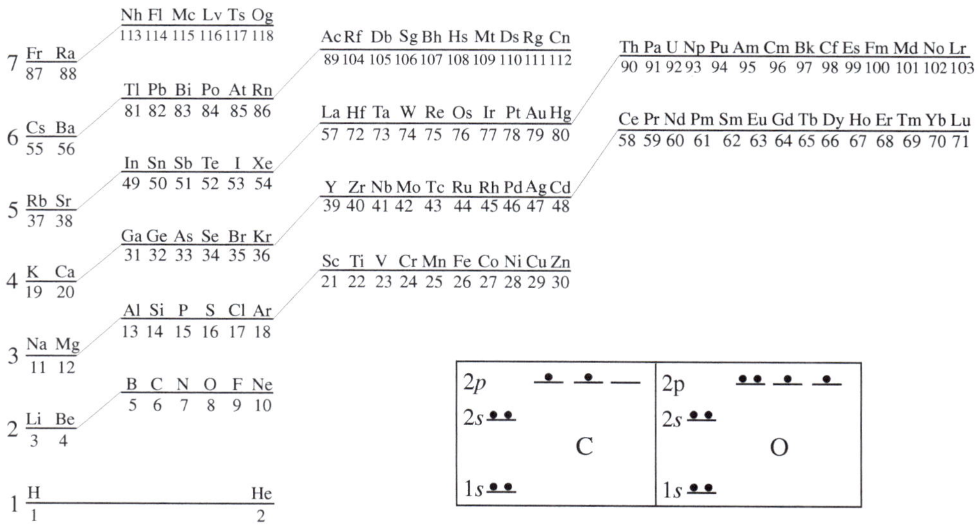

Figure 5.11 The periodic table of the elements. The elements are arranged in a scheme according to increasing atomic number Z; the scheme results from the energy scheme of the H atom, accounting for the orbital contraction increasing with increasing nuclear charge Ze and for the shielding of the nucleus when filling the p, d, and f levels. Hund's rule has to be considered (examples for C and O are shown in the boxes). Note: The periodic table given here is inverted relative to the usual representation to underscore the occupation of hydrogen atom like orbitals according to the Aufbau Principle.

to the relative energetics. The consideration on the shielding of $2s$ and $2p$ orbitals applies for the $3s$, $3p$, and $3d$ orbitals and the $4s$, $4p$, $4d$, and $4f$ orbitals as well. For a given principal quantum number n, we find

$$\text{orbital energy:} \quad ns < np < nd < nf$$

As one moves down a column in the periodic table the principal quantum number changes but orbitals of the same angular type are occupied in the same manner, with some notable exceptions for the transition metals. This feature is the primary reason that elements in a column, or "family," have similar chemical properties.

5.5.4 Periodic Properties of the Elements

In addition to understanding how the electronic structure of the atom corresponds to the arrangement of elements in the periodic table, we can understand how the electronic structure of atoms relates to trends in the properties of the atoms. Here we briefly review the trends in atomic ionization energies and atomic radii.

5.5.4.1 Ionization Energy

The first ionization energy is important because it is a measure of the binding energy of an atom's outermost valence electron. Of course it is the valence electrons and orbitals that are most likely to be involved in chemical reactions. The first ionization energy is

$$E_{\text{ion},1} = E^{+} - E$$

where E^{+} is the energy of the cation and E is the energy of the neutral. If the removal of an electron from orbital i does not affect the other $n-1$ orbitals in the atom, then the ionization energy represents the binding energy of the electron in that particular orbital.

$$E_{\text{ion},1} = -E_{\text{orbital}}$$

This is called *Koopmans Theorem*.

Remember that the SCF method computes the potential energy for each electron of an atom in the average field of the other electrons and converts the many-electron Schrödinger equation to a set of one-electron equations and one-electron energies. These one-electron energies and wavefunctions are the electron orbitals. The *Koopmans theorem* constraint, that the electrons do not reorganize during the ionization event, ensures that the average potential energy of the remaining electrons is not changed from that of the neutral atom.

Figure 5.12 plots the first ionization energy of the atomic elements, $E_{\text{ion},1}$, which provides the energy of the highest lying filled atomic orbital. For example, the large drop in ionization energy that occurs after the noble gas atoms arises because the principal quantum number must be increased for the next element in the plot. This necessity leads to a less stable valence electron. As one moves across a row in the periodic table however, the principal quantum number remains fixed and the nuclear charge increases. This behavior leads to a stabilization of the valence electrons. The finer variations in the ionization energy for a given principal quantum number can be rationalized by considering the effects of electrostatic repulsion as the orbitals are filled; see Exercises 5.9 and 5.10.

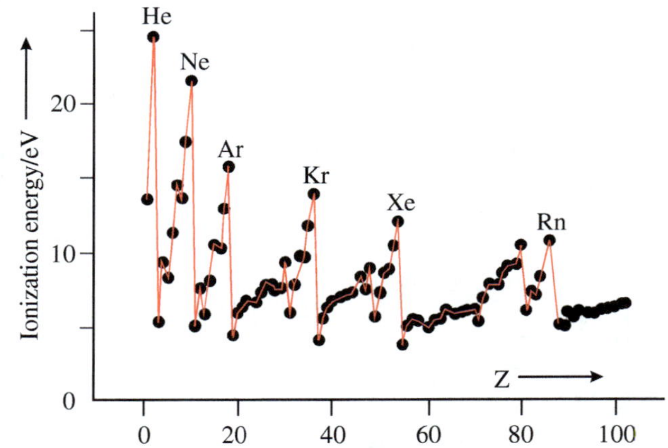

Figure 5.12 Plot of the elements' first ionization energies versus the atomic number Z (nuclear charge Ze).

5.5.4.2 Atomic Radii

Another important trend explained by consideration of the orbital energies is the change in atomic size through the periodic table. For example, as one moves from left to right across the periodic table the atom's size decreases for a fixed value of the principal quantum number. This decreasing size occurs because the nuclear charge is increasing, which stabilizes the orbitals causing them to contract. Figure 5.13 plots the electron density maps, obtained from SCF calculations, for the first 10 elements, and they clearly show this behavior. Note in particular how dramatically the $1s$ orbital contracts as one proceeds from H to Ne. Even though the total number of electrons about the nucleus increases as one proceeds across

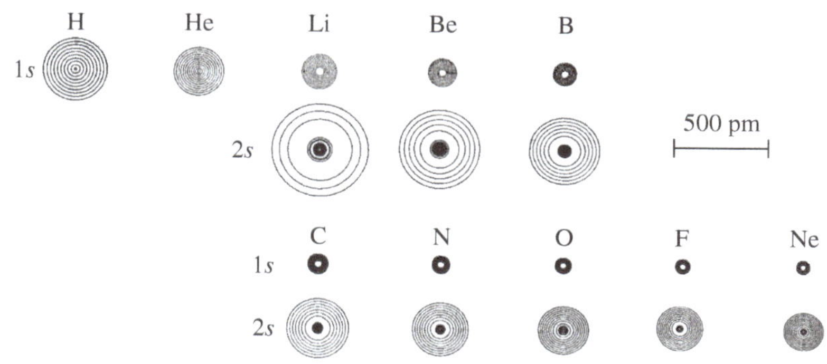

Figure 5.13 Plots of the SCF orbital's electron densities for the first 10 elements of the periodic table (1*s* and 2*s* orbitals). The innermost contour lines correspond to a density of $e \cdot 6800\text{nm}^{-3}$, except for the H atom where it is $e \cdot 1700\text{nm}^{-3}$. The lowest contour lines correspond to $e \cdot 0.41\text{nm}^{-3}$. The densities at adjacent contour lines differ by a factor of two.[8] These plots are taken from [8].

a row of the periodic table, the atomic size drops. As one moves down a column in the periodic table the size of the atom increases. This trend results from the increase in the number of electrons in the atom, and the fact that the outer electrons are in higher principal quantum number orbitals, which are found further from the nucleus on average.

5.6 An Electron Configuration Can Have Multiple Energy States

The specification of electronic configurations (e.g., C is $1s^2 2s^2 2p^2$) is ambiguous because it does not specify the k and m_s quantum numbers. In fact, a single electronic configuration may correspond to many different energy levels of the atom. How can we understand this fact? This splitting of the degeneracy (e.g., not all arrangements of the two *p*-electrons in the carbon atom having the same energy) arises from the multielectron nature of the atom and has two origins. The largest effect originates from electrostatic energy differences (energy range of 0.1–1 eV) and can be understood by considering the different orbital arrangements (see Section 5.5). The second effect originates from the magnetic interaction, or spin–orbit interaction, and depends on details of the electrons' orbital arrangement.

5.6.1 Angular Momentum and Vector Model of Atoms

We have seen in Section 4.3.5 that the degeneracy of 2*p* states in the H atom is removed by applying a magnetic field. In atoms with more than one electron, we find that some of these degeneracies are removed without applying an external magnetic field; namely, they arise from interactions among the electrons. In this case the energy depends on more than just the principal quantum number n. In the orbital picture, we find that the energies depend on the other quantum numbers: l, k, and m_s. By specifying these other quantum numbers of the atom, we are specifying its angular momentum state.

In the vector model of the atom each electron is assigned a hydrogenic orbital with an orbital angular momentum and a spin angular momentum. These angular momenta, which are vectors, must be added together to obtain a total angular momentum state for the atom, and these states are associated with particular energy levels. Although different "schemes" exist for combining these momenta, we will focus on the Russell–Saunders, or LS, coupling scheme which provides a realistic picture for atoms of lower atomic number. In this scheme the orbital angular momenta of the individual electrons are combined to give a total orbital angular momentum quantum number L, and the spin angular momenta of each of the electrons are combined to give a total spin angular momentum quantum number S. Lastly, these two angular momenta are combined to find a total angular momentum for the atom. A derivation of this coupling scheme from basic principles is beyond the scope of this book [9]. Rather we define and apply the scheme to develop a way of describing the electronic structure of atoms and their spectroscopic properties.

5.6.1.1 Orbital Angular Momentum

In a many-electron atom, the spatial distribution of electron density and the orbital angular momentum cause large differences in the energies, because of the electron–electron repulsion. The importance of this electron–electron repulsion can be seen from a simple consideration of two electrons moving in a circular orbit (Fig. 5.14). If the electrons are orbiting in the same direction, they will tend to stay further apart (on average) than if they are orbiting in opposite directions, because in the latter case they would cross every half-period.

8 Note that this is an exponential scale; therefore, the distances between adjacent contour lines are constant.

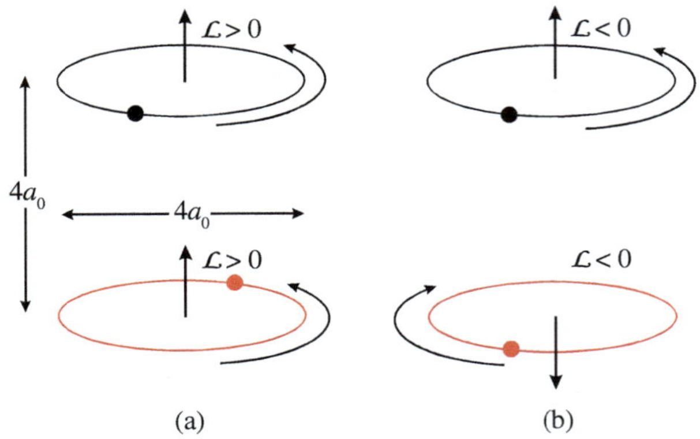

Figure 5.14 Repulsion of two electrons (a) orbiting in the same direction and (b) orbiting in opposite directions.

Example 5.4 *Estimate the electron–electron repulsion for two electrons in coaxial orbits that are displaced by a distance $4a_0$ and with orbital diameters of $4a_0$, typical for a 2p electron. Consider the two cases of co-circulating electrons and counter-circulating electrons and compare the results.*

Since we are asked to estimate the difference, we proceed with a rough calculation. For two electrons orbiting in the same direction, the electrostatic repulsion energy will be minimized if they are orbiting at the same speed but 180° phase shifted. In this case they are always at the maximum distance of $4a_0\sqrt{2}$, and their repulsion energy is

$$\frac{1}{4\pi\varepsilon_0}\frac{e^2}{\left(4a_0\sqrt{2}\right)} = 7.71 \times 10^{-19}\ \text{J} = 4.8\ \text{eV}$$

For two electrons orbiting in the opposite sense the average distance is smaller. We estimate the average repulsion energy by computing the average of their repulsion energy at the maximum distance $4\sqrt{2}a_0$ and at the minimum distance $4a_0$; thus we find

$$\frac{1}{2}\left[\frac{1}{4\pi\varepsilon_0}\frac{e^2}{\left(4\sqrt{2}a_0\right)} + \frac{1}{4\pi\varepsilon_0}\frac{e^2}{(4a_0)}\right] = 9.31 \times 10^{-19}\ \text{J} = 5.8\ \text{eV}$$

Comparison of these two estimates gives an energy difference of 1.0 eV, which is typical of the differences observed experimentally.

Although the model described by Figure 5.14 is greatly oversimplified, it illustrates the essence of how the electronic orbital angular momentum of a many-electron atom influences the atom's energy. For a many-electron atom, we take the vector sum of each electron's orbital angular momentum to obtain the total orbital angular momentum of the electrons. The total orbital angular momentum will depend on both the direction and magnitude of each electron's orbital vectors. For example, an electron configuration with two p-electrons has three possible total orbital angular momenta, depending on which of the different p-orbitals the two electrons occupy (see Foundation 5.5). We must also consider the spin angular momentum of the electron.

5.6.1.2 Spin Angular Momentum

As we saw with excited states of He, the spin angular momentum of the electrons (singlet state or triplet state) affects the symmetry of the spatial wavefunction (symmetric for the singlet and antisymmetric for the triplet), and this changes the electronic energy significantly. This same effect is found in many-electron atoms and causes large changes in the electronic energy.

To obtain the total spin angular momentum $\vec{S}$ of an atom's electrons, we perform a vector sum of the spin angular momentum from each electron, in direct analogy to the case for the orbital angular momentum.

5.6.1.3 Total Electron Angular Momentum: Spin–Orbit Coupling

In addition to the electrostatic contributions to the energy, the spin angular momentum influences the energy through an effect called spin–orbit coupling. In brief, the orbital motion of the electron(s) generates a magnetic field that interacts with the electron(s) intrinsic magnetic dipole moment. This effect becomes more important as the nuclear charge Ze of the atom increases. This dependence can be understood by considering the orbital motion from the reference frame of the electron. From this perspective, the nucleus is moving in an orbit around the electron and the magnitude of its orbital current increases with the charge on the nucleus. Because this orbital current generates the effective magnetic field, the increase in Z leads to an increase in the spin–orbit interaction energy.

The Russell–Saunders approach assumes that the electrostatic effect on the energy dominates and treats the spin–orbit energy effect as a perturbation. This assumption works well for low values of Z, typically $Z < 40$. For high atomic number atoms where the spin–orbit terms become more significant, an approach called $j - j$ coupling has been developed [10].

5.6.2 Atomic Term Symbols

An atomic term symbol is a shorthand notation for specifying the angular momentum of an atom's electronic energy level. A general form for the atomic term symbol is ${}^{2S+1}L_J$, where L specifies the total orbital angular momentum quantum number, S specifies the total spin angular momentum quantum number, and J specifies the total angular momentum quantum number. The $2S + 1$ term is called the multiplicity of the electronic state; e.g., the hydrogen atom has one unpaired electron so $S = 1/2$ and the multiplicity of the state is a doublet. If an atom's electronic state has $S = 1$ then it is a triplet. The value of L specifies the total orbital angular momentum for the electrons and is given a letter according to the following scheme

L	0	1	2	3	4	5
	S	P	D	F	G	H

This letter code is the same as that used for the atomic orbitals of different l; however, the letters are uppercase.

Table 5.3 provides the different spectroscopic term symbols that correspond to some of the common electron configurations. A procedure for determining the different possible term symbols from a given electronic configuration is developed in Foundation 5.5. Any filled subshell has a total angular momentum, total orbital angular momentum, and total spin angular momentum of zero, so it does not contribute to the overall angular momentum state of the atom. Hence we need only consider the partially filled subshells. The value of J is not given explicitly in the table. For each term symbol, J takes on values ranging from $|L + S|$ to $|L - S|$ in steps of one; i.e., $|L + S|, |L + S - 1|, \ldots, |L - S|$. Note that each of the term symbols (or terms) corresponds to a certain energy level, but can contain multiple states (have a degeneracy). From the table it is clear that a given electron configuration can correspond to many different electronic energy levels.

Table 5.3 Atomic Term Symbols Are Given for Different Electronic Configurations (n Denotes the Principal Quantum Number).

Ground States	Term Symbols	Excited State	Term Symbols
s	${}^2S$	$ns\ n's$	${}^1S,\ {}^3S$
s^2	${}^1S$	$ns\ np$	${}^1P,\ {}^3P$
p	${}^2P$	$ns\ nd$	${}^1D,\ {}^3D$
p^2	${}^1S,\ {}^1D,\ {}^3P$	$ns\ np^2$	${}^2S,\ {}^2P,\ {}^2D,\ {}^2S,\ {}^2P,\ {}^2D,\ {}^4S,\ {}^4P,\ {}^4D$
p^3	${}^4S,\ {}^2D,\ {}^2P$		
p^4	${}^1S,\ {}^1D,\ {}^3P$		
p^5	${}^2P$		
p^6	${}^1S$		

Taken from G. Herzberg *Atomic Spectra and Atomic Structure* (Prentice Hall, New York, 1937).

We could, in principle, compute the energy levels but instead we will use an empirical rule of thumb, known as Hund's rule, to determine the relative energy ordering of the energy levels which correspond to a given electron configuration (see Section 5.5.3.3).

- *The term with the largest value of S is the most stable, and the stability decreases with decreasing S.*
- *For terms with the same value of S, the term with the largest value of L is the most stable.*

Example 5.5 *Consider the case of the helium atom, for which the ground-state electronic configuration is $1s^2$ and has a term symbol of ${}^1S_0$. Find the term symbols for the first excited state configuration $1s2s$ and the second excited state configuration $1s2p$*

The first excited state configuration has two s-orbitals. From Table 5.3 we see that this configuration has two different energy levels available, the ${}^1S$ and the ${}^3S$. Because $L = 0$ and $S = 0$ for the ${}^1S$ term symbol it has $J = 0$, and we find ${}^1S_0$. Analogously for the ${}^3S$ we find $J = 1$ and the term symbol is ${}^3S_1$. According to Hund's rule we predict that the triplet level will be more stable. Hence, this single configuration gives two different energy levels, one which is a singlet and a second energy level which is a triplet (i.e., three-fold degenerate). This result is in accordance with the discussion in Section 5.4.3.3.

Following the same logic, we see that the next excited state configuration, $1s2p$, also has two electronic energy levels, which are a singlet ^1P and a triplet ^3P. In this case we find one J value for ^1P, $J = 1$, so that the singlet excited state term symbol is $^1P_1$. Analogously, we find three J values $J = 2, 1, 0$ for the triplet state and obtain the terms symbols $^3P_2$, $^3P_1$, and $^3P_0$. The different J values for the ^3P levels differ in energy because they have different spin–orbit energies; however, these energy differences are much smaller than those found for different L and S values (see Foundation 5.5). From the electron configuration $1s2p$ we expect to have 12 unique ways of placing the electrons into the atomic orbitals: the ^1P term has a degeneracy of three (the three orientations of the orbital angular momentum vector) and the ^3P term has a degeneracy of nine (for each of the three orientations of the orbital angular momentum it has three possible directions for its spin angular momentum).

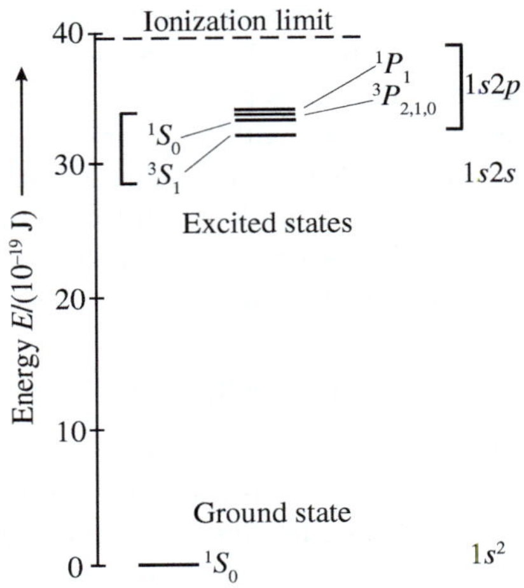

Figure 5.15 A plot of the first few electronic states of the helium atom is shown. The electronic configurations and the spectroscopic term symbols are given for the different energy levels.

Figure 5.15 shows the energy level diagram for the ground state ($^1S_0$), and for the excited states ($^3S_1$,$^1S_0$,$^3P_{2,1,0}$, and $^1P_1$) of helium. It is evident from this diagram that the electronic configuration change has the largest impact on the energy differences, but that the different terms in an electron configuration have significantly different energies. Note that the degenerate energy levels can be split into distinct electronic states by the application of an external perturbation, such as an electric or magnetic field.

5.7 Conclusion

This chapter discussed a number of important issues for understanding the electronic structure of atoms with more than one electron. In the beginning we found that the presence of more than one electron created a Schrödinger equation that could not be solved exactly. As a consequence we introduced the variational method as a way to calculate the electronic energies and wavefunctions for many-electron atoms. In the process of describing atoms with more than one electron, we required a number of new considerations and quantum mechanical postulates. In particular, we needed to maintain the indistinguishability of the electrons in our construction of the electronic wavefunction; we had to make sure that the total electronic wavefunction is antisymmetric, in accordance with the Pauli Exclusion Principle; and we had to postulate that electrons carry an intrinsic angular momentum (spin) which gives rise to a magnetic moment. Combining these new postulates with our earlier considerations, we are now able to understand the structure of the periodic table and the variation of atomic properties on both qualitative and quantitative levels.

5.8 Box 5.1. He Atom: Calculation of Energy

5.8.1 Energy in the Ground State

For a system with two electrons the coordinates of both electrons appear in the Hamiltonian operator:

$$\mathcal{H} = -\frac{\hbar^2}{2m_e}\left[\left(\frac{\partial^2}{\partial x_1^2}+\frac{\partial^2}{\partial y_1^2}+\frac{\partial^2}{\partial z_1^2}\right)+\left(\frac{\partial^2}{\partial x_2^2}+\frac{\partial^2}{\partial y_2^2}+\frac{\partial^2}{\partial z_2^2}\right)\right]+\frac{1}{4\pi\varepsilon_0}\cdot\left(-\frac{2e^2}{r_1}-\frac{2e^2}{r_2}+\frac{e^2}{r_{12}}\right)$$

$$= -\frac{\hbar^2}{2m_e}\left(\nabla_1^2+\nabla_2^2\right)+V_1+V_2+\frac{1}{4\pi\varepsilon_0}\frac{e^2}{r_{12}}$$

where

$$r_1 = \sqrt{x_1^2+y_1^2+z_1^2} \quad \text{and} \quad r_2 = \sqrt{x_2^2+y_2^2+z_2^2}$$

and

$$r_{12} = \sqrt{(x_1 - x_2)^2 + (y_1 - y_2)^2 + (z_1 - z_2)^2}$$

Because of the electron–electron repulsion term

$$\frac{1}{4\pi\varepsilon_0}\frac{e^2}{r_{12}}$$

it is not possible to separate the Schrödinger equation

$$\mathcal{H}\Psi(\vec{r}_1, \vec{r}_2) = E\Psi(\vec{r}_1, \vec{r}_2)$$

into two differential equations—one which depends on the coordinates of electron 1, $\vec{r}_1$, and one which depends on the coordinates of electron 2, $\vec{r}_2$. This complication makes the problem too difficult to solve exactly and we must proceed in an approximate manner. Note that the helium atom ground state is spherically symmetric, and we choose a trial wavefunction without any angle variables ϑ and φ.

We will proceed by ignoring the electron–electron repulsion term and finding an approximate ground-state wavefunction $\psi^{(0)}$, which we then use to calculate a correction term for the electron–electron repulsion term. Consider a simplified Schrödinger equation without the repulsion term:

$$\mathcal{H}_0\ \psi^{(0)} = E^{(0)} \cdot \psi^{(0)}$$

where

$$\mathcal{H}_0 = -\frac{h^2}{8\pi^2 m_e}\left(\nabla_1^2 + \nabla_2^2\right) + V_1 + V_2$$

We use the superscript (0) to distinguish the approximate form from the exact one. This equation can be separated into two independent He^+ Schrödinger equations, and it is solved by a product of two He^+ ground-state wavefunctions, one for each electron. For the ground state of He as a trial function ϕ, we use the lowest energy (ground state) solution of $\mathcal{H}_0$. This is the product of two 1s orbital wavefunctions Ψ_{He^+} of the He^+ ion, namely

$$\phi(r_1, r_2) = \Psi_{He^+}(r_1) \cdot \Psi_{He^+}(r_2) = \sqrt{\frac{8}{\pi a_0^3}}e^{-2r_1/a_0} \cdot \sqrt{\frac{8}{\pi a_0^3}}e^{-2r_2/a_0}$$

By inserting this wavefunction into the approximate Schrödinger equation we obtain the energy $E^{(0)}$ as the sum of two He^+ ground-state energies

$$E^{(0)} = 2E_{He^+} = 2\left(-\frac{1}{4\pi\varepsilon_0}\frac{2e^2}{a_0}\right) = -\frac{1}{4\pi\varepsilon_0}\frac{4e^2}{a_0} = -174.4 \times 10^{-19}\ \mathrm{J}$$

You are asked to show this result in Exercise 5.1.

To partially account for the mutual repulsion of the two electrons we use perturbation theory (see Foundation 5.2) with our trial wavefunction $\phi(r_1, r_2)$ to calculate the average repulsion energy $\overline{V}_e$. The probability density ρ is

$$\rho(r_1, r_2) = \phi^2(r_1, r_2) = \frac{8}{\pi a_0^3}e^{-4r_1/a_0} \cdot \frac{8}{\pi a_0^3}e^{-4r_2/a_0} = \rho_1(r_1) \cdot \rho_2(r_2)$$

Because $\rho_1 \cdot e \cdot d\tau_1$ is the charge of electron 1 in the volume element $d\tau_1$ and $\rho_2 \cdot e \cdot d\tau_2$ is the charge of electron 2 in the volume element $d\tau_2$, we obtain for $\overline{V}_e$

$$\overline{V}_e = \frac{1}{4\pi\varepsilon_0}e^2\int \rho_1\frac{1}{r_{12}}\rho_2 d\tau_1 d\tau_2$$

Using

$$\rho_1 = \frac{8}{\pi a_0^3}e^{-4r_1/a_0} \quad \text{and} \quad \rho_2 = \frac{8}{\pi a_0^3}e^{-4r_2/a_0}$$

we can solve this integral (see Justification 5.1) and find

$$\overline{V}_e = \frac{1}{4\pi\varepsilon_0}\frac{5}{4}\frac{e^2}{a_0} = 54.5 \times 10^{-19}\ \mathrm{J}$$

Now we add this correction term to the energy $E^{(0)}$ and find a corrected energy ε_{He} for the He atom ground state, namely

$$\varepsilon_{\text{He}} = E^{(0)} + \overline{V}_e = \frac{1}{4\pi\varepsilon_0}\left(-4\frac{e^2}{a_0} + \frac{5}{4}\frac{e^2}{a_0}\right) = -\frac{1}{4\pi\varepsilon_0}\frac{11}{4}\frac{e^2}{a_0} = -119.9 \times 10^{-19}\ \text{J}$$

This calculation can be improved by using better trial functions (see Equation (5.18) and Section 5.3.1.2). In these calculations the evasion of the electrons (correlation effect) is neglected; this effect can be taken into account by using two-electron trial functions which depend on the mutual distance r_{12} of the two electrons; see [10].

In the Schrödinger equation both electrons are indistinguishable. Thus we cannot attribute separate wavefunctions to each of the electrons, but only one wavefunction to both electrons as a whole. The indistinguishability of both electrons means that the function ψ^2 remains unchanged when the electrons are exchanged. This symmetry condition is fulfilled in our example.

5.8.2 Energy in the Excited State

We proceed in a similar way as for the ground state, assuming that electron 1 is in the ground state (1*s*-function) and that electron 2 is in the excited state (2*s*-function), thus using wavefunctions of the type

$$\phi_{\text{excited}}(r_1, r_2) = \Psi_{\text{He}^+_{1s}}(r_1) \cdot \Psi_{\text{He}^+_{2s}}(r_2) = \sqrt{\frac{8}{\pi a_0^3}}\,\text{e}^{-2r_1/a_0} \cdot \sqrt{\frac{8}{32\pi a_0^3}}(2 - r/a_0)\text{e}^{-r_2/a_0}$$

and introducing the condition that the final wavefunction must be symmetric or antisymmetric. In Problem 5.12, you show that the resulting energies are

$$\varepsilon_{\text{excited,sym}} = E_{1s} + E_{2s} + J_{12} + K_{12}$$

and

$$\varepsilon_{\text{excited,anti}} = E_{1s} + E_{2s} + J_{12} - K_{12}$$

where

$$J_{12} = \iint \varphi_{1s}^2(1) V_{12} \varphi_{2s}^2(2) \cdot \text{d}\tau_1 \text{d}\tau_2$$

$$K_{12} = \iint \left[\varphi_{1s}(1) \cdot \varphi_{2s}(1) \cdot V_{12} \cdot \varphi_{2s}(2)\varphi_{1s}(2)\right] \cdot \text{d}\tau_1 \text{d}\tau_2$$

are the Coulomb integral and the exchange integral. For the numerical calculation of the energies and of the integrals see Problem 5.12 and Justification 5.2.

5.9 Box 5.2. Current Loop in a Magnetic Field (Bohr Magneton)

Consider a current I flowing through a current loop (Fig. 5B.1) when applying a magnetic field of flux density B. A force f acts on a section Δy of the wire in the direction perpendicular to the direction of the current I and perpendicular to the direction of the magnetic flux density B, such that

$$f = I \cdot B \cdot \Delta y$$

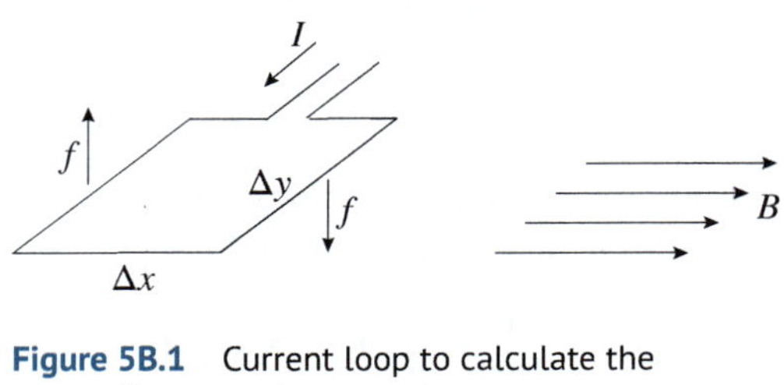

Figure 5B.1 Current loop to calculate the magnetic moment μ.

The torque τ, which acts on the loop, is

$$\tau = 2f \cdot \frac{\Delta x}{2} = 2I \cdot B \cdot \Delta y \cdot \frac{\Delta x}{2} = IA \cdot B = \mu B$$

where $A = \Delta x \cdot \Delta y$ is the area enclosed by the loop and $\mu = IA$ is called the *magnetic moment*. The magnetic moment μ is a property of the loop, being determined by its current and area, and is directed perpendicular to the plane of the loop. The direction that the magnetic moment points will depend on the sign of the current, that is whether it is flowing in a clockwise or counter-clockwise manner. These two directions have different energies because of the magnetic moment's interaction with the applied magnetic field. The energy ΔE needed to rotate μ by 180° (e.g., from being oriented parallel to

B to being oriented in the opposite direction) is

$$\Delta E = 2f \cdot \Delta x = 2I B \cdot \Delta x \cdot \Delta y = 2\mu B$$

Hence the energy splitting created by the application of a known magnetic field can be used to determine the magnetic moment, which is an intrinsic property of a system.

To estimate the magnitude of the magnetic moment of an electron, we consider the magnetic moment produced by an electron as described in Bohr's H-atom model.[9] In this model the electron moves with a speed

$$v = \frac{1}{4\pi\varepsilon_0} \cdot \frac{e^2}{\hbar}$$

on a circle of radius a_0

$$a_0 = 4\pi\varepsilon_0 \frac{\hbar^2}{m_e e^2}$$

Hence its orbital frequency is given by the ratio of the speed and the circle's circumference, so that

$$\nu = \frac{v}{2\pi a_0}$$

Using $A = \pi a_0^2$ for the area of the loop and $I = \nu e$ for the current, we obtain the magnetic moment

$$\mu_{\text{Bohr}} = I \cdot A = \nu e \cdot \pi a_0^2 = \frac{1}{2} e \cdot v a_0 = \frac{e\hbar}{2m_e}$$

which is known as the *Bohr magneton*.

5.10 Box 5.3. The Electron Spin Postulate and the Antisymmetry Postulate

5.10.1 The Electron Spin Postulate

The name spin derives from the classical picture of a spinning charge distribution (i.e., the electron or nucleus) that gives rise to an angular momentum. Like a current flowing in a loop (see Box 5.2), a spinning charge distribution will create a magnetic moment. The direction of spin (clockwise or counterclockwise) will affect the direction of the magnetic moment. This effect was proposed by Uhlenbeck and Goudsmit to explain the fine structure in atomic spectra; however, it is very nicely illustrated by the classic experiment of Stern and Gerlach.

In their experiment, Stern and Gerlach directed a beam of Ag atoms through an inhomogeneous magnetic field (see Fig. 5B.2). Because a Ag atom has one unpaired electron the magnetic field interacts with it. In contrast, atoms with no unpaired electrons, such as ground-state He, do not interact with a magnetic field. Our postulate states that the electron has one of two possible spin directions, corresponding to clockwise rotation or anti-clockwise rotation. These two spin orientations correspond to magnetic moments which are of equal magnitude but opposite direction and should experience opposite deflection angles in a magnetic field. If the electron had a continuous distribution of possible spin angular momenta, and hence magnetic moments, then the experiment would generate a continuous distribution of deflection angles. Stern and Gerlach observed discrete atom deflections (Fig. 5B.2) in their experiment. In general this effect can arise from orbital and/or spin angular momentum of the electron. Because the electron in a silver atom does not have any orbital angular momentum, this experiment probes the intrinsic magnetic moment of the electron that results from its spin.

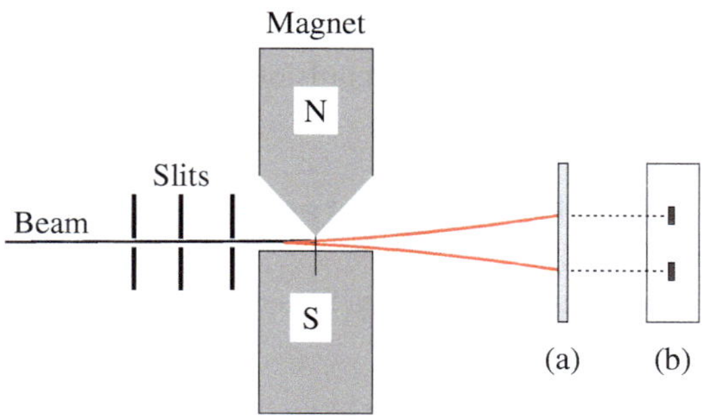

Figure 5B.2 Stern–Gerlach experiment. A beam of Ag atoms (expanding from a heated furnace) passes a series of slits before entering into an inhomogeneous magnetic field. When the magnetic field is switched on, the silver atoms in the beam are deflected as indicated in the picture, and silver is deposited on the surface of a glass plate. (a) side view and (b) top view. The filled rectangles in (b) represent Ag deposits.

9 See Pais [11] and Problem P4.1.

5.10.2 The Antisymmetry Postulate

The antisymmetry postulate states that electronic wavefunctions must be antisymmetric with respect to the exchange of two electrons. Here we illustrate the implications of this postulate by using one-dimensional box wavefunctions

$$\varphi(x_1) = \sqrt{\frac{2}{L}}\sin\left(\frac{\pi}{L}x_1\right) \qquad \varphi(x_2) = \sqrt{\frac{2}{L}}\sin\left(\frac{2\pi}{L}x_2\right)$$

to construct symmetrical and antisymmetrical wavefunctions:

$$\text{symmetric spatial function: } \phi(x_1, x_2) = \frac{1}{\sqrt{2}}\left[\varphi(x_1)\varphi(x_2) + \varphi(x_1)\varphi(x_2)\right] \tag{5B3.1}$$

and

$$\text{antisymmetric spatial function: } \phi(x_1, x_2) = \frac{1}{\sqrt{2}}\left[\varphi(x_1)\varphi(x_2) - \varphi(x_1)\varphi(x_2)\right] \tag{5B3.2}$$

According to the postulate, the symmetric spatial function must be multiplied by an antisymmetric spin function, so that the total wavefunction is antisymmetric. Correspondingly, the antisymmetric spatial function must be combined with a symmetric spin function. This symmetry constraint can have a strong impact on the energy of the states.

We can visualize the two spatial functions by drawing contour line plots in the x_1/x_2 space, see Fig. 5B.3. First consider the case in Fig. 5B.3a: symmetric spatial wavefunction combined with an antisymmetric spin function. The green line connects all points in the x_1/x_2-space where $x_1 = x_2$, that is, where the two electrons are at the same x-position. Along this line the wavefunction has finite values, that is, we have a high electron density along this line. This means that the probability of finding both electrons at the same place is high. Next consider the case in Fig. 5B.3b: antisymmetric spatial wavefunction combined with a symmetric spin function. In this case the wavefunction along the green line is zero, that is, the electron density is zero along this line. This means that the probability of finding both electrons at the same place is zero. Note that the difference between these two cases is a direct consequence of the antisymmetry postulate, and it gives rise to strong electrostatic energy differences for the electron–electron repulsion.

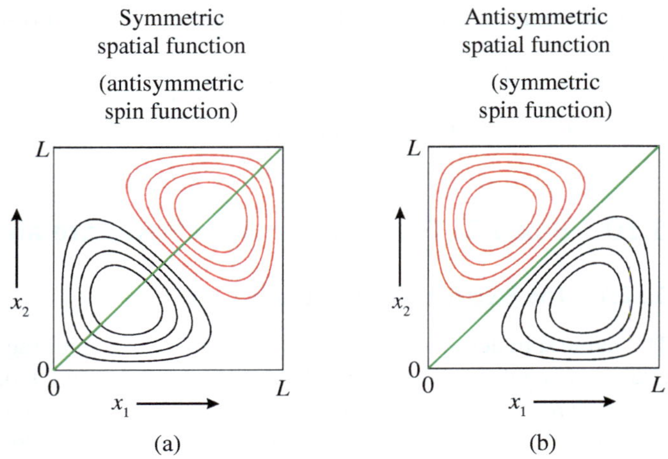

Figure 5B.3 Contour line plot of wavefunctions (5B3.1) (a) and (5B3.2) (b). The green lines indicate all points in the x_1/x_2-space where $x_1 = x_2$. Thus, the probability of $x_1 = x_2$ is zero for the wavefunction in panel (b) but is large for the wavefunction in panel (a).

5.11 Box 5.4. Slater Wavefunctions

To account for shielding, Slater suggested that we parameterize the radial part of an electron's orbital by an effective atomic number Z_{eff}:

$$Z_{\text{eff}} = Z - \sigma$$

where σ is called the shielding constant. In principle, σ should be determined for each electron by contributions of all the other electrons in the atom, for example see [12]. The shielding is dominated by electrons in the shells and subshells that are lower in energy, hence closer to the nucleus than the electron of interest, however. In Slater's scheme an approximate shielding value is assigned to each subshell: 1s; 2s, 2p; 3s, 3p, 3d; 4s, 4p, 4d, 4f; … as shown in Fig. 5B.4 for $Z = 20$ (Ca atom). The remaining electrons of the same shell as the selected electron contribute 0.35 to σ (an exception is 0.30 for the 1s shell). When they are located in the shell just below that of the selected electron, the d and f orbitals of a shell contribute 1.0 each to the shielding, and the s and p orbitals contribute 0.85 each to the shielding. Shells that are more than one principal quantum number below that of the selected electron completely shield the nucleus and contribute 1.0.

In the scheme of Fig. 5B.4 a 4s electron (Ca atom) is selected. Using the algorithm described above for the Ca atom we obtain

$$\sigma = 2 + 8 + 6.8 + 0.35 = 17.15$$

for the shielding constant, and this gives

$$Z_{\text{eff}} = 20 - 17.15 = 2.85$$

for the effective charge. In Table 5B.1, Z_{eff} is displayed for the outer-shell electrons of several elements:

Let us compare this method for estimating the effective charge to the experimental results for Li. For Li we find an effective atomic number of $Z_{\text{eff}} = 1.3$, which compares well to the value 1.26 which we found from the first ionization energy of Li. For $\text{Li}^+, Z_{\text{eff}} = 3 - 0.3 = 2.7$; thus $E_{\text{ion},2} = 21.8 \times 2.7^2 \times 10^{-19}\ \text{J} = 159 \times 10^{-19}\ \text{J}$. The experimental values are (see Table 5.1) $E_{\text{ion},2} = 121 \times 10^{-19}$ J and $E_{\text{ion},3} = 196 \times 10^{-19}$ J, and we see that the Slater value for the effective atomic number predicts the average of $E_{\text{ion},2}$ and $E_{\text{ion},3}$, namely 159×10^{-19} J, quite well.

Using these assumptions we obtain the same one-electron wavefunctions (orbitals) as discussed in Chapter 4 with the only difference that Z is substituted by Z_{eff}. As a further simplification Slater suggested that we use a simple power r^{n-1} instead of the polynomial of r that appears in the radial term of the wavefunctions ("*Slater functions*"), for example,

$$\psi_{2s,\text{Slater}} = A \cdot r \cdot \mathrm{e}^{-rZ_{\text{eff}}/(2a_0)}, \quad \psi_{3s,\text{Slater}} = A \cdot r^2 \cdot \mathrm{e}^{-rZ_{\text{eff}}/(3a_0)}$$

instead of

$$\psi_{2s} = A \cdot \left(r - \frac{2a_0}{Z_{\text{eff}}} \right) \cdot \mathrm{e}^{-rZ_{\text{eff}}/(2a_0)}$$

$$\psi_{3s} = A \cdot \left(r^2 - 9\frac{a_0}{Z_{\text{eff}}} r + \frac{27}{2} \left(\frac{a_0}{Z_{\text{eff}}} \right)^2 \right) \cdot \mathrm{e}^{-rZ_{\text{eff}}/(3a_0)}$$

where A is a constant. Only for ψ_{1s} and ψ_{2p} are the Slater functions hydrogen-like.

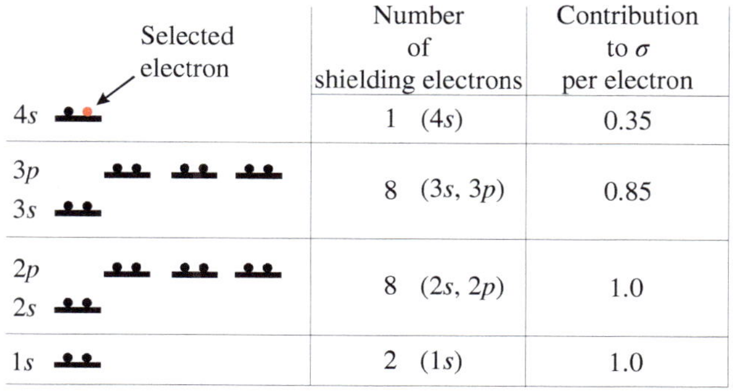

Figure 5B.4 Calculation of the shielding constant σ from contributions of different electrons for a Ca atom ($Z = 20$).

Table 5B.1 Calculation of some Z_{eff} values by the Slater method.

Atom	σ	Z_{eff}
He	$0.30 = 0.30$	$2 - 0.30 = 1.70$
Li	$2 \cdot 0.85 = 1.70$	$3 - 1.70 = 1.30$
Be	$0.35 + 2 \cdot 0.85 = 2.05$	$4 - 2.05 = 1.95$
B	$2 \cdot 0.35 + 2 \cdot 0.85 = 2.40$	$5 - 2.40 = 2.60$
C	$3 \cdot 0.35 + 2 \cdot 0.85 = 2.75$	$6 - 2.75 = 3.25$
N	$4 \cdot 0.35 + 2 \cdot 0.85 = 3.10$	$7 - 3.10 = 3.90$
O	$5 \cdot 0.35 + 2 \cdot 0.85 = 3.45$	$8 - 3.45 = 4.55$
F	$6 \cdot 0.35 + 2 \cdot 0.85 = 3.80$	$9 - 3.80 = 5.20$
Ne	$7 \cdot 0.35 + 2 \cdot 0.85 = 4.15$	$10 - 4.15 = 5.85$
Na	$8 \cdot 0.85 + 2 \cdot 1.00 = 8.80$	$11 - 8.80 = 2.20$
Mg	$0.35 + 8 \cdot 0.85 + 2 \cdot 1.00 = 9.15$	$12 - 9.15 = 2.85$
Ca	$0.35 + 8 \cdot 0.85 + 8 \cdot 1.0 + 2 = 17.15$	$20 - 17.15 = 2.85$

5.12 Justifications

5.12.1 Justification 5.1: He-Atom: Repulsion Energy in the Ground State

See the books website, www.wiley.com/go/Kuhn/physicalchemistry3e.

5.12.2 Justification 5.2: He-Atom: Repulsion Energy in the Excited State

See the books website, www.wiley.com/go/Kuhn/physicalchemistry3e.

5.12.3 Justification 5.3: Analytical Evaluation of Coulomb Integral

See the books website, www.wiley.com/go/Kuhn/physicalchemistry3e.

5.13 Foundations

5.13.1 Foundation 5.1: Proof of Variational Principle

Consider a normalized trial function ϕ_1, we want to show that the variational energy ε_1 is always greater than the true ground-state energy E_1; i.e.,

$$\varepsilon_1 = \int \phi_1 \mathcal{H} \phi_1 \cdot d\tau \geq E_1 \tag{5F1.1}$$

To begin, we assume that the true wavefunction solutions of the many electron Schrödinger equation $\mathcal{H}\psi = E\psi$ are ψ_1, $\psi_2, \ldots$, that they are orthogonal and normalized, and that they correspond to the energies $E_1, E_2, \ldots$. Next, we assume that the trial function ϕ_1 can be written as a linear combination of the exact wavefunctions, namely

$$\phi_1 = c_1\psi_1 + c_2\psi_2 + \cdots = \sum_i c_i\psi_i \tag{5F1.2}$$

This latter assumption can be justified rigorously, if the $\psi_1, \psi_2, \ldots$ form a complete set; see [13]. With these assumptions, the proof is straightforward.

We insert the expression for the trial wavefunction, Equation (5F1.2), into the integral for ε_1

$$\varepsilon_1 = \int \left(\sum_i c_i\psi_i\right) \mathcal{H} \left(\sum_i c_i\psi_i\right) \cdot d\tau = \sum_i c_i^2 \int \psi_i \mathcal{H} \psi_i \cdot d\tau + \sum_{j\neq i} c_i c_j \int \psi_i \mathcal{H} \psi_j \cdot d\tau \tag{5F1.3}$$

For the exact wavefunctions $\mathcal{H}\psi_i = E_i\psi_i$, so we can write

$$\varepsilon_1 = \sum_i c_i^2 \int \psi_i E_i \psi_i \cdot d\tau + \sum_{j\neq i} c_i c_j \int \psi_i E_j \psi_j \cdot d\tau = \sum_i c_i^2 E_i \int \psi_i \psi_i \cdot d\tau + \sum_{j\neq i} c_i c_j\ E_j \int \psi_i \psi_j \cdot d\tau \tag{5F1.4}$$

Because the exact wavefunctions are orthogonal and normalized, this simplifies further, and we find

$$\varepsilon_1 = \sum_i c_i^2 E_i \cdot (1) + \sum_{j\neq i} c_i c_j\ E_j \cdot (0) = \sum_i c_i^2 E_i \tag{5F1.5}$$

In the next step, we add and subtract $\sum_i c_i^2 E_1$ from the right-hand side of the equation (thus, it is like adding zero) and we find

$$\varepsilon_1 = \sum_i c_i^2 E_i + \sum_i c_i^2 E_1 - \sum_i c_i^2 E_1 = \sum_i c_i^2 \left(E_i - E_1\right) + E_1 \sum_i c_i^2 \tag{5F1.6}$$

Because the trial function ϕ_1 is normalized, we substitute our expansion for it in terms of the exact wavefunctions and write that

$$\int \phi_1^2 \cdot d\tau = 1 = \sum_i c_i^2 \int \psi_i^2 \cdot d\tau + \sum_{j\neq i} c_i c_j \int \psi_i \psi_j \cdot d\tau = \sum_i c_i^2 \tag{5F1.7}$$

Combining this result with our expression for the variational energy, we find that

$$\varepsilon_1 = E_1 + \sum_i c_i^2 \left(E_i - E_1\right) \tag{5F1.8}$$

Since E_1 is the ground state we know that $E_i \geq E_1$, so that the second term must be greater than or equal to zero. Hence we have shown that

$$\varepsilon_1 \geq E_1 \tag{5F1.9}$$

If ϕ_1 happens to be identical to ψ_1, then all coefficients except c_1 are zero and we have $\varepsilon_1 = E_1$.

In analogy, for the first excited state one can use a trial function ϕ_2 which is orthogonal to ψ_1. This means that $c_1 = 0$ and it can be shown in the same way that

$$\varepsilon_2 \geq E_2 \tag{5F1.10}$$

In practical cases ψ_1 is unknown, and one chooses ϕ_2 to be orthogonal to ϕ_1. Even in this case it can be shown that $\varepsilon_2 \geq E_2$, provided the trial function involves a sufficiently large number of variational parameters [14]. The same procedure can be applied for $\varepsilon_3, \varepsilon_4, \ldots$.

5.13.2 Foundation 5.2: First-Order Perturbation Theory

Consider the Schrödinger equation

$$\mathcal{H}\psi = E\psi \tag{5F2.1}$$

for a system that we wish to describe. For the application of perturbation theory, we write the Hamiltonian operator $\mathcal{H}$ as

$$\mathcal{H} = \mathcal{H}_0 + V \tag{5F2.2}$$

where $\mathcal{H}_0$ is the Hamiltonian for a Schrödinger equation

$$\mathcal{H}_0\psi^{(0)} = E^{(0)}\psi^{(0)} \tag{5F2.3}$$

that can be solved rigorously and V is a potential energy term that makes the system of interest different from that of the system for which we know the solution(s). Hence the wavefunctions ψ and energies E will be different from those of the zeroth-order Schrödinger equation, Equation (5F2.3).

We ask for an approximate solution of the Schrödinger equation (5F2.1). Following the considerations in Section 5.2 we can write the energy E as

$$E = \int \psi\mathcal{H}\psi \cdot d\tau = \int \psi\left(\mathcal{H}_0 + V\right)\psi \cdot d\tau = \int \psi\mathcal{H}_0\psi \cdot d\tau + \int \psi V\psi \cdot d\tau \tag{5F2.4}$$

In this expression, the exact wavefunction ψ is unknown, and therefore we cannot calculate the exact energy E. However, in a first attempt we can use the known wavefunction $\psi^{(0)}$ to calculate an approximate energy ε.

$$\boxed{\varepsilon = \int \psi^{(0)}\mathcal{H}_0\psi^{(0)} \cdot d\tau + \int \psi^{(0)}V\psi^{(0)} \cdot d\tau = E^{(0)} + \int \psi^{(0)}V\psi^{(0)} \cdot d\tau} \tag{5F2.5}$$

This approximation is called *first-order perturbation theory*(for a more rigorous development see Foundation 5.3).

How can we understand the physical meaning of the correction term? For simplicity, assume that the wavefunction ψ depends on the coordinates of one electron only. Then $\rho\, d\tau = \left|\psi^{(0)}\right|^2 d\tau$ is the probability of finding the electron in volume element $d\tau$. Thus $V \cdot \rho\, d\tau = V \cdot \left|\psi^{(0)}\right|^2 d\tau$ is the contribution of the electronic interaction energy in this volume element. To obtain the total potential energy change we have to integrate over space.

$$\text{change of potential energy} = \int_{\text{space}} V \cdot \rho d\tau = \int_{\text{space}} V \cdot \left|\psi^{(0)}\right|^2 d\tau \tag{5F2.6}$$

5.13.2.1 Example: Electron in a Potential Energy Well

As an example, we consider an electron enclosed in a one-dimensional box of length L (potential energy zero, Fig. 5F.1a).

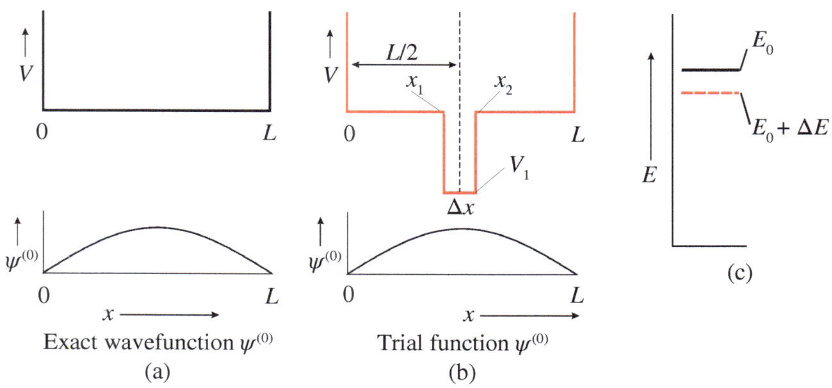

Figure 5F.1 Perturbation theory. (a) Particle in a box: potential energy $V_0 = 0$ and wavefunction $\psi_a = A\sin(\pi x/L)$. (b) Particle in a box with a potential energy well of depth V_1 and width Δx at the center of the box. (c) Energy E_a (case a) and estimated energy for case b, $E_a + \Delta E$, using ψ_a as a trial function.

Following the considerations in Chapter 3, the Hamiltonian operator is

$$\mathcal{H} = -\frac{h^2}{8\pi^2 m_e}\nabla^2 + V = \mathcal{H}_0 + V \tag{5F2.7}$$

Case (a) in Fig. 5F.1 is described by the Schrödinger equation

$$\mathcal{H}_0\psi^{(0)} = E^{(0)} \cdot \psi^{(0)}$$

with the solution

$$\psi^{(0)} = \sqrt{\frac{2}{L}} \sin\left(\frac{\pi x}{L}\right), \qquad E^{(0)} = \frac{h^2}{8mL^2} \tag{5F2.8}$$

for the lowest quantum state.

Next we introduce a potential well of depth V_1 and width Δx at the center of the box (Fig. 5F.1b). Using Equation (5F2.5) we approximate the energy as

$$\varepsilon = E^{(0)} + \int \psi^{(0)} V \psi^{(0)} \cdot \mathrm{d}x = E^{(0)} + \int V \cdot \left|\psi^{(0)}\right|^2 \mathrm{d}x = E^{(0)} + \int V\rho \cdot \mathrm{d}x \tag{5F2.9}$$

where $\rho \cdot \mathrm{d}x$ is probability of finding the electron between x and $x + \mathrm{d}x$. In the case of Fig. 5F.1b the potential energy V equals V_1 between $x = x_1$ and $x = x_2$; otherwise, V is zero. Then the approximate change in the electron's energy is

$$\Delta E = \int_{x_1}^{x_2} V_1 \rho \cdot \mathrm{d}x = \int_{x_1}^{x_2} V_1 \cdot \left|\psi^{(0)}\right|^2 \mathrm{d}x = \int_{x_1}^{x_2} V_1 \, \frac{2}{L}\sin^2\left(\frac{\pi x}{L}\right) \mathrm{d}x \tag{5F2.10}$$

In our case V_1 is a constant which can be taken out of the integral:

$$\Delta E = V_1 \frac{2}{L} \int_{x_1}^{x_2} \sin^2\left(\frac{\pi x}{L}\right) \cdot \mathrm{d}x \tag{5F2.11}$$

Special case 1: Δx in Fig. 5F.1b is small

To evaluate the integral we realize that the wavefunction is nearly constant at the center of the potential well, if $\Delta x = x_2 - x_1$ is sufficiently small. Then with $\sin\left(\frac{\pi x}{L}\right) \simeq 1$ we obtain

$$\Delta E = V_1 \frac{2}{L} \int_{x_1}^{x_2} \mathrm{d}x = V_1 \frac{2}{L}\left(x_2 - x_1\right) = V_1 \frac{2}{L}\Delta x \tag{5F2.12}$$

For example, for $\Delta x = L/8$ we find

$$\Delta E = V_1 \frac{2}{L}\frac{L}{8} = \frac{1}{4}V_1, \qquad \varepsilon = \frac{h^2}{8mL^2} + \frac{1}{4}V_1 \tag{5F2.13}$$

(dashed line in Fig. 5F.1c, note that V_1 is negative).

Special case 2: $\Delta x = L$

In this case the energy ε is, according to Equation (5F2.10),

$$\Delta E = V_1 \int_0^L \rho \cdot \mathrm{d}x = V_1 \tag{5F2.14}$$

where we used the fact that the wavefunction ψ is normalized.

General case:

For cases, where Δx in Fig. 5F.1b is not sufficiently small, we have to evaluate the integral in Equation (5F2.10) explicitly. From Appendix B, we know that $\int \sin^2 z \cdot \mathrm{d}z = \frac{1}{2}\left(z - \frac{1}{2}\sin(2z)\right)$. Hence we define $z = \pi x/L$ and write

$$\begin{aligned}
I &= \int_{x_1}^{x_2} \sin^2\left(\frac{\pi x}{L}\right) \cdot \mathrm{d}x = \frac{L}{\pi}\int_{z_1}^{z_2} \sin^2 z \cdot \mathrm{d}z \\
&= \frac{L}{\pi}\left|\frac{1}{2}\left(z - \frac{1}{2}\sin(2z)\right)\right|_{z_1}^{z_2} \\
&= \frac{L}{2\pi}\left(z_2 - z_1 - \frac{1}{4}\sin\left(2z_2\right) + \frac{1}{2}\sin\left(2z_1\right)\right) \\
&= \frac{1}{2}\left(x_2 - x_1\right) - \frac{L}{4\pi}\left(\sin\left(\frac{2\pi x_2}{L}\right) - \sin\left(\frac{2\pi x_1}{L}\right)\right)
\end{aligned}$$

According to Appendix B we know that

$$\sin\alpha - \sin\beta = 2\cos\left(\frac{\alpha+\beta}{2}\right)\cdot\sin\left(\frac{\alpha-\beta}{2}\right)$$

Thus we obtain

$$\begin{aligned}\sin\left(\frac{2\pi x_2}{L}\right) - \sin\left(\frac{2\pi x_1}{L}\right) &= 2\cos\left(\frac{\pi(x_2+x_1)}{L}\right)\cdot\sin\left(\frac{\pi(x_2-x_1)}{L}\right)\\ &= 2\cos\pi\cdot\sin\left(\frac{\pi(x_2-x_1)}{L}\right) = -2\sin\left(\frac{\pi(x_2-x_1)}{L}\right)\\ &= -2\sin\left(\frac{\pi\Delta x}{L}\right)\end{aligned}$$

where we expressed x_2 by x_1, according to Fig. 5F.1:

$$x_1 = L - x_2 \quad \text{or} \quad x_2 = L - x_1$$

Thus we find

$$\begin{aligned}I &= \frac{1}{2}\Delta x - \frac{L}{4\pi}\left(\sin\left(\frac{2\pi x_2}{L}\right) - \sin\left(\frac{2\pi x_1}{L}\right)\right)\\ &= \frac{1}{2}\Delta x + \frac{L}{2\pi}\sin\left(\frac{\pi\Delta x}{L}\right)\end{aligned}$$

and

$$\Delta E = V_1\frac{1}{L}\Delta x + V_1\frac{1}{\pi}\sin\left(\frac{\pi\Delta x}{L}\right)$$

For the special case 1 ($\pi\Delta x/L \ll 1$), $\sin\left(\frac{\pi\Delta x}{L}\right) \simeq \pi\Delta x/L$ so that

$$\Delta E = \frac{V_1\Delta x}{L} + \frac{V_1}{\pi}\frac{\pi\Delta x}{L} = 2\frac{V_1\Delta x}{L} \tag{5F2.15}$$

as we found in Equation (5F2.12). Also, for the special case 2 ($\Delta x = L$) we obtain

$$\Delta E = V_1\frac{1}{L}L + V_1\frac{1}{\pi}\sin\left(\frac{\pi L}{L}\right) = V_1$$

as we found in Equation (5F2.14).

Comparison with exact solution

For case 1 (Δx in Fig. 5F.1b is small) we compare the result of the perturbation method according to Equation (5F2.13) with the result of a numerical solution of the Schrödinger equation (exact solution)

$$\mathcal{H}\psi = (\mathcal{H}_0 + V)\psi = E\psi \tag{5F2.16}$$

with $\Delta x = L/8$ ($L = 1000$ pm) and different values of $V = V_1$.

	Perturbation		Exact	
V_1 10^{-19} J	ΔE 10^{-19} J	ε 10^{-19} J	ΔE 10^{-19} J	E 10^{-19} J
0.0	0.0	0.603	0.0	0.603
−0.5	−0.125	0.478	−0.128	0.475
−1.0	−0.250	0.353	−0.263	0.340
−2.0	−0.500	0.103	−0.558	0.045

Figure 5F.2 compares the exact wavefunctions.

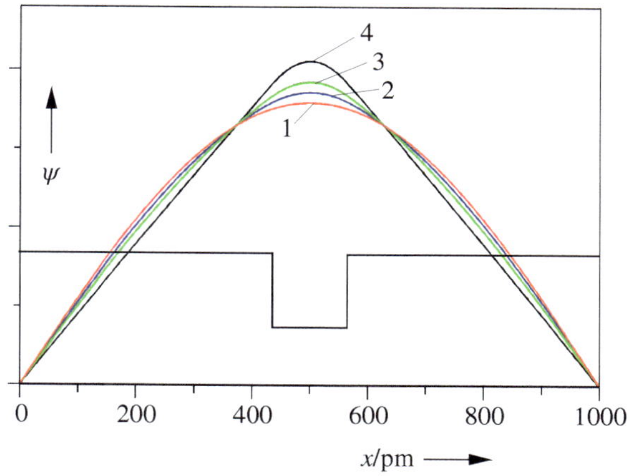

Figure 5F.2 Exact wavefunctions for the cases (1) $V_1 = 0$, (2) $V_1 = -0.5 \times 10^{-19}$ J, (3) $V_1 = -1.0 \times 10^{-19}$ J, and (4) $V_1 = -2.0 \times 10^{-19}$ J. The position of the potential well is indicated in the figure. The wavefunctions are obtained by numerical integration of the Schrödinger equation (5F2.16), see Appendix C.

5.13.3 Foundation 5.3: Perturbation Theory (Rigorous Treatment)

See the books website, www.wiley.com/go/Kuhn/physicalchemistry3e.

5.13.4 Foundation 5.4: The Hartree–Fock Self-Consistent Field Approximation

See the books website, www.wiley.com/go/Kuhn/physicalchemistry3e.

5.13.5 Foundation 5.5: Atomic Term Symbols

See the books website, www.wiley.com/go/Kuhn/physicalchemistry3e.

5.14 Exercises

E5.1 Consider the wavefunction ψ that is a product of two 1*s* electron orbitals, $\psi = \phi_{1s}(x_1, y_1, z_1) \cdot \phi_{1s}(x_2, y_2, z_2)$. Show that this wavefunction solves Equation (5.1) if you ignore the electron–electron repulsion term, $e^2/(4\pi\varepsilon_0 r_{12})$.

E5.2 Write the Hamiltonian for the lithium ion, Li^+ and for the lithium atom, Li. Identify the different terms in the Hamiltonian and discuss how the two Hamiltonians differ.

E5.3 The Schrödinger equation for He, given by Equation (5.1), does not include a kinetic energy operator for the He nucleus. Why is this reasonable? How can you account for the motion of the nucleus and the electrons?

E5.4 Apply the normalization condition to the trial wavefunctions in Equations (5.22) and (5.23) to find the constant *A*. You may use the fact that the one-electron wavefunctions are normalized already.

E5.5 Verify that the wavefunctions in Equations (5.31) and (5.32) satisfy the condition of indistinguishability.

E5.6 Show that the wavefunctions in Equations (5.29) and (5.30) do not satisfy the Pauli Exclusion principle.

E5.7 Calculate the ionization energy for Li^{2+} and compare it with the experimental value.

E5.8 Consider the 2*s* orbital of the atoms in the second row of the periodic table. How does the binding energy of the electron in this orbital change as one moves from Li to Ne? Contrast this behavior with that of the first ionization energies of the atoms.

E5.9 Consider the table of ionization potentials for the first 10 elements. Explain the trend in ionization potentials in the first column.

E5.10 Consider the table of ionization potentials for the first 10 elements. Explain the trend in ionization potential for the Be row.

E5.11 Compare the trend in atomic radii with the trend in ionization energies for the first 10 elements. Discuss why these trends might be correlated with each other.

E5.12 In Box 5.4 we computed an effective atomic number for Ca of 2.85. Use this effective atomic number to compute the ionization energy of a valence electron in the Ca atom. Compare your value with the literature value. Comment on your results.

E5.13 Consider the He^+ atomic ion. Show that the average distance of the electron from the nucleus is half the value found for the H atom; i.e., $\bar{r} = \frac{3}{4}a_0$ rather than $\bar{r} = \frac{3}{2}a_0$. Provide a physical explanation for this finding.

E5.14 In Table 5.1 two ionization energies are reported for He. Assume that the difference in ionization energies is caused solely by electron–electron repulsion and estimate the magnitude of the electron–electron repulsion energy. Use this repulsion energy to estimate the average distance between the two electrons. You need only perform simple calculations here and make rough estimates.

E5.15 Using a periodic table, identify the electron configurations of all the main group elements. Compare your identification to that you expect from analyzing Fig. 5.11.

E5.16 Excited sodium atoms gives rise to an intense yellow emission, called the sodium D-line, which results from an electronic transition from the $3p$ excited state configuration to the $3s$ ground-state configuration. What are the atomic term symbols for these configurations? The emission consists of two distinct lines at 589.0 and 589.6 nm. Use these emission lines to estimate the energy splitting that arises from spin–orbit interactions.

E5.17 In order to excite an electron in the $1s$ orbital of a hydrogen atom to the $3p$ orbital of the hydrogen atom, what wavelength of light is needed? What are the spectroscopic term symbols for the electron in each of these states?

E5.18 What is the ground electronic state configuration for the oxygen atom? What are the spectroscopic term symbols associated with this electronic configuration? Energy order these states using Hund's rules and state the rationale used to justify them.

E5.19 Write the ground electronic state's configuration for fluorine, F. Write the excited electronic state configuration which results from promoting the most weakly bound electron in the ground state to the next higher energy orbital in the atom. Determine the spectroscopic term symbols for your ground-state electron configuration and for your excited state electron configuration. If it is required, energy order the states that occur in each of these configurations.

E5.20 Write the ground electronic state's configuration for Mg. Write the excited electronic state configuration which results from promoting the most weakly bound electron in the ground state to the next higher energy orbital in the atom. Determine the spectroscopic term symbols for your ground-state electron configuration and for your excited state electron configuration. If it is required, energy order the states that occur in each of these configurations.

E5.21 In Section 5.6.2 the atomic term symbols for the excited states of He are discussed. Draw a vector model (analogous to Fig. 4.11) which shows the different orbital and spin angular momenta that are possible for the $1s2s$ configuration and for the $1s2p$ configuration.

5.15 Problems

P5.1 Given that the wavefunction of the lowest quantum state of the hydrogen atom is

$$\psi_1 = \frac{1}{\sqrt{\pi a_0^3}} \mathrm{e}^{-\frac{r}{a_0}}$$

use Equation (5.4) to calculate the hydrogen atom's ground-state energy. To evaluate the average potential energy $\overline{V_1}$ follow the example in Section 4.2.5. In order to find the average kinetic energy, you should use the fact that

$$\left(\frac{\partial^2}{\partial x^2} + \frac{\partial^2}{\partial y^2} + \frac{\partial^2}{\partial z^2}\right) \mathrm{e}^{-r/a_0} = -\frac{1}{a_0}\left(\frac{2}{r} - \frac{1}{a_0}\right) \cdot \mathrm{e}^{-r/a_0}$$

which is discussed in Section 4.2.2 and proven in Exercise 4.16.

P5.2 How does the electronic energy of a one-electron atom depend on the atomic number? Perform the calculation in Problem 5.1 for the case where the nucleus has a charge of Ze, so that

$$V(r) = \frac{-Ze^2}{4\pi\varepsilon_0 r}$$

and the ground-state wavefunction is given by

$$\psi(r) = \sqrt{\frac{Z^3}{\pi a_0^3}} \mathrm{e}^{-\frac{Zr}{a_0}}$$

Evaluate the energy for the case of $Z = 2$, He^+.

P5.3 Here you will use the variational method to calculate the ground-state electronic energy for the H-atom with two different trial functions.

Part 1) Use a normalized Gaussian trial function

$$\phi_{1G} = \left(\frac{2C_{11}}{\pi}\right)^{3/4} \exp(-C_{11}r^2) \quad \text{with} \quad C_{11} > 0$$

where C_{11} is an adjustable constant and

Given that the variational energy for the single Gaussian case is

$$E_{1G} = \int \phi_{1G} \mathcal{H} \phi_{1G} \mathrm{d}\tau = \frac{3h^2}{8\pi^2 m_e} C_{11} - \frac{e^2}{4\pi\varepsilon_0}\sqrt{\frac{8C_{11}}{\pi}}$$

find the value of C_{11} that minimizes the variational energy. Calculate the percent error in this energy, as compared to that for the exact solution. Make a plot of the optimized Gaussian wavefunction and the exact wavefunction on the same graph. Comment on your findings.

Part 2) Use a sum of two Gaussians

$$\phi_{2G} = \frac{\left(\frac{2C_{21}}{\pi}\right)^{3/4}\exp(-C_{21}r^2) + w\left(\frac{2C_{22}}{\pi}\right)^{3/4}\exp(-C_{22}r^2)}{\sqrt{1+w^2+2\cdot w\cdot S}} \quad \text{with } C_{21}, C_{22} > 0$$

where C_{21} and C_{22} are adjustable constants, the parameter w adjusts the relative statistical weight of the Gaussian functions, and the quantity S accounts for the overlap of the two Gaussians

$$S = \sqrt{8}\cdot\frac{\left(C_{21}C_{22}\right)^{3/4}}{\left(C_{21}+C_{22}\right)^{3/2}}$$

Use the MathCAD program "Problem5_3" to calculate the variational energy for different values of C_{21} and C_{22}, with $w = 1$ and $w = 0.2$, and report the results for three cases in which the variational energy is equal to or lower than that you found in part 1) for a single Gaussian. As in part 1) calculate the percent error in these energies as compared to that for the exact solution. Make plots of the two Gaussian wavefunctions and compare them to the exact solution. Comment on your findings and discuss how the wavefunction properties might affect the energy values you obtained.

P5.4 a) Insert the wavefunction given in Equation (5.23) into the Schrödinger equation for the helium atom and show that the energy is given by

$$E = E_{1s} + E_{2s} + \frac{e^2}{4\pi\varepsilon_0}\iint\frac{\varphi_{1s}^2\varphi_{2s}^2}{r_{12}}d\tau_1 d\tau_2$$

Provide a physical interpretation for each of the terms in the energy equation. State what is wrong with this choice of a wavefunction.

b) Perform the same analysis for the wavefunction in Equation (5.24) with $A = 1/\sqrt{2}$. You should find that

$$E = E_{1s} + E_{2s} + \frac{e^2}{4\pi\varepsilon_0}\iint\frac{\varphi_{1s}^2\varphi_{2s}^2}{r_{12}}d\tau_1 d\tau_2 + \frac{e^2}{4\pi\varepsilon_0}\iint\frac{\varphi_{1s}(1)\varphi_{2s}(2)\varphi_{1s}(2)\varphi_{2s}(1)}{r_{12}}d\tau_1 d\tau_2$$

Contrast your results in parts a) and b).

P5.5 Consider Equation (5.16) for the energy of a helium atom. Discuss how this energy expression would change if the nucleus had a charge of $Z = 1$, rather than $Z = 2$. Assuming that the electron–electron repulsion term is the same, would you expect the hydride ion H^- to be a stable ion? The hydride ion is known to have an ionization energy of 0.76 eV. Comment on this result, as compared to your rough estimations.

P5.6 Consider the Na atom, which has a single valence electron.

a) Use the procedure in Box 5.4 to find the effective charge Z_{eff}.

b) Compare this charge to the effective charge needed for the Na atom to reproduce its measured ionization energy.

Atom	**Ionization Energy**
H	13.60 eV
Na	5.39 eV

c) Use your effective charges to determine the root mean square distance $\sqrt{r^2}$ of the electron away from the nucleus. Comment on your results

P5.7 In this problem you will compare the Slater orbitals to the exact hydrogen atom orbitals. a) Plot the radial dependence of the Slater orbitals and hydrogen atom orbitals for the 2s orbitals on the same graph. Comment on how they compare. b) Perform the same procedure for the 1s orbitals. c) Show that the exact wavefunctions for the 1s and 2s orbitals are orthogonal to each other, but that the Slater orbitals are not orthogonal to each other.

P5.8 Consider the Li atom, which has a single valence electron. In the chapter we approximated this valence electron as being hydrogenic and found $Z_{eff} = 1.26$. Using this value for Z_{eff}, find the values of r at which the radial distribution function of a 2s orbital has its maxima.

P5.9 Consider the atomic and ionic radii given in the table. Use the electron shell structure of atoms to explain the trends in the radii. Look up the ionization energy for each of these species and plot it versus the radius. Identify any correlations between these data. Provide a physical explanation for the correlations you observe.

He, 128 pm	Li^+, 78 pm	Be^{2+}, 34 pm	B^{3+}, 23 pm
Ne, 160 pm	Na^+, 98 pm	Mg^{2+}, 78 pm	Al^{3+}, 57 pm
Ar, 174 pm	K^+, 133 pm	Ca^{2+}, 106 pm	Ga^{3+}, 62 pm
Kr, 198 pm	Rb^+, 149 pm	Sr^{2+}, 127 pm	In^{3+}, 92 pm
Xe, 218 pm	Cs^+, 165 pm	Ba^{2+}, 143 pm	Th^{3+}, 101 pm

P5.10 In Fig. 5F.1 we considered an electron in a potential $V = 0$ (electron in a box with box length L) and estimated its change in energy for a change in the potential well of V_1 over the distance Δx. Use the same method to calculate the change in the electron's energy for the following cases: a) V increases by V_1 over the range Δx in the box center, and b) V changes as $V_0 \sin(3\pi x/L)$.

P5.11 In this problem you use the perturbation theory method illustrated in Box 5.1 and Justification 5.1 to calculate the ground-state energy of the helium atom, using the trial wavefunction.

$$\phi(r_1, r_2) = \psi_{He^+}(r_1) \cdot \psi_{He^+}(r_2) = \sqrt{\frac{C^3}{\pi a_0^3}} e^{-Cr_1/a_0} \cdot \sqrt{\frac{C^3}{\pi a_0^3}} e^{-Cr_2/a_0}$$

Proceed in three steps. In the first step, follow the procedure of Problem 5.1 for the case where the nucleus has a charge of $2e$ and find the energy for an electron in each of the orbitals ψ_{He^+} while neglecting the electron–electron repulsion; i.e., solve the Schrödinger equation

$$-\frac{h^2}{8\pi^2 m_e}\nabla^2\psi - \frac{1}{4\pi\varepsilon_0} \cdot \frac{2e \cdot e}{r}\psi = E\psi$$

and show that

$$E(He^+) = -\frac{e^2}{4\pi\varepsilon_0} \frac{[4C - C^2]}{2a_0}$$

In the second step, use the results from Justification 5.1 to show that

$$\overline{V}_{12} = \frac{e^2}{4\pi\varepsilon_0 a_0} \cdot C \cdot \frac{5}{8}$$

You can begin with the result that

$$\overline{V}_{12} = \left(\frac{C^3}{\pi a_0^3}\right)^2 \cdot \frac{e^2}{4\pi\varepsilon_0} \cdot 8\pi^2 \cdot I$$

Rather than evaluate the integral

$$I = 2\int_{r_1=0}^{\infty} r_1^2 e^{-2Cr_1/a_0} \cdot \left[\frac{1}{r_1}\int_{r_2=0}^{r_1} r_2^2 e^{-2Cr_2/a_0}\, dr_2 + \int_{r_2=r_1}^{\infty} r_2 e^{-2Cr_2/a_0}\, dr_2\right] \cdot dr_1$$

explicitly, use the fact that the form of I that was found in Justification 5.1 corresponds to the case of $C = 2$. Thus, replacing a_0 in the original wavefunction by C/a_0 must result in a value of I which is inversely proportional to C^5 and you can write an expression for I directly.

In part three, you should combine the two results from parts 1 and 2 and minimize the variational energy by minimizing the parameter C. You should find that

$$E = -\frac{e^2}{4\pi\varepsilon_0 a_0}\left[\frac{(27)^2}{(16)^2}\right] = -124.2 \times 10^{-19}\ \mathrm{J}$$

Compare this result to the calculation in the text for case of $C = 2$, and to the best Hartree–Fock value of -124.8×10^{-19} J.

P5.12 Using the wavefunctions

$$\phi_2 = A \cdot [\varphi_{1s}(1) \cdot \varphi_{2s}(2) \pm \varphi_{1s}(2) \cdot \varphi_{2s}(1)]$$

calculate the energies for the symmetrical and antisymmetrical excited states of He.

References

1 C. Eckart, *Phys. Rev.*, The theory and calculation of screening constants, 1930, **36**, 878–892.

2 R.L. Snow and J.L. Bills, A simple illustration of the SCF-LCAO-MO method, *J. Chem. Educ.*, 1975, **52**, 506.

3 R.L. Snow and J.L. Bills, The Pauli principle and electronic repulsion in helium, *J. Chem. Educ.*, 1974, **51**, 585–586; D.A. Kohl, On the difference in energy between singlet and triplet states, *J. Chem. Phys.*, 1972, **56**, 4236–4238.

4 S.B. Crampton, D. Kleppner, and N.F. Ramsey, Hyperfine separation of ground-state atomic hydrogen, *Phys. Rev. Lett.*, 1963, **11**, 338–340.

5 D. R. Lide, ed *Handbook of Chemistry and Physics* (CRC Press, 1992).

6 R.K. Nesbet, Unrestricted Hartree–Fock then and now, *Int. J. Quant. Chem.*, 2005, **102**, 602–605; C.F. Fischer, *The Hartree-Fock Method for Atoms* (Wiley, New York, 1976).

7 R. Latter, Atomic energy levels for the Thomas-Fermi and Thomas-Fermi-Dirac potential, *Phys. Rev.*, 1955, **99**, 510. The values are taken from Table II.

8 A.C. Wahl, *Chemistry by Computer* (Scientific American, 1970, 54).

9 E.U. Condon and G.H. Shortley, *The Theory of Atomic Spectra* (Cambridge Press, 1979).

10 F.W. Byron, Jr. and C.J. Joachain, Correlation effects in atoms. I. Helium, *Phys. Rev.*, 1966, **146**, 1–8.

11 See S. Pais, *Niels Bohr's Times in Physics, Philosophy and Polity* (Clarendon Press, Oxford, 1991).

12 E. Clementi and D.I. Raimondi, Atomic screening constants from SCF functions, *J. Chem. Phys.*, 1963, **38**, 2686.

13 G. Arfken, *Mathematical Methods for Physicists* (Academic Press, New York, 1968).

14 J.K.L. MacDonald, Successive approximations by the Rayleigh-Ritz variation method, *Phys. Rev.*, 1933, **43**, 830.

6

A Quantitative View of Chemical Bonding

CHAPTER MENU

6.1 Introduction

In this chapter a quantitative quantum mechanical treatment of chemical bonding is developed for the H_2^+ molecular ion and the H_2 molecule. The quantitative rigor possible only for these simplest cases of chemical bonding is provided to help you appreciate the approximations required in the qualitative description of the more complex cases (see Chapters 7 and 8). At the same time you should gain some insight into the results of modern quantum chemical treatments of bonding.

The ground-state properties of H_2^+ are obtained by solving the Schrödinger equation exactly and also by using simple trial functions (particle-in-a-box wavefunction and linear combination of hydrogen atom wavefunctions). We show that the electron's kinetic energy increases when decreasing the distance between the protons, but this increase is counteracted by the decreasing potential energy, so that the net system is stable. Also we compare the energies and wavefunction properties for the approximate solutions to those of the exact solution. Comparison of these results allows us to assess the accuracy of the approximations.

The H_2 molecule shows the next level of complication, mutual repulsion of the electrons, and therefore the Schrödinger equation cannot be solved exactly. As with the He atom, the description of H_2 requires that we consider the indistinguishability of electrons and the Pauli Exclusion Principle. As a consequence we find that the electrons occupy the same spatial orbital as an electron pair (one spin up and the other spin down), in agreement with elementary ideas of electron pair bonds.

Our modern view of bonding has fundamental connections to our view of chemical reactions. The last section of the chapter discusses the relationship between bonding in H_2^+ at very long distances and electron transfer, redox reactions. In particular, we discuss the character of an electron in a chemical bond and its tunneling between an electron donor (reductant) and an electron acceptor (oxidant). To summarize, this chapter illustrates the rigorous approach to chemical bonding for the two simplest cases and provides a description of some important considerations we must employ to describe chemical bonding.

Principles of Physical Chemistry, Third Edition. Hans Kuhn, David H. Waldeck, and Horst-Dieter Försterling.

Companion website: www.wiley.com/go/Kuhn/physicalchemistry3e

6.2 H_2^+ Molecule Ion

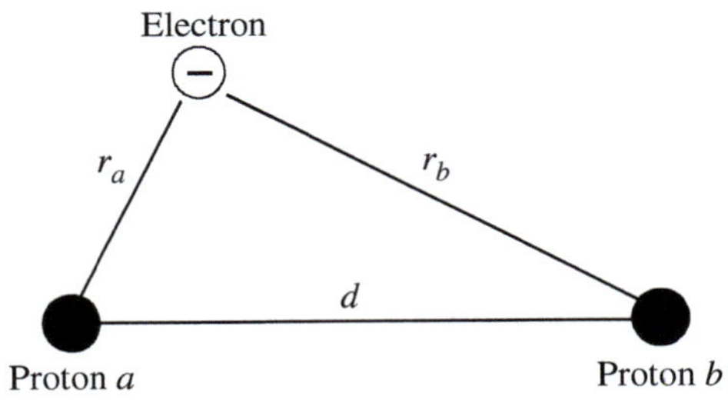

Figure 6.1 The H_2^+ molecular ion consists of a negatively charged electron in the field of two positively charged protons, which are separated by a distance d. r_a and r_b are the distances between the electron and each positive charge.

For our analysis, we consider H_2^+ to be made from an H atom and a proton (Fig. 6.1):

$$\text{H} + \text{proton} \rightarrow \text{H}_2^+$$

From the physical perspective, we can see that four kinds of energy terms will contribute to the total energy of this system. Two of these terms are potential energy terms. First is the electrostatic attraction between the electron and each of the protons a and b, given by

$$V_a = \frac{-e^2}{4\pi\varepsilon_0 r_a} \quad \text{and} \quad V_b = \frac{-e^2}{4\pi\varepsilon_0 r_b}$$

In addition to these stabilizing interactions, the electrostatic repulsion between the two protons contributes to the energy and is given by

$$V_{\text{repulsion}} = \frac{e^2}{4\pi\varepsilon_0 d}$$

The other two terms, which contribute to the total energy, are the kinetic energy of the electron, given by

$$T_{\text{op,electron}} = \frac{-\hbar^2}{2m_e}\nabla^2,$$

and the kinetic energy of the protons (described in Chapter 11). It is important and useful to remember the mathematical form of these different terms, since they will reappear for any problem of bonding.

Because its mass is almost 2000 times larger than that of the electron ($m_p \gg m_e$), the more massive protons have a slower speed than the electrons, at a given kinetic energy. This insight led Born and Oppenheimer to propose an approximate solution for the Schrödinger equation. On a physical level their approximation assumes that the electrons will instantaneously follow any motion of the protons (*Born–Oppenheimer approximation*). This approximation allows us to ignore the kinetic energy of the protons and solve the Schrödinger equation for the electrons at a fixed distance d between the protons. By solving this "electronic" Schrödinger equation at different distances d, we can determine how the electronic energy depends on the internuclear distance. This approximation is strongly coupled with our idea that the binding of the electrons to the protons forms the potential energy well of the molecule; the electrons are the "glue" that holds the protons (more generally the atomic cores) together as a molecule. By solving the electronic Schrödinger equation to find the binding energy as a function of the internuclear position(s), we will then use that electronic energy profile as a potential energy surface upon which we describe the internuclear motions (Chapter 11). The Born–Oppenheimer approximation underlies a chemist's view of bonding and, as we shall see, is intimately linked to a chemist's ideas of orbitals and potential energy curves.

6.2.1 Electron Described by Exact Wavefunction

Using the Born–Oppenheimer approximation and combining the energy terms described above, we can write an expression for the Hamiltonian operator of the H_2^+ molecular ion, namely

$$\mathcal{H} = -\frac{\hbar^2}{2m_e}\nabla^2 - \frac{e^2}{4\pi\varepsilon_0}\left(\frac{1}{r_a} + \frac{1}{r_b}\right) + \frac{e^2}{4\pi\varepsilon_0}\frac{1}{d} \tag{6.1}$$

where d is the distance between the protons and r_a and r_b are the distances between the electron and the protons a and b. With this Hamiltonian, the Schrödinger equation for the H_2^+ ion is

$$-\frac{\hbar^2}{2m_e}\left(\frac{\partial^2}{\partial x^2} + \frac{\partial^2}{\partial y^2} + \frac{\partial^2}{\partial z^2}\right)\psi + \frac{e^2}{4\pi\varepsilon_0}\left(-\frac{1}{r_a} - \frac{1}{r_b} + \frac{1}{d}\right)\psi = E\psi \tag{6.2}$$

where the ∇^2 operator has been replaced by its Cartesian form. x is the coordinate in the direction of the bond axis, and the coordinates y and z are perpendicular to the bond axis. The Schrödinger equation that results from the Born–Oppenheimer approximation is called the *electronic Schrödinger equation* because the only variables are the electron coordinates. This equation can be solved exactly (Foundation. 6.1 provides the detailed solution), and the wavefunction ψ and the energy E are found to depend on the interproton distance d (Fig. 6.2).

6.2.1.1 Electronic Energy as a Function of Distance *d*

Figure 6.2 plots the electronic energy E as a function of the distance d or H_2^+. As the figure shows, the electronic energy at very large distances corresponds to the energy of a separated hydrogen atom and a proton. Now let us consider the protons fixed at a distance d and the electron in the lowest energy state. Then the energy E has changed because of the electrostatic attraction of the electron in the hydrogen atom to the positive charge of the proton and the electrostatic repulsion of the two protons. At very large distances, this interaction "stretches" the wavefunction along the bond axis and causes a small decrease in the electron's average kinetic energy $\overline{T}$. On the other hand, the average potential energy $\overline{V}$ increases slightly with decreasing d. Both effects together result in a net stabilization, even at 300 pm, as can be seen from the curve for the energy E. At shorter distances between 150 and 50 pm, the electrostatic attraction between the electron and the two protons dominates and causes the large drop in the average potential energy. At the same time, this electrostatic attraction contracts the electron's wavefunction perpendicular to the direction of the interproton axis (x-axis) and causes the kinetic energy to rise above the value for the separated atom and proton. As the interproton distance becomes <50 pm, the electrostatic repulsion between the protons begins to gain importance and eventually dominates the interaction, leading to no net stabilization. Overall we see that the bond energy is a balance between the decrease in the average potential energy and the increase in the average kinetic energy, which arises from contraction of the wavefunction by the attractive forces. These features of bonding are general.

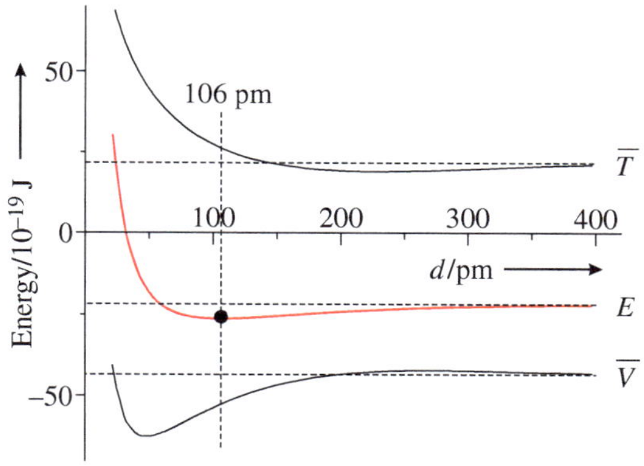

Figure 6.2 Energy E, average kinetic energy $\overline{T}$, and average potential energy $\overline{V}$ of the H_2^+ ion as a function of the interproton distance d. The dot denotes the energy minimum. The horizontal dashed lines correspond to the energies for an H atom and a proton, obtained as d increases to infinity (calculated from [1]).

6.2.1.2 Equilibrium Bond Length

We assign the equilibrium bond length d_{eq} to the interproton distance d at which the lowest energy is obtained in the energy well. The lowest total energy E is obtained for $d_{eq} = 106$ pm and is

$$E = -\frac{1}{4\pi\varepsilon_0} \cdot \frac{e^2}{2a_0} \times 1.203 = -26.28 \times 10^{-19}\ \text{J} \tag{6.3}$$

This energy is lower than the energy $E = -21.80 \times 10^{-19}$ J of the electron in an H atom, obtained in Chapter 4. For this reason, the reaction of an H atom with a proton releases energy, and the H_2^+ ion is a stable molecular ion. If we consider dissociating a H_2^+ molecular ion into a proton and an H atom, the electronic energy changes by $(26.28 - 21.80) \times 10^{-19}$ J = 4.48×10^{-19} J. This value is in complete agreement with experiment.[1]

The hydrogen molecular ion is the only molecular system for which the electronic Schrödinger equation can be solved exactly, and the description of chemical bonding in other molecules is an approximate one. This limitation makes the case of H_2^+ especially interesting because we can compare the exact solution to approximate descriptions that are based on the variational principle and gain insight into their limitations.

6.2.2 Electron Density and the Chemical Bond

6.2.2.1 Compression of Electron Cloud

The wavefunction of H_2^+ is shown in Fig. 6.3 in comparison to that of the H atom. The average electron density in the vicinity of the protons increases when we proceed from an H atom and a proton to H_2^+. The probability density of an electron in the H atom is $\rho(r_a = 0) = 1/\left(\pi a_0^3\right) = 0.32 a_0^{-3}$ and for a proton $\rho(r_b = 0) = 0$. On the average, the probability density is

$$\rho_{\text{average}} = \frac{1}{2}\left[\rho(r_a = 0) + \rho(r_b = 0)\right] = \frac{1}{2}(0.32 + 0)a_0^{-3} = 0.16 a_0^{-3} \tag{6.4}$$

From the exact calculation for the H_2^+ molecular ion, we find

$$\rho(r_a = 0) = \rho(r_b = 0) = 0.21 a_0^{-3} \tag{6.5}$$

1 As already noted in Chapter 2 the energy of an H_2^+ ion, as a result of the vibrational motion of the protons, lies about 0.2×10^{-19} J above the minimum of the electronic energy (see Section 11.3); the energy which is necessary to pull apart the bond must be smaller by this amount. Indeed, the experimental value for the bond energy is $\varepsilon_B = 4.2 \times 10^{-19}$ J.

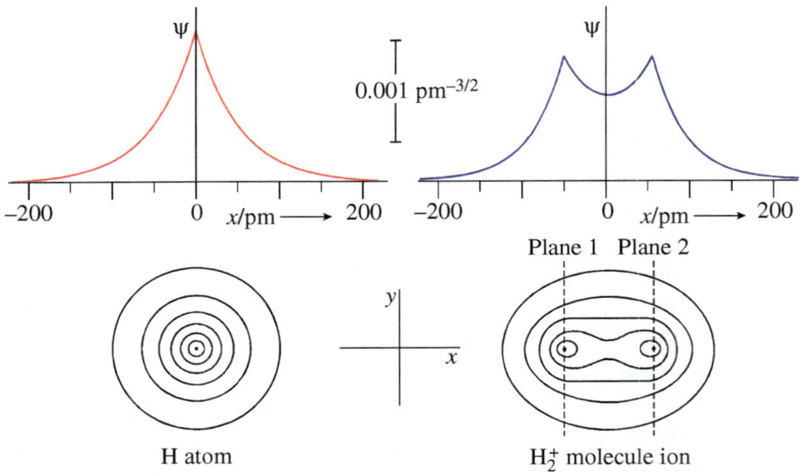

Figure 6.3 Graphs of the exact wavefunction, ψ, are shown for the H atom and H_2^+ ion along the x axis and as contour lines in the x, y plane ($z = 0$, x direction from left to right, y direction from bottom to top). The outermost contour line has the value $\psi = 2 \times 10^{-4}$ pm$^{-3/2}$, the next one has a value of 4×10^{-4}pm$^{-3/2}$, and so on. Planes 1 and 2 (yz planes) pass through protons a and b, respectively.

Thus the electron density at the protons in H_2^+ is about 30% larger than the average of the electron density at the proton of an H atom and that at an independent proton.

This increase of the electron density is caused by the compression of the electron cloud perpendicular to the interproton axis, i.e., the cylindrically symmetric electrostatic field of the two protons attracts the electron more strongly than a single proton does, because of their larger total positive charge. For this reason the electron density is increased along the interproton axis. Halfway between the protons the electron density is $\rho = 0.103\ a_0^{-3}$, that is 49% of the maximum value at the protons, and the electron is more likely found between the protons. Intuitively this makes sense, because the electron can be stabilized by attraction to both protons and can partially shield the proton–proton repulsion. The increased electron density between the protons acts to hold the molecular ion together by Coulomb forces and is a signature of chemical bonding.

6.2.2.2 Coulomb Attraction and Virial Theorem

In order to see more clearly how the two protons are held together by the electron cloud, we imagine the protons at a distance somewhat larger than the equilibrium separation. The two protons are pulled toward each other by the Coulomb attraction of the electron density between planes 1 and 2 (Fig. 6.3). This attractive force overcomes the forces acting in the opposite direction: (a) Coulomb repulsion of the protons and (b) Coulomb attraction of the protons toward the electron density outside the two planes 1 and 2. The protons approach each other, until these forces balance each other at the equilibrium distance. With a decrease of interproton distance a compression of the electron cloud occurs because of the increased field of the protons, and with it an increase in the kinetic energy of the electron. Hence the bond energy is smaller than one would expect from considering only the Coulomb energy. Lastly, we point out that the Coulomb attraction of the electron cloud and the protons accounts for the chemical bonding, the residual energy terms (proton–proton interaction, kinetic energy, and electron–electron interactions for molecules with more than one electron) are repulsive.

The balance between the average potential energy $\overline{V}$ and the average kinetic energy $\overline{T}$ is a fundamental feature of chemical bonding. For the equilibrium internuclear distance, $d_{eq} = 106$ pm, the average potential energy of the H_2^+ ion is $\overline{V} = -52.4 \times 10^{-19}$ J and the average kinetic energy is $\overline{T} = +26.2 \times 10^{-19}$ J, thus we find

$$\overline{V} = -2\overline{T} \quad \text{and} \quad \overline{E} = \overline{T} + \overline{V} = -\overline{T}$$

This relationship, which holds for the equilibrium distance, is called the Virial Theorem, see Foundation 4.3 and Section 4.2.5.

6.2.3 Electron Described by LCAO Wavefunction

In Chapter 2, we used particle-in-a-box trial functions to calculate the electronic energy of H_2^+; see Section 2.5. The variational principle discussed in Chapter 5 allows us to try out other functions in place of the box trial functions. A very common procedure for molecules is to imagine that the molecular wavefunctions are built from the wavefunctions of the constituent atoms. Approximating the electron system of a molecule by a linear combination of atomic orbitals (LCAO) is a very popular approach to describing chemical bonds. Therefore, using this approach for H_2^+ is of particular importance.

6.2.3.1 Procedure

To describe the ground state of the H_2^+ ion, we use two 1*s* hydrogen-atom wavefunctions

$$\varphi_a = \frac{1}{\sqrt{\pi a_0^3}} \cdot e^{-r_a/a_0} \quad \text{and} \quad \varphi_b = \frac{1}{\sqrt{\pi a_0^3}} \cdot e^{-r_b/a_0} \tag{6.6}$$

that are each centered on one of the protons. Our trial molecular wavefunction is

$$\phi = c \cdot (\varphi_a + \varphi_b) \tag{6.7}$$

where c is a normalization constant (see Fig. 6.4a). The method of proposing a trial function from a sum of atom functions is called the *LCAO method*. According to the variational principle we use this trial wavefunction to evaluate the energy integral

$$\varepsilon = \int \phi \mathcal{H} \phi \cdot d\tau \tag{6.8}$$

and we minimize its value with respect to any parameters. We see that this trial wavefunction does not have any parameters (c is determined by the normalization requirement), so our task is to evaluate Equation (6.8), using Equation (6.7) and the Hamiltonian, Equation (6.1).

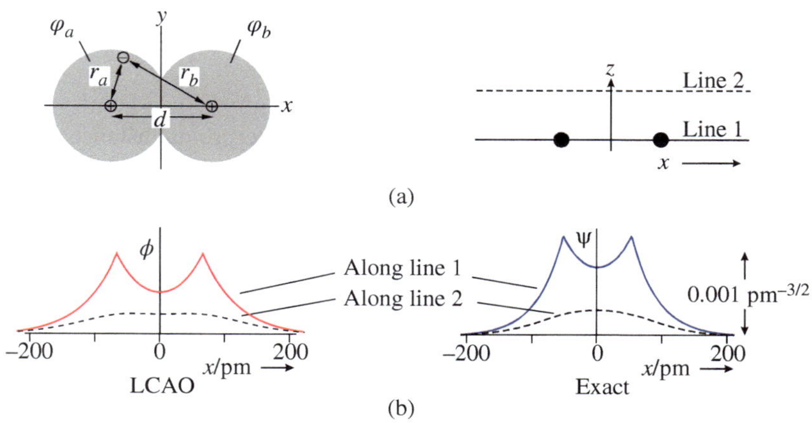

Figure 6.4 H_2^+ ion. (a) Atom functions φ_a and φ_b of two hydrogen atoms *a* and *b*, which are located at the equilibrium distance d_{eq} from each other. (b) Graph of the wavefunction along the internuclear axis (line 1) and along line 2 at a distance of 75 pm above the internuclear axis. Left curve: LCAO method (wavefunction ϕ; d_{eq} = 132 pm), right curve: exact (wavefunction ψ; d_{eq} = 106 pm).

First we evaluate the constant c by using the normalization condition

$$\int \phi^2 \cdot d\tau = c^2 \cdot \left(\int \varphi_a^2 \cdot d\tau + 2 \int \varphi_a \varphi_b \cdot d\tau + \int \varphi_b^2 \cdot d\tau \right) = 1 \tag{6.9}$$

Because of the normalization of the atomic functions we know that $\int \varphi_a^2 \cdot d\tau = \int \varphi_b^2 \cdot d\tau = 1$; and if we define the integral $\int \varphi_a \varphi_b \cdot d\tau$ by

$$S_{ab} = \int \varphi_a \varphi_b \cdot d\tau \tag{6.10}$$

we can solve Equation (6.9) for c to obtain

$$c = \frac{1}{\sqrt{2(1 + S_{ab})}} \tag{6.11}$$

(in Exercise 6.4 you are asked to show these steps in detail). The quantity S_{ab} is called the *overlap integral* and the product $\varphi_a \varphi_b$ is the *overlap density*. The overlap integral measures the change in electron density over that of the two independent atomic orbitals.

Next we insert the trial function (6.7) into the energy equation (6.8):

$$\varepsilon = \int \phi \mathcal{H} \phi \cdot d\tau = c^2 \int (\varphi_a + \varphi_b) \mathcal{H} (\varphi_a + \varphi_b) \cdot d\tau \tag{6.12}$$

$$= c^2 \left(\int \varphi_a \mathcal{H} \varphi_a \cdot d\tau + \int \varphi_a \mathcal{H} \varphi_b \cdot d\tau + \int \varphi_b \mathcal{H} \varphi_a \cdot d\tau + \int \varphi_b \mathcal{H} \varphi_b \cdot d\tau \right)$$

and find that it yields four energy integrals. Because the atomic functions φ_a and φ_b are identical we have

$$\int \varphi_a \mathcal{H} \varphi_b \cdot d\tau = \int \varphi_b \mathcal{H} \varphi_a \cdot d\tau$$

and

$$\int \varphi_a \mathcal{H} \varphi_a \cdot d\tau = \int \varphi_b \mathcal{H} \varphi_b \cdot d\tau$$

so that we need only evaluate two distinct integrals. We use a shorthand notation,

$$\int \varphi_a \mathcal{H} \varphi_a \cdot d\tau = \int \varphi_b \mathcal{H} \varphi_b \cdot d\tau = H_{aa}, \quad \int \varphi_a \mathcal{H} \varphi_b \cdot d\tau = \int \varphi_b \mathcal{H} \varphi_a \cdot d\tau = H_{ab} \tag{6.13}$$

The integrals H_{aa} are the average energy of the atomic orbitals, sometimes called the *self-energy* of the orbital, and the integral H_{ab} is the energy of the overlap density, sometimes called the *resonance integral*. From Equation (6.12) for the energy ε we obtain

$$\varepsilon = c^2 \left(2H_{aa} + 2H_{ab}\right)$$

and with Equation (6.11) for c, we find

$$\boxed{\varepsilon = \frac{H_{aa} + H_{ab}}{1 + S_{ab}}} \tag{6.14}$$

6.2.3.2 Results

The energy ε in Equation (6.14) depends on the interproton distance d, because H_{aa}, H_{ab}, and S_{ab} are functions of d. In Foundation 6.2 these terms are evaluated, and Fig. 6.5 shows $\overline{T}, \overline{V}$, and ε as a function of d. For the energy minimum we find

$$\overline{V} = -41.3 \times 10^{-19}\ \text{J}, \quad \overline{T} = 16.7 \times 10^{-19}\ \text{J}, \quad \varepsilon = -24.6 \times 10^{-19}\ \text{J}$$

and $d_{eq} = 132$ pm. Compared to the H atom, ε decreases by $(24.6 - 21.8) \times 10^{-19}\ \text{J} = 2.8 \times 10^{-19}\ \text{J} = 1.7$ eV. Using the box model wavefunction as a trial function (Section 2.4), we found $4.0 \times 10^{-19}\ \text{J} = 2.5$ eV. The wavefunction for the equilibrium distance $d = d_0$ is shown in Fig. 6.4b and compared to the exact solution of the Schrödinger equation.

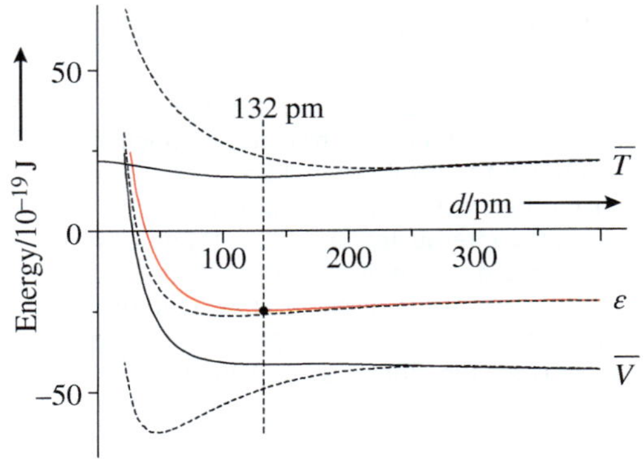

Figure 6.5 Energy ε, average kinetic energy $\overline{T}$, and average potential energy $\overline{V}$ of the H_2^+ ion are plotted as a function of internuclear distance d (solid lines: LCAO method; dashed lines: exact treatment, for comparison). The dot indicates the minimum of the energy ε for the LCAO model.

6.2.3.3 Failures

A closer inspection of this LCAO result reveals some problems. For example, the Virial Theorem is not fulfilled ($\varepsilon = -1.47 \times \overline{T}$ instead of $\varepsilon = -2.00 \times \overline{T}$). Thus, an important feature of chemical bonding, the balance of kinetic and potential energy, is not represented. Additionally, near the equilibrium distance the curve for $\overline{T}$ goes through a minimum; however, the exact result shows that $\overline{T}$ increases monotonically as d decreases. On the other hand, $\overline{V}$ increases with decreasing d, instead of passing through a minimum, as does the exact result. This treatment predicts a decrease of the electron's kinetic energy in forming the bond; in reality, however, the kinetic energy increases when proceeding from H to H_2^+, and the potential energy decreases. Lastly, the model predicts, erroneously, that charge is removed from the regions near the protons and brought into the overlap region. This tendency causes the electron density at the protons to decrease when forming the bond, although the exact solution shows that it increases because of the contraction of the electron cloud perpendicular to the bond axis. These failings result from the strict use of the 1s hydrogen atom wavefunction as the atomic orbitals and can be removed by allowing the atomic orbital wavefunctions to have more "flexibility."

6.2.3.4 Modifications

A common modification to the LCAO wavefunction is to add an adjustable parameter that changes with distance. With this modification, we can build an LCAO wavefunction that contracts perpendicular to the bond axis in shortening the bond, as does the exact solution. Consider the trial wavefunction

$$\phi = A\left[\exp\left(-\eta r_a/a_0\right) + \exp\left(-\eta r_b/a_0\right)\right] \tag{6.15}$$

where η is a variational parameter. Thus the spatial extent of the constituent atomic orbitals can change. In the simple LCAO treatment of H_2^+, η was equal to 1. The detailed solution for the energy ε is provided in Foundation 6.2 for the trial function (6.15), and ε is calculated for a given value of d and for different values of η.

At each fixed distance d we find the value η_{min} corresponding to the minimum of $\varepsilon(\eta)$. The curve for η_{min} as a function of distance d increases monotonically from the limiting case $\eta_{min} = 1$ at large distance d (for $d \to \infty$ the electron is in either one of two separate 1s orbitals of the H atoms) to the limiting case $\eta_{min} = 2$ at small distance d (for $d = 0$ the electron is in a 1s atomic orbital of the He^+-ion). Note that the curve $\eta_{min}(d)$ does not depend on the proton–proton repulsion term in $\varepsilon(d)$. The monotonic increase of $\eta_{min}(d)$ with decreasing distance d indicates that the electron cloud becomes more and more compressed toward the bond axis (Fig. 6.6a). We can use $\eta_{min}(d)$ to calculate the energies $\varepsilon(d)$, $\overline{T}(d)$, and $\overline{V}(d)$ for the H_2^+-ion (where, of course, the proton–proton repulsion term in $\varepsilon(d)$ is included). In this way we find the optimal distance d_{eq} of the two protons in the H_2^+-ion, corresponding to the minimum of $\varepsilon(d)$, see Fig. 6.6b. These energy curves are very close to the result of the exact calculation (Fig. 6.2), and the problems described above for the LCAO method are eliminated. In particular, the position of the energy minimum at $d_{eq} = 106$ pm is identical to the experimental value, and the kinetic energy increases with decreasing distance d. The LCAO method, with an insightful choice of atomic orbitals, provides a realistic description of bonding in H_2^+.

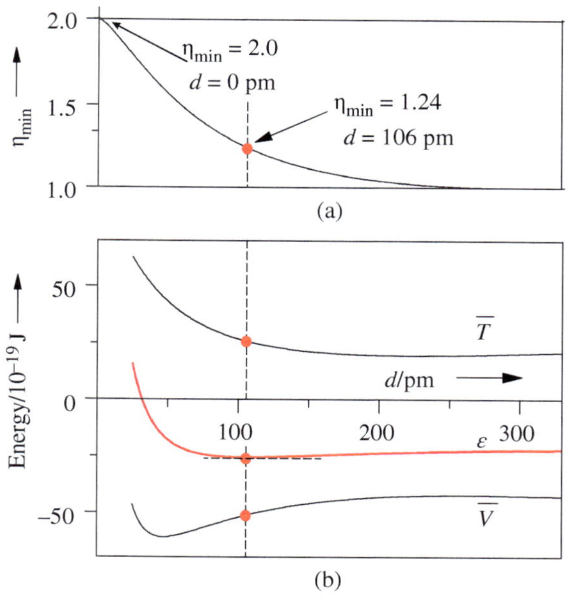

Figure 6.6 A variational LCAO solution for H_2^+ ion. (a) Parameter η_{min} versus distance d. (b) Energies, ε, $\overline{T}$, $\overline{V}$ and versus d. The circles denote the energy minimum at 106 pm.

6.2.4 Comparison of Box Model (Chapter 2) and LCAO Method

Insight into the appropriateness of the approximate solutions, LCAO method and box-wavefunctions, can be assessed by comparison to the exact solution.

In many ways the box wavefunctions (see Section 2.5) describe essential features of the bonding well. First, the negative charge between the protons, the glue that binds the protons by Coulomb forces, is well approximated, i.e., the density of the cloud is about the same as the actual density. Second the compression of the electron cloud when decreasing the distance between the protons because of the increased nuclear field and the corresponding decrease in potential energy is about the same as in reality, and third the balance of kinetic and potential energy is correctly described. So it is not surprising to find that the bond energy is well-approximated (it is almost 90% of the actual value). Other aspects of the electron wavefunctions of H_2^+ are badly represented by box-test functions. For example, the box wavefunction does not have peaks at the protons as do the actual wavefunctions. Moreover, the box wavefunction does not decay exponentially with increasing distance from the protons. Hence it is not surprising that the binding energy of the electron to the protons is much too small (only about 60% of the actual energy) and gives a bond length that is too large. In spite of the bad representation of the electron distribution by the box-model, it describes some essential features of the bonding correctly.

The LCAO method is widely applied and was of importance in the historical development of quantum chemistry. Application of this method with 1s orbitals gives a reasonable estimate of the total energy, but the bond energy is only about 60% of the actual value. This wavefunction generates electron density between the protons, which is needed for bonding, but it fails to contract the electron cloud perpendicular to the bond axis. Moreover it does not describe the balance between the kinetic and potential energy correctly (Fig. 6.7c), in contrast to the exact solution (Fig. 6.7a) and the box model (Fig. 6.7b) which are consistent with the Virial Theorem. If the 1s orbitals are modified by use of a variational parameter that adjusts the orbitals spatial extent, 85% of the true bond energy is reached (Fig. 6.7d). The qualitative power of the molecular orbital method and modern computational techniques has led to this method being widely used in chemistry.

The results of the exact calculation, of the box model, of the simple LCAO model, and of the modified LCAO model are summarized in Table 6.1.

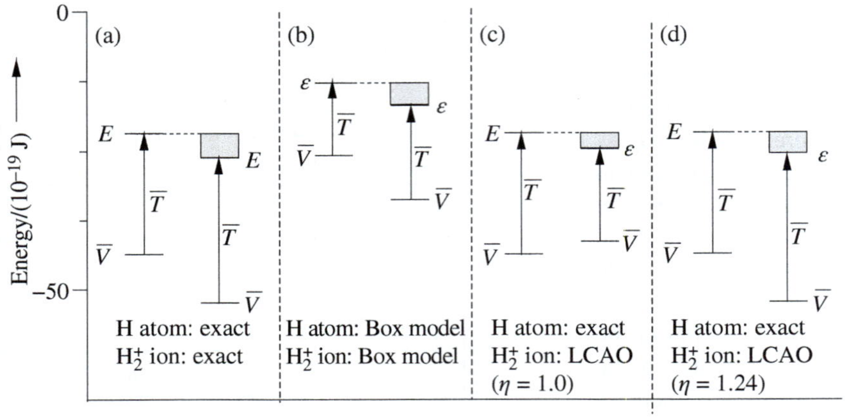

Figure 6.7 H atom (left) and H_2^+ ion (right); average potential energy $\overline{V}$, average kinetic energy $\overline{T}$, energy E, and ϵ. (a) Exact treatment (d_{eq} = 106 pm), (b) box model (d_{eq} = 152 pm, see Section 5.2), (c) H-atom: exact, H_2^+ ion: simple LCAO method (d_{eq} = 132 pm), (d) H-atom: exact, H_2^+ ion: LCAO method with $\eta = 1.24$ (d_{eq} = 106 pm). Note that in the LCAO model with $\eta = 1.0$ the Virial Theorem is not fulfilled. The energy is taken to be zero when the charges are separated (protons and electron removed infinitely far apart). In each case the bond energy is indicated by the height of the shaded area. Note that in the LCAO model the virial theorem is not fulfilled ($\overline{V}$ increases and $\overline{T}$ decreases in forming the bond).

Table 6.1 Comparison of Methods for Determining the Total Energy, the Electronic Contribution to the Bond Energy ε_{elec} and the Bond Length d for H_2^+ and H_2. Model A: Repulsion of the Two Electrons Neglected. Model AR: Repulsion Included, Evasion. Model ARE: Repulsion and Evasion Included. Kolos–Wolniewicz: Rigorous Treatment.

	Method	Energy / 10^{-19} J	ε_{elec} / 10^{-19} J	d_{eq} / pm	Parameters
H_2^+	Exact	−26.28	4.48	106	
	Box model	−16.9	4.0	152	L = 301 pm, b = 1.36
	LCAO	−24.6	2.8	132	$\eta = 1$
	LCAO, modified	−25.6	3.8	106	$\eta = 1.24$
	Experiment[2]	−26.28	4.4	106	
H_2	Kolos–Wolniewicz	−51.20	7.60	74.14	
	Model A	−80.8	(39.4)	52	
	Model AR	−48.4	4.8	74	
	Model ARE	−51.1	7.5	74	
	Experiment[3]	−51.20	7.6	74.16	

6.2.5 Excited State of H_2^+: Bonding and Antibonding Orbitals

Comparison of the first excited state of the H_2^+ molecular ion with its electronic ground state can also be made. Figure 6.8 plots the electronic energy versus the interproton distance, from the exact solution, for the ground and first excited state. The ground-state energy curve (energy E_1 versus bond distance d) has a well-defined minimum at d_{eq} = 106 pm (Fig. 6.8a, lower curve), and the corresponding wavefunction ψ_1 has a high electron density at the center between both nuclei (Fig. 6.8b, bottom). Therefore, the ground state is a bonding state, and its wavefunction is called a *bonding orbital*. In contrast, the energy curve of the first excited state of H_2^+ (energy E_2) does not show such a minimum (Fig. 6.8a, upper curve), and the corresponding wavefunction ψ_2 is zero at the center between both nuclei (Fig. 6.8b, top). Therefore, this

2 As already outlined in Chapter 2 the energy of an H_2^+ ion, as a result of the vibrational motion of the nuclei, lies about 0.2×10^{-19} J above the minimum of the electronic energy ε_{elec} (see Section 11.3); the energy which is necessary to pull apart the bond must be smaller by this amount. Indeed, the experimental value for the bond energy is $\varepsilon_B = 4.2 \times 10^{-19}$ J.

3 The energy of an H_2 molecule, as a result of the vibrational motion of the nuclei, lies about 0.4×10^{-19} J above the minimum of the electronic energy ε_{elec} (see Section 11.3); the energy which is necessary to pull apart the bond must be smaller by this amount. Indeed, the experimental value for the bond energy is $\varepsilon_B = 7.2 \times 10^{-19}$ J.

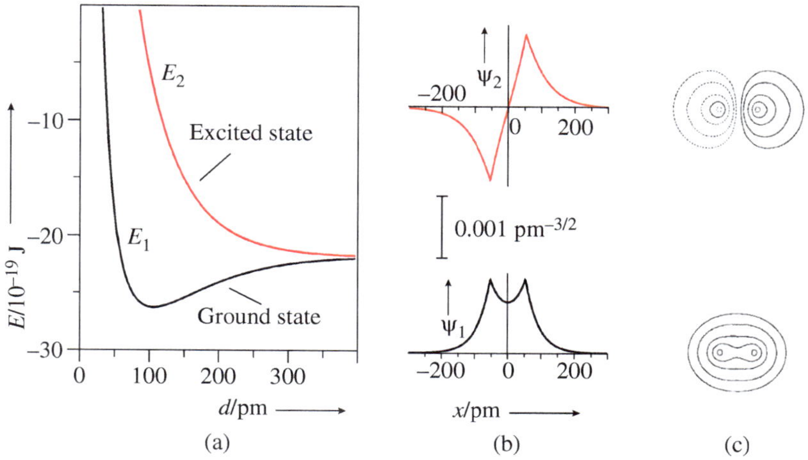

Figure 6.8 Comparison of ground state and excited state of the H_2^+ molecular ion. (a) Exact energies E_1 (ground state) and E_2 (excited state) versus interproton distance d. (b, c) Exact wavefunctions ψ_1 and ψ_2 along the x axis and as contour lines in the x, y plane ($z = 0$, x direction from left to right, y direction from bottom to top). The outermost contour line has the value $\psi = 2 \times 10^{-4} \text{pm}^{-3/2}$, the next one has a value of 4×10^{-4} pm$^{-3/2}$, and so on.

excited state does not contribute to bonding, and its wavefunction is called an *antibonding orbital*. Note that the bonding orbital has an antinode at the center of the bond, and the antibonding orbital has a node at the center of the bond.

The modified LCAO method makes a similar prediction. In Section 6.2.3 we used the ground-state trial wavefunction ϕ_1,

$$\phi_1 = c_1 \cdot (\varphi_a + \varphi_b) \quad \text{with} \quad c_1 = \frac{1}{\sqrt{2(1 + S_{ab})}} \tag{6.16}$$

and found the energy expression ε_1 is, according to Equation (6.14),

$$\varepsilon_1 = \frac{H_{aa} + H_{ab}}{1 + S_{ab}} \tag{6.17}$$

For describing the excited state, we use a trial function ϕ_2 which is orthogonal to the wavefunction ϕ_1, namely

$$\phi_2 = c_2 \cdot (\varphi_a - \varphi_b) \tag{6.18}$$

In Exercise 6.16 you show that

$$c_2 = \frac{1}{\sqrt{2(1 - S_{ab})}} \tag{6.19}$$

and that the energy ε_2 is

$$\varepsilon_2 = \frac{H_{aa} - H_{ab}}{1 - S_{ab}} \tag{6.20}$$

The wavefunction (6.16) has an antinode at the center of the bond, and this is a bonding orbital. In contrast, the wavefunction (6.18) has a node at the center of the bond, and this is an antibonding orbital. Foundation 6.2 shows that the energy curve for ε_2 does not have a minimum, in accordance with the exact wavefunction ψ_2 displayed in Fig. 6.8.

6.3 H_2: A Two-Electron System

The H_2 molecule consists of two protons and two electrons. The electron cloud of H_2 is similar to that of the H_2^+ ion; however, the internuclear distance of 74 pm is smaller than that of H_2^+. For H_2^+ we saw that the nuclei are attracted by the negatively charged electron density that lies in the region of space between the two nuclei and that this attraction is compensated by the repulsion between the nuclei. In H_2 this electron density is larger than that of H_2^+. Thus the distance between the nuclei becomes smaller. In the increased nuclear Coulomb field the average electron–electron distance decreases and causes the electron–electron repulsion to increase. These two effects compensate each other, so that the ground-state electron clouds in H_2^+ and H_2 are very similar

6.3.1 Rigorous Treatment

6.3.1.1 Procedure

The H_2 molecule represents the next step in difficulty above H_2^+ because it has two electrons, rather than one. This increase in complexity makes it no longer possible to find an exact solution to the electronic Schrödinger equation. This limitation arises because of a new term in the Schrödinger equation, the electrostatic repulsion between the electrons, $e^2/(4\pi\varepsilon_0 r_{12})$.

Using Fig. 6.9 we can write the electronic Hamiltonian for the H_2 molecule as

$$\mathcal{H} = -\frac{\hbar^2}{2m_e}\nabla_1^2 - \frac{\hbar^2}{2m_e}\nabla_2^2 - \frac{e^2}{4\pi\varepsilon_0}\left(\frac{1}{r_{1a}} + \frac{1}{r_{1b}} + \frac{1}{r_{2a}} + \frac{1}{r_{2b}}\right) + \frac{e^2}{4\pi\varepsilon_0}\frac{1}{d} + \frac{e^2}{4\pi\varepsilon_0}\frac{1}{r_{12}} \tag{6.21}$$

which has a kinetic energy term for each electron, an attraction term for each electron to each nucleus, the nuclear–nuclear repulsion term, and an electron–electron repulsion term (r_{12} is the distance between the electrons). By comparing this expression to that in Equation (6.1), you should be able to show that it consists of a sum of two H_2^+ Hamiltonians (see Problem 6.2) and the electron–electron repulsion term. The electron–electron repulsion term causes the electrons to avoid one another, even as they are attracted to the protons. As a consequence of this avoidance, the motion of electron 1 depends on the motion of electron 2, and vice versa (*correlation effect*). We saw the same phenomenon when discussing the electronic structure of the He atom in Chapter 5. This correlation in the electrons' motion remains a serious complication for the treatment of bonding in molecules. It is possible to obtain accurate energies by careful choice of variational wavefunctions and computational method, however. For H_2 this was accomplished many years ago by the clever choice of a wavefunction that possessed the correct limiting forms. In particular, Kolos and Wolniewicz [2] used H_2^+ molecular wavefunctions that were modified to look like 1*s* electron wavefunctions of two hydrogen atoms at long distance.

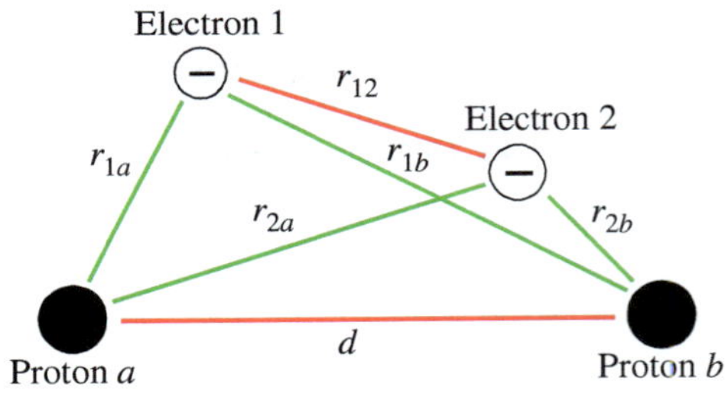

Figure 6.9 The H_2 molecule consists of two negatively charged electrons in the field of two positively charged protons, which are separated by a distance d. The green solid lines represent attractive Coulomb interaction and the red solid lines represent repulsive Coulomb interaction.

6.3.1.2 Results

A plot of the energy versus distance for this solution of the H_2 electronic Schrödinger equation, which leads to a very accurate energy curve,[4] is given in Fig. 6.10a. This electronic energy curve has a minimum at 74.14 pm and an electronic contribution to the bond energy that is $\varepsilon_{elec} = 7.60 \times 10^{-19}$ J. Both of these values are in excellent agreement with the experimental values of 74.16 pm and 7.60×10^{-19} J. A comparison of these values with the corresponding values for H_2^+ shows that the second electron in H_2 leads to a nearly factor of two increase in the bond energy and a decrease in the bond length (d-value that minimizes the energy). Figure 6.10b compares an electron density contour plot of H_2 to that of the hydrogen molecular ion.

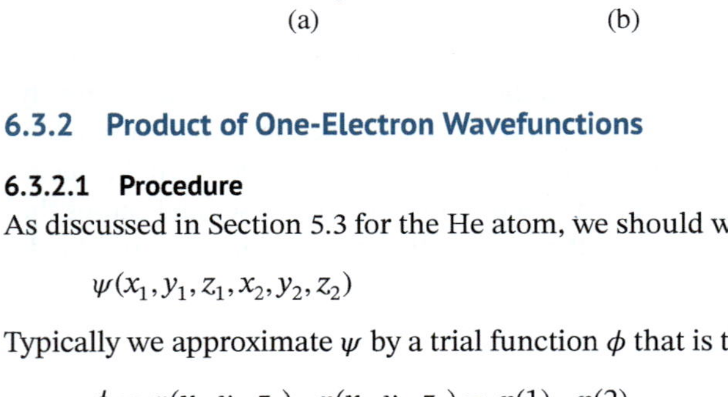

Figure 6.10 H_2 molecule. (a) Energy E, average kinetic energy $\overline{T}$, and average potential energy $\overline{V}$ as a function of the interproton distance d. The dashed lines correspond to the energies of two H atoms, obtained as d increases to infinity (Rigorous treatment by Kolos and Wolniewicz [2]). (b) Electron density contour plot of H_2 molecule, compared to that of H_2^+ molecular ion (nuclei in equilibrium distances). Within the gray zone, the electron density is greater than 1% of the density at the proton.

6.3.2 Product of One-Electron Wavefunctions

6.3.2.1 Procedure

As discussed in Section 5.3 for the He atom, we should write the wavefunction for both electrons 1 and 2 as a whole:

$$\psi(x_1, y_1, z_1, x_2, y_2, z_2) \tag{6.22}$$

Typically we approximate ψ by a trial function ϕ that is the product of two one-electron wavefunctions

$$\phi = \varphi(x_1, y_1, z_1) \cdot \varphi(x_2, y_2, z_2) = \varphi(1) \cdot \varphi(2) \tag{6.23}$$

---

4 We do not go into detail in describing the procedure of Kolos and Wolniewicz, but discuss a much simpler approach in Section 6.3.2 to focus on the essence.

and calculate a correction factor for the electron–electron repulsion. Such an approximation does not account for the correlated motion of the electrons, however. As we know the exact answer for the ground state of H_2^+ and the Hamiltonian for H_2 is so similar to H_2^+, we might choose the ground-state wavefunction of H_2^+ as our trial wavefunctions for $\varphi(1)$ and $\varphi(2)$ in the H_2 molecule. Attributing each electron to its own wavefunction (orbital, see Section 5.3.3) is called the *one-electron approximation*, and these one-electron wavefunctions are called *orbitals* (atomic or molecular orbitals). This choice of wavefunction is not yet complete because we have not considered the spin of the electrons (see Chapter 5). We delay that consideration to Section 6.3.3.

The exact Hamiltonian for the H_2 molecule in Equation (6.21), $\mathcal{H}(H_2)$, can be written in terms of the Hamiltonians for two H_2^+ molecular ions, $\mathcal{H}(H_2^+)$, according to Equation (6.1),

$$\mathcal{H}(H_2) = 2\mathcal{H}(H_2^+) - \frac{e^2}{4\pi\varepsilon_0}\frac{1}{d} + \frac{e^2}{4\pi\varepsilon_0}\frac{1}{r_{12}} \tag{6.24}$$

If we use the approximate ϕ from Equation (6.23), we find that the electronic energy of the H_2 molecule is given by

$$E(H_2) = 2E(H_2^+) - \frac{e^2}{4\pi\varepsilon_0}\frac{1}{d} + E_{\text{repulsion}} = E^{(0)}(H_2) + E_{\text{repulsion}} \tag{6.25}$$

where the superscript (0) indicates that we neglect the electron–electron repulsion. $E_{\text{repulsion}}$ is the average repulsion energy of the two electrons which is found by using the exact H_2^+ wavefunctions $\varphi(1)$ and $\varphi(2)$ and is given by

$$E_{\text{repulsion}} = \frac{e^2}{4\pi\varepsilon_0}\int \rho_1 \frac{1}{r_{12}} \rho_2 \, \mathrm{d}r_1 \, \mathrm{d}r_2 \tag{6.26}$$

where the probability densities ρ_1 and ρ_2 are $\rho_1 = \varphi(1)^2$ and $\rho_2 = \varphi(2)^2$. The energy being evaluated in Equation (6.26) is the mutual electrostatic repulsion of two electrons that have charge distributions of ρ_1 and ρ_2. The value of this electron–electron repulsion depends on the internuclear distance d because the wavefunctions, and hence the charge distributions, depend on d. This integral can be evaluated numerically as a function of the internuclear distance (see program "H2_molecule"). The d dependence of $E^{(0)}(H_2)$ is readily obtained from Equation (6.25) using the exact energy for H_2^+. Combining these results generates a curve of the electronic energy as a function of the internuclear distance d for H_2.

6.3.2.2 Results

Figure 6.11 shows the results of this calculation for H_2. Model **A** (**A** for *attraction*) shows the electronic energy as a function of distance for the case where the attraction of the electrons for the protons is included, but the electron–electron repulsion is excluded. This model greatly overestimates the bond energy and predicts an unrealistically short bond length (see Table 6.1). Model **AR** (**R** for *repulsion*) shows the electronic energy where the electron–electron repulsion has been incorporated according to Equation (6.26). The incorporation of the electron–electron repulsion in this way provides a correct bond length value and a physically reasonable, but inaccurate, bond energy (see Table 6.1). In approximating the two-electron wavefunction by the product of one-electron orbitals in Equation (6.23), the expansion of the electron cloud when proceeding from H_2^+ to H_2 is neglected (note that we are using H_2^+ one-electron wavefunctions). Furthermore, we have neglected evasion, namely, the fact that the electrons evade one another at short distances by their large Coulomb repulsion (by not using pair-wavefunctions). This evasion effect, which arises from the correlated motion of the electrons, is estimated in Box 6.1. Incorporation into the calculation results in model **ARE** (**E** for *evasion*) of Fig. 6.11.

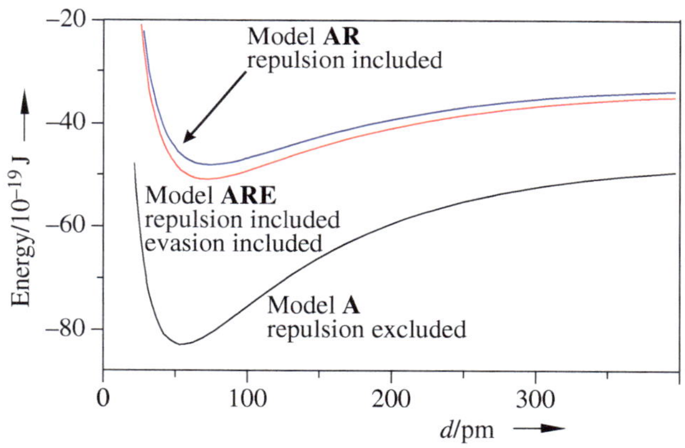

Figure 6.11 The electronic energy of the H_2 molecule is plotted as a function of the bond length d for the approximation of Equation (6.23), a trial function expressed as a product of two one-electron wavefunctions. The H_2^+ wavefunction is used as the one-electron trial orbital. Model A (black solid line): the calculation ignores the repulsion of the two electrons; model AR (blue solid line): electron–electron repulsion included, but the electron evasion effect is excluded; model ARE (red solid line): electron–electron repulsion and evasion effect included.

The calculated equilibrium distance is 74 pm, and the energy minimum is at -51.1×10^{-19} J. The contributions to the energy at the equilibrium distance $d = 74$ pm are:

E (**A**)	$E_{repulsion}$	E (**AR**) 10^{-19} J	$E_{evasion}$	E (**ARE**)
−80, 84	+32.45	−48.40	−2.7	−51.1

This approximate treatment is compared with the precise result of Kolos–Wolniewicz in Fig. 6.12. We calculate the electronic contribution to the bond energy ε_{elec} as the difference in electronic energy for the H_2 molecule and the electronic energy of two separated hydrogen atoms,

$$\begin{aligned} \varepsilon_{elec} &= E(H_2)_{minimum} - 2E(\text{H-atom}) \\ &= E(H_2)_{minimum} + 43.6 \times 10^{-19}\ \text{J} \end{aligned} \tag{6.27}$$

and obtain $\varepsilon_{elec} = 7.5 \times 10^{-19}$ J. A comparison of the results with the precise calculation of Kolos (see Table 6.1) demonstrates that the main features of the bonding are incorporated in the approximate approach. The approximation fails at larger internuclear distances, however. The curve does not go smoothly to the case of two independent hydrogen atoms. At long distance, the solution of the Schrödinger equation should have a wavefunction of the type $\psi(1,2) = \varphi_{1sa}(1)\varphi_{1sb}(2)$, where φ_{1sa} is a 1s electron orbital centered on nucleus a and φ_{1sb} is a 1s electron orbital centered on nucleus b. This wavefunction gives a limiting energy of

$$E(H_2, d \to \infty) = 2E_{1s} = 2(-21.80 \times 10^{-19})\ \text{J} = -43.60 \times 10^{-19}\ \text{J}$$

This limit is indicated in Fig. 6.12 by the dashed lines.

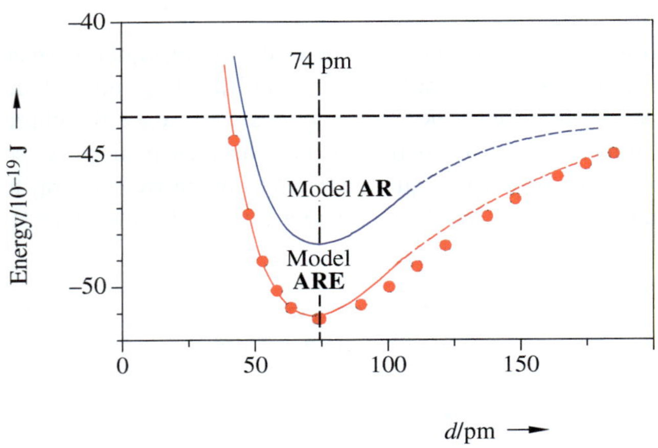

Figure 6.12 The electronic energy curves calculated for the H_2 molecule are plotted as a function of the bond length d. Blue solid line: model AR (repulsion included); red solid line: model ARE (repulsion and evasion included). The results for the approximate treatment (Equation (6.23) with the H_2^+ wavefunction used as the one-electron trial orbitals) are compared with the precise result of Kolos–Wolniewicz (full red circles). The H_2^+ wavefunctions do not well describe the electronic wavefunction at distances above 100 pm. The parts of the curves for $d > 100$ pm are obtained by requiring that the electronic energy extrapolates to that of two independent hydrogen atoms.

The first quantitative theory of the covalent chemical bond in terms of atomic orbitals was given by W. Heitler and F. London in 1927 [3]. For the ground state of H_2 (atoms a and b), they used the test function

$$\phi_1 = \varphi_{1sa}(1)\varphi_{1sb}(2) + \varphi_{1sa}(2)\varphi_{1sb}(1)$$

In this form the effect of the two electrons to evade each other is simply considered by electrons exchanging between the two atoms, the expansion of the electron cloud, however, is neglected.

6.3.3 Indistinguishability of Electrons

The indistinguishability of the electrons is accounted for in the ground state of H_2, approximated by wavefunction (6.23), because the electronic orbitals are the same. Changing the identity of the electron in the orbital has no effect on the overall wavefunction, so the electrons are not distinguishable.

To account for indistinguishability when two, or more, distinct orbitals are involved we proceed in a manner similar to that we used for describing the He atom. To begin, we create symmetric and antisymmetric superpositions of the spatial orbitals and the spin orbitals. For the first excited state of H_2 we use a product of a ground-state orbital φ, which has an antinode between the protons, and the first excited state orbital φ', which has a node between the protons. To fulfill the indistinguishability condition we create a linear combination of these wavefunction products to obtain the symmetric combination

$$\phi_{\mathrm{sym}} = \frac{1}{\sqrt{2}}\left[\varphi(2)\varphi'(1) + \varphi(1)\varphi'(2)\right] \tag{6.28}$$

and the antisymmetric combination

$$\phi_{\mathrm{asym}} = \frac{1}{\sqrt{2}}\left[\varphi(2)\varphi'(1) - \varphi(1)\varphi'(2)\right] \tag{6.29}$$

In the first case, Equation (6.28), the function remains unchanged when the two electrons' coordinates are exchanged and it is the symmetric combination. In the second case, exchange of the electrons' coordinates only changes the sign of the wavefunction and it is the antisymmetric combination. In both cases the probability density $\rho = \phi^2$ remains unchanged, and the electrons cannot be distinguished by their orbital location. For the spin part of the wavefunction, we find an antisymmetric combination

$$\frac{1}{\sqrt{2}}\{\alpha(1)\beta(2) - \alpha(2)\beta(1)\} \tag{6.30}$$

and three symmetric combinations

$$\begin{aligned} &\alpha(1)\alpha(2) \\ &\frac{1}{\sqrt{2}}\{\alpha(1)\beta(2) + \alpha(2)\beta(1)\} \\ &\beta(1)\beta(2) \end{aligned} \tag{6.31}$$

which correspond to a singlet state and triplet state, respectively. The total wavefunction of the H_2 molecule in the excited state is given by the product of the spatial wavefunctions (Equations (6.28) and (6.29)) and the spin wavefunctions (Equations (6.30) and (6.31)).

6.3.4 Pauli Exclusion Principle: Electronic Wavefunctions Must be Antisymmetric

With two spatial wavefunctions and four spin wavefunctions eight different possible products are possible, however only four of these satisfy the Pauli exclusion principle. They are

$$\frac{1}{2}\left\{\varphi(1)\varphi'(2) + \varphi(2)\varphi'(1)\right\} \cdot \{\alpha(1)\beta(2) - \alpha(2)\beta(1)\} \tag{6.32}$$

which uses a symmetric spatial wavefunction and antisymmetric spin wavefunction, and

$$\begin{aligned} &\frac{1}{\sqrt{2}}\left\{\varphi(1)\varphi'(2) - \varphi(2)\varphi'(1)\right\} \cdot \{\alpha(1)\alpha(2)\} \\ &\frac{1}{2}\left\{\varphi(1)\varphi'(2) - \varphi(2)\varphi'(1)\right\} \cdot \{\alpha(1)\beta(2) + \alpha(2)\beta(1)\} \\ &\frac{1}{\sqrt{2}}\left\{\varphi(1)\varphi'(2) - \varphi(2)\varphi'(1)\right\} \cdot \{\beta(1)\beta(2)\} \end{aligned} \tag{6.33}$$

which use an antisymmetric spatial function and symmetric spin functions. The combination of Equation (6.32) corresponds to a singlet state. The combinations in Equation (6.33) correspond to a triplet state, which has a degeneracy of three in the absence of a magnetic field, but has three different energy levels in an applied magnetic field, corresponding to each of the combinations.

In Fig. 6.13 the electronic energy curves, as calculated by Kolos and Wolniewicz, are shown for the singlet ground state, the first singlet excited state, and the triplet excited state. In the ground electronic state the spatial wavefunction (see Equation (6.23)) is symmetric; thus, the spin function must be antisymmetric. For the total wavefunction in the ground state we can write the expression

$$\frac{1}{\sqrt{2}}\varphi(1)\varphi(2) \cdot \{\alpha(1)\beta(2) - \alpha(2)\beta(1)\} \tag{6.34}$$

The energy difference between the singlet and triplet excited states is almost exclusively caused by Coulomb interaction effects and only very slightly by the magnetic interaction between the two electrons (see Section 5.4.4).

These three curves display some important general features. The ground electronic state and the first excited singlet electronic state both have minima in their electronic energy curves and form stable bonded species; however, the first triplet excited state has no minimum and does not form a stable molecule. Second, the two singlet states show minima with different well-depths. You will also notice that the deeper well (more stable chemical bond) is associated with a shorter equilibrium bond distance (value of d at the minimum) and is more strongly curved. These features often appear together.

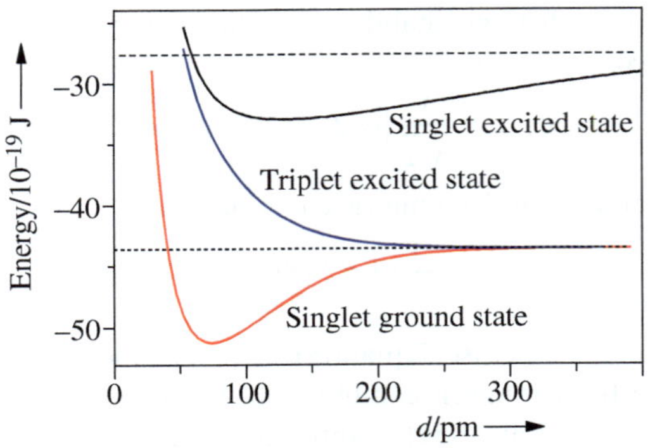

Figure 6.13 H_2 molecule, as calculated by Kolos and Wolniewicz [2]. The electronic energy E is plotted as a function of distance d for three different electronic states. The curves correspond to the ground state (singlet), and to the first two excited states: singlet state (stable) and triplet state (unstable). Lower dashed line: energy of two isolated H atoms in the ground state ($n = 1$), $E = 2\left(-21.8 \times 10^{-19}\text{ J}\right) = -43.6 \times 10^{-19}$ J. Upper dashed line: energy of one H atom in the ground state ($n = 1$) and one H atom in the first excited state ($n = 2$), $E = (-21.8 - 5.5) \times 10^{-19}\text{ J} = -27.3 \times 10^{-19}$ J.

6.4 Tunneling

6.4.1 Electron Oscillating Between Protons *a* and *b*

The view of bonding discussed above is "static," no time dependence included. H_2^+ provides a nice system for exploring a more "dynamic," time-dependent, view of electrons in molecules. Consider a hydrogen atom (nucleus a, electronic wavefunction φ_a) and a proton b at a distance d ($d \gg a_0$) at time $t = 0$. What is the probability of finding the electron at time t at atom b (wavefunction φ_b)?

6.4.2 Tunneling Barrier and Frequency of Oscillation

Figure 6.14 plots the potential energy that an electron charge experiences in the electrostatic field of the two protons a and b.

$$V = \frac{-e^2}{4\pi\varepsilon_0}\left(\frac{1}{r_a} + \frac{1}{r_b}\right)$$

where

$$r_a = \left|\frac{d}{2} + x\right| \qquad \text{and} \qquad r_b = \left|\frac{d}{2} - x\right|$$

The two horizontal lines show the energy level for the 1s electron in a hydrogen atom.

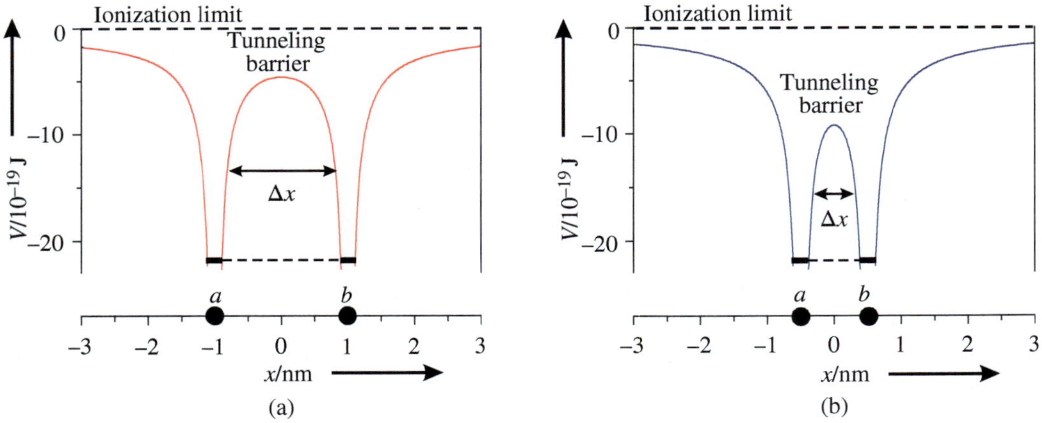

Figure 6.14 The potential energy V for an electron in the field of two protons a and b at a distance of $d = 2$ nm (a) and 1 nm (b). x is the position of the electron. Thick lines: energy of a bound 1s electron, $E = -21.80 \times 10^{-19}$ J. The height and the width of the tunneling barrier are $\Delta V = 17.2 \times 10^{-19}$ J and $\Delta x = 1.62$ nm for $d = 2$ nm and $\Delta V = 12.6 \times 10^{-19}$ J and $\Delta x = 0.63$ nm for $d = 1$ nm.

We demonstrate in Foundation 6.3 that we can describe the electron as oscillating between the protons a and b; that is, it moves back and forth through an energy barrier between the two bound energy levels on the protons. The energy barrier is 17.2×10^{-19} J = 10.7 eV for $d = 2$ nm (Fig. 6.14a) and 12.6×10^{-19} J = 7.8 eV for $d = 1$ nm (Fig. 6.14b). The period of oscillation depends strongly on the distance between the two protons: it is about one second for $d = 2$ nm and one nanosecond for $d = 1$ nm. As the interproton separation decreases the potential energy barrier that the electron moves through becomes smaller and narrower (see Fig. 6.14), hence the higher oscillation frequency. At closer distances, near the equilibrium bond length, the potential barrier drops below the 1s orbital energy and the electron moves freely between the protons.

6.4.3 Importance of Tunneling in Redox Chemistry

At large distances, the motion of an electron between the two Coulomb wells is in direct contradiction to what is possible for a particle that follows the laws of classical physics. In classical physics, a particle surmounts the barrier when it has sufficient energy, but it cannot overcome the barrier when its energy is smaller than that of the barrier. In quantum mechanics, this energy restriction is lifted. Instead, the electron is able to "tunnel" through the potential energy barrier. The present case of an electron in the potential well of a proton, when a second proton is nearby, can be viewed as a simple electron transfer reaction, in which the hydrogen atom is the reductant and the proton is the oxidant.

Electron tunneling is important for many processes in chemistry, biology, and physics. The tunneling effect is needed to understand many phenomena, such as the initial events in photosynthesis (see Chapter 28) and scanning tunneling microscopy. In these cases, the electron is removed in some other process after tunneling to the product state, rather than oscillating back and forth in the manner described here for H_2^+.

6.5 Conclusion

This chapter has used the two simplest molecules, H_2 and H_2^+, to illustrate the quantitative approach to chemical bonding. In both cases, the Born–Oppenheimer assumption was made and the electronic Schrödinger equation was solved to obtain the molecule's (or molecular ion's) electronic energy as a function of the internuclear distance. This essential approach is used to describe bonding throughout chemistry, and the dependence of the electronic energy on internuclear distances and angles underlies our structural view of molecules. The accuracy of these electronic energy curves is directly related to the reasonableness, perhaps cleverness, of our choice of a trial wavefunction. The wavefunction must account for the Pauli Exclusion Principle and allow the electrons to be indistinguishable in order to be physically reasonable. More than this we will choose it to most accurately represent the true molecular wavefunction and such a choice must be guided by our "chemical intuition," understanding of the nature of the interactions that are important. We have tried to illustrate some of those important interactions. In particular, the electronic energy represents a balance between the change in the average potential energy and the average kinetic energy, the extension of the electron cloud by the electron–electron repulsion effect must be included in any treatment of bonding, and the electrons' correlation should be addressed in truly accurate calculations of molecular properties.

Beyond providing us with an exposition on how to describe chemical bonds in a quantitative manner, this chapter has discussed many of the general features of chemical bonding. In each case a wavefunction with antinodes between the protons gives stable bonds. This feature of bonding is general. In both cases, the ground state (most stable state) corresponds to both electrons residing in a symmetric spatial orbital and an antisymmetric spin orbital (i.e., a singlet state); this is most common.

6.6 Box 6.1. Estimation of Evasion Energy (H_2 Ground State)

In Section 5.3 we treated the He atom, approximating the ground-state wavefunction as a product of two 1s He^+ wavefunctions. In the simple model the calculated energy minimum was at $E = -119.9 \times 10^{-19}$ J, whereas in the compressed orbital model the calculated energy minimum was at $E = -124.2 \times 10^{-19}$ J, compared to the experimental value of -126.6×10^{-19} J. We concluded that the difference is

$$\begin{aligned} E_{\text{evasion}}(\text{simple model}) &= E_{\text{experimental}} - E_{\text{calculated}} \\ &= (-126.6 \times 10^{-19}\ \text{J}) - (-119.9 \times 10^{-19}\ \text{J}) = 6.7 \times 10^{-19}\ \text{J} \end{aligned}$$

and

$$E_{\text{evasion}} \text{ (compressed orbital model)} = E_{\text{experimental}} - E_{\text{calculated}}$$
$$= (-126.6 \times 10^{-19}\ \text{J}) - (-124.2 \times 10^{-19}\ \text{J}) = -2.4 \times 10^{-19}\ \text{J}$$

arise from an overestimated electron–electron repulsion term in the model. The situation is analogous in the case of H_2 with the exception that the effective volume V of the electron cloud is larger; thus, the correction term is accordingly smaller. It is reasonable to assume that this term decreases with increasing volume V of the orbital.

For He we have $V_{\text{He}} = \frac{4}{3}\pi r^3$ where $r = 100$ pm is the approximate radius of the main portion of the spherical electron cloud (see Fig. 5.8). The volume is considered to be given by the gray zone in Fig. 5.8 for He and in Fig. 6.10 for H_2, and it is approximated in Fig. 6B.1.

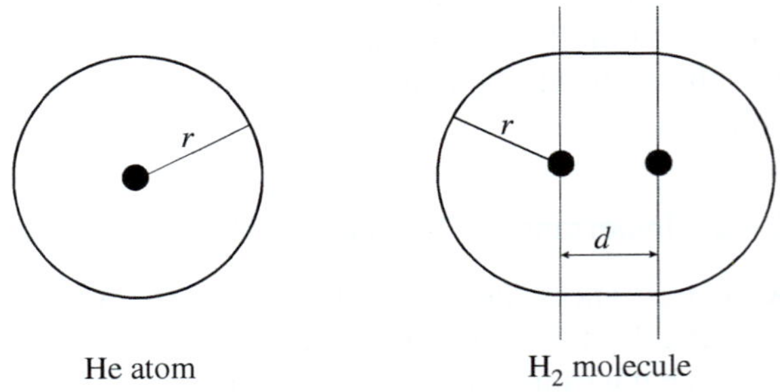

Figure 6B.1 Comparison of the volumes of electron clouds of He (a) and H_2 (b).

To estimate the volume V_{H_2} of a H_2 molecule we approximate H_2 as being constructed from two half spheres with radius r and a central cylinder of radius r and length d, where d is the bond distance (see Section 7.4.2)

$$V_{H_2} = 2 \times \frac{2}{3}\pi r^3 + \pi r^2 d = \frac{4}{3}\pi r^3 + \pi r^2 d$$

From V_{He} and V_{H_2} we calculate the evasion energy as

$$E_{\text{evasion},H_2} = E_{\text{evasion,He}} \cdot \frac{V_{\text{He}}}{V_{H_2}} = E_{\text{evasion,He}} \cdot \frac{\frac{4}{3}\pi r^3}{\frac{4}{3}\pi r^3 + \pi r^2 d}$$
$$= E_{\text{evasion,He}} \cdot \frac{1}{1 + \frac{3}{4}d/r} \qquad (6B.1)$$

where we use $E_{\text{evasion,He}} = -4.2 \times 10^{-19}$ J (approximately the mean value of both energies calculated above) and $r = 100$ pm. E_{evasion,H_2} depends on the bond distance d, see Fig. 6B.2. At the equilibrium distance of $d = 74$ pm the evasion energy is 2.70×10^{-19} J.

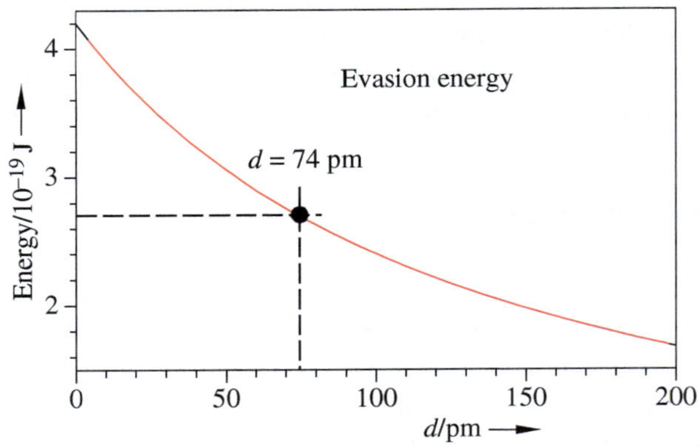

Figure 6B.2 Evasion energy of the two electrons in the H_2 molecule. At $d = 0$ the evasion energy is the evasion energy of helium, $E_{\text{evasion,He}} = -4.2 \times 10^{-19}$ J. E_{evasion,H_2} was calculated from Equation (6B.1) using $r = 100$ pm.

6.7 Foundations

6.7.1 Foundation 6.1: H_2^+ Ion Exact Wavefunction and Energy

See the books website, www.wiley.com/go/Kuhn/physicalchemistry3e.

6.7.2 Foundation 6.2: Evaluation of LCAO Integrals in H_2^+

See the books website, www.wiley.com/go/Kuhn/physicalchemistry3e.

6.7.3 Foundation 6.3: Oscillation of Electron Between Protons at Distance *d* (Tunneling)

See the books website, www.wiley.com/go/Kuhn/physicalchemistry3e.

6.8 Exercises

E6.1 Calculate the bond energy for H_2 by using the bond energy of H_2^+, the H-atom ionization potential, and the experimental ionization potential of H_2 (15.426 eV). Justify your calculation by building an energy cycle to determine the electronic energy released in breaking the bond.

E6.2 Use Fig. 6.1 and your knowledge of geometry to show that

$$r_a = \sqrt{(x + d/2)^2 + y^2 + z^2} \quad \text{and} \quad r_b = \sqrt{(x - d/2)^2 + y^2 + z^2}$$

E6.3 The particle-in-the-box wavefunctions are usually given as sine functions. In our analysis in Box 2.2, we used cosine functions for the box wavefunctions. Explain why this change is correct.

E6.4 Show in detail the mathematical steps needed to obtain Equation (6.11) from Equation (6.9).

E6.5 Show in detail the mathematical steps needed to obtain Equation (6.14) from Equation (6.12).

E6.6 Evaluate the average kinetic energy of an electron in the ground state of a three-dimensional box with $b = 1.36$ and $L = 301$ pm; see Section 2.5. Compare this value to the optimized energy, given in the text, and show that the Virial Theorem is satisfied.

E6.7 Write an explicit form for the electronic Schrödinger equation of HeH^{2+}. Discuss how it is different from that of H_2^+.

E6.8 Draw a figure like that of Fig. 6.9 for the molecular ion HeH^+. Use this diagram to write the Hamiltonian for HeH^+. Be explicit when you write the Hamiltonian and provide a physical interpretation of the different terms. Compare this Hamiltonian to that given for H_2.

E6.9 Show that the Hamiltonian in Equation (6.21) can be written as the sum of two Hamiltonians for an H_2^+ molecular ion and an electron–electron repulsion term. Derive Equation (6.24).

E6.10 State the underlying physical premise of the Born–Oppenheimer approximation. Write the complete Hamiltonian of H_2 and identify its different terms. Contrast this Hamiltonian with the electronic Hamiltonian of H_2.

E6.11 Using the fact that the molecular orbitals φ and φ' are orthogonal to one another and each normalized, show that the two-electron wavefunctions in Equations (6.28) and (6.29) are normalized.

E6.12 Show that the spin wavefunctions in Equation (6.31) are normalized, by using the fact that

$$\int \alpha(\zeta)\alpha(\zeta)\,\mathrm{d}\zeta = 1, \int \beta(\zeta)\beta(\zeta)\,\mathrm{d}\zeta = 1, \int \beta(\zeta)\alpha(\zeta)\,\mathrm{d}\zeta = 0$$

E6.13 In Section 6.2.3 an LCAO wavefunction with an effective charge η was used to describe the H_2^+ molecule's ground-state energy. Provide a physically based discussion of why a value between $\eta = 1$ and $\eta = 2$ is reasonable.

E6.14 Provide a physically based explanation for why the total energy, which is obtained for H_2^+, with the box wavefunction is in worse agreement with the exact result than that obtained from the modified LCAO calculation.

E6.15 Equation (2.35) gives an expression for the average potential energy of an electron. Show that it is equivalent to the expression

$$\overline{V} = \frac{-e^2}{4\pi\epsilon_0} \int \left[\frac{1}{r_a} + \frac{1}{r_b} - \frac{1}{d}\right] \varphi^2 \,\mathrm{d}x\,\mathrm{d}y\,\mathrm{d}z$$

E6.16 Starting with Equation (6.18) for the anti-bonding orbital trial function of H_2^+ derive the expression for its energy that is given in Equation (6.20).

E6.17 In Problem P6.10 you are asked to calculate the energy ε_1 for the excited state of the H_2^+ molecular ion as a function of bond distance d. As part of this calculation you must evaluate the expression

$$F(d) = \frac{\eta}{a_0} + \frac{1}{d}\left[1 - \left(1 + \eta\frac{d}{a_0}\right) \cdot \mathrm{e}^{-2\eta d/a_0}\right]$$

as the distance $d \to 0$. One must proceed carefully in the evaluation for small d in order to obtain a stable and accurate solution. a) Use L'Hopital's rule to show that $F(d) \to 2\eta/a_0$ as $d \to 0$. b) Expand the exponential to second order (using the Taylor's series method) and find a polynomial expression for $F(d)$ at small values of d.

6.9 Problems

P6.1 In Equations (6.4) and (6.5) we calculated the average charge density of an electron at the protons of H_2^+ using the exact solution. Perform this calculation by using the optimized particle-in-a-box wavefunction ($d = 152$ pm, $b = 1.36$, and $L = 301$ pm; see

Box 2.2). Also, perform this calculation using the optimized LCAO wavefunction $\phi(r) = [\varphi_a + \varphi_b] / \sqrt{2(1+S_{ab})}$ with $\eta = 1.24$ and $d = 106$ pm; from Foundation 6.2 use the facts that

$$\varphi_a = \sqrt{\frac{\eta^3}{\pi a_0^3}} \cdot e^{-\eta r_a/a_0}, \quad \varphi_b = \sqrt{\frac{\eta^3}{\pi a_0^3}} \cdot e^{-\eta r_b/a_0}$$

and

$$S_{ab} = \left(1 + \eta\frac{d}{a_0} + \frac{1}{3}\eta^2\frac{d^2}{a_0^2}\right) \cdot e^{-\eta d/a_0}$$

P6.2 In the first part of this problem, show that the Hamiltonian for the H_2 molecule's electronic Schrödinger equation can be written as a sum of two H_2^+ molecular ion Hamiltonians and an electron repulsion term. In the second part of this problem, use the wavefunction of Equation (6.23) in the electronic Schrödinger equation for H_2 and find the energy expression of Equation (6.25). This demonstration is formal, in the sense that you need not perform any derivatives explicitly. Rather substitute the expression for the wavefunction into the Schrödinger equation, and show that the equation can now be written as a sum of two terms that correspond to the Schrödinger equation for H_2^+ and a third term that is $-e^2/(4\pi\varepsilon_0 d)$.

P6.3 Figure 6.13 shows an electronic potential energy curve for the ground electronic state of H_2. The shape of this curve is characteristic of that for many diatomic molecules, and it is desirable to have an analytical expression that can reproduce such a shape. A widely used mathematical form for such electronic potential curves is the Morse potential $V(d - d_{eq})$ which is given by

$$V(d - d_{eq}) = \varepsilon_B[1 - \exp(-a(d - d_{eq}))]^2, \qquad a = \sqrt{\frac{k_f}{2\varepsilon_B}}$$

In this expression, ε_B is the well depth, d_{eq} is the equilibrium internuclear separation, and k_f is the force constant (see Chapter 7). Show that the Morse potential form is similar to that given in the figure by plotting it. For H_2 you should use $d_{eq} = 74.2$ pm, $\varepsilon_B = 7.60 \times 10^{-19}$ J, and $k_f = 575$ N/m.

P6.4 In Chapter 2 the mean kinetic energy of the H_2^+ molecular ion for the box wavefunctions was calculated as $\overline{T} = 16.9 \times 10^{-19}$ J. For a hypothetical H_2^+ ion use the same box length L_x, but the different lengths $L_y = L_z = 376$ pm to calculate the average kinetic energy. These perpendicular lengths correspond to the box lengths for the H atom. Explain your result in comparison to the optimized value for H_2^+ and that found for the hydrogen atom in Section 2.3.

P6.5 Consider the 1*s* orbital wavefunctions

$$\varphi_{1sa}(x,y,z) = \exp\left(-\eta\sqrt{(x-d/2)^2 + y^2 + z^2}\right)$$

and

$$\varphi_{1sb}(x,y,z) = \exp\left(-\eta\sqrt{(x+d/2)^2 + y^2 + z^2}\right)$$

Use a computer application to create a symmetric superposition of these two wavefunctions for $\eta = 1/a_0$ and $d = 74$ pm. Plot the superposition along the x axis ($y = 0, z = 0$). Also plot the value of the superposed wavefunction at $x = 0$ as a function of the internuclear distance d. Obtain an analytical expression that describes the d dependence you found.

P6.6 Perform Problem 6.5 for the antisymmetric combination of the 1*s* orbitals.

P6.7 For the hydrogen atom the energy of the electron is determined by the principal quantum number n. Hence the $n = 2$ energy level is four-fold degenerate because the 2*s* and the three 2*p* orbitals all have the same energy. Let us take a superposition of the 2*s* orbital and two of the 2*p* orbitals (the p_x and p_z). In particular we will restrict our discussion to the angular part of the wavefunctions and create three superposition wavefunctions that are each normalized and orthogonal to each other. These superposition states should have the angular shape of the sp^2 hybrid orbitals that you learned about in organic chemistry.
Given the sp^2 hybrid orbital

$$\psi_A = \sqrt{\frac{1}{3}}\psi_{2s} + \sqrt{\frac{2}{3}}\psi_{2p_z} = \sqrt{\frac{1}{3}}\psi_{2,0} + \sqrt{\frac{2}{3}}\psi_{2,1,0}$$

construct a second orbital of the form

$$\psi_B = \sqrt{\frac{1}{3}}\psi_{2s} + b_2\,\psi_{2p_z} + c_2\,\psi_{2px}$$

by requiring that ψ_A be orthogonal to ψ_B and that ψ_B be normalized. You should determine the numerical values for the b_2 and c_2 coefficients. Next, you should derive the third hybrid orbital, namely

$$\psi_C = a_3\,\psi_{2s} + b_3\,\psi_{2p_z} + c_3\,\psi_{2px}$$

(Hint: Each of these wavefunctions is orthogonal and normalized. If you use these properties, you will not need to perform an integration in solving this problem, rather you will be able to determine whether an integral is zero or unity by inspection.)

P6.8 Show that the trial wavefunctions for the H_2 molecule, ϕ_1 and ϕ_2 in Section 6.3.3, are orthogonal to each other.

P6.9 Using the trial function ϕ_2 for the excited state of the H_2^+ molecular ion in Section 6.2.5 derive Equations (6.19) and (6.20) for c_2 and for ε_2.

P6.10 In analogy to the procedure described in Foundation 6.2 for the calculation of the ground-state energy ε_0, calculate the energy ε_1 for the excited state of the H_2^+ molecular ion as a function of bond distance d.

P6.11 In Fig. 6.14 we consider the possibility that an electron in an H-atom can move to a proton over a distance d. Use the data given for the height and width of the barrier in Fig. 6.14 to calculate the probability $\mathcal{P}$ (see Section 3.2.2.4) for tunneling through the barrier.

P6.12 In this problem you will explore the Equation

$$E_{\text{kin,calc}} = -E - r \cdot \frac{\mathrm{d}E}{\mathrm{d}r} \qquad (4.30)$$

You should run the program "h2plus_exact." This program calculates the exact wavefunctions and energies for the H_2^+ ion. As an output you will find the following files:

Energy.dat: this is the total energy $E/10^{-19}$ J
Vpot.dat: this is the potential energy $V_{\text{pot}}/10^{-19}$ J
Ekin.dat: this is the kinetic energy $E_{\text{kin}}/10^{-19}$ J
Virial_correction.dat: this is the correction term $r \cdot \mathrm{d}E/\mathrm{d}r$ considered in Equation (4.30)

These energies are given as a function of the bond distance r in the range from 20 to 450 pm.
Select a number of data points and check if the prediction of Equation (4.30) is correct.

References

1 D.R. Bates, K. Ledsham, and A.L. Stewart, Wave functions of the hydrogen molecular ion, *Philos. Trans. R. Soc. London, Ser. A*, 1953, **246**, 215.
2 W. Kolos and L. Wolniewicz, Potential-energy curves for the $X^1\Sigma_g^+$, $b^3\Sigma_u^+$, and $C^1\Pi_u$ states of the hydrogen molecule, *J. Chem. Phys.*, 1965, **43**, 2429; *J. Chem. Phys.*, 1966, **45**, 509.
3 W. Heitler and F. London, Wechselwirkung neutraler Atome und homöopolare Bindung nach der Quantenmechanik, *Z Physik*, 1927, **44**, 455.

7

Bonding Described by Electron Pairs and Molecular Orbitals

CHAPTER MENU

7.1 Introduction

How can we understand the properties of a molecule composed of multiple atoms; in particular its connectivity pattern, geometry, energetics, and electron density profile? If we could solve the Schrödinger Equation ($\mathcal{H}\psi = E\psi$) to obtain ψ then we would know all that we could about the molecule and determine these characteristics. Unfortunately, we cannot solve this equation exactly and must resort to approximate methods. Although we cannot obtain a closed form analytical expression for the wavefunctions, we can numerically evaluate this equation to a high degree of accuracy for small molecules.

Despite this limitation, important aspects of molecular structure can be understood on a qualitative level without solving the Schrödinger equation. From Chapter 6, we found that the bond between two hydrogen atoms can be described by the overlap of atomic orbitals forming an electron pair bond. The assignment of electron pairs to distinct bonds is a very useful approximation for understanding the connectivity of atoms and the geometry of molecules, even though we cannot identify particular electrons with distinct bonds because electrons are indistinguishable. These elementary ideas should be familiar from introductory chemistry and organic chemistry courses; however, they are summarized here in Section 7.2. Although the electron pair description of bonding works well for many ground-state molecules and provides a reasonable description of the electron density distribution, a molecular orbital picture (using orbitals that are spread over the molecule) is superior because it meets these criteria and succeeds in situations where electron pair ideas fail (such as, electron-deficient systems, delocalized systems, and electronically excited states) and provides insight into chemical reactivity. Molecular orbitals are discussed in Sections 7.3 and 7.4. In Section 7.5, we discuss the forces that are required to change the distance between the nuclei of a molecule and to break a chemical bond. In this chapter, we aim to provide a qualitative and semi-quantitative guide to understanding the bonding in molecules.

7.2 Electron Pair Bonds (A Review)

Elementary ideas about the chemical bond rely on the concept of electron pair bonds and the "octet rule" proposed by G. N. Lewis. The occurrence of electrons in pairs for the formation of a chemical bond arises from the Pauli Exclusion

Principles of Physical Chemistry, Third Edition. Hans Kuhn, David H. Waldeck, and Horst-Dieter Försterling.

Companion website: www.wiley.com/go/Kuhn/physicalchemistry3e

principle, namely that no more than two electrons can occupy a given spatial orbital and they must do so with their spins paired (*electron pair bonds*), and the lowest energy orbitals must be occupied in the ground state (the lowest energy state). In this model each chemical bond comprises two electrons that are "shared" between two atoms, more precisely the atomic cores. Atoms tend to form filled electron shells, and for most main group elements this means an octet (four electron pairs). For first row elements this idea corresponds to two electrons, and for transition metals it is associated with the stability of 10 electrons in *d*-orbitals. Although this description makes the implicit assumption that particular electrons are localized between particular pairs of atoms and violates the indistinguishability of electrons, this idea has proven to be of great practical use as an elementary description.

7.2.1 Connectivity of Atoms in Three-Atom Molecules

The valency of an element is determined by the number of electrons needed to complete the shell, or subshell. Consider the case of a covalent electron pair bond, as in the H_2 molecule. In this case the bond is formed by the overlap of 1*s* orbitals of two H atoms. Since each atom contributes one electron to the bond and the nuclei have the same charge, the two electrons are shared in the bond as an electron pair. The valency of H is one since it can fill its shell by sharing its electron with an electron from another atom. Also, halogens have a valency of one because the atom has seven valence electrons and can fill an octet by forming a one-electron pair bond with another atom. This same logic follows for other elements, for example C (the carbon atom) has a valency of 4, N (the nitrogen atom) a valency of 3, and O (the oxygen atom) a valency of 2.

Example 7.1 *Rationalize the electron pair bonding in the F_2 molecule.*

The electron configuration of F (the fluorine atom) is

$$1s^2 2s^2 2p_y^2 2p_z^2 2p_x$$

The $2p_x$ orbital is occupied by an unpaired electron. The choice of p_x is arbitrary, of course, we could have chosen p_y or p_z as well. The *s* orbitals and the $2p_y$ and $2p_z$ orbitals are occupied by two electrons each. According to the Pauli Principle, they cannot contribute to bonding. The $2p_x$ orbitals overlap forming the F_2 molecule (Fig. 7.1). Thus the fluorine atom can form only one bond.

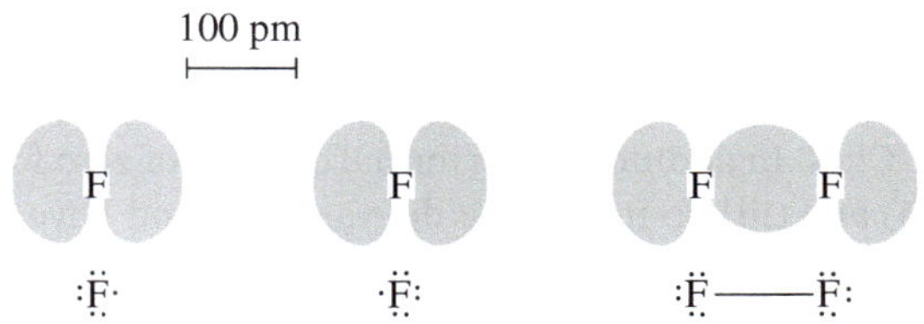

Figure 7.1 Electron pair bonding of F_2 (overlap of $2p_x$ orbitals). In the chemical formula representation a dot denotes a single electron; only the electrons of the outer shell are displayed. Inside the shaded area the electron density is larger than $e \cdot 0.02 \times 10^{30}$ m^{-3}. Electron clouds are depicted according to [1].

7.2.1.1 Double and Triple Bonds

In some circumstances more than one electron pair can be shared between two nuclei and a multiple bond is formed; numerous examples exist for double and triple bonds; however, quadruple and quintuple bonds have also been reported. The electron distribution in molecules can be schematically represented by bars and dots as indicated in the structural formulas in Figs. 7.1 and 7.2. Each bar corresponds to an electron pair, and a dot represents an unpaired electron.

The oxygen molecule O_2 is an example where a double bond is formed. The ground-state electron configuration of the O atom has four 2*p* and two 2*s* electrons in its valence shell and requires two more electrons to reach an octet. If each oxygen atom contributes two electrons to the bond then each atom has an octet, and this results in two-electron pair bonds between the nuclei and two non-bonding electron pairs at each atom, called *lone pairs*. Through analogous reasoning, you can predict that nitrogen N_2 has a triple bond.

7.2.1.2 Connectivity

These ideas are very useful for understanding the connectivity of atoms in larger molecules. By way of example, consider water H_2O. In general, we must consider two possible connectivities, H-H-O and H-O-H. Since O has a valency of two and

H has a valency of one, we see that the H-O-H connectivity pattern provides for electron pair bonds between each pair of atoms, a filled shell on each H atom, and a filled shell, octet, on the O atom. The H-O-H bonding pattern can be realized by the overlap of a $2p_z$ and $2p_x$ orbital of the oxygen atom with the $1s$ orbitals of the hydrogen atoms (Fig. 7.2). Accordingly the oxygen forms two bonds, and the H_2O molecule exists in a bent form. These simple considerations suggest a 90° bond angle;, however the average bond angle in the water molecule is 102°. These valency ideas must be enhanced to predict geometries, see Section 7.2.2.

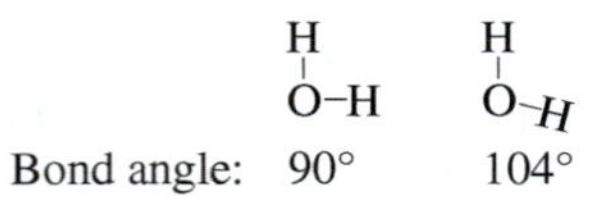

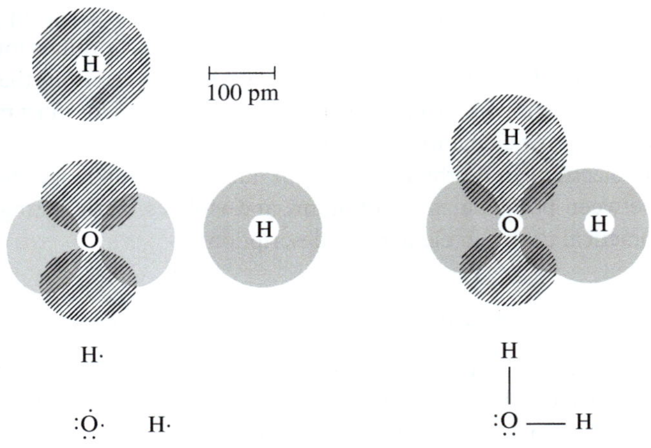

Figure 7.2 Formation of electron pair bonds in a H_2O molecule. The electron density in the shaded area is larger than $e \cdot 0.02 \times 10^{30}$ m^{-3}.

7.2.1.3 Formal Charge

It is often useful to extend these considerations by defining the formal charge of an atom. The formal charge on an atom is defined to be the difference between the number of electrons on the neutral atom and the number of electrons assigned to the atom in a molecular structure, assuming that lone pair electrons are attributed solely to the atom and half of each bonded electron is assigned to the atom. For the two water structures discussed above, we see that the H-O-H pattern has formal charges of zero on each atom center, whereas the pattern H-H-O has a zero formal charge on the terminal H atom, a formal charge $-e$ on the central H atom, and a formal charge $+e$ on the O atom. This latter structure is less stable because it contains more separated charge and because it has a positive formal charge on the more electronegative atom.

Example 7.2 *Consider the molecule FCN. Use the electron pair model to predict its most stable connectivity.*

The three possible connectivity patterns are (i) F–C–N, (ii) F–N–C, and (iii) N–F–C. The preferred valencies of the atoms are 1 for F, 4 for C, and 3 for N, which is satisfied for each atom if we have a single bond connecting F to C and a triple bond connecting C to N. Only pattern (i) gives this scenario F-C≡N. In addition it has formal charges of zero for each atom type. Patterns (ii) and (iii) do not satisfy the valencies and have significant formal charges on each of the atom types (see Fig. 7.3).

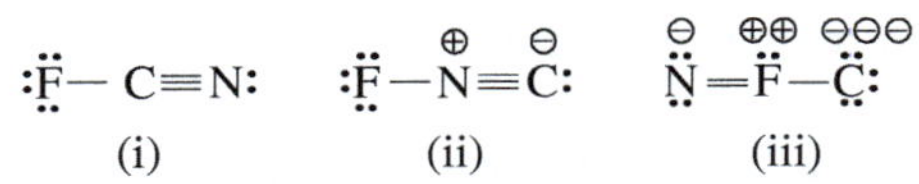

Figure 7.3 The Lewis dot structures for different bonding patterns of FCN, along with the formal charges on each atom center. Note that structure (iii) is unrealistic, because it implies that the molecule is twofold negatively charged, contrary to the fact that it is neutral.

7.2.1.4 Dative Bond

To form an electron pair bond, each bond partner must provide one suitable orbital (i.e., the corresponding energy levels must be sufficiently low) and two electrons are needed for the bond. Each partner can contribute one electron as discussed above, or both electrons can be contributed by one of the partners. In this case, the result bond is called a *dative bond*. Examples of dative bonds are found in NH_4^+ and $\overset{\oplus}{N}H_3 - \overset{\ominus}{B}(CH_3)_3$ (see Fig. 7.4). The partner which contributes the bonding electrons is called the donor (here the NH_3 molecule) and its partner ($B(CH_3)_3$) is called the acceptor. Because we associate one electron with each bond partner, the donor carries a formal charge of $+e$ while the acceptor carries a formal charge of $-e$.

7.2.1.5 Effective Charge

The formal charges should be distinguished from the effective charge, however. The bonding electrons are not uniformly spread over the bond partners, but are shifted toward the more electronegative atom. For example, the effective positive charge on the N atom in NH_4^+ is less than $+e$, and the valence electrons charge density is nearly equally distributed over the N atom and the H atoms (Fig. 7.4).

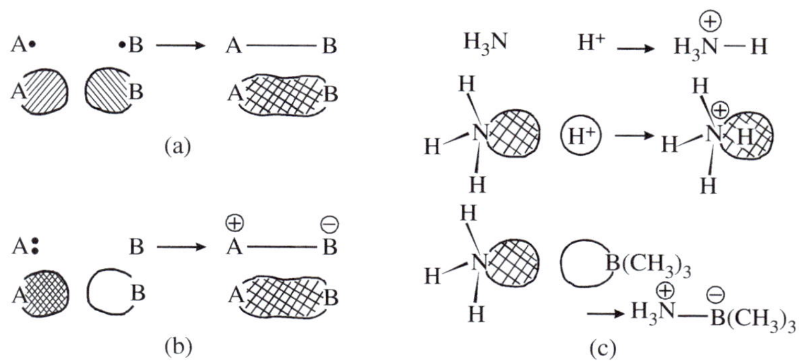

Figure 7.4 Formation of electron pair bonds. (a) Each of the two partners contributes one electron. (b) Bond partner A donates both electrons (donor), and bond partner B has an empty orbital (acceptor). (c) Examples of dative bonds.

7.2.2 Geometries for Electron Pair Bonds

7.2.2.1 VSEPR-Model

The Valence Shell Electron Pair Repulsion (VSEPR) model provides a general approach to predicting the geometries of electron pair bonds. Its primary assumption is that the electron pairs arrange around each atomic center to maximize the interpair distance, thereby minimizing the electron–electron repulsion. Because the electron clouds of bonded electron pairs are smaller than those of nonbonded electron pairs, the bonded electron pairs contribute less to the repulsion than do the nonbonded pairs. This simple model is surprisingly accurate and has been justified through more rigorous analysis [2].

Ignoring the finer differences between bonding and nonbonding electron pairs, a set of simple geometries can be found to minimize the electrostatic repulsion between electron pairs that are restricted to a spherical surface. In the case of two-electron pairs the maximum distance places them on opposite sides of the atom with a bond angle of 180°. Atoms with three electron pairs localized around them have a planar geometry with the electron pairs located at the vertices of a triangle. For equivalent electron pairs, the angle between any two electron pairs and the atomic center is 120°. For nonequivalent electron pairs this angle will change; e.g., one nonbonded pair and two bonded pairs have a bond angle that is less than 120° because the nonbonded pair is somewhat larger and acts to compress the bond angle. When four equivalent electron pairs are localized about an atomic center, the optimum geometry has electron pairs placed at the vertices of a tetrahedron. When five equivalent electron pairs are localized about an atomic center the optimum geometry places the electron pairs at the vertices of a trigonal bipyramid, and for six equivalent electron pairs they are localized at the vertices of an octahedron.

The key considerations in defining molecular geometries with this model are to first consider the electron pair number around a center and define the appropriate optimal geometry. With the geometry chosen, the electron pairs are divided into bonding and nonbonding sets and arranged to place the nonbonding pairs as far from each other as possible. When reporting the geometry, only the spatial arrangement of the atoms in space is specified.

Example 7.3 *Using the VSEPR model, predict the geometries of H_2O, NH_3, and CH_4.*

First consider H_2O. In this case O is the central atom and it has four valence electron pairs surrounding it, two bonding pairs and two nonbonding pairs. The optimal geometry in this case is tetrahedral of which two sites are filled by nonbonding electron pairs and two are filled by bonding electron pairs. Hence the molecular geometry is bent. For the case of ammonia, NH_3, the N atom has four electron pairs with an optimal geometry that is tetrahedral; however, three of the electron pairs are bonding and only one is nonbonding. Hence the molecular geometry is pyramidal. The final example is methane, CH_4, which also has four electron pairs about the central atom, so that its idealized geometry is tetrahedral. Since all of the electron pairs are bonding pairs, the molecular geometry is also tetrahedral.

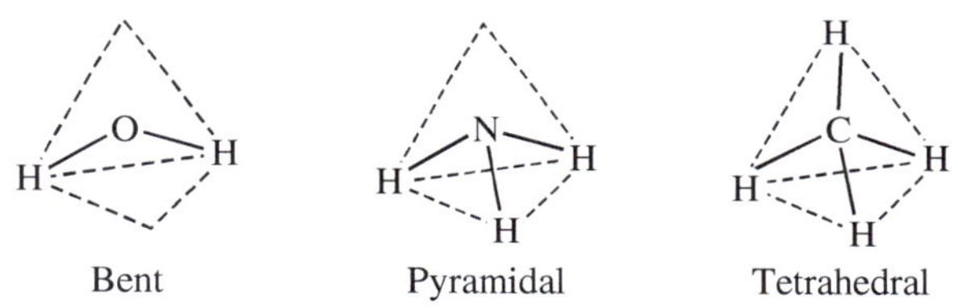

7.2.2.2 Hybrid Atomic Orbitals

Using s and p orbitals to describe the spatial electron pair bonds restricts the molecular geometries to bond angles of 90° and 180°. How can we retain the wave properties of the electrons and describe molecular geometries more accurately? In Chapters 3 and 4 we found that any linear combination of degenerate wavefunctions, corresponding to the same energy, is also a solution of the Schrödinger Equation. In a H-atom $2s$, $2p_x$, $2p_y$, and $2p_z$ wavefunctions are degenerate, and any linear combination is possible. By superposing s and p atomic orbitals on each atom (called *hybridization*), we can create atomic orbitals that are oriented at other angles. These orbital superpositions can be used as trial functions in a variational calculation to determine the superposition that gives the lowest energy. We expect that orbital superpositions which direct electron density between adjacent atoms will form the most stable chemical bonds.

Example 7.4 *Create a linear superposition of the ψ_{p_x} and ψ_{p_y} atomic wavefunctions*

$$\psi = a\psi_{p_x} + b\psi_{p_y} \tag{7.1}$$

where a and b are coefficients. Show that the orientation of these superpositions depends on the values of a and *b*.

The coefficients can be determined by requiring that the wavefunction is normalized ($\int \psi^2 \cdot d\tau = 1$) and that the hybrid orbitals, ψ_I and ψ_{II}, are orthogonal. For the present case of twofold degeneracy, the possible orthogonal sets of functions are given by

$$\psi_I = a \cdot \psi_{p_x} + \sqrt{1 - a^2} \cdot \psi_{p_y} \tag{7.2}$$

and

$$\psi_{II} = \sqrt{1 - a^2} \cdot \psi_{p_x} - a \cdot \psi_{p_y} \tag{7.3}$$

You are asked to derive these formulae in Exercise E7.8. Pairs of normalized functions with $a = 0$, $1/2$, and $1/\sqrt{2}$ are shown in Fig. 7.5. For $a = 0$ and $a = 1$ we obtain functions with horizontal and vertical nodal planes, respectively; for $a = 1/\sqrt{2}$ the nodal planes are diagonal. All the other cases fall in between. This superposition shows how the orientation of the atomic orbital's lobes can be changed by changing the relative magnitudes of the two basis wavefunctions, ψ_{p_x} and ψ_{p_y}.

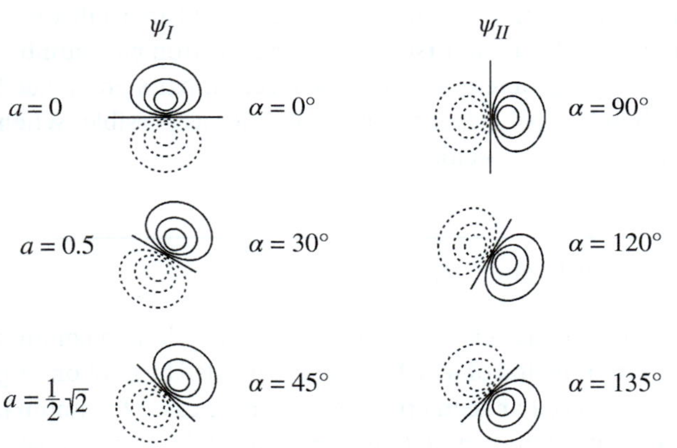

Figure 7.5 Linear combinations of wavefunctions $\psi_I = a\,\psi_{2p_x} + \sqrt{1-a^2}\,\psi_{2p_y}$ and $\psi_{II} = \sqrt{1-a^2}\,\psi_{2p_x} - a\,\psi_{2p_y}$, where ψ_{2p_x} and ψ_{2p_y} are the $2p_x$ and $2p_y$ wavefunctions of the H-atom. Wavefunctions in the xz plane are displayed by contour lines. α is the angle between the nodal plane and the x axis. Positive values: solid lines; negative values: dashed lines. Spacing of contour lines: $0.5 \times 10^{-4}\,\text{pm}^{-3/2}$. The parameter a is in the range from 0 to 1. For $a = 0$ we obtain $\psi_I = \psi_{2p_y}$ and $\psi_{II} = \psi_{2p_x}$.

These facts show that superpositions of atomic orbitals can be used to change orbital orientations and suggest that we can create superpositions of H-atom like atomic orbitals and use them as trial functions in a variational calculation. This idea is called hybridization (the new orbitals are called hybrid atomic orbitals). It is successful at predicting molecular geometries and retains the idea that a bond comprises electron pairs in two overlapping atomic orbitals.

Promotion, Strongest Overlap with Bond Partner For example, consider the carbon atom in methane. The ground-state electron configuration of the carbon atom is $1s^22s^22p^2$; however, to describe the presence of four equal bonds in methane we use a superposition of all four valence orbitals $2s$, $2p_x$, $2p_y$, and $2p_z$. In a many-electron atom, the $2s$ orbital has a slightly lower energy than the p-orbitals (see Fig. 5.11, inset). Therefore, it needs a certain energy (promotion energy) to create the electron configuration $2s2p_x2p_y2p_z$ from the ground-state configuration $2s^22p_x2p_y$. This energy is strongly overcompensated by forming the bonds. What orbitals are used for bonding? Those that have the strongest overlap with the orbitals of the bonding partners. The best set of hybrid orbitals is determined by a variational calculation to determine the lowest electronic energy for the molecule.

Hybridization in Carbon Compounds To see how this idea works, consider the hydrogen atom orbitals ψ_{2s}, ψ_{2p_x}, ψ_{2p_y}, and ψ_{2p_z} as trial functions, from which we create four linear combinations

$$\psi = a \cdot \psi_{2s} + b \cdot \psi_{2p_x} + c \cdot \psi_{2p_y} + d \cdot \psi_{2p_z} \tag{7.4}$$

corresponding to different sets of the coefficients a, b, c, and d. Out of the infinite number of possibilities to construct orthogonal sets of four hybrid functions ψ_I, ψ_{II}, ψ_{III}, and ψ_{IV} we discuss three important limiting cases.

Case 1 comprises two linear hybrid functions (sp hybrids). Each of the two orbitals consists of $\frac{1}{2}s$ and $\frac{1}{2}p$ ($s^{1/2}p^{1/2}$ orbitals). In this case we combine the functions ψ_{2s} and ψ_{2p_x} with $a = b = 1/\sqrt{2}$ (analogous to Equations (7.2) and (7.3)); the functions ψ_{2p_y} and ψ_{2p_z} remain unchanged. In Fig. 7.6 the two linear combinations ψ_I and ψ_{II} resulting from ψ_{2s} and ψ_{2p_x} are illustrated.

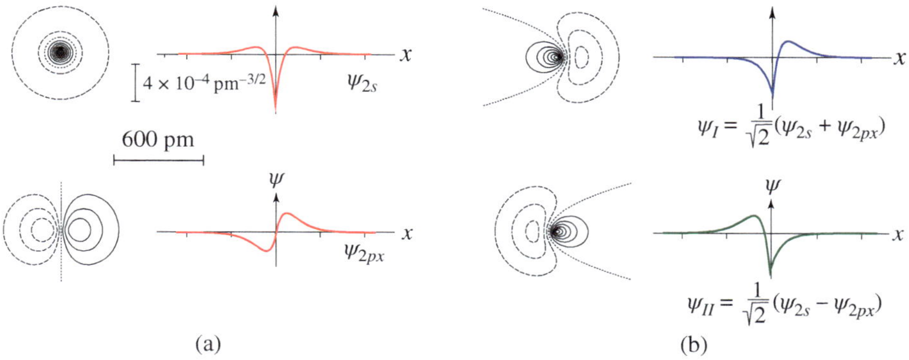

Figure 7.6 Linear sp-hybrid functions of the H atom. (a) Wavefunctions ψ_{2s} and ψ_{2px}. (b) Formation of the combinations ψ_I and ψ_{II} from ψ_{2s} and ψ_{2px}. (left) Wavefunctions in the *xz* plane displayed by contour lines. (right) Wavefunctions along the *x* axis. Positive values: solid lines; negative values: dashed lines; nodal lines: dotted lines. Spacing of contour lines: 0.5×10^{-4} pm$^{-3/2}$, the outermost contour line is at 0.5×10^{-4}pm$^{-3/2}$.

Case 2 comprises three trigonal planar hybrid functions (sp² hybrids). Each of the three orbitals consists of $\frac{1}{3}s$, $\frac{2}{3}p$ ($s^{1/3}p^{2/3}$ orbitals). The combinations ψ_I, ψ_{II}, and ψ_{III} are illustrated in Fig. 7.7. The symmetry axes of these functions are shifted by 120°, but all of them are in the *xy*-plane.

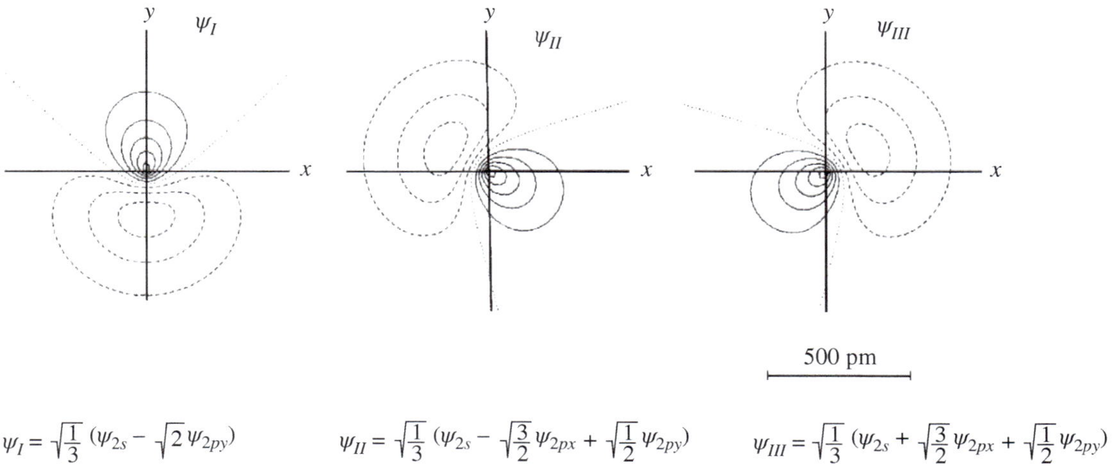

Figure 7.7 Trigonal planar sp²-hybrid functions ψ_I, ψ_{II}, and ψ_{III} of the H atom, displayed by contour lines. Positive values: solid lines; negative values: dashed lines; nodal lines: dotted lines. Spacing of contour lines: 0.5×10^{-4}pm$^{-3/2}$, the outermost contour line is at 0.5×10^{-4}pm$^{-3/2}$.

Case 3 includes four tetrahedral hybrid functions (sp^3 hybrids). Each of the four orbitals consists of $\frac{1}{4}s$, $\frac{3}{4}p$ ($s^{1/4}p^{3/4}$ orbitals). These combinations are illustrated in Fig. 7.8. Each of the major lobes of the hybrid points to the vertices of a tetrahedron and they generate a tetrahedral symmetry for the molecule.

The electron density in the overlap region is largest for sp^3 hybrids and decreases when proceeding from case 3 to case 2 to case 1 to p orbitals, as can be seen from Figs. 7.6–7.8. There are three contour lines in the lobes of the hybrid functions in Figs. 7.6 and 7.7, but four contour lines in the case of Fig. 7.8. Justification 7.1 shows that the overlap is strongest in the case of $s^{1/4}p^{3/4}$ hybrid orbitals.

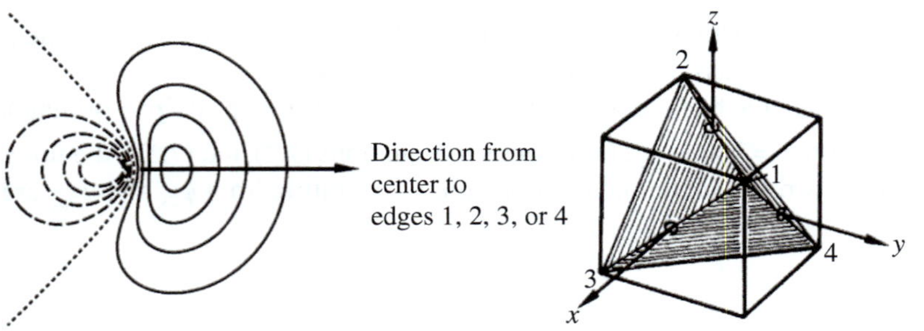

$$\psi_I = \frac{1}{2}(-\psi_{2s} + \psi_{2px} + \psi_{2py} + \psi_{2pz})$$
$$\psi_{II} = \frac{1}{2}(-\psi_{2s} - \psi_{2px} - \psi_{2py} + \psi_{2pz})$$

(a)

$$\psi_{III} = \frac{1}{2}(-\psi_{2s} + \psi_{2px} - \psi_{2py} - \psi_{2pz})$$
$$\psi_{IV} = \frac{1}{2}(-\psi_{2s} - \psi_{2px} + \psi_{2py} - \psi_{2pz})$$

(b)

Figure 7.8 Tetrahedral sp^3-hybrid functions of the H atom, for which one is displayed by contour lines (a). Positive values: solid lines; negative values: dashed lines; nodal lines: dotted lines. Spacing of contour lines: 0.5×10^{-4}pm$^{-3/2}$, the outermost contour line is at 0.5×10^{-4}pm$^{-3/2}$. (b) Direction of axis of hybrid function ψ_I to corner 1 of tetrahedron, axes of functions ψ_{II}, ψ_{III}, and ψ_{IV} accordingly to corners 2, 3, and 4.

Why is Methane Tetrahedral? Let us consider a C atom and four H atoms. What is the energetically lowest arrangement that is closest to reality, according to the variational principle? Figure 7.9 shows how a hybridization scheme predicts that the four hydrogen atoms in CH_4 tetrahedral hybrid functions are the most stable for forming the bonds. In case a, the bonding of two H atoms to a C atom using 2p orbitals does not require any promotion energy, but only two bonds are formed. In case b, four H atoms bind to a C atom using two linear hybrids and two 2p orbitals. This scheme requires a promotion energy of $\Delta E = \Delta E_1$ (where ΔE_1 is the energy difference between the 2s and the 2p orbitals), but four bonds are formed. However, the gain of energy by overlap of p orbitals and linear hybrid orbitals in bonding is small. In case c, four H atoms bind to a C atom using four tetrahedral hybrid orbitals. The promotion energy ΔE is the same as in case b, but the energy gain in

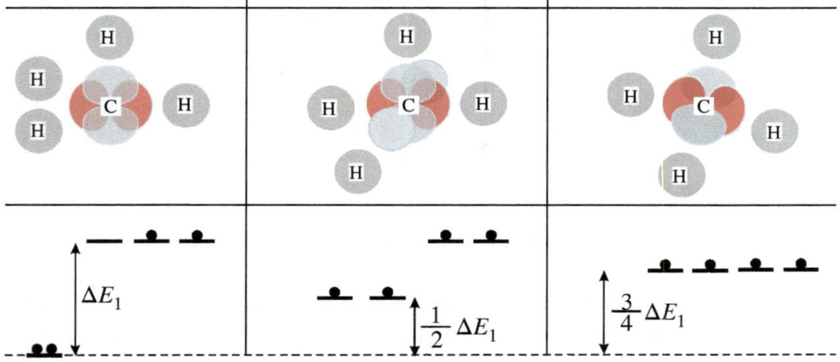

Figure 7.9 Four H atoms and one C atom. (a) Two H atoms form H_2, while the others form CH_2 (bonding through 2p orbitals of C). (b,c) four H atoms form CH_4 by (b) bonding through linear hybrid orbitals of C (x axis) and through 2p orbitals (y and z axes) and by (c) bonding through tetrahedral hybrid orbitals of C. Possibility (c) is favored. ΔE_1 is the energy difference between the 2s and the 2p orbitals. The promotion energies ΔE in (b) and (c) are equal, but the overlap with the 1s orbitals of the H atoms is better in case (c).

bonding is much larger due to the better orbital overlap. Hence this is the most stable bonding scheme according to the Variational principle, as proved in Justification 7.1.

Example 7.5 *Consider the three hybridizations sp, sp^2, and sp^3 for the oxygen on water. Which of these are the best bonding orbitals?*

Figure 7.10 provides an energy diagram that illustrates the energy penalty for hybridization in these different situations. On the left the valence electron configuration of ground-state oxygen is shown with the energy gap between the *s* and *p* orbitals given by ΔE_1. The hybrid orbital energies are intermediate between the *s*- and *p*-orbitals, with the value determined by the fraction of *p*-orbital in the hybrid. So, for an *sp* hybrid the energy is half way in the gap, for *sp*^{2} it is two-thirds of the way above the *s*-orbital energy, and for *sp*^{3} it is three-fourths of the way above the *s*-orbital energy. To create *sp* hybrids on the oxygen for bonding the energy cost is $\Delta E = 2 \cdot \frac{1}{2}\Delta E_1 = \Delta E_1$. For the *sp*^{2} hybrid it is $\Delta E = 2 \cdot \frac{2}{3}\Delta E_1 - 2 \cdot \frac{1}{3}\Delta E_1 = \frac{2}{3}\Delta E_1$, because there is an energy cost to place the *s* electrons into the *sp*^{2} hybrids and an energy gain to place two *p* electrons into the remaining *sp*^{2} hybrid as a lone pair. For the *sp*^{3} hybrid $\Delta E = 2 \cdot \frac{3}{4}\Delta E_1 - 4 \cdot \frac{1}{4}\Delta E_1 = \frac{1}{2}\Delta E_1$. This analysis shows that the *sp*^{3} hybrid has the lowest energy penalty, of the three cases considered.

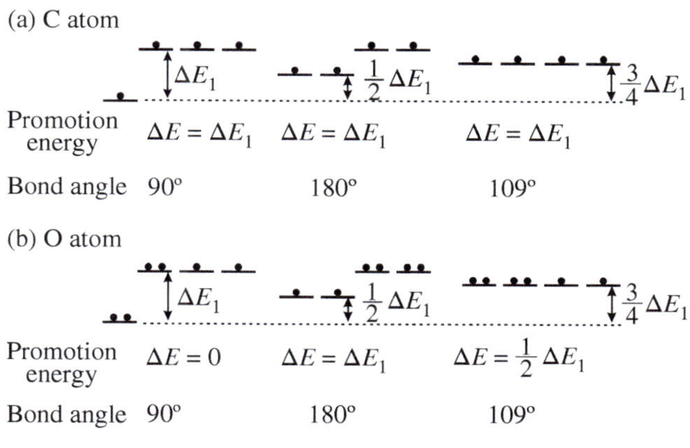

Figure 7.10 Promotion energy ΔE and bond angle in hybridization model. (a) C atom, (b) O atom. ΔE_1 is the energy difference between the 2*s* and the 2*p* orbitals.

In order to deduce the best hybridization for the oxygen atom, this energy penalty ΔE must be compared to the energy gain by bond formation with the H atoms. The optimal hybridization must be decided by a variational calculation. The change in the bonding with the angle has essentially three contributions. As the amount of *p* character in the hybrid decreases the energy penalty for hybridization increases, from pure *p* with $\Delta E = 0$ and an angle of 90° to pure *sp* with $\Delta E = \Delta E_1$ and an angle of 180°. The second contribution is the change in the overlap density of the orbitals with the amount of hybridization, for example *sp*^{3} hybrid functions have a larger overlap with the hydrogen 1*s* orbital than do the *p* orbitals. A third contribution is the electrostatic repulsion between the hydrogen atoms which decreases as the bond angle increases. A detailed calculation reveals that the electronic energy for an H_2O molecule is minimized near the experimental bond angle of 105° and the bonding is approximately $s^{1/8}p^{7/8}$ hybrid orbitals. This description of the bonding suggests that the OH bond energies will be the same and that the lone pair electrons on the oxygen atom have about the same energy. As discussed in Section 7.3.3.1 and illustrated in Fig. 7.21 these features are incorrect.

7.2.3 Limitations of Electron Pair Model

Although these elementary ideas are very powerful and comprise the initial stages of how we formulate our thinking about the structure of a molecule, they fail in important ways. A dramatic situation where it fails is for delocalized bonding, which is discussed at length in Chapter 8. Below, we give three other examples of how the electron pair model is deficient.

The electron pair approach does not adequately describe bonding in electron-deficient molecules, such as B_2H_6. The boron atom has three valence electrons and requires three covalent bonds and a dative bond to satisfy its octet, so B_2H_6 does not have enough electrons to form electron pair bonds between each pair of nuclei. Rather, B_2H_6 requires two electron pair bonds that are shared on three atomic centers if we insist on retaining this view of the bonding. Alternatively we can construct molecular orbitals from atomic orbitals centered on more than two nuclei and fill them with electron pairs to describe the bonding.

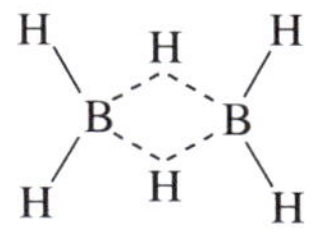

Another example where the electron pair model fails is predicting the relative strengths of bonds for molecules and their molecular ions. In the case of O_2 the electron pair model predicts a double bond between the atoms. If an electron is removed from the molecule to create O_2^+ it predicts that the bond will be weaker, or unchanged. In fact, the bond becomes stronger. The bond energy is 5.28 eV for O_2 and 6.71 eV for O_2^+. In addition, the electron pair bond model predicts that all the electrons are spin paired, but the paramagnetism of O_2 shows that this is not the case.

A third situation in which electron pair ideas fail are in describing electronically excited states of molecules and their electronic energies. The nature of a molecule's electronically excited state(s) can be very important for understanding its physicochemical properties and its reactivity.

These issues are discussed more below and in the next few chapters. By going to a molecular orbital picture we are able to eliminate these deficiencies and greatly improve our description of molecular electronic structure.

7.3 Molecular Orbitals

7.3.1 Homonuclear Diatomics

To illustrate the molecular orbital approach we first describe the bonding for O_2 in detail and then extend it to other diatomics.

7.3.1.1 Linear Combination of Atomic Orbitals (LCAO)

Let us consider the bonding in O_2. Each O atom has six valence electrons in the $n = 2$ shell. For O_2 we neglect any contribution of 1s electrons to the bonding because they do not overlap significantly with the orbitals on the neighboring atom. However, the 1s electrons affect the bonding indirectly by screening the valence electrons from the full nuclear charge. If we use the LCAO approximation we can construct molecular orbitals from the four valence atomic orbitals and perform a variational calculation to minimize the ground-state energy of the molecule.

Recall that H_2 and H_2^+ LCAO trial functions gave a reasonable description of the bonding, when appropriately parameterized (see Sections 6.2.3 and 6.2.5). In that case we used two 1s atomic orbitals (basis orbitals) to construct two molecular orbitals (symmetric and antisymmetric combinations of the atomic orbitals) as superpositions. The symmetric combination gave the lowest energy, and we filled it with two electrons (one spin up and one spin down) to generate the ground electronic state configuration. The energy was given by

$$\boxed{\varepsilon_\sigma = \frac{H_{aa} + H_{ab}}{1 + S_{ab}}} \tag{7.5}$$

where H_{aa} is the energy of the atomic orbitals, H_{ab} is the interaction energy between the two orbitals, and S_{ab} is the overlap integral. The subscript σ on the ground state is chosen to indicate that it is a bonding molecular orbital and has enhanced charge density between the nuclei. The antisymmetric combination corresponds to an excited state orbital, and we found its energy to be

$$\boxed{\varepsilon_{\sigma*} = \frac{H_{aa} - H_{ab}}{1 - S_{ab}}} \tag{7.6}$$

The subscript σ^* indicates an antibonding molecular orbital that has diminished electron density between the nuclei. In this picture, the atomic ground states of the two H-atoms are degenerate when the atoms are very far apart, but as they come closer together the orbital wavefunctions overlap and the atomic degeneracy is broken as two molecular states form, of which one is a bonding (symmetric) combination of the atomic orbitals and the other is an antibonding (antisymmetric) combination of the atomic orbitals. The amount of the energy splitting is determined by the interaction energy H_{ab} and the overlap S_{ab}.

Molecular Orbitals of O_2 (σ- and π-Orbitals) Figures 7.11 and 7.12 summarize the energies and LCAO molecular orbitals of O_2. One set of molecular orbitals results from combinations of the 2s-orbitals and the 2p_x-orbitals of atoms a and b, where x is the coordinate in the direction of the bond: these orbitals are commonly called *σ-orbitals*. A second set of orbitals results from combinations of the 2p_y- and 2p_z-orbitals of the atoms a and b, where y and z are the coordinates perpendicular to the direction of the internuclear axis; these orbitals are called π-orbitals.

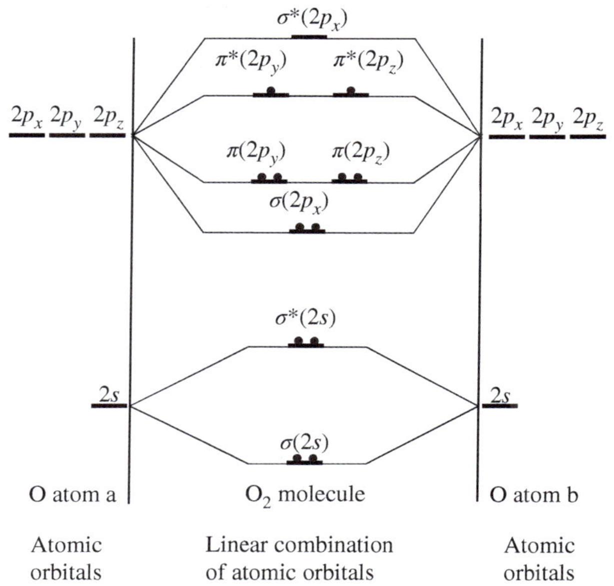

σ orbitals

$\sigma(2s)$: $\Psi_{2s,a} + \Psi_{2s,b}$

$\sigma^*(2s)$: $\Psi_{2s,a} - \Psi_{2s,b}$

$\sigma(2p_x)$: $\Psi_{2p_x,a} - \Psi_{2p_x,b}$

$\sigma^*(2p_x)$: $\Psi_{2p_x,a} - \Psi_{2p_x,b}$

π orbitals

$\pi(2p_y)$: $\Psi_{2p_y,a} + \Psi_{2p_y,b}$

$\pi(2p_z)$: $\Psi_{2p_z,a} + \Psi_{2p_z,b}$

$\pi^*(2p_y)$: $\Psi_{2p_y,a} - \Psi_{2p_y,b}$

$\pi^*(2p_z)$: $\Psi_{2p_z,a} - \Psi_{2p_z,b}$

Figure 7.11 Energy states of O_2. The eight lowest LCAO wavefunctions. The molecular $\sigma(2s)$ and $\sigma(2p_x)$ orbitals and the $\pi(2p_y)$ and $\pi(2p_z)$ orbitals have an antinode in the center of the molecule (bonding orbitals), whereas the molecular $\sigma^*(2s)$ and $\sigma^*(2p_x)$ orbitals and the $\pi^*(2p_y)$ and $\pi^*(2p_z)$ orbitals have a node (antibonding orbitals, marked by an asterisk).

The energy level diagram in Fig. 7.11 shows how the atomic orbitals interact with each other. In particular, the $2s$ orbitals form a bonding and anti-bonding combination; the $2p_x$ orbitals form a bonding and antibonding combination; and so forth.

The $2p$ orbital interactions are also governed by the extent of orbital overlap. Because the p_x orbitals are directed toward each other they have strong overlap and a strong interaction in forming bonding and antibonding combinations. In contrast the $2p_y$ and $2p_z$ orbitals which are oriented perpendicular to the bond axis have less overlap and therefore a weaker interaction. Because the $2p_y$ and $2p_z$ orbitals are oriented perpendicular to each other in space (i.e., the $2p_y$ orbital on one atom is perpendicular to the $2p_z$ orbital on the other atom), they do not bond with each other.

More generally, we can state: *If the orbitals on each atom (or molecular fragment) have different symmetries with respect to the internuclear axis, i.e., different numbers of nodal planes, then the net overlap of the orbitals will be zero (because of the sign changes in the wavefunctions).*

Electron Configuration and Bond Order Using the Aufbau principle we can proceed to fill the molecular orbitals in the molecular orbital energy diagram of Fig. 7.11. Each of the molecular orbitals is labeled by a symbol that identifies its symmetry. In analogy to the procedure with atoms, we specify the electron configuration of the molecule by indicating the electron populations in the molecular orbitals; for oxygen we find

$$\sigma(2s)^2\,\sigma^*(2s)^2\,\sigma(2p_x)^2\,\pi(2p_y)^2\,\pi(2p_z)^2\,\pi^*(2p_y)\,\pi^*(2p_z)$$

To characterize the bond properties we define the *bond order*, as

$$\text{bond order} = \frac{1}{2}\left[N_{\text{bonding}} - N_{\text{antibonding}}\right] \tag{7.7}$$

where N_{bonding} is the number of electrons in bonding orbitals, and $N_{\text{antibonding}}$ is the number of electrons in antibonding orbitals:

State	$\sigma(2s)$	$\sigma^*(2s)$	$\sigma(2p_x)$	$\pi(2p_y)$	$\pi(2p_z)$	$\pi^*(2p_y)$	$\pi^*(2p_z)$	$\sigma^*(2p_x)$
N_{bonding}	2		2	2	2			
$N_{\text{antibonding}}$		2				1	1	

For O_2 the total number of electrons in bonding orbitals is 8, the number of electrons in antibonding orbitals is 4; thus, the bond order in O_2 is two. Because the two most weakly bound electrons occupy a doubly degenerate molecular orbital, they are unpaired (following Hund's rule). This makes the ground electronic state of oxygen a triplet and the molecule is paramagnetic. The simple electron pair model discussed in Section 7.2.1 fails to predict this result.

Example 7.6 *Using the molecular orbital energy diagram in Fig. 7.11, determine the bond orders for O_2^+ and O_2^-. Compare the bonding to that of O_2 and describe the relative stabilities that you expect for these compounds. Contrast your conclusions with that of the electron pair bonding model.*

Using the Aufbau principle we see that the ground-state electronic configuration for the antibonding π orbitals of the oxygen cation is $\pi^*(2p_y)$, and it has a bond order of $2\frac{1}{2}$, whereas for the oxygen anion we have $\pi^*(2p_y)^2\pi^*(2p_z)$ and a bond order of $1\frac{1}{2}$. Hence, the cation should be the most stable, the anion the least stable, and the neutral oxygen molecule will be intermediate. In fact, their experimentally determined electronic binding energies are 6.60 eV for O_2^+, 5.17 eV for O_2, and 4 eV for O_2^-, in accordance with this prediction. In contrast to this result the electron pair model predicts that the removal of an electron from the neutral oxygen will cause either no change or a weakening of the bond strength between the oxygen atoms. The electron pair model would predict that the oxygen anion has a single bond and an unfilled octet on one of the atoms, making it significantly weaker than the other two compounds.

Figure 7.12 shows a molecular orbital energy diagram for O_2 in which the energies and wavefunction plots were calculated using the Hartree-Fock self-consistent field method (see Section 7.4.1). The molecular orbital plots in Figure 7.12 show the molecular orbital shapes and nodal planes.

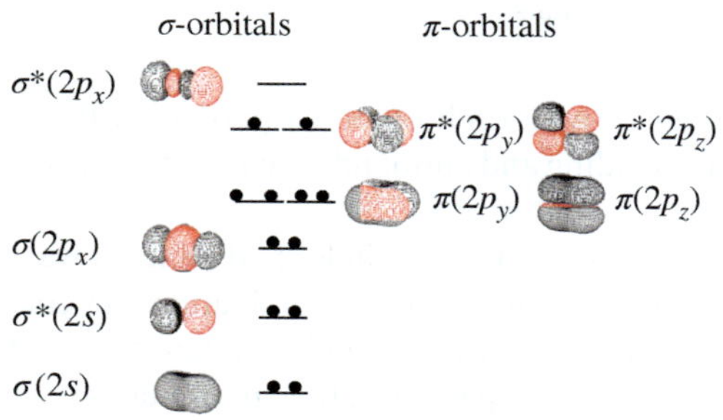

Figure 7.12 LCAO model for O_2 (Slater functions with effective charge number 4.55, internuclear distance 121 pm). Population scheme and illustration of the eight lowest orbitals. Left: σ orbitals, right: π orbitals. The contour lines connect points in space where the absolute value of the wavefunction is 2.5×10^{-4} pm$^{-3/2}$. Black lines: positive values, red lines: negative values.

The π-orbitals have a nodal plane that contains the internuclear axis (for example, the orbitals on the right-hand side in Fig. 7.12), and the σ-orbitals do not have a nodal plane containing the internuclear axis (for example, the orbitals on the left-hand side in Fig. 7.12).

Figure 7.13 shows the experimental photoelectron spectrum for O_2, in which the different peaks correspond to the binding energies of electrons in particular molecular orbitals (see the discussion of Koopmans theorem in Section 5.5.4). The valence electrons in O_2 have binding energies in the range of 10 to 50 eV, which is typical for valence electrons in atoms and molecules (Table 7.1).

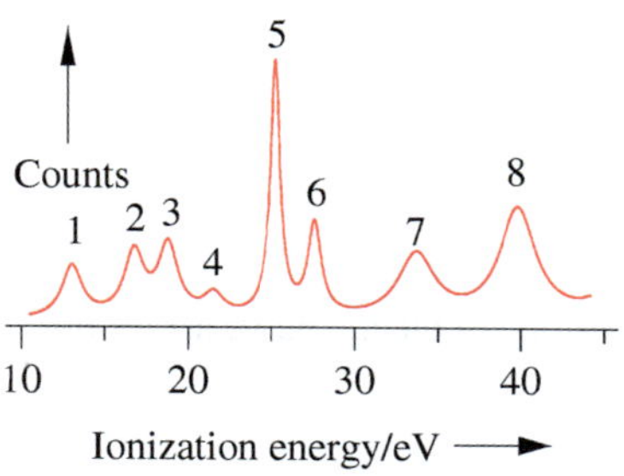

Figure 7.13 Photoelectron spectrum of O_2 molecule [3]. The peaks in this spectrum correspond to the following orbitals: 1: $\pi^*(2p)$, 2 and 3: $\pi(2p)$, 4: $\sigma(2p)$, 5 and 6: $\sigma^*(2s)$, 8: $\sigma(2s)$, see Table 7.1. Peak 7 is an artifact, not connected to the excitation from a molecular orbital of O_2. The method to obtain the experimental photoelectron spectrum is described in Chapter 11.

The ionization energies were calculated, using the energies displayed in Fig. 7.14, given by Equation (3.39) with $L_x = 450$ pm and $L_y = L_z = 300$ pm. We assume that the potential energy of the bottom of the box is −36.8 eV (adjustment to the experimental value of the ionization energy from the $\pi^*(2p)$ orbital, 13.1 eV).

7.3.1.2 Particle-in-Box Wavefunction Description for O_2

We can also describe small, symmetric molecules like O_2 with box model wavefunctions. As with the other treatments, we assume that the 1*s* electrons are localized on the nuclei and focus our attention on the valence electrons. As described earlier for H_2, we use the electron-in-a-box wavefunctions as trial functions in a variational calculation and minimize the energy to

Table 7.1 Ionization Energies of π Electrons in O_2 Orbitals as Measured by Photoelectron Spectrum in Fig. 7.13 and Calculated with the LCAO and Box Models.

Peak Assignment	1	2	3	4	5	6	8
Level Assignment	$\pi^*(2p)$	$\pi(2p)$		$\sigma(2p)$	$\sigma^*(2s)$		$\sigma(2s)$
Energies/eV							
Experiment	13.1	17.0	18.8	21.1	25.3	27.9	39.6
LCAO model[1]	12.2	17.2		20.6	24.5		
Box model[2]	13.1	19.3		21.3	30.7		36.8

find $L_x = 450$ pm and $L_y = L_z = 300$ pm. Figure 7.14 shows the energy level scheme and molecular wavefunctions obtained from this calculation. A comparison between this calculation and the LCAO calculation shows an excellent agreement between the two descriptions. Using the Aufbau principle, we obtain a bond order of two. The occupied molecular orbitals $2\sigma^*$, $3\pi^*$, $4\pi^*$, and $3\sigma^*$ have a node in the center of the bond and are antibonding, whereas the orbitals 1σ, 3σ, 1π, and 2π have an antinode in the center of the bond and are therefore bonding. In addition, predictions of the ground-state multiplicity, and relative stability of cation and anion species are the same as that of the LCAO description. Why do the simple box wavefunctions capture these essential features whereas the electron pair model fails? The box model applies the essential physical and symmetry properties that quantum mechanics requires to electron wavefunctions for describing chemical bonding.

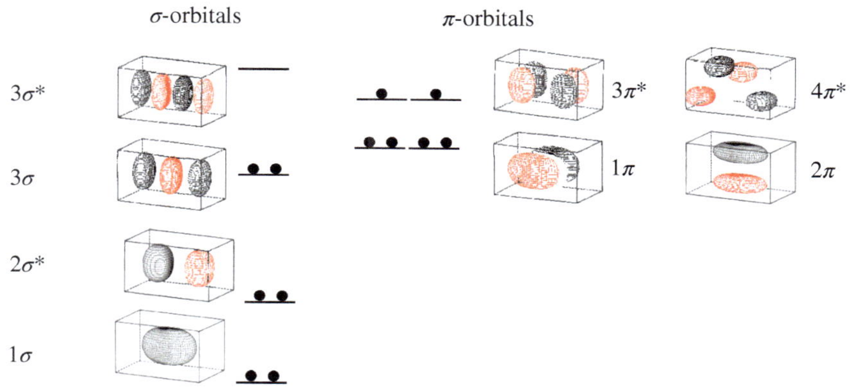

Figure 7.14 Box model for O_2 ($L_x = 450$ pm, $L_y = L_z = 300$ pm). Population scheme and illustration of the eight lowest box wavefunction electronic states. Left: σ orbitals, right: π orbitals. The contour lines connect points in space where the absolute value of the wavefunction is 2.5×10^{-4} pm$^{-3/2}$. Black lines: positive values, red lines: negative values.

7.3.1.3 The Second Row Diatomics

Figure 7.15 shows how the molecular orbital energies for the second row homonuclear diatomics change across the periodic table, from Li to F. Although the absolute scale on this plot is not meaningful, the relative energy ordering of the molecular orbitals as a function of atomic number is meaningful.

Example 7.7 *Use the molecular orbital energy diagram in Fig. 7.15 and the Aufbau principle to describe the bonding in the diatoms B_2, C_2, and N_2. In particular, determine the bond order and the electronic configuration of these diatoms.*

Following the Aufbau principle, we start at the lowest energy orbital and place two electrons in each orbital. In this manner we construct the electronic configurations that are appropriate to the molecules. It is clear from this diagram that in each case the σ orbital contributes a bond but that the σ^* orbital subtracts a bond from the bond order. In the case of B_2 the molecule has a bond order of 1 and an electron configuration of

$$\sigma(2s)^2\, \sigma^*(2s)^2\, \pi(2p_y)\, \pi(2p_z)$$

Following Hund's rule we place one electron in each degenerate orbital with their spins aligned before pairing the electrons.

---

1 Reference see Figure 7.13.
2 The ionization energies were calculated, using the energies displayed in Figure 7.14, given by Equation 3.39 with $L_x = 450$ pm and $L_y = L_z = 300$ pm. We assume that the potential energy of the bottom of the box is -36.8 eV (adjustment to the experimental value of the ionization energy from the $\pi^*(2p)$ orbital, 13.1 eV).

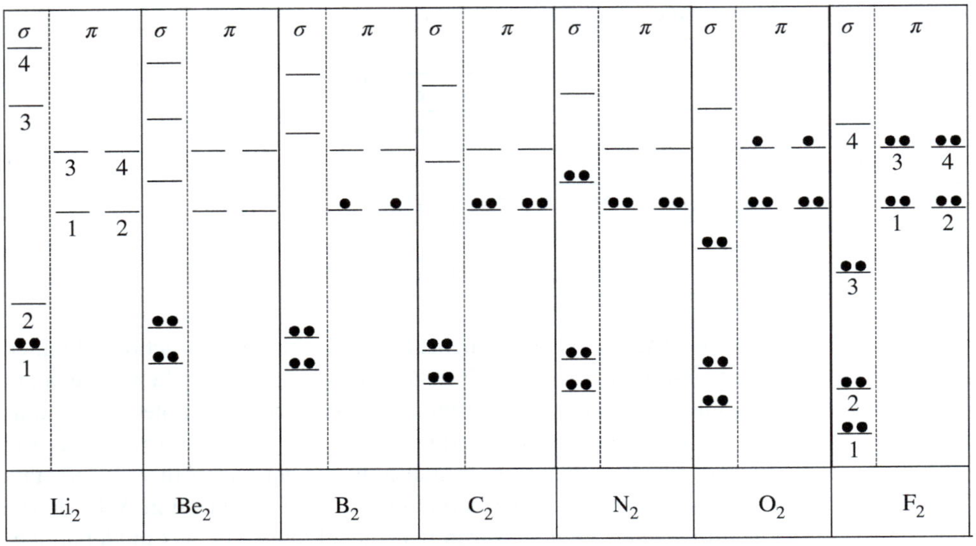

Figure 7.15 Molecular orbital occupation scheme for Li_2 to F_2 (relative positions of the energy levels, not to scale). Note that σ level 3, when proceeding from Li_2 to F_2 is continuously decreasing in energy relative to π levels 1 and 2, since the electrons in σ level 3 are less shielded from the nuclei and therefore more strongly stabilized by the increasing nuclear charge. According to Fig. 7.11, the σ and π orbitals correspond to : $1\sigma = \sigma(2s)$, $2\sigma = \sigma^*(2s)$, $3\sigma = \sigma(2p_x)$, $4\sigma = \sigma^*(2p_x)$, $1\pi = \pi(2p_y)$, $2\pi = \pi(2p_z)$, $3\pi = \pi^*(2p_y)$, $4\pi = \pi^*(2p_z)$.

The carbon diatom C_2 has two more electrons than B_2 and its last two electrons occupy the degenerate π orbitals. The bond order is two and the electron configuration is

$$\sigma(2s)^2\,\sigma^*(2s)^2\,\pi(2p_y)^2\,\pi(2p_z)^2$$

For N_2 the considerations are similar, but now we fill all of the π and σ bonding orbitals. The total bond order is three and the electron configuration is

$$\sigma(2s)^2\,\sigma^*(2s)^2\,\pi(2p_y)^2\,\pi(2p_z)^2\,\sigma(2p_x)^2$$

We predict that the bonding energy is largest for N_2 and weakest for B_2, in agreement with experiment.

Figure 7.16 shows electron density plots for the second row homonuclear diatomics. It is clear from these plots that the electrons are bound more tightly as the atomic charge of the nuclei increases. Consequently, the molecules are of smaller size. Note how the orbitals contract as the charge on the nuclei increases. As the orbitals contract the overlap decreases and the bonding energy decreases.

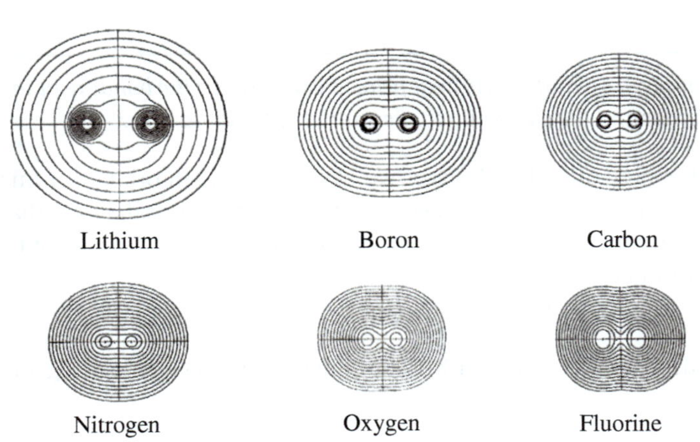

Figure 7.16 Contour plots of the total electron density of second row diatomic molecules Li_2, B_2, C_2, N_2, O_2, and F_2 (experimental bond lengths). Be_2 is not included because its ground state is unstable. The highest contour line has a value of $e \cdot 6800\ nm^{-3}$, the lowest contour line has a value of $e \cdot 0.41 nm^{-3}$. The densities at adjacent contour lines differ by a factor of two. Note that this is an exponential scale, and therefore the distances of adjacent contour lines are approximately constant (A.C. Wahl, Molecular Orbital Densities: Pictorial Studies, *Science*, 1966, **151**, 961).

Table 7.2 presents experimental data on the bond energy and the bond length of the homonuclear diatoms. Analyze the trends in these characteristics and rationalize them using the molecular orbital energy diagrams.

Table 7.2 Constants for Second Row Homonuclear Diatomics.[3]

	Bond Energy	Bond Length	Bond Order	Configuration
	E/eV	d/pm		
Li_2	1.046	267.3	1	$1\sigma^2$
Be_2	0.082	245	0	$1\sigma^2 2\sigma^{*2}$
B_2	3.02	159.0	1	$1\sigma^2 2\sigma^{*2} 1\pi 2\pi$
C_2	6.21	124.3	2	$1\sigma^2 2\sigma^{*2} 1\pi^2 2\pi^2$
N_2	9.759	109.8	3	$1\sigma^2 2\sigma^{*2} 1\pi^2 2\pi^2 3\sigma^2$
O_2	5.116	120.8	2	$1\sigma^2 2\sigma^{*2} 3\sigma^2 1\pi^2 2\pi^2 3\pi^* 4\pi^*$
F_2	1.602	141.2	1	$1\sigma^2 2\sigma^{*2} 3\sigma^2 1\pi^2 2\pi^2 3\pi^{*2} 4\pi^{*2}$
Ne_2	0.004	330	0	$1\sigma^2 2\sigma^{*2} 3\sigma^2 1\pi^2 2\pi^2 3\pi^{*2} 4\pi^{*2} 4\sigma^{*2}$

For the Configuration we are numbering the orbitals as in Fig. 7.15.

7.3.2 Heteronuclear Diatomics

The description of bonding in heteronuclear diatomics requires that we account for the different energies of the atoms' valence electrons. To simplify, we first consider each atom to have only one valence orbital, so the molecular wavefunction has the form

$$\phi = c_a\varphi_a + c_b\varphi_b$$

where φ_a and φ_b are the atomic orbitals on atoms a and b, and c_a, c_b are the corresponding coefficients. The values of c_a^2 and c_b^2 measure the contributions of the atomic orbitals a and b to the molecular orbital. For homonuclear diatomics these values are 0.5 in both molecular orbitals, leading to a symmetric charge distribution. Heteronuclear diatomics have orbital coefficients that are not the same and the charge distribution is asymmetric along the bond direction.

The charge density and extent of asymmetry must be quantified by performing a variational calculation; however, some insight is gained from elementary considerations. For homonuclear diatomics the bonding and antibonding levels have an energy difference of approximately $2H_{ab}$ (if we neglect the effect of S_{ab} in Equations (7.5) and (7.6)). Heteronuclear diatomics have different atomic orbital energies (see Fig. 7.17). As shown in Foundation 7.1, the bonding and antibonding molecular orbitals, which are formed in this case, are separated by an energy $2\sqrt{H_{ab}^2 + \frac{1}{4}(H_{aa} - H_{bb})^2}$. Hence, we find that the bonding molecular orbital lies closer in energy to the more electronegative atom's orbital and the antibonding molecular orbital lies closer in energy to the electropositive atom's orbital. Lastly, we find that the atomic orbital which lies closest in energy to the molecular orbital contributes more strongly (higher value of c^2) to the molecular orbital, as we can expect from perturbation theory; see Foundation 5.3.

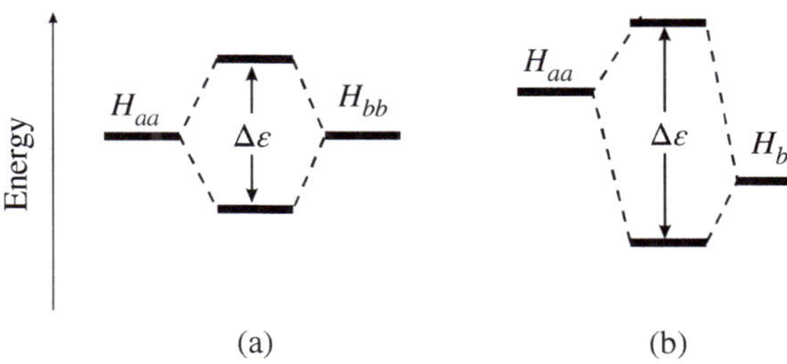

Figure 7.17 Comparison of homonuclear (a) and heteronuclear (b) generic molecular orbital pattern. (a) Homonuclear: $H_{aa} = H_{bb}$ is the energy of both atomic orbitals and $\Delta\varepsilon = 2H_{ab}$ is the energy difference between the ground state and the excited state (overlap integral neglected). (b) Heteronuclear: H_{aa} is the energy of the electropositive atom, H_{bb} is the energy of the electronegative atom. The energy difference between ground state and excited state is $\Delta\varepsilon = 2\sqrt{H_{ab}^2 + \frac{1}{4}(H_{aa} - H_{bb})^2}$ (overlap integral neglected).

7.3.2.1 Bonding in the HF Molecule

By way of illustration, consider the case of the heteronuclear diatomic molecule HF, for which the atoms have quite different electronic energies. The molecular orbital energy diagram in Fig. 7.18a,b uses the measured ionization energies of the F2s

3 Huber and Herzberg [4]. Be-dimer data from [5]. Ne-dimer data from [6].

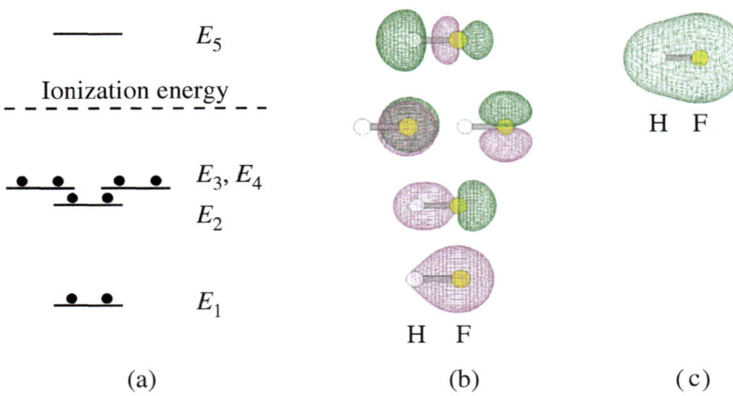

Figure 7.18 HF molecule. (a) Energy levels: $E_1 = -43.1$ eV, $E_2 = -20.3$ eV, $E_3 = E_4 = -17.1$ eV, $E_5 = +5.95$ eV. The dashed line denotes the zero energy level. (b) Corresponding orbitals. (c) Total electron density map.

orbital (40.2 eV), the F2*p* orbitals (18.6 eV), and the H atom (13.6 eV). The molecular orbital energies are determined by a variational calculation using the self-consistent field approach; see Section 7.14. The trial wavefunction in this particular calculation uses the 2*s* and 2*p* angular functions and the 6-31G basis set (see Foundation 7.2) for the fluorine valence electrons and the 1*s* function for the H atom. The molecular energy diagram plots the molecular orbital energies and shows their filling using the Aufbau principle. Because the 2*s* orbital of the F atom is so much lower in energy than the 2*p* orbital and the H atom's 1*s* orbital, it does not contribute strongly to the bonding. Rather the 2*p* orbital on the F atom that points toward the H atom ($2p_x$ in our convention) contributes the most to the bonding.

The asymmetry in the charge distribution becomes evident by analyzing the optimized wavefunctions that are found (Fig. 7.18b). From the MO energy diagram, we see that the bonding interaction arises primarily from the 1*s* orbital of the H atom and the $2p_x$ orbital of the F atom. The bonding orbital that is formed from these two atomic orbitals comprises 40% hydrogen atom 1s ($|c_H|^2 = 0.4$) and 60% the fluorine $2p_x$ ($|c_F|^2 = 0.6$). The antibonding orbital has the opposite distribution (i.e., more H atom character than F atom character). This asymmetry in the atomic orbital contributions to the molecular orbitals leads to an asymmetry in the electron density for the molecule (see Fig. 7.18c).

Example 7.8 *Describe the bonding in LiH. In particular, sketch a molecular orbital energy diagram, identify which atom is electropositive, and determine the bond order for LiH.*

The lithium atom has an ionization energy of 5.39 eV (see Table 5.1). Because the H atom's ionization energy is 13.6 eV, Li will be the electropositive atom and H will be the electronegative atom. The bonding occurs by interaction of the 2*s* orbital on the Li with the 1*s* orbital on the H atom; however, the bonding molecular orbital mostly comprises the H atom's 1*s* (a calculation indicates about 70%). Correspondingly, the antibonding orbital is mostly the Li atom's 2*s* orbital. From the filling pattern of the orbitals in the MO energy diagram (Fig. 7.19), it is evident that the bond order is one and that most of the electron density lies on the H atom. This result coincides with a description of the bonding as ionic bonding between Li^+ and H^-.

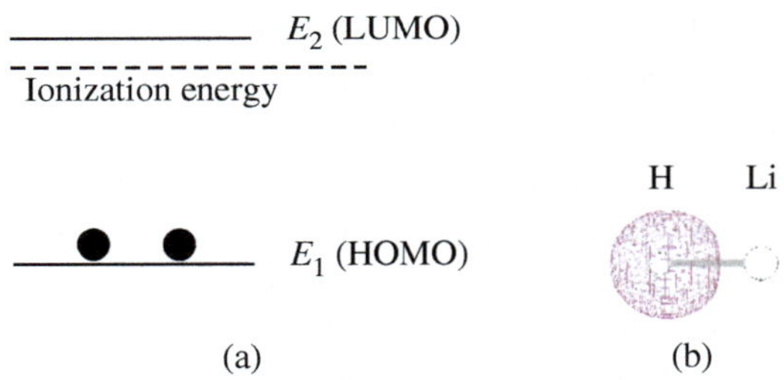

Figure 7.19 MO diagram for LiH. (a) Energy scheme: $E_1 = -8.1$ eV, $E_2 = +0.7$ eV. The dashed line denotes the zero energy level. (b) HOMO orbital showing that the electron is mostly located on the H atom.

7.3.3 Triatomics (H_2O)

The case of three atoms adds a qualitative change to the analysis because we need to predict more than a distance between the two nuclei, in particular the angle made by the atoms and the connectivity. For a molecule like water we need to predict whether the oxygen is bound to each hydrogen, HOH, or to only one, HHO, and whether it is linear or bent (a quantitative model should give a value for the angle which minimizes the energy). The electron pair model and the VSEPR model, as outlined above (see Example 7.3), can provide an excellent starting point for many molecules even if it fails in detail.

7.3.3.1 Structure and Bonding in H_2O

The molecular orbital picture provides a view of the bonding in water that is quite different from the electron pair model, because it allows electrons to delocalize over the entire molecule. To illustrate this difference, we consider a variational calculation that uses an LCAO trial function, which comprises valence orbitals of each atom. For water this involves six atomic orbitals (the 2*s* and three 2*p* orbitals on the oxygen and the two 1*s* orbitals on the hydrogens). Because of its small spatial extent, the 1*s* orbital on the oxygen atom does not contribute to bonding in a significant way and is neglected; shielding of the valence electrons from the full nuclear charge of the oxygen atom is included by using Slater-type orbitals with an effective charge [7].

Figure 7.20 shows these six molecular orbitals. By inspection you can convince yourself that the $2p_x$ orbital, which is oriented perpendicular to the plane of the molecule (defined by the three nuclei), has no net overlap with the hydrogen atom orbitals and is not involved in the bonding. This is orbital 4 in Fig. 7.20.

$$\psi_4 = \varphi_{2px}$$

The symmetric combination of the two H atom orbitals, $(\varphi_{1s_A} + \varphi_{1s_B})$, interacts to form superpositions with the oxygen's 2*s* and $2p_z$ orbitals in the form

$$\alpha \cdot (\varphi_{1s_A} + \varphi_{1s_B}) + \beta \cdot \varphi_{2s} + \gamma \cdot \varphi_{2p_z}$$

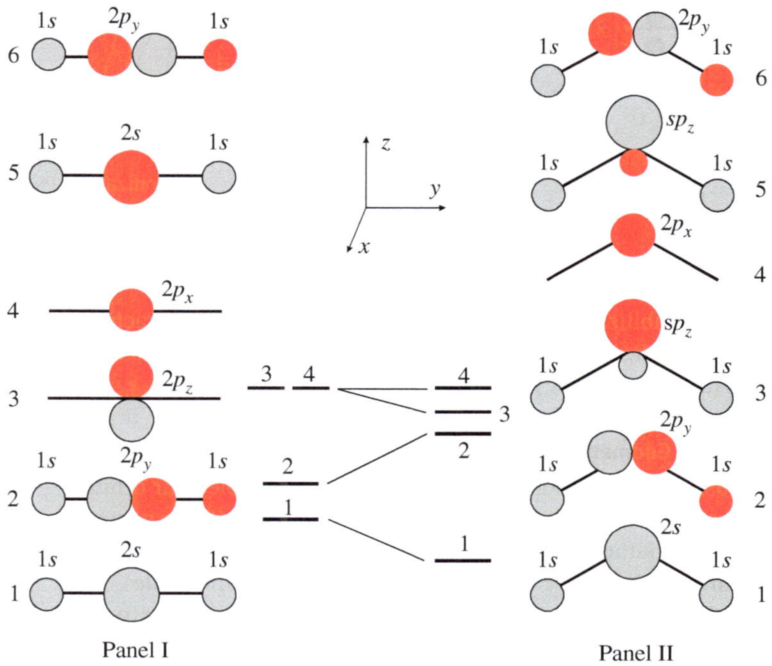

Figure 7.20 Water molecule, LCAO orbitals, and the four lowest (each doubly occupied) energy levels are shown for a linear structure, (a), and a bent structure, (b).

The values of the parameters α, β, and γ are obtained by minimizing the energy.[4] The variational energy calculation generates a minimum energy geometry that has a bond length of 98.6 pm (experimental is 95.8 pm) and bond angle of 100° (experimental is 104.5°). At this geometry, the most stable bonding orbital, labeled 1, is

$$\psi_1 = 0.107 \cdot (\varphi_{1s_A} + \varphi_{1s_B}) + 0.847 \cdot \varphi_{2s} - 0.124 \cdot \varphi_{2p_z}$$

A second bonding orbital with a symmetric combination of H atom orbitals is orbital 3, given by

$$\psi_3 = 0.208 \cdot (\varphi_{1s_A} + \varphi_{1s_B}) - 0.536 \cdot \varphi_{2s} - 0.758 \cdot \varphi_{2p_z}$$

which is orthogonal to ψ_1 (Fig. 7.20, orbital 3).[5] Accordingly, the antisymmetric combination of the H atom orbitals $(\varphi_{1s_A} - \varphi_{1s_B})$ interacts with φ_{2p_y} to form the bonding orbital

$$\psi_2 = 0.314 \cdot (\varphi_{1s_A} - \varphi_{1s_B}) - 0.616 \cdot \varphi_{2p_y}$$

(Fig. 7.20, orbital 2). Orbitals 1 to 4 are filled with the 8 electrons in the valence shell (remember that two electrons are in oxygen's 1*s* core). The remaining two unoccupied orbitals are formed by combining $(\varphi_{1s_A} + \varphi_{1s_B})$ with φ_{2s} and φ_{2pz} and $(\varphi_{1s_A} - \varphi_{1s_B})$ with φ_{2py} (Fig. 7.20, panel II, antibonding orbitals 5 and 6).

Figure 7.21a shows orbital diagrams and the energy levels of the four occupied orbitals. The calculated energies are compared to the peak positions in a photoelectron spectrum of H_2O (Fig. 7.21b). The variational calculation is performed at different geometries, and the bond length and bond angle which give the most stable energy are obtained. Some insight into the orbital interactions is gained by first considering the linear form of H_2O (Fig. 7.20a) and analyzing the changes which occur as the geometry is changed to the bent form. In the linear limit the sigma bonding is formed by the p_y orbital and the *s* orbital on the central atom with the 1*s* orbitals of H (orbitals 1 and 2). The other two *p* orbitals on the central atom do not contribute to the bonding, by symmetry (orbitals 3 and 4). The p_x orbital oriented out of the plane of the bent molecule will also not contribute to bonding as the molecule is bent (orbital 4). The p_z orbital in the plane of the molecule can contribute to the bonding in the bent molecule because of overlap with the H-atom's 1*s* orbitals. It is this orbital which stabilizes as the molecular geometry is bent, because of favorable overlap with the orbitals on the pendant H atoms (orbital

4 Pilar [8]. The calculation was performed using Slater-type orbitals and the SCF approximation described in Foundation 5.4.

5 Actually, in this calculation wavefunctions ψ_1 and ψ_3 have additional contributions of $-0.223\varphi_{1s}$ and $0.097\varphi_{1s}$, respectively, from the 1*s* orbital of the oxygen atom.

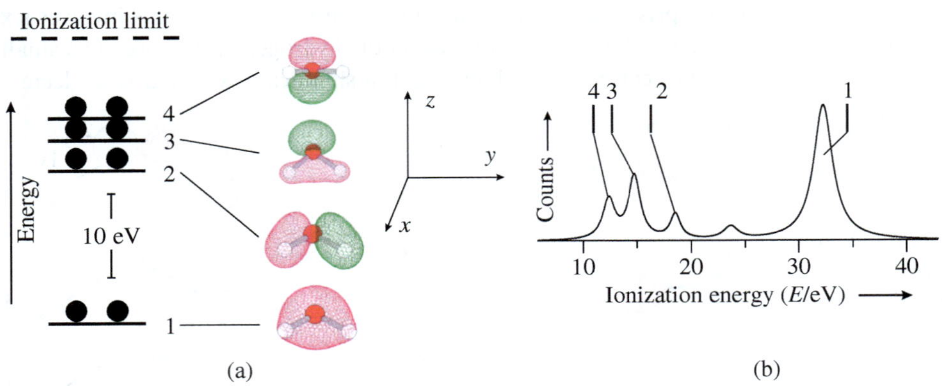

Figure 7.21 Water molecule. (a) Calculated molecular orbital diagrams and energy levels for the four occupied states. The calculated energies are $E_1 = -34.5$ eV, $E_2 = -16.3$ eV, $E_3 = -12.7$ eV, and $E_4 = -10.8$ eV. (b) Photoelectron spectrum (solid line, peak positions corresponding to orbitals 1 to 4 at 32.2, 18.4, 14.7, and 12.6 eV). The bars compare to the ionization energies calculated for orbitals 1 to 4.

3). The stabilization from this orbital is somewhat counteracted by a destabilization of the p_y bonding orbital 2, because its overlap decreases as the molecule bends. Figure 7.22 shows the energy shift of orbitals 1–4 when changing the bond angle (correlation diagram).

7.3.3.2 Geometry

By using a correlation diagram (Fig. 7.22), the molecular orbital description can be used to understand differences in the geometries of triatomic molecules. With modern computing power, quantum chemistry calculations on triatomic molecules are rapid. Nevertheless, it is important to develop intuition into the geometries and electronic structures of molecules.

Let us compare the molecular geometry in BeH_2 to that of H_2O. Both molecules have a B-A-B connectivity and access to the same valence orbitals. In the case of BeH_2 only four valence electrons are available, so that only orbitals 1 and 2 in Fig. 7.20 are filled. From the diagram we see that orbital 1 has a higher energy in the linear form than in the bent form. In contrast, orbital 2 has a lower energy in the linear form. Thus for a decision about the two possibilities we need a detailed calculation, which reveals that the linear form is the more stable one.

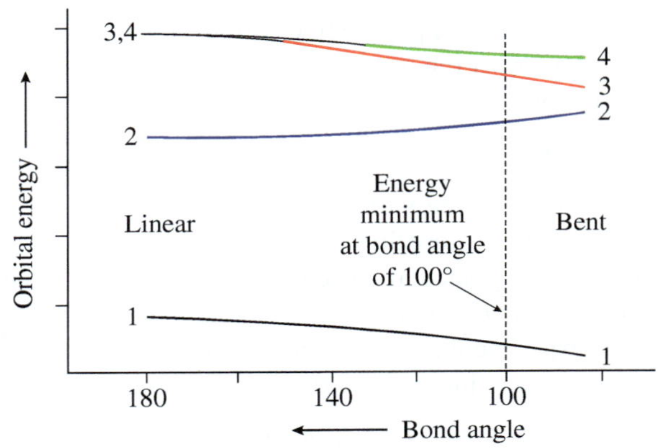

Figure 7.22 Dependence of orbital energies of H_2O (orbitals 1, 2, 3, and 4 in Fig. 7.20) on the bond angle (correlation diagram). The calculated energy minimum[6] is at a bond angle of 100° (R. J. Buenker, S. D. Peyerimhoff, *Chem. Rev.* 1974, **74**, 129). The experimental value is 104.52° (Tables of Interatomic Distances and Configurations in Molecules. Chem. Soc. Spec. Publ., No. 11 (1958); ibid., No. 18 (1965)).

Consider H_2O, which has eight valence electrons, so that orbitals 1 through 4 are filled for the ground-state electron configuration. From Fig. 7.20 we see that orbitals 1 and 3 are most stable in a bent geometry and orbital 2 is most stable in a linear geometry. In this case, we predict that the molecule will be bent at an angle which results from a balance of the destabilization of orbital 2 compared to the stabilization of orbitals 1 and 3. Exactly what angle requires a detailed calculation.

To use the correlation diagram we apply the Aufbau principle to fill the orbitals and then determine which geometry is more stable, linear or bent. For example, if the bonding orbitals 1 and 2 are occupied but the orbital 3 is not occupied, we expect the molecule to be linear (see Table 7.3 for examples). The presence of a single electron in orbital 3 causes the most stable geometry of the molecule to be bent, however. In contrast, the occupation of orbital 4 by electrons has little or no effect on the geometry, because this orbital does not contribute to bonding and its energy depends only weakly on angle. This logic may be extended further, and Table 7.3 reports some experimental geometries for different orbital occupations of AB_2 molecules.

6 The energy minimum follows from a SCF-calculation of the total energy

$$E_{\text{total}} = \sum \text{orbital energies} + V_{\text{nn}} + V_{\text{ee}}$$

as a function of the bond angle, where V_{nn} is the nuclear repulsion energy and V_{ee} is the electron repulsion energy.

Table 7.3 Geometries and Orbital Populations for AB_2 Molecules.

	ψ_4 (empty)	ψ_4 (1 electron)	ψ_4 (2 electrons)
ψ_3 (empty)	linear (BeH_2, MgH_2)	linear (BH_2*, AlH_2*)	
ψ_3 (1 electron)	BH_2 is 131°	triplet CH_2 is 136°	NH_2* is 144°
	AlH_2 is 119°		PH_2* is 123°
ψ_3 (2 electrons)	singlet CH_2 is 102°	NH_2 is 103°	H_2O is 105°
	SiH_2 is 92°	PH_2 is 92°	H_2S is 92°

An asterisk denotes an excited state of the molecule.

The considerations developed here are general and work equally well for predicting the geometries of excited electronic configurations. From a series of calculations it is known that the slope of the curve for orbital 3 in Fig. 7.22 is steeper for larger central atoms, leading to smaller bond angles. In addition it is found that the bond angle in an isoelectronic series depends on the charge state of the molecule, so that the bond angle of a cation is larger than that of the corresponding neutral species, which is larger than that of the corresponding anion. Similar plots, between other geometric shapes, can also be developed for other AB_n systems.

7.3.4 Polyatomics

Even for the simple example of the triatomic water molecule, H_2O, the differences between the electron pair description and the molecular orbital description become evident. The molecular orbital description of the bonding views the molecule as an entire entity and the bonding is delocalized over more than two atomic centers. In this description, the bond energy of OH is not directly linked to a particular electronic energy level of the molecule or pair of electrons. Nevertheless, a bond energy can be defined in a clear way (e.g., comparing the electronic energy of H_2O to that of OH and H) and calculations give good agreement with experiment. However, the delocalization of the wavefunctions in the molecular orbital description makes the calculations more difficult for very large molecules, such as proteins and even small peptides.

An approximate solution to this problem is suggested by the electron pair model and the trends in experimental bond energies. The simple electron pair description treats each electron bond as a separate and local entity that is largely unaffected by the presence of atoms not directly involved in the bond. For example, consider the change in OH bond energy through a homologous series of compounds, the n-alcohols. As we proceed from water to methanol, we expect that the OH bond will change somewhat and this can be attributed to changes in oxygen's hybridization because one of its OH bonds is replaced by a CO bond. If we proceed from methanol to ethanol, and higher n-alcohols, the electron pair model predicts that the OH bond energy is largely unchanged because the structural changes are more than one bond away from the OH moiety. In fact the OH bond energies change in the order: 8.29×10^{-19} J for H_2O, 7.26×10^{-19} J in CH_3OH, 7.29×10^{-19} J in CH_3CH_2OH, and 7.2×10^{-19} J in $CH_3(CH_2)_2OH$.

7.3.4.1 Divide and Conquer Strategy

The local behavior of bond properties suggests a "divide and conquer" strategy for describing the bonding in larger molecules. In this approach we divide the molecule into subunits, which are each described by a molecular orbital model, and treat the bonding between them in a local manner. The usefulness of this strategy relies on the accuracy of our approximation that the bonding between molecular subunits can be treated in a local way.

To illustrate this idea consider the bonding in acetonitrile, CH_3CN. In this case we treat the cyano radical CN as one subunit and consider its interaction with the methyl group CH_3. Figure 7.23 shows molecular orbital energy diagrams for CH_3CN. On the left we consider the molecular orbitals for the

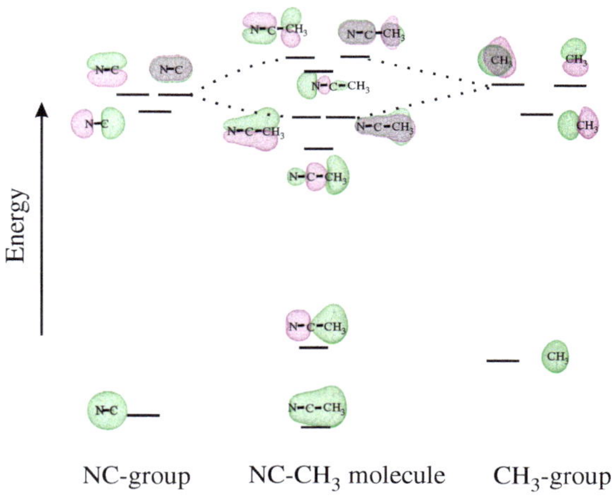

Figure 7.23 The molecular orbital energy diagram for filled levels of acetonitrile ($N{\equiv}C{-}CH_3$) is shown in the center. The relationship between these molecular orbitals and that of the fragments CH_3 (right) and NC (left) is indicated. In each case a probability density map of the orbital is shown near the energy level to which it corresponds.

CN radical and on the right are those of the CH_3 radical. We align the relative energies of these molecular orbitals and consider their interaction. As with bonding between atoms, the most weakly bound electrons (highest occupied molecular orbital, HOMO) on the molecular fragments are the outermost and interact most strongly with their counterparts of similar energy. In analogy to the orbital energy diagrams shown earlier (see Fig. 7.11) these "valence" orbitals interact to form bonding and antibonding combinations, which are illustrated in the diagram (center).

This local description assumes that the CN bond is largely unchanged by the interaction with a pendant group, e.g., methyl. This approximation is reasonable when the pendant moiety bonds primarily through a sigma bond to the carbon of the CN group. Table 7.4 shows CN bond properties for different molecules. These data reveal that the CN bond length is the same within 1% and that the force constant (see Section 7.5.3) is within 3% of the average value, suggesting that our approximation of local bonding is a good one. For pendant groups that interact strongly with the π orbitals of the CN unit the bond characteristics can change somewhat.

Table 7.4 Length d and Force Constant k_f for the C ≡ N Bond.[7]

Molecule	d/pm	k_f/N m^{-1}
HCN	115	1880
CH_3 CN	116	1770
CH_3CH_2 CN	116	1760
$CH_3(CH_2)_2$ CN	117	1760
FCN	116	1670
ClCN	116	1675
BrCN	116	1690
ICN	116	1670

7.3.4.2 Functional groups

In organic chemistry a functional group is defined as an atom or collection of atoms that determines the properties and identity of a family of organic compounds. This basic principle helps provide an underlying structure to much of synthetic organic chemistry. A first attempt to describing bonding and structure in complex organic compounds is to consider the molecular orbital description of the individual functional groups and provide a local description for their interaction with one another.

As a simple example, consider the family of alkanes which comprise two subunit types, the methyl group CH_3 and the methylene unit CH_2. Qualitatively, the bonding in ethane is reasonably described by local bonding of the sp^3-like hybrid orbitals of the carbons of the CH_3 units interacting to form a bonding and antibonding combination. Since each methyl radical contributes a single electron, a bond is formed between the two carbons. When describing a chain of methylene groups between two terminal methyl units, one uses two sp^3 orbitals on the CH_2 so that two carbon-carbon bonds can be made for each CH_2 group.

Other moieties, such as the carbonyl C=O, are more complex. The carbonyl group appears in a number of important functional groups, such as amides, aldehydes, ketones, and carboxylic acids. Figure 7.24 shows an orbital energy diagram for formaldehyde (H_2CO), the simplest carbonyl containing molecule. The relative orbital energy spacing used in this diagram were obtained from photoelectron spectra. If we replace one of the hydrogens with a methyl unit to make an acetaldehyde (CH_3CHO), the HOMO orbital retains its basic character but is destabilized by about 0.6 eV. Its energy changes further if we replace the methyl group by an ethyl group (another 0.4 eV) but remains fairly constant in energy with further increase in the alkane chain. Much of a molecule's chemistry is determined by the energetics and spatial arrangement of its most weakly bound (HOMO) electrons, so we expect that molecules of this type should have similar chemical properties. Although the alkyl aldehydes have similar chemistry and bond characteristics for the CO unit, other substitutions can provide more dramatic changes. For example the chemical behavior of an acetaldeyde is dramatically different from that of formic acid (HCOOH) or the amide functionality, RNHCOR', which is so important in biochemistry.

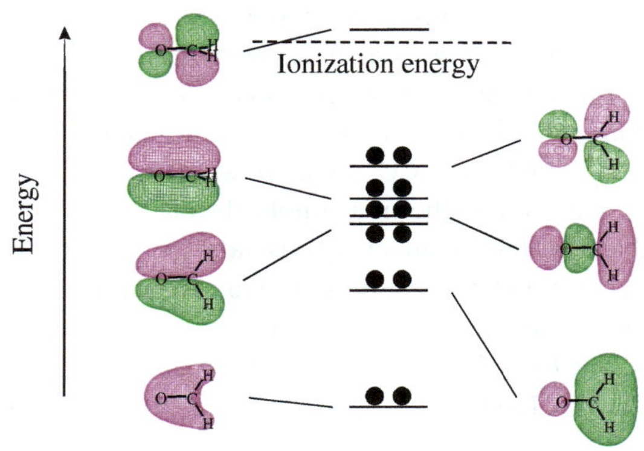

Figure 7.24 A molecular orbital energy diagram of formaldehyde ($H_2C{=}O$).

7 Tables of Interatomic Distances and Configurations in Molecules and Ions. Spec. Publ. 11 (1958); 18 (1965), The Chemical Society, London. Landolt–Börnstein: Zahlenwerte und Funktionen, 1. Bad, 2. Teil, Springer, Berlin 1950.

7.4 Quantum Chemistry Extension: Orbital-Based Numerical Methods

This section develops further the Hartree Fock/Self-Consistent Field ideas first described in Chapter 5 (Section 5.3 and Foundation 5.4). We proceed by considering the electronic Schrödinger equation for a molecule

$$\mathcal{H}^{(\text{el})}\Psi^{(\text{el})} = E^{(\text{el})} \cdot \Psi^{(\text{el})}$$

where the superscript (el) is used to emphasize that the Born–Oppenheimer approximation is made and the nuclear positions are fixed in a calculation. The wavefunction $\Psi^{(el)} = \Psi(\vec{r}_i; \vec{r}_a)$ is a function of the position $\vec{r}_i$ of each electron i, and it depends parametrically on the position $\vec{r}_a$ of each nucleus a. That is, we solve the Schrödinger equation for electrons as variables, but keep the nuclear positions fixed (this is the Born–Oppenheimer approximation). Separate calculations must be made to determine how the electronic energy changes for each choice of $\vec{r}_a$; see Chapter 6 The electronic Hamiltonian, $H^{(\text{el})}$, is

$$\mathcal{H}^{(\text{el})} = -\frac{\hbar^2}{2m_e}\sum_{i=1}^{N}\nabla_i^2 - \frac{e^2}{4\pi\varepsilon_0}\sum_{i=1}^{N}\sum_{a=1}^{M}\frac{Z_a}{r_{i,a}} + \frac{e^2}{4\pi\varepsilon_0}\sum_{j>i}^{N}\sum_{i=1}^{N}\frac{1}{r_{i,j}} + V_{NN}$$

The nuclear–nuclear repulsion term

$$V_{NN} = \frac{e^2}{4\pi\varepsilon_0}\sum_{a>b}^{M}\sum_{b=1}^{M}\frac{Z_aZ_b}{r_{ia,b}}$$

has a fixed value at any given choice for the geometry

$$\vec{R} = \left(\vec{r}_a, \vec{r}_b, \ldots \vec{r}_M\right)$$

of the M nuclei. The indices i and j denote the electrons, of which there are N, and the labels a and b denote the nuclei, of which there are M. Note that the sums with $j > i$ (and $a > b$) are used to ensure that the pairwise interactions between two different electrons (and nuclei) are not counted twice. Thus we must solve the equation

$$-\frac{\hbar^2}{2m_e}\sum_{i=1}^{N}\nabla_i^2\Psi^{(\text{el})} - \frac{e^2}{4\pi\varepsilon_0}\sum_{i=1}^{N}\sum_{a=1}^{M}\frac{Z_a}{r_{i,a}}\Psi^{(\text{el})} + \frac{e^2}{4\pi\varepsilon_0}\sum_{j>i}^{N}\sum_{i=1}^{N}\frac{1}{r_{i,j}}\Psi^{(\text{el})}$$
$$= \left(E^{(\text{el})} - V_{NN}\right)\Psi^{(\text{el})} = E \cdot \Psi^{(\text{el})} \tag{7.8}$$

In this form of the Schrödinger equation, the first term on the left corresponds to the sum of the kinetic energies of each electron i, the second term corresponds to the Coulomb attraction of each electron i to each nuclear charge Z_a, and the third term on the left corresponds to the electron–electron repulsion between each pair of electrons i and j. Note that the equation has been rearranged to place the nuclear–nuclear repulsion term, which is independent of the electron coordinates, on the right-hand side of the equation, along with the energy eigenvalue.

The solution to this equation follows in a manner which is similar to that used for the Hartree–Fock Self-Consistent Field (HF-SCF) method in Foundation 5.4, and we find

$$E^{(\text{el})} = 2\sum_{i=1}^{N/2}\varepsilon_i - \sum_{j=i}^{n/2}\sum_{i=1}^{n/2}\left(2 \cdot J_{ij} - K_{ij}\right) + V_{NN}$$

Foundation 7.2 and [8] for more discussion.

7.5 Polarity, Bond Length, Elasticity

7.5.1 Polarity of Bonds and Electronegativity

Whenever two different elements are bonded together an asymmetry in the charge distribution occurs and generates a bond dipole moment μ.

$$\mu = Q \cdot a$$

For neutral HF the center of negative charge is shifted toward the F atom and a partial negative charge $-Q$ is assigned to the fluorine atom ($-Q$ is the electron density on the F atom that is in excess of that found on a neutral F atom), and a partial positive charge $+Q$ at a distance a, is assigned to the hydrogen atom. Because the fluorine has a local excess of negative charge we say that it is *electronegative*. Correspondingly, the H atom is electropositive. The measured dipole moment for HF is 1.83 D, where D is the Debye unit. The conversion to SI units gives $1\text{D} = 3.33564 \times 10^{-30}$ C m.

The tendency of an atom to attract a bonding electron is quantified by the value of the electronegativity. A number of different electronegativity scales have been defined. Two of the more popular ones are the Pauling scale and the Mulliken scale. The Pauling scale uses the difference between the bond energy E of heteronuclear bonds and the arithmetic mean of the corresponding homonuclear bonds

$$\Delta = E_{A-B} - \frac{1}{2}\left[E_{A-A} + E_{B-B}\right]$$

and assumes that Δ is correlated with the relative tendency of the atoms to attract electrons. By analyzing a range of experimental data, Pauling proposed that $\sqrt{\Delta}$, not Δ itself, fulfills an additivity relation. Because its absolute scale is not determined, the electronegativities of F (the most electronegative atom) and C were fixed to be exactly 4.00 and 2.50. The other atoms are scaled accordingly, based on the proportionality of the difference in electronegativity with $\sqrt{\Delta}$ [9]. In contrast to this definition in terms of bonding properties of two atoms, Mulliken defined an electronegativity scale in terms of atom properties. In particular, he defined the average of an atom's ionization energy and its electron affinity as the electronegativity. A scaling factor can be applied to this definition in order to bring the two definitions into approximate agreement. Table 7.5 shows electronegativities for some common elements, in both the Pauling (P) and the Mulliken (M) scales.

Table 7.5 Electronegativities χ of Some Main Group Elements Are Reported for the Pauling[8] and Mulliken [11] Scales.

H						
2.20 P						
3.06 M						
Li	Be	B	C	N	O	F
0.98 P	1.57 P	2.04 P	2.55 P	3.04 P	3.44 P	4.00 P
1.28 M	1.99 M	1.83 M	2.67 M	3.08 M	3.22 M	4.43 M
Na	Mg	Al	Si	P	S	Cl
0.93 P	1.31 P	1.61 P	1.90 P	2.19 P	2.58 P	3.16 P
1.21 M	1.63M	1.37 M	2.03 M	2.39 M	2.65 M	3.54 M
K	Ca	Ga	Ge	As	Se	Br
0.82 P	1.00 P	1.81 P	2.01 P	2.18 P	2.55 P	2.96 P
1.03 M	1.30 M	1.34 M	1.95 M	2.26 M	2.51 M	3.24 M
Rb	Sr	In	Sn	Sb	Te	I
0.82 P	0.95 P	1.78 P	1.96 P	2.05 P	2.10 P	2.66 P
0.99 M	1.21 M	1.30 M	1.83 M	2.06 M	2.34 M	2.88 M

The electronegativity differences between atoms provide an excellent qualitative measure for the polarity of a bond. The larger the difference in electronegativity of the bonding atoms the larger is the bond polarity and the bond's dipole moment. In the homonuclear limit, when the electronegativity difference is zero, the electron density is symmetric and the bond is called "covalent." In the language of electron pair bonding, this limit (e.g., F_2) has equal sharing of the bonding electrons. As the electronegativity difference between the bonding partners increases, the electron density becomes more asymmetric and the bond is described as "polar covalent" (e.g., F-Cl). When the electronegativity difference between the atoms is so large that the electron pair is localized primarily around one atom and not the other, the bond is called "ionic" (e.g., F^-Li^+).

8 Allred [10]. These numbers differ slightly from those originally suggested by Pauling.

Example 7.9 *Use Table 7.5 to correlate the measured dipole moments of the hydrogen halides with the differences in the atomic electronegativities.*

Molecule	HF	HCl	HBr	HI
μ/D	1.826	1.109	0.827	0.448
$\Delta\chi$ (Pauling)	1.80	0.96	0.76	0.46

A plot of the experimental dipole moment versus the difference $\Delta\chi$ in the atomic electronegativities (see Fig. 7.25) shows a good linear correlation.

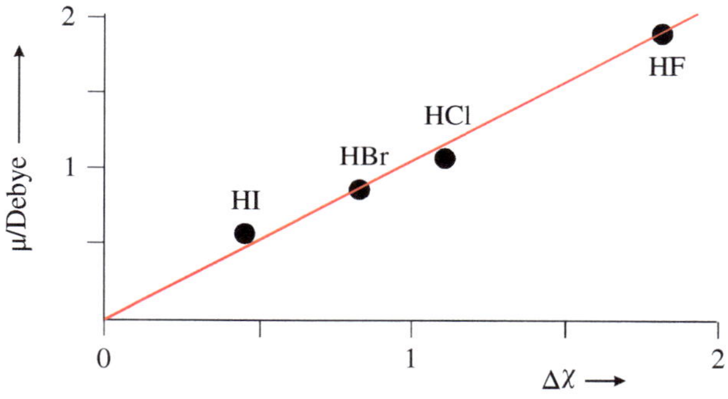

Figure 7.25 Plot of dipole moment versus electronegativity difference $\Delta\chi$ (Pauling electronegativities).

7.5.2 Covalent Radii and Van der Waals Radii

From the discussion of bonding, we see that the average distance between two nuclei is determined by the atom's size and the stabilizing interactions in the bond formation. Because only the valence electrons are involved in the bonding, we find that atoms appear to have somewhat regular volumes that are excluded to other atoms (the core electrons), even in bond formation. This feature allows us to define an effective atomic size for atoms in bonds, that is different from that of atoms which are not bonded.

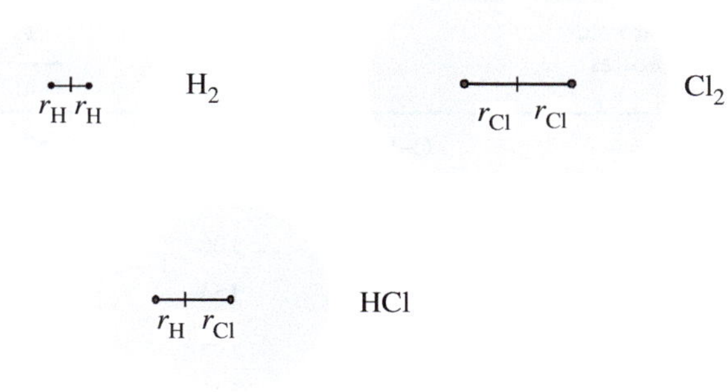

Figure 7.26 Calculating bond lengths from bond radii.

Table 7.6 provides some characteristic bond lengths for diatomic and polyatomic molecules. The bond lengths arise from two primary effects, the change in nuclear charge and the change in the principal quantum number of the valence electrons. For example, the bond length decreases from N – H (101 pm) to O – H (96 pm), because the increased nuclear charge of oxygen, compared to nitrogen, contracts the electron density around the atom. In contrast, the bond length increases from HF (92 pm) to HI (161 pm), even though I has a larger nuclear charge than F. This increase occurs because iodine uses a 5*p* orbital for bonding instead of the 2*p* orbital used by fluorine. Because the bond length depends upon the spatial extent of the atomic orbitals, we can approximately calculate the length of any chosen bond from atomic contributions. We assume that in a homonuclear diatomic molecule one-half the average length of a bond (Fig. 7.26) is the covalent radius of an atom. The covalent radii reported in Table 7.7 are obtained accordingly.

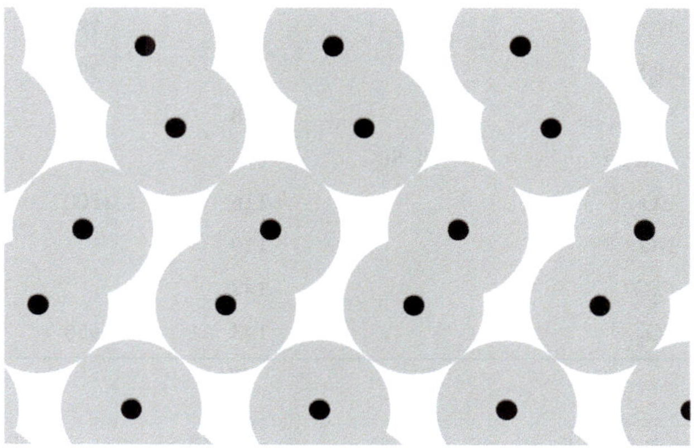

Figure 7.27 The arrangement of I_2 molecules in a crystal. The smallest distance between the nuclei of different molecules is much larger than the internuclear distance inside a molecule.

Another important measure of atomic size is the "contact" distance of atoms which are not bonded together, called "van der Waals" contact. For example, I_2 molecules in an iodine crystal pack at a distance of 400 pm between the nuclei of two adjacent I_2 molecules (Fig. 7.27), whereas the distance between the nuclei in one molecule is much smaller (267 pm, Table 7.6). Bonding orbitals overlap and bind; however, doubly occupied orbitals do not because of the Coulomb repulsion of the electron clouds and Pauli exclusion. Therefore covalent bonding radii and atomic radii must be distinguished; the latter are often called *van der Waals radii*. Table 7.7 compares these radii for some atom types.

Table 7.6 Bond Lengths d, Force Constants k_f, and Bond Energies ε_B for Some Diatomic[9,10] and Polyatomic[11] Molecules.

Diatomic Molecules	d_{eq} / pm	k_f / N m^{-1}	ε_B / 10^{-19} J	ε_B / eV	ε_{elec} / eV
H_2^+	105.7	160.2	4.247	2.651	2.795
H_2	74.1	575.9	7.175	4.478	4.751
O_2	121	1178	8.196	5.116	5.214
N_2	110	2298	15.636	9.759	9.905
F_2	141	470.9	2.566	1.602	1.659
Cl_2	199	323.2	3.972	2.479	2.514
Br_2	228	246.4	3.157	1.971	1.991
I_2	267	172.2	2.471	1.542	1.556
HF	95.7	486.3	9.403	5.869	6.055
HCl	127	517.0	7.104	4.434	4.619
HBr	141	412.1	6.021	3.758	3.922
HI	161	314.5	4.893	3.054	3.197
CO	113	1904	17.772	11.092	11.227
NO	115	1597	10.409	6.497	6.615

Polyatomic Molecules		d / pm	k_f / N m^{-1}	ε_B / 10^{-19} J	ε_B / eV
CH_4	C–H	109	480	7.28	4.54
C_2H_4	C–H	107		7.37	4.60
C_2H_2	C–H	106		9.17	5.72
CH_3-CH_3	C–C	154	434	6.24	3.90
Diamond	C–C	154			3.70
$CH_2=CH_2$	C=C	133	1080	11.96	7.46
$CH\equiv CH$	C≡C	120	1490	15.97	9.97
NH_3		101	635	7.46	4.66
H_2O		96	780	8.27	5.16
H_2S		134	390	6.33	3.95
SiH_4		148			3.92
Silicon	Si–Si				2.30
CO_2		116	1500	8.83	5.52
NO_2		119		5.07	3.16
F_2O		142			
SO_2		143	956	9.17	5.72

9 Experimental data for H_2^+: Chang et al. [12]: $d = 105.8$ pm. Pollard et al. [13]: $d = 105.6$ pm.
10 Theoretical data for H_2^+: Schaad and Hicks [14]: $d = 105.69$ pm.
11 Bond energies in polyatomic molecules are extracted from the energies of dissociation into atoms, which is assumed to be the sum of the energies for each bond.

These radii, covalent and van der Waals, can be used to construct simple mechanical models for a molecule. For a diatomic molecule, one cuts a hemisphere of van der Waals radius for each atom and puts them together at the correct internuclear distance. Such a "*space-filling model*" (after applying a magnification of about 1×10^8) is shown in Fig. 7.28. This approach can be used to model more complicated molecules as long as the internuclear distances and bond angles are known.

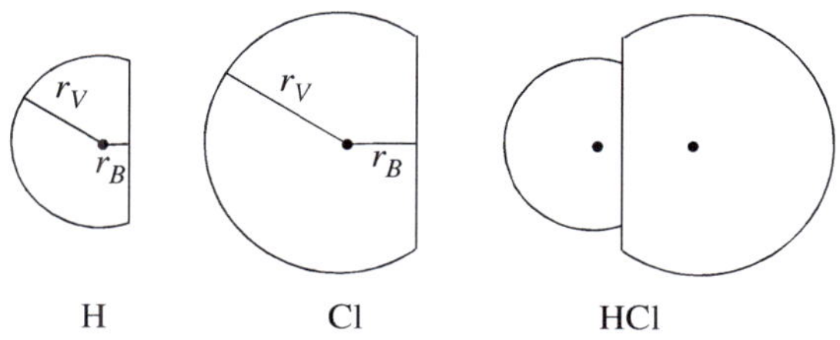

Figure 7.28 Space-filling models of H_2 and Cl_2 (r_B is the covalent bond radius, r_V is the van der Waals radius).

Example 7.10 *Use the covalent radii in Table 7.7 to predict the bond length and molecular size of the HCl molecule.*

The bond length of HCl is given by the sum of the H and Cl covalent radii,

$$d_{\mathrm{HCl}} = r_{\mathrm{H}} + r_{\mathrm{Cl}} = 31\ \mathrm{pm} + 102\ \mathrm{pm} = 133\ \mathrm{pm}$$

This distance is within 5% of the experimental value (127 pm). The overall length of the molecule is given by the sum of its bond length and the two van der Waals radii of the ending atoms,

$$\text{length} = d_{\mathrm{HCl}} + r_{\mathrm{H}} + r_{\mathrm{Cl}} = 136\ \mathrm{pm} + 120\ \mathrm{pm} + 180\ \mathrm{pm} = 436\ \mathrm{pm}$$

The molecule's cross-sectional diameter will range from about 240 pm on the hydrogen end to 360 pm on the chlorine end.

Table 7.7 Covalent Radii[12] and van der Waals Radii for Some Atoms.

Atom	Atom Radius/pm	
	Covalent	Van der Waals
F	57	135
Cl	102	180
Br	120	195
I	139	215
C	76	160
N	71	150
O	66	140
H	31	120

The discussion to this point has described the bonding in the molecule with regard to simple geometries and shapes, and it has considered the molecular structure in terms of its average molecular structure. The description of polyatomic molecules requires a more detailed description of the shape that is three-dimensional and properly includes the relative stability of different nuclear arrangements. In addition, molecules at thermal energies are dynamical entities, constantly in motion. In addition, to their overall translation and rotations in space, the molecules undergo internal motions. For a simple diatomic the motion corresponds to oscillations in its bond length. For more complex molecules, bending and torsional motions are important also.

12 Cordero et al. [15]. These covalent radii are averages of a large number of bond distances.

7.5.3 Stretching, Bending, and Torsion of Bonds

7.5.3.1 Stretching (Hooke's Law)

An increase or decrease in the distance between two nuclei changes a molecule's energy. A decrease in the bond length below the equilibrium bond length causes an increase in the repulsion between each atomic core's electron shells, and an increase of the distance beyond the equilibrium bond length causes a decrease in the electron overlap in the bonding region. We observed this dependence for the H_2^+ ion and the H_2 molecule (see Figs. 6.2, 6.8, and 6.10). Plots of the electronic energy with bond displacement are called "potential energy curves."

If the relative displacement of the nuclei is small enough, we can assume that the force f needed for an elongation x of the bond is proportional to x, such that

$$f = k_f \cdot x = k_f \cdot (d - d_{eq}) \tag{7.9}$$

This relation between the force and the displacement is the same as Hooke's Law for the deformation of a spring (k_f is the stretching force constant). From Equation (7.9) we can obtain the energy $\Delta\varepsilon$ that has to be supplied to elongate the bond by x

$$\Delta\varepsilon = \varepsilon - \varepsilon_{eq} = \int_0^x f \cdot \mathrm{d}x = \int_0^x k_f x \cdot \mathrm{d}x = \frac{1}{2} k_f \cdot x^2 \tag{7.10}$$

(ε is the energy at a distance d between the nuclei, ε_{eq} is the energy at the equilibrium distance d_{eq}). This relation looks qualitatively plausible, because the energy curves show a parabolic shape for small deflections. As an example, consider the energy curve for the H_2^+ ion (Fig. 7.29). Using the data for these potential energy curves around the minimum, we obtain the following values of k_f (see Problem P7.11)

$k_f = 164\,\text{N m}^{-1}$ (exact wavefunction)
$k_f = 60\,\text{N m}^{-1}$ (box model)
$k_f = 98\,\text{N m}^{-1}$ (LCAO model)

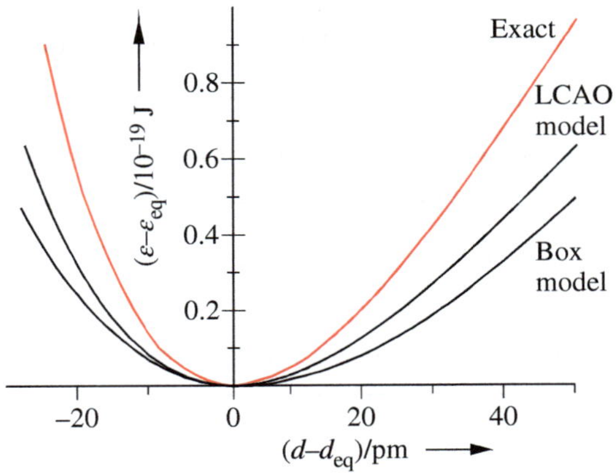

Figure 7.29 Energy $\varepsilon - \varepsilon_{eq}$ of an H_2^+ ion as a function of the elongation $d - d_{eq}$ of the nuclei (data see Problem P7.11); ε_{eq} is the energy at the equilibrium distance d_{eq}. The curves correspond to the exact calculation and to the box and LCAO models (see Chapter 6).

The experimental value for the H_2^+ ion is $k_f = 160\,\text{Nm}^{-1}$.

In Table 7.6 values of the force constant k_f are reported for bonds in some different molecules. For the C – C bond of ethane, $k_f = 434\,\text{Nm}^{-1}$. Hence, the application of a force of one nN, $f = 10^{-9}$ N, elongates the bond by two picometers; i.e.,

$$x = \frac{f}{k_f} = \frac{10^{-9}\ \text{N}}{434\text{N m}^{-1}} = 2 \times 10^{-12}\ \text{m} = 2\ \text{pm}$$

Figure 7.30 illustrates this process but exaggerates the bond length change, which really corresponds to an elongation of 1%.

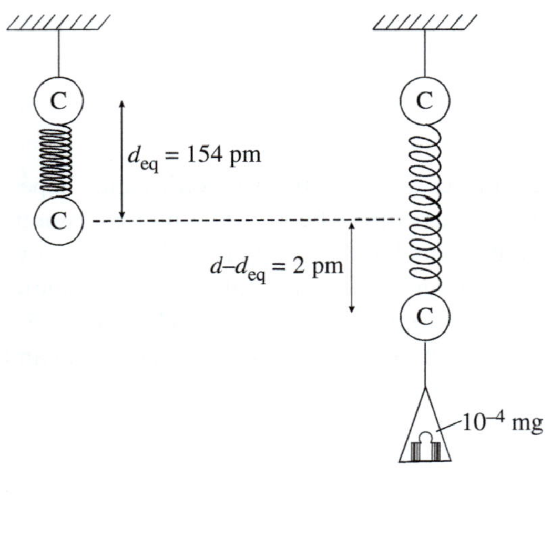

Figure 7.30 The C – C bond in an ethane molecule can be elongated 2 pm by applying a force of 10^{-9} N (this is the weight of about 10^{-4} mg or the weight of 10 lead cubes of 0.01 mm edge length). The drawing exaggerates the displacement.

Example 7.11 *Consider the stretching of the bond in HCl from the equilibrium distance of 127 pm to a distance of 133 pm (our estimated value in Example 7.10). Estimate the force which must be applied and the change in energy for the stretching.*

Using Hooke's law and the force constant $k_f = 517$ N/m, we find

$$f = k_f x = 517 \frac{\text{N}}{\text{m}} \cdot (133 - 127)\text{pm} = 3.10 \times 10^{-9} \text{ N}$$

for the force, and

$$\varepsilon - \varepsilon_{\text{eq}} = \frac{1}{2} k_f x^2 = \frac{517}{2} \frac{\text{N}}{\text{m}} \cdot (133 - 127)^2 \text{ pm}^2 = 9.29 \times 10^{-21} \text{ J} = 0.058 \text{ eV N}$$

This energy is 1 to 2 percent of a bond energy.

7.5.3.2 Stretching (Morse Potential)

Hooke's Law fails for larger forces because the energy curve is steeper for bond shortening than for bond lengthening; it is not parabolic. Figure 7.31 shows a more realistic shape. At large distances the energy ε becomes constant and equals the energy of the separated bond partners (e.g., H_2^+ dissociates into an H atom and a proton). The energy difference $(\varepsilon - \varepsilon_{\text{eq}})$ corresponds to the bond energy ε_{B}. The energy curve can be approximately described by

$$\varepsilon - \varepsilon_{\text{eq}} = \varepsilon_{\text{B}} \cdot \left(1 - e^{-a \cdot (d - d_{\text{eq}})}\right)^2 \tag{7.11}$$

with

$$a = \sqrt{\frac{k_f}{2\varepsilon_{\text{B}}}}$$

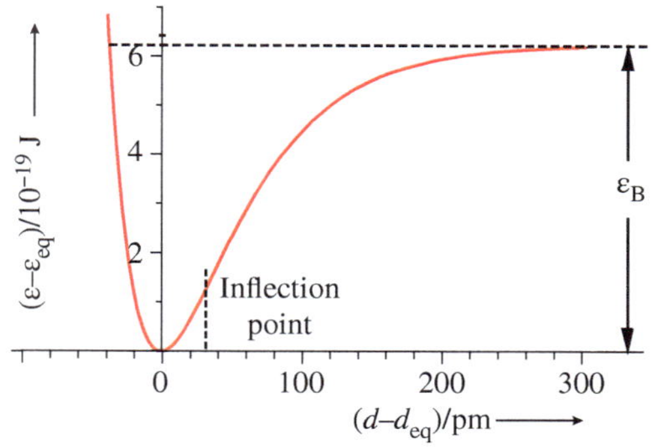

Figure 7.31 Energy of a diatomic molecule (Morse potential) according to Equation (7.11); the values used are $\varepsilon_{\text{B}} = 5.8 \times 10^{-19}$ J and $k_f = 434$ Nm^{-1} (corresponding to the C – C bond in ethane).

(called the *Morse potential*). Equation (7.11) places the energy minimum at $d = d_{\text{eq}}$, and for $d \to \infty$ the energy becomes equal to the bond energy ε_{B}. Near the minimum of the energy well (at small enough values of $x = d - d_{\text{eq}}$) Equation (7.11) is well approximated by Equation (7.10); you show this in Exercise 7.22.

The Morse potential has an inflection point at

$$x_{\text{inflection}} = \ln 2 \cdot \sqrt{\frac{2\,\varepsilon_B}{k_f}} \tag{7.12}$$

At this distance the force acting between the bonded atoms reaches a maximum

$$f_{\text{max}} = \sqrt{2k_f\,\varepsilon_B} \tag{7.13}$$

see Exercise E7.23. Once the applied force is larger than f_{max}, it should be possible to break the bond. For a C – C bond with $k_f = 434$ Nm^{-1} and $\varepsilon_B = 6.24 \times 10^{-19}$ J we obtain

$$f_{\text{max}} = 2.3 \times 10^{-8} \text{ N} \quad \text{at } x = 37 \text{ pm}$$

The force needed to break a C–C bond is about eight times larger that that for stretching the bond by 6 picometers. The bond breaks at an elongation of 37 pm, which is about 30% longer than the equilibrium bond length.

The stretching of chemical bonds can be experimentally performed by using an *atomic force microscope (AFM)*. Similar to a scanning tunneling microscope, an AFM holds a tip close to a surface; however, the tip is placed on the end of a cantilever and the tip position is changed by the attractive and repulsive forces between the tip and the surface. The deflection of the cantilever, which occurs upon displacement of the tip position, measures the forces. The forces are smaller than 10^{-8} N (often the forces are on the order of 10^{-11} N); thus, the tip does not damage the surface. The force between the surface and tip can be measured as a function of distance (attraction followed by repulsion when diminishing the distance between tip and surface). By moving the tip over the surface and recording the change in deflection of the cantilever the surface topography can be investigated. It is possible to shift the tip in such an accurate way that single molecules can be picked up by the tip and set down at a different location on the surface.

As an example we consider a single polysaccharide molecule with one end covalently bound to a substrate (glass) and the other end connected to the tip of an atomic force microscope (Fig. 7.32). In this case the bond breaks at $f_{max} = 2 \times 10^{-9}$ N. This value is smaller than $f_{max} = 5.6 \times 10^{-9}$ N calculated for a C – C bond. The breaking forces for C–Si are expected [16] to be about two times lower than those for C – C, suggesting that our simple estimation is reasonable.

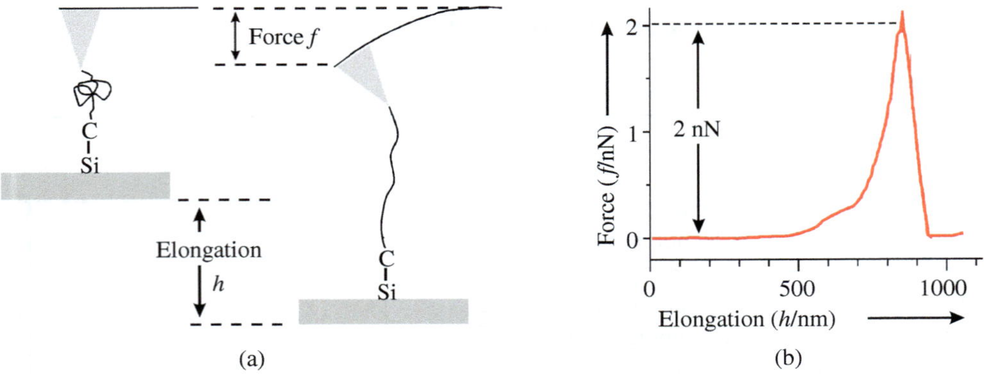

Figure 7.32 Single polysaccharide chain covalently attached to an AFM tip and to a glass surface. Individual chain stretched by a force f using an atomic force microscope. (a) Experimental setup. One end of the polymer chain is fixed to the AFM tip connected to a spring; the second end of the polymer is fixed to a movable plate. When the plate is moved downward by an elongation h, a force acts on the polymer chain which is measured by the bending of the spring. (b) Force acting on the polymer chain versus elongation h. The dashed line marks the breaking point [16].

Example 7.12 *Use the ε_B and k_f values for H_2, O_2, and N_2 to compute rupture forces for these bonds and compare them.*

The data are available in Table 7.6. For H_2 we find

$$f_{max} = \frac{1}{2}\sqrt{\frac{(7.17 \times 10^{-19})\,(576)}{2}}\,\text{N} = 7.18 \times 10^{-9}\ \text{N}$$

For O_2 we find $f_{max} = 11.0 \times 10^{-9}$ N, and for N_2 we find $f_{max} = 21.2 \times 10^{-9}$ N. The maximum rupture force increases systematically through the series H_2, O_2, and N_2 and correlates with the increasing bond order of the molecules.

7.5.3.3 Bending and Torsion

In addition to the stretching of a bond, a deformation of the bond angles occurs in molecules with three or more atoms. Because the change in overlap of the bonding orbitals is smaller for changing bond angles than it is for bond lengths, the force needed for bending a bond is much smaller than that for stretching. A small change $\Delta\alpha$ of the bond angle α causes a shift of the atom by (Fig. 7.33)

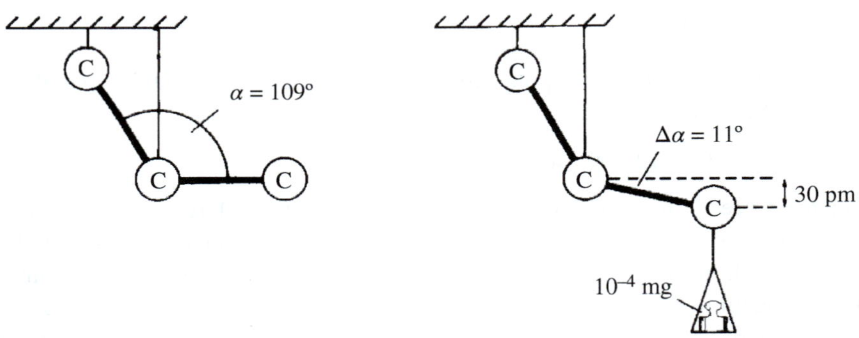

Figure 7.33 Bending a C – C bond.

$$x = d_{eq} \cdot \Delta\alpha \qquad (7.14)$$

The force required for this shift is

$$f = k_f' \cdot x \tag{7.15}$$

(k_f' = bending force constant). The experimental value for the bending force constant of the C − C − C bond angle in propane is $k_f' = 36\ \text{Nm}^{-1}$. According to Fig. 7.33, this corresponds to a shift of one C atom by

$$x = d_{\text{eq}} \cdot \Delta\alpha = \frac{f}{k_f'} = \frac{10^{-9}\ \text{N}}{36\text{N m}^{-1}}$$
$$= 30\ \text{pm}$$

at a force of 10^{-9} N. This shift is larger by a factor of about fifteen than for stretching the bond. This bending force can be directly measured using an atomic force microscope by fixing one end of a single chain molecule to the tip, fixing the other end to a surface, and then measuring the stretching force.

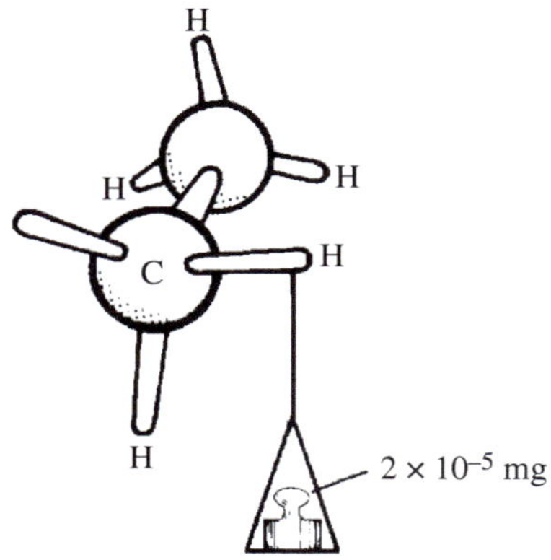

Figure 7.34 Free rotation (C—C bond in ethane).

In a similar way we can estimate the energy barrier for the rotation about a bond axis. For a methyl group around the C − C axis in ethane a force of 2×10^{-10} N is sufficient for overcoming the energy barrier against rotation of the methyl group (Fig. 7.34). Although rotation about single carbon-carbon bonds is relatively facile, the rotation about double-bonded systems is not because the overlap between the π-bonded atoms changes significantly with the rotation angle. Correspondingly, atoms that are part of a ring system do not allow rotation easily because it requires covalent bond rupture; hence, the axial and equatorial substituents on a cyclohexane ring are not easily interchanged.

7.6 Conclusion

Two different views of a molecule's electronic structure were described in this chapter. One view treats the interactions between atoms as local and describes the molecule's structure in terms of chemical bonds, defined as an electron pair localized between two nuclear centers. The second view treats the molecule as a single entity and describes the electronic structure in terms of molecular wavefunctions that are delocalized over the entire molecule. The second view provides a more accurate description, but it becomes very computationally demanding for large molecules (tens to hundreds of atoms). We discussed how the localized view can be combined with the molecular orbital view to build-up larger molecules from molecular subunits. A procedure that is qualitatively accurate and useful for rationalizing and/or predicting the molecular structures of large molecules.

Our discussion ended by considering some generic aspects of chemical bonds: asymmetry of the charge distribution (bond polarity), characteristic lengths of bonds for particular atom types, and the forces present as the nuclei in a molecule move away from the equilibrium geometry.

7.7 Justifications

7.7.1 Justification 7.1: Tetrahedral Hybrid Function

See the books website, www.wiley.com/go/Kuhn/physicalchemistry3e.

7.8 Foundations

7.8.1 Foundation 7.1: LCAO-Treatment of Heteronuclear Diatomic Molecules

Here we consider the interaction of two atomic orbitals of different energy, as shown in Fig. 7.17. Using the LCAO approximation introduced in Section 6.2.3, we represent the molecular orbitals of this diatomic molecule by the superposition of the normalized atomic orbitals φ_a and φ_b of atoms a and b. Consider the molecular orbital ϕ

$$\phi = c_a \cdot \varphi_a + c_b \cdot \varphi_b \tag{7F1.1}$$

where the square of the constants c_a and c_b (i.e., c_a^2 and c_b^2) gives the contribution of that atomic orbital to the molecular orbital.

We require that the trial function ϕ is normalized, i.e.,

$$\int \left(c_a\varphi_a + c_b\varphi_b\right)^2 = 1 \tag{7F1.2}$$

or

$$c_a^2 + 2c_ac_bS_{ab} + c_b^2 = 1 \tag{7F1.3}$$

where $S_{ab} = \int \varphi_a \cdot \varphi_b \cdot d\tau$ is the overlap integral. The variational energy is given by

$$\begin{aligned}\varepsilon &= \int \phi\mathcal{H}\phi \cdot d\tau = \int \left(c_A\varphi_A + c_B\varphi_B\right) \mathcal{H} \left(c_A\varphi_A + c_B\varphi_B\right)\ d\tau \\ &= c_a^2H_{aa} + 2c_ac_bH_{ab} + c_b^2H_{bb}\end{aligned} \tag{7F1.4}$$

where $H_{aa} = \int \varphi_aH\varphi_a \cdot d\tau$, $H_{bb} = \int \varphi_bH\varphi_b \cdot d\tau$, and $H_{ab} = \int \varphi_aH\varphi_b \cdot d\tau$. We do not change Equation (7F1.4) if we multiply ε by the left-hand side of Equation (7F1.3), so that

$$\varepsilon \cdot \left(c_a^2 + 2c_ac_bS_{ab} + c_b^2\right) = c_a^2H_{aa} + 2c_ac_bH_{ab} + c_b^2H_{bb}$$

We minimize the energy by variation of the coefficients c_a and c_b, so that

$$\frac{\partial\varepsilon}{\partial c_a} \cdot \left(c_a^2 + 2c_ac_bS_{ab} + c_b^2\right) + \varepsilon\left(2c_a + 2c_bS_{ab}\right) = 2c_aH_{aa} + 2c_bH_{ab}$$

and

$$\frac{\partial\varepsilon}{\partial c_b} \cdot \left(c_a^2 + 2c_ac_bS_{ab} + c_b^2\right) + \varepsilon\left(2c_aS_{ab} + 2c_b\right) = 2c_aH_{ab} + 2c_bH_{bb}$$

The minimum of the energy is reached for $\partial\varepsilon/\partial c_a = \partial\varepsilon/\partial c_b = 0$, from which we find

$$c_aH_{aa} + c_bH_{ab} - \varepsilon c_a - \varepsilon c_bS_{ab} = 0$$

$$c_aH_{ab} + c_bH_{bb} - \varepsilon c_aS_{ab} - \varepsilon c_b = 0$$

and

$$c_a\left(H_{aa} - \varepsilon\right) + c_bH_{ab} = 0 \tag{7F1.5}$$

$$c_aH_{ab} + c_b\left(H_{bb} - \varepsilon\right) = 0 \tag{7F1.6}$$

if we neglect the overlap integral S_{ab}. These two equations can be solved by the method of determinants. In this method the energy solutions are obtained by solving the determinant given by

$$\begin{vmatrix} (H_{aa} - \varepsilon) & H_{ab} \\ H_{ab} & (H_{bb} - \varepsilon) \end{vmatrix} = 0 \tag{7F1.7}$$

(see Appendix C for a discussion of the method of determinants). The diagonal terms in the determinant are associated with the self energy of the orbitals and the off-diagonal terms are associated with the interactions between orbitals. In particular, we find that

$$(H_{aa} - \varepsilon)(H_{bb} - \varepsilon) - H_{ab}^2 = 0$$

which rearranges to

$$H_{aa}H_{bb} - \varepsilon\left(H_{aa} + H_{bb}\right) + \varepsilon^2 = H_{ab}^2$$

or

$$\varepsilon^2 - \varepsilon\left(H_{aa} + H_{bb}\right) = H_{ab}^2 - H_{aa}H_{bb}$$

Using the quadratic equation, we find that the two roots $\varepsilon_\pm$ are

$$\begin{aligned}\varepsilon_\pm &= \frac{1}{2}\left(H_{aa}+H_{bb}\right) \pm \sqrt{H_{ab}^2 - H_{aa}H_{bb} + \frac{1}{4}\left(H_{aa}+H_{bb}\right)^2} \\ &= \frac{1}{2}\left(H_{aa}+H_{bb}\right) \pm \sqrt{H_{ab}^2 + \frac{1}{4}\left(H_{aa}-H_{bb}\right)^2}\end{aligned} \tag{7F1.8}$$

where $H_{aa} - H_{bb}$ is the energy difference between the atomic orbitals. Note that for homonuclear molecules $H_{aa} = H_{bb}$ and it follows that $\varepsilon_\pm = H_{aa} \pm H_{ab}$, as we found already in Equations (7.5) and (7.6) if we neglect the overlap integral[13]. Straightforward algebra gives the difference in energy between the two molecular orbitals, the bonding and antibonding states, to be

$$\varepsilon_+ - \varepsilon_- = 2\sqrt{H_{ab}^2 + \frac{1}{4}(H_{aa} - H_{bb})^2}$$

We can use these energy eigenvalues to determine the coefficients in the molecular orbital of the diatomic. Substituting the ε_+ result into Equation (7F1.5), we find

$$\begin{aligned}c_a &= -c_b \frac{H_{ab}}{(H_{aa} - (H_{aa}+H_{bb})/2 - \left|\sqrt{\Delta^2 + H_{ab}^2}\right|)} \\ &= -c_b \frac{H_{ab}}{(\Delta - \left|\sqrt{\Delta^2 + H_{ab}^2}\right|)} = c_b \frac{H_{ab}}{(\left|\sqrt{\Delta^2 + H_{ab}^2}\right| - \Delta)}\end{aligned} \tag{7F1.9}$$

where we have defined $\Delta = (H_{aa} - H_{bb})/2$ and the absolute value symbols ensure that we have chosen the positive root. Using this result for c_a in our wavefunction, we find

$$\psi_{\text{antibonding}} = c_b \left(\frac{H_{ab}}{(\left|\sqrt{\Delta^2 + H_{ab}^2}\right| - \Delta)} \varphi_a + \varphi_b \right) \tag{7F1.10}$$

Because H_{ab} is a negative quantity we see that this solution corresponds to destructive interference between the orbitals. A corresponding analysis for the ε_- root gives

$$c_a = -c_b \frac{H_{ab}}{(\left|\sqrt{\Delta^2 + H_{ab}^2}\right| + \Delta)} \tag{7F1.11}$$

and

$$\psi_{\text{bonding}} = c_b \left(\frac{-H_{ab}}{(\left|\sqrt{\Delta^2 + H_{ab}^2}\right| + \Delta)} \varphi_a + \varphi_b \right) \tag{7F1.12}$$

It is instructive to consider how the molecular orbitals change with the energy mismatch Δ between the atomic orbitals. To begin consider the homonuclear case for which $\Delta = 0$. In this limit we find that the antibonding orbital is higher in energy than the atomic orbitals, $\varepsilon_+ = H_{aa} + |H_{ab}|$, and the bonding orbital is lower in energy than the atomic orbitals, $\varepsilon_- = H_{aa} - |H_{ab}|$. The orbital coefficients indicate that each atomic orbital contributes equally to the molecular orbital in this limit; i.e., we find that $c_a = -c_b$ for the antibonding molecular orbital and $c_A = c_B$ for the bonding molecular orbital (Note: remember that H_{ab} is a negative quantity).

As the energy mismatch between the atomic orbitals becomes larger the atomic orbital contributions to the molecular orbitals become unequal. Consider the case in which $2\Delta = H_{aa} - H_{bb} > 0$. In the bonding case, ε_-, the coefficients have the relation $c_a < c_b$ and the difference becomes larger as Δ becomes larger. In addition, the energy ε_- approaches H_{bb} as Δ becomes larger. To summarize, the molecular orbital has more φ_b character and the energy approaches the atomic orbital's energy as Δ becomes larger. Correspondingly, the φ_a character of the ε_+ molecular orbital increases and its energy approaches that of the atomic orbital energy H_{aa}.

13 Equations (7.5) and (7.6) show that the neglect of the overlap integral results in an overestimation of the energy difference between ground state and excited state.

7.8.2 Foundation 7.2: Hartree–Fock SCF Method for Molecules

See the books website, www.wiley.com/go/Kuhn/physicalchemistry3e.

7.9 Exercises

E7.1 Use the Lewis electron pair idea to rationalize the triple bond in N_2.

E7.2 Use electron pair ideas to decide on the atom connectivity in the molecule NH_3. Explain why the structure you choose is expected to be the most stable, and comment on the geometry.

E7.3 *Extension of* Example 7.2: Analyze the bonding requirements for an FCN ring compound.

E7.4 *Extension of* Example 7.3: A perfect tetrahedron has angles between its diagonals (bond axes for methane) of 109.5°. Assume that nonbonded electron pairs take up more space than bonded electron pairs and rationalize why the bond angle for water is 104.0° and the bond angle in ammonia is 106.0°.

E7.5 Use the VSEPR method to predict the geometries of the molecules in Table 7.3.

E7.6 Consider Equation (7.1), which is a linear combination of two wavefunctions that are each solutions to a Schrödinger equation. Show that if these two wavefunctions correspond to the same energy for that Schrödinger equation then any linear combination of them is a solution to that Schrödinger equation also. Show that this does not hold for a linear combination of wavefunctions belonging to different energies.

E7.7 Show that the wavefunctions in Equations (7.2) and (7.3) form an orthonormal set of hybrid wavefunctions.

E7.8 In the Example 7.4 we considered the hybrid function

$$\psi = a\,\psi_{p_x} + b\,\psi_{p_y}$$

for which an infinite number of functions are possible, depending on the choice of a and b. By requiring that the two hybrid functions be orthogonal and normalized, show that we obtain

$$\psi_1 = a_1\psi_{p_x} + \sqrt{1-a_1^2}\psi_{p_y}, \quad \psi_2 = -\sqrt{1-a_1^2}\psi_{p_x} + a_1\psi_{p_y}$$

E7.9 Use hybridization ideas to predict the structure of $B(CH_3)_3$ and Hg(phenyl)$_2$, and O(phenyl)$_2$. For the case of Hg, only consider the $n = 6$ valence electrons.

E7.10 Qualitatively describe the nature of the bonding in amine oxide $(H_3)\overset{\oplus}{N} - \overset{\ominus}{\underline{\overline{O}}}|$.

E7.11 Why are the energy levels of the linear combinations of the $2p_x$ orbitals of the two atoms ($\sigma(2p_x)$ and $\sigma^*(2p_x)$) in Fig. 7.11 more strongly separated than those of the linear combinations of the $2p_y$ and $2p_z$ orbitals (π and π^* orbitals)?

E7.12 Make a molecular orbital energy diagram for the oxygen molecule in which the atomic orbitals used are considered to be the two equal *sp* hybrids, the p_y, and the p_z. Compare this energy diagram and its predictions to that considered in the text.

E7.13 Provide the molecular orbital energy diagrams for the three species C_2^+, NO, and Ne_2^+. Write out the electronic configurations for these diatomic molecules. Determine the bond order in each case.

E7.14 Draw a molecular orbital energy diagram for N_2^+. Use the valence shell electrons only (i.e., ignore the 1*s* electrons of each atom). Label the orbitals in your diagram. Using your diagram, determine the bond order of N_2^+. Write the electronic configuration of the ground state of N_2^+. Promote the most weakly bound electron to the next highest molecular orbital. What is the ion's bond order in this excited state configuration?

E7.15 Draw the molecular orbital energy diagrams for the three molecules C_2, N_2, and O_2 and use them to explain the variation in their dissociation energies which are 6.2 eV for C_2, 9.8 eV for N_2, and 5.1 eV for O_2. Sketch the first σ^* antibonding orbital and one of the π bonding orbitals.

E7.16 The dipole moment of an HCl molecule is 3.4×10^{-30} Cm = 1.03 Debye. Assume that the dipole moment can be represented by two partial charges, one on the H atom and one on the Cl atom, at a distance corresponding to the bond length 127 pm and determine the magnitude of the partial charges.

E7.17 Explain the trend in the electronegativities of atoms in the periodic table (see Table 7.5) in terms of the trends in the atoms' electronic properties.

E7.18 Consider the molecular orbital energy diagram for formaldehyde (Figure 7.24). Draw this diagram in a way that shows the origin of the molecular orbitals from the two fragments C=O and the two H-atoms, considered as a single unit.

E7.19 Use the data in Table 7.6 to compute the force required to stretch the X-H bond in the homologous series of HF, HCl, and HI by a similar displacement x.

E7.20 Given that the bending force constant for H_2O is 185 Nm^{-1}, that for H_2S is 94 Nm^{-1}, and that for H_2Se is 63 Nm^{-1} compute the change in H atom displacement when a one nanoNewton force is applied.

E7.21 Estimate the force needed to overcome the barrier to torsional motion of the methyl groups in ethane if the barrier height is 12 $kJmol^{-1}$. Perform a similar calculation for the torsion of the OH group in methanol using a 4.5 $kJmol^{-1}$ barrier. In each case assume that the potential energy varies sinusoidally with torsion angle and is three-fold symmetric for the methyl rotation and two-fold symmetric for the OH rotation.

E7.22 For small displacements $x = d - d_{eq}$, show that the a parameter in the Morse function

$$\varepsilon - \varepsilon_{eq} = \varepsilon_B \cdot \left(1 - e^{-a\cdot(d-d_{eq})}\right)^2$$

is given by $a = \sqrt{k_f/(2\varepsilon_B)}$, where k_f and ε_B are the parameters in the harmonic limit (see Section 7.5.3.2).

$$\Delta\varepsilon = \varepsilon - \varepsilon_{eq} = \int_0^x f \cdot dx = \int_0^x k_f x \cdot dx = \frac{1}{2} k_f \cdot x^2$$

E7.23 Using the Morse potential, derive the expression for f_{max} in Equation (7.13). Explain why the inflection point of the potential provides the maximum force.

7.10 Problems

P7.1 Plot the radial wavefunctions for the 1s, 2s, and 2p orbitals on oxygen, using the Slater wavefunctions in Box 5.4. What average distance would be needed to have an overlap integral that is 30% for the 1s orbital (see Foundation 6.2)? Explain how these findings impact the role of the 1s electrons in bonding.

P7.2 Show that for a hybrid function $\phi = a\psi_{2s} + b\psi_{2p_x} + c\psi_{2p_y} + d\psi_{2p_z}$ the expression $\varepsilon_{hybrid} = E_{2s} + \left(b^2 + c^2 + d^2\right) \cdot \Delta E$ is obtained; calculate ε_{hybrid} for the tetrahedral sp^3 hybrid state.

P7.3 In strongly acidic solution the carbonium ion CH_5^+ and the carbenium ion CH_3^+ are formed according to

$$CH_4 + H^+ \rightarrow CH_5^+ \rightarrow CH_3^+ + H_2$$

Discuss the orbital bonding of these ions.

P7.4 In Section 7.3.1.1 we discussed how symmetry considerations imply that the net overlap between σ and π orbitals is zero. Using $\varphi_a = 2p_y$ and $\varphi_b = 2s$ hydrogen atom wavefunctions, evaluate the overlap integral $S = \int \varphi_a \varphi_{b.} \cdot d\tau$, and demonstrate that it is zero. Consider the bond axis to be the z-axis.

P7.5 Our discussion of triatomic molecules considered hydride molecules, for which π-bonding is not important. Discuss how our bonding considerations must change to describe the bonding in CO_2. Sketch a molecular orbital energy diagram for $CO_{2.}$

P7.6 Make a qualitative molecular orbital energy diagram for methanol, CH_3OH. Sketch the molecular orbitals. Contrast this diagram with that of acetonitrile (Fig. 7.23).

P7.7 Discuss whether the electrostatic repulsion of the H atoms in the H_2O molecule is sufficient to cause a widening of the bond angle from 90° to 105°. The dipole moment of H_2O is $\mu = 6.16 \times 10^{-30}$ Cm = 1.85 Debye, the bending force constant k_f' is 68 N/m, and the bond length d is 96 pm.

P7.8 Use the Morse potential function in Equation (7.11) and determine how the force depends on distance. Plot the distance dependence of the force for the case of a hydrogen molecule, an oxygen molecule, and a nitrogen molecule. See Table 7.6 for values of the bond parameters.

P7.9 The figure below shows a photoelectron spectrum of H_2S. Construct a qualitative molecular orbital energy diagram for H_2S and use it to explain the structure in the photoelectron spectrum, which is provided in Figure P7.9.

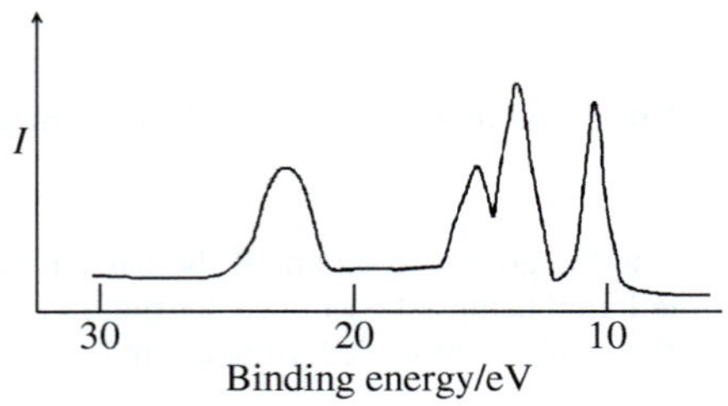

Figure P7.9 Photoelectron spectrum of H_2S.

P7.10 The table provides energy versus internuclear distance values that are calculated for HF at a sophisticated level of theory (see X. Li, J. Mol. Structure (Theochem) **2001**, 547, 69). Plot these data and then fit the Morse potential function to the data. Using your best fit parameters for the Morse potential calculate the force constant of the bond.

r/a_0	1.100	1.200	1.300	1.390	1.475	1.560	1.646	1.683
E/eV	0.920	−0.727	−3.080	−4.426	−5.237	−5.714	−5.954	−6.003
r/a_0	1.693	1.703	1.713	1.723	1.733	1.743	1.753	1.763
E/eV	−6.010	−6.016	−6.020	−6.022	−6.022	−6.020	−6.017	−6.013
r/a_0	1.773	1.790	1.870	1.950	2.030	2.110	2.166	2.310
E/eV	−6.006	−5.995	−5.883	−5.705	−5.486	−5.238	−5.051	−4.546
r/a_0	2.460	2.600	2.800	3.000	3.300	3.600	4.000	4.500
E/eV	−4.011	−3.524	−2.871	−2.287	−1.557	−1.004	−0.516	−0.199

P7.11 Values of the energy ε of the H_2^+ ion as a function of the distance d of the nuclei are given below (d_{eq} is the equilibrium distance, ε_{eq} is the energy at the equilibrium distance). Assume that the dependence of the energy on the displacement follows the cubic form

$$\varepsilon - \varepsilon_{\text{eq}} = \frac{1}{2} k_f \cdot x^2 + bx^3$$

where we have set $x = d - d_{eq}$. You should analyze the data for each of the three electronic structure models and extract a force constant, k_f, for the H_2^+ bond.

	$(\varepsilon - \varepsilon_{\text{eq}})/10^{-19}$ J		
$(d - d_{\text{eq}})$/pm	**Exact**	**Box Model**	**LCAO Model**
−20	0.44	0.15	0.25
−15	0.23	0.08	0.13
−10	0.09	0.03	0.05
−5	0.02	0.00	0.01
0	0	0	0

$\left(d - d_{eq}\right)$ /pm	$\left(\varepsilon - \varepsilon_{eq}\right)/10^{-19}$ J		
	Exact	Box Model	LCAO Model
5	0.02	0.01	0.01
10	0.07	0.03	0.05
15	0.15	0.07	0.10
20	0.24	0.12	0.16

a) Rearrange the equation to a linear form ($y = m \cdot x + b$), by showing that

$$y = \frac{\varepsilon - \varepsilon_{eq}}{x^2} = \frac{1}{2}k_f + bx$$

What is the slope and the intercept you expect?

b) Plot the data and extract a k_f value. Comment on the linearity of the data?
For each of the electronic energy curves, analyze the data for the case of $x \leq 0$ and for $x \geq 0$ and extract values for both b and k_f.
Compare and contrast the k_f values. Comment on the physical meaning of the parameter b.

P7.12 Consider the case of LiH which has an average bond distance of 159.5 pm. Given this constraint for the bond distance, evaluate the optimum geometry for the triatomic molecule Li_2H^+. More specifically, assume that the two LiH bond lengths are fixed and that the interactions between the ions Li^+ and H^- are strictly electrostatic. Vary the angle of the $[Li\text{-}H\text{-}Li]^+$. Make a plot of the change in the energy as a function of the angle and find the most stable (lowest energy) geometry. Provide a physical explanation for the answer that you find.

P7.13 Perform electronic structure calculations for hydrogen fluoride, HF, with the three different basis sets STO-3G, 3-21G, and 6-31G (see Foundation 7.2 for a discussion of these basis sets); and compare the results to experimental data. Many universities provide access to software programs for such calculations, but "freeware" versions also exist. The solutions given for this problem in the solution manual use the WebMO program, which you can access at https://www.webmo.net/demoserver/cgi-bin/webmo/login .cgi. If you use this application, select the GAMESS software package, which is free to users. You can obtain experimental values for HF properties from the USA government website: http://www.nist.gov/data.

a) Determine the equilibrium internuclear bond distance for HF and compare it to the experimental value.
b) Determine the dipole moment of HF and compare it to the experimental value.
c) Determine the molecular orbital energies of HF for the HOMO-2, HOMO-1, HOMO, and LUMO orbitals. (You may find the energy conversion factor of 1 Hartree = 27.2114 eV to be useful).
d) Plot the molecular orbital energy diagram that includes the valence electrons; i.e., make a plot to scale that shows the molecular orbital energies and the corresponding atomic orbital energies for the 6-31G basis set, which should be the most accurate one. There is no need to include the 1s orbital of the F or its corresponding molecular orbital). Take the atomic orbital energies to be −13.6 eV for the H atom, −40.2 eV for the 2*s* atomic orbital of F, and −18.6 eV for the 2*p* orbitals of F.

Finally, you should discuss and explain your results.

References

1 J.R. Wazar and I. Absar, *Electron Densities in Molecules and Molecular Orbitals* (Academic Press, New York 1975).

2 R.F.W. Bader, R.J. Gillespie, and P.J. MacDougall, A physical basis for the VSEPR model of molecular geometry, *J. Am. Chem. Soc.*, 1988, **110**, 7329; L.S. Bartell and Y.Z. Barshad, Valence-shell electron-pair repulsions: a quantum test of a naive mechanical model, *J. Am. Chem. Soc.*, 1984, **106**, 7799.

3 K. Siegbahn, C. Nordling, G. Johansson, J. Hedman, P.F. Hedén, K. Hamrin, U. Gelius, T. Bergmark, L.O. Werme, R. Manne, and Y. Baer, *ESCA Applied to Free Molecules* (North-Holland Publishing Co., Amsterdam, 1971).

4 K.P. Huber and G. Herzberg, *Molecular Spectra and Molecular Structure, IV. Constants of Diatomic Molecules* (D. van Nostarnd, New York, 1979).

5 V.E. Bondebey, Electronic structure and bonding of diatomic beryllium, *Chem. Phys. Lett.*, 1984, **109**, 436.

6 F.M. Tao and Y.K Pan, An accurate ab initio calculation of the Ne_2 potential, *Chem. Phys. Lett.*, 1992, **194**, 162; B. Ulrich, A. Vredenborg, A. Malakzadeh, L.Ph.H. Schmidt, T. Havermeier, M. Meckel, K. Cole, M. Smolarski, Z. Chang, T. Jahnke, and R. Dörner, Imaging of the structure of the argon and neon dimer, trimer, and tetramer, *J. Phys. Chem. A*, 2011, **115**, 6936.

7 R.J. Buenker and S.D. Peyerimhoff, Molecular geometry and the Mulliken-Walsh molecular orbital model. Ab initio study, *Chem. Rev.*, 1974, **74**, 129.

8 F.L. Pilar, *Quantum Chemistry* (Dover, New York, 1990).

9 L. Pauling, The nature of the chemical bond. IV. The energy of single bonds and the relative electronegativity of atoms, *J. Am. Chem. Soc.*, 1932, **54** (9), 3570–3582. doi: 10.1021/ja01348a011.

10 A.L. Allred, *J. Inorg. Nucl. Chem.*, Electronegativity values from thermochemical data, 1961, **17**, 215.

11 H.O. Pritchard and H.A. Skinner, The concept of electronegativity, *Chem. Rev.*, 1955, **55**, 745.

12 C. Chang, C.-Y. Ng, S. Stimson, M. Evans, and C.W. Hsu, Rotationally resolved vacuum ultraviolet pulsed field ionization-photoelectron vibrational bands for H_2^+ ($X^2\Sigma_g^+$, $v^+ = 0 - 18$), *Chin. J. Chem. Phys.*, 2007, **20**, 352.

13 J.E. Pollard, D.J. Trevor, J.E. Reutt, Y.T. Lee, and D.A. Shirley, Rotationally resolved photoelectron angular distributions for $H_2{}^+$ ($v' = 0$) at $\lambda = 584$ and 736 Å, *Chem. Phys. Lett.*, 1982, **88**, 434.

14 L.J. Schaad and W.V. Hicks, Equilibrium bond length in H_2^+, *J. Chem. Phys.*, 1970, **53**, 851.

15 B. Cordero, V. Gómez, A.E. Platero-Prats, M. Revés, J. Echeverria, E. Cremades, F. Baragan, and S. Alvarez, Covalent radii revisited, *Dalton Trans.*, 2008, 2832.

16 M. Grandbois, M. Beyer, M. Rief, H. Clausen-Schaumann, and H.E. Gaub, How strong is a covalent bond? *Science*, 1999, **283**, 1727.

8

Molecules with π-Electron Systems

CHAPTER MENU

8.1 Introduction

In cases where multiple bonds appear on adjacent atomic sites, it is often not possible to take a local view of the chemical bonding because the electrons (typically π-type) can delocalize across multiple atomic centers. Often these molecules are viewed as systems where π electrons move relatively freely and extend over large regions of the molecular skeleton, which is formed by σ electrons. Molecular orbital theories with their delocalized approach are natural for describing this behavior. It is fascinating and intuitively useful to explore how the basic features of quantum mechanics are manifest in the properties of π-electron systems.

Because of the poor orbital overlap between the σ-bonds and π-bonds in a molecule, it is common to view these two types of bonds independently. The σ and π networks can couple with each other however, even for planar geometries. For example, the charge distribution of the π electrons (given by the position of the antinodes of the wavefunctions) disturbs the molecular skeleton, and *vice versa*. Such changes in the geometry of the molecular skeleton will affect the σ-bonding. This interaction causes equal bond lengths in benzene and alternating short and long bonds in polyenes. Other than this coupling, the independence of the σ and π systems holds well for planar molecules.

Although realistic treatments are computationally accessible, it is beneficial to consider simple models that show the basic physics underlying properties of π-systems. This chapter discusses π-systems from the perspective of a simple free electron model and from a simple linear combination of atomic orbitals (LCAO) treatment, the Hückel molecular orbital (HMO) method. The discussion for the free electron model is organized by the geometric character of the π-systems, evolving from a linear connectivity, to ring systems, and then branched structures. The application of these models to the energy level structure of the molecules, the interatomic distances, and the charge density patterns for the π-structure are developed.

Principles of Physical Chemistry, Third Edition. Hans Kuhn, David H. Waldeck, and Horst-Dieter Försterling.

Companion website: www.wiley.com/go/Kuhn/physicalchemistry3e

8.2 Bonding Properties of π Electrons

So far we have concentrated on molecules with single bonds (one dash in the structural formula) or with multiple bonds (two or three dashes between a pair of atoms in the structural formula) localized between two or three nuclear centers. Now we consider some of the electronic properties of molecules with multiple bonds that can delocalize across multiple σ-bonded atoms in the molecular skeleton.

8.2.1 Sigma-Bonded Molecular Skeleton and π-Electrons

Let us begin by discussing ethene, C_2H_4; see Fig. 8.1. Each carbon atom in ethene has three bonding partners, but the bonding between the carbons involves four electrons (two electron pairs) rather than just two. In an electron pair description, the geometry about each carbon is trigonal planar and the C atoms are sp^2 hybridized (see Fig. 7.7). Each carbon atom forms three trigonal planar σ bonds, one for each of the H atoms and one for bonding to the other C atom. In addition to the σ bonds and perpendicular to the molecular frame (defined by the σ bonded framework), each carbon has a *p*-orbital and these overlap to form a π bond (Fig. 8.1), analogous to the multiple bonds we saw in diatomic molecules (see Section 7.3.1).

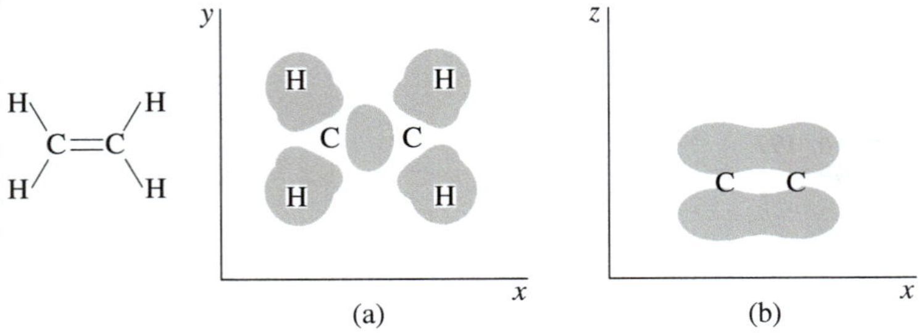

Figure 8.1 Ethene molecule. (a) Trigonal planar hybrid orbitals (view from above the *xy*-plane); (b) $2p_z$ orbitals (view from the side in the *xz*-plane). Inside the shaded area the charge density is larger than $e \cdot 0.02 \times 10^{30}\ \mathrm{m}^{-3}$.

Because ethene has two bonds instead of just one, we expect that the energy needed for breaking the carbon–carbon bond in ethene (bond energy[1]) must be larger than that for ethane, see Table 8.1. The bond energy increases substantially upon π-bond formation, nearly by a factor of two. Table 8.1 also gives data for the ethyne molecule, which has a triple bond. The bond energy of the C ≡ C bond in ethyne is even larger than for a C = C bond. In conjunction with this trend in bond energies, the force constant becomes systematically larger and the bond lengths become systematically shorter through the series (Table 8.1).

Table 8.1 Bond Energies and Bond Lengths of C–C Bonds (H_2^+ Ion and H_2 Molecule for Comparison).

		Bond Energy 10^{-19} J	Bond Length pm	Force Constant $\mathrm{N\ m^{-1}}$
Ethane	CH_3–CH_3	6.3	154	434
Ethene	CH_2=CH_2	12.2	133	1080
Ethyne	CH≡CH	16.0	120	1490
Hydrogen molecule ion	$(H–H)^+$	4.2	106	160
Hydrogen molecule	H–H	7.2	74	574

An important feature of multiple bonds in molecules is how they restrict the available molecular conformations. Because the rotation of the two parts of the ethene molecule around the carbon–carbon axis causes a strong change in the overlap of the *p*-orbitals (Fig. 8.2), the bond energy changes strongly with the rotation angle. In the planar geometry, good overlap and a bonding interaction occurs between the carbon *p*-orbitals. In contrast, no π-bonding occurs when the planes of the CH_2 units are perpendicular. The difference in energy for these two arrangements in ethene is about $1.1\ \mathrm{eV} = 1.8 \times 10^{-19}$ J. This value is almost 50 times larger than the available thermal energy (k_BT) at room temperature. The energy barrier inhibits

1 Bond energies in polyatomic molecules are extracted from the energies of dissociation into atoms. It is postulated that these energies are the sum of the energies for each bond. In this way, appropriate values for bond energies are obtained.

Figure 8.2 Decrease of the overlap of the π orbitals at a rotation of the ethene molecule.

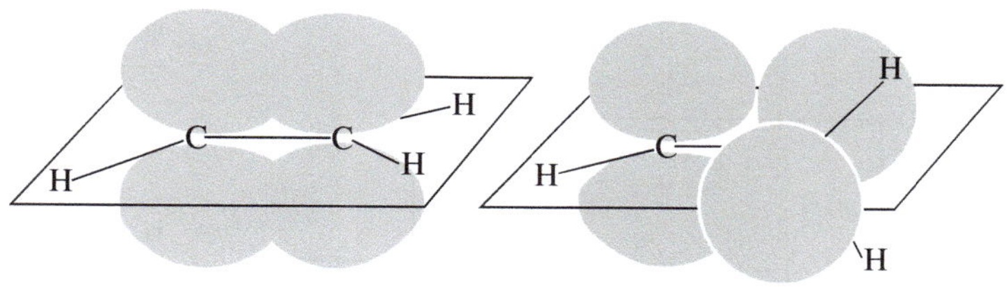

rotation about the double bond in ethene and restricts the internal rotation to small angular displacements around the planar configuration. For that reason *cis* and *trans* derivatives of ethene, e.g.,

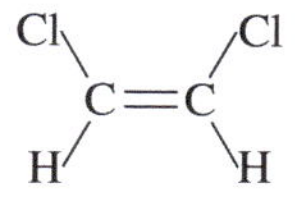

cis form

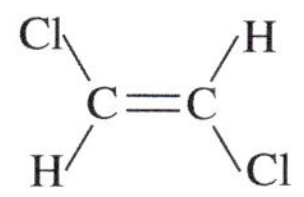

trans form

are stable and do not interconvert significantly at room temperature. In contrast, single carbon–carbon bonds do not have such a large energy barrier (ethane's barrier is about 0.125 eV $= 0.20 \times 10^{-20}$ J) and undergo rotation more freely.

8.2.2 Overview of Models for π-Systems

The weak coupling of the π and σ electrons suggests that we treat the π-electrons independently from the σ electrons; however, we must account for their mutual Coulomb interaction. To do this, we consider the π-electrons to interact with an effective electrostatic potential that is defined by the molecular skeleton of sigma bonded atoms. In addition, we consider that the π-electron to nuclear attraction is screened by the σ electrons; and we find solutions of the corresponding Schrödinger equation. In the following we approach solving this problem in two fundamentally different ways.

8.2.2.1 HMO Model

This method, which is discussed in Section 8.6, uses approximate molecular wavefunctions (i.e., LCAO model) and performs a variational calculation to find the energy. This calculation proceeds in the way described in Chapters 2, 6, and 7, but we make some important simplifications. In particular we describe the π-electrons by a linear combination of atomic $2p_z$ orbitals that are centered on the carbon atoms in the sigma framework. Rather than evaluating the energy integrals, we introduce two adjustable parameters α and β and determine them by comparison with experimental data. Hence the method is called *semiempirical*, in contrast to the ab initio Hartree–Fock approach described in Chapter 7. This method is called the *Hückel Molecular Orbital (HMO) model.*

8.2.2.2 BCD Model

This method (discussed in Sections 8.3 and 8.4) finds an approximate description for the electronic potential along the chain, an effective potential, and then solves the corresponding Schrödinger equation rigorously. The effective potential is obtained by assuming that the given π-electron finds itself in the field of the Slater effective nuclear charge of the nearest carbon or hetero-atom, and the charges of all other atoms are shielded by the other π-electrons. The wavefunction of the given π-electron is approximated by the product of $\psi(s)$ (where s is the coordinate along the molecular skeleton, see Fig. 8.3) and F (coordinate perpendicular to s). The potential energy function $V(s)$ is derived in Foundation 8.1. The wavefunctions $\psi(s)$ for this potential energy and the corresponding energies are obtained by solving the one-dimensional Schrödinger equation numerically (see Appendix C on the book's website). $V(s)$ and the solutions $\psi(s)$ are given in Fig. 8.3 for the three lowest states of butadiene. From the energies $E_2 = -109.5 \times 10^{-19}$ J and $E_3 = -101.3 \times 10^{-19}$ J for the excitation energy we obtain $\Delta E_{2\to3} = 8.2 \times 10^{-19}$ J, which differs by 1×10^{-19} J from the experimental value $\Delta E_{2\to3} = 9.2 \times 10^{-19}$ J.

In this approach the correlation between the measured bond length d and calculated π electron density in the middle of the bond is used, but no adjustable parameter is introduced. The π electron density can be seen as following from an ab initio calculation of the molecular skeleton. In this sense the approach differs from what is called semiempirical. This method is called the *bond length consistent with total π electron density (BCD) model.*

8.2.2.3 Step Potential Model

It is of interest to simplify the given potential $V(s)$ and to see in what way the wavefunctions change. We consider step potentials where the numerical evaluation is easy. From top to bottom Fig. 8.4a shows potentials of increasing simplicity,

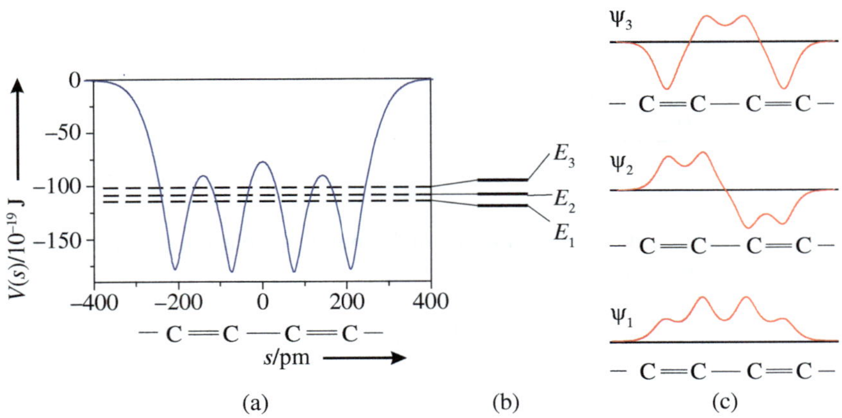

Figure 8.3 Wavefunctions $\psi(s)$ and energies E calculated for butadiene for an electron in an effective one-dimensional potential energy function $V(s)$. (a) $V(s)$ along the chain of carbon atoms, (b) energies $E_1 = -112, 8 \times 10^{-19}$ J, $E_2 = -109.5 \times 10^{-19}$ J, and $E_3 = -101.3 \times 10^{-19}$ J, and (c) wavefunctions ψ_1, ψ_2, and ψ_3. The energies and wavefunctions are obtained by solving the Schrödinger equation for $V(s)$ numerically (see Appendix C).

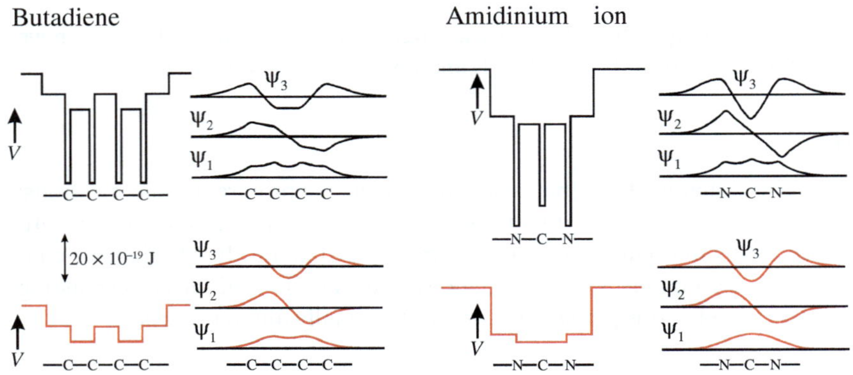

Figure 8.4a Approximate potential energy functions V and corresponding wavefunctions ψ for butadiene and amidinium ion. Top row: Atomic potential troughs and bond-length-dependent potential energies are considered. Bottom row: Atomic potential energy troughs are neglected, but bond-length-dependent potential energies are considered.

as well as their corresponding wavefunctions. In the case of Fig. 8.4a (top row) the wavefunctions represent the rigorous solution in Fig. 8.3 very well. In Fig. 8.4a (bottom row) the atomic contributions are neglected and the contribution depending on the bond length is accounted for by higher and lower steps at the center of the C–C and C=C bonds, respectively. The resulting ratio of 3.3 between the densities in C–C and C=C (see Table 8.2) is almost the same as in Fig. 8.3.

In Fig. 8.4b (top row) the contribution depending on the bond length is accounted for in the same way as in Fig. 8.4a (bottom row). However, it is assumed that the potential energy is infinite at both ends of the molecular skeleton. We call this the *step potential model*.

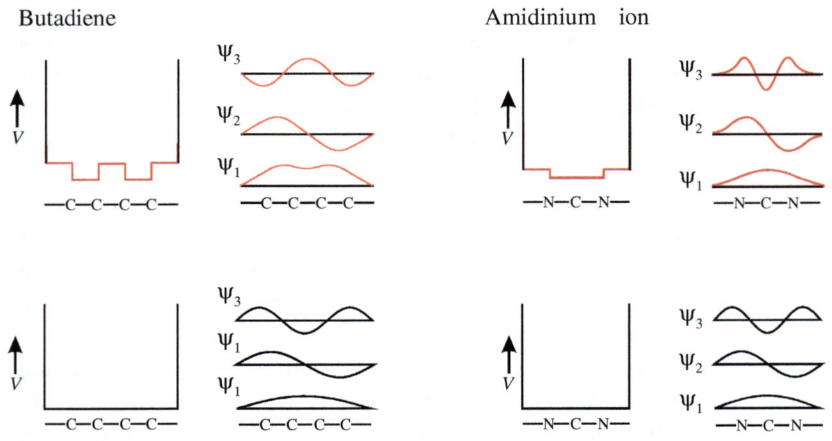

Figure 8.4b Approximate potential energy functions V and corresponding wavefunctions ψ for butadiene and amidinium ion. The potential at both ends of the molecular skeleton is assumed to be infinite. Top row: Bond-length-dependent potential energies are considered (step potential model). Bottom row: Bond-length-dependent potential energies are neglected (FEMO).

8.2.2.4 FEMO Model

Figure 8.4b (bottom row) shows a further simplification: the potential energy is assumed to be constant. This is the Free Electron Model (FEMO model). It still displays some essential features of the true wavefunctions. In Section 8.3 we explore this simple molecular orbital model for π-electrons.

The correspondence between the nodal pattern of the molecular orbitals in the LCAO model and the nodal pattern using a BCD, step potential, or FEMO model is general [1].

8.3 Free Electron Molecular Orbital (FEMO) Model

8.3.1 Ethene, Butadiene, Amidinium. π-Electron Chains

We approximate the π orbital electrons by one-dimensional waves along the bond axes. The de Broglie wavelengths (see Chapter 2) of the electrons are

$$\Lambda = \frac{2L}{n}, \quad n = 1, 2, 3 \ldots \tag{8.1}$$

where L is the overall length of delocalization. In this model the wavefunctions are given by

$$\psi_n = \sqrt{\frac{2}{L}} \sin\left(\frac{\pi n s}{L}\right), \quad n = 1, 2, 3 \ldots \tag{8.2}$$

where s is the displacement along the chain of carbon atoms. The corresponding energies E_n are

$$E_n = \frac{1}{2} m_e v^2 = \frac{h^2}{2m_e\Lambda^2} = \frac{h^2}{8m_eL^2} \cdot n^2 \tag{8.3}$$

where $n = 1, 2, 3 \ldots$ is a quantum number.

This treatment of the π electrons corresponds to a one-dimensional box model, in which the σ bonded framework defines the box length. The overall length of the box can be determined by summing up the carbon–carbon bond lengths and accounting for any extra length at the ends of the box. For carbon, the bond lengths in conjugated systems are in the range between 148 and 133 pm. As an average, we use $d_0 = 140$ pm. The extension of the electron wave beyond the last carbon nucleus in the chain can be obtained by comparison to a more exact calculation (see Section 8.4), and a value of 140 pm beyond the outermost carbon atoms is a good choice. Therefore, we determine the box length L for π electrons by the total length of the C–C bonds involved in the π-bonding and one bond length beyond each of the outermost carbon atoms.

Figure 8.5 shows the wavefunctions and the energy levels for ethene ($L = 3d_0 = 420$ pm) and for butadiene ($L = 5d_0 = 700$ pm) that are found using this model. Because the FEMO model only treats the π-electrons, the zero of the energy scale is shifted with respect to that normally chosen in electronic structure calculations. Nevertheless the energy spacings between π-electron levels in butadiene can be compared with the energy spacing measured in the photoelectron spectrum. The FEMO calculation for butadiene has the two π-orbitals separated by 3.07 eV (4.92×10^{-19} J), whereas the experimental value for the energy spacing is 2.5 eV.

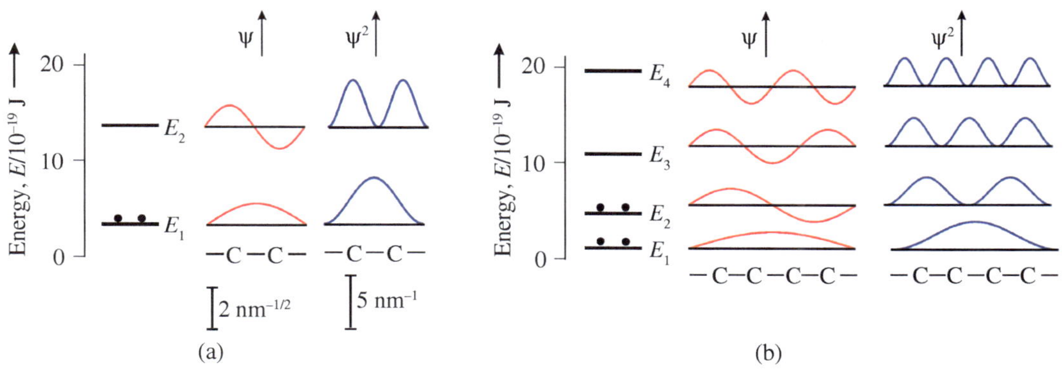

Figure 8.5 Energy E, wavefunction ψ, and square ψ^2 in the free-electron model. (a) Ethene ($L = 3 \cdot 140$ pm $= 420$ pm). (b) Butadiene ($L = 5 \cdot 140$ pm $= 700$ pm).

Example 8.1 *Use the FEMO model to discuss the π-bonding for the amidinium ion, described by resonance between the two structures*

$$H_2\overset{\oplus}{N}{=}CH{-}\overline{N}H_2 \longleftrightarrow H_2\overline{N}{-}CH{=}\overset{\oplus}{N}H_2$$

Using a CN bond length of 140 pm, determine the energies and write the wavefunctions for the occupied levels.

The box length in this case is $L = 4d_0 = 560$ pm (two bonds and two ends), and the system will have four π electrons occupying the two lowest energy orbitals, $n = 1$ and $n = 2$. Using Equation (8.3), we find that

$$E_n = \frac{h^2}{8m_eL^2} \cdot n^2 = \frac{(6.626 \times 10^{-34})^2\ \text{J}^2\text{-s}^2}{8(9.109 \times 10^{-31}\ \text{kg})(560 \times 10^{-12})^2\ \text{m}^2} \cdot n^2$$
$$= 1.92 \times 10^{-19} \cdot n^2\ \text{J}$$

so that $E_1 = 1.92 \times 10^{-19}$ J and $E_2 = 7.68 \times 10^{-19}$ J. The wavefunctions in these lowest two energy levels are given by the formulas

$$\psi_1 = \sqrt{\frac{2}{L}} \sin\left(\frac{\pi s}{L}\right) \quad \text{and} \quad \psi_2 = \sqrt{\frac{2}{L}} \sin\left(\frac{2\pi s}{L}\right)$$

where L is 560 pm and the position variable s ranges from zero to 560 pm.

In Exercise 8.21 you are asked to draw the energy level diagram and to plot the wavefunctions and probability densities for occupied levels.

8.3.2 Benzene. A π-Electron Ring

The benzene molecule, for which all six carbon atoms in the π-system have two nearest neighbor carbons, has a cyclic π system. Since the early days of organic chemistry, benzene's π-electron bonding has been described by resonance between the two structures

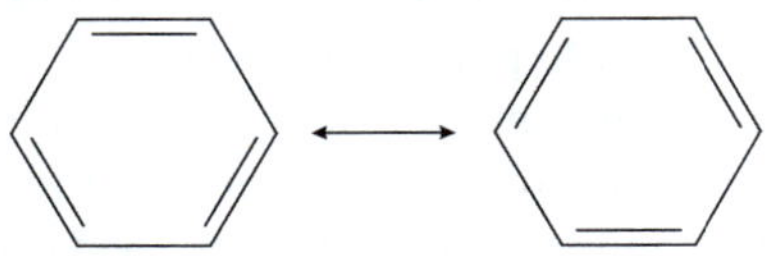

Benzene

In the free-electron model of benzene, we obtain the individual π-electron orbitals by investigating the stationary waves along a cyclic closed string (Fig. 8.6). Following our discussion in Section 3.5, we obtain the π electron energies and wavefunctions of this cyclic system with circumference L_c (where L_c is the sum of the six carbon–carbon bond lengths of benzene) by requiring that

$$\psi(s + L_c) = \psi(s) \quad \text{and} \quad \psi'(s + L_c) = \psi'(s) \tag{8.4}$$

where s is the displacement along the molecular skeleton and ψ' is the first derivative of ψ with respect to s. These boundary conditions for the electron wavefunction and its first derivative must be continuous at any point on the ring, ensure that the wavefunction is single-valued and that its second derivative (the kinetic energy operation) is defined. In the simplest case, this requirement is fulfilled for the wavefunction

$$\psi_0 = \frac{1}{\sqrt{L_c}} \qquad \Lambda_0 = \infty \tag{8.5}$$

which corresponds to a constant value of the wavefunction along the ring, that is, to a constant displacement of the cyclic string in Fig. 8.6 (the string is vibrating up and down as a whole) and to an infinite value of the de Broglie wavelength, $\Lambda_0 = \infty$. In contrast to the linear chains where the wavefunction must decay to zero beyond the last carbon atom, the ring structure has no last carbon atom and can have finite amplitude everywhere on the ring. This result for the wavefunction may seem strange; however, we have only considered the wave along the carbon–carbon ring of the molecule. The actual molecule has a three-dimensional wavefunction that has finite wavelengths along the other directions in space.

The next two wavefunctions fulfilling the boundary condition (8.4) are sine or cosine functions with de Broglie wavelengths of L_c and $L_c/2$, respectively (Fig. 8.6)

$$\psi_{1a} = \sqrt{\frac{2}{L_c}} \cdot \sin\left(\frac{2\pi s}{L_c}\right), \quad \psi_{1b} = \sqrt{\frac{2}{L_c}} \cdot \cos\left(\frac{2\pi s}{L_c}\right) \quad \Lambda_1 = L_c \tag{8.6}$$

$$\psi_{2a} = \sqrt{\frac{2}{L_c}} \cdot \sin\left(\frac{4\pi s}{L_c}\right), \quad \psi_{2b} = \sqrt{\frac{2}{L_c}} \cdot \cos\left(\frac{4\pi s}{L_c}\right) \quad \Lambda_2 = \frac{L_c}{2} \tag{8.7}$$

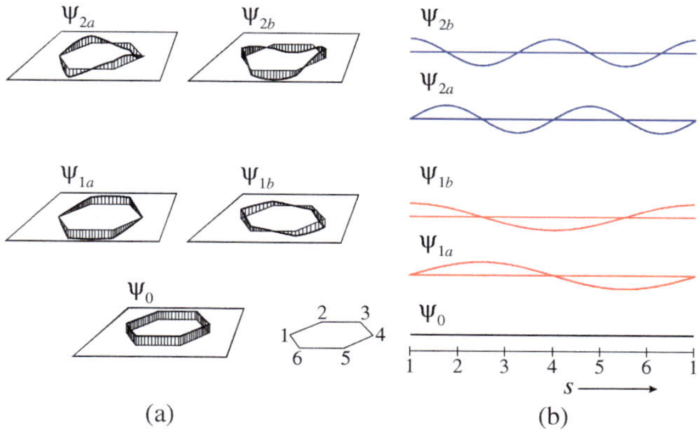

Figure 8.6 Vibrational states of a closed hexagonal string. (a) Maximum and minimum displacement of the string along the circumference of the hexagon. (b) Displacement along the coordinate s. These vibrations illustrate the standing de Broglie waves of the π electrons and should not be confused with vibrations of the molecular skeleton.

Generally, we can describe the de Broglie wavelength Λ by

$$\Lambda = \frac{L_c}{n}, \quad n = 0, 1, 2, \ldots \tag{8.8}$$

where the case $n = 0$ corresponds to infinite wavelength. The higher electronic states with $n \geq 1$ are twofold degenerate (e.g., ψ_{1a} and ψ_{1b}, Fig. 8.6). Equation (8.8) results in an integer number of wavelengths along the perimeter of the ring, for $n = 1, 2, 3, \ldots$.

We determine the energy level formula in analogy to Equation (8.3) by introducing the condition (8.8) instead of condition (8.1), and we find

$$E = \frac{h^2}{2m_e\Lambda^2} = \frac{h^2n^2}{2m_eL_c^2}, \quad n = 0, 1, 2, \ldots \tag{8.9}$$

For benzene we obtain $L_c = 6 \cdot d_0 = 840$ pm. The electron cloud and the energies of benzene are illustrated in Fig. 8.7. The energy separation found from the FEMO model, between the $n = 0$ and $n = 1$ levels, is 2.1 eV $= 3.42 \times 10^{-19}$ J, whereas the experimental energy separation is 2.9 eV.

The FEMO model describes the $n = 0$, 1, and 2 cases appropriately. For the $n = 3$ case the degeneracy is an artifact. Actually, the atomic potential troughs must be taken into account: the wavefunction with antinodes at each atom has a lower energy than that with nodes at the atoms. This energy difference is estimated in Problem 8.9.

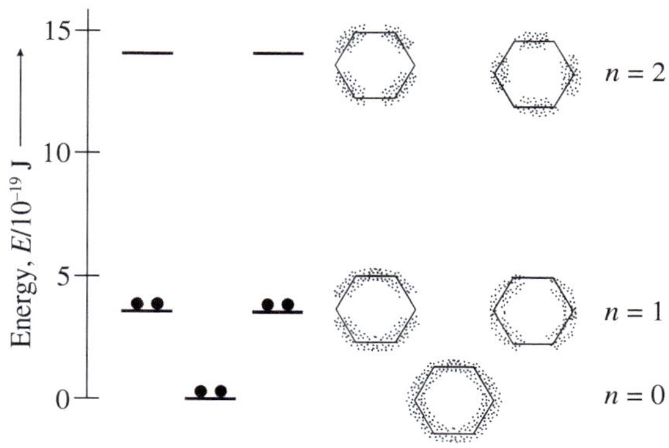

Figure 8.7 Energy E and cloud representation of the π-electron states of benzene in the free-electron model.

8.3.3 Charge Density dQ/ds

The charge density dQ/ds in the π-electron system is important to understand. For example, the variation in electron density along the π-bonds causes structural variations in the molecule, i.e., bond length changes. Also, the π-electrons often correspond to the most weakly bound electrons in a molecule so that their position (as manifest in the charge density) can influence what sites in a molecule are most reactive.

What is the electron distribution in a π-electron system? To answer this question we calculate the total probability density $\rho_{\text{total}}(s)$. The total probability density is given by

$$\rho_{\text{total}}(s) = \sum_k N_k \psi_k^2(s) \tag{8.10}$$

in which we multiply the probability densities $\psi_k^2(s)$ for each orbital k by the occupation number N_k of electrons in orbital k ($N_k = 0$: orbital empty; $N_k = 1$: orbital singly occupied; $N_k = 2$: orbital doubly occupied), and sum over all the π orbitals. From $\rho_{\text{total}}(s)$ we obtain the charge density dQ/ds as

$$\frac{dQ}{ds} = -e\,\rho_{\text{total}}(s) = -e\sum_k N_k\ \psi_k^2(s) \tag{8.11}$$

where the electron charge dQ is the charge between the positions s and $s + ds$.

In the case of ethene, we have $L = 3d_0$ and only one occupied orbital so that

$$\rho_{\text{total}}(s) = 2 \cdot \frac{2}{L} \cdot \sin^2\left(\frac{\pi s}{L}\right) = \frac{4}{3d_0} \cdot \sin^2\left(\frac{\pi s}{3d_0}\right)$$

and

$$\frac{dQ}{ds} = -e\,\rho_{\text{total}}(s) = -e\frac{4}{3d_0} \cdot \sin^2\left(\frac{\pi s}{3d_0}\right) \tag{8.12}$$

For butadiene $L = 5d_0$ and we must sum over both occupied orbitals, so that

$$\begin{aligned}\rho_{\text{total}}(s) &= 2 \cdot \psi_1^2 + 2 \cdot \psi_2^2 = 2 \cdot \frac{2}{L}\sin^2\left(\frac{\pi s}{L}\right) + 2 \cdot \frac{2}{L}\sin^2\left(\frac{2\pi s}{L}\right)\\ &= \frac{4}{5d_0}\sin^2\left(\frac{\pi s}{5d_0}\right) + \frac{4}{5d_0}\sin^2\left(\frac{2\pi s}{5d_0}\right)\end{aligned}$$

and

$$\frac{dQ}{ds} = -e\rho_{\text{total}}(s) = -e \cdot \frac{4}{5d_0} \cdot \left[\sin^2\left(\frac{\pi s}{5d_0}\right) + \sin^2\left(\frac{2\pi s}{5d_0}\right)\right] \tag{8.13}$$

In Fig. 8.8, ρ_{total} is displayed as a function of s for ethene, butadiene, amidinium ion, and benzene. For ethene we find a maximum of ρ_{total} at the center of the bond, which is consistent with our qualitative view of a double bond between two atoms. For butadiene ρ_{total} is larger in the centers of the formal double bonds than in the center of the formal single bond. Note that ρ_{total} of the single bond is greater than that for a single bond in ethane, and we say that the central bond in butadiene has some double bond "character." For benzene the total charge density is constant along the ring.

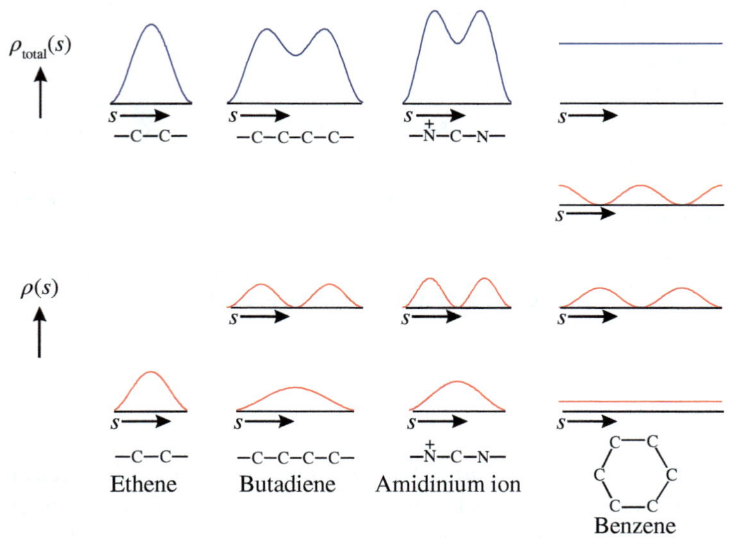

Figure 8.8 Probability densities $\rho(s)$ (red solid lines) and total probability densities $\rho_{\text{total}}(s)$ (blue solid lines) along the bond line, calculated from the FEMO model for ethene, butadiene, amidinium cation, and benzene.

Example 8.2 *Determine the total charge density profile for the amidinium ion. Compare the magnitude of the charge density at the carbon atom to that at the nitrogen atoms.*

Following Equation (8.11) and using the results of Example 8.1, with $L = 4d_0$ we find that

$$\begin{aligned}\rho_{\text{total}}(s) &= 2 \cdot \psi_1^2 + 2 \cdot \psi_2^2 = 2 \cdot \frac{2}{L}\sin^2\left(\frac{\pi s}{L}\right) + 2 \cdot \frac{2}{L}\sin^2\left(\frac{2\pi s}{L}\right)\\ &= \frac{4}{4d_0}\sin^2\left(\frac{\pi s}{4d_0}\right) + \frac{4}{4d_0}\sin^2\left(\frac{2\pi s}{4d_0}\right)\\ &= \frac{1}{d_0}\left[\sin^2\left(\frac{\pi s}{4d_0}\right) + \sin^2\left(\frac{2\pi s}{4d_0}\right)\right]\end{aligned}$$

and

$$\frac{dQ}{ds} = -e\rho_{\text{total}}(s) = -e\frac{1}{d_0} \cdot \left[\sin^2\left(\frac{\pi s}{4d_0}\right) + \sin^2\left(\frac{2\pi s}{4d_0}\right)\right]$$

The carbon atom is in the center of the molecule at the position $2d_0$. Evaluating our expression for (dQ/ds) at $s = 2d_0$ gives $(-e/d_0)$. The nitrogen atoms are located at the positions d_0 and $3d_0$. The value of (dQ/ds) at each of these positions is $(-\frac{3}{2}e/d_0)$. Clearly, we find a higher π-electron density near the nitrogen atoms. This charge density function is plotted in Fig. 8.8.

8.3.3.1 Total Probability Density $\rho_{\text{total}}(s)$ and the Bond Length

The larger the total probability density $\rho_{\text{total}}(s)$ between two nuclei, the stronger and shorter is the corresponding bond. Figure 8.9 attempts to illustrate the electrostatic origin of this effect. The uneven distribution of the electron charge causes an attraction of the nuclei to the positions where the electron cloud is accumulated. Consequently, we expect that the central carbon–carbon bond in butadiene will be longer than the other two bonds. Table 8.2 gives data for these bond lengths and shows that our expectation is met.

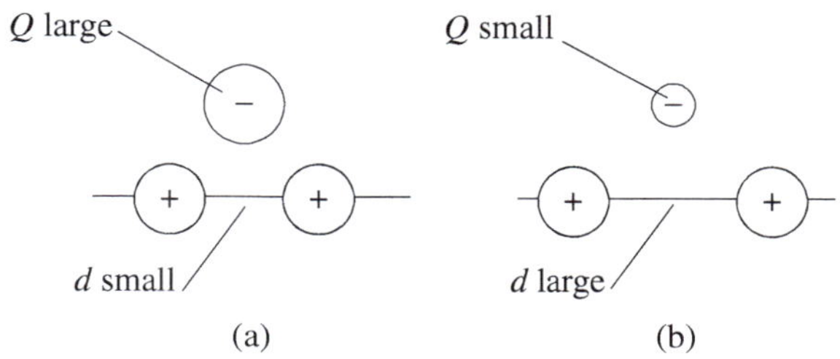

Figure 8.9 Attraction between a π electron cloud and two positive nuclear charges. (a) Electron density large: distance between nuclei small. (b) Electron density small: distance between nuclei large.

In contrast to the butadiene case, no bond alternation is found for benzene. In benzene the total probability density $\rho_{\text{total}}(s)$ is constant along the ring; i.e.,

$$\begin{aligned}\rho_{\text{total}}(s) &= 2 \cdot \psi_0^2 + 2 \cdot \psi_1^2 + 2 \cdot \psi_2^2 \\ &= 2 \cdot \frac{1}{L_c} + 2 \cdot \frac{2}{L_c} \cdot \left(\sin^2\left(\frac{2\pi s}{L_c}\right) + 2 \cdot \cos^2\left(\frac{2\pi s}{L_c}\right)\right) = \frac{6}{L_c}\end{aligned} \tag{8.14}$$

where L_c is the perimeter of the ring's carbon–carbon skeleton. Thus the electron charge is uniformly distributed along the bond axes. As in Section 8.3.1 we approximate the bond length in benzene as $d_0 = 140$ pm. Taking $L_c = 6 \times 140$ pm $= 6 \times 0.140$ nm, we obtain

$$\rho_{\text{total}}(s) = \frac{6}{6 \times 140 \text{ pm}} = 7.14 \text{ nm}^{-1} \tag{8.15}$$

Thus we conclude that the bars symbolizing the bonds in the structural formula of benzene do not correspond to the actual electron distribution. In Exercise 8.22 you extend this argument to the case of equal bond lengths in rings with $(4n + 2)$ carbon atoms: e.g., benzene ($n = 1$), [10] annulene (decapentaene, $n = 2$), [14] annulene ($n = 3$). This result is called the *Hückel rule*. However, a more sophisticated consideration in Section 8.4 proves that the validity of the Hückel rule is restricted to $n \leq 2$.

Table 8.2 FEMO and Step Potential Model Total Probability Density $\rho_{\text{total}}(s)$ in the Center of the Bond and Corresponding Experimental Bond Length for Some π Bonds.[2]

		$\rho_{\text{total}}(s)/\text{nm}^{-1}$		
Molecule	**Bond**	**FEMO**	**Step Potential**	$\frac{d_{\text{expt}}}{\text{pm}}$
Ethene	C=C	9.52	10.96	133.1
Butadiene	C–C	5.71	3.36	145.4
	C=C	8.91	10.87	133.8
Benzene	C$\overset{...}{-}$C	7.14	7.14	139.7

2 Ethene, Butadiene: Craig et al. [2]. Benzene: Baba et al. [3].

The correlation between the charge density and the bond length suggests that we can use the electron density profile of a molecule to predict bond length variations. Alternatively, we can directly probe the charge density along the chain by looking at measured bond lengths.

Example 8.3 *Derive the FEMO data for ethene, butadiene, and benzene in Table* 8.2.

We use Equation (8.12) to calculate $\psi^2_{\text{total}}(s)$ for ethene at the position $s = 3d_0/2$ using $d_0 = 140$ pm

$$\rho_{\text{total}}(s) = 2 \cdot \frac{2}{3d_0} \cdot \sin^2\left(\frac{\pi s}{3d_0}\right) = 2 \cdot \frac{2}{3d_0} \cdot \sin^2\left(\frac{\pi}{2}\right)$$
$$= \frac{4}{3d_0} = 9.52 \times 10^{-3}\ \text{pm}^{-1} = 9.52\ \text{nm}^{-1}$$

We use Equation (8.13) to calculate $\rho_{\text{total}}(s)$ for butadiene at the positions $s = 3d_0/2$ (double bond) and $s = 5d_0/2$ (single bond) using $d_0 = 140$ pm

$$\left[\rho_{\text{total}}(s)\right]_{\text{double bond}} = 2 \cdot \frac{2}{5d_0} \cdot \left[\sin^2\left(\frac{\pi s}{5d_0}\right) + \sin^2\left(\frac{2\pi s}{5d_0}\right)\right]$$
$$= 2 \cdot \frac{2}{5d_0} \cdot \left[\sin^2\left(\frac{3\pi}{2 \cdot 5}\right) + \sin^2\left(\frac{2 \cdot 3\pi}{2 \cdot 5}\right)\right]$$
$$= 2 \cdot \frac{2}{5d_0} \cdot [0.654 + 0.905] = 2 \cdot \frac{2}{5d_0} \cdot [0.654 + 0.905]$$
$$= 2 \cdot \frac{2}{5d_0} \cdot 1.559 = 8.91\ \text{nm}^{-1}$$

$$\left[\rho_{\text{total}}(s)\right]_{\text{single bond}} = 2 \cdot \frac{2}{5d_0} \cdot \left[\sin^2\left(\frac{5\pi}{2 \cdot 5}\right) + \sin^2\left(\frac{2 \cdot 5\pi}{2 \cdot 5}\right)\right]$$
$$= 2 \cdot \frac{2}{5d_0} \cdot [1.0 + 0] = 5.71\ \text{nm}^{-1}$$

Equation (8.15) calculates $\rho_{\text{total}}(s)$ for benzene.

$$\rho_{\text{total}}(s) = 7.14\ \text{nm}^{-1}$$

8.3.4 Branched Molecules

As an example of a branched π-electron system, we consider the guanidinium cation. It has six π electrons and is described by resonance between the three structures:

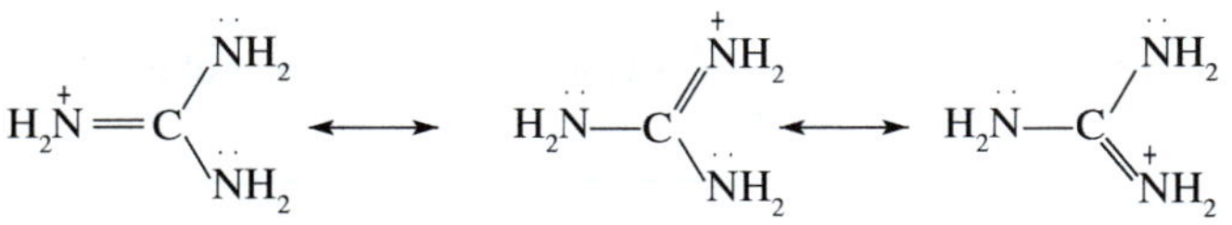

As for the linear and cyclic systems, the FEMO model describes the π electrons by the wavefunctions of an electron in a box, which extend over the sigma framework. The π-orbitals along the bond axes in a branched system can be approximated by the wavefunctions of an electron in an accordingly branched box. The wavefunctions can again be approximated by the product $\psi(s)$ and a function depending on coordinates perpendicular to s. $\psi(s)$ can be approximated by the amplitudes of the standing waves of a branched string, see Fig. 8.10.

8.3.4.1 Energies

We consider the de Broglie wavelength Λ in terms of the length L, where $L/2$ is the length of each branch. Using $d_0 =$ 140 pm for each carbon–nitrogen bond and adding d_0 for the end of a branch, we find that $L = 4d_0$. By inspection, we can classify the energies of the wavefunctions in Fig. 8.10 based on their de Broglie wavelengths. We see that orbital 1 has a de Broglie wavelength of $\Lambda = 2L$ and corresponds to the ground state. Also we see that orbital 4 has $\Lambda = 2L/3$ and corresponds to the highest energy state of the four shown here. Orbitals 2 and 3 belong to a doubly degenerate energy level ($\Lambda = L$), and form an orthogonal set.

According to Equation (8.3), the energies E result from the relation

$$E = \frac{1}{2}m_e v^2 = \frac{h^2}{2m_e\Lambda^2} \tag{8.16}$$

The energies for the states 1–4 are shown in Fig. 8.11. From the energy level diagram, we see that a seventh π-electron (for the neutral form of guanidinium) would have to be added to the anti-bonding π-orbital. The stability of the positively charged guanidinium ion relative to the neutral guanidine is reflected by the fact that guanidine is a very strong base. It easily binds a proton forming a guanidinium ion.

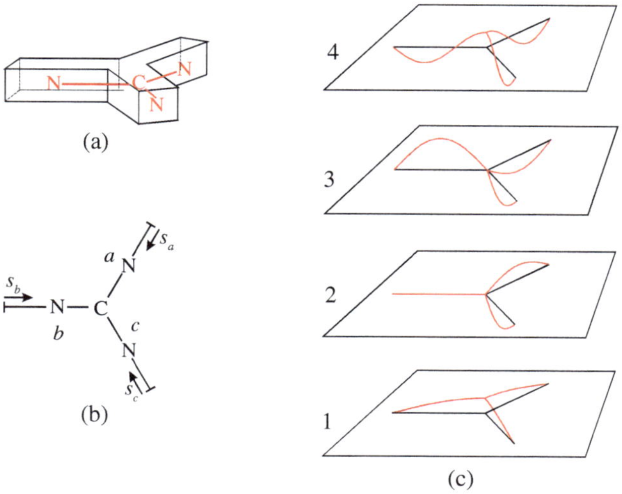

Figure 8.10 Guanidinium cation as an example for a branched π electron system. (a) Branched box. (b) Coordinates s_a, s_b, and s_c along the molecular skeleton. The electron gas extends by one bond length beyond the last atom. (c) Vibrational modes of a branched string correspond to the standing wave solutions for the guanidinium cation.

8.3.4.2 Wavefunctions

Orbitals 1 and 4 have an antinode at the branching point, whereas orbitals 2 and 3 have a node at the branching point. The wavefunctions for orbitals 2 and 3 may be somewhat puzzling for the beginner. Figure 8.12 shows the three lowest energy standing waves (A, B, and C) of a branched string that have a node at the branching point. In A the wave extends over branches b and c; branch a is at rest. Correspondingly, the branch b is at rest in B, and the branch c is at rest in C. Each of these waves has the same de Broglie wavelength, hence energy, and form a degenerate set.

These three degenerate wavefunctions ψ_A, ψ_B, and ψ_C are not independent of each other, however. For example, the integral $\int \psi_A \psi_B \cdot \mathrm{d}s \neq 0$ (the integrand is zero in branches a and b and has a finite value in branch c). However, by superimposing the degenerate functions ψ_A and ψ_B we obtain the wavefunction ψ_D (Fig. 8.12), that is orthogonal to ψ_B ((the integrand in $\int \psi_B \psi_D \cdot \mathrm{d}s$ is zero in branch b and equal in value, but opposite in sign in branches a and c; thus, the integral is zero). Thus we have an orthogonal two-fold set of degenerate wavefunctions ψ_B and ψ_D.

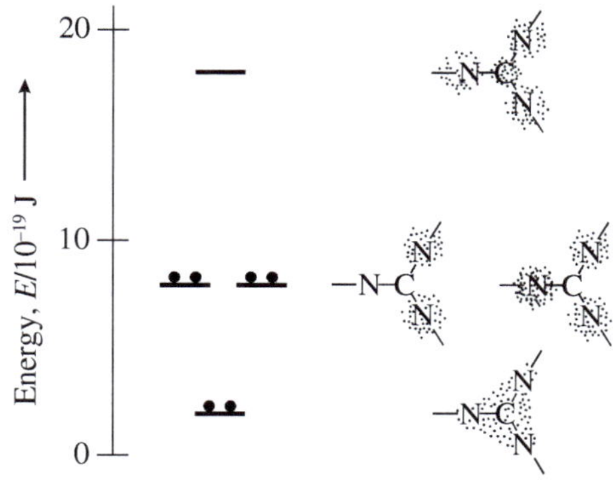

Figure 8.11 Guanidinium cation; energy scheme and charge distribution of the different states.

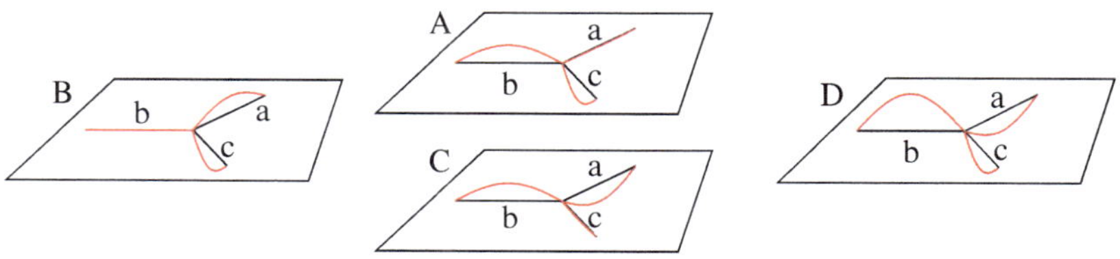

Figure 8.12 Some degenerate vibration modes of a string, analogous to the π-electron waves of the guanidinium ion. Mode B is orthogonal to mode D, which is a superposition of modes A and C. Modes A, B, and C are not mutually orthogonal.

8.3.4.3 π Electron Density

The three lowest energy levels in the guanidinium cation are occupied with six π electrons. The π electron density in each CN bond is the same, consistent with the three equivalent limiting structures. The charge distributions are

Orbital 1. The charge of 2 electrons is distributed equally over the three branches, giving $\frac{2}{3}e$ per branch.
Orbital 2. The charge of two electrons is equally distributed over branch a, and over branch c.
Orbital 3. The wavefunction ψ in branch b is twice the amplitude in branches a and c. Hence the electron density (proportional to ψ^2) in branch b is four times larger. Therefore, a charge of $\frac{4}{3}e$ is distributed over branch b, and the charge of $\frac{1}{3}e$ is distributed over each of branches a and c.

Summing over all three orbitals, we obtain

Branch a	charge of $\frac{2}{3}+1+\frac{1}{3}$	$=$ 2 electrons
Branch b	charge of $\frac{2}{3}+0+\frac{4}{3}$	$=$ 2 electrons
Branch c	charge of $\frac{2}{3}+1+\frac{1}{3}$	$=$ 2 electrons

The electron charge is equally distributed over the branches of the nuclear skeleton. The same charge density occurs for any orthogonal pair of wavefunctions that might be chosen for the degenerate energy level. The situation is the same as in benzene, where we can arbitrarily choose any point on the ring as the zero point of the coordinate s (the positions of the nodes and antinodes). This choice does not have any effect on the total charge distribution.

For larger molecules the energies and wavefunctions of arbitrary branched π-electron systems are obtained by applying numerical methods (see Foundations 8.1 and 8.2), making use of the program FEMO_HMO which is provided on the book's website.

8.4 BCD (Step Potential) Model Applied to Polyenes and Annulenes

8.4.1 Total Probability Density and Bond Length

In our considerations in Section 8.3.3 we observed that the π-electron probability density ρ_{total}, obtained by assuming equal carbon–carbon bond lengths, correlates with the bond lengths observed in experiment. However, we have neglected the resulting influence of bond length changes on the π-electron distribution. In a bond that is shortened by the Coulomb attraction between the π electron cloud and the adjacent nuclei (Fig. 8.9), the potential energy of the π electrons is decreased. This stabilization causes an additional accumulation of the π electron charge at the shortened bond. This effect is important for understanding the fact that in some π-electron systems (like the polyenes) the mutual interaction causes the formation of strongly alternating bonds and that no bond alternation occurs in polymethine chains or in benzene.

We use the step potential model to calculate the π-electron probability density ρ_{total} for ethene, butadiene, and benzene in the way illustrated by Fig. 8.4 using the experimentally determined values of the bond length (see Table 8.2). The plot in Fig. 8.13 shows a linear relationship between ρ_{total} and d. How can we understand this simple result of the computation? The nuclear cores, which are connected by a bond, are attracted by the π-electron density between them.

The data points are interpolated by a straight line:

$$\frac{\rho_{\text{total}}}{\text{nm}^{-1}} = 94.195 - 0.62416\,\frac{d}{\text{pm}} \tag{8.17}$$

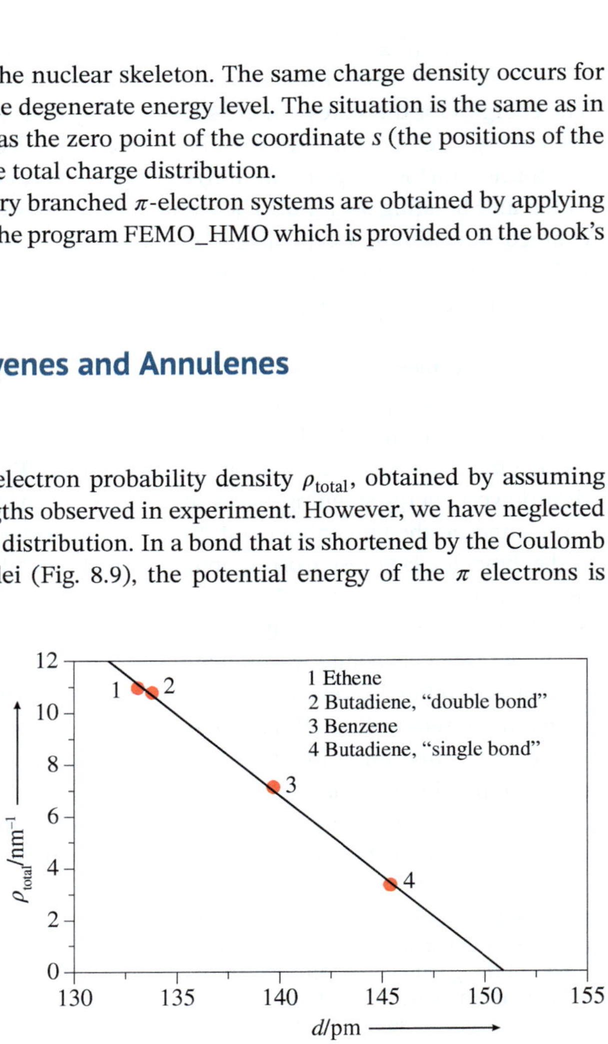

Figure 8.13 Total probability density $\rho_{\text{total}}(s)$ in the middle of a bond according to the step-potential model as a function of experimental bond length d for ethene, butadiene, and benzene (red full circles).

For determining the bond lengths of an arbitrary π-electron system, e.g., a polyene, we proceed in an iterative fashion. First we assume equal bond lengths for the polyene, that is, we choose the potential energy as $V(s) = 0$. Next we solve the Schrödinger equation with this potential energy to obtain the wavefunctions and the corresponding total probability density ρ_{total} at the middle of each bond. With this value we compute a new $V(s)$ and solve the Schrödinger equation again. This procedure is continued until the densities agree with those calculated with the new $V(s)$ (self-consistency). Finally the corresponding values for the bond lengths are obtained by using the correlation in Fig. 8.13.

This procedure is presented in Box 8.1 for hexatriene and benzene. For simplicity we use the approximation according to Fig. 8.4b (top, steps of heights given by bond lengths) rather than the more realistic model of Fig. 8.4a.

8.4.2 Energy Gap in a Long Chain Polyene

A step potential model treatment of polyenes of any number of CH units leads to bonds that alternate in length, as in butadiene. As a consequence, the π-electron distribution and the π-electron energies are strongly different from those obtained in a FEMO calculation. As an example, the energy scheme of a polyene with 20 carbon atoms is shown in Fig. 8.14. Note that the step potential model predicts an energy gap between the highest occupied (HOMO) and the lowest unoccupied (LUMO) orbital, a feature that is discussed more in Section 13.4. The program "step_model.exe" for the treatment of linear and annular molecules is provided on the book's website.

In the case of fullerene the step potential model calculation leads to alternating bonds of lengths 145 and 138 pm in agreement with experimental data resulting from X-ray diffraction data (143 and 139 pm), see Foundation 8.5.

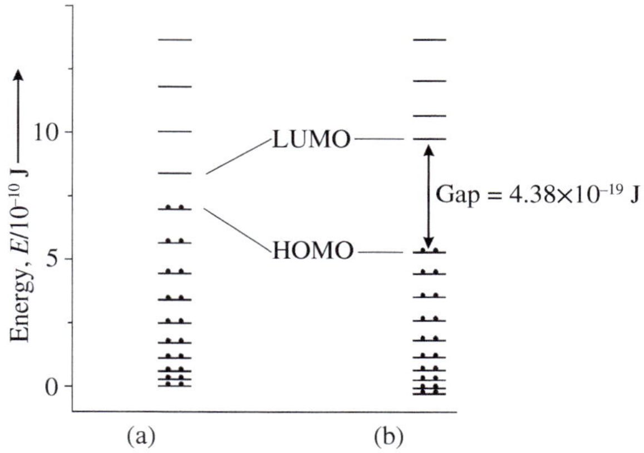

Figure 8.14 Energy scheme calculated for a polyene with 20 carbon atoms (10 doubly occupied energy levels). (a) FEMO calculation (assumption of equal bond lengths). (b) Step potential model calculation resulting in alternating double and single bonds and occurrence of an energy gap of 4.38×10^{-19} J between the HOMO and LUMO states.

8.4.3 Annulenes

In the case of π electron rings with $(4n + 2)$ carbon atoms, this treatment leads to equal bonds for benzene ($n = 1$) and for [10] annulene (polyene ring with 10 carbon atoms, $n = 2$). This is in agreement with experiment and in accordance with the Hückel Rule, which predicts that a ring with $4n + 2\pi$ electrons is aromatic. However, alternating bonds are obtained for $n > 2$, in agreement with experiment [4] and in contradiction to the Hückel Rule. This kind of behavior is expected; large rings must have alternating bonds like polyenes. The change over to equal bonds must take place at some distinct value of n. The critical value $n = 3$ ([14] annulene) results from a detailed calculation (Fig. 8.15. This result is not trivial, e.g., it is not obvious that benzene has equal bonds; it is the prediction of the step potential model calculation.

Figure 8.15 The change in bond length $\Delta d = d_{single} - d_{double}$, predicted from the step potential model, is plotted for the different [k] annulenes with $k = 4$ to 26. Red full circles: $4n + 2$ annulenes. Equal bond lengths for $n = 1$ and $n = 2$, according to the Hückel rule ($k = 6$ and 10, $\Delta d = 0$). Alternating bond lengths for $n > 2$, in disagreement with the Hückel rule ($k = 14, 18, 22, 26$). Black full circles: $4n$ annulenes. Dashed lines: Δd for butadiene ($d_{single} = 146$ pm, $d_{double} = 134$ pm, $\Delta d = 12$ pm) and polyacetylene ($d_{single} = 143$ pm, $d_{double} = 135$ pm, $\Delta d = 8$ pm), respectively.

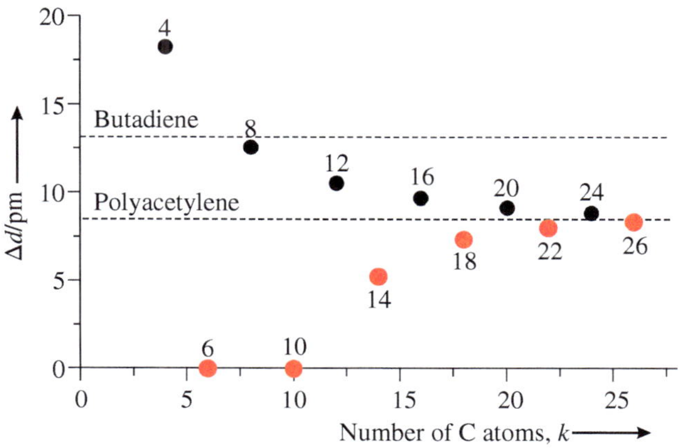

8.4.4 Neutral Soliton

A particular situation is expected in large rings with an odd number of carbon atoms. Although alternating bonds exist for the reason we have discussed, a smooth region extending over 15 CH units near to the odd electron (*neutral soliton*) has nearly equalized bond lengths. A band gap of 1.78 eV as in a long polyene, and an intergap state occupied by the odd electron appear (Fig. 8.16).

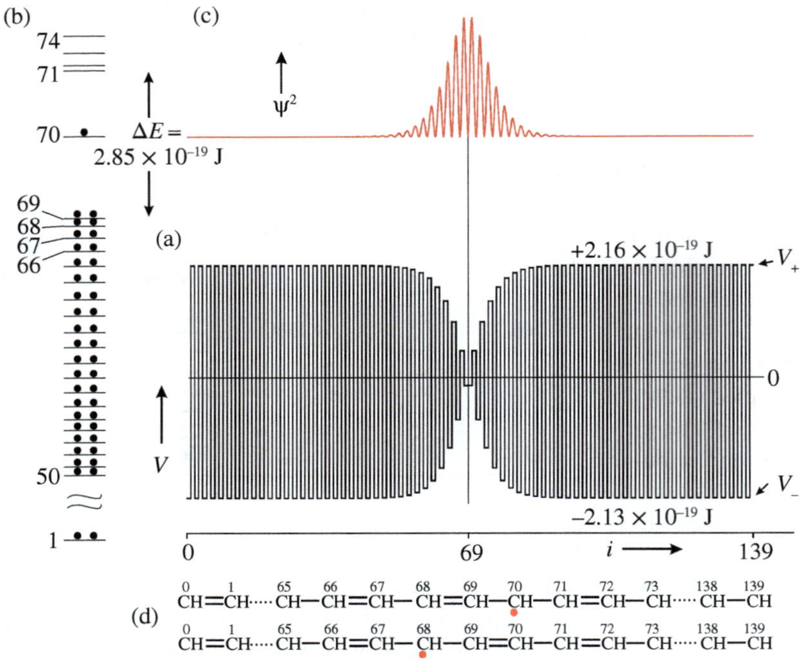

Figure 8.16 A large, odd numbered hydrocarbon ring must have a defect structure (called a soliton) that reverses the bond alternation pattern. The diagram corresponds to a hydrocarbon ring with 139 CH units and is cut at $i = 0$ which equals to $i = 139$; the index i numbers the CH units. The calculation is performed by the step potential model. (a) The bond potential energy V outside the soliton alternates between $V_+ = +1.35$ eV $= 2.16 \times 10^{-19}$ J (single bond) and $V_- = -1.44$ eV $= -2.31 \times 10^{-19}$ J (double bond). (b) The calculated energies give rise to a valence band (69 energy states, each state occupied by two electrons) and a conduction band (all states empty) with a band gap of $\Delta E = 1.78$ eV $= 2.85 \times 10^{-19}$ J. (c) The neutral soliton (centered at $i = 69$) with bond potential energy $V \approx 0$ eV (region of equal bonds) gives rise to a localized intergap state (occupied by one electron) with wavefunction ψ. (d) The soliton centered at site 69 is essentially described by two resonance structures.

8.5 Principles of Density Functional Theory (DFT)

The considerations on BCD are useful in developing an understanding of a rigorous approach to electronic structure calculations called Density Functional Theory, DFT. We have seen in Section 8.4 that the basic concepts in BCD are:

1) The total π electron density determines the ground state energy; it determines the elastic deformation of the σ bonded lattice in its equilibrium state, i.e., when the ground state of the system is reached.
2) The kinetic energy is obtained as a sum of orbital contributions.
3) The potential energy is described by the attraction of the π electron cloud to the atomic cores and by the repulsion between the π-electrons determined by their total density. The exchange-correlation is taken into account by the assumption that as long as a given pair of π electrons is in a given atom, no other π electron is in that same atom's orbital.
4) The total π electron density is obtained as a sum of orbital contributions.

In contrast to BCD, DFT is general and takes into account all $N_{\text{electrons}}$ electrons in the field of all N_{nuclei} nuclei. Systems in the ground state are considered in which the paired electrons have the same spatial one-electron orbitals.

1) according to the Hohenberg and Kohn theorem [5]: the total electron density ρ uniquely determines the ground state.
2) in the Kohn–Sham formalism [6] the kinetic energy is the sum of orbital contributions

$$T = -\frac{\hbar^2}{2m_e} \sum_{i=1}^{N_{\text{electrons}}} \int \psi_i(r) \nabla^2 \psi_i(r) \, dr \tag{8.18}$$

where $N_{\text{electrons}}$ is the number of electrons.

3) the potential energy is the sum of the energies of attraction of all electrons to all nuclei

$$V_{\text{attraction}} = -\sum_{l=1}^{N_{\text{nuclei}}} \int \frac{Z_l\ e^2}{4\pi\varepsilon_0\ r_l}\rho(r)\ \mathrm{d}r \tag{8.19}$$

(where N_{nuclei} is the number of nuclei) and of the mutual electron repulsion. The potential energy for repulsion is a sum of the Coulomb energy between the total charge distribution $\rho(r)$ and the exchange-correlation energy E_{ExCorr}

$$V = +\frac{1}{2}\int \frac{\rho(r_1)\rho(r_2)\cdot e^2}{4\pi\varepsilon_0 r_{12}}\mathrm{d}r_1\ \mathrm{d}r_2 + E_{\text{ExCorr}}(\rho) \tag{8.20}$$

E_{ExCorr} takes into account exchange as well as some electron correlation. This term is approximated by a number of different methods.

4) The total electron density is obtained as a sum of orbital contributions

$$\rho(r) = \sum_{i=1}^{N_{\text{electrons}}} \psi_i^2 \tag{8.21}$$

Finally, the ground-state energy is obtained by minimizing the total energy.

The Kohn–Sham DFT approach is analogous to the Hartree–Fock treatment; however, the calculation would be exact if the true effective potential energy

$$V_{\text{eff}} = V_{\text{attraction}} + \frac{1}{2}\int \frac{\rho(r_1)\,\rho(r_2)\cdot e^2}{4\pi\varepsilon_0 r_{12}}\mathrm{d}r_1\ \mathrm{d}r_2 + E_{\text{ExCorr}}(\rho)$$

were known. Because the true mathematical form of $E_{\text{ExCorr}}(\rho)$ is not known, practical DFT calculations proceed by choosing a model for $E_{\text{ExCorr}}(\rho)$. An important advantage of the DFT approach over the traditional HF-SCF method is that one can perform calculations on larger (i.e., more atoms) and extended systems, and this has opened up the application of quantum chemistry to understanding the properties of large molecules and of materials [7].

8.6 HMO Model

Although LCAO methods must be applied at a sophisticated level for quantitative accuracy, even very low level treatments provide a guide to π-bonding and orbital shapes. Here we present one approximate LCAO model, the Hückel model (named after its inventor Erich Hückel, 1896–1980). The Hückel model is of great historical importance to our understanding of bonding in aromatic molecules. The Hückel model need not be restricted to π-systems and has been used to describe σ-bonding in molecules, where it is called the *Extended Hückel Molecular Orbital (EHMO) method* [8], and bonding in covalent solids such as silicon. The Hückel model is simple enough that it allows for facile computational results to be achieved for even very large systems, for example, proteins and polymers.

8.6.1 Wavefunctions and Energies

The Hückel model describes π-electrons by an LCAO test function that is composed of $2p$ atomic orbitals. It is common to refer to these atomic orbitals as the basis orbitals, since they are superposed to construct the molecular orbitals. For carbon, the Slater-type p_z-orbital is given by

$$\varphi_{2p_z} = A\cdot z\cdot \mathrm{e}^{-3.25r/(2a_0)} \tag{8.22}$$

Let us begin with the simple example of ethene. For ethene the wavefunction ϕ_1 in the ground state is

$$\phi_1 = c_1\cdot(\varphi_a + \varphi_b) \tag{8.23}$$

where φ_a and φ_b are Slater atomic orbitals at the centers a and b, respectively; c_1 is a constant. By following a procedure like that used for the H_2^+ molecular ion in Section 6.2.3, we obtain,

$$\phi_1 = \frac{1}{\sqrt{2(1+S_{ab})}} \cdot (\varphi_a + \varphi_b) \quad \text{and} \quad \varepsilon_1 = \frac{H_{aa} + H_{ab}}{1 + S_{ab}} \tag{8.24}$$

for the energy and the wavefunction. Formally this result is like that for the H_2^+ molecular ion; however, the basis atomic orbitals are $2p_z$ rather than $1s$ and the effective nuclear charges (carbon core versus proton) are different.

The wavefunction ϕ_2 for the first excited state must be orthogonal to ϕ_1 and corresponds to the antisymmetric combination

$$\phi_2 = c_2 \cdot (\varphi_a - \varphi_b) \tag{8.25}$$

In Exercise 8.23 you show that the energy ε_2 is given by

$$\varepsilon_2 = \int \phi_2 H \phi_2 \cdot d\tau = 2c_2^2 \cdot (H_{aa} - H_{ab}) \tag{8.26}$$

By normalizing ϕ_2, we obtain

$$\phi_2 = \frac{1}{\sqrt{2(1-S_{ab})}} \cdot (\varphi_a - \varphi_b) \quad \text{and} \quad \varepsilon_2 = \frac{H_{aa} - H_{ab}}{1 - S_{ab}} \tag{8.27}$$

Hence the p-orbitals form bonding and antibonding pairs, in direct correspondence with the description of π-bonds for diatomic molecules in Chapter 7.

In order to simplify this approach so that it can be applied to more complex π electron systems, three common approximations are made.

1) As in the free-electron model the π electrons are considered to be in the field of a rigid molecular skeleton.
2) The integral S_{ab} which depends on the overlap of neighboring atomic orbitals φ_a and φ_b is neglected; that is, $S_{ab} = 0$.
3) H_{ab} is considered to be zero for atoms which are not directly linked by a σ bond. For atoms linked by a σ bond, H_{ab} depends on the nature of the atoms a and b and on their distance apart.

This last approximation restricts the interaction to nearest neighbors. The integrals H_{aa} (self-energy of atomic orbital φ_a) and H_{ab} (interaction energy between φ_a and φ_b) are not calculated explicitly but determined by comparison with experimental data. Hence we call this a *semi-empirical model*. Although the neglect of the overlap integral S_{ab} may seem strange in light of the role of overlap in bonding considerations, H_{ab} as well as S_{ab} depends on the overlap, and H_{ab} is the determining factor. The neglect of S_{ab} has considerable consequences, however, and this should be kept in mind when judging the results of HMO calculations (see Box 8.2).

Let us consider carbon π-systems, with equal bond lengths. In this case each carbon's p_z-electron has the same self-energy, H_{aa}, and the same interaction energy between each pair of neighboring carbon atoms, H_{ab}. Hence we can write that

$$\alpha = H_{aa}; \quad \beta = H_{ab} \tag{8.28}$$

where α is the energy of the electron in the $2p_z$ atomic orbital and β is the energy for the overlap density of two adjacent orbitals (α and β are negative quantities). With these definitions the result for the ethene molecule (Equations (8.24) and (8.27)) become

$$\phi_1 = \frac{1}{\sqrt{2}} \cdot (\varphi_a + \varphi_b); \quad \varepsilon_1 = \alpha + \beta \tag{8.29}$$

$$\phi_2 = \frac{1}{\sqrt{2}} \cdot (\varphi_a - \varphi_b); \quad \varepsilon_2 = \alpha - \beta$$

These energy levels and wavefunctions are plotted in Fig. 8.17 and compared to those calculated by the FEMO model. The molecular orbital ϕ_1 is a bonding orbital (antinode in the center of the bond), and ϕ_2 is an antibonding orbital (node in the center of the bond). The electronic stabilization that arises from the π-bond in ethene corresponds to 2β; i.e., the bonding energy level is more stable than the atomic orbital with energy α by the quantity β, and it is occupied with two electrons for a net stabilization of 2β.

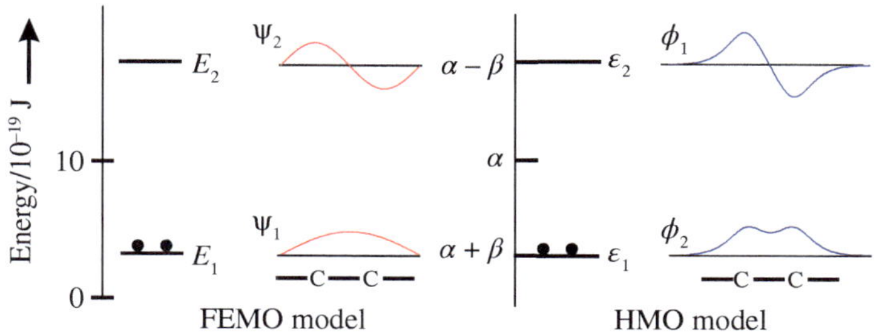

Figure 8.17 Energies and wavefunctions for ethene. (a) FEMO model. (b) HMO model. Wavefunction along the bond line 75 pm above the molecule plane.

8.6.2 Application of the HMO Method to π-Electron Chains

Now that we see the essential features of an HMO calculation, let us consider a more complicated system. Butadiene provides a good example because it can be compared to ethene in an instructive way, and it is complicated enough to illustrate the new mathematical tools we need to apply. The molecular structure of butadiene is $\overset{1}{CH_2} = \overset{2}{CH}–\overset{3}{CH} = \overset{4}{CH_2}$, where we label the carbon atoms 1 through 4.

Because our basis functions are four p orbitals (one on each carbon atom), butadiene will have four π-molecular orbitals (let us call them ϕ_1, ϕ_2, ϕ_3, and ϕ_4) and corresponding energies ($\varepsilon_1, \varepsilon_2, \varepsilon_3$, and ε_4). Following our prescription for constructing the LCAO molecular orbitals we can write ϕ_1 as

$$\phi_1 = c_1\varphi_1 + c_2\varphi_2 + c_3\varphi_3 + c_4\varphi_4 = \sum_{j=1}^{4} c_j\varphi_j$$

where φ_j is the $2p_z$ orbital centered on carbon j and the c_j are coefficients that provide the contribution of atomic orbital φ_j to molecular orbital ϕ_1. We proceed in a corresponding way to construct each molecular orbital, which requires a different set of coefficients. To find the energies and wavefunctions, we apply the variational method on each of the molecular orbitals under the constraint that they are mutually orthogonal. In Foundation 8.3, we show that this procedure generates a system of four algebraic equations, which need to be solved to find the energies ε and the wavefunctions. These equations are

$$\begin{aligned}
c_1 \cdot (\alpha - \varepsilon) + c_2 \cdot \beta &= 0 \\
c_1 \cdot \beta + c_2 \cdot (\alpha - \varepsilon) + c_3 \cdot \beta &= 0 \\
c_2 \cdot \beta + c_3 \cdot (\alpha - \varepsilon) + c_4 \cdot \beta &= 0 \\
c_3 \cdot \beta + c_4 \cdot (\alpha - \varepsilon) &= 0
\end{aligned}$$

We could proceed to solve these equations by the method of substitution; however, that method starts to become cumbersome for butadiene and is certainly so for larger systems. The method of determinants is easier to implement, especially for large systems, and it is amenable to effective numerical solution. Foundations 8.3 and 8.4 describe the basic features of this method. Using this approach we can determine the energy values and the coefficients by solving the determinant,

$$\begin{array}{c} \\ \mathbf{1} \\ \mathbf{2} \\ \mathbf{3} \\ \mathbf{4} \end{array}
\begin{array}{c} \begin{array}{cccc} \mathbf{1} & \mathbf{2} & \mathbf{3} & \mathbf{4} \end{array} \\
\begin{vmatrix}
(\alpha - \varepsilon) & \beta & 0 & 0 \\
\beta & (\alpha - \varepsilon) & \beta & 0 \\
0 & \beta & (\alpha - \varepsilon) & \beta \\
0 & 0 & \beta & (\alpha - \varepsilon)
\end{vmatrix} = 0 \end{array}$$

Often this quantity is called the *secular determinant*. The evaluation of the determinant to give the energies and coefficients is provided in Foundation 8.4. Table 8.3 summarizes the results where

$$x = \frac{\alpha - \varepsilon}{\beta} \quad \text{or} \quad \varepsilon = \alpha - x\beta$$

Figure 8.18 shows the energy level diagram for butadiene. We find that the lowest two energy levels are occupied, so that

Table 8.3 The HMO Energies and Coefficients for Butadiene.

Orbital	ϕ_1	ϕ_2	ϕ_3	ϕ_4
x	−1.618	−0.618	0.618	1.618
c_1	0.372	0.602	0.602	0.372
c_2	0.602	0.372	−0.372	−0.602
c_3	0.602	−0.372	−0.372	0.602
c_4	0.372	−0.602	0.602	−0.372

the total electronic energy that arises from the π-bonds is given by $2 \cdot 1.618 \cdot \beta + 2 \cdot 0.618 \cdot \beta = 4.472 \cdot \beta$. This bond energy is larger than that we would compute for two independent ethene bonds, $4 \cdot \beta$. The difference between these two values is the resonance energy associated with butadiene, see Section 8.7. The wavefunctions ϕ_1 to ϕ_4 for butadiene are illustrated in Fig. 8.18 and compared with the FEMO results. Comparison to the wavefunctions in Fig. 8.4 reveals that the HMO functions are more realistic than the FEMO functions.

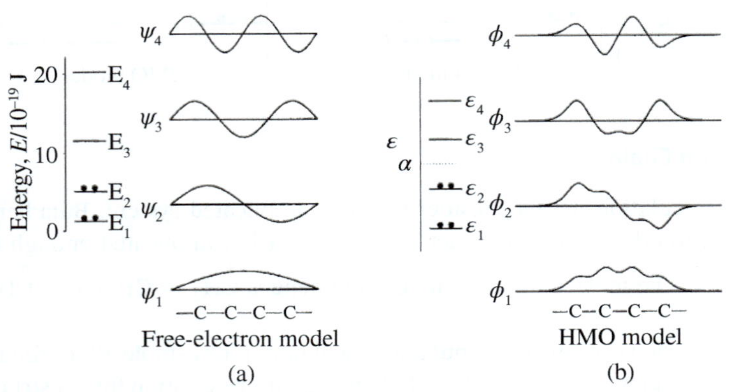

Figure 8.18 Energies and wavefunctions for butadiene. (a) FEMO model. (b) HMO model. Wavefunction along the bond line 75 pm above the molecule plane.

Example 8.4 *Consider the amidinium ion and use the HMO model to determine the energy level diagram of its π-system. To simplify the algebra, assume that the self energy of the nitrogen and carbon p-orbitals are the same ($\alpha_N = \alpha_C = \alpha$).*

The structural formula of the amidinium ion is $H_2\overset{1}{\underline{N}} = \overset{2}{C}H\text{–}\overset{3}{N}H_2^{\oplus}$ where the atoms that contribute p-orbitals are labeled 1, 2, and 3. This system is not entirely composed of carbon atoms; however, we simplify the procedure by treating the nitrogen atoms in the same way as we do for the carbon atoms. By inspection we can write the secular determinant as

$$\begin{vmatrix} \alpha - \varepsilon & \beta & 0 \\ \beta & \alpha - \varepsilon & \beta \\ 0 & \beta & \alpha - \varepsilon \end{vmatrix} = 0$$

Expanding the determinant, we find

$$\begin{aligned} 0 &= (\alpha - \varepsilon)\begin{vmatrix} \alpha - \varepsilon & \beta \\ \beta & \alpha - \varepsilon \end{vmatrix} - \beta \begin{vmatrix} \beta & \beta \\ 0 & \alpha - \varepsilon \end{vmatrix} \\ &= (\alpha - \varepsilon)\left[(\alpha - \varepsilon)(\alpha - \varepsilon) - \beta^2\right] - \beta^2(\alpha - \varepsilon) - 0 \\ &= (\alpha - \varepsilon)\left[\varepsilon^2 - (2\alpha)\varepsilon + \left(\alpha^2 - 2\beta^2\right)\right] \end{aligned}$$

By inspection we see that one of the roots for this equation is $\varepsilon = \alpha$, which corresponds to a nonbonding orbital. The other two roots are obtained by using the quadratic formula for the term in square brackets and yields $\varepsilon = \alpha \pm 2\sqrt{2}\beta$, which correspond to the bonding and anti-bonding orbitals.

8.6.3 Bond Length and HMO Bond Order

The HMO treatment can be used to obtain bond lengths by comparison to *HMO bond orders* and bond lengths of some reference compounds. The HMO bond order for bonded atoms i and j is defined by

$$\text{HMO bond order} = \sum_k N_k c_{ik} c_{jk} \tag{8.30}$$

where c_{ik} and c_{jk} are the HMO coefficients (see Foundation 8.3) of the atomic orbitals on atoms i and j in molecular orbital k. N_k is the number of π electrons in molecular orbital k. For ethene ($N_1 = 2$, $N_2 = 0$) we find

$$\text{HMO bond order} = 2 \cdot c_{11} \cdot c_{21} = 2 \cdot \frac{1}{\sqrt{2}} \cdot \frac{1}{\sqrt{2}} = 1$$

The HMO bond orders for ethene, butadiene, and benzene are given in Table 8.4. The data in Table 8.4 are plotted in Fig. 8.19. This plot can be used as a calibration curve to predict the bond length in arbitrary π-systems by calculating the HMO bond order.

Note that the HMO approach presented here leads to equal bond lengths in a long polyene chain since it neglects the increase of bond lengths by decreasing π-electron density. This is in contrast to the more rigorous approach in Section 8.4 (BCD). As the BCD model represents a perturbative improvement over the FEMO model, a perturbative treatment of the HMO bond order to the C–C bond length has been demonstrated also [9].

The HMO bond order is related to the π-electron density in the bond. The BCD π-electron density rho total decreases linearly with increasing bond length as does the HMO bond order.

Table 8.4 Comparison of the HMO Bond Order with the Experimental Bond Length.

	HMO Bond Order	Predicted Bond Length/pm	Exper. Bond Length/pm
Ethene	1.00	133.1	133.1
Butadiene	0.89	135.4	133.8
	0.45	144.6	145.4
Benzene	0.67	140.0	139.7

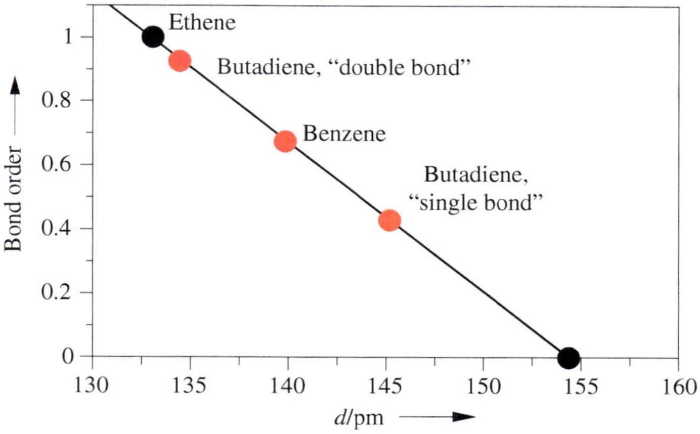

Figure 8.19 HMO bond order as a function of experimental bond length d for ethene, butadiene, and benzene. The solid line is defined by the data points for ethene (HMO bond order = 1, d = 133 pm) and ethane (HMO bond order = 0, d = 154 pm), black solid circles. Red solid circles: predicted data points for butadiene and benzene. For numerical data see Table 8.4.

8.6.4 Extensions of HMO Approach

The simplified HMO method described above can provide a useful guide to qualitative features of molecular orbital structure and bonding in π-systems; however, it lacks quantitative accuracy and is a semi-empirical theory. The HMO method has been improved to include effects of orbital overlap and to account for the differing self-energies of heteroatoms, as well as changes in π-electron density with bond length [10]. In Chapter 9, we discuss the effect of heteroatoms on the electron energetics of π-systems. Workers have generalized the HMO method to describe both the π- and σ-bonding frameworks of molecules, the EHMO method [8].

8.7 Resonance

From the discussion above, it is evident that the π electron distribution of many molecules is not well represented by a single structural formula. This limitation results because the structural formula has its origin in the electron pair model of bonding, which does not describe delocalized bonding well. To account for this limitation, we often describe such systems by resonance between two or more structures—*for example*,

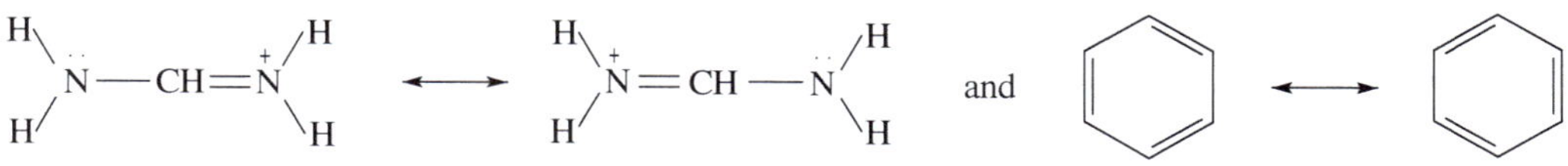

These individual structures represent limiting bonding motifs, rather than providing an accurate view of the bonding. Often we consider the average of these structures to provide a better description. The binding energy of the molecule for one of these limiting structures is less than that for the true molecule, and this additional binding energy is called the *resonance energy*.

For example, the molecule benzene has a lower π-orbital energy than comparable systems with the same number of isolated double bonds. To demonstrate this difference we consider the energies of hydrogenation of benzene and cyclohexene (Fig. 8.20). From calorimetric measurements (see Chapter 16) we obtain $\Delta E_{\text{benzene}} = -3.42 \times 10^{-19}$ J and $\Delta E_{\text{cyclohexene}} = -1.93 \times 10^{-19}$ J for the energy released upon hydrogenation of the double bonds, of which benzene has three formal double bonds and cyclohexene has one. If the double bonds in benzene were isolated, we expect that the hydrogenation energy of benzene would be three times the value of cyclohexene; however, it is smaller by (Fig. 8.20).

$$\varepsilon_R = \Delta E_{\text{benzene}} - 3 \times \Delta E_{\text{cyclohexene}} = 2.36 \times 10^{-19} \text{ J} \tag{8.31}$$

This means that the π electrons in benzene are more stable than would be expected from its formal number of double bonds.

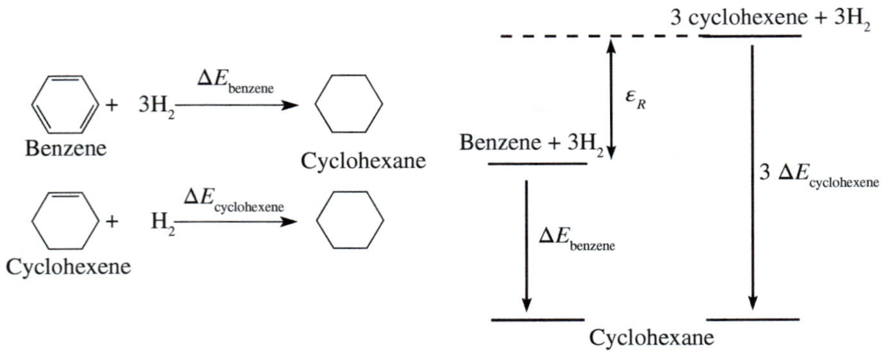

Figure 8.20 Energy change for the hydrogenation of benzene and cyclohexene. The energy diagram on the right shows the energy change for the hydrogenation of three cyclohexene molecules (i.e., $3 \times \Delta E_{\text{cyclohexene}}$), and compares it to the energy change released upon hydrogenation of benzene $\Delta E_{\text{benzene}}$ (i.e., three formal double bonds). The difference between these energy is the resonance energy ε_R of benzene.

The energy difference ε_R is the resonance energy of benzene. ε_R becomes large when the actual electron distribution in a conjugated system differs significantly from the structural formula of localized single and double bonds. In particular, benzene is a very stable system because the bonding orbitals with $n = 0$ and $n = 1$ are totally occupied, but the antibonding orbitals with $n = 2$ and higher are empty . The large resonance energy found for benzene is not unique. Such a stabilization is found for most cyclic systems with six π electrons—*for example*,

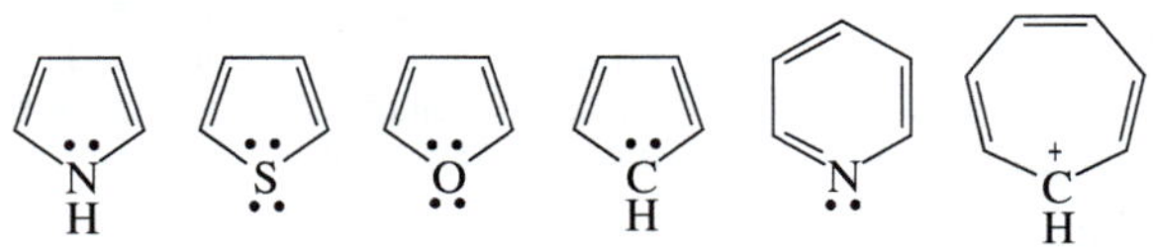

where the pair of dots below the heteroatoms symbolize lone pairs which do not contribute to the π electron system.

Example 8.5 *Given that the hydrogenation enthalpy of furan is* -2.54×10^{-19} *J and the hydrogenation enthalpy of dihydrofuran (DHF) is* -1.78×10^{-19} *J, calculate the resonance energy of furan.*

Furan Dihydrofuran

If we make the assumption that these hydrogenation enthalpies correspond to hydrogenation energies, we need only find the difference in these values. Since each DHF molecule has one double bond and furan has two, the resonance energy ε_R is

$$\varepsilon_R = \Delta E_{\text{furan}} - 2 \times \Delta E_{\text{DHF}}$$

Performing the arithmetic, we find that $\varepsilon_R = 1.02 \times 10^{-19}$ J $= 0.637$ eV, which is less than half the resonance stabilization found for benzene.

For larger ring structures with π-bonds, a significant resonance energy occurs when the electrons fulfill the *Hückel* $4n + 2$ *rule*. The case of 10 electrons corresponds to a filled $n = 2$ degenerate level, $4 \times 2 + 2 = 10$. Note that larger rings require special consideration (see Section 8.4).

8.8 Conclusion

We have seen in this chapter that π-electron systems are useful for testing simple quantum mechanical models by comparison with experimental data, such as bond lengths. It is fascinating to see that π-electrons modeled by standing de Broglie waves give a reasonable picture of the wavefunction's essential features. The success of the FEMO model reflects the importance of delocalization and the wavefunction's nodal character for π-systems. The HMO model provides a similar description of the nodal properties and delocalization as the FEMO model, but it uses atomic orbital basis functions on each atom site. FEMO assumes that the π electrons move in an appropriate electrostatic potential extending along the molecular skeleton. The orbital energies are obtained directly from this model.

The qualitative aspects (nodal structure, charge density profiles) of these models are in correspondence with more sophisticated methods. We introduced the BCD (bondlength consistent with total π electron density) method to illustrate one approach (a perturbative approach) to making simple models more realistic. Principles of a more profound molecular orbital model, the modern DFT method, are given.

8.9 Box 8.1. Self-Consistency in BCD (Step Potential) Method

Here we describe the BCD model in its simplest representation, the step potential model. We begin with an initial value and follow an iterative procedure to generate a sequence of solutions, for which the nth approximation is derived from the previous one. To show that the sequence converges to the final (self-consistent) result in a robust manner, we will proceed by choosing an initial value that is just the opposite of the presumed outcome. We examine the non-trivial question of localization and delocalization in π-electron systems by considering hexatriene and benzene as examples. We show that in hexatriene, starting with equal bond lengths, self-consistency converges to alternating bond lengths. Contrary, in benzene, starting with alternating bond lengths, self-consistency converges to equal bond lengths.

8.9.1 Bond Lengths in Hexatriene

First we assume equal bond lengths (potential energy $V(s) = 0$). The Schrödinger equation is solved, and the resulting total probability density $\rho_{\text{total}}(s)$ is calculated (Fig. 8B.1a, top). Note that this density already shows alternation at the middle of the bonds. Next we calculate a new potential energy from this density. We assume a linear relationship between V_i at bond i and the total probability density $\rho_{\text{total},i}$

$$\frac{V_i}{10^{-19}\ \text{J}} = c \cdot (7.143 - \rho_{\text{total},i}/\text{nm}^{-1}) \tag{8B.1}$$

where $7.143 = \rho_{\text{total,benzene}}/\text{nm}^{-1}$. We choose the benzene case as a reference for the potential energy variations because it corresponds to the case where each of the C-sites is equivalent. The proportionality factor c is determined by calibrating the result to the experimental values for ethene, butadiene, and benzene. First we calculate $\rho_{\text{total},i}$ from the experimental bond lengths using the relation 8.17. Next we make a guess for c in Equation (8B.1) and calculate a new V_i from the density. Then we solve the Schrödinger equation again and find a new value for $\rho_{\text{total},i}$. If the new $\rho_{\text{total},i}$ equals the old one, we know that we used the correct value for c. Otherwise, we have to continue with a new guess for c. Following this procedure, we find that a value of $c = 0.98097$ gives self-consistent results for the three molecules. Thus we obtain

$$\frac{V_i}{10^{-19}\ \text{J}} = 0.98097 \cdot (7.143 - \rho_{\text{total},i}/\text{nm}^{-1}) \tag{8B.2}$$

Using this relation and assuming that the bonds are all equal initially, we find the following values for the improved V_i in hexatriene

	First Cycle of Iteration	Bond 1	Bond 2	Bond 3	Bond 4	Bond 5
1)	Initial $V_i/10^{-19}$ J	0.0	0.0	0.0	0.0	0.0
2)	Calculated $\rho_{\text{total},i}/\text{nm}^{-1}$	8.79	6.00	8.14	6.00	8.79
3)	Improved $V_i/10^{-19}$ J	−1.62	+1.12	−0.98	+1.12	−1.62

Next, we solve the Schrödinger equation with the improved V_i. This process generates an improved $\rho_{\text{total},i}$ from which we can obtain a new improved V_i. The procedure is continued until self-consistency of the potential energy and the total probability density has been reached (Fig. 8B.1a, bottom). The self-consistent bond lengths are obtained by using Equation (8.17).

$$\frac{\rho_{\text{total}}}{\text{nm}^{-1}} = 94.195 - 0.62416\,\frac{d}{\text{pm}}$$

or

$$\frac{d}{\text{pm}^{-1}} = \frac{94.195 - \rho_{\text{total}}/\text{nm}^{-1}}{0.62416}$$

	Final Cycle of Iteration	Bond 1	Bond 2	Bond 3	Bond 4	Bond 5
1)	Self-consistent $V_i/10^{-19}$ J	−3.62	+3.37	−3.29	+3.37	−3.62
2)	Self-consistent $\rho_{\text{total},i}/\text{nm}^{-1}$	10.83	3.71	10.50	3.71	10.83
3)	Corresponding d_i/pm	133.6	145.0	134.2	145.0	133.6

In this case the mean value of $\Delta V = 7.72 \times 10^{-19}$ J is on the same order as that for butadiene ($\Delta V = 7.28 \times 10^{-19}$ J). Note that for the case of the single bonds represented by high potential steps in BCD, $\rho_{\text{total},i}$ is much smaller than according to the simple FEMO model (see Table 8.2).

8.9.2 Bond Lengths in Benzene

For the π-electron system in benzene, let us begin by assuming bond length alternation in the six-membered ring (Fig. B.1b, top). The table shows the results for the first iteration, and we see already that the bond alternation becomes smaller.

	First Cycle of Iteration	Bond 1	Bond 2	Bond 3	Bond 4	Bond 5	Bond 6
1)	Initial $V_i/10^{-19}$ J	−3.74	+3.74	−3.74	+3.74	−3.74	+3.74
2)	Calculated $\rho_{\text{total},i}/\text{nm}^{-1}$	10.36	4.21	10.36	4.21	10.36	4.21
3)	Improved $V_i/10^{-19}$ J	−3.18	+2.87	−3.18	+2.87	−3.18	+2.87

After proceeding through multiple cycles we find that self-consistency is reached when the bond lengths are equal (Fig. 8B.1b, bottom). Again, the self-consistent bond lengths are obtained by using Equation (8.17).

	Final Cycle of Iteration	Bond 1	Bond 2	Bond 3	Bond 4	Bond 5	Bond 6
1)	Self-consistent $V_i/10^{-19}$ J	0.00	0.00	0.00	0.00	0.00	0.00
2)	Self-consistent $\rho_{\text{total},i}/\text{nm}^{-1}$	0.74	0.74	0.74	0.74	0.74	0.74
3)	Corresponding d_i/pm	139.5	139.5	139.5	139.5	139.5	139.5

A program "step_Polyenes.exe" for linear and annular molecules is provided on the book's website.

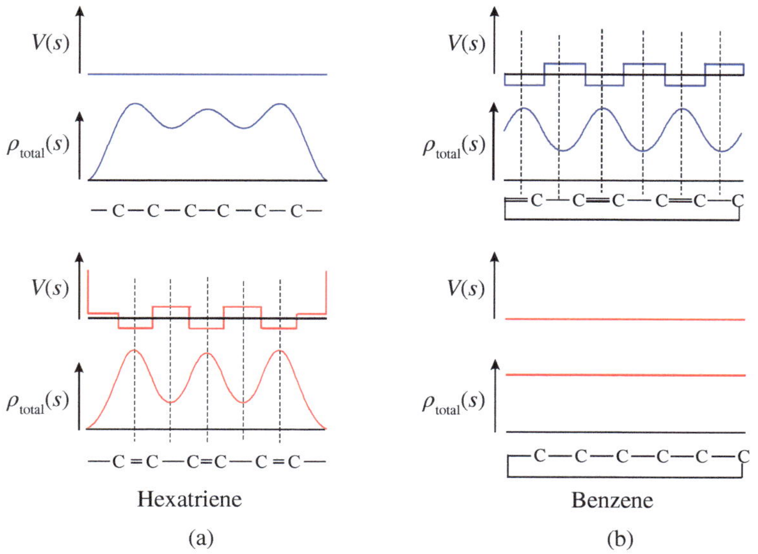

Figure 8B.1 Potential energy $V(s)$ and total probability density $\rho_{\text{total}}(s)$ of the π electrons in the step potential model. Comparison for hexatriene (a) and benzene (b). Top: Assuming equal bond lengths in hexatriene and alternating bond lengths in benzene. Bottom: Result after self-consistency has been reached.

8.10 Box 8.2. Neglect of the Overlap Integral S_{ab} in HMO

The implications of neglecting the overlap term S_{ab} in the HMO model can be assessed by considering the hydrogen molecular ion. For H_2^+ at its minimum energy internuclear distance, we calculated $S_{ab} = 0.46$, $H_{aa} = -21.4 \times 10^{-19}$ J, and $H_{ab} = -14.6 \times 10^{-19}$ J in Foundation 6.2. Using these values in Equations (8.24) and (8.27) we find that

$$\varepsilon_1 = \frac{H_{aa} + H_{ab}}{1 + S_{ab}} = -24.6 \times 10^{-19}\ \text{J};\quad \varepsilon_2 = \frac{H_{aa} - H_{ab}}{1 - S_{ab}} = -12.6 \times 10^{-19}\ \text{J}$$

Then the excitation energy is $\varepsilon_2 - \varepsilon_1 = 12.0 \times 10^{-19}$ J. The corresponding values with $S_{ab} = 0$ are

$$\varepsilon_1 = -36.0 \times 10^{-19}\ \text{J};\quad \varepsilon_2 = -6.8 \times 10^{-19}\ \text{J};\quad \varepsilon_2 - \varepsilon_1 = 29.2 \times 10^{-19}\ \text{J}$$

In general, the inclusion of the overlap term decreases the stabilization energy found for the bonding state and increases the destabilization found for the anti-bonding case. The effect is less important for ethene in which $S_{ab} = 0.17$, so that the bonding state's stabilization energy decreases by a factor of 0.86 and the anti-bonding state's destabilization increases by a factor of 1.2. Note that the HMO method uses experimental data to parameterize the H_{ab}, and this can mitigate some of these quantitative discrepancies.

8.11 Foundations

8.11.1 Foundation 8.1: Construction of $V(s)$ in Fig. 8.3

See the books website, www.wiley.com/go/Kuhn/physicalchemistry3e.

8.11.2 Foundation 8.2:b Free-Electron Model, Branched Systems

See the books website, www.wiley.com/go/Kuhn/physicalchemistry3e.

8.11.3 Foundation 8.3: Free-Electron Model, Representation by Determinants

See the books website, www.wiley.com/go/Kuhn/physicalchemistry3e.

8.11.4 Foundation 8.4: HMO Model

For N atoms, the trial molecular wavefunction is written as a linear combination of the atomic orbitals φ_i

$$\phi = \sum_{i=1}^{N} c_i \cdot \varphi_i$$

where the c_i are constants. This is the LCAO approximation. We calculate the energy ε according to

$$\varepsilon = \int \phi \mathcal{H} \phi \cdot d\tau = \int \left(\sum_{i=1}^{N} c_i \varphi_i \right) \mathcal{H} \left(\sum_{j=1}^{N} c_j \varphi_j \right) \cdot d\tau$$

$$= \sum_{i=1}^{N} c_i^2 \int \varphi_i \mathcal{H} \varphi_i \cdot d\tau + 2 \sum_{j>i} c_i c_j \int \varphi_i \mathcal{H} \varphi_j \cdot d\tau$$

Now we define the quantity α as

$$\alpha = \int \varphi_i \mathcal{H} \varphi_i \cdot d\tau \quad \text{(if only carbon atoms are in the } \pi\text{-system)}$$

which is the energy for an electron in atomic orbital i; we call it the self-energy. The first sum in the energy expression represents the total contribution of the atomic orbital self-energies to the total molecular orbital energy. If the molecule contains atoms other than carbon, we will have a distinct "self-energy" for each type of heteroatom. We also define the quantity β_{ij} as

$$\beta_{ij} = \int \varphi_i \mathcal{H} \varphi_j \cdot d\tau \quad \text{(atoms } i, j \text{ are linked by a } \sigma \text{ bond)}$$

If atoms i and j are direct neighbors, we set $\beta_{ij} = \beta$; otherwise, we set $\beta_{ij} = 0$. Using these definitions, we find that

$$\varepsilon = \sum_{i=1}^{N} c_i^2 \alpha + 2 \sum_{i>j} c_i c_j \beta_{ij}$$

8.11.4.1 Finding the Minimum of ε

Because of the normalization of the trial molecular wavefunction and the neglect of the overlap integrals we find

$$\int \left(\sum_{i=1}^{N} c_i \varphi_i \right)^2 \cdot d\tau = \sum_{i=1}^{N} c_i^2 = 1$$

Since this sum has the value one, we can multiply it by ε to obtain

$$\varepsilon = \varepsilon \cdot \sum_{i=1}^{N} c_i^2 = \sum_{i=1}^{N} c_i^2 \cdot \alpha + 2 \sum_{j>i}^{N} c_i c_j \cdot \beta_{ij}$$

Following the Variational Principle, we wish to minimize the energy with respect to the coefficients c_i. Hence we take the derivative of this equation with respect to each c_i and equate it to zero. For example, the result of differentiation with respect to c_1 is

$$\frac{\partial \varepsilon}{\partial c_1} \sum_{i=1}^{N} c_i^2 + \varepsilon \cdot 2c_1 = 2c_1 \alpha + 2 \sum_{j>1}^{N} c_j \cdot \beta_{1j}$$

The condition for the energy minimum is $\partial \varepsilon / \partial c_1 = 0$, so that

$$\varepsilon \cdot 2c_1 = 2c_1 \alpha + 2 \sum_{j>1}^{N} c_j \cdot \beta_{1j}$$

or

$$c_1 (\alpha - \varepsilon) + c_2 \beta_{12} + c_3 \beta_{13} + \cdots + c_N \beta_{1N} = 0$$

We proceed in a similar way for each of the coefficients and find similar equations. This procedure results in the following system of linear equations:

$$\begin{array}{lllllllll}
c_1(\alpha-\varepsilon) & + & c_2\beta_{12} & + & c_3\beta_{13} & + & \cdots & c_N\beta_{1N} & = 0\\
c_1\beta_{21} & + & c_2(\alpha-\varepsilon) & + & c_3\beta_{23} & + & \cdots & c_N\beta_{2N} & = 0\\
c_1\beta_{31} & + & c_2\beta_{32} & + & c_3(\alpha-\varepsilon) & + & \cdots & c_N\beta_{3N} & = 0\\
\cdots & + & \cdots & + & \cdots & + & \cdots & \cdots & = 0\\
c_1\beta_{N1} & + & c_2\beta_{N2} & + & c_3\beta_{N3} & + & \cdots & c_N(\alpha-\varepsilon) & = 0
\end{array}$$

For convenience, we use the abbreviation $x = (\alpha - \varepsilon)/\beta$ and $k_{ij} = \beta_{ij}/\beta$, where β is the non-diagonal element for a bond with a bond length of 140 pm. Thus we obtain

$$\begin{array}{llllllllll}
c_1x & + & c_2k_{12} & + & c_3k_{13} & + & \cdots & + & c_Nk_{1N} & = 0\\
c_1k_{21} & + & c_2x & + & c_3k_{23} & + & \cdots & + & c_Nk_{2N} & = 0\\
c_1k_{31} & + & c_2k_{32} & + & c_3x & + & \cdots & + & c_Nk_{3N} & = 0\\
\cdots & + & \cdots & + & \cdots & + & \cdots & + & \cdots & = 0\\
c_1k_{N1} & + & c_2k_{N2} & + & c_3k_{N3} & + & \cdots & + & c_Nx & = 0
\end{array}$$

Let us consider the simple case of an ethene molecule, assuming a bond length of 140 pm.

Example: Ethene $\overset{1}{\mathrm{C}}\mathrm{H}_2 = \overset{2}{\mathrm{C}}\mathrm{H}_2$

The number of atoms is $N = 2$, and the number of bonds is one; thus, $k_{12} = 1$, and we have two simultaneous equations,

$$\begin{array}{l}
c_1 \cdot x + \quad c_2 \quad = 0\\
\quad c_1 \quad + c_1 \cdot x = 0
\end{array}$$

These equations are easily solved by using substitution, and we find

$$x_1 = -1;\; c_1 = +c_2$$

$$x_2 = +1;\; c_1 = -c_2$$

Using our definition of x we find

$$\varepsilon_1 = \alpha + \beta;\; \varepsilon_2 = \alpha - \beta$$

Normalization of the molecular wavefunctions is

$$c_1^2 + c_2^2 = 1$$

whose solution gives $c_1 = 1/\sqrt{2}$, so that

$$\phi_1 = \frac{1}{\sqrt{2}} \cdot (\varphi_1 + \varphi_2) \quad \text{and} \quad \phi_2 = \frac{1}{\sqrt{2}} \cdot (\varphi_1 - \varphi_2)$$

This result corresponds with that found in Section 8.6.1.

8.11.4.2 Solving a System of *N* Linear Equations with Constant Coefficients

For more complicated cases it is useful to have a less tedious method. The determinant method for solving a system of linear equations is presented here. Appendix C provides mathematical background on this method.

Following the rules of matrix multiplication, we see that the system of linear equations, which is used to find the energies and wavefunctions, can be written as

$$\begin{bmatrix}
x & k_{12} & k_{13} & \dots & k_{1N}\\
k_{21} & x & k_{23} & \dots & k_{2N}\\
k_{31} & k_{32} & x & \dots & k_{3N}\\
\dots & \dots & \dots & \dots & \dots\\
k_{N1} & k_{N2} & k_{N3} & \dots & x
\end{bmatrix}
\begin{bmatrix} c_1\\ c_2\\ c_3\\ \dots\\ c_N \end{bmatrix} = 0$$

The eigenvalues x which satisfy this set of equations are obtained by evaluating the determinant

$$\begin{vmatrix} x & k_{12} & k_{13} & \dots & k_{1N} \\ k_{21} & x & k_{23} & \dots & k_{2N} \\ k_{31} & k_{32} & x & \dots & k_{3N} \\ \dots & \dots & \dots & \dots & \dots \\ k_{N1} & k_{N2} & k_{N3} & \dots & x \end{vmatrix} = 0$$

To find the coefficient for orbital j we generate the co-factor determinant from column j and evaluate it. For example c_1 is given by

$$c_1 = \begin{vmatrix} x & k_{23} & \dots & k_{2N} \\ k_{32} & x & \dots & k_{3N} \\ \dots & \dots & \dots & \dots \\ k_{N2} & k_{N3} & \dots & x \end{vmatrix}$$

We proceed in a similar way for each coefficient from c_1 to c_N.

8.11.5 Example: Butadiene

If we assume all bond lengths to be 140 pm (thus $k_{12} = k_{23} = k_{34} = 1$), and the secular determinant becomes

$$\begin{vmatrix} x & 1 & 0 & 0 \\ 1 & x & 1 & 0 \\ 0 & 1 & x & 1 \\ 0 & 0 & 1 & x \end{vmatrix} = 0$$

In this example, we have labeled the columns and rows so that we can identify each term with the carbon atoms to which it corresponds. To understand the utility of this approach, let us "walk across" the top row. The first term (first column, first row) contains the self-energy of the electron in the p-orbital of carbon 1 minus the energy of the molecular orbital ($x = (\alpha - \varepsilon)/\beta$). The second term contains the interaction energy β between carbon 1 and its bonding partner, carbon 2. The third term is zero, because carbon 1 is not bonded to carbon 3 (in the Hückel model only nearest neighbor carbons interact). The fourth term is zero for the same reason. If you proceed in a similar manner you should be able to relate the remaining terms in the determinant to the molecular structure. Once you gain some facility with the HMO method, you will be able to write the determinant directly by visual inspection of the molecule's connectivity pattern.

Following the method of determinants described in Appendix C, we find that

$$\begin{vmatrix} x & 1 & 0 & 0 \\ 1 & x & 1 & 0 \\ 0 & 1 & x & 1 \\ 0 & 0 & 1 & x \end{vmatrix} = x \cdot \begin{vmatrix} x & 1 & 0 \\ 1 & x & 1 \\ 0 & 1 & x \end{vmatrix} - \begin{vmatrix} 1 & 1 & 0 \\ 0 & x & 1 \\ 0 & 1 & x \end{vmatrix}$$

$$= x^2 \cdot \begin{vmatrix} x & 1 \\ 1 & x \end{vmatrix} - x \cdot \begin{vmatrix} 1 & 0 \\ 1 & x \end{vmatrix} - 1 \cdot \begin{vmatrix} x & 1 \\ 1 & x \end{vmatrix} + 1. \begin{vmatrix} 0 & 1 \\ 0 & x \end{vmatrix}$$

$$= x^4 - x^2 - x^2 - x^2 + 1 = x^4 - 3x^2 + 1 = 0$$

If we make the substitution $y = x^2$, we find

$$0 = y^2 - 3y + 1 \text{ so that} y = \frac{3}{2} \pm \frac{1}{2}\sqrt{9-4}$$

Since $x = \pm\sqrt{y}$, we find the four roots

$$x_1 = -1.618, \quad x_2 = -0.618, \quad x_3 = 0.618, \quad x_4 = 1.618$$

Using our definition of x, we find that the four energy values are

$$\begin{aligned} \varepsilon_1 &= \alpha + 1.618 \cdot \beta \\ \varepsilon_2 &= \alpha + 0.618 \cdot \beta \\ \varepsilon_3 &= \alpha - 0.618 \cdot \beta \\ \varepsilon_4 &= \alpha - 1.618 \cdot \beta \end{aligned}$$

We can also obtain the c_j coefficients and hence the molecular orbitals from this analysis. The coefficients for molecular orbital ϕ_1 are obtained by inserting the solution for x into the determinant and calculating the cofactor (or subdeterminant) corresponding to each column. For the first column we find

$$c_1 = \begin{vmatrix} x & 1 & 0 \\ 1 & x & 1 \\ 0 & 1 & x \end{vmatrix} = x \cdot (x^2 - 1) - x = x \cdot (x^2 - 2)$$

and for the second column we find

$$c_2 = - \begin{vmatrix} 1 & 0 & 0 \\ 1 & x & 1 \\ 0 & 1 & x \end{vmatrix} = -x^2 + 1$$

On grounds of symmetry, the two halves of the molecule must be equivalent, so that $c_3 = c_2$ and $c_4 = c_1$. For molecular orbital 1, we use $x = x_1 = -1.618$ to find that $c_1 = -1.000$ and $c_2 = -1.618$. Lastly we require that each molecular orbital be normalized; hence, the condition

$$c_1^2 + c_2^2 + c_3^2 + c_4^2 = 1$$

is used to determine the overall normalization factor for each molecular orbital. We find that the wavefunction ϕ_1 is

$$\phi_1 = 0.372\varphi_1 + 0.602\varphi_2 + 0.602\varphi_3 + 0.372\varphi_4$$

If we proceed through the series of molecular orbitals we find the information collected in Table 8.3.

8.11.6 Example: Benzene

To apply the HMO method to ring systems and branched structures, we proceed in a similar way. We illustrate the procedure for benzene. The carbons are numbered from 1 to 6.

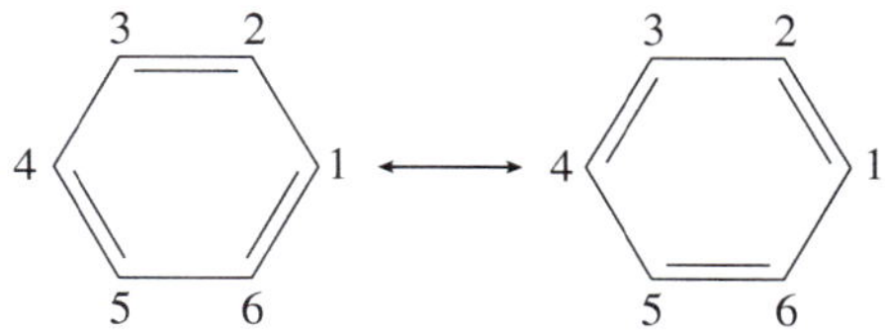

Benzene

In this case, the Hückel determinant is given by

$$\begin{vmatrix} x & 1 & 0 & 0 & 0 & 1 \\ 1 & x & 1 & 0 & 0 & 0 \\ 0 & 1 & x & 1 & 0 & 0 \\ 0 & 0 & 1 & x & 1 & 0 \\ 0 & 0 & 0 & 1 & x & 1 \\ 1 & 0 & 0 & 0 & 1 & x \end{vmatrix} = 0$$

This secular determinant has six solutions, which correspond to the energy eigenvalues. They are a nondegenerate ground state with an eigenvalue of $x_1 = -2.000$, a doubly degenerate level at $x_2 = x_3 = -1.000$, a doubly degenerate level at $x_4 = x_5 = 1.000$, and a nondegenerate level at $x_6 = 2.000$. Table 8F.2 shows the HMO energies x and the corresponding coefficients.

The HMO treatment only generates six molecular orbitals because it only has six atomic orbitals in its basis set. If we used more atomic orbitals (e.g., included 3d or 3p orbitals) more molecular orbitals could be generated. It is noteworthy that the FEMO model does not have this limitation because it solves the Schrödinger equation and finds molecular wavefunctions directly, without resorting to atomic orbitals. However, remember that a refinement of FEMO is needed to describe the wavefunctions with $n = 3$ appropriately (see end of Section 8.3.2 and Problem P8.9).

This procedure can be extended to arbitrary branched π electron systems by making use of the program "FEMO_HMO .exe" which is available electronically on the book's website.

8.11.7 Foundation 8.5: Fullerene

See the books website, www.wiley.com/go/Kuhn/physicalchemistry3e.

Table 8F.2 The HMO Energies and Coefficients for Benzene.

Orbital	1	2	3	4	5	6
x	−2.000	−1.000	−1.000	1.000	1.000	2.000
c_1	0.408	0.408	0.408	0.408	0.408	0.408
c_2	0.264	0.577	0.313	−0.264	−0.577	−0.313
c_3	−0.514	−0.029	0.485	0.514	0.028	−0.485
c_4	−0.284	0.577	−0.293	−0.284	0.577	−0.293
c_5	0.503	−0.005	−0.497	0.503	−0.005	−0.497
c_6	0.408	−0.408	0.408	−0.408	0.408	−0.408

8.12 Exercises

E8.1 Explain the data in Table 8.1. In particular, discuss a) the correlation between bond energy and bond length and b) the correlation between bond energy and force constant.

E8.2 Use Equation (8.3) to calculate the π orbital energy levels in the cases of ethene and butadiene.

E8.3 The experimental first ionization energy of ethene, 10.51 eV, is the energy required to remove a π-electron from the double bond, assuming Koopmans theorem applies: $IE = -E_{\text{orbital}}$, see Section 5.5.4. Use this binding energy value and the energy found from the FEMO model to estimate the average potential energy for the electron.

E8.4 Use the FEMO model to calculate the resonance energy ε_R for butadiene. Given that the hydrogenation energy of ethene is -2.16×10^{-19} J and the hydrogenation energy of butadiene is -3.72×10^{-19} J, compute an experimental resonance energy for butadiene and compare it to your predicted value.[3]

E8.5 Figure E8.5 shows the π-orbital energy levels of benzene and their orbital shapes in the LCAO approximation. Compare the π-electron nodal patterns for the LCAO and FEMO wavefunctions of benzene.

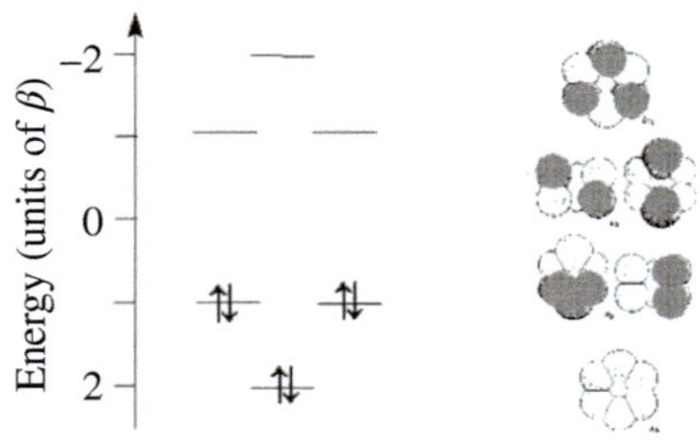

Figure E8.5 The LCAO energy levels and orbitals of benzene.

E8.6 In the FEMO model we confine the π-electrons for benzene to a circular box of circumference $6d_0$, where d_0 is the carbon–carbon bond length. Use a characteristic diameter for a carbon atom's $2p$ orbital of 400 pm and the experimental ionization energy of 9.40 eV $= 1.506 \times 10^{-18}$ J to estimate the extent to which the HOMO wavefunction tunnels outside of the box. You may wish to refer to Chapter 3.

E8.7 If we replace one of the carbon atoms (and its associated H atom) in benzene by a nitrogen atom, we obtain the molecule pyridine which is also an aromatic six-membered ring. Use Fig. 8.7 to describe how you expect the energy levels of pyridine to differ from those of benzene.

E8.8 Show that the two orthogonal wavefunctions ψ_B and ψ_D in Fig. 8.12 can be superposed to generate the wavefunctions ψ_A and ψ_C.

3 Hydrogenation enthalpies are taken from [11].

E8.9 For the guanidinium ion, the HOMO orbital is doubly degenerate, and we chose a particular combination of orthogonal wavefunctions. Perform the same analysis but choose a different orthogonal set of wavefunctions.

E8.10 Show that the energy expression in Equation (8.24) results from using the antisymmetric wavefunction $\phi_2 = c_2 \cdot (\varphi_a - \varphi_b)$ in the Schrödinger equation. Note: you may need to refer to Chapter 6.

E8.11 A HMO model treatment of naphthalene gives the π-electron energy level diagram shown in the table.

Orbital *i*	**1**	**2**	**3**	**4**	**5**	**6**	**7**
x_i	−2.30	−1.62	−1.30	−1.00	−0.62	0.62	1.00

Use these x_i values to find the ε_i value for each molecular orbital i. Compare the energy gaps between the occupied HMO orbitals to those you obtain from the first four ionization energies: 8.15, 8.87, 10.08, and 10.83 eV of naphthalene and calculate a value for β.[4]

E8.12 Use the FEMO model to determine the π-electron charge density of π electron rings with $(4n + 2)$ carbon atoms, $n = 1, 2, 3, \ldots$ (following the Hückel Rule). Compare to ring systems with $4n$ carbon atoms, $n = 1, 2, 3, \ldots$ Compare the degree of bond length alternation in these two cases.

E8.13 Consider the cyclopropene radical C_3H_3. Use the HMO method to find the π-electron energy level diagram. Determine the net binding energy in terms of β, for the species $C_3H_3^+$, C_3H_3, or $C_3H_3^-$. Which of these species has the largest net π-bonding and which has the least amount of net π-bonding?

E8.14 Analyze the π-structure of cyclopropene radical using the FEMO model and compare your results to those found with the HMO model in E8.13.

E8.15 Use the normalization condition to determine the normalization factor for the molecular wavefunctions ϕ_1 through ϕ_4 that we found for butadiene in Section 8.6.2 (Table 8.3).

E8.16 Verify that the four molecular wavefunctions found for butadiene in Section 8.6.2 (Table 8.3) are normalized and orthogonal to each other.

E8.17 Show that the energy ε_2 for the antibonding combination of the $2p_z$ orbitals in ethene (Equation (8.23)) can be written as

$$\varepsilon_2 = \int \phi_2 H \phi_2 \cdot d\tau = 2c_2^2 \cdot (H_{aa} - H_{ab})$$

E8.18 Using the FEMO model for the most weakly bound π-electron in butadiene, determine the uncertainty in the electron's momentum. Compare this uncertainty to that for a π electron in ethene. It may be helpful to refer to Section 3.5.5.

E8.19 In Exercise 8.4 you use the FEMO model to find the resonance energy of butadiene. Perform the same calculation using the HMO model of butadiene. In order to determine a value for the β parameter, compare the π-electron energy levels to the experimental ionization energies of butadiene's π-electrons.

E8.20 Using the bond orders in the HMO model and the bond lengths curve in Fig. 8.19, compute the bond lengths for anthracene (Fig. 8.20a). Compare to the experimental data.[5]

Bond	**Bond Order**	**Bond Length/pm (Experiment)**
1–2	0.736	137
1–11	0.539	144
2–3	0.582	142
11–12	0.486	143
9–11	0.606	140

4 The ionization energies are taken from [12].
5 The bond lengths are taken from [13].

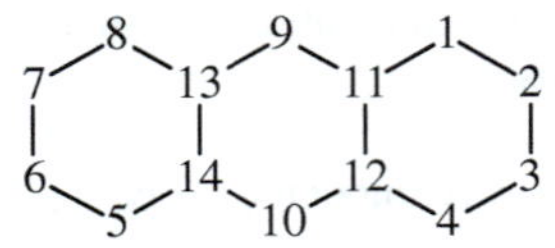

Figure E8.20a. The diagram shows the anthracene carbon framework and numbers the carbon atom sites.

E8.21 Example 8.1 provides the energy eigenvalues and the wavefunctions for the occupied π-electron levels of the amidinium ion. Draw the energy level diagram and plot the wavefunction amplitudes and the probability densities for each of the energy levels; i.e., make a figure like that of Fig. 8.5 but for the amidinium ion.

E8.22 (Hückel rule.) Extend the argument in Section 8.3.3.1 to the case of equal bond lengths in rings with $(4n + 2)$ carbon atoms: e.g., benzene ($n = 1$), [10] annulene (decapentaene, $n = 2$), [14] annulene ($n = 3$).

8.13 Problems

P8.1 Use the HMO method to calculate the energy levels for fulvene's π electrons. Make an orbital energy diagram and show its filling with electrons. You can use fulvene.mcdx to facilitate this calculation.

P8.2 Using the HMO coefficients listed below and the bond lengths curve in Fig. 8.19, calculate the bond lengths of fulvene from the HMO bond orders.

	c_1	c_2	c_3	c_4	c_5	c_6	Energy
ϕ_1	0.247	0.523	0.429	0.385	0.385	0.429	2.11β
ϕ_2	0.500	0.500	0.00	−0.500	−0.500	0.00	1.00β
ϕ_3	0.00	0.00	0.602	0.372	−0.372	−0.602	0.62β
ϕ_4	−0.749	0.190	0.351	−0.280	−0.280	0.351	-0.25β
ϕ_5	0.00	0.00	−0.372	0.602	−0.602	0.372	-1.62β
ϕ_6	0.357	−0.664	0.439	−0.153	−0.153	0.439	-1.86β

P8.3 Use the HMO method to calculate the π electron energies (in terms of β) for cyclobutadiene in a square geometry and in a rectangular geometry of the carbon ring. To model the rectangular geometry, assume that the nearest neighbor coupling is β for the short bond length and it is β' for the long bond length. Comment on which structure you find to be more reasonable for cyclobutadiene.

P8.4 Use the FEMO model to calculate the energies, wavefunctions, and charge densities in the centers of the bonds for cyclobutadiene. Perform your calculation for both a square and a rectangular shape of the molecule.

P8.5 Consider the molecular radical fluorocyclopropane (C_3H_2F) and describe the π-bonding in this system. Begin by assuming that each p_z-orbital has the same self-energy and calculate the energy level diagram and find the π molecular orbitals using Hückel theory; i.e., find the π-orbitals for cyclopropene. Next, you should describe (verbally) how you expect the fluorine to modify the molecular orbital energies.

P8.6 The description of the π-orbital structure of benzene with the FEMO model gives rise to a degenerate HOMO orbital. Rather than the standing wave solutions used in the text, describe the HOMO orbital of benzene with traveling waves, rotating clockwise and counterclockwise. Refer to Section 3.5. Given that the angular momentum operator is $\hat{l}_z = -i\hbar\frac{\partial}{\partial\varphi}$ where φ is the angle about the ring, evaluate the angular momentum of an electron in the $k = -1$ orbital, where $|k| = n$. Given this value for the angular momentum determine the magnetic moment.

P8.7 In Example 8.4 we solved for the HMO energies of the amidinium ion but we assumed that $\alpha_C = \alpha_N$ and $\beta = \beta_{CN}$. Find expressions for the HMO energies without making these assumptions.

P8.8 Use the MathCAD code "Problem8_8.mcd" to calculate the HMO energies for naphthalene. Make an energy level diagram for naphthalene. Determine the total stabilization energy from the π-system, in terms of β. Compare your result to that found from the energy level diagram in E8.11.

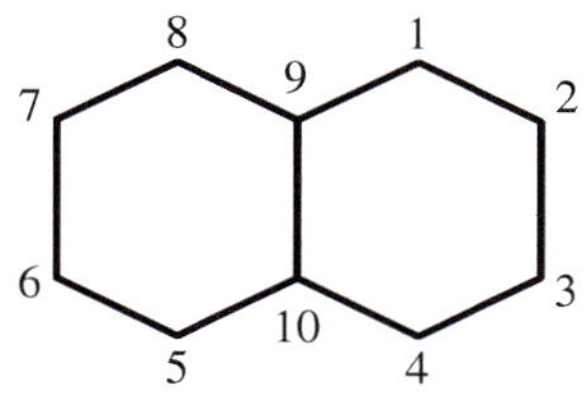

Figure P8.8a Molecular skeleton of naphthalene.

P8.9 Consider the FEMO model for benzene and compute the correction to the energies of the $n = 3$ states that arises from the carbon atom periodicity. Use a corrugated potential around the ring, where the attractive regions are centered on the atomic sites and their depths are $2b$.

P8.10 Make a rough estimate of the ΔV value found for the calculation in Box 8.1, by comparing the electrostatic energy of an electron in a $2p_z$ Slater-type orbital in a C=C double bond to that in the case of a C–C single bond. You can assume that the electron is at a distance $\delta = 75$ pm above the molecular plane, which corresponds to the center of the p_z orbital's probability density, that $Z_{\text{eff}} = 3.25$, and that the bond lengths are $d_{\text{C=C}} = 133$ pm and $d_{\text{C-C}} = 148$ pm.

References

1 A.A. Frost and B. Musulin, A mnemonic device for molecular orbital energies, *J. Chem. Phys.*, 1953, **21**, 572.

2 N.C. Craig, P. Groner, and D.C. McKean, Equilibrium structures for butadiene and ethylene: compelling evidence for π-electron delocalization in butadiene, *J. Phys. Chem. A*, 2006, **110**, 7461.

3 M. Baba, Y. Kowaka, U. Nagashima, T. Ishimoto, H. Goto, and N. Nakayama, Geometrical structure of benzene and naphthalene: ultrahigh-resolution laser spectroscopy and ab initio calculation, *J. Chem. Phys.*, 2011, **135**, 054305.

4 Ch.S. Wannere, K.W. Sattelmeyer, H.F. Schaefer III, and P. von Ragué Schleyer, Aromaticity: the alternating C–C bond length structures of [14]-, [18]-, and [22] annulen, *Angew. Chem.*, 2004, **116**, 4296.

5 P. Hohenberg and W. Kohn, Density functional theory (DFT), *Phys. Rev.*, 1964, **864**, B136.

6 W. Kohn and L.J. Sham, Self-consistent equations including exchange and correlation effects, *Phys. Rev.*, 1965, **1133**, A140.

7 K. Burke, Perspective on density functional theory, *J. Chem. Phys.*, 2012, **136**, 150901.

8 R. Hoffman, An extended Hückel theory. I. Hydrocarbons, *J. Chem. Phys.*, 1963, **39** (6), 1397–1412; ibid, 1964, **40**, 2245; ibid, 2474.

9 C.A. Coulson and A. Golebiewski, On perturbation calculations for the π-electrons and their application to bond length reconsiderations in aromatic hydrocarbons, *Proc. Phys. Soc.*, 1961, **78**, 1310.

10 K. Yates, *Huckel Molecular Orbital Theory* (Academic, New York, 1978).

11 M. Cowperthwaite and S.H. Bauer, Estimation of molecular parameters of C/H fragments, *J. Chem. Phys.*, 1962, **36**, 1743.

12 W. Schmidt, Photoelectron spectra of polynuclear aromatics. V. Correlations with ultraviolet absorption spectra in the catacondensed series, *J. Chem. Phys.*, 1977, **66**, 828.

13 D.W.J. Cruickshank and R.A. Sparks, Experimental and theoretical determinations of bond lengths in naphthalene, anthracene and other hydrocarbons, *Proc. R. Soc. London, Ser. A*, 1960, **258**, 270–285.

9

Absorption of Light

CHAPTER MENU

9.1 Introduction

In our earlier discussions we used photoelectron spectroscopy to provide experimental information on a molecule's electronic structure. One can also probe the electronic structure of atoms and molecules with ultraviolet and visible light that does not cause ionization, i.e., no electrons are ejected from the molecule. Light in the energy range of 1 to 6 eV excites valence electrons into higher excited states, but the electron remains bound to the molecule. Electronic spectroscopy is one of our primary experimental tools for investigating and characterizing molecules and materials. Beyond this, the interaction of light with the valence electrons of molecules is ubiquitous in the natural world. For example, our sense of sight and color relies on electronic excitations in molecules that are present in our eyeball. Of primary importance for life, the process of photosynthesis in plants begins with excitation of a π-electron, whose energy is later converted into useful chemical energy by a cascade of chemical transformations. Many important processes – photochemical, photobiological, and photoelectric – involve electronic excitations of molecules and materials.

In this chapter we discuss the interaction of light with molecules in which the π-electron system comprises the chromophore (i.e., the part of the molecule responsible for its color) and how its electronic structure can be used to explain molecular electronic spectra. For example, we show how a dye molecule's electronic structure determines the dye's color and explains the electronic transitions of light-absorbing biomolecules. In addition, we will understand the color changes that occur for indicator dyes and appreciate why they can directly probe the dye molecule's electron density distribution. We will see that the polarization properties of light can be used to probe the chirality of molecules. This sampling of phenomena is chosen to illustrate how quantum mechanics can enrich our understanding of molecules and nature.

Principles of Physical Chemistry, Third Edition. Hans Kuhn, David H. Waldeck, and Horst-Dieter Försterling.

Companion website: www.wiley.com/go/Kuhn/physicalchemistry3e

9.2 Excitation of π Electron Systems

9.2.1 Basic Experimental Facts

9.2.1.1 Transmittance and Absorbance

We can excite π-electrons in organic dye molecules by irradiating them with visible or ultraviolet light. Such dyes are widely used in man-made products, from your clothing to modern paints and beyond. Figure 9.1a shows the basic elements of an absorption spectrometer that can be used to measure the absorption spectrum of dye molecules. In this apparatus, a broad band source (e.g., a light bulb) shines on a solution containing the dye. Subsequently, a prism or a grating disperses the light into its different wavelength components and projects them onto a target, or screen. Without the dye molecules in the light path, the intensity distribution of the different wavelengths on the screen corresponds to the intensity distribution of the light emitted by the lamp. With the dye molecules in the light path, the light intensity's wavelength dependence is modulated by the dye molecule's absorption. It is this modulation that can be extracted to obtain the molecular absorption spectrum and analyzed to learn about the molecule's electronic structure.

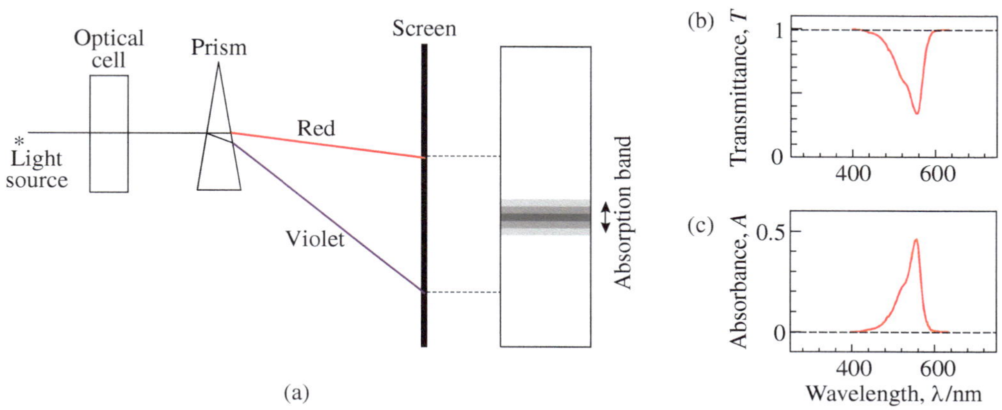

Figure 9.1 Light absorption of a molecular dye solution. (a) Experimental setup. (b) Transmittance $T = I/I_0$ spectrum. (c) Absorbance spectrum; i.e., $A = -\log(I/I_0)$ versus wavelength λ. The spectra shown are for a branched diethylthiacarbocyanine dye cation with $j = 1$ (see Table 9.1). The concentration is 6×10^{-6} mol/L, and the optical path length is 1 cm.

The intensity changes can be quantified by measuring the ratio

$$T = \frac{I}{I_0} \tag{9.1}$$

as a function of the wavelength. I is the light intensity with dye present, and I_0 is the light intensity without the dye. T is called the *transmittance*. At the wavelength λ where the dye absorbs the light most strongly ($\lambda = \lambda_{max}$), the transmittance is smallest. Typically, I and I_0 are measured quantitatively with a photodiode array or charge coupled device (or more traditionally with a photomultiplier tube). Hence we can quantify and plot the transmission spectrum, I/I_0 as a function of λ (see Fig. 9.1b). It is more common to convert the transmittance to an absorbance A through the definition

$$A = -\log \frac{I}{I_0} \tag{9.2}$$

Figure 9.1c plots the absorbance spectrum that is measured for a volume containing about 10^{13} cyanine dye molecules.

For samples that are not too concentrated (i.e., the concentration is such that the dye molecules do not interact with each other), the absorbance is proportional to the concentration c of the dye molecules in the optical cell. This proportionality is quantified by the Lambert–Beer law,

$$\boxed{A = \varepsilon\, c\, l} \tag{9.3}$$

where l is the length of the cell and ε is the molar decadic absorption coefficient (or extinction coefficient). Box 9.1 derives Equation (9.3) from simple considerations. The absorption coefficient ε depends on the nature of the absorbing molecules. Its magnitude and wavelength dependence provide the connection between the experimental measurement and the underlying electronic structure of the molecule.

Example 9.1 *The orange color in Orange Jello arises from the azo dye, Allura Red.*

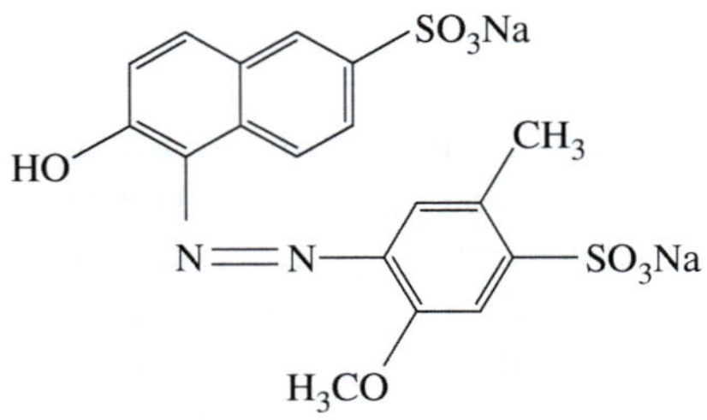

Allura_red

This dye molecule has a molar extinction coefficient of $\varepsilon = 2.5 \times 10^4\ M^{-1}\ cm^{-1}$ at 505 nm. If you prepare a bowl of orange Jello and then cut out a 1 cm thick slice, you will find that it has an absorbance of 0.1. Assuming all the absorbance comes from the Allura Red, what is its concentration in the orange jello? If you eat a 100 cm^3 portion of the orange jello, how many Allura Red molecules have you consumed?

Using the Lambert–Beer law, the concentration is given by

$$c = \frac{A}{\varepsilon l} = \frac{0.1}{(2.5 \times 10^4\ \text{mol}^{-1}\ \text{L cm}^{-1})\ (1\ \text{cm})} = 4.0 \times 10^{-6}\ \text{mol L}^{-1}$$

Given this concentration, the number of molecules in a 0.1 L volume is

$$N = \left(4.0 \times 10^{-6} \frac{\text{mol}}{\text{L}}\right)(0.1\ \text{L})\left(6.02 \times 10^{23}\ \frac{\text{molecules}}{\text{mol}}\right)$$
$$= 2.4 \times 10^{17}\ \text{molecules}$$

9.2.1.2 Dye Molecules

Much of our discussion will concern dye molecules that have a conjugated carbon chain whose π electrons can be excited. Many of these dyes comprise *polyene* chains with functional groups appended on each end. The structural formula for a polyene is

$$CH_2\ (=CH - CH)_k{=}CH_2, \quad k = 0, 1, 2, \ldots \tag{9.4}$$

The smallest polyene is ethene ($k = 0$), which absorbs in the far ultraviolet region of the spectrum.

By exchanging the terminal CH_2 groups in the polyenes for alkylamine groups

$$(CH_3)_2\overset{-}{N} - \quad \text{and} \quad = \overset{\oplus}{N}(CH_3)_2$$

we obtain the class of symmetric cyanine dye cations. These dyes have two equivalent resonance structures:

$$(CH_3)_2\overset{-}{N}\text{-}(CH = CH)_k\text{-}CH = \overset{\oplus}{N}(CH_3)_2, \quad k = 0, 1, 2, \ldots \tag{9.5}$$

and

$$(CH_3)_2\overset{\oplus}{N} = CH\text{-}(CH = CH)_k\text{-}\overset{-}{N}(CH_3)_2, \quad k = 0, 1, 2, \ldots \tag{9.6}$$

The shift of the absorption maximum of a cyanine dye cation compared with a polyene depends on the type of end group. The cyanine cations in (9.5) and (9.6) have a $(CH_3)_2\overset{\oplus}{N} =$ group on one side (electron acceptor) and a $(CH_3)_2\overset{-}{N}-$ group on the other side (electron donor). In the two resonance structures these groups are exchanged; hence, these are called *symmetric cyanine dyes.*

The end groups can also be asymmetric; for example, merocyanines have one amino group and one carbonyl group

$$(CH_3)_2\overline{N} - CH = CH - CH = \underline{O}|$$

Here nitrogen is an electron donor and oxygen is an electron acceptor; in the second resonance structure

$$(CH_3)_2\overset{\oplus}{N} = CH - CH = CH - \overset{\ominus}{\underline{\overline{O}}}|$$

the situation is reversed. The resonance structures are not equal in this case. Often a dye can be characterized by the structural formula

$$\text{donor} \quad -\mathrm{CH}=\mathrm{CH}-\mathrm{CH}=\mathrm{CH}-\mathrm{CH}= \quad \text{acceptor} \tag{9.7}$$

Cyanines Figure 9.2 shows absorption spectra for four related cationic dye molecules, the diethylthiacarbocyanines with $j = 0, 1, 2,$ and 3. For these molecules the absorption is caused by the excitation of a π electron in the dye molecule.

The position of the absorption maximum λ_{max} and the strength of the absorption band depend sensitively on the electronic structure. Qualitatively, the color of the dye solutions changes from yellow ($j = 0$) through red ($j = 1$) and blue ($j = 2$) to green ($j = 3$), as can be seen from Fig. 9.2 and Table 9.1. The wavelength λ_{max} increases systematically with the length of the conjugated carbon atom chain that links the two ring systems, by about 100 nm per ethylenic unit. The spectra in Fig. 9.2 clearly show that the molar decadic absorption coefficient ε increases as the length of the conjugated carbon chain increases, also.

How can these changes in the electronic spectrum be understood? In Section 9.2.2 we will see that these trends are well explained by simple considerations based on the FEMO model.

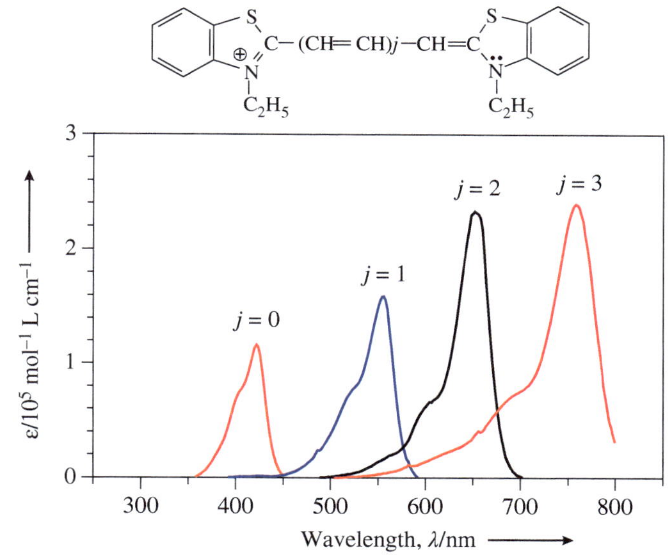

Figure 9.2 Shift of the absorption maximum of symmetric cyanine dye cations with increasing chain length (i.e., with increasing j, see Table 9.1). ε is the decadic absorption coefficient and λ is the wavelength.

Polyenes The changes in the absorption spectra of the cyanine dyes with carbon chain length are quite different from those found in the polyenes. Although the absorption maxima of the polyenes shift to lower energy as the carbon chain length increases (see Fig. 9.3a), the polyenes are colorless (i.e., do not absorb visible light) until $k > 8$. In addition, to this large shift in the energy of absorption, the polyenes approach a limiting value for λ_{max} with increasing chain length, whereas the λ_{max} of the cyanine cations do not.

How can these different behaviors result from a seemingly similar systematic trend in carbon chain length? In Section 9.2.5, we discuss a straightforward explanation that arises from considering the different character of the π-electrons in these two systems.

Table 9.1 Absorption of the Diethylthiacarbocyanine Dyes in Fig. 9.2 with Different Values of j.

j	λ_{max} /nm	Color of Absorbed Light	Color of Solution
0	420	Blue	Yellow
1	540	Green	Red
2	640	Orange	Blue
3	740	Red	Bluish green

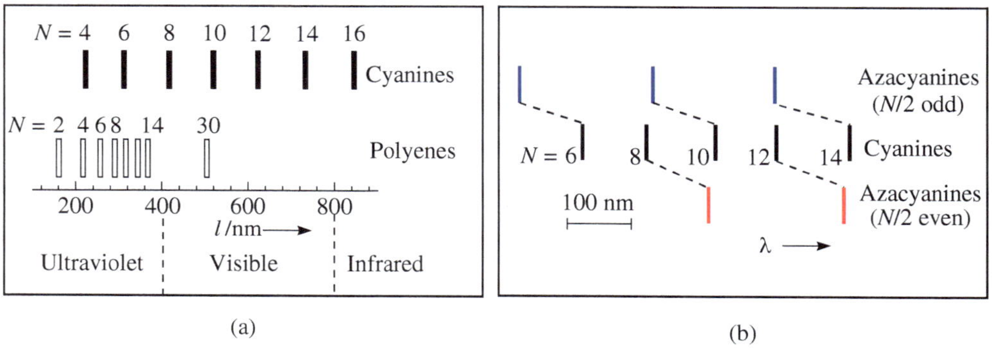

Figure 9.3 Absorption maxima λ_{max} of π-electron systems for different numbers N of π electrons. (a) Comparison of cyanine cations with polyenes. (b) Comparison of cyanine cations with azacyanines. For numerical data see Tables 9.1–9.4.

Azacyanines Another surprising observation occurs if we compare the absorption maxima of cyanine cations with those of the azacyanine cations. In an azacyanine cation the central CH group of a cyanine cation is replaced by an N atom, for example

$$(\mathrm{CH_3})_2\overline{\mathrm{N}} - \mathrm{CH} = \underline{\mathrm{N}} - \mathrm{CH} = \overset{\oplus}{\mathrm{N}}(\mathrm{CH_3})_2 \tag{9.8}$$

and

$$(\mathrm{CH_3})_2\overline{\mathrm{N}} - \mathrm{CH} = \mathrm{CH} - \underline{\mathrm{N}} = \mathrm{CH} - \mathrm{CH} = \overset{\oplus}{\mathrm{N}}(\mathrm{CH_3})_2 \tag{9.9}$$

This change causes the $\lambda_{\max}$ for compounds with an even number of occupied π electron energy states ($N/2$ even, e.g., dye represented by Equation (9.9), $N = 8$, $N/2 = 4$) to shift by about 100 nm to longer wavelengths. In contrast, the azacyanines with an odd number of occupied π electron states (e.g., dye 9.8), $N = 6$, $N/2 = 3$) shift by about 100 nm to shorter wavelengths (Fig. 9.3b). Neither the polyenes nor the cyanines display an odd-even effect of this sort. In Section 9.2.4 we see that this can be understood by considerations about the π-electron wavefunctions' nodal structure.

9.2.2 Absorption Maxima of Dye Molecules

How can we understand the shape and position of the absorption spectra? As discussed in Chapter 2 for atoms, the absorption of light energy can promote an electron from one electronic state to another. To understand the position of the spectrum, we need to relate the excitation energy to the electronic energy level diagram of the molecule. A photon is absorbed by the dye molecule when the resonance condition

$$h\nu = \Delta E \tag{9.10}$$

is fulfilled, where ν is the frequency of light and ΔE is an energy level spacing in the molecule. From ν we can obtain the wavelength λ

$$\lambda = \frac{c_0}{\nu} = \frac{hc_0}{\Delta E} \tag{9.11}$$

where c_0 is the speed of light. Since ΔE has only distinct values for a molecule, we expect sharp absorption lines in the spectrum, as we found for the spectra of atoms (see Chapter 4). The spectra are quite broad (*ca.* 0.5 eV), however.

9.2.2.1 Band Broadening

What is the reason for the broad absorption bands shown in Figs. 9.1 and 9.2? In fact, a number of different phenomena contribute to the band broadening.

1) For example, all electronically excited states have a finite lifetime for relaxing back to the ground electronic state. This intrinsic property causes them to have a finite spectral width; however, atoms also have this property and their bands are much narrower.
2) As we will discuss in Chapter 11, the motion of the nuclei can cause changes in the energy that is observed for the electronic transition and this contributes to the broadening.
3) A third important mechanism is the interaction between the molecule and its environment. Differences in how the chromophoric molecules interact with surrounding molecules can shift their electronic energy. Because the molecules experience a distribution of environments in the sample, the collection of molecules in the sample will have a distribution of energies. Typically, the absorption spectrum results from measurements on many molecules (10^{13} or more for the spectrum in Fig. 9.1), and the observed absorption band is a composite of all the distinct absorption lines of each molecule in the population. Hence we observe a broad absorption.

In some cases it is possible to distinguish particular subpopulations of the different molecules collected in a sample. For such cases the band is said to be inhomogeneously broadened. Figure 9.4 shows such a case for a sample of dye molecules embedded in a polymer. By exposing the sample to highly monochromatic light of high intensity, the dye molecules that absorb within the frequency range of the light beam are photochemically transformed to a species that does not absorb the light; they are "photobleached." Consequently, a narrow gap in the absorption band of the dye is observed (Fig. 9.4). This experiment shows that the molecules which were removed by the initial excitation light had an absorption

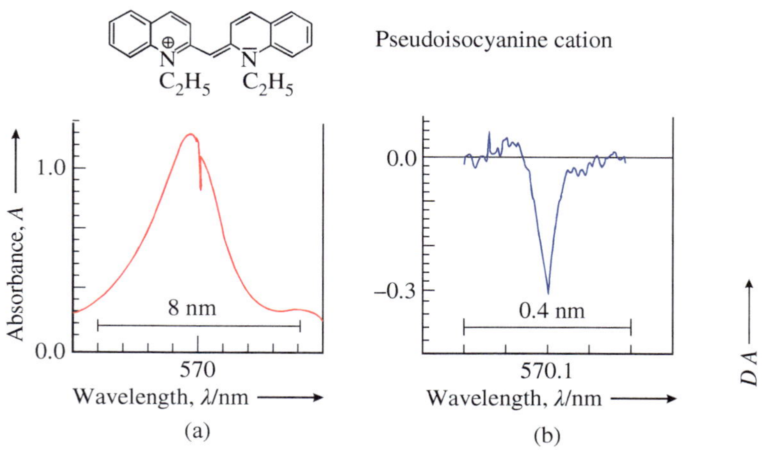

Figure 9.4 Spectral hole burning. A solution of the dye pseudoisocyanine bromide in polyvinylalcohol was cooled to the temperature of liquid helium (4.2 K). The dye was bleached by exposure to an intense light beam at a wavelength of 570.1 nm. (a) After bleaching, an absorption spectrum was taken showing a spectral hole at this wavelength. (b) The halfwidth of this hole is about 0.03 nm [1].

line just in the narrow range around the wavelength of the light that caused their bleaching. This experimental method is called *spectral hole burning*. As the thermal motion changes the individual molecules environment, the individual molecular absorption lines shift and the gap, or "hole" in the absorption profile.

9.2.2.2 Single-Molecule Absorption

Hole-burning experiments can be performed with sufficiently dilute samples so that only one molecule, which has its absorption line at the wavelength of the excitation light, is present. In such a case, the absorption line of that single molecule can be directly observed. Occasional shifts of this line by fluctuations in the individual molecule's environment appear. Also, the position of the absorption peak can be shifted by applying an electric field. The applied field changes the relative energy of the ground state and excited state since they have different electron distributions. Thus single-molecule spectroscopy allows direct access to an individual molecule's properties [2].

9.2.2.3 λ_{max} for Cyanine Dyes

When a dye molecule is irradiated with light, the molecule can take up energy from the electromagnetic field of the light source:

$$\text{molecule} + \text{photon} \rightarrow \text{molecule}^* \tag{9.12}$$

The asterisk indicates that the molecule is in an excited state. Typically, the excitation of the most weakly bound electron to the lowest energy unfilled electron orbital ($\Delta n = 1$) is responsible for the color of a dye molecule, so we focus on this transition. Other transitions will occur at higher energies. The accurate calculation of electronic spectra and the energies of electronically excited states of molecules remain a challenging task in computational chemistry. Instead of pursuing such a goal, we will use the simple FEMO model to identify important trends in the spectra and generate a "chemical intuition" about their origin in the chemical bonding.

The symmetric cyanine dye cation has two equivalent resonance structures (see structures (9.5) and (9.6)), so the carbon-carbon bonds of these molecules are all equal in length. Consequently, the potential energy $V(s)$ of a π electron is the same for all bonds. Hence this system should provide a good test case for the FEMO model of π-electrons.

Consider the cyanine cation

$$(CH_3)_2\bar{N} - CH = CH - CH = CH - CH = CH - CH = \overset{\oplus}{N}(CH_3)_2 \tag{9.13}$$

which has 10 π electrons. The FEMO model wavefunctions and energies for this molecule are shown in Fig. 9.5. The energies are

$$E = \frac{h^2}{8m_eL^2} \cdot n^2, \quad n = 1, 2, 3, \ldots \tag{9.14}$$

If we use the Aufbau principle, we obtain the ground electronic state configuration by distributing the π electrons over the lowest five energy levels. Level 5 is the highest occupied level (*HOMO* = *highest occupied molecular orbital*), and level 6 is the lowest unoccupied level (*LUMO* = *lowest unoccupied molecular orbital*). See Fig. 9.5.

Using the FEMO model, we calculate the excitation energy of the dye as the difference in energy between the HOMO and LUMO electronic states, each given by equation (9.14). We find that

$$\Delta E = \frac{h^2}{8m_e L^2} \cdot \left(n_{\text{LUMO}}^2 - n_{\text{HOMO}}^2\right) \tag{9.15}$$

where n_{LUMO} is the quantum number of the first unoccupied level, and n_{HOMO} is the quantum number of the highest occupied level. These numbers can be determined from the total number of π electrons, N. Since each level is occupied by two electrons, we find that

$$n_{\text{HOMO}} = \tfrac{1}{2}N, \quad n_{\text{LUMO}} = \tfrac{1}{2}N + 1 \tag{9.16}$$

and

$$\boxed{\Delta E = \frac{h^2}{8m_e L^2} \cdot (N+1) \qquad \text{(FEMO model)}} \tag{9.17}$$

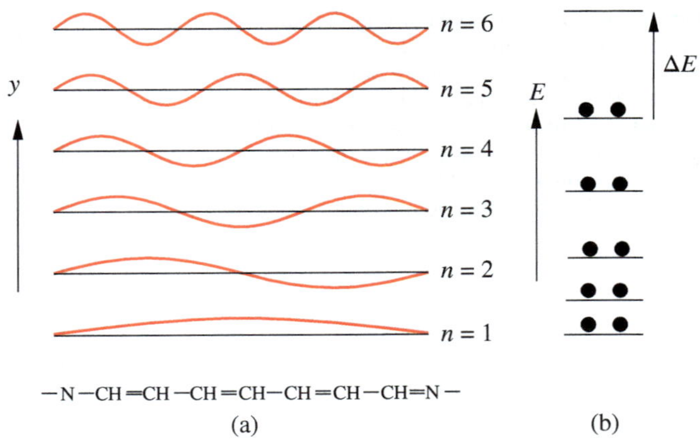

Figure 9.5 Wavefunctions (a) and energies (b) of the symmetrical cyanine cation in structure (9.56) according to the free-electron model.

Similar to our considerations in Chapter 8, the length L of the π system can be related to the number of π electrons N and the bond length d_0.

Example 9.2 *Use Equation (9.17) to determine the transition energy for the cyanine dye in (9.13).*

$$(CH_3)_2\overset{-}{N} - CH = CH - CH = CH - CH = CH - CH = \overset{\oplus}{N}(CH_3)_2$$

Assume that the characteristic bond length d_0 is 140 pm. Use the transition energy to determine the wavelength for the transition.

The dye molecule has eight bonds of length d_0, six C–C bonds and two C–N bonds, with a total length of $8d_0$. Now we add a length d_0 on each end of the chain for the terminal groups so that we find $L = 8d_0 + 2d_0 = 10d_0$ with $d_0 = 140$ pm. As discussed above the number of π-electrons is $N = 10$ for this dye molecule, so we have everything needed to evaluate Equation (9.17). We find that

$$\Delta E = \frac{h^2(N+1)}{8m \cdot (10d_0)^2} = \frac{h^2}{8m \cdot d_0^2}\frac{10+1}{100} = \frac{(6.626)^2 \times 10^{-68}\ \text{J}^2\ \text{s}^2 \times 11}{8 \cdot 9.109 \times 10^{-31}\ \text{kg} \cdot (140 \times 10^{-12})^2\ \text{m}^2}\frac{11}{100}$$
$$= 3.38 \times 10^{-19}\ \text{J} = 2.11\ \text{eV}$$

Using Equation (9.11), this energy corresponds to a photon wavelength of

$$\lambda = \frac{6.626 \times 10^{-34}\ \text{J s} \cdot 2.998 \times 10^{8}\ \text{m/s})}{3.38 \times 10^{-19}\ \text{J}}$$
$$= 5.88 \times 10^{-7}\ \text{m} = 588\ \text{nm}$$

This value is about 13% larger than the experimental value of 519 nm (see Table 9.2).

Although quantitative accuracy may be lacking, the free electron model can predict the linear relationship between λ_{max} and k for the cyanine dyes. Using the structural formula

$$(CH_3)_2\overset{\oplus}{N} = CH - (CH = CH-)_k\overset{-}{N}(CH_3)_2 \tag{9.18}$$

the length of the π electron system is given by

$$L = k \cdot 2\ \text{(C–C bonds)} + 2\ \text{(C–N bonds)} + 2\ \text{ends}$$
$$= k \cdot 2d_0 + 2d_0 + 2d_0 = 2(k+2) \cdot d_0$$

Using the fact that $N = 2(k + 2)$ and substituting the result for L into Equation (9.17) gives

$$\Delta E = \frac{h^2}{8m_e N^2 d_0^2} \cdot (N+1) = \frac{h^2}{8m_e d_0^2} \cdot \frac{N+1}{N^2} = 30.7 \times 10^{-19}\ \text{J} \cdot \frac{N+1}{N^2} \tag{9.19}$$

for the energy gap, where we chose $d_0 = 140$ pm. Combining this result with Equation (9.11), we find

$$\lambda_{\text{max}} = \frac{hc_0}{\Delta E} = \frac{8m_e d_0^2 c_0}{h} \cdot \frac{N^2}{N+1} = 64.7\ \text{nm} \cdot \frac{N^2}{N+1} \tag{9.20}$$

for the wavelength maximum. For cyanine dyes with a large number of $\pi-$ electrons ($N \gg 1$) we find that,

$$\frac{N^2}{N+1} \approx N \tag{9.21}$$

Thus λ_{max} is predicted to increase approximately linearly with N, in agreement with experiment and with the result of a more sophisticated calculation [3].

In Table 9.2 the calculated λ_{max} values are compared with experimental data. The experimental λ_{max} values increase linearly with the length of the chain. The increase of about 100 nm per π-electron pair (i.e., per double bond) is somewhat smaller than the 130 nm predicted by (9.20). Considering the very simple model used as a basis for our calculations, the agreement between theory and experiment is remarkably good.

Table 9.2 Calculated (FEMO) and Experimental Absorption Maxima λ_{max} and Oscillator Strengths f (see Equations (9.26) and (9.31)) of Symmetrical Cyanine Cations $(CH_3)_2\overset{\oplus}{N}=CH{-}(CH{=}CH{-})_k\overline{N}(CH_3)_2$.

k	*N*	λ_{max}/nm		Color[1]	*f*	
		Calculated	**Experiment**		**Calculated**	**Experiment**
0	4	207	224	Colorless	0.7	
1	6	332	313	Colorless	1.0	0.9
2	8	459	416	Yellow	1.3	1.0
3	10	588	519	Red	1.5	1.2
4	12	716	625	Blue	1.8	1.5
5	14	844	735	Green	2.1	2.0
6	16	973	848	Colorless	2.4	

The increase in λ_{max} (decrease in the HOMO–LUMO energy gap) is no longer valid for hypothetical cyanines with very long chains, however. Figure 9.6 shows the total probability density function for a cyanine with $k = 12$ (26 bonds), which is calculated by the step potential model, see Foundation 9.6.

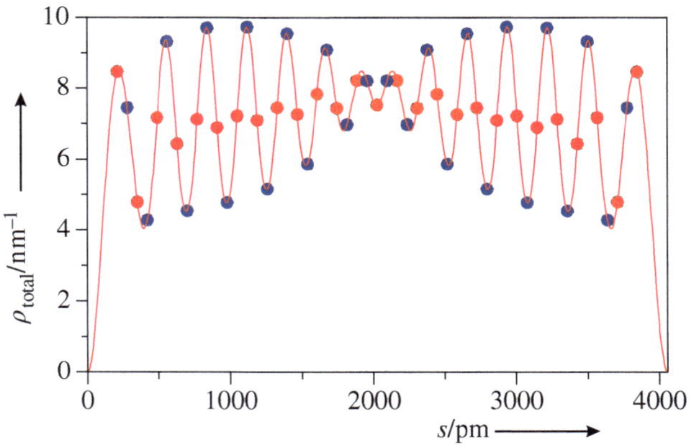

Figure 9.6 Total probability density ρ_{total} for a cyanine with $k = 12$ (26 bonds, structure see Equation (9.18)) versus the coordinate s along the molecular skeleton using the step potential model (red solid line). The blue solid circles indicate the locations for the probability density at the bonds, the red solid circles indicate the locations of the probability density at the nuclei. For details see Foundation 9.6.

1 The color of the solution of the dye is complementary to the color of the absorbed light; e.g., if red light is absorbed the color of the solution is green. The absorption maxima are calculated according to Equation (9.17), the oscillator strengths according to Equation (9.31) with $\hat{n} = 1.42$ (index of refraction of the solvent dichloromethane).

In this case the molecule forms regions of alternating single and double bonds and a region of CH units with diminished alternation in between (positive soliton, compare to neutral soliton discussed in Section 8.4), corresponding to the structural formula.[2]

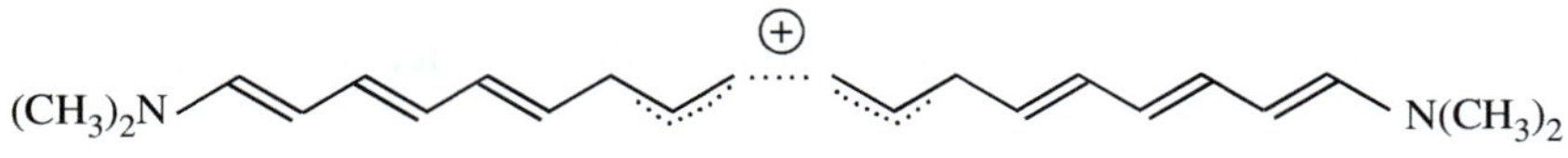

Figure 9.7 shows the result of a calculation of the absorption wavelengths λ_{max}, using the step potential model (for details see Foundation 9.6). Rather than the absorption wavelength λ_{max} increasing linearly with N, as supposed by the FEMO model, the step model predicts a limiting value of about 975 nm. This result is confirmed by independent calculations, see [4]. We can easily understand this phenomenon, because an infinitely long cyanine molecule cannot be distinguished from an infinitely long polyene molecule (see Section 9.2.5), for which the role of the end groups is greatly reduced.

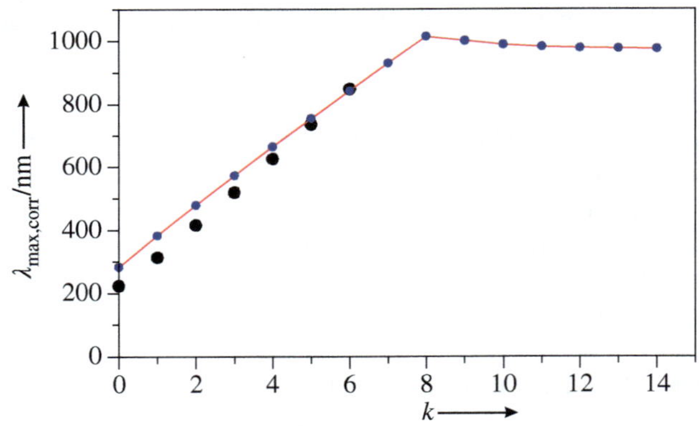

Figure 9.7 Absorbance maximum, λ_{max} calculated from the step potential model (blue circles) and experimental data (black circles) versus k (see Equation (9.18). For details, see Foundation 9.6.

9.2.3 Strength and Polarization of Absorption Bands

9.2.3.1 Interaction of Light With a Molecule

Although we have had some success in predicting spectral positions, we still need to understand the change in the strength of the transitions. This feature of the spectrum requires that we understand more about the mechanism by which the light energy couples into the molecules. We have seen in Chapter 1 that light has properties of a beam of particles and properties of a wave. Light's wave properties are well described by Maxwell's theory of an electromagnetic wave, in which an electric and a magnetic field are oscillating in space (associated with the light's wavelength) and time (associated with the light's frequency). The oscillating electric field of the light wave dominates the interaction with electrons in a molecule (see Exercise 9.20), and it creates a time-dependent force on the electrons in the molecule. When the characteristic frequency of the electric field's oscillation matches that of an electron's motion, a resonance occurs and energy is efficiently coupled between the light field and the electron. Let us proceed by describing a simple classical model of this effect, and then we will extend the discussion to a quantum mechanical description.

Electric Field Responsible for Light Absorption Figure 9.8 sketches a linearly polarized light field; that is, the electric field strength is constrained to oscillate in a single plane that contains its direction of travel. A variety of polarization conditions are possible for the light field. The "polarization" of light refers to the direction and behavior of its electric field strength vector (x- or y-direction). In general, the electric field strength vector must always be perpendicular to the direction of travel of the light beam (z-direction) but may lie anywhere in the plane perpendicular to that direction. The fields oscillate in time

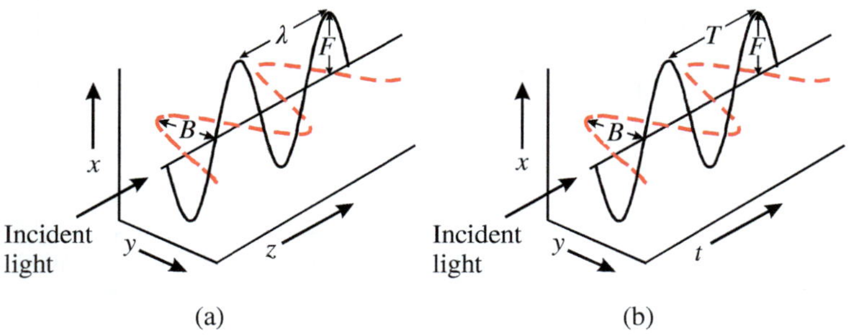

Figure 9.8 Linearly polarized electromagnetic wave. Electric field strength F (solid curve) and magnetic flux density B (dashed curve) (a) Dependence of field strength in space, light wave traveling in the z direction, distance of two adjacent maxima equals the wavelength λ, (b) dependence of field strength in time t, distance of two adjacent maxima equals the period T.

2 Here we consider the symmetric case only. The positive charge (positive soliton) can move along the molecular chain, leading to an asymmetric structure, depending on the position of the counterion.

as well as space. So we write the electric field strength $F(z, t)$ and the magnetic flux density $B(z, t)$

$$F(z,t) = F_0 \sin\left(\frac{2\pi z}{\lambda} + 2\pi \nu t\right) \quad \text{and} \quad B(z,t) = B_0 \sin\left(\frac{2\pi z}{\lambda} + 2\pi \nu t\right) \tag{9.22}$$

where ν is the frequency of the light field and λ is the wavelength. The electric (F_0) and magnetic (B_0) field amplitudes are written here as constants, but more generally can depend on x and y.

Classical Model Figure 9.9 illustrates the interaction of light with a molecule. The electron, which is excited by the light, is replaced by a particle of mass m_e and charge $-e$ attached to a spring (classical oscillator). When the light's electric field vector is pointing up, the electrons in the molecule experience a force driving them to the positive pole and the nuclei experience a force driving them to the negative pole. This picture can be simplified by considering the relative motion of the electrons with respect to the nuclei and modeling the process as a net negative charge that is displaced by the applied field but experiences a restoring force, illustrated by the spring. When the electric field is oscillating with a frequency that is much slower than the response time of the electron, then the electron's displacement will faithfully track the electric field (no phase shift). In contrast, when the electric field's oscillation frequency is much higher than the characteristic frequency of the electron, the electron's displacement is 180° out of phase from the applied light field, see Foundation 9.1 for a damped oscillator; that is, the electric field of the light points in the opposite direction before the electron moves appreciably. In this case, the electron experiences an average zero field.

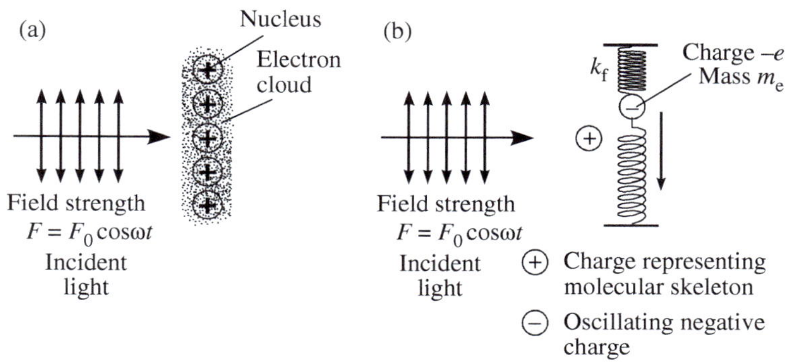

Figure 9.9 (a) Electron cloud of a molecule excited by an electric field with field strength $F = F_0 \cos \omega t$. (b) Motion of excited electron replaced by classical oscillator with force constant k_f, mass m_e, and charge $Q = -e$.

Interesting things happen in the intermediate timescale where the oscillation frequency of the electron and the light is similar. As the electron's displacement lags behind the field we see that it acts in the opposite direction to the light field. When the applied light field and the electron's displacement are phase shifted by 90° the force acting between them is the strongest and energy can be efficiently transferred between the electron and the light field. This coupling occurs when the frequency of the electron's oscillation and the frequency of the light's electric field match, the *resonance condition*. This picture is the same one used to describe the motion for a child on a swing, in which the child corresponds to the electron and the force applied by the parent pushing corresponds to the electric field. The child's motion is most efficiently generated when the frequency of the force applied is in synchrony with the frequency of motion; the resonance condition.

The absorption band, resulting from this classical model, becomes broader and weaker as we increase the damping of the oscillator, but the area covered by the absorption band, according to Foundation 9.1, is independent of the damping. For a classical oscillator with charge $-e$ and mass m_e absorbing light in the vacuum we find that the total integrated absorbance is

$$\left(\int_{\text{band}} \varepsilon \cdot d\nu\right)_{\text{classical }(-e,\, m_e),\text{vacuum}} = \frac{\mathbf{N}_A e^2}{4 \cdot \ln 10 \cdot c_0 \varepsilon_0 m_e} = 0.69 \times 10^{19}\ \text{mol}^{-1}\ \text{L cm}^{-1}\ \text{s}^{-1} \tag{9.23}$$

where $\mathbf{N}_A$ is the Avogadro constant (6.022×10^{23} mol^{-1}).

Quantum-Mechanical Description To describe the absorbing dye molecule we include a term in the Schrödinger equation to account for the energy of interaction between the applied light field $F = F_x^0 \cos(\omega \cdot t)$ and the induced charge displacement in the molecule (e.g., the dipole moment $e \cdot x$). For linear polarized light shining on the molecule, the time-dependent Schrödinger equation can be written as

$$i\hbar \frac{\partial \Psi}{\partial t} = \mathcal{H}_0 \Psi + e x F_x^0 \cdot \cos(\omega t) \cdot \Psi \tag{9.24}$$

where $\mathcal{H}_0$ is the Hamiltonian of the molecule in the absence of the light field and x is the electron's displacement. Usually the applied electric field is much weaker than the electric field that the electron experiences from the molecule (see Exercise E9.20). For this reason, we often use the energy states found from the Schrödinger equation $\mathcal{H}_0\Psi = E\Psi$ in the absence of the light field and consider the light to induce transitions between them. In Foundation 9.3, we show that the quantum mechanical description is mathematically equivalent to that for an electron on a spring (Fig. 9.9).

The probability of a transition occurring between two energy states is determined by the probability of inducing an electric dipole moment in that process. The absorption probability is related to the range of frequencies spanned by the absorption band (e.g., the marked area in Fig. 9.10, $\left(\int_{\text{band}} \varepsilon \cdot d\nu\right)_{\text{experimental}} = 0.9 \times 10^{19}\ \text{mol}^{-1}\ \text{L cm}^{-1}\ \text{s}^{-1}$). It is customary to relate the observed strength of an absorption band to the value expected for a classical oscillator of charge $-e$ and mass m_e. The ratio

$$f = \frac{\left(\int_{\text{band}} \varepsilon \cdot d\nu\right)_{\text{experimental}}}{\left(\int_{\text{band}} \varepsilon \cdot d\nu\right)_{\text{classical }(-e,\, m_e),\text{vacuum}}} \tag{9.25}$$

is called the *oscillator strength*.

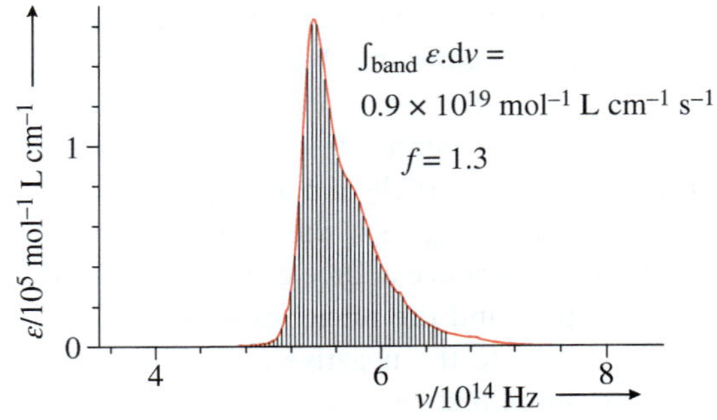

Figure 9.10 Evaluation of the oscillator strength f according to (9.26) for the cyanine dye cation with $j = 1$ in Fig. 9.2.

Inserting expression (9.23) into Equation (9.25) we obtain

$$f = \frac{4 \cdot \ln 10 \cdot c_0\varepsilon_0 m_e}{\mathbf{N}_A e^2}\left(\int_{\text{band}} \varepsilon \cdot d\nu\right)_{\text{experimental}} = \frac{\left(\int_{\text{band}} \varepsilon \cdot d\nu\right)_{\text{experimental}}}{0.69 \times 10^{19}\ \text{mol}^{-1}\ \text{L cm}^{-1}\ \text{s}^{-1}} \tag{9.26}$$

$$= 1.44 \times 10^{-19}\ \text{mol L}^{-1}\ \text{cm s} \cdot \left(\int_{\text{band}} \varepsilon \cdot d\nu\right)_{\text{experimental}}$$

For the absorption spectrum considered in Fig. 9.10 the oscillator strength is $f = 0.9 \times 10^{19}/0.69 \times 10^{19} = 1.3$. We show in Foundation 9.2, that f for a molecule aligned parallel to the electric field strength in the x direction can be calculated as

$$\boxed{f = \frac{8\pi^2 m_e}{h^2 e^2 \hat{n}} \cdot |M_x|^2 \cdot \Delta E} \tag{9.27}$$

where $\hat{n}$ is the index of refraction of the solvent.[3] The quantity M_x is the transition dipole moment between the states i and j

$$\boxed{M_x = e \cdot \int \psi_i \cdot x \cdot \psi_j \cdot d\tau} \tag{9.28}$$

For molecules that are statistically distributed in solution and have additional components of the transition moment in the y and x directions we show in Foundation 9.2, that the oscillatory strength is

$$f = \frac{1}{3} \cdot \frac{8\pi^2 m_e}{h^2 e^2 \hat{n}} \cdot |M|^2 \cdot \Delta E \quad \text{(molecules statistically distributed)} \tag{9.29}$$

with

$$|M|^2 = |M_x|^2 + |M_y|^2 + |M_z|^2 \tag{9.30}$$

(M_y and M_z written in a corresponding way as M_x). The direction of movement of the classical oscillator corresponds to the direction of the transition moment, and the magnitude of the oscillating charge corresponds to the value of M.

Experimental values for f of some symmetrical cyanine dyes are collected in Table 9.2 and compared to those calculated from Equation (9.29). In Foundation 9.2 we use the FEMO model and find that

$$f = \frac{1}{3} \cdot \frac{8\pi^2 m_e}{h^2 e^2 \hat{n}} \cdot |M|^2 \cdot \Delta E = \frac{2}{\pi^2 \hat{n}} \cdot (N+1) \qquad \text{(FEMO model)} \tag{9.31}$$

The calculated f-values are in good agreement with experiment (Table 9.2). They are on the order of 1, as for a classical oscillator with mass m_e and charge e in the vacuum.

Analysis of the spectra in Fig. 9.3 shows that the oscillator strength is larger for the longer chain cyanine dyes. This result reflects the dependence of the transition strength on the size of the transition dipole moment, which is larger for the longer dyes.

3 Here and in the following we assume that the π electrons are in a homogeneous medium of refraction index $\hat{n}$ (medium of σ electrons and solvent). Then the electric field of the light wave acting on the π electrons is given in Foundation 9.1 as $F_0^2 = 2\rho/(3\varepsilon_0\hat{n}^2)$, where ρ is the radiant energy density of the incident light.

9.2.3.2 Polarization of Absorption Bands

This simple classical model is also useful in describing a property called polarization of an absorption band. We consider a cyanine dye dissolved in a polymer (e.g., polyvinyl alcohol); when the polymer is stretched, the polymer chains and the rod-shaped dye molecules are oriented preferentially along the stretching direction. As a consequence light at wavelength $\lambda_{\max}$ polarized in the stretching direction (x-direction) is absorbed preferentially to that polarized in the y-direction (Fig. 9.11). In the present case

$$M_y = M_z = 0 \quad \text{and} \quad M_x = e \cdot \int \psi_n \cdot x \cdot \psi_{n+1} \cdot \mathrm{d}s$$

This means that only light polarized in the x-direction is absorbed.

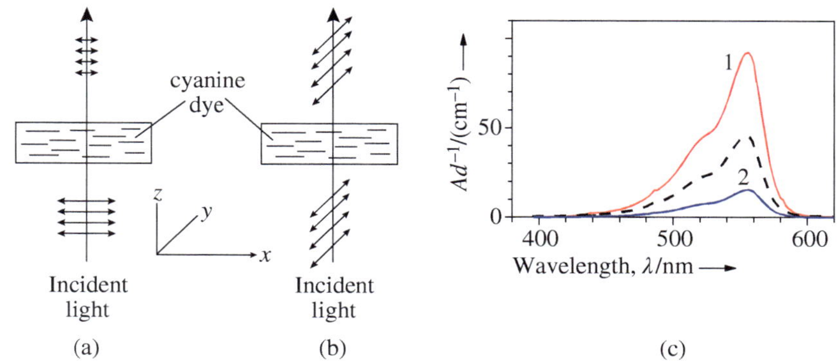

Figure 9.11 Absorption of light by a cyanine dye dissolved in a polymer. (a) Light polarized in the stretching direction of the polymer (x-direction). (b) Light polarized perpendicular to the stretching direction (y-direction). (c) Absorption spectrum of the diethylcarbocyanine with $j = 1$ (see Figure 9.2) dissolved in polyvinyl alcohol. (1) Polymer stretched in the direction of the electric field strength of the polarized light beam. (2) Polymer stretched perpendicular to the light's polarization direction. Dashed line: polymer unstretched.

9.2.4 Hetero-Atoms as Probes for Electron Distribution

Figure 9.3b reveals a surprising dependence of the absorption energy of azacyanine dyes on the chain length. To understand this behavior, we compare the cyanine cation

$$(CH_3)_2\bar{N} - CH = CH - CH = CH - CH = CH - CH = \overset{\oplus}{N}(CH_3)_2 \tag{9.32}$$

with the azacyanine cation

$$(CH_3)_2\bar{N} - CH = CH - CH = \underline{N} - CH = CH - CH = \overset{\oplus}{N}(CH_3)_2 \tag{9.33}$$

where the bonding about the N-atom is similar to that about the C-atoms. Rather than have a C–H bond in the molecular plane, the N-atom has a nonbonding electron pair in addition to its two C–N bonds. This orbital for the lone pair electrons lies in the molecular plane, and they are σ electrons. Consequently, azacyanine cations have the same number of π electrons as the corresponding cyanine cation. Although the free-electron model suggests that the electronic wavefunctions and energies are similar in these two cases, we must pay attention to the fact that an N-atom is more electronegative than a C-atom. Hence the π electrons are more strongly attracted by the core of the N-atom than by the core of a C-atom. It is this feature of the azacyanine dyes that causes their unusual behavior with increasing chain length, compared to the cyanine dyes.

For cyanine dyes the potential energy $V(s)$ along the chain of carbon atoms was approximated as constant because all the atoms in the chain are identical, Fig. 9.12a. The potential energy of a π-electron is strongly decreased at the nitrogen atom, however (see Fig. 9.12b). This extra stabilization causes the energies of all the orbitals of an azacyanine cation to be lower than those of the corresponding cyanine. The amount $\overline{V}$ by which the energy of a selected orbital is stabilized depends on the electron probability density near the N-atom, namely

$$\overline{V} = -a \int_{s_1}^{s_2} \psi^2 \cdot \mathrm{d}s = -a\left(\psi^2\right)_{\text{center of N}} \cdot d_0 \tag{9.34}$$

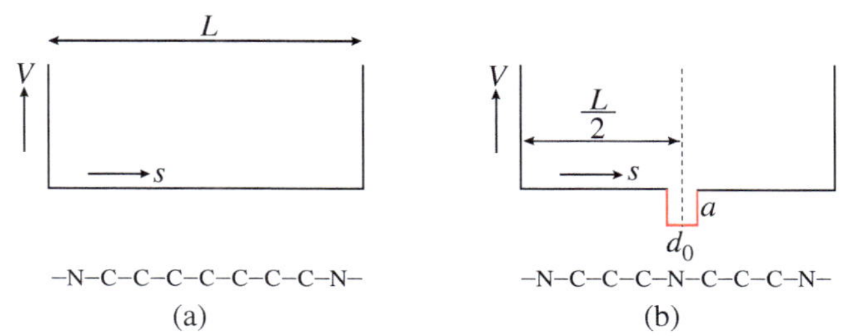

Figure 9.12 Potential energy $V(s)$ along the chain of a cyanine cation (a) and of an azacyanine cation (b). We assume a potential trough at the center of the molecule with depth a and width d_0 (the trough extends from center to center of the adjacent bonds).

where s_1 and s_2 are the coordinates in the region around the N-atom and a is the depth of the potential trough extending over the N-atom in Fig. 9.12. If ψ^2 is large at the position of the N-atom, then the decrease in energy will be large for this orbital. If ψ^2 has a node near the N-atom, then the probability density is low and the energy of the orbital remains unchanged.

Orbitals 5 and 6 of the cyanine (structure (9.13)) and azacyanine cations (9.33) are illustrated in Fig. 9.13 (HOMO and LUMO). ψ^2 has an antinode at the central N-atom in orbital 5, and it has a node at the central N-atom in orbital 6. Hence, orbital 5 in the azacyanine has a lower energy than in the corresponding cyanine, whereas the energy of orbital 6 changes little between the cyanine and azacyanine. This difference causes the excitation energy ΔE to increase for the azacyanine, and the dye (9.33) absorbs at shorter wavelengths than does the dye in structure (9.13).

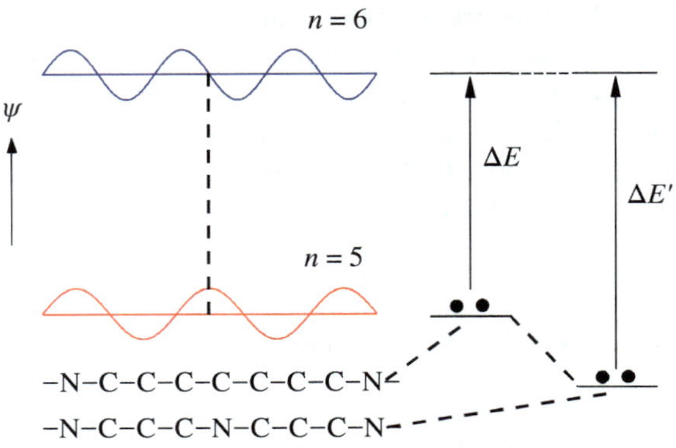

Figure 9.13 Wavefunctions and energies are shown for the cyanine cation in structure (9.32) and the azacyanine cation in structure (9.33) with $N = 10\pi$ electrons and $N/2 = 5$ (odd) occupied levels. Orbitals $n = 5$ (HOMO) and $n = 6$ (LUMO). Decrease of level $n = 5$ (HOMO) for the azacyanine: $\Delta E' > \Delta E$.

If we make both of the dye molecules shorter by two carbon atoms, e.g., the cyanine

$$(CH_3)_2\overline{N} - CH = CH - CH = CH - CH = \overset{\oplus}{N}(CH_3)_2 \quad (9.35)$$

and azacyanine

$$(CH_3)_2\overline{N} - CH = CH - \underline{N} = CH - CH = \overset{\oplus}{N}(CH_3)_2 \quad (9.36)$$

the situation is reversed. In this case the HOMO ($n = 4$) has a node at the N-atom and the LUMO ($n = 5$) has an antinode at the N-atom. Hence, the energy of the lowest unoccupied orbital decreases, and the energy of the highest occupied orbital remains unchanged. This aza dye then absorbs at longer wavelengths than does the corresponding cyanine, see Fig. 9.14.

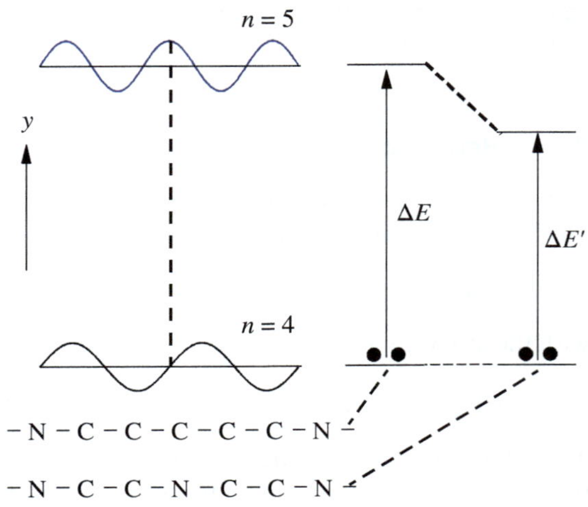

Figure 9.14 Wavefunctions and energies are shown for cyanine cation in structure (9.35) and azacyanine cation in structure (9.36) with $N = 8\pi$ electrons and $N/2 = 4$ (even) occupied levels. Orbitals $n = 4$ (HOMO) and $n = 5$ (LUMO). Decrease of level $n = 5$ (LUMO) for the azacyanine: $\Delta E' < \Delta E$.

Generally we find that λ_{max} is shifted to shorter wavelengths when the highest occupied orbital has an antinode in the center of the molecule. This situation always occurs for dyes with an odd number of occupied states ($N/2$ odd). For an even number of occupied states ($N/2$ even) we expect the opposite behavior, namely, a shift to longer wavelengths. Problem P9.10 calculates the aza shifts quantitatively, using perturbation theory discussed in Foundation 5.2:

$$\text{energy shift} = -a \cdot \frac{2}{N} \quad \text{for } \frac{N}{2} = 2, 4, 6, \ldots \qquad (9.37)$$
$$\text{energy shift} = +a \cdot \frac{2}{N} \quad \text{for } \frac{N}{2} = 3, 5, 7, \ldots$$

where $a = 3.5 \times 10^{-19}$ J. The result of this calculation is in good agreement with the experimental data displayed in Fig. 9.3b and Table 9.3. This good correspondence with experiment supports our considerations about the positions of nodes and antinodes of the wavefunctions. Furthermore, it shows how the model can be used to provide guidelines for designing a dye molecule's electronic characteristics.

In contrast to the potential trough of the N-atom in the center of the chain we have neglected the troughs of the N-atoms at the chain ends of cyanines and azacyanines. Is this justified? At the chain ends, the wavefunctions in the ground state and in the first excited state are similar; thus, both energies change by approximately the same value.

Table 9.3 Calculated and Experimental Aza Shifts for Some Cyanine Cations.[4]

	Shift $\Delta E/10^{-19}$ J			Shift $\Delta E/10^{-19}$ J	
$\frac{N}{2}$	Calculated	Experiment	$\frac{N}{2}$	Calculated	Experiment
2	−1.75		5	+0.70	+0.53
3	+1.17	+0.79	6	−0.58	−0.59
4	−0.88	−0.93	7	+0.50	+0.62

4 $N/2$ = number of occupied π electron states and $a = 3.5 \times 10^{-19}$ J.

Example 9.3 *The chromophore of vitamin B_{12} (an important compound in the biosynthesis of DNA) can be considered an azacyanine with a bent molecular structure (length of π system $L = 14d_0$). Use the free electron model to determine the energy gap for the HOMO to LUMO transition and for the transitions $6 \rightarrow 8$ and $7 \rightarrow 9$. For the characteristic bond length use $d_0 = 140$ pm and use $a = 3.5 \times 10^{-19}$ J to correct the orbital energy for the N-atom stabilization.*

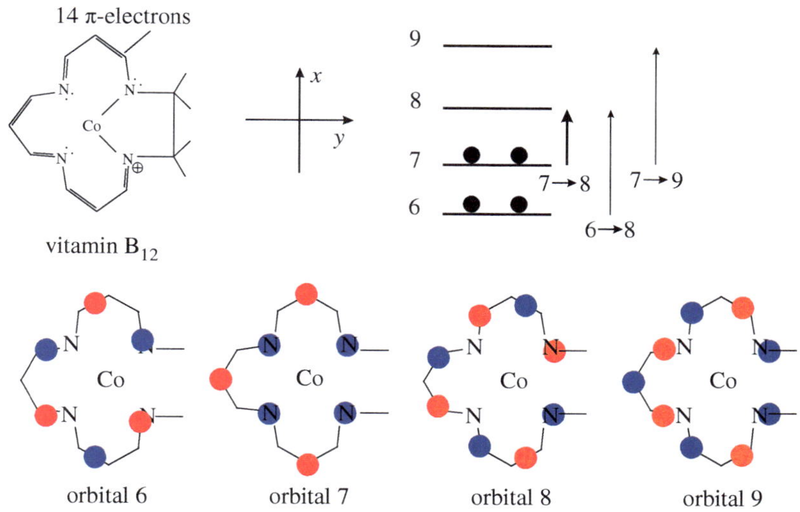

The HOMO to LUMO energy gap may be written as

$$\Delta E = \frac{h^2}{8m_e L^2} \cdot (N+1) + \Delta E_{\text{nitrogen}} \tag{9.38}$$

where $\Delta E_{\text{nitrogen}}$ is the change in ΔE that is caused by the lowering of the potential energy at the nitrogen atom sites. To calculate $\Delta E_{\text{nitrogen}}$ we use perturbation theory as discussed in Foundation 5.2. From the diagrams above, we see that the HOMO (orbital 7) is stabilized by $4 \cdot (2/L)ad_0$, because the four nitrogen atoms are at antinodes, and the LUMO (orbital 8) is stabilized by $2 \cdot (2/L)ad_0$, because two nitrogen atoms are at antinodes. The net change in the energy gap, which is caused by these shifts, is

$$\Delta E_{\text{nitrogen}} = 2 \cdot \frac{2}{L}\, a\, d_0 \tag{9.39}$$

With $L = 14d_0$ we can simplify Equation (9.39) to obtain

$$\Delta E_{\text{nitrogen}} = 2 \cdot \frac{2}{14\, d_0} \cdot a\, d_0 = \frac{2}{7}a$$

Now we can combine this result with Equation (9.38) to write

$$\Delta E_{7\rightarrow 8} = \frac{h^2}{8m_e L^2} \cdot (N+1) + \frac{2}{7}a = \\ = 2.35 \times 10^{-19}\ \text{J} + 1.00 \times 10^{-19}\ \text{J} = 3.35 \times 10^{-19}\ \text{J}$$

where we have used the fact that $N = 14$. This energy corresponds to a $\lambda_{\max} = 590$ nm, a value which is in fair agreement with experiment (560 nm). The transition is polarized in the x-direction. The transition moment is relatively small because of the bent form, leading to $f = 0.3$ in agreement with the experiment.

The transitions $7 \rightarrow 9$ ($\lambda_{\max} = 340$ nm) and $6 \rightarrow 8$ ($\lambda_{\max} = 420$ nm) are polarized in the y-direction. The transition moments M_y are relatively large because of the bent structure, in contrast to the linear cyanine dyes, where the corresponding absorption bands have negligible transition moments. The large M_y values in the present case lead to relatively large oscillator strengths. As a consequence of a coupling effect (see Section 9.2.7), the 420 nm band is essentially suppressed, and the 340 nm band becomes correspondingly stronger and is slightly shifted to shorter wavelengths; this is in accordance with the experiment (band at $\lambda_{\max} = 360$ nm with $f = 0.5$).

9.2.5 HOMO–LUMO Gap by Bond Alternation

Even though polyenes seem like simpler molecules than cyanine cations, their theoretical treatment is more complicated, because of the alternating single and double bonds (Fig. 9.15). This alternation means that the potential energy $V(s)$ along the chain of carbon atoms cannot be considered constant, as in a cyanine cation (Fig. 9.15a). As illustrated by Fig. 9.15b, $V(s)$ is lower at a double bond and higher at a single bond, because the positive charges of the nuclei are closer together at the double bond and the π electron is more strongly attracted. This alternation creates a "potential well" at a double bond and a "potential barrier" at a single bond.

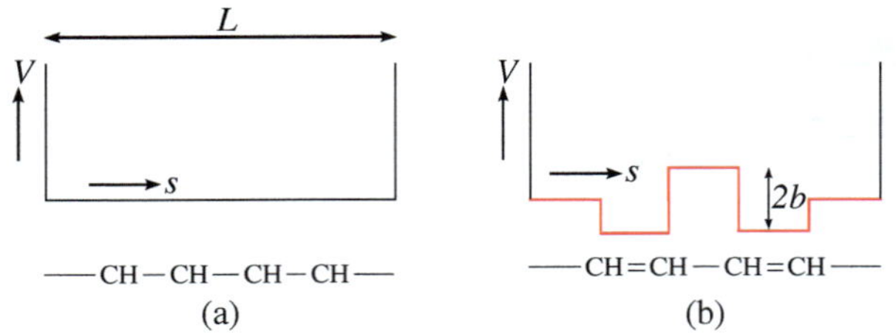

Figure 9.15 Potential energy $V(s)$ of a π electron along a chain of carbon atoms, case of butadiene. (a) All C–C bonds are assumed to be equally long (FEMO model). (b) Bond length alternation creates a potential well at a formal double bond and a potential barrier at a formal single bond.

In analogy to the considerations given for the azacyanine cations, we expect that the energy is decreased for orbitals with antinodes of the wavefunction at potential well sites, and increased for states with antinodes of the wavefunction at potential barrier sites (Fig. 9.16). We denote the difference in the potential energy at these sites by $2b$, where b is a parameter. We use perturbation theory, as discussed in Section 9.2.4 for the aza cyanines, to estimate the energy change $\overline{V}$ that arises from the difference b of the potential energy at a selected double bond site. In analogy to Equation (9.34), we find

$$\overline{V}_{\text{double}} = -b\int_{s_1}^{s_2} \psi^2 \cdot \mathrm{d}s = -b(\psi^2)_{\text{middle of the bond}} \cdot d_0 \tag{9.40}$$

at a selected double bond and

$$\overline{V}_{single} = +b\int_{s_1}^{s_2} \psi^2 \cdot \mathrm{d}s = +b(\psi^2)_{\text{middle of the bond}} \cdot d_0 \tag{9.41}$$

at a selected single bond. In Problem P9.11, you use these equations to show that the energy gap $\Delta E'$ of a polyene can be described approximately by

$$\Delta E' = \Delta E + \frac{6}{5}b \tag{9.42}$$

where, according to Equation (9.17),

$$\Delta E = \frac{h^2}{8m_e \cdot L^2} \cdot (N+1) = \frac{h^2}{8m_e d_0^2} \cdot \frac{1}{N+1} \tag{9.43}$$

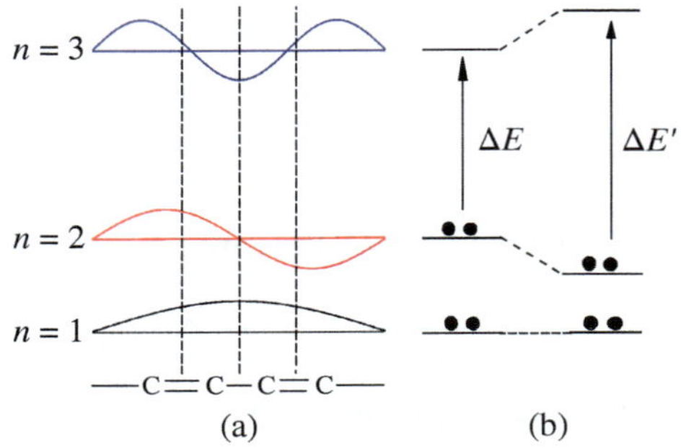

Figure 9.16 (a) FEMO wavefunctions and energy levels of butadiene assuming all C–C bonds are of equal length (potential energy $V(s)$ as in Fig. 9.15a). (b) A perturbation calculation for bond length alternation ($V(s)$ as in Fig. 9.15b) shows that level 2 (HOMO) is shifted downward, and level 3 (LUMO) is shifted upward.

with $L = (N+1) \cdot d_0$; N is the number of π– electrons, and d_0 is the average of the length of a C–C and a C = C bond ($d_0 = 140$ pm). For more details on perturbation theory see Foundation 5.2.

The parameter b can be obtained by comparison with the experimental value of the excitation energy of butadiene. We calculate this value from the measured absorption wavelength $\lambda_{\max}$ reported in Table 9.4, $\lambda_{\max} = 213$ nm. Thus from Equations (9.42) and (9.43) we obtain

$$b = \frac{5}{6}(\Delta E_{\text{experiment}} - \Delta E) = \frac{5}{6}\left(\Delta E_{\text{experiment}} - \frac{h^2}{8m_e d_0^2} \cdot \frac{1}{N+1}\right)$$

$$= \frac{5}{6}\left(\Delta E_{\text{experiment}} - \frac{h^2}{8m_e d_0^2} \cdot \frac{1}{4+1}\right) = \frac{5}{6}\left(\Delta E_{\text{experiment}} - 6.15 \times 10^{-19}\ \mathrm{J}\right)$$

with

$$\Delta E_{\text{experiment}} = \frac{hc_0}{\lambda_{\max}} = \frac{6.626 \times 10^{-34}\ \text{J s}\, 2.998 \times 10^{8}\ \text{m s-1}}{213 \times 10^{-9}\ \text{m}} = 9.33 \times 10^{-19}\ \text{J}$$

Table 9.4 Calculated (FEMO) Absorption Maxima λ_{max} of Polyenes with $N\pi$ Electrons ($N/2$ Double Bonds) According to Equations (9.42) and (9.43) with $b = 2.65 \times 10^{-19}$ J and Comparison with Experimental Data.[5,6]

Polyene	N	$\frac{\Delta E}{10^{-19}\ \text{J}}$	$\frac{\Delta E'}{10^{-19}\ \text{J}}$	$\frac{\lambda_{max}}{\text{nm}}$ Calculated	$\frac{\lambda_{max}}{\text{nm}}$ Experiment
Ethene	2	10.25	13.43	148	162
Butadiene	4	6.15	9.33	213	213
Hexatriene	6	4.39	7.57	262	254
Octatetraene	8	3.42	6.42	301	286
Decapentaene	10	2.79	5.97	333	312
Dodecahexaene	12	2.37	5.55	358	342
	14	2.05	5.23	340	368
	16	1.81	4.99	398	386
	18	1.62	4.80	414	413
	20	1.46	4.64	428	420
(β-Carotene)	22	1.34	4.52	440	453
(lycopene)	24	1.23	4.41	451	461
	26	1.14	4.32	460	471
	28	1.06	4.24	468	500
	30	1.0	4.17	476	504
Polyacetylene	∞	0	3.18	625	650

the result for b is

$$b = \frac{5}{6}(9.33 - 6.15) \times 10^{-19}\ \text{J} = 2.65 \times 10^{-19}\ \text{J}$$

The result of this calculation is in good agreement with the experiment (Table 9.4). Whereas the excitation energy ΔE of cyanines approaches zero with increasing chain length, the energy gap $\Delta E'$ of the polyenes approaches $6b/5 = 3.18 \times 10^{-19}$ J for polyacetylene, leading to $\lambda_{max} = 625$ nm; this is only a small difference of 4% compared to the experimental value. The large difference in the behavior of the cyanines and polyenes is connected with the nearly constant bond lengths in the cyanines as compared to the alternating bond lengths in the polyenes. Note that the HOMO–LUMO gap for the polyene with $N = 20$ ($\Delta E' = 4.64 \times 10^{-19}$ J) is only slightly different from that calculated from the BCD model ($\Delta E = 4.38 \times 10^{-19}$ J, see Fig. 8.14).

Example 9.4 *With reasoning analogous to that we used to explain the absorption spectra of polyenes and cyanine dyes, explain why the absorption spectrum of the indicator dye methyl orange changes in the manner shown by Fig. 9.17 for the alkaline and acid forms.*

Reaction of the neutral or alkaline form R^- (absorption maximum at $\lambda_{max} = 460$ nm) with H^+ ions leads to the acid form RH with an absorption maximum at $\lambda_{max} = 505$ nm (Fig. 9.17). The alkaline form has a structure like a polyene with

5 The experimental wavelengths are averages of the data reported in: Försterling et al. [5], Hudson et al. [6], and Schulten et al. [7].
6 Note that β-carotene is responsible for the yellow color of carrots and lycopene for the red color of tomatoes.

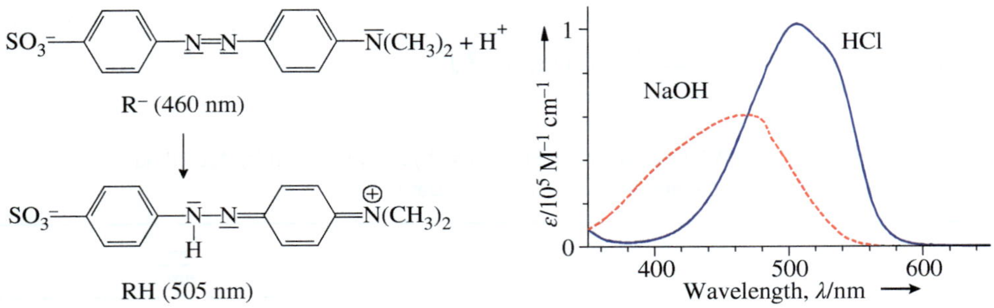

Figure 9.17 Absorption spectrum of solutions of the indicator dye methyl orange. Solid line: dissolved in 0.1 M HCl (cation R$^-$); dashed line: dissolved in 0.1 M NaOH (neutral RH).

considerable bond alternation, a double bond between the nitrogens and single bonds between the nitrogen atoms and the benzene ring. In contrast, the acidic form is more similar to the cyanines because it has two resonance structures – one with the positive charge at the nitrogen atom on the left and the other with the nitrogen atom on the right. Hence the acid form absorbs at longer wavelengths.

The distinction between the limiting cases of (a) symmetric cyanines (structures (9.5) and (9.6)) and of (b) polyenes (9.4) is crucial for understanding basic features of π-electron systems. The important features are equal bond lengths in case (a) and bond length alternation in case (b). Bond length alternation causes a gap between the HOMO and the LUMO. Therefore, the absorption band of a polyene is strongly shifted to shorter wavelengths relative to the absorption band of a cyanine of comparable size. A solution of the cyanine dye in structure (9.13) with $N = 10$ is red, whereas a solution of a polyene with $N = 10$ is colorless.

9.2.6 Cyclic π-Systems

Closed rings of π-electrons are an important motif in chemistry and biology. Here we discuss some electronic properties of these systems by discussing Cu^{2+} phthalocyanine as an example. This molecule is an important blue dye and is used as a coloring agent in industry; in addition, it plays a role as a catalyst in dehydrogenation reactions. The molecule's blue color is caused by an absorption band in the red part of the visible spectrum ($\lambda = 675$ nm, see Fig. 9.18).

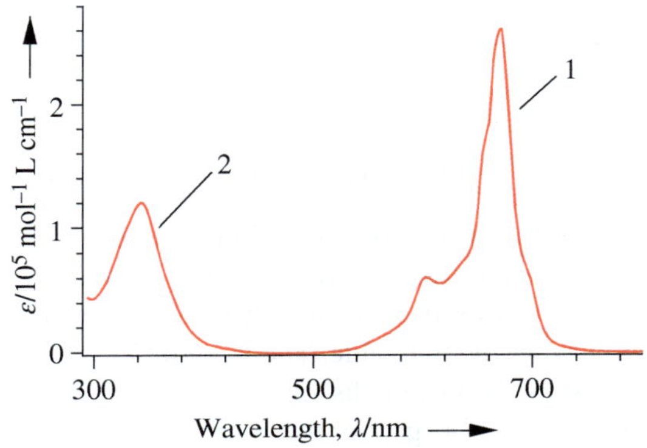

Figure 9.18 Absorption spectrum of Cu-phthalocyanine (solution of the tetrasulfonate-compound in a mixture of 67% ethanol and 33% water).

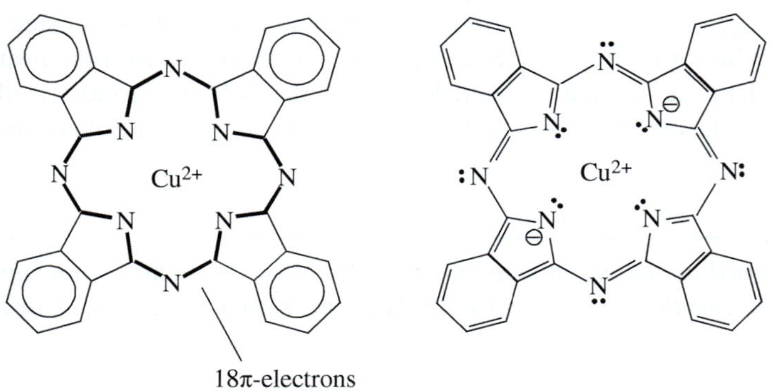

(a) (b)

9.2.6.1 Transition Energies

In structure (a) the benzene rings are connected to the inner π-electron system by single bonds, and we can consider the inner π-system, which comprises the chromophore, as being isolated from the benzene rings. In the inner ring (bold lines) each C and N atom donates one π electron; in addition, two electrons are donated by the Cu atom. Altogether the π-system has $16 + 2 = 18\pi$-electrons in the ring. Note that the structural formula (b) does not explain that the benzene branches are essentially isolated.

In analogy to our discussion of benzene (Chapter 8), we might write the energies of Cu^{2+} phthalocyanine as

$$E_n = \frac{h^2}{2m_e L_c^2} \cdot n^2, \quad n = 0, 1, 2, \ldots \tag{9.44}$$

where $L_c = 16d_0$ is the circumference of the inner ring. The energy levels and corresponding probability densities are plotted in Fig. 9.19a. According to this energy level diagram, the HOMO and LUMO orbitals are degenerate and four different transitions are possible, namely $4a \to 5a$, $4a \to 5b$, $4b \to 5a$, and $4b \to 5b$. However, we should account for the fact that the nitrogen atoms in the ring are more electronegative than the carbon atoms, and the potential energy is not constant along the ring (as for the azacyanine cations).

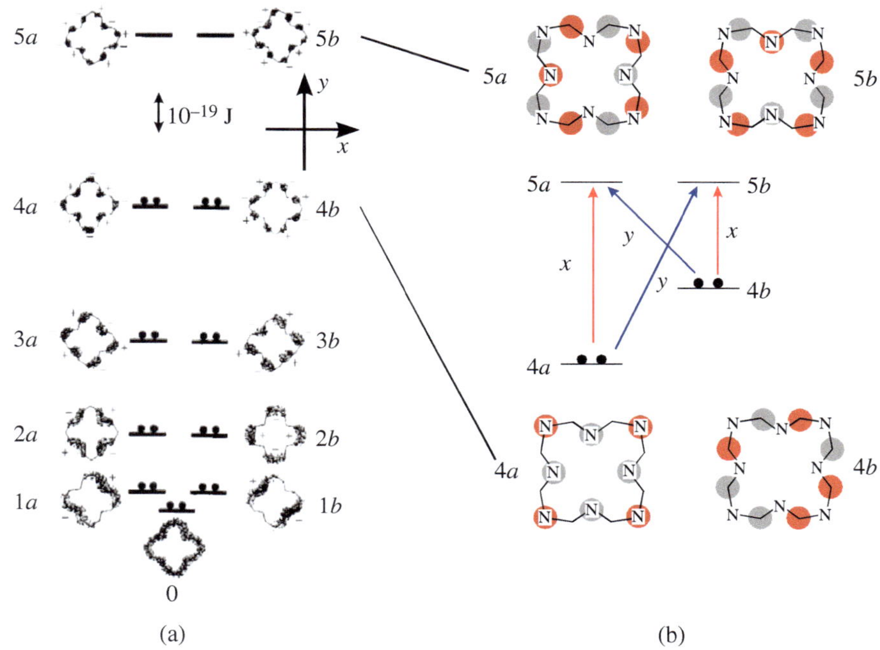

Figure 9.19 Phthalocyanine (benzene rings neglected). Charge densities of occupied and lowest unoccupied orbitals. The energy levels are calculated according to Equation (9.44) (the effect of the N atoms on the energy levels is neglected). (a) Orbitals $n = 0$ to $n = 5$. (b) Orbitals $4a$, $4b$, $5a$, and $5b$; location of the antinodes of the wavefunctions relative to the N-atoms; black circles: positive values of the wavefunction, red circles: negative values of the wavefunction. The energy shift in phthalocyanine caused by the electronegativities of the N-atoms is indicated.

Figure 9.19b shows the charge density pattern for orbitals 4 and 5 of the phthalocyanine. In orbital $4a$ the wavefunction has an antinode at each N-atom, and the energy of this state is decreased compared to the constant potential model. In contrast, orbital $4b$ remains unchanged in energy because it has a node at the positions of the N-atoms. The orbitals $5a$ and $5b$ are both decreased by the same amount and intermediate between the stabilization of the $4a$ orbital and the $4b$ orbital (see the energy diagram in Fig. 9.19b).

Example 9.5 *Calculate the transition energies for the transitions $4b \to 5b$ and $4a \to 5a$ of phthalocyanine. As for the azacyanine dyes take the extra stabilization energy at a nitrogen atom to be given by $-a \cdot d_0 \cdot \rho$, where ρ is the probability density at the atomic center, $d_0 = 140$ pm, and $a = 3.5 \times 10^{-19}$ J.*

We use perturbation theory as discussed in Foundation 5.2. In state $4a$ all eight N-atoms are placed at the antinodes of the wavefunction. According to the considerations in Section 9.2.4 on azacyanines, the average potential energy change at a nitrogen atom site is given by

$$\overline{V} = -a \int_{s_1}^{s_2} \psi^2 \cdot ds = -a \cdot (\psi^2)_{\text{at center}} \cdot d_0$$

For state $4a$ the wavefunction is

$$\psi_{4a} = \sqrt{\frac{2}{L_c}} \cdot \cos\left(4 \cdot \frac{2\pi s}{L_c}\right)$$

where L_c is the circumference and s is the coordinate along the cyclic system. The first antinode is at $s = 0$; thus, $(\psi^2)_{\text{at center}} = 2/L_c$ and

$$\overline{V} = -a \cdot \frac{2}{L_c} \cdot d_0$$

Because there are eight of these sites $\Delta E_{\text{nitrogen}} = 8\overline{V}$ and

$$E_{4a} = \frac{h^2}{2m_e L_c^2} \cdot 16 - \frac{16}{L_c} \cdot d_0 \cdot a$$

In state $4b$ all N-atoms are at nodes of the wavefunction, so that

$$E_{4b} = \frac{h^2}{2m_e L_c^2} \cdot 16$$

In state $5a$ the wavefunction is

$$\psi_{5a} = \sqrt{\frac{2}{L_c}} \cdot \cos\left(5 \cdot \frac{2\pi s}{L_c}\right)$$

Two N-atoms are at an antinode (at $s = 0$ and $s = L_c/2$), two are at a node (at $s = L_c/4$ and $s = 3L_c/4$), and four are adjacent to an antinode; for the first of the latter atoms, we have $s = L_c/8$ and ψ^2 is

$$\psi^2 = \frac{2}{L_c} \cdot \cos^2\left(5 \cdot \frac{2\pi \cdot L_c/8}{L_c}\right) = \frac{2}{L_c} \cdot \cos^2\left(\frac{5}{4}\pi\right) = \frac{1}{L_c}$$

so that

$$E_{5a} = \frac{h^2}{2m_e L_c^2} \cdot 25 - 2a \cdot \frac{2}{L_c} \cdot d_0 - 4a \cdot \frac{1}{L_c} \cdot d_0 = \frac{h^2}{2m_e L_c^2} \cdot 25 - \frac{8}{L_c} \cdot d_0 \cdot a$$

The same expression is obtained for state $5b$.

Thus with $L_c = 16 \cdot d_0 = 2240$ pm (with $d_0 = 140$ pm), we find that the transitions occurring from the $4b$ state occur at lower energy than those originating from the $4a$ state. The transition energies are

$$\Delta E_{4b\to 5b} = \Delta E_{4b\to 5a} = \frac{h^2}{512 m_e d_0^2} \cdot 9 - \frac{1}{2}a = (4.32 - 1.75) \times 10^{-19}\ \text{J} = 2.57 \times 10^{-19}\ \text{J}$$

$$\Delta E_{4a\to 5a} = \Delta E_{4a\to 5b} = \frac{h^2}{512 m_e d_0^2} \cdot 9 + \frac{1}{2}a = (4.32 + 1.75) \times 10^{-19}\ \text{J} = 6.07 \times 10^{-19}\ \text{J}$$

These energies correspond to $\lambda_{\max} = 772$ nm and $\lambda_{\max} = 327$ nm and are in reasonable agreement with experiment, see Table 9.5 and Fig. 9.18.

This analysis shows that the electronic transition responsible for the color originates in an orbital with nodes of the wavefunction at each N-atom, while the electronic transition of the second band (in the UV) originates from an orbital with antinodes at each N-atom. This is the reason for the strong separation of the two absorption bands.

9.2.6.2 Transition Polarization

The transitions identified by Fig. 9.20 show that the transition is polarized along the x or y direction because the transition moment M_x or M_y is nonzero, see Equation (9.28). We can see from Fig. 9.20 by symmetry considerations that in the case of transition $4a \to 5a$: $M_x \neq 0$ and $M_y = 0$. In more detail, from Fig. 9.20 and the scheme

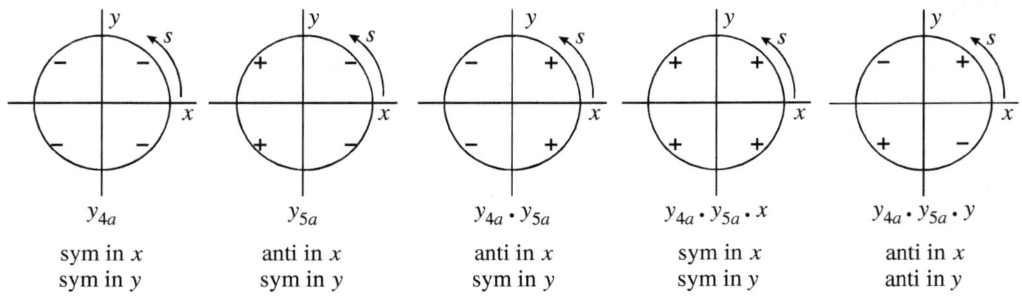

Table 9.5 Transitions in Cu^{2+} Phthalocyanine and in Zn^{2+} Porphyrin.

			Cu^{2+} Phthalocyanine			
			λ_{max} /nm	f	λ_{max} /nm	f
	Transitions	Polarization	Theory		Experiment	
1	$4b \rightarrow 5b, 4b \rightarrow 5a$	x, y	772	1.2	675	0.9
2	$4a \rightarrow 5a, 4a \rightarrow 5b$	x, y	327	2.8	360	1.7
					233	3.2
			Zn^{2+} Porphyrin			
			λ_{max} /nm	f	λ_{max} /nm	f
	Transitions	Polarization	Theory		Experiment	
1	$4b \rightarrow 5b, 4b \rightarrow 5a$	x, y	576	1.6	546	0.1
2	$4a \rightarrow 5a, 4a \rightarrow 5b$	x, y	382	2.4	422	2.0
					320	0.4

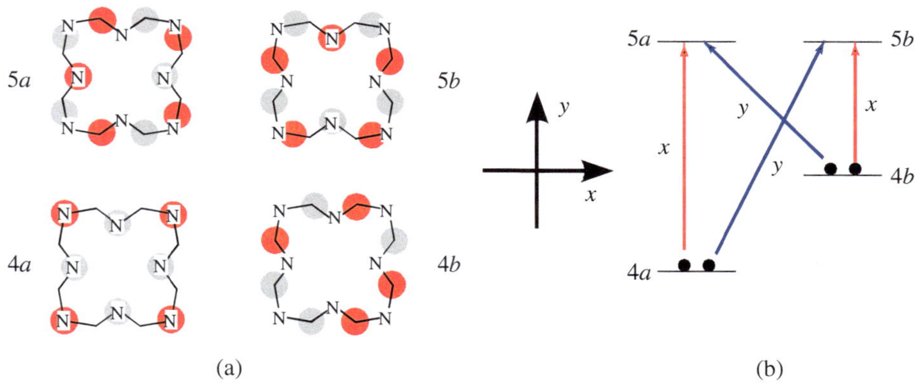

Figure 9.20 Phthalocyanine: (a) orbitals $4a$, $4b$, $5a$, and $5b$; location of the antinodes of the wavefunctions relative to the N atoms; red filled circles: positive values, gray filled circles: negative values of the wavefunction. (b) Energy shift in phthalocyanine caused by the electronegativities of the N atoms.

it is evident that the phthalocyanine wavefunctions have symmetry axes in the $x-$ and in the $y-$ directions (origin of the xy coordinate system at the center of the ring and the $\pm$ indicates the wavefunction sign). ψ_{4a} is symmetric in x whereas ψ_{5a} is antisymmetric in x. Since the coordinate x is antisymmetric, the function $\psi_{4a} \cdot \psi_{5a} \cdot x$ is symmetric in the $x-$ direction. Furthermore, $\psi_{4a} \cdot \psi_{5a} \cdot x$ is symmetric in the $y-$ direction, so the integration along the s-coordinate is nonzero. Thus M_x for transition $4a \longrightarrow 5a$ is nonzero. Correspondingly, the function $\psi_{4a} \cdot \psi_{5a} \cdot y$ is antisymmetric in the $x-$ and the $y-$ directions; thus, the integration along the s-coordinate is zero and M_y for transition $4a \longrightarrow 5a$ is zero.

A similar analysis can be applied to the transitions $4a \longrightarrow 5b$, $4b \longrightarrow 5a$, and $4b \longrightarrow 5b$. The result is collected in Table 9.6, where sym denotes symmetric and anti denotes antisymmetric. This means that out of the eight possible transitions two are polarized in the x-direction, two are polarized in the y-direction, and the other four transitions have transition dipole moments of zero and therefore are forbidden.

Because the transition dipole moment M_x for the $4a \longrightarrow 5a$ transition is nonzero, let us proceed to calculate it explicitly. To simplify the calculation we assume that the atoms in the molecule are arranged on a circle with radius $r = L_c/(2\pi)$. Then the coordinate s and the differential ds can be expressed by

$$\frac{2\pi s}{L_c} = \varphi, \quad \mathrm{d}s = \left(L_c/2\pi\right)\mathrm{d}\varphi$$

Table 9.6 Symmetry Considerations for Transition Moments in Phthalocyanine.

State i	State j	x-direction $\psi_i\psi_j$	x-direction $\psi_i\psi_j x$	y-direction $\psi_i\psi_j$	y-direction $\psi_i\psi_j y$	M_x	M_y
$4a$	$5a$	anti	sym	sym	anti	$\neq 0$	0
$4a$	$5b$	sym	anti	anti	sym	0	$\neq 0$
$4b$	$5a$	sym	anti	anti	sym	0	$\neq 0$
$4b$	$5b$	anti	sym	sym	anti	$\neq 0$	0

where φ is the angular displacement about the ring. The wavefunction for the $4a$ state has antinodes at each nitrogen atom and nodes at the intervening carbon atoms. Hence its de Broglie wavelength is $\Lambda = L_c/8 = 2d_0$ and its wavefunction is

$$\psi_{4a} = \sqrt{\frac{2}{L_c}} \cos\left(\frac{2\pi s}{L_c} \cdot 4\right) = \sqrt{\frac{2}{L_c}} \cos(\varphi \cdot 4)$$

In the state $5a$ the de Broglie wavelength is shorter, i.e., $\Lambda = L_c/10 = 8d_0/5$, so that

$$\psi_{5a} = \sqrt{\frac{2}{L_c}} \cos\left(\frac{2\pi s}{L_c} \cdot 5\right) = \sqrt{\frac{2}{L_c}} \cos(\varphi \cdot 5)$$

In the polar coordinates of this ring system, a displacement along x is given by $x = r\cos\varphi = (L_c/2\pi)\cos\varphi$. Using this substitution, the transition dipole moment integral M_x for the x transition 4a → 5a is given by

$$M_x = e\frac{2}{L_c}\left(\frac{L_c}{2\pi}\right)^2 \cdot \int_0^{2\pi} \cos(4\varphi)\cos\varphi\cos(5\varphi)\, \mathrm{d}\varphi \tag{9.45}$$

In Exercise 9.22 you show that the integrand in Equation (9.45) is symmetric with respect to the x and y directions (origin of the xy coordinate system in the center of the ring) and that the value of the integral is $\pi/2$. Hence we find that

$$M_x = e\frac{L_c}{2\pi^2} \cdot \frac{\pi}{2} = e\frac{L_c}{4\pi} = 2.9 \times 10^{-29}\ \text{C m} = 8.6\ \text{Debye} \tag{9.46}$$

where we used $L_c = 16 \cdot d_0 = 2240$ pm (with $d_0 = 140$ pm). Because this transition dipole moment has a non-zero value, the transition has a nonzero oscillator strength. One can show that the same value is obtained for M_y, the transition moment for the transition $4a \to 5\,b$ and, correspondingly, for the transitions $4b \to 5\,b$ and $4b \to 5\,a$.

For the oscillator strengths f, we obtain

$$f_{4a\to 5a,5b} = 2 \cdot \frac{8\pi^2 m_e}{h^2 e^2 \hat{n}} \cdot \left(|M_x|^2 + |M_y|^2\right) \cdot \Delta E_{4a\to 5a,5b} \tag{9.47}$$

and

$$f_{4b\to 5a,5b} = 2 \cdot \frac{8\pi^2 m_e}{h^2 e^2 \hat{n}} \cdot \left(|M_x|^2 + |M_y|^2\right) \cdot \Delta E_{4b\to 5a,5b} \tag{9.48}$$

from Equations (9.29) and (9.30). Here we apply a factor of two, because two degenerate electronic transitions are responsible for the experimentally observed absorption peak. Inserting the values for the excitation energies calculated in Example 9.5 and the transition moment $M_x = 2.9 \times 10^{-29}$ C m we obtain the data in Table 9.5, using $\hat{n} = 1.36$ for the index of refraction of the solvent.

9.2.7 Coupling of π Electrons

The spectroscopy of electronic excitations is often more complex than a simple HOMO to LUMO excitation of the electron. We consider the electronic excitation of porphyrin as an example to illustrate how electronic transitions can couple with each other. Because porphyrins are important biomolecules (see Chapter 28), it is intrinsically important to understand their absorption features.

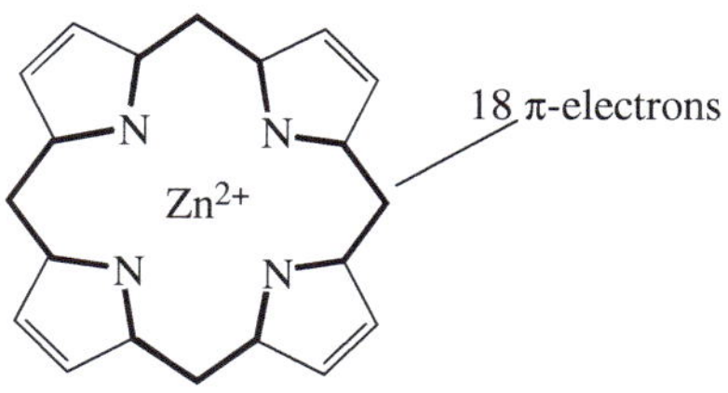

To a good approximation the porphyrin chromophore can be described as a 16 -membered ring with 18 π electrons (similar to phthalocyanine). The apparently small difference between porphyrin and phthalocyanine (four N-atoms instead of eight N-atoms in the inner ring) has considerable consequences, however. The separation of the absorption bands $4b \rightarrow 5a$, $4b \rightarrow 5b$, and $4a \rightarrow 5a$, $4a \rightarrow 5b$ is half as large in porphyrin as in phthalocyanine (see Exercise E9.22). This difference in peak positions agrees with experiment (Table 9.5, Fig. 9.21). The first transition at $\lambda = 550$ nm of porphyrin is related to the red color of blood. However, the oscillator strengths, which are calculated from Equation (9.29), are completely different from the experimental values (see Table 9.5). In particular, the oscillator strengths for the absorption band at long wavelengths (f_1) and for the absorption at short wavelengths (f_2) are predicted to be similar ($f_1 : f_2 = 0.7$), whereas the experimental ratio of these oscillator strengths is only $f_1 : f_2 = 0.06$.

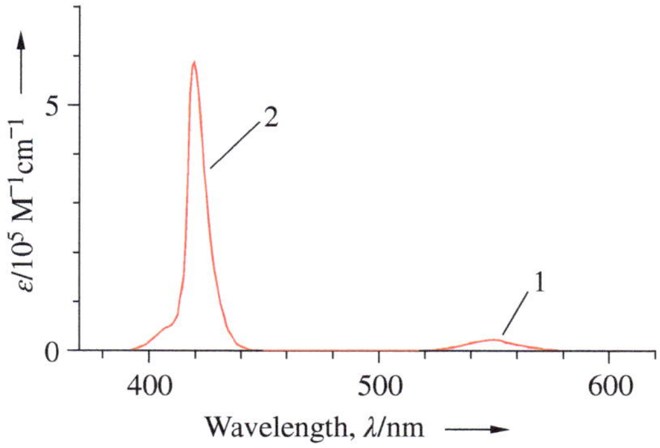

Figure 9.21 Absorption spectrum of Zn-tetraphenyl-porphyrin.

This change in the oscillator strength arises from a coupling between the two transitions. This coupling effect is caused by the repulsion of the electrons responsible for the two transitions. Foundations 9.4 and 9.5 show how a simple classical model can be used to understand this phenomenon. The electrons responsible for the two transitions are replaced by two classical oscillators (charge e) of frequencies

$$\nu_1 = \frac{hc_0}{\lambda_1} \quad \text{and} \quad \nu_2 = \frac{hc_0}{\lambda_2} \tag{9.49}$$

They are coupled by a force related to the repulsion of the electrons in the actual system (Fig. 9.20a), and the force is large when the frequencies ν_1 and ν_2 are similar. For the situation where the oscillators are nearly resonant, we must distinguish between the in-phase oscillation (Fig. 9.22b) and the out-of-phase oscillation (Fig. 9.22c) of the two oscillators. The in-phase oscillation corresponds to a strong absorption band because both dipoles are being driven by the light's electric field, and the dipoles interact constructively to increase the net charge displacement. In contrast, the out-of-phase oscillation corresponds to a weak absorption band. While one dipole is being accelerated by the light field, the other dipole is being decelerated, and this destructive interaction decreases the net charge displacement. As a consequence, the oscillator strength of transition 1 is decreased and that of transition 2 is increased. The quantitative application of this model (see Foundation 9.5) predicts an oscillator strength ratio of $f_1/f_2 = 0.05$, in good agreement with the experimental value (Table 9.5). For details see Problems P9.15 to P9.20.

Figure 9.22 Coupled oscillators. (a) Oscillators at rest. (b) In-phase oscillation. (c) Out-of-phase oscillation.

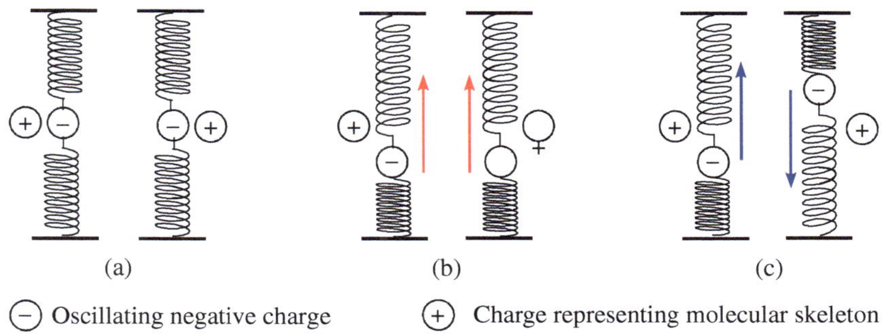

This coupling effect is also observed in the case of Cu^{2+} phthalocyanine. Because of the eight nitrogen atoms in the ring, however, the energy of orbital 4*a* is more strongly decreased than in porphyrin. Hence the transition energies are farther apart:

$$\Delta E_1 = 2.57 \times 10^{-19}\ \text{J}, \quad \Delta E_2 = 6.07 \times 10^{-19}\ \text{J} \tag{9.50}$$

You consider different cases of coupling in Problems P9.15 to P.21.

9.3 Optical Activity

Molecules that cannot be superimposed on their mirror image are called *chiral*. Chiral molecules play a crucial role in biology and biochemistry. It is well known that the highly specific interlocking of chiral molecules is an essential motif in the architecture of biomachinery and biochemical cycles. More recently, it has been shown that the chiral molecules display spin filtering properties that arise from the interaction between an electron's intrinsic magnetic moment and the effective magnetic field that is generated by a chiral molecule's electron cloud [8]. Here we consider the interaction between a chiral molecule and polarized light.

Chiral molecules have a "corkscrew" character in their electron clouds, and this feature generates some special considerations for their interaction with polarized light. Figure 9.8 diagrams the electric and magnetic field vectors for linearly polarized light as a function of space and time. Linearly polarized light can be viewed as a superposition of right and left circularly polarized light. Figure 9.23 shows the electric field vector for right circularly polarized light as a function of space. As the wave propagates, the electric field vector rotates in a clockwise manner. The figure illustrates this for the case of two linearly polarized electric field components (F_y and F_z) of equal amplitude that are propagating with a 90 degree phase shift, so that when the F_y component has zero amplitude the F_z component is at its maximum, and vice versa. This feature gives rise to a net electric field vector that rotates in space and in time. If the rotation direction of the light's electric field is in the same sense as that of the chiral molecule's electron cloud, we may expect a different interaction than if the rotation is in the opposite sense to that of the molecule's electron cloud.

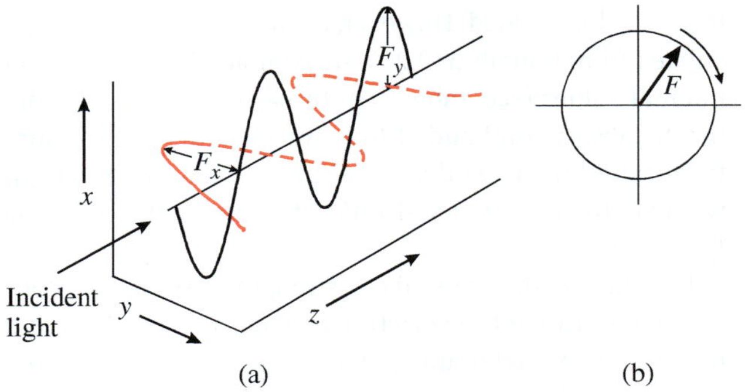

Figure 9.23 Circularly polarized light. (a) Light waves polarized in the *yz* plane (dashed curve) and in the *xz* plane (solid curve) with electric field strengths F_y and F_z, traveling in the *z* direction. (b) Vector of circular polarized light with electric field strength *F* rotating clockwise, imagined to be seen from the light source.

9.3.1 Rotatory Dispersion

When a beam of linearly polarized light travels through a solution of D-sucrose,

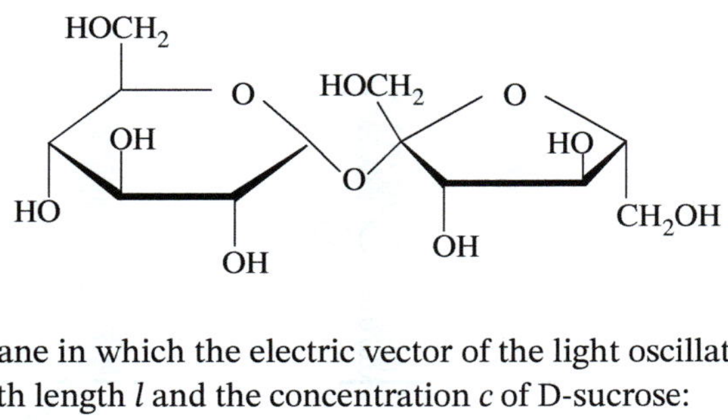

its plane of polarization (i.e., the plane in which the electric vector of the light oscillates) rotates by an angle α (Fig. 9.24). This angle is proportional to the path length l and the concentration c of D-sucrose:

$$\alpha = [\alpha] \cdot l \cdot c \tag{9.51}$$

and the proportionality constant, $[\alpha]$, is called the *molar rotation*. The angle α is assigned to be positive if the observer sees the plane of polarization rotated clockwise (as in Fig. 9.24). The molar rotation $[\alpha]$ is a property of the molecule (depending

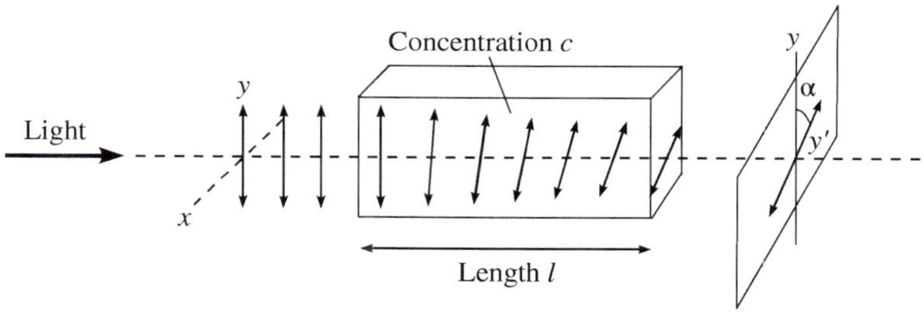

Figure 9.24 Optical activity. A light beam polarized in the y-direction passes through a solution of sucrose (concentration c, path length l). Thereby the plane of polarization is changed by an angle α into the y'-direction.

on the wavelength of the light) and can be determined by careful measurement of the optical rotation angle α as a function of the concentration c. For a wavelength $\lambda = 589$ nm (known as the Na-D line) we find $\alpha = +0.665°$ for a solution of natural sucrose in water at 20° C (concentration $c = 0.0292$ mol L^{-1}, path length $l = 0.10$ m). Then for the molar rotation we find

$$[\alpha]_D^{20} = +\frac{0.665 \text{ deg}}{0.1 \text{ m} \cdot 0.0292 \text{ mol L}^{-1}} = +227 \text{ deg m}^{-1} \text{ L mol}^{-1}$$

where we use the symbol deg instead of °, because it can be better identified in a product of units. If we perform the same experiment with L-sucrose, the magnitude of the effect is the same but its sign is opposite, that is the light is rotated counter-clockwise and $[\alpha]_L^{20} = -227$ deg m^{-1} L mol^{-1}.

The observed effect (*optical rotation*) is linked to the compound's chirality. We can understand this effect by describing the molecule's interaction with the incident light by two classical dipoles oscillating with frequency ν_0 that are coupled by a force related to the repulsion of the electrons in the actual system. In the case considered in Section 9.2.7 the dipoles were in the arrangements of Fig. 9.22. In the case of a chiral compound the oscillators replace the component chromophores (e.g., glucose and fructose in the case of sucrose). Let us consider two dipoles that are arranged in a right-hand screw geometry (Fig. 9.25a). The oscillators are coupled so that the displacement of one oscillator exerts a force on the second oscillator. When light travels through an assembly of such model oscillators, each oscillator acts as an antenna emitting a light wave, as shown in Foundation 10.1 for a single oscillator. The wave of the incident light and the waves scattered by each model oscillator interfere, resulting in a rotation of the plane of polarization. For the D-sucrose case considered above

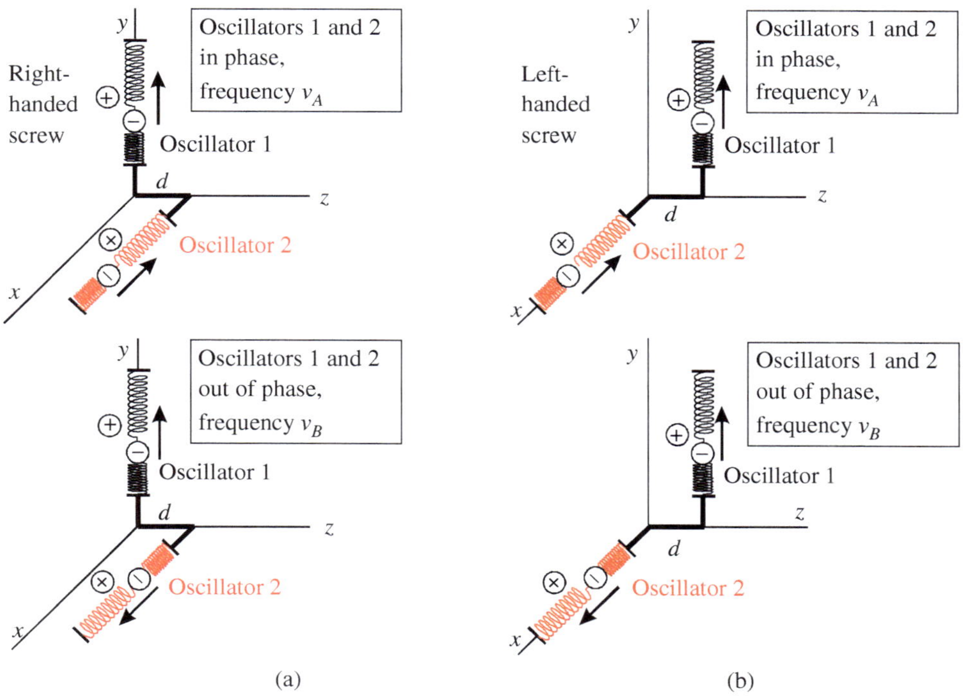

Figure 9.25 Model of a chiral molecule: Two coupled oscillators 1 and 2. When displacing oscillator 1, a force is exerted on oscillator 2 in the direction of oscillator 2. (a) Right-handed screw geometry, (b) Left-handed screw geometry. We distinguish between an oscillation of the two coupled oscillators at a frequency ν_A lower than the frequency of the uncoupled oscillators (in-phase oscillation) and an oscillation at higher frequency ν_B (out-of-phase oscillation).

the incident light frequency is much lower than the electronic absorption frequency, so that the scattered wave is in phase with the field of the incident wave and the net effect is to rotate the light polarization clockwise. If the coupled dipoles are constructed in a left-hand screw geometry (Fig. 9.25b), which is the mirror image of model a, the light polarization rotates in the opposite sense.

9.3.2 Ellipticity and Circular Dichroism

If the incident light frequency is in the range of a chiral molecule's absorption band, the absorption causes the relative intensity of the light field's left and right circularly polarized components to change because the molecule's absorption coefficients are different for the two light polarizations. Phenomenologically we observe that the transmitted light becomes elliptically polarized. For incident light with its electric field oriented in the *xz*-plane, the transmitted light has components in the *xz'*-plane and the *xy'*-plane, and it is elliptically polarized. This occurs because the amplitudes of the left and right circularly polarized light fields do not exactly cancel when oriented in opposing directions.

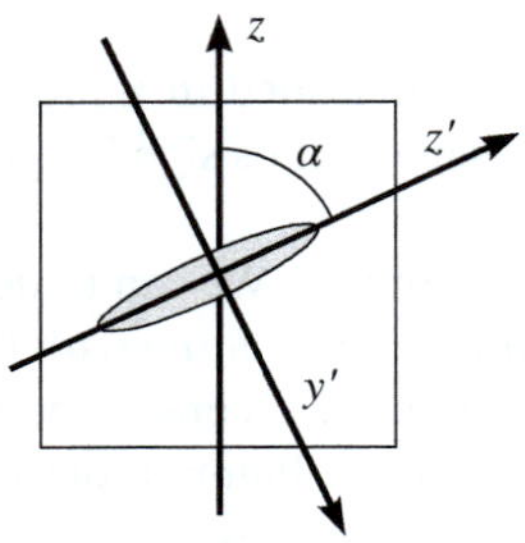

Plane of light rotated by an angle α

Long axis of ellipse = a

Short axis of ellipse = b

If we consider the elliptically polarized light as a superposition of a left-handed circularly polarized wave of amplitude C_1, with an electric field

$$F_{z',L} = C_1 \cos \omega t \quad \text{(component in the } z' \text{ direction, left polarized)}$$
$$F_{y',L} = C_1 \sin \omega t \quad \text{(component in the } y' \text{ direction, left polarized)} \tag{9.52}$$

and a right-handed circularly polarized wave of amplitude C_2, with an electric field

$$F_{z',R} = C_2 \cos \omega t \quad \text{(component in the } z' \text{ direction, right polarized)}$$
$$F_{y',R} = -C_2 \sin \omega t \quad \text{(component in the } y' \text{ direction, right polarized)} \tag{9.53}$$

then the electric field amplitudes trace out an ellipse in time, with a major axis $a = C_1 + C_2$ and a minor axis $b = C_1 - C_2$. The angle

$$\Theta = \arctan \frac{b}{a} \tag{9.54}$$

is called the *ellipticity*. The sign convention is that Θ is positive in the case of the scheme above — that is, when the rotation angle α is positive for $\nu < \nu_1$. Θ is proportional to l and c:

$$\Theta = [\Theta]\, l\ c \tag{9.55}$$

where $[\Theta]$ is the *molar ellipticity*. The molar ellipticity $[\Theta]$ depends on the molecular structure and on the wavelength of the light.

The ellipticity results because the molar decadic absorption coefficients ε_R and ε_L are different for an optically active compound. The difference

$$\Delta\varepsilon = \varepsilon_L - \varepsilon_R \tag{9.56}$$

is called *circular dichroism* (commonly abbreviated as "CD"). A detailed analysis in Box 9.2 shows that the molar ellipticity $[\Theta]$ is proportional to $\Delta\varepsilon$, so that

$$[\Theta] = \frac{2.303}{4}\Delta\varepsilon \tag{9.57}$$

We consider the ratio

$$g = \left(\frac{\Delta\varepsilon}{\varepsilon}\right)_{\text{extremum of } [\Theta]} \tag{9.58}$$

where

$$\varepsilon = \frac{\varepsilon_L + \varepsilon_R}{2} \tag{9.59}$$

g is called the *anisotropy factor* of the given absorption band.

9.3.3 Anisotropy Factor g for Model in Fig. 9.25

We calculate ε_L and ε_R for the model in Fig. 9.25a (oscillators 1 and 2 in phase). Let us first assume that the oscillators in the model interact with a beam of incident light of wavelength λ_{min}, at the minimum of circular dichroism, moving in the positive x direction. The phase of the electric field is delayed at oscillator 2 relative to the field at oscillator 1 by

$$\varphi = 2\pi \frac{d}{\lambda_{\text{min}}} \tag{9.60}$$

According to Fig. 9.25 the electric field strength driving oscillator 1 in the z-direction is

$$F_z = F_0 \cdot \cos \omega t \tag{9.61}$$

and the field strength driving oscillator 2 in the y-direction is

$$F_y = -F_0 \cdot \sin(\omega t - \varphi) \tag{9.62}$$

The displacements z_1 and $-y_2$ of the two oscillating charges Q_1 and Q_2 are equal because the fields F_z and F_y are equal but phaseshifted. We show in Justification 9.1 that

$$\varepsilon_L = const \cdot 2\,(1 + \sin \varphi) \tag{9.63}$$

Accordingly, for right circular polarized light, the forces act in the opposite sense so that

$$\varepsilon_R = const \cdot 2\,(1 - \sin \varphi) \tag{9.64}$$

We can calculate the mean value ε from ε_L and ε_R as

$$\varepsilon = \frac{\varepsilon_L + \varepsilon_R}{2} = 2 \cdot const \tag{9.65}$$

Hence we can write the anisotropy as

$$g = \frac{\Delta\varepsilon}{\varepsilon} = \frac{\varepsilon_L - \varepsilon_R}{2 \cdot const} = \frac{4 \cdot const \cdot \sin \varphi}{2 \cdot const} = 2 \sin \varphi \approx 2\varphi = \frac{4\pi d}{\lambda_{\text{min}}} \tag{9.66}$$

where we have made the approximation that $\sin \varphi \approx \varphi$ for small φ (alternatively $d << \lambda_{\text{min}}$).

For the present geometry (d in the direction of incident light: x-direction) the circular dichroism is strongest, and there is no effect when d is in the y- or z-direction. For the statistical average we obtain

$$g = \frac{\Delta\varepsilon}{\varepsilon} = \frac{1}{2} \cdot \frac{4\pi d}{\lambda_{\text{min}}} = \frac{2\pi d}{\lambda_{\text{min}}} \tag{9.67}$$

Generally:

	case (a) right handed screw	case (b) left handed screw
oscillators in phase	$g = -2\pi d/\lambda_{\text{min}}$	$g = +2\pi d/\lambda_{\text{min}}$
oscillators out of phase	$g = +2\pi d/\lambda_{\text{max}}$	$g = -2\pi d/\lambda_{\text{max}}$

(9.68)

where λ_{max} is the wavelength at the maximum of the circular dichroism.

9.3.4 Absolute Configuration of Chiral Molecules

In 1859, Louis Pasteur (1822–1895) proposed that the asymmetry of molecules is the reason for their optical activity. The basic question, how to relate chirality to optical activity, was open until Werner Kuhn (1899–1963) showed this in 1935 for the case of D-lactic acid [9], on the basis of the classical model discussed here. In the following, we consider a case of particular simplicity to demonstrate how to find the absolute configuration.

9.3.4.1 Circular Dichroism of Spirobisanthracene

In this molecule two anthracene subunits are connected in a chiral manner (see Fig. 9.25). Thus we consider the optical properties of anthracene first, and then we consider how their coupling can lead to a CD response.

Figure 9.26 shows the orbitals and energy levels that are obtained for anthracene within the free electron model. For the transition $3b \to 4a$ we calculate an absorption maximum of 495 nm and the transition is polarized in the y direction,
see considerations in Section 9.2.6. Experimentally this transition appears at 370 nm with the predicted polarization. A second transition $3a \to 4a$ (calculated wavelength 355 nm, experiment 260 nm) is polarized along the molecule's long axis, in the x direction. Its transition dipole moment is

$$M_{x,3a\to 4a} = e \int \psi_{3a} x \psi_{4a} \cdot \mathrm{d}\tau$$
$$= 2.21 \times 10^{-29}\ \mathrm{C\ m} \qquad (9.69)$$

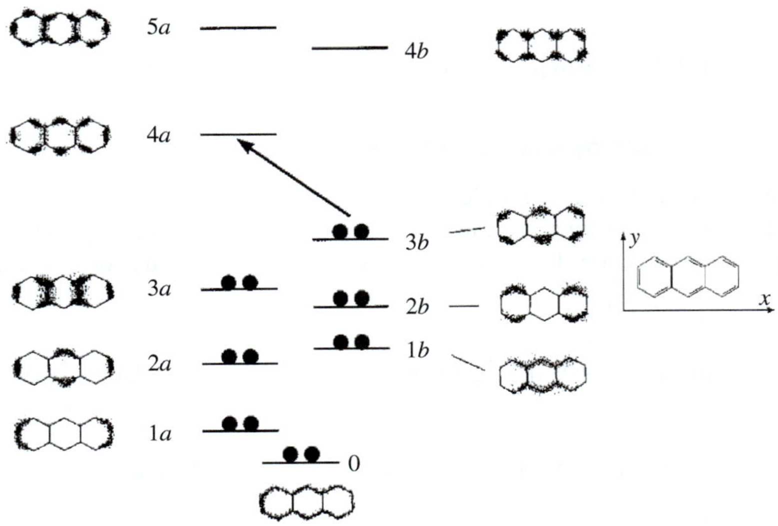

Figure 9.26 π-electron energies of anthracene. The calculated HOMO–LUMO excitation energy is $\Delta E_{3b\to 4a} = 4.01 \times 10^{-19}$ J corresponding to $\lambda = 495$ nm, followed by transition $3a \to 4a$ with $\Delta E_{3a\to 4a} = 5.59 \times 10^{-19}$ J and $\lambda = 355$ nm.

corresponding to an oscillator strength of $f = 1.2$ (calculated for the excitation of two electrons, using $\hat{n} = 1.3$).

The spirobisanthracene molecule has a strong absorption band at 290 nm (Fig. 9.27), which is not too different from the energy for the ($3a \to 4a$) absorption peak of independent anthracene molecules. The two anthracene units are arranged as shown in Fig. 9.28a. From this figure it is evident that in the case of the first transition ($3b \to 4a$) the transition dipoles of the two chromophores will be oriented parallel and no helicity is present. In contrast, the second transition ($3a \to 4a$) is polarized along the chromophore's long axes (x and z directions, respectively) and we expect a strong circular dichroism because the transition dipoles form a helical ("screw-like," Fig. 9.25) geometry (Fig. 9.28, right-handed screw).

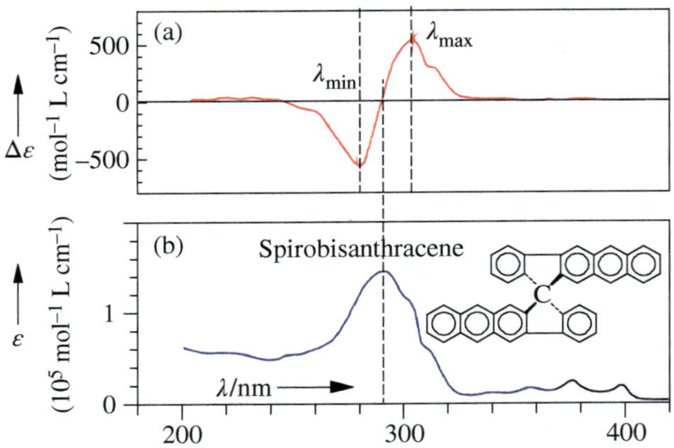

Figure 9.27 (a) CD-spectrum and (b) absorption spectrum of spirobisanthracene [10]. The in-phase oscillation of the two coupled oscillators at $\nu_{min} = c_0/\lambda_{min}$ and the out-of-phase oscillation at $\nu_{max} = c_0/\lambda_{max}$ are only weakly separated in this case. The two bands cannot be resolved in the absorption spectrum, but they appear in the CD spectrum.

Let us look at the coupling for this second transition $3a \to 4a$ more closely. If the chromophores did not interact with each other the transition on each chromophore would be at the same transition energy; however, their interaction causes a breaking of the degeneracy, or splitting. The change in the transition energy depends on details of the wavefunctions ψ_{3a} and ψ_{4a} and the geometry, but the essence of the interaction can be described by replacing each chromophore with a dipole (charges at distance $b = 480$ pm), so that its dipole moment is equal to the transition moment $M_x = 2.21 \times 10^{-29}$ C m. Thus a charge of $0.6 \times 10^{-19}\ \mathrm{C} \approx \frac{1}{2}e$ is at point 1, a charge of $-\frac{1}{2}e$ is at point 2, and so on, in Fig. 9.28. From the geometry it follows that the distances a and c in Fig. 9.28 are $a = 250$ pm and $c = 320$ pm.

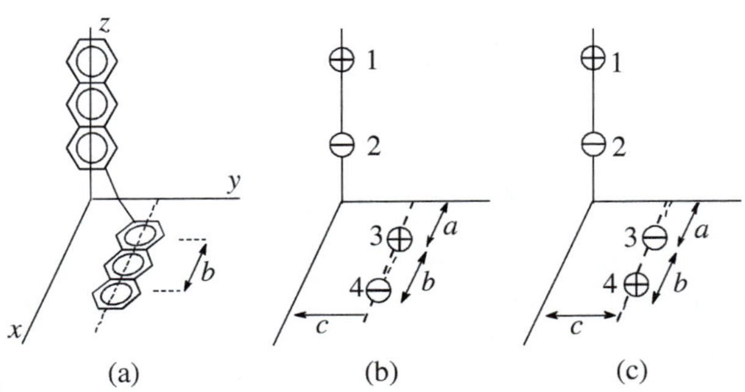

Figure 9.28 Arrangement of the anthracene units in spirobisanthracene. (a) Geometry, (b) in-phase coupling, (c) out-of-phase coupling in the oscillation.

When the dipoles are oscillating in-phase (Fig. 9.25a, top) they cause a shift to longer wavelengths for the oscillation.

When the dipoles are oscillating out-of-phase they cause a shift to shorter wavelengths for the oscillation (Fig. 9.25a, bottom). The direction of the energy shifts can be understood on a qualitative level. The in-phase oscillation shifts to lower energy because the attractive interaction between the dipoles helps to drive their oscillation; hence, less energy is needed from the light field for driving it. Correspondingly, the out-of-phase oscillation of the dipoles requires a higher energy light field because the dipoles have a repulsive interaction that inhibits their oscillation.

In Foundation 9.4 we show that the splitting ΔE of the absorption band in spirobisanthracene is

$$\Delta E = -2J_{12} \quad \text{(in-phase)} \quad \text{and} \quad \Delta E = +2J_{12} \quad \text{(out-of-phase)}$$

where J_{12} is the coupling integral. The interaction energy J_{12} between the two dipoles (dipoles $1 \rightarrow 2$ and $3 \rightarrow 4$ in Fig. 9.28b,c) is given by

$$J_{12} = \frac{(e/2)^2}{4\pi\varepsilon_0\varepsilon_r}\left[-\frac{1}{r_{23}} - \frac{1}{r_{14}} + \frac{1}{r_{24}} + \frac{1}{r_{13}}\right]$$

where $\varepsilon_r = 2.5$ is the effective relative permittivity of the molecule and r_{ij} = distance between charges i and j. Each term in this expression represents an electrostatic potential energy for a charge in one dipole with a charge in the neighboring dipole. Terms for each individual dipole are not included since they contribute to its "self-energy," not its interaction energy with the neighboring dipole. From the geometry in Fig. 9.28 we obtain

$$r_{23} = \sqrt{2a^2 + c^2}, \quad r_{14} = \sqrt{2(a+b)^2 + c^2}, \quad r_{24} = r_{13} = \sqrt{(a+b)^2 + a^2 + c^2}$$

Using the lengths given for a, b, and c, we obtain $r_{23} = 477$ pm, $r_{14} = 1080$ pm, $r_{24} = r_{13} = 835$ pm, and J_{12} is

$$J_{12} = 2.32 \times 10^{-29} \cdot (-6.24 \times 10^{8})\ \text{J} = 0.14 \times 10^{-19}\ \text{J} = 0.087\ \text{eV}$$

The energy splitting between the in-phase and out-of-phase transitions is

$$\Delta E = 4 \times J_{12} = 0.56 \times 10^{-19}\ \text{J} = 0.35\ \text{eV}$$

This splitting manifests in the circular dichroism spectrum of Fig. 9.27, for which the negative branch occurs at 278 nm ($\Delta E = 7.14 \times 10^{-19}$ J) and the positive branch occurs at 300 nm ($\Delta E = 6.62 \times 10^{-19}$ J). The experimental energy difference is 0.52×10^{-19} J = 0.33 eV in good agreement with the model.

Anisotropy Factor g Using Equation (9.68), $g = (2\pi d/\lambda_{min})$, with $d = c = 320$ pm and $\lambda = 278$ nm, we calculate $g = +7 \times 10^{-3}$ for the in-phase oscillation. The experimental CD curve gives the value

$$g = \left(\frac{\Delta\varepsilon}{\varepsilon}\right)_{min} = \frac{500}{10 \times 10^4} = 5 \times 10^{-3}$$

Correspondingly, for the out-of-phase oscillation we calculate $g = -7 \times 10^{-3}$, whereas the experimental value (maximum of the CD curve) is $g = -6.5 \times 10^{-3}$. The agreement in sign as well as in the magnitudes is unusually good.

9.3.4.2 Circular Dichroism of Chiral Cyanine Dye

The chiral cyanine dye

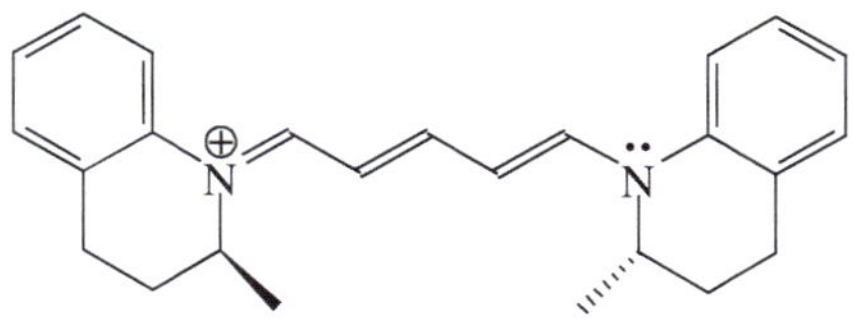

absorbs at $\lambda_{max} = 500$ nm, and the absorption coefficient is $\varepsilon = 8 \times 10^4$ mol^{-1} L cm^{-1}. This molecule displays circular dichroism [11] with $\Delta\varepsilon = \varepsilon_L - \varepsilon_R = 15$ mol^{-1} L cm^{-1}. In contrast to spirobisanthracene, the circular dichroism in this cyanine dye is not related to a coupling between two overlapping chromophores in the absorption band, rather it is caused by the left-handed screw like shape of the π-electron cloud between the N atoms. Consequently, the sign of the dichroism does not change within the absorption band.

The sign and the value of the quantity $\Delta\varepsilon/\varepsilon = 1.9 \times 10^{-4}$ can be readily estimated from a simple model. We replace the cyanine cation by a charge oscillating from the nitrogen atom on the left to the nitrogen atom on the right and back along the line indicated by the three arrows in the diagram.

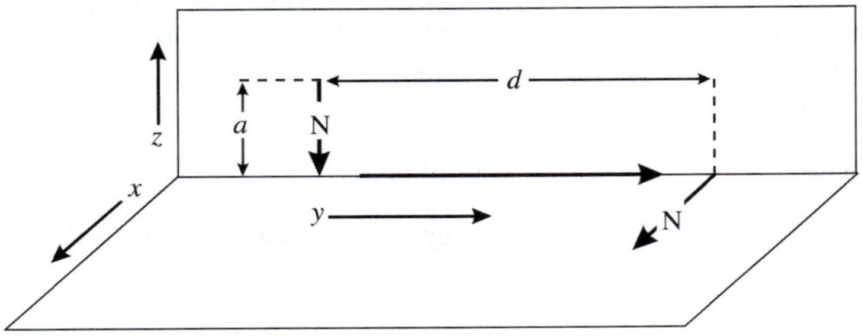

The transition dipole moment is related to the wavefunction's extent in each direction, so $M_x \sim d$, $M_y \sim a$, $M_z \sim a$, and

$$\varepsilon \sim \frac{1}{3}\left(d^2 + 2a^2\right)$$

For circularly polarized light moving in the x-direction, the signal results from the interaction of the light with M_y and M_z so that

$$\frac{\varepsilon_L - \varepsilon_R}{\varepsilon_{yz}} = \frac{2\pi d}{\lambda}$$

where

$$\varepsilon_{yz} \approx \frac{1}{3}\left(2a^2\right)$$

Given that

$$\frac{\varepsilon_{yz}}{\varepsilon} = \frac{2a^2}{d^2 + 2a^2}$$

we find

$$\frac{\Delta\varepsilon}{\varepsilon} = \frac{\varepsilon_L - \varepsilon_R}{\varepsilon} = \frac{2\pi d}{\lambda}\frac{2a^2}{d^2 + 2a^2}$$

For $\lambda = 500$ nm, $a = 0.1$ nm, $d = 0.73$ nm we obtain

$$\frac{\Delta\varepsilon}{\varepsilon} = \frac{2\pi \cdot 0.73\ \text{nm}}{500 \times 10^{-9}\ \text{m}}\frac{0.02\ \text{nm}^2}{\left(0.73^2 + 0.02\right)\ \text{nm}^2} = 3 \times 10^{-4}$$

This result agrees in sign and approximately in value with the experimental value $\Delta\varepsilon/\varepsilon = 1.9 \times 10^{-4}$.

9.3.5 Summary

Circular dichroism is very important in molecular biology, e.g., to detect α helical regions in proteins (see Chapter 27), and in synthetic chemistry aimed at the creation of bioactive agents (e.g., pharmaceuticals, agricultural chemicals). The mechanism for CD discussed in this section describes the response as arising from a coupling between electric dipole transitions; however, CD can also be described in terms of magnetic dipole transitions. By way of example, a chromophore like carbonyl has an n-to-π^* transition that is electric dipole forbidden but magnetic dipole allowed; i.e., the promotion of a lone pair electron on the oxygen to the π^* orbital of the C = O generates a magnetic dipole along the bond axis but does not generate an electric dipole. Moreover, the response in CD can depend sensitively on the properties of the molecular environment; see [12].

9.4 Conclusion

Investigating the interaction of light with molecules gives detailed information on their electronic structure and its quantum-mechanical basis.

In the first part of the chapter we used a prototypical example, the π electron system of dyes, to show how quantum-mechanical modeling can be used to elucidate relatively complex systems. By employing the simple model developed in Chapter 8 we saw how details in the shape of wavefunctions, which result from modifying the chemical structure, manifest as changes in the electronic absorption spectra. By comparing cyanine dyes and polyenes, we found that the FEMO model is appropriate for the cyanine dyes which have equal bond lengths, but that the understanding of the polyene spectra requires that the alternation of single and double bonds be adequately described.

The polarization of an absorption band could be understood on the basis of an even simpler model, replacing the molecule's electronic charge oscillation by a classical oscillating dipole. We showed that the quantum-mechanical treatment of a molecule in the electromagnetic field of a light wave is identical in its formalism with modeling the molecule as a classical oscillator. In many cases (e.g., considering the absorption of cyclic π electron systems or optical activity) the molecule could be represented by coupled classical dipole oscillators in order to model its quantum-mechanical properties.

Nevertheless, we must realize that a classical picture is ultimately limited, and a quantum mechanical description is required for a deeper understanding of electronic spectra. The more realistic calculation of the excited electronic states and electronic spectra by modern quantum chemical methods remains a current research challenge.

9.5 Box 9.1. Lambert–Beer Law

We consider a cell with a low concentration of absorbing molecules (i.e., the absorbing molecules can be treated as independent of each other), so that the light absorption is linearly proportional to the absorber concentration and the light intensity (Fig. 9B.1). In this limit, the differential change in the light intensity $\mathrm{d}I$ between a position x and $x + \mathrm{d}x$, is proportional to the incident light intensity I, and the sample concentration c, so that

$$\mathrm{d}I = -\alpha \cdot c \cdot I \cdot \mathrm{d}x$$

where α is the proportionality constant and is known as the absorption coefficient. To obtain the total light intensity at the right end of the cell (at the position $x = l$), we must integrate over the length of the optical cell, i.e.,

$$\int_{I_0}^{I} \frac{\mathrm{d}I'}{I'} = -\alpha \cdot c \cdot \int_0^l \mathrm{d}x$$

which gives

$$\ln \frac{I}{I_0} = -\alpha \cdot c \cdot l \quad \text{or} \quad I = I_0 \cdot \mathrm{e}^{-\alpha \cdot c \cdot l}$$

It is common to use the decadic logarithm, so we transform this result to find

$$\log \frac{I}{I_0} = -\alpha \cdot c \cdot l \cdot \log \mathrm{e}$$

and with

$$\varepsilon = \alpha \cdot \log \mathrm{e} \qquad \text{or} \qquad \alpha = \frac{1}{\log \mathrm{e}} \varepsilon$$

we obtain

$$\log \frac{I}{I_0} = -\varepsilon \cdot c \cdot l$$

where ε is the molar decadic absorption coefficient. We define the absorbance A as

$$A = -\log \frac{I}{I_0} = \varepsilon \cdot c \cdot l$$

and obtain a fundamental result, the *Lambert–Beer law*.

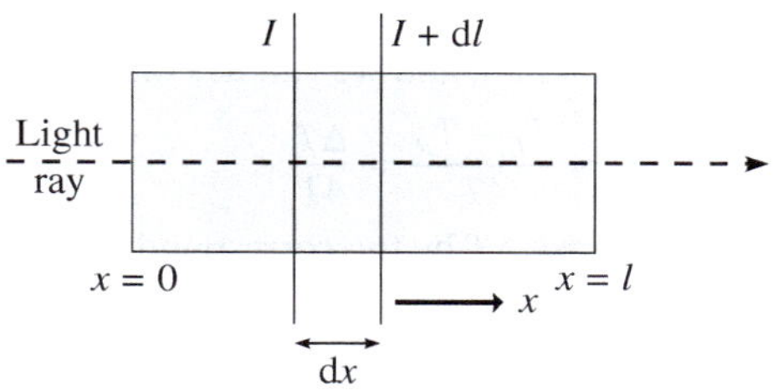

Figure 9B.1 Derivation of the law of Lambert–Beer. Note that I decreases with increasing x; thus, $\mathrm{d}I$ is negative.

9.6 Box 9.2. Relation Between Ellipticity Θ and $\Delta\varepsilon$

The transmission T is defined as I/I_0, where I_0 is the intensity of the incident light and I is the intensity of the light after passing through the sample. Because the intensity is proportional to the square of the electric field strength, $I \sim F^2$, we find that T is proportional to F^2. Then, according to Equations (9.52) and (9.53), for the transmissions T_L and T_R of the left and right polarized light we obtain

$$T_L \sim F_L^2 \sim C_1^2, \qquad T_R \sim F_R^2 \sim C_2^2$$

and

$$a = C_1 + C_2 \sim \sqrt{T_L} + \sqrt{T_R} \quad \text{and} \quad b = C_1 - C_2 \sim \sqrt{T_L} - \sqrt{T_R}$$

From Equation (9.54) we obtain

$$\tan\Theta = \frac{b}{a} = \frac{\sqrt{T_L} - \sqrt{T_R}}{\sqrt{T_L} + \sqrt{T_R}}$$

If we multiply the numerator and denominator by $\sqrt{T_L} + \sqrt{T_R}$, we find that

$$\tan\Theta = \frac{b}{a} = \frac{T_L - T_R}{T_L + 2\sqrt{T_L T_R} + T_R}$$

Because the difference between T_L and T_R is extremely small, we approximate the denominator by $4T$. Furthermore, Θ is extremely small, and we can use the approximation $\tan\Theta \simeq \Theta$ to write

$$\Theta = \frac{T_L - T_R}{4T} = \frac{\Delta T}{4T}$$

If we express ΔT by the corresponding change of absorbance, $\Delta A = A_L - A_R$, we can write

$$A_L = \log T_L = \frac{1}{2.303}\ln T_L, \quad A_R = \log T_R = \frac{1}{2.303}\ln T_R$$

so that

$$\Delta A = \frac{1}{2.303}\ln\frac{T_L}{T_R}$$

Because $T_L = T_R + \Delta T$, we find that

$$\Delta A = \frac{1}{2.303}\ln\left(\frac{T_R + \Delta T}{T_R}\right) = \frac{1}{2.303}\ln\left(1 + \frac{\Delta T}{T_R}\right)$$

In the limit that $\Delta T/T_R$ is small, we can use the relation $\ln(1 + x) \approx x$ for x small to write

$$\Delta A = \frac{1}{2.303}\frac{\Delta T}{T_R}$$

Thus we obtain

$$\Theta = \frac{\Delta T}{4T} = \frac{2.303 \cdot \Delta A}{4}$$

Because $\Delta A = \Delta\varepsilon \cdot lc$ and the definition $[\Theta] = \Theta/(lc)$ we find that

$$[\Theta] = \frac{2.303 \cdot \Delta A}{4lc} = \frac{2.303 \cdot \Delta\varepsilon}{4}$$

9.7 Justifications

9.7.1 Justification 9.1: Derivation of Equation (9.63) in Section 9.3.3

See the books website, www.wiley.com/go/Kuhn/physicalchemistry3e.

9.8 Foundations

9.8.1 Foundation 9.1: Integrated Absorption: Classical Oscillator

9.8.1.1 Absorbed Power in a Radiation Beam and Absorption Coefficient ε

Following Box 9.1 the Law of Lambert–Beer can be written in differential form as

$$\frac{\mathrm{d}I}{\mathrm{d}x} = -\ln 10 \cdot \varepsilon\, c\, I = -2.3026 \cdot \varepsilon\, c\, I \tag{9F1.1}$$

where c is the concentration of the dissolved molecules, and I is the intensity of the incident light. $\mathrm{d}I$ is the light's differential intensity change in a volume element $\mathrm{d}V$ with thickness $\mathrm{d}x$ ($\mathrm{d}V = A \cdot \mathrm{d}x$, A = cross section of the irradiated sample). Intensity is the mean power of the light wave divided by the area A of the light beam:

$$I = \frac{\text{mean power}}{A} \tag{9F1.2}$$

The number of absorbing molecules in this volume element is given by

$$\mathrm{d}N = \frac{N}{V} \cdot \mathrm{d}V = \frac{n\mathbf{N}_A}{V} \cdot \mathrm{d}V = \mathbf{N}_A \cdot c \cdot A \cdot \mathrm{d}x$$

where $\mathbf{N}_A$ is Avogadro's constant. If each molecule absorbs an energy $\mathrm{d}E$ in a time $\mathrm{d}t$ from the incident electromagnetic wave, then $\overline{\mathrm{d}E/\mathrm{d}t} \cdot \mathrm{d}N$ is the mean power absorbed by $\mathrm{d}N$ molecules. For this rate of power loss from the light field, the intensity change $\mathrm{d}I$ is

$$\mathrm{d}I = -\frac{\overline{\mathrm{d}E/\mathrm{d}t} \cdot \mathrm{d}N}{A} = -\overline{\left(\frac{\mathrm{d}E}{\mathrm{d}t}\right)} \cdot \mathbf{N}_A \cdot c \cdot \mathrm{d}x = -\ln 10 \cdot \varepsilon \cdot c \cdot I \cdot \mathrm{d}x$$

Comparison with Lambert–Beer law enables us to write

$$\ln 10 \cdot \varepsilon \cdot I = \overline{\left(\frac{\mathrm{d}E}{\mathrm{d}t}\right)} \cdot \mathbf{N}_A$$

or

$$\varepsilon = \overline{\left(\frac{\mathrm{d}E}{\mathrm{d}t}\right)} \cdot \frac{\mathbf{N}_A}{\ln 10} \cdot \frac{1}{I} \tag{9F1.3}$$

From physics,[7] we know that the intensity I of the incident light is proportional to the square of the amplitude F_0 of the electric field strength of the electromagnetic wave propagating in a medium with refractive index $\hat{n}$ and relative permittivity $\hat{\varepsilon}$ (c_0 = speed of light in the vacuum, $c_0/\hat{n}$ = speed of light in medium); such that:

$$I = \frac{1}{2} \cdot c_0 \cdot \varepsilon_0 \cdot \hat{n} \cdot F_0^2 \tag{9F1.4}$$

Here we assume that the π electrons are in a homogeneous medium of refraction index $\hat{n}$ (medium of σ electrons and solvent). Combining these latter two relations, we find that the absorption coefficient is

$$\varepsilon = \frac{2 \cdot \mathbf{N}_A}{\ln 10 \cdot c_0 \cdot \varepsilon_0 \cdot \hat{n}} \cdot \overline{\left(\frac{\mathrm{d}E}{\mathrm{d}t}\right)} \cdot \frac{1}{F_0^2} \tag{9F1.5}$$

9.8.1.2 Calculating the Absorption Coefficient ε Classically

First we calculate the mean power $\overline{\mathrm{d}E/\mathrm{d}t}$ for an isotropic damped classical oscillator of mass m and charge Q. The displacement ξ of the oscillator (change of its elongation) is obtained by solving the differential equation

$$m \cdot \frac{\mathrm{d}^2\xi}{\mathrm{d}t^2} = -k_f \cdot \xi - k_d \cdot \frac{\mathrm{d}\xi}{\mathrm{d}t} + Q \cdot F_0 \cdot \cos\omega t \tag{9F1.6}$$

where k_d is the damping constant and k_f is the force constant. This equation is Newton's law (force = mass × acceleration) for a spring system. The term on the left is the mass times the acceleration for the displacement and the first term on the right is the restoring force of the spring. If the equation only had these two terms it would be Hooke's law. The second term on the right is the friction term which accounts for energy exchange between the spring system and its environment. The last term

---

7 See D. Halliday, R. Resnick, and J. Walker. *Fundamentals of Physics*, 4th ed. (Wiley, NY, 1993).

in the equation is the driving force which corresponds to the electric field driving the electron's displacement. The solution to this equation is

$$\xi = \frac{Q}{m}F_0 \cdot \frac{1}{\sqrt{\left(\omega_0^2 - \omega^2\right)^2 + k_d^2\omega^2/m^2}} \cdot \cos(\omega t + \alpha) \tag{9F1.7}$$

with the resonant frequency $\omega_0 = \sqrt{k_f/m}$ and $\omega = 2\pi\nu$, where ν is the frequency of the electric field of the light wave. The phase shift α is obtained as

$$\alpha = \arctan\left[\frac{-k_d\omega}{m\left(\omega_0^2 - \omega^2\right)}\right] \tag{9F1.8}$$

You can verify this expression by substitution into the differential equation.

The energy $\mathrm{d}E$ absorbed is the product of the damping force $k_d \cdot \mathrm{d}\xi/\mathrm{d}t$ and the displacement $\mathrm{d}\xi$. Hence we can write that

$$\mathrm{d}E = k_d \cdot \left(\frac{\mathrm{d}\xi}{\mathrm{d}t}\right) \cdot \mathrm{d}\xi$$

To obtain the rate of energy absorption (power), we must determine the differential energy change $\mathrm{d}E$ in a time $\mathrm{d}t$, which may be written as

$$\frac{\mathrm{d}E}{\mathrm{d}t} = k_d \cdot \left(\frac{\mathrm{d}\xi}{\mathrm{d}t}\right)^2$$

Using the expression for ξ, we find that $\mathrm{d}\xi/\mathrm{d}t$ is

$$\frac{\mathrm{d}\xi}{\mathrm{d}t} = -\frac{Q}{m}F_0 \cdot \frac{\omega}{\sqrt{\left(\omega_0^2 - \omega^2\right)^2 + k_d^2\omega^2/m^2}} \cdot \sin(\omega t + \alpha)$$

and subsequently that the power of absorption is

$$\frac{\mathrm{d}E}{\mathrm{d}t} = k_d\frac{Q^2}{m^2}F_0^2 \cdot \frac{\omega^2}{\left(\omega_0^2 - \omega^2\right)^2 + k_d^2\omega^2/m^2} \cdot \sin^2(\omega t + \alpha)$$

To determine the average power we must integrate over the period T of the applied light field, and we find,

$$\begin{aligned}\overline{\left(\frac{\mathrm{d}E}{\mathrm{d}t}\right)} &= k_d\frac{Q^2}{m^2}F_0^2 \cdot \frac{\omega^2}{\left(\omega_0^2 - \omega^2\right)^2 + k_d^2\omega^2/m^2}\int_0^T \sin^2(\omega t + \alpha)\,\mathrm{d}t \\ &= k_d\frac{Q^2}{m^2}F_0^2 \cdot \frac{\omega^2}{\left(\omega_0^2 - \omega^2\right)^2 + k_d^2\omega^2/m^2} \cdot \frac{1}{2}\end{aligned} \tag{9F1.9}$$

Now we can substitute this result for the average power into our expression for the absorption coefficient ε, and we find that

$$\varepsilon = \frac{2\mathbf{N}_A}{\ln 10 \cdot c_0\ \varepsilon_0\ \hat{n}}k_d\frac{Q^2}{m^2} \cdot \frac{1}{2} \cdot \frac{\omega^2}{\left(\omega_0^2 - \omega^2\right)^2 + k_d^2\omega^2/m^2} \tag{9F1.10}$$

This classical expression provides the molar extinction coefficient in terms of fundamental constants and specific parameters of the spring system, in particular the magnitude of the displaced charge Q, the mass of the charge m, the damping constant k_d, and the spring's resonant frequency ω_0. These four parameters characterize the spring system.

9.8.1.3 Resonance Curve for ε: Calculating $\int_{\text{band}} \varepsilon \cdot \mathrm{d}\nu$

In Fig. 9F.1 the absorption coefficient ε is plotted versus ω for different values of the damping constant k_d. The absorption coefficient displays a clear peak at the resonance condition $\omega = \omega_0$. The half-width of the resonance curve depends on the value of k_d, so that the bandshape becomes broader as k_d increases.

Half-Width of Resonance Curve We write the resonance curve (9F1.10) for ε in the form

$$\varepsilon = const\frac{k_d\omega^2}{\left(\omega_0^2 - \omega^2\right)^2 + k_d^2\omega^2/m^2}$$

where

$$const = \frac{\mathbf{N}_A}{\ln 10 \cdot c_0 \varepsilon_0 \hat{n}} \frac{Q^2}{m^2} \qquad (9F1.11)$$

We find the maximum amplitude ε_{max} by setting $\omega = \omega_0$ to obtain

$$\varepsilon_{max} = const \cdot \frac{k_d \omega_0^2}{k_d^2 \omega_0^2 / m^2} = const \cdot \frac{m^2}{k_d} \qquad (9F1.12)$$

In order to find the frequency $\omega_{1/2}$ where ε is just one half of ε_{max}, we write

$$\frac{1}{2} \frac{m^2}{k_d} = \frac{k_d \omega_0^2}{\left(\omega_0^2 - \omega_{1/2}^2\right)^2 + k_d^2 \omega_0^2 / m^2}$$

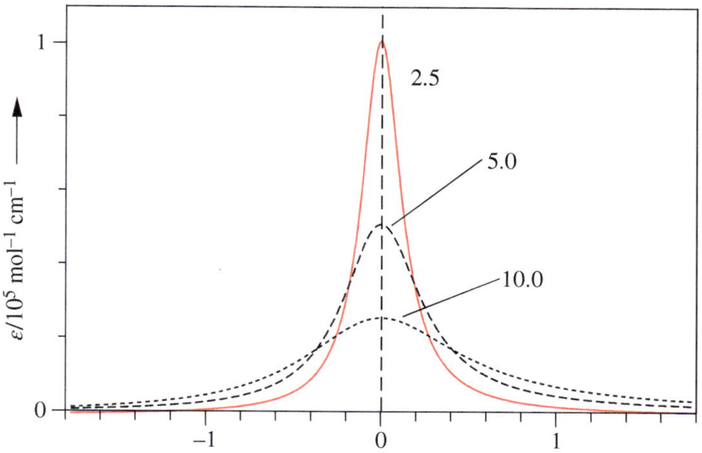

where we assumed that the resonance curve is so small that we can replace ω in the numerator and in the damping term in the denominator by ω_0. Then we obtain

$$2 \frac{k_d^2 \omega_0^2}{m^2} = \left(\omega_0^2 - \omega_{1/2}^2\right)^2 + k_d^2 \omega_0^2 / m^2$$

which rearranges to give

$$\frac{k_d^2 \omega_0^2}{m^2} = \left(\omega_0^2 - \omega_{1/2}^2\right)^2 = \left[\left(\omega_0 - \omega_{1/2}\right)\left(\omega_0 + \omega_{1/2}\right)\right]^2$$
$$= \left(\omega_0 - \omega_{1/2}\right)^2 \left(\omega_0 + \omega_{1/2}\right)^2$$

Figure 9F.1 Absorption coefficient ε versus $(\omega - \omega_0)$ in the classical oscillator model. Damping constants $k_d = 2.5$, 5.0, and 10.0×10^{-16} Nsm^{-1}; $\omega_0 = 3.77 \times 10^{15}$ s^{-1} (corresponding to $\lambda_{max} = 500$ nm). $Q = -e = -1.602 \times 10^{-19}$ C and $m = m_e = 9.109 \times 10^{-31}$ kg.

Using

$$\Delta\omega = \omega_0 - \omega_{1/2}$$

for small enough $\Delta\omega$ ($\Delta\omega \ll \omega_0$) we obtain

$$\frac{k_d^2 \omega_0^2}{m^2} = 4\omega_0^2 \Delta\omega^2$$

Hence we find that

$$\Delta\omega^2 = \frac{1}{4} \frac{k_d^2}{m^2} \quad \text{and} \quad \Delta\omega = \pm \frac{1}{2} \frac{k_d}{m} \qquad (9F1.13)$$

For example, with $k_d = 10 \times 10^{-16}$ Nsm^{-1} and $m = m_e = 9.109 \times 10^{-31}$ kg for the half width of the resonance curve we find $2\Delta\omega = 1 \times 10^{15}$ s^{-1} in agreement with the corresponding curve in Fig. 9F.1.

Area Under the Resonance Curve Although the maximum and the band width of the resonance curve depend on the damping constant k_d, the area under the curve is independent of k_d, however. This latter property can be proved by calculating the integral

$$\int_0^\infty \varepsilon \cdot d\omega = \frac{\mathbf{N}_A}{\ln 10 \cdot c_0 \varepsilon_0 \hat{n}} k_d \frac{Q^2}{m^2} \cdot \int_0^\infty \frac{\omega^2}{\left(\omega_0^2 - \omega^2\right)^2 + k_d^2 \omega^2 / m^2} \cdot d\omega \qquad (9F1.14)$$

and analyzing whether it depends on k_d.

Approximate Treatment

As a first step, we make an estimate by approximating the integral by the product of ε_{max} and the half-width of the absorption band, $2\Delta\omega$. Using Equations (9F1.11), (9F1.12), and (9F1.13) we find

$$\int_0^\infty \varepsilon \cdot d\omega \approx \varepsilon_{max} \cdot 2\Delta\omega = const \cdot k_d \frac{m^2}{k_d^2} \cdot \frac{k_d}{m}$$
$$= const \cdot \frac{1}{m}$$

where

$$const = \frac{\mathbf{N}_A}{\ln 10 \cdot c_0 \varepsilon_0 \hat{n}} \frac{Q^2}{m^2}$$

and for the integral we obtain

$$\int_0^\infty \varepsilon \cdot d\omega \approx \frac{\mathbf{N}_A}{\ln 10 \cdot c_0 \varepsilon_0 \hat{n}} \frac{Q^2}{m} \tag{9F1.15}$$

Rigorous Treatment

Now we perform the integration in Equation (9F1.14) rigorously and focus on the integral.

$$J = \int_{\omega=0}^\infty \frac{\omega^2}{\left(\omega_0^2 - \omega^2\right)^2 + k_d^2 \omega^2 / m^2} \cdot d\omega = \int_{\omega=0}^\infty \frac{\omega^2}{\left(\omega_0 - \omega\right)^2 \cdot \left(\omega_0 + \omega\right)^2 + k_d^2 \omega^2 / m^2} \cdot d\omega \tag{9F1.16}$$

From the form of the bandshape we see that most of the power absorption occurs near the resonance, so we replace the integral over the frequency ω by an integral over the deviation of the frequency from the resonant frequency. This variable change is given by $z = \omega - \omega_0$ and $dz = d\omega$. Performing this variable change, the integral becomes

$$J = \int_{z=-\omega_0}^\infty \frac{\left(\omega_0 + z\right)^2}{z^2 \cdot \left(2\omega_0 + z\right)^2 + k_d^2 \left(\omega_0 + z\right)^2 / m^2} \cdot dz$$

The integrand reaches a maximum for the resonance condition ($\omega_0 = \omega$ or $z = 0$):

$$\text{integrand } (z = 0) = \frac{\omega_0^2}{k_d^2 \omega_0^2 / m^2} = \frac{m^2}{k_d^2}$$

When z is significantly different from 0, the integrand becomes very small and does not contribute significantly to the integral (area under the curve). Consequently, $\omega_0 + z \approx \omega_0$ for all relevant values of z, and we can extend the integration limit of $-\omega_0$ to $-\infty$. With these approximations, the integral becomes

$$J = \int_{-\infty}^\infty \frac{1}{4z^2 + k_d^2 / m^2} \cdot dz = \frac{m^2}{k_d^2} \int_{-\infty}^\infty \frac{1}{(2mz/k_d)^2 + 1} \cdot dz \tag{9F1.17}$$

$$= -\frac{m}{2k_d} \cdot \left| \arctan \frac{2mz}{k_d} \right|_{-\infty}^\infty = \frac{m\pi}{2k_d}$$

Using this result for J, we find that

$$\int_0^\infty \varepsilon \cdot d\omega = \frac{\mathbf{N}_A}{\ln 10 \cdot c_0 \varepsilon_0 \hat{n}} k_d \frac{Q^2}{m^2} \cdot \frac{m\pi}{2k_d} = \frac{\mathbf{N}_A \, \pi}{2 \cdot \ln 10 \cdot c_0 \varepsilon_0 \hat{n}} \frac{Q^2}{m}$$

and does not depend on k_d. This result means that the integrated absorption is independent of the shape of the absorption band. Compared to our estimation in Equation (9F1.15) the exact solution has an additional factor of $\pi/2$.

Rather than measure the absorption coefficient as a function of angular frequency, it is commonly measured as a function of frequency. In this case we require the relation

$$\int_0^\infty \varepsilon \cdot d\nu = \frac{1}{2\pi} \int_0^\infty \varepsilon \cdot d\omega = \frac{\mathbf{N}_A}{4 \cdot \ln 10 \cdot c_0 \varepsilon_0 \hat{n}} \frac{Q^2}{m} \tag{9F1.18}$$

In the special case of a classical oscillator in vacuum with the charge and mass of an electron ($Q = -e$ and $m = m_e$), from Equation (9F1.18) we obtain

$$\left(\int_0^\infty \varepsilon \cdot d\nu \right)_{\text{classical } (-e,\, m_e),\text{vacuum}} = \frac{\mathbf{N}_A}{4 \cdot \ln 10 \cdot c_0 \varepsilon_0} \frac{e^2}{m_e} \tag{9F1.19}$$

9.8.2 Integrated Absorption Power

In Equation (9F1.9) we calculated the average power absorbed by one oscillator in an electric field oscillating with exactly one frequency ν.

$$\left(\frac{dE}{dt} \right)_\nu = \frac{1}{2} k_d \frac{Q^2}{m^2} F_0^2 \cdot \frac{\omega^2}{\left(\omega_0^2 - \omega^2\right)^2 + k_d^2 \omega^2 / m^2} \tag{9F1.20}$$

Now we calculate the average power absorbed by one oscillator in a distribution of electric fields oscillating at frequencies over the range from zero to infinity. To do this we have to sum up over the contributions of all frequencies. Instead of calculating the sum we consider the power absorbed in the frequency range dν and integrate over all frequencies. First, we replace F_0^2 by the radiant energy density ρ. According to Equation (9F1.28) (see below), we can write

$$F_0^2 = \frac{1}{3}\frac{2}{\varepsilon_0 \hat{n}^2}\rho \tag{9F1.21}$$

The factor $\frac{1}{3}$ accounts for the fact that in the case of an isotropic radiation distribution one-third of the radiant energy density contributes to the field in the x direction.

Now we consider the contribution of the electric field in a range of frequencies between ν and $\nu + \mathrm{d}\nu$; that is, we replace F_0^2 by $2/(3\varepsilon_0\hat{n}^2)\ \mathrm{d}\rho = 2/(3\varepsilon_0\hat{n}^2)\rho_\nu\ \mathrm{d}\nu$, where ρ_ν is the spectral radiant energy density. Then we perform the integration

$$\begin{aligned}\left(\frac{\overline{\mathrm{d}E}}{\mathrm{d}t}\right)_{\text{integrated}} &= \frac{1}{2}k_d\frac{Q^2}{m^2}\frac{2}{3\varepsilon_0\hat{n}^2}\rho_\nu \cdot \int_{\nu=0}^{\nu=\infty}\frac{\omega^2}{\left(\omega_0^2-\omega^2\right)^2 + k_d^2\omega^2/m^2}\mathrm{d}\nu \\ &= \frac{1}{2}k_d\frac{Q^2}{m^2}\frac{2}{3\varepsilon_0\hat{n}^2}\rho_\nu\frac{1}{2\pi} \cdot \int_{\omega=0}^{\omega=\infty}\frac{\omega^2}{\left(\omega_0^2-\omega^2\right)^2 + k_d^2\omega^2/m^2}\mathrm{d}\omega\end{aligned}$$

Here we have assumed that ρ_ν is constant within the frequency range of the absorption curve. The integral was found in Equation (9F1.17) as

$$\int_{\omega=0}^{\omega=\infty}\frac{\omega^2}{\left(\omega_0^2-\omega^2\right)^2 + k_d^2\omega^2/m^2}\mathrm{d}\omega = \frac{m\pi}{2k_d}$$

Thus for the integrated power we obtain

$$\left(\frac{\overline{\mathrm{d}E}}{\mathrm{d}t}\right)_{\text{integrated}} = \frac{1}{2}k_d\frac{Q^2}{m^2}\frac{2}{3\varepsilon_0\hat{n}^2}\rho_\nu\frac{1}{2\pi}\cdot\frac{m\pi}{2k_d} = \frac{Q^2}{m}\frac{1}{12\varepsilon_0\hat{n}^2}\rho_\nu \tag{9F1.22}$$

Note that this expression is independent of the damping constant k_d.

9.8.3 Derivation of Equation (9.77) for the Intensity I

According to Equation (9F1.2) the intensity I of a light wave is defined as

$$I = \frac{\text{mean power}}{A} = \frac{\overline{E}}{A \cdot t} \tag{9F1.23}$$

where $\overline{E}$ is the mean energy of the electric field of the light wave passing the cross section A in time t. $\overline{E}$ is related to the energy density ρ

$$\overline{E} = \rho \cdot A \cdot c \cdot t \tag{9F1.24}$$

where $c = c_0/\hat{n}$ is the speed of light in the medium of refractive index $\hat{n}$, thus

$$I = \rho\frac{c_0}{\hat{n}} \tag{9F1.25}$$

In what way is ρ related to the electric field strength F? Let us consider a parallel plate capacitor with the plates (area a) at distance d and charged to the voltage U. The electric field inside the plates is $F = U/d$, and the capacity is $C = \varepsilon_0\hat{\varepsilon}a/d$. The energy of the electric field is

$$\begin{aligned}E(U_1) &= \int_0^{U_1} Q\ \mathrm{d}U = \int_0^{U_1} CU\ \mathrm{d}U = C\int_0^{U_1} U\ \mathrm{d}U \\ &= C\frac{1}{2}U_1^2 = \frac{1}{2}C \cdot U_1^2\end{aligned}$$

Thus, omitting the index 1, we obtain

$$\rho = \frac{\frac{1}{2}CU^2}{ad} = \frac{\varepsilon_0\hat{\varepsilon}aU^2}{2ad^2} = \frac{\varepsilon_0\hat{\varepsilon}}{2}F^2 \tag{9F1.26}$$

In the case of a molecule placed in a radiation beam, $F(x,t)^2 = \left(F_0 \sin \omega t\right)^2$ is replaced by the mean value $\overline{F^2}$

$$\rho = \frac{\varepsilon_0 \hat{\varepsilon}}{2} \overline{F^2}$$

In a radiation beam the energy density is composed of an electric field component and an equally large magnetic component, thus

$$\rho = 2 \cdot \frac{\varepsilon_0 \cdot \hat{\varepsilon}}{2} \overline{F^2} = \varepsilon_0 \cdot \hat{\varepsilon} \cdot \overline{F^2}$$

For $\overline{F^2}$ we obtain

$$\overline{F^2} = \overline{\left(F_0 \cdot \cos 2\pi \nu t\right)^2} = \frac{1}{2} F_0^2 \tag{9F1.27}$$

where F_0 is the amplitude of the electric field. By using

$$\rho = \frac{1}{2} \varepsilon_0 \cdot \hat{\varepsilon} \cdot F_0^2$$

and the relation $\hat{\varepsilon} = \hat{n}^2$ we can write

$$\rho = \frac{1}{2} \varepsilon_0 \cdot \hat{n}^2 \cdot F_0^2 \tag{9F1.28}$$

and from Equation (9F1.25) it follows that

$$I = \frac{1}{2} \varepsilon_0 \cdot \hat{n}^2 F_0^2 \cdot \frac{c_0}{\hat{n}} = \frac{1}{2} \varepsilon_0 \cdot \hat{n} \cdot c_0 \cdot F_0^2 \tag{9F1.29}$$

if the light is polarized. This result is identical with Equation (9F1.4). If the light is unpolarized the result is

$$I = \varepsilon_0 \cdot \hat{n} \cdot c_0 \cdot F_0^2 \tag{9F1.30}$$

9.8.4 Foundation 9.2: Oscillator Strength in the Quantum Mechanical Treatment

9.8.4.1 Quantum Mechanical Expression for *f*

The time-dependent Schrödinger Equation for an electron in the field of the molecular skeleton and in the electric field of the light wave is, according to Equation (3.76),

$$\mathcal{H}\Psi = i\hbar \frac{\partial \Psi}{\partial t}$$

with

$$\mathcal{H} = \mathcal{H}_0 + e \cdot x \cdot F_0 \cdot \cos \omega t$$

$\mathcal{H}_0$ corresponds to the Hamiltonian operator (Equations (3.27), (3.28)) in the time-independent Schrödinger Equation (3.30). F_0 is the amplitude of the electric field of the incident light wave with frequency ν ($\omega = 2\pi\nu$), and $e \cdot x \cdot F_0 \cdot \cos \omega t$ is the potential energy of an electron in this field at time t in the x-direction, the direction of the transition moment of the molecule. From this time-dependent Schrödinger Equation we obtain the time-dependent part of the dipole moment of the molecule in the radiation field.

The time-dependent dipole moment can be modeled by considering the elongation of a classical oscillator (force constant k_f, mass m, charge Q) in the radiation field (Foundation 9.1). As we show in Foundation 9.3 (on the book's website) the quantum mechanical description of the interaction is equivalent to the classical description, if we set

$$\nu_0 = \frac{1}{2\pi} \sqrt{\frac{k_f}{m}} = \frac{\Delta E}{h} \quad \text{or} \quad \frac{k_f}{m} = \frac{4\pi^2 (\Delta E)^2}{h^2} \tag{9F2.1}$$

and

$$\frac{Q^2}{m} = \frac{8\pi^2}{h^2} \Delta E \cdot |M_x|^2 \tag{9F2.2}$$

where $\Delta E = E_B - E_A$ is the excitation energy from state A to state B and $M_x = e \cdot \int \psi_A \cdot x \cdot \psi_B \, d\tau$ is the x component of the transition moment M.

To connect with our discussion in Section 9.2.3, we begin by writing the oscillator strength f for a classical oscillator with charge Q and mass m in a medium with refractive index $\hat{n}$ is

$$f = \frac{\left(\int_{\text{band}} \varepsilon \cdot \mathrm{d}\nu\right)_{\text{classical }(Q,m),\text{medium}}}{\left(\int_{\text{band}} \varepsilon \cdot \mathrm{d}\nu\right)_{\text{classical }(-e,m_e),\text{vacuum}}} \tag{9F2.3}$$

with

$$\left(\int_0^\infty \varepsilon \cdot \mathrm{d}\nu\right)_{\text{classical }(Q,m),\text{medium}} = \frac{\mathbf{N}_A}{4 \cdot \ln 10 \cdot c_0 \varepsilon_0 \hat{n}} \frac{Q^2}{m} \tag{9F2.4}$$

(see Equation (9F1.18) in Foundation 9.1) and

$$\left(\int_0^\infty \varepsilon \cdot \mathrm{d}\nu\right)_{\text{classical }(-e,m_e),\text{vacuum}} = \frac{\mathbf{N}_A}{4 \cdot \ln 10 \cdot c_0 \varepsilon_0} \frac{e^2}{m_e} \tag{9F2.5}$$

(see Equation (9F1.19) in Foundation 9.1). Then

$$f = \frac{1}{\hat{n}} \cdot \frac{Q^2}{m} \cdot \frac{m_e}{e^2} \tag{9F2.6}$$

If we use the quantum mechanical expression (9F2.2) for Q^2/m, it follows that

$$f = \frac{8\pi^2 m_e}{h^2 e^2 \hat{n}} \cdot \Delta E \cdot |M_x|^2 \tag{9F2.7}$$

This consideration assumes that the molecule and the corresponding classical oscillator are aligned parallel to the electric field in the x direction. If we consider a statistically random distribution of molecular orientations in a solution, the absorption coefficient ε will be smaller. To simplify the consideration, we assume that one-third of the molecules have their transition dipole oriented in the direction of a linear polarized light field, and two-thirds of the molecules have their transition dipole oriented perpendicular to the light field. This assumption can be justified in analogy to the treatment of the motion of gas molecules, see Section 14.2.1. Hence, the calculated oscillator strength will decrease by a factor of three.

$$f = \frac{1}{3} \frac{8\pi^2 m_e}{h^2 e^2 \hat{n}} \cdot \Delta E \cdot |M_x|^2 \quad \text{(molecules statistically distributed)} \tag{9F2.8}$$

In the general case, the molecule can have additional components of the transition moment M_y and M_z in the y and z directions contributing to the oscillatory strength f. Then we obtain

$$f = \frac{1}{3} \frac{8\pi^2 m_e}{h^2 e^2 \hat{n}} \cdot \Delta E \cdot |M|^2 \quad \text{(molecules statistically distributed)} \tag{9F2.9}$$

with

$$|M|^2 = |M_x|^2 + |M_y|^2 + |M_z|^2 \tag{9F2.10}$$

Example 9.6 *Use the FEMO model to find an expression for the oscillator strength between two states i and j of a cyanine dye molecule. You may consider the transition moment to be oriented along the x coordinate in the molecular frame.*

As an example, we calculate the transition moment for a transition from orbital i to orbital j for a linear π-electron system with its molecular skeleton oriented in the x-direction:

$$M_x = e\frac{2}{L}\int_0^L \sin\left(\frac{\pi x \cdot i}{L}\right) \cdot x \cdot \sin\left(\frac{\pi x \cdot j}{L}\right) \cdot \mathrm{d}x$$

If we make the substitution $\xi = \pi x/L$, we find from Appendix B that

$$\begin{aligned} M_x &= e\frac{2}{L}\frac{L}{\pi}\frac{L}{\pi}\int_0^\pi \sin(\xi i) \cdot \xi \cdot \sin(\xi j) \cdot \mathrm{d}\xi \\ &= e\frac{2L}{\pi^2} \cdot \left[-\frac{1}{(i-j)^2} + \frac{1}{(i+j)^2}\right] \end{aligned}$$

for i even and j odd, or i odd and j even; otherwise, the transition moment is zero.

For the HOMO–LUMO transition ($j = i + 1$) we find that

$$M_x = e\frac{2L}{\pi^2} \cdot \left[-1 + \frac{1}{(2i+1)^2}\right] \approx -e\frac{2L}{\pi^2}$$

The excitation energy is (N = number of π electrons)

$$\Delta E = \frac{h^2}{8m_e L^2} \cdot (N+1)$$

and we obtain the following for the oscillator strength f:

$$f = \frac{1}{3}\frac{8\pi^2 m_e}{h^2 e^2 \hat{n}} \cdot \frac{h^2}{8m_e L^2} \cdot (N+1) \cdot e^2 \frac{4L^2}{\pi^4} = \frac{4}{3\pi^2 \hat{n}} \cdot (N+1)$$

This is the oscillator strength calculated for one absorbing electron. In the cyanine dyes the HOMO is occupied by two electrons, then

$$f = \frac{8}{3\pi^2 \hat{n}} \cdot (N+1)$$

We can easily account for the fact that the cyanine is a zigzag chain with an angle of 30° relative to the x-direction. Then the coordinate s along the chain is connected with x by

$$x = s \cdot \cos 30° = \frac{1}{2}\sqrt{3} \cdot s$$

and the oscillator strength is somewhat smaller:

$$f = \frac{8}{3\pi^2 \hat{n}} \cdot \frac{3}{4} \cdot (N+1) = \frac{2}{\pi^2 \hat{n}}(N+1)$$

9.8.5 Foundation 9.3: Classical and Quantum Mechanical Description of Light Absorption

9.8.5.1 Light Absorption Calculated from the Time-Dependent Schrödinger Equation

Let us perform a "thought experiment" for the excitation of a molecule with light. Initially, in the absence of light the molecule is in its ground state (wavefunction ψ_A). Imagine that we shine light on the molecule for a time t, and thereafter the light is switched off. Afterward the molecule can be found in the excited state (wavefunction ψ_B) with some probability.

We can describe the process by using the time-dependent wavefunction Ψ. We write Ψ as a linear combination of the wavefunctions ψ_A and ψ_B,

$$\Psi = \psi_A \cdot T_A + \psi_B \cdot T_B \tag{9F3.1}$$

where T_A and T_B are time-dependent coefficients. From our description of the experiment, we know that $T_A(t = 0) = 1$ and $T_B(t = 0) = 0$. We insert this wavefunction into the time-dependent Schrödinger equation for a molecule in the electric field of a light wave, namely

$$i\hbar\frac{\partial \Psi}{\partial t} = \mathcal{H}\Psi = \mathcal{H}_0\Psi + exF_x^0 \cdot \cos(\omega t) \cdot \Psi \tag{9F3.2}$$

Note that the time-independent functions ψ_A and ψ_B can be considered as real functions, but the time-dependent function Ψ and the functions T_A and T_A are complex (they are solutions of the time-dependent Schrödinger equation).

Upon substitution for Ψ, we find that

$$i\hbar\left(\psi_A\frac{\mathrm{d}T_A}{\mathrm{d}t} + \psi_B\frac{\mathrm{d}T_B}{\mathrm{d}t}\right) = T_A E_A \psi_A + T_B E_B \psi_B + exF_x^0 \cdot \cos(\omega t) \cdot (\psi_A \cdot T_A + \psi_B \cdot T_B)$$

For small enough intensity of the light wave, $\psi_A \cdot T_A \gg \psi_B \cdot T_B$. Hence we can neglect $\psi_B \cdot T_B$ in the term for the light field since it is a product of two small quantities $\psi_B \cdot T_B$ and F_x^0. This approximation gives

$$i\hbar\left(\psi_A\frac{\mathrm{d}T_A}{\mathrm{d}t} + \psi_B\frac{\mathrm{d}T_B}{\mathrm{d}t}\right) = T_A E_A \psi_A + T_B E_B \psi_B + exF_x^0 \cdot \cos(\omega t) \cdot \psi_A \cdot T_A$$

Now we multiply both sides of the equation by $\psi_A \,\mathrm{d}\tau$ and integrate. The integrals $\int \psi_B \psi_A \,\mathrm{d}\tau$ are zero because the wavefunctions are orthogonal, and we find

$$i\hbar\left(\frac{\mathrm{d}T_A}{\mathrm{d}t}\right) = T_A E_A + F_x^0 \cdot \cos(\omega t) \cdot T_A \int \psi_A ex \psi_A \,\mathrm{d}\tau$$

Correspondingly, for multiplication by $\psi_B\,\mathrm{d}\tau$ and integration, we obtain

$$i\hbar\left(\frac{\mathrm{d}T_B}{\mathrm{d}t}\right) = T_B E_B + F_x^0 \cdot \cos(\omega t) \cdot T_A \int \psi_A ex \psi_B\, d\tau$$

Rather than work with two differential equations, we find a single equation for the ratio T_B/T_A. If we divide both sides by T_A, we find

$$i\hbar\left(\frac{\mathrm{d}T_A/\mathrm{d}t}{T_A}\right) = E_A + F_x^0 \cdot \cos(\omega t) \cdot \int \psi_A ex \psi_A\, \mathrm{d}\tau \tag{9F3.3}$$

and

$$i\hbar\left(\frac{\mathrm{d}T_B/\mathrm{d}t}{T_A}\right) = \frac{T_B}{T_A}E_B + F_x^0 \cdot \cos(\omega t) \cdot M_x \tag{9F3.4}$$

where we introduce the abbreviation

$$M_x = \int \psi_A ex \psi_B\, \mathrm{d}\tau$$

From differential calculus, we recognize that the following relation is true

$$\frac{\mathrm{d}}{\mathrm{d}t}\left(\frac{T_B}{T_A}\right) = \frac{\mathrm{d}T_B/\mathrm{d}t \cdot T_A - \mathrm{d}T_A/\mathrm{d}t \cdot T_B}{T_A^2} = \frac{\mathrm{d}T_B/\mathrm{d}t}{T_A} - \frac{\mathrm{d}T_A/\mathrm{d}t}{T_A}\frac{T_B}{T_A}$$

Rearranging this result, we can write

$$\frac{\mathrm{d}T_B/\mathrm{d}t}{T_A} = \frac{\mathrm{d}}{\mathrm{d}t}\left(\frac{T_B}{T_A}\right) + \frac{\mathrm{d}T_A/\mathrm{d}t}{T_A}\frac{T_B}{T_A}$$

Combining this with Equation (9F3.4) gives

$$i\hbar\frac{\mathrm{d}}{\mathrm{d}t}\left(\frac{T_B}{T_A}\right) + i\hbar\frac{\mathrm{d}T_A/\mathrm{d}t}{T_A}\frac{T_B}{T_A} = \frac{T_B}{T_A}E_B + F_x^0 \cdot \cos(\omega t) \cdot M_x$$

Together with Equation (9F3.3)

$$i\hbar\frac{\mathrm{d}}{\mathrm{d}t}\left(\frac{T_B}{T_A}\right) - \frac{T_B}{T_A}\left[E_B - E_A + F_x^0 \cdot \cos(\omega t) \cdot \int \psi_A ex \psi_A \mathrm{d}\tau\right] = F_x^0 \cdot \cos(\omega t) \cdot M_x \tag{9F3.5}$$

We introduce the abbreviations

$$B = \frac{T_B}{T_A} \quad \text{and} \quad \Delta E = E_B - E_A \tag{9F3.6}$$

Furthermore, for low enough intensity of the light wave the expression in the square brackets equals ΔE. Then we obtain from (9F3.5)

$$i\hbar\frac{\mathrm{d}B}{\mathrm{d}t} - \Delta E \cdot B = F_x^0 \cdot \cos(\omega t) \cdot M_x \tag{9F3.7}$$

Equation (9F3.7) is a differential equation for the complex and time-dependent function B. We wish to obtain an equation for a real function, so we split B into its real and imaginary terms. Hence we define

$$B = b + i \cdot b' \tag{9F3.8}$$

and insert it into the differential equation for B (i is the $\sqrt{-1}$)

$$i\hbar\frac{\mathrm{d}b}{\mathrm{d}t} - \hbar\frac{\mathrm{d}b'}{\mathrm{d}t} - \Delta E \cdot b - \Delta E \cdot i \cdot b' = F_x^0 \cdot \cos(\omega t) \cdot M_x$$

We separate the real and imaginary parts in this equation

$$i\left[\hbar\frac{\mathrm{d}b}{\mathrm{d}t} - \Delta E \cdot b'\right] - \left[\hbar\frac{\mathrm{d}b'}{\mathrm{d}t} + \Delta E \cdot b + F_x^0 \cdot \cos(\omega t) \cdot M_x\right] = 0$$

This equation is valid only if the real and imaginary terms are each equal to zero

$$\hbar\frac{\mathrm{d}b}{\mathrm{d}t} - \Delta E \cdot b' = 0 \tag{9F3.9}$$

$$\hbar\frac{\mathrm{d}b'}{\mathrm{d}t} + \Delta E \cdot b + F_x^0 \cdot \cos(\omega t) \cdot M_x = 0 \tag{9F3.10}$$

We solve Equation (9F3.9) for b'

$$b' = \frac{\mathrm{d}b}{\mathrm{d}t}\frac{\hbar}{\Delta E} \tag{9F3.11}$$

and insert into Equation (9F3.10)

$$\frac{\mathrm{d}^2 b}{\mathrm{d}t^2} + \frac{\Delta E^2}{\hbar^2} \cdot b = -\frac{\Delta E}{\hbar^2} \cdot F_x^0 \cdot \cos(\omega t) \cdot M_x \tag{9F3.12}$$

This is a second-order differential equation for b.

We note that the differential equation for b (which is the real part of B) has the same structure as that for a classical oscillator. In this analogy, b corresponds to the elongation of the oscillator, and b' (which is the imaginary part of B) is proportional to the speed of the oscillator, according to Equation (9F3.11).

9.8.5.2 Power Absorbed by the Molecule

Now we will relate b to the absorbed power. The time-dependent energy of the system is

$$W(t) = \int \Psi \mathcal{H}_0 \Psi^* \mathrm{d}\tau$$

and thus for the power $\mathrm{d}W/\mathrm{d}t$ we obtain

$$\frac{\mathrm{d}W}{\mathrm{d}t} = \int \frac{\partial \Psi}{\partial t} \mathcal{H}_0 \Psi^* \, \mathrm{d}\tau + \int \Psi \mathcal{H}_0 \frac{\partial \Psi^*}{\partial t} \, \mathrm{d}\tau$$

Because of

$$\int \Psi \mathcal{H}_0 \Psi^* \, \mathrm{d}\tau = \int \Psi^* \mathcal{H}_0 \Psi \, \mathrm{d}\tau$$

it follows

$$\frac{\mathrm{d}W}{\mathrm{d}t} = \int \frac{\partial \Psi}{\partial t} \mathcal{H}_0 \Psi^* \, \mathrm{d}\tau + \int \frac{\partial \Psi^*}{\partial t} \mathcal{H}_0 \Psi \, \mathrm{d}\tau$$

From Equation (9F3.2) we obtain

$$\mathcal{H}_0 \Psi = i\hbar \frac{\partial \Psi}{\partial t} - exF_x^0 \cdot \cos(\omega t) \cdot \Psi$$

We multiply by $(\partial \Psi^* / \partial t)\, \mathrm{d}\tau$ and integrate over all space

$$\int \frac{\partial \Psi^*}{\partial t} \mathcal{H}_0 \Psi \, \mathrm{d}\tau = i\hbar \int \frac{\partial \Psi^*}{\partial t} \frac{\partial \Psi}{\partial t} \mathrm{d}\tau - \int \frac{\partial \Psi^*}{\partial t} \, exF_x^0 \cdot \cos(\omega t) \cdot \Psi^* \, \mathrm{d}\tau$$

Correspondingly, the result for the conjugate complex expression is

$$\int \frac{\partial \Psi}{\partial t} \mathcal{H}_0 \Psi^* \mathrm{d}\tau = -i\hbar \int \frac{\partial \Psi}{\partial t} \frac{\partial \Psi^*}{\partial t} \mathrm{d}\tau - \int \frac{\partial \Psi}{\partial t} \, exF_x^0 \cdot \cos(\omega t) \cdot \Psi \, \mathrm{d}\tau$$

Thus the power is

$$\begin{aligned} \frac{\mathrm{d}W}{\mathrm{d}t} &= -\int \frac{\partial \Psi}{\partial t} \, exF_x^0 \cdot \cos(\omega t) \cdot \Psi^* \, \mathrm{d}\tau - \int \frac{\partial \Psi^*}{\partial t} exF_x^0 \cdot \cos(\omega t) \cdot \Psi \, \mathrm{d}\tau \\ &= -\int \frac{\partial (\Psi \Psi^*)}{\partial t} \, exF_x^0 \cdot \cos(\omega t) \, \mathrm{d}\tau \end{aligned}$$

or

$$\frac{\mathrm{d}W}{\mathrm{d}t} = -F_x^0 \cdot \cos(\omega t) \cdot \int \frac{\partial (\Psi \Psi^*)}{\partial t} \, ex \, \mathrm{d}\tau \tag{9F3.13}$$

Now we insert the expression (9F3.1) for Ψ and calculate the product $\Psi\Psi^*$

$$\Psi \Psi^* = (\psi_A \cdot T_A + \psi_B \cdot T_B)(\psi_A \cdot T_A^* + \psi_B \cdot T_B^*)$$

If the intensity of the light wave is sufficiently small we can set $T_A = 1$, and T_B can be treated as a small quantity. Then for $\Psi\Psi^*$we obtain, together with Equation (9F3.6)

$$\Psi \Psi^* = \psi_A^2 + (T_B + T_B^*) \psi_A \psi_B = \psi_A^2 + (B + B^*) \psi_A \psi_B$$

Furthermore, from Equation (9F3.8) it follows that

$$B = b + ib' \; B^* = b - ib'$$

and we obtain

$$\Psi\Psi^* = \psi_A^2 + 2b \cdot \psi_A \psi_B \quad \text{and} \quad \frac{\partial\left(\Psi\Psi^*\right)}{\partial t} = 2\frac{\mathrm{d}b}{\mathrm{d}t} \cdot \psi_A \psi_B$$

Thus for the time-dependent power absorbed by the molecule we obtain from Equation (9F3.13) with the abbreviation $M_x = \int \psi_A e x \psi_B \, \mathrm{d}\tau$,

$$\frac{\mathrm{d}W}{\mathrm{d}t} = -2 \cdot F_x^0 \cdot \cos(\omega t) \cdot M_x \cdot \frac{\mathrm{d}b}{\mathrm{d}t} \tag{9F3.14}$$

9.8.5.3 Classical Oscillator

The differential equation (9F3.12) corresponds to the equation for the classical oscillator in the electric field of a light wave discussed in Foundation 9.1, if the damping constant k_d is set to zero and the field strength F_0 is replaced by F_x^0.

$$\frac{\mathrm{d}^2\xi}{\mathrm{d}t^2} + \frac{k_f}{m} \cdot \xi = -\frac{Q}{m} \cdot F_x^0 \cdot \cos(\omega t) \tag{9F3.15}$$

We calculate the power absorbed by the classical oscillator in the electric field of the light wave along the oscillator direction. The energy $\mathrm{d}W$ to move the charge Q in the direction of the electric field by the distance $\mathrm{d}\xi$ is the applied force times the displacement $\mathrm{d}\xi$

$$\mathrm{d}W = QF_x^0 \cdot \cos(\omega t) \cdot \mathrm{d}\xi$$

Thus for the absorbed power we obtain

$$\frac{\mathrm{d}W}{\mathrm{d}t} = QF_x^0 \cdot \cos(\omega t) \cdot \frac{\mathrm{d}\xi}{\mathrm{d}t} \tag{9F3.16}$$

9.8.5.4 Equivalence of Classical and Quantum Mechanical Treatment

In Equation (9F3.12) we found for the quantum mechanical treatment that

$$\frac{\mathrm{d}^2 b}{\mathrm{d}t^2} + \frac{\Delta E^2}{\hbar^2} \cdot b = -\frac{\Delta E}{\hbar^2} \cdot F_x^0 \cdot \cos(\omega t) \cdot M_x$$

In order to compare this result to the classical solution quantitatively we set $\xi = \alpha \cdot b$ and insert the result into Equation (9F3.15) to obtain

$$\frac{\mathrm{d}^2 b}{\mathrm{d}t^2} + \frac{k_f}{m} \cdot b = -\frac{1}{\alpha} \cdot \frac{Q}{m} \cdot F_x^0 \cdot \cos(\omega t) \tag{9F3.17}$$

Furthermore, we require both systems to absorb the same power, i.e., have the same absorption probability, so that

$$\alpha \cdot QF_0 \cdot \cos(\omega t) \cdot \frac{\mathrm{d}b}{\mathrm{d}t} = -2 \cdot F_x^0 \cdot \cos(\omega t) \cdot M_x \cdot \frac{\mathrm{d}b}{\mathrm{d}t}$$

which follows from Equations (9F3.14) and (9F3.16). Now we solve this equation for α and obtain

$$\alpha = -2M_x/Q$$

We insert this expression in Equation (9F3.17) and obtain

$$\frac{\mathrm{d}^2 b}{\mathrm{d}t^2} + \frac{k_f}{m} \cdot b = -\frac{Q}{2M_x} \cdot \frac{Q}{m} \cdot F_x^0 \cdot \cos(\omega t)$$

This equation becomes identical with the differential Equation (9F3.12) if we set

$$\boxed{\frac{k_f}{m} = \frac{\Delta E^2}{\hbar^2}, \qquad \frac{Q^2}{m} = 2\frac{M_x^2 \Delta E}{\hbar^2}} \tag{9F3.18}$$

We can understand the first relation in Equation (9F3.18) if we write the energy change ΔE as

$$\Delta E = h\nu_0$$

where ν_0 is the resonance frequency of a corresponding classical oscillator. Using

$$\nu_0 = \frac{1}{2\pi}\sqrt{\frac{k_f}{m}}$$

we find

$$\Delta E = h\frac{1}{2\pi}\sqrt{\frac{k_f}{m}} = \hbar\sqrt{\frac{k_f}{m}}$$

and

$$\frac{k_f}{m} = \left(\frac{\Delta E}{\hbar}\right)^2 = \frac{\Delta E^2}{\hbar^2}$$

Regarding the second relation in Equation (9F3.18), for a classical oscillator with the charge and the mass of an electron ($Q = e$ and $m = m_e$) we find

$$\frac{Q^2}{m} = \frac{e^2}{m_e} = 2.8 \times 10^{-8}\ \mathrm{C^2\ kg^{-1}}$$

The corresponding quantum mechanical result depends on the transition moment and on the excitation energy. This is the basis of the selection rules in spectroscopy. For reasons of symmetry, the transition moment can be zero; no transition occurs at all. Thus we can consider $|M_x|^2$ as a measure of the probability for a transition to occur. ΔE is the energy transferred from the light wave to the molecule by one photon; therefore, the power of absorption increases with ΔE.

For example, for an ethene molecule with two carbon atoms (transition moment $M_{x,1\to 2} = 1.12 \times 10^{-29}$ C m, excitation energy $\Delta E_{1\to 2} = 10.3 \times 10^{-19}$ J) the quantum mechanical result is

$$\frac{Q^2}{m} = 2\frac{|M_x|^2 \Delta E}{\hbar^2} = 2.2 \times 10^{-8}\ \mathrm{C^2\ kg^{-1}}$$

This value is on the same order as for the classical oscillator. However, for different classes of molecules the transition moment can differ over a wide range [13].

9.8.6 Foundation 9.4: Coupling Transitions with Parallel Transition Moments

9.8.6.1 Replacing Quantum Mechanical System by Coupled Classical Oscillators

We consider two π electrons giving rise to transitions polarized in the x-direction and assume that the incident light is polarized in the same direction. As in Foundation 9.2 we consider the time-dependent Schrödinger Equation for electron 1

$$\mathcal{H}_1\Psi_1 = i\hbar\frac{\partial\Psi_1}{\partial t}$$

and

$$\mathcal{H}_1 = \mathcal{H}_{10} + ex_1F_0 \cdot \cos(\omega t) + \mathcal{H}_{12}$$

where

$$\mathcal{H}_{12} = \int g(1,2) \cdot \Psi_2\Psi_2^* \cdot d\tau_2$$

$\mathcal{H}_{10}$ relates, as in Foundation 9.2, to the time-independent Schrödinger Equation for electron 1 in the field of the molecular skeleton. Correspondingly, $ex_1F_0 \cdot \cos(\omega t)$ is the potential energy of electron 1 in the radiation field. Ψ_2 is the time-dependent wavefunction of electron 2.

In addition to the case considered in Foundation 9.2, we have to include the repulsion energy of electron 1 in the field of electron 2. This corresponds to the term $\mathcal{H}_{12}$, where $\rho_2 = \Psi_2\Psi_2^*$ is the time-dependent probability density distribution of electron 2, and $g(1,2)$ is the repulsion energy of electrons 1 and 2 depending on the coordinates x_1 and x_2. Then $\mathcal{H}_{12}$ is the repulsion energy for a given coordinate x_1 of electron 1, averaged over the distribution of electron 2.

A corresponding time-dependent Schrödinger Equation holds for electron 2:

$$\mathcal{H}_2\Psi_2 = i\hbar\frac{\partial\Psi_2}{\partial t}$$

and

$$\mathcal{H}_2 = \mathcal{H}_{20} + ex_2F_0 \cdot \cos(\omega t) + \mathcal{H}_{21}$$

where

$$\mathcal{H}_{21} = \int g(1,2) \cdot \rho_1 \cdot d\tau_1$$

To obtain the time-dependent part of the dipole moment of the molecule in the radiation field we must solve these two coupled Schrödinger Equations. The time-dependent dipole moment can be modeled by considering the displacements ξ_1 and ξ_2 of two coupled classical oscillators (force constants $k_{f,1}$ and $k_{f,2}$, mass m, charges Q_1 and Q_2, damping constants $k_{d,1}$ and $k_{d,2}$) in the radiation field:

$$m \cdot \frac{d^2\xi_1}{dt^2} = -k_{f,1} \cdot \xi_1 - k_{d,1} \cdot \frac{d\xi_1}{dt} + Q_1 \cdot F_0 \cdot \cos(\omega t) - k_{f,12}\,\xi_2$$

$$m \cdot \frac{d^2\xi_2}{dt^2} = -k_{f,2} \cdot \xi_2 - k_{d,2} \cdot \frac{d\xi_2}{dt} + Q_2 \cdot F_0 \cdot \cos(\omega t) - k_{f,21}\,\xi_1$$

The coupling between the two oscillators with the frequencies

$$\nu_{10} = \frac{1}{2\pi}\sqrt{\frac{k_{f,1}}{m}} \quad \text{and} \quad \nu_{20} = \frac{1}{2\pi}\sqrt{\frac{k_{f,2}}{m}}$$

is included in the terms $k_{f,12}\xi_2$ and $k_{f,21}\xi_1$.

A detailed consideration [14] shows that the formal identity with the quantum mechanical treatment is obtained by the following relations (see Foundation 9.3):

$$\frac{k_{f,1}}{m} = \frac{4\pi^2}{h^2} \cdot \Delta E_1^2 \quad \text{and} \quad \frac{k_{f,2}}{m} = \frac{4\pi^2}{h^2} \cdot \Delta E_2^2 \tag{9F4.1}$$

where ΔE_1 and ΔE_2 are the transition energies for the two oscillators;

$$\frac{k_{f,12}}{m} = \frac{k_{f,21}}{m} = \frac{8\pi^2}{h^2} \cdot \sqrt{\Delta E_1 \cdot \Delta E_2} \cdot J_{12} \tag{9F4.2}$$

where J_{12} is the electron–electron interaction energy of the transition dipoles

$$J_{12} = \int \psi_{i,1}\psi_{j,1} \cdot g(1,2) \cdot \psi_{i,2}\psi_{j,2} \cdot d\tau_1 d\tau_2 \tag{9F4.3}$$

and

$$\frac{Q_1^2}{m} = \frac{8\pi^2}{h^2}\Delta E_1 \cdot |M_1|^2 \quad \text{and} \quad \frac{Q_2^2}{m} = \frac{8\pi^2}{h^2}\Delta E_2 \cdot |M_2|^2$$

The coupling of the two electrons causes a shift of the resonance frequencies of the two oscillators and a change in the oscillator strength of the two transitions. The uncoupled oscillators would be in resonance with the electric field of a light wave with frequency ν for $\nu = \nu_{10}$ and $\nu = \nu_{20}$, respectively. The system of two coupled oscillators, however, is in resonance at somewhat different frequencies ν_1 and ν_2, where ν_1 corresponds to an in-phase oscillation and ν_2 corresponds to an out-of-phase oscillation of both oscillators (Fig. 9.22). The calculation of these frequencies from the coupled differential equations is described in detail in Foundation 9.5.

In the out-of-phase oscillation, oscillator 1 is accelerated in the electric field of the light wave when oscillator 2 is retarded, and vice versa. As a consequence, the power absorbed from the electric field at ν_2 is small; the power is zero in the geometry of Fig. 9.22 (two identical oscillators). On the other hand, in the in-phase oscillation both oscillator 1 and oscillator 2 are accelerated in the electric field of the light wave, and the interaction is strong.

9.8.6.2 Doubly Occupied Orbitals

The equations given above are valid for transitions from singly occupied orbitals. For transitions from doubly occupied orbitals to empty orbitals the right side of the equations for Q_1, Q_2, and $k_{f,12}$ must be multiplied by a factor of 2:

$$\frac{Q_1^2}{m} = \frac{16\pi^2}{h^2}\Delta E_1 \cdot |M_1|^2 \quad \text{and} \quad \frac{Q_2^2}{m} = \frac{16\pi^2}{h^2}\Delta E_2 \cdot |M_2|^2$$

$$\frac{k_{f,12}}{m} = \frac{16\pi^2}{h^2} \cdot \sqrt{\Delta E_1 \cdot \Delta E_2} \cdot J_{12}$$

For the coupling of two electrons (each HOMO occupied with one electron) the shift is J_{12}, and for four electrons (each HOMO occupied with two electrons) it is $2 \cdot J_{12}$.

9.8.7 Foundation 9.5: Normal Modes of Coupled Oscillators

9.8.7.1 Resonance Frequencies of Two Coupled Oscillators

We consider the case of two coupled oscillators. In the absence of the field of a light wave and for sufficiently small damping constants k_d we obtain

$$m\frac{d^2\xi_1}{dt^2} + k_{f,1} \cdot \xi_1 + k_{f,12} \cdot \xi_2 = 0$$

$$m\frac{d^2\xi_2}{dt^2} + k_{f,2} \cdot \xi_2 + k_{f,12} \cdot \xi_1 = 0$$

With the assumption that both oscillators have the same phase factor α

$$\xi_1 = \gamma_1 \cdot \sin(\omega_0 t + \alpha) \quad \text{and} \quad \xi_2 = \gamma_2 \cdot \sin(\omega_0 t + \alpha)$$

the two differential equations reduce to a system of two linear homogeneous equations:

$$\gamma_1 \cdot (-m\omega_0^2 + k_{f,1}) + \gamma_2 \cdot k_{f,12} = 0$$

$$\gamma_1 \cdot k_{f,12} + \gamma_2 \cdot (-m\omega_0^2 + k_{f,2}) = 0$$

From the first two equations we obtain

$$\frac{\gamma_2}{\gamma_1} = \frac{m\omega_0^2 - k_{f,1}}{k_{f,12}} \qquad \frac{\gamma_2}{\gamma_1} = \frac{k_{f,12}}{m\omega_0^2 - k_{f,2}} \tag{9F5.1}$$

These equations can be solved by setting the determinant of the coefficients equal to zero (see Appendix C3):

$$\begin{vmatrix} (-m\omega_0^2 + k_{f,1}) & k_{f,12} \\ k_{f,12} & (-m\omega_0^2 + k_{f,2}) \end{vmatrix} = 0$$

By solving this determinant we obtain the resonance frequencies ω_0 belonging to the two different normal modes:

$$(-m\omega_0^2 + k_{f,1})(-m\omega_0^2 + k_{f,2}) = k_{f,12}^2$$

and

$$\omega_{0,\pm}^2 = \frac{1}{2}\frac{k_{f,1} + k_{f,2}}{m} \pm \frac{\sqrt{k_{f,12}^2 - k_{f,1}k_{f,2} + \frac{1}{4}(k_{f,1} + k_{f,2})^2}}{m}$$

$$= \frac{1}{2}\frac{k_{f,1} + k_{f,2}}{m} \pm \frac{\sqrt{k_{f,12}^2 + \frac{1}{4}(k_{f,1} - k_{f,2})^2}}{m}$$

$$\omega_{0,\pm} = \left| \sqrt{\frac{1}{2}\frac{k_{f,1} + k_{f,2}}{m} \pm \frac{\sqrt{k_{f,12}^2 + \frac{1}{4}(k_{f,1} - k_{f,2})^2}}{m}} \right| \tag{9F5.2}$$

9.8.7.2 Conversion of Resonance Frequencies into Excitation Energies

We express the classical quantities by their quantum mechanical counterparts. Using Equations (9F4.1) and (9F4.2)

$$\frac{k_{f,1}}{m} = \frac{4\pi^2}{h^2} \cdot \Delta E_1^2 \quad \text{and} \quad \frac{k_{f,2}}{m} = \frac{4\pi^2}{h^2} \cdot \Delta E_2^2,$$

$$\frac{k_{f,12}}{m} = \frac{8\pi^2}{h^2} \cdot \sqrt{\Delta E_1 \cdot \Delta E_2} \cdot J_{12}$$

in Foundation 9.4 and using the identity $\omega_0 = 2\pi\nu = 2\pi \cdot \Delta E_{\text{coupled}}/h$, we find that

$$\frac{2\pi \cdot \Delta E_{\text{coupled}}}{h} = \sqrt{\frac{1}{2}\frac{4\pi^2}{h^2}(\Delta E_1^2 + \Delta E_2^2) \pm \frac{4\pi^2}{h^2}\sqrt{4 \cdot \Delta E_1 \cdot \Delta E_2 \cdot J_{12}^2 + \frac{1}{4}(\Delta E_1^2 - \Delta E_2^2)^2}}$$

$$= \frac{2\pi}{h}\sqrt{\frac{1}{2}(\Delta E_1^2 + \Delta E_2^2) \pm \sqrt{4 \cdot \Delta E_1 \cdot \Delta E_2 \cdot J_{12}^2 + \frac{1}{4}(\Delta E_1^2 - \Delta E_2^2)^2}}$$

or

$$\Delta E_{\text{coupled}} = \sqrt{\frac{1}{2}\left(\Delta E_1^2 + \Delta E_2^2\right) \pm \sqrt{4 \cdot \Delta E_1 \cdot \Delta E_2 \cdot J_{12}^2 + \frac{1}{4}\left(\Delta E_1^2 - \Delta E_2^2\right)^2}} \tag{9F5.3}$$

where ΔE_1 and ΔE_2 are the transition energies for the two uncoupled oscillators, $\Delta E_{\text{coupled}}$ is the transition energy of the two coupled oscillators, and

$$J_{12} = \int \psi_{i,1}\psi_{j,1} \cdot g(1,2) \cdot \psi_{i,2}\psi_{j,2} \cdot d\tau_1 d\tau_2$$

is the coupling integral (9F4.3), which describes the repulsion energy of the two coupled electrons 1 and 2.

In the case of the coupling of two identical oscillators Equation (9F5.3) reduces to

$$\Delta E_{\text{coupled}} = \sqrt{\Delta E^2 \pm 2\Delta E \cdot J_{12}} \tag{9F5.4}$$

Below we show that in the case of the minus-sign the oscillatory strength of the oscillator is zero; thus, the only remaining solution is

$$\Delta E_{\text{coupled}} = \sqrt{\Delta E^2 + 2\Delta E \cdot J_{12}} \tag{9F5.5}$$

9.8.7.3 Oscillator Strength of Two Coupled Oscillators

We consider the system oscillating in a distinct normal mode (circular frequency ω_0, amplitude γ_1 of oscillator 1 with charge Q_1 and mass m, amplitude γ_2 of oscillator 2 with charge Q_2 and mass m). We replace the system by a single oscillator of mass m, circular frequency ω_0, and charge Q oscillating with amplitude γ. If this oscillator has the same dipole moment

$$Q \cdot \gamma = Q_1\gamma_1 + Q_2\gamma_2$$

and the same energy

$$\gamma^2 = \gamma_1^2 + \gamma_2^2$$

as the system of coupled oscillators, then it has the same oscillator strength f. Thus for Q we obtain

$$Q = \frac{Q_1\gamma_1 + Q_2\gamma_2}{\gamma} \quad \text{and} \quad Q^2 = \frac{\left(Q_1\gamma_1 + Q_2\gamma_2\right)^2}{\gamma^2} = \frac{\left(Q_1\gamma_1 + Q_2\gamma_2\right)^2}{\gamma_1^2 + \gamma_2^2}.$$

According to Foundation 9.2 the oscillator strength is

$$f = \frac{1}{3}\frac{Q^2}{m} \cdot \frac{m_e}{e^2} \cdot \frac{1}{\hat{n}} = \frac{1}{3}\frac{1}{m}\frac{m_e}{e^2} \cdot \frac{\left(Q_1\gamma_1 + Q_2\gamma_2\right)^2}{\gamma_1^2 + \gamma_2^2} \cdot \frac{1}{\hat{n}}$$

or

$$f = \frac{1}{3}\frac{m_e}{e^2 m}\frac{\left(Q_1 + Q_2\dfrac{\gamma_2}{\gamma_1}\right)^2}{1 + \left(\dfrac{\gamma_2}{\gamma_1}\right)^2} \cdot \frac{1}{\hat{n}}$$

where the ratio γ_2/γ_1 has been calculated above.

A special case is the coupling of two identical oscillators, where $Q_1 = Q_2 = Q$. According to Equation (9F5.1) we find that in this case $\gamma_2/\gamma_1 = \gamma_1/\gamma_2$ and therefore $\gamma_2/\gamma_1 = \pm 1$. For $\gamma_2/\gamma_1 = +1$ we obtain

$$f = \frac{1}{3}\frac{m_e}{e^2 m}\frac{(2Q)^2}{2} \cdot \frac{1}{\hat{n}} = 2 \cdot \frac{1}{3}\frac{m_e}{e^2 m}Q^2 \cdot \frac{1}{\hat{n}}$$

This is two times the oscillatory strength of one oscillator.

For $\gamma_2/\gamma_1 = -1$ we obtain

$$f = \frac{1}{3}\frac{m_e}{e^2 m}\frac{(Q - Q)^2}{2} \cdot \frac{1}{\hat{n}} = 0$$

Examples for coupled quantum mechanical systems are considered in Foundation 9.6 and in Problems 9.15 to 9.21.

9.8.8 Foundation 9.6: Cyanines in the FEMO and Step Model

See the books website, www.wiley.com/go/Kuhn/physicalchemistry3e.

9.9 Exercises

E9.1 For the molecules shown in Figure 9E.1 discuss whether the nitrogen atoms contribute to the number of π electrons in the conjugated system, and identify the number of π-electrons it contributes.

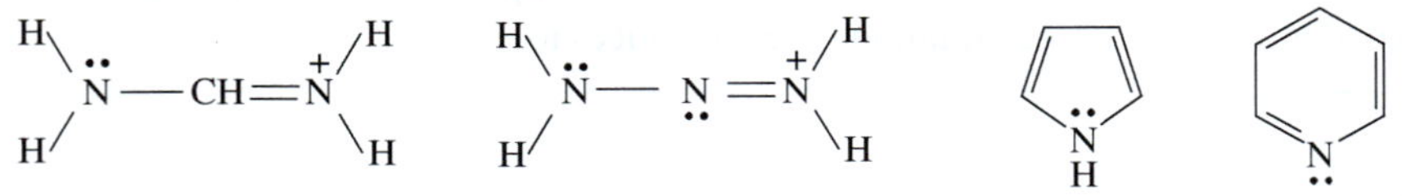

Figure 9E.1 Contribution of nitrogen atoms to the π-electron system.

E9.2 Discuss the color of the dyes in Table 9.1. Explain the shift in the absorbance maximum in terms of the FEMO model. Explain the changing color of the solution and how that relates to the molecule's absorption spectrum.

E9.3 Explain the shift in absorbance maxima for the different sets of dye molecules shown in Figure 9E.3.

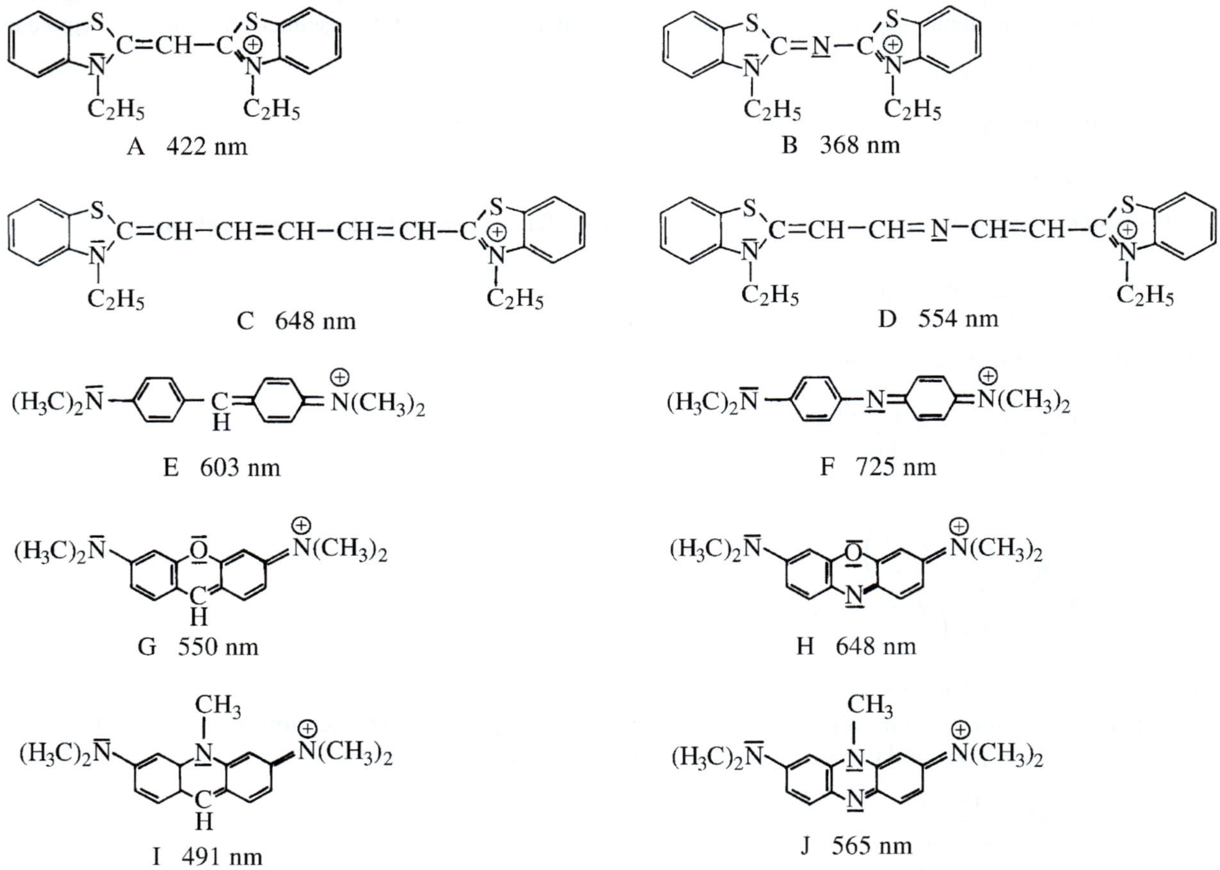

Figure 9E.3 Absorption maxima of different sets of dyes.

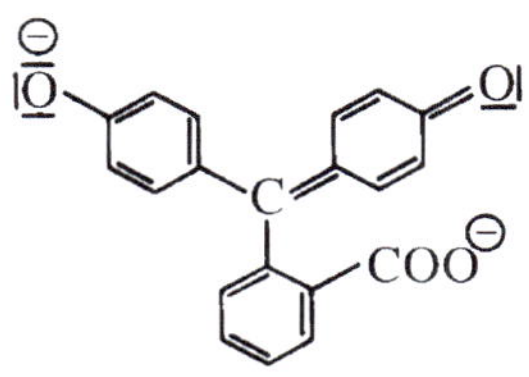

K 555 nm

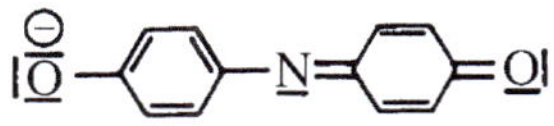

L 630 nm

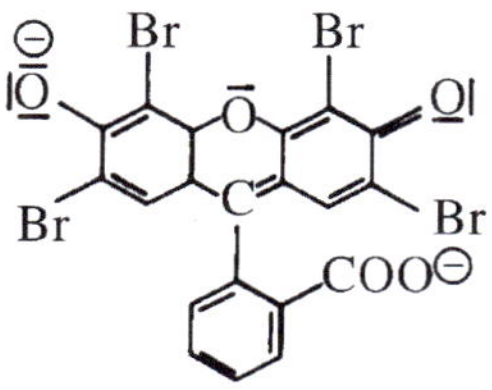

M 516 nm

N 609 nm

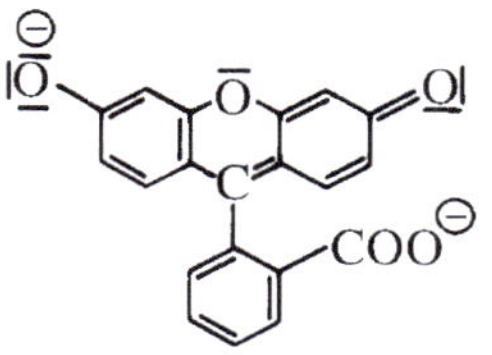

O 489 nm

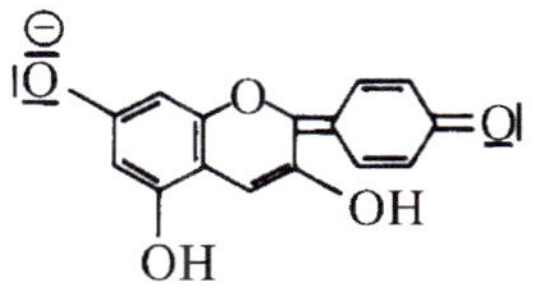

P 655 nm

Figure 9E.3 (*Continued*)

E9.4 Use the FEMO model to calculate the λ_{max} for the two dye molecules A and B shown in Figure 9E.4.

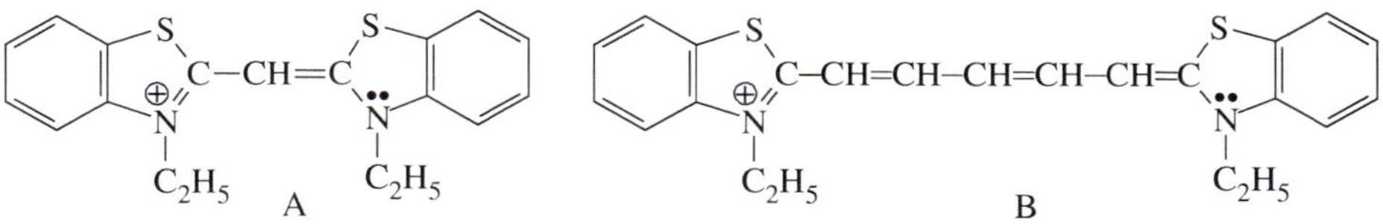

Figure 9E.4 Two different dye molecules A and B.

E9.5 Show that the energy gap calculated in Example 9.3 for Vitamin B12 corresponds to a λ_{max} of 580 nm.

E9.6 In Example 9.5, we calculated energy gaps for transitions in phthalocyanine. Use these energy gaps to calculate the λ_{max} values.

E9.7 Analyze the data in Table 9.4. Plot the energy of the experimental absorption maximum (taken as $\Delta E(\text{exp})$) versus $1/(N+1)$ and fit the data. Is it linear? How does the slope compare with the prediction of Equation (9.43)?

E9.8 In Example 9.5 we calculated the energies of the HOMO and LUMO electronic states in phthalocyanine using the FEMO model, and we corrected it for the self-energy of the N atom *p*-orbitals. Perform a similar calculation for porphyrin (see Figure 9E.8).

E9.9 A new spectroscopic technique (Cavity Ring Down Polarimetry) can be used to measure the optical rotation resulting from a gas of propylene oxide. Use the optical rotation data in the table to compute the molar rotation for propylene oxide. Consider the propylene oxide to behave like an ideal gas, and take the path length of the optical cell to be 1.6 m.

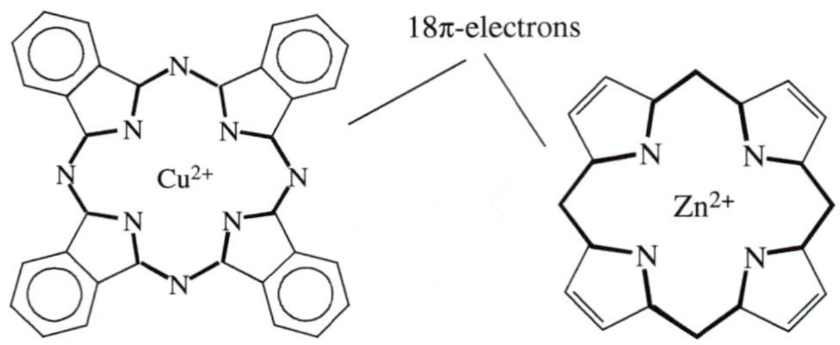

Figure 9E.8 Molecular skeleton of phthalocyanine and porphyrin.

Optical Rotation at T = 323 K and λ = 355 nm

P/bar	0.1	0.4	0.7	1.0
Rotation Angle α	0.0006	0.0025	0.0043	0.0062

E9.10 In the discussion of the azacyanines, we found the parameter $a = 3.5 \times 10^{-19}$ J to represent the difference in energy for a 2p-electron of a carbon atom versus a nitrogen atom. Compare this energy difference to the difference in first ionization potentials of carbon and nitrogen atoms (see Table 5.1).

E9.11 Figure 9.19 diagrams the FEMO energy of Cu phthalocyanine, excluding the difference in self-energy of nitrogen and carbon atoms. Calculate the difference in energy between the $n = 4$ and $n = 5$ states using this model.

E9.12 Imagine that you synthesize a free porphyrin which is supposed to have a molar extinction coefficient of ε = 50,000 M $^{-1}$ cm^{-1} at λ = 600 nm. Describe how you can use electronic spectroscopy to check the purity of your sample.

E9.13 Using a spreadsheet program plot the electric field found from Equations (9.61) and (9.62) 1 with $a = 1$ and $b = 0.2$. Your plot should show that it traces out an ellipse.

E9.14 The equation for a right circularly polarized field amplitude is $\vec{F}_{\text{right}} = F_x \sin(\omega t) \cdot \hat{x} + F_y \cos(\omega t) \cdot \hat{y}$ where F_x and F_y are the components along the directions x and y, respectively, and are equal to each other in magnitude. Plot F_{right} using a spreadsheet program and show that it traces out a circle by rotating in a clockwise fashion. What is the formula for F_{left}? Show that the sum of F_{right} and F_{left} gives linearly polarized light.

E9.15 Calculate the force on an electron in an electric field of $F = 10^6$ V/m (typical of an applied light field). Calculate the force on an electron in an applied magnetic field $B = F/c_0 = 3 \cdot 10^{-3}$ T, assuming that the force is given by $\mu B/a_0$ with a_0 the Bohr radius and μ the Bohr magneton. Compare the magnitudes of these two forces.

E9.16 An alternative description for the ORD signal is given by the difference of the index of refraction for left and right circularly polarized light, namely

$$\alpha = \frac{180 \cdot (n_l - n_r) \cdot l}{\lambda}$$

Using the difference in the rotation angle for saccharide, calculate the difference in refractive index.

E9.17 In Section 9.2.6 we calculated the transition moment $M_x = 8.5$ D $= 2.85 \times 10^{-29}$ C m for the transition between the 4a and 5a states of Cu phthalocyanine at λ = 345 nm. Use this transition moment value, $\hat{n} = 1.6$, and Equation (9.29) to calculate the transition's oscillator strength.

E9.18 Consider the molecule anthanthrene dissolved in benzene. The first electronic transition in this molecule occurs near an energy of 23, 090 cm^{-1}. A series of measurements of its absorbance A (at 23,090 cm^{-1}) as a function of concentration c gives the data that are reported and plotted in Figure 9E.18.

A	0.083	0.414	0.820	4.16
c/10^{-6} mol L^{-1}	1.0	5.0	10.	50.

Figure 9E.18 Absorbance of anthranthrene versus its concentration. The inset shows the C-skeleton of anthanthrene.

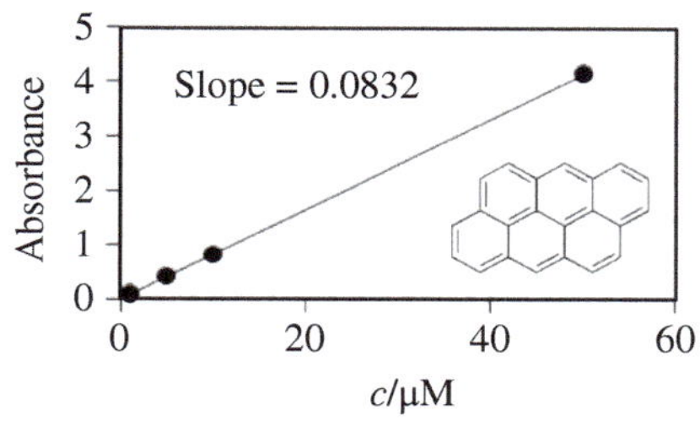

The path length of the optical cell was 1 cm. A) Determine the molar decadic absorption coefficient ε, in units of mol L^{-1} cm^{-1}, of the molecule at this wavelength. B) Determine the absorption cross-section σ of the molecule at this wavelength, where σ is given by

$$\sigma = \frac{\ln(10)}{\mathbf{N_A}} \cdot \varepsilon$$

C) Use bond increments and van der Waals radii to estimate the physical size of the molecule. Compare your absorption cross-section to the cross-sectional area of the molecule.

E9.19 Consider the Beer-Lambert law derivation in Box 9.1 for the case of a solid material rather than a liquid. In this case, we can combine the product $\alpha \cdot c$ into a single parameter that is indicative of the solid; let us call it α'. In this case, show that

$$I = I_0 \exp(-\alpha' x)$$

where x is the distance into the solid sample. The table gives α' values for pure silicon at 300 K and four different photon energies; use these data to compute the distance $x_{1\%}$ at which 99% of the incident light would be absorbed.

α'/cm^{-1}	10^3	4×10^3	6×10^4	1×10^6
$h\nu/\text{eV}$	1.6	2.0	3.0	4.0

In your calculations, you can ignore the contribution of reflection at the surface on your results; however, you should comment on its effect.

E9.20 Using the fundamental relation between the intensity I of a light beam and its electric field strength F, namely $I = c_0 \cdot \frac{1}{2}\varepsilon_0 F^2$, show that the force acting on a valence electron in a carbon atom (use a screened charge of $q = 3.25e$ and a distance $d = 150$ pm for the electron from the carbon nucleus) is much larger than that from a light beam of intensity 100 W/cm^2.

E9.21 Evaluate the integral for the transition moment in Equation (9.45) by approximating the phthalocyanine ring by a circle with a radius of r.

E9.22 Explain why the energy shift between the two low lying π-electron transitions in porphyrin is smaller than that in phthalocyanine. Using logic like that used in Section 9.2.6 for phthalocyanine estimate the absorption maxima for the porphyrin transitions.

9.10 Problems

P9.1 Show that the energy shift $\Delta E'$, when proceeding from a cyanine cation to the corresponding azacyanine cation, is given by $\Delta E' = \Delta E - a \cdot 2/N$ for $N/2 = 2, 4, 6, \ldots$ and $\Delta E' = \Delta E + a \cdot 2/N$ for $N/2 = 3, 5, 7, \ldots$ where a is a constant and N is the number of π-electrons. Use perturbation theory as discussed in Foundation 5.2 (on the book's website). Proceed by calculating the mean potential energy of a π electron in the electric field of the nitrogen atom's core using

$$\overline{V} = \int_0^L V_N \, \psi^2 \cdot ds$$

V_N is the potential energy of a π electron in the field of the N atom (more precisely, the difference between the potential energies of N and C). As shown in Fig. 9.13 assume that $V_N = -a$ for $s_1 < s < s_2$ and $V_N = 0$ for s elsewhere.

P9.2 a) Calculate how the HOMO to LUMO gap energy shifts when proceeding from a hypothetical butadiene with equal bond lengths to a butadiene with alternating single and double bonds. In analogy to the considerations in Problem P9.1 the energy in orbital n should be

$$E' = E - 2 \cdot b \cdot \psi^2_{n,\text{C=C}} \cdot d_0 + b \cdot \psi^2_{n,\text{C-C}} \cdot d_0$$

where b is a constant, $\psi_{n,\text{C=C}}$ is the value of the wavefunction in orbital n at the place of the double bond, $\psi_{n,\text{C-C}}$ is the corresponding value at the place of the single bond, and d_0 is the bond length (see Fig. 9.16). You should find that $\Delta E' = \Delta E + \frac{6}{5}b$.

b) For longer polyenes explain why the difference between $\Delta E'$ and ΔE is approximately the same, independent of length.

P9.3 a) For the porphyrin transitions in Exercise E9.8, calculate the transition dipole matrix elements $M_{x,\,4a\to 5a}$ and $M_{x,\,4b\to 5b}$ by using the free electron model and taking the porphyrin ring to be a circle with circumference $L_c = 16d_0$.

b) b) Use the facts $M_x = M_y$, $M_z = 0$, and your result in part a to evaluate the oscillator strength f where

$$f = \frac{8\pi^2 m_e}{3h^2 e^2 \hat{n}} \cdot M^2 \cdot \Delta E$$

and $\hat{n} = 1.5$.

P9.4 Fit the data in Table 9.2 to a more general model of the energy shift, in which the box length at the ends of the π-system is an adjustable parameter γ and the bond length d_0 is an adjustable parameter. In particular analyze the data by

$$\Delta E = \frac{h^2}{8m_e d_0^2} \frac{2(k+2)+1}{4(k+1+\gamma)^2}$$

Comment on your findings.

P9.5 Using the particle-in-a-box model, calculate the transition dipole moment for the excitation of an electron from the $n = 1$ level to the $n' = 2$ level. a) Determine the dependence of the transition dipole moment on the box length. b) Determine the transition dipole moment between an electron in the ground state $n = 1$ and an arbitrary excited sate n'. What general conclusion can you draw from this calculation; i.e., what is the selection rule?

P9.6 Explain why the terminal nitrogen atoms do not modulate the lowest energy electronic transitions of a cyanine dye, such as

$$(CH_3)_2\bar{N} - CH = CH - CH = CH - CH = \overset{\oplus}{N}(CH_3)_2$$

as do nitrogen atoms in the center of the molecule, such as

$$(CH_3)_2\bar{N} - CH = CH - \underline{N} = CH - CH = \overset{\oplus}{N}(CH_3)_2$$

P9.7 (a) Show that the integrand in the transition moment integral, Equation (9.45), for the $4a \to 5a$ transition in phthalocyanine is symmetric with respect to the x and y direction (origin of the coordinate system at the center of the ring). Hint: plot the integrand as a function of the angle φ on the circumference of the ring. (b) Solve the transition moment integral, Equation (9.45).

P9.8 Show that the equation

$$\xi = \frac{Q}{m}F_0 \cdot \frac{1}{\sqrt{\left(\omega_0^2 - \omega^2\right)^2 + k_d^2\omega^2/m^2}} \cdot \cos(\omega t + \alpha)$$

in Foundation 9.1 is a solution of the differential equation

$$m \cdot \frac{d^2\xi}{dt^2} = -k_f \cdot \xi - k_d \cdot \frac{d\xi}{dt} + Q \cdot F_0 \cdot \cos(\omega t)$$

for a damped oscillator in an external electric field. Hint: In a first step solve the differential equation by using the general solution $\xi = A\cos(\omega t) + B\sin(\omega t)$ and obtain expressions for A and B. In a second step use the relation

$$\cos\alpha\cos\beta - \sin\alpha\sin\beta = \cos(\alpha + \beta)$$

with $\cos\alpha = A/x_0$, $\sin\alpha = -B/x_0$, and $\beta = \omega t$ and show that the general solution for ξ is equivalent to $\xi = x_0\cos(\omega t + \alpha)$. Finally, express x_0 and α in terms of the expressions for A and B obtained in the first step.

P9.9 According to Foundation 9.1, and derived in Problem P9.8, the phase shift α and the amplitude ξ of a damped driven oscillator are given by

$$\alpha = \arctan\left[-\frac{k_d\omega/m}{\omega_0^2 - \omega^2}\right]$$

and

$$\xi = \frac{Q}{m}F_0 \cdot \frac{1}{\sqrt{\left(\omega_0^2 - \omega^2\right)^2 + k_d^2\omega^2/m^2}} \cdot \cos(\omega t + \alpha)$$

where m is the mass of the charge, Q is the displaced charge, and ω_0 is the resonant frequency of the oscillator. The driving field has an amplitude F_0 and a frequency ω. Plot the amplitude ξ and the phase shift α of the damped oscillator versus the frequency ω of the external electric field, for $\omega_0 = 3.77 \times 10^{15}\ \text{s}^{-1}$, $m = m_e = 9.109 \times 10^{-31}$ kg, and damping constants $k_d = 2.5$, 5.0, and $20.0 \times 10^{-16}\ \text{Nsm}^{-1}$.

P9.10 Here you are asked to estimate the mean potential energy $\overline{V}$ of a π electron in the electric field of the charge of the core of the aza nitrogen atom, by using the first-order perturbation theory result

$$\overline{V} = \int_0^L V_N\ \psi^2 \cdot ds$$

where V_N is the potential energy of a π electron in the field of the N atom (more precisely: the difference between the potential energies of N and C). You can approximate V_N by assuming that the potential energy of an electron around a N atom is lower by a constant amount a compared to the constant potential energy of a chain of carbon atoms. You should show that

$$\overline{V} = -a \cdot \frac{2}{N}$$

Using this result explain the even–odd effect in the absorption spectra for the azacyanine dyes.

P9.11 In this problem you should calculate the HOMO to LUMO energy gap for polyenes, in which you include the effects of bond alternation. Use perturbation theory (in a manner analogous to that for Problem 9.10) to show that the HOMO-to-LUMO energy gap $\Delta E'$ in butadiene can be written as

$$\Delta E' = \Delta E + \frac{6}{5}b$$

where ΔE is the energy gap for the unperturbed FEMO model, $2b$ is the energy difference between the "double bond" and "single bond" sites (see Fig. 9.13), and d_0 is the bond length in the unperturbed model. Explain why you would expect this result to hold for longer polyenes.

P9.12 Discuss the absorption spectrum of β-Carotene, distinguishing between an all-trans and a cis-configuration.

P9.13 Solve the coupling problem discussed for the cis-peak of β-carotene in Problem 9.12.

P9.14 Using the numerical data for the experimental absorbance curves for the carbocyanines with $j = 0, j = 1, j = 2$, and $j = 3$ in Fig. 9.2 evaluate the oscillator strengths f. You can find the numerical data for these curves (decadic absorbance coefficient $\varepsilon(\lambda)$ versus wavelength λ) in Appendix A.6. The wavelengths are given in nanometers (in steps of $\Delta\lambda = 2$ nm) and the decadic absorbance coefficients ε in $10^5\ \text{Lmol}^{-1}\ \text{cm}^{-1}$.

P9.15 Resonance frequency of two identical coupled oscillators. In Foundation 9.5, we found the resonance frequency $\omega_{0,}$ of two coupled oscillators

$$\omega_{0,\pm} = \left|\sqrt{\frac{1}{2}\frac{k_{f,1} + k_{f,2}}{m} \pm \frac{\sqrt{k_{f,12}^2 + \frac{1}{4}\left(k_{f,1} - k_{f,2}\right)^2}}{m}}\right|$$

where $k_{f,1}$ and $k_{f,2}$ are the force constants and m is the mass of the oscillators, and $k_{f,12}$ is the coupling constant. Consider the case that both oscillators are identical, and show that two new oscillation frequencies, at higher and lower frequencies, are generated

$$\omega_{0,1} = \sqrt{\frac{k_f}{m} + \frac{k_{f,12}}{m}} \quad \text{and} \quad \omega_{0,2} = \sqrt{\frac{k_f}{m} - \frac{k_{f,12}}{m}}$$

and use the results of Foundation 9.4 to show that the transition energies can be written as

$$\Delta E_{\text{coupled},1} = \sqrt{\Delta E^2 + 2\Delta E \cdot J_{12}} \qquad \Delta E_{\text{coupled},2} = \sqrt{\Delta E^2 - 2\Delta E \cdot J_{12}}$$

P9.16 Oscillator strength of two identical coupled oscillators. According to Foundation 9.3 the oscillator strength of a classical oscillator is

$$f = \frac{1}{3}\frac{m_e}{e^2 m} \cdot Q^2 \cdot \frac{1}{\hbar}$$

and according to Foundation 9.5 the oscillator strength of two coupled classical oscillators is

$$f = \frac{1}{3}\frac{m_e}{e^2 m} \cdot \frac{(Q_1\gamma_1 + Q_2\gamma_2)^2}{\gamma_1^2 + \gamma_2^2} \cdot \frac{1}{\hbar}$$

where Q_1 and Q_2 are the electrical charges of both oscillators and the amplitudes γ_1 and γ_2 of the oscillators are connected by

$$\frac{\gamma_2}{\gamma_1} = \frac{m\omega_0^2 - k_{f,1}}{k_{f,12}} \qquad \frac{\gamma_2}{\gamma_1} = \frac{k_{f,12}}{m\omega_0^2 - k_{f,1}}$$

Consider the case that both oscillators are identical, and find expressions for the oscillator strength of the two different transitions. Discuss your result in terms of what you expect to observe in the experimental absorption spectrum.

P9.17 Coupling Effect in Cyanines. The HOMO energy level in a cyanine molecule is occupied by two electrons which can be excited to the LUMO energy level. Consider the transitions of both HOMO electrons to the LUMO as two coupled classical oscillators. Calculate the excitation energy considering the coupling effect.

P9.18 Electron pairs in porphyrin. Calculate the shift of the transition energy for both separate transitions considering the coupling between the two electrons in each transition, in analogy to the cyanine considered in Problem P9.17.

P9.19 Coupling of the two electron pairs in porphyrin. Calculate the coupling between the two electron pairs (coupling integral $J_{12} = 0.6 \times 10^{-19}$ J), following the treatment in Foundation 9.4:

$$\frac{k_{f,\text{pair1}}}{m} = \frac{4\pi^2}{h^2}\Delta E_{\text{pair1}}^2, \quad \frac{k_{f,\text{pair2}}}{m} = \frac{4\pi^2}{h^2}\Delta E_{\text{pair2}}^2, \quad \text{and} \quad \frac{k_{f,12}}{m} = \frac{16\pi^2}{h^2}\sqrt{\Delta E_{\text{pair1}}\Delta E_{\text{pair2}}} \cdot J_{12}$$

P9.20 Oscillator strength f of porphyrin absorption bands. Calculate f for the coupled transitions using the transition moments $M_{x,1} = M_{x,2} = 2.85 \times 10^{-29}$ C m of transitions 1 and 2 and the excitation energies $\Delta E_1 = 3.92 \times 10^{-19}$ J, $\Delta E_2 = 5.68 \times 10^{-19}$ J.

P9.21 Coupling of transitions in phthalocyanine. Perform the same calculations as in Problems 9.18 to 9.20 for phthalocyanine. From the calculated absorption maxima $\lambda_1 = 772$ nm and $\lambda_2 = 327$ nm of phthalocyanine that are reported in Table 9.5 (uncoupled transitions), we obtain $\Delta E_1 = 2.57 \times 10^{-19}$ J, $\Delta E_2 = 6.07 \times 10^{-19}$ J. The coupling integrals for the electrons in the electron pairs ($J_{12} = 0.5 \times 10^{-19}$) J and between the two electron pairs ($J_{12} = 0.6 \times 10^{-19}$ J) and the transition moments ($M_{x,1} = M_{x,2} = 2.85 \times 10^{-29}$ C m) are the same as those for porphyrin. Thus for the transition energies of the electron pairs we obtain

$$\Delta E_1 = 3.03 \times 10^{-19}\ \text{J}, \quad \Delta E_2 = 6.55 \times 10^{-19}\ \text{J}$$

P9.22 Using the step model and the program Step_Cyanines_Polyenes.nb (this is a Mathematica application) show that in a long cyanine cation alternating bond lengths are obtained.

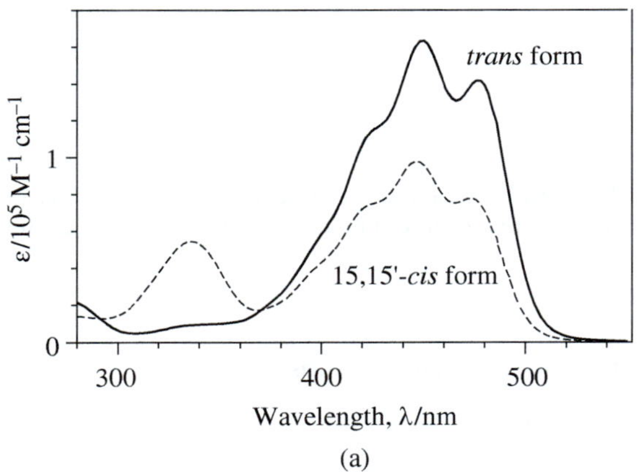

Figure 9P.22 (a) Absorption spectrum of β-Carotene. Solid line: trans-form, dashed line: cis-form.

Figure P9.22b (b) Structural formulae of β-carotene. (b) all-Trans-form, (c) 15, 15′-cis-form. The energy level diagram shows the 10 → 12 and 11 → 13 transitions.

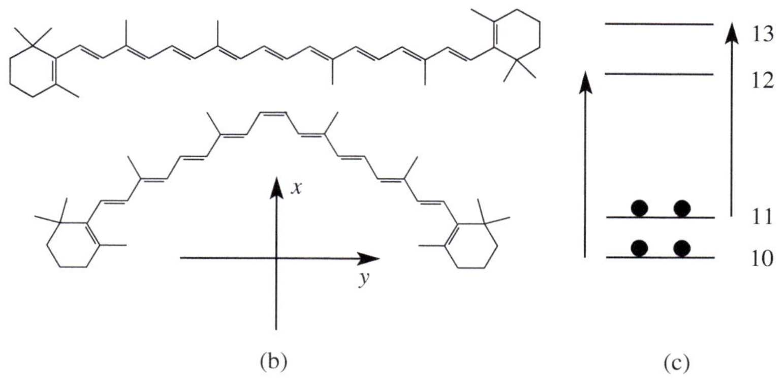

References

1 K. Misawa and T. Kobayshi, Hierarchical Structure in Oriented J-Aggregates, in: *J-Aggregates*, ed. T. Kobayashi (World Scientific, 1996, p. 50).

2 T. Basche, W.E. Moerner, M. Orrit, and V.P. Wild, eds. *Single Molecule Optical Detection, Imaging and Spectroscopy* (Wiley-VCH, Weinheim, 1998).

3 M. Schreiber, V. Buß, and M.P. Fülscher, The electronic spectra of symmetric cyanine dyes: a CASPT2 study, *Phys. Chem. Chem. Phys.*, 2001, **3**, 3906.

4 M. Eskandari, J.C. Roldao, J. Cerezo, and B. Milin-Medina, Counterion-mediated crossing of the cyanine limit in crystals and fluid solution: bond length alternation and spectral broadening unveiled by quantum chemistry, *J. Am. Chem. Soc.*, 2020, **142**, 2835; S. Pascal, A. Haefele, C. Monnereau, A. Charf-Eddin, D. Jaquemin, B. Le Guennic, O. Maury, and C. Andraud, On the versatility of electronic structures in polymethine dyes, *Conference paper in Proceedings of SPIE-The Int. Soc. for Opt. Eng.*, September 2014. DOI: 10.1117/12.2072624; L.M. Tolbert and X. Zhao, Beyond the cyanine limit: peierls distortion and symmetry collapse in a polymethine dye, *J. Am. Chem. Soc.*, 1997, **119**, 3253; W. Xu, E. Leary, S. Sangtarash, M. Jirasek, M.T. González, K.E. Christensen, L.A. Vicente, N. Agraït, S.J. Higgins, R.J. Nichols, C.J. Lambert, and H.L. Anderson, A Peierls transition in long polymethine molecular wires: evolution of molecular geometry and single-molecule conductance, *J. Am. Chem. Soc.*, 2021, **143** (48), 20472, DOI: 10.1021/jacs.1c10747.

5 H.D. Försterling, W. Huber, H. Kuhn, H.H. Martin, A. Schweig, F.F. Seelig, and W. Stratmann, Neuere Verfahren zur Behandlung von π-Elektronensystemen, in: *Optische Anregung organischer Systeme*, Hrsg. W. Foerst (Verlag Chemie, Weinheim, 1966).

6 B.S. Hudson, B.E. Kohler, and K. Schulten, in: *Excited States*, ed. E.C. Lim (Academic Press, New York 1982, pp. 1–95).

7 K. Schulten, I. Ohmine, and M. Karplus, Correlation effects in the spectra of polyenes, *J. Chem. Phys.*, 1976, **64**, 4422.

8 R. Naaman and D.H. Waldeck, Chiral-induced spin selectivity effect, *J. Phys. Chem. Lett.*, 2012, **3**, 2178–2187; R. Naaman, Y. Paltiel, and D.H. Waldeck, Chiral molecules and the spin selectivity effect, *J. Phys. Chem. Lett.*, 2020, **11**, 3660–3666.

9 W. Kuhn, Absolute Konfiguration der Milchsäure, *Z. Phys. Chem.*, 1935, **B31**, 23.

10 N. Harada, H. Ono, T. Nishiwaki, and H. Uda, Synthesis, circular dichroism and absolute stereochemistry of chiral spiroaromatic compounds. 9,9′-Spirobifluorene derivatives, *J. Chem. Soc., Chem. Commun.*, 1991, 1753.

11 V. Buß, D. Ulbrich, and Ch. Reichardt, Chiral tri-and pentamethinium cyanine dyes with 1,2,3,4-tetrahydro-6-methylquinolyl end groups: UV/Vis and CD spectroscopy and structure correlations, *Liebigs Ann.*, 1996, **1996**, 1823.

12 B. Norden, A. Rodger, and T. Dafforn, *Linear Dichroism and Circular Dichroism* (RSC, Cambridge, 2010); L. Barron, *Molecular Light Scattering and Optical Activity* (Cambridge, New York, 2004) 2nd edition.

13 W. Huber, G. Simon, and H. Kuhn, Analogiebetrachtungen und Analogrechner zur Behandlung der Korrelation von π-Elektronen (1), *Z. Naturforsch.*, 1962, **17a**, 99

14 W. Huber, G. Simon, and H. Kuhn, Analogiebetrachtungen und Analogrechner zur Behandlung der Korrelation von π-Elektronen (1), *Z. Naturforsch.*, 1962, **17a**, 99; Ch. Kuhn and H. Kuhn, Considerations on correlation effects in π-electron systems, *Synth. Metals*, 1995, **68**, 173.

10

Emission of Light

CHAPTER MENU

10.1 Introduction

After a molecule in solution is electronically excited, e.g., by the absorption of a photon, the molecule does not remain excited forever. A variety of different processes can occur to remove the energy from the excited molecule, including chemical reaction, conversion of the excess energy into heat, conformational rearrangement of the molecule, and emission of light (spontaneous emission). What occurs in any particular case depends on the molecule's properties and its environment. In this chapter we discuss the emission of light by electronically excited molecules. We will discuss fluorescence emission spectroscopy, which is so sensitive that we are able to detect and measure the electronic spectra of individual molecules. In addition, we will discuss stimulated emission, which is the working principle of lasers, and we will study the operation of dye lasers.

10.2 Spontaneous Emission

We consider dye molecules that relax back to their ground electronic state, either in a radiationless process (energy loss by collisions with solvent molecules) or in a radiative process (emission of light). In particular, we focus our discussion on the radiative process called *spontaneous emission*. Spontaneous emission takes place without external influence and is of two primary types: fluorescence and phosphorescence.

Figure 10.1 shows an experimental arrangement for recording the spectral distribution of the emitted light. The apparatus is similar to that used for absorption studies; however, the light emitted by the sample is monitored instead of the excitation light which is transmitted through the sample. Monochromators are optical devices that select particular light wavelengths for the excitation and/or emission. One can also include imaging optics that allow for the selective excitation and or detection of particular regions of a sample. Fluorescence microscopy has wide use in both fundamental studies (e.g., single molecule spectroscopy) and applications (e.g., biomedical assays, purity of materials).

Figure 10.2 provides an example of a molecule (carbocyanine dye cation) that displays both fluorescence and phosphorescence. The figure shows the absorption spectrum (black solid line, λ_{max} = 420 nm) and the corresponding emission spectrum (blue solid line, λ_{max} = 470 nm, corresponding to blue light) of this dye at room temperature. This emission is called fluorescence, and its maximum is shifted to longer wavelengths (lower energy) than that of the absorption band (in this case by about 50 nm). If one excites the solution with a very short burst of light (10^{-12} seconds) and monitors the light emission

Principles of Physical Chemistry, Third Edition. Hans Kuhn, David H. Waldeck, and Horst-Dieter Försterling.

Companion website: www.wiley.com/go/Kuhn/physicalchemistry3e

as a function of time, it decays in intensity over a time period of 10^{-9} seconds. Fluorescence occurs rapidly and is relatively efficient at room temperature. At low temperatures (e.g., 80 K) an additional emission with λ_{max} = 540 nm, corresponding to yellow light, is observed (see red solid line in Fig. 10.2). Typically, it is also shifted to longer wavelengths, even to longer wavelengths than the fluorescence and it has a much longer decay time, typically > 10^{-3} seconds. This slowly decaying emission is called *phosphorescence.*

How can we understand these fascinating phenomena? What is the reason for the tremendous difference in the decay time of fluorescence and phosphorescence? Why does the phosphorescence spectrum disappear as the molecule is warmed to room temperature?

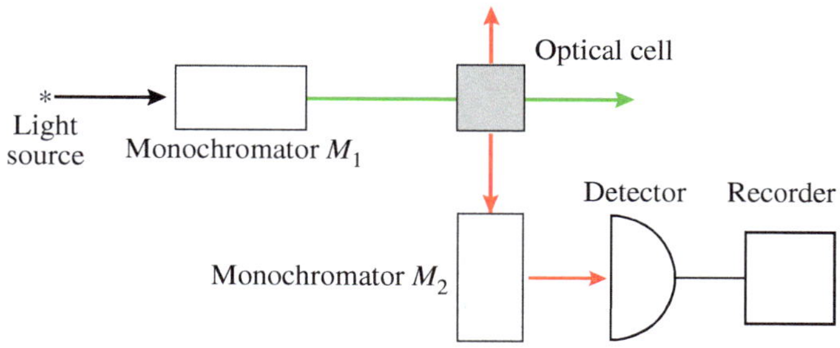

Figure 10.1 Experimental setup for the measurement of fluorescence and phosphorescence. Light from a source, selected for a wavelength $\lambda_{excitation}$ by a monochromator M_1, excites dye molecules in an optical cell. The light emitted from the excited dye molecules and selected for a wavelength $\lambda_{emission}$ by a monochromator M_2 is measured by a detector (e.g., a photomultiplier tube or a photodiode).

10.2.1 Fluorescence

In principle, we should solve the time-dependent Schrödinger equation for the excited molecule to calculate its emission rate. In the spirit of our earlier discussion, we use the oscillating dipole model for the molecule and calculate the rate that the dipole emits radiation (see Foundation 10.1). In this process the oscillating charge density associated with the transition moment between the excited state and ground state generates an oscillating electric field that can propagate away from the molecule, resulting in a net conversion of energy from the molecule's excited state into light. We find that the intensity of emission at time t is given by

$$I(t) = I_0\, e^{-t/\tau_0}$$

where I_0 is the initial intensity and τ_0 is a characteristic decay time. This decay time τ_0 is given by (see Foundation 10.1)

$$\tau_0 = \frac{3m_e\varepsilon_0 c_0^3}{2\pi e^2 \nu_0^2} \cdot \frac{1}{f} \tag{10.1}$$

and is called the *radiative lifetime.* τ_0 depends on the dipole's resonant frequency ν_0, its oscillator strength f, and on the universal constants m_e, ε_0, and e. The rate of de-excitation increases (decay time decreases) strongly with the frequency ν_0 of the oscillator and with the magnitude of the oscillating charge Q, which is determined by the oscillator strength f.

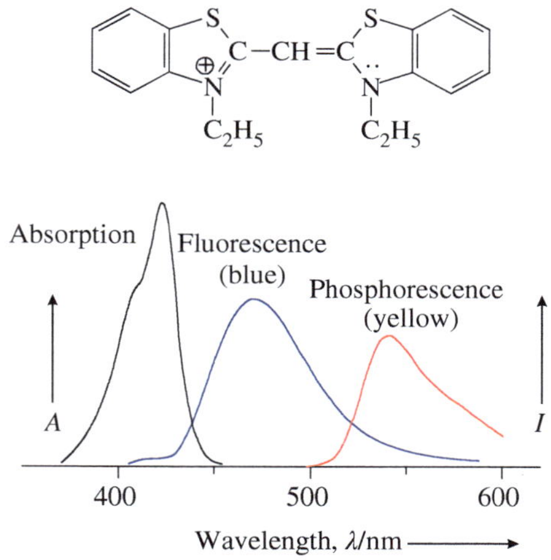

Figure 10.2 Absorption and emission spectra of a carbocyanine dye cation in methanol solution at room temperature. Molecular structure of the dye molecule is shown at the top. Absorbance A (black solid curve, left-hand scale), fluorescence intensity I (blue solid curve, right-hand scale) and phosphorescence intensity I (red solid curve, right-hand scale).[1]

Example 10.1 *Calculate the radiative lifetime of the carbocyanine dye cation considered above. Use the fact that* λ_{max} = *470 nm, and* f = 0.67.

First we use λ_{max} to estimate the resonance frequency ν_0, by $\nu_0 \sim c_0/\lambda_{max}$. We find that $\nu_0 = 6.38 \times 10^{14}$ s^{-1}. Using this result in Equation (10.1) with the other parameters we find that

$$\tau_0 = \frac{3 \cdot 9.109 \times 10^{-31}\ \text{kg} \cdot 8.854 \times 10^{-12}\ \text{C}^2\ \text{J}^{-1}\ \text{m}^{-1} \cdot (3.00 \times 10^8)^3}{2\pi \cdot (1.602 \times 10^{-19})^2 \cdot 0.67 \cdot (6.38 \times 10^{14}\ \text{s}^{-1})^2}$$
$$= 15 \times 10^{-9}\ \text{s} = 15\ \text{ns}$$

---

1 The phosphorescence was measured at liquid nitrogen temperature (dye adsorbed on the surface of a filter paper). To separate the phosphorescence (decay time ≈ 1 seconds) from the fluorescence (decay time ≈ 1 ns), rotating shutters in the illuminating and the emission light beams, which can be phase shifted by 180°, were used.

In Equation (9.26) we found a relationship between the absorption spectrum of a molecule and its oscillator strength.

$$f = \frac{4 \cdot \ln\ 10 \cdot c_0 \varepsilon_0 m_e}{\mathbf{N}_A\, e^2} \int_{\text{band}} \varepsilon \cdot \mathrm{d}\nu$$

where N_A is the Avogadro constant (see Chapter 14). Combining that result with Equation (10.1),we can relate the radiative lifetime to the absorption spectrum of a molecule:

$$\frac{1}{\tau_0} = \ln 10 \cdot \frac{8\pi\nu_0^2}{3c_0^2\mathbf{N}_A} \cdot \int_{\text{band}} \varepsilon \cdot \mathrm{d}\nu \tag{10.2}$$

The finite time for the existence of the electronically excited state gives rise to a width in its emission spectrum because the excited state is not a stationary state; see Section 3.5.3. As shown in Foundation 10.1, the natural spectral line width $2 \cdot \Delta\nu$ and τ_0 are related by

$$\text{natural spectral line width} = 2 \cdot \Delta\nu = \frac{1}{2\pi\tau_0} \tag{10.3}$$

where $2 \cdot \Delta\nu$ is the frequency difference between the two sides of the spectral line where the intensity is half of its maximum value. The observed fluorescence band (Fig. 10.2, $\Delta\nu \approx 1 \times 10^{14}$ Hz) is by far broader than the natural band width, however. The reason for this discrepancy is the same as discussed in Section 9.2.2.1 for the broadening of the absorption band. The energy of the fluorescence of a selected molecule depends on its surroundings and internal nuclear motions. The observed fluorescence band is a superposition of the bands of all molecules in the solution that are emitting light into the collection angle of the detector.

10.2.1.1 Einstein Theory of Emission and Absorption of Radiation

In Foundation 10.1 we derive Equation (10.1) for the decay time τ_0 by studying the emission of radiation by an oscillating classical dipole and making an analogy between the classical and quantum-mechanical. We will use this result with Einstein's steady-state rate analysis to construct a quantum mechanical description of light absorption and emission; see Box 10.1.

To illustrate this analysis, we assume that molecules can exist in states a and b with energy difference ΔE_{ab}, a two-level model of the molecule. The molecules are irradiated by an electromagnetic field of spectral radiant energy density ρ_ν

$$\rho_\nu(\nu_{ab}) = \frac{\mathrm{d}\rho}{\mathrm{d}\nu}, \quad \text{with} \quad \rho = \frac{E_{\text{rad}}}{V} \tag{10.4}$$

where E_{rad} is the energy of radiation, V is the volume, and $\nu_{ab} = \Delta E_{ab}/h$. For a particular frequency ν_{ab}, $\rho = h\nu_{ab} \cdot N_{\text{photon}}/V$, where N_{photon} is the number of photons of frequency ν_{ab} in volume V. Next, we consider molecules M that absorb photons to generate excited molecules M*.

$$\mathrm{M} + h\nu_{ab} \rightarrow \mathrm{M}^*$$

The rate that molecules M in the lower state a will absorb a quantum of energy and undergo a transition to state b is

$$\left(\frac{\mathrm{d}N}{\mathrm{d}t}\right)_{a\to b} = B_{a\to b} \cdot \rho_\nu(\nu_{ab}) \cdot N_a \tag{10.5}$$

where N_a is the number of molecules M in state a, and $B_{a\to b}$ is called the Einstein coefficient *of absorption*.

Accordingly, in the process where molecules M* in the upper state b undergo a transition to the lower state a Einstein has assumed that the rate is

$$\left(\frac{\mathrm{d}N}{\mathrm{d}t}\right)_{b\to a} = B_{b\to a} \cdot \rho_\nu(\nu_{ab}) \cdot N_b + A_{b\to a} \cdot N_b \tag{10.6}$$

where N_b is the number of molecules M* in state b. $B_{b\to a}$ is called *Einstein's coefficient of stimulated emission*, and $A_{b\to a}$ is called *Einstein's coefficient of spontaneous emission*. By considering the limit in which the radiation field and the molecules are under steady-state conditions, Einstein showed that these coefficients are related by the equations,

$$B_{a\to b} = B_{b\to a}, \quad A_{b\to a} = \frac{8\pi h\nu_{ab}^3}{c_0^3} \cdot B_{b\to a} \tag{10.7}$$

see Problem P10.5. These considerations are important for an understanding of laser action discussed in Section 10.3. Einstein's postulates can be related to the strength of the optical transitions; see Foundations 9.1 and 10.1.

10.2.2 Single Molecule Emission

Because fluorescence detection is so sensitive it is possible to probe the fluorescence emission of a single molecule. Initial studies by physical chemists in the late 1980s and 1990s measured the spectra and fluorescence decay properties of single molecules for the first time. These developments were recognized by award of the 2014 Nobel Prize in Chemistry to W. E. Moerner. Although fundamental studies continue and much remains to be discovered, fluorescence of single molecules is now being used to probe the dynamics of biomolecules and biological machines, see Chapter 28.

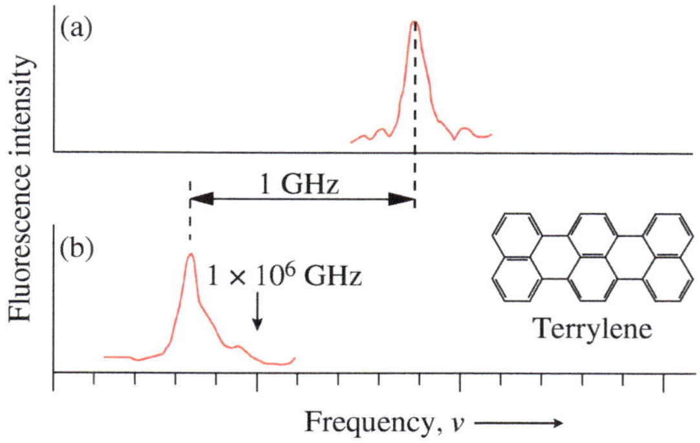

Figure 10.3 Fluorescence excitation spectra of a single terrylene molecule (highly diluted solution in a polyethylene glass at a temperature of 2 K) taken within a time span of several minutes. Occasionally the spectrum changes, all of a sudden, from (a) to (b). This behavior shows that the local environment of the terrylene molecule changes in a discrete manner, causing the spectral shift. On the ordinate is the fluorescence intensity, on the abscissa the frequency ν of the exciting laser beam.

As an example consider using a nearly monochromatic light source to excite a terrylene molecule [1] in a highly diluted polyethylene glass at a temperature of 2 K. Figure 10.3a shows the molecule's fluorescence intensity as a function of the excitation light's frequency. This plot is called a *fluorescence excitation spectrum*, and it is very similar to the absorption spectrum. These data are for one molecule, however. The bandwidth of 0.1 GHz is 10^6 times narrower than the width of the overall absorption band of the sample, which is a sum of the spectra for each molecule in the sample. In this particular experiment, the single molecule spectrum is unchanged for about a minute, after which it "jumps" to a new place in the spectrum (see Fig. 10.3b). The subsequent jumping back and forth between the two spectra on the timescale of minutes indicates a flip-flop of this individual molecule between two distinct energy environments of the surrounding matrix. Such phenomena are observed in many experiments on single molecules, and their origin is not always clear.

10.2.3 Fluorescence Optical Microscopy

Figure 10.4a shows the scheme of a fluorescence optical microscope. The sample is excited by the light of a laser, and the fluorescence light passes through a spatial aperture (e.g., pinhole) and a spectral filter that blocks the laser light frequency. In this way only a very small area of the sample is in the focus of the microscope, according to the diameter of the pinhole. In order to obtain an image, the sample can be moved stepwise in the x- and y-directions, and a spatial map of the fluorescence intensity across the surface can be constructed. The equipment necessary to realize this technique requires specialized optics for the microscope and is called confocal fluorescence microscopy. The spatial resolution with this technique is limited by the diffraction limit of the light, which depends on the light photon wavelength and the optical arrangement (typically it is 250–500 nm for visible light).

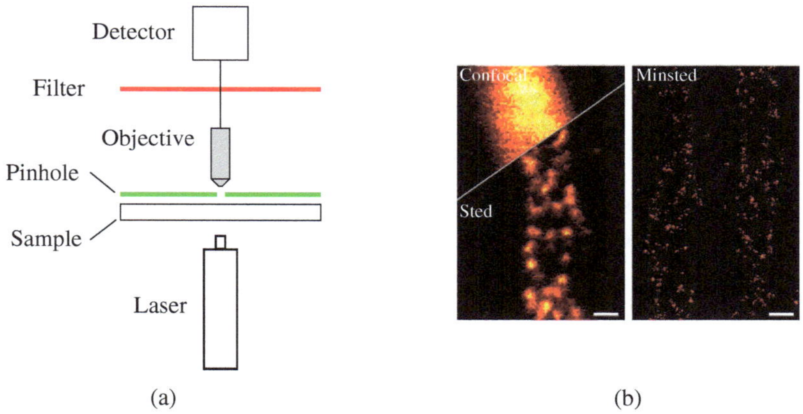

Figure 10.4 (a) Scheme of a fluorescence optical microscope. The sample is excited by a beam of laser light. The fluorescence light passes a pinhole and a filter cutting the contribution of the laser light. The fluorescence intensity is measured by the detector. The sample can be moved in the x- and y directions; in this way the surface can be scanned. (b) Improvement of the resolution of a sample from confocal microscopy (upper left) to STED microscopy (lower left) and to MINSTED microscopy (right). The length of the scale bars is 200 nm. (Weber et al. 2021/With permission of Springer Nature/CC BY 4.0.)

More sophisticated techniques have been developed to improve the spatial resolution well below the diffraction limit. In one such technique only a very small part of the fluorescence light passing the pinhole reaches the detector and is called STimulated Emission Depletion (STED) microscopy. It increases the spatial resolution by a factor of about 10 below the diffraction limit (ca. 20–30 nm). A still more advanced technique is MINSTED, which increases the resolution by an additional factor of 10 (on the order of 1–2 nm). These techniques are described in detail in Foundation 10.3. The STED microscopy was first realized in 1999 by Stefan Hell and Thomas Klar; for his work Stefan Hell won the Nobel prize in Chemistry in 2014. The MINSTED microscopy was developed by Stefan Hell in 2021 [2].

Figure 10.4b compares the resolution of a sample, using confocal microscopy, STED microscopy, and MINSTED microscopy. The sample consists of the mitochondrial inner membrane protein Mic60 in chemically fixed human cells. The chromophore silicon rhodamine [3]

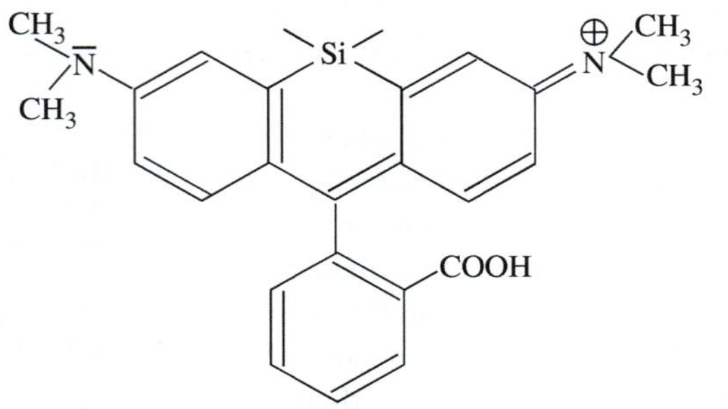

was fixed to the protein and used as a fluorescence marker. Compared to rhodamine B (containing an oxygen bridge),the fluorescence wavelength is shifted from 627 to 775 nm. The resolution in the MINSTED experiment was about 5 nm.

10.2.4 Phosphorescence and Triplet States

Phosphorescence happens when radiative transitions between triplet and singlet states occur. Important evidence for this assignment arises from studies of fluorescence and phosphorescence as a function of an external magnetic field [4]. Whereas fluorescence does not depend on the magnetic field, the phosphorescence of a molecule changes dramatically in a magnetic field. Here we discuss the red shift of the phosphorescence, compared to the fluorescence, and its longer radiative lifetime.

10.2.4.1 Energy Shift

We proceed in a similar way to that used in Sections 5.3 and 5.4 for He. To better understand this phenomenon let us approximate the carbocyanine dye cation in Fig. 10.2 by the symmetric cyanine cation

$$(CH_3)_2\overset{\oplus}{N}{=}CH{-}CH{=}CH{-}\overline{N}(CH_3)_2 \tag{10.8}$$

and evaluate its fluorescence and phosphorescence energies. This molecule has six π electrons occupying three energy levels in the ground state (Fig. 10.5a). Irradiation with light at 313 nm excites the electron in the HOMO to the LUMO (from level 3 to level 4 in Fig. 10.5b), while the electron distribution in levels 1 and 2 remains unchanged. Thus, we restrict our discussion to the two most weakly bound π electrons, neglecting the interaction with the electrons in levels 1 and 2. In analogy to the treatment of He, the two-electron wavefunction for the ground state (singlet state S_0) is

$$\psi_{S_0} = \phi_3(1) \cdot \phi_3(2) \tag{10.9}$$

where ϕ_3 is the one-electron wavefunction in orbital 3. After excitation the system can be found in the singlet state S_1. In these considerations, we omit the electron spin contribution to the energy from its magnetic moment because it is a small correction.

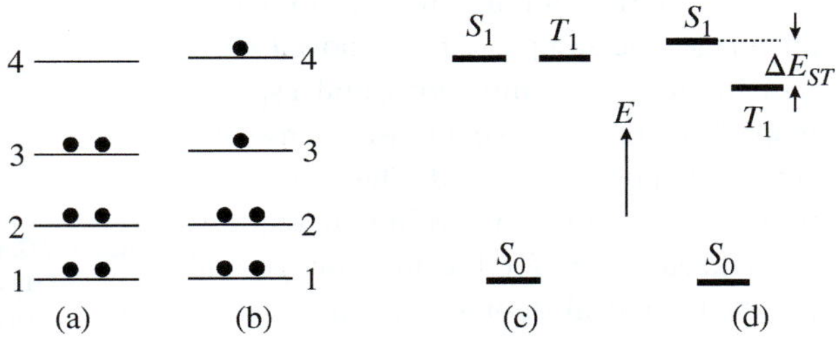

Figure 10.5 (a,b) One-electron energy levels of the cyanine cation in structure (10.8) in the ground state (a) and in the excited state (b). (c,d) Energy scheme of a two-electron system in the ground state S_0 and in the first excited singlet state S_1 and triplet state T_1. (c) Electron repulsion neglected. (d) Electron repulsion included.

By approximating the one-electron wavefunctions with those from the free-electron model, we can evaluate the energies for the singlet and triplet states. The free electron wavefunctions are

$$\phi_3(1) = \sqrt{\frac{2}{L}} \cdot \sin\left(\frac{3\pi s_1}{L}\right)$$

$$\phi_4(1) = \sqrt{\frac{2}{L}} \cdot \sin\left(\frac{4\pi s_1}{L}\right)$$

where s_1 is the coordinate of electron 1 along the bond axis in the cyanine cation (10.8). In the ground-state electron configuration $\phi_3(1) \cdot \phi_3(2)$, the energy of the two electrons is

$$E'_{S_0} = \frac{h^2}{8m_e L^2}(3^2 + 3^2) = \frac{h^2}{8m_e L^2} \cdot 18 = 7.69 \times 10^{-19}\ \mathrm{J} \tag{10.10}$$

In the excited state configuration, the spatial orbital wavefunctions are either a symmetric superposition

$$\psi_{S_1} = \frac{1}{\sqrt{2}}\left[\phi_3(1) \cdot \phi_4(2) + \phi_3(2) \cdot \phi_4(1)\right]$$

which is multiplied by the antisymmetric spin combination (singlet state) or an antisymmetric superposition

$$\psi_{T_1} = \frac{1}{\sqrt{2}}\left[\phi_3(1) \cdot \phi_4(2) - \phi_3(2) \cdot \phi_4(1)\right]$$

which is multiplied by the symmetric spin combinations (triplet state). Within the FEMO model, the energy for these two spatial orbitals is the same

$$E'_{S_1} = E'_{T_1} = \frac{h^2}{8m_e L^2}(3^2 + 4^2) = \frac{h^2}{8m_e L^2} \cdot 25 = 13.66 \times 10^{-19}\ \mathrm{J} \tag{10.11}$$

where the prime in E' indicates that the Coulomb repulsion between the two electrons is neglected in the energy calculation. These energies are shown in Fig. 10.5c.

To make our energy calculation more realistic, we must add the Coulomb repulsion energy between the electrons. Following in the spirit of the He atom calculation (see Foundation 10.2 on the Book's website for details of this calculation), we find

$$\overline{V}_{e,S_0} = \frac{e^2}{4\pi\varepsilon_0\varepsilon}\int\int\frac{\psi^2_{S_0}}{r_{12}}\,ds_1\,ds_2 = 3.6 \times 10^{-19}\ \mathrm{J}$$

$$\overline{V}_{e,S_1} = \frac{e^2}{4\pi\varepsilon_0\varepsilon}\int\int\frac{\psi^2_{S_1}}{r_{12}}\,ds_1\,ds_2 = 4.2 \times 10^{-19}\ \mathrm{J}$$

$$\overline{V}_{e,T_1} = \frac{e^2}{4\pi\varepsilon_0\varepsilon}\int\int\frac{\psi^2_{T_1}}{r_{12}}\,ds_1\,ds_2 = 2.9 \times 10^{-19}\ \mathrm{J}$$

where ε is the relative permittivity attributed to the medium of the σ-electrons. Now we use these electron–electron repulsion energies to correct the energies in Equations (10.10) and (10.11), and find that

$$E_{S_0} = 11.3 \times 10^{-19}\ \mathrm{J}, \quad E_{S_1} = 17.9 \times 10^{-19}\ \mathrm{J}, \quad \text{and} \quad E_{T_1} = 16.4 \times 10^{-19}\ \mathrm{J}$$

These energy levels are displayed in Fig. 10.5d. The energy difference between the singlet and triplet state is

$$\Delta E_{ST} = E_{S_1} - E_{T_1} = 1.4 \times 10^{-19}\ \mathrm{J}$$

Note that this energy difference arises from electron–electron repulsion differences that originate from the different spatial orbitals used for the singlet and triplet states, not from magnetic effects.

The corresponding calculation for the branched cyanine cation in Fig. 10.2 yields $\Delta E_{ST} = 0.9 \times 10^{-19}$ J. This smaller value arises from the larger extent of the π-electron system in the branched cyanine, resulting in a smaller electron–electron repulsion. From the experimental values $\lambda_S = 470$ nm and $\lambda_T = 540$ nm (Fig. 10.2) we obtain

$$\Delta E_{ST,\mathrm{exp}} = 4.2 \times 10^{-19}\ \mathrm{J} - 3.7 \times 10^{-19}\ \mathrm{J} = 0.5 \times 10^{-19}\ \mathrm{J}$$

within a factor of 3 of our rough calculation.

10.2.4.2 Lifetime

The transition $S_0 \to T_1$ requires an inversion of the spin of an electron; thus, a transition by absorption of light is not possible, in contrast to the transition $S_0 \to S_1$ (see Problem 10.4). For the same reason the transition $T_1 \to S_0$ by emission of light is not allowed. However, we can observe this transition in the experiment (yellow emission) with a lifetime of about one second. From this very long lifetime (it is larger than for the $S_1 \to S_0$ transition by nine orders of magnitude) we conclude that the probability for this transition is extremely small. In the theoretical treatment we can account for this small probability by assuming that the excited state is not a pure triplet state, but is in fact a mixture of a triplet and a tiny contribution of singlet wavefunctions, which arises from spin–orbit coupling [5].

10.2.5 Relative Energetics Fluorescence, Phosphorescence, and Absorption

10.2.5.1 Shift of Wavelengths

Why is the fluorescence band shifted to longer wavelengths relative to the absorption band (Fig. 10.2)? Light absorption is accompanied by a change of the π-electron distribution between the ground and excited electronic states. Hence the newly created electronic state has a different set of bond lengths and bond angles that minimize its energy. Because the electronic excitation occurs more rapidly than the characteristic time for the nuclei to rearrange, they are no longer in their equilibrium positions immediately after excitation. As time progresses the nuclei move in the new force field of the excited electronic state, i.e., the molecular skeleton starts to oscillate. This energy is transferred among the internal motions of the molecule and to the surroundings by collisions with the solvent molecules. Furthermore, the new charge distribution of the molecule causes a change in the interaction forces with the solvent molecules. Hence the solvent molecules reorganize, and this process leads to an additional energy decrease for the dye molecule. Similar processes occur when an excited state molecule relaxes back down to the ground state (Fig. 10.6).

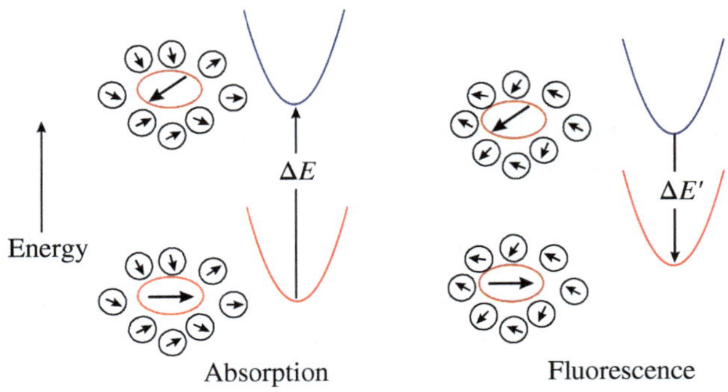

Figure 10.6 This diagram illustrates the change in solvation about different electronic states of a solute molecule (red ellipse with an arrow indicating its dipole moment) and its impact on the energy gap between the two states. The solvent molecules are represented by circles and an arrow indicating their dipole moment direction.

The time-dependent shift in the fluorescence spectra of numerous dye molecules have been measured. The spectral shift arises from internal nuclear relaxation of the molecule (see Chapter 11) and rearrangement of the solvent molecules about the solute. In order to quantify the time dependence of the solvent relaxation, we can define the function $S_\nu(t)$,

$$S_\nu(t) = \frac{\nu(t) - \nu(\infty)}{\nu 0) - \nu(\infty)} \tag{10.12}$$

where $\nu(t)$ is the frequency of the peak emission intensity at time t, $\nu(0)$ is the hypothetical frequency of the peak emission intensity at early time after vibrational relaxation has occurred but before solvent relaxation proceeds, and $\nu(\infty)$ is the frequency of the peak emission intensity for the spectrum at times long compared to the nuclear rearrangements of the molecule and the solvent. Figure 10.7 shows the time profile of $S_\nu(t)$ for the dye molecule coumarin 153 in both acetonitrile and methanol.

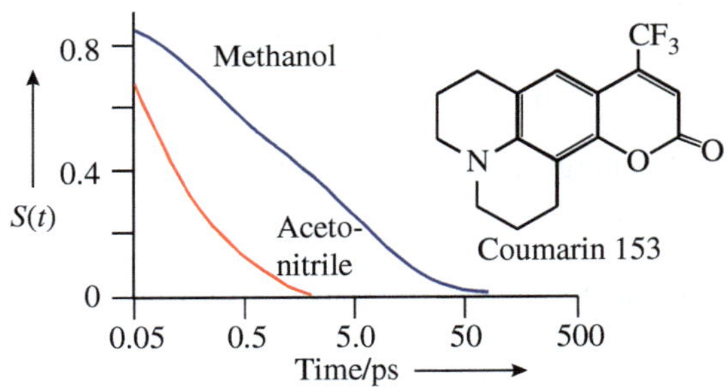

Figure 10.7 The dynamic Stokes shift of coumarin 153 is shown for acetonitrile and methanol solvents [6].

$S_\nu(t)$ changes from one to zero as a function of time, and its characteristic decay time measures the time for the solvation energy to reach equilibrium. The decay curves in Fig. 10.7 show that in the polar aprotic solvent acetonitrile the relaxation is complete in a few picoseconds (a characteristic decay time of <300 fs), whereas the similarly sized, but hydrogen bonding solvent, methanol evolves for nearly 50 ps (a characteristic decay time of about 5 ps). This large difference in relaxation time is linked to the structural features of the hydrogen bonding solvent. Characteristic solvation times range from less than a picosecond to hundreds of picoseconds in hydrogen bonding solvents (like alcohols), and even longer in glasses and other slowly relaxing materials.

10.2.5.2 Jablonski-Diagram

Figure 10.8 illustrates these phenomena with an energy level scheme called a *Jablonski diagram*. In this figure three electronic states S_0, S_1, and T_1 are shown. Within each of these electronic states a number of distinct energy levels are drawn with narrower horizontal lines. These energy levels are associated with different possible nuclear arrangements of the molecules and its surroundings. The nature of these different vibrational states of the electronically excited molecules and the vibrational dynamics will be discussed in Chapter 11. Typically the energy spacing between these levels is small enough that the molecule will fluctuate among them as energy is exchanged between internal motions of the molecule and the surroundings; however, most molecules will occupy the lower energy states within an electronic state as the molecule equilibrates. The characteristic relaxation rates within a single electronic state are about 10^{12}–10^{13}s^{-1}.

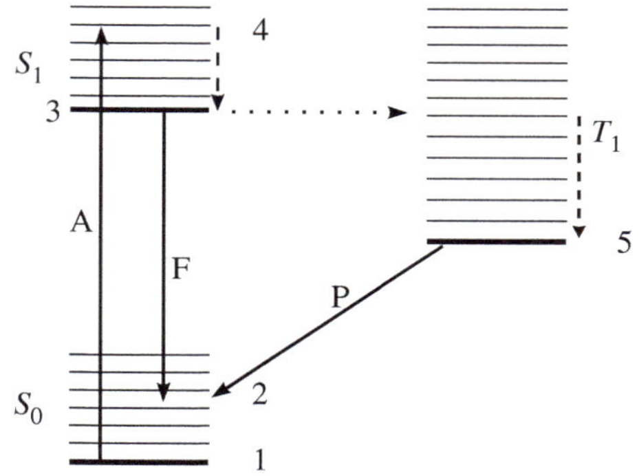

Figure 10.8 Energy states of a dye molecule in solution. 1: Lowest vibronic level of electronic ground state S_0. 4: Vibronically excited level of electronic singlet state S_1 reached immediately after light absorption (process A). 3: Lowest vibronic level of electronic state S_1 reached by the molecule in level 4 after transfer of vibrational energy to the solvent and solvent reorganization. 2: Vibronically excited level of electronic ground state S_0 reached by emission of fluorescence (process F) and by phosphorescence (process P). 5: Lowest vibronic level of electronic triplet state T_1.

The molecule can remain in level 3 for many nanoseconds (10^{-9} seconds) before it relaxes to the ground state by fluorescence or relaxes to the triplet state. The timescale for the molecule to remain excited depends strongly on its molecular structure and the environment in which it is placed; for example, triphenylmethane dyes are known to relax nonradiatively to the ground electronic state (a process called internal conversion) in a few picoseconds (10^{-12} seconds) whereas cyanine dyes with rigid π-frameworks relax primarily by light emission on a nanoseconds timescale, $>10^{-9}$ seconds. Even upon fluorescence and relaxation to the ground electronic state, the molecule finds itself in a high energy arrangement of nuclear coordinates. Once again it relaxes by thermal exchange with the surroundings. As a consequence of these nuclear rearrangements, the fluorescence emission occurs at a lower energy than the absorption.

An alternative relaxation pathway for the π system in level 3 is to cross over into the triplet state T_1 (dotted arrow in Fig. 10.8); a process called *intersystem crossing*. The term intersystem reflects a distinction between the system (or manifold) of singlet states in a molecule from the system (or manifold) of triplet states. Once the π system converts to the excited levels of the triplet state T_1, vibrational relaxation can occur rapidly down to level 5, the lowest vibronic level. From level 5 a transition into level 2 by emission of light (phosphorescence) occurs, albeit with low probability. Just as emission from the triplet to the singlet state (phosphorescence) is a forbidden process, the direct absorption of light to populate T_1 from S_0 is forbidden (highly improbable). Thus it is common for triplet states to be populated optically by excitation of the S_1 state which then undergoes intersystem crossing to generate the T_1 state, or to generate triplet states by nonradiative processes, such as chemical reactions or energy transfer.

10.2.6 Quenching of Fluorescence

The decay of an excited electronic singlet state of a molecule A by spontaneous emission of light

$$A^* \rightarrow A + h\nu \tag{10.13}$$

is a first-order process (see Chapter 24) and can be described by a lifetime τ. The rate at which a population N^* of excited state molecules will decay is given by

$$\frac{\mathrm{d}N^*}{\mathrm{d}t} = -k_{\mathrm{fl}} \cdot N^* \tag{10.14}$$

where the rate constant k_{fl} is connected with the natural lifetime τ_0 by

$$k_{\mathrm{fl}} = \frac{1}{\tau_0} \tag{10.15}$$

In most cases radiation is not the only deexcitation process. Other processes, such as thermal deactivation or a chemical reaction of the molecule in the excited state, can occur. Since these processes reduce the number of excited states and they do not cause light emission, they reduce the amount of fluorescence, or *quench the fluorescence*. In the given environment of the molecule we can characterize these processes by a rate constant k', and the rate equation becomes

$$\frac{\mathrm{d}N^*}{\mathrm{d}t} = -(k_{\mathrm{fl}} + k') \cdot N^* \tag{10.16}$$

The total rate constant $(k_{\mathrm{fl}} + k')$ with the observed lifetime τ by

$$k_{\mathrm{fl}} + k' = \frac{1}{\tau} \tag{10.17}$$

For $k' = 0$ the only deexcitation process is fluorescence emission and a maximum fluorescence intensity I_0 is observed. In contrast, when $k' \gg k_{\mathrm{fl}}$ the competitive process dominates and no fluorescence light appears. In between these extremes a fluorescence intensity I is observed which is smaller than I_0.

The ratio $\phi = I/I_0$ is called the *quantum yield of fluorescence*, and it is related to the rate constants by

$$\phi = \frac{I}{I_0} = \frac{k_{\mathrm{fl}}}{k_{\mathrm{fl}} + k'} = \frac{1}{1 + k'/k_{\mathrm{fl}}} \tag{10.18}$$

(known as the *Stern–Volmer Equation*). Correspondingly, the lifetime τ of the excited state is

$$\tau = \frac{1}{k_{\mathrm{fl}} + k'} = \tau_0 \cdot \phi \tag{10.19}$$

Note that the analysis performed here is an example of that discussed for first-order chemical kinetics in Chapter 24.

Example 10.2 *Consider the molecule trans-1,2-diphenylethylene (trans-stilbene) in the solvent hexane at 298 K. After photoexcitation with 330 nm light this molecule relaxes to its ground state by two primary pathways, fluorescence emission or a conformational change to generate the cis isomer. Experimental measurements give 80 ps for the excited state lifetime and 0.04 for the fluorescence quantum yield. Use these values to determine the radiative and nonradiative rate constants.*

By rearrangement of Equation (10.19) we can determine the radiative lifetime of the molecule by

$$\tau_0 = \frac{\tau}{\phi} = \frac{80 \times 10^{-12}\ \mathrm{s}}{0.04} = 2 \times 10^{-9}\ \mathrm{s}$$

which gives a radiative rate constant k_{fl} of $5 \times 10^8\ \mathrm{s}^{-1}$. Now we can use Equation (10.17) to find the nonradiative rate constant k',

$$k' = \frac{1}{\tau} - k_{\mathrm{fl}} = \frac{1}{80 \times 10^{-12}\ \mathrm{s}} - 5 \times 10^8\ \mathrm{s}^{-1} = 1.1 \times 10^{10}\ \mathrm{s}^{-1}$$

10.2.7 Absorption from Excited States

By using time-resolved spectroscopies, we can probe optical transitions between excited states, and this allows us to probe features of molecular excited states. In Fig. 10.9, we generalize the energy scheme in Fig. 10.8 to include higher excited singlet and triplet states. For example, time-resolved methods allow us to probe triplet states, even though direct transitions between singlet and triplet states are extremely weak. This limitation arises because the light wave's electric field does not act on the spin state of the electron; hence, the spin wavefunction does not change. In one approach to probing triplet states, we can excite a molecule to a singlet state in which spin-orbit coupling (see Chapter 5) may occur, and the singlet state may convert to a triplet state. Once in the triplet state T_1, a transition from T_1 to T_2, or higher excited triplet states, can be induced by a light wave with appropriate wavelength (*triplet–triplet absorption*). We can also perform time-resolved experiments with singlet states; i.e., populate an excited singlet state with one light source and then probe its absorption spectrum with a second light source (see Fig. 10.10). Essential features to successful studies of this sort are to have an initial light source that is intense enough to create a large population of the excited states and a second light source that probes this excited state population before it decays away. The development of laser methods has made studies like this relatively common for both singlet and triplet states.

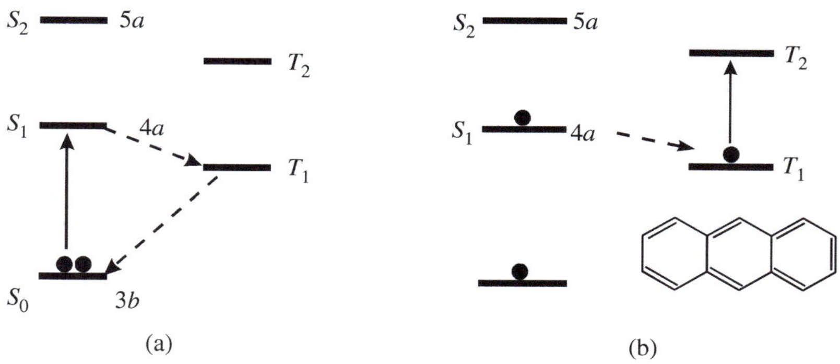

Figure 10.9 Triplet–triplet absorption in anthracene. Energy scheme for the anthracene as a two-electron system including higher singlet (S_2) and triplet (T_2) electronic states (for the notation of the states see Fig. 9.26). Panel (a) shows promotion of an electron from the ground-state singlet to the first excited singlet state, S_1, which can then undergo intersystem crossing to generate the triplet state T_1. Panel (b) shows the light-induced transition from T_1 to T_2, which can be approximated by the energy difference between 4*a* and 5*a* (Fig. 9.26). The calculated excitation energy is $\Delta E_{4a\to 5a} = 3.9 \times 10^{-19}$ J corresponding to $\lambda = 510$ nm; in the experiment we find $\lambda = 424$ nm (see Fig. 10.10).

10.2.7.1 Anthracene: $T_1 \to T_2$ Absorption

As an example, we consider anthracene. In the free-electron model the orbitals and energy levels in Fig. 9.26 are obtained. For the transition from the S_0 ground state to the S_1 first excited state ($3b \to 4a$), we calculate an absorption maximum at 496 nm and the transition is polarized in the *y*-direction. Experimentally this transition appears at 363 nm [7] with the predicted polarization. We can induce this transition by a short pulse of light. Subsequent to the initial excitation, the triplet state T_1 is populated by intersystem crossing (Fig. 10.9).

When simultaneously illuminating the sample with a second source of light with a wavelength of 424 nm, a transition from T_1 to T_2 can be observed, and it corresponds to the promotion of an electron from the orbital 4*a* to the orbital 5*a*. Note that the lower transition $4a \to 4b$ (see Fig. 9.24) is forbidden, because the transition moments are zero: $M_x = M_y = M_z = 0$. This result can be seen from the symmetry of the wavefunctions ψ_{4a} and ψ_{4b}. The lifetime of the T_1 state is on the order of some seconds, so that the population of molecules in the T_1 state diminishes over this time interval and the absorbance decreases accordingly (Fig. 10.10b).

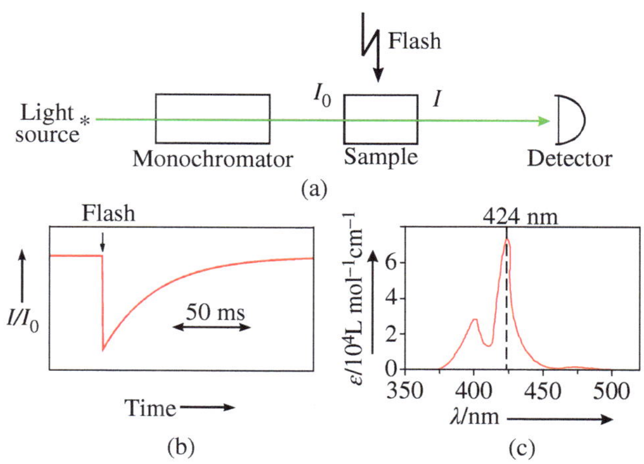

Figure 10.10 Triplet–triplet absorption of anthracene. (a) Experimental setup. (b) Intensity of light ($\lambda = 424$ nm) passing the sample. By starting the flash, T_1 is occupied and a strong $T_1 \to T_2$ absorption is observed. According to the deexcitation of T_1 the $T_1 \to T_2$ absorbance decreases exponentially (increase of the transmittance I/I_0) with time. (c) Absorption spectrum for the $T_1 \to T_2$ transition. The experiment was carried out with anthracene in a matrix of polymethylmethacrylate in order to minimize radiationless deexcitations.

Actually, the considered transition occurs from the T_1 state (energy 2.92×10^{-19} J) to the T_3 state (energy 7.61×10^{-19} J), which is only slightly higher that the S_2 state (energy 7.17×10^{-19} J). The energy difference $\Delta E = E_{T_3} - E_{T_1} = 4.69 \times 10^{-19}$ J corresponds to the absorption wavelength of 424 nm, which is used in the experiment in Fig. 10.10; see [8]. Figure 10.14c shows the absorption spectrum for this transition. To obtain the spectrum, the experiment in Fig. 10.14a,b is repeated by measuring I/I_0 in Fig. 10.14b for different wavelengths.

10.2.7.2 Anthracene: $S_1 \rightarrow S_2$ Absorption

Instead of following the $T_1 \rightarrow T_2$ absorption, the $S_1 \rightarrow S_2$ absorption can be measured, if a fast laser pulse (a time duration much shorter than the lifetime of S_1, about 10^{-9} seconds) is used for excitation, and a second fast laser pulse is used to measure the $S_1 \rightarrow S_2$ absorption (singlet–singlet absorption).

10.3 Stimulated Emission and Laser Action

In the preceding sections our discussion of different processes for the interaction of light with molecules has emphasized spontaneous emission and absorption. Now we turn our attention to stimulated emission which is the third process in Einstein's description for the emission and absorption of radiation:

$$B_{a\rightarrow b} = B_{b\rightarrow a}$$

where $B_{a\rightarrow b}$ is the Einstein coefficient of absorption and $B_{b\rightarrow a}$ is the Einstein coefficient for stimulated emission, see Section 10.2.1.1. In spontaneous emission an excited molecule relaxes to a lower energy state (usually the ground state) and emits a photon of light. This process is not driven by an externally applied field and is called *spontaneous* (Fig. 10.11a). In absorption we apply an external light field, and a ground-state molecule can be energized to an electronically excited state (Fig. 10.11c). In fact, it would be appropriate to describe this process as stimulated absorption, because the applied light field drives the electronic transition in the molecule. In stimulated emission, we apply a light field to an excited molecule and cause it to relax to a lower energy state by emitting a photon (Fig. 10.11b). Stimulated emission is just the reverse of absorption. For the absorption process the intensity of a light wave decreases when passing through a sample containing molecules in the ground state (Fig. 10.11c), whereas in stimulated emission the intensity of the light increases when passing through a sample containing molecules in the excited state (Fig. 10.11b). Hence a molecule's relaxation by stimulated emission generates a new photon for the applied light field. More than this, the new photon is an exact replica of the photon that stimulated the emission; i.e., it has the same phase and the same propagation direction as the incident photon.

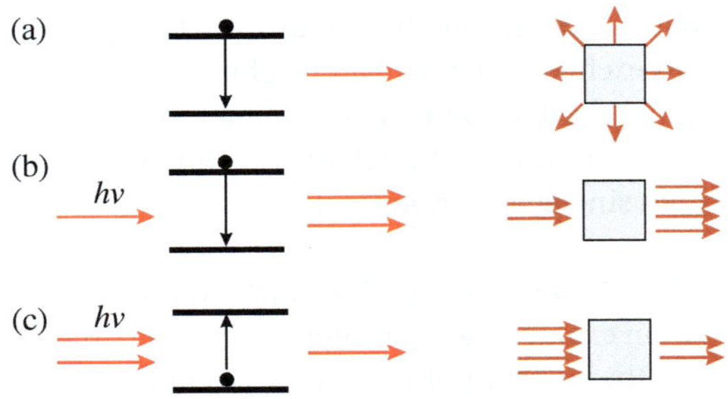

Figure 10.11 Emission and absorption of light. (a) Spontaneous emission, (b) stimulated emission, and (c) absorption.

The stimulated emission phenomenon is the fundamental physical process underlying the operation of a laser. In fact the word laser, originated as an acronym (Light Amplification by Stimulated Emission of Radiation). The early development of laser devices largely occurred through the efforts of physicists and physical chemists, and lasers have had a profound impact on a chemist's ability to study and manipulate the behavior of atoms and molecules. Beyond pure science, laser devices are increasingly used in medicine, communications, information technology, and sensing.

10.3.1 Inversion of Population

To observe stimulated emission in a solution of dye molecules, more molecules must be in the excited state than in the ground state (contrast Figure 10.12a,b). Otherwise the reverse process (absorption) predominates. In fact, the probability that a light field of intensity I_0 induces absorption in a ground-state molecule is identical to the probability of that same light field inducing emission in the same molecule when it is in an excited state. Which process predominates is determined by the relative populations of molecules in the ground states and excited states in the sample.

The situation required for observing stimulated emission can be reached by preparing the sample in an excited state. This preparation can take many forms, ranging from electrical current to chemical reactions. For a solution of dye molecules, one can apply a strong light field, or flash of light, that excites the dye molecules to state 4 in Fig. 10.12a. The molecules relax by evolving heat (e.g., transferring energy to vibrational motions of nuclei) until they reach energy level 3, where they remain for a while, typically nanoseconds. Thus for a short time more electronically excited molecules are present than in the lower energy state 2 to which the dye molecule relaxes by light emission. Hence the population is inverted, more molecules are in the upper (electronically excited) level 3 than in the lower state 2. If photons of appropriate frequency $\nu = (E_3 - E_2)/h$ impinge on the dye solution within the very short time interval during which the population is inverted, they will induce, by stimulated emission, the creation of more photons (with frequency ν and traveling in the same direction) and the number of photons (light intensity) will grow. Although some absorption of the light is also possible, the population inversion biases the dye solution so that stimulated emission is more probable than absorption.

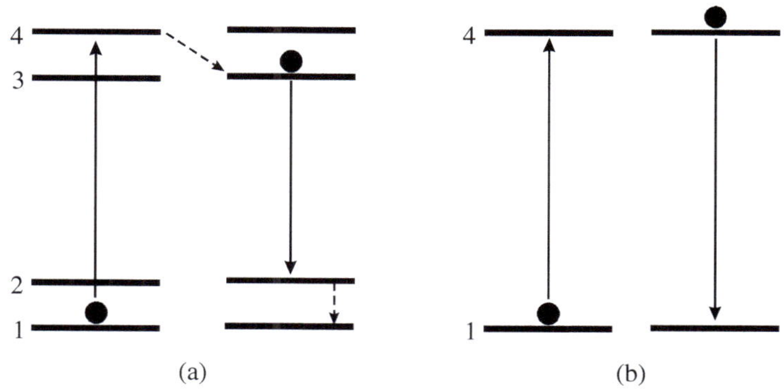

Figure 10.12 Comparison of four-level system (a) and two-level system (b). Dye laser operation is possible in the four-level system, but not in the two-level system.

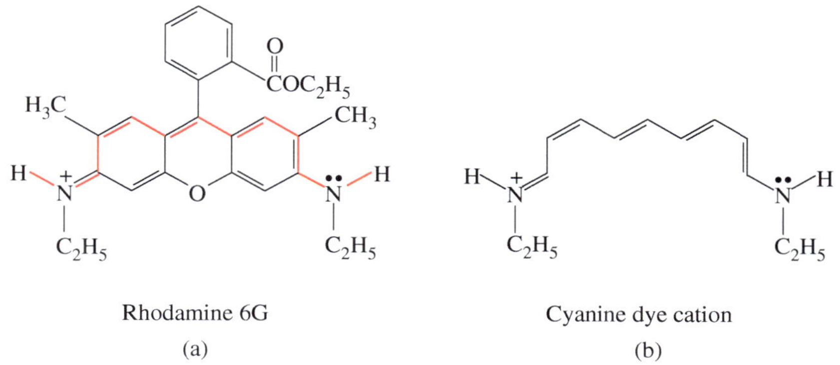

10.3.2 Dye Laser Operation

A very common dye used in dye lasers is the dye cation rhodamine 6G. Its π-electron chromophore is similar to a cyanine cation (red outlined region in rhodamine 6G structure) with 12 π electrons. According to Table 9.2, we expect an absorption maximum at 625 nm. Because of rhodamine 6G's branched structure it absorbs at somewhat shorter wavelengths (experimental absorption maximum at 522 nm). In contrast to the cyanine dye molecule, rhodamine 6G has a rigid structure, so that the lifetime of the excited state is long enough to allow efficient emission of the excitation energy. The fluorescence quantum yield of rhodamine 6G is nearly one, and the emission maximum is at 556 nm in methanol (see Fig. 10.13).

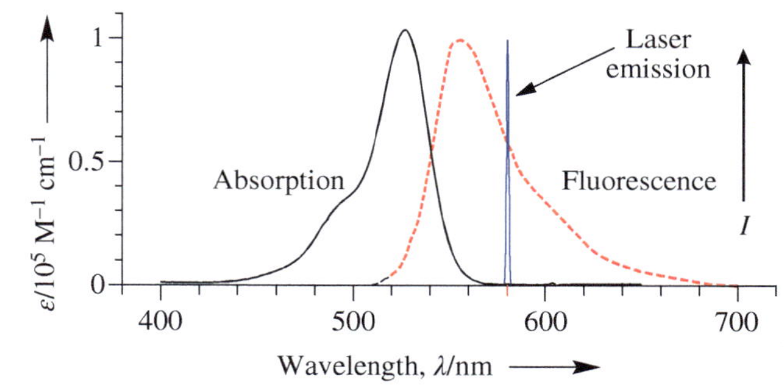

Figure 10.13 Rhodamine 6G as an example for a laser dye. Absorption spectrum (black solid line, left-hand scale), fluorescence spectrum (red dashed line, maximum at 556 nm), and laser emission spectrum (blue solid line, maximum at 580 nm), intensity I in arbitrary units. The concentration of the dye is 1×10^{-5} M, dissolved in methanol (optical path length 5 cm).

10.3.2.1 Pumping

A dye laser can be constructed by positioning a solution of dye molecules in an optical cell between two parallel mirrors and illuminating the solution with a strong flash of light (Fig. 10.14). Some of the light is absorbed by the dye molecules, and electronically excited dye molecules are created. These excited molecules shift their energy by thermal relaxation processes; consequently, the emission intensity no longer overlaps with the absorption. If enough of these excited state molecules are created, then stimulated emission predominates. The excited dye solution is called the *gain medium* because it leads to amplification, or gain, of the light field.

A number of different energy level schemes can be used for the gain medium of a laser. The dye laser is an example of a four-level gain medium (Fig. 10.12a). The light flash pumps the dye molecules from the ground state (level 1) to an excited state (level 4). The dye molecules lose energy as heat to the surroundings and relax to level 3, which is still in the electronically excited state. Subsequently, the dye molecule undergoes stimulated emission and relaxes from level 3 to level 2, which is in the ground electronic state but at a slightly higher internal energy (i.e., an excited vibrational state of the ground electronic state, see Chapter 11).

10.3.2.2 Laser Action

In the initial stages of the laser's operation, the gain medium emits a photon of frequency ν by spontaneous emission. This photon acts as a seed that stimulates the emission of more photons (avalanche effect). This increase in the light intensity is called gain, and the collection of excited dye molecules is called the dye medium. Only the photons which happen to be moving perpendicular to the mirrors are reflected back and forth through the dye solution. This feedback between the light field in the laser cavity and the gain medium creates a "survival of the fittest" mechanism, in which a subset of the photons, initially generated by spontaneous emission, grow in population to the exclusion of others. After the photons make 10 to a 100 transits back and forth between the mirrors, an intense narrow beam of light, which is moving perpendicular to the mirrors, is created. If one of the mirrors (mirror 2 in Fig. 10.14) is partially transparent (typically about 1% to 10%), then a narrow beam of coherent light perpendicular to the mirrors leaves the sample (or laser cavity).

The wavelength of the emitted laser light is determined by the spontaneous emission that initiates the process and whether it overlaps with the wavelength λ_m of one of the standing waves supported by the laser cavity. Although the energy of the spontaneous emission and stimulated emission may be spread over a range of energies, the transitions near the fluorescence maximum are preferentially amplified in the laser. Once the feedback process takes hold, the laser light emission occurs in a narrow range of wavelengths near the fluorescence maximum. For example, dye laser light has a much narrower bandwidth than does spontaneous fluorescence: 2 nm compared with 40 nm for the case shown in Fig. 10.13.

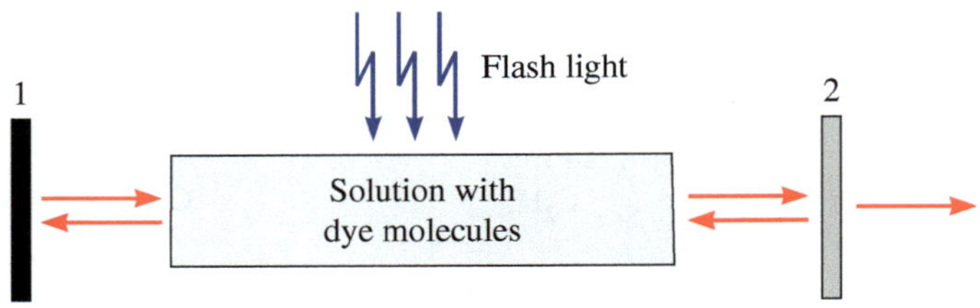

Figure 10.14 Experimental setup of a dye laser. The dye (e.g., rhodamine 6G) dissolved in methanol is in an optical cell which is placed between two parallel mirrors 1 and 2, where mirror 2 is partially transparent (1%). The dye molecules are excited by means of a strong light flash. The space of length L, which is defined by the mirrors 1 and 2, is called the laser cavity, and it supports light wavelengths $\lambda_m = 2L/m$ where the mode number m is an integer $1, 2, 3, \ldots$.

In Fig. 10.13 it can be seen that the laser emission occurs at longer wavelengths relative to the maximum of the fluorescence band. What is the reason for this behavior? It is the overlap of fluorescence and absorption bands. Part of the emitted fluorescence light can be reabsorbed by surrounding molecules which are in the electronic ground state. The reabsorption is strongest at the maximum of the absorption band. But even at the fluorescence maximum the reabsorption can be strong enough to cause a decrease in the gain of the laser. Hence the laser emission is more probable at longer wavelengths where the overlap of fluorescence and absorption is much smaller.

10.3.2.3 Tuning

Although the laser will evolve to a steady state with a preferred wavelength of the emitted light, it is possible to bias the device so that it operates at a desired wavelength. For example, one can use the reabsorption of emitted photons to shift the frequency of the light. The reabsorption increases with increasing concentration of the dye molecules, and this can shift the laser emission to longer wavelengths with increasing dye concentration. The laser emission at 580 nm indicated in Fig. 10.13 occurs at a dye concentration of 10^{-5} mol L^{-1}; however, the emission wavelength is 585 nm at 10^{-4} mol L^{-1} and 600 nm at 10^{-3} mol L^{-1}. Thus changing the dye concentration can be used to tune the laser wavelength within the range of the fluorescence band.

It is more common however to add one or more tuning elements (such as a diffraction grating or a prism) in the laser. For example, one of the mirrors in Fig. 10.14 can be replaced by a grating. A grating reflects only light in a very small wavelength range back into the direction of the incident laser beam. By rotating the grating, the laser light can be tuned to any wavelength in the range of the fluorescence band (tunable dye laser) and the spectral band width can be further decreased. Using dyes with various positions of the fluorescence maximum, laser light can be produced at any wavelength between 300 and 900 nm.

10.3.3 Excimer and Exciplex Laser

In the ultraviolet region, a common type of laser gain medium uses excimers (*exc*ited d*imers*) or exciplexes (*exc*ited com*plexes*). These excited complexes are only stable in the excited state. The simplest example of an excimer is the He_2^* molecule. From a simple molecular orbital picture we predict that helium dimer will not form (see Chapter 7); however, that same model predicts that the excited state species He_2^* is bound (see Exercise 10.10) and exists as a stable excited state species for milliseconds. Common exciplex-based lasers generate excited state complexes between noble gas atoms and halogen atoms with an electrical discharge. The exciplex emits ultraviolet light during its transition back to the ground state. Because the ground state is unstable, the ground electronic state complex dissociates into the component atoms within a time of 1 ps. Consequently, it is easy to maintain the population inversion (more molecules in the excited state than in the ground state). The wavelength range of exciplex lasers extends from 193 nm (ArF laser) to 352 nm (XeF laser).

10.4 Conclusion

We calculated the fluorescence lifetime using the classical oscillating dipole model introduced in the previous chapter. We studied the Einstein postulates relating absorption, spontaneous emission, and stimulated emission, and we used this classical model to show their plausibility. These phenomena provide a basis for understanding the operation of lasers. Our considerations have shown that simple models, which can guide our evaluation of theory and experiment in proper limits, are of great benefit.

10.5 Box 10.1. Einstein Coefficients

In Foundation 9.1, we showed that the average integrated power absorbed by one molecule is

$$\left(\frac{\overline{\mathrm{d}E}}{\mathrm{d}t}\right)_{\text{per molecule}} = \frac{Q^2}{m}\frac{1}{12\varepsilon_0\hat{n}^2}\rho_\nu$$

and in Foundation 9.2 we derived the relationship

$$\frac{Q^2}{m} = \hat{n}\frac{e^2}{m_e}f$$

where f is the oscillatory strength of the transition. Combining these results, we can write that the integrated power absorbed by N_a molecules is

$$\left(\frac{\overline{\mathrm{d}E}}{\mathrm{d}t}\right)_{a\to b} = \frac{Q^2}{m}\frac{1}{12\varepsilon_0}\rho_\nu(\nu_{ab})\cdot N_a = \frac{e^2}{12\varepsilon_0 m_e}f\cdot\rho_\nu(\nu_{ab})\cdot N_a \tag{10B1.1}$$

for vacuum conditions ($\hat{n} = 1$). Accordingly, in the quantum-mechanical view, the rate is

$$\left(\frac{\mathrm{d}N}{\mathrm{d}t}\right)_{a\to b} = \left(\frac{\overline{\mathrm{d}E}}{\mathrm{d}t}\right)_{a\to b}\frac{1}{\Delta E} = \frac{e^2}{12\varepsilon_0 m_e h\nu_{ab}}f\cdot\rho_\nu(\nu_{ab})\cdot N_a \tag{10B1.2}$$

with $\Delta E = h\nu_{ab}$. By comparing this result with Equation (10.5), we find that the Einstein coefficient $B_{a\to b}$ as

$$B_{a\to b} = \frac{e^2}{12\varepsilon_0 m_e h\nu_{ab}}f \tag{10B1.3}$$

On the other hand, the coefficient $A_{b\to a}$ for the spontaneous emission rate must be equal to the inverse of the natural lifetime τ_0 given by Equation (10.1)

$$A_{b\to a} = \frac{1}{\tau_0} = \frac{2\pi e^2\nu_{ab}^2}{3m_e\varepsilon_0 c_0^3}f \tag{10B1.4}$$

(with the notation ν_{ab} instead of ν_0). If we divide Equation (10B1.4) by (10B1.3), we find that

$$\frac{A_{b\to a}}{B_{a\to b}} = \frac{2\pi e^2\nu_{ab}^2}{3m_e\varepsilon_0 c_0^3}f\cdot\frac{12\varepsilon_0 m_e h\nu_{ab}}{e^2}\frac{1}{f} = \frac{8\pi h\nu_{ab}^3}{c_0^3} \tag{10B1.5}$$

Using Einstein's postulate that

$$B_{b\to a} = B_{a\to b} \tag{10B1.6}$$

which is closely related to the classical theory according to which an oscillator interacting with an electromagnetic wave can either absorb energy from the field or lose energy to it, and substituting into Equation (10B1.5) for $B_{a\to b}$, we find Einstein's relation (10.7) between $A_{b\to a}$ and $B_{b\to a}$.

In Equation (9.27) we found a relation between the oscillatory strength f and the transition moment M

$$f = \frac{8\pi^2 m_e}{h^2 e^2} \cdot M^2 \cdot \Delta E \tag{10B1.7}$$

This relation follows from the time-dependent Schrödinger equation (9.24), where we added a first-order perturbation term including the electric field strength F of the light wave. If we substitute this expression for f into the expressions for the Einstein coefficients B and A we obtain

$$B_{a\to b} = \frac{2\pi^2}{3h^2\varepsilon_0}M^2 \quad \text{and} \quad A_{b\to a} = \frac{16\pi^3\nu_{ab}^3}{3\varepsilon_0 c_0^3 h}M^2$$

10.6 Foundations

10.6.1 Foundation 10.1: Fluorescence Lifetime

10.6.1.1 Antenna Equation of Hertz

A classical oscillator (charge Q, displacement ξ) emits electromagnetic waves. Heinrich Rudolf Hertz (1847–1894) first solved this problem for the case of radiowave transmission. In this treatment a dipole is oscillating and generating an electromagnetic wave. We can use the rate of light emission from a dipole as a model to calculate the rate of fluorescence emission from a transition dipole.

We consider the dipole to be located at the center of a sphere with radius r. According to the wave equations developed by Hertz [9] the electric field at a point P (Fig. 10F.1) has a component F_a along the direction $\vec{r}$,

$$F_a = \frac{2Q\cos\vartheta}{4\pi\varepsilon_0}\left\{\frac{1}{c_0 r^2}\frac{\mathrm{d}\xi}{\mathrm{d}t} + \frac{1}{r^3}\xi\right\} \tag{10F1.1}$$

and a component F_b perpendicular to the direction $\vec{r}$

$$F_b = \frac{Q\sin\vartheta}{4\pi\varepsilon_0}\left\{\frac{1}{c_0^2 r}\frac{\mathrm{d}^2\xi}{\mathrm{d}t^2} + \frac{1}{c_0 r^2}\frac{\mathrm{d}\xi}{\mathrm{d}t} + \frac{1}{r^3}\xi\right\} \tag{10F1.2}$$

where ϑ is the angle between the direction of the dipole orientation and the position vector $\vec{r}$ and

$$\xi = \xi_0\cos(\omega t - \gamma), \quad \gamma = \frac{\omega r}{c_0} = \frac{2\pi r}{\lambda} \tag{10F1.3}$$

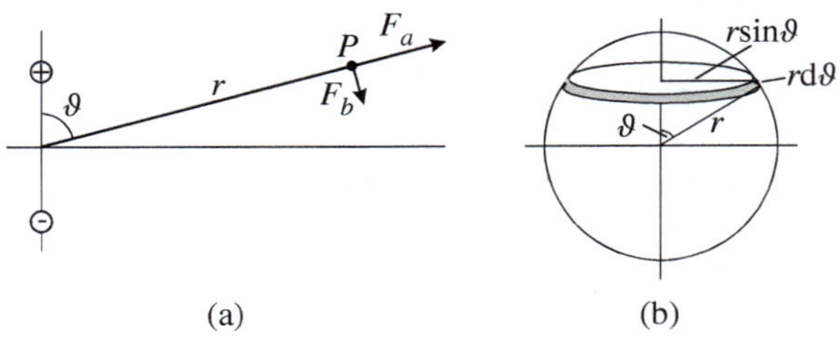

Figure 10F.1 (a) Electric field strength at point P which is at a distance r from an oscillating dipole: components F_a (in direction of $\vec{r}$) and F_b (perpendicular to the direction of $\vec{r}$). (b) Area element $2\pi r^2 \sin\vartheta \cdot \mathrm{d}\vartheta$ on the surface of a sphere (shaded area). In both (a) and (b) the z-direction is vertical, along the dipole axis.

is the displacement of the oscillator. Inserting these expressions into Equations (10F1.1) and (10F1.2), we obtain

$$F_a = \frac{2Q\xi_0\cos\vartheta}{4\pi\varepsilon_0}\xi_0\left\{-\frac{\omega}{c_0 r^2}\sin(\omega t - \gamma) + \frac{1}{r^3}\cos(\omega t - \gamma)\right\} \tag{10F1.4}$$

$$F_b = \frac{Q\xi_0\sin\vartheta}{4\pi\varepsilon_0}\xi_0 \cdot \left\{-\frac{\omega^2}{c_0^2 r}\cos(\omega t - \gamma) - \frac{\omega}{c_0 r^2}\sin(\omega t - \gamma) + \frac{1}{r^3}\cos(\omega t - \gamma)\right\} \tag{10F1.5}$$

The energy $\mathrm{d}E$ emitted by the oscillator in time $\mathrm{d}t$ can be determined by finding the light intensity I on a spherical surface of large radius r. We choose the sphere's radius to be large because we only wish to know the electric field components that escape from the dipole and propagate away into free space. The emitted power is

$$\text{Emitted power} = -\frac{\mathrm{d}E}{\mathrm{d}t} = \int_{\text{surface of sphere}} I \cdot \mathrm{d}A \tag{10F1.6}$$

where $\mathrm{d}A = r^2 \sin\theta \,\mathrm{d}\theta\,\mathrm{d}\varphi$ is the surface area element on the sphere. The intensity I can be expressed by the amplitude F_0 of the electric field (see Foundation 9.1):

$$I = \frac{c_0 \varepsilon_0}{2} F_0^2 \tag{10F1.7}$$

This expression holds for vacuum conditions (refractive index $\hat{n} = 1$). The amplitude F_0 is determined from the expressions for F_a and F_b. At large distances we can neglect the terms with $1/r^2$ and $1/r^3$, because they will be small. In this case we find that

$$F_a = 0,\; F_b = -\frac{Q\xi_0 \sin\vartheta}{4\pi\varepsilon_0}\frac{\omega^2}{c_0^2 r} \cdot \cos(\omega t - \gamma) \tag{10F1.8}$$

so that the amplitude is

$$F_0 = \frac{Q\xi_0}{4\pi\varepsilon_0}\frac{\omega^2}{c_0^2 r} \sin\vartheta \tag{10F1.9}$$

Thus the emitted light has the same intensity in all directions in the xy plane and no light is emitted in the z direction:

$$I = \frac{c_0\varepsilon_0}{2}\left(\frac{Q\xi_0}{4\pi\varepsilon_0}\frac{\omega^2}{c_0^2 r}\sin\vartheta\right)^2 \tag{10F1.10}$$

$$= \frac{c_0\varepsilon_0}{2}\frac{1}{\left(4\pi\varepsilon_0\right)^2}\frac{\left(Q\xi_0\right)^2\omega^4}{c_0^4 r^2}\sin^2\vartheta$$

Using this form for I, we find that the emitted power is

$$-\frac{\mathrm{d}E}{\mathrm{d}t} = \frac{Q^2\xi_0^2}{(4\pi\varepsilon_0)^2}\frac{\omega^4}{c_0^4}\frac{c_0\varepsilon_0}{2}\int\frac{\sin^2\vartheta}{r^2}\cdot 2\pi r^2 \sin\vartheta \cdot \mathrm{d}\,\vartheta = \frac{Q^2\xi_0^2}{4\pi\varepsilon_0}\frac{\omega^4}{4c_0^3}\cdot\frac{4}{3} \tag{10F1.11}$$

or

$$-\frac{\mathrm{d}E}{\mathrm{d}t} = \frac{1}{4\pi\varepsilon_0}\frac{Q^2\omega^4}{3c_0^3}\xi_0^2 \tag{10F1.12}$$

With

$$E = \frac{k_f}{2}\xi_0^2 \quad \text{and} \quad v_0^2 = \frac{1}{4\pi^2}\frac{k_f}{m} \tag{10F1.13}$$

(k_f is the force constant of the oscillator), we obtain

$$-\frac{\mathrm{d}E}{\mathrm{d}t} = \frac{1}{4\pi\varepsilon_0}\frac{2}{3}\frac{Q^2\omega^2}{mc_0^3}\cdot E = \frac{1}{\tau_0}\cdot E \tag{10F1.14}$$

10.6.1.2 Natural Lifetime

Correspondingly, for an excited molecule emitting a light quantum the average lifetime of the excited state equals τ_0. In this case Q^2/m can be replaced by

$$\frac{Q^2}{m} = f\cdot\frac{e^2}{m_e} \tag{10F1.15}$$

where f is the oscillator strength (see Foundation 9.2). $1/\tau_0$ is the radiative rate constant and τ_0 is called the *natural lifetime*.

$$\tau_0 = \frac{3m\varepsilon_0 c_0^3}{2\pi Q^2 v_0^2} \quad \text{or} \quad \tau_0 = \frac{3m_e\varepsilon_0 c_0^3}{2\pi e^2 v_0^2}\cdot\frac{1}{f} \tag{10F1.16}$$

10.6.1.3 Lifetime and Line Width

The displacement x of a damped classical oscillator is determined by the differential equation

$$\frac{\mathrm{d}^2 x}{\mathrm{d}t^2} + \frac{k_d}{m}\frac{\mathrm{d}x}{\mathrm{d}t} + \frac{k_f}{m}x = 0$$

where k_d is the damping constant, k_f is the force constant, and m is the oscillator's mass. In the weak damping limit (k_d small), we can solve this differential equation for the displacement x of the oscillator and obtain

$$x = Ce^{-(k_d/2m)t} \cdot \cos(\omega_0 t), \quad \omega_0 = \sqrt{\frac{k_f}{m}}$$

where C is a constant and the change of the resonance frequency by the damping force is neglected. In Exercise 10.10, you show that this expression for x solves the equation. Thus the amplitude x_0 of the oscillator is

$$x_0 = Ce^{-(k_d/2m)t}$$

Because the intensity I of the radiation emitted from an oscillating dipole is proportional to the square of the amplitude we obtain

$$I = C'e^{-(k_d/m)t} = C'e^{-t/\tau_0}$$

where $\tau_0 = \frac{m}{k_d}$ is the lifetime.

In Foundation 9.1 we considered a damped classical oscillator excited by an external electric field, and we calculated the resonance curve for the absorption coefficient ε. We found in Equation (9.86) that the amplitude is one-half of the maximum value of the resonance curve (circular frequency ω_0) at

$$\Delta\omega = \pm\frac{1}{2}\frac{k_d}{m}$$

where $\Delta\omega = \omega_0 - \omega$. Thus for the line width $2\Delta\omega$ we obtain

$$2\Delta\omega = \frac{k_d}{m}$$

This relation is also valid for the spectral band width of the damped oscillator; thus, we find that

$$2\Delta\omega = \frac{1}{\tau_0}$$

or, in terms of the frequency ν,

$$\text{natural spectral line width} = 2\Delta\nu = \frac{1}{2\pi}\frac{1}{\tau_0}$$

10.6.2 Foundation 10.2: Calculation of Repulsion Integrals

See the books website, www.wiley.com/go/Kuhn/physicalchemistry3e.

10.6.3 Foundation 10.3: Fluorescence Microscopy: STED, MINSTED

See the books website, www.wiley.com/go/Kuhn/physicalchemistry3e.

10.7 Exercises

E10.1 The molecule 9,10-diphenylanthracene is a commonly used quantum yield standard in fluorescence spectroscopy. Its quantum yield is measured to be $\phi = 0.95$ and its fluorescence lifetime τ is 8.2 ns. Determine the radiative lifetime, the radiative rate constant, and the nonradiative rate constant for this molecule.

E10.2 Single molecule spectroscopy relies on the ability to detect single photons. If a photomultiplier tube has a gain of 10^6 (i.e., generates a million electrons for each photon impinging on it), how many photons per second are needed to generate an average current of 10 pA from the photomultiplier tube. If a single photon strikes the photomultiplier tube it generates a short burst of electrons from the tube, estimate how short this burst of electrons is if it has an average current of 1 nA.

E10.3 A He–Ne laser generates 1 mW of laser light at a wavelength of 632.8 nm. Calculate the number of photons per second that are being emitted by the laser.

E10.4 Using the results in Foundations 9.1 and 10.1, relate the radiative rate constant to the molecules absorption spectrum; derive Equation 10.2, i.e.,

$$\frac{1}{\tau_0} = \ln\ 10 \cdot \frac{8\pi\nu_0^2}{3c_0^2\mathbf{N}_A} \cdot \int_0^\infty \varepsilon \cdot d\nu$$

E10.5 The fluorescence excitation spectrum for terrylene (see Fig. 10.3) shows an intrinsic molecular linewidth of 0.1 GHz. Use this linewidth value to calculate the radiative lifetime of the molecule.

E10.6 Consider a 100 mW laser source that has a circular beam cross-section with a diameter d of 3 mm and a wavelength λ of 740 nm. Consider the light beam to consist of 10^8 light pulses per second and the light pulses to have a characteristic time width of 10^{-13} seconds (you may assume that the pulses have a square profile in time). a) Calculate the peak intensity of the light field and compare it to the corresponding instantaneous intensity you would find if the light beam were continuous wave (not pulsed). b) Justify why it is reasonable to assume that the peak field strength of the light pulse is small compared to a typical molecular field strength of 10^{11} V/m. (Hint: Use the fact that $I = c_0 \cdot \epsilon_0 \cdot \hat{n} \cdot F_0^2$).

E10.7 In E10.6) you calculated the peak intensity of a 100 fs light pulse, assuming a diameter d of 3 mm which is typical of an unfocused laser source. a) Using the same parameters as in E10.6) calculate the intensity for a pulse that is focused to a 30 μm diameter spot. b) Calculate the intensity for the case where the beam is focused to near its diffraction limit, taken to be $\lambda/2$. b) Assuming that you can focus the beam to the diffraction limit and that you can change the laser power without changing its other parameters, calculate the power you would need for the peak field strength of the light pulse to be comparable to a typical molecular field strength of 10^{11} V/m.

E10.8 Near U.V. chromophores are commonly found in detergents, because they make the clothing "brighter" and "whiter."Chemical derivatives of the stilbene chromophore (1,2-diphenylethylene) are one such fluorescent chromophore found in "whitening agents."The molecule *trans*-stilbene has a radiative rate constant $k_{\text{rad}} = 1/\tau = 3.6 \times 10^8\ \text{s}^{-1}$. a) If radiative emission determines the lifetime of the excited state, what do you expect for the frequency width of the excited state (i.e., full width of the spectral line at half its maximum)? Assume you have a Lorentzian lineshape; namely

$$I(\omega - \omega_0) = \frac{I_0}{(\omega - \omega_0)^2 + (\gamma/2)^2}$$

and show that $\gamma = 1/\tau$. b) Take the transition energy of stilbene (i.e., from the ground vibrational, ground electronic state to the ground vibrational, excited electronic state) to be $32,234.74 \pm 0.02\ \text{cm}^{-1}$, what precision must your spectrometer have in order to determine the bandwidth of the transition? Give your answer in terms of 1 part in 10^n, where you will determine n.

E10.9 Using a qualitative molecular orbital energy diagram, explain why you do not expect helium to form a stable diatomic molecule (He_2) in its electronic ground state, but that it could form stable bond in an excited electronic state.

E10.10 In Foundation 10.1, we write the equation of motion for a damped harmonic oscillator with a force constant k_f, a mass m, and a damping constant k_d as

$$\frac{d^2x}{dt^2} + \frac{k_d}{m}\frac{dx}{dt} + \frac{k_f}{m}x = 0$$

and claim that it has the solution

$$x(t) = Ce^{-(k_d/2m)t} \cdot \cos(\omega_0 t), \quad \omega_0 = \sqrt{\frac{k_f}{m}}$$

in the underdamped limit. More specifically, the underdamped limit corresponds to the case where the frictional decay rate k_d/m is small compared to the oscillation frequency ω_0. Show that $x(t)$ is a solution of the equation.

E10.11 Compare Equation (10B1.2) to Equation (10.5) and show that

$$B_{a\to b} = \frac{e^2}{12\varepsilon_0 m_e h\nu_{ab}} f$$

10.8 Problems

P10.1 a) The intensity of a light beam I is the product of the light field's energy density and the velocity of light, namely

$$I = c_0 \cdot \rho_{\text{energy}} = c_0 \cdot \frac{1}{2}\varepsilon_0 F^2$$

where F is the light's electric field strength. Calculate the electric field strength for an intensity of 100 W/cm^2.

b) The electric field experienced by a valence electron in a molecule can be estimated as

$$F_{\text{valence}} = -\frac{1}{4\pi\varepsilon_0}\frac{q}{d^2}$$

where d is the distance from the nucleus and q is the shielded nuclear charge, which the valence electron experiences. Use typical values for a π-electron on a carbon atom (let $d = 75$ pm and $q = -3.25\ e$; see Box 8.1, which are appropriate for the p_z-orbital on carbon) to calculate the field strength.

c) Compare your two field strengths. Is it reasonable to assume that the light field is small compared to the molecular field?

P10.2 Calculate the electron–electron repulsion energy for S_1 and T_1, which are discussed in Section 10.2.2. You may use the MathCAD code which is provided, or you can write your own algorithm along the lines given in Foundation 10.2.

P10.3 Using the Einstein A coefficient (Equation (10B1.4)) and our expression for the oscillator strength (Equation (10B1.7)) calculate the lifetime for a $2p_z$ excited state of an H-atom, by assuming that it decays by the electronic transition for an electron in the $2p_z$ orbital of an H-atom to the $1s$ orbital of the H-atom.

P10.4 Show that the transition moment integral between the singlet

$$\psi_{S_1} = \frac{1}{\sqrt{2}}\left[\phi_3(1)\cdot\phi_4(2) + \phi_3(2)\cdot\phi_4(1)\right]$$

and triplet states

$$\psi_{T_1} = \frac{1}{\sqrt{2}}\left[\phi_3(1)\cdot\phi_4(2) - \phi_3(2)\cdot\phi_4(1)\right]$$

given in Section 10.2.2 is zero; thus, the transition is "forbidden."

P10.5 Consider a two-state system (levels a and b) that is in thermal equilibrium with a radiation field having a spectral energy density of $\rho_\nu\left(\nu_{ab}\right)$ and undergoes the three transitions absorption, stimulated emission, and spontaneous emission (see Fig. P10.5).

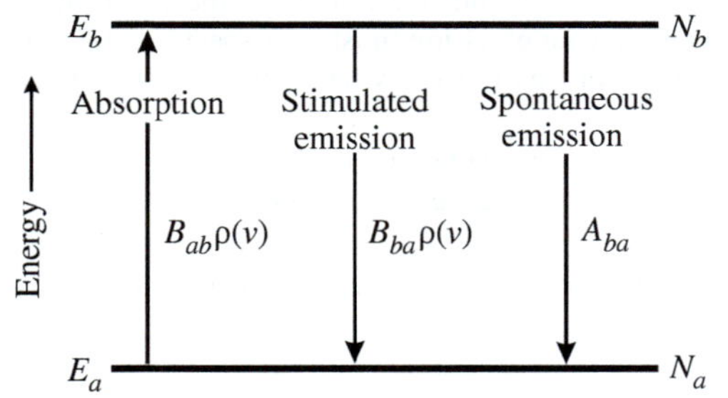

Figure P10.5 Absorption, spontaneous and stimulated emission in a two-level energy system with particle numbers N_a and $N_b = N_a \exp\left[-h\nu/(kT)\right]$.

First, show that the equilibrium condition implies that

$$\rho\left(\nu_{ab}\right) = \frac{A_{ba}}{\left(B_{ab}\cdot\exp\left(h\nu_{ab}/kT\right) - B_{ba}\right)}$$

Second, require that $\rho(\nu)$ be equal to the Planck distribution law ($\rho(\nu) = \left(8\pi\nu^2/c^3\right)(h\nu/(\exp(h\nu/kT) - 1))$) and $B_{ab} = B_{ba}$, and then show that this leads to

$$A_{ba} = \frac{8\pi h\nu_{ab}^3}{c^3}B_{ba}$$

References

1 T. Basche and C. Bräuchle, Single molecule spectroscopy in solids: quantum optics and spectral dynamics, *Ber. Bunsenges. Phys. Chem.*, 1996, **100**, 1269.

2 M. Weber, M. Leutenegger, S. Stoldt, S. Jakobs, T.S. Mihaila, A. N. Butkevich, and S.W. Hell, MINSTED fluorescence localization and nanoscopy, *Nat. Photon.*, 2021, **15**, 361–366.

3 Y. Xiao and X. Qian, Substitution of oxygen with silicon: a big step forward for fluorescent dyes in life science, *Coord. Chem. Rev.*, 2020, **423**, 213513.

4 G.N. Lewis and M. Calvin, Paramagnetism of the phosphorescent state, *J. Am. Chem. Soc.*, 1945, **67**, 1232; M. Kasha, Phosphorence and the role of the triplet state in the electronic excitation of complex molecules, *J. Am. Chem. Soc.*, 1947, **69**, 401–419.

5 G. Baryshnikov, B. Minaev, and H. Ågren, Theory and calculation of the phosphorescence phenomenon, *Chem. Rev.*, 2017, **117**, 6500–6537.

6 M.L. Horng, J.A. Gardecki, A. Papazyan, and M. Maroncelli, Subpicosecond measurements of polar solvation dynamics: coumarin 153 revisited, *J. Phys. Chem.*, 1995, **99**, 17311.

7 A. Staicu, G. Rouille, O. Sukhorukov, Th. Henning, and F. Huisken, Cavity ring-down laser absorption spectroscopy of jet-cooled anthracene, *Mol. Phys.*, 2004, **102**, 1777–1783.

8 G. Porter and M.W. Windsor, The triplet state in fluid media, *Proc. R. Soc. London, Ser. A, Math. Phys. Sci.*, 1958, **245**, 238.

9 See P. Lorrain and D.R. Corson, *Electromagnetic Fields and Waves* (Freeman, New York, 1970); S. Flügge, *Lehrbuch der theoretischen Physik*, Vol. **3** (Springer-Verlag, Berlin, 1961, p. 154).

11

Nuclei: Particle and Wave Properties

CHAPTER MENU

11.1 Introduction

We have seen that the behavior of electrons is governed by their wave–particle duality; however, we have treated nuclei as fixed point charges that define the molecular skeleton. Although their masses are much larger than the electron mass, nuclei have wave–particle properties also, and we explore their consequences here. The nuclear motions can be divided into three types: the overall translation motion of a molecule's center-of-mass, the rotational motion of a molecule about its center of mass, and the relative internal displacement of nuclei with respect to each other (vibrational motion). Although the rotational and vibrational motions of a molecule's nuclei are coupled with each other, they often occur on very different timescales. The internuclear vibrations occur at frequencies of 10^{13} to 10^{14} s^{-1}, whereas the rotations occur at frequencies from 10^{11} s^{-1} to 10^{12} s^{-1}. This timescale separation allows us to approximate the motions as independent.

The translational motions can be described classically. This conclusion is evident if we estimate the de Broglie wavelength, using $\Lambda = h/mv$ where m is the molecular mass. Even for a very light system, such as He or H_2, the de Broglie wavelength is very small, about 20 pm. Because this size is much smaller than the typical volume confining a molecule, it behaves more like a particle than it does like a wave for its translational motion. Hence, we will treat translation as classical motion of the molecule's center of mass, which is discussed in Chapter 14. Currently, we restrict our analysis to the rotational and vibrational motions.

For isolated molecules in the gas phase, the rotational motion must be treated quantum-mechanically. As we will see, this gives rise to rotational periods of about a picosecond and a discrete set of energy levels. The energy level spacings can be probed using absorption spectroscopy; however, the wavelength of the light will correspond to microwaves rather than the ultraviolet or visible light that is typically needed to excite electrons in molecules (see Chapter 9).

During vibrational motion, bond lengths and bond angles of a molecule change, and the electron cloud deforms. In the Born–Oppenheimer treatment, the electron distortion is accounted for by the molecule's electronic energy well which confines the nuclear motion and gives rise to discrete vibrational energy states. This confinement causes the vibrational

Principles of Physical Chemistry, Third Edition. Hans Kuhn, David H. Waldeck, and Horst-Dieter Försterling.

Companion website: www.wiley.com/go/Kuhn/physicalchemistry3e

motions to have larger energy level spacings than the rotational energy level spacings. The absorption of light in the infrared spectral region can be used to probe the vibrational energy level structure of a molecule. We will see how such absorption spectra provide detailed information on vibrational and rotational states and thus detailed knowledge of the molecule's structure.

In the preceding chapters we discussed the position, strength, and polarization of absorption and emission bands on the basis of quantum mechanics, restricting our consideration to electronic wavefunctions. Now we include the vibration of the molecule's nuclei when electronically exciting a molecule. In particular, we investigate light scattering by the molecules (Raman effect) in Section 11.4 and the detailed structure of electronic absorption bands (vibrational structure) in Section 11.5.

11.2 Rotational Motion of Molecules

11.2.1 Diatomic Quantum-Mechanical Rotator

As an example consider the HCl molecule. The Cl atom is much heavier than the H atom ($m_{\mathrm{Cl}} = 35 \cdot m_{\mathrm{H}}$) so that the molecule's center of mass lies very close to the Cl nucleus. For a bond length of $d = 128$ pm, the center of mass lies 3.6 pm from the chlorine nucleus; hence, let us approximate the motion as the H atom circling around the Cl atom on the surface of a sphere of radius d (Fig. 11.1a). From a classical view, the H atom can circle with any speed; however, experiments establish the astonishing fact that the system exists in discrete rotational energy states. Rotation does not contribute to the potential energy, so these energy states correspond to the angular kinetic energy of the molecule and the molecule has discrete angular velocities. The kinetic energy of the lowest rotational quantum state is zero. How can we understand these facts?

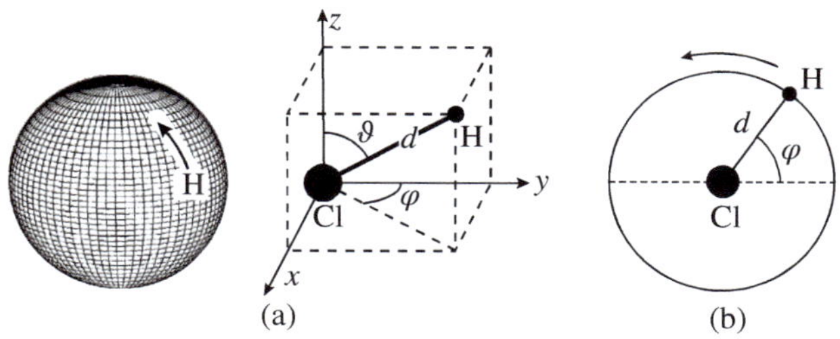

Figure 11.1 Rotational motion of an HCl molecule. The mass of the Cl atom is 35 times larger than that of H, so the Cl atom is practically at rest and the H atom circles at a distance d. (a) Rotation in space: the H atom circles on the surface of a sphere, (b) rotation in a plane.

11.2.1.1 Simplified Model

A simplified model which retains the essence of the nuclear motion's quantization is described here. Consider an H atom moving with a speed v on the circumference of a circle with radius d (Fig. 11.1b). Its kinetic energy is

$$E = \frac{1}{2} m_{\mathrm{H}} v^2 \tag{11.1}$$

Because of wave–particle duality (see Chapter 1), we describe the H nucleus by a standing wave of wavelength Λ along a circle of circumference $2\pi d$. The requirements that the wavefunction be single-valued and continuous gives rise to waves with an integer number of wavelengths on the circle. For Λ we obtain

$$\Lambda = \frac{2\pi d}{n}, \quad n = 0, 1, 2, \ldots \tag{11.2}$$

Because $\Lambda = h/(m_{\mathrm{H}} v)$, the energy E is

$$E = \frac{h^2}{8\pi^2 m_{\mathrm{H}} d^2} \cdot n^2 = B \cdot n^2, \quad n = 0, 1, 2, \ldots \tag{11.3}$$

where we used Equations (11.1) and (11.2). The energy levels are quantized, and the quantity

$$B = \frac{h^2}{8\pi^2 m_{\mathrm{H}} d^2}$$

is called the *rotational constant*. The wavefunctions in the simple rotator are

$$\psi_0 = \frac{1}{\sqrt{2\pi}} \quad \text{for} \quad n = 0 \quad (\Lambda = \infty)$$

$$\psi_{1a} = \frac{1}{\sqrt{\pi}} \sin\varphi, \quad \psi_{1b} = \frac{1}{\sqrt{\pi}} \cos\varphi, \quad \text{for} \quad n = 1 \quad (\Lambda = 2\pi d)$$

$$\psi_{2a} = \frac{1}{\sqrt{\pi}} \sin 2\varphi, \quad \psi_{2b} = \frac{1}{\sqrt{\pi}} \cos 2\varphi, \quad \text{for} \quad n = 2 \quad (\Lambda = \pi d)$$

This simplified model focuses on the heart of the issue by viewing the de Broglie wave interference to generate a standing wave. The first few energy levels and wavefunctions are shown in Fig. 11.2. These wavefunctions are standing waves; however, the solutions can also be expressed as traveling waves corresponding to a clockwise or counter-clockwise rotation. This view was discussed in Sections 3.5 and 4.2.

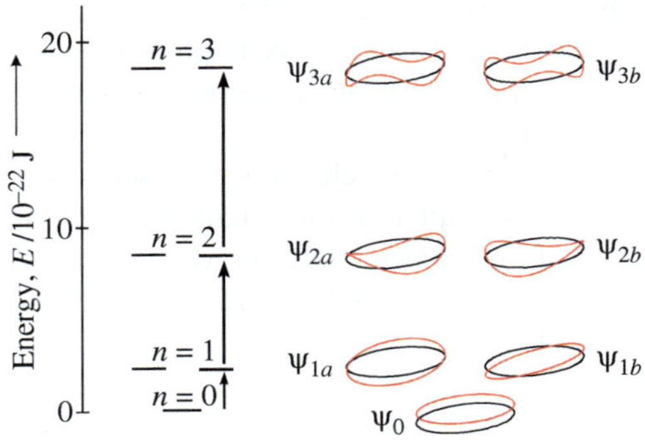

Figure 11.2 Energy levels and wavefunctions of the rotating H nucleus in the HCl molecule (simplified model: rotation in a plane). The arrows indicate the allowed transitions ($\Delta n = 1$).

11.2.1.2 Rigorous Solution

We have to solve the Schrödinger Equation for the H-atom moving about the Cl-atom at a constant potential energy. This motion corresponds to a particle of mass m_H moving on the surface of a sphere with radius d. The Schrödinger equation in this case

$$-\frac{h^2}{8\pi^2 m_H}\left(\frac{\partial^2\psi}{\partial x^2} + \frac{\partial^2\psi}{\partial y^2} + \frac{\partial^2\psi}{\partial z^2}\right) = E\psi$$

with the condition

$$x^2 + y^2 + z^2 = d^2$$

In Foundation 11.1 we solve this Schrödinger equation (using spherical coordinates instead of Cartesian coordinates) and find that the molecule's energy is quantized with energy levels

$$\boxed{E = B \cdot J(J+1), \quad J = 0, 1, 2, \ldots} \tag{11.4}$$

where J is a quantum number and

$$B = \frac{h^2}{8\pi^2 m_H d^2} = \frac{h^2}{8\pi^2 I} \tag{11.5}$$

is the rotational constant. This relation specifies the rotational energy levels (see Fig. 11.3). It tells us that the rotational energy depends on the nuclear mass m and the distance d between the nuclei through the product $I = m_H d^2$, which is called the molecule's *moment of inertia*.

The wavefunctions $\psi(\vartheta, \varphi)$ are polynomials of the sine and cosine functions of the angles ϑ and φ. For the case of a diatomic molecule these functions have the same dependence on angle as the hydrogen atom orbitals (see Foundation 11.1). The first few energy levels and wavefunctions are shown in Fig. 11.3. We obtain one wavefunction for $J = 0$, three wavefunctions for $J = 1$, and five wavefunctions for $J = 2$. The wavefunctions have the functional forms:

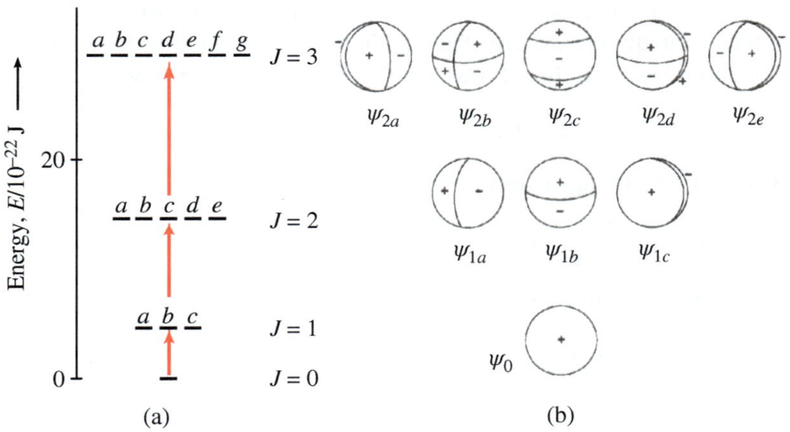

Figure 11.3 Rotational motion of the H nucleus in the HCl molecule (rotation in space). (a) Energy levels for $J = 0$, 1, 2, and 3. The allowed transitions are indicated by arrows. (b) Wavefunctions (indicated by the closed nodal lines and the signs of the wavefunction on the surface of the sphere) for $J = 0$, 1, and 2.

$J = 0$:

$$\psi_0 = \frac{1}{\sqrt{4\pi}} \tag{11.6}$$

$J = 1$:

$$\psi_{1a} = \sqrt{\frac{3}{4\pi}} \cdot \sin\vartheta \cdot \cos\varphi, \quad \psi_{1b} = \sqrt{\frac{3}{4\pi}} \cdot \cos\vartheta, \quad \psi_{1c} = \sqrt{\frac{3}{4\pi}} \cdot \sin\vartheta \cdot \sin\varphi$$

$J = 2$:

$$\psi_{2a} = \sqrt{\frac{15}{16\pi}} \cdot \sin^2\vartheta \cos 2\varphi, \quad \psi_{2b} = \sqrt{\frac{15}{4\pi}} \cdot \sin\vartheta \cos\vartheta \cos\varphi$$

$$\psi_{2c} = \sqrt{\frac{5}{16\pi}} \cdot (3\cos^2\vartheta - 1), \quad \psi_{2d} = \sqrt{\frac{15}{4\pi}} \cdot \sin\vartheta \cos\vartheta \sin\varphi$$

$$\psi_{2e} = \sqrt{\frac{15}{16\pi}} \cdot \sin^2\vartheta \sin 2\varphi$$

In general the degeneracy g_J of energy level j is given by

$$g_J = 2J + 1 \tag{11.7}$$

From the plots in Fig. 11.3 it is evident that the number of nodal lines in the wavefunction increases with the quantum number J ($J = 0$ has no nodes, $J = 1$ has one nodal line which is a great circle, $J = 2$ has two nodal lines, and so forth). This nodal structure corresponds to a decrease in the de Broglie wavelength with the quantum number J, hence a larger angular momentum and a larger angular kinetic energy.

Example 11.1 *Calculate the energy difference between the $J = 0$ and $J = 3$ level of the HCl molecule (bond length 128 pm). Compare this energy to that for a typical electronic transition energy* (5.0×10^{-19} J).

Using Equation (11.4), the energy change between the $J = 0$ and the $J = 3$ level is

$$\Delta E = B \cdot 3 \cdot 4 - 0 = 12B$$

We calculate B according to Equation (11.5)

$$B = \frac{(6.626 \times 10^{-34}\ \text{J s})^2}{8\pi^2(1.674 \times 10^{-27}\ \text{kg})(128 \times 10^{-12}\ \text{m})^2} = 2.03 \times 10^{-22}\ \text{J}$$

Then we obtain $\Delta E = 2.43 \times 10^{-21}$ J. ΔE is much smaller than the energy difference between electronic states which is on the order of 1×10^{-19} J or more.

11.2.2 Polyatomic Molecules

The case of a diatomic molecule is a particularly simple example of rotational motion in which we need only specify one axis for the rotation. Polyatomic molecules can have different inertial moments I for rotation about different molecular axes. The moments of inertia for a rigid body in three dimensions depend on the direction in space and are determined by the molecule's geometry.

11.2.2.1 Linear Polyatomic Molecules

For linear polyatomic molecules (like O = C = O or H − C ≡ C − H) only one rotational axis must be specified, as for diatomic molecules. Remember that the moment of inertia, I, of a linear array of masses m_i at distances r_i from the center of mass is defined as

$$I = \sum_i m_i r_i^2 \tag{11.8}$$

For example, in the case of CO_2 (Figure 11.4a) the central carbon atom remains at rest during rotation; thus, the moment of inertia is that of two oxygen atoms, each of mass m_O, rotating at

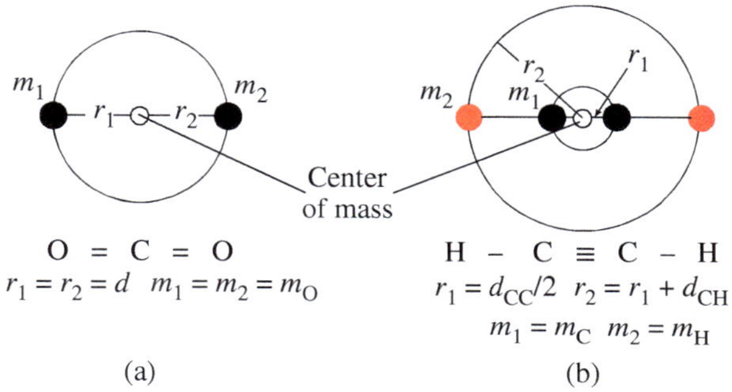

Figure 11.4 Calculation of moment of inertia for CO_2 (a) and ethyne (b).

the distance $r_1 = r_2 = d$ from the center of mass,

$$I = m_1 r_1^2 + m_2 r_2^2 = 2m_O d^2$$

Example 11.2 *Calculate the moment of inertia I and the rotational constant B for the ethyne molecule, H–C≡C–H (Figure 11.4b).*

In this case two bond lengths have to be considered: $d_{CH} = 106$ pm and $d_{CC} = 120$ pm. The moment of inertia is that for two carbon atoms ($m_1 = m_C$) rotating at a distance $r_1 = d_{CC}/2 = 60$ pm plus that for two H atoms ($m_2 = m_H$) rotating at a distance $r_2 = d_{CC}/2 + d_{CH} = 166$ pm from the center of mass. Thus

$$\begin{aligned} I &= 2\left(r_1^2 m_C + r_2^2 m_H\right) \\ &= 2\left(3.60 \times 10^{-21} \cdot 1.99 \times 10^{-26} + 2.76 \times 10^{-20} \cdot 1.67 \times 10^{-27}\right)\ \text{kg-m}^2 \\ &= 2.35 \times 10^{-46}\ \text{kg m}^2 \end{aligned}$$

Using this value for the inertial moment, we find

$$B = \frac{h^2}{8\pi^2 I} = \frac{\left(6.626 \times 10^{-34}\right)^2\ \text{J}^2\ \text{s}^2}{8\pi^2\left(2.35 \times 10^{-46}\ \text{kg m}^2\right)} = 2.37 \times 10^{-23}\ \text{J}$$

11.2.2.2 Nonlinear Polyatomic Molecules

Rigid nonlinear polyatomic molecules can be assigned three axes of rotation; thus, we have to deal with three inertial moments and correspondingly with three rotational constants. If all three are equal, the rotator is called a spherical top. If two inertial moments are equal, the rotator is called a symmetric top. If all three moments are different, the rotator is called an asymmetric top. In the following we restrict our consideration to spherical and symmetric top molecules, for which the Schrödinger equation can be solved analytically.

For a spherical top, all moments of inertia are the same. For the energy we obtain (see Foundation 11.1)

$$E_{rot} = B \cdot J(J+1),\ J = 0, 1, 2, \ldots$$

as for a linear molecule, and for the degeneracy we obtain

$$g_J = (2J+1)^2$$

in contrast to a linear molecule where $g_J = (2J+1)$.

For a symmetric top molecule we have two different moments of inertia, I_A corresponding to the axis of the top (the unique axis), and I_B corresponding to the two axes perpendicular to the unique axis. If $I_A > I_B$ then the body rotates like a disk or hockey puck, and is called an *oblate rotor*, scheme (a). Molecules of this type are benzene and nitrate ion. The other case, $I_A < I_B$, corresponds to a body that rotates like a rod or pencil, and is called a *prolate rotor*, scheme (b). Molecules of this type are ethane and methyl chloride.

For symmetric top molecules, two quantum numbers J and K are needed to specify the energy levels of the molecule. $J = 0, 1, 2, \ldots$ corresponds to the total angular momentum (Figure 11.5). The quantum number K corresponds to rotation about the unique axis of the symmetric rotor. In Foundation 11.1, the energy level formula is shown to be

$$E_{J,K} = B \cdot J(J+1) + (A - B) \cdot K^2 \tag{11.9}$$

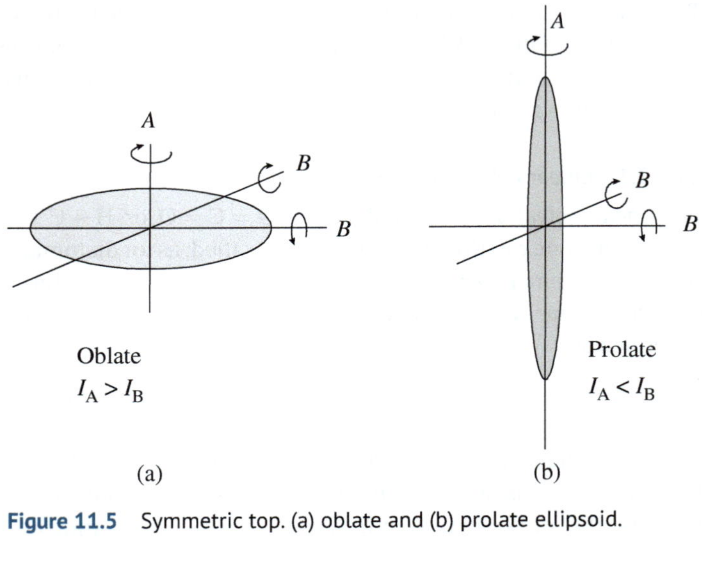

Figure 11.5 Symmetric top. (a) oblate and (b) prolate ellipsoid.

where $|K| \leq J$, $K = 0, \pm1, \pm2, \ldots$. A and B are the rotational constants

$$A = \frac{h^2}{8\pi^2 I_A}, \quad B = \frac{h^2}{8\pi^2 I_B} \tag{11.10}$$

The energy level diagrams for the two symmetric top limits are shown in Fig. 11.6. Each energy level has a degeneracy of two when $K \neq 0$ and is non-degenerate when $K = 0$. In the special case $A = B$ (spherical top, e.g., methane) the second term in Equation (11.9) is zero, and the energy levels are the same as for diatomic and linear polyatomic molecules; they do not depend on K.

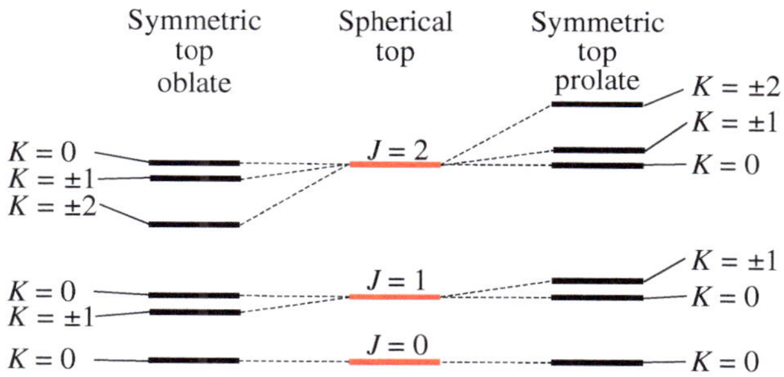

Figure 11.6 Rotational energy levels of symmetric top (oblate), spherical top, and symmetric top (prolate). Note that for the oblate case we have $I_A > I_B$, thus $A - B < 0$; correspondingly, for the prolate case we have $A - B > 0$.

11.2.3 Rotational Spectra

Absorption spectroscopy can be used to probe transitions between rotational energy states of molecules; just as we saw for electronic states in Chapter 9. The photon energy for such transitions is only about 10^{-21} J, and the light's wavelength is in the microwave region of the electromagnetic spectrum. Because microwaves span the frequency range from 10^{12} Hz to 10^9 Hz, the applied electric field is changing slowly in time compared to the internal motions of a molecule's electrons. Consequently, the electrons can be considered to be distributed in a cloud.

Microwave spectroscopy provides a sensitive method for determining the geometries and dipole moments of gas-phase molecules. As an example, consider the HCl molecule. To measure an absorption spectrum of HCl, we irradiate an optical cell containing HCl gas with microwaves and measure the transmitted intensity as a function of the wavelength (see Box 10.1). For HCl, such an experiment yields the spectrum in Fig. 11.7.

The position and intensity of the lines in the spectrum are determined by the energy level diagram and the molecule's transition dipole moment. The transitions only occur if the transition moment integral, which in this case corresponds to the permanent dipole moment, is nonzero. Analysis of this condition is described in Foundation 11.4, and it gives rise to selection rules for the transitions; namely the change in quantum number must be $\Delta J = \pm 1$ and the molecule must have a permanent dipole moment. The dependence on the permanent dipole moment arises from the nature of the interaction between the molecule and the light field.

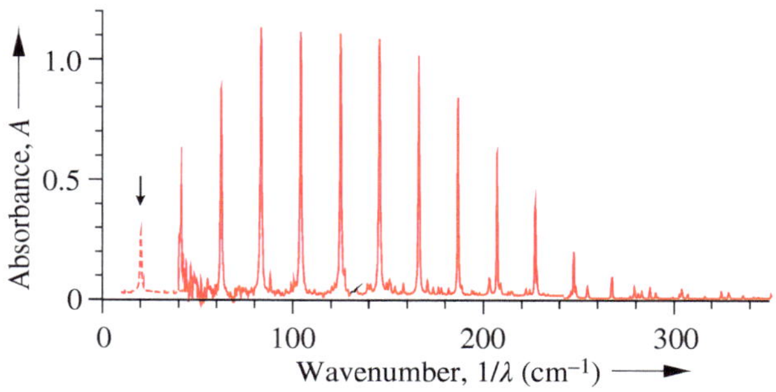

Figure 11.7 Absorption spectrum of HCl gas in the far infrared spectral region. The absorbance, A is plotted *versus* the wavenumber $1/\lambda$ (defined as $1/\lambda = \nu/c_0$), which is directly proportional to the energy of the transition. Solid line: experimental spectrum starting at 40 cm^{-1}. Dashed line: estimated location of first absorption band (arrow). (Source: Frank Weller, Philipps-Universität Marburg, private communication).

11.2.3.1 Spectra of Diatomic Molecules

Position of Absorption Lines If all the HCl molecules were in the lowest energy level ($J = 0$), then only one transition would be possible with the excitation energy (see Equations (11.4) and (11.5))

$$\Delta E = B \cdot [1 \cdot (1 + 1) - 0] = 2B \tag{11.11}$$

with

$$B = \frac{h^2}{8\pi^2 I} = \frac{h^2}{8\pi^2 m_H d^2} \tag{11.12}$$

and only one single line would appear in the spectrum. Yet we observe a sequence of equally spaced lines. How can this be? At room temperature, the molecules are not all in the ground rotational state. In contrast to electronic energy level spacings, the spacing of the rotational levels is smaller than the energy available in typical collisions with other molecules. Hence, a molecule can accumulate enough energy through collisions to populate excited rotational energy levels. Consequently, many transitions, originating from different J states, with $\Delta J = +1$ are possible (arrows in Fig. 11.3).

Let us determine the energies for the rotational transitions of a diatomic molecule. Using Equation (11.4), we find that the energy difference between a state with a quantum number J and one with a quantum number $(J+1)$ is

$$\begin{aligned}\Delta E &= B\cdot[(J+1)\cdot(J+2)-J\cdot(J+1)] \\ &= 2B\cdot(J+1), \quad J=0,1,2,\ldots\end{aligned} \tag{11.13}$$

This result shows that ΔE increases linearly with J, in agreement with Fig. 11.7. In other words, the spacing of the lines in the spectrum, $\Delta(\Delta E)$, is a constant:

$$\Delta(\Delta E) = \Delta E_{J+1} - \Delta E_J = 2B \tag{11.14}$$

This spacing of $2B$ provides a direct determination of the molecule's moment of inertia, which depends on the mass m and on the bond distance d. Because the atomic masses are usually known, the rotational constant can be used to determine bond distances. The success of Equation (11.13) can be shown by assigning the peaks in Fig. 11.5 to particular transition frequencies and finding their energy differences, see Table 11.1.

Example 11.3 *Use the microwave absorption spectrum of $H^{35}Cl$ (Fig. 11.7) and the numerical data in Table 11.1 to determine the molecule's rotational constant and bond length.*

First we assign each peak in the spectrum to a wavenumber $1/\lambda$ (column 2 of Table 11.1). Next we convert each of these wavenumbers $\tilde{\nu} = 1/\lambda$ into an excitation energy ΔE using

$$\Delta E = h\nu = hc_0\frac{1}{\lambda}$$

Now we evaluate the difference in these energy changes (column 4 of Table 11.1). We obtain $\Delta(\Delta E) = 40.58\times10^{-23}$ J as an average, and the rotational constant is $B = 20.29\times10^{-23}$ J, according to Equation (11.14). From this value and Equation (11.12) for B, we can calculate the average bond distance d of HCl

$$d = \sqrt{\frac{h^2}{8\pi^2 m_H}\cdot\frac{1}{B}} = 128 \text{ pm}$$

Table 11.1 Evaluation of the Absorption Lines in the Rotational Spectrum of HCl in Fig. 11.7 According to Equations (11.13) and (11.14).

J	$1/\lambda$ (cm^{-1})	ΔE (10^{-23} J)	$\Delta(\Delta E)$ (10^{-23} J)
0	(20.93)	(41.58)	
1	41.78	82.99	41.4
2	62.60	124.3	41.3
3	83.35	165.6	41.3
4	104.02	206.6	41.0
5	124.74	247.8	41.2
6	145.36	288.7	40.9
7	165.83	329.4	40.7
8	186.25	370.0	40.6
9	206.54	410.3	40.3
10	226.71	450.3	40.0
11	246.82	490.3	40.0
12	266.70	529.8	39.5
13	286.50	569.1	39.3

At closer inspection we see that the energy difference $\Delta(\Delta E)$ decreases slightly with increasing quantum number. This occurs because the centrifugal force, which increases with the rotational frequency, increases the bond distance. As we show in Foundation 11.5 this leads to a correction term in the expression for the rotational energy:

$$E = B\cdot J(J+1) - \frac{2B^2}{k_f\cdot d_0^2}\cdot J^2\,(J+1)^2 \tag{11.15}$$

(k_f = force constant of bond, d_0 = bond distance for quantum number $J=0$). In Problem 11.12 you show that the observed decrease in $\Delta\,(\Delta E)$ is in quantitative accordance with this equation.

In our treatment of the HCl molecule we assumed that the Cl atom is at rest. This is an approximation, of course. In a rigorous treatment, the center of mass is fixed and we must treat the rotation of the molecule about the center of mass. In Exercise 11.21, you show that the moment of inertia is given by

$$I = \mu\cdot d^2 = \left(\frac{m_A\cdot m_B}{m_A+m_B}\right)\cdot d^2 \tag{11.16}$$

where μ is called the reduced mass, and you use this more exact formula to calculate a more precise bond distance d for the HCl molecule.

The choice of the center-of-mass does not change the relationship between the rotational constant and the inertial moment, but it does change the expression for the inertial moment and thus the bond length determination; the rotational constant B for a diatomic molecule A–B with arbitrary masses m_A and m_B is

$$B = \frac{h^2}{8\pi^2 I} = \frac{h^2}{8\pi^2 \mu d^2} \tag{11.17}$$

instead of Equation (11.5).

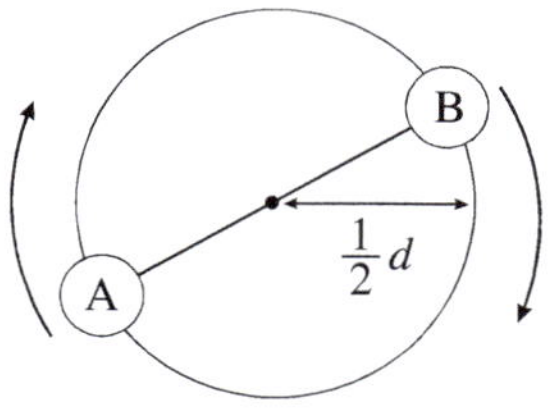

Figure 11.8 Rotational motion of a diatomic molecule with nuclei A and B of identical mass.

For example, if the masses of the atoms A and B in a diatomic molecule are identical ($m_A = m_B$), both nuclei rotate about the center of the bond (Fig. 11.8). In this case,

$$I = \mu d^2 = \frac{m_A^2}{2m_A} d^2 = \frac{1}{2} m_A d^2 \tag{11.18}$$

in accordance with Equations 11.17 and 11.16. The kinetic energy is the same as for rotation of the mass $(m_A + m_B) = 2m_A$ on a circle of radius $d/2$.

Example 11.4 *Given that the rotational constant B for $^{12}C^{16}O$ is 1.931 cm^{-1} , calculate the bond length of the molecule in two ways. In the first case, make the approximation that the center-of-mass lies on the O atom; and in the second case use the more exact treatment with the reduced mass for the inertial moment calculation. Comment on the result.*

To begin, we use the rotational constant to find the inertial moment. Using $B = 1.931\ \text{cm}^{-1} = 3.836 \times 10^{-23}$ J, we find that

$$I = \frac{h^2}{8\pi^2} \cdot \frac{1}{B} = \frac{\left(6.626 \times 10^{-34}\right)^2}{8\pi^2 \left(3.836 \times 10^{-23}\right)}\ \text{kg m}^2 = 1.449 \times 10^{-46}\ \text{kg m}^2$$

In the approximate treatment, which places the center of mass on the O atom, we find that

$$d = \sqrt{\frac{I}{m_C}} = \sqrt{\frac{1.449 \times 10^{-46}\ \text{kg m}^2}{1.9926 \times 10^{-26}\ \text{kg}}} = 85.3\ \text{pm}$$

For the more exact treatment, we use $\mu = m_O \cdot m_C / \left(m_O + m_C\right) = (16/28) \cdot m_C$ and find that

$$d = \sqrt{\frac{I}{\mu}} = \sqrt{\frac{28}{16} \frac{1.449 \times 10^{-46}\ \text{kg m}^2}{1.9926 \times 10^{-26}\ \text{kg}}} = 112.8\ \text{pm}$$

The latter value agrees well with the value reported in the literature, whereas the approximate method gives a value that is 25% too small.

Intensity of Absorption Lines Initially, the absorbances of the spectral lines in the HCl spectrum (see Fig. 11.7) increase with increasing wavenumber, that is with increasing quantum number J; however, after passing a maximum the absorbances decrease rapidly. We show in Foundation 11.4 that the absorbance of a spectral line with quantum number J is

$$\text{absorbance} = const' \cdot \mu_{\text{perm}}^2 \cdot (J+1)^2 \cdot e^{-BJ(J+1)/(kT)} \tag{11.19}$$

where μ_{perm} is the molecule's permanent dipole moment, k is the Boltzmann constant, and T is the temperature (see Chapter 14). The term $(J+1)^2 \cdot \exp(-BJ(J+1)/(kT))$ originates from the population number of the energy level with quantum number J (see Chapter 15). The factor $(J+1)^2$ is responsible for the increase of the absorbance at low J values, and the decrease of the absorbance arises from the exponential term at high J values.

11.2.3.2 Spectra of Polyatomic Molecules

The methods developed above for diatomic molecules can be generalized to the case of polyatomic molecules. As described in Section 11.2.2, nonlinear polyatomic molecules can display more complex rotational spectra, which arises from the need to account for their three-dimensional mass distribution (moment of inertia tensor). The solution of the Schrödinger equation for the rigid rotator of nonlinear molecules is discussed in Foundation 11.1, on Wiley's website; we show the formal solution for a symmetric top molecule and we apply it to the case of NH_3, which we model as a pyramidal molecule, in order to illustrate how one can extract bond lengths and bond angles from the rotational spectrum of small polyatomic molecules.

Lastly, we note that the observation of well-defined rotational transitions (see Fig. 11.7) requires that the rotational angular momentum of the molecule be well defined over the time period of the rotational motion; and this is often only true for molecular gases. For liquids (at room temperature), the collisional frequency between molecules randomizes the angular momentum, and it is more reasonable to treat its rotational motion classically.

11.3 Vibrational Motion of Molecules

11.3.1 Classical Oscillator

In Chapters 6 and 7 we found that atoms bonded together have a minimum energy at a particular separation distance, i.e., an electronic energy well for the molecule. A small displacement of the nuclei away from the energy minimum (Fig. 11.9a, example of an HCl molecule) increases the molecule's energy, and the nuclei experience a restoring force which pushes them back toward the minimum. If we limit ourselves to a small enough deflection x, then the energy increase is proportional to x^2. In this limit the relative motion of the nuclei corresponds to the motion of two point masses m_H and m_{Cl} connected by a spring obeying Hooke's law (Fig. 11.9b). Its vibrational motion is harmonic, and a system with this potential energy is referred to as a *harmonic oscillator*. The vibrational frequency (Boxes 11.1 and 11.2)

$$\nu_0 = \frac{1}{2\pi} \cdot \sqrt{\frac{k_f}{\mu}} \tag{11.20}$$

is the same as that of a mass

$$\mu = \frac{m_H m_{Cl}}{m_H + m_{Cl}} \tag{11.21}$$

(reduced mass) connected to a spring which is fixed to a solid wall (Fig. 11.9c, Box 11.1).

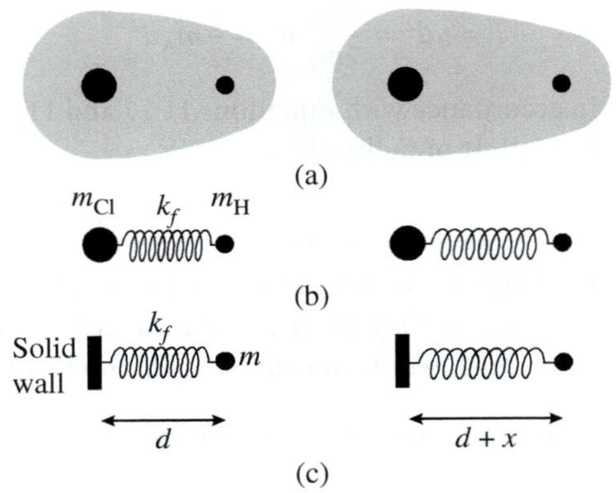

Figure 11.9 The deviation of the bond distance in a HCl molecule by x from the equilibrium distance d. (a) Deformation of the electron cloud during the deviation. (b) Spring model of the HCl molecule, vibration frequency $\nu_0 = \frac{1}{2\pi}\sqrt{\frac{k_f}{\mu}}$, where $\mu = \frac{m_H m_{Cl}}{m_H + m_{Cl}}$ is the reduced mass and k_f is the force constant. If we approximate the mass of the H atom as $m_H \ll m_{Cl}$ the reduced mass becomes equal to $\mu = m_H$. (c) Equivalent spring system with the same frequency as in (b): mass μ connected to a spring which is fixed to a solid wall (force constant k_f).

Example 11.5 *If the force constant of $H^{35}Cl$ is 480 N m^{-1} (see Example 11.6), calculate the vibrational period of the oscillator and compare it to the rotational period of a HCl molecule (bond length d = 128 pm) in quantum state J = 1.*

The vibrational period T_{vib} of the oscillator is given by

$$T_{vib} = \frac{1}{\nu_0} = 2\pi\sqrt{\frac{\mu}{k_f}}$$

To evaluate this expression we must know the reduced mass, which is $\mu = 1.61 \times 10^{-27}$ kg. For the vibrational period we find

$$T_{vib} = 2\pi\sqrt{\frac{1.61 \times 10^{-27}}{480}}\,\text{s} = 1.15 \times 10^{-14}\ \text{s}$$

To estimate the rotational period, we use the rotational kinetic energy formula, $E_{rot} = B \cdot J(J+1) = \frac{1}{2}\mu v^2$, and find that the speed v is

$$v = \sqrt{\frac{2B \cdot J(J+1)}{\mu}}$$

In Example 11.1 we calculated $B = 2.03 \times 10^{-22}$ J. Then for $J = 1$ we obtain $v = 7.10 \times 10^2$ m s^{-1}. The period T_{rot} is the time needed for one revolution

$$T_{rot} = \frac{2\pi r}{v} = \frac{\pi d}{v} = \frac{\pi \cdot 128 \times 10^{-12}\ \text{m}}{7.10 \times 10^2\ \text{m s}^{-1}} = 56.6 \times 10^{-14}\ \text{s}$$

Hence the HCl molecule undergoes about fifty vibrational periods in the time it takes to rotate through a single cycle.

11.3.2 Quantum-Mechanical Harmonic Oscillator

Classically a harmonic oscillator can assume any energy; however, in quantum mechanics the energy states are quantized. The experiments discussed in Section 11.3.3 reveal this quantization. Let us proceed by discussing HCl (Fig. 11.9), as an example.

11.3.2.1 Simplified Model

A simple model that accounts for the de Broglie wave character of nuclei can be used to appreciate the quantization of the nuclear energy levels in the electronic energy well

$$V = \frac{1}{2}k_f x^2 \tag{11.22}$$

We approximate the oscillations of the HCl molecule by considering the oscillations of a mass μ connected to a spring which is fixed to a solid wall (Fig. 11.9c). We begin by using particle-in-a-box wavefunctions as trial functions in a variational calculation where the length L in the wavefunction is considered as a variational parameter which is optimized for each trial function (Fig. 11.10). Two trial functions are

$$\phi_0 = \sqrt{\frac{2}{L_0}} \cos\left(\frac{\pi x}{L_0}\right) \quad \text{and} \quad \phi_1 = \sqrt{\frac{2}{L_1}} \sin\left(\frac{2\pi x}{L_1}\right) \tag{11.23}$$

ϕ_0 and ϕ_1 are given by Equation (11.23) within the limits $-L_0/2$ and $+L_0/2$ and $-L_1/2$ and $+L_1/2$, respectively; outside these limits ϕ_0 and ϕ_1 are equal to zero. In Problem P11.3 you calculate the energies ε_0 and ε_1 and the lengths L_0 and L_1 as

$$\varepsilon_0 = 0.57 \cdot h\nu_0, \quad L_0^2 = 2.77 \cdot \frac{h}{\sqrt{k_f \mu}}$$

and

$$\varepsilon_1 = 1.67 \cdot h\nu_0, \quad L_1^2 = 3.76 \cdot \frac{h}{\sqrt{k_f \mu}}$$

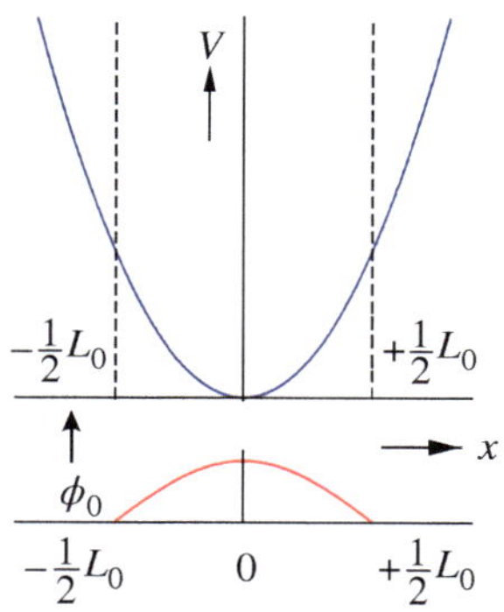

Figure 11.10 Potential energy $V = \frac{1}{2}k_f x^2$ of a harmonic oscillator and trial function ϕ_0 to estimate the ground-state energy level ε_0. L_0 is varied to find the minimum variational energy according to the trial function (11.23). The dashed lines mark the turning points of the corresponding classical oscillator with the same energy.

The energies ε_0, ε_1 and the wavefunctions ϕ_0, ϕ_1 are shown in Fig. 11.11a. In this model, the lowest energy state of the oscillator is not zero, in contrast to the case of rotational motion. Thus the oscillator has a zero-point energy. This finding reflects the fact that the oscillator's wavefunction must have a finite de Broglie wavelength, which results from the confinement by the potential energy well. Second, as the number of nodes in the wavefunction increases, so does the energy of the state. Lastly, the spatial extent of the wavefunction increases as the energy of the state increases, i.e., L becomes larger, which means that the amplitude of the vibrational motion increases.

11.3.2.2 Rigorous Solution

The Schrödinger Equation for a particle of reduced mass μ moving in the x-direction with the potential energy $V(x) = \frac{1}{2}k_f x^2$ is

$$-\frac{h^2}{8\pi^2\mu} \cdot \frac{d^2\psi}{dx^2} + \frac{1}{2}k_f x^2 \psi = E\psi \tag{11.24}$$

in which the first term corresponds to the kinetic energy and the second term corresponds to the potential energy. The mathematical solution of this differential equation is provided in Foundation 11.2, on Wiley's website. For quantum numbers n the exact solutions for the energy are

$$\boxed{E = h\nu_0 \cdot \left(n + \frac{1}{2}\right), \quad n = 0, 1, 2, \ldots.} \tag{11.25}$$

The energy levels for the vibrational motion are equally spaced by $h\nu_0$ and the ground state has the energy $h\nu_0/2$.

The wavefunctions are a product of a polynomial in x and a Gaussian function, $e^{-ax^2/2}$ where a is a constant. The exact solutions of the Schrödinger Equation for the two lowest energy states are

$$E_0 = \frac{1}{2}h\nu_0, \quad \psi_0 = \sqrt[4]{\frac{a}{\pi}} \cdot e^{-ax^2/2}$$

$$E_1 = \frac{3}{2}h\nu_0, \quad \psi_1 = \sqrt[4]{\frac{a}{\pi}}\sqrt{2a} \cdot x \cdot e^{-ax^2/2}$$

with

$$a = \frac{2\pi}{h}\sqrt{k_f\mu}, \quad \nu_0 = \frac{1}{2\pi}\sqrt{\frac{k_f}{\mu}} \tag{11.26}$$

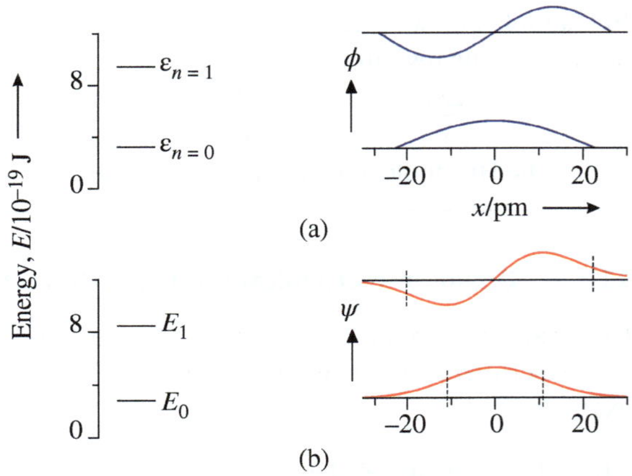

Figure 11.11 Energies and wavefunctions of a quantum-mechanical oscillator (HCl molecule: $m_H = 1.67 \times 10^{-27}$ kg, $k_f = 480$ N m^{-1}). (a): Box model (energies ε_0 and ε_1, wavefunctions ϕ_0 and ϕ_1, $L_0 = 45.3$ pm, $L_1 = 52.8$ pm). (b): Exact (energies E_0 and E_1, wavefunctions ψ_0 and ψ_1). Dashed lines: positions of the inflection points of the wavefunctions.

These energies and wavefunctions are shown for the HCl molecule in Fig. 11.11b. The energies are only slightly different from the energies calculated using the box wavefunctions. As shown in the previous section the energies of the first and second quantum states are 0.57 $h\nu_0$ and 1.67 $h\nu_0$ for the box wavefunction, as compared to 0.5 $h\nu_0$ and 1.5 $h\nu_0$ in the exact treatment. Whereas the box wavefunctions are zero for $x > L/2$ or $x < -L/2$ the exact wavefunctions approach the baseline exponentially. The distance of the inflection points (see Fig. 11.11) of the wavefunctions increases with increasing quantum number.

The energies and wavefunctions are displayed in Fig. 11.12 for $n = 0$ to $n = 5$. According to Equation (11.25), the energy E increases in proportion to the quantum number n. This is in contrast to a particle in a box, where E increases with n^2. However, in our approximation of the oscillator wavefunctions by box trial functions, the box length L is not a constant, but increases with increasing n, thus decreasing the mean kinetic energy. Thus, the energy increases with n, but not as strongly as n^2.

The wavefunctions in Fig. 11.12 have a number of interesting features. First you should notice that the number of nodes in the wavefunction matches the quantum number for the energy state. Second you should notice that the wavefunctions are either even or odd with respect to the origin, where the origin is chosen to be the equilibrium position, $x = 0$. Third you will notice that the ground state wavefunction $n = 0$ has its maximum at $x = 0$. The maxima and minima of the wavefunction with $n \geq 3$ increase in value with increasing distance from $x = 0$ (Fig. 11.12). In our discussion below the significance of these properties will become more apparent.

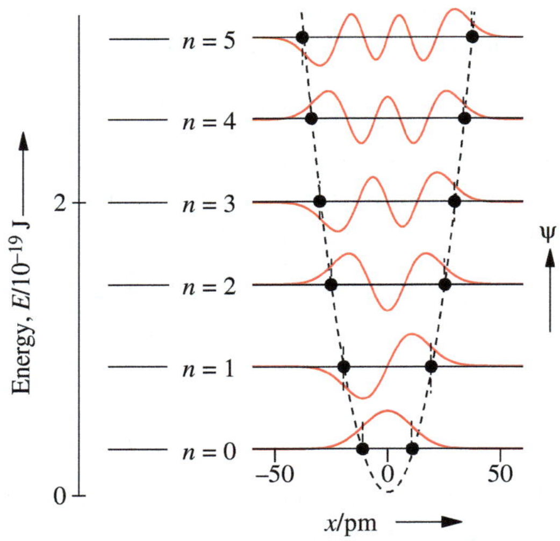

Figure 11.12 Energies and wavefunctions of a quantum-mechanical oscillator for the quantum numbers $n = 0$ to $n = 5$ (HCl molecule: reduced mass $\mu = 1.627 \times 10^{-27}$ kg, $k_f = 480$ N m^{-1}). The full circles mark the turning points (coordinates x_0 and $-x_0$) of the corresponding classical oscillator with the same energy. The dashed curve shows the potential energy $V = \frac{1}{2}k_f x^2$.

11.3.2.3 Comparison of Quantum-Mechanical Oscillator with Classical Oscillator

An important difference between the classical and quantum-mechanical oscillators is nuclear tunneling. To understand the tunneling, we analyze the wavefunction's inflection points, identified as the coordinate x_0 (Fig. 11.12). From calculus we recall that an inflection point occurs when the second derivative of the wavefunction is zero:

$$\left(\frac{d^2\psi}{dx^2}\right)_{x_0} = 0 \tag{11.27}$$

Inserting this condition into the Schrödinger Equation (11.24), we find that

$$E = \frac{1}{2}k_f x_0^2 \tag{11.28}$$

That is, the kinetic energy is zero at the position x_0, and the total energy is equal to the potential energy. For a classical oscillator, this position corresponds to the turning point, the position where the oscillation stops before moving back in the opposite direction. A classical oscillator cannot reach any point beyond x_0, because it does not have enough energy to climb farther up the potential energy well. For the quantum-mechanical oscillator, however, the wavefunction is nonzero even at $x > x_0$, or at $x < -x_0$, and we can find the oscillator at displacements where the classical oscillator can never be found.

Let us calculate the probability $\mathcal{P}$ to find the oscillator between x and $x + \mathrm{d}x$ for both the quantum-mechanical and the classical oscillator. In the classical description, the probability $\mathrm{d}\mathcal{P}_{\text{class}}$ of finding the oscillator between x and $x + \mathrm{d}x$ is proportional to the time it spends between x and $x + \mathrm{d}x$. Hence it is inversely proportional to its classical speed v

$$\mathrm{d}\mathcal{P}_{\text{class}} \sim \frac{1}{v} \cdot \mathrm{d}x \tag{11.29}$$

In Exercise E11.23 you show that

$$\frac{\mathrm{d}\mathcal{P}_{\text{class}}}{\mathrm{d}x} = \frac{1}{\pi x_0} \cdot \frac{1}{\sqrt{1 - (x/x_0)^2}} \tag{11.30}$$

The classical model predicts a minimum of $\mathrm{d}\mathcal{P}/\mathrm{d}x$ at $x = 0$, independent of the energy. This behavior arises because the classical oscillator swaps its kinetic and potential energy during the oscillation. When the oscillator nears its maximum displacement, most of its energy has been converted into potential energy, but as it moves back toward the minimum displacement (at $x = 0$) that potential energy is converted into kinetic energy. Consequently, the classical oscillator is moving more rapidly near $x = 0$ and spends less time near its minimum displacement than near its maximum displacement.

In the quantum-mechanical description, the probability of finding the oscillator between x and $x + \mathrm{d}x$ is given by

$$\mathrm{d}\mathcal{P} = \psi^2(x) \cdot \mathrm{d}x \tag{11.31}$$

In Fig. 11.13 the probability densities

$$\frac{\mathrm{d}\mathcal{P}}{\mathrm{d}x} = \psi^2(x) \tag{11.32}$$

for the quantum numbers $n = 0$ and $n = 50$ are displayed and compared with the classical prediction. For both cases we see a substantial "tunneling" of the quantum-mechanical oscillator's probability density into the nonclassical region of displacement. Also we obtain a maximum at the equilibrium displacement for the $n = 0$ state, in contrast to the classical result. As n increases the probability density for the quantum-mechanical model oscillates more rapidly in space. When the energy of the quantum-mechanical oscillator is very high (see the curve for $n = 50$), the quantum-mechanical and the classical oscillator correspond much better.[1] This is in accordance with Bohr's Correspondence Principle discussed in Chapter 3.

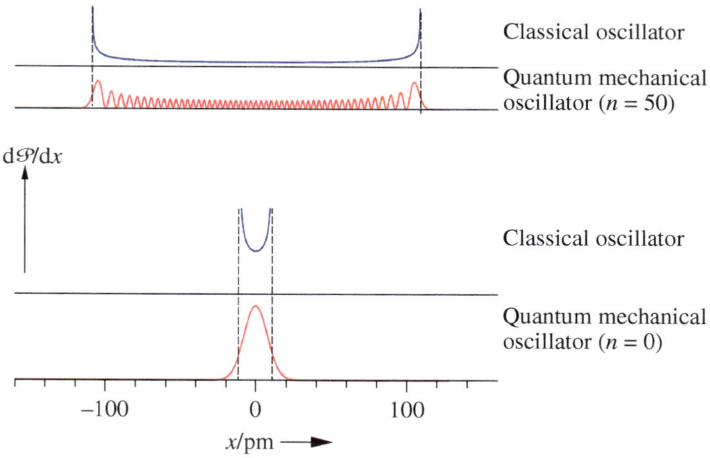

Figure 11.13 Probability density $\mathrm{d}\mathcal{P}/\mathrm{d}x$ of a quantum-mechanical oscillator for the quantum numbers $n = 0$ ($E_0 = 2.83 \times 10^{-20}$ J) and $n = 50$ ($E_{50} = 2.86 \times 10^{-18}$ J) and of a classical oscillator with the same energies ($k_f = 480$ N m^{-1}, $\mu = 1.627 \times 10^{-27}$ kg).

11.3.3 Vibrational–Rotational Spectra

The vibrational energy level structure can be probed by absorption spectroscopy using infrared light. As we have discussed (Chapters 9 and 10), a spectroscopic transition requires the resonance condition and a nonzero transition dipole moment integral. Figure 11.14 provides a classical view of how the applied electric field causes a vibrational transition in a diatomic molecule. The sequence of nuclear displacements show "snapshots" along a single vibrational period of a molecule, and the sinusoidal curve shows the time-dependent dipole moment of the molecule. When an applied electric field oscillates at a frequency that matches the natural oscillation frequency of the nuclei, energy couples efficiently

1 We consider the case $n = 50$ for pedagogical reasons, to compare the classical and quantum-mechanical treatment. The potential energy well of typical molecules would not be harmonic (or bound) at such high vibrational excitations.

between the molecule and the applied field. For vibrational motions of a molecule these frequencies lie in the infrared region of the electromagnetic spectrum, from 10^{14} Hz to 10^{12} Hz.

For an infrared transition the selection rules require that the vibrational motion causes a change in the dipole moment of the molecule and that the change in quantum number is ± 1; i.e., $\Delta n = \pm 1$. The origin of these selection rules is demonstrated in Foundation 11.6. The restriction to $\Delta n = \pm 1$ only applies to a potential energy well that is strictly harmonic. When the potential energy curve deviates from this ideal limit, transitions with $\Delta n = \pm 1, \pm 2, \pm 3, \ldots$ can occur and can be used to determine properties of the potential energy curve (see Section 11.3.4).

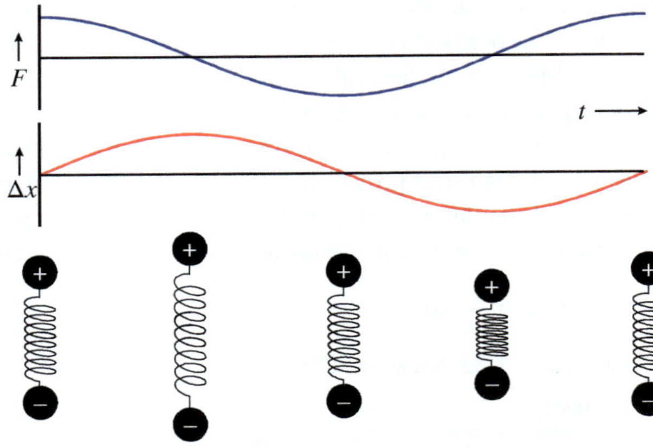

Figure 11.14 Interaction of the applied electric field (field strength F, top) with the vibrational motion of a diatomic molecule (elongation Δx, bottom), in the classical view. The elongation is phase shifted relative to the electric field by 90°.

As we discussed in Section 11.2.3, many rotational energy states are occupied at room temperature (for details see Chapter 15). In contrast, the energy spacing between the vibrational energy levels of HCl is 2990 cm^{-1} which is ten to fifteen times larger than typical thermal energies. Therefore, at room temperature an HCl molecule is usually found in the lowest vibrational state.

11.3.3.1 Spectra of Diatomic Molecules

Irradiation of a molecule with infrared light can excite it from the ground state ($n = 0$) to the next vibrational state ($n = 1$). From Equation (11.25), we see that this change in vibrational energy must be $h\nu_0$. During such an excitation, however, the molecule's rotational energy can change, and the rotational excitation follows its own set of selection rules (Fig. 11.15). These facts lead to a complex pattern of transitions that provide information on the vibrational mode (force constant and reduced mass) and the rotational parameters (inertial moment) in each vibrational state. In principle, we can determine how the rotational properties of the molecule might change with vibrational state.

The selection rules for the transitions and the properties of the molecule determine the pattern of transitions. For HCl, the vibrational selection rule is $\Delta n = \pm 1$ and the rotational selection rule is $\Delta J = \pm 1$. Assuming we start in the ground vibrational state and the average bond length does not change with excitation, $\Delta n = 1$ and $\Delta J = 1$ give the transition energies

$$\Delta E = h\nu_0 + 2B, \quad h\nu_0 + 4B, \quad h\nu_0 + 6B, \ldots \tag{11.33}$$

and correspond to the transitions from $(n = 0, J = 0)$ to $(n = 1, J = 1)$, from $(n = 0, J = 1)$ to $(n = 1, J = 2)$, etc., referred to as the R branch, see Fig. 11.16. In contrast, a change in rotational quantum number of minus one $\Delta J = -1$ gives

$$\Delta E = h\nu_0 - 2B, \quad h\nu_0 - 4B, \quad h\nu_0 - 6B, \ldots \tag{11.34}$$

and correspond to the transitions $(n = 0, J = 1)$ to $(n = 1, J = 0)$, from $(n = 0, J = 2)$ to $(n = 1, J = 1)$, etc., referred to as the P branch, see Fig. 11.16. From this analysis we expect a system of equally spaced absorption lines around the frequency ν_0, or wavenumber $\tilde{\nu} = \nu_0/c_0$. According to Table 11.1 the lines should be spaced by 41 cm^{-1}. The high-resolution infrared absorption spectrum for HCl (see Fig. 11.16) largely agrees with this prediction.

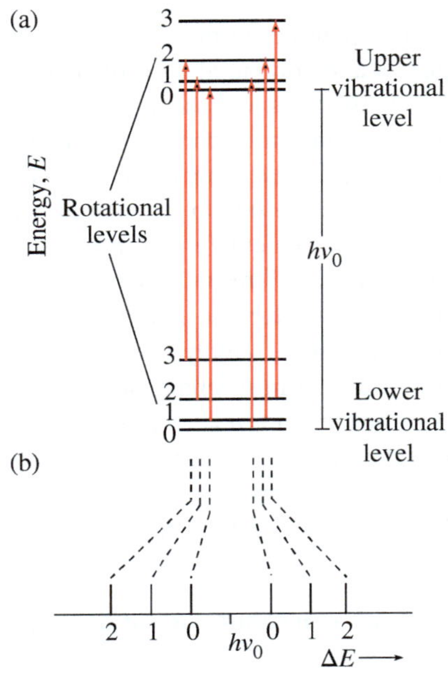

Figure 11.15 Vibrational and rotational motion of a diatomic molecule. (a) Energy scheme and transitions that can be excited by irradiating with infrared light. (b) Expected absorption lines in the infrared spectrum.

The structure of the spectrum can be analyzed to determine molecular properties. In particular, we can calculate the force constant of the molecule and its inertial moment, hence its bond length.

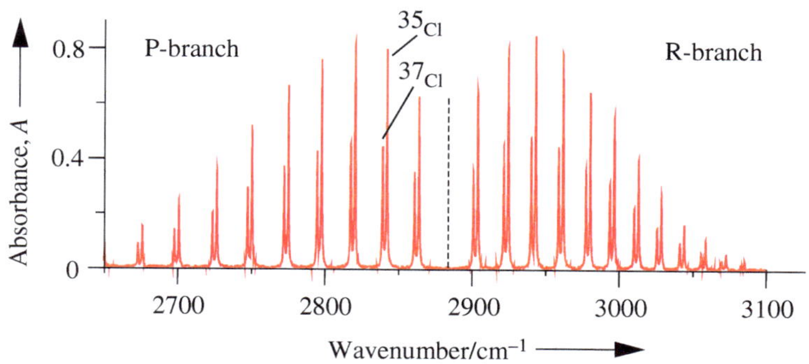

Figure 11.16 Infrared spectrum of HCl gas. Absorbance A versus wavenumber $\widetilde{\nu} = \nu/c_0$. The line system on the left of the dashed line is called the P branch; the line system on the right is called the R branch. Optical path length 10 cm, temperature 25 °C, HCl pressure, 65 mbar. (Source: D. Schiöberg, Marburg, private communication).

Example 11.6 *Use the center frequency of the high-resolution infrared spectrum to calculate the force constant of the HCl vibration.*

From Equations (11.33) and (11.34), we see that the wavenumber $\widetilde{\nu} = 2885\ \text{cm}^{-1}$ at the center of the line system in the HCl spectrum corresponds to $h\nu_0$. We calculate

$$\Delta E = h\nu_0 = hc_0\widetilde{\nu} = 0.5731 \times 10^{-19}\ \text{J}$$

and

$$\nu_0 = c_0\widetilde{\nu} = 8.65 \times 10^{13}\ \text{s}^{-1}$$

From the frequency ν_0 we can use Equation (11.26) to calculate the force constant k_f

$$k_f = 4\pi^2\nu_0^2\mu_{\text{HCl}} = 4\pi^2 \cdot (8.65 \times 10^{13}\ \text{s}^{-1})^2 \cdot 1.627 \times 10^{-27}\ \text{kg} = 480\ \text{N m}^{-1}$$

where $\mu_{\text{HCl}} = m_{\text{H}} \cdot m_{\text{Cl}}/\left(m_{\text{H}} + m_{\text{Cl}}\right) = (35/36) \cdot m_{\text{H}} = 1.627 \times 10^{-27}$ kg is used for the reduced mass.

A number of important features of the spectrum are not accounted for in this simple analysis.

^{35}Cl, ^{37}Cl Isotope Effect As in the case of the rotational spectrum, the strengths of the absorption lines reflect the population of the energy levels at room temperature. This explains why the intensity pattern falls away on the wings; however, it does not explain the presence of the two peaks at each position. Each line is accompanied by a weaker line. This occurs because the sample is a mixture of chlorine isotopes whose different mass gives rise to slightly different inertial moments and hence transition frequencies. The stronger lines belong to the ^{35}Cl isotope and the weaker lines belong to ^{37}Cl.

Deviation from Harmonic Oscillator Model From Table 11.1 for the energy spacing between two rotational lines of HCl we calculated $\Delta\Delta E = 41 \times 10^{-23}$ J, corresponding to a spacing of 20.6 cm^{-1} in the absorption spectrum. However, the absorption lines in Fig. 11.16 are not exactly spaced by this amount: the distance between adjacent lines is larger than 20.6 cm^{-1} at small wavenumbers and it is smaller than 20.6 cm^{-1} at large wavenumbers. This behavior results from the deviation of the potential curve of HCl from that of a harmonic oscillator (for details see Section 11.3.4). As a consequence, the mean bond distance, d_1, in vibrational state $n = 1$ is larger than the mean bond distance d_0 in vibrational state $n = 0$. In Problem P11.13 you show that this affects the spacing of the absorption lines: in the P branch the spacing is significantly larger than in the R branch. Note that this effect on the spacing of the absorption lines is different from the centrifugal distortion effect included in Equation (11.15). First, the centrifugal distortion should decrease the spacing of the absorption lines in both the P and the R branches. Second, the centrifugal distortion effect is much smaller than the shifts observed in Fig. 11.16.

Pure Vibrational Transition Requires Unpaired Electron The center of both line systems (Fig. 11.16, indicated by the dashed line) corresponds to the energy ΔE for the pure vibrational transition, which is forbidden in the case of most diatomic molecules, like HCl (the rotational quantum number J cannot remain unchanged). Diatomic molecules with an unpaired electron (and thus with an uncompensated electron spin), e.g., NO, have this transition ($\Delta n = 1, \Delta J = 0$) appear.[2]

2 When the molecule's electron cloud has a net angular momentum, it couples with the rotational angular momentum of the nuclei, and $\vec{J}$ is no longer perpendicular to the internuclear axis. Instead it lies at an angle somewhat smaller than 90 degrees and nutates around the normal. This fact leads to an equation like the symmetric top (see Foundation 11.1) and the addition of another quantum number K. Transitions are allowed for $\Delta J \neq 0, \Delta K = 0$ or $\Delta J = 0, \Delta K \neq 0$ and this effect explains the 0/0 transition in the NO spectrum. See G. Herzberg [1] for a detailed discussion.

11.3.3.2 Spectra of Polyatomic Molecules

Number of Vibrational Modes In polyatomic molecules, different types of vibrations are possible (*vibrational modes*). A molecule with N_{atom} atoms possesses $3N_{\text{atom}}$ degrees of freedom for the motions of the atoms in space; three for each atom in the molecule. Because the molecule is bound together, three of these motions, or degrees of freedom, correspond to its overall translation in space, i.e., its center of mass motion. In addition, a linear molecule has 2 degrees of freedom for the rotation corresponding to the two equivalent axes of rotation. The number of remaining motions is

$$3N_{\text{atom}} - 5 \quad \text{(linear molecule)} \tag{11.35}$$

and correspond to the internal nuclear motions, vibrational modes, of the molecule. If the molecule is nonlinear, it has three axes of rotation; hence, the number of vibrational modes is

$$3N_{\text{atom}} - 6 \quad \text{(nonlinear molecule)} \tag{11.36}$$

Linear Molecule, CO_2 Consider CO_2 as an example of a linear polyatomic molecule. In this case we have $3 \times 3 - 5 = 4$ vibrational modes. These vibrational modes are illustrated in Fig. 11.17. They consist of a twofold degenerate bending mode (*xy*- and *xz*-plane), a symmetric stretching mode, and an asymmetric stretching mode. Each of these vibrational modes has its own resonant frequency. The amplitudes and resonant frequencies of these vibrational modes are derived in Box 11.2.

These vibrational modes are called *normal modes*. The use of normal modes simplifies the mathematical description of the nuclear displacements in a molecule and provides a direct connection between the measured vibrational frequencies and the particular modes. Often the harmonic approximation provides a useful description of the molecule's dynamics and spectrum.

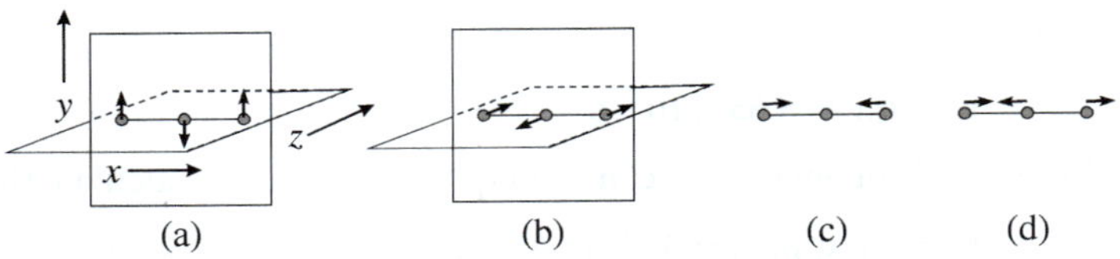

Figure 11.17 Vibrational modes of a CO_2 molecule. (a,b) Bending mode; this vibration mode can occur in two perpendicular planes and is therefore twofold degenerate. (c) Symmetric stretching mode. (d) Asymmetric stretching mode.

In the case of the CO_2 molecule the quantum numbers n_{sym} and n_{anti} correspond to the symmetric and asymmetric stretching modes, respectively; the bending mode is twofold degenerate and therefore has to be characterized by two quantum numbers ($n_{\text{bending,1}}$ and $n_{\text{bending,2}}$). Figure 11.18 shows the energy levels corresponding to the different vibrational modes.

The total vibrational energy of CO_2 is the sum of the contributions of the four vibrational modes:

$$\begin{aligned} E = h\nu_{0,\text{bending}} \cdot \left(\tfrac{1}{2} + n_{\text{bending,1}} + \tfrac{1}{2} + n_{\text{bending,2}}\right) \\ + h\nu_{0,\text{sym}} \cdot \left(\tfrac{1}{2} + n_{\text{sym}}\right) + h\nu_{0,\text{anti}} \cdot \left(\tfrac{1}{2} + n_{\text{anti}}\right) \end{aligned} \tag{11.37}$$

At low enough temperature, only the lowest energy levels in each vibration mode are occupied. Infrared radiation excites the CO_2 molecules to higher energies (arrows in Fig. 11.18). As derived in Foundation 11.6, transitions in the infrared are only possible if the vibration of the molecule is connected with a periodically changing dipole moment. For diatomic molecules this means that they must also have a permanent dipole moment (like HCl and CO); in polyatomic molecules, however, a dipole moment can occur in certain vibrational modes, even if the permanent dipole moment of the molecule is zero.

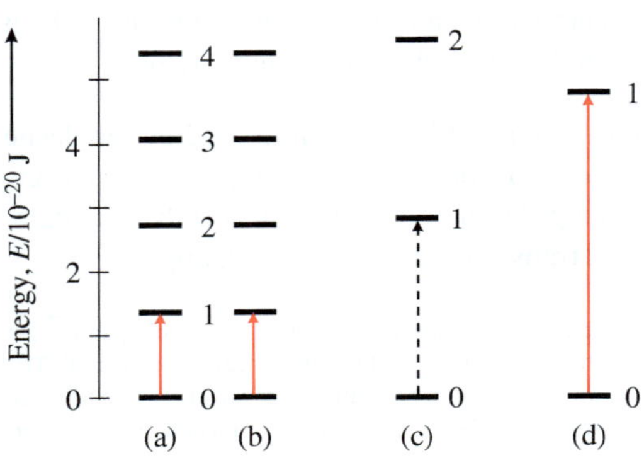

Figure 11.18 Energy scheme of the different vibrational modes of a CO_2 molecule. (a,b) Bending modes. (c) Symmetric stretching mode. (d) Asymmetric stretching mode. The illustration is based on the vibrational frequencies: $\nu_{0,\text{bending}} = 0.20 \times 10^{14}$ Hz, $\nu_{0,\text{sym}} = 0.42 \times 10^{14}$ Hz, and $\nu_{0,\text{anti}} = 0.71 \times 10^{14}$ Hz. These values are obtained from the vibrational spectrum of CO_2. Solid arrows: transitions from the lowest quantum state that can be observed in the IR spectrum; dashed line: this transition does not appear in the IR spectrum but appears, instead, in the Raman spectrum.

CO_2 is a linear molecule, and because of its symmetry the permanent dipole moment is zero. Nevertheless CO_2 absorbs infrared light and undergoes vibrational excitation. Both the bending mode (667 cm^{-1}) and the asymmetric stretching mode (2349 cm^{-1}) generate an oscillating dipole moment when they are excited (Fig. 11.19a). The infrared spectrum of CO_2 is displayed in Fig. 11.19b; the two strong transitions are called fundamentals because they arise by exciting the vibration from its $n = 0$ ground state to its $n = 1$ excited state. Force constants can be derived from these main absorption bands, as shown in Foundation 11.6. Excitation of the symmetric stretching mode at 1388 cm^{-1} does not appear in the infrared spectrum, because the displacements of the nuclei do not generate a dipole moment. This vibrational motion can be excited by a Raman process, however (see Section 11.4).

In addition to the two strong transitions in the infrared spectrum at 667 cm^{-1} and 2349 cm^{-1}, a number of weaker features are apparent in the infrared spectrum (see table in Fig. 11.19). These features arise from exciting harmonics and combinations of the molecules' fundamental vibrations. For example the transition at 3609 cm^{-1} arises from excitation of the molecule from its ground vibrational level to an excited state that has two vibrational quanta in the bending mode and one vibrational quantum in the asymmetric stretching mode. Such excitations arise because the vibrational energy profile of the molecule is not strictly harmonic, thus allowing coupling between the different vibrational motions.

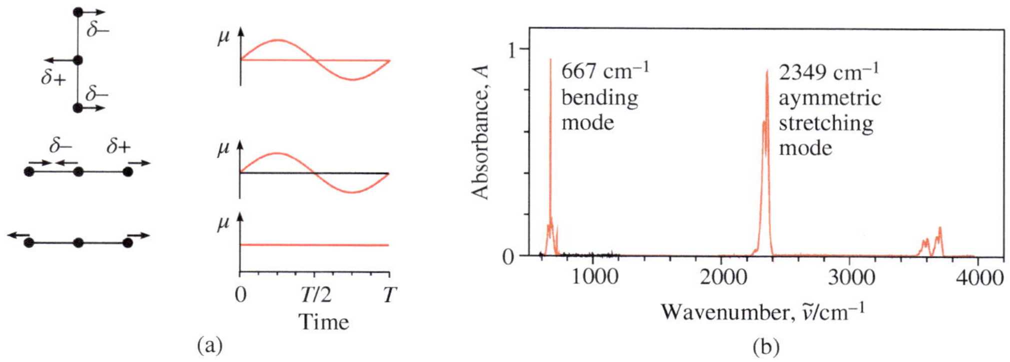

No.	$\tilde{\nu}$/cm^{-1}	$n_{bending}$	n_{sym}	n_{anti}		$n_{bending}$	n_{sym}	n_{anti}
1	720	1	0	0	→	0	1	0
2	3609	0	0	0	→	2	0	1
3	3716	0	0	0	→	0	1	1

Figure 11.19 Infrared absorption of CO_2. (a) Dipole moment μ of a CO_2 molecule during vibration. *Top*: Bending mode (direction of μ perpendicular to the molecular axis). *Center*: Asymmetric stretching mode. *Bottom*: Symmetric stretching mode. (b) Infrared spectrum. The absorbance $A = -\ln I/I_0$ is plotted versus the wavenumber $\tilde{\nu} = \nu/c_0$. Optical path length 10 cm, temperature 298 K, CO_2 pressure 100 mbar. The bending mode and the asymmetric stretching mode give rise to strong absorption bands ($n_{bending}$: 0 → 1, n_{anti} : 0 → 1). The table lists three weak absorption bands, which correspond to transitions where different quantum numbers change simultaneously (for more details see Box 11.2).

Nonlinear Molecule, H_2O For a nonlinear triatomic, like water, with $N_{atom} = 3$, we have $3N_{atom} - 6 = 3$ vibrational degrees of freedom. An analysis of the vibrations identifies a bending motion, a symmetric stretch, and an asymmetric stretch, with which all the vibrational motions of the molecule can be described. The nuclear displacements for these vibrational motions are shown in Fig. 11.20.

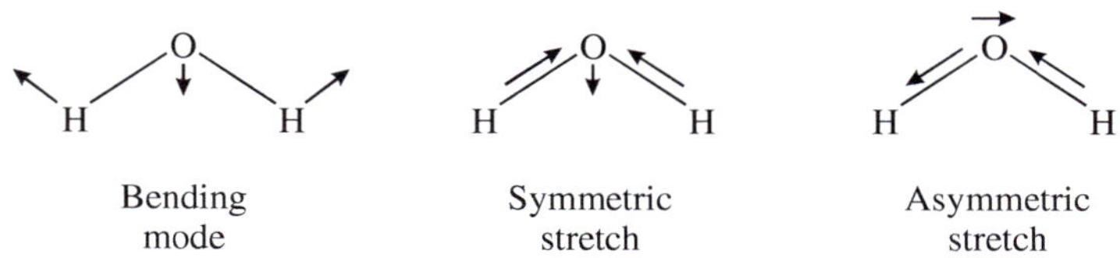

Figure 11.20 Normal modes of the water molecule.

The highest frequency mode is the asymmetric stretch with a fundamental frequency of 3756 cm^{-1}, and the second highest frequency mode is the symmetric stretch with a fundamental frequency of 3657 cm^{-1}. These motions are directly analogous to the symmetric and asymmetric stretch motions in CO_2. The bending motion has the lowest fundamental frequency 1595 cm^{-1}, but only one such mode exists, rather than the two found for CO_2.

Figure 11.21 shows an energy level diagram and infrared absorption spectrum for water vapor. The fundamental transitions are indicated by the vertical arrows, and they are the most intense ones observed.

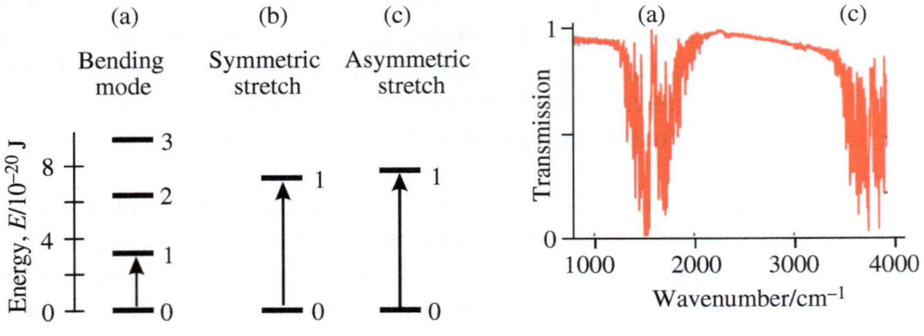

Figure 11.21 Left: Energy scheme of the different vibrational modes of a H_2O molecule. (a) Bending mode. (b) Symmetric stretching mode. (c) Asymmetric stretching mode. The illustration is based on the vibrational frequencies: $\nu_{0,\text{bending}} = 0.48 \times 10^{14}$ Hz, $\nu_{0,\text{sym}} = 1.09 \times 10^{14}$ Hz, and $\nu_{0,\text{anti}} = 1.13 \times 10^{14}$ Hz, corresponding to the wavenumbers 1595, 3657, and 3756 cm^{-1}. Arrows: transitions from the lowest quantum state. Right: Infrared spectrum of H_2O (transmission versus wavenumber). Note that the symmetric stretch cannot be seen in the infrared spectrum, because the dipole moment does not change during vibration.

Normal Modes and Chemical Bonds It is evident from the discussion of CO_2 and H_2O that the vibrational motion of a polyatomic molecule is much more complex to describe than that of a diatomic molecule. The symmetric and asymmetric stretching motions of each of the triatomic molecules involve simultaneous changes in each of the bond lengths. In general the displacements of any atom will be linked with the displacement of the other atoms (see the diagram in Fig. 11.22).

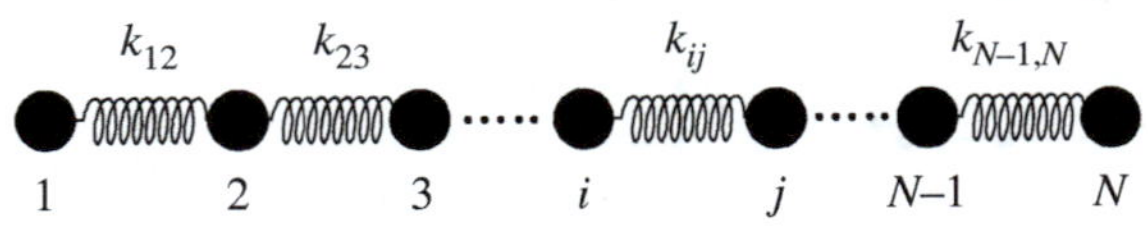

Figure 11.22 Coupled spring system.

Oftentimes a molecule's vibrational mode may have its dominant contribution from the displacement of atoms in a particular functional group or bond. Such modes are called "local modes." This approximation holds in many cases, but it is not generally true; for example, the breathing mode of a benzene ring involves the concerted motion of the carbon and hydrogen atoms away from the center of the molecule. Table 11.2 shows some characteristic frequencies for particular functionalities and more detailed tables are available [2]. Insight into the frequencies of these modes and its origin in the functional group composition may be obtained by using Equation (11.26), which connects the frequency of a harmonic oscillator to its force constant and reduced mass. For example, notice that the functionalities with H atom vibrations have the highest frequencies, ca. 2800 to 3700 cm^{-1}. This high frequency arises from the small reduced mass for the vibration. Notice that the vibrational frequencies of functionalities with two heavy atoms are significantly lower, reflecting the larger reduced mass.

The presence of local modes underlies the use of vibrational spectroscopy for the identification of functional groups in a polyatomic molecule. This spectroscopy is not selective enough to determine the composition and structure of a compound by itself, but it is very useful in combination with other methods; for example, NMR and mass spectrometry. When using the vibrational spectrum to identify a molecule, it is best to begin with the high frequency modes that are more characteristic of particular functionalities and correspond to bond stretching motions. Then proceed to the lower frequency modes which arise from twisting motions, wagging motions, bending motions and torsional processes.

Table 11.2 Some Characteristic Stretching Frequencies of Bonds Commonly Found in Organic Molecules (Taken from Harris and Bertolucci, Symmetry and Spectroscopy).

Functionality	Type of Compound	Wavenumber/ cm^{-1}	IR Intensity
–C–H	Alkane	2800–3000	Strong
=C–H	Alkene or arene	3000–3100	Medium
≡C–H	Alkyne	3300	Strong
C=C	Alkene	1620–1680	Variable
C≡C	Alkyne	2100–2260	Variable
–C≡N	Nitrile	2200–2300	Variable
C=O	Ketones, aldehydes	1700–1750	Strong
C=O	Acids, esters	1700–1750	Strong
–O–H	Alcohols	3590–3650	Variable, sharp
–O–H	H-bonded alcohols	3200–3400	Strong, broad
–O–H	H-bonded acids	2500–3000	Variable, broad
>N–H	Amines	3300–3500	Medium

11.3.4 Quantum-Mechanical Anharmonic Oscillator

So far we have treated the molecular vibrations as harmonic, that is, the restoring force follows Hooke's law and the potential energy V of the vibrating molecule is

$$V = \frac{1}{2} k_f x^2 \tag{11.38}$$

where $x = d - d_{eq}$ is the elongation from the equilibrium distance. However, we found in Chapter 7 that the electronic energy curves for diatomic molecules do not follow this simple relationship. This is shown in Fig. 11.23 for the electronic energy of H_2^+. Figure 11.23a shows $V = E - E_{eq}$ (calculated from the energy E and the energy E_{eq} at the equilibrium distance, using the exact solution for H_2^+) in a small range of elongations from $x = -30$ pm to $+40$ pm (red solid line). It is evident that the curve corresponding to Equation (11.38) is too high at large values of x and it is too low at low values of x (dashed line), although there is a small region where the harmonic curve is a good fit to the exact curve.

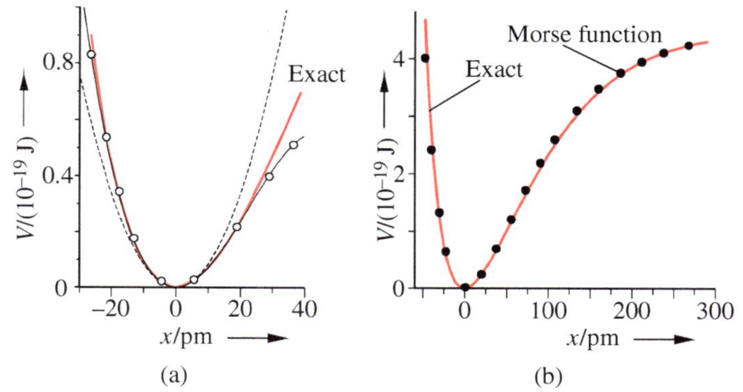

Figure 11.23 Potential energy V of H_2^+ versus elongation $x = d - d_{eq}$. Solid red curve: exact calculation, see Chapter 6. (a) Range of x from −30 to +40 pm. The dashed line shows the $V = a \cdot x^2$. curve which is too high for elongation and too low for compression larger than 10 pm. The open circles show the $V = a \cdot x^2 + b \cdot x^3$ curve, which is appropriate up to an elongation of 30 pm (with $a = 84.5$ N m^{-1} and $b = -126 \times 10^{10}$ N m^{-2}). (b) Range of x from −50 to +300 pm. The full circles show the Morse function calculated with $k_f = 160$ N m^{-1} and $\varepsilon_B = 4.48 \times 10^{-19}$ J.

It is common to account for this deviation by including anharmonic terms in the potential curve. Rather than use a parabola to mimic the intramolecular potential energy curve we can include higher order terms in the displacement, i.e.,

$$V(x) = \frac{1}{2} k_f x^2 + a x^3 + b x^4 + \cdots \tag{11.39}$$

where the parameters a, b, etc. are determined by comparison with experiment. Figure 11.23a compares this polynomial approximation (open circles) with the harmonic approximation and the exact potential energy curve for H_2^+. It is evident from the plot that the polynomial better mimics the molecule's potential energy curve at large displacements. Only near the bottom of the potential energy well does the harmonic approximation show good agreement.

Instead of using the polynomial approach it is a good choice to replace the actual electronic energy curve of a diatomic molecule by the Morse function, see Chapter 7.

$$V = \varepsilon_B \left[1 - e^{-ax}\right]^2 \quad \text{with} \quad a = \sqrt{\frac{k_f}{2\varepsilon_B}} \tag{11.40}$$

We discussed in Chapter 7 that this function becomes identical with the harmonic potential energy (11.38) for small enough elongations. In Fig. 11.23b we compare the exact potential curve for the H_2^+ molecular ion (red solid line) with the Morse function (filled circles) obtained with $k_f = 160$ N m^{-1} (see Table 7.6) and $\varepsilon_B = 4.48 \times 10^{-19}$ J (recall that this value is the difference between the minimum of the electronic energy of H_2^+, -26.276×10^{-19} J, and the energy of a hydrogen atom, -21.799×10^{-19} J).

11.3.4.1 Schrödinger Equation for the Anharmonic Oscillator

To obtain the energies and the wavefunctions of an anharmonic oscillator we proceed in analogy to the discussion of the harmonic oscillator, Equation (11.24). We solve the Schrödinger equation

$$-\frac{h^2}{8\pi^2\mu} \cdot \frac{d^2\psi}{dx^2} + V\psi = E\psi \tag{11.41}$$

where V is given by the Morse function (11.40). Instead of solving this equation analytically, we integrate this second-order differential equation numerically, using the program morse.exe (provided on the book's website).

As an example, we choose the HCl molecule with the reduced mass $\mu = 1.627 \times 10^{-27}$ kg (see Section 11.3.3), the force constant $k_f = 516$ N m^{-1} and $\varepsilon_B = 8,85 \times 10^{-19}$ J. Table 11.3 and Fig. 11.24a,b show the calculated energies in comparison to those expected for a harmonic potential. The energies for the anharmonic oscillator are lower than those for the harmonic oscillator.

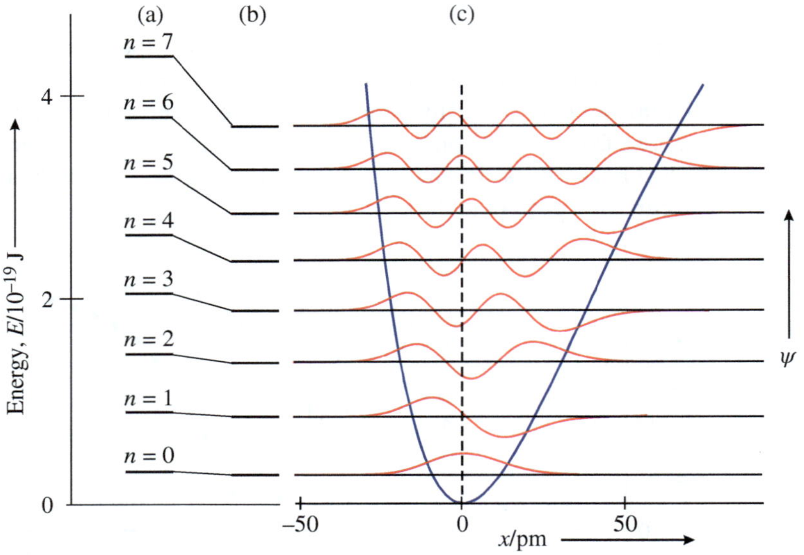

Figure 11.24 Energies and wavefunctions of quantum-mechanical oscillator (HCl molecule, reduced mass $\mu = 1.627 \times 10^{-27}$ kg, force constant $k_f = 516$ N m^{-1}, and bond energy $\varepsilon_B = 8.58 \times 10^{-19}$ J). (a) Energies E in the harmonic oscillator approximation ($n = 0$ to $n = 7$), (b) energies E and (c) wavefunctions ψ in the anharmonic oscillator model, potential energy V described by the Morse function. x is the elongation of the molecule from the equilibrium distance.

Table 11.3 HCl Molecule. Calculated Energies E for the Harmonic and the Anharmonic Oscillator Model for Quantum Numbers $n = 0$ to $n = 8$. The Harmonic Energies are Calculated According to Equation (11.25), the Anharmonic Energies are Calculated Numerically, see text; the Data Used are $\mu = 1.627 \times 10^{-27}$ kg, $k_f = 516$ Nm^{-1}, and $\varepsilon_B = 8.58 \times 10^{-19}$ J.

	n	0	1	2	3	4	5	6	7	8
Harmonic	$E/10^{-20}$ J	2.97	8.91	14.85	20.79	26.73	32.66	38.60	44.54	50.48
Anharmonic	$E/10^{-20}$ J	2.94	8.68	14.21	19.53	24.64	29.56	34.26	38.76	43.06

The origin of this shift can be traced to the anharmonic potential, which is steeper for compressions (x negative) and flatter under elongation (positive x). Because of the flatness of the potential curve at positive elongations, the restoring force on the mass is smaller than assumed for a mass oscillating in a harmonic potential, and the oscillation frequency becomes accordingly smaller. With increasing quantum number this effect becomes more and more apparent.

Figure 11.24c shows the corresponding wavefunctions. Compared to the harmonic wavefunctions in Fig. 11.12 the anharmonic wavefunctions are asymmetric with respect to the position $x = 0$. The extension of the wavefunctions at positive values of x is greater than for negative values of x, and this tendency increases with increasing quantum number. That is, the probability of finding the particle at positive x values is higher than that for negative x values, and the average bond distance $\overline{d}$ becomes larger than the equilibrium distance d_{eq}.

The asymmetry of the wavefunctions has direct consequences for the transition moments $M_{0\to n} = \int \psi_0 x \psi_n \cdot \mathrm{d}x$. We found in Section 11.3.3 that for the harmonic oscillator the only allowed transition is $0 \to 1$ because the transition moments for higher transitions are zero. Because of the asymmetry of the anharmonic wavefunctions the transition moments for higher transitions ($0 \to 2$, $0 \to 3$, etc.) do not vanish, and corresponding absorption bands are observed; however, the intensity of these absorption bands decreases with increasing excitation energy.

11.3.4.2 Evaluation of k_f from Spectroscopic Data

Besides solving the Schrödinger equation for the Morse function numerically, it is also possible to find an approximate analytical solution [3] which is precise enough for describing the vibrational energies of diatomic molecules [4].

$$E = h\nu_0 \left(\frac{1}{2} + n\right) - C \cdot \left(\frac{1}{2} + n\right)^2 \tag{11.42}$$

with

$$C = \frac{(h\nu_0)^2}{4\varepsilon_B} \tag{11.43}$$

Using this expression, we can calculate the excitation energies from the ground state ($n = 0$) to a state with quantum number n as

$$\begin{aligned}\Delta E_{0\to n} &= \left[h\nu_0\left(\frac{1}{2}+n\right) - C\left(\frac{1}{2}+n\right)^2\right] - \left[h\nu_0\left(\frac{1}{2}\right) - C\left(\frac{1}{2}\right)^2\right] \\ &= h\nu_0 \cdot n - C \cdot n(n+1)\end{aligned} \tag{11.44}$$

Rearranging this expression, we find

$$\frac{\Delta E_{0\to n}}{n} = h\nu_0 - C\cdot(n+1) \tag{11.45}$$

This relation enables us to evaluate k_f and C for a diatomic molecule from the experimental excitation energies.

In Fig. 11.25 we use experimental data for HCl (Table 11.4) and plot $\Delta E_{0\to n}/n$ versus $(n+1)$. We find an excellent linear relationship. According to Equation (11.45), from the intercept with the ordinate and from the slope we obtain

$$h\nu_0 = 5.938\times10^{-20}\ \text{J} \quad \text{and} \quad C = 0.1027\times10^{-20}\ \text{J}$$

which gives a $\nu_0 = 8.962\times10^{13}\ \text{s}^{-1}$, corresponding to a wavenumber of $\tilde{\nu} = 2989\ \text{cm}^{-1}$. From Equation (11.20) for the force constant of HCl we obtain $k_f = (2\pi\nu_0)^2\mu = 516\ \text{N m}^{-1}$, using $\mu = 1.627\times10^{-27}$ kg. Note that this value is larger than $k_f = 480\ \text{N m}^{-1}$ calculated in Example 11.6. This method is important for deriving force constants from spectroscopic measurements. All force constants for the diatomic molecules shown in Table 7.6 have been obtained using this method. In Exercise E11.22 you repeat the procedure in Equation (11.45), Table 11.4, and Fig. 11.25 using the calculated data in Table 11.3 for the anharmonic oscillator and compare the result with that obtained from the experimental data in Table 11.4.

Accordingly, using Equation (11.43), for the bond energy of HCl we find $\varepsilon_B = (h\nu_0)^2/(4\cdot C) = 8.58\times10^{-19}$ J. However, this value differs significantly from the value 7.17×10^{-19} J given in Table 7.5. This discrepancy results from the fact that the Morse function is only an approximation to the exact potential energy curve, and the small remaining deviations are

Table 11.4 HCl Molecule, Transitions $0\to1$ (Fundamental Absorption Band), $0\to2$, $0\to3$, and $0\to4$ (Overtone Bands). Measured [5] Wavenumbers $\tilde{\nu}$ and Corresponding Frequencies ν and Excitation Energies $\Delta E_{0\to n}$.

n	$\tilde{\nu}/\text{cm}^{-1}$	$\nu/10^{13}\ \text{s}^{-1}$	$\Delta E_{0\to n}/10^{-20}$ J	$(n+1)$	$\frac{\Delta E_{0\to n}/10^{-20}\ \text{J}}{n}$
1	2886	8.652	5.723	2	5.733
2	5668	16.99	11.259	3	5.630
3	8347	25.02	16.581	4	5.527
4	10923	32.75	21.698	5	5.425
5	13397	40.16	26.611	6	5.322

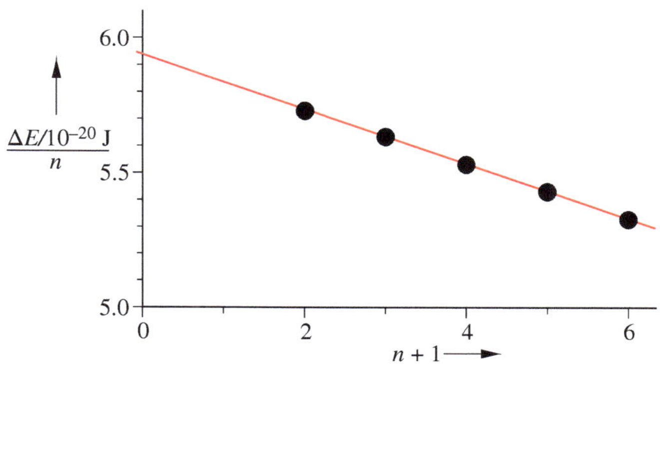

Figure 11.25 Evaluation of force constant k_f and bond energy ε_B of HCl from vibrational spectra. The excitation energies ΔE from state $n = 0$ to states $n = 1, 2, 3, 4$ are obtained from the fundamental and the overtone bands. $\Delta E/n$ is plotted versus $n + 1$.

sufficient to make an extrapolation to the bond energy inaccurate. On the other hand, in most cases the two terms in Equation (11.42) give a good fit for the excitation energies of the overtone bands, if C is treated as an adjustable parameter.

11.3.5 Multidimensional Infrared Spectroscopy

Improvements in laser and detector technologies have enabled the development of multidimensional optical spectroscopies; i.e., the examination of how optical transitions between quantum states are correlated with each other and are affected by their environment. Over the past twenty years, two-dimensional infrared (2D-IR) spectroscopy has proven useful for asking incisive questions about molecular structures and solute molecule interactions with solvent. An example for this method is given in Foundation 11.8.

11.4 Raman Spectra

The Raman effect is often used to obtain the same type of molecular parameters as found from infrared spectra (bond lengths, force constants, geometry of molecules). In a Raman spectrophotometer the molecules are irradiated with light in the visible or ultraviolet, e.g., the green light ($\lambda = 514$ nm) of an argon ion laser. The laser light is scattered by the molecules, and the scattered light is detected after passing through a monochromator (Fig. 11.26a). To understand the Raman effect we first consider elastic light scattering, or Rayleigh scattering, in which the scattered light has the same frequency as the incident light. Then, we consider Raman transitions for which the scattered laser light has a different frequency than the incident laser light, arising from changes in the internal energy of the molecules.

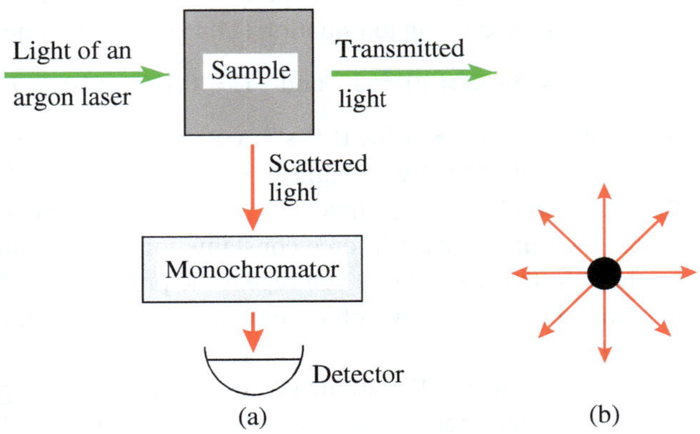

Figure 11.26 Raman effect: (a) Experimental setup with the scattered light collected at 90° from the incident light; i.e., the incident light propagates from left to right, and the scattered light propagates from top to bottom. After passing a monochromator its intensity is measured by a photoelectric device (e.g., a photomultiplier). (b) The black full circle indicates a scattering molecule. The intensity of the scattered light is uniformly distributed in the directions perpendicular to the direction of its propagation.

11.4.1 Rayleigh Scattering

We begin by using a classical model. The electric field (amplitude F_0) of the incident light can distort the electron cloud of a molecule and create a dipole moment μ_{ind}. This induced dipole moment in the molecule is given by

$$\mu_{\text{ind}} = \alpha \cdot F_0 \cdot \cos\left(\omega_{\text{laser}} t\right) \tag{11.46}$$

where α is the polarizability of the molecule (see Section 13. Induction forces) and $\nu_{\text{laser}} = \omega_{\text{laser}}/(2\pi)$ is the frequency of the applied light field (Fig. 11.27). The induced dipole, oscillating with the frequency ν_{laser}, can be viewed as a radiating antenna that emits light in different directions. This process does not change the energy of the molecule or the photon (it is elastic) and is called *Rayleigh scattering*. In this case the wavelength of the scattered light is identical with the wavelength of the incident light.

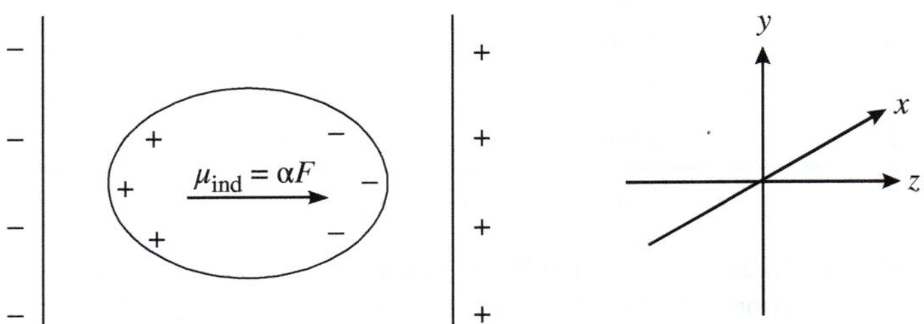

Figure 11.27 Induced dipole moment $\mu_{\text{ind}} = \alpha \cdot F$ of a molecule in an applied electric field of field strength F where α is the polarizability of the molecule.

In Foundation 10.1 we showed that the light intensity $I_{\text{scattered}}$ from an oscillating dipole is

$$I_{\text{scattered}} = \frac{c_0 \varepsilon_0}{2} \frac{1}{\left(4\pi\varepsilon_0\right)^2} \frac{\left(Q\xi_0\right)^2}{c_0^4 d^2} \omega^4 \cdot \sin^2\vartheta \tag{11.47}$$

where ξ_0 is the amplitude of oscillation of charge Q, ω is the angular frequency, and d is the distance from the dipole. ϑ is the angle between the direction of the dipole (the z direction) and the direction of d. Because of the sine function, the scattered light has the same intensity in the xy plane and no light is emitted in the z direction. In the following we restrict our consideration to the case $\vartheta = 90°$.

To describe Rayleigh scattering, we replace the displaced charge of the oscillator $Q\xi_0$ with the magnitude of the induced dipole αF_0. Together with the relation

$$I_{\text{incident}} = \frac{c_0\varepsilon_0}{2}F_0^2 \tag{11.48}$$

where I_{incident} is the intensity of the incident laser light wave, we obtain

$$(Q\xi_0)^2 = \alpha^2 F_0^2 = \alpha^2 \frac{2}{\varepsilon_0 c_0} I_{\text{incident}} \tag{11.49}$$

Then

$$\begin{aligned} I_{\text{scattered}} &= \frac{1}{(4\pi\varepsilon_0)^2}\frac{\omega_{\text{laser}}^4}{c_0^4 d^2}\alpha^2 \cdot I_{\text{incident}} \\ &= \frac{1}{(4\pi\varepsilon_0)^2}\frac{\alpha^2}{d^2}\left(\frac{2\pi}{\lambda_{\text{laser}}}\right)^4 \cdot I_{\text{incident}} \end{aligned} \tag{11.50}$$

where ω_{laser} is the angular frequency and λ_{laser} is the wavelength of the excitation laser beam used in the experiment. Note the difference between Rayleigh scattering (molecule emitting light because it is polarized in the field of the incident light) and fluorescence (molecule emitting light by transitioning between an excited state and a lower state). Rayleigh scattering is discussed more in Chapter 26.

Example 11.7 *The sunlight impinging on the atmosphere is scattered by air molecules before reaching the planet's surface. Compare the scattered intensity for 400 nm light to that for 700 nm light. Describe how this effect can explain the color of the sky.*

To compare the scattered intensity at the two different wavelengths we take their ratio, and find that

$$\frac{I_{\text{scattered}}(400\text{ nm})}{I_{\text{scattered}}(700\text{ nm})} = \left(\frac{700}{400}\right)^4 \cdot \frac{I_{\text{incident}}(400\text{ nm})}{I_{\text{incident}}(700\text{ nm})}$$

If the incident intensity at the two wavelengths is the same, then the intensity of the scattered 400 nm light is almost 10 times larger than that of the 700 nm light. This comparison shows that the shorter wavelength (blue) light is scattered more strongly than the longer wavelength (red) light. Consequently, the longer wavelength incident light is more likely to travel a straight trajectory to the planet's surface, but the shorter wavelength (blue) incident light is more likely to be scattered in the air and to your eye. At sunset and sunrise the path length of the sun light through the atmosphere is extremely long, therefore the sun shows a red or orange color because most of the shorter wavelength colors are depleted and only the longer wavelengths colors propagate far enough to provide detectable light.

11.4.2 Raman Spectroscopy

If light from an argon ion laser at 514.5 nm irradiates a sample of molecules, some light is scattered elastically, giving rise to a Rayleigh line, but spectral lines are also observed at the detector with wavelengths different from the laser wavelength. These spectral features correspond to inelastic scattering (processes that exchange energy between the incident photon and the molecule). The most common inelastic scattering process is called *Raman scattering*. In the case of a rotational Raman spectrum the lines are very close (within 1 nm) to the Rayleigh line at 514 nm and are caused by rotational excitation (or deexcitation) of the molecules. In the case of a vibrational–rotational Raman spectrum the molecules take up or release vibrational energy, in addition to the rotational energy. Then the spectral lines are shifted by about ± 20 nm from the Rayleigh line. Although less common, it is also possible to perform Raman scattering with the excitation of electronic states.

Because Raman scattering occurs by inducing a dipole moment in a molecule, it does not require that the molecule have an intrinsic dipole moment (as does microwave spectroscopy) nor does it require that the vibrational motion change a molecule's dipole moment (as does infrared spectroscopy). Instead, Raman spectra require a change in the molecule's polarizability, α. For these reasons, Raman spectroscopy can provide an important and complementary method for probing the nuclear geometry and motions of molecules.

Because the incident light beam does not need to be resonant with the transition, the Raman method is particularly useful if the surrounding medium absorbs infrared light and interferes with the infrared measurement. Furthermore, the selection rules in Raman spectroscopy are different from those in an infrared spectrum, so that transitions which are not observed in the infrared spectrum often can be observed in the Raman spectrum.

11.4.3 Rotational Raman Spectra of Heteronuclear Diatomic Molecules

For a nonspherical molecule, the polarizability α depends on the direction because of the electron clouds shape. Typically we expect that the induced dipole moment μ_{ind} will be larger along the long axis of the molecule than the short axis. The induced dipole moment becomes small and large two times during a rotation; hence, α changes at twice the rotation frequency $\nu_{\text{rot}} = \omega_{\text{rot}}/(2\pi)$ (Fig. 11.28). We can write the time-dependent polarizability as

$$\alpha = \alpha_0 + \Delta\alpha \cdot \cos\left(2\omega_{\text{rot}} t\right) \tag{11.51}$$

If we insert this result into Equation (11.46), we find that the induced dipole moment is given by

$$\begin{aligned} \mu_{\text{ind}} &= (\alpha_0 + \Delta\alpha \cdot \cos\left(2\omega_{\text{rot}} t\right)) \cdot F_0 \cdot \cos\left(\omega_{\text{laser}} t\right) \\ &= \alpha_0 \cdot F_0 \cdot \cos\left(\omega_{\text{laser}} t\right) + (\Delta\alpha \cdot F_0 \cdot \cos\left(2\omega_{\text{rot}} t\right)) \cdot \cos\left(\omega_{\text{laser}} t\right) \end{aligned}$$

By taking the product of the cosine functions (see Appendix B.4), we can write

$$\begin{aligned} \mu_{\text{ind}} &= \alpha_0 \cdot F_0 \cdot \cos\left(\omega_{\text{laser}} t\right) \\ &\quad + \frac{1}{2}\Delta\alpha \cdot F_0 \cdot \left[\cos\left((\omega_{\text{laser}} + 2\omega_{\text{rot}})t\right) + \cos\left((\omega_{\text{laser}} - 2\omega_{\text{rot}})t\right)\right] \end{aligned} \tag{11.52}$$

Equation (11.52) shows that a molecule rotating with frequency ν_{rot} scatters light at the frequencies ν_{laser}, $\nu_{\text{laser}} + 2\nu_{\text{rot}}$, and $\nu_{\text{laser}} - 2\nu_{\text{rot}}$. The light at the laser frequency ν_{laser} corresponds to the elastically (Rayleigh) scattered light, the light at the lower frequency $\nu_{\text{laser}} - 2\nu_{\text{rot}}$ is the Raman scattered light (sometimes called *Stokes scattering*), and the light at higher frequency $\nu_{\text{laser}} + 2\nu_{\text{rot}}$ is the Raman scattered light (called *anti-Stokes scattering*).

Figure 11.29 shows the rotational Raman spectrum of CO gas, for which the rotational energy of a molecule is restricted to discrete energy states, Equation (11.4). The transitions $\nu_{\text{laser}} - 2\nu_{\text{rot}}$ correspond to the case where the rotational quantum number increases by two (see Foundation 11.7):

$$\begin{aligned} \Delta E_{J\to J+2} &= B \cdot [(J+2)(J+3) - J(J+1)] \\ &= 2B \cdot (2J+3), \quad J = 0, 1, 2, \ldots \end{aligned} \tag{11.53}$$

The lines at lower frequencies relative to the Rayleigh line correspond to a change of quantum numbers $0 \to 2$, $1 \to 3$, … (Stokes lines); the lines at higher frequencies relative to the Rayleigh line correspond to changes $2 \to 0$, $3 \to 1$, …

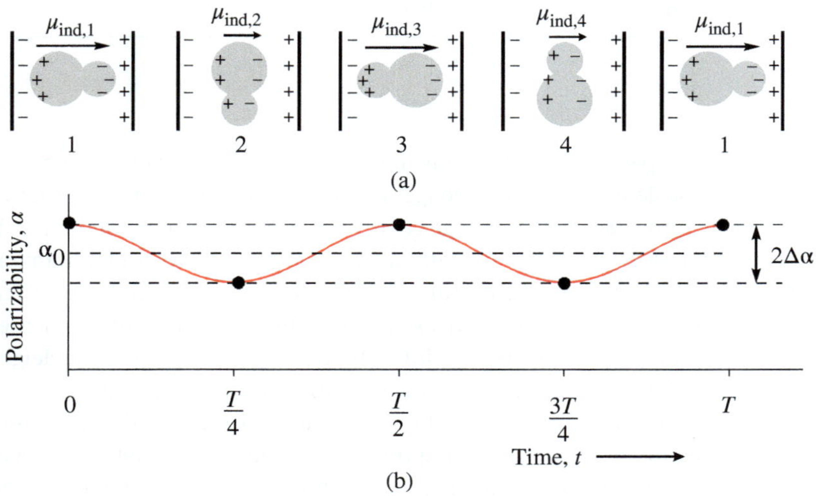

Figure 11.28 Change of the polarizability α for a rotating molecule in the classical view. (a) Five stages of rotation inside a capacitor replacing the x-component of the electric field strength $F_0 \cos \omega_{\text{laser}} t$ of the exciting laser beam. (b) Polarizability α versus time t during one rotation cycle (period T): $\alpha = \alpha_0 + \Delta\alpha \cos 2\omega_{\text{rot}} t$, with $\alpha_0 = \frac{1}{2}(\alpha_1 + \alpha_2)$, $\Delta\alpha = \frac{1}{2}(\alpha_1 - \alpha_2)$, and $\omega_{\text{rot}} = 2\pi/T$. The filled circles indicate stages 1 to 4 in Fig. 11.28a.

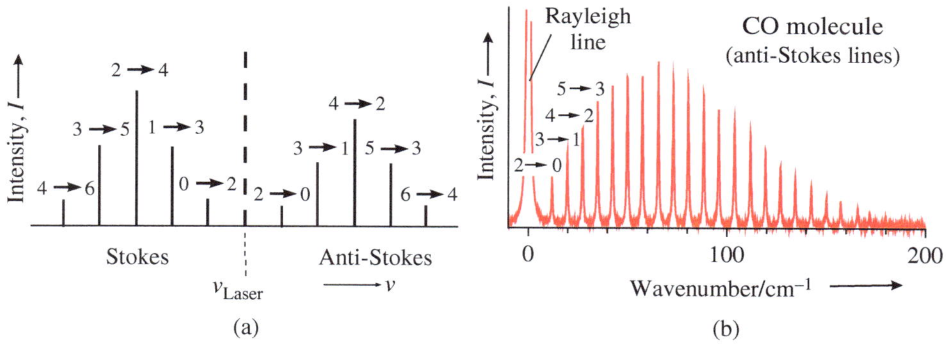

Figure 11.29 (a) Intensity I expected in a Raman experiment with rotational excitation of diatomic molecules (schematic). ν_{laser}: Rayleigh line; the lines at lower frequencies $\nu_{laser} - 2\nu_{rot}$ relative to the Rayleigh line correspond to a change in quantum numbers $0 \to 2, 1 \to 3, \ldots$ (Stokes lines). The transitions $\nu_{laser} + 2\nu_{rot}$ correspond to a change in quantum numbers $2 \to 0, 3 \to 1, \ldots$ (*anti*-Stokes lines). (b) Rotational Raman spectrum of CO (*anti*-Stokes lines). Intensity I versus wavenumber $(\tilde{\nu} - \tilde{\nu}_{laser}) = (\nu - \nu_{laser})/c_0$. (Source: P. Dechant, Philipps-University Marburg, private communication).

(anti-Stokes lines). The spacing between two adjacent lines

$$\Delta(\Delta E) = \Delta E_{n+1 \to n+3} - \Delta E_{n \to n+2} = 4B = \frac{h^2}{2\pi^2 \mu d^2} \tag{11.54}$$

is just twice the spacing of the rovibrational lines in the infrared spectrum. The Raman lines in Fig. 11.29b are spaced at equal frequencies, as expected from (11.54). As in the case of the high-resolution infrared spectrum, the intensities I of the Raman lines reflect the population of the rotational energy levels at room temperature. It is shown in Foundation 11.7 that

$$I = const' \cdot (\Delta\alpha)^2 \cdot \frac{3(J+1)(J+2)}{2(2J+3)} \cdot e^{-BJ(J+1)/(kT)} \tag{11.55}$$

Example 11.8 *Calculate the bond length of CO from the rotational Raman Spectrum*

The spacing between two adjacent rotational Raman lines in Fig. 11.29b is $\tilde{\nu} = 7.75\ \text{cm}^{-1}$. This corresponds to

$$\Delta(\Delta E) = \frac{hc_0}{\lambda} = hc_0\tilde{\nu} = 1.54 \times 10^{-22}\ \text{J}$$

Using Equation (11.54) with the reduced mass

$$\mu = \frac{m_C m_O}{m_C + m_O} = 1.14 \times 10^{-26}\ \text{kg}$$

we find

$$d = \sqrt{\frac{h^2}{2\pi^2 \mu} \cdot \frac{1}{\Delta(\Delta E)}} = 113\ \text{pm}$$

Raman spectra of homonuclear diatomic molecules will be discussed in Chapter 12.

11.4.3.1 Vibrational–Rotational Raman Spectra of Diatomic Molecules

In the classical picture, the polarizability α becomes periodically smaller and larger with the same frequency as the vibrational motion of a diatomic molecule (Fig. 11.30). Consequently, the induced dipole moment is given by

$$\begin{aligned} \mu_{ind} &= \alpha(t) \cdot F(t) \\ &= (\alpha_0 + \Delta\alpha \cdot \cos(\omega_{vib} t)) \cdot F_0 \cdot \cos(\omega_{laser} t) \end{aligned} \tag{11.56}$$

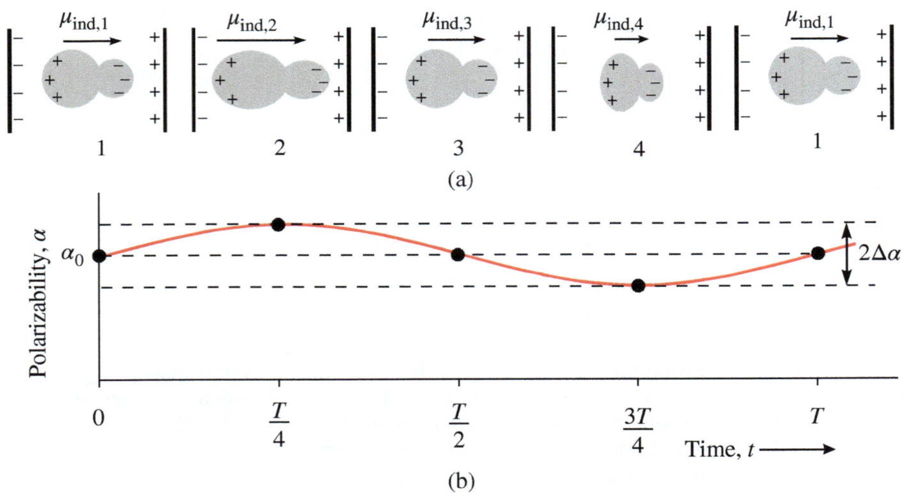

Figure 11.30 Polarizability α for vibrational motion of a diatomic molecule in the classical view. (a) Five stages of vibration. (b) Polarizability α versus time t during one vibration cycle (period T). The filled circles indicate stages 1 to 4 in Fig. 11.30a.

where $\Delta\alpha$ is the change in polarizability that results from the molecule's vibrational motion. Multiplying these terms out we find that

$$\mu_{ind} = \alpha_0 F_0 \cdot \cos(\omega_{laser} t) + \frac{1}{2}\Delta\alpha F_0 \left[\cos\left((\omega_{vib} + \omega_{laser})\, t\right) + \cos\left((\omega_{vib} - \omega_{laser})\, t\right)\right]$$

The first term in this formula gives rise to Rayleigh scattering at the angular frequency ω_{laser}. The other two terms in this expansion give the sum and difference frequencies, $\nu_{laser} + \nu_{vib}$ (anti-Stokes) and $\nu_{laser} - \nu_{vib}$ (Stokes), for the vibrational Raman transitions. The selection rules for vibrational Raman transitions of diatomic molecules require one quantum transitions. Importantly, these transitions can be driven for molecules that do not have a dipole moment.

As in the case of the infrared spectrum, the vibrational transitions are accompanied by rotational transitions. The quantum-mechanical treatment (Foundation 11.7) reveals that the rotational quantum number can change by +2, −2, or 0 in a vibrational transition (see Fig. 11.31). Because all of the transitions where the rotational quantum number remains unchanged appear at the same place in the Raman spectrum (at ν_{vib}), we expect a strong line corresponding to the pure vibrational transition and many weak lines at both sides of the strong line. Parts of the vibrational–rotational Raman spectrum of CO are shown in Fig. 11.32. The lines correspond to transitions from the vibrational ground state (Stokes lines). Because of the large energy difference between the vibrational ground state and the first vibrational excited state, the excited state is only slightly occupied at room temperature. Therefore, the anti-Stokes lines are significantly weaker than the Stokes lines.

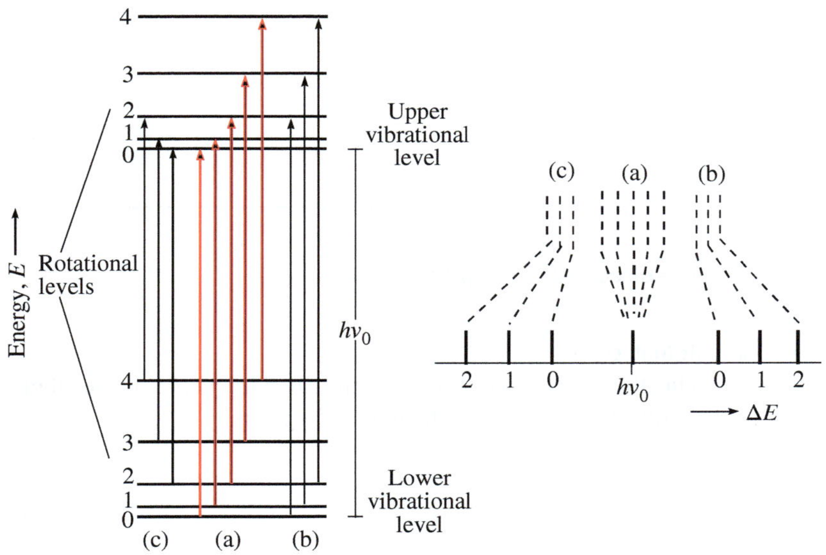

Figure 11.31 Left: Transitions in a vibrational–rotational Raman spectrum. $h\nu_0$ is the energy for the pure vibrational transition. (a) No change of the rotational quantum number. (b,c) Rotation quantum number changing by +2 and −2, respectively. Right: corresponding lines in the Raman spectrum.

Example 11.9 *Use the vibrational Raman spectrum of CO to calculate the force constant of its bond.*

According to Fig. 11.32, the pure vibrational transition of CO is at $\tilde{\nu} = 2142\ \text{cm}^{-1}$. This corresponds to $\Delta E = 4.25 \times 10^{-20}$ J. In analogy to Section 11.3.3.1 it follows that

$$k_f = 4\pi^2 \mu \cdot \left(\frac{\Delta E}{h}\right)^2 = 1850\ \text{Nm}^{-1}$$

if $\mu = 1.14 \times 10^{-26}$ kg.

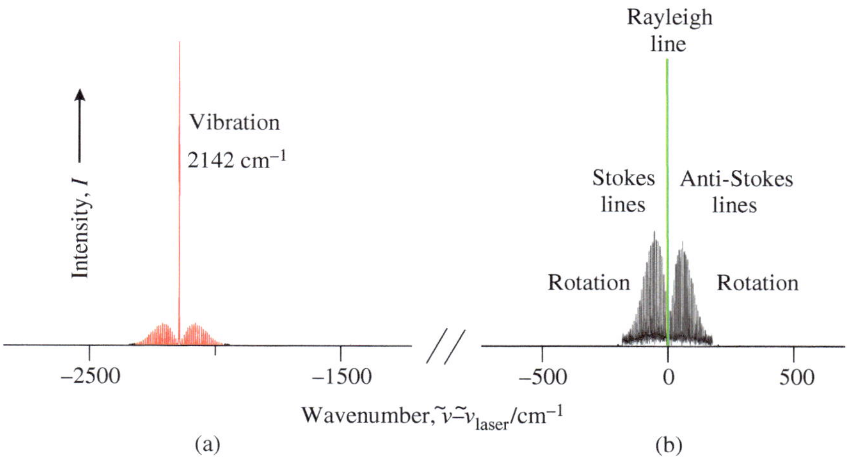

Figure 11.32 Complete Raman spectrum of CO. Intensity I versus wavenumber $\tilde{\nu} - \tilde{\nu}_{\text{laser}} = (\nu - \nu_{\text{laser}})/c_0$. (a): Vibrational–rotational lines. Note the strong line corresponding to the $0 \to 0$ transition. (b): Rotational lines (Stokes and *anti*-Stokes lines, see Fig. 11.29 for better spectral resolution).(Source: P. Dechant, Philipps-University marburg, private communication).

11.4.4 Raman Spectra of Polyatomics

11.4.4.1 Raman Spectrum of CO_2

Here we discuss CO_2 as an example for a polyatomic molecule. In Fig. 11.33 the polarizability α during the vibration of CO_2 is displayed for the three different vibrational modes. In Fig. 11.33a α is changing with twice the vibration frequency

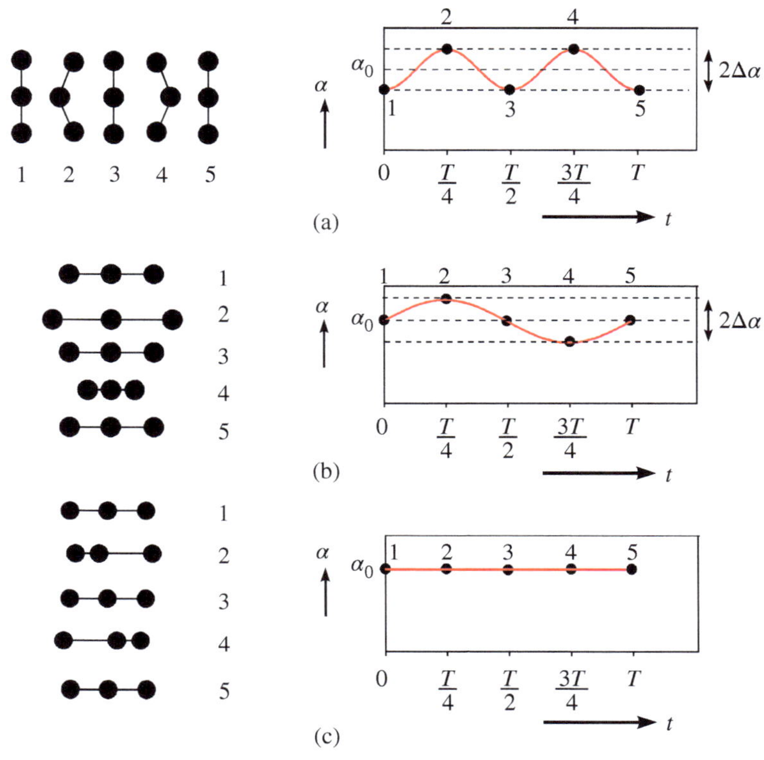

Figure 11.33 Vibrational modes for CO_2: Raman activity. Different stages of vibration (*left*) and corresponding polarizability α versus time t (*right*) during one vibrational cycle. (a) Bending mode. (b) Symmetric stretching mode. (c) Asymmetric stretching mode. The bending, compression, and expansion of atoms are strongly exaggerated.

$\tilde{\nu}$/cm^{-1}	($n_{bending}$, n_{sym}, n_{anti})		($n_{bending}$, n_{sym}, n_{anti})
1264	(1,0,0)	→	(3,0,0)
1369	(0,1,0)	→	(2,1,0)
1409	(1,0,0)	→	(1,1,0)

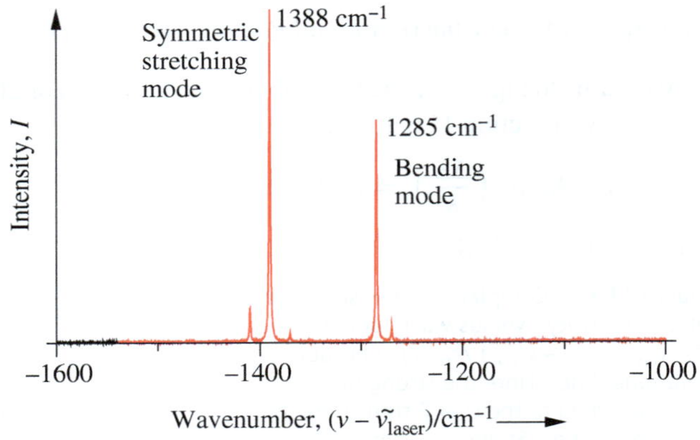

Figure 11.34 Vibrational Raman spectrum for CO_2. Intensity I versus wavenumber $(\tilde{\nu} - \tilde{\nu}_{laser}) = (\nu - \nu_{laser})/c_0$. The strong Raman lines correspond to an excitation of the bending mode ($n_{bending}$: 0 → 2) and of the symmetric stretching mode (n_{sym}: 0 → 1). Additionally, three weak absorption bands occur, which correspond to the transitions shown in the table (for more details see Box 11.2). (source: P. Dechant, Philipps-University Marburg, private communication).

for the bending mode, and a Raman line corresponding to twice the vibration frequency is expected. In Fig. 11.33b α is changing periodically with the same frequency as the vibration for the symmetric stretching mode: A Raman line corresponding to the vibration frequency is expected. In Fig. 11.33c α is constant for the asymmetric stretching mode: no Raman line is expected. Consequently, we find two strong lines in the Raman spectrum of CO_2 (Fig. 11.34): one corresponds to the symmetric stretch fundamental and the second corresponds to the overtone (two quanta) of the bending motion.

11.4.4.2 Complementarity of IR and Raman Spectra

Comparison of the IR and Raman spectrum of CO_2 reveals their complementarity. In the IR spectrum the fundamental (single quantum excitations from $n = 0$ to $n = 1$) transitions which appear are the bending mode and the asymmetric stretching mode, but not the symmetric stretching mode. In the Raman spectrum, the fundamental transition which appears is the symmetric stretching motion but not the bending or asymmetric stretching motion. The bending excitation that appears in the Raman spectrum is a $\Delta n = 2$ transition, not the fundamental. It can be proven that any molecule that possesses a center of symmetry, as does CO_2, will show this complementarity.

11.4.4.3 IR and Raman Spectra of CH_4

Because not all vibrational modes can be excited by infrared absorption (or Raman excitation), it is often desirable to obtain both the infrared absorption spectrum and the Raman spectrum of a molecule. Figure 11.35 shows the infrared

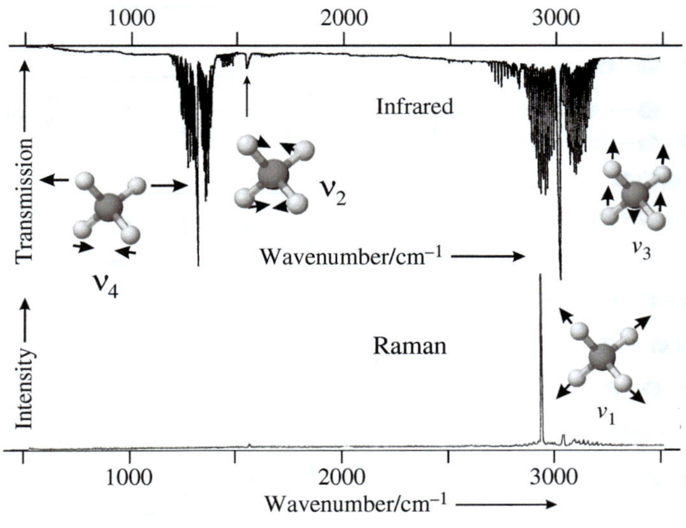

Figure 11.35 Infrared absorption (top) and the Raman spectrum (bottom) of gaseous methane. Note the rotational structure in the spectra. The schematic of the molecule illustrates the type of nuclear displacement associated with the normal modes, which have fundamental frequencies of $\nu_1 = 2914$ cm^{-1}; $\nu_2 = 1526$ cm^{-1}, doubly degenerate (weak transition); $\nu_3 = 3020$ cm^{-1}, triply degenerate; and $\nu_4 = 1306$ cm^{-1}, triply degenerate. (Taken from Schrader [6]).

absorption spectrum and the Raman spectrum of methane gas. The vibrational spectra of methane can be understood using four vibrational fundamentals. Examples of each of these fundamentals are shown in Fig. 11.35. The modes ν_3 and ν_4 generate dipole moments and are infrared active, whereas the ν_1 which corresponds to the symmetric stretch does not generate a dipole moment but is Raman active. The fourth fundamental with frequency ν_2 has a weak oscillator strength. It can be observed near 1526 cm^{-1} in the infrared spectrum, and its first overtone can be observed near 3072 cm^{-1} in the Raman spectrum. These four fundamentals correspond to the nine normal modes of the methane molecule, for which one is doubly degenerate and two are triply degenerate. Although it is easier to observe the rotational structure in the infrared spectrum shown here, it is also present in the Raman spectrum. The assignment of the rovibrational structure allows the determination of important methane structural properties, e.g., the rotational constant is 1.04×10^{-22} J, hence a moment of inertia of 5.33×10^{-47} kg m^2. This inertial moment corresponds to a bond length of 109.7 pm for the CH bonds in this tetrahedral molecule.

11.5 Vibrational Structure of Electronic Spectra

When a molecule is electronically excited the nature of its bonding changes and its geometry (bond angles and bond lengths) can change. For the simple case of an electronically excited diatomic molecule the equilibrium bond length d^* is usually different from that in the ground state. Often the equilibrium bond length in the excited state is larger than in the ground state; however, this need not always be the case. The different bonding in the excited electronic state causes significant vibrational rearrangement and gives rise to vibrational substructure in the electronic spectrum, called vibronic structure.

11.5.1 Franck–Condon Principle

From our earlier discussions it is evident that electrons have a higher velocity than nuclei, so we might expect the electron cloud to rearrange more rapidly than the nuclei during an electronic transition. The approximation that the electronic transition occurs with the nuclei frozen in their initial state configuration is called the *Franck–Condon principle*. After the electronic cloud has adjusted to its final state arrangement, the nuclei move.

To simplify we consider the case of a diatomic molecule. Figure 11.36 shows a ground electronic state whose equilibrium bond length is d and an excited electronic state curve with an equilibrium bond length of d^*. Applying the Franck–Condon approximation, an electronic transition occurs without a change of nuclear position. Electronic transitions of this sort are called *vertical transitions*. A vertical transition has the nuclear configuration fixed, and the intensity profile of the electronic transition is determined by the overlap of the vibrational wavefunctions in the different states.

Let us analyze this last statement in more detail for the case of a diatomic molecule. As the electronic motion is fast compared to the motion of the nuclei, the molecule, immediately after excitation, is in the geometry of the ground state but in the electronically excited state. In a classical description, the molecule starts to oscillate with amplitude $d^* - d$. The vibrational frequency ν_0^* and the vibrational energy E^* are

$$\nu_0^* = \frac{1}{2\pi}\sqrt{\frac{k_f^*}{m_r}} \quad \text{and} \quad E^* = \frac{1}{2}k_f^* \cdot (d^* - d)^2 \tag{11.57}$$

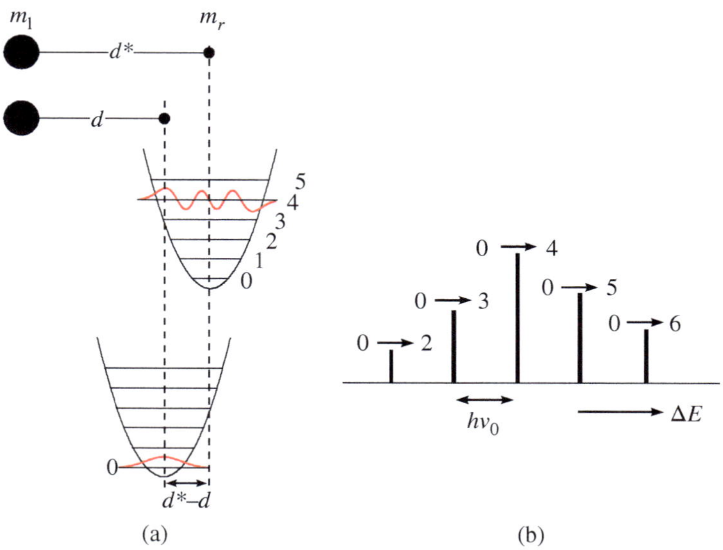

Figure 11.36 Vibrational structure of electronic transitions in a diatomic molecule according to the Franck–Condon Principle. (a) Scheme of potential curves in the electronic ground state (equilibrium distance d) and in the electronically excited state (equilibrium distance d^*). In this scheme the transition from vibrational quantum number 0 in the electronic ground state to the excited state with quantum number 4 has maximum probability because of the large overlap of the vibrational wave functions. (b) Corresponding intensity distribution of the absorption bands in this scheme.

According to quantum mechanics any vibrational energy

$$E_n^* = h\nu_0^* \cdot (\tfrac{1}{2} + n) \tag{11.58}$$

can be taken up in the electronic excitation, but the quantum state with energy E_n^* closest to E^* is reached with highest probability ($n = 4$ in the case of Fig. 11.36). Although states with smaller and larger values of n are populated, those transitions are weaker.

The excitation of the different vibrational sublevels in the excited electronic state gives rise to a series of absorption lines in the electronic absorption band; this substructure is called *vibronic structure*. The origin of the vibronic structure and the relative intensity changes may be understood, by analyzing the electronic transition moment. Consider the transition dipole moment integral

$$M_x = e \int \psi_i x \psi_j \cdot d\tau$$

where x is the electron displacement. Although we could write analogous expressions for the transition moments M_y and M_z, we simplify the discussion by only considering M_x. In the Born–Oppenheimer approximation we can write the total wavefunction ψ_i as a product of an electronic wavefunction and a nuclear wavefunction,

$$\psi_i = \psi_{\text{el},i}(x, y, z; \xi) \cdot \psi_{\text{vib},i}(\xi)$$

where x, y, z are the electron coordinates and ξ is the bond stretching coordinate. The electronic wavefunction depends on the bond distance as a parameter (see Chapter 6). Accordingly, for the excited state we may write ψ_j as

$$\psi_j = \psi_{\text{el},j}(x, y, z; \xi) \cdot \psi_{\text{vib},j}(\xi)$$

and the transition moment integral is

$$M_x = e \int \psi_{\text{el},i} \cdot \psi_{\text{vib},i} \cdot x \cdot \psi_{\text{el},j} \cdot \psi_{\text{vib},j} \cdot dx\, dy\, dz \cdot d\xi$$

The Franck–Condon approximation allows us to separate the electronic and nuclear integrals so that

$$M_x = e \left[\int \psi_{\text{el},i} \cdot x \cdot \psi_j \cdot dx\, dy\, dz \right] \cdot \left[\int \psi_{\text{vib},i} \cdot \psi_{\text{vib},j} \cdot d\xi \right]$$
$$= M_{x,el} \cdot \int \psi_{\text{vib},i} \cdot \psi_{\text{vib},j} \cdot d\xi$$

where $M_{x,el}$ is evaluated by integrating over the electron displacement, at a fixed nuclear geometry. This means that the total transition moment is a product of a purely electronic transition moment $M_{x,el}$ and an integral that gives the overlap of the vibrational wavefunctions. This overlap term is called the Franck–Condon factor, and it is a sensitive function of the nuclear displacement and the force constants in the electronic states.

Figure 11.37 shows an electronic spectrum of HCl. The overlap between the initial state and the $n^* = 0$ final state is small because the wavefunctions are displaced. The maximum overlap occurs for the case of $n^* = 1$, because $\psi_{n^*=1}$ has its inflection point nearest to the equilibrium bond length of the ground state (i.e., the vertical transition from the ground vibrational state is nearest to the $n^* = 1$ vibrational level of the excited electronic state). The spectral lines shown in the figure are part of the vibronic spectrum. The energy spacings between the absorption lines in the spectrum correspond to the vibrational energy spacing in the excited electronic state. Hence, vibrational structure in an absorption spectrum can be used to learn about the force constants and vibrational frequencies of a molecule's excited electronic state.

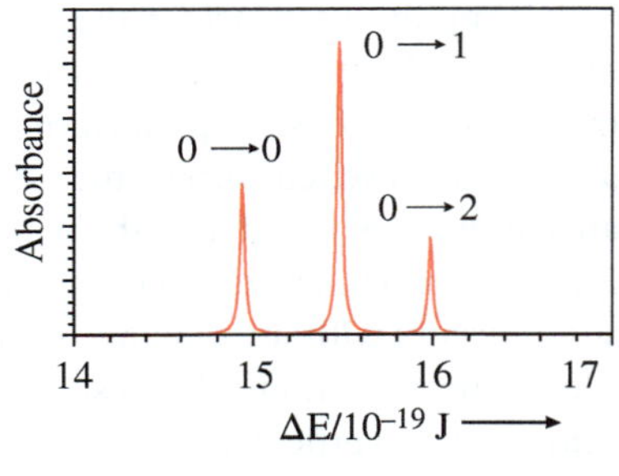

Figure 11.37 Experimental electronic spectrum of HCl.[3] The spectrum shows transitions from a vibrational quantum state with $n = 0$ in the electronic ground state to vibrational quantum states with $n^* = 0$, $n^* = 1$, and $n^* = 2$ in the electronically excited state. The transition energies are $\Delta E_{0\to0} = 14.94 \times 10^{-19}$ J, $\Delta E_{0\to1} = 15.48 \times 10^{-19}$ J, and $\Delta E_{0\to2} = 15.99 \times 10^{-19}$ J.

11.5.1.1 Types of Vibronic Spectra for Diatomics

Vibronic spectra can be classified into three different types. One type is that discussed above for HCl in which the initial and final states are both bound and have well-defined vibrational substates, and the resulting absorption spectrum has vibrational structure (Fig. 11.38a).

3 Redrawn from Figure 2 in Cheng et al. [7].

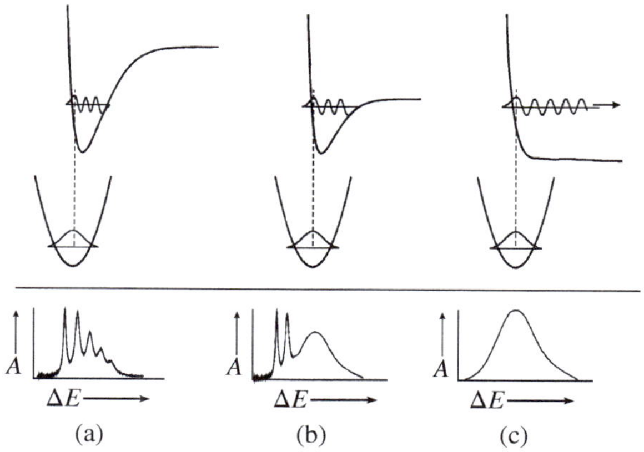

Figure 11.38 Types of vibronic spectra. Top: potential curves of the electronic ground state and the first electronic excited state. The dashed line corresponds to the transition with highest probability. Bottom: corresponding absorption spectra. The curves for the excited state differ in the depth of the potential well. (a) Deep well, bound states are reached, full vibrational structure in the spectrum. (b) Medium well depth, bound states are only reached at low excitation energy, limited vibrational structure. (c) No well, no bound states can be reached, no vibrational structure.

The second type corresponds to a situation in which the transition spans the dissociation threshold in the excited electronic state (Fig. 11.38b). In the higher energy region transitions occur into nonbound states; hence, the spectrum in this region is unstructured. But in the lower energy part of the spectrum one observes spectral lines associated with particular vibronic transitions. A situation of this sort is very useful, because the structure of the molecule in the excited state, the depth of its energy well, and its dissociation can be studied. An example is the spectrum of O_2.

A third type of transition is like that shown in Fig. 11.38c. The vertical transition occurs from the ground electronic state to a position on the excited electronic state's potential energy surface that lies above the dissociation energy for the electronically excited state. In this case a continuum of energy levels are available in the excited state, and a discrete set of lines is not observed. Nevertheless the Franck–Condon principle applies, and the intensity of the absorption changes with the excitation energy because the overlap of the initial state and final state wavefunctions changes. Because the excited state is not bound, a continuum of final states can be reached (i.e., not discrete) and this gives rise to a broad and unstructured spectrum. In this transition the molecule directly dissociates into atoms. An example is the spectrum of Cl_2.

In addition to the three basic types discussed here, molecules often have more than one electronic state that exists over a similar energy range (recall that a single excited state can be formed by promotion of an electron from any one of the filled molecular orbitals and an unfilled molecular orbital). Thus electronic spectra can become congested by multiple possible transitions, especially as the excitation increases. The electronic states can even overlap, or "cross," and this gives rise to additional complexity [8].

11.5.2 Photoelectron Spectroscopy

An interesting version of absorption spectroscopy is the ionization of molecules by monochromatic UV radiation and the measurement of the kinetic energy of the emitted electrons (*photoelectron spectroscopy*). The kinetic energy E_{kin} is measured by deflecting the electrons in a magnetic or electric field (Fig. 11.39).

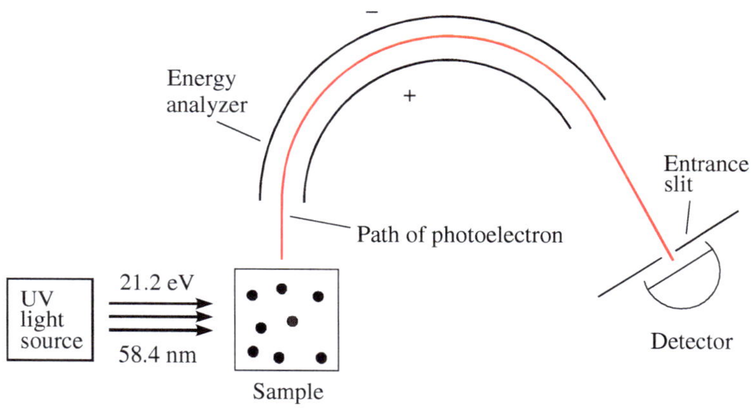

Figure 11.39 Experimental setup of a photoelectron spectrophotometer. The sample molecules are irradiated with UV light of 58.4 nm and photoelectrons are ejected from the molecules. The kinetic energy of the photoelectrons is the difference of the excitation energy and the energy required to transform a molecule into the corresponding ion. By using a charged capacitor and an entrance slit in front of the detector, photoelectrons in a small energy range are selected.

We consider a gas of H_2 molecules. The H_2 molecules are excited by using a He discharge tube that produces ultraviolet light of 58.4 nm wavelength (corresponding to an energy of 21.2 eV). This photon energy is high enough that H_2^+ ions are generated. From the measured kinetic energy E_{kin} and using energy conservation, we can obtain the ionization energy E_{Ion} as

$$E_{Ion} = 21.2\ \text{eV} - E_{kin}$$

In Fig. 11.40 the counting rate at the detector is shown as a function of E_{Ion} (photoelectron spectrum). In this ionization process H_2^+ is produced in various vibrational states ($n_{vib} = 0, 1, 2, \ldots$):

$$H_2(n'_{vib} = 0) \xrightarrow{h\nu=21.2\ \text{eV}} H_2^+(n_{vib}) + e^-(E_{kin})$$

In the case $H_2^+(n_{vib} = 0)$ the electron arrives at the detector with the largest energy ($E_{kin} = 5.77$ eV) corresponding to the $0 \to 0$ transition and an energy difference

$$\Delta E_{0\to 0} = (21.22 - 5.77)\ \text{eV} = 15.45\ \text{eV} \tag{11.59}$$

$\Delta E_{0\to 0}$ is the difference between the energy of $H_2(n'_{vib} = 0)$ and $H_2^+(n_{vib} = 0)$. In a corresponding way, the energy for transitions to other vibrational states of the H_2^+ ion is

$$\Delta E_{0\to n_{vib}} = 15.45\ \text{eV} + n_{vib} h\nu_0$$

where $h\nu_0$ is the distance between subsequent vibrational levels in H_2^+. The vibronic structure for this spectrum is shown in Fig. 11.40. The difference between the bond lengths of H_2 (74 pm) and H_2^+ (106 pm) determines the intensities of the vibrational transitions through the Franck–Condon factors. The most probable transition occurs for $0 \to 2$, $n'_{vib} = 0 \to n_{vib} = 2$.

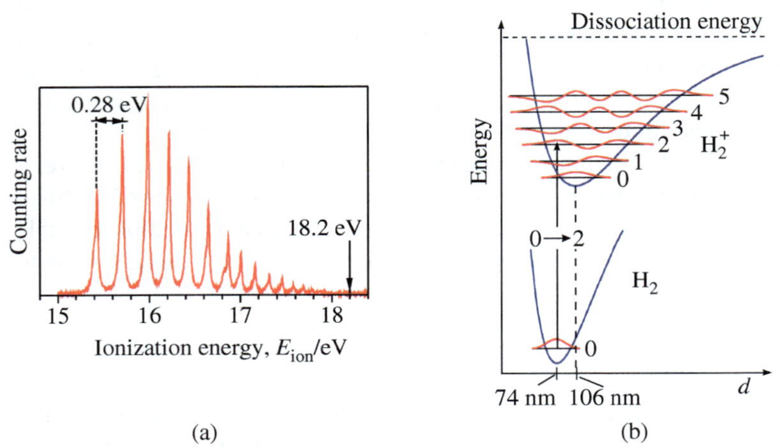

Figure 11.40 Photodissociation of H_2. (a) Photoelectron spectrum [8] of H_2. Counting rate of the detector versus ionization energy E_{Ion}. The arrow denotes the dissociation limit of the H_2^+ ion at 18.2 eV. The energy difference of 0.28 eV between the first and second bands corresponds to the change of vibration energy of H_2 for the $0 \to 1$ transition. (b) Potential energy curves of H_2 and H_2^+. In the photoelectron spectroscopy experiment transitions from the lowest vibrational level in H_2 into different vibrational levels in H_2^+ are induced. Transition $0 \to 2$ has maximum transition probability in accordance with the experimental spectrum.

Example 11.10 *Obtain the force constant for the bond in H_2^+ from the photoelectron spectrum in Fig. 11.40.*

The difference between the two lines on the left in Fig. 11.40 is $\Delta E = 0.28$ eV. This energy corresponds to the vibrational excitation energy $h\nu_0$. Using the reduced mass

$$\mu = \frac{m_H m_H}{m_H + m_H} = 8.31 \times 10^{-28}\ \text{kg}$$

we obtain

$$k_f = 4\pi^2 \mu \cdot \left(\frac{\Delta E}{h}\right)^2 = 150\ \text{N m}^{-1}$$

This is very close to the value as we calculated in Chapter 7 ($k_f = 163\ \text{N m}^{-1}$).

11.5.2.1 Evaluation of Bond Energy

The energy of the $0 \to 0$ transition provides a way to determine the bond energy of the molecular ion, if we know the bond energy of the neutral H_2 molecule. The energy level diagram in Fig. 11.41 corresponds to an energy cycle that allows the determination of the bond energy of H_2^+, which corresponds to the energy consumed in the process

$$H_2^+ \rightarrow H + H^+$$

The transition energy in the photoelectron spectrum tells us that the energy for $H_2 \rightarrow H_2^+ + e^-(E_{kin} = 0)$ is 15.5 eV. Hence, if we can determine the energy for the process

$$H_2 \rightarrow H + H^+ + e^-$$

then the difference should give the bond energy of H_2^+. If we take the dissociation energy of H_2 to be 4.5 eV and use the ionization energy of the H atom (13.6 eV), this process requires the energy 18.1 eV. Hence

$$E(H_2^+) = (18.1)\text{ eV} - 15.5\text{ eV} = 2.6\text{ eV} \tag{11.60}$$

which is in good agreement with the value $\varepsilon_B = 4.247 \times 10^{-19}$ J = 2.65 eV in Table 7.6.

Photoelectron spectroscopy has been applied to many molecules; it gives important information on their electronic structure. The photoelectron spectra of O_2 and ethene were discussed in Chapters 7 and 8, respectively.

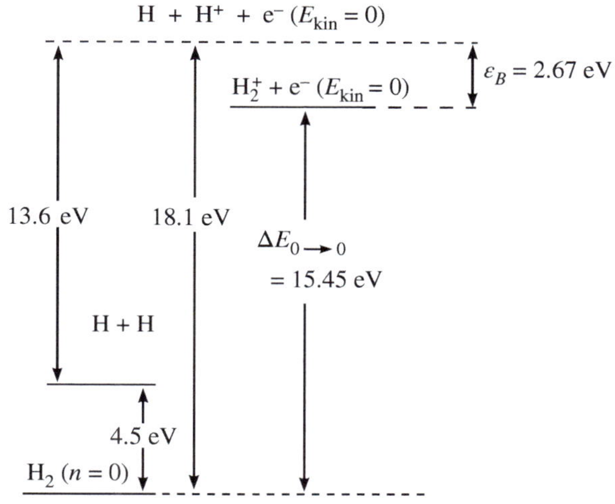

Figure 11.41 Energy scheme for H_2 and H_2^+ (photo ionization of H_2); calculation of $\Delta E_{0\rightarrow 0}$ in Equation (11.59). The bond energies of H_2 and H_2^+ are 4.2×10^{-19} J = 2.6 eV and 7.1×10^{-19} J = 4.4 eV, respectively (see Chapter 2).

11.5.3 Polyatomic Molecules

In polyatomic molecules the situation is similar; however, their electronic spectra can be quite complicated. One origin of this complexity is the large number of spectral transitions (or lines in the spectrum). For example the molecule naphthalene is known to have about 10^4 spectral transitions in the 10 cm^{-1} region centered around 32433 cm^{-1}. In dense gases and liquids, collisional processes broaden each of these lines significantly and they can overlap. This broadening and congestion makes it difficult to discern the individual quantum states.

11.5.3.1 Vibrational Structure in Naphthalene Spectrum

Figure 11.42a shows part of the absorption spectrum of naphthalene in isooctane near the molecule's lowest energy transition. As one excites higher energy transitions in the molecule the congestion problem becomes worse. By placing the molecule in a special environment, such as a supersonic jet expansion[4] which reduces the collisions and cools the molecule, the spectrum can be greatly simplified. Fig. 11.42b shows an absorption spectrum (more precisely, an excitation-emission spectrum) for naphthalene over the same wavelength region as in Fig. 11.42a. Note the change in scale of the abscissa. The spectrum in Fig. 11.42b shows clear evidence of having been simplified. It is indeed possible to assign each of the transitions in this spectrum to transitions involving specific vibrationally excited levels of the naphthalene species. However, a great amount of spectral information is hidden inside each of these narrow lines. Figure 11.42c shows an expanded view around the region of the strongest line in Fig. 11.42b.

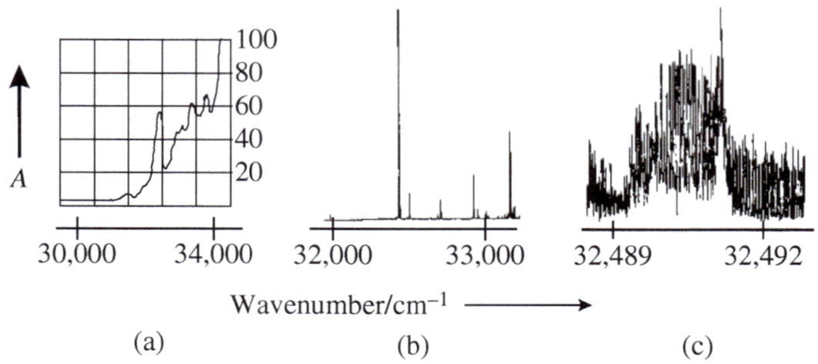

Figure 11.42 Vibronic absorption spectra for naphthalene at different resolutions. (a) Naphthalene dissolved in isooctane.[5] (b) Spectrum taken after supersonic jet expansion. (c) Same spectrum, but expanded for the strongest line in (b).[6]

4 In a supersonic jet expansion a gas is very rapidly expanded from a high pressure into vacuum, thereby "cooling down" such that the thermal motions become frozen. In the process, the energy stored in thermal motions is transferred to directed translational motion of the gas molecules at high (supersonic) average speeds. See Chapter 14 for more discussion.

5 Taken from J.B. Birks [9].

6 The high resolution spectra are given as a courtesy by D.W. Pratt, Univ. of Pittsburgh.

The large number of lines in Fig. 11.42c corresponds, primarily, to transitions involving the different rotational energy levels of the naphthalene molecule.

11.5.3.2 Vibrational Structure in Cyanine Dye and Polyene Spectrum

Even though the vibronic spectrum of a polyatomic molecule in a liquid may be broadened and congested, vibronic structure can still be observed and provides useful information. Here we compare the vibronic structure for the solution spectrum of a cyanine dye to that of a polyene.

Cyanine Dye For a dye molecule, such as a cyanine, the electron density distribution is only slightly changed when exciting the molecule. This occurs because the π-electron density is distributed over many atoms. The $\pi \rightarrow \pi^*$ transition changes the bonding along the C-C backbone and a normal mode corresponding to C-C stretching dominates the band's vibronic structure. Because the displacement is small for the cyanine dye, the bond length changes very little. The electronic absorption peak corresponding to a transition into the ground vibrational level of the C-C stretch in the electronically excited state is strongest; increasingly weaker bands corresponding to higher vibrational states appear at energy levels higher by $h\nu_0$, $2h\nu_0, \ldots$ ($\nu_0 = 4.2 \times 10^{13}$ s^{-1}, corresponding to $h\nu_0 = 0.17$ eV). This vibrational structure is shown in Fig. 11.43, left panel.

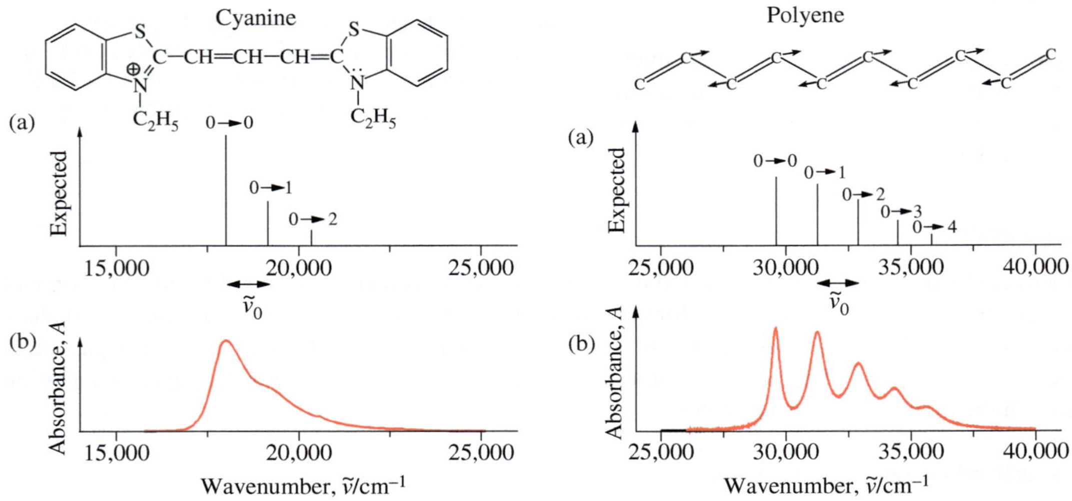

Figure 11.43 Vibrational structure of the electronic absorption band according to the Franck–Condon Principle. Left: Cyanine dye, right: polyene. (a) Expected intensity distribution for different vibrational transitions. (b) Experimental absorption spectrum. Vibrational modes in polyene indicated [10].

Polyene Polyenes constitute a somewhat different case because of the strong bond alternation. This bond alternation is caused by the fact that the electrons in the highest occupied π orbital have antinodes at the double bonds (Section 8.4). When exciting one of these electrons into the next orbital the bond lengths change significantly (Fig. 11.43, right panel). From the Franck–Condon principle, the vibrational structure according to Fig. 11.43a is expected (strongest transition for $n_{\text{vib}} = 1$) and observed (Fig. 11.43b). Again absorption bands at distance $\nu_0 = 4.2 \times 10^{13}$ s^{-1}, corresponding to $h\nu_0 = 0.17$ eV, occur. This agreement, and the strong differences between the vibrational structure of a cyanine dye and a polyene is a nice verification of the fact that polyenes, in contrast to cyanines, have bonds alternating between 134 pm and 146 pm.

11.5.3.3 Jablonski Diagram

Nonradiative processes are quite important in polyatomic molecules. Figure 11.44 shows an energy diagram (called a *Jablonski diagram*), that indicates various radiative and nonradiative processes. In this diagram the horizontal lines represent energy levels of the molecule. The subscript on the letter numbers the electronic states (thick lines) from lowest energy to highest energy. The horizontal lines within a particular electronic state (thin lines) indicate the presence of excited vibrational levels of the molecule. Rotational levels are not explicitly shown in the Jablonski diagram, for reasons of graphical clarity. However, they are present and can be quite important in various nonradiative, and radiative, processes.

The arrows in Fig. 11.44 illustrate the various radiative and nonradiative processes which may occur. The full vertical lines indicate radiative processes such as absorption (1), fluorescence (2), and phosphorescence (3). The radiative processes conserve energy by the emission, or absorption, of a photon whose energy matches that of the transition. The dashed lines indicate nonradiative processes such as dissociation, predissociation, and ionization (4), vibrational relaxation (5), internal conversion (6), and intersystem crossing (7). These processes conserve energy by exchange with a thermal medium (or bath), typically through collisions between the molecule of interest and surrounding molecules. Whenever a molecule experiences collisions and the density of states is high, these processes can be quite rapid.

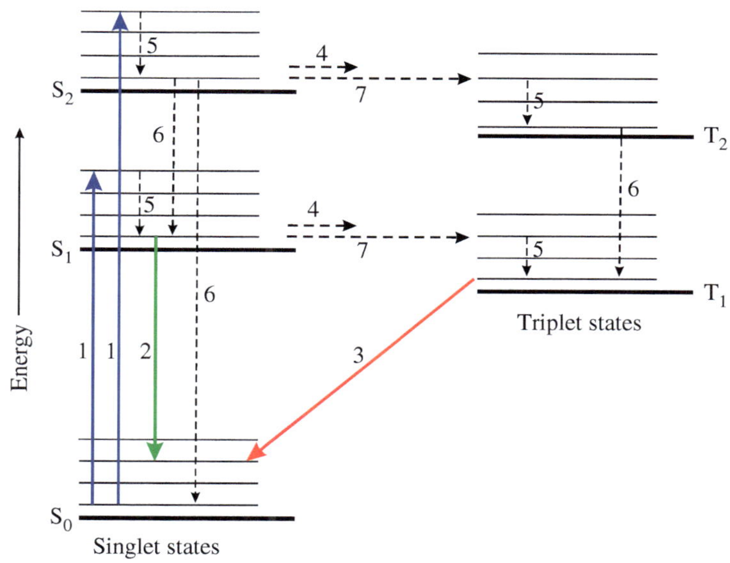

Figure 11.44 Jablonski diagram. Radiative processes (full lines). Blue: absorption (1), green: fluorescence (2), red: phosphorescence (3). Nonradiative processes (dashed lines): dissociation, predissociation, ionization (4), vibrational relaxation (5), internal conversion (6), intersystem crossing (7). S_0, S_1, and S_2 are singlet states, T_1 and T_2 are triplet states. The bold lines indicate the electronic energy levels.

11.5.3.4 Time Scales of Processes

Many of the processes indicated here have characteristic time scales associated with them. It is quite useful to be aware of these timescales. First, let us consider the radiative processes. The absorption event is typically treated as instantaneous by chemists. Its timescale is finite (typically of the order of 10^{-15} s) but shorter than those of relevance to most chemistry. The relaxation of the excited state by fluorescence (a singlet to singlet radiative transition) may occur on timescales that range from about 10^{-9} s to 10^{-7} s. The relaxation of triplet states by phosphorescence (a singlet to triplet radiative transition) is typically of much longer timescale (ranging from 10^{-6} s to seconds).

Nonradiative processes can have a wide range of timescales also, and these processes are under active investigation in chemistry. For example, all of the processes lumped together under process 4 constitute important chemical changes; such as the dissociation of a molecule or its isomerization to a new form. These processes range from timescales of 10^{-13} s (sometimes shorter) to many seconds. The process labeled as internal conversion can also be important, and sometimes may lead to new molecular conformations or isomers. Internal conversion is a nonradiative relaxation pathway between electronic states of the same spin multiplicity. Internal conversion processes can also occur over a wide range of timescales, but they are typically rapid (ca. 10^{-12} s) for higher lying energy states in a manifold (i.e., S_2 and T_2). Intersystem crossing (processes 7) is a nonradiative transition between electronic states of different spin multiplicity and typically occurs on timescales longer than 10^{-10} s. The process called vibrational relaxation may also have a wide range of timescales, and this range of timescales depends strongly on the internal energy of the molecule and the molecule's environment. This process occurs within an electronic state and is discussed more below.

11.5.3.5 Vibrational Dynamics of Anthracene

To illustrate vibrational dynamics/relaxation of a molecule, let us consider the case of anthracene. If the molecule anthracene is isolated in a supersonic jet expansion, then one can eliminate the effects of collisions and considerably reduce the contribution of "hot" transitions in the spectrum (see Fig. 11.45). Compare the spectrum shown here on the right with that found for the anthracene in solution at room temperature (on the left side of Fig. 11.45).

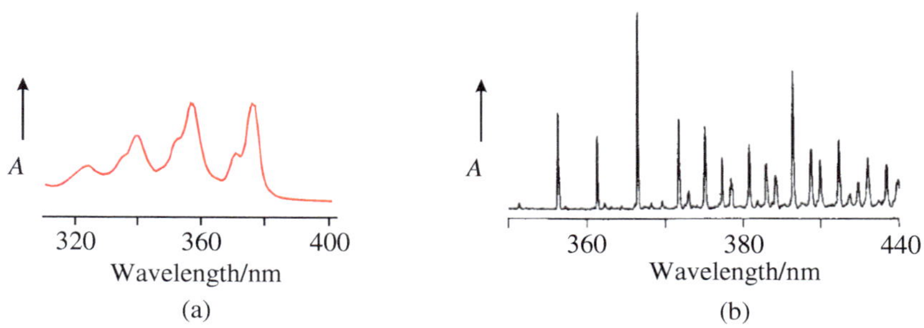

Figure 11.45 Absorption spectra of anthracene. (a) Spectrum in solution,[7] (b) Spectrum of isolated molecule.[8]

7 Taken from Birks [9].
8 From Felker and Zewail [11].

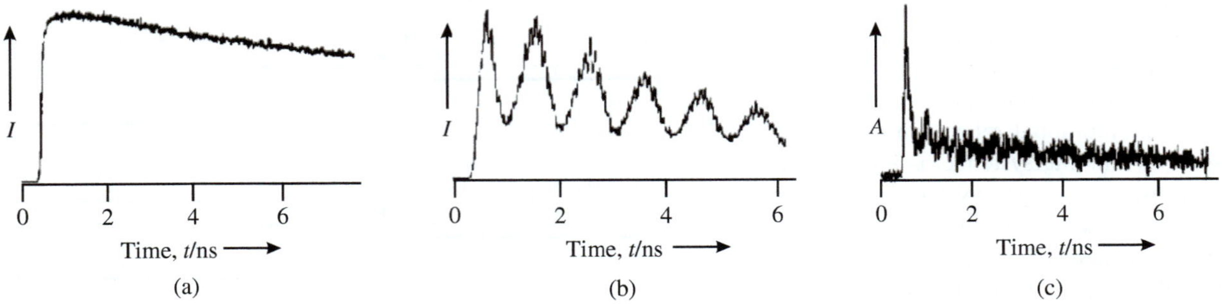

Figure 11.46 Time evolution of the fluorescence intensity I from an anthracene molecule with different amounts of excess vibrational energy. (a) $E_{vib} = 1.54 \times 10^{-20}$ J, (b) $E_{vib} = 2.82 \times 10^{-20}$ J, and (c) $E_{vib} = 3.56 \times 10^{-20}$ J. The fluorescence was measured after supersonic jet expansion to avoid collisions.[9]

Low Excess Energy One can study the excited state relaxation processes of anthracene in a supersonic jet expansion by using a pulse of light and examining how the molecule's fluorescence emission changes in time. Because the molecules are in a supersonic jet expansion, they are mostly in their ground vibrational levels and very low rotational energy levels before excitation with light. The panels in Fig. 11.46 show three examples of how the time-dependent fluorescence emission intensity changes as the amount of vibrational (and rotational) energy in the molecule changes. When the amount of vibrational and rotational energy in the excited electronic state is low, the molecule's geometry does not change significantly and the time dependence of the light emission is determined by the radiative decay rate of the molecule back down to the ground electronic state at a geometry very close to the starting geometry. Figure 11.46a shows this situation, in which the decay profile of the molecule is nearly exponential with a decay time of many tens of nanoseconds.

High Excess Energy Figure 11.46c shows an example where the molecule is excited with a pulse containing shorter wavelength light so that the same electronic state is populated but more vibrational (and rotational) energy is imparted to the molecule; the excess energy (energy above the 00 transition of the molecule) is high. Initially, the molecule's geometry is near that of the ground electronic state and fluorescence can be observed (see sharp peak near early time); however, as the molecule starts to relax and change its geometry the vibrational energy can flow between rotational and vibrational modes of the molecule (anthracene has 66 vibrational normal modes). As a consequence the fluorescence decay curve decreases. This decrease is not caused by energy relaxation however, because the molecule is isolated and the energy cannot flow out of it. Rather, the energy flows into vibrational modes that have very small Franck–Condon factors, so the fluorescence intensity drops.

Intermediate Excess Energy: Quantum Beats Figure 11.46b shows the situation for an intermediate value of the vibrational (and rotational) energy imparted to the molecule. Because the molecules of the sample were excited with a short burst of light in this experiment, they begin their nuclear motion on the excited state surface at nearly the same time. In the example shown here it appears that two vibrational modes of the electronically excited molecule dominate the time-dependent behavior. One of the vibrational modes has a large Franck–Condon factor ("bright") and the other mode has a small Franck–Condon factor ("dark") so that the fluorescence emission intensity of the molecule changes as the vibrational energy moves between the two vibrational modes. Because the molecules begin their nuclear motion at nearly the same time and they are initially prepared in the "bright" vibrational mode (it has the higher oscillator strength), the different molecules vibrational motion proceeds in phase (they have the same nuclear displacements at the same time). Hence the whole collection of molecules in the sample becomes more, or less, fluorescent as the energy moves between the two modes. The movement of the vibrational energy proceeds in an oscillatory manner so that oscillations in the fluorescence signal are observed. Of course, the overall fluorescence intensity decays as the molecules relax to the ground electronic state. The oscillations in the fluorescence profile are called *quantum beats*.

9 Adapted from Felker and Zewail [11], with permission.

11.6 Conclusion

This chapter discusses the wave–particle properties of nuclei. As with electrons, we see that the wave–particle duality of nuclei leads to quantized energies for the vibrational and rotational nuclear motions. The energy level structure can be probed by spectroscopies that drive transitions between the energy levels. Rotational transitions have energies on the order of 10^{-21} J and occur in the microwave region of the electromagnetic spectrum. Vibrational transitions occur at larger energies, on the order of 10^{-20} J and occur in the infrared region of the electromagnetic spectrum. We also showed how electronic and nuclear spectroscopy could be combined to examine properties of molecular excited states. In this chapter we showed how these spectroscopies can be used to determine bond lengths, bond angles, bond force constants, and the connectivity pattern of molecules. For these reasons, the spectroscopies discussed in this chapter are very useful in science and technology, but especially for chemists.

11.7 Box 11.1. Reduced Mass of Diatomic Molecules and Frequency of Vibration

11.7.1 Reduced Mass

11.7.1.1 Vibration

We consider the vibration as a motion of the masses m_A and m_B relative to the center of mass (Fig. 11B.1a). Then atom A is an oscillator with mass m_A and force constant k_{fA}. Because the force constant is inversely proportional to the length of the spring representing the bond, we have

$$k_{fA} = k_f \cdot \frac{d}{a} \quad \text{and} \quad k_{fB} = k_f \cdot \frac{d}{b}$$

(k_f = force constant of the spring). During vibration the center of mass stays at rest, and we can use the facts $m_A \cdot a = m_B \cdot b$ and $a + b = d$ to obtain a and b in terms of d and the atomic masses; i.e.

$$a = d \cdot \frac{m_B}{m_A + m_B} \quad \text{and} \quad b = d \cdot \frac{m_A}{m_A + m_B}$$

Then the frequency of oscillator A is

$$\nu_0 = \frac{1}{2\pi} \cdot \sqrt{\frac{k_{fA}}{m_A}} = \frac{1}{2\pi} \cdot \sqrt{\frac{k_f d}{a m_A}} = \frac{1}{2\pi} \cdot \sqrt{k_f \frac{m_A + m_B}{m_A m_B}} = \frac{1}{2\pi} \cdot \sqrt{\frac{k_f}{\mu}} \quad \text{with} \quad \mu = \frac{m_A m_B}{m_A + m_B}$$

The same expression holds for oscillator B:

$$\nu_0 = \frac{1}{2\pi} \cdot \sqrt{\frac{k_{fB}}{m_B}} = \frac{1}{2\pi} \cdot \sqrt{\frac{k_f d}{b m_B}} = \frac{1}{2\pi} \cdot \sqrt{k_f \frac{m_A + m_B}{m_A m_B}} = \frac{1}{2\pi} \cdot \sqrt{\frac{k_f}{\mu}}$$

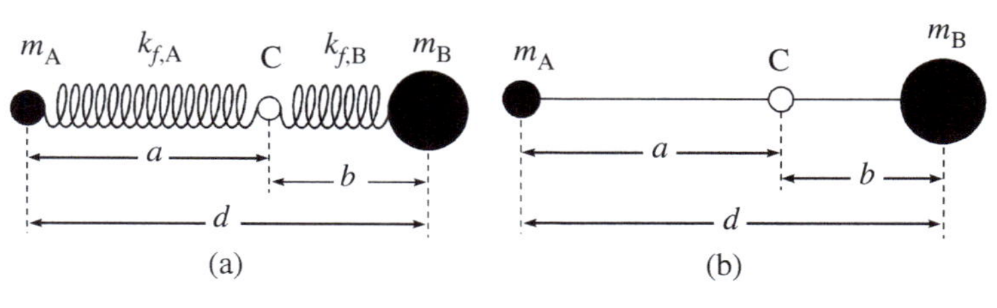

Figure 11B.1 Diatomic molecule. How to find the center of mass C. (a) vibration, (b) rotation.

11.7.1.2 Rotation

Next we consider the rotation of a diatomic molecule (Fig. 11B.1b) about its center of mass. Using the expressions for a and b from above, we can write the moment of inertia of the molecule as in the case of vibration

$$a m_A = b m_B \quad \text{and} \quad a + b = d$$

$$a = d \frac{m_B}{m_A + m_B} \quad \text{and} \quad b = d \frac{m_A}{m_A + m_B}$$

$$I = m_A a^2 + m_B b^2 = m_A d^2 \left[\frac{m_B}{m_A + m_B}\right]^2 + m_B d^2 \left[\frac{m_A}{m_A + m_B}\right]^2$$

$$= d^2 \left[m_A \frac{m_B^2}{(m_A + m_B)^2} + m_B \frac{m_A^2}{(m_A + m_B)^2}\right] = d^2 \left[m_A m_B \frac{m_B}{(m_A + m_B)^2} + m_B m_A \frac{m_A}{(m_A + m_B)^2}\right]$$

$$= d^2 \left[m_A m_B \frac{m_B + m_A}{(m_A + m_B)^2}\right] = d^2 \frac{m_A m_B}{m_A + m_B} = d^2 \mu$$

11.7.2 Frequency of Vibration

The force acting on a reduced mass μ connected to a spring with force constant k_f and moving in the x-direction is

$$\mu \frac{d^2x}{dt^2} = -k_f x$$

This differential equation for $x(t)$ is solved by

$$x = A \cdot \sin(2\pi\nu_0 t)$$

where A is a constant. Insertion into the differential equation yields

$$-A \cdot 4\pi^2\nu_0^2 \cdot \sin(2\pi\nu_0 t) = -\frac{k_f}{\mu} \cdot A \cdot \sin(2\pi\nu_0 t)$$

and we find that

$$4\pi^2\nu_0^2 = \frac{k_f}{\mu} \quad \text{and} \quad \nu_0 = \frac{1}{2\pi}\sqrt{\frac{k_f}{\mu}}$$

The total energy of the oscillator is fixed; however, the kinetic energy and potential energy change with time. The potential energy is

$$E_{pot} = \frac{1}{2}k_f x^2 = \frac{A^2}{2}k_f \cdot \sin^2(2\pi\nu_0 t)$$

and the kinetic energy is

$$E_{kin} = \frac{1}{2}m\frac{d^2x}{dt^2} = \frac{A^2}{2}m(4\pi^2\nu_0^2) \cdot \cos^2(2\pi\nu_0 t) = \frac{A^2}{2}k_f \cos^2(2\pi\nu_0 t)$$

The total energy is

$$E_{total} = E_{kin} + E_{pot} = \frac{1}{2}A^2 k_f$$

The potential and kinetic energy vary sinusoidally and exactly out of phase so that the kinetic energy is at a minimum when the potential energy is at a maximum and vice versa.

11.8 Box 11.2. Oscillator: Normal Modes of a Triatomic Linear Molecule

We restrict our discussion to CO_2 as an example for a triatomic linear molecule.

11.8.1 Reduced Mass μ

11.8.1.1 Symmetric Stretching Mode (Fig. 11B.2b)

The carbon atom is at the center of gravity and both oxygen atoms are moving with the same amplitude, but in opposite directions.

$$x_2 = 0 \quad \text{and} \quad x_1 = -x_3$$

Then

$$m_O \cdot \frac{d^2x_1}{dt^2} = -k_f \cdot x_1 \quad \text{and} \quad \nu_{0,sym} = \frac{1}{2\pi} \cdot \sqrt{\frac{k_f}{\mu_{sym}}}, \quad \mu_{sym} = m_O$$

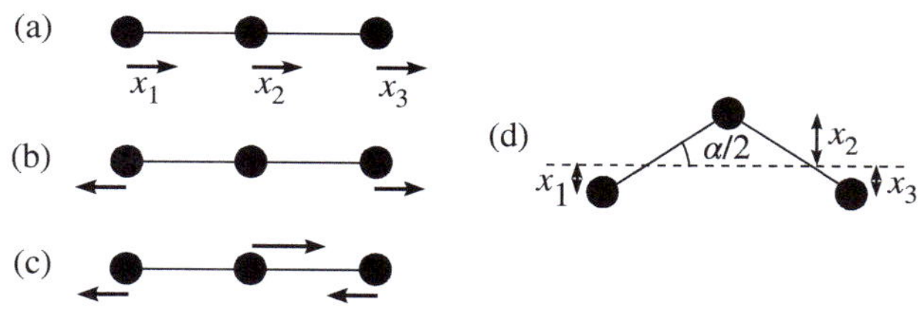

Figure 11B.2 Linear triatomic molecule (carbon dioxide). (a) How to calculate the moment of inertia from the displacements x_1, x_2, and x_3 for the symmetric (b) and asymmetric (c) stretching mode. Corresponding procedure for the bending mode (d).

11.8.2 Asymmetric Stretching Mode (Fig. 11B.2c)

In this case the oxygen atoms are again moving with the same amplitude, but in the same direction

$$x_1 = x_3$$

In order to keep the center of mass fixed in space, the carbon atom moves in the opposite direction

$$m_C \cdot x_2 = -2 \cdot m_O \cdot x_1, \quad x_2 = -\frac{2m_O}{m_C} \cdot x_1$$

Then

$$m_O \cdot \frac{d^2x_1}{dt^2} = -k_f \cdot (x_1 - x_2) = -k_f \cdot \left(1 + \frac{2m_O}{m_C}\right) \cdot x_1$$

and

$$\nu_{0,\text{anti}} = \frac{1}{2\pi} \cdot \sqrt{\frac{k_f}{\mu_{\text{anti}}}}, \quad \mu_{\text{anti}} = \frac{m_C m_O}{m_C + 2m_O}$$

11.8.2.1 Bending Mode (Fig. 11B.2d)

Both oxygen atoms are moving in the same direction, but opposite to the direction of the carbon atom. During vibration, the bond length d is constant, and the bond angle α is changing periodically. Then for a fixed center of mass we obtain

$$2x_1 \cdot m_O = x_2 \cdot m_C \quad \text{or} \quad \frac{x_1}{x_2} = \frac{m_C}{2m_O}$$

Furthermore,

$$x_1 + x_2 = d \sin\frac{\alpha}{2} \approx \frac{\alpha}{2} \cdot d$$

Thus

$$x_1 = d \cdot \frac{m_C}{2(2m_O + m_C)} \cdot \alpha$$

The force acting on an oxygen atom is

$$m_O \cdot \frac{d^2x_1}{dt^2} = -k_f' \cdot d \cdot \alpha$$

where k_f' is the bending force constant, see Chapter 7. Thus

$$\frac{d^2\alpha}{dt^2} = -\frac{k_f'}{\mu_{\text{bending}}} \cdot \alpha$$

and

$$\nu_{0,\text{bending}} = \frac{1}{2\pi} \cdot \sqrt{\frac{k_f'}{\mu_{\text{bending}}}}, \quad \text{with} \quad \mu_{\text{bending}} = \frac{m_C m_O}{2(2m_O + m_C)}$$

11.8.2.2 Force Constants

For k_f we obtain:

$$k_f = \mu\left(2\pi\nu_0\right)^2 = 4\pi^2c^2 \cdot \mu \cdot \tilde{\nu}^2$$

Asymmetric stretching mode: $\mu = 7.25 \times 10^{-27}$ kg, $\widetilde{\nu} = 2349$ cm^{-1} (Fig. 11.19); thus $k_f = 1420$ N m^{-1}.
Symmetric stretching mode: $\mu = 26.6 \times 10^{-27}$ kg, $\widetilde{\nu} = 1388$ cm^{-1} (Fig. 11.34); thus $k_f = 1820$ N m^{-1}.
Bending mode: $\mu = 3.62 \times 10^{-27}$ kg, $\widetilde{\nu} = 667$ cm^{-1} (Fig. 11.19); thus $k_f' = 57$ N m^{-1}.

We see that k_f' is much smaller than the force constant k_f for the stretching modes: The deformation of bond angles requires less force than that of bond lengths.

Why is the force constant k_f for the symmetric stretching mode larger than that for the asymmetric stretching mode? This discrepancy arises from the simple spring model used to describe bond length deformations (harmonic oscillator). Actually, when lengthening one bond in the CO_2 molecule, the electron density in the second bond will be affected and this results in a change of the force constant attributed to this bond. Therefore, only a rough value of the force constant in CO_2 can be estimated ($k_f \approx 1500$ N m^{-1}, see Table 7.6).

For a quantitative treatment of vibrations of molecules nonlinear terms must be added in the expression (11.42) for the vibrational energy (correction for *anharmonicity*). Furthermore, in polyatomic molecules coupling can occur between different vibrational modes, in a similar way as coupling of electronic transitions occurs (see Chapter 9); this type of coupling is called *Fermi resonance*. These are the reasons why the Raman line corresponding to the bending mode of CO_2 is found at 1285 cm^{-1} (Fig. 11.34) instead of 2×667 cm^{-1} = 1334 cm^{-1}, as expected from the harmonic oscillator model. The same is true for the weak transitions observed in Figs. 11.19b and 11.34.

11.9 Foundations

11.9.1 Foundation 11.1: Rigid Rotator Solution of the Schrödinger Equation

See the books website, www.wiley.com/go/Kuhn/physicalchemistry3e.

11.9.2 Foundation 11.2: Harmonic Oscillator: Quantum-Mechanical Treatment

See the books website, www.wiley.com/go/Kuhn/physicalchemistry3e.

11.9.3 Foundation 11.3: Hermitian Polynomials

See the books website, www.wiley.com/go/Kuhn/physicalchemistry3e.

11.9.4 Foundation 11.4: Absorption Spectra, Selection Rules for Rotation of Linear Molecules

11.9.4.1 Transition Moment

We consider a linear molecule with the permanent dipole moment μ_{perm}. In analogy to the light absorption of an electron (Foundation 9.2), we find for the transition moment

$$M^2 = M_x^2 + M_y^2 + M_z^2 \tag{11F4.1}$$

with

$$M_x = \int \psi_i \, \mu_x \, \psi_j \cdot \mathrm{d}\tau \quad (M_y,\ M_z \text{ correspondingly}) \tag{11F4.2}$$

μ_x is the x-component of μ_{perm}. Only transitions with a nonzero value of the transition moment are allowed.

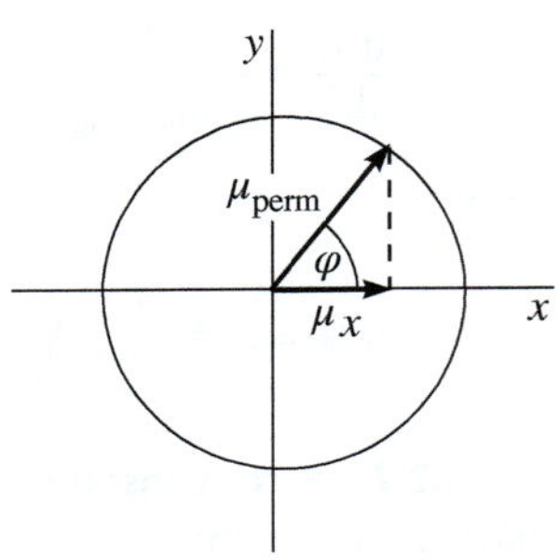

Figure 11F.4 Permanent dipole moment μ and its x-component μ_x.

Example: Rotation in the xy Plane (Fig. 11F.4).
In this case the wavefunctions depend on φ only and for μ_x we find (Fig. 11F.4)

$$\mu_x = \mu_{\text{perm}} \cdot \cos\varphi \tag{11F4.3}$$

and

$$M_x = \mu_{\text{perm}} \int_0^{2\pi} \psi_i \cos\varphi \, \psi_j \cdot \mathrm{d}\varphi \tag{11F4.4}$$

From this relation we conclude that the rotator can only be excited by light if it possesses a permanent dipole moment μ_{perm} (e.g., HCl, NO, CO). We can rewrite Equation (11F4.4) by using the identity

$$x = d\cos\varphi \quad \text{or} \quad \cos\varphi = \frac{x}{d}$$

where d is the bond length. We find that

$$M_x = \mu_{perm}\frac{1}{d}\int \psi_i \cdot x \cdot \psi_j \cdot d\varphi \tag{11F4.5}$$

We consider transitions from the ground state ($J = 0$) and from the states with $J = 1$ (Fig. 11F.5) to higher states. From the symmetries of the corresponding wavefunctions and the symmetries of the product functions in the transition integral, we see that only transitions are allowed for $\Delta J = \pm 1$. Particularly, we consider the transition $0 \to 1b$. The wavefunction ψ_0 is symmetric, and ψ_{1b} is antisymmetric with respect to the x-coordinate. Consequently, the product $(\psi_0\psi_{1b}) \cdot x$ is symmetric and the integral is nonzero. Now we consider the transition $0 \to 1a$. The wavefunction ψ_0 is symmetric, and ψ_{1a} is symmetric about x (note it is antisymmetric about y). Hence the product $(\psi_0\psi_{1a}) \cdot x$ is antisymmetric about x and the integral is zero. Similar arguments can be made to show that the transitions $0 \to 2b$ and $0 \to 2a$ are zero (Fig. 11F.5).

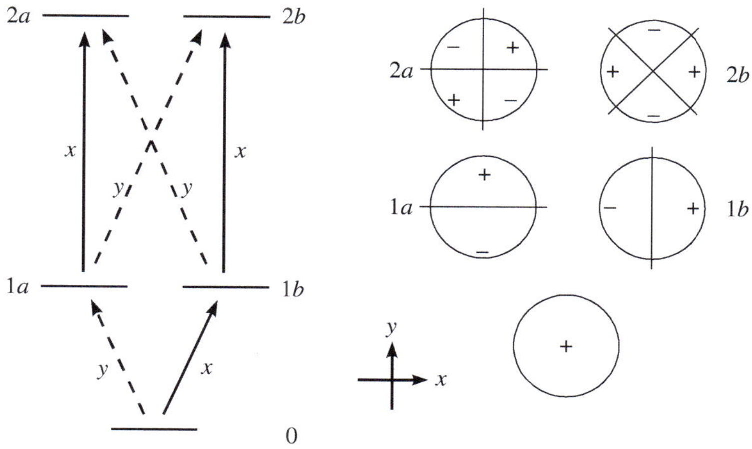

Figure 11F.5 Transitions between rotational states (rotation in a plane). Left: Energy scheme. Right: Symmetry of the wave functions.

One can extend these considerations to the rotator in three-dimensional space, considering the symmetry properties of the wavefunctions ψ shown in Fig. 11.2b. Accordingly, in this general case the selection rule is $\Delta J = \pm 1$ and $\mu_{perm} \neq 0$.

11.9.4.2 Intensity Distribution of Absorption Lines (Rotator in Space)

According to Chapter 9 the oscillator strength for a transition from quantum state J to quantum state $J + 1$ is proportional to the square of the transition moment M and to the excitation energy ΔE. In addition to the oscillator strength, the observed intensity of an absorption line depends on the number N_J of molecules in quantum state J.

First we calculate the oscillator strength, f,

$$f = const.\Delta E \cdot M^2 \tag{11F4.6}$$

with

$$\Delta E = 2B(J+1), \qquad M^2 = M_x^2 + M_y^2 + M_z^2 \tag{11F4.7}$$

(see Chapter 9). As an example, we consider the change from $J = 0$ to $J = 1$, where we have three different transitions ($0 \to 1a$, $0 \to 1b$, and $0 \to 1c$, see Fig. 11.3). By making symmetry arguments analogous to those made in the previous section, these transitions can be shown to occur with polarizations along the x-, y-, and z-directions, respectively. For the transition moment in the x-direction, M_x, we find

$$M_x = \int \psi_0\, \mu_x\, \psi_{1a} \cdot d\tau \tag{11F4.8}$$

where μ_x is the x-component of the permanent dipole moment of the molecule (Fig. 11F.6)

$$\mu_x = \mu_{perm} \cdot \sin\vartheta \cdot \cos\varphi \tag{11F4.9}$$

The rotator wavefunctions for $J = 0$ and $J = 1$ are

$$\psi_0 = \frac{1}{\sqrt{4\pi}}, \qquad \psi_{1a} = \sqrt{\frac{3}{4\pi}} \cdot \sin\vartheta\cos\varphi \tag{11F4.10}$$

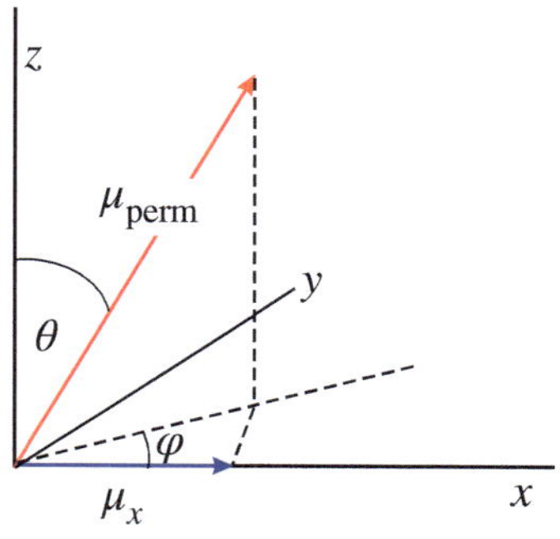

Figure 11F.6 The diagram shows the component μ_x of permanent dipole moment μ_{perm}.

Using

$$d\tau = \sin\vartheta \cdot d\vartheta \cdot d\varphi \tag{11F4.11}$$

we obtain

$$\begin{aligned} M_{x,0\to 1} &= \frac{1}{\sqrt{4\pi}}\sqrt{\frac{3}{4\pi}}\mu_{\text{perm}} \int (\sin^2\vartheta \cdot \cos^2\varphi) \cdot \sin\vartheta \cdot d\vartheta \cdot d\varphi \\ &= \frac{1}{4\pi}\sqrt{3}\mu_{\text{perm}} \int_0^{\pi} \sin^3\vartheta \cdot d\vartheta \cdot \int_0^{2\pi} \cos^2\varphi \cdot d\varphi \\ &= \frac{1}{4\pi}\sqrt{3}\mu_{\text{perm}} \cdot \frac{4}{3}\pi = \frac{1}{\sqrt{3}}\mu_{\text{perm}} \end{aligned} \tag{11F4.12}$$

and

$$M_{0\to 1}^2 = 3 \cdot M_{x,0\to 1}^2 = \mu_{\text{perm}}^2 \tag{11F4.13}$$

Generally, for a transition $J \to J+1$ we find

$$M_{J\to J+1}^2 = \frac{J+1}{2J+1} \cdot \mu_{\text{perm}}^2 \tag{11F4.14}$$

Then for the oscillator strength we obtain from Equations (11F4.6) and (11F4.7)

$$\begin{aligned} f_{J\to J+1} &= const.2B(J+1) \cdot \frac{J+1}{2J+1} \cdot \mu_{\text{perm}}^2 \\ &= const.2B \cdot \frac{(J+1)^2}{2J+1} \cdot \mu_{\text{perm}}^2 \end{aligned} \tag{11F4.15}$$

This result shows that f increases linearly with J as J becomes large.

Next we account for the number of molecules N_J in quantum state J. In Chapter 15 we will derive the formula for N_J and find that it is

$$N_J = const_1.(2J+1)e^{-BJ(J+1)/(kT)} \tag{11F4.16}$$

where k is the Boltzmann constant and T is the temperature. Thus for the absorbance of a spectral line we obtain (collecting all constants in *const′*)

$$\text{absorbance} = const' \cdot \mu_{\text{perm}}^2 \cdot (J+1)^2 \cdot e^{-BJ(J+1)/(kT)} \tag{11F4.17}$$

11.9.5 Foundation 11.5: Nonrigid Rotator, Centrifugal Effect on Energy

See the books website, www.wiley.com/go/Kuhn/physicalchemistry3e.

11.9.6 Foundation 11.6: Absorption Spectra, Selection Rules for Vibration

11.9.6.1 Diatomic Molecules

We calculate the transition moment M_x

$$M_x = \int \psi_i \mu_x \psi_j \cdot d\tau \tag{11F6.1}$$

where μ_x is the dipole moment of the molecule and is oriented along the bond axis in the x-direction. Now we consider the fact that the change in the bond length of a heteronuclear diatomic molecule will change its dipole moment, because the asymmetric charge distribution will have a different separation distance between the nuclei.

In a first approximation, we assume that the charges on the atoms remain constant during vibration, but their distance changes proportional to the elongation x of the bond. Then we obtain

$$\mu_x = \mu_{\text{perm}} + ax \tag{11F6.2}$$

where $x = d - d_{\text{eq}}$ (d is the bond distance and d_{eq} is the equilibrium bond distance), μ_{perm} is the molecule's permanent dipole moment, and the change in dipole moment with bond length is accounted for through the parameter a. Using a Taylor series expansion, you can show that

$$\mu_x = (\mu_x)_{x=0} + \left(\frac{d\mu_x}{dx}\right)_{x=0} \cdot x + \dots .$$

so that

$$(\mu_x)_{x=0} = \mu_{\text{perm}} \quad \text{and} \quad \left(\frac{d\mu_x}{dx}\right)_{x=0} = a$$

in agreement with Equation (11F6.2). Thus a corresponds to the gradient of the dipole moment with bond length.

The transition moment M_x for a molecule vibrating in the x direction is

$$M_x = \int_{-\infty}^{\infty} \psi_i \cdot \mu_x \cdot \psi_j \cdot dx \tag{11F6.3}$$

$$= \mu_{\text{perm}} \int_{-\infty}^{\infty} \psi_i \psi_j \cdot dx + a \int_{-\infty}^{\infty} \psi_i x \psi_j \cdot dx$$

Because of the orthogonality of the wavefunctions ψ_i and ψ_j the first integral is zero and for M_x we obtain

$$M_x = a \int_{-\infty}^{\infty} \psi_i x \psi_j \cdot dx \tag{11F6.4}$$

The transition is only possible if a is nonzero. For a diatomic molecule this is only possible if the molecule has a permanent dipole moment. This requirement is fulfilled for most heteronuclear diatomic molecules (exceptions are molecules containing nuclei with the same electronegativity, e.g., HD).

If we consider harmonic oscillator functions the functions are either symmetric (quantum numbers 0, 2, …) or antisymmetric (quantum numbers 1, 3, …) in the x direction (Fig. 11.12). If the product function $\psi_i\psi_j$ is symmetric the integrand becomes antisymmetric (because x is an antisymmetric function) and M_x is zero. If the product function $\psi_i\psi_j$ is antisymmetric the integrand becomes symmetric and M_x is nonzero (Fig. 11F.7). This means a transition is allowed if the quantum number changes by one, and it is forbidden if the quantum number changes by two.

In principle, it might be possible that for a change of the quantum number by three (e.g., $n = 0 \rightarrow n = 3$) or some other odd value M_x becomes nonzero; however, this is not the case in the harmonic oscillator approximation. In Problem P11.14, you prove this by using the orthonormality property of the harmonic oscillator wavefunctions, and you demonstrate that the selection rule for light absorption of a vibrating molecule is $\Delta n = \pm 1$.

If anharmonic oscillator functions are taken into account, the situation is different: these functions do not have a symmetry element. Nevertheless it can be easily seen that transitions with $\Delta n = \pm 1$ are allowed as for the harmonic oscillator. However, transitions with $\Delta n = \pm 2$ or $\Delta n = \pm 3$ can occur, although the transition moments are much smaller than that for $\Delta n = \pm 1$.

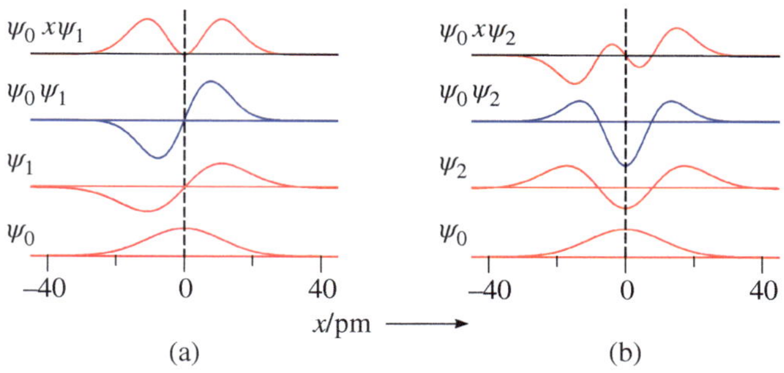

Figure 11F.7 The diagrams show how one can analyze the symmetry of the integrand in the transition moment integral M_x and qualitatively discern whether a vibrational transition is allowed. x is the displacement of the harmonic oscillator. (a) For the transition $n = 0 \rightarrow n = 1$, M_x is nonzero. (b) For the transition $n = 0 \rightarrow n = 2$, $M_x = 0$.

11.9.6.2 Polyatomic Molecules

As an example, we consider the normal modes of vibration of the CO_2 molecule. This molecule does not have a permanent dipole moment. However, we see in Fig. 11.17 that a changing dipole moment can occur during vibration in the bending mode and in the antisymmetric stretching mode, that is the constant a in Equation (11F6.4) is nonzero. In both cases the transition moment is nonzero for $\Delta n = \pm 1$, as for diatomic molecules. For the symmetric stretching mode no dipole moment arises during vibration; thus, $a = 0$ and $M_x = 0$.

11.9.7 Foundation 11.7: Raman Spectra, Selection Rules

See the books website, www.wiley.com/go/Kuhn/physicalchemistry3e.

11.9.8 Foundation 11.8: Multidimensional Infrared Spectroscopy

See the books website, www.wiley.com/go/Kuhn/physicalchemistry3e.

11.10 Exercises

E11.1 Calculate the center of mass position in an HCl molecule, an HF molecule, and an H_2 molecule. Comment on the trend.

E11.2 Given that the rotational constant of H_2 is $B = 1.209 \times 10^{-21}$ J, calculate the bond length of the molecule. If you use the mass of a hydrogen atom rather than the reduced mass, how large is the error in your calculation?

E11.3 Given that the rotational constant of $^{12}C^{16}$ O is $B = 3.84 \times 10^{-23}$ J, calculate the bond length of the molecule. Compare your result to that found in Example 11.8.

E11.4 Show that the wavefunction ψ_0 in Equation (11.26) satisfies the Schrödinger equation for the harmonic oscillator Equation (11.24).

E11.5 Calculate the energy difference between the $H^{35}Cl$ and the $H^{37}Cl$ line in the high-resolution infrared spectrum of HCl (Fig. 11.16). Do you expect the energy spacing to change with the J quantum number?

E11.6 Use the data given for O_2 and its ions to determine the force constants of their bonds, in the harmonic oscillator approximation. Explain the trend in the value of the force constant in terms of the bonding in the different species.

Molecular Species	$^{16}O_2$	$^{16}O_2^-$	$^{16}O_2^+$
Vibrational Frequency $\nu_0/10^{13}$ s^{-1}	4.74	3.27	5.71

E11.7 The rotational period of a diatomic molecule is the time it takes for it to rotate through an angle of 2π; i.e., one full revolution. Consider the two diatomic molecules in the table below and the data associated with them. What is the rotational period of these two molecules, when they have an angular momentum of $\sqrt{2}\hbar$? What is the rotational period when they have an angular momentum of $\sqrt{110}\hbar$?

Molecule	$^1H_2$	$^{127}I_2$
I/kg m^2	4.59×10^{-48}	7.54×10^{-45}

E11.8 Consider the largest value of the rotational period found in E11.7. How many revolutions does this molecule make in one second? If we consider that human history has only been recorded for 5000 years, how many seconds in recorded history? Are there more seconds in recorded history, or are there more revolutions of a molecule in one second?

E11.9 Consider the two potential energy curves (A and B) in Fig. E11.9 for a diatomic molecule.
(Assume the same atomic nuclei for both curves). Which curve corresponds to the larger internuclear distance? Which curve corresponds to the higher moment of inertia? Which curve has a higher binding energy? Which curve has a higher vibrational energy level spacing? Which curve has a higher rotational energy level spacing?

E11.10 Given that the vibrational frequency of H_2 is $\nu_0 = 1.321 \times 10^{14}$ s^{-1}, determine the vibrational frequencies of HD and D_2.

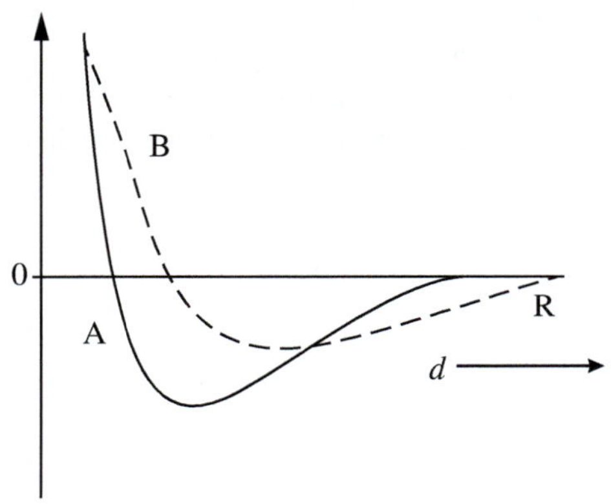

Figure E11.9 Potential energy curves. Energy V versus bond length d.

E11.11 Plot the data in Table 11.1 and determine the best value of the parameter B using Equation (11.15). Use this B value to determine the HCl molecule's bond length. Compare your result with a value obtained from the literature.

E11.12 Using a force constant of 516 N m^{-1} calculate the vibrational frequencies for different isotopes of HCl, namely $D^{37}Cl$, $H^{37}Cl$, and $H^{35}Cl$. Comment on your findings.

E11.13 In Fig. 11.35 identify the P, Q, and R branches for the rotational fine structure in the infrared spectrum of methane.

E11.14 Energy order the following spectroscopic transitions for the molecule CO: vibrational excitation of the CO bond, electronic excitation of CO (from HOMO to LUMO), rotational transition of CO, and photoionization of CO. Also, state the order of magnitude of the energy in each case

E11.15 Show that the kinetic energy of a classical oscillator is at a maximum when its potential energy is at a minimum and vice versa (see Box 11.1).

E11.16 Use Fig. 11.14 to provide a physical explanation for how an infrared light wave excites the vibrational motion of a chemical bond.

E11.17 Explain why the bending motion and the asymmetric stretching motion of the CO_2 molecule generate a dipole moment but the symmetric stretching motion does not.

E11.18 Compare the vibrational wavefunctions for a Morse potential to those for a harmonic oscillator potential by exploring the MathCAD exercise morse.xmcd.

E11.19 Use the MathCAD exercise FCfactors_displacement.xmcd to explore the vibronic structure of electronic transitions and its dependence on the displacement of the ground and excited electronic states. In this MathCAD excercise you assume that the ground-state electronic surface is parabolic with an energy

$$E_{\text{gnd}} = \frac{1}{2}k_f x^2$$

and the excited state surface is parabolic and has the same force constant as the ground state, namely

$$E_{\text{exc}} = \frac{1}{2}k_f(x-\delta)^2 + E_{00}$$

You are asked to evaluate the Franck–Condon factors (FCF) as a function of the displacement δ between the minima of the two parabolic curves. The absorption probability $\mathcal{P}$ in a spectroscopy experiment is proportional to

$$\mathcal{P} = \left|\int_{-\infty}^{\infty} \psi_g \psi_e \, \mathrm{d}\tau\right|^2 = |M_{x,el}|^2 \cdot |\int \psi_{\text{vib},g} \cdot \psi_{\text{vib},e} \cdot \mathrm{d}x|^2$$

where $|M_{x,el}|^2$ is the square of the electronic transition dipole moment integral and the vibrational overlap term is the FCF; namely

$$FCF = |\int \psi_{\text{vib},g} \cdot \psi_{\text{vib},e} \cdot \mathrm{d}x|^2$$

You are asked to plot the FCF values as a function of the vibrational quantum number in each state and to rationalize the trends you observe.

E11.20 In this exercise you extend the MathCAD-based calculations of Franck–Condon factors (see E11.19) to include changes in the force constants between the two harmonic oscillators, in addition to their displacement from each other. The calculations are a straightforward generalization of those used in E11.19, in which the ground-state vibrational energy spacing is given by

$$E_{\text{gnd},n} = \left(n + \frac{1}{2}\right) h\nu_g = \left(n + \frac{1}{2}\right) \frac{h}{2\pi}\sqrt{\frac{k_{fg}}{\mu}}$$

where n is the vibrational quantum number. The excited state surface, which is displaced by the electronic energy, E_{00}, has the vibrational spacing of

$$E_{\text{exc},n'} = \left(n' + \frac{1}{2}\right) h\nu_e + E_{00} = \left(n' + \frac{1}{2}\right) \frac{h}{2\pi}\sqrt{\frac{k_{fe}}{\mu}} + E_{00}$$

where n' is the vibrational quantum number in the excited state. In this case you should explore how the vibronic structure (FCFs) in the transition depends on the force constant differences. You are asked to plot the FCF values as a function of the vibrational quantum number in each state and to rationalize the trends you observe, for both the absorption spectrum and the fluorescence spectrum.

E11.21 Using the diagram in Box 11.1 for a diatomic molecule, deduce the center of mass and show that the formula for the moment of inertia becomes $I = d^2 \cdot \mu$, where μ is the reduced mass ($\mu = m_A m_B / (m_A + m_B)$). Use this more exact formula for I to determine the bond length of the $H^{35}Cl$ molecule, which has a rotational constant of $B = 10.439\ \text{cm}^{-1}$.

E11.22 Repeat the procedure in Equation (11.45), Table 11.4, and Fig. 11.25 using the calculated data in Table 11.3 for the anharmonic oscillator and compare the result with that obtained from the experimental data in Table 11.4.

E11.23 Given that the probability $d\mathcal{P}_{\text{class}}$ of finding a classical oscillator between x and $x + dx$ is inversely proportional to its classical speed v

$$d\mathcal{P}_{\text{class}} \sim \frac{1}{v} \cdot dx$$

(i.e., the time it spends in the region of x to $x + dx$), show that

$$\rho = \frac{d\mathcal{P}_{\text{class}}}{dx} = \frac{1}{x_0 \cdot \pi} \cdot \frac{1}{\sqrt{1 - (x/x_0)^2}}$$

Proceed by using the fact that the time-dependent displacement $x(t)$ of a classical oscillator can be written as $x(t) = x_0 \cdot \sin(2\pi\nu_0 t)$, where ν_0 is the frequency of oscillation and x_0 is the amplitude of the oscillator.

E11.24 Consider the water molecule in its gaseous state. Given the experimental values for $I_B = 1.023 \times 10^{-47}$ kg m^2 and $I_C = 2.945 \times 10^{-47}$ kg m^2, [12] calculate the bond angle and the bond length, in analogy to the consideration in Foundation 11.1 for the NH_3 molecule.

E11.25 Given that the rotational constant of CH_4 is $B = 5.336 \times 10^{-47}$ kg m^2, calculate the bond distance R of the molecule.

11.11 Problems

P11.1 Use Equation (11.19) to calculate the intensity pattern of the rotational absorption spectrum of HCl at 298 K. Compare your result with the experimental pattern shown in Fig. 11.7.

P11.2 Show that the minimum of a potential energy curve is always harmonic. Consider an arbitrary potential energy function $V(x)$ and show that the leading term in its Taylor series expansion about the minimum is quadratic.

P11.3 In Section 11.3.2 particle-in-a-box wavefunctions are used as trial functions for solving the Schrödinger equation of a harmonic oscillator. Use the variational principle with Equations (11.22) and (11.23) to calculate the energies ε_0 and ε_1.

P11.4 Given the bond length $d = 156 \times 10^{-12}$ m and force constant $k_f = 250$ N m^{-1} for ^{6}LiF, use the rigid rotator/harmonic oscillator approximation to construct, to scale, an energy level diagram for the first four rotational levels in both the $n = 0$ and $n = 1$ vibrational states (i.e., a total of eight states). Indicate the allowed transitions in an infrared absorption experiment. Clearly label which transitions belong to the R branch and which belong to the P branch.

P11.5 From the different masses of ^{35}Cl and ^{37}Cl calculate the line splitting in the HCl spectrum shown in Fig. 11.16. Note that this problem is very similar to E11.5.

P11.6 Consider a vibrational transition from quantum number $n = 0$ to $n = 3$. Using harmonic oscillator wavefunctions show that the corresponding transition moment M_x is zero. Use Equation (11.81); i.e.,

$$M_x = a \int_{-\infty}^{\infty} \psi_i x \psi_j \cdot dx$$

from Foundation 11.6.

P11.7 Consider the radical species OH and OD. Their vibrational energy level spectrum is given by Equation (11.42):

$$E_n = h\nu_0(n + \frac{1}{2}) - C(n + \frac{1}{2})^2 \quad \text{with} \quad C = \frac{(h\nu_0)^2}{4\varepsilon_B}$$

with the following parameters

Species	d/pm	ε_B/eV	ν_0/c_0 (cm^{-1})	$\frac{\nu_0}{c_0}C$ (cm^{-1})
OH	97.06	4.35	3735.2	82.8
OD	96.99	4.39	2720.9	44.2

a) Graph the positions of the fundamental vibration and the first four overtones of OH; i.e., the infrared transitions. b) On the same graph plot the fundamental and the first four overtones of OD. c) Why is the anharmonic term ($\nu_0 C$) smaller for OD than OH? d) Sketch the vibrational spectrum you would expect to observe for OH.

P11.8 The table provides transition wavelengths and absorption cross-sections $\sigma = 2.303\varepsilon/\mathbf{N}_A$, where ε is the molar decadic extinction coefficient, for vibronic transitions in the first allowed electronic transition of CO. Using these data make a qualitative sketch of the ground and excited state potential energy curves. Discuss how your potential energy curves can explain the relative transition intensities and energy spacings. Explain how you can determine the force constant for the excited state potential energy surface and do so for the lowest energy and highest energy spacings.

Ground State (X)	Excited State (A)	λ/nm	$\sigma/10^{-18}$cm ^{2}
$n = 0$	$n = 0$	154.45	350 ± 35
$n = 0$	$n = 1$	150.98	680 ± 68
$n = 0$	$n = 2$	147.77	820 ± 82
$n = 0$	$n = 3$	144.74	700 ± 70
$n = 0$	$n = 4$	141.91	460 ± 46
$n = 0$	$n = 5$	139.26	280 ± 28
$n = 0$	$n = 6$	136.77	172 ± 17
$n = 0$	$n = 7$	134.43	95 ± 10
$n = 0$	$n = 8$	132.22	45 ± 5
$n = 0$	$n = 9$	130.15	25.4 ± 3
$n = 0$	$n = 10$	128.19	9.5 ± 1.2
$n = 0$	$n = 11$	126.35	4.0 ± 0.4

P11.9 a) Illustrate the Franck–Condon principle using the potential energy curves of ^{40}CaH that are given in Fig. P11.9a. Sketch a vibrationally resolved absorption spectrum associated with transitions from the ground rovibronic state of ^{40}CaH to the energy levels of the excited electronic state of ^{40}CaH, which is given in the diagram. Describe the important features of your sketch. b) Sketch the absorption spectrum you would expect to observe for transitions between the ground vibrational state of this molecule's ground electronic state to the first excited vibrational state of this molecule's ground electronic state. This transition is centered at 1299 cm^{-1}. Indicate clearly any features in your spectrum. c) State four molecular parameters that can be determined from the spectra in parts a and b. Describe in a detailed manner how you would go about determining them.

Figure P11.9a Potential energy curves of ^{40}CaH. This image is redrawn from the curve reported by Herzberg [1].

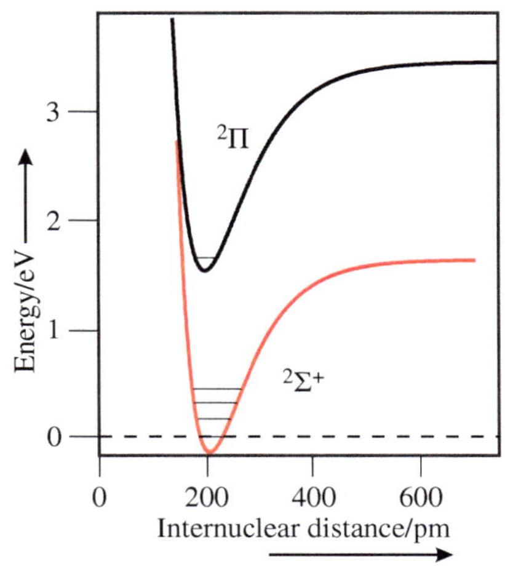

P11.10 For the CO molecule the following wavenumbers for the fundamental and the overtone bands were measured: $0 \to 1 : 2144\ \text{cm}^{-1}$, $0 \to 2 : 4266\ \text{cm}^{-1}$, $0 \to 3 : 6357\ \text{cm}^{-1}$. Evaluate the force constant k_f and the bond energy ε_B, in analogy to the treatment of HCl in Section 11.3.4.

P11.11 Consider the $^{12}C^{14}N$ molecule. The potential energy curves of the first two electronic states are shown in Fig. P11.11a and the vibrational energy levels in each state are given by $E_n = h\nu_0(n + \frac{1}{2}) - C(n + \frac{1}{2})^2$. The molecular constants are given in the table. B is the rotational constant for the equilibrium bond length.

State	d/pm	$\left[\frac{B}{hc_0}\right]$/cm^{-1}	$\frac{\nu_0}{c_0}$ (cm^{-1})	$\frac{\nu_0}{c_0}C$ (cm^{-1})
X	117.18	1.8996	2068.71	13.14
A	123.27	1.7165	1814.43	12.83

a) Sketch the rovibrational spectrum of the molecule in the X state. Be sure to label the lines as quantitatively as possible. Assume the molecule is initially in $n = 0$. Perform a similar sketch for the A state. In what ways do these two sketches differ? b) Give an explicit expression for the Franck–Condon factor involved in the lowest energy allowed electronic transition between the X and A state. Assume harmonic oscillator wavefunctions are valid. Evaluate the integral. (Note that the X state is $^2\Sigma^+$ and does not have electronic angular momentum, whereas the A state is $^2\Pi$ and does have angular momentum.)

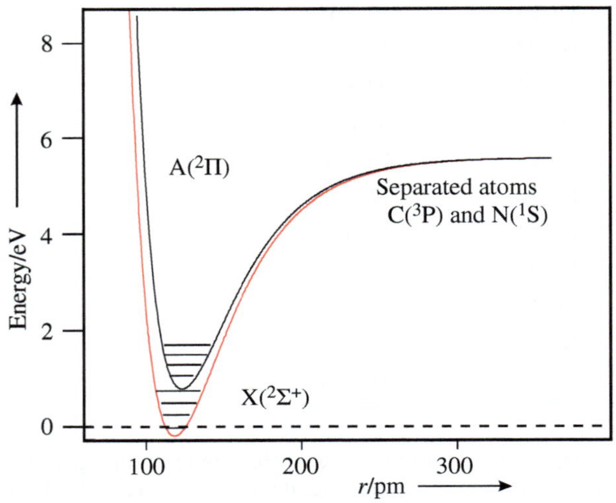

Figure P11.11a Potential energy curves of the first two electronic states of $^{12}C^{14}N$. This image is redrawn from the curve reported by Herzberg [1].

P11.12 Use Equation (11.15) to fit the experimental data in Table 11.1 for the infrared absorption lines in the rotational spectrum of HCl. Compare your result with the result from the rigid rotator calculation that is given in the text.

P11.13 Show that the anharmonicity of the potential energy curve causes the R-branch lines of HCl to come closer together with increasing J whereas those for the P-branch move farther apart. Begin with the energy expressions for the two vibrational quantum states of

$$E_J(v = 0) = B_0 J(J + 1) + \frac{1}{2}h\nu_0, \quad J = 0, 1, 2, \ldots.$$

$$E_J(v = 1) = B_1 J(J + 1) + \frac{3}{2}h\nu_0, \quad J = 0, 1, 2, \ldots.$$

where $E_J(v = 0)$ is the energy of states originating from the lowest vibrational state and $E_J(v = 1)$ is the energy of states originating from the first excited vibrational state. Derive expressions for the difference in energy corresponding to the R-branch lines (a $\Delta E(J)$) and then an expression for the difference in energy of these ΔE (thus a $\Delta\Delta E$). Then perform a corresponding calculation for the P-branch lines and compare the formulas to each other.

P11.14 Use the general form of the harmonic oscillator wavefunctions from Foundation 11.4, namely

$$\psi_n(q) = N_n \cdot H_n(q) \cdot \exp(-q^2/2)$$

to derive the selection rule $\Delta n = \pm 1$ for a vibrational transition; i.e., show that

$$M_x = \int \psi_{n'}(q) \cdot q \cdot \psi_n(q) \mathrm{d}q$$

is only nonzero for $n' = n \pm 1$.

P11.15 Use the infrared absorption spectrum of CO_2 in Fig. 11.19 to estimate the CO_2 absorbance in the atmosphere. The current pressure of CO_2 on sea level is about $P_{\text{atmosphere}} = 400$ ppm ×1 bar = 400×10^{-6} bar. Assume that the atmosphere reaches up to 15 km (troposphere). Compare your result to the CO_2 concentration of 280×10^{-6} bar in the pre-industrial period.

References

1 G. Herzberg, *Molecular Spectra and Molecular Structure: I Spectra of Diatomic Molecules* (Van Nostrand, New York, 1950).

2 N.B. Colthrup, Spectra-structure correlations in the infra-red region, *J. Opt. Soc. Am.*, 1950, **40**, 397; J.R. Dyer, *Applications of Absorption Spectroscopy of Organic Compounds* (Prentice-Hall, 1965).

3 P.M. Morse, Diatomic molecules according to the wave mechanics. II. Vibrational levels, *Phys. Rev.*, 1929, **34**, 57.

4 D. ter Haar, The vibrational levels of an anharmonic oscillator, *Phys. Rev.*, 1946, **70**, 222.

5 G.M. Barrow, *Introduction to Molecular Spectroscopy* (McGraw-Hill, New York, 1962).

6 B. Schrader, *Raman/Infrared Atlas of Organic Compounds* (Wiley-VCH, New York, 1989).

7 B.M. Cheng, C.Y. Chung, Y.P. Lee, and L.C. Lee, Quantitative spectral analysis of HCl and DCl in 120–220 nm: effects of singlet–triplet mixing, *J. Chem. Phys.*, 2002, **117**, 4293.

8 N. Knöpfel, Th. Olbricht, and A. Schweig, in: *Physikalische Methoden in der Chemie*, eds., B. Schröder and J. Rudolph (VCH Weinheim, 1985, p. 205).

9 J.B. Birks, *Photophysics of Aromatic Molecules* (John Wiley & Sons, New York, 1970).

10 F. Sondheimer, D.A. Ben-Efraim, and R. Wolovsky, Unsaturated macrocyclic compounds. XVII. ^ 1 The prototropic rearrangement of linear 1,5-enynes to conjugated polyenes. The synthesis of a series of vinylogs of butadiene, *J. Am. Chem. Soc.*, 1961, **83**, 1675.

11 P.M. Felker and A.H. Zewail, Dynamics of intramolecular vibrational-energy redistribution (IVR). II. Excess energy dependence, *J. Chem. Phys.*, 1985, **82**, 2975.

12 D.W. Posener, Microwave spectrum of the water molecule, Technical Report No. 255, May 14, 1953,Massachusetts Institute of Technology, Research Laboratory of Electronics.

12

Nuclear Spin

CHAPTER MENU

12.1 Introduction

Nuclei have an intrinsic angular momentum (spin). This fact leads to important symmetry consequences for some molecules: the nuclear spin wavefunction of a molecule is included in the total wavefunction, and the total wavefunction must obey symmetry rules, for example, the Pauli exclusion principle. We will show fascinating effects, exemplifying this astonishing feature. Another important consequence of nuclear spin is the energy splitting of degenerate nuclear spin states under an external magnetic field. It is possible to drive transitions between different spin states by applying radio-frequency (RF) electromagnetic radiation. The local field experienced by the nuclei, hence the resonant frequencies, depends on details of the molecular structure. This phenomenon is the reason for the importance of nuclear magnetic resonance (NMR) spectroscopy in chemistry and biology and underlies the use of magnetic resonance imaging (MRI) in medicine.

12.2 Nuclear Spin: Fundamentals

Here we discuss some experiments with H_2 that display the quantum nature of nuclei, arising from the Antisymmetry Postulate. In Section 5.4 we discussed how an electron exists in a state with a spin function α or β. As for the case of electrons, a property called nuclear spin, can be attributed to the nuclei and their energies are sensitive to an external magnetic field.

12.2.1 Spin of Protons in H_2

A proton has a spin, and it can exist in two nuclear spin states (with spin quantum number $I = \frac{1}{2}$ and z-components $m_I = \pm\frac{1}{2}$). In an H_2 molecule the spins of the two protons can be parallel (*ortho*- H_2, analogous to a triplet) or antiparallel (*para*-H_2, analogous to a singlet). We can symbolically describe these nuclear spin states by α and β. This leads to symmetric and antisymmetric spin combinations with respect to the exchange of protons 1 and 2, in direct analogy to the

Principles of Physical Chemistry, Third Edition. Hans Kuhn, David H. Waldeck, and Horst-Dieter Försterling.

Companion website: www.wiley.com/go/Kuhn/physicalchemistry3e

case of electrons (see Section 5.4). The symmetric combinations (*ortho*-hydrogen) are

$$\alpha(1)\alpha(2), \quad \beta(1)\beta(2), \quad \frac{1}{\sqrt{2}}[\alpha(1)\beta(2)+\beta(1)\alpha(2)] \tag{12.1}$$

and the antisymmetric combination (*para*-hydrogen) is

$$\frac{1}{\sqrt{2}}[\alpha(1)\beta(2)-\beta(1)\alpha(2)] \tag{12.2}$$

Therefore we expect that hydrogen gas is a mixture of *ortho* and *para* forms. In equilibrium, these forms occur in a ratio of 3 : 1. This conclusion will be discussed in more detail in Chapter 15 and compared to experiment.

Ortho-H_2 and *para*-H_2 can be separated experimentally. As shown below in Section 12.2.2.4 *ortho*-H_2 and *para*-H_2 exist only in rotational states with odd and even quantum numbers J, respectively. Therefore, at very low temperature, *para*-H_2 ($J = 0$) is the stable form. *Para*-H_2 is produced from the natural gas mixture at low enough temperature by contacting the gas with charcoal, catalyzing the conversion of nuclear spins into the antisymmetric state. *Para*-H_2 is metastable when heating the gas up to room temperature (after removing the catalyst).

12.2.2 Antisymmetry of Total Wavefunction of a Molecule Including Electrons and Nuclei

How can we understand the astonishing fact that *ortho*-H_2 and *para*-H_2 exist only in rotational states with odd and even quantum numbers J, respectively? The complete wavefunction of a molecule results as a solution of the Schrödinger equation for all the electrons and nuclei. Making use of the *Born–Oppenheimer Approximation*, we can write the wavefunction as a product of the electronic wavefunction (depending on the electron coordinates) and a nuclear wavefunction

$$\psi_{\text{total}} = \psi_{\text{electronic}} \cdot \psi_{\text{nuclear}} \tag{12.3}$$

In addition, we approximate the nuclear wavefunction by a product of rotational, vibronic, and spin wavefunctions:

$$\psi_{\text{nuclear}} = \psi_{\text{rotation}} \cdot \psi_{\text{vibration}} \cdot \psi_{\text{nuclear spin}} \tag{12.4}$$

To analyze the symmetry properties of hydrogen's total wavefunction with respect to the exchange of its nuclei, we analyze the symmetry properties of its individual components and put them together as a product. First we consider the ground state of the molecule.

12.2.2.1 Ground State of H_2: *Para*-H_2

We restrict our consideration to closed shell molecules, where the electronic ground state wavefunction $\psi_{\text{electronic}}$ is symmetric with respect to space, that is, exchanging protons 1 and 2 does not change $\psi_{\text{electronic}}$.

$\psi_{\text{vibration}}$ depends on the distance between the protons and the electronic energy well. Thus exchanging protons 1 and 2 does not change $\psi_{\text{vibration}}$.

ψ_{rotation} is a constant in the ground state ($J = 0$, see Fig. 11.3). Thus, exchanging protons 1 and 2 does not change ψ_{rotation}.

Because the ground-state wavefunction is symmetric for the electronic, vibrational, and rotational modes, the $\psi_{\text{nuclear spin}}$ must be antisymmetric for the total wavefunction to be antisymmetric. For *para*-H_2, exchanging protons 1 and 2 changes the sign of $\psi_{\text{nuclear spin}}$. In contrast, the nuclear spin wavefunction of *ortho*-H_2 is symmetric with respect to proton exchange and its combination with the ground rotational, vibrational, and electronic wavefunctions does not satisfy the Pauli exclusion principle. Thus the ground state of the H_2 molecule is the *para*-H_2 molecule.

12.2.2.2 First Excited Rotational State of H_2: *Ortho*-H_2

What happens for excited states of the molecule? For the first excited rotational state ($J = 1$), ψ_{rotation} changes sign through the nodal plane (Fig. 11.3). Thus exchanging protons 1 and 2 changes the sign of ψ_{rotation}. Since the electronic and vibrational wavefunctions remain in the ground state, they are symmetric. In order for the total wavefunction to be antisymmetric, the nuclear spin wavefunction must be symmetric. Hence the first excited rotational state ($J = 1$) has the symmetric nuclear spin combination and is *ortho*-H_2. The *para*-H_2 cannot exist in the $J = 1$ rotational energy state.

This argument can be applied to each of the rotational energy levels of hydrogen in its ground vibrational, ground electronic state. All of the odd (J is an odd number and the rotational wavefunction is antisymmetric) rotational states are *ortho*-H_2 (symmetric nuclear spin wavefunction), and all of the even (J is an even number and the rotational wavefunction is symmetric) rotational states are *para*-H_2 (antisymmetric nuclear spin wavefunction).

12.2.2.3 No Distinction Between Identical Particles

This discussion illustrates a strange quantum feature: It is impossible to distinguish between identical particles. This holds for protons as well as for electrons. Amazingly, these symmetry constraints on the wavefunction are experimentally manifest and apply for any homonuclear diatomic and polyatomic molecules with a center of symmetry. No such effect occurs for the HD molecule because the nuclei are distinguishable and no symmetry restriction exists.

12.2.2.4 Experimental Proof of Indistinguishability by Rotational Raman Spectrum of H_2

The phenomenon of indistinguishability can be directly demonstrated by the rotational Raman spectrum (Fig. 12.1). From the spacing of the rotational lines from Equation (11.54) we obtain the rotational constant $B = 1.19 \times 10^{-21}$ J. Then from Equation (11.55) we calculate the relative intensities of the spectral lines for $T = 298$ K, first proceeding as for the heteronuclear case (Table 12.1). We see immediately that this intensity distribution differs significantly from the experimental one. Next we repeat the calculation assuming that H_2 is a natural mixture of *para*- and *ortho*-hydrogen in the ratio 1:3 (Table 12.2). These values agree reasonably with experiment. This means that the transitions $0 \rightarrow 2$ (i.e., transition from $J = 0$ to $J = 2$) and $2 \rightarrow 4$ (i.e., transition from $J = 2$ to $J = 4$) arise from *para*-hydrogen and the transitions $1 \rightarrow 3$ and $3 \rightarrow 5$ arise from *ortho*-hydrogen molecules. Therefore only two lines are expected for pure *para*-hydrogen in the wavenumber range in Fig. 12.1.

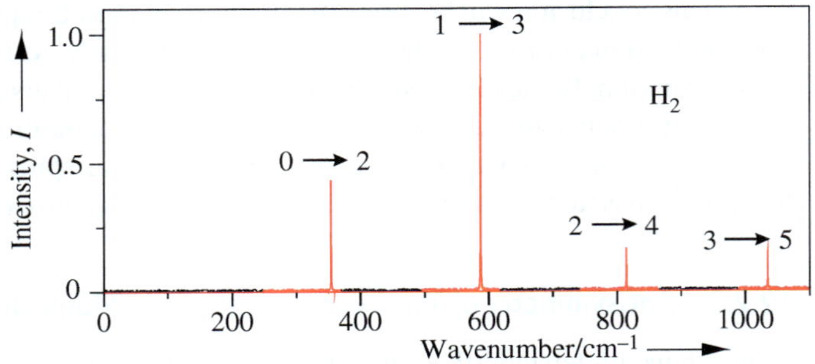

Figure 12.1 Rotational Raman spectrum of H_2. Excitation with an argon laser at 514.5 nm, recording time 12.5 h. The Rayleigh line at the origin is not shown. The intensity scale is arbitrarily chosen such that the highest peak has a value of 1. (Data from [1].)

In Section 16.3.5 we show that the symmetry properties of the nuclear wavefunctions are manifest in the temperature dependence of the heat capacity for hydrogen gas, also.

Table 12.1 Intensity of Spectral Line in the Raman Spectrum of H_2 Calculated Under the Assumption that H_2 can be Treated in the Same Way as a Heteronuclear Molecule.

				Relative Intensity	
J	Transition	$\frac{3(J+1)(J+2)}{2(2J+3)}$	$e^{-BJ(J+1)/kT}$	Calculated from Equation (11.55)	Experiment
0	$0 \rightarrow 2$	1	1	1	1.0
1	$1 \rightarrow 3$	1.8	0.56	1.0	2.3
2	$2 \rightarrow 4$	2.57	0.18	0.46	0.4
3	$3 \rightarrow 5$	3.33	0.031	0.10	0.3

Table 12.2 Intensity of Spectral Line in the Raman Spectrum of H_2 Calculated from the Contributions of *Para*- and *Ortho*-H_2.

		Relative Intensity				
J	Transition	From Equation (11.55)	*Para*	*Ortho*	Total	Experiment
0	$0 \rightarrow 2$	1	1		1.0	1.0
1	$1 \rightarrow 3$	1.0		3.0	3.0	2.3
2	$2 \rightarrow 4$	0.46	0.46		0.46	0.4
3	$3 \rightarrow 5$	0.10		0.30	0.30	0.3

12.2.3 Nuclei with Half-Integer Spin (Antisymmetric Total Wavefunction), Nuclei with Integral Spin (Symmetric Total Wavefunction)

Similar to protons, nuclei with an odd number of protons and neutrons have half-integer spin (in units of $\hbar$). Nuclei with an even number of protons and neutrons have integral spin. Reality is described by postulating that the total wavefunction is antisymmetric in the first case and symmetric in the second case.

12.2.3.1 Fermions and Bosons

Generally, particles with a total wavefunction that is antisymmetric with respect to the exchange of identical particles are called *fermions*. Particles with a total wavefunction that is symmetric with respect to the exchange of identical particles are called *bosons*. This is expressed by Pauli in the following form:

> The total wavefunction of fermions (particles with noninteger spin) is antisymmetric. The total wavefunction of bosons (particles with integer spin) is symmetric with respect to the exchange of identical particles.

It is straightforward to generalize the discussion of Section 5.4 and deduce that the antisymmetry principle implies that fermions must obey the Pauli Exclusion Principle. As a special case, electrons have half-integer spin and can be considered as fermions: electronic wavefunctions must be antisymmetric. In contrast, the wavefunction of bosons is symmetric with respect to exchange of two particles. Hence two bosons could have the identical wavefunction (in an orbital picture, the same orbital) and satisfy the symmetry requirement. To conclude, two fermions cannot have the same wavefunction, but any number of bosons could have the same wavefunction.

Some examples for each type of symmetry, fermion and boson nuclei, are given in Table 12.3. The spins of different nuclei can be understood using the nuclear shell model [2]: odd mass number nuclei have half-integral spin, odd–odd nuclei with odd numbers of both protons and neutrons have integral spin, even–even nuclei with even numbers of both protons and neutrons have zero spin.

From Table 12.3 we see that D_2 (or $^2H_2$), N_2, and O_2 consist of bosons, and these molecules must be treated according to the symmetry principle instead of the antisymmetry principle. Furthermore, the total nuclear spin which is 1/2 in the case of ^1H can be different for different nuclei.

Table 12.3 Examples for Fermions and Bosons. *I* Is the Nuclear Spin Quantum Number.

Fermions				Bosons			
Nucleus	Protons	Neutrons	*I*	Nucleus	Protons	Neutrons	*I*
^1H	1	0	$\frac{1}{2}$	^2H	1	1	1
^3H	1	2	$\frac{1}{2}$	^4He	2	2	0
^3He	2	1	$\frac{1}{2}$	^{12}C	6	6	0
^7Li	3	4	$\frac{3}{2}$	^{14}N	7	7	1
^9Be	4	5	$\frac{3}{2}$	^{16}O	8	8	0
^{11}B	5	6	$\frac{3}{2}$	^{20}Ne	10	10	0
^{13}C	6	7	$\frac{1}{2}$				
^{19}F	9	10	$\frac{1}{2}$				

12.2.3.2 *Ortho*-D_2 and *Para*-D_2

For example, for D_2 the ^2H-nucleus has a total spin of 1. This means the nucleus can assume the spin values −1, 0, and +1. Accordingly, in this case we have three spin functions α, β, and γ which combine in the D_2 molecule in the way shown in Table 12.4.

Altogether, D_2 has 6 symmetric and 3 antisymmetric nuclear spin combinations. Since the total wavefunction must by symmetric, the symmetric spin functions combine with the symmetric rotational functions (J even) and the antisymmetric spin functions combine with the antisymmetric rotational functions (J odd). Therefore the rotational states with $J = 0,2, \ldots$ (*ortho*-D_2) are sixfold degenerate and the states with $J = 1,3, \ldots$ (*para*-D_2) are threefold degenerate, that is they are in a 2:1 ratio. In this way we understand

Table 12.4 Combinations of Nuclear Spin Functions α, β, and γ for the D_2 Molecule.

Spin Functions	
Symmetric	Antisymmetric
$\alpha(1)\alpha(2)$	
$\beta(1)\beta(2)$	
$\gamma(1)\gamma(2)$	
$\alpha(1)\beta(2) + \alpha(2)\beta(1)$	$\alpha(1)\beta(2) - \alpha(2)\beta(1)$
$\alpha(1)\gamma(2) + \alpha(2)\gamma(1)$	$\alpha(1)\gamma(2) - \alpha(2)\gamma(1)$
$\beta(1)\gamma(2) + \beta(2)\gamma(1)$	$\beta(1)\gamma(2) - \beta(2)\gamma(1)$

Table 12.5 Intensity of Spectral Line in the Rotational Raman Spectrum of D_2 Calculated from the Contributions of *Para*- and *Ortho*-D_2 (with $B = 0.594 \times 10^{-21}$ J and $T = 298$ K).

		Relative intensity				
J	Transition	From Equation (11.55)	*Ortho*	*Para*	Total	Experiment
0	$0 \to 2$	1	2		2.0	2.0
1	$1 \to 3$	1.35		1.35	1.35	1.4
2	$2 \to 4$	1.08	2.16		2.16	2.2
3	$3 \to 5$	0.59		0.59	0.59	0.6
4	$4 \to 6$	0.23	0.46		0.46	0.4
5	$5 \to 7$	0.06		0.06	0.06	0.06

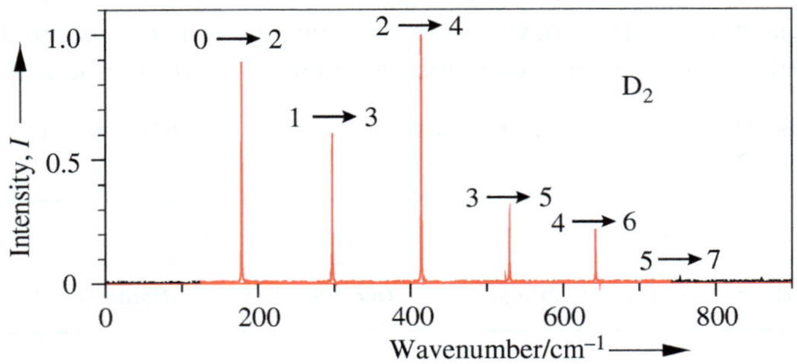

Figure 12.2 Rotational Raman spectrum of D_2. Excitation with an argon laser at 514.5 nm, recording time 12.5 h. The Rayleigh line at the origin is not shown. The intensity scale is arbitrarily chosen such that the highest peak has a value of 1. (Data from [1].)

immediately that the second line in the Raman spectrum of D_2 is weaker than the first one. In Table 12.5 we calculate the intensities of the Raman lines for D_2, in analogy to the case of H_2. They are in excellent agreement with the experiment (Fig. 12.2).

12.2.3.3 Raman Spectrum of N_2: Alternating Intensities

^{14}N has a total spin of 1, that is, N_2 can be treated in the same way as D_2. Justification 12.1 calculates the Raman intensities of N_2 and compares it with the experimental spectrum in Fig. 12.3.

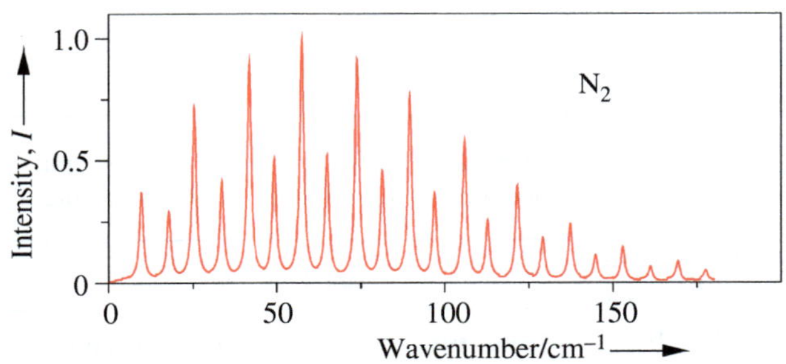

Figure 12.3 Rotational Raman spectrum of N_2. The Rayleigh line at the origin is not shown. The intensity scale is arbitrarily chosen such that the highest peak has a value of 1. Excitation with a 300 mW laser at 488 nm, spectral resolution 3 cm^{-1}, recording time 20 min. (Courtesy to P. Dechant, Marburg, private communication).

12.2.3.4 Raman Spectrum of O_2: Every Second Line Is Missing

^{16}O has a total spin of zero, so that only one spin function α is available. Hence, we have one symmetric nuclear spin function and no antisymmetric spin function for O_2; and a dramatic effect is observed: every second line is missing in the Raman spectrum. Justification 12.2 calculates the spacing of two adjacent Raman lines in O_2 and compares it with the spectrum shown in Fig. 12.4. The same effect is expected in the rotational Raman spectrum of CO_2 (during rotation, the carbon atom remains at rest and the two oxygen atoms rotate at a distance of $2d$). CO_2 is discussed in Justification J12.2.

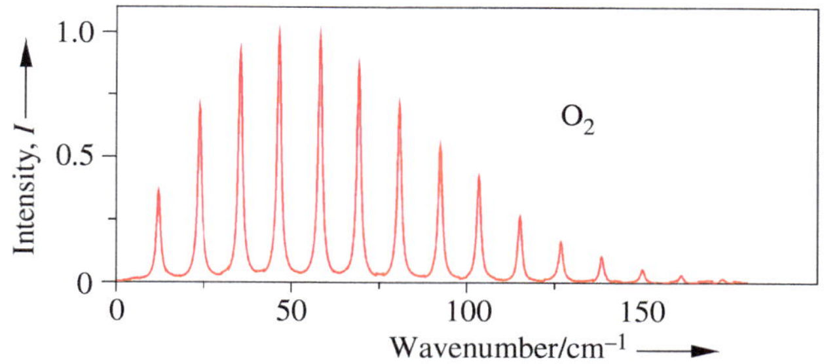

Figure 12.4 Rotational Raman spectrum of O_2. The Rayleigh line at the origin is not shown. The intensity scale is arbitrarily chosen such that the highest peak has a value of 1. Excitation with an 300 mW laser at 488 nm, spectral resolution 3 cm^{-1}, recording time 20 min. (Courtesy to P. Dechant, Marburg, private communication).

12.3 Nuclear Magnetic Resonance (NMR)

NMR is based on the nuclear magnetic moment, which is associated with nuclear spin. When a nucleus is exposed to an external magnetic field, the interaction energy between the applied field and the nuclear magnetic moment causes the degeneracy of the nuclear spin states (α and β for a proton) to be broken, giving rise to different energy levels. By applying radiofrequency waves the nuclear spin orientation can be turned (or "flipped"); i.e., transitions between the nuclear spin states occur. The magnetic field at the nucleus depends on the external magnetic field and a local magnetic field, which is determined by the electronic structure near the nucleus. For this reason, the resonance condition depends on details of the molecule's bonding, and NMR is a general probe of molecular structure.

12.3.1 Fundamentals

To simplify, we consider the atom ^1H (spin α or β) in a magnetic field with magnetic flux density B. By analogy to the case of electrons (Section 5.4), the nuclear magnetic moment of ^1H (specified by the nuclear quantum number I and the z-component m_I) is oriented in two different directions relative to the magnetic field. We measure the nuclear magnetic moment in terms of the nuclear magneton μ_N

$$\mu_N = \frac{eh}{4\pi m_p} = 5.051 \times 10^{-27}\ \mathrm{J\,T^{-1}}$$

where m_p is the mass of the proton. μ_N is smaller than the Bohr magneton (Section 5.4) by a factor of 2000 because of the larger proton mass. We can write

$$\text{nuclear magnetic moment} = g_N\ \mu_N\ m_I$$

g_N is called the nuclear g-factor. For a magnetic field oriented in the z-direction we can write that the energy of the nucleus is

$$E = -g_N\ \mu_N\ m_I\ B$$

A proton, $m_I = \pm 1/2$, can assume the energies

$$E_1 = -g_N\ \mu_N \frac{1}{2} B \quad \text{and} \quad E_2 = g_N\ \mu_N \frac{1}{2}\ B$$

Thus the energy difference between these two energy states is

$$\Delta E = E_2 - E_1 = g_N\ \mu_N\ B\left(\frac{1}{2} + \frac{1}{2}\right) = g_N\ \mu_N\ B \tag{12.5}$$

The spectroscopic technique of inducing transitions between different nuclear spin states is called Nuclear Magnetic Resonance (NMR) spectroscopy (Fig. 12.5).

The nucleus can be excited by irradiation with an electromagnetic field with a frequency of

$$\nu = \frac{\Delta E}{h} = \frac{g_N \mu_N B}{h} \tag{12.6}$$

As an example, ^1H atoms in a magnetic field with a magnetic flux density of $B = 2.3488$ T are in resonance with the exciting electric field at a frequency of 100 MHz, this is in the radio frequency range. From this experimental result we can calculate the nuclear g-factor of the proton in ^1H as

$$g_N = \frac{h}{\mu_N B}\nu = \frac{6.626 \times 10^{-34}\ \mathrm{J\,s}}{5.051 \times 10^{-27}\ \mathrm{J\,T^{-1}} 2.3488\,\mathrm{T}} 100 \times 10^6\ \mathrm{s^{-1}} = 5.585$$

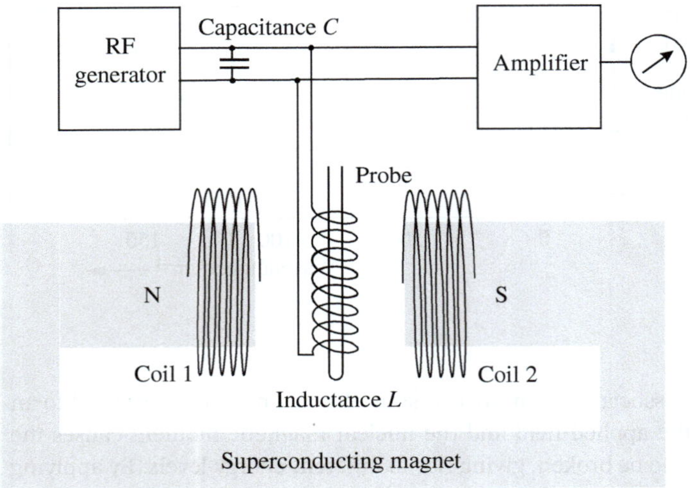

Figure 12.5 Continuous wave NMR spectrometer. The sample is located inside a glass tube, surrounded by a coil with inductance L, and is connected to a capacitance C. The system of inductance and capacitance acts as a resonant circuit. The signal of an RF generator (fixed radio frequency) is fed into this circuit, and the signal is measured by an amplifier. The external magnetic field is that of a superconducting magnet and can be changed by a current applied to the coils 1 and 2. When changing the magnetic field to a value for which the sample absorbs energy from the RF field the resonant circuit becomes damped and the signal measured by the amplifier decreases.

It is common to express the resonance frequency ν by the gyromagnetic ratio γ

$$\nu = \gamma \frac{1}{2\pi} B \quad \text{with} \quad \gamma = \frac{2\pi g_N \, \mu_N}{h} \tag{12.7}$$

For ^{1}H we have

$$\gamma = \frac{2\pi \cdot 5.585 \cdot 5.051 \times 10^{-27} \text{ J T}^{-1}}{6.626 \times 10^{-34} \text{ J s}} = 2.675 \times 10^8 \text{ T}^{-1} \text{ s}^{-1}$$

Besides the proton (^{1}H), many other nuclei can be excited by the NMR technique, see Table 12.6.

Table 12.6 Nuclear Quantum Number I, Nuclear g-Factor g_N, and Gyromagnetic Ratio γ for Some Nuclei.

Nucleus	I	g_N	γ	ν at $B = 2.3488$ T
			10^8 T^{-1} s^{-1}	MHz
^{1}H	1/2	5.585	2.675	100
^{2}H	1	1.714	0.411	15.35
^{13}C	1/2	1.405	0.673	25.14
^{14}N	1	0.808	0.193	7.22
^{19}F	1/2	5.257	2.517	94.08
^{31}P	1/2	2.264	1.083	40.48

Adapted from https://en.wikipedia.org/wiki/Gyromagnetic_ratio and https://en.wikipedia.org/wiki/G-factor_(physics).)

A difficulty with this spectroscopy is that ΔE is extremely small compared to the thermal energy, and the population of the ground state exceeds the population of the excited state only by one part in 10^6 (see Chapter 15). Consequently, the probability of absorption is almost the same as that for stimulated emission, and the net absorption is very small. By using very strong superconducting magnets ($B > 10$ T) the energy splitting ΔE, and thus the difference in the population between ground state and excited state, can be substantially increased. In addition to increasing the population difference between the states, sensitive detection methods have been developed to make NMR a very powerful spectroscopy. Instead of scanning the magnetic field in the region of resonance, a short high frequency pulse of electromagnetic radiation is applied to the sample. In this way all spins in the sample can be excited simultaneously. By measuring their relaxation as a function of time, the entire NMR spectrum can be collected in a matter of seconds. The signal-to-noise ratio can be significantly improved by the accumulation of a large number of measurements. Numerical methods are used to convert these time-dependent data into frequency-dependent data for interpretation.

12.3.2 Chemical Shift

The effect of the molecule's local magnetic field on the resonance frequency of a nucleus' transitions is characterized by a "chemical shift" parameter. When a magnetic field is applied to the molecule, the electrons in the molecule interact with the field in such a way that the local magnetic field at the nucleus B_{nucleus} is less than the externally applied magnetic field B

$$B_{\text{nucleus}} = B \cdot (1 - \sigma) \tag{12.8}$$

where σ is a shielding parameter. The shielding parameter depends on a number of factors, including the molecule's electron distribution near the nucleus whose spin is being probed, the orientation of the molecule relative to the direction of the magnetic field, and coupling with spins of solvent molecules. We restrict our considerations to molecules in a liquid or gas, where the measured shielding is an average over all possible orientations, and we only consider proton NMR.

Instead of measuring the absolute value of σ it is customary to measure the difference of σ relative to a reference. The difference

$$\delta = (\sigma_{\text{ref}} - \sigma) \times 10^6 \tag{12.9}$$

is called the chemical shift. Tetramethylsilane (TMS, $(CH_3)_4Si$) is a common reference for proton NMR spectra.

In Equation (12.6) we calculated the resonance frequency for the unshielded case as

$$\nu_{\text{unshielded}} = \frac{g_N \, \mu_N \, B}{h}$$

Accounting for shielding, the resonance frequencies of the sample are

$$\nu = \nu_{\text{unshielded}} \cdot (1 - \sigma) = A \cdot (1 - \sigma)$$

with

$$A = \frac{g_N \cdot \mu_N}{h} \cdot B \tag{12.10}$$

for each proton in the sample. Correspondingly, the resonance frequency of the reference is $\nu_{\text{ref}} = A \cdot (1 - \sigma_{\text{ref}})$. Thus for the shielding parameters we obtain

$$\sigma = 1 - \frac{\nu}{A} \qquad \sigma_{\text{ref}} = 1 - \frac{\nu_{\text{ref}}}{A}$$

and for the chemical shift (Equation (12.9)) we find that

$$\delta = (\sigma_{\text{ref}} - \sigma) \times 10^6 = \left(-\frac{\nu_{\text{ref}}}{A} + \frac{\nu}{A}\right) \times 10^6 = (\nu - \nu_{\text{ref}}) \times 10^6 \times \frac{1}{A} \tag{12.11}$$

As the shielding constant σ is very small (on the order of 10^{-6}), we can approximate ν_{ref} as

$$\nu_{\text{ref}} = A(1 - \sigma_{\text{ref}}) \approx A$$

and replace A in Equation (12.11) by ν_{ref} to obtain

$$\boxed{\delta = \frac{\nu - \nu_{\text{ref}}}{\nu_{\text{ref}}} \times 10^6} \tag{12.12}$$

The chemical shift increases with decreasing electron density at the proton. Thus δ increases if the neighboring atoms of the proton are more electronegative. Some common chemical shifts are reported in Table 12.7. The protons in tetramethylsilane (TMS) are more shielded than most protons in organic compounds. Using TMS as a reference in proton NMR results in positive values for δ in the range from 0 to 12.

The chemical shift is exceptionally large in aromatic compounds: $\delta = 7$ for benzene, compared to $\delta = 5$ for ethene and $\delta = 1$ in methane. To explain this observation, we analyze the π

Table 12.7 Typical Chemical Shift Values for Protons in Organic Molecules.

		δ
Methyl proton	$-C\mathbf{H}_3$	1
Ethene proton	$CH_2-C\mathbf{H}_2$	5
Phenyl proton	$-C_6\mathbf{H}_6$	7
Carboxyl proton	$CH_3-COO\mathbf{H}$	9.5
Carboxyl proton	$CCl_3-COO\mathbf{H}$	12

electron structure of a cyclic π-system. As we discussed in Chapter 8, the π-electrons can be described as de Broglie waves moving along the carbon–carbon perimeter. For the isolated benzene whether the electron waves propagate clockwise or counter-clockwise they have the same energy. Application of an external magnetic field can generate preference for motion in one direction (break the degeneracy) and cause a ring current (see Fig. 12.6). The additional magnetic field generated by the ring current enlarges the magnetic field outside the ring—that is, at the positions of the protons. Consequently, the shielding effect becomes smaller and δ becomes larger.

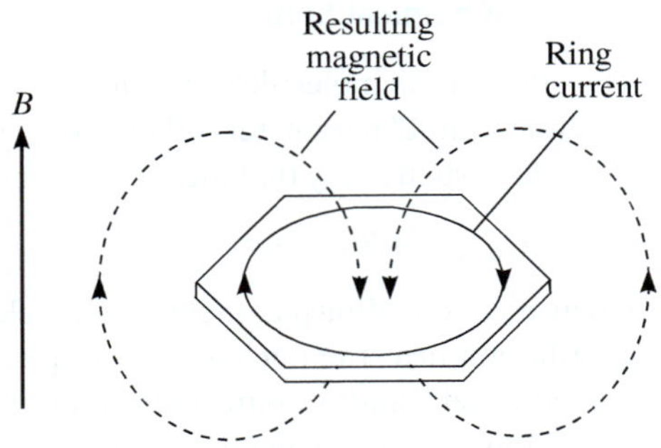

Figure 12.6 Ring current in benzene.

Example 12.1 *Figure 12.7 shows a low-resolution NMR spectrum of ethanol. Assign the proton peaks in the spectrum.*

Three distinct peaks are evident in the spectrum and the molecule has three different types of protons: the OH proton, the CH_2 protons, and the CH_3 protons. Since oxygen is more electronegative than carbon, we expect that the OH proton is more deshielded than a CH proton and has the highest δ value. The methyl protons should be shielded the most since they are bonded to a C atom and are furthest away from the O atom. We expect the methylene protons to be intermediate between the other two types of protons because the protons are bonded to a C atom that is directly bonded to the O atom. This assignment of the peaks is further substantiated by the relative intensities of the transitions, which corresponds to the 1:2:3 population of the proton types.

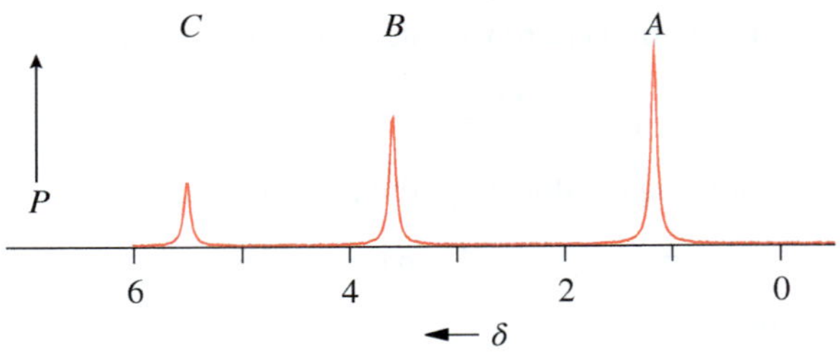

Figure 12.7 Low resolution ^{1}H-NMR spectrum of ethanol $\overset{C}{CH_3}-\overset{B}{CH_2}-\overset{A}{OH}$. P is the power absorbed by the molecule. (Data from [3].)

The position of the peaks in an NMR spectrum can be used to identify the type of protons in a molecule and their relative abundance. This feature of NMR spectroscopy makes it useful for identifying the types of chemical functionalities in a molecule.

12.3.3 Fine Structure of NMR Spectra

12.3.3.1 Coupling of Two Protons

The magnetic moment of a proton can couple with another proton in its neighborhood, and this coupling can make the NMR spectrum more complex. If two protons 1 and 2 are far enough away from each other in a molecule, they do not interact significantly and observe two single peaks in the NMR spectrum, provided that the protons are not equivalent (transitions in Fig. 12.8, top). As the protons become closer (only one bond removed) their coupling increases and the energy levels are split into two doublets (Fig. 12.8, bottom), because the local magnetic field experienced by proton 1 depends on whether it is aligned parallel or

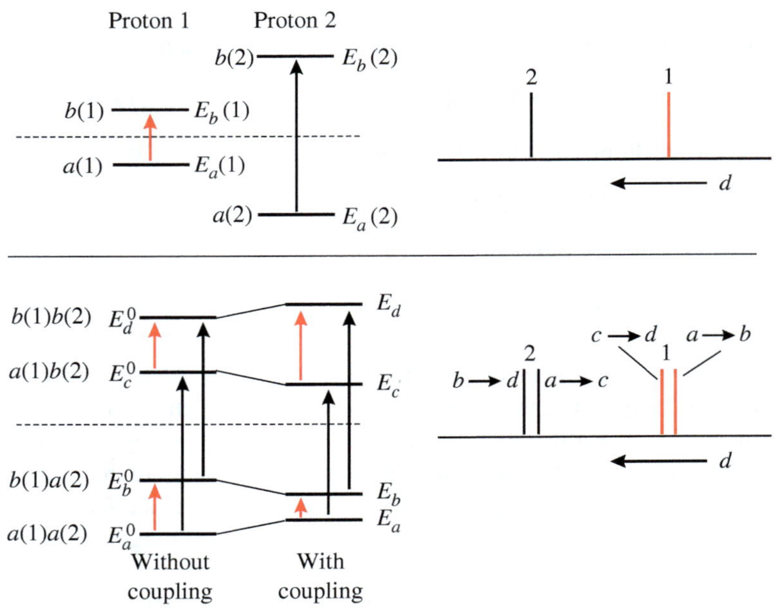

Figure 12.8 Splitting scheme for two nonequivalent protons (spins α and β, respectively). Top: protons 1 and 2 considered as separate systems (two single lines in the NMR spectrum). Bottom: both protons considered as one system. The left energy scheme is for the case that the nuclear spins are not coupled. The right energy scheme is for the case of coupling (splitting of the NMR lines into two doublets). Dashed line: zero point of energy (nuclei in no external magnetic field).

antiparallel to the nuclear spin of proton 2. In this situation two different transitions, in which the spin of proton 2 remains unchanged, are possible for proton 1. In this way we observe two close lying transitions, a doublet, for proton 1. Through a corresponding argument, we deduce that proton 2 should also show a doublet.

(1) *No coupling*. In detail, protons 1 and 2, if treated as separate systems, have energies $E_\alpha(1)$, $E_\beta(1)$, $E_\alpha(2)$, and $E_\beta(1)$, according to their spin functions $\alpha(1)$, $\beta(1)$, $\alpha(2)$, and $\beta(2)$, in the external magnetic field (Fig. 12.8, top). If we treat both protons as one system but independently, their spin functions can be described as $\alpha(1)\alpha(2)$, $\beta(1)\alpha(2)$, $\alpha(1)\beta(2)$, and $\beta(1)\beta(2)$. In analogy to the case of the electron spin wavefunction of He (see Chapter 5) the corresponding energies are sums of the energies of the two isolated protons, if spin–spin coupling is neglected (note that $E_\alpha(1)$ and $E_\alpha(2)$ are negative, and $E_\beta(1)$ and $E_\beta(2)$ are positive):

Spin Function	Energies (Coupling Neglected)
$\alpha(1)\alpha(2)$	$E_a^0 = E_\alpha(1) + E_\alpha(2)$
$\beta(1)\alpha(2)$	$E_b^0 = E_\beta(1) + E_\alpha(2)$
$\alpha(1)\beta(2)$	$E_c^0 = E_\alpha(1) + E_\beta(2)$
$\beta(1)\beta(2)$	$E_d^0 = E_\beta(1) + E_\beta(2)$

When exciting the system, transitions are allowed in which the spin orientation of only one proton changes. That is, the allowed transitions are $a \to b$, $c \to d$, $a \to c$, and $b \to d$ as indicated by arrows in Fig. 12.8, bottom (left scheme). Note that for the excitation energies we obtain $\Delta E_{a\to b} = \Delta E_{c\to d}$ and $\Delta E_{a\to c} = \Delta E_{b\to d}$. Thus the system of two protons shows only two lines in the NMR spectrum, if coupling is neglected.

(2) *Coupling*. If the two protons couple with each other, the position of the energy levels shift, as indicated in Fig. 12.8, bottom (right scheme). Now we find $\Delta E_{a\to b} < \Delta E_{c\to d}$ and $\Delta E_{a\to c} < \Delta E_{b\to d}$. As a consequence, the NMR lines split into two doublets (Fig. 12.8, bottom).

Here we have considered coupling of two non-equivalent protons. Justification 12.3 shows that no splitting into doublets occurs in the case of coupling of two equivalent protons. This result is not trivial and is a consequence of the indistinguishability of such protons.

12.3.3.2 Coupling of Three Nonequivalent Protons

The coupling between nuclei is a strong function of distance. For protons on nuclei that are directly bonded together this coupling can be significant, but if the nuclei are more than one bond away the coupling is much weaker. This feature provides a method for experimentally determining the connectivity of a molecule by determining the couplings, or splittings, in the NMR spectra. We illustrate this analysis for the ethanol molecule.

In an ethanol molecule we have three nonequivalent types of protons (*A*, *B*, and *C*) which give rise to three NMR peaks. Because of coupling between the nuclei, fine structure is observed (Fig. 12.9).

$$\overset{A}{CH_3}-\overset{B}{CH_2}-\overset{C}{OH}$$

Let us start with the protons *A*, which are coupled to protons *B*. Protons *B* have four possible spin orientations (Fig. 12.10), of

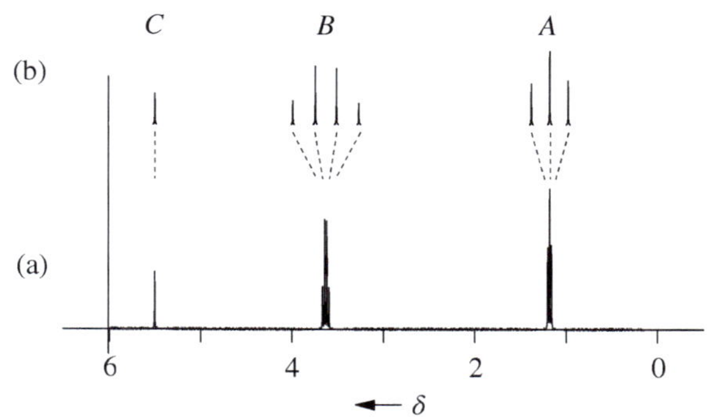

Figure 12.9 High resolution ^{1}H-NMR spectrum of ethanol $\overset{C}{CH_3}-\overset{B}{CH_2}-\overset{A}{OH}$. (a) shows the high resolution spectrum and (b) shows the spectral assignments.

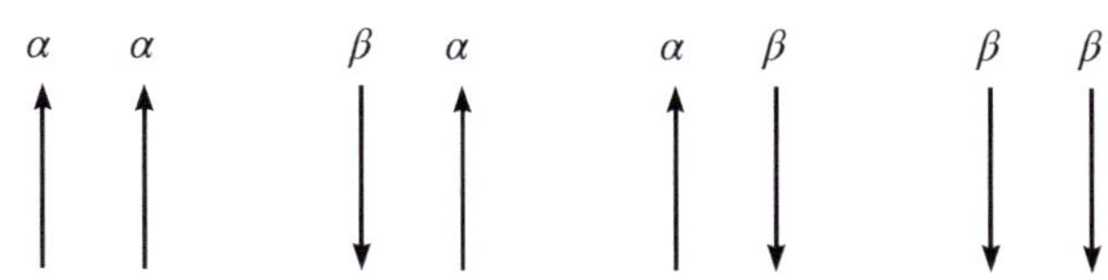

Figure 12.10 Combinations of spins α and β of protons *B* in the ethanol molecule.

which two are equivalent, and give rise to three different local magnetic fields at protons *A*. The spectral line for protons A is split into three lines, a triplet, with intensities 1 : 2 : 1 (where the two arises from the two equivalent spin orientations).

The *A* protons also modify the local magnetic field of the *B* protons. The three spins of protons *A* have eight possible arrangements; however, three possibilities with two spins up and one spin down and three possibilities with one spin up and two spins down are equivalent. Thus only four different local magnetic fields are generated at protons *B*, and we expect the signal of protons *B* to split into a quadruplet with intensities 1 : 3 : 3 : 1.

In addition to the coupling of the *A* and *B* protons, we expect a coupling of protons *B* with proton *C*. Including the effect of the spin orientation of the *C* proton, which is independent of the *A* proton orientations, would further modify the local magnetic field about the *B* protons, generating an octuplet for protons *B*. Indeed this fine structure can be observed in experiments with properly controlled conditions.

Oftentimes, this additional splitting is not observed because the OH proton is exchanging rapidly with protons in other molecules.

12.3.3.3 Exchange and Relaxation Effects

For solvent conditions that promote proton exchange the expected fine structure disappears. For ethanol, the exchange of the OH proton is an acid-catalyzed process, and a very small concentration of acid is enough for the exchange. More precisely the protons *B* experience an average coupling with the spin magnetic moment of proton *C*, if the exchange of proton *C* with the solvent is rapid, as compared to the timescale of the nuclear spin transitions (typically 10^{-6} s). In this case the proton being removed and then replaced on the hydroxyl group is not necessarily the same proton and does not necessarily have the same spin orientation, so that (on average) the magnetic moment arising from proton *C* is zero. This leads to no net coupling and thus no spectral splitting.

Why are protons from the solvent not spin correlated with those on the solute molecule? As we saw in Chapter 11, the timescales for the vibrational motions (10^{-14} to 10^{-12} s) and rotational motions (10^{-12} to 10^{-10} s) are rapid, and these motions give rise to local magnetic field fluctuations which cause the proton magnetic moments to change direction. Thus, while the external magnetic field does orient the spins of the protons in the entire sample, the thermal fluctuations of the solvent molecules cause relaxation between the spin sublevels; i.e., cause α spins to flip to β spins and vice versa. This dynamic equilibrium means that any given spin will spend part of the time oriented "up" and part oriented "down," with a small bias associated with the energy level difference.

12.3.4 Nuclear Overhauser Effect (NOE) and 2D Spectroscopy

So far our discussion has imagined experiments in which an applied radiofrequency is tuned or an applied magnetic field is scanned until it comes into resonance with the sample and then the loss or gain is measured. In contrast to this "frequency domain" approach, we can perform a time-dependent experiment, in which a radiofrequency (RF) pulse impinges on the sample. The RF pulse comprises many different frequencies which are superposed to create a localized RF pulse—in analogy to the superposition of electron waves discussed in Section 3.5.5 and Problem 3.10. After the RF pulse passes through the sample, we can decompose it again into its frequency components and determine what frequencies were either absorbed or stimulated to emit light. This strategy allows us to collect the whole spectrum at once and improve the signal-to-noise ratio, which is critical in making NMR spectroscopy broadly useful.

In addition to its signal-to-noise benefits, the pulsed approach provides a mechanism to create non-equilibrium states of a molecule's spin subsystem and to manipulate it through a sequence of RF pulses and time intervals between pulses. This technology has made NMR spectroscopy a very powerful tool for examining the three-dimensional structure of molecules. While conventional NMR spectra, discussed above, report on the connectivity of a molecule (i.e., the bonding pattern), time-resolved measurements provide information on the "through space" interaction of nuclear spins and can be used to infer a three-dimensional structure of a molecule. The rise of these methods over the past 50 years has made NMR one of the most powerful spectroscopic methods in chemistry, and the active transfer of these methods to IR and UV-Vis spectroscopies is an active frontier in molecular spectroscopy.

12.3.4.1 Nuclear Overhauser Effect

Consider a two-spin system like that shown in the bottom of Fig. 12.8, but not directly bonded, rather they are near each other in space because of their three dimensional arrangement. Imagine applying an RF pulse for a long enough duration that it saturates the transitions for the spins on proton 2. That is, we apply RF radiation of sufficient intensity and duration that we create equal populations of α and β spins on proton 2. This irradiation shifts the population distribution of the spin states. If we then perform a measurement of the NMR signal for the spin on proton 1, we find that its intensity is changed.

That is, pumping the transition on spin 2 can effect the intensity of the transition on spin 1, if they lie close enough to each other in space. This phenomenon is called the nuclear Overhauser effect, and it can be understood through consideration of the excitation and relaxation kinetics of the spin system.

How do the spins interact when the atoms are not directly bonded together? The magnetic dipoles associated with the nuclear spins interact directly through space. The dipole of a magnetic moment on a nucleus gives rise to a local magnetic field B_{local} where

$$B_{\text{local}} = \frac{\gamma \mu_0}{4\pi r^3}\left(1 - 3 \cdot \cos^2\theta\right) m_I \hbar$$

m_I is the z-component of the nuclear spin, θ is the azimuthal angle, and r is the distance from the dipole. Consider a proton (1) with an α spin ($m_I = 1/2$) aligned along z (i.e., $\theta = 0$), which generates a magnetic field

$$B_1 = -\frac{\gamma \mu_0}{4\pi r^3}\hbar,$$

and interacts at a distance r with a second proton (2) which has the magnetic field

$$B_2 = \frac{\gamma \mu_0}{4\pi r^3}\left(1 - 3 \cdot \cos^2\theta\right) m_I \hbar$$

In this case, the interaction energy is

$$E_{\text{interaction}} = \frac{\vec{B}_1 \bullet \vec{B}_2}{2\mu_0} = -\frac{\mu_0}{2}\left(\frac{\gamma}{4\pi r^3}\right)^2 \left(1 - 3 \cdot \cos^2\theta\right) m_I \hbar^2$$
$$= -\frac{\mu_0}{2}\left(\frac{g_N\, \mu_N}{4\pi r^3}\right)^2 \left(1 - 3 \cdot \cos^2\theta\right) m_I$$

Because the interaction energy becomes much weaker as the distance increases; i.e.,

$$E_{\text{interaction}} \propto \frac{1}{r^6}$$

only the spins which are very close to each other in space will have a significant dipole–dipole interaction.

This interaction can drive spin relaxation and its strong dependence on the distance between the two nuclei makes it very local. Thus if we excite a spin its relaxation can be affected by a nearby spin which may not necessarily be bonded; that is, it is affected by through space interaction.

Consider the case of two coupled protons, 1 and 2. We are going to pump/saturate the transition to flip the spin of proton 2, but not that of spin 1. The nonradiative relaxation among the spin sublevels gives rise to population changes that affect the strength of the transition for proton 1. Figure 12.11 illustrates the energy level diagram for the spin substates of the two spin system in which the RF pulse is driving the $\alpha(1)\alpha(2)$ to the $\alpha(1)\beta(2)$ transition, represented by a thick double headed blue arrow. Because the RF pulse is longer than the nonradiative spin relaxation rates between the spin sublevels (dashed arrows in the figure), the population of the spin sublevels achieve a steady-state population distribution that is different from the thermal distribution without the RF pumping pulse. Note that a corresponding energy level scheme must be considered for the $\beta(1)\alpha(2)$ to the $\beta(1)\beta(2)$ transition. This population shift causes the signal intensities to be different in the NMR spectrum. It is possible to have either a positive or a negative Nuclear Overhauser Effect, but the main thing is that the intensity of the transition for proton 1 changes.

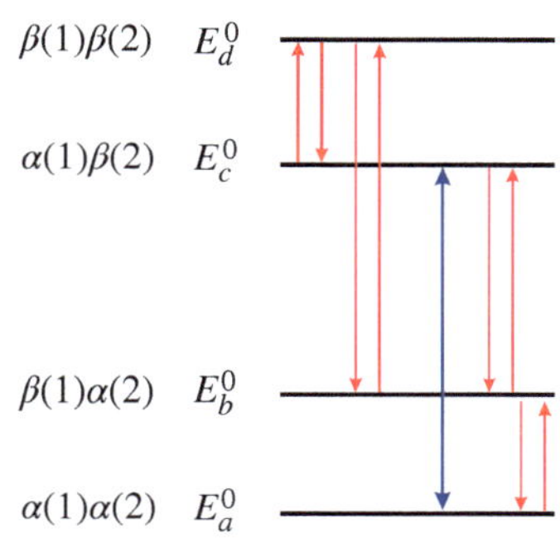

Figure 12.11 Energy level diagram for a two spin system in which the $\alpha(1)\alpha(2) \rightarrow \alpha(1)\beta(2)$ transition is being saturated by an RF pulse (thick blue arrow). The red lines illustrate the dipolar relaxation processes that occur to create a new population of the energy levels under irradiation by the RF pulse.

If the proton 1 intensity changes, then it tells us that the proton 2 spin is nearby in space. This feature along with the "through space" nature of the dipole–dipole coupling allows one to gain information about the three-dimensional structure of molecules rather than their bond connectivity. Because the saturation of a particular spin, teaches us what other spins it is nearby, we can perform an experiment in which we scan the frequency and saturate the transitions of each of the different spins in the molecule and examine how the other spins respond. This experiment allows us to create a map of the "through space" arrangement of the spins in a molecule. That is we plot the NMR signal intensity versus the transition frequency probed and versus the frequency of the saturation ("pump") pulse; a three-dimensional graph. From the information this experiment provides, we can infer the arrangement of the protons in three-dimensional space. A number of different methods based on this idea have been developed, one very popular version is called NOESY (Nuclear Overhauser Effect Spectroscopy) [4].

Example 12.2 *Consider the two-dimensional NMR spectrum shown in Fig. 12.12 for the molecule dimethyl acetamide* $(CH_3)_2N\text{–}C(O)CH_3$ *at 303 K. Assign the peaks along the diagonal in this spectrum and rationalize the origin of the two "cross-peaks."*

The peaks along the diagonal correspond to the frequency positions (chemical shifts) for the different protons of dimethylacetamide. Because of the deshielding effect of the carbonyl group, we expect that the peak near 50 Hz corresponds to the protons on the methyl group which is directly bonded to the carbonyl unit (protons c). The other two peaks must then correspond to the protons on the methyl groups which are bonded to the nitrogen atom. These protons have different NMR shifts, which arise from whether the methyl group is cis or trans to the methyl group that is directly bonded to the carbonyl unit.

The cross-peaks arise because the molecule can rotate about the N–C bond, and this exchanges the position of the methyl groups bonded to the N with respect to the methyl bonded to the carbonyl unit. This changes the local field of the protons because of the through space dipole–dipole coupling and gives rise to the cross-peaks. While the internal rotation frequency is 100–200 Hz at this temperature, it is slow enough on the timescale of the NMR experiment that distinct cross-peaks are evident.

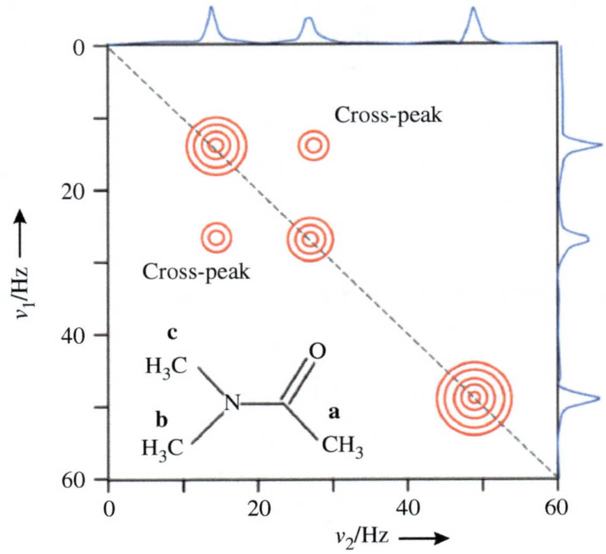

Figure 12.12 Two-dimensional proton NMR spectrum, in which the circular contours represent peaks in the NMR spectrum for the molecule dimethylacetamide at 30 K. The one-dimensional NMR spectrum is plotted along the outer axes of the box. The image is adapted from,[1] for which a 60 MHz NMR spectrometer was used.

12.3.4.2 NMR Spectroscopy for Determination of Protein Structures in Solution

An extremely important method that allows for the determination of protein structures in solution is based on the spin relaxation involving $^1H\text{–}^1H$ dipolar interaction using the Nuclear Overhauser effect. It allows to relate signals by moving along the peptide backbone and in this way to obtain details of the spatial structure [6].

12.3.5 Magnetic Resonance Imaging (MRI)

By applying an inhomogeneous magnetic field (a magnetic field gradient), one can change the nuclear splittings by different amounts throughout a sample, making it possible to identify a position with a resonant frequency and image the sample. This method allows one to investigate metabolic processes in soft tissues and is important in medicine [7].

12.4 Conclusion

In this chapter we discussed how the spin properties of nuclei place important symmetry restrictions on the types of nuclear states that are possible, and we discussed how to use the nuclear spin states of a molecule to probe the structure of molecules. If a molecule is placed in an external magnetic field, the nuclear spin states have different energies, and spectroscopic transitions between these spin states are possible. Although the energy splittings are small, the probing of such transitions for common nuclei (protons and carbon 13) provides a selective and important probe of molecular structure.

1 The spectrum shown here is adapted from the report by Jeener et al. [5].

12.5 Justifications

12.5.1 Justification 12.1: Raman Spectrum of N_2

See the books website, www.wiley.com/go/Kuhn/physicalchemistry3e.

12.5.2 Justification 12.2: Spacing of Raman Lines in O_2 and CO_2

See the books website, www.wiley.com/go/Kuhn/physicalchemistry3e.

12.5.3 Justification 12.3: Coupling of Two Equivalent Protons

See the books website, www.wiley.com/go/Kuhn/physicalchemistry3e.

12.6 Exercises

E12.1 Show that the spacing of the lines in the Raman spectrum of O_2 (Fig. 12.4) is $8B$ instead of $2B$ as for heteronuclear diatomics. Use 120.752 pm for the bond length of O_2.

E12.2 Calculate the relative intensity of the Raman lines in the O_2 spectrum for the first eight transitions.

E12.3 Following the procedure for D_2 explain the intensity distribution of the Raman lines of N_2 (Fig. 12.3).

E12.4 Estimate the spacing of the spectral lines in the Raman spectrum of CO_2. Use this spacing to calculate the C–O bond length. Note that carbon dioxide will have a rotational spectrum like that found for oxygen (Fig. E12.4).

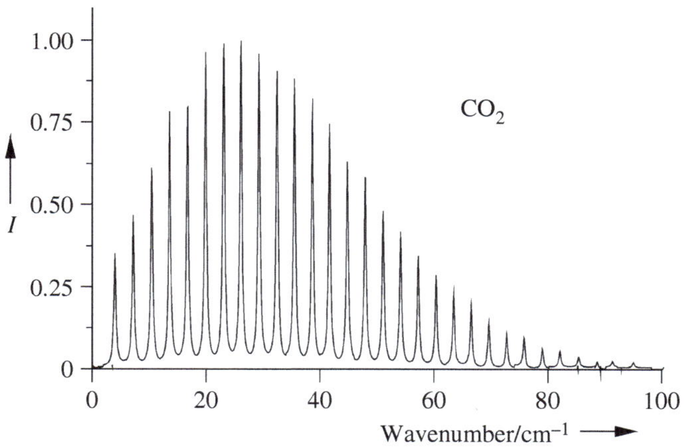

Figure E12.4 Rotational Raman spectrum of CO_2.

E12.5 Calculate the magnetic field strength that satisfies the nuclear resonance condition for protons in a radiofrequency field of i) 60 MHz; ii) 300 MHz, and iii) 500 MHz.

E12.6 Calculate the nuclear resonance frequency for a ^{13}C nucleus in a 10 T magnetic field.

E12.7 For the ethanol molecule draw the eight different spin combinations for the methyl protons (A) and draw a splitting pattern for the methylene protons (B) that arises from the methyl protons.

E12.8 Some of the fine structure in the NMR spectrum of ethanol (Fig. 12.9) is lost because of rapid OH proton exchange. Draw the type of spectrum you would expect if the rapid OH proton exchange could be stopped.

E12.9 Consider the cyanine cation

$$\overset{+}{-N} = \overset{1}{C}H-\overset{2}{C}H=\overset{3}{C}H-\overset{4}{C}H=\overset{5}{C}H-\overset{6}{C}H=\overset{7}{C}H-\overset{-}{N}-$$

The measured ^{13}C shielding parameters σ (see Equation (12.8)) at the C atoms 1–7 are [8]

C Atom	C_1	C_2	C_3	C_4	C_5	C_6	C_7
σ/ppm	163.2	100.0	150.4	120.8	150.4	100.0	163.2

Use the FEMO model to compare these values to the total probability density at these atoms.

12.7 Problems

P12.1 Energy order the following spectroscopic transitions for the molecule NO: vibrational excitation of the NO bond, nuclear spin "flip" of O nucleus, electronic excitation of NO (from HOMO to LUMO), electron spin "flip," rotational transition of NO, and photoionization of NO. Also, state the order of magnitude of the energy in each case.

P12.2 In Section 12.3.3 we have considered the splitting of the NMR lines of two non-equivalent protons into two doublets. Perform a corresponding treatment for two equivalent protons (as the protons in dichloromethane). Hint: equivalent protons cannot be distinguished; therefore, their spin functions must be either symmetric or antisymmetric. Transitions can only occur from a symmetric state to a symmetric state or from an antisymmetric state to an antisymmetric state.

References

1 D.E. Jennings, A. Weber, and J.W. Brault, FTS-Raman flame spectroscopy of high-*J* lines in H_2 and D_2, *J. Mol. Spectrosc.*, 1987, **126**, 19.
2 I. Talmi, "Simple Models of Complex Nuclei", in: *Contempory Concepts in Physics*, Volume **7** (Harwood Academic, Chur, Switzerland, 1993).
3 W.A. Guillory, *Introduction to Molecular Structure and Spectroscopy* (Allyn and Bacon, Boston, MA, 1977).
4 M.H. Levitt, *Spin Dynamics* (Wiley, New York, 2009).
5 J. Jeener, B.H. Meier, P. Bachmann, and R.R. Ernst, Investigation of exchange processes by two-dimensional NMR spectroscopy, *J. Chem. Phys.*, 1979, **71**, 4546.
6 K. Wüthrich, *NMR of proteins and nucleic acids* (Wiley, New York, 1986).
7 W.A. Gibby, Basic principles of magnetic tresonance imaging, *Neursurg. Clin. N. Am.*, 2005, **16**, 1–64.
8 H. Mustroph, Cyanine dyes, *Phys. Sci. Rev.*, 2020, **5**, 145.

13

Solids and Intermolecular Forces

CHAPTER MENU

13.1 Introduction

Solids are often categorized by their ability or inability to conduct electricity. Metals show high conduction (low resistance) to electron current and the conductivity decreases as the temperature increases. Semiconductors have a conductivity that is low, but strongly dependent on introducing distinct impurities (called *doping*), and the conductivity increases as the temperature increases. Insulators have very small conductivities and are relatively insensitive to the temperature.

An alternative classification scheme distinguishes solids and clusters by the nature of the interaction forces that cause the atoms and molecules to aggregate. The structure of ionic crystals can be described as rigid ions that arrange to create a neutral structure which minimizes the overall electrostatic energy. In this case the electrons are localized and bound tightly to the individual ions. In contrast, metals are often described as atomic ions embedded in a cloud of free electrons. Their structure results from a balance between the stabilizing effect of Coulomb attraction and the destabilization arising from the electrons' increased kinetic energy. Covalent solids lie between these two limits. In this latter case localized bonding interactions occur between adjacent atoms, but electron wavefunctions can delocalize through the solid's lattice.

Also, we consider molecular crystals and aggregates. The aggregation of neutral molecules is governed by dispersion forces, dipole forces, and induction forces that are considerably weaker than the electrostatic forces in ionic crystals. Although weak, understanding these forces is extremely important for understanding the structure of biological machines and other "soft" materials. For example, the special properties of liquid water arise from the importance of hydrogen bonds.

Albeit incomplete, this chapter describes some important models for describing bonding in condensed matter. We discuss some typical examples, using simple models that focus on the essentials and illustrate foundations. We extend this approach in Chapters 27 and 28 where molecular assemblies of quite different nature are considered. Our aim is to stimulate your interest in future supramolecular design, to assist you in developing your facility to create new materials, and to excite you about research.

Principles of Physical Chemistry, Third Edition. Hans Kuhn, David H. Waldeck, and Horst-Dieter Försterling.

Companion website: www.wiley.com/go/Kuhn/physicalchemistry3e

13.2 Ionic Crystals

Ionic crystals are formed by the binding of ions through electrostatic attraction. These solids have tightly bound electron clouds around each ion. Because the electron clouds only overlap weakly, a classical model based on Coulomb's law provides a good description of the crystal structure. This model can be viewed as an extension of that we used in Section 7.3.2 to describe LiH, which is bound by Coulomb forces

$$f = \frac{-1}{4\pi\varepsilon_0} \cdot \frac{e^2}{d^2} \tag{13.1}$$

between two oppositely charged ions (charges $+e$ and $-e$). The distance between the ions is d, and ε_0 is the permittivity of vacuum.

13.2.1 Bond Energy of an Ion Pair

As the distance between two ions decreases, the electrostatic attraction stabilizes the ion pair until a repulsive force, resulting from overlap of the ions' electron clouds, becomes dominant. The energy of attraction between two singly charged ions at a distance d is

$$E_{\text{attraction}} = -\frac{1}{4\pi\varepsilon_0} \cdot \frac{e^2}{d} \tag{13.2}$$

As the ions get very close a repulsive interaction occurs, because the filled electron clouds of the two ions cannot exist in the same region of space (Pauli Exclusion Principle). Hence the electron clouds must distort as they begin to overlap, and this process is energetically unfavorable. Because it is difficult to describe the repulsion energy of the electron clouds theoretically, we use the empirical formula

$$E_{\text{repulsion}} = C \cdot \frac{1}{d^n} \tag{13.3}$$

where C and n are constants. A range for n from $n = 6$ to $n = 12$ can be estimated from the compressibility of ionic crystals.

Using this approximation, the electrostatic bond energy of an ion pair may be written as

$$E = E_{\text{attraction}} + E_{\text{repulsion}} = -\frac{1}{4\pi\varepsilon_0} \cdot \frac{e^2}{d} + C \cdot \frac{1}{d^n} \tag{13.4}$$

Equation (13.4) is a special case of the Mie-potential [1]

$$E_{\text{Mie}} = const\left[\left(\frac{C}{d}\right)^n - \left(\frac{1}{d}\right)^m\right]$$

if we use $m = 1$ for the electrostatic attraction. At large distances the energy E is zero, and at small distances the repulsion energy dominates so that E becomes positive. At intermediate distances the attraction energy dominates and E is negative. For a distinct distance d_{eq} (equilibrium distance) E must go through a minimum. We can find this distance by setting the derivative of E with respect to d equal to zero:

$$\left(\frac{\mathrm{d}E}{\mathrm{d}d}\right)_{d=d_{\text{eq}}} = \left(\frac{1}{4\pi\varepsilon_0} \cdot \frac{e^2}{d^2} - C \cdot n \cdot \frac{1}{d^{n+1}}\right)_{d=d_{\text{eq}}} = 0 \tag{13.5}$$

Rearranging this equation we find that

$$C = \frac{e^2}{4\pi\varepsilon_0}\frac{1}{n}d_{\text{eq}}^{n-1}$$

and from Equations (13.3) and (13.4) it follows that

$$E_{\text{repulsion}} = \frac{e^2}{4\pi\varepsilon_0}\frac{1}{n}d_{\text{eq}}^{n-1} \cdot \frac{1}{d^n} \tag{13.6}$$

and

$$E = -\frac{1}{4\pi\varepsilon_0} \cdot \frac{e^2}{d} \cdot \left[1 - \frac{1}{n}\left(\frac{d_{\text{eq}}}{d}\right)^{n-1}\right] \tag{13.7}$$

As an example, we consider the ion pair Na^+ and Cl^-. Treating the ions as hard spheres (the cation Na^+ with radius $r_+ = 102$ pm and the anion Cl^- with radius $r_- = 181$ pm) that touch each other at the equilibrium distance d_{eq}, results in $d_{eq} = r_+ + r_- = 283$ pm. The energy E is plotted in Fig. 13.1 versus the distance d with $d_{eq} = 283$ pm and $n = 8$ together with the energies $E_{attraction}$ and $E_{repulsion}$. At the equilibrium distance $d = d_{eq}$ we find

$$E_{eq} = -\frac{1}{4\pi\varepsilon_0} \cdot \frac{e^2}{d_{eq}} \cdot \left[1 - \frac{1}{n}\right] \tag{13.8}$$

As n takes values between 6 and 12, $1/n$ is only a small correction in comparison to 1; this means that only a minor deformation of the electron clouds takes place when an ion pair is formed. Consequently, the energy needed for separating the ions is mainly determined by the Coulomb attraction. In Problem P13.13 you check this.

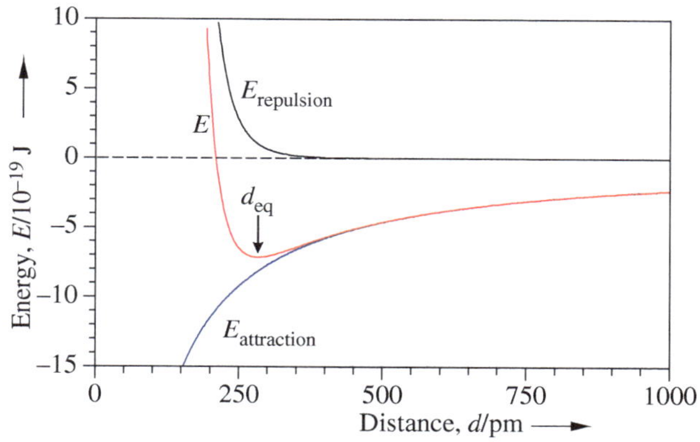

Figure 13.1 The graph plots $E_{attraction}$, $E_{repulsion}$, and the total energy E of the ion pair Na^+ and Cl^- versus the distance d between the ions according to Equations (13.2), (13.6), and (13.7) with $n = 8$ and $d_{eq} = 283$ pm.

13.2.2 Lattice Types

Ion pairs can aggregate, and an energetically favorable arrangement is reached when a three-dimensional lattice has been formed (for details, see Foundation 13.1). By way of example, consider common table salt, NaCl, for which the Na^+ and Cl^- ions form a lattice like that shown in Fig. 13.2. The cations occupy the corners and the centers of the faces of cubes; and the anions form a similar interpenetrating set of cubes. Each Na^+ ion is surrounded by six Cl^- ions as nearest neighbors. Correspondingly, each anion is surrounded by six cations. We describe this fact by introducing the coordination number as

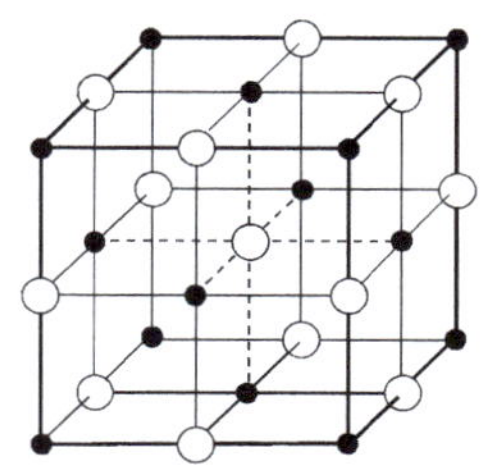

Figure 13.2 Arrangement of the Na^+- and Cl^- ions in the sodium chloride lattice.

$$\begin{aligned}\text{coordination number } &= \text{Number of anions surrounding a cation} \\ &\quad \text{or Number of cations surrounding an anion}\end{aligned} \tag{13.9}$$

Thus the coordination number in a NaCl lattice is six. The crystals grow in cubes as a result of the arrangement of Na^+ and Cl^- in the lattice, and therefore the cleavage of NaCl crystals results in cubic pieces (Fig. 13.3).

The crystal structure is energetically more stable than the isolated ion pair, because each ion is adjacent to six ions with opposite charge. In Box 13.1 we show that the energy E' per ion pair in the NaCl crystal is 1.748 times larger than the energy E of an isolated ion pair, so that the equilibrium bond energy is

$$E'_{eq,NaCl} = -\frac{1}{4\pi\varepsilon_0} \cdot \frac{e^2}{d_{eq}} \cdot 1.748 \cdot \left[1 - \frac{1}{n}\right] \tag{13.10}$$

Figure 13.3 The cleavage of a NaCl crystal results in cubic pieces.

The parameter n can be obtained from measurements of the compressibility of the NaCl crystal, leading to the empirical value $n = 8$ for NaCl. The factor of 1.748 is called the *Madelung constant* for the NaCl lattice. The Madelung constant depends upon the geometry of the crystal. The sodium chloride lattice is especially favorable for the Na^+ and Cl^- ions because the ions just touch each other (Fig. 13.4a).

The relative sizes of the ions influence the structure of the lattice which is formed. Consider the BeO ion pair in which the O^{2-} anions are significantly larger than the Be^{2+} cations. If we assume that the ions are arranged in a sodium chloride lattice, the anions touch each other before the anions and cations contact one another, because the cations are so small (see Fig. 13.4b). The electrostatic energy can be decreased if the cations and anions are closer to each other. For BeO this occurs in a *Wurtzite lattice*, where each cation is tetrahedrally surrounded by anions (Fig. 13.5).

$$E'_{\text{eq,BeO,Wurtzit}} = -\frac{1}{4\pi\varepsilon_0}\cdot\frac{(2e)^2}{d_{\text{eq}}}\cdot 1.639\cdot\left[1-\frac{1}{n}\right] \quad (13.11)$$

Note that the charge of Be^{2+} is $2e$ and the charge of O^{2-} is $-2e$. The Madelung constant is smaller than for the sodium chloride lattice (1.639 instead of 1.748), but the cations and anions get closer together which helps stabilize the structure. The increase in the stabilization from the smaller d_{eq} more than compensates for the smaller Madelung constant.

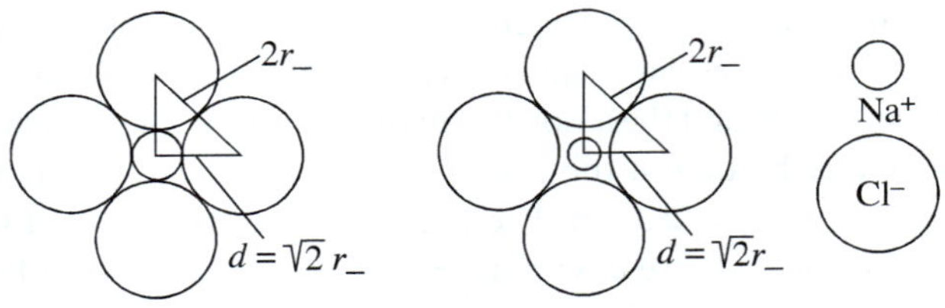

Figure 13.4 The sodium chloride lattice. (a) Realized case: the anions (Cl^-, with radius r_-) touch each other and the cations (Na^+, with radius r_+) fill the space in between, touching the anions. Thus the distance between the ions in the ion pair equals $d = d_{\text{eq}}$ (where $d = \sqrt{2r_-}$ by geometry and $d_{\text{eq}} = r_+ + r_-$ as defined in Fig. 13.1). (b) Hypothetical case: the anions touch each other, but the cations do not fill the space in between (no contact between anions and cations). In this case the distance between the ions in the ion pair is $d < d_{\text{eq}}$.

Example 13.1 *Calculate the electrostatic energy for an ionic crystal of BeO in two cases, the Sodium-Chloride lattice and the Wurtzite lattice, and determine which is more stable.*

The ionic radii of Be^{2+} and O^{2-} are $r_+ = 27$ pm and $r_- = 140$ pm. In the Wurtzite lattice with $d_{\text{eq}} = r_+ + r_- = 167$ pm we obtain

$$E'_{\text{eq,BeO}} = -\frac{1}{4\pi\varepsilon_0}\cdot\frac{4e^2}{167\ \text{pm}}\cdot 1.639\cdot\left[1-\frac{1}{n}\right] \quad (13.12)$$

For a crystal in a sodium chloride lattice, the distance d_{eq} would be, according to Fig. 13.5b, $d_{\text{eq}} = 140 \times \sqrt{2}$ pm $= 198$ pm and

$$E'_{\text{eq,BeO}} = -\frac{1}{4\pi\varepsilon_0}\cdot\frac{4e^2}{198\ \text{pm}}\cdot 1.748\cdot\left[1-\frac{1}{n}\right] \quad (13.13)$$

The energy gain for forming the Wurtzite lattice is larger by 11%,

$$\frac{1.639 \times 198}{167 \times 1.748} = 1.11$$

Indeed, BeO crystallizes in the Wurtzite lattice. The empirical factor n is obtained from a measurement of the compressibility of the BeO crystal.

The effect of the ionic size on the type of lattice that is formed is well known and general. As shown for the sodium chloride lattice in Fig. 13.5a, contact between anions and cations is possible for

$$r_+ + r_- \geq \sqrt{2}\cdot r_- \quad (13.14)$$

resulting in a ratio

$$\frac{r_+}{r_-} \geq \sqrt{2} - 1 = 0.414 \quad \text{(sodium chloride lattice)} \quad (13.15)$$

for the radii of cation and anion (in the case of the > sign the cations touch the anions, but the anions do not touch each other). If this ratio is smaller than 0.414, the anions and cations cannot touch each other and the ions should crystallize in the Wurtzite lattice. If the ratio is larger than 0.732 a cation can be surrounded by eight nearest neighbors. Then a cesium

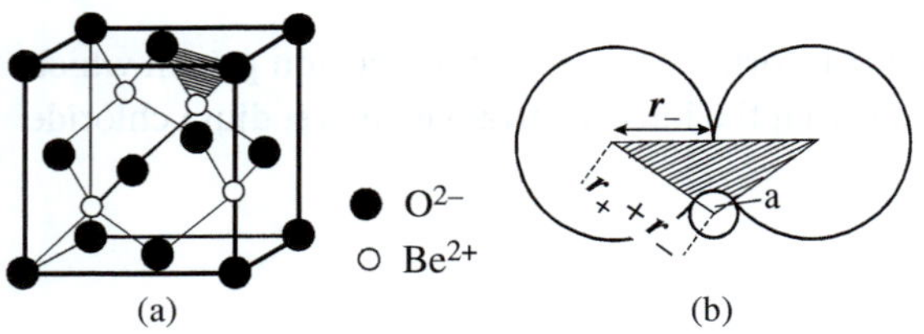

Figure 13.5 (a) Three-dimensional arrangement of Be^{2+} and O^{2-} ions in a Wurtzite lattice. (b) Cross section through the plane shown in (a) when anions and cations touch each other. The ratio of the radii is obtained from $r_-/(r_+ + r_-) = \sin \alpha/2$, where $\alpha = 109.47°$ is the tetrahedral angle:$r_+/r_- = 0.225$.

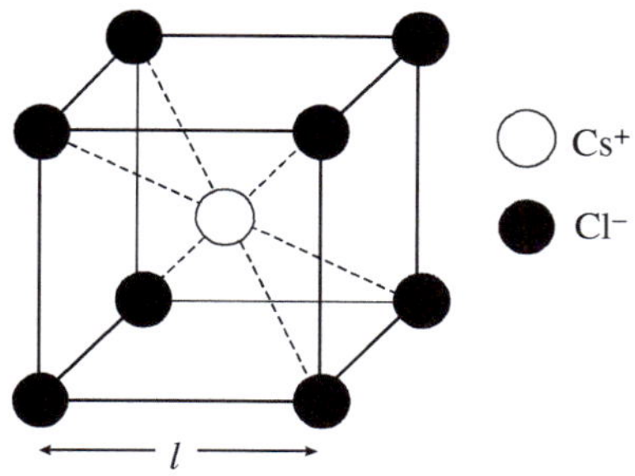

Figure 13.6 CsCl lattice. The distance between a Cs^+ and the Cl^- is $(\frac{1}{2})\sqrt{3} \cdot l$, where l is the length of the cube. $N_A \cdot l^3$ is the volume of 1 mol CsCl; thus, $l = \sqrt[3]{m_{CsCl}/\rho}$, where m_{CsCl} is the mass of CsCl and ρ is the density of the crystal.

Table 13.1 Lattice Types and Radii Ratios r_+/r_- of Different Crystals of Type AB.[1]

Lattice Type	Wurtzite		NaCl		CsCl	
Coordination Number	4		6		8	
Madelung Constant	1.638		1.748		1.763	
	ZnO	0.43	NaF	0.77	CsCl	0.91
	ZnS	0.35	NaCl	0.56	CsBr	0.83
	BeO	0.20	NaBr	0.52	CsI	0.77
	BeS	0.16	NaI	0.46	RbCl	0.84
	BeSe	0.15	KF	1.04	TlCl	0.83
	MgTe	0.28	KCl	0.76	TlI	0.67

chloride lattice (Fig. 13.6) is formed with a larger Madelung constant (1.763). Table 13.1 shows data for three lattice types and demonstrates the correlation of the ratio r_+/r_- with the type of lattice formed.

So far we have considered only crystals consisting of ions with the same charge number; the smallest electrically neutral unit is then an ion pair AB. In components with different charge numbers (units A_nB_m) we can proceed similarly. The resulting structures may be more complicated, however. The considerations become even more complicated when ions cannot be approximated as spheres (e.g., CH_3COO^-, CN^-, CO_3^{2-}). In these cases, detailed aspects of a molecular ion's geometry can influence the structure. Nevertheless all possible crystalline solids fall into one of 14 lattice types (see Foundation 13.1).

13.3 Metals

To illustrate the model for bonding in metals, consider a monatomic metal like lithium which crystallizes in a body-centered cubic lattice (see Foundation 13.1). In the solid each Li atom releases one electron, so that the crystal is composed of Li^+ ions and valence electrons (Fig. 13.7). The Li^+ ions are embedded in a gas, or cloud, of valence electrons; and the attractive Coulomb forces between the negatively charged electrons and the Li^+ ion cores stabilize the lattice. As the crystal volume decreases, the kinetic energy of the electron gas increases because the de Broglie wavelengths of the electrons decrease. An energy minimum is reached at a certain volume which results as a compromise between the potential energy and kinetic energy.

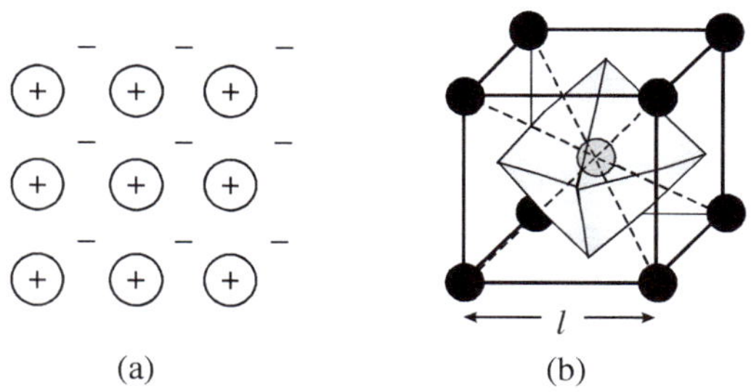

Figure 13.7 (a) Metal lattice of positively charged nuclei (e.g., Li^+ ions) surrounded by valence electrons. (b) Li lattice arrangement (body-centered), and polyhedron around a lithium atom.

13.3.1 Free Electron Model for Conduction Electrons

Rather than provide a detailed treatment of all the metal states, we proceed by treating the conduction ("valence") electrons as moving in an effective potential of the metal nuclei and their core electrons. Furthermore we consider elemental solids

1 See [2]. Note that the calculation uses ionic radii for S^{2-}, Se^{2-}, and Te^{2-} which are coordination number 6; however, the radii of anions change weakly with coordination number.

and approximate the effective potential as a constant. In this approximation the electronic structure of the metal corresponds to that of N electrons in a box potential.

13.3.1.1 Particle-in-a-Box Model

Recall that the energy levels for the particle in a three-dimensional box are given by

$$E = \frac{h^2}{8m_e}\left(\frac{n_x^2}{L_x^2} + \frac{n_y^2}{L_y^2} + \frac{n_z^2}{L_z^2}\right) \tag{13.16}$$

and the wavefunctions are given by

$$\psi(x,y,z) = \sqrt{\frac{8}{L_x L_y L_z}} \cdot \sin\left(\frac{n_x \pi x}{L_x}\right) \cdot \sin\left(\frac{n_y \pi y}{L_y}\right) \cdot \sin\left(\frac{n_z \pi z}{L_z}\right) \tag{13.17}$$

We assume that the selected electron is in the field of the closest ion and that the electrostatic fields of the other ions are compensated by the other electrons. Furthermore, we assume that this effective field acting on the selected electron is constant, as we did in our analysis of π electrons (see Fig. 8.4, bottom). Then the electrons occupy the energy levels according to Equation (13.16) subject to the Pauli Exclusion principle. For a metal of macroscopic size the lengths L_x, L_y, and L_z are large, and the energy spacing between quantum states is small enough that their distribution can be approximated as continuous.

13.3.1.2 Fermi Energy

For an alkali metal each atom contributes one electron from its valence shell so that the total number of "free" electrons in the metal equals the number N of atoms of the metal. In Box 13.2 we show that the kinetic energy for the highest occupied energy level is

$$E_F = \left(\frac{3}{\pi}\right)^{2/3} \cdot \frac{h^2}{8m_e} \cdot \left(\frac{N}{V}\right)^{2/3} \tag{13.18}$$

where E_F is called the *Fermi energy* (Fig. 13.8) and $V = L^3$ is the volume of the considered cube of metal. The number density N/V can be calculated from the density ρ of the metal and the mass m_{atom} of the atoms in the metal

$$\frac{N}{V} = \frac{\rho}{m_{\text{atom}}} \tag{13.19}$$

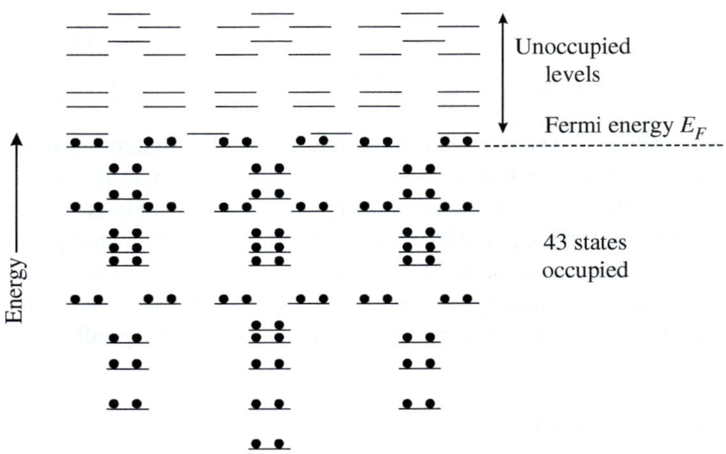

Figure 13.8 Distribution of freely moving electrons on the different orbitals (quantum numbers n_x, n_y, and n_z: 111, 112, 121, 112, 113, 131, 311, ...) in a cube of metal. Actually, in a cube of volume 1 cm^3 there are about 10^{20} states occupied.

13.3.1.3 Li Metal: Number *g* of Electronic States Between *E* and *E* + Δ*E*

For Li metal we have $\rho_{\text{Li}} = 0.534$ g cm^{-3} and $m_{\text{Li}} = 1.151 \times 10^{-26}$ kg; thus, $N/V = 4.64 \times 10^{28}$ m^{-3} and $E_F = 7.54 \times 10^{-19}$ J. For a metal the Fermi energy corresponds to the energy of the most weakly bound electron, at zero temperature (see Fig. 13.9). Because the energy that is needed to excite an electron to a higher energy state is very small compared to typical thermal energies, the number of accessible electronic states in an energy interval around the Fermi energy is very high. In Box 13.2 we show that the number g of electronic states per energy interval between the energies E and $E + \Delta E$ is given by

$$g = \frac{\pi}{4}\left(\frac{8m_e L^2}{h^2}\right)^{3/2} E^{1/2} \cdot \Delta E = \frac{\pi}{4}\left(\frac{8m_e}{h^2}\right)^{3/2} V\ E^{1/2} \cdot \Delta E \tag{13.20}$$
$$= 5.31 \times 10^{55}\ \text{J}^{-3/2}\ \text{m}^{-3} \cdot V \sqrt{E} \cdot \Delta E$$

where V is the crystal's volume. In an energy interval ΔE of 4×10^{-21} J (this corresponds to the thermal energy at room temperature, see Chapter 14) near the Fermi energy of Li ($E_F = 7.54 \times 10^{-19}$ J) the number of electronic states for $V = 1$ cm^3 is $g = 1.8 \times 10^{20}$.

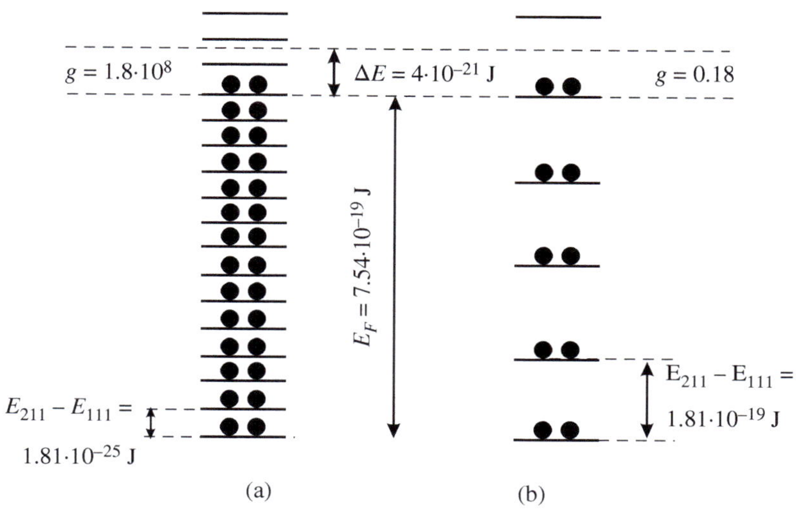

Figure 13.9 Energy levels of electrons in a small block of Li metal, schematic. (a) cube with 1 μm side length. In the energy range $\Delta E = 4 \times 10^{-21}$ J the number of electronic states is $g = 1.8 \times 10^8$. (b) cube with 1 nm side length. The number of electronic states is $g = 0.18$; this means that the distance between the energy levels is so large that the probability of finding a level in the energy range $\Delta E = 4 \times 10^{-21}$ J is very low.

These two features, the small energy spacing between the filled and unfilled energy levels and the large density of electronic states, account for many of the electrical and optical characteristics of bulk metals. Because the electrons are easy to excite in a metal, a small applied electric field causes electron motion. For visible light and other high-frequency AC electric fields this means that the electrons can respond readily to the applied field and shield it, giving rise to the high reflectivity and silvery lustre of metals.

Example 13.2 *Li-chunk. Consider cubic blocks of Li of 1 μm³ and of 1 nm³. Calculate the number of electronic states g for the given energy interval $\Delta E = 4 \times 10^{-21}$ J at the Fermi energy level $E_F = 7.54 \times 10^{-19}$ J.*

We use Equation (13.20) and find for the 1 μm³ chunk that the number of electronic states is

$$g = 5.31 \times 10^{55}\ \mathrm{J^{-3/2}\ m^{-3}} \cdot (10^{-6}\ \mathrm{m})^3\ \sqrt{E_F} \cdot 4 \times 10^{-21}\ \mathrm{J} = 2.12 \times 10^{17}\ \mathrm{J^{-1/2}} \sqrt{E_F} = 1.8 \times 10^8$$

if we identify the energy E with the Fermi energy $E_F = 7.54 \times 10^{-19}$ J.

The corresponding calculation for the 1 nm³ block of lithium (about 40–50 Li atoms) gives

$$g = 5.31 \times 10^{55}\ \mathrm{J^{-3/2}\ m^{-3}} \cdot (10^{-9}\ \mathrm{m})^3\ \sqrt{E_F} \cdot (4 \times 10^{-21}\ \mathrm{J}) = 2.12 \times 10^{8}\ \mathrm{J^{-1/2}} \sqrt{E_F} = 0.18$$

(Fig. 13.9b). Hence quantization of the electronic energy will play an important role for metal particles of nanometer size; however, particles of micron size have a large density of states and the properties are like those of the bulk metal.

It is of interest to compare the given energy interval $\Delta E = 4 \times 10^{-21}$ J with the energy spacing between the most stable energy level and the next energy level. From Equation (13.16), we see that the energy difference between the lowest electronic state and the next state is

$$\begin{aligned} E_{211} - E_{111} &= \frac{h^2}{8m_eL^2}[(4+1+1) - (1+1+1)] \\ &= \frac{3 \cdot (6.63 \times 10^{-34})^2}{8 \cdot 9.11 \times 10^{-31} \cdot (10^{-2})^2} \mathrm{J} = 1.81 \times 10^{-31}\ \mathrm{J} \end{aligned}$$

for a 1 cm³ cube, and it is 1.81×10^{-25} J for a 1 μm³ cube. These ΔE values are small compared to the thermal energy at 300 K, which is about 4×10^{-21} J. In contrast, we obtain $E_{211} - E_{111} = 1.81 \times 10^{-19}$ J for a 1 nm³ cube, which is much larger than the thermal energy at 300 K.

13.3.2 Cohesion Energy of Metals

The ductility of Li can be understood by considering the forces that bind the atoms together into a solid. A metal's total cohesion energy comprises two parts: the kinetic energy and the potential energy, which we consider here.

The particle-in-a-box energies (Equation (13.16)) can be used to determine the total average kinetic energy of the electrons in the metal. The N electrons in the metal are distributed in pairs over the lowest energy orbitals (Fig. 13.9), following the Aufbau principle. The average kinetic energy is given by

$$\overline{E}_{\text{kin}} = \frac{\sum_{\text{occupied}} E(n_x, n_y, n_z)}{N/2}$$

If we approximate the distribution of occupied states as continuous, we can replace the sum over the occupied energy levels with an integral. In Box 13.3 we show that this expression for the average kinetic energy can be integrated to yield

$$\overline{E}_{\text{kin}} = \frac{3}{5}E_F = \frac{3}{5}\left(\frac{3}{\pi}\right)^{3/2}\frac{h^2}{8m_e}\cdot\left(\frac{N}{V}\right)^{2/3} \tag{13.21}$$

For Li we obtain

$$\overline{E}_{\text{kin}} = 4.5\times 10^{-19}\text{J} = 2.8\text{ eV}$$

The average potential energy is determined by the array of positive atomic cores interacting with the continuous distribution of negative charge, the electrons. In the case of Li, the solid forms a body-centered cubic lattice. We can divide the solid into polyhedra that are centered on the Li^+ cores. The polyhedra will have eight faces, each of which are perpendicular to the axis connecting the atom of interest with a nearest neighbor (see Fig. 13.8). Because the polyhedra are neutral, their interaction energy is small. If we assume that this interaction energy can be neglected, then the average potential energy becomes N times the average potential energy within a single polyhedron.

To determine the electrostatic energy in a single polyhedron, we further simplify by approximating the polyhedron as a sphere and ignoring the volume of the atomic core compared to the volume of the polyhedron. In this case, the electrostatic energy is that for an electron interacting with a positive point charge in a sphere of radius r, so that

$$\overline{E}_{\text{pot}} = -\frac{e^2}{4\pi\varepsilon_0}\cdot\overline{\left(\frac{1}{r}\right)} \tag{13.22}$$

For a sphere with radius r (volume V/N)

$$\frac{V}{N} = \frac{4}{3}\pi r^3\text{ , thus }\frac{1}{r} = \left(\frac{N}{V}\frac{4\pi}{3}\right)^{1/3}$$

and

$$\overline{E}_{\text{pot}} = -\frac{e^2}{4\pi\varepsilon_0}\cdot\left(\frac{4\pi}{3}\right)^{1/3}\left(\frac{N}{V}\right)^{1/3} \tag{13.23}$$

where N/V is the number density. For Li, $N/V = 4.63\times 10^{28}\text{ m}^{-3} = 46.3\text{ nm}^{-3}$ and we find that

$$\overline{E}_{\text{pot}} = -11.7\times 10^{-19}\text{ J}$$

The cohesive energy is given by

$$E = \overline{E}_{\text{pot}} + \overline{E}_{\text{kin}} = -A\cdot\left(\frac{N}{V}\right)^{1/3} + B\cdot\left(\frac{N}{V}\right)^{2/3} \tag{13.24}$$

where

$$A = \frac{e^2}{4\pi\varepsilon_0}\cdot\left(\frac{4\pi}{3}\right)^{1/3} = 3.27\times 10^{-19}\text{ J nm} \tag{13.25}$$

$$B = \frac{3}{5}\left(\frac{3}{\pi}\right)^{3/2}\frac{h^2}{8m_e} = 3.50\times 10^{-20}\text{ J nm}^2$$

With the number density for Li, $N/V = 46.4\text{ nm}^{-3}$, we find

$$E = (-11.7 + 4.5)\times 10^{-19}\text{ J} = -7.2\times 10^{-19}\text{ J}$$

This means that an energy 7.2×10^{-19} J has to be supplied to convert a Li atom in a Li crystal into a Li^+ ion and an electron. This predicted value is reasonably close to the experimental value of 11×10^{-19} J (the sum of the heat of sublimation, 2.32×10^{-19} J, and the ionization energy, 8.64×10^{-19} J).

Instead of using the measured number density to estimate the cohesion energy we may estimate the number density by minimizing E with respect to the number density. As shown in Exercise 13.21 $(N/V)_{\text{min}} = 101.9\text{ nm}^{-3}$,

$\overline{E}_{\text{pot}} = -15.3 \times 10^{-19}$ J, $\overline{E}_{\text{kin}} = 7.6 \times 10^{-19}$ J and $E = -7.7 \times 10^{-19}$ J. The typical radius about an atom in the metal is

$$l = \left(\frac{N}{V}\right)_{\text{min}}^{-1/3} = \left(101.9\ \text{nm}^{-3}\right)^{-1/3} = 0.214\ \text{nm} = 214\ \text{pm}$$

This result is on the order of the experimental value $l = 280$ pm.

Example 13.3 *Use the density, sublimation energy, and ionization energy for the alkali metals in Table 13.1 to quantitatively analyze the predictions of Equation (13.24).*

Table 13.2 shows the density data and the atomic mass data for the alkali metals. These data can be used to calculate the number density of atoms. By way of example, for Li we find

$$\frac{N}{V} = \frac{\rho}{m_{\text{atom}}} = \frac{\rho N_A}{M}$$
$$= \frac{534\ \text{kg m}^{-3}\ 6.022 \times 10^{23}\ \text{mol}^{-1}}{6.941\ \text{g mol}^{-1}}$$
$$= 46.3\ \text{nm}^{-3}$$

The sublimation energy per atom and the first ionization energy of the element can be used to compute the cohesive energy E. Figure 13.10 compares the experimental cohesion energies to those calculated from Equation (13.24) with the parameters $A = 4.50 \times 10^{-19}$ J nm and $B = 3.92 \times 10^{-20}$ J nm^2.

Table 13.2 Cohesion Energy E and Number Density N/V of the Alkali Metals.

Metal	ρ kg m^{-3}	M g mol^{-1}	N/V nm^{-3}	E_{sub} (10^{-19} J)	$E_{\text{ionization}}$ (10^{-19} J)	E (10^{-19} J)
Li	534	6.94	46.3	−2.54	−8.52	−11.06
Na	971	22.99	25.4	−1.81	−8.23	−10.04
K	862	39.10	13.3	−1.57	−6.95	−8.52
Rb	1532	85.47	10.79	−1.31	−6.69	−8.00
Cs	1873	132.9	8.48	−1.31	−6.24	−7.55

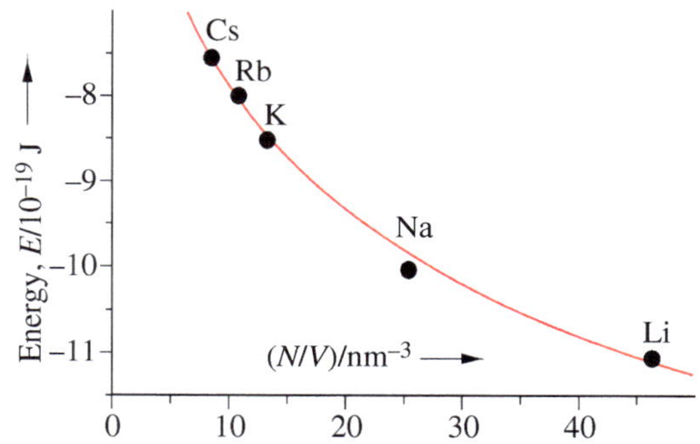

Figure 13.10 Cohesion energy E for the alkali metals versus number density N/V. Dots: experiment, data see Table 13.2. Solid line: calculated according to Equation (13.24) with $A = 4.50 \times 10^{-19}$ J nm and $B = 3.92 \times 10^{-20}$ J nm^2.

13.3.3 Quantum Wires and Nanostructure

What happens as we constrict the diameter of a metal wire, making it thinner and thinner until it reaches the limit of a single chain of metal atoms? Such experiments have been realized. Figure 13.11 is taken from a study by H. Ohnishi et al. [3], who measured the current through a single atom wide wire of Au atoms. These workers constructed a scanning tunneling microscope on the sample stage of an electron microscope so that they could simultaneously measure the current–voltage curve and image the Au "wire." They find that the conductance of a metal wire increases in a stepwise manner with the thickness of the wire. Such studies have been verified in a number of different experiments and demonstrate that the electrical conductance is quantized.

How can we understand this phenomenon? The quantization of conductance can be understood by considering the free electron model for the motion of electrons in a "quantum wire." The electron wavelengths perpendicular to a quantum wire's length are short and the corresponding energy spacings are large, but along its length (that is, in the z direction) the energy spacings are small and can be approximated as continuous. Imagine two identical large pieces of metal with bulk properties that are connected by a chain of atoms (the quantum wire). For simplicity, we assume that the junction between our quantum wire and the two macroscopic slabs is perfect and every electron passing through this boundary is transmitted

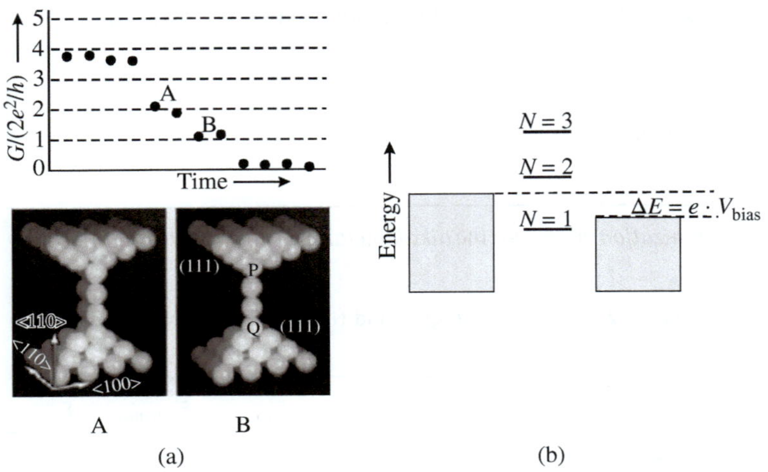

Figure 13.11 (a) some details of the conductance measurement through an atomic chain of Au atoms. In the beginning of the experiment an Au STM tip is moved into contact with an Au crystal, then the conductance G of the metal–metal contact is measured as the tip is slowly pulled away. The bottom image shows a model of the contact, indicated by two Au strands for conductance in region A and one Au strand for conductance in region B. (b) The energetics of the electronic states for the two bulk metal regions and the Au chain, which consists of subbands $N = 1$, $N = 2$, etc. [3].

with a unit probability. In the ground energy state of the whole system the energy levels of the quantum wire are filled up to the Fermi energy, determined by the large metal slabs on the ends. Figure 13.11b illustrates this structure for the case of a small bias voltage V_{bias}, which shifts the Fermi levels by $\Delta E = eV_{\text{bias}}$.

What causes the current to jump by discrete amounts as the Au chain's thickness increases? When the chain is one atom thick the electrons can only propagate through electronic states with $n_x = 1, n_y = 1$ ($N = 1$ in Fig. 13.11b); i.e., the ground state. As the chain thickness increases, from a single strand to a double strand as shown in the figure, a second set of electronic states, $n_x = 1, n_y = 2$ ($N = 2$ in Fig. 13.11b), become available because the energy spacing decreases as $1/L^2$. Hence an additional "channel" for electron flow opens up. Because the current flow through the excited state acts independently of that through the ground state, the conductance increases (the resistance decreases). In the example below, we show that the increase is equal to $2e^2/h$. This feature of the quantum wire gives rise to steps in the current–voltage curve, unlike that found for a traditional metal wire. As the wire becomes larger in the x and y directions, more energy states become available and the discrete nature of the increase in current is no longer evident.

Example 13.4 *In Problem 13.12 you show that the electron current moving through the quantum wire in Fig. 13.11 can be written as*

$$I = e\sum \frac{v_z}{L} = e\sum \frac{hn_z}{2m_eL^2} \tag{13.26}$$

Given this result show that the conductance is $2e^2/h$; i.e., the current is related to the voltage bias by

$$I = \frac{2e^2}{h} \cdot V_{\text{bias}}$$

To relate the current to the voltage, we find how the voltage affects the number of states in the summation in Equation (13.26). For the case where the Fermi level changes by one quantum state, $g = 1$, in the z-direction, the energy difference ΔE is

$$\Delta E = \frac{h^2}{8m_eL^2}\left[(n_z+1)^2 - n_z^2\right] = \frac{h^2}{8m_eL^2}2n_z$$

Accordingly, if the Fermi level changes by g quantum states, the energy difference ΔE must be larger by a factor of g:

$$\Delta E = \frac{h^2}{8m_eL^2}2n_z \cdot g$$

For the case of a small constant bias voltage, the energy difference between the two sides of the quantum wire is given by the charge on an electron times the bias voltage, $\Delta E = eV_{\text{bias}}$. Substituting this result into the above expression, we can rearrange and relate g, the number of quantum states accessed along z, to V_{bias}, the voltage drop along z. We find that

$$g = \frac{8m_eL^2}{2h^2n_z} \cdot \Delta E = \frac{8m_eL^2}{2h^2n_z} \cdot eV_{\text{bias}}$$

Using this result for the number of states to replace the sum in Equation (13.26), we find that

$$I = e\sum \frac{hn_z}{2m_eL^2} = \frac{ehn_z}{2m_eL^2}\cdot g = \frac{ehn_z}{2m_eL^2}\frac{8m_eL^2}{2h^2n_z}\cdot eV_{\text{bias}} \tag{13.27}$$

and cancelling terms we obtain

$$I = \frac{2e^2}{h}\cdot V_{\text{bias}}$$

Thus the quantized conductance $G = I/V_{\text{bias}}$ of the metal is equal to

$$G = \frac{2e^2}{h} \tag{13.28}$$

Because the conductance is just the inverse of resistance R, we see that the quantized resistance is

$$R = \frac{h}{2e^2} = 12.9\ k\Omega$$

Equation (13.28) is the basic equation for the electric conductivity in nanotechnology, first derived by Landauer [4].

13.4 Semiconductors

In this section we discuss semiconductors for which the bonding between atoms is covalent, e.g., Si. In these materials electrons must be excited through a significant energy gap in order for electrical conduction to occur. Materials with an energy gap that ranges from a few tenths of an eV to a few eV are called *semiconductors*. The energy gap must be sufficiently small and the density of available excited states must be sufficiently large in order to achieve substantial conductivity.

13.4.1 Organic Semiconductors: Polyacetylene

In Section 8.4 we found that for a polyene a row of equal bonds is unstable because the π electron density is alternating in the middle of the bonds—in contrast to a cyanine dye. This alternation leads to alternating "double" bonds and "single" bonds in butadiene. This effect is called *Peierls distortion* [5], but it was proposed and investigated earlier in the case of polyenes [6]. This effect leads to a gap in the energy spectrum between occupied and unoccupied orbitals. With increasing chain length and increasing number of orbitals this energy gap decreases in magnitude. The occupied states below the energy gap form what is called the *valence band*, and the unoccupied states above the energy gap form the *conduction band*. The gap between these bands reaches a limit of 1.78 eV in the step potential model discussed in Chapter 8. The value obtained experimentally in polyacetylene (upper limit of absorption band) is 1.5 eV.

In Section 8.4 we considered large rings with an odd number of carbon atoms and showed that a defect in the row of alternating single and double bonds introduced by an extra electron extends over about 15 C atoms. The same effect takes place in the case of a *trans*-polyacetylene when a small portion is oxidized (p-doping). Again the defect introduced in the chain extends over about 15 CH units. This quasi-particle (a positively charged soliton, see Section 8.4) can move along the chain and act as an electron conductor under an applied electric field. The conduction of *trans*-polyacetylene can be increased from 10^{-5} S cm^{-1} to 10^5 S cm^{-1} by exposing it to an oxidant (iodine); for comparison, the conductivity of Cu metal is 5.8×10^5 S cm^{-1}. The excitation of an electron in pure *trans*-polyacetylene is not very likely. When the polyacetylene is partially oxidized, however, empty states are created in the valence band and electrical conduction of the material can occur without the need to excite an electron through the energy gap. This effect was discovered in 1977 by MacDiarmid, Shirakava, and Heeger who were later awarded the 2000 Nobel Prize in Chemistry for it. Developments with conjugated polymers have spawned a subfield of materials science and applications, called organic electronics.

13.4.2 Silicon

Inorganic semiconductors, in particular silicon, dominated twentieth century electronics. When silicon atoms interact, the bonding between orbitals on adjacent atoms is strong and directed; i.e., the energy splitting between bonding and antibonding orbitals of a silicon dimer is much larger than the energy splitting caused by their aggregation.

In a description with electron pair bonds the Si atoms, like the carbon atoms in diamond, are connected by tetrahedral sp^3 hybrid orbitals arranged in an infinite tetrahedral pattern. Why is diamond considered an insulator, but silicon a semiconductor? The C–C bond energy is 2.54 eV (see Chapter 7), whereas the Si–Si bond energy is only 1.84 eV. The splitting between bonding and antibonding orbitals should be about two times as large, that is 5.1 and 3.7 eV, respectively. The measured band gap is 5.5 eV in diamond, but it is only 1.1 eV in silicon (Fig. 13.12). Thus the case of diamond is largely consistent with a simplified view of the bonding as localized between the C–C centers, but the case of Si is not; the estimate differs from the reality by more than three times. In the MO description bonding and antibonding orbitals split into bands, the valence band (filled orbitals at 0 K) and the conduction band (empty orbitals at 0 K). Because the Si sp^3 hybrids are not as tightly bound as those in diamond, they delocalize more in the extended structure and give rise to a band gap that is much smaller than that of the dimer (Fig. 13.12a, center), whereas the diamond band gap is similar to that of the dimer. Because of the smaller band gap, silicon is a semiconductor at room temperature; it is an insulator at 0 K. In contrast, diamond acts as an insulator at common temperatures.

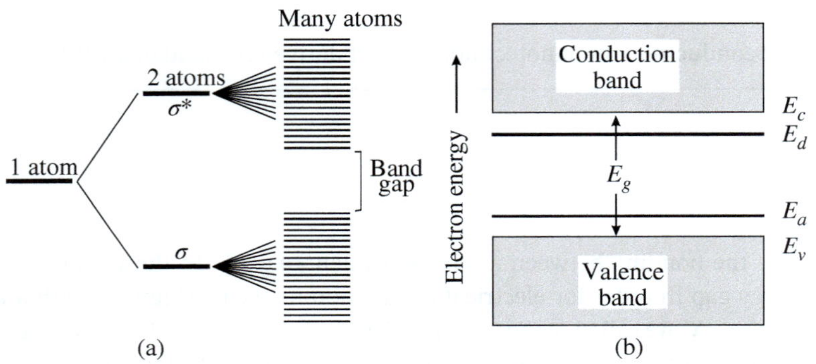

Figure 13.12 (a) Band formation in silicon (schematic). The horizontal lines correspond to the energy levels. Left: single Si atom. Center: σ and σ^* levels of two bound Si atoms. Right: σ and σ^* levels of many bound Si atoms. (b) Band diagram that illustrates dopant levels E_a and E_d. E_c and E_v are the energies of the valence and the conduction band, respectively.

13.4.3 Semiconductor Bandgap

The energy gap between the filled valence band and the empty conduction band of a pure silicon crystal is an essential feature that makes this material so useful in solid-state electronics. Because the bandgap in Si is 1.1 eV, it is possible to thermally excite a small but significant number of electrons from the valence band to the conduction band. Once an electron is excited to the conduction band a large number of unoccupied electronic states are available in the solid and it can move "freely," as with a metal. The empty state in the valence band also contributes to the conductivity because other valence band electrons can occupy it, leaving a new unoccupied state which may then be filled by another electron, and so on. Rather than considering the correlated motions of many different electrons, it is simpler to consider the empty site as a "quasi-particle," called a hole, that moves through the valence band and carries a net positive charge (opposite to that of the electron). The number of conduction band electrons, and valence band holes, that are created increases with increasing temperature; hence, the conductivity of the semiconductor increases with temperature.

13.4.3.1 Doping by a Donor

If the crystalline lattice is imperfect or the material has impurity atoms, energy states can exist between the conduction band edge, E_c and the valence band edge, E_v (Fig. 13.12b), as has been shown already in the case of a neutral soliton in Fig. 8.16. If an atom with one more valence electron than the silicon atom (e.g., a phosphorus atom in silicon) is placed in the lattice, its extra electron would not be used in bonding. Instead this electron would be localized on the impurity atom with an energy, E_d. If E_d lies close enough to the conduction band, the electron can be thermally excited into the conduction band. This type of impurity is called a *donor*. With the correct choice of dopant atom, this state lies very close to the conduction band edge and can dramatically modify the semiconductor's conductivity. Chemical control over the type of impurity atoms, concentration of impurities, and their placement in the semiconductor crystal are used to control and manipulate the crystal's electrical properties.

13.4.3.2 Doping by Acceptor

Analogous to the donor impurities, an impurity atom with one less valence electron than the semiconductor atom (e.g., aluminum atom in silicon) provides a vacant state which can be filled with an electron at an energy level, E_a, just above the valence band. This type of impurity is called an *acceptor*.

13.4.3.3 Bandgap Emission

In addition to thermal excitation, an electron can be excited from the valence band to the conduction band by absorption of a photon. Correspondingly, an excited electron in the conduction band can combine with a hole in the valence bond to generate a photon of light, called *bandgap emission*. These processes correspond directly to the optical absorption and emission that occurs between electronic states in molecules (see Chapter 10). Although the excited electron is confined in a molecular excited state, the excited electron can move very far in a bulk semiconductor; e.g., millimeters. This property underlies the use of semiconductor materials as photoconductors, photovoltaics, and photodiodes. The recombination of electrons and holes to generate light is used to construct light emitting diodes (LEDs) and diode lasers.

Example 13.5 *You plan to use a semiconductor detector to monitor the output of an erbium fiber laser operating at 1.55 μm. Is it better to use a silicon-based detector (1.12 eV bandgap) or a germanium detector (0.72 eV bandgap)?*

The energy of the a 1.55 μm photon is

$$E = \frac{hc}{\lambda} = 1.28 \times 10^{-19}\ \text{J} = 0.80\ \text{eV}$$

This energy is large enough to excite an electron from the valence band to the conduction band in Ge but not in Si, making Ge the better choice.

13.4.4 Semiconductor Quantum Dots

Semiconductors reveal a dramatic dependence of their electrical and optical properties on the size of the crystallites. Figure 13.13 shows the change in band gap energy as a function of crystallite size for CdS. The figure plots the absorbance versus the photon energy for samples of different size. The absorption peak marks the band gap energy, and it shifts to higher energy as the size of the crystallite decreases. This shift arises from "quantum confinement" of the electron in the crystallite. Such crystallites are called *quantum dots*.

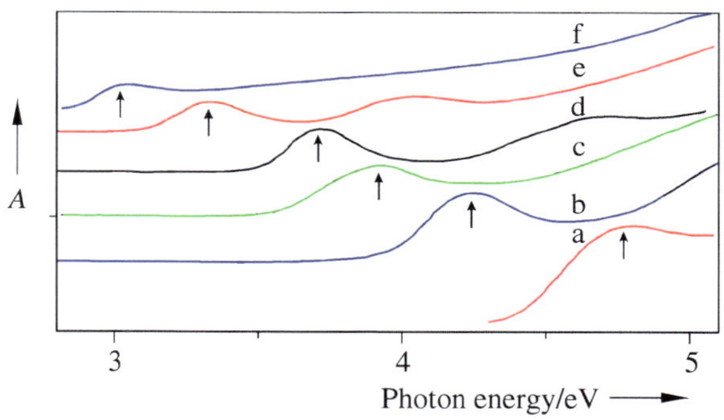

Figure 13.13 Absorption spectra of CdS crystallites of different sizes. The curves a–f correspond to the different lengths L in Table 13.3 (a: 1.28 nm, b: 1.44 nm, c: 1.60 nm, d: 1.86 nm, e: 2.32 nm, f: 3.88 nm). The absorption peaks (arrows) mark the band gap. (Adapted from [7].)

The evolution of a semiconductor's absorption energy with size can be understood [8] by considering how the energy of the excited electron differs from that of the bulk semiconductor E_{bulk}. In a bulk semiconductor, the band gap is the amount of energy needed to remove the most weakly bound electron from the valence band and place it in the conduction band, thereby generating a hole in the valence band. To simplify the analysis of the semiconductor crystallite, we approximate its electronic structure by a particle-in-a-box model with side length L.

For a semiconductor crystallite the electronic energy is modified from its bulk value in three important ways. First, the kinetic energy of the excited electron changes between the bulk and the crystallite (approximated by a cubic box). If we account for the additional kinetic energy of the electron from confinement (zero point energy) we find from Chapter 2 (energy of a particle in a cubic box)

$$T_{\text{electron}} = \frac{3h^2}{8m_eL^2} \tag{13.29}$$

A corresponding kinetic energy correction appears for the hole

$$T_{\text{hole}} = \frac{3h^2}{8m_hL^2} \tag{13.30}$$

The third correction term arises from the electrostatic attraction between the electron and the hole. This correction is approximately

$$V_{e-h} = -\frac{e^2}{4\pi\varepsilon_0\varepsilon} \cdot \frac{1}{L/2} = -\frac{2e^2}{4\pi\varepsilon_0\varepsilon L} \tag{13.31}$$

Combining these results we write the bandgap energy of the crystallite as

$$E_{gap} = E_{bulk} + \frac{3h^2}{8L^2}\left(\frac{1}{m_e} + \frac{1}{m_h}\right) - \frac{2e^2}{4\pi\varepsilon_0\varepsilon L} \tag{13.32}$$

Example 13.6 *Consider the shift of absorption energy with particle size that is found for CdS in Fig. 13.13. Compare this measured E_{gap} to the prediction of Equation (13.32). The bulk value of CdS energy gap is 2.5 eV.*

We can linearize Equation (13.32) by plotting $(E_{gap} - E_{bulk}) \cdot L$ versus $1/L$. Using the reported values of the crystallite size and the gap energy (see Table 13.3), the plot in Fig. 13.14 is generated. Fitting these data to a line gives an intercept of 0.61 eV nm and a slope of 2.8 eV nm^2, if we exclude the anomalous point for sample f. Equation (13.32) predicts an intercept of 0.50 eV nm and a slope of 2.3 eV nm^2, if we use the mass of the free electron for m_e and m_h.

Table 13.3 E_{gap} and L for Crystallites of Different Sizes.

Sample	E_{gap} (eV)	L (nm)	Sample	E_{gap} (eV)	L (nm)
a	4.78	1.28	e	3.32	2.32
b	4.22	1.44	f	3.05	3.88
c	3.89	1.60	g	2.68	5.60
d	3.72	1.86	h	2.60	9.60

In addition to their optical absorption spectra, a number of other properties of semiconductor nanocrystals change with the particle size, for example the fluorescence emission wavelength shifts with particle size and the conductivity of the quantum dot is unusual. Their optical properties make these materials interesting as fluorophores and optical sensitizers in nanostructured and microstructured devices. For their pioneering work on quantum dots M. G. Bawendi, L. E. Brus and A. Yekimov were awarded the 2023 Nobel prize in Chemistry.

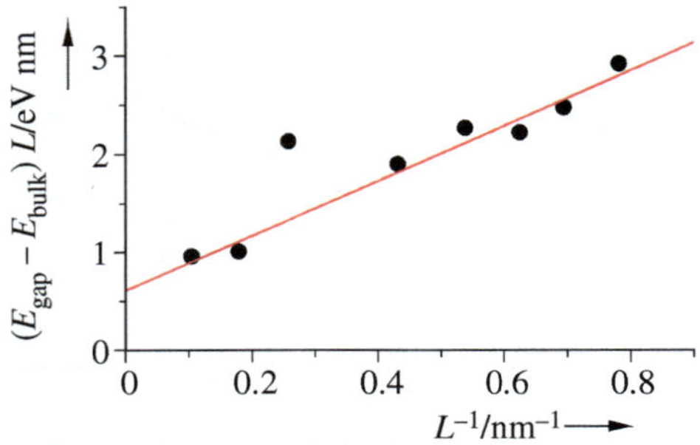

Figure 13.14 Evaluation of the data from Fig. 13.13 according to Equation (13.32). $(E_{gap} - E_{bulk}) \cdot L$ is plotted as a function of $1/L$ with E_{bulk} = 2.5 eV. Data are taken from Table 13.3.

13.5 Molecular Crystals and Intermolecular Forces

The energy stabilization in forming a molecular crystal is much smaller than that for metallic, semiconducting, and ionic crystals. The smaller bond energy reflects the fact that bonding and ion–ion interactions do not contribute to formation of the solid, rather the binding is driven by dispersion forces, induction forces, dipole–dipole forces, and higher multipole electrostatic forces (such as quadrupole and octupole).

Consider the case of two interacting ammonia molecules. As the molecules approach each other in space but are far enough away to not have significant overlap of their electron clouds, they interact electrostatically through their charge distributions. Because ammonia has a dipole moment, the most important term arises from the interaction of the molecular dipole moments with each other (see Section 13.5.1). Also contributing to the interaction are the induction force, which arises from the dipole on one NH_3 molecule polarizing the electron cloud of the other molecule (see Section 13.5.3), and the dispersion force, which arises from the polarizable electron clouds interacting with each other (see Section 13.5.4). Figure 13.15 shows how each of these interactions depends on the distance between ammonia molecules. When two NH_3 molecules are nearly in "contact" (distance d = 341 pm), we obtain

$$E = E_{dipole} + E_{ind} + E_{disp} = -(0.11 + 0.0062 + 0.045) \times 10^{-19}\ \text{J} = -0.16 \times 10^{-19}\ \text{J}$$

At short distances the overlap of the filled electron shell on each molecule causes a repulsive force that balances the attractive forces. Microwave spectra show that the ammonia dimer has an experimental attraction energy of -0.19×10^{-19} J and an equilibrium distance of 341 pm [9].

According to Equation (13.7) the energy of an ion pair with the same distance $d = 300$ pm is $E = -12 \times 10^{-19}$ J; a factor of 60 larger. Therefore, an NH_3 crystal is much less stable than a NaCl crystal; consequently, NH_3 is a gas at room temperature. The polarizability α increases strongly with the size of a molecule and many organic molecules (e.g., naphthalene and anthracene) form stable crystals at room temperature, even though they do not possess a net charge or permanent dipole moment. A complex composed of a molecule with electron donor properties and an electron acceptor molecule (e.g., quinhydrone, a molecular crystal of quinone and hydroquinone) can be additionally stabilized by partial charge transfer (charge transfer complex). Although small, the intermolecular forces that give rise to binding in molecular crystals are very important for understanding the interactions between molecules, the solvation forces in solutions and molecular assemblies, and biological structures—hence they are of great importance in chemistry and biology.

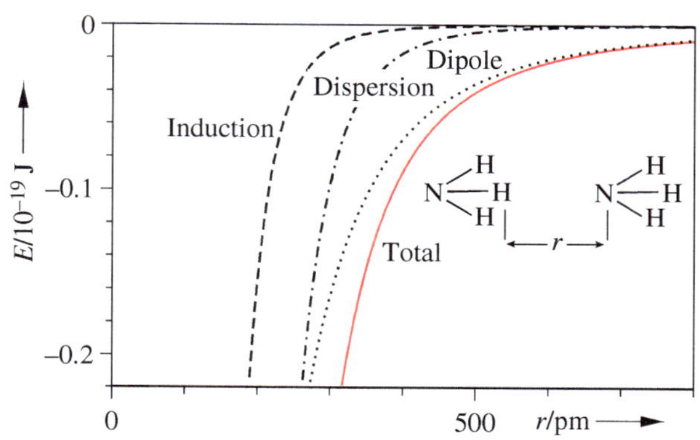

Figure 13.15 Attraction energy as a function of the distance r for NH_3 molecules; comparison of the contributions of the dipole energy, induction energy, and dispersion energy (dashed and dotted lines). Solid line: total attraction energy. The two NH_3 molecules are considered to approach each other in the geometry shown in the figure.

13.5.1 Electrostatic Forces, the Dipole

Consider two dipoles arranged according to Fig. 13.16a. What is their dipole–dipole attraction energy? This energy is the sum of all attraction and repulsion energies between the point charges Q_1 and Q_2 on each molecule

$$E = \frac{1}{4\pi\varepsilon_0} \cdot \left[\frac{Q_1Q_2}{r} + \frac{Q_1Q_2}{r} - \frac{Q_1Q_2}{(r+a)} - \frac{Q_1Q_2}{(r-a)}\right]$$
$$= \frac{1}{4\pi\varepsilon_0} \cdot Q_1Q_2 \cdot \left[\frac{2}{r} - \frac{1}{(r+a)} - \frac{1}{(r-a)}\right] \tag{13.33}$$

In Exercise E13.22, you consider the limit where $a \ll r$, and show that the energy is

$$E = -\frac{1}{4\pi\varepsilon_0} \cdot \frac{2\mu_1\mu_2}{r^3} \tag{13.34}$$

where $\mu_1 = Q_1 a$ and $\mu_2 = Q_2 a$ are the dipole moments in the point dipole approximation. The energy between two dipoles decreases much more rapidly with distance r than does the energy between two point charges. The force exerted between the dipoles is

$$f = \frac{dE}{dr} = \frac{1}{4\pi\varepsilon_0} \cdot \frac{6\mu_1\mu_2}{r^4} \tag{13.35}$$

in the point dipole approximation.

Dipole moments are often reported in units of Debye (D) rather than SI units. One Debye unit is equal to 3.33564×10^{-30} C m.

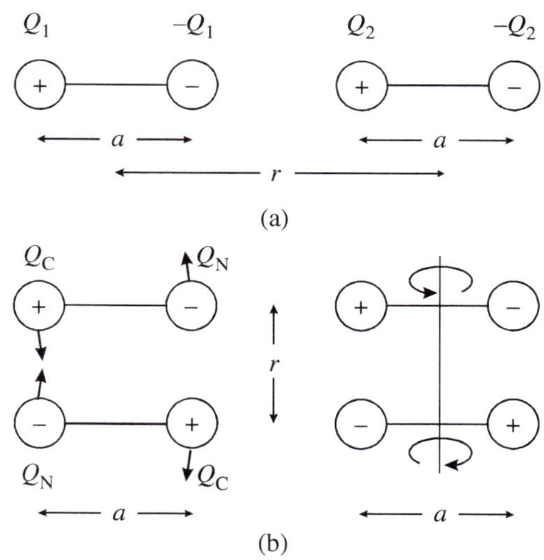

Figure 13.16 Attraction of two dipoles (charge Q, dipole length a). (a) Dipoles arranged linearly. (b) Dipoles arranged parallel, example of the C≡N bond in acetonitrile, CH_3CN (Q_C = partial charge on carbon atom, Q_N = partial charge on nitrogen atom); (left) rocking mode, (right) twisting mode.

Example 13.7 *Calculate the dipole–dipole interaction energy for two acetonitrile molecules that are antiparallel but displaced by a distance of $r = 550$ pm, in the point dipole approximation and in the extended dipole calculation. Approximate the molecule as having its dipole moment localized on the CN bond with partial charges of $Q_C = 0.704e$ and $Q_N = -0.704e$, and a CN bond length of $d = 116$ pm.*

Figure 13.16b shows the geometry for this particular acetonitrile dimer. The partial charges and CN bond length give a dipole moment of $\mu = 160\text{ pm} \cdot 0.704 \cdot 1.602 \times 10^{-31}\text{ C} = 1.31 \times 10^{-19}$ C m, which is 3.92 D. Using the result in Box 13.4

(with $\beta = 90°$), we find

$$E = -\frac{1}{4\pi\varepsilon_0}\frac{\mu^2}{r^3} = 8.99\times10^{9}\ \mathrm{C^{-2}\ J\ m}\cdot\frac{(1.31\times10^{-29}\ \mathrm{C\ m})^2}{(550\times10^{-12}\ \mathrm{m})^3}$$
$$= -0.927\times10^{-20}\ \mathrm{J}$$

The distance 550 pm is fairly typical of that found for liquid acetonitrile, and the interaction energy is significant (two to three times thermal energy at room temperature).

Now we compare this point dipole result to that found by considering the discrete charges in our model geometry. In analogy to Equation (13.33), we sum the interaction between the point charges on each site of the dipole with those on the neighboring dipole. According to Fig. 13.16 we find

$$E = \frac{1}{4\pi\varepsilon_0}\left[2\cdot\frac{Q_N Q_C}{r} + \frac{Q_C Q_C}{\sqrt{r^2+a^2}} + \frac{Q_N Q_N}{\sqrt{r^2+a^2}}\right]$$

With the values $Q_C = 0.704e = 1.14\times10^{-19}$ C, $Q_N = -0.704e = -1.14\times10^{-19}$ C, $r = 550\times10^{-12}$ m, and $a = d = 116\times10^{-12}$ m, we obtain

$$E = 8.99\times10^{9}\ \mathrm{C^{-2}\ J\ m}\cdot\left[-4.73\times10^{-29} + 2\times2.31\times10^{-29}\right]\mathrm{C^2\ m^{-1}}$$
$$= -0.989\times10^{-20}\ \mathrm{J}$$

which is 10% larger than the point dipole value.

The energy of interaction between two CH_3CN molecules is large enough that they arrange as a pair when trapped in a solid argon matrix at 14 K. The molecules are no longer independent, and the interaction strength is large enough that the frustrated translational and rotational motions give rise to quantized vibrational motions (shown in Fig. 13.16b). Infrared absorption bands, corresponding to these vibrations, which change the total dipole moment, are observed [10] at $1/\lambda = 80$ and 130 $\mathrm{cm^{-1}}$. Because these quantized motions are physically related to the relative rotation or translation of the individual molecules, they are often referred to as *librations*, rather than *vibrations*.

In the case of the acetonitrile example, the distance between the dipoles is about five times the dipole length. Therefore, it is not surprising that the point dipole model gives nearly the correct result for the dipole–dipole energy. In the following example the situation is quite different; namely, the distance between the dipoles is about one-half of the dipole length.

Example 13.8 *Consider the dye molecule in Fig. 13.17a. It has a dipole moment $\mu = 2.0$ Debye $= 6.67\times10^{-30}$ C m. We approximate the dipole of the molecule by two charges $Q_1 = Q$ and $Q_2 = -Q$ at a distance $a = 5\times140$ pm $=$ 700 pm Fig. (13.17b). From $\mu = Qa$ we find $Q = \mu/a = 6.67\times10^{-30}$ C m/700 pm $= 9.53\times10^{-21}$ C. Calculate the dipole–dipole energy of two such dye molecules in contact at $r_1 = 400$ pm Fig. (13.17c), using the point dipole approximation and the extended dipole approximation.*

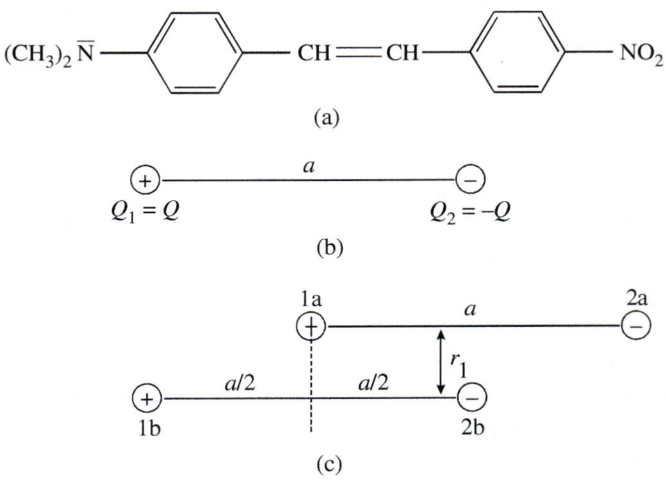

Figure 13.17 Interaction of extended dipoles. (a) Structural formula of dipole molecule. (b) Molecule modeled as an extended dipole. (c) Two extended dipoles at a distance r_1.

Point dipole approximation: from Fig. 13.18 we calculate the angle β. Because of $r_1 = r\cdot\sin\beta$ and $a/2 = r\cdot\cos\beta$ we obtain $\tan\beta = 2r_1/a = 800/700 = 1.14$ and $\beta = 48.7°$. From Box 13.4 it follows that

$$E = -\frac{1}{4\pi\varepsilon_0}\frac{\mu^2}{r^3}\left(3\cos^2\beta - 1\right)$$
$$= -6.24\times10^{-21}\cdot(+0.307)\ \mathrm{J}$$
$$= -1.92\times10^{-21}\ \mathrm{J}$$

Extended dipole approximation:

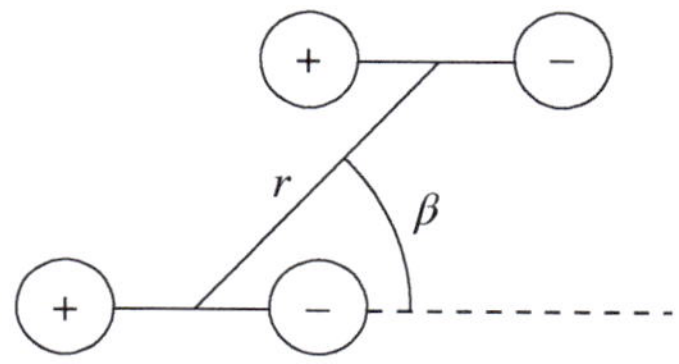

Figure 13.18 Orientation of two dipoles with arbitrary angle β.

$$
\begin{aligned}
E &= \frac{1}{4\pi\varepsilon_0}\sum_i\sum_j \frac{Q_j Q_i}{r_{ij}} \\
&= \frac{1}{4\pi\varepsilon_0}\left[\frac{Q_1Q_1}{r_{1a,1b}} + \frac{Q_1Q_2}{r_{1a,2b}} + \frac{Q_2Q_1}{r_{2a,1b}} + \frac{Q_2Q_1}{r_{1a,2b}}\right] \\
&= \frac{1}{4\pi\varepsilon_0}Q^2\left[+\frac{1}{r_{1a,1b}} - \frac{1}{r_{1a,2b}} - \frac{1}{r_{2a,1b}} + \frac{1}{r_{1a,2b}}\right]
\end{aligned}
$$

From the geometry in (c) it follows that $r_{1a,1b} = r_{1a,2b}$, $r_{2a,1b} = \sqrt{(1.5a)^2 + r_1^2} = 1124$ pm, and $r_{1a,2b} = \sqrt{(0.5a)^2 + r_1^2} =$ 532 pm. Then

$$
\begin{aligned}
E &= 8.17 \times 10^{-31}\ \mathrm{J\,m} \cdot \left[-8.90 \times 10^8 + 18.8 \times 10^8\right]\mathrm{m}^{-1} \\
&= +0.81 \times 10^{-21}\ \mathrm{J}
\end{aligned}
$$

Note that in this case the point dipole model gives a completely wrong result predicting that the sign of the interaction energy is negative (attraction of the dipoles), in contrast to the positive sign (repulsion of the dipoles) calculated from the extended dipole model.

For large polyatomic molecules packed closely in a solid, the point dipole or molecular dipole approximation is not sufficiently accurate. Rather, one must compute the electrostatic energy as a sum over the Coulomb interaction between each of the atomic partial charges of the interacting molecules; i.e.,

$$
E_{\text{electrostatic}} = \frac{1}{4\pi\varepsilon_0}\sum_i\sum_j \frac{Q_j Q_i}{r_{ij}}
$$

With modern computational tools, such summations can be readily performed.

13.5.2 Hydrogen Bonds

Hydrogen bonds are particularly important in processes where the formation and disruption of weak bonds is crucial, including fundamental processes in biosystems. Hydrogen bonds are believed to be essential for base pairing in DNA (Fig. 13.19).

Figure 13.19 Base pairing in nucleotides. The hydrogen bonds for the base pairs are indicated by dashed lines. (a) Uracil/adenine pair. (b) Cytosine/guanine pair.

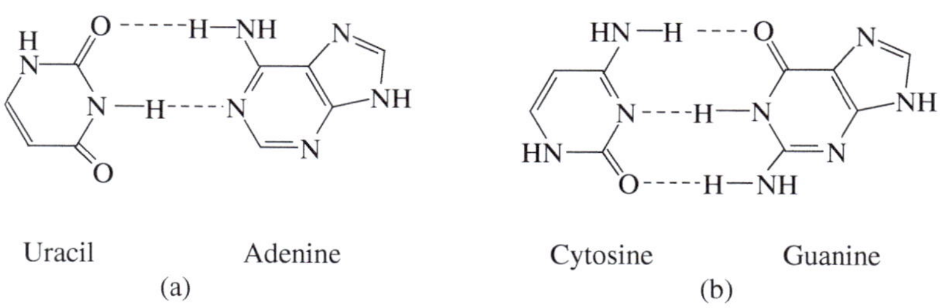

In a very first approximation hydrogen bonds in molecules with hydrogen atoms bound to strongly electronegative atoms (e.g., NH_3 and H_2O) can be treated as being caused by electrostatic interaction. In general, the shift of electronic charge from the donor to the acceptor makes a considerable contribution to the bond energy and a treatment on a fundamental basis (DFT or MO theory) is required for an understanding.

First we consider an interaction of HF and H_2O molecules by using an electrostatic model. In such systems, the electronegative atoms have a high negative charge, and the van der Waals radius of the hydrogen atom is small. These two features cause such a strong dipole interaction that the H-atom of one molecule is preferentially oriented toward an unshared electron pair of a second molecule (hydrogen bond). For example, F–H forms such a strong hydrogen bond with F^- (bond energy 4×10^{-19} J) that in the energy minimum the H nucleus is at the center between the F nuclei. Thus this aggregate can vibrate at frequencies comparable with frequencies of molecules with covalent bonds (see Exercise E13.23) [11].

Hydrogen bonds (Fig. 13.20) are very important for understanding the structure of liquid water, hence much aqueous chemistry and biology. In the water dimer the OH distance between the oxygen on one water and the H atom to which it hydrogen bonds is 180 pm [12]. This distance is much smaller than the sum of the van der Waals radii of O and H (140 +120 pm = 260 pm), but larger than the covalent O–H bond distance (97 pm). The electrostatic energy gain in this example is about 0.5×10^{-19} J. Typically, the energy of a hydrogen bond is $\lesssim 1 \times 10^{-19}$ J, less than 10% of typical covalent bond energies (Table 7.1).

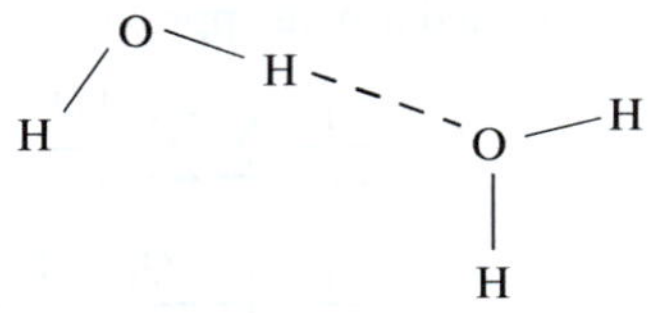

Figure 13.20 Hydrogen bond in the water dimer.

Hydrogen bonds affect many properties of water. The directional bonding and the tendency of H_2O molecules to aggregate in a tetrahedral arrangement generate its open structure. The low density of ice and its high melting point are caused by the stability of the three-dimensional tetrahedral network. When ice melts, some of these bonds are broken, and a denser packing of the H_2O molecules is possible. As a result, the density of water is higher than that of ice at the normal freezing point, and it further increases after melting up to a temperature of 4 °C. The strong attractive interactions between water molecules give rise to the high boiling temperature and heat of vaporization for water.

Example 13.9 *Gas phase spectra of the hydronium ion H_3O^+ find a symmetric structure with HOH bond angles of 113.6° and OH bond lengths of 97.45 pm. Use this geometry with the partial charges in Fig. 13.21 to calculate the electrostatic binding energy of the proton to the water molecule. Compare this result to that found for binding of the proton to the water dipole of 1.84 D.*

Using the distances and partial charges in Fig. 13.21, the electrostatic bond energy of the proton to the water molecule is

$$E = \frac{e^2}{4\pi\epsilon_0}\left[\frac{-0.72}{97.4 \times 10^{-12}} + 2\frac{0.36}{163.0 \times 10^{-12}}\right]$$
$$= -6.86 \times 10^{-19}\ \text{J} = -4.28\ \text{eV}$$

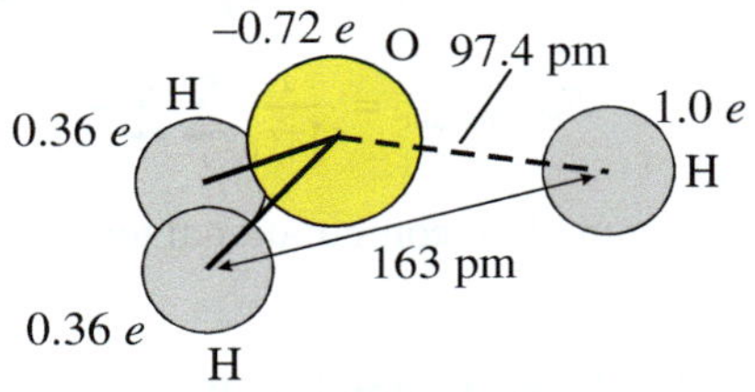

Figure 13.21 Distances in a hydronium ion H_3O^+. Yellow circle: Oxygen atom, gray circles: hydrogen atoms.

For comparison, consider the charge–dipole interaction between a point dipole μ and a charge Q is (see Box 13.4)

$$E = \frac{1}{4\pi\epsilon_0}\frac{Q\mu}{r^2}\cos\theta \tag{13.36}$$

where θ is the angle between the dipole and the line connecting it to the charge and r is their distance. For the proton bound to the water molecule we find $\theta = 149.9°$ and estimate $r \approx 120$ pm to find

$$E = \frac{1.602 \times 10^{-19} \cdot 1.84 \cdot 3.3356 \times 10^{-30} \cdot \cos(149.9)}{4\pi \cdot 8.854 \times 10^{-12} \cdot \left(120 \cdot 10^{-12}\right)^2}$$
$$= -5.31 \times 10^{-19}\ \text{J} = -3.31\ \text{eV}$$

Although our calculation does not compare the bond energy to a water molecule in its equilibrium geometry and charge distribution, comparison with the experimental value for the proton affinity of a water molecule in the gas phase (7.16 eV = 691 kJ mol^{-1}) shows a qualitative agreement; i.e., the bond energy is significant, 5–10 eV.

An example for a density functional theory (DFT) treatment is given in Fig. 13.22. Systems a–f are considered where the donor is identical, but the acceptor differs in the substituent. In this way, the charge surrounding the oxygen atom is manipulated.[2] The H-bond energies calculated by a DFT method are plotted versus the charge on the oxygen. A linear relationship is observed: the bond energy increases in proportion to the charge surrounding the oxygen.

2 This follows from ESCA measurements of the 1s orbital of oxygen. The energy to remove an electron from this orbital depends on the environment. This method is a direct measurement of the oxygen's charge.

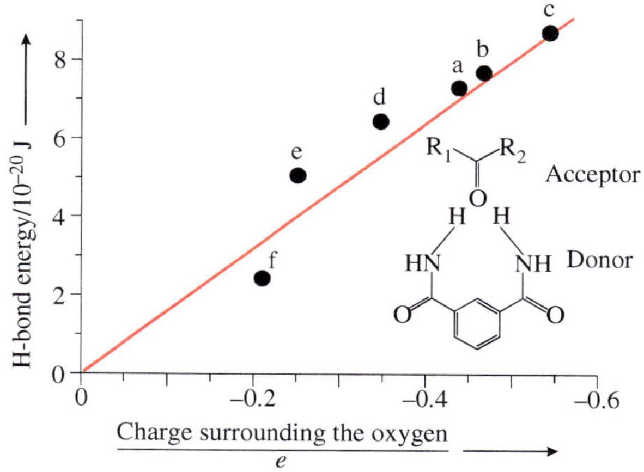

Figure 13.22 H-bond energy versus charge surrounding the oxygen atom of the acceptor molecule. (a) R_1=NH_2, R_2=H, (b) R_1=$(CH_3)_2N$, R_2=H, (c) R_1=$(CH_3)_2N$, R_2=$N(CH_3)_2$, (d) R_1=CH_3, R_2=CH_3, (e) R_1=R_2=H, (f) R_1=R_2=CF_3. (Data from Ref. [12] (bonding energies from Table 1, column 1, charges from Table 3, column 1).)

13.5.3 Induction Forces

Attractive forces, called induction forces, can occur when a permanent dipole on one molecule induces a dipole moment μ_{ind} on a second molecule, or atom (Fig. 13.23). When the electric field from a dipole moment acts on a polarizable electron cloud, it shifts the center of negative charge relative to the nuclear charge (see Fig. 13.24, left). This asymmetry gives rise to an induced dipole moment.

If F is the strength of the electric field (originating from a permanent dipole) at the position of a polarizable molecule, then

$$\mu_{\text{ind}} = \alpha \cdot F \tag{13.37}$$

where α is the *polarizability* of the molecule. This relation defines the polarizability. The polarizability describes an electron cloud's ability to deform when an electric field F acts on it.

The value found experimentally for the polarizability of an H-atom is

$$\alpha = 4\pi\varepsilon_0 \cdot 0.67 \times 10^{-24}\ \text{cm}^3 = 4\pi\varepsilon_0 \cdot 4.5\ a_0^3 \tag{13.38}$$

As a point of reference, we can compare this polarizability to that expected for an idealized metal sphere of a radius like the H-atom. Box 13.5 shows that a metal sphere of radius r has a polarizability of

$$\alpha = 4\pi\varepsilon_0 \cdot r^3 \tag{13.39}$$

Using the size of an H-atom ($r = 125$ pm $= 2.4a_0$, obtained from the collision diameter, see Table 11.1), we find

$$\alpha = 4\pi\varepsilon_0 \cdot (2.4a_0)^3 = 4\pi\varepsilon_0 \cdot 14a_0^3 \tag{13.40}$$

This value is about three times larger than the measured polarizability of an H-atom, a moderate difference. An He atom is smaller than an H-atom (Fig. 2.13), so its polarizability is smaller. In general, the polarizability increases with increasing size of the atom or molecule (see Table 13.4). This observation is consistent with Equation (13.39), which predicts that α is proportional to the volume.

To calculate the energy caused by the attraction of a polarizable molecule to a permanent dipole, we assume that the strength of the electric field F at a distance r from a point charge Q is

$$F = \frac{1}{4\pi\varepsilon_0} \cdot \frac{Q}{r^2} \tag{13.41}$$

Thus the field strength of the dipole in Fig. 13.23 at the position of the polarizable molecule is

$$F = \frac{1}{4\pi\varepsilon_0} \cdot \left[\frac{+Q}{(r-a/2)^2} + \frac{-Q}{(r+a/2)^2}\right] \tag{13.42}$$

For $a \ll r$ (point dipole approximation) this expression simplifies to (see Box 13.4)

$$F = \frac{1}{4\pi\varepsilon_0} \cdot \frac{2\mu}{r^3} \tag{13.43}$$

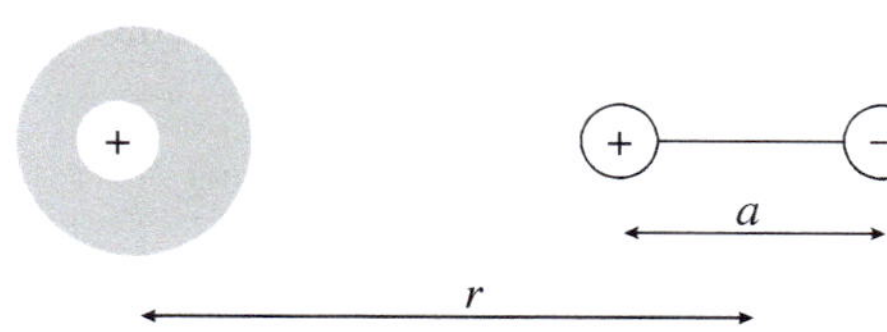

Figure 13.23 Interaction between a dipole and a polarizable molecule.

Using Equations (13.35) and (13.37), the attraction force f between the permanent dipole and the polarizable molecule with $\mu_1 = a \cdot Q$ and $\mu_2 = \mu_{\text{ind}}$ is

$$f = \frac{1}{4\pi\varepsilon_0} \cdot \frac{6\mu \cdot \mu_{\text{ind}}}{r^4} = \frac{1}{(4\pi\varepsilon_0)^2} \cdot \frac{12\mu^2 \cdot \alpha}{r^7} \tag{13.44}$$

The energy E is obtained by integration:

$$E = \int_{\infty}^{r} f \cdot \mathrm{d}r = -\frac{1}{(4\pi\varepsilon_0)^2} \cdot \frac{2\mu^2 \cdot \alpha}{r^6} \tag{13.45}$$

Example 13.10 *Calculate the induction energy for an ammonia dipole interacting with a neighboring ammonia molecule at the distances of 341 and 682 pm.*

The dipole moment of an ammonia molecule is 1.47 D and its polarizability is about 2.5×10^{-40} C^2 m^2 J^{-1}. Using Equation (13.45), we find

$$E = \frac{2 \times (1.47 \times 3.3356 \times 10^{-30})^2 \times 2.5 \times 10^{-40}}{(4\pi \times 8.854 \times 10^{-12})^2 \times r^6} \text{J m}^6$$
$$= 9.7 \times 10^{-79}/r^6 \text{ J m}^6$$

which gives 6.2×10^{-22} J = 3.9 meV at 341 pm and 9.6×10^{-24} J at 1800 pm. These energies are quite small. For large organic molecules with polarizabilities that are 20–50 times larger, these interactions can be important.

13.5.4 Dispersion Forces

From our considerations it is clear that two HCl molecules or an HCl molecule and an H_2 molecule attract each other; however, it is not evident that two He atoms will attract each other. Experimentally we do find that two He atoms attract each other, even though a permanent dipole moment is not present and the charge distribution is spherically symmetric.

How can we explain this fact? Both electrons of the He-atom occupy a spherically symmetric 1*s* orbital; thus, the center of positive charge and the center of negative charge coincide, and no permanent dipole manifests. In slow processes (like the shift of nuclei by a distance of the size of an atom), the electrons behave as if they were distributed over a cloud (see Chapter 2). If, however, we consider fast processes (like the shift of two electrons toward each other), the momentary position of the electrons becomes important and can explain the attraction between two He atoms.

To simplify the analysis, consider the model in Fig. 13.24. This diagram shows a "snapshot" of atom 1 with an asymmetric arrangement of the electrons and nucleus that generates a momentary dipole moment (dipole moment $\mu_{\text{mom},1}$). This instantaneous dipole polarizes the electron cloud of atom 2, inducing a dipole moment $\mu_{\text{mom},2}$ and generating an attraction force between the two atoms. Panels (a) and (b) show that an attraction results for either orientation of the instantaneous dipole.

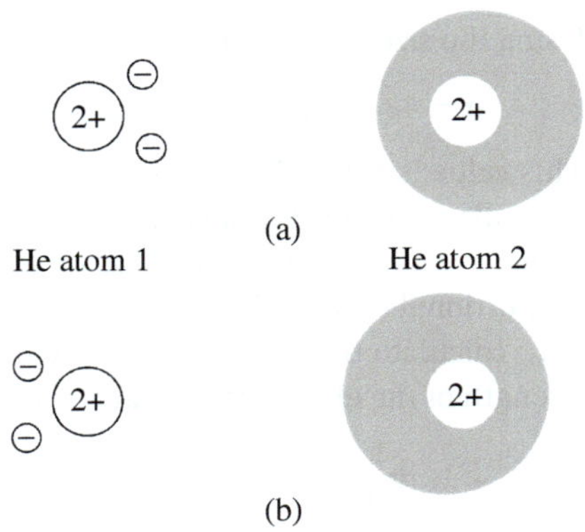

Figure 13.24 . He atom 2 in different instantaneous dipole fields of He atom 1; i.e., the drawings in panels (a) and (b) represent two different fluctuations of the charge density.

Now we use the model to estimate the interaction energy between two He atoms. Let us assume that the momentary dipole moment of He atom 1 is on the order of

$$\mu_{\text{mom},1} = e \cdot a_0$$

Using this dipole moment with Equation (13.44) we find the induction force it creates with He atom 2 to be

$$f = \frac{1}{(4\pi\varepsilon_0)^2} \cdot \frac{12\,\mu_{\text{mom},1}^2 \cdot \alpha_2}{r^7} \tag{13.46}$$

where α_2 is the polarizability of He atom 2.

In order to generalize later, it is advantageous to express the momentary dipole moment in terms of the atom's polarizability and its ionization energy. We take the polarizability of a helium atom to be

$$\alpha_1 = 4\pi\varepsilon_0 \cdot a_0^3$$

which is smaller than that of an H-atom. According to Section 5.3, Equation (5.17), the ionization energy is

$$E_{\mathrm{Ion},1} = \frac{1}{4\pi\varepsilon_0} \cdot \frac{3}{4} \cdot \frac{e^2}{a_0}$$

Combining these two results we see that $\alpha_1 E_{\mathrm{Ion},1} = 3e^2 a_0^2/4$, and we can write

$$\mu^2_{\mathrm{mom},1} = 4\alpha_1 E_{\mathrm{Ion},1}/3 \tag{13.47}$$

Using this result, the force f in Equation (13.46) can be written as

$$f = \frac{1}{(4\pi\varepsilon_0)^2} \cdot \frac{9\alpha_1 E_{\mathrm{Ion},1} \cdot \alpha_2}{r^7} \tag{13.48}$$

We can integrate this force to determine the energy for this interaction and find that

$$E_{\mathrm{disp}} = \int_{\infty}^{r} f \cdot \mathrm{d}r = -\frac{1}{(4\pi\varepsilon_0)^2} \cdot \frac{3}{2} \cdot \frac{\alpha_1 \cdot \alpha_2}{r^6} E_{\mathrm{Ion},1} \tag{13.49}$$

A more rigorous treatment for the dispersion energy [13] results in the expression

$$E_{\mathrm{disp}} = -\frac{1}{(4\pi\varepsilon_0)^2} \cdot \frac{3}{2} \cdot \frac{\alpha_1 \cdot \alpha_2}{r^6} \frac{E_{\mathrm{Ion},1} E_{\mathrm{Ion},2}}{E_{\mathrm{Ion},1} + E_{\mathrm{Ion},2}} \tag{13.50}$$

Since $\alpha_1 \cdot \alpha_2$ appears in the expression for the dispersion energy, this energy is much smaller for the He atom than for H_2 molecules, even though the ionization energy is twice as large. Indeed He, in contrast to H_2, does not solidify even at the lowest possible temperatures under atmospheric pressure, and it has a unique structure at low temperatures. Figure 13.25 shows the dispersion energy of H, He, and Xe as a function of the distance r (data see Table 13.4). This energy decreases very rapidly with increasing distance.

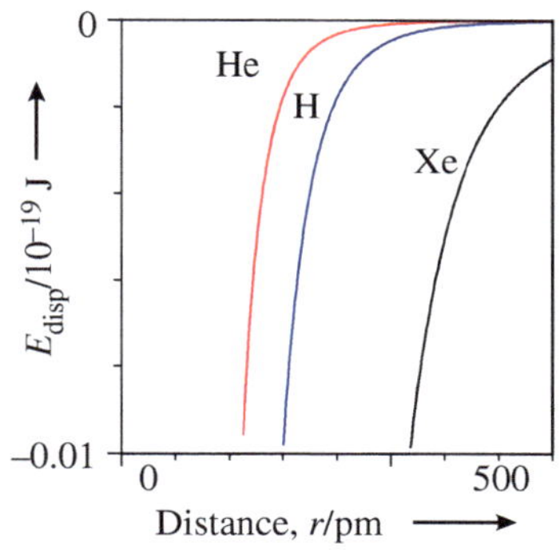

Figure 13.25 Dispersion energy E_{disp} as a function of the distance r.

Table 13.4 Dipole Moment μ, Polarizability α, Ionization Energy E_{Ion}, and Collision Diameters of Some Atoms and Molecules.[3]

	μ				
	10^{-30} C m	Debye	$\alpha/(4\pi\varepsilon_0)$ 10^{-30} m³	E_{Ion} 10^{-19} J	Collision Diameter/pm
H	0	0	0.67	21.8	256
He	0	0	0.21	39.2	220
Ne	0	0	0.39	41.2	260
Ar	0	0	1.63	28.0	370
Kr	0	0	2.46	23.5	420
Xe	0	0	4.0	19.5	495
H_2	0	0	0.81	23.2	276
HCl	3.42	1.03	2.63	21.5	450
HI	1.26	0.38	5.4	16.8	530
NH_3	4.97	1.5	2.24	18.7	450
H_2O	6.11	1.84	1.48	28.8	462

Example 13.11 *Calculate the dispersion energy between two ammonia molecules at the distance of 341 pm.*

Using Equation (13.50) and the values in Table 13.4, we find

$$E_{\mathrm{disp}} = -\frac{3 \cdot (2.24 \times 10^{-30})^2}{2 \cdot (341 \times 10^{-12})^6} \frac{18.7 \times 10^{-19}}{2} \mathrm{J}$$

$$= -\frac{3 \cdot (1.12)^2 \cdot 18.7}{(3.41)^6} \times 10^{-19}\ \mathrm{J} = 4.5 \times 10^{-21}\ \mathrm{J}$$

3 The dipole moments are given in SI units as well as in Debye units (1 Debye = $3.33.10^{-30}$ Cm)

In Fig. 13.15 the contributions of dipole energy (going over to hydrogen bond energy at small distances), induction energy, and dispersion energy are shown for NH_3 molecules. Even though the dipole moment of NH_3 is large, the dispersion energy is about as large as the dipole energy at small distances. At larger distances, only the dipole forces play a role.

According to our considerations the polarizability α increases strongly with the size of the molecule; this means that large organic molecules (e.g., naphthalene, anthracene) can readily form crystals at room temperature, even if they do not possess a permanent dipole moment.

13.6 Conclusion

The diverse forms and properties of solids may be understood by considering the covalent and electrostatic bond energies between atoms and molecules. We began our discussion with the case of atomic solids. For closed shell ions the solid takes on a form that optimizes the electrostatic attraction between the cations and anions. Because the electrons are strongly bound about each ion and not delocalized, these materials are hard and insulating. Next we discussed the opposite limit in which the valence electrons of atoms are delocalized throughout the solid, which gives rise to the ductility and conductivity of metals. Lastly we discussed the intermediate case of covalent, elemental solids (carbon and silicon) for which the electrons are bound strongly between the atomic cores but have some delocalization throughout the solid. A model in which a band of molecular orbitals is formed in the solid explains the electronic properties of these interesting materials very well.

In the latter part of the chapter, we discussed noncovalent interactions between atoms and molecules; in particular dipole–dipole, dipole-induced dipole, and dispersion forces were discussed. These forces are very important in understanding how two molecules interact with one another and form complexes. The assembly of molecular crystals and "soft" materials (membranes, micelles, etc.) arises from these forces and can be controlled by manipulating the properties of the molecules.

13.7 Box 13.1. Madelung Constant

Consider an infinite one-dimensional arrangement of NaCl ion pairs, where d is the distance between two adjacent ions. At one end we add a new ion pair (Fig. 13B.1). First we add an Na^+ at the end of the infinite chain. The energy required is the energy of the Na^+ in the electric field of the ions already present in the arrangement

$$E_+ = \frac{1}{4\pi\varepsilon_0}\frac{e^2}{d} \cdot \left[-1 + \frac{1}{2} - \frac{1}{3} + \frac{1}{4} - \frac{1}{5} + \cdots\right]$$

To evaluate the infinite sum in the brackets we consider the power series

$$\ln(1+x) = x - \frac{1}{2}x^2 + \frac{1}{3}x^3 - \cdots$$

For $x = 1$ we obtain

$$\ln 2 = 1 - \frac{1}{2} + \frac{1}{3} - \frac{1}{4} + \frac{1}{5} + \cdots$$

Then

$$E_+ = -\frac{1}{4\pi\varepsilon_0}\frac{e^2}{d} \cdot \ln 2 = -\frac{1}{4\pi\varepsilon_0}\frac{e^2}{d} \cdot 0.693$$

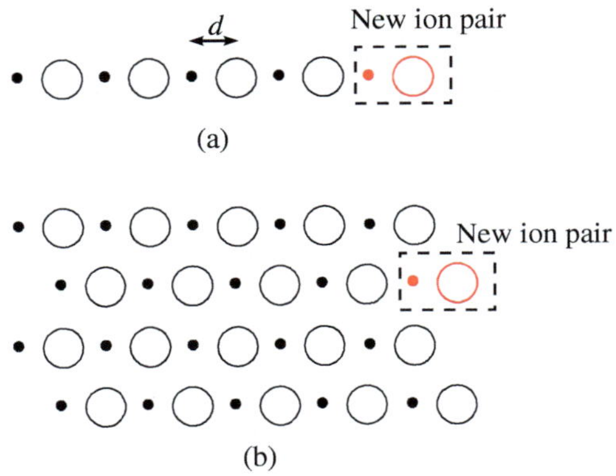

Figure 13B.1 Calculation of the Madelung constant. (a) One-dimensional arrangement of positively and negatively charged ions, where d is the distance between two adjacent ions. (b) Two-dimensional arrangement.

Adding the Cl^- in a second step requires the same energy. Then the energy to add one ion pair is

$$E = E_+ + E_- = -\frac{1}{4\pi\varepsilon_0}\frac{e^2}{d} \cdot 1.386$$

Correspondingly, we proceed for a two-dimensional and a three-dimensional arrangement: In these cases the sums are evaluated numerically. Instead of the factor 1.386 we obtain 1.614 (two-dimensional) and 1.748 (three-dimensional).

13.8 Box 13.2. Number and Density of Energy States of a Metal's Electron Gas

For simplicity we consider a cube of volume L^3. The energy levels are described by the equation

$$E_{n_x,n_y,n_z} = \frac{h^2}{8mL^2}\left(n_x^2 + n_y^2 + n_z^2\right) \tag{13B2.1}$$

The quantum numbers (n_x, n_y, n_z) specify a single quantum state. How many quantum states have energies below a certain energy E? All states with

$$R^2 = n_x^2 + n_y^2 + n_z^2 = E \cdot \frac{8m_eL^2}{h^2} \tag{13B2.2}$$

belong to this group.

In Fig. 13B.2 each of these states is represented by the vertices of the cubes, and the corresponding energies are given by the square of length of the vector connecting each vertex with the origin. For all quantum states under consideration this length is smaller than or equal to R. That is, the total number of the quantum states under consideration, g_{total}, is given by the number of vertices within the volume of $\frac{1}{8}$ of the sphere of radius R (the fraction $\frac{1}{8}$ arises because the quantum numbers are all positive and the spherical shell only occupies one-eighth of the whole sphere). Thus

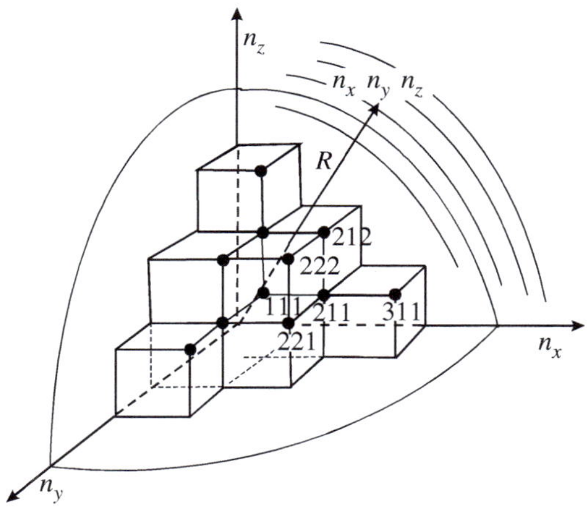

Figure 13B.2 Counting the number g of energy levels. The numbers on the cubes denote the quantum numbers n_x, n_y, and n_z.

$$g_{\text{total}} \approx \frac{1}{8}\frac{4\pi R^3}{3} = \frac{1}{8}\frac{4\pi}{3}\left(E\frac{8m_eL^2}{h^2}\right)^{3/2} = \frac{\pi}{6}\left(\frac{8m_eL^2}{h^2}\right)^{3/2} E^{3/2} \tag{13B2.3}$$

g_{total} is very large and therefore can be considered as a continuous function of E. For example, for a cube of volume $L^3 = 1\ \text{cm}^3$ from Equation (13B2.3) we obtain

$$g_{\text{total}} = 3.5 \times 10^{49}\ \text{J}^{-3/2}\ E^{3/2}$$

Now we ask for the number g of quantum states in an energy interval between E and $E + \Delta E$. Since g_{total} can be treated as a continuous function we obtain by differentiation:

$$\frac{dg_{\text{total}}}{dE} = \frac{\pi}{4}\left(\frac{8m_eL^2}{h^2}\right)^{3/2} E^{1/2} \tag{13B2.4}$$

Then for an interval ΔE_i the number g_i of states is

$$g_i = \frac{\pi}{4}\left(\frac{8mL^2}{h^2}\right)^{3/2} E^{1/2}\ \Delta E_i \tag{13B2.5}$$

The relations (13B2.3) and (13B2.5) are illustrated in Fig. 13B.3. The points in Fig. 13B.3b represent g_i obtained from the actual number of states in the five lowest energy intervals ($\Delta E_i = 0.72 \times 10^{-32}$ J, Fig. 13B.3a). Figure 13B.3c represents the total number g_{total} of states, summed up over the five lowest intervals.

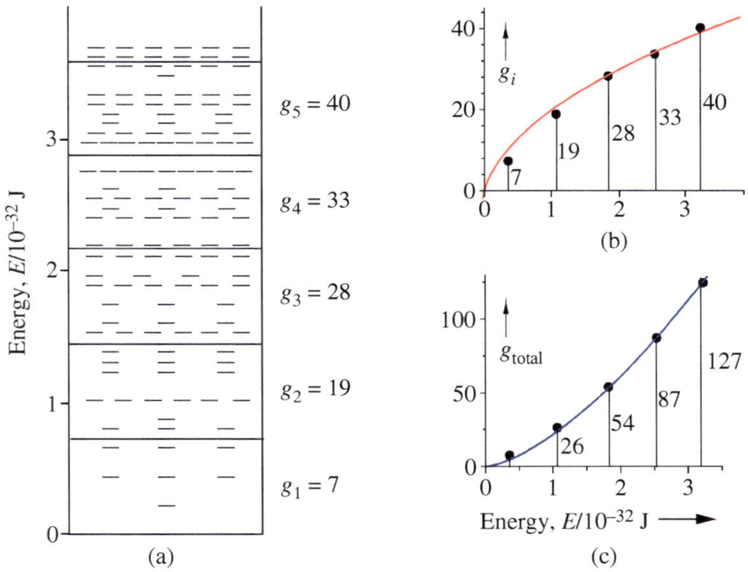

Figure 13B.3 Number g_{total} of energy levels (between $E = 0$ und E) and number g_i of energy levels in the interval ΔE_i (electron in a cube of volume $V = 1\ \text{cm}^3$). (a) Energy scheme (corresponding to Fig. 3.7b). (b) Number g_i as a function of E (for $\Delta E_i = 0.72 \times 10^{-32}$ J); the dots correspond to the numbers 7, 19, 28, 33, and 40 in the five lowest intervals. (c) Total number g_{total} as a function of E; the dots correspond to the numbers 7, 7 + 19 = 26, 7 + 19 + 28 = 54 and so on.

13.9 Box 13.3. Fermi Energy and Mean Energy of Electrons in a Metal

13.9.1 Fermi Energy

The N electrons of a metal occupy $N/2$ orbitals since each orbital can be occupied with two electrons each. We approximate the total number of orbitals of an electron moving in a cube with side length L in the x-, y-, and z-directions for energies between 0 and E by Equation (13B2.3)

$$g_{\text{total}} = \frac{\pi}{6}\left(\frac{8m_e L^2 E}{h^2}\right)^{3/2}$$

Then we can write that

$$\frac{N}{2} = \frac{\pi}{6}\left(\frac{8m_e L^2 E_F}{h^2}\right)^{3/2}$$

where E_F is the energy of the highest occupied energy level (*Fermi energy*). By solving this equation for E_F we find

$$E_F = \left(\frac{3}{\pi}\right)^{2/3} \cdot \frac{h^2}{8m_e} \cdot \left(\frac{N}{V}\right)^{2/3} \tag{13B3.1}$$

where $V = L^3$ is the volume of the considered cube.

We can substitute the volume V of the crystal by the density ρ of the crystal and the mass of the corresponding atom, which is m_{Li} in the case of Li metal. $N \cdot m_{\text{Li}}$ is the mass of all Li atoms in the volume V; thus $(N \cdot m_{\text{Li}})/V = \rho$ and for E_F we obtain

$$E_F = \left(\frac{3}{\pi}\right)^{2/3} \cdot \frac{h^2}{8m_e} \cdot \left(\frac{\rho}{m_{\text{Li}}}\right)^{2/3} \tag{13B3.2}$$

13.9.2 Calculation of the Mean Energy $\overline{E}$ of the Electron Gas

To calculate the mean energy $\overline{E}$, we must sum the energy for each electron and divide by the total number of electrons in the system, namely

$$\overline{E} = \frac{1}{N}\sum N_i E_i$$

The number of particles, N_i, in the energy interval ΔE_i is $2g_i$, for all levels below the Fermi energy, and the total number N is $N = \sum_{i=1}^{i_F} 2g_i$. Thus

$$\overline{E} = \frac{\sum_{i=1}^{i_F} 2g_i E_i}{\sum_{i=1}^{i_F} 2g_i}$$

According to Equation (13B2.5) for g_i, we find $g_i = const \cdot E_i^{1/2} \cdot \Delta E_i$. Thus

$$\overline{E} = \frac{\sum_{i=1}^{i_F} E_i^{3/2} \cdot \Delta E_i}{\sum_{i=1}^{i_F} E_i^{1/2} \cdot \Delta E_i}$$

Since the energy levels are close to each other we can replace the sums by integrals

$$\overline{E} = \frac{\int_0^{E_F} E^{3/2} \cdot \text{d}E}{\int_0^{E_F} E^{1/2} \cdot \text{d}E} = \frac{\frac{2}{5}(E_F)^{5/2}}{\frac{2}{3}(E_F)^{3/2}} = \frac{3}{5}E_F$$

or

$$\overline{E} = \frac{3}{5}\left(\frac{3}{\pi}\right)^{2/3} \cdot \frac{h^2}{8m_e} \cdot \left(\frac{N}{V}\right)^{2/3} = \frac{3}{5}\left(\frac{3}{\pi}\right)^{2/3} \cdot \frac{h^2}{8m_e} \cdot \left(\frac{\rho}{m_{\text{Li}}}\right)^{2/3}$$

13.10 Box 13.4. Dipole and Induction Energies

Here we present the results for energies and field strengths arising from the interaction of a dipole with a dipole or with a point charge, or of an induced dipole with a point charge. For the detailed calculation see Foundation 13.2.

13.10.1 Dipole Energy (Dipoles at an Angle β)

For two dipoles arranged as in Fig. 13B.4 the energy is.

$$E = -\frac{1}{4\pi\varepsilon_0} \cdot \frac{\mu_1\mu_2}{r^3} \cdot \left(3\cos^2\beta - 1\right)$$

with the definitions $\mu_1 = aQ_1$ and $\mu_2 = aQ_2$. This expression holds for an arbitrary angle β.

For the case of $\beta = 0$ we find

$$E = -\frac{1}{4\pi\varepsilon_0} \cdot \frac{2\mu_1\mu_2}{r^3} \qquad \text{(dipoles in line)}$$

and $\beta = 90°$ we find

$$E = \frac{1}{4\pi\varepsilon_0} \cdot \frac{\mu_1\mu_2}{r^3} \qquad \text{(dipoles parallel)}$$

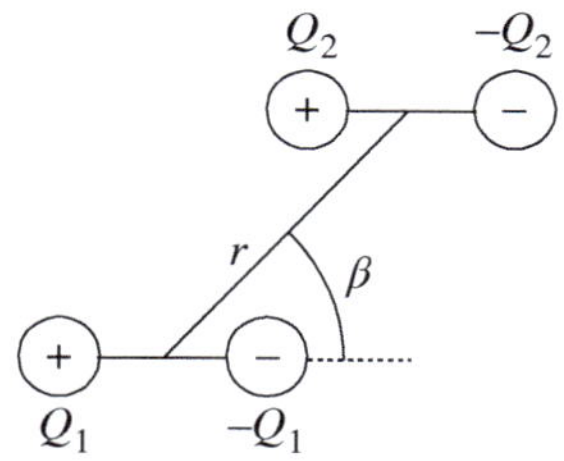

Figure 13B.4 Orientation of two dipoles (charges Q_1, $-Q_1$ and Q_2, $-Q_2$) with arbitrary angle β.

13.10.2 Dipole in the Field of a Point Charge

According to Fig. 13B.5 the energy of a point charge in the field of a dipole is with $\mu = aQ_1$

$$E = \frac{1}{4\pi\varepsilon_0}\frac{\mu}{r^2}Q_2 \cdot \cos\beta$$

In the special cases $\beta = 0$ and $\beta = 90°$ this relation simplifies to

$$E = \frac{1}{4\pi\varepsilon_0}\frac{\mu}{r^2}Q_2 \qquad \text{and} \qquad E = 0$$

respectively.

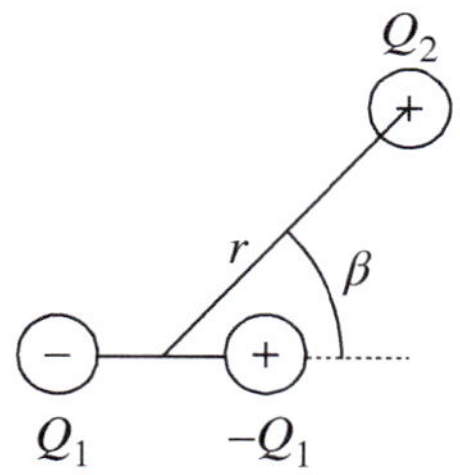

Figure 13B.5 Dipole (charges Q_1, $-Q_1$) in the field of a point charge (charge Q_2) in distance r at an angle β.

13.10.3 Electric Field Strength of Dipole

13.10.3.1 Case 1: Field Strength at Point *P* in Distance *r* Along the Direction of the Dipole (Fig. 13B.6)

For the electric field strength F it follows that

$$\boxed{F = -\frac{1}{4\pi\varepsilon_0} \cdot \frac{2Qa}{r^3} = -\frac{1}{4\pi\varepsilon_0} \cdot \frac{2\mu}{r^3}}$$

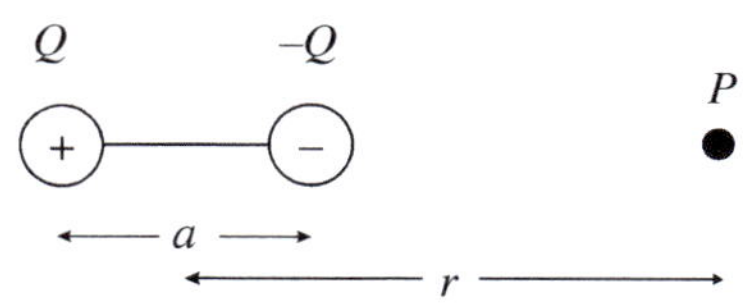

Figure 13B.6 Field strength at point *P* along the direction of a dipole.

13.10.3.2 Case 2: Field Strength in Distance *r* Perpendicular to the Direction of the Dipole (Fig. 13B.7)

The component of the electric field strength in direction of r is zero at point P (the contributions of the positive and of the negative charge cancel each other). Here we calculate the component of the electric field strength that is perpendicular to the direction of r (Fig. 13B.8).

For the component F_v of the field strength we find that

$$\boxed{F_v = 2F_{+,v} = \frac{Qa}{4\pi\varepsilon_0} \cdot \frac{1}{r^3} = \frac{1}{4\pi\varepsilon_0} \cdot \frac{\mu}{r^3}}$$

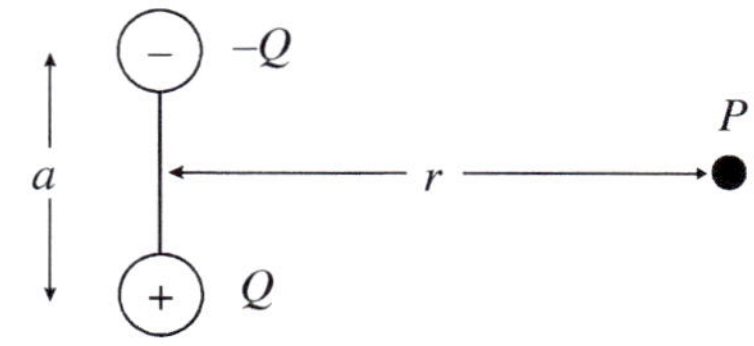

Figure 13B.7 Field strength at point *P* perpendicular to the dipole.

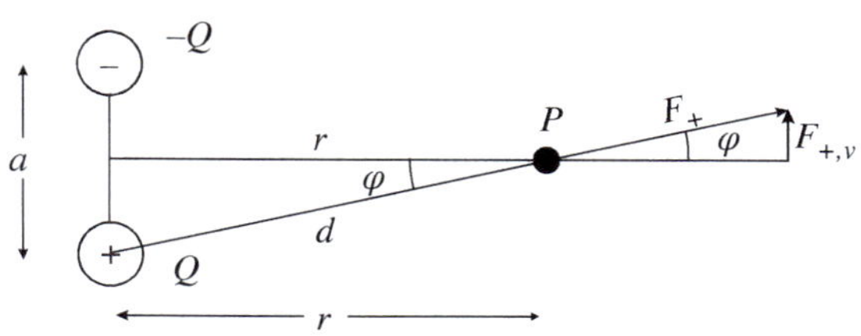

Figure 13B.8 Component of the electric field strength perpendicular to the direction of r.

13.11 Box 13.5. Polarizability of a Conducting Plate and of a Conducting Sphere

We consider a capacitor with area A and plate distance l (Fig. 13B.9). Its capacitance is

$$C = \frac{Q}{V} = \varepsilon_0 \frac{A}{l} \tag{13B5.1}$$

where Q is the charge on the plates, and V is the voltage. Now we place a metallic plate of the same area (thickness d) inside the capacitor. Then the charge Q is induced on the plate and the dipole moment of the plate is

$$\mu_{\text{ind}} = Q \cdot d = \varepsilon_0 \frac{A}{l} V \cdot d = \varepsilon_0 Ad \cdot F \tag{13B5.2}$$

where Ad is the volume of the plate and $F = V/l$ is the electric field strength. Then the polarizability of the plate is

$$\alpha = \frac{\mu_{\text{ind}}}{F} = \varepsilon_0 \cdot Ad = \varepsilon_0 \cdot volume \quad \text{(plate)} \tag{13B5.3}$$

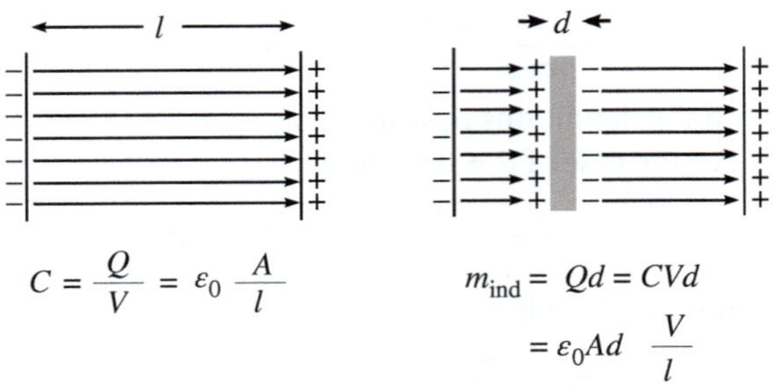

Figure 13B.9 Polarization of a metallic plate inside a capacitor.

In Fig. 13B.10 we compare the polarizations of a cube and of a sphere. Apparently, Equation (13B5.3) derived for the polarizability of a plate holds also approximately for the polarizability of a cube: the electric field lines are mainly parallel, if we neglect the edge effects (Fig. 13B10a). In the case of a sphere Fig. 13B.10b shows that the electric field lines are no longer parallel because of the condition that they must be perpendicular to the surface of the sphere. Consequently, the induced charge density on the surface of the sphere is larger than that on the surface of the cube. From Fig. 13B.10b we can estimate that the one-dimensional density is larger by a factor of about two; thus, the two-dimensional density is larger by a factor of four. A detailed treatment[4] shows that α of a conducting sphere is larger by a factor of three compared to that of a plate

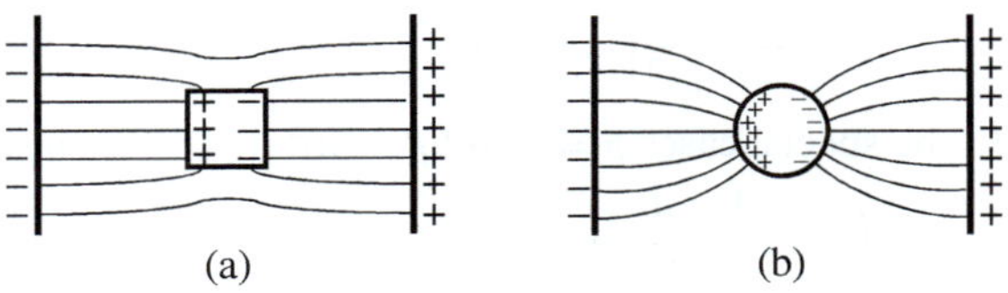

Figure 13B.10 Polarization of a metallic cube (a) and of a metallic sphere with the same volume (b).

$$\boxed{\alpha = 3\varepsilon_0 \cdot \text{volume} \quad \text{(sphere)}} \tag{13B5.4}$$

13.12 Foundations

13.12.1 Foundation 13.1: Some Features of Crystal Structures and Lattices

See the books website, www.wiley.com/go/Kuhn/physicalchemistry3e.

13.12.2 Foundation 13.2: Calculation of Dipole and Induction Energies

See the books website, www.wiley.com/go/Kuhn/physicalchemistry3e.

13.13 Exercises

E13.1 Derive Equation (13.5) from Equation (13.4), and use this result to derive a formula for the energy of an ion pair in terms of the equilibrium bond distance.

E13.2 Plot Equation (13.4) over the range of 190–1000 pm for the case where $n = 10$, $C = e^2 d_{\text{eq}}^9/(40\pi\varepsilon_0)$, and $d_{\text{eq}} = 250$ pm. Find the minimum and verify that it occurs at the distance 250 pm.

E13.3 Use the experimental lattice energy of NaCl (equal to -769 kJ mol^{-1}) to estimate the equilibrium distance d_{eq}. Compare your result with that found from the ionic radii.

4 The detailed solution can be found in standard textbooks on electricity and magnetism. For examples, see [14] or [15].

E13.4 Use the experimental lattice energy of CsCl (equal to -657 kJ mol^{-1}) to estimate the equilibrium distance d_{eq}. Compare your result with that found from the ionic radii.

E13.5 Use the ionic radii of Na^+ and Br^- to estimate the equilibrium bond distance d_{eq} and calculate the lattice energy. Compare your result to the experimental value of -732 kJ mol^{-1}.

E13.6 Use the ionic radii of Li^+ and H^- to estimate the equilibrium bond distance d_{eq} and calculate the lattice energy. Compare your result to the experimental value of -858 kJ mol^{-1}.

E13.7 Calculate the electrostatic force that the Li^+ cation exerts on the H^- anion at their equilibrium bond distance, $d_{eq} = 209$ pm. Compare this force to the Coulomb force experienced by an electron that is one Bohr radius from a proton.

E13.8 Calculate the binding energy of MgO for a Wurtzite structure and compare it to that for a cubic structure. In nature, MgO forms a cubic structure with an energy of $-3795 \text{ kJ mol}^{-1}$. The ionic radius of O^{2-} is 140 pm and the ionic radius of Mg^{2+} is 72 pm.

E13.9 Using the most common valency for the metals listed, calculate the Fermi energy for metallic Na and Al.

E13.10 Given that the work function φ of a metal (see Chapter 1) is the energy difference between the Fermi energy and the energy of a free electron (with zero kinetic energy), explain the trend in work function for the alkali metals (Li (2.9 eV), K (2.3 eV), and Cs (2.1 eV)) in terms of the metals' electronic structure.

E13.11 For an energy difference ΔE between the two sides of a quantum wire and a change of g quantum states, show that $\Delta E = g \cdot (2n_z h^2)/(8m_e L^2)$.

E13.12 The bulk solids, Si, Ge, and C (diamond), each have tetrahedral (sp^3) bonding. Why is diamond transparent, whereas Si and Ge have a silvery luster (much like a metal)?

E13.13 Using a semiconductor band diagram explain how impurity atoms in silicon (band gap of 1.12 eV) with an $E_{acceptor}$ energy 0.05 eV above the valence band edge can effect the conductivity. Use the fact that the probability to excite an electron from one electronic state to another (energy difference of ΔE) can be written as a probability $\sim \exp(-\Delta E/(kT))$.

E13.14 Use Equation (13.35) to calculate the force between the two dipoles in Example 13.7.

E13.15 In Example 13.7 we considered two anti-parallel dipoles as a model for the acetonitrile dimer. Use the measured far-infrared bands at (80 and 130 cm^{-1}) and calculate the force constant for the vibration in the harmonic oscillator approximation.

E13.16 Calculate the attraction energy between two ICl dipoles arranged in the collinear geometry of Fig. 13.16a and the antiparallel geometry of Fig. 13.16b, with separation distances of 1.0 nm. The bond length of ICl is 232 pm and its dipole moment is 0.65 D.

E13.17 Calculate the dispersion energy between two lithium atoms (take their polarizability to be $24.3 \times 10^{-30}\ \text{m}^3 = \alpha/(4\pi\varepsilon_0)$ and their ionization energy to be 5.39 eV) at distances of 350 pm (that in the solid) and 700 pm.

E13.18 Consider the dipole–dipole interaction between two ammonia molecules ($\mu = 1.5$ D) at 400 pm. In the point dipole approximation, what angle between the two dipoles has the most stable energy? How does this geometry compare to that you expect for hydrogen bond formation between the two ammonia molecules? (Refer to Box 13.4).

E13.19 Consider an HCl molecule ($\mu = 1.18$ Debye) that is 400 pm from an Ar atom in a collinear geometry (Ar–HCl). Calculate the induced dipole moment in the Ar atom. Use the induced dipole to estimate the charge displacement (assume a unit charge) in units of the atom's radius.

E13.20 In Example 13.9 we calculated the bond energy of a proton with a water molecule. Perform a similar calculation for a sodium ion ($d_{eq} = 288$ pm).

E13.21 In the text we calculated the cohesion energy of Li metal by combining the experimental number density of Li atoms in the metal with the energy expression (Equations (13.24) and (13.25)) for the cohesive energy:

$$E = \overline{E}_{pot} + \overline{E}_{kin} = -A \cdot \left(\frac{N}{V}\right)^{1/3} + B \cdot \left(\frac{N}{V}\right)^{2/3}$$

where

$$A = \frac{e^2}{4\pi\varepsilon_0} \cdot \left(\frac{4\pi}{3}\right)^{1/3} = 3.27 \times 10^{-19}\ \text{J nm} \quad \text{and} \quad B = \frac{3}{5}\left(\frac{3}{\pi}\right)^{3/2} \frac{h^2}{8m_e} = 3.50 \times 10^{-20}\ \text{J nm}^2$$

In this problem, you are asked to estimate the optimum number density by minimizing E with respect to the number density ($x = N/V$), and then use this value to calculate the cohesion energy.

E13.22 Consider the case of two electric dipoles arranged in a line (see Fig. E13.22), for which the electrostatic energy is

$$E = \frac{1}{4\pi\varepsilon_0} \cdot \left[\frac{Q_1Q_2}{r} + \frac{Q_1Q_2}{r} - \frac{Q_1Q_2}{(r+a)} - \frac{Q_1Q_2}{(r-a)}\right]$$

Show that the energy expression simplifies to $2a^2/r^3$ in the limit where $a \ll r$.

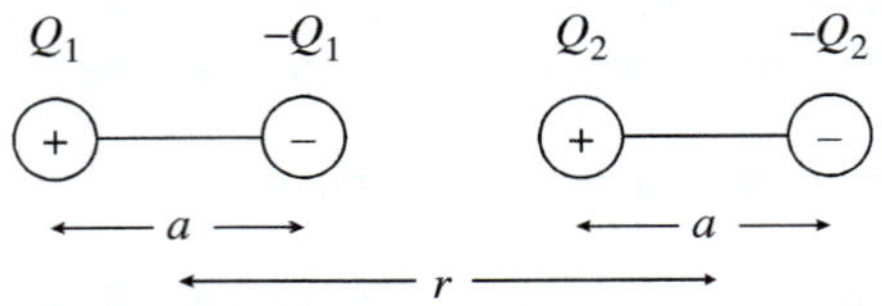

Figure E13.22 Two dipoles arranged in line.

E13.23 Consider the hydrogen bonded aggregate

$(\text{F} \ldots \text{H} \ldots \text{F})^-$

The H nucleus can be considered as a harmonic oscillator, which can be excited by infrared light. The infrared absorption bands for this aggregate occur at $1/\lambda = 1364\ \text{cm}^{-1}$ (excitation in the direction of the bond line: antisymmetric vibration mode) and at $1217\ \text{cm}^{-1}$ (excitation perpendicular to the bond line: bending mode). Using a harmonic oscillator approximation for the vibrations, calculate the wavenumber shift in these transitions if the H atom is replaced by a deuterium atom. Compare your answer to the experimental value of 969 and 880 cm^{-1}.

E13.24 Beginning with the energy expression for two dipoles at an angle ß, given in Box 13.4, namely

$$E = \frac{Q_1Q_2}{4\pi\varepsilon_0} \cdot \left[\frac{2}{r} - \frac{1}{r}\frac{1}{\sqrt{1 + \frac{a^2}{r^2} + 2\frac{a}{r}\cos\beta}} - \frac{1}{r}\frac{1}{\sqrt{1 + \frac{a^2}{r^2} - 2\frac{a}{r}\cos\beta}}\right]$$

expand each of the square root terms in brackets in a Taylor series and demonstrate that

$$E = \frac{1}{4\pi\varepsilon_0}\frac{Q_1Q_2a^2}{r^3} \cdot (1 - 3\cos^2\beta)$$

up to second order in a/r.

E13.25 Beginning with the expressions

$$r_{23}^2 = r^2\left(1 - \frac{a}{r}\cos\beta + \frac{a^2}{4r^2}\right) \quad \text{and} \quad r_{13}^2 = r^2\left(1 + \frac{a}{r}\cos\beta + \frac{a^2}{4r^2}\right)$$

show that

$$\frac{1}{r_{13}} - \frac{1}{r_{23}} = -\frac{a}{r^2}\cos\beta$$

by expanding the term $1/r$ in a Taylor series and collecting terms up to second order in a/r.

E13.26 Describe an experiment that demonstrates the wave nature of matter. Make a sketch that illustrates your observations; i.e., measured data. Draw a sketch of an experimental apparatus that shows the essential components needed in making such a measurement. Explain in words how this experiment demonstrates the wave nature of matter.

13.14 Problems

P13.1 Estimate how much energy has to be supplied in order to increase or decrease the distance d of an ion pair by 1% and 10% ($d_{eq} = 100$ pm, $n = 10$).

P13.2 The lattice energies and equilibrium ion distances for some simple halide salts (MCl) are provided by the table shown in problem P13.3. Using these data (taken from CRC handbook), calculate the best value of n for Equation (13.9).

P13.3 Using the crystal data for these different salts, which form cubic crystals with the sodium chloride lattice, and the fact that the radius of I^- is much greater than that of Li^+, determine "best values" for the ionic radii. Describe your reasoning.

Salt	d/pm	Salt	d/pm	Salt	d/pm	Salt	d/pm
LiF	201	LiCl	257	LiBr	275	LiI	300
NaF	231	NaCl	282	NaBr	297	NaI	323
KF	266	KCl	314	KBr	329	KI	353
RbF	282	RbCl	329	RbBr	343	RbI	363

P13.4 The semiconductor Si has a bulk bandgap of 1.1 eV and a dielectric constant of 11.8. Graph the bandgap you expect for this material over a range of crystal sizes between 1 and 100 nm. At what size do you expect the material to have a bandgap of 2.5 eV (use $m_{e,\text{eff}} = 0.98m_e$ and $m_{h,\text{eff}} = 0.16m_e$ where $m_e = 9.109 \times 10^{-31}$ kg)?

P13.5 A simple model for an impurity atom in a semiconductor is to approximate the impurity site as an effective hydrogen atom. Consider a phosphorous atom in silicon to use four of its valence electrons to participate in the sp^3 bonding of the silicon crystal and have its fifth valence electron to be loosely bound to the phosphorous cation. Approximate the interaction of this electron with the phosphorous cation as a hydrogen atom in a medium with the relative permittivity of silicon ($\varepsilon = 11.8$), and calculate 1) the energy of the ground electronic state and 2) the energy required to ionize this effective hydrogen atom. Also calculate and plot the ground state wavefunction of the electron.

P13.6 The resistance of a semiconductor decreases with increasing temperature because it is possible to excite electrons through the bandgap; hence, the temperature dependence of the resistance R can be used to determine the bandgap of the material. In order to eliminate the effects of a solid's size one usually finds the resistivity $\rho = 1/\kappa$ reported, where $\kappa = l/(RA)$ is the conductivity. Thus $\rho = R \cdot A/l$ with A equal to the cross-sectional area of the sample and l equal to the sample length. Assume that the resistivity of germanium is given by

$$\rho = C \cdot \exp\left(E_{\text{gap}}/(2kT)\right)$$

where C is a constant and use the data in the table to determine its bandgap energy. Compare your result to the experimental value of 0.66 eV.

T/K	ρ/Ω cm	T/K	ρ/Ω cm
833	5.00×10^{-3}	357	4.55
700	1.24×10^{-2}	333	10.0
554	5.56×10^{-2}	312	25.0
500	1.43×10^{-1}	300	50.0
385	1.82	278	167

P13.7 Consider two HCl molecules that are in a collinear geometry (Fig. 13.16a) and separated by a distance of 600 pm. Plot the dipole–dipole interaction energy as a function of the relative orientation of the dipoles; i.e., rotate one of the dipoles about its center of charge and compute the interaction energy for a series of angles.

P13.8 Consider two HCl molecules in the geometry of Fig. 13.16b. Plot the dipole–dipole energy, the induction energy, and the dispersion energy as a function of the distance between the HCl centers, ranging from van der Waals contact to 5 nm. (Take van der Waals contact to be 450 pm, the dipole moment to be 1.03 D, the polarizability α to be $4\pi\varepsilon_0\left(2.63\ \text{Å}^3\right) = 2.93 \times 10^{-40}\ \text{C}^2\ \text{J}^{-1}\ \text{m}^{-1}$, and the ionization energy to be 21.5×10^{-19} J).

P13.9 Calculate the polarizability of an electrically conducting plate. Imagine a plate in a homogeneous electric field and calculate the induced dipole moment of the plate. Compare with the polarizability of a sphere of the same volume.

P13.10 Derive Equation (13.36), the attraction energy between a charge Q and a point dipole (dipole moment μ) for $\theta = 0°$.

P13.11 We can use Equation (13.24) for the cohesion energy to find the number density for a metal. Calculate the minimum of the energy with respect to the number density by taking its derivative and setting it equal to zero.

P13.12 Show that electron current traveling in the z-direction through a quantum wire in the $n_x = n_y = 1$ energy level is given by the formula

$$I = e\sum \frac{v_z}{L} = e\sum \frac{hn_z}{2m_e L^2}$$

P13.13 Using Equation (13.7)

$$E = -\frac{1}{4\pi\varepsilon_0}\cdot\frac{e^2}{d}\cdot\left[1-\frac{1}{n}\left(\frac{d_{eq}}{d}\right)^{n-1}\right]$$

calculate the energy E for an ion pair of NaCl for different values of the parameter n.

P13.14 Often when constructing a theory, we begin with an idealization of the realistic system to see if the essential ideas apply. In this problem we consider the diffraction of X-ray light wavelengths from a lattice of atoms. Let us view the crystalline solid as being a two-dimensional array of atoms as in Fig. P13.14a.

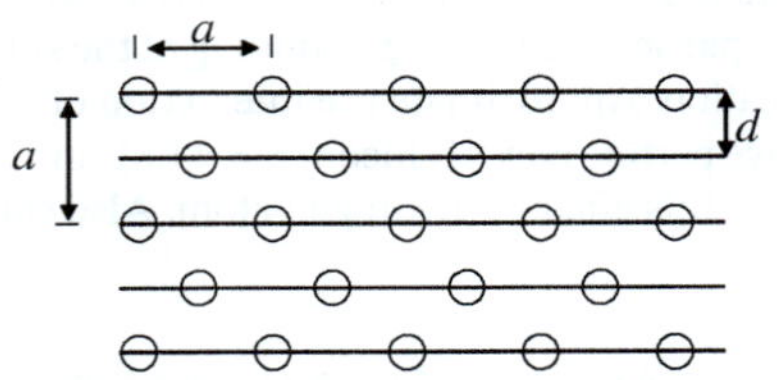

Figure P13.14a Two-dimensional lattice.

The distance a is called the lattice parameter, and let us take its value to be that of Au, $a = 408$ pm. If X-rays of wavelength λ are incident on this array of atoms from the top at an angle θ, it will be diffracted at some angle. You are asked to find this angle.

a) Consider the incident light field to be described by a sine wave and consider each layer of atoms (indicated by the dashed lines) to act as a partial mirror that reflects the X-rays. Using the superposition of light waves, show that the reflected waves add constructively when $n\lambda = 2d \cdot \sin(\theta)$ in which n is an integer and d is the distance between diffraction planes. This is called the Bragg condition.
b) One common source produces X-rays at a mean wavelength of 154 pm; using this value and the lattice constant given above determine the angles that satisfy the Bragg condition for $n = 1$ and $n = 2$.
c) The material C_{60} (fullerene) forms cubic crystals with a lattice constant of 1470 pm near room temperature. What value of the diffraction angles would you expect to find in this case?
d) Consider forming a solid lattice from micron-sized particles (sometimes called a colloidal crystal) that has a lattice parameter of 1.0 μm. What wavelength of light will your crystal diffract in first order ($n = 1$) at 15°? Diffraction peaks will appear at points where the wave amplitudes add in phase, and this occurs when the distances traveled by the different parts of the wave (e.g., that reflected from one layer of atoms versus that from the adjacent layer) differ by a multiple of the wavelength. Figure P13.14b shows that the difference in distance traveled by two light rays (red and black) is $2l$, where $l = d \cdot \sin(\theta)$.

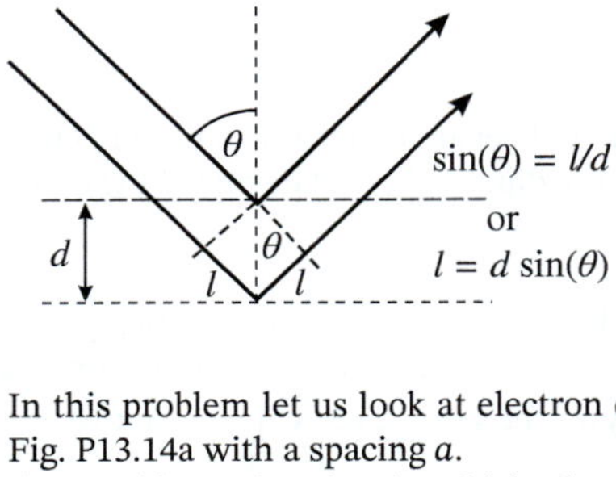

Figure P13.14b Diffraction: traveling distance.

P13.15 In this problem let us look at electron diffraction and consider the diffraction of electrons just from the top row of atoms in Fig. P13.14a with a spacing a.

a) Consider a situation in which electron waves of wavelength Λ are incident on the row of atoms at an angle normal to the surface and are scattered from the atoms at an angle θ. Show that the electron waves superpose constructively when $n\Lambda = a\ \sin(\theta)$, where n is an integer called the order of diffraction.
b) Consider the case when the row of atoms has a lattice parameter $a = 408$ pm (that of Au), what wavelength of electrons is needed for the first-order diffraction peak to appear at 15°? What must the kinetic energy of the electrons be for them to have the required wavelength?

c) A number of important experimental methods use electron diffraction to investigate the structural features of solid surfaces. In LEED (Low Energy Electron Diffraction) spectroscopy the energy of the electrons ranges from about 10 to 500 eV. What range of electron wavelengths are accessed in LEED spectroscopy and what characteristic structural features can you expect to study with LEED?
d) Another electron diffraction method, called RHEED (Reflection High-Energy Electron Diffraction), uses electron energies in the range of 5–50 keV. What range of electron wavelengths are accessed in this method?

References

1 G. Mie, Zur kinetischen Theorie der einatomigen Körper, *Ann. Phys.*, 1903, **11**, 657–697.
2 R.D. Shannon, Revised effective ionic radii and systematic studies of interatomic distances in halides and chalcogenides, *Acta Crystallogr.*, 1976, **A32**, 751–767; R.D. Shannon, Effective ionic radii in oxides and fluorides, *Acta Crystallogr.*, 1969, **B25**, 925.
3 H. Ohnishi, Y. Kondo, and K. Takayanagi, Quantized conductance through individual rows of suspended gold atoms, *Nature*, 1998, **395**, 780.
4 Y. Imry and R. Landauer, Conductance viewed as transmission, *Rev. Mod. Phys.*, 1999, **71**, S306–S312.
5 R. Peierls, *Quantum Theory of Solids* (Oxford University Press, New York, 1955).
6 H. Kuhn, A quantum-mechanical theory of light absorption of organic dyes and similar compounds, *J. Chem. Phys.*, 1949, **17**, 1198; *Chimia*, 1950, **9**, 204.
7 T. Vossmeyer, L. Katsikas, M. Giersig, I.G. Popovic, K. Diesner, A. Chemseddine, A. EychmiUler, and H. Weller, CdS nanoclusters: synthesis, characterization, size dependent oscillator strength, temperature shift of the excitonic transition energy, and reversible absorbance shift, *J. Phys. Chem.*, 1994, **98**, 7665.
8 L. Brus, Electron–electron and electron-hole interactions in small semiconductor crystallites: The size dependence of the lowest excited electronic state, *J. Chem. Phys.*, 1984, **80**, 4403.
9 G.T. Fraser, D.D. Nelson Jr., A. Charo, and W. Klemperer, Microwave and infrared characterization of several weakly bound NH3 complexes, *J. Chem. Phys.*, 1985, **82**, 2535.
10 E. Knözinger and D. Leutloff, Far infrared spectra of strongly polar molecules in solid solution. I. Acetonirile, *J. Chem. Phys.*, 1981, **74**, 4812.
11 B. Dereka, Q. Yu, N.H.C. Lewis, W.B. Carpenter, J.M. Bowman, and A. Tokmakoff, Crossover from hydrogen to chemical bonding, *Science*, 2021, **371**, 160–164; W. Reckien and S.D. Peyerimhoff, Twofold hydrogen bridges as observed in amide-templated rotaxanes, *J. Phys. Chem. A*, 2003, **107**, 9634–9640.
12 See J.A. Odutola and T.R. Dyke, Partially deuterated water dimers: microwave spectra and structure, *J. Chem. Phys.*, 1980, **72**, 5062.
13 F. London, Zur theorie und systematik der molekularkräfte, *Z. Phys.*, 1930, **63**, 245; *Z. Phys. Chem.*, 1930, **B11**, 222.
14 The detailed solution can be found in standard textbooks on electricity and magnetism. For examples, see P. Lorrain and D.R. Corson, *Electromagnetic Fields and Waves* (Freeman, New York, 1970).
15 J.D. Jackson, *Classical Electrodynamics* (Wiley, New York, 1962).

14

Thermal Motion of Molecules

CHAPTER MENU

14.1 Introduction

Our earlier discussions concentrated on understanding structural and energy level features of molecules and molecular assemblies. Although such knowledge is essential for a molecular-based picture of chemistry and matter, it is not complete. It is just as important to understand the distribution of energy and its flow within molecules and among molecules in assemblies.

What is temperature, and how is this phenomenological quantity related to molecular processes? In this chapter we consider some general features about the thermal motion of atoms and molecules. We begin with the case of a collection of independent molecules in a gas and ask how the energy may be distributed. In particular, we consider a container of hot gas that is brought into thermal contact with a container of a cold gas. We show in this chapter that the hot gas comprises fast moving molecules, whereas the cold gas consists of slow moving molecules. Bringing the containers into contact causes them to equilibrate so that the fast molecules become slower and the slow molecules become faster, on average. The temperature of the gas measures the average translational energy of the molecules. Through statistical considerations it is possible to relate the temperature directly to the average speed of the gas molecules.

Later in the chapter we consider other phenomena that reveal the motion of molecules in relation to temperature and heat, in particular the number of collisions with other molecules, the length of the path between collisions, and how far a molecule moves as it diffuses. These considerations are basic for appreciating the nature of thermal molecular assemblies, understanding chemical reactions, and describing how molecules interact and interlock to form increasingly complex molecular assemblies.

Principles of Physical Chemistry, Third Edition. Hans Kuhn, David H. Waldeck, and Horst-Dieter Försterling.

Companion website: www.wiley.com/go/Kuhn/physicalchemistry3e

14.2 Kinetic Gas Theory and Temperature

Consider a collection of identical particles (atoms or molecules) that are so far away from each other that their mutual attraction forces are negligible. In this limit, the total energy is composed of translational energy and of the internal energy of each particle (rotational, vibrational, and electronic energy). In this chapter we focus our discussion on the translational kinetic energy.

In an exact description the motion of these particles follows the laws of quantum mechanics; however, we consider the situation where the de Broglie wavelength is small compared to the typical distance a molecule travels between subsequent collisions (10^{-7} m, see Section 14.3.2). By way of example, consider hydrogen gas. The mass of a hydrogen molecule is $m = 3 \times 10^{-27}$ kg, and its speed v is on the order of 10^3 m s^{-1} (see Section 14.3.1). In the quantum mechanical treatment we would assign each molecule a de Broglie wavelength

$$\Lambda = \frac{h}{mv} = 2 \times 10^{-10} \text{ m} \tag{14.1}$$

according to Equation (1.7). Consequently, the molecules can be well described as particles that obey the laws of classical mechanics.

As an aside, we note that it is more conventional to first consider the macroscopic properties of matter, without taking account of molecular structure ("classical thermodynamics"). However, we follow a different pathway that proceeds systematically from the postulates of quantum mechanics and develops our understanding of increasingly complex molecular assemblies. This approach provides a unified view for understanding chemistry as a consequence of the basic principles of quantum mechanics. In the present chapter tremendous simplification is allowed because the translational motion of gases is well described classically, under most conditions. In addition, we proceed by constructing highly simplified models because it allows us to avoid considerable mathematical "baggage" and to illustrate the essential features of a molecular assembly's thermal properties.

14.2.1 Thermal Motion and Pressure

The gas molecules collide with the walls of their container, and these collisions exert a force on the walls. The force per unit area is the pressure of the gas. What is the pressure exerted on the walls by the gas? Can we relate this macroscopic property of pressure to some microscopic property of the gas?

In the following we will proceed in different steps of rigor.

1) First, we consider a simplified model for the calculation of the pressure, assuming that all molecules have the same speed and that one-sixth of all molecules are moving perpendicular toward one of the six walls of a cubic container. In spite of its crudeness, this model contains the essential physical elements needed for an understanding of the behavior of the gas molecules.
2) Next, we take account of the fact that the molecules are moving statistically into all directions of space. This refinement of the model does not change the result for the gas pressure obtained in the simplified model, but has consequences when discussing the number of collisions with a wall, or plate.
3) Finally, we take account of the fact that the speeds of the molecules are distributed according to the Maxwell–Boltzmann law. Regarding the gas pressure in our model, this has the only effect, that the speed of the gas molecules has to be replaced by their mean speed. We will show this explicitly in Chapter 15 where the Maxwell–Boltzmann law is derived. Here we restrict our considerations in order to introduce the main points.

14.2.1.1 Calculation of Pressure

Simplified Model First we calculate the pressure that gas molecules exert on the walls of the container in which they are enclosed. For simplicity we neglect intermolecular forces, attribute the same speed v to all the molecules, describe the molecules as spheres, and assume elastic collisions of the molecules with the walls of the container. We consider N molecules in a cubic container of side edge L (Fig. 14.1a). In our model $\frac{1}{6}N$ molecules are moving with speed v in the x direction, perpendicular to the wall on the right in Fig. 14.1b (gray zone, area L^2) and exert a force f at their impact with the wall.

The pressure P on a container wall is then

$$P = \frac{\text{force}}{\text{area}} = \frac{f}{L^2} \tag{14.2}$$

Using Newton's law for the calculation of the force exerted upon collision of the molecules with the wall, we can write

$$\text{force} = \frac{\text{momentum change}}{\Delta t} = m_{\text{gas}} \frac{\text{velocity change}}{\Delta t} \tag{14.3}$$

where m_{gas} is the total mass of all the gas molecules that collide with the wall within the time interval Δt. In our model the molecules move with the same speed after the collision (i.e., their kinetic energy is unchanged) but into the opposite direction; that is, their velocity change is

$$\text{velocity change} = v_{\text{final}} - v_{\text{initial}} = (-v) - v = -2v \tag{14.4}$$

(Fig. 14.2). As the force f exerted on the wall is opposite to the force exerted on the particles, we obtain

$$f = -m_{\text{gas}} \frac{-2v}{\Delta t} = m_{\text{gas}} \frac{2v}{\Delta t} \tag{14.5}$$

If ΔN gas particles reach the wall in time Δt, then

$$m_{\text{gas}} = m \cdot \Delta N \tag{14.6}$$

where m is the mass of one gas particle.

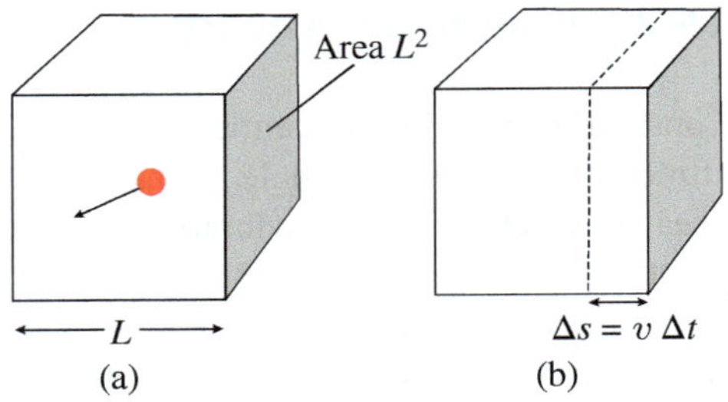

Figure 14.1 (a) Gas particles in a cubic container of length L. (b) Volume $L^2\Delta s = L^2 v\Delta t$. In our model one-sixth of the gas molecules in that volume move perpendicular toward the right wall and reach this wall in time Δt.

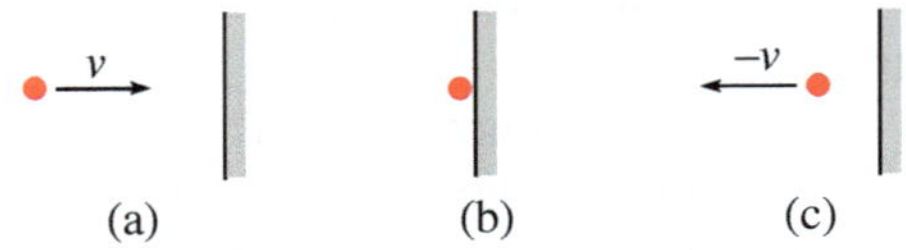

Figure 14.2 Elastic collisions of a particle with the wall (a) before the collision, (b) at the moment of the collision, and (c) after the collision.

In our simplified model we assume that one-sixth of the gas particles are moving perpendicular toward the wall. Then in time Δt only one-sixth of all particles that are at a distance smaller than $\Delta s = v\Delta t$ (Fig. 14.1b) will reach the wall. These particles must be in a volume ΔV where

$$\Delta V = L^2 \Delta s = L^2 v \Delta t \tag{14.7}$$

The number of gas particles in this volume element is given by $(N/V) \cdot \Delta V$, where N/V is the number density. Hence, the number of gas particles that will strike the wall perpendicularly is given by

$$\Delta N = \frac{1}{6}\frac{N}{V}\Delta V = \frac{1}{6}\frac{N}{V}L^2 v \Delta t \tag{14.8}$$

Combining Equations (14.2), (14.5), (14.6), and (14.8) for the pressure P we find that

$$P = \frac{f}{L^2} = m \cdot \Delta N \cdot \frac{2v}{\Delta t}\frac{1}{L^2} = \frac{1}{3}\frac{N}{V} m v^2 \tag{14.9}$$

Since the translational kinetic energy of a gas particle is

$$E_{\text{trans}} = \frac{1}{2} m v^2 \tag{14.10}$$

we can write the pressure as

$$P = \frac{2N}{3V} \cdot \frac{mv^2}{2} = \frac{2}{3}\frac{N}{V} E_{\text{trans}} \tag{14.11}$$

or

$$PV = \frac{2}{3} N \cdot E_{\text{trans}} \tag{14.12}$$

This means that the product PV is constant for a given translational energy.

Furthermore, we note that the number N of particles can be expressed by the amount of substance **n** and Avogadro's constant $\mathbf{N}_A$ (see Box 14.1)

$$N = \mathbf{N}_A \cdot \mathbf{n} \tag{14.13}$$

This relationship allows us to write Equation (14.12) as

$$PV = \frac{2}{3} N \cdot E_{\text{trans}} = \frac{2}{3} \mathbf{n}\mathbf{N}_A \cdot E_{\text{trans}} \tag{14.14}$$

Rigorous Result The same result for the product PV is obtained when we more exactly account for the particles' motion in different directions. A derivation of the pressure expression, which accounts for this fact, is presented in Box 14.2.

In our simplified model we have assumed that the molecules are moving with the same speed. Actually, the molecules in a gas do not have the same speed. In Chapter 15 we will discuss the distribution of speeds in a gas and include this distribution in the calculation of the gas pressure. It turns out that Equations (14.10) and (14.14) remain essentially valid, but υ^2 and E_{trans} have to be replaced by their averages, $\overline{\upsilon^2}$ and $\overline{E}_{\text{trans}} = \frac{1}{2}m\overline{\upsilon^2}$. Thus, we can write that

$$\boxed{PV = \frac{2}{3}N \cdot \overline{E}_{\text{trans}} = \frac{2}{3}\mathbf{n}\mathbf{N}_A \cdot \overline{E}_{\text{trans}}} \tag{14.15}$$

This equation relates the pressure P and the volume V of a gas ($N = \mathbf{n}\mathbf{N}_A$ molecules) to the average translational energy $\overline{E}_{\text{trans}}$ of each molecule.

Lastly, we note that for the number of collisions ΔN with a wall, in a more detailed consideration (see Box 14.2), instead of Equation (14.8) we obtain

$$\Delta N = \frac{1}{4}\frac{N}{V}L^2\overline{\upsilon} \cdot \Delta t \tag{14.16}$$

On first sight, it may seem strange that Equation (14.14) for the gas pressure remains the same as in the simplified model, although the number of collisions ΔN with the wall is larger by a factor of 1.5 and therefore the pressure on the wall should be accordingly higher. However, many of these collisions have only a small velocity component perpendicular to the wall; thus, they transfer only small amounts of momentum to the wall, and this fact counteracts the effect of the increase in collision number (see the derivation in Box 14.2).

14.2.1.2 Comparison with Experiment

Consider a cylinder filled with exactly 1 mol of hydrogen gas (i.e., with 2.016 g H_2) at a pressure of exactly[1] $P_1 = 1$ bar $= 1 \times 10^5$ Pa $= 1 \times 10^5$ N m^{-2}. The cylinder is placed in a bath with melting ice (Fig. 14.3). In the experiment we find that the volume of the gas is $V_1 = 22.7119$ L. A pressure change to $P_2 = 2$ bar causes a change in the enclosed gas volume to $V_2 = \frac{1}{2}V_1$. Correspondingly, a pressure change to $P_3 = \frac{1}{2}$ bar causes a change in the enclosed gas volume to $V_3 = 2V_1$ bar (Fig. 14.3). More generally we find (Fig. 14.4) that

$$PV = 22.7119 \text{ bar L (1 mol gas, melting ice)} \tag{14.17}$$

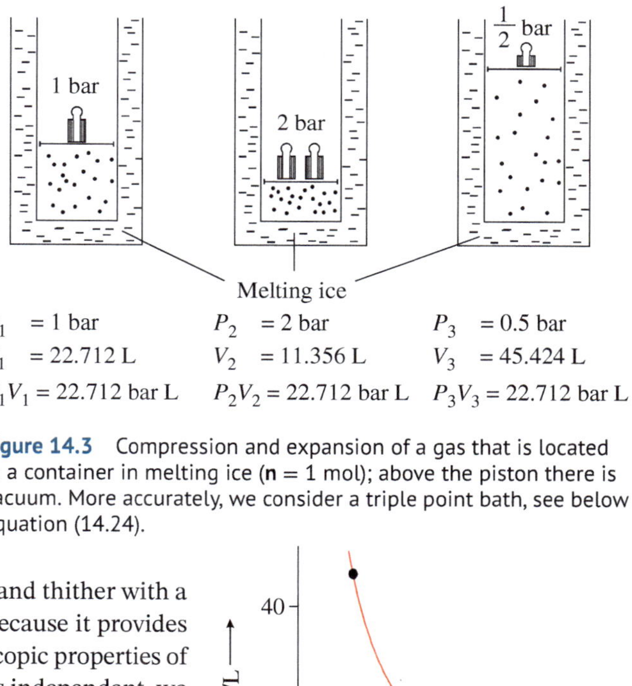

Figure 14.3 Compression and expansion of a gas that is located in a container in melting ice ($\mathbf{n}$ = 1 mol); above the piston there is vacuum. More accurately, we consider a triple point bath, see below Equation (14.24).

The volume V of the gas is inversely proportional to its pressure P (Fig. 14.4), known as *Boyle's law* (1664: also known as Mariotte's law).

In 1738, Daniel Bernoulli demonstrated that this early finding can be explained by considering the container to consist of "very minute corpuscles, which are driven hither and thither with a very rapid motion." The kinetic theory of gases is important because it provides a connection between macroscopic measurements and microscopic properties of molecules. Using the PV data and treating the gas molecules as independent, we can calculate the average translational energy of a hydrogen molecule in the gas container. According to Equation (14.15), we find

$$\overline{E}_{\text{trans}} = \frac{PV}{\frac{2}{3}\mathbf{n}\mathbf{N}_A} = \frac{22.7119 \text{ bar L}}{\frac{2}{3} \text{ mol } 6.022 \times 10^{23} \text{ mol}^{-1}} = 5.657 \times 10^{-21} \text{ J} \tag{14.18}$$

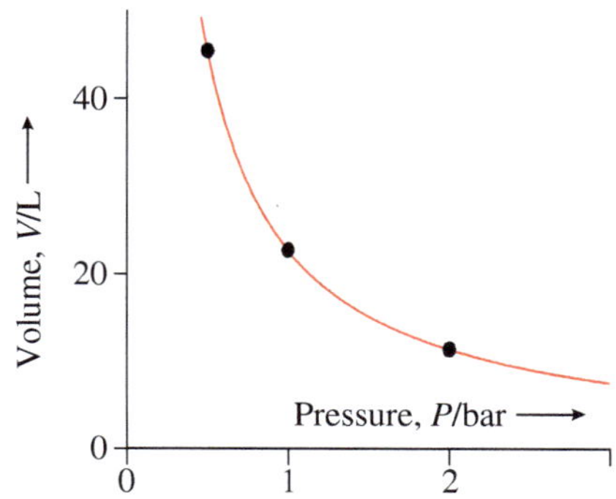

Figure 14.4 The volume of a gas which is in a container in melting ice as a function of pressure ($\mathbf{n}$ = 1 mol).

1 One bar is approximately one atmosphere pressure.

(with 1 bar L = 105 N m^{-2} L = 100 N m = 100 J). In addition, we can use Equation (14.10) to calculate the root mean square speed of the H_2 molecule. By rearranging Equation (14.10) we find

$$v_{\text{rms}} = \sqrt{\overline{v^2}} = \sqrt{\frac{2\overline{E}_{\text{trans}}}{m}} = \sqrt{\frac{2 \times 5.66 \times 10^{-21}\ \text{J}}{3.32 \times 10^{-27}\ \text{kg}}} = 1850\ \text{m s}^{-1}$$

where we use a mass of $m = 3.32 \times 10^{-27}$ kg for the H_2 molecule.

14.2.2 Avogadro's Law

If we measure the volume of 1 mol of gas at a pressure of 1 bar in a container that is immersed in an ice water bath, we always obtain the same value, independent of the type of gas molecules (e.g., N_2, CO_2, O_2, H_2). This result is known as *Avogadro's law* (Avogadro, 1776–1856). In other words, Avogadro's and Boyle's Laws can be expressed by stating that Equation (14.17), obtained from measurements with H_2, is true for all gases.

Can we understand this surprising result on the basis of the kinetic theory of gases? To do this we consider again a container with H_2 molecules (mass m_1, speed v_1) in a bath with melting ice, but now we imagine adding a CO_2 molecule (mass m_2) which is initially at rest (Fig. 14.5). As the H_2 molecules collide with the CO_2 molecule, energy is transferred to the CO_2 molecule, and its speed v_2 increases at the expense of the speed of the H_2 molecules. This process continues until a steady state is reached, in which all the molecule types have the same average translational energy, namely

$$\tfrac{1}{2}m_1\overline{v_1^2} = \tfrac{1}{2}m_2\overline{v_2^2} \tag{14.19}$$

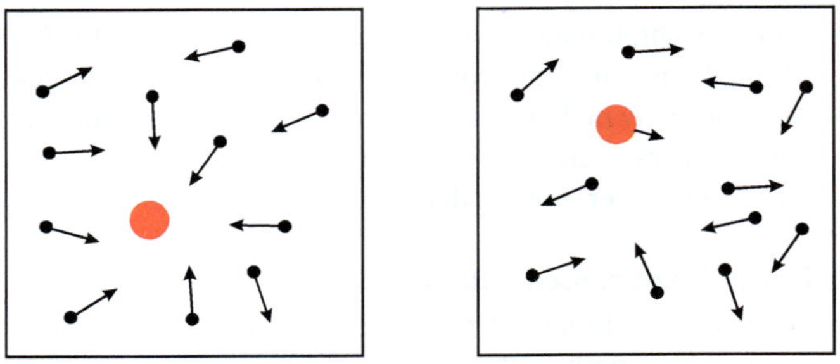

(a) (b)

Figure 14.5 A CO_2 molecule (red) in between H_2 gas molecules (black). (a) At the beginning the CO_2 molecule is at rest. (b) The motion of the CO_2 molecule after thermal equilibrium has been reached.

According to Equation (14.19), particles with a small mass move faster than particles with a larger mass, on average. The exact proof for Equation (14.19) is involved, and was first given by James Clerk Maxwell (1831–1879) [1].

Rather than prove Equation (14.19), we give some plausibility considerations. Consider the central ('head-on') collision of two particles with very different masses, but equal speed v. After the collision the speed of the heavy particle is practically unchanged, whereas the lighter particle moves in the opposite direction at a three times larger speed. This result can be understood by realizing that the relative speed of the two particles is $2v$ just before the collision. In the approximation that the heavy particle is much more massive than the lighter particle, it is like the collision with a wall (see Fig. 14.2) and the light particle has a speed $2v$ relative to the massive particle after the collision. Because the massive particle is moving at a speed v, in the laboratory coordinate frame, the speed of the light particle must be $3v$ in the laboratory frame. This fact tends to increase the average speed of the lighter particles, as compared to the heavier particles.

On the other hand, two particles with masses m_1 and m_2 do not change their speeds v_1 and v_2 in a central collision when the ratio of these speeds is inversely proportional to the ratio of their masses. If we take the particles' speeds after the collision to be u_1 and u_2, the conservation of momentum requires that[2]

$$m_1\overline{v_1} + m_2\overline{v_2} = m_1\overline{u_1} + m_2\overline{u_2}$$

For the case where

$$m_1\overline{v_1} = -m_2\overline{v_2}$$

it follows that

$$\frac{u_1}{u_2} = \frac{m_2}{m_1} = \frac{v_1}{v_2}$$

2 The arrow over the symbol, v or u, indicates that it is a velocity, whereas the absence of an arrow indicates that it is a speed.

This result demonstrates that for the speeds of two particles to be unchanged after a collision, the heavier particle must be slower than the lighter particle by their mass ratio. These considerations clearly show that the average speeds should be different for two molecule types of different mass. For more details on the collision of two particles see Foundation 14.4.

Equation (14.19) states that the mean translational energies of the different kinds of molecules in a gas mixture at equilibrium are equal. The conclusion is the same when we consider two separate containers with different kinds of molecules in a bath with melting ice (Fig. 14.6). In this case also, the translational energies of both kinds of molecules must be equal. Combining this result with Equation (14.12) we find that

$$\frac{1}{2}m_1\overline{v_1^2} = \frac{1}{2}m_2\overline{v_2^2} = \frac{3}{2}\frac{P_1V_1}{N_1} = \frac{3}{2}\frac{P_2V_2}{N_2} \tag{14.20}$$

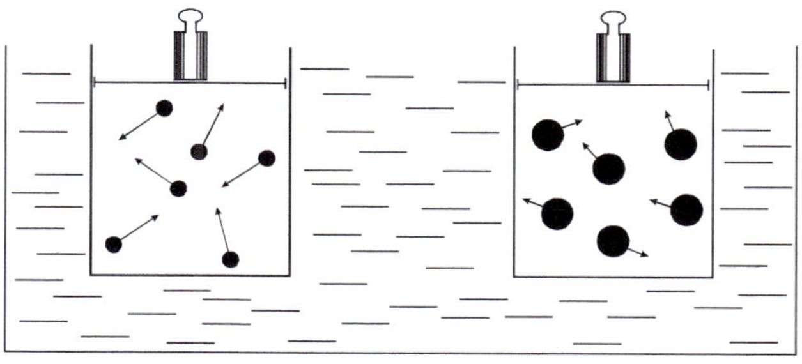

Figure 14.6 Two containers with different kinds of molecules in a bath of melting ice.

In this way we can understand *Avogadro's law* on the basis of molecular motion. Two different gases with the same number of molecules ($N_1 = N_2$) under the same pressure ($P_2 = P_1$) in a container at the same temperature have the same volume.

Example 14.1 *Consider a mole of N_2 gas under the same conditions as the H_2 gas discussed above. Compute the root mean square speed v_{rms} of an N_2 gas molecule and compare it to the case of H_2.*

Following along the lines of the H_2 calculation above we can write

$$v_{rms} = \sqrt{\overline{v^2}} = \sqrt{\frac{2\overline{E}_{trans}}{m}} = \sqrt{\frac{2 \times 5.66 \times 10^{-21}\ \text{J}}{4.605 \times 10^{-26}\ \text{kg}}} = 496\ \text{m s}^{-1}$$

where we use the mass of N_2, $m = 4.605 \times 10^{-26}$ kg, and the mean translational energy of 5.66×10^{-21} J. Because the average kinetic energy of a hydrogen molecule is the same as that of a nitrogen molecule and the mass of a nitrogen molecule is 14 times larger than that of a hydrogen molecule, the root mean square speed of the nitrogen molecules is smaller than that of the hydrogen molecules by a factor of $1/\sqrt{14}$.

14.2.3 Thermal Equilibration and Heat

Consider what happens if we immerse a small container with gas molecules into a huge container filled with gas molecules. The large container acts as a bath for the small container. Initially, if the molecules in the small container move slowly:

$$\overline{E}_{\text{trans, small container}} < \overline{E}_{\text{trans, bath}}$$

the molecules in the bath will transfer energy to the small container by colliding with the wall of the small container. After some time the mean translational energy of the molecules in the small container equals the mean translational energy in the bath:

$$\overline{E}_{\text{trans, small container}} = \overline{E}_{\text{trans, bath}}$$

This result can be related to a well-known fact in everyday life. If no other work is done when a hot body is brought into thermal contact with a cold body, the hot body cools down and the cold body warms up until both bodies are equal in temperature (*thermal equilibration*). In other words, heat flows from the hot body to the cold body. This energy transfer (or heat) from the assembly with high thermal energy to the assembly with low thermal energy occurs spontaneously and proceeds until an equilibrium (or balance) is reached.

Heat going from a hot to a cold body when contacting the bodies can be understood as energy being transferred from the assembly with high thermal energy to the assembly with low thermal energy. In molecular terms, the observation that net energy transfer occurs only from a hot to a cold body without performing work results from the establishment of a dynamic equilibrium.

14.2.4 Ideal Gas Law and the Definition of Absolute Temperature

14.2.4.1 Temperature Measures the Average Translational Energy

When we transfer a cylinder filled with 1 mol of a gas at a pressure of $P = 1$ bar from a bath of melting ice (Fig. 14.3) to a bath with boiling water, the gas volume increases from 22.7119 to 31.0256 L (Fig. 14.7). This means that the change from an ice bath to a bath with boiling water changes the product PV of the gas, so that

$$(PV)_{\text{boiling water}} = (PV)_{\text{melting ice}} \cdot \frac{31.0256}{22.7119} \tag{14.21}$$

A higher value of PV (or a warmer bath) corresponds to a higher average translational energy of the gas particles. Consequently, the molecules move more rapidly as the bath is warmed up (Fig. 14.7).

From these considerations we see that the motion of molecules can be used to decide if the bath is cold or warm, and vice versa. Therefore, we define a quantity, the thermodynamic temperature T of the bath, which is proportional to the mean translational energy of the gas molecules, namely

$$\overline{E}_{\text{Trans}} = C \cdot T \tag{14.22}$$

where C is a constant. From Equation (14.15) it follows that

$$\begin{aligned} PV &= \tfrac{2}{3}N \cdot \overline{E}_{\text{Trans}} = \tfrac{2}{3}NC \cdot T \\ &= \tfrac{2}{3}\mathbf{nN}_A \cdot C \cdot T \end{aligned} \tag{14.23}$$

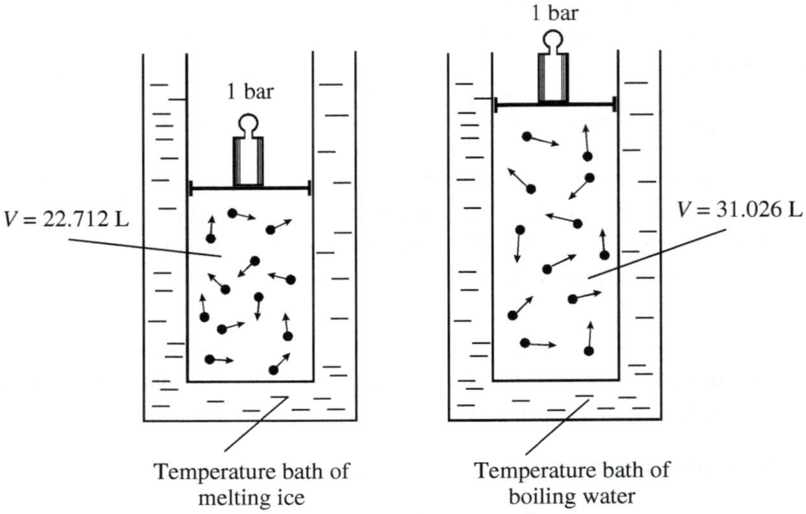

Figure 14.7 Expansion of a gas when transferring the gas from a bath with melting ice to a bath with boiling water.

This definition enables us to measure the temperature T by measuring PV.

We define the value of the temperature by setting the water's triple point temperature to $T = 273.160$ temperature units. We call this temperature unit, Kelvin (K), and the temperature T is measured by

$$\boxed{T = 273.160\ \text{K} \cdot \frac{(PV)_T}{(PV)_{\text{triple point bath of water}}}} \tag{14.24}$$

T is called the absolute temperature because it is not conceivable that PV can become less than zero. A triple point bath of water is an ice bath prepared from pure water without dissolved air, so the three phases of water (liquid, solid, and gas) are in equilibrium under a pressure of 6.133 mbar. For more details see Chapter 19 (Figs. 19.4 and 19.5).

We choose the temperature scale so that the temperature difference between a bath with melting ice[3] and boiling water[4] is 100 K. These two temperatures are the fixed points of the *Celsius scale* (temperature t), see Fig. 14.8. The Celsius and Kelvin scales are related by

$$\frac{t}{°\text{C}} = \frac{T - 273.15\ \text{K}}{\text{K}}$$

Accordingly, 273.15 K corresponds to 0 °C and 373.15 K corresponds to 100 °C. The triple point temperature of water (273.160 K) is then 0.01 °C.

Relation (14.24) tells us how to measure the absolute temperature T. We compare the pressure of a gas in a container at constant volume

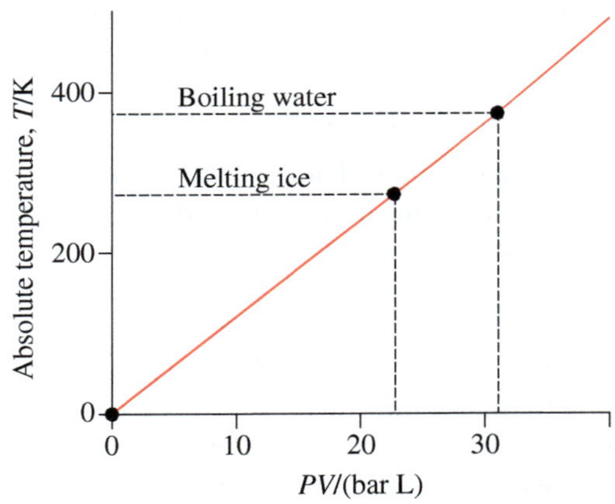

Figure 14.8 The absolute temperature T as a function of PV.

3 Ice bath saturated with air at 1 atm = 1.013 bar, temperature 273.15 K.
4 Boiling water at 1 atm, temperature 273.160 K × 31.0256/22.7119 = 373.15 K.

and temperature to its pressure when putting the container at constant volume into a temperature bath at the triple point of water. Generally, a temperature bath is defined by the property that the heat exchanged between the system and the bath does not change the bath's temperature by a measurable amount. In the following, the temperature bath is indicated in all figures only schematically.

14.2.4.2 Ideal Gas Law

With this definition of the temperature, we are able to specify the constant C in Equation (14.22). Using $\overline{E}_{\text{trans}}$ from (14.18) we obtain

$$C = \frac{\overline{E}_{\text{trans}}}{T} = \frac{5.66 \times 10^{-23}\ \text{bar L}}{273.160\ \text{K}} = 2.072 \times 10^{-25}\ \text{bar L K}^{-1}$$

It is customary to use the expression $\frac{3}{2}k$ instead of C in (14.22):

$$\boxed{\overline{E}_{\text{trans}} = \frac{3}{2}kT} \qquad (14.25)$$

We rewrite Equation (14.23) as

$$\boxed{PV = NkT = \mathbf{n}\mathbf{N}_A kT = \mathbf{n}\mathbf{R}T} \qquad (14.26)$$

The quantity k is called the *Boltzmann constant:*

$$k = \frac{2}{3} \cdot C = \frac{2}{3} 2.072 \times 10^{-25}\ \text{bar L K}^{-1} = 1.381 \times 10^{-25}\ \text{bar L K}^{-1}$$
$$= 1.381 \times 10^{-23}\ \text{J K}^{-1}\ (\text{more accurately: } k = 1.380658 \times 10^{-23}\ \text{J K}^{-1})$$

The constant

$$\mathbf{R} = \mathbf{N}_A k = 6.022 \times 10^{23} \cdot 1.381 \times 10^{-25}\ \text{bar L K}^{-1}\ \text{mol}^{-1}$$
$$= 8.314 \times 10^{-2}\ \text{bar L K}^{-1}\ \text{mol}^{-1} = 8.314\ \text{J K}^{-1}\ \text{mol}^{-1}$$

is called the *gas constant* (more accurately: $\mathbf{R} = 8.314510\ \text{J K}^{-1}\ \text{mol}^{-1}$).

Equation (14.26) is called the *ideal gas law* (Fig. 14.9). An ideal gas is a hypothetical assembly of particles that do not interact by intermolecular forces. Real gases in high dilution behave approximately like an ideal gas. The ideal gas law can also be written in the form

$$\boxed{PV_m = \mathbf{R}T} \qquad (14.27)$$

where

$$V_m = \frac{V}{\mathbf{n}} \qquad (14.28)$$

is the *molar volume.*

Equations (14.25) and (14.26) are fundamental in physical chemistry. Equation (14.25) gives the mean translational energy E_{trans} of a molecule as a function of temperature T. Equation (14.26) connects pressure P, volume V, and temperature T for an ideal gas.

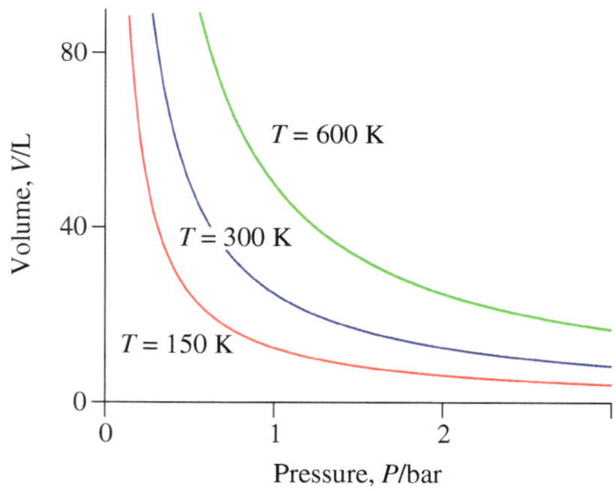

Figure 14.9 Ideal gas. V as a function of P (isotherms) for $\mathbf{n} = 1$ mol and different temperatures T according to (14.28).

14.2.4.3 Extensive and Intensive Variables, Molar Quantities

The volume is called an *extensive variable* because it depends on the amount of substance. In contrast, an *intensive variable* does not depend on the amount. For a substance, an extensive variable can be made intensive by dividing it by the amount of substance. For example, the volume V is an extensive variable, and the molar volume V_m is an intensive variable. Molar quantities such as V_m are represented by the index "m" throughout this book (see Box 14.1).

14.2.4.4 Measuring Temperature

At this point we describe more precisely our procedure of measuring the temperature. When plotting PV_m at constant temperature as a function of P according to Equation (14.26), a line parallel to the P-axis is expected. In reality, however, deviations occur at higher pressure (Fig. 14.10). These deviations result from the interaction forces between the gas molecules and from the fact that the volume of the molecules is not small enough to be neglected completely. At $P = 1$ bar the deviations from the ideal gas law are on the order of 1% for air. However, as $P \to 0$ all the curves in Fig. 14.10 meet at the same point independent of the nature of the gas. Following this observation we replace PV by the limit of PV for $P \to 0$ in the ideal gas law. Then instead of Equation (14.26) we have

$$\lim_{P\to 0} PV = \mathbf{n}\mathbf{R}T \tag{14.29}$$

Only this form enables us to define the temperature T exactly:

$$T = \frac{\lim_{P\to 0} PV}{\lim_{P\to 0} (PV)_{\text{triple point bath of water}}} \times 273.160\ \text{K} \tag{14.30}$$

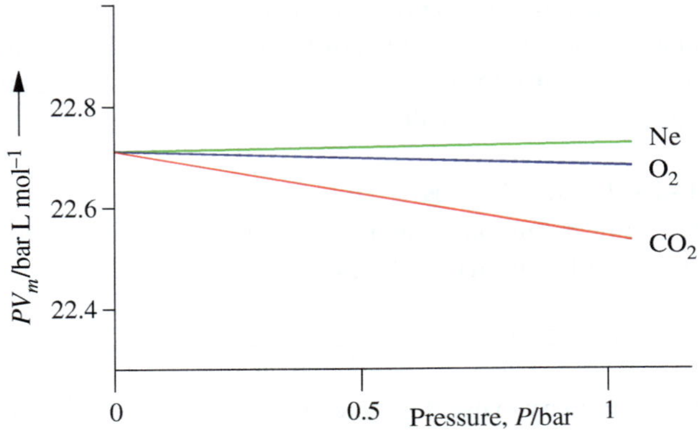

Figure 14.10 PV_m as a function of pressure P at $T = 273.16$ K. This figure shows that at $P = 1$ bar the deviation from the ideal gas law is less than 1%.

14.2.5 Law of Partial Pressures

The analysis presented here leads naturally to the observation that a gas comprising different types of molecules has a total pressure that is a sum of the "partial pressures" of the individual types. For example, air is a mixture of different molecules and the total pressure for a container of air is a sum of a partial pressure for each molecule type

$$P = P_{N_2} + P_{O_2} + P_{Ar} + P_{CO_2} + P_{Ne} + P_{He} + P_{Kr} + P_{Xe} + P_{CH_4} + P_{H_2}$$

If we assume that the gas molecules do not interact with each other, then we can treat each gas type independently. Thus

$$P = \sum_i P_i \tag{14.31}$$

and

$$P_i = x_i P$$

where

$$x_i = \frac{\mathbf{n}_i}{\mathbf{n}}, \ \mathbf{n} = \sum_i \mathbf{n}_i$$

is called the *mole fraction* of gas i. P_i is called the partial pressure of gas i. We can write Equation (14.31) in the form

$$P = \sum_i P_i = \sum_i x_i P \tag{14.32}$$

This result is called the *Law of partial pressures*.

14.3 Speed and Collisions of Gas Molecules

14.3.1 Average Speed of Molecules in a Gas

From the average translational energy we can calculate a speed for the gas molecules. According to Equations (14.10) and (14.25) we have

$$\overline{v^2} = \frac{2\overline{E}_{\text{trans}}}{m} = \frac{3kT}{m} = \frac{3\mathbf{R}T}{\mathbf{M}} \tag{14.33}$$

Here the mass m of a molecule is expressed by the molar mass $\mathbf{M}$ and Avogadro's constant

$$\mathbf{M} = m \cdot \mathbf{N}_A \tag{14.34}$$

The speed is obtained by taking the square root of $\overline{v^2}$

$$v_{\text{rms}} = \sqrt{\overline{v^2}} = \sqrt{\frac{3\mathbf{R}T}{\mathbf{M}}} \tag{14.35}$$

Often, it is more useful to calculate $\overline{v}$, the mean speed

$$\overline{v} = \int v \cdot \mathrm{d}\mathcal{P}$$

where $\mathrm{d}\mathcal{P}$ is the probability that the speed is in the interval between v and $v + \mathrm{d}v$. When $\mathrm{d}\mathcal{P}$ is described by the Maxwell–Boltzmann distribution (considered in Chapter 15) we find that

$$\overline{v} = \sqrt{\frac{8kT}{\pi m}} = \sqrt{\frac{8\mathbf{R}T}{\pi\mathbf{M}}} \tag{14.36}$$

The mean speed $\overline{v}$ is smaller than v_{rms} by 8%; when averaging the speeds, small and large values of v contribute equally, whereas when averaging the squares, larger values of v contribute more, biasing the average to larger values.

Example 14.2 *Calculate the mean speed of an H_2 molecule at $T = 300$ K and two lower temperatures. Compare your result to the root mean square speed.*

Using $\mathbf{M} = 2.016$ g mol^{-1} we find

$$\overline{v} = \sqrt{\frac{8 \times 8.314\ \text{J K}^{-1}\ \text{mol}^{-1} \times 300\ \text{K}}{\pi \times 2.016 \times 10^{-3}\ \text{kg mol}^{-1}}} = \sqrt{3.15 \times 10^6\ \text{J kg}^{-1}}$$
$$= 1.77 \times 10^3\ \text{m s}^{-1} = 1.77\ \text{km/s}$$

The H_2 molecules move with a very high speed at room temperature. Performing the calculation at different temperatures 100 K and 25 K gives the results below. The v_{rms} is calculated according to the procedure in Example 14.1.

T/K	$\overline{v}$/(m s^{-1})	v_{rms}/(m s^{-1})
300	1770	1921
100	1022	1109
25	511	555

If we cool the gas, the average speed and the root mean square speed of the molecules become lower. The root mean square speed is consistently larger than the mean speed, $\overline{v} = 0.92 \cdot v_{\text{rms}}$.

According to Equation (14.33) $\overline{v^2}$ is inversely proportional to the mass of the molecules, heavy molecules move slower than light molecules. For example, the speed of Br_2 molecules ($\mathbf{M} = 159$ g mol^{-1}) at 300 K is $\overline{v} = 203$ m s^{-1}, which is one-tenth of the speed of H_2 molecules. A number of methods have been devised to measure the speed of molecules. Figure 14.11 shows a simple method for measuring the speed of atoms or molecules. In this method we fire a stream of particles (e.g., Na atoms) at two discs, which are rotating on the same axis inside a vacuum chamber (Fig. 14.11). The first of these discs has a slit through which the

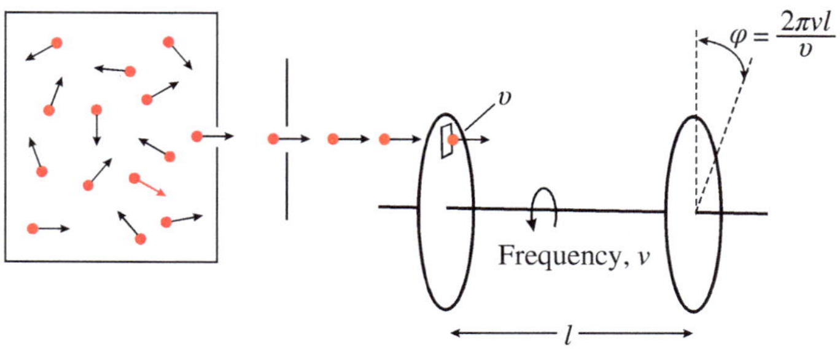

Figure 14.11 Apparatus for measuring the speed of molecules.

molecules can pass. The particles deposit on the surface of the second disc and are detected as a small spot. Because the molecules require a finite time to reach the second disc, it will have rotated by an additional angle φ before the particles strike it. The speed can be calculated from the angle φ, the frequency ν of rotation, and the distance l between the two discs.

14.3.1.1 Effusion

Consider a gas that is confined in a container with a very small hole (diameter less than the mean free path, Section 14.3.2) connected to a vacuum line. The number of molecules hitting the wall of the container is, according to Equation (14.16), proportional to their speed v. Therefore, the probability that a molecule enters the hole and escapes is also proportional to v. Thus the rate of effusion is proportional to the speed. Because the speed, on the average, is inversely proportional to the square root of the molecular mass (*Graham's Law*), this container allows us to separate isotopes. Light isotopes escape the container faster than heavy isotopes.

Example 14.3 *Separation of isotopes by effusion. A mixture of* $^{235}UF_6$ *(0.1%) and* $^{238}UF_6$ *(99.9%) is allowed to flow from a container through a small hole into vacuum for a short time. By how much is the lighter isotope enriched as a consequence of its larger velocity?*

According to Equation (14.36) we have $\overline{v_1}/\overline{v_2} = \sqrt{m_2/m_1} = 1.004$. If we attribute v_1, m_1 to the compound $^{235}UF_6$ and v_2, m_2 to the compound $^{238}UF_6$, then the lighter molecules are faster by 0.4%. This causes an enrichment from 0.1% to 0.1004%. The process must be repeated many times to obtain an appreciable separation.

14.3.2 Mean Free Path and Number of Collisions

Even though a molecule in a gas moves at a high speed, it may not move very far before its trajectory is changed by collisions with other molecules (Fig. 14.12). It is of general interest to know the average path length of a molecule between subsequent collisions (*mean free path*). We will see that the frequency of collisions is related to the mean free path and to the speed $\overline{v}$. The collision frequency is of particular interest when considering the rate of chemical reactions (Chapter 24).

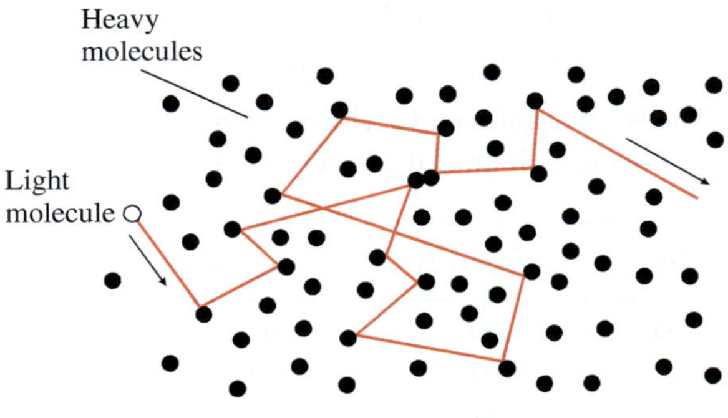

Figure 14.12 A zigzag path of a light molecule in a gas of heavy molecules.

14.3.2.1 Mean Free Path

The mean free path of a molecule will depend on its speed and the density of molecules surrounding it. Before calculating the mean free path let us first determine the average distance a between gas molecules. To simplify, imagine that the molecules are evenly distributed in a cube of length L (Fig. 14.13); then

$$L^3 = N \cdot a^3$$

where N is the total number of molecules in the cube and a is the average distance of two nearest neighbors. Solving for a we find

$$a = \sqrt[3]{\frac{L^3}{N}} = \sqrt[3]{\frac{V}{N}} = \sqrt[3]{\frac{kT}{P}}$$

where we have used the ideal gas law for the last equality. At $P = 1$ bar and $T = 273.16$ K, the average distance is $a = 3.3 \times 10^{-9}$ m $= 3.3$ nm.

The free path λ (the distance a molecule traverses between collisions) is much longer than the average distance a because molecules can pass many neighboring molecules between collisions (Fig. 14.12). To determine the mean free path $\overline{\lambda}$, we make some simplifying assumptions.

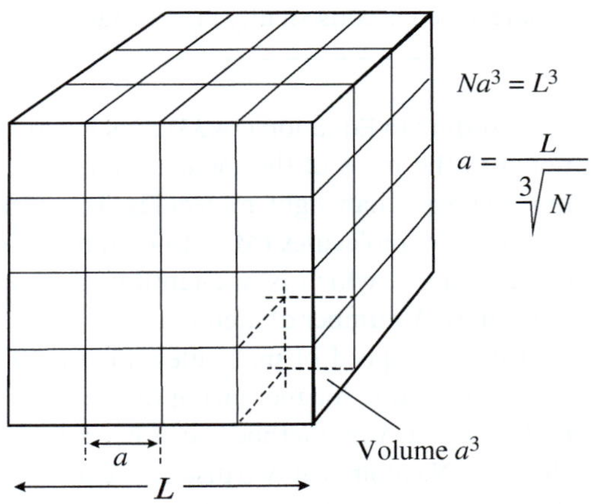

Figure 14.13 The average distance a of two molecules in a gas is estimated by assuming, for simplicity, a regular arrangement of gas molecules in a cube.

We approximate the molecules as hard spheres, and we consider, a particle of type 1 moving among many particles of type 2 that are so massive that they can be considered nearly stationary. A collision occurs each time a type 1 particle with radius r_1 touches a type 2 particle with radius r_2 (Fig. 14.14). As the type 1 molecule moves through space it sweeps out a cylindrical volume. A collision with a type 2 particle will only occur if its center of mass lies inside the surface of a cylinder of diameter $2(r_1 + r_2)$. Type 2 particles that have their centers of mass inside or on the surface of the considered cylinder collide with the light particle, and all those outside of this cylinder do not.

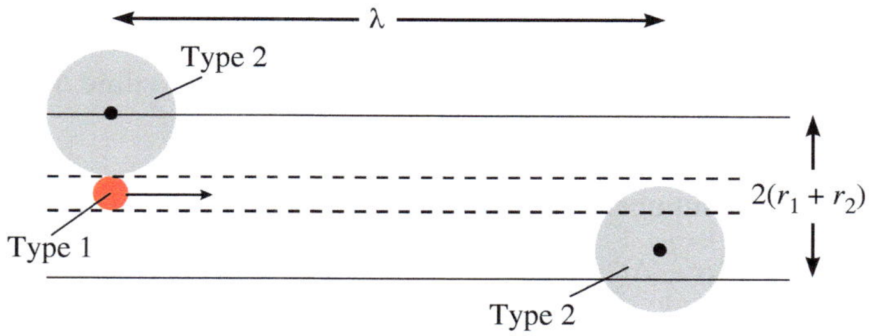

Figure 14.14 Calculation of the mean free path λ. A collision occurs when the colliding molecules of type 1 and type 2 touch each other.

According to Fig. 14.14, a collision occurs when the type 1 particle has traveled a distance λ. The volume V_{cyl} that is swept out before the collision is

$$V_{cyl} = \pi(r_1 + r_2)^2 \lambda \tag{14.37}$$

If there are N type 2 particles in the total volume V of the gas, then the average volume element for a type 2 particle is V/N and the average cylinder volume $\overline{V}_{cyl}$ is

$$\overline{V}_{cyl} = \frac{V}{N} = \pi(r_1 + r_2)^2 \,\overline{\lambda} \tag{14.38}$$

where $\overline{\lambda}$ is the mean free path. Thus we obtain

$$\overline{\lambda} = \frac{V}{\pi(r_1 + r_2)^2 N} \tag{14.39}$$

and can write

$$\boxed{\overline{\lambda} = \frac{V}{\sigma N} \quad \text{(one light particle among } N \text{ heavy particles)}} \tag{14.40}$$

where we define the collision cross section σ as

$$\sigma = \pi(r_1 + r_2)^2 \tag{14.41}$$

In our consideration we assumed that the heavy molecules are practically at rest. In the general case (mean free path $\overline{\lambda}_1$ of a particle with mass m_1 moving among particles with mass m_2) we obtain [2]

$$\boxed{\overline{\lambda}_1 = \frac{V}{\sigma N} \cdot \frac{1}{\sqrt{1 + m_1/m_2}} \quad \text{(one particle of mass } m_1 \text{ among } N \text{ particles of mass } m_2)} \tag{14.42}$$

For $m_1 \ll m_2$ this leads to Equation (14.40), and for $m_1 = m_2$ we obtain

$$\boxed{\overline{\lambda} = \frac{V}{\sqrt{2}\sigma N} \quad \text{(one particle of mass } m \text{ among } N \text{ particles of mass } m)} \tag{14.43}$$

In the case $m_1 \gg m_2$ (one very heavy particle moving among light particles) the mean free path approaches zero; many collisions with the light particles occur, while the heavy particle travels only a very small distance.

Example 14.4 *Determine the mean free path for an H_2 molecule (r_{H_2} = 138 pm) moving among Br_2 (r_{Br_2} = 312 pm) molecules at P = 1 bar and T = 350 K.*

In this case a light particle is moving among heavy particles. With $(r_1 + r_2) = (138 + 312)$ pm = 450 pm, where the index 1 refers to the H_2 molecule and the index 2 refers to the Br_2 molecules, we obtain $\sigma = 6.36 \times 10^{-19}$ m^2. Under these conditions, we have the number density $N/V = P/kT = 2.07 \times 10^{25}$ m^{-3}. Using Equation (14.40) we find that

$$\overline{\lambda}_{H_2} = \frac{1}{6.36 \times 10^{-19}\ \text{m}^2 \cdot 2.07 \times 10^{25}\ \text{m}^{-3}} \frac{1}{\sqrt{1 + 2.02/159.80}} = 75.5\ \text{nm}$$

$\overline{\lambda}$ is then about 20 times larger than the average distance 3.6 nm between the gas molecules.

14.3.2.2 Number of Collisions

From $\bar{\lambda}$ and the average speed $\bar{v}$ we can calculate an average number z_{coll} of collisions that occur for a molecule in a time interval Δt. The total distance s that is traveled in the time Δt can be written as

$$s = \bar{\lambda} z_{coll} = \bar{v} \Delta t \tag{14.44}$$

and for the number of collisions we obtain

$$z_{coll} = \frac{\bar{v}}{\bar{\lambda}} \Delta t \tag{14.45}$$

Example 14.5 *Calculate the number of collisions in* $\Delta t = 1\,s$ *for* H_2 *molecules* ($r_{H_2} = 138\,pm$) *at* $T = 300\,K$ *and* $P = 1\,bar$.

From Example 14.2 we know that the average speed is $\bar{v} = 1.77 \times 10^3$ m s^{-1}. Using Equation (14.43), we calculate the mean free path to be

$$\bar{\lambda} = \frac{V}{\sqrt{2}\sigma N} = \frac{kT}{\sqrt{2}\sigma P}$$
$$= \frac{1.380658 \times 10^{-23} \cdot 300\ \text{J}}{\sqrt{2}\pi \cdot (276 \times 10^{-12})^2 \cdot 10^5\ \text{m}^2\ \text{Pa}} = 122\ \text{nm}$$

Now we can evaluate the number of collisions z_{coll}

$$z_{coll} = \frac{1.77 \times 10^3\ \text{m s}^{-1}}{122 \times 10^{-9}\ \text{m}} = 1.45 \times 10^{10}\ \text{s}^{-1} \cdot \Delta t$$

Thus in every second a molecule collides about 10^{10} times with other molecules in the gas.

14.3.3 Diffusion

From the above discussion it is clear that a gas molecule does not move freely from one side of a container to the other, at atmospheric pressure and common temperatures. Instead the molecule collides many times with other molecules as it moves through space. This mechanism of translational motion is operative for both gases and liquids and is called diffusion. Figure 14.15 illustrates a diffusive trajectory for a gas molecule.

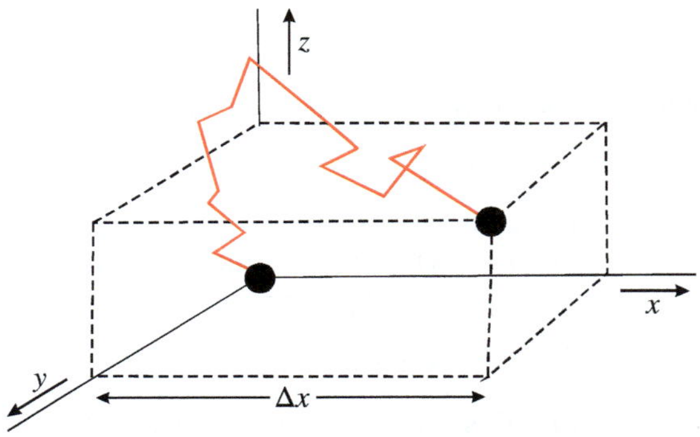

Figure 14.15 A zigzag path of a molecule. After time Δt the molecule traverses a distance Δx in the x-direction.

14.3.3.1 Mean Displacement in the x-Direction (Random Walk Model)

We ask, "How far can a gas molecule travel from its initial position in a time Δt?" Again we simplify by considering a light molecule moving in a gas of heavy molecules. The light molecule, after collision, moves in a random direction as illustrated for three typical cases (Fig. 14.16). The solid arrows denote the initial velocity of a light particle, the dashed arrows denote the final velocity of the light particle after collision. In contrast to the light particle, the motion of the heavy particle is essentially unaffected, it takes many collisions to change its motion gradually into a random one.

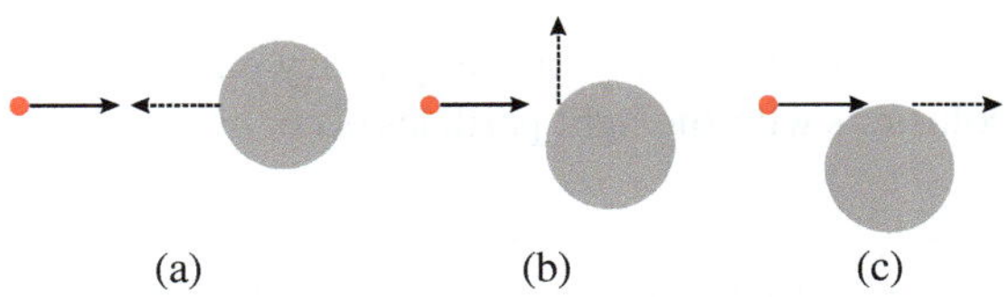

Figure 14.16 Three collision scenarios: (a) a central ("head on") collision, (b) a collision at 45°, (c) a glancing collision.

To simplify, we first consider the one-dimensional case and assume that the light molecule, after collision, moves a distance λ with equal probability in the positive or negative x direction (*random walk model*). Thus we neglect the probability distribution of λ and set $\lambda = \bar{\lambda}$.

Imagine the particle starts at the origin and proceeds a distance λ in the positive x-direction. At the new position the particle has an equal probability of continuing in the positive x-direction during the next time interval or proceeding in the reverse direction. The next step in the particle's trajectory will either return it back to its original starting point (the

$-\lambda$ event), or it will proceed forward to 2λ. If diffusion has been proceeding for some time the distribution broadens but its average value remains at zero. In order to calculate the spread in the distribution we compute $\overline{x^2}$ where $x = 0$ is the position of the particle at the beginning.

In Box 14.3 we use this one-dimensional model to find the mean square displacement $\overline{x^2}$,

$$\boxed{\overline{x^2} = j \cdot \lambda^2} \tag{14.46}$$

Of course any particular trajectory may be different from this average. For example, after $j = 7$ collisions on the path, the net displacement could be

$$x = \lambda + \lambda + \lambda - \lambda + \lambda - \lambda + \lambda = 3\lambda$$

(five times by λ in the $+x$-direction, and twice in the $-x$-direction), and we obtain $x^2 = 9 \cdot \lambda^2$. Alternatively the particle may take one step in the $+x$-direction and six steps in the $-x$-direction.

$$x = \lambda - \lambda - \lambda - \lambda - \lambda - \lambda - \lambda = -5\lambda$$

so that $x = -5\lambda$ and $x^2 = 25\lambda^2$. On average, according to Equation (14.46), we expect

$$\overline{x^2} = 7 \cdot \lambda^2 \tag{14.47}$$

The root mean square displacement x_{rms} is

$$\boxed{x_{\text{rms}} = \sqrt{\overline{x^2}} = \sqrt{j} \cdot \lambda}$$

Distribution Function of Diffusion Path The evolution of the diffusion path is shown in Table 14.1 for $j = 1$ to $j = 8$. We start at $x = 0$. After one step the particle can be at $x = -\lambda$ or at $x = +\lambda$ with equal probability. Then the mean square of the displacement after this step is

$$\overline{x^2} = \frac{1}{2}\left(\lambda^2 + \lambda^2\right) = \lambda^2$$

After two steps the particle can be found at $x = -2\lambda$ (one possibility), at $x = 0$ (two possibilities), or at $x = +2\lambda$ (one possibility), and we obtain

$$\overline{x^2} = \frac{1}{4}\left(4\lambda^2 + 0 + 4\lambda^2\right) = 2\lambda^2$$

Table 14.1 Number $N_j(x_i)$ of Possibilities to Find a Particle at Position $x_i = i \cdot \lambda$ After Taking Sequential Steps in the x and $-x$-Direction.

	Positions x_i in Units of λ																
	To the Left												To the Right				
i	−8	−7	−6	−5	−4	−3	−2	−1	0	1	2	3	4	5	6	7	8
j																	
0									1								
1								1		1							
2							1		2		1						
3						1		3		3		1					
4					1		4		6		4		1				
5				1		5		10		10		5		1			
6			1		6		15		20		15		6		1		
7		1		7		21		35		35		21		7		1	
8	1		8		28		56		70		56		28		8		1

The distributions and the mean square displacements $\overline{x^2}$ for the first eight steps are summarized in Tables 14.1 and 14.2 (see also the data points in Fig. 14.17).

In the general case, $N_j(x_i) = N_j(i\lambda)$ is the number of different ways (trajectories) to reach position $x_i = i \cdot \lambda$, where $i = -j, (-j+2), \ldots +j$ and j is the number of steps. With these definitions, the mean square displacement after j diffusion steps, $\overline{x_j^2}$, is given by

$$\overline{x_j^2} = \frac{\sum_i x_i^2}{\sum_i N_j(x_i)} = \frac{\sum_i N_j(i\lambda) \cdot i^2 \cdot \lambda^2}{\sum_i N_j(i\lambda)}$$

Table 14.2 confirms the result of Equation (14.46), $\overline{x^2} = j \cdot \lambda^2$, and Figure 14.17 plots the number $N_j(x_i)$ of possibilities to find a particle versus the displacement $x_i = i \cdot \lambda$ for $j = 8$ diffusion steps (dots). $N_j(x)$ can be described by a Gaussian function (solid line in Fig. 14.17, omitting the index i in x_i)

$$N_j(x) = const \cdot e^{-bx^2} \tag{14.48}$$

with $b = 1/(16\lambda^2)$. The mean square displacement is then

$$\overline{x^2} = \frac{\int_{-\infty}^{\infty} x^2 \cdot e^{-bx^2} \cdot dx}{\int_{-\infty}^{\infty} e^{-bx^2} \cdot dx} = \frac{1}{b}\frac{\int_{-\infty}^{\infty} z^2\, e^{-z^2} \cdot dz}{\int_{-\infty}^{\infty} e^{-z^2} \cdot dz} = \frac{1}{b}\frac{\frac{1}{2}\sqrt{\pi}}{\sqrt{\pi}}$$

$$= \frac{1}{2b} = 8 \cdot \lambda^2$$

as expected from Equation (14.46). That is, we obtain

$$\overline{x^2} = j\lambda^2 = \frac{1}{2b} \quad \text{or} \quad b = \frac{1}{2j\lambda^2}$$

and the distribution function is well described by

$$\mathcal{P}_j(x) = \frac{N_j(x)}{N} = const \cdot \exp\left[-\frac{1}{2j\lambda^2}x^2\right] \tag{14.49}$$

where $N = \sum_x N_j(x)$. $\mathcal{P}_j(x) \cdot dx$ is the probability of finding a particle between x and $x + dx$ after j diffusion steps.

Table 14.2 Mean Square Displacements $\overline{x_j^2}$ in the Model of Table 14.1 After j Sequential Steps.

Number j of Steps	$\sum_i x_i^2$	Number $\sum_i N_j(x_i)$ of Possibilities	$\overline{x_j^2}$
1	$2 \cdot \lambda^2$	2	$1 \cdot \lambda^2$
2	$8 \cdot \lambda^2$	4	$2 \cdot \lambda^2$
3	$24 \cdot \lambda^2$	8	$3 \cdot \lambda^2$
4	$64 \cdot \lambda^2$	16	$4 \cdot \lambda^2$
5	$160 \cdot \lambda^2$	32	$5 \cdot \lambda^2$
6	$384 \cdot \lambda^2$	64	$6 \cdot \lambda^2$
7	$896 \cdot \lambda^2$	128	$7 \cdot \lambda^2$
8	$2048 \cdot \lambda^2$	256	$8 \cdot \lambda^2$

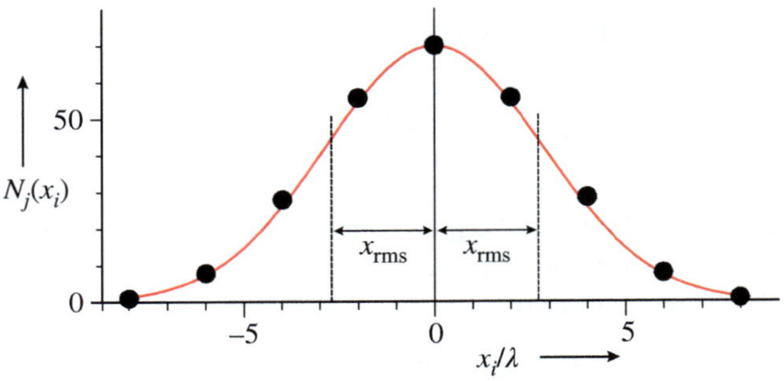

Figure 14.17 Distribution of one-dimensional diffusion path. Number $N_j(x_i)$ of possibilities to find a diffusing particle at displacement x_i after $j = 8$ diffusion steps of size λ. Dots: data according to Table 14.1. Solid line: Gaussian function, according to Equation (14.48) with $b = 1/(16\lambda^2)$. x_{rms} is the root mean square displacement, $x_{rms} = \sqrt{\overline{x^2}}$.

14.3.3.2 Mean Square Displacement in Space

For three-dimensional motion of the molecules, only every third collision, in our approximation, leads to a displacement in the x-direction. Hence the mean square displacement in the x-direction is only one-third of the value calculated by Equation (14.46), namely

$$\boxed{\overline{x^2} = \frac{1}{3} j \cdot \lambda^2} \tag{14.50}$$

(see Foundation 14.1 for a proof). We obtain a corresponding result for displacements in the y-direction and z-direction, so that

$$\overline{r^2} = \overline{x^2} + \overline{y^2} + \overline{z^2} = j \cdot \lambda^2$$

A more rigorous analysis should account for the distribution of the free path lengths (Box 14.4). In this case the result is similar, however we must replace λ^2 by $\overline{\lambda^2}$ in Equation (14.50), and we find

$$\boxed{\overline{x^2} = \tfrac{1}{3} j \cdot \overline{\lambda^2}} \tag{14.51}$$

Distribution Function in Space The probability of finding a particle in a volume element dx dy dz after j diffusion steps is

$$\mathcal{P}_j(x, y, z) \cdot \mathrm{d}x\ \mathrm{d}y\ \mathrm{d}z = const \cdot \mathrm{e}^{-bx^2}\ \mathrm{e}^{-by^2}\ \mathrm{e}^{-bz^2} \cdot \mathrm{d}x\ \mathrm{d}y\ \mathrm{d}z \tag{14.52}$$

$$= const \cdot \exp\left[-\frac{3}{2j\overline{\lambda^2}}\left(x^2 + y^2 + z^2\right)\right] \cdot \mathrm{d}x\ \mathrm{d}y\ \mathrm{d}z$$

$$= const \cdot \exp\left[-\frac{3}{2j\overline{\lambda^2}} r^2\right] \cdot \mathrm{d}x\ \mathrm{d}y\ \mathrm{d}z$$

(with $r^2 = x^2 + y^2 + z^2$), and the relation for $\overline{x^2}$ is

$$\overline{x^2} = \frac{1}{3}\ j\overline{\lambda^2} = \frac{1}{2b} \qquad \text{or} \qquad b = \frac{3}{2j\overline{\lambda^2}}$$

The probability of finding the particle in a distance between r and $r + \mathrm{d}r$ (in a spherical shell of volume $4\pi r^2 \mathrm{d}r$) after j diffusion steps is

$$\mathcal{P}_j(r) \cdot \mathrm{d}r = const \cdot 4\pi r^2 \exp\left[-\frac{3}{2j\overline{\lambda^2}} r^2\right] \cdot \mathrm{d}r \tag{14.53}$$

where *const* is given by the normalization condition $\int_{r=0}^{\infty} \mathcal{P}_j(r) \cdot \mathrm{d}r = 1$.

14.3.3.3 Mean Square Displacement as a Function of Time

Because $j\overline{\lambda}$ is the total path length traversed by the molecule in time Δt, we can write

$$j\overline{\lambda} = \overline{v}\Delta t \tag{14.54}$$

which allows us to transform Equation (14.51) into

$$\overline{x^2} = \frac{1}{3} \cdot \frac{\overline{\lambda^2}}{\overline{\lambda}} \cdot \overline{v}\Delta t \tag{14.55}$$

The average of the square $\overline{\lambda^2}$ is larger than the square of the average $\overline{\lambda}^2$, because in averaging the square's large values contribute more. In Box 14.4 we show that

$$\overline{\lambda^2} = 2 \cdot \left(\overline{\lambda}\right)^2 \tag{14.56}$$

Thus, we can cast our result for the mean square displacement in the x-direction as

$$\boxed{\overline{x^2} = \frac{2}{3}\ \overline{\lambda}\ \overline{v} \cdot \Delta t \quad \text{(light particle among heavy particles)}} \tag{14.57}$$

$\overline{x^2}$ increases proportionally to Δt. This result means that the root mean square displacement increases as the square root of time ($\sqrt{\overline{x^2}} = x_{\mathrm{rms}} \propto \sqrt{\Delta t}$), not linearly in time as we would expect from Newton's law of motion for a free particle. For example, increasing the time by a factor of four results in an increase of the displacement by a factor of two only.

With the relations (14.36) and (14.40) we can rewrite Equation (14.57) as

$$\overline{x^2} = \frac{2}{3} \cdot \frac{V}{\sigma \cdot N} \cdot \sqrt{\frac{8kT}{\pi m}} \cdot \Delta t \tag{14.58}$$

This equation relates $\overline{x^2}$ to the number density N/V of the heavy particles, to the mass m of the light particle, and to the collision cross section σ. In the general case of a particle of mass m_1 moving among particles of mass m_2 one obtains a similar result, except that the mass m has to be replaced by the reduced mass $\mu = (m_1 m_2/(m_1 + m_2))$, so that

$$\boxed{\overline{x^2} = \frac{2}{3} \cdot \frac{V}{\sigma \cdot N} \cdot \sqrt{\frac{8kT}{\pi \mu}} \cdot \Delta t} \tag{14.59}$$

For $m_1 \ll m_2$ this leads to Equation (14.58) with $\mu \approx m_1$, the mass of the light particle moving among heavy particles. For $m_1 \gg m_2$ (heavy particle moving among light particles) we obtain $\mu \approx m_2$, which is again the mass of the light particle moving among heavy particles.[5]

Example 14.6 *Compute the root mean square displacement of Br_2 molecules in air after a time of 15 s. Compare your result to the root mean square displacement of H_2 molecules in air. Use $P = 1$ bar and $T = 300$ K.*

We consider a Br_2 molecule moving in air at atmospheric pressure and room temperature. If we approximate air as nitrogen gas, the collision diameter is $r_1 + r_2 = (307 + 187)$ pm $= 494$ pm. Thus we obtain $\sigma = 7.67 \times 10^{-19}$ m^2. Because $m_{Br_2} = 26.4 \times 10^{-26}$ kg and $m_{N_2} = 4.65 \times 10^{-26}$ kg, the reduced mass is

$$\mu = \frac{26.4 \cdot 4.65}{26.4 + 4.65} \times 10^{-26} \text{ kg} = 3.95 \times 10^{-26} \text{ kg}$$

We obtain $\overline{v} = \sqrt{8kT/(\pi\mu)} = 517$ m s^{-1} and the reciprocal number density of $V/N = kT/P = 4.14 \times 10^{-26}$ m^3 at 1 bar. Combining these values in Equation (14.59) we find

$$\overline{x^2} = \frac{2}{3}\frac{4.14 \times 10^{-26}\text{m}^3}{7.67 \times 10^{-19}\text{ m}^2} \cdot 517 \text{ m s}^{-1} \cdot \Delta t = 1.86 \times 10^{-5} \text{ m}^2 \text{ s}^{-1} \, \Delta t$$

In $\Delta t = 15$ s the Br_2 molecule travels the distance $x_{rms} = \sqrt{1.86 \times 10^{-5} \times 15 \text{ m}^2} = 1.67$ cm.

The corresponding calculation for H_2 diffusing in nitrogen ($r_1 + r_2 = (137 + 187)$ pm $= 325$ pm; $\sigma = 3.31 \times 10^{-19}$ m^2; $\mu = 3.10 \times 10^{-27}$ kg) gives $\sqrt{8kT/\pi\mu} = 1845$ m s^{-1} and $x_{rms} = \sqrt{\overline{x^2}} = 4.80$ cm. In spite of its enormous speed, the root mean square displacement of the H_2 molecule is small because of the large number of collisions leading to a random walk involving both forward and backward motions.

The result for Br_2 can be checked experimentally by transferring a few drops of liquid bromine to the bottom of a cylinder filled with air and following the advancement of the brown-colored zone (Fig. 14.18a). When we plot the squares of the measured displacement versus the time Δt, we obtain the straight line displayed in Fig. 14.18b with a slope of 1.6×10^{-5} m^2 s^{-1} in good agreement with the calculation.

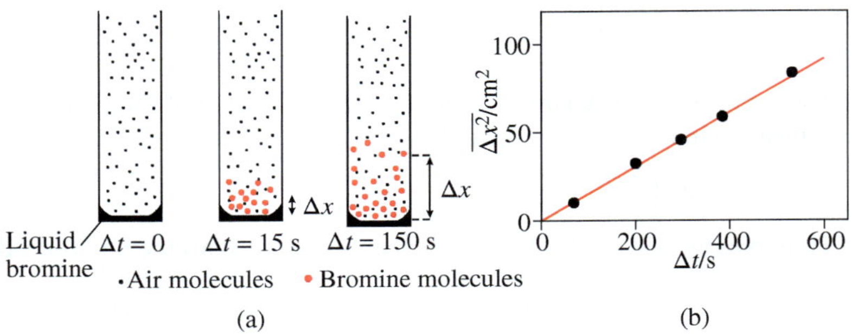

Figure 14.18 (a) The advancing of the brown color of Br_2 gas in a cylinder filled with air. (b) Mean square displacement $\overline{\Delta x^2}$ for the diffusion of Br_2 in air.

14.3.3.4 Equation of Einstein and Smoluchowski

We have shown that the mean square displacement $\overline{x^2}$ is proportional to the time Δt allowed for diffusion. Usually this proportionality is written as

$$\boxed{\overline{x^2} = \frac{1}{3}\overline{\Delta r^2} = 2D \cdot \Delta t} \tag{14.60}$$

and called the *Einstein and Smoluchowski* equation. D is called the *diffusion coefficient*. This equation is of particular importance when considering Brownian motion of particles in solution (see Section 14.4.4).

To describe molecules diffusing in a gas we combine this result with Equation (14.59) and obtain

$$D = \frac{1}{2}\frac{\overline{x^2}}{\Delta t} = \frac{1}{3} \cdot \frac{V}{\sigma \cdot N} \cdot \sqrt{\frac{8kT}{\pi\mu}}, \quad \text{where} \quad \mu = \frac{m_1 m_2}{m_1 + m_2} \tag{14.61}$$

5 Our analysis does not explicitly include the dynamics of the motion; i.e., we assumed the particle proceeds at a fixed rate rather than actually solve Newton's equation and treat the collisions. A more general treatment that describes the particle's motion correctly, but assumes that the collisions occur randomly, produces a similar result. This more general treatment shows that at early times the particles displacement grows linearly with time (according to Newton's equation) but at longer times when many collisions have occurred the displacement grows as the square root of the time (according to the random walk model).

for the diffusion coefficient. Equations (14.60) and (14.61) allow us to calculate the time Δt required for a particle to move a certain distance by diffusion.

According to Equations (14.57) and (14.60), in the special case of light particles moving among heavy particles, Equation (14.61) reduces to

$$D = \frac{\overline{\Delta x^2}}{2\Delta t} = \frac{1}{3}\overline{\lambda} \cdot \overline{v} \tag{14.62}$$

For bromine molecules diffusing through nitrogen gas at 300 K and 1 bar, we find with the data in Example 14.6

$$D = \frac{1}{3} \cdot \frac{4.14 \times 10^{-26}\ \text{m}^3}{7.92 \times 10^{-19}\ \text{m}^2} \cdot 517\ \text{m s}^{-1} = 0.90 \times 10^{-5}\ \text{m}^2\ \text{s}^{-1}$$

(which is about 75% of the experimental value, 1.2×10^{-5} m^2 s^{-1}). Correspondingly, for the diffusion of H_2 molecules through a gas of nitrogen molecules we obtain

$$D = \frac{1}{3} \cdot \frac{4.14 \times 10^{-26}\ \text{m}^3}{3.31 \times 10^{-19}\ \text{m}^2} \cdot 1845\ \text{m s}^{-1} = 7.68 \times 10^{-5}\ \text{m}^2\ \text{s}^{-1}$$

(which is about 90% of the experimental value, 8.3×10^{-5} m^2 s^{-1}).

In the case of self-diffusion (e.g., the diffusion of He atoms in a He gas) the reduced mass in Equation (14.61) is $\mu = \frac{1}{2}m$; thus, for the diffusion coefficient we obtain

$$D_{\text{self-diffusion}} = \frac{1}{3} \cdot \sqrt{2}\frac{V}{\sigma \cdot N}, \quad \text{where} \quad \overline{v} = \sqrt{\frac{8kT}{\pi m}} \tag{14.63}$$

In Table 14.3 $D_{\text{self-diffusion}}$ is calculated for some gases and compared to experiment.

Table 14.3 Diffusion Coefficients for Self-Diffusion of Some Gases Calculated from Equation (14.63) and Compared to Experiment ($T = 273$ K, $P = 1$ bar, $V/N = 3.77 \times 10^{-26}$ m^3).

Gas	$\frac{2r}{\text{pm}}$	$\frac{m}{10^{-27}\ \text{kg}}$	$\frac{v}{\text{m s}^{-1}}$	$\frac{\sigma}{10^4\ \text{pm}^2}$	$\frac{D_{\text{Calculated}}}{10^{-5}\ \text{m}^2\ \text{s}^{-1}}$	$\frac{D_{\text{Experiment}}}{10^{-5}\ \text{m}^2\ \text{s}^{-1}}$
He	218	6.65	1202	14.91	14.30	14.03
Ne	259	33.51	535	21.03	4.51	4.52
Ar	364	66.34	380	41.71	1.62	1.56
Kr	416	139.16	263	54.44	0.86	0.84
Xe	489	218.02	210	75.25	0.49	0.48
H_2	273	3.35	1693	23.42	12.82	11.9
CO_2	462	73.08	362	67.10	0.96	0.97
N_2	375	46.53	454	44.19	1.82	1.85

The Diameters $2r$ are Taken from Table 14.4.
Data taken from Landolt–Börnstein, 5. Teil, Bandteil a.

As expected, the diffusion coefficients decrease significantly from He to Xe, in accordance with experiment.

Example 14.7 *It is remarkable that single nitrogen atoms absorbed on a ruthenium single crystal surface can be observed by using a scanning tunneling microscope (STM, see Chapter 3). Each atom changes place after about 50 s and can be seen in a sequence of STM pictures. The displacement x of each single atom was measured as a function of time and can be analyzed (see Fig. 14.19).*

In Fig. 14.19 the average mean square displacement $\overline{x^2}$ is plotted versus time t. The straight line indicates that diffusive motion applies even here and is not restricted to the thermal motion of molecules in a gas [3]. The slope of the line corresponds to the diffusion coefficient of the N atom on the surface, namely

$$D = \frac{1}{2}\frac{\overline{x^2}}{\Delta t} = 3 \times 10^{-22}\ \text{m}^2\ \text{s}^{-1} = 300\ \text{pm}^2\ \text{s}^{-1}$$

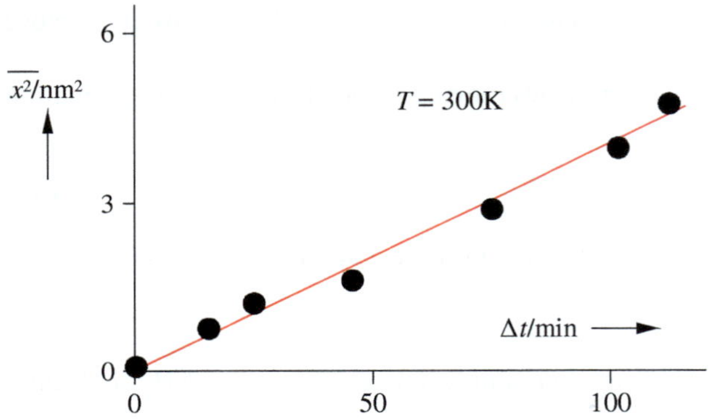

Figure 14.19 Nitrogen atoms adsorbed on a surface. Average of the square displacement $\overline{x^2}$ of selected nitrogen atoms versus time at $T = 300$ K. For the slope we find $\overline{x^2}/\Delta t = 6 \times 10^{-22}$ m² s⁻¹.

14.3.3.5 Fick's First Law

Although it is interesting to calculate the average behavior of a molecule, experimentally we most often deal with collections of molecules, and it is useful to calculate the behavior of these collections. Hence, we commonly ask a question of the type, "What amount of bromine vapor disappears at the top of the cylinder in Fig. 14.20 in time Δt, if the cylinder has a cross section A and height H?"

We will perform the calculation under steady-state conditions so that we do not have to worry about the depletion of the bromine gas and we will assume that molecules reaching the top are irreversibly consumed. These conditions can be met experimentally by placing some liquid bromine at the bottom of the cylinder, so that the bromine gas is continuously generated by vaporizing the liquid bromine, and by placing a silver foil at the top to trap the bromine molecules (the $Br_2(g)$ reacts with Ag(s) to make AgBr(s)).

Under these conditions a linear concentration gradient is established in the cylinder from the bottom to the top. Through any cross section of the cylinder the same amount Δn of Br_2 diffuses in time Δt (Fig. 14.20). The concentration at the bottom ($x = H$) is c_H and at the top it is $c(h = 0) = 0$. The concentration $c(h)$ at a distance h from the top (Fig. 14.20b) is

$$c(h) = c_H \cdot \frac{h}{H} \tag{14.64}$$

We consider the top layer next to the silver foil between $h = 0$ and $h = 0 + \Delta h$. The amount $\Delta \mathbf{n}$ of Br_2 in this layer moving in the direction of h is

$$\Delta \mathbf{n} = -\frac{1}{2} c(\Delta h) \cdot \Delta h \cdot A$$

and $\frac{1}{2} c(\Delta h)$ is the average concentration between $c = 0$ and $c = c(\Delta h)$. Using $c(\Delta h) = c_H \cdot \frac{\Delta h}{H}$, we find

$$\Delta \mathbf{n} = -\frac{1}{2} c_H \cdot \frac{(\Delta h)^2}{H} \cdot A \tag{14.65}$$

The negative sign indicates that the particles are moving opposite to the direction of h, from bottom to top. According to Equation (14.60) the mean square displacement $\overline{x^2}$ can be expressed by the time Δt needed for the bromine molecules in the top layer to be trapped. For simplicity we assume that the diffusion path Δx in time Δt is always Δh, i.e., we replace $x_{rms} = \sqrt{\overline{x^2}}$ by Δh:

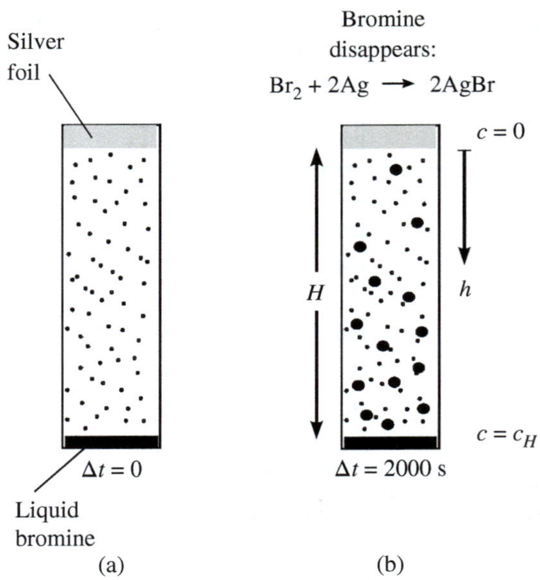

Figure 14.20 Diffusion of bromine vapor through a cylinder filled with air. Evaporation of liquid bromine on the bottom provides a constant flow of Br_2 gas; the bromine reaching the top is trapped by a silver foil. A linear concentration gradient is formed inside the cylinder; the concentration decreases from c_H on the bottom to zero on top. A stationary state is reached after a time of 2000 s. This time is large compared to the time a bromine molecule needs to diffuse from the bottom to the top (see below).

$$(\Delta h)^2 = \overline{x^2} = 2D \cdot \Delta t \tag{14.66}$$

With Equation (14.65) we obtain

$$\Delta \mathbf{n} = -c_H \cdot \frac{D \cdot \Delta t}{H} \cdot A \tag{14.67}$$

or

$$\boxed{\frac{\Delta \mathbf{n}}{\Delta t} = -A \cdot D \cdot \frac{c_H}{H}} \tag{14.68}$$

The same amount of bromine passes each cross sectional area A in time Δt. Using this relation we can calculate the amount of substance passing the opening of the cylinder in time Δt.

Example 14.8 *Calculate the amount of bromine gas (with diffusion coefficient[6] $D = 0.93 \times 10^{-5}$ m^2 s^{-1} and vapor pressure 220 mbar) that passes through a cylinder of length 10 cm and cross-sectional area 1.0 cm^2. Use a temperature of $T = 300$ K.*

If we use the diffusion coefficient and set $x_{\text{rms}} = \sqrt{\overline{x^2}} = 10$ cm (the length of the cylinder), we can estimate the time for the apparatus to reach steady-state. Using Equation (14.66) we calculate a time of

$$\Delta t = \frac{H^2}{2D} = \frac{100 \times 10^{-4}\ \text{m}^2}{2 \cdot 0.93 \times 10^{-5}\ \text{m}^2\ \text{s}^{-1}} = 538\ \text{s}$$

for the system to reach steady state (estimated by the time for a bromine molecule to diffuse the full length of the cylinder), about 10 minutes.

The amount $\Delta \mathbf{n}$ of bromine molecules passing the opening with cross section $A = 1$ cm^2 in the direction of h is given by

$$\Delta \mathbf{n} = -AD\,\frac{c_H}{H} \cdot \Delta t$$

Using the vapor pressure of bromine $P_{\text{vap}} = 220$ mbar and the ideal gas law, its concentration is

$$c_H = \frac{\mathbf{n}}{V} = \frac{N}{V}\frac{1}{\mathbf{N}_A} = \frac{P_{\text{vap}}}{\mathbf{N}_A kT}$$

$$= \frac{2.20 \times 10^4\ \text{N m}^{-2}}{6.02 \times 10^{23} \cdot 1.38 \times 10^{-23} \cdot 300\ \text{J}} = 8.82 \times 10^{-3}\ \text{mol L}^{-1}$$

For a time interval $\Delta t = 60$ s, we obtain

$$\Delta \mathbf{n} = -8.2 \times 10^{-8}\ \text{mol s}^{-1} \cdot \Delta t = -4.9 \times 10^{-6}\ \text{mol}$$

This corresponds to a mass of $\Delta m = \Delta \mathbf{n} \cdot 159.8$ g mol^{-1} = 0.78 mg. This means that in about 10 minutes only 8 mg of liquid bromine disappears; it takes 20 hours $\approx$ 1 day for the disappearance of 1 g of liquid bromine. This sort of time is observed in an experiment.

The quantity c_H/H in Equation (14.64) is the linear concentration gradient that we assumed for the diffusing particles inside the cylinder. In the general case the concentration gradient can be written as $\mathrm{d}c/\mathrm{d}h$. For sufficiently small time changes, $\Delta \mathbf{n}/\Delta t$ can be written as $\mathrm{d}\mathbf{n}/\mathrm{d}t$, and we obtain *Fick's First Law*:

$$\boxed{\frac{\mathrm{d}\mathbf{n}}{\mathrm{d}t} = -AD \cdot \frac{\mathrm{d}c}{\mathrm{d}h}} \tag{14.69}$$

According to this relation the amount of substance $\mathrm{d}\mathbf{n}$ passing any cross-sectional area of the cylinder in time $\mathrm{d}t$ is proportional to the cross-sectional area A and the concentration gradient $\mathrm{d}c/\mathrm{d}h$. The proportionality constant D increases with the average molecular speed $\overline{v}$ and the mean free path $\overline{\lambda}$, see Equation (14.62).

14.3.3.6 The Influence of Gravity

In the above consideration the influence of gravity on the molecules has been neglected. Is this reasonable? If the molecules are under a gravitational field, e.g., that of the Earth, then a concentration gradient will be established. How significant is this concentration gradient?

In a gravitational field g each molecule is pulled down by a force $f = mg$ (m = mass of a molecule; $g = 9.81$ m s^{-2} on Earth), but diffuses up as a result of thermal motion. At a certain height d the potential energy in the gravitational field $f \cdot d$ equals the thermal energy kT:

$$fd = mgd = kT$$

6 D is calculated from Equation (14.60) using $\overline{x^2} = 1.86 \times 10^{-5}$ m^2 $s^{-1} \Delta t$ calculated in Example 14.6.

We can use this relationship to estimate the characteristic length of the gravitational field's influence. Solving for d, we find

$$d = \frac{kT}{mg}$$

For bromine gas ($m = 26.4 \times 10^{-26}$ kg), $g = 9.81\ \mathrm{m\ s^{-2}}$, and $T = 300$ K, so that we obtain $d = 1.6$ km. Hence it is reasonable to ignore these effects for the 10 cm long cylinder discussed above.

In Box 14.5, we show that the concentration profile in the Earth's gravitational field is

$$c(h) = c_0 \cdot \mathrm{e}^{-mgh/(kT)} = c_0 \cdot \mathrm{e}^{-h/d} \tag{14.70}$$

Thus d can be identified with the height h at which the concentration has dropped to c_0/e. For the Earth's atmosphere d is about 9 km. This effect explains why the air gets "thinner" as you climb a mountain.

In an artificial gravity field of an ultracentrifuge (e.g., $g_{\mathrm{centrifuge}} = 10^6\ \mathrm{m\ s^{-2}}$), d can be brought to laboratory dimensions ($d \approx 10$ cm) and the dependence of d on the mass of the molecules can be used to separate isotopes.

14.3.4 Viscosity Arising from Collisions of Molecules

So far our discussion has viewed the role of collisions in the mechanism of gas molecule motion through space. Now we wish to change our perspective and view how collisions inhibit motion. This inhibition will be described by a friction or viscosity.

14.3.4.1 Viscous Flow

Consider a square plate of area A that is moved in a gas parallel between two large plates at rest, each at distance d from the moving plate (Fig. 14.21). When the gas molecules bounce off the moving plate they are accelerated in the direction of the plate's motion and the plate is decelerated. This effect increases with increasing molecular speed and increasing molecular mass, because more momentum is transferred in those collisions, and it increases with increasing mean free path; thus, it decreases with increasing collision diameter.

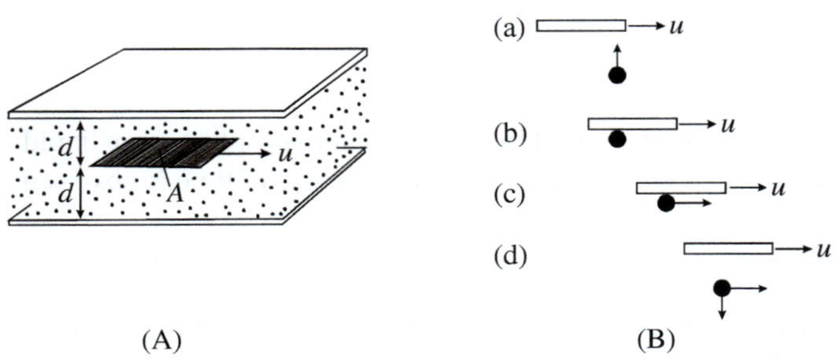

Figure 14.21 (A) A plate moving with speed u in a gas between two resting plates. (B) Processes responsible for the viscosity of a gas: A molecule bouncing back and forth between the moving plate and the stationary walls of the container is accelerated at the moving plate. After the collision (a, b) the particle is fixed for a short while on the surface of the plate and brought to velocity u (c); after leaving the plate (d) it keeps this velocity component.

In order to move the plate at a constant speed u we have to apply a constant force f that is balanced by the frictional drag in the gas. If u is sufficiently small the frictional force is proportional to u. If the motion of the plate is slow enough (conditions of laminar flow) then the friction is also proportional to the area of the plate (A, surface above and surface below) and reciprocally proportional to the distance d of the plate from the immobile plates. In this case

$$f = \eta \frac{A}{d} \cdot u \tag{14.71}$$

where the proportionality factor η is called the *coefficient of shear viscosity*.

14.3.4.2 Calculating the Viscosity Coefficient

For a plate that is moving slow enough (velocity u), the change in velocity between the moving plate and the wall, which is at rest (velocity of zero), can be approximated as linear. When traveling between the resting plate and the moving plate, gas molecules collide with other molecules and exchange their momentum with these molecules. So even molecules which do not collide with the moving plate obtain an additional velocity in the direction of the moving plate (x-direction). This additional velocity decreases linearly when proceeding from the moving plate to the wall (from a speed of u to a speed of zero, see Fig. 14.22). Hence the velocity gradient normal to the plate (call it the y-direction), is

$$\frac{\text{change in velocity}}{\text{distance}} = \frac{u - 0}{d} = \frac{u}{d} \tag{14.72}$$

The change in velocity between the layers in the gas (see Fig. 14.22) occurs because of momentum transfer in the collisions. This transfer of momentum from the motion of the plate along the x-direction into the perpendicular direction y is the microscopic mechanism of the shear viscosity.

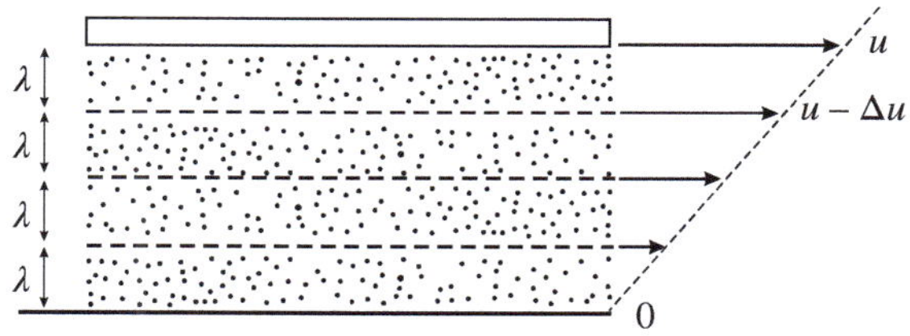

Figure 14.22 Moving plate at a distance $d = 4\lambda$ from the stationary container wall; the additional velocity in the different gas layers decreases linearly from u to zero, because the momentum transferred from one layer of thickness λ to the next layer is the same for all layers in the stationary state considered here.

To elucidate this phenomenon we use a simple model, in which the mean free path is the same for all molecules (mean path = $\overline{\lambda}$) and all molecules have the same speed (average speed $\overline{v}$). Furthermore, as in Section 14.2.1, we assume that $\frac{1}{6}$ of all molecules move perpendicular toward the moving plate. On average, only molecules in a distance smaller than the mean free path $\overline{\lambda}$ can reach the moving plate without a collision with other gas molecules. For a linear velocity gradient (Fig. 14.22), the velocity difference between the moving plate and the molecules hitting the plate is

$$\Delta u = u \cdot \frac{\overline{\lambda}}{d} \tag{14.73}$$

The force acting on the plate by one molecule reaching the plate in time $\Delta t = \overline{\lambda}/\overline{v}$ is

$$\text{force} = \frac{\text{momentum change}}{\Delta t} = \frac{m \cdot \Delta u}{\Delta t} = \frac{m \cdot u \cdot \overline{\lambda}/d}{\overline{\lambda}/\overline{v}} = \frac{m \cdot u \cdot \overline{v}}{d} \tag{14.74}$$

To obtain the total force acting on the plate, we multiply by the number ΔN of all molecules reaching the plate in time Δt, which in our simple model is

$$\Delta N = \frac{1}{6} \frac{N}{V} \overline{\lambda} A \tag{14.75}$$

where N/V is the number density (ΔN is the number of all molecules in distance $\overline{\lambda}$ below or above the moving plate, moving perpendicular toward the plate). Thus the force exerted on the plate by all molecules reaching the plate in time Δt is

$$f = \Delta N \cdot \frac{m \cdot u\, \overline{v}}{d} = \frac{1}{6} \frac{N}{V} \overline{\lambda} A \cdot \frac{m \cdot u\, \overline{v}}{d} = \frac{1}{6} \frac{N}{V} m \cdot \overline{v}\, \overline{\lambda} \cdot u\, \frac{A}{d} \tag{14.76}$$

By comparing Equation (14.76) with our requirement of steady motion, given by Equation (14.71)

$$f = \eta \cdot u\, \frac{A}{d}$$

we can find the shear viscosity of the gas

$$\eta = \frac{1}{6} \frac{N}{V} m \cdot \overline{v}\, \overline{\lambda} \tag{14.77}$$

In this simplified model the force acting on the plate is underestimated. As we discussed already in Section 14.2.1, the number of molecules reaching the plate is larger by a factor of 1.5, if the distribution of their directions in space is taken into account, according to Equation (14.16). From the distribution of λ it follows that the contribution of the molecules with larger λ is more effective than that estimated from the average $\overline{\lambda}$. Also, molecules can reach more distant layers after noncentral collisions. A rigorous calculation [4] leads to a shear viscosity η that is $15\pi/16$ times larger than given by Equation (14.77):

$$\boxed{\eta = \frac{5\pi}{32} \frac{N}{V} m \cdot \overline{v}\, \overline{\lambda}} \tag{14.78}$$

14.3.4.3 Dependence of η on Temperature

From Equation (14.78) we see that the viscosity of a gas increases with increasing temperature.

$$\eta = \frac{5\pi}{32} \frac{N}{V} m\, \overline{\lambda} \cdot \overline{v} = \frac{5\pi}{32} \frac{N}{V} m\, \overline{\lambda} \cdot \sqrt{\frac{8kT}{\pi m}} \tag{14.79}$$

Because the average speed $\overline{v}$ of the molecules increases with temperature, more particles collide with the moving plate (Fig. 14.22) in a given time and the frictional drag is higher. This behavior contrasts with that of a liquid, where η decreases with increasing temperature (see Section 14.4.1). From Equation (14.79) we expect that the mean speed $\overline{v}$ and thus the viscosity η increase proportional to the square root of the temperature T. However, as shown in Foundation 14.2, the situation is more complicated when attractive forces act between the molecules.

14.3.4.4 Dependence of η on Pressure

Despite the form of (14.78), η does not depend on the number density N/V or on the pressure

$$P = \frac{NkT}{V} = \frac{N}{V} \cdot kT$$

This result is caused by a compensation between P and $\overline{\lambda}$. When P decreases, the number of collisions decreases; however, the momentum transferred per collision increases because of the increase in the mean free path $\overline{\lambda}$

$$\text{momentum} = m \cdot u\,\frac{\overline{\lambda}}{d} \quad \text{for} \quad \overline{\lambda} < d$$

see Equation (14.73). At small enough pressures, so that $\overline{\lambda} > d$, the mean free path $\overline{\lambda}$ in Equation (14.78) has to be exchanged by the distance d; thus, we obtain

$$\eta = \frac{5\pi}{32}\frac{N}{V} m \cdot \overline{v}\, d$$

In this limit we expect that the viscosity becomes proportional to the pressure P.

14.3.4.5 Calculating Collision Diameters

The shear viscosity of a gas can be experimentally determined by measuring its flow rate through a capillary tube (Fig. 14.23). Consider a capillary tube of length l and an inner radius R. If P_2 is the pressure on the right end of the capillary tube and P_1 is the pressure on the left of the capillary tube, then the volume ΔV of the gas flowing through the capillary tube from left to right in time Δt is

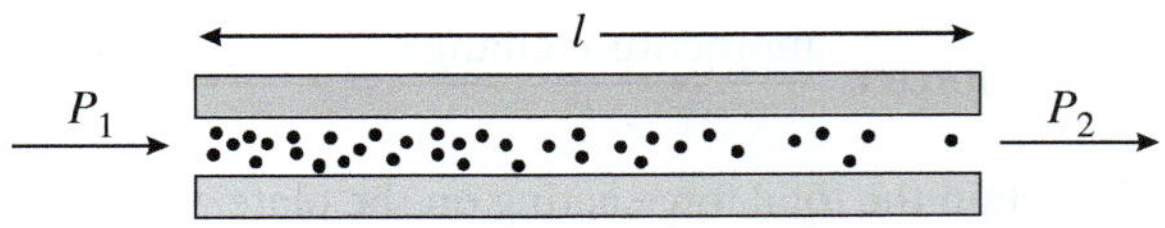

Figure 14.23 The flow of a gas through a capillary tube.

$$\Delta V = \frac{1}{\eta}\frac{\pi\,(P_2 - P_1)\,(P_1 + P_2)\,R^4}{16\,l P_1}\Delta t \tag{14.80}$$

(*Hagen–Poiseuille's Law*, derived in Foundation 14.3). The volume of the gas flowing out on the right-hand side of the capillary is larger because of the smaller pressure P_2. Thus one can determine η for the gas by measuring ΔV versus Δt as a function of the pressures, for a given capillary tube.

From the measured shear viscosity η, the mean free path $\overline{\lambda}$ and the collision diameters of gas molecules can be calculated (Table 14.4) according to Equations (14.78), (14.43), (14.41), and (14.36):

$$\overline{\lambda} = \frac{32}{5\pi}\frac{V}{N}\frac{1}{m\,\overline{v}}\eta, \quad \sigma = \frac{5\pi}{32\sqrt{2}}\frac{m\,\overline{v}}{\eta}, \quad 2r = \sqrt{\frac{\sigma}{\pi}}, \quad \overline{v} = \sqrt{\frac{8kT}{\pi m}} \tag{14.81}$$

14.4 Thermal Motion in Liquids

14.4.1 Collisions in Liquids

At a given temperature, a molecule in a liquid has the same mean translational energy $\frac{3}{2}kT$ and the same mean speed as in a gas. Because the molecules are packed more closely together in a liquid than in a gas, the idea of independent collisions of two molecules may not always be valid, however.

According to Section 14.3.2 the molecules in a gas at 1 bar and 273.16 K are spaced 3.3 nm apart, on average. If we suppose that 1 cm^3 of water has $\frac{1}{18}$ mol$\cdot\mathbf{N}_A = 3.3 \times 10^{22}$ molecules, the average distance between molecular centers of mass is 0.3 nm. This intermolecular distance is on the order of the molecular diameter, and a particle does not move freely between collisions, as in a gas. Because of the high density, the mean free path $\overline{\lambda}$ in liquids is smaller than the average distance between the molecular centers; $\overline{\lambda}$ is on the order of 0.1 nm.

For these reasons, the molecules in a liquid collide more frequently than do the molecules in a gas. If we use Equation (14.45) to calculate the number Z_{coll} of collisions in time Δt, we find

$$z_{\text{coll}} = \frac{\overline{v}}{\overline{\lambda}}\Delta t = \frac{10^3\ \text{m s}^{-1}}{10^{-10}\ \text{m}}\Delta t = 10^{13}\ \text{s}^{-1} \cdot \Delta t$$

Table 14.4 Experimental Viscosity η, Calculated Mean Speed $\overline{v}$, Mean Free Path $\overline{\lambda}$ at $P = 1$ bar, and Collision Diameter $2r$ for Some Gases, using (14.81).

	$\frac{T}{\text{K}}$	$\frac{\eta}{10^{-6}\ \text{g cm}^{-1}\ \text{s}^{-1}}$	$\frac{M}{\text{g mol}^{-1}}$	$\frac{\overline{v}}{\text{m s}^{-1}}$	$\frac{\overline{\lambda}}{\text{nm}}$	$\frac{2r}{\text{pm}}$
H-atom[7]	273	69	1.01	2395	132	253
H_2	273	84	2.02	1691	114	273
CO_2	273	137	44.01	362	40	462
H_2O	373	125	18.00	662	66	418
O_2	273	192	32.00	425	65	360
N_2	273	166	28.02	454	60	375
He	273	186	4.003	1202	179	218
Ne	273	296	20.18	535	127	259
Ar	273	210	39.95	380	64	364
Kr	273	233	83.80	263	49	416
Xe	273	211	131.29	210	35	489
Br_2	286	152	159.80	195	24	613
Ethene	339	117	28.04	506	47	472
Butadiene	339	97	54.09	364	28	611

That is, $z_{\text{coll}} = 10^{13}$ for $\Delta t = 1$ s, whereas the corresponding value calculated for gases (Section 14.3.2) is only $z_{\text{coll}} = 10^{10}$. We define the quantity

$$\nu_{\text{coll}} = \frac{z_{\text{coll}}}{\Delta t} \tag{14.82}$$

as the *collision frequency*.

14.4.2 Diffusion Coefficient *D* of a Liquid

As a result of this difference in collision frequencies, it takes much more time for a molecule in a liquid to traverse a certain distance by diffusion than it does in a gas. The molecule collides many times with other molecules in the liquid before it changes its position (Fig. 14.24).

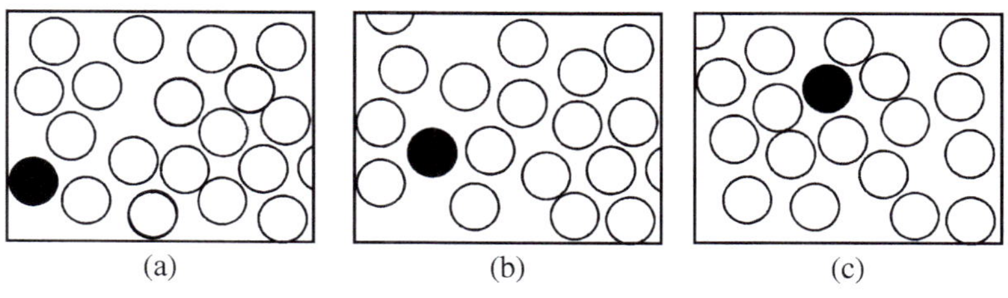

Figure 14.24 A change of position of a molecule diffusing in a liquid: (a) starting position of molecule, (b) one site change of the molecule, and (c) second site change of the molecule.

We assume that a molecule in a liquid changes its position if it travels the mean distance between molecules. We call this distance a. If the position is changed after every ξth collision, then $\nu_{\text{coll}}\Delta t/\xi$ is the number of position changes in time Δt. According to Equation (14.51), the mean square displacement $\overline{x^2}$ is

$$\overline{x^2} = \frac{1}{3} j \cdot \overline{\lambda^2} = \frac{1}{3}\frac{\nu_{\text{coll}}\ \Delta t}{\xi} \cdot a^2 = \frac{1}{3}\frac{\nu_{\text{coll}}\ a^2}{\xi} \cdot \Delta t$$

By comparison with the equation of Einstein and Smoluchowski (14.60), we obtain

$$D = \frac{\overline{x^2}}{2\ \Delta t} = \frac{1}{6}\frac{\nu_{\text{coll}}\ a^2}{\xi} \tag{14.83}$$

7 The viscosities of H atoms can be measured in gas discharge tubes filled with dihydrogen gas.

For a liquid such as water, $D = 1 \times 10^{-5}$ cm^2 s^{-1}. Using $a = 0.3$ nm we calculate the number ξ of collisions to be

$$\xi = \frac{1}{6}\frac{v_{\text{coll}}\, a^2}{D} = \frac{1}{6}\frac{10^{13}\ \text{s}^{-1} \cdot \left(0.3 \times 10^{-9}\ \text{m}\right)^2}{1 \times 10^{-5}\ \text{cm}^2\ \text{s}^{-1}} = 150$$

About 150 collisions occur before a position change occurs.

The frequency of position changes depends on the intermolecular forces which are determined by the specific surroundings of the molecule. Therefore, the quantity D has specific values for liquids, in contrast to gases, where D depends only on the mass and cross section of the isolated gas molecules.

Example 14.9 *Brownian Motion. By microscopic observation the x- and y-coordinates of a small sphere of diameter 2 μm, in water ($\eta = 8.9 \times 10^{-4}$ kg s^{-1} m^{-1}) were determined at room temperature in the beginning of the experiment, after 30 s, 60 s, and so on for 135 min = 8100 s (Fig. 14.25). Compare the observed value of the square of the displacement at 135 min with the mean square displacement calculated using the diffusion coefficient $D = 2.5 \times 10^{-13}$ m^2 s^{-1}, which is obtained from the Stokes–Einstein equation (14.85).*

The observed value of the square of the displacement is $x^2 = (40\ \mu\text{m})^2 = 1.6 \times 10^{-9}$ m^2. Using the diffusion coefficient and Equation (14.83) we obtain $\overline{\Delta x^2} = 2D\Delta t = 2 \cdot 2.5 \times 10^{-13} \cdot 8100$ m$^2 = 4 \times 10^{-9}$ m^2.

We see that the observed square of the displacement is about one-third of the calculated mean square displacement. This deviation is not unreasonable, because the data in Fig. 14.25 represent an arbitrary single trajectory. To obtain the experimental mean square displacement, the experiment has to be repeated for a large number of cases and a statistical average over all the possible trajectories has to be taken.

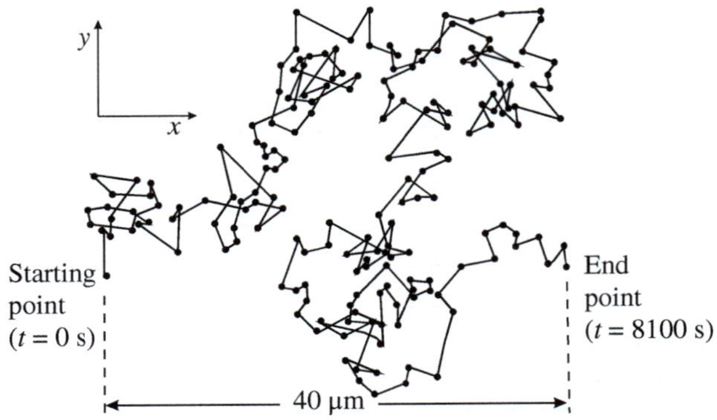

Figure 14.25 Brownian motion. In an experiment the x- and y-coordinates of a particle with diameter 2 μm, moving in water, are observed every 30 s for 8100 s. The distance in the x-direction between starting point and end point is 40 μm. For comparison the mean displacement in the x direction, as calculated from Equations (14.60) and (14.85), is

$$x_{\text{rms}} = \sqrt{\overline{x^2}} = 63\ \mu\text{m}.$$

14.4.3 Viscosity of a Liquid

It is also true that the viscosity is much larger for liquids than for gases. Primarily, this increase is caused by the close packing of the molecules in a liquid. Intermolecular forces play a significant role in determining a liquid's viscosity, also. At 300 K $\eta = 0.0089$ g s^{-1} cm^{-1} for water, but for glycerin it is $\eta = 9.54$ g s^{-1} cm^{-1}, more than a thousand times larger.

In contrast to gases, the viscosity of liquids decreases with increasing temperature. This dependence results because higher temperatures cause a decrease in the liquid's density and make it easier for the molecules to overcome the energy barrier associated with positional changes.

14.4.4 Stokes–Einstein Equation

To estimate the diffusion coefficient D of a molecule in a liquid, we approximate the dissolved molecule as a sphere of radius r. At steady-state the force f that acts on the molecule to pull it through the liquid with a constant speed u is balanced by the frictional drag $-\zeta u$, where ζ is called the friction coefficient. This force is given by *Stokes Law,* [5]

$$f = \zeta u = 6\pi\eta r u \tag{14.84}$$

Stokes law is derived for a macroscopic sphere moving in a fluid and considers the solvent as a continuum characterized by its viscosity coefficient η, neglecting its molecular structure. In this approximation D is given by the *Stokes–Einstein Equation*

$$\boxed{D = \frac{kT}{6\pi\eta r}} \tag{14.85}$$

or by the more general Einstein relation

$$D = \frac{kT}{\zeta} \tag{14.86}$$

(see Box 14.6). These equations are very useful for estimating diffusion coefficients of a particle in a liquid from the viscosity of the liquid and the average size of the particle. For example, the diffusion coefficient for a molecule with $r = 200$ pm diffusing in water ($\eta = 0.0089$ g s^{-1} cm^{-1}) becomes $D = 1.2 \times 10^{-5}$ cm^2 s^{-1} in reasonable agreement with experimental data.

Equation (14.85) is justified for sufficiently large solute particles. In this limit the solvent molecules are small enough that their detailed structural features (or that of their clusters) are not so important. Amazingly, the equation can also be used to estimate D for molecules which are only a few times larger than the solvent molecules.

Example 14.10 *A bacteriophage attacks a bacterium (diameter 1 μm) in water. After 20 min the bacterium is destroyed, and the bacteriophage has produced 200 daughter phages. How long does it take before another bacterium at a distance of 5 μm is infected?*

We approximate the bacteriophage as a sphere of radius $r = 100$ nm (Fig. 14.26). From Equation (14.85) and $\eta = 0.0089$ g s^{-1} cm^{-1} we calculate a diffusion coefficient of

$$D = \frac{(1.38 \times 10^{-23})(300)\ \text{J}}{6\pi(8.9 \times 10^{-4})(100 \times 10^{-9})\ \text{kg/s}} = 2.5 \times 10^{-8}\ \text{cm}^2\ \text{s}^{-1}$$

To have a mean square displacement of 5 μm, requires an average time Δt

$$\Delta t = \frac{\overline{x^2}}{2D} = \frac{\left(5 \times 10^{-6}\right)^2\ \text{m}^2}{2 \cdot 2.5 \times 10^{-8}\ \text{cm}^2\ \text{s}^{-1}} = 5\ \text{s}$$

Because the bacterium covers an area of about 1 μm^2, the probability of finding a bacterium on the surface of a sphere with radius 5 μm is

$$\frac{1\ \mu\text{m}^2}{4\pi \cdot (5\ \mu\text{m})^2} = \frac{1}{300}$$

Thus about one of the 200 phages reaches the actual position of the bacterium in 5 s. This time is negligibly small compared to the time of 20 min needed to destroy the bacterium.

Figure 14.26 Bacteriophage approximated by a sphere.

14.5 Conclusion

In this chapter we introduced the concepts of temperature and heat in association with molecular assemblies, and we considered thermal equilibration in molecular terms.

We began with a discussion of gas molecules that were treated as independent spherical particles with no internal structure. This approximation allowed us to focus the discussion on the thermal properties that arise from collisions and molecular transport. We inferred that heat is a form of energy that is transferred from an assembly at higher temperature to an assembly at lower temperature and that it does not proceed in the reverse direction without performing work. This feature is fundamental and will be discussed more thoroughly in the following chapters. We used this simple model to understand the ideal gas law and the creation of an absolute temperature scale. Importantly this model provides a bridge between macroscopic properties (pressure, temperature, shear viscosity, diffusion) and average microscopic properties of the molecules (momentum transfer, average speed, mean free path, collision frequency).

This chapter illustrates that molecules are constantly undergoing motion, and this feature is essential to understanding molecular assemblies. For example, the assemblage of functional units, which is the basic process in molecular biology and supramolecular engineering, relies on processes in which molecular components diffuse as a result of their thermal motion. The components that fit together find each other and bind, whereas the poorly fitting ones separate, disperse, and are replaced by better fitting components. In this way, complex systems of molecular components can interlock, analogous to pieces in a jigsaw-puzzle, and form supramolecular structures with properties and function that arise from the assembly (see Chapters 28 and 29).

14.6 Box 14.1. Molar Quantities

The number N of particles in a molecular assembly can be expressed by the amount of substance $\mathbf{n}$. The amount of substance $\mathbf{n} = 1$ mol is defined as an assembly of atoms or molecules that contains exactly $6.02214076 \times 10^{23}$ entities.

For example, from this definition we obtain the amount of substance in a reservoir of hydrogen gas when we examine the chemical reaction

$$2H_2 + C \rightarrow CH_4$$

We know that 12 g of carbon ^{12}C corresponds to an amount of substance of $n = 1$ mol. It is found experimentally that 12 g of the carbon isotope ^{12}C react with 4.032 g H_2. Since for each ^{12}C atom two H_2 molecules are consumed, the amount of substance $\mathbf{n}_{H_2} = 1$ mol corresponds to 2.016 g H_2.

The quantity

$$\mathbf{M} = \frac{\text{mass of the molecular assembly}}{\text{amount of substance}} = \frac{N \cdot m}{\mathbf{n}}$$

(N = number of molecules, m = mass of one molecule) is called the *molar mass*; for ^{12}C we get $\mathbf{M} = 12$ g mol^{-1}, for H_2 we get $\mathbf{M} = 2.016$ g mol^{-1}. For other molecules $\mathbf{M}$ is calculated in a corresponding manner. The number of particles N is, on the other hand, proportional to the amount of substance $\mathbf{n}$:

$$N = \mathbf{N}_A \cdot \mathbf{n}$$

The proportionality factor $\mathrm{N}_A = 6.02214076 \times 10^{23}$ mol^{-1} is called the Avogadro constant. The value obtained for $\mathbf{N}_A$ is

$$\mathbf{N}_A = 6.02214 \times 10^{23}\ \mathrm{mol}^{-1}$$

Thus the molar mass $\mathbf{M}$ is related to its molecular mass m by

$$\mathbf{M} = \mathbf{N}_A \cdot m$$

For example, for ^{12}C we have $\mathbf{M} = 12$ g mol^{-1} and $m = (12\ \mathrm{g\ mol}^{-1})/(6.022 \times 10^{23}\ \mathrm{mol}^{-1}) = 1.99 \times 10^{-26}$ kg.

Furthermore,

$$c = \frac{\mathbf{n}}{V} \quad \text{and} \quad V_m = \frac{V}{\mathbf{n}} = \frac{1}{c}$$

are the concentration and the molar volume, respectively.

Note that in the following molar quantities will be represented by the index "m."

14.7 Box 14.2. Pressure on the Wall of a Container and Number of Collisions with a Wall

14.7.1 Pressure on the Wall

The simplified analysis assumes that one-sixth of all particles are moving perpendicular to the wall. Actually, the particles are moving with arbitrary angles ϑ relative to the wall (see Fig. 14B.1). The force on the wall is proportional to the particles' velocity component that is perpendicular to the wall. Hence v in Equation (14.5) must be replaced by $v_\perp = v \cdot \cos\vartheta$. To calculate the number of particles reaching the wall in time Δt, we replace $v/6$ in Equation (14.8) by $v \cdot \cos\vartheta$. The force $f(\vartheta)$ exerted on the wall is

$$f(\vartheta) = \left(\frac{N}{V} L^2 v\ \cos\vartheta\ \Delta t\right)\left(\frac{2v\cos\vartheta}{\Delta t}\right)$$

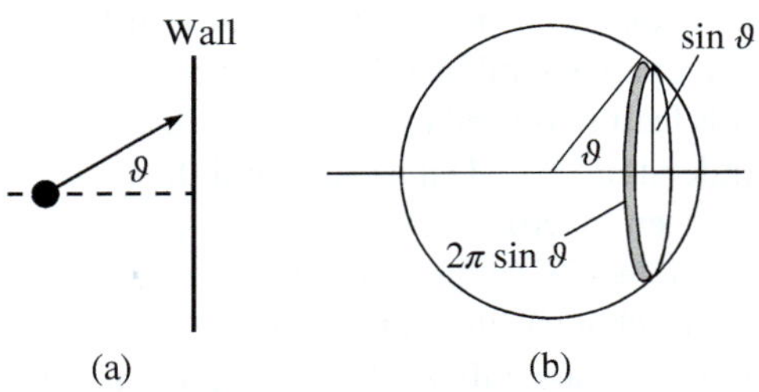

Figure 14B.1 Panel (a) shows a particle incident on a wall at an angle ϑ. Panel (b) illustrates the angular element between ϑ and $\vartheta + \mathrm{d}\vartheta$.

The probability of finding a particle moving in the differential angular element between ϑ and $\vartheta + \mathrm{d}\vartheta$ is (see Fig. 14B.1)

$$\frac{2\pi \sin\vartheta \cdot \mathrm{d}\vartheta}{4\pi} = \frac{1}{2}\sin\vartheta \cdot \mathrm{d}\vartheta$$

To obtain the total average force $\overline{f}$ we average the angular dependent force over the angles and find

$$\overline{f} = \int_0^{\pi/2} 2\frac{N}{V}L^2 mv^2 \cos^2\vartheta \cdot \frac{1}{2}\sin\vartheta \cdot \mathrm{d}\vartheta$$
$$= \frac{N}{V}L^2 mv^2 \int_0^{\pi/2} \cos^2\vartheta \cdot \sin\vartheta \cdot \mathrm{d}\vartheta$$

The integral is (see Appendix A.5)

$$\int_0^{\pi/2} \cos^2\vartheta \cdot \sin\vartheta \cdot \mathrm{d}\vartheta = \frac{1}{3}, \quad \text{so that} \quad \overline{f} = \frac{1}{3}\frac{N}{V}L^2 mv^2$$

We find the pressure P by

$$P = \frac{\overline{f}}{L^2} = \frac{1}{3}\frac{N}{V}mv^2$$

This is the same result as obtained in the simplified model, Equation (14.9).

14.7.2 Number of Collisions with a Wall

In a similar way, we proceed with the number of collisions with the wall. In this case $\frac{1}{6}v$ in Equation (14.8) must be replaced by $v \cdot \cos\vartheta$.

$$\frac{N}{V}L^2 v \cos\vartheta \; \Delta t$$

To obtain the number ΔN of particles reaching the wall we average over the angles and find

$$\Delta N = \int_0^{\pi/2} \frac{N}{V}L^2 v \cos\vartheta \; \Delta t \cdot \frac{1}{2}\sin\vartheta \cdot \mathrm{d}\vartheta$$
$$= \frac{N}{V}L^2 v \frac{1}{2}\Delta t \int_0^{\pi/2} \cos\vartheta \cdot \sin\vartheta \cdot \mathrm{d}\vartheta$$

The integral is (see Appendix A.5)

$$\int_0^{\pi/2} \cos\vartheta \cdot \sin\vartheta \cdot \mathrm{d}\vartheta = \frac{1}{2}$$

and we find for the number ΔN of particles reaching the wall in time Δt

$$\Delta N = \frac{1}{4}\frac{N}{V}L^2 v \Delta t$$

14.8 Box 14.3. Averaging the Displacement x^2

We consider the displacement x of a particle moving randomly in the x-direction (one-dimensional diffusion); at each collision, x can change by $+\lambda$ or by $-\lambda$. Let us assume that after j collisions the displacement is x_j. We can express x_j in relation to the displacement x_{j-1} (displacement after $j-1$ collisions). If in the last collision the change in x was $+\lambda$, then

$$(x_j)^2 = (x_{j-1} + \lambda)^2 = (x_{j-1})^2 + 2x_{j-1}\lambda + \lambda^2$$

If in the last collision the change in x was $-\lambda$, then

$$(x_j)^2 = (x_{j-1} - \lambda)^2 = (x_{j-1})^2 - 2x_{j-1}\lambda + \lambda^2$$

Since the particle is moving randomly, both cases are equally probable; thus on average we obtain

$$\overline{(x_j)^2} = \overline{(x_{j-1})^2} + \lambda^2$$

because, when averaging, the terms $+2x_{j-1}\lambda$ and $-2x_{j-1}\lambda$ occur statistically and cancel each other. In a corresponding way we obtain

$$\overline{(x_{j-1})^2} = \overline{(x_{j-2})^2} + \lambda^2$$

that is,

$$\overline{(x_j)^2} = \overline{(x_{j-2})^2} + 2\lambda^2 = \overline{(x_{j-3})^2} + 3\lambda^2 = \cdots = \overline{(x_1)^2} + (j-1) \cdot \lambda^2$$

Because

$$\overline{(x_1)^2} = \lambda^2$$

we finally obtain

$$\overline{(x_j)^2} = j \cdot \lambda^2$$

Thus we have proved Equation (14.46).

14.9 Box 14.4. Averaging the Free Path λ

We consider N_0 particles that can collide with other particles. In a range of the free path between λ and $\lambda + d\lambda$ the number of particles without any collisions changes by

$$\mathrm{d}N = -a \cdot N \cdot \mathrm{d}\lambda$$

(a = constant). We integrate to find the number N of particles traveling a distance λ without any collision after a path λ to be

$$N = N_0 \mathrm{e}^{-a\lambda}$$

This shows that N decreases exponentially with λ. Thus the number of particles colliding in a range between λ and $\lambda + \mathrm{d}\lambda$ is

$$\mathrm{d}N = aN_0 \mathrm{e}^{-a\lambda} \cdot \mathrm{d}\lambda$$

This is the distribution function of the free path.

If we calculate $\overline{\lambda}$, we find that

$$\overline{\lambda} = \frac{1}{N_0} \int_0^\infty \lambda \cdot \mathrm{d}N = \frac{1}{N_0} aN_0 \int_0^\infty \lambda \mathrm{e}^{-a\lambda} \cdot \mathrm{d}\lambda = a\frac{1}{a^2} \int_0^\infty z \cdot \mathrm{e}^{-z} \cdot \mathrm{d}z = \frac{1}{a}$$

From this result we see that the distribution can be written as $N = N_0 \exp\left(-\lambda/\overline{\lambda}\right)$. Next we calculate $\overline{\lambda^2}$ and find

$$\overline{\lambda^2} = \frac{1}{N_0} \int_0^\infty \lambda^2 \cdot \mathrm{d}N = \frac{1}{N_0} aN_0 \int_0^\infty \lambda^2\, \mathrm{e}^{-a\lambda} \cdot \mathrm{d}\lambda = a\frac{1}{a^3} \int_0^\infty z^2 \cdot \mathrm{e}^{-z} \cdot \mathrm{d}z = 2\frac{1}{a^2}$$

Thus we obtain

$$\overline{\lambda^2} = 2 \cdot \left(\overline{\lambda}\right)^2$$

and have proved Equation (14.56).

14.10 Box 14.5. Atmospheric Distribution

By accounting for the gravitational force acting on a volume of gas, we deduce the distribution function for the change in gas pressure with height above the Earth. Consider a cylinder (cross section A) filled with a gas (molecules with mass m) as displayed in Fig. 14B.2. We select a layer of thickness Δh (volume $V = A \cdot \Delta h$). The number N of molecules in this layer is related to the pressure P by the ideal gas law, so that

$$N = \frac{PV}{kT} = \frac{PA \cdot \Delta h}{kT}$$

The force of gravity acting on these N molecules is

$$f = N \cdot mg$$

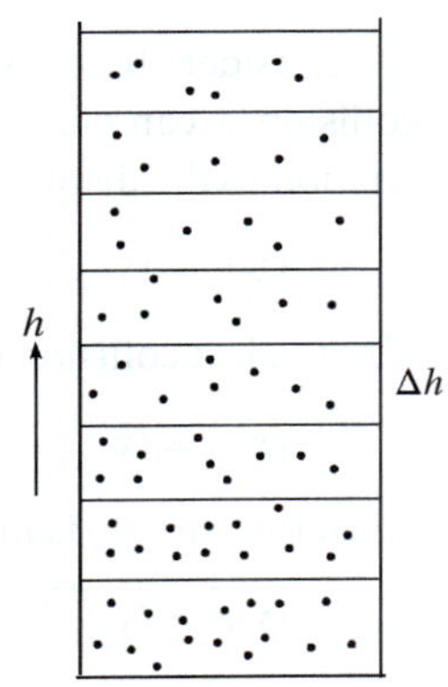

Figure 14B.2 Layers of thickness Δh (atmospheric distribution).

(g = gravitational acceleration). This force causes a pressure change $\Delta P = P_{h+\Delta h} - P_h$ in the gas

$$\Delta P = -\frac{f}{A} = -\frac{N \cdot mg}{A} = -\frac{P\,\Delta h \cdot mg}{kT}$$

By replacing the differences by differentials, we find

$$\mathrm{d}P = -P\,\frac{mg}{kT} \cdot \mathrm{d}x \quad \text{or} \quad \frac{\mathrm{d}P}{P} = -\frac{mg}{kT} \cdot \mathrm{d}h$$

Integration gives the atmospheric distribution function

$$P = P_0 \cdot \mathrm{e}^{-mgh/(kT)} \quad \text{or} \quad c = c_0 \cdot \mathrm{e}^{-mgh/(kT)}$$

where P_0 is the pressure at $h = 0$ and c_0 is the concentration at $h = 0$. Thus we have proved Equation (14.70).

14.11 Box 14.6. Derivation of the Stokes–Einstein Equation

Consider a dissolved particle in a force field, like the artificial gravitational field of an ultracentrifuge. Each molecule is pulled down with the force f; however, the thermal motion of the particles causes a distribution of particle positions to be established.

To find this distribution, we consider that the particles' motion downwards under the action of the force f is balanced by the motion upwards under diffusion, at the steady state. The particles move in the force field at a velocity

$$u = \frac{f}{\zeta}$$

where ζ is the friction coefficient. Consequently, the amount of substance moving downward through a cross section A at height x in time Δt is

$$\Delta \mathbf{n}_f = -c \cdot A \cdot u\,\Delta t = -cA\,\frac{f}{\zeta}\Delta t$$

where the subscript f indicates the force field and c denotes the concentration. At the same time, however, particles are diffusing upward according to the concentration gradient $\mathrm{d}c/\mathrm{d}x$. The amount of substance moving upward is, according to Equation (14.67),

$$\Delta \mathbf{n}_d = -A\,D\,\frac{\mathrm{d}c}{\mathrm{d}x}\,\Delta t$$

where the subscript d indicates diffusion. In the stationary state we have $\Delta n_f + \Delta n_d = 0$, and therefore

$$D\,\frac{\mathrm{d}c}{\mathrm{d}x} = -\frac{f\,c}{\zeta} \quad \text{or} \quad \frac{\mathrm{d}c}{c} = -\frac{f}{\zeta \cdot D} \cdot \mathrm{d}x$$

Integration yields

$$c(x) = c_0 \cdot \mathrm{e}^{-fx/(\zeta \cdot D)}$$

On the other hand, we can apply (14.70) to the same problem:

$$c(x) = c_0 \cdot \mathrm{e}^{-fx/(kT)}$$

Equating the exponentials yields the *general Einstein equation* of diffusion

$$D = \frac{kT}{\zeta}$$

If we describe the diffusing particle as a sphere with radius r in a solvent with viscosity η we can apply Stokes Law

$$f = \zeta \cdot u = 6\pi\eta r \cdot u \qquad \zeta = 6\pi\eta r$$

to find that

$$D = \frac{kT}{6\pi\eta r}$$

which is the *Stokes–Einstein equation.*

14.12 Foundations

14.12.1 Foundation 14.1: The Random Walk in Three Dimensions

See the books website, www.wiley.com/go/Kuhn/physicalchemistry3e.

14.12.2 Foundation 14.2: Intermolecular Forces Affecting the Mean Free Path

See the books website, www.wiley.com/go/Kuhn/physicalchemistry3e.

14.12.3 Foundation 14.3: Law of Hagen–Poiseuille

See the books website, www.wiley.com/go/Kuhn/physicalchemistry3e.

14.12.4 Foundation 14.4: Central Collision of Particles

We consider the central elastic collision between two particles with masses m_1, m_2 and speeds v_1, v_2. To analyze for the final particle speeds u_1 and u_2 after the collision, we apply the laws of conservation of energy and of momentum:

$$\frac{1}{2}m_1 v^2 + \frac{1}{2}m_2 v^2 = \frac{1}{2}m_1 u_1^2 + \frac{1}{2}m_2 u_2^2 = a$$

$$m_1 v_1 - m_2 v_2 = m_1 u_1 + m_2 u_2 = b$$

We rearrange these equations and find

$$u_1^2 = \frac{a}{m_1} - \frac{m_2}{m_1} u_2^2 \tag{14F4.1}$$

$$u_1 = \frac{b}{m_1} - \frac{m_2}{m_1} u_2 \tag{14F4.2}$$

and

$$u_1^2 = \left(\frac{b}{m_1}\right)^2 - 2\frac{b}{m_1}\frac{m_2}{m_1} u_2 + \left(\frac{m_2}{m_1}\right)^2$$

From Equations (14F4.1) and (14F4.2) we eliminate the variable u_1 and find an equation for u_2.

$$\frac{a}{m_1} - \frac{m_2}{m_1} u_2^2 = \left(\frac{b}{m_1}\right)^2 - 2\frac{b}{m_1}\frac{m_2}{m_1} u_2 + \left(\frac{m_2}{m_1}\right)^2$$

or

$$u_2^2 c - d u_2 = e \tag{14F4.3}$$

where we are using the abbreviations

$$c = \left(\frac{m_2}{m_1}\right)^2 + \frac{m_2}{m_1};\ d = 2\frac{b}{m_1}\frac{m_2}{m_1};\ e = \frac{a}{m_1} - \left(\frac{b}{m_1}\right)^2$$

Equation (14F4.3) is a second-order equation with the solutions

$$u_2 = \frac{1}{2}\frac{d}{c} \pm \sqrt{\frac{e}{c} + \frac{1}{4}\frac{d^2}{c^2}} \tag{14F4.4}$$

u_1 is obtained by insertion of u_2 into Equation (14F4.2)

$$u_1 = \frac{b}{m_1} - \frac{m_2}{m_1}\left[\frac{1}{2}\frac{d}{c} \pm \sqrt{\frac{e}{c} + \frac{1}{4}\frac{d^2}{c^2}}\right] \tag{14F4.5}$$

Because of the complexity of these general expressions it is difficult to discuss them analytically in detail. Instead, we provide a program "collision.exe" where the student can enter arbitrary values for the masses m_1 and m_2 and for the initial speeds, and the program returns the final speeds calculated from Equations (14F4.4) and (14F4.5) together with a check for the conservation of energy and momentum.

14.13 Exercises

E14.1 In Section 14.2 we calculated the de Broglie wavelength of H_2. Perform a similar calculation for N_2, CO_2, and He and compare to the H_2 calculation. As the molecular mass increases do you expect a molecule's translational motion to behave more classically or less classically? Of the five gases which would you expect to reveal quantum effects first?

E14.2 Consider a "head-on" collision (one-dimensional) between a CO_2 molecule and an H_2 molecule that are each traveling at 500 m s^{-1} in opposite directions. Use the constraints that the linear momentum is conserved and the kinetic energy is conserved to determine the final speed of the two molecules.

E14.3 Two spheres (masses m_1 and m_2; initial speeds υ_1 and υ_2) collide head-on (central collision). Find the final speed for each of the particles if the initial conditions are $m_1 = m_2 = m$; $\upsilon_1 = 0$, $\upsilon_2 = \upsilon$.

E14.4 Two spheres (masses m_1 and m_2; initial speeds υ_1 and υ_2) collide head-on (central collision). Take the displacement along the collision direction to be x and consider particle 1 to be moving in the positive x-direction. Find the final speed for each of the particles if the initial conditions are $m_1 \gg m_2$; $\upsilon_1 = \upsilon$, $\upsilon_2 = -\upsilon$.

E14.5 Calculate the mean free path for an H_2 molecule in a container of hydrogen gas at a pressure of 1 bar and a temperature of 300 K. Approximate the H_2 molecules as spheres of radius 138 pm.

E14.6 Calculate the mean free path for a Br_2 molecule moving among a collection of N_2 molecules at a pressure of 1 bar and a temperature of 300 K. Approximate the molecules as spheres of radius 312 pm for Br_2 and 190 pm for N_2.

E14.7 Consider a container of I_2 gas molecules at a temperature of 298 K. What is the average translational energy of an I_2 molecule in the container? Consider two I_2 molecules, X and Y, to have the average energy. If molecule X collides with molecule Y and transfers all of its translational energy to molecule Y, is it possible for molecule Y to become vibrationally excited? If so, what is the maximum vibrational energy level to which Y could be excited? The I_2 molecule has a fundamental vibrational frequency ν_0 of 6.42×10^{12} s^{-1}.

E14.8 For practice you should compute the root mean square speeds of some gases; e.g., He, N_2, O_2, Ar, and Xe at $T = 298$ K and compare the results you find with the speed of sound in air, which is about 340 m s^{-1} at sea level.

E14.9 Calculate the collision frequency of a nitrogen molecule (the dominant component of air) at 1 bar pressure and 300 K. Compute the number of collisions that occur in one second. Compare this number to the number of seconds in recorded history, which you can estimate to be 5000 years.

E14.10 Compute the mean free path and average speed of a He atom under the following conditions: a) 1 bar pressure and 300 K (typical of Earth surface), b) 0.01 bar pressure and 250 K (typical of Mars surface), c) 90 bar pressure and 700 K (typical of Venus surface), d) 1.4×10^{-5} bar pressure and 40 K (typical surface values for Triton, a moon of Neptune).

E14.11 Consider a gas of nitrogen at 273 K. Calculate the average time it takes for a nitrogen molecule to travel a distance of 5585 km (about the distance from London, UK, to New York, NY, USA). Compare this time to average airline flight time of about 8 hours.

E14.12 Use the atmospheric distribution function in Box 14.5 to calculate the atmospheric pressure at the top of Mount Everest. Take the height of Mount Everest to be 8882 m. Use a temperature of $T = 273$ K as an average of the temperature at sea level $x = 0$ and at $x = 8882$ m.

E14.13 Analyze the following viscosity data as a function of temperature for Ar at 1 bar pressure ($\eta = 1.5951 \times 10^{-5}$ Pa s at 200 K, 2.2676×10^{-5} Pa s at 300 K, and 2.8627×10^{-5} Pa s at 400 K). Determine the effective diameter of an argon atom from these data. Comment on any temperature dependence you observe for the diameter that is obtained.

E14.14 In Section 14.3.3.5 we discussed the diffusion of Br_2 from a graduated cylinder and approximated the concentration gradient as linear. Why is this approximation reasonable?

E14.15 Here you consider the removal of excess thiosulfate from a photographic emulsion by diffusion. How long does the process take if the thickness of the emulsion is $d = 0.2$ mm? Consider the emulsion to be a gel with $\eta = 0.0089$ g s^{-1} cm^{-1} and the thiosulfate to be a sphere of radius $r = 120$ pm.

E14.16 Consider the formation of a bubble of methane gas (diameter 1 cm) at the bottom of a lake (depth 10 m, $P = 2$ bar and $T = 278$ K). What is the diameter of the bubble when it reaches the surface? Consider the lake to have a surface temperature of 298 K and a surface pressure of 1 bar.

E14.17 The air inside a refrigerator (volume 100 L) is chilled from 20 °C to 0 °C after closing the door. What is the force needed to open the refrigerator door (area 1 m^2)? Suppose the refrigerator is totally airtight and neglect the friction between the door seal and the refrigerator.

E14.18 In a previous form of an automobile airbag (in use until 1995), NaN_3 and KNO_3 react to make metal oxides and nitrogen gas (1.6 moles for each mole of reactant). Consider the typical airbag to be 60 L in volume and fill to a pressure of 1.5 bar. How many grams of NaN_3 will be needed?

E14.19 In obtaining the rate of escape for effusion, we assume that the gas remains at equilibrium and the hole is smaller than the mean free path in the gas. Discuss how the flow of molecules from the container changes when the hole is larger than the mean free path.

E14.20 Consider a standard mixture of air (78.084% N_2, 20.9476% O_2, 0.934% Ar, 0.0314% CO_2, 0.001818% Ne, 0.000524% He, 0.000114% Kr, 0.0000087% Xe, 0.0002% CH_4 and 0.00005% H_2) at 1 bar pressure. Calculate the partial pressure of each of the atomic gases. What is the partial pressure arising from all of the atomic gases? What is the partial pressure arising from all of the molecular gases?

E14.21 Use Equation (14.52) to derive Equation (14.53).

14.14 Problems

P14.1 A gas thermometer problem. The following molar volumes V_m are measured for SO_2 when the gas is examined at different pressures:

P/bar	0.253	0.506	0.660	1.013
V_m/(L mol^{-1})	89.13	44.30	33.83	21.89

What is the temperature of the gas?

P14.2 In Box 14.5 we considered the pressure P of a gas in height h above the bottom of a container

$$P = P_0 \cdot \exp\left[-mgh/(kT)\right]$$

where P_0 is the pressure at the bottom, m is the mass of a gas molecule, k is the Boltzmann constant, and T is the temperature. mgh is the potential energy of the molecule and kT is the thermal energy. At $T = 0$ all gas molecules are at the bottom, but with increasing temperature more and more molecules have the chance to take up energy from the temperature bath to reach a height h above the bottom. We also see that molecules with a lighter mass can reach greater heights than heavier molecules can. In principle we can use this effect to separate isotopes. However, the effect is too small in the gravity field, and an ultracentrifuge is used to apply a much greater force on the molecules.

Consider a gaseous mixture of $^{235}UF_6$ and $^{238}UF_6$ in an ultracentrifuge cylinder rotating with a speed of $v = \omega r = 350$ m s^{-1} at a radius $r = 15$ cm. The force f acting on a particle of mass m in the centrifuge is given by $f = m\omega^2 r$. Consider that the partial pressure of $^{235}UF_6$ is 0.7% of the partial pressure of $^{238}UF_6$. What will be the enrichment of $^{235}UF_6$ in this process?

P14.3 For a one-dimensional diffusion path (random walk with steps of $+\lambda$ and $-\lambda$) calculate the distribution function (i.e., probability versus position) after 10 diffusion steps; i.e., extend the results given in Section 14.3. Use your distribution function to calculate the average displacement of the particle and its mean square displacement.

P14.4 Our development of the diffusion process is called the "random walk" model. In the limit that we allow the diffusion steps to be continuous, rather than discrete, we obtain a continuous distribution function which corresponds to a Gaussian distribution

$$\frac{dN}{dx} = a\, e^{-bx^2}$$

If you have N_0 total particles and a root mean square displacement of 4λ, determine the constants a and b in the distribution function.

P14.5 Consider a measuring cylinder (inner diameter 3 cm, height 15 cm) with a layer (height 3 cm) of water on the bottom. Above the water surface gaseous water is in equilibrium with liquid water (vapor pressure of 30 mbar at room temperature). The water molecules escape from the cylinder by diffusion. How long does it take until 1 ml of liquid water has disappeared by diffusion? How long does it take if the cylinder height is 5 cm?

P14.6 It is possible to view small spherical particles undergoing Brownian motion on the surface of a liquid with a video camera and analyze the mean square displacement of the particles. The table contains data for such an experiment that was performed on 107 polystyrene spheres of 1.02 μm diameter. Analyze these data by plotting the root mean square Δx_{rms} displacement versus time t and also versus $\sqrt{t}$. After demonstrating that the motion is diffusive, determine the diffusion coefficient. From your diffusion constant and the size of the spherical particle determine the effective viscosity of the air–water interface at which the spheres are diffusing.

$\Delta R^2/\mu m^2$	t/s	$\Delta R^2/\mu m^2$	t/s	$\Delta R^2/\mu m^2$	t/s	$\Delta R^2/\mu m^2$	t/s
0	0	6.9	3.6	13.6	7.2	20.8	10.8
1.25	0.4	7.7	4	13.6	7.6	22.1	11.2
1.8	0.8	8.2	4.4	15	8	22.3	11.6
2.5	1.2	8.75	4.8	15.2	8.4	22.5	12
2.9	1.6	9.8	5.2	16.25	8.8		
4.2	2	10.8	5.6	17.9	9.2		
5	2.4	11.7	6	19.0	9.6		
6.0	2.8	11.9	6.4	19.0	10		
6.5	3.2	12.4	6.8	19.7	10.4		

P14.7 Use a statistical model to estimate the time it takes for an enzyme to assemble and interlock with a substrate of radius 1 nm that is present in a concentration of one molecule in a cubic micron of volume V. An enzyme **E** reacts with a substrate **S** if the point **s** on the substrate has approached point **e** on the enzyme within about d = 100 pm (see Fig. P14.7). Then point **s′** must find point **e′**; this process, however, requires a short time compared to the first process. Analyze this phenomenon by considering

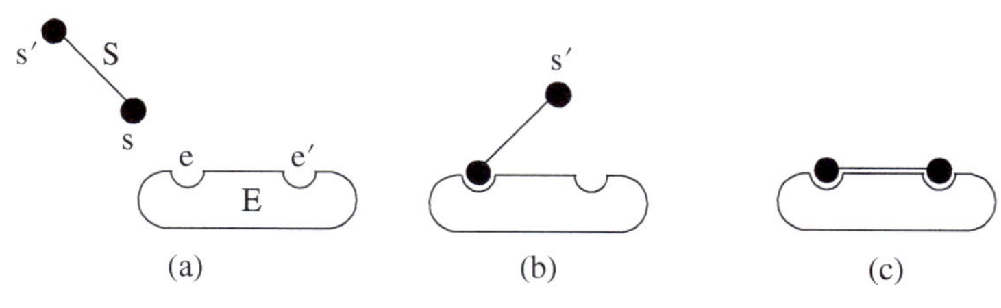

Figure P14.7 Enzyme interlocking with a substrate: (a) Substrate S approaching the enzyme E; (b) Substrate docked on the enzyme with one end; (c) Substrate docked on the enzyme with both ends.

the enzyme molecule to be immobilized and divide up the volume V into cells with an edge length of d = 100 pm. Consider point **s** of the substrate to be in one of these cells and diffuse from cell to cell until it reaches the cell with point **e**. Find the average time $\bar{t}$ required for this process. Why is the time you find so much longer than the time it takes the substrate to diffuse over a length of 1 μm?

P14.8 Consider the following data for water vapor as a function of temperature at 1 bar pressure. Compare the experimental values for the shear viscosity and the average kinetic energy with that you calculate using the kinetic theory of gases. Compare the molar volume with the value you calculate from the ideal gas law. Discuss any differences in terms of the molecular properties of water.

T/K	V_m/l/mol	E_{kin}/kJ/mol	η/µ Pa s
376.11	30.814	10.467	12.381
397.36	32.67	10.814	13.184
416.97	34.36	11.252	13.941
436.58	36.036	11.75	14.712
456.19	37.702	12.287	15.494
477.44	39.499	12.903	16.350
497.05	41.152	13.494	17.149
518.29	42.939	14.157	18.020
539.53	44.723	14.837	18.896
559.15	46.367	15.481	19.709
578.76	48.009	16.137	20.523
600.00	49.786	16.862	21.407

P14.9 A Fortran program is provided for performing computer simulations of diffusion of a particle. In the calculation the direction and the mean free path are chosen from a pseudorandom number. Use the program to run 200 simulations for the diffusion of a particle undergoing 200 collisions. Plot the endpoints of these trajectories and determine a) the average displacement, b) the mean square displacement, and c) the diffusion coefficient of the particle.

P14.10 Consider a plastic balloon that you fill to a pressure of 1.1 bar with 5 L of He gas at 300 K. Eventually the He escapes from the balloon through tiny "holes" between the polymer chains of the plastic. For a balloon that deflates to half its volume over a time of 24 hours, determine the effective area of the tiny "holes" (or microleaks) in the balloon. Model the escape using a model of effusion through the holes. You may neglect the presence of He outside the balloon and assume that the pressure in the balloon is 1.1 bar throughout its deflation. Perform your calculation by neglecting the rate of effusion of air into the balloon.

References

1 J.C. Maxwell, V. Illustrations of the dynamical theory of gases.—Part I. On the motions and collisions of perfectly elastic spheres, *Philos. Mag.*, 1860, **19**, 19.

2 R.D. Present, *Kinetic Theory of Gases* (McGraw-Hill, New York, 1958, p. 55).

3 G. Ertl, Dynamik der Wechselwirkung von Molekülen mit Oberflächen, *Ber. Bunsen. Phys. Chem.*, 1995, **99**, 1282.

4 See R.D. Present, *Kinetic Theory of Gases* (McGraw-Hill, New York, 1958, p. 222).

5 J. Happel and H. Brenner, *Low Reynolds Number Hydrodynamics* (Kluwer, Dordrecht, 1991).

15

Energy Distribution in Molecular Assemblies

CHAPTER MENU

15.1 Introduction

What happens inside a molecule when it collides with another molecule in a gas? In Chapter 14, we assumed that all the collisions are elastic, i.e., no energy is transferred to form internally excited states of the molecule. Energy is only swapped among the translational motions of the gas molecules. As our earlier analysis showed the temperature determines the average molecular speed and the average energy in a collision. If a collision is energetic enough, however, it is possible to excite the molecule's rotational, vibrational, and electronic energy levels. In this chapter, we will generalize our considerations about collisions and their effect on a molecule's internal energy.

In this chapter we discuss how the energy is distributed among the molecules in an assembly and the molecules' different internal energy states. The Boltzmann distribution describes the occupation probability $\mathcal{P}_i$ of quantum state i with energy E_i as a function of temperature:

$$\mathcal{P}_i \propto \mathrm{e}^{-E_i/(kT)}$$

and we find that it can be used to describe the energy distribution in molecules and their assemblies at equilibrium. The Maxwell–Boltzmann distribution of molecular speeds v

$$\frac{\mathrm{d}N}{\mathrm{d}v} \sim v^2 \cdot \mathrm{e}^{-mv^2/2kT}$$

is a special case of this energy distribution and describes the distribution of translational speeds as a function of temperature for a collection of gas molecules. The Maxwell–Boltzmann distribution can be determined by considering the

Principles of Physical Chemistry, Third Edition. Hans Kuhn, David H. Waldeck, and Horst-Dieter Försterling.

Companion website: www.wiley.com/go/Kuhn/physicalchemistry3e

translational quantum states of gas molecules or by treating the translational energy classically, so it provides an example of the correspondence between classical and quantum mechanical considerations. The essence and characteristics of the Boltzmann distribution are straightforward to understand, but its formal derivation from statistical considerations requires deeper reflection. These considerations are given in Section 15.8.

15.2 The Boltzmann Distribution Law

15.2.1 System Consisting of Two Quantum States

Consider an ideal gas of atoms (i.e., we neglect forces acting between the gas atoms: model of independent particles), in which each atom has two internal quantum states, in contact with a temperature bath (temperature T). We consider a simple model for the atoms where the ground state has an energy E_1, and the excited state has an energy E_2 (Fig. 15.1).

For N gas particles we have

$$N_1 + N_2 = N \tag{15.1}$$

with N_1 particles in quantum state 1 and N_2 particles in quantum state 2. At low enough temperature, all the particles are in the ground state (lower energy state) and we have

$$N_2 = 0, \quad N_1 = N \quad \text{(low temperature)}$$

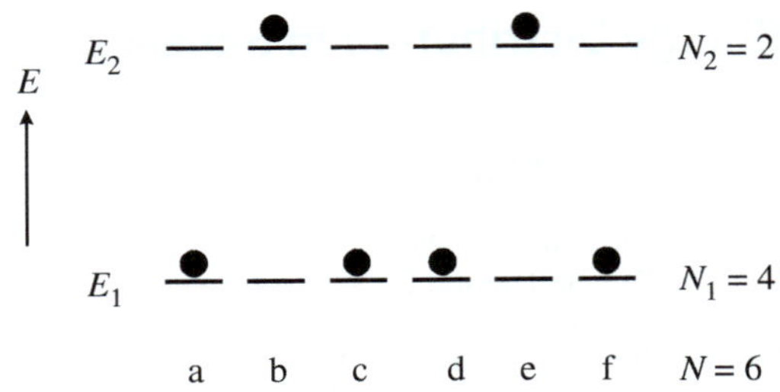

Figure 15.1 Ground state E_1 and electronically excited state E_2 (two quantum states) of $N = 6$ atoms labeled a to f. In the present case atoms a, c, d, and f ($N_1 = 4$) are in the ground state E_1, atoms b and e ($N_2 = 2$) are in the excited state E_2 ($N_1 + N_2 = N$).

With increasing temperature the population number N_2 in the excited state increases because of collisions and energy exchange with the temperature bath. When the mean translational energy of the gas particles is similar to the energy difference between the two quantum states, that is

$$\Delta E = E_2 - E_1 \approx \tfrac{3}{2}kT \quad \text{(medium temperature)},$$

a considerable fraction of the molecules will be in quantum state 2. At very high temperature it is equally probable that a particle in the ground state is excited in a collision or that a particle in the excited state is de-excited. This means that

$$N_2 = N_1 = \tfrac{1}{2}N \quad \text{(high temperature)}$$

and N_2 does not exceed N_1.

As a simple model, we imagine a container with a bottom made of two steps (Fig. 15.2a). On the bottom we place steel spheres. As long as the container is at rest, all the spheres will be on the lower step at lowest potential energy (Fig. 15.2b). Now we shake the container, thus providing the steel spheres with kinetic energy. Weak shaking corresponds to low kinetic energy – that is, to low temperature. On a time average, most of the spheres are on the lower step and only a few are on the upper step (Fig. 15.2c). The stronger we shake, the more spheres will reach the upper step, but the spheres will not accumulate on the upper step. When very strongly shaking the container, the spheres do not recognize any difference between the two parts of the container, and therefore they will be equally distributed (Fig. 15.2d).

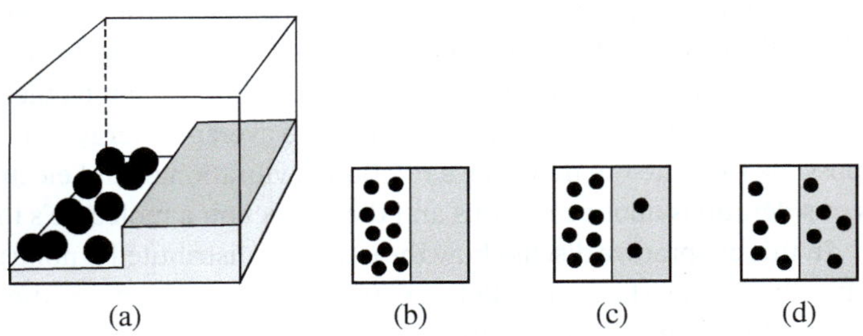

Figure 15.2 Model experiment for the demonstration of the distribution of molecules on two energy levels. (a) Experimental setup. (b) Box at rest, (c) box weakly shaken, (d) box vigorously shaken.

In a similar way, the number of gas atoms in an internal quantum state depends on the energy of the state compared to the thermal energy kT; in Section 15.8, we show that

$$N_1 = C \cdot e^{-E_1/(kT)} \quad \text{and} \quad N_2 = C \cdot e^{-E_2/(kT)} \tag{15.2}$$

in our model. We shall give a rigorous proof for this important relation in Section 15.8. With $\Delta E = E_2 - E_1$, the ratio of gas atoms in each internal energy state is given by

$$\frac{N_2}{N_1} = e^{-\Delta E/(kT)} \tag{15.3}$$

Combining this result with the conservation of particles, Equation (15.1), it follows that

$$N_1 = N \cdot \frac{1}{1 + e^{-\Delta E/(kT)}} \quad \text{and} \quad N_2 = N \cdot \frac{1}{1 + e^{+\Delta E/(kT)}} \tag{15.4}$$

Equation (15.4) satisfies our earlier considerations. As $T \to 0$, we obtain $N_1 = N$ and $N_2 = 0$. All the atoms are in the ground state (Fig. 15.3a). As $T \to \infty$ we find that $N_1 = N_2 = N/2$, so that half the gas atoms are in the ground state and half are in the excited state (Fig. 15.3c). When $\Delta E = kT$, $N_1 = 0.73 \cdot N$, and $N_2 = 0.27 \cdot N$ (Fig. 15.3b).

In Fig. 15.3d, the population fractions N_1/N and N_2/N (see Equation (15.4)) in states 1 and 2 are plotted as a function of T. For this excitation energy, which is typical for electronic states of atoms and molecules, the temperature must be about 10,000 K to excite 10% of the particles to state 2.

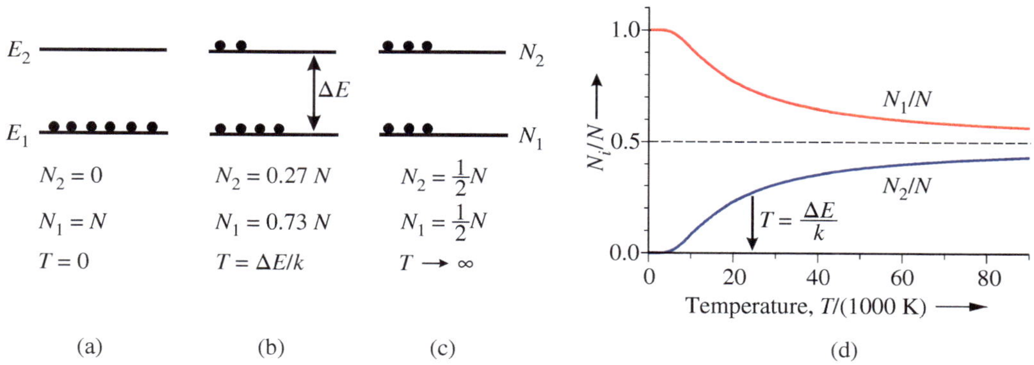

Figure 15.3 Population of ground state E_1 and electronically excited state E_2 for $N = N_1 + N_2 = 6$ atoms at different temperatures (each atom has two quantum states with a separation of $\Delta E = E_2 - E_1$). (a) $T = 0$, (b) $T = \Delta E/k$, (c) $T \to \infty$. N_1 and N_2 are calculated according to Equation (15.4). Lines for states E_1 and E_2, respectively, are drawn continued to indicate that we count the number of atoms in each energy level. (d) N_1/N_2 as a function of T according to Equation (15.4) with $\Delta E = 3.37 \times 10^{-19}$ J (excitation energy of a Na atom for $3s \to 3p$).

15.2.2 Systems Consisting of Many Quantum States

Now we proceed from particles with only two internal quantum states to molecules with many internal quantum states. In Section 15.8 we show that for independent particles the number N_i of the particles in a certain quantum state i is given by the *Boltzmann distribution* (first derived by Ludwig Boltzmann, 1844–1906), as a generalization of Equation (15.2):

$$N_i = C \cdot e^{-E_i/(kT)}, \quad i = 1, 2, 3, \ldots \tag{15.5}$$

where E_i is the energy of quantum state i and C is a constant. This relation corresponds to what is expected from classical considerations, see the atmospheric distribution discussed in Section 14.3.3.6 and in Box 14.5. Next we evaluate the constant C in Equation (15.5). Using the condition that the sum over all population numbers N_i equals the total number N of particles

$$N_1 + N_2 + N_3 + \cdots = N \tag{15.6}$$

and substituting Equation (15.5) for N_i we obtain

$$N = \sum_{i=1}^{\infty} N_i = \sum_{i=1}^{\infty} C \cdot e^{-E_i/(kT)} = C \cdot \sum_{i=1}^{\infty} e^{-E_i/(kT)} = C \cdot Z \tag{15.7}$$

The quantity

$$\boxed{Z = \sum_{i=1}^{\infty} e^{-E_i/(kT)}} \tag{15.8}$$

is called the *molecular partition function* and the index i identifies the individual quantum states. Z is a function of the temperature and the molecular energy states. Using Equation (15.7), we can write the Boltzmann distribution (Equation (15.5)) as

$$\boxed{N_i = \frac{N}{Z} \cdot e^{-E_i/(kT)}} \tag{15.9}$$

For the probability $\mathcal{P}_i$ of finding a particle in a quantum state of energy E_i at temperature T we obtain

$$\boxed{\mathcal{P}_i = \frac{N_i}{N} = \frac{1}{Z} \cdot e^{-E_i/(kT)}} \tag{15.10}$$

These are the basic equations for calculating the distribution of N molecules over all quantum states i in equilibrium with a temperature bath of temperature T.

Example 15.1 *Consider a three-state system at a temperature T with an energy spectrum of $E_1 = 0$, $E_2 = a$, and $E_3 = 2a$. Calculate the partition function Z and the probabilities $\mathcal{P}_1$, $\mathcal{P}_2$, and $\mathcal{P}_3$ of occupying the three energy levels for the cases $a = 0.1kT$, $a = kT$, and $a = 10kT$.*

From Equation (15.8) for the partition function we obtain

$$Z = 1 + e^{-a/(kT)} + e^{-2a/(kT)}$$

and from Equation (15.10) for the probabilities it follows that

$$\mathcal{P}_1 = \frac{1}{Z}, \quad \mathcal{P}_2 = \frac{1}{Z}e^{-a/(kT)}, \quad \mathcal{P}_3 = \frac{1}{Z}e^{-2a/(kT)}$$

These expressions are evaluated for the three cases in the following table:

a	Z	$\mathcal{P}_1$	$\mathcal{P}_2$	$\mathcal{P}_3$
$0.1kT$	2.724	0.367	0.332	0.301
kT	1.503	0.665	0.245	0.090
$10kT$	1.000	1.000	0.000	0.000

We see that for small values of a the three levels are nearly equally occupied, whereas for large a only the lowest level is occupied. The value of Z provides a statistical measure of the number of quantum states which are available to the system at temperature T.

Below we show how the molecular partition function Z can be used to compute the average energy of molecules at thermal equilibrium.

15.2.3 Internal Energy *U*

The energy of a collection of molecules, consisting of translational, rotational, vibrational, and electronic contributions is called the *internal energy*, U. For independent particles, U can be calculated by adding up the energies E_i of the N_i molecules in the quantum states i, namely

$$U = \sum_{i=1}^{\infty} N_i E_i = \frac{N}{Z} \sum_{i=1}^{\infty} e^{-E_i/(kT)} \cdot E_i \tag{15.11}$$

with

$$Z = \sum_{i=0}^{\infty} e^{-E_i/(kT)} \tag{15.12}$$

We can write U directly in terms of Z. First, we realize that the temperature derivative of the molecular partition function is

$$\frac{dZ}{dT} = \sum_{i=0}^{\infty} e^{-E_i/(kT)} \cdot \left(\frac{E_i}{kT^2}\right) = \frac{1}{kT^2} \cdot \sum_{i=0}^{\infty} e^{-E_i/(kT)} \cdot E_i$$

Next we combine this result with Equation (15.11) and write

$$U = \frac{N}{Z} \cdot kT^2 \frac{dZ}{dT} = NkT^2 \cdot \left(\frac{1}{Z}\frac{dZ}{dT}\right) \quad \text{or} \quad U = NkT^2 \cdot \frac{d\ln Z}{dT} \tag{15.13}$$

This equation is very useful because it relates the internal energy U directly to the molecular partition function Z. Once the molecular partition function Z is known, U can be easily calculated.

If the particles (molecules or atoms) are independent of each other, then U can be related to the average energy $\overline{E}$ of a particle as

$$U = N\overline{E}$$

Although each particle has a specific energy at any given time, it is common to speak of a particle's average energy. Over an appropriately long span of time, during which a particular particle exchanges energy with other particles and the temperature bath, it will sample many different available energies.

Often, it is useful to approximate a molecule's individual degrees of freedom (electronic, vibrational, rotational, and translational) as independent. As we show in Foundation 15.5, in this case the internal energy U can be written as a sum of contributions of the electronic, vibrational, rotational, and translational energy of the particles:

$$U = U_{el} + U_{vib} + U_{rot} + U_{trans} \tag{15.14}$$

and these contributions are given by

$$\boxed{U_{el} = kT^2 \cdot \frac{d\ln Z_{el}}{dT}\ ; \ U_{vib} = kT^2 \cdot \frac{d\ln Z_{vib}}{dT}} \tag{15.15}$$

$$\boxed{U_{rot} = kT^2 \cdot \frac{d\ln Z_{rot}}{dT}\ ; \ U_{trans} = kT^2 \cdot \frac{d\ln Z_{trans}}{dT}} \tag{15.16}$$

This analysis shows that we can write the internal energy of the macroscopic system as a sum of terms for the different degrees of freedom. In the next sections, we proceed to find these contributions (electronic, vibrational, rotational, and translational).

15.3 Electronic Energy

The energy spacing between electronic states is typically so large that the thermal energy at room temperature is not sufficient to populate a significant number of them (see Example 15.1). Since almost all N molecules are in the electronic ground state with energy E_1, the electronic contribution to the internal energy U_{el} is well approximated by

$$U_{el} = N \cdot E_1$$

In special cases, the population of electronically excited levels plays a role. One such example is the NO molecule. The NO molecule is a stable radical, i.e., has one unpaired electron. Because it has an unpaired electron it forms a doublet state whose magnetic moment couples to the magnetic moment generated by its orbital motion (it is in a π-orbital). This spin–orbit interaction generates a splitting of the lowest energy level into two energy levels separated by 0.015 eV (2.4×10^{-21} J). The lower level (energy E_1) is occupied at low temperature, and the upper level (energy E_2) becomes gradually populated as the temperature increases. Because the next highest electronic state lies 7.6×10^{-19} J higher in energy, we need not consider its population at the temperatures of interest here. Hence we can treat NO approximately as a system with two electronic energy levels.

According to Equation (15.4) we obtain[1]

$$U_{\mathrm{el}} = N_1E_1 + N_2E_2 = N \cdot \left[\frac{E_1}{1 + \mathrm{e}^{-\Delta E/(kT)}} + \frac{E_2}{1 + \mathrm{e}^{\Delta E/(kT)}}\right] \tag{15.17}$$

and

$$U_{\mathrm{el,m}} = \frac{U_{\mathrm{el}}}{\boldsymbol{n}} = \boldsymbol{N}_{\mathrm{A}} \cdot \left[\frac{E_1}{1 + \mathrm{e}^{-\Delta E/(kT)}} + \frac{E_2}{1 + \mathrm{e}^{\Delta E/(kT)}}\right] \tag{15.18}$$

For simplicity, we choose E_1 as the zero point of the electronic energy. Then for a gas of NO molecules we calculate $U_{\mathrm{el,m}}$ with $E_1 = 0$ and $\Delta E = E_2 = 2.4 \times 10^{-21}$ J. $U_{\mathrm{el,m}}$ is displayed in Fig. 15.4 as a function of temperature. At absolute zero, $U_{\mathrm{el}} = 0$ (since $N_1 = N$, $N_2 = 0$); at high enough temperature $U_{\mathrm{el}} = \frac{1}{2}NE_2$ (since $N_1 = N_2 = N/2$).

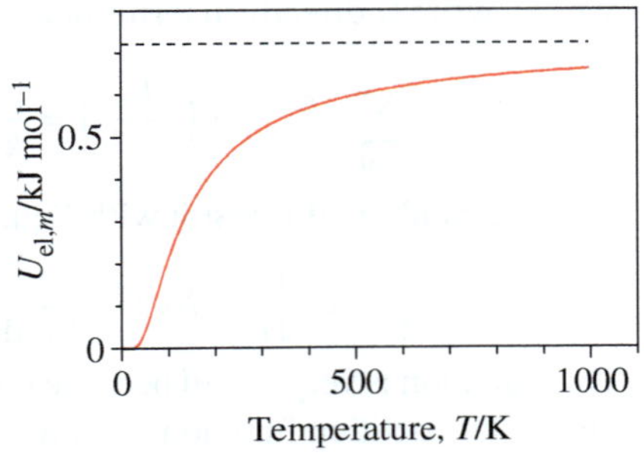

Figure 15.4 $U_{\mathrm{el,m}}$ calculated for $E_1 = 0$, $E_2 = 2.4 \times 10^{-21}$ J (NO molecule). The limit as $T \to \infty$ is $0.5 \cdot \boldsymbol{N}_{\mathrm{A}}E_2 = 0.72$ kJ mol^{-1} (dashed line).

15.4 Vibrational Energy

Consider a molecule such as I_2 in the vapor phase. The molecule can be in the vibrational ground state or in a vibrational excited state. How can we determine the fraction of molecules, N_i/N, in each of these different vibrational quantum states? If we know the energy level spectrum of the vibrational states (e.g., see Table 11.4 of Chapter 11), then we can use the vibrational energies to evaluate Equation (15.12) for Z and Equation (15.10) $P_i = N_i/N$. If the energy spectrum is known, the numerical evaluation of the partition function is straightforward, see Chapter 16. Often however, these data may not be available and in these cases it can be helpful to make an estimate by using the harmonic oscillator approximation for the vibration.

Following the analysis in Chapter 11, we approximate the vibrational mode of the molecule (e.g., iodine) as harmonic. The vibrational energy E_n of a harmonic oscillator with quantum number n is

$$E_n = \left(\frac{1}{2} + n\right) \cdot h\nu_0, \quad n = 0, 1, 2, 3, \ldots \tag{15.19}$$

where

$$\nu_0 = \frac{1}{2\pi}\sqrt{\frac{k_f}{\mu}} \tag{15.20}$$

μ is the reduced mass and k_f is the force constant. The corresponding quantum states are shown in Fig. 15.5.

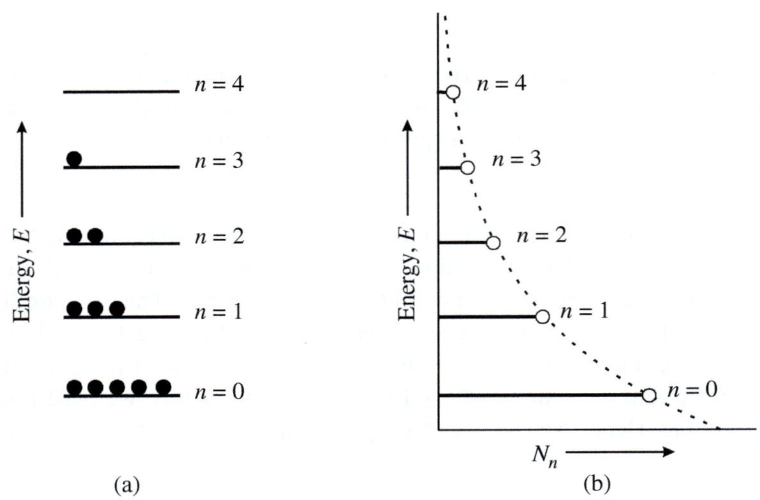

Figure 15.5 Population numbers N_n as a function of the energy E_n at constant temperature. (a) Population scheme, (b) illustration of the Boltzmann distribution (15.9). The population scheme is shown for the distribution of vibrational energy of a molecule in the harmonic oscillator approximation (equal distances of energy levels); n is the vibrational quantum number.

1 Actually,each of the two NO energy levels has two states (doubly degenerate). In this case each of the two lower states is occupied by $N_1/2$ molecules and each of the two higher states is occupied by $N_2/2$ molecules. Thus the result for the internal energy remains unchanged.

15.4.1 Population Number N_n of Harmonic Oscillator

Equation (15.9) for the population numbers refer to the number N_i of molecules in quantum state i ($i = 1, 2, 3, \ldots$) for a collection of N molecules. Because of the dependence of the vibrational energy on the quantum number n it is more convenient to consider the number N_n of molecules in a state with quantum number n ($n = 0, 1, 2, \ldots$). Then from Equation (15.9) for the population number N_n of molecules we obtain

$$N_n = \frac{N}{Z_{\text{vib}}} e^{-E_n/(kT)} = \frac{N}{Z_{\text{vib}}} \exp\left[-\frac{\left(\frac{1}{2}+n\right) h\nu_0}{kT}\right], \quad n = 0, 1, 2, 3, \ldots \tag{15.21}$$

The molecular partition function for vibration, Z_{vib} is

$$Z_{\text{vib}} = \sum_{n=0}^{\infty} e^{-E_n/(kT)} = \sum_{n=0}^{\infty} \exp\left[-\frac{\left(\frac{1}{2}+n\right) h\nu_0}{kT}\right] \tag{15.22}$$

With the abbreviation

$$x = \frac{h\nu_0}{kT}$$

we can write

$$Z_{\text{vib}} = e^{-x/2} \cdot \sum_{n=0}^{\infty} e^{-xn} = e^{-x/2} \cdot \sum_{n=0}^{\infty} \left(e^{-x}\right)^n = e^{-x/2} \cdot \sum_{n=0}^{\infty} a^n$$

where $a = e^{-x} \leq 1$. In this case the sum corresponds to that of an infinite geometrical series so that

$$\sum_{n=0}^{\infty} a^n = a^0 + a^1 + a^2 + \cdots = \frac{1}{1-a};$$

see Appendix B5. Using this result we find the molecular partition function to be

$$Z_{\text{vib}} = e^{-h\nu_0/(2kT)} \cdot \frac{1}{1 - e^{-h\nu_0/(kT)}} \tag{15.23}$$

Combining this result with Equation (15.21) it follows that

$$N_n = \frac{N}{Z_{\text{vib}}} e^{-E_n/(kT)} = N\left(1 - e^{-h\nu_0/(kT)}\right) \cdot e^{-nh\nu_0/(kT)} \tag{15.24}$$

Using this result, we can calculate the population numbers for each of the vibrational quantum states of I_2 at any given temperature (Table 15.1 and Fig. 15.6). For I_2 we use $k_f = 173$ N m^{-1} and $\mu = m_I \cdot m_I/(m_I + m_I) = \frac{1}{2} m_I = 1.05 \times 10^{-25}$ kg to

Table 15.1 Population Probability $\mathcal{P}_n = N_n/N$, Molecular Partition Function Z_{vib} and Internal Energy U_{vib} for I_2 ($h\nu_0 = 4.27 \times 10^{-21}$ J).

		$\mathcal{P}_n = N_n/N$			
n	$E_n/10^{-20}$ J	50 K $Z_{\text{vib}} = 0.04$	250 K $Z_{\text{vib}} = 0.75$	500 K $Z_{\text{vib}} = 1.58$	1000 K $Z_{\text{vib}} = 3.20$
0	0.21	0.92	0.71	0.46	0.27
1	0.64	0.08	0.21	0.25	0.20
2	1.07		0.06	0.13	0.14
3	1.49		0.02	0.07	0.11
4	1.92			0.04	0.08
5	2.35			0.02	0.06
			$U_{\text{vib},m}$/kJ mol^{-1}		
		1.30	2.34	4.29	8.38

$U_{\text{vib},m} = U_{\text{vib}}/\boldsymbol{n}$ is the molar internal energy of vibration where $\boldsymbol{n}$ is the amount of substance.

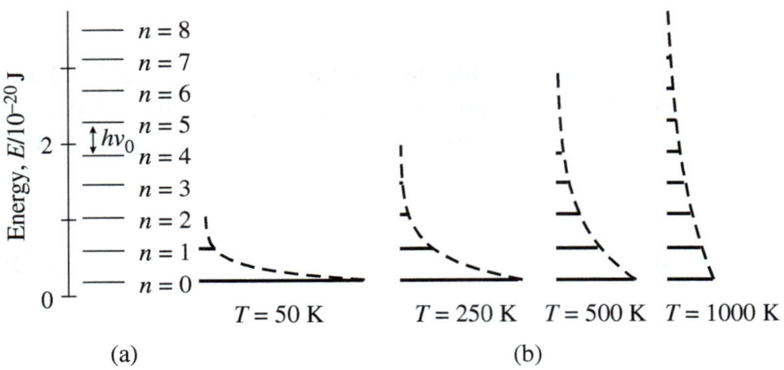

Figure 15.6 Vibrational quantum states of the I_2 molecule ($h\nu_0 = 4.3 \times 10^{-21}$ J). (a) Energy of lowest quantum state with quantum number $n = 0$ ($E_0 = \frac{1}{2}h\nu_0$) and equally spaced energy levels $E_n = (\frac{1}{2} + n) \cdot h\nu_0$ up to quantum number $n = 8$. (b) Populations of quantum states according to Equations (15.21) and (15.23) at various temperatures T. The length of a line of an energy level is proportional to the number of particles that occupy that level.

obtain $h\nu_0 = 4.27 \times 10^{-21}$ J. At low enough temperature, all N molecules are in the lowest quantum state. As the temperature increases however, the percentage of molecules in excited vibrational levels increases. At 1000 K we find that some I_2 molecules will be in very high vibrationally excited states ($n = 5$ and larger).

15.4.2 Internal Energy U_{vib} of Diatomic Molecules

The vibrational contribution to the internal energy U_{vib} can be calculated by adding up the vibrational energies E_n of the N_n molecules in each quantum state with quantum number n. Using the sum over the quantum number n, we obtain

$$U_{vib} = \sum_{n=0}^{\infty} N_n E_n = \frac{N}{Z_{vib}} \sum_{n=0}^{\infty} e^{-E_n/(kT)} \cdot E_n \tag{15.25}$$

with

$$Z_{vib} = e^{-h\nu_0/(2kT)} \cdot \frac{1}{1 - e^{-h\nu_0/(kT)}}$$

Equation (15.15) relates U_{vib} to the partition function Z_{vib}

$$U_{vib} = NkT^2 \frac{d \ln Z_{vib}}{dT}$$

Using

$$\ln Z_{vib} = -h\nu_0/(2kT) - \ln\left(1 - e^{-h\nu_0/(kT)}\right)$$

we can evaluate the temperature derivative to find

$$\frac{d \ln Z_{vib}}{dT} = \frac{h\nu_0}{2kT^2} + \frac{e^{-h\nu_0/(kT)}}{1 - e^{-h\nu_0/(kT)}} \frac{h\nu_0}{kT^2} = \frac{h\nu_0}{kT^2}\left(\frac{1}{2} + \frac{1}{e^{h\nu_0/(kT)} - 1}\right)$$

With this result, we find that

$$U_{vib} = Nh\nu_0 \left(\frac{1}{2} + \frac{1}{e^{h\nu_0/(kT)} - 1}\right) \tag{15.26}$$

or

$$U_{vib,m} = \frac{U_{vib}}{\boldsymbol{n}} = \boldsymbol{N}_A h\nu_0 \left(\frac{1}{2} + \frac{1}{e^{h\nu_0/(kT)} - 1}\right) \tag{15.27}$$

where $U_{vib,m}$ is the molar internal energy of vibration.

Figure 15.7 shows $U_{vib,m}$ for I_2, HD, and H_2 as a function of temperature. The temperature behavior of U_{vib} depends strongly on the force constant and the mass of the harmonic oscillator. At sufficiently low temperatures all of the molecules occupy the ground vibrational state, zero point energy level, and the vibrational contribution to the internal energy goes to

$$U_{vib} = \frac{1}{2} Nh\nu_0$$

Therefore the internal vibrational energy differences for the three diatoms in Fig. 15.7 at $T = 0$ result from their different zero point vibrational energies.

In contrast to the two-level system, U_{vib} continuously increases with temperature, without reaching a limiting value. This occurs because more and more energy levels become available.

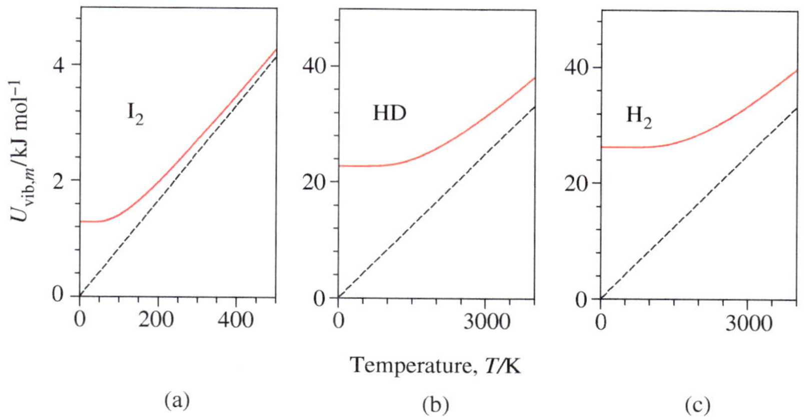

Figure 15.7 Molar vibrational energy $U_{vib,m} = U_{vib}/n$ as a function of the temperature T. (a) I_2 gas ($h\nu_0 = 4.27 \times 10^{-21}$ J), (b) HD gas ($h\nu_0 = 75 \times 10^{-21}$ J), (c) H_2 gas ($h\nu_0 = 87 \times 10^{-21}$ J). The dashed lines correspond to the classically expected curve ($U_{vib} = NkT$). $U_{vib,m}$ was calculated according to (15.27).

At high temperatures, Equation (15.27) simplifies. Then the first term in Equation (15.27) is much smaller than the second term and we can write

$$U_{vib} \approx Nh\nu_0 \frac{1}{e^{h\nu_0/(kT)} - 1} \quad \text{at high enough temperature}$$

We define $x = h\nu_0/(kT)$. Because x is small at high temperature, we can expand the exponential function in a series

$$e^{h\nu_0/(kT)} - 1 = e^x - 1 = 1 + x + \frac{1}{2}x^2 + \ldots - 1 \approx x$$

so that

$$U_{vib} \approx Nh\nu_0 \cdot \frac{1}{x}$$

Substituting for x, we find that

$$U_{vib} = NkT \quad \text{(for large enough } T\text{)} \tag{15.28}$$

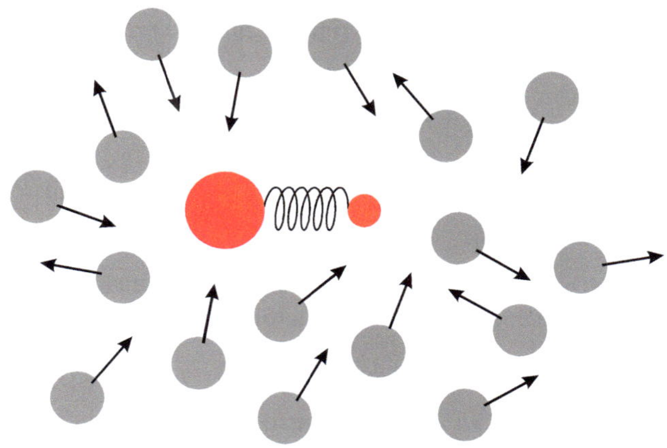

Figure 15.8 Oscillator (red) in a temperature equilibrium with gas molecules (gray).

At high enough temperature U_{vib} increases linearly with T (see dashed lines in Fig. 15.7). Equation (15.28) is identical with the expression obtained when the oscillator is considered as a classical oscillator in equilibrium with a temperature bath of temperature T (Equipartition Principle, Fig. 15.8, Box 15.1).

15.4.3 Internal Energy U_{vib} of Polyatomic Molecules

To find the vibrational internal energy for polyatomic molecules we must consider all of the vibrational motions of the molecules. If we use the normal modes for a polyatomic molecule (see Chapter 11), we can treat each normal mode independently of the others and write the vibrational energy as a sum over that in each normal mode. If we make the harmonic oscillator approximation for each normal mode, then we can use Equation (15.27) to describe the internal energy for each normal mode.

For example, in the case of CO_2 we have four normal modes. U_{vib} is calculated as a sum of the contributions of these different normal modes; thus, we obtain

$$U_{vib} = U_{vib,1} + U_{vib,2} + U_{vib,3} + U_{vib,4} \tag{15.29}$$

where

$$U_{vib,1} = Nh\nu_{0,1}\left(\frac{1}{2} + \frac{1}{e^{h\nu_{0,1}/(kT)} - 1}\right) \tag{15.30}$$

is the contribution of normal mode 1 with frequency $\nu_{0,1}$ ($U_{vib,2}$, $U_{vib,3}$, and $U_{vib,4}$ are written in a corresponding manner). For CO_2, the frequencies $\nu_{0,1} = \nu_{0,2}$ are those of the bending modes, $\nu_{0,3}$ is the frequency of the symmetric stretching mode, and $\nu_{0,4}$ is the frequency of the antisymmetric stretching mode.

As outlined in Box 15.1, at high enough temperature each normal mode of vibration contributes NkT to the internal energy U_{vib}. Some examples are

molecule	number of vibration modes	U_{vib} (classical limit)
CO_2	4	$4NkT$
H_2O	3	$3NkT$
CH_4	9	$9NkT$

This limit is rarely achieved for all the modes in a polyatomic molecule.

15.5 Rotational Energy

As found in Chapter 11 for a rigid rotator, the rotational energy E_J of a diatomic molecule is characterized by the quantum number J

$$E_J = B \cdot J(J+1), \quad J = 0, 1, 2, \ldots \tag{15.31}$$

with

$$B = \frac{h^2}{8\pi^2 \mu d^2} \tag{15.32}$$

where B is called the *rotational constant*. In contrast to the oscillator where only one wavefunction belongs to each quantum number, the energy levels of a rotator are

$$g_J = 2J + 1 \quad \text{-fold degenerate.} \tag{15.33}$$

In Sections 15.5.1 and 15.5.2 we restrict our treatment to heteronuclear diatomic and polyatomic molecules. As outlined in Chapter 11 special quantum conditions exist for the population of rotational quantum states of homonuclear diatomic molecules. This fascinating quantum effect will be discussed for the H_2 molecule in Section 15.5.3.

If we consider HD as an example for a heteronuclear diatomic molecule, then $\mu = 1.12 \times 10^{-27}$ kg and $d = 74$ pm so that $B = 9.07 \times 10^{-22}$ J. The corresponding quantum states are shown in Fig. 15.9a.

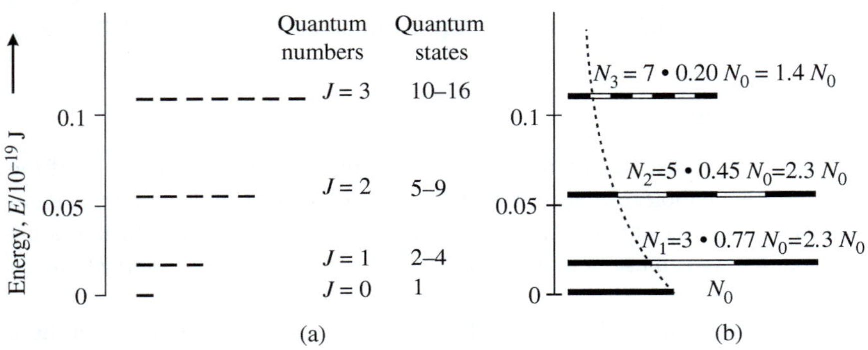

Figure 15.9 Rotational quantum states (HD molecule) (a) Energy scheme. (b) Population scheme at $T = 100$ K. The lengths of the horizontal lines on the left from the dashed curve correspond to the number of particles per quantum state, the total lengths correspond to the number of particles per quantum number. N_0 is the number of particles in the lowest energy state: $J = 0$ ($i = 1$). $N_1 = 3 \cdot 0.74\ N_0 = 2.22\ N_0$: number of particles in energy level $J = 1$ (quantum states $i = 2, 3, 4$). $N_2 = 5 \cdot 0.40 \cdot N_0 = 2.0\ N_0$: $J = 2$ ($i = 5, 6, 7, 8, 9$). $N_3 = 7 \cdot 0.16 \cdot N_0 = 1.12\ N_0$: $J = 3$ ($i = 10, 11, 12, 13, 14, 15, 16$).

15.5.1 Population Number N_J of Rigid Rotator (Heteronuclear Molecules)

Because of the energy level degeneracy, we consider the population number N_J of molecules in an energy level with quantum number J ($J = 0, 1, 2, \ldots$), rather than the number N_i of molecules in quantum state i.

Equation (15.9) for the population of molecules in rotational quantum state i is

$$\boxed{N_i = \frac{N}{Z_{\text{rot}}} \cdot e^{-E_i/(kT)}} \tag{15.34}$$

Each of the $(2J+1)$ quantum states of energy level E_J has the same population so that the total population in energy level with quantum number J is

$$N_J = \frac{N}{Z_{\text{rot}}} g_J \cdot e^{-E_J/(kT)} = \frac{N}{Z_{\text{rot}}}(2J+1) \cdot \exp\left[-\frac{B \cdot J(J+1)}{kT}\right], \quad J = 0, 1, 2, 3, \ldots \tag{15.35}$$

where g_J is the degeneracy. Correspondingly, we find that the molecular partition function for rotation is

$$Z_{\text{rot}} = \sum_{J=0}^{\infty} g_J \cdot e^{-E_J/(kT)} = \sum_{J=0}^{\infty} (2J+1) \cdot \exp\left[-\frac{B \cdot J(J+1)}{kT}\right] \tag{15.36}$$

where the sum is taken over the quantum number J.

At most temperatures, the population number of rotational energy levels has a maximum for $J \neq 0$. The exponential term in Equation (15.35) causes N_J to decrease with increasing J, and the degeneracy term, $2J+1$, causes N_J to increase with J. These two competing effects create a maximum in N_J at a nonzero value of J. The population numbers N_J for HD are calculated numerically (Table 15.2) and displayed in Fig. 15.9b for $T = 100$ K. The population increases, reaching a maximum at $J = 1$; then, it decreases with increasing J.

Table 15.2 Rigid Rotator: Population Probability $\mathcal{P}_j = N_J/N$, Molecular Partition Function Z_{rot}, and Internal Energy U_{rot} calculated numerically for HD using the data: $\mu = 1.116 \times 10^{-27}$ kg, $d = 74$ pm, and $B = 0.00910 \times 10^{-19}$ J. $U_{\text{rot},m} = U_{\text{rot}}/\boldsymbol{n}$ is the molar internal energy of rotation where $\boldsymbol{n}$ is the amount of substance.

		$\mathcal{P}_j = N_J/N$			
J	$E_J/10^{-19}$ J	10 K $Z_{\text{rot}} = 1.00$	50 K $Z_{\text{rot}} = 1.22$	100 K $Z_{\text{rot}} = 1.89$	500 K $Z_{\text{rot}} = 7.91$
0	0	1.00	0.82	0.53	0.13
1	0.0182	0.00	0.18	0.42	0.29
2	0.0547		0.00	0.05	0.29
3	0.109			0.00	0.18
4	0.182				0.08
5	0.274				0.03
6	0.383				0.00
		$U_{\text{rot},m}$/kJ mol^{-1}			
		0.00	0.20	0.66	3.83

15.5.1.1 High Temperature Limit

Unlike the harmonic oscillator case, the molecular partition function (15.36) does not have a simple analytical solution at all temperatures. If the temperature is large enough, compared to the rotational energy spacing, an excellent approximation is possible, however. In this limit the summation in Equation (15.36) can be approximated by an integral; i.e.,

$$Z_{\text{rot}} = \int_0^{\infty} (2J+1) \cdot \exp\left[-\frac{B \cdot J(J+1)}{kT}\right] \mathrm{d}J \tag{15.37}$$

With the substitution

$$x = \frac{BJ(J+1)}{kT} = \frac{B}{kT} \cdot (J^2 + J) \quad \text{and} \quad \mathrm{d}x = \frac{B}{kT} \cdot (2J+1) \cdot \mathrm{d}J$$

we obtain

$$Z_{\text{rot}} = \frac{kT}{B} \int_0^{\infty} (2J+1) \cdot e^{-x} \cdot \frac{1}{2J+1} \cdot \mathrm{d}x = \frac{kT}{B} \int_0^{\infty} e^{-x} \cdot \mathrm{d}x$$

which can be evaluated to give

$$Z_{\text{rot}} = \frac{kT}{B} \quad \text{(high temperature)} \tag{15.38}$$

This approximation is very commonly applied.

15.5.2 Internal Energy U_{rot} of Rotation (Heteronuclear Molecules)

15.5.2.1 Diatomic and Linear Polyatomic Molecules

To calculate the energy U_{rot} of N rotators we modify Equation (15.11) to perform a sum over quantum numbers rather than a sum over states, and we find

$$U_{\text{rot}} = \sum_{J=0}^{\infty} N_J E_J = \frac{N}{Z_{\text{rot}}} \sum_{J=0}^{\infty} g_J \, e^{-E_J/(kT)} \cdot E_J \tag{15.39}$$

or

$$U_{\text{rot},m} = \frac{U_{\text{rot}}}{\boldsymbol{n}} = \frac{\boldsymbol{N}_{\text{A}}}{Z_{\text{rot}}} \sum_{J=0}^{\infty} g_J \, e^{-E_J/(kT)} \cdot E_J \tag{15.40}$$

for the molar internal energy of rotation. As for the population numbers this expression is evaluated numerically (Table 15.2), and the result is shown in Fig. 15.10 for the heteronuclear diatomic molecules IBr (which has a relatively small B) and for HD (which has a large B).

At $T = 0$ K we obtain $U_{\text{rot}} = 0$ (all molecules occupy the lowest energy state with $E_0 = 0$). As the temperature increases the rotational energy U_{rot} increases. At high enough temperatures, the molecular partition function for rotation can be approximated by the high temperature limit, Equation (15.38). We use Equation (15.13) to calculate the high temperature limit of U_{rot}.

$$U_{\text{rot}} = NkT^2 \cdot \frac{\mathrm{d} \ln Z_{\text{rot}}}{\mathrm{d}T} = NkT^2 \cdot \frac{1}{T} = NkT \tag{15.41}$$

This is the same expression as expected from the Equipartition Principle (Box 15.1).

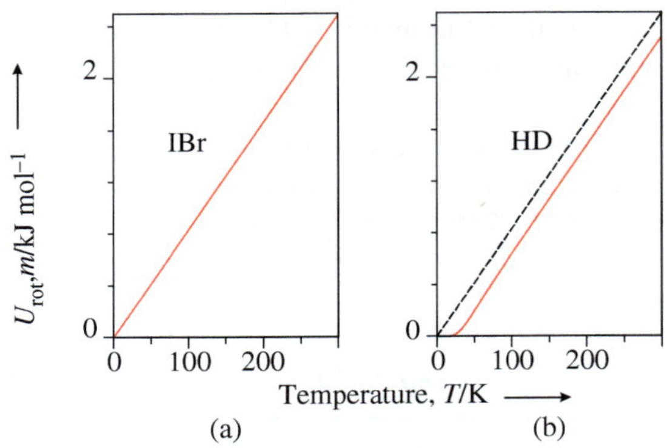

Figure 15.10 Molar rotational energy $U_{\text{rot},m}$ as a function of temperature (solid lines). (a) IBr gas ($B = 0.11 \times 10^{-23}$ J), (b) HD gas ($B = 0.91 \times 10^{-21}$ J). In the case of HD, the dashed line corresponds to the curve expected from the equipartition principle ($U_m = \boldsymbol{N}_{\text{A}} kT$).

Because the spacing between the rotational energy levels is small enough, the rotational energy of gas molecules at room temperature is often well described by the classical relation (15.41). From Fig. 15.10a we find that the classical value NkT for U_{rot} is achieved for IBr at nearly all temperatures. In contrast U_{rot} for HD is only 80% of the classical value at 100 K and only reaches 92% at 300 K (see Fig. 15.10b).

Example 15.2 *Use the average rotational energy for an IBr molecule at $T = 300$ K to compute the molecule's average rotational frequency (revolutions per second).*

At 300 K the rotational energy of IBr can be well approximated by the classical limit. To find the average rotational energy per molecule, we can write that

$$\frac{U_{\text{rot}}}{N} = kT = \frac{1}{2} I \omega^2$$

where $I = \mu d^2$ is the inertial moment ($\mu = 0.82 \times 10^{-25}$ kg, $d = 267$ pm, $I = 5.8 \times 10^{-45}$ kg m^2), $\omega = 2\pi\nu$, and ν is the rotational frequency. For IBr at $T = 300$ K we find

$$\omega = \sqrt{\frac{2kT}{I}} = 1.2 \times 10^{12}\,\text{s}^{-1} \quad \text{or} \quad \nu = 1.9 \times 10^{11}\,\text{Hz}$$

This frequency corresponds to a rotational period ($T = 1/\nu$) of 5.3 ps.

15.5.2.2 Nonlinear Polyatomic Molecules

Nonlinear polyatomic molecules have three axes of rotation, and the molecular partition function for rotation is evaluated in a similar manner. In the classical limit (kT large compared to the rotational energy level spacing), rotation about each of the principal inertial axes has an angular kinetic energy of $I\omega^2/2$. According to the equipartition principle (Box 15.1), rotation about each axis has an average energy of $kT/2$. Hence the rotational internal energy for N molecules is

$$U_{\text{rot}} = N\frac{3}{2}kT \tag{15.42}$$

Problem P15.17 shows that the same result is obtained by taking the high temperature limit of the quantum mechanical form.

15.5.3 Homonuclear Diatomic Molecules: Ortho- and Para-Hydrogen

As outlined in Chapter 12, H_2 is a 3:1 mixture of *ortho*- and *para*-H_2. *Para*-H_2 exists in rotational states with even quantum numbers ($J = 0, 2, 4, \ldots$). Hence the excitation energy from the lowest energy state ($J = 0$) to the next higher available rotational energy level ($J = 2$) is

$$\Delta E_{0\to 2} = B\cdot[2(2+1)-0] = 6B \qquad (\textit{para}\text{-H}_2) \tag{15.43}$$

where B is the rotational constant. Similarly, *ortho*-H_2 only exists in states with odd quantum numbers ($J = 1, 3, 5, \ldots$), for which the lowest energy transition is

$$\Delta E_{1\to 3} = B\cdot[3(3+1)-1(1+1)] = 10B \quad (\textit{ortho}\text{-H}_2) \tag{15.44}$$

The energy gap is almost two times larger for *ortho*-H_2 than for *para*-H_2.

In analogy to Equations (15.35) and (15.36) the population numbers and partition functions of the ortho- and para-hydrogen molecules are calculated by

$$N_{para,J} = \frac{N}{Z_{\text{rot,para}}} g_J\cdot e^{-E_J/(kT)}, \quad Z_{\text{rot,para}} = \sum_{J=0}^{\infty} g_J\cdot e^{-E_J/(kT)}, \quad J = 0, 2, 4, \ldots \tag{15.45}$$

$$N_{ortho,J} = \frac{N}{Z_{\text{rot,ortho}}} g_J\cdot e^{-E_J/(kT)}, \quad Z_{\text{rot,ortho}} = \sum_{J=1}^{\infty} g_J\cdot e^{-E_J/(kT)}, \quad J = 1, 3, 5, \ldots \tag{15.46}$$

In analogy to Equation (15.40) we calculate the molar internal energies of para- and ortho-hydrogen.

$$U_{\text{rot,para},m} = \frac{\boldsymbol{N}_{\text{A}}}{Z_{\text{rot,para}}}\sum_{J=0}^{\infty} g_J\cdot e^{-E_J/(kT)}\cdot E_J, \quad J = 0, 2, 4, \ldots \tag{15.47}$$

$$U_{\text{rot,ortho},m} = \frac{\boldsymbol{N}_{\text{A}}}{Z_{\text{rot,ortho}}}\sum_{J=1}^{\infty} g_J\cdot e^{-E_J/(kT)}\cdot E_J, \quad J = 1, 3, 5, \ldots \tag{15.48}$$

For the natural mixture of para- and ortho-hydrogen in the ratio of 1:3 the molar internal energy is

$$U_{\text{rot},m} = \frac{1}{4}\left(U_{\text{rot,para},m} + 3\cdot U_{\text{rot,ortho},m}\right) \tag{15.49}$$

As in the case of HD the molecular partition functions $Z_{\text{rot,para}}$ and $Z_{\text{rot,ortho}}$ and the molar internal energies $U_{\text{rot,para},m}$, $U_{\text{rot,ortho},m}$, and $U_{\text{rot},m}$ are calculated numerically.

The result for the internal energies is shown in Fig. 15.11. Whereas the molar internal energy for para-H_2 is zero at $T = 0$ K, we observe a zero point energy of $2B\boldsymbol{N}_{\text{A}}$ for ortho-H_2 (at $T = 0$ K all ortho-H_2 molecules occupy the energy level with $J = 1$). For natural H_2 the molar zero point energy is just 3/4 of that for ortho-H_2. Compared to the corresponding curves of HD the increase of the internal energy with temperature

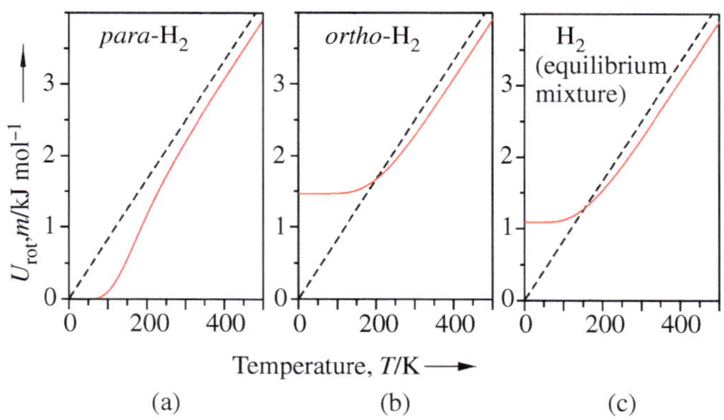

Figure 15.11 Molar rotational energy $U_{\text{rot},m}$ as a function of temperature of hydrogen gas ($B = 1.22\times 10^{-21}$ J, solid lines). (a) para-H_2, (b) ortho-H_2, (c) natural mixture (para : ortho = 1 : 3). The dashed lines correspond to the curves expected from the equipartition principle ($U_m = \boldsymbol{N}_{\text{A}}kT$).

occurs at higher temperatures (25 K for HD, 100 K for para-H $_2$, 150 K for ortho-H_2), corresponding to the larger ΔE values calculated in Equations (15.43) and (15.44).

At high enough temperature, we can approximate the molecular partition functions as

$$Z_{\text{rot,para}} = Z_{\text{rot,ortho}} = \frac{1}{2}Z_{\text{rot}} \quad \text{(high temperature limit)} \tag{15.50}$$

where

$$Z_{\text{rot}} = \sum_{J=0}^{\infty} g_J \cdot e^{-E_J/(kT)}, \qquad J = 0, 1, 2, 3, 4, \ldots \tag{15.51}$$

In Exercise E15.32, you show that this leads to

$$U_{\text{rot}} = NkT \quad \text{(high temperature limit)} \tag{15.52}$$

That is, for sufficiently high temperature the expression for the internal energy is the same as for heteronuclear diatomic molecules. HD and H_2 are the only molecular gases where the quantum nature in the rotational motion becomes important, because of their exceptionally large rotational constants and of their low boiling temperatures. In all other cases the gases follow closely to what is expected from the equipartition principle at all accessible temperatures. We will compare these predictions to experimental data in Chapter 16.

15.6 Translational Energy

In Chapter 14 we assumed that the translational motion of a molecule can be described by classical mechanics. The de Broglie wavelength of an H_2 molecule with velocity $v = 10^3$ m s^{-1} is $\Lambda = 200$ pm; thus, Λ is small compared to the mean free path. The quantum mechanical description should, therefore, lead to the same result as the classic consideration. Since H_2 has one of the lowest masses, hence large Λ, we expect that the classical approximation will work for all molecular gases if it works for H_2.

Is this really the case? According to Equation 14.25 the mean translational energy of a molecule at temperature T is

$$\overline{E}_{\text{trans}} = \tfrac{3}{2}kT \tag{15.53}$$

Can we reproduce this classical result by quantum mechanical considerations?

15.6.1 Internal Energy U_{trans} of Translation According to Quantum Mechanics

According to quantum theory (Chapter 3) the energy of a particle of mass m in a cubic box of length L is

$$E_{n_x, n_y, n_z} = \frac{h^2}{8mL^2}\left(n_x^2 + n_y^2 + n_z^2\right), \quad n_x, n_y, n_z = 1, 2, 3, \ldots \tag{15.54}$$

In Chapter 3 we used the same relation with the mass m_e of an electron, now m is the mass of the molecule. The probability $\mathcal{P}_{n_x,n_y,n_z}$ of finding a molecule in a state characterized by the quantum numbers n_x, n_y, n_z is

$$\mathcal{P}_{n_x,n_y,n_z} = \frac{1}{Z_{\text{trans}}} \cdot e^{-E_{n_x,n_y,n_z}/(kT)} \tag{15.55}$$

with

$$Z_{\text{trans}} = \sum_{n_x,n_y,n_z=1}^{\infty} e^{-E_{n_x,n_y,n_z}/(kT)} \tag{15.56}$$

Then, using Equation (15.11), we can write that

$$U_{\text{trans}} = \frac{N}{Z_{\text{trans}}} \sum_{n_x,n_y,n_z=1}^{\infty} e^{-E_{n_x,n_y,n_z}/(kT)} \cdot E_{n_x, n_y, n_z} \tag{15.57}$$

for the internal translational energy. In Box 13.2 we calculated g_{total}, the number of translational quantum states in a container of macroscopic size (side length L);

$$g_{\text{total}} = \frac{\pi}{6} \cdot \left(\frac{8m}{h^2}\right)^{3/2} \cdot VE^{3/2} \tag{15.58}$$

where $V = L^3$ is the volume of the cubic box. For $L = 1$ cm, $m = 3.32 \times 10^{-27}$ kg (mass of an H_2 molecule) and $E = kT = 4 \times 10^{-21}$ J at 300 K this number is on the order of 10^{24}. Therefore it is reasonable to replace the summation in Equation (15.56) by an integral. Justification 15.1 shows this analysis in detail and finds

$$Z_{\text{trans}} = V \cdot \left(\frac{2\pi mk}{h^2}\right)^{3/2} \cdot T^{3/2} \tag{15.59}$$

where $V = L^3$ is the volume of the cubic box.

With this analytical expression for Z_{trans} in hand, we can use Equation (15.16) for the average translational energy $\overline{E}_{\text{trans}}$

$$U_{\text{trans}} = NkT^2 \frac{\mathrm{d} \ln Z_{\text{trans}}}{\mathrm{d}T}$$

and find that

$$U_{\text{trans}} = NkT^2 \cdot \frac{3}{2}\frac{1}{T} = \frac{3}{2}kT \tag{15.60}$$

Correspondingly, the molar internal translational energy $U_{\text{trans},m}$ of a molecular gas is

$$U_{\text{trans},m} = \frac{3}{2}\mathbf{N}_{\text{A}}kT = \frac{3}{2}\boldsymbol{R}T \tag{15.61}$$

This result is identical to that found by classical considerations (Chapter 14).

Example 15.3 *Assume that all translational quantum states below $E = kT$ are occupied with equal probability and all quantum states above $E = kT$ are empty, and calculate the number of occupied translational quantum states. Compare your result to the expression for the translational partition function, Equation (15.59).*

From Equation (15.58) we obtain

$$g_{\text{total}} = \frac{\pi}{6} \cdot V\left(\frac{8mk}{h^2}\right)^{3/2} \cdot T^{3/2} = \frac{4}{3\sqrt{\pi}} \cdot V\left(\frac{2\pi mk}{h^2}\right)^{3/2} \cdot T^{3/2}$$

In this approximation (except for a small numerical factor),

$$g_{\text{total}} \approx Z_{\text{trans}} \quad \text{for } E = kT$$

Thus Z_{trans} provides a measure of the number of available translational quantum states at temperature T.

Here we note an important caveat concerning the independence of the translational energy states for the gas molecules. Whereas the electronic, vibrational, and rotational quantum states correspond to individual molecules, the translational quantum states belong to the entire gas. If a gas particle occupies one of the box's quantum states that might block another gas particle from occupying the state. For most molecular and atomic gases of chemical interest the probability P_{n_x,n_y,n_z} of finding a molecule in a particular translational quantum state n_x, n_y, n_z is quite low, and these considerations are not important. This issue is discussed more fully in Section 15.10.

15.6.2 Maxwell–Boltzmann Distribution of Speeds

15.6.2.1 Classical Treatment

In Chapter 14 our simple model assumed that the molecules in a gas are moving with the same speed; however, this cannot be true. To see this, assume that at some initial time all molecules have the same speed v. When the molecules begin to collide with each other, they cause the speeds to change (become unequal) in a way that depends on the collision geometry. Only in the case of centered collisions do the speeds of identical molecules remain equivalent (Chapter 14). This is not the case for a noncentered collision, however (Fig. 15.12). For example, consider a collision like that in Fig. 15.12c, molecule 2 is at rest, and the resulting speed of molecule 1 is $v_1' = \sqrt{2} \cdot v$. Noncentered collisions occur frequently. Hence, even if we start with gas molecules that all have the same speed, many of the molecules will have speeds higher or lower than the average speed after collisions start to occur.

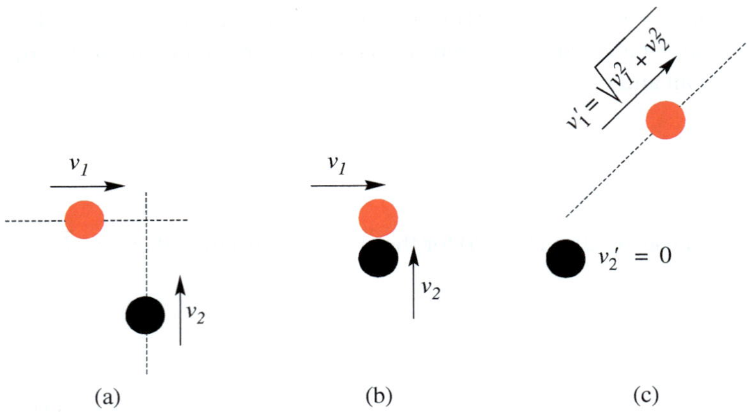

Figure 15.12 Collision of two identical molecules moving in directions perpendicular to each other. Special case when particle 2 collides centrally with particle 1 and particle 1 collides tangentially to particle 2. (a) Velocities v_1 and v_2 shortly before the collision, (b) velocities at collision, (c) velocities $v_1' = \sqrt{v_1^2 + v_2^2}$ and $v_2' = 0$ shortly after the collision. In the special case $v_1 = v_2 = v$ we obtain $v_1' = \sqrt{2}v$.

From classical considerations, Maxwell derived the distribution of molecular speeds. He found that the number of molecules with components of velocity in the interval between v_x and $v_x + dv_x$ is given by

$$dN\left(v_x, v_x + dv_x\right) = const. \cdot \exp\left(-\frac{mv_x^2}{2kT}\right) \cdot dv_x \tag{15.62}$$

You can find a derivation of this equation in Foundation 15.1. We ask for the number of molecules with velocities v_x, v_y, and v_z in the interval between v_x and $v_x + dv_x$, v_y and $v_y + dv_y$, and v_z and $v_z + dv_z$

$$\begin{aligned} &dN\left(v_x, v_x + dv_x, v_y, v_y + dv_y, v_z, v_z + dv_z\right) \\ &= const. \cdot \exp\left(-\frac{mv_x^2}{2kT}\right) \cdot \exp\left(-\frac{mv_y^2}{2kT}\right) \cdot \exp\left(-\frac{mv_z^2}{2kT}\right) \cdot dv_x\, dv_y\, dv_z \\ &= const. \cdot \exp\left(-m\frac{v_x^2 + v_y^2 + v_z^2}{2kT}\right) \cdot dv_x\, dv_y\, dv_z \end{aligned} \tag{15.63}$$

The number of molecules with speed

$$v = \sqrt{v_x^2 + v_y^2 + v_z^2}$$

between v and $v + dv$ is then obtained by replacing the volume element $dv_x dv_y dv_z$ in the velocity space by the corresponding element between spheres of radius v and $v + dv$, which is $4\pi v^2 \cdot dv$. Thus from Equation (15.63) we obtain

$$dN(v, v + dv) = const. \cdot 4\pi v^2 \cdot \exp\left(-\frac{mv^2}{2kT}\right) \cdot dv$$

The constant is determined by the condition

$$\int_0^\infty dN(v, v + dv) = N$$

and we find

$$\boxed{dN = N \cdot 4\pi \cdot \left(\frac{m}{2\pi kT}\right)^{3/2} \cdot v^2 \cdot e^{-mv^2/2kT} \cdot dv} \tag{15.64}$$

Figure 15.13 plots this *Maxwell–Boltzmann distribution of speeds* together with experimental data. This distribution is found to correspond to a thermally equilibrated gas. Small speeds are rare, as are extremely high speeds.

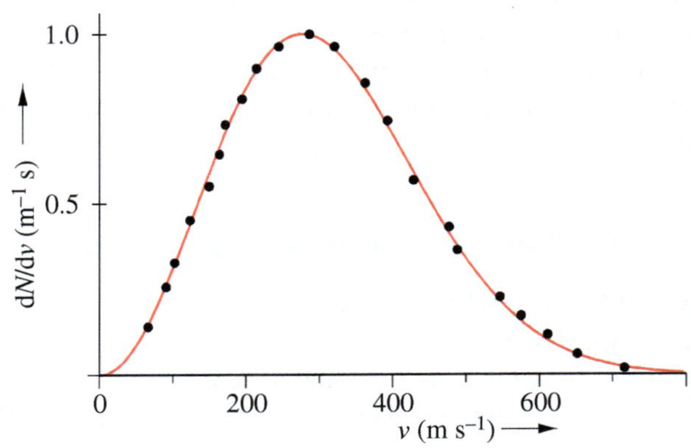

Figure 15.13 Distribution of speeds of thallium atoms in the gas state at 944 K; the ordinate scale is arbitrarily chosen such that the maximum is 1.0 m^{-1} s. Dots: experiment [1]; the experimental points were extracted from the measured intensity data. Solid line: calculation according to (15.64).

15.6.2.2 Quantum Mechanical Treatment

Can we obtain the Maxwell–Boltzmann equation (15.64) while beginning with quantum mechanical considerations? If we view the gas molecules as a collection of particles in a very large box, we can arrive at the Maxwell–Boltzmann distribution of molecular speeds from the Boltzmann distribution.

Consider N molecules of mass m in a cubic box of side length L and at a temperature T. The probability of occupying a quantum state may be calculated from Equations (15.54) and (15.55). As in our analysis for the rotational energy distribution, we wish to find the probability of a molecule having a particular energy and must account for the energy level degeneracy. Because of the high number of quantum states available in a box of macroscopic size, we will determine the number of quantum states g_i in a small energy interval E_i and $E_i + \Delta E_i$ and use it for the degeneracy.

Figure 15.14a shows the lowest 130 energy levels obtained from Equation (15.54) for H_2 molecules in a cubic box of volume 1 cm^3 and reports the number of states g_i in energy intervals of $\Delta E_i = 1.66 \times 10^{-36}$ J. Figure 15.14b plots g_i in these energy intervals as a function of the energy E_i (dots). The number of translational quantum states g_i increases with the energy E_i. In Box 13.2 we show that the number of quantum states g_i in the energy interval from E_i to $E_i + \Delta E_i$ is

$$g_i = \frac{\pi}{4} \cdot \left(\frac{8mL^2}{h^2}\right)^{3/2} \sqrt{E_i}\,\Delta E \qquad (15.65)$$

for the particle in a box model, in accordance with Fig. 15.14b.

The number ΔN of molecules with a translational energy between E_i and $E_i + \Delta E_i$ is proportional to the Boltzmann factor and proportional to g_i

$$\Delta N_{E,E+\Delta E} = C \cdot e^{-E_i/(kT)} \sqrt{E_i} \cdot \Delta E \qquad (15.66)$$

where the constants are combined in C. For small enough ΔE, we can write Equation (15.66) in differential form as

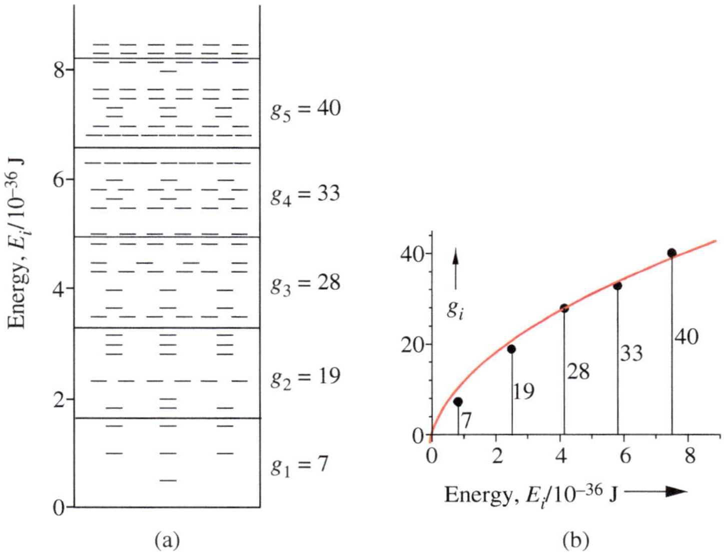

Figure 15.14 Number g_i of quantum states in the interval ΔE with increasing energy (lowest quantum state of H_2 in a cube of volume $V = 1$ cm^3, energies calculated according to Equation (15.54). (a) Energy scheme (corresponding to Fig. 3.7b); (b) g_i as a function of E_i (for $\Delta E = 1.66 \times 10^{-36}$ J).

$$dN_{E,E+dE} = C \cdot e^{-E/(kT)} \sqrt{E} \cdot dE \qquad (15.67)$$

Equation (15.67) is the distribution function of the translational energies in a gas. In Justification 15.1, we show that

$$C = N \cdot \frac{2}{(kT)^{3/2}\sqrt{\pi}}$$

To obtain the distribution of speeds, we realize that the energy E of the gas is purely kinetic energy, so we can write

$$E = \frac{1}{2}mv^2, \quad dE = mv \cdot dv \qquad (15.68)$$

Substituting these expressions for E into Equation (15.67), we find

$$dN_{v,v+dv} = C' \cdot v^2 \cdot e^{-mv^2/2kT} \cdot dv \qquad (15.69)$$

where the constants are combined into the parameter C'. In Exercise E15.10 you show that

$$C' = 4\pi N \cdot \left(\frac{m}{2\pi kT}\right)^{3/2}$$

Then from Equation (15.69) we obtain the Maxwell–Boltzmann distribution of speeds (15.64)

$$\boxed{dN = N \cdot 4\pi \cdot \left(\frac{m}{2\pi kT}\right)^{3/2} \cdot v^2 \cdot e^{-mv^2/2kT} \cdot dv}$$

Figure 15.15 plots the Maxwell–Boltzmann distribution of speeds for He gas ($m = 6.65 \times 10^{-27}$ kg) at two different temperatures, 100 K and 600 K. Two important changes are evident. First the maximum of the distribution shifts to higher speeds with increasing temperature – it increases as $\sqrt{T}$. Second, the width of the distribution broadens as the temperature increases.

15.6.2.3 Most Probable Speed, Mean Speed, Root Mean Square Speed

Rather than use the entire distribution it is common to use different "characteristic speeds" as a measure of the gas particles' behavior.

The location of the maximum of the distribution curve provides the most probable speed v_P. To find the maximum we set the derivative of the distribution function with respect to speed equal to zero and evaluate it for the speed; i.e.,

$$\frac{\mathrm{d}}{\mathrm{d}v}\left(v^2 \cdot \mathrm{e}^{-mv^2/2kT}\right)_{v=v_P} = 0$$

or

$$\mathrm{e}^{-mv_P^2/2kT} \cdot 2v_P \left[1 - \frac{mv_P^2}{2kT}\right] = 0$$

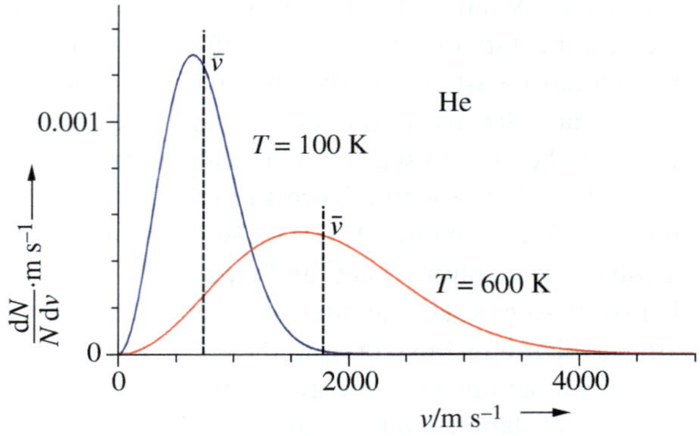

Figure 15.15 Distribution $dN/(N\,dv)$ (see Equation (15.64)) of speeds for He atoms ($m = 6.65 \times 10^{-27}$ kg) at 100 K and 600 K. The mean speed is obtained by averaging the speeds.

By solving, we find

$$v_P = \sqrt{\frac{2kT}{m}} \qquad \text{(most probable speed)} \tag{15.70}$$

Note that the most probable speed and thus the position of the distribution maximum increases as $\sqrt{T}$. The mean speed $\overline{v}$ is an important measure for the gas and is used to model reaction rates and other transport properties of gases. The mean speed is defined by

$$\overline{v} = \int_{v=0}^{\infty} v \cdot \mathrm{d}\mathcal{P}$$

with $\mathrm{d}\mathcal{P} = \mathrm{d}N/N$. Using the distribution function we find

$$\overline{v} = \int_0^{\infty} v \cdot 4\pi \cdot \left(\frac{m}{2\pi kT}\right)^{3/2} \cdot v^2 \cdot \mathrm{e}^{-mv^2/2kT} \cdot \mathrm{d}v$$

The integral evaluation is given as Exercise E15.16, and one finds

$$\overline{v} = \sqrt{\frac{8kT}{\pi m}} \qquad \text{(mean speed)} \tag{15.71}$$

$\overline{v}$ is higher than the most probable speed v_P because speeds $v > v_P$ are more likely than speeds $v < v_P$; thus, they contribute more to the average.

The root mean square speed v_{rms} is also an important measure for the gas particle speed and is defined by

$$v_{rms} = \sqrt{\int_{v=0}^{\infty} v^2 \cdot \mathrm{d}\mathcal{P}}$$

First we evaluate v_{rms}^2, which may be written as

$$v_{rms}^2 = \overline{v^2} = \int_{v=0}^{\infty} v^2 \cdot \mathrm{d}\mathcal{P} = \frac{2}{m}\int_{v=0}^{\infty} E_{trans} \cdot \mathrm{d}\mathcal{P} = \frac{2}{m}\overline{E}_{trans}$$

Hence the mean square speed is easily obtained from the average translational energy of a gas particle. Using Equation (15.60)

$$\overline{E}_{trans} = \frac{3}{2}kT$$

we find that

$$v_{rms} = \sqrt{\overline{v^2}} = \sqrt{\frac{3kT}{m}} \qquad \text{(root mean square speed)} \tag{15.72}$$

In the case of the He gas considered in Fig. 15.15 ($m = 6.65 \times 10^{-27}$ kg) the following values are obtained:

T K	v_P	$\bar{v}$ m s^{-1}	v_{rms}
100	645	728	790
600	1579	1782	1934

15.6.2.4 Supersonic Expansion

In a supersonic expansion of a molecular gas, the internal energy of the molecules (and atoms) is largely converted into forward translational energy. This process acts to reduce the populations of excited rotational and vibrational energy levels of gas molecules and to define better the gas molecules translational velocity, which are useful for studies of collisions and chemical reaction dynamics. The pressure difference between the two sides of the orifice (small hole) and the orifice diameter determine whether the gas expands supersonically or effusively. For effusion the diameter of the orifice is smaller than the average distance a molecule travels between its collision with another molecule (mean free path discussed in Chapter 14). In this case no collisions occur between the gas molecules as they move through the orifice and the flow is "molecular."

In the supersonic expansion the diameter of the hole is larger than the average distance between collisions; hence, many collisions between the gas molecules can occur during the expansion through the hole. The collisions act to transfer energy from the random atomic and molecular motions of the gas into the directed flow of the gas molecules along the expansion direction. This means that the gas moves as a whole; i.e., most of the molecules have the same speed in the direction of the flow, and the speed of the gas molecules relative to each other becomes very small. Because the relative speed of the gas molecules determines the temperature of the gas, its temperature becomes very low; the gas cools down.

According to Equation (15.70) the most probable speed of an ideal gas is

$$v_P = \sqrt{\frac{2kT}{m}}$$

For a gas of argon atoms ($m = 0.03995$ kg mol$^{-1}/N_A = 6.64 \times 10^{-26}$ kg) at a temperature of $T = 300$ K we find $v_P = 353$ m s^{-1}. This value corresponds to the maximum of the Maxwell–Boltzmann distribution (Equation (15.64)) in Fig. 15.16 (curve a, effusive source). We calculate the half width w of this distribution numerically: at $v_{left} = 170.5$ m s^{-1} and at $v_{right} = 578.5$ m s^{-1} the curve reaches one-half of its maximum, resulting in a value of $w = v_{right} - v_{left} = 408.0$ m s^{-1}.

In an actual supersonic expansion experiment (Ref. [2]), Ar gas was found to expand with a most probable speed of 553 m s^{-1}. The distribution of speed is measured in a manner analogous to that described in Chapter 14 (see Fig. 14.11) and is displayed in Fig. 15.16 (curve b). The half width of this distribution is 63 m s^{-1}. This means that the speeds of the atoms in the supersonic gas are very near together, that is, the atoms in the expanded gas essentially move as a whole. A Maxwell–Boltzmann distribution with this half width corresponds to a temperature of $T = 7.0$ K. This means that the argon gas expands from an initial temperature of $T = 300$ K to a final temperature of $T = 7.0$ K., see Fig. 15.16 (curve b).

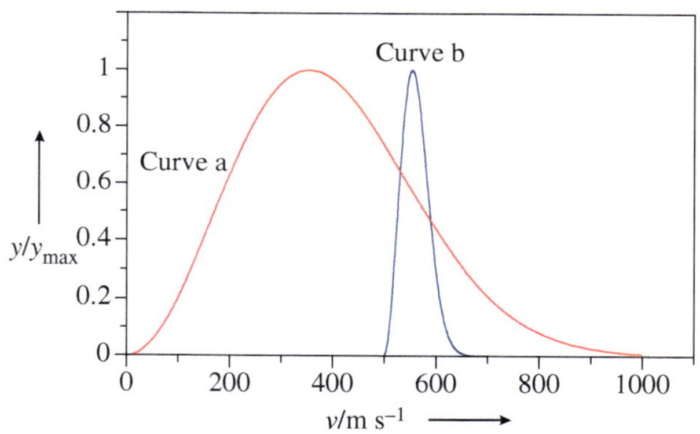

Figure 15.16 The distribution of molecular speeds $y/y_{max} = dN/dv/(dN/dv)_{max}$ for Ar gas from an effusive source (curve a, red) is compared to that from a supersonic source (curve b, blue). The temperature of the reservoir is 300 K. The speed of the supersonic gas atoms moving at a whole is 553 m s^{-1}. (Data taken from [2], Figure 6.)

15.7 Characteristic Temperature

In the limit of $T = 0$ K the molecules are in their lowest quantum state (Fig. 15.3, left). As energy is added to the assembly, molecules will not be excited unless it is large enough to excite the molecule (or atom) through the energy gap $E_2 - E_1$ to the next higher energy level (Fig. 15.3, center). This energy requirement can be quantified by a *characteristic temperature* θ, defined by

$$\theta = \frac{E_2 - E_1}{k} = \frac{\Delta E}{k} \tag{15.73}$$

At $T = \theta = \Delta E/k$ the ratio of the number of molecules in state 2 to those in state 1, N_2/N_1, is

$$\frac{N_2}{N_1} = \mathrm{e}^{-\Delta E/(kT)} = \mathrm{e}^{-1} = 0.3679$$

that is, the number of molecules in the excited (upper) state is about one-third the number in the ground (lower) state, according to the Boltzmann law for non-degenerate energy levels.

The characteristic temperature is a property of the molecule and provides a useful guide to the importance of the energy level quantization. For vibrational motion the characteristic temperature is (with $\Delta E = h\nu_0$)

$$\theta_{\text{vib}} = \frac{\Delta E}{k} = \frac{h\nu_0}{k} \tag{15.74}$$

For rotational motions the first excited energy level is three-fold degenerate; thus,

$$\frac{N_2}{N_1} = 3\,\mathrm{e}^{-\Delta E/(kT)}$$

In this case the ratio $N_2/N_1 = 0.3679$ is reached at a temperature of $T = \Delta E/(2.099k)$. Therefore it is reasonable to define the characteristic temperature θ_{rot} for rotation as

$$\theta_{\text{rot}} = \frac{\Delta E}{2k} = \frac{B}{k} \tag{15.75}$$

where B is the rotational constant and $\Delta E = 2B$. In this case $k\theta_{\text{rot}}$ is one-half of the lowest energy ($J = 0$ to $J = 1$) transition.

For HD we obtain $\theta_{\text{rot}} = 65.6$ K and $\theta_{\text{vib}} = 5480$ K. Consequently we expect that quantum effects from rotational energy levels of HD will be important unless the temperature is above 100 K. Correspondingly the importance of vibrational excitation of HD will not be significant until thousands of degrees Kelvin.

For translational motions the energy difference between the two lowest quantum states is, according to Equation (15.54),

$$\Delta E = \frac{h^2}{8mL^2}\left[(2^2 + 1 + 1) - (1 + 1 + 1)\right] = \frac{3h^2}{8mL^2}$$

For a gas of H_2 molecules ($m = 3.3 \times 10^{-27}$ kg) in a cubic box with a side length of $L = 10$ cm, $\Delta E = 5 \times 10^{-39}$ J. At all accessible temperatures ΔE is by far smaller than kT and quantum effects will not play a role when discussing the translational contribution to the gas molecules' internal energy.

15.8 Proving the Boltzmann Distribution for Distinguishable Particles

In this section, we describe the statistical considerations that underlie our use of the Boltzmann distribution for energy level populations.

15.8.1 Distribution of N Particles Among 3 Levels with Given Internal Energy

To illustrate the analysis, we first ask "How many possibilities exist to distribute a distinct energy $4a$ over six independent particles, each of which can occupy three equally spaced energy states of spacing a?" Figure 15.17a sketches this situation. Our analysis of the possible arrangements of the energy must consider that only six particles are present, i.e.,

$$N_1 + N_2 + N_3 = 6 \tag{15.76}$$

and that the internal energy is $4a$, that is,

$$N_1E_1 + N_2E_2 + N_3E_3 = 4a \tag{15.77}$$

Because of these constraints on the arrangement of particle energies, only the three population types A, B, and C shown in Fig. 15.17b are possible. Which of these population types is the most probable one? Let us assume that we can distinguish among the six particles. For example, in a crystal we can assign site numbers to the different particles (1 to 6) and distinguish

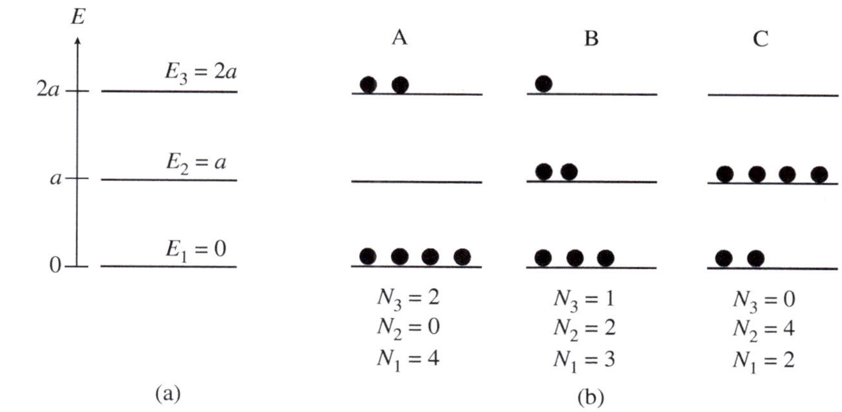

Figure 15.17 System with three equidistant quantum states. (a) Energy scheme. (b) Six particles, internal energy $4a$. Possible population types A, B, and C, considering the conditions (15.76) and (15.77).

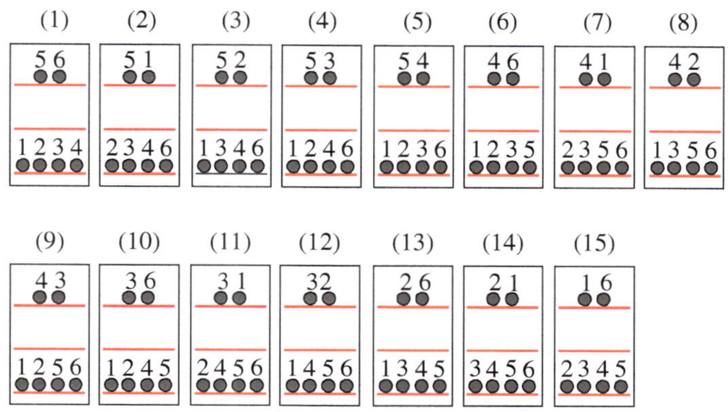

Figure 15.18 Distinguishable Particles. The 15 possibilities of distributing particles 1 to 6 over quantum states 1 and 3, according to population type A in Fig. 15.17: $N_1 = 4$, $N_2 = 0$, and $N_3 = 2$.

between molecules that occupy different lattice sites. As demonstrated in Fig. 15.18, there are $\omega_A = 15$ ways to realize population type A by exchanging particles between the energy levels 1 and 3. In the same way it can be shown that there are $\omega_B = 60$ ways to represent population type B and $\omega_C = 15$ ways to represent population type C.

We postulate that each of these possibilities has the same probability to be realized.

15.8.1.1 Number of Representations of a System with Given Internal Energy

How can we mathematically describe and generalize this result? For population type A we have six particles (1 to 6) arranged in groups of four (corresponding to the population in level 1) and two (corresponding to the population in level 3). We ask how many unique arrangements of these particles in the energy level distribution can be made.

Given six particles we might initially expect that there are 6! ways of arranging them (i.e., six ways to choose the first particle we place in an energy level, five ways to choose the second particle we place in an energy level, and so forth). If the particles were each in a different energy level, this would indeed be the correct way of counting; however, we have an energy level with four particles in it and another one with two particles. Because the individual arrangement of particles in a given energy level is not distinguishable (i.e., it does not matter whether we assign particle 4 an energy of zero first or last; all that matters is whether it has the energy zero or the energy $2a$), we have overcounted the number of distinguishable arrangements.

For the ground state energy level, which has four particles, 4! arrangements for the particles, which each satisfy the conditions, are possible. Hence we must divide 6! by 4!. A corresponding argument leads to the conclusion that energy level $2a$, which has two particles, is overcounted and we need to divide by 2!. Thus the number of distinguishable representations ω_A for population type A is given by

$$\omega_A = \frac{6!}{4!2!} = 15 \quad \text{(type A)} \tag{15.78}$$

In general, the number of arrangements for placing N distinguishable particles in i energy levels is

$$\omega = \frac{N!}{N_1! \cdot N_2! \cdot N_3! .. \cdot N_i!} \tag{15.79}$$

Example 15.4 *Given the general expression for the number of arrangements of distinguishable particles, determine which of the population types in Fig. 15.17 is the most probable*

We already calculated that $\omega_A = 15$. For the population types B and C we find

$$\omega_B = \frac{6!}{3!2!1!} = 60 \quad \text{and} \quad \omega_C = \frac{6!}{2!4!} = 15$$

Hence the total number of different representations that satisfy our initial conditions is

$$\Omega = \omega_A + \omega_B + \omega_C = 15 + 60 + 15 = 90$$

Since each representation satisfies our knowledge about the system, we consider each of them to be equally probable. Therefore, the probability that population type A is realized is $15/90 = \frac{1}{6}$; the probability of population type B occurring is $\frac{2}{3}$, and the probability of population type C occurring is $\frac{1}{6}$. We conclude that the most probable population type is B; but that A and C can also occur (Fig. 15.19).

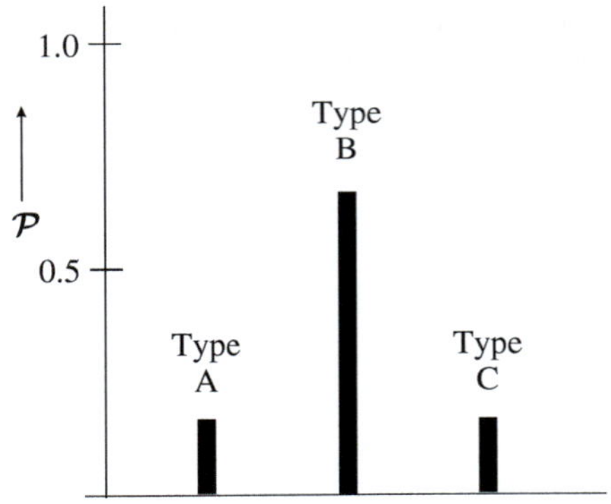

Figure 15.19 Probability $\mathcal{P}$ that a population type A, B, or C in Fig. 15.18 is realized. $\omega_A = 15$. $\omega_B = 60$. $\omega_C = 15$. $\Omega = \omega_A + \omega_B + \omega_C = 90$. $\mathcal{P} = \omega/\Omega$.

When we repeat this calculation for 100 particles possessing an energy of $40a$ we find 21 different population types (Table 15.3). This condition can be realized for $N_1 = 80, 79, \ldots, 60$. In the case A ($N_1 = 80, N_2 = 0, N_3 = 20$), we find that $\omega_A = 2.4 \times 10^{22}$. In case B ($N_1 = 79, N_2 = 2, N_3 = 19$), we find that $\omega_B = 1.9 \times 10^{25}$. From the Table 15.3 it is evident that of 21 possible population types the most probable type M is more than ten times larger than all but seven other ones.

Table 15.3 Population Types for a 3 Level System According to Fig. 15.19 with $N = 100$ Particles Possessing the Energy $U = 40a$.

	A	B	C	D	E	F	G	H	I	J
N_3	20	19	18	17	16	15	14	13	12	11
N_2	0	2	4	6	8	10	12	14	16	18
N_1	80	79	78	77	76	75	74	73	72	71
$\omega/10^{32}$	0	0	0	0	0	0	0.1	0.4	1.5	4.3

	K	L	M	N	O	P	Q	R	S	T	U
N_3	10	9	8	7	6	5	4	3	2	1	0
N_2	20	22	24	26	28	30	32	34	36	38	40
N_1	70	69	68	67	66	65	64	63	62	61	60
$\omega/10^{32}$	8.8	13.4	15.0	12.6	7.8	3.6	1.2	0.3	0	0	0

Altogether there are

$$\begin{aligned}\Omega &= \omega_A + \omega_B + \ldots + \omega_U \\ &= \begin{pmatrix} 0+0+0+0+0+0+0.1+0.4+1.5+4.3+8.8 \\ +13.4+15.0+12.6+7.8+3.6+1.2+0.3+0+0 \end{pmatrix} \times 10^{32} \\ &= 69.0 \times 10^{32}\end{aligned}$$

representations. 80% of all representations belong to types K to O, 60% to types L, M, and N and 20% belong to type M (Fig. 15.20). In the present case (energy $40a$ distributed among 100 particles), it is most probable that $N_1 = 68$ particles

Figure 15.20 Probability $\mathcal{P}$ that a population type A, B, C, … U in Table 15.3 (in the order of decreasing number of particles in energy level 3) is realized. $\mathcal{P}$ is given by the number of representations of population types A to U. The maximum of $\mathcal{P}$ is reached for population type M ($N_1 = 68, N_2 = 24, N_3 = 8$).

are in the lowest energy level, $N_2 = 24$ particles are in level 2, and $N_3 = 8$ particles are in the highest energy level. The population types L and N have slightly different population numbers from that of M.

The larger the number of particles under consideration, the more closely concentrated are the population types around the most probable one. That is, the distribution becomes more narrow (in relative terms) as the particle number increases. Thus the distribution of the N particles over the given energy states at a fixed value of the internal energy is practically determined by the most probable distribution $N_1, N_2, N_3, \dots$.

Our considerations have assumed that the energy U of the system is fixed. For a system in contact with a heat bath the internal energy can take on different values. Nevertheless, the deviation of the system's energy from its average energy is extremely small when the number of particles N is large, and our analysis can be reasonably applied to systems in thermal equilibrium (see Box 15.2 for more discussion of this point).

15.8.1.2 Calculating Population Numbers N_i

In Fig. 15.21a the population numbers N_i corresponding to type M (Table 15.3) are plotted as a function of the energy E_i — that is, for $E_1 = 0$, $E_2 = a$, and $E_3 = 2a$. The points can be well described by the function

$$N_i = C \cdot e^{-\beta E_i} \tag{15.80}$$

if we set $C = 68$ and $\beta = 1.04/a$ (solid line in Fig. 15.21a).

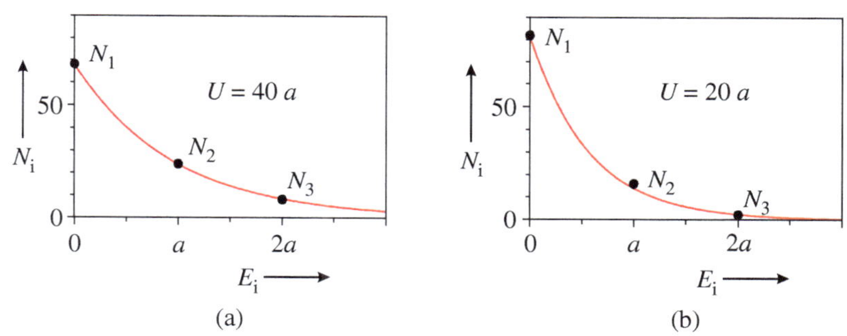

Figure 15.21 Population N_i of an arbitrary quantum state i as a function of its energy E_i. Dots: most probable population type (N_1, N_2, N_3) of 3-level system with $E_1 = 0, E_2 = a, E_3 = 2a$, number of particles $N = N_1 + N_2 + N_3 = 100$ and given internal energy U. Solid lines: calculated from Equation (15.80) with optimized values C and β. (a) $U = 40a$ ($N_1 = 68, N_2 = 24, N_3 = 8$, from population type M in Fig. 15.20 and Table 15.3), $C = 68, \beta = 1.04a$. (b) $U = 20a, C = 83, \beta = 1.76a$.

If we decrease the internal energy in our example from $40a$ to $20a$, then the same exponential form occurs but with $C = 83$ and $\beta = 1.76/a$ (Fig. 15.21b). The constant β increases with decreasing energy of the system, i.e., with decreasing temperature. The quantity β is therefore a measure for the temperature of the system. At higher temperatures (i.e., with increasing energy uptake), the higher levels become more populated and β becomes correspondingly smaller. At high enough temperature, according to the consideration in Section 15.2.1, all levels are equally populated, leading to a value of $\beta = 0$. This means that β cannot assume negative values.[2]

15.8.2 Rigorous Treatment

As discussed in Section 15.8.1, the distribution of population types becomes very narrow as the number of particles increases. Hence we make the approximation that we can replace the distribution of population types by the most probable

2 This statement is valid for systems in thermal equilibrium. For systems that are not in thermal equilibrium, it is possible that the population number in a high energy state can become greater than that in a low energy state. For example, particles in a low energy state can be excited by a laser pulse to create a population inversion (see Chapter 10). In such a case the parameter β is negative.

one; i.e., with the maximum ω. This distribution is the Boltzmann distribution. In practice, it is more convenient to search for the maximum of $\ln \omega$ instead of the maximum of ω, and we proceed in this manner. In Foundation 15.2 we derive that

$$\left(\frac{\partial \ln \omega}{\partial N_i}\right) - \alpha - \beta E_i = 0, \quad (i = 1, 2, 3, \ldots) \quad \text{(condition for the maximum of } \ln \omega) \tag{15.81}$$

in which α and β are parameters that must be determined.

15.8.2.1 How to Calculate the Population Numbers N_i

For distinguishable particles we calculate ω according to Equation (15.79)

$$\omega = \frac{N!}{N_1!N_2!N_3!\ldots} \quad \text{(distinguishable particles)}$$

According to Stirling's formula (see Appendix A.10) we have

$$\ln N! = N \cdot \ln N - N$$

and we can write $\ln \omega$ as

$$\ln \omega = N \cdot \ln N - N - N_1 \cdot \ln N_1 + N_1 - \ldots \tag{15.82}$$

Evaluating its derivative with respect to N_1, we find that

$$\frac{\partial \ln \omega}{\partial N_1} = -\ln N_1 - N_1 \frac{1}{N_1} + 1 = -\ln N_1 \tag{15.83}$$

Inserting this result into the condition (15.81) for the maximum of ω we obtain

$$-\ln N_1 - \alpha - \beta E_1 = 0 \tag{15.84}$$

The solution for N_1 is

$$N_1 = e^{-\alpha - \beta E_1} = e^{-\alpha} \cdot e^{-\beta E_1}$$

We can proceed in a like manner for any arbitrary N_i and find that

$$\boxed{N_i = C \cdot e^{-\beta E_i}} \tag{15.85}$$

where $C = e^{\alpha}$ is a constant.

15.9 Proving the Boltzmann Distribution for Indistinguishable Particles (High Temperature)

In counting the number of representations we have assumed that the particles are distinguishable. This assumption was reasonable for particles occupying different lattice sites which can be distinguished by using different site numbers. The assumption is not correct when considering identical particles moving freely in a gas. This can be understood by recognizing that the rotational, vibrational, and electronic energy states are associated with individual molecules, whereas the translational energy states of the container are shared among all the gas molecules.

15.9.1 Number of Available Quantum States

First we show that the number of available translational quantum states is much larger than the number of molecules if the temperature is sufficiently high. In Box 13.2 we have calculated the total number of available quantum states, g_{total}, for a particle of mass m in a cubic box of volume V as

$$g_{\text{total}} = \frac{\pi}{6} \cdot \left(\frac{8m}{h^2}\right)^{3/2} V E^{3/2} \tag{15.86}$$

Then the number g_{total} of available quantum states between $E = 0$ and $E = \frac{3}{2}kT$ is:

$$g_{\text{total}} = \frac{\pi}{6} \cdot \left(\frac{8m}{h^2}\right)^{3/2} V \left(\frac{3}{2}kT\right)^{3/2} \tag{15.87}$$

Thus for a helium gas at $T = 300$ K in a cubic box with side length $L = 10$ cm ($V = 1000$ mL)

$$g_{\text{total}} = 1.1 \times 10^{28}$$

If the helium gas is at a pressure of 1 bar the number of molecules in 1000 mL is

$$N = 2.4 \times 10^{22} \tag{15.88}$$

and there are $g_{\text{total}}/N = 5 \times 10^5$ quantum states available per molecule, on average. That is, the average occupation probability is $N/g_{\text{total}} = 2 \times 10^{-6}$. In addition, for the lowest quantum state at $T = 300$ K the occupation probability does not exceed 3×10^{-6} (see Problem 15.18). This result is valid for helium; however, for gas molecules with masses larger than the helium mass the average occupation probability at $T = 300$ K is even smaller than 2×10^{-6}.

If the number g_i of quantum states in an energy interval ΔE is much larger than the number of particles in this interval, quantum states can be occupied by one particle, or they are unoccupied (Fig. 15.22). Then we obtain the number of representations ω_i by counting the possibilities to exchange empty and singly occupied quantum states.

15.9.2 Number of Representations and Boltzmann Law

If g_i quantum states are available for one particle, we obtain $\omega_i = g_i^1$; for n_i particles in the energy interval, $\omega_i = g_i^{n_i}$. However, this calculation includes all the possibilities where two particles are interchanged, and there are $n_i!$ possibilities. For example, in the case $n_i = 3$ (as shown in Fig. 15.22) we have $n_i! = 6$ possibilities (123, 132, 213, 231, 312, 321) of interchanging. Therefore

$$\omega_i = \frac{g_i^{n_i}}{n_i!} \tag{15.89}$$

The total number of representations ω is given by the product

$$\omega = \frac{g_1^{n_1} \cdot g_2^{n_2} \cdot g_3^{n_3} \cdots}{n_1! \cdot n_2! \cdot n_3! \cdots}$$
(indistinguishable particles) (15.90)

because the number of representations at energy E_i is independent of the number of representations at other energies.

By using Stirling's formula and Equation (15.90), we can show that

$$\frac{\partial \ln \omega}{\partial n_1} = \ln \frac{g_1}{n_1} \tag{15.91}$$

Substituting this result in Equation (15.81), we find

$$\ln \frac{g_1}{n_1} - \alpha - \beta E_1 = 0 \quad \text{and} \quad \frac{n_1}{g_1} = \mathrm{e}^{-\alpha} \cdot \mathrm{e}^{-\beta E_1} \tag{15.92}$$

This relation holds for any particle i, so that

$$\boxed{n_i = g_i \cdot \mathrm{e}^{-\alpha} \cdot \mathrm{e}^{-\beta E_i}} \tag{15.93}$$

The number N_i of particles in a given quantum state at energy E_i is

$$N_i = \frac{n_i}{g_i} = C \cdot \mathrm{e}^{-\beta Ei} \tag{15.94}$$

where n_i is the number of particles in the energy interval ΔE (containing g_i quantum states) and N_i is the number of particles in quantum state i, and $C = \mathrm{e}^{-\alpha}$ is a constant given by Equation (15.6).

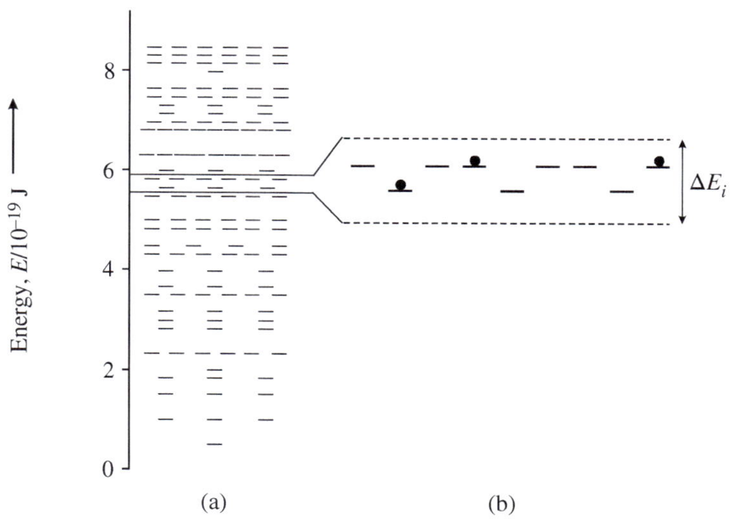

Figure 15.22 (a) Translational quantum states, conditions as in Fig. 15.14. (b) Selected interval between energies E_i and $E_i + \Delta E_i$ with g_i quantum states. At high enough temperature only a few quantum states are occupied by one particle each, and most quantum states are empty.

On first sight, Equation (15.90) looks quite different from the corresponding Equation (15.79) for distinguishable particles. However, Justification 15.2 shows that we can rewrite Equation (15.79) for the case of g-fold degenerate energy levels (e.g., for rotational motion) as

$$\omega = N! \cdot \frac{g_1^{n_1} \cdot g_2^{n_2} \cdot g_3^{n_3} \cdots}{n_1! \cdot n_2! \cdot n_3! \cdots} \tag{15.95}$$

In the case of non-degenerate systems ($g_i = 1$, $n_i = N_i$) this equation becomes identical with Equation (15.79). Thus in the case of indistinguishable particles the values of ω calculated above for distinguishable particles must be divided by $N!$. A division by a constant, however, does not change the shape of the distribution function; hence, our study of how energy is distributed is not affected. We shall see that indistinguishability plays a very important role in other properties of gases (see Chapter 17).

15.9.3 Calculating the Internal Energy *U* of an Ideal Atomic Gas

First we find an expression[3] for the constant C in Equation (15.94); in analogy to the procedure in Section 15.2.2:

$$N_i = C \cdot e^{-\beta E_i} = \frac{N}{\sum_i e^{-\beta E_i}} \cdot e^{-\beta E_i} = \frac{N}{Z'} \cdot e^{-\beta E_i} \quad \text{where} \quad Z' = \sum_i e^{-\beta E_i} \tag{15.96}$$

Next we find a relation between U and Z', in analogy to Section 15.2.3:

$$U = \sum_i N_i E_i = \frac{N}{Z'} \cdot \sum_i E_i \, e^{-\beta E_i} \tag{15.97}$$

so that

$$\frac{dZ'}{d\beta} = -\sum_i E_i \, e^{-\beta E_i} = -U \frac{Z'}{N} \tag{15.98}$$

and

$$U = -N \frac{dZ'/Z'}{d\beta} = -N \frac{d \ln Z'}{d\beta} \tag{15.99}$$

We obtain Z'_{trans} in a way as outlined for the calculation of Z_{trans} in Justification 15.1:

$$Z'_{\text{trans}} = \sum_{n_x, n_y, n_z} e^{-a(n_x^2 + n_y^2 + n_z^2)} = \left(\frac{1}{\sqrt{a}} \frac{1}{2} \sqrt{\pi} \right)^3, \quad a = \frac{h^2}{8mL^2} \cdot \beta$$

Then

$$Z'_{\text{trans}} = \left(\frac{2\pi m L^2}{h^2} \right)^{3/2} \cdot \beta^{-3/2}, \quad \frac{d \ln Z'_{\text{trans}}}{d\beta} = -\frac{3}{2} \frac{1}{\beta}$$

and for U_{trans} we obtain

$$U_{\text{trans}} = \frac{3}{2} \frac{N}{\beta} \tag{15.100}$$

15.9.4 Relation Between β and T

According to Chapter 14, the internal energy of translation, U_{trans}, is related to the temperature T by

$$U_{\text{trans}} = \frac{3}{2} NkT \tag{15.101}$$

We compare this result with the expression obtained for U_{trans} from Equation (15.100). and find that

$$\beta = \frac{1}{kT} \tag{15.102}$$

Then we can write the Boltzmann distribution (15.94) as

$$N_i = \frac{N}{Z} \cdot e^{-E_i/(kT)} \tag{15.103}$$

Accordingly, we replace β by $1/(kT)$ in Equation (15.85) for distinguishable particles.

3 Note that in the following we are using the notation Z' instead Z, because we do not yet know the parameter β.

15.10 Energy Distribution of Fermions and Bosons Among Quantum States

In Sections 15.8 and 15.9 we restricted our consideration to systems of distinguishable particles and to systems with indistinguishable particles at high temperature. Now we discuss systems with indistinguishable particles at low temperature, where the number of quantum states is comparable to the number of particles.

We consider helium gas as an example. In a volume V we have, according to the ideal gas law,

$$N = \frac{PV}{kT} \tag{15.104}$$

particles. In Equation (15.87) we have calculated the total number of available quantum states, g_{total}, for a particle of mass m in a three-dimensional box of length L in the energy range between $E = 0$ and $E = \frac{3}{2}kT$:

$$g_{\text{total}} = \frac{\pi}{6} \cdot V\left(\frac{8m}{h^2}\right)^{3/2}\left(\frac{3}{2}kT\right)^{3/2} \tag{15.105}$$

where we have introduced $V = L^3$. Then for the ratio g_{total}/N we obtain

$$\frac{g_{\text{total}}}{N} = \frac{\pi}{6} \cdot \left(\frac{8m}{h^2}\frac{3}{2}\right)^{3/2}(kT)^{5/2} \cdot \frac{1}{P} \tag{15.106}$$

If we decrease the temperature, according to Equation (15.106), the ratio g_{total}/N decreases (Table 15.4) dramatically.

Table 15.4 Ratio g_{total}/N for He Gas at Different Temperatures ($P = 1$ bar).

T/K	300	100	50	10	5
g_{total}/N	4.5×10^5	0.3×10^5	1.0×10^3	91	9.5

If g_{total} is on the order of N or even smaller we are faced with a completely different situation compared to Section 15.9, because our assumption that the number of quantum states in an energy interval ΔE is much larger than the number n_i of particles is no longer correct. This very different situation is realized already at room temperature for an electron gas in a metal (see Chapter 13). For atomic and molecular gases this approximation can break down at temperatures near to the zero point of T. In this case (number of quantum states comparable with the number of particles) we have to distinguish between two different situations.

Fermions:

Fermions are particles with half-integer spin such as electrons or protons ^{1}H (spin up and spin down, see Chapter 12). Each quantum state can only be occupied by one fermion. We consider n_i indistinguishable particles distributed over g_i quantum states (energy level i). Then the number of distinguishable representations of these particles, as outlined in Foundation 15.3, is

$$\omega_i = \frac{g_i!}{n_i!\,(g_i - n_i)!} \qquad \text{(Fermions)} \tag{15.107}$$

Bosons:

Bosons are particles with integer spin such as ^{2}H (see Chapter 12). Each quantum state can be occupied by any number of bosons. Again we consider n_i indistinguishable particles distributed over g_i quantum states (energy level i). Then the number of distinguishable representations of these particles, as outlined in Foundation 15.3, is

$$\omega_i = \frac{(g_i + n_i - 1)!}{n_i!\,(g_i - 1)!} \qquad \text{(Bosons)} \tag{15.108}$$

Example: Consider the case $n_i = 3$ and $g_i = 3$. In this case, fermions have only one representation

$$\omega_i = \frac{3!}{3!0!} = 1$$

whereas bosons have 10 representations

$$\omega_i = \frac{5!}{3!2!} = 10$$

These representations are shown in Fig. 15.23.

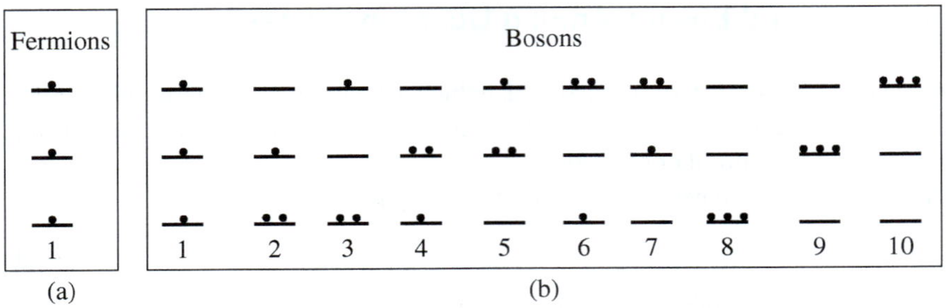

Figure 15.23 Distribution of 3 particles on 3 quantum states. (a) Fermions, one possible distribution. (b) Bosons, 10 possible distributions. Note that we do not imply the condition that all distributions have the same internal energy.

In Foundation 15.3 we calculate the distribution functions for fermions and bosons, as we earlier did for the Boltzmann distribution. The result is

$$N_i = \frac{n_i}{g_i} = \frac{1}{\mathrm{e}^{\alpha}\,\mathrm{e}^{E_i/(kT)} + 1} \quad \text{(distribution function of fermions)} \tag{15.109}$$

and

$$N_i = \frac{n_i}{g_i} = \frac{1}{\mathrm{e}^{\alpha}\,\mathrm{e}^{E_i/(kT)} - 1} \quad \text{(distribution function of bosons)} \tag{15.110}$$

where N_i is the number of particles per quantum state and e^{α} is a function of T, given by Equation (15.6). Note that N_i, in the present case, is small compared to one, it denotes the probability that quantum state i is occupied by a particle. As we have seen in Section 15.9, at sufficiently high temperature we have $n_i \ll g_i$, that is, the denominator in Equations (15.109) and (15.110) becomes large and we can neglect the terms +1 or −1. Then these equations become identical with the Boltzmann distribution for indistinguishable particles, Equation (15.94).

15.10.1 Photons Obey Boson Statistics: Planck's Radiation Law

Consider a heated body, for example the filament of an incandescent light bulb. With increasing temperature the body emits electromagnetic radiation. At low temperature the radiation is mostly at infrared wavelengths (you recognize it as heat), at higher temperatures the spectrum shifts to shorter visible wavelengths (the filament's color changes from red to white). This phenomenon can be explained by assuming that photons are generated by the filament and their energy distribution is determined by the filament's equilibrium temperature.

Photons can be considered as a special kind of bosons. Note that the factor e^{α} in Equation (15.110) is determined by the condition that the sum of the occupation numbers equals the total number of particles. Usually, we can count the number of particles, but in the case of photons this is impossible: the number of photons increases with temperature, and it is zero at the absolute zero of temperature. Because the photon number is not fixed, the distribution function 15.110 for photons becomes simplified as

$$N_i = \frac{n_i}{g_i} = \frac{1}{\mathrm{e}^{E_i/(kT)} - 1} \quad \text{(distribution function of photons)} \tag{15.111}$$

Starting with this equation, namely,

$$n_i = \frac{1}{\mathrm{e}^{E_i/(kT)} - 1} g_i$$

we show in Foundation F15.6 that the number $\mathrm{d}n$ of photons in the frequency range from ν to $\nu + \mathrm{d}\nu$ is described as

$$\mathrm{d}n = \frac{8\pi}{c_0^3} V \frac{\nu^2}{\exp(h\nu/(kT)) - 1}\,\mathrm{d}\nu \tag{15.112}$$

where V is the volume of the heated body. Each photon contributes an energy $h\nu$ to the radiation that is determined by its frequency ν (or wavelength λ). With this result we can write the differential energy $\mathrm{d}E$ as

$$\mathrm{d}E = \mathrm{d}n\,\mathrm{h}\nu = \frac{8\pi h}{c_0^3} V \frac{\nu^3}{\exp(h\nu/(kT)) - 1}\,\mathrm{d}\nu \tag{15.113}$$

Finally we define the spectral energy density ρ as

$$\rho = \frac{\mathrm{d}E}{\mathrm{d}\nu}\frac{1}{V} = \frac{8\pi h}{c_0^3}\frac{\nu^3}{\exp(h\nu/(kT)) - 1} \tag{15.114}$$

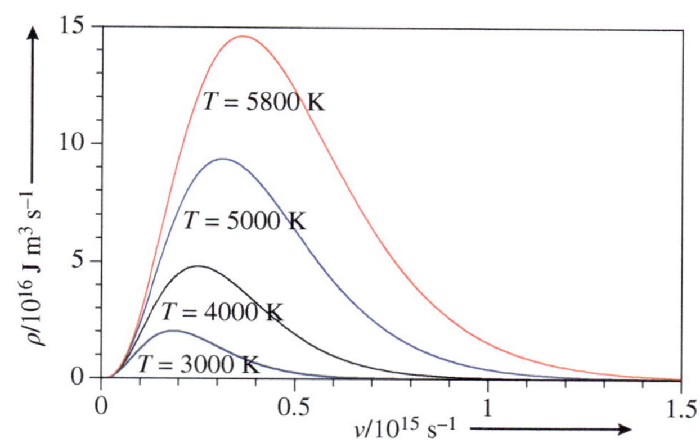

Figure 15.24 Planck's radiation law: spectral energy density ρ versus the frequency ν of the emitted radiation for different temperatures (5800 K is similar to the temperature at the sun's surface).

This is Planck's radiation law. Figure 15.24 plots ρ as a function of the frequency ν for different temperatures. Note that the maximum of the curve is shifted to higher frequencies with increasing temperature.

As an application, we compare the spectral energy density according to the Planck radiation law to experimental data on the solar spectrum. For a better comparison, we consider the spectral density as a function of the wavelength rather than as function of the frequency.

$$\rho_\lambda = -\frac{\mathrm{d}E}{\mathrm{d}\lambda}\frac{1}{V}$$

Note that the minus-sign ensures that the spectral energy density is a positive quantity. Using Equation (15.113) we find with $\nu = c_0/\lambda$ and $\mathrm{d}\nu = -(c_0/\lambda^2)\,\mathrm{d}\lambda$ that;

$$\begin{aligned}\mathrm{d}E &= \frac{8\pi h}{c_0^3}V\frac{\nu^3}{\exp(h\nu/(kT))-1}\mathrm{d}\nu \\ &= -\frac{8\pi h}{c_0^3}V\frac{c_0^3/\lambda^3}{\exp(hc_0/(\lambda kT))-1}\frac{c_0}{\lambda^2}\mathrm{d}\lambda \\ &= -\frac{8\pi hVc_0}{\exp(hc_0/(\lambda kT))-1}\frac{1}{\lambda^5}\mathrm{d}\lambda\end{aligned}$$

and

$$\rho_\lambda = \frac{8\pi hc_0}{\exp(hc_0/(\lambda kT))-1}\frac{1}{\lambda^5} \qquad (15.115)$$

In Fig. 15.25 we plot $\rho_\lambda/\rho_{\lambda,\max}$ according to Equation (15.115) (solid line) in comparison to experimental data of the solar spectrum above the atmosphere (dots). Apparently, theory and experiment agree very well at wavelengths longer than 500 nm. However, the experimental peak is slightly higher than calculated, and in the ultraviolet the experimental data are significantly smaller than calculated. The reason is that the sun is not an ideal black body. Also, the black body radiation analysis assumes thermal equilibrium over the complete surface of the body; however, this is not the case for the sun.

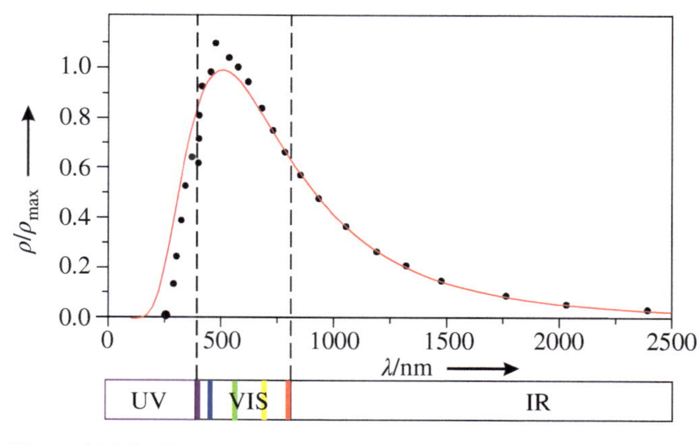

Figure 15.25 Experimental solar spectrum (dots, taken from above the atmosphere), compared to the Planck radiation function for $T = 5778$ K (red solid line). Experimental data taken from[4].

15.11 Conclusion

We have considered thermally excited molecular assemblies. Although any particular molecule has a well-defined energy at a particular time, the energy contained in the thermal collection of molecules is distributed statistically. The

4 https://en.wikipedia.org/wiki/Sunlight.

statistical probability $\mathcal{P}_i$ of a molecule being in energy state E_i is proportional to $\exp(-E_i/(kT))$, called the Boltzmann distribution, namely

$$\mathcal{P}_i = \frac{N_i}{N} = \frac{1}{Z} \cdot e^{-E_i/(kT)}$$

where Z is the molecular partition function. This distribution law holds well for most chemically interesting systems.

In addition to this distribution law we assumed that the internal energy associated with a molecule's different types of motions can be treated independently. This assumption allows us to treat the energy distributions for each type of motion and to write the molecules' internal energy as a sum of components for the different types of motions

$$U = U_{\text{trans}} + U_{\text{rot}} + U_{\text{vib}} + U_{\text{el}}$$

If we know the quantum states and their populations, then we can determine the internal energy directly by performing the sum over states.

From the molecular perspective we see that the internal energy is determined in two ways, namely, by the positions of the energy levels and by the population numbers.

15.12 Box 15.1. Equipartition Principle in the Classical Limit

The internal energy of a molecule's translational and rotational degrees of freedom (motions), is often well described by classical physics; i.e., the high temperature limit.

15.12.1 Translational Degrees of Freedom

The translational motion of a gas molecule has three components (x-, y-, and z-) for the kinetic energy, and each of these contributes $\frac{1}{2}kT$ to the molecule's average energy. Consequently the translational contribution to the internal energy of N molecules is

$$U_{\text{trans}} = \frac{3}{2}NkT$$

15.12.2 Rotational Degrees of Freedom

The overall rotational motion of molecules in a gas contains kinetic energy. Rotation about each of the inertial axes contributes $\frac{1}{2}kT$ to the molecule's average internal energy. For a linear molecule, a diatomic or linear polyatomic, the average rotational kinetic energy is kT because it only has two possible axes of rotation, each contributing $\frac{1}{2}kT$ to the molecule's internal energy. The rotational contribution to the internal energy of N linear molecules is

$$U_{\text{rot,linear}} = NkT$$

The rotation of a nonlinear molecule contributes with three terms of $\frac{1}{2}kT$ corresponding to three possible axes of rotation, so that its rotational contribution to the internal energy of N molecules is

$$U_{\text{rot,nonlinear}} = \frac{3}{2}NkT$$

15.12.3 Vibrational Degrees of Freedom

At typical Earth temperatures (few hundred Kelvin), the quantization of vibrational motion must be used to assess the internal energy of molecules. Nevertheless it can be useful to compare such energies to that expected from a classical description.

Let us consider a molecule like HCl in which the center of mass lies close to the heavy nucleus (Fig. 11.9). In the classical picture the light atom has, on average, the same kinetic energy $\frac{3}{2}kT$ as the gas molecules that hit it. The energy is equally distributed in the direction of the bond between the atoms and in the two directions perpendicular to it. On average, $\frac{1}{2}kT$ is stored as the kinetic energy of vibration. Additionally, the oscillator stores potential energy (stretching of the spring), which, on average, equals the kinetic energy, $\frac{1}{2}kT$. Consequently the vibrational contribution to the internal energy of a gas of N diatomic molecules is

$$U_{\text{vib,classic}} = U_{\text{vib,kin}} + U_{\text{vib,pot}} = \frac{1}{2}NkT + \frac{1}{2}NkT = NkT$$

At high enough temperature the quantum mechanical consideration approaches the classical one; according to Fig. 15.8 this happens for HD at temperatures much higher than 5000 K. At lower temperatures the deviations from classical behavior are considerable.

In polyatomic molecules with N_{atom} atoms several vibration modes are possible, as shown in Equations (11.35) and (11.36):

$(3N_{\text{atom}} - 5)$ vibration modes (linear molecules)

$(3N_{\text{atom}} - 6)$ vibration modes (nonlinear molecules)

In the classical limit each of the vibrational modes contribute kT to the average energy of a molecule.

15.12.4 Summary

In the classical limit each of these terms contributes $\frac{1}{2}NkT$ to the internal energy: $3 \times \frac{1}{2}NkT$ for translation in the x-, y-, and z-direction, $2 \times \frac{1}{2}NkT$ for rotation of a diatomic or a linear molecule, $3 \times \frac{1}{2}NkT$ for rotation of a nonlinear molecule, $2 \times \frac{1}{2}NkT$ for each vibrational normal mode ($\frac{1}{2}NkT$ for kinetic energy, $\frac{1}{2}NkT$ for potential energy). This statement is called the *Equipartition Principle*, and each term is called a *degree of freedom*.

15.13 Box 15.2. Canonical and Microcanonical Ensembles

In deriving the Boltzmann law we have considered isolated systems; that is, the internal energy U of the system has a fixed value, and the number of representations Ω is an integer (microcanonical ensemble). A different situation arises if we consider a closed system in contact with a temperature bath (canonical ensemble), because energy can be exchanged between the temperature bath and the system. Hence the energy of the system, and accordingly Ω, fluctuate.

Consider the case discussed in Fig. 15.21, where $N = 100$ particles occupy energy levels 0, a, and $2a$ with the internal energy fixed at $40a$, but now we place the system in a temperature bath of temperature T. For the temperature bath we choose a mean energy of

$$\overline{E}_{\text{trans}} = \frac{3}{2}kT = \frac{40a}{N} \quad \text{or} \quad T = \frac{2}{3}\frac{40a}{Nk}$$

In this case the internal energy of the system is no longer fixed to $U = 40a$. In principle, it can assume other values $\ldots .38a, 39a, 40a, 41a, 42a, \ldots$ Imagine that sometime after the system and temperature bath are at equilibrium, we isolate the system from the temperature bath. It is most probable that its internal energy is $U = 40a$, but any different energy $\ldots .38a, 39a, 40a, 41a, 42a, \ldots$ is also possible.

For a large enough number of particles N the distribution of energies approaches a Gaussian distribution (see Foundation 15.4). Then the mean deviation $\overline{\Delta U}$ from $\overline{U}$ is

$$\overline{\Delta U} = \pm \overline{U}\frac{1}{\sqrt{N}}$$

on the average. $\overline{\Delta U}/\overline{U}$ approaches zero as N becomes large. For 100 particles, according to this equation, $\overline{\Delta U}$ is 10% of $\overline{U}$, for 10,000 particles we obtain $\overline{\Delta U}/\overline{U} = \pm 0.01$. For a mole of gas atoms $N = 6.0 \times 10^{23}$ and $\overline{\Delta U}/\overline{U} = 1.3 \times 10^{-12}$. This means that for a very large number N the energy U is given to a high degree of approximation by $\overline{U}$.

15.14 Justifications

15.14.1 Justification 15.1: Molecular Partition Function Z_{trans} of Translation

See the books website, www.wiley.com/go/Kuhn/physicalchemistry3e.

15.14.2 Justification 15.2: Distinguishable Particles Occupying Degenerate Energy Levels

See the books website, www.wiley.com/go/Kuhn/physicalchemistry3e.

15.15 Foundations

15.15.1 Foundation 15.1: Classical Derivation of the One-Dimensional Maxwell–Boltzmann Distribution of Speeds

We assume that the speeds of molecules are uniformly distributed over space. In particular, the distribution function $f_x(\upsilon_x)$ in the x direction does not depend on the distribution functions $f_y(\upsilon_y)$ and $f_z(\upsilon_z)$ in the y and z directions. Then the three-dimensional distribution function $F(\upsilon)$ can be written as

$$F(\upsilon) = f_x(\upsilon_x) \cdot f_y(\upsilon_y) \cdot f_z(\upsilon_z) \tag{15F1.1}$$

where

$$\upsilon^2 = \upsilon_x^2 + \upsilon_y^2 + \upsilon_z^2 \quad \text{or} \quad \upsilon = \sqrt{\upsilon_x^2 + \upsilon_y^2 + \upsilon_z^2} \tag{15F1.2}$$

Next we consider the logarithms of this expression

$$\ln F(\upsilon) = \ln f_x(\upsilon_x) + \ln f_y(\upsilon_y) + \ln f_z(\upsilon_z) \tag{15F1.3}$$

and calculate the partial derivative with respect to υ_x:

$$\frac{\partial \ln F(\upsilon)}{\partial \upsilon_x} = \frac{1}{f_x}\frac{df_x}{d\upsilon_x} \tag{15F1.4}$$

We rewrite this expression according to the chain rule

$$\frac{\partial \ln F(\upsilon)}{\partial \upsilon}\frac{\partial \upsilon}{\partial \upsilon_x} = \frac{1}{f_x}\frac{df_x}{d\upsilon_x} \tag{15F1.5}$$

where, according to Equation (15F1.2),

$$\frac{\partial \upsilon}{\partial \upsilon_x} = \frac{2\upsilon_x}{2\sqrt{\upsilon_x^2 + \upsilon_y^2 + \upsilon_z^2}} = \frac{\upsilon_x}{\upsilon} \tag{15F1.6}$$

Then

$$\frac{\partial \ln F(\upsilon)}{\partial \upsilon}\frac{1}{\upsilon} = \frac{1}{f_x}\frac{df_x}{d\upsilon_x}\frac{1}{\upsilon_x} \tag{15F1.7}$$

Corresponding expressions are obtained for the y and z directions, resulting in

$$\frac{\partial \ln F(\upsilon)}{\partial \upsilon}\frac{1}{\upsilon} = \frac{1}{f_x}\frac{df_x}{d\upsilon_x}\frac{1}{\upsilon_x} = \frac{1}{f_y}\frac{df_y}{d\upsilon_y}\frac{1}{\upsilon_y} = \frac{1}{f_z}\frac{df_z}{d\upsilon_z}\frac{1}{\upsilon_z} = -A \tag{15F1.8}$$

Since the terms for x, y, and z are independent from each other, each term must be equal to a common constant $-A$. By way of example we solve Equation (15F1.8) for the x direction,

$$\frac{df_x}{f_x} = -A\upsilon_x \cdot d\upsilon_x \tag{15F1.9}$$

and integration yields

$$\ln f_x = -A\int \upsilon_x d\upsilon_x = -A\frac{1}{2}\upsilon_x^2 + \ln B \tag{15F1.10}$$

or

$$f_x = const \cdot \exp\left(-\frac{A\upsilon_x^2}{2}\right) \tag{15F1.11}$$

A must be a positive quantity, because f_x must become zero for $\upsilon_x \to \infty$. This is the Maxwell–Boltzmann distribution, except that the constant A is still unknown.

A corresponding analysis can be used for v_y and v_z; hence, we find that

$$F(v) = const \cdot \exp\left(-\frac{A\left(v_x^2 + v_y^2 + v_z^2\right)}{2}\right)$$
$$= const \cdot \exp\left(-\frac{Av^2}{2}\right)$$

15.15.1.1 Calculation of Constant A

From Chapter 14 we know that the mean square speed in the x direction is $\overline{v_x^2} = kT/m$. Thus

$$\overline{v_x^2} = \frac{\int_0^\infty v_x^2 \cdot f_x \cdot dv_x}{\int_0^\infty f_x \cdot dv_x} = \frac{\int_0^\infty v_x^2 \cdot e^{-Av_x^2/2} \cdot dv_x}{\int_0^\infty e^{-Av_x^2/2} \cdot dv_x} = \frac{kT}{m} \tag{15F1.12}$$

With the substitution

$$Av_x^2/2 = w^2, \quad v_x^2 = 2w^2/A, \quad v_x = w\sqrt{2/A}, \quad \text{and} \quad dv_x = dw\sqrt{2/A} \tag{15F1.13}$$

we obtain

$$\int_0^\infty v_x^2 \cdot e^{-Av_x^2/2} \cdot dv_x = \left(\frac{2}{A}\right)^{3/2} \int_0^\infty w^2 \cdot e^{-w^2} \cdot dw = \left(\frac{2}{A}\right)^{3/2} \cdot \frac{\sqrt{\pi}}{4} \tag{15F1.14}$$

and

$$\int_0^\infty e^{-Av_x^2/2} \cdot dv_x = \left(\frac{2}{A}\right)^{1/2} \int_0^\infty e^{-w^2} \cdot dw = \left(\frac{2}{A}\right)^{1/2} \cdot \frac{\sqrt{\pi}}{2} \tag{15F1.15}$$

Then

$$\overline{v_x^2} = \frac{\left(\frac{2}{A}\right)^{3/2} \cdot \frac{\sqrt{\pi}}{4}}{\left(\frac{2}{A}\right)^{1/2} \cdot \frac{\sqrt{\pi}}{2}} = \frac{1}{A} = \frac{kT}{m} \quad \text{and} \quad A = \frac{m}{kT} \tag{15F1.16}$$

and

$$f_x = const \cdot \exp\left(-\frac{mv_x^2}{2kT}\right) \tag{15F1.17}$$

15.15.2 Foundation 15.2: How to Find the Maximum of ln ω

In Section 15.8.2 we considered the number of arrangements ω for distinguishable particles

$$\omega = \frac{N!}{N_1! \cdot N_2! \cdot N_3! .. \cdot N_i!}$$

and asked for the maximum of ω or for the maximum of $\ln \omega$, respectively. ω is a function that depends on very many variables $N_1, N_2, \ldots$ The condition for the $\ln \omega$ to be an extremum is

$$\delta \ln \omega = \left(\frac{\partial \ln \omega}{\partial N_1}\right)_{N_2, N_3, \ldots} \cdot dN_1 + \left(\frac{\partial \ln \omega}{\partial N_2}\right)_{N_1, N_3, \ldots} \cdot dN_2 + \left(\frac{\partial \ln \omega}{\partial N_3}\right)_{N_1, N_2, \ldots} \cdot dN_3 + \ldots = 0 \tag{15F2.1}$$

In addition we must ensure that the conditions

$$N_1 + N_2 + N_3 + \cdots = N \quad \text{or} \quad dN = 0 = \sum dN_i \qquad \text{(condition 1)}$$
$$E_1 N_1 + E_2 N_2 + E_3 N_3 + \cdots = U \quad \text{or} \quad \sum E_i \cdot dN_i = 0 \qquad \text{(condition 2)}$$

hold[5] as we perform the variation of $\ln \omega$. These conditions are called *constraints*. For example, condition 1 means that the particle numbers $N_1, N_2, \ldots$ are not independent of each other, because the total number of particles in the system is fixed. Hence the maximum of $\ln \omega$ is not simply obtained by setting the partial derivatives equal to zero.

5 Note that condition 2 is valid only for a gas of independent particles.

We can include the conditions 1 and 2 in the expression for $\delta \ln \omega$ by using a mathematical trick, known as the method of *Lagrange's multipliers*. We begin by multiplying the expressions in conditions 1 and 2 with arbitrary constants α and β, so that the conditions become

$$\alpha \cdot \sum \mathrm{d}N_i = 0 \quad \text{and} \quad \beta \cdot \sum E_i \cdot \mathrm{d}N_i = 0 \tag{15F2.2}$$

Now we subtract both expressions from $\delta \ln \omega$ (we are allowed to do this, because effectively we subtract zero):

$$\delta \ln \omega = \left[\left(\frac{\partial \ln \omega}{\partial N_1}\right) - \alpha - \beta E_1\right] \cdot \mathrm{d}N_1 + \left[\left(\frac{\partial \ln \omega}{\partial N_2}\right) - \alpha - \beta E_2\right] \cdot \mathrm{d}N_2 + \left[\left(\frac{\partial \ln \omega}{\partial N_3}\right) - \alpha - \beta E_3\right] \cdot \mathrm{d}N_3 + \cdots = 0$$

Because the constants α and β are arbitrary, we can choose any value for them we like. Arbitrarily, we define these constants by setting

$$\left(\frac{\partial \ln \omega}{\partial N_1}\right) - \alpha - \beta E_1 = 0 \quad \text{and} \quad \left(\frac{\partial \ln \omega}{\partial N_2}\right) - \alpha - \beta E_2 = 0 \tag{15F2.3}$$

Combining these results with the original expression for $\delta \ln \omega$, we find

$$\delta \ln \omega = \left[\left(\frac{\partial \ln \omega}{\partial N_3}\right) - \alpha - \beta E_3\right] \cdot \mathrm{d}N_3 + \left[\left(\frac{\partial \ln \omega}{\partial N_4}\right) - \alpha - \beta E_4\right] \cdot \mathrm{d}N_4 + \cdots = 0 \tag{15F2.4}$$

In this expression the remaining variables $N_3, N_4, N_5, \ldots$ are independent of each other, and $\delta \ln \omega$ is zero for

$$\left(\frac{\partial \ln \omega}{\partial N_i}\right) - \alpha - \beta E_i = 0, \quad (i = 1, 2, 3, \ldots) \quad \text{(condition for the maximum of } \ln \omega)$$

15.15.3 Foundation 15.3: Energy Distribution of Fermions and Bosons

See the books website, www.wiley.com/go/Kuhn/physicalchemistry3e.

15.15.4 Foundation 15.4: Canonical and Microcanonical Ensembles

As an example for a canonical system we consider a container with molecules immersed in a temperature bath (Fig. 15F.3). The molecules in the container have an average internal energy $\overline{U}$, depending on the temperature of the bath. However, the energy can be distributed over the system in a great number of ways, leading to deviations ΔU from the internal energy.

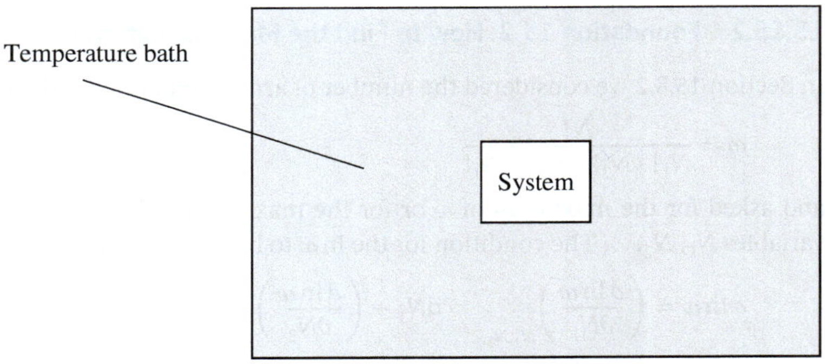

Figure 15F.3 Canonical system. System immersed in a temperature bath.

We imagine that the system consists of N molecules distributed over three energy levels with energies $E_1 = 0$, $E_2 = a$, and $E_3 = 2a$ as in Fig. 15.19, and we assume that the mean internal energy is $\overline{U} = 0.4 \cdot N \cdot a$, corresponding to the distribution M in Table 15.3. For example, for $N = 100$ we have $N_1 = 68$, $N_2 = 24$, and $N_3 = 8$, and $\overline{U} = 40a$.

At equilibrium with the temperature bath the average internal energy of the system $\overline{U}$ is well defined, but fluctuations can occur, leading to deviations ΔU of the system from the mean value $\overline{U}$. Because of the fluctuations the value of U may differ from $\overline{U}$ at any particular time but the time average of U is $\overline{U}$. In order to calculate the probability of a fluctuation, we replace the time average by the ensemble average. For simplicity, we divide the system of N molecules into two subsystems

a and b with $N/2$ molecules each. We allow system a to increase its internal energy if system b decreases its internal energy by the same amount. In this way we ensure that the average internal energy of the total system remains constant. To obtain the probability of finding the system in a state with $\overline{U} \pm \Delta U$ we calculate the number of representations ω_a and ω_b of the subsystems,

$$\omega_a = \frac{\left(\frac{N}{2}\right)!}{N_{a1}!N_{a2}!N_{a3}!}, \quad \omega_b = \frac{\left(\frac{N}{2}\right)!}{N_{b1}!N_{b2}!N_{b3}!} \tag{15F4.1}$$

where N_{a1}, N_{a2}, N_{a3} are the occupation numbers in the energy levels 1, 2, and 3 of subsystem a and N_{b1}, N_{b2}, N_{b3} are the occupation numbers in the energy levels 1, 2, and 3 of subsystem b. Note that

$$\sum_{i=1}^{i=3} N_{ai} = \frac{N}{2} \qquad \sum_{i=1}^{i=3} N_{bi} = \frac{N}{2}$$

Thus the number ω of the total system is

$$\omega = \omega_a \cdot \omega_b \tag{15F4.2}$$

Finally we calculate the total number of representations Ω as the sum over all possible ω values to obtain the probability $\mathcal{P}$ of finding the system in a state with a given energy distribution.

$$\mathcal{P} = \frac{\omega_a \cdot \omega_b}{\Omega} \tag{15F4.3}$$

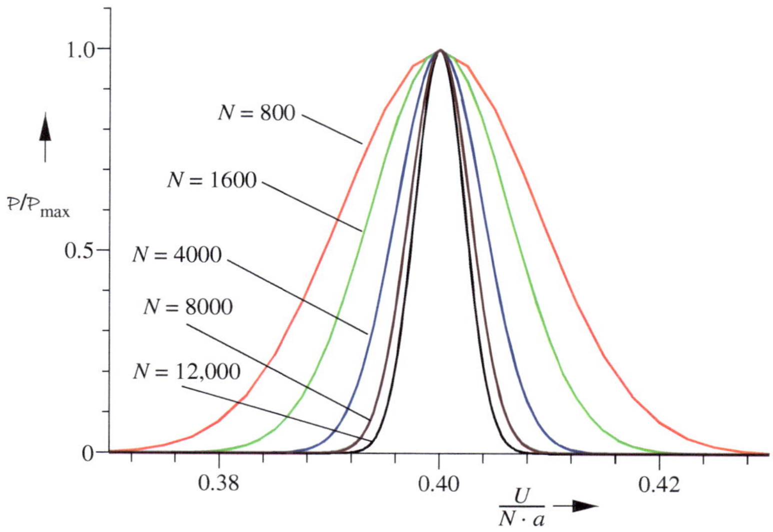

Figure 15F.4 $N = 9$ molecules distributed on energy levels $E_1 = 0$, $E_2 = a$, and $E_3 = 2a$.

Figure 15F.4 shows the distribution function $\mathcal{P}$ as a function of $U/(N \cdot \Delta E)$ for $N = 800$, 1600, 4000, 8000, and 12, 000.

The figure shows that we obtain a Gaussian distribution for $\mathcal{P}$ with a maximum at $\overline{U}$; the half-width of the distribution decreases with an increasing number of N. We identify one-half of the half-width with the mean deviation $\overline{\Delta U}$ (this corresponds approximately to the standard deviation σ). These data are collected in Table 15F.1 and displayed in Fig. 15F.5 (filled circles).

Table 15F.1 $\overline{\Delta U}/\overline{U}$ as a Function of the Number N of Particles and Calculation of the Constant *const.* in Equation (15F4.4).

N	200	400	800	1600	2000	4000	8000	12,000
$\frac{\overline{\Delta U}}{\overline{U}}$	0.0210	0.0149	0.0105	0.00739	0.00661	0.00466	0.00329	0.00269

From these data we conclude that $\overline{\Delta U}/\overline{U}$ is inversely proportional to $\sqrt{N}$:

$$\left(\frac{\overline{\Delta U}}{\overline{U}}\right)_{\text{theoretical}} = const. \frac{1}{\sqrt{N}} \tag{15F4.4}$$

Comparison with the data in Table 15F.1 allows us to fix the constant as *const.* = 2.96. In Fig. 15F.5 we compare the values taken from Table 15F.1 to the values calculated from Equation (15F4.4). Both curves are in excellent agreement. These considerations show that for sufficiently large values of N (in practical cases we have $N \approx 10^{20}$) the two subsystems have the same internal energy.

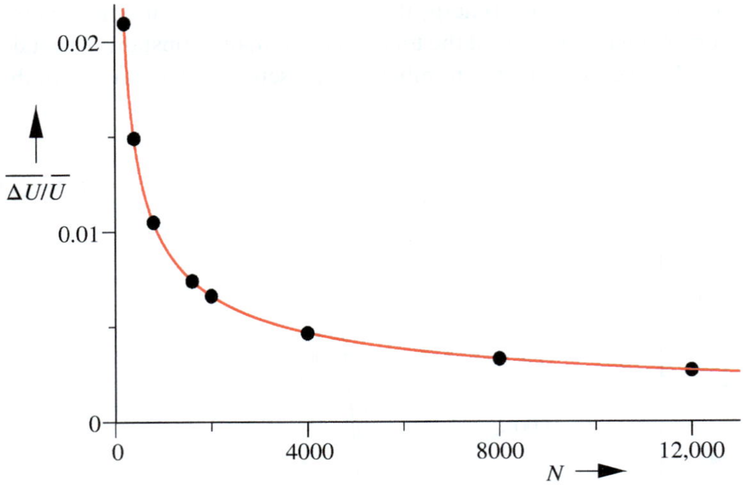

Figure 15F.5 $\overline{\Delta U}/\overline{U}$ as a function of the number N of molecules distributed on the three energy levels. (a) Calculated from the halfwidths in Fig. 15F.4, filled circles, (b) calculated according to Equation (15.141) with *const.* = 2.96, solid line.

15.15.5 Foundation 15.5: Internal Energy U as Sum of Contributions U_{el}, U_{vib}, U_{rot}, and U_{trans}

Often, it is useful to approximate a molecule's individual degrees of freedom (electronic, vibrational, rotational, and translational) as independent. In this case, the energy $E_{k,l,m,n}$ of a particle which is in electronic quantum state k, vibrational quantum state l, rotational quantum state m, and translational quantum state n is

$$E_{k,l,m,n} = E_{el,k} + E_{\text{vib},l} + E_{\text{rot},m} + E_{\text{trans},n}$$

Accordingly, the molecular partition function Z can be written in terms of molecular partition functions for the electronic, vibrational, rotational, and translational contributions; i.e.,

$$\begin{aligned} Z &= \sum_{k,l,m,n=1}^{\infty} \mathrm{e}^{-E_{k,l,m,n}/(kT)} \\ &= \sum_{k,l,m,n=1}^{\infty} \left(\mathrm{e}^{-E_{el,k}/(kT)} \cdot \mathrm{e}^{-E_{\text{vib},l}/(kT)} \cdot \mathrm{e}^{-E_{\text{rot},m}/(kT)} \cdot \mathrm{e}^{-E_{\text{trans},n}/(kT)}\right) \\ &= \sum_{k=1}^{\infty} \mathrm{e}^{-E_{el,k}/(kT)} \cdot \sum_{l=1}^{\infty} \mathrm{e}^{-E_{\text{vib},l}/(kT)} \cdot \sum_{m=1}^{\infty} \mathrm{e}^{-E_{\text{rot},m}/(kT)} \cdot \sum_{n=1}^{\infty} \mathrm{e}^{-E_{\text{trans},n}/(kT)} \end{aligned}$$

or

$$Z = Z_{\text{el}} \cdot Z_{\text{vib}} \cdot Z_{\text{rot}} \cdot Z_{\text{trans}} \tag{15F5.1}$$

Using Equation (15.14) for the internal energy we obtain

$$U = NkT^2 \frac{\mathrm{d}\ln Z}{\mathrm{d}T} = NkT^2 \frac{\mathrm{d}\ln\left(Z_{\text{el}} \cdot Z_{\text{vib}} \cdot Z_{\text{rot}} \cdot Z_{\text{trans}}\right)}{\mathrm{d}T} \tag{15F5.2}$$

$$= NkT^2 \left[\frac{\mathrm{d}\ln\left(Z_{\text{el}}\right)}{\mathrm{d}T} + \frac{\mathrm{d}\ln\left(Z_{\text{vib}}\right)}{\mathrm{d}T} + \frac{\mathrm{d}\ln\left(Z_{\text{rot}}\right)}{\mathrm{d}T} + \frac{\mathrm{d}\ln\left(Z_{\text{trans}}\right)}{\mathrm{d}T}\right] \tag{15F5.3}$$

$$= U_{\text{el}} + U_{\text{vib}} + U_{\text{rot}} + U_{\text{trans}} \tag{15F5.4}$$

where U_{el}, U_{vib}, U_{rot}, and U_{trans} are the contributions of electronic, vibrational, rotational, and translational energy of the particle, and are given by

$$\boxed{U_{\text{el}} = NkT^2 \cdot \frac{\mathrm{d}\ln Z_{\text{el}}}{\mathrm{d}T}; \quad U_{\text{vib}} = NkT^2 \cdot \frac{\mathrm{d}\ln Z_{\text{vib}}}{\mathrm{d}T}}$$

$$\boxed{U_{\text{rot}} = NkT^2 \cdot \frac{\mathrm{d}\ln Z_{\text{rot}}}{\mathrm{d}T}; \quad U_{\text{trans}} = NkT^2 \cdot \frac{\mathrm{d}\ln Z_{\text{trans}}}{\mathrm{d}T}}$$

15.15.6 Foundation 15.6: Planck's Radiation Law

See the books website, www.wiley.com/go/Kuhn/physicalchemistry3e.

15.16 Exercises

E15.1 Calculate the characteristic vibrational temperature for HD ($\tilde{\nu} = 3813$ cm^{-1}), HCl ($\tilde{\nu} = 2991$ cm^{-1}), and CO ($\tilde{\nu} = 2169$ cm^{-1}), where $\tilde{\nu} = 1/\lambda$ is the wavenumber. Over what temperatures are quantum effects important? At what temperatures is a classical description appropriate?

E15.2 Calculate the characteristic rotational temperature for HD ($d = 74$ pm), HCl ($d = 127$ pm), and CO ($d = 113$ pm). Over what temperatures are quantum effects important? At what temperatures is a classical description appropriate?

E15.3 Consider the I_2 molecule, with $\mu = 1.05 \times 10^{-25}$ kg and $d = 267$ pm, calculate the characteristic rotational temperature. Calculate the rotational energy level that has the highest population at $T = 300$ K.

E15.4 Obtain the fundamental vibrational frequencies of H_2, HI, and I_2; compute the energy difference between the ground vibrational state and the first excited vibrational state. Compute the ratio of this energy to $k_B T$ for temperatures of $T = 300$ K and $T = 1000$ K. Comment on the molecules' abilities to store thermal energy in their vibrational degrees of freedom at these temperatures.

E15.5 Obtain the rotational constants for H_2, HI, I_2 from the literature and compute the energy difference between the ground rotational state and the first excited rotational state. Compute the ratio of this energy to kT for temperatures of $T = 10$ K, $T = 100$ K, and $T = 300$ K. Comment on the molecules' abilities to store thermal energy in their rotational degrees of freedom at these temperatures.

E15.6 Derive an expression for the internal energy U_{rot} of a molecule rotating in two dimensions, i.e., rotating on a circle (see the discussion for Fig. 11.2).

E15.7 Use the harmonic oscillator model to describe the vibrational energy levels of HCl (the fundamental vibration occurs at $\tilde{\nu} = 2991$ cm^{-1}). What fraction of HCl molecules are in the 3rd excited state level at $T = 1000$ K? What fraction of HCl molecules are in the ground energy level at $T = 1000$ K?

E15.8 Consider the diatomic molecules H_2, Cl_2, and I_2. At what temperature does the value of kT equal the energy level spacing between the lowest two energy levels for the vibrational levels, the rotational levels, and the translational levels in a cubic box of volume 1 mL? Compare trends between these three molecules and between the types of motions. Find literature values for the molecular parameters. (Note: Although these molecules are homonuclear you can neglect symmetry constraints arising from nuclear spin considerations for the purposes of this problem; i.e., treat the molecules as if they were heteronuclear diatomics.)

E15.9 Using the harmonic oscillator approximation, compute the fraction of O_2 molecules ($\theta_{vib} = 2230$ K) in the first excited vibrational state at $T = 298$ K. Compute the fraction of O_2 molecules in all excited vibrational states at 298 K. Perform the same calculations for H_2 ($\theta_{vib} = 6332$ K). The bond strengths of these molecules are similar (in fact, the O_2 bond is a little stronger than H_2). Why is the H_2 population in the first excited level lower than that of O_2?

E15.10 Determine the constants C and C' in Equations (15.67) and (15.69).

$$C = \frac{2N}{\sqrt{\pi}(kT)^{3/2}} \qquad C' = \frac{4N}{\sqrt{\pi}}\left(\frac{m}{2kT}\right)^{3/2}$$

which are the values given in the text.

E15.11 Calculate the root mean square speed v_{rms} and the average speed v for Ar gas at $T = 300$ K. Perform the same calculation for CH_4 gas at $T = 300$ K. Discuss your results.

E15.12 Derive Equation (15.14) using the approximations described in Section 15.2.3.

E15.13 The lowest electronically excited energy level of O_2 is 1.0 eV $= 1.602 \times 10^{-19}$ J above the ground state. At what temperature $T_{1\%}$ would the population of excited molecules be 1.0% of that in the ground state? Take the degeneracy of the ground level to be $g_0 = 3$ and that of the excited levels to be $g_{i>0} = 1$. Assume that the energy of the ground state is $E_0 = 0$. Compute the electronic partition function for O_2 at the temperature $T_{1\%}$.

E15.14 Figure 15.18 shows that the distribution of translational speeds broadens with temperature. Quantify how the width of the distribution changes with temperature by finding an expression for the standard deviation in the translational speeds; i.e., $\sigma_v = \sqrt{\overline{v^2} - \overline{v}^2}$.

E15.15 Analyze the expression for the most probable speed v_P of the Maxwell–Boltzmann distribution and determine how it changes with temperature. How does it change with molar mass?

E15.16 Use the three-dimensional Maxwell–Boltzmann distribution of molecular speeds to find an expression for the average speed $\overline{v}$, Equation (15.71).

E15.17 Sketch the form of the three-dimensional Maxwell–Boltzmann distribution law for molecular speeds. Perform this sketch for two different values of the temperature. On your graph mark the most probable speed, the average speed, and the root mean square speed of the gas particle.

E15.18 Sketch the Maxwell–Boltzmann distribution of molecular speeds for two monatomic gases of mass m_1 and m_2, at the same temperature. Assume that $m_1 > m_2$. Explain your sketch in words.

E15.19 An example of a two-state system is the case of a proton in an external magnetic field (as created in an NMR spectrometer). The energy splitting between the spin states of an H atom nucleus in a magnetic field is given by Equation (12.5)

$$\Delta\varepsilon = g_N \cdot \mu_N \cdot B$$

where B is the magnetic field strength. $g_N = 5.586$ is the nuclear g -factor for the proton and $\mu_N = 5.051 \times 10^{-27}$ J T^{-1} is the nuclear magneton. Ignore the translational degrees of freedom and compute the molecular partition function for the case of an H atom nucleus in a 12 T magnetic field and $T = 77$ K. Also compute the relative populations of the two spin energy levels and their average energy in the magnetic field.

E15.20 A molecular radical (e.g., methyl radical) has an unpaired electron. In an external magnetic field the ground electronic energy level splits into two different energy states associated with the two possible spin orientations. The energy splitting between the spin states of an unpaired electron in a magnetic field (without orbital angular momentum) is given by

$$\Delta\varepsilon = g \cdot \mu_{\text{Bohr}} \cdot B$$

where $g = 2.002$ is the g-factor of the electron and $\mu_{\text{Bohr}} = 9.274 \times 10^{-24}$ J T^{-1} is the Bohr magneton. Compute the molecular partition function for the two different electron spin states in a 1.2 Tesla magnetic field at $T = 77$ K. Compute the relative populations of the two energy levels and the average energy of the electron in the magnetic field.

E15.21 Use the MathCAD exercise, "distribution.mcdx," to explore some fundamental properties of two different probability distributions: the Poisson distribution,

$$f(\lambda, k) = \exp(-\lambda) \cdot \frac{\lambda^k}{k!}$$

which often applies when the number of particles (or events) is small, and the Gaussian distribution,

$$f(x) = \frac{1}{\sigma\sqrt{2\pi}} \cdot \exp\left(-\frac{1}{2}\left(\frac{x-\mu}{\sigma}\right)^2\right)$$

which is very widely used and is appropriate when the particle (or event) number is large. In the first part of the exercise you are asked to plot the distributions with various parameter values and for various particle numbers N, to understand the meaning of the parameters for the distributions and to assess how the number of particles affects the shape of the distribution. In the second part of the exercise, you are asked to consider the probability distribution for the energy of a one-dimensional ideal gas, which is well described as a Gaussian distribution under common conditions. You should examine how the energy distribution for the gas changes with the number of particles N. In the last question for this part, you are given the relationship between the standard deviation σ of the distribution and the constant volume heat capacity C_V of the gas (i.e., $\sigma = \sqrt{kT^2 C_V}$), and you are asked to comment on its implications for how the energy distribution must change near a phase transition.

E15.22 Plot the following three functions on a single graph: $f_1(J) = 2J + 1$; $f_2(J) = \exp(-E_J/kT)$ with $E_J = 10^{-22}$ J; and their product $f_3(J) = f_1(J) \cdot f_2(J)$. Discuss your result in terms of the population distribution of rotational energy levels.

E15.23 Consider a two-state system with a ground-state energy of $E_1 = 0$ and an excited state energy of $E_2 = 1.6 \times 10^{-21}$ J. Write an expression for the molecular partition function and evaluate it at the temperatures 1 K, 10 K, 100 K, 1000 K, and 10,000 K. What value does the molecular partition function take as $T \to 0$? What value does the partition function take as $T \to \infty$? How does the partition function reflect the population of the states (refer to Fig. 15.3)?

E15.24 A simple model for the lowest excited states of the Na atom is a ground energy level that is doubly degenerate ($E_1 = 0$, and $g_1 = 2$) and an excited energy level that is six-fold degenerate ($E_2 = 3.37 \times 10^{-19}$ J, $g_2 = 6$). (Note: the excited state configuration [Na]$3p^1$ is actually split into sublevels; however, they lie very close in energy as compared to the energy gap $E_2 - E_1$ so we ignore this difference.) Calculate the fraction of atoms in each of these levels at 1000 and 5000 K. For 0.1 mol of Na atoms, how many are in the excited level under these conditions?

E15.25 The molecular partition function Z_{trans} for a monatomic ideal gas is given by (see Equation (15.59))

$$Z_{\text{trans}} = V\left(\frac{2\pi mkT}{h^2}\right)^{3/2}$$

Evaluate Z_{trans} for a He atom at a temperature of 298 K and in a container of volume i) 1 cm^3; ii) 1 μm^3; iii) 10 nm^3.

E15.26 Calculate the molecular partition function for a mole of He atoms and for a mole of Ar atoms at 298 K and 1 bar. Assuming that Z_{trans} measures the number of available quantum states, calculate the average number of states available for each gas atom in the two cases. Compare your results with each other.

E15.27 Show that for He in a volume of $V = 1000$ mL at $T = 300$ K and $P = 1$ bar the probability to find an atom in the lowest translational quantum state is 3×10^{-6}.

E15.28 Consider a two-state system with an excitation energy of $\Delta E = 3.37 \times 10^{-19}$ J (e.g., the $3s$ to $3p$ electron promotion in Na is close to this value). Calculate the fraction of excited states at $T = 300$ K, 1000 K, and 2000 K.

E15.29 The fraction of molecules in a particular vibrational state depends on the energy spacing of the vibrational levels, as well as the temperature. Use the data in Table E15.29 to plot the population probabilities $\mathcal{P}_n$ of molecules in vibrational level n for two different diatomic halogens at 300 K.

Table E15.29 Vibrational Parameters and Population Probabilities $\mathcal{P}_0$ and $\mathcal{P}_1$ for Some Interhalogens at 300 K.

System	k_f (N/m)	$\frac{1}{\lambda}$ (cm^{-1})	$\Delta E/10^{-21}$ J	$\mathcal{P}_0$	$\mathcal{P}_1$
$^{19}F^{35}Cl$	449	786	15.6	0.977	0.023
$^{19}F^{79}Br$	407	671	13.3	0.960	0.039
$^{127}I^{35}Cl$	239	384	7.63	0.842	0.133
$^{127}I^{79}Br$	208	269	5.34	0.725	0.200

E15.30 Calculate the fraction of molecules that are in excited vibrational states at 300 K for HD, $h\nu_0 = 75 \times 10^{-21}$ J, and I_2, $h\nu_0 = 4.27 \times 10^{-21}$ J.

E15.31 Plot the fraction of molecules in each rotational energy level $\mathcal{P}_J$ as a function of the quantum number J for two interhalogen compounds $^{19}F^{35}Cl$ ($B = 1.02 \cdot 10^{-23}$ J; $r = 163$ pm) and $^{127}I^{79}Br$ ($B = 0.11 \cdot 10^{-23}$ J; $r = 247$ pm). Comment on your findings.

E15.32 Assuming that you can approximate the rotational partition functions for para and ortho hydrogen by

$$Z_{\text{rot,para}} = Z_{\text{rot,ortho}} = \frac{1}{2}Z_{\text{rot}} = \frac{1}{2}\sum_{J=0}^{\infty} g_J \cdot e^{-E_J/(kT)}, \quad J = 0, 1, 2, 3, 4, ..$$

at high temperature, show that $U_{\text{rot}} = NkT$.

15.17 Problems

P15.1 Perform an analysis like that in Section 15.8 for a system with $N = 100$ and $U = 20a$. Calculate C and β in Equation (15.80).

P15.2 (adapted from Physical Chemistry, W. Moore) Consider the molecule N_2 ($\theta_{\text{vib}} = 3397$ K). A spectroscopic measurement of the population of molecules in excited vibrational states is

n	0	1	2	3
$\mathcal{P}_n$	0.75	0.19	0.048	0.012

where $\mathcal{P}_n$ is the fraction of molecules in energy level n. Describe how you would measure these populations spectroscopically; i.e., what spectroscopic method would you use and why (refer to Chapter 9)? Show that the gas is in thermodynamic equilibrium with respect to the distribution of vibrational energy. What is the temperature of the gas?

P15.3 Use a computer to directly evaluate the rotational partition function Z_{rot} for HD and I_2 at 100 K, 300 K, and 1000 K. Compare your value to that obtained by considering the classical limit, for which $Z_{\text{rot}} = kT/B$.

P15.4 Figure 11.16 shows an image of a high-resolution infrared spectrum for HCl. Calculate the populations of HCl molecules in the different rotational levels of the ground vibrational state and use them to explain the intensity profile in the spectrum.

P15.5 Given that

$$U = NkT^2 \frac{\mathrm{d} \ln Z}{\mathrm{d}T}$$

Derive an expression for the internal energy U_{rot} of a nonlinear polyatomic molecule in the high temperature limit, using the rotational energy relation for a spherical top molecule

$$E_{\text{rot}} = B \cdot J(J+1),\ J = 0, 1, 2, ..$$
$$g_J = (2J+1)^2$$

and for a symmetric top molecule:

$$E_{\text{rot}} = B \cdot J(J+1) + (A - B) \cdot K^2$$
$$J = 0, 1, 2, \ldots;\ K = 0, \pm 1, \pm 2, \ldots, \pm J$$

Compare your result with the equipartition principle. (Note: In your analysis, you can approximate the nuclei as having no spin. In the high temperature limit spin symmetry considerations will affect the partition function and the entropy, but not the energy. See Sections 15.5 and 17.8 of the text)

P15.6 Consider an atomic gas, which is constrained to move in two dimensions and deduce an expression for the normalized Maxwell–Boltzmann distribution of speeds. You should find that

$$\boxed{\mathrm{d}N = N \cdot \frac{m}{kT} \cdot v \cdot \mathrm{e}^{-mv^2/(2kT)} \cdot \mathrm{d}v}$$

Use the two-dimensional form of the Maxwell–Boltzmann distribution to show that the average speed of a gas atom is given by

$$\overline{|v|} = \sqrt{\frac{\pi kT}{2m}}$$

that the root mean square speed is given by

$$\sqrt{\overline{v^2}} = \sqrt{2kT/m}$$

and the most probable speed is given by

$$v_p = \sqrt{kT/m}$$

In addition, show that the average energy of a gas atom is given by $\overline{E} = kT$ and that the internal energy of a mole of that gas is given by $U_{\text{trans},m} = \boldsymbol{R}T$. (Hint: For the normalization use polar coordinates $\mathrm{d}v_x\, \mathrm{d}v_y = v\, \mathrm{d}v\, \mathrm{d}\varphi$).

P15.7 Use a spreadsheet program (e.g., Excel...) or MathCAD to plot the Maxwell–Boltzmann distribution for N_2 gas. Perform the plot for temperatures of 77 K, 300 K and 1000 K. On the 300 K graph mark $\overline{v}$, v_{rms}, and v_{max}.

P15.8 Plot the Maxwell–Boltzmann distribution of speeds for He, Ne, and Ar at a temperature of 300 K on the same graph. Perform a corresponding plot at a temperature of 1200 K. Discuss your findings.

P15.9 Find an expression for the number of available translational quantum states for a one-dimensional model gas. Compare your result with that we found for the corresponding three-dimensional gas (see Example 15.3).

P15.10 Compare the vibrational energy levels for I_2 in the table [3] to those obtained from the harmonic oscillator approximation, with a characteristic frequency of $\nu_0 = 6.43 \times 10^{12} s^{-1}$, by plotting them on a graph. Calculate the vibrational partition function and the average vibrational energy of an iodine molecule at $T = 300$ K by using both the harmonic oscillator approximation and the values in the table. Comment on the differences between the two calculations.

n	$\tilde{\nu}$ (cm^{-1})	n	$\tilde{\nu}$ (cm^{-1})	n	$\tilde{\nu}$ (cm^{-1})	n	$\tilde{\nu}$ (cm^{-1})
0	107.120	6	1368.379	11	2384.577	16	3367.822
1	320.435	7	1574.194	12	2583.905	17	3560.397
2	532.515	8	1778.731	13	2781.904	18	3751.586
3	743.356	9	1981.980	14	2978.564	19	3941.377
4	952.952	10	2183.932	15	3173.874	20	4129.756
5	1161.296					21	4316.709

P15.11 Using a spreadsheet program (such as Excel) evaluate and plot the molar vibrational internal energy for CO_2 as a function of temperature from 10 K to 2000 K (see Equations (15.29) and (15.30) and Chapter 11). What is the molar vibrational internal energy at 300 K? at 2000 K? Compare these values to that expected in the high temperature, classical, limit.

P15.12 In Chapter 14 (Box 14.2) we calculated the pressure of a gas and the number of collisions with a wall by including the different directions of motion of the gas molecules, but we assumed that the molecules are moving with equal speeds. Following the derivation in Box 14.2 and using the Maxwell–Boltzmann distribution (Equation (15.64)) show that in the expression for the pressure the square of the speed v^2 has to be replaced by its mean value $\overline{v^2}$ and in the expression for the number of collisions the speed v has to be replaced by its mean value $\overline{v}$.

P15.13 Figure 15.4 plots the temperature dependence of the electronic internal energy of NO, using the two-state approximation. Actually the ground and first excited state levels are each doubly degenerate, perform the analysis for NO using this fact and show that the formulas for N_1, N_2, and U_{el} are the same as those given in the text. Calculate and plot the populations for the ground state N_1 and the excited state N_2 over the temperature range of 0 K to 1000 K for NO; i.e., a plot analogous to that shown for Na in Fig. 15.3. Perform a similar calculation for a system which is non-degenerate in the ground state, but three-fold degenerate in the excited state.

P15.14 Show that the expression for ω in Equation (15.79) (written in terms of the quantum states i) can be written for the rotational motion in terms of the quantum number J, namely

$$\omega = N! \cdot \frac{g_1^{n_1} \cdot g_2^{n_2} \cdot g_3^{n_3} \cdots}{N_1! \cdot N_2! \cdot N_3! \cdots}$$

where N is the total number of particles, g_J is the degeneracy of energy level with quantum number J, and N_J is the number of particles with quantum number J.

P15.15 Equation (15.112) is written in terms of frequencies. Rewrite the equation in terms of wavelengths.

P15.16 Generalize the one-dimensional wave equation (Equation 3.6) to three dimensions, in analogy to the Schrödinger Equation (3.31). Using this equation, derive the equation for the frequency ν of a three-dimensional standing wave in a cubic box with side length L,

$$\nu = \frac{v}{2L}\sqrt{n_x^2 + n_y^2 + n_z^2}$$

P15.17 Use Equation (11.9) in Section 11.2.2 for the rotational energy levels of symmetric top polyatomic molecules.

$$E_{J,K} = B \cdot J(J+1) + (A - B) \cdot K^2$$

where $|K| \leq J$, $K = 0, \pm1, \pm2, .. \pm J$; and A and B are the rotational constants

$$A = \frac{h^2}{8\pi^2 I_A}, \quad B = \frac{h^2}{8\pi^2 I_B}$$

to derive a formula for the rotational partition function Z_{rot} in the high temperature limit and show that the internal energy $U_{rot} = N\frac{3}{2}kT$.

P15.18 Calculate the occupation probability for He gas in the lowest translational quantum state of a 1 L cubic box at $T = 300$ K at $P = 1$ bar.

References

1 R.C. Miller and P. Kusch, Velocity distribution in potassium and thallium atomic beams, *Phys. Rev.*, 1955, **99**, 1314.

2 J.B. Anderson and J.B. Fenn, Velocity distributions in molecular beams from nozzle sources, *Phys. Fluids*, 1965, **8**, 780.

3 Values taken from R.J. LeRoy, Molecular constants and internuclear potential of ground-state molecular iodine, *J. Chem. Phys.*, 1970, **52**, 2683.

16

Work w, Heat q, and Internal Energy U

CHAPTER MENU

16.1 Introduction

In Chapters 14 and 15 we discussed the thermal excitation of molecular assemblies and the distribution of energy over the quantum states of the molecules forming a gas. Now we relate this molecular description to concepts and quantities describing thermal properties of macroscopic systems, traditionally called thermodynamics. We define a thermodynamic system as that part of the universe which is of interest and which is separated from its surroundings, the remainder of the universe. The walls belonging to the surroundings constrain what external forces we can apply on the system. An equilibrium state of such a system is a state with well-defined macroscopic properties, such as composition, volume, temperature, and internal energy.

In Chapter 15 we have considered the internal energy, U, of an assembly. In the following U is defined as the internal energy of a thermodynamic system, q is defined as heat transfer to a thermodynamic system, and w is defined as work done on the system. Of particular interest is the internal energy of the system in thermal equilibrium (that is, after the system has reached a uniform temperature). If we apply a stress (e.g., a temperature or pressure jump) and wait for thermal equilibration, the change in U is determined solely by the difference between its initial value (before the stress is applied) and its final value (after the stress is applied). From the macroscopic perspective ΔU is the sum of the work done on the system (w) and the heat transferred from the surroundings into the system (q):

$$\Delta U = q + w$$

In contrast to the internal energy, the heat q and the work w are not determined solely by the thermodynamic state. The heat transfer to the system and the work done on the system are energy fluxes and depend on details of how the system changes between thermodynamic states rather than on the ending states themselves (e.g., a body can be heated by contacting it with a hot body or by doing work on it).

Principles of Physical Chemistry, Third Edition. Hans Kuhn, David H. Waldeck, and Horst-Dieter Försterling.

Companion website: www.wiley.com/go/Kuhn/physicalchemistry3e

16.2 Thermodynamic Systems

16.2.1 Defining System and Surroundings

In the macroscopic description of the behavior of molecular assemblies we must clearly distinguish between the designations "system" and "surroundings." A system is a collection of matter that is specified by the walls, or boundaries, that separate it from the rest of the universe (e.g., the molecules enclosed in the container in Fig. 16.1). The part of the universe outside of the system including the walls is called *surroundings* (e.g., the temperature bath, the piston, and the weight on the piston in Fig. 16.1b).

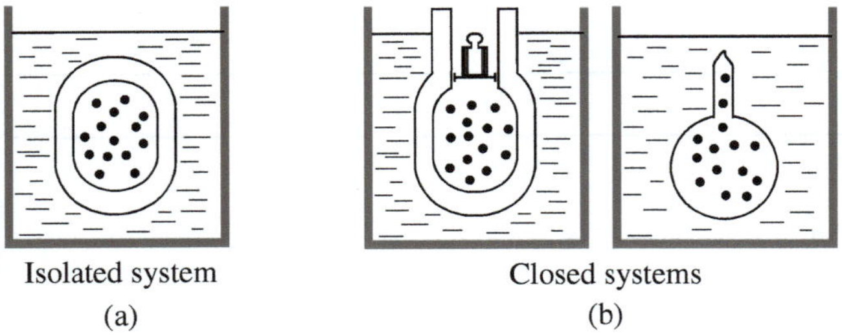

Figure 16.1 Isolated and closed systems. (a) Isolated system: heat cannot be transferred (heat isolation is indicated by double wall), matter cannot be transferred (closed container), and work cannot be done on the system. (b) Closed systems: No exchange of matter (closed container), but work can be done on the system (left) or heat can be transferred (right).

Not only do the *walls* specify what belongs to the system and to the surroundings, but they determine the *constraints* on the system. For example, a system may be enclosed by heat isolating walls or the walls may allow heat flow. In addition to walls that control the heat flow, we can design the walls to control work. *Rigid* walls do not allow the system to expand so that the pressure–volume work is zero, whereas *moveable* walls allow such work.

Commonly, we classify thermodynamic systems into one of three types: isolated systems, closed systems, and open systems. An *isolated system* does not exchange either energy (heat and work) or matter with its surroundings (Fig. 16.1a). Because any change of state in an isolated system occurs without exchange of heat q and work w with the surroundings ($q = 0, w = 0$), we can say

$$\Delta U = 0 \quad \text{(isolated system)} \tag{16.1}$$

The properties of isolated systems are useful to understand because any system together with devices for the storage of energy (e.g., temperature baths, springs, weights) can be considered as a new system, which is an isolated system (Fig. 16.2). A *closed system* exchanges energy with the surroundings (exchange of work or heat, see Fig. 16.1b), but does not exchange matter. In this case, the mass of the system is fixed. We will consider many examples of closed systems. An *open system* exchanges both energy and matter with its surroundings. This type of system is very important for understanding chemical processes in biology and engineering.

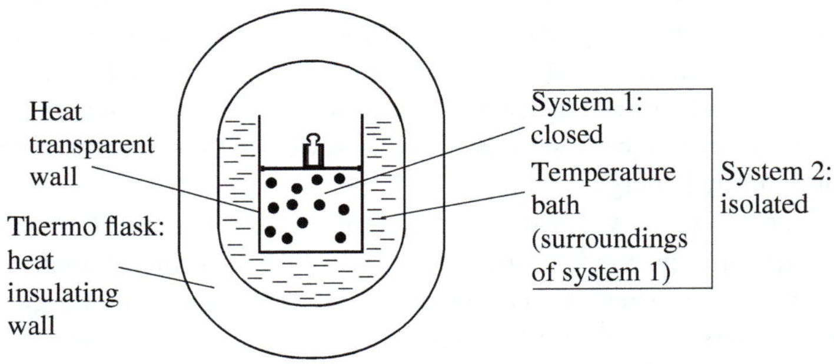

Figure 16.2 System 2, composed of system 1 and surroundings, constitutes an isolated system.

16.2.2 Defining Thermodynamic States

Often we will discuss a thermodynamic state, or simply state, of a thermodynamic system. Such states do not necessarily correspond to a single quantum mechanical state of the system; typically many different quantum states (or representations) can be compatible with a single thermodynamic state (see Chapter 15 for the internal energy U).

Thermodynamic states are characterized by their internal energy U, their volume V, and the amounts of substance $\boldsymbol{n}_1, \boldsymbol{n}_2 \ldots$ of their chemical components. For example, the system displayed in Fig. 16.3 can be described by the following information: $\boldsymbol{n}_{O_2} = 1$ mol, $T = 273.15$ K, $P = 1$ bar, and $V = 22.71$ L. We can also say that the system is in a certain state, where the state is defined by the thermodynamic variables (composition, volume, internal energy). This set of variables define the thermodynamic state of the system, they are independent of how the state has been reached, and they are called *state variables*.

The internal energy U is the energy of the molecular assembly arising from the energies of its constituent molecules (electronic, vibrational, rotational, and translational energies) and their mutual interaction energies. $U(T, V, \boldsymbol{n}_i, \ldots)$ is called a *state function*. Because its value depends only on the state variables, the change in its value depends only on the difference between the initial state U_{initial} and the final state U_{final}.

A complete description of the state of the O_2 gas in Fig. 16.3 is possible by specifying the amount of substance and two other variables — for example, by

$\boldsymbol{n}$, T, and P or $\boldsymbol{n}$, T, and V or

$\boldsymbol{n}$, V, and U

All other state variables are then uniquely defined by an equation of state—for example, the ideal gas equation ($PV = \boldsymbol{nR}T$). Note that the state variables exactly specify the state of a pure substance.

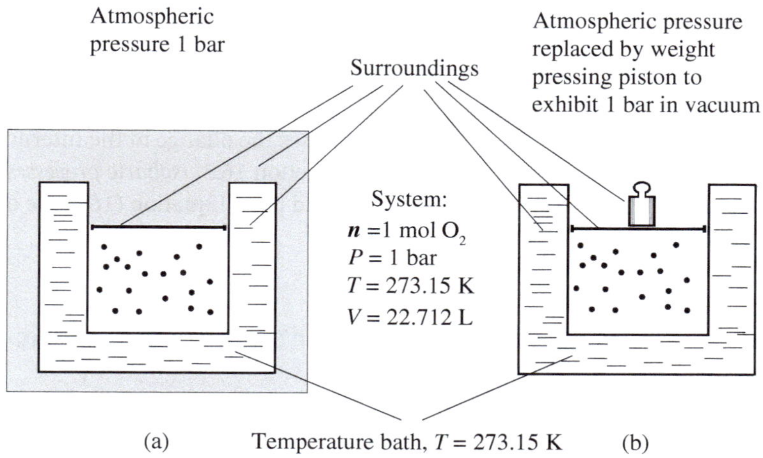

Figure 16.3 A system and its surroundings. Description of a state by composition, pressure, temperature, and volume. (a) Pressure (1 bar) exerted by the atmosphere (gray); (b) pressure (1 bar) exerted by a weight.

Whereas the change of a state variable is defined by the difference between the initial state's value (1) and the final state's value (2), for example,

$$\Delta T = T_2 - T_1, \quad \Delta P = P_2 - P_1,$$
$$\Delta U = U_2 - U_1$$

other quantities of interest depend upon how the change of state is performed. Importantly, the heat q or the work w supplied to a system during a change of state depend on how the change of state is performed (see Box 16.1). As a consequence, we must be careful to control how a process occurs if we wish to know the heat flow.

16.2.3 Change of State

Let us consider a closed system. The work w done on the system and the heat q flowing into the system in a distinct change of state increases the internal energy U of the system, so that

$$\Delta U = q + w \tag{16.2}$$

Since work and heat are forms of energy, Equation (16.2) is a special form of the law of Conservation of Energy.

The property of U as a state function is particularly useful for deducing energy changes of chemical systems under extraordinary conditions, by analyzing cyclic processes. In a *cyclic process* the system is in the same state at the end as in the beginning. Because U is a state function, for a cyclic process

$$\Delta U = 0 \quad \text{(cyclic process)} \tag{16.3}$$

Because of Equations (16.2) and (16.3) we obtain

$$w = -q \quad \text{(cyclic process)} \tag{16.4}$$

Note that this relation is general, not restricted to processes with ideal gases. The work done on the system (w) equals the heat flowing from the system to the surroundings ($-q$) if the system goes through a cyclic process. Chemical cycles are discussed more fully in Sections 16.5.1 and 16.5.7.

Processes in which one, or more, state variables are held constant can greatly simplify the analysis of molecular assemblies. Of particular utility are

(1) *Isothermal* processes: processes at constant temperature (gas enclosed in *diathermal*, i.e., heat conducting walls)
(2) *Isochoric* processes: processes at constant volume
(3) *Isobaric* processes: processes at constant pressure
(4) *Adiabatic* processes: processes with no heat flow (gas enclosed in *adiabatic*, i.e., heat insulating walls)

Examples for *isothermal processes* are shown in Box 16.1, Fig. 16B.1: the system is in contact with a temperature bath. In process (a) heat q is supplied to the system, and the system does work (w is negative). In process (b) heat q is supplied to the system, no work is done. In both processes the change of the internal energy, ΔU, is the same.

Isochoric processes will be discussed in Section 16.3, *isobaric processes* will be discussed in Section 16.4. In an *adiabatic process* no heat is supplied to the system, and from Equation (16.2) we obtain $\Delta U = w$ (adiabatic process). Such processes will be considered in Section 17.2.2.

16.3 Change of State at Constant Volume (Isochoric Process)

Consider a system which is composed of gas molecules that do not interact with each other. To begin, consider a gas consisting of N hydrogen molecules in a closed container that has rigid walls (constant volume) and exchanges heat with a bath of temperature T (Fig. 16.4a). In the approximation that we can ignore interactions between the gas molecules, the internal energy of the gas is the sum of the electronic, translational, rotational, and vibrational energies of the individual H_2 molecules (Fig. 16.4b):

$$U = U_{el} + U_{trans} + U_{rot} + U_{vib} \tag{16.5}$$

In Chapter 15, we saw how these different contributions to the internal energy can be determined and combined.

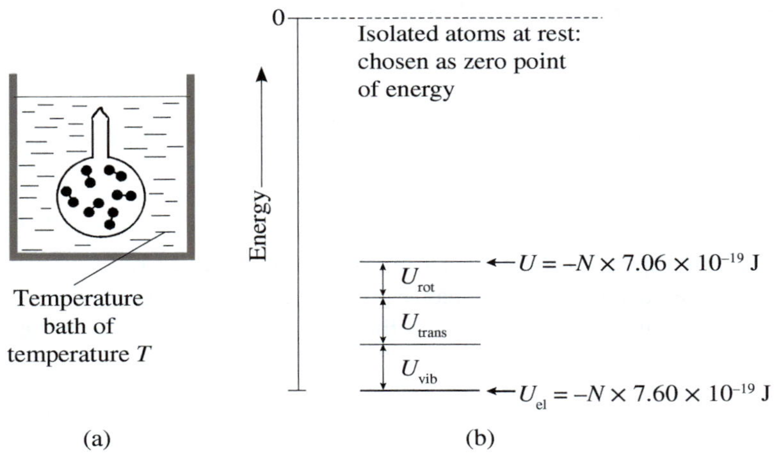

Figure 16.4 (a) Hydrogen gas in a closed container at temperature T. (b) Contributions to the internal energy U of N molecules of H_2 at $T = 300$ K. The electronic energy is strongly negative, if we choose isolated H atoms at rest as the zero point of energy. The contributions of vibration ($U_{vib} = N \cdot 4.37 \times 10^{-20}$ J, zero point energy), rotation ($U_{rot} = N \cdot 4.14 \times 10^{-21}$ J), and translation ($U_{rot} = N \cdot 6.21 \times 10^{-21}$ J) are slightly positive; thus, the internal energy $U = -N \cdot 7.06 \times 10^{-19}$ J is still strongly negative.

16.3.1 Change of Internal Energy ΔU and the Heat q

What happens if we change the temperature of the gas in Fig. 16.4a? In this case the gas is transformed from its original thermodynamic state into a different thermodynamic state. Such processes are called *changes of thermodynamic state*. In this section we consider processes at constant volume (*isochoric processes*). Because the volume of the gas does not change, it does not perform any work and $\Delta U = q$. If we warm the gas but do not change its molecular structure (i.e., no chemical reactions occur) and the volume of the container is fixed, then we change the average translational energy of the molecules and the populations of the rotational and vibrational quantum states, but we do not change the energy levels themselves.

16.3.1.1 Atomic Gas

To simplify the discussion, we consider an atomic gas, such as He, for which $U_{rot} = U_{vib} = 0$ (remember that atoms cannot store rotational or vibrational energy). As we increase the temperature of the gas from T_1 to T_2 at constant volume, the gas atoms speed up and their translational energy increases (Fig. 16.5). Following the discussion in Section 15.6.1 their internal energy

$$U = U_{trans} = \frac{3}{2}NkT \tag{16.6}$$

changes by

$$\Delta U = \Delta U_{trans} = U_{trans,2} - U_{trans,1}$$
$$= \frac{3}{2}Nk \cdot (T_2 - T_1) \tag{16.7}$$

where N is the number of molecules. In this process, heat q is taken from the bath at temperature T_2 and supplied to the gas.

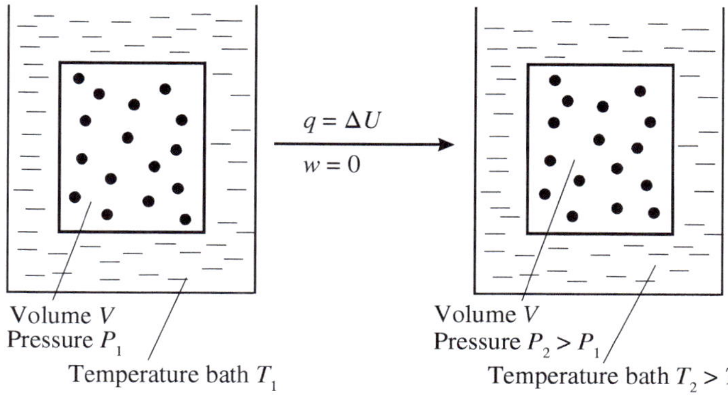

Figure 16.5 Heating a gas from temperature T_1 to T_2 at constant volume.

16.3.1.2 Diatomic Molecules at Moderate Temperature

When we analyze a molecular gas instead of an atomic gas, we must include the contributions of rotational and vibrational motions to the internal energy. As with the atomic gas we assume that the temperature is low enough that only the ground electronic state is populated, and we choose its energy as the zero. As discussed in Section 15.4 we expect that diatomic molecules remain in the lowest vibrational state at moderate temperatures; thus, $\Delta U_{vib} = \frac{1}{2}Nh\nu_0$. On the other hand, the rotational energy level spacings are assumed to be small compared to kT, and the rotational energy will be close to its value in the high temperature limit

$$U_{rot} = \frac{2}{2}NkT \tag{16.8}$$

Thus we obtain

$$U = U_{trans} + U_{rot} + U_{vib} = \frac{5}{2}NkT + \frac{1}{2}Nh\nu_0 \tag{16.9}$$

and its internal energy change is

$$\Delta U = U(T_2) - U(T_1) = \frac{5}{2}Nk \cdot (T_2 - T_1) \tag{16.10}$$

A larger amount of heat q must be transferred from the temperature bath for the molecular gas than for the atomic gas in order to achieve the same temperature rise.

16.3.1.3 Conservation of Energy

In all cases, the energy corresponding to the heat q is stored in the gas molecules, thus increasing its internal energy U. Because energy is conserved, we can write that

$$\boxed{\Delta U = q \quad \text{(heating at constant volume)}} \tag{16.11}$$

It is important to appreciate that heat is a flux of energy between the temperature bath (or temperature reservoir) and the gas. The heat is used to energize the gas molecules by changing the population distribution over the energy levels.

For the calculations discussed above, the electronic energy levels are assumed to be not excited; i.e., excited electronic states were neglected.

16.3.2 Heat Capacity C_V: Definition

It is useful to define a quantity called the *heat capacity*. It quantifies how much the system changes temperature as it is heated. This will be important in studying the energetics of chemical reactions later in this chapter.

If we transfer the rigid container with the gas from a bath of temperature T into a bath of temperature $T + \Delta T$, then heat $q_{T,T+\Delta T}$ flows from the bath into the container and increases the internal energy U of the gas by ΔU. In the limit $\Delta T \to 0$, we define the ratio $q_{T,T+\Delta T}/\Delta T$ as the *heat capacity C_V at constant volume*; i.e., the heat capacity C_V is defined by

$$C_V = \lim_{\Delta T \to 0} \left(\frac{q_{T,T+\Delta T}}{\Delta T} \right)_{V=\text{const}} \tag{16.12}$$

Because of Equation (16.11) we can write

$$C_V = \lim_{\Delta T \to 0} \left(\frac{\Delta U}{\Delta T} \right)_{V=\text{const}} = \left(\frac{\partial U}{\partial T} \right)_V \tag{16.13}$$

where the symbol ∂ denotes a partial differential (see Appendix C.1.3). C_V is an important measure for the ability of a substance to absorb heat.

Because U is an extensive quantity, C_V is also extensive. Often it is useful to define a *molar internal energy* and a *molar heat capacity at constant volume* by

$$U_m = \frac{U}{\boldsymbol{n}}, \quad C_{V,m} = \frac{C_V}{\boldsymbol{n}} \tag{16.14}$$

where $\boldsymbol{n}$ is the amount of substance.

16.3.2.1 Atomic Gas

We can use our results for the internal energy of the atomic and molecular gases to determine the heat capacities at constant volume. From Equations (16.6) and (16.13), the heat capacity C_V of the atomic gas He is

$$C_V(\text{He}) = \left(\frac{\partial U}{\partial T} \right)_V = \tfrac{3}{2} Nk \quad \text{(atomic gas)} \tag{16.15}$$

Because of

$$Nk = \boldsymbol{n}\boldsymbol{N}_A k = \boldsymbol{n}\boldsymbol{R} \tag{16.16}$$

the molar heat capacity for He is

$$C_{V,m}(\text{He}) = \tfrac{3}{2}\boldsymbol{R} \tag{16.17}$$

16.3.2.2 Diatomic Molecules at Moderate Temperature

Correspondingly, for the molecular gas H_2 from Equation (16.9) we obtain in the high temperature limit of rotation, but at a temperature when vibration can still be neglected,

$$C_V(\text{H}_2) = \tfrac{5}{2} Nk \quad \text{(diatomic gas)} \tag{16.18}$$

and the molar heat capacity for H_2 is

$$C_{V,m}(\text{H}_2) = \tfrac{5}{2}\boldsymbol{R} \tag{16.19}$$

16.3.2.3 Generalization (Full Range of Temperature)

These considerations are valid for a range of temperature where the rotational contribution reaches the value expected from the equipartition principle and where the vibrational contribution can be neglected. In the following sections we discuss the full range of temperature, starting at $T = 0$ and increasing T until the vibrations become important. The experimental molar heat capacity $C_{V,m}$ of gases measured at different temperatures is shown in Fig. 16.6 for He (atomic gas), HD (diatomic gas),[1] and CO_2 (polyatomic gas). How can we understand this dependence of C_V on temperature?

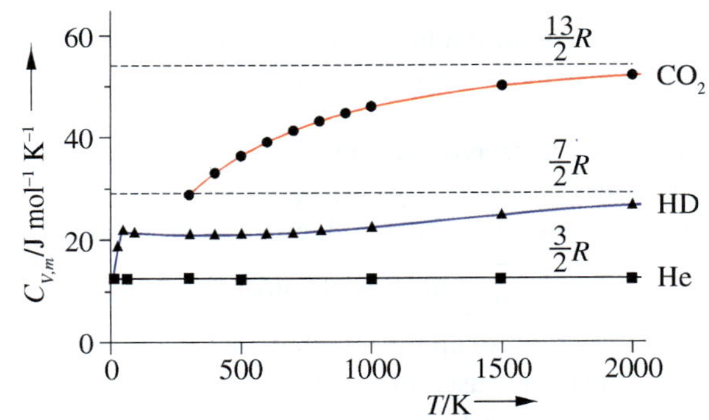

Figure 16.6 Experimental heat capacity $C_{V,m}$ for several gases as a function of temperature. Full circles, triangles, and squares: experiment. Solid lines: interpolated curves. Dashed lines: limiting values for translation ($\frac{3}{2}\boldsymbol{R}$) and translation + rotation + vibration ($\frac{7}{2}\boldsymbol{R}$ for HD, $\frac{13}{2}\boldsymbol{R}$ for CO_2).

16.3.3 Translational Contribution to C_V of a Gas

The experimental heat capacity for He is $C_{V,m} = \frac{3}{2}\boldsymbol{R}$. From Equation (16.17) and the preceding discussion, we see that it arises from the translational motion of the He atoms. Since every type of gas will have translational kinetic energy available to the particles, we expect that the translational contribution to the heat capacity will be

$$C_{V,\text{trans},m} = \tfrac{3}{2}\boldsymbol{R} = 12.47 \text{ J K}^{-1} \text{ mol}^{-1} \tag{16.20}$$

1 Here we use HD instead of H_2 as an example because in homonuclear diatomic molecules a particular quantum effect leads to a more complicated behavior in the low temperature range (see Section 12.2). H_2 will be discussed in Section 16.3.5.

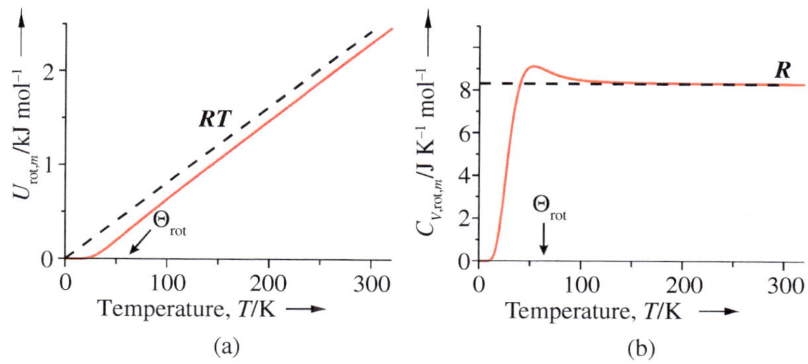

Figure 16.7 Calculated rotational contribution to U and C_V for HD versus temperature (solid lines). (a) $U_{rot,m}$ taken from Fig. 15.13, (b) $C_{V,rot,m}$ obtained from $U_{rot,m}$ by numerical differentiation. Dashed lines: classical limit. $\theta_{rot} = 65.6$ K is the characteristic temperature for the rotational motion of HD.

When decreasing the temperature, $C_{V,m}$ of gases does not decrease below $\frac{3}{2}\boldsymbol{R}$; this holds even for the lowest temperature at which gases can be examined experimentally.

16.3.4 Rotational and Vibrational Contributions to C_V of a Gas

16.3.4.1 Diatomic Molecule: HD

The experimental value $C_{V,m} = \frac{5}{2}\boldsymbol{R}$ for HD in the temperature range from 200 to 600 K (Fig. 16.6) arises from the contributions of the molecules' translational and rotational degrees of freedom. At a low enough temperature, however, the contribution of the rotational motion becomes zero. This is in accordance with the theoretical $C_{V,rot,m}$ curve for HD in Fig. 16.7b, which is obtained by calculating the slope of the theoretical curve for $U_{rot,m}$ (Fig. 16.7a). It can be seen from Fig. 16.7b that the contribution of the rotation to $C_{V,m}$ is zero up to 10 K; after which it increases strongly with temperature. At high enough temperature (above 200 K, after passing a maximum at 50 K), we have

$$U_{rot,m} = \boldsymbol{R}T \quad \text{and} \quad C_{V,rot,m} = \boldsymbol{R} \tag{16.21}$$

Above 600 K, $C_{V,m}$ increases further (Fig. 16.6). This increase arises from the contribution of the vibrational motion. In Section 15.4.2 we calculated $U_{vib,m}$ as

$$U_{vib,m} = \boldsymbol{N}_A h\nu_0 \cdot \left(\frac{1}{2} + \frac{1}{e^{h\nu_0/(kT)} - 1}\right) \tag{16.22}$$

where ν_0 is the vibrational frequency. Combining this result with Equation (16.13) we obtain

$$C_{V,vib,m} = \left(\frac{\partial U_{vib,m}}{\partial T}\right)_V = \boldsymbol{N}_A h\nu_0 \frac{h\nu_0 \cdot e^{h\nu_0/(kT)}}{kT^2\left(e^{h\nu_0/(kT)} - 1\right)^2} \tag{16.23}$$

This equation is plotted in Fig. 16.8b together with the temperature dependence of $U_{vib,m}$ (Fig. 16.8a), and compared with the experimental values (see Table 16.1). At high enough temperature we expect

$$U_{vib,m} = \boldsymbol{R}T \quad \text{and} \quad C_{V,vib,m} = \boldsymbol{R} \tag{16.24}$$

Because of the large vibrational frequency of HD, the classical limit

$$C_{V,m} = \frac{3}{2}\boldsymbol{R} + \boldsymbol{R} + \boldsymbol{R} = \frac{7}{2}\boldsymbol{R} = 29.10 \text{ J K}^{-1} \text{ mol}^{-1}$$

is not reached until the temperature is well above 2000 K (see Section 15.4.2).

Example 16.1 *Consider 1 mol of the diatomic molecule iodine in the ground electronic state, I_2, at $T = 500$ K and $P = 1$ bar. Assume it behaves as an ideal gas and determine its molar heat capacity at constant volume, $C_{V,m}$.*

The translational motion contributes $\frac{3}{2}\boldsymbol{R}$ to the molar heat capacity. Because the rotational energy level spacings of I_2 are quite small (a result of iodine's large moment of inertia), the rotational motion contributes $\boldsymbol{R}$ to the molar heat capacity.[2] Then with $C_{V,vib,m}$ from Equation (16.23) we obtain

$$C_{V,m} = \frac{5}{2}\boldsymbol{R} + \boldsymbol{N}_A h\nu_0 \cdot \frac{h\nu_0}{kT^2} \cdot \frac{e^{h\nu_0/(kT)}}{\left(e^{h\nu_0/(kT)} - 1\right)^2} \tag{16.25}$$

---

2 As outlined in Section 15.5.3 this statement is valid even for homonuclear diatomic molecules like I_2 at high enough temperature.

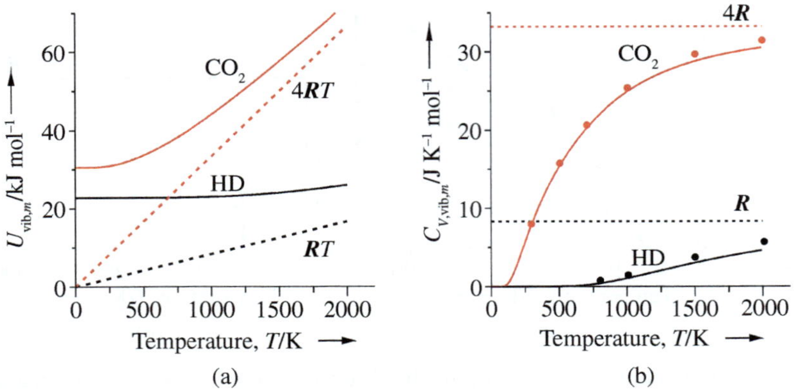

Figure 16.8 Calculated vibrational contribution to $U_{vib,m}$ and $C_{V,vib,m}$ for HD and CO_2 versus temperature (solid lines). For data see Table 16.1. (a) $U_{vib,m}$ calculated from Equation (16.22) and (b) $C_{V,vib,m}$ calculated from Equation (16.23). Dashed lines: classical limits. Dots: experimental data (experimental $C_{V,m}$ minus the contributions of translation and rotation).

Using $h\nu_0 = 4.27 \times 10^{-21}$ J for I_2, we find for $T = 500$ K

$$C_{V,m} = [20.79 + 8.05]\ \text{J K}^{-1}\ \text{mol}^{-1} = 28.84\ \text{J K}^{-1}\ \text{mol}^{-1}$$

This value is in excellent agreement with the experimental value of 29.2 J K^{-1} mol^{-1}. For iodine, which has a relatively low vibrational frequency ($1/\lambda = 214$ cm^{-1}), the vibrational motion contributes significantly to the heat capacity at constant volume.[3]

16.3.4.2 Polyatomic Molecules: CO_2

A gas consisting of polyatomic molecules, CO_2, has an even larger molar heat capacity (see Fig. 16.6). Because CO_2 has a larger moment of inertia than HD, we expect that the classical limit of $\boldsymbol{R}$ for the rotational heat capacity is reached at moderately low temperatures, below 20 K. Because of the four different vibrational modes of CO_2 (see Chapter 11), the classical limit for their contribution to the total heat capacity is $C_{V,vib,m} = 4 \cdot \boldsymbol{R}$; see Box 15.1. Hence, the molar heat capacity would be[4]

$$C_{V,m} = \tfrac{3}{2}\boldsymbol{R} + \boldsymbol{R} + 4\boldsymbol{R} = \tfrac{13}{2}\boldsymbol{R} = 54.04\ \text{J K}^{-1}\ \text{mol}^{-1}$$

if all of the vibrational modes contribute fully. Clearly this limit is not reached until after 2000 K.

The calculated $C_{V,vib,m}$ curves for HD and CO_2 (Table 16.1) are compared with experiment in Fig. 16.8b and found to be in excellent agreement. We find that the lower frequency vibrational modes start to contribute to the heat capacity at lower

Table 16.1 Contributions of $C_{V,vib,m}$ (in J K^{-1} mol^{-1}) for HD and CO_2.

		CO_2			
T/K	**HD**	**Bending**	**Symmetric**	**Antisymmetric**	**Total**
200	0.0	3.2	0.0	0.0	3.2
300	0.0	7.5	0.5	0.0	8.0
400	0.0	10.5	1.4	0.1	12.0
600	0.1	13.5	3.6	1.0	18.1
800	0.4	14.8	5.1	2.2	22.1
1000	1.1	15.4	6.0	3.4	24.8
1500	3.0	16.1	7.2	5.5	28.8
2000	4.6	16.3	7.7	6.6	30.6

According to Equation (16.23). HD: $\nu_0 = 1.14 \times 10^{14}$ s^{-1}. CO_2: $\nu_{bending} = 0.20 \times 10^{14}$ s^{-1}; $\nu_{symmetric} = 0.42 \times 10^{14}$ s^{-1}; $\nu_{antisymmetric} = 0.71 \times 10^{14}$ s^{-1}. Because the Bending Mode Is Twofold Degenerate, It Contributes Twice.

3 The experimental value of $C_{V,m}$ is larger than the calculated value, by 0.4 J K^{-1} mol^{-1}, because of anharmonicity in the molecule's vibrational motion, which is not accounted for in this calculation.

4 Regarding its rotational motion CO_2 must be treated in the same way as a homonuclear diatomic molecule. However, at high enough temperature, the relation $U_{rot,m} = \boldsymbol{RT}$ holds.

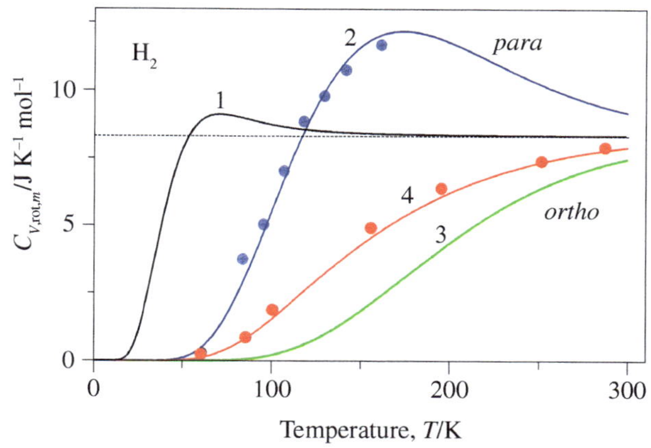

Figure 16.9 Rotational contribution to $C_{V,m}$ for H_2 versus temperature. Solid lines: calculation. Curve 1: all rotational levels accessible (as for HD). Curve 2: only even rotational levels accessible ($n = 0, 2, \ldots$, *para*-H_2). Curve 3: only odd rotational levels accessible ($n = 1, 3, \ldots$, *ortho*-H_2). Curve 4: natural mixture of *ortho*- and *para*-H_2 (3:1). Dots: experimental data for curves 2 and 4; pure *ortho*-H_2 cannot be isolated; thus, experimental data for curve 3 are not available.

temperatures than do the higher frequency modes. A closer inspection of the experimental and theoretical data reveals that the experimental values are slightly larger than the theoretical ones at high temperatures (by 5% for HD and by 1% for CO_2 at 2000 K). This discrepancy occurs because the harmonic oscillator model used in the calculation is not strictly valid for highly excited vibrational states, see Section 11.3.4.

16.3.5 *Ortho*- and *Para*-H_2: Fascinating Quantum Effects on C_V

In Fig. 16.9 the heat capacity $C_{V,\text{rot}}$ for H_2 is displayed. The theoretical curve calculated in the corresponding way as for HD (curve 1) is in strong disagreement with the experiment (dots in curve 4). How can we understand this discrepancy?

As outlined in Chapter 15, H_2 must be considered as a 3:1 mixture of *ortho*- and *para*-H_2. In Section 15.5.3 we have calculated the internal energies U_rot of rotation for *ortho*- and *para*-H_2 as a function of temperature. When warming up the gases from $T = 0$, *para*-H_2 is more strongly excited at lower temperatures than is *ortho*-H_2. By calculating the slopes of the curves for U_rot in Fig. 15.11 we obtain the corresponding curves for $C_{V,\text{rot,ortho}}$ (Fig. 16.9, curve 3) and $C_{V,\text{rot,para}}$ (Fig. 16.9, curve 2). At higher temperature, $C_{V,\text{rot,para}}$ exhibits a maximum at $T = 168$ K, whereas $C_{V,\text{rot,ortho}}$ approaches the classical value Nk monotonically. The calculated $C_{V,\text{rot}}$ curve of H_2 (a 3:1 mixture of *ortho*-H_2 and *para*-H_2) is in excellent agreement with the experiment (Fig. 16.9, curve 4). This is a fascinating macroscopic manifestation of a most astonishing quantum feature discussed in Section 15.5.3, the indistinguishability of protons.

16.3.6 Electronic Contribution to C_V ($C_{V,\text{el}}$)

For most molecules at moderate temperatures, electronic excitation does not contribute significantly to the heat capacity. In some cases a molecule or atom may have important low-lying energy levels and then electronic excitation can become important.

In Chapter 15 we calculated the electronic energy U_el for a two-level model of the NO molecule as

$$U_{\text{el},m} = \boldsymbol{N}_\text{A} \cdot \frac{E_1\,\mathrm{e}^{-E_1/(kT)} + E_2\,\mathrm{e}^{-E_2/(kT)}}{\mathrm{e}^{-E_1/(kT)} + \mathrm{e}^{-E_2/(kT)}}$$

or with $\Delta E = E_2 - E_1$

$$U_{\text{el},m} = \boldsymbol{N}_\text{A} \cdot \frac{E_1 + E_2\mathrm{e}^{-\Delta E/(kT)}}{1 + \mathrm{e}^{-\Delta E/(kT)}} \tag{16.26}$$

We can obtain the contribution of this electronic degree of freedom to C_V by direct differentiation. In Exercise E16.22, you show that

$$C_{V,\text{el},m} = \left(\frac{\partial U_{\text{el},m}}{\partial T}\right)_V = \boldsymbol{N}_\text{A}\frac{\Delta E}{kT^2}\frac{\Delta E \cdot \mathrm{e}^{-\Delta E/(kT)}}{\left(1 + \mathrm{e}^{-\Delta E/(kT)}\right)^2} \tag{16.27}$$

Figure 16.10 compares the experimental determination of $C_{V,\text{el}}$ for NO to that calculated from Equation (16.27) with $\Delta E = 2.40 \times 10^{-21}$ J. The calculated curve for $C_{V,\text{el}}$ exhibits a maximum at $T = 72$ K and decreases to $C_{V,\text{el}} = 0$ as $T \to \infty$.

How can we understand this behavior? At low temperature $\Delta E \gg kT$, and it is very unlikely for the molecules to occupy energy level E_2 and $C_{V,\text{el}} = 0$ (similar to what was observed for rotational and vibrational excitations at low temperature). At intermediate temperatures where $\Delta E \approx kT$ the energy level E_2 becomes more and more populated; thus, the internal energy U increases with temperature and $C_{V,\text{el}}$ increases as well. At sufficiently high temperature, however, the energy levels E_1 and E_2 are equally populated, and the internal energy reaches the constant value $U_{\text{el}} = \frac{1}{2}N\left(E_1 + E_2\right)$. Because U_{el} is temperature independent in this regime, $C_{V,\text{el}}$ goes to zero.

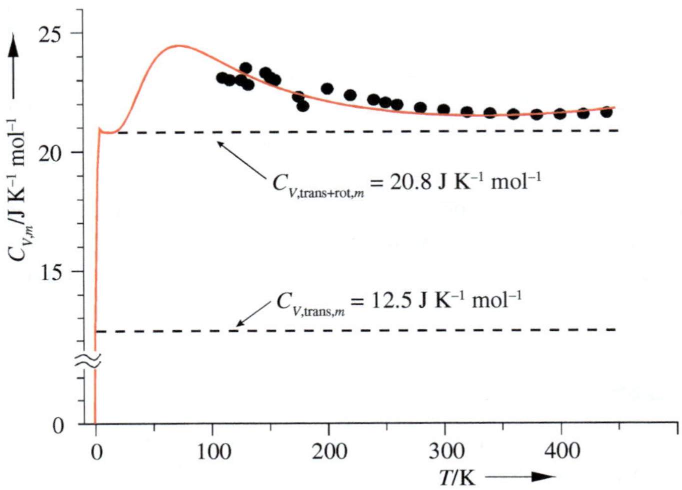

Figure 16.10 Molar heat capacity $C_{V,m}$ of NO versus temperature. Full circles: experiment (no experimental data are available below the boiling point of NO). Solid line: theory. The calculated maximum at $T = 72$ K is due to the electronic contribution to $C_{V,m}$ (excitation energy 2.4×10^{-21} J). The increase of $C_{V,m}$ above $T = 350$ K is due to the contribution of vibration (excitation energy 3.8×10^{-20} J). Dashed lines: classical limiting values for contributions of translation and of rotation to $C_{V,m}$.

16.3.7 Heat Capacity and Characteristic Temperature

The fact that C_V decreases as the temperature decreases results directly from the quantum nature of matter, and it may be understood by comparing the temperature to the characteristic temperatures, θ, of the different quantized modes of the molecules (see Section 15.7). For example, we found that the characteristic rotational temperature for HD is $\theta_{\text{rot}} = 65.6$ K, and its characteristic vibrational temperature is $\theta_{\text{vib}} = 5480$ K. Consequently, at temperatures below 50 K the energy is not effectively stored in rotational and vibrational degrees of freedom; only translational degrees of freedom can be excited.

For CO_2 we obtain $\theta_{\text{rot}} = 0.56$ K and $\theta_{\text{vib}} = 960$ K for the bending mode, 1890 K for the symmetric stretch, and 3360 K for the asymmetric stretch. Clearly, we do not expect that rotational quantum effects will be important for CO_2 unless the temperatures are extremely close to zero Kelvin. In contrast to HD, we expect that the excitation of bending vibrations could be an important contributor to the heat capacity at temperatures of a few hundred Kelvin, as seen in Fig. 16.6.

It is remarkable that the heat capacity of a macroscopic collection of molecules is determined by the wave-particle duality of atomic nuclei.

Example 16.2 *Like carbon dioxide (CO_2), sulfur dioxide (SO_2) is a triatomic molecule; however, it is not linear. In what ways do you expect its heat capacity to be similar to that of CO_2 and it what ways might it be different?*

The linear CO_2 molecule has two identical axes for rotational motion, which can be characterized by $\theta_{\text{rot}} = 0.56$ K, and four vibrational modes (see above). The nonlinear SO_2 molecule has three different rotational axes, which each have a different characteristic temperature $\theta_{\text{rot},1} = 0.42$ K, $\theta_{\text{rot},2} = 0.50$ K, and $\theta_{\text{rot},3} = 2.92$ K, and three vibrational modes with characteristic temperatures $\theta_{\text{vib}} = 750$ K for the bending mode, 1660 K for the symmetric stretch, and 1960 K for the asymmetric stretch. In the temperature range where the vibrational modes are not strongly excited, the heat capacities of these molecules will be different. For CO_2 we expect the rotational motion to contribute $\boldsymbol{R}$ to the molar heat capacity and for SO_2 the rotational motion will contribute $\frac{3}{2}\boldsymbol{R}$ ($\frac{1}{2}\boldsymbol{R}$ for motion about each rotational axis). Hence CO_2 will have a molar heat capacity of $\frac{5}{2}\boldsymbol{R}$ plus a small amount from partial excitation of the doubly degenerate bending mode, whereas the SO_2

will have a molar heat capacity of 3$\boldsymbol{R}$ with a small amount from partial excitation of its bending mode. As the temperature increases the importance of the vibrational motions to the heat capacity will increase and the molecules' heat capacities will become more similar.

16.3.8 C_V of Solids

Here we consider the heat capacity of solids that arises from the motion of the nuclei in space, and we ignore the contribution from electronic excitations. The atoms in a solid vibrate in space (in the x-, y-, and z-directions), but they do not translate or rotate. Hence the heat capacity of a solid can be modeled as that of a molecule with 3N vibrational modes.

The molar internal energy U_m of the solid, which arises from its 3N normal modes, is

$$U_m = \frac{1}{\boldsymbol{n}} \sum_{n=1}^{n=3N} \left(\frac{1}{2} h\nu_n + \frac{h\nu_n}{e^{h\nu_n/(kT)} - 1} \right) \tag{16.28}$$

This approximation reduces the problem of finding the heat capacity of a solid to that of finding its normal modes.

16.3.8.1 Einstein Model

In *Einstein's model of a solid* it is assumed that each of the solid's normal modes oscillates with the same frequency ν_0 (Fig. 16.11a). In this case, each term in the summand of Equation (16.28) is identical, and the molar internal energy U_m of the solid is

$$U_m = 3 \cdot \boldsymbol{N}_A h\nu_0 \cdot \left(\frac{1}{2} + \frac{1}{e^{h\nu_0/(kT)} - 1} \right) \tag{16.29}$$

Using Equation (16.13) gives

$$C_{V,m} = 3 \cdot \boldsymbol{R} \left(\frac{h\nu_0}{kT} \right)^2 \cdot \frac{e^{h\nu_0/(kT)}}{\left(e^{h\nu_0/(kT)} - 1\right)^2} \tag{16.30}$$

for the molar constant volume heat capacity $C_{V,m}$. The temperature dependence of the Einstein solid's heat capacity is the same curve as in Fig. 16.8 for the vibrational contribution for gases, but $C_{V,m}$ is three times larger than for a linear oscillator.

This model is in reasonable agreement with the experimental data. Figure 16.11b compares the model to experimental heat capacity data for lead, aluminum, and diamond. A limiting value is reached, which is shifted to higher temperatures with increasing vibrational frequency of the atoms, as in the cases of the gas molecules discussed above. At high enough temperature

$$\lim_{T \to \infty} C_{V,m} = 3\boldsymbol{R} \tag{16.31}$$

Because many metals have low vibration frequencies, the high temperature limit is achieved at moderate temperatures and we can understand the *rule of Dulong and Petit*, which states that the molar heat capacity $C_{V,m}$ of such metals at room temperature (e.g., lead in Fig. 16.11b) is about 3$\boldsymbol{R}$.

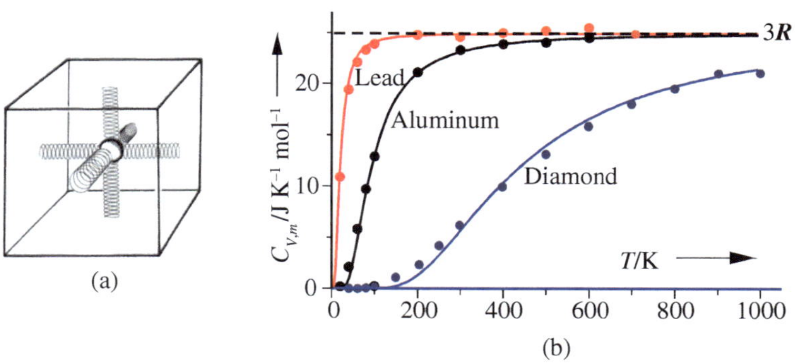

Figure 16.11 (a) Three-dimensional oscillator. (b) Heat capacity C_V of solids versus temperature. Filled circles: experiment. Solid lines: calculation according to Equation (16.30). The vibrational frequencies are adjusted to best fit the measured points: $\nu_0 = 1.33 \times 10^{12}$ s^{-1} (lead), $\nu_0 = 6.00 \times 10^{12}$ s^{-1} (aluminum), $\nu_0 = 28.3 \times 10^{12}$ s^{-1} (diamond). Dashed line: classical limit. It must be noted that for solids the C_V data cannot be measured directly, but they can be obtained from measurements under constant pressure, C_p data; see Section 16.4.2.

16.3.8.2 Debye Model

Although the Einstein model captures the essence of the temperature behavior (approaching zero at low T and 3$\boldsymbol{R}$ at high T), it is not quantitatively accurate (see Fig. 16.12 for the case of diamond). *Debye's model of a solid* constructs a more realistic

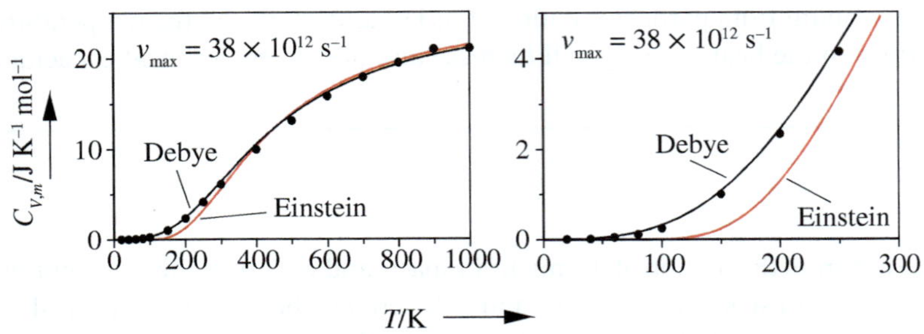

Figure 16.12 Heat capacity C_V of diamond versus temperature. Filled circles: experiment. Solid lines: calculation according to the Einstein model (red), Equation (16.30), vibrational frequency $\nu_0 = 28.3 \times 10^{12}\ s^{-1}$ and according to the Debye model (black), Equation (16.33), vibrational frequency $\nu_{max} = 38 \times 10^{12}\ s^{-1}$. Left: Temperature range 0–1000 K, Right: temperature range 0–300 K to show the T^3 law at low temperature.

description for the normal mode frequencies, ν_k, in Equation (16.28). The frequency spectrum of the normal modes is assumed to be quasi-continuous and to increase in number as ν^2 until a "cut-off" frequency ν_{max} (see Foundation 16.1). In this case

$$U_m = U_{0,m} + \frac{9h\boldsymbol{N}_A}{\nu_{max}^3} \cdot \int_0^{\nu_{max}} \frac{\nu^3}{e^{h\nu/(kT)} - 1}\, d\nu \tag{16.32}$$

where $U_{0,m} = \frac{9}{8}\boldsymbol{N}_A h\nu_{max}$ corresponds to the molar zero point energy of the oscillators. The molar heat capacity $C_{V,m}$ is obtained by differentiation of Equation (16.32)

$$C_{V,m} = \frac{9h\boldsymbol{N}_A}{\nu_{max}^3} \cdot \int_0^{\nu_{max}} \frac{h\nu}{kT^2} \frac{\nu^3\, e^{h\nu/(kT)}}{\left(e^{h\nu/(kT)} - 1\right)^2}\, d\nu \tag{16.33}$$

(*Debye function of heat capacity*). In Problem P16.8, you show that Equation (16.33) has the limits $C_{V,m} = 0$ for $T \to 0$ and $C_{V,m} = 3\boldsymbol{R}$ for $T \to \infty$, in accordance with the Einstein model. Also it is shown that for small T

$$C_{V,m} = const. \cdot T^3 \tag{16.34}$$

in contrast to the Einstein model which predicts an exponential increase. Debye's T^3 law is in excellent agreement with experimental data.[5]

The results of the Einstein and the Debye models are compared for diamond in Fig. 16.12. According to Equation (16.29) the molar zero point energy in the Einstein model is $\frac{3}{2}\boldsymbol{N}_A h\nu_0$, compared to $\frac{9}{8}\boldsymbol{N}_A h\nu_{max}$ in the Debye model; thus, we expect the relation $\nu_{max} = \frac{4}{3}\nu_0$. Using $\nu_0 = 28.3 \times 10^{12}\ s^{-1}$ we obtain $\nu_{max} = 37.7 \times 10^{12}\ s^{-1}$ which is the value used in calculating the Debye curve in Fig. 16.12. The frequency ν_0 in the Einstein model can be interpreted as a mean value of the frequencies ν appearing in the Debye model. Whereas in the Debye model enough low frequency oscillations are available which can be excited at low temperature (reproducing the T^3 dependence) much higher temperatures must be applied to excite the oscillators with the single frequency ν_0 in the Einstein model. At high enough temperature both models lead to the same prediction.

16.4 Change of State at Constant Pressure (Isobaric Process)

16.4.1 Change of Internal Energy ΔU; Heat q and Work w

Consider a change of state at constant pressure (*isobaric process*). Figure 16.13 illustrates the heating of a gas from T_1 to T_2 at constant pressure. Because the internal energy U of an ideal gas depends only on the temperature T and not on the volume (see Section 14.2.4.2), the change ΔU is the same as for heating the gas at constant volume. However, more heat is taken from the bath with temperature T_2 because the gas must lift the weight on the piston as it expands.

To lift the weight by a distance ds, the gas exerts a force f on the piston. The differential work is

$$f \cdot ds = \left(\frac{f}{\text{area of piston}}\right) \cdot \text{area of piston} \cdot ds = P \cdot dV \tag{16.35}$$

where dV is the volume change and P is the pressure of the gas. The total work done by the expanding gas under constant pressure is

$$\int_{V_1}^{V_2} P \cdot dV = P \int_{V_1}^{V_2} dV = P(V_2 - V_1) = P \cdot \Delta V \tag{16.36}$$

5 In metals, at low enough temperature, an additional contribution of the electron gas to C_V has to be taken into account.

where the pressure P can be removed from the integrand because it is constant. Alternatively, we may say that the work w done on the system is

$$w = -P\,\Delta V \qquad (16.37)$$

The internal energy change of the gas is the heat flow plus the work performed on the system

$$\boxed{\begin{array}{c}\Delta U = q + w = q - P\,\Delta V \\ \text{(heating at constant pressure)}\end{array}} \qquad (16.38)$$

and for the heat q we obtain

$$q = \Delta U + P\,\Delta V \quad \text{(heating at constant pressure)} \qquad (16.39)$$

Note that we attribute the positive sign to heat flowing to the system and to work done on the system. In the example of Fig. 16.13 q is positive and w is negative.

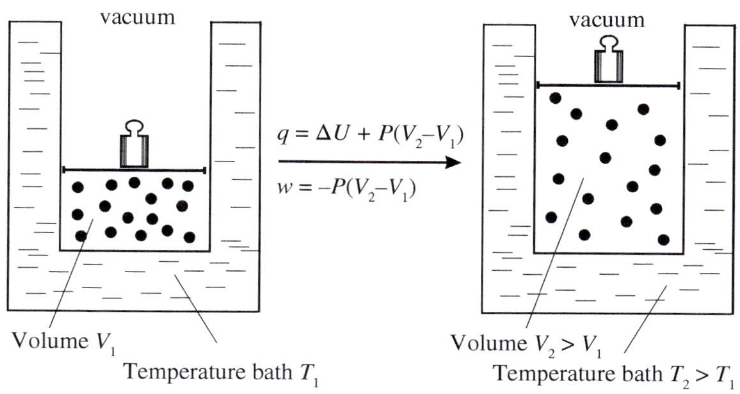

Figure 16.13 Heating a gas from T_1 to T_2 at constant pressure. External pressure P replaced by weight pressing piston to exhibit pressure P.

Example 16.3 *How much heat is needed to change the temperature of* 1 *mol of He gas from* 300 *to* 400 *K at* 1 *bar pressure? (Assume that the ideal gas equation of state holds for He).*

From the change in temperature we can determine the change in internal energy by

$$\Delta U = \tfrac{3}{2}Nk\Delta T$$
$$= \left(\tfrac{3}{2}\cdot 6.02\times 10^{23}\cdot 1.38\times 10^{-23}\times 100\right)\ \text{J} = 1.25\ \text{kJ}$$

Because $PV_1 = NkT_1$ and $PV_2 = NkT_2$, we obtain

$$V_1 = \frac{NkT_1}{P} = 24.9\ \text{L},\quad V_2 = \frac{NkT_2}{P} = 33.2\ \text{L}$$

and find that the work done on the system is

$$w = -P\,\Delta V = (-1\ \text{bar})\,(8.3\ \text{L}) = -0.83\ \text{kJ}$$

Combining these results, the heat flow into the system is

$$q = \Delta U - w = (1.25 + 0.83)\ \text{kJ} = 2.08\ \text{kJ}$$

The heat supplied to the gas (2.08 kJ) is used to increase the internal energy of the gas (1.25 kJ) and to perform work on the surroundings by displacing the piston (0.83 kJ). This example shows how we can exploit the state function property of U and use the constraint of constant P to evaluate the work, and hence obtain the heat.

Equation (16.38) means that the energy flow to the gas (work and heat) is stored by the molecules as internal energy, just as in the corresponding isochoric process. In this example, ΔU is less than q, since w is negative; work is done by the system. Clearly this case is different from that of constant volume because of the work which is performed.

16.4.2 Enthalpy *H* and Heat Capacity C_p: Definition

It is often more convenient to perform chemical transformations under constant pressure rather than constant volume conditions. Also, it is of practical interest to know the heat produced in a chemical reaction (such as the combustion of coal), as well as the temperatures which can be reached as a result of chemical reactions. Rather than explicitly account for the work each time we perform such a process, we define a new energy function, called the enthalpy H, that is directly related to the heat flow for a constant pressure. As shown below the enthalpy reports on the populations of the molecular energy levels and is directly related to the system's internal energy.

16.4.2.1 Defining the Enthalpy *H*

The enthalpy H is defined by

$$H \equiv U + PV \tag{16.40}$$

Using this definition and Equation (16.38), the enthalpy change, $\Delta H = H_2 - H_1$, for a constant pressure process is

$$\Delta H = \Delta U + P\,\Delta V = (q - P\,\Delta V) + P\,\Delta V = q \quad \text{(at constant pressure)} \tag{16.41}$$

That is, the change of the enthalpy ΔH equals the heat q transferred in a process at constant pressure, like that in Example 16.3. In addition, the enthalpy is a state function, because the internal energy, the pressure, and the volume are state functions. This state function property and its direct connection to the heat in a constant pressure process makes the enthalpy a very useful and important quantity.

16.4.2.2 Defining the Heat Capacity C_P

In analogy to the constant volume heat capacity C_V, we can define a *constant pressure heat capacity* C_P by

$$C_P = \lim_{\Delta T \to 0}\left(\frac{q_{T,T+\Delta T}}{\Delta T}\right)_{P=\text{const.}} = \lim_{\Delta T \to 0}\left(\frac{\Delta H}{\Delta T}\right)_{P=\text{const.}} = \left(\frac{\partial H}{\partial T}\right)_P \tag{16.42}$$

16.4.2.3 Relation Between C_V and C_P

According to Equation (16.40) we have

$$C_P = \left(\frac{\partial H}{\partial T}\right)_P = \left(\frac{\partial U}{\partial T}\right)_P + P\left(\frac{\partial V}{\partial T}\right)_P \tag{16.43}$$

In the special case of an ideal gas, U does not depend on pressure or volume, so

$$\left(\frac{\partial U}{\partial T}\right)_P = \left(\frac{\partial U}{\partial T}\right)_V = C_V$$

and from the ideal gas law

$$PV = NkT \quad \text{or} \quad V = \frac{NkT}{P}$$

we can show

$$\left(\frac{\partial V}{\partial T}\right)_P = \frac{Nk}{P}$$

Hence we find for an ideal gas that

$$C_P = C_V + Nk \quad \text{or} \quad C_{P,m} = C_{V,m} + \boldsymbol{R} \tag{16.44}$$

C_P is larger than C_V because part of the heat is used to expand the gas, performing work.

Example 16.4 *Assuming that HD behaves as an ideal gas, use its constant pressure heat capacity to determine its enthalpy change upon heating from* 200 *to* 400 K *at constant pressure.*

From the definition of the enthalpy, we find that

$$\Delta H_m = C_{P,m} \cdot \Delta T = (C_{V,m} + \boldsymbol{R}) \cdot \Delta T$$

Because the vibrational frequency of HD is sufficiently high, we can use $C_{V,m} = 2.5\,\boldsymbol{R}$ for HD, and we find

$$\Delta H_m = 3.5\,\boldsymbol{R} \cdot \Delta T = 29.1(400 - 200)\ \text{J mol}^{-1} = 5.82\ \text{kJ mol}^{-1}$$

16.4.2.4 General Relation Between C_V and C_P

Although the relationship between the constant pressure heat capacity C_P and the constant volume heat capacity C_V is simple for an ideal gas, the relationship is more complicated for the general case.

In Section 19.2.5, we derive the general relation

$$C_P - C_V = T \cdot V \frac{\alpha_V^2}{\kappa_T} \tag{16.45}$$

where α_V is called the *cubic expansion coefficient* and κ_T is called the *isothermal compressibility*. Equation (16.45) is useful because α_V and κ_T can be obtained from experiments. The cubic expansion coefficient is defined by the relation

$$\alpha_V = \frac{1}{V}\left(\frac{\partial V}{\partial T}\right)_P \tag{16.46}$$

It measures how much the volume of a substance changes upon heating at constant pressure. The isothermal compressibility is defined by the relation

$$\kappa_T = -\frac{1}{V}\left(\frac{\partial V}{\partial P}\right)_T \tag{16.47}$$

The isothermal compressibility quantifies the dependence of a system's volume on the pressure applied.

In the special case of an ideal gas we have $(\partial V/\partial T)_P = \boldsymbol{nR}/P$ and $(\partial V/\partial P)_T = -\boldsymbol{nR}T/P^2$, so that $\alpha_V = \boldsymbol{nR}/(PV) = 1/T$ and $\kappa_T = \boldsymbol{nR}T/(P^2V) = 1/P$. Thus, we obtain $C_P - C_V = \boldsymbol{nR}$, which agrees with Equation (16.44).

Figure 16.11b plots $C_{V,m}$ for some solids (lead, aluminum, diamond) as a function of temperature. In this case only the $C_{P,m}$ data can be measured directly, and the corresponding $C_{V,m}$ data are obtained by using Equation (16.45). As an example, consider metallic lead at $T = 600$ K. Using the experimental data $C_{P,m} = 29.4$ J K^{-1} mol^{-1}, $\alpha_V = 105 \times 10^{-6}$ K^{-1}, $\kappa_T = 3.1 \times 10^{-6}$ bar^{-1}, and $V_m = 18.4$ cm^3 mol^{-1}, we obtain

$$\begin{aligned} C_{P,m} - C_{V,m} &= 600\ \text{K} \cdot 18.4\ \text{cm}^3\ \text{mol}^{-1} \frac{\left(105 \times 10^{-6}\ \text{K}^{-1}\right)^2}{3.1 \times 10^{-6}\ \text{bar}^{-1}} \\ &= 3.92 \times 10^{-2}\ \text{bar L K}^{-1}\ \text{mol}^{-1} = 3.92\ \text{J K}^{-1}\ \text{mol}^{-1} \end{aligned}$$

The resulting value for $C_{V,m}$ is (29.4 − 3.92) J K^{-1} mol^{-1} = 25.5 J K^{-1} mol^{-1}. Data for lead and aluminum in the temperature range considered in Fig. 16.11b are given in Table 16.2, data for some selected compounds for $T = 298$ K are collected in Table 16.3.

Table 16.2 Experimental Data for the Conversion of C_P into C_V for Pb and Al in Fig. 16.11.

	T	C_P	α_V	κ_T	V_m	$T \cdot V \frac{\alpha_V^2}{\kappa_T}$	C_V
	K	J K^{-1} mol^{-1}	10^{-6} K^{-1}	10^{-6} bar^{-1}	cm^3 mol^{-1}	J K^{-1} mol^{-1}	
Pb	20	11.0	70	1.9	17.5	0.09	10.9
	40	19.6	72	1.9	17.5	0.18	19.4
	60	22.4	73	2.0	17.5	0.27	22.1
	100	24.4	75	2.0	17.6	0.49	23.9
	200	25.9	81	2.1	17.7	1.10	24.8
	400	27.5	93	2.5	18.1	2.49	25.0
	600	29.4	105	3.1	18.4	3.91	25.5
Al	20	0.2	10	1.0	9.6	0.00	0.2
	40	2.1	16	1.0	9.6	0.01	2.1
	60	5.8	22	1.1	9.6	0.03	5.8
	80	9.7	29	1.1	9.6	0.06	9.7
	100	13.0	36	1.2	9.9	0.10	12.9
	200	21.6	60	1.3	9.9	0.54	21.1
	400	25.6	80	1.4	10.1	1.84	23.8
	600	28.1	95	1.5	10.3	3.71	24.4

Table 16.3 Experimental Data α_V, κ_T, V_m, and $C_{P,m}$, for Some Compounds at 298 K. $C_{V,m}$ Is Calculated According to Equation (16.45).

	α_V 10^{-6} K^{-1}	κ_T 10^{-6} bar^{-1}	V_m mL mol^{-1}	$C_{P,m}$	$\frac{\alpha_V^2 V}{\kappa_T}T$	$C_{V,m}$
				J K^{-1} mol^{-1}		
Benzene	1240	94	89.4	136.3	43.6	92.7
Ethanol	1120	111	58.5	112.3	23.3	89.0
Water	210	45.8	18	75.3	5.2	70.1
Lead	86.7	2.42	17.9	26.4	1.76	24.6
Aluminum	69	1.3	9.9	24.4	1.08	23.3
Diamond	3.5	0.2	3.4	6.2	0.005	6.2

16.5 Heat Exchange and Chemical Reactions

16.5.1 Reaction at Constant Volume: ΔU

Consider a closed container that is filled with a stoichiometric mixture of hydrogen and oxygen and is immersed in a bath at a temperature T (Fig. 16.14). Ignition of the mixture, e.g., by a spark, causes an explosive chemical reaction in the container

$$2\,H_2 + O_2 \rightarrow 2\,H_2O \tag{16.48}$$

In this reaction bonds are broken and new bonds are formed. The water molecules formed as products possess a lower internal energy than the starting materials; i.e., the change of internal energy is $\Delta U < 0$. Since the reaction is performed at constant volume, no work is done on the system, and

$$\boxed{q = \Delta U \quad \text{(reaction at constant volume)}} \tag{16.49}$$

In this case the heat q supplied to the system (that is, to the gas mixture) is negative. This means that heat is transferred to the bath (an *exothermic reaction*).

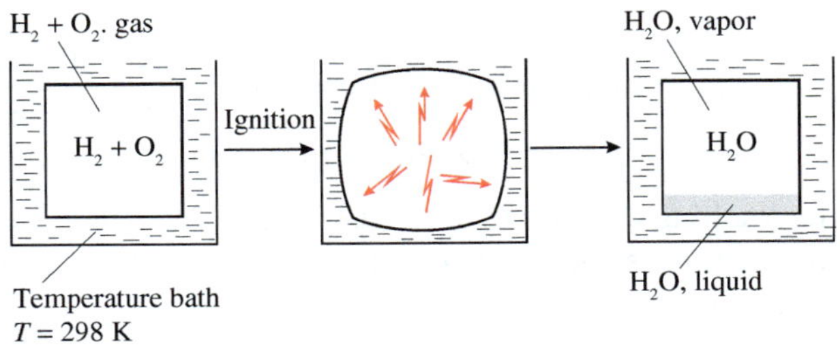

Figure 16.14 Hydrogen and oxygen gas reaction, reaction at $T = const$ and $V = const$. Immediately after ignition the temperature in the container increases and the container is being deformed, but after equilibration with the temperature bath, T and V are the same as in the beginning. System: $H_2 + O_2$ (in gas phase), H_2 and O_2 in a stoichiometric mixture.

This heat can be measured with a device called an adiabatic bomb calorimeter, Fig. 16.15. In this device the reaction occurs inside of a chamber that is enclosed by rigid, diathermal walls. The walls of the reaction chamber allow heat transfer between a temperature bath (typically water) that is enclosed in a set of adiabatic walls, hence no heat exchange between the bath and the external surroundings. By measuring the temperature rise of the water and knowing the heat capacity of the container (calorimeter), the amount of heat released in the reaction can be determined. For the stoichiometric mixture

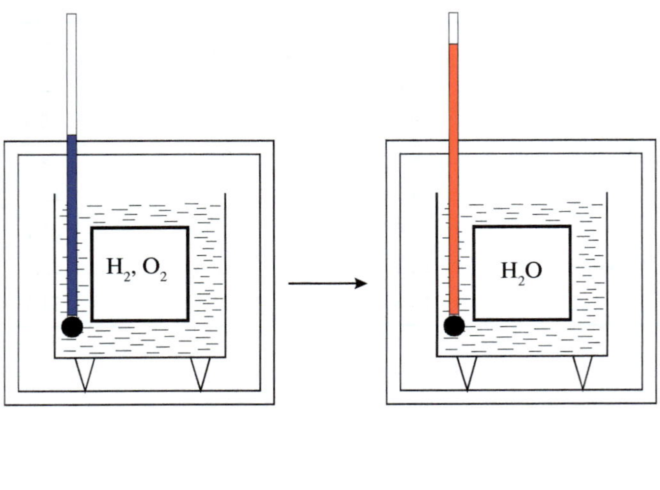

Figure 16.15 Adiabatic bomb calorimeter to measure the heat transferred into the temperature bath. The small increase of the temperature T of the bath is measured using a very sensitive thermometer, then the heat can be calculated from ΔT and the heat capacity of the calorimeter.

of hydrogen and oxygen we find that

$$\Delta U = -564.4 \text{ kJ mol}^{-1} \cdot \boldsymbol{n}_{O_2} \quad (T = 298 \text{ K})$$

where $\boldsymbol{n}_{O_2}$ is the amount of reacting O_2.

Not all chemical reactions can be directly investigated in a calorimeter. In such cases, we can use the state function property of the internal energy and the idea of thermodynamic cycles to study the reactions indirectly. If a reaction $1 \rightarrow 2$ has to be investigated, two indirect reactions $1 \rightarrow 3$ and $2 \rightarrow 3$ can be used:

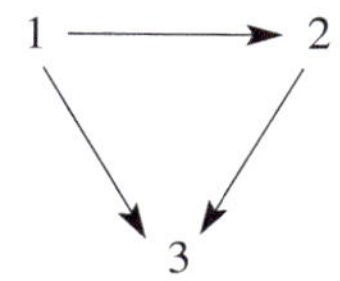

According to Equation (16.3), in a cyclic process the internal energy of a system does not change.

$$\Delta U_{1\rightarrow 2} + \Delta U_{2\rightarrow 3} + \Delta U_{3\rightarrow 1} = 0$$

Then, as long as the initial and final states are the same, we can calculate $\Delta U_{1\rightarrow 2}$ as

$$\Delta U_{1\rightarrow 2} = -\Delta U_{3\rightarrow 1} - \Delta U_{2\rightarrow 3} = \Delta U_{1\rightarrow 3} - \Delta U_{2\rightarrow 3} \tag{16.50}$$

The fact that the internal energy change of a reaction is obtained by adding the internal energy changes of different reaction steps is called *Hess' law*.

Example 16.5 *Consider the hydrogenation of ethene*

$$1 \rightarrow 2: \; C_2H_4 + H_2 \rightarrow C_2H_6 \tag{16.51}$$

$\Delta U_{1\rightarrow 2}$ can be obtained indirectly by measuring the heat of oxidation of C_2H_4, H_2, and C_2H_6 at constant volume. Use these oxidation reactions to construct a cycle and evaluate the internal energy change for the hydrogenation of ethene $\Delta U_{1\rightarrow 2}$ at 298 K.

The oxidation reactions are rapid and clean if a large excess of pure oxygen is used:

$$1 \rightarrow 3 \qquad C_2H_4 + H_2 + \tfrac{7}{2}O_2 \rightarrow 2CO_2 + 3H_2O \qquad \Delta U_{1\rightarrow 3} = -1688.1 \,\frac{\text{kJ}}{\text{mol}} \cdot \boldsymbol{n}_{C_2H_4}$$

$$2 \rightarrow 3 \qquad C_2H_6 + \tfrac{7}{2}O_2 \rightarrow 2CO_2 + 3H_2O \qquad \Delta U_{2\rightarrow 3} = -1554.0 \,\frac{\text{kJ}}{\text{mol}} \cdot \boldsymbol{n}_{C_2H_6}$$

These indirect reactions are chosen so that the products are the same. The difference of the reactions $1 \rightarrow 3$ and $2 \rightarrow 3$ yields the reaction $1 \rightarrow 2$, which is of interest. With $\boldsymbol{n}_{C_2H_4} = \boldsymbol{n}_{C_2H_6}$ we obtain

$$\begin{aligned} \Delta U_{1\rightarrow 2} &= \Delta U_{1\rightarrow 3} - \Delta U_{2\rightarrow 3} = (-1688.1 + 1554.0) \text{ kJ mol}^{-1} \cdot \boldsymbol{n}_{C_2H_4} \\ &= -134.1 \text{ kJ mol}^{-1} \cdot \boldsymbol{n}_{C_2H_4} \end{aligned}$$

This macroscopic analysis does not require any detailed information about the molecular properties of the reactants and products. In one sense, this makes it powerful because it does not rely on our ability to accurately model the detailed molecular properties. In another sense, this feature is a weakness because it does not tell us how the difference in internal energy is composed of electronic, vibrational, rotational, and translational degrees of freedom. Rather we must combine the ΔU determination with other experiments and information to address such questions.

16.5.2 Reaction at Constant Pressure: ΔH

Often chemical reactions are not carried out in closed containers, but in open vessels (e.g., test tubes, Erlenmeyer flasks) in which the reactants are under external pressure. In such cases, the pressure remains constant during the reaction, but the volume of the reaction mixture changes (*reaction at constant pressure*).

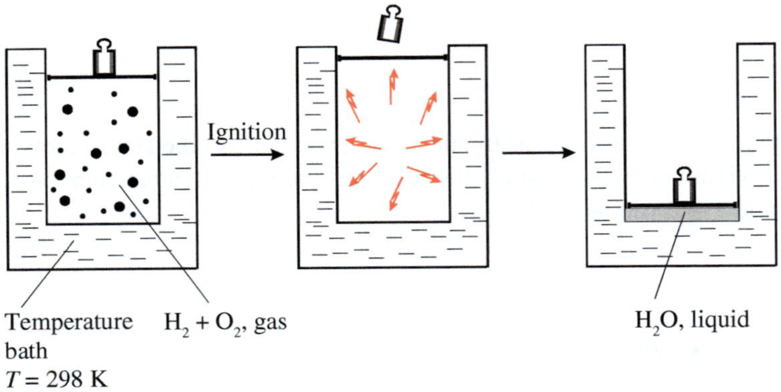

Figure 16.16 Hydrogen and oxygen gas reaction, reaction at $T = const$ and $P = const$. Immediately after ignition the temperature in the container increases and the container is being deformed, but after equilibration with the temperature bath, T and P are the same as in the beginning. The initial system is $H_2 + O_2$ (in gas phase), H_2 and O_2 in a stoichiometric mixture.

Figure 16.16 illustrates such a process for the explosive reaction of hydrogen and oxygen, where the external air pressure is represented by a weight on the piston. At 298 K the reaction of gaseous H_2 and O_2 forms liquid water and the molar volume of the reaction mixture decreases as the products are generated and cool. The piston falls, and work is done on the system. The change of internal energy is

$$\Delta U = q + w = q - P\Delta V \qquad \text{(reaction at constant pressure)} \tag{16.52}$$

so that the heat q transferred into the system is

$$q = \Delta U + P\Delta V = \Delta U + P(V_{H_2O} - V_{H_2} - V_{O_2}) \tag{16.53}$$

For this reaction q is smaller than ΔU, because ΔV is negative (the volume of liquid water is much smaller than the volumes of H_2 and O_2). Whereas in a reaction at constant volume the heat q equals the change ΔU of the internal energy, the heat transferred at constant pressure is $\Delta U + P\Delta V$.

Since the reaction is performed at constant pressure, it is useful to evaluate the enthalpy, according to Equation (16.41)

$$\Delta H = \Delta U + P\Delta V = q \quad \text{(reaction at constant pressure)} \tag{16.54}$$

which is directly linked to the heat. The change in enthalpy ΔH is equal to the heat transferred into the system at constant pressure. The heat of reaction is converted to internal energy of the system and to work done by the system against the external pressure.

In our example of a hydrogen and oxygen explosion the volume of the product liquid water is much smaller than the volumes of H_2 and O_2. By neglecting the volume of the liquid and treating H_2 and O_2 as ideal gases we can calculate the work $P\Delta V$. We find that

$$P\Delta V = -\boldsymbol{RT}\cdot(\boldsymbol{n}_{H_2} + \boldsymbol{n}_{O_2}) = -\boldsymbol{RT}\cdot(2\boldsymbol{n}_{O_2} + \boldsymbol{n}_{O_2}) = -3\boldsymbol{RT}\cdot\boldsymbol{n}_{O_2} \tag{16.55}$$

If we combine this result with Equation (16.53), we find that at $T = 298$ K

$$\Delta H = \Delta U - 3\boldsymbol{RT}\cdot\boldsymbol{n}_{O_2} = (-564.4 - 7.4)\ \text{kJ mol}^{-1}\cdot\boldsymbol{n}_{O_2}$$
$$= -571.8\ \text{kJ mol}^{-1}\cdot\boldsymbol{n}_{O_2}$$

Example 16.6 *The heat produced in burning H_2 ($\Delta H_m = 285.9$ kJ mol^{-1}) is used to heat liquid water from room temperature (20 °C) to the boiling point (100 °C, $C_{P,m} = 75.3$ J K^{-1} mol^{-1}). How many moles of water can we heat with 1 kg of hydrogen gas? If the heat produced by 1 mol CH_4 is 890 kJ, how many kg of methane are needed to perform the same boiling process?*

For heating 1 mol (18 g) of water we need $q = \boldsymbol{n}C_{P,m}\Delta T = 1\ \text{mol} \times 75.3\ \text{J K}^{-1}\ \text{mol}^{-1} \times 80\ \text{K} = 6.02$ kJ. Thus 6.02 kJ/285.9 kJ mol^{-1} = 0.021 mol of hydrogen (corresponding to 42 mg) is needed to heat 1 mol of water from 20 to 100 °C. This means that 1 kg of H_2 (500 mol) can be used to heat 24,000 mol (430 kg) of water to boiling.

If we use methane, we will need 6.02 kJ/890 kJ mol^{-1} = 0.00676 mol methane per mol of water. To heat 24,000 mol of water, we need 0.00676·24,000 = 160 mol of methane, which is 2.6 kg of methane. Although hydrogen releases less heat per mole of substance combusted, it produces more heat per gram of substance.

16.5.3 Temperature Dependence of ΔU and ΔH

Consider a chemical reaction performed at the temperature T_1 which has an internal energy change of ΔU_1. What will the change ΔU_2 be if the same reaction is performed at a temperature T_2? The electronic energy of bonding and the energies of translation, rotation, and vibration contribute to the internal energy U of a molecular assembly. Increasing the temperature of the system causes an increase in U, and we expect that the internal energy of the reactants might depend differently on the temperature than those of the products. According to Equation (16.13) the change in the internal energy with temperature is given by the constant volume heat capacity,

$$C_V = \left(\frac{\partial U}{\partial T}\right)_V \quad \text{or} \quad \mathrm{d}U = C_V\, \mathrm{d}T$$

By integrating $\mathrm{d}U$ from T_1 to T_2 we find that

$$U(T_2) = U(T_1) + \int_{T_1}^{T_2} C_V \cdot \mathrm{d}T \tag{16.56}$$

which for a temperature independent C_V reduces to $\Delta U = C_V\,\Delta T$.

We can use the cycle in Fig. 16.17 to quantify how the internal energy of the reaction changes with temperature. To answer this question we imagine the reaction carried out at temperature T_1; after that the products are heated from T_1 to T_2. For this process the change of internal energy is

$$\Delta U_{T_1} + \int_{T_1}^{T_2} C_{V,\text{products}} \cdot \mathrm{d}T$$

On the other hand, if the reactants are first heated from T_1 to T_2 and the reaction is carried out at temperature T_2, the change of internal energy is

$$\Delta U_{T_2} + \int_{T_1}^{T_2} C_{V,\text{reactants}} \cdot \mathrm{d}T$$

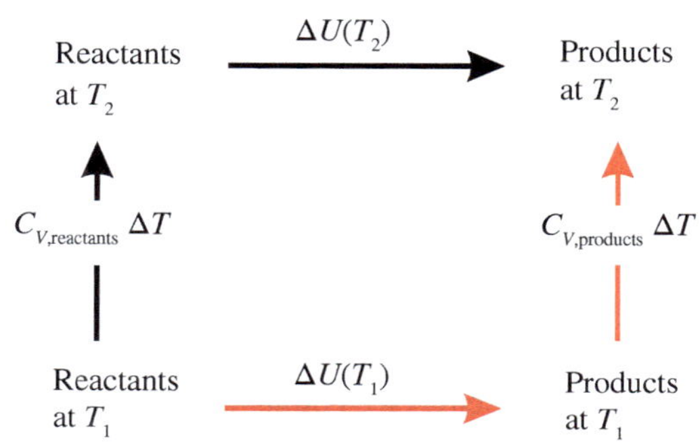

Figure 16.17 The cycle shows the connection between a reaction performed at two different temperatures, T_1 and T_2.

Both these changes of internal energy must be the same, so that

$$\Delta U_{T_1} + \int_{T_1}^{T_2} C_{V,\text{products}} \cdot \mathrm{d}T = \Delta U_{T_2} + \int_{T_1}^{T_2} C_{V,\text{reactants}} \cdot \mathrm{d}T \tag{16.57}$$

or

$$\boxed{\Delta U_{T_2} = \Delta U_{T_1} + \int_{T1}^{T2} \Delta C_V \cdot \mathrm{d}T} \tag{16.58}$$

where

$$\Delta C_V = C_{V,\text{products}} - C_{V,\text{reactants}} \tag{16.59}$$

A similar analysis applies for reactions under constant pressure conditions. The change in a system's enthalpy with temperature is

$$H(T_2) = H(T_1) + \int_{T_1}^{T_2} C_P \cdot \mathrm{d}T \tag{16.60}$$

in which C_P is the constant pressure heat capacity. For a chemical reaction performed at constant pressure we obtain

$$\boxed{\Delta H_{T_2} = \Delta H_{T_1} + \int_{T_1}^{T_2} \Delta C_P \cdot \mathrm{d}T} \tag{16.61}$$

with

$$\Delta C_P = C_{P,\text{products}} - C_{P,\text{reactants}} \tag{16.62}$$

Example 16.7 *Calculate ΔH at 350 K for the reaction $2\,H_2 + O_2 \rightarrow 2\,H_2O$. The molar heat capacities $C_{P,m} = C_P/\boldsymbol{n}$ of H_2, O_2, and liquid H_2O are given in Table 16.4 at 298 and 350 K and 1 bar pressure.*

For these substances the heat capacities do not change much with temperature, and we approximate $C_{P,m}(T)$ by the average $\overline{C}_{P,m}$. Now Equation (16.61) can be simplified to give

$$\Delta H_{T_2} = \Delta H_{T_1} + \Delta\overline{C}_P \cdot (T_2 - T_1)$$

with

$$\begin{aligned}\Delta C_P &= \boldsymbol{n}_{H_2O} \cdot \overline{C}_{P,m}(H_2O) - \boldsymbol{n}_{H_2} \cdot \overline{C}_{P,m}(H_2) \\ &\quad -\boldsymbol{n}_{O_2} \cdot \overline{C}_{P,m}(O_2) \\ &= \boldsymbol{n}_{O_2} \cdot \left[2 \cdot \overline{C}_{P,m}(H_2O) - 2 \cdot \overline{C}_{P,m}(H_2) - \overline{C}_{P,m}(O_2)\right] \\ &= \boldsymbol{n}_{O_2} \cdot 63.6\ \text{J K}^{-1}\ \text{mol}^{-1}\end{aligned}$$

Table 16.4 Experimental Molar Heat Capacity $C_{P,m}$. Data from Table 16F.1 in Foundation 16.2.

	$C_{P,m}$/J K^{-1} mol^{-1}		
Compound	**At 298 K**	**at 350 K**	**Average**
H_2O(liquid)	75.3	75.6	75.5
H_2(gas)	28.8	29.0	28.9
O_2(gas)	29.4	29.7	29.6

Then with $T_2 - T_1 = 373 - 298$ K $= 52$ K we obtain

$$\Delta H_{350} = (-571.8 \times 10^3 + 63.6 \cdot 52)\ \text{J mol}^{-1} \cdot \boldsymbol{n}_{O_2} = -568.6\ \text{kJ mol}^{-1} \cdot \boldsymbol{n}_{O_2}$$

The heat produced when burning H_2 at 350 K is about 1% smaller than the heat produced when burning H_2 at 298 K. This difference occurs because the heat capacity of liquid water exceeds the heat capacities of H_2 and O_2. The high heat capacity of liquid water is caused by the important hydrogen bonding interactions between water molecules and their dissociation upon heating.

Box 16.2 shows how we can use the ΔH of a combustion reaction to estimate the temperature of a flame. For large changes of the temperature and/or strong temperature dependence of the heat capacities, the integral in Equation (16.61) must be evaluated explicitly (see Foundation 16.2).

16.5.4 Molar Enthalpies of Formation from Elements $\Delta_f H^\ominus$

The enthalpy change for the formation of a compound from its elements under standard conditions (compounds separated and under standard pressure $P^\ominus = 1$ bar) is useful for obtaining enthalpy changes in arbitrary reactions. The reaction of H_2 and O_2 to form H_2O

$$H_2 + \frac{1}{2}O_2 \rightarrow H_2O$$

is a reaction of this type. In this case ($T = 298$ K)

$$\Delta H = -571.8\ \text{kJ mol}^{-1} \cdot \boldsymbol{n}_{O_2} = -285.9\ \text{kJ mol}^{-1} \cdot \boldsymbol{n}_{H_2O}$$

The *molar enthalpy of formation under standard conditions*, i.e., the enthalpy of formation of one mol of H_2O from its elements under standard conditions is

$$\begin{aligned}\Delta_f H^\ominus(H_2O) &= \frac{\Delta H\ \text{(formation from elements, separated, 1 bar)}}{\boldsymbol{n}_{H_2O}} \\ &= -285.9\ \text{kJ mol}^{-1}\ \text{(for 298 K)}\end{aligned}$$

The symbol $^\ominus$ indicates standard conditions (reactants separated, standard pressure $P^\ominus = 1$ bar). A subscript at the symbol Δ designates a molar quantity; additionally, subscript f indicates formation of a substance from elements; other subscripts will be used to indicate chemical reactions (subscript r), vaporization (subscript v), melting (subscript fus), sublimation (subscript sub).

For selected compounds the molar enthalpies of formation $\Delta_f H^\ominus_{298\ K}$ are listed in Appendices A.1 and A.2.

16.5.5 Molar Enthalpy of Reaction $\Delta_r H^\ominus$

The change in enthalpy of a reaction under standard conditions (species separated, standard pressure $P^\ominus = 1$ bar) at temperature T can be obtained for any reaction from the enthalpies of formation of the reactants and products. Because the enthalpy is a state function, we can imagine a process in which the reactants are converted into their elemental components and then reformed as the products. The enthalpy and internal energy changes in this latter process must be the same as that for the direct reaction.

Figure 16.18 illustrates how this cycle works for understanding the energetics for the oxidation of CO to produce carbon dioxide.

$$CO + \frac{1}{2}O_2 \rightarrow CO_2$$

In step 1 the reactants CO and oxygen are converted into their constituent elements under standard conditions, and the enthalpy change is minus the enthalpy of formation for the reactants. In step 2 the elements are converted into the product CO_2 for which the enthalpy change is the enthalpy of formation. Because we define the molar enthalpy of reaction under standard conditions as the enthalpy change for the reaction of one mol of CO and one-half mol of O_2 to form one mol of CO_2 (species separated, pressure 1 bar), we find that

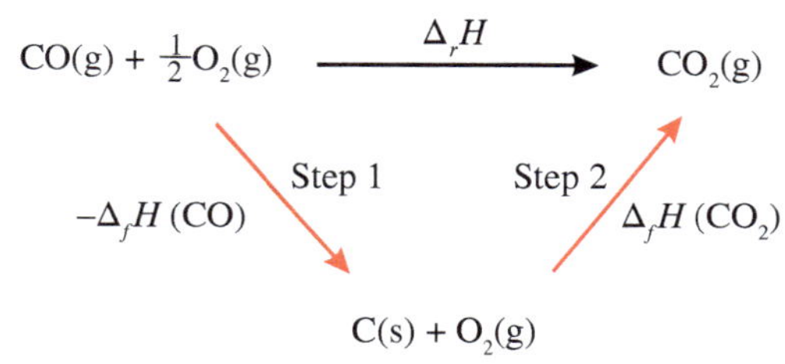

Figure 16.18 A cycle showing the relation between the enthalpy of reaction, $\Delta_r H$, and the enthalpies of formation from the elements, $\Delta_f H$.

$$\Delta_r H^{\ominus}(\text{Eqn 1}) = \Delta_f H^{\ominus}(CO_2) - \Delta_f H^{\ominus}(CO) - \frac{1}{2}\Delta_f H^{\ominus}(O_2)$$
$$= \left(-393.5 + 110.5 + \frac{1}{2}\cdot 0\right)\ \text{kJ mol}^{-1} = -283.0\ \text{kJ mol}^{-1}\ (\text{for } 298\ \text{K})$$

If we write the same reaction with twice the number of reactants,

$$2\ CO + O_2 \rightarrow 2\ CO_2$$

then $\Delta_r H^{\ominus}$ refers to two molecules CO reacting with one molecule O_2, and

$$\Delta_r H^{\ominus} = 2\cdot\Delta_f H^{\ominus}(CO_2) - 2\cdot\Delta_f H^{\ominus}(CO) - \Delta_f H^{\ominus}(O_2)$$
$$= 2\cdot(-283.0\ \text{kJ mol}^{-1}) = -566.0\ \text{kJ mol}^{-1}\ (\text{for } 298\ \text{K})$$

Thus, the value of the molar reaction enthalpy will depend on the way in which the reaction equation is written.

The quantities ΔU, ΔH, ΔV are *extensive* and depend on the amounts of material. In contrast, $\Delta_f H^{\ominus}$, $\Delta_r H^{\ominus}$ are molar quantities; hence, *intensive* quantities like the pressure P or the temperature T and do not depend on the size of the system.

Generalization

This result can be generalized for the reaction

$$\alpha A + \beta B + \ldots \cdots \rightarrow \gamma C + \delta D + \cdots \tag{16.63}$$

Then we obtain $\Delta_r H^{\ominus}$, the molar enthalpy change for the reaction of α mol A, β mol B … to γ mol C and δ mol D …, under standard conditions:

$$\Delta_r H^{\ominus} = \gamma\cdot\Delta_f H^{\ominus}(C) + \delta\cdot\Delta_f H^{\ominus}(D) - \alpha\cdot\Delta_f H^{\ominus}(A) - \beta\cdot\Delta_f H^{\ominus}(B) \tag{16.64}$$

16.5.6 Bond Enthalpies and Bond Energies

Enthalpy changes in chemical reactions are related to the bond energies. To illustrate this relationship, we consider the reaction

$$C(s) + 2H_2(g) \rightarrow CH_4(g)$$

in which graphitic carbon and hydrogen combine to form methane. The enthalpy change in this reaction is the molar enthalpy of formation of methane from the elements,

$$\Delta_r H^{\ominus}_{298} = \Delta_f H^{\ominus}_{298}(CH_4) = -74.8\ \text{kJ mol}^{-1}$$

The average bond energy of C–H is considered to be $\frac{1}{4}$ of the molar enthalpy change of the reaction

$$C(g) + 4\ H(g) \rightarrow CH_4(g) \qquad \text{at } T = 0\ \text{K},\ \Delta_r H^{\ominus}_0$$

To relate $\Delta_r H_0^\ominus$ to $\Delta_r H_{298}^\ominus$ we consider the following scheme:

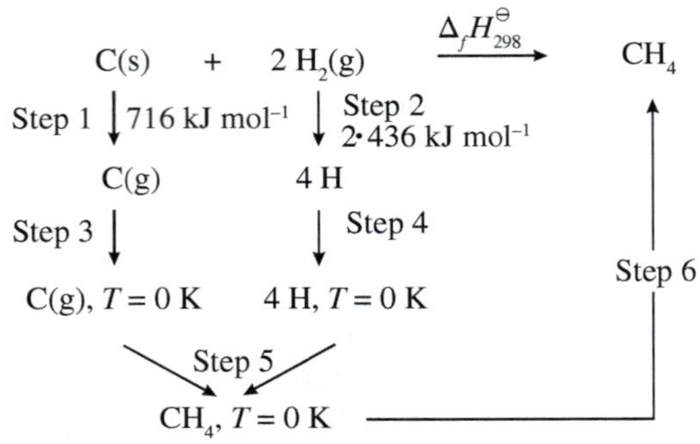

First we must atomize C(s) and H_2(g) These processes occur in step 1 (sublimation of C, $\Delta H_m^\ominus = 716$ kJ mol^{-1}) and step 2 (dissociation of 2 H_2, $\Delta H_m^\ominus = 2 \cdot 436$ kJ mol^{-1}). Then C(g) and H must be cooled down to $T = 0$ K. This occurs in step 3 ($\Delta H_m^\ominus = -\frac{3}{2}\boldsymbol{R}T$) and step 4 ($\Delta H_m^\ominus = -4 \cdot \frac{3}{2}\boldsymbol{R}T$). CH_4 is formed in step 5 at $T = 0$ K ($\Delta_r H_0^\ominus$) and heated to 298 K in step 6 ($\Delta H_m^\ominus = \frac{5}{2}\boldsymbol{R}T$). The contributions of steps 3, 4, and 6 are small ($-5\boldsymbol{R}T = -12.4$ kJ mol^{-1}). Combining these enthalpy changes for the cycle we can write

$$(716 + 2 \cdot 436 - 12) \text{ kJ mol}^{-1} + \Delta_r H_0^\ominus - \Delta_f H_{298}^\ominus(CH_4) = 0$$

and with $\Delta_f H_{298}^\ominus(CH_4) = -74.8$ kJ mol^{-1}

$$\Delta_r H_0^\ominus = -74.8 \text{ kJ mol}^{-1} - (716 + 2 \cdot 436 - 12) \text{ kJ mol}^{-1} = -1651 \text{ kJ mol}^{-1}$$

Finally, for the average bond enthalpy of C–H in CH_4 we obtain $\frac{1}{4}\Delta_r H_0^\ominus = \frac{1}{4}(-1651 \text{ kJ mol}^{-1}) = -413$ kJ mol^{-1}.

As this example shows the dominant contribution to the enthalpy change in chemical reactions is often associated with the cleavage of chemical bonds. The contributions from rotational and translational degrees of freedom play a small role, unless the enthalpy change in a reaction is near zero. Hence it can often be reasonable to use enthalpy changes in chemical reactions as a measure of chemical bond energies: $\varepsilon_{B,m} = +413$ kJ mol^{-1} for the CH bond in CH_4. This value is about 5% smaller than $\varepsilon_{B,m} = +432$ kJ mol^{-1} obtained from direct spectroscopic experiments [1]. However, $\varepsilon_{B,m}$ is not the same for different compounds. For example, $\varepsilon_{B,m} = +365$ kJ mol^{-1} for C–H in the CH_3 group of toluene and $\varepsilon_{B,m} = +417$ kJ mol^{-1} for C–H in ethane. Some mean bond enthalpies are shown in Tables 16.5 and 16.6.

Table 16.5 Some Mean Single Bond Energies ($\varepsilon_{B,m}$/kJ mol^{-1}) [2].

	H	C	N	O	F	S	P
H	436						
C	413	348					
N	391	305	163				
O	463	358	201	146			
F	567	489	278	193	159		
S	367	272			327	255	
P	322	264		335	503		172

Table 16.6 Some Mean Double and Triple Bond Dissociation Energies (kJ mol^{-1}).

C=C	N=N	C=N	O=O	C=O	C=S	C≡C	N≡N	C≡N
614	418	615	498	745	536	839	945	891

16.5.7 Enthalpies and Reaction Cycles

Enthalpies have been measured for many different chemical and physical processes. By combining enthalpies together in thermodynamic cycles, we can deduce enthalpy changes for processes in which it is not possible, or very difficult, to measure the enthalpy reliably. By way of example, consider the reaction of sodium (Na) with chlorine (Cl_2) gas to make ordinary table salt (NaCl). This reaction is very rapid and exothermic, and can generate a lot of heat and intense yellow light emission from excited sodium atoms.

At large separations the energy of the neutral atoms is lower than that of the cation and anion. As the distance between the atoms decreases the Coulomb attraction between a Na^+ and Cl^- stabilizes more rapidly than that for the neutral pair. The atoms remain neutral until the energy of the neutral atom pair is equal to that for the ion pair. At this distance an electron is transferred from the Na atom to the Cl atom to generate the ion pair. The ions are then drawn toward each other by the Coulomb force until their electron clouds touch.

How can we determine the enthalpy of reaction to form the salt from the ions? The molar enthalpy of formation under standard conditions for NaCl(s) is $\Delta_f H^{\ominus}_{298} = -441$ kJ mol^{-1}, but this is for the formation of NaCl from the elements, not from the ions. The scheme

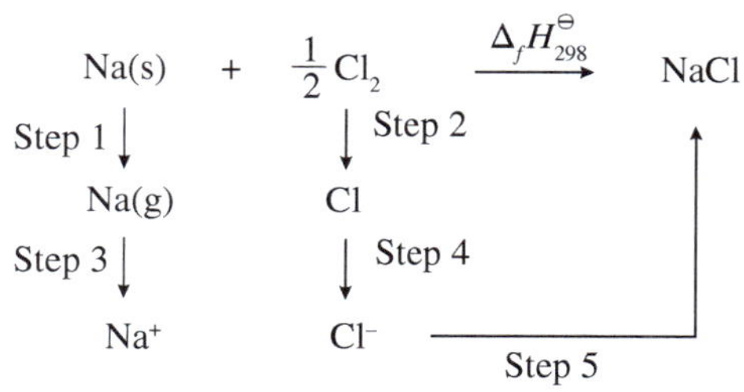

illustrates a cycle that allows us to deduce the reaction enthalpy for the ions from enthalpies of other processes. Step 1 is the enthalpy for the sublimation (92 kJ mol^{-1}) of sodium. Step 2 is the dissociation of chlorine molecule into chlorine atoms (121 kJ mol^{-1}). Step 3 is the ionization energy of sodium atoms (496 kJ mol^{-1}), and step 4 is the electron affinity of chlorine atoms (−349 kJ mol^{-1}). Lastly, step 5 is the enthalpy of interest and is determined by requiring that the enthalpy change for the cycle be zero; namely

$$\Delta_f H^{\ominus}_{298} + \Delta H^{\ominus}_{m,\text{step }1} + \Delta H^{\ominus}_{m,\text{step }2} + \Delta H^{\ominus}_{m,\text{step }3} + \Delta H^{\ominus}_{m,\text{step }4} + \Delta H^{\ominus}_{m,\text{step }5} = 0$$

Then

$$\Delta H^{-}_{m,\text{step }5} = (-441 - 92 - 121 - 496 + 349) = 801 \text{ kJ mol}^{-1}$$

Enthalpy changes of many processes can be determined by building cycles that combine enthalpies from known processes. Because experimental enthalpies have error, it is often desirable to use cycles for which enthalpies can be reliably measured and to use cycles that are direct so that errors have less of an impact on the final result.

16.6 Conclusions

We considered a system changing from one state to another one in thermal equilibrium. Of particular interest was the change in the internal energy U that is determined by the macroscopic variables ($\boldsymbol{n}$, T, and V). We have shown that ΔU, for chemical and physical processes, can generally be obtained from calorimetric data. In simple cases, as shown in Chapter 15, ΔU can be calculated from atomic and molecular data, and these values agree well with those determined calorimetrically. An important quantity in the analysis of the calorimetric data is the heat capacity, quantifying how much the system changes temperature as it is heated. Highlights are the analysis of macroscopic quantum effects (*ortho*- and *para*-hydrogen, Einstein- and Debye models of solids).

For chemical reactions performed at constant pressure it is convenient to have a state function that is in a direct relation to the heat released or absorbed by the system under constant pressure conditions, the enthalpy H. Its change ΔH between two states, both in thermal equilibrium, is identical with the heat supplied to the system in this special case. By absorbing this heat the system's internal energy increases by ΔU and the system does work against the external pressure, $-w = P\,\Delta V$. Thus $\Delta H = \Delta U + P\,\Delta V$.

16.7 Box 16.1. ΔU, q, and w in a Change of State

We consider a change from state 1 to state 2 carried out in two different ways (Fig. 16B.1). First, we evaporate water in a cylinder connected to a piston (Fig. 16B.1a, green line) from state 1 ($T = 373$ K, $P = 1$ bar, $V = 0.018$ L; the water is liquid) to state 2 ($T = 373$ K, $P = 1$ bar, $V = 31.2$ L; the water is a saturated vapor). After opening the valve (red bar), the water evaporates and lifts the weight on the piston (force f) in height h above the valve in state 2. The weight exerts a pressure P on the vapor, where

$$P = \frac{f}{\text{area of piston}} \tag{16B1.1}$$

Then the work w_a supplied to the system is

$$w_a = -f \cdot h \tag{16B1.2}$$

and we obtain

$$w = w_a = -P \cdot \text{area of piston} \cdot h = -P \cdot \Delta V \tag{16B1.3}$$

with $\Delta V = V_2 - V_1 = V_{\text{vapor}} - V_{\text{liquid}}$. ΔV is positive; thus, w_a is negative.

$$w_a = -P \cdot \Delta V = -1 \text{ bar} \cdot (31.2 - 0.018) \text{ L} = -3.12 \text{ kJ}$$

In a second experiment, we evaporate the water into a vacuum (Fig. 16B.1b). No weight is lifted, and therefore

$$w_b = 0$$

On the other hand, the change ΔU of the internal energy is the same in both cases:

$$\Delta U = U_2 - U_1 = q_a + w_a = q_b + w_b$$

The heat supplied in route (a) is larger than that in route (b), because part of the energy is used for external work:

Route (a)	**Route (b)**
$w_a = -P \cdot \Delta V$	$w_b = 0$
$q_a = q_{\text{vap}}$	$q_b = q_{\text{vap}} - P \cdot \Delta V$
$\Delta U = w_a + q_a = -P \cdot \Delta V + q_{\text{vap}}$	$\Delta U = w_b + q_b = -P \cdot \Delta V + q_{\text{vap}}$

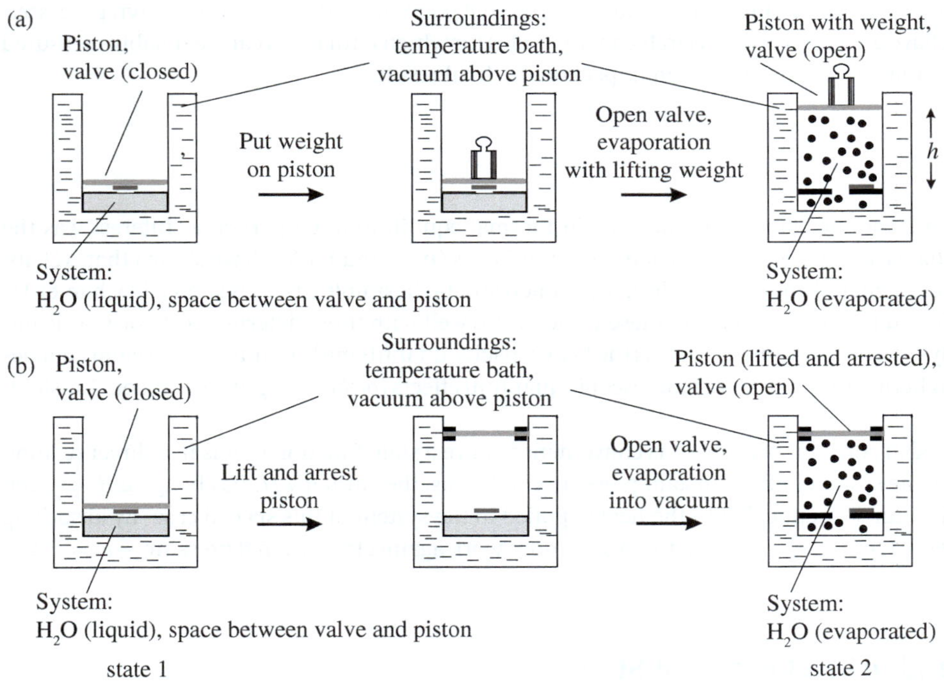

Figure 16B.1 Transition from state 1 ($n = 1$ mol H_2O at $P = 1$ bar and $T = 373$ K, $V = 0.018$ L) to state 2 ($n = 1$ mol H_2O at $P = 1$ bar and $T = 373$ K, $V = 31.2$ L) of a system by different routes (a) and (b). In route (b) the difference is that no weight is lifted, but the piston (green bar) is arrested at a position equal to the position in the end of route (a). ΔU is the same in both cases (U is a state function). The heat supplied in route (a) is larger than that in route (b), because part of the energy is used for external work. Note that piston and weight belong to the surroundings.

16.8 Box 16.2. Temperature of a Flame

What is the temperature of the flame of a Bunsen burner (Fig. 16B.2) fed with a mixture of hydrogen and air? In the flame the detonating gas reaction

$$2\,H_2 + O_2 \rightarrow 2\,H_2O$$

takes place. Contrary to our previous considerations, the reaction occurs adiabatically, and the reaction heat is used to warm up the reaction products from the initial temperature T_1 to the final temperature T_2. In a first approach, we neglect the temperature dependence of the heat capacities and start at $T_1 = 373$ K, where H_2O is in the gaseous state.

We imagine that H_2 and O_2 react at the starting temperature, immediately after ignition. The reaction enthalpy $\Delta_r H^{\ominus}_{373}$ is used to increase the temperature of the reaction products, and we obtain

$$\Delta_r H^{\ominus}_{373} = -C^{\ominus}_{P,m}(\text{products at 373 K}) \cdot \Delta T$$

where $C^{\ominus}_{P,m}$ (products at 373 K) is the heat capacity of the products. Then for the increase of temperature, ΔT, we obtain

$$\Delta T = \frac{-\Delta_r H^{\ominus}_{373}}{C^{\ominus}_{P,m}(\text{products at 373 K})}$$

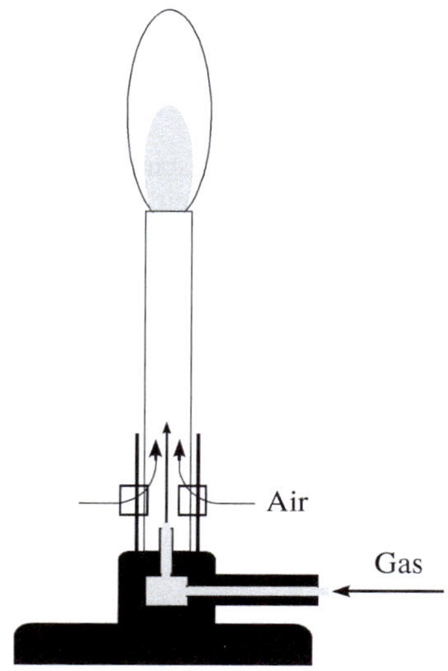

Figure 16B.2 Schematic diagram of a Bunsen burner.

Because the Bunsen burner operates with air (4:1 mixture of N_2 and O_2), nitrogen must be warmed up also. Then

$$C^{\ominus}_{P,m}(\text{products at 373 K}) = 2 \cdot C^{\ominus}_{P,m}(H_2O) + 4 \cdot C^{\ominus}_{P,m}(N_2)$$

In analogy to Example 16.7 we obtain $\Delta_r H^{\ominus}_{373}$ for water in the gas state if we add the heat of evaporation, 87.8 kJ mol^{-1}. Then

$$\Delta_r H^{\ominus}_{373} = (-567.0 + 87.8)\ \text{kJ mol}^{-1} = -479.2\ \text{kJ mol}^{-1}$$

Using the experimental $C_{P,m}$ data in Foundation 16.2 data, $C_{P,m}(H_2O,\ 373\ \text{K}) = 34.0\ \text{J K}^{-1}\ \text{mol}^{-1}$ and $C_{P,m}$ $(N_2, 373\ \text{K}) = 29.2\ \text{J K}^{-1}\ \text{mol}^{-1}$, we obtain

$$C^{\ominus}_{P,m}(\text{products at 373 K}) = (2 \times 34 + 4 \times 29.2)\ \text{J K}^{-1}\ \text{mol}^{-1} = 185\ \text{J K}^{-1}\ \text{mol}^{-1}$$

Then

$$\Delta T = \frac{479.2\ \text{kJ mol}^{-1}}{185\ \text{J K}^{-1}\ \text{mol}^{-1}} = 2590\ \text{K}$$

Actually, the heat capacities are larger at higher temperature (see Fig. 16F.1 and Table 16F2.1); using the mean values 40 and 32 J K^{-1} mol^{-1} for H_2O and N_2, that is $C^{\ominus}_{P,m}$ (products at 373 K) = 208 J K^{-1} mol^{-1}, we obtain $\Delta T = 2304$ K. In reality the process is not completely adiabatic, because it cannot be avoided that the air surrounding the flame is heated as well.[6] In practice, temperatures around 1800 K can be reached with a Bunsen burner.

16.9 Foundations

16.9.1 Foundation 16.1: Deriving the Debye Function (C_V of Solids)

See the books website, www.wiley.com/go/Kuhn/physicalchemistry3e.

16.9.2 Foundation 16.2: How to Calculate $\int_{T_1}^{T_2} \Delta C^{\ominus}_{P,m} \cdot dT$

See the books website, www.wiley.com/go/Kuhn/physicalchemistry3e.

16.10 Exercises

E16.1 Calculate the relative populations of the ground electronic energy level and the first excited electronic energy level for He ($\Delta E = E_2 - E_1 = 31.7 \times 10^{-19}$ J) at the temperatures 300 and 1000 K. (Note: You may find it helpful to use the fact that $e^{-a} = 10^{-a/\ln 10}$, rather than your calculator directly.)

E16.2 Calculate the relative populations of the ground electronic state and the first excited singlet electronic state for H_2 ($\Delta E = 18.3 \times 10^{-19}$ J) at the temperatures 300 and 1000 K.

6 Our assumption that the reaction is complete at this temperature will be justified in Chapter 18.

E16.3 Consider the NO radical whose ground electronic state and first excited electronic state are separated by $\Delta E = 2.4 \times 10^{-21}$ J. How much does the inclusion of electronic excitation modify the heat capacity of NO from that for a diatomic molecule that does not include electronic excitation, at 300 and 1000 K? Compute a percentage error.

E16.4 The molar constant volume heat capacities $C_{V,m}$ of Ne and Xe are the same at $T = 300$ K. What is the value of $C_{V,m}$? Why are their heat capacities the same?

E16.5 The molar heat capacity of a diatomic gas is always higher than that of a monatomic gas. Why is the molar heat capacity of a diatomic gas higher and what range of values may it take?

E16.6 The heat capacity of molecules is temperature dependent, even for an ideal gas of molecules. Explain why this is so.

E16.7 [Adapted from [3]]. *The Setting*: You have two identical containers and they each contain 1 mol of a gas at the same pressure and temperature. In one of the containers you know that the gas is nitrogen, which behaves ideally. In addition, you know that the second container has a pure substance X which behaves as an ideal gas, but you need to determine its identity.

The Clues: a) At 100 K, when vibrational modes are not excited both nitrogen and the unknown gas have the same heat capacity. b) At very high temperatures (when vibrational modes are fully excited) it is found that the molar heat capacity of nitrogen is $C_{V,m} = 29.1\ \mathrm{J\ K^{-1}\ mol^{-1}}$, whereas that of unknown gas is $C_{V,m} = 54.0\ \mathrm{J\ K^{-1}\ mol^{-1}}$.

The Question: What can you say about the molecular structure and properties of unknown gas? Explain your reasoning.

E16.8 Consider the molecule HF which has a characteristic vibrational frequency of $1.24 \times 10^{14}\ \mathrm{s^{-1}}$ and a rotational constant $B = 4.2 \times 10^{-22}$ J. Sketch the constant pressure molar heat capacity $C_{P,m}$ versus temperature of this gas from a temperature of 200 K to a temperature of 1000 K. On the same graph sketch the temperature dependence of the heat capacity of Ne. Explain your reasoning.

E16.9 Why is the constant volume heat capacity always smaller than the constant pressure heat capacity?

E16.10 Consider the series of diatomic molecules HI, ICl, and I_2. Describe how the temperature-dependent constant volume heat capacities of these molecules differ under conditions where they behave as ideal gases. The vibrational frequencies are $\nu_0 = 6.92 \times 10^{13}\ \mathrm{s^{-1}}$ for HI, $\nu_0 = 1.15 \times 10^{13}\ \mathrm{s^{-1}}$ for ICl, and $\nu_0 = 0.64 \times 10^{13}\ \mathrm{s^{-1}}$ for I_2; the rotational constants are $B = 128 \times 10^{-24}$ J for HI, $B = 2.26 \times 10^{-24}$ J for ICl, and $B = 0.74 \times 10^{-24}$ J for I_2. Use isotope 35 for Cl and 127 for I in your calculations.

E16.11 Consider the molecules CH_4, CH_2Cl_2, and CCl_4. Describe how the temperature-dependent constant pressure heat capacities of these molecules differ under conditions where they behave as ideal gases.

E16.12 Using $\Delta_f H^\ominus$ values at $T = 298$ K from Appendix A.1, calculate $\Delta_r H^\ominus$ for the combustion of diamond and of graphite. Use your result to determine the $\Delta_r H^\ominus$ for the formation of diamond from graphite. Perform the same calculation directly by using the heats of formation and compare your answers.

E16.13 Using $\Delta_f H^\ominus$ at $T = 298$ K calculate $\Delta_r H^\ominus$ for the hydrogenation of benzene and cyclohexene to cyclohexane. Use your result to determine the $\Delta_r H^\ominus$ for the hydrogenation of benzene to cyclohexene. Perform the same calculation directly by using the heats of formation and compare your answers.

E16.14 Following an argument like that used for the internal energy, derive Equation (16.61) for the temperature dependence of the enthalpy.

E16.15 Use the bond enthalpies in Tables 16.5 and 16.6 to compute the heat of isomerization of cyclopropane to propylene. Comment on any difference between this value and the one found in E16.19.

E16.16 Using $\Delta_f H^\ominus$ values at $T = 298$ K from Appendix A.2, calculate $\Delta_r H^\ominus$ for the combustion of methane and propane. Which substance generates the most heat per mol? Which generates the most heat per gram?

E16.17 Using $\Delta_f H^\ominus$ values at $T = 298$ K from Appendix A.2, calculate $\Delta_r H^\ominus$ for the combustion of *n*-butane and isobutane. Which substance generates the most heat per mol? Which generates the most heat per gram?

E16.18 Using $\Delta_f H^\ominus$ values at $T = 298$ K from Appendix A.1, calculate the heat of combustion for graphite. How many grams of graphite must be burned in order to warm the air in a 100 $\mathrm{m^3}$ room from 0 to 25 °C. You may assume that the air has a heat capacity like that of pure nitrogen.

E16.19 The heat of combustion of cyclopropane (C_3H_6), graphite (C(s)), and H_2 are −2092 kJ mol^{-1}, −393.3 kJ mol^{-1}, and −285.83 kJ mol^{-1}, respectively. In addition, the heat of formation of propylene (CH_3–CH=CH_2) is $\Delta H_f^{\ominus} = 20.5$ kJ mol^{-1}. Calculate the heat of formation of cyclopropane. Calculate the heat of isomerization of cyclopropane to propylene [Adapted from [3]].

E16.20 Compute the average translational kinetic energy of an argon atom at 300 K. If one converted the argon atom's kinetic energy to gravitational potential energy, how high could one raise the argon atom in the Earth's gravitational field (remember that $g = 9.81$ m s^{-2})? Perform the same calculation for a polymer molecule which has a molar mass of 100 kg mol^{-1}. Comment on your result.

E16.21 Make a pressure versus volume (PV) diagram that corresponds to the state change represented in Fig. 16.13. Provide a graphical interpretation of the work.

E16.22 From the expression for the molar internal energy in Equation (16.26) calculate $C_V = \left(\mathrm{d}U_{\mathrm{el},m}/\mathrm{d}T\right)_V = \mathrm{d}U_{\mathrm{el},m}/\mathrm{d}T$ ($U_{\mathrm{el},m}$ does not depend on V).

E16.23 Using the equation of state for an ideal gas, show that the general expression for $C_P - C_V$, Equation (16.45), reduces to the simple form of Equation (16.44).

E16.24 Using Equation 16.45 calculate $C_{V,m}$ from $C_{P,m}$ data at 298 K for the data in the table; i.e., fill in the missing columns in the table. Comment on your results.

	α_V	κ_T	V_m	$C_{P,m}$	$\frac{\alpha_V^2 V}{\kappa_T} T$	$C_{V,m}$
	10^{-6} K^{-1}	10^{-6} bar^{-1}	mL mol^{-1}	J K^{-1} mol^{-1}		
Benzene	1240	94	89.4	136.3		
Ethanol	1120	111	58.5	112.3		
Water	210	45.8	18	75.3		
Lead	86.7	2.42	17.9	26.4		
Aluminum	69	1.3	9.9	24.4		
Diamond	3.5	0.2	3.4	6.2		

E16.25 What is the isobaric expansion coefficient of an ideal gas? Compare it to those reported for the substances in Table 16.3.

E16.26 What is the isothermal compressibility of an ideal gas? Compare it to those reported for the substances in Table 16.3.

16.11 Problems

P16.1 The MathCAD exercise "workandheat.mcdx" asks you to explore our distinction of the heat and work from an atomistic perspective. You will use a one-dimensional particle in a box model for the energy levels of an ideal monatomic gas (pure translational energy), and you will examine how an adiabatic change in the length of the box affects the internal energy of the gas (this corresponds to "work") and compare that to a change in the energy that can be realized at fixed length by changing the population of the energy levels (this corresponds to "heat"). For the "work" analysis, you are asked to show how the energy level structure changes for three different box lengths, to discuss the relationship between your calculated energy diagram and "macroscopic" energy for 10 gas atoms (assuming the populations of the energy levels is not changing), and to discuss how your model calculations are like performing an adiabatic compression (or expansion). For the "heat" analysis, you are asked to keep the box length fixed and to calculate the distribution of energy level populations; in particular, you compare three different temperatures (100, 400, and 1300 K) to the distribution at 300 K; and in the second part of this part you are asked to describe how the distributions change and what its implications might be. In the last part of the exercise, you are asked to calculate the total internal energy numerically, and to compare it with the value that you would find by integrating over the entire distribution.

P16.2 An ideal gas ($T = 300$ K, $P = 10$ bar, $V_1 = 1$ L) expands isothermally (see Fig. P16.2) by lifting a weight on a piston corresponding to the pressure $P_2 = 1$ bar. What is the volume V_2 of the gas after thermal equilibration, and what is the work w done on the gas and the heat q flowing in the gas during the process?

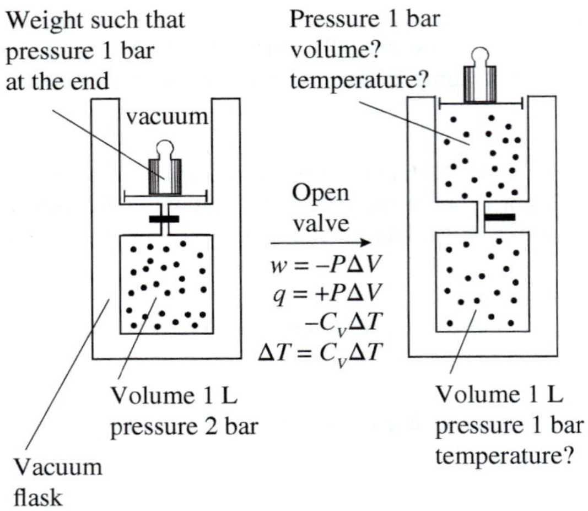

Figure P16.2 Isothermal expansion of an ideal gas.

P16.3 An ideal gas ($T_1 = 300$ K, $P_1 = 10$ bar, $V_1 = 1$ L) expands adiabatically to the pressure $P_2 = 1$ bar. Calculate T_2, V_2, and w.

P16.4 What is the flame temperature of a Bunsen burner that operates with a methane–air mixture? Compare your result with that found for a hydrogen–air mixture in Box 16.2.

P16.5 Consider the values for the experimental constant pressure heat capacity data of nitrogen as a function of temperature, given in the table. Use a spreadsheet program to plot these data. On the same graph plot the translational, the rotational, and the vibrational contribution to the heat capacity, and the sum of these three contributions (use the molecular parameters $k_f = 2296$ Nm^{-1}, $d_0 = 110$ pm).

T/K	300	400	500	600	700	800	900	1000	1100
$C_{P,m}$/J K^{-1} mol^{-1}	28.88	29.35	29.91	30.47	1.02	31.55	32.06	32.54	32.99

P16.6 Consider pure gases of each of the diatomic molecules H_2 ($k_f = 574$ Nm^{-1}), Cl_2 ($k_f = 328$ Nm^{-1}) and I_2 ($k_f = 173$ Nm^{-1}). Calculate the ratio of the number of molecules in the first excited vibrational level to the ground-state vibrational energy level for each of these molecules at 298 K. Do you expect the vibrational contribution to the heat capacity of the gas to be large or small at this temperature?

P16.7 Calculate the vibrational contribution to the heat capacity of Cl_2 gas at 100, 298, 500, 1000, and 5000 K. Plot these values on a graph. What heat capacity would be expected if the vibrational mode was fully excited? [Adapted from [3]].

P16.8 Consider the Debye function for the molar heat capacity $C_{V,m}$ of a solid, Equation (16.33). Find limiting expressions for this function as $T \rightarrow 0$, $T \rightarrow \infty$, and for small T.

P16.9 Consider the isomerization reaction of acetonitrile ($H_3C{-}C{\equiv}N$) to isocyanomethane ($H_3C{-}N{\equiv}C$). The enthalpy of formation of acetonitrile is $\Delta_f H^{\ominus}_{298} = 31.4$ kJ mol^{-1} and the enthalpy of formation of isocyanomethane is $\Delta_f H^{\ominus}_{298} = 149.0$ kJ mol^{-1}. Compute the enthalpy change for the isomerization reaction at 298 K. If the heat capacity of acetonitrile is $C_{P,m} = 39.66$ J K^{-1} mol^{-1}+ 0.242 J K^{-2} mol^{-1} $\cdot\, T$ and the heat capacity of isocyanomethane is $C_{P,m} = 51.05$ J K^{-1} mol^{-1} + 0.201 J K^{-2} mol^{-1} $\cdot\, T$, find the enthalpy change for the isomerization reaction at 750 K.

P16.10 The heat of formation of gaseous isoprene, C_5H_8, is $\Delta_f H^{\ominus}_{298} = 75.5$ kJ mol^{-1}. Compute this heat of formation by using the bond enthalpies (the heat of atomization of graphite is 718 kJ mol^{-1}). Compare the heat of formation you calculate with the value given above. Use this energy difference to estimate the resonance energy of isoprene.

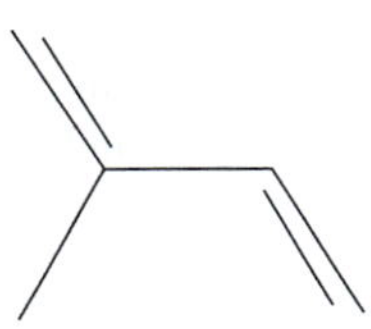

P16.11 Consider the exchange reaction

$$Cl_2\ (g) + 2\ HBr\ (g) \rightleftarrows Br_2\ (g) + 2\ HCl\ (g)$$

a) Given that the heat of formation of HBr is $\Delta_f H^{\ominus}_{298} = -36.40$ kJ mol^{-1} and the heat of formation of HCl is $\Delta_f H^{\ominus}_{298} = -92.31$ kJ mol^{-1}, find the enthalpy change for this reaction at 298 K.
b) Find an expression for the enthalpy change in this exchange reaction in terms of its molecular constants. You may use the high temperature limit expressions for the rotational degrees of freedom and the translational degrees of freedom. For the vibrational partition functions of HBr and HCl you can assume that $\theta_{\text{vib}} \gg T$, but for Br_2 ($\theta_{\text{vib}} \sim 467$ K) and Cl_2 ($\theta_{\text{vib}} \sim 804$ K) you should include the vibrational partition function.
c) How would you expect the $\Delta_r H$ to change as the temperature increases from 298 to 500 K?

P16.12 Consider the Otto Cycle (that used for the internal combustion engine) with the ideal gas as medium. The four steps in the Otto cycle are sketched in Figure P16.12 and may be stated as 1: adiabatic compression stage involving the compression and the expulsion of the gas, 2: isochoric heating at V_2 stage involving the ignition of the fuel, 3: adiabatic expansion stage involving the expansion of the piston, and 4: isochoric cooling at V_1 stage involving cooling at the end of the piston stroke.

Figure P16.12 Otto cycle.

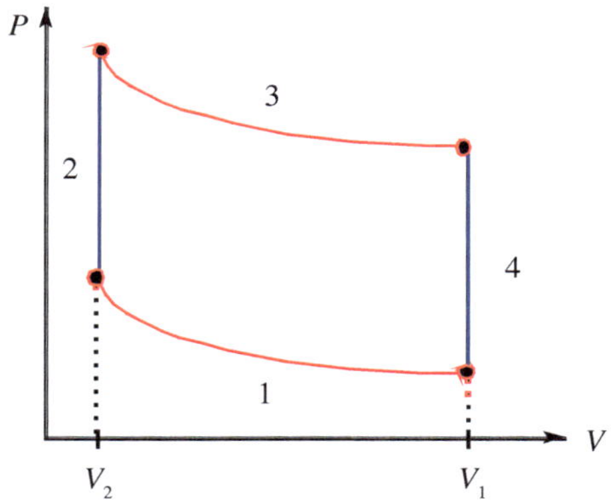

Assume that this cycle operates on a gas that follows the ideal gas law. Consider the gases formed in the combustion to have a heat capacity of $C_{V,m} = 2.5\boldsymbol{R}$. Take the gas which is combusted to be 0.001 moles of pure octane, such that

$$C_8H_{18} + 12.5O_2 \rightleftarrows 8CO_2 + 9H_2O$$

Assume your piston has a $V_2 = 50.0$ mL and a $V_1 = 500$ mL. Assume that your engine operates between 50 °C at the low end and the gas has a temperature of 1200 °C at the high end. Assume that the gas has a temperature of 490 °C after it is compressed but before it is ignited, and that the gas has a temperature of 310 °C after it drives the expansion of the piston but before it is cooled. Find the work performed in each step of the cycle and the total work.

References

1 J. Berkowitz, G.B. Ellison, and D. Gutman, Three methods to measure RH bond energies, *J. Phys. Chem.*, 1994, **98**, 2744.
2 G.H. Aylward and T.J.V. Findlay, *SI Chemical Data* (John Wiley & Sons, New York, 1971) 2nd edition. Adapted from Table 7.2 of Winn, Physical Chemistry.
3 R.E. Dickerson, *Molecular Thermodynamics* (Benjamin, 1969).

17

Reversible Work w_{rev}, Reversible Heat q_{rev}, and Entropy S

CHAPTER MENU

17.1 Introduction

Chapters 15 and 16 have discussed the energetics of molecular assemblies and the energy changes associated with transitions between thermodynamic states. We ask, "Why does a system transform from one equilibrium state to another?" For example, a gas expands into a vacuum spontaneously, but the reverse process does not happen. We will see that although the reverse process is possible, it is extraordinarily improbable.

A macroscopic system in a given thermodynamic state can be represented by a number of different quantum states, called the *number of representations* Ω. We will consider a system that changes from an initial state 1 (Ω_1 representations) to a final state 2 (Ω_2 representations) spontaneously, when the process is not hindered kinetically. This change occurs spontaneously when $\Omega_2 > \Omega_1$. The reverse process is not impossible, but the probability $\mathcal{P} = \Omega_1 / \left(\Omega_1 + \Omega_2\right)$ of its occurrence is often extremely small, while the probability $1 - \mathcal{P} = 1 - \Omega_1 / \left(\Omega_1 + \Omega_2\right)$ of the change from state 1 to state 2 is nearly one. Spontaneously occurring processes are called *(thermodynamically) irreversible*. A reversal of the process can only occur when $\Omega_1 \approx \Omega_2$, that is, when the number of representations does not change significantly; such a process is called a *(thermodynamically) reversible process*.

In order to judge whether a process can occur spontaneously, it is of fundamental importance to know how Ω changes in proceeding from one thermodynamic state to another. We discuss how the number Ω is calculated for simple cases. We will find that it is convenient to work with the logarithm of Ω, so we define a quantity S by

$$S \equiv k \cdot \ln \Omega$$

S is called the *entropy* of the system and k is the Boltzmann constant.

For processes in an isolated system the entropy S always increases, because the number of representations Ω increases. The relation

$$\Delta S \geq 0 \qquad \text{(isolated system)}$$

Principles of Physical Chemistry, Third Edition. Hans Kuhn, David H. Waldeck, and Horst-Dieter Försterling.

Companion website: www.wiley.com/go/Kuhn/physicalchemistry3e

is the fundamental condition for a spontaneous process (e.g., a chemical reaction that proceeds spontaneously), and therefore it is of great practical importance to know the entropies of reacting species in order to calculate entropy changes. S can be directly related to calorimetric properties of a system. This feature allows us to find S for systems that may be too complex to calculate Ω accurately from first principles. Before proceeding with our exposition on spontaneity and number of representations in a system, we discuss some basic aspects of thermodynamic changes.

17.2 Irreversible and Reversible Changes of State

Before proceeding with our exposition on spontaneity and the number of representations in a system, we discuss some basic aspects of thermodynamic changes.

17.2.1 Irreversible Changes

In Section 14.2, we considered the thermal equilibration of two bodies at different temperatures and in diathermal contact. From our experience, we know that heat flows from the higher temperature body to the lower temperature body until they have the same temperature. In addition, the reverse process does not occur; that is, two bodies with the same temperature do not evolve spontaneously to a state where they have different temperatures.

No net heat flows spontaneously from a colder to a warmer body, even if energy is conserved. This statement, established empirically, was also deduced by statistical considerations of large numbers of molecules. How can we understand this phenomenon for a container of gas molecules? The fast molecules in the hot container (Fig. 17.1a) exchange energy with the slow molecules in the cold container. The reverse process, in which one body gets warmer and the other one gets colder is extremely improbable. The greater the number of particles in the two containers, the more improbable it is that the temperature in one container exceeds the temperature in the other, unless we interfere with the equilibration in some way. For macroscopic systems we always have such large numbers of molecules that we can view the reverse of temperature equilibration as practically impossible. These considerations are discussed more in Section 17.5.1 and reveal the statistical nature of heat flow.

Processes which occur by themselves only in a certain direction are called *irreversible*, because they cannot be reversed without external interaction. Examples of irreversible processes are temperature equilibration, mixing of two gases, expansion of a gas into a vacuum, and evaporation in vacuum (see Fig. 17.1).

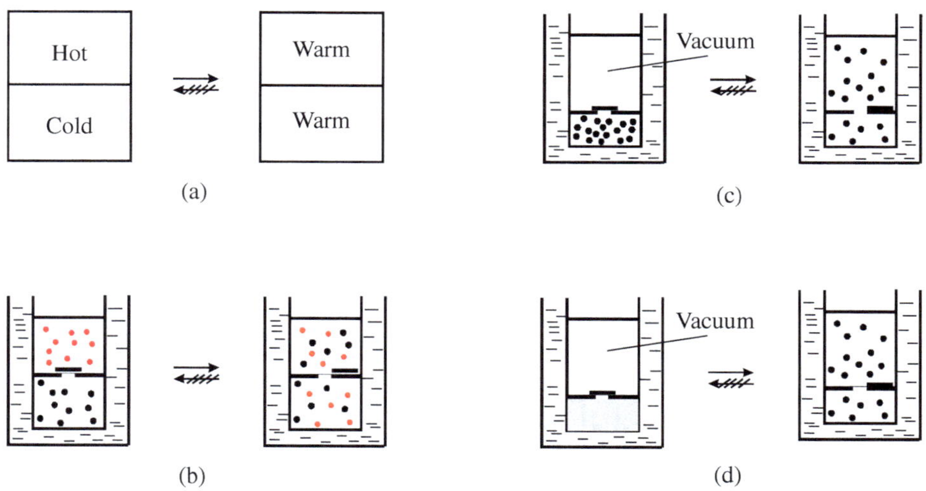

Figure 17.1 Examples of irreversible processes. (a) Temperature equilibration of two closed systems in diathermal contact. (b) Mixing of two gases. (c) Expansion of a gas into a vacuum. (d) Evaporation into a vacuum.

17.2.2 Reversible Changes

The change of state for each case in Fig. 17.1 can also be carried out in a more controlled way so that the corresponding system is close to equilibrium at each instant. For example, consider a fluid with equilibrium vapor pressure P_0 in a temperature bath as shown in Fig. 17.2 (State 1). If we apply an external pressure that is greater than the vapor pressure by a very small amount ΔP, then the vapor will condense and thermal equilibrium will be maintained by heat exchange with

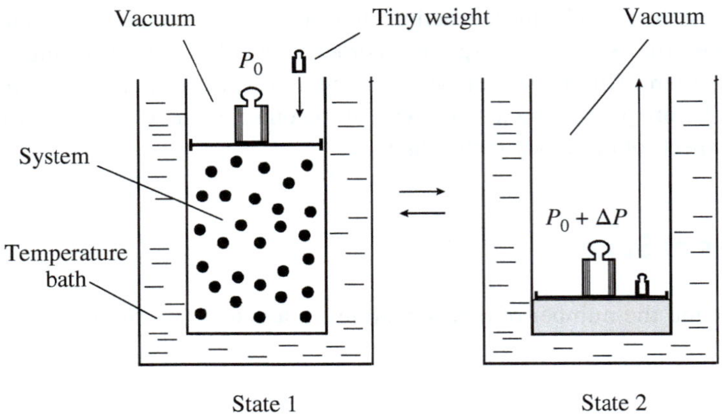

Figure 17.2 Condensation and evaporation of a liquid in a reversible path (e.g., water at 373 K, $P_0 = 1$ bar). State 1: liquid completely evaporated under the equilibrium vapor pressure P_0. State 2: liquid completely condensed under the pressure $P_0 + \Delta P$. Process state 1 → state 2: pressure minimally increased by adding a tiny weight. Process state 2 → state 1: pressure minimally decreased by taking off the tiny weight.

the bath (Fig. 17.2, State 2). If we remove the tiny additional weight, then the liquid again will slowly evaporate completely (Fig. 17.2, State 1). This process can be reversed at any time by adding the tiny additional weight. Because the additional weight can be made arbitrarily small, a negligibly small increment of work from the outside is expended consecutively in reversing the evaporation process.

Processes which occur in the manner described above are called reversible, because their direction can be reversed through an arbitrarily small change of a state variable (in this case the pressure). Below we discuss two important types of reversible processes (isothermal processes and adiabatic processes) more quantitatively, and we show how they can be combined to create a cyclic process.

17.2.2.1 Reversible Isothermal Expansion of an Ideal Gas

How can the expansion of a gas (Fig. 17.1c) be carried out on a reversible path? Consider the reversible isothermal expansion of an ideal gas from pressure P_1 to pressure P_2. We imagine the pressure decreases in small steps ΔP as we remove small weights, one after the other, so that the gas pressure matches the externally applied pressure (Fig. 17.3). We do this very slowly to allow heat exchange between the system and temperature bath so that the temperature of the system is constant during the process.

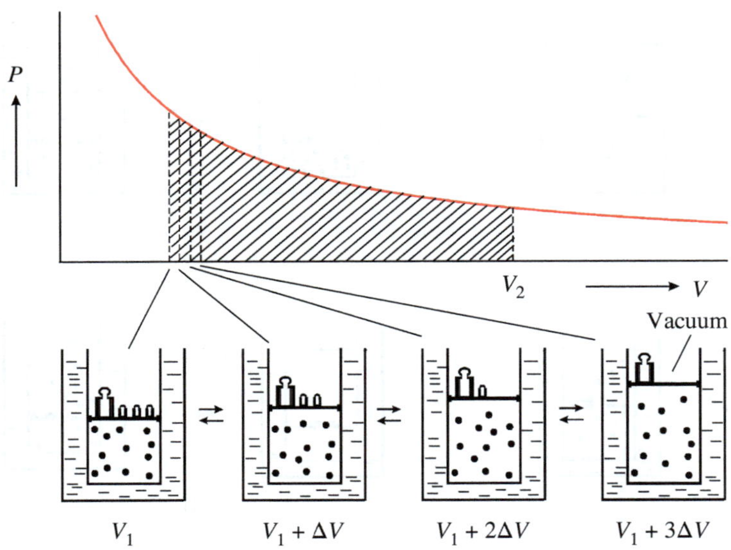

Figure 17.3 Reversible isothermal expansion of an ideal gas. Expansion from volume V_1 to volume V_2. Isothermal process in contact with a temperature bath. Reversible process: decrease of the pressure by stepwise removal of (infinitesimal) tiny weights. Calculation of reversible work w_{rev} (the striped area in the *PV* diagram corresponds to $\int_{V_1}^{V_2} P \cdot dV = \boldsymbol{nRT} \cdot \ln \frac{V_2}{V_1}$, according to Equation (17.2)).

If we perform the process at an infinitesimal rate, then the work in each step is $-P \cdot dV$. The total work in the expansion is then

$$w_{rev} = -\int_{V_1}^{V_2} P \cdot dV = -\int_{V_1}^{V_2} \boldsymbol{nRT} \cdot \frac{dV}{V} \tag{17.1}$$

where we have used the ideal gas law. Because the temperature is constant in the process we can take $\boldsymbol{nR}T$ outside the integral, and write

$$\boxed{w_{\text{rev}} = -\boldsymbol{nR}T \int_{V_1}^{V_2} \frac{\mathrm{d}V}{V} = -\boldsymbol{nR}T \ln \frac{V_2}{V_1}} \tag{17.2}$$

This integral corresponds to the striped area in Fig. 17.3.

Because the temperature is constant and the gas is ideal, from the law of conservation of energy we obtain

$$\Delta U = q_{\text{rev}} + w_{\text{rev}} = 0 \tag{17.3}$$

in the process. Consequently, the heat that is transferred from the temperature bath to the gas exactly compensates the work, and we find

$$\boxed{q_{\text{rev}} = -w_{\text{rev}} = \boldsymbol{nR}T \cdot \ln \frac{V_2}{V_1}} \tag{17.4}$$

Example 17.1 *Calculate the reversible work to expand $\boldsymbol{n} = 0.2$ mol of an ideal gas isothermally from $V_1 = 7$ L to $V_2 = 20$ L at $T = 400$ K. What are the initial and final pressures P_1 and P_2?*

Using Equation (17.4), we obtain

$$w_{\text{rev}} = -\boldsymbol{nR}T \cdot \ln \frac{V_2}{V_1} = -0.2\ \text{mol} \cdot \boldsymbol{R} \cdot 400\ \text{K} \cdot \ln \frac{20\ \text{L}}{7\ \text{L}} = -698\ \text{J}$$

and $q_{\text{rev}} = +698$ J. The energy transferred from the temperature bath to the system (heat q_{rev}) when isothermally expanding a gas is converted into mechanical energy (work $-w_{\text{rev}}$ done by the system). The initial pressure of the gas is

$$P_1 = \frac{\boldsymbol{nR}T}{V_1} = 0.95\ \text{bar}$$

and its final pressure is

$$P_2 = \frac{\boldsymbol{nR}T}{V_2} = 0.33\ \text{bar}$$

In any irreversible process leading from state 1 to state 2 the work $-w$ done by the system is smaller than $-w_{\text{rev}}$. This is seen in performing the thought experiment in Fig. 17.3, but now removing stepwise weights that are not infinitesimally small. Again, we wait for temperature equilibration in each step. Then the work $-w$, given by the gray area in Fig. 17.4, is smaller than the work $-w_{\text{rev}}$, shown in Fig. 17.3 (striped area):

$$(-w_{\text{rev}}) \geq (-w) \tag{17.5}$$

Because of

$$q_{\text{rev}} = -w_{\text{rev}} \tag{17.6}$$

heat flowing from the temperature bath to the system is largest when using a reversible pathway.

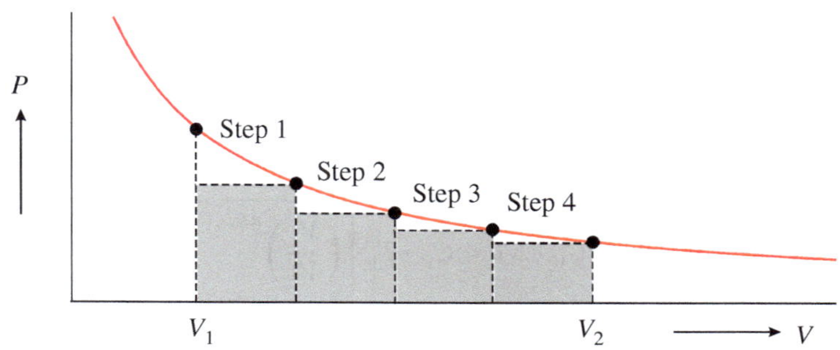

Figure 17.4 Isothermal expansion of an ideal gas as in Fig. 17.3, but the pressure decreases in large steps from volume V_1 to volume V_2. The gray area in the PV diagram corresponds to $\int_{V_1}^{V_2} P \cdot \mathrm{d}V = P_1 \cdot \Delta V + P_2 \cdot \Delta V + P_3 \cdot \Delta V + P_4 \cdot \Delta V$, where $\Delta V = (V_2 - V_1)/4$. The gray area shown here is smaller than the striped area in Fig. 17.3.

17.2.2.2 Reversible Adiabatic Expansion of an Ideal Gas

How much mechanical work w_{rev} is done on the system when an ideal gas is expanded from volume V_1 (at temperature T_1) to volume V_2 (at temperature T_2) in a heat insulated cylinder (thus adiabatically, see Fig. 17.5a) by a reversible path?

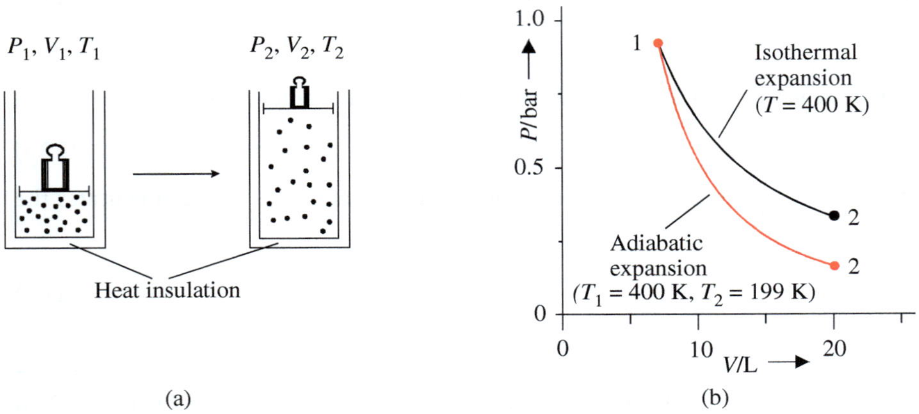

Figure 17.5 Reversible adiabatic expansion of an ideal gas ($n = 0.2$ mol). (a) Experimental setup. The gas is in a heat insulated container. The expansion is performed in small steps, as indicated in Fig. 17.3. (b) *PV* diagram (red solid curve). The gas is expanded from $V_1 = 7$ L and $T_1 = 400$ K ($P_1 = 0.95$ bar) to a final volume of $V_2 = 20$ L. The final temperature is $T_2 = 199$ K, the final pressure is $P_2 = 0.16$ bar. For comparison, the curve for a reversible isothermal expansion is shown (black solid line). For the calculation of the solid red line see Equation 17.16 (with $V_1 = 7$ L, $P_1 = 0.95$ bar, and $C_P/C_V = 5/3$) and Example 17.2.

For an adiabatic change of state, $q = 0$ and from the law of conservation of energy we obtain $\Delta U = w$. Then, for the reversible process,

$$w_{\mathrm{rev}} = \Delta U \tag{17.7}$$

If the temperature changes by $\mathrm{d}T$, the change of the internal energy is

$$\mathrm{d}U = C_V \cdot \mathrm{d}T \tag{17.8}$$

The work done on the system in this step is

$$-P \cdot \mathrm{d}V = -\frac{\boldsymbol{nR}T}{V} \cdot \mathrm{d}V \tag{17.9}$$

Then, because of $-P \cdot \mathrm{d}V = \mathrm{d}U$,

$$C_V \cdot \mathrm{d}T = -\frac{\boldsymbol{nR}T}{V} \cdot \mathrm{d}V \quad \text{or} \quad \frac{\mathrm{d}T}{T} = -\frac{\boldsymbol{nR}}{C_V} \cdot \frac{\mathrm{d}V}{V} \tag{17.10}$$

and by integration we obtain

$$\ln\frac{T_1}{T_2} = -\frac{\boldsymbol{nR}}{C_V} \cdot \ln\frac{V_1}{V_2} = \ln\left(\frac{V_1}{V_2}\right)^{-\boldsymbol{nR}/C_V} \tag{17.11}$$

and

$$T_2 = T_1 \cdot \left(\frac{V_1}{V_2}\right)^{\boldsymbol{nR}/C_V} \tag{17.12}$$

or

$$\boxed{\Delta T = T_2 - T_1 = T_1\left[\left(\frac{V_1}{V_2}\right)^{\boldsymbol{nR}/C_V} - 1\right]} \tag{17.13}$$

This equation relates the change in volume from V_1 to V_2 to the change in temperature from T_1 to T_2 for an adiabatic process. Now we are able to calculate the reversible work w_{rev}, according to Equation (17.7), by

$$\boxed{w_{\mathrm{rev}} = \Delta U = C_V \cdot \Delta T = C_V \cdot T_1\left[\left(\frac{V_1}{V_2}\right)^{\boldsymbol{nR}/C_V} - 1\right]} \tag{17.14}$$

(assuming that C_V does not change in the temperature range from T_1 to T_2).

To calculate the PV diagram for the reversible adiabatic expansion we combine Equation (17.12) with the ideal gas law

$$P_1 V_1 = \boldsymbol{nR}T_1, \qquad P_2 V_2 = \boldsymbol{nR}T_2$$

$$\left(\frac{V_1}{V_2}\right)^{\boldsymbol{nR}/C_V} = \frac{T_2}{T_1} = \frac{P_2 V_2}{P_1 V_1}$$

and

$$\left(\frac{V_1}{V_2}\right)^{\boldsymbol{nR}/C_V} \frac{V_1}{V_2} = \frac{P_2}{P_1} \quad \text{or} \quad \left(\frac{V_1}{V_2}\right)^{\boldsymbol{nR}/C_V+1} = \frac{P_2}{P_1} \tag{17.15}$$

For an ideal gas we can write

$$\frac{\boldsymbol{nR}}{C_V} + 1 = \frac{\boldsymbol{nR} + C_V}{C_V} = \frac{C_P}{C_V}$$

Furthermore, when starting the expansion at volume V_1, relation (17.15) holds for any final volume V. Then for the PV diagram we obtain

$$\boxed{P = P_1 \cdot \left(\frac{V_1}{V}\right)^{C_P/C_V}} \tag{17.16}$$

This diagram is plotted for $V_1 = 7$ L, $P_1 = 0.95$ bar, and $C_P/C_V = 5/3$ in Figure 17.5 (red solid line).

Example 17.2 *Calculate the reversible work arising from the adiabatic expansion of* $\boldsymbol{n} = 0.2$ *mol of an ideal atomic gas from* $V_1 = 7$ *L* ($T_1 = 400$ *K) to a volume* $V_2 = 20$ *L. What is the final temperature* T_2*? What are the initial and final pressures* P_1 *and* P_2*? Compare to a reversible isothermal process at 400 K.*

For $V_1 = 7$ L, $V_2 = 20$ L V_1, $T_1 = 400$ K, and $C_V = \frac{3}{2}\boldsymbol{nR}$ (atomic gas) from Equation (17.13) we obtain

$$\Delta T = 400 \text{ K} \cdot \left[\left(\frac{7}{20}\right)^{2/3} - 1\right] = -400 \text{ K} \cdot 0.503 = -201 \text{ K}$$

and $T_2 = 199$ K. Equation (17.14) gives

$$w_{\text{rev}} = \tfrac{3}{2}\boldsymbol{nR} \cdot \Delta T = \tfrac{3}{2} 0.2 \text{ mol} \cdot \boldsymbol{R} \cdot (-201 \text{ K}) = -502 \text{ J}$$

The pressures P_1 and P_2 are

$$P_1 = \frac{\boldsymbol{nR}T_1}{V_1} = \frac{0.2 \text{ mol} \cdot \boldsymbol{R} \cdot 400 \text{ K}}{7 \text{ L}} = 0.95 \text{ bar}$$

and

$$P_2 = \frac{\boldsymbol{nR}T_2}{V_2} = \frac{0.2 \text{ mol} \cdot \boldsymbol{R} \cdot 289 \text{ K}}{20 \text{ L}} = 0.16 \text{ bar}$$

The corresponding PV diagram calculated from Equation (17.16) is shown in Fig. 17.5b, lower curve.

In Example 17.1 we showed that the corresponding reversible isothermal expansion gave a reversible work $w_{\text{rev}} = -698$ J and a final pressure of $P_2 = 0.33$ bar (Fig. 17.5b, upper curve). P_2 in the isothermal process is larger than in the adiabatic process, due to the higher final temperature of the gas. Correspondingly, the reversible work done by the system is larger in the isothermal process, as can be seen from the areas under both curves in Fig. 17.5b.

17.2.2.3 Carnot Cycle

We consider the special case of a reversible cyclic process, namely, a cyclic process with an ideal gas connected to two temperature baths (*Carnot cycle*). In Section 17.6 we will use the Carnot cycle as an important construct to derive basic relations between entropy change and reversible heat transfer. Here we consider changes of state for an ideal gas, and in Section 17.6 we generalize for arbitrary processes. The ideal gas is first in contact with bath T_2 and is expanded (isothermal expansion) from volume V_a to volume V_b (Fig. 17.6a). In this process, heat $q_{\text{rev},2}$ is transferred from bath T_2 to the gas in order to maintain its temperature. Finally the gas is removed from the bath and expanded adiabatically (no heat exchange

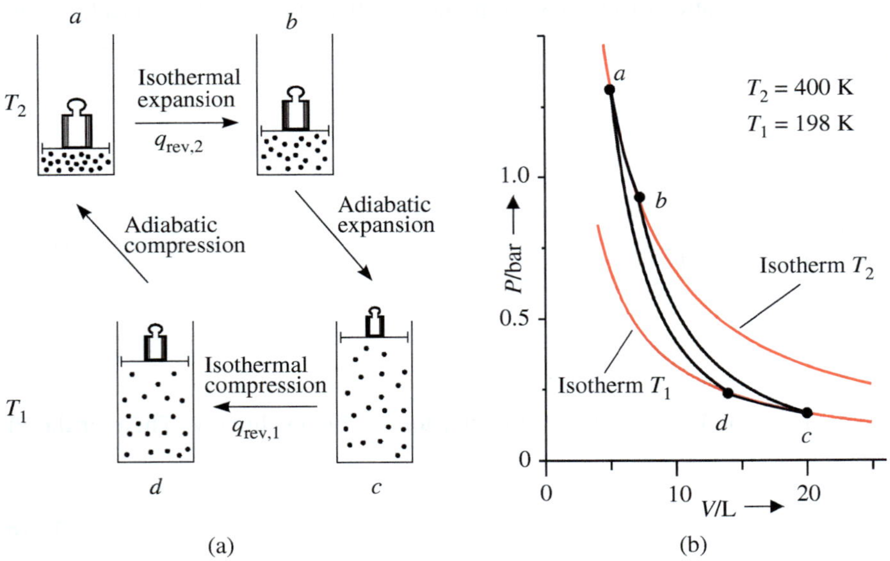

Figure 17.6 Carnot process. (a) Experimental setup. (b) *PV* diagram for the cycle. The isotherms ($a \to b$, $c \to d$, ideal gas law) and the adiabats ($b \to c$, $d \to a$, see Section 17.2.2.2) are calculated for an ideal atomic gas ($\boldsymbol{n} = 0.2$ mol, $C_P/C_V = 5/3$) for $T_2 = 400$ K and $T_1 = 198$ K. Points a ($V_a = 5$ L), b ($V_b = 7$ L), c ($V_c = 20$ L), and d ($V_d = 14.3$ L) correspond to the locations of the piston in (a).

with the bath) until it is cooled down to temperature T_1 (volume V_c). Next it is contacted with bath T_1 and compressed from volume V_c to volume V_d. Heat $-q_{rev,1}$ is transferred from the gas to bath T_1. Finally the gas is removed from the bath and further compressed adiabatically (no heat exchange with the bath) until it is heated up to temperature T_2 and has reached its initial state (volume V_a).

The reversible work done on the system in the cycle is

$$w_{rev} = w_{rev,a \to b} + w_{rev,b \to c} + w_{rev,c \to d} + w_{rev,d \to a} \tag{17.17}$$

where

$$w_{rev,a \to b} = -\boldsymbol{nR}T_2 \cdot \ln \frac{V_b}{V_a}, \qquad w_{rev,c \to d} = -\boldsymbol{nR}T_1 \cdot \ln \frac{V_d}{V_c} \tag{17.18}$$

and, according to Equation (17.14),

$$w_{rev,b \to c} = C_V \left(T_1 - T_2\right), \qquad w_{rev,d \to a} = C_V \left(T_2 - T_1\right) \tag{17.19}$$

thus

$$w_{rev} = -\boldsymbol{nR}T_2 \cdot \ln \frac{V_b}{V_a} - \boldsymbol{nR}T_1 \cdot \ln \frac{V_d}{V_c} \tag{17.20}$$

On the other hand, from Equation (17.12) it follows that

$$\left(\frac{V_b}{V_c}\right)^{\boldsymbol{nR}/C_V} = \frac{T_1}{T_2} \quad \text{and} \quad \left(\frac{V_d}{V_a}\right)^{\boldsymbol{nR}/C_V} = \frac{T_2}{T_1}$$

Then we obtain

$$\left(\frac{V_b}{V_c}\right)^{\boldsymbol{nR}/C_V} = \left(\frac{V_a}{V_d}\right)^{\boldsymbol{nR}/C_V} \quad \text{and} \quad \frac{V_b}{V_c} = \frac{V_a}{V_d}$$

or

$$\frac{V_b}{V_a} = \frac{V_c}{V_d} \tag{17.21}$$

Combining this result with Equation (17.20) we obtain

$$\boxed{w_{rev} = -\boldsymbol{nR}\left(T_2 - T_1\right) \cdot \ln \frac{V_b}{V_a}} \tag{17.22}$$

The reversible heat q_{rev} supplied to the system in the cycle is the sum of the heat supplied at temperature T_2 ($q_{rev,2} = q_{rev,a \to b}$) and the heat supplied at temperature T_1 ($q_{rev,1} = q_{rev,c \to d}$)

$$q_{rev} = q_{rev,2} + q_{rev,1} \tag{17.23}$$

where

$$q_{\text{rev},2} = +\boldsymbol{nR}T_2 \cdot \ln \frac{V_b}{V_a}, \qquad q_{\text{rev},1} = +\boldsymbol{nR}T_1 \cdot \ln \frac{V_d}{V_c} \tag{17.24}$$

or, together with Equation (17.21),

$$\boxed{q_{\text{rev},2} = +\boldsymbol{nR}T_2 \cdot \ln \frac{V_b}{V_a}, \qquad q_{\text{rev},1} = -\boldsymbol{nR}T_1 \cdot \ln \frac{V_b}{V_a}} \tag{17.25}$$

Then

$$q_{\text{rev}} = +\boldsymbol{nR}\left(T_2 - T_1\right) \cdot \ln \frac{V_b}{V_a} \tag{17.26}$$

that is, $w_{\text{rev}} + q_{\text{rev}} = 0$, as required for a cyclic process where $\Delta U = 0$.

Example 17.3 *Consider a Carnot cycle operating between temperatures $T_1 = 198$ K and $T_2 = 400$ K with $\boldsymbol{n} = 0.2$ mol of an ideal atomic gas (Fig. 17.6b). Start at point a ($V_a = 5$ L) and expand isothermally to point b ($V_b = 7$ L). What are the volumes at points c and d? Calculate the reversible work and the reversible heat in one cycle.*

The adiabatic process from point b to point c is the same as described in the previous example; thus, $V_c = 20$ L. To obtain the volume at point d, we use Equation (17.12) with $C_V = \frac{3}{2}nR$

$$V_d = V_a \left(\frac{T_2}{T_1}\right)^{C_V/(\boldsymbol{nR})} = 5\ \text{L}\left(\frac{400}{198}\right)^{3/2} = 14.3\ \text{L}$$

According to Equation (17.22)

$$w_{\text{rev}} = -\boldsymbol{nR}\left(T_2 - T_1\right) \cdot \ln \frac{V_b}{V_a} = -0.2\ \text{mol} \cdot \boldsymbol{R}(400\ \text{K} - 198\ \text{K}) \cdot \ln \frac{7}{5}$$
$$= -113\ \text{J}$$

According to Equation 17.25, in one cycle the heat

$$q_{\text{rev},a\to b} = +\boldsymbol{nR}T_2 \cdot \ln \frac{V_b}{V_a} = 223.8\ \text{J}$$

is taken from temperature bath T_2 and the heat $-q_{\text{rev},c\to d} = 110.8$ J is supplied to the temperature bath T_1. The difference (223.8 J – 110.8 J) = 113 J is transformed into work ($w_{\text{rev}} = -113$ J).

17.2.2.4 Efficiency of a Carnot Engine

The Carnot cycle can be considered as a *heat engine*. The engine is connected to two temperature baths at temperatures T_1 and T_2. In one cycle of the engine, part of the heat supplied to the system from the warmer bath ($q_{\text{rev},2}$ at temperature T_2) is transferred into reversible work w_{rev}, the remaining heat ($-q_{\text{rev},1}$) is supplied to the colder temperature bath (Fig. 17.7a). The *efficiency* η of the engine is defined by

$$\eta = \frac{-w_{\text{rev}}}{q_{\text{rev},2}} \tag{17.27}$$

where $-w_{\text{rev}}$ is the reversible work done by the engine. Using Equations (17.22) and (17.25) we obtain

$$\boxed{\eta = \frac{\boldsymbol{nR}\left(T_2 - T_1\right) \cdot \ln \frac{V_b}{V_a}}{\boldsymbol{nR}T_2 \cdot \ln \frac{V_b}{V_a}} = \frac{T_2 - T_1}{T_2} = 1 - \frac{T_1}{T_2}} \tag{17.28}$$

Moreover, it follows that

$$\boxed{\frac{q_{\text{rev},1}}{T_1} + \frac{q_{\text{rev},2}}{T_2} = 0 \qquad \text{(cyclic process)}} \tag{17.29}$$

This is a remarkable result, derived for a Carnot engine working with ideal gases. We will discuss it further in Section 17.6.

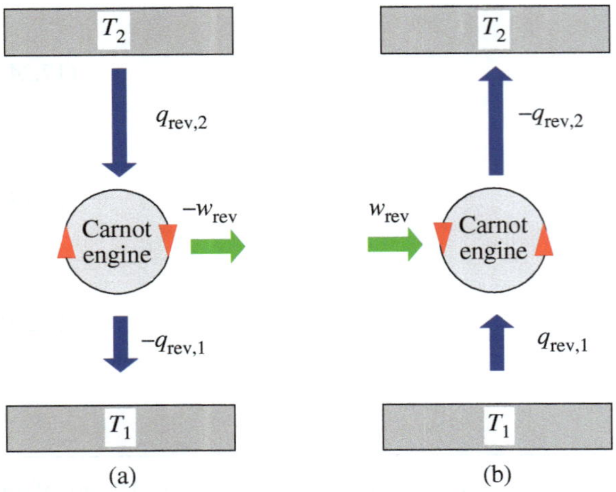

Figure 17.7 Flow of heat and work in the Carnot engine (a) Carnot engine working as a heat engine; (b) Carnot engine working as a heat pump. The arrows in the circles indicate the direction of the cycle: clockwise as in Fig. 17.6 ($a \to b \to c \to d \to a$), or anticlockwise ($a \to d \to c \to b \to a$).

Example 17.4 *What is the efficiency of a Carnot engine working between temperatures $T_2 = 373$ K and $T_1 = 273$ K?*

According to Equation (17.28) we obtain

$$\eta = 1 - \frac{T_1}{T_2} = 1 - \frac{273}{373} = 0.27$$

This means that only 27% of the energy supplied by the warmer temperature bath can be transferred into mechanical work.

So far we considered a Carnot engine running in the same direction as indicated in Fig. 17.6 ($a \to b \to c \to d \to a$). If the cycle is carried out in the reverse direction ($a \to d \to c \to b \to a$, Fig. 17.7b), mechanical work is supplied to the system, and heat is transferred from the colder bath into the warmer bath. Then only 27% of the transferred heat is taken from the mechanical energy, and 73% is taken from the colder bath. This is the working principle of a *heat pump*.

17.3 Counting the Number of Representations of a Thermodynamic State, Ω

In Chapter 15 we introduced the quantity Ω for the number of ways to represent a population of N particles in a given volume, distributed between distinct energy states at constant total energy, see Section 15.8. Now we generalize the quantity Ω to describe the number of quantum states representing a given thermodynamic state. To better understand the origin of spontaneity we must consider how Ω changes as the thermodynamic system evolves. We will see that spontaneous processes occur by proceeding to the thermodynamic state with the largest value of Ω, the largest number of representations.

17.3.1 Distribution Possibilities

Finding Ω is involved, so let us begin by considering the relatively simple case of the irreversible expansion of a classical ideal gas (Fig. 17.8) under isothermal conditions. We begin in state 1 which has a volume V_1. Next we open a valve that connects the chamber containing the gas to a second chamber which has been evacuated of gas molecules. The final state 2 has a larger volume V_2 available to the molecules. This process cannot be run in the opposite direction by a minimal change of the conditions. Although it is not absolutely impossible that the molecules transform from state 2 to state 1 spontaneously, it is extraordinarily improbable.

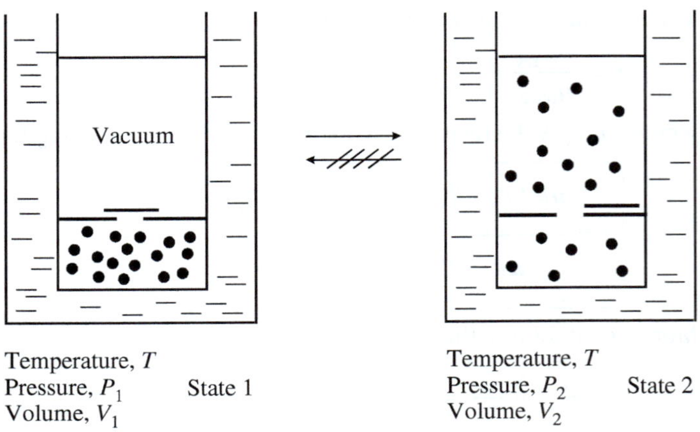

Figure 17.8 Expansion of an ideal gas into a vacuum.

The probability $\mathcal{P}$ that all molecules appear in subvolume V_1 at the same time by chance can be calculated in a simple way when the molecules are considered as noninteracting particles. To simplify the argument we consider the case where

$$V_1 = \frac{1}{2}V_2 \tag{17.30}$$

and the molecules do not interact with each other (i.e., an ideal gas). Imagine that only one particle is present in the container. On average, it is equally likely that the particle is found in the upper chamber as in the lower chamber (Fig. 17.9a). There are two possibilities for the particle to stay in V_2, and one possibility to stay in V_1. So the probability $\mathcal{P}$ that the particle is in the lower subvolume is

$$\mathcal{P} = \tfrac{1}{2} \tag{17.31}$$

For two particles, four a priori equally probable possibilities exist, as shown in Fig. 17.9b. Only in one case are both particles in the lower subvolume. So the probability of finding both particles in the lower subvolume is

$$\mathcal{P} = \tfrac{1}{4} = \left(\tfrac{1}{2}\right)^2 \tag{17.32}$$

With three particles eight possibilities exist because two possibilities exist for adding the third particle; hence, $2 \times 4 = 8$ total possibilities. The probability that all three particles are found in the lower subvolume is

$$\mathcal{P} = \tfrac{1}{8} = \left(\tfrac{1}{2}\right)^3 \tag{17.33}$$

The argument is readily extended to more particles, and for N particles the probability of finding all the particles in the bottom chamber is

$$\mathcal{P} = \left(\tfrac{1}{2}\right)^N \tag{17.34}$$

In cases of practical interest, N is extraordinarily large. If we imagine 1 mol of a gas in the container, then N is of the order of 10^{23}. The probability that all the molecules are in the lower half of the container is then

$$\mathcal{P} = \left(\tfrac{1}{2}\right)^{10^{23}} = \frac{1}{2^{(10^{23})}} \approx \frac{1}{10^{(10^{22})}} = \frac{1}{10^{10000000000000000000000}}$$

This probability is extremely small. To appreciate how small, we compare this value with the probability that every human on earth ($N = 10^{10}$) dies of a heart attack within 1 second of each other, purely by chance. The life expectancy of a human is about 10^2 years $\approx 10^9$ seconds, and heart attacks are the cause of death for about a tenth of the population. So the probability that a person chosen at random will suffer a fatal heart attack within a certain given time interval of 1 second equals $0.1 \times 10^{-9} = 10^{-10}$. The probability for everyone to have a heart attack within the same second is then

$$\mathcal{P} = \left(10^{-10}\right)^N = \left(10^{-10}\right)^{10^{10}} = \frac{1}{\left(10^{10}\right)^{(10^{10})}} = \frac{1}{10^{(10^{11})}}$$

This probability is greater by an enormous factor than the probability that 10^{23} molecules will be found, by chance, in the volume V_1.

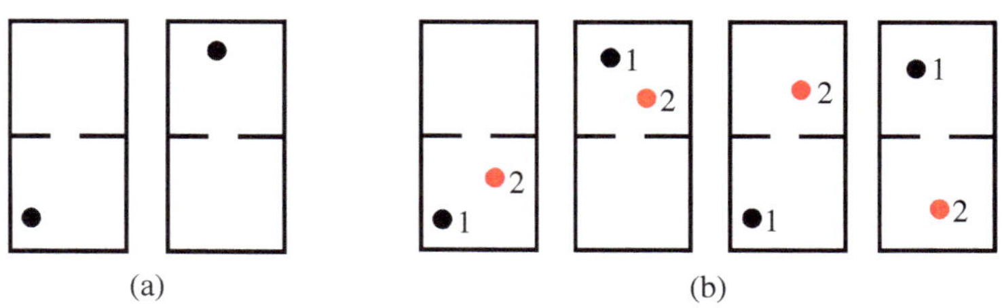

Figure 17.9 (a) A particle can be found in the upper or the lower subvolume, V_1, but is always in the container of volume, V_2. (b) Possibilities to distribute two particles 1 and 2 in two equal subvolumes.

We can understand spontaneity as the tendency of a thermodynamic system to evolve to the most probable representation that is consistent with the macroscopic state. Hence the gas molecules change from state 1 to state 2 (Fig. 17.8), because state 2 has a much greater number of possibilities for the molecular arrangements. Once the system reaches state 2, the probability $\mathcal{P}$ that it returns to state 1 through random fluctuations is very small, practically impossible if the system is large enough.

In Box 17.1 we show that for a gas the probability of finding all the gas particles in a subvolume V_1 is

$$\mathcal{P} = \left(\frac{V_1}{V_2}\right)^N \tag{17.35}$$

where V_2 is the total volume. So the smaller the value of V_1/V_2, the smaller is $\mathcal{P}$. Consequently, the probability $\mathcal{P}$ of finding all the gas molecules in subvolume V_1 decreases as a gas expands in an irreversible state change.

Example 17.5 *Consider the mixing of two gases as illustrated in Fig. 17.1b. In State 1 all the molecules of the first type (black dots) are concentrated in the lower chamber, and all the molecules of the second type (red dots) are concentrated in the upper chamber. When the partition is opened, the molecules distribute over the whole volume until state 2 (fully mixed) is reached. For the case of 4 particles (2 red and 2 black), show that the number of distribution possibilities increases in this process.*

Before mixing (state 1), particles 1 and 2 are in the upper chamber, and particles 3 and 4 are in the lower. We denote this case symbolically as:

1 2 | 3 4

Allowing mixing (state 2) we have 16 distinguishable distribution possibilities (both chambers can be accessed by all particles):

1 2 | 3 4	1 3 | 2 4	1 4 | 2 3	2 3 | 1 4
2 4 | 1 3	3 4 | 1 2	1 | 2 3 4	2 | 1 3 4
3 | 1 2 4	4 | 1 2 3	1 2 3 | 4	1 2 4 | 3
1 3 4 | 2	2 3 4 | 1	1 2 3 4 |	| 1 2 3 4

The number of distribution possibilities has increased from 1 to 16 when proceeding from state 1 (molecules unmixed) to state 2 (molecules mixed). That is, the probability $\mathcal{P}$ of finding the system in state 1 is

$$\mathcal{P} = \frac{1}{16} = 0.0625$$

17.3.2 Number of Representations Ω of an Atomic Gas in a Given Thermodynamic State

We start with a simple model to show the essentials, and then we compare to a rigorous treatment.

17.3.2.1 Approximate Treatment

Let us calculate the number of representations Ω for a gas of N particles that are contained in a cubic volume $V = L^3$ at the temperature T. The average translational kinetic energy of a particle is

$$\overline{E}_{\text{trans}} = \tfrac{3}{2}kT \tag{17.36}$$

Most of the particles are distributed over quantum states between $E = 0$ and an energy which is about twice the value of the average translational energy ($E = 2\overline{E}_{\text{trans}} = 3kT$). In Box 13.2 we calculate the number of quantum states g_{total} between $E = 0$ and an energy E; for $E = 3kT$ we have

$$g_{\text{total}} = \frac{\pi}{6}\left[\frac{8mL^2}{h^2}\right]^{3/2} E^{3/2} = \frac{\pi}{6}\left[\frac{8mL^2}{h^2}\right]^{3/2} (3kT)^{3/2} \tag{17.37}$$

Rearranging the variables, we obtain

$$\begin{aligned} g_{\text{total}} &= \frac{4}{3}\pi 6^{3/2}\frac{(km)^{3/2}}{h^3}L^3 \cdot T^{3/2} \\ &= \frac{4}{3}\pi 6^{3/2}\frac{(km)^{3/2}}{h^3}V \cdot T^{3/2} \end{aligned} \tag{17.38}$$

To simplify the analysis, assume that all g_{total} quantum states are occupied with the same probability and that all states with higher energy are empty (in reality, the occupation probability decreases exponentially with $E/(kT)$). Now we calculate the number of representations, Ω, by proceeding in two steps. First we treat the gas molecules as if they are distinguishable. Then we account for the indistinguishability of the gas particles, starting with the expression for Ω obtained in the first step.

Treatment as Distinguishable Particles Here we treat the gas molecules as if they were distinguishable. Also, we assume that any number of particles can occupy the same quantum state (actually, this is true for bosons, but not for fermions, see Chapter 15). We start with one particle occupying any of the g_{total} quantum states, then $\Omega = g_{\text{total}}$. In the case of two particles, there are g_{total} representations of particle 2 for each representation of particle 1; thus, $\Omega = g_{\text{total}}^2$. We generalize for N particles,

$$\Omega = g_{\text{total}}^N$$

For example, for $g_{\text{total}} = 3$ and $N = 2$, there are altogether $\Omega = 3^2 = 9$ representations (Fig. 17.10).

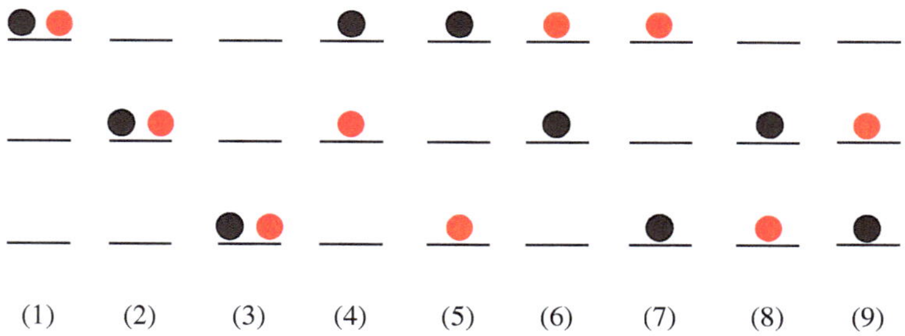

Figure 17.10 Nine possible representations (1)–(9) for two distinguishable particles (red and black filled circles) occupying three quantum states ($N = 2, g = 3$).

Treatment as Indistinguishable Particles Now we take account of the fact that the N particles of our gas are indistinguishable. Hence any possibilities which are formed through exchange of particles should be included only once in the calculation of Ω (see Chapter 15). In our numerical example ($N = 2$, Fig. 17.10), three representations are formed through exchange of particles (6 from 4, 7, from 5, and 9 from 8), so that $\Omega = 6$ instead of $\Omega = 9$ for indistinguishable molecules.

How can we obtain Ω in the general case of N particles? We have proved in Section 15.9 that in the case of a gas at high enough temperature, g_{total} is much greater than N. In this limit, we can easily calculate Ω because most quantum states are empty, a few are singly occupied, and states with more than one particle play practically no role. Therefore, Ω is obtained by dividing the number of representations for distinguishable particles by the number of possibilities to interchange N particles with each other. This number is $N!$, also called the number of permutations. For example, if $N = 3$, then $3! = 6$ permutations of the particle numbers are possible: 123, 132, 213, 231, 312, 321. Thus we find

$$\boxed{\Omega = \frac{g_{\text{total}}^N}{N!} \quad \text{for } g_{\text{total}} \gg N} \tag{17.39}$$

Note that the condition $g_{\text{total}} \gg N$ holds for most cases of chemical interest.

If we approximate $N!$ by Stirling's formula (see Appendix C.3)

$$N! \approx N^N \cdot \mathrm{e}^{-N} = \left(N \cdot \frac{1}{\mathrm{e}}\right)^N \tag{17.40}$$

then from Equation (17.38) we obtain

$$\Omega = \left[\frac{g_{\text{total}} \cdot \mathrm{e}}{N}\right]^N = \left[\frac{4}{3}\pi 6^{3/2}\mathrm{e} \cdot \frac{(km)^{3/2}}{h^3}\frac{V}{N}T^{3/2}\right]^N \tag{17.41}$$

or

$$\ln \Omega = N \cdot \ln \frac{g_{\text{total}} \cdot \mathrm{e}}{N} = N \cdot \ln \left[\frac{4}{3}\pi 6^{3/2}\mathrm{e} \cdot \frac{(km)^{3/2}}{h^3}\frac{V}{N}T^{3/2}\right] \tag{17.42}$$

17.3.2.2 Rigorous Treatment

Foundation 17.1 provides a rigorous derivation for $\ln \Omega$ of an atomic gas and finds that

$$\ln \Omega = N \ln Z_{\text{trans}} + \frac{U}{kT} + (N - N \ln N) \tag{17.43}$$

where Z_{trans} is the molecular partition function for translation. For high enough temperature, the internal energy of translation is $U = \frac{3}{2}NkT$, and we can simplify Equation (17.43) as

$$\begin{aligned} \ln \Omega &= N \ln Z_{\text{trans}} + \frac{3}{2}N + (N - N \ln N) \\ &= N\left(\ln \frac{Z_{\text{trans}}}{N} + \frac{5}{2}\right) = N \ln\left(\frac{Z_{\text{trans}}}{N}\mathrm{e}^{5/2}\right) \end{aligned}$$

In Chapter 15 we found that the molecular translational partition function is

$$Z_{\text{trans}} = \left(\frac{2\pi kTmL^2}{h^2}\right)^{3/2}$$

Substituting this result into the expression for ln Ω gives

$$\ln \Omega = N \ln \left[\left(\frac{2\pi kTmL^2}{h^2}\right)^{3/2} \frac{e^{5/2}}{N}\right] = N \ln \left[e^{5/2}(2\pi)^{3/2}\frac{(km)^{3/2}}{h^3}\frac{V}{N}T^{3/2}\right] \tag{17.44}$$

The dependence of ln Ω on m, V, and T is the same as in Equation (17.42), only the numerical factor is $e^{5/2}(2\pi)^{3/2} = 191.9$ instead of $\frac{4}{3}\pi 6^{3/2}\,e = 167.3$.

ln Ω depends on properties of the particle (its mass m), the number of particles N, and the thermodynamic variables V and T. ln Ω increases with the volume V of the gas because the spacing between neighboring energy levels decreases as the side length L increases, i.e., the volume V increases. For a given internal energy U, this means that more quantum states are available (see Fig. 17.11). ln Ω increases with the mass m of the gas particles, because the spacing of the energy levels decreases with increasing mass. Lastly, ln Ω increases with the temperature T of the gas. For a given volume V, an increase in T means that the energy distribution of gas particles is broader, and more high energy quantum states are populated than at lower temperatures.

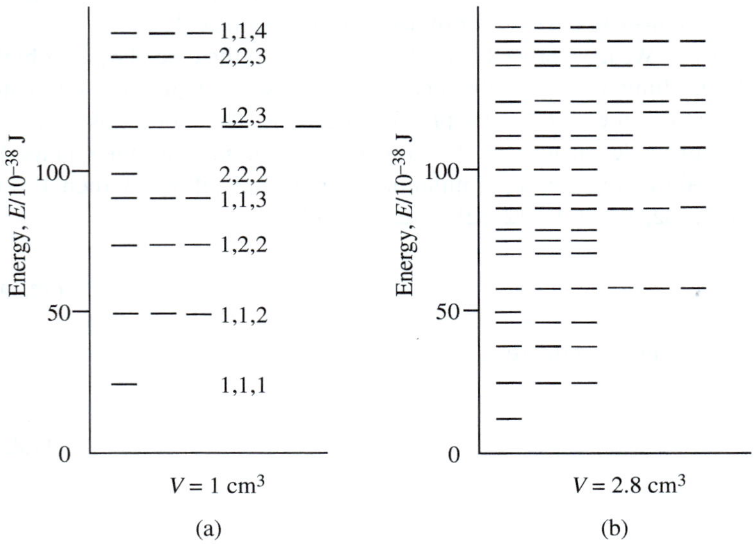

Figure 17.11 The diagrams show the first few translational quantum states for He in a cube of volume (a) $V = 1\ \text{cm}^3$ and (b) $V = 2.8\ \text{cm}^3$. The figure shows that the number of available quantum states increases with increasing volume. The quantum numbers in (a) correspond to n_x, n_y, n_z for the different energy states. The degeneracies occur because of the interchange of n_x, n_y, and n_z leading to the same energy, e.g., (1,1,2), (1,2,1), (2,1,1). Note the lower energies of the quantum states in (b).

Example 17.6 *Consider a gas that has expanded from the volume V_1 to the volume V_2 at constant temperature. Use the result for Ω to evaluate the probability of finding all of the gas molecules in the subvolume V_1.*

Using our expression for Ω, we can calculate the probability $\mathcal{P}$ for finding the gas in state 1. The result is

$$\mathcal{P} = \frac{\Omega_1}{\Omega_2} = \frac{\left[e^{5/2}(2\pi)^{3/2}\frac{(km)^{3/2}}{h^3}\frac{1}{N}T^{3/2}\right]^N}{\left[e^{5/2}(2\pi)^{3/2}\frac{(km)^{3/2}}{h^3}\frac{1}{N}T^{3/2}\right]^N} \cdot \frac{V_1^N}{V_2^N} = \left(\frac{V_1}{V_2}\right)^N$$

This finding agrees with the result (17.35) from our previous approach. Using the volumes $V_1 = 1.0$ L, $V_2 = 3.0$ L, and $N = 1.2 \times 10^{23}$ atoms, we find $\ln \mathcal{P} = -1.3 \times 10^{23}$ or $\mathcal{P} < 10^{-100000000000000000000000}$-extremely unlikely.

Note that Equation (17.44) is valid only under the condition that the number of quantum states is much larger than the number of gas particles, see Equation (17.39). This condition requires that the gas be dilute enough; i.e., the particle number density, N/V, must be sufficiently small.

17.4 The Entropy: $S = k \ln \Omega$

In practice it is convenient to use the quantity

$$\boxed{S = k \cdot \ln \Omega} \tag{17.45}$$

rather than the number of representations Ω for a system of N particles in a given volume with given internal energy U. The quantity S is called the entropy. Because Ω depends only on the thermodynamic state of the system and not on the way in which the thermodynamic state is reached, S is a state variable.

17.4.1 Entropy of Subsystems

Because Ω is an extensive quantity (that is, it depends on the number of molecules in the system), the entropy S is also extensive. For extensive quantities like the internal energy U we showed that U of a large system is the sum of contributions of its parts. Does this statement hold for the entropy S as well?

To clarify this additivity, we consider two subsystems separated by a wall (Fig. 17.12) and assume that the particles in Subsystem I are distinguishable from the particles in Subsystem II (e.g., the molecules in a solid). We let Ω_1 be the number of representations for Subsystem I, and we assume the subsystems are independent. If for each representation in Subsystem I there are Ω_{II} representations for Subsystem II, then the total system has

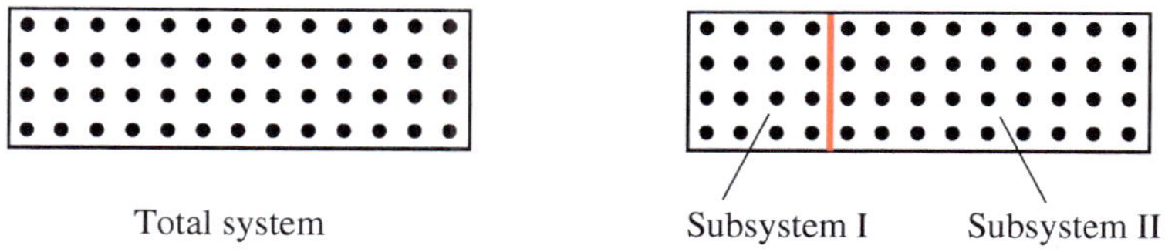

Figure 17.12 Entropy of subsystems in case of a system with a lattice structure: $S_{total} = S_I + S_{II}$, because $\Omega_{total} = \Omega_1 \cdot \Omega_2$.

$$\Omega = \Omega_I \cdot \Omega_{II} \tag{17.46}$$

representations. According to Equation (17.45), we have

$$S_I = k \cdot \ln \Omega_I \quad \text{and} \quad S_{II} = k \cdot \ln \Omega_{II} \tag{17.47}$$

and find that

$$\boxed{S = k \cdot \ln \Omega = k \cdot \ln\left(\Omega_I \cdot \Omega_{II}\right) = k \cdot \left(\ln \Omega_I + \ln \Omega_{II}\right) = S_I + S_{II}} \tag{17.48}$$

Does Equation (17.48) apply to the case when the particles in Subsystem I are indistinguishable from the particles in Subsystem II (e.g., the molecules in a gas)? This situation is more complicated. Consider an atomic gas (N particles) in a container with volume V and divide the container into two subsystems, each with volume $\frac{1}{2}V$ (so $\frac{1}{2}N$ particles are in each subsystem), by inserting a wall. This process produces a completely new situation, because the spacing of the translational energy levels in the smaller partial containers is larger than the spacing of the energy levels in the whole container. This change means that the number of representations in both subsystems together is smaller than that of the original system. Nevertheless, in Box 17.2 we show that Equation (17.48) still holds even in this case, provided that N is large enough.

17.4.2 Entropy of Atomic Gas: Sackur–Tetrode Equation

From Equations (17.45) and (17.44) for the entropy of an atomic gas we obtain

$$S = k \ln \Omega = Nk \ln \left[e^{5/2} (2\pi)^{3/2} \frac{(km)^{3/2}}{h^3} \frac{V}{N} T^{3/2} \right] \tag{17.49}$$

This relation is called the *Sackur–Tetrode equation*. The entropy increases with the volume V of the gas, with its temperature T, and with the mass m of the gas particles, corresponding to the increase of $\ln \Omega$ already discussed with regard to Equation (17.44).

Example 17.7 *Calculate the entropy of 1 mol of Ar at $P = 1$ bar (standard condition) and $T = 298.15$ K. Compare this entropy to that you calculate for 1 mol of Xe under the same conditions.*

For argon gas $m = 6.63 \times 10^{-26}$ kg and the number density $N/V = 2.43 \times 10^{25}$ m^{-3} at $T = 298.15$ K and $P = 1$ bar. Using these facts with Equation (17.49), the molar entropy under standard conditions, $S_m^{\ominus} = S^{\ominus}/\boldsymbol{n}$, is

$$S^{\ominus}_{m,298} = \boldsymbol{R} \cdot \ln \left[e^{5/2}(2\pi)^{3/2} \frac{(km)^{3/2}}{h^3} \frac{V}{N} T^{3/2} \right] = 154.8 \text{ J K}^{-1} \text{ mol}^{-1}$$

with $\boldsymbol{n} = N/\boldsymbol{N}_A$. Because of its larger mass, xenon ($m = 21.8 \times 10^{-26}$ kg) has a larger entropy ($S^{\ominus}_{m,298} = 169.5$ J K^{-1} mol^{-1}) than argon. The difference in entropy is a direct consequence of the different energy level spacings for the translational quantum states. Increasing the mass m decreases the spacing of the energy levels; thus, more energy levels are available for the Xe atoms and the number of representations Ω increases.

17.5 Entropy Change ΔS

Because the entropy S is a state variable, the entropy change in a transformation from a state 1 to a state 2,

$$\Delta S = S_2 - S_1 \tag{17.50}$$

is independent of the path. The entropy change for an arbitrary process is

$$\Delta S = k \ln \Omega_2 - k \ln \Omega_1 = k \cdot \ln \frac{\Omega_2}{\Omega_1} \tag{17.51}$$

Consider the case of an ideal gas. Expanding an ideal gas at constant temperature increases its entropy. Using Equation (17.49), we find that the entropy of the gas changes by

$$\Delta S = Nk \cdot \ln \left(\frac{const \cdot V_2}{const \cdot V_1} \right) = Nk \cdot \ln \frac{V_2}{V_1} = Nk \cdot \ln \frac{P_1}{P_2} \tag{17.52}$$

S increases with increasing volume or decreasing pressure ($P_2 < P_1$) reflecting the increasing number of representations for the gas.

Heating an ideal gas at constant volume increases its entropy also. Using Equation (17.49), the heating of an atomic gas from temperature T_1 to T_2 at constant volume increases the entropy by

$$\Delta S = Nk \cdot \ln \left(\frac{const \cdot T_2^{3/2}}{const \cdot T_1^{3/2}} \right) = \frac{3}{2} Nk \cdot \ln \frac{T_2}{T_1} \tag{17.53}$$

This increase arises because the probability of occupying the higher energy levels increases with temperature.

Finally, if the ideal gas changes its temperature from T_1 to T_2 and at the same time its volume is changed from V_1 to V_2, from Equation (17.49) we obtain

$$\Delta S = Nk \cdot \ln \left(\frac{const \cdot V_2 T_2^{3/2}}{const \cdot V_1 T_1^{3/2}} \right) = Nk \cdot \ln \left[\frac{V_2}{V_1} \left(\frac{T_2}{T_1} \right)^{3/2} \right] \tag{17.54}$$

17.5.1 Temperature Equilibration: Entropy Increase

17.5.1.1 Calculating the Entropy Change

We can use the entropy to analyze the process of temperature equilibration illustrated by Fig. 17.1a. For the case of N atoms of an atomic gas in the upper container at temperature T_2 and N atoms of an atomic gas in the lower container at temperature T_1, the final temperature will be

$$T_{eq} = \frac{T_1 + T_2}{2} \tag{17.55}$$

because the heat capacities of the two gases are the same. The entropy change in this process is the sum of the entropy changes in each subsystem, so that

$$\begin{aligned} \Delta S &= \Delta S_{upper} + \Delta S_{lower} \\ &= \frac{3}{2} Nk \cdot \ln \frac{T_{eq}}{T_2} + \frac{3}{2} Nk \cdot \ln \frac{T_{eq}}{T_1} = \frac{3}{2} Nk \cdot \ln \frac{T_{eq}^2}{T_1 T_2} \end{aligned} \tag{17.56}$$

For $T_1 < T_2$, the entropy in the upper container decreases, while the entropy in the lower container increases; the entropy of both containers as a whole increases.

For example, if we choose $T_1 = 200$ K and $T_2 = 600$ K, then $T_{eq} = 400$ K, and we obtain

$$\Delta S = \tfrac{3}{2}Nk \cdot \ln \frac{400^2}{200 \times 600} = \tfrac{3}{2}Nk \times 0.29$$

The overall entropy increases, hence Ω increases, for this irreversible process.

17.5.1.2 Calculating the Change of the Number of Representations

Figure 17.13 illustrates the temperature equilibration in a model system with only two quantum states ($E_{low} = 0, E_{high} = a$) available to each molecule. In State 1, left, we have all four particles in the lower level, and in State 1, right, we have two particles in the lower level and two particles in the higher level (internal energy $U_1 = U_{1,left} + U_{1,right} = 0 + 2a = 2a$). Thus there are

$$\Omega_1 = \Omega_{1,left} \cdot \Omega_{1,right} = \frac{4!}{4!} \cdot \frac{4!}{2!2!} = 1 \cdot 6 = 6$$

representations (Fig. 17.13, State 1).

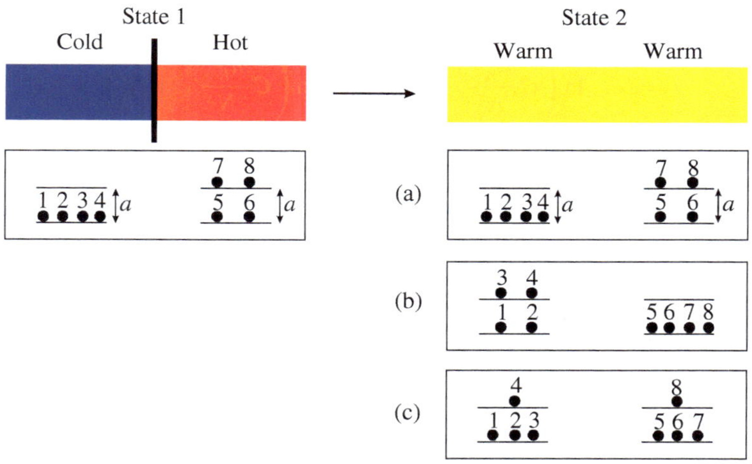

Figure 17.13 Description of temperature equilibration in a model system with only two quantum states (energies $E_{low} = 0$, $E_{high} = a$) available to each molecule. State 1: cold body (left, one representation) and hot body (right, six representations) thermally isolated (black bar); the total number of representations is six. State 2: both bodies in thermal contact, after temperature equilibration: (a) has six representations, (b) has six representations, and (c) has 16 representations; the total number of representations is 28.

In State 2 both subsystems are at the same intermediate temperature. Now we have three possibilities to maintain the internal energy ($U_2 = U_1 = 2a$):

a) In State 2, left, the four particles are in their ground state; and in State 2, right, two particles are in their ground state and two particles are in their excited state; then

$$\Omega_{2,a} = \frac{4!}{4!} \cdot \frac{4!}{2!2!} = 1 \cdot 6 = 6$$

b) In State 2, left, two particles are in their ground state and two particles are in their excited state; and in State 2, right, all four particles are in their ground state; then

$$\Omega_{2,b} = \frac{4!}{2!2!} \cdot \frac{4!}{4!} = 6 \cdot 1 = 6$$

c) In State 2, left, three particles are in their ground state and one particle is in its excited state; in State 2, right, three particles are in their ground state and one particle is in its excited state; then

$$\Omega_{2,c} = \frac{4!}{3!1!} \cdot \frac{4!}{3!1!} = 4 \cdot 4 = 16$$

Altogether the number of representations Ω_2 is

$$\Omega_2 = \Omega_{2,a} + \Omega_{2,b} + \Omega_{2,c} = 6 + 6 + 16 = 28$$

(Fig. 17.13, State 2). Thus, the number of representations increases during temperature equilibration.

The finding that the temperature equilibrium corresponds to the largest number of representations can be generalized to quantum-mechanical systems with any sequence of energy levels. Thus, the fact that heat can flow from a hot body to a cold body and not from a cold body to a hot body is explained through considerations of statistical quantum mechanics.

17.5.2 Mixing: Entropy Increase

17.5.2.1 Mixing of Two Different Gases

Here we consider the entropy change when two different gases mix, which is important for understanding chemical reactions.

Consider two separated containers A and B (each of volume V) with different ideal gases A and B at pressure P and temperature T (process shown in Fig. 17.1b). We can calculate the entropy change in this irreversible mixing process by counting the number of representations before and after mixing. Using Equation (17.49) we can write the entropy of the total system before mixing S_1 as the sum of the entropies of the two parts, so that

$$S_1 = Nk \cdot \ln\left(C\frac{m_A^{3/2}}{N} \cdot V \right) + Nk \cdot \ln\left(C\frac{m_B^{3/2}}{N} \cdot V \right)$$

where $N = N(A) = N(B)$ is the number of A atoms and of B atoms, respectively, and

$$C = e^{5/2}(2\pi)^{3/2}\frac{(k)^{3/2}}{h^3}T^{3/2}$$

After mixing, both gases occupy a volume $2V$, and thus the entropy is

$$\begin{aligned} S_2 &= Nk \cdot \ln\left(C\frac{m_A^{3/2}}{N} \cdot 2V \right) + Nk \cdot \ln\left(C\frac{m_B^{3/2}}{N} \cdot 2V \right) \\ &= Nk \cdot \ln\left(C\frac{m_A^{3/2}}{N} \cdot V \right) + Nk \cdot \ln 2 + Nk \cdot \ln\left(C\frac{m_B^{3/2}}{N} \cdot V \right) + Nk \cdot \ln 2 \\ &= Nk \cdot \ln\left(C\frac{m_A^{3/2}}{N} \cdot V \right) + Nk \cdot \ln\left(C\frac{m_B^{3/2}}{N} \cdot V \right) + 2Nk \cdot \ln 2 \end{aligned}$$

For the entropy change we obtain

$$\Delta S = S_2 - S_1 = 2Nk \cdot \ln 2$$

so that the entropy increases after mixing.

17.5.2.2 Mixing of Identical Gases

Consider the case where the molecules A and B in Fig. 17.1b are identical particles, their masses are equal, $m = m_A = m_B$. Before mixing, N particles are in the upper volume V and N particles are in the lower volume V, so that

$$\begin{aligned} S_1 &= Nk \cdot \ln\left(C\frac{m^{3/2}}{N} \cdot V \right) + Nk \cdot \ln\left(C\frac{m^{3/2}}{N} \cdot V \right) \\ &= 2Nk \cdot \ln\left(C\frac{m^{3/2}}{N} \cdot V \right) \end{aligned}$$

After mixing $2N$ particles occupy the total volume $2V$, so that

$$S_2 = 2Nk \cdot \ln\left(C\frac{m^{3/2}}{2N} \cdot 2V \right) = 2Nk \cdot \ln\left(C\frac{m^{3/2}}{N} \cdot V \right)$$

Thus the entropy remains unchanged:

$$\Delta S = S_2 - S_1 = 0$$

Note that a tiny difference between the two kinds of gas molecules (e.g., the tiny difference in mass between the isotopes $^{235}UF_6$ and $^{237}UF_6$ molecules) is sufficient to generate an entropy effect on mixing. It does not matter how much the two kinds of gas molecules differ from each other, just that they are distinguishable.

17.5.3 Entropy Cannot Decrease for an Isolated System

As we discussed previously, the most probable thermodynamic state for an isolated system is that with the maximum possible value of Ω. Changes of thermodynamic state can only occur if the system transforms from an initial state 1 to a final state 2 which has a greater (or equal) number of representations, that is

$$\boxed{\Omega_2 > \Omega_1 \text{ (isolated system)}} \tag{17.57}$$

Correspondingly, the probability that the isolated system transforms from state 2 to state 1 is negligibly small for a system with a large number of particles.

If Ω is at a maximum, then the entropy is at a maximum. From Equations (17.51) and (17.57) we have

$$\boxed{\Delta S \geq 0 \text{ (isolated system)}} \tag{17.58}$$

The equal sign is valid for the limiting case of a reversible process. Equation (17.58) is called the principle of entropy increase.

Principle of entropy increase

For an isolated system the entropy must increase or remain the same; it cannot decrease.

If the entropy of an isolated system is at a maximum, then the system is in thermal equilibrium. Thermal equilibrium differs from static equilibrium in mechanical systems because molecular fluctuations still occur (*dynamic equilibrium*). For example, the temperature equilibration discussed above leads to a constant T_{eq}, but energy is still exchanged between the two systems.

17.5.4 Entropy Can Decrease in Closed Systems

Let us consider Fig. 17.2 as an example. The closed system can be combined with the temperature bath, thus constituting a new system which is isolated. Figure 17.2 illustrates, for a reversible process, an entropy increase in one part of the isolated system and an entropy decrease in another part. In the process from left to right the molecules of the gas condense to a liquid and Ω decreases in the closed subsystem. However, the heat of condensation is transferred to the temperature bath, and thus the number of representations in the bath increases. The total number of representations (in the isolated system consisting of the closed subsystem and the temperature bath) and thus the entropy of the isolated system remains unchanged. In the reverse process (from right to left) the number of representations in the closed subsystem increases and that in the temperature bath decreases (heat flows into the closed subsystem during evaporation).

17.6 Heat and Entropy Change

Now we explore the relationship between the change of entropy, ΔS, and the reversible heat, q_{rev}, by considering an arbitrary reversible process between an initial state 1 and a final state 2 of a substance. The purpose of this section is to demonstrate that

$$\Delta S = k \ln \Omega_2 - k \ln \Omega_1$$

can be generally expressed in terms of calorimetric data by

$$\Delta S = \sum \frac{q_{rev,i}}{T_i}$$

First, we consider our model system of an ideal gas. By considering a volume change at constant temperature and a temperature change at constant volume, we can combine the two reversible processes to change between any two arbitrary states of an ideal gas. Later, we generalize our considerations to arbitrary changes of any substance.

17.6.1 Heat and Entropy Change for an Ideal Gas

17.6.1.1 Expansion at Constant Temperature

We consider the isothermal expansion of an ideal atomic gas from state 1 to state 2 (Fig. 17.14). The change of entropy in this process is, according to Equation (17.52),

$$\Delta S = Nk \cdot \ln \frac{V_2}{V_1} = \boldsymbol{nR} \cdot \ln \frac{V_2}{V_1} \tag{17.59}$$

If the isothermal expansion is carried out reversibly (Fig. 17.14b), then the heat supplied to the system is, according to Equation (17.4),

$$q_{rev} = \boldsymbol{nR}T \cdot \ln \frac{V_2}{V_1} \tag{17.60}$$

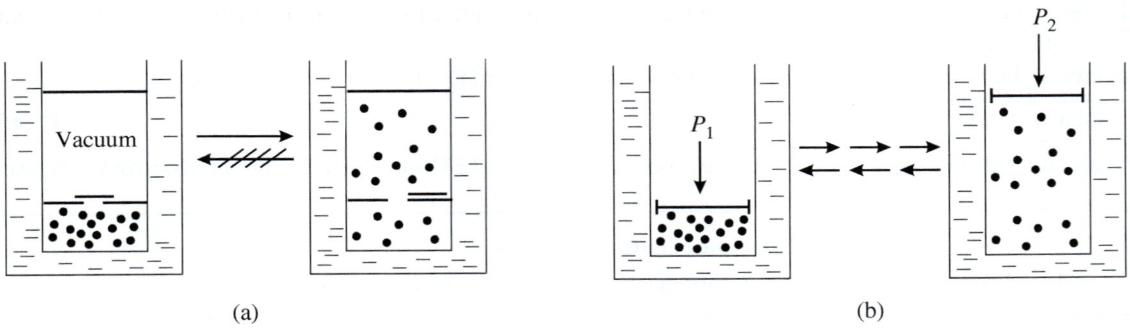

Figure 17.14 Isothermal expansion of an ideal gas. (a) Irreversible (expansion into a vacuum) and (b) reversible (the pressure decreases in small steps from the starting value P_1 to the final value P_2), see Fig. 17.3.

Combining these two results, we find that

$$\boxed{\Delta S = \frac{q_{\text{rev}}}{T} \quad \text{(isothermal change of state)}} \tag{17.61}$$

if the change of state occurs at constant temperature.

17.6.1.2 Heating at Constant Volume

To relate the heat and the entropy, we will compare the entropy change with the heat q_{rev} flowing into the system for a reversible process.

We contact a gas with a series of temperature baths that differ in temperature by a small amount ΔT. Thus in each step the temperature of the bath is only slightly different from the temperature of the gas (Fig. 17.15). Each step can be considered as reversible. The heat $q_{\text{rev},i}$ transferred to the gas in step i is related to the heat capacity C_V at temperature T_i, as discussed in Chapter 16. C_V has been defined by Equation (16.12)

$$C_V = \lim_{\Delta T \to 0} \left(\frac{q_{T,T+\Delta T}}{\Delta T} \right)_{V=const}$$

For small enough ΔT we can write

$$q_{\text{rev},i} = C_V \cdot \Delta T \tag{17.62}$$

for the heat supplied to the gas. We are interested in the quantity

$$\frac{q_{\text{rev},i}}{T_i} = \frac{C_V \cdot \Delta T}{T_i} \tag{17.63}$$

For the total process of heating, we sum up the contributions in each step:

$$\sum_{T_1}^{T_2} \frac{q_{\text{rev},i}}{T_i} = \sum_{T_1}^{T_2} \frac{C_V \cdot \Delta T}{T_i} \tag{17.64}$$

If our temperature increments are small enough, we can replace the summation by an integration and find

$$\sum_{T_1}^{T_2} \frac{C_V \cdot \Delta T}{T_i} = \int_{T_1}^{T_2} \frac{C_V \cdot \mathrm{d}T}{T} \tag{17.65}$$

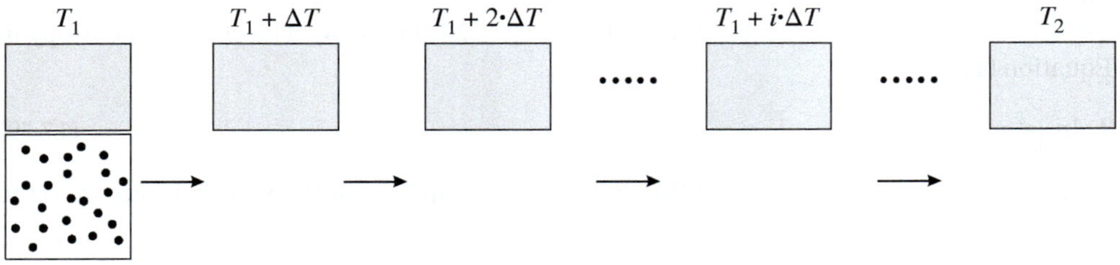

Figure 17.15 Reversible heating of an ideal gas at constant volume in small steps from the starting temperature T_1 to the final temperature T_2. The gas container (below left) is slowly shifted from temperature bath T_1 to bath $T_1 + \Delta T$, to bath $T_1 + 2\Delta T$, etc.

Now we restrict the discussion to an atomic ideal gas, for which the heat capacity C_V is

$$C_V = \tfrac{3}{2}Nk \tag{17.66}$$

Making this substitution for the heat capacity, the integral in Equation (17.65) is straightforward and we find

$$\sum_i \frac{q_{\text{rev},i}}{T_i} = \frac{3}{2}Nk \int_{T_1}^{T_2} \frac{\mathrm{d}T}{T} = \frac{3}{2}Nk \cdot \ln\frac{T_2}{T_1} \tag{17.67}$$

According to Equation (17.53) the entropy change when heating an atomic gas from T_1 to T_2 at constant volume is

$$\Delta S = \frac{3}{2}Nk \cdot \ln\frac{T_2}{T_1} \tag{17.68}$$

By comparing Equations (17.67) and (17.68) we find

$$\boxed{\Delta S = \sum_i \frac{q_{\text{rev},i}}{T_i} \quad \text{(isochoric change of state)}} \tag{17.69}$$

17.6.1.3 Generalization to an Arbitrary Process

Here we show that Equation (17.61) for an isothermal process and Equation (17.69) for an isochoric process can be generalized and applied to any reversible process; i.e.,

$$\boxed{\Delta S = \sum_i \frac{q_{\text{rev},i}}{T_i}} \tag{17.70}$$

This relation is valid for any change of state of an ideal atomic gas, because any change of state for an atomic gas can be realized by a combination of constant volume and constant temperature processes.

Consider the diagram in Fig. 17.16, which shows the pathways of two different reversible processes between the states A and B of a gas. Since A and B are each states which have a well-defined T, V, and S, we can build a reversible pathway ($A \to C \to B$) for proceeding between the two states. The pathway $A \to C \to B$ contains a change at constant volume for $A \to C$ (the temperature increasing from T_A to T_C) leading to an entropy change of

$$\Delta S_{A\to C} = \sum_{T_A}^{T_B} \frac{q_{\text{rev},i}}{T_i}$$

where C denotes the ending point of this first part of the pathway. The second part of the pathway is an isothermal volume change from the intermediate volume V_C to the final volume V_B. The total entropy change is

$$\Delta S_{A\to B} = \sum_{T_A}^{T_B} \frac{q_{\text{rev},i}}{T_i} + \frac{q_{\text{rev},C\to B}}{T_B} = \sum_i \frac{q_{\text{rev},i}}{T_i}$$

Because the states A and B are arbitrary, we can build a reversible pathway between any two states for our gas. Any reversible pathway from A to B can be considered as a sum of steps consisting of a process at constant volume and a process at constant temperature. An example is the reversible pathway in Fig. 17.16 leading from A to B via steps 1, 3, 5, 7 at constant volume and steps 2, 4, 6 at constant temperature.

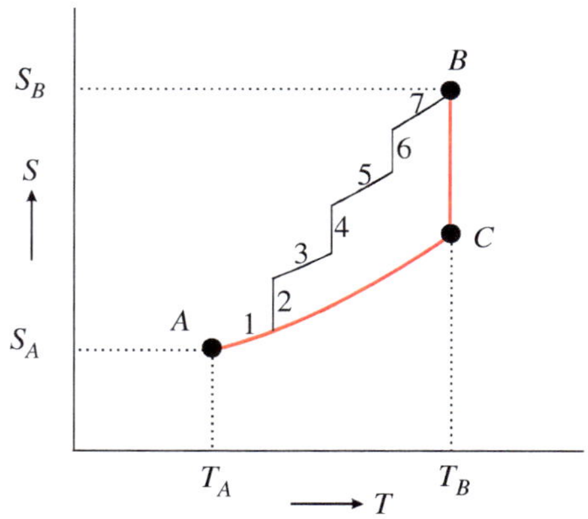

Figure 17.16 Sketch of S versus T for two different reversible pathways between the starting and ending points (A and B) of the change of state. The first process has a step $A \to C$ at constant volume (with increasing temperature and entropy) and a step $C \to B$ at constant temperature (with increasing volume and entropy). The second process has the steps 1, 3, 5, 7 at constant volume and the steps 2, 4, 6 at constant temperature.

Example 17.8 *Using Equation (17.70), show that the entropy change ΔS from state 1 (temperature T_1, volume V_1) to state 2 (temperature T_2, volume V_2) of an ideal atomic gas can be calculated from process a (heating at constant volume), followed by process b (isothermal expansion). Compare this result with the ΔS obtained from the Sackur–Tetrode equation.*

First we heat the gas at constant volume, for which

$$q_{\text{rev},i} = C_v \cdot \Delta T = \frac{3}{2}Nk \cdot \Delta T$$

As with Equation (17.67), we consider infinitesimally small steps and replace the sum in Equation (17.70) with an integral to find

$$\Delta S_{\text{step 1}} = \sum_i \frac{3}{2} Nk \cdot \frac{\Delta T_i}{T_i} = \int_{T_1}^{T_2} \frac{3}{2} Nk \cdot \frac{\mathrm{d}T}{T}$$
$$= \frac{3}{2} Nk \cdot \ln \frac{T_2}{T_1} = Nk \cdot \ln \left(\frac{T_2}{T_1} \right)^{3/2}$$

After performing this step the volume of the gas is still V_1. In the second step we expand the gas isothermally at temperature T_2 to the volume V_2. Then from Equations (17.60) and (17.61) we obtain

$$\Delta S_{\text{step 2}} = \frac{q_{\text{rev}}}{T_2} = Nk \cdot \ln \frac{V_2}{V_1}$$

For the total process we find

$$\Delta S = \Delta S_{\text{step 1}} + \Delta S_{\text{step 2}} = Nk \cdot \ln \left(\frac{T_2}{T_1} \right)^{3/2} + Nk \cdot \ln \left[\frac{V_2}{V_1} \right]$$
$$= Nk \cdot \ln \left[\frac{V_2}{V_1} \cdot \left(\frac{T_2}{T_1} \right)^{3/2} \right]$$

The same relation is directly obtained by using the Sackur–Tetrode equation (17.49).

17.6.2 Cyclic Processes ($\Delta S = 0$) and Processes in Isolated Systems ($\Delta S \geq 0$)

Because the entropy S is a state function, the change in entropy ΔS for a cyclic process of a system is zero

$$\Delta S = 0 \quad \text{(cyclic process)}$$

For an atomic gas we showed that

$$\boxed{\Delta S = \sum_i \frac{q_{\text{rev},i}}{T_i} = 0 \qquad \text{(cyclic process)}} \tag{17.71}$$

For an arbitrary process in an isolated system we found, according to Equation (17.58),

$$\Delta S \geq 0 \quad \text{(isolated system)}$$

Hence for a process in an isolated atomic gas, we find

$$\boxed{\sum_i \frac{q_{\text{rev},i}}{T_i} \geq 0 \qquad \text{(process in an isolated system)}} \tag{17.72}$$

Note that relations (17.71) and (17.72) have so far been proved for an ideal atomic gas only.

17.6.3 Heat and Entropy Change in Arbitrary Processes

Equation (17.70) is of great importance because ΔS determines the direction of a process in an isolated system (whether it can be spontaneous), and the connection to heat provides a way to measure entropy changes. Thus, we ask: "How general is the remarkable relation between ΔS and $\sum_i q_{\text{rev},i}/T_i$?"

We cannot prove this result directly by counting the number of representations Ω_1 and Ω_2 for any change of state, because the calculation of Ω is practically impossible for complex systems. Rather, we proceed by showing that the quantity

$$\sum_i \frac{q_{\text{rev},i}}{T_i} \tag{17.73}$$

has the same properties as the entropy change:

$$\Delta S = 0 \quad \text{(cyclic process)}, \qquad \Delta S \geq 0 \quad \text{(isolated system)}$$

To accomplish this task, we must prove the following two proposals.

Proposal 1: The quantity

$$\sum_i \frac{q_{\text{rev},i}}{T_i}$$

is independent of the path leading from an initial state to a final state. Hence it is the change of a state variable, and

$$\boxed{\sum_i \frac{q_{\text{rev},i}}{T_i} = 0 \quad \text{(cyclic process)}} \tag{17.74}$$

Proposal 2: For an arbitrary process

$$\boxed{\sum_i \frac{q_{\text{rev},i}}{T_i} \geq 0 \quad \text{(process in an isolated system)}} \tag{17.75}$$

A general proof of proposals 1 and 2, based on a consideration of the efficiency of heat engines not restricted to ideal gases as a working medium, is given in Foundation 17.1.

17.7 Thermodynamic Temperature Scale and Cooling

17.7.1 Thermodynamic Temperature Scale

We can define a *thermodynamic temperature scale*; i.e., a temperature scale that does not require any special properties of the "thermometer." Our temperature should be defined for any equilibrium state of the system, so we must be able to access any equilibrium state of the system.

To begin we first consider the special case of a reversible cyclic process, namely, the Carnot cycle discussed in Section 17.2.2.3, see Fig. 17.6. According to Equation (17.29), we obtain

$$\frac{q_{\text{rev},2}}{T_2} + \frac{q_{\text{rev},1}}{T_1} = 0 \tag{17.76}$$

This is identical with what we expect from Equation (17.74)

$$\sum_i \frac{q_{\text{rev},i}}{T_i} = 0 \quad \text{(cyclic process)}$$

In Foundation 17.1 we prove that Equation (17.76) is valid for any reversible cyclic process, not only for processes with ideal gases.

Rearrangement of Equation (17.76) shows that

$$\boxed{\frac{T_2}{T_1} = -\frac{q_{\text{rev},2}}{q_{\text{rev},1}}} \tag{17.77}$$

This relation holds independently of the details of the system; it relies only on the reversibility of the processes and the heat exchange at the two well-defined temperatures. Because the points a and c in Fig. 17.6 are chosen arbitrarily, the cycle could be built between any of the equilibrium states on these two isotherms. Furthermore, the adiabatic steps could be chosen to correspond to any pair of isotherms. Hence Equation (17.77) provides a recipe to measure the temperature between any two equilibrium states.

If we define temperature T_1 arbitrarily (e.g., the triple point of water), then we can determine T_2 by measuring $q_{\text{rev},1}$ and $q_{\text{rev},2}$. In Exercise E17.27 you show that Equation (17.77) reduces to the corresponding relation 14.24 for the gas thermometer. The recipe (17.77) to measure the temperature can still be used in temperature regions where working with ideal gases is impossible.

17.7.2 How Low Can Temperature Go?

Temperatures below 1 K can be reached by using a substance M containing atoms with unpaired electron spins (e.g., gadolinium sulfate) or with large nuclear spins (e.g., copper). First, the body M is placed in a low temperature bath and exposed to a strong magnetic field orienting the spins. The magnetic moments originally aligned with the field are stabilized further. Those originally oriented against the field (in an excited state) align with the field and release their excess energy as heat to the low temperature bath (Fig. 17.17, step 1 → 2). In the next step of the process, M is isolated from the bath and the magnetic field is slowly reduced to zero (*adiabatic demagnetization*). Because the field is no longer applied, the magnetic moments can change their direction; however, they must absorb energy from other modes of the substance M; e.g., vibrational motions (Fig. 17.17, step 2 → 3). In the next step a strong magnetic field is applied to a section

M_1 of M. The residual part of M constitutes the low temperature bath allowing essentially isothermal magnetization of M_1 (Fig. 17.17, step 3 → 4). Then M_1 is separated from the residual part of M and the magnetic field is reduced to zero, so further cooling of M_1 is reached (Fig. 17.17, step 4 → 5).

By using the orientation of electron spins, temperatures of 2×10^{-3} K can be reached, and temperatures of 10^{-5} K have been reached by using nuclear spins. Lower and lower temperatures can be reached by repeating the process of cooling. To reach absolute zero, an infinite number of steps would be necessary. The lowest reported temperature, 5×10^{-10} K for a gas of ^{23}Na atoms, has been realized through somewhat different processes [1].

It is possible to reach temperatures in the laboratory that are lower than the average temperature of interstellar space. For comparison, the microwave background radiation corresponds to a temperature of 3.5 K of the universe considered as a temperature bath [2].

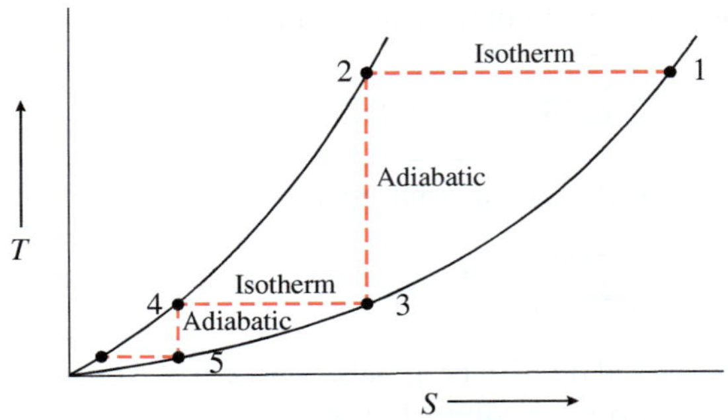

Figure 17.17 Cooling by adiabatic demagnetization. Step 1 → 2: magnetic field on, alignment of the spins (system in contact with temperature bath, decrease of entropy). Step 2 → 3: magnetic field off, equilibration of the spins (system isolated from temperature bath, decrease of temperature). The procedure is repeated in Steps 3 → 4, 4 → 5, etc.

17.8 Entropies of Substances

In simple cases the entropy of a substance can be obtained by finding the number of representations that are available (see Section 17.4). When this is not practical, entropies can be determined experimentally through the relationship

$$\boxed{\Delta S = \sum_i \frac{q_{\text{rev},i}}{T_i}} \tag{17.78}$$

In the following examples we compare simple models for the entropy of a gas to experimentally determined (calorimetric) ΔS values.

17.8.1 Entropy of an Atomic Gas

As an example of an atomic gas, we consider the substance argon (Ar) which has a melting point at 83.9 K and a boiling point at 87.3 K under a pressure of 1 bar (Fig. 17.18a).

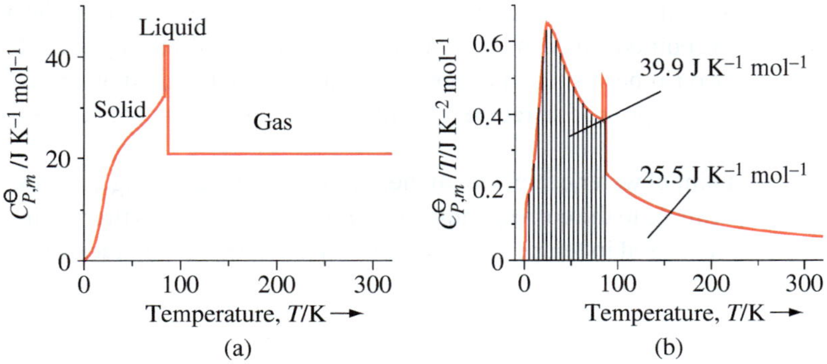

Figure 17.18 Evaluation of the entropy of Ar by plotting $C^{\ominus}_{P,m}$ (a) and $C^{\ominus}_{P,m}/T$ (b) versus temperature T (for data see Table 17.1). The shaded area corresponds to the measured entropy of the gas at the boiling point (83.9 K).

To begin, we imagine a perfect crystalline solid of Ar whose temperature is arbitrarily close to absolute zero. At this temperature no atoms are electronically excited, and each of the 3N vibrational modes of the crystal lattice is in their zero point energy level. Hence only one quantum state of the crystal is possible, the vibrational ground state, so that $\Omega = 1$ and $S_{m,0} = 0$. Next we heat the sample in small steps $q_{\text{rev},i}$ from zero Kelvin to the desired temperature at the pressure $P = 1$ bar (under standard conditions).

$$S^{\ominus}_{m,298} = \frac{1}{\boldsymbol{n}} \sum_{T_i=0}^{298\ \text{K}} \frac{q_{\text{rev},i}}{T_i}$$

As we begin to heat the solid the vibrational energy levels of the crystal become excited. For heating at constant pressure from T_1 to T_2, the entropy change is

$$\Delta S = \int_{T_1}^{T_2} \frac{C_P}{T} \cdot \mathrm{d}T \tag{17.79}$$

as long as it is warmed slowly enough (small enough temperature increments). For argon solid, we find that the molar entropy just below the melting point, T_{mp}, is

$$S^{\ominus}_{m,T_{\text{mp}}} = \int_0^{T_{\text{mp}}} \frac{C^{\ominus}_{P,m} \cdot \mathrm{d}T}{T}$$

At the melting point ($T_{\text{mp}} = 83.9$ K), the temperature stops increasing as we continue heating. Rather, the energy added to the system causes a phase change from the solid to the liquid at a constant temperature. For a process at constant temperature T, the change in entropy is

$$\Delta_{\text{fus}} S^{\ominus}_m = \frac{\Delta_{\text{fus}} H^{\ominus}}{T_{\text{mp}}} \tag{17.80}$$

where $\Delta_{\text{fus}} H = 1.18$ kJ mol^{-1} is the molar enthalpy of fusion ("melting" = "fusion").

Proceeding in a like manner, we continue heating the liquid argon until it reaches the boiling point ($T_{\text{bp}} = 87.3$ K). At the boiling point the added heat converts the liquid to a gas at constant temperature, and the entropy increases (molar enthalpy of evaporation $\Delta_{\text{vap}} H^{\ominus} = 6.52$ kJ mol^{-1}). Lastly, we warm the argon gas up to $T_3 = 298.15$ K. Combining these terms we obtain

$$S^{\ominus}_{m,298} = \int_0^{T_{\text{mp}}} \frac{C^{\ominus}_{P,m} \cdot \mathrm{d}T}{T} + \frac{\Delta_{\text{fus}} H^{\ominus}}{T_{\text{mp}}} + \int_{T_{\text{mp}}}^{T_{\text{bp}}} \frac{C^{\ominus}_{P,m} \cdot \mathrm{d}T}{T} + \frac{\Delta_{\text{vap}} H^{\ominus}}{T_{\text{bp}}} + \int_{T_{\text{bp}}}^{T_3} \frac{C^{\ominus}_{P,m} \cdot \mathrm{d}T}{T} \tag{17.81}$$

We can evaluate this formula for the entropy at 298 K, by measuring the enthalpy changes and heat capacities in each step. For the phase changes, we find

$$\frac{\Delta_{\text{fus}} H^{\ominus}}{T_{\text{mp}}} + \frac{\Delta_{\text{vap}} H^{\ominus}}{T_{\text{bp}}} = \left(\frac{1.18}{83.9} + \frac{6.52}{87.3}\right) \text{ kJ K}^{-1} \text{ mol}^{-1} = 88.7 \text{ J K}^{-1} \text{ mol}^{-1}$$

The entropy changes for the solid and the liquid are calculated by plotting the measured $C^{\ominus}_{P,m}/T$ versus T and numerically evaluating the area under the curve (Table 17.1 and Fig. 17.18b, shaded area)

$$\int_0^{T_{\text{mp}}} \frac{C^{\ominus}_{P,m} \cdot \mathrm{d}T}{T} + \int_{T_{\text{mp}}}^{T_{\text{bp}}} \frac{C^{\ominus}_{P,m} \cdot \mathrm{d}T}{T} = 39.9 \text{ J K}^{-1} \text{ mol}^{-1}$$

In the last integral $C^{\ominus}_{P,m}$ is the molar heat capacity of argon gas at constant pressure ($C^{\ominus}_{P,m} = \frac{5}{2}\boldsymbol{R}$); thus,

$$\int_{T_{\text{bp}}}^{298} \frac{C^{\ominus}_{P,m} \cdot \mathrm{d}T}{T} = \frac{5}{2}\boldsymbol{R} \int_{T_{\text{bp}}}^{298} \frac{\mathrm{d}T}{T} = \frac{5}{2}\boldsymbol{R} \cdot \ln \frac{298 \text{ K}}{T_{\text{bp}}} = 25.5 \text{ J K}^{-1} \text{ mol}^{-1}$$

The result for the molar entropy $S^{\ominus}_{m,298}$ at $P = 1$ bar is

$$S^{\ominus}_{m,298} = 154.1 \text{ J K}^{-1} \text{ mol}^{-1}$$

Table 17.1 $C^{\ominus}_{P,m}$ for Argon as a Function of Temperature.

T/K	20	40	60	80	83.85	83.85	87.29	87.29	100	300
$\frac{C^{\ominus}_{P,m}}{\text{J K}^{-1}\text{ mol}^{-1}}$	11.8	22.1	26.6	32.1	33.3	42.1	42.1	20.8	20.8	20.8
$\frac{C^{\ominus}_{P,m}/T}{\text{J K}^{-2}\text{ mol}^{-1}}$	0.588	0.552	0.443	0.401	0.397	0.501	0.482	0.238	0.208	0.069
$\frac{S^{\ominus}}{\text{J K}^{-1}\text{ mol}^{-1}}$					39.5	53.6	54.0	128.7		154.8
	Solid					Liquid		Gas		

This value is in excellent agreement with the value $S_{m,298} = 154.8$ J K^{-1} mol^{-1} calculated in Section 17.4.2 by using the Sackur–Tetrode equation.

A fascinating fact: The entropy, which is obtained by summing up calorimetrically measured data, agrees exactly with the entropy obtained from N/V, m, T, Planck's constant h, and the Boltzmann constant k. Note that for the calculation based on the Sackur–Tetrode equation we used only data about the final state of the argon gas; we did not need any information about the behavior of Ar at low temperatures. Neither do we need the enthalpy changes for the phase transitions. These observations underscore the fact that the entropy is a state function and the quantum nature of matter is essential for understanding macroscopic properties. The distribution of the translational quantum states of the gas particles is decisive for the evaluation of the entropy. According to the Sackur–Tetrode equation the entropy increases with increasing mass of the atomic gas. Thus the entropy of He ($S^{\ominus}_{m,298} = 126$ J K^{-1} mol^{-1}) is smaller than the entropy of argon, and the entropy of Xe ($S^{\ominus}_{m,298} = 170$ J K^{-1} mol^{-1}) is larger, in complete agreement with the entropies obtained from calorimetric data (see Table 17.2).

17.8.2 Entropy of Diatomic Gases

To evaluate the entropy for a molecular gas, we must extend our derivation for the Sackur–Tetrode equation. As with our consideration of the energy distribution of gases, it is often reasonable to consider the entropy contributions from different motions independently. By approximating these motions as independent we can consider them separate subsystems, so that Equation (17.48) applies and we find

$$S = S_{\text{trans}} + S_{\text{rot}} + S_{\text{vib}} + S_{\text{electr}} \tag{17.82}$$

For diatomic gases (e.g., HD, HCl, CO), the entropy S is often dominated by contributions from the translational, rotational, and vibrational motions. As long as the energy gap between the ground electronic state and higher excited states is large enough and the ground state has a degeneracy of one, the entropy contribution from the electronic internal motions is zero. The contribution of translation is, according to Equation (17.43),

$$S_{\text{trans}} = k \ln \Omega = Nk \ln Z_{trans} + k\,(N - N \ln N) + \frac{U_{\text{trans}}}{T} \tag{17.83}$$

resulting in the Sackur–Tetrode equation (17.49) for S_{trans}.

As outlined in Chapter 15 rotators and oscillators can be treated as distinguishable particles. In Foundation 17.2 we show that

$$\ln \Omega = N \ln Z + \frac{U}{kT} \tag{17.84}$$

for N independent and distinguishable particles, so that

$$S = k \ln \Omega = Nk \ln Z + \frac{U}{T} \tag{17.85}$$

where Z is the molecular partition function and U is the internal energy.

17.8.2.1 Rotation of Diatomic Molecules

For high enough temperature ($T >> \theta_{\text{rot}}$), the internal energy of rotation of a diatomic molecule is $U_{\text{rot}} = NkT$ and the molecular partition function is, according to Section 15.5,

$$Z_{\text{rot}} = \frac{kT}{B} \quad \text{with} \quad B = \frac{h^2}{8\pi^2 \mu d^2} = \frac{h^2}{8\pi^2 I} \tag{17.86}$$

where $I = \mu d^2$ is the inertial moment, d is the average bond length, and μ is the reduced mass. As outlined in Chapters 15 and 16 the rotational motions of hetero- and homonuclear diatomic molecules must be treated separately. In Foundation 17.3 we show that the rotational contribution S_{rot} for N independent diatomic molecules is

$$S_{\text{rot}} = Nk \left[\ln \frac{Z_{\text{rot}}}{\sigma} + 1 \right] = Nk \cdot \left[\ln \left(\frac{8\pi^2 IkT}{\sigma h^2} \right) + 1 \right] \tag{17.87}$$

in the high temperature limit, where σ is called the *symmetry factor* ($\sigma = 1$ for a heteronuclear diatomic and $\sigma = 2$ for a homonuclear diatomic). The contribution S_{rot} increases with decreasing spacing of the energy levels — that is, with increasing moment of inertia I.

17.8.2.2 Vibration of Diatomic Molecules

According to Section 15.4, the internal energy and the molecular partition function for N independent harmonic oscillators is

$$U_{\mathrm{vib}} = Nh\nu_0 \left(\frac{1}{2} + \frac{1}{\mathrm{e}^{h\nu_0/(kT)} - 1}\right) \tag{17.88}$$

and

$$Z_{\mathrm{vib}} = \mathrm{e}^{-h\nu_0/(2kT)} \frac{1}{1 - \mathrm{e}^{-h\nu_0/(kT)}} \tag{17.89}$$

Using Equation (17.85), the vibrational contribution to the entropy S_{vib} is

$$\begin{aligned} S_{\mathrm{vib}} &= Nk \ln Z_{\mathrm{vib}} + \frac{U_{\mathrm{vib}}}{T} \\ &= Nk \cdot \left[-\ln\left(1 - \mathrm{e}^{-h\nu_0/(kT)}\right) + \frac{h\nu_0}{kT} \cdot \frac{1}{\mathrm{e}^{h\nu_0/(kT)} - 1}\right] \end{aligned} \tag{17.90}$$

S_{vib} follows from an exact analytical treatment in the harmonic oscillator approximation. In the limit of high temperature ($h\nu_0/(kT) \ll 1$) we can approximate Equation (17.90) by

$$S_{\mathrm{vib}} = Nk \cdot \left[\ln\left(\frac{kT}{h\nu_0}\right) + 1\right] \tag{17.91}$$

In this case it is evident that the entropy increases with decreasing spacing of the energy levels – that is, with decreasing vibrational frequency ν_0.

Example 17.9 *Calculate the molar entropy of molecular HD at 298 K and 1 bar pressure.*

We will calculate the different contributions and combine them to obtain the total entropy. To begin, we calculate the translational part of the entropy using the Sackur–Tetrode equation (17.49). For HD $m = m_H + m_D = (1.678 + 3.339) \times 10^{-27}$ kg $= 5.017 \times 10^{-27}$ kg and the number density is $N/V = 2.43 \times 10^{25}$ m^{-3} at $T = 298$ K and $P = 1$ bar. With these facts we obtain

$$\begin{aligned} S^{\ominus}_{m,\mathrm{trans},298} &= \boldsymbol{R} \cdot \ln\left[\mathrm{e}^{5/2}(2\pi)^{3/2}\frac{(km)^{3/2}}{h^3}\frac{V}{N}T^{3/2}\right] \\ &= 14.75\ \boldsymbol{R} = 122.6\ \mathrm{J\ K^{-1}\ mol^{-1}} \end{aligned}$$

To calculate the rotational part of the entropy from Equation (17.87) we need the reduced mass $\mu = \frac{m_H \cdot m_D}{m_H + m_D} = 1.117 \times 10^{-27}$ kg, and the bond length $d = 74$ pm. With these facts we find $Z_{\mathrm{rot}} = 4.52$ and

$$S^{\ominus}_{m,\mathrm{rot},298} = \boldsymbol{R} \cdot [\ln 4.52 + 1] = 2.51\ \boldsymbol{R} = 20.9\ \mathrm{J\ K^{-1}\ mol^{-1}}$$

To calculate the vibrational part of the entropy from Equation (17.90) we need $\nu_0 = \frac{1}{2\pi}\sqrt{\frac{k_f}{\mu}}$ where k_f is the force constant, $k_f = 574$ N m^{-1}. Hence $\nu_0 = 1.14 \times 10^{14}$ s^{-1} and we find

$$\begin{aligned} S^{\ominus}_{m,\mathrm{vib},298} &= \boldsymbol{R} \cdot \left[-\ln\left(1 - \mathrm{e}^{-h\nu_0/(kT)}\right) + \frac{h\nu_0}{kT} \cdot \frac{1}{\mathrm{e}^{h\nu_0/(kT)} - 1}\right] \\ &= 2.0 \times 10^{-7}\ \boldsymbol{R} = 1.7 \times 10^{-6}\ \mathrm{J\ K^{-1}\ mol^{-1}} \end{aligned}$$

We assume that all the HD molecules are in the electronic ground state, which is a singlet, $S^{\ominus}_{m,\mathrm{elec},298} = 0$. Combining these results we find that

$$S^{\ominus}_{m,\mathrm{total},298} = S^{\ominus}_{m,\mathrm{trans},298} + S^{\ominus}_{m,\mathrm{rot},298} + S^{\ominus}_{m,\mathrm{vib},298} = 143.5\ \mathrm{J\ K^{-1}\ mol^{-1}}.$$

This result is in excellent agreement with the value $S^{\ominus}_{m,\mathrm{total},298} = 143.8$ J K^{-1} mol^{-1} obtained from thermochemical measurements. In Foundation 17.3 we calculate the entropy of H_2 as $S^{\ominus}_{m,\mathrm{total},298} = 130.3$ J K^{-1} mol^{-1}. This result is smaller than that for HD, because of the smaller nuclear mass (D versus H) and the symmetry factor $\sigma = 2$.

17.8.3 Entropy of Polyatomic Gases

To evaluate the entropy for a polyatomic molecular gas, we can use Equation (17.82) as well. Considerations of the translational and electronic contributions to the entropy proceed in a manner analogous to that used for diatomic molecules. We must modify our calculations for the rotational and vibrational contributions to the entropy, however.

17.8.3.1 Rotation of Polyatomic Molecules

As with the discussion of diatomic molecules we consider the high temperature (classical) limit for the rotational motion. The molecular partition function for a linear polyatomic is exactly like that of a diatomic

$$Z_{\text{rot}} = \frac{kT}{B} \quad \text{with} \quad B = \frac{h^2}{8\pi^2 I} \tag{17.92}$$

where I is the inertial moment. Using Equation (17.87), the rotational contribution S_{rot} is

$$S_{\text{rot}} = Nk\left[\ln\frac{Z_{\text{rot}}}{\sigma} + 1\right] = Nk \cdot \left[\ln\frac{8\pi^2 IkT}{\sigma h^2} + 1\right], \quad \text{(linear molecule)} \tag{17.93}$$

The symmetry factor is $\sigma = 1$ for a molecule without a center of symmetry (e.g., HCN) and $\sigma = 2$ for a molecule with a center of symmetry (e.g., CO_2).

For high enough temperature ($T >> \theta_{\text{rot}}$), the internal energy of a nonlinear polyatomic molecule is $U_{\text{rot}} = \frac{3}{2}NkT$. For a nonlinear polyatomic the molecular partition function is

$$Z_{\text{rot}} = \sqrt{\pi}\left(\frac{8\pi^2 kT}{h^2}\right)^{\frac{3}{2}} \sqrt{I_A I_B I_C} \tag{17.94}$$

where I_A, I_B, and I_C are the inertial moments corresponding to the three principal rotation axes (asymmetric top). In analogy to the derivation of Equation (17.93), the rotational contribution S_{rot} is

$$S_{\text{rot}} = Nk\left[\ln\frac{Z_{\text{rot}}}{\sigma} + \frac{3}{2}\right] \tag{17.95}$$

or

$$S_{\text{rot}} = Nk \cdot \left\{\ln\left[\frac{\sqrt{\pi}}{\sigma}\left(\frac{8\pi^2 kT}{h^2}\right)^{\frac{3}{2}} \sqrt{I_A I_B I_C}\right] + \frac{3}{2}\right\} \tag{17.96}$$

(nonlinear molecule, asymmetric top)

The symmetry factor σ depends on the symmetry of the molecule: H_2O has two indistinguishable orientations; thus, $\sigma = 2$. NH_3 has $\sigma = 3$, CH_4 has $\sigma = 12$.[1]

17.8.3.2 Vibration of Polyatomic Molecules

To determine the total vibrational contribution to the entropy, we sum up the contribution from each of the molecule's normal modes, which can be treated independently. In the approximation that each normal mode i can be described by the harmonic oscillator model, the energy will be

$$U_{\text{vib},i} = Nh\nu_{0,i}\left(\frac{1}{2} + \frac{1}{e^{h\nu_{0,i}/(kT)} - 1}\right) \tag{17.97}$$

and the partition function will be

$$Z_{\text{vib},i} = e^{-h\nu_{0,i}/(2kT)} \frac{1}{1 - e^{-h\nu_{0,i}/(kT)}} \tag{17.98}$$

Using Equation (17.85), the vibrational contribution to the entropy from mode i, $S_{\text{vib},i}$, is

$$\begin{aligned} S_{\text{vib},i} &= Nk\ln Z_{\text{vib},i} + \frac{U_{\text{vib},i}}{T} \\ &= Nk \cdot \left[-\ln\left(1 - e^{-h\nu_{0,i}/(kT)}\right) + \frac{h\nu_{0,i}}{kT} \cdot \frac{1}{e^{h\nu_{0,i}/(kT)} - 1}\right] \end{aligned} \tag{17.99}$$

Evidently the vibrational contribution to the entropy will be larger for the low frequency modes. The total vibrational contribution to the entropy S_{vib} is found by summing over the contributions from each mode.

1 The factor 12 arises because each of the 4 CH bonds in methane has a three-fold axis of symmetry; rotation by 120 degrees swaps the H-atom positions and is indistinguishable.

17.8.4 Entropy of Different Substances

In analogy to the case of argon and hydrogen we can obtain the entropies for a number of different gases. Table 17.2 compares entropy data for some different monatomic and diatomic gases. In each case the agreement between the spectroscopic and calorimetric data is very good (within a few percent).

The table also shows how different motions contribute to the overall entropy of the substance. In every case it is clear that the translational motion of the molecules makes the largest contribution. The difference in the translational entropy values is determined solely by the mass of the molecule or atom. The diatomic molecules have contributions to their entropy from rotational motion which is significant. For the diatomic molecules in Table 17.2 the vibrational contribution to the entropy is very small because most of these molecules are in the ground vibrational state at 298.15 K.

Table 17.2 Comparison of Entropies Calculated with Spectroscopic Data to Those from Heat Capacity C_p Data.[2]

		$S^{\ominus}_{m,298}/(\text{J K}^{-1}\text{ mol}^{-1})$					
		From Quantum Mechanics					
Molecule	σ	**Trans**	**Rot**	**Vib**	**Total**	**From $C_{p,m}$ Data**	**Difference**
He	1	126.0			126.0	124.7	−1.3
Ar	1	154.8			154.8	154.1	−0.7
Xe	1	169.5			169.5	169.2	−0.3
HD	1	122.6	20.9	2×10^{-6}	143.5	143.8	+0.3
HCl	1	153.7	33.0	7×10^{-5}	186.6	185.7	−0.9
CO	1	150.4	47.2	3×10^{-3}	197.6	192.4	−5.2
H_2	2	117.6	12.7	1×10^{-7}	130.3	130.6	+0.3
D_2	2	126.1	18.4	3.8×10^{-5}	144.5		
N_2	2	150.4	41.2	1×10^{-3}	191.6	190.6	−1.0
F_2	2	154.2	48.0	0.54	202.7	202.7	0.0
Cl_2	2	162.0	58.8	2.2	222.9	223.1	+0.2
Br_2	2	172.1	67.8	5.4	245.3	245.4	+0.1
I_2	2	177.9	74.3	8.4	260.5	260.7	+0.2

According to Table 17.2, the calorimetric value of S for CO is too small by 5 J K^{-1} mol^{-1}. This discrepancy can be explained by considering the dipole moment of CO and its impact on the entropy of solid CO (see Fig. 17.19). Because the energy difference between the two possible orientations of CO is small compared with kT, the molecules do not arrange in the most favorable orientation when forming a crystal at the freezing point of 68 K. Once the crystal is formed, a CO molecule cannot easily rotate, and the random arrangement of the CO dipoles remains. For this reason the calorimetric data do not sample this part of the entropy. In contrast, the dipole moment of HCl is large, and in forming a crystal the orientations of the molecules are fixed and ordered in the ground state.

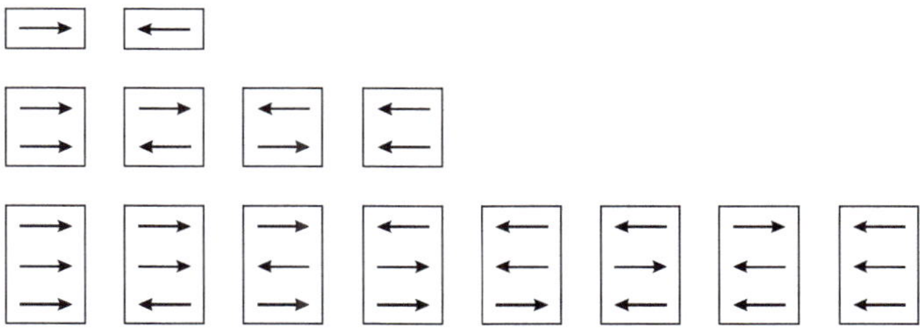

Figure 17.19 Distribution possibilities for CO molecules in a crystal. The direction of the arrows symbolizes the orientations CO and OC, respectively. The number of representations is $\Omega = 2$ for $N = 1$, $\Omega = 2^2$ for $N = 2$, and $\Omega = 2^3$ for $N = 3$.

2 The values correspond to room temperature (298.15 K) and atmospheric pressure (1.013 bar).

Example 17.10 *Determine the number of representations for a CO crystal and show that it can explain the discrepancy between the spectroscopic and calorimetric determinations of the entropy for gaseous CO*

If we assume that one CO molecule can exist in two equally probable representations in the crystal (CO or OC), then for N molecules

$$\Omega = 2^N$$

(Fig. 17.19). With these representations, the entropy of CO near absolute zero is

$$S_m = \frac{k}{\boldsymbol{n}} \cdot \ln \Omega = \frac{k}{\boldsymbol{n}} \cdot \ln\left(2^N\right) = \frac{Nk}{\boldsymbol{n}} \cdot \ln 2 = \boldsymbol{R} \cdot \ln 2 = 5.6 \text{ J K}^{-1} \text{ mol}^{-1}$$

If we add this entropy to the calorimetric value, we obtain (5.6 + 192.4) J K^{-1} mol^{-1} = 198 J K^{-1} mol^{-1} in good agreement with the spectroscopic value 197.6 J K^{-1} mol^{-1}.

Although our discussions have emphasized the case of gases, many of the important results apply also to liquids and solids. Table 17.3 provides entropy data for some different substances. If you focus on homologous series of compounds, you can identify trends in the data. How does the entropy depend on the molecule's mass, or its vibrational frequency?

Table 17.3 Molar Entropies Under Standard Conditions for Some Substances at 298 K.

Gas Substance	$S^{\ominus}_{m,298}$ J K^{-1} mol^{-1}	Liquid Substance	$S^{\ominus}_{m,298}$ J K^{-1} mol^{-1}	Solid Substance	$S^{\ominus}_{m,298}$ J K^{-1} mol^{-1}
He	126.0	H_2O	76.02	C (graphite)	5.74
Ne	146.2	CH_3OH	126.8	C (diamond)	2.4
Ar	154.8	C_2H_5OH	160.7		
Kr	164.0			$AlCl_3$	110.7
Xe	170.0	C_6H_{14}	204.3	Al_2O_3	50.9
		C_6H_{18}	361.1		
F_2	202.7			CaF_2	68.9
Cl_2	223.0			$CaCl_2$	104.6
Br_2	245.4	Br_2	152.2	$CaBr_2$	130
I_2	260.6				
CO_2	213.8	HCOOH	129.0		
H_2O	188.8	CH_3COOH	159.8		
NH_3	192.8				
CH_4	186.3				

17.9 Laws of Thermodynamics

In Chapters 15–17 we have introduced U and S as useful quantities that correspond to molecular assemblies which are in thermal equilibrium. Historically U and S were introduced without reference to the atomistic structure of matter. Based on experimental observations, the thermal properties of macroscopic systems can be deduced from the following postulates (the laws of classical thermodynamics):

- *Zeroth Law.* "A state function exists, the temperature T, which has the same value in all systems which are in thermal equilibrium"
- *First Law.* Three possible statements of the First Law of Thermodynamics are
  "The internal energy U is a function of the state variables ($T, V, N_i, \ldots$). The change in its value, ΔU, from an initial state to a final state is always the same, independent of the path used in making the transition."

"The internal energy of an isolated system is constant, or $\Delta U = 0$ (law of conservation of energy)."

A device that could increase the internal energy of a closed system is called a "*perpetuum mobile of the first kind*." The first law can be stated by saying that a perpetuum mobile of the first kind is not possible.

- *Second Law*. Three possible statements of the Second Law of Thermodynamics are

  It can be expressed in several ways:

  1) *The principle of Clausius*. "It is impossible by a cyclic process to transfer heat from a colder to a warmer reservoir without changes in other bodies."
  2) *The principle of William Thomson*. "It is impossible by a cyclic process to take heat from a reservoir and convert it into work without, in the same operation, transferring heat from a hot to a cold reservoir."

     In other words: The work $-w$ done by any system when proceeding on an arbitrary path 1 from a state A to a state B of the same temperature, cannot be larger than the work $-w_{\mathrm{rev}}$ done by the system when taking a reversible path 2 from A to B.

     $$(-w)_{\mathrm{path\ 1}} \leq (-w_{\mathrm{rev}})_{\mathrm{path\ 2}} \tag{17.100}$$

     Otherwise, in a cyclic process at constant temperature (path 1 from A to B and back on reversed path 2 from B to A), continuously repeated, work would be done by the system continuously by extracting heat from the environment. A device that could operate in this way is called a "*perpetuum mobile of the second kind*." The second law can be stated by saying that a perpetuum mobile of the second kind is not possible.
  3) As shown in Section 17.6.2 the second law can be expressed as:

     In an isolated system

     $$\sum_i \frac{q_{\mathrm{rev},i}}{T_i} \geq 0$$

     Since ΔS, in classical thermodynamics, is defined as

     $$\Delta S = \sum_i \frac{q_{\mathrm{rev},i}}{T_i}$$

     this statement is equivalent to

     $$\Delta S \geq 0 \quad \text{for an isolated system}$$

     Sometimes called the "law of increasing entropy":

     Each of these three statements of the Second Law can be shown to be equivalent to one another.

- *Third Law*. "The entropy of each pure substance[3] in a perfect crystalline form is zero at the absolute zero point of temperature."

  A consequence of this statement is the fact that $T = 0$ cannot be reached. This statement is an alternative version of the Third Law.

The approach in this chapter does not rest on the postulates (laws) of thermodynamics. However, it does rest on the assumptions underlying statistical mechanics. We started by introducing temperature as a measure for the mean kinetic energy of molecules in a molecular aggregate at thermodynamic equilibrium. Then we considered the establishment of temperature equilibrium when two systems with different temperatures come into contact. For statistical reasons the molecules of the system with the higher temperature transfer energy to the molecules of the system with the lower temperature (Fig. 17.1a). Temperature appears as a state variable (it has the same value for systems which are in thermal equilibrium), and heat appears as a form of energy change. Thus, the statement that energy is being conserved (this is valid for mechanical systems) can immediately be used for processes in which mechanical energy is converted to heat and vice versa.

We defined the entropy by

$$S = k \cdot \ln \Omega$$

The law of increasing entropy results immediately from a probabilistic analysis of the change in the number of representations. Later we found that the quantity

$$\sum_i \frac{q_{\mathrm{rev},i}}{T_i}$$

has all the properties of the change of entropy ΔS where S is defined as $k \cdot \ln \Omega$; thus, $\sum_i q_{\mathrm{rev},i}/T_i$ can be identified with ΔS.

3 In this context, "pure" means not only chemically pure, but also isotopically pure.

From the fact that a perfect crystal (no lattice disorder) has only one possible representation ($\Omega = 1$) at the zero point of temperature, we concluded that its entropy must be zero.

17.10 Conclusion

Macroscopic assemblies of molecules are called thermodynamic systems. A thermodynamic system that is isolated from the rest of the universe evolves spontaneously toward a thermodynamic state that has the largest number of different quantum states (representations Ω), if the process is not kinetically hindered. It is useful to define the entropy S of a thermodynamic state of the system, $S = k \ln \Omega$, because its change ΔS is directly related to the heat q_{rev} supplied to the system. The entropy S is a state function, and it can be used to predict how a thermodynamic system can change. The entropy of a substance depends on molecular properties. In simple cases S can be obtained from quantum-mechanical considerations, and generally it can be obtained from experimental data.

We first discussed reversible and irreversible changes of state and studied their relation to changes in the number of representations and in entropy. We showed that the entropy change in a process that goes from an initial state 1 to a final state 2, $\Delta S_{1\to 2}$, is given by

$$\Delta S_{1\to 2} = \sum_i \frac{q_{\text{rev},i}}{T_i}$$

if the heat transfer is performed in a reversible manner. The excellent agreement of the experimentally determined entropy of a gas, obtained by summing up calorimetric data as a function of temperature, with the quantum mechanically calculated entropy, is an important demonstration that the statistical model of the thermal properties of molecular assemblies is appropriate. Entropies of some substances are reported in Tables 17.2, 17.3, and in Appendix A.1.

17.11 Box 17.1. Derivation of Equation (17.35)

First we consider the example $V_1 = \frac{2}{3}V_2$. In this case we divide the volume V_2 into three subvolumes (Fig. 17B.1). Three possibilities exist for finding a particle in one of these three subvolumes and the probability $\mathcal{P}$ of finding the particle in the subvolume V_1 is $\mathcal{P} = \frac{2}{3}$. In the general case we divide the volume V_2 into subvolumes according to the ratio V_1/V_2 (e.g., 11 subvolumes for $V_1 = \frac{7}{11}V_2$, or, more generally, r subvolumes for $V_1 = \frac{q}{r}V_2$ with q and r integers and $q < r$). This choice gives $\mathcal{P} = V_1/V_2$ for one particle. In analogy to the case $V_1 = \frac{1}{2}V_2$ we obtain Equation (17.35) for N particles.

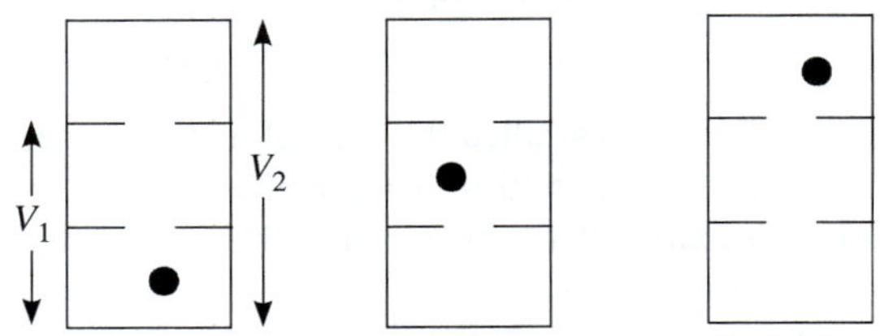

Figure 17B.1 A particle can be found in one of three subvolumes.

17.12 Box 17.2. Indistinguishable Particles: Change of ΔS When Dividing a System into Subsystems

To elucidate the issue, we consider the simple case of $N = 2$ indistinguishable particles which are distributed over four quantum states. When considered as one system, we have $\Omega_1 = 10$ representations available (Fig. 17B.2a, representations 1-10). After division into two equal subsystems I and II, we have $\Omega_2 = 4$ representations (Fig. 17B.2b, representations 1–4). Clearly, the creation of subsystems changes Ω by restricting the particles' access to quantum states.

17.12.1 2 Particles Occupying g Quantum States

In the general case, we consider g quantum states available for the two particles. The system will have g representations with doubly occupied quantum states and $\frac{1}{2}g(g-1)$ representations with singly occupied quantum states. Hence the total number of indistinguishable representations is

$$\Omega_1 = g + \tfrac{1}{2}\left(g^2 - g\right) = \tfrac{1}{2}g\left(g+1\right)$$

After division into two equal subsystems the number of possible representations is

$$\Omega_2 = \Omega_\mathrm{I} \cdot \Omega_\mathrm{II} = \binom{g}{2} \cdot \binom{g}{2} = \binom{g}{2}^2$$

where Ω_I and Ω_II are the numbers of representations of the subsystems. If we evaluate the ratio of Ω_2 and Ω_1, we find that

$$\frac{\Omega_2}{\Omega_1} = \frac{\frac{1}{4}g^2}{\frac{1}{2}g\,(g+1)} = \frac{g}{2(g+1)}$$

cannot become greater than 0.5 as g increases. This result means that the number of representations in the two partial systems is much smaller than the number of representations in the whole system.

However, if we compare the logarithms of the numbers of representations

$$\frac{\ln \Omega_2}{\ln \Omega_1} = \frac{\ln\left(\frac{1}{4}g^2\right)}{\ln\left(\frac{1}{2}g(g+1)\right)} = \frac{\ln \frac{1}{4} + 2\ln g}{\ln \frac{1}{2} + \ln g + \ln(g+1)}$$

we see that this ratio converges to 1 at large g. Thus for a large number of quantum states we have

$$\boxed{\ln \Omega_1 = \ln \Omega_2 = \ln \Omega_\mathrm{I}\Omega_\mathrm{II} = \ln \Omega_\mathrm{I} + \ln \Omega_\mathrm{II}}$$

and

$$\boxed{S_1 = S_\mathrm{I} + S_\mathrm{II}}$$

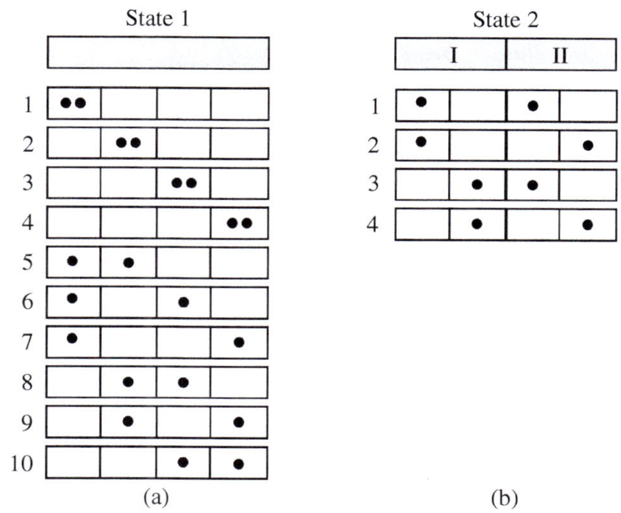

Figure 17B.2 Illustration of the representations for the case $N = 2$ indistinguishable particles and $g = 4$ quantum states. (a) System with four quantum states available to both particles (10 representations). (b) System of four quantum states divided in two subsystems I and II. Only two quantum states available to each particle (four representations).

17.12.2 *N* Particles Occupying *g* Quantum States ($g \gg N$)

We can generalize this consideration to a large number N of particles occupying g quantum states, provided that $g \gg N$ and N is large enough that Stirling's formula can be applied. For two equal subvolumes

$$\Omega_1 = \frac{g^N}{N!} \quad \text{and} \quad \Omega_2 = \Omega_\mathrm{I} \cdot \Omega_\mathrm{II} = \frac{(g/2)^{N/2}}{(N/2)!} \cdot \frac{(g/2)^{N/2}}{(N/2)!} = \frac{(g/2)^N}{(N/2)!\,(N/2)!} = \frac{g^N}{2^N\,(N/2)!\,(N/2)!}$$

For the logarithms we obtain, using Stirling's formula (see Appendix C.3),

$$\ln \Omega_1 = N\ln g - \ln N! = N\ln g - N\ln N + N$$

and

$$\begin{aligned}\ln \Omega_2 &= N\ln g - N\ln 2 - 2\frac{N}{2}\ln\frac{N}{2} + 2\frac{N}{2}\\ &= N\ln g - N\ln 2 + N\ln 2 - N\ln N + N\\ &= N\ln g - N\ln N + N\end{aligned}$$

That is, for $g \gg N$ and the validity of Stirling's formula, we find

$$\boxed{\ln \Omega_1 = \ln \Omega_2 = \ln \Omega_\mathrm{I} + \ln \Omega_\mathrm{II}} \quad \text{and} \quad \boxed{S_1 = S_\mathrm{I} + S_\mathrm{II}}$$

independent of the absolute number of particles or quantum states.

17.13 Foundations

17.13.1 Foundation 17.1: The Relationship Between Entropy and Heat

17.13.1.1 Proof of Proposal 1: $\sum_i \frac{q_{\mathrm{rev},i}}{T_i} = 0$ (Cyclic Process)

Arbitrary Reversible Heat Engine (Two Temperature Baths Involved) Here we prove that $\sum_i \frac{q_{\mathrm{rev},i}}{T_i}$ is the change of a state variable. We consider an arbitrary reversible heat engine and show that its efficiency must be the same as the efficiency

$$\eta = 1 - \frac{T_1}{T_2}$$

of the Carnot engine, see Section 17.2.2.4. This means that the relation

$$\frac{q_{\mathrm{rev},1}}{T_1} + \frac{q_{\mathrm{rev},2}}{T_2} = 0 \quad \text{(cyclic process)} \tag{17F1.1}$$

derived for the Carnot engine, Equation (17.29), must hold for any arbitrary reversible heat engine. We proceed by using "proof by contradiction."

What is the efficiency η of an arbitrary reversible heat engine? We imagine, besides the Carnot engine, a second reversible heat engine (fictional engine) with an efficiency

$$\eta^* = \frac{-w^*_{\mathrm{rev}}}{q^*_{\mathrm{rev},2}} > \eta \tag{17F1.2}$$

This means that the efficiency η^* is larger than the efficiency η for the Carnot engine. We couple this engine with a Carnot engine by using the Carnot engine as a heat pump. The two engines are coupled such that the mechanical energy supplied by the fictional engine is used up by the Carnot engine to pump heat (Fig. 17F.1). Then the total work done on the surroundings in a cycle is zero.

However, because $\eta^* > \eta$, more heat is pumped by the Carnot engine into the warmer bath than taken away from it by the fictional engine. All together, in a cycle (i.e., when both engines are included in the system) heat is transferred from the colder bath to the warmer bath without any external work being done. Such a result, however, contradicts the statement that heat can flow only from a warmer to a colder body. This statement is a result of statistical mechanics, as we have shown in Section 17.5.1.

What happens if $\eta^* < \eta$ and the fictional engine is working reversibly? Then again we can transfer heat from a colder body to a warmer body without supplying mechanical energy by using the fictional engine as a heat pump and the Carnot machine as a heat engine.

From these considerations it follows that no reversibly working heat engine can exist with an efficiency different from the Carnot efficiency. In this way we have proved that $\sum_i \frac{q_{\mathrm{rev},i}}{T_i}$ is the change of a state function for a system connected to two temperature baths.

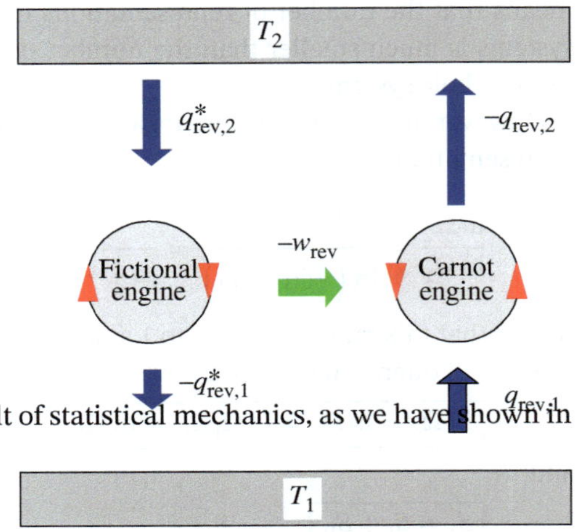

Figure 17F.1 Carnot engine coupled to a reversibly working fictional engine with $\eta^* > \eta$. The arrows in the circles indicate the direction of the cycle: clockwise as in Fig. 17.6 ($a \to b \to c \to d \to a$), or anticlockwise ($a \to d \to c \to b \to a$).

Arbitrary Heat Engine (Two Temperature Baths Involved) Any practically realizable heat engine works nonreversibly. This means that such an engine yields less work than a Carnot engine when consuming the same amount of heat from bath T_2. Used as a heat pump, it needs more external mechanical energy to transfer a given amount of heat to bath T_2 than a Carnot engine (Fig. 17F.2). Thus, the relation

$$\eta_{\mathrm{realistic}} < \eta \tag{17F1.3}$$

must generally hold. That is, all reversible Carnot cycles operating between the same initial and final temperatures must have the same efficiency. Since the cycles are reversible, this efficiency is the maximum possible.

From Equation (17.27)

$$\eta = \frac{-w_{\mathrm{rev}}}{q_{\mathrm{rev},2}}$$

and the relation

$$w_{\mathrm{rev}} + q_{\mathrm{rev},1} + q_{\mathrm{rev},2} = 0$$

for a cyclic process (see Fig. 17.7) it follows that

$$\frac{q_{\mathrm{rev},1}}{q_{\mathrm{rev},2}} = \eta - 1$$

Accordingly,

$$\frac{q_1}{q_2} = \eta_{\mathrm{realistic}} - 1 < \frac{q_{\mathrm{rev},1}}{q_{\mathrm{rev},2}}, \text{ where } \frac{q_{\mathrm{rev},1}}{q_{\mathrm{rev},2}} = -\frac{T_1}{T_2}$$

Then

$$\frac{q_1}{q_2} < -\frac{T_1}{T_2}$$

and

$$\frac{q_1}{T_1} + \frac{q_2}{T_2} < 0 \qquad (17F1.4)$$

must generally hold for a realistic heat engine operating in a cyclic process. The heat flowing into the system is less than in the reversible limit because more work is done on the system than in the reversible limit.

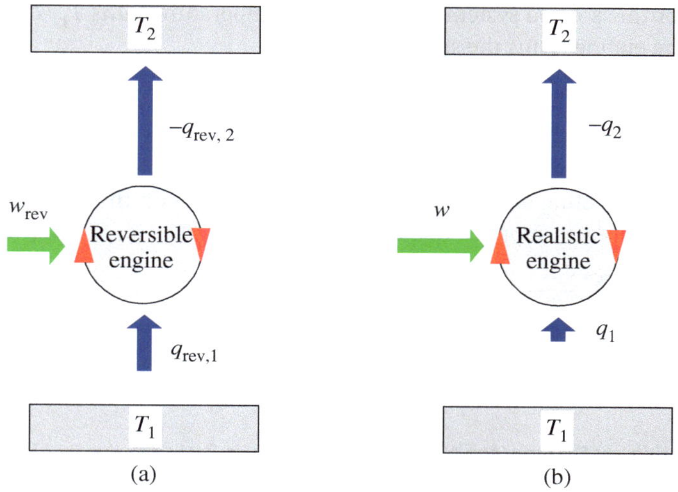

Figure 17F.2 Comparison of a reversible engine with a realistic engine ($\eta_{\text{realistic}} < \eta$). The colored arrows indicate the direction of the cycle: clockwise as in Fig. 17.6 ($a \to b \to c \to d \to a$).

17.13.1.2 Clausius Inequality

We apply Equation (17F1.4) to a cyclic process at constant temperature ($T = T_1 = T_2$). First we consider an irreversible process from State 1 to State 2 with $q_{1\to2}$. Next we perform a reversible process from State 2 to State 1 with $q_{\text{rev},2\to1}$. Then

$$\frac{q_{1\to2}}{T} + \frac{q_{\text{rev},2\to1}}{T} < 0 \quad \text{or} \quad \frac{q_{1\to2}}{T} - \frac{q_{\text{rev},1\to2}}{T} < 0$$

Then with $q_{\text{rev},1\to2}/T = \Delta S_{1\to2}$ we obtain

$$\frac{q_{1\to2}}{T} - \Delta S < 0 \quad \text{and} \quad \boxed{\Delta S_{1\to2} > \frac{q_{1\to2}}{T}}$$

This relation is called the *Clausius inequality*.

Example 17.11 *Compare the entropy change ΔS in an isothermal expansion to q/T.*

Let us consider Fig. 17.14a, where the gas is expanded irreversibly from volume V_1 to volume V_2 (no work supplied to the system). In this case $q = 0$. In Fig. 17.14b the gas is expanded reversibly. In this case $q_{\text{rev}} = \boldsymbol{nR}T\ln(V_2/V_1)$ and $\Delta S = q_{\text{rev}}/T > q/T$, in accordance with the Clausius inequality relation.

17.13.1.3 Proof of Proposal 2: $\sum_i \frac{q_{\text{rev},i}}{T_i} \geq 0$ (Isolated System)

Arbitrary Process in an Isolated System (Two Temperature Baths Involved)

Here we prove that $\sum_i \frac{q_{\text{rev},i}}{T_i} \geq 0$ for an isolated system. Let us consider the isolated system in Fig. 17F.3 consisting of a realistic heat engine, two temperature baths (one at T_1, one at T_2), and a device for exchanging mechanical energy between the engine and the outside. The baths are so large that the addition or removal of heat does not change their temperature. We consider an engine acting as a heat pump. The engine consumes heat q_1 from temperature bath T_1 and produces heat at temperature T_2. This heat $-q_2$ is transferred to bath T_2. This transfer is considered to be a reversible process as well as the removal of heat from bath T_1.

We are interested in the quantity

$$\sum_i \frac{q_{\text{rev},i}}{T_i}$$

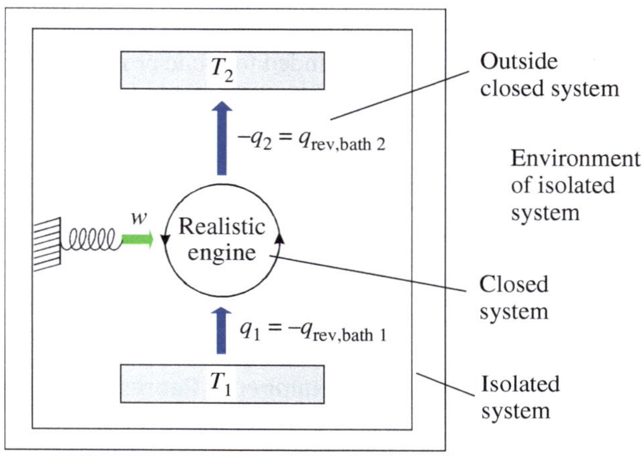

Figure 17F.3 Realistic engine and temperature baths T_1 and T_2 considered as an isolated system. Work is done on the realistic engine (symbolized by a relaxing spring) and heat is transferred to the high temperature bath.

for the isolated system (consisting of temperature baths T_1, T_2, and a heat engine) when a cyclic process is carried out by the engine. Thus the sum is

$$\sum_i \frac{q_{\text{rev},i}}{T_i} = \frac{\text{heat supplied to bath } T_2}{T_2} + \frac{\text{heat supplied to bath } T_1}{T_1} = \frac{-q_2}{T_2} + \frac{-q_1}{T_1} = -\left(\frac{q_2}{T_2} + \frac{q_1}{T_1}\right)$$

In the case of a reversible engine performing a cyclic process we have $q_2/T_2 + q_1/T_1 = 0$. In the present case of a realistic engine acting as heat pump, more work is done on the engine to remove a certain amount of heat from the bath T_1, and more heat is supplied to bath T_2; therefore $q_2/T_2 + q_1/T_1 < 0$, see Equation (17F1.4). Then we obtain

$$\sum_i \frac{q_{\text{rev},i}}{T_i} \geq 0 \quad \text{(isolated system)} \tag{17F1.5}$$

where the equals sign corresponds to a reversible process. In this way we have proved for a system connected to two temperature baths that $\sum_i \frac{q_{\text{rev},i}}{T_i} \geq 0$ for an isolated system.

Arbitrary Process (Any Number of Temperature Baths Involved) Our considerations can be extended to processes in which the changes of state occur at any chosen number of temperatures. We consider a cyclic process involving three temperatures (Figure 17F.4) We imagine this process is composed of the two partial processes I ($a \to b \to c \to g \to a$) and II ($g \to d \to e \to f \to g$). The sum of the heats supplied to the system in the partial isothermal steps is equal to that of the total system. The steps $a \to b \to c \to d$ and $e \to f$ occur as well in the partial processes as in the total process and the contributions of steps $c \to g$ (of I) and $g \to c$ (of II) cancel each other. For the partial processes we obtain

$$\frac{q_{\text{rev}}(a \to b)}{T_1} + \frac{q_{\text{rev}}(c \to g)}{T_2} = 0 \quad \text{(2 temperature baths involved)}$$

$$\frac{q_{\text{rev}}(e \to f)}{T_1} + \frac{q_{\text{rev},2}(g \to d)}{T_2} = 0 \quad \text{(2 temperature baths involved)}$$

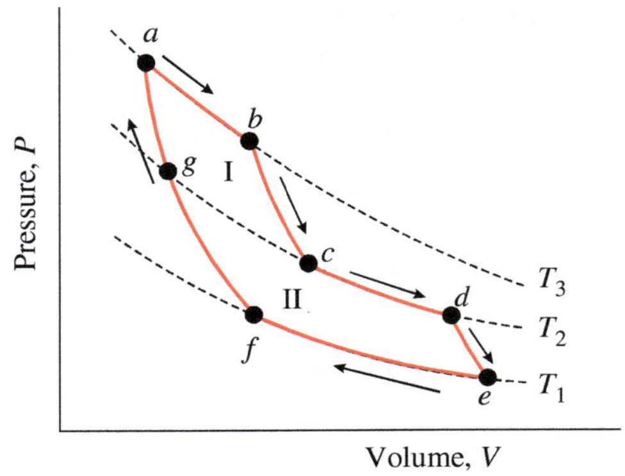

Figure 17F.4 Cyclic process involving three temperatures.

From

$$q_{\text{rev}}(g \to c) + q_{\text{rev}}(c \to d) = q_{\text{rev}}(g \to d)$$

$$q_{\text{rev}}(g \to c) = -q_{\text{rev}}(c \to g)$$

it follows

$$q_{\text{rev}}(c \to g) + q_{\text{rev}}(g \to d) = q_{\text{rev}}(c \to d)$$

For the total process with three temperature baths involved, we find

$$\frac{q_{\text{rev}}(a \to b)}{T_3} + \frac{q_{\text{rev}}(c \to d)}{T_2} + \frac{q_{\text{rev}}(e \to f)}{T_1} = 0$$

This consideration can be extended to cyclic processes with 4, 5,... or any number of temperature baths. Then

$$\sum_i \frac{q_{\text{rev},i}}{T_i} = 0 \quad \text{(cyclic process)} \tag{17F1.6}$$

and

$$\sum_i \frac{q_{\text{rev},i}}{T_i} \geq 0 \quad \text{(process in an isolated system)} \tag{17F1.7}$$

Thus we have proved relations (17.74) and (17.75) for a system with an arbitrary number of temperature baths.

17.13.2 Foundation 17.2: Number of Representations Ω from Molecular Partition Function *Z*

17.13.2.1 Distinguishable Particles

For rotational and vibrational motions a molecule's center of mass remains unchanged. Thus we can attribute site numbers to any molecule, and the rotational and vibrational motions are distinguishable, in contrast to the translational motions of molecules. According to Section 15.8.1 the number of representations ω for a particular population type is

$$\omega = \frac{N!}{N_1!N_2!\ldots} \tag{17F2.1}$$

and the total number of representations Ω is

$$\Omega = \omega_1 + \omega_2 + \omega_3 + \cdots \tag{17F2.2}$$

where $\omega_1, \omega_2, \omega_3, \ldots$ are the numbers of representations belonging to different population types.

***Relation Between* $\ln \Omega$ *and* $\ln \omega_{max}$** As an example we consider the situation described in Section 15.8.1, where each of N particles has three energy levels. The table compares Ω and ω_{max} (ω_{max} corresponds to the population type with the maximum number of representations) for different numbers, N, of particles.

N	Ω	$\ln \Omega$	ω_{max}	$\ln \omega_{max}$	ω_{max}/Ω	$\ln \omega_{max} / \ln \Omega$
6	90	4.5	60	4.09	0.67	0.90
100	69×10^{32}	77.9	15×10^{32}	76.4	0.22	0.98
1000	16×10^{349}	806.5	1×10^{349}	803.7	0.063	0.997
10,000	164×10^{3503}	8072.5	3.6×10^{3503}	8068.7	0.022	0.9995

The ratio ω_{max}/Ω becomes smaller with increasing N and converges to zero. In contrast, the ratio $\ln \omega_{max} / \ln \Omega$ converges to 1 with increasing N. For N large enough, we can make the approximation

$$\boxed{\ln \Omega = \ln \omega_{max}} \tag{17F2.3}$$

Since we often deal with large values of N this approximation is excellent.

***Calculating* $\ln \omega$** Using Equation (17F2.1), we calculate $\ln \omega$ by writing

$$\ln \omega = \ln N! - \ln N_1! - \ln N_2! - \cdots \tag{17F2.4}$$

Using Equation (17F2.4) and Stirling's approximation (see Appendix C.3), we find

$$\ln \omega = N \cdot \ln N - N - N_1 \cdot \ln N_1 + N_1 - N_2 \cdot \ln N_2 + N_2 - \cdots \tag{17F2.5}$$

Calculating* $\ln \omega_{max}$ *by Introducing the Boltzmann Distribution We showed in Section 15.8.2 that the Boltzmann distribution corresponds to ω_{max}. Hence we can find an expression for $\ln \omega_{max}$ by expressing the population numbers N_i in Equation (17F2.5) by the Boltzmann distribution

$$N_i = \frac{N}{Z} \cdot \mathrm{e}^{-E_i/(kT)} \tag{17F2.6}$$

(Z is the molecular partition function). Then we find that

$$\ln \omega_{max} = N \cdot \ln N - \sum_{i=1}^{\infty} \left(\frac{N}{Z}\mathrm{e}^{-E_i/(kT)} \cdot \ln \left[\frac{N}{Z}\mathrm{e}^{-E_i/(kT)} \right] \right)$$

where we have also used the fact that $N_1 + N_2 + \cdots = N$. Because of

$$\ln \left[\frac{N}{Z}\mathrm{e}^{-E_i/(kT)} \right] = \ln \frac{N}{Z} - \frac{E_i}{kT}$$

we obtain

$$\ln \omega_{max} = N \cdot \ln N - \frac{N}{Z} \cdot \ln \frac{N}{Z} \sum_{i=1}^{\infty} \left(\mathrm{e}^{-E_i/(kT)} \right) + \frac{1}{kT} \sum_{i=1}^{\infty} \left(\frac{N}{Z}\mathrm{e}^{-E_i/(kT)} \cdot E_i \right) \tag{17F2.7}$$

Introducing the Partition Function The first sum in Equation (17F2.7) is equal to the molecular partition function Z and the second sum is equal to the internal energy U. Using Equation (17F2.3), we find

$$\begin{aligned}\ln \Omega = \ln \omega_{\max} &= N \cdot \ln N - \frac{N}{Z} \cdot \ln \frac{N}{Z} \cdot Z + \frac{1}{kT} U \\ &= N \cdot \ln N - N \cdot \ln \frac{N}{Z} + \frac{1}{kT} U \\ &= N \cdot \ln Z + \frac{U}{kT}\end{aligned} \tag{17F2.8}$$

Using the fact that $S = k \ln \Omega$, we see that

$$S = Nk \cdot \ln Z + \frac{U}{T} \tag{17F2.9}$$

for distinguishable particles.

17.13.2.2 Indistinguishable Particles

Calculating ln ω In Section 15.9.2 we derived the number of representations ω for indistinguishable particles and found

$$\omega = \frac{g_1^{n_1}}{n_1!} \cdot \frac{g_2^{n_2}}{n_2!} \cdot \frac{g_3^{n_3}}{n_3!} \cdots \tag{17F2.10}$$

where g_i is the number of states and n_i is the number of particles in an energy interval ΔE. This relation holds for $g_i \gg n_i$, that is for gas particles at sufficiently high temperature. We calculate $\ln \omega$ by writing

$$\ln \omega = n_1 \ln g_1 - \ln n_1! + n_2 \ln g_2 - \ln n_2! + \cdots \tag{17F2.11}$$

and applying Stirling's formula to obtain

$$\begin{aligned}\ln \omega &= n_1 \ln g_1 - n_1 \ln n_1 + n_1 + n_2 \ln g_2 - n_2 \ln n_2 + n_2 + \cdots \\ &= \sum_i (n_i \ln g_i - n_i \ln n_i + n_i) = N - \sum_i \left(n_i \ln \frac{n_i}{g_i} \right)\end{aligned} \tag{17F2.12}$$

Calculating ln $\omega_{\max}$ by Introducing the Boltzmann Distribution The n_i in Equation (17F2.12) are given by the Boltzmann law in the form:

$$n_i = \frac{N}{Z} g_i \mathrm{e}^{-E_i/(kT)} \tag{17F2.13}$$

Now we proceed in analogy to the case of distinguishable particles. By using the Boltzmann distribution for n_i, we obtain

$$\begin{aligned}\ln \omega_{\max} &= N - \frac{N}{Z} \cdot \sum_{i=1}^{\infty} g_i \,\mathrm{e}^{-E_i/(kT)} \ln \left(\frac{N}{Z} \mathrm{e}^{-E_i/(kT)} \right) \\ &= N - \frac{N}{Z} \cdot \sum_{i=1}^{\infty} g_i \,\mathrm{e}^{-E_i/(kT)} \left(\ln \frac{N}{Z} - \frac{E_i}{kT} \right) \\ &= N - \frac{N}{Z} \ln \frac{N}{Z} \cdot \sum_{i=1}^{\infty} g_i \,\mathrm{e}^{-E_i/(kT)} + \frac{1}{kT} \cdot \sum_{i=1}^{\infty} \frac{N}{Z} E_i \, g_i \,\mathrm{e}^{-E_i/(kT)}\end{aligned} \tag{17F2.14}$$

Introducing the Partition Function The first sum in Equation (17F2.14) is equal to the molecular partition function Z and the second sum is equal to the internal energy U. Using Equation (17F2.3), we find

$$\ln \Omega = \ln \omega_{\max} = N - N \ln \frac{N}{Z} + \frac{U}{kT} = N \ln Z + \frac{U}{kT} + (N - N \ln N) \tag{17F2.15}$$

and

$$S = k \ln \Omega = Nk \ln Z + \frac{U}{T} + k\,(N - N \ln N)$$

This expression differs from the corresponding Equation (17F2.8) for distinguishable particles by the additional term $(N - N\ln N)$. Alternatively, we can apply Stirling's approximation (in reverse) and write

$$\ln \Omega = \ln \frac{Z^N}{N!} + \frac{U}{kT} \tag{17F2.16}$$

and for the entropy we find

$$S = k \ln \Omega = Nk \ln Z + \frac{U}{T} - k \ln N! \tag{17F2.17}$$

for indistinguishable particles.

17.13.3 Foundation 17.3: Entropy of Homonuclear Diatomic Gases

See the books website, www.wiley.com/go/Kuhn/physicalchemistry3e.

17.14 Exercises

E17.1 Consider the reversible isothermal expansion of 1 mol of an ideal gas at $T = 300$ K from a volume V_1 to a volume $V_2 = 2V_1$, and calculate the work performed. Using this value of the work, how many meters can this expansion lift a weight of 1 kg in earth's gravitational field 9.81 m s^{-2}?

E17.2 Calculate the work for an isothermal expansion of a gas from V_1 to $V_2 = 2V_1$ against a constant pressure P. Compare the work found in this case to that found for the isothermal expansion in E17.1.

E17.3 In Section 17.5.2 we computed the entropy change for the mixing of two gases. Find this entropy by performing the reverse process. Compute the probability $\mathcal{P}$ that the mixture separates into gases A and B, and the change of the entropy.

E17.4 Perform an analysis for heat flow at constant volume for a system like that in Fig. 17.1a. Derive a mathematical expression for the entropy change in the upper container and for the final temperature in terms of the two bodies heat capacities, C_{upper} and C_{lower}, and initial temperatures, T_{upper} and T_{lower}. You may assume that the constant volume heat capacities are temperature independent. Evaluate the final temperature for the initial conditions of $T_{\text{lower}} = 273$ K, $T_{\text{upper}} = 373$ K, and 1 mol of a monatomic gas. Calculate how much the final temperature changes if the upper chamber has 2 mol of gas, rather than 1 mol.

E17.5 Consider an isothermal and reversible expansion of an ideal gas to twice its initial volume. Determine whether the work w is positive, negative, or zero; the heat q is positive, negative, or zero; the internal energy change ΔU is positive, negative, or zero; and the entropy change ΔS is positive, negative, or zero.

E17.6 In Equation (17.56) we calculated the entropy change for temperature equilibration as

$$\Delta S = \frac{3}{2} Nk \cdot \ln\left(\frac{T_{\text{eq}}^2}{T_1 T_2}\right) \quad \text{with } T_{\text{eq}} = \frac{T_1 + T_2}{2}$$

Show that ΔS is positive for any pair of temperatures T_1 and T_2.

E17.7 Consider an ideal gas at 10 bar pressure and a temperature of 298 K in an isolated system. While remaining isolated, the gas is expanded into a vacuum from a volume $V_0 = 1$ L to a volume $V_f = 3$ L. Calculate ΔS for this process and the final temperature of the gas. Using your result, determine the heat q_{rev} that would be needed to perform this process reversibly.

E17.8 A piece of metal with heat capacity of 8 kJ K^{-1} and an initial temperature $T_1 = 273$ K is placed on top of another piece of metal with a heat capacity of 4 kJ K^{-1} and an initial temperature $T_2 = 373$ K. Assume that the two blocks are isolated from the environment and determine the final temperature and the total entropy change in the temperature equilibration between the two metal pieces.

E17.9 Engineering tables list values of S for water. At 0 °C, $S^{\ominus}_{273} = 3.9$ kJ K^{-1} for 1 kg of water. What is the number of representations Ω_{273} for water at 0 °C and atmospheric pressure? A kilogram of liquid water at 100 °C and atmospheric pressure has the entropy $S^{\ominus}_{373} = S^{\ominus}_{273} + 1300$ J K^{-1}. Calculate the number of representations Ω_{373} for the water at 100 °C. Compare the two values.

E17.10 Consider the three-level system of energy $U = 40a$ in Table 15.3. What is the entropy of the system?

E17.11 In Exercises E15.19 and E15.20 you computed the average energy and populations for spin sublevels in an applied magnetic field. Compute the entropy of each of those systems, for the cases of $N = 1$, $N = 10$, and $N = 100$.

E17.12 Consider a gas that has expanded from the volume V_1 to the volume V_2 in Example 17.1. For the gas at V_2 calculate the probability of finding all of the gas molecules in the subvolume V_1. Calculate the entropy change for the gas.

E17.13 Calculate the entropy for argon gas at $T = 300$ K and $P = 1$ bar in a container of volume V, where V is a) one milliliter, b) one microliter, and c) one nanoliter; assume that the gas behaves ideally.

E17.14 Calculate the translational, rotational, and vibrational contributions to U, S, and C_V for 1 mol of ICl molecules at 300 and 1000 K. The fundamental vibrational frequency of ICl is $\nu_0 = 1.15 \times 10^{13}$ s^{-1} and the rotational constant of ICl is $B = 2.27 \times 10^{-24}$ J. You may consider the gas to behave ideally and have a pressure of 1 bar.

E17.15 At 1 bar pressure and $T = 194.65$ K solid CO_2 sublimates to form CO_2 vapor. Compare the molar entropy of the solid (68.3 J mol^{-1} K^{-1}) and vapor (199.1 J mol^{-1} K^{-1}) forms. Explain which motions contribute to each and how it explains the relative size of the entropies.

E17.16 Calculate the molar entropies of HF and HI from spectroscopic parameters (for HF $g_{el} = 1$, $\sigma = 1$, $\theta_{vib} = 5954$ K, $\theta_{rot} = 29.4$ K, and for HI $g_{el} = 1$, $\sigma = 1$, $\theta_{vib} = 3266$ K, $\theta_{rot} = 9.1$ K). Compare the entropy values and explain why they are different. You may assume that the gases behave ideally and are at standard conditions of 1 bar and 298.15 K.

E17.17 Calculate the molar entropies of CO and N_2 from their spectroscopic parameters (for CO $g_{el} = 1$, $\sigma = 1$, $\theta_{vib} = 3103$ K, $\theta_{rot} = 2.8$ K, and for N_2 $g_{el} = 1$, $\sigma = 2$, $\theta_{vib} = 3374$ K, $\theta_{rot} = 2.9$ K). Note that the difference in the symmetry factors σ is based on the fact that CO is a heteronuclear molecule and N_2 is homonuclear. Compare the entropy values and explain why they are different. You may assume that the gases behave ideally and are at standard conditions of 1 bar and 298.15 K.

E17.18 For a given substance, the entropy of the solid is less than that of the liquid which is less than that of the gas. Discuss how the different motions (translation, rotation, vibration, and electronic) contribute to each and explain this trend.

E17.19 Consider the absolute entropies for the gases in Table 17.3. Explain the trend in the entropies for the monatomic series of gases. Explain the trend in the entropy for the diatomic gases. Explain the difference between the two sets of gases.

E17.20 Consider the absolute entropies for the three gases Ne ($S^{\ominus}_{m,298} = 146.2$ J mol^{-1} K^{-1}), HF ($S^{\ominus}_{m,298} = 173.8$ J mol^{-1} K^{-1}), and D_2O ($S^{\ominus}_{m,298} = 198.3$ J mol^{-1} K^{-1}). These gases all have about the same total mass of 20 g mol^{-1}. Explain the difference in the entropies of the gases in terms of the gas molecule's molecular characteristics. Also consider the molecule ND_3, where do you expect its absolute entropy to lie with respect to Ne, HF, and D_2O?

E17.21 In Example 17.8 we compute the entropy change of an ideal gas in a two step process: 1) heating at constant volume and 2) expansion at constant temperature. Find the entropy change between the same initial and final states by using a different two-step process: 1) heating at constant pressure and 2) expansion at constant temperature.

E17.22 Consider two CO dipoles that are arranged side by side (take the CO dipole moment to be 0.1 D, its bond length to be 113 pm, and its diameter to be 350 pm). Using information from Chapter 13 calculate the difference in the dipole–dipole interaction energy for the antiparallel arrangement of the dipoles and for the parallel arrangement of the dipoles. At what temperature is this energy difference equal to kT?

E17.23 Two equal containers (each of volume V) contain two different ideal gases A and B (pressure P and temperature T). By connecting the containers the gases are mixed. What is the mixing entropy and what is the reversible work necessary to separate the gases?

E17.24 Consider the evaporation of a liquid at constant pressure (as shown in Fig. 17.2). The entropy change for the process is

$$\Delta S = \frac{(q_{rev})_P}{T} = \frac{\Delta H}{T}$$

Given that the molar enthalpy of vaporization for water is 44.1 kJ mol^{-1} and its boiling point is 373 K at 1 bar pressure, calculate the molar entropy change for water when it vaporizes.

E17.25 In our considerations of the entropy of subsystems, we wrote the number of representations of the total system as a product of that for the subsystems. Explain why this procedure is appropriate.

E17.26 Equation (17.12) relates the change in temperature for a reversible adiabatic expansion in terms of the ratio V_1/V_2. Rewrite the equation in terms of P_1/P_2. If you expand the gas from $P_1 = 10$ bar and $T_1 = 300$ K to $P_2 = 1$ bar, what is the final temperature?

17.15 Problems

P17.1 Figure 17.13 illustrates a process of temperature equilibration. Calculate the probability $\mathcal{P}$ that, by chance, the molecules are distributed as in State 1.

P17.2 Consider an ideal atomic gas ($C_V = \frac{3}{2}Nk$) whose volume changes from V_1 to $V_2 = 2V_1$ in two different ways. In the first case the gas expansion proceeds isothermally and reversibly at the temperature $T_1 = 300$ K, and in the second case it proceeds adiabatically and reversibly. Find expressions for the work w, heat q, internal energy change ΔU, and entropy change ΔS for these two different processes and compare them.

P17.3 The molecular partition function Z for a monatomic ideal gas is given by

$$Z = V\left(\frac{2\pi mkT}{h^2}\right)^{3/2}$$

Use this expression for the molecular partition function to derive an expression for the internal energy U and for the entropy of an ideal monatomic gas.

P17.4 Consider the following heat capacity data for the gases neon (Ne), water (H_2O), and hydrogen fluoride (HF). Plot these data with a spreadsheet program, such as Excel. Compare the data for the different substances and explain the different behavior from a molecular perspective.

$\frac{T}{\text{K}}$	$\frac{C^{\ominus}_{P,m}(\text{Ne})}{\text{J K}^{-1}\text{ mol}^{-1}}$	$\frac{C^{\ominus}_{P,m}(\text{HF})}{\text{J K}^{-1}\text{ mol}^{-1}}$	$\frac{C^{\ominus}_{P,m}(\text{H}_2\text{O})}{\text{J K}^{-1}\text{ mol}^{-1}}$
400	20.786	29.180	34.261
600	20.786	29.205	36.322
800	20.786	29.582	38.723
1000	20.786	30.142	41.627
1200	20.786	30.934	43.769
1400	20.786	31.832	46.055
1600	20.786	32.666	48.050
1800	20.786	33.396	49.747
2000	20.786	34.030	51.185

Use these data to find the enthalpy change at constant pressure for each gas upon heating it from $T = 400$ K to $T = 2000$ K. Use these data to find the entropy change at constant pressure for each gas upon heating it from 400 to 2000 K.

P17.5 Consider a mole of CO_2 gas at 1 bar pressure and 194.67 K. The entropy of this gas was determined by measuring its heat capacity as a function of temperature and found to be 199.12 J mol^{-1} K^{-1} (see [3]). Compare this calorimetric value for the entropy to that you would calculate using the CO_2 molecule's spectroscopic parameters. Consider the molecule to be linear with $\theta_{rot} = 0.561$ K and $\sigma = 2$. Given that the molecule's normal modes have characteristic temperatures of $\theta_{vib} = 3360$ K, $\theta_{vib} = 1890$ K, $\theta_{vib} = 954$ K, the vibrational contribution to the entropy is small, 0.732 J mol^{-1} K^{-1}. Compute the percentage difference between the thermodynamic and spectroscopic values of the entropy.

P17.6 Consider a three-level system like that discussed in Section 15.8, except it has a fixed temperature of $10a/k$ (a is the energy spacing and k is the Boltzmann constant), rather than a fixed energy. What is the average energy of the system? What is the entropy of the system?

P17.7 Use the following calorimetric data for the molar constant pressure heat capacity of O_2 and oxygen phase transitions to determine the molar entropy of oxygen at 298 K and 1 bar pressure.

Solid I		Solid II		Solid III	
$\frac{T}{K}$	$\frac{C_{P,m}}{J\ mol^{-1}\ K^{-1}}$	$\frac{T}{K}$	$\frac{C_{P,m}}{J\ mol^{-1}\ K^{-1}}$	$\frac{T}{K}$	$\frac{C_{P,m}}{J\ mol^{-1}\ K^{-1}}$
12.97	4.60	25.02	22.68	45.90	46.13
14.14	6.36	26.75	24.07	47.76	46.34
15.12	7.07	28.0	25.32	48.11	46.08
15.57	7.49	29.88	27.67	48.97	46.00
16.66	9.75	30.63	29.05	50.55	46.08
16.94	9.42	33.05	31.48	51.68	46.17
18.13	11.18	34.41	33.82	52.12	46.29
18.45	11.68	35.77	35.54		
19.34	12.85	37.85	38.17		
20.26	14.65	38.47	41.02		
20.85	15.07	39.99	41.02		
21.84	17.58	40.67	41.52		
22.24	18.42	42.21	44.91		

Liquid		Gas	
$\frac{T}{K}$	$\frac{C_{P,m}}{J\ mol^{-1}\ K^{-1}}$	$\frac{T}{K}$	$\frac{C_{P,m}}{J\ mol^{-1}\ K^{-1}}$
56.95	53.41	91	30.75
57.95	53.24	100	29.93
60.97	53.20	140	29.50
65.57	53.20	180	29.30
69.12	53.37	220	29.25
74.95	53.79	260	29.30
81.13	53.91	298	29.44
84.79	54.12		
86.97	54.08		
90.33	54.37		

Phase transitions occur at a) $T = 23.66$ K with $\Delta H_m = 93.84$ J mol^{-1}, b) $T = 43.76$ K with $\Delta H_m = 743.4$ J mol^{-1}, c) $T = 54.39$ K with $\Delta H_m = 444.9$ J mol^{-1}, and d) $T = 90.13$ K with $\Delta H_m = 6817.6$ J mol^{-1}. The transitions at a) and b) are solid-to-solid phase transitions. The transition at c) is the solid-to-liquid phase transition and the transition at d) is the liquid-to-gas phase transition.

P17.8 In Problem P17.7 you should find that the calorimetrically determined entropy for diatomic oxygen is 203.6 J K^{-1} mol^{-1}. Calculate the entropy of oxygen from spectroscopic parameters and compare it to the calorimetric value. Note: In addition to the translational, rotational, and vibrational contributions to the entropy, you will also need to include the electronic contribution for oxygen. At room temperature the electronic ground state of diatomic oxygen may be considered to be a triplet, $g = 3$.

P17.9 We have calculated the rotational contribution S_{rot} for HD at 298 K with the assumption that the molecular partition function Z_{rot} can be evaluated by integration ($Z = kT/B$) and the internal energy by the high temperature value $U_{rot} = NkT$. We learned in Chapter 15 that the internal energy U_{rot} for HD at 298 K is lower than the classical value (see Fig. 15.10). The numerical values at $T = 298$ K are $Z_{\text{rot}} = 4.875$ and $U_{\text{rot},m} = 2.295$ kJ mol^{-1}, whereas the integrated values used in Example 17.10 are $Z_{\text{rot}} = 4.52$ and $U_{\text{rot},m} = 2.48$ kJ mol^{-1}. Calculate S_{rot} for both data sets, using Equation (17.87).

P17.10 From Equation (17.77) show that the thermodynamic definition of temperature is equivalent to Equation (14.24), using the gas thermometer. Use the Carnot cycle with an ideal gas as a working medium.

P17.11 Reversible adiabatic compression. Consider a process with an ideal monatomic gas, which is thermally isolated from the surroundings. The gas (amount of substance $\boldsymbol{n} = 0.2$ mol) is reversibly compressed (*reversible adiabatic compression*) from volume $V_1 = 10$ L to volume $V_2 = 1$ L, while its temperature changes from $T_1 = 300$ K to T_2. Using the considerations on a reversible adiabatic expansion derive the final temperature T_2 and the reversible work w_{rev}.

P17.12 In Section 17.8 we calculated the rotational contribution of the entropy S_{rot} of HD at 298 K and $P = 1$ bar from the molecular partition function Z_{rot}. Perform a similar calculation by describing S_{rot} as

$$S_{\text{rot}} = \int_0^{298} \frac{C_{P,\text{rot}}}{T} \text{d}T$$

Use the methods of Chapters 15 and 16 to calculate the values of $C_{P,\text{rot}}$ by direct summation. Compare the results of both treatments.

P17.13 Consider a reversible adiabatic compression of an ideal gas from volume V_1 and temperature T_1 to volume V_2 and temperature T_2. Treating the gas as particles in a cubic box of side length L, show that the spacing of the energy levels increases, but the population probabilities remain the same.

P17.14 In this problem you find expressions for the internal energy and the entropy for a gas of N photons. Use the distribution function for photons in Equation (15.112) to show that the number N of photons is given by

$$N = \frac{8\pi}{c_0^3} V \left(\frac{kT}{h}\right)^3 \int_{x=0}^{\infty} \frac{x^2}{\text{e}^x - 1} \text{d}x = \frac{8\pi^4}{12.897 c_0^3} V \left(\frac{kT}{h}\right)^3$$

where $x = h\nu/kT$. The last equality results from evaluating the integral (see Appendix B.3) which is

$$\int_{x=0}^{\infty} \frac{x^2}{\text{e}^x - 1} \text{d}x = 2.4041 = \frac{\pi^3}{12.897}$$

Next, you use the definition of the internal energy $U \equiv \int_{x=0}^{\infty} h\nu \, \text{d}N$, to show that

$$U = \frac{8\pi}{c_0^3} V \left(\frac{kT}{h}\right)^4 \int_{x=0}^{\infty} \frac{x^3}{\text{e}^x - 1} \text{d}\nu = \frac{8\pi}{c_0^3} V \left(\frac{kT}{h}\right)^4 \frac{\pi^4}{15}$$

In the last part of the problem, you will derive an expression for the entropy. Comparison of the distribution function for photons in the limit that $\exp(E_i/(kT)) \gg 1$, namely

$$N_i = \frac{g_i}{\exp(E_i/(kT)) - 1} \approx g_i \text{e}^{-E_i/(kT)}$$

with our Equation (15.9)[4]

$$N_i = \frac{N}{Z} g_i \text{e}^{-E_i/(kT)}$$

implies that

$$Z = N$$

Use this equality, and the general result that

$$S = Nk \ln Z + Nk(1 - \ln N) + \frac{U}{T}$$

to show that entropy S for a photon gas is

$$S = Nk + \frac{8\pi}{c_0^3} V \left(\frac{k}{h}\right)^4 \frac{\pi^4}{15} T^3$$

Using these formulae calculate the photon density (N/V), the energy density U/V, and the entropy S of the photon gas under standard conditions of 298 K and 1 atmosphere.

---

4 Note that in Equation (15.9) the index i is related to the energy state i. In our procedure the index i is related to the quantum state i; therefore, we have to add the factor g_i.

References

1 A.E. Leanhardt, T.A. Pasquini, M. Saba, A. Schirotzek, Y. Shin, D. Kielpinski, D.E. Pritchard, and W. Ketterle, *Science*, 2003, **301**, 1513.

2 S. Weinberg, *The First Threee Minutes: A Modern View of the Origin of the Universe*, Perseus Books Group, 1993.

3 W.F. Giaque and C.J. Egan, Carbon dioxide. The heat capacity and vapor pressure of the solid. The heat of sublimation. Thermodynamic and spectroscopic values of the entropy, *J. Chem. Phys.*, 1937, **5**, 45.

18

General Conditions for Spontaneity and its Application to Equilibria of Ideal Gases and Dilute Solutions

CHAPTER MENU

18.1 Introduction

We begin our discussion of chemical reactions and chemical equilibria, by asking the question: What is the condition for a reaction to be spontaneous? In answering this question, we introduce the Gibbs energy $G = H - TS$ and the change of the molar Gibbs energy $\Delta_r G$ in a reaction (the subscript r symbolizes "reaction" and "per mole"). We show that a chemical reaction at constant pressure is only spontaneous if the condition

$$\Delta_r G < 0$$

is fulfilled. In the limiting case of $\Delta_r G = 0$ an equilibrium is established.

A useful quantity to be introduced is the molar Gibbs energy of formation from the elements under standard conditions, $\Delta_f G^{\ominus}$. The Gibbs energy of a reaction at constant pressure and constant temperature, $\Delta_r G$ is composed of the Gibbs energies of formation of each constituent compound.

To clearly display the fundamental considerations we restrict our discussion to the case of ideal gases and dilute solutions, but extend these methods to nonideal systems in Chapters 19–21. As an example, we discuss the dissociation of water vapor into H_2 and O_2 at high temperature

$$H_2O \rightarrow H_2 + \tfrac{1}{2}O_2$$

The equilibrium state ($\Delta_r G = 0$) has certain partial pressures $P_{H_2O,\,eq}$, $P_{H_2,\,eq}$, and $P_{O_2,\,eq}$ for each of the substances in the container. What is the mixing ratio of H_2O, H_2, and O_2 in the equilibrium state? How does this ratio change with temperature and total pressure? These questions can be answered by the fundamental equilibrium relation

$$\Delta_r G^{\ominus} = -\boldsymbol{R}T \cdot \ln K$$

where K is the equilibrium constant

$$K = \frac{\widehat{P}_{H_2,\,eq} \cdot \widehat{P}^{1/2}_{O_2,\,eq}}{\widehat{P}_{H_2O,\,eq}}$$

Principles of Physical Chemistry, Third Edition. Hans Kuhn, David H. Waldeck, and Horst-Dieter Försterling.

Companion website: www.wiley.com/go/Kuhn/physicalchemistry3e

$\widehat{P}_{H_2,eq}$ is the numerical value of the partial pressure of H_2 in the equilibrium state measured in units of the standard pressure $P^{\ominus} = 1$ bar: $\widehat{P}_{H_2,eq} = P_{H_2,eq}/P^{\ominus}$. $\widehat{P}_{O_2,eq}$ and $\widehat{P}_{H_2O,eq}$ in the equilibrium mixture are defined accordingly. $\Delta_r G^{\ominus}$ is the change of the molar Gibbs energy in the reaction under standard conditions (initial state: one mol of water vapor at pressure $P^{\ominus} = 1$ bar and final state: one mol of H_2 and one-half mol of O_2, each in separate containers at pressure $P^{\ominus} = 1$ bar). The fundamental equilibrium relation enables us to calculate the equilibrium constant K directly from $\Delta_r G^{\ominus}$:

$$\ln K = \frac{-\Delta_r G^{\ominus}}{\boldsymbol{R}T}$$

K increases with increasing $-\Delta_r G^{\ominus}$, thereby shifting the equilibrium in favor of the products. $-\Delta_r G^{\ominus}$ is often called the *driving force* for the reaction.

These results can be immediately extended to dilute solutions. If the concentration is sufficiently low, the interaction of the particles of the solute (molecules or ions) with each other can be neglected, like molecules in an ideal gas. In the expression for the equilibrium constant the gas pressure P has to be replaced by the osmotic pressure ^{osm}P (the pressure on a piston between solution and solvent which is permeable to the solvent and impermeable to the solute).

18.2 General Conditions for Spontaneity

A process that occurs immediately or after ignition is called a *spontaneous process* (for example, see the expansion of an ideal gas into the vacuum, Fig. 18.1). The gas-phase reaction

$$H_2 + \frac{1}{2}O_2 \rightarrow H_2O$$

takes place spontaneously after ignition. In contrast, the opposite reaction

$$H_2O \rightarrow H_2 + \frac{1}{2}O_2$$

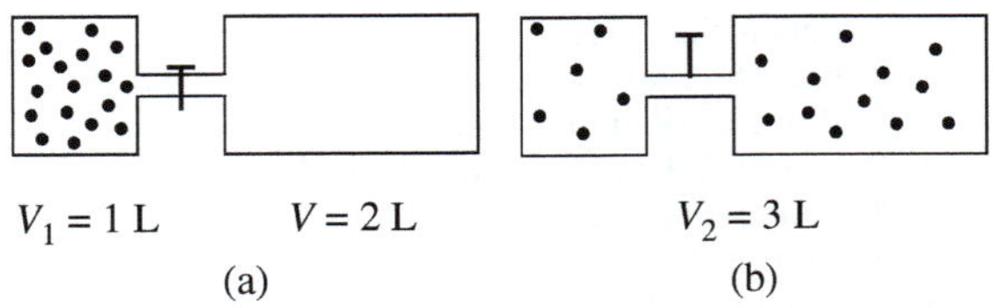

Figure 18.1 The expansion of a gas at constant temperature. (a) Before expansion ($V_1 = 1$ L), (b) after expansion ($V_2 = 3$ L).

does not occur under usual conditions, such as 1 bar pressure and room temperature. A reaction that occurs immediately or after ignition is called a *spontaneous reaction.*

Under what conditions is a given reaction spontaneous? As we discussed in Section 17.5, the entropy for a spontaneous process must increase in an isolated system, that is,

$$\Delta S_{\text{isolated}} \geq 0$$

The process stops when the entropy reaches its maximum value, because once the entropy reaches a maximum any further change must be negative or zero.

To better understand the condition for a spontaneous reaction we simplify the analysis by constraining some of the thermodynamic variables. For example, we can place the container with the reaction mixture of H_2 and O_2 in a temperature bath of temperature T and wait for equilibration of the temperature. Then we can ignite the reaction mixture and wait again until the gas mixture has reached the temperature T of the bath. This process, as mentioned in Chapter 16, is called a *reaction at constant volume and at constant temperature.* In spite of the fact that the temperature of the system changes during reaction, the temperature of the initial and final states is the same; hence, the change in state functions can be treated as if the reaction occurs *at constant temperature.* If the volume and the temperature, under the given conditions, are the same at the end of the reaction as before ignition, the process is said to be *at constant volume* and *at constant temperature.* Correspondingly, if the final pressure and the temperature are the same as before ignition, the process is said to occur *at constant pressure* and *at constant temperature.*

Because the system and temperature bath together can be considered as an isolated system, we see that

$$\Delta S_{\text{isolated}} = \Delta S_{\text{system}} + \Delta S_{\text{temperature bath}} \geq 0 \tag{18.1}$$

In the limit that the temperature bath is much larger than the system, the temperature of the bath changes only by an infinitesimally small amount, and this temperature change can be neglected. Because the temperature of the bath and the temperature of the system are the same within an infinitesimally small amount, the process is reversible. For the reversible transfer of heat q from the system into a temperature bath we can write

$$\Delta S_{\text{temperature bath}} = -\frac{q}{T} \tag{18.2}$$

Note that q is defined as the heat transferred into the system from the surroundings, see Chapter 16. q is negative for an exothermic reaction; thus, in this case $\Delta S_{\text{temperature bath}}$ is positive.

Defining $\Delta S \equiv \Delta S_{\text{system}}$, and substituting with Equation (18.2), we can write Equation (18.1) as

$$\boxed{\Delta S - \frac{q}{T} \geq 0 \quad (T = const)} \tag{18.3}$$

which is the key condition for a spontaneous reaction (or any spontaneous process). Equation (18.3) can be written in the form

$$q - T\Delta S \leq 0 \quad (T = const) \tag{18.4}$$

If the volume and the temperature, under the given conditions, are the same at the end of the reaction as before ignition, the process can be analyzed as if it occurs at constant volume and at constant temperature. In the case of a reaction at constant V we have $q = \Delta U$, so that Equation (18.4) becomes

$$\boxed{\Delta U - T\Delta S \leq 0 \quad (T = const \text{ and } V = const)} \tag{18.5}$$

If the final pressure and the temperature, under the given conditions, are the same at the end of the reaction as before ignition, the process can be treated as if it occurs at constant pressure and at constant temperature. In the case of a reaction at constant P we have $q = \Delta H$, so that

$$\boxed{\Delta H - T\Delta S \leq 0 \quad (T = const \text{ and } P = const)} \tag{18.6}$$

Many chemical reactions take place at atmospheric pressure (the earth's atmosphere acts as a large pressure reservoir); thus, this latter case is of particular interest.

Equations (18.5) and (18.6) are the fundamental criteria for the conditions under which a reaction occurs spontaneously, provided that the reaction is not hindered kinetically. They are expressions based on the fact that the whole reaction from the initial state to the final state must lead to an increase of entropy of the system and the temperature bath.

18.2.1 Helmholtz Energy *A*

For processes at constant volume and constant temperature, it is useful to define an energy function, called the *Helmholtz energy A* by

$$\boxed{A = U - TS} \tag{18.7}$$

A is a state variable because its value is determined by the values of U, T, and S, which are state variables. With this definition we see that a reaction at constant temperature has a Helmholtz energy change of

$$\boxed{\Delta A = \Delta U - T\Delta S \quad (T = const)} \tag{18.8}$$

Thus we can write the condition (18.5), which is restricted to $T = const$ and $V = const$, in the form

$$\boxed{\Delta A \leq 0 \quad (T = const \text{ and } V = const)} \tag{18.9}$$

By directly computing the Helmholtz energy change for a process at constant T and V, we can determine whether it is spontaneous.

In Chapter 17, we showed that w_{rev} at $T = const$ is greater than the work w done by a system in any corresponding non-reversible process. Here we show that the Helmholtz energy is directly related to the reversible work, w_{rev}; hence, it provides an upper bound on the work accessible from a process at constant T. Conservation of energy requires

$$\Delta U = q_{\text{rev}} + w_{\text{rev}} = T\Delta S + w_{\text{rev}} \quad \text{or} \quad w_{\text{rev}} = \Delta U - q_{\text{rev}} \tag{18.10}$$

and with

$$q_{\text{rev}} = T\Delta S \qquad (T = const) \tag{18.11}$$

we obtain

$$w_{\text{rev}} = \Delta U - T\Delta S \quad (T = const) \tag{18.12}$$

Comparing this result with Equation (18.8), we find that

$$\boxed{\Delta A = w_{\text{rev}} \quad (T = const)} \tag{18.13}$$

The change in the Helmholtz energy ΔA is equal to the reversible work for an isothermal process.

According to Equation (18.9), the reversible work done on the system in a change of state at constant temperature and at constant volume (no PV work occurs) must be

$$w_{\text{rev}} \leq 0 \qquad (T = const,\ V = const) \tag{18.14}$$

i.e., the system must be able to do work ($-w_{\text{rev}}$) on a reversible path.

18.2.2 Gibbs Energy *G*

For processes at constant pressure and temperature, it is useful to define the *Gibbs energy G*, namely

$$\boxed{G = H - TS} \tag{18.15}$$

G is a state variable because its value is determined by the values of H, T, and S, which are state variables. For a reaction at constant temperature we find that

$$\boxed{\Delta G = \Delta H - T\Delta S \quad (T = const)} \tag{18.16}$$

and we can write condition (18.6) in the form

$$\boxed{\Delta G \leq 0 \quad (T = const \text{ and } P = const)} \tag{18.17}$$

Because chemical reactions are often carried out at constant pressure and constant temperature, Equations (18.16) and (18.17) are of particular interest and will be used throughout this book. As demonstrated by the logic given above, this criterion reflects the condition that the total entropy change of the system and surroundings must be positive.

Like the Helmholtz energy, the Gibbs energy can be related to the reversible work. Consider a process at $T = const$ and $P = const$, so that, according to Equation (18.10),

$$\Delta H = \Delta U + P\Delta V = T\Delta S + w_{\text{rev}} + P\Delta V$$

or

$$w_{\text{rev}} + P\Delta V = \Delta H - T\Delta S \quad (T = const \text{ and } P = const)$$

and with Equation (18.16)

$$\boxed{\Delta G = w_{\text{rev}} + P\Delta V \quad (T = const \text{ and } P = const)} \tag{18.18}$$

This result means that

$$\Delta G = \begin{matrix}\text{work done on}\\ \text{the system}\\ (w_{\text{rev}})\end{matrix} \; - \; \begin{matrix}\text{pressure–volume work}\\ \text{done on the system}\\ (-P\Delta V)\end{matrix}$$

According to Equation (18.17), the reversible work done on the system in a change of state at constant temperature and constant pressure must be

$$w_{\text{rev}} + P\Delta V \leq 0 \;\; (T = const,\ P = const) \tag{18.19}$$

or

$$-w_{\text{rev}} \geq P\Delta V \tag{18.20}$$

for a spontaneous process, i.e., in a change of state at constant pressure, $-w_{\text{rev}}$, the reversible work done by the system, must be larger than the PV work done by the system.

Relations (18.9) and (18.17) are fundamental conditions which must be fulfilled for a reaction, or any process, to proceed spontaneously. Whether or not the reaction occurs can depend upon kinetic factors, and in many cases the reaction has to be started by an initiation event (such as ignition, adding a catalyst, or a change of conditions).

18.3 ΔG and its Dependence on Temperature

18.3.1 Molar Gibbs Energy of Formation from Elements $\Delta_f G^\ominus$

According to Equation (18.16), the Gibbs energy change is

$$\Delta G = \Delta H - T\Delta S \quad (T = const)$$

In the following we use ΔG for the Gibbs energy change of a system consisting of an amount $\boldsymbol{n}$ of substance, and $\Delta_f G$ for the Gibbs energy change per mole in the formation of a substance from the elements. For example, the Gibbs energy change to form $\boldsymbol{n} = 5$ mole of water is

$$\Delta G^\ominus(H_2O) = \boldsymbol{n} \cdot \Delta_f G^\ominus(H_2O) = -5 \cdot 237.2 \text{ kJ mol}^{-1} = -1186 \text{ kJ}$$

For the formation of a substance from elements in molar quantity under standard conditions the molar Gibbs energy of formation $\Delta_f G^\ominus$ is

$$\Delta_f G^\ominus = \Delta_f H^\ominus - T\Delta_f S^\ominus \quad (T = const) \tag{18.21}$$

where $\Delta_f S^\ominus$ is the molar change in entropy when forming one mol of substance from elements under standard pressure $P^\ominus = 1$ bar (again, the subscript f symbolizes "formation" and "molar"). Accordingly, $\Delta_f H^\ominus$ is the enthalpy of formation (see Chapter 16). Values for these quantities at 298 K are included in Appendix A.1.

18.3.2 Molar Gibbs Energy of Reaction $\Delta_r G^\ominus$

From Equation (18.16) the molar change of the Gibbs energy in a reaction $\Delta_r G^\ominus$ under standard conditions (species separated, under standard pressure $P^\ominus = 1$ bar) at temperature T is calculated as

$$\boxed{\Delta_r G^\ominus = \Delta_r H^\ominus - T \cdot \Delta_r S^\ominus} \tag{18.22}$$

(the subscript r symbolizes "reaction" and "molar"). $\Delta_r H^\ominus$ is obtained from the enthalpies of formation, $\Delta_f H$, see Section 16.5. $\Delta_r S^\ominus$ is obtained from the molar entropies $S_m^\ominus$ of the species involved in the reaction. The entropy and enthalpy values for selected compounds are listed in Appendix A.1. If we consider a chemical reaction

$$\alpha A + \beta B \rightarrow \gamma C + \delta D \tag{18.23}$$

then $\Delta_r G^\ominus$ is related to the reaction of α moles of A with β moles of B.

The relation (18.22) is of great importance. It allows us to calculate the Gibbs energy change for any suggested reaction provided the data for the reactants and products can be found in a table.

Example 18.1 *Determine the molar Gibbs energy change $\Delta_r G^\ominus$ for the oxidation of CO at 298 K and 1 bar. Is this reaction spontaneous?*

We consider the reaction

$$2\,CO + O_2 \rightarrow 2\,CO_2$$

at a temperature of 298 K. According to Section 16.5, the molar reaction enthalpy under standard conditions at 298 K is

$$\Delta_r H^\ominus = -566.0 \text{ kJ mol}^{-1}$$

Note that in this case "molar reaction enthalpy" means the enthalpy change per two moles of CO and one mole of O_2, according to the given reaction equation.

From the absolute molar entropies under standard conditions given in Appendix A1 we find

$$\Delta_r S^\ominus = 2S_m^\ominus(CO_2) - 2S_m^\ominus(CO) - S_m^\ominus(O_2) = -173.6 \text{ J mol}^{-1}$$

Combining these results for the molar Gibbs energy of the reaction under standard conditions we find

$$\Delta_r G^\ominus = -566.0 \text{ kJ mol}^{-1} - 298 \cdot \left(-173.6 \text{ J mol}^{-1}\right) = -514.3 \text{ kJ mol}^{-1}$$

Even though the entropy of the products is smaller than the reactants (entropy change is negative), the enthalpy change is very strongly negative so that the reaction Gibbs energy change is negative. Since $\Delta_r G^\ominus < 0$, the reaction is spontaneous, provided that the reaction is not hindered kinetically.

18.3.3 Temperature Dependence of $\Delta G^\ominus$

By using the data in Appendix A.1 we are able to calculate $\Delta G^\ominus(298\text{ K})$, the change of the Gibbs energy under standard conditions and at $T_1 = 298$ K. How can we calculate $\Delta G^\ominus$ at another temperature T_2?

Once again, we use the general result (Equation (18.16)) to calculate $\Delta G_T^\ominus$, namely

$$\Delta G^\ominus(T) = \Delta H^\ominus(T) - T \cdot \Delta S^\ominus(T) \tag{18.24}$$

Using Equation (16.61) in Section 16.5, the enthalpy change at an arbitrary temperature T_2 is

$$\boxed{\Delta H^\ominus(T_2) = \Delta H^\ominus(298\text{ K}) + \int_{298\text{ K}}^{T_2} \Delta C_P^\ominus \cdot dT \qquad (P = const)} \tag{18.25}$$

Because the process occurs at constant pressure, we can relate the reversible heat (and hence the entropy change) to the constant pressure heat capacity, so that

$$dS = \frac{dq_{rev}}{T} = \frac{C_P \cdot dT}{T} \tag{18.26}$$

and for $\Delta S^\ominus(T_2)$ we obtain

$$\boxed{\Delta S^\ominus(T_2) = \Delta S^\ominus(298\text{ K}) + \int_{298\text{ K}}^{T_2} \frac{\Delta C_P^\ominus}{T} \cdot dT \qquad (P = const)} \tag{18.27}$$

This treatment assumes that no phase transitions (melting, evaporation) occur during heating. Otherwise, as shown in the examples in Foundation 18.1, the heat of the phase transitions (e.g., $\Delta_{vap}H$ in the case of evaporation) must be added to $\Delta H_T^\ominus$ and the entropy change of the phase transitions (e.g., $\Delta_{vap}H/T_{vap}$ in the case of evaporation) must be added to ΔS_T. We find that

$$\Delta G^\ominus(T_2) = \Delta H^\ominus(T_2) - T_2 \cdot \Delta S^\ominus(T_2) \tag{18.28}$$

Example 18.2 *Consider the decomposition of limestone, $CaCO_3$ (Fig. 18.2). How does this reaction's Gibbs energy change with temperature? At what temperature is the Gibbs energy change zero?*

We consider the reaction (Fig. 18.2a)

$$CaCO_3 \rightarrow CaO + CO_2$$

From the data in Appendix A.1, we find that

$$\Delta_r H^\ominus = 177.4\text{ kJ mol}^{-1},$$
$$\Delta_r S^\ominus = 160.4\text{ J K}^{-1}\text{ mol}^{-1}$$

and

$$\begin{aligned}\Delta_r G^\ominus = \Delta_r H^\ominus - T\Delta_r S^\ominus &= 177.4\text{ kJ mol}^{-1} \\ &+ 298 \cdot 160.4\text{ J K}^{-1}\text{ mol}^{-1} \\ &= (177.4 - 47.8)\text{ kJ mol}^{-1} \\ &= 129.6\text{ kJ mol}^{-1}\end{aligned}$$

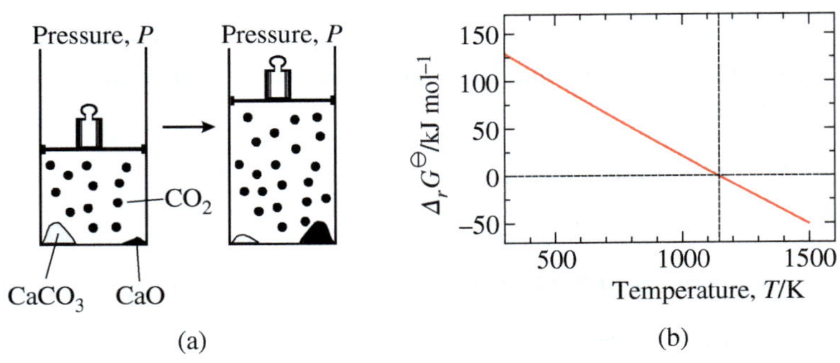

Figure 18.2 Formation of CO_2 and CaO from $CaCO_3$ (decomposition of limestone). (a) Experimental setup. (b) $\Delta_r G^\ominus$ for the reaction $CaCO_3 \rightarrow CaO + CO_2$ as a function of temperature at an external pressure of $P = 1$ bar. At $T = 1145$ K we obtain $\Delta_r G^\ominus = 0$ (dashed lines).

for $T = 298$ K. $\Delta_r G^\ominus$ is positive at 298 K and the reaction does not proceed spontaneously under standard conditions.

What is $\Delta_r G^\ominus$ at $T = 1200$ K? As shown in Foundation 18.1, we can use Equations (18.24) to (18.27) to find

$$\Delta_r G^\ominus = -8.0\text{ kJ mol}^{-1} \quad \text{at } T = 1200\text{ K}$$

$\Delta_r G^{\ominus}$ is negative at 1200 K and the reaction proceeds spontaneously. The decomposition of limestone is a well-known process which needs a minimum temperature to proceed at a CO_2 pressure of 1 bar. This minimum temperature is determined by requiring that

$$\Delta_r G^{\ominus} = \Delta_r H^{\ominus} - T \cdot \Delta_r S^{\ominus} = 0$$

By calculating $\Delta_r G^{\ominus}$ for different temperatures (Fig. 18.2b), we find $T_{minimum} = 1145$ K, which is in agreement with the experimentally determined temperature for the decomposition of limestone.

Although the procedure described above is always appropriate, it is often more convenient to use the direct general relationship between the temperature dependence of the Gibbs energy change and the entropy change in a reaction. In Box 18.1 we show that the general temperature dependence of the molar Gibbs energy change is

$$\left(\frac{\partial \Delta_r G}{\partial T}\right)_P = -\Delta_r S \tag{18.29}$$

at constant pressure. Then

$$\Delta_r G_{T+\mathrm{d}T} = \Delta_r G_T - \Delta_r S \cdot \mathrm{d}T \tag{18.30}$$

18.4 Pressure Dependence of ΔG In Ideal Gases

Let us first consider the pressure dependence of ΔG in an ideal gas. How does ΔG change if we change the external pressure from the standard pressure $P^{\ominus}$ to an arbitrary pressure P? As for the temperature dependence of ΔG, we have to consider the changes in ΔH and ΔS. In order to illustrate the principles, we restrict our considerations to ideal gases and treat solids and liquids as incompressible in the following discussion. In Chapters 19-21 we will remove this restriction.

The enthalpy change can be found by considering the basic relation

$$H = U + PV$$

For an ideal gas U depends on the temperature and not on the pressure, i.e., $U = U(T)$. Combining this result with the ideal gas law, the enthalpy is given by

$$H = U + \boldsymbol{nR}T$$

It is a function of temperature and independent of pressure.

In contrast to the enthalpy, the entropy of gases depends on the pressure in our approximation. The volume of an ideal gas decreases with increasing pressure P and its entropy changes strongly. Using Equation (17.52) we can write

$$S_T = S_T^{\ominus} + Nk \cdot \ln \frac{P^{\ominus}}{P} = S_T^{\ominus} - \boldsymbol{nR} \cdot \ln \frac{P}{P^{\ominus}}$$

or

$$\boxed{\Delta S = S_T - S_T^{\ominus} = -\boldsymbol{nR} \cdot \ln \frac{P}{P^{\ominus}} \qquad \text{(ideal gas)}} \tag{18.31}$$

where S_T is the entropy at temperature T and pressure P, and $S_T^{\ominus}$ is the entropy at temperature T and standard pressure $P^{\ominus} = 1$ bar of the ideal gas. In the approximation that solids and liquids are not compressible, their volume does not change with pressure; hence, their entropy does not change with pressure.

Using Equation (18.16),

$$\Delta G = \Delta H - T\Delta S \qquad (T = const)$$

we can calculate the Gibbs energy at a particular temperature and pressure. For a solid or liquid substance, $\Delta G(T, P) = \Delta G^{\ominus}(T)$ in this approximation. For a gaseous substance

$$\boxed{\Delta G = \Delta G^{\ominus} + \boldsymbol{nR}T \cdot \ln \frac{P}{P^{\ominus}} \qquad \text{(ideal gas)}} \tag{18.32}$$

which explicitly shows that the Gibbs energy is a function of T and P. Using the abbreviation

$$\hat{P} = \frac{P}{P^{\ominus}} \tag{18.33}$$

and making use of $\Delta_r G = \Delta G / \boldsymbol{n}$ we can rewrite Equation (18.32) as

$$\Delta_r G = \Delta_r G^{\ominus} + \boldsymbol{R} T \cdot \ln \hat{P} \qquad \text{(ideal gas)} \tag{18.34}$$

Below we apply these results to the formation of a gas from a physical transformation (evaporation of a liquid; sublimation of a solid) and from a chemical transformation (decomposition of $CaCO_3$).

18.4.1 Vapor Pressure (Clausius–Clapeyron Equation)

Both sublimation and evaporation processes can be treated in the same manner; hence, we provide a detailed presentation for evaporation and simply state the result for sublimation.

18.4.1.1 Evaporation of a Liquid

An interesting special case is the evaporation of a liquid, e.g., of water:

$$H_2O \text{ (liquid)} \rightarrow H_2O \text{ (vapor)}$$

In a closed container, the liquid will come to equilibrium with its vapor, so that $\Delta_r G = 0$. At equilibrium, we can use Equation (18.32) to write

$$\ln \hat{P} = -\frac{\Delta_{\text{vap}} G^{\ominus}}{\boldsymbol{R} T} \tag{18.35}$$

with $\hat{P} = P/P^{\ominus}$ where P is the vapor pressure. $\Delta_{\text{vap}} G^{\ominus}$ is the molar Gibbs energy of evaporation under standard pressure.

Example 18.3 *Using $\Delta_{\text{vap}} G^{\ominus}$ values from Appendix A.1 and Equation (18.35), calculate the vapor pressure of water at 25 °C.*

With $\Delta_{\text{vap}} G^{\ominus} = \Delta_f G^{\ominus}(H_2O\text{,vapor}) - \Delta_f G^{\ominus}(H_2O\text{,liquid}) = (-228.57 + 237.13)\ \text{kJ mol}^{-1} = 8.56\ \text{kJ mol}^{-1}$ we obtain

$$\ln \hat{P} = -\frac{8.56\ \text{kJ mol}^{-1}}{8.314\ \text{J K}^{-1}\ \text{mol}^{-1} \cdot 298\ \text{K}} = -3.46$$

and $P = 0.0314$ bar $= 31.4$ mbar.

If we approximate the liquid as incompressible and the vapor as an ideal gas, we can derive an expression for the temperature dependence of the vapor pressure. Taking the derivative of Equation (18.35), we find that

$$\frac{\mathrm{d} \ln \hat{P}}{\mathrm{d}T} = -\frac{1}{\boldsymbol{R}} \frac{\mathrm{d}}{\mathrm{d}T} \left[\frac{\Delta_{\text{vap}} G^{\ominus}}{T} \right] = -\frac{1}{\boldsymbol{R}} \left[\frac{1}{T} \frac{\mathrm{d}\Delta_{\text{vap}} G^{\ominus}}{\mathrm{d}\,T} - \frac{\Delta_{\text{vap}} G^{\ominus}}{T^2} \right] \tag{18.36}$$

Using the result from Box 18.1 and substituting for $\Delta_{\text{vap}} G^{\ominus}$ from Equation (18.16), we can write that

$$\frac{\mathrm{d}\Delta_{\text{vap}} G^{\ominus}}{\mathrm{d}T} = \left(\frac{\partial \Delta_{\text{vap}} G}{\partial T} \right)_{P=P^{\ominus}} = -\Delta_{\text{vap}} S^{\ominus} \quad \text{and} \quad \frac{\Delta_{\text{vap}} G^{\ominus}}{T^2} = \frac{\Delta_{\text{vap}} H^{\ominus}}{T^2} - T \frac{\Delta_{\text{vap}} S^{\ominus}}{T^2} \tag{18.37}$$

In our approximation $\Delta_{\text{vap}} H$ does not depend on pressure, and our result simplifies to

$$\boxed{\frac{\mathrm{d} \ln \hat{P}}{\mathrm{d}T} = \frac{\Delta_{\text{vap}} H}{\boldsymbol{R} T^2}} \tag{18.38}$$

This result is called the *Clausius–Clapeyron equation*.

Because of $\Delta_{\text{vap}} H > 0$ the vapor pressure is strongly increasing with temperature. In many cases it is useful to estimate the vapor pressure at a higher temperature, while avoiding a rigorous treatment. Using Equation (18.35) we can write

$$\ln \hat{P}_1 = -\frac{\Delta_{\text{vap}} G_1^{\ominus}}{\boldsymbol{R} T_1} = -\frac{\Delta_{\text{vap}} H_1^{\ominus}}{\boldsymbol{R} T_1} + \frac{\Delta_{\text{vap}} S_1^{\ominus}}{\boldsymbol{R}} \tag{18.39}$$

for a starting temperature T_1. If we neglect the temperature dependence of $\Delta H^\ominus$ and $\Delta S^\ominus$, then it follows for any temperature T that

$$\ln \widehat{P} = -\frac{\Delta_{\text{vap}} H_1^\ominus}{\boldsymbol{R}T} + \frac{\Delta_{\text{vap}} S_1^\ominus}{\boldsymbol{R}} \tag{18.40}$$

and, considering that the enthalpy of vaporization does not depend on pressure,

$$\ln \widehat{P} = \ln \widehat{P}_1 - \frac{\Delta_{\text{vap}} H}{\boldsymbol{R}} \left(\frac{1}{T} - \frac{1}{T_1} \right) \tag{18.41}$$

Example 18.4 *A common demonstration experiment for introductory chemistry courses is to boil water at room temperature. Given that the enthalpy of vaporization for water is $\Delta_{\text{vap}} H = 44.0$ kJ mol^{-1}, calculate the pressure of a vacuum that is needed to boil water at 298 K.*

At the boiling point the vapor pressure of a liquid is equal to the ambient pressure. Given that the vapor pressure of water is $P_1 = 1$ bar at 373 K, we can use Equation (18.41) to find the vapor pressure P of water at 298 K. We find

$$\ln \frac{P}{P_1} = -\frac{44.0 \text{ kJ mol}^{-1}}{8.314 \text{ J mol}^{-1} \text{ K}^{-1}} \left(\frac{1}{298 \text{ K}} - \frac{1}{373 \text{ K}} \right) = -3.57$$

so that $P = 0.028$ bar $= 28$ mbar. We need to reduce the pressure to 28 mbar or less to vaporize liquid water at 298 K.

The calculated pressure is smaller by 10% compared to the calculation in Example 18.3. This is because the integrated Equation (18.41) is an approximation, neglecting the temperature dependence of $\Delta_r H$.

18.4.1.2 Evaporation of a Solid

Correspondingly, Equations (18.38) and (18.41) can be applied to the vapor pressure above a solid, for example in the equilibrium

$$H_2O \text{ (solid)} \rightarrow H_2O \text{ (vapor)}$$

Then $\Delta_{\text{vap}} H$ has to be replaced by the molar enthalpy of sublimation, $\Delta_{\text{sub}} H$ (i.e., the heat to vaporize the solid),

$$\Delta_{\text{sub}} H = \Delta_{\text{fus}} H + \Delta_{\text{vap}} H$$

where $\Delta_{\text{fus}} H$ is the molar enthalpy of melting. Because of $\Delta_{\text{sub}} H > \Delta_{\text{vap}} H$ the vapor pressure above a solid is more strongly increasing with temperature than the vapor pressure above the corresponding liquid. In Chapter 19, Fig. 19.8, the vapor pressure for liquid and solid water is shown as a function of temperature.

18.4.2 Chemical Evolution of a Gas

For a chemical reaction, the Gibbs energy difference ΔG between the products and the reactants is given by the difference in the Gibbs energy of the products and the Gibbs energy of the reactants. At equilibrium, we have the condition that $\Delta G(T, P) = 0$. In this situation, the forward and reverse reactions are in balance, and the chemical system behaves reversibly. For a reaction with a single product and at equilibrium ($\Delta G = 0$), we find from Equation (18.32) that

$$\Delta G^\ominus(T) = -\boldsymbol{nR}T \cdot \ln \frac{P_{\text{eq}}}{P^\ominus} \qquad \text{(ideal gas)}$$

or

$$\Delta_r G^\ominus(T) = -\boldsymbol{R}T \cdot \ln \frac{P_{\text{eq}}}{P^\ominus} \qquad \text{(ideal gas)} \tag{18.42}$$

where P_{eq} refers to the equilibrium pressure of the gaseous product.

Example 18.5 *Analyze how the external pressure affects the decomposition of limestone (Fig. 18.3).*

In Example 18.2, we calculated $\Delta_r G^\ominus = -8.0$ kJ mol^{-1} at $T = 1200$ K for the reaction

$$CaCO_3 \rightarrow CaO + CO_2$$

at 1 bar pressure. Because $CaCO_3$ and CaO are solids we will assume that they are incompressible and their Gibbs energy does not change with pressure. Hence, we can compute the molar change of Gibbs energy for this reaction as a function of pressure.

$$\Delta_r G = \Delta_r G^\ominus + \boldsymbol{R}T \cdot \ln \frac{P_{CO_2}}{P^\ominus} \quad \text{at} \quad T = 1200\ \text{K}$$

$\Delta_r G$ increases with increasing pressure and eventually switches from a negative value to a positive value (see Fig. 18.3). With $\Delta_r G = 0$, we can solve for the equilibrium pressure of CO_2:

$$\ln\left(\frac{P_{CO_2,\text{eq}}}{P^\ominus}\right) = -\frac{\Delta_r G^\ominus}{\boldsymbol{R}T} = \frac{8.0\ \text{kJ mol}^{-1}}{\boldsymbol{R} \cdot 1200\ \text{K}} = 0.802$$

so that

$$P_{CO_2,\text{eq}} = 2.2\ \text{bar} \quad \text{at} \quad T = 1200\ \text{K}$$

This means that the decomposition of limestone at 1200 K is spontaneous up to a CO_2 pressure of 2.2 bar.

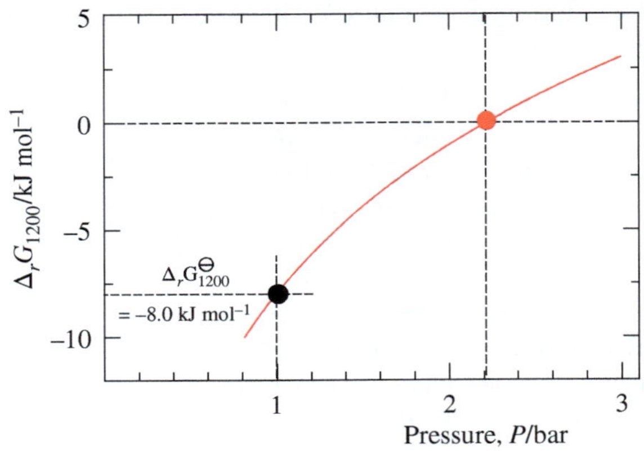

Figure 18.3 $\Delta_r G_{1200}$ for the reaction $CaCO_3 \rightarrow CaO + CO_2$ (decomposition of limestone) as a function of external pressure P. At $T = 1200$ K and $P = 1$ bar the change in Gibbs energy is negative ($\Delta_r G_{1200} = \Delta_r G^\ominus_{1200} = -8.0$ kJ mol^{-1}, black solid circle), at $P = 2.2$ bar we obtain $\Delta_r G_{1200} = 0$ (red solid circle).

Our discussion in this chapter will make the approximation that solids and liquids are incompressible. In this approximation their volume does not change with pressure; hence, neither their enthalpy nor their entropy change with pressure. For a solid or liquid substance, $\Delta G = \Delta G^\ominus(T)$ in this approximation. In Chapter 19 we will remove this restriction.

18.5 ΔG and Chemical Equilibrium in Ideal Gases

Now we discuss the spontaneity of chemical reactions and chemical equilibria. First we treat the chemical equilibrium and the mass action law in a formal way. Because it is important to see the basic features of chemical equilibrium, we also consider the mass action law by omitting any formalism; rather we use the fact that the work $-w$ done by the system when changing from one equilibrium state to another at constant temperature has a maximum for a reversible change.

18.5.1 Formal Derivation of Mass Action Law

Consider a chemical reaction

$$\alpha\text{A} + \beta\text{B} \rightarrow \gamma\text{C} + \delta\text{D} \tag{18.43}$$

where A, B, C, and D are gaseous reactants, which are treated as ideal gases. We calculate ΔG for this reaction from the Gibbs energies of the different species.

$$\begin{aligned}\Delta G &= G_\text{C} + G_\text{D} - G_\text{A} - G_\text{B} \\ &= G_{m,\text{C}} \cdot \boldsymbol{n}_\text{C} + G_{m,\text{D}} \cdot \boldsymbol{n}_\text{D} - G_{m,\text{A}} \cdot \boldsymbol{n}_\text{A} - G_{m,\text{B}} \cdot \boldsymbol{n}_\text{B}\end{aligned} \tag{18.44}$$

where G_m is the molar Gibbs energy and $\boldsymbol{n}$ is the amount of substance. We rewrite Equation (18.44) in molar terms

$$\Delta_r G = G_{m,\text{C}} \cdot \gamma + G_{m,\text{D}} \cdot \delta - G_{m,\text{A}} \cdot \alpha - G_{m,\text{B}} \cdot \beta \tag{18.45}$$

$\Delta_r G$ is the change in Gibbs energy when α moles of A and β moles of B react to form γ moles of C and δ moles of D. With Equation (18.32)

$$G_m = G_m^\ominus + \boldsymbol{R}T \cdot \ln \frac{P}{P^\ominus} \quad \text{(ideal gas)} \tag{18.46}$$

and the abbreviation

$$\hat{P} = \frac{P}{P^\ominus} \tag{18.47}$$

for ideal gases, we obtain

$$\Delta_r G = \left(G^\ominus_{m,\mathrm{C}} \cdot \gamma + G^\ominus_{m,\mathrm{D}} \cdot \delta - G^\ominus_{m,\mathrm{A}} \cdot \alpha - G^\ominus_{m,\mathrm{B}} \cdot \beta\right) + \boldsymbol{R}T \cdot \ln \frac{\hat{P}^\gamma_\mathrm{C} \cdot \hat{P}^\delta_\mathrm{D}}{\hat{P}^\alpha_\mathrm{A} \cdot \hat{P}^\beta_\mathrm{B}} \tag{18.48}$$

We call the expression

$$Q = \frac{\hat{P}^\gamma_\mathrm{C} \cdot \hat{P}^\delta_\mathrm{D}}{\hat{P}^\alpha_\mathrm{A} \cdot \hat{P}^\beta_\mathrm{B}} \tag{18.49}$$

the *reaction quotient*. With

$$\Delta_r G^\ominus = G^\ominus_{m,\mathrm{C}} \cdot \gamma + G^\ominus_{m,\mathrm{D}} \cdot \delta - G^\ominus_{m,\mathrm{A}} \cdot \alpha - G^\ominus_{m,\mathrm{B}} \cdot \beta$$

where $\Delta_r G^\ominus$ is the Gibbs energy of reaction under standard conditions (reactants and products separated and under standard pressure $P^\ominus$). We obtain

$$\Delta_r G = \Delta_r G^\ominus + \boldsymbol{R}T \cdot \ln Q \qquad \text{(ideal gases)} \tag{18.50}$$

If reaction (18.43) is carried out at constant temperature T and constant pressure (i.e., reactants and products are under the same pressure P) we can apply the equilibrium condition, $\Delta_r G = 0$. In this case the reaction quotient Q becomes identical with the *equilibrium constant* K

$$K = \frac{\hat{P}^\gamma_\mathrm{C,eq} \cdot \hat{P}^\delta_\mathrm{D,eq}}{\hat{P}^\alpha_\mathrm{A,eq} \cdot \hat{P}^\beta_\mathrm{B,eq}} \tag{18.51}$$

where we have used the abbreviation

$$\hat{P}_\mathrm{eq} = \frac{P_\mathrm{eq}}{P^\ominus} \tag{18.52}$$

and the index *eq* denotes the equilibrium state. Thus from Equation (18.50) we obtain

$$\boxed{\Delta_r G^\ominus = -\boldsymbol{R}T \cdot \ln K \qquad \text{(ideal gases)}} \tag{18.53}$$

and we can rewrite Equation (18.50) as

$$\boxed{\Delta_r G = -\boldsymbol{R}T \cdot \ln K + \boldsymbol{R}T \cdot \ln Q \qquad \text{(ideal gases)}} \tag{18.54}$$

Equation (18.51) is called the *Law of Mass Action*. Equation (18.54) has three interesting cases.

case 1:	$Q < K$	$\Delta_r G < 0$	reaction occurs spontaneously
case 2:	$Q > K$	$\Delta_r G > 0$	reverse reaction occurs spontaneously
case 3:	$Q = K$	$\Delta_r G = 0$	equilibrium

The equilibrium case can be realized in various ways. Whenever the partial pressures of the reacting species fulfill the mass action law, relation (18.51), the reacting species are in equilibrium regardless of the particular choice of P_A, P_B, P_C, or P_D. For example, P_C does not change if we arbitrarily increase the partial pressures of D and of B by the same factor.

Example 18.6 *Consider the reaction* $H_2O \rightarrow H_2 + \frac{1}{2}O_2$ *at* $T = 1500\,\mathrm{K}$ *(under conditions where* H_2O *is a gas) and give expressions for K and Q.*

In this case from Equations (18.49) and (18.51) we obtain

$$K = \frac{\hat{P}_{\mathrm{H_2,eq}} \cdot \hat{P}^{1/2}_{\mathrm{O_2,eq}}}{\hat{P}_{\mathrm{H_2O,eq}}} \quad \text{and} \quad Q = \frac{\hat{P}_{\mathrm{H_2}} \cdot \hat{P}^{1/2}_{\mathrm{O_2}}}{\hat{P}_{\mathrm{H_2O}}}$$

Again we consider the following three cases:

Case 1. $Q < K$. In this case the pressure of H_2O is higher than in equilibrium, and the reaction will proceed toward the products H_2 and O_2 to reach equilibrium.

Case 2. $Q > K$. In this case the pressure of H_2O is lower than in equilibrium, and the reaction will proceed toward the reactant H_2O to reach equilibrium

Case 3. $Q = K$. In this case the system is in equilibrium.

In Foundation 18.1 we calculate the change of Gibbs energy for this reaction; $\Delta_r G^{\ominus}_{1500} = +158$ kJ mol^{-1}. Using this result with Equation (18.53), we can calculate the equilibrium constant K, via

$$\ln K = -\frac{\Delta_r G^{\ominus}(1500\text{ K})}{RT} = \frac{-158\text{ kJ mol}^{-1}}{8.3145\text{ J K}^{-1}\text{ mol}^{-1}\cdot 1500\text{ K}} = -12.67$$

so that

$$K = \exp(-12.67) = 3.15\times 10^{-6}$$

During the dissociation of H_2O molecules, H_2 and O_2 are formed in a 2:1 ratio; thus, $\widehat{P}_{H_2,eq} = 2\times \widehat{P}_{O_2,eq}$ and

$$K = \frac{\widehat{P}_{H_2,eq}\cdot \widehat{P}^{1/2}_{O_2,eq}}{\widehat{P}_{H_2O,eq}} = \frac{\left(2\widehat{P}_{O_2,eq}\right)\cdot \widehat{P}^{1/2}_{O_2,eq}}{\widehat{P}_{H_2O,eq}} = \frac{2\cdot \widehat{P}^{3/2}_{O_2,eq}}{\widehat{P}_{H_2O,eq}}$$

so that

$$\widehat{P}_{O_2,eq} = \left(\frac{1}{2}K\cdot \widehat{P}_{H_2O,eq}\right)^{2/3}$$

If we assume the H_2O gas to be under a pressure of 1 bar, then we obtain $\widehat{P}_{O_2,eq} = 1.35\times 10^{-4}$ and $\widehat{P}_{H_2,eq} = 2.70\times 10^{-4}$; that is, the water dissociates to a very small extent.

18.5.2 Mass Action Law – A Direct Consequence of the Relation Between Work and Heat

The mass action law is so important in chemistry that a direct physical understanding without any abstracting formalism is important. In Chapter 17 we showed that the work $(-w)$ done by the system when proceeding on an arbitrary path from state A to state B at constant temperature cannot be larger than the reversible work $(-w_{rev})$ done by the system when taking a reversible path from A to B. Thus, according to Equation (17.5),

$$(-w) \le (-w_{rev}) \tag{18.55}$$

Otherwise, moving back on a reversible path, heat would be taken from the temperature bath and transferred into work in a cyclic process. This is not possible (second law of thermodynamics).

In the special case of a chemical reaction in a closed container at constant V and T no work is done by the system; thus, we have $w = 0$ and therefore

$$\boxed{w_{rev} \le 0\ (T = const \text{ and } V = const)} \tag{18.56}$$

That is, the reaction at constant temperature and constant volume can only proceed if the reversible work w_{rev} is negative; in other words, the system must do work when proceeding from A to B on a reversible path.

18.5.2.1 Reaction in a Closed Container at Constant T and V

How can we imagine that a chemical reaction can be carried out reversibly? As an example, we choose the detonating gas reaction.

Consider a mixture of H_2, O_2, and gaseous H_2O with partial pressures of P_{H_2}, P_{O_2}, and P_{H_2O} at fixed volume and temperature (Fig. 18.4, container 1). Now we consider the reaction

$$H_2O \rightarrow H_2 + \frac{1}{2}O_2 \tag{18.57}$$

under conditions where H_2O is a gas. We imagine that a small amount of H_2O (n_{H_2O}) reacts to form H_2 and O_2, according to (18.57). Provided that the reaction is not hindered kinetically, can this process occur spontaneously?

In a thought experiment, we imagine a reversible pathway leading to the same final state. To perform the reaction (18.57) reversibly we consider a second container where H_2, O_2, and H_2O are at the same temperature as in the first container, but under their equilibrium partial pressures $P_{H_2,eq}$, $P_{O_2,eq}$, and $P_{H_2O,eq}$ (Fig. 18.4, container 2). We take a small amount of H_2O from container 1 and transfer it reversibly into container 2; at the same time we transfer the products H_2 and O_2 from container 2 to container 1, in the proper stoichiometric amounts. In order to transfer H_2O from container 1 to container 2 reversibly, the partial pressure of H_2O must be changed reversibly from P_{H_2O} to $P_{H_2O,\,eq}$. In the ideal gas approximation the reversible work needed for this process can be obtained[1] by using Equation (17.2) in Section 17.2.2,

$$w_{rev} = -\boldsymbol{nR}T \cdot \ln\frac{V_2}{V_1} = \boldsymbol{nR}T \cdot \ln\frac{P_2}{P_1}$$

as

$$w_{rev,H_2O} = \boldsymbol{n}_{H_2O}\boldsymbol{R}T \cdot \ln\frac{P_{H_2O,\,eq}}{P_{H_2O}} \tag{18.58}$$

Proceeding in a like manner for H_2 and O_2, we find that w_{rev} for the whole process is

$$w_{rev} = \boldsymbol{n}_{H_2O}\boldsymbol{R}T \cdot \ln\frac{P_{H_2O,\,eq}}{P_{H_2O}} + \boldsymbol{n}_{H_2}\boldsymbol{R}T \cdot \ln\frac{P_{H_2}}{P_{H_2,\,eq}} + \boldsymbol{n}_{O_2}\boldsymbol{R}T \cdot \ln\frac{P_{O_2}}{P_{O_2,\,eq}} \tag{18.59}$$

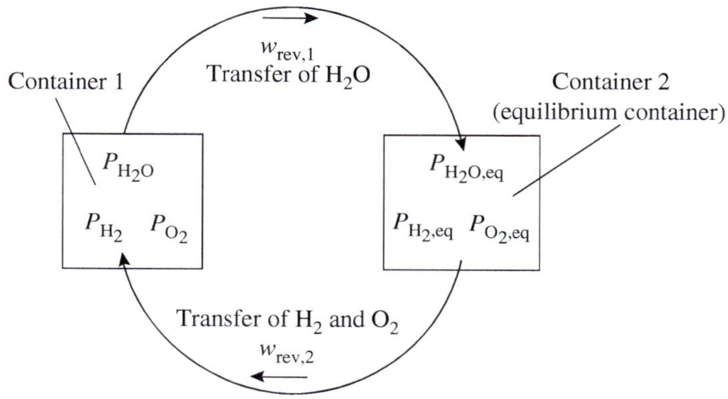

Figure 18.4 Performing the detonating gas reaction reversibly at $T = const$ and $V = const$. Reacting species under arbitrary partial pressures are in container 1 and those under equilibrium pressures are in container 2 (equilibrium container).

Equation (18.59) is our result, but it is useful to modify its form. First, we use the dimensionless pressure defined in Equations (18.47) and (18.52)

$$\widehat{P} = \frac{P}{P^\ominus} \quad \text{and} \quad \widehat{P}_{eq} = \frac{P_{eq}}{P^\ominus} \tag{18.60}$$

Second, the amounts of substance are determined by the reaction's stoichiometry, so that

$$\boldsymbol{n} = \boldsymbol{n}_{H_2O} = \boldsymbol{n}_{H_2} = 2\boldsymbol{n}_{O_2} \tag{18.61}$$

Substituting these relationships, Equation (18.59) becomes

$$w_{rev} = \boldsymbol{nR}T \cdot \ln\frac{\widehat{P}_{H_2O,\,eq}}{\widehat{P}_{H_2O}} + \boldsymbol{nR}T \cdot \ln\frac{\widehat{P}_{H_2}}{\widehat{P}_{H_2,\,eq}} + \frac{1}{2}\boldsymbol{nR}T \cdot \ln\frac{\widehat{P}_{O_2}}{\widehat{P}_{O_2,\,eq}} \tag{18.62}$$

Now we use the properties of logarithms and obtain

$$\begin{aligned} w_{rev} &= \boldsymbol{nR}T \cdot \left[\ln\frac{\widehat{P}_{H_2O,\,eq}}{\widehat{P}_{H_2,\,eq}\widehat{P}_{O_2,\,eq}} + \ln\frac{\widehat{P}_{H_2}\widehat{P}_{O_2}}{\widehat{P}_{H_2O}}\right] \\ &= \boldsymbol{nR}T \cdot \left[-\ln\frac{\widehat{P}_{H_2,\,eq}\widehat{P}_{O_2,\,eq}}{\widehat{P}_{H_2O,\,eq}} + \ln\frac{\widehat{P}_{H_2}\widehat{P}_{O_2}}{\widehat{P}_{H_2O}}\right] \end{aligned}$$

and

$$\boxed{w_{rev} = -\boldsymbol{nR}T \cdot \ln K + \boldsymbol{nR}T \cdot \ln Q} \tag{18.63}$$

where K is the equilibrium constant and Q is the reaction quotient

$$K = \frac{\widehat{P}_{H_2,\,eq} \cdot \widehat{P}^{1/2}_{O_2,\,eq}}{\widehat{P}_{H_2O,\,eq}} \quad \text{and} \quad Q = \frac{\widehat{P}_{H_2} \cdot \widehat{P}^{1/2}_{O_2}}{\widehat{P}_{H_2O}}$$

As in Example 18.6 Equation (18.63) has three interesting cases.

Case 1. $Q < K$, then $w_{rev} < 0$ (spontaneous reaction toward products). In this case the pressure of H_2O is higher than in equilibrium, and the reaction will proceed toward the products H_2 and O_2 to reach equilibrium.

1 H_2O must be taken out from container 1 separately in a reversible manner. To overcome this problem we imagine a filter which is permeable to H_2O molecules, but not to O_2 and H_2. In principle it should be possible to separate any gas with its particular semipermeable membrane, even if not practically realizable. In our considerations, which are typical for a "thought experiment," we are interested in the realization possibility as a matter of principle, and not in the practical realization. Thought experiments play an important role in physical chemistry.

Case 2. $Q > K$, then $w_{rev} > 0$ (spontaneous reaction toward reactants). In this case the pressure of H_2O is lower than in equilibrium, and the reaction will proceed toward the reactant H_2O to reach equilibrium.

Case 3. $Q = K$, then $w_{rev} = 0$ and the system is in equilibrium.

18.5.2.2 Reaction in a Closed Container at Constant *T* and *P*

We have assumed that the reacting gases in container 1 (Fig. 18.4) are at constant volume. Now we consider the same reaction, under conditions where the reactants are at constant pressure P. In this case, the volume changes when the reaction occurs, and we obtain an additional contribution $-P\Delta V$ to w_{rev}:

$$w_{rev} = -\boldsymbol{n}\boldsymbol{R}T \cdot \ln K + \boldsymbol{n}\boldsymbol{R}T \cdot \ln Q - P\Delta V \tag{18.64}$$

or

$$\boxed{w_{rev} + P\Delta V = -\boldsymbol{n}\boldsymbol{R}T \cdot \ln K + \boldsymbol{n}\boldsymbol{R}T \cdot \ln Q} \tag{18.65}$$

In this case the reaction is possible for

$$w_{rev} + P\Delta V \leq 0 \tag{18.66}$$

Assuming ideal gas behavior, the volume increases for this reaction by

$$\Delta V = \boldsymbol{n}_{O_2}\frac{\boldsymbol{R}T}{P} + \boldsymbol{n}_{H_2}\frac{\boldsymbol{R}T}{P} - \boldsymbol{n}_{H_2O}\frac{\boldsymbol{R}T}{P} \tag{18.67}$$

Using Equation (18.61)

$$P\Delta V = \boldsymbol{R}T \cdot \left(\frac{1}{2} + 1 - 1\right)\boldsymbol{n} = \frac{1}{2}\boldsymbol{n}\boldsymbol{R}T \tag{18.68}$$

Using Equation (18.18) we express $w_{rev} + P\Delta V$ in Equation (18.65) by the change of Gibbs energy, ΔG and find from Equation (18.65) that

$$\boxed{\Delta G = -\boldsymbol{n}\boldsymbol{R}T \cdot \ln K + \boldsymbol{n}\boldsymbol{R}T \cdot \ln Q} \tag{18.69}$$

or in molar form,

$$\Delta_r G = -\boldsymbol{R}T \cdot \ln K + \boldsymbol{R}T \cdot \ln Q \tag{18.70}$$

where $\Delta_r G$ is the molar change in Gibbs energy under the given conditions.

If the reactant and product species are separated and under standard pressure $P^\ominus$ (see Fig. 18.5 for reaction scheme A + B → C + D), the reaction quotient becomes $Q = 1$ and from Equation (18.70) it follows that

$$\Delta_r G^\ominus = -\boldsymbol{R}T \cdot \ln K \tag{18.71}$$

These results are the same as we obtained in Section 18.5.1 using the formal derivation.

Inventing thought experiments can deepen our understanding of scientific problems under investigation and is enormously important and challenging in doing successful research. The formal procedure in Section 18.5.1, rewards us by its beauty. However, analyzing this important subject through basic principles is crucial for developing an intuitive understanding.

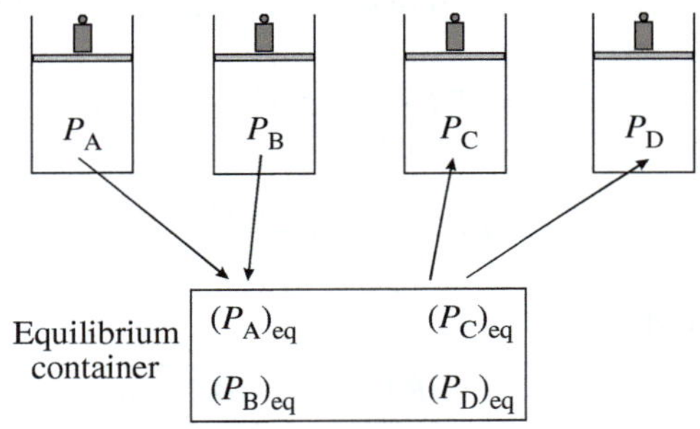

Figure 18.5 Performing reaction A + B → C + D reversibly at $T = const$ and $P = const$. Reacting species under standard pressure $P^\ominus = 1$ bar in separate reservoirs.

18.5.3 Applying the Mass Action Law

18.5.3.1 Dimerization Equilibrium For NO_2

We consider the equilibrium

$$N_2O_4 \leftrightharpoons 2NO_2 \quad \text{and} \quad K = \frac{\widehat{P}^2_{NO_2,\,eq}}{\widehat{P}_{N_2O_4,\,eq}} = 0.15 \text{ at } 298 \text{ K}$$

We determine how the Gibbs energy changes as the proportions of the reactant and product change, and we show that the equilibrium state corresponds to a minimum in the Gibbs energy.

Dependence of $\Delta_r G$ on Degree of Dissociation ξ We define a degree of dissociation parameter ξ, with the range $0 < \xi < 1$. At the value zero only dimer is present, and at the value one the dimer is fully dissociated. We assume that if none of the dimers are dissociated, we have an amount $\boldsymbol{n}$ of gas molecules N_2O_4. Hence the amount of dimer is $\boldsymbol{n}(1-\xi)$ and the amount of NO_2 is $2\xi\boldsymbol{n}$. In an arbitrary state of the mixture the total amount of gas is

$$\boldsymbol{n}_{\text{total}} = \boldsymbol{n}_{N_2O_4} + \boldsymbol{n}_{NO_2} = \boldsymbol{n}(1-\xi) + 2\boldsymbol{n}\xi = \boldsymbol{n}(1+\xi) \tag{18.72}$$

For example, for $\xi = 0$ (only the dimer is present) we have $\boldsymbol{n}_{\text{total}} = \boldsymbol{n}$; for $\xi = 1$ (only the monomer is present) we have $\boldsymbol{n}_{\text{total}} = 2\boldsymbol{n}$.

To evaluate Q and $\Delta_r G$ we want expressions for the partial pressures of the constituent gas molecules. Using Equation (18.72) and the ideal gas equation of state, we find that

$$P = P_0 \cdot (1-\xi) + P_0 \cdot 2\xi = P_0 \cdot (1+\xi) \tag{18.73}$$

where P is the total pressure for dissociation parameter ξ and P_0 is the initial, $\xi = 0$, pressure; note that for $\xi = 0$ the total pressure is $P = P_0$ and that for $\xi = 1$ the total pressure is $P = 2\,P_0$. Rearranging Equation (18.73) we find that

$$P_0 = P\frac{1}{1+\xi}$$

and for the partial pressures $P_{N_2O_4}$ and P_{NO_2} we obtain

$$P_{N_2O_4} = P_0(1-\xi) = \frac{1-\xi}{1+\xi}P \quad \text{and } P_{NO_2} = P_0 \cdot 2\xi = \frac{2\xi}{1+\xi}P \tag{18.74}$$

Using these expressions for the partial pressure we may write the reaction quotient as

$$Q = \frac{\widehat{P}^2_{NO_2}}{\widehat{P}_{N_2O_4}} = \frac{4\xi^2}{1-\xi^2}\widehat{P}$$

Next we set $P = 1\text{ bar} = P^{\ominus}$ that is, $\widehat{P} = P/P^{\ominus} = 1$, to obtain

$$Q = \frac{\widehat{P}^2_{NO_2}}{\widehat{P}_{N_2O_4}} = \frac{4\xi^2}{1-\xi^2}$$

Using these results with Equation (18.54), we find

$$\begin{aligned}\Delta_r G &= -\boldsymbol{R}T \cdot \ln K + \boldsymbol{R}T \cdot \ln Q \\ &= 4.7\text{ kJ mol}^{-1} + 2.478 \cdot \ln\left(\frac{4\xi^2}{1-\xi^2}\right)\text{ kJ mol}^{-1}\end{aligned}$$

We can evaluate the change of the molar Gibbs energy $\Delta_r G$ for different values of ξ, see Fig. 18.6a. At $\xi = 0.190$ we find $\Delta_r G = 0$, that is, system is in equilibrium. In the limit $\xi \to 0$, $\Delta_r G$ becomes extremely negative, and in the limit $\xi \to 1$, it becomes extremely positive.

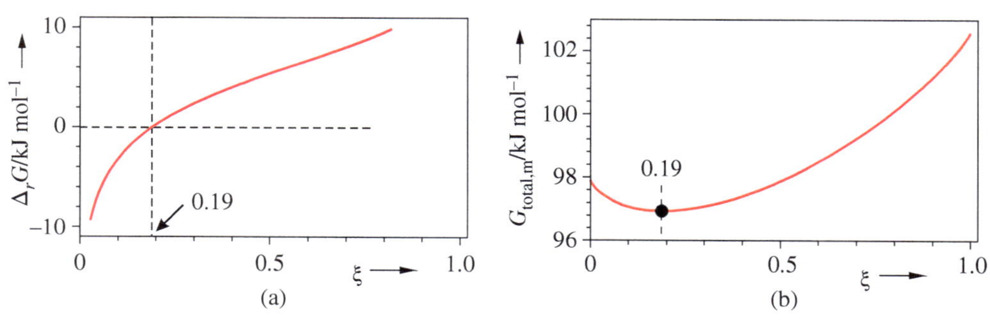

Figure 18.6 Reaction $N_2O_4 \to 2\,NO_2$. (a) $\Delta_r G$ as a function of the dissociation parameter, ξ. At $\xi = 0.19$ we find $\Delta_r G = 0$ (equilibrium). (b) $\Delta G_{\text{total,m}}$ as a function of ξ (solid line). At $\xi = 0.19$ we find a minimum for $\Delta G_{\text{total,m}}$ (equilibrium, filled circle).

Dependence of G_{total} on Degree of Dissociation ξ From Appendix A.1 we see that the Gibbs energy of formation for NO_2 is $\Delta_f G^{\ominus}_{NO_2} = 51.5\text{ kJ mol}^{-1}$ and that for N_2O_4 is $\Delta_f G^{\ominus}_{N_2O_4} = 97.9\text{ kJ mol}^{-1}$. If we consider the total Gibbs energy $G^{\ominus}_{\text{total}}$ for the relative amounts of the pure substances at standard pressure, we find

$$\begin{aligned} G^{\ominus}_{\text{total}} &= \boldsymbol{n}_{\text{N}_2\text{O}_4}\ \Delta_f G^{\ominus}_{\text{N}_2\text{O}_4} + 2\boldsymbol{n}_{\text{NO}_2} \Delta_f G^{\ominus}_{\text{NO}_2} \\ &= \boldsymbol{n}(1-\xi)\ \Delta_f G^{\ominus}_{\text{N}_2\text{O}_4} + 2\boldsymbol{n}\xi\ \Delta_f G^{\ominus}_{\text{NO}_2} \end{aligned} \tag{18.75}$$

and

$$G^{\ominus}_{\text{total,m}} = (1-\xi)\ \Delta_f G^{\ominus}_{\text{N}_2\text{O}_4} + 2\xi\ \Delta_f G^{\ominus}_{\text{NO}_2} \tag{18.76}$$

To obtain the total Gibbs energy G_{total}, where N_2O_4 and NO_2 are under their partial pressures $P_{N_2O_4}$ and P_{NO_2} (total pressure = $P^{\ominus} = 1$ bar), we have to change the pressures from $P^{\ominus}$ to $P_{N_2O_4}$ and P_{NO_2}, respectively. Then the total molar Gibbs energy is

$$\begin{aligned} G_{\text{total,m}} &= (1-\xi)\ \Delta_f G^{\ominus}_{N_2O_4} + 2\xi\ \Delta_f G^{\ominus}_{NO_2} \\ &\quad + (1-\xi)\ \boldsymbol{R}T \cdot \ln\left(\frac{P_{\text{N}_2\text{O}_4}}{P^{\ominus}}\right) + 2\xi\ \boldsymbol{R}T \cdot \ln\left(\frac{P_{NO_2}}{P^{\ominus}}\right) \end{aligned} \tag{18.77}$$

or with Equation (18.74)

$$\begin{aligned} G_{\text{total,m}} &= (1-\xi)\ \Delta_f G^{\ominus}_{N_2O_4} + 2\xi\ \Delta_f G^{\ominus}_{NO_2} \\ &\quad + \boldsymbol{R}T\left((1-\xi)\ \cdot \ln\frac{1-\xi}{1+\xi} + 2\xi\ \cdot \ln\frac{2\xi}{1+\xi}\right) \end{aligned} \tag{18.78}$$

The total molar Gibbs energy $G_{\text{total,m}}$ displays a minimum at $\xi = 0.190$, see Fig. 18.6b (solid line). This minimum is at the same place we found for the equilibrium condition $\Delta_r G = 0$ in Fig. 18.6a.

18.5.3.2 Dissociation of Water at Different Temperatures

We compare the equilibrium constant for the dissociation of water vapor at 1500 K and 1 bar pressure to that at 4000 K and 1 bar pressure.

For the reaction R1

$$H_2O \rightarrow H_2 + \frac{1}{2}O_2 \qquad \text{R1}$$

the equilibrium constant is, according to Equation (18.53),

$$\ln K = -\frac{\Delta_r G^{\ominus}}{\boldsymbol{R}T}$$

In Example 18.6 we used $\Delta_r G^{\ominus}(1500\text{ K}) = +158\text{ kJ mol}^{-1}$ and found $K(1500\text{ K}) = 3.15 \times 10^{-6}$. This result shows that for $T = 1500$ K and $\hat{P}_{H_2O} = 1$ the equilibrium is strongly shifted toward the reactant H_2O.

On the other hand, if the reaction is carried out at $T = 4000$ K, $\Delta_r G^{\ominus}$ becomes negative, $\Delta_r G^{\ominus}(4000\text{ K}) = -1.3\text{ kJ mol}^{-1}$, and the equilibrium is shifted toward the products:

$$\ln K(4000\text{ K}) = \frac{1.3\text{ kJ mol}^{-1}}{\boldsymbol{R}T} = 0.0391, \quad \text{or} \quad K(4000\text{ K}) = \frac{\hat{P}_{\text{H}_2,\text{eq}} \cdot \hat{P}^{1/2}_{\text{O}_2,\text{eq}}}{\hat{P}_{\text{H}_2\text{O,eq}}} = 1.04$$

Following the considerations in Example 18.6 and assuming that the equilibrium pressure of the H_2O gas is 1 bar, we find that

$$\hat{P}_{\text{O}_2,\text{eq}} = \left(\frac{1}{2}K \cdot \hat{P}_{\text{H}_2\text{O, eq}}\right)^{2/3} = 0.65$$

In this case the total pressure is $1.00 + 0.65 + 1.30$ bar $= 2.95$ bar. This means that the H_2O in the equilibrium mixture is only $1.00/2.95 = 34\%$, that is, the H_2O is mainly dissociated.

The equilibrium constant K depends on how the reaction is written. If we consider the reverse of reaction R1 (detonating gas reaction),

$$H_2 + \frac{1}{2}O_2 \rightarrow H_2O \qquad \text{(R-1)}$$

we find $\Delta_r G^{\ominus}(1500\text{ K}) = -158\text{ kJ mol}^{-1}$. Then

$$K(1500\text{ K}) = \frac{\hat{P}_{\text{H}_2\text{O, eq}}}{\hat{P}_{\text{H}_2,\text{ eq}} \cdot \hat{P}^{1/2}_{\text{O}_2,\text{ eq}}} = \frac{1}{3.14 \times 10^{-6}} = 3.19 \times 10^5$$

For the reaction R2

$$2H_2 + O_2 \rightarrow 2H_2O \qquad \text{(R2)}$$

we find $\Delta_r G^\ominus(1500\text{ K}) = -316\text{ kJ mol}^{-1}$, and the equilibrium constant is

$$K(1500\text{ K}) = \frac{\hat{P}^2_{H_2O,eq}}{\hat{P}^2_{H_2,eq} \cdot \hat{P}_{O_2,eq}} = \left(3.19 \times 10^5\right)^2 = 1.02 \times 10^{11}$$

at 1500 K.

Note that the equilibrium constant is always defined by the corresponding reaction equation. Thus the equilibrium constant for one and the same equilibrium depends on the way the reaction is written. In the case of reaction R1, $\Delta_r G$ is related to the dissociation of one mole of H_2O; $-R_1$ is related to the reaction of one mole of H_2 and one-half mole of O_2. R2 is related to the reaction of two moles of H_2 and one mole of O_2.

18.5.4 Temperature Dependence of Equilibrium Constant *K*

According to Equation (18.53) the equilibrium constant K is determined by $\Delta_r G^\ominus$

$$\ln K = -\frac{\Delta_r G^\ominus}{\boldsymbol{R}T} \tag{18.79}$$

Changing the temperature changes $\Delta_r G^\ominus$ and thus K. To determine the temperature dependence of K over a small temperature interval we evaluate the temperature derivative of Equation (18.79). In analogy to our procedure in deriving the Clausius–Clapeyron equation in Section 18.4.1, we find

$$\frac{\mathrm{d}\ln K}{\mathrm{d}T} = -\frac{1}{\boldsymbol{R}}\left(\frac{1}{T}\frac{\mathrm{d}\Delta_r G^\ominus}{\mathrm{d}T} - \Delta_r G^\ominus \frac{1}{T^2}\right) \tag{18.80}$$

Using the result of Box 18.1, namely $\mathrm{d}\Delta_r G^\ominus/\mathrm{d}T = -\Delta_r S^\ominus$, we can write that

$$\frac{\mathrm{d}\ln K}{\mathrm{d}T} = -\frac{1}{\boldsymbol{R}}\left[-\frac{1}{T}\Delta_r S^\ominus - \left(\Delta_r H^\ominus - T\Delta_r S^\ominus\right)\frac{1}{T^2}\right] = -\frac{1}{\boldsymbol{R}}\left[-\Delta_r H^\ominus \frac{1}{T^2}\right]$$

or

$$\boxed{\frac{\mathrm{d}\ln K}{\mathrm{d}T} = \frac{\Delta_r H^\ominus}{\boldsymbol{R}T^2}} \tag{18.81}$$

which is known as the *van't Hoff equation*.

This relation enables us to judge whether K will increase or decrease with temperature by knowing whether the reaction is endothermic or exothermic. In addition, this equation is useful to obtain the enthalpy change of a reaction by measuring the equilibrium constant at different temperatures. If $\Delta H^\ominus$ is positive (endothermic reaction), then $\mathrm{d}(\ln K)/\mathrm{d}T$ is positive and K increases with temperature, that is, the equilibrium is shifted toward the products.

This result is in accordance with *Le Chatelier's Principle of minimum constraint*:

> When an independent variable of a system at equilibrium is changed, the equilibrium shifts in the direction that tends to reduce the effect of this change. For example, when pressure is increased, the equilibrium shifts in the direction to reduce the number of molecules. If the temperature is increased in an endothermic reaction, the equilibrium shifts in the direction that absorbs the heat – toward the products.

18.5.4.1 Estimation of $K = f(T)$

As in the case of the temperature dependence of the vapor pressure, it is useful to estimate K at a higher temperature, while avoiding a rigorous treatment. Using Equation (18.79) we can write

$$\ln K_1 = -\frac{\Delta_r G_1^\ominus}{\boldsymbol{R}T_1} = -\frac{\Delta_r H_1^\ominus}{\boldsymbol{R}T_1} + \frac{\Delta_r S_1^\ominus}{\boldsymbol{R}} \tag{18.82}$$

for a starting temperature T_1. If we neglect the temperature dependence of $\Delta H^\ominus$ and $\Delta S^\ominus$, then it follows for any other temperature T that

$$\ln K = -\frac{\Delta_r H_1^\ominus}{\boldsymbol{R}T} + \frac{\Delta_r S_1^\ominus}{\boldsymbol{R}} \tag{18.83}$$

and

$$\ln K = \ln K_1 - \frac{\Delta_r H_1^\ominus}{\boldsymbol{R}} \left(\frac{1}{T} - \frac{1}{T_1} \right) \tag{18.84}$$

Example 18.7 *Use Equation (18.84) to estimate the change in equilibrium constant with temperature from 298 K to 700 K for the ammonia synthesis reaction*

$$N_2 + 3\,H_2 \rightarrow 2\,NH_3$$

The standard enthalpy change and equilibrium constant can be found from the data in Appendix A.1

$$\Delta H_{298}^\ominus = -2 \cdot 45.9 \text{ kJ mol}^{-1} = -91.8 \text{ kJ mol}^{-1}$$
$$\Delta S_{298}^\ominus = (2 \cdot 192.5 - 3 \cdot 130.6 - 191.5) \text{ J K}^{-1} \text{ mol}^{-1}$$
$$= -198.3 \text{ J K}^{-1} \text{ mol}^{-1}$$
$$\Delta G_{298}^\ominus = -91.8 \text{ kJ mol}^{-1} - 298 \text{ K} \cdot (-198.3 \text{ J K}^{-1} \text{ mol}^{-1})$$
$$= -32.7 \text{ kJ mol}^{-1}$$

Then

$$\ln K_{298} = -\frac{-32.7 \text{ kJ mol}^{-1}}{8.314 \text{ J K}^{-1} \text{ mol}^{-1} \cdot 298 \text{ K}} = 13.2$$

so that

$$K_{298} = 5.40 \times 10^5$$

Using Equation (18.84) we can write

$$\ln K_{700} = 13.2 - \frac{-91.8 \text{ kJ mol}^{-1}}{8.314 \text{ J K}^{-1} \text{ mol}^{-1}} \left(\frac{1}{700 \text{ K}} - \frac{1}{298 \text{ K}} \right)$$
$$= 13.2 - 21.3 = -8.10$$

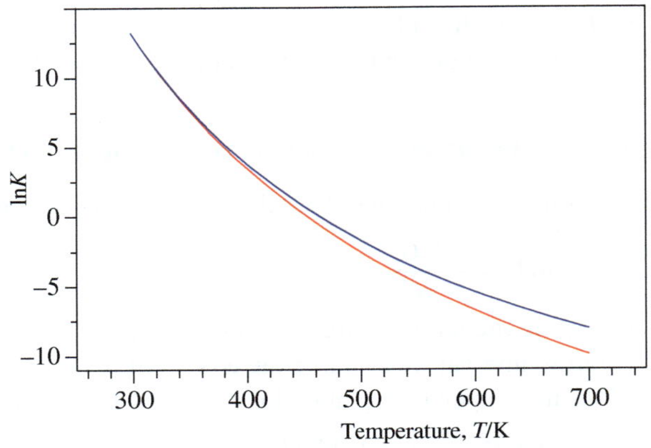

Figure 18.7 $\ln K$ for the formation of NH_3 from the elements H_2 and N_2 versus temperature under standard conditions. Solid red line: rigorous calculation (from Equation (18.79), including the temperature dependence of $\Delta_r G^\ominus$, see Foundation 18.1). Solid blue line: $\ln K$ approximated by Equation (18.84).

The resulting estimate $K_{700} = 3.03 \times 10^{-4}$ is in reasonable agreement with the value 4.88×10^{-5} obtained from the rigorous treatment (based on Equation (18.79), including the temperature dependence of $\Delta_r G^\ominus$) and the experimental value 4.37×10^{-5} (see Table 20.3); note that the change in K is about nine orders of magnitude. Figure 18.7 compares the rigorous analysis and the approximate analysis for $\ln K$ over the temperature range $T = 300$ K to $T = 700$ K. You can follow the calculation by running the program "ammonia_equilibrium.exe" included on Wiley's website.

18.5.4.2 ΔH and ΔS from Measured K

In the approximation (18.84), $\ln K$ depends linearly on $1/T$. Thus plotting experimental $\ln K$ data versus $1/T$ should result in a straight line with slope $-\Delta_r H^\ominus/\boldsymbol{R}$. With the enthalpy change in hand it is straightforward to calculate the entropy change from $\Delta_r G^\ominus = \Delta_r H^\ominus - T\Delta_r S^\ominus$.

Example 18.8 *Use the temperature dependence of the equilibrium constant for the Iodine-Hydrogen reaction to determine the reaction's enthalpy and entropy changes.*

Experimental data for K of the reaction

$$H_2 + I_2 \rightarrow 2\,HI$$

are listed in Table 18.1. In Fig. 18.8a, $\ln K$ is plotted versus T for this equilibrium; note that the line connecting the data points is slightly curved. Figure 18.8b shows a plot of $\ln K$ versus $1/T$. Fitting the data to a straight line (see Equation (18.84)) yields a slope from which we find $\Delta_r H^\ominus = -13.0$ kJ mol^{-1}.

Using the values at 667 K, we find that

$$\Delta_r S^{\ominus} = \frac{\Delta_r H^{\ominus} - \Delta_r G^{\ominus}}{T} = \frac{\Delta_r H^{\ominus}}{T} + \boldsymbol{R} \ln K$$

$$= \frac{-13.0 \text{ kJ mol}^{-1}}{667 \text{ K}} + \left(8.314 \text{ J K}^{-1} \text{ mol}^{-1}\right) \cdot 4.12 = 14.8 \text{ J K}^{-1} \text{ mol}^{-1}$$

This reaction has a slight (about $2\boldsymbol{R}$) increase in entropy.

Table 18.1 Experimental Equilibrium Constant $K = \widehat{P}^2_{HI}/(\widehat{P}_{H_2} \cdot \widehat{P}_{I_2})$ at Different Temperatures.

T (K)	K	$\ln K$	$\frac{1}{T}$ (K^{-1})
667	61.0	4.12	0.00150
699	55.2	4.02	0.00143
731	49.9	3.92	0.00137
764	45.9	3.84	0.00132

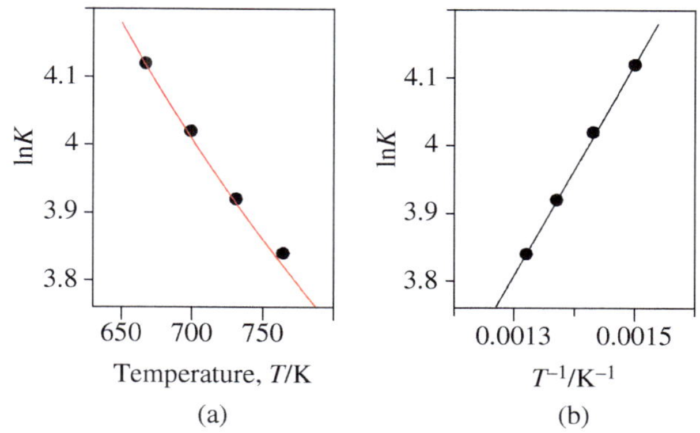

Figure 18.8 Evaluation of $\Delta_r H^{\ominus}$ for the reaction $I_2 + H_2 \rightarrow 2$ HI at $T = 700$ K. For data see Table 18.1. (a) $\ln K$ versus T, (b) $\ln K$ versus $1/T$.

18.5.5 Pressure Changes and Equilibrium

The equilibrium constant is only a function of temperature, not the pressure. This fact is evident from Equation (18.53) which relates the equilibrium constant directly to the Gibbs energy under standard conditions. Nevertheless the pressure can affect the amounts of species in a reaction mixture under equilibrium conditions.

As an example of the pressure effect, consider the dimerization reaction

$$N_2O_4 \leftrightarrows 2NO_2$$

In Section 18.5.3, we found that the reaction quotient is

$$Q = \frac{4\xi^2}{1 - \xi^2} \widehat{P}$$

where ξ is the degree of dissociation and $\widehat{P}$ is the total pressure in units of the standard pressure. Then the equilibrium constant K is given by

$$K = \frac{4\xi^2_{eq}}{1 - \xi^2_{eq}} \widehat{P}$$

Solving this expression for ξ_{eq}, we find

$$\xi_{eq} = \sqrt{\frac{K}{K + 4\widehat{P}}}$$

At 298 K, $K_{298} = 0.15$, and for 1 bar pressure we find that $\xi_{eq} = 0.190$; i.e., 19% of the dimers are dissociated. A similar calculation at a pressure of 10 bar results in a decrease of the dissociation to 6%; i.e., an increase in pressure displaces the equilibrium to the reactant side, which has fewer molecules. As the pressure increases ξ_{eq} approaches 0 (no dissociation); as the pressure decreases ξ_{eq} approaches 1 (all N_2O_4 is dissociated).

As for the effect of temperature on the equilibrium, these observations coincide with LeChatelier's principle.

Example 18.9 *How does the pressure effect the equilibrium for the formation of NH_3 from its elements at 700 K?*

$$3H_2 + N_2 \rightarrow 2\ NH_3$$

In Example 18.7 we showed that $K(700\ K) = 9.3 \times 10^{-5}$; hence, only a very small fraction of H_2 and N_2 can be converted into NH_3. We expect the percentage conversion into NH_3 to depend strongly on the total pressure because of the different amounts of substance between the products ($n_{products} = 2$ mol) and reactants ($n_{reactants} = 4$ mol). We expect that a higher pressure shifts the equilibrium towards the products. What is the yield of NH_3 at different total pressures $P_{total} = P_{H_2} + P_{N_2} + P_{NH_3}$? We restrict our discussion to the case where N_2 and H_2 are in the stoichiometric ratio 1:3. In the beginning of the reaction the pressures of N_2 and H_2 are P_{N_2} and $P_{H_2} = 3P_{N_2}$ and $P_{total} = 4P_{N_2}$. At equilibrium the partial pressures of the gases are

$$P_{N_2,eq} = P_{N_2} - x, \quad P_{H_2,eq} = 3P_{N_2} - 3x, \quad P_{NH_3,eq} = 2x$$

and the total pressure is

$$P_{total,eq} = 4P_{N_2} - 2x$$

If we reference our pressures to the standard state pressure, $\hat{x} = x/P^{\ominus}$, we obtain

$$K = \frac{4\hat{x}^2}{27\left(\hat{P}_{N_2} - \hat{x}\right)^4} \quad \text{or} \quad \frac{\hat{x}}{\left(\hat{P}_{N_2} - \hat{x}\right)^2} = \sqrt{\frac{27}{4}K}$$

This quadratic equation can be solved to find

$$\hat{x} = \hat{P}_{N_2} + \frac{1}{2a} - \sqrt{\frac{\hat{P}_{N_2}}{a} + \frac{1}{4a^2}} \quad \text{with} \quad a = \sqrt{\frac{27}{4}K}$$

For different pressures P_{N_2} the equilibrium pressures in Table 18.2 are obtained.

Table 18.2 Equilibrum Pressures for the $3H_2 + N_2 \rightarrow 2\ NH_3$ Equilibrium at $T = 700$ K ($K = 9.3 \times 10^{-5}$).[2]

$\hat{P}_{N_2}$	$\hat{P}_{total}$	$\hat{x}$	$\hat{P}_{total,eq}$	$\hat{P}_{NH_3,eq}$	$\hat{P}_{H_2,eq}$	$\hat{P}_{N_2,eq}$	$\frac{\hat{P}_{NH_3,eq}}{\hat{P}_{total,eq}}$
1	4	0.023	3.95	0.05	2.93	0.98	0.013
30	120	9.95	100	19.9	60.2	20.1	0.20
100	400	53.5	293	107	140	47	0.36
200	800	130	540	260	210	70	0.48

Increasing P_{total} to 400 bar ($P_{total,eq}$ to 293 bar) increases the ratio $P_{NH_3,eq} : P_{total,eq}$ to 0.36 for a stoichiometric mixture of H_2 and N_2. The strong shift of the equilibrium with pressure to favor the product NH_3 occurs because the number of molecules decreases (three molecules of H_2 react with one molecule of N_2 to form two molecules of NH_3) and is consistent with the Principle of Le Chatelier.

These considerations are important for the industrial synthesis of NH_3 from the elements. Only at high temperature is a catalyst available to accelerate the reaction. Thus the reaction must be performed at high temperature (700 K) and high pressure (400 – 600 bar): *Haber–Bosch process.*

18.5.6 Reactions Involving Gases and Immiscible Condensed Species

18.5.6.1 Equilibrium Constant

If we assume that solid and liquid reactants are incompressible, then the internal energy and the entropy of these compounds do not depend on pressure (see Section 18.4). Consider the decomposition of limestone

$$CaCO_3\ (\text{solid}) \rightarrow CaO\ (\text{solid}) + CO_2$$

2 At the high pressures used in this table the gases might not behave ideally. See Chapter 20 for corrections in the equilibrium constant.

In Example 18.5 we showed that

$$\Delta_r G = \Delta_r G^{\ominus} + \boldsymbol{R}T \cdot \ln \hat{P}_{CO_2}$$

for this reaction. At equilibrium $\hat{P}_{CO_2} = \hat{P}_{CO_2,\,eq}$, so that

$$\Delta_r G^{\ominus} = -\boldsymbol{R}T \cdot \ln \hat{P}_{CO_2,\,eq}$$

and the result for K is very simple:

$$K = \hat{p}_{CO_2,\,eq}$$

In this approximation, solids and liquids do not contribute to the equilibrium constant K.

Example 18.10 *Write the equilibrium constant for the combustion of benzene at 298 K*

$$C_6H_6\ (liquid) + \frac{15}{2} O_2 \rightarrow 6\ CO_2 + 3\ H_2O\ (liquid)$$

Assume that the liquids are incompressible and immiscible with the other reactants and that the gases behave ideally.[3]

Using this approximation, the pressure change from standard pressure to equilibrium pressure of these liquids does not lead to a contribution to ΔG. Thus these species do not appear in the expression for K and we find

$$\Delta_r G^{\ominus} = -\boldsymbol{R}T \cdot \ln K, \qquad K = \frac{\hat{P}^{6}_{CO_2,\,eq}}{\hat{P}^{15/2}_{O_2,\,eq}}$$

18.5.7 Equilibrium Constant from Molecular Properties

The equilibrium constant is an important property for chemical systems, and its value is determined by the molecular characteristics of the reactants and products as well as the external conditions. The connection of the equilibrium constant to the molecular partition function $Z = Z_{trans} \cdot Z_{rot} \cdot Z_{vib} \cdot Z_{electr}$ is direct. In Foundation 18.2, we show that for an equilibrium

$$\alpha A + \beta B + \ldots \rightleftarrows \gamma C + \delta D + \cdots \tag{18.85}$$

the equilibrium constant K can be expressed as

$$K = \frac{\left(\dfrac{Z_C^{\ominus}}{\sigma_C N_C}\right)^{\gamma} \cdot \left(\dfrac{Z_D^{\ominus}}{\sigma_D N_D}\right)^{\delta} \cdot \ldots}{\left(\dfrac{Z_A^{\ominus}}{\sigma_A N_A}\right)^{\alpha} \cdot \left(\dfrac{Z_B^{\ominus}}{\sigma_B N_B}\right)^{\beta} \cdot \ldots} \tag{18.86}$$

where σ is the symmetry factor defined in Chapter 17 (see Sections 17.8.2 and 17.8.3, e.g., $\sigma = 1$ for heteronuclear diatomic molecules and $\sigma = 2$ for homonuclear diatomic molecules) and N is the particle number.

18.5.7.1 Isomerization Reaction *i*-Butane ⇄ *n*-Butane

To illustrate this connection, we consider the isomerization reaction between *i*-butane and *n*-butane:

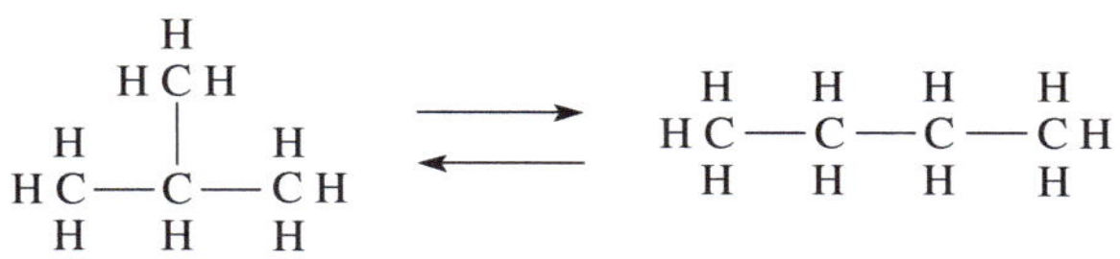

i-Butane *n*-Butane

3 See Chapter 20 for corrections to this approximation.

In this case the equilibrium constant K depends strongly on temperature. At low temperature the equilibrium is completely on the reactant side, and with increasing temperature it is shifted to the product side. The connection between K and the molecular partition function provides insight into this behavior.

According to Equation (18.86) the equilibrium constant is (setting $\sigma_{i\text{-butane}} = \sigma_{n\text{-butane}} = 1$)

$$K = \frac{P_{n\text{-butane}}}{P_{i\text{-butane}}} = \frac{Z^{\ominus}_{n\text{-butane}}/N_{n\text{-butane}}}{Z^{\ominus}_{i\text{-butane}}/N_{i\text{-butane}}} = \frac{Z^{\ominus}_{n\text{-butane}}}{Z^{\ominus}_{i\text{-butane}}} \tag{18.87}$$

because the number of molecules does not change in the reaction. Since the mass of the product molecules equals the mass of the reactant molecules, the translational contributions to Z are the same. Because the moment of inertia does not change significantly, the rotational contributions to Z are approximately equal and Equation (18.87) simplifies to

$$K = \frac{(Z_{\text{vib}} Z_{\text{electr}})_{n\text{-butane}}}{(Z_{\text{vib}} Z_{\text{electr}})_{i\text{-butane}}} \tag{18.88}$$

If both molecules are in the electronic ground state, E_1, then

$$Z_{\text{electr}} = \sum_{i=1}^{\infty} \mathrm{e}^{-E_i/(kT)} \approx \mathrm{e}^{-E_1/(kT)} \tag{18.89}$$

and

$$K = \frac{(Z_{\text{vib}})_{n\text{-butane}}}{(Z_{\text{vib}})_{i\text{-butane}}} \cdot \mathrm{e}^{-\Delta E/(kT)} \tag{18.90}$$

where ΔE is the electronic energy difference between n-butane and i-butane. The electronic energy of i-butane is lower than that for n-butane. Thus ΔE is positive. Isobutane is more rigid than n-butane in particular, n-butane has low frequency torsional modes that interchange different conformers (trans and gauche forms) giving rise to a larger density of vibrational energy levels.

At low temperature, the exponential factor is small and K becomes small (all molecules accumulate in the i-butane ground state, see Fig. 18.9, left). At high temperature the exponential factor increases; moreover, because of the smaller energy spacing of the vibrational states, Z_{vib} becomes larger for n-butane than for i-butane, resulting in an increase of K: more molecules are found in the n-butane states than in the i-butane states, see Fig. 18.9, right.

Figure 18.10 presents different characteristic energy level structures that may occur. In the butane case the electronic energy is higher for the product ($\Delta H > 0$), but the product has a larger density of states ($\Delta S > 0$). Thus $K < 1$ at low temperature, and $K > 1$ at high temperature (Fig. 18.10a). Now imagine an equilibrium for which $\Delta H > 0$, but $\Delta S < 0$ (Fig. 18.10b). In this case the reactant species is preferred at low temperature and at high temperature. The equilibrium constant K can never reach one at high temperature.

Figure 18.10c,d shows the cases $\Delta H < 0$, $\Delta S > 0$ and $\Delta H < 0$, $\Delta S < 0$ also.

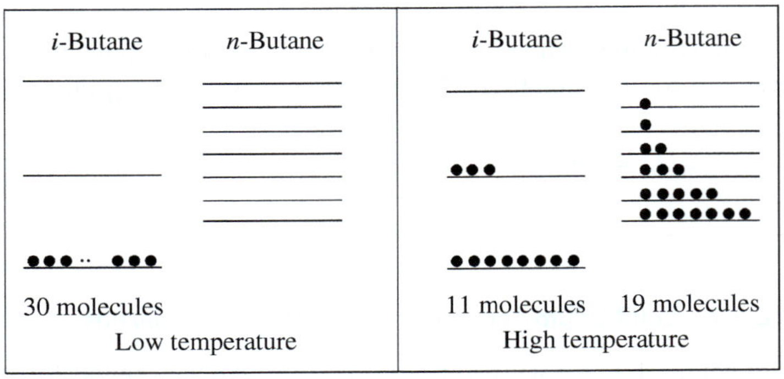

Figure 18.9 Schematic diagram of the electronic and vibrational energy levels of i-butane and n-butane. Population at low and high temperature. Low temperature: all molecules are in the i-form. High temperature: most molecules are in the n-form.

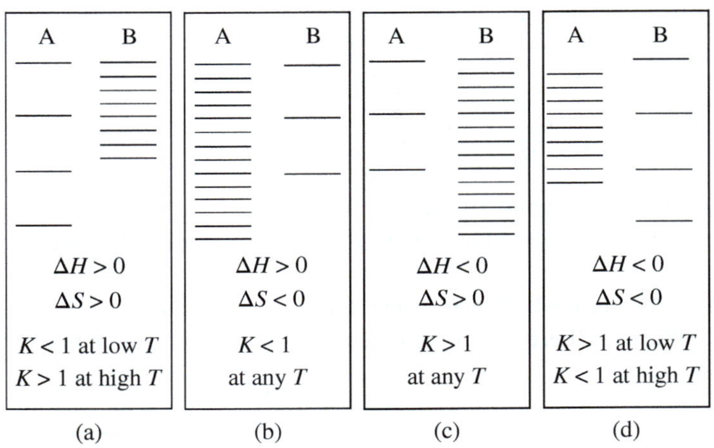

Figure 18.10 Equilibrium constant K of the equilibrium A $\rightleftharpoons$ B for A and B. Panel (a) shows the case where species A is more stable and species B has a higher density of state; panel (b) shows the case where species A is more stable and has a higher density of states; panel (c) shows the case where species B is more stable and has a higher density of states; and panel (d) shows the case where B is more stable and species A has a higher density of states.

Example 18.11 *Determine the equilibrium constant for the exchange reaction H_2+D_2 ⇌ 2 HD at 383 K and show how it depends on the masses of the reactant and product molecules.*

We start with Equation (18.53)

$$\ln K = \frac{-\Delta_r G^\ominus}{RT}, \quad \Delta_r G^\ominus = \Delta_r H^\ominus - T\Delta_r S^\ominus = \Delta_r U^\ominus + P^\ominus \Delta_r V - T\Delta_r S^\ominus$$

In our case the volume does not change during the reaction, thus

$$\Delta_r G^\ominus = \Delta_r U^\ominus - T\Delta_r S^\ominus$$

To determine $\Delta_r U^\ominus$ we consider how each degree of freedom contributes to it. The translational energy $\frac{3}{2}NkT$ and the rotational energy $\frac{2}{2}NkT$ do not change between the reactants and products, as long as the reaction is isothermal. In the approximation that the electronic energies do not change with the isotope, $\Delta_r U^\ominus$ is given by the difference in vibrational energies

$$\Delta_r U^\ominus = 2U^\ominus_{m,vib,HD} - U^\ominus_{m,vib,H_2} - U^\ominus_{m,vib,D_2}$$

In Chapter 15 we found that HD is in its vibrational ground state up to about 1000 K. Hence we approximate $U_{\mathrm{vib,HD}}$ by

$$U^\ominus_{m,vib,HD} = \mathbf{N}_A \frac{1}{2} h\nu_{0,HD} = \mathbf{N}_A \frac{1}{2} h \cdot \frac{1}{2\pi}\sqrt{\frac{k_f}{\mu_{\mathrm{HD}}}} = 26.30 \text{ kJ mol}^{-1}$$

where we have used $k_f = 574$ N m^{-1} for the force constant and $\mu_{\mathrm{HD}} = m_{\mathrm{H}} m_{\mathrm{D}}/(m_{\mathrm{H}} + m_{\mathrm{D}}) = 1.11 \times 10^{-27}$ kg for the reduced mass of HD. Assuming the same force constant in H_2 and D_2 we obtain the $U^\ominus_{m,vib,H_2}$, $U^\ominus_{m,vib,D_2}$, and $\Delta_r U^\ominus$ values in Table 18.3. The molar entropies $S^\ominus_m$ and the change $\Delta_r S^\ominus$ in Table 18.3 are calculated from the corresponding partition functions, as outlined in Chapter 17 (Table 17.2). The differences $\Delta_r U^\ominus$ and $\Delta_r S^\ominus$ are determined by the different masses of the reactants: $\Delta_r U^\ominus = 0.604$ kJ mol^{-1} and $\Delta_r S^\ominus = +12.09$ J K^{-1} mol^{-1}.

Using these results, we find that $\Delta_r G^\ominus = -4.03$ kJ mol^{-1} and $K = 3.54$ at 383 K. This result agrees with the experimental value [1] $K = 3.46$.

Table 18.3 Calculation of $\Delta_r U^\ominus$ and $\Delta_r S^\ominus$ for the Reaction $H_2 + D_2 \rightleftarrows$ 2HD at $T = 383$ K from Quantum Mechanical Data.

	$U^\ominus_{m,vib}$/(kJ mol^{-1})	$S^\ominus_m$/(J K^{-1} mol^{-1})
H_2	26.30	137.59
D_2	18.67	151.84
HD	22.79	150.76

In Foundation 18.3 the equilibrium constant K, for the H_2 +D_2 ⇌ 2 HD reaction, is expressed in terms of the masses and the vibrational frequency of the reactants, starting from Equation (18.86):

$$K = \frac{2\left(m_{\mathrm{H}} + m_{\mathrm{D}}\right)}{\sqrt{m_{\mathrm{H}} m_{\mathrm{D}}}} \cdot \exp\left[-\frac{h\nu_{0,\mathrm{HD}}}{kT}\left(1 - \frac{\sqrt{m_{\mathrm{H}}} + \sqrt{m_{\mathrm{D}}}}{\sqrt{2\left(m_{\mathrm{H}} + m_{\mathrm{D}}\right)}}\right)\right] \tag{18.91}$$

This calculation demonstrates that in simple cases the partition function for a substance can be calculated on the basis of quantum mechanics, without using thermochemical data. Such calculations provide insight into how the molecular structure and properties influence the chemical equilibrium. In problem P18.11 you calculate $K = 3.18$ for $T = 273$ K and $K = 3.45$ for $T = 383$ K from Equation (18.91) in accordance with the experimental data $K = 3.19$ and $K = 3.46$.

18.6 ΔG and Equilibrium in Dilute Solution

To this point our analysis has treated the reactants and products as ideal gases; however, this analysis may be extended to describe chemical equilibria in dilute solutions. If the solution is dilute enough the interactions between solute molecules can be neglected. Moreover, under these conditions the solute–solvent interactions do not depend on the concentration of the solute, even though the solute molecules may interact strongly with the solvent. In this way, the solute molecules in an ideal dilute solution are similar to the particles of an ideal gas.

18.6.1 Osmotic Pressure and Concentration

As an example we consider a dilute solution of sugar in water. The solution is in a container connected with a piston that allows water molecules to pass through, but not sugar molecules (Fig. 18.11a). The sugar solution becomes diluted as water molecules (solvent) diffuse through the membrane, but the upper container remains pure water.

Because a net flow of solvent molecules into the lower chamber occurs, the solution exerts a pressure (force) on the piston. This force can be counteracted by placing a weight on the piston to stop the influx of solvent. The pressure exerted on the piston is called the osmotic pressure ${}^{\mathrm{osm}}P$. If we decrease the weight, the piston is pushed up, and the solution is diluted further. This process takes place until the pressure exerted by the diluted solution corresponds to the smaller weight.

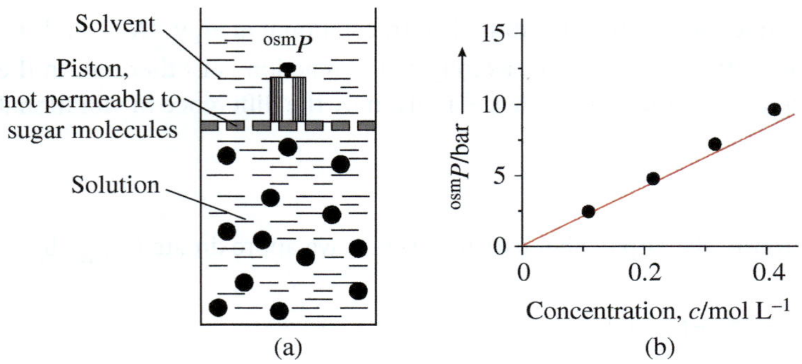

Figure 18.11 Osmotic pressure. (a) Experimental setup for measuring ${}^{\mathrm{osm}}P$ of a solution of sugar in water. The semipermeable piston is a membrane with small pore size, permeable for water and impermeable for the solute (e.g., a porous glass disk treated with Cu(II)-hexacyanoferrate(II)). (b) ${}^{\mathrm{osm}}P$ of a solution of sugar in water versus the concentration c (temperature $T = 298$ K). Solid line: calculated from Equation (18.98), dots: experiment [2].

We consider the solution, the solvent, and the semipermeable membrane as the system; and we consider the weight, the wall, and the temperature bath as the surroundings. When we increase the weight in Fig. 18.11a slightly, the piston moves downward by a distance ds. In this process, reversible work $-{}^{\mathrm{osm}}P \cdot \mathrm{d}V$ is done on the system. Because the process is isothermal

$$-{}^{\mathrm{osm}}P \cdot \mathrm{d}V = \mathrm{d}U - T \cdot \mathrm{d}S \tag{18.92}$$

Then we find that

$${}^{\mathrm{osm}}P = -\frac{\mathrm{d}U}{\mathrm{d}V} + T\frac{\mathrm{d}S}{\mathrm{d}V} \tag{18.93}$$

In the limit that the solute molecules are in low concentration, they do not interact with each other. The internal energy U does not change with the volume V, because i) the surroundings of a solvent molecule, when being shifted from the pure solvent to the dilute solution does not change significantly if the solution is sufficiently dilute, and ii) the interaction of the solvent with the solute is not changed if the solution is dilute enough; so that

$${}^{\mathrm{osm}}P = T\frac{\mathrm{d}S}{\mathrm{d}V} \quad (T = const) \tag{18.94}$$

Since ${}^{\mathrm{osm}}P$ and T are positive quantities, Equation (18.94) requires that the entropy S decreases (d$S < 0$) if the volume V decreases (d$V < 0$).

We can understand this entropy decrease by considering the number of representations Ω_V in the beginning and the number of representations $\Omega_{V+\Delta V}$ at the end of the process (where ΔV is smaller than zero),

$$\Delta S = k \ln \frac{\Omega_{V+\Delta V}}{\Omega_V}$$

The number of representations of a solvent molecule does not change for the same reasons that ΔU does not change, namely i) the surroundings of a solvent molecule, when being shifted from the pure solvent to the solution, is unchanged if the solution is sufficiently dilute, and ii) the interaction of the solute with the solvent remains unchanged if the solution is dilute enough. On the other hand, the number of representations of a solute molecule decreases for a decrease in volume. Thus, the ratio $\Omega_{V+\Delta V}/\Omega_V$ is given by the probability that all solute molecules in volume V appear in the volume $V + \Delta V$ (where ΔV is smaller than zero)

$$\frac{\Omega_{V+\Delta V}}{\Omega_V} = \mathcal{P} = \left(\frac{V + \Delta V}{V}\right)^N$$

where N is the number of molecules of the solute.

This ratio for the number of representations, $\Omega_{V+\Delta V}/\Omega_V$, is given by the same factor $\mathcal{P}$ as in the case of an ideal gas (see Section 17.3.1, Equation (17.34)). We should anticipate this correspondence from the similarity of the two models. For the ideal gas we ignore interactions between the gas molecules and assume that a gas molecule can occupy any part of the container's volume with equal probability. Correspondingly, we neglect interactions between solute molecules, we assume there are no changes in the solvent properties with solute concentration, and we assume that the solute molecule can occupy any part of the solvent volume with equal probability.

The change of entropy in the process in Fig. 18.11a is

$$\Delta S = k \ln \frac{\Omega_{V+\Delta V}}{\Omega_V} = k \ln \left(\frac{V+\Delta V}{V}\right)^N = Nk \ln \left(1 + \frac{\Delta V}{V}\right) \tag{18.95}$$

Now we can obtain a relation for the osmotic pressure. For the case when x is small enough $\ln(1+x) \approx x$; hence, for small volume changes ($\Delta V/V$ small) Equation (18.95) becomes

$$\Delta S \approx Nk \cdot \frac{\Delta V}{V}$$

For an infinitesimally small change, we find that

$$dS = Nk \cdot \frac{dV}{V} \tag{18.96}$$

or

$$\frac{dS}{dV} = Nk \cdot \frac{1}{V}$$

Combining this result with Equation (18.94), we find

$${}^{\text{osm}}P = T\frac{dS}{dV} = NkT\frac{1}{V} = \boldsymbol{R}T\frac{\boldsymbol{n}}{V} \tag{18.97}$$

or

$$\boxed{{}^{\text{osm}}P = \boldsymbol{R}T \cdot c \quad \text{with} \quad c = \frac{\boldsymbol{n}}{V}} \tag{18.98}$$

(c = concentration of the sugar solution). The osmotic pressure increases linearly with the concentration c (Fig. 18.11b, solid line). Equation (18.97) is analogous to the ideal gas equation of state, in which the pressure P of the gas is replaced by the osmotic pressure ${}^{\text{osm}}P$. Equation (18.98) is called the *van't Hoff equation for the osmotic pressure of dilute solutions*.

Example 18.12 *Calculate the osmotic pressure for an aqueous solution of sucrose with a concentration of* $c = 0.1$ *mol* L^{-1} *at 298 K.*

Using Equation (18.98) and $\boldsymbol{R} = 0.08314$ bar LK^{-1} mol^{-1}, we obtain

$${}^{\text{osm}}P = 2.48 \text{ bar}$$

This value is in good agreement with the experimental value 2.63 bar. Figure 18.11b compares calculated values (solid line) to the experimental ones (filled circles) up to a concentration of 0.4 mol L^{-1}. At higher concentration, however, the agreement becomes worse. We will discuss this fact in Chapter 21.

18.6.1.1 Reversible Change of Concentration

By changing the weight on the piston in the arrangement in Fig. 18.11 we can concentrate or dilute the solution reversibly, in a similar way as for the compression or expansion of a gas. This result allows us to extend the considerations on gases to dilute solutions. What we must do is substitute the osmotic pressure for the gas pressure.

The isothermal reversible work w_{rev} to concentrate a sugar solution from a concentration c_1 to a concentration c_2 is

$$w_{\text{rev}} = \boldsymbol{nR}T \cdot \ln\left(\frac{{}^{\text{osm}}P_2}{{}^{\text{osm}}P_1}\right) = \boldsymbol{nR}T \cdot \ln \frac{c_2}{c_1} \tag{18.99}$$

Because $\Delta U = 0$ (the solvation energy of the solute molecules is the same in the starting and ending states), the reversible heat transfer is

$$q_{\text{rev}} = -\boldsymbol{nR}T \cdot \ln \frac{c_2}{c_1} \tag{18.100}$$

and the entropy change is

$$\Delta S = \frac{q_{\text{rev}}}{T} = -\boldsymbol{nR} \cdot \ln \frac{c_2}{c_1} \tag{18.101}$$

Equation (18.101) can also be derived directly from Equation (18.96):

$$\Delta S = \int_{V_1}^{V_2} Nk \cdot \frac{\mathrm{d}V}{V} = Nk \cdot \ln \frac{V_2}{V_1} = -\boldsymbol{nR} \cdot \ln \frac{c_2}{c_1}$$

18.6.2 Depression of Vapor Pressure (Raoult's Law)

Consider a dilute solution of a nonvolatile substance, say sucrose in water. If a part of the solvent evaporates, then the dissolved molecules are confined to a smaller volume; the concentration increases during evaporation of the solvent. As a consequence, it is more difficult to evaporate the solution than the pure solvent. The vapor pressure P above the solution is lower than the vapor pressure P_s of the solvent. This lowering of the vapor pressure may be intuited by considering a container of the pure solvent in equilibrium with its vapor pressure P_s. Now imagine adding a nonvolatile solute to the liquid. The entropy of the solution will increase if the molecules of the solute could occupy a larger space. Hence some of the solvent vapor condenses into the solution and increases the solution's entropy at the expense of decreasing the entropy of the vapor. This process proceeds until an equilibrium is reached at a lower vapor pressure P.

18.6.2.1 Nonvolatile Solute and Volatile Solvent

How does the vapor pressure P above the solution depend on the concentration of the dissolved species? In Box 18.2 we use a thought experiment and our results for the osmotic pressure to show that

$$\frac{P_s - P}{P_s} = x \quad \text{or} \quad \frac{P}{P_s} = 1 - x = x_s \tag{18.102}$$

where P_s is the vapor pressure of the pure solvent and x is the mole fraction of the dissolved species

$$x = \frac{\boldsymbol{n}}{\boldsymbol{n} + \boldsymbol{n}_s} \approx \frac{\boldsymbol{n}}{\boldsymbol{n}_s}, \qquad x_s = \frac{\boldsymbol{n}_s}{\boldsymbol{n} + \boldsymbol{n}_s} = 1 - x \tag{18.103}$$

($\boldsymbol{n}$ = amount of substance of the dissolved species, $\boldsymbol{n}_s$ = amount of substance of the solvent). Equation (18.102) is called *Raoult's law*.

Example 18.13 *What is the vapor pressure of a* 0.2 *M sucrose solution in water at* 273 *K?*

The vapor pressure P_s of pure water at 273 K is 0.0061 bar. For a dilute solution of sucrose in water (0.2 mol of sucrose dissolved in 1000 mL of water) we obtain

$$x = \frac{0.2 \text{ mol}}{(55.5 + 0.2) \text{ mol}} = 0.0036, \quad x_s = 0.996$$

Then $P = 0.996 \cdot P_s = 0.00607$ bar, this is a change by 0.4%. The experimental value is $P = 0.00608$ bar.

18.6.2.2 Two Volatile Components

According to Equation (18.102) the vapor pressure P above a solution is

$$P = x_s P_s$$

A binary solution of volatile components that follows Raoult's law is called an *ideal binary solution*. In this case the two components A and B do not interact with each other, so that the vapor pressure P of the solution is

$$P = x_{\text{A}} P_{\text{A,pure}} + x_{\text{B}} P_{\text{B,pure}} \tag{18.104}$$

Figure 18.12 shows vapor pressure data for ethylene bromide (A) and dibromopropane (B), for which the total vapor pressure of the solution changes linearly with the mole fraction of ethylene bromide.

For an ideal solution, the composition of the vapor is determined in a direct way by the composition of the liquid. Using the fact that $x_B = 1 - x_A$, we find

$$P = x_A P_{A,pure} + (1 - x_A) P_{B,pure} \tag{18.105}$$

We can solve this equation for x_A and obtain

$$x_A = \frac{P - P_{B,pure}}{P_{A,pure} - P_{B,pure}}$$

From this formula we can see that measurement of the total vapor pressure P (and knowledge of the vapor pressures for the pure components) can be used to determine the mole fraction of the two components.

From Equation (18.105), it is evident that the vapor phase will have a larger mole fraction of the more volatile component than does the condensed phase. This behavior is in accord with intuition, and such enrichment in the vapor phase is the physical premise upon which the purification of substances by distillation is based.

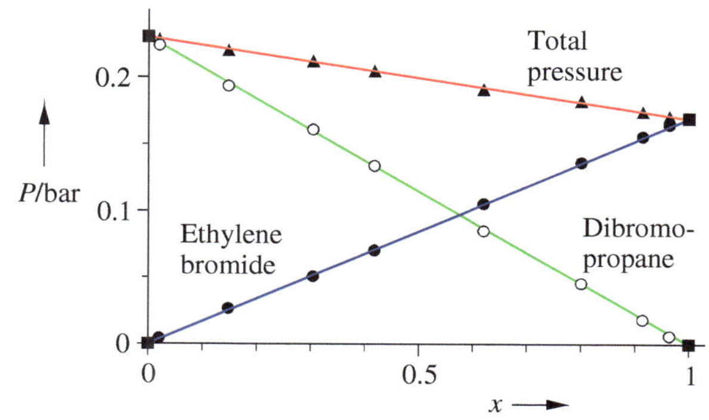

Figure 18.12 The vapor pressure of an ethylene bromide/dibromopropane mixture at $T = 358$ K is plotted versus the mole fraction x of ethylene bromide. The total vapor pressure (filled triangles), ethylene bromide partial pressure (filled circles), and dibromopropane partial pressure (open circles) are shown. (Data taken from International Critical Tables III, page 287 (1923).)

Example 18.14 *Given that the vapor pressure of pure ethylene bromide is 172.6 mbar and the vapor pressure of pure dibromopropane is 127.2 mbar at 358.2 K, calculate the mole fraction of each component in the solution when the total pressure is* $P = 150.0$ *mbar.*

Because the mixture behaves ideally, we can write

$$x_A = \frac{P - P_{B,pure}}{P_{A,pure} - P_{B,pure}} = \frac{(150.0 - 127.2)\ \text{mbar}}{(172.6 - 127.2)\ \text{mbar}} = 0.502$$

and hence $x_B = 1 - 0.502 = 0.498$.

18.6.3 Elevation of Boiling Point and Depression of Melting Point

The elevation of the boiling point and the depression of the freezing point of the solution as compared to the pure solvent are immediate consequences of the depression of the vapor pressure (Fig. 18.13). We presume that crystallization of the solvent takes place when decreasing the temperature of the solution below the freezing point, which is generally the case for dilute solutions.

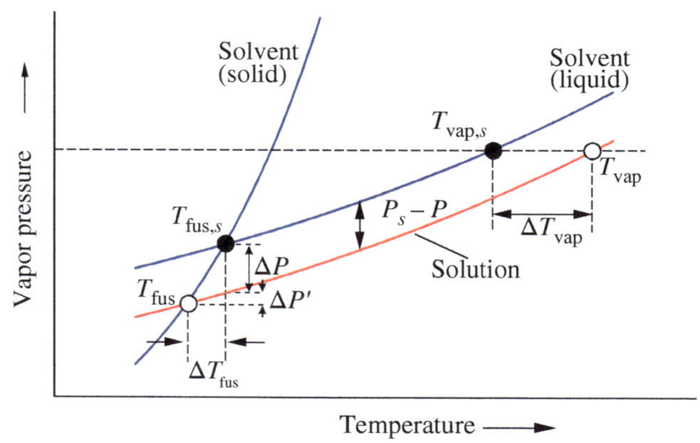

Figure 18.13 Depression ($P_s - P$) of vapor pressure, elevation ΔT_{vap} of boiling point, and depression ΔT_{fus} of melting point. Note that $\Delta T_{vap} = T_{vap} - T_{vap,s}$ is positive, while $\Delta T_{fus} = T_{fus} - T_{fus,s}$ is negative.

18.6.3.1 Elevation of Boiling Point

From Fig. 18.13, using Raoult's law, Equation (18.102), and the Clausius–Clapeyron equation, Equation (18.38), we calculate the elevation of boiling point, $\Delta T_{vap} = T_{vap} - T_{vap,s}$ (where T_{vap} is the boiling point of the solution and $T_{vap,s}$ is the boiling point of the pure solvent), as a function of mole fraction x of the solute.

From Fig. 18.13 we wish to obtain the temperature T_{vap} at which the vapor pressure of the solution is equal to that for the pure solvent at its boiling point $T_{\text{vap},s}$. If the temperature shift is small enough that the vapor pressure curve is approximately linear, we can write that

$$P_s - P = \frac{dP}{dT} \cdot \left(T_{\text{vap}} - T_{\text{vap},s}\right) \tag{18.106}$$

where P_s is the vapor pressure of the pure solvent and P is the vapor pressure of the solution (both at the boiling temperature of the pure solvent) and dP/dT is the slope of the vapor pressure/temperature curve. Rearranging this result, we find that

$$\frac{\Delta P}{\Delta T_{\text{vap}}} = \frac{dP}{dT} \tag{18.107}$$

where $\Delta P = P_s - P$ and $\Delta T_{\text{vap}} = T_{\text{vap}} - T_{\text{vap},s}$. According to the Clausius–Clapeyron equation,

$$\frac{d \ln \hat{P}}{dT} = \frac{\Delta_{\text{vap}} H}{RT^2} \quad \text{or} \quad \frac{dP}{dT} = P \frac{\Delta_{\text{vap}} H}{RT^2} \tag{18.108}$$

so that

$$\frac{\Delta P}{\Delta T_{\text{vap}}} = P \frac{\Delta_{\text{vap}} H}{RT^2} \tag{18.109}$$

Using Raoult's law,

$$\Delta P = P_s \cdot x \tag{18.110}$$

we can write

$$\frac{P_s \cdot x}{\Delta T_{\text{vap}}} = P \frac{\Delta_{\text{vap}} H}{RT^2} \tag{18.111}$$

or

$$\Delta T_{\text{vap}} = \frac{RT^2}{\Delta_{\text{vap}} H} \cdot \frac{P_s}{P} x \tag{18.112}$$

If ΔP is small enough, then $P_s/P \approx 1$, and we obtain

$$\boxed{\Delta T_{\text{vap}} = \frac{RT^2}{\Delta_{\text{vap}} H} \cdot x} \tag{18.113}$$

18.6.3.2 Depression of Melting point

In a similar way, we can find the depression of the melting point (see Fig. 18.13) as a function of mole fraction x of the solute. In Problem P18.11 you show that

$$\boxed{\Delta T_{\text{fus}} = -\frac{RT^2}{\Delta_{\text{fus}} H} \cdot x} \tag{18.114}$$

Example 18.15 *What is the elevation of the boiling point and the depression of the melting point for a 0.2 M solution of sucrose in water?*

We consider a dilute solution of sucrose in water (0.2 mol of sucrose in 1000 mL of water), corresponding to a mole fraction of

$$x = \frac{0.2 \text{ mol}}{(55.5 + 0.2)} = 0.0036$$

For water the enthalpy of fusion and the enthalpy of vaporization are $\Delta_{\text{fus}} H = 6.0 \text{ kJ mol}^{-1}$ and $\Delta_{\text{vap}} H = 44.1 \text{ kJ mol}^{-1}$. Then, according to Equation (18.113), for the change of the boiling point with $T = 373$ K we obtain

$$\Delta T_{\text{vap}} = \frac{RT^2}{\Delta_{\text{vap}} H} \cdot x = \frac{8.314 \text{ J K}^{-1} \text{ mol}^{-1} \cdot 1.39 \times 10^5 \text{ K}^2}{44.1 \text{ kJ mol}^{-1}} \cdot x$$
$$= 26.2 \text{ K} \cdot x = 0.094 \text{ K}$$

(elevation of boiling point, experimental value $\Delta T_{\text{vap}} = 0.098$ K) and the change of the melting point with $T = 273$ K is, according to Equation (18.114),

$$\Delta T_{\text{fus}} = \frac{-\boldsymbol{R}T^2}{\Delta_{\text{fus}}H} \cdot x = \frac{-8.314 \text{ J K}^{-1} \text{ mol}^{-1} \cdot 7.45 \times 10^4 \text{ K}^{\ 2}}{6.0 \text{ kJ mol}^{-1}} \cdot x$$
$$= -103.2 \text{ K} \ \cdot x = -0.37 \text{ K}$$

(depression of melting point, experimental value $\Delta T_{\text{fus}} = -0.38$ K).

Note that the effect on the melting point is much larger than that on the boiling point, because $\Delta_{\text{fus}}H$ is much smaller than $\Delta_{\text{vap}}H$.

Note that Equations (18.113) and (18.114) are restricted to low enough concentrations of the solute. Solutions of higher concentration are considered in Chapter 21.

18.6.4 Mass Action Law (Solutions of Neutral Particles)

In analogy to the gas reaction (18.43) we consider a chemical reaction

$$\alpha\text{A} + \beta\text{B} + \ldots \rightarrow \gamma\text{C} + \delta\text{D} + \ldots \tag{18.115}$$

at constant temperature in aqueous solution. How can we perform this reaction reversibly?

To answer this question we assume that before and after the reaction the reacting species are in aqueous solution at standard concentrations $c^{\ominus} = 1$ mol L^{-1} in separate reservoirs. We take certain amounts $\boldsymbol{n}_{\text{A}}$, $\boldsymbol{n}_{\text{B}}$,... of species A, B, ... of the starting materials from their reservoirs and transfer them reversibly to an equilibrium container (equilibrium concentrations c_{A}, c_{B}, ...). After reaction the products C, D, ... are reversibly transferred to the corresponding reservoirs. In analogy to a gas reaction we obtain

$$\Delta_r G^{\ominus} = -\boldsymbol{R}T \cdot \ln K, \qquad \Delta_r G = \Delta_r G^{\ominus} + \boldsymbol{R}T \cdot \ln Q \tag{18.116}$$

with

$$K = \frac{\hat{c}^{\gamma}_{\text{C,eq}} \cdot \hat{c}^{\delta}_{\text{D,eq}}}{\hat{c}^{\alpha}_{\text{A,eq}} \cdot \hat{c}^{\beta}_{\text{B,eq}}} \qquad Q = \frac{\hat{c}^{\gamma}_{\text{C}} \cdot \hat{c}^{\delta}_{\text{D}}}{\hat{c}^{\alpha}_{\text{A}} \cdot \hat{c}^{\beta}_{\text{B}}} \tag{18.117}$$

In analogy to our definition of a dimensionless pressure $\hat{P}$ for the gas reaction, we define a dimensionless concentration $\hat{c}$ by

$$\hat{c} = \frac{c}{c^{\ominus}} \tag{18.118}$$

where $c^{\ominus}$ is the concentration in the standard state. All species in the beginning and at the end of the process are separated and at concentrations of $c^{\ominus} = 1$ mol L^{-1}. $\Delta_r G^{\ominus}$ is the change in Gibbs energy when α mol of species A and β mol of species B react to form γ mol of species C and δ mol of species D.

These relations are only valid if the mutual interactions of the solute molecules can be neglected – that is, for small enough concentrations of the dissolved species. This treatment will be extended in Chapter 21 for concentrated solutions.

Example 18.16 *Iodine forms a complex with dioxane in a solution of n-heptane. The equilibrium constant at $T = 298$ K is measured by monitoring the electronic absorbance of the complex and is found to be*

$$K = \frac{\hat{c}_{\text{complex}}}{\hat{c}_{I_2} \cdot \hat{c}_{\text{dioxane}}} = 9.3$$

What is the change of Gibbs energy for this reaction? If the initial iodine concentration is 0.01 M and the initial dioxane concentration is 0.1 M, what is the concentration of the complex at equilibrium?

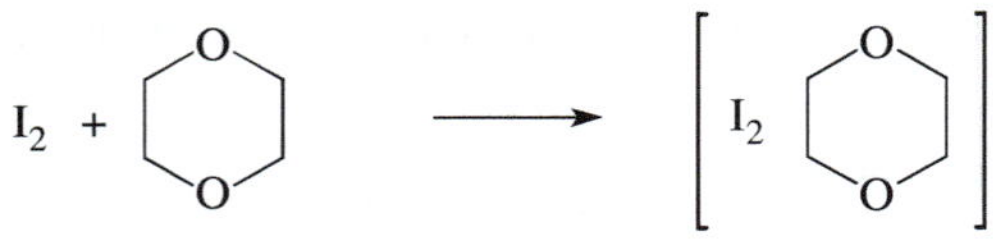

Using Equation (18.116), the change of Gibbs energy is directly calculated

$$\Delta_r G^{\ominus}(298\ \text{K}) = -(8.314) \cdot 298 \cdot \ln(9.3)\ \text{J mol}^{-1} = -5.5\ \text{kJ mol}^{-1}$$

We define the concentration of the complex as $x = \widehat{c}_{\text{complex}}$. From the reaction stoichiometry, the concentrations of iodine and dioxane at equilibrium are

$$\widehat{c}_{I_2} = 0.01 - x \quad \text{and} \quad \widehat{c}_{\text{dioxane}} = 0.1 - x$$

Hence the equilibrium constant is

$$K = \frac{x}{(0.01 - x) \cdot (0.1 - x)} = 9.3$$

which gives a quadratic equation for x

$$x^2 - x\left(0.11 + \frac{1}{9.3}\right) + 0.001 = 0$$

Using the value given for K we find two roots for x (0.00470 and 0.213) of which $x = \widehat{c}_{\text{complex}} = 0.00470$ is the physically realizable one; that is, about one-half of the I_2 is bound in the complex. (Note: we exclude the value $x = 0.213$ because it would exceed the total initial concentration of the complex and of the iodine.)

18.6.5 Mass Action Law (Solutions of Charged Particles)

Let us discuss the reaction (HAc = acetic acid, NaAc = sodium acetate)

$$\text{HAc(aq)} + \text{NaCl(aq)} \rightarrow \text{NaAc(aq)} + \text{HCl(aq)} \qquad (18.119)$$

where the subscript aq indicates that the corresponding species are dissolved in water. In this case NaCl, NaAc, and HCl are strong electrolytes and dissociate into ions. Because the concentrations of Na^+ and Cl^- do not change in the reaction, we can write a net reaction for (18.119) as

$$\text{HAc} \rightarrow \text{Ac}^- + \text{H}^+ \qquad (18.120)$$

Can we still apply the mass action law,

$$K = \frac{\widehat{c}_{\text{H}^+,\text{eq}} \cdot \widehat{c}_{\text{Ac}^-,\text{eq}}}{\widehat{c}_{\text{HAc,eq}}} \qquad (18.121)$$

to reactions of ions in solution? The mass action law applies and it can be shown directly using the formal, Gibbs energy, approach. Because of the condition of electroneutrality, our thought experiment for placing each species in a separate container will not work, however. Nevertheless, the thought experiment can be modified, although a tricky operation is needed for the proof (See Foundation 18.4).

18.6.6 Gibbs Energy of Formation in Aqueous Solution

In Section 18.3.1 we defined the Gibbs energy of formation with respect to the Gibbs energy of the elements in their standard state. Although this choice is convenient for gas reactions, this choice of a reference state is not convenient for reactions in solution because it would require evaluating the solvation energy of the species by the solvent. For reactions in solution, it is convenient to define the Gibbs energy of formation with respect to a reference state of the solutes (reactants and products) at a particular concentration.

18.6.6.1 Solutions of Neutral Particles

Here we define a Gibbs energy of formation in aqueous solution of each species, $\Delta_f G^{\ominus}_{\text{aq}}$ (species separated, in water at $c^{\ominus} = 1\ \text{mol L}^{-1}$).

$\Delta_f G_{\text{aq}}^{\ominus}$ is obtained from $\Delta_f G^{\ominus}$ by adding the molar Gibbs energy required to transfer the species from a reservoir with the pure substance into an aqueous solution of concentration $c^{\ominus} = 1$ mol L^{-1} (Box 18.3). Values of $\Delta_f G_{\text{aq}}^{\ominus}$ are listed in Appendix A.1 for selected compounds.

Example 18.17 *Calculate $\Delta_r G_{\text{aq}}^{\ominus}$ for the reaction*

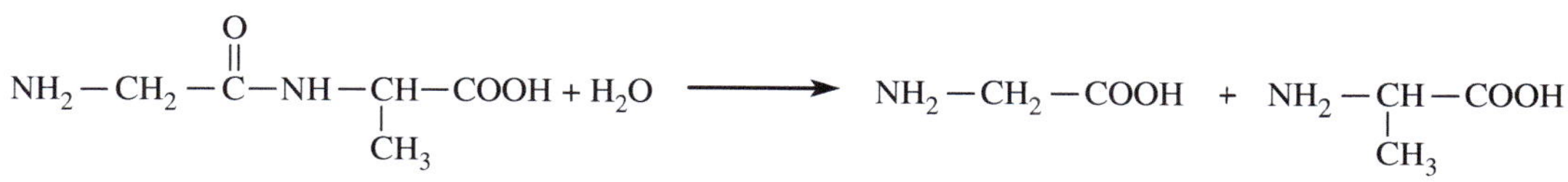

Glycylalanine — Glycine — Alanine

in aqueous solution.

From Appendix A.4 we obtain

$$\Delta_r G_{\text{aq}}^{\ominus} = (-371.3 - 379.9 + 733.9)\ \text{kJ mol}^{-1} = -17.3\ \text{kJ mol}^{-1}$$

The hydrolysis of glycylalanine under standard conditions and at $T = 298$ K is a spontaneous reaction ($K = 1.1 \times 10^3$). Actually, this reaction is kinetically hindered. This fact is of general importance since proteins, chains of amino acids linked by peptide bonds, would otherwise hydrolyze spontaneously.

18.6.6.2 Solutions of Charged Particles

This case requires special attention, since a solution of isolated ions cannot be prepared. In spite of this we can define values of $\Delta_f G_{\text{aq}}^{\ominus}$ for each ionic species in the following sense.

Consider the solution of a salt like NaCl. $\Delta_f G_{\text{aq}}^{\ominus}$ is obtained from $\Delta_f G^{\ominus}$ of crystalline NaCl as described in Foundation 18.4. $\Delta_f G_{\text{aq}}^{\ominus}$ can be considered as the sum of contributions of the corresponding ions:

$$\Delta_f G_{\text{aq}}^{\ominus}(\text{NaCl}) = \Delta_f G_{\text{aq}}^{\ominus}(\text{Na}^+) + \Delta_f G_{\text{aq}}^{\ominus}(\text{Cl}^-) \tag{18.122}$$

Similarly for HCl, we find

$$\Delta_f G_{\text{aq}}^{\ominus}(\text{HCl}) = \Delta_f G_{\text{aq}}^{\ominus}(\text{H}^+) + \Delta_f G_{\text{aq}}^{\ominus}(\text{Cl}^-) \tag{18.123}$$

It is useful to have a Gibbs energy for individual ions, however. To this end we define the Gibbs energies of formation of single ions by arbitrarily setting the Gibbs energy of the aqueous proton to zero, i.e.,

$$\boxed{\Delta_f G_{\text{aq}}^{\ominus}(\text{H}^+) = 0} \tag{18.124}$$

Then we obtain

$$\Delta_f G_{\text{aq}}^{\ominus}(\text{Cl}^-) = \Delta_f G_{\text{aq}}^{\ominus}(\text{HCl}), \quad \Delta_f G_{\text{aq}}^{\ominus}(\text{Na}^+) = \Delta_f G_{\text{aq}}^{\ominus}(\text{NaCl}) - \Delta_f G_{\text{aq}}^{\ominus}(\text{HCl}) \tag{18.125}$$

The values of $\Delta_f G_{\text{aq}}^{\ominus}$ that are listed in Appendix A.1 for selected ions are obtained in this way. This allows us to calculate $\Delta_r G_{\text{aq}}^{\ominus}$ and thus, in analogy to Equation (18.71),

$$\Delta_r G_{\text{aq}}^{\ominus} = -\boldsymbol{R}T \cdot \ln K \tag{18.126}$$

Example 18.18 *Use the values of $\Delta_f G_{\text{aq}}^{\ominus}$ to determine $\Delta_r G_{\text{aq}}^{\ominus}$ and K for the oxidation of the Fe^{2+} ion by the Cu^{2+} ion.*

$$Fe^{2+} + Cu^{2+} \rightarrow Fe^{3+} + Cu^{+}$$

From Appendix A.1 we obtain

$$\Delta_r G_{\text{aq}}^{\ominus} = (50.2 - 10.5 + 84.9 - 65)\ \text{kJ mol}^{-1} = 59.6\ \text{kJ mol}^{-1}$$

so that the equilibrium constant is

$$\ln K = -\frac{\Delta_r G^{\ominus}_{\text{aq}}}{RT} = -24.1, \quad K = \frac{\hat{c}_{\text{Fe}^{3+}} \cdot \hat{c}_{\text{Cu}^{1+}}}{\hat{c}_{\text{Fe}^{2+}} \cdot \hat{c}_{\text{Cu}^{2+}}} = 3.5 \times 10^{-11}$$

This means that in a solution of Fe^{2+} and Cu^{2+} a tiny amount of Fe^{3+} and Cu^{1+} are present, formed by electron exchange. For $c_{\text{Fe}^{2+}} = c_{\text{Cu}^{2+}} = 1$ mol L^{-1} we obtain

$$c_{\text{Fe}^{3+}} = c_{\text{Cu}^{1+}} = \sqrt{3.5 \times 10^{-11}} \text{ mol L}^{-1} = 6 \times 10^{-6} \text{ mol L}^{-1}$$

18.6.7 Part of Reactants or Products in Condensed or Gaseous State

We have seen in Section 18.5.6 that immiscible condensed species do not appear in the expression for the equilibrium constant for gas reactions. For the same reason condensed species do not appear in the expression for the equilibrium constant for reactions in solution, for example:

$$\text{AgCl (solid)} \rightarrow \text{Ag}^+\text{(aq)} + \text{Cl}^-\text{ (aq)} \qquad K = \hat{c}_{\text{Ag}^+,\text{eq}} \cdot \hat{c}_{\text{Cl}^-,\text{eq}}$$

$$\text{H}_2\text{O (liquid)} \rightarrow \text{H}^+\text{(aq)} + \text{OH}^-\text{ (aq)} \qquad K = \hat{c}_{H^+,\text{eq}} \cdot \hat{c}_{OH^-,\text{eq}}$$

$$\text{Zn (solid)} + \text{Cu}^{2+}\text{ (aq)} \rightarrow \text{Zn}^{2+}\text{ (aq)} + \text{Cu (solid)} \qquad K = \frac{\hat{c}_{Zn^{2+},\text{eq}}}{\hat{c}_{\text{Cu}^{2+},\text{eq}}}$$

$$\text{Hg (liquid)} + 2\,\text{Ag}^+\text{ (aq)} \rightarrow \text{Hg}^{2+}\text{ (aq)} + 2\,\text{Ag (solid)} \qquad K = \frac{\hat{c}_{Hg^{2+},\text{eq}}}{\left(\hat{c}_{\text{Ag}^+,\text{eq}}\right)^2}$$

Example 18.19 *Does Zn dissolve in an acidic medium?*

We consider the reaction

$$\text{Zn (solid)} + 2\,\text{H}^+\text{ (aq)} \rightarrow \text{Zn}^{2+}\text{(aq)} + \text{H}_2\text{ (gas)}$$

According to Appendix A.1 at 298 K we have

$$\Delta_f G^{\ominus}(\text{Zn}) = 0, \quad \Delta_f G^{\ominus}(\text{H}_2) = 0, \quad \Delta_f G^{\ominus}_{\text{aq}}(\text{H}^+) = 0$$

Then we obtain

$$\Delta_r G^{\ominus} = \Delta_f G^{\ominus}_{\text{aq}}(\text{Zn}^{2+}) = -147.2 \text{ kJ mol}^{-1}$$

$\Delta_r G^{\ominus}$ is negative under standard conditions; that is, Zn dissolves in acids at a concentration $c^{\ominus} = 1$ mol L^{-1} and at an external pressure $P^{\ominus} = 1$ bar. The equilibrium constant is

$$K = \frac{\hat{c}_{Zn^{2+},\text{eq}} \cdot \hat{P}_{H_2,\text{eq}}}{\hat{c}^2_{H^+,\text{eq}}}, \qquad \text{where} \qquad \ln K = -\frac{\Delta_r G^{\ominus}}{RT} = 59.3, \qquad K = 6 \times 10^{25}$$

18.7 Conclusion

The application of entropy ideas to the behavior of chemical reactions and chemical equilibria enables us to predict how a chemical system evolves to its final, or equilibrium, state. We proceed by defining two new energy functions ($A = U - TS$ and $G = H - TS$). The change in these energy functions corresponds to the reversible work between the initial and final states. For a reaction at constant volume and constant temperature ΔA equals w_{rev}, and for a reaction at constant pressure and constant temperature ΔG equals w_{rev}, diminished by the pressure volume work, $-P\Delta V$.

The Gibbs energy G is particularly useful because many natural and industrial chemical processes occur under fixed pressure and/or temperature conditions. We showed how the change in Gibbs energy can be used to predict the direction in which a system will evolve spontaneously, and we applied it to both physical transformations (phase transitions) and chemical transformations (reactions and equilibria). In particular, we derived the law of mass action. Under equilibrium conditions, $\Delta G = 0$, we showed that the standard Gibbs energy change $\Delta G^{\ominus}$ determines the equilibrium constant. We investigated the dependence of the equilibrium constant on the temperature and discussed its relation to the energy level structure of the molecules. We derived the Clausius–Clapeyron equation; and, using Raoult's law, we calculated the depression of the vapor pressure above a solution, the elevation of the boiling temperature, and the depression of the freezing temperature.

In this chapter we restricted our considerations to ideal gases and ideally diluted solutions. In the following chapters we will generalize these findings to real gases and real solutions.

18.8 Box 18.1. Temperature Dependence of ΔG

In the general case

$$\Delta G = \Delta H - T \cdot \Delta S$$

so that the temperature derivative of ΔG at constant pressure is

$$\left(\frac{\partial \Delta G}{\partial T}\right)_P = \left(\frac{\partial \Delta H}{\partial T}\right)_P - T \cdot \left(\frac{\partial \Delta S}{\partial T}\right)_P - \Delta S \tag{18B1.1}$$

where we have used the chain rule. From Section 16.5.3 we know that

$$\Delta H(T_2) = \Delta H(T_1) + \int_{T_1}^{T_2} \Delta C_P \, \mathrm{d}T, \text{ thus } \left(\frac{\partial \Delta H}{\partial T}\right)_P = \Delta C_P \tag{18B1.2}$$

On the other hand, from Equation (17.70) in Section 17.6.1 the entropy change for heating at constant pressure is

$$\mathrm{d}S = \frac{C_P}{T} \, \mathrm{d}T$$

and we obtain

$$\left(\frac{\partial \Delta S}{\partial T}\right)_P = \frac{\Delta C_P}{T} \tag{18B1.3}$$

Combining these results Equation (18B1.1) becomes

$$\left(\frac{\partial \Delta G}{\partial T}\right)_P = \Delta C_P - T \cdot \frac{\Delta C_P}{T} - \Delta S = -\Delta S \tag{18B1.4}$$

18.9 Box 18.2. Raoult's Law

We imagine a container A with a dilute solution of concentration c and a container B with the pure solvent (Fig. B18.1). We transfer a small amount of solvent ($\mathrm{d}n_s$) reversibly at constant temperature T from A to B by pressing the solvent through a semipermeable membrane. The reversible work for this process is

$${}^{\mathrm{osm}}P \cdot \mathrm{d}V = {}^{\mathrm{osm}}P \cdot V_{s,m} \cdot \mathrm{d}\boldsymbol{n}_s$$

($V_{s,m}$ = molar volume of the solvent). Next we transfer the same amount of vapor reversibly from B to A. The reversible work for this process is

$$\mathrm{d}\boldsymbol{n}_s \cdot \boldsymbol{R}T \cdot \ln \frac{P}{P_s}$$

Both processes together constitute a cyclic process, thus

$${}^{\mathrm{osm}}P \cdot V_{s,m} \cdot \mathrm{d}\boldsymbol{n}_s = -\mathrm{d}\boldsymbol{n}_s \cdot \boldsymbol{R}T \cdot \ln \frac{P}{P_s}$$

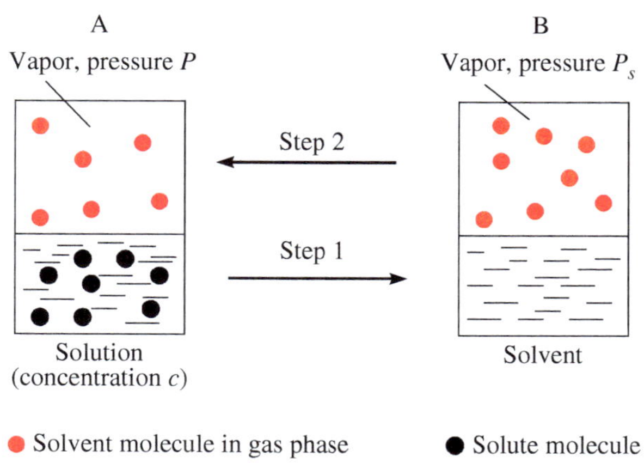

Figure B18.1 How to calculate the vapor pressure above a solution. Solvent from the solution (concentration c) in container A is reversible transferred to the solvent in container B (step 1). Vapor from container B (vapor pressure P_s of the pure solvent) is reversibly transferred to container A, where the vapor is under pressure P (step 2).

Because of

$$^{\text{osm}}P = c \cdot \boldsymbol{R}T$$

and

$$\ln \frac{P}{P_s} = \ln \left(1 - \frac{P_s}{P_s} + \frac{P}{P_s}\right) = \ln \left(1 - \frac{P_s - P}{P_s}\right) \approx -\frac{P_s - P}{P_s}$$

we obtain

$$\frac{P_s - P}{P_s} = V_{s,m} \cdot c \quad \text{or} \quad \frac{P}{P_s} = 1 - V_{s,m} \cdot c$$

For dilute solutions the mole fraction x and the concentration c are

$$x = \frac{\boldsymbol{n}}{\boldsymbol{n} + \boldsymbol{n}_s} \approx \frac{\boldsymbol{n}}{\boldsymbol{n}_s}, \quad c = \frac{\boldsymbol{n}}{V_{s,m} \cdot \boldsymbol{n}_s}$$

Then

$$c = \frac{x}{V_{s,m}}$$

and we obtain *Raoult's law*

$$\frac{P_s - P}{P_s} = x, \quad \frac{P}{P_s} = 1 - x$$

18.10 Box 18.3. How to Obtain $\Delta_f G^{\ominus}_{\text{aq}}$ from $\Delta_f G^{\ominus}$

To calculate the Gibbs energy for dissolving a given species in water we imagine a container filled with a saturated solution of that species (saturation concentration c, Fig. B18.2). We add a certain amount of the pure species to this solution; because the solution is saturated, the additional amount remains as a precipitate on the bottom, and no work is performed on the system in this process.

Now we apply a semipermeable piston with a weight corresponding to the osmotic pressure of the saturated solution and then add water. The water moves through the membrane into the solution, and the piston is slowly moved up until the precipitate has dissolved ($\Delta G = 0$). Finally, the saturated solution is reversibly diluted to the standard concentration $c^{\ominus}$. For this step we have

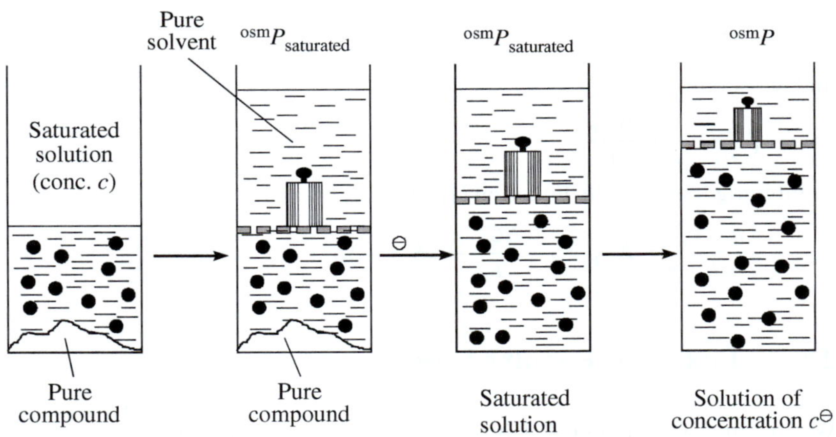

Figure B18.2 Isothermal reversible dissolution of a pure solid compound in a solvent. First, the pure compound is added to a saturated solution of concentration c. Finally, a solution of concentration $c^{\ominus}$ is obtained.

$$\Delta G = \boldsymbol{nR}T \cdot \ln \frac{c^{\ominus}}{c} = -\boldsymbol{nR}T \cdot \ln \hat{c}$$

if we neglect the interactions between the dissolved particles ($\boldsymbol{n}$ = amount of solute). Thus, we find that the molar Gibbs energy of formation in aqueous solution is

$$\Delta_f G^{\ominus}_{\text{aq}} = \Delta_f G^{\ominus} - \boldsymbol{R}T \cdot \ln \hat{c}$$

For a saturated solution of glucose in water, $\hat{c} = 3.1$ at $T = 298$ K and we obtain

$$\Delta_f G^{\ominus}_{\text{aq}} = -910.6 \text{ kJ mol}^{-1} - \boldsymbol{R}T \cdot \ln 3.1 = -913.4 \text{ kJ mol}^{-1}$$

This is only slightly different from the value -914.5 kJ mol^{-1} in Appendix A.2, where the deviation from ideal behavior has been taken into account.

18.11 Foundations

18.11.1 Foundation 18.1: To Calculate ΔG_{T_2} from ΔG_{T_1}

In order to calculate ΔG at an arbitrary temperature according to

$$\Delta G = \Delta H - T \cdot \Delta S$$

we must find the temperature dependence of ΔH and ΔS. In Foundation 16.2 we demonstrated how $\Delta H(T_2)$ can be calculated from $\Delta H(T_1)$:

$$\Delta H(T_2) = \Delta H(T_1) + \int_{T_1}^{T_2} \Delta C_P \cdot \mathrm{d}\, T \tag{18F1.1}$$

In analogy we calculate ΔS by

$$\Delta S(T_2) = \Delta S(T_1) + \int_{T_1}^{T_2} \frac{\Delta C_P}{T} \cdot \mathrm{d}T \tag{18F1.2}$$

see Box 18.1. As in Foundation 16.2 we approximate $C^{\ominus}_{P,m}$ by a power series

$$C^{\ominus}_{P,m} = c_1 + c_2 T + c_3 T^2$$

For the general reaction

$$\alpha \mathrm{A} + \beta \mathrm{B} + \ldots \rightarrow \gamma \mathrm{C} + \delta \mathrm{D} + \ldots$$

we obtain

$$\begin{aligned} &\int_{T_1}^{T_2} \frac{\Delta C^{\ominus}_{P,m}}{T} \cdot \mathrm{d}T \qquad (18F1.3)\\ &= \int_{T_1}^{T_2} \left(\frac{\Delta c_1}{T} + \Delta c_2 + \Delta c_3 \cdot T \right) \cdot \mathrm{d}T \\ &= \Delta c_1 \cdot \ln\left(\frac{T_2}{T_1}\right) + \Delta c_2 \cdot (T_2 - T_1) + \frac{1}{2} \Delta c_3 \cdot (T_2^2 - T_1^2) \end{aligned}$$

with

$$\Delta c_1 = \gamma c_1(\mathrm{C}) + \delta c_1(\mathrm{D}) + \ldots - \alpha c_1(\mathrm{A}) - \beta c_1(\mathrm{B}) - \ldots$$

(Δc_2, Δc_3 correspondingly).

Example 18.20 *Calculate the Gibbs energy change for the detonating gas reaction at 1500 K*

$$2\,\mathrm{H_2} + \mathrm{O_2} \rightarrow 2\,\mathrm{H_2O}$$

First we find the entropy change for performing this reaction at 1500 K. Above 373 K H_2O is in the gaseous state. Then

$$\Delta S^{\ominus}_m(1500\ \mathrm{K}) = \Delta S^{\ominus}_m(298\ \mathrm{K}) + \int_{298}^{373} \frac{\Delta C^{\ominus}_{P,m}(\text{liquid})}{T} \cdot \mathrm{d}T + \frac{\Delta_{\text{vap}} H^{\ominus}}{T_{\text{vap}}} + \int_{373}^{1500} \frac{\Delta C^{\ominus}_{P,m}(\text{gas})}{T} \cdot \mathrm{d}T$$

The entropy change $\Delta S^{\ominus}_m(298\ \mathrm{K})$ is

$$\Delta S^{\ominus}_m(298\ \mathrm{K}) = -326.4\ \mathrm{J\ K^{-1}\ mol^{-1}}$$

Using the results $\Delta c_1 = 96.3\ \mathrm{J\ mol^{-1}\ K^{-1}}$, $\Delta c_2 = -0.204\ \mathrm{J\ mol^{-1}\ K^{-2}}$, and $\Delta c_3 = 31 \times 10^{-5}\ \mathrm{J\ mol^{-1}\ K^{-3}}$ from Foundation 16.2 for liquid H_2O as a product, we obtain

$$\begin{aligned} &\int_{298}^{373} \frac{\Delta C^{\ominus}_{P,m}(\text{liquid})}{T} \cdot \mathrm{d}T \\ &\Delta c_1 \cdot \ln\frac{373}{298} + \Delta c_2 \cdot (373 - 298)\ \mathrm{K} + \frac{1}{2}\Delta c_3 \cdot (373^2 - 298^2)\ \mathrm{K}^2 \\ &= 21.6\ \mathrm{J\ mol^{-1}\ K^{-1}} - 15.3\ \mathrm{J\ mol^{-1}\ K^{-1}} + 7.8\ \mathrm{J\ mol^{-1}\ K^{-1}} \\ &= 14.1\ \mathrm{J\ mol^{-1}\ K^{-1}} \end{aligned}$$

Using the results $\Delta c_1 = -23.39$ J mol^{-1} K^{-1}, $\Delta c_2 = 0.0084$ J mol^{-1} K^{-2}, and $\Delta c_3 = 0.185$ J mol^{-1} K^{-3} from Foundation 16.2 for gaseous H_2O as a product, we obtain

$$\int_{373}^{1500} \frac{\Delta C_{P,m}^{\ominus}(\text{liquid})}{T} \cdot dT$$
$$= \Delta c_1 \cdot \ln \frac{1500}{373} + \Delta c_2 \cdot (1500 - 373)\ \text{K} + \frac{1}{2} \Delta c_3 \cdot (1500^2 - 373^2)\ \text{K}^2$$
$$= -32.6\ \text{J mol}^{-1}\ \text{K}^{-1} + 9.5\ \text{J mol}^{-1}\ \text{K}^{-1} + 2.0\ \text{J mol}^{-1}\ \text{K}^{-1} = -21.1\ \text{J mol}^{-1}\ \text{K}^{-1}$$

With $\Delta_{\text{vap}} H^{\ominus} = 83.5$ kJ at $T = 373$ K we obtain

$$\frac{\Delta_{\text{vap}} H^{\ominus}}{T_{\text{vap}}} = 223.9\ \text{J K}^{-1}\ \text{mol}^{-1}$$

Thus for $\Delta S_m^{\ominus}(1500\ \text{K})$ we find

$$\Delta S_m^{\ominus}(1500\ \text{K}) = -326.4\ \text{J K}^{-1}\ \text{mol}^{-1} + (14.1 - 21.1 + 223.9)\ \text{J K}^{-1}\ \text{mol}^{-1} = -109.6\ \text{J K}^{-1}\ \text{mol}^{-1}$$

In Foundation 16.2, we showed that $\Delta_r H^{\ominus}(1500\ \text{K}) = -480.8$ kJ mol^{-1}; hence, we can write

$$\Delta_r G^{\ominus}(1500\ \text{K}) = (-480.8 \times 10^3 + 1500 \cdot 109.6)\ \text{J mol}^{-1} = -316\ \text{kJ mol}^{-1}$$

For the reverse reaction

$$2\,H_2O \rightarrow 2\,H_2 + O_2$$

we obtain $\Delta_r G^{\ominus}(1500\ \text{K}) = 316$ kJ mol^{-1}.

18.11.2 Foundation 18.2: Relationship between the Molecular Partition Function and the Equilibrium Constant

See the books website, www.wiley.com/go/Kuhn/physicalchemistry3e.

18.11.3 Foundation 18.3: Isotope Exchange Equilibrium

See the books website, www.wiley.com/go/Kuhn/physicalchemistry3e.

18.11.4 Foundation 18.4: How to Determine $\Delta G^{\ominus}$ for Ions

As in Section 18.6.4 we determine $\Delta G^{\ominus}$ for the reaction

$$HAc_{aq} + NaCl_{aq} \rightarrow NaAc_{aq} + HCl_{aq}$$

The reactants are in reservoirs at standard concentration $c^{\ominus} = 1$ mol L^{-1} (Fig. 18F.1).

We transfer small amounts of HAc and NaCl reversibly into the equilibrium container and return NaAc and HCl reversibly into the corresponding reservoirs. Now the following problem arises: Na^+ and Cl^- have the same concentrations $c^{\ominus}$ in the reservoirs, but different concentrations c_{Na^+} and c_{Cl^-} in the equilibrium container. Because of the condition of electroneutrality, we cannot transfer the single ions Na^+ and Cl^- separately, but we have to transfer NaCl as a whole.

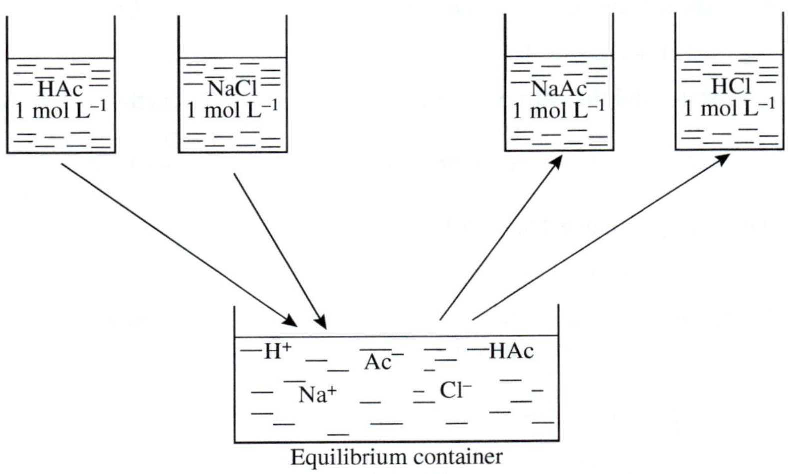

Figure 18F.1 Performing the reaction $HAc \rightarrow Ac^- + H^+$ under standard conditions (reacting species in separate reservoirs at standard concentration $c^{\ominus} = 1$ mol L^{-1}). The species are at equilibrium concentrations in the equilibrium container.

How can we overcome this difficulty? The trick is to transfer NaCl (amount n) from the standard concentration to the concentration

$$c = \sqrt{c_{\mathrm{Na}^+} \cdot c_{\mathrm{Cl}^-}}$$

The work needed in this step is

$$\begin{aligned} w_{\mathrm{rev},1} &= \boldsymbol{nRT} \cdot \ln \frac{c}{c^{\ominus}} + \boldsymbol{nRT} \cdot \ln \frac{c}{c^{\ominus}} \\ &= \boldsymbol{nRT} \cdot \ln \frac{c^2}{(c^{\ominus})^2} = \boldsymbol{nRT} \cdot \ln \frac{c_{\mathrm{Na}^+} \cdot c_{\mathrm{Cl}^-}}{(c^{\ominus})^2} \\ &= \boldsymbol{nRT} \cdot \ln \widehat{c}_{\mathrm{Na}^+} + \boldsymbol{nRT} \cdot \ln \widehat{c}_{\mathrm{Cl}^-} \end{aligned}$$

Then this solution is transferred into the equilibrium container by using a piston and a membrane permeable to Na^+ and Cl^- (Fig. 18F.2).

The work needed for this step is

$$\begin{aligned} w_{\mathrm{rev},2} &= \boldsymbol{nRT} \cdot \ln \frac{c_{\mathrm{Na}^+}}{c} + \boldsymbol{nRT} \cdot \ln \frac{c_{\mathrm{Cl}^-}}{c} - {}^{\mathrm{osm}}P\Delta V \\ &= \boldsymbol{nRT} \cdot \ln \frac{c_{\mathrm{Na}^+} \cdot c_{\mathrm{Cl}^-}}{c^2} - {}^{\mathrm{osm}}P\Delta V \end{aligned}$$

Because of our special choice of c we obtain

$$w_{\mathrm{rev},2} = -{}^{\mathrm{osm}}P\Delta V$$

If, for example, $c_{\mathrm{Na}^+} < c_{\mathrm{Cl}^-}$, then the work to transfer Na^+ from c to c_{Na^+} is negative, whereas the work to transfer Cl^- from c to c_{Cl^-} is positive; both contributions cancel each other. This means that by supplying or removing a tiny additional weight on the piston, the process can proceed in the forward or in the reverse direction. The total work required to transfer NaCl from its reservoir into the equilibrium reservoir is

$$w_{\mathrm{rev}} = \boldsymbol{nRT} \cdot \ln(\widehat{c}_{\mathrm{Na}^+} \cdot \widehat{c}_{\mathrm{Cl}^-}) - {}^{\mathrm{osm}}P\Delta V$$

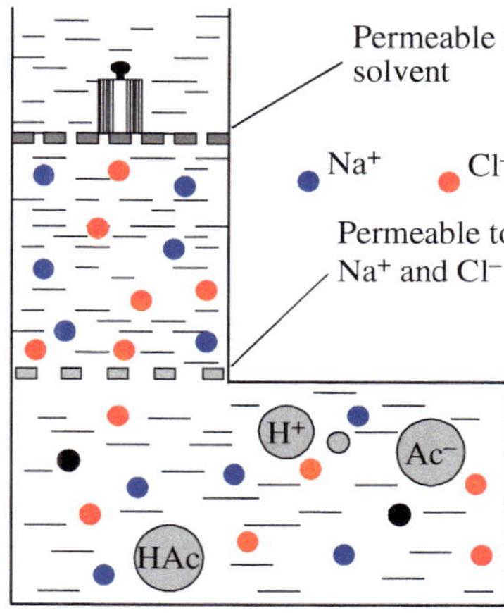

Figure 18F.2 Transfer of NaCl (concentration c) into the equilibrium container (concentrations c_{Na^+}, c_{Cl^-}). The process is reversible, if $c = \sqrt{c_{\mathrm{Na}^+} \cdot c_{\mathrm{Cl}^-}}$.

The same consideration holds for NaAc and HCl. Thus

$$\begin{aligned} \Delta G^{\ominus} &= \boldsymbol{nRT} \cdot \left[\ln(\widehat{c}_{\mathrm{Na}^+} \cdot \widehat{c}_{\mathrm{Cl}^-}) + \ln \widehat{c}_{\mathrm{HAc}} - \ln(\widehat{c}_{\mathrm{Na}^+} \cdot \widehat{c}_{\mathrm{Ac}^-}) \right. \\ &\quad \left. - \ln(\widehat{c}_{\mathrm{H}^+} \cdot \widehat{c}_{\mathrm{Cl}^-})\right] \\ &= -\boldsymbol{nRT} \cdot \ln \frac{\widehat{c}_{\mathrm{H}^+} \cdot \widehat{c}_{\mathrm{Ac}^-}}{\widehat{c}_{\mathrm{HAc}}} = -\boldsymbol{nRT} \cdot \ln K \end{aligned}$$

This shows that the mass action law is indeed valid.

Many of the exercises and problems in this section require that you look up thermodynamic quantities of the substances. These facts may be found in Appendix A.

18.12 Exercises

Some of the exercises for this chapter require that you look up thermodynamic quantities of the substances. These facts may be found in Appendix A.

E18.1 Calculate the Gibbs energy change and the equilibrium constant K for the reaction: $n\text{-}C_4H_{10} \rightarrow i\text{-}C_4H_{10}$ under standard conditions. Estimate the equilibrium constant at 1500 K. How does your estimated value compare to the more rigorous result of $K = 0.55$?

E18.2 Consider the decomposition reaction for TNT (2,4,6-trinitrotoluene, $C_7H_5N_3O_6$).

$$2\ \mathrm{TNT} \rightarrow 3\ \mathrm{N_2(g)} + 7\ \mathrm{CO(g)} + 5\ \mathrm{H_2\ O(g)} + 7\ \mathrm{C(s)}$$

Calculate the enthalpy change for this reaction at 298K and 1 bar pressure. Compare your value to the enthalpy of combustion for TNT, which is -3410 kJ mol^{-1}. Use the value $\Delta_f H^{\ominus} = -63.2$ kJ mol^{-1} for TNT.[4]

4 These values are taken from the NIST Chemistry WebBook, 2008; http://webbook.nist.gov/.

E18.3 Calculate the Gibbs energy change and the equilibrium constant for the combustion of methane:

$$CH_4 + 2O_2 \rightarrow CO_2 + 2H_2O$$

Estimate the temperature at which this reaction is no longer spontaneous.

E18.4 Calculate the equilibrium constant for the reaction:

$$I_2(g) \rightarrow 2\,I\,(g)$$

at 1298 K. What is the fraction of iodine atoms at total pressures of 0.1, 1.0, and 10 bar?

E18.5 Calculate the equilibrium constant for the reaction:

$$H_2(g) + I_2\,(g) \rightarrow 2HI\,(g)$$

at 700 K. What is the fraction of HI at total pressures of 0.1, 1.0, and 10 bar?

E18.6 Use the standard enthalpy and absolute entropies for the reactions and products in the hydrogen and oxygen explosion to compute the $\Delta_r G$ for the reaction. Determine the temperature at which the reaction is no longer spontaneous.

E18.7 Consider the molecular energy levels for the reactants and products in $2\,H_2 + O_2 \rightarrow 2\,H_2O$. Discuss how the reaction's enthalpy and entropy changes can be understood in terms of the molecular properties.

E18.8 Do the reactions written in Table E18.8 occur spontaneously under standard conditions (reactants and products in their standard states)?

Table E18.8 List of Chemical Reactions

	Reaction
1	$Fe + S \rightarrow FeS$
2	$2\,AgI + Pb \rightarrow PbI_2 + 2\,Ag$
3	$2\,Al + Fe_2O_3 \rightarrow Al_2O_3 + 2\,Fe$
4	$C_2H_4 \rightarrow$ polyethylene
5	$CaCO_3 \rightarrow CaO + CO_2$
6	$H_2O\,(l) \rightarrow H_2O\,(g)$
7	$2Ag + PbI_2 \rightarrow 2AgI + Pb$

E18.9 What is the minimum temperature that is required for the decomposition of limestone on the top of Mount Everest (atmospheric pressure 0.34 bar) to be spontaneous?

E18.10 In the isotope exchange reaction in Example 18.11,

$$H_2 + D_2 \rightleftarrows 2HD$$

we considered the limit where the vibrational energy spacing is large compared to kT. Discuss how you expect the analysis to change as this condition is relaxed. In particular, comment on which motions contribute to the enthalpy and entropy changes in the reaction.

E18.11 Consider the gas reaction $H_2O \rightarrow H_2 + \frac{1}{2}O_2$. Following Example 18.6 and Foundation 18.1 determine K at 1500 and 4000 K.

E18.12 Consider Example 18.9 for the ammonia synthesis. At each of the pressures in this example calculate the mole fraction of each component. Also calculate the entropy change that arises from the mixing of the different substances. Assume that the enthalpy change for mixing is zero, and use your entropy change to calculate the Gibbs energy change.

E18.13 Calculate the partial pressure of ammonia for the ammonia synthesis reaction at $T = 298$ K and $P_{H_2} = P_{N_2} = 1$ bar. Perform the same calculation for $P_{H_2} = P_{N_2} = 100$ bar and compare your results.

E18.14 By using the heat capacity data given in Foundation 16.2 for the limestone decomposition reaction, calculate the Gibbs energy change for the decomposition of limestone $(CaCO_3 \rightarrow CaO + CO_2)$ at 1200 K.

E18.15 Consider the reaction

$$CH_3CH_2OH \ (l) + CH_3COCl \ (l) \rightarrow CH_3CH_2O\text{-}C(O)CH_3 \ (l) + HCl \ (g)$$

What is the equilibrium constant for this reaction at 298 K? Estimate the equilibrium constant at 370 K. How would the addition of a base, such as pyridine, affect this equilibrium?

	CH_3CH_2OH	CH_3COCl	$CH_3CH_2O(CO)CH_3$	HCl
$\Delta_f G^\ominus$ (kJ mol^{-1})	−174.8	−208.0	−337.2	−92.3
$\Delta_f H^\ominus$ (kJ mol^{-1})	−277.7	−273.8	−479.3	−95.3

E18.16 Temperature dependence of ΔS. Use Equation (18.26) to derive Equation (18.27). Proceed in a manner similar to that we used in Section 16.5.3 for the temperature dependence of ΔU and ΔH.

E18.17 Reverse Osmosis. Consider sea water with a salt concentration of $c_1 = 1$ mol L^{-1} below the piston in Fig. 18.11. By exerting an external pressure on the piston, the sea water is concentrated, and pure water appears above the piston (reverse osmosis). What is the reversible work required to extract 0.5 L of pure water from 1 L of sea water at a temperature of 298 K?

E18.18 Temperature Dependence of the Equilibrium Constant. Begin with Equation (18.53) for the equilibrium constant and derive the van't Hoff equation

$$\boxed{\frac{d \ln K}{dT} = \frac{\Delta_r H^\ominus}{RT^2}}$$

E18.19 Calculate the standard Gibbs energy change for the precipitation of calcium carbonate, Ca^{2+} (aq) + CO_3^{2-} (aq) → $CaCO_3$ (s). What is the concentration of ions in a saturated solution?

E18.20 From Equation (18.91) calculate the equilibrium constant K for the $H_2 + D_2 \rightleftharpoons 2\,HD$ equilibrium at $T = 83$ K, 195 K, and 383 K. Compare your result with the corresponding experimental values $K = 2.20$, 2.88, and 3.46.

E18.21 Calculate the autoprotolysis constant K_w of water.

E18.22 Zn^{2+} forms a complex with cyanide: $Zn^{2+} + 4\,CN^- \rightarrow [Zn(CN)_4]^{2-}$. For concentrations of Zn^{2+} and CN^- of 1×10^{-3} mol L^{-1}, calculate the equilibrium concentration of the complex.

E18.23 Using the Clausius–Clapeyron equation in its integrated form, calculate the vapor pressure above liquid and solid water (a) in the range of $T = 350$ K to $T = 500$ K and (b) in the range of $T = 271$ K to $T = 275$ K. The molar enthalpies of evaporation and sublimation are $\Delta_{vap}H = 44.1$ kJ mol^{-1}, and $\Delta_{sub}H = 50.1$ kJ mol^{-1}. The vapor pressure at $T_1 = 373$ K is $P_{vap,1} = 1.013$ bar, the vapor pressure at the triple point, $T_1 = 273.16$ K, is $P_{vap,1} = 6.11 \times 10^{-3}$ bar.

E18.24 The vapor pressure of water at 301 K is 0.0031 bar. Compute the vapor pressure at 301 K of a solution containing 68.0 grams of sucrose, $C_{12}H_{22}O_{11}$ in 1000 grams of water. Assume the solution is ideal.

E18.25 Given that the vapor pressure of pure water is $P_s = 0.0031$ bar at 298 K, and the vapor pressure of water above a sucrose solution is $P = 0.0030$ bar, determine the concentration of the sucrose solution.

E18.26 Given that its vapor pressure is 0.790 bar, determine the change of Gibbs energy for the evaporation of liquid Ar at 85.0 K.

E18.27 In Example 18.8 we found a positive entropy change for the reaction of one mole of H_2 with one mole of I_2 to form two moles of HI. Rationalize this entropy change from a molecular perspective.

E18.28 Show that the Clausius–Clapeyron Equation (18.38) can be derived as a special case of the van't Hoff Equation (18.81)

.

18.13 Problems

Some of the problems for this chapter require that you look up thermodynamic quantities of the substances. These facts may be found in Appendix A.

P18.1 Find an expression for the equilibrium constant of the reaction: I_2 (g) → 2 I (g), in terms of the molecular and atomic parameters. Use the following data for I_2: $\theta_{vib} = 308$ K; $\theta_{rot} = 0.0537$ K; $\sigma = 2$; $\varepsilon_B = 2.51 \cdot 10^{-19}$ J and the ground electronic state of the I atom is $^2P_{3/2}$. Given your expression for the equilibrium constant, describe how this equilibrium will be affected by temperature. Evaluate the equilibrium constant at 1298 K and compare it to your result in E18.4.

P18.2 In Example 18.7, we estimated the $\Delta_r G^{\ominus}$ and K for the reaction $3H_2 + N_2 \rightarrow 2\ NH_3$, at 700 K. Using the heat capacity data in the table, perform a rigorous calculation of the equilibrium constant's temperature dependence between 298 K and 1000 K. The parameters in the table should be used in the formula

$$C^0_{P,m} = c_1 + c_2 \cdot T + c_3 \cdot T^2$$

Molecule	c_1 J/(K mol)	c_2 J/(K^2 mol)	c_3 J/(K^3 mol)
N_2	27.20	0.00535	−0.0121
H_2	29.04	−0.0008	0.197
NH_3	24.15	0.0392	−0.783

P18.3 Temperature Regulation in the Human Body. A daily intake of 300 g glucose is sufficient to cover the human energy demand.
a) What is the heat produced by the oxidation of 300 g glucose to CO_2 and H_2O?
b) What would be the increase of temperature in the body in one day in the case of complete heat isolation? For simplicity, use the heat capacity of water.
c) The temperature of the body is kept constant by evaporation of water; what amount of water is necessary?
d) The temperature of the body ($T_2 = 100$ °F = 311 K) is kept constant by thermal conduction to the surroundings (temperature T_1) through clothing. Model the clothing by a bolster of air of thickness 1 cm (thermal conductivity $\kappa = 25 \times 10^{-5}$ J K^{-1} cm^{-1} s^{-1}). Calculate T_1.
e) In analogy to part (d) calculate the demand of glucose for the case that $T_1 = -40$ °F = 233 K.
f) What power P, wattage, of a continuously burning electrical bulb could be operated by the energy equivalent of 300 g of glucose per day?

P18.4 Phosphorus pentachloride dissociates according to the reaction

$$PCl_5\ (g) \rightarrow PCl_3\ (g) + Cl_2\ (g)$$

Find an expression for the equilibrium constant in terms of the fraction ξ of PCl_5 that is dissociated at equilibrium and the total pressure P. Given that the equilibrium constant is 1.78 bar at 523 K, calculate ξ for total pressures of 0.01, 0.10, and 1.00 bar. Calculate and plot $\Delta_r G$ as a function of ξ at a pressure of 1 bar and a temperature of 523 K. (Adapted from Dickerson, Molecular Thermodynamics 1969).

P18.5 Consider that 0.050 mol I_2 vapor is combined with 0.050 mol of Br_2(g) in a high temperature 10.0L reaction vessel, and the equilibrium for the reaction is

$$I_2\ (g) + Br_2\ (g) \rightarrow 2\ IBr\ (g)$$

The equilibrium is studied by extracting gas samples and using gas chromatography. At 1000 K the mole ratio of $\boldsymbol{n}_{IBr}/\boldsymbol{n}_{Br_2}$ is found to be 5.41, while at 1400 K it is 4.62. a) Determine the $\Delta_r G^{\ominus}$ for the reaction at these two temperatures, assuming that the gases behave ideally. b) Find $\Delta_r H^{\ominus}$ and $\Delta_r S^{\ominus}$ for the reaction, assuming that they are independent of temperature. Check the sign of your answer by using LeChatelier's principle. c) Provide a justification for the assumption that the enthalpy change is independent of temperature. Describe in detail what molecular properties will contribute to the temperature dependence of the enthalpy and which molecules will contribute most to the enthalpies temperature dependence. d) Combine the result of part B with the following data to estimate the bond energy of IBr. e) How would decreasing the volume of the reaction vessel have affected the results of the equilibrium measurements?

Substance	Br_2 (g)	I_2 (g)	Br (g)	I (g)
$\Delta_f H^{\ominus}$ (kJ mol^{-1})	30.7	62.2	111.8	106.6

P18.6 Consider the isomerization reaction of methylisocyanide to acetonitrile,

$$CH_3NC \rightleftarrows CH_3CN$$

a) Given the following thermochemical data, compute the $\Delta_r G^\ominus$ and the equilibrium constant K for this reaction at 298 K. Take both compounds to be in the gas phase.

Molecule	$\Delta_f H^\ominus$ (kJ mol^{-1})	$\Delta_f G^\ominus$ (kJ mol^{-1})	$S_m^\ominus$ (J mol^{-1} K^{-1})
CH_3NC	149.0	165.7	246.9
CH_3CN	64.3	81.7	245.1

b) Given the following spectroscopic data, compute the $\Delta_r G^\ominus$ and the equilibrium constant K for this reaction at 298 K.

Molecule	CH_3NC	CH_3CN
$\varepsilon_{B,m}$ (kJ mol^{-1})	2346	2476
$\theta_{rot,A}$ (K)	7.2	7.2
$\theta_{rot,B}$ (K)	0.48	0.45
$\theta_{rot,C}$ (K)	0.48	0.45

For this analysis, ignore the contribution associated with the thermal population of all vibrational modes with wavenumbers above 1000 cm^{-1}. The vibrational modes with wavenumbers less than 1000 cm^{-1} are 444 cm^{-1} (2) for acetonitrile (corresponding to a characteristic vibrational temperature of $\theta_{vib} = 639$ K) and 343 cm^{-1} (2) for methylisocyanide (corresponding to a characteristic vibrational temperature of $\theta_{vib} = 343$ K). The number in parentheses after the vibrational wavenumber is the degeneracy of the normal mode.

c) Provide a justification for why it is reasonable to not consider the vibrational modes with wavenumbers of 1000 cm^{-1} and higher.

P18.7 Consider the decomposition of hydrogen peroxide

$$H_2O_2(l) \rightarrow H_2O(l) + \frac{1}{2}O_2(g)$$

where $\Delta_r G^\ominus = -123.1$ kJ mol^{-1}. Calculate $\Delta_r G$ under atmospheric conditions, $\hat{P}_{O_2} = 0.209$, and decide whether the reaction is spontaneous. Calculate the equilibrium pressure of O_2 in a closed system held at 323 K and fixed volume, which initially contained only $H_2O_2(l)$. Account for the temperature dependence of $\Delta_r H^\ominus$ by using $\Delta_r H^\ominus = -98.2$ kJ mol^{-1} at 298 K, and the heat capacities $C_{P,m}(H_2O_2(l)) = 89.1$ J K^{-1} mol^{-1}, $C_{P,m}(H_2O(l)) = 75.3$ J K^{-1} mol^{-1} and $C_{P,m}(O_2(g)) = 29.4$ J K^{-1} mol^{-1}. Assume the heat capacities are independent of temperature over the temperature range of interest.

P18.8 Consider the following reaction:

$$2\,XO\,(g) \rightarrow 2\,X\,(g) + O_2\,(g)$$

where X is a halogen: Cl, Br, or I. Relevant information is given here for the different XO species. Find the other information you need in the Appendix.

Table of data for the XO species.

Species XO	$\Delta_f H^\ominus$ kJ mol^{-1}	$S_m^\ominus$ J mol^{-1} K^{-1}	$\varepsilon_{B,m}$ kJ mol^{-1}	θ_{vib} K	θ_{rot} K
ClO	101.84	226.52	264.90 ± 0.12	1431	2.8
BrO	125.77	237.44	232.6 ± 2.4	1025	1.9
IO	175.06	245.39	176 ± 20	980	0.65

The ground electronic state for these XO species is a $^2\Pi_{3/2}$ (a degeneracy of 2) and the symmetry parameter σ is equal to 1 since they are heteronuclear diatomics. The ground electronic state of O_2 is a $^3\Sigma_g$ (a degeneracy of 3) and its symmetry parameter σ is equal to 2 since it is a homonuclear diatomic.

First, use thermochemical data to compute the $\Delta_r G^\ominus$ for these reactions at 298 K.

Second assume that the gases behave ideally and use the molecular parameters given above to find an expression for the $\Delta_r G^{\ominus}$ of these reactions and for their equilibrium constants. Evaluate your $\Delta_r G^{\ominus}$ expression at the temperature 298 K and compare your results with the thermochemical ones. Analyze your expression for $\Delta_r G^{\ominus}$ and explain (verbally) how you expect it to change with increasing temperature. Explain what terms contribute to the change in $\Delta_r G^{\ominus}$ through this series of three reactions. Which terms are most important and which terms are least important?

P18.9 Consider the following data for the osmotic pressure of aqueous solutions of the protein α-chymotrypsin (298 K and pH = 8.25). Analyze these two datasets and determine the molecular weight of the protein. Data are taken from [3].

	0.1 M K_2SO_4		0.1 M Na_3PO_4
m/V	${}^{osm}p$	m/V	${}^{osm}p$
g/L	Pa	g/L	Pa
0.902	63.6	0.957	75.80
1.811	125.6	2.209	167.2
3.659	259.6	5.677	269.9
5.391	377.4	7.209	416.1
7.185	501.1	9.366	513.3
8.932	625.3		

P18.10 Calculate the enthalpy change, the entropy change, the Gibbs energy change, and the equilibrium constant for the autoionization of water, $H_2O \rightarrow H^+\ (aq) + OH^-\ (aq)$, at 298 K. Using a molecular analysis, estimate how the equilibrium constant changes upon replacing the protons by deuterons (i.e., the autoionization of D_2O). Compare your result to the experimental value of 1.0×10^{-15} M. Lastly, comment on the relative importance of enthalpic and entropic terms in the isotope effect.

P18.11 Depression of Melting point. Using Fig. 18.13 of the text, show that the depression of melting point as a function of mole fraction x of the solute is given by

$$\Delta T_{\text{fus}} = -\frac{RT^2}{\Delta_{\text{fus}} H} \cdot x$$

References

1 A.J. Gould, W. Bleakney, and H.S. Taylor, The inter-relations of hydrogen and deuterium molecules, *J. Chem. Phys.*, 1934, **2**, 362.

2 H. Landolt and R. Börnstein, *Landolt-Börnstein*, **II**. Band, 2. Teil (Springer, 1960, p. 933).

3 C.A. Haynes, K. Tamura, H.R. Kórfer, H.W. Blanch, and J.M. Prausnitz, Thermodynamic properties of aqueous α-chymotrypsin solution from membrane osmometry measurements, *J. Phys. Chem.*, 1992, **96**, 905.

19

Formal Thermodynamics and Its Application to Phase Equilibria

CHAPTER MENU

19.1 Introduction

The laws of classical thermodynamics are stated in Chapter 17 and were deduced from statistical considerations. In the previous chapters we have used statistical ideas and our understanding of the energy level structure of molecules to describe how energy is distributed in assemblies of non-interacting molecules. For an isolated system, these considerations led to the conservation of the total internal energy U of assemblies and to the idea that the most probable distribution for the assembly is the one that maximizes the entropy S. Historically U and S were introduced without reference to the atomistic structure of matter and were based upon observations of macroscopic systems (classical thermodynamics).

In this chapter we introduce classical thermodynamics in a mathematically rigorous way, and we illustrate its use by the example of phase equilibria. Much of the power of thermodynamics lies with the relationships between different macroscopic properties that may be found using purely mathematical considerations. Because no assumptions about the molecular nature of the substance(s) are needed, these relationships are general. They are especially useful when they relate thermodynamic properties that are not directly measurable (e.g., entropy) to properties that are directly measurable (e.g., heat capacity). Although these relationships are quite general, thermodynamics is mute on the origins of these properties in atomic and molecular structure of substances.

19.2 Internal Energy, Enthalpy, Work, and Heat

In Chapter 16 we showed that the change in internal energy U for a cycle is zero. Mathematically, we can write this fact as

$$\oint \mathrm{d}U = 0 \tag{19.1}$$

where the integral sign with a circle indicates that the integration is performed over a closed path. This condition corresponds to the law of conservation of energy. That is, the integral over the closed cycle is zero because energy cannot be either created or destroyed.

Principles of Physical Chemistry, Third Edition. Hans Kuhn, David H. Waldeck, and Horst-Dieter Försterling.

Companion website: www.wiley.com/go/Kuhn/physicalchemistry3e

Because energy is conserved, the change in a system's internal energy must equal the energy flux of heat q and work w between the system and its surroundings, namely

$$\Delta U = q + w + w' + w'' + \cdots \tag{19.2}$$

where w, w', w'' denote different forms of work (the term w represents the energy term we call PV work; other work terms are possible, see Table 19.1). By convention the sign of these energy changes is positive when energy is added to the system and negative when energy leaves the system.

The fact that the internal energy U is a state function places a strict requirement on its differential change dU; namely, dU must be an exact differential. What is an exact differential? It means that the cross-derivatives of the function are equal. As an example, consider U to depend on the two independent variables T and V, then its differential $dU(T, V)$ is written as

$$dU = \left(\frac{\partial U}{\partial T}\right)_V dT + \left(\frac{\partial U}{\partial V}\right)_T dV \tag{19.3}$$

For dU to be exact, its second cross-derivatives must be equal (*Euler criterion*); i.e., the condition

$$\left[\frac{\partial}{\partial V}\left(\frac{\partial U}{\partial T}\right)_V\right]_T = \left[\frac{\partial}{\partial T}\left(\frac{\partial U}{\partial V}\right)_T\right]_V \quad \text{or} \quad \frac{\partial^2 U}{\partial V \partial T} = \frac{\partial^2 U}{\partial T \partial V} \tag{19.4}$$

holds. This equality means that the differential change in the value of U does not depend on the order of moving between two equilibrium points; whether we first move along V and then along T, or whether we first move along T and then along V. If you are unfamiliar with these concepts, you are strongly encouraged to review Appendix C.1.3.

In Chapters 16 and 17 we found that work and heat are not changes of state functions, because they depend on the path of the state change. Also, we cannot attribute work or heat to a system. In other words, a system has internal energy, but it cannot have heat or work. Heat and work are energy fluxes that are supplied to (or extracted from) the system in a change of equilibrium state. Consequently, differentials of heat and work are not exact differentials; they need not fulfill a condition corresponding to Equation (19.4). To distinguish the inexact differentials "dq" and "dw," we will denote small amounts of heat or work by the notation "δq" and "δw." With this convention, condition (19.2) may be written in differential form as

$$dU = \delta q + \delta w + \delta w' + \delta w'' + \cdots \tag{19.5}$$

Equation (19.5) is a mathematical statement of the *First Law of Thermodynamics.*

19.2.1 Work

When a macroscopic system performs work (or work is being performed on it), its properties change in a macroscopically observable manner. The primary areas of classical physics (mechanics, electricity and magnetism, etc.) study these properties, or mechanical modes of the system. Energy that is transferred to a mechanical mode of a system is called *mechanical work*. Some different types of work terms are listed in Table 19.1.

Table 19.1 A Listing of Some Work Terms.

Type of Work	δw	Comment
Volume change	$-P\,dV$	P is the external pressure
Tension	$f\,dl$	f is tension
Surface tension	$\gamma\,dA$	γ is surface tension
Electrical	$\phi\,dQ$	ϕ is electrical potential
Gravitational	$m \cdot g\,dz$	$m \cdot g$ is the gravitational force on earth

19.2.2 Heat

Here we discuss some quantitative aspects of heat. By combining Equations (19.5) and (19.3), we can write

$$dU = \left(\frac{\partial U}{\partial T}\right)_V dT + \left(\frac{\partial U}{\partial V}\right)_T dV = \delta q + \delta w = \delta q - PdV \tag{19.6}$$

where we have simplified the work term by considering only PV work. This expression may be rearranged to give an expression for the heat δq

$$\begin{aligned}\delta q = dU + PdV &= \left(\frac{\partial U}{\partial T}\right)_V \cdot dT + \left(\frac{\partial U}{\partial V}\right)_T \cdot dV + PdV \\ &= \left(\frac{\partial U}{\partial T}\right)_V \cdot dT + \left[P + \left(\frac{\partial U}{\partial V}\right)_T\right] \cdot dV\end{aligned} \tag{19.7}$$

If we use the definition of the constant volume heat capacity (Section 16.3.2), $C_V = (\partial U/\partial T)_V$, we find that

$$\delta q = C_V \,\mathrm{d}T + P\mathrm{d}V + \left(\frac{\partial U}{\partial V}\right)_T \mathrm{d}V \tag{19.8}$$

The first term on the right side of this equation represents the flow of energy into the system from temperature change, and the second term on the right side of Equation (19.8) corresponds to the heat used by the system to perform PV work on the surroundings. Both of these terms should be familiar from the discussions in Chapters 16 and 17. The last term on the right is new; it represents the change of the internal energy that arises from volume change, but is not PV–work. Our earlier discussions did not include this last term because U for an ideal gas depends only on T, not V. For interacting systems (e.g., a gas of dipolar molecules) this term will not necessarily be zero. How we might measure such a quantity is discussed in Section 19.2.3.

Example 19.1 *Use Equation (19.7) to verify that the differential heat δq is not an exact differential; i.e., heat is not a state function.*

If we write the differential for $q(T, V)$, we find

$$\delta q = \left(\frac{\partial q}{\partial T}\right)_V \mathrm{d}T + \left(\frac{\partial q}{\partial V}\right)_T \mathrm{d}V$$

Comparison with Equation (19.7) shows that

$$\left(\frac{\partial q}{\partial T}\right)_V = \left(\frac{\partial U}{\partial T}\right)_V, \qquad \left(\frac{\partial q}{\partial V}\right)_T = \left[P + \left(\frac{\partial U}{\partial V}\right)_T\right]$$

Next we calculate the second cross-derivatives

$$\frac{\partial^2 q}{\partial V \partial T} = \frac{\partial^2 U}{\partial V \partial T}, \qquad \frac{\partial^2 q}{\partial T \partial V} = \frac{\partial P}{\partial T} + \frac{\partial^2 U}{\partial T \partial V}$$

and take their difference

$$\frac{\partial^2 q}{\partial T \partial V} - \frac{\partial^2 q}{\partial V \partial T} = \frac{\partial^2 U}{\partial T \partial V} + \left(\frac{\partial P}{\partial T}\right)_V - \frac{\partial^2 U}{\partial V \partial T}$$

Because U is a state function we can apply Equation (19.4) and find

$$\frac{\partial^2 q}{\partial T \partial V} - \frac{\partial^2 q}{\partial V \partial T} = \left(\frac{\partial P}{\partial T}\right)_V$$

P and T are connected by the equation of state, e.g., for an ideal gas we have

$$P = \frac{\boldsymbol{nRT}}{V} \qquad \text{and} \qquad \left(\frac{\partial P}{\partial T}\right)_V = \frac{\boldsymbol{nR}}{V}$$

Because $(\partial P/\partial T)$ is not zero, the second cross-derivatives must be different:

$$\frac{\partial^2 q}{\partial T \partial V} \neq \frac{\partial^2 q}{\partial V \partial T}$$

and δq is not an exact differential.

19.2.3 Combined Form of the First and Second Laws

A very important relationship between the internal energy and the entropy is called the combined form of the first and second laws of thermodynamics. Recall from Chapter 17, that the second law of thermodynamics concerns the state function S and places constraints on whether a process can proceed spontaneously. It is expressed mathematically using the Clausius inequality

$$\mathrm{d}S \geq \frac{\delta q}{T} \qquad \text{or} \qquad \Delta S \geq \int_{\text{initial}}^{\text{final}} \frac{\delta q}{T}$$

Here we consider the case of a reversible process between equilibrium states, for which $\delta q_{\text{rev}} = T\mathrm{d}S$. Let us consider a closed system (no exchange of mass) that undergoes a reversible process from an initial state to a final state. Then we can write the first law of thermodynamics, Equation (19.5), as

$$\mathrm{d}U = \delta q_{\text{rev}} + \delta w_{\text{rev}}$$

Because the heat flow is reversible, we can use the second law of thermodynamics to write that $\delta q_{\text{rev}} = T\mathrm{d}S$, and in the limit of only PV work, $\delta w_{\text{rev}} = -P\,\mathrm{d}V$, we can write that

$$\boxed{\mathrm{d}U = T\mathrm{d}S - P\,\mathrm{d}V} \tag{19.9}$$

Equation (19.9) is called the combined form of the first and second laws. By rearranging Equation (19.9) we find

$$\mathrm{d}S = \frac{1}{T}\cdot \mathrm{d}U + \frac{P}{T}\cdot \mathrm{d}V \tag{19.10}$$

19.2.4 Relating U and H to Measurable Quantities

In part, thermodynamics is useful because it provides a rigorous connection between thermodynamic properties (e.g., the internal energy U and the entropy S). By expressing hard-to-measure properties, such as U, in terms of directly measurable properties, we can build a framework that allows the experimental determination of thermodynamic properties without relying on approximate models. In this section, we show how to express two important state functions, U and H, in terms of measurable thermodynamic properties.

19.2.4.1 dU in Terms of Measurable Quantities

Using, Equation (19.3) and the definition of the constant volume heat capacity $C_V = (\mathrm{d}U/\mathrm{d}T)_V$, we can write the differential of U as

$$\mathrm{d}U = C_V\,\mathrm{d}T + \left(\frac{\partial U}{\partial V}\right)_T \mathrm{d}V \tag{19.11}$$

In this equation, the constant volume heat capacity C_V is directly measurable; see Chapter 16. Thus, we need only find an expression for $(\partial U/V)_T$ in terms of measurable quantities, and we will have dU in terms of measurable quantities.

We begin by considering the combined form of the first and second laws, Equation (19.10), and substituting in the formal expression for dU (Equation (19.3)), to write dS as

$$\mathrm{d}S = \frac{1}{T}\,\mathrm{d}U + \frac{P}{T}\,\mathrm{d}V = \frac{1}{T}\left[\left(\frac{\partial U}{\partial T}\right)_V \mathrm{d}T + \left(\frac{\partial U}{\partial V}\right)_T \mathrm{d}V\right] + \frac{P}{T}\cdot \mathrm{d}V$$

or

$$\mathrm{d}S = \frac{1}{T}\left(\frac{\partial U}{\partial T}\right)_V \cdot \mathrm{d}T + \frac{1}{T}\left[\left(\frac{\partial U}{\partial V}\right)_T + P\right]\cdot \mathrm{d}V \tag{19.12}$$

In Equation (19.12), the first term corresponds to $(\partial S/\partial T)_V$ and the second term corresponds to $(\partial S/\partial V)_T$; i.e.,

$$\left(\frac{\partial S}{\partial T}\right)_V = \frac{1}{T}\left(\frac{\partial U}{\partial T}\right)_V, \qquad \left(\frac{\partial S}{\partial V}\right)_T = \frac{1}{T}\left[P + \left(\frac{\partial U}{\partial V}\right)_T\right] \tag{19.13}$$

Next, we use the fact that S is a state function, so that dS must satisfy the conditions of an exact differential. If we evaluate the cross derivatives, we obtain

$$\frac{\partial^2 S}{\partial V \partial T} = \frac{1}{T}\left(\frac{\partial^2 U}{\partial V \partial T}\right) \tag{19.14}$$

$$\frac{\partial^2 S}{\partial T \partial V} = -\frac{1}{T^2}\left[P + \left(\frac{\partial U}{\partial V}\right)_T\right] + \frac{1}{T}\left[\left(\frac{\partial P}{\partial T}\right)_V + \frac{\partial^2 U}{\partial T \partial V}\right]$$

Because the cross derivatives must be equal, their difference is zero so that

$$\frac{\partial^2 S}{\partial V \partial T} - \frac{\partial^2 S}{\partial T \partial V} = \frac{1}{T}\left(\frac{\partial^2 U}{\partial V \partial T}\right) + \frac{1}{T^2}\left[P + \left(\frac{\partial U}{\partial V}\right)_T\right] - \frac{1}{T}\left[\left(\frac{\partial P}{\partial T}\right)_V + \frac{\partial^2 U}{\partial T \partial V}\right] = 0 \tag{19.15}$$

By using Equation (19.4), this result simplifies to

$$\frac{1}{T^2}\left[P + \left(\frac{\partial U}{\partial V}\right)_T\right] - \frac{1}{T}\left(\frac{\partial P}{\partial T}\right)_V = 0$$

or

$$\boxed{\left(\frac{\partial U}{\partial V}\right)_T = T\left(\frac{\partial P}{\partial T}\right)_V - P} \tag{19.16}$$

This relation connects $(\partial U/\partial V)_T$ to the measurable quantities $(\partial P/\partial T)_V$ and P, and is called the *Thermodynamic Equation of State*.

Using this relation in Equation (19.11), we can write the differential of the internal energy as

$$dU = C_V dT + \left[T\left(\frac{\partial P}{\partial T}\right)_V - P\right] dV \tag{19.17}$$

This equation relates the differential change in the internal energy to measurable properties of the system, namely the constant volume heat capacity, the temperature dependence of the pressure, the pressure, and the temperature. No assumptions about the nature of the substance have been made; e.g., it could be a gas, a solid, or a liquid. We have assumed that the system is closed and exists in an equilibrium state so that the quantities in this expression are well-defined. Generalization of this expression to consideration of other types of work terms is possible.

19.2.4.2 *dH* in Terms of Measurable Quantities

Recall that the enthalpy is defined as

$$H = U + PV \tag{19.18}$$

and we can write dH as

$$\begin{aligned} dH &= d(U + PV) = dU + d(PV) \\ &= dU + PdV + VdP \end{aligned} \tag{19.19}$$

By substituting the combined form of the first and second laws, $dU = TdS - PdV$, for dU, we can write the differential enthalpy as

$$\boxed{dH = TdS + VdP} \tag{19.20}$$

where we are only considering PV work. Rearranging this equation gives

$$dS = \frac{1}{T} \cdot dH + \frac{V}{T} \cdot dP$$

Next we write the formal expression for dH to be a function of T and P,

$$dH = \left(\frac{\partial H}{\partial T}\right)_P dT + \left(\frac{\partial H}{\partial P}\right)_T dP \tag{19.21}$$

and substitute it into the equation for dS, giving

$$\begin{aligned} dS &= \frac{1}{T}\left(\frac{\partial H}{\partial T}\right)_P dT + \frac{1}{T}\left(\left(\frac{\partial H}{\partial P}\right)_T + V\right) dP \\ &= \frac{1}{T}C_P\, dT + \frac{1}{T}\left(\left(\frac{\partial H}{\partial P}\right)_T + V\right) dP \end{aligned}$$

To find $(\partial H/\partial P)_T$ in terms of measurable properties, we can proceed in a manner much like that used above for finding the Thermodynamic Equation of State. In Problem P19.10 you show that

$$\boxed{\left(\frac{\partial H}{\partial P}\right)_T = -T\left(\frac{\partial V}{\partial T}\right)_P + V} \tag{19.22}$$

This relation connects $(\partial H/\partial P)_T$ to the measurable quantities $(\partial V/\partial T)_P$ and V.

By substituting Equation (19.22) into Equation (19.21) for dH, with $(\partial H/\partial T)_P = C_P$ we find that

$$dH = C_P \cdot dT + \left[V - T\left(\frac{\partial V}{\partial T}\right)_P\right] \cdot dP \tag{19.23}$$

which expresses the differential enthalpy in terms of measurable properties of the system.

Example 19.2 *Given the mean value of the constant pressure heat capacity of NH_3 at 1 bar pressure to be $\overline{C}_{p,m}$ = 40.45 $J\ K^{-1}\ mol^{-1}$, and using a mean value of the molar volume, $\overline{V}_m = 1.35\ L\ mol^{-1}$, and a mean value of $\left(\overline{dV_m/dT}\right)_P$ = 0.0038 $L\ K^{-1}\ mol^{-1}$ at 26 bar as estimates of the NH_3 gas fluid properties, calculate the change in molar enthalpy of NH_3 for a change of state from 1 bar and 400 K to 51 bar and 500 K.*

Because the constant pressure heat capacity is specified at the lower pressure, we will calculate the enthalpy change by a two-step process. In the first step, we will heat the gas at constant pressure from 400 to 500 K to find ΔH_1, and in the second step we will perform an isothermal compression of the liquid from 1 bar to 51 bar at 500 K to find ΔH_2.

Step 1: Because the first step occurs at constant pressure $(\mathrm{d}H)_P = C_P \cdot \mathrm{d}T$, so that

$$\Delta H_{1,m} = \int_{400\ \mathrm{K}}^{500\ \mathrm{K}} \overline{C}_{P,m} \cdot \mathrm{d}T = \overline{C}_P \cdot 100\ \mathrm{K} = 40.45\ \mathrm{J\ K^{-1}\ mol^{-1}} \cdot 100\ \mathrm{K} = 4.05\ \mathrm{kJ\ mol^{-1}}$$

Step 2: Because this step occurs at constant temperature, $(\mathrm{d}H) = \left[V - T\left(\frac{\partial V}{\partial T}\right)_P\right]\ \mathrm{d}P$, so that

$$\begin{aligned}\Delta H_{2,m} &= \overline{V}_m \int_{1\ \mathrm{bar}}^{51\ \mathrm{bar}} \mathrm{d}P - T\overline{\left(\frac{\partial V_m}{\partial T}\right)_P} \int_{1\ \mathrm{bar}}^{51\ \mathrm{bar}} \mathrm{d}P \\ &= 1.35\ \mathrm{L\ mol^{-1}} \cdot 50\ \mathrm{bar} - 500\ \mathrm{K}\ \cdot 0.0038\ \mathrm{L\ K^{-1}\ mol^{-1}} \cdot 50\ \mathrm{bar} \\ &= 67.5\ \mathrm{bar\ L\ mol^{-1}} - 95.0\ \mathrm{bar\ L\ mol^{-1}} = -27.5\ \mathrm{bar\ L\ mol^{-1}} \\ &= -2.75\ \mathrm{kJ\ mol^{-1}}\end{aligned}$$

Combining these two contributions, we find that the total enthalpy change $\Delta H_{\mathrm{total}}$ is

$$\begin{aligned}\Delta H_{\mathrm{total},m} &= \Delta H_{1,m} + \Delta H_{2,m} \\ &= 4.05\ \mathrm{kJ\ mol^{-1}} - 2.75\ \mathrm{kJ\ mol^{-1}} = 1.30\ \mathrm{kJ\ mol^{-1}}\end{aligned}$$

This value is about one-half of the experimentally determined value of 2.81 kJ mol^{-1}. However, here we are using mean values of $C_{P,m}$, V_m, and $\mathrm{d}V_m/\mathrm{d}T$ instead of using the actual values. In Problem P19.13, you use the actual values for this calculation, and the result is $\Delta H_{\mathrm{total},m} = 2.28$ kJ mol^{-1}.

19.2.5 An Important Application: Calculating $C_P - C_V$

In Chapter 16 we considered the heat capacities, C_P and C_V, as

$$C_P = \left(\frac{\partial H}{\partial T}\right)_P \quad \text{and} \quad C_V = \left(\frac{\partial U}{\partial T}\right)_V$$

As we discussed at that time it is sometimes more convenient to measure one of these properties as opposed to the other; hence, it is useful to relate them to each other.

To relate these quantities, we consider the exact differential $\mathrm{d}H$ for a change at $P = const$. According to Equations (19.19) and (19.3), we can write

$$\begin{aligned}\mathrm{d}H = \mathrm{d}U + P\mathrm{d}V &= \left(\frac{\partial U}{\partial T}\right)_V \cdot \mathrm{d}T + \left(\frac{\partial U}{\partial V}\right)_T \cdot \mathrm{d}V + P\mathrm{d}V \\ &= \left(\frac{\partial U}{\partial T}\right)_V \cdot \mathrm{d}T + \left[\left(\frac{\partial U}{\partial V}\right)_T + P\right] \cdot \mathrm{d}V \qquad \text{(constant } P\text{)}\end{aligned}$$

Note that the term in square brackets corresponds to $(\partial H/\partial V)_T$.

Next we calculate $(\partial H/\partial T)_P$ as[1]

$$\left(\frac{\partial H}{\partial T}\right)_P = \left(\frac{\partial U}{\partial T}\right)_V + \left[\left(\frac{\partial U}{\partial V}\right)_T + P\right] \cdot \left(\frac{\partial V}{\partial T}\right)_P$$

to obtain

$$\left(\frac{\partial H}{\partial T}\right)_P - \left(\frac{\partial U}{\partial T}\right)_V = \left[\left(\frac{\partial U}{\partial V}\right)_T + P\right] \cdot \left(\frac{\partial V}{\partial T}\right)_P$$

Substituting by the definitions of the heat capacities, we can write

$$C_P - C_V = \left[\left(\frac{\partial U}{\partial V}\right)_T + P\right] \cdot \left(\frac{\partial V}{\partial T}\right)_P$$

---

1 Given the expression

$$\mathrm{d}f = \left(\frac{\partial f}{\partial x}\right)_y \mathrm{d}x + \left(\frac{\partial f}{\partial y}\right)_x \mathrm{d}y$$

we find that (see Appendix C3.1.5)

$$\left(\frac{\partial f}{\partial x}\right)_g = \left(\frac{\partial f}{\partial x}\right)_y + \left(\frac{\partial f}{\partial y}\right)_x \left(\frac{\partial y}{\partial x}\right)_g$$

Next we replace $(\partial U/\partial V)_T$ by the expression in Equation (19.16), the Thermodynamic Equation of State.

$$\left(\frac{\partial U}{\partial V}\right)_T = T\left(\frac{\partial P}{\partial T}\right)_V - P$$

to find

$$\begin{aligned} C_P - C_V &= \left[T\left(\frac{\partial P}{\partial T}\right)_V - P + P\right] \cdot \left(\frac{\partial V}{\partial T}\right)_P \\ &= T\left(\frac{\partial P}{\partial T}\right)_V \cdot \left(\frac{\partial V}{\partial T}\right)_P \end{aligned} \tag{19.24}$$

Rather than use this result directly, it is common to rearrange it so that it is written in terms of substance properties that are commonly reported in thermophysical tables. Next, we find a modified expression for $(\partial P/\partial T)_V$. Because of

$$\mathrm{d}V = \left(\frac{\partial V}{\partial T}\right)_P \cdot \mathrm{d}T + \left(\frac{\partial V}{\partial P}\right)_T \cdot \mathrm{d}P$$

for a process at $V = const$ we have $\mathrm{d}V = 0$, thus

$$\left(\frac{\partial V}{\partial T}\right)_P \cdot \mathrm{d}T = -\left(\frac{\partial V}{\partial P}\right)_T \cdot \mathrm{d}P$$

or

$$\left(\frac{\partial P}{\partial T}\right)_V = -\frac{\left(\frac{\partial V}{\partial T}\right)_P}{\left(\frac{\partial V}{\partial P}\right)_T}$$

Using these facts, Equation (19.24) becomes

$$\begin{aligned} C_P - C_V &= T\left(\frac{\partial P}{\partial T}\right)_V \cdot \left(\frac{\partial V}{\partial T}\right)_P = -T\frac{\left(\frac{\partial V}{\partial T}\right)_P}{\left(\frac{\partial V}{\partial P}\right)_T} \cdot \left(\frac{\partial V}{\partial T}\right)_P \\ &= -T\left(\frac{\partial V}{\partial T}\right)_P^2 \cdot \frac{1}{(\partial V/\partial P)_T} \end{aligned}$$

Finally we express the quantities $(\partial V/\partial T)_P$ and $(\partial V/\partial P)_T$ by the cubic expansion coefficient α_V and the isothermal compressibility κ_T.

$$\kappa_T = -\frac{1}{V}\left(\frac{\partial V}{\partial P}\right)_T \qquad \alpha_V = \frac{1}{V}\left(\frac{\partial V}{\partial T}\right)_P \tag{19.25}$$

Both of these quantities are commonly reported in data tables (e.g., the *CRC Handbook of Chemistry and Physics*). Thus

$$\left(\frac{\partial V}{\partial P}\right)_T = -V\kappa_T \qquad \left(\frac{\partial V}{\partial T}\right)_P = V\alpha_V$$

We insert this relation in Equation (19.24) to obtain:

$$C_P - C_V = -T\left(V\alpha_V\right)^2\frac{1}{\left(-V\kappa_T\right)} = T \cdot V\frac{\alpha_V^2}{\kappa_T} \tag{19.26}$$

Because the expansion coefficient and the compressibility are positive for a stable system, the result demonstrates that $C_P > C_V$. Equation (19.26) was already used in Chapter 16, to convert measured values of C_p into values of C_V, see Table 16.2.

19.2.6 Maxwell Relations

Up to this point we have assumed that the enthalpy was written as a function of T and P; however, this is not a requirement. We can also write the enthalpy as a function of the two thermodynamic variables S and P, namely $H(S, P)$. In this case the total differential of H is

$$\mathrm{d}H = \left(\frac{\partial H}{\partial S}\right)_P \mathrm{d}S + \left(\frac{\partial H}{\partial P}\right)_S \mathrm{d}P \tag{19.27}$$

By comparing this result to Equation (19.20), $\mathrm{d}H = T\mathrm{d}S + V\mathrm{d}P$, we find that

$$\left(\frac{\partial H}{\partial S}\right)_P = T \quad \text{and} \quad \left(\frac{\partial H}{\partial P}\right)_S = V \tag{19.28}$$

Because H is a state function, its differential is exact and the cross-derivatives must be equal. Hence we require that

$$\left(\frac{\partial}{\partial P}\left(\frac{\partial H}{\partial S}\right)_P\right)_S = \left(\frac{\partial}{\partial S}\left(\frac{\partial H}{\partial P}\right)_S\right)_P \tag{19.29}$$

and substituting for the first derivatives gives

$$\left(\frac{\partial T}{\partial P}\right)_S = \left(\frac{\partial V}{\partial S}\right)_P \tag{19.30}$$

A number of relationships between the cross-derivatives of thermodynamic energy functions, called *Maxwell Relations*, can be derived (see Foundation 19.1), for example

$$\left(\frac{\partial V}{\partial T}\right)_P = -\left(\frac{\partial S}{\partial P}\right)_T \quad \left(\frac{\partial T}{\partial V}\right)_S = -\left(\frac{\partial P}{\partial S}\right)_V$$

$$\left(\frac{\partial P}{\partial T}\right)_V = \left(\frac{\partial S}{\partial V}\right)_T \quad \left(\frac{\partial T}{\partial P}\right)_S = \left(\frac{\partial V}{\partial S}\right)_P$$

The Maxwell relations are useful because they relate the entropy S, which is not directly measurable, to quantities that are directly measurable.

Example 19.3 *Use the Maxwell relation,* $\left(\frac{\partial P}{\partial T}\right)_V = \left(\frac{\partial S}{\partial V}\right)_T$, *to find how the entropy of an ideal gas changes upon isothermal expansion, and compare your result to the case of a van der Waals gas, for which* $\left(P + \frac{a}{V_m^2}\right) \cdot (V_m - b) = \boldsymbol{R}T$

Using the ideal gas equation of state, we can write that

$$\left(\frac{\partial S}{\partial V}\right)_T = \left(\frac{\partial P}{\partial T}\right)_V = \frac{\boldsymbol{R}}{V_m} \quad \text{or} \quad (\mathrm{d}S)_T = \boldsymbol{R}\left(\frac{\mathrm{d}V_m}{V_m}\right)_T$$

Upon integration from state 1 to state 2, we find that

$$\Delta S_m = \frac{S_2 - S_1}{\boldsymbol{n}} = \boldsymbol{R} \cdot \ln\left(\frac{V_{m,2}}{V_{m,1}}\right)$$

which agrees with that found in Section 17.5 for an ideal gas. For a van der Waals gas (see Chapter 20 for a discussion)

$$\left(P + \frac{a}{V_m^2}\right)(V_m - b) = \boldsymbol{R}T$$

or

$$P = \frac{\boldsymbol{R}T}{V_m - b} - \frac{a}{V_m^2}$$

we can write that

$$\left(\frac{\partial S}{\partial V}\right)_T = \left(\frac{\partial P}{\partial T}\right)_V = \frac{\boldsymbol{R}}{V_m - b} \quad \text{or} \quad (\mathrm{d}S)_T = \boldsymbol{R}\left(\frac{\mathrm{d}V}{V_m - b}\right)_T$$

Upon integration from state 1 to state 2, we find that

$$\Delta S_m = \frac{S_2 - S_1}{\boldsymbol{n}} = \boldsymbol{R} \cdot \ln\left(\frac{V_{m,2} - b}{V_{m,1} - b}\right)$$

Thus we see that the entropy change has a similar mathematical form to that for an ideal gas, but the terms used in the volume ratio for the van der Waals gas are corrected by the empirical factor b. As discussed in Chapter 20, b accounts for the finite size of the gas molecules.

Note that this procedure can be used to calculate the entropy change in an isothermal expansion for any pure substance if we know its equation of state (P, V, T relationship). The equation of state can be determined empirically for a substance—gases, liquids, and solids.

19.3 Spontaneity and Free Energy

In the previous section, we discussed properties of equilibrium states and reversible processes using the First and Second Laws of Thermodynamics. In this section we extend these discussions to include irreversible processes and the directionality a system takes in approaching equilibrium. The Second Law of Thermodynamics,

$$dS \geq \frac{\delta q}{T} \tag{19.31}$$

expresses the increase of entropy principle; for reversible processes $TdS = \delta q_{\text{rev}}$ and for irreversible processes $TdS > \delta q_{\text{irr}}$. Combining this result with the First Law of Thermodynamics, we find that

$$dU = \delta q + \delta w \leq TdS + \delta w \tag{19.32}$$

or, in the special case where only PV work occurs,

$$dU \leq TdS - PdV$$

For the reversible case we find the same result as in Equation (19.9),

$$\boxed{dU = TdS - PdV \qquad \text{(reversible processes)}} \tag{19.33}$$

and for the irreversible case

$$\boxed{dU < TdS - PdV \qquad \text{(irreversible processes)}} \tag{19.34}$$

This result has some important implications. For example, we find that pressures and temperatures equalize (or become uniform) as a system moves toward equilibrium in a spontaneous process. Example 4 shows this for the case of pressure.

Example 19.4 *Figure 19.1 shows a piston separating two chambers of an isolated system. Initially the pressure and volume of the gases are different, but when the piston is released the system evolves to an equilibrium state. Use the maximum entropy principle to show that the irreversible process in Fig. 19.1 proceeds to an equilibrium state for which the pressure on the two sides of the piston is equal.*

At equilibrium the entropy is a maximum, so that $dS = dS_1 + dS_2 = 0$ and we can write that

$$dS_1 = -dS_2$$

The total entropy change will be the sum of the differential changes in the left chamber and in the right chamber. We consider differential changes about the equilibrium state that are small enough to be reversible. Hence we can use Equation (19.33) to write

$$\frac{dU_1}{T} + \frac{P_1}{T}\,dV_1 = -\frac{dU_2}{T} - \frac{P_2}{T}\,dV_2$$

where T is the final temperature. Given that the system is isolated, $dU = 0 = dU_1 + dU_2$ and $dU_1 = -dU_2$, so that

$$P_1\,dV_1 = -P_2\,dV_2$$

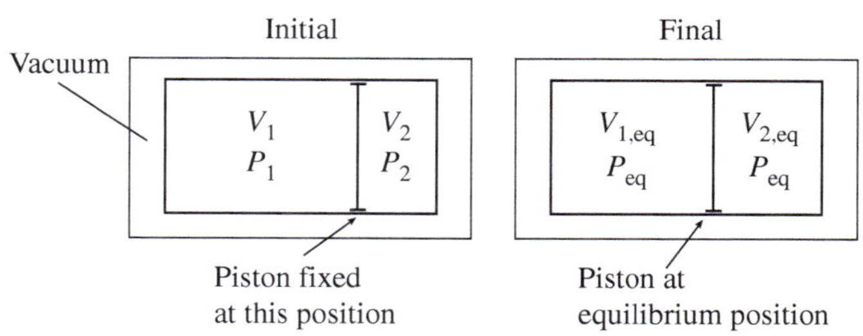

Figure 19.1 The diagram shows the case of an adiabatic enclosure containing two subchambers that are connected by a moveable piston. Initially the piston is fixed (arbitrary volumes V_1 and V_2 and pressures P_1 and P_2). When it is released the piston can move to reach the equilibrium position (equilibrium pressure P_{eq} and equilibrium volumes $V_{1,\text{eq}}$ and $V_{2,\text{eq}}$).

Because the system's total volume is fixed, $dV = dV_1 + dV_2 = 0$ and $dV_1 = -dV_2$. Hence, we must conclude that $P_1 = P_2 = P_{\text{eq}}$ at equilibrium.

19.3.1 Helmholtz Energy *A*

In Chapter 18 (Section 18.2) we defined the Helmholtz energy (also called the *free energy*) as

$$A = U - TS \tag{19.35}$$

Hence the differential Helmholtz energy $\mathrm{d}A$ is

$$\mathrm{d}A = \mathrm{d}U - T\mathrm{d}S - S\mathrm{d}T \tag{19.36}$$

If we substitute this expression into Equation (19.32), we find that

$$\mathrm{d}A \leq -S\mathrm{d}T + \delta w \tag{19.37}$$

For the special case of only PV work ($\delta w = -P\mathrm{d}V$) and a reversible process, we have

$$\mathrm{d}A = -S\mathrm{d}T - P\mathrm{d}V \tag{19.38}$$

For isothermal ($T = const.$) processes, Equation (19.37) reduces to

$$\mathrm{d}A \leq \delta w$$

For a reversible isothermal process, the equality holds and we can write

$$\mathrm{d}A = \delta w_{\mathrm{rev}} \quad \text{(reversible change at } T = const.\text{)} \tag{19.39}$$

The differential Helmholtz energy is equal to the reversible work for an isothermal process. For an irreversible process, the inequality holds and we can write

$$\mathrm{d}A < \delta w \text{ (irreversible change at } T = const.\text{)} \tag{19.40}$$

This result agrees with our findings in Section 18.2. The Helmholtz energy corresponds to the maximum amount of work that can be performed by the system in an isothermal process. If the process is reversible, then the equality holds and the maximum possible work is performed. If the process is irreversible, then less than the maximum possible work is performed.

In the limit of only PV work, Equation (19.37) becomes

$$\mathrm{d}A \leq -S\mathrm{d}T - P\mathrm{d}V$$

so that for a process at constant temperature and constant volume

$$\mathrm{d}A \leq 0 \qquad (T = const.,\ V = const.)$$

This result is the differential version of our earlier (see Section 18.2) result

$$\Delta A \leq 0 \qquad (T = const.,\ V = const.)$$

and means that a spontaneous process, occurring at constant T and V, must have a negative change in the Helmholtz energy.

19.3.2 Gibbs Energy G

In Chapter 18 we defined the Gibbs energy (also called the *free enthalpy*) as

$$G = H - TS \tag{19.41}$$

Introducing Equations (19.18) and (19.35), we obtain

$$G = (U + PV) - TS = (U - TS) + PV = A + PV \tag{19.42}$$

Hence the differential Gibbs energy $\mathrm{d}G$ is

$$\mathrm{d}G = \mathrm{d}A + P\mathrm{d}V + V\mathrm{d}P \tag{19.43}$$

Using Equation (19.37), $\mathrm{d}A \leq -S\mathrm{d}T + \delta w$, we find that

$$\mathrm{d}G \leq -S\mathrm{d}T + \delta w + P\mathrm{d}V + V\mathrm{d}P \tag{19.44}$$

For the special case of only PV work ($\delta w = -P\mathrm{d}V$), we find

$$\mathrm{d}G = -S\mathrm{d}T + V\mathrm{d}P \tag{19.45}$$

for a reversible process, and

$$\mathrm{d}G < -S\mathrm{d}T + V\mathrm{d}P \tag{19.46}$$

for an irreversible process.

For processes that occur at constant temperature (isothermal) and constant pressure (isobaric), we find that

$$dG \leq \delta w + PdV \qquad (T = const.,\ P = const.) \tag{19.47}$$

If we write δw as a sum of PV work and non-PV work ($\delta w = \delta w_{\text{nonPV}} - PdV$), we find that

$$dG \leq \delta w_{\text{nonPV}} \qquad (T = const.,\ P = const.) \tag{19.48}$$

The Gibbs energy function provides a bound on the maximum non-PV work performed by a system. In the reversible limit, we find that

$$dG = \delta w_{\text{nonPV,rev}} \qquad (T = const.,\ P = const.) \tag{19.49}$$

This feature is extremely useful in a number of important situations. For example, the design and construction of electrochemical cells (batteries, fuel cells, etc.) can be analyzed by considering the Gibbs energy changes and allows us to focus on the useful (electrical) work rather than the PV work which the device might perform.

In the special case where only PV work is performed, Equation (19.47) simplifies to

$$dG \leq 0 \qquad (T = const.,\ P = const.) \tag{19.50}$$

Using the definition (19.41), $G = H - TS$, we see that for an isothermal process

$$dG = dH - TdS \qquad (T = const.) \tag{19.51}$$

These differential expressions correspond directly to our integral expressions from Section 18.2,

$$\Delta G \leq 0 \qquad (T = const.,\ P = const.) \tag{19.52}$$

and

$$\Delta G = \Delta H - T\Delta S \qquad (T = const.) \tag{19.53}$$

According to Equation (19.50) the Gibbs energy must decrease for a spontaneous process in which T and P are constant. The Gibbs energy may be treated as a potential function (or thermodynamic potential) for the case of constant temperature and pressure processes. These processes are very common in chemistry and this potential function is quite useful.

19.4 Phase Equilibria and Phase Transitions

In order to see the usefulness of purely thermodynamic considerations, we consider the case of first-order phase transitions of a substance.[2] From experience we know that a substance can exist in different phases. A substance has one vapor phase, one liquid phase, and one or more solid phases (corresponding to different crystal structures). Although thermodynamics cannot predict the properties, or even existence of these phases, thermodynamics can predict which phase is the most stable under a given set of conditions. Thermodynamically, each particular phase of a substance is defined by its $G(T, P)$ for that phase. If two phases have the same value of $G(T, P)$, e.g., the surfaces for each phase can cross or touch each other, then G for some values of T and P could correspond to one or more phases and a phase equilibrium occurs.

To illustrate this abstract thermodynamic view more physically, consider the effect of temperature on water at 1 bar pressure. Near zero Kelvin, the water is solid and the water's enthalpy is determined by the binding energy of the water molecules in the solid with little excitation of internal modes. We expect that the ordered solid form, ice, will have a lower enthalpy than a form which is disordered, e.g., frozen into a liquid-like structure, because the ordered array of the water molecules in the crystal optimize the binding forces (hydrogen bonds). Moreover, the Gibbs energy G and the enthalpy H for the solid are the same as T approaches zero, because $G = H - TS$ and S decreases as T decreases. As the temperature increases, the water molecule's internal excitation and motion increases; the vibrational motions store energy in the crystal, and the enthalpy increases. As the ice is warmed the enthalpy increases, but we also find that the number Ω of configurations increases, hence the entropy S increases. The net effect causes the Gibbs energy to decrease; see Figure 19.2.

2 Here we discuss phase transitions in which the Gibbs energy curves cross, which gives rise to a discontinuous change in the enthalpy and entropy at the crossing point. It is also possible that the Gibbs energy curves can touch, but not cross. Such a situation gives rise to a critical point and is discussed more below.

Figure 19.2 shows the Gibbs energy change for a solid and a hypothetical liquid form below the freezing point. As for the case of the solid, the Gibbs energy and enthalpy of the liquid are equal near $T = 0$. The enthalpy and Gibbs energy of the liquid are higher than that of the solid at $T = 0$ K; the solid is the more stable phase at low temperature. Because the liquid phase has a higher entropy than the solid (the liquid can access translational and rotational motions as well as vibrational ones), its Gibbs energy decreases more rapidly with increasing temperature than that of the solid; i.e., recall that $(\partial G/\partial T)_P = -S$. At some temperature T_{fus} the Gibbs energy curves can cross.[3]

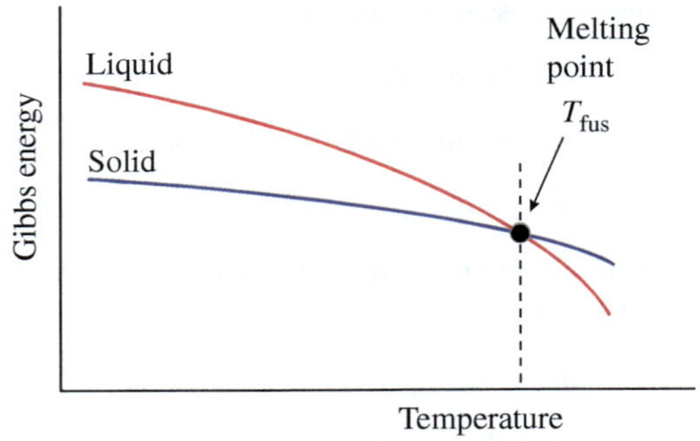

Figure 19.2 Temperature effect on the Gibbs energy of a solid (blue line) and a liquid (red line) at constant pressure. Melting occurs at the temperature where the two Gibbs energy curves cross.

At the particular temperature T_{fus}, the molar Gibbs energy of the liquid and solid forms is equal (i.e., $\Delta_{fus}G = G_{m,\text{liquid}} - G_{m,\text{solid}} = 0$) and the solid ice begins to melt. Although the Gibbs energies of the two phases are the same, their molar enthalpies and molar entropies are different. As we continue to add heat to the system the temperature does not change, rather the heating excites the translational and rotational motions of the water molecules, melts the ice. This process continues until all of the ice is melted. At this point the system has increased its molar enthalpy by $\Delta_{fus}H = H_{m,\text{liquid}} - H_{m,\text{solid}}$ and increased its molar entropy by $\Delta_{fus}S = S_{m,\text{liquid}} - S_{m,\text{solid}}$, but its total Gibbs energy has not changed. Because the process can be performed reversibly, and the temperature is fixed at T_{fus}, we find according to Equation (17.61)

$$\Delta S = \frac{\Delta H}{T_{fus}} \quad \text{or} \quad \Delta_{fus}S = \frac{\Delta_{fus}H}{T_{fus}} \tag{19.54}$$

Conditions in which both phases coexist are called biphasic.

19.4.1 Solid–Liquid Equilibria

While thermodynamics is mute about the mechanism of phase transitions, it can make useful predictions about how they depend on the system variables, T and P. For example, the melting temperature T_{fus} of water ice changes with the pressure, from 0 °C at 1 bar pressure to −1.5 °C for a pressure of 200 bar. Usually the melting temperature of a substance increases with pressure; however, solid water (ice) is less dense at 1 bar than the coexisting liquid water (ice floats) and, as a consequence, the melting point decreases with pressure.

Melting of a solid is a physical transformation represented by

$$\text{solid} \rightarrow \text{liquid} \tag{19.55}$$

For equilibrium between the solid and liquid phases,[4]

$$\Delta_{fus}G = G_{m,\text{liquid}} - G_{m,\text{solid}} = 0 \quad \text{and} \quad G_{m,\text{liquid}} = G_{m,\text{solid}}$$

At constant temperature, the differential Gibbs energy is, according to Equation (19.45),

$$\left(\mathrm{d}G_m\right)_T = V_m \cdot \mathrm{d}P$$

This equation shows that a change in the pressure can shift the equilibrium position if the solid and liquid forms have different molar volumes. From a more physical perspective, the reversible work performed on the system is larger for the phase (solid or liquid) that has the larger molar volume.

In Box 19.1 we show that the temperature and pressure dependence of the solid–liquid phase equilibrium is described by

$$\boxed{\frac{\mathrm{d}P}{\mathrm{d}T} = \frac{\Delta_{fus}S}{\Delta_{fus}V} = \frac{\Delta_{fus}H}{T_{fus}\Delta_{fus}V}} \tag{19.56}$$

3 We note that thermodynamics does not state whether the solid or liquid form is more stable; within thermodynamics the enthalpy and entropy must be measured. It is through our molecular understanding of solids and liquids that we can provide the rationale for their enthalpy and entropy values.

4 It is important to use molar quantities here because the amounts of solid and liquid change during the melting process. Although the total Gibbs energy $G_{total} = G_{liquid} + G_{solid}$ is a constant, the magnitude of $G_{solid} = n_{solid}G_{m,\text{solid}}$ decreases and $G_{liquid} = n_{liquid}G_{m,\text{liquid}}$ increases as the solid melts.

where $\Delta_{fus}S = S_{m,liquid} - S_{m,solid} = \Delta_{fus}H/T_{fus}$, $\Delta_{fus}H$ is the molar enthalpy of melting, and T_{fus} is the temperature of the melting point. Equation (19.56) is called the *Clapeyron equation* for the solid-to-liquid transition.

Example 19.5 *What is the change in the temperature of ice in equilibrium with liquid water if we apply an external pressure of 200 bar?*

To determine the effect of pressure on the melting temperature T_{fus}, we apply the Clapeyron equation (19.56)

$$dT = \frac{\Delta_{fus}V}{\Delta_{fus}S} \cdot dP$$

If we assume that the molar volume and molar entropy of the two phases do not change much with pressure, we can use

$$\Delta_{fus}V = -1.64 \text{ cm}^3 \text{ mol}^{-1}, \quad \Delta_{fus}S = \frac{\Delta_{fus}H}{T_{fus}} = \frac{6.01 \text{ kJ mol}^{-1}}{273.16 \text{ K}} = 22.0 \text{ J K}^{-1} \text{ mol}^{-1}$$

for these values at 1 bar, see Appendix A.[5] Then for $\Delta P = 200$ bar we find

$$\Delta T_{fus} = \frac{-1.64 \text{ ml mol}^{-1}}{22.0 \text{ J K}^{-1} \text{ mol}^{-1}} 200 \text{ bar} = \frac{-1.64}{22.0} 10^{-6} \frac{\text{m}^3}{\text{N m}} 200 \frac{\text{N}}{\text{m}^2} 10^5 \text{ K} \approx -1.49 \text{ K}$$

This means that the freezing point of ice at a pressure of 200 bar is about 1.5° lower than the freezing point at atmospheric pressure. For comparison, the pressure exerted on ice by a skater is on the order of 80 bar if the weight of the skater is 800 N and the area of the skate in contact with the ice is 1 cm^2. Note that this pressure is too small to reduce the melting point significantly. Why is skating on ice possible even at temperatures far below the freezing point? The reason is that near to the surface of solid ice a thin layer of liquid water exists, even at temperatures below −150 °C, see Hong, J., Tian, Y., Liang, T. et al. Imaging surface structure and premelting of ice Ih with atomic resolution. Nature (2024). https://doi.org/10.1038/s41586-024-07427-8.

19.4.1.1 Solid–Liquid Phase Coexistence

Figure 19.3 illustrates the pressure effect on a generic solid–liquid phase transition, by sketching how the Gibbs energies of the solid and liquid change at two different pressures. For both phases we expect that the Gibbs energy of the substance will increase as we increase the external pressure, because we are performing work on the system. The increase in the Gibbs energy is determined by the molar volume change of the phase when increasing the pressure. For most substances the solid is more dense and its molar volume changes less than that of its liquid form. Hence the Gibbs energy of the liquid increases more with an increase in pressure than does that of the solid.

Because of their different molar volumes, the temperature at which the Gibbs energy curves cross is shifted; see Fig. 19.3 For most liquids the shift is to higher temperature with higher pressure. A plot of the pressures and temperatures at which the two phases have the same Gibbs energy comprises the coexistence curve; i.e., it shows the (T, P) conditions needed for biphasic equilibrium. The Clapeyron equation, Equation (19.56), states that the slope of the coexistence curve is determined by the ratio of the molar entropy change in the phase transition to the molar volume change in the phase transition. Values of pressure

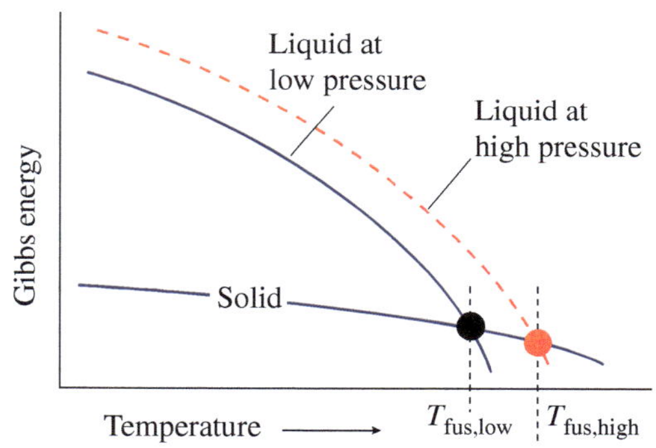

Figure 19.3 Temperature dependence of the Gibbs energy for a solid and a liquid at two different pressures. Because the liquid is more compressible than the solid, its Gibbs energy changes more, thereby shifting the temperature of the crossing point and the temperature of the phase transition. $T_{fus,low}$ (black circle) and $T_{fus,high}$ (open circle) are the melting points at high and low pressure, respectively. In this example, the solid's Gibbs energy is assumed to not change.

---

5 $\Delta_{fus}V$ is calculated from the densities $\rho_{liquid} = 0.999818$ g/ml for liquid water and $\rho_{solid} = 0.9168$ g/ml for solid water at 0 °C.

$$\begin{aligned}\Delta_{fus}V &= (18.02 \text{ g mol}^{-1}/\rho_{solid}) - (18.02 \text{ g mol}^{-1}/\rho_{liquid}) \\ &= (18.02 - 19.66) \text{ ml mol}^{-1} \\ &= -1.64 \text{ ml mol}^{-1}\end{aligned}$$

and temperature to the left of the coexistence curve are conditions under which the solid is the most stable phase (lowest Gibbs energy), and pressure and temperature values to the right of the line are conditions under which the liquid is the most stable phase (lowest Gibbs energy). See Section 19.4.3 for more discussion.

19.4.2 Liquid–Gas and Solid–Gas Equilibria

After all of the ice has melted in our thought experiment, the continued application of heat acts to warm the liquid. As the liquid water is heated the average translational kinetic energy of the water molecules increases. Correspondingly, the percentage of molecules with enough kinetic energy to escape the liquid increases and the vapor pressure of the liquid increases (see Section 18.4). When the temperature becomes high enough that the vapor pressure of the liquid equals the ambient pressure, the molecules in the liquid begin to vaporize

$$H_2O(\text{liquid}) \rightarrow H_2O(\text{vapor}) \tag{19.57}$$

This temperature is the boiling point temperature. At the boiling point temperature, additional heating of the liquid converts more water molecules from the liquid phase into vapor; and the temperature does not change until all of the water molecules are vaporized. This behavior is similar to that found for the liquid-to-solid transition and we can write the Clapeyron equation (19.56) as

$$\boxed{\frac{dP}{dT} = \frac{\Delta_{vap}S}{\Delta_{vap}V} = \frac{\Delta_{vap}H}{T_{vap} \cdot \Delta_{vap}V}} \tag{19.58}$$

where $\Delta_{vap}S = \Delta_{vap}H/T_{vap}$ where T_{vap} is the boiling temperature.

Because the vapor has a much larger molar volume than the liquid ($V_{vapor} \gg V_{liquid}$), it is often a good approximation to let $\Delta_{vap}V \approx V_{m,vapor}$. If we further assume that the vapor behaves like an ideal gas we can write that

$$\Delta_{vap}V = V_{m,vapor} = \frac{\boldsymbol{RT}}{P} \tag{19.59}$$

Substituting this expression into the Clapeyron equation (19.58) we obtain

$$\frac{dP}{dT} = P \cdot \frac{\Delta_{vap}H}{\boldsymbol{R}T_{vap}^2} \quad \text{or} \quad d\ln\widehat{P} = \frac{\Delta_{vap}H}{\boldsymbol{R}T_{vap}^2}\, dT \tag{19.60}$$

This result is the Clausius–Clapeyron equation which we discussed in Chapter 18.

Example 19.6 *At 501 K and 26.9 bar water boils and its molar enthalpy of vaporization is $\Delta_{vap}H = 32.83$ kJ mol^{-1}. Under these conditions, the liquid water has a molar volume of $V_{m,liquid} = 0.0217$ L mol^{-1} and the gaseous water (steam) has a molar volume of $V_{m,gas} = 1.340$ L mol^{-1}. Use these data to determine the slope of the vapor pressure curve. Compare your finding to the value you obtain using the more approximate Clausius–Clapeyron equation.*

Substituting these data into Equation (19.58) we find

$$\frac{dP}{dT} = \frac{32.83 \times 10^3 \text{ J mol}^{-1}}{(501 \text{ K})(1.328 \times 10^{-3} \text{ m}^3 \text{ mol}^{-1})} = 0.493\frac{\text{bar}}{\text{K}}$$

If we use the Clausius–Clapeyron equation, which assumes that the water vapor behaves ideally, we find

$$\frac{dP}{dT} = (26.9 \text{ bar})\frac{32.83 \cdot 10^3 \text{ J mol}^{-1}}{(8.3145 \text{ J mol}^{-1} \text{ K}^{-1})\,(500 \text{ K})^2} = 0.423\frac{\text{bar}}{\text{K}}$$

The Clausius–Clapeyron equation underestimates the slope by 14%.

19.4.2.1 Liquid–Gas and Solid–Gas Phase Coexistence

The Clapeyron equation gives the slope of the coexistence curve between the liquid and gas phases of a substance on a P versus T diagram—as we discussed for the solid–liquid coexistence curve. In contrast to the case of the solid–liquid coexistence curves, the solid–gas and liquid–gas coexistence curves always have a positive slope, see Figure 19.4. The slope is positive because the entropy of the gas is always higher than that of the liquid or solid and the molar volume of the gas is always higher than that of the liquid or solid.

Thermodynamics provides important definite statements about the nature of the phases and their coexistence, but it does not make predictions about the value of the molar entropy changes or the molar volume changes. These quantities have to be measured. In contrast, molecular-based models in combination with statistical considerations allow us to predict these quantities and hence the coexistence curves; see Foundation 20.1.

19.4.3 Phase Diagrams and Phase Rule

19.4.3.1 *P–T* Phase Diagrams

Figure 19.4 shows a generic phase diagram for a substance in the P/T representation (volume V of the system kept constant). Three different single phases which correspond to the solid, the liquid, and the vapor are evident in the diagram and are defined by the corresponding (P, T) conditions. Three biphasic lines which correspond to the solid–liquid, solid–gas, and gas–liquid are shown in the diagram also. One finds that the solid–gas coexistence curve and the liquid–gas coexistence curve always have a positive slope. At the triple point three phases (solid, liquid, vapor) coexist. A special point (the *critical point*) on this diagram occurs at (P_c, V_c, T_c). At temperatures higher than the critical temperature T_c the phase transition between the liquid and gas is no longer apparent; i.e., liquid–gas phase line terminates at a particular pressure and temperature, the critical point. The nature of the critical point is discussed more below.

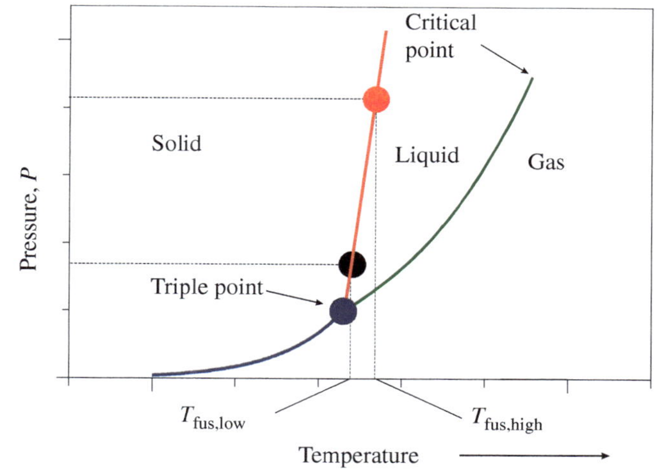

Figure 19.4 P/T phase diagram. The black and red circles indicate the positions of the melting points corresponding to Fig. 19.3. The blue circle indicates the triple point.

The geometric structure of a PT phase diagram can be understood from fundamental considerations. The Gibbs energy G of a substance can be specified by specifying T, P, and the amount of substance (or V). In the PT diagram the volume is fixed so that the amount of substance is fixed. Hence two variables (P and T) are needed to specify the state of the substance, and this is why the single-phase regions are areas (two dimensional). If two phases I and II coexist, then equilibrium requires that $G_I = G_{II}$. This condition means that the pressure and temperature are not independent; they are related through the Clapeyron equation. Hence the two-phase condition only has one independent variable and appears as a line (one dimensional). For the case of three phases coexisting, two equilibrium conditions apply ($G_I = G_{II} = G_{III}$) so that the system has no independent variables. Hence the three-phase condition corresponds to a point (zero dimensional).

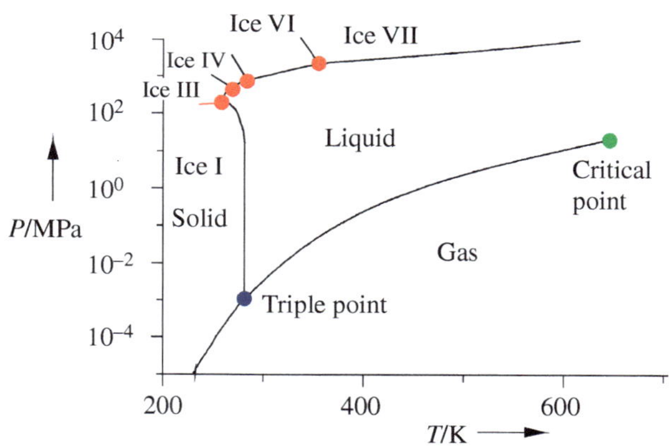

Figure 19.5 P/T phase diagram for water,[6] illustrating the numerous solid forms of ice. Note the logarithmic pressure axis. The triple point is at a temperature of 273.16 K and a pressure of 6.117 mbar.

While Fig. 19.4 shows essential elements of solid:liquid:gas phase equilibria, phase diagrams of substances often show a much richer behavior. Figure 19.5 shows an experimentally determined P versus T phase diagram for water. The diagram[7] shows that water has multiple solid phases, each of which corresponds to a distinct crystal structure, at low temperatures and or high pressures. The liquid and vapor phases of water are realized at high temperatures and or low pressures. Above the critical pressure of 218 bar and/or 647.30 K the fluid water no longer undergoes a liquid to vapor phase transition. Under these conditions the water is called "*supercritical*."

6 Redrawn from [1].
7 Other phase diagrams may be found in the International Critical Tables, or [2].

19.4.3.2 *P–V* Phase Diagram

Although somewhat less common than a P/T phase diagram, the two-dimensional projection of the three-dimensional (P, V, T) diagram of a substance onto the P/V plane can be useful also.

Figure 19.6 shows a P/V phase diagram for water (curves at constant temperature: isotherms). Consider the isotherm for 573 K. At point A on the isotherm shown, the substance is at low pressure and large volume and is a gas. As the gas is compressed (move to smaller volume) isothermally, point B is reached at a pressure of 84.5 bar and a molar volume of, $V_m = 0.39$ L mol^{-1}. This point corresponds to the saturated vapor. Further compression (move to smaller volume) causes the vapor to condense. Neither the pressure nor the temperature change, rather the relative amounts of the vapor and the liquid change along the isotherm. In fact, the position along the isotherm (i.e., system volume) may be used to determine the relative amounts of vapor and liquid. Upon compression all the way to point C (molar volume of, $V_m = 0.025$ L mol^{-1}), all of the vapor has been converted to the liquid form (this point is called the *saturated liquid*). Further compression causes a large rise in pressure (the compressibility of a liquid is much smaller than its vapor) and the system's thermodynamic state moves into the single phase liquid region.

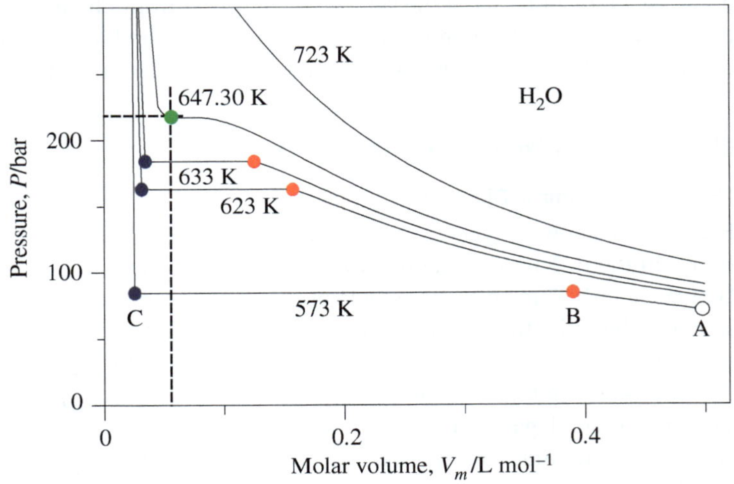

Figure 19.6 P/V phase diagram for water. Isotherms at different temperatures (573, 623, 633, 647.30, and 723 K).[8] The open circle (point A) indicates the gaseous region. The filled circles indicate the limits of the line segment marking the gas–liquid coexistence region—where the right-hand side (points B, red) is gas and the left-hand side (points C, blue) is liquid. The green full circle indicates the critical point.

19.4.3.3 Critical Point

Inspection of the P/V diagram reveals that the difference in density between gas (saturated vapor) and coexisting liquid decreases with increasing temperature. In Fig. 19.6 the horizontal line segments (biphasic regions) become shorter as T increases (compare the curves for 623 and 633 K). In the isotherm for 647.30 K this line segment shrinks to a point. We can no longer distinguish between vapor and liquid. This point is the critical point ($V_m = 0.0563$ L mol^{-1}, $P = 218$ bar).

At temperatures above the critical point temperature, compressing the dilute gas (first behaving as an ideal gas) proceeds continuously into a liquid (curve for 723 K); no biphasic region exists. If we then cool, this liquid, it will continuously develop its specific cluster structure. This fact means that it is possible to change the system from a gas to a liquid without passing through a phase transition; stated more abstractly, the molar Gibbs energy surface for the liquid and the gas are the same surface however it has a complicated shape and intersects with itself.

How can we distinguish between phases? The distinction between a liquid, a solid, and a gas may be illustrated by the difference in their density functions. In general the density of a substance will depend on the distance in some way. Let us consider a monatomic substance and choose our origin at the center of one of the atoms. Figure 19.7a shows the dependence of the density of a gas away from the center of the gas atom and compares it to that of a liquid. Within the radius of the atom $\sigma/2$, all other gas atoms are excluded and the density of other atoms here is zero. Once at the radius of the atom we see that density shows a value that is greater than the average value ρ_0 but monotonically

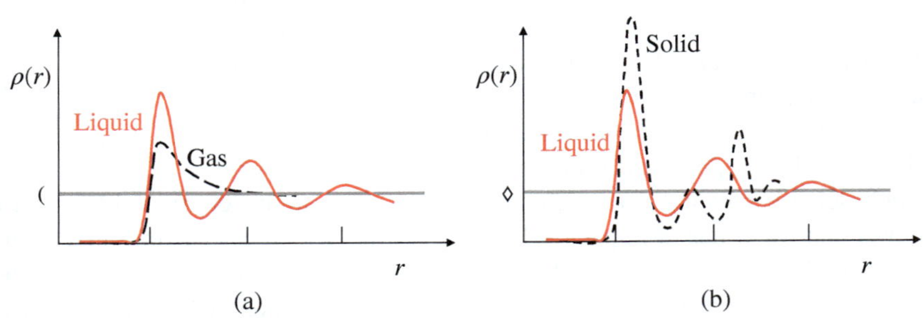

Figure 19.7 Density function $\rho(r)$. (a) Comparison between gas and liquid and (b) comparison between solid and liquid.

8 Experimental data from [3].

decays to this value with distance. In contrast the density function of a solid in Fig. 19.7b is highly structured. The peaks in this curve represent the positions of atoms in the solid's lattice. The slight broadening of the peaks is meant to illustrate the fact that the atoms oscillate about their equilibrium position.

The density function for a liquid lies somewhere between that of the solid and the gas. The features in the density illustrate that structure exists in the liquid, because of attractive interactions and packing forces. As with the gas it is clear that the other atoms in the liquid are excluded by a strong repulsive force associated with their electron clouds and Pauli repulsion. Again we see a density greater than the average just outside this radius, and it proceeds for a few atomic diameters. This distance dependence in the density represents the solvation shells of other atoms in the liquid about the atom of interest, eventually however (as the distance increases) these structural features decay away and the density approaches its average "long distance" value. To summarize, a solid is highly ordered (structured), a gas is not ordered, and a liquid has some local order.

A second distinction between these phases is the relative importance of interatomic interactions and the kinetic energy of the constituent atoms. In gases, we find that the interaction between molecules is quite limited so that an accurate treatment of the kinetic energy of the particles, the ideal gas approximation, is often sufficient. Corrections to this approximation are discussed in the next chapter. As demonstrated above, solids lie at the other extreme. For the case of the solid the potential energy interactions dominate and the kinetic energy terms are least important. In fact, our earlier discussion of the heat capacity of a solid approximated it as one large molecule. Liquids are more challenging to describe because both the kinetic energy terms and the potential energy terms are important. Hence neither can be neglected.

19.4.3.4 Vapor Pressure

The phase diagrams provide the vapor pressure curve of a liquid or solid. For example, the liquid–gas coexistence curve in the P/T diagram corresponds to the vapor pressure of the liquid substance. A plot of the saturation pressure versus the saturation temperature is the vapor pressure curve. Figure 19.8 shows this curve for water (red solid curves). A similar argument applies for the gas–solid equilibrium (blue solid curves in Fig. 19.8).

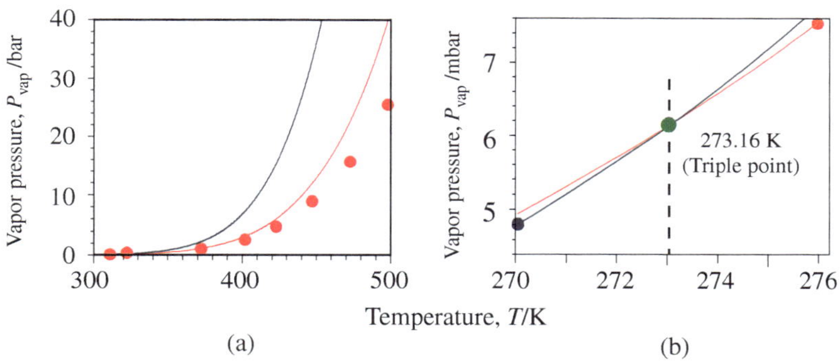

Figure 19.8 Vapor pressure P_{vap} of water as a function of temperature calculated from the Clausius–Clapeyron equation (19.61), using $P_{vap,1} = 6.115$ mbar at 273 K, $\Delta_{vap}H_1 = 44.1$ kJ mol^{-1} and $\Delta_{sub}H_1 = 50.1$ kJ mol^{-1}. Red solid lines: vapor pressure above liquid water, blue solid lines: vapor pressure above solid water. (a) Temperature range 300–500 K, filled red circles: experiment. (b) Temperature range 270–276 K. Filled circles: experiment: green, triple point (273.16 K, 6.117 mbar); blue, vapor pressure above solid water (270.15 K, 4.76 mbar); red, vapor pressure above liquid water (276.15 K, 5.69 mbar). Experimental data see Appendix A.5.

As discussed above the coexistence curve for the liquid–vapor (or solid–vapor) is described by the Clapeyron equation. In the approximation that the gas behaves ideally, we can use the Clausius–Clapeyron equation (18.41)

$$\ln \hat{P}_{vap} = \ln \hat{P}_{vap,1} - \frac{\Delta_{vap}H}{\boldsymbol{R}}\left(\frac{1}{T} - \frac{1}{T_1}\right) \tag{19.61}$$

This equation enables us to calculate the vapor pressure P_{vap} at temperature T from the vapor pressure $P_{vap,1}$ at temperature T_1. In this derivation the molar heat of vaporization was assumed to be independent of temperature. In Problem P19.12 you are asked to derive a corresponding expression, taking into account the temperature dependence of $\Delta_{vap}H$. The result is

$$\ln \hat{P}_{vap} = \ln \hat{P}_{vap,1} - \frac{\Delta_{vap}H_1}{\boldsymbol{R}}\left(\frac{1}{T} - \frac{1}{T_1}\right) + \frac{\Delta C_{P,m}}{\boldsymbol{R}}\left[\ln \frac{T}{T_1} + T_1\left(\frac{1}{T} - \frac{1}{T_1}\right)\right] \tag{19.62}$$

where $\Delta C_{P,m}$ is the difference of the molar constant pressure heat capacities of vapor and liquid ($\Delta C_{P,m}$ is assumed to be independent of temperature). This equation is a commonly used mathematical form for describing the vapor pressure curves of substances.

Example 19.7 *Compare the result of Equations (19.61) and (19.62) for the vapor pressure of water at $T = 500$ K, using $P_{\text{vap},1} = 1$ bar at $T_1 = 373$ K.*

Under standard conditions, for water we have $\Delta_{\text{vap}}H^\ominus = (-241.8 + 285.9)$ kJ mol^{-1} = 44.1 kJ mol^{-1} and $\Delta C_P = (33.6 - 75.3)$ J K^{-1} mol^{-1} = -41.7 J K^{-1} mol^{-1} see data in Appendix A.

Then from Equation (19.61) we obtain

$$\ln \widehat{P}_{\text{vap}} = -\frac{44.1 \text{ kJ mol}^{-1}}{8.314 \text{ J K}^{-1} \text{ mol}^{-1}} \left(\frac{1}{500 \text{ K}} + \frac{1}{373 \text{ K}} \right) = 3.61 \quad \text{or} \quad P_{\text{vap}} = 36.9 \text{ bar}$$

To compare to Equation (19.62) we have to add the additional term

$$\frac{\Delta C_{P,m}}{\boldsymbol{R}} \left[\ln \frac{T}{T_1} - T_1 \left(\frac{1}{T} - \frac{1}{T_1} \right) \right]$$

$$= \frac{-41.7 \text{ J K}^{-1} \text{ mol}^{-1}}{8.314 \text{ J K}^{-1} \text{ mol}^{-1}} \left[\ln \frac{500 \text{ K}}{373 \text{ K}} + 373 \text{ K} \left(\frac{1}{500 \text{ K}} - \frac{1}{373 \text{ K}} \right) \right]$$

$$= -5.02 \cdot [0.293 - 0.254] = -0.20$$

Then

$$\ln \widehat{P}_{\text{vap}} = 3.61 - 0.20 = 3.41 \quad \text{or} \quad P_{\text{vap}} = 30.3 \text{ bar}$$

This is a change by 18%.

See Problem P19.11 for a calculation of the vapor pressure in the range 300–500 K.

19.5 Conclusion

In this chapter we developed classical thermodynamics for a substance and applied it to phase equilibria of substances. Although thermodynamics provides an accurate description of substances, it requires empirical measurement of a substance's properties, such as heat capacity and compressibility. In the next chapter we will develop this description further for the case of real gases, for which we can replace the empirical measurement with equations of state and/or with molecular-based models.

19.6 Box 19.1. Clapeyron Equation

We consider the melting of a solid

$$\text{solid} \rightarrow \text{liquid}$$

At equilibrium, the molar Gibbs energy of the two phases is equal, so that

$$\Delta G = G_{m,\text{liquid}} - G_{m,\text{solid}} = 0 \quad \text{and} \quad G_{m,\text{liquid}} = G_{m,\text{solid}}$$

For a differential change that maintains the equilibrium (e.g., for differential changes of pressure and temperature)

$$\text{d}G_{m,\text{liquid}} = \text{d}G_{m,\text{solid}} \tag{19B1.1}$$

Using our basic equation (19.45) for the differential Gibbs energy

$$\text{d}G_m = V_m \text{d}P - S_m \text{d}T$$

we find that

$$V_{m,\text{liquid}} \cdot \mathrm{d}P - S_{m,\text{liquid}} \cdot \mathrm{d}T = V_{m,\text{solid}} \cdot \mathrm{d}P - S_{m,\text{solid}} \cdot \mathrm{d}T \tag{19B1.2}$$

which rearranges to the form

$$\frac{\mathrm{d}P}{\mathrm{d}T} = \frac{\Delta_{\text{fus}}S}{\Delta_{\text{fus}}V}$$

with

$$\Delta_{\text{fus}}S = S_{m,\text{liquid}} - S_{m,\text{solid}} \quad \text{and} \quad \Delta_{\text{fus}}V = V_{m,\text{liquid}} - V_{m,\text{solid}}$$

19.7 Exercises

E19.1 Show that the relations

$$\Delta G = \int V\,\mathrm{d}P \qquad \text{and} \qquad \Delta G = w_{\text{rev}} = -\int P\,\mathrm{d}V$$

are equivalent for the isothermal expansion of an ideal gas.

E19.2 Use Equation (19.3) and the combined form of the first and second laws to find a general expression for the differential entropy in terms of directly measurable quantities.

E19.3 a) What is the criterion for a spontaneous process at constant T and P? b) What is the criterion for a spontaneous process at constant S and V?

E19.4 Beginning with the Sackur–Tetrode Equation (17.49) and $U = C_V T$, show that the internal energy is

$$U(S,V) = c_1 \cdot V^{-R/C_V} \exp\left(S/C_V\right)$$

for an ideal gas. In the first part of your derivation, eliminate T from the Sackur–Tetrode equation by replacing it with U, and in the second part of the equation rearrange your results to write U as a function of S and V. Using your knowledge of thermodynamics and this $U(S,V)$, derive a) the ideal gas equation of state and b) show that the internal energy U is a function of T only.

E19.5 In Section 19.2.6 we derived the Maxwell relation

$$\left(\frac{\partial T}{\partial P}\right)_S = \left(\frac{\partial V}{\partial S}\right)_P$$

Derive the other three Maxwell relations

$$\left(\frac{\partial V}{\partial T}\right)_P = -\left(\frac{\partial S}{\partial P}\right)_T \quad \left(\frac{\partial P}{\partial T}\right)_V = \left(\frac{\partial S}{\partial V}\right)_T \quad \left(\frac{\partial T}{\partial V}\right)_S = -\left(\frac{\partial P}{\partial S}\right)_V$$

from their corresponding energy functions: G, A, and U.

E19.6 Imagine a vertical tube filled with mercury (height $h = 10$ m). What is the freezing point of the mercury at the bottom when the freezing point at the top is 234.3 K?

E19.7 What is the boiling point of water on the top of Mount Everest (8882 m above sea level and a pressure of 0.34 bar)? You may take the enthalpy change for the vaporization of water $\Delta_{\text{vap}}H$ to be 32.83 kJ mol^{-1}.

E19.8 Sketch and label a P versus T phase diagram for a pure substance which has a liquid phase at $P = 1$ bar. Indicate the normal melting point, boiling point, triple point, critical point, and label the phases. Using the triple point (T_3, P_3) as a reference, give equations describing each of the phase boundaries, assuming ideal behavior for the gas phase and neglecting the temperature dependence of the enthalpies.

E19.9 Using the following experimental data, make a rough sketch of the phase diagram of acetic acid. Solid acetic acid may exist in two phases, α and β. The two solid phases α and β are in equilibrium at a temperature of 328 K and a pressure of 200 bar. The high pressure form β is more dense than α, and both solids are more dense than the liquid. The normal boiling point of the liquid is 391 K. The low pressure form α melts at 289 K under its own vapor pressure of 0.012 bar. The critical pressure and critical temperature are 57.9 bar and 593 K. (Adapted [4]).

E19.10 Sketch the Gibbs energy for a gas and a liquid versus temperature to explain the phenomenon of boiling. What is the slope of the Gibbs energy curve with temperature of a substance in a given state? How does this slope change with temperature?

E19.11 At 298 K and 1 bar, the molar entropies of liquid and gaseous water are 69.9 and 188.8 J K^{-1} mol^{-1}, respectively. The enthalpy of vaporization for water $\Delta_{vap}H^{\ominus}$ is 44.0 kJ mol^{-1} at 298 K. a) Will the phase change, $H_2O(l) \rightarrow H_2O(g, 1\ bar)$, occur spontaneously at 298 K? b) Will the phase change, $H_2O(l) \rightarrow H_2O(g, 0.01\ bar)$, occur spontaneously at 298 K? c) Estimate the normal boiling point of water from the data given above. State what assumptions or approximations you make. [Adapted from [4]].

E19.12 Take the molar volume of solid benzene to be 74.4 cm^3 mol^{-1} at 70.5 MPa and 298. K, its melting temperature. The molar volume of the liquid at this temperature and pressure is 83.4 cm^3 mol^{-1}. If the melting temperature changes to 313 K at 128 MPa, calculate the molar enthalpy and entropy of fusion of the solid.

E19.13 Using the Clausius–Clapeyron equation in its integrated form calculate and plot the vapor pressure above liquid and solid water (a) in the range of $T = 350$ K to $T = 500$ K and (b) in the range of $T = 271$ K to $T = 275$ K. The molar enthalpies of evaporation and sublimation are $\Delta_{vap}H = 44.1$ kJ mol^{-1}, and $\Delta_{sub}H = (6.0 + 44.1)$ kJ mol^{-1} = 50.1 kJ mol^{-1}. The vapor pressure at 373 K is 1.013 bar, and the vapor pressure at 273.16 K (triple point) is 6.11×10^{-3} bar.

E19.14 Using Equation (19.12) and the Thermodynamic Equation of State, show that

$$dS = C_V \frac{dT}{T} + \left(\frac{\partial P}{\partial T}\right)_V dV \tag{19.61}$$

E19.15 A butane cigarette lighter contains a biphasic (liquid–gas) mixture of butane. The normal boiling point of butane occurs at 272.6 K and 1 bar. The enthalpy of vaporization of butane is 23.8 kJ mol^{-1} K^{-1}. Use the Clausius–Clapeyron equation to estimate the vapor pressure of the gaseous butane in the cigarette lighter at 300 K.

E19.16 Use the equation ($dH = TdS + VdP$) to derive the equation

$$dH = C_P \cdot dT + \left[V - T\left(\frac{\partial V}{\partial T}\right)_P\right] \cdot dP$$

by writing dS as a function of T and P and using a Maxwell relation.

E19.17 Use the fact that dV/V is an exact differential to relate the pressure dependence of the expansivity to the temperature dependence of the compressibility, namely prove that

$$\left(\frac{\partial \alpha_V}{\partial P}\right)_T = -\left(\frac{\partial \kappa_T}{\partial T}\right)_P$$

E19.18 For a first-order phase transition, the molar entropy, the molar enthalpy, and the molar volume are discontinuous. What happens to the molar internal energy of the system at such a phase transition?

E19.19 Prove that the volume dependence of the heat capacity at constant volume is equal to the second derivative of the pressure, namely

$$\left(\frac{\partial C_V}{\partial V}\right)_T = T\left(\frac{\partial^2 P}{\partial T^2}\right)_V$$

Use this result to show that the constant volume heat capacity of an ideal gas does not depend on the volume.

E19.20 The table contains vapor pressure data for water, taken from [5]. Plot these data. Compare the data to the result of the Clausius–Clapeyron equation and of Equation (19.62). Use the values $\Delta_{vap}H^{\ominus} = 44.1$ kJ mol^{-1}, $\Delta C_P^{\ominus} = -41.7$ J K^{-1} mol^{-1}, $T_1 = 373$ K, and $P_{vap,1} = 1$ bar.

P/bar	0.005948	0.022762	0.030844	0.071851	0.19406	0.46130	0.98678	1.9334	3.5189
T/K	273.15	293.15	298.15	313.15	333.15	353.15	373.15	393.15	413.15
P/bar	6.0189	9.7653	15.1451	22.592	32.6051	45.7124	62.5133	83.6708	109.9467
T/K	433.15	453.15	473.15	493.15	513.15	533.15	553.15	573.15	593.15

E19.21 Using Equation (19.12) and the Thermodynamic Equation of State, show that

$$dS = C_V \frac{dT}{T} + \left(\frac{\partial P}{\partial T}\right)_V dV$$

19.8 Problems

P19.1 The following data are found for Cl_2:

T/K	P_{vap}/bar	ρ_{liquid}/g cm^{-3}	ρ_{vapor} /g cm^{-3}
273	3.636	1.468	0.0128
283	4.993	1.438	0.0175
293	6.706	1.408	0.0226

Calculate the enthalpy of vaporization of Cl_2 at 283 K. If Cl_2 were assumed to behave as an ideal gas and the volume of the liquid is neglected, how would it change your results? (Adapted from [4]).

P19.2 Figure P19.2 shows a sketch of the phase diagram of iodine. Some data on I_2 are a) the normal boiling point of iodine is 456 K; b) the vapor pressure of liquid I_2 at 390 K is 100 torr; c) the molar enthalpy of sublimation is 61.05 kJ mol^{-1}; d) the vapor pressure of solid I_2 is 1.00 torr at 312 K.

Figure P19.2 Sketch of phase diagram of I_2.

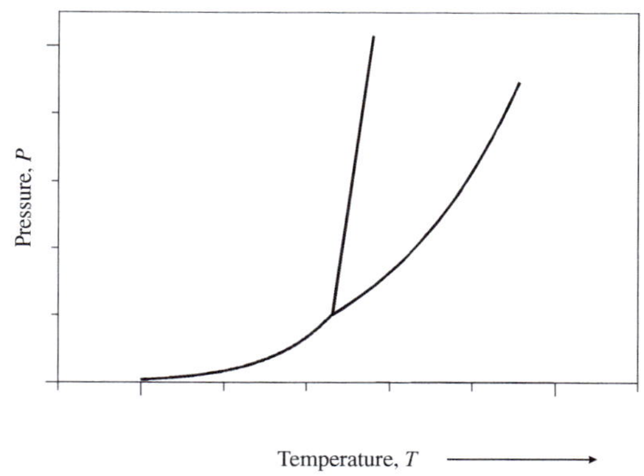

Identify the three phases on the diagram and label the points indicated with the appropriate values.

Using the data given above and your knowledge of phase transitions, calculate the molar enthalpy of vaporization of iodine.
Using the data given above and your knowledge of phase transitions, calculate the triple point temperature and pressure.

Sketch the Gibbs energy curves for gaseous, liquid, and solid I_2 as a function of temperature and at the triple point pressure. Label any of the temperatures you know.

On a different graph sketch the Gibbs energy curves for gaseous, liquid, and solid I_2 as a function of temperature and at a pressure of 1 bar. Label any temperatures that you know.

P19.3 The table gives phase equilibrium data for CO_2.

Sublimation	Curve								
T/K	130	140	155	170	185	194.7	205	216.658	
P/bar	0.00032	0.00187	0.01674	0.0987	0.4402	1.013	2.271	5.185	
Vaporization	Curve								
T/K	216.658	220	230	240	250	260	280	300	304.14
P/bar	5.185	5.991	8.929	12.825	17.85	24.19	41.61	67.13	73.75

Liquid–Solid	Curve	
T/K	216.55	236.45
P/bar	5.185	1013

a) Use the data in this table to make a P versus T plot that shows the different phases of CO_2. The critical point occurs at $T = 304.14$ K and $P = 73.75$ bar, and the triple point occurs at $T = 216.658$ K and $P = 5.185$ bar. Mark the different phase regions on the graph and indicate the position of the triple pint and critical point.
b) Use these data to determine $\Delta_{vap}H$ and $\Delta_{sub}H$ for CO_2 by making the assumption that $V_{gas} \gg V_{liq}$ and $V_{gas} \gg V_{solid}$.
c) Use the density data given below for the liquid and gas[9] to determine $\Delta_{vap}H$ using the Clapeyron equation but accounting for $\Delta_{vap}V$ in a more realistic manner.

T/K	233.15	243.15	253.15	263.15	273.15	283.15	293.15
ρ_{liquid}/kg m^{-3}	1115.9	1075.0	1031.0	980.97	925.07	859.99	772.02
ρ_{gas}/kg m^{-3}		36.78	52.03	72.31	97.46	132.	190.

Compare the enthalpies of vaporization with each other and with the literature value.

P19.4 You observe that a certain liquid has a vapor pressure of 1 bar at 300 K; i.e., it boils. In addition the liquid has a molar volume of 0.150 L mol^{-1}, the compression factor of the vapor above the liquid is $\Phi = pV/\boldsymbol{R}T = 0.95$ and $dP/dT = 0.026$ bar K^{-1}. Determine the enthalpy of vaporization by assuming that the volume of the liquid may be neglected. Determine the enthalpy of vaporization of the liquid without any assumptions.

P19.5 You plan to measure the spectrum of I_2 vapor in order to calibrate your spectrometer. You wish to have an absorbance of 1.0 or less for the I_2 vapor in your optical cell. If the maximum absorption cross-section[10] of I_2 is $\sigma = 4.84 \cdot 10^{-18}$ cm^2 and your optical cell's path length is $d = 10$ cm, what number density of I_2 will you need to have? Assuming the gas is ideal, what pressure of I_2 will you need? Given that I_2 is a solid under standard conditions, what temperature is necessary to provide your desired vapor pressure ($\Delta_f G(I_2(g)) = 19.3$ kJ mol^{-1} and $\Delta_{sub}H = 62.4$ kJ mol^{-1} at 298 K and 1 bar)? Assume that I_2 vapor behaves ideally and that $\Delta_f H$ is independent of T.

P19.6 Use a Maxwell relation to show that a phase transition that has a discontinuity in the volume must also have a discontinuity in its entropy, and hence have a latent heat.

P19.7 Because the enthalpies $H_{m,c}$ and $H_{m,v}$ of two phases are state functions, their difference is also. Use this fact to show that the heat of vaporization of a substance may be written as

$$\frac{d\left(\Delta_{vap}H\right)}{dT} = \Delta C_{P,m} + \frac{\Delta_{vap}H}{T} - \frac{\Delta_{vap}H}{\Delta V_m}\left(\frac{\partial \Delta V_m}{\partial T}\right)_P$$

where $\Delta V_m = V_{v,m} - V_{c,m}$ and $\Delta C_{P,m} = C_{P,m,v} - C_{P,m,c}$

P19.8 Explain what happens to the constant pressure heat capacity of a system undergoing a first-order phase transition. Provide graphs of the Gibbs energy versus T (as in Fig. 19.2), H versus T and C_P versus T and a verbal explanation.

P19.9 Use the fundamental expression for the Helmholtz free energy (Equation (19.34))

$$A = U - TS$$

and the results for U and S from Problem 17.14 to show that A for a photon gas is

$$A = -\frac{8\pi^4}{12.897}\frac{Vk^4}{c_0^3h^3}T^4$$

Then use this expression for A to find the radiation pressure P of the photon gas and deduce the equation of state for the photon gas. Lastly, calculate the radiation pressure created by an electric filament (incandescent light bulb filament) at $T = 1000$ K.

P19.10 Beginning with the definition of the enthalpy and the combined form of the first and second laws of thermodynamics, derive an expression for $(\partial H/\partial P)_T$.

P19.11 Calculate the vapor pressure of water in the range of $T = 373$ to 500 K using the refined Equation (19.62) (data see Example 19.6)

9 The density data are from the International Critical Tables, Volume 3 (McGraw Hill, 1928).
10 This cross-section is determined for a wavelength of 533 nm (a Fourier spectrometer measurement at a resolution of 0.1 nm) and is taken from [6].

P19.12 Derive Equation (19.62) for the vapor pressure curve from the Clausius–Clapeyron equation (19.61) in Section 19.4.2 by accounting for the temperature dependence of $\Delta_{vap}H$. In particular, assume that the $\Delta_{vap}H$ at any temperature T is described by the linear equation

$$\Delta_{vap}H = \Delta_{vap}H_1 + \Delta C_P \cdot (T - T_1)$$

where $\Delta_{vap}H_1$ is the molar enthalpy of vaporization at the temperature T_1 and is the difference in the molar constant pressure heat capacities at T and T_1.

P19.13 In Example 19.2 we calculate the change in enthalpy for the ammonia gas undergoing a transition between the equilibrium state of 1 bar and 400 K to 51 bar and 500 K. In that calculation we used the mean values of the heat capacity, molar volume, and temperature dependence of the molar volume. In this problem you will account for the full temperature and pressure dependence of those quantities to find the enthalpy change. Over this pressure range, the molar volume can be described by

$$\overline{V}_m = \frac{41.573}{P}\ \text{L mol}^{-1}\ \text{bar} - 0.0638\ \text{L mol}^{-1}$$

and its temperature dependence by

$$\left(\mathrm{d}\overline{V}_m/\mathrm{d}T\right)_P = \frac{0.0831}{P}\ \text{L mol}^{-1}\ \text{bar K}^{-1} + 5.83 \times 10^{-4}\ \text{L K}^{-1}\ \text{mol}^{-1}$$

The temperature dependence of the constant pressure heat capacity can be described by a linear equation from 400 to 500 K.

$$C_{P,m} = 26.661\ \text{J mol}^{-1}\ \text{K}^{-1} + 0.0307 \cdot T\ \text{J mol}^{-1}\ \text{K}^{-2}$$

Compare your result to those reported in Example 19.2.

References

1 W. Wagner and A. Pruß, The IAPWS formulation 1995 for the thermodynamic properties of ordinary water substance for general and scientific use, *J. Phys. Chem. Ref. Data*, 2002, **31**, 387.

2 D.A. Young, *Phase Diagrams of the Elements* (U.C. Press, Berkeley, 1991).

3 H. Landolt and R. Börnstein, *Zahlenwerte und Funktionen*, 2. Band, 4. Teil (Springer-Verlag, Berlin, 1961).

4 R.E. Dickerson, *Molecular Thermodynamics* (Benjamin, 1969).

5 O.C. Bridgeman and E.W. Aldrich, Vapor Pressure Tables for Water, *J. Heat Transfer*, 1964, **86**, 279.

6 A. Saiz-Lopez, R.W. Saunders, D.M. Joseph, S.H. Ashworth, and J.M.C. Plane, Absolute absorption cross-section and photolysis rate of I_2, *Atmos. Chem. Phys. Discuss.*, 2004, **4**, 2379.

20

Real Gases

CHAPTER MENU

20.1 Introduction

In this chapter, we extend our discussion of gas behavior to real gases; i.e., gases for which intermolecular interactions are important. To this point in our discussions we have treated the molecules in a gas by the ideal gas approximation, in which the gas molecules do not interact with each other and their finite size is neglected. This approximation is not always reasonable. In addition we note that the ideal gas model cannot account for the condensation of the gas into a liquid; the existence of the vapor–liquid transition requires attractive interactions between gas molecules and a mechanism of forming molecular aggregates (clusters). Hence it is important to develop a more realistic description for gases and their P, V, T behavior.

We analyze the properties of gases, which contain interacting molecules, and contrast them with those of ideal gases. In particular, we will define the fugacity for a gas. The fugacity reduces to the pressure in the limit of an ideal gas. We will also introduce a molecular-based model for real gases and evaluate how the interactions between molecules affect the thermodynamic properties.

20.2 ΔG for Real Gases and the Fugacity

From the basic Equation (19.45)

$$dG = -S\,dT + V\,dP$$

we see that the PV behavior in an isothermal process ($T = const.$) is directly related to the Gibbs energy change,

$$\Delta G = \int_{G_1}^{G_2} dG = \int_{P_1}^{P_2} V \cdot dP \tag{20.1}$$

For an ideal gas, for which the equation of state is

$$V_{\text{ideal gas}} = \frac{NkT}{P} \tag{20.2}$$

Principles of Physical Chemistry, Third Edition. Hans Kuhn, David H. Waldeck, and Horst-Dieter Försterling.

Companion website: www.wiley.com/go/Kuhn/physicalchemistry3e

we obtain

$$\Delta G_{\text{ideal gas}} = \int_{P_1}^{P_2} V_{\text{ideal gas}} \cdot dP = \int_{P_1}^{P_2} \frac{NkT}{P} dP = NkT \cdot \ln \frac{P_2}{P_1} \quad (20.3)$$

In fact it is possible to define an ideal gas by this relationship; specifically we define an ideal gas as a substance whose Gibbs energy is given by

$$G(T,P) = G^{\ominus}(T) + NkT \cdot \ln \hat{P} \quad (20.4)$$

where $G^{\ominus}(T)$ is the Gibbs energy under standard conditions and $\hat{P} = P/P^{\ominus}$. In Problem P20.10 you show that this definition and Equation (19.45) imply that $PV = NkT$, which is our previous definition.

For a real gas the integrand in Equation (20.1) depends in a more complicated way on P. Because the ideal gas model describes the properties of a real gas at very low pressures, our model for a real gas should coincide with the ideal gas model at very low pressures. Thus, we maintain the structure of Equation 20.3 for the description of a real gas, but we replace the pressure P by the fugacity

$$f = \phi \cdot P \quad (20.5)$$

where ϕ is called the fugacity coefficient. Then Equation (20.1) becomes

$$\Delta G = NkT \cdot \ln \frac{f_2}{P_1} = NkT \cdot \ln \frac{\phi \cdot P_2}{P_1} \quad (20.6)$$

provided that P_1 is so small that the ideal gas law (20.2) is fulfilled. This choice places the effects of the gases nonideality into the fugacity f. By defining the fugacity in this way the thermodynamic relations for ΔG remain unchanged if we replace the pressures by fugacities. If $\phi = 1$, then the gas behaves ideally; otherwise, it does not.

The fugacity coefficient ϕ can be calculated from measured PV data. Using Equations (20.1), (20.6), and (20.5) we can write that

$$\int_{P_1}^{P_2} V_m \cdot dP = \boldsymbol{R}T \cdot \ln \frac{P_2}{P_1} + \boldsymbol{R}T \cdot \ln \phi$$
$$= \int_{P_1}^{P_2} V_{m,\text{ideal}} \cdot dP + \boldsymbol{R}T \cdot \ln \phi$$

where P_1 is a small enough pressure such that $\phi_0 = 1$ and V_m is the molar volume. Rearranging this result, we find

$$\ln \phi = \frac{1}{\boldsymbol{R}T} \int_{P_1}^{P_2} \left(V_m - V_{m,\text{ideal}}\right) \cdot dP \quad (20.7)$$

where

$$V_{m,\text{ideal}} = \frac{\boldsymbol{R}T}{P}$$

The molar volume V_m is obtained from experimental PV data. Table 20.1 provides PV_m data for CO_2 at $T = 313$ K and gives values of f and ϕ. from these data and Equations (20.5) and (20.7).

In Fig. 20.1 $V_m - V_{m,\text{ideal}}$ (data from Table 20.2, column 4) is plotted versus the pressure P_2. The area between the curve and the abscissa for $P_2 = 101$ bar is -13.6 bar L mol^{-1}; i.e.,

$$\int_{0 \text{ bar}}^{101 \text{ bar}} \left(V_m - V_{m,\text{ideal}}\right) \cdot dP = -13.6 \text{ bar L mol}^{-1}$$

Then for $P_2 = 101$ bar we obtain

$$\ln \phi = -\frac{1}{\boldsymbol{R}T} \cdot 13.6 \text{ bar L mol}^{-1} = -0.52, \quad \phi = 0.59$$

The fugacity coefficients and fugacities for CO_2 in Table 20.1, columns 5 and 6, are obtained accordingly.

Table 20.1 Calculation of the Fugacity Coefficient ϕ and the fugacity f of CO_2 ($T = 313$ K) from Experimental PV Data, Using Equation (20.7) with $P_1 = 0$ bar.

P_2	V_m	$V_{m,\text{ideal}}$	$\left(V_m - V_{m,\text{ideal}}\right)$		f_2
bar	L mol^{-1}			ϕ	bar
0	–	–	0	1.0	0
1.013	25.60	25.67	−0.070	1.0	1.013
20	1.190	1.310	−0.120	0.93	18.6
28.6	0.786	0.909	−0.123	0.90	25.6
42.4	0.469	0.613	−0.127	0.84	35.5
50.6	0.380	0.513	−0.133	0.80	40.6
75.9	0.184	0.343	−0.159	0.70	52.9
101	0.069	0.257	−0.188	0.59	59.9

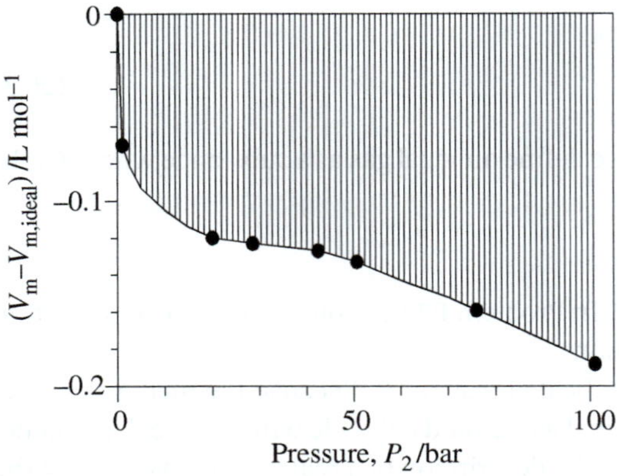

Figure 20.1 $V_m - V_{m,\text{ideal}}$ versus pressure P_2 for the data in Table 20.2. The shaded area corresponds to −13.6 bar L mol^{-1}.

Example 20.1 *What is ΔG for the compression of 1 mol CO_2 at 313 K from $P_1 = 1$ bar to $P_2 = 66$ bar?*

With $\phi_1 = 1$ and $\phi_2 = 0.73$ (interpolation in Table 20.2) we obtain $f_1 = 1$ bar, $f_2 = 48$ bar, and

$$\Delta G = \boldsymbol{n}\boldsymbol{R}T \cdot \ln \frac{f_2}{f_1} = 10 \text{ kJ}$$

This result is 8% lower than that predicted by making an ideal gas approximation.

20.3 Equations of State for Real Gases

Figure 20.2 plots the compression factor Φ, defined as

$$\Phi = \frac{PV}{\boldsymbol{n}\boldsymbol{R}T} \tag{20.8}$$

as a function of pressure. For an ideal gas the compression factor should be $\Phi = 1.0$ at all pressures and temperatures. Clearly this approximation does not hold, and both positive and negative deviations from unity are observed. How can we understand this behavior? Here we discuss improvements to the ideal gas equation of state that account for intermolecular forces. In all cases we assume that the interaction can be treated classically.

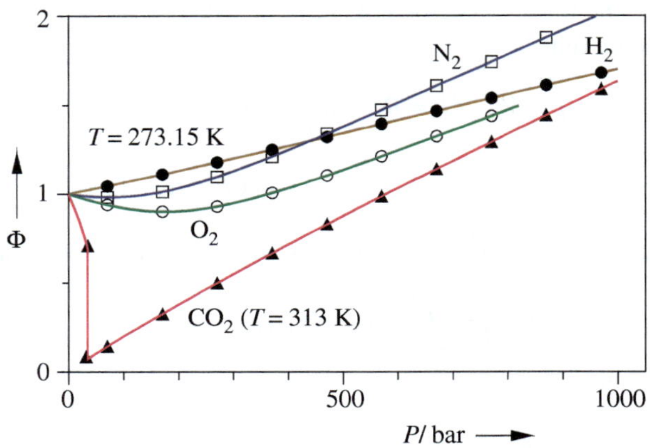

Figure 20.2 The compression factor Φ is plotted versus pressure for four different gases (H_2, N_2, O_2 at 273.15 K, and CO_2 at 313 K). The special behavior of CO_2 at low pressure arises from its much stronger attraction forces compared to H_2, N_2, and O_2. The ideal gas behavior would be represented by a horizontal line at $\Phi = 1$. Experimental data see [1].

20.3.1 Hard Sphere Gas

One obvious limitation in the ideal gas model is its failure to account for the finite volume occupied by the molecules (or atoms). Because of their size, molecules occupy part of the volume of the gas container and the total volume of the container is not accessible to a given molecule; i.e., part of the volume is excluded because of the intermolecular repulsion at very close distances between molecules (Fig. 20.3). If the total volume which is excluded by the gas molecules is V_1, then the accessible volume in a container of volume V is $V - V_1$. If we substitute this accessible volume into the ideal gas equation of state for the volume, we find

$$P(V - V_1) = \boldsymbol{n}\boldsymbol{R}T \tag{20.9}$$

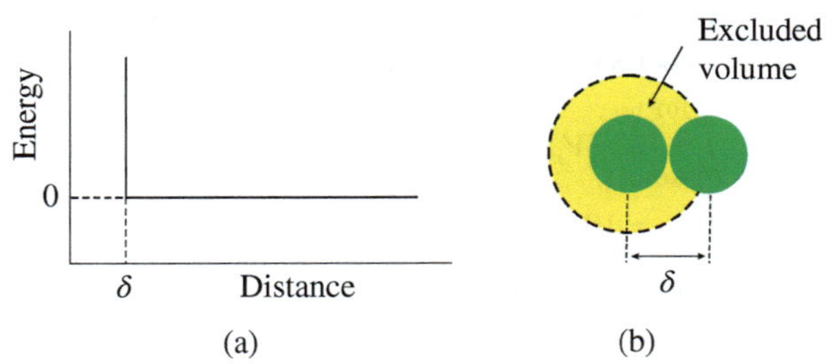

Figure 20.3 Hard sphere model. (a) Energy versus distance, (b) excluded volume and distance δ.

$V - V_1$ is the volume available for the molecules to move freely, and Equation (20.9) is called the *Hard Sphere Gas Equation of State.*

The inclusion of V_1 acts to increase the pressure of the gas, compared to the pressure of an ideal gas. The volume V_1 can be estimated as (Box 20.1)

$$V_1 = 4.2 \cdot V_{\text{molecule}} \boldsymbol{N}_\text{A} \cdot \boldsymbol{n} = b \cdot \boldsymbol{n} \tag{20.10}$$

where V_{molecule} is the volume of one molecule. The parameter $b = V_1/\boldsymbol{n}$ has units of volume per mole, and we expect that it should correspond to the volume that is unavailable (excluded) because of the finite size of gas molecules.

Example 20.2 *Use the hard sphere gas equation of state with $V_{\text{molecule}} = 0.10\ nm^3$ to determine the compression factor. Contrast your result with that of an ideal gas.*

We can rearrange Equation (20.9) to obtain

$$\Phi = \frac{PV}{\boldsymbol{n}\boldsymbol{R}T} = 1 + \frac{PV_1}{\boldsymbol{n}\boldsymbol{R}T} = 1 + \frac{Pb}{\boldsymbol{R}T}$$

This result shows us that Φ for the hard sphere gas increases linearly with pressure. According to Equation (20.10)

$$b = 4.2 \cdot V_{\text{molecule}} \boldsymbol{N}_\text{A}$$

b is a positive constant and the deviation from the ideal gas behavior, $\Phi = 1$, is always positive. For 10 bar pressure and a temperature of 298 K, for $V_{\text{molecule}} = 0.10\ \text{nm}^3$ we find $b = 0.253\ \text{L mol}^{-1}$; then

$$\Phi = 1 + \frac{10\ \text{bar}\ 0.253\ \text{L mol}^{-1}}{0.08314\ \text{bar L K}^{-1}\ \text{mol}^{-1}\ 298\ \text{K}} = 1.10$$

which is a 10% correction. The magnitude of the deviation will increase with gas pressure and molecular size.

This analysis shows that repulsive forces, due to the excluded volume, account for positive deviations from the ideal gas law, but do not account for negative deviations.

20.3.2 Van der Waals Equation

In contrast to the repulsive forces, attractive forces between molecules decrease the pressure. This decrease results because the attraction of a selected molecule by its neighbors diminishes the momentum transferred to the wall of the container when the molecule strikes it. To take account of the decrease of the pressure P we can extend Equation (20.9) to

$$(P + P_1) \cdot (V - V_1) = \boldsymbol{n}\boldsymbol{R}T \tag{20.11}$$

in which P_1 is an empirically determined parameter. For molar quantities we find

$$(P + P_1) \cdot (V_m - b) = \boldsymbol{R}T \tag{20.12}$$

What form does P_1 have? The attractive force between a selected molecule and its neighbors is considered to be proportional to the number of molecules in the close neighborhood (see Fig. 20.4), i.e., proportional to the number density of molecules, thus proportional to $1/V_m$:

$$\textit{force} \propto \frac{1}{V_m}$$

Because the number of molecules hitting the wall is also proportional to $1/V_m$, the total force, and P_1, is proportional to $1/V^2$. Hence we postulate that

$$P_1 = \frac{a}{V_m^2} = a \cdot \frac{\boldsymbol{n}^2}{V^2} \tag{20.13}$$

where the constant a depends on the nature of the gas.

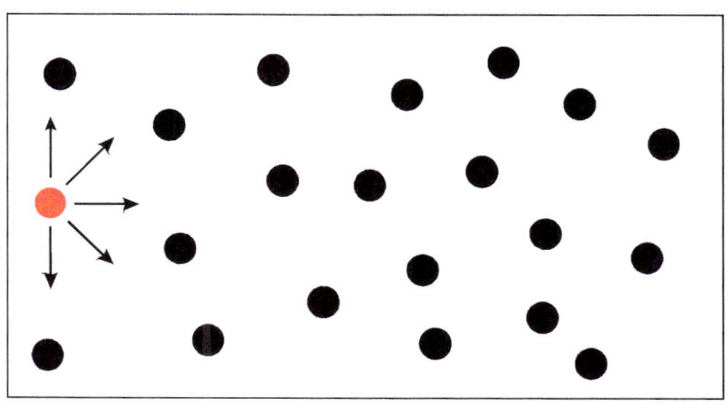

Figure 20.4 Attractive forces acting on a particle near the wall of a container (red solid circle).

Equation (20.12) can be written as

$$\boxed{\left(P + \frac{a}{V_m^2}\right) \cdot (V_m - b) = \boldsymbol{R}T} \tag{20.14}$$

or

$$\boxed{P = \frac{\boldsymbol{R}T}{V_m - b} - \frac{a}{V_m^2}} \tag{20.15}$$

in which a and b are constants. b accounts for the excluded volume and a accounts for attractive forces between the molecules. Equations (20.14) and (20.15) are called the *van der Waals Equation of State.*

Example 20.3 *Using the van der Waals equation of state, calculate the compression factor for CO_2 as a function of pressure at 500 K. Compare your result with the experimental data in the table. For CO_2 $a = 3.66$ bar L^2 mol^{-2}* and $b = 0.043$ *L mol^{-1}*.

P (bar)	1	50	100	150	200	250	300	400	500	600	700	800	900	1000
Φ	0.999	0.965	0.936	0.914	0.900	0.896	0.901	0.935	0.990	1.06	1.13	1.21	1.29	1.37

Figure 20.5 plots the data as a function of pressure. The solid curve is calculated using the van der Waals equation with the parameters given above. It is evident from the plot that the qualitative behavior of the compressibility factor is accounted for by the van der Waals equation but that a quantitative comparison shows a 15% deviation at the highest pressures plotted.

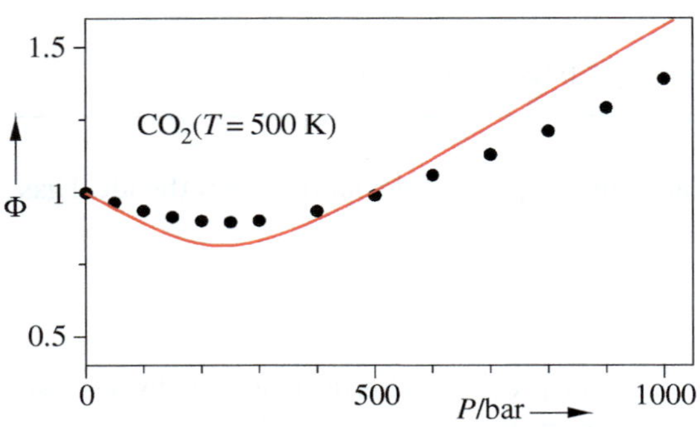

Figure 20.5 The compressibility factor Φ for CO_2 as a function of pressure P at 500 K. Dots: experiment. Red solid line: calculated from the van der Waals equation with $a = 3.66$ bar L^2 mol^{-2} and $b = 0.043$ L mol^{-1}. In detail, in a first step P is calculated for different values of V_m using Equation (20.15); from these P/V_m data, the product PV_m and the compressibility factor $PV_m/(\boldsymbol{R}T)$ are calculated and Φ is plotted versus P. Note that in this figure the temperature is much higher ($T = 500$ K) than in the corresponding case of Fig. 20.2 ($T = 313$ K). Experimental data see [1].

20.3.3 Critical Point and Van der Waals Constants

Figure 20.6 shows isotherms for CO_2. The solid curves in Fig. 20.6a show the isotherms calculated by the van der Waals equation, using Equation (20.15) with

$$a = 3.66 \text{ bar L}^2 \text{ mol}^{-2}, \quad b = 0.043 \text{ L mol}^{-1}$$

at $T = 273$ and 323 K. These curves are compared with the experimental data in Fig. 20.6b. For comparison, the corresponding curve for the ideal gas at 323 K is included (Fig. 20.6a, dashed line). At high enough temperature and large enough volume we find reasonable agreement with the ideal gas law. At $T = 323$ K the experimental data are well reproduced by Equation (20.15). However, at lower temperature, the Van der Waals equation fails, see 273 K isotherms in Fig. 20.6.

Consider the P/V diagram for CO_2 displayed in Fig. 19.6, especially the curve for $T = 647.30$ K with an inflection point at $V_m = 0.04$ L mol^{-1} (green filled circle in Fig. 19.6) indicating the critical point. This special point is determined by the conditions

$$\left(\frac{\partial P}{\partial V}\right)_T = 0 \quad \text{and} \quad \left(\frac{\partial^2 P}{\partial V^2}\right)_T = 0 \tag{20.16}$$

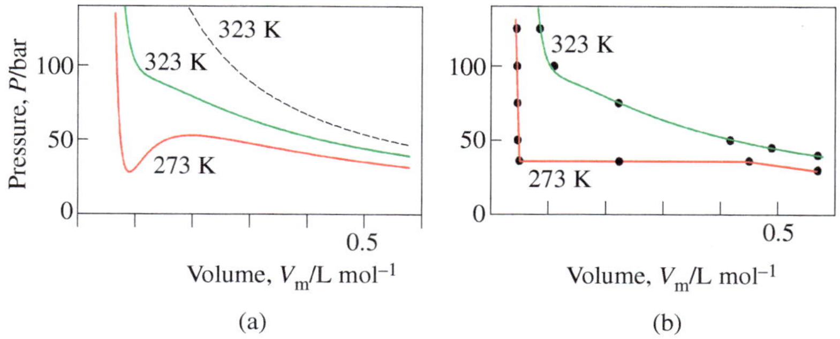

Figure 20.6 Isotherms of CO_2 for $T = 273$ K (red solid lines) and $T = 323$ K (green solid lines). (a) Calculation. Solid lines: van der Waals equation with $a = 3.66$ L^2 bar mol^{-2} and $b = 0.043$ L mol^{-1}. Dashed line: ideal gas law. (b) Experiment (dots) showing the condensation of the gas into a liquid at $T = 273$ K.

In Problem P20.11, you use the van der Waals equation of state and Equation (20.16) to show that

$$T_c = \frac{8 \cdot a}{27 \cdot \boldsymbol{R} \cdot b}, \quad P_c = \frac{a}{27 \cdot b^2}, \quad V_{m,c} = 3 \cdot b \tag{20.17}$$

where T_c, P_c, and $V_{m,c}$ are the temperature, the pressure, and the molar volume at the critical point.

Example 20.4 *Evaluate the a and b van der Waals parameters from the critical data for CO_2, namely $T_c = 304$ K, $P_c = 73.6$ bar, and $V_{m,c} = 0.095$ L mol^{-1}*

From Equation (20.17) for T_c and P_c we find that

$$b = \frac{T_c \boldsymbol{R}}{8\, P_c} \quad \text{and} \quad a = \frac{27\, T_c^2\, \boldsymbol{R}^2}{64\, P_c} \tag{20.18}$$

which gives

$$a = 3.66 \text{ bar L}^2 \text{ mol}^{-2}, \quad b = 0.043 \text{ L mol}^{-1}$$

With these parameters, we calculate that $V_{m,c} = 3 \cdot 0.043$ L mol^{-1} $= 0.129$ L mol^{-1} which is about 35% larger than the measured $V_{m,c} = 0.095$ L mol^{-1}.

From Equation (20.17) we find

$$\frac{V_{m,c}\, P_c}{T_c} = \frac{3 \cdot \boldsymbol{R}}{8} = 3.13 \text{ J K}^{-1}$$

This means that the ratio $V_{m,c}\, P_c/T_c$ should be independent of the nature of the gas (*principle of corresponding states*). In the experiment, values between 2.28 and 2.54 are found (see Table 20.2).

Below the critical temperature, the fluid's isotherms show a discontinuity in the volume. The situation in this region is not correctly described by the van der Waals Equation (20.14), which predicts a maximum and a minimum for P, contrary to the experiment (Fig. 20.6a, $T = 273$ K). In this region the gas condenses to a liquid, something not considered in our simple construction of the van der Waals Equation. These ideas have provided a starting point for describing liquids, however (see [3]).

20.3.4 The Vapor–Liquid Phase Change?

How can we understand the phase change from a liquid to a gas (vaporization of a liquid)?. The molecules possess thermal energy kT, as pointed out in Chapter 14. Depending on their average kinetic energy (temperature), molecules at the surface of a liquid are able to escape from it (high density region) into the gas (low density region). At equilibrium, the flux of molecules leaving the liquid surface (thus entering into the gas state) is matched by the flux of molecules returning from the gas state into the liquid structure. On average, the density of molecules above the liquid (in the gas phase) is stable and a vapor pressure is established. With increasing temperature a larger percentage of molecules have enough kinetic energy

Table 20.2 Table of Critical Parameters $V_{m,c}$, T_c, and P_c, Second Virial Coefficient B_2 (see Equation (20.21), at 298.15 K), and Van der Waals Constants a and b for Some Gases.[1]

Gas	$V_{m,c}$ cm³ mol⁻¹	T_c K	P_c MPa	$V_{m,c}P_c/T_c$ J K⁻¹	B_2 cm³ mol⁻¹	a L² bar mol⁻²	b L mol⁻¹
He	57	5.19	0.227	2.5	12	0.035	0.024
Ar	75	150.87	4.898	2.4	−16	1.36	0.032
Xe	118	289.73	5.84	2.4	−130	4.19	0.052
H_2	65	32.97	1.293	2.5	15.4	0.25	0.027
N_2	90	126.21	3.39	2.4	−4	1.37	0.039
O_2	73	154.59	5.043	2.4	−16	1.38	0.032
H_2O	56	647.14	22.06	1.9	−1158	5.54	0.031
CO_2	94	304.14	7.375	2.3	−127	3.66	0.043
CH_4	99	190.53	4.604	2.4	−43	2.30	0.043
CF_4	140	227.6	3.74	2.3	−88	4.04	0.063

to escape and enter the gas state, and the vapor pressure increases accordingly. Above a high enough temperature, that is, the critical temperature, the density of molecules is uniform throughout the sample (no distinct liquid and gas states exist); and the substance is called a supercritical vapor.

How can we understand the phase change from the gas state to the liquid state (condensation of a gas into a liquid)? Given the weak attraction forces acting between gas molecules, their formation into a liquid may seem surprising. From a molecular point of view, however, imagine that by chance two gas molecules can come very near to each other after a collision. Because of their small distance, the force acting between them is much stronger than that acting between them in the gas phase at large distance. Provided that the relative kinetic energy of the two molecules is small enough, they can form a dimer. The dimer can decompose, or additional molecules can be added to form larger aggregates, i.e., a trimer, tetramer, and so on. We call such aggregates clusters. We can estimate the stability of a cluster by regarding the change of the enthalpy ΔH and of the entropy ΔS in forming the cluster. On the one hand, we expect that ΔH is negative because of the attractive forces between the molecules; on the other hand, the molecules in the cluster have fewer configurations available than they did as individual molecules, that is, the entropy change ΔS is positive. With increase of the size of the cluster the contribution of ΔH eventually dominates and $\Delta G = \Delta H - T\Delta S$ becomes negative. This means a stable cluster has formed. Once one cluster has formed, it can act as a seed that facilitates the formation of additional clusters. Thus we can understand a liquid as composed of a great number of clusters: we have an ordered region in the near vicinity of a molecule (that is, inside a cluster), but we have disorder far away from the molecule, in contrast to a solid with a long-range order. This is the reason for the difference in the density function $\rho(r)$ for solids and liquids discussed in Section 19.4.3.3, Fig. 19.7. For more details of cluster nucleation theory, see [4].

20.3.5 A Molecular Model for the Thermodynamics of Gas Condensation

Developing a molecular-based model for a gas-to-liquid phase transition is an important area of research; however, it is complicated by the challenge of describing the molecular thermodynamics of a liquid. The condensation of a gas to a solid, however, allows for simplified statistical thermodynamic models of the gas and the solid to be employed. In this section we outline the condensation of Ar gas to Ar solid. To describe the thermodynamics of the gas-to-solid phase transition for Ar, we first make simplified models for calculating the Gibbs energy of the solid and that of the gas; and then we equate them to find the coexistence curve.

For solid Ar we assume that it forms into a macroscopic single crystal so that it has only electronic and vibrational degrees of freedom that we need to describe. We describe the vibrational energy of the solid by using the Einstein model and neglect excited electronic states. We assume that the electronic energy of Ar in the solid $E_{\text{electronic}}$ arises from van der

1 Critical constants data are from [2]; Van der Waals constants a and b are calculated from critical constants.

Waals interactions that we write as

$$E_{\text{electronic}} = N\phi/2$$

where ϕ is the interaction energy for a pair of argon atoms. Using Equation (16.29) for the vibrational energy (Einstein model), we write that

$$E_{\text{vib}} = 3Nk\left[\frac{\theta_E}{2} + \frac{\theta_E}{\exp\left(-\frac{\theta_E}{T}\right) - 1}\right]$$

where $\theta_E = h\nu_0/k$ and the subscript E denotes *Einstein*. In Foundation 20.1 (on Wiley's website) we show that this model for the Gibbs energy of the solid gives

$$G_{\text{solid}} = \left(\frac{3Nk\,\theta_E}{2} + \frac{N\phi}{2}\right) + 3NkT\left[\ln\left(1 - \exp\left(-\frac{\theta_E}{T}\right)\right)\right]$$

where the first two terms describe G_{solid} for $T = 0$:

$$G_{\text{solid,T=0}} = \frac{3Nk\theta_E}{2} + \frac{N\phi}{2} = E_{\text{vib},T=0} + E_{\text{electronic}}$$

For gaseous Ar, we make the ideal gas approximation (see Foundation 20.1 (on Wiley's website)) and write its Gibbs energy as

$$G_{\text{gas}} = \frac{5}{2}NkT + NkT\ln\frac{P^{\ominus}}{a\cdot T^{5/2}} + NkT\ln\hat{P}$$

with $\hat{P} = P/P^{\ominus}$, where $P^{\ominus}$ is the standard pressure, and the constant a is defined as

$$a = e^{5/2}(2\pi)^{3/2}\frac{(k\cdot m)^{3/2}}{h^3}k$$

For phase coexistence, the Gibbs energies of the gas and that of the solid must be the same; i.e., $G_{\text{gas}} = G_{\text{solid}}$. By setting these two equations equal to each other, we can solve for the vapor pressure and find

$$\ln\hat{P} = \frac{3}{2}\frac{\theta_E}{T} + 3\ln\left[1 - \exp\left(-\frac{\theta_E}{T}\right)\right] + \frac{\phi}{2kT} - \frac{5}{2} - \ln\frac{P^{\ominus}}{a\cdot T^{5/2}} \tag{20.19}$$

By comparing this result with the Clausius–Clapeyron equation, where $\Delta_{\text{sub}}H^{\ominus}$ is the enthalpy of sublimation,

$$\frac{\mathrm{d}\ln\hat{P}}{\mathrm{d}T} = \frac{\Delta_{\text{sub}}H^{\ominus}}{\boldsymbol{R}T^2}$$

one can show that

$$\Delta_{\text{sub}}H^{\ominus} = \frac{5}{2}\boldsymbol{R}T_{\text{sub}} - \frac{3}{2}\boldsymbol{R}\theta_E\frac{\exp\left(-\theta_E/T_{\text{sub}}\right)}{\left(1 - \exp\left(-\theta_E/T_{\text{sub}}\right)\right)} - \boldsymbol{R}\left(\frac{\phi}{2k} + \frac{3}{2}\theta_E\right) \tag{20.20}$$

See Foundation 20.1 for this derivation.

The enthalpy of sublimation is determined by the difference between the gases enthalpy ($5\boldsymbol{R}T/2$) and that of the solid, which is composed of the binding energy of the solid and the energy stored in vibrational modes. For an Einstein temperature $\theta_E = 30$ K, $T_{\text{sub}} = 83.8$ K (triple point temperature of argon) and $\Delta_{\text{sub}}H_{\text{triple point}} = 7.694$ kJ mol^{-1} we find that $\phi = -0.239\times10^{-19}$ J. Figure 20.7a plots $\Delta_{\text{sub}}H^{\ominus}$ versus the temperature. In Exercise 20.21, you use $\theta_E = 30$ K, $P_0 = 1$ bar, and $\phi = -0.239\times10^{-19}$ J to calculate the vapor pressure curve. Using these parameters, Figure 20.7b plots the vapor pressure above solid argon according to Equation (20.19). The calculated vapor pressure at the triple point is 0.64 bar, slightly less than the experimental value of 0.68 bar, see[2].

This simple model treats the lattice vibrations only crudely, as a single effective vibrational frequency ν_E. A molecular solid can have vibrational modes corresponding to the individual molecules as well as a distribution of frequencies for the motions between molecules. Hence we expect both lower frequency modes (50 cm^{-1} – 200 cm^{-1}) and high frequency modes associated with individual molecules to be important. In addition, this model does not include excitation of electronic states. The heat capacity of a metal will have an important contribution from its electrons when the temperature becomes low.

In Exercise E20.22 you use this expression to graph the gas–solid coexistence line for Ar and for Mg.

---

2 On the Vapor Pressure of Solid Argon, [5].

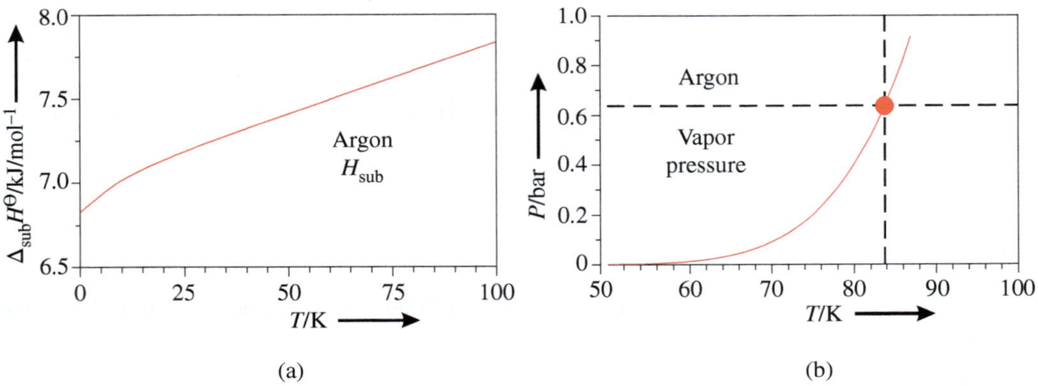

Figure 20.7 (a) $\Delta_{sub}H^{\ominus}$ for argon plotted versus temperature according to Equation (20.20), using $\theta_E = 30$ K and $\phi = -0.239 \times 10^{-19}$ J. (b) Vapor pressure above solid argon calculated from Equation (20.19) using $\theta_E = 30$ K, $P_0 = 1$ bar, and $\phi = -0.239 \times 10^{-19}$ J. The red filled circle denotes the triple point ($T_{triple} = 83.8$ K).

20.3.6 Virial Equation of State

In this last approach, we assume that the pressure of the gas can be written as a polynomial in the gas density N/V, so that

$$\frac{P}{kT} = \frac{N}{V} + B_2(T)\frac{N^2}{V^2} + B_3(T)\frac{N^3}{V^3} + \cdots \tag{20.21}$$

where the coefficients $B_2(T)$, $B_3(T)$ are taken to be functions of temperature only. At very low densities the quadratic, cubic, and higher terms are not important, and the ideal gas law is obtained. As the density of the gas becomes higher, however, interactions between molecules (e.g., clusters of molecules) become important and the higher order density terms must be included to accurately model the equation of state for the gas. This expansion is called a virial expansion and the coefficients are called virial coefficients; $B_2(T)$ is the second virial. Table 20.2 contains some representative values for the second virial coefficient.

This description for the equation of state is particularly important because the virial coefficients are directly related to the intermolecular forces. In Foundation 20.2 (on Wiley's website) we show that for spherical gas molecules (e.g., methane).

$$B_2(T) = \frac{-2\pi}{V}\int_0^{\infty} r^2 \left[\exp\left(-E(r)/kT\right) - 1\right] \, \mathrm{d}r \tag{20.22}$$

where $E(r)$ is the intermolecular potential between the molecules.

Example 20.5 *Show that the second virial coefficient for a hard sphere gas is equal to $2\pi\delta^3/3$ where the parameter δ is the contact distance defined in Fig. 20.3. Afterward find an expression for the hard sphere equation of state parameter V_1 in terms of the hard sphere potential's δ parameter.*

According to Fig. 20.3 for a hard sphere gas the intermolecular potential is infinite ($E(r < \delta) = \infty$) when the intermolecular distance r is less than the contact distance δ, hence impossible to realize. The hard sphere potential is zero when the gas atoms are farther/apart ($E(r > \delta) = 0$), and not touching. With this form for the potential, the integral in Equation (20.22) separates into two pieces, namely

$$\begin{aligned} B_2(T) &= -2\pi\int_0^{\delta} r^2 \cdot (-1) \, \mathrm{d}r - 2\pi\int_{\delta}^{\infty} r^2 \cdot (0) \, \mathrm{d}r \\ &= 2\pi\int_0^{\delta} r^2 \cdot \mathrm{d}r = 2\pi\left|\frac{1}{3}r^3\right|_0^{\delta} = \frac{2\pi\delta^3}{3} \end{aligned}$$

This virial coefficient is temperature independent and it has a value that is half the volume of a sphere of radius δ. When the atoms collide it occurs at their surfaces, hence a spherical volume of radius δ is excluded to one atom by the other.

For this model, we can relate the virial expansion to the hard sphere gas equation of state. To second order in the density, the virial equation is

$$\frac{P}{kT} = \frac{N}{V} + \frac{2\pi\delta^3}{3}\frac{N^2}{V^2} = \frac{N}{V}\left(1 + \frac{2\pi\delta^3}{3}\frac{N}{V}\right)$$

which rearranges to

$$\frac{PV}{kT} = N\left[1 + \frac{2\pi\delta^3}{3}\frac{N}{V}\right] = N\left[1 + \frac{2\pi\delta^3}{3}\frac{P}{kT}\right]$$

where we made the approximation $N/V = P/(kT)$ in the correction term. Next we multiply through by kT

$$PV = NkT + N\frac{2\pi\delta^3}{3}P$$

which rearranges to

$$P\left(V - \frac{2\pi\delta^3}{3}N\right) = NkT = \boldsymbol{nR}T$$

Comparison with Equation (20.9), gives $V_1 = N \cdot 2\pi\delta^3/3$.

Although the hard sphere gas potential is easy to implement in Equation (20.22), it does not include attractive interactions. A more realistic description of the interaction between nonpolar molecules (discussed in Section 13.5) is the Lennard–Jones potential, given by

$$V(r) = 4\varepsilon \cdot \left[\left(\frac{\delta}{r}\right)^{12} - \left(\frac{\delta}{r}\right)^{6}\right] \tag{20.23}$$

see Fig. 20.8. The attractive part of the intermolecular potential decreases as the sixth power of the distance, and the repulsive part of the intermolecular potential is modeled with a dependence to the twelfth power of the distance. With this potential function the second virial coefficient is

$$B_2 = -\frac{3}{\delta^3}\int_0^\infty r^2\left[\exp\left(-\frac{4\varepsilon}{kT}\left(\left(\frac{\delta}{r}\right)^{12} - \left(\frac{\delta}{r}\right)^{6}\right)\right) - 1\right]\mathrm{d}r \tag{20.24}$$

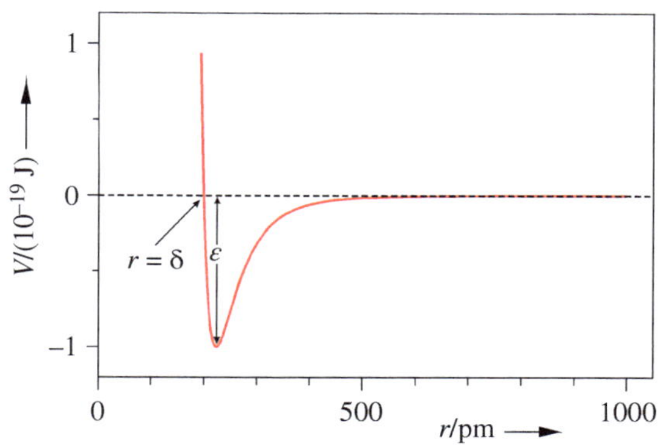

Figure 20.8 Lennard–Jones potential ($\delta = 200$ pm, $\varepsilon = 1.0 \times 10^{-19}$ J).

The integral in Equation (20.24) must be evaluated numerically.

We can express the integral in dimensionless variables by defining $x = \delta/r$ and $T^* = kT/\varepsilon$, and we find that

$$B_2^* - 3\int_0^\infty x^2\left[\exp\left(-\frac{4}{T^*}\left(x^{12} - x^6\right)\right) - 1\right]\,\mathrm{d}x \tag{20.25}$$

Expressed in this form we see that the integral can be evaluated once and applied to gases that have different values for δ and ε but the same shape for the intermolecular potential energy. Equation (20.25) provides a connection between the parameters of the intermolecular potential and the gas properties. As an example, we write the following expressions for the rescaled temperature T^*, volume V^*, and pressure P^*.

$$kT/\varepsilon = T^*, \quad V/(N\sigma^3) = V^*, \quad \text{and} \qquad P\sigma^3/\varepsilon = P^*$$

Table 20.3 reports values of the critical temperature T_c, critical volume V_c, and critical pressure P_c, and the Lennard–Jones parameters δ and ε for some nonpolar gases. Table 20.3 also provides values for the critical parameters when they are rescaled by the Lennard–Jones parameters. The rescaled values for the critical pressure, temperature, and volume hardly change through the series of molecules.

Expressed in this form, the integral can be evaluated once and applied to gases that have different values for δ and ε but the same shape for the intermolecular potential energy. Figure 20.9 plots the reduced second Virial coefficient as a

Table 20.3 Critical Constants for Some Nonpolar Gases and Their Lennard-Jones Potential Rescalings.

Gas	T_c/K	V_c/cm^3	P_c/atm	T_c^*	V_c^*	P_c^*	$\varepsilon/k_$K	$\delta\ 10^{-8}$ cm
Ne	44.5	41.7	25.9	1.25	3.3	0.111	35.60	2.75
Ar	151	75.2	48	1.26	3.16	0.116	120	3.41
Xe	289.81	120.2	57.89	1.31	2.90	0.132	229	4.06
N_2	126.1	90.1	33.5	1.33	2.96	0.131	95.1	3.70
O_2	154.4	74.4	49.7	1.31	2.69	0.142	118	3.58
CH_4	190.7	99.0	45.8	1.29	2.96	0.126	149	3.78

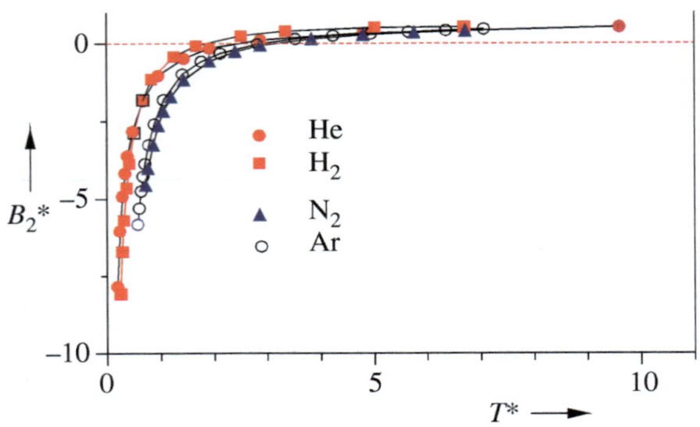

Figure 20.9 Reduced Virial coefficient B_2^* for four gases. (The data were taken from [6].) For He and H_2 quantum effects cause a deviation from the data for other nonpolar gases, which is significant at low temperatures. Otherwise, the data fall on a single curve.

function of the reduced temperature for some different gases. This scaling of the thermal properties is called the *Law of Corresponding States.*

20.4 Change of State for Real Gases

In contrast to an ideal gas, a real gas has an internal energy which depends on both the temperature and the volume. This distinction has both practical and historical importance. Historically, the definition of an ideal gas was linked to the experimental observations of Joule who measured the temperature of a gas which expands into a vacuum, and we discuss this case first. On a practical level, the dependence of a gas's energy on volume can be exploited in refrigeration and in the liquefaction of gases. We discuss this feature in terms of the Joule–Thomson coefficient.

20.4.1 Adiabatic Expansion into Vacuum

Consider an adiabatic expansion of a gas into vacuum (Fig. 20.10). Because the system is isolated, no heat or work is supplied to the system:

$$q = 0, \quad w = 0 \tag{20.26}$$

From the First Law of Thermodynamics, the internal energy change is

$$\Delta U = w + q = 0 \tag{20.27}$$

Because the internal energy of an ideal gas is only a function of the temperature, the temperature of an ideal gas does not change in the experiment (after thermal equilibration).

In a van der Waals gas the molecules attract each other. After an expansion, the molecules are at a larger average distance from each other, and during the expansion they must move against the attractive forces. As a consequence, their average speed is smaller after the expansion, and the temperature of the gas has decreased (after thermal equilibration).

What is the decrease in temperature for a van der Waals gas that expands adiabatically from a volume of V_1 to V_2 into a vacuum? In a thought experiment we divide the overall process in Fig. 20.10a into two subprocesses (Fig. 20.10b). First we expand the gas to its final volume isothermally (temperature T_1); in this subprocess the internal energy changes by ΔU_a (gas at the same temperature, but molecules at larger average distance). In a second step we cool the gas to the temperature T_2 at constant volume. In this subprocess the internal energy changes by ΔU_b (molecules at the same average distance, but at lower temperature). Because the initial and final states are the same as in the adiabatic process, we obtain

$$\Delta U = \Delta U_a + \Delta U_b = 0 \quad \text{or} \quad \Delta U_a = -\Delta U_b \tag{20.28}$$

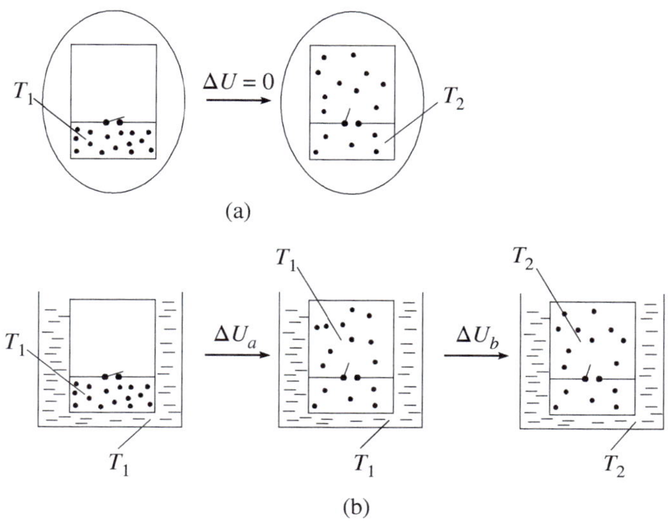

Figure 20.10 Adiabatic expansion of a real gas into vacuum. (a) Direct process: the temperature decreases from T_1 to T_2. (b) Indirect process: isothermal expansion at T_1 (ΔU_a) followed by cooling down to T_2 (ΔU_a) at constant volume. $\Delta U_a + \Delta U_b = 0$.

Step b is a cooling at constant volume, so that

$$\Delta U_b = C_V \cdot (T_2 - T_1) \tag{20.29}$$

Step a is an isothermal work expansion. Using Equation (19.16)

$$\left(\frac{\partial U}{\partial V}\right)_T = T\left(\frac{\partial P}{\partial T}\right)_V - P$$

we have

$$\mathrm{d}U_a = \left[T\left(\frac{\partial P}{\partial T}\right)_V - P\right]\,\mathrm{d}V$$

Using the van der Waals equation of state (20.15) we find

$$\mathrm{d}U_a = \left[T\frac{\boldsymbol{nR}}{V - \boldsymbol{nb}} - \frac{\boldsymbol{nR}T}{V - \boldsymbol{nb}} + \frac{\boldsymbol{n}^2 a}{V^2}\right]\,\mathrm{d}V = \frac{\boldsymbol{n}^2 a}{V^2}\,\mathrm{d}V$$

By integration we obtain

$$\Delta U_a = \int_{V_1}^{V_2} a\boldsymbol{n}^2 \cdot \frac{1}{V^2} \cdot \mathrm{d}V = -a\boldsymbol{n}^2 \cdot \left(\frac{1}{V_2} - \frac{1}{V_1}\right) \tag{20.30}$$

Combining Equations (20.29) and (20.30) we find

$$C_V \cdot (T_2 - T_1) = a\boldsymbol{n}^2 \cdot \left(\frac{1}{V_2} - \frac{1}{V_1}\right)$$

and

$$T_2 - T_1 = \frac{a}{C_{V,m}} \cdot \left(\frac{1}{V_{m,2}} - \frac{1}{V_{m,1}}\right) \tag{20.31}$$

Example 20.6 *Using the van der Waals parameter a for CO_2 ($a = 3.66$ bar L^2 mol^{-2}) and $C_{V,m} = 2.5\,\boldsymbol{R}$ (neglecting the vibrational contribution to $C_{V,m}$), calculate the temperature change that results from the adiabatic expansion from $V_{m,1} = 2.27$ L mol^{-1} to $V_{m,2} = 22.7$ L mol^{-1}.*

Using Equation (20.31), we obtain

$$T_2 - T_1 = \frac{3.66\ \text{bar L}^2\ \text{mol}^{-2}}{2.5 \cdot 0.08314\ \text{bar L K}^{-1}\ \text{mol}^{-1}} \cdot \left(\frac{1}{22.7} - \frac{1}{2.27}\right)\ \text{mol L}^{-1} = -7\ \text{K}$$

The temperature of the gas decreases by 7 K. This should be contrasted to the prediction of the ideal gas, for which the temperature change is expected to be zero.

20.4.2 Joule–Thomson Effect

The cooling of a gas upon expansion at constant pressure is called the Joule–Thomson effect and is of practical utility in the liquefaction of gases. Figure 20.11 illustrates a Joule–Thomson expansion. In this process the gas flows through a throttle valve (typically a porous frit) and is adiabatically expanded from a pressure P_1 (volume V_1) to a pressure P_2 (volume V_2). Because the process is adiabatic ($\delta q = 0$), the change in internal energy, ΔU, is equal to the work w

$$\Delta U = w$$

By means of piston 1, the work $P_1 V_1$ is supplied to transfer the gas through the throttle valve from left to right (Fig. 20.11a). On the right side, the work $-P_2 V_2$ is supplied to move piston 2 accordingly (Fig. 20.11b). The total work done on the system is

$$w = -P_2 V_2 + P_1 V_1 = U_2 - U_1 \tag{20.32}$$

Rearrangement of this equation shows that

$$U_2 + P_2 V_2 = U_1 + P_1 V_1 \quad \text{or} \quad H_2 = H_1 \tag{20.33}$$

The enthalpy does not change in this process ($\mathrm{d}H = 0$).

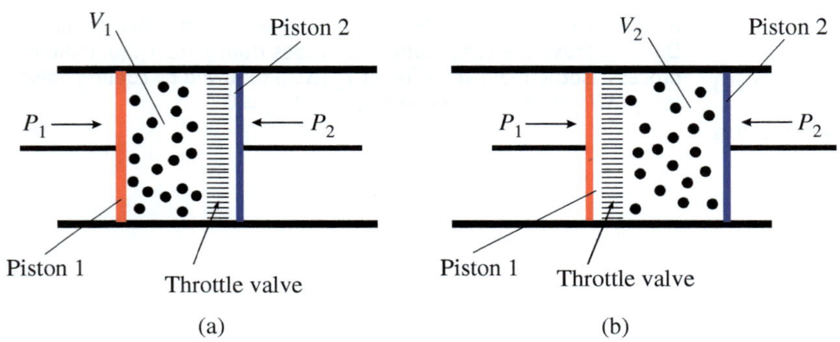

Figure 20.11 Joule–Thomson experiment. Adiabatic expansion of a gas from pressure P_1 and volume V_1 (a) to pressure P_2 and volume V_2 (b) using a throttle valve.

20.4.2.1 Joule–Thomson Coefficient

Using Equation (19.23) for the differential enthalpy change, we find

$$\mathrm{d}H = C_P \mathrm{d}T + \left(-T\left(\frac{\partial V}{\partial T}\right)_P + V\right) \mathrm{d}P = 0 \tag{20.34}$$

We resolve this equation for

$$\mu_{\mathrm{JT}} = \left(\frac{\mathrm{d}T}{\mathrm{d}P}\right)_H \tag{20.35}$$

where μ_{JT} is called the *Joule–Thomson coefficient*.

$$\mu_{\mathrm{JT}} = \left(\frac{\mathrm{d}T}{\mathrm{d}P}\right)_H = \frac{1}{C_P}\left(T\left(\frac{\partial V}{\partial T}\right)_P - V\right) \tag{20.36}$$

For an ideal gas, $V = NkT/P$ and $(\partial V/\partial T)_P = Nk/P$, thus $\mu_{\mathrm{JT}} = 0$, and the temperature does not change for a change in pressure.

For a real gas the temperature does change. In Problem P20.12 you show that the van der Waals equation of state predicts a Joule–Thomson coefficient of

$$\mu_{\mathrm{JT}} = \frac{1}{C_{P,m}} \cdot \left[\frac{2a}{\boldsymbol{R}T} - b - \frac{3ab}{\boldsymbol{R}^2 T^2} \cdot P\right] \tag{20.37}$$

For small enough pressure P this result can be approximated as

$$\mu_{\mathrm{JT}} \approx \frac{1}{C_{P,m}} \cdot \left(\frac{2a}{\boldsymbol{R}T} - b\right) \tag{20.38}$$

At low enough temperature T and pressure P the Joule–Thomson coefficient is positive: the gas cools upon expansion. However, at low enough pressure and sufficiently high temperature T, the Joule–Thomson coefficient becomes equal to $-b$ so that the gas heats as it expands.

Example 20.7 *Approximate CO_2 ($C_{P,m} = \frac{7}{2}\boldsymbol{R}$) as a van der Waals gas ($a = 3.66$ bar L^2 mol^{-2}, $b = 0.043$ L mol^{-1}) and find the temperature change for a Joule–Thomson expansion with $\Delta P = P_2 - P_1 = -10$ bar and $T_1 = 300$ K, using Equation (20.38).*

In a small enough pressure range we can replace the differentials dT and dP in Equation (20.35) by the differences $\Delta T = T_2 - T_1$ and $\Delta P = P_2 - P_1$, that is $(dT/dP)_H \sim (\Delta T/\Delta P)$, and we obtain

$$T_2 - T_1 = \mu_{JT} \cdot (P_2 - P_1)$$

Then with

$$\mu_{JT} = \frac{1}{C_{P,m}} \cdot \left(\frac{2a}{\boldsymbol{R}T} - b\right)$$
$$= \frac{2}{7 \cdot \boldsymbol{R}} \cdot \left(\frac{2 \cdot 3.66 \text{ bar L}^2 \text{ mol}^{-2}}{0.08314 \cdot 300 \text{ bar L mol}^{-1}} - 0.043 \text{ L mol}^{-1}\right) = 0.86 \text{ K bar}^{-1}$$

and for $P_2 - P_1 = -10$ bar, we obtain $T_2 - T_1 = -8.6$ K. The experimental value is -10 K.

20.4.2.2 Inversion Temperature T_i

Inspection of Equations (20.37) and (20.38) shows that the Joule–Thomson coefficient can take both positive (cooling) and negative (heating) values. The temperature at which the Joule–Thomson coefficient changes sign is called the *inversion temperature* T_i, and its value depends on the starting pressure. In Problem P20.12, you show that T_i is

$$T_i = \frac{1}{b\boldsymbol{R}}\left[a \pm \sqrt{a^2 - 3ab^2 \cdot P}\right] \tag{20.39}$$

for a van der Waals equation of state. Figure 20.12 plots the inversion temperature of nitrogen as a function of the pressure; i.e., the *inversion curve*. At pressures and temperatures within the shaded area of the inversion curve, the gas characteristics are dominated by molecules that are interacting in the attractive part of the intermolecular potential. When the gas expands, the molecules are further apart on average and have a higher average potential energy. This increase in the average potential energy leads to a decrease in the kinetic energy of the gas atoms and hence the temperature of the gas. μ_{JT} is positive because we must do work against the attractive well.

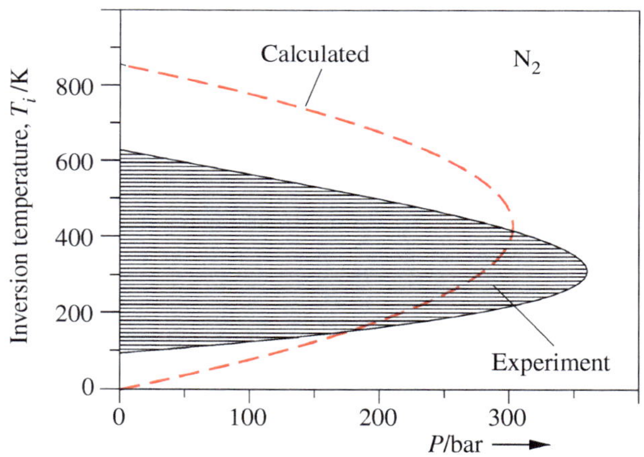

Figure 20.12 Inversion temperature T_i for N_2 plotted against P. Solid line: experiment. Inside the shaded area a cooling effect occurs, and outside the shaded area the gas is warmed up. Red dashed line: calculated from Equation 20.39, using the van der Waals model with $a = 1.390$ L^2 bar mol^{-2} and $b = 0.0391$ L mol^{-1}. Experimental data from [7]

The sensitivity to the attractive forces is a function of temperature; as the temperature increases the average kinetic energy increases and the attractive forces, which are not temperature dependent, become relatively less important. The other regime (corresponding to high pressures and/or high temperatures) is the case in which the initial state of the gas is dominated by the repulsive part of the intermolecular potential and the final state is dominated by interactions lower down on the intermolecular potential. In this case, the expansion causes a decrease in the average potential energy, and hence an increase in the gases kinetic energy (i.e., temperature). In this case, μ_{JT} is less than zero because potential energy is converted into kinetic energy and the temperature increases.

To cool a gas using the Joule–Thomson effect it is important that the starting temperature lies below the inversion temperature of the gas. For example, to liquify helium, the helium gas must be cooled below 43 K before starting the Joule–Thomson process. Liquid N_2 (boiling point 80 K) and liquid He (boiling point 4 K) are produced in industry by using the Linde process based on the Joule–Thomson effect.

Example 20.8 *Use Equation (20.38) to calculate the inversion temperature for helium ($a = 0.035$ bar L^2 mol^{-2}, $b = 0.024$ L mol^{-1}), which is reported to be 43 K* [8].

From Equation (20.38), we see that the Joule–Thomson coefficient is zero when

$$\frac{2a}{\boldsymbol{R}T} = b \tag{20.40}$$

so that the inversion temperature is

$$T_i = \frac{2a}{b \cdot \boldsymbol{R}} \tag{20.41}$$

For He we obtain

$$T_i = \frac{2 \cdot 0.035}{0.024 \cdot 0.08314}\ \text{K} = 35\ \text{K}$$

which is 22% smaller than the measured value 43 K.

20.5 Chemical Equilibria Involving Real Gases

In Chapter 18 we discussed chemical equilibria of gases in the approximation that they behave ideally, namely that

$$PV = \boldsymbol{nR}T \quad \text{and} \quad P_i = x_i P \tag{20.42}$$

where P_i is the partial pressure and x_i is the mole fraction of component i in a gas mixture. Here we discuss how the chemical equilibria of gas mixtures can be described more realistically using the fugacity and Equation (20.6).

Following our development in Section 18.5, we write the Gibbs energy for a reaction, such as

$$\alpha A + \beta B \rightarrow \gamma C + \delta D \tag{20.43}$$

as

$$\Delta_r G = G_{m,C} \cdot \gamma + G_{m,D} \cdot \delta - G_{m,A} \cdot \alpha - G_{m,B} \cdot \beta \tag{20.44}$$

The Gibbs energy expression for an ideal gas is

$$G_m = G_m^{\ominus} + \boldsymbol{R}T \cdot \ln \frac{P}{P^{\ominus}} \tag{20.45}$$

Within the ideal gas approximation it is possible to replace the pressure of the pure substance i with its partial pressure in the equation for G_m to obtain an expression for the molar Gibbs energy of component i, $G_{m,i}$, in the mixture. Rather than the ideal gas expression, we use our result for a real gas with fugacity f, namely

$$G_m = G_m^{\ominus} + \boldsymbol{R}T \cdot \ln \frac{f}{f^{\ominus}} = G_m^{\ominus} + \boldsymbol{R}T \cdot \ln \widehat{f} \tag{20.46}$$

and we assume that the fugacity of the pure species i is the same as its fugacity in the mixture; i.e., $f_i = x_i f$, so that

$$G_{m,i} = G_{m,i}^{\ominus} + \boldsymbol{R}T \cdot \ln \frac{f_i}{f_i^{\ominus}} \tag{20.47}$$

Using this result in our equation for $\Delta_r G$, we find the analogue for Equation (18.50)

$$\Delta_r G = \Delta_r G^{\ominus} + \boldsymbol{R}T \cdot \ln \frac{\widehat{f}_{\mathrm{C}}^{\gamma} \cdot \widehat{f}_{\mathrm{D}}^{\delta}}{\widehat{f}_{\mathrm{A}}^{\alpha} \cdot \widehat{f}_{\mathrm{B}}^{\beta}} \tag{20.48}$$

At equilibrium $\Delta_r G = 0$, and we obtain

$$\Delta_r G^{\ominus} = -\boldsymbol{R}T \cdot \ln K = -\boldsymbol{R}T \cdot \ln \frac{\widehat{f}_{\mathrm{C,eq}}^{\gamma} \cdot \widehat{f}_{\mathrm{D,eq}}^{\delta}}{\widehat{f}_{\mathrm{A,eq}}^{\alpha} \cdot \widehat{f}_{\mathrm{B,eq}}^{\beta}} \tag{20.49}$$

which defines the equilibrium constant K.

As an example we consider the equilibrium constant for the ammonia equilibrium

$$N_2 + 3\ H_2 \ \rightarrow 2\ NH_3$$

where we write the equilibrium constant as

$$K = \frac{f^2_{NH_3,\,eq}}{f_{N_2,\,eq} \cdot f^3_{H_2,\,eq}}$$

using the fugacities $f = \widehat{P}\phi$ instead of the pressures $\widehat{P} = P/P^{\ominus}$. Then

$$K = \frac{\left(\widehat{P}\phi\right)^2_{NH_3,\,eq}}{\left(\widehat{P}\phi\right)_{N_2,\,eq}\left(\widehat{P}\phi\right)^3_{H_2,\,eq}} = \frac{\widehat{P}^2_{NH_3,\,eq}}{\widehat{P}_{N_2,\,eq} \cdot \widehat{P}^3_{H_2,\,eq}} \cdot \frac{\phi^2_{NH_3}}{\phi_{N_2} \cdot \phi^3_{H_2}} = K_{ideal} \cdot \frac{\phi^2_{NH_3}}{\phi_{N_2} \cdot \phi^3_{H_2}}$$

In Table 20.4 (column 5), K_{ideal} is calculated for different total pressures up to 1013 bar, using the experimentally[3] measured equilibrium pressures P_{eq}. The corresponding values for K are given in column 6, using the experimentally determined fugacity coefficients, ϕ. For example, at $T = 723$ K and $P = 304$ bar we find $K_{ideal} = 7.76 \times 10^{-5}$. The fugacity coefficients ϕ of the reactants depend on pressure. For $P = 304$ bar, they are $\phi(N_2) = 1.14$, $\phi(H_2) = 1.09$, and $\phi(NH_3) = 0.91$ (see Section 20.2 for the determination of fugacity coefficients). Thus we calculate

$$K = 7.76 \times 10^{-5} \cdot \frac{0.91^2}{1.14 \cdot 1.09^3} = 7.76 \times 10^{-5} \cdot 0.561 = 4.35 \times 10^{-5}$$

Table 20.4 Calculation of K for the Synthesis of NH_3 from the Elements at $T = 723$ K.

$\widehat{P}_{total}$	$\widehat{P}_{NH_3,\,eq}$	$\widehat{P}_{N_2,\,eq}$	$\widehat{P}_{H_2,\,eq}$	$K_{ideal}/10^{-5}$	$K/10^{-5}$
10	0.21	2.48	7.44	4.48	4.37
30	1.78	7.15	21.46	4.49	4.17
51	4.63	11.50	34.51	4.54	4.03
101	16.64	21.16	63.49	5.11	4.02
304	108.9	48.8	146.3	7.76	4.35
608	326.5	70.3	211.0	16.13	4.10
1013	706.0	76.8	230.3	53.17	9.78

Whereas K_{ideal} is constant only at low pressures (serious deviations occur above 100 bar; see Table 20.4, column 5). The equilibrium constant K (partial pressures replaced by fugacities) is constant up to 600 bar. The remaining deviation at 1013 bar arises from interactions between different gases in the mixture. Our use of the fugacity for the pure gases cannot account for this effect. To improve the agreement we would need to measure the fugacities in the mixture.

20.6 Conclusion

In this chapter we advanced our considerations about the thermochemical properties of gases. Intermolecular forces give rise to important effects, such as the existence of phases. To describe the gas behavior, we exploited the mathematical framework that was developed for the case of ideal gases, but we replaced the pressure P with the fugacity f. The fugacity became the essential variable in determining the Gibbs energy, and it can be obtained directly from experimental measurements or by theoretical models.

We developed models for the *PVT* behavior of gases, the equations of state. Knowledge of the equation of state for a substance allows us to calculate its thermodynamic properties. We discovered that both repulsive and attractive interactions are important to consider when describing gases at high densities and/or high pressures.

3 Larson [9] The data in Table 20.4 are calculated from the data given in Table III.

20.7 Box 20.1. Estimation of V_1 in the Hard Sphere Model

For a hard sphere gas the intermolecular force is zero when greater than some distance δ, but is infinitely repulsive at the contact distance δ. This model corresponds to ping pong balls zooming around in a box. In this case the potential energy is written as

$$E_{\text{pot}}(r) = \infty \text{ at } r < \delta \quad \text{and} \quad E_{\text{pot}}(r) = 0 \text{ at } r > \delta$$

where δ is the diameter of the hard sphere. See Fig. 20.3.

To approximate the increase in pressure by the volume V_1 of the molecules, consider the diagram in Fig. 20B.1, which shows two colliding molecules. Because of their finite size, the molecules (diameter $2r$) reach the collision sooner than they would if they were infinitesimally small. For a given collision, this additional distance is $2r \cdot \cos\alpha$, and on average over all collisions the distance is

$$2r\overline{\cos\alpha} = 2r\frac{\int_0^{\pi/2}\cos\alpha\,d\alpha}{\int_0^{\pi/2}d\alpha} = 2r\frac{|\sin\alpha|_0^{\pi/2}}{\pi/2} = \frac{4}{\pi}r \approx r$$

In each collision the molecules achieve this additional distance. Effectively, the molecules collide more often, namely, by a factor

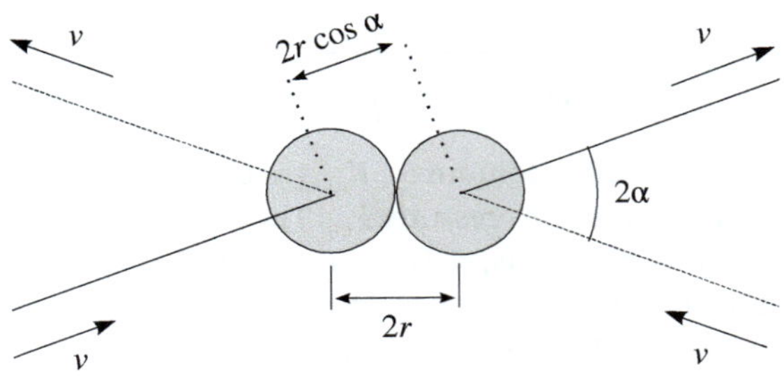

Figure 20B.1 Collision frequency increases with increasing diameter of the molecules.

$$\frac{\overline{\lambda}+r}{\overline{\lambda}} = 1 + \frac{r}{\overline{\lambda}}$$

where $\overline{\lambda}$ is the mean free path; i.e., the frequency of collisions is increased by this factor. Then instead of $PV = \boldsymbol{n}\boldsymbol{R}T$, we obtain a pressure that is higher by a factor of $r/\overline{\lambda}$

$$P = \left(1 + \frac{r}{\overline{\lambda}}\right)\cdot\frac{\boldsymbol{n}\boldsymbol{R}T}{V}$$

If $r \ll \overline{\lambda}$, then $1 + r/\lambda \approx 1/(1 - r/\lambda)$ and we find that

$$P = \frac{\boldsymbol{n}\boldsymbol{R}T}{\left(1 - r/\overline{\lambda}\right)\cdot V} = \frac{\boldsymbol{n}\boldsymbol{R}T}{V - rV/\overline{\lambda}} = \frac{\boldsymbol{n}\boldsymbol{R}T}{V - V_1}$$

Now we use Equation (14.43), $\lambda = V/(\sqrt{2}N\pi(2r)^2)$, to obtain

$$V_1 = \frac{r\cdot V}{\overline{\lambda}} = r\cdot\sqrt{2}\cdot\pi(2r)^2\cdot N = 4\sqrt{2}\pi r^3\cdot N$$

The volume of a molecule, approximated as a sphere, is

$$V_{\text{molecule}} = \frac{4}{3}\pi r^3$$

so that

$$V_1 = 3\sqrt{2}\cdot V_{\text{molecule}}\cdot N = 4.2V_{\text{molecule}}\cdot N = 4.2V_{\text{molecule}}\cdot \boldsymbol{N}_{\text{A}}\cdot\boldsymbol{n}$$

20.8 Foundations

20.8.1 Foundation 20.1: Molecular Perspective on Solid–Gas Equilibria

See the books website, www.wiley.com/go/Kuhn/physicalchemistry3e.

20.8.2 Foundation 20.2: Virial Equation of State

See the books website, www.wiley.com/go/Kuhn/physicalchemistry3e.

20.9 Exercises

E20.1 Use the Thermodynamic Equation of State,

$$\boxed{\left(\frac{\partial U}{\partial V}\right)_T = T\left(\frac{\partial P}{\partial T}\right)_V - P},$$

to show that the internal energy of a hard sphere gas depends only on T. Discuss your result and compare it to that for an ideal gas.

E20.2 Integrate the differential form of the entropy which you found in E19.2 and show that the entropy change of a monatomic hard sphere gas ($C_V = 1.5\ \boldsymbol{nR}$) is

$$\Delta S = \frac{3}{2}\boldsymbol{nR}\ln\left(\frac{T_2}{T_1}\right) + \boldsymbol{nR}\ln\left[\frac{(V_{m,2} - b)}{(V_{m,1} - b)}\right]$$

E20.3 Given that the molar Helmholtz energy of a gas is

$$A_m = -\frac{a}{V_m} - \boldsymbol{R}T\ln\frac{(V_m - b)}{V_m^{\ominus}} + f(T)$$

where V_m is the molar volume of the gas, $V_m^{\ominus}$ is the molar volume of the gas under standard state conditions, $\boldsymbol{R}$ is the gas constant (8.314 J mol^{-1} K^{-1}), a and b are empirical constants, and $f(T)$ is an arbitrary function of T, find the equation of state for the gas. Interpret the parameters a and b. What features of the intermolecular interactions do they capture?

E20.4 Use the van der Waals equation of state $P = \frac{\boldsymbol{R}T}{V_m - b} - \frac{a}{V_m^2}$ to find an expression for dU of a gas. Contrast your result with that for an ideal gas.

E20.5 Consider 1 mol of Ar gas in a volume $V = 1$ L. Compute the pressure P in the ideal gas approximation and the hard sphere gas approximation at 298 K. Perform the same calculation for methane. Comment on your findings.

E20.6 Using the hard sphere gas equation of state, $P(V_m - b) = \boldsymbol{R}T$ and assuming that $C_V = \frac{3}{2}\ \boldsymbol{nR}$, find mathematical expressions for the internal energy, the entropy, and the enthalpy of the gas.

E20.7 Extend the calculations in E20.6 to determine the change in Gibbs energy and Helmholtz energy of the gas along an isotherm

E20.8 Consider an isothermal and reversible expansion of a hard sphere gas from an initial volume V_0 to a final volume V_f. Find expressions for the work and the heat in the process. Compare the reversible work to the change in Helmholtz energy between the initial and final states.

E20.9 Show that the change in internal energy with an isothermal volume change is $\boldsymbol{n}^2 a/V^2$ for a van der Waals gas. Evaluate the reversible work to compress CO_2 isothermally from 1 bar to 66 bar.

E20.10 Show that the equation

$$G(T, P) = G^{\ominus}(T) + NkT \cdot \ln\left(\frac{P}{P^{\ominus}}\right)$$

can be used as an alternative definition of the ideal gas, by showing that it implies (a) $PV = NkT$ and (b) U is a function of only T.

E20.11 What is the Law of Corresponding States?

E20.12 Show that the Joule–Thomson coefficient for an ideal gas is zero.

E20.13 Find an expression for the Joule–Thomson coefficient of a hard sphere gas. Assume the gas is monatomic and that $C_{P,m} = \frac{5}{2}\boldsymbol{R}$.

E20.14 In Example E20.7 we have calculated the temperature jump when expanding a gas in the Joule–Thomson experiment from pressure P_1 to P_2, using Equation (20.38) for the Joule–Thomson coefficient. Repeat the calculation by using the more general Equation (E20.37).

E20.15 Use the fact that dU is an exact differential to show that the constant volume molar heat capacity of a van der Waals gas depends only on the temperature.

E20.16 Using the data provided in Table 20.1, determine the fugacity of CO_2 at 101 bar and 50.6 bar; i.e., perform the analysis of the text.

E20.17 Analyze the PV_m data [12] for hydrogen at 273 K to obtain values for the fugacity at 200 bar and 1000 bar.

P (bar)	1	50	100	200	300	400	600	800	1000
V_m (L mol^{-1})	22.712	0.46818	0.24146	0.12844	0.09096	0.072300	0.053670	0.044327	0.038670

E20.18 Analyze the data given for nitrogen at 273.15 K [13] to obtain values for the fugacity at 202.65 bar and 1013.25 bar (Φ is the compression factor). To proceed, assume that the gas behaves ideally (i.e., the compression factor $\Phi = 1$, see Equation (20.8)) for pressures at 1.01325 bar and lower.

P (bar)	1.01325	50.6625	101.325	202.650	303.975	405.300	607.95	810.6	1013.25
Φ	1.0000	0.9839	0.9840	1.0330	1.1332	1.2540	1.5211	1.7985	2.0659

Use your result to determine the change in Gibbs energy of the gas between 202.65 bar and 1013.25 bar. Compare your result to that expected for an ideal gas.

E20.19 In Fig. 20.5 we compared the pressure dependence of the compression factor for CO_2 at 500 K to the prediction by the van der Waals equation of state model. Perform a similar comparison using the Beattie–Bridgman equation of state, which is

$$P = \frac{\boldsymbol{R}T \cdot (1 - c/(V_m T^3))}{V_m^2}(V_m + B_0(1 - b/V_m)) - \frac{A_0 \cdot (1 - a/V_m)}{V_m^2}$$

where the parameters are $A_0 = 5.072836$ bar L^2 mol^{-2}, $a = 0.07132$ L mol^{-1}, $B_0 = 0.10476$ L mol^{-1}, $b = 0.07235$ L mol^{-1}, and $c = 6.600 \cdot 10^5$ K^3 L mol^{-1}. The data for CO_2 are

P (bar)	1	50	100	150	200	250	300
V_m (L mol^{-1})	41.55	0.8091	0.389	0.253	0.187	0.149	0.125
Φ_{expt}	0.999	0.965	0.936	0.914	0.900	0.896	0.901

P (bar)	400	500	600	700	800	900	1000
V_m (L mol^{-1})	0.0972	0.0824	0.0733	0.0673	0.0630	0.0597	0.0571
Φ_{expt}	0.935	0.990	1.06	1.13	1.21	1.29	1.37

E20.20 The Dieterici equation of state is

$$P(V_m - b) \cdot \exp\left(\frac{a}{V_m \boldsymbol{R}T}\right) = \boldsymbol{R}T$$

where a and b are constants (parameters). a) Show that the Dieterici equation of state reduces to the ideal gas equation of state as the temperature T becomes large and the molar volume V_m becomes large. b) Use this equation of state and find an expression for the volume dependence of the gas's internal energy.

E20.21 A simple application of thermodynamic equilibria to chemically significant systems is the description of the phase transitions which pure substances undergo. In this MathCAD worksheet (sublimation.mcdx) you investigate the biphasic equilibrium between a solid and its monatomic gas, see Foundation 20.1. To construct the phase diagram you will calculate the Gibbs free energy for the solid ($G_s(T)$), by using an Einstein model (i.e., one normal mode frequency) and for the gas ($G_g(T)$), by using an ideal gas approximation. You will find the temperature and pressures at which the Gibbs energies of the two phases are equal and plot the coexistence curve.

20.10 Problems

P20.1 Consider hydrogen to be a hard sphere gas [$P(V - \boldsymbol{n}b) = \boldsymbol{nR}T$] with constant $b = 0.02661$ L mol^{-1}. a) Show that $dH = C_P\, dT + \left(V - T\left(\frac{\partial V}{\partial T}\right)_P\right) dP$. b) Provide a molecular-based explanation as to why it is reasonable to estimate the molar heat capacity at constant pressure for hydrogen to be 29.100 J mol^{-1} K^{-1}. c) Use the information given in parts a) and b) to determine the enthalpy change for 10 mol of hydrogen that is compressed from 1 bar pressure at 298 K to 500 bar pressure at 298 K. d) Use the enthalpy expression in part a) to derive an expression for the internal energy of the hydrogen gas. Evaluate the internal energy change for compressing the gas from 1 bar to 500 bar at 298 K.

P20.2 The differential Gibbs energy change of a gas is

$$dG = V\,dP = \boldsymbol{nR}T\,\frac{df}{f}$$

under isothermal conditions. P is the pressure and f is the fugacity. a) Show that the fugacity is related to the compressibility factor Φ of the gas by

$$\ln\left(\frac{f}{P}\right) = \int_{P_0}^{P} dP' \left(\frac{\Phi - 1}{P'}\right)$$

b) Using the Berthelot equation of state for a gas, namely

$$PV_m = \boldsymbol{R}T + \frac{9\boldsymbol{R}T_c P}{128P_c}\left(1 - 6\frac{T_c^2}{T^2}\right)$$

find an expression for the fugacity of the gas. Note that T_c is the critical temperature and P_c is the critical pressure of the gas. c) Evaluate the fugacity of carbon dioxide at 298 K and 1 bar pressure, using your result from b) and the facts $T_c = 304$ K and $P_c = 74$ bar for CO_2.

P20.3 Consider the thermodynamic properties of a "real" gas with a molar Gibbs energy of

$$G_m = G^{\ominus}(T) + \boldsymbol{R}T\ln\left(P/P^{\ominus}\right) + bP$$

where $G^{\ominus}(T)$ is the standard state value of the Gibbs energy, i.e., at 1 bar pressure. $\boldsymbol{R}$ is the gas constant, and b is an empirical constant. a) Find the equation of state for the gas. Interpret the parameter b. b) Assume that $G^{\ominus}(T) = c_1 T \cdot \ln(c_2 T)$, where c_1 and c_2 are constants, and find an expression for the entropy of the gas. c) Using your results from parts a) and b), write expressions for the internal energy of the gas, the enthalpy of the gas, and the Helmholtz energy of the gas. d) Find an expression for the maximum work that can be performed by isothermally expanding the gas from an initial volume $V_{m,1}$ to a final volume $V_{m,2}$.

P20.4 In problem P20.2, you showed that the fugacity f is related to the compression factor of the gas by

$$\ln\left(\frac{f}{P}\right) = \int_{P_0}^{P} dP' \left(\frac{\Phi - 1}{P'}\right)$$

Find the fugacity and the compression factor for a hard sphere gas whose equation of state is given by $P(V - \boldsymbol{n}b) = \boldsymbol{nR}T$. Sketch the dependence of Φ on P and the dependence of f on P at three different temperatures. Lastly, sketch the dependence of Φ on P for an ideal gas.

P20.5 Consider our definition of the isothermal compressibility κ_T. Derive a mathematical expression for this quantity in the case of an ideal gas equation of state. Derive a mathematical expression for the isothermal compressibility for a hard sphere gas equation of state $P(V - \boldsymbol{n}b) = \boldsymbol{nR}T$. Compare your results and find an expression for the deviation $\delta\kappa_T$ in the isothermal compressibility, defined as $\delta\kappa_T = \kappa_T(\text{ideal}) - \kappa_T(\text{hard sphere})$. Provide a physical interpretation for why the isothermal compressibility differs in the way it does.

P20.6 The molar Helmholtz free energy of one mole of a van der Waals gas may be written as

$$A_m = -\frac{a}{V_m} - \boldsymbol{R}T\ln\left[(V_m - b)/V_m^{\ominus}\right] - \boldsymbol{R}T - 3\boldsymbol{R}T\ln\left(C_1 T^{1/2}\right)$$

where a, b, and C_1 are constants. a) Use this expression to derive an expression for the molar entropy of the van der Waals gas. b) Use the previous results to derive an expression for the molar internal energy of the van der Waals gas, U_m. c) Compare your expression for the molar internal energy of the van der Waals gas to that for one mole of a monatomic ideal gas. Determine how they differ mathematically and provide a physically based explanation for this difference. d) Find an expression for the molar constant volume heat capacity $C_{V,m}$ of the van der Waals gas. Compare your finding to that of a monatomic ideal gas and provide a physically based explanation for your findings.

P20.7 If the equation of state for a liquid is

$$V_m = V_{m,0}\left[1 + a \cdot T - b \cdot P\right]$$

where P is the pressure in bar, $V_{m,0}$ is the molar volume of the liquid at 273 K and 1 bar pressure, and a and b are empirical constants, show that

$$\left(\frac{\partial S}{\partial P}\right)_T = -aV_{m,0}\,; \quad \left(\frac{\partial S}{\partial V_m}\right)_T = -\frac{a}{b}\,; \text{ and } \left(\frac{\partial H}{\partial P}\right)_T = V_{m,0}\,(1 - bP)$$

Calculate ΔS and ΔH when 1 mol of the liquid is compressed at 293 K from 1 bar to 25 bar. Assume that $a = 2.1 \times 10^{-4}$ K^{-1} and $b = 49 \times 10^{-6}$ bar^{-1}. (adapted from R. E. Dickerson, Molecular Thermodynamics).

P20.8 In the MathCAD exercise, sublimation.mcdx, you study the phase equilibria of Ar between its solid and its vapor. For the molar Gibbs energy of Ar gas you use

$$G_{m,\text{gas}}(T,P) = \boldsymbol{R}T\ln(P/P^{\ominus}) - \boldsymbol{R}T\ln\left[kT\left(\frac{2\pi mkT}{h^2}\right)^{1/2}\right]$$

and for the molar Gibbs energy of Ar solid you use

$$G_{m,\text{solid}}(T,P) = \frac{3}{2}\boldsymbol{R}\theta + \frac{\phi}{2} + \frac{3}{2}\boldsymbol{R}T\ln\left(1 - e^{-\theta/T}\right)$$

where θ is 30 K and ϕ is -147 kJ mol^{-1}. Use the Gibbs energy expression to determine the molar entropy of the gas, $S_{m,\text{gas}}$. Given that the entropy of the solid is

$$S_{m,\text{solid}} = -\frac{3}{2}\boldsymbol{R}\ln\left[1 - e^{-\theta/T}\right] + \frac{3}{2}\boldsymbol{R}\frac{\theta}{T\,(\exp(\theta/T) - 1}$$

find an expression for the entropy change in the phase transition. Use these entropy expressions and the Gibbs energy expressions to determine the enthalpy change in the phase transition. Simplify your answer and evaluate it for $T = 298$ K. Use the Gibbs energy expression to determine the volume change in the phase transition. Combine these results with the Clapeyron equation to find $\mathrm{d}P/\mathrm{d}T$.

P20.9 Given the data in the table and that the constant pressure heat capacity at 5 bar follows the empirical equation $C_{P,m} = a \cdot T^2 - b \cdot T + c$ with $a = 1.680 \cdot 10^{-4}$ J mol^{-1} K^{-3}, $b = 0.1043$ J mol^{-1} K^{-2}, and $c = 91.458$ J mol^{-1} K^{-1}, calculate the change in enthalpy of 1 mol of liquid water if it is compressed from 5 bar and 300 K to 100 bar and 400 K.

P/bar	ρ/kg m^{-3}	$V_m/10^{-5}$ m^3 mol^{-1}	$\frac{\partial V_m}{\partial T}/10^{-8}$ m^3 mol^{-1} K^{-1}
5	937.617	1.92139	1.140
7.5	937.745	1.92113	1.139
10	937.873	1.92086	1.138
20	938.385	1.91982	1.136
30	938.895	1.91877	1.134
40	939.404	1.91773	1.131
50	939.91	1.9167	1.129
60	940.415	1.91567	1.127
70	940.919	1.91465	1.124
80	941.42	1.91363	1.122
100	942.418	1.9116	1.118

P20.10 In Section 20.2 we claim that an ideal gas can be defined as a substance whose Gibbs energy is given by

$$G(T,P) = G^{\ominus}(T) + NkT \cdot \ln \hat{P} \tag{20.50}$$

where $G^{\ominus}(T)$ is the Gibbs energy under standard conditions and $\hat{P} = P/P^{\ominus}$. Using your knowledge of thermodynamics, prove that this is true by deriving an expression for the ideal gas equation of state and showing that the internal energy depends only on T.

P20.11 Find mathematical expressions (Equation (20.18)) for the van der Waals parameters a and b in terms of the critical constants, T_c, P_c, and V_c, of a gas.

P20.12 Derive an expression for the Joule–Thomson coefficient, and its inversion temperature, of a van der Waals gas. To simplify the analysis, you should write the van der Waals equation of state in the approximate form

$$V_m \simeq \frac{\boldsymbol{R}}{P}T + b - \frac{a}{\boldsymbol{R}T} + \frac{abP}{\boldsymbol{R}^2T^2}$$

Explain the reasoning behind this approximate form.

References

1 E.W. Washburn, *International Critical Tables*, Vol. **III** (McGraw Hill, New York, 1928, pp. 11–12).

2 D.R. Lide, ed., *CRC Handbook of Chemistry and Physics* (CRC Press, 1994) 75th edition.

3 J.P. Hansen and I.R. McDonald, *Theory of Simple Liquids* (Academic Press, New York, 1986).

4 B.E. Wyslouzil and J. Wölk, Overview: Homogeneous nucleation from the vapor phase—the experimental science, *J. Chem. Phys.*, 2016, **145**, 211702; I.J. Ford, Statistical mechanics of nucleation: a review, *Proc. Inst. Mech. Eng., Part C: J. Mech. Eng. Sci.*, 2004, **218**, 883–899; M. Yarom and A. Marmur, Vapor–liquid nucleation: the solid touch, *Adv. Colloid Interface Sci.*, 2014, http://dx.doi.org/10.1016/j.cis2014.0711; R. Dettmer, Thermodynamische Betrachtungen zur Kondensation, Dissertation Marburg, 1997.

5 H.H. Chen, R.A. Aziz, and C.C. Lim, On the vapor pressure of solid argon, *Can. J. Phys.*, 1971, **49**, 1569–1581.

6 G.C. Maitland, M. Rigby, E.B. Smith, and W.A. Wakeham, *Intermolecular Forces* (Oxford, New York, 1981).

7 J.R. Roebuck and H. Osterberg, The Joule-Thomson effect in nitrogen, *Phys. Rev.*, 1935, **48**, 450.

8 R.D. McCarty, Thermodynamic properties of helium 4 from 2 to 1500 K at pressures to 10^8 Pa, *J. Phys. Chem. Ref. Data*, 1973, **2**, 923.

9 A.T. Larson, The ammonia equilibrium at high pressures, *J. Am. Chem. Soc.*, 1924, **46**, 367.

10 D.A. McQuarrie, *Statistical Mechanics* (Harper, 1976).

11 R.K. Pathria, *Statistical Mechanics* (Pergamon, 1972).

12 E.W. Lemmon, M.O. McLinden, and D.G. Friend, Thermophysical Properties of Fluid Systems, in *NIST Chemistry WebBook, NIST Standard Reference Database Number 69*, eds. P.J. Linstrom and W.G. Mallard (National Institute of Standards and Technology, Gaithersburg MD, 2010), http://webbook.nist.gov.

13 E.P. Bartlett, The compressibility isotherms of hydrogen, nitrogen and mixtures of these gases at 0 and pressures to 1000 atmospheres, *J. Am. Chem. Soc.*, 1927, **49**, 687.

21

Real Solutions

CHAPTER MENU

21.1 Introduction

In Chapter 18 we discussed chemical equilibria in ideally diluted solutions. Here we discuss the physical properties and equilibria of solutions more generally, including ideal and non-ideal solutions. We introduce the concept of activity, which is analogous to fugacity. In an ideal solution the state variables are obtained by adding the corresponding contributions of solvent and solute together. In contrast, an ideally diluted solution is any solution (ideal or nonideal) at small enough concentration, for which the equilibrium properties can be described in terms of the concentration rather than the activities. We restrict our considerations on solutions to cases where the solvent is a liquid, typically water (aqueous solution). For ions dissolved in water, the intermolecular forces are particularly strong; however, we will see that a relatively simple model (Debye–Hückel theory) can guide our understanding.

21.2 Partial Molar Quantities and Thermodynamics of Multicomponent Systems

21.2.1 Ideal and Non-Ideal Solution

For an ideal solution, state variables such as V, U, H, and G are obtained by adding the corresponding contributions of solvent (A) and solute (B). For example the volume V_{AB} of a solution is written as

$$V_{\text{AB}} = \boldsymbol{n}_{\text{A}} V_{m,\text{A}} + \boldsymbol{n}_{\text{B}} V_{m,\text{B}} \tag{21.1}$$

where $V_{m,\text{A}}$ and $V_{m,\text{B}}$ are the molar volumes of A and B. This simple relation is not always fulfilled however; e.g., V_{AB} may depend on $\boldsymbol{n}_{\text{A}}$ and $\boldsymbol{n}_{\text{B}}$ in a complicated manner. For instance, a dilute solution of $MgSO_4$ in water has a smaller volume than the water used to prepare the solution (Fig. 21.1). This reduction in molar volume (increase in density) arises from the attractive forces between the water molecules and the Mg^{2+} ions. Such a solution is considered non-ideal, which should not be confused with an ideally diluted solution, see introduction.

Principles of Physical Chemistry, Third Edition. Hans Kuhn, David H. Waldeck, and Horst-Dieter Försterling.

Companion website: www.wiley.com/go/Kuhn/physicalchemistry3e

21.2.2 Partial Molar Volume

How can we describe the volume V_{AB} of a real solution? We consider a solution of a given composition $\boldsymbol{n}_B/\boldsymbol{n}_A$, and we ask, "How much does the volume V_{AB} of the solution change if we add a small amount $d\boldsymbol{n}_B$ of solute and a small amount $d\boldsymbol{n}_A$ of solvent?" From calculus we can write the differential change dV_{AB} in volume by

$$dV_{AB} = \left(\frac{\partial V}{\partial \boldsymbol{n}_A}\right)_{\boldsymbol{n}_B} \cdot d\boldsymbol{n}_A + \left(\frac{\partial V}{\partial \boldsymbol{n}_B}\right)_{\boldsymbol{n}_A} \cdot d\boldsymbol{n}_B \qquad (21.2)$$

The partial derivatives are called *partial molar volumes*, and we write them in lower case font as

$$\upsilon_A = \left(\frac{\partial V}{\partial \boldsymbol{n}_A}\right)_{\boldsymbol{n}_B} \quad \text{and} \quad \upsilon_B = \left(\frac{\partial V}{\partial \boldsymbol{n}_B}\right)_{\boldsymbol{n}_A}$$

so that

$$dV_{AB} = \upsilon_A \cdot d\boldsymbol{n}_A + \upsilon_B \cdot d\boldsymbol{n}_B \qquad (21.3)$$

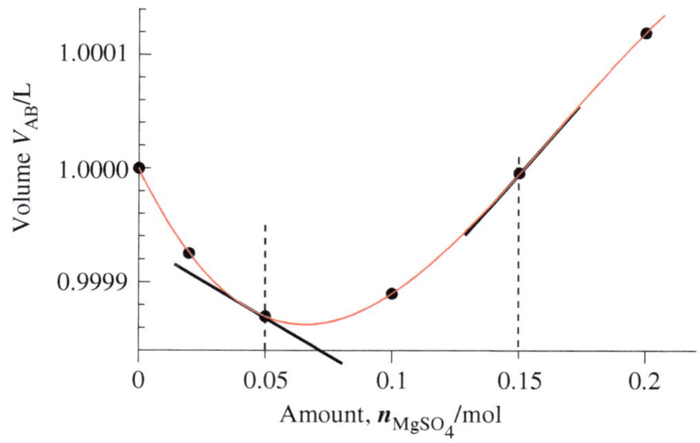

Figure 21.1 Volume V_{AB} of a solution of $MgSO_4$ in 1000 mL water as a function of the amount $\boldsymbol{n}_{MgSO_4}$ at 298 K. The slopes of the curve at $\boldsymbol{n}_{MgSO_4} = 0.05$ mol and 0.15 mol are indicated.

The partial molar volumes υ can depend on the composition of the solution. For example, for a solution of $MgSO_4$ in water, υ_B is negative at small concentrations and positive at large concentrations, although the volume is always a positive quantity.

Example 21.1 *From the slope of the curve in Fig. 21.1 (volume V_{AB} versus the amount $\boldsymbol{n}_B$ of $MgSO_4$) calculate υ_B for the amounts $\boldsymbol{n}_B = 0.05$ mol and $\boldsymbol{n}_B = 0.15$ mol of $MgSO_4$.*

As indicated in the figure, we obtain

$$\upsilon_B = \left(\frac{\partial V}{\partial \boldsymbol{n}_B}\right)_{\boldsymbol{n}_A} = -0.67 \times 10^{-3}\ \text{L mol}^{-1} \ \text{for } \boldsymbol{n}_B = 0.05\ \text{mol}$$

and

$$\upsilon_B = \left(\frac{\partial V}{\partial \boldsymbol{n}_B}\right)_{\boldsymbol{n}_A} = +2.24 \times 10^{-3}\ \text{L mol}^{-1} \ \text{for } \boldsymbol{n}_B = 0.15\ \text{mol}$$

The partial molar volume is an intensive property and depends on the composition of the solution, $\boldsymbol{n}_A/\boldsymbol{n}_B$, not the amount of each species. If we prepare a solution by adding, alternately in many small steps, small amounts of solvent and solute in the fixed ratio $\boldsymbol{n}_A/\boldsymbol{n}_B$, then all changes of volume are equal, and the total volume is given by

$$V_{AB} = \upsilon_A \cdot \boldsymbol{n}_A + \upsilon_B \cdot \boldsymbol{n}_B \qquad (21.4)$$

This equation corresponds to Equation (21.1), except that the molar volumes $V_{m,A}$ and $V_{m,B}$ are replaced by the partial molar volumes υ_A and υ_B. Taking the differential of V_{AB}

$$dV_{AB} = d(\boldsymbol{n}_A\upsilon_A + \boldsymbol{n}_B\upsilon_B) = d\boldsymbol{n}_A \cdot \upsilon_A + d\upsilon_A \cdot \boldsymbol{n}_A + d\boldsymbol{n}_B \cdot \upsilon_B + d\upsilon_B \cdot \boldsymbol{n}_B \qquad (21.5)$$

and combining this result with Equation (21.3) reveals that

$$\boldsymbol{n}_A \cdot d\upsilon_A + \boldsymbol{n}_B \cdot d\upsilon_B = 0 \qquad (21.6)$$

This result shows that the partial molar volume changes in a binary solution are not independent, but rather that

$$d\upsilon_A = -\frac{\boldsymbol{n}_B}{\boldsymbol{n}_A}\, d\upsilon_B = -\frac{x_B}{x_A}\, d\upsilon_B$$

where $x_i = \boldsymbol{n}_i/\boldsymbol{n}_{total}$ is the mole fraction of component i. These considerations are not restricted to the molar volume as a state variable but may be applied to all the state variables.

21.2.3 Chemical Potential

For chemical systems the partial molar change of the Gibb's energy is an extremely important quantity. The partial molar change in the Gibb's energy is called the *chemical potential*. The change $\mathrm{d}G_{AB}$ of the Gibbs energy at constant temperature and constant pressure may be written as

$$\mathrm{d}G_{AB} = \mu_A \cdot \mathrm{d}\boldsymbol{n}_A + \mu_B \cdot \mathrm{d}\boldsymbol{n}_B \tag{21.7}$$

where

$$\mu_A = \left(\frac{\partial G}{\partial \boldsymbol{n}_A}\right)_{T,P,\boldsymbol{n}_B} \quad \text{and} \quad \mu_B = \left(\frac{\partial G}{\partial \boldsymbol{n}_B}\right)_{T,P,\boldsymbol{n}_A} \tag{21.8}$$

are partial molar Gibbs energies; i.e., the chemical potentials. Similar to Equations (21.4) and (21.6) we find

$$G_{AB} = \mu_A \cdot \boldsymbol{n}_A + \mu_B \cdot \boldsymbol{n}_B \tag{21.9}$$

and

$$\boldsymbol{n}_A \cdot \mathrm{d}\mu_A + \boldsymbol{n}_B \cdot \mathrm{d}\mu_B = 0 \tag{21.10}$$

This relation is called the *Gibbs–Duhem Equation* (for the special case of a two component system at constant temperature). The Gibbs–Duhem equation is general and applies for an arbitrary number of components. In general, we have that

$$\sum_i \boldsymbol{n}_i \cdot \mathrm{d}\mu_i = 0 \tag{21.11}$$

where $\boldsymbol{n}_i$ is the amount of component i and μ_i is the chemical potential of component i.

In order to make the physical meaning of the chemical potential more clear, consider a pure substance where

$$G = G_m \cdot \boldsymbol{n} \qquad \text{(pure substance)} \tag{21.12}$$

is proportional to the amount of substance, $\boldsymbol{n}$. Then μ is the molar Gibbs energy,

$$\mu = \left(\frac{\partial G}{\partial \boldsymbol{n}}\right) = G_m \quad \text{(pure substance)} \tag{21.13}$$

Also for an ideally diluted solution (see Section 18.7), the Gibbs energy G_i of a dissolved compound i is proportional to its amount $\boldsymbol{n}_i$,

$$G_i = G_i^{\ominus} + \boldsymbol{n}_i \boldsymbol{R}T \cdot \ln \hat{c}_i \qquad \text{(ideally diluted solution)} \tag{21.14}$$

and the chemical potential μ_i is identical with the partial molar Gibbs energy

$$\mu_i = \left(\frac{\partial G}{\partial \boldsymbol{n}_i}\right)_{T,P} = \frac{G}{\boldsymbol{n}_i} = G_{m,i}^{\ominus} + \boldsymbol{R}T \cdot \ln \hat{c}_i \qquad \text{(ideally diluted solution)} \tag{21.15}$$

For an ideally diluted solution, we find that

$$\mu_i = \mu_i^{\ominus} + \boldsymbol{R}T \cdot \ln \hat{c}_i \quad \text{with} \quad \mu_i^{\ominus} = G_{m,i}^{\ominus} \tag{21.16}$$

where $\mu_i^{\ominus}$ is the chemical potential at standard pressure. Note that $\mu_i^{\ominus}$ depends on temperature, and that μ_i depends on temperature and concentration.

21.2.4 Thermodynamic Relations

Our previous development of thermodynamics focused on the description of a pure substance, and here we generalize it to describe multicomponent systems. In this description, the extensive properties (e.g., S, G, and U) depend on the amounts of each component. For example the internal energy is a function of S, V, and the amount $\boldsymbol{n}_i$ of each component i; i.e., $U(S, V, \boldsymbol{n}_1, \boldsymbol{n}_2, \boldsymbol{n}_3, \ldots.)$ or more compactly $U(S, V, \boldsymbol{n}_i)$. For a system with N components, the differential of U is

$$\mathrm{d}U = \left(\frac{\partial U}{\partial S}\right)_{V,\boldsymbol{n}_i} \mathrm{d}S + \left(\frac{\partial U}{\partial V}\right)_{S,\boldsymbol{n}_i} \mathrm{d}V + \sum_{j=1}^{N} \left(\frac{\partial U}{\partial \boldsymbol{n}_j}\right)_{S,V,\boldsymbol{n}_i} \mathrm{d}\boldsymbol{n}_j \tag{21.17}$$

and the combined form of the first and second laws becomes

$$\mathrm{d}U = T\mathrm{d}S - P\mathrm{d}V + \sum_{j=1}^{N} \mu_j \, \mathrm{d}\boldsymbol{n}_j \tag{21.18}$$

where μ_j is the chemical potential of component j. From this result we see that it is the differential change in internal energy of the system with a differential change in the amount of component j at fixed entropy, volume, and amounts of other components. Corresponding equations can be written for the other important energy functions H, A, G, as well as the entropy S and the volume V (as shown above).

It is worthwhile to note that the results we derived previously for the thermodynamics of a pure substance hold also for the case of a solution at fixed composition. To illustrate this fact and to show some new relations, let us consider the differential Gibbs energy for a two-component system

$$dG = -SdT + VdP + \mu_A\, d\boldsymbol{n}_A + \mu_B\, d\boldsymbol{n}_B \tag{21.19}$$

Our previous relationships,

$$\left(\frac{\partial G}{\partial T}\right)_{P,\boldsymbol{n}_A,\boldsymbol{n}_B} = -S \quad \text{and} \quad \left(\frac{\partial G}{\partial P}\right)_{T,\boldsymbol{n}_A,\boldsymbol{n}_B} = V \tag{21.20}$$

still hold for the case of fixed composition and total amount $\boldsymbol{n}$. Rather than generating one Maxwell relation, we see that a number of Maxwell relations are possible. For example, we find that

$$\left(\frac{\partial \mu_A}{\partial P}\right)_{T,\boldsymbol{n}_A,\boldsymbol{n}_B} = \left(\frac{\partial V}{\partial \boldsymbol{n}_A}\right)_{P,T,\boldsymbol{n}_B} = v_A \tag{21.21}$$

which relates the partial molar volume of component A to the pressure dependence of its chemical potential. Similarly we find that the partial molar entropy of component A is related to the temperature dependence of the chemical potential.

$$\left(\frac{\partial \mu_A}{\partial T}\right)_{P,\boldsymbol{n}_A,\boldsymbol{n}_B} = -\left(\frac{\partial S}{\partial \boldsymbol{n}_A}\right)_{P,T,\boldsymbol{n}_B} = -s_A \tag{21.22}$$

These relations are analogues to those found for the total entropy and total volume in Chapter 19. A totally new relationship is

$$\left(\frac{\partial \mu_A}{\partial \boldsymbol{n}_B}\right)_{T,P,\boldsymbol{n}_A} = \left(\frac{\partial \mu_B}{\partial \boldsymbol{n}_A}\right)_{P,T,\boldsymbol{n}_B} \tag{21.23}$$

which relates the composition dependence of the chemical potentials to each other.

Example 21.2 *Using Equation (21.23), show that*

$$x_A\left(\frac{\partial \mu_A}{\partial x_B}\right) = x_B\left(\frac{\partial \mu_B}{\partial x_A}\right)$$

where $x_A = \boldsymbol{n}_A/(\boldsymbol{n}_A + \boldsymbol{n}_B)$ *and* $x_B = \boldsymbol{n}_B/(\boldsymbol{n}_A + \boldsymbol{n}_B)$.

To perform the change of variables, we write that

$$\left(\frac{\partial \mu_A}{\partial \boldsymbol{n}_B}\right) = \left(\frac{\partial \mu_A}{\partial x_B}\right)\left(\frac{\partial x_B}{\partial \boldsymbol{n}_B}\right)$$

Substituting for x_B, we find

$$\left(\frac{\partial \mu_A}{\partial \boldsymbol{n}_B}\right) = \left(\frac{\partial \mu_A}{\partial x_B}\right)\left[\frac{1}{(\boldsymbol{n}_A + \boldsymbol{n}_B)} - \frac{\boldsymbol{n}_B}{(\boldsymbol{n}_A + \boldsymbol{n}_B)^2}\right]$$

$$= \left[\frac{1 - x_B}{(\boldsymbol{n}_A + \boldsymbol{n}_B)}\right]\left(\frac{\partial \mu_A}{\partial x_B}\right) = \frac{x_A}{(\boldsymbol{n}_A + \boldsymbol{n}_B)}\left(\frac{\partial \mu_A}{\partial x_B}\right)$$

Similarly we find that

$$\left(\frac{\partial \mu_B}{\partial \boldsymbol{n}_A}\right) = \frac{x_B}{(\boldsymbol{n}_A + \boldsymbol{n}_B)}\left(\frac{\partial \mu_B}{\partial x_A}\right)$$

Substituting into our original relation, we find that

$$x_A\left(\frac{\partial \mu_A}{\partial x_B}\right) = x_B\left(\frac{\partial \mu_B}{\partial x_A}\right)$$

21.3 Activities and Activity Coefficient for Real Solutions

For gases, we replaced pressures by fugacities when we needed to account for nonideality. We use a similar approach for non-ideal solutions, by replacing the concentration with the *activity*. In treating real solutions it is convenient to consider the activity in terms of the mole fraction x_s of the solvent, a_s. In Problem P21.7 you show that the chemical potential of the solvent for an ideal solution is, according to Equation (21.16),

$$\mu_s = \mu_s^{\ominus} + \boldsymbol{R}T \cdot \ln x_s \qquad \text{(ideal solution)} \tag{21.24}$$

and for a real solution, we obtain

$$\mu_s = \mu_s^{\ominus} + \boldsymbol{R}T \cdot \ln a_s \tag{21.25}$$

with

$$a_s = x_s \cdot \gamma_s \tag{21.26}$$

where a_s is the activity of the solvent and γ_s is the activity coefficient. Note that in this case the standard state is the pure liquid ($x_s = 1$, $a_s = 1$ and $\gamma_s = 1$).

21.3.1 Activity from Vapor Pressure Above a Solution

Consider a solution of a nonvolatile solute in a liquid, e.g., sucrose in water. At equilibrium the water vapor above the solution is in equilibrium with the liquid water in the solution; hence, the chemical potential of the vapor, $\mu_{s,\text{vap}}$, must be the same as the chemical potential of the liquid water, $\mu_{s,\text{liq}}$ in the solution; i.e.,

$$\mu_{s,\text{vap}} = \mu_{s,\text{liq}} \tag{21.27}$$

where the subscript s is for the solvent. Using Equation (21.25), we know that

$$\mu_{s,\text{liq}} = \mu_{s,\text{liq}}^{\ominus} + \boldsymbol{R}T \cdot \ln a_s \tag{21.28}$$

and with Equation (21.27) we can write that

$$\mu_{s,\text{vap}} = \mu_{s,\text{liq}}^{\ominus} + \boldsymbol{R}T \cdot \ln a_s \tag{21.29}$$

This equation tells us how the activity is related to the chemical potential of the vapor.

Here we show how to relate the activity of the solvent to its vapor pressure. The differential of Equation (21.29) is

$$\mathrm{d}\mu_{s,\text{vap}} = \mathrm{d}\mu_{s,\text{liq}}^{\ominus} + \boldsymbol{R}T \cdot \mathrm{d}\ln a_s \tag{21.30}$$

at constant temperature. For pure substances Equation (21.13) applies, so that,

$$\mathrm{d}\mu = \left(\frac{\partial \mu}{\partial P}\right)_T \cdot \mathrm{d}P = \left(\frac{\partial G_m}{\partial P}\right)_T \cdot \mathrm{d}P = V_m \cdot \mathrm{d}P \tag{21.31}$$

and Equation (21.30) becomes

$$V_{m,s,\text{vap}} \cdot \mathrm{d}P = V_{m,s,\text{liq}} \cdot \mathrm{d}P + \boldsymbol{R}T \cdot \mathrm{d}\ln a_s \tag{21.32}$$

or

$$\boxed{\mathrm{d}P = \frac{\boldsymbol{R}T}{V_{m,s,\text{vap}} - V_{m,s,\text{liq}}} \cdot \mathrm{d}\ln a_s \qquad \text{(constant temperature)}} \tag{21.33}$$

This is a general expression for the dependence of the vapor pressure on the activity of the solvent.

We can simplify this equation further, if we approximate the vapor as an ideal gas and neglect the molar volume of the liquid. Then we can use $V_{m,s,\text{vap}} - V_{m,s,\text{liq}} = \boldsymbol{R}T/P$ and find

$$\frac{\mathrm{d}P}{P} = \mathrm{d}\ln a_s \tag{21.34}$$

for the relative change of the vapor pressure. On the left side, we integrate from the vapor pressure $P = P_s$ of the pure solvent to the vapor pressure P_1 of the solution corresponding to the activity $a_{s,1}$ of the solvent, and on the right side from the activity $a_s = 1$ of the pure solvent ($\ln a_s = 0$) to the activity $a_{s,1}$.

$$\int_{P=P_s}^{P=P_1} \frac{dP}{P} = \int_{\ln a_s=0}^{\ln a_{s,1}} d\ln a_s \qquad \text{or} \qquad \ln \frac{P_1}{P_s} = \ln a_{s,1} \tag{21.35}$$

Thus we obtain, omitting the index 1,

$$\boxed{\frac{P}{P_s} = a_s \qquad \text{or} \qquad \frac{\Delta P}{P_s} = a_s - 1} \tag{21.36}$$

where $\Delta P = P - P_s$ is the change of the vapor pressure. For sucrose in water $a_s < 1$ and the change ΔP is negative: depression of vapor pressure.

This result is a generalization of Raoult's law discussed in Chapter 18, Equation (18.102), where we used the mole fraction x of the solute instead of the activity a_s.

$$\frac{P_s - P}{P_s} = x \quad \text{or} \quad \frac{\Delta P}{P_s} = -x = x_s - 1 \qquad \text{(Raoult's law)} \tag{21.37}$$

Example 21.3 *For solutions of sucrose in water calculate the vapor pressure depression from Raoult's law, Equation (21.37). Compare the result to the experimental values and evaluate the activities and the activity coefficients of the solvent, using Equation (21.36).*

First we use the simplified Equation (21.37) to calculate $\Delta P/P_s$ for some solutions with the mole fractions x of the solute shown in column 1 of Table 21.1. Using the experimental data for $\Delta P/P_s$ in column 4 we calculate the activity of water, a_s, from Equation (21.36), $a_s = (\Delta P/P_s) + 1$ and derive the activity coefficients from Equation (21.26), $\gamma_s = a_s/x_s$.

Table 21.1 Relative Change of Vapor Pressure $\Delta P/P_s$ for Solutions of Sucrose in Water at 0° C. x is the Mole Fraction of Sucrose, x_s is the Mole Fraction of Water.

Calculated				Calculated		
Equation (21.37)			Experiment[a)]	Equation (21.36)		Equation (21.38)
x	$x_s = 1 - x$	$\frac{\Delta P}{P_s}$	$\frac{\Delta P}{P_s}$	a_s	γ_s	γ
0.000	1.000	0.000	0.000	1.000	1.000	1.00
0.0036	0.996	−0.0036	−0.00372	0.996	1.000	1.00
0.0089	0.991	−0.0089	−0.00945	0.991	1.000	1.00
0.0177	0.982	−0.0177	−0.0197	0.981	0.998	1.12
0.059	0.941	−0.059	−0.0837	0.920	0.978	1.91
0.075	0.925	−0.075	−0.1125	0.894	0.966	2.25
0.083	0.917	−0.082	−0.1275	0.880	0.960	2.42
0.098	0.903	−0.098	−0.1554	0.856	0.949	2.74
0.112	0.888	−0.112	−0.1841	0.832	0.937	3.01

a) The data are taken from *The International Critical Tables, III*. New York: McGraw Hill Book Co.; 1928, p. 293.

From the activities a_s of the solvent we can obtain the activities a of the solute. In Foundation 21.1 we use the Gibbs–Duhem Equation (21.10) to show that

$$\ln \gamma_1 = -\int_{\ln \gamma_s=0}^{\ln \gamma_{s,1}} \frac{x_s}{1 - x_s} \cdot d\ln \gamma_s \tag{21.38}$$

where γ_1 is the activity coefficient of the solute with mole fraction x_s of the solute. The integral may be evaluated by numerical integration. In Problem P21.3 you perform this integration for the data in Table 21.1 (column 7).

21.3.2 Activity from Osmotic Pressure

In this case (see Fig. 18.11) the pure solvent is under standard pressure $P^\ominus$ and in equilibrium with the solvent in the solution under a pressure $P^\ominus + {}^{\text{osm}}P$, where ${}^{\text{osm}}P$ is the osmotic pressure. At equilibrium

$$\mu_{s,\text{pure}} = \mu_{s,\text{solution}} \tag{21.39}$$

For the chemical potential of the pure solvent and of the solvent in the solution we have

$$\mu_{s,\text{pure}} = \mu^\ominus_{s,\text{pure}}, \qquad \mu_{s,\text{solution}} = \mu^*_{s,\text{solution}} + \boldsymbol{R}T \ln a_s \tag{21.40}$$

Here we use the superscript * for the standard state of the solvent in the solution to indicate that it is under a different pressure. Then

$$\mu^\ominus_{s,\text{pure}} - \mu^*_{s,\text{solution}} = \boldsymbol{R}T \ln a_s \tag{21.41}$$

The difference in the reference state chemical potentials may be written as

$$\mu^\ominus_{s,\text{pure}} - \mu^*_{s,\text{solution}} = \int_{P^\ominus + {}^{\text{osm}}P}^{P^\ominus} \left(\frac{\partial \mu_s^\ominus}{\partial P}\right)_T \cdot \mathrm{d}P = \int_{P^\ominus + {}^{\text{osm}}P}^{P^\ominus} V^\ominus_{m,s} \cdot \mathrm{d}P \tag{21.42}$$

If we approximate the liquid as incompressible, then $V^\ominus_{m,s} = V_{m,s}$ and

$$\int_{P^\ominus + {}^{\text{osm}}P}^{P^\ominus} V^\ominus_{m,s} \cdot \mathrm{d}P = V_{m,s}\left(P^\ominus - P^\ominus - {}^{\text{osm}}P\right) = -V_{m,s} \cdot {}^{\text{osm}}P \tag{21.43}$$

Combining this result with Equation (21.41) we obtain

$$-V_{m,s} \cdot {}^{\text{osm}}P = \boldsymbol{R}T \cdot \ln a_s \tag{21.44}$$

or

$$\boxed{{}^{\text{osm}}P = -\frac{\boldsymbol{R}T}{V_{m,s}} \cdot \ln a_s} \tag{21.45}$$

For the limiting case of ideally diluted solutions we have $a_s = x_s = 1 - x$, where x is the mole fraction of the solute, so that $\ln a_s = \ln(1 - x) \approx -x$ for small x and.

$${}^{\text{osm}}P = \frac{\boldsymbol{R}T}{V_{m,s}} \cdot x \tag{21.46}$$

Because of

$$x = \frac{\boldsymbol{n}}{\boldsymbol{n}_s + \boldsymbol{n}} \approx \frac{\boldsymbol{n}}{\boldsymbol{n}_s} \quad \text{and} \quad V_{m,s} \cdot \boldsymbol{n}_s = V_s \tag{21.47}$$

we can write Equation (21.46) as

$${}^{\text{osm}}P = \frac{\boldsymbol{R}T}{V_{m,s}} \cdot \frac{\boldsymbol{n}}{\boldsymbol{n}_s} = \frac{\boldsymbol{R}T}{V_s} \cdot \boldsymbol{n} \tag{21.48}$$

or

$$\boxed{{}^{\text{osm}}P = \boldsymbol{R}T \cdot c \qquad \text{(ideally dilute solution)}} \tag{21.49}$$

where c is the concentration of the solute. Equation (21.49) is identical with Equation (18.98), derived in Chapter 18.

Example 21.4 *Calculate the osmotic pressure of a solution of sucrose in water and compare to the experiment.*

Table 21.2 shows the concentration c, the mole fraction x of the solute, the mole fraction x_s of the solvent, the activity coefficients γ_s (taken from Table 21.1), and the activities a_s of the solvent. The osmotic pressure is calculated according to

Table 21.2 Osmotic Pressure ${}^{osm}P$ for Solutions of Sucrose in Water at 0° C. The Osmotic Pressure is Calculated According to Equations (21.49) and (21.45) and Compared to the Experiment.[1]

					${}^{osm}P$ bar		
c mol L^{-1}	x	x_s	γ_s	a_s	Equation (21.49)	Equation (21.45)	Experiment
0.098	0.00180	0.99820	1.000	0.99820	2.22	2.27	2.49
0.192	0.00359	0.99641	1.000	0.99641	4.36	4.54	4.78
0.370	0.00716	0.99284	1.000	0.99284	8.40	9.06	9.57
0.453	0.00893	0.99107	0.999	0.99008	10.28	12.57	12.05
0.610	0.01246	0.98754	0.999	0.98656	13.85	17.07	17.11
0.757	0.01596	0.98404	0.998	0.98207	17.18	22.81	22.41
0.825	0.0177	0.98230	0.998	0.98034	18.73	25.04	25.15

Equations (21.49) and (21.45). The experimental values for the osmotic pressure are shown in column 8. They are in good agreement with the calculations according to Equation (21.45), whereas the simplified Equation (21.49) works only in the low concentration range.

21.4 Phase Transitions of Solutions

21.4.1 Elevation of Boiling Point and Depression of Melting Point

We consider the liquid/vapor equilibrium:

$$\mu_{s,\mathrm{vap}} = \mu_{s,\mathrm{liq}} \tag{21.50}$$

where $\mu_{s,\mathrm{vap}}$ is the chemical potential of the solvent s in the vapor state and $\mu_{s,\mathrm{liq}}$ is the chemical potential of the solvent in the liquid state (i.e., in the solution). We may write that

$$\frac{\mu_{s,\mathrm{vap}}}{T} = \frac{\mu_{s,\mathrm{liq}}}{T} \tag{21.51}$$

or for a differential change of the chemical potential

$$\mathrm{d}\left(\frac{\mu_{s,\mathrm{vap}}}{T}\right) = \mathrm{d}\left(\frac{\mu_{s,\mathrm{liq}}}{T}\right) \tag{21.52}$$

For a solution we have

$$\mu_{s,\mathrm{liq}} = \mu^{\ominus}_{s,\mathrm{liq}} + \boldsymbol{R}T \cdot \ln a_s \quad \text{or} \quad \frac{\mu_{s,\mathrm{liq}}}{T} = \frac{\mu^{\ominus}_{s,\mathrm{liq}}}{T} + \boldsymbol{R} \cdot \ln a_s \tag{21.53}$$

where $a_s = x_s \cdot \gamma_s$ is the activity of the solvent in the solution (x_s = mole fraction of the solvent, γ_s is the activity coefficient of the solvent) and the standard state is defined for $a_s = 1$ (pure solvent). Thus we obtain from Equation (21.53)

$$\mathrm{d}\left(\frac{\mu_{s,\mathrm{vap}}}{T}\right) = \mathrm{d}\left(\frac{\mu^{\ominus}_{s,\mathrm{liq}}}{T}\right) + \mathrm{d}\left(\boldsymbol{R}\ln a_s\right) \tag{21.54}$$

For a pure substance $\mu = G_m = H_m - TS_m$ and

$$\begin{aligned}\frac{\partial(\mu/T)}{\partial T} &= \frac{\partial\left(H_m T^{-1} - S_m\right)}{\partial T} = \frac{\partial H_m}{\partial T}\frac{1}{T} - H_m\frac{1}{T^2} - \frac{\partial S_m}{\partial T} \\ &= \frac{C_P}{T} - \frac{H_m}{T^2} - \frac{C_P}{T} = -\frac{H_m}{T^2}\end{aligned} \tag{21.55}$$

1 Landolt–Börnstein.

Then

$$-\frac{H_{m,s,\mathrm{vap}}}{T^2}\,\mathrm{d}T = -\frac{H^{\ominus}_{m,s,\mathrm{liq}}}{T^2}\,\mathrm{d}T + \mathrm{d}\left(\boldsymbol{R}\ln a_s\right) \tag{21.56}$$

where $H^{\ominus}_{m,s,\mathrm{liq}}$ is the molar enthalpy of the pure liquid solvent in its standard state and $H_{m,s,\mathrm{vap}}$ is the molar enthalpy of the pure solvent in the vapor state. Then $\Delta_{\mathrm{vap}}H = H_{m,s,\mathrm{vap}} - H^{\ominus}_{m,s,\mathrm{liq}}$ is the molar enthalpy of vaporization of the pure solvent, and we obtain

$$\boxed{-\frac{\Delta_{\mathrm{vap}}H}{\boldsymbol{R}T^2}\,\mathrm{d}T = \mathrm{d}\left(\ln a_s\right)} \tag{21.57}$$

This relation is exact.

For practical use we have to integrate this equation. Then

$$-\int_{T=T_s}^{T_1} \frac{\Delta_{\mathrm{vap}}H}{\boldsymbol{R}T^2}\,\mathrm{d}T = \int_{\ln a_s=0}^{\ln a_{s,1}} \mathrm{d}\left(\ln a_s\right) \tag{21.58}$$

where T_s is the boiling temperature of the pure solvent and T_1 is the boiling temperature of the solution with activity $a_{s,1}$. As an approximation we assume that i) $\Delta_{\mathrm{vap}}H$ and ii) that the activity coefficient γ_s does not depend on temperature. Often these assumptions are appropriate for the small range of temperature between the boiling temperature of the pure solvent and that of the solution. Then we find that

$$\left|\frac{\Delta_{\mathrm{vap}}H}{\boldsymbol{R}T}\right|_{T_s}^{T_1} = \left|\ln a_s\right|_0^{\ln a_{s,1}} \tag{21.59}$$

and, omitting the subscript 1, we obtain

$$\frac{\Delta_{\mathrm{vap}}H}{\boldsymbol{R}}\left(\frac{1}{T} - \frac{1}{T_s}\right) = \ln a_s \tag{21.60}$$

In Exercise E21.21 you show that for sufficiently small temperature shifts this expression reduces to

$$-\frac{\Delta_{\mathrm{vap}}H}{\boldsymbol{R}}\left(\frac{\Delta T}{T_s^2}\right) = \ln a_s \tag{21.61}$$

with $\Delta T = T - T_s$. Solving for ΔT we find

$$\boxed{\Delta T = -\frac{\boldsymbol{R}T_s^2}{\Delta_{\mathrm{vap}}H}\cdot\ln a_s \qquad \text{(elevation of boiling point)}} \tag{21.62}$$

Since a_s is smaller than one, $\ln a_s$ is negative and ΔT is positive: elevation of boiling point.

A similar derivation can be used to find an expression for the depression of the freezing point. In Exercise E21.14, you show that

$$\frac{\Delta_{\mathrm{fus}}H}{\boldsymbol{R}}\left[\frac{1}{T_s} - \frac{1}{T}\right] = \ln a_s \tag{21.63}$$

and for small enough ΔT

$$\boxed{\Delta T = \frac{\boldsymbol{R}T_s^2}{\Delta_{\mathrm{fus}}H}\cdot\ln a_s \quad \text{(depression of freezing point)}} \tag{21.64}$$

Since a_s is smaller than one, $\ln a_s$ is negative and ΔT is negative, i.e., depression of freezing point.

In the special case of ideally dilute solutions the activity coefficient is $\gamma_s = 1$; thus, $\ln a_s = \ln x_s$. Because the mole fraction of the solute is $x = 1 - x_s$ we obtain

$$\ln a_s = \ln x_s = \ln(1 - x) \approx -x \tag{21.65}$$

Then for the boiling point and for the freezing point we obtain

$$\boxed{\Delta T = \frac{\boldsymbol{R}T_s^2}{\Delta_{\mathrm{vap}}H}\cdot x \qquad \text{and} \qquad \Delta T = -\frac{\boldsymbol{R}T_s^2}{\Delta_{\mathrm{fus}}H}\cdot x \quad \text{(ideally dilute solution)}} \tag{21.66}$$

These are the same expressions which we derived already in Chapter 18.

Example 21.5 *Consider a solution of* 2.116 *mol sucrose in* 1000 *kg of water at* 298 *K. Calculate the depression of the freezing point and compare it to the experimental [1] value* $\Delta T = -4.9$ *K.*

This molality corresponds to a mole fraction of $x = 0.0367$ and $x_s = 1 - x = 0.963$. If we interpolate the activity coefficient at 0 °C from the data in Table 21.2, we find that $\gamma_s = 0.986$, so that the activity of the solvent is $a_s = 0.963 \cdot 0.986 = 0.950$. With $\Delta_{\text{fus}}H = 6.0$ kJ mol^{-1} and $T_s = 273$ K from Equation (21.64) we obtain

$$\Delta T = \frac{8.314 \text{ J K}^{-1} \text{ mol}^{-1} \cdot 273^2 \text{ K}^2}{6.0 \text{ kJ mol}^{-1}} \cdot \ln 0.950 = -5.3 \text{ K}$$

The deviation between the experimental value and that of Equation (21.64) is 8%. If we use Equation (21.66) for the ideally diluted solution, we find that

$$\Delta T = \frac{8.314 \text{ J K}^{-1} \text{ mol}^{-1} \cdot 273^2 \text{ K}^2}{6.0 \text{ kJ mol}^{-1}} \cdot 0.0367 = -3.8 \text{ K}$$

This is about 22% less than the experimental value.

21.4.2 Solutions with Two Volatile Components

Figure 21.2 shows vapor pressure plots for water/ethanol and for acetone/chloroform mixtures. Neither solution follows Raoult's law (see Chapter 18 and Fig. 18.12), in which the total vapor pressure P depends linearly on the mole fractions x_A of compound A and x_B of compound B in the binary mixture

$$P = x_A P_{A,\text{pure}} + x_B P_{B,\text{pure}} \qquad \text{(ideal solution)} \tag{21.67}$$

$P_{A,\text{pure}}$ and $P_{B,\text{pure}}$ are the vapor pressures of the pure compounds A and B.

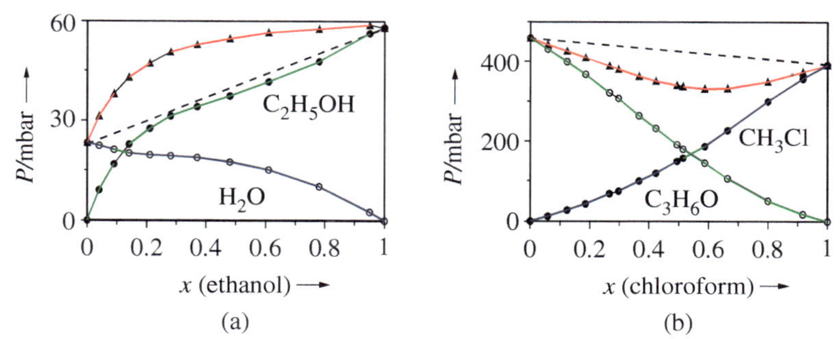

Figure 21.2 Vapor pressure data for (a) ethanol (C_2H_5OH, green solid line)/water (H_2O, blue solid line) mixtures (red solid line) at 293 K and (b) chloroform (CH_3Cl, blue solid line)/acetone (C_3H_6O, green solid line) mixtures (red slid line) at 308 K. The dashed lines correspond to the total pressure expected if Raoult's law is valid. (Data are taken from the International Critical Tables III, pages 286 and 290 (1923).)

In the case of Fig. 21.2 the deviations from Raoult's law are negative for the water/ethanol mixture (the total pressure is above the dashed line) and positive for the acetone/chloroform mixture (the total pressure is below the dashed line). What are the activities for the water/ethanol mixtures? At equilibrium the chemical potential of the ethanol in the vapor phase is equal to the chemical potential of the ethanol in the solution

$$\mu_{\text{EtOH,vap}} = \mu_{\text{EtOH,liq}}$$

so that

$$\mu^{\ominus}_{\text{EtOH,vap}} + \boldsymbol{R}T \cdot \ln \widehat{P}_{\text{EtOH}} = \mu^{\ominus}_{\text{EtOH,liq}} + \boldsymbol{R}T \cdot \ln a_{\text{EtOH}} \tag{21.68}$$

where $\mu^{\ominus}_{\text{EtOH,vap}}$ is the standard chemical potential of ethanol vapor, P_{EtOH} is the vapor pressure, $\mu^{\ominus}_{\text{EtOH,liq}}$ is the chemical potential of pure liquid ethanol ($a_{\text{EtOH}} = 1$), and a_{EtOH} is the activity of ethanol in the solution. For the special case of pure ethanol, we find that

$$\ln \widehat{P}_{\text{EtOH,pure}} = \frac{\mu^{\ominus}_{\text{EtOH,liq}} - \mu^{\ominus}_{\text{EtOH,vap}}}{\boldsymbol{R}T}$$

Combining this result with Equation (21.68), we may write that

$$\ln \widehat{P}_{\text{EtOH}} = \ln \widehat{P}_{\text{EtOH,pure}} + \ln a_{\text{EtOH}}$$

which simplifies to

$$a_{\mathrm{EtOH}} = \frac{P_{\mathrm{EtOH}}}{P_{\mathrm{EtOH,pure}}} \tag{21.69}$$

This result tells us that the activity of ethanol in the solution can be obtained by measuring the partial pressure of ethanol in the vapor P_{EtOH} and comparing it to the vapor pressure of pure ethanol $P_{\mathrm{EtOH,pure}}$.

Example 21.6 *Using the data for the ethanol/water mixture in Fig. 21.2 at a mole fraction of 0.48 ethanol and $T = 352$ K (the vapor pressure of the water is 200 mbar and the vapor pressure of the ethanol is 140 mbar), find the activities and activity coefficients of the water and the ethanol.*

Because the vapor pressure of pure water is 460 mbar, we find

$$a_{H_2O} = \frac{200}{460} = 0.44 \text{ and } \gamma_{H_2O} = \frac{a_{H_2O}}{x_{H_2O}} = \frac{0.435}{0.52} = 0.84$$

Because the vapor pressure of pure ethanol is 390 mbar, we find

$$a_{\mathrm{EtOH}} = \frac{140}{390} = 0.36 \text{ and } \gamma_{\mathrm{EtOH}} = \frac{a_{\mathrm{EtOH}}}{x_{\mathrm{EtOH}}} = \frac{0.359}{0.48} = 0.75$$

21.5 Mass Action Law For Reactions in Solution

To account for interactions in real solutions we replace the concentration c of the solute by the activity a_{solute} such that

$$a_{\mathrm{solute}} = c \cdot \gamma_{\mathrm{solute}} \tag{21.70}$$

where γ_{solute} is the activity coefficient. In the limit that $c \to 0$, we require that $\gamma_{\mathrm{solute}} \to 1$ and hence $a_{\mathrm{solute}} \to c$. If we transfer reversibly a small amount $\boldsymbol{n}$ of a solute from an ideally diluted solution ($\gamma_{\mathrm{solute}} = 1$ at concentration c_0) to a solution with concentration c, then the change in Gibbs energy ΔG is

$$\Delta G = \boldsymbol{nR}T \cdot \ln \frac{\gamma_{\mathrm{solute}} \cdot c}{c_0} = \boldsymbol{nR}T \cdot \ln \frac{a_{\mathrm{solute}}}{c_0} \tag{21.71}$$

This equation defines the activity a_{solute} and the activity coefficient γ_{solute}. Then instead of Equation (21.16) for the chemical potential μ we can write

$$\mu = \mu^{\ominus} + \boldsymbol{R}T \cdot \ln \frac{a_{\mathrm{solute}}}{c^{\ominus}} \tag{21.72}$$

In this case the standard state is defined as a hypothetical state of the solute at standard concentration $c^{\ominus}$, behaving as a solute in an ideally diluted solution.

We consider a chemical reaction in solution

$$\alpha\mathrm{A} + \beta\mathrm{B} \;\rightarrow\; \varepsilon\mathrm{E} + \delta\mathrm{d} \tag{21.73}$$

We imagine the reactants reversibly transferred from the containers with the standard concentrations into the equilibrium container; correspondingly, the products are transferred from their equilibrium concentrations to standard concentration. Then

$$\mathrm{d}G = 0 \tag{21.74}$$

for this process (equilibrium condition). Using (21.7) we calculate $\mathrm{d}G$ from the chemical potentials of the different species.

$$\mathrm{d}G = \mu_{\mathrm{E}} \cdot \mathrm{d}\boldsymbol{n}_{\mathrm{E}} + \mu_{\mathrm{D}} \cdot \mathrm{d}\boldsymbol{n}_{\mathrm{D}} - \mu_{\mathrm{A}} \cdot \mathrm{d}\boldsymbol{n}_{\mathrm{A}} - \mu_{\mathrm{B}} \cdot \mathrm{d}\boldsymbol{n}_{\mathrm{B}} \tag{21.75}$$

With

$$\mathrm{d}\boldsymbol{n} = \frac{\mathrm{d}\boldsymbol{n}_{\mathrm{E}}}{\varepsilon} = \frac{\mathrm{d}\boldsymbol{n}_{\mathrm{D}}}{\delta} = \frac{\mathrm{d}\boldsymbol{n}_{\mathrm{A}}}{\alpha} = \frac{\mathrm{d}\boldsymbol{n}_{\mathrm{B}}}{\beta} \tag{21.76}$$

we obtain

$$dG = \left(\mu_E \cdot \varepsilon + \mu_D \cdot \delta - \mu_A \cdot \alpha - \mu_B \cdot \beta\right) \cdot d\boldsymbol{n} = 0 \tag{21.77}$$

Using Equation (21.72) for the chemical potential,

$$\mu = \mu^{\ominus} + \boldsymbol{R}T \cdot \ln \widehat{a} \tag{21.78}$$

with

$$\widehat{a} = \frac{a}{c^{\ominus}} \tag{21.79}$$

we obtain

$$\left(\mu_E^{\ominus} \cdot \varepsilon + \mu_D^{\ominus} \cdot \delta - \mu_A^{\ominus} \cdot \alpha - \mu_B^{\ominus} \cdot \beta\right) + \boldsymbol{R}T \cdot \ln \frac{\widehat{a}_E^{\varepsilon} \cdot \widehat{a}_D^{\delta}}{\widehat{a}_A^{\alpha} \cdot \widehat{a}_B^{\beta}} = 0 \tag{21.80}$$

With

$$\Delta_r G^{\ominus} = \mu_E^{\ominus} \cdot \varepsilon + \mu_D^{\ominus} \cdot \delta - \mu_A^{\ominus} \cdot \alpha - \mu_B^{\ominus} \cdot \beta \tag{21.81}$$

and

$$K = \frac{\widehat{a}_E^{\varepsilon} \cdot \widehat{a}_D^{\delta}}{\widehat{a}_A^{\alpha} \cdot \widehat{a}_B^{\beta}} \tag{21.82}$$

this leads to

$$\boxed{\Delta_r G^{\ominus} = -\boldsymbol{R}T \cdot \ln K} \tag{21.83}$$

This is the same result as obtained in Chapter 18, except that we have to replace the concentrations c by the activities a.

Example 21.7 *Consider the equilibrium*

$$AgCl(solid) \leftrightarrows Ag^+ + Cl^-$$

which has the solubility product $K_{sp} = \widehat{a}_{Ag^+} \cdot \widehat{a}_{Cl^-} = 1.77 \times 10^{-10}$ in water. Calculate the concentration of Ag^+ by assuming that the solution behaves ideally. For AgCl in a 0.020 mol L^{-1} solution of KNO_3 the observed concentration of Ag^+ is 1.5×10^{-5} mol L^{-1}. Calculate the activity coefficient for this case.

At equilibrium

$$\Delta_r G^{\ominus} = -\boldsymbol{R}T \cdot \ln K_{sp} = -\boldsymbol{R}T \cdot \ln\left(\widehat{a}_{Ag^+} \cdot \widehat{a}_{Cl^-}\right)$$

and

$$\widehat{a}_{Ag^+} \cdot \widehat{a}_{Cl^-} = \gamma_{Ag^+}\widehat{c}_{Ag^+} \cdot \gamma_{Cl^-}\widehat{c}_{Cl^-} = K_{sp}$$

Because of $\widehat{c}_{Ag^+} = \widehat{c}_{Cl^-}$, we write

$$\gamma_{Ag^+} \cdot \gamma_{Cl^-} \cdot \left(\widehat{c}_{Ag^+}\right)^2 = K_{sp}$$

In the first case, we assume the system is ideal ($\gamma_{Ag^+} = \gamma_{Cl^-} = 1$), and we find that

$$\left(\widehat{c}_{Ag^+}\right)^2 = K_{sp} \quad \text{and} \quad \widehat{c}_{Ag^+} = \sqrt{K_{sp}} = \sqrt{1.77 \times 10^{-10}} = 1.33 \times 10^{-5}$$

For the case where an inert salt (KNO_3) is present, we find that

$$\gamma_{Ag^+} \cdot \gamma_{Cl^-} = \frac{K_{sp}}{\left(\widehat{c}_{Ag^+}\right)^2} = \frac{1.77 \times 10^{-10}}{(1.5 \times 10^{-5})^2} = 0.79$$

From this example we see that the activity coefficients for cations γ_+ and for anions γ_- cannot be measured separately. For this reason, we define the mean activity coefficient $\gamma_\pm$

$$\gamma_\pm = \left(\gamma_+^{z_+} \cdot \gamma_-^{z_-}\right)^{1/(z_+ + z_-)} \tag{21.84}$$

where z_+e is the charge of the cation and $-z_-e$ is the charge of the anion. For example, in the case of AgCl dissociating into Ag^+ and Cl^- ions, we have $z_+ = z_- = 1$ and $\gamma_\pm = \sqrt{\gamma_+ \cdot \gamma_-} = \sqrt{\gamma_{Ag^+} \cdot \gamma_{Cl^-}} = \sqrt{0.79} = 0.89$. This value is significantly different from unity.

21.6 Electrolyte Solutions and the Debye–Hückel Theory

From the previous discussions it is evident that the nonideality of the solution is correlated with the interaction strength between the solute(s) and the solvent. A simple statistical model captures this feature and relates the activity coefficient to the intermolecular interactions, albeit crudely. Electrolyte solutions have important long range Coulomb interactions and here we discuss the Debye–Hückel theory, which has been developed to account for them.

21.6.1 Debye–Hückel Theory for Electrolytes

Understanding the properties of aqueous electrolyte solutions requires that we treat long range interactions because the Coulomb force between two charges is proportional to $1/r^2$. How can we estimate the activity coefficient for this case?

21.6.1.1 The Activity Coefficient Depends on r_D

First we imagine what happens when the electrostatic interaction between the ions is neglected. In this case the reversible work to transform the ionic solution from c_0 to c is

$$w_{\text{rev,ideal}} = \boldsymbol{nRT} \cdot \ln \frac{c}{c_0} \tag{21.85}$$

The difference between w_{rev} and $w_{\text{rev,ideal}}$ arises because a selected ion (charge e in the case of a singly charged ion) in the concentrated solution is preferentially surrounded by oppositely charged ions (i.e., counterion). In contrast, the ions in an ideally dilute solution are statistically distributed.

In Box 21.1 we show that, according to the Debye–Hückel theory, the additional reversible work is

$$w_{\text{rev}} - w_{\text{rev,ideal}} = -\frac{1}{2} \cdot \frac{e^2}{4\pi\varepsilon_0\hat{\varepsilon} \cdot r_D} \cdot N \tag{21.86}$$

when the counterions are at an average distance r_D from the selected ion; $\hat{\varepsilon}$ is the relative permittivity, ε_0 is the vacuum permittivity, and N is the number of ions.

If we transfer reversibly a small amount $\boldsymbol{n}$ of a solute from an ideally diluted solution ($\gamma_{\text{solute}} = 1$ at concentration c_0) to a solution with concentration c, then the change in Gibbs energy ΔG is, according to Equation (21.71)

$$\Delta G = \boldsymbol{nRT} \cdot \ln \frac{\gamma \cdot c}{c_0}$$

Let us approximate ΔG in our model by w_{rev} (neglecting the volume change of the solution). Then from

$$w_{\text{rev}} = \boldsymbol{nRT} \cdot \ln \frac{\gamma \cdot c}{c_0}$$

and Equation (21.86), we obtain

$$w_{\text{rev}} - w_{\text{rev,ideal}} = \boldsymbol{nRT} \cdot \ln \gamma = -\frac{1}{2} \cdot \frac{e^2}{4\pi\varepsilon_0\hat{\varepsilon} \cdot r_D} \cdot N \tag{21.87}$$

or

$$\ln \gamma = \frac{-e^2}{2kT \cdot 4\pi\varepsilon_0\hat{\varepsilon} \cdot r_D} \tag{21.88}$$

This relation can be generalized to ions with arbitrary charge ze, so that

$$\boxed{\ln \gamma = \frac{-(ze)^2}{2kT \cdot 4\pi\varepsilon_0\hat{\varepsilon} \cdot r_D}} \tag{21.89}$$

According to Equation (21.89) the activity coefficient γ decreases with decreasing distance r_D (i.e., with increasing concentration of the solution).

21.6.1.2 Calculating the Average Distance r_D

In Foundation 21.2 we use the model of Guoy and Chapman to find the average distance r_D,

$$r_D = R + \frac{1}{\kappa} \tag{21.90}$$

where R is the smallest possible distance between an ion and a counter ion, and $1/\kappa$ is called the *Debye screening length*. It measures the spatial extent of the region around an ion (called the ionic atmosphere) that is preferentially occupied by counter ions. The parameter κ has units of inverse length and is given by

$$\kappa = \sqrt{\frac{2e^2\mathbf{N}_A}{\varepsilon_0\hat{\varepsilon}kT}} \cdot \sqrt{I} \tag{21.91}$$

where

$$I = \frac{1}{2} \cdot \sum_m c_m z_m^2 \tag{21.92}$$

is the ionic strength, c_m and $z_m\, e$ are the concentration and the charge of ion species m. As the ionic strength I of the solution increases the value of κ increases, which contracts the ionic atmosphere. Hence, the charge on an ion is better screened.

By substituting our results for r_D into Equation (21.89) and performing some algebraic rearrangements, we obtain the *Debye–Hückel law*

$$\boxed{\ln\gamma = -\frac{A \cdot z^2\sqrt{I}}{1 + B \cdot R\sqrt{I}}} \tag{21.93}$$

with the abbreviations

$$A = \frac{e^3}{(4\pi\varepsilon_0)^{3/2}} \cdot \sqrt{\frac{2\pi\mathbf{N}_A}{k^3}} \cdot \frac{1}{\left(\hat{\varepsilon}T\right)^{3/2}} \tag{21.94}$$

$$B = \frac{2e}{(4\pi\varepsilon_0)^{1/2}} \cdot \sqrt{\frac{2\pi\mathbf{N}_A}{k}} \cdot \frac{1}{\left(\hat{\varepsilon}T\right)^{1/2}} \tag{21.95}$$

In the case of water ($\hat{\varepsilon} = 78.3$) at 298 K we find $A = 1.175\ \mathrm{L}^{1/2}\ \mathrm{mol}^{-1/2}$ and $B = 0.330 \times 10^{10}\ \mathrm{m}^{-1}\ \mathrm{L}^{1/2}\ \mathrm{mol}^{-1/2}$.

For sufficiently small concentrations (i.e., $1 \gg B \cdot R\sqrt{I}$), Equation (21.93) simplifies to

$$\ln\gamma = -A \cdot z^2\sqrt{I} \tag{21.96}$$

This equation is called the *Debye–Hückel limiting law.*

Example 21.8 *Calculate the activity coefficient of sodium ions in a 0.01 M NaCl solution for the case where $R = 0$ and $R = 500$ pm. Compare your result with a similar calculation for $MgSO_4$.*

First we calculate the ionic strength I

$$I = \frac{1}{2} \cdot \left(c_{Na^+} + c_{Cl^-}\right) = 0.01\ \mathrm{mol\ L^{-1}}$$

Assuming $R = 0$, from (21.93) we obtain

$$\ln\gamma = -1.175 \cdot \sqrt{0.01} = -0.1175 \text{ and } \gamma = 0.89$$

Using $R = 500$ pm (approximately accounting for the hydration sphere of an ion), we obtain

$$\begin{aligned} 1 + BR\sqrt{I} &= 1 + 0.330 \times 10^{10} \cdot 500 \times 10^{-12} \cdot \sqrt{0.01} \\ &= 1 + 0.17 = 1.17 \end{aligned}$$

and

$$\ln\gamma = -0.1175/1.17 = -0.100 \text{ and } \gamma = 0.90$$

For 0.01 molar $MgSO_4$ we have

$$I = \frac{1}{2} \cdot (0.01 \cdot 4 + 0.01 \cdot 4)\ \mathrm{mol\ L^{-1}} = 0.04\ \mathrm{mol\ L^{-1}}$$

With $R = 0$ we obtain

$$\ln \gamma = -1.175 \cdot \sqrt{0.04} \cdot 4 = -0.94 \text{ and } \gamma = 0.39$$

and with $R = 500$ pm we have

$$1 + BR\sqrt{I} = 1.34, \quad \ln \gamma = -0.70, \quad \gamma = 0.49$$

Note that the Coulomb interaction increases quadratically with the ionic charge!

21.6.1.3 Mean Activity Coefficient

Figure 21.3 plots the mean activity coefficient $\gamma_\pm = \sqrt{\gamma_+ \cdot \gamma_-}$ for HCl as a function of the concentration. Figure 21.3a shows concentrations from $0 - 0.15$ mol L^{-1}, for which the experimental values[2] (full circles) agree well with those calculated from the Debye–Hückel Equation (21.93) up to $c = 0.1$ M (solid line). The validity of the Debye–Hückel limiting law (21.96) is restricted to concentrations lower than 0.01 mol L^{-1}. Figure 21.3b shows the range of concentrations of $0 - 2$ mol L^{-1}. It is evident in this graph that the activity coefficient does not decrease monotonically with the electrolyte concentration. In fact, it reaches $\gamma_\pm = 1$ at $c = 2$ mol L^{-1} and may even exceed unity at higher concentrations, e.g., a 4 molar HCl solution has a mean activity coefficient of 1.76.

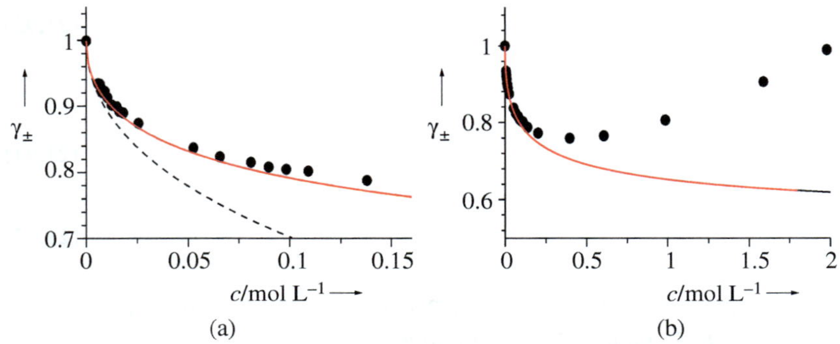

Figure 21.3 Mean activity coefficient $\gamma_\pm$ of an aqueous HCl solution as a function of the concentration c of HCl at $T = 308$ K. Filled circles: experiment, solid line: Debye–Hückel Equation (21.93), calculated with $R = 500$ pm, dashed line: Debye–Hückel limiting law (21.96). (a) Concentration range $0 - 0.15$ mol L^{-1}, (b) concentration range $0 - 2$ mol L^{-1}.

21.7 Conclusion

The description of realistic chemical solutions with more than one component and/or phase may be realized through modification of the essential thermodynamical relationships that we found for ideal gases and solutions. We introduced the chemical potential and the activity as useful quantities to describe equilibria in solution. Activities of the solvent can be deduced from measurements of the vapor pressure above a solution and of the osmotic pressure. Using these activities, we show that the elevation of the boiling point and the depression of the freezing point are in good agreement with experimental data. The activities of the dissolved electrolytes in dilute solution can be quantitatively understood by applying the Debye–Hückel theory.

21.8 Box 21.1. Electrical Work to Transfer Ions into Concentrated Solution

To calculate the electrical work in (21.86) we consider transferring an ion reversibly from an ideally diluted solution to a concentrated solution. In a thought experiment we first remove the charge from the ion and transfer the neutral particle into the concentrated solution. Next we charge the particle from $Q = 0$ to $Q = e$. On charging the ion to Q the counterions in the concentrated solution will arrange in the neighborhood of the ion such that an opposite charge $-Q$ at an average (or effective) distance r_D from the ion results. The work to add a small charge dQ to Q is then

$$-\frac{1}{4\pi\varepsilon_0\varepsilon} \cdot \frac{Q \cdot dQ}{r_D}$$

---

2 Harned and Ehlers [2] For an evaluation of the activity coefficients see Chapter 23, Section 23.5.2.

To find the total work to charge the ion we integrate the differential work expression from $Q = 0$ to $Q = e$ and obtain

$$-\frac{1}{4\pi\epsilon_0\epsilon\cdot r_D}\cdot\int_{Q=0}^{Q=e} Q\cdot dQ = -\frac{1}{4\pi\epsilon_0\epsilon\cdot r_D}\cdot\left|\frac{1}{2}Q^2\right|_0^e = -\frac{1}{2}\cdot\frac{e^2}{4\pi\epsilon_0\epsilon\cdot r_D}$$

Multiplication with the number N of ions leads to the expression in (21.86).

21.9 Foundations

21.9.1 Foundation 21.1: Activity Coefficient of Solute From Activity Coefficient of Solvent

Here we derive an expression for the activity coefficient of the solute in terms of the solvent's activity coefficient. We begin with the Gibbs–Duhem Equation (21.10) in Section 21.2.3 in the form

$$\boldsymbol{n}\cdot d\mu + \boldsymbol{n}_s\cdot d\mu_s = 0 \qquad (21F1.1)$$

where $\boldsymbol{n}$ is the amount of substance of the solute and $\boldsymbol{n}_s$ is the amount of substance of the solvent. Dividing the equation by the total amount $\boldsymbol{n}+\boldsymbol{n}_s$ we obtain

$$x\cdot d\mu + x_s\cdot d\mu_s = 0 \qquad (21F1.2)$$

where x and x_s are the mole fractions. With Equation (21.25) of Section 21.3, namely

$$\mu = \mu^{\ominus} + \boldsymbol{R}T\cdot\ln a \qquad \text{and} \qquad \mu_s = \mu_s^{\ominus} + \boldsymbol{R}T\cdot\ln a_s \qquad (21F1.3)$$

it follows that

$$x\cdot\boldsymbol{R}T\cdot d\ln a + x_s\cdot\boldsymbol{R}T\cdot d\ln a_s = 0 \qquad (21F1.4)$$

where we have used the fact that the standard state chemical potentials do not depend on the composition of the system. Rearrangement of this expression gives

$$d\ln a = -\frac{x_s}{1-x_s}\, d\ln a_s \qquad (21F1.5)$$

Equation (21F1.5) provides the essential connection between the activity of the solute and that of the solvent.

To find the connection between the activity coefficients, we write the activities in terms of the activity coefficients γ and γ_s.

$$d\ln(x\gamma) = -\frac{x_s}{1-x_s}\, d\ln(x_s\gamma_s) \qquad (21F1.6)$$

and differentiate by x_s to obtain

$$\frac{d\ln(x\gamma)}{dx_s} = -\frac{x_s}{1-x_s}\,\frac{d\ln(x_s\gamma_s)}{dx_s} \qquad (21F1.7)$$

The logarithms can be written as

$$\ln(x\gamma) = \ln x + \ln\gamma \qquad \text{and} \qquad \ln(x_s\gamma_s) = \ln x_s + \ln\gamma_s$$

and we obtain

$$\frac{d\ln x}{dx_s} + \frac{d\ln\gamma}{dx_s} = -\frac{x_s}{1-x_s}\,\frac{d\ln x_s}{dx_s} - \frac{x_s}{1-x_s}\,\frac{d\ln\gamma_s}{dx_s} \qquad (21F1.8)$$

Because of

$$\frac{d\ln x}{dx_s} = \frac{d\ln(1-x_s)}{dx_s} = -\frac{1}{1-x_s} \qquad \text{and} \qquad \frac{d\ln x_s}{dx_s} = \frac{1}{x_s} \qquad (21F1.9)$$

we find that

$$-\frac{1}{1-x_s} + \frac{d\ln\gamma}{dx_s} = -\frac{x_s}{1-x_s}\,\frac{1}{x_s} - \frac{x_s}{1-x_s}\,\frac{d\ln\gamma_s}{dx_s}$$

and Equation (21F1.8) simplifies to

$$d\ln\gamma = -\frac{x_s}{1-x_s}\cdot d\ln\gamma_s \qquad (21F1.10)$$

By integration we obtain

$$\ln \gamma_1 = -\int_{\ln \gamma_s=0}^{\ln \gamma_{s,1}} \frac{x_s}{1-x_s} \cdot d\ln \gamma_s \tag{21F1.11}$$

which is Equation (21.38) of the text.

21.9.2 Foundation 21.2: Distribution of Ions in Solution

See the books website, www.wiley.com/go/Kuhn/physicalchemistry3e.

21.10 Exercises

E21.1 Consider a 1.50 molal aqueous HCl solution and assume that the HCl is not volatile. When the external pressure is 355.1 Torr, the enthalpy of vaporization $\Delta_{vap}H$ for water is 40.7 kJ mol^{-1} and its entropy of vaporization is $\Delta_{vap}S$ is 115 J mol^{-1} K^{-1}. Find the boiling point of pure water and of the HCl solution under these conditions. Provide a physical explanation for the boiling point elevation, or vapor pressure lowering, of the water. In your explanation discuss the Gibbs energy and provide a sketch of its concentration dependence.

E21.2 The vapor pressure of pure water at 298 K is 23.76 Torr. Assume that the water vapor above an HCl solution acts ideally and compute the activity of the water in solution. If the vapor pressure of water above the solution at 298 K is 22.28 Torr, what is its activity?

E21.3 Charge Distribution $\rho(r)$: Prove that the total charge of the cloud of counterions in the treatment of a spherical distribution equals $-e$. Refer to Foundation 21.2.

E21.4 Consider an ideal mixture of two miscible liquids, and derive a formula for the composition that results in the largest entropy of mixing. Now consider a non-ideal mixture of two miscible liquids and derive a formula for the composition that results in the largest entropy of mixing. Assume that $a_i = \gamma_i x_i$ where γ_i is a constant.

E21.5 Starting with Equation (21.60), namely

$$\frac{\Delta_{vap}H}{\boldsymbol{R}}\left(\frac{1}{T}-\frac{1}{T_s}\right) = \ln a_s$$

derive the approximate relation

$$\boxed{\Delta T = -\frac{\boldsymbol{R}T_s^2}{\Delta_{vap}H} \cdot \ln a_s \qquad \text{(elevation of boiling point)}}$$

for ΔT.

E21.6 An aqueous solution of KCl at 298 K, which is 1.2455 molar, has a freezing point depression of 4.048°. Use these data to calculate the activity coefficient for the water.

E21.7 The data in the table show the volume of a KCl solution as a function of its molality at 298 K and 1 bar pressure. Plot these data and use it to determine the partial molar volume of KCl at 0.5 mol of KCl. (Note that the molality is moles of solute per kilogram of solvent.)

m_{KCl}, molal	0.0	0.1656	0.3342	0.6689	1.6694	3.3481
V, mL	1000.	998.318	996.781	993.967	986.791	977.973

E21.8 The data in the table show the measured activity coefficient for HNO_3 as a function of molality. Plot these data versus the solution molality and fit it to the Debye–Hückel model. Comment on the quality of the fit and your best fit value for R.

m, mol kg^{-1}	0.001	0.002	0.005	0.01	0.02
$\gamma_\pm$	0.965	0.951	0.927	0.902	0.871

m, mol kg^{-1}	0.05	0.1	0.2	0.5	1.0
$\gamma_\pm$	0.823	0.785	0.748	0.715	0.720

E21.9 The data in the table [3] show the measured activity coefficient for Na_2SO_4 as a function of concentration. Plot these data versus the solution molality and fit it to the Debye–Hückel model. Comment on the quality of the fit and your best fit value for R.

m, mol kg^{-1}	0.001	0.002	0.005	0.01	0.02
$\gamma_\pm$	0.8846	0.8505	0.7808	0.6889	0.6468

m, mol kg^{-1}	0.05	0.1	0.2	0.5	1.0
$\gamma_\pm$	0.5439	0.4342	0.3923	0.2996	0.2335

E21.10 Beginning with Equation (21.83) prove that the van't Hoff equation,

$$\frac{\mathrm{d}(\ln K)}{\mathrm{d}T} = \frac{\Delta_r H^\ominus}{\boldsymbol{R}T^2}$$

holds; i.e., it is a general result that does not depend on approximations about a system's ideality.

E21.11 In Example 21.7 the K_{sp} of AgCl in water are given to be $K_{sp} = 1.77 \cdot 10^{-10}$. Using this knowledge and the Debye–Hückel model predict what the solubility of AgCl will be in a 0.100 molar solution of KCl.

E21.12 By far the major component of seawater is NaCl and it is about 0.55 molar. If sea water were an ideal solution with this concentration of NaCl, what would be the vapor pressure of the water in sea water at 298 K? Compare your value for the vapor pressure of seawater to that of pure water at 298 K. State whether it is higher or lower and explain why. If the actual vapor pressure of such a solution is 0.027 bar, what is the activity of the water? Consider the vapor pressure of pure water to be 0.035 bar at 298 K, and take the density of 0.55 molar salt water to be $\rho = 0.98$ g mL^{-1}.

E21.13 Use Equations (21.8) and (21.22) to write a general expression for the partial molar enthalpy of a component in a solution. Subsequently, use your enthalpy expression to show that

$$\left(\frac{\partial \ln \hat{a}_i}{\partial T}\right)_{P,\boldsymbol{n}_{i\neq i}} = \frac{h_i^\ominus - h_i}{\boldsymbol{R}T^2}$$

E21.14 Derive the expression for the depression of the freezing point, namely

$$\boxed{\Delta T = \frac{\boldsymbol{R}T_s^2}{\Delta_{\text{vap}}H} \cdot \ln a_s \qquad \text{(depression of freezing point)}}$$

E21.15 Consider a solution of two volatile components, A and B, at fixed composition in equilibrium with its vapor at a total pressure P of 1 bar. By assuming that the vapor behaves like an ideal gas and that $\Delta_{\text{vap}}H$ is independent of temperature, derive the relationship

$$\ln\left(\frac{y_\text{A}}{\hat{a}_\text{A}}\right) = \frac{\Delta_{\text{vap}}H_\text{A}}{\boldsymbol{R}}\left(\frac{1}{T_{\text{BP,A}}} - \frac{1}{T}\right)$$

where $T_{\text{BP,A}}$ is the boiling temperature, y_A is the mole fraction of A in the vapor, and a_A is the activity of A. (Hint: Refer to the derivation in Section 21.4.)

E21.16 Derive the general relationship

$$\left(\frac{\partial \ln \hat{a}_i}{\partial P}\right)_{T,\boldsymbol{n}_j} = \frac{v_i}{\boldsymbol{R}T}$$

which describes the pressure dependence of the activity for component i.

E21.17 Show that

$$\left(\frac{\partial \ln \hat{a}_i}{\partial T}\right)_{p,n_i,n_{j\neq i}} = \frac{\Delta H_{m,i}}{RT^2}$$

where $\Delta H_{m,i}$ is the difference in enthalpy between the solution and the standard state.

E21.18 The table shows data for the solubility x_{O_2} of oxygen in water at 1.013 bar pressure. Use these data to calculate the enthalpy of solution for O_2 in water.

x_{O_2}	$2.756 \cdot 10^{-5}$	$2.501 \cdot 10^{-5}$	$2.293 \cdot 10^{-5}$	$2.122 \cdot 10^{-5}$	$1.982 \cdot 10^{-5}$
T/K	288.15	293.15	298.15	303.15	308.15

E21.19 The table shows data for the lowering of the vapor pressure of water as a function of NaCl concentration at a temperature of 373 K and pressure of 760 Torr. Using these data calculate how much NaCl must be added to 1 L of water in order to raise its boiling point by 5 °C.

c_{NaCl}/mol L^{-1}	0.50	1.00	2.00	3.00	4.00	5.00	6.00
ΔP/torr	12.3	25.2	52.1	80.0	111.0	143.0	176.5

E21.20 Consider the Debye–Hückel equation for the mean activity coefficient of a 1:1 electrolyte (e.g., KCl) and show that it may be written in terms of the concentration c as

$$\ln \gamma_{\pm} = -\frac{A\sqrt{c}}{1 + B \cdot R\sqrt{c}}$$

For the salts KCl and NaCl estimate the value of R from the average of their ionic radii and calculate the concentration c at which the two terms in the denominator are equal. Determine the molarity at which the denominator is 1.01. Under what conditions would you expect to be able to use the more approximate equation

$$\ln \gamma_{\pm} = -A\sqrt{c}$$

E21.21 Starting with Equation (21.60), show that it reduces to Equation (21.62) for a small temperature shift.

21.11 Problems

P21.1 The data in the table report masses and volumes for preparing NaCl solutions with water. In each case 14.881 g NaCl with a volume of 6.902 mL was used.

m mol/kg	Mass$_{solution}$ g	Mass$_{H_2O}$ g	$V_{solution}$ mL	V_{H_2O} mL
5.1420	58.836	43.955	49.504	44.084
2.5638	108.842	93.961	99.286	94.237
1.0273	257.213	242.332	247.779	243.045
0.5110	506.986	492.105	498.163	493.551
0.2543	1008.572	993.691	1001.094	996.610

Fit these data and write an empirical expression for the partial molar volume over this range of concentration. Evaluate your expression for the case of 2 molal NaCl solution.

P21.2 Consider the following set of data for mixtures of acetone and water.

$x_{acetone}$	T = 298 K		T = 333 K	
	$\frac{P_{water}}{Torr}$	$\frac{P_{acetone}}{Torr}$	$\frac{P_{water}}{Torr}$	$\frac{P_{acetone}}{Torr}$
0.0	23	0	149	0
0.0333	27	28	149	190
0.0720	27	77	143	342
0.117	25	105	134	443
0.171	24	125	145	495
0.236	22	146	129	553
0.318	28	152	126	588
0.420	23	164	116	624
0.554	16	182	102	672
0.737	17	192	97	711
1.00	0	229	0	860

Plot the vapor pressures of each component and the total vapor pressure as a function of the mole fraction. Perform this plot for each temperature. Explain, in words, how the observed vapor pressure data differs from that which one would predict for an ideal solution.

Compute the activity of each component at each of the mole fractions and temperatures. Present a table of the activities and give a representative calculation in detail. Use the activities to determine the $\Delta_{mix}G$ at each of the intermediate mole fractions and for each temperature. Compute the $\Delta_{mix}G$ that you would expect if acetone and water formed an ideal solution. Present your values in tabular form and plot the data. Compare your two results and try to explain it in terms of the intermolecular interactions between acetone and water.

P21.3 Using Equation (21.38) and the data for the activity coefficients of the solvent γ_s in Table 21.1, calculate the corresponding activity coefficients γ of the solute.

P21.4 Consider a solution of glycol $(CH_2OH)_2$ in water (glycol is used in the water cooling system of cars). Calculate the mole fraction x of glycol that is necessary to decrease the freezing point by 3, 10, and 20 K. How much glycol must be added to 1 L of water?

P21.5 Lead iodide, PbI_2, has a solubility product of 9.49×10^{-9}. Calculate the solubility of Pb^{2+} in a 0.01 molar KNO_3 solution, in a 0.01 molar KI solution, and a 0.1 molar KI solution. You can assume that the KNO_3 and KI are strong electrolytes and infinitely soluble. Comment on your findings.

P21.6 One important feature of a biological membrane is its ability to maintain concentration gradients of ions. Consider the idealized situation in which a membrane separates two solutions, one with a concentration of 0.15 molal KCl and the other with a concentration of 5×10^{-3} molal KCl, typical differences for the inside and outside of a cell. Assume that both solutions have a density of $\rho = 1$ g mL^{-1} and are at 298 K. a) Find the molar Gibbs energy difference between the two sides of the membrane, assuming that the solutions are ideal. b) Using the Debye–Hückel limiting law given by

$$\ln \gamma_{\pm} = -1.141 \cdot z_+ z_- \sqrt{I} \quad \text{with} \quad I = \frac{1}{2}\sum_i \hat{c}_i z_i^2$$

find the activity for each of the two solutions. c) Using your activities from part b) find the molar Gibbs energy difference between the two sides. d) Compute the percentage error in the two molar Gibbs energy differences that you found in parts a) and c).

P21.7 Show that the expression for the chemical potential of an ideal solution in terms of the mole fraction (e.g., Equation (21.24) of the text) and the expression in terms of the concentration (e.g., Equation (21.16) of the text) differ from each other only in terms of the reference state.

P21.8 Show that

$$\phi(x) = \frac{1}{2}a_0\,(e^{\kappa x} + e^{-\kappa x}) + \frac{1}{2}a_1\frac{1}{\kappa}\,(e^{\kappa x} - e^{-\kappa x})$$

is a solution of the equation

$$\frac{d^2\phi(x)}{dx^2} = 2n\frac{e^2}{\varepsilon_0 \varepsilon kT} \cdot \phi(x)$$

in Foundation 21.2, and that

$$\phi(x) = a_0 + a_1 \cdot x \qquad \text{for small values of } x$$

References

1 R.C. Weast, *CRC Handbook of Physics and Chemistry* (The Chemical Rubber Co., Cleveland, OH, 1970/71) 51st edition.

2 H.S. Harned and R.W. Ehlers, The thermodynamics of aqueous hydrochloric acid solutions from electromotive force measurements, *J. Am. Chem. Soc.*, 1933, **55**, 2179.

3 D. Zaytsev and G.G. Aseyev, eds., *Properties of Aqueous Solutions of Electrolytes I* (CRC, Boca Raton, FL, 1992).

22

Reaction Equilibria in Aqueous Solutions and Biosystems

CHAPTER MENU

22.1 Introduction

A great number of compounds dissolve in water where they interact in various ways. Of particular importance in chemistry and biology are acid–base reactions and redox reactions. The energetic considerations in understanding these reactions are crucial for understanding the interplay of biomolecules and the operation of biological machinery.

How can we understand reactions of acids and bases? Acid–base reactions in water are strongly affected by solvation and the autoionization of water. Water autoionizes to produce a hydronium ion (H_3O^+), the protonated form of H_2O, and hydroxide ion (OH^-), the deprotonated form of H_2O; i.e.,

$$H_2O + H_2O \rightleftarrows H_3O^+ + OH^- \qquad K_w = \hat{c}_{H_3O^+} \cdot \hat{c}_{OH^-} = 1.0 \times 10^{-14} \text{ at 298 K}$$

In this reaction one water molecule acts as an acid and another as the base; i.e., water is amphoteric. Because water is the solvent and has this equilibrium, the strongest acid that can exist in water is H_3O^+ since any acid stronger than this would react with a water molecule to produce H_3O^+. Correspondingly, the strongest base is OH^-. Because an acid and its conjugate base have different net charges, their relative solvation energies can be very different.

Reduction and oxidation (redox) reactions are ubiquitous in nature, ranging from the initial chemical steps in photosynthesis to the rusting of iron. Unlike the proton in acid–base reactions (where the solvent H_2O is the proton carrier), the electron is not commonly captured by a solvent molecule before reacting with the oxidant. How can we understand reactions of reductants and oxidants? Although redox reactions can be strongly affected by solvent, the electron transfer equilibria are controlled by the energetics of the oxidant and reductant species.

Acid–base and redox reactions are both important in biochemistry and the bioenergetics of organisms. To illustrate such processes we introduce group transfer potentials and apply them to the synthesis and combustion of glucose.

22.2 Proton Transfer Reactions: Dissociation of Weak Acids

A very common reaction type is the Brønsted–Lowry acid–base reaction

$$HA + B \rightleftarrows A^- + HB^+$$

Principles of Physical Chemistry, Third Edition. Hans Kuhn, David H. Waldeck, and Horst-Dieter Försterling.

Companion website: www.wiley.com/go/Kuhn/physicalchemistry3e

which is written here for the case of neutral reactants. Because a proton is transferred from the acid HA to the base B, it is called a proton transfer reaction. The species A^- is called the conjugate base of the acid A, and the species HB^+ is called the conjugate acid of the base B.

A simple example of such an equilibrium is a solution of vinegar (acetic acid solution). In this case, we write the equilibrium as

$$CH_3COOH + H_2O \rightarrow CH_3COO^- + H_3O^+$$

With the intermediate

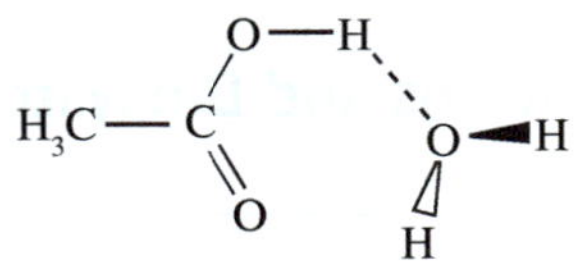

The conjugate base to acetic acid is the acetate ion. Water acts as the base and the hydronium ion is its conjugate acid.

22.2.1 Henderson–Hasselbalch Equation

A short notation for the acid dissociation reaction is

$$HA \rightarrow A^- + H^+ \tag{22.1}$$

in which the proton is transferred to the solvent (H_2O in this case). The equilibrium constant K_a for this reaction is called the *acid dissociation constant* and given by

$$K_a = \frac{\hat{a}_{A^-} \cdot \hat{a}_{H^+}}{\hat{a}_{HA}} \tag{22.2}$$

where $\hat{a} = a/c^{\ominus}$, $c^{\ominus}$ is the standard concentration ($c^{\ominus} = 1$ mol L^{-1}) and a is the activity. It is very common to discuss acid–base chemistry by consideration of the pK_a of a molecule, where pK_a is defined as

$$\boxed{pK_a = -\log K_a} \tag{22.3}$$

From the conditions of equilibrium, we see that for reaction (22.1)

$$\Delta_r G^{\ominus} = 2.303 \cdot \boldsymbol{R}T \cdot \ pK_a$$

so that a discussion of pK_a values is equivalent to a discussion of $\Delta_r G^{\ominus}$.

To simplify our considerations we approximate the solution as ideally diluted, so that

$$K_a = \frac{\hat{c}_{A^-} \cdot \hat{c}_{H^+}}{\hat{c}_{HA}} \tag{22.4}$$

Furthermore, we define the quantity

$$\boxed{pH = -\log\left(\hat{c}_{H^+}\right)} \tag{22.5}$$

Combining Equations (22.3), (22.4), and (22.5) yields

$$pK_a = -\log\left(\frac{\hat{c}_{A^-} \cdot \hat{c}_{H^+}}{\hat{c}_{HA}}\right) = -\log\left(\frac{\hat{c}_{A^-}}{\hat{c}_{HA}}\right) + pH \tag{22.6}$$

or

$$\boxed{pH = pK_a + \log\left(\frac{\hat{c}_{A^-}}{\hat{c}_{HA}}\right)} \tag{22.7}$$

which is called the *Henderson–Hasselbalch equation.*

In the special case of a solution of a weak acid (e.g., acetic acid) and its salt (e.g., sodium acetate) in a 1:1 ratio we obtain

$$\hat{c}_{A^-} = \hat{c}_{HA} \tag{22.8}$$

(the acid is almost undissociated and the salt is completely dissociated). In this case

$$pH = pK_a \tag{22.9}$$

At sufficiently high concentration the pH of such solutions is relatively insensitive to the addition of strong acids or bases and are called *buffer solutions*. The pH of a buffer solution depends primarily on the dissociation constant of the weak acid. It does not change, or only slightly, upon the addition of small amounts of strong acids or bases.

Example 22.1 *Change of pH: A 1 mL aliquot of 1 M HCl is added to 99 mL of water (case 1) and to 99 mL of a buffer solution (case 2). The buffer is a 1:1 mixture of acetic acid (HAc, pK_a = 4.62) and sodium acetate (NaAc), each of concentration c_0 = 0.10 M. Assume that the sodium acetate (a strong electrolyte) and hydrochloric acid (a strong acid) are completely dissociated. What is the change of pH in both cases?*

Case 1: The initial concentration c_a of HCl is $c_a = 0.010$ M, so that the concentration of H^+ is equal to $c_a = 0.010$ M and this results in a pH = 2. The change in pH from pure water is $\Delta pH = 2 - 7 = -5$, a drastic change.

Case 2: To calculate the change of pH in the buffer solution we must analyze how its equilibrium changes as we add HCl:

$$K_a = \frac{\hat{c}_{H^+} \cdot \hat{c}_{Ac^-}}{\hat{c}_{HAc}}$$

From the stoichiometry of the acetic acid dissociation we know that $\hat{c}_{HAc} + \hat{c}_{Ac^-} = 2\,\hat{c}_0$, and from electroneutrality we know that $\hat{c}_{H^+} + \hat{c}_{Na^+} = \hat{c}_{OH^-} + \hat{c}_{Cl^-} + \hat{c}_{Ac^-}$. Furthermore, the conservation of mass requires that $\hat{c}_{Na^+} = \hat{c}_0$ and $\hat{c}_{Cl^-} = \hat{c}_a$.

Neglecting $\hat{c}_{H^+}$ and $\hat{c}_{OH^-}$ relative to $\hat{c}_0$ and $\hat{c}_a$, we obtain

$$\hat{c}_{Ac^-} = \hat{c}_0 - \hat{c}_a, \quad \hat{c}_{HAc} = 2\hat{c}_0 - \hat{c}_0 + \hat{c}_a = \hat{c}_0 + \hat{c}_a$$

so that

$$\hat{c}_{H^+} = K_a \frac{\hat{c}_{HAc}}{\hat{c}_{Ac^-}} = K_a \cdot \frac{\hat{c}_0 + \hat{c}_a}{\hat{c}_0 - \hat{c}_a}, \qquad pH = pK_a - \log \frac{\hat{c}_0 + \hat{c}_a}{\hat{c}_0 - \hat{c}_a}$$

For $\hat{c}_a = 0$ (i.e., for the pure buffer) we have pH = pK_a = 4.62.

For $\hat{c}_a = 0.010$ and $\hat{c}_0 = 0.10$ we have pH = 4.62 – log(0.11/0.09) = 4.53, thus ΔpH= –0.09. We see that the pH of the buffer solution remains nearly unchanged.

From pH = 4.53 we calculate $\hat{c}_{H^+} = 10^{-4.53} = 3 \times 10^{-5}$ and $\hat{c}_{OH^-} = 1 \times 10^{-14}/3 \times 10^{-5} = 3 \times 10^{-10}$. Thus our assumption that $\hat{c}_{H^+}$ and $\hat{c}_{OH^-}$ are small relative to $\hat{c}_0$ and $\hat{c}_a$ is justified.

How much of a strong acid can be dissolved in a buffer without appreciably changing the pH? This question requires an analysis as a function of the concentrations $\hat{c}_a$ and $\hat{c}_0$, see Problem P22.9.

22.2.2 Degree of Dissociation in Aqueous Solution

If we dissolve a weak acid HA in water at a concentration c_{total}, what is the fraction of HA molecules forming A^- and H^+? We start with Equations (22.1) and (22.4)

$$HA \rightarrow A^- + H^+, \qquad K_a = \frac{\hat{c}_{A^-} \cdot \hat{c}_{H^+}}{\hat{c}_{HA}}$$

If we can neglect the autoionization of water, then

$$c_{A^-} = c_{H^+} \quad \text{and} \quad c_{HA} = c_{total} - c_{H^+} \tag{22.10}$$

K_a becomes

$$K_a = \frac{\hat{c}^2_{H^+}}{\hat{c}_{total} - \hat{c}_{H^+}} \tag{22.11}$$

Hence the proton concentration is

$$\hat{c}_{H^+} = -\frac{1}{2}K_a + \sqrt{\hat{c}_{total}K_a + \tfrac{1}{4}K_a^2} \tag{22.12}$$

The fraction of dissociated HA molecules is called the *degree of dissociation* α, and it is given by

$$\alpha = \frac{c_{A^-}}{c_{total}} = \frac{c_{H^+}}{c_{total}} = \frac{\hat{c}_{H^+}}{\hat{c}_{total}} \tag{22.13}$$

Combining Equations (22.12) and (22.13) we can write the degree of dissociation as

$$\boxed{\alpha = \frac{-\frac{1}{2}K_a + \sqrt{\widehat{c}_{\text{total}} \cdot K_a + \frac{1}{4}K_a^2}}{\widehat{c}_{\text{total}}}} \tag{22.14}$$

For a large enough total concentration, Equation (22.14) simplifies to

$$\alpha \approx \frac{\sqrt{\widehat{c}_{\text{total}} \cdot K_a}}{\widehat{c}_{\text{total}}} = \sqrt{\frac{K_a}{\widehat{c}_{\text{total}}}}$$

and α approaches zero with increasing $\widehat{c}_{\text{total}}$. For a sufficiently small total concentration

$$\sqrt{\widehat{c}_{\text{total}} \cdot K_a + \frac{1}{4}K_a^2} = \frac{1}{2}K_a \cdot \sqrt{1 + \frac{4 \cdot \widehat{c}_{\text{total}}}{K_a}}$$
$$\approx \frac{1}{2}K_a \cdot \left(1 + \frac{2 \cdot \widehat{c}_{\text{total}}}{K_a}\right)$$

and Equation (22.14) may be approximated by

$$\alpha \approx \frac{-\frac{1}{2}K_a + \frac{1}{2}K_a + \widehat{c}_{\text{total}}}{\widehat{c}_{\text{total}}} = 1$$

The dissociation is complete; α approaches unity (Fig. 22.1).

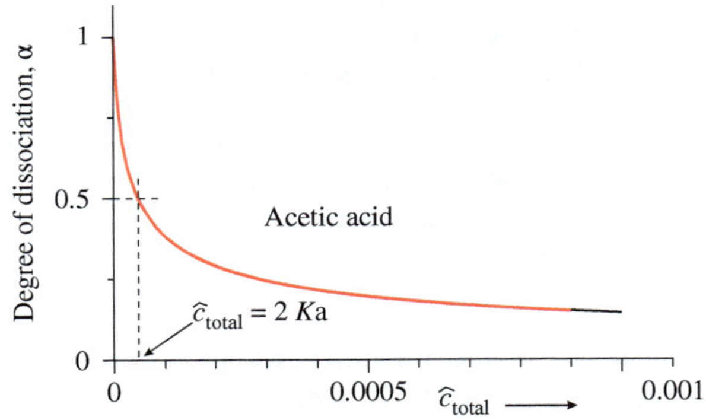

Figure 22.1 Degree of dissociation α for acetic acid versus the total concentration, $\widehat{c}_{\text{total}}$. Calculated from Equation (22.14) with $K_a = 2.4 \times 10^{-5}$. For $\widehat{c}_{\text{total}} = 2K_a = 4.8 \times 10^{-5}$ we find $\alpha = \left(-\frac{1}{2}K_a + \frac{3}{2}K_a\right)/(2K_a) = 0.5$.

22.2.3 Degree of Dissociation in a Buffer Solution

If we dissolve a weak acid in a buffer solution with a given pH, what is the degree of dissociation α? Again we start with Equations (22.1) and (22.4)

$$\text{HA} \rightarrow \text{A}^- + \text{H}^+, \qquad K_a = \frac{\widehat{c}_{\text{A}^-} \cdot \widehat{c}_{\text{H}^+}}{\widehat{c}_{\text{HA}}}$$

In this case c_{H^+} is given by the buffer. Because of the stoichiometry,

$$\widehat{c}_{\text{HA}} = \widehat{c}_{\text{total}} - \widehat{c}_{\text{A}^-} \tag{22.15}$$

and

$$K_a = \frac{\widehat{c}_{\text{A}^-} \cdot \widehat{c}_{\text{H}^+}}{\widehat{c}_{\text{total}} - \widehat{c}_{\text{A}^-}} \tag{22.16}$$

We can solve Equations (22.15) and (22.16) for $\widehat{c}_{\text{A}^-}$ and $\widehat{c}_{\text{HA}}$ to find

$$\widehat{c}_{\text{A}^-} = \widehat{c}_{\text{total}} \frac{K_a}{\widehat{c}_{\text{H}^+} + K_a}, \qquad \widehat{c}_{\text{HA}} = \widehat{c}_{\text{total}} \frac{\widehat{c}_{\text{H}^+}}{\widehat{c}_{\text{H}^+} + K_a} \tag{22.17}$$

The degree of dissociation is

$$\alpha = \frac{\widehat{c}_{\text{A}^-}}{c_{\text{total}}} = \frac{K_a}{\widehat{c}_{\text{H}^+} + K_a} = \frac{1}{1 + \widehat{c}_{\text{H}^+}/K_a} \tag{22.18}$$

Using the definitions in Equation (22.5)

$$\widehat{c}_{\text{H}^+} = 10^{-\text{pH}} \quad \text{and} \quad K_a = 10^{-\text{p}K_a}$$

we can rearrange the equation for α and find

$$\alpha = \frac{1}{1 + 10^{\text{p}K_a - \text{pH}}} \tag{22.19}$$

In Fig. 22.2a α is shown as a function of the buffer pH. Figure 22.2b shows a corresponding diagram for the concentrations c_{A^-} and c_{HA}, according to Equation (22.17). In this case the degree of dissociation α is independent of the total concentration c_{total}; it is determined by the pK_A and the pH. For pH = pK_a, $\alpha = \frac{1}{2}$.

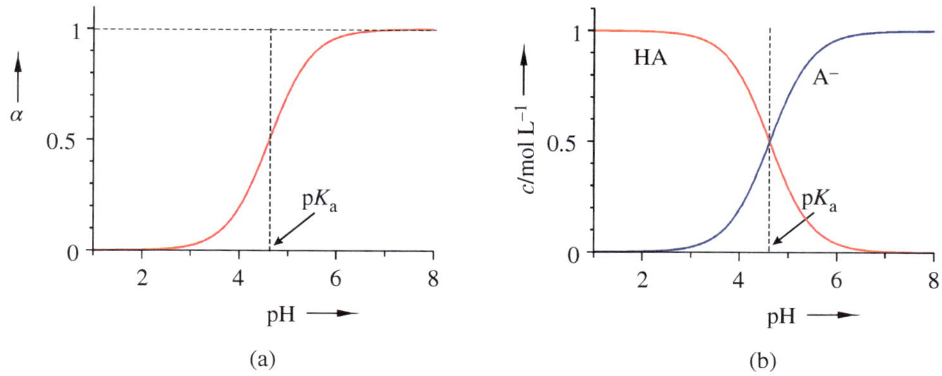

Figure 22.2 Dissociation of acetic acid (c_{total} = 1 mol L^{-1}, pK_a = 4.62). (a) Degree of dissociation α versus (pH)$_{buffer}$, calculated from Equation (22.19). (b) Concentrations c_{HA} and c_{A^-} versus (pH)$_{buffer}$, calculated from Equation (22.17).

22.2.4 Titration Curve of a Weak Acid

The addition of NaOH to a solution of a weak acid causes an increase of the pH, because the protons generated by the weak acid's dissociation react with the OH^- ions to form water:

$$H^+ + OH^- \rightleftharpoons H_2O \tag{22.20}$$

How can we quantify the pH change as more and more NaOH is added to the solution?

To calculate the pH of the solution at different amounts of added NaOH we consider the dissociation of the acid

$$HA \rightarrow H^+ + A^-$$

As in Section 22.2.3, we calculate the concentrations $\hat{c}_{A^-}$ and $\hat{c}_{HA}$. Furthermore, from electroneutrality we know that

$$c_{H^+} + c_{Na^+} = c_{OH^-} + c_{A^-} \tag{22.21}$$

Finally we can relate the hydroxide concentration c_{OH^-} to the proton concentration through the autoionization equilibrium of water. The dissociation constant K_w of water is

$$K_w = \hat{c}_{H^+} \cdot \hat{c}_{OH^-} \tag{22.22}$$

Since Na^+ does not undergo any further reaction after the dissociation of NaOH, c_{Na^+} is known from the amount $\boldsymbol{n}_{NaOH}$ of NaOH added to the solution. The only remaining quantity c_{total} is given by the amount $\boldsymbol{n}_{total}$ of the acid at the start of the titration ($c_{total} = \boldsymbol{n}_{total}/V$, where V is the volume of the solution). Hence we have four algebraic equations for the four unknowns $\hat{c}_{H^+}$, $\hat{c}_{OH^-}$, $\hat{c}_{A^-}$, and $\hat{c}_{HA}$. The solution of these equations is discussed in Foundation 22.1 and the following relation between $\boldsymbol{n}_{NaOH}$ and $\hat{c}_{H^+}$ is obtained

$$\boldsymbol{n}_{NaOH} = Vc^{\ominus}\left(\frac{K_w}{\hat{c}_{H^+}} - \hat{c}_{H^+}\right) + \boldsymbol{n}_{total} \cdot \frac{K_a}{\hat{c}_{H^+} + K_a} \tag{22.23}$$

(with the standard concentration $c^{\ominus}$ = 1 mol L^{-1}).

Figure 22.3 shows the change in solution pH for an acetic acid-acetate buffer as a function of $\boldsymbol{n}_{NaOH}$ (for c_{total} = 1.0 M, $\boldsymbol{n}_{total}$ = 1 mol); i.e., the titration curve. For small amounts of NaOH added to the solution the pH changes little. This region, centered around the pH = pK_a point is called the *buffer region*. Near the condition where the amount of added strong base is equivalent to the amount of weak acid (called the *equivalence point*), the pH changes sharply. This behavior results because the acetic acid is totally converted to the acetate by the base and the weak acid's equilibrium is overwhelmed. The pH changes rapidly to that expected for the strong base whose equilibrium now controls the solution pH.

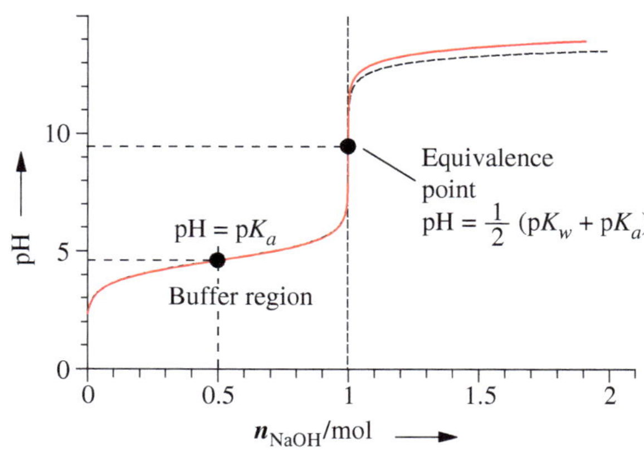

Figure 22.3 Titration curve for acetic acid (K_a = 2.4 × 10^{-5}, pK_a = 4.62) according to Equation (22.23). Initial concentration of acetic acid 1 mol L^{-1}, amount $\boldsymbol{n}_{total}$ = 1 mol dissolved in V = 1 L. pH versus the added amount $\boldsymbol{n}_{NaOH}$ of NaOH. Solid line: volume of the solution assumed as constant; dashed line: volume V increasing with added solution of NaOH (c_{NaOH} = 1 mol L^{-1}). For numerical data see Foundation 22.1.

22.3 Stepwise Proton Transfer

22.3.1 Diprotic Acid

Some acids can donate more than one proton to an acceptor. For example, consider the dissociation of a diprotic acid

$$H_2A \rightarrow HA^- + H^+, \qquad HA^- \rightarrow A^{2-} + H^+ \tag{22.24}$$

with

$$K_{a1} = \frac{\widehat{c}_{H^+} \cdot \widehat{c}_{HA^-}}{\widehat{c}_{H_2A}} \quad \text{and} \quad K_{a2} = \frac{\widehat{c}_{H^+} \cdot \widehat{c}_{A^{2-}}}{\widehat{c}_{HA^-}} \tag{22.25}$$

In this case the deprotonation processes occur in a stepwise fashion but the equilibria are coupled with each other. The concentrations of $\widehat{c}_{H_2A}$, $\widehat{c}_{HA^-}$, and $\widehat{c}_{A^{2-}}$ are calculated in a manner analogous to that used for acetic acid (see Foundation 22.2 for the derivation), and we find

$$\widehat{c}_{A^{2-}} = \widehat{c}_{total} \frac{K_{a1}K_{a2}}{\widehat{c}^2_{H^+} + K_{a1}\widehat{c}_{H^+} + K_{a1}K_{a2}} \tag{22.26}$$

$$\widehat{c}_{HA^-} = \widehat{c}_{total} \frac{K_{a1}\,\widehat{c}_{H^+}}{\widehat{c}^2_{H^+} + K_{a1}\widehat{c}_{H^+} + K_{a1}K_{a2}} \tag{22.27}$$

$$\widehat{c}_{H_2A} = \widehat{c}_{total} \frac{\left(\widehat{c}_{H^+}\right)^2}{\left(\widehat{c}_{H^+}\right)^2 + K_{a1}\widehat{c}_{H^+} + K_{a1}K_{a2}} \tag{22.28}$$

Figure 22.4a plots the concentrations as a function of pH for the case of p tolyl-phosphonic acid (H_2A = tolyl-H_2PO_4) with

$$K_{a1} = 2.82 \times 10^{-2} \quad \text{and} \quad K_{a2} = 8.32 \times 10^{-8}$$

Only at very low pH (<3) will we find a significant quantity of the diprotic species available. At physiological pH values (near 7.0) we expect a mixture of the tolyl-HPO_4^- and tolyl-PO_4^{2-} species to be found.

The titration curve of a diprotic acid is derived in Foundation 22.2. By making use of the equilibrium concentrations in Equations (22.26) to (22.28), the neutrality condition

$$\widehat{c}_{H^+} + \widehat{c}_{Na^+} = \widehat{c}_{OH^-} + \widehat{c}_{HA^-} + 2\widehat{c}_{A^{2-}} \tag{22.29}$$

and the autoprotolysis reaction of water (Equation (22.22)), the following relation between $\boldsymbol{n}_{NaOH}$ and $\widehat{c}_{H^+}$ is obtained

$$\boldsymbol{n}_{NaOH} = Vc^{\ominus}\left(\frac{K_w}{\widehat{c}_{H^+}} - \widehat{c}_{H^+}\right) + \boldsymbol{n}_{total} \cdot \frac{K_{a1}\widehat{c}_{H^+} + 2K_{a1}K_{a2}}{\widehat{c}^2_{H^+} + K_{a1}\widehat{c}_{H^+} + K_{a1}K_{a2}} \tag{22.30}$$

(with the standard concentration $c^{\ominus} = 1$ mol L^{-1} and $\boldsymbol{n}_{total} = \widehat{c}_{total}Vc^{\ominus}$). The titration curve is plotted in Fig. 22.4b.

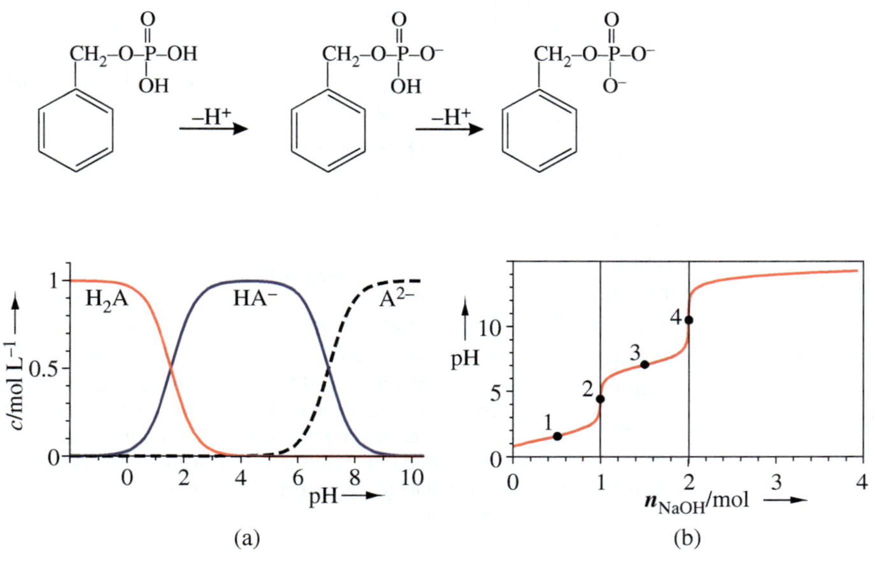

Figure 22.4 Diprotic acid p-tolylphosphonic acid ($c_{total} = 1$ mol L^{-1}, $\boldsymbol{n}_{total} = 1$ mol,. $K_{a1} = 2.82 \times 10^{-2}$, $K_{a2} = 8.32 \times 10^{-8}$, volume of the solution $V = 1$ L). (a) Concentrations c_{H_2A} (red solid line), c_{HA^-} (blue solid line), and $c_{A^{2-}}$ (dashed line, right) as a function of the solution pH. (b) Titration curve. Point 1: $\boldsymbol{n}_{NaOH} = 0.5$ mol, pH = 1.58, point 2: $\boldsymbol{n}_{NaOH} = 1.0$ mol, pH = 4.30 (first equivalence point), point 3: $\boldsymbol{n}_{NaOH} = 1.5$ mol, pH = 7.10, point 4: $\boldsymbol{n}_{NaOH} = 2.0$ mol, pH = 10.55 (second equivalence point). Calculation according to Equations (22.26) to (22.30), see Foundation 22.2.

Example 22.2 *The partial pressure of CO_2 in the earth's atmosphere was 3.89×10^{-3} bar = 394 Pa in the year 1990. At a partial pressure of 394 Pa, CO_2 has a solubility in water of 1.36×10^{-4} mol L^{-1}. Given that the equilibrium between CO_2 and H_2CO_3 has an equilibrium constant of 0.031, compute the pH of one liter of pure water that is exposed to the atmosphere.*

Using the reaction for carbonic acid formation,

$$H_2O + CO_2 \rightleftarrows H_2CO_3$$

we find a carbonic acid concentration of $c_{\text{total}} = 0.031 \cdot 1.36 \times 10^{-4}$ M $= 4.2 \times 10^{-6}$ M. Because pure water has a pH of 7 and H_2CO_3 is an acid, we should expect the pH of the solution to be reduced somewhat. Carbonic acid has two possible equilibria,

$$H_2CO_3 \rightleftharpoons H^+ + HCO_3^- \quad \text{with} \quad K_{a1} = \frac{\hat{c}_{H^+} \cdot \hat{c}_{HCO_3^-}}{\hat{c}_{H_2CO_3}} = 4.47 \times 10^{-7}$$

and

$$HCO_3^- \rightleftarrows H^+ + CO_3^{2-} \quad \text{with} \quad K_{a2} = \frac{\hat{c}_{H^+} \cdot \hat{c}_{HCO_3^{2-}}}{\hat{c}_{HCO_3^-}} = 4.68 \times 10^{-11}$$

Because the second equilibrium constant is far smaller than the first,[1] let us neglect it for the moment. From the first equilibrium we can determine the proton concentration by using Equation (22.12)

$$\hat{c}_{H^+} = -\tfrac{1}{2}K_{a1} + \sqrt{\hat{c}_{\text{total}}K_{a1} + \tfrac{1}{4}K_{a1}^2}$$

With $\hat{c}_{\text{total}} = 4.2 \times 10^{-6}$ we obtain $\hat{c}_{H^+} = 1.22 \times 10^{-6}$; this value gives a pH of 5.9, a nontrivial shift from that of pure water.

Now we use these values in the equilibrium constant expression for the second equilibrium,

$$K_{a2} = \frac{\hat{c}_{H^+} \cdot \hat{c}_{CO_3^{2-}}}{\hat{c}_{HCO_3^-}}$$

Approximating the proton and bicarbonate concentrations as equal, we find $\hat{c}_{CO_3^{2-}} \simeq K_{a2} = 10^{-10.33}$. Because the carbonate concentration is significantly smaller than the bicarbonate concentration we see that this equilibrium has little impact and can be neglected.

22.3.2 Amino Acids

In the case of an amino acid two different proton transfer reactions are possible, one at the carboxyl group and another at the amino group. The figure illustrates these equilibria for glycine.

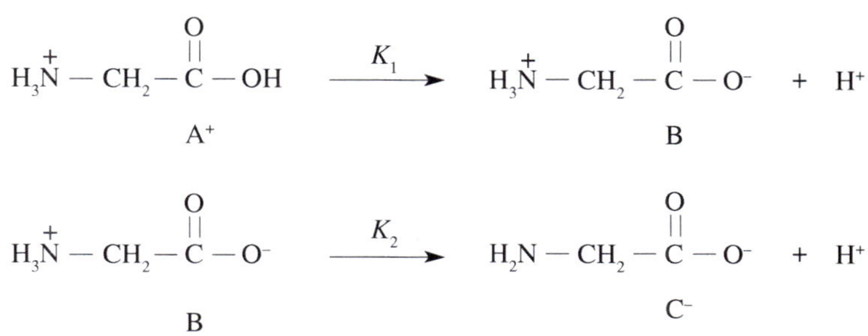

Species A^+ is stable in acidic solution, whereas species C^- is stable in alkaline solution. Species B (*zwitterion*) exists at an intermediate pH (*isoelectric point*).

1 K_a values are from [1].

What is the pH of an amino acid in water? We start with the equilibrium constants

$$K_1 = \frac{\hat{c}_{H^+} \cdot \hat{c}_B}{\hat{c}_{A^+}}, \quad K_2 = \frac{\hat{c}_{H^+} \cdot \hat{c}_{C^-}}{\hat{c}_B} \tag{22.31}$$

and include the condition for mass balance

$$c_{A^+} + c_B + c_{C^-} = c_{total} \tag{22.32}$$

From these equations, we obtain the concentrations c_{A^+}, c_B, and c_{C^-}at any pH, in analogy to the procedure for a diprotic acid. In order to calculate the titration curve of the amino acid, we use the electroneutrality condition

$$c_{H^+} + c_{A^+} + c_{Na^+} = c_{C^-} + c_{OH^-} \tag{22.33}$$

Note that this relation is different from Equation (22.29) for a diprotic acid, because A^+ is positively charged, B is uncharged, and C^- is negatively charged. Instead of Equation (22.30) we obtain

$$\boldsymbol{n}_{NaOH} = Vc^{\ominus} \left(\frac{K_w}{\hat{c}_{H^+}} - \hat{c}_{H^+} \right) + \boldsymbol{n}_{total} \cdot \frac{K_1 K_2 - \hat{c}^2_{H^+}}{\hat{c}^2_{H^+} + K_1 \hat{c}_{H^+} + K_1 K_2} \tag{22.34}$$

Figure 22.5 shows the titration curve for glycine with $K_1 = 3.98 \times 10^{-3}$ and $K_2 = 1.58 \times 10^{-10}$ ($pK_1 = 2.4$, $pK_1 = 9.8$) with $\boldsymbol{n}_{total} = 0.1$ mol and $V = 1$ L. For $\boldsymbol{n}_{NaOH} = 0$ the result for the proton concentration is pH = 6.1 (the initial pH of an aqueous solution of glycine).

For a better understanding of these equations we make two simplifying assumptions. First, we realize that the dissociation constants for Equations (22.31) are much larger than those of water and we neglect water's autoionization. Hence the proton concentration is determined primarily by the formation of C^- and A^+, so that $c_{H^+} = c_{C^-} - c_{A^+}$. Second, we realize that the predominant form is B (i.e., K_1 is about 0.01 and K_2 is small) so that only a small amount of B is converted into A^+ or C^-; i.e.,

$$c_B \simeq c_{total} \tag{22.35}$$

With these simplifications, we can recast Equations (22.31) as

$$K_1 = \frac{\hat{c}_{H^+} \cdot \hat{c}_{total}}{\hat{c}_{A^+}}, \quad K_2 = \frac{\hat{c}_{H^+} \cdot (\hat{c}_{H^+} + \hat{c}_{A^+})}{\hat{c}_{total}} \tag{22.36}$$

which simplifies to

$$(\hat{c}_{H^+})^2 = K_1 K_2 \cdot \frac{\hat{c}_{total}}{\hat{c}_{total} + K_1} \tag{22.37}$$

For sufficiently large $\hat{c}_{total}$ this result simplifies further so that

$$(\hat{c}_{H^+})^2 = K_1 K_2 \quad \text{or} \quad pH = \tfrac{1}{2}(pK_1 + pK_2) \tag{22.38}$$

Using the values for glycine, $pK_1 = 2.4$ and $pK_2 = 9.8$, we obtain

$$pH = \tfrac{1}{2}(2.4 + 9.8) = 6.1$$

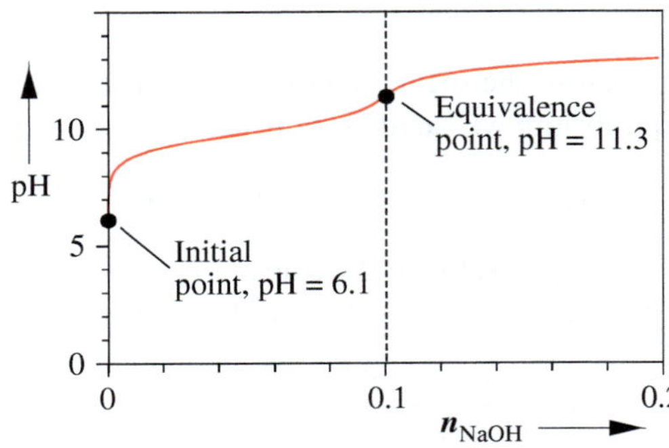

Figure 22.5 A titration curve for a solution of the amino acid glycine ($pK_{a1} = 2.4$, $pK_{a2} = 9.8$, $V = 1$ L, $\boldsymbol{n}_{total} = 0.1$ mol), calculated according to Equation (22.34).

as already discussed above.

In Problem P22.8, you show that for $c_{total} = 0.1$ mol L^{-1} the pH = 5.4 is obtained if the reaction from B to A is neglected, and the pH = 7.7 if the reaction from B to C is neglected. This behavior demonstrates the significance of the coupling of both equilibria.

22.4 Electron Transfer Reactions

Electron transfer from one reactant species to another is an important elementary chemical transformation. Examples of such reactions are found broadly in chemistry, biology, and engineering applications.

22.4.1 Electron Transfer from Metal to Proton: Dissolution of Metals in Acid

When we insert a zinc rod into a beaker with dilute acid (Fig. 22.6), the reaction

$$\tfrac{1}{2}\,\mathrm{Zn} + \mathrm{H}^+ \rightarrow \tfrac{1}{2}\,\mathrm{Zn}^{2+} + \tfrac{1}{2}\,\mathrm{H}_2 \tag{22.39}$$

occurs spontaneously under standard conditions ($\widehat{c}_{\mathrm{Zn}^{2+}} = \widehat{c}_{\mathrm{H}^+} = 1$, $\widehat{P}_{\mathrm{H}_2} = 1$), with $\Delta_r G^{\ominus}_{298} = -73.6$ kJ mol^{-1} (see Appendix A.3).[2] In this reaction the metallic zinc transfers electrons to the protons in the solution, and it is useful to think of the process as two half reactions

$$\tfrac{1}{2}\,\mathrm{Zn} \rightarrow \tfrac{1}{2}\,\mathrm{Zn}^{2+} + \mathrm{e}^-, \qquad \mathrm{e}^- + \mathrm{H}^+ \rightarrow \tfrac{1}{2}\,\mathrm{H}_2 \tag{22.40}$$

Reactions of this sort are called *electron transfer reactions* or *redox reactions*. Zn is the reductant (it donates electrons) and H^+ is the oxidant (it accepts electrons).

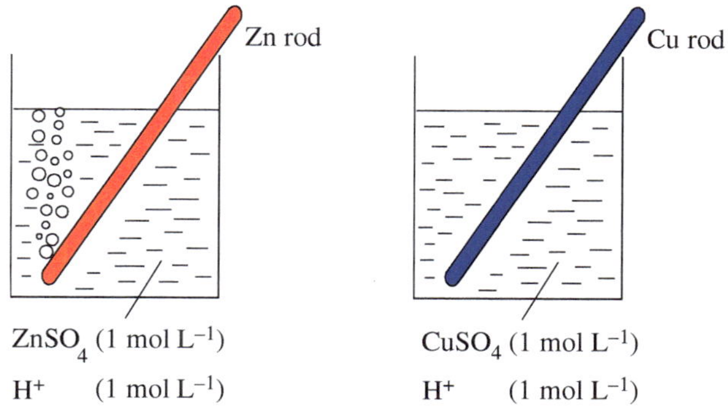

Figure 22.6 Different behavior of zinc and copper when immersed in an acidic solution of zinc sulfate and copper sulfate, respectively.

What happens if the concentrations and the pressure are different from the standard state values? According to Section 18.5 the change of Gibbs energy is

$$\Delta_r G = \Delta_r G^{\ominus} + \boldsymbol{R}T \ln Q = \Delta_r G^{\ominus} + \boldsymbol{R}T \cdot \ln\left(\frac{\left(\widehat{c}_{\mathrm{Zn}^{2+}}\right)^{1/2} \cdot \left(\widehat{P}_{\mathrm{H}_2}\right)^{1/2}}{\widehat{c}_{\mathrm{H}^+}}\right) \tag{22.41}$$

$\Delta_r G$ is the change in Gibbs energy when one-half mol Zn reacts with one mol H^+ (concentration c_{H^+}) forming one-half mol Zn^{2+} (concentration $c_{\mathrm{Zn}^{2+}}$) and one-half mol H_2 (pressure P_{H_2}).

Consider immersing the zinc rod in a large volume of a solution which is one molar in H^+ and Zn^{2+}; then

$$\widehat{c}_{\mathrm{Zn}^{2+}} = \widehat{c}_{\mathrm{H}^+} = 1 \quad \text{and} \quad \Delta_r G = \Delta_r G^{\ominus} + \boldsymbol{R}T \cdot \ln\left(\widehat{P}_{\mathrm{H}_2}\right)^{1/2}$$

As the pressure of H_2 increases, $\Delta_r G$ becomes larger; eventually $\Delta_r G$ switches from a negative value to a positive value. We can calculate the equilibrium pressure $\left(\widehat{P}_{\mathrm{H}_2}\right)_{\mathrm{eq}}$ by setting $\Delta_r G = 0$. In this case

$$0 = \Delta_r G^{\ominus} + \boldsymbol{R}T \cdot \ln\left(\widehat{P}_{\mathrm{H}_2}\right)^{1/2}_{\mathrm{eq}}, \qquad \left(\widehat{P}_{\mathrm{H}_2}\right)_{\mathrm{eq}} = 6 \times 10^{25}\ \mathrm{bar}$$

This means the reaction proceeds spontaneously unless H_2 reaches the incredible pressure of $\sim 10^{26}$ bar. Clearly, this reaction is likely to be stopped by the consumption of Zn or H^+ before it is significantly affected by the partial pressure of H_2.

Example 22.3 *Consider the reaction*

$$\tfrac{1}{2}\,Cu + H^+ \rightarrow \tfrac{1}{2}\,Cu^{2+} + \tfrac{1}{2}\,H_2 \tag{22.42}$$

with $\Delta_r G^{\ominus} = +32.5$ kJ mol^{-1}. Does this reaction occur spontaneously under standard conditions?

The positive value of $\Delta_r G^{\ominus}$ means that the reaction is not spontaneous and copper does not dissolve in water under standard conditions ($c_{\mathrm{Cu}^{2+}} = c_{\mathrm{H}^+} = 1$ mol L^{-1}, $P_{\mathrm{H}_2} = 1$ bar).

22.4.2 Electron Transfer from Metal 1 to Metal 2 Ion: Coupled Redox Reactions

Consider a zinc rod immersed in a CuSO_4 solution, for which the reaction

$$\tfrac{1}{2}\,\mathrm{Zn} + \tfrac{1}{2}\,\mathrm{Cu}^{2+} \rightarrow \tfrac{1}{2}\,\mathrm{Zn}^{2+} + \tfrac{1}{2}\,\mathrm{Cu} \tag{22.43}$$

2 The value $\Delta_r G^{\ominus}_{298} = -73.6$ kJ mol^{-1} is half the value reported in Appendix A.1 because that value corresponds to one mol Zn, whereas this reaction considers one-half mol Zn.

occurs. This reaction can be considered as the difference between the reactions (22.40) and (22.42). Then $\Delta_r G$ is the difference between

$$\Delta_r G_1 = \Delta_r G_1^\ominus + \boldsymbol{RT} \cdot \ln \left(\frac{\left(\widehat{c}_{\mathrm{Zn}^{2+}}\right)^{1/2} \cdot \left(\widehat{P}_{\mathrm{H}_2}\right)^{1/2}}{\widehat{c}_{\mathrm{H}^+}} \right) \tag{22.44}$$

and

$$\Delta_r G_2 = \Delta_r G_2^\ominus + \boldsymbol{RT} \cdot \ln \left(\frac{\left(\widehat{c}_{\mathrm{Cu}^{2+}}\right)^{1/2} \cdot \left(\widehat{P}_{\mathrm{H}_2}\right)^{1/2}}{\widehat{c}_{\mathrm{H}^+}} \right) \tag{22.45}$$

Thus

$$\Delta_r G = \Delta_r G_1 - \Delta_r G_2 = \Delta_r G_1^\ominus - \Delta_r G_2^\ominus + \boldsymbol{RT} \cdot \frac{1}{2} \ln \left(\frac{\widehat{c}_{\mathrm{Zn}^{2+}}}{\widehat{c}_{\mathrm{Cu}^{2+}}} \right) \tag{22.46}$$

During the reaction Zn^{2+} is formed and Cu^{2+} is consumed. What are the equilibrium concentrations of both species? At equilibrium ($\Delta_r G = 0$) we obtain

$$\ln \left(\frac{\widehat{c}_{\mathrm{Zn}^{2+}}}{\widehat{c}_{\mathrm{Cu}^{2+}}} \right) = -2 \frac{\Delta_r G_1^\ominus - \Delta_r G_2^\ominus}{\boldsymbol{RT}} = 85.6 \quad \text{and} \quad \frac{\widehat{c}_{\mathrm{Zn}^{2+}}}{\widehat{c}_{\mathrm{Cu}^{2+}}} = 10^{37}$$

This reaction continues until practically all the Cu^{2+} has been consumed.

Example 22.4 *The oxidation of iron (Fe) to rust (Fe_2O_3) occurs spontaneously upon exposure of iron to the air and/or moisture. To inhibit rusting of iron nails, they are often galvanized (coated with zinc). Compare the reaction Gibbs energies for oxidation of pure iron*

$$2Fe + \frac{3}{2} O_2 \rightleftarrows Fe_2O_3$$

and zinc

$$Zn + \frac{1}{2} O_2 \rightleftarrows ZnO$$

to explain how galvanizing inhibits rust formation.

Using the data in Appendix A.1 the standard Gibbs energy for iron oxidation is $\Delta_r G^\ominus_{298,\mathrm{Fe_2O_3}} = -739.9\ \mathrm{kJ\ mol^{-1}}$ and for zinc oxidation is $\Delta_r G^\ominus_{298,\mathrm{ZnO}} = -318.6\ \mathrm{kJ\ mol^{-1}}$. Relevant for the galvanized nail is the equilibrium

$$3\mathrm{Zn} + \mathrm{Fe_2O_3} \rightleftarrows 3\ \mathrm{ZnO} + 2\mathrm{Fe}$$

which has a standard Gibbs energy $\Delta_r G^\ominus_{298} = 3 \times (-318.6\ \mathrm{kJ\ mol^{-1}}) - (-739.9\ \mathrm{kJ\ mol^{-1}}) = -215.9\ \mathrm{kJ\ mol^{-1}}$ so that the zinc is oxidized and the iron not.

How does oxidizing the zinc, and not the iron, protect the nail? The advantage of oxidizing the Zn in preference to the Fe arises from the different properties of their oxides. Zinc oxide forms a compact film that adheres to the metal surface and inhibits access of water and oxygen to the underlying pure metal layers, whereas the iron oxide does not adhere well to the metal surface and allows more facile diffusion of water and oxygen to the underlying metal. In this way ZnO forms a barrier to inhibit further corrosion, whereas Fe_2O_3 does not.

22.4.3 Electron Transfer to Proton at pH 7: $\Delta G^{\ominus\prime}$

As outlined in the previous sections, $\Delta G^\ominus$ values for reactions of the general scheme

$$\text{species} + \mathrm{H}^+ \rightarrow \text{product} + \tfrac{1}{2}\,\mathrm{H}_2 \tag{22.47}$$

are useful for determining ΔG of redox reactions between different species. Such values are listed in Appendix A.3 for selected compounds. Many reactions occur in buffered solutions with constant pH; hence, it is convenient to choose a standard state for which $c_{H^+} = 10^{-7}$ mol L^{-1}; so that, $\hat{c}_{H^+} = c_{H^+}/(10^{-7}$ mol L^{-1}). Such conditions are important in technological applications (e.g., batteries) and in biochemistry where reactions occur in a buffer solution with pH = 7.

For the reaction

$$\alpha A + \beta B + ... + H^+ \rightarrow \gamma C + \delta d + \cdots + \tfrac{1}{2} H_2 \tag{22.48}$$

we can write that

$$K = \frac{\hat{c}^{\gamma}_{C,eq} \cdot \hat{c}^{\delta}_{D,eq} \cdot \dots \cdot \hat{P}^{1/2}_{H_2,eq}}{\hat{c}^{\alpha}_{A,eq} \cdot \hat{c}^{\beta}_{B,eq} \cdot \dots \cdot \hat{c}_{H^+}} = K' \cdot \frac{1}{\hat{c}_{H^+}} \tag{22.49}$$

where

$$K' = \frac{\hat{c}^{\gamma}_{C,eq} \cdot \hat{c}^{\delta}_{D,eq} \cdot \dots \cdot \hat{P}^{1/2}_{H_2,eq}}{\hat{c}^{\alpha}_{A,eq} \cdot \hat{c}^{\beta}_{B,eq} \cdot \dots} \tag{22.50}$$

Here we use the notation K' for the equilibrium constant in buffered solution at pH = 7. The corresponding molar change of the Gibbs energy is $\Delta_r G'$. Using

$$\Delta_r G^{\ominus\prime} = -\boldsymbol{R}T \cdot \ln K' \quad \text{with} \quad K' = K \cdot \hat{c}_{H^+} \tag{22.51}$$

and

$$\Delta_r G^{\ominus} = -\boldsymbol{R}T \cdot \ln K$$

we obtain

$$\Delta_r G^{\ominus\prime} = -\boldsymbol{R}T \cdot \ln K - \boldsymbol{R}T \cdot \ln \hat{c}_{H^+}$$

or

$$\Delta_r G^{\ominus\prime} = \Delta_r G^{\ominus} - \boldsymbol{R}T \cdot \ln \hat{c}_{H^+} \tag{22.52}$$

For the case of a pH = 7 solution and $T = 298$ K, $\hat{c}_{H^+} = 10^{-7}$ and we obtain

$$\Delta_r G^{\ominus\prime} = \Delta_r G^{\ominus} - \boldsymbol{R}T \cdot \ln\, 10^{-7} = \Delta_r G^{\ominus} + 40.0 \text{ kJ mol}^{-1} \tag{22.53}$$

Values of $\Delta_r G^{\ominus\prime}$ for selected compounds are included in Appendix A.3. For the reaction

$$\tfrac{1}{2}Cu + H^+ \rightarrow \tfrac{1}{2}\, Cu^{2+} + \tfrac{1}{2}\, H_2$$

no change in the sign from $\Delta_r G^{\ominus}$ to $\Delta_r G^{\ominus\prime}$occurs because $\Delta_r G^{\ominus} = 32.8$ kJ mol^{-1}; hence, $\Delta_r G^{\ominus\prime} = 72.8$ kJ mol^{-1} at 298 K. Other reactions, such as

$$\tfrac{1}{2}Cd + H^+ \rightarrow \tfrac{1}{2}\, Cd^{2+} + \tfrac{1}{2}\, H_2$$

do change sign (e.g., $\Delta_r G^{\ominus} = -38.8$ kJ mol^{-1} and $\Delta_r G^{\ominus\prime} = +1.2$ kJ mol^{-1} at 298 K). Hence we see that for $\hat{c}_{Cd^{2+}} = 1$ and $\hat{P}_{H_2} = 1$, cadmium dissolves in acid solution at $\hat{c}_{H^+} = 1$ (pH = 0) but not in a pH = 7 solution (that is, in pure water). However, sodium dissolves in water

$$Na + H^+ \rightarrow Na^+ + \tfrac{1}{2} H_2$$

because $\Delta_r G^{\ominus} = -261.6$ kJ mol^{-1} and $\Delta_r G^{\ominus\prime} = -221.6$ kJ mol^{-1} at 298 K.

22.4.4 Photoinduced Electron Transfer

An important class of electron transfer reactions are those initiated by the excitation of one of the reactants by a photon. For example, 9,10-dimethylanthracene and 1,4-dicyanobenzene are two molecules which do not react in their ground electronic states; however, when the anthracene is photoexcited the following reaction occurs.

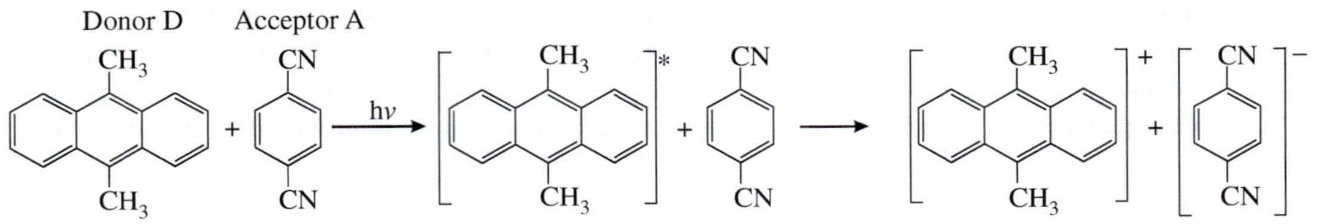

Even though this system is time-dependent (a kinetic process), it is often reasonable to apply equilibrium concepts to the process. This approximation is useful because the excited state of the anthracene molecule equilibrates rapidly with the surrounding solvent, compared to its excited state lifetime, and it is possible to approximate its energetics. The common approach to such processes was developed by Rehm and Weller [2].

Figure 22.7 shows an energy diagram for a photoinduced electron transfer reaction of D* (where D is the electron donor, e.g., the excited anthracene) and A (where A is the electron acceptor, e.g., dicyanobenzene). The Gibbs energy change for the reaction has four major contributions and is given by

$$\Delta_r G = \Delta_r G^{A/A^-} + \Delta_r G^{D/D^+} - \Delta_r G_{0,0} - \frac{\boldsymbol{N}_A e^2}{4\pi\varepsilon\varepsilon_0 R}$$

One is the Gibbs energy change for the reduction of the acceptor ($\Delta_r G^{A/A^-} = 158.2$ kJ mol^{-1} for 1,4-dicyanobenzene) and one is the Gibbs energy change for the oxidation of the donor ($\Delta_r G^{D/D^+} = 90.7$ kJ mol^{-1} for 9,10-dimethylanthracene). These two contributions provide the Gibbs energy change for the reaction in the ground electronic state. The third contribution is the energy difference between the ground electronic state and the excited electronic state ($\Delta_r G_{0,0} = \boldsymbol{N}_A \Delta E_{0,0} = 6.03 \times 10^{23} \cdot 4.97 \times 10^{-19}$ J mol^{-1} = 299.7 kJ mol^{-1} for 9,10-dimethylanthracene in acetonitrile). This term accounts for the extra destabilization of the donor by the electronic excitation. The last term accounts for the Coulomb stabilization between the monoanion and the monocation, $e^2/(4\pi\hat{\varepsilon}\varepsilon_0 R)$, when located at the distance R from one another. In the limit that R becomes large, this contribution to the energy change becomes negligible. Using the values reported above for the case of dimethylanthracene and dicyanobenzene, we find

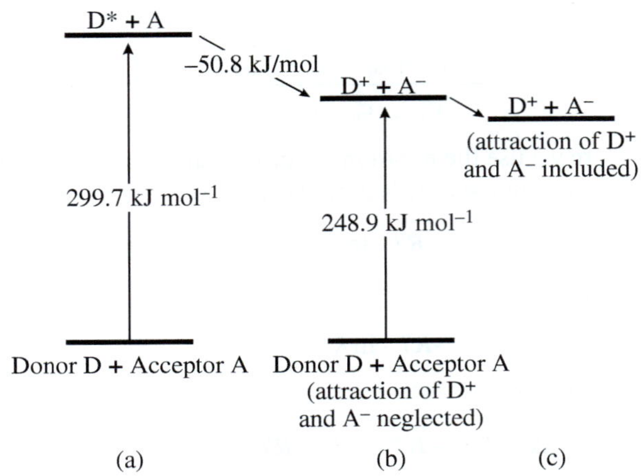

Figure 22.7 Photoinduced electron transfer from Donor D (9,10-dimethylanthracene) to acceptor A (1,4-dicyanobenzene). (a) Photoexcitation of D into D*. (b) Electron transfer reaction from the electronic ground state of D (D^+ and A^- assumed in distance $R = \infty$): $\Delta_r G$ is positive, this reaction cannot occur spontaneously. Because the energy of the excited donor in (a) is higher than the energy of $D^+ + A^-$ in (b), spontaneous electron transfer from the electronically excited donor is possible. (c) The energy of the system $D^+ + A^-$ decreases if the distance of the product molecules changes from ∞ to R.

$$\Delta_r G = (158.2 + 90.7 - 299.7)\ \text{kJ mol}^{-1} - \frac{\boldsymbol{N}_A e^2}{4\pi\hat{\varepsilon}\varepsilon_0 R} = -50.8\ \text{kJ mol}^{-1} - \frac{\boldsymbol{N}_A e^2}{4\pi\hat{\varepsilon}\varepsilon_0 R}$$

so that the photoinduced reaction is spontaneous at all values of R. For the ground-state electron transfer process the change of Gibbs energy, $\Delta_r G^{A/A^-} + \Delta_r G^{D/D^+} = +248.9$ kJ mol^{-1}, is positive.

Example 22.5 *Calculate the Gibbs energy change for the photoinduced electron transfer between 9.10-dimethylanthracene and 4′-methylacetophenone ($\Delta_r G^{A/A^-} = 211.3$ kJ mol^{-1}) at a contact distance of $R = 0.7$ nm in acetonitrile (relative permittivity $\hat{\varepsilon} = 37$). At what distance is $\Delta_r G = 0$?*

First we calculate $\Delta_r G$ for a distance R.

$$\Delta_r G = \Delta_r G^{A/A-} + \Delta_r G^{D/D+} - \Delta_r G_{0,0} - \frac{\boldsymbol{N}_A e^2}{4\pi\hat{\varepsilon}\varepsilon_0 R}$$

$$= 211.3\ \text{kJ mol}^{-1} + 90.7\ \text{kJ mol}^{-1} - 299.7\ \text{kJ mol}^{-1} - \frac{\boldsymbol{N}_A e^2}{4\pi\hat{\varepsilon}\varepsilon_0 R}$$

$$= 2.3\ \text{kJ mol}^{-1} - \frac{3.75 \times 10^{-6}\ \text{m}}{R}\ \text{J mol}^{-1}$$

For a contact distance $R = 0.7$ nm we obtain

$$\Delta_r G = 2.3 \text{ kJ mol}^{-1} - 5.36 \text{ kJ mol}^{-1} = -3.1 \text{ J mol}^{-1}$$

The distance at which $\Delta_r G = 0$ is found by

$$2.3 \text{ kJ mol}^{-1} - \frac{3.75 \times 10^{-6} \text{ m}}{R} \text{ J mol}^{-1} = 0$$

or

$$R = \frac{3.75 \times 10^{-6} \text{ m}}{2300} = 1.6 \text{ nm}$$

Hence the molecules must be within 1.5 nm, nearly touching, before the reaction is spontaneous. Although this latter calculation accounts for the change in Coulomb interaction between the ions, it is approximate and ignores possible changes in solvation of the ions with distance.

22.5 Electron Transfer Coupled with Proton Transfer

Proton coupled electron transfer reactions are an important reaction type and of current research interest. By way of example, an important step in photosynthesis involves the oxidation of a tyrosine amino acid which releases its phenol proton to a nearby base. Hammarstrom and coworkers [3] have studied a model system to better understand this process. One part of this system is the amino acid tyrosine, after esterification with ethyl alcohol:

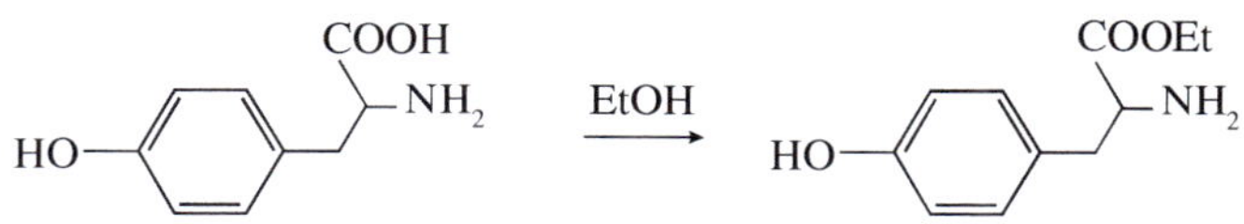

The second part of the system is the *tris*-dipyridyl complex of ruthenium, and both parts are connected by a peptide bond.

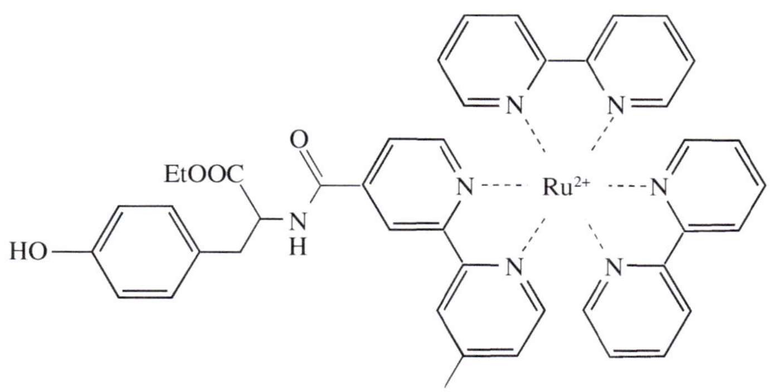

The photoexcitation of the ruthenium complex by irradiation with light of $\lambda = 450$ nm initiates an electron transfer and proton transfer reaction. In a first step an electron from the excited ruthenium complex is transferred to an external electron acceptor (not shown here), oxidizing the ruthenium into the Ru^{3+} state. In the next step the tyrosine part of the molecule donates an electron back to the ruthenium complex, and the OH group is deprotonated. Through careful kinetic studies Hammarstrom showed that this reaction proceeds by a concerted mechanism (both proton transfer and electron transfer proceed simultaneously):

$$\text{Ru}^{3+} - \text{Tyr} - \text{OH} \rightarrow \text{Ru}^{2+} - \text{Tyr} - \text{O} \cdot + \text{H}^{+}$$

In their system they can rule out that the proton transfer and electron transfer proceed sequentially, e.g.,

$$\text{Ru}^{3+} - \text{Tyr} - \text{OH} \rightarrow \text{Ru}^{3+} - \text{Tyr} - \text{O}^{-} + \text{H}^{+} \rightarrow \text{Ru}^{2+} + \text{Tyr} - \text{O} \cdot + \text{H}^{+}$$

or

$$\text{Ru}^{3+} - \text{Tyr} - \text{OH} \rightarrow \text{Ru}^{2+} - \text{Tyr} - \text{OH}^{\cdot +} \rightarrow \text{Ru}^{2+} + \text{Tyr} - \text{O} \cdot + \text{H}^{+}$$

Many electron transfer reactions coupled with proton transfer involve more than one electron and/or proton. Although these systems are less well understood in detail, it is likely that sequential reaction steps are important to consider.

An important example of a reaction involving two electrons and two proton transfers is the oxidation of hydroquinone. Consider the reaction

$$\tfrac{1}{2}\ H_2Q + Fe^{3+} \rightarrow \tfrac{1}{2}\ Q + Fe^{2+} + H^+ \qquad (22.54)$$

where H_2Q is hydroquinone and Q is quinone

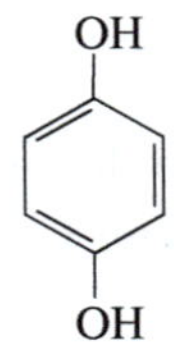

Hydroquinone H_2Q

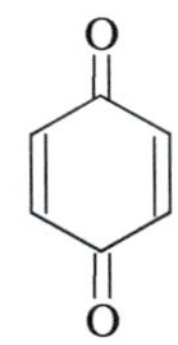

Quinone Q

Reaction (22.54) can be treated as the sum of the reactions

$$\tfrac{1}{2}\ H_2Q \rightarrow \tfrac{1}{2}\ Q + \tfrac{1}{2}\ H_2, \quad \Delta_r G_1^\ominus = 33.8\ \text{kJ mol}^{-1} \qquad (22.55)$$

and

$$Fe^{3+} + \tfrac{1}{2}\ H_2 \rightarrow Fe^{2+} + H^+, \quad \Delta_r G_2^\ominus = -74.48\ \text{kJ mol}^{-1} \qquad (22.56)$$

so that

$$\Delta_r G = \Delta_r G_1^\ominus + \Delta_r G_2^\ominus + \boldsymbol{RT} \cdot \ln \frac{(\hat{c}_Q)^{1/2} \cdot \hat{c}_{Fe^{2+}} \cdot \hat{c}_{H^+}}{\left(\hat{c}_{H_2Q}\right)^{1/2} \cdot \hat{c}_{Fe^{3+}}} \qquad (22.57)$$

At equilibrium ($\Delta_r G = 0$) we have

$$\ln \frac{\left(\hat{c}_{Q,eq}\right)^{1/2} \cdot \hat{c}_{Fe^{2+},eq} \cdot \hat{c}_{H^+,eq}}{\left(\hat{c}_{H_2Q,eq}\right)^{1/2} \cdot \hat{c}_{Fe^{3+},eq}} = -\frac{\Delta_r G_1^\ominus + \Delta_r G_2^\ominus}{\boldsymbol{RT}} = 16.3, \quad K = e^{16.3} = 1.2 \times 10^7$$

In this case the ratio $c_{Fe^{2+}}/c_{Fe^{3+}}$ depends strongly on the acidity of the solution. For example, with $\hat{c}_Q = \hat{c}_{H_2Q}$ and $\hat{c}_{H^+} = 1$ we obtain $c_{Fe^{2+}}/c_{Fe^{3+}} = 1.2 \times 10^7$; for $\hat{c}_Q = \hat{c}_{H_2Q}$ and $\hat{c}_{H^+} = 10^{-7}$ (neutral solution) the corresponding value is 1.3×10^{14}.

Example 22.6 *The reaction*

$$2FeS_2 + \frac{15}{2}O_2 + H_2O \rightleftarrows 2Fe^{3+} + 4SO_4^{2-} + 2H^+$$

often occurs in abandoned coal mines (iron sulfide ores (e.g., pyrite) are often found near coal seams) and generate acid water drainage that damages ecosystems. At what pH does the reaction stop, if the other components are present in standard concentrations?

This overall reaction involves multiple electron and proton transfer. Here we consider it as a sum of the two reactions, an electron transfer reaction involving a change in the iron oxidation

$$2Fe^{2+} + \frac{1}{2}O_2 + 2H^+ \rightleftarrows 2Fe^{3+} + H_2O, \ \Delta_r G_{A,298}^\ominus = -325.8\ \text{kJ mol}^{-1}$$

and an acid–base reaction involving the pyrite, in which the sulfur is oxidized

$$2FeS_2 + 7O_2 + 2H_2O \rightleftarrows 2Fe^{2+} + 4SO_4^{2-} + 4H^+\ ,\ \Delta_r G_{B,298}^\ominus = -2227.8\ \text{kJ mol}^{-1}$$

The reaction stops being spontaneous when

$$\ln \frac{\hat{c}^4_{SO_4^{2-}} \cdot \hat{c}^2_{Fe^{3+}} \cdot \hat{c}^2_{H^+}}{\hat{P}^{15/2}_{O_2}} = -\frac{\Delta_r G_{A,298}^\ominus + \Delta_r G_{B,298}^\ominus}{\boldsymbol{RT}}$$

For standard concentrations of SO_4^{2-}, Fe^{3+}, and standard pressure of oxygen we find that

$$\widehat{c}_{H^+} = \exp\left(-\frac{\Delta_r G_A^\ominus + \Delta_r G_B^\ominus}{2\boldsymbol{R}T}\right) = \exp\left(\frac{-2553.6}{4955.1}\right) = \exp(-0.515) = 0.59$$

That is, the reaction stops at a hydrogen concentration of 0.59 mol L^{-1}, corresponding to pH = 2.2.

22.6 Group Transfer Reactions in Biochemistry

A similar reaction motif, in which a chemical unit is transferred between reactants, is important in biochemistry. Let us consider the reaction of adenosine triphosphate (ATP) to yield adenosine diphosphate (ADP) and phosphoric acid:

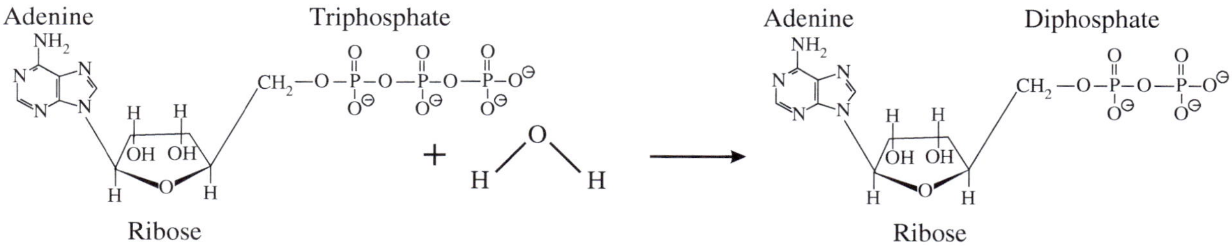

$$ATP^{4-} + H_2O \rightarrow ADP^{3-} + HPO_4^{2-} + H^+ \tag{22.58}$$

This reaction equilibrium is used to store energy in biological systems and later extract it for useful processes.

We perform the reaction in a buffer solution at pH 7 and use the shorter notation

$$ATP \rightarrow ADP + P \tag{22.59}$$

According to (22.51) the Gibbs energy $\Delta G^{\ominus\prime}$ is

$$\Delta_r G^{\ominus\prime} = -\boldsymbol{R}T \cdot \ln K' \tag{22.60}$$

and the equilibrium constant is

$$K' = \frac{\widehat{c}_{ADP,eq} \cdot \widehat{c}_{P,eq}}{\widehat{c}_{ATP,eq}} \tag{22.61}$$

$\Delta_r G^{\ominus\prime}$ is the molar change in Gibbs energy of the reaction under standard physiological conditions ($c_{ATP} = c_{ADP} = c_P =$ 1 M; species separated at pH= 7 in buffer solution). With $\Delta_r G^{\ominus\prime} = -30.5$ kJ mol^{-1} we find

$$K' = 2.22 \times 10^5$$

The equilibrium (22.59) is completely shifted to the right for $\widehat{c}_P = 1$; therefore, ATP is called an energy-rich compound.

$\Delta_r G^{\ominus\prime}$ for the hydrolysis of selected substances important in biochemistry at pH 7 are listed in Appendix A.4. Using these values, one can predict whether or not biochemical reactions can occur spontaneously.

Example 22.7 *Because phosphorylated compounds are important intermediates in biochemical syntheses and degradation processes, it is interesting to know whether a substance can be phosphorylated by ATP. Let us ask whether the reaction*

$$\textit{ATP} + \textit{glucose} \rightarrow \textit{ADP} + \textit{glucose-6-phosphate} \tag{22.62}$$

can occur spontaneously in aqueous solution at pH 7.

To calculate $\Delta G^{\ominus\prime}$ for this reaction we combine reaction (22.59) with the reaction (P = phosphate)

$$\text{glucose-6-phosphate} + H_2O \rightarrow \text{glucose} + P \tag{22.63}$$

($\Delta_r G^{\ominus\prime} = -13.8$ kJ mol^{-1}). Hence the net reaction (22.62) has a Gibbs energy of

$$\Delta_r G^{\ominus\prime} = (-30.5 + 13.8)\ \text{kJ mol}^{-1} = -16.7\ \text{kJ mol}^{-1}$$

and occurs spontaneously under standard conditions. The equilibrium constant is

$$K' = \mathrm{e}^{-\Delta_r G^{\ominus\prime}/RT} = 845$$

In the reaction of Example 22.7 phosphate group is transferred from ATP to glucose; such reactions are called *group transfer reactions*.

22.6.1 Group Transfer Potential

Other compounds, besides ATP, can transfer phosphate. The larger the value of $-\Delta_r G^{\ominus\prime}$ for the cleavage of the phosphate group, the higher the thermodynamic driving force for group transfer to other species. Therefore $-\Delta_r G^{\ominus\prime}$ is called the *group transfer potential*. The group transfer potential of ATP is $-\Delta_r G^{\ominus\prime} = 30.5$ kJ mol^{-1}.

By way of example, consider the hydrolysis of acetylphosphate

$$\text{acetylphosphate} + H_2O \rightarrow \text{acetic acid} + \text{phosphate}$$

for which $\Delta_r G^{\ominus\prime} = -43.9$ kJ mol^{-1}. Hence the group transfer potential of acetylphosphate is 43.9 kJ mol^{-1}. Because its group transfer potential is larger than that of ATP, acetylphosphate can be used to convert ADP to ATP.

22.6.2 Coupled Reactions in Biology

Enzymes can couple specific reactions for which $\Delta G > 0$ with other reactions for which $\Delta G < 0$. Consider the oxidation of 3-phosphoglycerolaldehyde R-CHO to phosphoglyceric acid R–COOH, which is a reaction that plays an important role in the degradation of carbohydrates. Direct combustion of the 3-phosphoglycerolaldehyde with O_2

$$R - CHO + \frac{1}{2}O_2 \rightarrow R - COOH \tag{22.64}$$

results in a large negative value of $\Delta_r G^{\ominus\prime}$ (-263 kJ mol^{-1}) and release of heat. In the biological process, most of the Gibbs energy is kept in the biological system rather than being released to the environment as heat. This retention is realized by coupling the oxidation process (aldehyde to carboxyl) to a reduction process, namely the formation of NADH from NAD^+ ($\Delta_r G^{\ominus\prime} = 220$ kJ mol $^{-1}$). In the same process, ATP is formed from ADP, the reverse of reactions (22.58) and (22.59), with $\Delta_r G^{\ominus\prime} = 30.5$ kJ mol^{-1}:

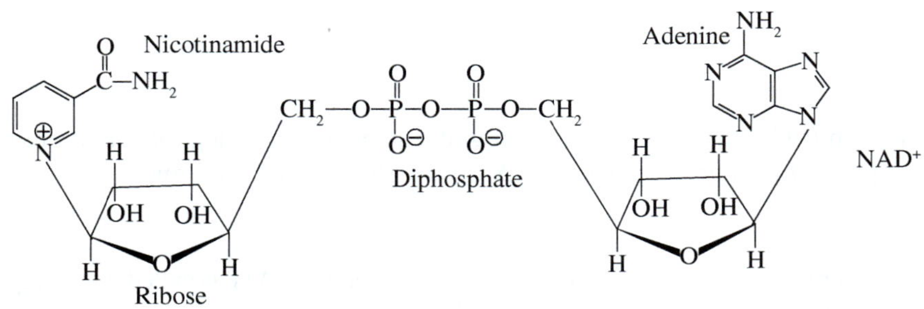

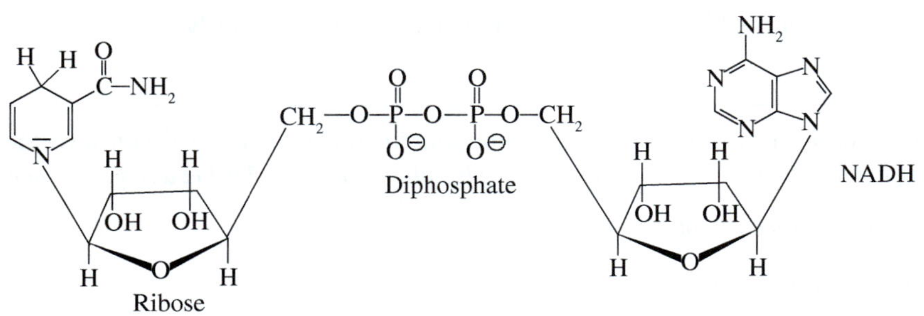

$$NAD^+ + H_2O \rightarrow NADH + H^+ + \tfrac{1}{2}\,O_2 \tag{22.65}$$

$$ADP + P \rightarrow ATP \tag{22.66}$$

The overall enzymatic conversion is the sum of reactions (22.64), (22.65), and (22.66):

$$\text{R-CHO} + \text{NAD}^+ + \text{H}_2\text{O} + \text{ADP} + \text{PO}_4 \rightarrow \text{R-COOH} + \text{NADH} + \text{ATP} + \text{H}^+ \tag{22.67}$$

Then $\Delta_r G^{\ominus\prime}$ for reaction (22.67) is

$$\Delta_r G^{\ominus\prime} = (-263 + 220 + 30.5\ )\ \text{kJ mol}^{-1} = -13\ \text{kJ mol}^{-1}$$

The first enzyme enables oxidation of the aldehyde to the acid, and subsequent phosphorylation of the acid in position 1. A second enzyme transfers a phosphate group to ADP, and ATP is formed.

The net Gibbs energy decreases by much less than in reaction (22.64). The products NADH and ATP serve the organism in many other reactions, NADH as a reducing agent and ATP as an energy carrier. The processes that define the biochemical energy balance are based on such couplings of reactions in which the Gibbs energy decreases only slightly in each reaction step.

22.7 Bioenergetics

In biological systems four types of polymers (proteins, nucleic acids, polysaccharides, and lipids) must be synthesized. The formation of a biopolymer from monomers is accompanied by an increase in the molecule's Gibbs energy, by 15 to 20 kJ mol^{-1} per bond. For example, the formation of a peptide bond between alanine and glycine requires a $\Delta_r G^{\ominus\prime} =$ 17.3 kJ mol^{-1} (Section 18.6.6). Also the synthesis of lipids from sugar requires energy.

The monomer units described above are complex molecules, however. The primary compounds that biosystems can use for biosynthesis are mostly CO_2, H_2O, N_2, and inorganic phosphate. Hence the monomer units must be constructed as well. Here we describe the energetic considerations for the formation and degradation of the important molecule, glucose. Glucose can be used as a monomer unit in the synthesis of biopolymers, and it is also used to store chemical energy in biosystems.

The Gibbs energy suppliers in biosystems are 1) NADPH formed from NADP$^+$ and 2) ATP formed from ADP and phosphate, with the help of sunlight. NADPH is used for the reduction of CO_2 (and of N_2 in some bacteria), for which ATP is needed as an additional supplier of energy. Energy is stored in the form of glucose and can be used on demand by formation of ATP from ADP.

22.7.1 Synthesis of Glucose

The production of glucose from CO_2,

$$6\ \text{CO}_2 + 6\ \text{H}_2\text{O} \rightarrow \text{C}_6\text{H}_{12}\text{O}_6 + 6\ \text{O}_2 \tag{22.68}$$

has a $\Delta_r G^{\ominus\prime} = 2872$ kJ mol^{-1} and does not occur spontaneously. Energy must be supplied to construct glucose from carbon dioxide and water.

22.7.1.1 Energy Supplied by NADPH and ATP

In biological organisms the net reaction (22.68) occurs by coupling it to reactions involving NADPH and ATP. By coupling the reaction

$$12\ \text{NADPH} + 12\ \text{H}^+ + 6\ \text{O}_2 \rightarrow 12\ \text{NADP}^+ + 12\ \text{H}_2\text{O} \tag{22.69}$$

($\Delta_r G^{\ominus\prime} = -2640$ kJ mol^{-1}, see Table A.3) with reaction (22.68) with $\Delta_r G^{\ominus\prime} = 2872$ kJ mol^{-1}, we find

$$6\ \text{CO}_2 + 12\ \text{NADPH} + 12\ \text{H}^+ \rightarrow \text{C}_6\text{H}_{12}\text{O}_6 + 12\ \text{NADP}^+ + 6\ \text{H}_2\text{O} \tag{22.70}$$

for which

$$\Delta_r G^{\ominus\prime} = (2872 - 2640)\ \text{kJ mol}^{-1} = 232\ \text{kJ mol}^{-1}$$

Reaction (22.70) does not occur spontaneously under standard conditions. Indeed, the reaction is driven by converting 18 ATP molecules in 18 enzymatic reactions to ADP and phosphate, transferring their energy to intermediates. According to Section 22.6, $\Delta_r G^{\ominus\prime}$ for this reaction sequence is

$$\Delta_r G^{\ominus\prime} = 18 \cdot (-30.5)\ \text{kJ mol}^{-1} = -549\ \text{kJ mol}^{-1}$$

For the total reaction

$$6\,CO_2 + 12\,NADPH + 18\,ATP + 12\,H^+ \rightarrow C_6H_{12}O_6 + 12\,NADP^+ + 18\,ADP + 18\,P + 6\,H_2O \qquad (22.71)$$

we obtain

$$\Delta_r G^{\ominus\prime} = (232 - 549)\ \text{kJ mol}^{-1} = -317\ \text{kJ mol}^{-1}$$

This process can occur spontaneously.

22.7.1.2 Sunlight-Induced NADPH and ATP Production

The synthesis of ATP and NADPH takes place in the photosynthetic system of plants. For the formation of 18 ATP and of twelve NADPH molecules ($\Delta_r G^{\ominus\prime} = 3189$ kJ mol^{-1}) about 50 light quanta have to be absorbed. Each light quantum has an energy of about 2.8×10^{-19} J (corresponding to the absorption maximum of chlorophyll at 700 nm), so that 50 light quanta supply about 140×10^{-19} J per glucose molecule (corresponding to 8500 kJ mol^{-1}). Only one-third of this amount is used to increase the Gibbs energy of the system. The low energy yield in the primary process of the photosynthesis arises from the difficulty of keeping the electrical charges separated (see Chapter 28).

22.7.2 Combustion of Glucose

The combustion of glucose with O_2 to produce CO_2 and H_2O does not proceed in one step ($\Delta_r G^{\ominus\prime} = -2872$ kJ mol^{-1}), but in many small steps which are coupled, each with a small decrease in $\Delta_r G$. We have considered one of these steps in the previous section (the oxidation of phosphoglycerolaldehyde to 3-phosphoglyceric acid (22.67)). Through enzymatic coupling with reactions in which ATP is synthesized from ADP and phosphate, about half of the Gibbs energy is recovered. Note that the Gibbs energy initially available from the glucose (2872 kJ mol^{-1}) would be sufficient for the formation of ninety-four molecules of ATP from ADP and phosphate; however, only 38 molecules are formed.

22.7.3 Energy Balance of Formation and Degradation (Combustion) of Glucose

Figure 22.8 shows a schematic illustration of the energy conversion processes in the formation and degradation of glucose. To initiate the formation of a molecule of glucose in plants about 50 light quanta are required, corresponding to 8500 kJ mol^{-1} (see Section 22.7.1.2): Fig. 22.8, left. The system transforms from state 1 (ground state) to the photo-excited state 2.

In step 2 → 3 18 ATP molecules from 18 ADP and 12 NADPH molecules from 12 $NADP^+$(reactants in the photo-excited state) are formed. The Gibbs energy required to form the same products from the reactants in their ground state is $\Delta_r G^{\ominus\prime} = (18 \times 30.5 + 2640)$ kJ mol^{-1} = 3189 kJ mol^{-1}, see Equations (22.59) and (22.69). In this step the Gibbs energy of the system decreases by (8500 – 3189) kJ mol^{-1} = 5311 kJ mol^{-1} (loss of Gibbs energy by transfer into heat).

In step 3 → 4 glucose is formed (as an energy reservoir) from 12 NADPH and 18 ATP molecules. The Gibbs energy required to form glucose from 6 CO_2 and 6 H_2O is $\Delta_r G^{\ominus\prime} = 2872$ kJ mol^{-1}, see Section 22.7.1. In this step the loss of Gibbs energy of the system is (3189 – 2872) kJ mol^{-1} = 317 kJ mol^{-1}

In step 4 → 5 one molecule of glucose is used to form 38 molecules ATP (combustion of glucose). The Gibbs energy required to form 38 ATP from 38 ADP molecules is $\Delta_r G^{\ominus\prime} = 38 \times 30.5$ kJ mol^{-1} = 1159 kJ mol^{-1}. In this step the loss of Gibbs energy of the system is (2872 – 1159) kJ mol^{-1} = 1713 kJ mol^{-1}.

In step 5 → 1 the 38 ATP molecules formed in step 4 → 5 (Gibbs energy $\Delta_r G^{\ominus\prime} = 1159$ kJ mol^{-1}) are used for driving bioreactions. For details see [4].

22.8 Conclusion

Thermodynamic considerations were discussed for chemical equilibria of two fundamental reaction types, proton transfer and electron transfer. These reaction types are found in a wide range of chemical processes: inorganic, organic, and biological. Illustrations of these reaction types were discussed for some important biological processes. The considerations discussed here, for electron transfer, proton transfer, and group transfer, can be adapted to describe any homogeneous equilibria.

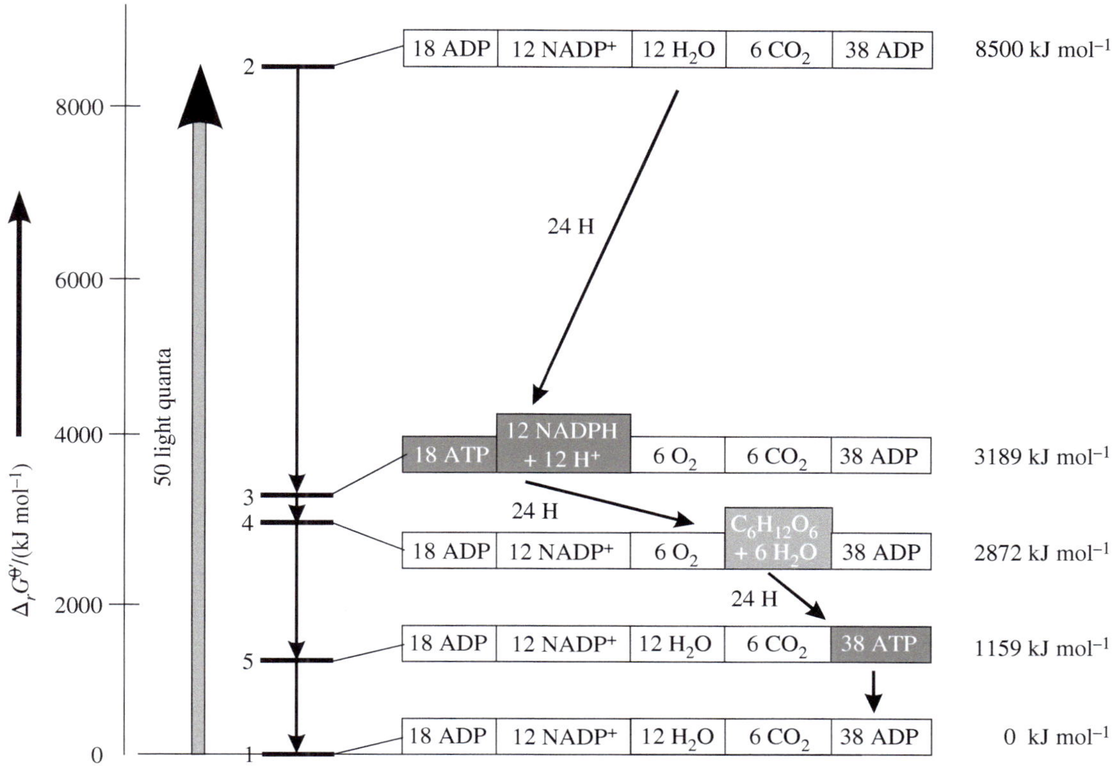

Figure 22.8 Biosynthesis of glucose and its degradation. Transfer of the absorbed light energy (step 1 → 2) in chemical energy (ATP, NADPH, step 2 → 3). Formation of glucose as an energy reservoir (step 3 → 4). Degradation of glucose forming ATP (step 4 → 5). ATP (supplier of energy for the formation of biopolymers and maintenance of life processes) is converted into ADP (step 5 → 1). Shaded areas: high-energy compounds occurring in the sequence of reactions.

Processes in complex systems, such as biological organisms and the environment, involve coupled equilibria and reaction sequences. Biological systems have evolved to perform coupled sets of chemical transformations in a manner useful to the organism.

22.9 Foundations

22.9.1 Foundation 22.1: Titration of Acetic Acid by NaOH

Combining the equilibrium expression

$$K_a = \frac{\hat{c}_{A^-} \cdot \hat{c}_{H^+}}{\hat{c}_{HA}} \tag{22F1.1}$$

and the conservation of mass

$$\hat{c}_{total} = \hat{c}_{A^-} + \hat{c}_{HA} \tag{22F1.2}$$

results in

$$K_a = \hat{c}_{H^+} \cdot \frac{\hat{c}_{A^-}}{\hat{c}_{total} - \hat{c}_{A^-}} \tag{22F1.3}$$

and for the concentration c_{A^-} we obtain

$$\hat{c}_{A^-} = \hat{c}_{total} \frac{K_a}{\hat{c}_{H^+} + K_a} \tag{22F1.4}$$

From the condition of electroneutrality

$$c_{H^+} + c_{Na^+} = c_{OH^-} + c_{A^-} \tag{22F1.5}$$

and the dissociation equilibrium of water

$$K_w = \hat{c}_{H^+} \cdot \hat{c}_{OH^-} \tag{22F1.6}$$

we obtain

$$\hat{c}_{H^+} + \hat{c}_{Na^+} = \hat{c}_{OH^-} + \hat{c}_{A^-} = \frac{K_w}{\hat{c}_{H^+}} + \hat{c}_{A^-} \tag{22F1.7}$$

or

$$\hat{c}_{H^+} + \hat{c}_{Na^+} = \frac{K_w}{\hat{c}_{H^+}} + \hat{c}_{total}\frac{K_a}{\hat{c}_{H^+} + K_a} \tag{22F1.8}$$

This equation is cubic in the proton concentration $\hat{c}_{H^+}$. Therefore we proceed by solving for c_{Na^+}. We obtain

$$\hat{c}_{Na^+} = \frac{K_w}{\hat{c}_{H^+}} - \hat{c}_{H^+} + \hat{c}_{total}\frac{K_a}{\hat{c}_{H^+} + K_a} \tag{22F1.9}$$

At the beginning of the titration, $\hat{c}_{Na^+} = 0$, thus:

$$0 = \frac{K_w}{\hat{c}_{H^+}} - \hat{c}_{H^+} + \hat{c}_{total}\frac{K_a}{\hat{c}_{H^+} + K_a} \tag{22F1.10}$$

This is an equation for the proton concentration of the pure weak acid dissolved in water. Usually, in this case $(K_w/\hat{c}_{H^+}) \ll \hat{c}_{H^+}$, and we obtain

$$\hat{c}_{H^+} = -\frac{1}{2}K_a + \sqrt{\hat{c}_{total}K_a + \frac{1}{4}K_a^2} \tag{22F1.11}$$

In the course of the titration $\hat{c}_{Na^+}$ increases. $\boldsymbol{n}_{NaOH}$, the amount of added NaOH, equals the amount of Na^+ ions in the solution:

$$\boldsymbol{n}_{NaOH} = \boldsymbol{n}_{Na^+} = Vc_{Na^+} = Vc^{\ominus}\hat{c}_{Na^+} \tag{22F1.12}$$

with V= volume of the solution and $c^{\ominus} = 1\ \text{mol L}^{-1}$ (standard concentration). Then with $\boldsymbol{n}_{total} = c_{total}V = \hat{c}_{total}Vc^{\ominus}$ ($\boldsymbol{n}_{total}$ = total amount of acid at the start of the titration)

$$\boxed{\boldsymbol{n}_{NaOH} = Vc^{\ominus}\left(\frac{K_w}{\hat{c}_{H^+}} - \hat{c}_{H^+}\right) + \boldsymbol{n}_{total}\frac{K_a}{\hat{c}_{H^+} + K_a} = AV + B} \tag{22F1.13}$$

with

$$\boxed{A = \left(\frac{K_w}{\hat{c}_{H^+}} - \hat{c}_{H^+}\right)c^{\ominus} \qquad B = \boldsymbol{n}_{total}\frac{K_a}{\hat{c}_{H^+} + K_a}} \tag{22F1.14}$$

In Table 22B.1 $\boldsymbol{n}_{NaOH}$ is displayed for different values of $\hat{c}_{H^+}$ for acetic acid as an example (dissociation constant $K_a = 2.4 \times 10^{-5}$, initial amount of acetic acid $\boldsymbol{n}_{total} = 1$ mol). It is assumed that the volume V ($V = 1$ L) of the solution does not change significantly during the titration (this is possible by using NaOH in a high enough concentration). By plotting pH versus $\boldsymbol{n}_{NaOH}$ the titration curve in Fig. 22.3 (solid line) is obtained.

At the beginning of the titration ($\boldsymbol{n}_{NaOH} = 0$), $\hat{c}_{H^+}$ is much larger than $K_w/\hat{c}_{H^+}$, and $\hat{c}_{H^+}$ is much larger than K_a:

$$\boldsymbol{n}_{NaOH} \approx \frac{K_a \cdot \boldsymbol{n}_{total}}{\hat{c}_{H^+}} - Vc^{\ominus}\hat{c}_{H^+} = 0$$

Then

$$\hat{c}_{H^+} = \sqrt{K_a\hat{c}_{total}} \quad \text{and} \quad \text{pH} = 2.31 \text{ for } \hat{c}_{total} = 1$$

For $\boldsymbol{n}_{NaOH} = 0.5$ mol (one-half of the acid has been titrated) we can neglect the term AV; then

$$\frac{\boldsymbol{n}_{NaOH}}{\boldsymbol{n}_{total}} \approx \frac{K_a}{\hat{c}_{H^+} + K_a} = \frac{1}{2} \quad \text{and} \quad \hat{c}_{H^+} = K_a,\ \text{pH} = 4.62$$

Table 22B.1 Titration of Acetic Acid ($\hat{c}_{\text{total}} = 1$, $n_{\text{total}} = 1$ mol, $V = 1$ L) with NaOH.[3]

pH	$\hat{c}_{H^+}$	*B*/ mol	*A*/mol	n_{NaOH}/mol
2.31	4.9×10^{-3}	0.005	4.9×10^{-3}	0.000
4	1.0×10^{-4}	0.193	1.0×10^{-4}	0.193
4.62	2.3×10^{-5}	0.500	2.4×10^{-5}	0.500
5	1.0×10^{-5}	0.705	1.0×10^{-5}	0.705
9.31	4.9×10^{-10}	1.000	-1.0×10^{-5}	1.000
11	1.0×10^{-11}	1.000	-1.0×10^{-3}	1.001
12	1.0×10^{-12}	1.000	-1.0×10^{-2}	1.010
13	1.0×10^{-13}	1.000	-1.0×10^{-1}	1.100
13.75	1.8×10^{-14}	1.000	−0.555	1.555
14	1.0×10^{-14}	1.000	−1.000	2.000

At the equivalence point ($n_{\text{NaOH}} = 1.0$ mol) with $n_{\text{total}} = 1$ mol, $V = 1$ L we obtain

$$1 = \frac{K_a}{\hat{c}_{H^+} + K_a} - \hat{c}_{H^+} + \frac{K_w}{\hat{c}_{H^+}}$$

$$\frac{\hat{c}_{H^+}}{\hat{c}_{H^+} + K_a} = -\hat{c}_{H^+} + \frac{K_w}{\hat{c}_{H^+}}$$

With the approximation $\hat{c}_{H^+} \ll K_a$ and $\hat{c}_{H^+} \ll K_w/\hat{c}_{H^+}$ we calculate

$$\text{pH} = \frac{1}{2}\left(\text{p}K_w + \text{p}K_a\right) = 9.31.$$

Finally, for $n_{\text{NaOH}} = 2.0$ mol there is an excess of NaOH by 1 mol L^{-1}, thus pH = 14.

22.9.1.1 Effect of Increasing Volume

Actually, the volume V of the solution increases during the titration:

$$V = V_0 + n_{\text{NaOH}} \cdot \frac{1}{c_{\text{NaOH}}} \tag{22F1.15}$$

(V_0 = starting volume) and we obtain

$$n_{\text{NaOH}} = AV + B = A\left(V_0 + n_{\text{NaOH}} \cdot \frac{1}{c_{\text{NaOH}}}\right) + B \tag{22F1.16}$$

and

$$n_{\text{NaOH}} = \frac{AV_0 + B}{1 - A/c_{\text{NaOH}}} \tag{22F1.17}$$

The corresponding values of n_{NaOH} are plotted in Fig. 22.3 (dashed line); the effect of dilution is a less steep increase of pH at the end of the titration.

22.9.2 Foundation 22.2: Two Coupled Chemical Equilibria

See the books website, www.wiley.com/go/Kuhn/physicalchemistry3e.

3 Calculation of added amounts n_{NaOH} for various values of pH using $K_a = 2.4 \times 10^{-5}$ and $K_w = 10^{-14}$. n_{NaOH} in column 5 is calculated for constant volume.

22.10 Exercises

E22.1 The pK_a of benzoic acid in water at 298 K is 4.213. Calculate the pH of a 1:1 mixture of benzoic acid (C_6H_5COOH, HA) and benzoate ion ($C_6H_5COO^-$, A^-). Use the pK_a to calculate the Gibbs energy change for the acid ionization reaction.

E22.2 Determine the pK_a of lactic acid by using Gibbs energies of formation data and compare it to the experimental value of 3.86 at $T = 298$ K. Calculate the pH of a 1:1 mixture of lactic acid ($CH_3C(OH)HCOOH$) and lactate ion ($CH_3C(OH)HCOO^-$).

E22.3 Calculate the pH of acetic acid CH_3COOH ($K_a = 2.5 \times 10^{-5}$), a weak acid, at different total concentrations of 0.010 mol L^{-1}, 1.0×10^{-3} mol L^{-1}, and 1.0×10^{-4} mol L^{-1}.

E22.4 Test the approximation about complete dissociation of the acid HCl by calculating its acid dissociation constants (using Gibbs energies of formation) and comparing it to K_w. Contrast your result to that of acetic acid with its pK_a of 4.62.

E22.5 In Section 22.3.2 the pH of amino acid solutions is calculated. Equation (22.37) was simplified to Equation (22.38) by considering large enough c_{total}. Can the dissociation of water and K_1 be neglected?

E22.6 In Section 22.3.2 the pH of a glycine solution is calculated. First calculate the pH by assuming that the reaction from B to A^+ does not happen. Second, calculate the pH by assuming that the reaction from B to C^- does not take place.

E22.7 The pK_a values for the dissociation of H_2O and D_2O at two different temperatures are:

T(K)	pK_a(H_2O)	pK_a(D_2O)
283	14.528	15.527
313	13.542	14.462

Determine $\Delta_r H$ and $\Delta_r S$ for the autoionization of H_2O and D_2O. Provide a molecular level explanation for any differences you find between the enthalpies and entropies.
Provide a molecular level explanation for any differences you find between the enthalpies and entropies.

E22.8 In analogy to K_a

$$HA \rightleftarrows H^+ + A^- \quad \text{for which } K_a = \frac{\hat{c}_{A^-}\,\hat{c}_{H^+}}{\hat{c}_{HA}},$$

we can define an equilibrium constant K_b for the reaction of an acid's conjugate base

$$H_2O + A^- \rightleftarrows HA + OH^- \quad \text{for which } K_b = \frac{\hat{c}_{HA}\,\hat{c}_{OH^-}}{\hat{c}_{A^-}}$$

Show that $K_a K_b = K_w$.

E22.9 Using the information in E22.1 and E22.2, write the equilibria for benzoate, the conjugate base of benzoic acid, and lactate, the conjugate base of lactic acid. Determine their pK_b values.

E22.10 Calculate the pH of an equimolar mixture of NH_4NO_3 and NH_3 which acts as a basic buffer. What relative concentrations are needed to generate a pH of 9.50? Use a value of 9.25 for the pKa of ammonia at 298 K.

E22.11 Human blood plasma is a complex mixture of many important biochemicals; however, it also contains simple buffering compounds (HCO_3^-/H_2CO_3 and $HPO_4^{2-}/H_2PO_4^-$) that maintain its pH at 7.4. Calculate the ratio of concentrations of HCO_3^-/H_2CO_3 that create a pH of 7.4. Also perform the calculation for $HPO_4^{2-}/H_2PO_4^-$. Use pK_a values of 6.35 and 10.33 for carbonic acid at 298 K, and use pK_a values of 2.16, 7.21, and 12.32 for phosphoric acid.

E22.12 Perform the Gibbs energy calculation for the reaction of Equation (22.39) at a pH of 7 rather than the 1 molar solution.

E22.13 The amino acid alanine has a pK_a of 2.3 for its carboxylic acid group. What is the pH of a solution that is 0.1 molar HCl and 0.1 molar alanine?

E22.14 The amino acid alanine (NH_2-$CH(CH_3)CO_2$ H) has a p$K_\mathbf{a}$ of 9.9 for its amino group and a pK_a of 2.3 for its acid group. What is the pH of a solution that is 0.1 molar NaOH and 0.1 molar alanine?

E22.15 In Foundation 22.2 we derived equations for the concentrations c_{H_2A}, c_{HA^-}, and $c_{A^{2-}}$ for the coupled equilibria in Equation (22.24). Show that these equations simplify to Equation (22.17) for $K_{a2} = 0$.

E22.16 Determine $\Delta_r G$ for the reaction

$$NADH \rightarrow NAD^+ + H^+ + 2e^-$$

under standard conditions at pH = 7, at pH = 5, and at pH = 9. Use $\Delta_r G^\ominus = 21.8$ kJ mol^{-1} for the standard conditions of $\hat{c}_{H^+} = 1$ (pH = 0).

E22.17 Use Equations (22.15) and (22.16) to derive an expression for $\hat{c}_{HA}$ in terms of $\hat{c}_{total}$, $\hat{c}_{H^+}$, and K_a. Use your result to rationalize Fig. 22.2.

E22.18 Use the half reactions

$$I_2 + 2e^- \rightarrow 2I^- \qquad E^\ominus = 0.6197 \text{ V}$$
$$I_3^- + 2e^- \rightarrow 3I^- \qquad E^\ominus = 0.5355 \text{ V}$$

to determine ΔG and K for the equilibrium

$$I_2 + I^- \rightarrow I_3^-$$

Is the net reaction a redox reaction? If so, can you identify the reductant and oxidant?

E22.19 The formation of a peptide bond is an essential step in protein synthesis. What is the ΔG for the reaction

$$\text{alanine} + \text{glycine} \rightarrow \text{alylglycine}$$

under standard aqueous conditions? What is the minimum number of ATP molecules that must be converted into ADP molecules to provide the necessary Gibbs energy to drive this reaction?

E22.20 Calculate the Gibbs energy change for the reactions

$$\text{a}: \ 2\,H_2O \rightarrow O_2 + 2H_2$$
$$\text{b}: \ CH_3COOH + Zn + H_2 \rightarrow CH_3CHO + Zn^{2+} + H_2O$$
$$\text{c}: \ 2\,H_2O + 2Li \rightarrow 2\,OH^- + H_2 + 2\,Li^+$$

Determine whether these are redox reactions and identify the reductant and oxidant if they are. You may use the value $\Delta_f G^\ominus_{aq}(CH_3CHO) = -139.24$ kJ mol^{-1}; the other values may be found in the appendix.

E22.21 In Example 22.6 the oxidation of FeS_2 to ferric ion Fe^{3+} and sulfate SO_4^{2-}, important in the generation of acid mine drainage, was discussed. Streambeds suffering from this form of pollution have a red/orange color caused by ferric hydroxide

$$Fe^{3+}(aq) + 3H_2O \rightleftarrows Fe(OH)_3\,(s) + 3\,H^+(aq)$$

How does the concentration of the ferric ion to its hydroxide depend on a solution's pH, under standard conditions? (Take the K_{sp} of $Fe(OH)_3$ to be 4×10^{-38}.)

E22.22 Alcohol dehydrogenase removes ethanol from your bloodstream and generates acetaldehyde. In simplified form this reaction is

$$NAD^+ + C_2H_5OH \rightarrow CH_3C(O)H + NADH$$

What is the Gibbs energy of this reaction under standard conditions? What is the Gibbs energy of this reaction at a typical body temperature 37 C?

	NAD^+	C_2H_5OH	$CH_3C(O)H$	NADH[4]
$\Delta_f G^{\ominus\prime}$ (kJ mol^{-1})	1059.1	63.0	24.1	1120.1
$\Delta_f H^{\ominus\prime}$ (kJ mol^{-1})	−10.3	−290.8	−213.6	−41.4

4 These data are taken from [5].

E22.23 In Section 22.4.1 we discussed the reaction of a Zn rod immersed in a solution which is 1 M in H^+ and in Zn^{2+}. Perform the same calculation if the Zn rod is immersed in pure water.

E22.24 One of the steps in the chain of coupled bio-reactions for respiration is the transfer of an electron from cytochrome b [cyt b(Fe^{2+})] to the oxidized cytochrome c_1 [cyt c_1(Fe^{3+})]. For the reaction

$$\text{cyt } c_1(Fe^{2+}) + H^+ \rightarrow \text{cyt } c_1(Fe^{3+}) + \tfrac{1}{2}H_2$$

$\Delta_r G^{\ominus\prime} = 21\ \text{kJ mol}^{-1}$ and for the reaction

$$\text{cyt b}(Fe^{2+}) + H^+ \rightarrow \text{cyt b}(Fe^{3+}) + \tfrac{1}{2}H_2$$

$\Delta_r G^{\ominus\prime} = 4\ \text{kJ mol}^{-1}$. Assuming that one molecule of ATP is produced by coupling to the oxidation of three cytochrome c_1 molecules in this way, calculate the $\Delta_r G^{\ominus\prime}$ for the formation of ATP.

E22.25 Consider the electron transfer reaction between N,N-dimethylaniline ($\Delta_r G(D/D^+) = 51.1$ kJ mol^{-1}) and 9,10-dimethylanthracene ($\Delta_r G(A/A^-) = 185.3$ kJ mol^{-1}) in acetonitrile. Is this reaction spontaneous in the ground state? Is it spontaneous upon photoexcitation of the 9,10-dimethylanthracene?

E22.26 Show by insertion into Equation (22.14), that the degree of dissociation of a weak acid is $\alpha = 0.618$ for $\hat{c}_{\text{total}} = K_a$ and $\alpha = 0.5$ for $\hat{c}_{\text{total}} = 2K_a$.

22.11 Problems

P22.1 The combustion of fossil fuels contributes to the buildup of CO_2 and oxides of sulfur and nitrogen in the atmosphere. As we saw in Example 22.2 CO_2 in water can modify the pH. In an analogous fashion the dissolution of SO_2 can lead to the formation of sulfonic acid H_2SO_3. If the partial pressure of SO_2 in the atmosphere is 2×10^{-8} bar, and you expose a solution of pure water to this partial pressure of SO_2, the solution becomes acidic. Use the equilibria below to calculate the pH, [HSO_3^-] and [SO_3^{2-}] in the solution.

$$H_2O\ (l) + SO_2\ (g) \rightleftarrows H_2SO_3\ (aq) \qquad pK = -0.097$$

$$H_2SO_3\ (aq) \rightleftarrows H^+\ (aq) + HSO_3^-\ (aq) \qquad pK_{a1} = 1.79$$

$$HSO_3^-\ (aq) \rightleftarrows H^+\ (aq) + SO_3^{2-}\ (aq) \qquad pK_{a2} = 9.00$$

P22.2 In Example 22.2 we computed the pH of pure water when exposed to a partial pressure of 3.89×10^{-3} bar in CO_2. If you add an excess of $CaCO_3$ (s) to the water solution (the solubility of $CaCO_3$ in water is $4.0 \times 10^{-9}(\text{mol L}^{-1})^2$) to set the carbonate concentration, what will the new pH of the solution be?

P22.3 The sulfurous smell that arises from wastewater in sewers is often created by the reduction of sulfur oxides to hydrogen sulfide (H_2S) by bacteria. Using the equilibria

$$H_2S \rightarrow HS^- + H^+, \qquad HS^- \rightarrow S^{2-} + H^+$$

$$K_{a1} = \frac{\hat{c}_{HS^-}\hat{c}_{H^+}}{\hat{c}_{H_2S}} = 8.9 \times 10^{-8}, \quad K_{a2} = \frac{\hat{c}_{S^{2-}}\hat{c}_{H^+}}{\hat{c}_{HS^-}} = 1 \times 10^{-19}$$

calculate the concentrations of H_2S, HS^-, and S^{2-} as a function of pH. Plot the fraction of each species present over a pH range from 5 to 10. What can you say about the pH of solution from which you can smell H_2S?

P22.4 Consider the acid dissociation

$$CH_3COOH \rightleftarrows CH_3COO^- + H^+$$

and take $\hat{c}_{CH_3COO^-} + \hat{c}_{CH_3COOH} = 0.01$. Calculate the $\Delta_r G^{\ominus}$ for this reaction under standard conditions. Compute the concentrations of acetate ions and protons assuming that the solution is ideally diluted. Account for nonideality of the solution by using the extended Debye–Hückel model for the activity coefficients of the ions and compute corrected ion concentrations. How much does the solution's pH change between your two estimates?

P22.5 Consider the dissociation of HF, HCl, and HI into protons and anions in the gas phase. What determines the trend in acidity? Contrast this result with the dissociation constants of these molecules in water. How does the acidity change and what determines the trend in acidity? Some of the facts you need can be found in the appendix; you may also use the fact that the electron affinity is $EA = 349$ kJ mol^{-1} for Cl, $EA = 328$ kJ mol^{-1} for F, and $EA = 324.7$ kJ mol^{-1} for Br and that the ionization energy of H is $IP = 1312.0$ kJ mol^{-1}. (You may also need to use the *CRC Handbook of Chemistry and Physics* for some facts.)

P22.6 Consider the acid–base equilibria for the amino acid glycine, for which $K_1 = 3.98 \times 10^{-3}$ and $K_2 = 1.58 \times 10^{-10}$. Given a total concentration of 0.10 mol L^{-1}, plot the concentrations for each of glycine's forms as a function of the solution pH. (Hint: Refer to Foundation 22.2).

P22.7 In Fig. 22.5 the titration curve of glycine ($n_{total} = 0.1$ mol) with a NaOH solution is shown, calculated according to Equation (22.34). In analogy, calculate the titration curve of glycine with an HCl solution. Make the corresponding plot to Fig. 22.5.

P22.8 Consider the equilibria of the amino acid glycine,

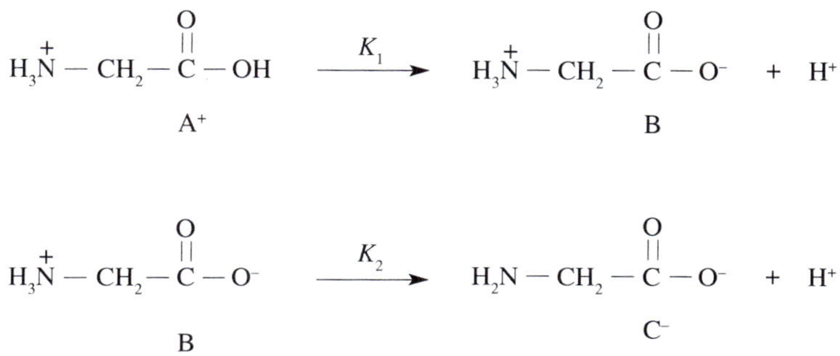

which was discussed in Section 22.3.2., and calculate the pH for two limiting cases: 1) Equilibrium $A^+ \rightleftharpoons B$ is neglected and 2) Equilibrium $B \rightleftharpoons C^-$ is neglected.

P22.9 pH change in buffer solutions: In Example 22.1 in Section 22.2.1 we derived the equation

$$\mathrm{pH} = \mathrm{pK} - \log \frac{\hat{c}_0 + \hat{c}_a}{\hat{c}_0 - \hat{c}_a} \tag{22.55}$$

for the pH as a function of the concentration c_a of added acid for different buffer concentrations c_0. In this problem you are asked to extend the calculations for the example of HCl and Acetic acid: Acetate buffer given in Example 22.1; for four different values of $\hat{c}_0$ (0.1, 0.2, 0.3, and 0.4) Please check whether "0.," can be changed to "0.3,". and five different values of $\hat{c}_a$ (0.00, 0.01, 0.02, 0.03, and 0.04). Plot the pH change as a function of $\hat{c}_a$ for each of the buffer concentrations $\hat{c}_0$. Comment on your findings.

P22.10 Consider the acid dissociation of oxalic acid

$$\text{COOH–COOH} \rightleftarrows \text{COOH–COO}^- + H^+$$

and take $\hat{c}_{\text{COOH–COOH}} = 0.10$. The p$K_a$ value for this reaction is p$K = 1.23$. Neglect the second dissociation step and compute the concentrations of oxalic acid ions and protons assuming that the solution is ideally diluted. Account for nonideality of the solution by using the extended Debye–Hückel model for the activity coefficients of the ions and compute corrected ion concentrations. How much does the solution's pH change between your two estimates?

References

1 A.F. Hollemann, E. Wiberg, and N. Wiberg, *Lehrbuch der Anorganischen Chemie* (Walter der Gruyter, Berlin, 2007); Also, see F.A. Cotton and G. Wilkerson, *Advanced Inorganic Chemistry* (Wiley, New York, 1980) 4th edition.
2 D. Rehm and A. Weller, Kinetics of fluorescence quenching by electron and H-atom transfer, *Israel J. Chem.*, 1970, **8**, 259.
3 M. Sjödin, S. Styring, B. Akermark, L. Sun, and L. Hammarström, Proton-coupled electron transfer from tyrosine in a tyrosine-ruthenium-tris-bipyridine complex: comparison with tyrosine$_z$ oxidation in photosystem II, *J. Am. Chem. Soc.*, 2000, **122**, 3932.
4 C.S. Silva, W.D. Seider, and N. Lior, Exergy efficiency of plant photosynthesis, *Chem. Eng. Sci.*, 2015, **130**, 151.
5 R.A. Alberty, Effect of temperature on standard transformed Gibbs energies of formation of reactants at specified pH and ionic strength and apparent equilibrium constants of biochemical reactions, *J. Phys. Chem. B*, 2001, **105**, 7865.

23

Chemical Reactions in Electrochemical Cells

CHAPTER MENU

23.1 Introduction

The conversion of electrical energy into chemical energy (electrolysis) and the conversion of chemical energy into electrical energy (batteries, fuel cells) are important in science and technology. These processes often occur in electrochemical cells, in which two or more electron transfer reactions are coupled together.

What happens in an electrochemical cell? A chemical reaction takes place in the cell, producing or consuming electrical energy, and the cell has a distinct electrical potential E (also called the *electromotive force* or *emf*). If we allow a chemical reaction to occur spontaneously ($\Delta G < 0$), it can generate a current that flows through an external electrical circuit and can be used to perform electrical work. Alternatively, we can apply an external potential $\phi > E$ to make the reverse reaction occur. In the limiting case when ϕ is only slightly smaller than E the reaction is reversible. In either case, the change in Gibbs energy is

$$\Delta G = -E \cdot I \cdot t$$

where $I \cdot t$ (current × time) is the amount of charge that passes through the circuit in time t.

ΔG for chemical reactions can be easily and accurately obtained from electrochemical measurements by using this relation. The ease and precision of these measurements make electrochemistry useful for quantifying thermodynamic properties. The electronics revolution in the latter twentieth century has made electrochemical measurements of broad utility in quantifying the properties of complex systems in chemistry and biology, as well as the environment and medicine.

23.2 ΔG and Potential E of an Electrochemical Cell

Figure 23.1 diagrams an electrochemical cell in which the reaction

$$AgI(s) + \frac{1}{2}\,Pb(s) \rightarrow Ag(s) + \frac{1}{2}\,PbI_2(s) \tag{23.1}$$

occurs. A lead electrode and a silver electrode are immersed in a beaker which contains a KI solution, solid PbI_2, and solid AgI. The Pb and PbI_2 are in one part of the beaker and are separated from the Ag and AgI in the other part, so that reaction (23.1) cannot occur directly in solution but must proceed through the electrically connected solid electrodes.

Principles of Physical Chemistry, Third Edition. Hans Kuhn, David H. Waldeck, and Horst-Dieter Försterling.

Companion website: www.wiley.com/go/Kuhn/physicalchemistry3e

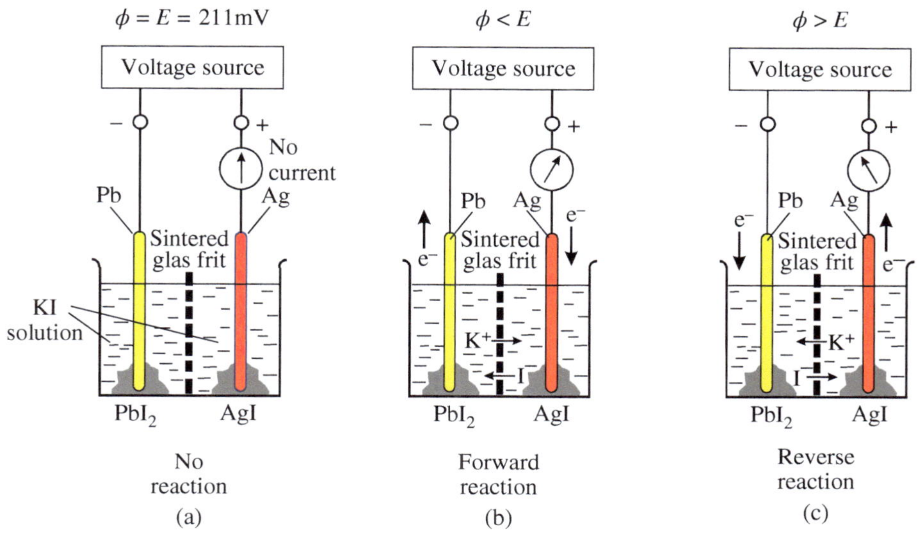

Figure 23.1 Electrochemical cell in which reaction (23.1) can occur reversibly at atmospheric pressure and $T = 298$ K. Dashed line: sintered glass frit preventing the mixing of the solutions in the half-cells by convection or diffusion. KI solution is in great excess. (a) No reaction of Equation (23.1), the external voltage ϕ is $\phi = E$ ($E = 211$ mV). (b) Forward reaction of Equation (23.1), $\phi < E$ (discharging the cell). (c) Reverse reaction of Equation (23.1), $\phi > E$ (charging the cell).

Electrode reactions: At the silver electrode (cathode, i.e., the electrode where reduction takes place) silver ions are reduced

$$Ag^+ + e^- \rightarrow Ag(s) \tag{23.2}$$

and at the lead electrode (anode, i.e., the electrode where oxidation takes place) lead atoms are oxidized

$$\tfrac{1}{2}\,Pb(s) \rightarrow \tfrac{1}{2}\,Pb^{2+} + e^- \tag{23.3}$$

Thus the silver electrode is positively charged and the lead electrode is negatively charged.
Succeeding reactions: The Pb^{2+} ions which are formed at the anode react with I^- ions to generate solid PbI_2

$$\tfrac{1}{2}\,Pb^{2+} + I^- \rightarrow \tfrac{1}{2}\,PbI_2(s) \tag{23.4}$$

Correspondingly, solid AgI dissolves to generate silver ions:

$$AgI(s) \rightarrow Ag^+ + I^- \tag{23.5}$$

All four reactions (23.2)–(23.5) result in the net reaction (23.1). Using the data in Appendix A.1, $\Delta_r G^\ominus$ for this reaction is negative at 298 K; hence, it occurs spontaneously.

If the silver and lead electrodes are not electrically connected, the external circuit is "open." Under these conditions reaction (23.2) cannot proceed because the electrons are not delivered from the silver electrode; the same phenomenon stops reaction (23.3) because the electrons are not removed from the lead electrode. In this situation a potential difference (called an *open circuit voltage*) of $E = 0.211$ V arises across the electrodes. The same condition can be realized when the electrodes are connected to an external voltage source with a voltage $\phi = 0.211$ V (Fig. 23.1a): the ammeter shows zero current.

If we decrease ϕ below E (e.g., $\phi = 0.1$ V), then electrons can move through the external circuit from the lead electrode to the silver electrode (Fig. 23.1b, discharging the cell). Reaction (23.1) occurs and the system (the electrochemical cell) supplies electrical work to the surroundings (the external voltage source). On the other hand, if we increase ϕ above E (e.g., $\phi = 0.3$ V), the reverse of reaction (23.1) occurs (Fig. 23.1c, charging the cell); the surroundings supply electrical work to the system. Even small changes of the external potential ϕ around 0.211 V enable us to reverse the direction of the chemical reaction; thus, it is reversible when applying a voltage $\phi = E$, and we can relate the cell potential E to the change in Gibbs energy of the reaction, ΔG.

What is the relationship between ΔG and E? Consider the case where the external potential ϕ is only slightly different from E, such that reaction (23.1) can occur, but the current I is extremely small. The electrical energy that is supplied by the system in a time t is

$$\text{electrical energy supplied by the system} = E \cdot I \cdot t \tag{23.6}$$

Because the cell is under atmospheric pressure, the volume of the solution in the cell can change by ΔV, and additional work $P\Delta V$ is supplied by the system. Consequently the total reversible work $-w_{rev}$ done by the system is

$$\begin{aligned} -w_{rev} &= (\text{electrical energy supplied by the system}) + P\Delta V \\ &= E \cdot I \cdot t + P\Delta V \end{aligned} \tag{23.7}$$

or

$$w_{rev} + P\Delta V = -E \cdot I \cdot t \tag{23.8}$$

For a change of state at constant temperature and constant pressure the change of Gibbs energy is, according to Section 18.2.2,

$$\Delta G = w_{rev} + P\Delta V \tag{23.9}$$

hence

$$\Delta G = -E \cdot I \cdot t \tag{23.10}$$

$I \cdot t$ is the total charge that passes through the external circuit during time t. If $\boldsymbol{n}$ is the corresponding amount of electrons then the total charge is

$$\text{total charge} = I \cdot t = \boldsymbol{n}\boldsymbol{N}_A \cdot e \tag{23.11}$$

and we obtain

$$\boxed{\Delta G = -E \cdot \boldsymbol{n}\boldsymbol{N}_A \cdot e = -\boldsymbol{n}\boldsymbol{F} \cdot E} \tag{23.12}$$

or

$$\boxed{\Delta_r G = \frac{\Delta G}{\boldsymbol{n}} = -\boldsymbol{F} \cdot E} \tag{23.13}$$

where

$$\boldsymbol{F} = \boldsymbol{N}_A \cdot e = 96{,}485.309 \text{ C mol}^{-1} \tag{23.14}$$

is called the *Faraday constant*.

With Equation (23.13) we can use the measured cell potential $E = 0.211$ V to calculate $\Delta_r G$ for reaction (23.1) at 298 K and find

$$\Delta_r G = -96{,}485.309 \text{ C mol}^{-1} \cdot 0.211 \text{ V} = -20.4 \times 10^3 \text{ C V mol}^{-1} = -20.4 \text{ kJ mol}^{-1}$$

(Note that 1 C V = 1 J.). In Exercise E23.18, you compare this value to the value for $\Delta_r G$ from $\Delta_f G^{\ominus}$. That is, we can measure the Gibbs energies of redox reactions by measuring their electrochemical cell potentials. Note that Equation (23.13) is valid only if we write the reaction (23.1) such that just one electron is involved in the reaction, see Equations (23.2) and (23.3); i.e., $\Delta_r G$ is the change of Gibbs energy when one mol of electrons is involved.

In addition to $\Delta_r G$, electrochemical potential measurements provide a route to determining $\Delta_r H$ and $\Delta_r S$ for reactions. By measuring the cell potential E at various temperatures, we can measure the entropy change $\Delta_r S$ for a reaction. According to Equation (18.29)

$$\Delta_r S = -\left(\frac{\partial \Delta_r G}{\partial T}\right)_P \tag{23.15}$$

By substituting for $\Delta_r G$ with $-E\boldsymbol{F}$ we obtain

$$\Delta_r S = \boldsymbol{F} \cdot \left(\frac{\partial E}{\partial T}\right)_P \tag{23.16}$$

Finally, from $\Delta_r G$ and $\Delta_r S$ we can obtain $\Delta_r H$ via

$$\Delta_r H = \Delta_r G + T \cdot \Delta_r S$$

Example 23.1 *Table 23.1 shows the change in potential with temperature for the electrochemical cell, $AgI + \frac{1}{2}\,Pb \rightarrow Ag + \frac{1}{2}PbI_2$. Determine the entropy and enthalpy change for this reaction.*

Analyzing the data in Table 23.1 (see Fig. 23.2), we find for the slope $(\partial E/\partial T)_P = -0.169$ mV K^{-1}. Then

$$\Delta_r S = \boldsymbol{F} \cdot \left(\frac{\partial E}{\partial T}\right)_P = \boldsymbol{F} \cdot \left(-0.169 \text{ mV K}^{-1}\right) = -16.3 \text{ J K}^{-1} \text{ mol}^{-1}$$

$$\Delta_r H = \Delta_r G + T\Delta_r S = \left(-20.4 \times 10^3 - 298 \times 16.3\right) \text{ J mol}^{-1}$$
$$= -25.3 \text{ kJ mol}^{-1}$$

Because reaction takes place at a pressure of 1 bar and the species are separated at the beginning and at the end of the reaction, we can identify $\Delta_r H$ and $\Delta_r S$ with $\Delta_r H^{\ominus}$ and $\Delta_r S^{\ominus}$.

The negative sign of ΔS means that heat

$$-q_{\text{rev}} = -T\Delta S > 0$$

flows from the cell to the surroundings even when the reaction is carried out reversibly. When the current flow is reversed (by slightly increasing ϕ) the reverse reaction takes place; i.e., heat flows from the surroundings into the cell (isothermal process), or the cell cools down when it is not connected with the heat bath (adiabatic process). Because $\Delta H < 0$, heat is transferred to the surroundings when the cell is short-circuited.

Table 23.1 Potential E of the Cell in Fig. 23.1 for Different Temperatures T.

T/K	298	320	350	370	390
E/mV	211	208	201	199	196

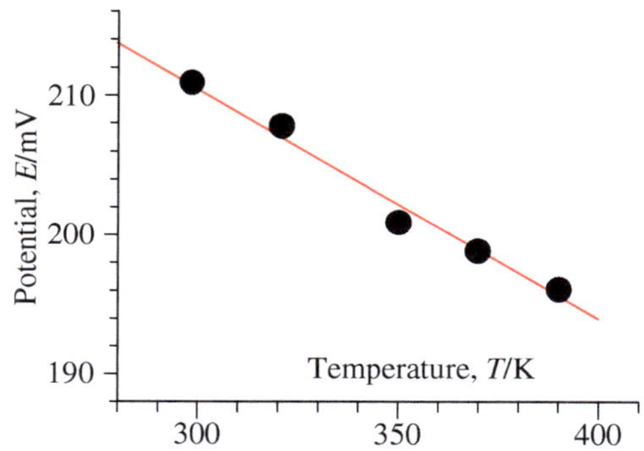

Figure 23.2 Potential E of the electrochemical cell in Fig. 23.1 as a function of temperature T.

23.3 Simple Cells and Nernst Equation

To simplify, we consider very dilute solutions of electrolytes and use concentrations instead of activities. To generalize we can always replace concentrations by activities, as outlined in Chapter 21 and Section 23.5.

23.3.1 Metal/Metal Ions

Figure 23.3a diagrams a concentration cell, in which two silver electrodes are immersed in $AgNO_3$ solutions of concentrations c_1 and c_2. As in Fig. 23.1 the solutions are separated by a sintered glass frit in order to avoid a direct mixing of the solutions in the two compartments of the cell or the diffusion of ions from one side to the other; however, it allows for ion passage under an electric field generated by an external voltage. It is convenient to describe this cell arrangement by the short notation

$$\text{Ag} \mid \text{Ag}^+(c_1) \mid \text{Ag}^+(c_2) \mid \text{Ag}$$

where Ag corresponds to the two solid electrodes and Ag^+ corresponds to the silver ions in solution. A solid vertical line (|) indicates a phase boundary, a dashed vertical line (¦) indicates a sintered glass frit (diaphragm). Two dashed lines (¦¦) indicate a salt bridge (see below).

For $c_1 > c_2$, the net flux moves Ag^+ from the left (c_1) to the right (c_2) and reaction

$$\text{Ag}^+(c_1) \rightarrow \text{Ag}^+(c_2) \qquad (23.17)$$

can occur spontaneously. Assuming that the sintered glass frit inhibits diffusion of Ag^+ ions from left to right (i.e., makes it sufficiently slow), the process (23.17) occurs indirectly by the electrode reactions

$$\text{Ag}^+(c_1) + \text{e}^- \rightarrow \text{Ag (left)} \qquad (23.18)$$

$$\text{Ag} \rightarrow \text{Ag}^+(c_2) + \text{e}^-\text{(right)} \qquad (23.19)$$

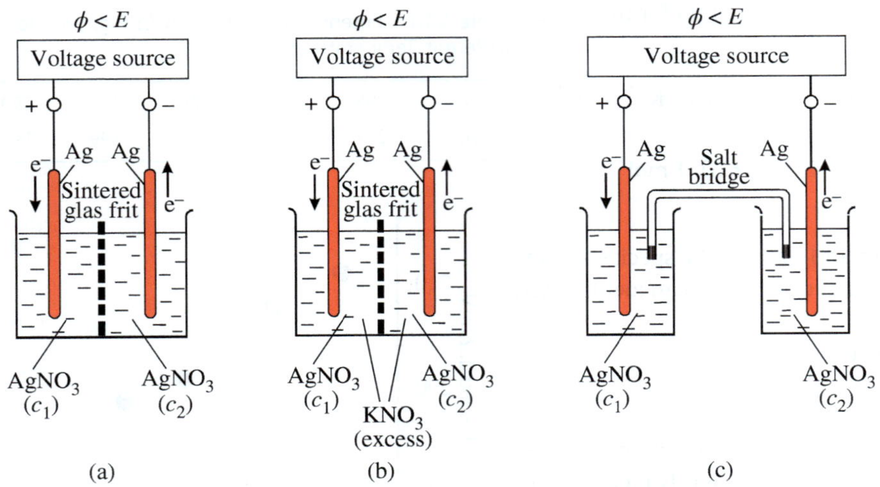

Figure 23.3 Concentration cell. The polarity of the cell is for $c_1 > c_2$ and the external voltage is $\phi < E$. The direction of the current is such that reaction (23.17) takes place. Ag^+ ions disappear in the left-hand half-cell and appear in the right-hand half-cell at lower concentration. The cell acts as a battery. Transport of ionic charge occurs between the half-cells to maintain electroneutrality. Charge balance is maintained by the flow of Ag^+ and NO_3^- ions in case (a); by the flow of K^+ and NO_3^- ions in excess and in equal concentration in both compartments in case (b); and by a salt bridge filled with a concentrated solution of KNO_3 in case (c). The function of the sintered glass frit in (a), (b) and of the salt bridge in (c) is to avoid a direct mixing of the solutions in the two compartments of the cell or the diffusion of ions from one side to the other; however, ions can pass through the frit under an electric field generated by an external voltage.

Consequently, the silver electrode on the right-hand side becomes negatively charged, relative to the silver electrode on the left-hand side.

By applying a voltage ϕ in the external circuit which is slightly lower than the cell potential E we can achieve a practically reversible electron transport between the two electrodes from left to right. What will happen in the two-electrode compartments? Ag^+ ions will be consumed (forming solid Ag according to Equation (23.18)) in the left compartment, and Ag^+ ions will be formed from solid Ag in the right compartment. Thus the overall reaction is (23.17), a change in concentration on the two sides.

An additional process occurs at the same time, however. Because the solution is part of the electrical circuit, NO_3^- ions pass from left to right, away from the negative electrode to the positive electrode, to maintain electroneutrality between the two sides. Correspondingly, Ag^+ ions move from right to left through the sintered glass frit, thus counteracting the flow due to the electrode reaction. To avoid this complication we add an inert salt (e.g., KNO_3) to the solution such that its concentration is equal in both compartments and much larger than c_1 or c_2. Consequently, the ion transport is mainly determined by this indifferent salt, see Fig. 23.3b. Because the inert salt is at a high concentration, the ion transport is rapid and the two compartments maintain electroneutrality.[1]

If we deal with ideally dilute solutions, then we can use the approximations considered in Sections 18.5 and write

$$\Delta_r G = \boldsymbol{R}T \cdot \ln\left(\frac{c_2}{c_1}\right) \tag{23.20}$$

and

$$E = -\frac{\Delta_r G}{\boldsymbol{F}} = -\frac{\boldsymbol{R}T}{\boldsymbol{F}} \cdot \ln\left(\frac{c_2}{c_1}\right) \tag{23.21}$$

Thus the cell potential E is determined by the ratio of the two concentrations. At $T = 298$ K

$$\frac{\boldsymbol{R}T}{\boldsymbol{F}} = 25.7 \times 10^{-3}\ \mathrm{J\,C^{-1}} = 25.7\ \mathrm{mV}$$

and

$$E = -25.7\ \mathrm{mV} \cdot \ln\left(\frac{c_2}{c_1}\right) = -59.1\ \mathrm{mV} \cdot \log\left(\frac{c_2}{c_1}\right) \tag{23.22}$$

1 In the arrangement of Fig. 23.1 the KI solution has the same purpose.

23.3.1.1 Salt Bridge

To avoid complications by the transport of charge through the sintered glass frit we added an indifferent salt in excess (Fig. 23.3b). Another possibility is to use two separate containers for the reaction solutions (which are called *half-cells*) and to connect them by a *salt bridge*. A salt bridge is a glass tube filled with a concentrated solution of an electrolyte, the ions of which move with similar speed and do not react with the ions of the solution at the electrodes, see Fig. 23.3c. The potentials E measured with such an arrangement usually differ only by a few millivolts from the values expected according to Equations (23.12) and (23.21); such an error can be neglected in most cases.

Example 23.2 *Construct an electrochemical cell that measures the solubility product for AgCl.*

If we add enough KCl into the right compartment of the concentration cell, Ag^+ ions will precipitate as AgCl (Fig. 23.4). For an excess of KCl (concentration of Cl^- ions 0.1 mol L^{-1} after the AgCl has precipitated) in the right compartment and $c_1 = 0.1$ mol L^{-1} in the left compartment, the measured cell potential is $E = 0.461$ V. Using Equation (23.22),

$$\log\left(\frac{c_2}{c_1}\right) = -\frac{E}{59.1\ \text{mV}} \quad \text{or} \quad \frac{c_2}{c_1} = 10^{-E/59.1\ \text{mV}}$$

the concentration c_2 of Ag^+ ions in equilibrium with Cl^- and AgCl is obtained as

$$c_2 = 0.1\ \text{mol L}^{-1} \cdot 10^{-(0.461/0.0591)} = 1.6 \times 10^{-9}\ \text{mol L}^{-1}$$

Even though practically all the Ag^+ ions have precipitated as AgCl, the tiny amount of still dissolved Ag^+ can be measured very accurately. Using the ion concentrations, we can determine the solubility product of AgCl as

$$K_{sol} = \hat{c}_{Ag^+} \cdot \hat{c}_{Cl^-} = 1.6 \times 10^{-10}$$

Remember that $\hat{c} = \frac{c}{c^{\ominus}}$, where $c^{\ominus} = 1$ mol L^{-1}.

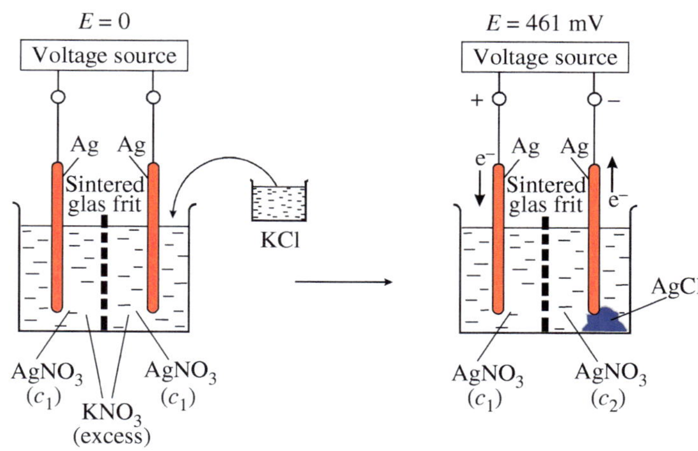

Figure 23.4 Measuring the Ag^+ concentration in the concentration cell in Fig. 23.3 after adding a KCl solution to the half-cell on the right ($c_1 = 0.1$ mol L^{-1}, chloride concentration after the formation of AgCl: $c_{Cl^-} = 0.1$ mol L^{-1}). The external voltage is $\phi = E$ (left) and $\phi < E$ (right). Transport of charge between the half-cells through porous wall by K^+ ions (from right to left) and NO_3^- ions (from left to right) in excess.

23.3.2 Gas Electrodes

Consider the electrochemical cell in Fig. 23.5, which shows two half-cells with *hydrogen electrodes* connected by a salt bridge. On the left a platinum electrode is immersed in acid solution (HCl solution with a high H^+ concentration c_1) and on the right a platinum electrode is immersed in a basic solution (NaOH solution with a low H^+ concentration c_2). In each compartment hydrogen gas is bubbled through the solutions (saturation with H_2). Again we assume $c_1 > c_2$, and the reaction

$$H^+(c_1) \rightarrow H^+(c_2) \tag{23.23}$$

can occur spontaneously. At the surface of the platinum electrodes the reactions

$$H^+(c_1) + e^- \rightarrow \tfrac{1}{2}\ H_2(\text{Pt}) \qquad \text{(left)} \tag{23.24}$$

$$\tfrac{1}{2}\ H_2(\text{Pt}) \rightarrow H^+(c_2) + e^- \quad \text{(right)} \tag{23.25}$$

occur. The right electrode becomes negatively charged relative to the left one.

What processes occur at the platinum electrodes? At the right electrode H_2 molecules diffuse to the platinum surface and transfer electrons to the metal, converting one H_2 molecule into two H^+ ions. The H^+ ions escape from the surface by diffusing into the solution. The reverse process occurs at the left electrode, H^+ ions convert into H_2 molecules. Both processes are in a fast dynamic equilibrium. Although platinum is practically inert, it works as a catalyst.

Example 23.3 *We consider the gas cell in Fig. 23.5 with concentrations* $c_{\mathrm{HCl}} = c_{\mathrm{NaOH}} = 0.01$ *mol* L^{-1}. *The electrical potential measured with this cell is* $E = 0.590$ *V. What is the proton concentration* c_2 *in the right compartment?*

From Equation (23.22) we find

$$\frac{c_2}{c_1} = 10^{-E/59.1\ \mathrm{mV}} = 1.04 \times 10^{-10}$$

The HCl solution in the left compartment is completely dissociated into H^+ and Cl^- ions (strong acid), thus $c_1 = 0.01$ mol L^{-1}. The right compartment, with 0.01 molar NaOH solution, has a proton concentration of

$$c_2 = 0.01 \times 1.04 \times 10^{-10}\ \mathrm{mol\ L^{-1}} = 1.04 \times 10^{-12}\ \mathrm{mol\ L^{-1}}$$

Because the concentration of the OH^- ions in the NaOH solution is 0.01 mol L^{-1}, we can calculate the ionic product of water as

$$K_w = \hat{c}_{\mathrm{H^+}} \cdot \hat{c}_{\mathrm{OH^-}} = 0.01 \times 1.04 \times 10^{-12} = 1.04 \times 10^{-14}$$

This example shows clearly that extremely small ion concentrations can be determined by measuring cell potentials.

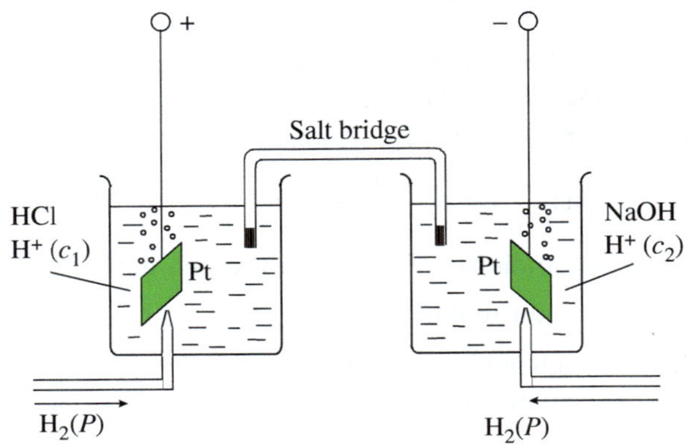

Figure 23.5 Concentration cell with gas electrodes (hydrogen electrodes). The half-cells are connected by a salt bridge.

23.3.3 Nernst Equation and Standard Potential

The effect of concentration on the potential of an electrochemical cell is quantified through the *Nernst equation*.

Consider a hydrogen electrode connected to a copper electrode immersed in a Cu^{2+} solution (Fig. 23.6). The cell reaction

$$\tfrac{1}{2}\ \mathrm{Cu^{2+}} + \tfrac{1}{2}\ \mathrm{H_2} \rightarrow \tfrac{1}{2}\ \mathrm{Cu} + \mathrm{H^+} \tag{23.26}$$

occurs spontaneously under standard conditions (Cu^{2+} is reduced to Cu by H_2). In this case the copper electrode becomes positively charged relative to the hydrogen electrode. The change of Gibbs energy ΔG for this reaction is, according to Section 18.5,

$$\Delta_r G = \Delta_r G^{\ominus} + \boldsymbol{RT} \cdot \ln Q \quad \text{with} \quad Q = \frac{\hat{c}_{\mathrm{H^+}}}{\left(\hat{c}_{\mathrm{Cu^{2+}}}\right)^{1/2} \cdot \left(\hat{P}_{\mathrm{H_2}}\right)^{1/2}} \tag{23.27}$$

where Q is the reaction quotient. The potential E is

$$E = -\frac{\Delta_r G}{\boldsymbol{F}} = -\frac{\Delta_r G^{\ominus}}{\boldsymbol{F}} - \frac{\boldsymbol{RT}}{\boldsymbol{F}} \cdot \ln Q \tag{23.28}$$

The quantity

$$E^{\ominus} = -\frac{\Delta_r G^{\ominus}}{\boldsymbol{F}} \tag{23.29}$$

is called the *standard potential*. With this definition, Equation (23.28) becomes

$$\boxed{E = E^{\ominus} - \frac{\boldsymbol{RT}}{\boldsymbol{F}} \cdot \ln Q} \tag{23.30}$$

This relation is called the *Nernst Equation*.

Note that $\Delta_r G^{\ominus}$ is the change of Gibbs energy when the equation for the chemical reaction is written such that one electron is involved; for example, see reaction (23.26).

Example 23.4 *Determine the standard potential $E^\ominus$ for the cell in Fig. 23.6, which has $T = 298$ K, $E = 425$ mV for $c_{Cu^{2+}} = 0.001$ mol L^{-1}, $P_{H_2} = 1$ bar, and $c_{H^+} = 0.001$ mol L^{-1}.*

From Equation (23.30) we obtain

$$E^\ominus = E + \frac{RT}{F} \cdot \ln Q = E + 25.7\ \text{mV} \cdot \ln Q$$

We calculate Q as

$$Q = \frac{\hat{c}_{H^+}}{\left(\hat{c}_{Cu^{2+}}\right)^{1/2} \cdot \left(\hat{P}_{H_2}\right)^{1/2}} = \frac{0.001}{\sqrt{0.001} \times \sqrt{1}} = 0.0316$$

so that

$$E^\ominus = [425 + 25.7 \cdot \ln(0.0316)]\ \text{mV} = 336\ \text{mV}$$

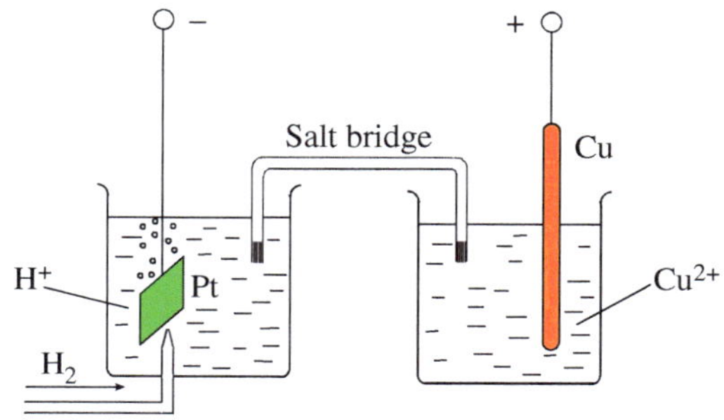

Figure 23.6 Potential of a Cu electrode in Cu^{2+} solution measured relative to a hydrogen electrode.

23.4 Standard Potential $E^\ominus$ and Reference Electrodes

It is conventional to define the standard potential $E^\ominus$ for a reaction in a cell where one half-cell contains the hydrogen electrode and the second half-cell contains the electrode where the reaction under consideration takes place. The cell reaction is written such that H_2 appears on the left-hand side and H^+ on the right-hand side of the equation, see Equation (23.26)

$$\tfrac{1}{2}\ Cu^{2+} + \tfrac{1}{2}\ H_2 \rightarrow \tfrac{1}{2}\ Cu + H^+$$

This choice means that the half-cell under consideration is a reduction reaction. For the copper reduction we write

$$\tfrac{1}{2}\ Cu^{2+} + e^- \rightarrow \tfrac{1}{2}\ Cu, \quad E^\ominus_{Cu^{2+}/Cu} = 336\ \text{mV} \tag{23.31}$$

Equation (23.31) is called a *half-cell reaction*. Standard potentials have been measured for many reduction processes at $T = 298$ K, and some are listed in Table A.3. By combining two half-cell reactions, we can predict the standard potential for any electrochemical cell.

Generally the standard potentials $E^\ominus$ can be obtained from the measurement of the cell potential for any concentrations in both half-cells, as demonstrated above. The measurement can be simplified by using a hydrogen electrode with $P_{H_2} = 1$ bar and $c_{H^+} = 1$ mol L^{-1}. Then formally $\hat{P}_{H_2}$ and $\hat{c}_{H^+}$ are equal to one, and these quantities can be omitted in writing the Nernst equation. This statement is not completely correct, because $c_{H^+} = 1$ mol L^{-1} is far from an ideally dilute solution. However, the cell potential of a cell with a hydrogen electrode immersed in a HCl solution of $c_{HCl} = 1.184$ mol L^{-1} and under a H_2 pressure of 1 bar is just the same as expected from Equation (23.30) with $\hat{P}_{H_2} = 1$ and $\hat{c}_{H^+} = 1$. Such an electrode is commonly used as a reference electrode and is called a *standard hydrogen electrode* (SHE) or *normal hydrogen electrode* (NHE). Its standard potential is chosen to be exactly 0 V.

Example 23.5 *Determine the operating potential for a quinhydrone electrode (Fig. 23.7) in which*

$$\tfrac{1}{2}\ Q + H^+ + e^- \rightleftharpoons \tfrac{1}{2} H_2Q \tag{23.32}$$

where the half-cell contains an equimolar mixture of quinone (Q) and hydroquinone (H_2Q), see Section 22.5, at pH = 5 (acetic acid/acetate buffer).

For Fig. 23.7, the cell reaction is

$$\tfrac{1}{2}\ Q + H^+ + \tfrac{1}{2}\ H_2(\text{SHE}) \rightarrow \tfrac{1}{2}\ H_2Q + H^+(\text{SHE}) \tag{23.33}$$

so that a proton is consumed in the Q/H_2Q half-cell and produced at the hydrogen electrode. The cell potential is

$$E = E^\ominus - \frac{RT}{F} \cdot \ln\left(\frac{\left(\hat{c}_{H_2Q}\right)^{1/2}}{\left(\hat{c}_Q\right)^{1/2} \cdot \hat{c}_{H^+}}\right) \tag{23.34}$$

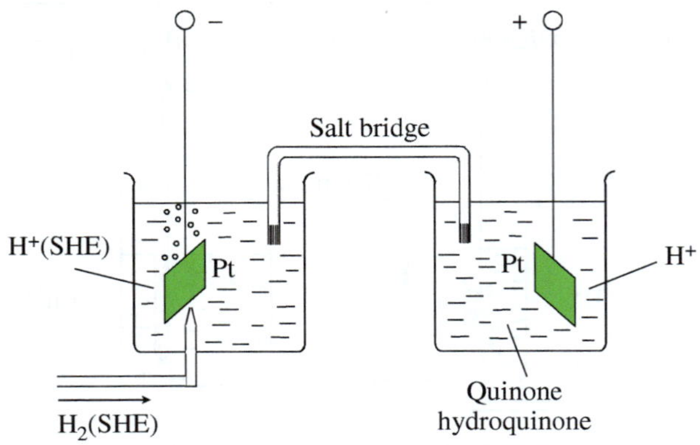

Figure 23.7 Quinone/hydroquinone electrode measured against a standard hydrogen electrode (SHE).

where c_{H^+} is the H^+ concentration in the Q/H_2Q half-cell. From Table A.3 the standard potential $E^\ominus$ is 700 mV. Using $c_Q = c_{H_2Q}$ and $c_{H^+} = 1 \times 10^{-5}$ mol L^{-1}, the measured cell potential is

$$E = 700 \text{ mV} - 25.7 \text{ mV} \cdot \ln\left(10^5\right) = +404 \text{ mV}$$

The cell potential changes with the solution pH. At pH 7, the cell potential is +286 mV, and at pH 9 the cell potential is +167 mV. It changes by 59.2 mV per pH unit.

23.4.1 Practical Determination of $E^\ominus$

For the determination of $E^\ominus$ we used electrochemical cells with a hydrogen electrode. Because working with such electrodes can be cumbersome, it is common to use other reference electrodes which are more convenient. Figure 23.8 shows a silver chloride electrode (AgCl).

The half-cell reaction in the AgCl electrode is

$$Ag^+ + e^- \rightarrow Ag \tag{23.35}$$

and the potential relative to the standard hydrogen electrode is

$$E_{AgCl} = E^\ominus - \frac{\boldsymbol{RT}}{\boldsymbol{F}} \cdot \ln \frac{1}{\hat{c}_{Ag^+}} \tag{23.36}$$

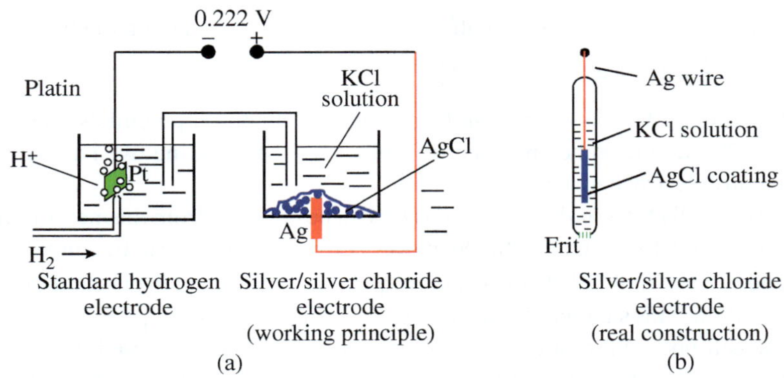

Figure 23.8 Ag/AgCl electrode (reference electrode) measured against a standard hydrogen electrode. (a) Working principle and (b) real construction: a silver wire is coated by a thin layer of AgCl and immersed in a KCl solution.

Because the electrode contains an excess of chloride ions, the concentration of Ag^+ is determined by the equilibrium

$$Ag^+ + Cl^- \rightleftharpoons AgCl \tag{23.37}$$

$$\hat{c}_{Ag^+} \cdot \hat{c}_{Cl^-} = K_s \tag{23.38}$$

and

$$E_{AgCl} = E^\ominus - \frac{\boldsymbol{RT}}{\boldsymbol{F}} \cdot \ln \frac{\hat{c}_{Cl^-}}{K_s} \tag{23.39}$$

For a AgCl electrode with $\hat{c}_{Cl^-} = 1$ (1M AgCl electrode) at $T = 298$ K, we use $E^\ominus = 800$ mV and $K_s = 2.0 \times 10^{-10}$ and calculate

$$E_{AgCl,\,1M} = 800 \text{ mV} - 25.68 \text{ mV} \cdot \ln \frac{1.0}{2.0 \times 10^{-10}} = (800 - 574) \text{ mV} = 226 \text{ mV}$$

Actually, the voltage is 222 mV. The small difference results from the fact that the calculated value is valid only for ideally diluted solutions.

23.4.2 Absolute Electrode Potential

In Chapter 13 we discussed the work function of a solid and chose the energy zero (vacuum scale) with reference to the energy of a free electron having "zero" kinetic energy. Hence it is useful to relate the electrochemical potential of a redox couple on the SHE scale to its value on the vacuum scale.

23.4.2.1 Absolute Potential of the Hydrogen Electrode

Figure 23.9 shows a thermodynamic cycle that relates the half-cell potential of the SHE to the vacuum potential. Solving for the molar Gibbs energy change $\Delta_r G^\ominus(H_2/H^+)$ under standard conditions for the oxidation of hydrogen gas to protons

$$\frac{1}{2}H_2 \rightarrow H^+ + e^- \tag{23.40}$$

we find that

$$\Delta_r G^\ominus(H_2/H^+) = \frac{1}{2}\Delta_{diss} G^\ominus(H_2) + \boldsymbol{N}_A \cdot E_{Ion}(H) + \Delta_{solv} G^\ominus(H^+) - \boldsymbol{N}_A \cdot W\,(Pt) \tag{23.41}$$

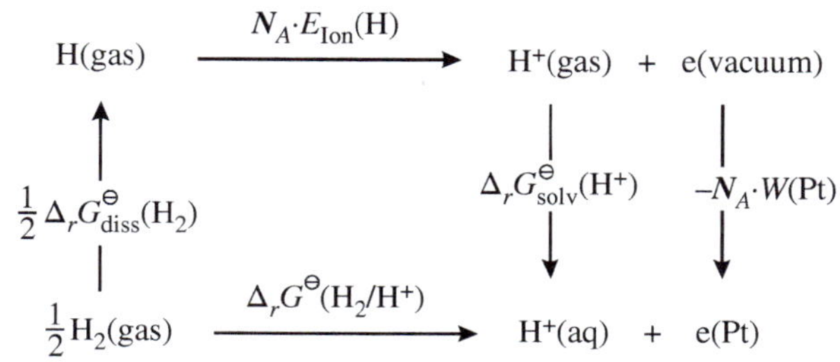

Figure 23.9 Thermodynamic cycle to relate the electrode potential on the SHE scale to that on the vacuum scale.

where $\Delta_{diss} G^\ominus(H_2)$ and $\Delta_{solv} G^\ominus(H^+)$ are the molar changes of Gibbs energy for the dissociation of H_2 into hydrogen atoms and for the solvation of the proton, respectively, and $E_{Ion}(H)$ is the ionization energy of the hydrogen atom. $W(Pt)$ is the work function for an electron in platinum; however, the platinum is taken to have its surface in contact with the liquid. This latter requirement is needed to account for the fact that a dipole layer (e.g., cations may adsorb at the surface of the metal) forms at the interface of the phases and contributes to the measured work function. Thus $W(Pt)$ is smaller than the work function in the vacuum case.

Using Equation (23.13), $\Delta_r G = -\boldsymbol{F}E$, we can write

$$E^\ominus(H_2/H^+) = -\frac{1}{\boldsymbol{F}}\left[\frac{1}{2}\Delta_{diss} G^\ominus(H_2) + \boldsymbol{N}_A \cdot E_{Ion}(H) + \Delta_{solv} G^\ominus(H^+)\right] + \frac{\boldsymbol{N}_A \cdot W(Pt)}{\boldsymbol{F}} \tag{23.42}$$

This expression for $E^\ominus(H_2/H^+)$ depends on intrinsic properties of the redox couple and the work function of the platinum electrode. In order to create a reference hydrogen potential that is independent of the electrode material, we define the *absolute electrode potential* $E^\ominus_{abs}(H_2/H^+)$ as

$$E^\ominus_{abs}(H_2/H^+) = E^\ominus(H_2/H^+) - \frac{\boldsymbol{N}_A \cdot W\,(Pt)}{\boldsymbol{F}} = -\frac{1}{\boldsymbol{F}}\left[\frac{1}{2}\Delta_{diss} G^\ominus(H_2) + \boldsymbol{N}_A \cdot E_{Ion}(H) + \Delta_{solv} G^\ominus(H^+)\right] \tag{23.43}$$

Hence $E^\ominus_{abs}(H_2/H^+)$ depends on the dissociation energy of the hydrogen, the ionization energy of the H atom, and the solvation energy of the proton.

A number of different studies using a range of methods have been used to relate the SHE half-cell potential to the vacuum scale, and the values found for $\boldsymbol{N}_A \cdot W(Pt)/\boldsymbol{F}$ range from 4.4 to 4.8 V. Gomer[2] found 4.73 V, IUPAC [2] recommends a value of 4.44 ± 0.02 V. This is 1.21 V below the voltage we expect from the work function of platinum under vacuum conditions (5.65 V). Thus we obtain

$$E^\ominus_{abs}(H_2/H^+) = E^\ominus(H_2/H^+) - 4.44\ \text{V} = -4.44\ \text{V} \tag{23.44}$$

with an uncertainty of ± 0.02 V.

23.4.2.2 Solvation Energy of the Proton

Equation (23.43) can be used to determine the molar solvation energy of the proton.

$$\Delta_{solv} G^\ominus(H^+) = -\boldsymbol{F} \cdot E^\ominus_{abs}(H_2/H^+) - \frac{1}{2}\Delta_{diss} G^\ominus(H_2) - \boldsymbol{N}_A \cdot E_{Ion}(H)$$

2 See [1].

Using the value $E^{\ominus}_{abs}(H_2/H^+) = -4.44$ V and the measured values of $\Delta_{diss}G^{\ominus}(H_2) = 431.9$ kJ mol^{-1} (taken from G. Herzberg, Spectra of Diatomic Molecules) and $\boldsymbol{N}_A E_{Ion}(H) = 1312.75$ kJ mol^{-1} calculated in Chapter 2, we obtain

$$\Delta_{solv}G^{\ominus}(H^+) = (428.4 - 215.9 - 1312.8)\ \text{kJ mol}^{-1} = -1100.3\ \text{kJ mol}^{-1}$$

This value is in good agreement with the value of −1104.5 kJ mol^{-1} obtained by Coe and coworkers [3] by using cluster ion data.

23.4.2.3 Absolute Potential of Arbitrary Electrodes $E^{\ominus}_{abs}(M^+/M)$

The thermodynamic cycle and analysis used here is not restricted to the case of the hydrogen electrode. In fact, we can define the absolute potential for half-cell reactions in this way. The absolute potentials have the same relationship with each other as the more commonly used electrochemical potentials. These potentials are defined by considering the cell reaction

$$M^+ + \frac{1}{2}H_2 \rightarrow M + H^+$$

with the cell voltage of

$$E^{\ominus} = E^{\ominus}(M^+/M) + E^{\ominus}(H_2/H^+)$$

The cell voltage is the same if we refer the potentials to the potential of the absolute hydrogen electrode

$$E^{\ominus} = E^{\ominus}_{abs}(M^+/M) + E^{\ominus}_{abs}(H_2/H^+)$$

Thus

$$E^{\ominus}(M^+/M) + E^{\ominus}(H_2/H^+) = E^{\ominus}_{abs}(M^+/M) + E^{\ominus}_{abs}(H_2/H^+)$$

so that, setting $E^{\ominus}(H_2/H^+) = 0$, we can define the absolute potential $E^{\ominus}_{abs}(M^+/M)$ as

$$E^{\ominus}_{abs}(M^+/M) = E^{\ominus}(M^+/M) - E^{\ominus}_{abs}(H_2/H^+) \tag{23.45}$$

with $E^{\ominus}_{abs}(H_2/H^+) = -4.44$ V.

For example, $E^{\ominus}(Ag^+/Ag) = 0.800$ V and $E^{\ominus}_{abs}(Ag^+/Ag) = 0.800\ \text{V} - (-4.44\ \text{V}) = +5.24$ V. Thus we see that the standard potential for the Ag^+ reduction half-reaction is 0.800 V more positive than the standard potential for the H^+ reduction half-reaction.

23.4.2.4 Solvation Energies of Metal Ions

Again we consider the reaction

$$M^+ + \frac{1}{2}H_2 \rightarrow M + H^+$$

If this reaction is carried out nonelectrically, we calculate $\Delta_r G$ as

$$\begin{aligned}\Delta_r G = &\left[\frac{1}{2}\Delta_{diss}G^{\ominus}(H_2) + \boldsymbol{N}_A \cdot E_{Ion}(H) + \Delta_{solv}G^{\ominus}(H^+)\right]\\ &- \left[\Delta_{sub}G^{\ominus}(M) + \boldsymbol{N}_A \cdot E_{Ion}(M) + \Delta_{solv}G^{\ominus}(M^+)\right]\end{aligned} \tag{23.46}$$

The terms in the first brackets are in analogy to the corresponding terms in Equation (23.41), except that the work function term of the electron does not occur. The corresponding terms in the second brackets describe the processes of sublimation and ionization of the metal M and the solvation of M^+.

According to Equation (23.43) the terms in the first brackets are connected to the absolute potential of the hydrogen electrode. Thus

$$\begin{aligned}\Delta_r G = &-\boldsymbol{F}E^{\ominus}_{abs}(H_2/H^+)\\ &- \left[\Delta_{sub}G^{\ominus}(M) + \boldsymbol{N}_A \cdot E_{Ion}(M) + \Delta_{solv}G^{\ominus}(M^+)\right]\end{aligned} \tag{23.47}$$

On the other hand, we can express $\Delta_r G$ by the voltage of an electrochemical cell consisting of a hydrogen electrode and a metal electrode,

$$\Delta_r G = -\boldsymbol{F}\left[E^{\ominus}(M^+/M) + E^{\ominus}(H_2/H^+)\right] \tag{23.48}$$

where $E^{\ominus}(M^+/M)$ is the standard potential of the M^+/M redox couple. Thus we obtain, using the definition of the absolute potential of the redox couple M^+/M, Equation (23.45),

$$\begin{aligned}\Delta_r G + \boldsymbol{F}E^{\ominus}_{\text{abs}}(H_2/H^+) &= -\boldsymbol{F}\left[E^{\ominus}(M^+/M) + E^{\ominus}(H_2/H^+) - E^{\ominus}_{\text{abs}}(H_2/H^+)\right]\\ &= -\boldsymbol{F}\left[E^{\ominus}(M^+/M) - E^{\ominus}_{\text{abs}}(H_2/H^+)\right]\\ &= -\boldsymbol{F}E^{\ominus}_{\text{abs}}(M^+/M)\end{aligned}$$

and

$$\Delta_{\text{sub}}G^{\ominus}(M) + \boldsymbol{N}_A \cdot E_{\text{Ion}}(M) + \Delta_{\text{solv}}G^{\ominus}(M^+) = \boldsymbol{F}E^{\ominus}_{\text{abs}}(M^+/M) \tag{23.49}$$

It is interesting to solve for the solvation energy $\Delta_{\text{solv}}G^{\ominus}(M^+)$, which cannot be measured directly.

$$\Delta_{\text{solv}}G^{\ominus}(M^+) = -\Delta_{\text{sub}}G^{\ominus}(M) - \boldsymbol{N}_A \cdot E_{\text{Ion}}(M) + \boldsymbol{F} \cdot E^{\ominus}_{\text{abs}}(M^+/M) \tag{23.50}$$

In Table 23.2 $\Delta_{\text{solv}}G^{\ominus}(M^+)$ for the series of alkali metals is calculated from this equation with $E^{\ominus}_{\text{abs}}(M^+/M) = E^{\ominus}(M^+/M) - E^{\ominus}_{\text{abs}}(H_2/H^+) = E^{\ominus}(M^+/M) + 4.44$ V.

Table 23.2 Calculation of the Solvation Energy $\Delta_{\text{solv}}G^{\ominus}(M^+)$ According to Equation (23.50) with $E^{\ominus}_{\text{abs}}(M^+/M) = E^{\ominus}(M^+/M) + 4.44$ V. $\Delta_{\text{sub}}G^{\ominus}(M)$ is the Sublimation Energy of the Metal, E_{Ion} Is Its Ionization Energy. r_{crystal} Is the Ionic Radius.[3]

Reaction	r_{crystal} pm	$E^{\ominus}(M^+/M)$ V	$\Delta_{\text{sub}}G^{\ominus}(M)$ kJ mol^{-1}	$N_A E_{\text{Ion}}(M)$ kJ mol^{-1}	$\Delta_{\text{solv}}G^{\ominus}(M^+)$ kJ mol^{-1}
$Cs^+ + e^- \rightarrow Cs$	184	−3.02	49.12	375.7	−287.9
$Rb^+ + e^- \rightarrow Rb$	163	−2.98	53.06	403.0	−315.2
$K^+ + e^- \rightarrow K$	152	−2.931	60.59	418.8	−333.8
$Na^+ + e^- \rightarrow Na$	116	−2.710	76.76	495.8	−405.6
$Li^+ + e^- \rightarrow Li$	88	−3.05	126.7	513.3	−505.8

23.4.2.5 Born Model

The Born model approximates $\Delta_{\text{solv}}G^{\ominus}(M^+)$ by the reversible work to transfer the ion M^+ (effective ionic radius a) from the vacuum into the solvent, which is modeled as a continuum fluid with a relative permittivity $\hat{\varepsilon}_s$. Hence, we imagine a process in which we discharge the sphere in vacuum and charge up the sphere in a dielectric. In vacuum, the electrostatic potential ϕ at a distance a from the center of a charged sphere is $\phi = Q/(4\pi\varepsilon_0 a)$ where Q is the charge. Hence, the energy to change Q by dQ is

$$\frac{Q}{4\pi\varepsilon_0 a} \cdot \mathrm{d}Q$$

and the reversible work w_{rev} to discharge the sphere from $Q = e$ is

$$w_{\text{rev}}(\text{vacuum}) = \int_e^0 \frac{Q \cdot \mathrm{d}Q}{4\pi\varepsilon_0 a} = -\frac{e^2}{8\pi\varepsilon_0 a}$$

The reversible work to charge a sphere in a dielectric of relative permittivity $\hat{\varepsilon}_s$ proceeds in a similar way (using $\varepsilon_0\hat{\varepsilon}_s$ instead of ε_0) and yields

$$w_{\text{rev}}(\text{dielectric}) = \int_0^e \frac{Q \cdot \mathrm{d}Q}{4\pi\varepsilon_0\hat{\varepsilon}_s a} = \frac{e^2}{8\pi\varepsilon_0\hat{\varepsilon}_s a}$$

3 r_{crystal} is the ionic radius. Source: [4]. In this work two sets of radii for a coordination number of six are presented: set 1 is based on the radius of the F^- ion, set 2 is based on the radius of the O^{2-} ion. Both sets differ by a constant value of 14 pm. Here we use set 1.

Combining these results, we find that the reversible work to transfer one ion is

$$w_{\text{rev}}(\text{vacuum}) + w_{\text{rev}}(\text{dielectric}) = -\frac{e^2}{8\pi\varepsilon_0 a}\left(1 - \frac{1}{\hat{\varepsilon}_s}\right)$$

Then for $\Delta_{\text{solv}}G^{\ominus}(\text{M}^+)$ we find

$$\begin{aligned}\Delta_{\text{solv}}G^{\ominus}(\text{M}^+) &= \boldsymbol{N}_\text{A} \cdot \left[w_{\text{rev}}(\text{vacuum}) + w_{\text{rev}}(\text{dielectric})\right] \\ &= -\boldsymbol{N}_\text{A}\frac{e^2}{8\pi\varepsilon_0 a}\left(1 - \frac{1}{\hat{\varepsilon}_s}\right)\end{aligned} \tag{23.51}$$

The Born model predicts that a plot of $\Delta_r G_{\text{solv}}(\text{M}^+)$ versus $1/a$, where a is the effective ionic radius, is a straight line ($\hat{\varepsilon}_s = 80$ for the solvent water). Because of the shell of solvent molecules, the effective ionic radius a is larger than the crystal radius r_{crystal} (Table 23.2, column 2). It is assumed that a corresponds to the distance from the center of the ion to the center of the dipoles constituting the solvent and a is calculated as

$$a = r_{\text{crystal}} + 54 \text{ pm}$$

The values calculated from Equation (23.50) are in very good agreement with the prediction of the Born model, Equation (23.51), see Fig. 23.10.

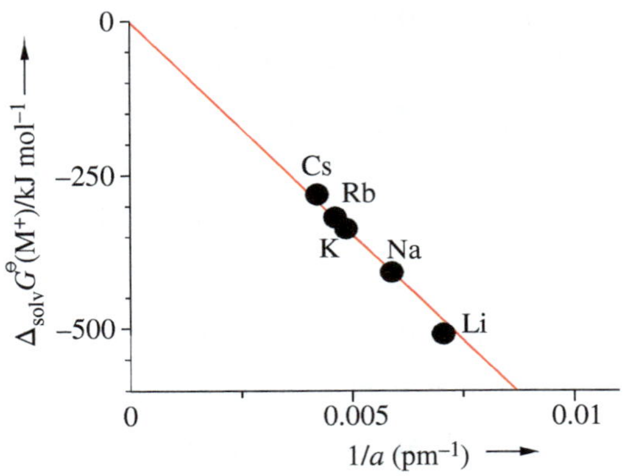

Figure 23.10 $\Delta_{\text{solv}}G^{\ominus}(M^+)$ versus $1/a$, where $a = r_{\text{crystal}} + 54$ pm is the effective radius of the ion, for the series of the alkali ions in Table 23.2. Filled circles: calculated from Equation (23.50) and solid line: calculated from the Born model (23.51).

23.5 Use of Electrochemical Cells for Thermodynamic Measurements

The above discussion has illustrated the relationship between the cell potential and the Gibbs energy, and it demonstrates that the Gibbs energy of an electrochemical cell is conveniently measured. In Example 23.1 we used the dependence of the cell potential on temperature to measure the enthalpy and entropy changes in a reaction. Furthermore we saw that the electrode potential changes of the Ag/AgCl half-reaction depend on the solubility product of AgCl (Example 23.2) and that the quinone/hydroquinone (Q/H_2Q) electrode potential changes with pH (Example 23.5). Here we provide other examples of interesting thermodynamic properties that can be measured electrochemically.

23.5.1 pH Electrodes

Two half-cells filled with solutions containing H^+ ions (one with a known concentration c_1, and the second with an unknown concentration c_2) are connected by a membrane permeable only to H^+ ions. Such membranes can be made from properly treated glass.[4] In each half-cell a AgCl electrode is immersed (Fig. 23.11a). We consider the case $c_1 < c_2$. Since both silver chloride electrodes are identical, the cell potential is

$$E = \frac{\boldsymbol{R}T}{\boldsymbol{F}} \cdot \ln\left(\frac{c_2}{c_1}\right) \tag{23.52}$$

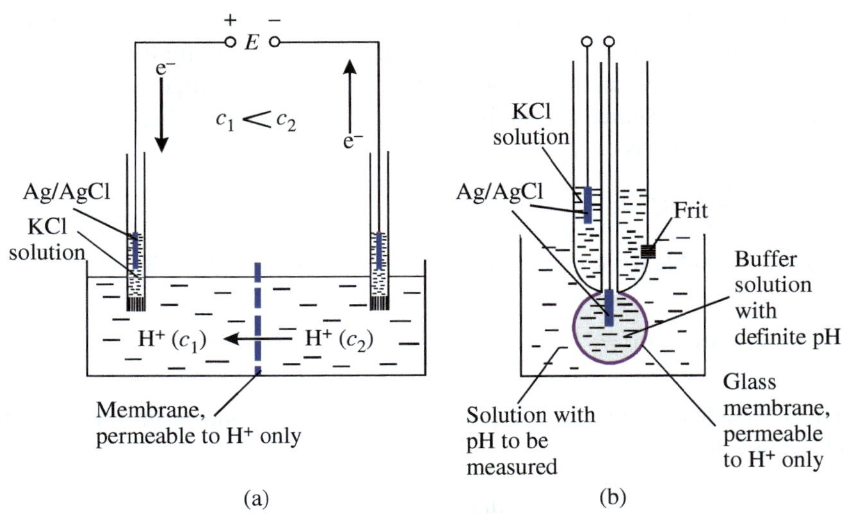

Figure 23.11 Glass electrode. (a) Fundamental setup and (b) real construction.

4 Actually, the glass membrane is not semipermeable for H^+-ions; the transport of charge inside the membrane occurs by Na^+-ions, not by H^+-ions. At the surface of the glass membrane Na^+-ions are exchanged by H^+-ions. In this way H^+-ions disappear in the solution on the outer side of the membrane, and H^+-ions enter the solution on the inner side of the membrane. In spite of this situation, Equation (23.52) is still valid, because the mechanism of the charge transport does not play a role in its derivation.

In the real construction (Fig. 23.11b) a glass ball is filled with a buffer solution of known pH, namely, $pH_{buffer} = -\log \hat{c}_1$. This solution is then connected to a silver chloride electrode. A second silver chloride electrode is inserted into the same glass tube. The hydration of the outer surface of the glass ball generates a junction potential (see Section 23.5.4) across the surface that is proportional to the pH difference. If c_2 is the concentration of the solution to be examined, then $pH = -\log \hat{c}_2$ and according to Equation (23.52) we find

$$\boxed{pH = pH_{buffer} - \frac{\boldsymbol{F}}{2.303 \cdot \boldsymbol{RT}} \cdot E} \tag{23.53}$$

and the pH value is a linear function of the cell voltage.

Equation (23.53) is the practical definition of pH; usually a KH_2PO_4/Na_2HPO_4 buffer with 0.1 mol L^{-1} of NaCl is used as a standard (pH = 6.865 at T = 298 K) [5].

Example 23.6 *To determine the concentration of acetic acid in vinegar you titrate a 25 mL vinegar solution (unknown molarity of acetic acid in water) with a 0.100 M NaOH solution and monitor the pH change. Using the data in the table, calculate the concentration of acetic acid in vinegar.*

pH	2.4	4.2	4.62	5.1	5.5	5.6	7.9	9.31	10.8	12.1	12.7	13.3
NaOH/mL	0.0	10.0	20.0	30.0	35.0	39.5	39.9	40.0	40.1	40.5	41.0	45.0

Figure 23.12 plots the pH of the solution as a function of the volume of NaOH. The equivalence point (see Chapter 22) occurs at pH = 9.31 where the volume of NaOH added is 40.0 mL, or 0.040 × 0.100 = 0.0040 moles. Hence the solution has 0.0040 moles of acetic acid, so that its concentration must be 0.0040/0.025 = 0.16 M or 9.6 g/1000 ml; i.e., the mass concentration of acetic acid in the vinegar is about 1%.

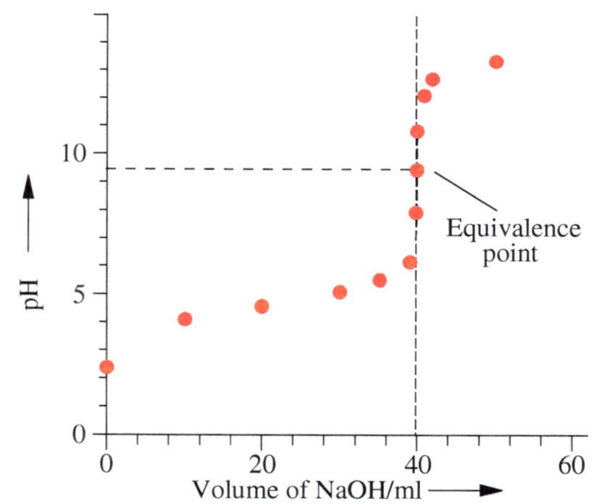

Figure 23.12 The potentiometric titration of a vinegar solution (V = 25 ml) with NaOH (solution of concentration 0.1 mol L^{-1}) is shown.

23.5.2 Measurement of Mean Activity Coefficient $\gamma_\pm$

More generally, the Nernst equation and the reaction quotient, described in Section 23.3.3, should be written in terms of activities. In our discussion of reference electrodes in Section 23.4.1 we found that the potential at high concentrations deviated from that we calculated using the formulae for ideally diluted solutions. Here we show how to use the measured potential to determine the activity coefficients.

Consider a hydrogen electrode in contact with a silver–silverchloride electrode. The cell potential E of the cell

Pt | H_2(gas) | HCl(aq) | AgCl(solid) | Ag

is measured for different concentrations c of HCl at 298 K (columns 1 and 2 in Table 23.3).

The reaction in the cell is

$$\frac{1}{2}\,H_2 + Ag^+ \rightarrow H^+ + Ag \tag{23.54}$$

and the resulting cell potential is given by

$$E = E^{\ominus}_{Ag^+/Ag} - \frac{\boldsymbol{R}T}{\boldsymbol{F}} \cdot \ln\left(\frac{\hat{a}_{H^+}}{\hat{P}^{1/2}_{H_2} \cdot \hat{a}_{Ag^+}}\right)$$

For $P_{H_2} = 1$ bar this result reduces to

$$E = E^{\ominus}_{Ag^+/Ag} - \frac{\boldsymbol{R}T}{\boldsymbol{F}} \cdot \ln\left(\frac{\hat{a}_{H^+}}{\hat{a}_{Ag^+}}\right) \tag{23.55}$$

Here we have used the activities to include any nonidealities in our treatment. Because

$$\hat{a}_{Ag^+} \cdot \hat{a}_{Cl^-} = K_s \tag{23.56}$$

we obtain

$$\begin{aligned}\frac{\hat{a}_{H^+}}{\hat{a}_{Ag^+}} &= \frac{\hat{a}_{H^+} \cdot \hat{a}_{Cl^-}}{K_s} \\ &= \frac{\hat{c}_{H^+}\gamma_+ \cdot \hat{c}_{Cl^-}\gamma_-}{K_s} = \frac{\hat{c}_{H^+} \cdot \hat{c}_{Cl^-} \cdot \gamma_\pm^2}{K_s} = \frac{\hat{c}^2 \cdot \gamma_\pm^2}{K_s}\end{aligned} \tag{23.57}$$

Remember that $\gamma_\pm$ is the mean activity coefficient (see Chapter 21), in this case defined by

$$\gamma_\pm = \sqrt{\gamma_+ \cdot \gamma_-}$$

By inserting Equation (23.57) into Equation (23.55), we find

$$E = E^{\ominus}_{Ag^+/Ag} + \frac{\boldsymbol{R}T}{\boldsymbol{F}} \cdot \ln K_s - 2\frac{\boldsymbol{R}T}{\boldsymbol{F}} \cdot \ln \hat{c} - 2\frac{\boldsymbol{R}T}{\boldsymbol{F}} \cdot \ln \gamma_\pm \tag{23.58}$$

By defining

$$E' = E^{\ominus}_{Ag^+/Ag} + \frac{\boldsymbol{R}T}{\boldsymbol{F}} \cdot \ln K_s \tag{23.59}$$

we obtain the following for the activity coefficient

$$\ln \gamma_\pm = \frac{\boldsymbol{F}}{2\boldsymbol{R}T}\left(E' - E\right) - \ln \hat{c} \tag{23.60}$$

If E' is known, then measuring E as a function of $\hat{c}$ allows us to determine the activity coefficient $\gamma_\pm$. To determine E' we rearrange this expression:

$$y = E + 2\frac{\boldsymbol{R}T}{\boldsymbol{F}} \ln \hat{c} = E' - 2\frac{\boldsymbol{R}T}{\boldsymbol{F}} \ln \gamma_\pm \tag{23.61}$$

For small enough concentration $\hat{c}$ the activity coefficient reaches the limiting value $\gamma_\pm = 1$, i.e.,

$$\lim_{\hat{c}\to 0} y = \lim_{\hat{c}\to 0}\left(E + 2\frac{\boldsymbol{R}T}{\boldsymbol{F}} \ln \hat{c}\right) = E' \tag{23.62}$$

E' can be obtained by plotting $y = (E + 2\boldsymbol{R}T/\boldsymbol{F} \cdot \ln \hat{c})$ versus c and extrapolating for $c \rightarrow 0$. However, y is not a linear function of c (Fig. 23.13a), and therefore it is difficult to perform the extrapolation. On the

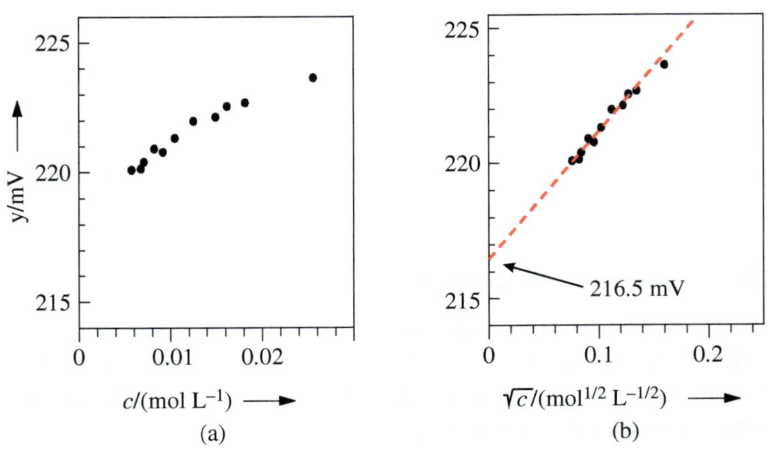

Figure 23.13 Evaluating E' by extrapolating the expression $y = E + 2(\boldsymbol{R}T/\boldsymbol{F}) \cdot \ln \hat{c}$ for $\hat{c} \rightarrow 0$, where c is the concentration of HCl, according to Equations (23.61) and (23.62). Filled circles: experiment; dashed line: extrapolation. (a) y plotted versus c and (b) y plotted versus $\sqrt{c}$.

other hand, according to the Debye–Hückel limiting law, $\ln \gamma_\pm$ decreases proportional to $\sqrt{\hat{c}}$; thus plotting y versus $\sqrt{\hat{c}}$ results in a straight line (Fig. 23.13b) and from the intersection with the ordinate we obtain $E' = 0.2165$ V.

Using $E' = 0.2165$ V, we calculate $\gamma_\pm$ from Equation (23.60). These experimental values are provided in Table 23.3, column 4. From the Debye–Hückel law, Equation (21.93),

$$\ln \gamma = -\frac{A \cdot z^2 \sqrt{I}}{1 + B \cdot R\sqrt{I}} \tag{23.63}$$

with $A = 1.175$ $\mathrm{L^{1/2}\ mol^{-1/2}}$, $B = 0.330 \times 10^{10}$ $\mathrm{m^{-1}\ L^{1/2}\ mol^{-1/2}}$, and $R = 500$ pm we calculate the mean activity coefficients in Table 23.3, column 5, which are in excellent agreement with the experiment up to a concentration of 0.1 mol $\mathrm{L^{-1}}$. In column 6 the result of the Debye–Hückel limiting law, Equation (21.96),

$$\ln \gamma = -A \cdot z^2 \sqrt{I} \tag{23.64}$$

is shown; these values are satisfactory up to a concentration of 0.01 mol $\mathrm{L^{-1}}$.

Table 23.3 Mean Activity Coefficient $\gamma_\pm$ of HCl Solutions at 308 K from the Measured Cell Potential E. The Experimental [6] Values of $\gamma_\pm$ (Column 4) Are Compared with the Result of the Debye–Hückel Law (Equation (23.63), Column 5) and with that of the Debye–Hückel Limiting Law (Equation (23.64), Column 6).

				$\gamma_\pm$ (calc.)	
$\hat{c}$	E (mV)	$\left(E + 2\frac{RT}{F} \cdot \ln \hat{c}\right)$ (mV)	$\gamma_\pm$ (exp.)	(Equation (23.63))	(Equation (23.64))
0.00573	494.0	220.1	0.93	0.93	0.92
0.00675	485.5	220.1	0.93	0.93	0.91
0.00707	483.2	220.4	0.93	0.93	0.91
0.00820	475.9	220.9	0.92	0.92	0.90
0.00914	470.0	220.8	0.92	0.92	0.90
0.01045	463.4	221.3	0.91	0.91	0.89
0.01249	454.6	222.0	0.90	0.91	0.88
0.01492	445.3	222.1	0.90	0.90	0.87
0.01617	441.5	222.6	0.89	0.90	0.87
0.01814	435.5	222.7	0.89	0.90	0.86
0.02558	418.2	223.6	0.87	0.88	0.84
0.05254	382.3	225.9	0.84	0.85	0.77
0.1091	345.8	228.2	0.80	0.82	0.69
0.3964	280.2	231.1	0.76	0.77	0.49
0.9862	228.7	227.9	0.81	0.74	0.33

23.5.3 Measurement of Equilibrium Constant

If we can measure the activities of electrolytes, then we can measure the equilibrium constants for weak electrolytes and weak acids or bases from the potential of electrochemical cells containing those electrolytes. This ability was demonstrated in Example 23.2 for the solubility product of AgCl, and in Section 23.3.2 for the ionization constant of water.

If the equilibrium constant for an electrode cell reaction is K, the standard Gibbs energy change for the cell is

$$\Delta G^\ominus = -\boldsymbol{nR}T \ln K$$

Combining this relationship at equilibrium with the Nernst equation shows that

$$E^\ominus = \frac{-\Delta G^\ominus}{\boldsymbol{nF}} = \frac{\boldsymbol{R}T}{\boldsymbol{F}} \ln K$$

23.5.4 Liquid–Liquid Junctions

When two immiscible electrolyte solutions are brought into electrical contact an electrical potential, called a *liquid junction potential*, is created between them. In this way, liquid junctions are useful for relating the ionic activities of ions in one solvent (e.g., water) to that in other solvents.[5] The potential difference between two liquids is important itself, and the same phenomenon applies to any two solutions that are separated by a boundary, such as the interior and exterior of a biological cell, lysosome, or micelle.

The difference in potential between two liquids (1 and 2) that contain the same ion is determined by equating their electrochemical potentials. The analysis proceeds in a similar way to that for the metal and liquid phases, and we find

$$E_2 - E_1 = E_2^{\ominus} - E_1^{\ominus} - \frac{RT}{F}\ln\left(\frac{a_2}{a_1}\right)$$

The difference in their standard potentials is

$$E_2^{\ominus} - E_1^{\ominus} = -\frac{\Delta_{\text{transfer}}G^{\ominus}}{F}$$

where $\Delta_{\text{transfer}}G^{\ominus}$ is called the *molar standard Gibb's energy of transfer* for the ion. A difficulty arises in assigning a Gibbs energy to a single ion, and a consistent standard state must be chosen. It is common to choose the Gibbs energy of transfer for the tetraphenylarsonium cation ($TPAs^+$) and the tetraphenylborate anion (TPB^-) to be equal for any pair of liquids.

Determination of the Gibbs energy of transfer determines the relative solvation energy for an ion in different liquids. Table 23.4 gives some values for the transfer of ions from liquid water to nitrobenzene. These data show that the inorganic ions with large charge to size ratios are more stable in water and the organic ions are more stable in nitrobenzene.

Table 23.4 Standard Gibbs Energy of Transfer for Some Ions from Water to Nitrobenzene.

Ion	$\Delta_{\text{transfer}}G^{\ominus}$/kJ mol^{-1}	Ion	$\Delta_{\text{transfer}}G^{\ominus}$/kJ mol^{-1}
Li^+	38.2	$(C_2H_5)_4N^+$	−5.7
Na^+	34.2	$(C_4H_9)_4N^+$	−24.0
K^+	23.4		
NH_4^+	26.8	$TPAs^+$	−35.9
H^+	32.5	TPB^-	−35.9

23.5.4.1 Ion Selective Electrodes

Ion selective electrodes exploit this phenomenon. Such an electrode is constructed much like a pH electrode except the glass membrane is replaced by a liquid or polymer membrane which is chemically designed to selectively transfer the ion of interest.

An electrode that is selective for calcium ions can be created by sealing the end of a glass tube with a cellulose membrane and filling the tube with a 0.1 M solution of a calcium salt of didecylphosphoric acid in di-*n*-octylphenylphosphonate (inner solution), which is electrically connected to a Ag/AgCl electrode through a salt bridge, see Fig. 23.14. When the electrode is immersed in an aqueous solution of calcium ions (outer solution) a liquid junction is formed between the outer solution and the organic buffer solution (inner solution) at the membrane.

A potential difference is generated at the membrane and its magnitude is controlled by the relative activity of the ions in the two liquids. The membrane potential, E_{membrane}, depends on the calcium ion activity as

$$E_{\text{membrane}} = const + \frac{RT}{F}\ln\left(\frac{a^{1/2}_{Ca^{2+},\text{solution}}}{a^{1/2}_{Ca^{2+},\text{buffer}}}\right) = const + \frac{RT}{2F}\ln\left(\frac{a_{Ca^{2+},\text{solution}}}{a_{Ca^{2+},\text{buffer}}}\right) \quad (23.65)$$

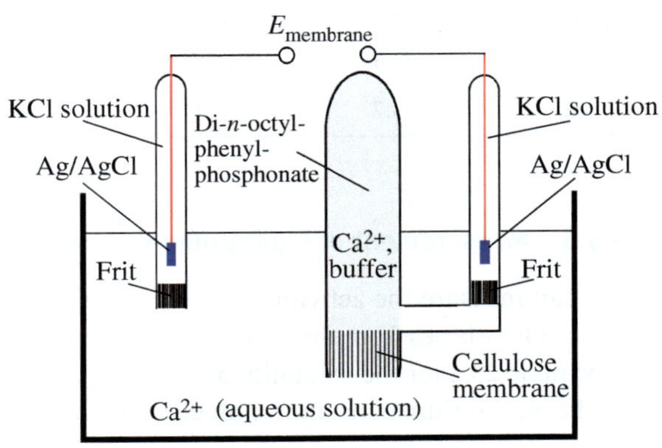

Figure 23.14 Scheme of a calcium selective electrode.

Figure 23.15 shows data for the measured potential E of a Ca^{2+} selective electrode as a function of the logarithm of $\gamma\hat{m}_{Ca^{2+}}$ of a $CaCl_2$ solution, where γ is the activity coefficient and $\hat{m}_{Ca^{2+}}$ is the molality divided by 1 mol kg^{-1}.[6]

---

5 Also, a liquid junction potential is created for the same solutions on both sides of the junction if the ion concentrations are different. In Section 23.2 we discussed this latter case with regard to the different potentials on the metal electrodes; however, the same analysis shows that an electrical potential between the two sides of the frit (or membrane) is present.

6 The data are taken from J. W. Ross, *Calcium-Selective Electrode with Liquid Ion Exchanger*, Science 1967, **156**, 1378.

Example 23.7 *The slope of the straight line through the data points in Fig. 23.15 is 36.5 mV. Show that this value is what we expect for a $CaCl_2$ solution.*

We have to consider that, for reason of electroneutrality, chloride ions contribute to the potential in addition to the Ca^{2+} ions. Because for each Ca^{2+} two Cl^- are transferred, the potential for this system is

$$E_{\text{membrane}} = const' + \frac{\boldsymbol{RT}}{2\boldsymbol{F}} \ln\left(\frac{a_{\text{Ca}^{2+}} \cdot a^2_{\text{Cl}^-}}{a_{\text{Ca}^{2+},\text{buffer}} \cdot a^2_{\text{Cl}^-,\text{buffer}}}\right)$$

Using the relations

$$a_{\text{Cl}^-} = 2a_{\text{Ca}^{2+}} \qquad \text{and} \qquad a_{\text{Ca}^{2+}} = \gamma m_{\text{Ca}^{2+}}$$

where $m_{\text{Ca}^{2+}}$ is the molality of Ca^{2+} and γ is the mean activity coefficient we obtain

$$\begin{aligned} E_{\text{membrane}} &= const' + \frac{\boldsymbol{RT}}{2\boldsymbol{F}} \ln\left(\frac{4a^3_{\text{Ca}^{2+}}}{4a^3_{\text{Ca}^{2+},\text{buffer}}}\right) \\ &= const' + \frac{3\boldsymbol{RT}}{2\boldsymbol{F}} \ln\left(\frac{\gamma m_{\text{Ca}^{2+}}}{\gamma m_{\text{Ca}^{2+},\text{buffer}}}\right) \end{aligned}$$

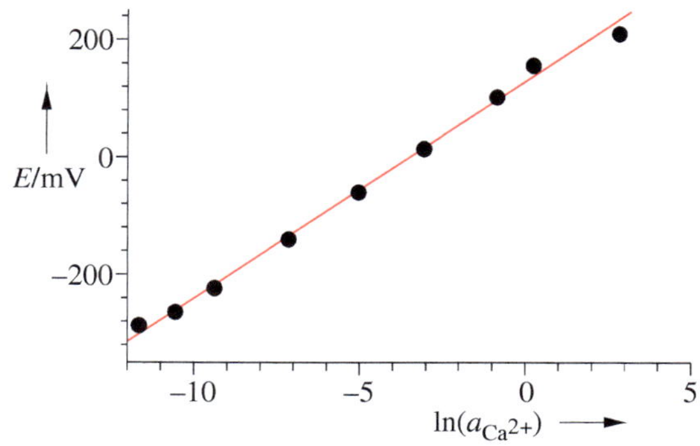

Figure 23.15 The measured potential for the Ca^{2+} selective electrode is plotted as a function of the natural logarithm of the ion activity.

Finally we collect the contribution of the buffer solution in the constant *const''*, and find

$$\begin{aligned} E_{\text{membrane}} &= const'' - \frac{3\boldsymbol{RT}}{2\boldsymbol{F}} \ln \gamma \hat{m}_{\text{Ca}^{2+},\text{buffer}} + \frac{3\boldsymbol{RT}}{2\boldsymbol{F}} \ln \gamma \hat{m}_{\text{Ca}^{2+}} \\ &= const'' + \frac{3\boldsymbol{RT}}{2\boldsymbol{F}} \ln \gamma \hat{m}_{\text{Ca}^{2+}} \end{aligned}$$

with $\hat{m}_{\text{Ca}^{2+}} = m_{\text{Ca}^{2+}}/(1 \text{ mol kg}^{-1})$. The expected slope for $T = 298$ K is $3\boldsymbol{RT}/(2\boldsymbol{F}) = 38.5$ mV.

23.6 Applications of Electrochemical Cells

A common use of electrochemical cells is to convert chemical energy into electrical energy or convert electrical energy into chemical energy (electrolysis).

23.6.1 Primary Batteries and Rechargeable Batteries

A primary battery uses chemical reactions between its components to generate electricity.

23.6.1.1 Leclanché Element (Zn/Graphite)

The half cell reactions in the Leclanché cell are (Fig. 23.16a)

$$\tfrac{1}{2}\,\text{Zn} \rightarrow \tfrac{1}{2}\,\text{Zn}^{2+} + \text{e}^-, \qquad \text{H}^+ + \text{e}^- \rightarrow \tfrac{1}{2}\,\text{H}_2 \tag{23.66}$$

The reaction product Zn^{2+} is complexed by NH_4Cl into $[Zn(NH_3)_4]^{2+}$. A graphite rod surrounded by MnO_2 (mixed with graphite for electrical conductivity) is used as a counterelectrode; MnO_2 oxidizes H_2:

$$2\,\text{MnO}_2 + \text{H}_2 \rightarrow 2\,\text{MnO(OH)} \tag{23.67}$$

This cell can be considered as a Zn electrode combined with a hydrogen electrode. For a Zn^{2+} concentration of 1 mol L^{-1} we expect a voltage of 762 mV (this is the standard potential of Zn). Because of the complex formation

$$\text{Zn}^{2+} + 4\text{NH}_4\text{Cl} \rightarrow [\text{Zn(NH}_3)_4]^{2+} + 4\text{HCl}$$

the concentration of Zn^{2+} is much smaller and the voltage becomes much larger, according to the Nernst equation (23.30). The actual cell voltage is 1.6 V. Solid zinc is used in the form of a beaker containing NH_4Cl and the counterelectrode (Fig. 23.16a).

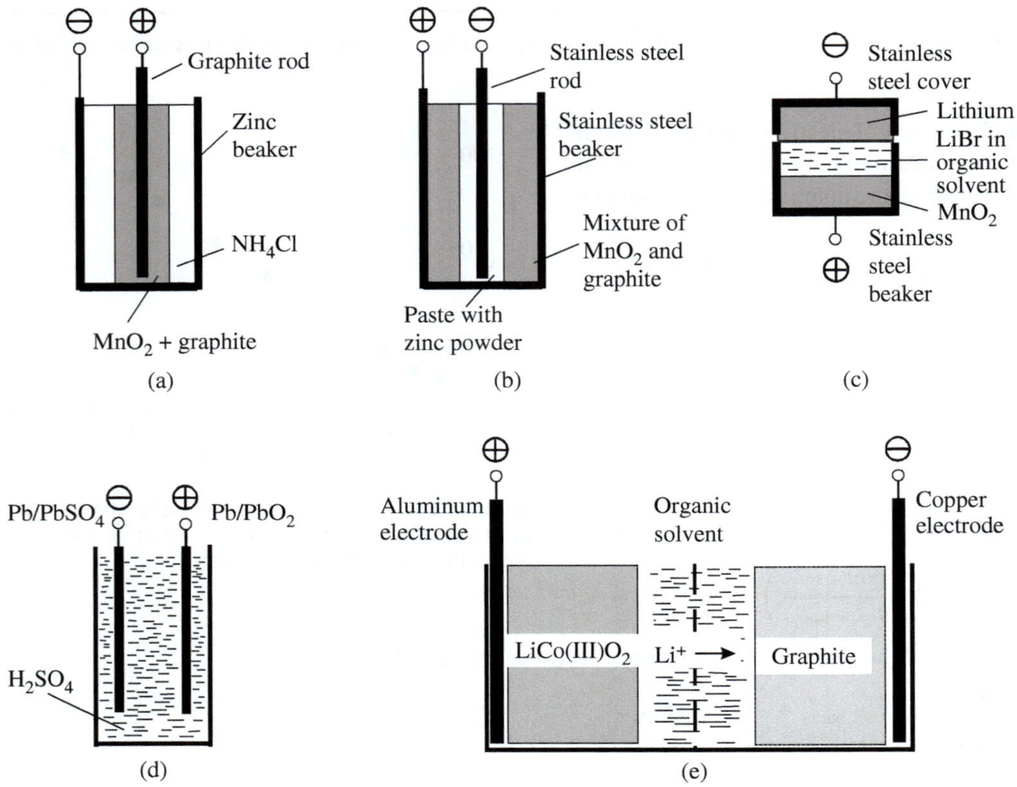

Figure 23.16 Primary and rechargeable batteries. (a) Leclanché battery. (b) Alkaline manganese battery. (c) Lithium battery. (d) Lead battery. (e) Lithium ion battery.

23.6.1.2 Alkaline Manganese Battery

This cell is a variation of the Leclanché element, where NH_4Cl is replaced by KOH (Fig. 23.16b). The electrode half cell reactions are

$$\tfrac{1}{2}\,\mathrm{Zn} + \mathrm{OH}^- \rightarrow \tfrac{1}{2}\,\mathrm{Zn(OH)_2} + \mathrm{e}^-, \qquad \mathrm{H}^+ + \mathrm{e}^- \rightarrow \tfrac{1}{2}\,\mathrm{H}_2 \tag{23.68}$$

The hydrogen is oxidized by MnO_2 in the same way as in the Leclanché cell. In this case, zinc is used in the form of powder in the inner part of the cell; the zinc powder is surrounded by alkaline MnO_2 placed in a beaker of stainless steel (Fig. 23.16b). In this case the concentration of Zn^{2+} is reduced by the formation of $Zn(OH)_2$, leading to a similar cell voltage as for the Leclanché element. This is the most common cell used nowadays.

23.6.1.3 Lithium Battery

Lithium batteries (Fig. 23.16c) are desired for their high cell voltage of about 3 V, long lifetime, high energy content, and broad temperature range. The reaction at the lithium anode is

$$\mathrm{Li} \rightarrow \mathrm{Li}^+ + \mathrm{e}^- \tag{23.69}$$

A layer of MnO_2 is used to remove electrons from the counterelectrode (stainless steel):

$$\mathrm{MnO_2} + \mathrm{e}^- - \rightarrow [\mathrm{MnO_2}]^-$$

Because Li reacts directly with water, a mixture of propylene carbonate and dimethoxyethane is used as a solvent for the electrolyte (LiBr, $LiAlCl_4$).

The overall cell reaction is

$$\mathrm{Li} + \mathrm{MnO_2} \rightarrow (\mathrm{Li}^+)\,(\mathrm{MnO_2})^-$$

From the standard voltage of Li^+ we expect a cell voltage of 3024 mV. The actual voltage of this cell is $E = 3.5$ V.

Example 23.8 *Besides MnO_2, a number of different compounds are used as electron removing agents in lithium batteries. If the cathode is a mixture of solid I_2 and poly-2-vinylpyridine (P2VP) and the solid electrolyte is LiI, what is the voltage of this cell?*

In this case the overall cell reaction is

$$2\ \text{Li} + \text{I}_2 \rightarrow 2\ \text{LiI}$$

Using the table of redox potentials (Appendix A.3), we expect a cell voltage of 3.045 + 0.535 = 3.580 V.

23.6.1.4 Rechargeable Lead Battery

Electrical energy can be stored in rechargeable batteries.

In loading a lead battery (Fig. 23.16d), Pb^{2+} (in the form of solid $PbSO_4$) is reduced to Pb at the left electrode

$$\text{PbSO}_4 + 2\text{e}^- \rightarrow \text{Pb} + \text{SO}_4^{2-} \qquad (23.70)$$

and oxidized to PbO_2 at the right electrode

$$\text{PbSO}_4 + 2\text{H}_2\text{O} \rightarrow \text{PbO}_2 + \text{H}_2\text{SO}_4 + 2\text{H}^+ + 2\text{e}^-$$

leading to the total reaction

$$2\text{PbSO}_4 + 2\text{H}_2\text{O} \rightarrow \text{Pb} + \text{PbO}_2 + 2\text{H}_2\text{SO}_4 \qquad (23.71)$$

In order to obtain the voltage of this battery, we calculate $\Delta_r G^\ominus_{298}$ for reaction (23.71). Using the data in Appendix A.1., we obtain (see Exercise E23.19) $\Delta_r G^\ominus_{298} = +397.5$ kJ mol^{-1}. Hence, we see that the reverse of reaction (23.71), the formation of $PbSO_4$, occurs spontaneously, and the left electrode becomes negative because of Equation (23.70).

According to Equation (23.13) we find that the voltage of the cell is $E = -\Delta_r G^\ominus_{298}/\boldsymbol{F}$, if one electron is needed for this process. In our case we need two electrons, thus

$$E = -\frac{397.5 \text{ kJ mol}^{-1}}{2 \cdot 96{,}485 \text{ C}} = -2.06 \text{ V}$$

Because water can be split into H_2 and O_2 when applying a voltage above 1.23 V, one may wonder why this process does not occur in the lead accumulator. This is because the production of H_2 at the negative electrode and of O_2 at the positive electrode is kinetically hindered, leading to an overvoltage, see Section 23.6.4. The overvoltage of H_2 at the lead electrode is 0.67 V, whereas there is little to no overvoltage for H_2 at a platinum electrode.

The fact that there is an overvoltage for H_2 at a lead electrode is crucial for the realization of a lead battery using an aqueous electrolyte. Without that overvoltage it would not work at all. In practice, traces of platinum (e.g., from analytical grade sulfuric acid produced in platinum vessels) are sufficient to catalyze the decomposition of water in a lead battery.

Lithium Ion battery Contrary to the case of the lithium battery discussed in Fig. 23.16c, no elemental lithium is used in a rechargeable lithium ion battery. Instead, lithium is used in the form of a solid lithium metal oxide, such as lithium cobalt oxide, LiCo(III)O_2 (Fig. 23.16e). In the charging process

$$\text{LiCo(III)O}_2 \rightarrow \text{Li}^+ + \text{Co(IV)O}_2 + \text{e}^-$$

lithium ions are formed and cobalt is oxidized to Co(IV) at the left electrode. The lithium ions move through a nonaqueous electrolyte (for example $LiBF_4$ in propylenecarbonate) to the counterelectrode consisting of graphite. Graphite (C_n) forms an intercalation compound with Li^+, LiC_n at the right electrode.

$$\text{Li}^+ + \text{C}_n + \text{e}^- \rightarrow \text{LiC}_n$$

The total process is

$$\text{LiCo(III)O}_2 + \text{C}_n \rightarrow \text{Co(IV)O}_2 + \text{LiC}_n$$

The actual voltage of a lithium ion battery is about 3.6 V, comparable to that of a lithium battery.

23.6.2 Fuel Cells

A fuel cell is a galvanic device that oxidizes a fuel directly into electrical energy. It differs from a battery in that the fuel is supplied continuously from an external source (open system) rather than being an integral part of the device (closed system), as with a battery. Figure 23.17a shows a fuel cell that operates by converting hydrogen and oxygen into water,

$$2\ H_2 + O_2 \rightarrow 2\ H_2O \qquad (23.72)$$

In such fuel cells, porous nickel electrodes (Raney nickel)[7] are used. These electrodes have large pore size toward the gas phase (allowing good contact with the gases) and small pore size toward the liquid phase (avoiding mixing of gas and liquid), see Fig. 23.17b. The hydrogen is oxidized at the positive electrode and converted into protons which diffuse through the proton exchange membrane to the cathode where they participate in the reduction at the cathode. The electrons generated at the anode flow through an external circuit, where they can perform work, and travel to the cathode where they act to reduce the oxygen to water.

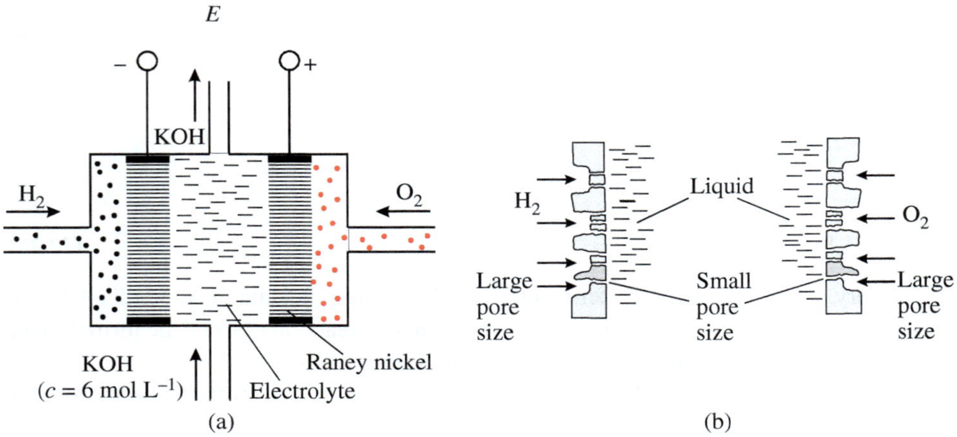

Figure 23.17 Fuel cell. (a) Setup of a hydrogen/oxygen fuel cell using electrodes of Raney nickel. The electrolyte KOH is pumped through the cell in a cycle to maintain the concentration of 6 mol L^{-1}. (b) Double layer structure at the Raney nickel electrodes: large pore size toward the gas and small pore size toward the liquid. Left: hydrogen electrode, right: oxygen electrode.

The voltage of an idealized electrochemical cell in which reaction (23.72) is carried out can be calculated from Equation (23.12)

$$E = -\frac{\Delta G}{\boldsymbol{n}F}$$

where $\boldsymbol{n}$ is the amount of electrons involved in the reaction; in our case $\boldsymbol{n} = 4\boldsymbol{n}_{O_2}$.For the fuel cell, the work done is $-\Delta_r G^{\ominus}(T = 298\ \text{K}) = 474.4\ \text{kJ mol}^{-1} \cdot \boldsymbol{n}_{O_2}$, then

$$E = \frac{474.4\ \text{kJ mol}^{-1} \cdot \boldsymbol{n}_{O_2}}{4\boldsymbol{n}_{O_2} \cdot 9.6484 \times 10^4\ \text{C mol}^{-1}} = 1.229\ \text{V}$$

(with V C = V A s = J). This fuel cell's voltage is 0.73 V, about 60% of the theoretical value of 1.23 V.

This fuel cell has a higher efficiency than a heat engine. From the considerations in Chapter 17 we know that it is impossible to convert heat completely into mechanical work, and hence to convert chemical energy completely into work by means of a heat engine. According to Section 17.2.2.4, the work done by an idealized heat engine is

$$w_{\text{heat engine}} = \eta \cdot q_{\text{rev},2} = \left(1 - \frac{T_1}{T_2}\right) \cdot q_{\text{rev},2} \qquad (23.73)$$

where T_1 and T_2 are the working temperatures of the heat engine, and $q_{\text{rev},2}$ is the heat supplied by the combustion of the fuel at temperature T_2. For the combustion of hydrogen gas

$$2\ H_2 + O_2 \rightarrow 2\ H_2O$$

7 See [7].

in a heat engine operating at the temperatures $T_1 = 300$ K and $T_2 = 800$ K we obtain

$$q_{rev,2} = -\Delta H^{\ominus}(T = 800 \text{ K}) = 533 \text{ kJ mol}^{-1} \cdot \boldsymbol{n}_{O_2}$$

and the work done by the engine is

$$\left(1 - \frac{300}{800}\right) \times (533) \text{ kJ mol}^{-1} \cdot \boldsymbol{n}_{O_2} = 333 \text{ kJ mol}^{-1} \cdot \boldsymbol{n}_{O_2}$$

$\boldsymbol{n}_{O_2}$ is the amount of oxygen being consumed. This value is only about 70% of the idealized electrochemical cell.

23.6.3 Electrolysis and Electrosynthesis

Instead of converting chemical energy into electrical energy or storing electrical energy in chemical compounds, we can perform chemical reactions by supplying electrical energy. As an example, we examine the electrolysis of an HCl solution in which hydrogen and chlorine are formed. When the reaction

$$HCl(aq) \rightarrow \tfrac{1}{2}\, H_2 + \tfrac{1}{2}\, Cl_2 \qquad (23.74)$$

takes place in an electrochemical cell (hydrogen electrode connected to a chlorine electrode in an HCl solution, Fig. 23.18a), we expect, according to Appendix A.3, a voltage of 1.36 V. Applying a somewhat higher voltage enables us to force the reaction (23.74) to occur: H_2 and Cl_2 are formed at atmospheric pressure. The formation of the gases, as bubbles, can be observed in the electrolysis cell (Fig. 23.18b).

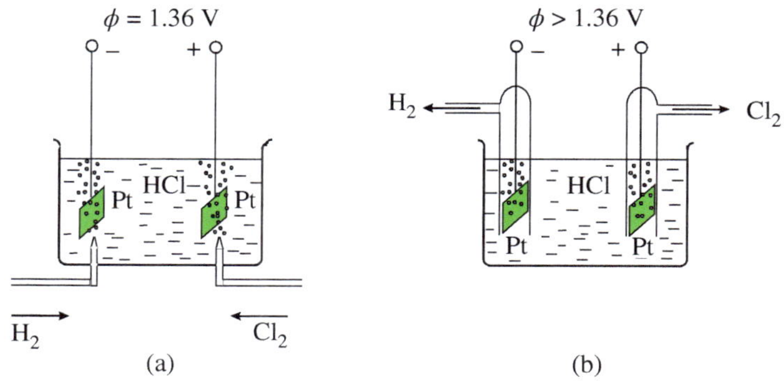

Figure 23.18 Electrochemical cell for reaction (23.74). (a) Cell working as a voltage source and (b) electrolysis cell.

By plotting the electrical current in the electrolysis cell as a function of the applied potential ϕ, we obtain the curve displayed in Fig. 23.19. Above 1.36 V the current increases linearly with ϕ, because the migration velocities of H^+ and Cl^- increase linearly with ϕ; that is, in the same time more ions are discharged at the electrodes. Below 1.36 V, H^+ ions are discharged at the negative electrode and Cl^- ions at the positive electrode, until equilibrium concentrations of dissolved H_2 and Cl_2 at the electrodes have been established. If the concentrations of the gases H_2 and Cl_2 were able to reach their equilibrium values, then the current would become zero. Because the H_2 and Cl_2 molecules diffuse away from the electrode surface and eventually escape from the liquid, their concentrations are smaller than the equilibrium value, and a small current is maintained.

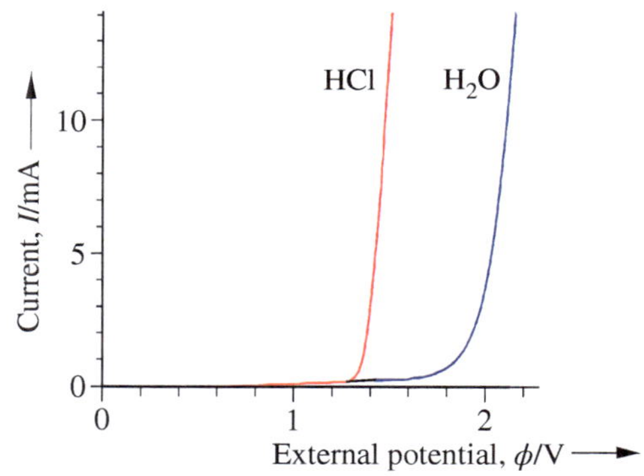

Figure 23.19 Electrolysis current I versus the external potential ϕ. The HCl electrolysis is carried out in an HCl solution (concentration 1 mol L^{-1}), and the H_2O electrolysis is carried out in an H_2SO_4 solution (concentration 1 mol L^{-1}).

Example 23.9 *The Hall–Heroult process, in which alumina (Al_2O_3) is reduced to metallic aluminum, underlies the multi-billion dollar aluminum industry. Performed in a molten salt of Na_3AlF, the major reaction that proceeds is*

$$2\, Al_2O_3 + 3\, C \rightarrow 4\, Al + 3\, CO_2$$

Assume that the reaction vessel operates with an efficiency of 50%, and calculate how many Joules of electrical energy are required to produce 100 kg of aluminum.

The $\Delta_r G^{\ominus}$ for this reaction is

$$\Delta_r G^{\ominus} = 3 \text{ mol} \cdot \Delta_f G^{\ominus}(CO_2) - 2 \text{ mol} \cdot \Delta_f G^{\ominus}(Al_2O_3)$$
$$= [3 \text{ mol} \cdot (-394.4) - 2 \text{ mol} \cdot (-1581.9)] \text{ kJ mol}^{-1} = +1980.6 \text{ kJ}$$

and the reaction is not spontaneous. However, it can be carried out in an electrochemical cell with the half cell reactions

$$4\, Al^{3+} + 12\, e^- \rightarrow 4\, Al, \qquad 6\, O^{2-} \rightarrow 6\, O + 12\, e^-$$

followed by a reaction of the carbon with the oxygen to form carbon dioxide. What must be the voltage of the electrochemical cell to make the reaction spontaneous? From Equation (23.13) we obtain

$$E = \frac{\Delta_r G^\ominus}{\boldsymbol{n}F} = \frac{+1980.6 \text{ kJ}}{12 \text{ mol} \cdot 96{,}485 \text{ C mol}^{-1}} = 1.71 \text{ V}$$

where $\boldsymbol{n}$ is the amount of electrons.

From the reaction, each atom of Al requires three electrons. 100 kg of aluminum corresponds to an amount of $\boldsymbol{n} = 100 \text{ kg}/\boldsymbol{M} = 100 \text{ kg}/(26.9815 \text{ g mol}^{-1}) = 3.706 \times 10^3$ mol. Then for the charge Q needed to produce 100 kg of aluminum we find

$$Q = 3\boldsymbol{n}F = 3 \cdot 3.706 \times 10^3 \text{ mol} \cdot 96{,}485 \text{ C mol}^{-1} = 1.1 \times 10^9 \text{ C}$$

At 1.71 V, the energy is $E \cdot Q = 1.88 \times 10^9$ J or 522 kWh. Actually, the electrolysis is performed at a voltage of 5 V, and the energy becomes 1500 kWh. This value is about one-fourth that used annually by an average US household in 1993.

23.6.4 Overvoltage

If we apply a sufficiently high voltage across two platinum electrodes immersed in an aqueous solution, we observe that hydrogen gas is formed at the negative electrode, and oxygen gas is formed at the positive electrode according to the electrode reactions:

$$H^+ + e^- \rightarrow \frac{1}{2}H_2 \text{ (negative electrode)}$$

$$OH^- \rightarrow e^- + \frac{1}{4}O_2 + \frac{1}{2}H_2O \quad \text{(positive electrode)}$$

resulting in the overall reaction

$$\frac{1}{2}H_2O \rightarrow \frac{1}{2}H_2 + \frac{1}{4}O_2 \tag{23.75}$$

The molar Gibbs energy of formation of H_2O under standard conditions is $\Delta_f G^\ominus = -237.2$ kJ mol^{-1}; thus for reaction (23.75) the change of Gibbs energy is $\Delta_r G^\ominus = \frac{1}{2} \cdot 237.2$ kJ mol^{-1} = 118.6 kJ mol^{-1}. According to Equation (23.29), the standard potential is $E^\ominus = \Delta_r G^\ominus/\boldsymbol{F} = 1.23$ V.

This voltage is smaller by 0.13 V than the voltage of 1.36 V which is necessary for the formation of H_2 and Cl_2 when electrolyzing the HCl solution (concentration 1 mol L^{-1}) considered in Section 23.6.3. For this reason we should expect that in the experiment shown in Fig. 23.18b we do not observe the formation of H_2 and Cl_2, but instead the formation of H_2 and O_2. However, in the actual experiment, the formation of H_2 and Cl_2 takes place. How can we understand this discrepancy?

We perform an electrolysis experiment with a solution of H_2SO_4 (concentration 1 mol L^{-1}). In this case the products of the electrolysis are H_2 and O_2. In Fig. 23.19 it is shown that the formation of these gases starts at a voltage of 1.7 V, instead of 1.23 V as expected theoretically. From this result we conclude that in the case of formation of oxygen at a platinum electrode our assumption that the electrode processes occur almost reversibly is not fulfilled. Actually, the formation of oxygen is kinetically hindered, and the electrical energy which is supplied to the system is larger than the electrical energy supplied in a reversible process. Accordingly, a higher voltage is needed. The difference between the measured voltage and the theoretically calculated voltage is called the *overvoltage*. Overvoltages for H_2 and O_2 at different electrode materials are collected in Table 23.5.

Table 23.5 Overvoltages of Hydrogen and Oxygen for Different Electrode Materials.

	Hydrogen		Oxygen	
Electrode	**Electrolyte**	**Overvoltage/V**	**Electrolyte**	**Overvoltage/V**
Pt (platinum black)	HCl	<0.01		
Pt	HCl	0.09	H_2SO_4	0.29
Pd	H_2SO_4	<0.01	H_2SO_4	0.30
Au	H_2SO_4	0.08	H_2SO_4	0.45
Pb	HCl	0.67	H_2SO_4	0.64
Ni	HCl	0.33		
Fe	HCl	0.40		
Ag	HCl	0.44		
Hg	HCl	1.04		

23.7 Conductivity of Electrolyte Solutions

So far the discussion has emphasized the electron transfer process at electrodes, and the subsequent flow of electrons through an external circuit. Current must flow through the solution in order to complete the electrical circuit, however. The current flow in the solution occurs by the transport of ions, which is discussed here.

23.7.1 Mobility of Ions

Consider a solution containing positive and negative ions with charges ze, e.g., a solution of NaCl in water. In an electric field the ions (Na^+, Cl^-) are accelerated; thus, a current flows through the solution. First we restrict our consideration to the case that only one kind of ion contributes to the current (we assume that the counterions are immobilized).

Two electrodes are immersed in the solution (Fig. 23.20). For a potential ϕ across the electrodes, we approximate the electrical field strength between the electrodes by

$$F = \frac{\phi}{l} \tag{23.76}$$

where l is the distance between the electrodes. The electrical field exerts a force f_{el} on the ions

$$f_{el} = ze \cdot \frac{\phi}{l} \tag{23.77}$$

Thus the positive ions are accelerated toward the negative electrode, and the negative ions are accelerated toward the positive electrode. On the other hand, the ions are decelerated by the force of friction f_{fric}. To estimate this friction, we approximate the ions as spheres moving in a viscous fluid and apply Stokes' law, see Section 14.4.4,

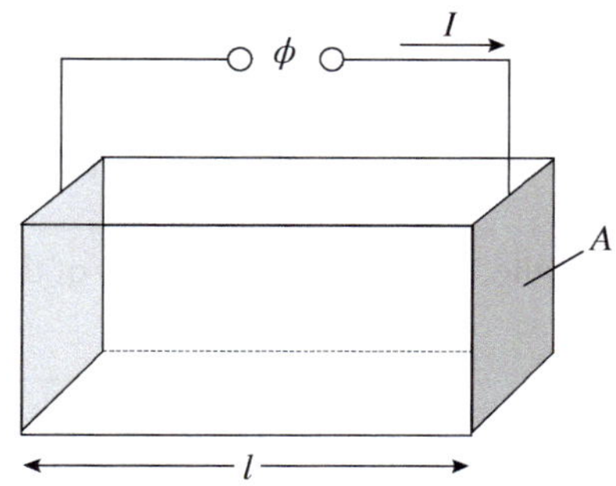

Figure 23.20 Conductivity cell.

$$f_{fric} = -6\pi\eta r v \tag{23.78}$$

where η is the shear viscosity of the solvent, r is the radius of the sphere, and v is the speed of the ions. The speed v becomes constant when the net force on the ion becomes zero:

$$f_{el} + f_{fric} = 0 \tag{23.79}$$

Then

$$v = \frac{ze}{6\pi\eta r} \cdot \frac{\phi}{l} \tag{23.80}$$

As a consequence of the movement of the ions, a current I can be measured in the external circuit.

$$I = \frac{dQ}{dt} \tag{23.81}$$

where dQ is the charge of the ions passing the cross section A of the container in time dt.

During time dt the ions travel the distance

$$dx = v \cdot dt \tag{23.82}$$

and the number of ions passing the cross section A in time dt is

$$dN = c \cdot \mathbf{N}_A A \cdot dx = c \cdot \mathbf{N}_A A v \cdot dt \tag{23.83}$$

where c is the concentration of the ions. Hence we obtain the ion current

$$I = \frac{dQ}{dt} = ze \cdot \frac{dN}{dt} = ze \cdot c \cdot \mathbf{N}_A A v \tag{23.84}$$

If we define the mobility u by

$$v = u\frac{\phi}{l} \tag{23.85}$$

we obtain

$$I = \boldsymbol{F} c z u \frac{A}{l} \cdot \phi \tag{23.86}$$

where $\boldsymbol{F} = e\mathbf{N}_A$. This expression can be simplified by introducing the conductivity κ

$$\boxed{I = \kappa \cdot \frac{A}{l}\phi, \quad \kappa = \boldsymbol{F} \cdot c \cdot z \cdot u} \tag{23.87}$$

23.7.2 Generalization

In this consideration we restricted our discussion to the case where only one kind of ion is responsible for the current I; this is the case when the mobility of the counterions is sufficiently small. Generally, if cations (charge z_+e, concentration c_+, mobility u_+) and anions (charge $-z_-e$, concentration c_-, mobility u_-) contribute to the current, we obtain

$$\kappa = \boldsymbol{F} \cdot \left(c_+z_+u_+ + c_-z_-u_-\right) \tag{23.88}$$

Because of electroneutrality,

$$c_+z_+ = c_-z_- \tag{23.89}$$

and thus

$$\kappa = \boldsymbol{F}c_+z_+ \cdot \left(u_+ + u_-\right) \tag{23.90}$$

The quantity

$$\boxed{\Lambda = \frac{\kappa}{c_+z_+} = \boldsymbol{F} \cdot \left(u_+ + u_-\right)} \tag{23.91}$$

is called the *molar conductivity* of the solution.

Example 23.10 *Speed of Ions. What is the speed of Na^+ ions in water that move from an anode to a cathode with a separation distance $l = 1$ cm and a potential difference of $\phi = 10$ V, at 25 °C and at 100 °C?*

According to Equation (23.80) the speed is

$$v = \frac{e}{6\pi\eta r} \cdot F = \frac{e}{6\pi\eta r} \cdot \frac{\phi}{l}$$

The viscosity of water at 25° C is $\eta = 0.89 \times 10^{-2}$ g s^{-1} cm^{-1} and with $r = 200$ pm (accounting for the hydration sphere of the ion) we obtain

$$v = \frac{1.60 \times 10^{-19}\ \text{C}}{6\pi \times 0.89 \times 10^{-3}\ \text{kg s}^{-1}\ \text{m}^{-1} \cdot 200 \times 10^{-12}\ \text{m}} \cdot \frac{10\ \text{V}}{10^{-2}\ \text{m}} = 48\ \mu\text{m s}^{-1}$$

The viscosity of water at 100°C is $\eta = 0.28 \times 10^{-2}$ g s^{-1} cm^{-1}; then $v = 153$ µm s^{-1}. These values are in accordance with the results of direct experiments on the speed of ions. The speed of the ions is remarkably low. The speed increases significantly with temperature because of the decreasing viscosity of the solvent. The agreement of the calculated values with experiment is remarkable.

Using the result for 25 °C, the displacement of an Na^+ ion within a time of $t = 1000$ s is 48 µm s^{-1} · 1000 s = 48 mm. We compare this displacement to that obtained by diffusion of the particle, using the Einstein and Smoluchowski equation and the Stokes–Einstein equation (see Sections 14.3.3 and 14.4.4)

$$\overline{\Delta x^2} = 2D \cdot t, \qquad D = \frac{kT}{6\pi\eta r}$$

Thus we obtain

$$D = \frac{1.38 \times 10^{-23}\ \text{J K}^{-1} \cdot 298\ \text{K}}{6\pi \times 0.89 \times 10^{-3}\ \text{kg s}^{-1}\ \text{m}^{-1} \cdot 200 \times 10^{-12}\ \text{m}} = 1.23 \times 10^{-9}\ \text{m}^2\ \text{s}^{-1}$$

and the mean displacement within 1000 s is

$$\text{mean displacement} = \sqrt{2D \cdot t} = \sqrt{2 \cdot 1.23 \times 10^{-9}\ \text{m}^2\ \text{s}^{-1} \cdot 1000\ \text{s}} = 1.6\ \text{mm}$$

Hence the distance traveled by diffusion is only 3% of that traveled in an electric field of $F = 10$ V cm^{-1}.

Using Equation (23.80), we can calculate radii r of the ions from the mobility u (Table 23.6). The radii r are much larger than the ionic radii discussed in Chapter 13, because of the solvation shell formed by the solvent molecules. The solvation shell increases with increasing field strength at the surface of the considered ion. Therefore Li^+ has the lowest mobility for the alkali ions, and Rb^+ has a much higher mobility.

23.7.3 Mobility of H^+

The data in Table 23.6 show that the mobility of an H^+ ion is about seven times higher than that for Na^+. Because of its hydration sphere, the radius r cannot be significantly smaller than 200 pm. What causes the high mobility of H^+?

Water molecules are connected by hydrogen bonds. If an H^+ is attached to an H_2O molecule, an additional bond is formed:

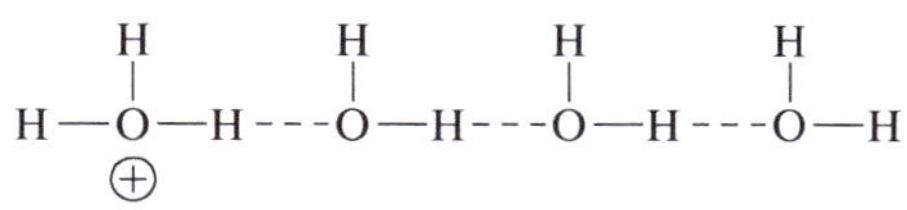

Because of the hydrogen bonding network, a positive charge can move to the adjacent water molecule by exchanging one OH bond for a hydrogen bond and vice versa

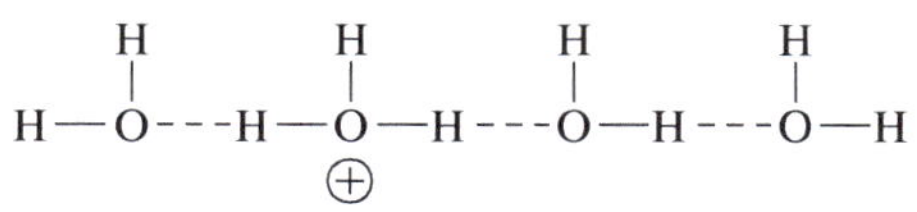

Table 23.6 Mobility u and Radius r for Some Ions in Aqueous Solution.

Ion	$u/10^{-8}$ m^2 s^{-1} V^{-1}	r/pm
Li^+	4.01	238
Na^+	5.19	184
Rb^+	7.92	120
Cs^+	7.85	122
F^-	5.70	168
Cl^-	7.91	121
I^-	7.96	120
H^+	36.2	26
OH^-	20.6	46

This process (*Grotthus mechanism*) continues along the chain of hydrogen-bonded water molecules. It is much faster than the migration of a proton. This mechanism was postulated by Theodor von Grotthus (1785–1822) in the beginning of the nineteenth century. Recently direct experiments, using ultrashort laser pulses of 150 fs, have revealed that this mechanism operates for proton transport [8].

23.7.4 Ion Transport Through Membranes

Biological cells must transport material between the inside and the outside as a way to sustain their viability and to communicate with the rest of the organism (or environment). Signals between nerve and muscle cells are mediated by changes in ion currents; ions such as Na^+, K^+, Ca^{2+}, and Cl^- which have different activities between the inside and outside of a cell membrane. B. Katz proposed the existence of "ion channels," the smallest unit of conductivity in biological membranes in the early 1970s, and in 1976 Neher and Sakmann used a "patch- clamp" electrode to measure the ion currents through single channels in a cell membrane of muscle fibers. These experiments provided the first proof that ion channels through a membrane exist. Since that time ion channels (proteins that span the membrane) have been extensively characterized, structurally and kinetically.

In a *patch clamp* experiment (Fig. 23.21a–c) an extracellular pipette filled with a Ringer's solution (0.73% NaCl, 0.02% KCl, 0.01% $CaCl_2$, 0.02% $NaHCO_3$) is held on a membrane. It is connected to a metal electrode at a voltage bias with respect to a second, intracellular electrode. The extracellular pipette has a very small opening (about 1 μm). The opening is pressed

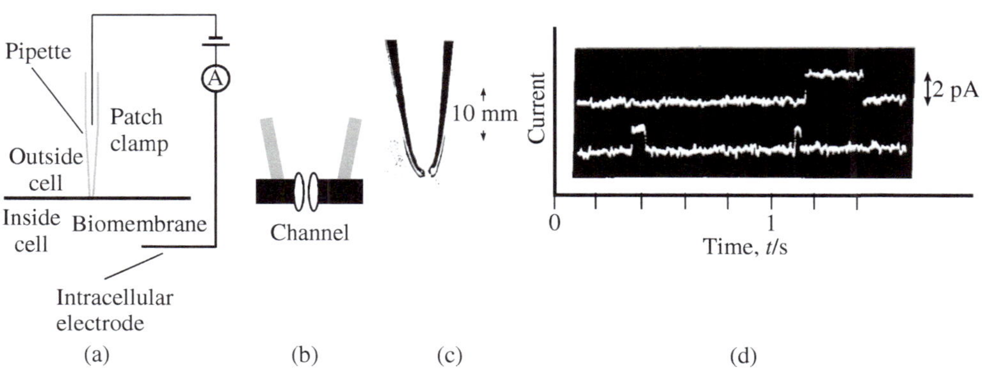

Figure 23.21 Biomembrane: single-channel recording. (a) Electric circuit using the patch clamp technique. (b) Enlargement of the channel in the membrane. (c) photograph of the tip of a capillary. (d) Two traces showing the electric current measured on opening and closing a single channel. (Adapted from Ref. [9]).

against the membrane. To avoid leakage current, the contact between the pipette and the membrane must be carefully controlled. By making the pipette opening very small and creating a good seal the membrane stray ion currents can be eliminated and a single ion channel can be measured. Figure 23.21d shows some of the data. One can see jumps in the current from the baseline value. These jumps correspond to individual openings of the ion channel. The length of the openings can be seen to vary from trace to trace. A current of about 10^{-12} A is observed during the opening of the channel for about 10^{-4} s. Then a number of about $(10^{-4} \cdot 10^{-12})/(1.6 \times 10^{-19}) \approx 500$ ions pass the channel. This is a powerful method to investigate the properties of the channels, and how they can be affected by an external influence—for example, by drugs.

23.8 Conclusion

Electrochemical cells are of fundamental and practical significance. On a practical level, electrochemical cells are very important as devices for the interconversion of electrical and chemical energy and as devices for the sensing and quantification of chemicals. On a more fundamental level, electrochemical cells often provide a convenient method for determining important properties of molecular ensembles, such as activities, equilibrium constants, and Gibbs energy changes. In a heuristic sense, they demonstrate the coupling of chemical reactions to create more complex systems and functionality.

The discussion in this chapter has addressed equilibrium properties of electrochemical cells. This feature simplifies our analysis and allows us to analyze the features of the system while ignoring details of the transport of ions, or electrons, from one phase of the system to another. We need only know the initial and final states.

23.9 Exercises

E23.1 Prove that the $\Delta G^{\ominus}$ calculated from the cell potential of the electrochemical cell in Fig. 23.1 are the same as calculated from $\Delta H^{\ominus}$, $\Delta S^{\ominus}$, and $\Delta G^{\ominus}$ in Table A.1 for 298 K.

E23.2 Consider the reaction

$$2\ \text{Ag(s)} + \text{Cl}_2\text{(g)} \rightleftarrows 2\ \text{Ag}^+\text{(aq)} + 2\ \text{Cl}^-\text{(aq)}$$

which may be used to construct an electrochemical cell. The electrical potential of this cell is 1.1372 V at 298 K, and its potential has a temperature dependence of -0.595 mV K^{-1}. What is the Gibbs energy change associated with this cell when 1 mol of electrons have passed the cell by the conversion of reactants into products? What is the change in enthalpy and the change in entropy for the reaction?

E23.3 Using a standard potential of $E^{\ominus} = -0.113$ V versus NHE for the half-cell reaction

$$\text{NAD}^+ + 2\text{e}^- + \text{H}^+ \rightarrow \text{NADH}$$

calculate $\Delta_r G^{\ominus\prime}$(pH= 7) for the reaction

$$\text{NAD}^+ + \text{H}_2 \rightarrow \text{NADH} + \text{H}^+$$

E23.4 The redox protein Cytc_1 functions by oxidation and reduction of an iron ion which is bound by a heme unit of the protein. Use its reduction potential of $E^{\ominus} = 0.220$ V to calculate $\Delta_r G^{\ominus\prime}$ for the reaction

$$\text{Cytc}_1\ \text{Fe(II)} + \text{H}^+ \rightarrow \text{Cytc}_1\ \text{Fe(III)} + \frac{1}{2}\text{H}_2$$

E23.5 In the cell in Fig. 23.5, we used different H^+ concentrations in the two containers, but the hydrogen is under the same pressure P. Alternatively, we can choose equal H^+ concentrations, but different H_2 pressures P_1 and P_2. This can be achieved by bubbling H_2 at atmospheric pressure (P_1) on the left side and, say, a 1:9 mixture of H_2 and argon (as an inert gas) on the right side; then $P_1/P_2 = 10$. Derive an expression for the cell potentials in terms of the partial pressure of hydrogen on each side.

E23.6 Write the chemical reaction for an electrochemical cell consisting of an SHE connected to a half-cell with a platinum electrode immersed in a solution containing a mixture of Fe^{2+} and Fe^{3+}. Write a mathematical expression for the standard potential of this cell and determine its value for $c_{Fe^{2+}} = 0.1 \cdot c_{Fe^{3+}}$.

E23.7 Write the Nernst equation for the electrochemical cell reaction

$$\tfrac{1}{2}\ \text{Cu}^{2+} + \tfrac{1}{2}\ \text{H}_2 \rightarrow \tfrac{1}{2}\text{Cu} + \text{H}^+$$

in which $c_{HCl} = 1.184$ mol L^{-1}, $P_{H_2} = 1$ bar (i.e., conditions for an SHE). Determine the cell potential for $c_{Cu^{2+}}$ of 1.0×10^{-3} mol L^{-1}, 1.0×10^{-6} mol L^{-1}, and 1.0×10^{-9} mol L^{-1}.

E23.8 As for the quinone/hydroquinone reaction the cell potential for the reaction

$$\frac{1}{2}\text{NAD}^+ + \frac{1}{2}\text{H}^+ + \frac{1}{2}\text{H}_2(\text{SHE}) \rightarrow \frac{1}{2}\text{NADH} + \text{H}^+(\text{SHE})$$

depends on pH. Using $E^\ominus = -0.113$ V, evaluate the cell potential at pH values of 5.0, 7.0, and 9.0. Comment on the change in cell potential and compare it to the pH dependence of the quinone/hydroquinone cell potential.

E23.9 Use the data point at $c = 0.00573$ mol L^{-1} in Table 23.3 to calculate the mean activity coefficient of the ions. Compare this experimental result to the prediction of the Debye–Hückel law and the Debye Hückel limiting law, Equations (23.63) and (23.64).

E23.10 Calculate the cell potential of a calomel electrode. Compare this calculated value with the experimental value of 281 mV versus NHE and determine the activity coefficient of the chloride ion.

E23.11 Determine the ideal cell potential for a Mallory Cell which operates with the cell reaction

$$\text{Zn} + \text{HgO} \rightarrow \text{ZnO} + \text{Hg}$$

(Hint: Identify the two half-cell reactions and find their potentials).

E23.12 Upon loading the Ni/Cd accumulator Ni^{3+} is reduced to Ni^{2+}, and Cd is oxidized to Cd^{2+} under basic conditions. For the Ni half-cell reaction use

$$2\text{NiO(OH)} + 2\text{H}_2\text{O} + 2\,e^- \rightarrow 2\text{Ni(OH)}_2 + 2\text{OH}^- \qquad E^\ominus(\text{Ni}) = 0.48 \text{ V}$$

Propose a reaction for the Cd half-cell and use it to compute a cell voltage. Compare your result to the 1.29 V EMF of the battery. (Note: You can find the relevant half-cell reactions in the *CRC Handbook of Chemistry and Physics*).

E23.13 One can build an oxygen electrode by bubbling oxygen over a platinum foil and coupling it to a standard hydrogen electrode. Write a cell reaction for such an electrochemical cell. Using your result, write an expression for the electrode potential in terms of the pH and the oxygen pressure. What is the Gibbs energy change for this cell? What is its standard potential?

E23.14 Acceleration of Na^+ Ions. What is the time required for Na^+ ions to reach a constant speed after switching on an electrical field strength of 1 V cm^{-1} ?

E23.15 Plot the molar conductivity Λ data in the table for KCl solutions at 298 K to determine the limiting conductance (at infinite dilution) Λ_0. Show that these data agree with Kohlrausch's law, namely $\Lambda = \Lambda_0 - A \cdot \sqrt{c}$ in which A is an empirical constant and c is the concentration.[8]

$c/10^{-4}$ mol L^{-1}	0.32576	2.6570	3.5217	6.0895	11.321	20.568	32.827
Λ/cm^2 Ω^{-1} mol^{-1}	149.33	148.38	148.12	147.52	146.76	145.71	144.64

E23.16 The limiting conductance for a salt is equal to the sum of the conductance for each ion; i.e., for a 1:1 salt we have $\Lambda_0 = \lambda_+^0 + \lambda_-^0$. If the limiting conductance of chloride is $\lambda_{Cl^-}^0 = 76.35$ cm^2 Ω^{-1} mol^{-1} at 298 K, determine the limiting conductance for potassium and sodium ions from the data: Λ_0(NaCl) = 126.45 cm^2 Ω^{-1} mol^{-1} and Λ_0(KCl) = 149.85 cm^2 Ω^{-1} mol^{-1}.

E23.17 Explain how you might use conductivity data for a weak acid (e.g., acetic acid) to determine the acid's pK_a.

E23.18 In Section 23.1 we considered the reaction

$$\text{AgI} + \tfrac{1}{2}\text{ Pb} \rightarrow \text{Ag} + \tfrac{1}{2}\text{ PbI}_2$$

and used the electrochemical cell properties to find its $\Delta_r G^\ominus$, $\Delta_r H^\ominus$, and $\Delta_r S^\ominus$. Use the thermodynamic data in Appendix A.1 to calculate $\Delta_r G^\ominus$ for the reaction and compare them to the values found electrochemically.

E23.19 The chemical reaction

$$2\text{PbSO}_4 + 2\text{H}_2\text{O} \rightarrow \text{Pb} + \text{PbO}_2 + 2\text{H}_2\text{SO}_4$$

occurs when charging a lead rechargeable battery. Calculate $\Delta_r G^\ominus$ for this reaction.

8 Data are taken from [10].

23.10 Problems

P23.1 The cells discussed in the chapter rely on activity differences or chemical differences to generate the Gibbs energy difference that drives the reaction, hence the EMF of the cell. An electrochemical cell can also be generated by mechanical energy differences between the two half-reactions. Consider the case of two Hg electrodes, of which one of the mercury columns has a 10 cm height and the other has a 110 cm height (in the gravitational field of the earth, 9.8 m s^{-2}), in contact with the same solution of 0.1M KI. In this arrangement the liquid mercury in the 110 cm column dissolves in the solution as Hg_2^+ and diffuses to the 10 cm column where it is reduced to Hg. The process proceeds until the columns have equal height. Determine the cell potential for this mechanical system (you may assume the KI behaves ideally and the cross-sectional areas A of the columns are 1 cm^2; $\rho(Hg) = 13546$ kg m^{-3}).

P23.2 Consider the experiment depicted by Fig. 23.4. a) How does E change if KI is added to the $AgNO_3$ solution instead of KCl? The solubility product of AgI is $K_s = 1.6 \times 10^{-16}$. b) What will E be when a KCl solution (0.1 mol L^{-1}) is added to the left compartment and a KCN solution to the right compartment? The complex formation constant of $[Ag(CN)_2]^-$ is $K = 10^{21}$. In all cases the concentration of excess reagent is 0.1 mol L^{-1}.

P23.3 Compare the calculated values for the saturated calomel and the silver chloride electrode with the experimental values and determine the activity coefficients of the ions. Compare this experimental activity coefficient to that you calculate from the extended Debye–Hückel model.

P23.4 The temperature dependence of the standard potential for a silver-silver chloride cell (versus the SHE) was determined by Harned and Ehlers [6]. Use these data to determine the enthalpy and entropy change for this cell.

T/K	273	283	293	303	313	323	333
$E^\ominus$/mV	236.34	231.26	225.51	219.12	212.00	204.37	196.20

P23.5 In the same study as that cited in P23.4, the workers measured the cell potential for a silver-silver chloride cell (versus the SHE) as a function of HCl concentration, at 298 K. Analyze these data to determine the activity of HCl and compare your result with the predictions of the Debye–Hückel model.

$\hat{m}$	E/V	$\hat{m}$	E/V
0.0001	0.69620	0.02	0.43022
0.0002	0.66083	0.05	0.38587
0.0005	0.61421	0.1	0.35239
0.001	0.57912	0.2	0.31871
0.002	0.54421	0.5	0.27229
0.005	0.49841	1	0.23328
0.01	0.46416	2	0.18634

P23.6 A variation of the silver-silver chloride electrode (Problem P23.4 and 23.5) can be used to determine the dissociation constant of acetic acid. In this case known amounts of CH_3COOH (molality $\hat{m}_1$, HAc), CH_3COONa (molality $\hat{m}_2$, NaAc), and NaCl (molality $\hat{m}_3$) are used for the acid solution, and the cell potential is measured as a function of concentration and temperature. For this case, show that the cell potential may be written as

$$E - E^\ominus + \frac{RT}{F}\ln\left(\frac{\hat{m}_{\mathrm{HAc}}\hat{m}_{\mathrm{Cl}}}{\hat{m}_{\mathrm{Ac}}}\right) = -\frac{RT}{F}\ln\left(\frac{\gamma_{\mathrm{HAc}}\gamma_{\mathrm{Cl}}}{\gamma_{\mathrm{Ac}}}\right) - \frac{RT}{F}\ln K_a$$

where K_a is the acid dissociation constant of acetic acid. At low ionic strength (infinite dilution) the ratio of the activity coefficients on the right-hand side goes to unity. Use the data in the table to determine the acid dissociation constant of acetic acid by extrapolating to the case of infinite dilution. Perform the analysis for each of the three temperatures (278, 293, and 308 K) and then analyze the enthalpy and entropy change for the ionization of acetic acid. Use the values $E^\ominus$(278 K) = 0.23386 V; $E^\ominus$(293 K) = 0.2250 V; and $E^\ominus$(308 K) = 0.21591 V. Compare your result to literature values.

$\hat{m}_1$	$\hat{m}_2$	$\hat{m}_3$	I	$E_{T=278}$	$E_{T=293}$	$E_{T=308}$
mol kg^{-1}				V		
0.004779	0.004599	0.004896	0.00951	0.62392	0.63580	0.64722
0.012035	0.011582	0.012426	0.02403	0.60183	0.61241	0.62264
0.021006	0.020216	0.021516	0.04175	0.58855	0.59840	0.60792
0.04922	0.04737	0.05042	0.09781	0.56833	0.57699	0.58529
0.08101	0.07796	0.08297	0.16095	0.55667	0.56456	0.57213
0.09056	0.08716	0.09276	0.17994	0.55397	0.56171	0.56917

P23.7 According to Equation (23.73) the work done by an idealized heat engine is

$$w_{\text{heat engine}} = \eta \cdot q_{\text{rev},2} = \left(1 - \frac{T_1}{T_2}\right) \cdot q_{\text{rev},2}$$

where T_1 and T_2 are the working temperatures of the heat engine, and $q_{\text{rev},2}$ is the heat supplied by the combustion of the fuel at temperature T_2. Compare $w_{\text{heat engine}}$ to the work done by performing the same chemical reaction in a fuel cell.

P23.8 In Section 23.5.4, we discussed ion selective electrodes. A common issue with such devices is that they can respond to the presence of ions other than the desired one, i.e., the analyte. A common way to account for this behavior is through the empirical relationship

$$E = E^{\ominus} - \frac{RT}{F} \ln\left(a_1 + K \cdot a_2\right)$$

where a_1 is the activity of the analyte ion, a_2 is the activity of some other ion to which the electrode responds, and K is an empirical constant called a selectivity ratio. Provide a rationalization for this equation.

An iodide selective electrode, constructed using a silicone rubber membrane that is soaked in silver iodide solution, also responds to other anions in test solutions. The data in the table reports the change in electrode response between a 0.1 molar solution of KI and the corresponding 0.1 molar solution of some other anion. Use these data to determine the selectivity ratio for the electrode against the anions.

Anion	Br^-	Cl^-	SO_4^{2-}	PO_4^{3-}
ΔE/V	−0.140	−0.314	−0.451	−0.341

P23.9 Consider the experimental reduction potentials and electron affinities for the aromatic hydrocarbons in the table. Determine if the observed reduction potential E (actually a half-wave potential measured in dimethylformamide) correlates with the electron affinity E_A. Use a thermodynamic cycle to explain your findings and discuss any approximations that are important for your analysis. Values are taken from the NIST Webbook and from [11].

Molecule	E/V, versus SCE	E_A/eV
Pyrene	−2.09	0.58
Anthracene	−1.95	0.52
1,2-Benzanthracene	−2.0	0.63
Perylene	−1.67	0.97
Tetracene	−1.58	1.07

P23.10 Use the conductance data in the Table to determine the limiting conductances of KCl and NaCl at T = 298 K.

$c/10^{-3}$ mol L^{-1}	0.50	1.0	2.0	5.0	10.0
Λ/cm^2 Ω^{-1} mol^{-1} for KCl	147.79	146.95	145.79	143.59	141.30
Λ/cm^2 Ω^{-1} mol^{-1} for NaCl	124.54	123.77	122.69	120.67	118.55

P23.11 Analyze the Gibbs energy for ion solvation of the alkali metals in water by using the Born model with different measures of the ion's radius. In the first case, determine if the ion solvation energy is linear with $1/r_{crystal}$ and compare your best fit slope to the value predicted by the Born model (Section 23.4.2). In the second case, determine if the solvation energy is linear with $1/r_{ion\text{-}H_2O}$ and compare your best fit slope to the value predicted by the Born model. The Born model treats the solvent as dielectric continuum with a relative permittivity of $\hat{\varepsilon}_s$. Discuss this approximation in terms of the two different cavity radii used in your analysis and contrast it with what you would expect from a more physical/molecular perspective.

Ion	Cs^+	Rb^+	K^+	Na^+	Li^+
$r_{crystal}$/pm	184	163	152	116	88
$r_{ion\text{-}H_2O}$/pm[9]	314	289	280	236	208
$\Delta_{solv}G^{\ominus}(M^+)$/kJ mol^{-1}	−278.2	−315.2	−333.8	−405.6	−505.8

Lastly, analyze the ion solvation energy by assuming that the Born model applies with an effective radius for each ion of $r_{crystal} + C$, where C is an adjustable parameter.

P23.12 In Exercise E23.14 we found that the speed of an ion depends inversely on η, the viscosity of the solvent. Use the experimental data for Λ_0 for KCl (molar conductivity at infinite dilution) and η for water to prove this dependency.

T/K	288	298	308	318
Λ_0/cm^2 Ω^{-1} mol^{-1}	120.8	149.6	181.9	214.0
η/g s^{-1} cm^{-1}	1.138×10^{-2}	0.890×10^{-2}	0.719×10^{-2}	0.596×10^{-2}

References

1 R. Gomer and G. Tryson, An experimental determination of absolute half-cell emf's and single ion free energies of solvation, *J. Chem. Phys.*, 1977, **66**, 4413.

2 S. Trasatti, The absolute electrode potential: an explanatory note (Recommendations 1986), *Pure Appl. Chem.*, 1986, **58**, 955.

3 M.D. Tissandier, K.A. Cowen, W.Y. Feng, E. Gundlach, M.H. Cohen, A.D. Earhart, J.V. Coe, and T.R. Tuttle, Jr., The proton's absolute aqueous enthalpy and Gibbs free energy of solvation from cluster-ion solvation data, *J. Phys Chem.*, 1998, **A102**, 7787.

4 R.D. Shannon and C.T. Prewitt, Effective ionic radii in oxides and fluorides, *Acta Crystallogr.*, 1969, **B 25**, 925.

5 A.K. Covington, R.G. Bates, and R.A. Durst, Definition of pH scales, standard reference values, measurement of pH and related terminology, *Pure Appl. Chem.*, 1985, **57**, 531.

6 H.S. Harned and R.W. Ehlers, The thermodynamics of aqueous hydrochloric acid solutions from electromotive force measurements, *J. Am. Chem. Soc.*, 1933, **55**, 2179.

7 D. Linden and T.B. Reddy eds., *Handbook of Batteries* (McGraw Hill, New York, 2002) 3rd Edition.

8 O.F. Mohammed, D. Pines, J. Dreyer, E. Pines, and E.T.J. Nibbering, Sequential proton transfer through water bridges in acid-base reactions, *Science*, 2005, **310**, 83.

9 E. Neher, B. Sakmann, and J.H. Steinbach, The extracellular patch clamp: a method for resolving currents through individual open channels in biological membranes, *Pflügers Arch.*, 1978, **375**, 219.

10 T. Shedlovsky, The electrolytic conductivity of some uni-univalent electrolytes in water at 25 deg, *J. Am. Chem. Soc.*, 1932, **54**, 1411.

11 S.L. Murov, I. Carmichael, and G.L. Hug, *Handbook of Photochemistry* (Marcel Dekker, New York, 1993) 2nd Edition.

12 Y. Marcus, Ionic radii in aqueous solutions, *Chem. Rev.*, 1988, **88**, 1475.

9 $r_{ion\text{-}H_2O}$ is an empirically determined mean distance between the metal ion and the first shell of the water molecules; see [12].

24

Chemical Kinetics

CHAPTER MENU

24.1 Introduction

In the previous chapters we discussed the conditions for a chemical reaction equilibrium to change spontaneously and move toward a new equilibrium state. We did not ask, however, how much time it takes until the equilibrium has been established. Often, this question is of great importance. For instance, the condition $\Delta G \leq 0$ is fulfilled for the oxidation of organic compounds in air at standard pressure; however, the reaction is extremely slow, and most organic compounds are stable in air at room temperature.

A reaction rate is the speed at which the reactant(s) convert into the product(s). This speed or rate is written in terms of the time derivative of the reactant, or product, concentration. For the reaction

$$\mathrm{A} + \mathrm{B} \rightarrow \mathrm{C} \tag{24.1}$$

we define the reaction rate as

$$\boxed{Rate = \frac{\mathrm{d}c_\mathrm{C}}{\mathrm{d}t} = -\frac{\mathrm{d}c_\mathrm{A}}{\mathrm{d}t}} \tag{24.2}$$

where c is the concentration. The minus sign accounts for the fact that the concentration of the reactants decreases ($\mathrm{d}c_\mathrm{A}/\mathrm{d}t$ is negative), and the rate of reaction is considered to be a positive quantity.

For reactions whose stoichiometry is not as simple as that indicated here, we must account for the scaling of the concentration changes by the reaction's stoichiometry. Our general definition of the reaction rate for reaction

$$\alpha\mathrm{A} + \beta\mathrm{B} + \cdots \rightarrow \gamma\mathrm{C} + \delta\mathrm{D} + \cdots \tag{24.3}$$

is

$$Rate = \frac{1}{\gamma}\frac{\mathrm{d}c_\mathrm{C}}{\mathrm{d}t} = \frac{1}{\delta}\frac{\mathrm{d}c_\mathrm{D}}{\mathrm{d}t} = -\frac{1}{\alpha}\frac{\mathrm{d}c_\mathrm{A}}{\mathrm{d}t} = -\frac{1}{\beta}\frac{\mathrm{d}c_\mathrm{B}}{\mathrm{d}t} \tag{24.4}$$

Principles of Physical Chemistry, Third Edition. Hans Kuhn, David H. Waldeck, and Horst-Dieter Försterling.

Companion website: www.wiley.com/go/Kuhn/physicalchemistry3e

The reaction rate depends on a number of system parameters and properties. Some system variables that can modify the reaction rate are temperature, pressure, and concentration. In addition to the system variables, the reaction rate may depend on the transport of the reactants, the coupling of reactions with each other, and the concentration of inert species (i.e., molecules which do not transform into a different chemical form) such as catalysts.

This chapter discusses the observed phenomena (primarily rate laws), their behavior, and their determination. The discussion begins by using a simple kinetic theory for gas molecule collisions and reaction. Then it proceeds to describing elementary reactions and their combination into forming more complex reaction mechanisms. The chapter ends by discussing experimental methods that are used to monitor chemical reaction rates.

24.2 Collision Theory for Gas Reactions

24.2.1 Counting the Number of Collisions

In Chapter 14 we considered the mean speed of a molecule in a gas, and calculated its mean free path and the frequency of collisions. These properties are very important for understanding reactions because they allow us to calculate how often the reactants are likely to encounter each other.

Imagine a molecule A moving in a gas of $N(\mathrm{B})$ molecules of type B. According to Equation (14.45), the number z_{coll} of collisions of one molecule A with molecules B in time Δt is

$$z_{\mathrm{coll}} = \frac{\overline{v} \cdot \Delta t}{\overline{\lambda}} \tag{24.5}$$

where, according to Equation (14.36), $\overline{v}$ is the mean speed

$$\overline{v} = \sqrt{\frac{8kT}{\pi m_{\mathrm{A}}}} \tag{24.6}$$

and $\overline{\lambda}$ is the mean free path of molecule A. The total number Z_{coll} of collisions of $N(\mathrm{A})$ molecules A in a gas of $N(\mathrm{B})$ molecules B is

$$Z_{\mathrm{coll}} = z_{\mathrm{coll}} \cdot N(\mathrm{A}) = \frac{\overline{v} \cdot \Delta t}{\overline{\lambda}} \cdot N(\mathrm{A}) \tag{24.7}$$

If we assume that molecules B have a much larger mass (mass m_{B}) than molecules A (mass m_{A}), then according to Equation (14.40) we can write that

$$\overline{\lambda} = \frac{V}{\sigma N(\mathrm{B})} = \frac{V}{\pi (r_{\mathrm{A}} + r_{\mathrm{B}})^2 N(\mathrm{B})} \tag{24.8}$$

where V is the volume of the gas, $\sigma = \pi (r_{\mathrm{A}} + r_{\mathrm{B}})^2$ is the collision cross section, and r_{A}, r_{B} are the radii of the colliding molecules (molecules considered as spheres). Combining Equations (24.7) and (24.8), we find

$$Z_{\mathrm{coll}} = \sqrt{\frac{8kT}{\pi m_{\mathrm{A}}}} \cdot \pi (r_{\mathrm{A}} + r_{\mathrm{B}})^2 \frac{1}{V} \cdot N(\mathrm{A}) \cdot N(\mathrm{B}) \cdot \Delta t \tag{24.9}$$

In the general case when the masses m_{A} and m_{B} are arbitrary, according to Equation (14.42) the mean free path is

$$\overline{\lambda} = \frac{V}{\pi (r_{\mathrm{A}} + r_{\mathrm{B}})^2 N(\mathrm{B})} \frac{1}{\sqrt{1 + m_{\mathrm{A}}/m_{\mathrm{B}}}}$$

and instead of Equation (24.9) we obtain

$$Z_{\mathrm{coll}} = \sqrt{\frac{8kT}{\pi m_{\mathrm{A}}}} \cdot \sqrt{1 + m_{\mathrm{A}}/m_{\mathrm{B}}} \cdot \pi (r_{\mathrm{A}} + r_{\mathrm{B}})^2 \frac{1}{V} \cdot N(\mathrm{A}) \cdot N(\mathrm{B}) \cdot \Delta t$$

or

$$Z_{\mathrm{coll}} = \sqrt{\frac{8kT}{\pi \mu}} \cdot \pi (r_{\mathrm{A}} + r_{\mathrm{B}})^2 \frac{1}{V} \cdot N(\mathrm{A}) \cdot N(\mathrm{B}) \cdot \Delta t \tag{24.10}$$

where

$$\mu = \frac{m_{\mathrm{A}} m_{\mathrm{B}}}{m_{\mathrm{A}} + m_{\mathrm{B}}} \tag{24.11}$$

is the reduced mass.

Example 24.1 *For the Diels–Alder reaction of ethylene with butadiene (C_4H_6) to form cyclohexene (C_6H_{10})*

$$H_2C{=}CH_2 + H_2C{=}CH\text{-}CH{=}CH_2 \rightarrow C_6H_{10}$$

butadiene and ethylene molecules must collide. Determine the number of collisions of butadiene molecules B with ethene molecules A, under the condition that A particles are at 298 K and under a partial pressure of 1 bar in a volume $V = 1\ cm^3$.

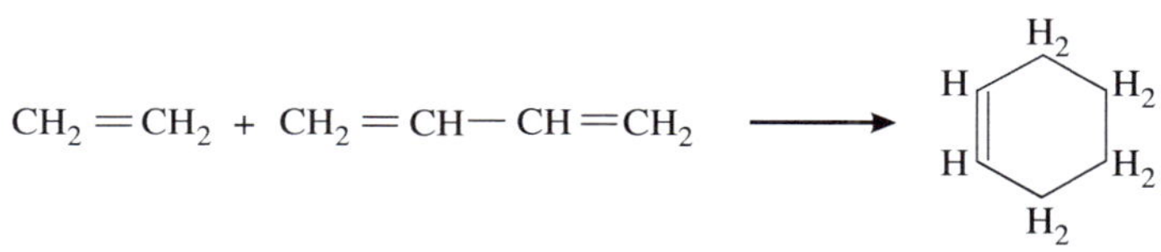

In this case $m_B = 8.98 \times 10^{-26}$ kg and $m_A = 4.66 \times 10^{-26}$ kg so that $\mu = 3.07 \times 10^{-26}$ kg. Using the collision radii of C_6H_{10} and C_2H_4 displayed in Table 14.4 ($r_{\text{ethene}} = 236$ pm, $r_{\text{butadiene}} = 306$ pm) we estimate $r_A + r_B = 542$ pm. For $V = 1\ \text{cm}^3 = 1 \times 10^{-6}\ \text{m}^3$ and $N(\text{A}) = 2.43 \times 10^{19}$ (this is the number of A particles in a gas of volume 1 cm^3 at 298 K and under a partial pressure of 1 bar) the number of collisions Z_{coll} is, according to Equation (24.10),

$$Z_{\text{coll}} = 5.84 \times 10^2\ \text{m s}^{-1} \cdot \pi \cdot (542 \times 10^{-12}\ \text{m})^2 \cdot \frac{2.43 \times 10^{19}}{10^{-6}\ \text{m}^3} \cdot N(\text{B}) \cdot \Delta t$$
$$= 1.3 \times 10^{10}\ \text{s}^{-1} \cdot N(\text{B}) \cdot \Delta t$$

This means that one molecule B undergoes 1.3×10^{10} collisions with molecules B in one second.

The large collision frequency arises, in part, from the large number of molecules in the volume. To see this, consider one molecule A and one molecule B in a volume of 1 cm^3. The number of collisions is only 1.5×10^{-9} in one second. On average, this means that it takes 46 years for the two molecules to collide with each other once. This provides a picture of the smallness of the molecules.

24.2.2 Activation

Two molecules, as in Example 24.1, do not always react upon collision. Oftentimes, they must collide with enough energy to reorganize the bonding structure of the atoms, and this involves an energy threshold. Often the energy in the molecules is statistically distributed (see Chapter 15), and we can calculate the fraction which have enough energy to undergo reaction.

For example, consider the simple case of a deuterium atom reacting with a hydrogen molecule

$$\text{D} + \text{H}_2 \rightarrow \text{HD} + \text{H} \tag{24.12}$$

In a collision, the D atom approaches the H_2 molecule until the repulsion of the electrons becomes important. On approaching further, energy must be supplied to deform the electron clouds of the reacting particles (Fig. 24.1a,b). This condition is achieved only by means of the kinetic energy of the colliding particles (translational energy in the direction of the collision). For a large enough collision energy, a configuration with equal bond lengths can be obtained (Fig. 24.1c). Finally, a transition to the final state occurs (Fig. 24.1d,e). The various stages of the reaction can be characterized by the distances d_1 and d_2 between adjacent nuclei.

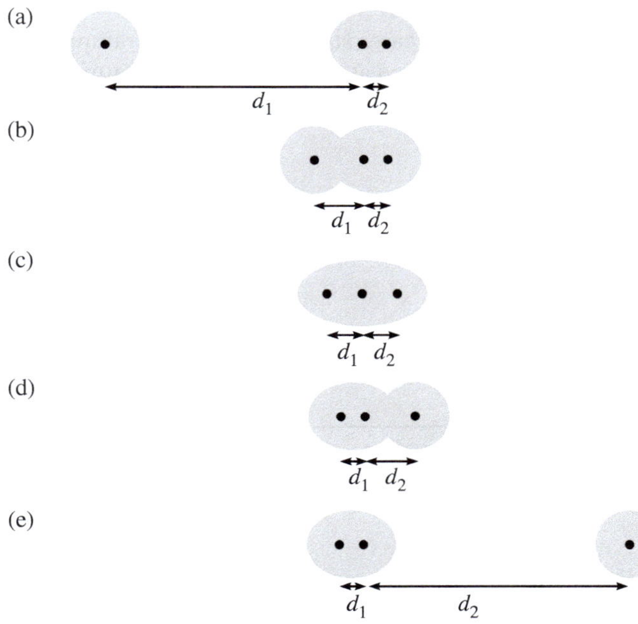

Figure 24.1 Reaction $D + H_2 \rightarrow HD + H$ via activated complex (transition state, stage c).

Figure 24.2 shows how the electronic energy E of the system changes with the different distances d_1 and d_2 corresponding to the stages (a) to (e) in Fig. 24.1. The difference between the energy at the maximum

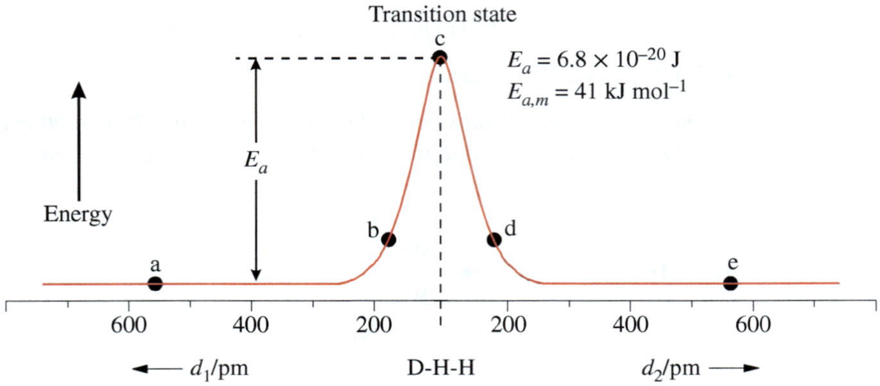

Figure 24.2 Reaction $D + H_2 \rightarrow HD + H$. Electronic energy of the system in the various stages of the reaction (see Fig. 24.1). The curve was drawn from numerical data [1].

of the curve (point c, at $d_1 = d_2 = 100$ pm) and that at point (a) is the activation energy,[1] $E_a = 6.8 \times 10^{-20}$ J, of the reaction (activated complex of transition state); the molar activation energy is $E_{a,m} = E_a \cdot \boldsymbol{N}_A = 41$ kJ mol^{-1}. This value is much smaller than the energy required for breaking the H – H bond (Section 6.3.1.2): 7.6×10^{-19} J $\cdot$ 6.02×10^{23} mol^{-1} = 457 kJ mol^{-1}).

24.2.2.1 Accumulation of Energy

For many reactions the molar activation energy $E_{a,m}$ is on the order of 50 kJ mol^{-1}, whereas the average molar kinetic energy at room temperature is, according to Chapter 14, only $\frac{3}{2}\boldsymbol{R}T = 3.7$ kJ mol^{-1}. Of course, this energy is not enough to surmount the activation barrier. From the distribution of molecular speeds (see Chapter 15), we can see that a small but finite number of molecules have enough energy to surmount the activation barrier (distribution of energies shown in Fig. 24.3 for $T = 300$ K).

The reaction rate will be proportional to the probability $\mathcal{P}$ that the energy exchanged between the collision partners is larger than or equal to E_a. This energy is the contribution of the kinetic energy of the collision partners that is transformed into potential energy during the collision and then available for initiating the chemical reaction. In general, this energy is the sum of the contributions of both partners. In the special case of a collision between a light and a heavy particle the contribution of the heavy particle is negligible, the light particle must have enough energy to surmount the activation barrier. The probability $\mathcal{P}$ for accumulating this energy is very low (see Fig. 24.3). In Box 24.1 it is shown that the result for $\mathcal{P}$ is

$$\mathcal{P} = e^{-E_a/kT} = e^{-E_{a,m}/\boldsymbol{R}T} \tag{24.13}$$

The probability $\mathcal{P}$ increases strongly with temperature because the average collision energy increases with temperature. In Fig. 24.4, $\mathcal{P}$ is displayed as a function of temperature.

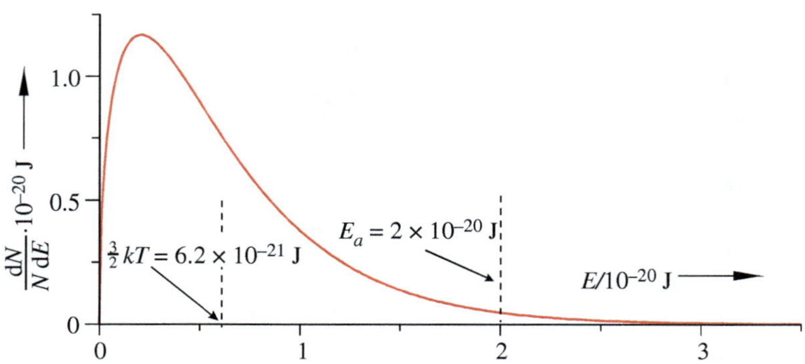

Figure 24.3 Maxwell–Boltzmann distribution of translational energy, according to Equation (15.62), for $T = 300$ K. The dashed lines correspond to the average kinetic energy $E_{kin} = \frac{3}{2}kT = 6.2 \times 10^{-21}$ J (equivalent to $E_{kin,m} = \frac{3}{2}\boldsymbol{R}T = 3.7$ kJ mol^{-1}) and to an activation energy of $E_a = 2 \times 10^{-20}$ J (equivalent to $E_{a,m} = 12$ kJ mol^{-1}).

1 In this consideration we have assumed that the activation energy depends only on the electronic energy of the system. Actually, H_2 as well as the activated complex DHH are in their vibrational ground states. Thus the vibrational energies change in the reaction, and this leads to a smaller value of the activation energy ($E_{a,m} = 31.8$ kJ mol^{-1} instead of 41 kJ mol^{-1}).

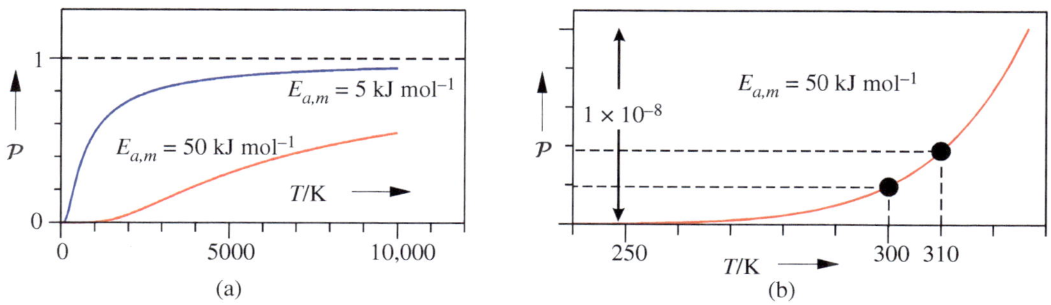

Figure 24.4 Probability $\mathcal{P}$ for a particle to surmount the activation barrier at temperature T. (a) Temperature range 0–10,000 K, $E_{a,m} = 5$ kJ mol^{-1} (blue curve) and $E_{a,m} = 50$ kJ mol^{-1} (red curve); (b) temperature range 240–330 K for $E_{a,m} = 50$ kJ mol^{-1}. Increasing the temperature from 300 to 310 K results in an increase of $\mathcal{P}$ from 2×10^{-9} to 4×10^{-9}.

24.2.3 Steric Factor

Having sufficient energy to overcome the activation barrier of a reaction is one critical condition for the reaction to proceed successfully. In addition to this condition, the reactants must collide in the correct geometric arrangement for the reaction to be successful. To account for the correct orientation of the molecules in the collision initiating the chemical reaction, we introduce the steric factor α. In this case the number of successful collisions is

$$Z_{\text{coll,successful}} = Z_{\text{coll}} \cdot \mathcal{P} \cdot \alpha = Z_{\text{coll}} \cdot e^{-E_{a,m}/\boldsymbol{R}T} \cdot \alpha \tag{24.14}$$

The steric factor is discussed more in Section 24.3.1 and treated as an empirical parameter. A first principles prediction of α depends on details of the reactant (and product) molecules. These aspects are discussed in Chapter 25.

24.3 Rate Equation for Elementary Bimolecular Reactions

Here we discuss a simple model for two reactant molecules colliding and generating a single product, namely

$$\text{A} + \text{B} \rightarrow \text{C} \tag{24.15}$$

An example is the Diels–Alder reaction in which ethylene and butadiene (C_4H_6) combine to form cyclohexene (C_6H_{10})

$$H_2C{=}CH_2 + H_2C{=}CH{-}CH{=}CH_2 \rightarrow C_6H_{10}$$

discussed in Example 24.1. This simplified treatment illustrates the essential elements of a reaction rate law and relates the number of successful collisions to the reaction rate constant.

24.3.1 Rate Constant and Frequency Factor for a Gas-Phase Reaction

In the chemical reaction (24.15), one molecule A and one molecule B disappear after every successful collision, and the change $\Delta N(\text{A})$ in the number of molecules A is

$$\Delta N(\text{A}) = -Z_{\text{coll,successful}} = -Z_{\text{coll}} \cdot e^{-E_{a,m}/\boldsymbol{R}T} \cdot \alpha \tag{24.16}$$

Combining this result with Equation (24.10) for the collision frequency, we find

$$\Delta N(\text{A}) = e^{-E_{a,m}/\boldsymbol{R}T} \cdot \alpha \cdot \sqrt{\frac{8kT}{\pi\mu}} \cdot \pi(r_\text{A} + r_\text{B})^2 \frac{1}{V} \cdot N(\text{A})N(\text{B}) \cdot \Delta t$$

If we convert the particle numbers $N(\text{A})$ and $N(\text{B})$ into concentrations c_A and c_B, and $\Delta N(\text{A})$ into Δc_A:

$$c_\text{A} = \frac{N(\text{A})}{V \cdot \boldsymbol{N}_\text{A}}, \quad c_\text{B} = \frac{N(\text{B})}{V \cdot \boldsymbol{N}_\text{A}}, \quad \Delta c_\text{A} = \frac{\Delta N}{V \cdot \boldsymbol{N}_\text{A}} \tag{24.17}$$

($\boldsymbol{N}_\text{A}$ = Avogadro's constant), we find

$$\Delta c_\text{A} = e^{-E_{a,m}/\boldsymbol{R}T} \cdot \alpha \cdot \sqrt{\frac{8kT}{\pi\mu}} \cdot \pi(r_\text{A} + r_\text{B})^2 \boldsymbol{N}_\text{A} \cdot c_\text{A} \cdot c_\text{B} \cdot \Delta t$$

Now we replace the differences by differentials and find that

$$\boxed{\frac{dc_A}{dt} = -k_r \cdot c_A \cdot c_B} \tag{24.18}$$

with

$$\boxed{k_r = A_f \cdot e^{-E_{a,m}/RT}} \tag{24.19}$$

and

$$\boxed{A_f = \sqrt{\frac{8kT}{\pi \mu}} \cdot \pi \cdot \left(r_A + r_B\right)^2 \cdot \boldsymbol{N}_A \cdot \alpha} \quad \text{(gas reaction through activated complex)} \tag{24.20}$$

This relation applies to the case of a simple bimolecular gas reaction that proceeds through an activated complex. In our example of the Diels–Alder reaction of ethene and butadiene, Equation (24.18), becomes

$$\frac{dc_{C_2H_4}}{dt} = -k_r \cdot c_{C_2H_4} \cdot c_{C_4H_6}$$

The constant k_r is called the *rate constant*. Equation (24.19) was first suggested by Arrhenius as an empirical equation and is called the *Arrhenius Equation*. The constant A_f depends on the frequency of collisions and is called the *frequency factor*. The term $\sqrt{8kT/(\pi\mu)}$ is the average relative speed of the gas molecules and $\pi\left(r_A + r_B\right)^2$ is a cross-sectional area, see Fig. 14.14. In this simplified treatment, it is assumed that the collision partners can be approximated by spheres of radii r_A and r_B. A collision occurs when the distance between the atomic centers (called the *impact parameter*) is smaller than the sum of their radii $r_A + r_B$. The factor $e^{-E_a/RT} \cdot \alpha$ is the probability that a collision proceeds successfully to reaction; i.e., the energy is sufficient to surmount the activation barrier and the collision is sterically appropriate.

In special cases, when $E_a = 0$, every collision of the two partners has enough energy for reaction, so that

$$\boxed{k_r = \sqrt{\frac{8kT}{\pi \mu}} \cdot \pi\left(r_A + r_B\right)^2 \cdot \boldsymbol{N}_A \cdot \alpha} \tag{24.21}$$

(gas reaction, diffusion controlled)

This relation applies to the case of a simple bimolecular gas reaction where the gas molecules A and B react instantaneously when they come into contact with each other on their diffusion paths. Such a reaction is called *diffusion controlled*.

Example 24.2 *For the Diels–Alder reaction*

$$H_2C{=}CH_2 + H_2C{=}CH{-}CH{=}CH_2 \rightarrow C_6H_{10}$$

the activation energy is $E_{a,m} = 115$ *kJ mol*$^{-1}$. *What is the probability* $\mathcal{P}$ *of successful collisions and the rate constant* k_r *at* 800 *K? What is the steric factor* α*?*

At $T = 800$ K we obtain, according to Equation (24.13),

$$\mathcal{P} = e^{-E_{a,m}/RT} = 3.1 \times 10^{-8}$$

for the probability $\mathcal{P}$. About 31 million collisions have to occur before two colliding molecules have enough energy to react with one another. Using Equations (24.19) and (24.20) and the numerical data in Example 24.1, we obtain

$$\begin{aligned} A_f &= 9.57 \times 10^2 \text{ m s}^{-1} \cdot \pi \cdot (542 \times 10^{-12} \text{ m})^2 \cdot 6.02 \times 10^{23} \text{ mol}^{-1} \cdot \alpha \\ &= 5.32 \times 10^8 \text{ m}^3 \text{ mol}^{-1} \text{ s}^{-1} \cdot \alpha = 5.32 \times 10^{11} \text{ L mol}^{-1} \text{ s}^{-1} \cdot \alpha \\ k_r &= 5.32 \times 10^{11} \text{ L mol}^{-1} \text{ s}^{-1} \cdot \alpha \cdot 3.1 \times 10^{-8} = 1.6 \times 10^4 \text{ L mol}^{-1} \text{ s}^{-1} \cdot \alpha \end{aligned}$$

The experimental value [2] is $k_r = 6.29$ L mol^{-1} s^{-1}, which implies a steric factor of $\alpha = 6.29/1.6 \times 10^4 = 3.9 \times 10^{-4}$.

Given the expectation that the collision must occur in a geometry for the ethylene bond and the butadiene π-system to align and form the six-membered ring, we expect the steric factor to be much smaller than one. Models for the steric factor in collisions are discussed in Chapter 25.

24.3.1.1 Reactants A and B are Identical Molecules

If the reactants are identical molecules (e.g., dimerization, disproportionation reactions), such as

$$2\,\mathrm{A} \rightarrow \mathrm{C} \qquad \text{or} \qquad 2\,\mathrm{A} \rightarrow \mathrm{C} + \mathrm{D} \tag{24.22}$$

then two molecules of A disappear in one successful collision

$$\frac{\mathrm{d}c_\mathrm{C}}{\mathrm{d}t} = -\frac{1}{2}\frac{\mathrm{d}c_\mathrm{A}}{\mathrm{d}t} = k_r\, c_A^2$$

and the rate of disappearance of A is

$$\frac{\mathrm{d}c_\mathrm{A}}{\mathrm{d}t} = -2k_r\, c_\mathrm{A}^2 \tag{24.23}$$

The rate constant k_r is obtained from Equations (24.19) and (24.20) by setting[2] $r_\mathrm{A} = r_\mathrm{B}$.

Example 24.3 *The recombination of ethyl radicals*

$$^\bullet CH_2\text{–}CH_3 \ + {^\bullet}CH_2\text{–}CH_3 \rightarrow CH_3CH_2CH_2CH_3$$

is diffusion-controlled ($E_{a,m} = 0$). Use the simple collision theory to calculate the rate constant at 298 K.

With $\mu = 24 \times 10^{-27}$ kg and $(r_\mathrm{A} + r_\mathrm{B}) = 500$ pm from Equation (24.21) we obtain the rate constant for $T = 298$ K

$$k_r = 3 \times 10^{11}\ \mathrm{L\ mol^{-1}\ s^{-1}} \cdot \alpha$$

The experimental value is $k_r = 1.6 \times 10^{11}\ \mathrm{L\ mol^{-1}\ s^{-1}}$, corresponding to $\alpha = 0.5$.

24.3.2 Rate Constant and Frequency Factor for a Reaction in Solution

24.3.2.1 Reaction Through Activated Complex

Let us consider a reaction of two particles A and B in a liquid, like iodide and chloromethane dissolved in methanol and reacting to form chloride and iodomethane.

$$\mathrm{I^-} + \mathrm{ClCH_3} \rightarrow \mathrm{Cl^-} + \mathrm{ICH_3} \tag{24.24}$$

In contrast to a gas reaction, A collides with particles B as well as with solvent molecules. The mean free path is much shorter in a liquid than in a gas; thus, it takes longer for a particle to diffuse and collide with a reaction partner. However, when the reactants collide (called an encounter pair) they can collide many times with each other; thus, the likelihood that an encounter pair proceeds to product successfully is much higher than in a gas. Surprisingly, these two effects largely compensate, and the reaction rates are similar for a reaction in solution and in a gas reaction.

In analogy to Equation (24.5), the number of collisions z'_coll of one particle A with particles B and with solvent molecules in time Δt can be written as

$$z'_\mathrm{coll} = \nu \cdot \Delta t \tag{24.25}$$

where ν is the collision frequency of particle A. What is the number z_coll of collisions of particle A with particles B? Apparently, z_coll must be smaller than z'_coll. To calculate z_coll we consider the probability $\mathcal{P}$ to find particle A at the site of a particle B. To simplify the calculation, we assume that a solvent molecule as well as a molecule A can be described by a sphere of radius r_A. After a certain number of collisions, particle A exchanges its place with a neighboring solvent molecule (Fig. 24.5). A favorable site is reached if molecules A and B come into contact (Fig. 24.5b). Then

$$\mathcal{P} = \frac{\text{number of favorable sites}}{\text{total number of sites}} = \frac{\text{volume of favorable sites}}{\text{volume of solution}} \tag{24.26}$$

$$= \frac{N(\mathrm{B}) \cdot \frac{4}{3}\pi r_\mathrm{A}^3}{V}$$

---

2 Note that this identity is not trivial. On the one hand, $-\Delta N_\mathrm{A}$ is twice the number of successful collisions; on the other hand, applying the consideration to the present case, each collision is counted twice, thus $Z_\mathrm{coll,successful}$ must be divided by a factor of two. Thus k_r has to be identified with one-half of the rate constant calculated from the expressions (24.19) and (24.20).

and the number of collisions of a molecule A with molecules B becomes

$$z_{\text{coll}} = z'_{\text{coll}} \cdot \mathcal{P} = \frac{4}{3}\pi r_{\text{A}}^3 \cdot \frac{N(\text{B})}{V} \cdot \nu \cdot \Delta t \tag{24.27}$$

Following the same arguments as in the derivation of the relations (24.18)–(24.20) for reactions in the gas phase we obtain

$$\boxed{\frac{\mathrm{d}c_{\text{A}}}{\mathrm{d}t} = -k_r \cdot c_{\text{A}} \cdot c_{\text{B}}} \tag{24.28}$$

with

$$\boxed{k_r = A_f \cdot \mathrm{e}^{-E_a/\boldsymbol{R}T}} \tag{24.29}$$

$$\boxed{A_f = \nu \cdot \frac{4}{3}\pi r_{\text{A}}^3 \cdot \boldsymbol{N}_{\text{A}} \cdot \alpha} \tag{24.30}$$

(reaction in solution through activated complex)

Figure 24.5 Reaction in solution (solvent methanol): $I^- + CH_3Cl \rightarrow Cl^- + CH_3I$. Red: I^- and CH_3Cl, blue: solvent molecules. (a) Reactants separated by solvent molecules. (b) Favorable position of I^- to react with CH_3Cl.

What is the difference between Equation (24.30) and the corresponding expression (24.20) for a gas reaction? We can write the frequency as

$$\nu = \frac{\overline{v}}{\overline{\lambda}_{\text{solution}}} \tag{24.31}$$

where the mean speed $\overline{v}$ is given by Equation (24.6). The mean free path $\overline{\lambda}_{\text{solution}}$ is much smaller than the diameter of a solvent molecule. With the estimate

$$\overline{\lambda}_{\text{solution}} = \frac{1}{3} r_{\text{A}} \tag{24.32}$$

we find

$$\nu = \frac{3\overline{v}}{r_{\text{A}}} \tag{24.33}$$

and

$$A_f = \frac{3\overline{v}}{r_{\text{A}}} \cdot \frac{4}{3}\pi r_{\text{A}}^3 \cdot \boldsymbol{N}_{\text{A}} \cdot \alpha = \overline{v} \cdot \pi \cdot 4 r_{\text{A}}^2 \cdot \boldsymbol{N}_{\text{A}} \cdot \alpha \tag{24.34}$$

This corresponds to Equation (24.20) for $r_{\text{A}} = r_{\text{B}}$. Although the mean free path is much smaller in solution than in a gas and the collision frequency ν is accordingly higher, the number of successful collisions is comparable in both cases.

24.3.2.2 Diffusion Controlled Reaction in Solution

Often, it is useful to divide the reaction sequence in solution into two stages: formation of the encounter complex AB and its subsequent transformation into the product

$$\text{A} + \text{B} \rightleftarrows \text{AB} \rightarrow \text{product}$$

If the second stage is slow compared to the first stage, then the reaction is not affected by formation of the encounter pair. Rather, it is determined by the intrinsic rate of conversion to the product. If the first stage is slow compared to the second state, then the molecules A and B react when they touch each other, or reach a critical distance, and the reaction is *diffusion controlled*.

For a diffusion controlled reaction, the number $Z_{\text{coll,successful}}$ of successful collisions equals the number of encounter complexes AB formed in time Δt. To model the reaction rate, we consider the diffusion of A to the location of B. If a molecule A exchanges its location in the liquid in time τ, then it moves a distance

$$d = 2 \cdot r_{\text{A}} \tag{24.35}$$

where d is the distance between two adjacent locations (solvent molecule and molecule A described by a sphere of radius r_A). By modeling the molecular movement as a diffusion process, we can apply the Einstein–Smoluchowski relation, Equation (14.60), so that

$$\tau = \frac{x^2}{2D} = \frac{d^2}{2D} = \frac{4r_A^2}{2D} \tag{24.36}$$

where D is the diffusion coefficient. Let us first assume that molecule B reacts with molecule A when A is just at the right site. In our present case the reaction occurs whenever the molecules A and B touch each other. If both molecules are considered as spheres of radius r_A in closest packing (molecule B surrounded by 12 molecules of A) there are 12 right sites. Then the probability $\mathcal{P}$ of finding a molecule A at a right site, touching a molecule B, is

$$\mathcal{P} = \frac{\text{volume of all molecules B}}{\text{volume of solution}} \times 12 = \frac{N(\text{B})\frac{4\pi}{3}r_A^3}{V} \times 12$$

Hence the number of successful collisions of $N(\text{A})$ molecules A with molecules B in time Δt is

$$Z_{\text{coll,successful}} = N(\text{A}) \cdot \frac{\Delta t}{\tau} \cdot \mathcal{P} \tag{24.37}$$

$N(\text{A}) \cdot \Delta t/\tau$ is the number of exchanges of location of molecules A taking place during time Δt. In this way we obtain

$$Z_{\text{coll,successful}} = \frac{N(\text{A})\,N(\text{B})}{V} \cdot 4\pi \cdot 2r_A \cdot D \cdot \Delta t \tag{24.38}$$

Additionally, we have to consider that the molecules B are not at rest, but change their locations by diffusion. Thus D must be replaced by $D_A + D_B$ (diffusion coefficients of molecules A and B). We obtain (in agreement with the rigorous treatment [3])

$$Z_{\text{coll,successful}} = \frac{N(\text{A})\,N(\text{B})}{V} \cdot 4\pi \cdot (r_A + r_B) \cdot (D_A + D_B) \cdot \Delta t \tag{24.39}$$

Then

$$\boxed{\frac{\text{d}c_A}{\text{d}t} = -k_r c_A c_B} \tag{24.40}$$

with

$$\boxed{k_r = 4\pi \cdot (r_A + r_B) \cdot (D_A + D_B) \cdot \boldsymbol{N}_A} \qquad \text{(diffusion-controlled reaction in solution)} \tag{24.41}$$

The reaction rate depends on the rate of diffusion of the reaction partners (diffusion coefficients $D_A + D_B$) and on the distance $(r_A + r_B)$ at which the reactants touch and are able to react.

Example 24.4 *Consider the neutralization reaction: $H^+ + OH^- \rightarrow H_2O$. Because of their hydration spheres the radii of H^+, r_A, and OH^-, r_B, are on the order of 250 pm. Using this information and the measured diffusion coefficients: $D_{H^+} = 9 \times 10^{-5}$ cm^2 s^{-1} and $D_{OH^-} = 5 \times 10^{-5}$ cm^2 s^{-1}, calculate the diffusion limited rate constant.*

To calculate the rate constant, we use Equation (24.41) so that

$$k_r = 4\pi \cdot (500\ \text{pm}) \cdot \left(14 \times 10^{-5}\ \text{cm}^2\ \text{s}^{-1}\right) \cdot \boldsymbol{N}_A$$
$$= 5 \times 10^{10}\ \text{L mol}^{-1}\ \text{s}^{-1}$$

The experimental value is $k_r = 15 \times 10^{10}$ L mol^{-1} s^{-1}. Given the simplicity of the model, this difference is not significant; however, it might arise because the reactants have opposite charges and attract each other.

24.4 Rate Laws

We have considered the bimolecular reaction

$$\text{A} + \text{B} \rightarrow \text{C} \tag{24.42}$$

According to Equation (24.18) the rate for this reaction is

$$\frac{dc_A}{dt} = -k_r \cdot c_A \cdot c_B \tag{24.43}$$

In the following we denote this equation as a *rate law*. We discuss for some important cases how we can obtain rate constants from experimental data, using the corresponding rate laws. In particular, we are interested in describing the course of the overall reaction.

This goal must be distinguished from our goal in Sections 24.2.1 and 24.3.1, where we have attempted to understand the basic mechanisms of some elementary reactions. The simple *bimolecular* reactions discussed so far correspond to second-order rate equations; i.e., the reaction rate depends quadratically on concentration. Oftentimes reactions, such as decomposition or isomerization, depend only on the concentration to the first power and are called *unimolecular reactions*. In most cases rate laws are more complicated because back reactions or side reactions may occur.

24.4.1 What Is a Rate Law?

A rate law is the observed dependence of a reaction's rate on the concentrations of the reacting species, typically the reactants but sometimes the products also. In this expression for the rate, the concentrations of the species are raised to a power, which may be positive, negative, zero, or fractional. For the simple elementary reactions discussed in Section 24.3 the power is determined by the reaction stoichiometry. For complex reaction mechanisms, this relationship does not necessarily hold however.

For the case of A + B → C, the rate law may be written as

$$Rate = k\ c_A^a\ c_B^b$$

where k is the rate constant, the parameters a and b are the powers, and we have assumed that the concentration of the product does not affect the rate law. The order of the reaction is defined as the sum of these powers; i.e., $a + b$. The order of the rate law for A is a, and the order of the rate law for B is b. Knowledge of the rate law is important for two primary reasons. First, it allows us to predict the rate of the reaction and hence optimize it for our needs. Second, it places constraints on the mechanism of the reaction that can help us understand its molecular features.

Below we first discuss simple rate laws and later we look at how they can be combined to generate more complicated rate behavior. Mathematical expressions for the simple rate laws are compiled in Table 24.1 at the end of this section. A combination of elementary reactions that are linked is called the reaction mechanism (see Sections 24.6 and 24.7). Here we discuss, elementary reactions of order zero (rare), one and two (both of which are fairly common), and three (rare). An example for a rate law with non-integer reaction order is discussed in Section 24.7.3.

Table 24.1 Summary of Rate Laws and Half-Lives for Zero-, First-, and Second-order Reactions. $c_{A,0}$ and $c_{B,0}$ Are the Initial Concentrations of Species A and B.

Order	Rate Law	Units of k_r	Integrated Rate Law	Half-Life $\tau_{1/2}$
0th	$\frac{dc_A}{dt} = -k_r$	mol L^{-1} s^{-1}	$c_A = c_{A,0} - k_r t$	$\frac{c_{A,0}}{2k_r}$
1st	$\frac{dc_A}{dt} = -k_r \cdot c_A$	s^{-1}	$c_A = c_{A,0} \cdot e^{-k_r t}$	$\frac{\ln 2}{k_r}$
2nd	$\frac{dc_A}{dt} = -k_r \cdot c_A^2$	L mol^{-1} s^{-1}	$c_A = \frac{1}{1/c_{A,0} + k_r t}$	$\frac{1}{k_r c_{A,0}}$
2nd	$\frac{dc_A}{dt} = -k_r \cdot c_A c_B$	L mol^{-1} s^{-1}	$c_A = \frac{c_{B,0} - c_{A,0}}{\frac{c_{B,0}}{c_{A,0}} \cdot e^{(c_{B,0}-c_{A,0})k_r t} - 1}$	$\frac{\ln(2 - c_{A,0}/c_{B,0})}{k_r (c_{B,0} - c_{A,0})}$

24.4.2 Zero-Order Rate Law

The simplest rate law to integrate is one that is zero order. For the decay of species A with a zero-order rate law, we may write

$$\frac{dc_A}{dt} = -k_r \quad \text{or} \quad c_A = c_{A,0} - k_r t \tag{24.44}$$

where $c_{A,0}$ is the initial concentration of species A (c_A restricted to be a nonnegative value: $k_r \cdot t \le c_{A,0}$).

The time when c_A decreases to one-half of its original value is called the *half-life* $\tau_{1/2}$. For this rate law, the *half-life* $\tau_{1/2}$ is:

$$\frac{1}{2}c_{A,0} = c_{A,0} - k_r\tau_{1/2}, \quad \tau_{1/2} = \frac{c_{A,0}}{2k_r} \tag{24.45}$$

The half-life $\tau_{1/2}$ for a zero-order reaction depends on the initial concentration $c_{A,0}$, and it increases with increasing initial concentration.

Reactions with this rate law are rare. One example is the bromination of malonic acid at high malonic acid initial concentration, see Section 24.7.1, Equation (24.102). Other examples are found in heterogeneous or surface reactions in which a catalytic site is typically involved. In such a case, the concentration of the reactant species in solution will not affect the reaction rate when the catalytic sites are saturated with reactant species.

24.4.3 First-Order Rate Law

A number of photochemical reactions are unimolecular, such as the *cis* to *trans* isomerization of retinal (see Chapter 28). Reactant A is electronically excited by a light flash forming species A*: $A \xrightarrow{h\nu} A^*$, and in a second step A* forms product B.

$$A^* \xrightarrow{k_r} B \tag{24.46}$$

In this case the reaction rate is proportional to the concentration of one reacting species, A*, and the rate law may be written as

$$\frac{dc_{A^*}}{dt} = -k_r \cdot c_{A^*} \tag{24.47}$$

Such rate laws are called *first-order rate laws*.[3]

If we integrate Equation (24.47) from time zero (initial concentration $c_{A_0^*}$) to time t (concentration c_{A^*}) we find

$$\int_{c_{A_0^*}}^{c_{A^*}} \frac{dc_{A^*}}{c_{A^*}} = -k_r \int_0^t dt,$$

which gives

$$\ln \frac{c_{A^*}}{c_{A_0^*}} = -k_r t \tag{24.48}$$

This equation may be rearranged to give

$$c_{A^*} = c_{A_0^*} \cdot e^{-k_r t} \tag{24.49}$$

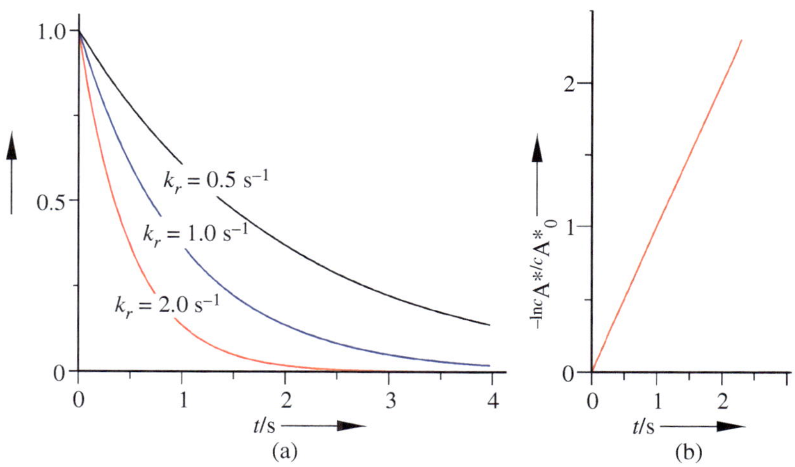

Figure 24.6 Reaction A* → B, Equation (24.46), with initial concentration $c_{A_0^*}$ of reactant A*. (a) Exponential decay of $c_{A^*}/c_{A_0^*}$ according to Equation (24.49) with $k_r = 0.5$, 1, and 2 s^{-1}. (b) Plot of $-\ln(c_{A^*}/c_{A_0^*})$ versus time, according to Equation (24.48) for $k_r = 1\ s^{-1}$.

Figure 24.6a plots this expression for three different values of k_r. Figure 24.6b shows that plots of $\ln(c_{A^*}/c_{A_0^*})$ versus time are linear. This property can be used to extract rate constants from experimental data.

For the half-life $\tau_{1/2}$ we obtain

$$\frac{1}{2}c_{A_0^*} = c_{A_0^*} \cdot e^{-k_r\tau_{1/2}},$$

so that

$$\tau_{1/2} = \frac{\ln 2}{k_r} \tag{24.50}$$

In the case of a first-order rate law the half-life $\tau_{1/2}$ does not depend on the initial concentration $c_{A_0^*}$.

A common case occurs when one of the reactants (e.g., B) in a bimolecular reaction (24.42)

$$A + B \rightarrow C$$

3 Note that we have to distinguish between the order of the reaction and the order of the differential equation. All rate laws are first-order differential equations, containing only the first derivative dc/dt.

is in large excess over the other reactant, A. Then c_B is practically constant during the reaction and the rate may be written as

$$\frac{dc_A}{dt} = -k_r \cdot c_B \cdot c_A = -k_r' \cdot c_A \quad \text{with } k_r' = k_r \cdot c_B \tag{24.51}$$

(*pseudo first-order* reaction).

Example 24.5 *Consider the hydrolysis of the triphenylmethane dye cation R^+ to form ROH,*

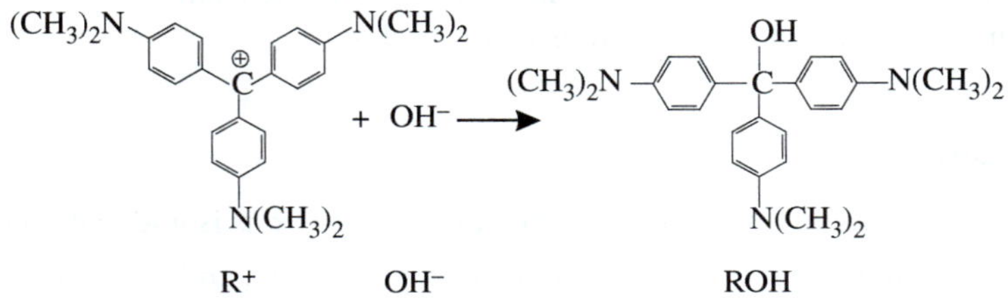

at 296 K with OH^- in large excess ($c_{OH^-} = 0.2$ mol L^{-1}, $c_{R^+} = 5 \times 10^{-5}$ mol L^{-1}). Use the experimental data in the table to evaluate the pseudo first-order rate constant

Concentration of the Cation $\boldsymbol{R}^+$ as a Function of Time.

t/s	0	19	39	59	79	98	118	137	157
$c_{R^+}/10^{-5}$ mol L^{-1}	5.00	3.31	2.19	1.45	0.94	0.61	0.39	0.24	0.15

The dye cation R^+ is trigonal planar, and the positive charge is uniformly distributed over the molecular skeleton; therefore, R^+ behaves like a cyanine cation, absorbing light at about 590 nm. The structure of the reaction product ROH is tetrahedral, and the π electrons are confined to the isolated benzene rings; thus, ROH does not absorb light in the visible (resulting in a bleaching of the dye cation R^+). The concentration c_{R^+} (measured by following the absorbance of the solution at 590 nm) is listed as a function of time in the table and plotted in Fig. 24.7a. From the logarithmic plot in Figure 24.7b we obtain $k_r' = 0.0214\ s^{-1}$ and $k_r = k_r'/c_{OH^-} = 0.11$ L mol s^{-1}.

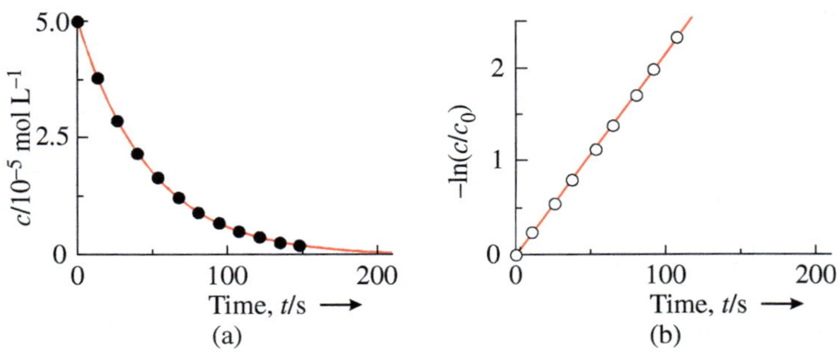

Figure 24.7 First-order reaction: hydrolysis of triphenylmethane dye cation in 0.2 mol L^{-1} aqueous NaOH solution at 296 K. (a) Concentration versus time plot. Filled circles: experimental data points. (b) Evaluation of the data. Open circles: $-\ln(c/c_0)$ versus time, calculated from the experimental data points; red line: regression line.

24.4.4 Second-Order Rate Law

Now we consider a bimolecular reaction

$$A + B \rightarrow C \tag{24.52}$$

in which the rate is second order in the concentrations of the reactant species. If we start the reaction with equal initial concentrations of the reactants A and B

$$c_{A,0} = c_{B,0}$$

then the concentrations are equal, $c_A = c_B$, at any time, because they are consumed at an equal rate. Following the analysis for Equation (24.43) we obtain the rate law

$$\frac{dc_A}{dt} = -k_r\, c_A^2 \tag{24.53}$$

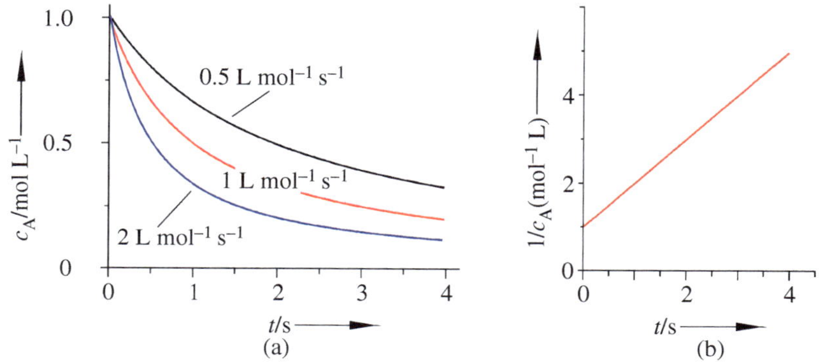

Figure 24.8 Reaction A + B → C, Equation (24.52), with initial concentration $c_{A,0} = 1$ mol L^{-1} of reactant A and concentration $c_{B,0} = c_{A,0}$ of reactant B. (a) Second-order decay of concentration c_A according to Equation (24.55) with $k_r = 0.5$ L mol^{-1} s^{-1}, 1 L mol^{-1} s^{-1}, and 2 L mol^{-1} s^{-1}. (b) Plot of $1/c_A$ versus time, according to Equation (24.54) for $k_r = 1$ L mol^{-1} s^{-1}.

For Equation (24.53) the rate is proportional to the square of the concentration of one reacting species, A (*second-order rate law*). We can integrate this equation and find

$$\int_{c_{A,0}}^{c_A} \frac{dc_A}{c_A^2} = -k_r \int_0^t dt, \qquad \frac{1}{c_A} - \frac{1}{c_{A,0}} = k_r t \tag{24.54}$$

and

$$c_A = \frac{1}{1/c_{A,0} + k_r t} \tag{24.55}$$

This equation is plotted in Figure 24.8a. Compared to the first-order reaction, the reactant concentration falls off more slowly, for a given k_r, because as the concentration drops the collision frequency decreases and the reaction rate slows. Figure 24.8b shows that plots of $1/c_A$ versus t are linear in time. This property can be used to extract rate constants from experimental data.

For the *half-life* $\tau_{1/2}$ we obtain:

$$\frac{1}{2}c_{A,0} = \frac{1}{1/c_{A,0} + k_r \tau_{1/2}}, \quad \tau_{1/2} = \frac{1}{k_r c_{A,0}} \tag{24.56}$$

The half-life $\tau_{1/2}$ for a second-order reaction depends on the initial concentration $c_{A,0}$ and decreases with increasing initial concentration.

In the general case of reaction A + B → C the reaction rate depends on the concentration of both species, namely

$$\frac{dc_A}{dt} = -k_r\, c_A\, c_B \tag{24.57}$$

In this case the integration is more involved. In Problem P24.8, you show that c_A is

$$c_A = \frac{c_{B,0} - c_{A,0}}{\frac{c_{B,0}}{c_{A,0}} \cdot e^{(b-a)k_r t} - 1} \tag{24.58}$$

and $\tau_{1/2}$ is

$$\tau_{1/2} = \frac{\ln\left(2 - c_{A,0}/c_{B,0}\right)}{k_r\left(c_{B,0} - c_{A,0}\right)} \quad \text{for } c_{A,0} \le c_{B,0} \tag{24.59}$$

where $c_{A,0}$ and $c_{B,0}$ are the initial concentrations of the species A and B, respectively. In Problem P24.9, you show that Equation (24.55) is a special case of the general solution (24.58), and you plot these equations for different initial concentrations $c_{A,0}$ and $c_{B,0}$.

Example 24.6 *From the experimental data in the table evaluate the second-order rate constant for the disproportionation of $HBrO_2$. The concentration c was measured by following the absorbance of the solution at 240 nm.*

t/s	0	7.66	15.3	23.0	30.7	38.3	46.0	53.7	61.3	69.0	76.7	84.3
$c/10^{-5}$ mol L^{-1}	2.00	0.84	0.53	0.39	0.31	0.25	0.22	0.19	0.17	0.15	0.14	0.12

We consider the disproportionation of $HBrO_2$ at 293 K (initial concentration $c_{A,0} = 2 \times 10^{-5}$ mol L^{-1}) in aqueous acidic solution

$$2\,HBrO_2 \rightarrow HOBr + HBrO_3$$

In this case two identical molecules react and, according to Equation (24.23),

$$\frac{dc}{dt} = -2k_r \cdot c^2$$

where c is the concentration of $HBrO_2$. In analogy to Equation (24.54) we find

$$\frac{1}{c} = \frac{1}{c_0} + 2k_r t \quad \text{or} \quad \frac{c_0}{c} - 1 = 2c_0 k_r t$$

where c_0 is the initial concentration. From a plot of $y = c_0/c - 1$ versus time (Fig. 24.9b) we obtain $k_r = 4500$ L mol^{-1} s^{-1}.

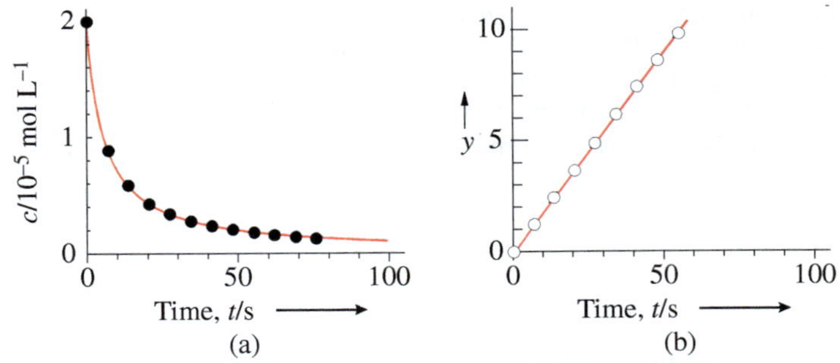

Figure 24.9 Second-order reaction: Disproportionation of $HBrO_2$ in aqueous acidic solution at 293 K. (a) Concentration versus time plot. Filled circles: measured data points. (b) Evaluation of the data. Open circles: $y = (c_0/c - 1)$ versus time, calculated from the experimental data points, dashed line: regression line.

This reaction is an important step in the Belousov–Zhabotinsky reaction discussed in Section 24.7.7. Note that the half-life of this reaction is 5 s, in spite of the high value of the rate constant. This long time constant arises because of the low initial concentration c_0 used in the experiment. If we increase c_0 to 2×10^{-2} mol L^{-1} the half-life would be only 5 ms.

24.5 Activation Energy and Frequency Factor

24.5.1 Arrhenius Plot

According to Equations (24.19) and (24.29)

$$k_r = A_f \cdot e^{-E_{a,m}/RT} \tag{24.60}$$

the rate constant depends strongly on temperature. By rearranging Equation (24.60) we obtain

$$\boxed{\ln\left(\frac{k_r}{k_{r,1}}\right) = -\frac{E_{a,m}}{\boldsymbol{R}}\left(\frac{1}{T} - \frac{1}{T_1}\right)} \tag{24.61}$$

where $k_{r,1}$ is the rate constant at temperature T_1. We can obtain $E_{a,m}$ by measuring k_r at different temperatures and plotting $\ln(k_r/k_{r,1})$ versus $1/T$; we obtain $E_{a,m}$ from the slope of the resulting straight line. The frequency factor A_f can be evaluated by inserting $E_{a,m}$ into Equation (24.60) and evaluating it at a particular temperature. A plot of $\ln(k_r/k_{r,1})$ versus $1/T$ is called an *Arrhenius plot*.

Example 24.7 *The hydrolysis of the triphenylmethane dye cation R^+ (see Example 24.5)*

$$R^+ + OH^- \rightarrow ROH \tag{24.62}$$

was measured at different temperatures (see Table 24.2). Evaluate the activation energy $E_{a,m}$ and estimate the frequency factor A_f.

Table 24.2 Rate Constants $k_r' = k_r \cdot c_{OH^-}$ for Reaction (24.62) at Different Temperatures T; $c_{OH^-} = 0.2$ mol L^{-1}.

T/K	293	296	299	301	303	307	311
k_r'/s^{-1}	0.0170	0.0214	0.0272	0.0325	0.0389	0.0540	0.0780

In Fig. 24.10a, $k_r' = k_r \cdot c_{OH^-}$ is plotted versus T. In Fig. 24.10b $\ln(k_r'/k_{r,1}')$ is plotted versus $1/T$; from the slope −7640 K of the straight line we obtain $E_{a,m} = \boldsymbol{R} \cdot 7640\ \text{K} = 63.5$ kJ mol^{-1}. The frequency factor A_f is calculated from Equation (24.60) as $A_f = 1.8 \times 10^{10}$ L mol^{-1} s^{-1}.

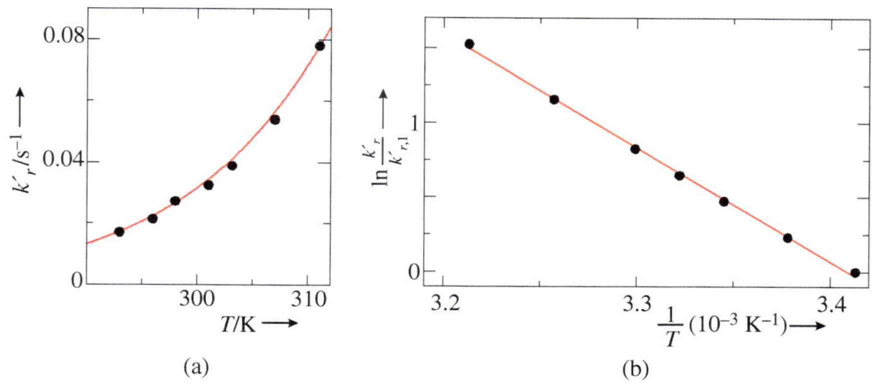

Figure 24.10 Reaction (24.62) measured at different temperatures. (a) Rate constant k_r' versus temperature T. (b) $\ln\left(k_r'/k_{r,1}'\right)$ versus $1/T$ (Arrhenius plot). For data see Table 24.2.

From Equation (24.34) we can estimate A_f by using $r_A = r_{OH^-} = 250$ pm, and $\mu \approx m_{OH^-} = 3 \times 10^{-26}$ kg. Then $\bar{v} = 609$ m s^{-1} at 298 K and $A_f = 1.73 \times 10^{10}$ L mol^{-1} s^{-1} $\cdot\, \alpha$. By comparison with the experiment we find $\alpha = 0.06$.

24.5.2 The Chemist's Rule of Thumb

For many reactions in solution that are important for the organic chemist the molar activation energy is on the order of 50 kJ mol^{-1}. By increasing the temperature, these reactions can be considerably accelerated. According to (24.60) an increase of temperature by ΔT results in a change of the rate constant from $k_r(T)$ to $k_r(T + \Delta T)$:

$$\frac{k_r(T+\Delta T)}{k_r(T)} = \frac{e^{-E_{a,m}/R(T+\Delta T)}}{e^{-E_{a,m}/RT}} \approx \exp\left(\frac{E_{a,m}\Delta T}{RT^2}\right) \tag{24.63}$$

if we neglect the small temperature dependence of the number of collisions. In this approximation, changing the temperature from 300 to 310 K ($\Delta T = 10$ K, this is a change by 3%) and using $E_{a,m} = 50$ kJ mol^{-1} we obtain

$$k_r(310 \text{ K}) = 2 \cdot k_r(300 \text{ K})$$

This is in accordance with the chemist's rule of thumb that an increase of temperature by 10 K increases the reaction rate by a factor of two.

24.6 Combinations of Elementary Reactions

This section describes particular combinations of elementary reactions that are commonly observed in chemical systems. They are the kinetics of reactions leading to equilibrium, reactions that operate in parallel, and reactions that are coupled serially.

24.6.1 Reactions Leading to Equilibrium

Oftentimes reactions lead to an equilibrium with measurable amounts of both reactants and products. This means that not only the reactants combine to make product molecules

$$A + B \xrightarrow{k_1} C + D \tag{24.64}$$

but the products can react with each other to form reactant molecules

$$C + D \xrightarrow{k_{-1}} A + B \tag{24.65}$$

The forward reaction decreases the concentration of A and the backward reaction increases the concentration of A. Hence, the net rate of formation of molecules A is

$$\frac{dc_A}{dt} = -k_1 \cdot c_A \cdot c_B + k_{-1} \cdot c_C \cdot c_D \tag{24.66}$$

To simplify the analysis, we assume that the concentrations of molecules A and B are equal ($c_{A,0} = c_{B,0}$) at time $t = 0$ and that no molecules C and D are present ($c_{C,0} = c_{D,0} = 0$). Thus, at time t: $c_C = c_D = c_{A,0} - c_A$ and

$$\frac{dc_A}{dt} = -k_1 \cdot c_A^2 + k_{-1} \cdot (c_{A,0} - c_A)^2 \tag{24.67}$$

In the special case $k_1 = k_{-1} = k_r$, this differential equation can be integrated (see Exercise E24.4) to find

$$\boxed{c_A = c_B = \frac{1}{2} c_{A,0} \cdot \left(1 + e^{-2k_r c_{A,0} t}\right)} \tag{24.68}$$

Correspondingly

$$\boxed{c_C = c_D = \frac{1}{2} c_{A,0} \cdot \left(1 - e^{-2k_r c_{A,0} t}\right)} \tag{24.69}$$

The concentrations c_A and c_C are displayed in Fig. 24.11a as a function of time. At very long times $\left(\text{i.e., } t \gg 1/\left(2k_r c_{A,0}\right)\right)$ the equilibrium concentrations

$$c_{A,\infty} = c_{C,\infty} = \frac{1}{2} c_{A,0} \tag{24.70}$$

are established.

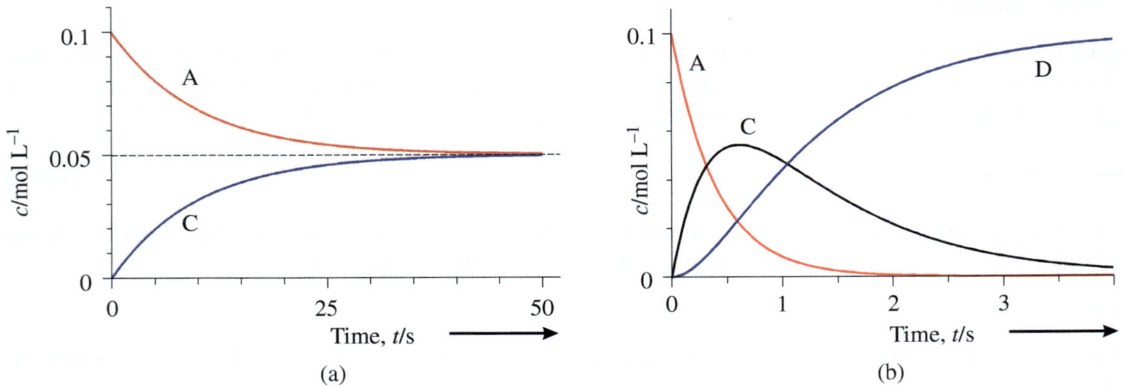

Figure 24.11 (a) Chemical reaction leading to equilibrium, Equations (24.68) and (24.69): A + B ⇄ C + D. Rate constants $k_1 = k_{-1} = 0.5\ M^{-1}\ s^{-1}$; initial concentrations $c_{A,0} = c_{B,0} = 0.1\ M^{-1}$, $c_{C,0} = c_{D,0} = 0$. (b) Consecutive reaction (24.79) with B in large excess: A + B ⇄ C + D, C + B ⇄ D. Concentrations c_A, c_C, and c_d versus time t according to Equations (24.83), (24.85), and (24.86). Initial concentrations $c_{A,0} = 0.1\ M^{-1}$, $c_{C,0} = c_{D,0} = 0$. Rate constants $k_1' = 2.5\ s^{-1}$, $k_2' = 1\ s^{-1}$.

In the general case (arbitrary rate constants k_1 and k_2, different initial concentrations of $c_{A,0}$ and $c_{B,0}$) the integration of the differential equation (24.66) is no longer so easy. Therefore, we restrict consideration to the calculation of the equilibrium concentrations. After an infinite time the concentrations of the reactants and of the products do not change; thus, $dc_A/dt = 0$ and

$$-k_1 \cdot c_A c_B + k_2 \cdot c_C c_D = 0 \tag{24.71}$$

leading to

$$\frac{c_C \cdot c_D}{c_A \cdot c_B} = \frac{k_1}{k_{-1}} \tag{24.72}$$

Because the expression on the left-hand side represents the equilibrium constant K

$$K = \frac{\hat{c}_C \cdot \hat{c}_D}{\hat{c}_A \cdot \hat{c}_B} = \frac{c_C \cdot c_D}{c_A \cdot c_B} \tag{24.73}$$

we can relate K to the ratio of the rate constants k_1 and k_{-1}:

$$\boxed{K = \frac{k_1}{k_{-1}}} \tag{24.74}$$

Example 24.8 *For the iodine hydrogen equilibrium.* $H_2 + I_2 \rightarrow 2\,HI$ *determine the equilibrium constants at* $T = 667$ *K and* $T = 699$ *K by using the rate constants* k_1 *and* k_{-1} *of the forward and reverse reaction and compare them with the constants obtained from equilibrium measurements (use the data in Table 24.3).*

Using Equation (24.74), we find

$$K(667\ \text{K}) = \frac{1.56 \times 10^{-2}}{2.59 \times 10^{-4}} = 60.4$$

which compares favorably with the experimentally determined equilibrium constant, $K = 61.0$. A similar calculation gives $K(699\ \text{K}) = 54.0$.

Table 24.3 Rate Constants k_1, k_{-1}, and Equilibrium Constant K (from Measurement of Equilibrium) for the $H_2 + I_2 \rightleftharpoons 2HI$ Reaction.

$\frac{T}{\text{K}}$	$\frac{k_1}{\text{L mol}^{-1}\ \text{s}^{-1}}$	$\frac{k_{-1}}{\text{L mol}^{-1}\ \text{s}^{-1}}$	K
667	1.56×10^{-2}	2.59×10^{-4}	61.0
699	6.70×10^{-2}	12.4×10^{-4}	55.2

24.6.2 Caveat on the Interpretation of Kinetic Data

From this discussion it is evident that a reaction model involving a bimolecular reaction leads to a second-order rate law. However, the converse need not hold; i.e., a second-order rate law does not imply that a reaction is a simple bimolecular process.

The reaction of H_2 and I_2 provides an example of this. Because the kinetics are second-order, it was believed for a long time that the iodine hydrogen reaction is based on direct collisions between H_2 and I_2 molecules. It is now known that this reaction is initiated by dissociation of iodine:

$$I_2 \rightleftharpoons 2\,I$$

which can be described by the dissociation constant

$$K' = \frac{\hat{c}_I^2}{\hat{c}_{I_2}}$$

Subsequently I atoms react with H_2

$$H_2 + 2I \xrightarrow{k_3} 2HI$$

and we obtain

$$\frac{\mathrm{d}c_{HI}}{\mathrm{d}t} = k_3 \cdot c_{H_2} \cdot c_I^2 = k_3 \cdot c_{H_2} \cdot K' \cdot \hat{c}_{I_2} = k_3 K' \frac{1}{c^{\ominus}} \cdot c_{H_2} \cdot c_{I_2}$$

In spite of the complicated kinetics and a termolecular step in the mechanism [4], the rate equation for the forward reaction is formally second-order with $k_1 = k_3 K'/c^{\ominus}$.

24.6.3 Parallel Reactions

24.6.3.1 Same Reaction Order in Both Reactions

We consider a main reaction between molecule A and molecule B to yield C:

$$A + B \xrightarrow{k_1} C \tag{24.75}$$

At the same time a parallel reaction occurs, forming a side product D:

$$A + B \xrightarrow{k_2} D \tag{24.76}$$

The rate law for this pair of reactions is

$$\frac{\mathrm{d}c_A}{\mathrm{d}t} = -k_1 \cdot c_A\, c_B - k_2 \cdot c_A\, c_B = -(k_1 + k_2) \cdot c_A\, c_B \tag{24.77}$$

In this case, A decreases according to a second-order reaction, but the effective rate constant is $k_1 + k_2$.

24.6.3.2 Different Reaction Orders in Both Reactions

The situation is more complicated when the side reaction follows a different reaction order than the main reaction:

$$\begin{aligned} \mathrm{A} + \mathrm{B} &\rightarrow \mathrm{C} \quad \text{(main reaction)} \\ 2\,\mathrm{A} &\rightarrow \mathrm{D} \quad \text{(side reaction)} \end{aligned} \tag{24.78}$$

To simplify, we assume that B is present in large excess over A; so that the main reaction is *quasi*-first-order, and the side reaction is second-order. In this case, the overall rate equation for A is

$$\frac{dc_A}{dt} = -k_1' \cdot c_A - 2k_2 \cdot c_A^2$$

with $k_1' = k_1 c_{B,0}$ (B in great excess). By integration (see Exercise E24.10) we find

$$c_A = c_{A,0} \cdot \frac{k_1' \cdot e^{-k_1' t}}{k_1' + 2k_2 c_{A,0} \cdot [1 - \exp(-k_1' t)]}$$

Example 24.9 *Consider the formation of 2-formyl-methyl-norbornene (C) from cyclopentadiene (A) and crotonaldehyde (B)*

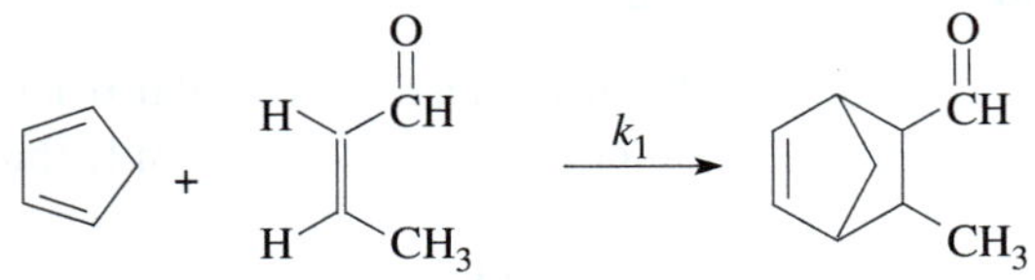

and account for the dimerization of cyclopentadiene as a side reaction

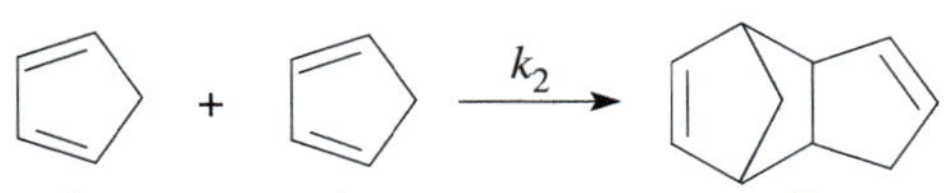

What must be the excess of B over A to obtain a yield of 80% when $k_1 = 6.6 \times 10^{-6}$ L mol^{-1} s^{-1} and $k_2 = 16.8 \times 10^{-6}$ L mol^{-1} s^{-1} at 70 °C?

To obtain a yield for the reaction we need to find the concentration of the product. Hence, we analyze the rate equation for C, namely

$$\frac{dc_C}{dt} = k_1' \cdot c_A = c_{A,0} k_1' \cdot \frac{k_1' \cdot e^{-k_1' t}}{k_1' + 2k_2 c_{A,0} \cdot [1 - \exp(-k_1' t)]}$$

We can integrate this rate equation by using the substitution $z = 1 - \exp(-k_1' t)$, $dz = k_1' \cdot \exp(-k_1' t) \cdot dt$. We find that

$$c_C = \frac{k_1'}{2k_2} \cdot \ln\left[1 + 2c_{A,0} \cdot \frac{k_2}{k_1'} \cdot (1 - \exp(-k_1' t))\right]$$

The yield of C at the end of the reaction ($t \rightarrow \infty$) is

$$\frac{c_{C,\infty}}{c_{A,0}} = \frac{k_1'}{2k_2 c_{A,0}} \cdot \ln\left(1 + \frac{2c_{A,0} k_2}{k_1'}\right) = \frac{\ln(1 + x)}{x}$$

with

$$x = \frac{2c_{A,0} k_2}{k_1'} = \frac{2c_{A,0} k_2}{k_1 c_{B,0}}$$

For $c_{C,\infty}/c_{A,0} = 0.8$ we obtain $x = 0.54$. Then

$$\frac{c_{B,0}}{c_{A,0}} = \frac{2k_2}{k_1} \cdot \frac{1}{x} = \frac{33.6}{6.6} \cdot \frac{1}{0.54} = 9.4$$

This means that to obtain 80% of the main product, reactant B must be in an almost tenfold excess over A.

24.6.4 Consecutive Reactions

We imagine molecules A reacting with B to form an intermediate C; in a consecutive reaction, the product D is formed from C:

$$A + B \xrightarrow{k_1} C, \quad C + B \xrightarrow{k_2} D \tag{24.79}$$

We restrict consideration to the case where B is in large excess. In this limit the rate equations for this sequence of reactions are

$$\frac{dc_A}{dt} = -k_1' c_A \quad \text{(disappearance of A)} \tag{24.80}$$

$$\frac{dc_C}{dt} = k_1' c_A - k_2' c_C \quad \text{(formation of C)} \tag{24.81}$$

$$\frac{dc_D}{dt} = k_2' c_C \quad \text{(formation of D)} \tag{24.82}$$

with $k_1' = k_1 c_{B,0}$ and $k_2' = k_2 c_{B,0}$.

Equation (24.80) is a first-order reaction step and can be integrated as

$$\boxed{c_A = c_{A,0} \cdot e^{-k_1' t}} \tag{24.83}$$

By substituting this result into Equation (24.81), we find

$$\frac{dc_C}{dt} = k_1' \, c_{A,0} \cdot e^{-k_1' t} - k_2' \, c_C \tag{24.84}$$

This differential equation is solved in Problem 24.6, to give

$$\boxed{c_C = c_{A,0} \frac{k_1'}{k_2' - k_1'} \left(e^{-k_1' t} - e^{-k_2' t} \right)} \tag{24.85}$$

Substituting this result into Equation (24.82) allows the time-dependent concentration of D to be found by integration, namely

$$\boxed{c_D = \frac{c_{A,0}}{k_2' - k_1'} \left[k_1' \left(e^{-k_2' t} - 1 \right) - k_2' \left(e^{-k_1' t} - 1 \right) \right]} \tag{24.86}$$

This solution is illustrated in Fig. 24.11b. As expected, the concentration of the intermediate C increases initially (its formation is faster than its decomposition); it reaches a maximum and finally it decreases.

In Problem 24.7 you show that the maximum concentration of the intermediate is reached at the time

$$t_{max} = \frac{\ln \left(k_2' / k_1' \right)}{k_2' - k_1'} \tag{24.87}$$

and in the special case of $k_1' = k_2' = k'$, you show that $t_{max} = 1/k'$.

Example 24.10 *Consider the special case in reaction (24.79) when both rate constants are equal: $k_1' = k_2' = k'$. In this special case the concentration c_C and the time t_{max} from Equations (24.85) and (24.87) are*

$$c_C = c_{A,0} k' \cdot t \cdot e^{-k' \cdot t} \quad \text{and} \quad t_{max} = \frac{1}{k'} \tag{24.88}$$

Use these results to calculate the maximum yield of the intermediate C.

Combining these results, we find

$$c_{C,max} = c_{A,0} k' \cdot t_{max} \cdot e^{-k' \cdot t_{max}} = c_{A,0} \cdot e^{-1} = \frac{c_{A,0}}{2.718} = 0.37 \cdot c_{A,0}$$

Thus the yield of the intermediate C ($c_{C,max}/c_{A,0}$) cannot exceed 37%.

24.7 Complex Reactions

24.7.1 Approximation Methods

As the number of elementary steps in a reaction sequence increases, the complexity can increase significantly. In many cases, however, the overall reaction may appear simple. In such cases, approximate treatments of the rate process are adequate.

24.7.1.1 Approximation by Rate-Determining Step

If one reaction step in a sequence of consecutive reactions is much slower than the other reactions, then it can be considered as rate-determining and the analysis can be simplified. Consider the consecutive reactions in Equation (24.79)

$$\mathrm{A + B} \xrightarrow{k_1} \mathrm{C}, \quad \mathrm{C + B} \xrightarrow{k_2} \mathrm{D}$$

and assume that B is in large excess over A. If $k_1' = k_1 c_\mathrm{B}$ is much smaller than $k_2' = k_2 c_\mathrm{B}$ then the rate-determining step is the decay of A. Soon after reactant C forms it reacts to form the product D, and the concentration of C remains small all the time. In this approximation, we have

$$c_\mathrm{C} \approx 0, \quad \text{and} \quad \frac{\mathrm{d}c_\mathrm{D}}{\mathrm{d}t} = -\frac{\mathrm{d}c_\mathrm{A}}{\mathrm{d}t} = k_1' c_\mathrm{A} \tag{24.89}$$

Because the formation of A is quasi-first order, $c_\mathrm{A} = c_{\mathrm{A},0} \cdot \mathrm{e}^{-k_1' t}$, we can integrate Equation (24.89) to find c_D

$$c_\mathrm{D} = c_{\mathrm{A},0}\, k_1' \int_0^t \mathrm{e}^{-k_1' t}\, \mathrm{d}t = -c_{\mathrm{A},0}\left(\mathrm{e}^{-k_1' t} - \mathrm{e}^{-0}\right) = c_{\mathrm{A},0}\left(1 - \mathrm{e}^{-k_1' t}\right) \tag{24.90}$$

This result is the same as that we found from Equation (24.86) by considering the limit where $k_1' \ll k_2'$.

Example 24.11 *In the Belousov–Zhabotinsky reaction (see Section 24.7.7) the inhibitor bromide is removed by the reaction*

$$5\,Br^- + BrO_3^- + 6\,H^+ \rightarrow 3\,Br_2 + 3\,H_2O \tag{24.91}$$

The rate equation for this reaction is

$$\frac{dc_{\mathrm{Br^-}}}{dt} = -k_r \cdot c_{\mathrm{BrO_3^-}} \cdot c_{\mathrm{Br^-}} \cdot c_{\mathrm{H^+}}^2 \tag{24.92}$$

and the reaction proceeds in three consecutive steps:

$$Br^- + BrO_3^- + 2\,H^+ \xrightarrow{k_1} HOBr + HBrO_2 \tag{24.93}$$

$$Br^- + HBrO_2 + H^+ \xrightarrow{k_2} 2\,HOBr \tag{24.94}$$

$$Br^- + HOBr + H^+ \xrightarrow{k_3} Br_2 + H_2O \tag{24.95}$$

Assume that reaction (24.93) is the slow step in the mechanism ($k_1 \ll k_2, k_3$) and find an expression for k_r in terms of the rate constants for the elementary reactions.

From Equations (24.93)–(24.95), the rate law for the disappearance of bromide is

$$\frac{dc_{\mathrm{Br^-}}}{dt} = -k_1 c_{\mathrm{BrO_3^-}} c_{\mathrm{Br^-}} c_{\mathrm{H^+}}^2 - k_2 c_{\mathrm{HBrO_2}} c_{\mathrm{Br^-}} c_{\mathrm{H^+}} - k_3 c_{\mathrm{HOBr}} c_{\mathrm{Br^-}} c_{\mathrm{H^+}}$$

Because the first step is rate determining, the HOBr and the $HBrO_2$ created in reaction (24.93) will be rapidly consumed in the subsequent reactions. Each HOBr created in reaction (24.93) consumes another Br^-, and each $HBrO_2$ created in Equation (24.93) reacts to consume three Br^-. Hence for each bromide that reacts with bromate, four additional bromide ions react with the products HOBr and $HBrO_2$. Then the rate law can be approximated by

$$\frac{dc_{\mathrm{Br^-}}}{dt} = -5k_1 \cdot c_{\mathrm{BrO_3^-}} \cdot c_{\mathrm{Br^-}} \cdot c_{\mathrm{H^+}}^2 \tag{24.96}$$

and we can identify k_r with $5k_1$.

24.7.1.2 Steady-State Approximation

In many cases one intermediate in a consecutive reaction appears only in small concentration. Consequently, the change of this concentration is also small, and it can be approximated as zero, thereby simplifying the analysis. This approximation is called the *steady-state approximation*.

Again consider the consecutive reactions in Equation (24.79) with B in large excess over A and assume $k'_1 \ll k'_2$. In this limit, the intermediate C is in a small concentration and its change, dc_C/dt, can be approximately set to zero; hence, we write

$$\frac{dc_C}{dt} = k'_1 c_A - k'_2 c_C = 0 \tag{24.97}$$

Solving Equation (24.97) for the concentration of the intermediate C gives the steady-state concentration $c_{C,ss}$ as

$$c_{C,ss} = c_A \frac{k'_1}{k'_2} = \frac{k'_1}{k'_2} c_{A,0}\, e^{-k'_1 t} \tag{24.98}$$

Substituting this result into Equation (24.82), for the product D gives

$$\frac{dc_D}{dt} = k'_2 c_C = k'_1 c_{A,0}\, e^{-k'_1 t} \tag{24.99}$$

and by integration we obtain

$$c_D = c_{A,0}\left(1 - e^{-k'_1 t}\right)$$

This is the same result as we obtained for the rate-determining step approximation.

Example 24.12 *As an example of applying the steady-state approximation, consider the bromination of malonic acid (HOOC–CH$_2$–COOH) in the Belousov–Zhabotinsky reaction (see Section 24.7.7). This reaction proceeds via the enol form (MAE) of the acid*

$$HOOC\text{–}CH = C(OH)_2$$

which is in equilibrium with the acid in its keto form

$$MA \underset{k_{-1}}{\overset{k_1}{\rightleftharpoons}} MAE \tag{24.100}$$

$$HOOC\text{–}CH = C(OH)_2 + Br_2 \xrightarrow{k_2} HOOC\text{–}CHBr\text{–}COOH + HBr \tag{24.101}$$

Use the steady-state approximation to find the rate law when the malonic acid is present in great excess.

The rate for the consumption of bromine is

$$\frac{dc_{Br_2}}{dt} = -k_2 \cdot c_{MAE} \cdot c_{Br_2}$$

and the rate for the formation and consumption of the intermediate MAE is

$$\frac{dc_{MAE}}{dt} = k_1 c_{MA} - k_{-1} \cdot c_{MAE} - k_2 \cdot c_{MAE} \cdot c_{Br_2}$$

The enol form MAE appears as an intermediate in a small concentration, so we apply the steady-state approximation

$$\frac{dc_{MAE}}{dt} = 0$$

If we assume that MA exists in great excess over bromine ($c_{MA} = c_{MA,0}$) we obtain

$$c_{MAE} = \frac{k_1 c_{MA,0}}{k_{-1} + k_2 c_{Br_2}}$$

for the enol concentration. By inserting this result into the rate equation for bromine, we find the rate law

$$\frac{dc_{Br_2}}{dt} = -\frac{k_1 k_2 \cdot c_{MA,0} \cdot c_{Br_2}}{k_{-1} + k_2 \cdot c_{Br_2}}$$

Now we consider two limiting cases.

Case 1: For high enough Br_2 concentration, $k_2 c_{Br_2} \gg k_{-1}$ and

$$\frac{dc_{Br_2}}{dt} = -k_1 \cdot c_{MA,0} \quad \text{and} \quad c_{Br_2} = c_{Br_2,0} - k_1 \cdot c_{MA,0} \cdot t \tag{24.102}$$

This means that Br_2 disappears linearly with time (zero-order kinetics, Fig. 24.12a).

Case 2: For low enough Br_2 concentration, $k_2 c_{Br_2} \ll k_{-1}$ and

$$\frac{dc_{Br_2}}{dt} = -\frac{k_1 k_2}{k_{-1}} \cdot c_{MA,0} \cdot c_{Br_2} \quad \text{and} \quad c_{Br_2} = c_{Br_2,\,0} \cdot e^{-k_r' t} \tag{24.103}$$

with $k_r' = k_1 k_2 c_{MA,0} / k_{-1}$. This means that Br_2 disappears exponentially with time (first-order kinetics, Fig. 24.12b).

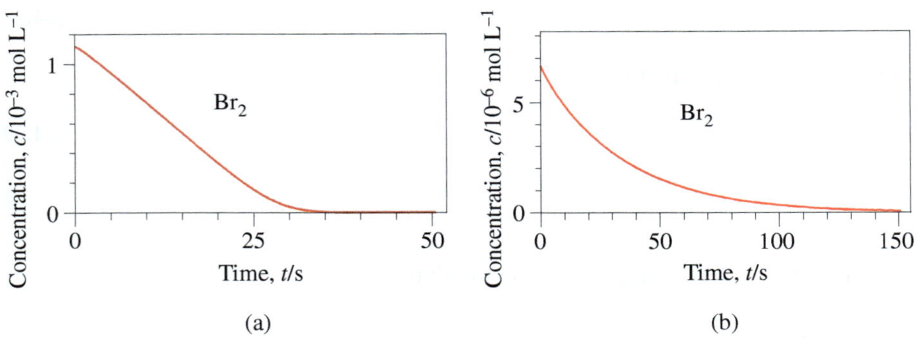

Figure 24.12 Bromination of malonic acid via enol form at 20 °C in acidic medium ($c_{H^+} = 1.29$ mol L^{-1}): reactions (24.101) and (24.100). Bromine is injected into a malonic acid solution and the concentration c of bromine is measured [5]. (a) High bromine concentration ($c_{\text{malonic acid},0} = 0.016$ mol L^{-1}, $c_{Br_{2,0}} = 1.1 \times 10^{-3}$ mol L^{-1}): linear decay of Br_2. (b) Low bromine concentration ($c_{\text{malonic acid},0} = 0.001$ mol L^{-1}, $c_{Br_{2,0}} = 6.5 \times 10^{-6}$ mol L^{-1}): exponential decay of Br_2.

24.7.2 Lindemann–Hinshelwood Mechanism

Many reactions appear to be first order in the concentration of a single reactant species. Common examples of such processes are isomerization reactions and decomposition reactions. Reactions of this type may be initiated by the absorption of light or other forms of energy. However in the case of thermal reactions, the molecule gains enough energy to react by collision with surrounding atoms and molecules. Because collisions are necessary to activate the molecule, we might expect such reactions to be second order, rather than the observed first-order behavior. This apparent paradox was explained by Lindemann and Hinshelwood and is called the Lindemann–Hinshelwood mechanism.

We consider the isomerization of 3-methyl-cyclobutene into 4-methyl-butadiene:

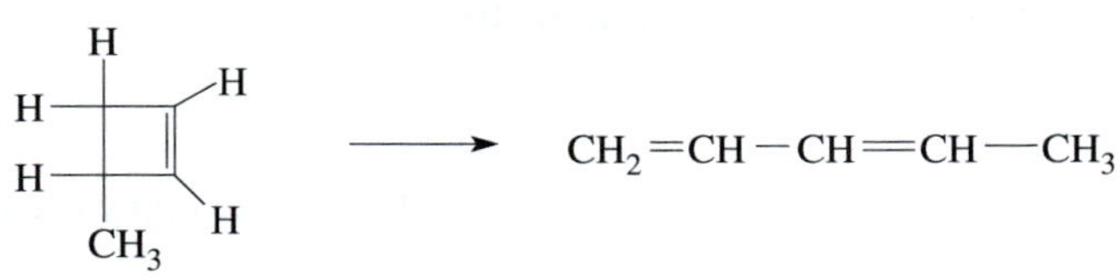

A B

Following the Lindemann–Hinshelwood mechanism of unimolecular reactions the first step in this reaction is a collision between two molecules A, leading to one molecule A in a vibrationally excited state, A*

$$A + A \underset{k_{-1}}{\overset{k_1}{\rightleftharpoons}} A^* + A \tag{24.104}$$

In A* the weakest bond breaks, forming the product B after reorganization

$$A^* \xrightarrow{k_2} B \tag{24.105}$$

We have to take into account that collisions between A and B can produce A* as well

$$A + B \underset{k_{-3}}{\overset{k_3}{\rightleftharpoons}} A^* + B \tag{24.106}$$

The rate equations for B and for A* are)

$$\frac{dc_B}{dt} = k_2 \cdot c_{A^*} \tag{24.107}$$

$$\frac{dc_{A^*}}{dt} = k_1 \cdot c_A^2 - k_{-1} \cdot c_A \cdot c_{A^*} - k_2 \cdot c_{A^*} + k_3 \cdot c_A c_B - k_{-3} \cdot c_B \cdot c_{A^*} \tag{24.108}$$

Because the intermediate A* occurs in small concentration, we apply the steady-state approximation

$$\frac{dc_{A^*}}{dt} = 0$$

To further simplify, we assume that $k_1 \approx k_3$ and $k_{-1} \approx k_{-3}$ which we can justify because A and B in our example are of comparable complexity and of equal mass. We find

$$c_{A^*} = \frac{k_1 \cdot c_A \cdot (c_A + c_B)}{k_{-1} \cdot (c_A + c_B) + k_2} \tag{24.109}$$

If we choose the condition of only A present initially, then $c_A + c_B = c_{A,0}$ and $c_A = c_{A,0} - c_B$. Hence, we obtain

$$\frac{dc_B}{dt} = k_2 \cdot c_{A^*} = \frac{k_1 k_2 \cdot c_{A,0}}{k_{-1} \cdot c_{A,0} + k_2} \cdot (c_{A,0} - c_B) \tag{24.110}$$
$$= k_r' \cdot (c_{A,0} - c_B)$$

Thus the product B is formed according to first-order kinetics, and the first-order rate constant k_r' is

$$k_r' = \frac{k_1 k_2 \cdot c_{A,0}}{k_{-1} \cdot c_{A,0} + k_2} = \frac{(k_1 k_2 / k_{-1}) \cdot c_{A,0}}{c_{A,0} + k_2 / k_{-1}} \tag{24.111}$$

The rate constant in Equation (24.111) depends on the initial concentration $c_{A,0}$. At low $c_{A,0}$ we expect $k_r' = k_1 \cdot c_{A,0}$, i.e., k_r' increases linearly with $c_{A,0}$. At high $c_{A,0}$ we expect $k = k_1 k_2 / k_{-1}$, i.e., k_r' becomes independent of the initial concentration $c_{A,0}$.

The dependence on initial concentration can be understood by considering the relative time scales associated with the collision frequency and the molecule A's energy relaxation rate. At low $c_{A,0}$ fewer collisions occur, and the formation of A* is rate determining. This rate increases with increasing initial concentration $c_{A,0}$. At high $c_{A,0}$ the average time between collisions of A* and A or B becomes shorter than the lifetime of A*, and the isomerization reaction (24.105) becomes rate determining.

The prediction of the theory is compared with the experimental data in Fig. 24.13. The experimental points are fairly reproduced, although the theory predicts values that are somewhat too low for small initial concentrations. Moreover, measurements at higher initial concentrations reveal that the theory needs improvement. Essentially, one must account for how the unimolecular reaction step depends on the specific structure of A*: such improvements were introduced by Hinshelwood, Slater, Rice–Ramsberger–Kassel (RRK-Theory), and Rice–Ramsberger–Kassel–Marcus (RRKM-Theory).

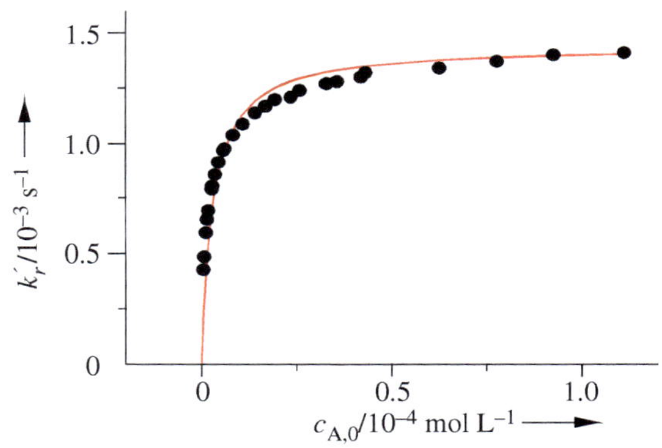

Figure 24.13 Experimental data [6] for the isomerization of 3-methyl-cyclobutene at 148.5 °C (dots) compared with the prediction of the Lindemann mechanism (solid line). The calculation was carried out with the values $k_1 k_2 / k_{-1} = 1.44 \times 10^{-3}$ s^{-1} and $k_2 / k_{-1} = 2.80 \times 10^{-6}$ mol L^{-1}.

24.7.3 Chain Reactions

In the iodine hydrogen reaction, iodine radicals are important as intermediates (Example 24.8); however, the formation of H radicals is not significant (see Problem P24.12). In contrast, reactions involving hydrogen with bromine, and hydrogen with chlorine, involve the formation of both hydrogen radicals and halogen radicals. These reactions proceed via a chain mechanism.

In the bromine hydrogen reaction

$$Br_2 + H_2 \rightarrow 2\,HBr \tag{24.112}$$

both hydrogen radicals and bromine radicals are formed. The elementary steps in the reaction mechanism are:

a) Production of radicals

$$\text{Initiation:}\quad Br_2 \xrightarrow{k_1} 2\,Br \tag{24.113}$$

b) Conservation of radicals

$$\text{Propagation:}\quad Br + H_2 \xrightarrow{k_2} HBr + H \tag{24.114}$$

$$\text{Propagation:}\quad HBr + H \xrightarrow{k_{-2}} Br + H_2 \tag{24.115}$$

$$\text{Propagation:}\quad H + Br_2 \xrightarrow{k_3} HBr + Br \tag{24.116}$$

c) Removal of radicals

$$\text{Termination:}\quad 2\,Br \xrightarrow{k_4} Br_2 \tag{24.117}$$

This mechanism has three characteristic reaction types: a) the generation of radicals, b) the conservation of radicals, and c) the removal of radicals. Such a mechanism is called a *chain reaction*. The H and Br radicals are called *chain carriers*, because they allow the process to continue. Br is first formed in the chain-initiating step (a). Br and H are conserved in the *chain-propagation* steps (b), until they are removed by a *chain-termination* or *chain-breaking* step (c).

In Box 24.2 we solve for the rate of these reactions and obtain the following rate equation for the overall process

$$\frac{dc_{HBr}}{dt} = \frac{A}{c_{Br_2} + B \cdot c_{HBr}} \cdot \left(c_{Br_2}\right)^{3/2} \cdot c_{H_2} \tag{24.118}$$

with

$$A = 2k_2\sqrt{\frac{k_1}{k_4}} \quad \text{and} \quad B = \frac{k_{-2}}{k_3}$$

This result is in agreement with the empirical rate law.

24.7.3.1 Branching Chain Reactions

If the chain-propagating steps generate more radicals than they consume, then the reaction is called a *branched chain reaction*. An example is the detonating gas reaction

$$2\,H_2 + O_2 \rightarrow 2\,H_2O \tag{24.119}$$

which follows the mechanism:

$$\text{Initiation:}\quad H_2 + O_2 \rightarrow 2\,OH \tag{24.120}$$

$$\text{Propagation:}\quad OH + H_2 \rightarrow H_2O + H \tag{24.121}$$

$$\text{Branching:}\quad H + O_2 \rightarrow OH + O \tag{24.122}$$

$$\text{Branching:}\quad O + H_2 \rightarrow OH + H \tag{24.123}$$

$$\text{Termination:}\quad H \rightarrow \text{wall} \tag{24.124}$$

$$\text{Termination:}\quad H + O_2 + M \rightarrow HO_2 + M^* \tag{24.125}$$

where M is a third body acting as an inert collision partner. The overall reaction in the propagation and branching steps is

$$2\,H_2 + O_2 + OH \rightarrow H_2\,O + H + 2\,OH \tag{24.126}$$

In Equation (24.126) each OH radical produces two OH radicals and an H radical. This feature causes the number of H and OH radicals to increase exponentially with time. The result is an explosion.

The explosion is sensitive to the total pressure of gases. At low pressure the reaction is terminated by collisions with the wall, see Equation (24.124), and this reaction sets a lower explosion limit. At high pressure the reaction is terminated by formation of unreactive hydroperoxide radicals HO_2, which diffuse to the wall, see Equation (24.125), and sets an *upper explosion limit*. At very high pressures the gas mixture becomes again explosive: HO_2 radicals react with H_2 and form OH radicals and H_2O before reaching the wall.

24.7.4 Enzyme Reactions (Michaelis–Menten Mechanism)

Enzymes act as catalysts and can increase reaction rates by many orders of magnitude. The original definition of catalysis is attributed to Ostwald, "A catalyst is a substance that changes the velocity of a chemical reaction without itself appearing in the end products." He showed that a catalyst does not change the position of the equilibrium, rather it changes the rate at which equilibrium is attained. Thus a catalyst changes the forward and reverse rates in the same proportion.

24.7.4.1 Hydrolysis of Urea by the Enzyme Urease

As an example consider the hydrolysis of urea by using the enzyme urease

$$\mathrm{O{=}C(NH_2)_2 + H_2O \xrightarrow{\text{enzyme urease}} (NH_4)_2CO_3} \tag{24.127}$$

In a first step a complex UE between urea (U) and the enzyme (E) is formed:

$$\mathrm{U + E} \underset{k_{-1}}{\overset{k_1}{\rightleftharpoons}} \mathrm{UE} \tag{24.128}$$

In a second step the complex hydrolyzes into the product P and the enzyme E:

$$\mathrm{UE + H_2O} \xrightarrow{k_2} \mathrm{P + E} \tag{24.129}$$

With this mechanism, the rate equations for U, E, UE, and P are

$$\frac{dc_U}{dt} = -k_1 \cdot c_U \cdot c_E + k_{-1} \cdot c_{UE} \tag{24.130}$$

$$\frac{dc_{UE}}{dt} = k_1 \cdot c_U \cdot c_E - k_{-1} \cdot c_{UE} - k_2 \cdot c_{UE} \tag{24.131}$$

$$\frac{dc_E}{dt} = -\frac{dc_{UE}}{dt} \tag{24.132}$$

$$\frac{dc_P}{dt} = k_2 c_{UE} \tag{24.133}$$

The reactant which is consumed, urea in this case, is called the *substrate* upon which the enzyme acts.

The rate equations (24.130)–(24.133) cannot be integrated directly. Therefore we apply the steady-state approximation for the complex UE (UE is present in a very low concentration):

$$\frac{dc_{UE}}{dt} = k_1 \cdot c_U \cdot c_E - k_{-1} \cdot c_{UE} - k_2 \cdot c_{UE} = 0 \tag{24.134}$$

We can eliminate c_E from this equation by using the fact that the enzyme is conserved, namely

$$c_E + c_{UE} = c_{E,0} \tag{24.135}$$

where $c_{E,0}$ is the initial concentration of the enzyme. Solving Equation (24.134) for c_{UE}, we obtain,

$$c_{UE} = \frac{k_1 \cdot c_U \cdot c_{E,0}}{k_1 c_U + k_{-1} + k_2} \tag{24.136}$$

Now we substitute this expression for the UE concentration into Equation (24.133) and find

$$\boxed{\frac{dc_P}{dt} = k_2 \cdot c_{UE} = \frac{k_2 \cdot c_U}{c_U + K_M} \cdot c_{E,0}} \quad \text{with} \quad K_M = \frac{k_{-1} + k_2}{k_1} \tag{24.137}$$

This rate equation is called the *Michaelis–Menten Equation* and K_M is called the Michaelis constant.

The reaction rate increases linearly with the concentration of the enzyme but has a more complicated dependence on the urea concentration. The limiting conditions for Equation (24.137) are instructive. If we define the initial rate by

$$v_0 = \left(\frac{dc_P}{dt}\right)_{t=0}$$

then at low initial concentrations of urea Equation (24.137) is approximately

$$v_0 = \frac{k_2}{K_M} \cdot c_{U,0} \cdot c_{E,0} \quad (c_U \ll K_M) \tag{24.138}$$

The initial rate v_0 increases linearly with the initial urea concentration. At high initial concentrations of urea we can neglect K_M in the denominator of Equation (24.137):

$$v_0 = k_2 \cdot c_{E,0} \quad (c_U \gg K_M) \tag{24.139}$$

and v_0 is independent of c_U. It becomes independent because the equilibrium (24.128) is completely shifted to the right-hand side (the maximum amount of UE has been formed). By plotting v_0 versus $c_{U,0}$ (Fig. 24.14a) we can evaluate k_2/K_M from the initial slope and k_2 from the final value of the curve. Alternatively, we can rearrange Equation (24.137):

$$\frac{1}{v_0} = \frac{c_{U,0} + K_M}{k_2 \cdot c_{U,0} \cdot c_{E,0}} = \frac{1}{k_2 \cdot c_{E,0}} + \frac{K_M}{k_2 \cdot c_{E,0}} \cdot \frac{1}{c_{U,0}} \tag{24.140}$$

By plotting $1/v_0$ versus $1/c_{U,0}$ (Fig. 24.14b) we obtain K_M/k_2 from the slope and k_2 from the intercept with the ordinate (*Lineweaver–Burk diagram*).

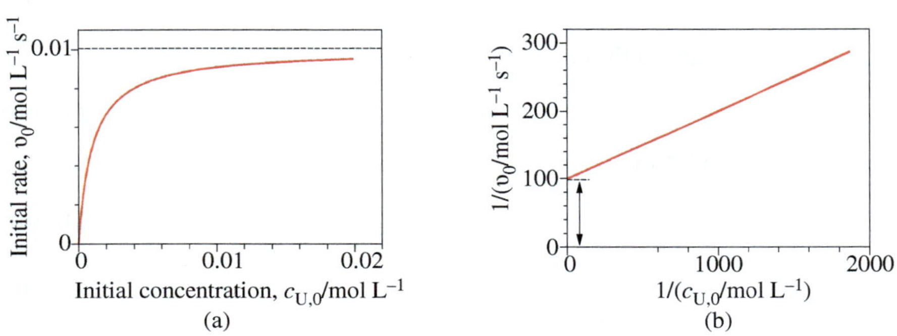

Figure 24.14 Enzyme reaction: hydrolysis of urea by the enzyme urease. (a) Initial rate v_0 versus initial concentration $c_{U,0}$ of urea ($c_{E,0} = 1 \times 10^{-5}$ mol L^{-1}, $K_M = 1 \times 10^{-3}$ mol L^{-1}, $k_2 = 1000$ s^{-1}). (b) Lineweaver–Burk diagram. Ordinate intercept $= 1/(k_2 \cdot c_{E,0})$, slope $= K_M/(k_2 \cdot c_{E,0})$.

24.7.4.2 Inhibition of Enzyme Activity

The binding between an enzyme and a substrate is crucial for the specificity of enzyme reactions. Molecules which reduce the activity of an enzyme are called *inhibitors*, and they can play a fundamental role in triggering the interplay of enzymes, determining their functional network [7]. Several kinds of inhibitors are known; however, we only consider *competitive inhibitors*. In many cases a reaction product acts as a competitive inhibitor and is an important form of regulation: the enzymatic reaction is turned off when sufficient product for further requirements is present.

A competitive inhibitor is a molecule that resembles the substrate and can bind reversibly to the active site of the enzyme, thereby competing with the substrate. The inhibition can be reversed by diluting the inhibitor or by increasing the concentration c_S of the substrate. The initial rate v_0 of the reaction is diminished by the inhibitor, but its influence is less at high substrate concentration $c_{S,0}$.

To account for inhibition we must include the reactions

$$I + E \underset{k'_{-1}}{\overset{k'_1}{\rightleftharpoons}} IE \tag{24.141}$$

in addition to the Michaelis–Menten scheme

$$S + E \underset{k_{-1}}{\overset{k_1}{\rightleftharpoons}} SE \xrightarrow{k_2} P + E \tag{24.142}$$

where I is the inhibitor and IE is the inhibitor–enzyme complex. The equilibrium between I, E, and IE follows from the relation

$$K_I = \frac{k'_{-1}}{k'_1} = \frac{c_I \cdot c_E}{c_{IE}} \tag{24.143}$$

Accounting for the conservation of enzyme

$$c_E + c_{SE} + c_{IE} = c_{E,0} \tag{24.144}$$

we can solve the rate equations by making the steady-state approximation. We find that the concentration of SE is

$$c_{SE} = \frac{k_1 \cdot c_S \cdot c_{E,0}}{k_1 c_S + (k_{-1} + k_2)(1 + c_I/K_I)} \tag{24.145}$$

and the reaction rate is

$$\frac{dc_P}{dt} = \frac{k_2 \cdot c_S}{c_S + K'_M} \cdot c_{E,0} \quad \text{with} \quad K'_M = K_M \cdot \left(1 + \frac{c_I}{K_I}\right) \tag{24.146}$$

Example 24.13 *Malonic Acid as Competitive Inhibitor. The enzyme succinic acid dehydrogenase catalyzes the oxidation of succinic acid to form fumaric acid:*

$$HOOC–CH_2–CH_2\ –COOH \rightarrow HOOC\ –CH{=}CH\ –COOH$$

Malonic acid $HOOC–CH_2–COOH$ acts as a competitive inhibitor since its structure resembles that of succinic acid. It can bind to the active site of the enzyme with $K_I = 10^{-5}$ mol L^{-1}, but it cannot act as an alternative substrate. How much does malonic acid affect the initial reaction rate for oxidation of succinic acid?

At a small concentration c_S, of malonic acid, the initial rate v_0 is

$$\frac{dc_P}{dt} = \frac{k_2 \cdot c_S}{K'_M} \cdot c_{E,0} \quad \text{with} \quad K'_M = K_M \cdot \left(1 + \frac{c_I}{K_I}\right)$$

When the inhibitor is not present, $c_I = 0$, the rate is $\left[k_2 \cdot c_S/K_M\right] \cdot c_{E,0}$, and when $c_I = 10^{-5}$ M the initial rate is $\left[k_2 \cdot c_S \,/\, (2K_M)\right] \cdot c_{E,0}$; i.e., the rate is one-half of the value without inhibitor.

24.7.5 Autocatalytic Reactions

In many processes (e.g., oscillating chemical reactions) autocatalysis plays an important role. In an autocatalytic reaction a reactant appears also as a product of the reaction:

$$A + B \xrightarrow{k_r} C + 2\,B \tag{24.147}$$

If the reaction is started with a very low concentration of B, then the reaction is very slow in the beginning. As the reaction proceeds, the concentration of B increases and the reaction becomes faster and faster. Because of the consumption of reactant A, however, the reaction rate reaches a maximum and then decreases. The rate equation for B is

$$\frac{dc_B}{dt} = k_r \cdot c_A \cdot c_B \tag{24.148}$$

For each molecule of B that is formed, one molecule of A has disappeared, and the concentration c_A can be written as

$$c_A = c_{A,0} - (c_B - c_{B,0}) \tag{24.149}$$

where $c_{A,0}$ and $c_{B,0}$ are the initial concentrations of A and B. Substituting into Equation (24.148), we find

$$\frac{dc_B}{dt} = k_r \cdot c_B \cdot (c_{A,0} + c_{B,0} - c_B) \tag{24.150}$$

The solution of this differential equation is (see Problem P24.10):

$$c_B = \frac{c_{A,0} + c_{B,0}}{1 + (c_{A,0}/c_{B,0}) \cdot e^{-(c_{A,0}+c_{B,0})k_r t}} \tag{24.151}$$

Correspondingly,

$$c_A = (c_{A,0} + c_{B,0}) - c_B = \frac{c_{A,0} + c_{B,0}}{1 + (c_{B,0}/c_{A,0}) \cdot e^{+(c_{A,0}+c_{B,0})k_r t}} \tag{24.152}$$

The concentrations of c_A and c_B are plotted in Fig. 24.15 and display the S-shape that is typical for autocatalytic behavior.

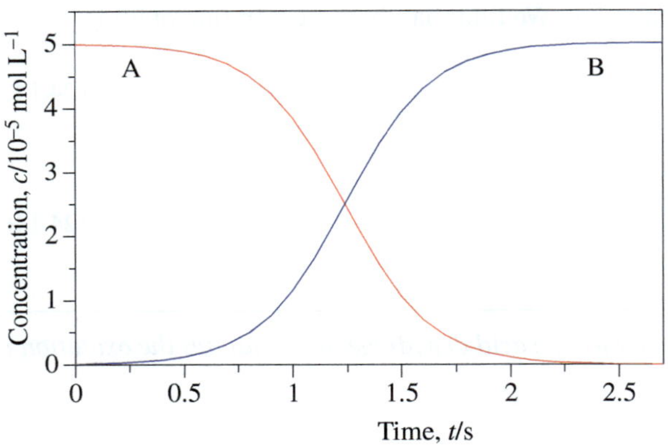

Figure 24.15 Autocatalytic reaction (24.147). Concentrations c_A (red solid line) and c_B (blue solid line) calculated from Equations (24.151) and (24.152), $k_r = 1 \times 10^5$ L mol^{-1} s^{-1}. Initial conditions: $c_{A,0} = 5 \times 10^{-5}$ mol L^{-1}, $c_{B,0} = 1 \times 10^{-7}$ mol L^{-1}.

Example 24.14 *Relate the autocatalytic formation of trypsin to the mechanism in Equation (24.147).*

The formation of the enzyme trypsin from its precursor trypsinogen is catalyzed by trypsin:

trypsinogen + trypsin → C + 2 trypsin

where the other product C is a hexapeptide. In this case we identify the reactants in Equation (24.147) as A = trypsinogen, B = trypsin, and C = other products. From the experimental data in the table below (dots in Fig. 24.16), we evaluate the rate constant k_r, using Equation (24.151) and treat k_r as a fitting parameter. Figure 24.16 (solid line) shows the calculated curve for $k_r = 1.55 \times 10^3$ L min $^{-1}$ mol^{-1}.

t/min	4.5	15.8	24.8	36.0	45.0	54.0	74.3	83.3	94.5	103.5
$c/10^{-5}$ mol L^{-1}	0.020	0.030	0.051	0.081	0.142	0.203	0.386	0.508	0.508	0.935

t/min	115	123	135	144	153	164	173	184	200	214	230	295
$c/10^{-5}$ mol L^{-1}	1.20	1.40	1.62	1.83	1.95	2.03	2.15	2.28	2.34	2.34	2.42	2.50

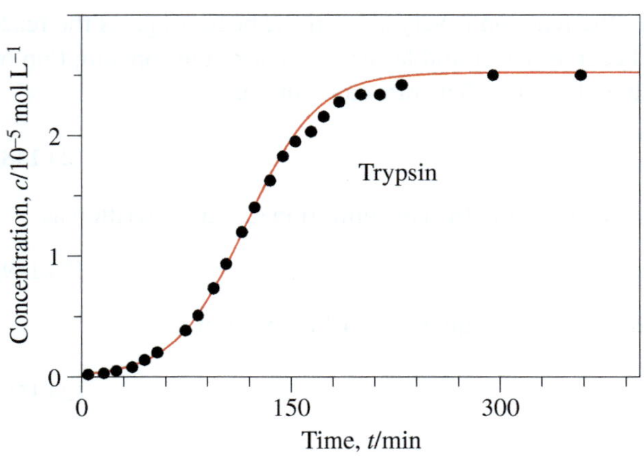

Figure 24.16 Autocatalytic formation of trypsin from trypsinogen (at pH 8, 298 K in 0.1 M Tris/HCl buffer + 0.05 M $CaCl_2$). The concentration c of trypsin is displayed. Initial concentrations: trypsin: 2.5×10^{-7} M, trypsinogen: 2.5×10^{-5} M. Black filled circles: experiment (H. Lachmann, private communication); red solid line: calculated from Equation (24.151) with rate constant $k_r = 1.55 \times 10^3$ L min $^{-1}$ mol^{-1}.

24.7.5.1 Autocatalytic Oxidation of Ru^{2+}

In most cases, autocatalytic reactions proceed in several steps. One important example is the oxidation of Ru^{2+} by bromate dissolved in diluted sulfuric acid.[4] This reaction is one of the key reactions in the Belousov–Zhabotinsky reaction (see Section 24.7.7).

4 Ru^{2+} is used in the form of its trisbipyridyl complex: $[Ru(bipy)_3]^{2+}$.

Consider an aqueous solution of Ru^{2+} and bromate in 1 molar sulfuric acid containing traces (10^{-7} mol L^{-1}) of $HBrO_2$; bromate and H^+ are in large excess over Ru^{2+}.[5] BrO_3^- reacts with $HBrO_2$ forming BrO_2 which is immediately removed by reaction with Ru^{2+}:

$$BrO_3^- + HBrO_2 + H^+ \xrightarrow[\text{slow}]{k_2} 2\ BrO_2 + H_2O \tag{24.153}$$

$$BrO_2 + Ru^{2+} + H^+ \xrightarrow[\text{fast}]{k_1} Ru^{3+} + HBrO_2 \tag{24.154}$$

By investing one $HBrO_2$ in the first reaction, two $HBrO_2$ and two Ru^{3+} are produced in the second reaction. The $HBrO_2$ molecules thus formed enter into the first reaction, leading to the overall reaction

$$2\ Ru^{2+} + BrO_3^- + HBrO_2 + 3\ H^+ \rightarrow 2\ Ru^{3+} + 2\ HBrO_2 + H_2O \tag{24.155}$$

In this way an exponential increase in the amount of $HBrO_2$ and Ru^{3+} takes place until the Ru^{2+} is consumed, stopping the process.

The corresponding rate equations are (with $k_1' = k_1 \cdot c_{H^+}$ and $k_2' = k_2 \cdot c_{BrO_3^-} c_{H^+}$)

$$\frac{dc_{Ru^{2+}}}{dt} = -k_1' \cdot c_{Ru^{2+}} \cdot c_{BrO_2} \tag{24.156}$$

$$\frac{dc_{BrO_2}}{dt} = -k_1' \cdot c_{Ru^{2+}} \cdot c_{BrO_2} + 2k_2' \cdot c_{HBrO_2} \tag{24.157}$$

$$\frac{dc_{HBrO_2}}{dt} = +k_1' \cdot c_{Ru^{2+}} \cdot c_{BrO_2} - k_2' \cdot c_{HBrO_2} \tag{24.158}$$

Example 24.15 *Find an approximate solution of the rate equations (24.156)–(24.158) for autocatalysis. It is not possible to integrate these rate equations rigorously in analytical form. Because reaction (24.154) is much faster than reaction (24.153), the BrO_2 is in a low concentration. Apply the steady-state approximation and find the time dependence of the Ru^{3+} concentration.*

Using the steady-state approximation

$$\frac{dc_{BrO_2}}{dt} = 0, \qquad \text{and we obtain} \quad c_{BrO_2} = \frac{2k_2' \cdot c_{HBrO_2}}{k_1' \cdot c_{Ru^{2+}}}$$

from Equation (24.157). From Equation (24.158) we have

$$\frac{dc_{HBrO_2}}{dt} = +k_2' \cdot c_{HBrO_2}$$

so that the $HBrO_2$ concentration grows exponentially:

$$c_{HBrO_2} = c_{HBrO_2,0} \cdot e^{k_2' t}$$

Using Equation (24.156), we obtain

$$\frac{dc_{Ru^{2+}}}{dt} = -2k_2' \cdot c_{HBrO_2} = -2c_{HBrO_2,0} \cdot k_2' \cdot e^{k_2' t}$$

for the change in Ru^{2+} concentration with time and by integration

$$c_{Ru^{2+}} = c_{Ru^{2+},0} - 2c_{HBrO_2,0} \cdot (k_2')^2 \cdot \left(e^{k_2' t} - 1\right) \qquad (\text{for } c_{Ru^{2+}} > 0)$$

with the restriction that $c_{Ru^{2+}}$ is a positive quantity.

Figure 24.17 (curve 1) plots the decay of the Ru^{2+} concentration obtained in this approximation together with the experimental data. The decay is somewhat slower in the experiment than in the calculation. This discrepancy arises because reaction (24.153) is reversible and $HBrO_2$ disproportionates according to

$$2\ HBrO_2 \rightarrow HOBr + BrO_3^- + H^+ \tag{24.159}$$

Including these reactions and solving the differential equations numerically (see Table 24.4, reactions 6–8 from top) we obtain curve 2 in Fig. 24.17, which is in agreement with the experiment.

5 Note that Ru^{2+} and Ru^{3+} denote the trisbipyridyl complexes of ruthenium.

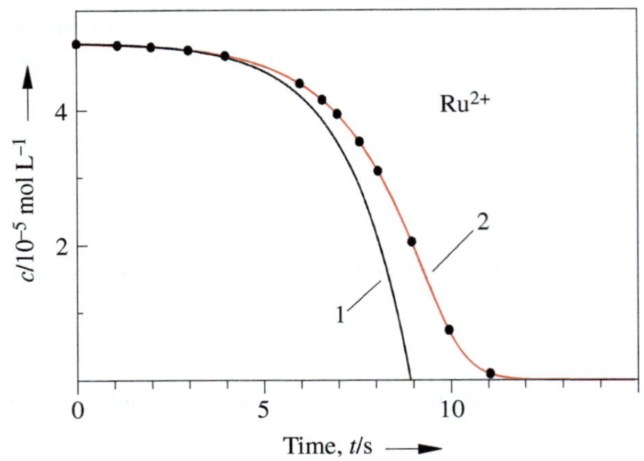

Figure 24.17 Autocatalytic oxidation of Ru^{2+} by bromate in 1 molar sulfuric acid solution at 20 °C. Initial concentrations: $c_{BrO_3^-} = 0.01$ mol L^{-1}, $c_{Ru^{2+}} = 5 \times 10^{-5}$ mol L^{-1}, $c_{H^+} = 1.29$ mol L^{-1}. Black filled circles: experiment. Solid lines: calculation; curve 1 (black): calculated from Equations (24.156)–(24.158) with the approximation in Example 24.15; curve 2 (red): calculated by including reversibility of reaction (24.153) and of disproportionation of $HBrO_2$ (24.159), rate equations solved numerically. Initial concentration of $HBrO_2$ in the calculation: 1×10^{-7} mol L^{-1}. For rate constants used in the calculation see Table 24.4, reactions 1–5 from top. Experimental data from [8].

Table 24.4 Rate Constants of Reactions Which Are Important for the Autocatalytic Formation of Ru^{3+} and for Its Inhibition by Bromide. (Temperature 20 °C).[6]

Reaction			$k_{forward}$	$k_{reverse}$
$Ru^{2+} + BrO_2 + H^+$	→	$Ru^{3+} + HBrO_2$	4×10^6	0
$HBrO_2 + HBrO_3$	→	$Br_2O_4 + H_2O$	4.8×10^1	3.2×10^3
Br_2O_4	→	$2\ BrO_2$	7.5×10^4	1.4×10^9
$HBrO_2 + H^+$	→	$H_2BrO_2^+$	2×10^6	1×10^8
$H_2BrO_2^+ + HBrO_2$	→	$HOBr + HBrO_3$	1.7×10^5	0
$Br^- + HBrO_2 + H^+$	→	$2\ HOBr$	2.5×10^6	2×10^{-5}
$Br^- + HOBr + H^+$	→	$Br_2 + H_2O$	8×10^9	80
$Br^- + HBrO_3 + H^+$	→	$HOBr + HBrO_2$	0.12×10^1	3.2

Note that Ru^{2+} and Ru^{3+} denote the trisbipyridyl complexes of ruthenium.

24.7.5.2 Inhibition of Autocatalysis

Reactants that remove intermediates of the autocatalytic process inhibit the onset of the autocatalysis. For example, adding Br^- to the aqueous solution of Ru^{2+} and bromate considered in Example 24.15 inhibits the autocatalysis of HOBr. The bromide reacts with the intermediate, $HBrO_2$, in a fast process:

$$HBrO_2 + Br^- + H^+ \xrightarrow{k_3} 2\ HOBr \tag{24.160}$$

Consequently the rate of reaction (24.153) in the oxidation of Ru^{2+} is slowed down: less $HBrO_2$ is oxidized to BrO_2.

Figure 24.18 illustrates the effect of bromide added in a small concentration (2% of the initial Ru^{2+} concentration) on the Ru^{2+} concentration. The bromide concentration decreases because of reaction (24.160) with $HBrO_2$ and of direct reaction with bromate (see Section 24.6.4). As soon as the bromide concentration falls below a critical limit (c_{crit} in Fig. 24.18) the autocatalytic process starts. As shown in Problem P24.11, the critical bromide concentration is given by

$$c_{crit} = k_2'/k_3' \tag{24.161}$$

with $k_2' = 48 \times 0.01\ s^{-1} = 0.48\ s^{-1}$ and $k_3' = 2.5 \times 10^6$ L mol^{-1} s^{-1} we obtain $c_{crit} = 2 \times 10^{-7}$ mol L^{-1}.

During the inhibitory phase of the reaction the Ru^{2+} concentration is practically constant; then it decreases abruptly. This behavior can be considered as a *clock reaction*, because the time where the reaction mixture crosses c_{crit} can be controlled by controlling the Br^- concentration.

6 The units of the rate constants are s^{-1}, if one reactant appears in the chemical reaction, L mol^{-1} s^{-1} for two reactants and L^2 mol^{-2} s^{-1} for three reactants. The solvent water is included in the rate constant k_8. Y. Gao, H.D. Försterling, J. Phys. Chem. 99, 8638 (1995).

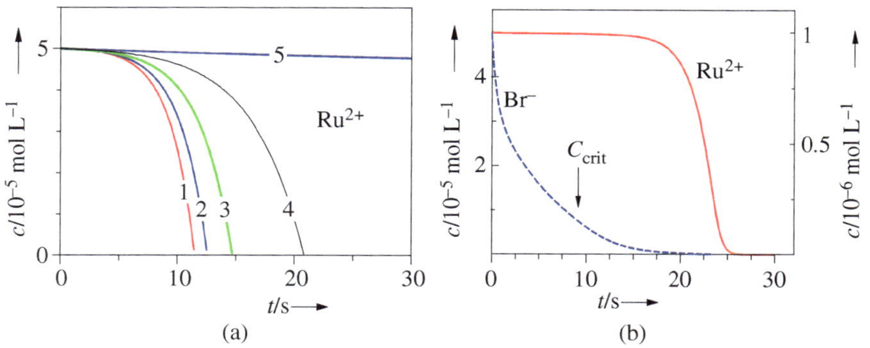

Figure 24.18 Inhibition of autocatalytic oxidation of Ru^{2+} by addition of bromide. Conditions as in Fig. 24.17. (a) Concentration of Ru^{2+} versus time curves for constant bromide concentration, calculated as described in Problem 24.11. Bromide concentrations for curves 1–5: 0, 0.2, 0.5, 1.0, 2.0×10^{-7} mol L^{-1}. (b) Concentrations of Ru^{2+} (left-hand scale) and Br^- (right-hand scale) for initial bromide concentration of 1×10^{-6} mol L^{-1} (bromide reacts with bromate, thus its concentration decreases with time). For rate constants used in the calculation see Table 24B.1, reactions 1–7 from top. The concentration of Ru^{2+} is shown in Figure 24.18a for $k_2' = 48 \times 0.01\ s^{-1} = 0.48\ s^{-1}$, $k_3' = 2.5 \times 10^6$ L mol^{-1} s^{-1}, $c_{HBrO_2,0} = 1 \times 10^{-7}$ mol L^{-1}, $c_{Ru^{2+},0} = 5 \times 10^{-5}$ mol L^{-1}, and for different bromide concentrations: $c_{Br^-} = 0$, 0.2×10^{-7}, 0.5×10^{-7}, 1×10^{-7}, and 2×10^{-7} mol L^{-1}.

24.7.6 Bistability

When the autocatalytic reaction is performed in a flow reactor instead of in a batch reactor, instabilities can occur; i.e., different steady-states can be established, depending on the flow conditions. This behavior is called *bistability*. A flow reactor (Fig. 24.19a) consists of a well-stirred reaction vessel with a continuous inlet of reactants and a continuous outlet of products. The steady-state concentrations of the reacting species can be established inside the reactor. If we feed a flow reactor with Ru^{2+} and bromate, the steady-state concentrations of Ru^{2+} and Ru^{3+} depend on the flow rate. At low flow rate there is enough time for the autocatalytic reaction to convert all Ru^{2+} into Ru^{3+}. At high flow rate Ru^{3+} disappears fast through the outlet of the reactor, and the steady-state concentration of Ru^{3+} is small.

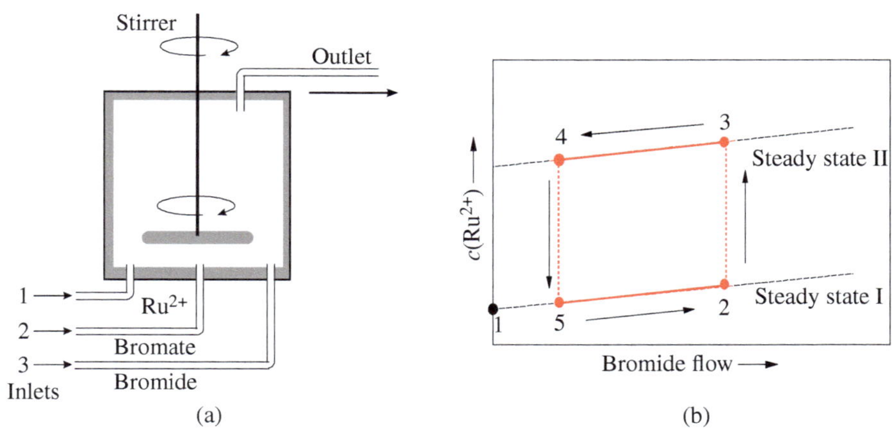

Figure 24.19 Reaction in a flow reactor. (a) Flow reactor with three inlets for reactants and one outlet for products. Inlet 1, Ru^{2+}; inlet 2, bromate; inlet 3, bromide.
(b) Bistability in the Ru^{2+}-bromate-bromide system. Steady-state Ru^{2+} concentration $c_{Ru^{2+}}$ at different inlet concentrations of bromide. The starting point is 1 (no bromide flow); then the bromide concentration in the inlet flow is increased up to point 2 and decreased again from points 3 to 4.

Consider a medium flow rate and imagine adding bromide continuously to inlet 3. The situation is similar to that discussed in Section 24.7.5.2, but now we have a steady state in which Ru^{3+} is removed and Ru^{2+} is supplied. For $c_{Br^-} = 0$, the steady-state concentration of Ru^{2+} corresponds to point 1 in Fig. 24.19 (steady-state I; the concentration of Ru^{2+} is relatively low because the autocatalytic formation of Ru^{3+} with $HBrO_2$ as an intermediate is effective). If we increase c_{Br^-} in the inlet flow, the Ru^{2+} concentration changes only slightly. At point 2 a critical bromide concentration is reached, and it inhibits the autocatalytic reaction by removal of $HBrO_2$ according to reaction (24.160). The Ru^{3+} present in the solution disappears by the outlet flow; at the same time the Ru^{2+} concentration increases by the inlet flow. The system switches from steady-state I to steady-state II (point 3). Note that a high concentration of bromide is necessary for the inhibition at point 2 because the rate of formation of $HBrO_2$ by the autocatalytic process is high (corresponding to the strong decay of Ru^{2+} in the final part of curves 1 and 2 in Fig. 24.17).

If we decrease c_{Br^-} slightly, the system will not switch back to steady-state I. This hysteresis occurs because the rate of formation of $HBrO_2$ by the autocatalytic process is low (corresponding to the very slow decay of Ru^{2+} at the starting point of curves 1, 2 in Fig. 24.17), and the autocatalytic process cannot start. When we decrease c_{Br^-} further, the rate of formation of $HBrO_2$ can compete with the rate of consumption by bromide, and the autocatalytic reaction starts again. The system switches back to steady-state I at point 5.

The system is bistable because two different steady states occur for identical bromide flow rates. In steady state I, Ru^{2+} is rapidly oxidized to Ru^{3+} in the autocatalytic sequence (24.153), (24.154); so that $HBrO_2$ appears immediately in such an amount that its removal by reaction (24.160) is negligible. In steady state II the inhibitor Br^- removes the intermediate $HBrO_2$ by reaction (24.160) before the autocatalytic sequence (24.153), (24.154) can become effective. How can we reach steady state I at a given moderate Br^- flow? We have to increase the flow, starting from a value where autocatalysis is not suppressed (e.g., from point 1 in Fig. 24.19b). Accordingly, if we start from a value where autocatalysis is suppressed (from top right in Fig. 24.19b), steady state II is reached.

24.7.7 Oscillating Reactions

24.7.7.1 Lotka–Volterra Mechanism

Periodic changes in population dynamics are well known, e.g., the famous case of the two kinds of fish: large fish (predators) and small fish (prey). The small fish live in an unlimited supply of food, and the large fish eat the small ones. The number of predator fish, initially small, increases by eating more and more prey fish. As a consequence, the number of prey fish decreases accordingly. Because of a lack of food, the number of predator fish decreases some time later. Subsequently, the population of the prey fish can increase again because the predator population has decreased. This cycle is repeated periodically.

The dynamics of this problem can be described by the *Lotka–Volterra mechanism*, assuming autocatalytic growth of species A and B, if we denote A = prey fish, B = predator fish:

$$A \xrightarrow{k_1} 2A$$

$$A + B \xrightarrow{k_2} 2B$$

$$B \xrightarrow{k_3} C$$

The last process describes the natural death of B. The corresponding differential equations are:

$$\frac{dN(A)}{dt} = k_1 \cdot N(A) - k_2 \cdot N(A) \cdot N(B) \quad (24.162)$$

$$\frac{dN(B)}{dt} = k_2 \cdot N(A) \cdot N(B) - k_3 \cdot N(B) \quad (24.163)$$

where $N(A)$ and $N(B)$ are the population numbers of A and B. A numerical solution of these differential equations is shown in Fig. 24.20. Can such a periodic behavior be realized by means of chemical reactions?

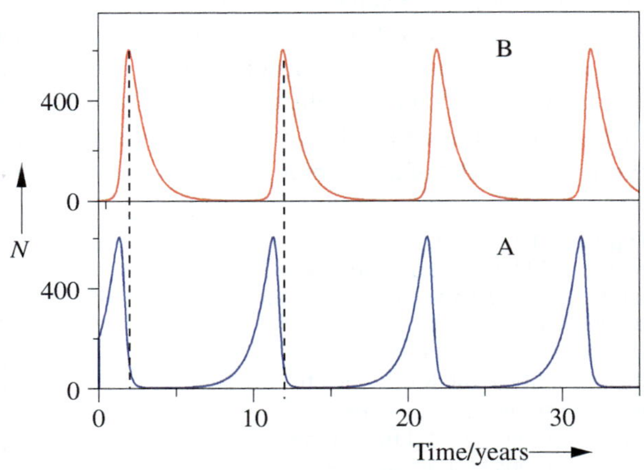

Figure 24.20 Lotka–Volterra model. Populations of prey fishes A (blue solid line) and of predator fishes B (red solid line) versus time. Initial conditions: $N(A)_0 = 200$, $N(B)_0 = 2$. The rate constants are $k_1 = 1$ year^{-1}, $k_2 = 0.01$ year^{-1}, $k_3 = 1$ year^{-1}. The rate equations (24.162) and (24.163) are solved numerically.

24.7.7.2 Belousov–Zhabotinsky Reaction

According to Section 24.7.6 the oxidation of Ru^{2+} to Ru^{3+} in the presence of excess bromate in dilute sulfuric acid solution is an autocatalytic process showing bistability in a flow reactor. By adding an agent which is able (1) to reduce Ru^{3+} to Ru^{2+} and (2) to generate a product inhibiting the autocatalytic reaction, it is possible to reach bistability in a batch reactor. Consequently, the system switches periodically from steady-state I to steady-state II and vice versa.

This oscillation occurs in the *Belousov–Zhabotinsky reaction*, where bromomalonic acid (BrMA; COOH–CHBr–COOH) is initially present with Ru^{2+}, BrO_3^-, and traces of BrO_2. First Ru^{2+} is oxidized to Ru^{3+} according to (24.154) and (24.153):

$$Ru^{2+} + BrO_2 + H^+ \rightarrow Ru^{3+} + HBrO_2$$

$$HBrO_2 + BrO_3^- + H^+ \rightarrow 2\,BrO_2 + H_2O$$

Ru^{3+} reacts with bromomalonic acid to form bromomalonyl radicals (BrMA•; $COOH-\dot{C}Br-COOH$)

$$Ru^{3+} + BrMA \rightarrow Ru^{2+} + BrMA \bullet + H^{+} \quad (24.164)$$

These radicals recombine, forming the inhibitor bromide and an organic product P_1

$$2\,BrMA\bullet \rightarrow Br^{-} + P_1 + H^{+} \quad (24.165)$$

(Fig. 24.21). At the same time, the bromomalonyl radicals remove BrO_2 from the autocatalytic process (second inhibition reaction):

$$BrMA \bullet + BrO_2 \rightarrow P_2 \quad (24.166)$$

P_2 reacts further by

$$P_2 \rightarrow Br^{-} + P_3$$

to form the inhibitor bromide.

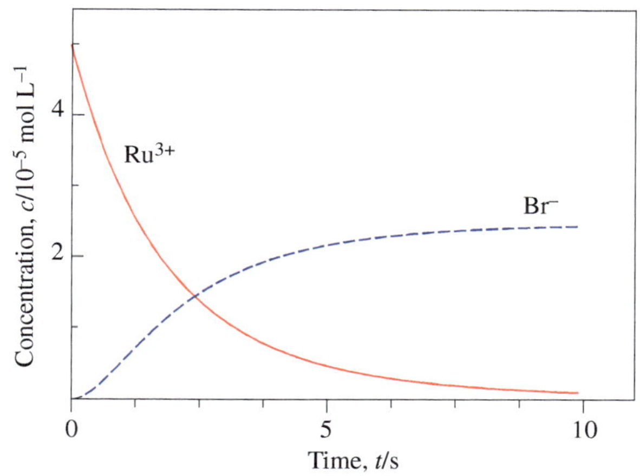

Figure 24.21 Reaction of Ru^{3+} with excess bromomalonic acid according to Equation (24.164). Initial conditions: $c_{BrMA} = 0.01$ mol L^{-1}, $c_{Ru^{3+}} = 5 \times 10^{-5}$ mol L^{-1}. Decay of Ru^{3+} (red solid line) and formation of bromide (blue dashed line). Note that Ru^{3+} denotes the trisbipyridyl complexes of ruthenium.

The formation of BrO_2 occurs via the precursor Br_2O_4, and the disproportionation of bromous acid occurs via a protonated form. The intermediate P_2 disappears by two parallel reactions. The abbreviations are: BrMA: bromomalonic acid; BrMA•: bromomalonyl radical, BrTA: bromotartronic acid, MOA: mesoxalic acid, BrMAE: enol form of bromomalonic acid. P_1 can be identified as bromoethenetricarboxylic acid. The identity of P_3 is not yet completely clear. For a discussion of possible pathways see [9, 10].

The calculated oscillatory behavior of this system is displayed in Fig. 24.22. For reactions and rate constants used in the calculation see Tables 24.4 and 24.5, The rate equations are solved numerically. The reaction is started by an injection of Ru^{2+} into a solution of bromate and bromomalonic acid in 1 molar sulfuric acid. Since no inhibitor is present, the autocatalytic reaction starts immediately; thus, Ru^{2+} is consumed and Ru^{3+} is formed. Because of reactions (24.165) and (24.166), bromide and bromomalonyl radicals BrMA• are formed, thereby inhibiting the autocatalysis (arrows on bottom in Fig. 24.22a). As a consequence, Ru^{3+} is reduced to Ru^{2+}; the bromide formed during this process disappears by its reaction with bromate (Fig. 24.22b). When the bromide concentration is below a critical value the autocatalysis starts again (arrows on top of Fig. 24.22a). This process continues periodically until the bromomalonic acid has been consumed (Fig. 24.23a, experiment, Fig. 24.23b, theory).

Oscillations can also be obtained by starting the reaction with malonic acid instead of bromomalonic acid. In this case Ru^{3+} reacts with malonic acid instead of bromomalonic acid, and no production of bromide is possible. However, HOBr, the final product of the autocatalytic reaction, brominates malonic acid. After a short induction time enough bromomalonic acid has accumulated to start oscillations (Fig. 24.24a). The amplitude of these oscillations increases with time because

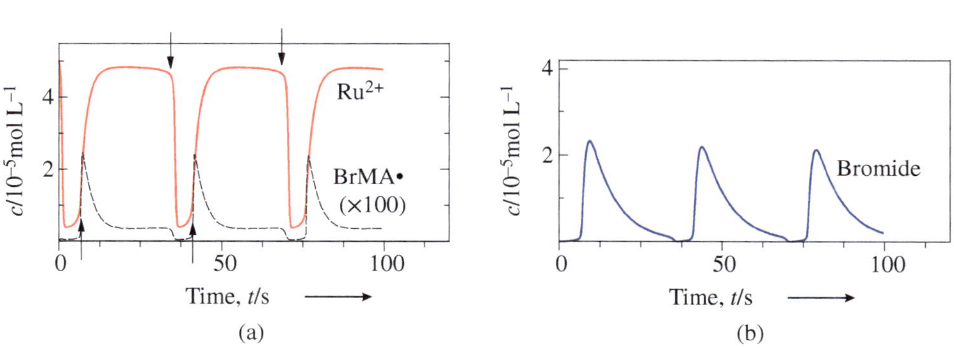

Figure 24.22 Ruthenium-bromate-bromomalonic acid (BrMA) oscillator. Calculated concentrations c of Ru^{2+} (red solid line) and BrMA• (bromomalonyl radical, black dashed line) (a) and of bromide (blue solid line) (b). Initial conditions: $c_{bromate} = 0.1$ mol L^{-1}, $c_{BrMA} = 0.01$ mol L^{-1}, $c_{Ru^{2+}} = 5 \times 10^{-5}$ mol L^{-1}, $c_{H^{+}} = 1.29$ mol L^{-1}, and $c_{HBrO_2} = 1 \times 10^{-6}$ mol L^{-1}. Arrows on bottom of part (a): inhibition of autocatalytic reaction by bromide and bromomalonyl radicals. Arrows on top of part (a): onset of autocatalytic reaction, when all bromide has been consumed. For reactions and rate constants used in the calculation see Tables 24.4 and 24.5. The rate equations are solved numerically. Note that Ru^{2+} denotes the trisbipyridyl complexes of ruthenium.

Table 24.5 Rate Constants of Reactions of Ru^{3+} With Bromomalonic Acid BrMA (Temperature 20 °C).[7]

Reaction			$k_{forward}$	$k_{reverse}$
Ru^{3+} + BrMA	→	Ru^{2+} + BrMA • +H^+	5.5×10^1	1.3×10^5
2 BrMA•	→	Br^- + P_1 + H^+ + CO_2	1×10^8	0
BrMA • +BrO_2	→	P_2	4×10^9	0
P_2	→	Br^- + P_3	6.2×10^{-1}	0
P_2	→	BrTA + $HBrO_2$	4.6×10^{-1}	0
BrTA	→	Br^- + MOA	0.15×10^1	0
BrMA	→	BrMAE	1.2×10^{-2}	800
BrMAE + Br_2	→	Br_2MA + Br^- + H^+	3.5×10^6	0
BrMAE + HOBr	→	Br_2MA + H_2O	6.6×10^4	0

Note that Ru^{2+} and Ru^{3+} denote the trisbipyridyl complexes of ruthenium.

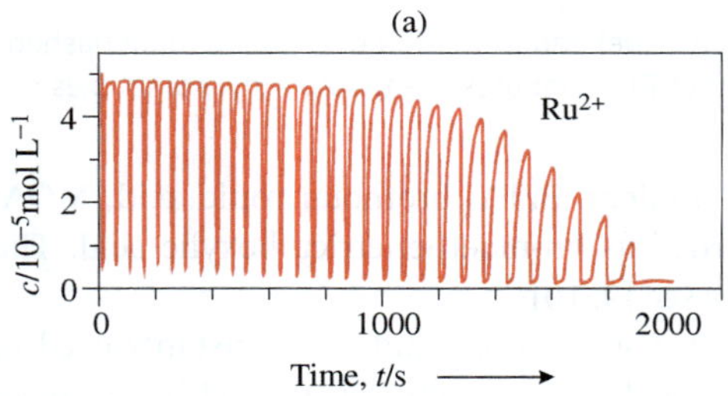

Figure 24.23 Oscillatory system as in Fig. 24.22, but the concentration c of Ru^{2+} is followed for 2000 s. (a) Experiment; Ru^{2+} is measured by its absorbance at 450 nm. (b) Theory. For reactions and rate constants used in the calculation see Table 24.4. The oscillations continue until most of the bromomalonic acid has been consumed. Note that Ru^{2+} denotes the trisbipyridyl complexes of ruthenium.

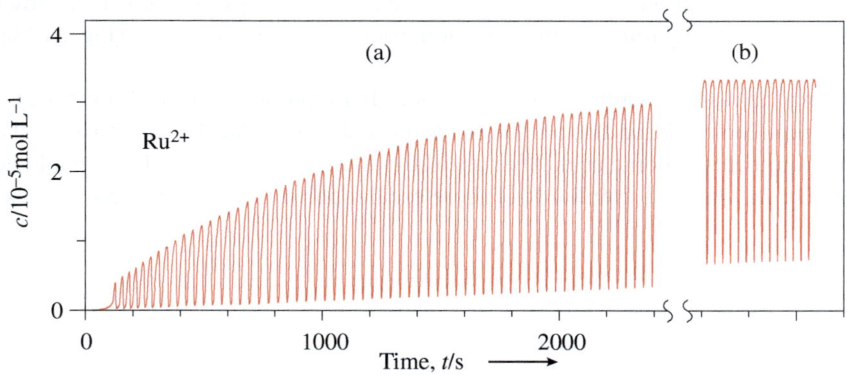

Figure 24.24 Oscillatory system as in Fig. 24.22, but the reaction was started with an addition of malonic acid ($c = 0.1$ mol L^{-1}) into a mixture of bromate (0.1 M) and Ru^{2+} (5×10^{-5} M). (a) $t = 0–2400$ s. (b) $t = 4000–4500$ s. At $t = 0$ all ruthenium salt is in the oxidized form (Ru^{3+}). The oscillations disappear after about 5 hours. Note that Ru^{2+} and Ru^{3+} denote the trisbipyridyl complexes of ruthenium.

of the increasing concentration of bromomalonic acid. After a longer time the BrMA concentration is kept on a nearly constant level and nearly undamped oscillations occur (Fig. 24.24b). Because of the high concentration of malonic acid the oscillations continue for about 5 hours, in contrast to Fig. 24.23, for which the oscillations disappear already after 2000 s.

24.7.8 Chemical Waves

24.7.8.1 Circular Waves

First, let us consider a prairie fire, as an analogy for a chemical wave. We assume a large area covered with dry grass. Thus the grass is in an excitable state: a single spark is enough to start burning at a certain spot. A circular firefront propagates, in

7 Units of the rate constants see Table 24.4.

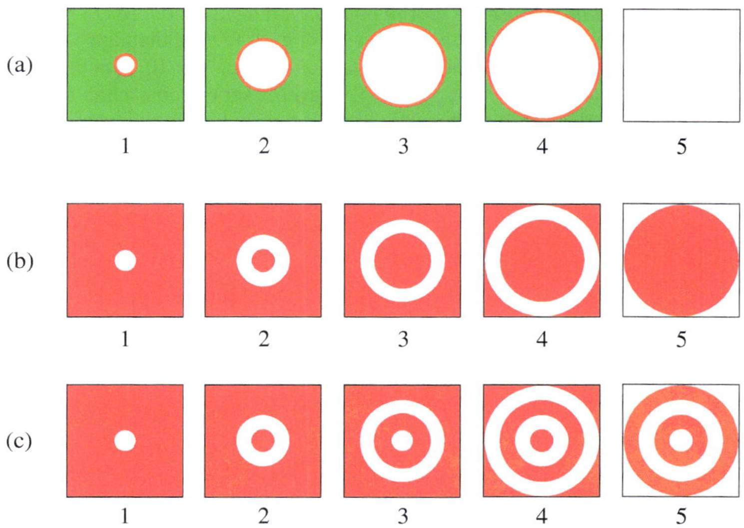

Figure 24.25 Propagation of waves in an excitable medium. (a) Prairie fire. Dry grass (green area) is burned by a spark at a small spot (1). The fire front propagates from inside to outside, leaving burned grass behind (white area, 2–4). Finally, the fire front reaches solid walls and is extinguished; the full area, originally covered with dry grass, is burned out (5). (b,c) Chemical wave. An excitable chemical medium (Ru^{2+}, bromate, and bromomalonic acid, orange area) is excited at a small spot by touching with a silver wire (1). (b) The wavefront (Ru^{3+} at the front) propagates from the inside to the outside, leaving the medium in a non-excitable state (white area, no Ru^{2+} available to start the autocatalytic reaction). The Ru^{3+} behind the wavefront is reduced by bromomalonic acid, and after a short while the medium is again in an excitable state. Finally, the wave reaches the solid walls and is extinguished. (c) Same as in (b), but in positions 3 and 5 new waves are induced by touching again with a silver wire. Thus a series of waves propagates simultaneously (3–5). Finally, the first wave reaches the solid walls and is extinguished. Note that Ru^{2+} and Ru^{3+} denote the trisbipyridyl complexes of ruthenium.

analogy to a water wave. The fire is fed by dry grass at its front leaving burned grass behind it. Hence the medium behind the firefront is no longer in an excitable state, and a second firefront cannot start from the same spot, in contrast to a water wave. Finally, when the firefront runs against a solid wall it is extinguished, in contrast to a water wave showing reflection at the wall. These stages are illustrated in Fig. 24.25a. Two colliding firefronts extinguish each other, in contrast to two water waves which would display interference phenomena. After a recovery time, new grass grows, dries out in the sun, and the dry grass is ready to generate a new firefront. Hence, this phenomenon shows periodicity in time and space.

Now let us consider a chemical wave. As in the example of the prairie fire we need an excitable chemical medium. Such a medium can be found in the Belousov–Zhabotinsky reaction discussed in Section 24.7.7. We prepare a thin layer of the reactive solution (mixture of bromomalonic acid, bromate, and ruthenium complex in sulfuric acid solution) in a Petri dish. Such a system is usually self-oscillatory; however, under appropriate conditions, the inhibitor bromide can be kept above its critical value so that oscillations cannot start spontaneously. Decreasing the bromide concentration at a small spot for a short time (e.g., by touching with a silver wire, which produces Ag^+ ions that remove Br^-) can start the autocatalytic process. Once the reaction has been started a circular chemical wavefront propagates, just as in the case of the firefront.

In the beginning, the thin layer is characterized by a uniformly distributed concentration of Ru^{2+} (orange color). When the chemical wave propagates a circular ring of Ru^{3+} is formed, running from the inside to the outside, and it is easily detectable by eye because of the color change from orange to colorless. The system inside the circular wavefront recovers because Ru^{3+} is reduced by excess bromomalonic acid. Typically, the layer is ready for the start of a second wavefront before the first wavefront has reached the borders of the excitable medium (Fig. 24.25b). As a consequence, the concentration profile of Ru^{3+} in the wave is steep in its front region (due to the fast autocatalytic process) and it is flat behind the front (due to the slow process of reduction of Ru^{3+} by bromomalonic acid).

It is possible to establish a high bromide concentration in the medium by illumination of the layer containing the BZ medium with visible light. Ru^{2+} in the excited state is able to oxidize bromomalonic acid, producing additional bromide and preventing the onset of the autocatalytic process. A small part in the center of the layer is kept in the dark (Fig. 24.26). Thus a wavefront can start spontaneously in the central part of the Petri dish, reach the illuminated part, and start a wavefront there. In Fig. 24.27 a sequence of several wavefronts is shown following each other with a period of 30 s.

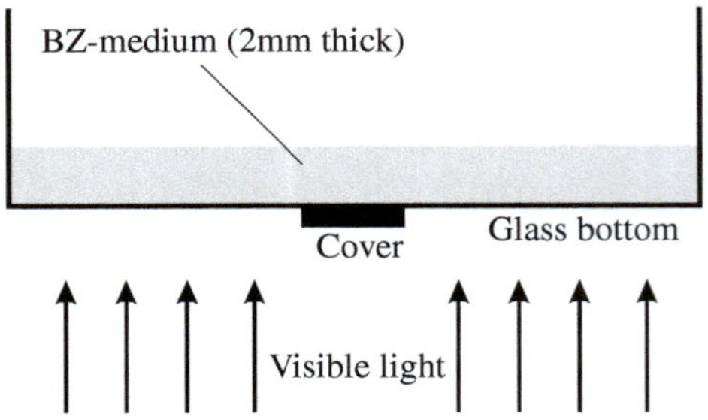

Figure 24.26 Experimental setup for the observation of self-induced chemical waves. On the bottom of a Petri dish a thin layer of excitable BZ medium (bromate 0.8 mol L^{-1}, bromomalonic acid 0.12 mol L^{-1}, Ru^{2+} 6×10^{-4} mol L^{-1}, dissolved in H_2SO_4 (0.8 mol L^{-1})) is placed. The layer is kept in an excitable, but not self-oscillatory state by shining visible light. If the center of the layer is covered such that it cannot be reached by the light, then this part of the medium is self-oscillatory and waves begin to propagate [11].

24.7.8.2 Chemical Pinwheel

We consider an excitable Belousov–Zhabotinsky (BZ) medium confined to a ring-shaped area, and a circular chemical wave propagating in this medium (Fig. 24.28a,b. White zone: excitable medium; gray inner circle: no excitable medium). In the wavefront the catalyst (e.g., the ruthenium complex) is in the oxidized state, due to the fast running autocatalytic process. After reaching the outer and inner boundaries of the excitable medium the upper and the lower parts of the wavefront are extinguished. Thus the circular wave splits into two arms propagating clockwise and counterclockwise around the ring (Fig. 24.28c). We expect that, after a rotation by 180°, they meet at the opposite side of the ring and extinguish each other. This can be avoided by killing off one of the two arms before they meet: touching the wavefront with an iron wire (e.g., an iron paperclip) is sufficient to bring the catalyst back to the reduced state. Then, the remaining arm rotates permanently around the ring, like the arm of a pinwheel (Fig. 24.28d). In the same manner, a second and a third arm can be introduced, which, after equilibration, rotate at equal distances from each other (Fig. 24.29).

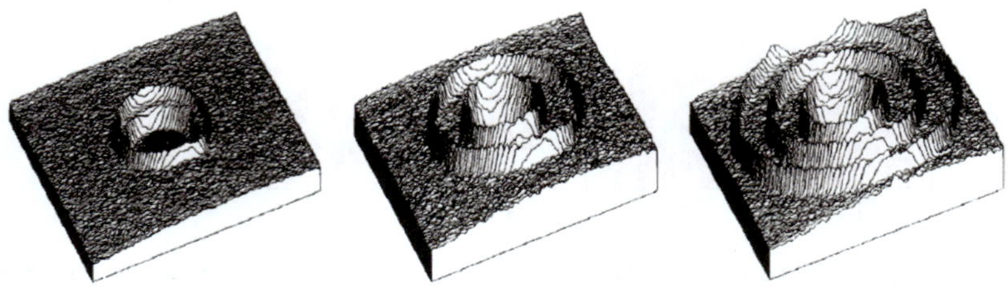

Figure 24.27 Propagation of Ru^{3+} waves in a thin layer of BZ medium (see Fig. 24.26). The Ru^{3+} concentration is measured by recording the transmitted light intensity using a video camera. The transmittance is converted into Ru^{3+} concentration which is shown as a 3D representation. The three pictures are taken every 30 s.

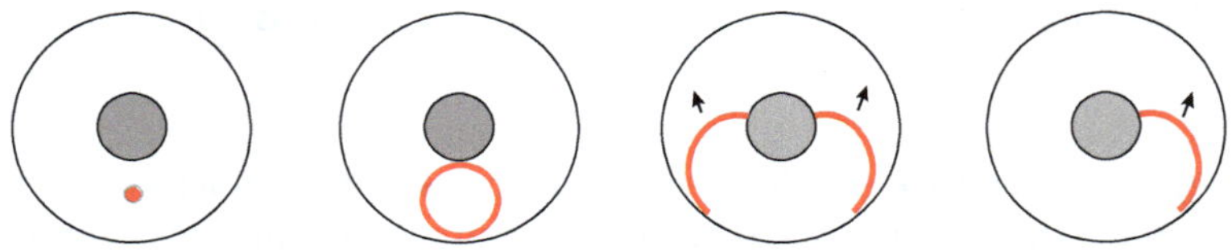

Figure 24.28 Formation of a pinwheel in a ring-shaped excitable medium (white area). The boundaries of the outer circle and of the inner circle (gray zone: no excitable medium) act as solid walls, extinguishing the waves. (a) Starting a circular wave. (b) The circular wave reaches the boundaries, (c) leading to two pinwheels rotating in opposite directions. (d) Remaining one-armed pinwheel rotating counterclockwise.

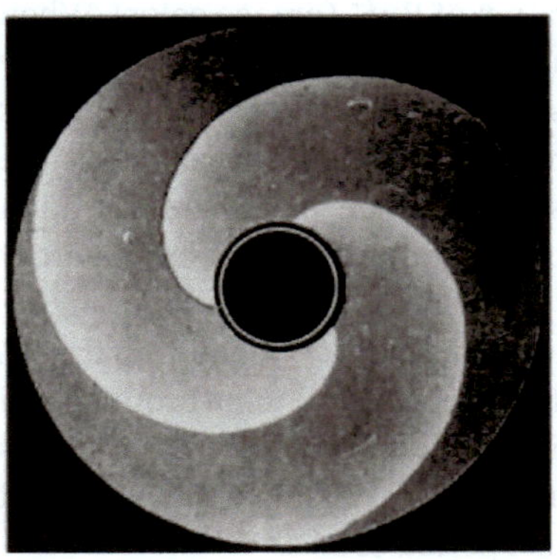

Figure 24.29 Three-armed chemical pinwheel rotating counterclockwise. The time needed for one rotation is on the order of one hour. Composition of the excitable medium: bromate 0.03 mol L^{-1}, bromomalonic acid 0.03 mol L^{-1}, malonic acid 0.08 mol L^{-1}; ferroin (trisphenanthroline complex of Fe^{2+}) is added as a catalyst instead of the ruthenium complex used in the experiment in Figs. 24.22–24.24. The reaction takes place on the surface of a circular polysulfone membrane (pore size 0.45 μm, outer diameter 40 mm) with a hole cut in the center (diameter 10 mm). The membrane is in contact with a solution containing bromate, bromomalonic acid, and malonic acid; the ferroin is adsorbed at the surface of the membrane. Thus the excitable areas on the membrane are red (strong absorption of ferroin), and at the locations of the wavefronts (which are colorless because of the weak absorption of ferroin), the white color of the membrane appears. ([12]. Attila Lázár et al. 1997/Reproduced from AIP Publishing.)

24.8 Experimental Methods

A reaction's rate (or its progress) can be monitored in a variety of different ways. The unifying aspect of such measurements is the determination of the reactant or product concentration as a function of time. One particularly useful method for monitoring rates is by spectroscopy (UV/visible, infrared, and NMR are all commonly used) because it allows for very good temporal resolution and molecular specificity. Other experimental probes of reactions are pH, conductivity, pressure, redox potential, and chromatographic methods (GC/MS and others). An essential feature of the property monitored is that it be sensitive to the formation of product and/or the decay of reactant. Below, we outline a few of the methodologies used to monitor reaction rates.

The most detailed way to determine the rate of a reaction would be to follow the decay and formation of each of the chemical species as a function of time. This is not always possible, or practical, and approximate methods are commonly used for determining rate laws.

24.8.1 Monitor Reaction Progress and Sampling

The real-time measurement of a reaction's progress is very common and quite useful for reactions that proceed on longer time scales. In this case one prepares the reaction mixture and then monitors some variable that is proportional to the reaction's progress. The time resolution of this approach is determined by the inherent time limit of the monitoring method and/or the ability to keep the reaction well mixed during the course of the experiment. A possibility to monitor the reaction progress is by sampling; the reaction mixture at different time points i.e., you take an aliquot, or sample, from the reacting mixture and analyze it. One very popular method to analyze samples is thin layer chromatography of a reaction mixture to follow a reaction's progress, qualitatively. A more sophisticated version of this method uses gas chromatography coupled with mass spectrometry to identify the species present.

Only for very slow reactions can the concentrations of the reactants be analyzed by taking samples from the reaction mixture. For fast reactions the concentration is measured by selecting a physical quantity proportional to the concentration. By way of example consider the catalyzed reaction of nitrogen and hydrogen to make ammonia,

$$N_2 + 3H_2 \rightleftarrows 2NH_3$$

The progress of this reaction can be followed by monitoring the total pressure of the reactant mixture. In such experiments and experimental design, one must be sure that the reactants are well mixed and one must minimize the available dead space in the apparatus. This method has the virtue of being straightforward experimentally; however, it is not sensitive to chemical identity.

24.8.2 Flow Methods

Flow methods initiate reactions by mixing the reactants in a chamber and follow the time evolution through a physical change in the sample that arises from product formation.

In the stopped flow method, reactants A and B are in two cylinders. By means of two pistons they are rapidly transferred through a mixing chamber and a measuring cell into a third cylinder (Fig. 24.30). When the piston in that third cylinder reaches the stop face, the flow is stopped abruptly. Then the concentration in the measuring cell is followed by selecting an appropriate physical property (such as absorbance, electrical conductivity, optical rotation, fluorescence). Figure 24.30 shows the case in which the sample in the monitoring cell is followed spectroscopically; C is a computer or some other device for recording the signal.

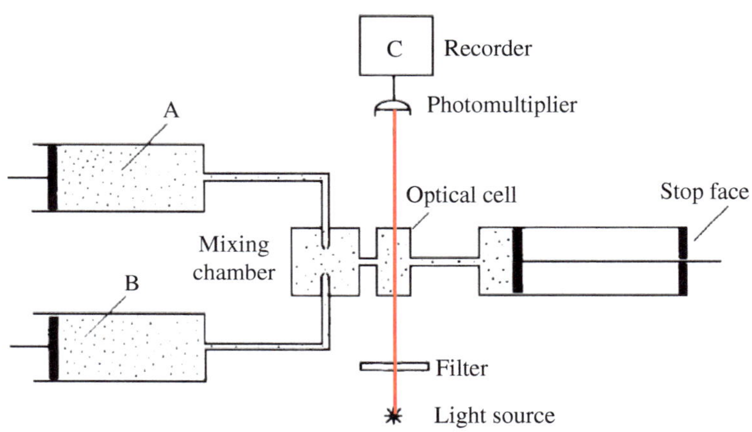

Figure 24.30 Stopped flow apparatus. The diagram shows the vessels for manipulating the reactants in time, and it depicts how the reaction mixture can be followed with optical spectroscopy. See text for a detailed description.

With this method, reactions with a half-life longer than 1 ms can be investigated [13].

Instead of stopping the flow, we can apply a continuous flow (velocity v) and measure the steady-state concentration along a flow tube (Fig. 24.31). Because the reactants at point $x = v \cdot t$ along the tube have concentrations corresponding to the time t after mixing, the measurement of the reaction mixture composition along the flow tube can be used to infer the reaction-time profile.

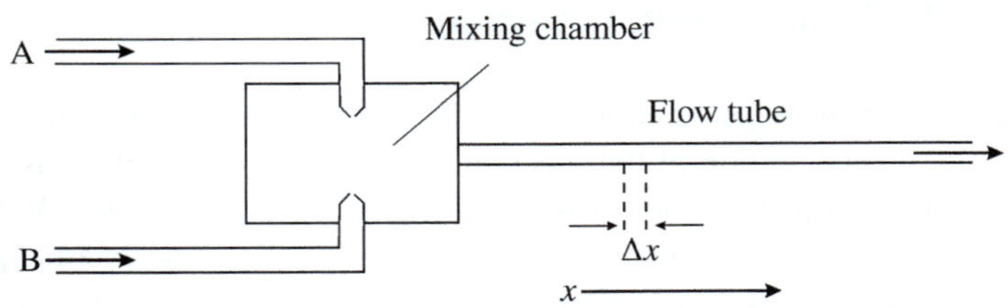

Figure 24.31 Flow method. Under a constant flow speed, the time course of the reaction can be inferred by measuring the composition of the mixture at different spatial positions along the flow tube.

24.8.3 Quenching Methods

Quenching methods can be used over a wide range of time scales but do not directly monitor the progress of a reaction in time. In the quenching method one adds a chemical species to the reaction that deactivates one or more of the reactants and stops the reaction. Alternatively, one might cool the system rapidly to turn off the reaction. Once the reaction is stopped the concentrations of the reactants may be determined.

Another variation on this method is to add a chemical species (call it species X) to the reacting mixture, which reacts with one of the reactants at a known rate. In this case the added chemical species does not totally inhibit the reaction, but it affects the yield of the reaction. By changing the concentration of species X, we can vary the reaction rate and change the yield of the reaction being studied. For example, imagine the reaction

$$\mathrm{A} + \mathrm{B} \rightarrow \mathrm{AB}, \qquad \frac{dc_\mathrm{B}}{dt} = -k_r(\mathrm{B}) \cdot c_\mathrm{A} \cdot c_\mathrm{B} \tag{24.167}$$

and a competitive reaction whose rate we know

$$\mathrm{A} + \mathrm{X} \rightarrow \mathrm{AX}, \qquad \frac{dc_\mathrm{X}}{dt} = -k_r\,(\mathrm{X}) \cdot c_\mathrm{A} \cdot c_\mathrm{X} \tag{24.168}$$

then we can write

$$c_\mathrm{A} = -\frac{1}{k_r(\mathrm{B})}\frac{dc_\mathrm{B}}{dt}\frac{1}{c_\mathrm{B}} \quad \text{or} \quad c_\mathrm{A} = -\frac{1}{k_r(\mathrm{B})}\frac{d\left(c_\mathrm{B}/c_{\mathrm{B},0}\right)}{dt}\frac{1}{\left(c_\mathrm{B}/c_{\mathrm{B},0}\right)} \tag{24.169}$$

An analogous equation can be written for reaction (24.168), hence we find that

$$c_\mathrm{A} = -\frac{1}{k_r(\mathrm{B})}\frac{d\ln\left(c_\mathrm{B}/c_{\mathrm{B},0}\right)}{dt} = -\frac{1}{k_r(\mathrm{X})}\frac{d\ln\left(c_\mathrm{X}/c_{\mathrm{X},0}\right)}{dt} \tag{24.170}$$

where $c_{\mathrm{B},0}$ and $c_{\mathrm{X},0}$ are the initial concentrations of A and X. Upon integration, we see that

$$k_r(\mathrm{B}) = k_r(\mathrm{X}) \cdot \frac{\ln\left(c_\mathrm{B}/c_{\mathrm{B},0}\right)}{\ln\left(c_\mathrm{X}/c_{\mathrm{X},0}\right)} = k_r(\mathrm{X}) \cdot ratio \tag{24.171}$$

We must choose AX so that it does not react with B. By adjusting the rate of the second reaction we can modify the yield of AB. This procedure allows us to use the rate of the second reaction as a "clock" to measure the rate of the first reaction.

Example 24.16 *OH radicals are produced by photolysis of water at a wavelength of 185 nm. The radicals react with a mixture of ethane and propane; i.e.,*

$$\cdot OH + CH_3\text{–}CH_3 \rightarrow CH_3\text{–}CH_2\cdot + H_2O \quad and \quad \cdot OH + CH_3\text{–}CH_2\text{–}CH_3 \rightarrow CH_3\text{–}CH_2\text{–}CH_2\cdot + H_2O$$

The rate constant for the reaction of OH radicals with ethane is $k_r(ethane) = 1.54 \times 10^8\ L\ mol^{-1}\ s^{-1}$ at $T = 298$ K. To determine the reaction rate of OH radicals with propane, you used the stopped-flow method to monitor the depletion of the hydrocarbons in a mixture of ethane and propane and you found that the ratio of the propane and ethane hydroxyl radical reaction rates is 4.394. Use this information to determine the rate constant k_r for the reaction of OH radicals with propane.

Using Equation (24.171), we calculate

$$k_r(\text{propane}) = k_r(\text{ethane}) \cdot \frac{\ln\left(c_{\text{propane}}/c_{\text{propane},0}\right)}{\ln\left(c_{\text{ethane}}/c_{\text{ethane},0}\right)}$$
$$= 1.54 \times 10^8 \text{ L mol}^{-1} \text{ s}^{-1} \cdot 4.394 = 6.76 \times 10^8 \text{ L mol}^{-1} \text{ s}^{-1}$$

24.8.4 Flash Photolysis

In the flash photolysis method, a reaction is initiated by a short burst of light and then the reacting species (reactants and/or products) are monitored in time spectroscopically. In its original realization [14, 15], the method used a light flash from a Xe discharge tube (1–50 μs flash); however, this approach has been extended through laser technology into the femtosecond (10^{-15} s) time range. Such short bursts of light allow one to initiate and monitor chemical reactions on timescales that are shorter than individual vibrational periods, thereby providing the ability to probe the elementary physical motions of the atoms during a chemical reaction (see Chapter 25).

Figure 24.32a diagrams a flash photolysis apparatus. For example, one may photodissociate the molecule I_2 using visible light, such that

$$I_2 + h\nu \rightarrow I + I$$

The recombination of I radicals

$$2\,I \rightarrow I_2$$

can be followed spectroscopically through the recovery of the electronic absorbance of I_2 (Fig. 24.32b for recombination in the gas phase). This method probes the time evolution for processes that are slower than the decay time of the flash pulse.

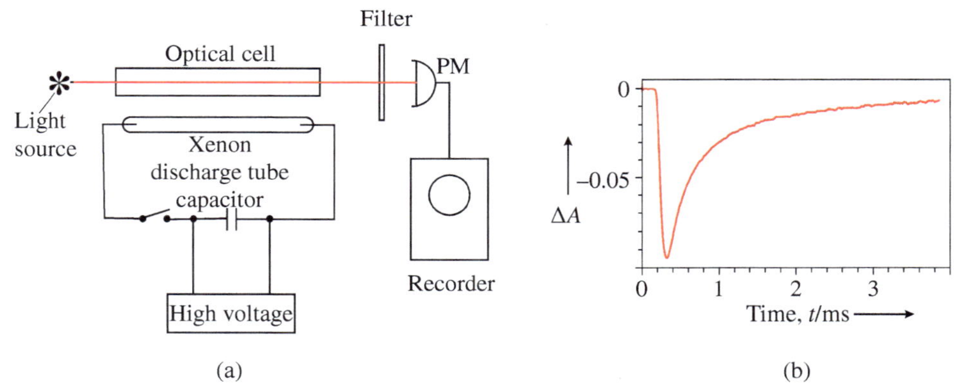

Figure 24.32 Flash photolysis. (a) Flash photolysis apparatus. (b) Change ΔA of the absorbance of I_2 molecules after flash photolysis of gaseous I_2 at 25 °C ($\lambda = 540$ nm, optical path length $d = 15$ cm). Immediately after the flash ΔA is negative because I_2 molecules disappear. Then ΔA increases because of the recombination of I atoms.

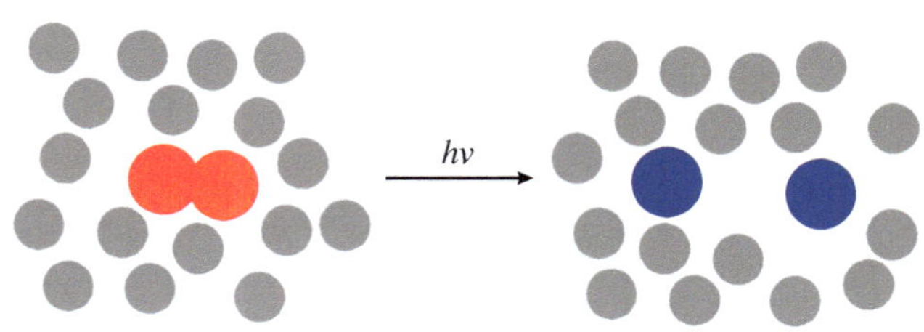

Figure 24.33 Photodissociation of iodine in solution: $I_2 + h\nu \rightarrow I + I$. Red: I_2 molecule, blue: I atoms. Gray: solvent molecules.

The recombination of I atoms can also be followed in solution, and it occurs on timescales of hundreds of nanoseconds and longer. When the light flash photolyzes the I_2 molecules, two things can happen. A fraction of the atoms escape the "solvent cage" in which they are produced (see Fig. 24.33) and diffuse in solution until they encounter one another and recombine. Not all of the I atoms escape from the solvent shell around the I_2, however. Although the initial photodissociation generates two distinct I atoms, some fraction of these "geminate" (twins) atoms are trapped together by the surrounding shell of solvent molecules. These caged atoms can recombine rapidly, on the order of picoseconds. This latter process is called *geminate recombination*. The branching between the two processes depends upon a number of parameters, such as the excess energy in the photodissociation and the properties of the solvent.

24.8.5 Relaxation Method

In the relaxation method, a reaction mixture is perturbed away from equilibrium and the recovery back to the new equilibrium state is followed in time. This can be realized by changing temperature, pressure, or electric field strength (*temperature, pressure, or field jump method*, see Fig. 24.34). We consider the equilibrium

$$A + B \underset{k_{-1}}{\overset{k_1}{\rightleftharpoons}} C \tag{24.172}$$

According to the van't Hoff Equation (18.81)

$$\frac{d \ln K}{dT} = \frac{\Delta_r H^\ominus}{RT^2} \tag{24.173}$$

we can estimate the relative change of the equilibrium constant K for a change of temperature. Because of $d \ln K = dK/K$ we obtain

$$\frac{dK}{K} = \frac{\Delta_r H^\ominus}{RT^2} \cdot dT \tag{24.174}$$

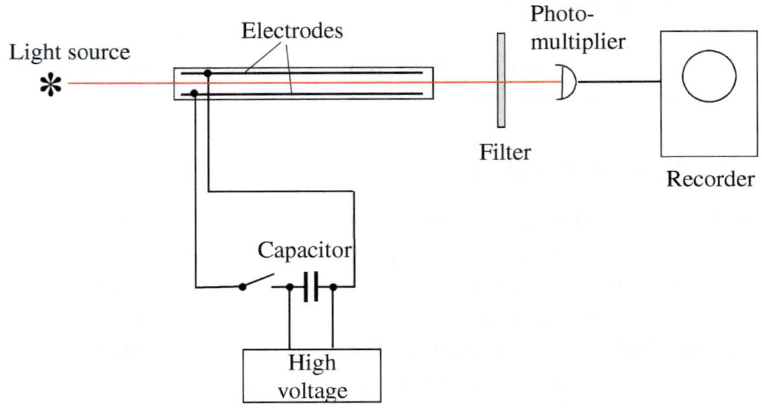

Figure 24.34 Temperature jump apparatus. The capacitor is discharged through the solution with the reactants. The equilibrium is shifted by the resulting temperature jump caused by the discharge.

We can use Equation (24.174) to estimate the relative change in the equilibrium constant K for a small enough change in temperature T. For a small temperature jump of 5 K, $\Delta_r H^\ominus = 10$ kJ mol^{-1}, and $T = 298$ K, we estimate a 7% relative change in K. This is only a small change, but it is enough to follow the relaxation of the system into the new equilibrium by using sensitive detection methods (Fig. 24.35).

Let us denote the concentrations in the new equilibrium with $c_{A,\infty}$, $c_{B,\infty}$, and $c_{C,\infty}$. Then

$$\frac{dc_C}{dt} = k_1 c_{A,\infty} c_{B,\infty} - k_{-1} c_{C,\infty} = 0 \tag{24.175}$$

After the temperature jump the concentrations are

$$c_{A,\infty} + x, \quad c_{B,\infty} + x, \quad \text{and} \quad c_{C,\infty} - x$$

where x is a measure of the deviation of the system from the new equilibrium ($x = 0$ for $t \to \infty$). The rate of concentration change of molecules C is

$$\begin{aligned}\frac{dc_C}{dt} &= k_1 (c_{A,\infty} + x) \cdot (c_{B,\infty} + x) - k_{-1}(c_{C,\infty} - x) \\ &= k_1 c_{A,\infty} c_{B,\infty} - k_{-1} c_{C,\infty} + k_1 x \left(c_{A,\infty} + c_{B,\infty}\right) + k_{-1} x + k_1 x^2\end{aligned} \tag{24.176}$$

By inserting the equilibrium condition (24.175), neglecting the term with x^2 and using the equation

$$\frac{dc_C}{dt} = -\frac{dx}{dt} \tag{24.177}$$

we obtain

$$\frac{dx}{dt} = -k_r \cdot x \quad \text{with} \quad k_r = k_1 \cdot (c_{A,\infty} + c_{B,\infty}) + k_{-1} \tag{24.178}$$

so that

$$x = x_0 \cdot e^{-k_r t} \tag{24.179}$$

x decreases exponentially with time, where $x_0 = c_{A,0} - c_{A,\infty} = c_{C,0} - c_{C,\infty}$ ($c_{A,0}$ and $c_{C,0}$ are the equilibrium concentrations of A and C before the temperature jump) is the deviation from the equilibrium concentration immediately after the temperature jump.

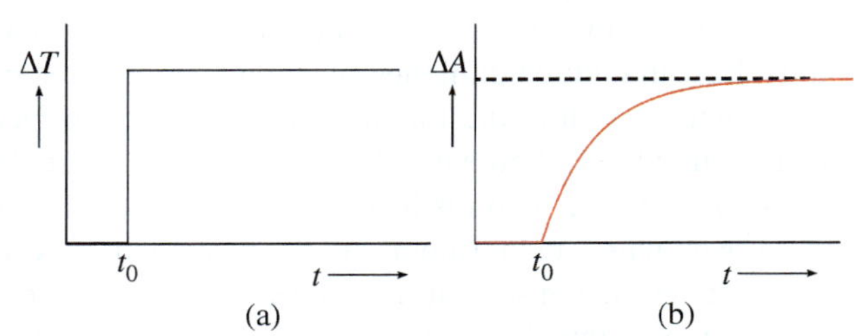

Figure 24.35 Temperature jump method. Time course of temperature change ΔT (a) and of absorbance change ΔA (b), which is proportional to $x_0\left(1 - \exp(k_r t)\right)$, according to Equation (24.180).

We consider the case where the change ΔA of the absorbance is proportional to the change of the concentration of reactant C:

$$\Delta A \sim \Delta c_{\mathrm{C}} = c_{\mathrm{C}} - c_{\mathrm{C},0} = \left(c_{\mathrm{C},\infty} - x\right) - \left(c_{\mathrm{C},\infty} - x_0\right) = x_0\left(1 - \mathrm{e}^{-k_r t}\right) \tag{24.180}$$

By measuring ΔA as a function of time (Fig. 24.35b) we can obtain k_r. To evaluate k_1 and k_2, the experiment is repeated for different values of $c_{\mathrm{A},\infty}$ and $c_{\mathrm{B},\infty}$. Then k_r is plotted versus $(c_{\mathrm{A},\infty} + c_{\mathrm{B},\infty})$. According to Equation (24.178), k_1 is obtained from the slope and k_{-1} from the intercept of the straight line. Another possibility is to make use of the equilibrium constant

$$K_1 = \frac{\hat{c}_{\mathrm{C}}}{\hat{c}_{\mathrm{A}} \cdot \hat{c}_{\mathrm{B}}} = \frac{k_1}{k_{-1}} \cdot c^{\ominus} \quad \text{or} \quad k_{-1} = \frac{k_1 \cdot c^{\ominus}}{K_1} \tag{24.181}$$

Then using Equation (24.178) we obtain

$$k_r = k_1 \cdot (c_{\mathrm{A},\infty} + c_{\mathrm{B},\infty}) + \frac{k_1 \cdot c^{\ominus}}{K_1}$$

from which it follows that

$$k_1 = \frac{k_r}{(c_{\mathrm{A},\infty} + c_{\mathrm{B},\infty}) + c^{\ominus}/K_1}, \quad k_{-1} = \frac{k_1}{K_1} \cdot c^{\ominus} \tag{24.182}$$

Example 24.17 *Consider the reaction of the indicator dye alizarin yellow R (A, yellow) with OH^- to its deprotonated form (C, orange-red).*[8]

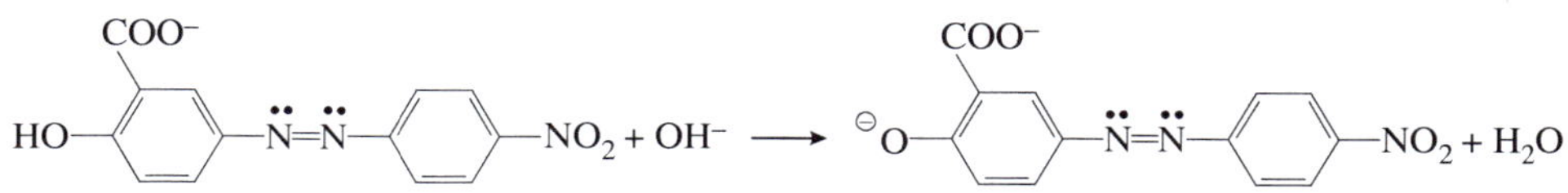

or

$$\mathrm{A} + \mathrm{OH^-} \underset{k_{-1}}{\overset{k_1}{\rightleftharpoons}} \mathrm{C} + \mathrm{H_2O}$$

Use the observed relaxation rate constants k_r to determine the elementary rate constants k_1 and k_{-1}.

The table gives the measured k_r at 298 K with $c_{\mathrm{A}} = 1 \times 10^{-4}$ mol L^{-1} as a function of pH.

pH	10.30	10.49	10.60	10.70
$(c_{\mathrm{A}} + c_{\mathrm{OH^-}})$ /mol L^{-1}	3.26×10^{-4}	4.57×10^{-4}	5.71×10^{-4}	7.02×10^{-4}
k_r/s^{-1}	4.43	4.60	5.46	5.92

The data are plotted in Fig. 24.36. From the slope of the straight line we obtain $k_1 = 4.0 \times 10^7$ L mol^{-1} s^{-1} and from the intercept we obtain $k_{-1} = 3.2 \times 10^4$ s^{-1}.

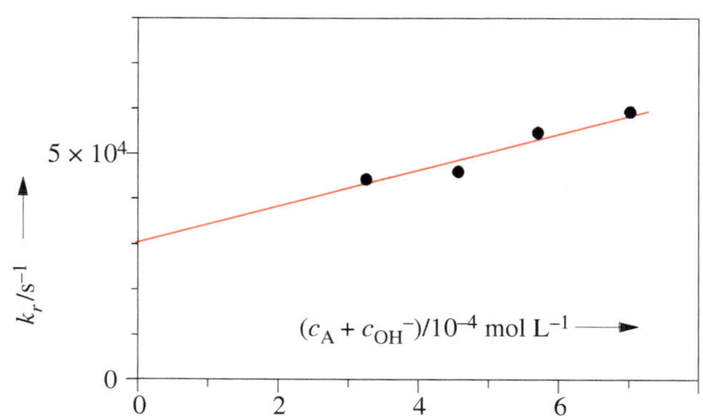

Figure 24.36 Reaction of alizarin yellow (A) with OH^-. The rate constant k_r is plotted versus $(c_{\mathrm{A}} + c_{\mathrm{OH^-}})$ [16]. The red line shows a best fit to the data and is extrapolated to the limit of zero concentration.

---

8 In this case, in contrast to the indicator methylorange (see Fig. 9.17) the alkaline form absorbs at longer wavelengths compared with the acidic form, in accordance with the considerations in Chapter 9.

24.9 Conclusion

Our discussion in this chapter began with elementary chemical reactions in which one or more molecules react to form new molecules. We deduced that the rate for that process depends on the concentrations of the molecular species and a reaction rate constant. The reaction rate constant depends on the collision cross sections of the reacting molecules and on temperature. We showed how to quantify reaction rates, analyze reaction mechanisms, and experimentally probe a reaction's progress. Our discussion in this chapter shows that the coupling of multiple elementary chemical reactions leads to a wide and rich range of phenomena, which can change in both time and space.

24.10 Box 24.1. Activation Factor

For simplicity we consider a light particle A (mass m) colliding with a much heavier particle B. According to Section 15.6.2, the probability $\mathrm{d}\mathcal{P}_s$ of finding a light particle with a velocity component, in a given direction s (Fig. 24B.1), between v_s and $v_s + \mathrm{d}v_s$ is

$$\mathrm{d}\mathcal{P}_s = C \cdot \mathrm{e}^{-mv_s^2/(2kT)} \cdot \mathrm{d}v_s = f(v_s) \cdot \mathrm{d}v_s$$

where C is a proportionality factor. Additionally, we have to consider that fast moving particles collide more often than slowly moving particles: the number of collisions is proportional to the velocity v_s. Then the probability $\mathrm{d}\mathcal{P}_{\mathrm{coll}}$ that a light particle hits a heavy particle is

$$\mathrm{d}\mathcal{P}_{\mathrm{coll}} = const. \cdot v_s \cdot \mathrm{d}\mathcal{P}_s = const. \cdot v_s f(v_s) \cdot \mathrm{d}v_s$$

In a collision, a particle's kinetic energy $\frac{1}{2}mv_s^2$ can be transformed into potential energy (Fig. 24B.1). We are interested in the probability $\mathcal{P}$ of a collision in which the kinetic energy is

$$\frac{1}{2}m \cdot v_s^2 \geq E_a$$

and therefore available for initiating the chemical reaction.

$$\mathcal{P} = \frac{\int_{v_{s,a}}^{\infty} \mathrm{d}\mathcal{P}_{\mathrm{coll}}}{\int_0^{\infty} \mathrm{d}\mathcal{P}_{\mathrm{coll}}} = \frac{\int_{v_{s,a}}^{\infty} v_s \cdot f(v_s) \cdot \mathrm{d}v_s}{\int_0^{\infty} v_s \cdot f(v_s) \cdot \mathrm{d}v_s} = \frac{\int_{v_{s,a}}^{\infty} v_s \cdot \mathrm{e}^{-mv_s^2/(2kT)} \cdot \mathrm{d}v_s}{\int_0^{\infty} v_s \cdot \mathrm{e}^{-mv_s^2/(2kT)} \cdot \mathrm{d}v_s}$$

where

$$v_{s,a} = \sqrt{\frac{2E_a}{m}}$$

The integrals may be solved by the substitution

$$x = \frac{mv_s^2}{2kT}, \qquad \mathrm{d}x = \frac{m}{kT} v_s \cdot \mathrm{d}v_s$$

so that

$$\int v_s \cdot \mathrm{e}^{-mv_s^2/(2kT)} \cdot \mathrm{d}v_s = \frac{kT}{m} \cdot \int \mathrm{e}^{-x} \cdot \mathrm{d}x = -\frac{kT}{m} \cdot \mathrm{e}^{-x}$$

and with $x_a = E_a/(kT)$ for $v_s = v_{s,a}$ we obtain

$$\mathcal{P} = \frac{(\mathrm{e}|_{x_a}^{-x\infty}}{(\mathrm{e}|_0^{-x\infty}} = \mathrm{e}^{-x_a} = \mathrm{e}^{-E_a/(kT)}$$

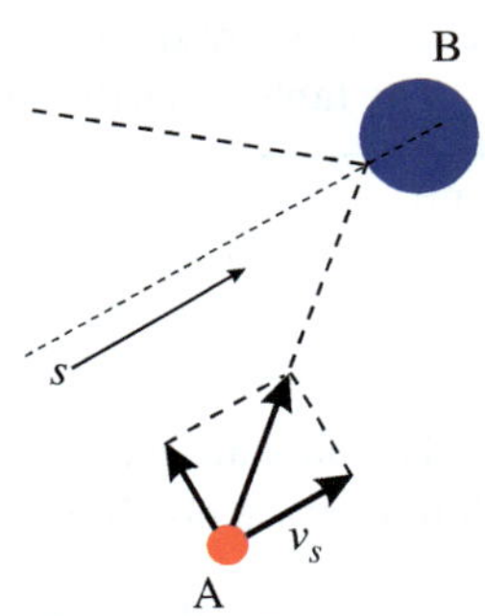

Figure 24B.1 Light particle A colliding with heavy particle B. v_s is the velocity component of particle A in the s-direction (in direction of the center of mass of particle A).

24.11 Box 24.2. Rate Law for a Chain Reaction

To solve the rate equations

$$\frac{\mathrm{d}c_{\mathrm{HBr}}}{dt} = k_2 \cdot c_{\mathrm{Br}} \cdot c_{\mathrm{H_2}} - k_{-2} \cdot c_{\mathrm{HBr}} \cdot c_{\mathrm{H}} + k_3 \cdot c_{\mathrm{H}} \cdot c_{\mathrm{Br_2}}$$

$$\frac{dc_{Br}}{dt} = 2k_1 \cdot c_{Br_2} - k_2 \cdot c_{Br} \cdot c_{H_2} + k_{-2} \cdot c_{HBr} \cdot c_H + k_3 \cdot c_H \cdot c_{Br_2} - 2k_4 \cdot c_{Br}^2$$

and

$$\frac{dc_H}{dt} = k_2 \cdot c_{Br} \cdot c_{H_2} - k_{-2} \cdot c_{HBr} \cdot c_H - k_3 \cdot c_H \cdot c_{Br_2}$$

we apply the steady-state approximation. For the Br and H radicals, we assume that

$$\frac{dc_{Br}}{dt} = \frac{dc_H}{dt} = 0$$

so that

$$2k_1 c_{Br_2} - k_2 c_{Br} c_{H_2} + k_{-2} c_{HBr} c_H + k_3 c_H c_{Br_2} - 2k_4 c_{Br}^2 = 0$$

$$k_2 c_{Br} c_{H_2} - k_{-2} c_{HBr} c_H - k_3 c_H c_{Br_2} = 0$$

By adding these equations, we obtain

$$2k_1 c_{Br_2} - 2k_4 c_{Br}^2 = 0$$

and

$$c_{Br} = \sqrt{\frac{k_1}{k_4}} \cdot \sqrt{c_{Br_2}}$$

We insert this expression in the second equation and solve for c_H

$$c_H = \frac{k_2 c_{H_2}}{k_{-2} c_{HBr} + k_3 c_{Br_2}} \cdot \sqrt{\frac{k_1}{k_4}} \cdot \sqrt{c_{Br_2}}$$

Furthermore, from the second equation we obtain

$$k_2 c_{Br} c_{H_2} - k_{-2} c_{HBr} c_H = k_3 c_H c_{Br_2}$$

By inserting this relation in the first rate equation we obtain

$$\frac{dc_{HBr}}{dt} = 2k_3 c_H c_{Br_2}$$

$$= \frac{2k_2 \sqrt{k_1/k_4}}{c_{Br_2} + (k_{-2}/k_3) c_{HBr}} \cdot \sqrt{c_{Br_2}^3} \cdot c_{H_2}$$

24.12 Exercises

E24.1 Many organic reactions, which are used in laboratory synthesis, have activation barriers around 50 kJ mol^{-1}. Use the Arrhenius equation with a frequency factor of 10^{13} s^{-1} and a temperature of 300 K, to estimate the rate of reaction for a few different activation barriers: 30, 50, 70, and 90 kJ mol^{-1}. Which of these reactions would be most convenient to operate?

E24.2 The molar activation energy for the Diels–Alder reaction (ethylene reacting with 1,3-butadiene) is $E_{a,m} = 115$ kJ mol^{-1}

$$H_2C{=}CH_2 + H_2C{=}CH{-}CH{=}CH_2 \rightarrow C_6H_{10}$$

What is the probability of the reactants exceeding this energy at $T = 760$ K? Calculate the rate constant using the collision theory model. Using the experimental value of $k_r = 0.384$ L mol^{-1} s^{-1}, calculate the steric factor.

E24.3 Perform the analysis of the Lindemann–Hinshelwood mechanism but remove the restriction that $k_1 \approx k_3$ and $k_{-1} \approx k_{-3}$.

E24.4 Reaction Leading to Equilibrium. For the special case of $k_1 = k_{-1}$, integrate the differential equation (24.67)

$$\frac{dc_A}{dt} = -k_1 \cdot c_A^2 + k_{-1} \cdot (c_{A,0} - c_A)^2$$

E24.5 Derive Equation (24.146) for the inhibited enzyme reaction.

E24.6 Consider the reaction

$$2NO + 2H_2 \rightarrow N_2 + 2H_2O$$

The initial rate of this reaction was measured by following the total pressure change of the system [see [17]]. Use the data below (taken at 1099 K) for this reaction to determine the rate law.

$P_{H_2}(t = 0)/\text{atm}$	0.380	0.270	0.193	0.526	0.526	0.526
$P_{NO}(t = 0)/\text{atm}$	0.526	0.526	0.526	0.472	0.395	0.200
Initial rate $/10^{-4}$ atm s^{-1}	2.11	1.45	1.04	1.97	1.36	0.33

E24.7 Consider the reaction

$$O + HCl \rightarrow OH + Cl$$

Measured rate constants for this reaction are $k_r = 7.83 \times 10^4$ L mol^{-1} s^{-1} at 293 K and $k_r = 1.00 \times 10^6$ L mol^{-1} s^{-1} at 403 K [see [18]]. Perform an Arrhenius analysis and find the activation energy and the pre-exponential factor for the reaction. Use your Arrhenius parameters to compute the rate constant at 298 K.

E24.8 Use the simple collision theory to calculate the frequency factor and rate constant for the reaction

$$D + H_2 \rightarrow HD + H$$

at 298 K. The measured [19] molar activation energy is $E_{a,m} = N_A \cdot 5.28 \times 10^{-20}$ J $= 31.8$ kJ mol^{-1}.

E24.9 For the reaction

$$H^+ + OH^- \underset{k_{-1}}{\overset{k_1}{\rightleftharpoons}} H_2O$$

a value $k_r = 2.9 \times 10^4$ s^{-1} is obtained from a temperature jump experiment with $c_{H^+} = c_{OH^-} = 10^{-7}$ mol L^{-1} at $T = 298$ K. Given that the ionization constant of water is

$$K_w = \hat{c}_{H^+} \cdot \hat{c}_{OH^-} = 1.0 \times 10^{-14}$$

determine the forward and back reaction rate constants.

E24.10 Parallel reactions:

$$A + B \rightarrow C \qquad \text{main reaction}$$
$$2A \rightarrow d \qquad \text{side reaction}$$

where B is assumed to be in large excess. Solve the differential rate equation for A

$$\frac{dc_A}{dt} = -k_1' \cdot c_A - 2k_2 \cdot c_A^2$$

that is found below Equation (24.78).

E24.11 Autocatalytic Reaction. Solve the differential equation (24.150)

$$\frac{dc_B}{dt} = k_r \cdot c_B \cdot (c_{A,0} + c_{B,0} - c_B)$$

E24.12 Consider the reaction

$$N_2O_5 \rightarrow 2NO_2 + \frac{1}{2}O_2$$

for the thermal decomposition of N_2O_5 [20] which is first order in the N_2O_5 concentration. The temperature dependence of the rate constant is given by the following data:

$k_r/10^{-5}$ s^{-1}	0.0787	3.46	49.8	487
T/K	273	298	318	338

Determine the Arrhenius parameters for this reaction.

E24.13 Consider the gas-phase thermal reaction

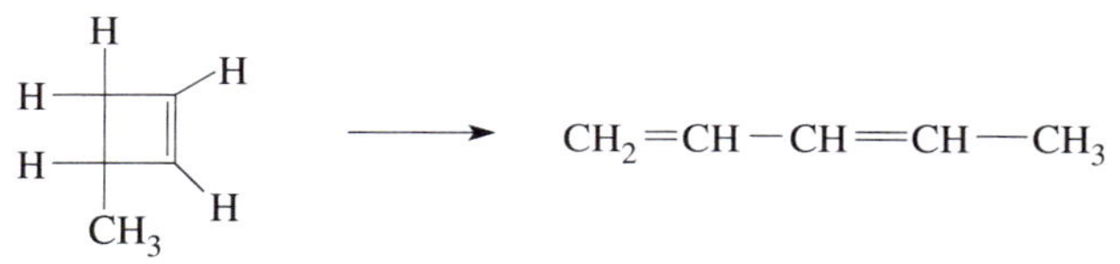

A B

for which the rate constant at different initial concentrations of A and at 396.6 K is

$c_{A,0}/10^{-3}$ mol L^{-1}	4.03	8.06	21.8	41.1	78.6	105	203
$k_r/10^{-4}$ s^{-1}	0.48	0.57	0.79	0.92	1.03	1.10	1.17

a) Plot these data and comment on how the rate constant changes with the initial concentration of A. b) Use the Lindemann–Hinshelwood mechanism to describe this reaction and to explain the dependence of the measured rate constant on the initial concentration of species A.

E24.14 The reaction between hypochlorite ion and iodide ion in alkaline solution,

$$ClO^-\ (aq) + I^-\ (aq) \rightarrow Cl^-\ (aq) + IO^-\ (aq)$$

has been studied by Chia and Connick [21]. They found the following initial rate data for different initial concentrations of the reactants.

	$c_{OH^-,0}$/mol L^{-1}	$c_{ClO^-,0}$/mol L^{-1}	$c_{I^-,0}$/mol L^{-1}	Rate/mol L^{-1} s^{-1}
A	1.0×10^{-5}	4.7×10^{-3}	2.2×10^{-3}	62.0
B	1.0×10^{-5}	2.5×10^{-3}	4.7×10^{-3}	70.5
C	1.0×10^{-5}	1.9×10^{-3}	2.1×10^{-3}	23.9
D	4.5×10^{-6}	1.9×10^{-3}	2.1×10^{-3}	53.2

Use these data to determine the rate law and the rate constants. Describe an experimental method that you would use to measure the initial rate in this reaction.

E24.15 Consider the electron transfer reaction

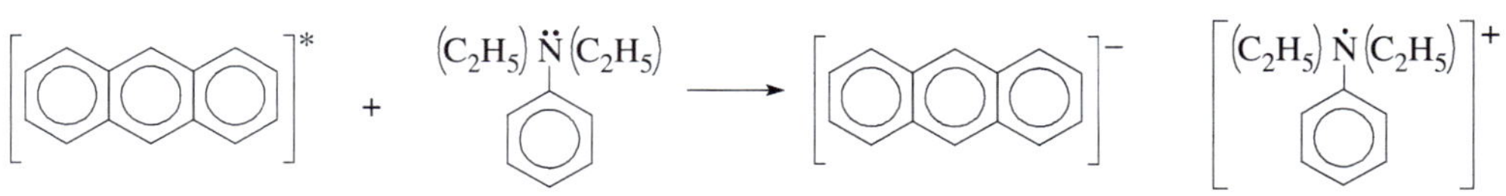

Anthracene molecules (molecules A) dissolved in acetonitrile are excited by light (the asterisk symbolizes the excited state). They react with diethylaniline molecules (molecules B). Estimate the sizes of the two molecules, r_A and r_B. Combine your estimate with the measured value $(D_A + D_B) = 3.8 \times 10^{-5}$ cm^2 s^{-1}, and calculate the rate constant from Equation (24.41). Compare your result to the experimental value $k_r = 2.0 \times 10^7$ m^3 mol^{-1} s^{-1} [22].

E24.16 Consider the equilibrium of the indicator dye alizarin yellow (A) with OH^- to make the deprotonated form (C) which was studied using a temperature jump method.

$$A(aq) + OH^-\ (aq) \rightarrow C(aq).$$

A C

At a pH of 10.30 and $c_{A,0} = 1.26 \times 10^{-4}$ mol L^{-1} the relaxation rate constant is $k_r = 4.43$ s^{-1}, and at a pH of 10.70 and $c_{A,0} = 2.01 \times 10^{-4}$ mol L^{-1} the relaxation rate constant is $k_r = 5.92$ s^{-1}. Use these data to determine k_1 and k_{-1}.

E24.17 In Fig. 24.10a, $k_r' = k_r \cdot c_{OH^-}$ is plotted versus T and Fig. 24.10b plots $\ln(k_r'/k_{r,1}')$ versus $1/T$ for reaction 24.62 (experimental data points see Table 24.2). Compare the experimental frequency factor to that you estimate from Equation (24.34). Using Fig. 24.10b the slope −7640 K of the straight line gives $E_{a,m} = \boldsymbol{R} \cdot 7640$ K $= 63.5$ kJ mol^{-1} and the intercept gives a frequency factor $A_f = 1.8 \times 10^{10}$ L mol^{-1} s^{-1}.

E24.18 Consider the recombination reaction of two separated iodine atoms to make diatomic iodine

$$I + I \rightleftarrows I_2$$

Find an expression for the rate law of iodine recombination, by using the following mechanism

$$I + I \underset{k_{-1}}{\overset{k_1}{\rightleftarrows}} I_2^*$$

$$I_2^* + Ar \xrightarrow{k_2} I_2 + Ar$$

Using your expression for the rate law, describe, verbally, how the formation rate of I_2 will depend on the pressure of Ar.

E24.19 In Equation (24.32), we estimate the mean free path for a molecule in a liquid to be one-third the radius of the molecule. Given that the density of water is $\rho = 1.00$ g cm^{-3} and a water molecule's van der Waals radius is approximately $r_W = 0.15$ nm radius, estimate the distance between two water molecules in solution. Comment on the validity of our estimate.

E24.20 Stern–Volmer Kinetics. An excited electronic state of a molecule can be relaxed (quenched) by collisions with appropriate reaction partners. Often this process leads to de-excitation of the excited state by transfer of the energy to the acceptor molecule. For example, electronically exciting tryptophan molecules in solution leads to a long lived excited state with an average lifetime of 2.5 ns. By increasing the concentration of oxygen in solution one can reduce the molecule's excited state lifetime and quench the fluorescence. Assume that the reaction mechanism is

$$\text{tryptophan}^* + O_2 \rightarrow \text{tryptophan} + O_2^*$$

and show that the fluorescence intensity decreases linearly with the concentration of O_2.

E24.21 Temperature jump experiments on the reaction

$$Br^- + HOBr + H^+ \underset{k_{-1}}{\overset{k_1}{\rightleftharpoons}} Br_2 + H_2O$$

give the overall rate constant values $k_1 = 1.61 \times 10^{10}$ L^2 mol^{-2} s^{-1} and $k_{-1} = 110$ s^{-1} at 20 °C [23]. Assume that bromine is injected into pure water so that the pH and the relative concentrations of Br^-, HOBr are determined by the chemical equilibrium, that is

$$(c_{HOBr})_\infty = (c_{H^+})_\infty = (c_{Br^-})_\infty$$

and take the steady-state concentration of bromine to be $(c_{Br_2})_\infty = 2 \times 10^{-3}$ mol L^{-1}. What is the observed relaxation rate constant in the measurement?

E24.22 Use the simple collision theory to calculate the frequency factor and rate constant for the reaction

$$D + H_2 \rightarrow HD + H$$

at 298 K. The measured [18] molar activation energy is $E_{a,m} = \boldsymbol{N}_A \cdot 5.28 \times 10^{-20}$ J $= 31.8$ kJ mol^{-1}. You can take the radii of D and H_2 to be $r_D = 128$ pm and $r_{H_2} = 138$ pm. Compare your result with the experimental value of 1.68×10^5 L mol^{-1} s^{-1} and calculate a steric factor for the reaction.

24.13 Problems

P24.1 Jeong and Grady [24] investigated the enzyme kinetics of dihydrofolate reductase (DHFR) with the substrate 6,8-dimethylpterin (S). Both the enzyme DHFR and the substrate S have a carboxy group that is involved in the reaction. They used the Michaelis–Menten mechanism to describe the reaction

$$DHFR + S \underset{k_{-a}}{\overset{k_a}{\rightleftharpoons}} (DHFR\text{-}S) \xrightarrow{k_b} DHFR + P$$

where

$$\frac{dc_P}{dt} = \frac{k_b \cdot c_{DHFR,0} \cdot c_S}{K_M + c_S}$$

By controlling the reaction conditions and measuring the initial rate of the reaction, they were able to determine the rate constant k_b. Using the rate law, show mathematically and explain verbally how this is possible. In their conclusions they state: "The active complex in the reduction of the mechanism-based substrate 6,8-diMe-Pterin by DHFRs with NADPH is characterized by a deprotonated enzyme carboxy group which binds protonated 6,8-diMe-Pterin. The rate-limiting step for the reaction is hydride-ion transfer, with no perturbation on the catalytic rate by product release."
Based on this statement, describe how you think the rate constant might change with pH.

P24.2 Consider the unimolecular decomposition of di-*t* -butylperoxide in the gas phase,

$$(CH_3)_3COOC(CH_3)_3 \rightarrow 2(CH_3)_2CO + C_2H_6$$

The following table gives the data for the total pressure of the system versus time:

t/min	0	6	12	18
P_{total}/Torr	173.5	211.3	244.4	273.9

Find a mathematical expression for the time dependence of the total pressure in terms of the initial pressure, the time, and the rate constant. Make a plot which shows that the data follow your mathematical expression. In particular, rearrange the expression and perform a plot, which should be linear. Obtain a value for the rate constant by analyzing your plot. Use the value of your rate constant to determine the half-life for the reaction.

P24.3 Campbell and Plane [25] studied the gas-phase recombination reaction kinetics between Pd and O_2. The Pd atoms were generated by photodissociation of a palladium(II) trifluoroacetate compound in the gas phase. The concentrations of O_2 and gaseous Pd were followed spectroscopically. The bimolecular recombination reaction may be written as

$$Pd + O_2 \rightarrow PdO_2$$

where the observed second-order rate constant k_{obs} is defined through the relation

$$\frac{dP_{O_2}}{dt} = -k_{obs}\, P_{O_2} \times P_{Pd}$$

The table summarizes their findings for the observed rate constant as a function of the pressure and temperature of an argon buffer gas. Explain why the observed rate constant k_{obs} depends on the presence of a buffer gas. Propose a set of elementary steps for the reaction mechanism and explain how your mechanism accounts for the observed pressure dependence at 294 K.

Second-Order Rate Constants for Pd Reacting with O_2.

T/K	P/Torr	$k_{obs}\left(10^8 \frac{L}{mol\ s}\right)$	T/K	P/Torr	$k_{obs}\left(10^8 \frac{L}{mol\ s}\right)$
294	5.0	0.8	294	20.0	3.7
294	10.0	1.5	348	20.0	2.5
294	15.0	3.3	373	20.0	2.2
294	40.0	8.4	398	20.0	1.8
294	60.0	13	423	20.0	1.3
294	80.0	16	448	20.0	1.1
294	100.0	20	473	20.0	0.72
294	150.0	30	498	20.0	0.54
294	200.0	36	523	20.0	0.45
294	250.0	45			
294	300.0	49			

The temperature dependence for this reaction is unusual. Using the temperature dependence of the k_{obs} data at 20 Torr, make an Arrhenius plot for the rate constant and comment on the result. By analyzing their data with a detailed molecular model, they found that the rate constant at low Ar pressure $k_{rec,0}$ was given by $2.71 \times 10^{18}/(T/K)^{2.76}$ $L^2\ mol^{-2}\ s^{-1}$ and that the rate constant at high Ar pressures was given by

$$k_{rec,\infty} = 1.52 \times 10^{11} \exp\left[-69.3/(T/K)\right]\ L\ mol^{-1}\ s^{-1}.$$

Discuss these qualitatively different temperature dependences in terms of your proposed mechanism.

P24.4 Consider the self-exchange reaction

$$^{55}Fe^{2+} + {}^{56}Fe^{3+} \rightarrow {}^{55}Fe^{3+} + {}^{56}Fe^{2+}$$

which occurs spontaneously and can be followed by chemically identifying a particular oxidation state and monitoring the radioactive decay of the ^{55}Fe nuclide (see [26]). For example, one would begin an experimental run with $^{55}Fe^{2+}$ and no $^{55}Fe^{3+}$ present. The formation of $^{55}Fe^{3+}$ would reflect the reaction rate. From the following rate data determine the rate law for the reaction, and find a value for the rate constant, k_r.

Run	A	B	C	D	E	F	G	H
$c_{^{55}Fe^{3+},0}/10^{-4}$ mol L^{-1}	1.03	1.05	1.06	1.08	2.10	3.11	4.18	4.24
$c_{^{55}Fe^{2+},0}/10^{-4}$ mol L^{-1}	0.998	1.97	2.94	3.92	3.94	3.95	3.92	3.85
Rate/10^{-8} mol L^{-1} s^{-1}	1.38	2.85	4.08	5.63	11.17	16.46	23.10	21.87

Given the results of your determination for the rate law, state an alternative experimental test you could perform to determine the rate law. Be specific about how you would proceed with an experimental procedure.

P24.5 The rate for the thermal decomposition of methyl ethyl ether has been determined using the method of initial rates by Steacie [27]. Figure P24.5 plots some of their data at three temperatures as a function of pressure. Postulate a reaction mechanism for the decomposition of methyl ethyl ether. Why does the rate increase as the pressure of the gas increases?

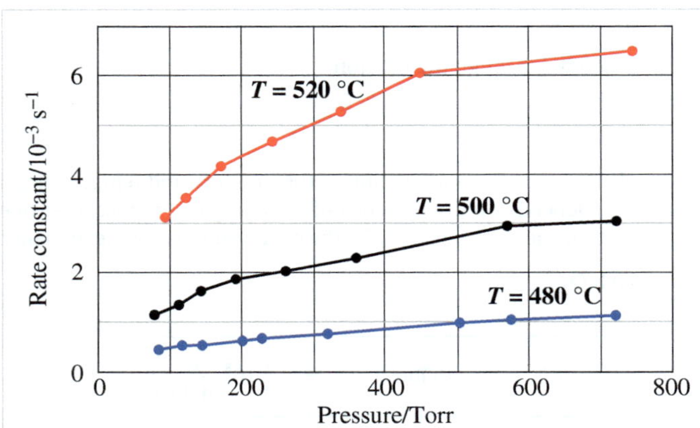

Figure P24.5 Dependence of the first-order rate constant for the thermal decomposition of methylethylether on the pressure.

P24.6 Consecutive reaction:

$$A + B \xrightarrow{k_1} C \quad C + B \xrightarrow{k_2} d$$

where B is assumed to be in large excess. Solve the corresponding differential equation (24.84)

$$\frac{dc_C}{dt} + k_2' c_C = k_1' c_{A,0} \cdot e^{-k_1' t}$$

P24.7 Given the expression for the concentration of C in the reaction sequence, $A + B \xrightarrow{k_1} C, C + B \xrightarrow{k_2} D$

$$c_C = -\frac{k_1' c_{A,0}}{k_1' - k_2'} \cdot \left(e^{-k_1' t} - e^{-k_2' t}\right),$$

which you show in P24.6), determine the time t_{max} required to reach the maximum concentration of c_C. What is the result for c_C and for t_{max} in the special case $k'_1 = k'_2 = k'$?

P24.8 General Form of Second-Order Reaction. Consider the general case of the bimolecular reaction

$$\mathrm{A} + \mathrm{B} \rightarrow \mathrm{C} \qquad \text{with the rate} \qquad \frac{\mathrm{d}c_A}{\mathrm{d}t} = -k_r \cdot c_A \cdot c_B$$

in which $c_{A,0}$ and $c_{B,0}$ can be different. Step 1: Using the definitions $c_A = c_{A,0} - x$, and $c_B = c_{B,0} - x$ where x is the amount of reacted species, show that the integrated rate law can be written as

$$\int \frac{\mathrm{d}x}{(c_{A,0} - x) \cdot (c_{B,0} - x)} = k_r \int \mathrm{d}t$$

Step 2: Show that the integrand can be written as

$$\frac{1}{(c_{A,0} - x) \cdot (c_{B,0} - x)} = \frac{A}{(c_{A,0} - x)} + \frac{B}{(c_{B,0} - x)} \quad \text{with} \quad A = -B = 1/\left(c_{B,0} - c_{A,0}\right)$$

Step 3: Using the result in Step 2, perform the integral. Apply the boundary condition of $x = 0$ at $t = 0$ and show that

$$\frac{c_{B,0} - x}{c_{A,0} - x} = \frac{c_{B,0}}{c_{A,0}} \cdot \mathrm{e}^{(c_{B,0} - c_{A,0})k_r t}$$

Step 4: Use the definition $x = c_{A,0} - c_A$ to write the rate law in terms of c_A and show that the half-life is given by the expression

$$\tau_{1/2} = \frac{\ln\left(2 - \dfrac{c_{A,0}}{c_{B,0}}\right)}{\left(c_{B,0} - c_{A,0}\right) k_r} \qquad \text{for } c_{A,0} \leq c_{B,0}$$

P24.9 In this problem you show that the general expression for the second-order rate law where $c_{B,0} \neq c_{A,0}$,

$$c_A = \frac{c_{B,0} - c_{A,0}}{\dfrac{c_{B,0}}{c_{A,0}} \cdot \mathrm{e}^{(c_{B,0} - c_{A,0})k_r t} - 1}$$

reduces to the expression

$$c_A = \frac{1}{\dfrac{1}{c_{A,0}} + k_r t} \qquad \text{for } c_{B,0} = c_{A,0}.$$

and to the quasi-first-order expression

$$c_A = c_{A,0} \cdot \mathrm{e}^{-c_{B,0} k_r t} \qquad \text{for } c_{B,0} \gg c_{A,0}$$

Make plots of c_A versus time t plots of the three expressions using $k_r = 2\ \mathrm{L\ mol^{-1}\ s^{-1}}$ and $c_{A,0} = 0.1\ \mathrm{mol\ L^{-1}}$ for Curve 1: $c_{B,0} = 0.15\ \mathrm{mol\ L^{-1}}$. Curve 2: $c_{B,0} = 0.10\ \mathrm{mol\ L^{-1}}$. Curve 3: $c_{B,0} = 1.0\ \mathrm{mol\ L^{-1}}$.

(Hint: To show the simplification of the general expression to the special case of $c_{B,0} = c_{A,0}$, you must be careful to avoid divergences. Use l'Hopital's rule, and proceed by defining $c_{B,0} = c_{A,0} + z$ and taking the limit as z approaches zero.)

P24.10 In Equation (24.150) in Section 24.7.5, we found

$$\frac{\mathrm{d}c_B}{\mathrm{d}t} = k_r \cdot c_B \cdot (c_{A,0} + c_{B,0} - c_B) \tag{24.183}$$

for the differential equation for an autocatalytic reaction. Solve this equation and show that

$$c_B = \frac{c_{A,0} + c_{B,0}}{1 + \dfrac{c_{A,0}}{c_{B,0}}\ \mathrm{e}^{-(c_{A,0} + c_{B,0})k_r t}}$$

P24.11 Critical Bromide Concentration. Estimate c_{crit} in the inhibition of the autocatalytic oxidation of Ru^{2+} by considering the rate equations (24.156)–(24.158) for the autocatalytic process and include the inhibition process (24.160). Assume that the pH is low, so that $k'_3 = k_3 \cdot c_{H^+}$ and use the steady-state approximation to find an expression for c_{BrO_2}. Combine this result with the results in Example 24.15 to find an expression for $c_{Ru^{2+}}$. If we define the critical bromide concentration to be the transition point between an exponentially decreasing $c_{Ru^{2+}}$ and an exponentially increasing $c_{Ru^{2+}}$, show that $c_{crit} = k'_2/k'_3$.

P24.12 The $H_2 + I_2 \rightarrow 2\ HI$ reaction proceeds by the formation of H_2I adducts, rather than through a chain mechanism [4]. Using the reaction mechanism

$$I_2 \underset{k_{-1}}{\overset{k_1}{\rightleftharpoons}} 2\ I \qquad R1 \tag{24P12.1}$$

$$H_2 + I \underset{k_{-2}}{\overset{k_2}{\rightleftharpoons}} H_2I \qquad R2 \tag{24P12.2}$$

$$H_2I + I \xrightarrow{k_3} 2\ HI \qquad R3 \tag{24P12.3}$$

show that the rate law is bimolecular and can be written as

$$\frac{dc_{HI}}{dt} = k_{eff} \cdot c_{I_2} \cdot c_{H_2}$$

Find an expression for k_{eff} in terms of the rate constants given in the fundamental steps of the reaction mechanism.

References

1 B. Liu, Ab initio potential surface for linear H_3, *J. Chem. Phys.*, 1973, **58**, 1925, Table 10.

2 D. Rowley and H. Steiner, Kinetics of diene reactions at high temperatures, *Discuss. Faraday Soc.*, 1951, **10**, 198.

3 J.W. Moore and R.G. Pearson, *Kinetics and Mechanism* (John Wiley & Sons, New York, 1981, p. 238).

4 a) J.H. Sullivan, Mechanism of the "bimolecular" hydrogen—iodine reaction, *J. Chem. Phys.*, 1967, **46**, 73; b) J.B. Anderson, The Hydrogen-Iodine Reactions: 100 Years Later. In: Wolfrum, J., Volpp, H.R., Rannacher, R., and Warnatz, J. (eds) *Gas Phase Chemical Reaction Systems*. Springer Series in Chemical Physics, 1996, **61**, 167. (Springer, Berlin).

5 A. Sirimungkala, H.D. Försterling, V. Dlask, and R.J. Field, Bromination reactions important in the mechanism of the Belousov-Zhabotinsky system, *J. Phys. Chem.*, 1999, **A103**, 1038.

6 H.M. Frey and D.C. Marshall, Thermal unimolecular isomerization of cyclobutenes. Part 5.—The isomerization of 3-methylcyclobutene at low pressures, *Trans. Faraday Soc.*, 1965, **61**, 1715.

7 G. Michel, *Biochemical Pathways* (John Wiley & Sons, 1999, pp. 9–12).

8 Y. Gao and H.D. Försterling, Oscillations in the bromomalonic acid/bromate system catalyzed by $[Ru(bipy)_3]_2^+$, *J. Phys. Chem.*, 1995, **99**, 8638.

9 L. Hegedüs, H.D. Försterling, E. KoKai, K. Pelle, G. Taba, M. Wittmann, and Z. Noszticius, Chemical mechanism of the radical feedback loop in the classical BZ reaction. Malonyl bromite and oxalic acid as flow-through intermediates, *Phys. Chem. Chem. Phys.*, 2000, **2**, 4023;

10 L. Hegedüs, M. Wittmann, Z. Noszticius, S. Yan, A. Sirimungkala, H.D. Försterling, and R.J. Field, HPLC analysis of complete BZ systems. Evolution of the chemical composition in cerium and ferroin catalysed batch oscillators: experiments and model calculations, *Faraday Discuss.*, 2001, **120**, 21.

11 T. Akagi and T. Matsumura-Inoue, Dynamic photochemical response of ruthenium(II)/(III) redox chemical waves in Belousov-Zhabotinsky reaction, *Chem. Lett.*, 1996, **25**, 557.

12 A. Lazar, H.D. Försterling, H. Farkas, P. Simon, A. Volford, and Z. Noszticzius, Waves of excitation on nonuniform membrane rings, caustics, and reverse involutes, *Chaos*, 1997, **7**, 731.

13 H. Strehlow, W. Knoche, *Fundamentals of Chemical Relaxation* (Wiley-VCH, Weinheim, New York, 1977).

14 G. Porter, Flash photolysis and some of its applications, *Science*, 1968, **160**, 1299–1307.

15 R.G.W. Norrish and G. Porter, Chemical reactions produced by very high light intensities, *Nature*, 1949, **164**, 658–659.

16 M.C. Rose and J. Struehr, Kinetics of proton transfer reactions in aqueous solution. III. Rates of internally hydrogen-bonded systems, *J. Am. Chem. Soc.*, 1968, **90**, 7205.

17 C.N. Hinshelwood and T.E. Green, CII.—The interaction of nitric oxide and hydrogen and the molecular statistics of termolecular gaseous reactions, *J. Chem. Soc.*, 1926, **129**, 730.

18 R.H. Brown and I.W.M. Smith, Absolute Rate Constants for the Reactions $O(^3P)$ Atoms with HCI and HBr, *Int. J. Chem. Kinet. VII*, 1975, **7**, 301.

19 J.P. Toennies, J. Arnold, and J. Wolfrum, H + H_2: the current status, *Ber. Bunsenges Phys. Chem.*, 1990, **94**, 1231.

20 F. Daniels and E.H. Johnston, The thermal decomposition of gaseous nitrogen pentoxide. A monomolecular reaction, *J. Am. Chem. Soc.*, 1921, **43**, 53.

21 Y.T. Chia and R.E. Connick, The rate of oxidation of I^- to hypoiodite ion by hypochlorite ion, *J. Phys. Chem.*, 1959, **63**, 1518.

22 D. Rehm and A. Weller, Kinetik und Mechanismus der Elektronübertragung bei der Fluoreszenzlöschung in Acetonitril, *Ber. Bunsenges. Phys. Chem.*, 1969, **73**, 834.
23 M. Eigen and K. Kustin, The kinetics of halogen hydrolysis, *J. Am. Chem. Soc.*, 1962, **84**, 1355.
24 S.S. Jeong and J.E. Grady, pH dependence of enzyme reaction rates and deuterium isotope effects on the reduction of a new mechanism-based substrate by dihydrofolate reductase (DHFR), *Biochemistry*, 1995, **34**, 3734.
25 M.L. Campbell and J.M.C. Plane, Kinetic study of the recombination reaction of gas phase Pd(aS_0) with O_2 from 294 to 523 K, *J. Phys. Chem. A*, 2003, **107**, 3747.
26 J. Silverman and R.W. Dodson, The exchange reaction between the two oxidation states of iron in acid solution, *J. Am. Chem. Soc.*, 1952, **56**, 846.
27 E.W.R. Steacie, The kinetics of the thermal decomposition of methyl ethyl ether, *J. Chem. Phys.*, 1933, **1**, 618.

25

Transition States and Chemical Reactions

CHAPTER MENU

25.1 Introduction

In Chapter 24 we discussed the rate of a chemical reaction, using the simple collision theory. However, many reactions are extremely slow and the resulting steric factor is 1×10^{-5} or even less. In such cases the collision theory becomes unreasonable, and it is advisable to consider structural details of the transition state. This chapter focuses on molecular aspects of elementary reactions, in particular how the reaction rate constant depends on molecular properties.

We begin our discussion with the fundamentals of the transition state theory, which is a statistical model for reactions; it provides a quantitatively accurate description for many reactions. For many reactions this "statistical" view of the reaction is inadequate, however. In these cases details of how the molecules encounter each other are required for describing the reaction. Such "dynamical" issues were at the frontier of physical chemistry research in the latter part of the twentieth century.

25.2 Transition State in a Statistical View

Most measurements of reaction rates are performed on large collections of molecules, and statistical averaging over the many different molecular possibilities provides a convenient approach for describing the average reaction rate constant. In this model we assume knowledge of the potential energy surface along the reaction coordinate; i.e., we assume that we can compute the potential energy of the molecule(s) as a function of the displacements of the atoms, which lead from reactant to product[1].

Figure 25.1 shows a contour plot for part of the potential energy surface for an atom exchange reaction

$$A + BC \rightarrow AB + C$$

A and BC are the reactants, AB and C are the products of the reaction.To simplify the discussion we constrain the atoms to be collinear and the diagram shows the potential energy as a function of the distances between the atoms A and B (abscissa)

1 Commonly, a reaction coordinate is a one-dimensional coordinate that indicates progress of a reaction along a reaction pathway. The path leads from the configuration of the reactants through that of the transition state to the configuration of the products. For example, see Fig. 24.2.

Principles of Physical Chemistry, Third Edition. Hans Kuhn, David H. Waldeck, and Horst-Dieter Försterling.

Companion website: www.wiley.com/go/Kuhn/physicalchemistry3e

and between the atoms B and C (ordinate). The distances in the transition state are much smaller in the case of low energy (blue line in Fig. 25.1) than those in the case of high energy (red line in Fig. 25.1).

	Blue line	Red line
Initial state:	A........BC	A........BC
Transition state:	ABC	A...B...C
Final state:	AB........C	AB........C

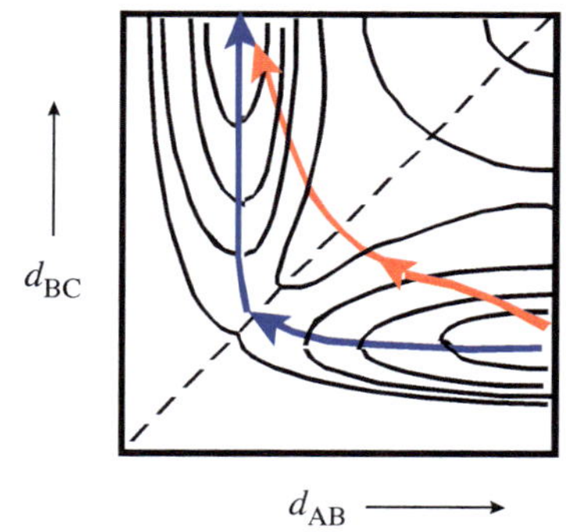

Figure 25.1 Potential energy surface (two-dimensional contour) for the A + BC → AB + C reaction. Two possible reactive trajectories are sketched: one with a low energy of the activated complex (blue line) and one with a high energy (red line). d_{BC} is the distance of the atoms B and C, and d_{AB} is the distance of the atoms A and B; and the contour lines increase monotonically in energy from bottom to top along the dashed line. The dashed line indicates the transition state.

A particular reactive encounter follows a particular trajectory (or path) along this potential energy surface. The transition state is the set of distances and angles (nuclear configuration) at the top of the energy barrier through which the system passes as it proceeds from reactants to products. For the two-dimensional situation given here, the transition state appears as a line, or ridge, through the surface. In the classical limit the reactive trajectories pass through the minimum in the ridge, and under normal circumstances this trajectory dominates the process because it is favored energetically.

In general, for a $3N$ dimensional potential energy surface the transition state is a surface of dimension $3N - 1$. If this is so, how can we use the simple one-dimensional plots of potential energy versus reaction coordinate? In the transition state theory, we replace a detailed consideration of all the degrees of freedom by a thermal average, via the partition functions. A partition function ratio is used to statistically weight the relative importance of different reactive trajectories along the potential energy surface.

25.2.1 Transition State Theory for Bimolecular Reactions

In Section 24.2.2 we introduced the terms “activated complex” and “transition state” for the reaction

$$D + H_2 \rightarrow H + DH \tag{25.1}$$

(Figs. 24.1 and 24.2) and considered this reaction, using collision theory. That simple treatment does not consider that the activated complex may have many different realizations, however, the transition state theory does.

We represent the reaction mechanism as

$$A + B \overset{K^{\ddagger}}{\rightleftharpoons} M \overset{k^{\ddagger}}{\rightarrow} C \tag{25.2}$$

The activated complex M can dissociate, forming either reaction product C or the reactants A and B. For the rate of product formation we can write

$$\frac{dc_C}{dt} = k^{\ddagger} \cdot c_M \tag{25.3}$$

Assuming an equilibrium between the activated complex and the reactants, we obtain

$$\frac{\hat{c}_M}{\hat{c}_A \cdot \hat{c}_B} = K^{\ddagger} \tag{25.4}$$

and, because of the definition $\hat{c} = c/c^{\ominus}$,

$$c_M = \hat{c}_M \cdot c^{\ominus} = K^{\ddagger} \cdot \hat{c}_A \cdot \hat{c}_B \cdot c^{\ominus} = K^{\ddagger} \cdot c_A\, c_B \cdot \frac{1}{c^{\ominus}} \tag{25.5}$$

where $c^{\ominus} = 1$ mol L^{-1} is the standard concentration. Using this result, Equation (25.3) becomes

$$\frac{dc_C}{dt} = k_r \cdot c_A\, c_B \quad \text{with} \quad k_r = k^{\ddagger} \cdot K^{\ddagger} \cdot \frac{1}{c^{\ominus}} \tag{25.6}$$

and we must find $k^{\ddagger}$ and $K^{\ddagger}$ to build a model for obtaining k_r.

In our example the activated complex M is triatomic and linear[2]

$$\mathrm{D \cdots\cdot H \cdots\cdot H}$$

Thus, besides three translational and two rotational degrees of freedom it possesses four vibrational modes (asymmetric, symmetric, and a twofold degenerate bending mode). From Fig. 25.1 we see that along the minimum energy path in the transition state region, one of the bond lengths decreases as the other increases; i.e., the reaction coordinate has the features of an asymmetric stretching vibration

$$\overrightarrow{\mathrm{D}} \cdots\cdot \overleftarrow{\mathrm{H}} \cdots\cdot \overrightarrow{\mathrm{H}} \qquad \text{or} \qquad \overleftarrow{\mathrm{D}} \cdots\cdot \overrightarrow{\mathrm{H}} \cdots\cdot \overleftarrow{\mathrm{H}}$$

This vibration mode is stable only if the molecule is in a potential well (see Chapter 11). However, there is no potential well along this normal coordinate on top of the activation barrier. Thus, when the vibration starts, the complex dissociates. There are two possibilities:

i) It dissociates to the right on the reaction coordinate, i.e., in direction of the products

$$\overrightarrow{\mathrm{D}} \cdots\cdot \overleftarrow{\mathrm{H}} \cdots\cdot \overrightarrow{\mathrm{H}} \quad \rightarrow \quad \mathrm{DH + H}$$

or ii) it dissociates to the left on the reaction coordinate, i.e., in direction of the reactants

$$\overleftarrow{\mathrm{D}} \cdots\cdot \overrightarrow{\mathrm{H}} \cdots\cdot \overleftarrow{\mathrm{H}} \quad \rightarrow \quad \mathrm{D + H_2}$$

Hence it is reasonable to treat the asymmetric vibrational mode of the activated complex as a translational motion along the reaction coordinate.

If the activated complex M has an extent for a distance δ along the coordinate in Fig. 25.2 (*reaction coordinate*), we can define a characteristic time τ for the motion across the barrier top as

$$\tau = \frac{\delta}{\overline{v}} \tag{25.7}$$

where $\overline{v}$ is the average speed of the complex along the reaction coordinate, i.e.,

$$\overline{v} = \frac{\int_0^\infty v \cdot \exp(-m^\ddagger v^2/(2kT)) \cdot \mathrm{d}v}{\int_0^\infty \exp(-m^\ddagger v^2/(2kT)) \cdot \mathrm{d}v} = \sqrt{\frac{2kT}{\pi m^\ddagger}} \tag{25.8}$$

($m^\ddagger$ is the effective mass along the coordinate of the motion for the activated complex). Thus the rate of formation of the product C is, according to Equation (25.3),

$$\frac{\mathrm{d}c_C}{\mathrm{d}t} = k^\ddagger \cdot c_M = \frac{1}{2}\frac{1}{\tau} \cdot c_M \tag{25.9}$$

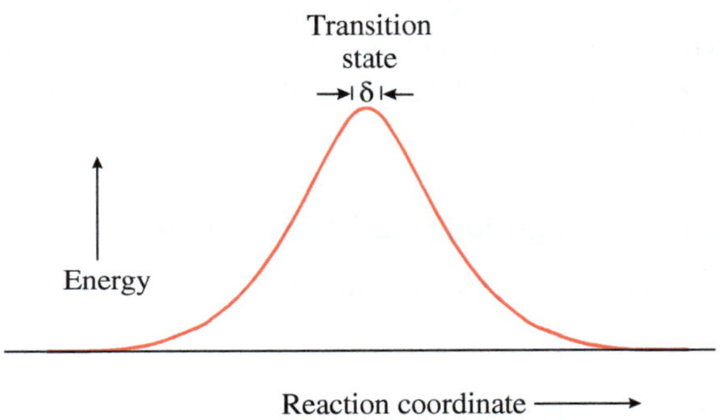

Figure 25.2 One-dimensional potential energy plot (energy versus reaction coordinate) for the reaction A + B → products.

and

$$k^\ddagger = \frac{1}{2}\frac{1}{\tau} = \frac{1}{2}\frac{\overline{v}}{\delta} = \frac{1}{\delta} \cdot \sqrt{\frac{kT}{2\pi m^\ddagger}} \tag{25.10}$$

where the factor of one-half accounts for the fact that only one-half of the activated complex molecules move toward the products, on average. Combining this result with Equation (25.6), we obtain

$$k_r = \frac{1}{c^\ominus} \cdot k^\ddagger \cdot K^\ddagger = \frac{1}{c^\ominus} \cdot \frac{1}{\delta} \cdot \sqrt{\frac{kT}{2\pi m^\ddagger}} \cdot K^\ddagger \tag{25.11}$$

25.2.1.1 Rough Estimation of Rate Constant k_r

Here we give a simple consideration to estimate $k^\ddagger$ from Equation (25.10). The de Broglie wavelength for the motion of the activated complex with velocity v along the reaction coordinate is

$$\Lambda = \frac{h}{m^\ddagger v} \tag{25.12}$$

---

2 If the complex is triatomic and nonlinear, the consideration is the same except that we have three rotational and three vibrational modes.

For our estimation, we replace υ by the average speed $\overline{\upsilon}$ and postulate that the characteristic distance δ is approximately $\Lambda/2$, namely $\delta \sim \Lambda/2$. In the particle in a box model, this length corresponds to the smallest possible energy state; we neglect higher energy states. Thus

$$k^{\ddagger} = \frac{1}{2}\frac{\overline{\upsilon}}{\delta} \sim \frac{\overline{\upsilon}}{\Lambda} = \overline{\upsilon} \cdot \frac{m^{\ddagger}\overline{\upsilon}}{h} = \frac{m^{\ddagger}\left(\overline{\upsilon}\right)^2}{h} \tag{25.13}$$

Using Equation (25.8), we find that

$$k^{\ddagger} = \frac{m^{\ddagger}}{h}\left(\frac{2kT}{\pi m^{\ddagger}}\right) = \frac{2}{\pi}\frac{kT}{h} \tag{25.14}$$

and for the rate constant from Equations (25.6) and (25.11) we obtain

$$k_r = \frac{1}{c^{\ominus}} \cdot k^{\ddagger} \cdot K^{\ddagger} = \frac{1}{c^{\ominus}} \cdot \frac{2}{\pi}\frac{kT}{h} \cdot K^{\ddagger} \tag{25.15}$$

25.2.1.2 Rigorous Calculation of Rate Constant k_r

A rigorous derivation is given in Box 25.1, leading to an only slightly different expression for the rate constant

$$\boxed{k_r = \frac{1}{c^{\ominus}} \cdot \frac{kT}{h} \cdot K^{\ddagger} \qquad \text{(bimolecular reaction)}} \tag{25.16}$$

Following the consideration in Section 18.5.7, Equation (18.86), we can express the equilibrium constant in Equation (25.4) by the molecular partition functions of the reactants.

$$K^{\ddagger} = \frac{\dfrac{Z_{\mathrm{M}}^{\ominus}}{\sigma_{\mathrm{M}} \cdot N(\mathrm{M})}}{\dfrac{Z_{\mathrm{A}}^{\ominus}}{\sigma_{\mathrm{A}} \cdot N(\mathrm{A})} \cdot \dfrac{Z_{\mathrm{B}}^{\ominus}}{\sigma_{\mathrm{B}} \cdot N(\mathrm{B})}} \tag{25.17}$$

where $Z^{\ominus}$ is the molecular partition function under standard conditions, $N(\mathrm{A})$, $N(\mathrm{B})$, and $N(\mathrm{M})$ are the number of particles A, B, and M, and σ is the symmetry number. Because one molecule of A reacts with one molecule of B to form one molecule of M, we have

$$N(\mathrm{A}) = N(\mathrm{B}) = N(\mathrm{M}) \tag{25.18}$$

and therefore

$$K^{\ddagger} = N(\mathrm{A}) \cdot \frac{\dfrac{Z_{\mathrm{M}}^{\ominus}}{\sigma_{\mathrm{M}}}}{\dfrac{Z_{\mathrm{A}}^{\ominus}}{\sigma_{\mathrm{A}}} \cdot \dfrac{Z_{\mathrm{B}}^{\ominus}}{\sigma_{\mathrm{B}}}} \tag{25.19}$$

Using Equation (25.16) we obtain

$$k_r = \frac{1}{c^{\ominus}} \cdot \frac{kT}{h} \cdot N(\mathrm{A}) \cdot \frac{\dfrac{Z_{\mathrm{M}}^{\ominus}}{\sigma_{\mathrm{M}}}}{\dfrac{Z_{\mathrm{A}}^{\ominus}}{\sigma_{\mathrm{A}}} \cdot \dfrac{Z_{\mathrm{B}}^{\ominus}}{\sigma_{\mathrm{B}}}} \tag{25.20}$$

where $N(\mathrm{A})$ is the number of molecules A. Note that the partition function for the activated complex has the same features as a typical molecular partition function, except that the asymmetric vibration mode is excluded because it has been treated explicitly (see Box 25.1).

As an example, k_r is calculated in Box 25.2 for reaction (25.1),

$$\mathrm{D} + \mathrm{H}_2 \rightarrow \mathrm{D}--\mathrm{H}--\mathrm{H} \rightarrow \mathrm{DH} + \mathrm{H}$$

the reaction of a D atom with an H_2 molecule, where $E_{a,m}$ is the molar activation energy. We find that

$$k_r = \sqrt{\frac{8kT}{\pi\mu}} \cdot \pi d^2 \cdot \boldsymbol{N}_{\mathrm{A}} \cdot \left[\frac{\mu^{\ddagger}}{\mu}\left(\frac{d^{\ddagger}}{d}\right)^2 \frac{2h^2}{8\pi^2 kT\, I_{H_2}}\right] \cdot \exp\left(-\frac{E_{a,m}}{\boldsymbol{RT}}\right) \tag{25.21}$$

where $d = r_D + r_{H_2}$ is the sum of the collision radii of D and H_2, μ is the reduced mass, $\mu = m_D m_{H_2}/(m_D + m_{H_2})$, and I_{H_2} is the moment of inertia of H_2. The moment of inertia for the activated complex I_{DHH} is modeled by a linear rotator

$$I_{DHH} = \mu^{\ddagger}(d^{\ddagger})^2$$

where $\mu^{\ddagger}$ is the reduced mass and $d^{\ddagger}$ is the bond distance in the activated complex (D–H and H–H, respectively).

According to Equations (24.19) and (24.20) the corresponding expression derived from the simple collision theory is

$$k_r = \sqrt{\frac{8kT}{\pi\mu}} \cdot \pi d^2 \cdot \boldsymbol{N}_A \cdot \alpha \cdot \exp\left(-\frac{E_{a,m}}{\boldsymbol{R}T}\right) \tag{25.22}$$

Thus, we can identify the steric factor α as

$$\alpha = \left[\frac{\mu^{\ddagger}}{\mu}\left(\frac{d^{\ddagger}}{d}\right)^2 \frac{2h^2}{8\pi^2 kT\, I_{H_2}}\right] \tag{25.23}$$

Example 25.1 *Use Equation (25.23) to calculate the steric factor for the D + H_2 reaction and to compare it with the experimentally derived value $\alpha = 0.2$, which is found in Exercise E24.8 of Chapter 24. For H_2 use the parameters $d_{H_2} = 74.14$ pm and $I_{H_2} = 4.59 \times 10^{-48}$ kg m^2 and take the bond distance in the activated complex to be $d^{\ddagger} = 100$ pm (see Fig. 24.2).*

Using the parameters given we find

$$\frac{2h^2}{8\pi^2 kT\, I_{H_2}} = 0.59 \quad \text{at} \quad 298\ \text{K}$$

Using molecular and atomic radii given in Chapter 14, the collision diameter d is $d = r_D + r_{H_2} = r_H + r_{H_2} = (126 + 136)$ pm $= 262$ pm . Thus

$$\left(\frac{d^{\ddagger}}{d}\right)^2 = \left(\frac{100}{262}\right)^2 = 0.15$$

The reduced mass μ is

$$\mu = \frac{m_D m_{H_2}}{m_D + m_{H_2}} = m_H \frac{4}{4} = m_H$$

We estimate the reduced mass $\mu^{\ddagger}$ for the rotational motion of the activated complex

D–H–H

$\mu^{\ddagger}$ is in between the reduced masses of D–H–D and H–H–H:

$$I_{DHD} = 2m_D \cdot (d^{\ddagger})^2, \quad \text{thus } \mu^{\ddagger} = 2m_D = 4m_H$$

$$I_{HHH} = 2m_H \cdot (d^{\ddagger})^2, \quad \text{thus } \mu^{\ddagger} = 2m_H$$

thus approximately

$$\mu^{\ddagger} = \frac{1}{2} m_H \cdot (4 + 2) = 3m_H$$

In Exercise E25.20 you calculate the reduced mass $\mu^{\ddagger}$ more rigorously and find $\mu^{\ddagger} = \frac{11}{4} m_H = 2.75 m_H$. Thus, we obtain

$$\alpha = \frac{\mu^{\ddagger}}{\mu}\left(\frac{d^{\ddagger}}{d}\right)^2 \frac{2h^2}{8\pi^2 kTI_{H_2}} = 3 \cdot 0.15 \cdot 0.59 = 0.24$$

This calculated value agrees well with $\alpha = 0.2$ calculated in Exercise 24.8 in Chapter 24.

The transition state theory provides a theoretical basis for the calculation of the steric factor. In Equation (25.23), α depends inversely on the inertial moment of H_2. If we consider the reaction

$$H + Cl_2 \rightarrow HCl + Cl$$

then the inertial moment is $I_{Cl_2} = 1.15 \times 10^{-45}$ kg m^2 and for α we obtain $\alpha = 0.003$. Furthermore, if we extend our consideration to more complex reactants (e.g., two reacting diatomic molecules and two reacting polyatomic molecules) α becomes smaller by many orders of magnitude, in accordance with experiment.

25.2.1.3 Activation Enthalpy and Activation Entropy

In our consideration we have assumed an equilibrium between reactants and the activated complex. With this connection, we can use the relations

$$\Delta_r G^\ominus = \Delta_r H^\ominus - T\Delta_r S^\ominus$$

and

$$\Delta_r G^\ominus = -\boldsymbol{R}T \cdot \ln K, \qquad K = \exp\left(-\frac{\Delta_r G^\ominus}{\boldsymbol{R}T}\right)$$

(see Chapter 18) to determine the equilibrium constant for the activated complex $K^\ddagger$. Using Equation (25.16), we find that

$$k_r = \frac{1}{c^\ominus}\frac{kT}{h} \cdot K^\ddagger = \frac{1}{c^\ominus}\frac{kT}{h} \cdot \exp\left(-\frac{\Delta_r G^{\ddagger\ominus}}{\boldsymbol{R}T}\right) \tag{25.24}$$

Then with

$$\Delta_r G^{\ddagger\ominus} = \Delta_r H^{\ddagger\ominus} - T\Delta_r S^{\ddagger\ominus}$$

where $\Delta_r H^{\ddagger\ominus}$ is the activation enthalpy and $\Delta_r S^{\ddagger\ominus}$ is the activation entropy. Under standard conditions, we obtain

$$\boxed{k_r = A_f \cdot \mathrm{e}^{-\Delta_r H^{\ddagger\ominus}/(\boldsymbol{R}T)}} \tag{25.25}$$

with

$$\boxed{A_f = \frac{1}{c^\ominus} \cdot \frac{kT}{h} \cdot \mathrm{e}^{\Delta_r S^{\ddagger\ominus}/\boldsymbol{R}}} \tag{25.26}$$

for the bimolecular transition state theory rate constant. This "thermodynamic" version of the expression for the rate constant is called the *Eyring equation*.

Compared with the result of the collision theory, the dependence of the rate constant on temperature is the same,[3] if we neglect the small difference in the temperature dependence of A_f and replace $E_{a,m}$ by $\Delta_r H^{\ddagger\ominus}$. As outlined in Chapters 16 and 17 the quantities $\Delta_r H^{\ddagger\ominus}$ and $\Delta_r S^{\ddagger\ominus}$ are related to the Boltzmann distribution of the energy levels of the reactants, A and B, and the activated complex, M.

Example 25.2 *Dimerization of cyclopentadiene*

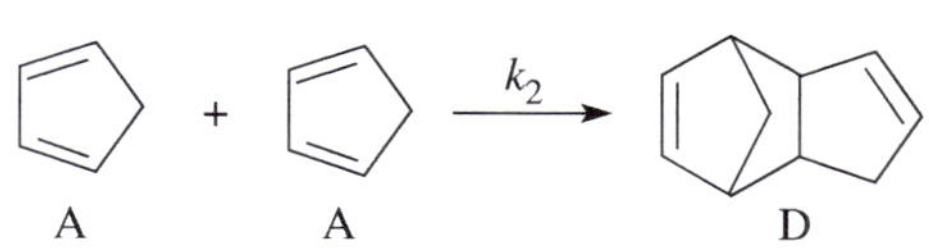

This reaction *has the rate constant* $k_r = 16.8 \times 10^{-6}$ *L mol*$^{-1}$ s^{-1} *at* $T = 343$ *K, and an activation energy of* $E_{a,m} =$ 62 *kJ mol*$^{-1}$ *obtained from the temperature dependence of* k_r. *According to the simple collision theory, the steric factor is extremely small, namely,* 1×10^{-6}. *Determine the activation entropy for the reaction and discuss it.*

Using Equation (25.25) and setting $\Delta_r H^{\ddagger\ominus} = E_{a,m} = 62$ kJ mol^{-1}, the frequency factor for this reaction is

$$A_f = \frac{k_r}{\mathrm{e}^{-\Delta_r H^{\ddagger\ominus}/\boldsymbol{R}T}} = \frac{16.8 \times 10^{-6}\ \mathrm{L\ mol^{-1}\ s^{-1}}}{\exp\left[-62\ \mathrm{kJ\ mol^{-1}}/(8.314\ \mathrm{J\ K^{-1}\ mol^{-1}} \cdot 343\ \mathrm{K})\right]} = 4.7 \times 10^4\ \mathrm{L\ mol^{-1}\ s^{-1}}$$

From Equation (25.26), we find

$$\exp\left(\frac{\Delta_r S^{\ddagger\ominus}}{\boldsymbol{R}}\right) = \frac{A_f\, c^\ominus\, h}{kT}$$

3 More rigorously, we have to consider the change of volume so that $\Delta_r H^{\ddagger\ominus} = E_{a,m} + P \cdot \Delta V_m^{\ddagger\ominus}$. Often the difference between $\Delta_r H^{\ddagger\ominus}$ and $E_{a,m}$ is small.

and

$$\Delta_r S^{\ddagger\ominus} = \boldsymbol{R} \cdot \ln\left(\frac{A_f\, c^{\ominus}\, h}{kT}\right) = -157 \text{ J K}^{-1} \text{ mol}^{-1}$$

$\Delta_r S^{\ddagger\ominus}$ is negative indicating that the number of configurations of the activated complex is much smaller than that for the reactants. This result is consistent with the small steric factor obtained from the simple collision theory. The large negative entropy displayed by this reaction suggests that well-defined orientations of the reactants are needed to proceed successfully to product.

25.2.2 Transition State Theory for Unimolecular Reactions

Similar considerations may be applied to unimolecular reactions. We consider, for example, the reaction of *cis*-dichloroethene to trans-dichloroethene

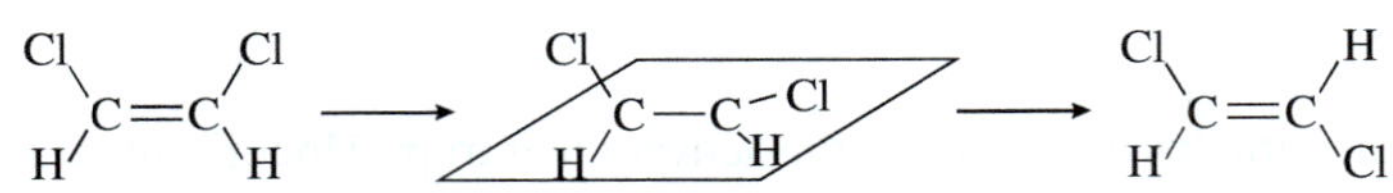

cis-Form   Transition state   *trans*-Form

Between the reactant and product wells, one finds a barrier to the reaction. The geometry of the molecule at the top of the barrier is the activated complex (or transition state) geometry, and we write the reaction mechanism as

$$\mathrm{A} \rightleftarrows \mathrm{M} \xrightarrow{k^{\ddagger}} \mathrm{B} \tag{25.27}$$

Assuming an equilibrium condition between A and M, we can write

$$K^{\ddagger} = \frac{c_M}{c_A} \tag{25.28}$$

Using these assumptions, we can write an expression for the formation rate of product as

$$\frac{dc_{\mathrm{B}}}{dt} = k^{\ddagger}\, c_{\mathrm{M}} = k^{\ddagger}\, K^{\ddagger}\, c_{\mathrm{A}} \tag{25.29}$$

The considerations here are identical to those to find Equation (25.16), except that we must replace $K^{\ddagger}/c^{\ominus}$ by $K^{\ddagger}$, so that

$$\boxed{k_r = \frac{kT}{h} \cdot K^{\ddagger} \qquad \text{(unimolecular reaction)}} \tag{25.30}$$

Accordingly, instead of Equations (25.25) and (25.26) for the bimolecular reaction we find

$$\boxed{k_r = A_f \cdot \mathrm{e}^{-\Delta_r H^{\ddagger\ominus}/(\boldsymbol{R}T)}} \tag{25.31}$$

with

$$\boxed{A_f = \frac{kT}{h} \cdot \mathrm{e}^{\Delta_r S^{\ddagger\ominus}/\boldsymbol{R}} \qquad \text{(unimolecular reaction)}} \tag{25.32}$$

Expressing the equilibrium constant, $K^{\ddagger}$, in terms of the partition functions (Equation (25.17), using $N(\mathrm{A}) = N(\mathrm{M})$, and setting $\sigma_{\mathrm{A}} = \sigma_M = 1$, we find

$$k_r = \frac{kT}{h} \cdot \frac{Z_{\mathrm{M}}^{\ominus}}{Z_{\mathrm{A}}^{\ominus}} \tag{25.33}$$

$Z_{\mathrm{M}}^{\ominus}$ and $Z_{\mathrm{A}}^{\ominus}$ are the activated complex and reactant partition functions in the standard state. In general, this equation is difficult to evaluate theoretically, because it requires knowledge of the activated complexes energetics and geometry.

Example 25.3 *Consider the cis to trans isomerization of dichloroethene and find an explicit expression for the rate constant in terms of the molecular parameters. To simplify your considerations, assume that the vibrational coordinates, other than the torsional motion of isomerization, have the same characteristic frequencies in the reactant well as they do at the transition state.*

Using Equation (25.33), we can write that

$$k_r = \frac{kT}{h} \cdot \frac{Z^{\ominus}_{\text{M,trans}}}{Z^{\ominus}_{\text{A,trans}}} \cdot \frac{Z_{\text{M,rot}}}{Z_{\text{A,rot}}} \cdot \frac{Z_{\text{M,vib}}}{Z_{\text{A,vib}}} \cdot \frac{Z_{\text{M,electr}}}{Z_{\text{A,electr}}} \tag{25.34}$$

where the index A refers to the *cis*-form and the index M refers to the transition state. Because the reaction is unimolecular, the translational partition functions are identical. In addition we are ignoring any changes in the inertial moments and in the vibrational partition functions, except that we take into account the zero point energy of the vibrational torsion mode of reactant A, $\frac{1}{2}h\nu_{\text{torsion}}$ (this vibrational mode does not occur for the transition state). Hence

$$k_r = \frac{kT}{h} \cdot \frac{1}{Z_{\text{A,torsion}}} \cdot \frac{Z_{\text{M,electr}}}{Z_{\text{A,electr}}} \tag{25.35}$$

According to Section 15.4.2 we find

$$Z_{\text{A,torsion}} = \frac{1}{1 - \exp\left(-h\nu_{\text{torsion}}/kT\right)} \exp\left(-\frac{h\nu_{\text{torsion}}}{2kT}\right)$$

so that

$$k_r = \frac{kT}{h}\left(1 - e^{-h\nu_{\text{torsion}}/kT}\right) \cdot \exp\left(-\frac{E_{a,m}}{\boldsymbol{R}T}\right) \tag{25.36}$$

where the molar activation energy $E_{a,m} = \Delta E^{\ddagger}_{\text{electronic,m}} - \frac{1}{2}\boldsymbol{N}_{\text{A}}h\nu_{\text{torsion}}$ is the energy difference between the zero point energy level of the reactant state and the zero point energy level of the transition state.

For the case when $h\nu_{\text{torsion}} \gg kT$, most of the molecules A are in the torsional ground state, and $\exp(-h\nu_{\text{torsion}}/kT) \ll 1$, so that

$$k_r = \frac{kT}{h} \cdot \exp\left(-\frac{E_{a,m}}{\boldsymbol{R}T}\right) \qquad (\text{case } h\nu_{\text{torsion}} \gg kT\ ) \tag{25.37}$$

In the opposite limit of $h\nu_{\text{torsion}} \ll kT$, with $(1 - \exp\left(-h\nu_{\text{torsion}}/kT\right)) \approx h\nu_{\text{torsion}}/kT$, we find

$$k_r = \nu_{\text{torsion}} \cdot \exp\left(-\frac{E_{a,m}}{\boldsymbol{R}T}\right) \qquad (\text{case } h\nu_{\text{torsion}} \ll kT\ ) \tag{25.38}$$

Hawton and Semeluk [1] measured the rate constant of the thermal trans to *cis* isomerization of 1,2-dichloroethene in the gas phase from 806 to 846 K. They were able to fit their data to the Arrhenius expression: $A \cdot \exp(-E_{a,m}/\boldsymbol{R}T)$, with $E_{a,m} =$ 231 kJ mol^{-1} and $A = 4.8 \times 10^{12}$ s^{-1}, corresponding to a rate constant of $k_r = 3.96 \times 10^{-3}$ s^{-1} at $T = 800$ K. The vibrational spectrum of *trans*-dichloroethene gives a frequency of $\nu_{\text{torsion}} = 227$ cm^{-1} $\cdot$ 3.00×10^{10} cm s^{-1} $= 6.8 \times 10^{12}$ s^{-1}, thus $h\nu_{\text{torsion}} = 0.45 \times 10^{-20}$ J . From Equation (25.36) for $T = 800$ K ($kT = 1.10 \times 10^{-20}$ J) we obtain

$$A = \frac{kT}{h}\left(1 - e^{-h\nu_{\text{torsion}}/kT}\right) = \frac{kT}{h} \cdot 0.33 = 5.5 \times 10^{12}\ \text{s}^{-1}$$

in good agreement with the experiment.

Note that the activation barrier $E_{a,m} = 231$ kJ mol^{-1} is about 100 times larger than the available thermal energy $\boldsymbol{R}T =$ 2.49 kJ mol^{-1} at 300 K, and the trans and *cis* forms of dichloroethene are quite stable at room temperature.

25.2.3 More on Activation Energy

In Chapter 24, Equation (24.60) we write the rate constant as

$$k_r = A_f \cdot \exp\left(-\frac{E_{a,.m}}{\boldsymbol{R}T}\right)$$

where A_f is the preexponential factor. For practical use, we consider the ratio $k_r/k_{r,1}$ instead of k_r where k_r is the rate constant at temperature T and $k_{r,1}$ is the rate constant at a selected temperature T_1. Thus we write that

$$\frac{k_r}{k_{r,1}} = \frac{A_f}{A_{f,1}} \frac{\exp\left(-\frac{E_{a,m}}{\boldsymbol{R}T}\right)}{\exp\left(-\frac{E_{a,m}}{\boldsymbol{R}T_1}\right)}$$

Using this form we calculate the logarithm as

$$\ln\left(\frac{k_r}{k_{r,1}}\right) = \ln\left(\frac{A_f}{A_{f,1}}\right) - \frac{E_{a,m}}{\boldsymbol{R}}\left(\frac{1}{T} - \frac{1}{T_1}\right)$$

If we consider A_f to be temperature independent, then this equation simplifies to

$$\ln\left(\frac{k_r}{k_{r,1}}\right) = -\frac{E_{a,m}^{\text{empirical}}}{\boldsymbol{R}}\left(\frac{1}{T} - \frac{1}{T_1}\right) \tag{25.39}$$

The slope of a plot of $\ln\left(k_r/k_{r,1}\right)$ versus $1/T$ defines $E_{a,m}^{\text{empirical}}/\boldsymbol{R}$, where $E_{a,m}^{\text{empirical}}$ is an empirical activation energy.

The assumption that A_f is independent of temperature does not always hold true, however. Moreover, in collision theory A_f is proportional to $\sqrt{T}$, and in transition state theory it is proportional to T. Thus we have a difficulty when we want to extract $E_{a,,m}$ (collision theory) or $\Delta_r H^{\ddagger\ominus}$ (transition state theory) from experimental k_r data. In these case, we have to account explicitly for the temperature dependence of A_f. For example, in transition state theory A_f depends linearly on the temperature, so that

$$A_f = A_{f,0} \cdot T$$

Then we have, according to Equations (25.25) and (25.26)

$$k_r = A_{f,0} \cdot T \cdot \exp\left(-\frac{\Delta_r H^{\ddagger\ominus}}{\boldsymbol{R}T}\right) \quad \text{and} \quad k_{r,1} = A_{f,0} \cdot T_1 \cdot \exp\left(-\frac{\Delta_r H^{\ddagger\ominus}}{\boldsymbol{R}T_1}\right)$$

so that

$$\frac{k_r}{k_{r,1}} = \frac{T}{T_1} \cdot \frac{\exp\left(-\frac{\Delta_r H^{\ddagger\ominus}}{\boldsymbol{R}T}\right)}{\exp\left(-\frac{\Delta_r H^{\ddagger\ominus}}{\boldsymbol{R}T_1}\right)}$$

Lastly, we take the logarithm of both sides and find

$$\begin{aligned}\ln\left(\frac{k_r}{k_{r,1}}\right) &= \ln\left(\frac{T}{T_1}\right) - \frac{\Delta_r H^{\ddagger\ominus}}{\boldsymbol{R}T} + \frac{\Delta_r H^{\ddagger\ominus}}{\boldsymbol{R}T_1} \\ &= \ln\left(\frac{T}{T_1}\right) - \frac{\Delta_r H^{\ddagger\ominus}}{\boldsymbol{R}}\left(\frac{1}{T} - \frac{1}{T_1}\right)\end{aligned}$$

Reorganization of the terms yields

$$\ln\left(\frac{k_r}{k_{r,1}}\right) - \ln\left(\frac{T}{T_1}\right) = -\frac{\Delta_r H^{\ddagger\ominus}}{\boldsymbol{R}}\left(\frac{1}{T} - \frac{1}{T_1}\right) \tag{25.40}$$

If we plot the quantity

$$y = \ln\left(\frac{k_r}{k_{r,1}}\right) - \ln\left(\frac{T}{T_1}\right)$$

versus $1/T$, then we expect a straight line with slope $-\Delta_r H^{\ddagger\ominus}/\boldsymbol{R}$.

How is this slope related to the empirical activation energy? Using Equations (25.39) and (25.40) for the activation energies, we find that

$$\frac{\mathrm{d}\ln(k_r/k_{r,1})}{\mathrm{d}(1/T)} = -\frac{E_{a,m}^{\text{empirical}}}{\boldsymbol{R}}$$

and

$$\begin{aligned}\frac{\mathrm{d}\left[\ln(k_r/k_{r,1}) - \ln(T/T_1)\right]}{\mathrm{d}(1/T)} &= \frac{\mathrm{d}\ln(k_r/k_{r,1})}{\mathrm{d}(1/T)} - \frac{\mathrm{d}\left[\ln(T/T_1)\right]}{\mathrm{d}(1/T)} \\ &= -\frac{E_{a,m}^{\text{empirical}}}{\boldsymbol{R}} - \frac{\mathrm{d}\left[\ln(T/T_1)\right]}{\mathrm{d}(1/T)} = -\frac{\Delta_r H^{\ddagger\ominus}}{\boldsymbol{R}}\end{aligned}$$

Thus we obtain

$$\frac{\Delta_r H^{\ddagger\ominus}}{\boldsymbol{R}} = \frac{E_{a,m}^{\text{empirical}}}{\boldsymbol{R}} + \frac{\mathrm{d}\,[\ln(T/T_1)]}{\mathrm{d}(1/T)}$$

$$= \frac{E_{a,m}^{\text{empirical}}}{\boldsymbol{R}} + \frac{\mathrm{d}\,[\ln(T/T_1)]}{\mathrm{d}T} \cdot \frac{\mathrm{d}T}{\mathrm{d}(1/T)}$$

$$= \frac{E_{a,m}^{\text{empirical}}}{\boldsymbol{R}} + \frac{1}{T} \cdot \frac{\mathrm{d}T}{\mathrm{d}(1/T)}$$

We calculate the last term by setting $x = 1/T$:

$$\frac{\mathrm{d}T}{\mathrm{d}(1/T)} = \frac{\mathrm{d}(1/x)}{\mathrm{d}x} = -\frac{1}{x^2} = -T^2$$

to obtain

$$\frac{\Delta_r H^{\ddagger\ominus}}{\boldsymbol{R}} = \frac{E_{a,m}^{\text{empirical}}}{\boldsymbol{R}} - T \quad \text{or} \quad \Delta_r H^{\ddagger\ominus} = E_{a,m}^{\text{empirical}} - \boldsymbol{R}T \tag{25.41}$$

In analyzing experimental data, workers often assume a model and account for the correction explicitly or they extract the slope from k_r values measured over a range of temperatures, starting with T_{start} to T_{end}. In the latter case, one can replace T in Equation (25.41) by the mean value,

$$T_{\text{mean}} = \frac{1}{2}(T_{\text{end}} + T_{\text{start}})$$

In Exercise E25.10 you analyze this correction term and show that it works well for a temperature range of about 50 K. In Exercise E25.22, you perform a similar analysis with the rate constant expression from the collision theory and find that $E_{a,m} = E_{a,m}^{\text{empirical}} - \frac{1}{2}\boldsymbol{R}T_{\text{mean}}$.

25.2.4 Applications of Transition State Theory

Transition state theory makes some important predictions about how rates can be modified by molecular and atomic characteristics.

25.2.4.1 Isotope Effects

For bond cleavage reactions and reactions involving changes in bond lengths and frequencies, the isotope of the atoms in the bonds can have an important effect on the reaction rate. This effect is shown most dramatically for the case of the H atom because the deuterium isotope has twice the mass of the H atom itself. The data in Table 25.1 show the isotope effect on the rate constant for the H atom transfer reaction

$$Cl + H_2 \rightarrow HCl + H \tag{25.42}$$

Table 25.1 Isotope Dependence [2] of Reactions Described by Equation (25.42) (the Data Are Averages).

T/ K	295.2	384.7	498.6
$k_r(H_2)/k_r(D_2)$	9.0 ± 0.4	6.0 ± 0.5	4.0 ± 0.3

Here we illustrate how the zero point energy shift changes the overall activation energy of the reaction. In Fig. 25.3 we assume that the isotope effect on the electronic energy of the transition state, $\Delta E_0^{\ddagger}$, can be neglected, but that we must account for the zero point energy shifts in the reactant well and the transition state. This effect accounts for most of the reaction rate constant's change. The activation energy $E_a(H_2)$ for reaction (25.42) is

$$E_a\,(H_2) = \Delta E_0^{\ddagger} - \Delta E_0\,(H_2) + \Delta E_0\,(\text{Cl–H–H})$$

where $\Delta E_0^{\ddagger}$ is the electronic contribution to the activation energy, $\Delta E_0(H_2)$ is the zero point energy of vibration of H_2, and ΔE_0(Cl–H–H) is the zero point energy of vibration of the activated complex Cl–H–H. Accordingly, for the reaction

$$Cl + D_2 \rightarrow DCl + D \tag{25.43}$$

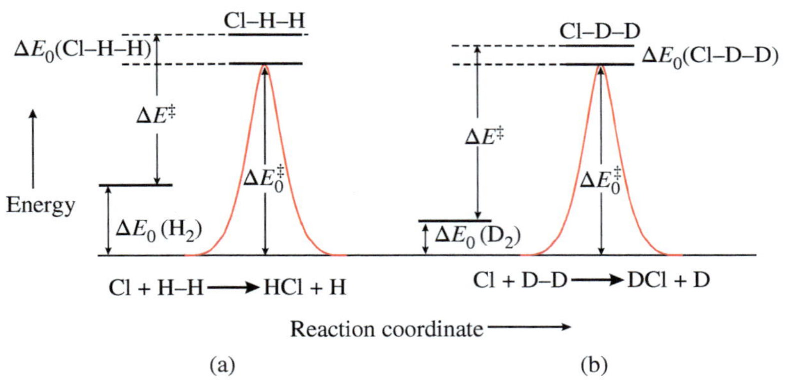

Figure 25.3 Isotope effect in the reactions a) $Cl + H_2 \rightarrow HCl + H$ and b) $Cl + D_2 \rightarrow DCl + D$. The potential energy curve is shown to illustrate how replacing H_2 by D_2 can modify the activation energy $\Delta E^‡$. In both cases $\Delta E_0^‡$ is the same electronic contribution to the activation energy. Because of the greater zero point energy of vibration, the total activation energy $\Delta E^‡$ is smaller in case a) compared to case b). Thus reaction (a) is faster than reaction (b).

we obtain

$$E_a(D_2) = \Delta E_0^‡ - \Delta E_0(D_2) + \Delta E_0\ (\text{Cl–D–D})$$

In this approximation, the isotope effect should be

$$\begin{aligned}\frac{k_r\ (H_2)}{k_r(D_2)} &= \frac{\exp\left[-E_a\ (H_2)/(kT)\right]}{\exp\left[-E_a\ (D_2)/(kT)\right]} \\ &= \frac{\exp\left[\left(-\Delta E_0^‡ + \Delta E_0\ (H_2) - \Delta E_0\ (\text{Cl–H–H})\right)/(kT)\right]}{\exp\left[\left(-\Delta E_0^‡ + \Delta E_0(D_2) - \Delta E_0\ (\text{Cl–D–D})\right)/(kT)\right]} \\ &= \exp\left[\left(\Delta E_0\ (H_2) - \Delta E_0\ (\text{Cl–H–H}) - \Delta E_0(D_2) + \Delta E_0\ (\text{Cl–D–D})\right)/(kT)\right]\end{aligned} \quad (25.44)$$

Example 25.4 *From Equation (25.44) rationalize the observed isotope ratios in Table 25.1.*

For the zero point energies $\frac{1}{2}h\upsilon_0(H_2)$ and $\frac{1}{2}h\upsilon_0(D_2)$ we obtain

$$\Delta E_0\ (H_2) = \frac{1}{2}h\nu_0\ (H_2) = \frac{1}{2}h\frac{1}{2\pi}\sqrt{\frac{k_f}{\mu\ (H_2)}} = 4.39 \times 10^{-20}\ \text{J}$$

$$\Delta E_0(D_2) = \frac{1}{2}h\nu_0\ (H_2) = \frac{1}{2}h\frac{1}{2\pi}\sqrt{\frac{k_f}{\mu(D_2)}} = 3.10 \times 10^{-20}\ \text{J}$$

where $k_f = 574\ \text{N m}^{-1}$ is the force constant of H_2 and D_2. $\mu(H_2) = 0.83 \times 10^{-27}$ kg and $\mu(D_2) = 1.66 \times 10^{-27}$ kg are the reduced masses of H_2 and D_2, respectively.

In reactions (25.42) and (25.43) the transition states are Cl–H–H and Cl–D–D, respectively. For the zero point energies of the vibrational bending and stretching modes calculated data [3] are available:

	Zero point energy/10^{-20} J	
	Cl–H–H	Cl–D–D
Bending mode	0.58	0.41
stretching mode	1.34	0.98
total	1.92	1.38

This means that the zero point energy of the two transition states is

$$\Delta E_0\ (\text{Cl–H–H}) = 1.92 \times 10^{-20}\ \text{J}$$

and

$$\Delta E_0\ (\text{Cl–D–D}) = 1.38 \times 10^{-20}\ \text{J}$$

Thus, according to Equation (25.44), the ratio of the rate constants at 295.2 K is

$$\frac{k_r\,(H_2)}{k_r(D_2)} = \exp\left(\frac{[4.39 - 1.92 - 3.10 + 1.38] \times 10^{-20}\ \text{J}}{kT}\right)$$
$$= \exp\left(\frac{0.75 \times 10^{-20}\ \text{J}}{kT}\right) = 6.3$$

The corresponding values for 384.7 and 498.6 K are 4.1 and 3.0, respectively.

These values differ from the experimental data by about 50%. In general we must also consider how the isotopes affect the ratio of the rotational and translational partition functions.

This calculation is done in Problem P25.10, and an improved agreement with the experimental values is reached.

This example shows that detailed features of the transition state structure can strongly influence the magnitude of the isotope effect (typical effects range from 1.1 to a factor of 15), and isotope studies provide an important tool for elucidating reaction mechanisms and transition state properties.

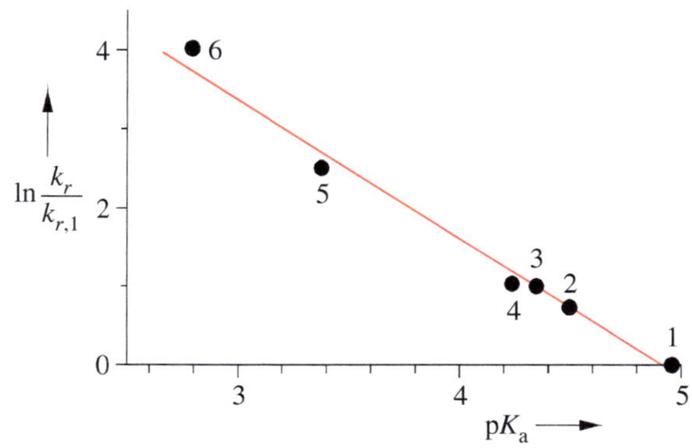

Figure 25.4 The rate constant k_r for the nucleophilic substitution reaction (25.45) is shown ($k_{r,1}$ is the rate constant for substituent 1) to correlate with the pK_a of R-COOH for different substituents R (1: acetate, 2: p-toluate, 3: benzoate, 4: o-toluate, 5: o-chlorobenzoate, 6: o-nitrobenzoate). Full circles: experiment, $T = 373$ K; solid line: regression line.

25.2.4.2 Linear Gibbs Energy Relationship (Hammett Equation)

Reactions of a similar type and involving largely similar compounds show a general correlation of the reaction rate with the overall Gibbs energy change in the reaction. The origin of this dependence can be understood from the perspective of transition state theory. An example of this effect is shown in Fig. 25.4, where the rate of a nucleophilic substitution reaction

$$\text{R-COOCH}_3 + \text{N(CH}_3)_3 \rightarrow \text{RCOO}^- + \text{N(CH}_3)_4^+ \qquad (25.45)$$

displays a linear correlation with the pK_a of the corresponding acid R-COOH.

This correlation can be rationalized by considering how the reaction driving force changes and can affect the activation barrier. Figure 25.5 shows plots for a reactant and two possible products, one with a reaction Gibbs energy of $\Delta_r G_1$ and the other with a reaction Gibbs energy of $\Delta_r G_2$. In addition, the diagram shows the two Gibbs energies of activation. As long as the product and reactant curves have the same shape for the two reactions, then we can write that

$$\Delta_r G_2^\ddagger - \Delta_r G_1^\ddagger = \alpha \cdot (\Delta_r G_2 - \Delta_r G_1) \qquad (25.46)$$

where α is a proportionality factor. Hence the change in the Gibbs energy of activation between the two reactions is proportional to the change in their reaction Gibbs energy.

From Equation (25.24), we can write that

$$\ln\left(\frac{k_r}{const}\right) = -\frac{\Delta_r G^{\ddagger\ominus}}{\boldsymbol{R}T} \qquad (25.47)$$

and from the law of mass action, we can write that

$$\ln K = -\frac{\Delta_r G^\ominus}{\boldsymbol{R}T}$$

Substituting these expressions into Equation (25.46), we find that

$$\ln\left(\frac{k_{r2}}{k_{r1}}\right) = -\alpha \cdot (\Delta_r G_2 - \Delta_r G_1)$$
$$= \alpha \cdot \ln\left(\frac{K_{r2}}{K_{r1}}\right) \qquad (25.48)$$

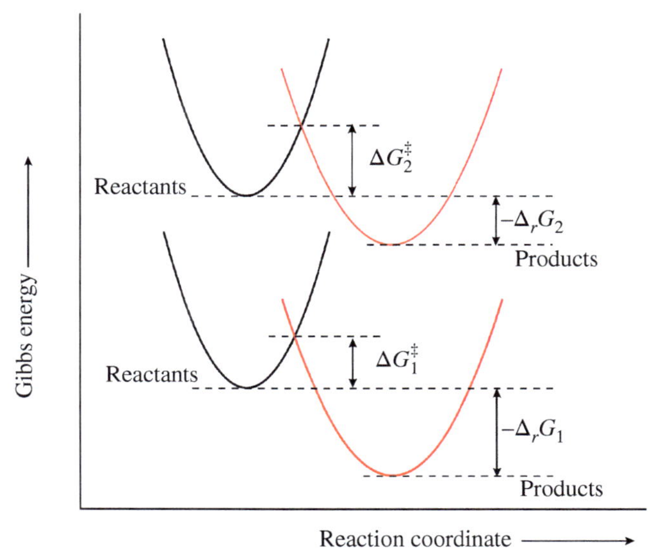

Figure 25.5 This diagram shows how a change in the exoergocity of a reaction, $\Delta_r G$, can correlate with the change in the activation energy $\Delta G^\ddagger$ and hence the reaction rate.

This correlation of the reaction rate with the standard Gibbs energy change (hence the equilibrium constant) relies on a similarity in the reaction mechanism and general features of the reaction coordinate for the different reactions – except for an overall shift in the relative Gibbs energy of the reaction.

The correlation for the data in Fig. 25.4 can be understood from this proportionality. The pK_a of the acid R-COOH corresponds to the Gibbs energy change for the acid dissociation reaction,

$$\text{R-COOH} \rightarrow \text{R-COO}^- + \text{H}^+$$

hence the stability of the carboxylate $RCOO^-$, namely

$$\frac{\Delta_r G^\ominus}{RT} = -\ln K = +2.303 \cdot \text{p}K_a \tag{25.49}$$

Assuming that reactions of a similar type proceed by the same mechanism and that the change in the stability of the ester is proportional to that of the acid, we can write that

$$\ln\left(\frac{k_{r2}}{k_{r1}}\right) = -2.303\alpha \cdot \left(\text{p}K_{a2} - \text{p}K_{a1}\right)$$

or, generalizing for arbitrary k_r,

$$\ln\left(\frac{k_r}{k_{r1}}\right) = -2.303\alpha \cdot \left(\text{p}K_a - \text{p}K_{a1}\right) = -2.303\alpha \cdot \text{p}K_a + 2.303\alpha \cdot \text{p}K_{a1}$$

which is an equation for a straight line with slope $-2.303 \cdot \alpha$.

Linear Gibbs energy relationships are used quite extensively in physical organic chemistry. It is found that these correlations hold for a wide range of reactions and can be used to rationalize substituent effects on reaction rates in a semiquantitative manner.[4]

25.2.4.3 Limits of Transition State Theory

The transition state theory describes chemical reaction rates in terms of structural (static) details of the reactants and the activated complex. Knowledge of the activated complex's structure remains a very challenging task, however. By knowing the molecular properties (hence partition functions) of the stable reactants and the activated complex, we can compute the temperature-dependent rate constant for a reaction. Hence we need not know detailed features of the multidimensional potential energy surface; rather, we need only know the potential energy surface in the region around the transition state and in the region of the reactant well. This great simplification arises from the assumption of a Maxwell–Boltzmann distribution for the occupation of the system's quantum states and the assumption that the reaction proceeds through the lowest energy region of the potential energy surface (see Fig. 25.1).

In many reactions the transition state theory model does not accurately predict the reaction rate. Such deviations arise from limitations, or breakdowns, in the assumptions used in deriving transition state theory. For example, we assumed that once the system moved beyond the transition state to the product side of the barrier that it could not return to the reactant side of the barrier. If this assumption is not correct, then effects occur which are called "recrossings," and they can be important for reducing the rate constant below that predicted theoretically; particularly for condensed phase reactions in which collisions with solvent molecules take place. Another feature of the transition state theory is that it requires the reaction to proceed by going "over" the barrier; however, reactions with light atoms (e.g., H) can display significant quantum mechanical tunneling effects.

The difference between the transition state theory result and the observed reaction rate k_{observed} is accounted for by defining a transmission coefficient κ, where

$$k_{\text{exp}} = \kappa \cdot k_r \tag{25.50}$$

Usually κ is on the order of one. However, in some cases κ becomes significantly different (larger or smaller than one). Table 25.2 gives some examples of transmission coefficients for κ. Reactions 3 and 4 in the table show large deviations from $\kappa = 1$, because in these cases tunneling of the proton occurs which is not accounted for in our derivation of transition state theory, see Section 25.4.3.

---

4 For a detailed discussion of these issues, see the text Grunwald and Leffler [4].

Table 25.2 Transmission Coefficients for Some Thermal Gas-Phase Reactions.[5]

	Reaction	T/K	k_r/L mol^{-1} s^{-1} exp.	calc.	κ
1	$D + H_2 \rightarrow DH + H$	298	1.7×10^5	2.5×10^5	0.7
2	$CH_4 + Cl \rightarrow HCl + CH_3$	400	2.5×10^7	5.7×10^7	0.44
3	$OH + H_2 \rightarrow H_2O + H$	298	3.7×10^6	6.9×10^5	5.3
4	$CF_3 + CH_4 \rightarrow CF_3H + CH_3$	400	890	90	9.9
5	$Cl^- + CH_3Br \rightarrow ClCH_3 + Br^-$	300	1.4×10^{10}	1.4×10^{10}	1.0
6	$CH_4 + OH \rightarrow CH_3 + H_2O$	300	4.1×10^6	3.8×10^6	1.1

25.3 Transition State in a Dynamical View

Reactions. are intrinsically time dependent and dynamic; in general, we should expect to solve the quantum mechanical equations of motion (time-dependent Schrödinger equation) to provide the most accurate picture of a reactive collision. More qualitatively we expect that the probability for a reaction to occur in a particular collision will depend on the amount of energy with which the particles collide (i.e., the collision energy or relative translational kinetic energy), the geometry of the collision and reactant angular momenta, and the population of the internal energy states of the reactants.

25.3.1 State-to-State Reaction Rates

To improve our description, we consider the quantum states of the reactants and products in a reaction and the details of how the reactants interact with each other, e.g., their relative velocity or whether a collision is "head-on" or glancing. To this end, we will generalize our concept of a reaction rate constant k_r.

As an example, let us consider the reaction scheme

$$A + BC\ (n, J) \rightarrow AB\ (n', J') + C$$

and define a rate constant for reactions involving particular quantum states $k_r(E, n, J, n', J')$, where E specifies the collision energy, n and n' are vibrational quantum numbers, and J and J' are rotational quantum numbers of the reactants and products, respectively. In principle, we make a specification for each reactant and product. The development of this view stands as a major accomplishment in physical chemistry in the latter half of the twentieth century.

The development of molecular (and atomic) beams and laser spectroscopy methods, among others, has enabled the study of chemical reactions as a function of collision energy, angle, and quantum state. Such studies have provided new and detailed insights into the nature of chemical reactions. For example, Fig. 25.6 shows two sources, which generate atomic (or molecular) beams of the reactants. The two beams of reactants are aimed so that they collide with each other inside of a vacuum chamber, and the products can be detected as a function of angle. One can also prepare the reactants in particular excited initial states (e.g., laser excitation of the reactant beams) and detect particular quantum states (typically through spectroscopy) of the products. In more sophisticated apparatuses the reactants undergo a supersonic expansion (see Section 15.6.2.4), so that the translational

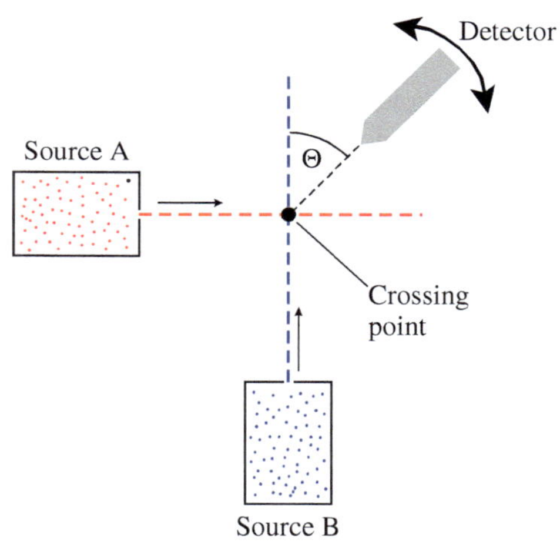

Figure 25.6 Schematic diagram of a crossed molecular beam apparatus. Molecules in heated containers A and B escape through a small hole, forming a molecular beam. Both beams collide at the crossing point. The detector may be positioned in various orientations; i.e., it is rotatable.

5 Reaction 1: See Example 25.1.
Reaction 2: Corchado et al. [5]
Reaction 3: Isaacson and Truhlar [6].

velocity distribution becomes very narrow but of high velocity, and the other energy states (rotational, vibrational, and electronic energy levels) have population distributions that correspond to those of very low temperature, typically just a few Kelvin. In summary, it is possible to measure the state-to-state reaction rate constant, and compare it with theoretical calculations[6].

By way of example, Fig. 25.7 shows molecular beam data for the reaction

$$H + D_2\ (n = 0,\ \ J = 0\ ) \rightarrow HD\ (n' = 0,\ \ J') + D$$

for a collision energy of $E = 1.29$ eV [8]. In this experiment the collision energy is high enough that the HD product can be excited both rotationally and vibrationally. The data in Fig. 25.7 show the translational energy of the D-atom measured at an angle of $\Theta = 11°$. This translational energy spectrum shows discrete peaks whose spacing corresponds to the energy spacing between vibrational n' and rotational J' states of the HD product (the assignments are shown by the bar graph); the HD distribution can be measured in a separate experiment. Through experiments of this sort, one can map out the entire state-to-state reaction probability for a reaction.

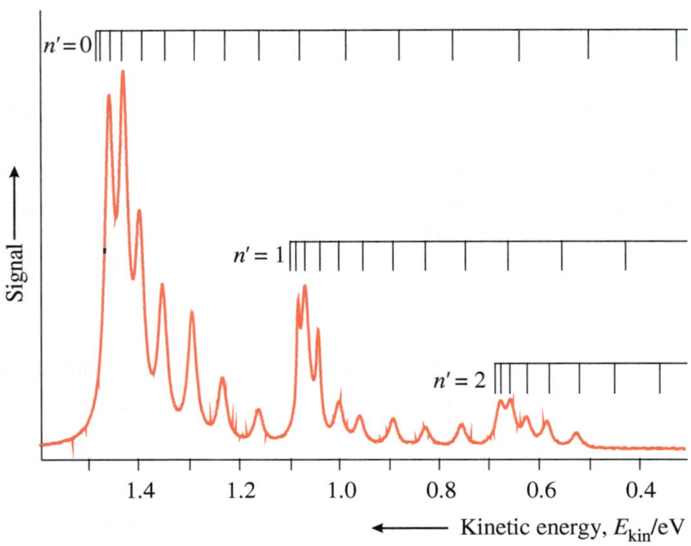

Figure 25.7 A source of H-atoms (source A in Fig. 25.6) crosses with a source of D_2-molecules (source B in Fig. 25.6) at 90°. The product D-atoms are detected at an angle θ from the direction of the H-atom source (see Fig. 25.6) and their kinetic energy E_{kin} is resolved. The energy distribution of the D-atoms that are detected at an angle of $\Phi = 11°$ for a collision energy of 1.29 eV is shown (signal as a function of E_{kin}). The bar graphs show the assignments of the peaks which correspond to the product state distribution of HD (rotational energies for vibrational quantum numbers $n' = 0$, 1, and 2). These data are taken from [8] and are adapted from work of others.

In principle, the state-to-state reaction rate constants can be directly determined by quantum mechanical calculation of the nuclear motion on the potential energy surface(s) of the reactants; referred to as trajectory calculations. This task is very challenging, because it requires the calculation of the potential energy surface(s) for the reactants (e.g., for the simple case of A + BC this surface has at least three dimensions, two bond lengths and one angle) and the calculation of the time-dependent Schrödinger equation for the nuclear motion on this potential energy surface. This challenge has been met for the $H + H_2$ reaction (and its isotopomers).

Even though quantitatively accurate potential surfaces and quantum calculations may not be accessible, important insights into dynamical features of reactions can be obtained by classical (or quasi-classical) trajectory calculations on qualitatively accurate potential energy surfaces. To illustrate, Fig. 25.8 shows four different trajectories for the H atom exchange reaction

$$H + H_2 \rightarrow H_2\ + H$$

In this case the H_3 bond angle is taken to be 180°; i.e., we have a collinear geometry. Panel (a) shows a reactive trajectory, in which the H atom strikes an H_2 molecule (in its ground vibrational state) with a relative speed of 7.5 km s^{-1}. The collision energy is large enough that reaction occurs (the potential energy barrier is surmounted). The product is formed in its ground vibrational state as well. The oscillations in the trajectory result from the zero point vibrational motion of H_2. Panel (b) shows a nonreactive collision; however, energy is transferred in this collision. Initially the H_2 molecule is vibrationally excited but after the collision it is in the ground vibrational state and the particles are moving apart with higher kinetic energy. This energy exchange is apparent from the decreased amplitude of the vibrational oscillations after collision. Panel

6 For more detailed discussion of these issues see R.B. Bernstein [7].

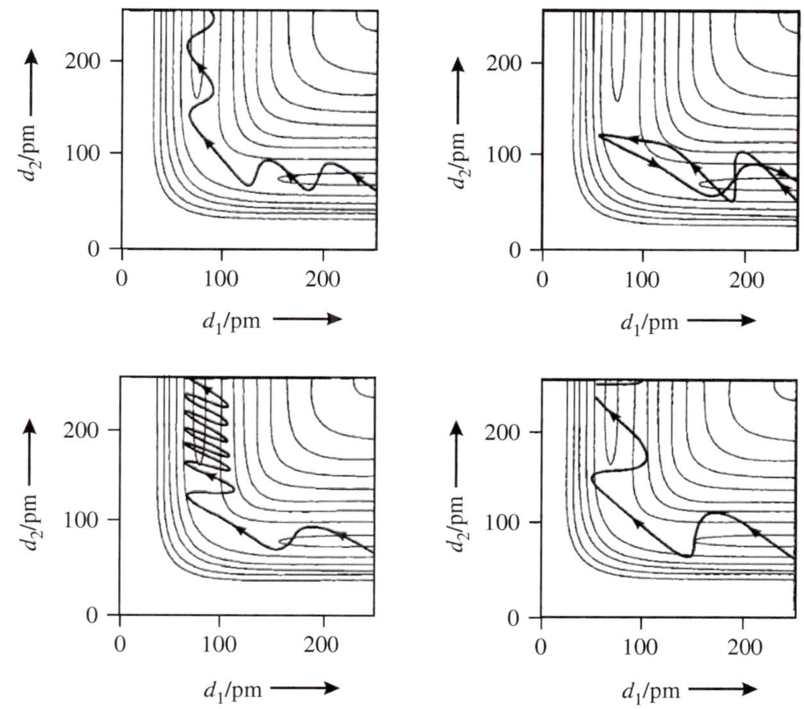

Figure 25.8 Trajectories for the H atom exchange reaction $H + H_2 \rightarrow H_2 + H$. d_1 is the distance between the reactants, and d_2 is the distance between the products of the reaction. (a) Reactive trajectory with no vibrational excitation ($\upsilon = 0$), (b) nonreactive trajectory with a single quantum of vibrational excitation in the entrance channel ($\upsilon = 1$), (c) reactive trajectory that results in vibrational excitation of the product ($\upsilon = 1$), (d) reactive trajectory in which the vibrational excitation is the same for the reactant and product states ($\upsilon = 1$). (From [9]. With permission of Harper Collins.)

(c) shows a reactive encounter in which the H_2 molecule is initially in its ground vibrational state, but the collision energy has been increased enough that the product is formed in a vibrationally excited state. Note that the particles are now moving apart more slowly than they came together. Panel (d) shows a trajectory which has high internal energy (vibrationally hot) and large translational energy. Given that we have restricted these collisions to be collinear, imagine the added richness which occurs by considering noncollinear collisions and allowing for the excitation of rotational energy levels of the diatom, as well as vibrational energy levels.

25.3.2 Transition State Spectroscopy

Although the idea of a transition state for describing reactions is very powerful, one might ask: "Is the transition state more than a theoretical construct?" Beginning in the 1980s this question was answered by directly probing the time evolution of molecules moving through transition states.

As an example, consider the photodissociation of the molecule ICN. The absorption of a photon from a femtosecond laser pulse can excite the molecule from its stable electronic ground state to an excited electronic state that is directly dissociative; i.e., the iodine to carbon bond is broken in the excited state. Hence the reactant molecules ICN* are prepared in a condition which corresponds to the transition state.

$$\text{I-CN} + h\nu \rightarrow (\text{I-CN})^* \rightarrow (\text{I} \cdots \text{CN})^* \rightarrow \text{I} + \text{CN} + E_{\text{trans}}$$

The progress of the molecule's fragmentation into the products I and CN is followed by probing the absorption of the CN radical as a function of time. By probing at different wavelengths one sees different features of the dynamics (Fig. 25.9). The 389.7 nm probing wavelength (top) monitors the formation of the final CN product state; hence, the absorbance increases with time to a constant value. The risetime of the signal corresponds to the reaction rate. At longer wavelengths, such as 389.8 and 390.4 nm (center), one sees a short transient and then a decay to a plateau value. In this case the signal has two components. It contains the formation of CN radical (with a weaker absorbance) as before and an absorption from CN moieties in proximity to the I atom which causes a red shift in the absorbance. This latter component decays away as the molecule falls apart; hence, it is short lived (about 200 fs). The bottom transient shows the most extreme case (391.4 nm) in which the product absorption is now absent, and one only sees the absorption of the transition state (I···CN).

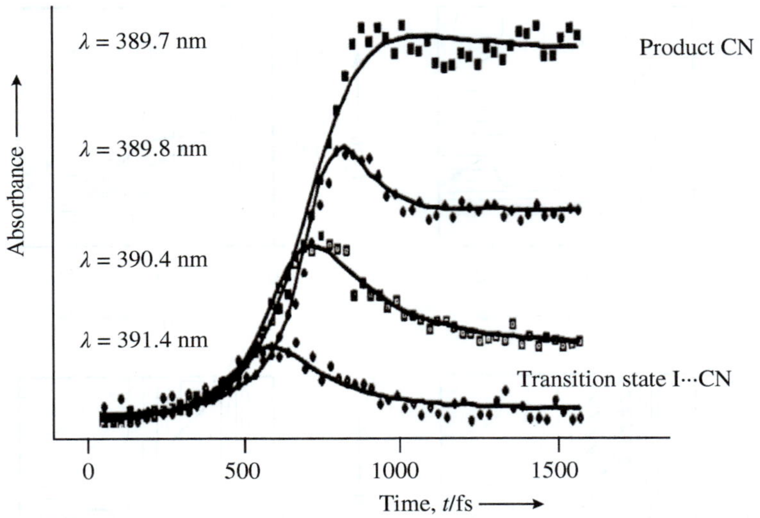

Figure 25.9 Femtosecond transition state spectra for the photodissociation reaction.

$$\mathrm{I-CN} \rightarrow (\mathrm{I-CN})^* \rightarrow (\mathrm{I}\cdots\mathrm{CN})^* \rightarrow \mathrm{I} + \mathrm{CN}$$

The reaction is followed at $\lambda = 389.7$ nm (top, kinetics of product CN), $\lambda = 389.8$ nm (center), and $\lambda = 390.4$ nm (bottom, kinetics of transition state I$\cdots$CN). (Adapted from [10], with permission.)

25.3.3 Reaction Constant $k_r(E)$ from Reaction Cross-Sections

Given that we can measure or calculate the rate constant between each of the initial states and final states for a reaction, how can we calculate the overall thermal rate constant? In Section 25.3.1 we discussed rate constants $k_r(E, n, J, n', J')$ depending on the collision energy E and the vibrational and rotational quantum numbers of the reactants and products. From an appropriate set of these rate constants we can calculate a mean value $k_r(E)$, depending only on the collision energy E, by averaging over all quantum states (see Foundation 25.1, on Wiley's website).

In molecular beam scattering experiments, data are often reported in terms of cross-sections or differential cross-sections, rather than rate constants. The rate constant $k_r(E)$ is defined for a bimolecular reaction as

$$k_r(E) = \mathbf{N}_\mathrm{A} \cdot v_{rel} \cdot \sigma_r(E) \tag{25.51}$$

where

$$v_\mathrm{rel}(E) = \sqrt{\frac{2E}{\mu}} \tag{25.52}$$

is the relative velocity of the reactants (μ = reduced mass) and $\sigma_r(E)$ is the reactive cross section. Note that the activation term $\exp(-E_{a,m}/\boldsymbol{R}T)$ and the steric factor α are not present here, because we are considering the reaction to occur at a fixed collision energy, rather than a fixed temperature, and we will consider collisions occurring with different geometries (head on, glancing, etc.).

In the following, we develop a model for $\sigma_r(E)$, where we account for the change in geometry of a collision; for example, a head-on collision should have a higher reaction probability than a glancing collision. To begin, we define the impact parameter b. If we have two hard spheres (see inset of Fig. 25.10) with a closest distance of approach given by $d = r_A + r_B$, then the spheres will not collide for $b > d$ and they will collide for impact parameters b less than d; i.e., $0 \le b \le d$. When b is zero, a head-on collision occurs and the reaction cross-section will take its highest value. The reaction cross-section will decrease as the impact parameter b increases, eventually becoming zero for $b > d$.

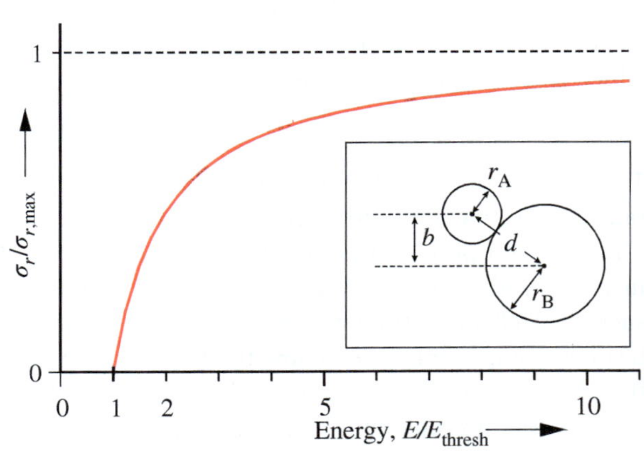

Figure 25.10 Two different approximations for the energy-dependent reactive cross-section σ_r: a step function (dashed line) and the "line-of-centers" cross-section (solid line). The inset defines the impact parameter b.

In this model we require that the collision energy exceeds a certain threshold E_thresh along the direction connecting the two centers of mass of the bodies. Thus for a central (head-on) collision with $E = E_\mathrm{thresh}$ a reaction will occur, but no reaction will occur at this energy for a glancing collision. Rather, E must be larger than E_thresh for a reaction to occur with other geometries.

Because there is only a small probability for a central collision, the reaction cross section is very small for $E = E_{\text{thresh}}$ and it increases with increasing E. In Foundation 25.1 we show that the quantitative result for $\sigma_r(E)$ is

$$\sigma_r(E) = 0 \qquad \text{for } E \le E_{\text{thresh}}$$
$$\sigma_r(E) = \sigma_{r,\max} \cdot \left(1 - \frac{E_{\text{thresh}}}{E}\right) \qquad \text{for } E > E_{\text{thresh}} \tag{25.53}$$

This equation is illustrated in Fig. 25.10 (solid curve).

Example 25.5 *The reaction H +D_2 → HD + D has a reaction cross-section that depends on the collision energy. Workers*[7] *have measured the rate constant at different collision energies E to determine the total reaction cross section σ_r (with an uncertainty of about 10%) as a function of the energy:*

E/eV	0.87	0.9	1.57	1.6	1.97	2.0	2.7
$\sigma_r/10^4$ pm^2	0.88	0.97	1.28	1.23	1.27	1.28	1.25

and other workers [12] have computed the reaction cross-sections theoretically, using three-dimensional quantum mechanical scattering calculations, to find

E/eV	0.45	0.48	0.67	0.85	1.29	2.0
$\sigma_r/10^4$ pm^2	0.075	0.16	0.52	0.81	1.20	1.35

The estimated error in the scattering calculations is $\pm 10^3$ pm^2. Compare these data with each other. Compare both data sets to Equation (25.53) and obtain the parameters $\sigma_{r,\max}$ and E_{thresh}.

From Equation (25.54), we see that

$$\sigma_r E = \sigma_{r,\max} \cdot E - \sigma_{r,\max} \cdot E_{\text{thresh}}$$

Thus a plot of $\sigma_r \cdot E$ versus E should be linear, with a slope of $\sigma_{r,\max}$ and an intercept of $E_{\text{thresh}} \cdot \sigma_{r,\max}$. Figure 25.11 shows this plot for the experimental and computational data reported in the tables. The best fit parameter values give $\sigma_{r,\max} = 1.57 \times 10^4$ pm^2 and an $E_{\text{thresh}} = 0.38$ eV.

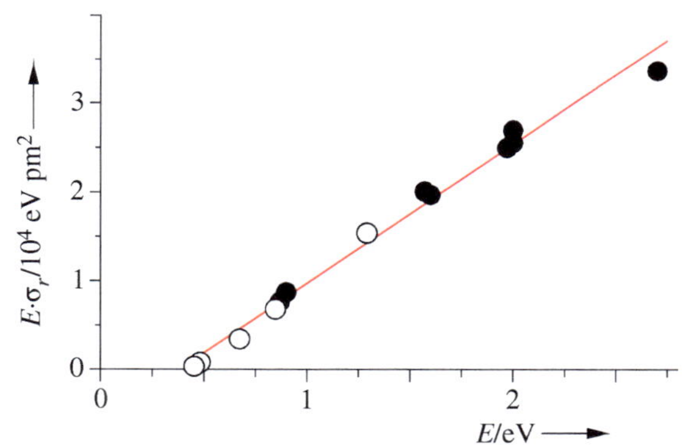

Figure 25.11 Plot of $E \cdot \sigma_r$ as a function of the collision energy E. Evaluation of $\sigma_{r,\max}$ and E_{thresh} from the slope and intercept of the regression line. Filled circles: experimental, open circles: theoretical, from scattering calculations.

Using Equation (25.51) with Equations (25.52) and (25.53) we can write an expression for the rate constant $k_r(E)$ of the H + D_2 → HD + D reaction, and we obtain

$$k_r(E) = \mathbf{N}_A \cdot \sqrt{\frac{2E}{\mu}} \cdot \sigma_{r,\max} \cdot \left(1 - \frac{E_{\text{thresh}}}{E}\right) \tag{25.54}$$

with $\mu = m_H m_{D_2}/(m_H + m_{D_2}) = 1.34 \times 10^{-27}$ kg, $E_{\text{thresh}} = 0.38$ eV, and $\sigma_{r,\max} = 1.57 \times 10^4$ pm^2. This calculation gives the following result for $k_r(E)$ values

E/eV	0.38	0.39	0.40	0.50	1.0	1.5	2.0
$k_r(E)/10^4$ L mol^{-1} s^{-1}	0.000	0.023	0.046	0.248	0.906	1.336	1.674

---

7 Taken from [15].

25.3.4 Relation Between $k_r(E)$ and the Rate Constant $k_r(T)$

To relate $k_r(E)$ with the thermal rate constant $k_r(T)$, we explicitly perform the average over the thermal distribution. Using Equation (25.54) for $k_r(E)$ we find

$$k_r(T) = \frac{\mathbf{N}_A}{N_{total}} \int_{N=0}^{N_{total}} \sqrt{\frac{2E}{\mu}} \cdot \sigma_{r,max} \left(1 - \frac{E_{thresh}}{E}\right) \mathrm{d}N \tag{25.55}$$

where N_{total} is the total number of reacting molecules. We use the Maxwell–Boltzmann distribution from Section 15.6.2

$$\mathrm{d}N = N_{total} \sqrt{\frac{4}{\pi(kT)^3}} \sqrt{E} \cdot \exp\left(-\frac{E}{kT}\right) \cdot \mathrm{d}E \tag{25.56}$$

and obtain

$$k_r(T) = \mathbf{N}_A \sqrt{\frac{8}{\pi\mu(kT)^3}} \int_0^{\infty} E \cdot \sigma_{r,max} \left(1 - \frac{E_{thresh}}{E}\right) \cdot \exp\left(-\frac{E}{kT}\right) \cdot \ \mathrm{d}E \tag{25.57}$$

Equation (25.57) relates the thermal rate constant to the microscopic energy-dependent reactive cross-section. This equation may be integrated to give

$$k_r(T) = \mathbf{N}_A \sqrt{\frac{8kT}{\pi\mu}} \sigma_{r,max} \exp\left(-\frac{E_{thresh}}{kT}\right) \tag{25.58}$$

which matches the collision state rate constant (see Chapter 24, Equations (24.19) and (24.20))

$$k_r(T) = \alpha \mathbf{N}_A \sqrt{\frac{8kT}{\pi\mu}} \left[\pi\left(r_A + r_B\right)^2\right] \exp\left(-\frac{E_{a,m}}{\mathbf{R}T}\right) \tag{25.59}$$

with $E_{thresh} = E_a$ and $\sigma_{r,max} = \alpha\pi\left(r_A + r_B\right)^2$. The experimental rate constant [13] for the $H + D_2 \rightarrow HD + D$ reaction at 300 K is $k_r = 1.56 \times 10^4$ L mol^{-1} s^{-1}, and the activation barrier is $E_a = 0.40$ eV , in good agreement with the threshold energy of $E_{thresh} = 0.38$ eV found in Example 25.5. If we model the H atom and the H_2 molecule as hard spheres we predict a collision cross section of $\alpha\pi\left(r_A + r_B\right)^2 = \alpha\pi(126 + 136)^2$ pm^2 $= \alpha \cdot 2.16 \times 10^5$ pm^2, and we can find a value for the steric factor of $\alpha = 1.57 \times 10^4 / 2.16 \times 10^5 = 0.07$ whereas the experimental value is $\alpha = 0.3$.

Note that the observed thermal rate constant k_r at 300 K is many orders of magnitude smaller than the $k_r(E)$ values calculated in Example 25.5, because only a very few reactant molecules have collision energies as high as assumed in the example [14].

25.4 Transition State Theory and Reactions in Solution

Important differences exist between chemical transformations that occur in the gas phase and those that occur in fluid solutions. Two of the primary effects are that the electronic states of the molecules; i.e., potential energy surfaces, are modified by solvation and the reactants in solution experience collisions with solvent molecules during reaction. Solvation refers to the stabilization in energy of a molecule that arises from intermolecular forces; such as dispersion forces, dipole–dipole forces, ion–dipole forces, and others. The extent of this interaction depends on the nature of both the solvent and the solute. Furthermore the magnitude of the solvation energy may be quite different for the reactants than it is for the transition state, thereby causing significant changes in the activation barrier for the reaction. In a liquid solution the number and strength of collisions is quite high. In fact, the solute is in such intimate contact with the solvent that it can become difficult to define what is meant by a collision. It is however clear that the magnitude of the collision frequency is on the order of 10^{13} s^{-1} or even higher. The large number and strength of collisions means that energy is exchanged with the solvent on short time scales. Hence the approximations concerning thermal equilibration, which lie in the transition state theory model, are likely good, but friction and dissipation, which are not included in the transition state theory, may become important.

25.4.1 Unimolecular Reactions and Frictional Coupling

According to Equations (25.31) and (25.32) for unimolecular reactions, we can write the rate constant k_r as

$$k_r = \frac{kT}{h} \cdot \exp\left(-\Delta_r S^{\ddagger\ominus}/\mathbf{R}\right) \cdot \exp\left(-\Delta_r H^{\ddagger\ominus}/\mathbf{R}T\right) \tag{25.60}$$

The symbols in this equation need to be interpreted in a manner somewhat different from that for gas-phase reactions, however. The activation enthalpy represents not an intrinsic barrier of the molecule, but a barrier energy for a solvated molecule. If the solvation for the transition state is different from that for the reactant state, this barrier could be quite different in solution than it is in the gas phase.

25.4.1.1 Solvation and Reaction Rates

As a molecule undergoes reaction its charge distribution is likely to change. If the solvent molecules are able to respond to the changing charge distribution in order to maintain its optimal solvation, then we can model the effect of solvation by plotting a "solvated" reaction coordinate. If the solvent molecules respond slowly to the reactant's changing charge distribution, then we cannot use such a simple treatment and must account for the nonequilibrium between the reactant and the solvent polarization field.

Figure 25.12 sketches potential energy surfaces for the low lying singlet states of stilbene (1,2-diphenylethene), as a function of the torsional angle θ about the ethylenic moiety. This molecule possesses a reaction barrier of 14.6 kJ mol^{-1} for the *trans* to *cis* isomerization in the first excited electronic state in the gas phase. In alkane solvents this molecule has the same barrier as the gas-phase case – solvation appears to not affect the barrier height so much. In alkane nitrile solvents however, the activation barrier is lowered to a value near 10.5 kJ mol^{-1} by solvation. In *n*-alkyl alcohol solvents, this barrier is lowered even further (to a value near 4 kJ mol^{-1} or even less) [16].

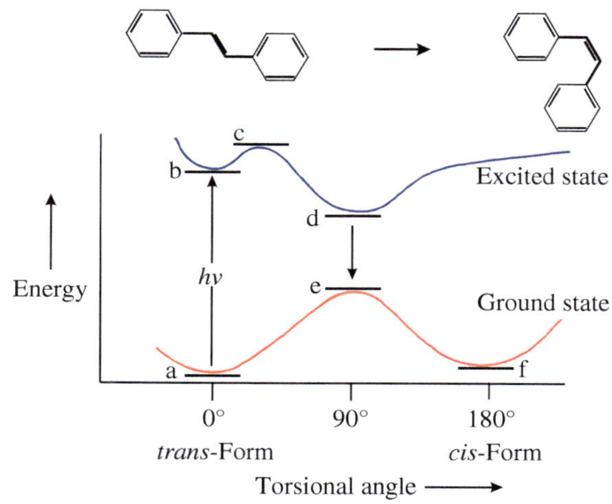

Figure 25.12 Schematic of the potential energy surface for the *trans* to *cis* isomerization of stilbene. Lower curve: ground singlet state S_0, upper curve: excited singlet state S_1. This shape of the reaction coordinate is appropriate for stilbene in the gas phase and in nonpolar solvents.

25.4.1.2 Friction and Reaction Rates

Studies have shown that after excitation by light to the lowest lying excited state (state b in Fig. 25.12), trans-stilbene may either relax back to the ground state *trans* form or it may "twist" along the excited state potential energy surface. At state c the transition state is reached at a twisting angle of about 30°, and then a 90° twisted geometry forms (state d in Fig. 25.12) which relaxes very rapidly (less than 200 fs) to the ground-state potential energy surface where it ends up at the top of the ground state reaction barrier (state e). Subsequently the molecule relaxes toward the *cis* form (state f) or the *trans* form (state a).

The rate-limiting step in the *trans* to *cis* isomerization is the initial barrier crossing from the excited *trans* form in state b to point c. This step can be followed experimentally by measuring the lifetime τ of the excited state b. According to Section 10.2.6, the rate constant for the isomerization reaction can be obtained by

$$k_{\text{iso}} = \frac{1}{\tau} - \frac{1}{\tau_0}$$

where τ_0 (see Example 10.2) is the pure radiative lifetime of the excited stilbene molecule (fluorescence lifetime when the isomerization reaction is prohibited) and $\tau < \tau_0$ is the total lifetime. The large amplitude motion during the isomerization reaction requires some displacement of solvent molecules in the solution and leads to a dependence of the rate constant on the solute–solvent frictional coupling (i.e., in a simplified picture the phenyl rings are like paddlewheels which displace solvent upon proceeding from a trans to cis configuration).

Fleming and coworkers[8] reported data on the lifetime τ in a series of alkane solvents with different viscosities η. These data are shown in Table 25.3 together with the calculated rate constants k_{iso}. It can be seen that k_{iso} decreases through the series of alkanes and correlates with the solvent viscosity η. By using a homologous series of solvent (see Table 25.3), it was possible to keep the solvation energetics of the stilbene relatively constant and extract an activation barrier of $E_{a,m}$ of 14.6 ± 4 kJ mol^{-1} for process b→c. This activation barrier is in good experimental agreement with that determined for the isomerization of stilbene in the gas phase.

One can describe their data by the empirical equation

$$k_{\text{iso}} = F(\zeta) \cdot \exp\left(-E_{a,m}/(\boldsymbol{R}T)\right) \tag{25.61}$$

8 [17] The data for τ are taken from Table 1, column 5 (τ_F values).

where $F(\zeta)$ represents a pre-exponential factor that changes with the solute–solvent friction coefficient, ζ, which is defined in Equation (14.84):

$$f = \zeta \cdot u$$

where f is the force needed to move particles through a viscous medium at a speed u. In the special case when the particles are spheres of radius r and the medium is considered as continuous, the force is (Stokes law)

$$f = 6\pi\eta r \cdot u \qquad \text{or} \qquad \zeta = 6\pi\eta r$$

where η is the viscosity. In Chapter 14, we discussed the Stokes–Einstein equation Equation (14.85),

$$D = \frac{kT}{6\pi\eta r}$$

where D is the diffusion coefficient and the solvent is treated as a continuous medium with a viscosity η.

In their analysis, Fleming and coworkers used two ways to model the friction coefficient ζ : i) the continuum model with $\zeta = 6\pi\eta r$ and ii) experimental measurement of the rotational diffusion constant D_{rot} of trans-stilbene and using the more general Einstein relation

$$D_{\text{rot}} = \frac{kT}{\zeta}$$

Table 25.3 shows the isomerization rate constant k_{iso} (column 6) and the preexponential factor F (column 7) for different values of the different solvents.

Table 25.3 Isomerization Rate and Diffusion Data for Stilbene in Alkane Solvents at Room Temperature.[9]

Solvent	T (K)	η/cP	$D_{\text{rot}}/10^9$ s^{-1}	τ/ps	$k_{\text{iso}}/10^{10}$ s^{-1}	$F(\zeta)/10^{12}$ s^{-1}
n-Pentane	302	0.22	14	61	1.59	5.33
n-Hexane	304	0.29	11	64	1.51	4.88
n-Octane	304	0.49	6.7	74	1.30	4.20
n-Decane	304	0.79	4.8	83	1.15	3.73
n-Dodecane	303	1.27	3.5	98	0.97	3.19
n-Tetradecane	305	1.84	2.1	104	0.91	2.89
n-Hexadecane	303	2.77	1.8	119	0.79	2.60

Source: Data from[9]

The transition state theory, presented in Section 25.2, does not account for the dependence of the unimolecular reaction rate on viscosity. In 1940 Kramers [18] proposed a model for unimolecular chemical reactions that accounts for solute–solvent frictional coupling. In this treatment, he approximates the reactant molecule by an effective particle that is diffusing along the reaction coordinate and reaction occurs when it crosses through the transition state. Thus, the reactant is performing a "random walk" along the reaction coordinate but rather than the probability of moving forward or backward being equally likely (see Chapter 14), the statistical weight for the direction is determined by the potential energy along the reaction coordinate. Kramers showed that the rate constant was similar to the Arrhenius form, but has a preexponential factor $F(\zeta)$ that depends on the solute–solvent frictional coupling as well as parameters of the potential energy surface; namely

$$F(\zeta) = \frac{\sqrt{k_0} \cdot \zeta}{4\pi\mu\sqrt{k_C}} \left(\sqrt{1 - \frac{4\mu \cdot k_C}{\zeta^2}} - 1 \right) \tag{25.62}$$

where μ is the reduced mass for the twisting motion of the molecule in the potential. k_0 is the curvature of the potential well near its minimum point (see Fig. 25.12, point b), and $-k_C$ is the curvature of the potential barrier at the transition state (see Fig. 25.12, point c):[10]

$$k_0 = \frac{d^2V}{dx^2} \text{ (potential well)} \qquad k_C = -\frac{d^2V}{dx^2} \text{ (potential barrier)}$$

where x denotes the reaction coordinate. For details see Foundation 25.2.

9 [17] The data for τ are taken from Table 1, column 5 (τ_F values).
10 Because the potential barrier is approximated by an inverted parabola at the barrier top, d^2V/dx^2 is negative, but k_C is positive.

In contrast to our treatment of the motion through the transition state for a gas-phase reaction (Section 25.1), in a liquid collisions are occurring between the solute and solvent molecules. The likelihood of a collision occurring during a reactant molecule's motion through the transition state depends on the magnitude of the frictional coupling and the time the molecule spends in the transition state region. When the reaction barrier is sharp (i.e., k_C is large) then there is less chance of solute–solvent collisions. In Foundation 25.2, we show that Kramers equation reduces to the transition state theory result in this limit. In contrast, a broad reaction barrier (i.e., k_C is small) leads to a larger probability of collisions occurring during the reactant molecule's motion through the transition state region. In this limit (i.e., when the friction coefficient is large), we show that Kramers expression simplifies to

$$F(\zeta) = \frac{\sqrt{k_0 \cdot k_C}}{2\pi\zeta} \qquad \text{high friction limit} \tag{25.63}$$

which is called the *Smoluchowski limit* of the Kramers equation.

Here we use two different methods to model the solution solvent frictional coupling for the stilbene rate data in Table 25.3. In the first case, we use Stokes law, $\zeta = 6\pi\eta r$, and find

$$F(\eta) = C_1 \cdot \eta \left[\sqrt{1 + C_2/\eta^2} - 1\right] \tag{25.64}$$

which becomes $F(\eta) = const./\eta$ at high friction. Panel a) of Fig. 25.13 compares the rate data in Table 25.3 with Equation (25.64) for the reaction's dependence on the frictional coupling. In the second case, we assume that the rotational diffusion coefficient is inversely proportional to the friction coefficient,

$$D_{\text{rot}} = \frac{kT}{\zeta}$$

(the Einstein relation for rotational diffusion), and we find that

$$F(D_{\text{rot}}) = \frac{C_1'}{D_{\text{rot}}} \left[\sqrt{1 + C_2' \cdot D_{\text{rot}}^2} - 1\right] \tag{25.65}$$

which becomes $F(D_{\text{rot}}) = const. \cdot D_{\text{rot}}$ at high friction.[11] Panel b) in Fig. 25.13 compares the rate data in Table 25.3 with Equation (25.65) for the reaction's dependence on the frictional coupling. These comparisons show that Equation (25.64), which uses Stokes law to model the solute-solvent frictional coupling (panel a), gives a somewhat worse agreement with the experimental data than does Equation (25.65), which uses the experimentally determined D_{rot} to model the friction coefficient in the Kramers expression.

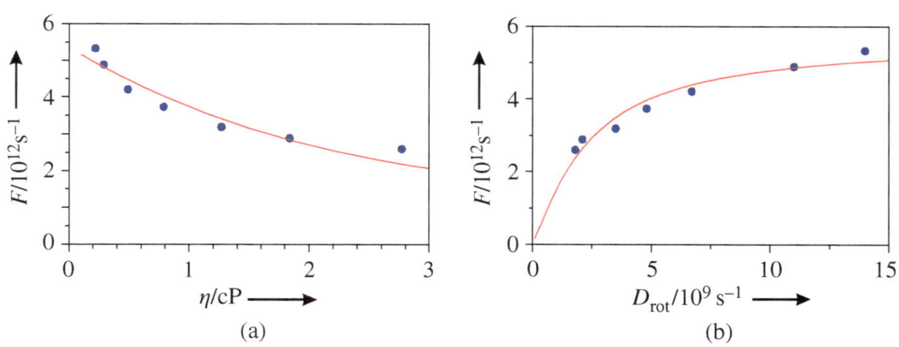

Figure 25.13 Isomerization reaction of *trans*-stilbene in alkanes. a) Preexponential factor F as a function of the viscosity η. b) Preexponential factor F as a function of the rotational diffusion coefficient D_{rot}. Blue filled circles: experiment (see Table 25.3), red solid lines: regression curves, using Equation (25.64) with $C_1 = 1.02 \times 10^{13}\ \text{s}^{-1}$ and $C_2 = 3.1 \times 10^{17}\ \text{s}^2$ and Equation (25.65) with $C_1' = 10.25 \times 10^{21}\ \text{s}^{-1}$ and $C_2' = 0.3108 \times 10^{-18}\ \text{s}^2$. For a comparison of these fitting parameters with the features of the intermolecular potential, see [17].

11 The rotational diffusion coefficient is an approximation for the diffusion associated with the isomerization motion. The rotational diffusion constant offers the advantage of a simple substitution and is affected by solute–solvent coupling on timescales somewhat closer to that of the chemical reaction than would the translational diffusion coefficient. A few workers have used the translational diffusion coefficient of benzene in combination with a mechanical model for the benzene moieties movement around an axle (the ethylenic unit).

25.4.2 Dissociation Reactions

For bimolecular reactions, we must consider the transport of the reactants to contact (formation of an "encounter pair complex") and the unimolecular rate for that complex. In Chapter 24, we discussed the limit in which the unimolecular reaction step is fast compared to the encounter pair formation, then the reaction proceeds at a rate constant determined by the diffusion and is called a diffusion limited reaction. What about the opposite limit, in which the diffusion of the reactant species is fast compared to the unimolecular reaction step? To circumvent the limitations of diffusion and to focus our attention on the dynamics of a reactive encounter pair, we consider the creation of iodine atoms I by photodissociating a stable molecule, I_2, namely

$$I_2 + h\nu \rightarrow I + I$$

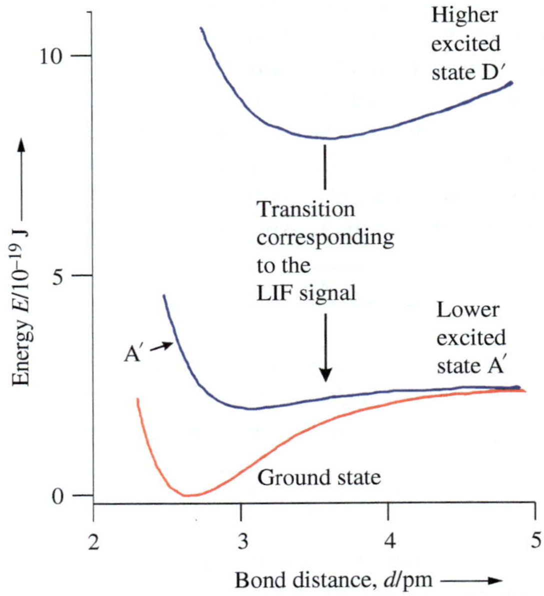

Figure 25.14 Potential curves for I_2. Ground state, lower excited state A′, and higher excited state D′. For the LIF (Laser-Induced Fluorescence) detection method, the molecule is excited to D′ and the fluorescence corresponding to the transition D′ → A′ is measured as a function of time.

Figure 25.14 shows the potential energy surfaces for the ground state and excited electronic states of the iodine diatom. By exciting the iodine diatom with light in the wavelength region of 590 to 710 nm, the I_2 can be excited directly to the A′ state above the dissociation limit, causing direct dissociation of I_2 molecules in the gas phase. In the condensed phase the atoms can collide with solvent molecules and are "caged" so that they recombine to form molecular iodine. To observe this caging phenomenon, Zewail and coworkers directly photodissociated I_2 by a 614 nm, 100 fs laser pulse (an energy above A′ state dissociation limit) and followed the excited state evolution by excitation to the higher D′ state with 307 nm light and monitoring its fluorescence (*Laser-Induced Fluorescence*, LIF).

Figure 25.15 (data from [19]) shows data from these studies for $I_2 \cdot Ar_{44}$ clusters at 30 K. The time dependence of the photodissociation is followed in time using the LIF, which has its maximum strength (Franck–Condon overlap, see Section 11.5) near the well region of the A′ state. The data show a prompt signal near time zero (arrow 1) which corresponds to the initially excited I_2 molecule after its creation on the repulsive wall of the A′ state. Interestingly the signal decreases to a minimum near 600 fs, suggesting that the I_2 molecule is dissociating and moving out of the well region, and then it recovers because the I atoms 'bounce off' the Ar atom cage and return to the well region of the A′ state. Because they retain their excess energy they once again climb the A′ state repulsive wall and then start to dissociate again (arrow 2). Because the collisions with the cage are not wholly elastic the I_2 molecules can lose their excess energy and become trapped in the A′ state well. The signal rises strongly at later times because the I_2 molecule is thermalizing in the well region of the A′ state, and these vibronic levels have stronger Franck–Condon overlap; hence, a higher signal is seen (arrows 3–5). On longer time timescales than shown in this figure (tens of picoseconds), the signal reaches a plateau; the vibrational relaxation is complete and the molecules have geminately recombined.

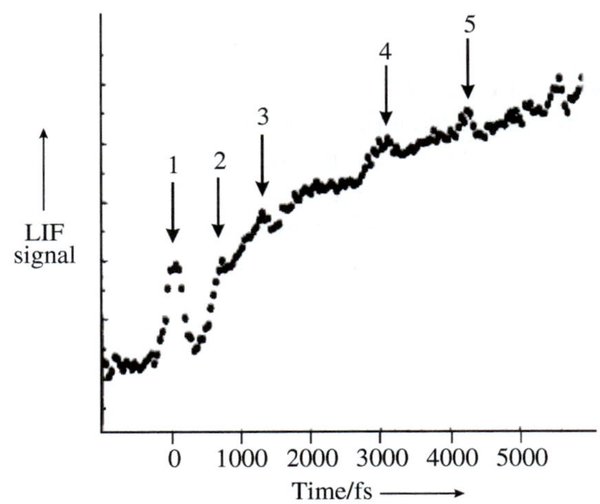

Figure 25.15 I_2 molecule. LIF signal during the transition from the higher excited state D′ to the lower excited state A′. Discontinuities in the signal are marked by the arrows. (From [18], with permission.) Note that 1 fs = 10^{-15} s = 10^{-3} ps.

Example 25.6 *Given that the excess energy (energy above the dissociation limit) for I_2 in Ar clusters is about $E = 3.8 \times 10^{-20}$ J when excited with 614 nm light, estimate the maximum displacement of the I_2 atoms from each other by assuming that their motion in the 660 fs recurrence is classical.*

After the dissociation of the I_2 molecule the motion of the two separated I atoms can be very different. Let us assume that both atoms move at the same speed, but in opposite directions. Then the excess energy is equally distributed to both atoms and their speeds are

$$v_1 = \sqrt{\frac{E}{m_I}} = \sqrt{\frac{3.8 \times 10^{-20}\ \mathrm{J}}{2.05 \cdot 10^{-25}\ \mathrm{kg}}} = 430\ \mathrm{m\ s^{-1}}$$

$$v_2 = -v_1 = -430\ \mathrm{m\ s^{-1}}$$

Thus the relative speed is $v_{\mathrm{relative}} = v_1 - v_2 = -860\ \mathrm{m\ s^{-1}}$.

Assuming that the 660 fs recurrence corresponds to the period for a full cycle (I atoms move apart, reflect from the solvent cage, and proceed back to the repulsive wall of the potential where they started), we find that the maximum displacement will occur at about 330 fs, so that

$$d_{\max} = d_{eq} + \left(330 \cdot 10^{-15}\ \mathrm{s}\right)(860\ \mathrm{m\ s^{-1}}) = 267\ \mathrm{pm} + 284\ \mathrm{pm} = 551\ \mathrm{pm}$$

Hence the atoms move by about an atomic radius before striking the Ar atom cage; a distance which is about twice as large as the displacement you would expect for a ground electronic state I_2 molecule that has this excess energy in vibrational excitation.

These experiments reveal important dynamical features of the I_2 dissociation and recombination at early times; however, the A state is not the ground electronic state of the molecule. In a different set of studies Paige and Harris [19] probed the photodissociation and geminate recombination to the ground electronic state of I_2 in liquid Xe. In this case, the I_2 was photodissociated by exciting with 590 nm light and the recovery of the ground electronic state was monitored by probing its absorption at six different probe wavelengths (500, 632, 710, 760, 860, and 950 nm). By using these different wavelengths, they were able to probe the time profile of the vibrationally excited ground electronic state I_2, which relaxes to a thermal equilibrium condition by transferring excess vibrational energy to translational motion of the Xe liquid. They found that the vibrational relaxation near the top of the X-state well is rapid (<150 ps) but that near the bottom of the well it is long (>3 ns). This disparity in the relaxation rate arises from changes in the vibrational energy level spacing for the I_2 diatomic.

25.4.3 Proton Transfer Reactions

Tunneling can play an important role in proton transfer reactions, and these reactions can be much faster than expected from transition state theory. For example, we consider the case of 1-naphthol. The pK_a of the hydroxyl proton of 1-naphthol changes from 9.4 in its ground electronic state to about −0.2 in its excited state [21]. If 1-naphthol is placed in a solvent such as ammonia, the following reaction occurs

$$\text{1-naphthol} + NH_3 + h\nu \rightarrow \text{1-naphthol*} + NH_3 \rightarrow \text{naphtholate} + NH_4^+$$

Kelley performed measurements on this photoinduced proton transfer in Ar matrices doped with 1-naphthol and ammonia and found that the rate constant ratio for different isotopes was $k_r(\mathrm{H})/k_r(\mathrm{D}) > 200$. Studies [21] of the proton transfer in small clusters of $\left[\text{1-naphthol:3}\left(NH_3\right)\right]$ display an isotope ratio ranging from 3 to 20. Although this value is somewhat smaller than that reported for the matrix studies, the conditions of the reaction (an Ar matrix, versus small molecular cluster) are quite different.

Example 25.7 *When excited by light, molecule 1 transforms into molecule 2. Then molecule 2 rearranges by shifting a proton by a distance of* $r = 200$ *pm (formation of molecule 3). At high temperatures the rate constant* k_r *for the proton shift follows the Arrhenius equation (*$E_{a,m} = 50\ \mathrm{kJ\ mol^{-1}}$*); the result is the same if we replace the proton by a deuteron. At low temperatures, however,* k_r *becomes constant. By extrapolation for* $T \rightarrow 0$ *it is found that* k_r *is* 10^8 *times larger for a shift of H than for a shift of D [23]. Provide an estimate of this large isotope effect.*

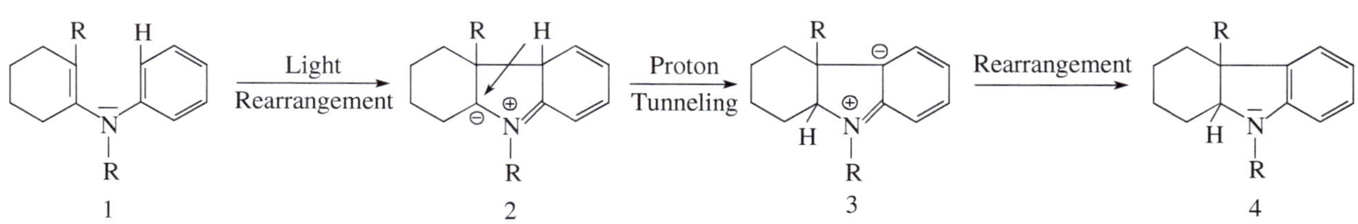

This tremendous difference in reaction rate can be understood by considering tunneling of the proton and of the deuteron through the activation barrier. The rate constant k_r is proportional to the probability density ψ^2 of the proton and of the deuteron, respectively, at distance r

$$k_r = a\psi^2$$

We describe the proton as a particle (mass m, energy E) in a potential trough of depth V_0. According to Foundation 3.1 and Fig. 3.2 the wavefunction outside the trough is

$$\psi \sim \mathrm{e}^{-br} \quad \text{with} \quad b = \frac{2\pi\sqrt{2m}}{h} \cdot \sqrt{V_0 - E}$$

$V_0 - E$ corresponds to the activation barrier $E_a = E_{a,m}/\boldsymbol{N}_\mathrm{A} = 8.31 \times 10^{-20}$ J , and we obtain

$$\psi^2 \sim \exp\left(-C\sqrt{m}\right), \quad C = \frac{4\pi}{h}\sqrt{2E_a} \cdot r$$

Then the ratio of the rate constants k_H/k_D is

$$\frac{k_r\ (\mathrm{H})}{k_r\ (\mathrm{D})} = \frac{\exp\left(-C\sqrt{m_\mathrm{H}}\right)}{\exp\left(-C\sqrt{m_\mathrm{D}}\right)} = \exp\left(-C \cdot \left[\sqrt{m_\mathrm{H}} - \sqrt{m_\mathrm{D}}\right]\right)$$
$$= \exp\left(-C\sqrt{m_\mathrm{H}} \cdot \left[1 - \sqrt{2}\right]\right)$$

With $E_a = 8.31 \times 10^{-20}$ J and $r = 200$ pm, we obtain

$$C \cdot \sqrt{m_\mathrm{H}} = 7.73 \times 10^{24}\ \mathrm{J}^{1/2}\ \mathrm{s}^{-1} \cdot 4.09 \times 10^{-14}\ \mathrm{kg}^{-1/2} \cdot r$$
$$= 3.16 \times 10^{11}\ \mathrm{m}^{-1} \cdot r = 63.2$$

and the result for $k_r(\mathrm{H})/k_r(\mathrm{D})$ is

$$\frac{k_r\ (\mathrm{H})}{k_r\ (\mathrm{D})} = \exp\left(-63.4 \times \left[1 - \sqrt{2}\right]\right) = 2 \times 10^{11}$$

This estimate shows that tunneling of the H nucleus is many orders of magnitude more probable than tunneling of the heavier D nucleus, in agreement with the experiment ($k_r(\mathrm{H})/k_r(\mathrm{D}) = 10^8$).

25.5 Conclusion

The molecular details of reactants and reaction conditions are of paramount importance to understanding and predicting the products formed in a chemical reaction. The challenge of predicting the outcome and rate of reactions from first principles is a daunting one; however, the discussion in Section 25.3 shows that this challenge is being addressed and for the simple H plus H_2 reaction, it has been largely met [15]. While it is important to understand reactions at such detail, it is often acceptable to understand and predict the rate constant for a thermal assembly of reactants. Through a statistical treatment of the reactants and transition state (activated complex), the transition state theory model for the rate constant provides an important connection between the nature of reactions on an elementary state-to-state level and the thermal rate constant. In many cases, in particular for gas-phase reactions, the transition state theory rate constant provides a good description; however, it is limited in quantitative accuracy. Two important limitations to the transition state theory description were discussed; recrossings at the transition state are important in solution-phase reactions and tunneling is important in proton transfer (and electron transfer) reactions.

25.6 Box 25.1. Rigorous Derivation of Rate Equation (25.16)

According to Equation (25.6) the rate constant k_r for a bimolecular reaction is

$$k_r = k^\ddagger \cdot K^\ddagger \cdot \frac{1}{c^\ominus} \tag{25.66}$$

and in Equation (25.10) we found

$$k^{\ddagger} = \frac{1}{\delta}\sqrt{\frac{kT}{2\pi m^{\ddagger}}} \tag{25.67}$$

Now we proceed to find an expression for $K^{\ddagger}$, which tells us the ratio of transition state species to reactants. From Chapter 18, we can write the equilibrium constant $K^{\ddagger}$ in terms of the partition functions. For example, for the bimolecular reaction of Equation (25.2) the expression for $K^{\ddagger}$ would be

$$K^{\ddagger} = \frac{\left(\dfrac{Z_{\mathrm{M}}^{\ominus}}{\sigma_{\mathrm{M}}\, N(\mathrm{M})}\right)}{\left(\dfrac{Z_{\mathrm{A}}^{\ominus}}{\sigma_{\mathrm{A}}\, N(\mathrm{A})}\right)\cdot\left(\dfrac{Z_{\mathrm{B}}^{\ominus}}{\sigma_{\mathrm{B}}\, N(\mathrm{A})}\right)} \tag{25.68}$$

where $Z^{\ominus}$ is the molecular partition function under standard conditions, $N(\mathrm{A})$, $N(\mathrm{B})$, and $N(\mathrm{M})$ are the number of particles A, B, and M, and σ is the symmetry number. As outlined in the text we assume that in the activated complex the asymmetric vibration mode is replaced by a translational motion along the reaction coordinate. The contribution of this motion to the partition function of the complex is

$$Z_{\mathrm{M,\,reaction\ coordinate}} = \delta\cdot\sqrt{\frac{2\pi m^{\ddagger}kT}{h^2}} \tag{25.69}$$

This expression differs slightly from that provided in Chapter 15, because we are only considering translation in one-dimension, rather than three. Thus we obtain

$$K^{\ddagger} = \frac{\left(\dfrac{Z_{1,\mathrm{M}}^{\ominus}}{\sigma_{\mathrm{M}}N(\mathrm{M})}\right)}{\left(\dfrac{Z_{\mathrm{A}}^{\ominus}}{\sigma_{\mathrm{A}}N(\mathrm{A})}\right)\cdot\left(\dfrac{Z_{\mathrm{B}}^{\ominus}}{\sigma_{\mathrm{B}}N(\mathrm{B})}\right)}\cdot\delta\left(\frac{2\pi m^{\ddagger}kT}{h^2}\right)^{1/2} = K_1^{\ddagger}\cdot\delta\sqrt{\frac{2\pi m^{\ddagger}kT}{h^2}} \tag{25.70}$$

where the index 1 in $Z_{1,\mathrm{M}}^{\ominus}$ and $K_1^{\ddagger}$ denotes the remaining part of the partition function for M (asymmetric vibration mode excluded).

Using Equations (25.66) and (25.67) we can now combine these results to find the reaction rate constant k_r

$$k_r = k^{\ddagger}\cdot K^{\ddagger}\cdot\frac{1}{c^{\ominus}} = \frac{1}{\delta}\sqrt{\frac{kT}{2\pi m^{\ddagger}}}\cdot K_1^{\ddagger}\cdot\delta\cdot\sqrt{\frac{2\pi m^{\ddagger}kT}{h^2}}\cdot\frac{1}{c^{\ominus}}$$

which simplifies to

$$k_r = \frac{1}{c^{\ominus}}\cdot\frac{kT}{h}\cdot K_1^{\ddagger} \qquad \text{(bimolecular reaction)}$$

Note that in this consideration the extension δ cancels. It is common to omit the index 1, and thus we arrive at Equation (25.16).

25.7 Box 25.2. Rate Constant from Transition State Theory

By way of example, we calculate the transition state theory rate constant for the reaction

$$\mathrm{D} + \mathrm{H\text{-}H} \rightarrow \mathrm{D}\cdots\mathrm{H}\cdots\mathrm{H} \rightarrow \mathrm{D\text{-}H} + \mathrm{H}$$

and compare it to that we found using the simple kinetic theory.

The D atom has translational and electronic contributions to the molecular partition function, the H_2 molecule has translational, rotational, vibrational, and electronic contributions. If the molecules are in the vibrational and electronic ground state, the partition functions $Z_{\mathrm{D}}^{\ominus}$ and $Z_{\mathrm{H}_2}^{\ominus}$ are

$$Z_{\mathrm{D}}^{\ominus} = V^{\ominus}\left(\frac{2\pi m_{\mathrm{D}}k}{h^2}\right)^{3/2}\cdot T^{3/2}\cdot \mathrm{e}^{-E_{\mathrm{D},0}/(kT)}$$

$$Z_{\mathrm{H}_2}^{\ominus} = V^{\ominus}\cdot\left(\frac{2\pi m_{\mathrm{H}_2}k}{h^2}\right)^{3/2}\cdot T^{3/2}\cdot\left(\frac{8\pi^2kT}{h^2}I_{\mathrm{H}_2}\right)\cdot \mathrm{e}^{-\frac{1}{2}h\nu_{\mathrm{H}_2}}\cdot \mathrm{e}^{-E_{\mathrm{H}_2,0}/(kT)}$$

where I is the inertial moment, ν is the vibration frequency, and $E_{\mathrm{D},0}$ and $E_{\mathrm{H_2},0}$ are the electronic energies of the D atom and the H_2 molecule in their ground states. Note that we treat the rotational partition function in the high temperature limit ($\theta_{\mathrm{rot}} \ll T$), and the vibrational partition function in its low temperature limit ($\theta_{\mathrm{vib}} \gg T$).

The activated complex has the structure of a triatomic molecule. The motion along the reaction coordinate (asymmetric stretch motion) is treated explicitly; that is, we include only the symmetric stretch and the bending mode in the vibrational partition function. Then we find that

$$\begin{aligned} Z^{\ominus}_{\mathrm{DHH}} &= Z^{\ominus}_{\mathrm{DHH,trans}} \cdot Z_{\mathrm{DHH,rot}} \cdot Z_{\mathrm{DHH,vib}} \cdot Z_{\mathrm{DHH,electr}} \\ &= V^{\ominus}\left(\frac{2\pi\left(m_{\mathrm{D}}+2m_{\mathrm{H}}\right)k}{h^2}\right)^{3/2} \cdot T^{3/2} \cdot \left(\frac{8\pi^2 kT}{h^2} I_{\mathrm{DHH}}\right) \\ &\quad \cdot \mathrm{e}^{-\frac{1}{2}h(\nu_2+\nu_3)_{\mathrm{DHH}}} \cdot \mathrm{e}^{-E^{\ddagger}_{\mathrm{DHH},0}/(kT)} \end{aligned}$$

where ν_2 and ν_3 are the frequencies of two normal modes of HDD and $E^{\ddagger}_{\mathrm{DHH},0}$ is the electronic energy of HDD in its ground state. We substitute these partition functions into Equation (25.20). Setting $\sigma_{\mathrm{H_2}} = 2$, $\sigma_{\mathrm{D}} = 1$ and $\sigma_{\mathrm{DHH}} = 1$, we find that

$$k_r = \frac{1}{c^{\ominus}} \cdot \frac{kT}{h} \cdot N(A) \cdot \frac{1}{V^{\ominus}}\left(\frac{m_{\mathrm{D}}+m_{\mathrm{H_2}}}{m_{\mathrm{D}}m_{\mathrm{H_2}}}\right)^{3/2} \frac{\sqrt{8\pi I_{\mathrm{DHH}}}h}{\sqrt{kT}} \cdot \left(\frac{2h^2}{8\pi^2 kT I_{\mathrm{H_2}}}\right) \cdot \exp\left(-\frac{E_a}{kT}\right)$$

where the activation energy is $E_a = \left[E^{\ddagger}_{\mathrm{DHH},0} + \frac{1}{2}h(\nu_2+\nu_3)_{\mathrm{,DHH}}\right] - E_{\mathrm{D},0} - \left(E_{\mathrm{H_2},0} + \frac{1}{2}h\nu_{\mathrm{H_2}}\right)$. Using the relationship $N(\mathrm{A})/V^{\ominus} = \boldsymbol{N}_{\mathrm{A}}c^{\ominus}$, we find that

$$k_r = \boldsymbol{N}_{\mathrm{A}}\sqrt{kT} \cdot \left(\frac{m_{\mathrm{D}}+m_{\mathrm{H_2}}}{m_{\mathrm{D}}m_{\mathrm{H_2}}}\right)^{3/2} \sqrt{8\pi I_{\mathrm{DHH}}} \cdot \left(\frac{2h^2}{8\pi^2 kT I_{\mathrm{H_2}}}\right) \cdot \exp\left(-\frac{E_{a,m}}{\boldsymbol{R}T}\right)$$

where $E_{a,m}$ is the molar activation energy. Now we set

$$\frac{m_{\mathrm{D}}+m_{\mathrm{H_2}}}{m_{\mathrm{D}}m_{\mathrm{H_2}}} = \frac{1}{\mu} \quad \text{and} \quad I_{\mathrm{DHH}} = \mu^{\ddagger}(d^{\ddagger})^2$$

where $\mu^{\ddagger}$ is the reduced mass (see Section 11.2.3, Equation (11.16)) and $d^{\ddagger}$ is the bond distance of the activated complex D–H–H (in our case, the distances D–H and H–H are identical). We can rewrite the expression for I_{DHH} as

$$I_{\mathrm{DHH}} = \mu^{\ddagger}(d^{\ddagger})^2 = \frac{\mu^{\ddagger}}{\mu}\left(\frac{d^{\ddagger}}{d}\right)^2 \cdot \mu d^2$$

where $d = r_{\mathrm{D}} + r_{\mathrm{H_2}}$ is the collision diameter of the reacting molecules. Then we obtain

$$k_r = \sqrt{\frac{8kT}{\pi\mu}} \cdot \pi d^2 \cdot \boldsymbol{N}_{\mathrm{A}} \cdot \left[\frac{\mu^{\ddagger}}{\mu}\left(\frac{d^{\ddagger}}{d}\right)^2 \frac{2h^2}{8\pi^2 kT I_{\mathrm{H_2}}}\right] \cdot \exp\left(-\frac{E_{a,m}}{\boldsymbol{R}T}\right)$$

25.8 Foundations

25.8.1 Foundation 25.1: Thermal Rate Constants

See the books website, www.wiley.com/go/Kuhn/physicalchemistry3e.

25.8.2 Foundation 25.2: Reaction Rate and Solvent Friction

See the books website, www.wiley.com/go/Kuhn/physicalchemistry3e.

25.9 Exercises

E25.1 The gas-phase reaction between F_2 and SF_5 is first order in each of the reactants. At 298 K the rate constant is 4.39×10^5 L mol^{-1} s^{-1} and the empirically determined activation energy is 20.29 kJ mol^{-1}. Use the definition of the empirical activation energy $E^{\mathrm{empirical}}_{a,m}$ given in the text and the transition state theory expression for a bimolecular gas-phase rate constant, namely

$$k_{\mathrm{TST}} = \left(\frac{kT}{h}\right)\left(\frac{\boldsymbol{R}T}{P^{\ominus}}\right) \cdot \exp\left(\frac{\Delta_r S^{\ddagger\ominus}}{\boldsymbol{R}}\right) \cdot \exp\left(-\frac{\Delta_r H^{\ddagger\ominus}}{\boldsymbol{R}T}\right),$$

to calculate the entropy of activation and the enthalpy of activation at 298 K.

E25.2 Compare the unimolecular transition state theory expression for the rate constant to the Arrhenius expression for the rate constant ($k_r = A \cdot \exp(-E_{a,m}/\boldsymbol{R}T)$) and show that

$$E_{a,m} = \Delta_r H^{\ddagger\ominus} + \boldsymbol{R}T$$

E25.3 For the reaction in Exercise E24.12, perform a transition state theory analysis, using the bimolecular expression for the rate constant, to determine $\Delta_r H^{\ddagger\ominus}$ and $\Delta_r S^{\ddagger\ominus}$. Comment on the sign of your value for $\Delta_r S^{\ddagger\ominus}$ and on its magnitude.

E25.4 Consider the transition state theory expression for the rate constant in a unimolecular reaction. Using a potential energy diagram, describe the primary assumptions used in deriving this expression and provide the important highlights of the derivation.

E25.5 Consider the thermodynamic form of the transition state theory expression for a unimolecular reaction. Describe how each of the different degrees of freedom of the reactant contribute to the $\Delta_r S^{\ddagger\ominus}$ term and the $\Delta_r H^{\ddagger\ominus}$ term. Be explicit and mathematically exact. Provide an example of a unimolecular reaction in which you expect $\Delta_r S^{\ddagger\ominus}$ to be positive. Provide an example of a unimolecular reaction in which you expect $\Delta_r S^{\ddagger\ominus}$ to be negative.

E25.6 Consider a potential energy surface for a C–H bond cleavage and a C–D bond cleavage – of some large organic molecule. If we take a typical C–H vibrational frequency to be 3500 cm^{-1}, what is the zero point energy shift between the two bonds? Assume that the zero point energy shift at the transition state can be neglected. How would you justify this assumption? Ignoring any other differences in the two molecules' structures and molecular parameters, estimate the isotope effect on the bond cleavage; i.e., determine $k_r(\text{CH})/k_r(\text{CD})$ using the zero point energy shifts.

E25.7 Consider the reaction

$$H_2 + {}^{79}\text{Br} \rightarrow \text{H}^{79}\text{Br} + \text{H}$$

This reaction is quite sensitive to the isotope of hydrogen used. The data plotted in Fig. E25.7 are measured reaction rate constants for this reaction with different isotopes of hydrogen; XY indicates HD, HT, and D_2. Note that the data are plotted as $\ln\left[k_r\,(\text{H}_2)/k_r\,(\text{XY})\right]$. Use transition state theory to explain the trend in the rate constant ratios. The H_2 vibrational wavenumber is 4401 cm^{-1}.

E25.8 Consider the bimolecular reaction

$$\text{OH (g)} + \text{H}_2\text{(g)} \rightarrow \text{H}_2\text{O(g)} + \text{H (g)}$$

and take the Arrhenius parameters for this reaction to be $A = 3.0 \times 10^9\ \text{L mol}^{-1}\ \text{s}^{-1}$ and $E_{a,m} = 16.5\ \text{kJ mol}^{-1}$ [24]. Find the values of $\Delta_r H^{\ddagger}$ and $\Delta_r S^{\ddagger}$ at 298 K. Use a standard state of $P^{\ominus} = 1$ bar in the expression for the transition state theory rate constant of this gas-phase reaction, rather than the $c^{\ominus} = 1\ \text{mol L}^{-1}$ used for the solution rate constant expression that is given in the text.

E25.9 Consider the recombination of ethyl radicals in the gas phase. The reaction

$$^{\bullet}\text{CH}_2 - \text{CH}_3 + {}^{\bullet}\text{CH}_2 - \text{CH}_3 \rightarrow \text{CH}_3\text{CH}_2\text{CH}_2\text{CH}_3$$

is diffusion-controlled ($E_a = 0$). Estimate the rate constant at $T = 298$ K . Given that the experimental rate constant is $1.6 \times 10^{11}\ \text{L mol}^{-1}\ \text{s}^{-1}$ calculate the steric factor α.

Figure E25.7 Plot of $\ln\left[k_r\,(\text{H}_2)/k_r\,(\text{XY})\right]$ versus $1/T$ for the reaction 25.75 with different isotopes.

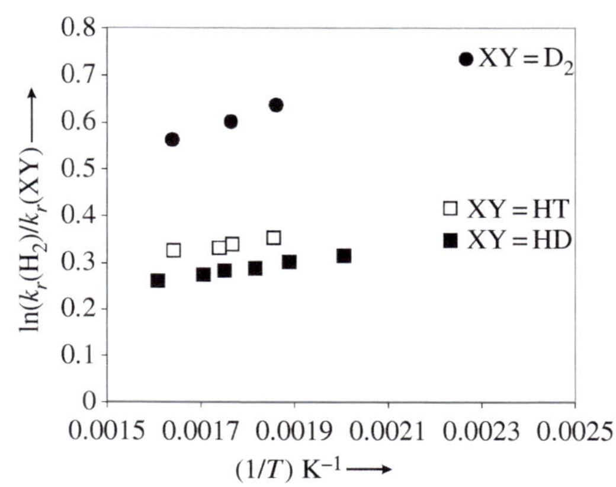

E25.10 Consider the artificial data given in the table, which is calculated from the relation

$$k_r = A_{f,0} \cdot T \cdot \exp\left(-\frac{\Delta_r H^{\ddagger\ominus}}{\boldsymbol{R}T}\right)$$

using $\Delta_r H^{\ddagger\ominus} = 50$ kJ mol^{-1}, $A_{f,0} = 1 \times 10^7$ s^{-1}, and different ranges of the temperature.

T/ K	300	305	310	320	340	350	400	450	500
k_r/s^{-1}	5.90	8.34	11.6	22.0	70.7	1.21×10^2	1.18×10^3	7.07×10^3	2.99×10^4

Perform an Arrhenius analysis for these data and compare the result to Equation (25.41) in which you use the mean value,

$$T_{\text{mean}} = \frac{1}{2}(T_{\text{end}} + T_{\text{start}})$$

To observe a significant trend in how the difference changes, consider temperature ranges of 10, 50, 100, and 200 K.

E25.11 Consider the reaction: $O(^3P) + Cl_2 \rightarrow OCl + Cl$, where $O(^3P)$ means that the oxygen atom is in its triplet state rather than its singlet state. Consider the potential energy surface shown in Fig. E25.11. The energies assigned to the contours are labeled in kJ mol^{-1}, and the surface contours assume a linear collision geometry. Provide a verbal description of what is happening for the trajectory drawn in each diagram – as the system moves from points A to B to C to D. (Figure E25.11 is sketched from that given by Kosmas [25]).

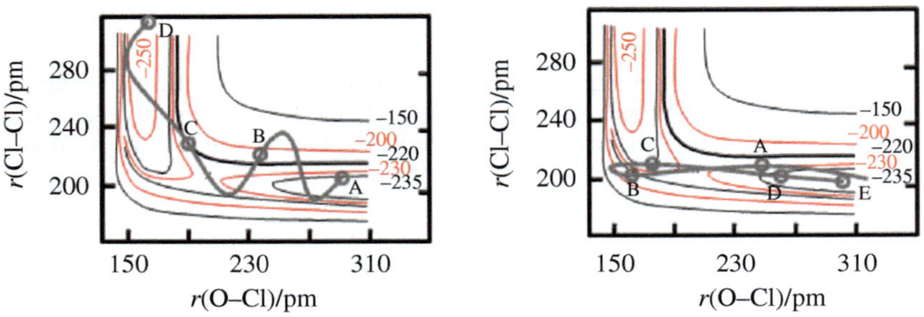

Figure E25.11 Potential energy surface for the reaction of O with Cl_2.

E25.12 Consider the reaction: $O(^3P) + Cl_2 \rightarrow OCl + Cl$, where $O(^3P)$ means that the oxygen atom is in its triplet state rather than its singlet state. Figure 25.12 shows two different potential energy contours. Describe how you expect the distribution of energy in the final products to change as the input energy changes. In particular, the distribution of energy in the translational, rotational, and vibrational degrees of freedom. (Note the shift in the transition state's location, labeled by a cross.) (Figure E25.12 is sketched from those given by Kosmas [25].)

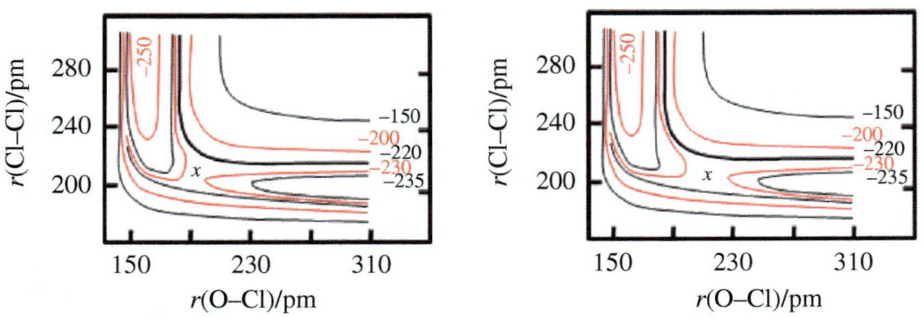

Figure E25.12 Potential energy surface for the reaction of O with Cl_2.

E25.13 Given that $A = 7.0 \times 10^{10}$ L mol^{-1} s^{-1} for the reaction

$$D + H_2 \rightarrow HD + H$$

under standard conditions. Determine $\Delta_r S^{\ddagger\ominus}$.

E25.14 The excited molecule ICN* is created with an internuclear distance of about 275 pm and the molecule is fully dissociated at about 400 pm. Use these distances and the 200 fs risetime for the CN formation (Fig. 25.9) to estimate the relative velocity of the dissociating fragments.

E25.15 The formation of the CN radical from the bond dissociation in Fig. 25.9 occurs in 200 fs. Compare this time to the vibrational period of the CN radical which has a vibrational frequency of $\nu = 6.2 \times 10^{13}\ s^{-1}$. Compare this time to the thermally averaged rotational period of the ICN molecule ($T = 298$ K and $I = 2.6 \times 10^{-45}$ kg m^2).

E25.16 In the gas-phase reaction

$$K + Br_2 \rightarrow KBr + Br$$

$E_{a,m} = 0$ and the steric factor is $\alpha = 4.8$. The large α is explained by a harpooning mechanism, which can be represented by

$$K + Br_2 \rightarrow K^+ + Br_2^- \rightarrow KBr + Br$$

When a K atom approaches a Br_2 molecule and a critical distance $\boldsymbol{R}$ is reached, an electron can be transferred from K to Br_2. Then the Coulomb attraction initiates a fast reaction of both ions, and the effective collision cross section is $\pi \boldsymbol{R}^2$ instead of $\pi\left(r_K + r_{Br_2}\right)^2$, where $r_K + r_{Br_2} = 400$ pm. Use the steric factor to estimate the distance of approach at which the electron "jumps" from the potassium atom to the bromine molecule. What is the Coulomb force between the two ions at this distance?

E25.17 In contrast to the slow vibrational relaxation for the ground electronic state of I_2 in Xe (> 3 ns), the vibrational relaxation is nearly complete within 200 ps in CCl_4. Provide an explanation for this difference.

E25.18 In Section 25.2.1 we discussed the transition state theory for the

$$D + H_2 \rightarrow H + DH$$

reaction and identified the asymmetric stretch motion with the reaction coordinate. Why is this an appropriate choice? Why is the symmetric stretch an inappropriate choice? Cast your discussion in terms of the energy profile for the reaction in Fig. 25.1.

E25.19 In Example 25.5 we found that $\sigma_{r,\max} = 1.57 \times 10^4$ pm^2 and $E_{\text{thresh}} = 0.38$ eV. Use these values in Equation (25.57) to compute the thermal rate constant.

E25.20 Calculate the moment of inertia I_{DHH} of the transition state complex D–H–H that is discussed in the text and extract an expression for the reduced mass $\mu^{\ddagger}$.

E25.21 Assume that a gas-phase I_2 molecule has an excess vibrational energy of $E = 3.80 \times 10^{-20}$ J and is in its ground electronic state with a characteristic vibrational frequency of $\nu_0 = 6.45 \times 10^{12}\ s^{-1}$. Given that the equilibrium bond distance is $x_{\text{eq}} = 266$ pm, estimate the maximum distance of the I atoms from each other during the molecule's vibrational motion. Compare this value to the maximum displacement that was estimated in Example 25.6.

E25.22 In the collision theory model, the rate constant is described by

$$\boxed{k_r = A_f \cdot e^{-E_{a,m}/RT}} \tag{25.71}$$

and

$$\boxed{A_f = \sqrt{\frac{8kT}{\pi\mu}} \cdot \pi \cdot \left(r_A + r_B\right)^2 \cdot \boldsymbol{N}_A \cdot \alpha} \tag{25.72}$$

Describe how to rigorously extract the activation energy $E_{a,m}$ from experimental rate constant data.

25.10 Problems

P25.1 Consider the photodissociation of the CF_3I molecule into a CF_3 radical and an I* radical (the * indicates that the I atom is electronically excited) with a 275 nm photon. Given that the binding energy (ε_B) of the C–I bond corresponds to a wavenumber of 18,600 cm^{-1}, and the energy difference between I and I* is 7603 cm^{-1}, compute the excess energy left in the CF_3 and I* after the photodissociation. *In part 1*: Assume that all the excess energy goes into the relative translational motion of the CF_3 and I* fragments and find the relative velocity of the two fragments. Then determine how long it takes for them to reach a separation distance of 840 pm – assuming they began at the equilibrium bond distance of 210 pm. *In part 2*: Assume that five quanta of the CF_3 umbrella motion (580 cm $^{-1}$) are excited and the rest of the excess energy is statistically distributed among the translational and rotational degrees of freedom of the fragments, estimate the relative translational velocity of the CF_3 and I* fragments and compute the time it takes for the I* atom and the CF_3 fragment to reach a distance of 840 pm, Provide a verbal explanation for why you might expect the umbrella motion of CF_3 to be highly excited.

P25.2 You wish to follow the thermal isomerization reaction

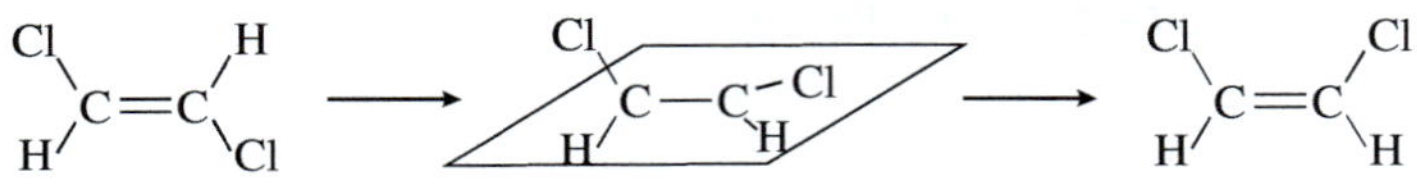

trans-Form Transition state *cis*-Form

This unimolecular reaction proceeds with a rate constant of about 2 s^{-1} at 500 K and 43 s^{-1} at 600 K. Describe in detail how you would characterize the kinetics of this reaction over the temperature range of 500 to 600 K. In particular, describe the experimental method that you would use to determine the rate law, including a) what quantity, or quantities, you would measure and b) how you would use these quantities to determine the rate law. Provide a detailed sketch of your experimental apparatus(es). Indicate clearly how you would perform the measurement and control the system's properties.

P25.3 (Adapted from [27]) Consider the reaction

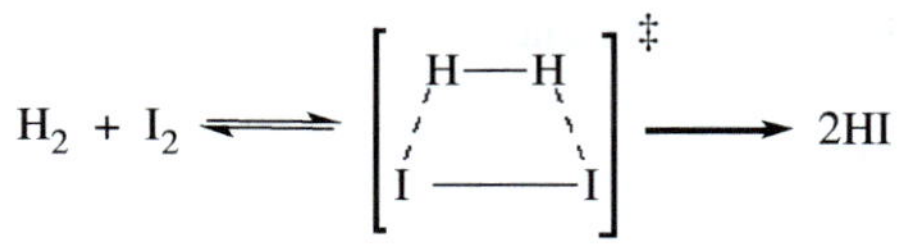

Assume that the transition state is planar and has an axis of symmetry. Show using transition state theory that the rate constant may be written as

$$k_r = \frac{\boldsymbol{N}_A h^3}{4\pi^2 kT}\left(\frac{m^\ddagger}{m_{H_2} m_{I_2}}\right)^{3/2}\left(\frac{\sqrt{I_A I_B I_C}}{I_{H_2} I_{I_2}}\right)\left[\frac{\left(1-\exp\left(-\frac{h\nu_{H_2}}{kT}\right)\right)\left(1-\exp\left(-\frac{h\nu_{I_2}}{kT}\right)\right)}{\prod_{i=1}^{5}\left(1-\exp\left(-\frac{h\nu_i^\ddagger}{kT}\right)\right)}\right]\exp\left(-\frac{E^\ddagger_{0,m}}{\boldsymbol{RT}}\right)$$

Note that you should assume that $\sigma_{H_2} = \sigma_{I_2} = \sigma^\ddagger = 2$. Now use the following parameters to estimate the reaction rate constant. a) $I_{I_2} = 7.485 \times 10^{-45}$ kg^{-1} m^2; $\nu_{I_2} = 214.57$ cm^{-1}. b) $I_{H_2} = 4.56 \times 10^{-48}$ kg^{-1} m^2; $\nu_{H_2} = 4395.2$ cm^{-1}. c) The bond distances in the activated complex are estimated to be H–H: 97 pm, I–I: 295 pm, and H–I: 175 pm. d) The normal mode wavenumbers of the activated complex are estimated to be 994 cm^{-1}, 86 cm^{-1}, 1280 cm^{-1}, $965^\ddagger$ cm^{-1}, 1400 cm^{-1}, and 1730 cm^{-1}. The reaction coordinate is associated with the 965 cm^{-1} mode. e) The activation barrier is $E^\ddagger_{0,m} = 170.5$ kJ mol^{-1} at a temperature of 600 K. Compare your calculated rate constant value to the experimental value of 9.0×10^{-6} M^{-1} s^{-1}.

P25.4 Consider the reaction

$$HO(g) + HO(g) \rightarrow [\ HO\text{–}HO]^\ddagger \rightarrow H_2O(g) + O(g)$$

The measured thermal rate constant for this reaction is 8.7×10^8 L mol^{-1} s^{-1} at 298 K [27]. Take the energy barrier $\Delta E^\ddagger_{0,m}$ for this reaction to be 2.0 kJ mol^{-1} and consider the transition state to be a linear geometry, as shown above. Using the data given below and assuming that the frequency for the reaction coordinate mode is $\nu_0 = 9.37 \times 10^{13}$ s^{-1} and $\kappa = 1$, compute the transition state theory rate constant for this reaction at 298 K.

Species	σ	θ_{rot} (K)	θ_{vib} (K)
OH	1	27	5384
HO—HO	1	5	5400(ν_s), 4500(ν_a), 1500, 1100, 800, 100, 100

Compare your calculated value of the rate constant to the experimental value.

P25.5 Consider the reaction

$$F + H_2 \rightarrow HF + H$$

which has been studied by Polanyi and coworkers [28]. Two observations are noteworthy. First, it is possible to determine the relative populations of the vibrational states formed for HF in this reaction; see the table. The vibrational wavenumber of HF is 4138 cm^{-1} and of H_2 is 4401 cm^{-1}.

Table: Vibrational Energy Distribution of Product HF at 298 K.

Vibrational quantum number	1	2	3
Fraction of molecules	0.174	0.562	0.264

Second, this reaction is exothermic and the energy is partitioned into the products as 66% vibration, 8% rotation, and 26% translation. Use the potential energy surface in Fig. P25.5 to explain these two observations.

Figure P25.5 This surface is sketched from that given by Schaeffer [29]. The contour labels have units of kcal/mol.

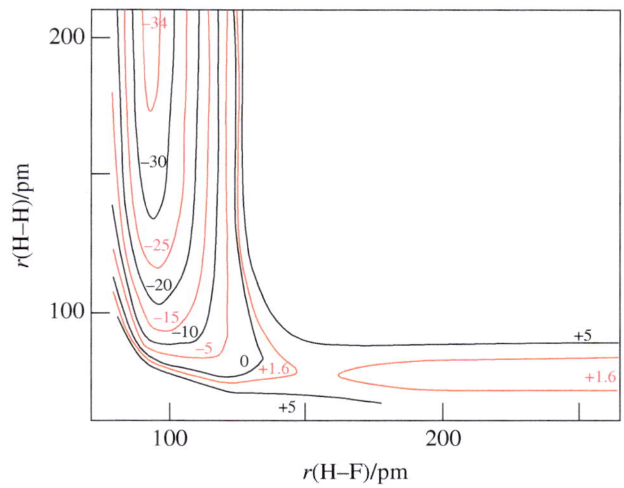

Describe an experimental arrangement that you might use to investigate the vibrational distribution of the HF product. Provide a sketch of an apparatus. Describe in words your experimental method.

P25.6 Consider the reaction

$$H_2 + I_2 \underset{k_{-1}}{\overset{k_1}{\rightleftarrows}} 2\ HI$$

The forward rate constant is $k_1 = 10^{11} \exp\left(-E_{1,m}/\boldsymbol{R}T\right)$ L mol^{-1} s^{-1} where $E_{1,m} = 165$ kJ mol^{-1}, and the backward rate constant is $k_{-1} = 6 \times 10^{10} \exp\left(-E_{2,m}/\boldsymbol{R}T\right)$ L mol^{-1} s^{-1} where $E_{2,m} = 185$ kJ mol^{-1}. What is the equilibrium constant for a mixture of H_2, I_2, and HI at 1000 K? What is the standard Gibbs energy change at this temperature? Determine the activation entropy and the activation enthalpy for the forward reaction at 1000 K. Perform the same calculation for the backward reaction rate constant. Sketch the Gibbs energy versus the reaction coordinate and label any of the energies that you know.

P25.7 The enol compound E was investigated in a hydrocarbon glass at low temperature [30]. When exciting by light, E goes over into the excited singlet state of the keto form K, then into the triplet state of K, into the triplet state of E and finally back into the ground state of E:

$$E \xrightarrow{\text{light}} {}^1K^* \rightarrow {}^3K^* \rightarrow {}^3E^* \rightarrow E$$

Below 25 K, the rate of reaction ${}^3K^* \rightarrow {}^3E^*$ is independent of temperature. Explain why replacing the proton by a deuteron reduces the rate by 2000 times. Use a simple calculation to estimate the change in rate constant.

P25.8 Consider the H atom transfer reaction

$$F + HI \rightarrow FHI \rightarrow HF + I \quad \text{or} \quad F + DI \rightarrow FDI \rightarrow DF + I$$

which shows an isotope effect on the rate constant that is $k_r(HI)/k_r(DI) = 1.29$ at $T = 298$ K. In their paper [31], Mei and Moore use a bent transition state to calculate this isotope ratio. Perform a related calculation but assume that the transition state is linear. Compare your calculated isotope ratio to that of Moore and Mei's for the bent transition state. Comment on which geometry you think is more reasonable.

P25.9 Recently Baldwin and Chapman [32] studied 1,5 sigmatropic shift of H atoms for 1,3-cyclohexadienes by synthesizing the monodeuterated molecule and studying its interconversion between the three forms:

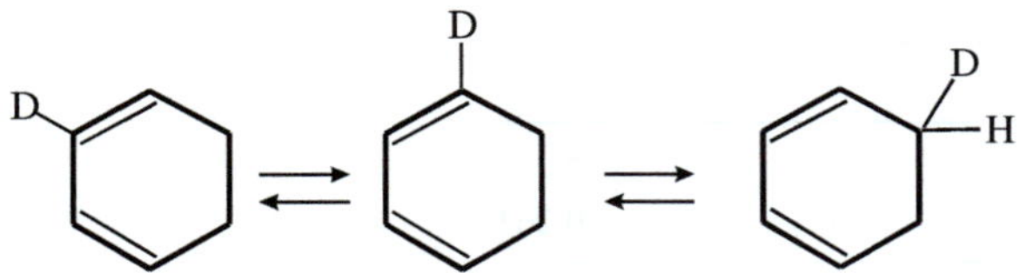

In their study they measured the time for the system to equilibrate between the three forms, and they found the following temperature dependence for this reaction

T/K	527	557
k/s^{-1}	0.394×10^{-4}	3.00×10^{-4}

a) From their data, determine the activation enthalpy for this reaction. Comment on the size of the activation enthalpy you find; i.e., is it what you would expect and why (or why not)? b) From their data, determine the activation entropy for the reaction. Comment on the sign of the activation entropy; i.e., is it what you expect and why (or why not)? c) In their analysis they assume that the H atoms move but that the D atom stays in place. Is this a good assumption? Justify your answer with an estimate of the relative rates in the two cases.

P25.10 Use transition state theory to rationalize the observed isotope ratios in Table 25.1.

P25.11 Consider the reaction of H + D_2 and assume that the reaction barrier has a functional form of

$$E_a(\gamma) = E_a + E_a'(1 - \cos\gamma)$$

where γ is the angle that the H-atom trajectory makes with the D_2 molecule's bond axis.

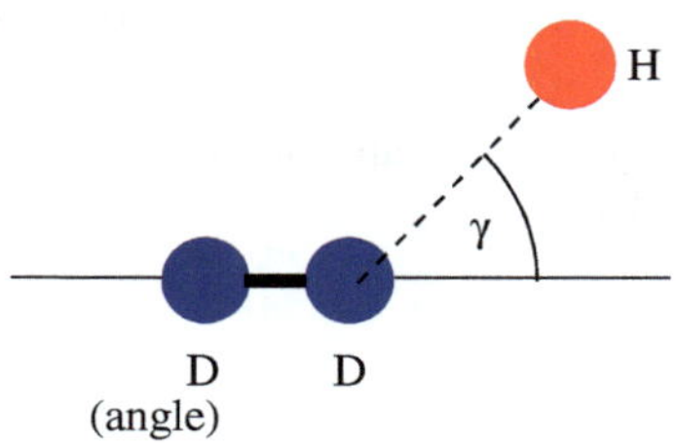

Combine this expression with the angle-dependent cross-section

$$\sigma_r(E,\gamma) = 0 \qquad \text{for } E < E_a(\gamma)$$

$$\sigma_r(E,\gamma) = \pi d^2 \cdot \left(1 - \frac{E_a(\gamma)}{E}\right) \qquad \text{for } E > E_a(\gamma)$$

to obtain an explicit expression for the dependence of $\sigma_r(E,\gamma)$ on γ. Compare your result to the angle dependence of the cross-section found from quasiclassical trajectory calculations at $E = 0.55$ eV (see Table below) by plotting the computational data [33] and comparing it to the model.

$\sigma_r(E = 0.55\text{ eV})/\text{Å}^2$	1.1	0.80	0.56	0.22	0.02	< 0.01	< 0.01
$\cos\gamma$	0.96	0.84	0.74	0.64	0.54	0.45	0.35

Lastly, perform an average over the angle cos(γ), to show that

$$\langle \sigma_r(E) \rangle_\gamma = \pi d^2 \cdot \frac{(E - E_a)^2}{4EE_a}$$

References

1 L.D. Hawton and G.P. Semeluk, THE Thermal unimolecular isomerization of 1,2-dichloroethylene, *Can. J. Chem.*, 1966, **44**, 2143.

2 J.C. Miller and R.J. Gordon, Kinetics of the Cl–H_2 system. III. The deuterium isotope effect in Cl+H_2, *J. Chem. Phys.*, 1983, **79**, 1252.

3 T.C. Allison, G.C. Lynch, D.G. Truhlar, and M.S. Gordon, An improved potential energy surface for the H_2Cl system and its use for calculations of rate coefficients and kinetic isotope effects, *J. Phys. Chem.*, 1996, **100**, 13575.

4 E. Grunwald and J.E. Leffler, *Rates and Equilibria of Organic Reactions* (Dover, New York, 1963).

5 J.C. Corchado, D.G. Truhlar, and J. Espinosa-García, Potential energy surface, thermal, and state-selected rate coefficients, and kinetic isotope effects for $Cl^+CH_4 \rightarrow$ HC l^+CH_3, *J. Chem. Phys.*, 2000, **112**, 9375.

6 A.D. Isaacson and D.G. Truhlar, Polyatomic canonical variational theory for chemical reaction rates. Separable-mode formalism with application to $OH^+H_2 \rightarrow H_2O^+H$, *J. Chem. Phys.*, 1982, **76**, 1380.

7 R.B. Bernstein, *Chemical Dynamics via Molecular Beam and Laser Techniques*, (Oxford Press, New York, 1982).

8 P. Casavecchia, Chemical reaction dynamics with molecular beams, *Rep. Prog. Phys.*, 2000, **63**, 355–414.

9 J.S. Winn, *Physical Chemistry* (Harper Collins, New York, 1995).

10 A.H. Zewail, Laser femtochemistry, *Science*, 1988, **242**, 1645.

11 G.W. Johnston, B. Katz, K. Tsukiyama, and R. Bersohn, Isotopic variants of the H + H_2 reaction. 1. Total reaction cross sections of the H + D_2 and H + HD reactions as a function of relative energy, *J. Phys. Chem.*, 1987, **91**, 5445.

12 R.A. Brownsword, M. Hillenkamp, T. Laurent, H.-R. Volpp, J. Wolfrum, R.K. Vatsa, and H.-S. Yoo, Excitation function and reaction threshold studies of isotope exchange reactions: $H + D_2 \rightarrow D + HDandH + D_2O \rightarrow D + HOD$, *J. Phys. Chem. A*, 1997, **101**, 6448.

13 D.M. Charutz, I. Last, and M. Baer, The Toeplitz approach to treating three-dimensional reactive exchange processes: quantum mechanical cross sections and rate constants for the D+H_2 and H+D_2 reactions, *J. Chem. Phys.*, 1997, **106**, 7654–7661.

14 A.A. Westenberg and N. de Haas, Atom—molecule kinetics using ESR detection. II. Results for $D^+H_2 \rightarrow HD^+H$ and $H^+D_2 \rightarrow HD^+D$, *J. Chem. Phys.*, 1967, **47**, 1393.

15 See S.L. Mielke, K.A. Peterson, D.W. Schwenke, B.C. Garrett, D.G. Truhlar, J.V. Michael, M.-C. Su, and J.W. Sutherland, H+H_2 thermal reaction: a convergence of theory and experiment, *Phys. Rev. Lett.*, 2003, **91**, 063201-1.

16 D.H. Waldeck, Photoisomerization dynamics of stilbenes, *Chem. Rev.*, 1991, **91**, 415.

17 S.H. Courtney, S.K. Kim, S. Canonica, and G.R. Fleming, Rotational diffusion of stilbene in alkane and alcohol solutions, *J. Chem. Soc., Faraday Trans. 2*, 1986, **82**, 2065.

18 H.A. Kramers, Brownian motion in a field of force and the diffusion model of chemical reactions, *Physica*, 1940, **7**, 284–304.

19 Q. Liu, J.K. Wang, and A.H. Zewail, Solvation ultrafast dynamics of reactions. 10. Molecular dynamics studies of dissociation, recombination, and coherence, *J. Phys. Chem.*, 1995, **99**, 11321.

20 M.E. Paige and C.B. Harris, Ultrafast studies of chemical reactions in liquids: validity of gas phase vibrational relaxation models and density dependence of bound electronic state lifetimes, *Chem. Phys.*, 1990, **149**, 37.

21 M. Premont-Schwarz, T. Barak, D. Pines, E.T.J. Nibbering, and E. Pines, Ultrafast excited-state proton-transfer reaction of 1-naphthol-3,6-disulfonate and several 5-substituted 1-naphthol derivatives, *J. Phys. Chem. B*, 2013, **117**, 4594–4603.

22 J.A. Syage, Ultrafast measurements of chemistry in clusters: excited-state proton transfer, *J. Phys. Chem.*, 1995, **99**, 5772.

23 K.H. Grellmann, U. Schmitt, and H. Weller, Tunnel effect on the kinetics of a sigmatropic shift, *Chem. Phys. Lett.*, 1982, **88**, 40.

24 A.R. Ravishankara, J.M. Nlcovich, R.L. Thompson, and F.P. Tully, Kinetic study of the reaction of hydroxyl with hydrogen and deuterium from 250 to 1050 K, *J. Phys. Chem.*, 1981, **85**, 2498–2503.

25 A.M. Kosmas, Dynamical trajectory study of the reaction $O(^3P) + Cl_2 \rightarrow OCl + Cl$, *J. Chem. Soc., Faraday Trans.*, 1993, **89**, 2999.

26 S. Benson, *Foundations of Chemical Kinetics*, McGraw Hill, New York, 1960.

27 Y. Bedjanian, G. LeBras, and G. Poulet, Kinetic study of OH+ OH and OD+ OD reactions, *J. Phys. Chem. A*, 1999, **103**, 7017.

28 D.H. Maylotte, J.C. Polanyi, and K.B. Woodall, Energy distribution among reaction products. IV. X+HY (X≡C1, Br; Y≡Br, I), Cl+DI, *J. Chem. Phys.*, 1972, **57**, 1547.

29 H.F. Schaeffer III, The atomic fluorine+ molecular hydrogen potential energy surface: the ecstasy and the agony, *J. Phys. Chem.*, 1985, **89**, 5336.

30 B. Nickel, K.H. Grellmann, J.S. Stephan, and P.J. Walla, Keto-enol tautomerism in the triplet state of hydroxyphenylbenzoxazoles in an alkane glass: hydrogen tunneling and isotope effects down to 2 K, *Ber. Bunsenges.*, 1998, **102**, 436.

31 C.C. Mei and C.B. Moore, Thermal rate constants, energy dependence, and isotope effect for halogen–hydrogen halide reactions, *J. Chem. Phys.*, 1979, **70**, 1759.

32 J.E. Baldwin and B.R. Chapman, Gas-phase kinetics and activation parameters for thermal 1,5-hydrogen shifts interconverting monodeuterium-labeled 1,3-cyclohexadienes, *J. Org. Chem.*, 2005, **70**, 377.

33 N.C. Blais, R.B. Bernstein, and R.D. Levine, Orientation dependence of the atomic hydrogen+ molecular deuterium reaction cross section: steric model vs. trajectory calculations, *J. Phys. Chem.*, 1985, **89**, 10.

26

Macromolecules

CHAPTER MENU

26.1 Introduction

The principles of self-organizing molecular structures are needed for understanding the occurrence of complexity in natural and artificial supramolecular systems. Some of the factors that control these assemblies are already manifest in the structure of macromolecules—comprising many covalently appended molecular subunits that form a large molecule, greater than a few nm in diameter. Many kinds of macromolecules exist, including artificial polymers and biopolymers, and they can display a variety of properties and functions, e.g., complex nano-structured polymers can act as optical sensors [1] and many proteins act as catalysts or enzymes with functions and properties distinct from their monomeric subunits. The huge variety of properties is related to particular features of the macromolecular structure. In this chapter we focus on polymers forming coils and illustrate how physicochemical principles, which we have discussed already, can be used and developed to understand their change of shape caused by external forces, uncoiling and recoiling, winding through mesh works, etc.

26.2 Random Coil

We first consider a homopolymer, which is a macromolecule consisting of identical monomers in a chain such as

Polyethylene $-CH_2-CH_2-CH_2-CH_2-CH_2-CH_2-$

Polystyrene

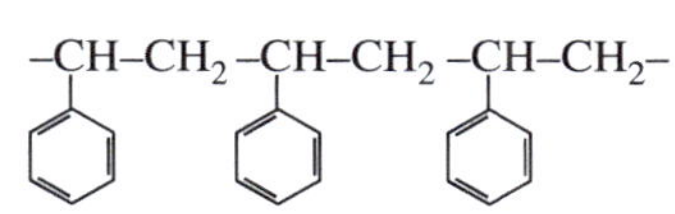

Principles of Physical Chemistry, Third Edition. Hans Kuhn, David H. Waldeck, and Horst-Dieter Försterling.

Companion website: www.wiley.com/go/Kuhn/physicalchemistry3e

Cellulose

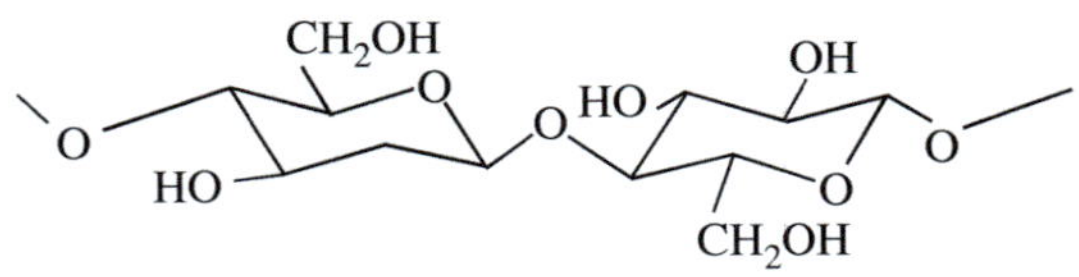

Because of synthetic limitations, man-made polymer materials often have a distribution of monomer units in their composition. To simplify the analysis, we will assume that the polymer materials comprise chains that all have the same number of monomer units. In our treatment the molar mass $\boldsymbol{M}$ of a polymer chain is

$$\boldsymbol{M} = N_{\text{monomer}} \cdot \boldsymbol{M}_{\text{monomer}} \tag{26.1}$$

where $\boldsymbol{M}_{\text{monomer}}$ is the molar mass of the monomer unit and N_{monomer} is the number of monomers in the chain.

The shape of macromolecules in dilute solution and related properties (such as viscosity, flow birefringence, light scattering) is subject to statistical considerations. To see this, we first consider the small molecule butane that can exist in three different conformations, corresponding to rotation about C–C bonds (one *trans* and two *gauche* conformations, Fig. 26.1a).

The number of conformations of equal or nearly equal energy increases with the number of carbon atoms in the chain. If we define the $-CH_2-$ units as monomers, and neglect steric restrictions and the energy difference between *trans* and *gauche* forms, then the number of equally probable conformations for a hydrocarbon chain is

$$\text{number of conformations} = 3^{N_{\text{monomer}}-3},$$

with $N_{\text{monomers}} \geq 3$. In our example of butane $N_{\text{monomer}} = 4$ and thus the number of conformations is 3. In the case of propane ($N_{\text{monomer}} = 3$) only one conformation is possible. For a polymer with 100 carbon atoms the number of conformations is 3^{97}. Only one of these corresponds to the fully extended (all *trans*) conformation, and the rest correspond to an immense number of coiled conformations.

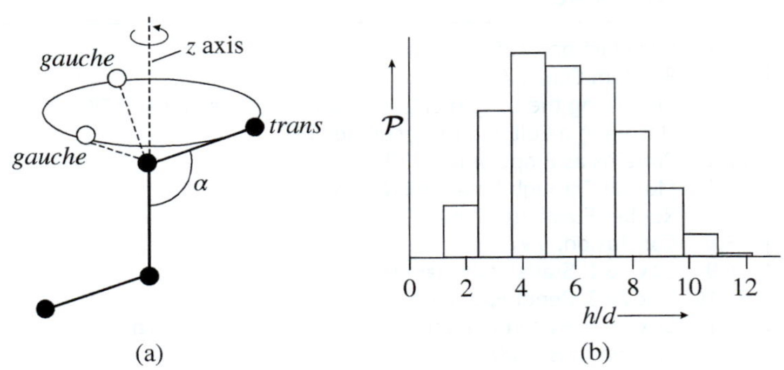

Figure 26.1 (a) Conformations of butane (one *trans*- and two *gauche*-conformations). In the *trans* conformation the four carbon atoms (full circles) are in the paper-plane, in the *gauche* conformations the carbon atom on top is above or below the paper plane (open circles). The bond angles are 109°28′ (the tetrahedral angle) and the number of conformations is $3^{N_{\text{monomer}}-3} = 3^1 = 3$, where N_{monomer} is the number of CH_2 groups ($N_{\text{monomer}} = 4$ in this case). Definition of z axis and bond angle α. (b) The calculated distribution function $\mathcal{P}$ of the end-to-end distance h is shown for a hydrocarbon with $N_{\text{monomer}} = 17$, assuming equal probabilities of all $3^{17-3} = 3^{14} = 4{,}782{,}969$ *trans* and *gauche* conformations. The resulting average of h^2 is $\overline{h^2} = 30.52 \cdot d^2 = 1.79 \cdot N_{\text{monomer}} \cdot d^2$.

Because a polymer chain in solution has many molecular conformations available at any given time, we characterize a polymer solution by the distribution in size and shape of the polymer chains and their average properties. To illustrate, we consider the average of the square of the end-to-end distance h of a polymer chain, see Fig. 26.2.

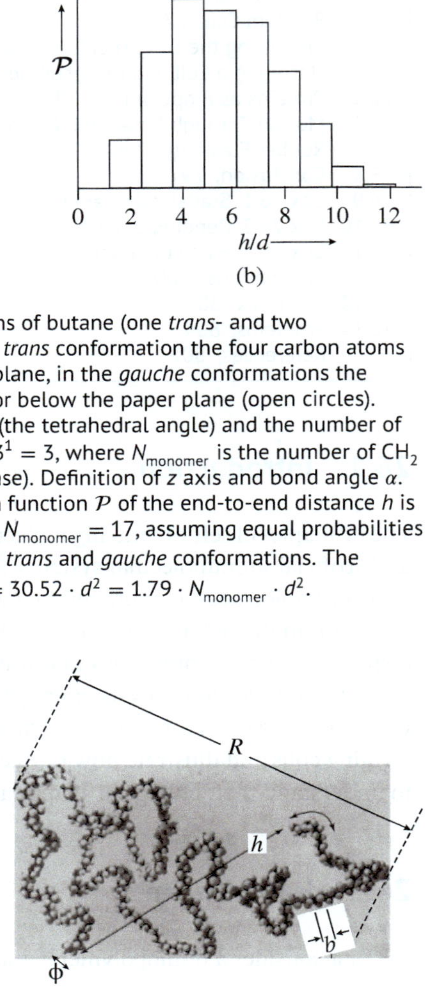

Figure 26.2 Parameters of a coil. R is the extension of the coil, h is the distance between the chain ends, b is the length of a monomer unit, and ϕ is the cross section diameter of the chain. The length L of the chain is $L = bN_{\text{monomer}}$, and the molar mass of the polymer is $M_m = N_{\text{monomer}}\boldsymbol{M}_{m,\text{monomer}}$, where N_{monomer} is the number of monomer units and $\boldsymbol{M}_{m,\text{monomer}}$ is the molar mass of the monomer.

Consider the case of a hydrocarbon chain and assume that the *trans*- and *gauche*-conformations are equally probable, the bond angle α (Chapter 7, tetrahedral angle $\alpha = 109°28'$) is fixed, and rotation about the z axis is free (see Fig. 26.1a). Then the end-to-end distance h of a given chain for each conformation can be calculated numerically, and we obtain the distribution function $\mathcal{P}(h)$. The result for the case $N_{\text{monomer}} = 17$ is shown in Fig. 26.1b. For the average of h^2 we find $\overline{h^2} = 30.52 \cdot d^2 = 1.79 \cdot N_{\text{monomer}} \cdot d^2$, where d is the bond length of the monomer.

Generally, for sufficiently large N_{monomer}, it can be shown [2] that

$$\overline{h^2} = 2 \cdot N_{\text{monomer}} \cdot d^2 \qquad (N_{\text{monomer}} \gg 1) \tag{26.2}$$

Actually, the *gauche* conformations are sterically hindered. A more quantitative treatment, which accounts for the tetrahedral geometry of the methylene units and includes the energy barrier for rotation around the C–C bonds, leads to a more extended chain (larger $\overline{h^2}$). In addition to these features arising from the bond connectivity of the chain, one must also account for a steric factor (i.e., a region of space already occupied by part of the polymer chain is not accessible to another part of the chain, like an excluded volume); see [3]. These considerations show that the *trans* conformation is about three times more probable than each of the *gauche* conformations at room temperature, and it follows that

$$\overline{h^2} = 7.1 \cdot N_{\text{monomer}} \cdot d^2 \tag{26.3}$$

26.2.1 A Chain of Statistical Chain Elements

In a more general description we ask for the average size and for the distribution in size and shape of a polymer. Two quantities are important to approach this task in a simplified manner, the *contour length* and the *statistical chain length*.

26.2.1.1 Contour Length

The contour length L is the length of the path from head to tail along the polymer chain

$$L = N_{\text{monomer}} \cdot b \tag{26.4}$$

where b is the contribution of a monomer to the contour length. In the case of a hydrocarbon (polyethylene) $b = d \cdot \sin(\alpha/2) = 154\text{ pm} \sin(54.5°) = 125\text{ pm}$.

26.2.1.2 Length of Statistical Chain Elements

The length l of a statistical chain element is an abstraction that we use to describe the size and shape of actual polymer chains in terms of the *random coil model*. In this description, we replace the chain of N_{monomer} monomer units of length b by a number N_{stat} of straight elements of length l, such that

$$L = N_{\text{stat}} \cdot l \tag{26.5}$$

Each of the N_{stat} elements is directed at random, i.e., each direction in space has an a priori equal probability. These elements are called *statistical chain elements*, see Fig. 26.3.

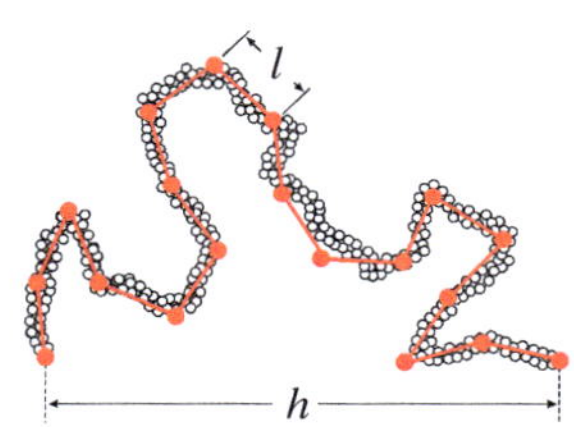

Figure 26.3 Polymer coil (total chain length L) replaced by a chain of $N_{\text{stat}} = L/l$ statistical chain elements of length l. Each element is directed at random. l depends on the nature of the monomer. For a hydrocarbon, $l = 1$ nm.

In Box 26.1 we show that the probability $\mathcal{P}(h) \cdot \mathrm{d}h$ of finding the end-to-end distance between h and $h + \mathrm{d}h$ equals

$$\mathcal{P}(h) \cdot \mathrm{d}h = const \cdot 4\pi h^2 \cdot \exp\left(-\frac{3h^2}{2N_{\text{stat}} \cdot l^2}\right) \cdot \mathrm{d}h \tag{26.6}$$

for the random coil model and that the mean square of h is

$$\overline{h^2} = N_{\text{stat}} \cdot l^2 \tag{26.7}$$

Using Equation (26.5) we rewrite Equation (26.7) as

$$\overline{h^2} = L \cdot l \tag{26.8}$$

Hence, we find that

$$l = \frac{\overline{h^2}}{L} \quad \text{and} \quad N_{\text{stat}} = \frac{L}{l} = \frac{L^2}{\overline{h^2}}$$

The statistical chain length l and the number N_{stat} of statistical chain elements can be derived from the contour length L (given by the molecular structure) and $\overline{h^2}$ (experimentally obtained from optical or hydrodynamic properties of polymer solutions, see below).

Example 26.1 *Calculate the statistical chain element in a hydrocarbon chain, with the bond length $d = 154$ pm and $b = 125$ pm.*

The parameter b is the contribution of a monomer unit to the contour length L, see Equation (26.4). According to Equation (26.3),

$$\overline{h^2} = 7.1 \cdot N_{\text{monomer}} \cdot d^2$$

and 26.8

$$\overline{h^2} = L \cdot l = b \cdot N_{monomer} \cdot l$$

where l is the length of the statistical chain element. Combining these results, we obtain

$$l = \frac{\overline{h^2}}{b \cdot N_{monomer}} = \frac{7.1 \cdot N_{monomer} \cdot d^2}{b \cdot N_{monomer}} = \frac{7.1 \cdot d^2}{b}$$

Using $d = 154$ pm and $b = 125$ pm for a methylene chain, we obtain $l = 1.3$ nm. This is in good agreement with the experimentally determined length of the statistical chain element [4] of $l = 1.1$ nm.

26.2.1.3 Probability Distribution of Coil Diameter

The picture of a random coil can be extended to cases such as DNA double helices where the molecular strand is bent by thermal fluctuations and its elasticity, not conformational changes. As outlined in Box 26.2, a bending of 90° occurs over a section of 100 nm, on average, along the strand at room temperature. That is, in this section of the DNA double helix the memory about the direction of the strand's contour is effectively lost, and the polymer is said to be "worm-like." In modeling DNA we replace this section by a statistical chain element, i.e., we describe the worm-like DNA double helix as a random coil consisting of statistical chain elements of about 100 nm length.

The prediction of the model can be checked for DNA by labeling it with a fluorescent dye molecule. Using fluorescence microscopy, the molecule can be directly observed. The extension of the coil, R (defined in Fig. 26.2 for the case of a hydrocarbon) can be measured, and its probability distribution is obtained (Fig. 26.4, dots). The solid line in Fig. 26.4 gives the theoretical distribution function $\mathcal{P}(R)$ for $l = 100$ nm. The value agrees well with the value of l estimated in Box 26.2.

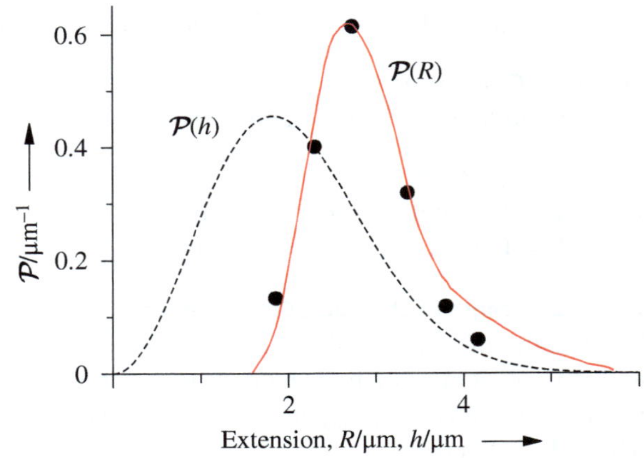

Figure 26.4 DNA double helix with $N_{monomer} = 1.5 \times 10^5$ base pairs ($l = 100$ nm; $L = 50$ μm, $N_{stat} = 500$). Distribution $\mathcal{P}(R)$ of maximum extension R (solid line). The distribution of R has been measured on fluorescence-labeled strands observed [5] in a microscope (dots) and compared with the theoretical [6] distribution. $\mathcal{P}(R) \cdot dR$ is the probability of finding the maximum extension of the coil between R and $R + dR$. On average, R is $1.4\sqrt{L \cdot l}$, this is somewhat larger than $\sqrt{\overline{h^2}}$. For comparison, the distribution $\mathcal{P}(h)$ of the end-to-end distance h, according to Equation (26.6), is shown (dashed line).

26.3 Measuring the Length of Statistical Chain Elements

A number of experimental methods are used to measure the average size and shape of polymer chains. Here we discuss how light scattering, diffusion, sedimentation, and viscosity measurements can be used to quantify $\overline{h^2}$ and hence l.

26.3.1 Light Scattering

Solutions that are created by dissolving macromolecules in a solvent are turbid; i.e., they strongly scatter light and have a milky appearance. By quantifying the turbidity we can quantify the molecular size of the macromolecule.

Case of $\sqrt{\overline{h^2}} \ll \lambda$: Imagine illuminating a dilute solution of macromolecules with light polarized in the z-direction (Fig. 26.5). First we consider solute molecules that are small compared to the wavelength λ of light (Fig. 26.5a), and we assume that the light is not resonant with any transitions in the molecule. In this case the applied light field polarizes the electron cloud of the molecule and Rayleigh scattering (see Chapter 11) occurs. Each molecule can be considered as a dipole oscillator, oscillating with the frequency of the incident light, that emits light of the same intensity and wavelength in all directions in the xy-plane. No light is emitted in the z-direction, as predicted by the Hertz antenna equation [7].

In Section 11.4.1, Equation (11.50), we found that the intensity $I_{scattered}$ of the scattered light of one molecule is

$$I_{scattered} = I_{incident} \cdot \left(\frac{1}{4\pi\varepsilon_0}\right)^2 \frac{\alpha^2\omega^4}{c_0^4 d^2} \cdot \sin^2\vartheta$$

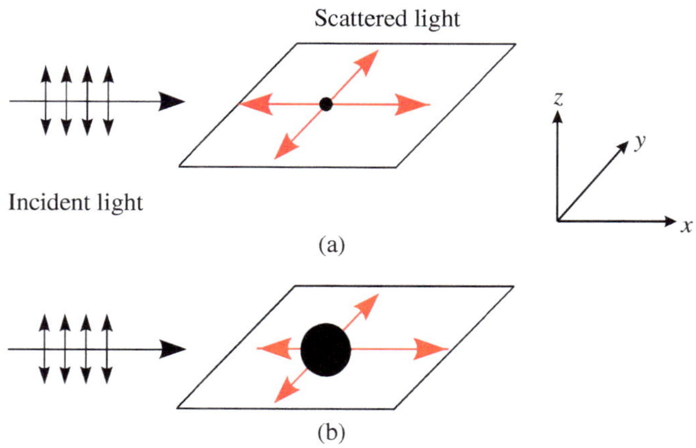

Figure 26.5 Light scattering by a polymer coil. The incident light wave is propagating in the x-direction and linearly polarized in the z direction. The coil is represented as a sphere of radius $r = \frac{1}{2}\sqrt{\overline{h^2}}$. (a) $\sqrt{\overline{h^2}} \ll \lambda$: equal intensity of forward and back scattering. (b) $\sqrt{\overline{h^2}} \geq \lambda/4$: intensity of forward scattering is about twice that of back scattering. This difference arises from interference effects, see Fig. 26.6a and Exercise 26.19.

where d is the distance between the sample and the observer, α is the polarizability of the scattering molecule, ω is the angular frequency, I_{incident} is the intensity of the incident light, and ϑ is the angle between the direction of the electric field strength of the incident light (z direction in Fig. 26.5a) and the xy plane. In the following we restrict our consideration to the case $\vartheta = 90°$, that is, $\sin^2\vartheta = 1$. Then the intensity $I_{\text{scattered}}$ of N molecules is

$$I_{\text{scattered}} = I_{\text{incident}} \cdot N \cdot \left(\frac{1}{4\pi\varepsilon_0}\right)^2 \frac{\alpha^2\omega^4}{c_0^4 d^2}$$

For molecules in a liquid solution, α is the difference between the polarizability of the macromolecule and that of the solvent molecules displaced by the macromolecule. The polarizability α is obtained by measuring the refractive index $\hat{n}_{\text{solution}}$ of the dilute solution of macromolecules and the refractive index $\hat{n}_{\text{solvent}}$ of the pure solvent. We show in Foundation 26.1 that

$$\alpha = \frac{\varepsilon_0 V \cdot 2\hat{n}_{\text{solution}} \left(\hat{n}_{\text{solution}} - \hat{n}_{\text{solvent}}\right)}{N}$$

and $I_{\text{scattered}}$ becomes

$$I_{\text{scattered}} = I_{\text{incident}} \cdot \left(\frac{1}{4\pi\varepsilon_0}\right)^2 \cdot \left[\varepsilon_0 V \cdot 2\hat{n}_{\text{solution}} \left(\hat{n}_{\text{solution}} - \hat{n}_{\text{solvent}}\right)\right]^2 \cdot \frac{\omega^4}{c_0^4 d^2} \cdot \frac{1}{N}$$

Finally, we replace N and ω by

$$N = \frac{m_{\text{total}}}{m}, \quad \text{and} \quad \omega = 2\pi\nu = \frac{2\pi c_0}{\lambda}$$

where m_{total} is the total mass of the dissolved polymer molecules in volume V and m is the molecular mass. Then

$$I_{\text{scattered}} = I_{\text{incident}} \cdot 4\pi^2 \cdot (\hat{n}_{\text{solution}} - \hat{n}_{\text{solvent}})^2 \cdot \hat{n}_{\text{solution}} \frac{V^2}{m_{\text{total}} d^2} \frac{m}{\lambda^4} \tag{26.9}$$

The strong dependence of $I_{\text{scattered}}$ on λ arises from the strongly increasing field strength of an oscillating dipole with increasing frequency.

This method allows us to determine the molar mass $\boldsymbol{M} = \boldsymbol{N}_{\text{A}} \cdot m$ from the scattering intensity. In the present case this method does not provide information on the size of the particles.

Case of $\sqrt{\overline{h^2}} \geq \lambda/4$: If the solute particle is no longer small compared to the wavelength of the incident light, then the intensity of the light scattered in the forward direction is higher than the intensity in the backward direction (Fig. 26.5b) and conclusions about the size of the particle can be drawn.

To understand how the scattering depends on molecular size consider the solute particle to have two scattering centers A and B at a distance a from each other (Fig. 26.6a). For light scattered in the direction of the incident light (forward scattering), the delay of the wave emitted from center B relative to center A is compensated by the shorter path; thus, both beams arrive at the observer (point P_2) with nearly the same phase. For the opposite direction (backward scattering), the light scattered from point B travels an extra distance $2a$ and is phase shifted by $2a/\lambda$. Superposition of the two backward scattered wave amplitudes generates a smaller intensity at point P_1. As derived in Foundation 26.1 the intensity ratio for the forward and backward waves is

$$\frac{I_{\text{backward}}}{I_{\text{forward}}} = 1 - \left(\frac{2\pi a}{\lambda}\right)^2 = 1 - \frac{\pi^2 \overline{h^2}}{\lambda^2} \tag{26.10}$$

This ratio is 1 for $a \ll \lambda$ and 0.38 for $a = \lambda/8$. By measuring this ratio the root mean square diameter of the polymer coil,

$$\sqrt{\overline{h^2}} \approx 2a$$

can be obtained, thereby allowing us to determine the length l of the statistical chain element, according to Equation (26.8).

For a coil of larger size, the model of two scattering centers becomes unrealistic. Rather we must include multiple scattering centers in the macromolecule. We can proceed by identifying each statistical element of the polymer chain as a scattering center, and perform a summation over the contributions of each chain element's scattered wave to find the total scattered intensity. For incident light polarized in the z-direction, this summation gives the expression [8]

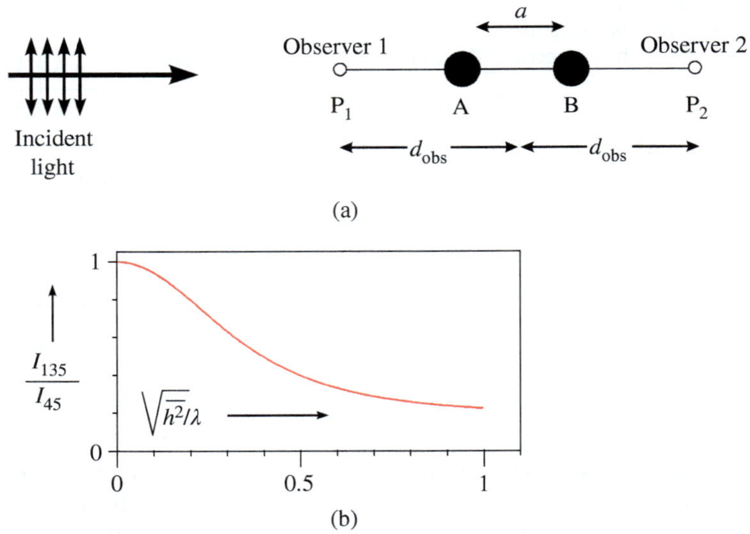

Figure 26.6 Light scattering on macromolecules. (a) Light scattered from points A and B reach point P_2 with equal phase. In contrast, point P_1 is reached by light scattered at points A and B with phase shift $2a/\lambda$, where $a = \frac{1}{2}\sqrt{\overline{h^2}} = r_{\text{coil}}$. (b) Scattering function I_{135}/I_{45} versus $\sqrt{\overline{h^2}}/\lambda$ according to Equation (26.11).

$$\frac{I_\varphi}{I_{\text{forward}}} = \frac{2}{C^2}\left(e^{-C} - 1 + C\right) \quad \text{with} \quad C = \frac{8\pi^2}{3}\frac{\overline{h^2}}{\lambda^2}\sin^2\left(\frac{\varphi}{2}\right) \tag{26.11}$$

I_φ is the intensity of the light scattered in the xy-plane in a direction forming an angle φ with the direction of the incident beam ($I_{\varphi=0} = I_{\text{forward}}$). When the parameter C is smaller than unity, Equation (26.11) approximates Equation (26.10), see Exercise E26.19.

Experimentally, it is often easier to measure light scattering at 45° and 135° rather than at 0° and 180°. The ratio $I_{135°}/I_{45°}$ obtained from Equation (26.11) is displayed in Fig. 26.6b as a function of $\sqrt{\overline{h^2}}/\lambda$, see Exercise E26.2.

Example 26.2 *Consider a solution of polystyrene showing a scattering ratio* $I_{135°}/I_{45°} = 0.5$ *at a wavelength of* $\lambda = 500$ *nm. Determine the coil diameter* $\sqrt{\overline{h^2}}$.

Figure 26.6b shows how the scattering function I_{135}/I_{45} changes with $\sqrt{\overline{h^2}}/\lambda$. Reading from the graph, we find $\sqrt{\overline{h^2}}/\lambda = 0.4$ for $I_{135°}/I_{45°} = 0.5$ nm. Then $\sqrt{\overline{h^2}} = 0.4 \cdot \lambda = 0.4 \cdot 500\ \text{nm} = 0.2\ \mu\text{m}$.

26.3.2 Hydrodynamics: Coil Approximated as a Sphere

We begin with a simple model describing the behavior of very large coils consisting of many statistical chain elements. In this limit of large N_{stat}, the chain can be viewed in a first approximation as spheres of radius r_{coil}, where we set $r_{\text{coil}} = \frac{1}{2}\sqrt{\overline{h^2}}$. The description for the diffusion of colloid particles (Chapter 14) and our treatment for the mobility of ions (see Section 23.7) are based on the same idea, considering the solvent as a continuum of macroscopic viscosity. Then, according to Equation (26.8),

$$r_{\text{coil}} = \frac{1}{2}\sqrt{\overline{h^2}} = \frac{1}{2}\sqrt{Ll} \tag{26.12}$$

We consider the coil to be "undrained" when moving through the solvent, i.e., any solvent molecules that are inside the coil remain so, and the solvent can be represented by a homogeneous medium of macroscopic viscosity η. Thus the steady-state speed v of the particle under a constant applied force f can be calculated through Stokes' law

$$f = \zeta \cdot v = 6\pi\eta \cdot r_{\text{coil}} \cdot v \tag{26.13}$$

where ζ is the friction coefficient.

26.3.2.1 Diffusion

In this approximation, we can substitute the hydrodynamic expression for the friction coefficient into Einstein's relation for the diffusion coefficient, $D = kT/\zeta$, and obtain the Stokes–Einstein expression for the diffusion coefficient D of the polymer coil (see Chapter 14), which is

$$D = \frac{kT}{6\pi\eta \cdot r_{\text{coil}}} \tag{26.14}$$

Because

$$L = N_{\text{monomer}} b, \quad (b = \text{length of a monomer unit}) \tag{26.15}$$

we can use Equation (26.12) to write

$$2r_{\text{coil}} = \sqrt{L \cdot l} = \sqrt{N_{\text{monomer}} bl} \tag{26.16}$$

Substituting this result in our expression for D, we find

$$\boxed{D = \frac{kT}{3\pi\eta} \cdot \frac{1}{\sqrt{b \cdot l}} \frac{1}{\sqrt{N_{\text{monomer}}}}} \tag{26.17}$$

Thus, if we know the number N_{monomer} and length b of the monomer units, we can determine the statistical length l of the polymer by measuring its diffusion coefficient.

26.3.2.2 Sedimentation

In the artificial gravity field of an ultracentrifuge, a polymer molecule experiences a force f. In vacuum, this force would be

$$f = m \cdot g_{\text{centrifuge}} \tag{26.18}$$

where m is the mass of the polymer molecule and $g_{\text{centrifuge}}$ is the acceleration in the ultracentrifuge:

$$g_{\text{centrifuge}} = \omega^2 r_{\text{axis}} \tag{26.19}$$

(ω = angular speed, r_{axis} = distance from axis of revolution). Typically, $g_{\text{centrifuge}}$ is on the order of $10^5\ \text{m s}^{-2}$, that is 10,000 times that of earth's gravitational field.

In a solvent of density ρ the force must be corrected for buoyancy, and it becomes

$$f = m \cdot g_{\text{centrifuge}} \cdot \left(1 - \rho \cdot v_{\text{part}}\right) \tag{26.20}$$

v_{part} is the partial specific volume of solute, which is given by

$$v_{\text{part}} = \frac{V_{\text{solution}} - V_{\text{solvent}}}{m_{\text{solute}}} \tag{26.21}$$

If V_{solution} can be considered as $V_{\text{solution}} = \left(V_{\text{solvent}} + V_{\text{solute}}\right)$, we obtain

$$v_{\text{part}} \approx \frac{V_{\text{solute}}}{m_{\text{solute}}} = \frac{1}{\rho_{\text{solute}}}$$

The molecule is accelerated in the centrifuge until the force acting on it is matched by the friction at a certain speed v and reaches a steady state. If the friction is given by Stokes' law, then with Equations (26.12) and (26.4), we can write

$$\begin{aligned} f &= 6\pi\eta \cdot r_{\text{coil}} \cdot v = 6\pi\eta \cdot \frac{1}{2}\sqrt{L \cdot l} \cdot v \\ &= 3\pi\eta\sqrt{b \cdot l}\sqrt{N_{\text{monomer}}} \cdot v \end{aligned} \tag{26.22}$$

Upon equating the two force expressions we find

$$3\pi\eta\sqrt{b \cdot l}\sqrt{N_{\text{monomer}}} \cdot v = m \cdot g_{\text{centrifuge}} \cdot \left(1 - \rho \cdot v_{\text{part}}\right)$$

By replacing m by m_{monomer} we obtain

$$3\pi\eta\sqrt{b \cdot l}\sqrt{N_{\text{monomer}}} \cdot v = m_{\text{monomer}} \cdot N_{\text{monomer}} \cdot g_{\text{centrifuge}} \cdot \left(1 - \rho \cdot v_{\text{part}}\right)$$

This equation rearranges to give the ratio

$$\boxed{s = \frac{v}{g_{\text{centrifuge}}} = \frac{m_{\text{monomer}} \cdot (1 - v_{\text{part}}\rho)}{3\pi\eta \cdot \sqrt{bl}} \sqrt{N_{\text{monomer}}}} \tag{26.23}$$

where s is called the *sedimentation constant*. In typical cases, the sedimentation speed v is very small: about 1 cm per day for $g_{\text{centrifuge}} = 10^5$ m s^{-2}.

Molecular Mass of Any Macromolecule from D and s By combining Equations (26.17) and (26.23) we obtain

$$\boxed{s = \frac{m \cdot D}{kT}(1 - v_{\text{part}}\rho)} \tag{26.24}$$

showing that we can determine m from D and s. Equation (26.24) is not restricted to statistical coils, but holds for any small particle. This relation is important for determining the molecular mass $\boldsymbol{M}$ of macromolecules.

26.3.2.3 Viscosity

If macromolecules are added to a solvent, the hydrodynamic properties of the solution are different from the pure solvent. Here we discuss the change in the shear viscosity.

When small spheres are suspended in a liquid, the viscosity increases (Fig. 26.7). At high dilution, when the interaction between the spheres can be neglected, we assume that the viscosity increase is proportional to the number of particles per unit volume, thus proportional to the volume fraction x_{volume}, where

$$x_{\text{volume}} = \frac{\text{volume of spheres in suspension}}{\text{volume of suspension}} = \frac{V_{\text{spheres}}}{V} \tag{26.25}$$

In this limit, the viscosity of the suspension η is given by

$$\eta = \eta_0 \cdot (1 + const \cdot x_{\text{volume}}) \tag{26.26}$$

where η_0 is the viscosity of the pure solvent (for definition of viscosity see Chapter 14) and *const* is a constant.

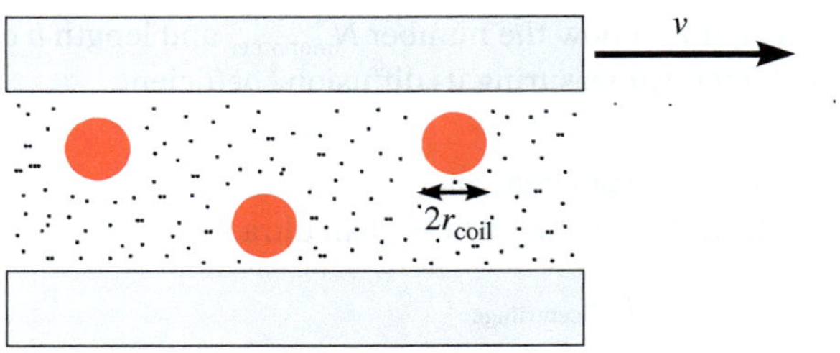

Figure 26.7 Increase in the viscosity by suspending spheres (filled circles) in a solvent (indicated by small dots). Measurement of the viscosity by fixing the solution between two plates, the upper plate moving at a speed v relative to the lower plate, see Fig. 14.21.

If the macromolecular coil including solvent is undrained when applying a flow gradient, it behaves roughly like a sphere of radius r_{coil}. The effective volume occupied by all of the spheres is

$$V_s = N\frac{4}{3}\pi r_{\text{coil}}^3 \tag{26.27}$$

where N is the number of polymer coils. Using Equations (26.25) and (26.26) we obtain

$$x_{\text{volume}} = \frac{N}{V}\frac{4}{3}\pi r_{\text{coil}}^3 = \frac{1}{const} \cdot \left(\frac{\eta}{\eta_0} - 1\right) \tag{26.28}$$

and

$$\left(\frac{\eta}{\eta_0} - 1\right) = const \cdot \frac{N}{V}\frac{4}{3}\pi r_{\text{coil}}^3 \tag{26.29}$$

The number N of coils can be expressed by

$$N = \frac{m_{\text{total}}}{m} \tag{26.30}$$

where m_{total} is the mass of the polymer dissolved in volume V and m is the molecular mass.

We make the connection that

$$2r_{\text{coil}} = \sqrt{\overline{h^2}} \tag{26.31}$$

to obtain

$$\left(\frac{\eta}{\eta_0} - 1\right) = const \cdot \frac{N}{V}\frac{4}{3}\pi \cdot \frac{1}{8}\left(\overline{h^2}\right)^{3/2} \tag{26.32}$$

Einstein used hydrodynamics and calculated the energy dissipation in the liquid around a sphere in a flow gradient to find the value *const* = 2.5 [9]. Using that result, we obtain

$$\left(\frac{\eta}{\eta_0}-1\right) = 2.5\cdot\frac{m_{\text{total}}}{V}\frac{4}{3}\pi\cdot\frac{1}{8}\left(\overline{h^2}\right)^{3/2}\cdot\frac{1}{m} = 1.31\cdot\left(\overline{h^2}\right)^{3/2}\frac{1}{m}\cdot\frac{m_{\text{total}}}{V}$$

or

$$\frac{\dfrac{\eta}{\eta_0}-1}{m_{\text{total}}/V} = 1.31\cdot\left(\overline{h^2}\right)^{3/2}\frac{1}{m}$$

Finally we consider the limiting value of this expression for $m_{\text{total}}/V \to 0$, and find

$$\boxed{[\eta] = \lim_{m_{\text{total}}/V\to 0}\left(\frac{\dfrac{\eta}{\eta_0}-1}{m_{\text{total}}/V}\right) = 1.31\cdot\left(\overline{h^2}\right)^{3/2}\cdot\frac{1}{m}} \tag{26.33}$$

$[\eta]$ is called the *Staudinger index*. It is a measure of the relative viscosity increase by the dissolved polymer in the limit of high dilution.

26.3.3 Hydrodynamics: Macroscopic Modeling

Our assumption in Section 26.3.2 that the coils consisting of many statistical chain elements are spheres and that they are undrained when moving through the solvent or exposing them to a flow gradient is a severe simplification. The coil becomes increasingly drained as the length of the chain decreases. Finding the resistance of a random coil to flow is a formidable problem because of the coil's complicated, continuously changing shape. How can we take account of that effect? If we keep the approximation of the solvent as a continuum of macroscopic viscosity, hydrodynamics tells us that the resistance (friction) to flow is still proportional to the viscosity even though the particle is not a sphere. However, the numerical factor in $\zeta = 6\pi\eta\cdot r_{\text{coil}}$ is no longer 6π; rather, it depends on the shape of the particle. We call this numerical factor a form factor.

Our task is to find the form factor for coils of given shape. Solving this kind of hydrodynamic problem remains challenging; however, advancements in algorithms and computer hardware make it possible to numerically test and assess different models for the hydrodynamic friction and its change with the properties of a macromolecule (e.g., drained versus undrained and shape). Here we bypass this formal approach by performing experiments with macroscopic models of coils. The situation is similar to investigating airplanes by performing experiments with much smaller models in wind tunnels. In the following we describe the result of many experiments on macroscopic models. The data obtained on models that correspond to different conformations of a given macromolecule are averaged (Monte Carlo method). A sufficient number of experiments had to be performed to get a reproducible average and corresponding to different values of the number of statistical chain elements, N_{stat}, the length of a statistical chain element, l, and the chain diameter, ϕ. These experiments give rise to an empirical formula for the hydrodynamic friction and is in good agreement with more sophisticated and modern calculations [10].

Models of polymer coils are made by bending wires in a random fashion to statistical coils in the range from $N_{\text{stat}} = 1$ to 100. The models are made by dividing the contour length L into L/l_{segment} equal segments of length l_{segment} and by bending each segment randomly by an angle between 0° and 180°. This kind of model of random coils (Fig. 26.8a) fits better to the contour line of the actual molecule than the model made of stretched segments (Fig. 26.3). It can be demonstrated that the length l_{segment} of a

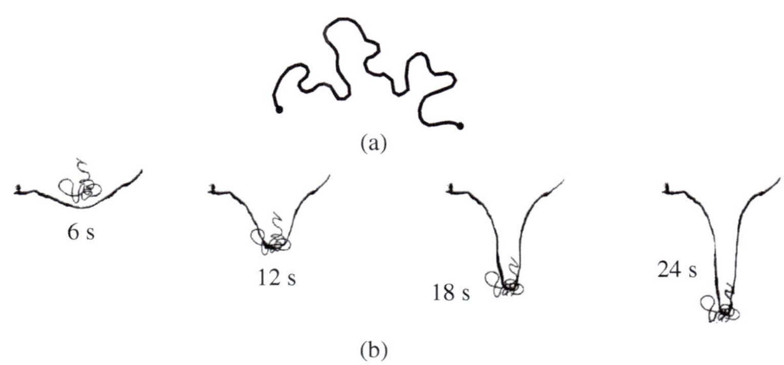

Figure 26.8 Macroscopic models of random coil. (a) Model made of wire by bending segments of length $l_{\text{segment}} = 0.65l$ in a random fashion. (b) Model falling through a viscous liquid. A horizontal strip of ink in the viscous liquid is carried along with the model showing that the coil is essentially undrained. (From Ref. [12], with permission.)

statistical segment must be made by a factor $1/1.53 = 0.65$ shorter than the length l of a statistical chain element [11] in order to correspond to the same value of $\overline{h^2}$ and to the same contour length L. The wire models are dropped in a viscous liquid and the speed of fall is measured (*translational motion*); furthermore, they are rotated about their axes of gyration to search for the behavior of macromolecules in a flow gradient (*rotational motion*).

Figure 26.8b shows such a model during its fall through a viscous liquid (translational motion). The limited draining is seen by the deformation of the mark—a horizontal strip of ink in the viscous liquid. For hydrodynamic reasons the solvent flows through the coil at its border and the coil is essentially undrained in its center. From experiments with many models a statistical average can be obtained. The result of these measurements can be represented by simple mathematical expressions. On average, the force f to move a model of a random coil with speed v through the solvent is given by

$$f = C_{\text{transl}} \cdot \eta \cdot r_{\text{coil}} \cdot v \tag{26.34}$$

where C_{transl} is called the *translational form factor*. Equation (26.34) follows from hydrodynamics: for any object's shape the drag force is proportional to its speed, the linear dimensions of the body and the shear viscosity of the medium through which the body moves. Over a broad range of values for l, N_{monomer}, and ϕ the translational form factor C_{transl} can be represented by

$$\boxed{C_{\text{transl}} = \frac{15}{1 + B/\sqrt{N_{\text{monomer}}}} \quad \text{with} \quad B = 0.5\sqrt{\frac{l}{b}} \cdot \ln\left(\frac{l}{\phi}\right)} \tag{26.35}$$

According to Equation (26.35), the translational form factor C_{transl}, in the limiting case of a large number of statistical chain elements (undrained coils, N_{monomer} large), is $C_{\text{transl}} = 15$, and this is close to the value expected when approximating the coil as a sphere (Section 26.3.2, form factor $= 6\pi = 18.8$).

In the case of molecules consisting of a small number of statistical chain elements (drained chain) with $r_{\text{coil}} = \frac{1}{2}\sqrt{Ll}$, Equation (26.12), and $N_{\text{monomer}} b = L$, Equation (26.4), we obtain

$$f = \frac{15}{0.5\sqrt{l/b} \cdot \ln(l/\phi)} \sqrt{N_{\text{monomer}}} \eta r_{\text{coil}} \cdot v = \frac{15}{\ln(l/\phi)} \cdot L\eta \cdot v \tag{26.36}$$

If we use $l = 100$ nm and $\phi = 3$ nm for DNA, then $\ln(l/\phi) = 3.5$, and the numerical factor in Equation (26.36) is about 4.2. The proportionality of f with L is expected from a simple approximation (necklace model), that will be used in Section 26.4.2, Equation (26.47).

$$f = \frac{3}{2}\pi\eta L v = 4.7 \cdot L\eta \cdot v$$

26.3.3.1 Diffusion

To introduce the translational form factor C_{transl} into the expression for the diffusion coefficient, we replace 6π in Stokes law with C_{transl}. Thus we find the ratio

$$\frac{6\pi}{C_{\text{transl}}} = \frac{3\pi}{7.5} \cdot \left(1 + B/\sqrt{N_{\text{monomer}}}\right) = \frac{1}{0.8} \cdot \left(1 + \frac{B}{\sqrt{N_{\text{monomer}}}}\right)$$

and then we multiply it by our expression for the diffusion coefficient, Equation (26.17), to obtain

$$\boxed{D = \frac{kT}{3\pi\eta} \cdot \frac{1}{\sqrt{bl}} \frac{1}{\sqrt{N_{\text{monomer}}}} \times \frac{1}{0.8}\left(1 + \frac{B}{\sqrt{N_{\text{monomer}}}}\right)} \tag{26.37}$$

with

$$B = 0.5\sqrt{\frac{l}{b}} \cdot \ln\frac{l}{\phi}$$

Example 26.3 *Use Equation (26.37) to analyze the diffusion data for cellulose acetate in acetone [13] at $T = 300$ K. Use the fact that X-ray diffraction data gives a monomer unit length of $b = 0.515$ nm and that $\eta = 0.3 \times 10^{-2}$ Poise $= 0.3 \times 10^{-3}$ N m^{-2} s (viscosity of acetone). Determine the statistical chain element length l and the diameter ϕ of the chain.*

$N_{monomer}$	$\sqrt{N_{monomer}}$	D 10^{-7} cm² s⁻¹	$D \cdot N_{monomer}$ 10^{-5} cm² s⁻¹
40	6.33	20.7	8.3
196	14.0	7.7	15.1
550	23.5	4.0	22.0
746	27.3	3.4	25.4

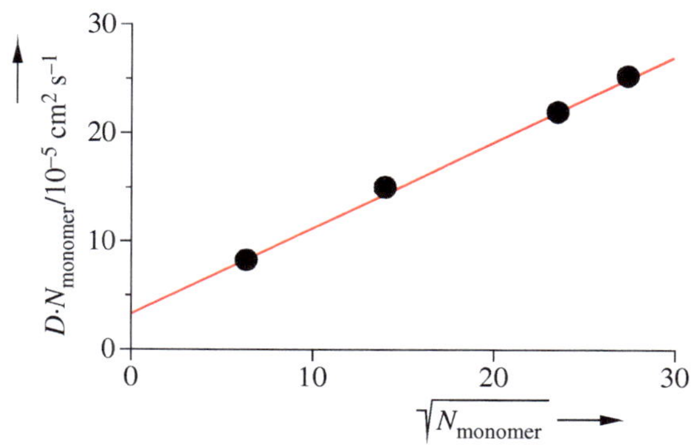

Figure 26.9 Diffusion of cellulose acetate in acetone; $D \cdot N_{monomer}$ versus $\sqrt{N_{monomer}}$. Dots: experiment. Solid line: regression line with slope 0.80×10^{-5} cm² s⁻¹ and y-intercept 3.3×10^{-5} cm² s⁻¹. By comparison with Equation (26.38) the length of the statistical chain element ($l = 10$ nm) and the diameter of the chain ($\phi = 1.5$ nm) are obtained. (Experimental data from Ref. [13]).

We rearrange Equation (26.37) as

$$D \cdot N_{monomer} = \frac{kT}{3\pi\eta} \cdot \frac{1}{\sqrt{bl}} \cdot \frac{1}{0.8}\left(\sqrt{N_{monomer}} + B\right) \qquad (26.38)$$

In Fig. 26.9, $D \cdot N_{monomer}$ is plotted versus $\sqrt{N_{monomer}}$. A fit of the experimental points to a straight line gives a slope 0.80×10^{-5} cm² s⁻¹ and a y-intercept of 3.3×10^{-5} cm² s⁻¹. From Equation (26.38) and the slope we find

$$\frac{kT}{3\pi\eta} \cdot \frac{1}{\sqrt{bl}} \cdot \frac{1}{0.8} = 0.80 \times 10^{-5}\ \text{cm}^2\ \text{s}^{-1}$$

so that $bl = 5.23 \times 10^{-18}$ m² and $l = 5.23 \times 10^{-18}\ \text{m}^2/0.515 \times 10^{-9}\ \text{m} = 10$ nm. From the intercept and the slope, we find

$$B = 0.5\sqrt{\frac{l}{b}} \cdot \ln\left(\frac{l}{\phi}\right) = \frac{3.3 \times 10^{-5}\ \text{cm}^2\ \text{s}^{-1}}{0.80 \times 10^{-5}\ \text{cm}^2\ \text{s}^{-1}} = 4.13$$

Using $l = 10$ nm, we obtain $\ln(l/\phi) = 2B/\sqrt{l/b} = 1.88$ and $\phi = 1.5$ nm. The actual thickness of the chain is about 1.3 nm, indicating that the hydrodynamic approximation is appropriate.

26.3.3.2 Sedimentation

Accordingly, from Equation (26.24) and using the expression in Equation (26.37), we obtain the sedimentation constant with $m = N_{monomer} \cdot m_{monomer}$:

$$s = \frac{m \cdot (1 - v_{part}\rho)}{3\pi\eta \cdot \sqrt{b \cdot l}\sqrt{N_{monomer}}} \times \frac{1}{0.8}\left(1 + B\frac{1}{\sqrt{N_{monomer}}}\right) \qquad (26.39)$$

and

$$B = 0.5\sqrt{\frac{l}{b}} \cdot \ln\left(\frac{l}{\phi}\right)$$

Example 26.4 *Use the data for the sedimentation of cellulose acetate in acetone to determine the statistical chain element length l [13].*

$N_{monomer}$	40	196	550	746
$\sqrt{N_{monomer}}$	6.33	14.0	23.5	27.3
$s\ 10^{-13}$ s	4.07	7.5	10.9	12.5

We rearrange Equation (26.39) as

$$s = \frac{m \cdot (1 - v_{part}\rho)}{3\pi\eta \cdot \sqrt{b \cdot l}} \times \frac{1}{0.8}\left(\sqrt{N_{monomer}} + B\right) \qquad (26.40)$$

According to Fig. 26.10 and Equation (26.40) the expected linear dependence when plotting the sedimentation constant s versus $\sqrt{N_{monomer}}$ is observed; the slope is 0.39×10^{-13} s and the intercept is 2×10^{-13} s. With

$m = \mathbf{M}_{\text{monomer}}/\mathbf{N}_{\text{A}} = 260$ g mol^{-1}/(6.02 × 10^{23} mol^{-1}) = 4.32 × 10^{-25} kg, $b = 0.515$ nm, $\rho = 0.79$ g cm^{-3}, $\eta = 0.3 \times 10^{-2}$ Poise $= 0.3 \times 10^{-3}$ N m^{-2} s (viscosity of acetone) and $v_{\text{part}} = 0.68$ cm^3 g^{-1} we obtain $bl = 5.13 \times 10^{-18}$ m^2. Then $l = 5.13 \times 10^{-18}$ m^2/0.515 × 10^{-9} m = 10 nm. On the other hand, $\ln(l/\phi) = 2B/\sqrt{l/b} = 2.33$ and $\phi = 1.0$ nm. Note that s is extremely small.

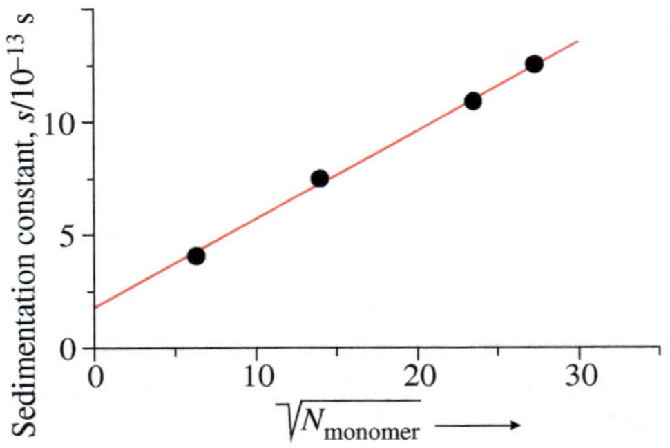

Figure 26.10 Sedimentation of cellulose acetate in acetone, sedimentation constant s versus $\sqrt{N_{\text{monomer}}}$. Filled circles: experiment, same reference as for Fig. 26.9. Solid line: regression line with slope 0.39×10^{-13} s and y-intercept 2×10^{-13} s. By comparison with Equation (26.39) the length of the statistical chain element (l = 10 nm) and the diameter of the chain (ϕ = 1 nm) are obtained. (Experimental data from Ref. [13].)

Our modeling of these macroscopic experiments and its application to polymer solutions provides an avenue to account for the complicated shape and the draining of the solvent from the macromolecules. These studies reveal that replacing the factor 6π in the Stokes law, Equation (26.13), by the translational form factor C_{transl}, Equation (26.35), provides realistic values of the diffusion and sedimentation constants.

26.3.3.3 Viscosity

In order to determine the shear viscosity of a dilute solution of random coils, models made of wire were rotated in a viscous liquid by applying a torque and measuring the angular speed. The result of a great number of such measurements was averaged in a manner similar to the translational experiments. As shown in Foundation 26.2, on the book's website, it follows that Equation (26.33) should be replaced by

$$\boxed{[\eta] = 0.45\left(\overline{h^2}\right)^{3/2}\frac{1}{m}\cdot\frac{1}{1+\beta/\sqrt{N_{\text{monomer}}}}} \qquad (26.41)$$

with

$$\beta = \sqrt{\frac{l}{b}}\cdot\left(\ln\left(\frac{l}{\phi}\right) - 1.5\right) \qquad (26.42)$$

For $N_{\text{monomer}} \gg 1$ this equation reduces to

$$[\eta] = 0.45\cdot\left(\overline{h^2}\right)^{3/2}\frac{1}{m} \qquad (26.43)$$

Equation (26.43) differs from Equation (26.33) by the proportionality factor 0.45, instead of 1.31. Using the molar mass $M = N_A \cdot m$ we obtain

$$[\eta] = 0.45\cdot\mathbf{N}_A\cdot\left(\overline{h^2}\right)^{3/2}\frac{1}{\mathbf{M}} = 2.71\times10^{23}\cdot\left(\overline{h^2}\right)^{3/2}\frac{1}{\mathbf{M}}\text{mol}^{-1} \qquad (26.44)$$

This expression agrees with an empirical equation called the *Flory Equation* [14].

Equations (26.41) and (26.43) result from an assumption that the resistance of the random coil can be approximated by considering the solvent as a continuum with the macroscopic viscosity and using simulations with macroscopic bodies. The agreement of the result of this simulation with the Flory equation suggests that the assumption of using the shear viscosity and hydrodynamics to model solutions of random coil polymers is reasonable.

With $\overline{h^2} = lL = lbN_{\text{monomer}}$ and $m = m_{\text{monomer}}\cdot N_{\text{monomer}}$ we can rewrite Equation (26.41) as

$$[\eta] = 0.45\cdot(lb)^{3/2}\cdot\frac{1}{m_{\text{monomer}}}\cdot\frac{N_{\text{monomer}}}{\sqrt{N_{\text{monomer}}}+\beta} \qquad (26.45)$$

with

$$\beta = \sqrt{\frac{l}{b}} \cdot \left(\ln \frac{l}{\phi} - 1.5 \right)$$

Example 26.5 *From Equation (26.45) calculate the Staudinger index* $[\eta]$ *as a function of* N_{monomer} *for cellulose acetate in acetone, using the values of* l *and* ϕ *obtained in Example 26.4 from sedimentation experiments (*$l = 10$ *nm,* $\phi = 1.0$ *nm), and compare the result with experimental data [13]*

N_{monomer}	30	80	120	200	250	300	500	610	860	1050
$[\eta]$/cm^3 g^{-1}	20	50	70	140	140	190	270	250	310	320

When introducing the values for l and ϕ and the value for b (0.515 nm) in Equation (26.45) we obtain

$$[\eta] = 12 \cdot \frac{N_{\text{monomer}}}{\sqrt{N_{\text{monomer}}} + 3.1} \text{cm}^3 \text{ g}^{-1}$$

(curve in Fig. 26.11). The agreement between the experimental data and the model is excellent and suggests that continuum hydrodynamics provides a reasonable first description of the viscosity for polymer solutions.

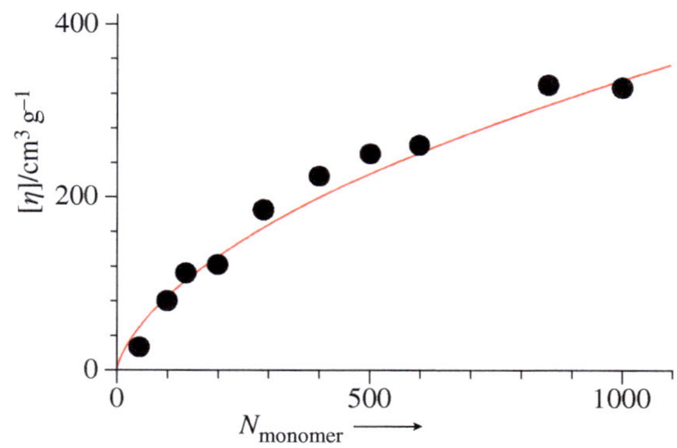

Figure 26.11 Staudinger index $[\eta]$ of cellulose acetate in acetone, $[\eta]$ versus N_{monomer}. Filled circles: experiment. Solid line: calculated according to Equation (26.45) with $l = 10$ nm, $\phi = 1.0$ nm (these values were obtained from the sedimentation experiment in Example 26.4). (Experimental data from Ref. [13].)

The description of a macromolecule as a random coil by Werner Kuhn (1899–1963) in 1934 was of importance to the early development of molecular biology. This was the first quantitative analysis explaining the particular properties of macromolecules as a result of molecular shape and dynamics [15]. The replacement of a molecular chain by a sequence of elements of length l, or, as well, the division into segments of length l_{segment} has been introduced in connection with constructing macroscopic models [16]. Another way to describe a random coil is by dividing the chain into sections called *persistence length* (this is exactly one-half of the length of a statistical chain element) [17].

26.4 Uncoiling a Coil and Its Recoiling

The coiling and uncoiling, or folding and unfolding, of polymer strands is extremely important in biology (e.g., DNA replication, protein function) and materials science (e.g., elasticity). Here we discuss fundamental considerations for describing the coiling and uncoiling of polymer molecules.

26.4.1 Unraveling a Coil by Applying a Force at the Chain Ends

The force to unravel a coil can be measured by fixing one end to a substrate, attaching a magnetic bead (iron grain in latex particles of 200 nm diameter) to the other end and using a magnetic field to unravel the coil [18]. A force of 100 pN unravels a DNA double helix (L = 50 μm). For comparison, 1000 pN is the force to stretch a C – C bond by 1% (see Chapter 7). The breaking point for a DNA strand is reached when stretching the chain with a force of 500 pN. This value corresponds to a force per unit area of 10^8 N m^{-2}; this breaking limit is similar to that of steel.

26.4.1.1 Small Forces: Length of a Statistical Chain Element

First we consider small forces f, when the chain is largely coiled, i.e., the distance h between chain ends is small compared to the contour length L. In Box 26.3 we assume that f is proportional to the distance h between the chain ends (Fig. 26.12), and find that the average force f is

$$f = \frac{3kT}{l \cdot L} h \tag{26.46}$$

From Equation (26.46) we can calculate the statistical length $l = 3kT \cdot h/(fL)$ if we know the other parameters. For DNA the ratio $f/h = 4 \times 10^{-9}$ N m^{-1} was measured for a DNA duplex in water (L = 33 μm; T = 300 K), so that l = 94 nm. This finding is in agreement with the value found in Box 26.2.

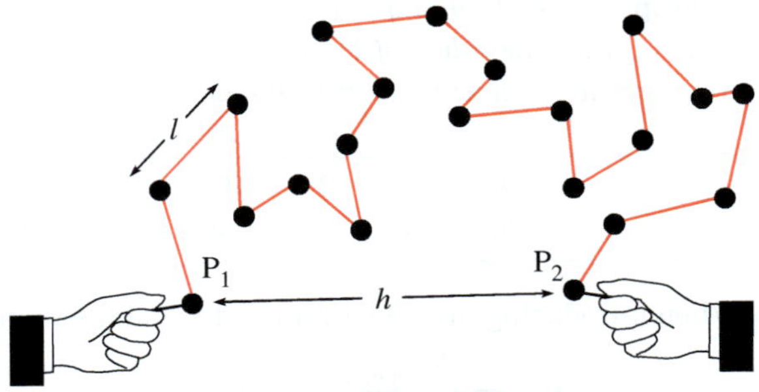

Figure 26.12 Random coil held tight at chain ends. Contracting force due to thermal fluctuation.

26.4.1.2 Large Forces: Elasticity of Stretched Chain

At large forces, when the chain is uncoiled, it extends slightly in proportion to the increasing force. For a DNA double helix with L = 33 μm, an extension of Δx = 1 μm is found at a force of Δf = 15 pN. In this elastic limit, with r = 1.5 nm we can calculate Young's modulus A_{Young} for stretching

$$A_{\text{Young}} = \frac{L \cdot \Delta f}{\pi r^2 \cdot \Delta x} = 7 \times 10^7 \text{ N m}^{-2}$$

For comparison, Young's modulus for stretching for steel is $A_{\text{Young}} = 10^{11}$ N m^{-2} and that for rubber is $A_{\text{Young}} = 10^6$ N m^{-2}. That is, Young's moduli for stretching of rubber and DNA differ by a factor of ~70.

Example 26.6 *Consider a single chain (L = 160 nm) of polyethylene glycol HO–(–CH_2–CH_2–O–=)$_n$–H that bonds to an AFM tip on one end and to a solid surface on its other end. Figure 26.13 shows the measured force as a function of the elongation h (see Chapter 7). At small forces f the ratio $f/h = 1 \times 10^{-4}$ N m^{-1} was measured, and at large forces an increase in Δf causes an additional elongation Δh so that $\Delta f = \Delta h \cdot 0.4$ N m^{-1}. For small forces calculate the length l of the statistical chain element. Calculate the force increase Δf needed for a 1% increase in elongation at large forces.*

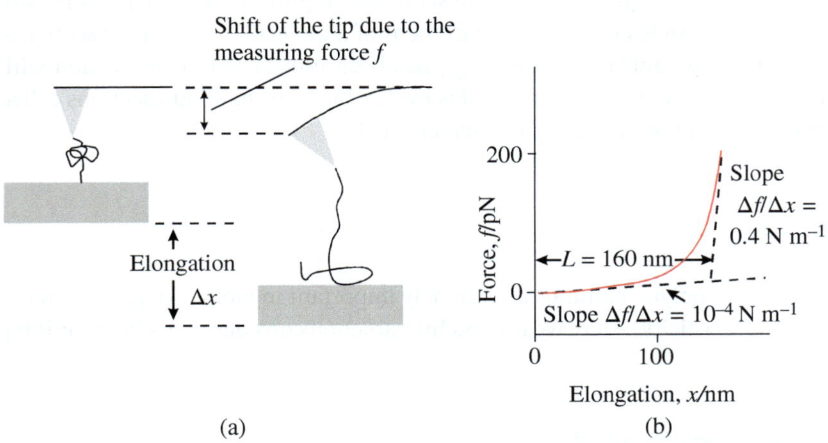

Figure 26.13 Polyethylene glycol in hexane. Individual chain stretched by force f using an atomic force microscope. x is the elongation of the chain [19]. (a) Experimental setup. One end of the polymer chain is fixed to a tip connected to a spring; the second end of the polymer is fixed to a movable plate. When the plate is moved downward by an elongation h, a force acts on the polymer chain which is measured by the bending of the spring. (b) Force acting on the polymer chain versus elongation x.

At small forces Equation (26.46) applies and

$$l = \frac{h}{f} \cdot \frac{3kT}{L} = \frac{1}{1 \times 10^{-4} \text{ N m}^{-1}} \cdot \frac{3 \cdot 4.2 \times 10^{-21} \text{ N m}}{160 \times 10^{-9} \text{ m}} = 1 \text{ nm}$$

in agreement with the value obtained from viscosity measurements.

At large forces f, a 1% increase in length is $0.01 \cdot L = 1.6$ nm and

$$\Delta f = 1.6\ \text{nm} \cdot 0.4\ \text{N m}^{-1} = 6 \times 10^{-10}\ \text{N}$$

This value is similar to the force required to extend a hydrocarbon chain by 1% via C–C–C bond angle distortions, namely

$$\Delta f = 0.01 \cdot k_f' \cdot d_{eq} \frac{2 \sin(\alpha/2)}{\cos^2(\alpha/2)} = 3 \times 10^{-10}\ \text{N}$$

where $k_f' = 36$ N m^{-1}, $d_{eq} = 154$ pm, and $\alpha = 109°$ (see Section 7.5).

26.4.2 Fully Unraveling a Coil in a Flowing Medium

We consider a coil fixed at one end (Fig. 26.14a), and we flow a fluid past the coil. In a weak current the coil is slightly deformed, but as the current increases the coil begins to unravel. The unraveling begins at the fixed end because the strain is strongest there. This occurs because the entire chain, resisting against the current, pulls on that end. At a sufficiently large current and time the molecule becomes fully uncoiled. What minimum speed v_c of the flow is needed to unravel the coil? In the steady state and a flow speed $v < v_c$, what is the length of the unraveled section of the chain?

26.4.2.1 Minimum Speed v_c to Unravel a Coil

To fully unravel a coil, the energy to orient a given statistical chain element by the stream surrounding it must be greater than the energy of thermal fluctuations. This must be the case even for the statistical element constituting the open end (Fig. 26.14b). The energy E to turn the element oriented in the flow by 180° (to move the center by the distance l against the current) must be greater than kT. How can we approximate the friction of the element and estimate the energy needed?

We build a model by replacing the chain (diameter ϕ) by a row of spheres of radius $r = \phi$, whose centers are spaced at a distance of 4ϕ (*necklace model*, see Fig. 26.14c). Using Stokes law (see Chapter 14) we can estimate the hydrodynamic

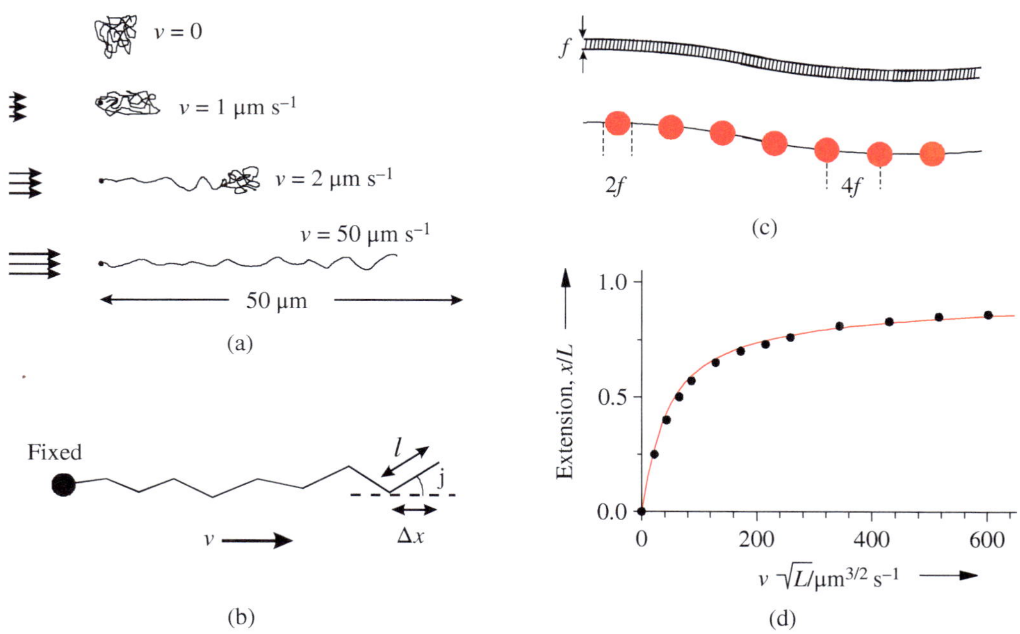

Figure 26.14 (a) A DNA double helix, which is fixed at one end, is uncoiled by a flow of water with a speed v larger than $v_c = 50$ μm s^{-1}; $L = 50$ μm; $l = 100$ nm; $N_{stat} = 500$. The complete and partial unraveling of the DNA duplex strand carrying a fluorescent dye can be seen in an optical microscope. The molecule is in slightly moving water and one end is fixed [20]. (b) Critical current for uncoiling. The molecule is described as a chain of statistical chain elements. The frictional force of flowing solvent counteracts thermal fluctuations. $\Delta x = l \cdot \cos \varphi$ is the extension of the statistical chain element in the x direction. (c) The DNA chain (top, diameter ϕ) is replaced by a necklace model (bottom, spheres of diameter 2ϕ and distance 4ϕ). (d) DNA double helix. The molecular chain end is bound to a small bead of 0.3 μm diameter. The bead is pulled by a laser beam and moved through the solvent with constant speed. The force pulling the bead arises from the polarization of the bead in the electric field of light; the bead is driven toward the point of maximum field. The graph plots the fractional extension x/L versus $v\sqrt{L}$ for samples with $L = 50$ μm. Dots: experiment [21], solid line: theoretical curve according to Equation (26.51) and Box 26.3 with $l = 100$ nm, $N_{stat} = 500$, $\eta = 1 \times 10^{-3}$ N m^{-2} s, and $T = 300$ K.

resistance of each sphere and the resistance of any given section of the chain. The force of friction f_{sphere} of one sphere is

$$f_{\text{sphere}} = 6\pi\eta\phi v$$

In our model, a chain element of statistical length l consists of $l/(4\phi)$ spheres, so that the force of friction f of the chain element is

$$f = 6\pi\eta\phi v \cdot \frac{l}{4\phi} = \frac{3}{2}\pi\eta l v \tag{26.47}$$

This approximation is substantiated by measurements on macroscopic models.

The energy E to turn the element against the flow is $E = f \cdot l$, where l is the average distance that a sphere moves (the sphere at the end moves a distance of $2l$, the sphere at the beginning of the element does not move, see Fig. 26.14b, right chain element). By comparing the E in this approximation to the available thermal energy kT, the condition for fully unraveling the coil is

$$E = f \cdot l = \frac{3}{2}\pi\eta l v \cdot l = \frac{3}{2}\pi\eta l^2 v \geq kT \tag{26.48}$$

and the flow speed must be

$$v \geq v_{c,} \quad v_c = \frac{2kT}{3\pi\eta l^2} \tag{26.49}$$

Example 26.7 *Flow Required to Unravel a Coil. Calculate the critical flow speed needed to unravel a DNA strand in water ($l = 100$ nm and $\eta = 0.01$ poise $= 1 \times 10^{-3}$ N m^{-2} s).*

Substituting values into Equation (26.49) we obtain

$$v_c = \frac{2 \cdot 1.38 \times 10^{-23} \cdot 300 \text{ J}}{3\pi \cdot 0.001 \text{ N m}^{-2} \text{ s } 1 \times 10^{-14} \text{ m}^2} = 100 \text{ µm s}^{-1}$$

This is in accordance with experiment (Fig. 26.14d). Thus a tiny current is sufficient, and the unraveling of a coil can be a very sensitive sensor for currents.

26.4.2.2 Force Acting on Fixed End

What is the force acting on the fixed end at the speed v_c? Each statistical chain element exerts a friction force $6\pi\eta\frac{1}{4}lv_c$, and the force acting on the fixed end is the sum of the resistance forces of all N_{stat} statistical elements

$$\boxed{f = 6\pi\eta\frac{1}{4}lv_c \cdot N_{\text{stat}}} \tag{26.50}$$

Using the expression for v_c in Equation (26.49) we obtain

$$f \approx \frac{kT}{l}N_{\text{stat}} = 2 \times 10^{-11} \text{ N} = 20 \text{ pN}$$

for $N_{\text{stat}} = 500$, $l = 100$ nm, and $T = 300$ K.

When attaching a magnetic bead to the end of the chain of a DNA double helix dissolved in water ($L = 40$ µm) and pulling the bead with a magnetic field, the force f and the speed v can be measured. When applying a force of 30 pN a speed of 100 µm s^{-1} is observed (fully uncoiled chain [22]). Thus we obtain

$$\frac{f}{v} = 0.3 \text{ pN µm}^{-1} \text{ s}$$

On the other hand, the simple model discussed in Equation (26.50) predicts

$$\frac{f}{v_c} = \frac{3\pi}{2}\eta L = 0.2 \text{ pN µm}^{-1} \text{ s}$$

in good agreement with the measurement.

26.4.3 Partially Unraveling a Coil in a Flowing Medium

The unraveling of a DNA strand, which is labeled with a fluorescent dye molecule, has been monitored in an optical microscope (Fig. 26.14a). The molecule is in slowly moving water (at a speed up to 50 µm s^{-1}), and one end is fixed.

What flow speed v keeps a section unraveled or loosely coiled? To address this question consider a statistical chain element number n (numbering from the fixed end). In Box 26.3 we show that the chain segment between element n and the fixed end exerts a force f

$$f = \frac{kT}{2l}\left[\frac{1}{\left(1 - \overline{\Delta x}_n/l\right)^2} - 1 + \frac{4\overline{\Delta x}_n}{l}\right]$$

where $\overline{\Delta x}_n$ is the average extension of the chain element n in the x direction (Fig. 26.14b). On the other hand, the chain segment between element n and the open end acts on element n with a hydrodynamic force f'. For simplicity we consider this segment as undrained, so that

$$f' = 6\pi\eta r_{\text{coil}} \cdot v \qquad \text{with} \qquad r_{\text{coil}} = \frac{l}{2}\sqrt{N_{\text{stat}} - n}$$

At steady state the forces f and f' balance, so that

$$\frac{kT}{2l}\left[\frac{1}{\left(1 - \overline{\Delta x}_n/l\right)^2} - 1 + \frac{4\overline{\Delta x}_n}{l}\right] = 6\pi\eta \cdot l \cdot \frac{1}{2}\sqrt{N_{\text{stat}} - n} \cdot v \tag{26.51}$$

This relation provides the average contribution $\overline{\Delta x}_n$ of element n to the extension x of the molecule in the direction of the current. In Problem P26.10. you show that the fractional extension x/L can be obtained from this relation. The fractional extension x/L is essentially a function of $v\sqrt{l}$. It is plotted versus $v\sqrt{L}$ in Fig. 26.14d (solid line); the experiment (dots in Fig. 26.14d) is in good agreement with the model.

Also, it is possible to estimate the extension for small speeds when the coil is deformed but not unraveled. In this limit we assume that the hydrodynamic force f' is acting on the open end, where

$$f' = 6\pi\eta \cdot r_{\text{coil}} \cdot v$$

With

$$r_{\text{coil}} = \frac{l}{2}\sqrt{N_{\text{stat}}} = \frac{1}{2}\sqrt{\overline{h^2}} = \frac{1}{2}\sqrt{l \cdot L}$$

we find that

$$f' = 6\pi\eta \cdot \frac{1}{2}\sqrt{l \cdot L} \cdot v$$

Also, the open end is attracted toward the fixed end at distance x by a force f, given on average, by Equation (26.46)

$$f = \frac{3kT}{l \cdot L} \cdot x \quad \text{for} \qquad x \ll L$$

Obviously, f increases with T and with x. At steady state the forces balance, namely

$$\frac{3kT}{l \cdot L} \cdot x = 6\pi\eta \cdot \frac{1}{2}\sqrt{l \cdot L} \cdot v$$

and for the fractional extension we obtain

$$\frac{x}{L} = \frac{\pi\eta l^{3/2}}{kT} v\sqrt{L} \tag{26.52}$$

With $l = 100$ nm, $\eta = 1$ cP $= 1 \times 10^{-3}$ N m^{-2} s, and $T = 300$ K, the fractional extension is

$$\frac{x}{L} = \frac{1}{41\ \mu\text{m}^{3/2}\ \text{s}^{-1}} v\sqrt{L}$$

Unlike the model used here, in which the initial slope is independent of L, the measured initial slope depends slightly on L. It is found that the numerical factor of 41 μm$^{3/2}$ s^{-1} has to be replaced by 58 μm$^{3/2}$ s^{-1} for $L = 22$ μm and by 55 μm$^{3/2}$ s^{-1} f or $L = 84$ μm.

In the refined model the factor 6π in Equation (26.51) is replaced by the form factor, Equation (26.35). With $l = 100$ nm, $b = 0.3$ nm, and $\phi = 3$ nm the translational form factor becomes $C_{\text{transl}} = 13.4$ for $L = 22$ μm and $C_{\text{transl}} = 14.15$ for $L = 84$ μm. The ratio of the two slopes should then be

$$\frac{C_{\text{trans}}(L = 22\ \mu\text{m})}{C_{\text{transl}}(L = 84\ \mu\text{m})} = \frac{13.4}{14.15} = 0.95$$

in accordance with the experimental ratio $55/58 = 0.95$.

26.4.4 Restoring a Coil

Now let us stop the flow that has unraveled the coil and ask, "How long before the coil is restored?" An initial coil develops at the open end and grows with an average speed v toward the fixed end, coiling the unraveled portion (length x), see Fig. 26.15a. As shown in Problem P26.11, the decrease in x with time t is proportional to $t^{2/3}$:

$$x = L - \left(\frac{kT}{2\pi\eta}\right)^{2/3} \frac{1}{l} \cdot t^{2/3} \tag{26.53}$$

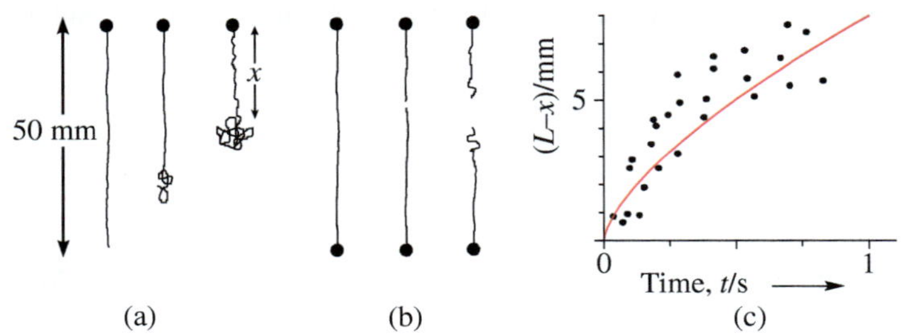

Figure 26.15 Restoring a coil from a fully uncoiled chain. (a) Shape of coil fixed at one end at different times t after stopping the flow. (b) Coils fixed at both ends and cut by UV radiation. (c) Theory (solid line) according to (26.53) compared with experiment (filled circles) observing single fluorescence-labeled DNA molecules. (Experimental data from [5].)

An experimental realization of this phenomenon uses an extended strand of a DNA double helix that is labeled with a fluorescent dye and fixed at both ends. Upon illumination, the strand occasionally breaks (a light quantum initiates a photochemical reaction). Subsequently, the growth of the two coils and their motion to the fixed-points can be observed [5] (Fig. 26.15b). The measured decrease of length x with time (experimental points in Fig. 26.15c) follows the theoretical relation (26.53) with $\eta = 1 \times 10^{-2}$ Poise $= 10^{-3}$ N m^{-2} s, $l = 100$ nm, and $T = 300$ K,

$$L - x = \left(\frac{kT}{2\pi\eta}\right)^{2/3} \frac{1}{l} \cdot t^{2/3} = 7.6\ \mu\text{m} \cdot \left(\frac{t}{\text{s}}\right)^{2/3}$$

Example 26.8 *In another experiment [23] the chain is bound to a small bead. The bead is pulled by a laser beam until the chain is fully uncoiled; then the pulling is stopped and the evolution of the shape of the molecule is observed. Given the parameters $\eta = 1 \times 10^{-2}$ Poise $= 10^{-3}$ N m^{-2} s, $l = 100$ nm, $T = 300$ K, and $L = 100$ μm, how long does it take for the strand to coil up to 51 μm?*

In this case Equation (26.53) may be written as

$$t = (L - x)^{3/2} \left(\frac{2\pi\eta}{kT}\right) l^{3/2}$$
$$= (51\ \mu\text{m})^{3/2} \cdot \left(\frac{2\pi 10^{-3}\ \text{N m}^{-2}\ \text{s}}{1.38 \cdot 10^{-23}\ \text{J K}^{-1} \cdot 300\ \text{K}}\right) \cdot (100\ \text{nm})^{3/2} = 17\ \text{s}$$

According to Equation (26.53) the time to restore a coil of contour length $L - x$ is independent of the total chain length L; this result is justified by the experiment. However, when the fractional extension becomes small, (26.53) is no longer valid.

26.5 Proteins as Biopolymers

A protein macromolecule differs from the polymer molecules we have discussed in a number of important ways. For example, it is a heteropolymer consisting of a particular number of different amino acid groups that are covalently linked in a particular sequence. This sequence, called the *primary structure* of the protein, specifies the protein. The interaction between the amino acid side groups on the peptide chain and with the solvent (or surrounding biological medium) determines the local conformation of the backbone, and is called the *secondary structure*. Particular conformational motifs, such as an α-helix and a β-sheet, are evident in the secondary structure of different proteins. In contrast to the synthetic polymers discussed above, the overall three-dimensional shape of a protein molecule can be well defined, and is called the *folded structure* of the protein. Below we describe how these features arise.

A protein is a polymer composed of monomer units that are selected from the 20 possible amino acids. The backbone of the polymer strand comprised peptide bonds, which are formed from the condensation reaction of two amino acids, e.g.,

$$R_1\text{–COOH} + H_2N\text{–}CH_2\text{–}R_2 \ \rightarrow R_1\text{–CO–NH–}CH_2\text{–}R_2 + H_2O$$

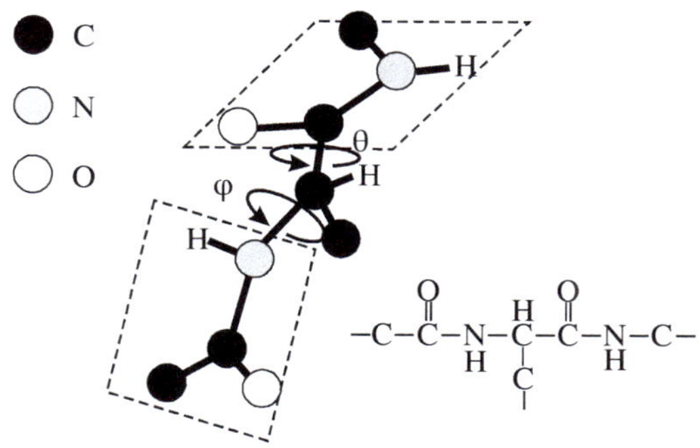

The peptide bond has a significant amount of double bond character and the five highlighted atoms are largely in the same plane. Two conformations (*trans* and *cis*) are possible for this unit; however, the other CN bond and the CC bond have a number of different conformations possible. Even a conservative estimate of 6 conformations per amino acid gives rise to 6^{100} conformations for a 100 amino acid polypeptide chain, which would be considered a small protein.

Because of its well-defined folded structure, a globular protein cannot be considered as a random coil. Therefore we approximate the protein as a compact sphere of radius $\bar{r}$. The volume V of the sphere is $(4\pi/3)\bar{r}^3$ and its mass is $m = \rho V$, where ρ is the density, considered as that of water, $\rho = 1\ \text{g cm}^{-3}$. Thus for the molar mass $\boldsymbol{M}$ we obtain

$$\boldsymbol{M} = m \cdot \boldsymbol{N}_A = \rho \frac{4}{3}\pi \bar{r}^3 \cdot \boldsymbol{N}_A \tag{26.54}$$

and

$$\bar{r} = \left(\frac{3}{4\pi\rho \boldsymbol{N}_A}\right)^{1/3} \cdot \boldsymbol{M}^{1/3} = 0.0736\ \text{nm} \cdot \left(\boldsymbol{M}/\text{g mol}^{-1}\right)^{1/3} \tag{26.55}$$

Figure 26.16 plots the mean radius $\bar{r}$, which is calculated from the measured diffusion coefficient D using Equation (26.14), versus the molar mass $\boldsymbol{M}$ (filled circles) for 27 different globular proteins. The solid line shows the theoretical prediction of Equation (26.55). The experimental values are only slightly larger than the calculated ones, indicating that our treatment of the protein as a compact sphere is essentially correct.

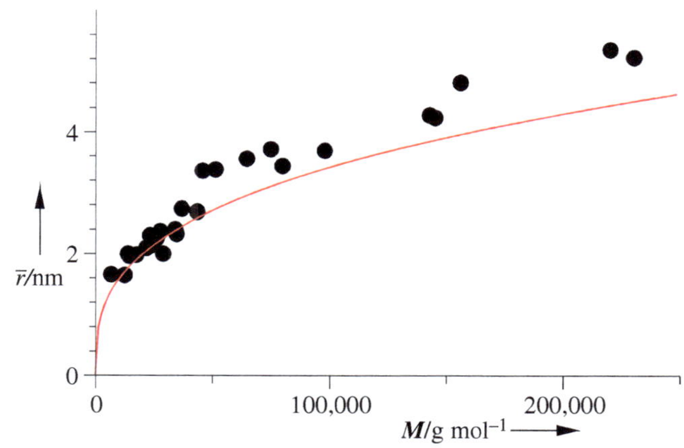

Figure 26.16 Mean radius $\bar{r}$ of proteins as a function of their molar mass $\boldsymbol{M}$. Filled circles: experiment and solid line: calculated from Equation (26.55).

The native structure of a globular protein results from a number of different interactions; however, some primary principles are the solvation energy differences for the side chains, the propensity to form particular conformations, and specific interactions between remote groups. The amino acids (see Table 26.1) can be divided into subclasses based on the solvation properties of their side groups: hydrophilic versus hydrophobic. The decrease in Gibbs energy that arises from presenting the hydrophilic groups to the water and the hydrophobic groups to the interior of the protein is important for creating the compact structures in water. We will see in Chapter 27 that this "hydrophobic force" is also important in the assembly of supramolecular structures. In addition to this somewhat nonspecific hydrophobic force, the conformational energetics of the amino acid residues on the peptide chain give rise to a propensity to form a particular subset of the φ and θ angles (see peptide bond structure above). These preferences and their association with particular structures (α-helices, β-sheets, etc.) have been widely studied [24]. Lastly, the directed interaction between chemical functionalities on particular side groups (hydrogen bonds, disulfide bonds) that are remote from one another on the peptide chain can be important contributors to the overall folded structure of the protein.

Table 26.1 Amino Acids and Some Characteristics. The Groups R Refer to the General Structure: R–CH(NH_2)–COOH.[1]

Amino Acid	Symbol	R	M/g mol^{-1}	V/nm^3	
Alanine	Ala	$-CH_3$	71	0.067	Nonpolar
Arginine	Arg	$-(CH_2)_3NHC(NH_2)_2$	156	0.148	Polar
Asparagine	Asn	$-CH_2CONH_2$	114	0.096	Polar
Aspartic acid	Asp	$-CH_2CO_2H$	115	0.091	Polar
Cysteine	Cys	$-CH_2SH$	103	0.086	Polar
Glutamine	Gln	$-CH_2CH_2CONH_2$	128	0.114	Polar
Glutamic acid	Glu	$-CH_2CH_2CO_2H$	129	0.109	Polar
Glycine	Gly	$-H$	57	0.048	
Histidine	His	$-CH_2(C_3N_2H_3)$	137	0.118	Polar
Isoleucine	Ile	$-CH(CH_3)C_2H_5$	113	0.124	Nonpolar
Leucine	Leu	$-CH_2CH(CH_3)_2$	113	0.124	Nonpolar
Lysine	Lys	$-(CH_2)_4NH_2$	128	0.135	Polar
Methionine	Met	$-CH_2CH_2SCH_3$	131	0.124	Nonpolar
Phenylalanine	Phe	$-CH_2(C_6H_5)$	147	0.135	Nonpolar
Proline	Pro	$-CH_2CH_2CH_2-$	97	0.090	Nonpolar
Serine	Ser	$-CH_2OH$	87	0.073	Nonpolar
Threonine	Thr	$-CH(CH_3)OH$	101	0.093	Nonpolar
Tryptophan	Trp	$-CH_2(C_8NH_6)$	186	0.163	Nonpolar
Tyrosine	Tyr	$-CH_2(C_6H_4OH)$	163	0.141	Nonpolar
Valine	Val	$-CH(CH_3)_2$	119	0.105	Nonpolar

26.6 Motion Through Entangled Polymer Chains

The previous discussion considered polymer molecules at sufficiently high dilution that they do not directly interact. At higher concentrations, the interactions between polymer molecules give rise to a number of interesting effects, and some of these have important technological and practical applications. For example, biopolymers can be separated by their different rates of transport through entangled polymer meshes.

For isolated polymer molecules, the sedimentation rate in a centrifuge depends on the polymer's chain length, but for entangled polymers it may not. For example, a polymer of polystyrene with $N_{\text{monomer}} = 2300$ has a sedimentation rate that is about 10 times faster than that of a polymer with $N_{\text{monomer}} = 26$ when the solution is very dilute. This difference arises because the sedimentation constant s is approximately proportional to $\sqrt{N_{\text{monomer}}}$, as discussed in Section 26.3.3. At a dilution of 1 volume % of polymer however, the chains penetrate each other, and the sedimentation speed is equal for these two molecular weights (Fig. 26.17). In this

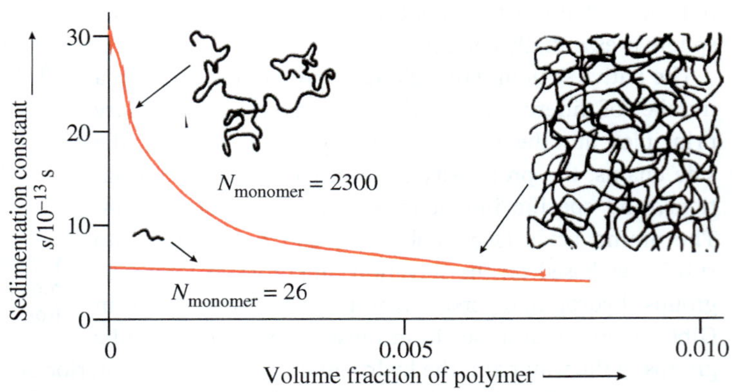

Figure 26.17 Sedimentation constant of isolated molecules versus the volume fraction of polymer. In the case of an entanglement the solvent must be pressed through the network: slow sedimentation. (Adapted from [25].)

1 This residue van der Waals volume V is taken from [24].

limit, the sedimentation speed is determined by the resistance against the draining of solvent through the entangled polymer molecules.

The motion of a polymer chain through an entangled mesh of polymer depends on its size. This can be demonstrated by observing duplex DNA, which is labeled with a fluorescent dye and bound to a bead at one end, in a solution of other unlabeled chains (Fig. 26.18). The bead is pulled by a laser beam and moved along a curved path sufficiently fast to uncoil the chain bound to the bead.[2] The whole chain follows the trajectory of the bead because it is entangled in the meshwork of the polymer.

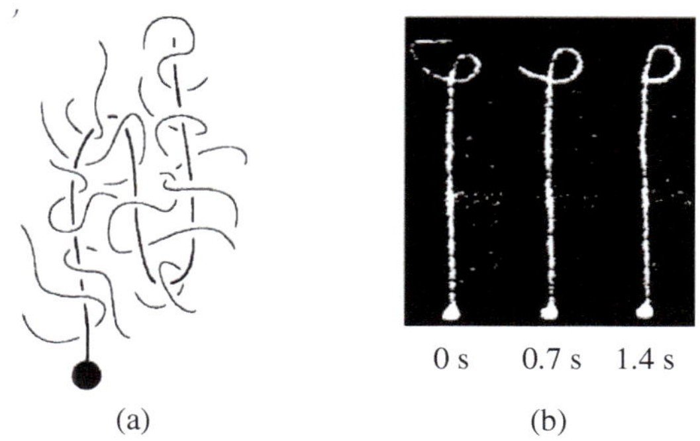

Figure 26.18 Fluorescently stained DNA strand driven through entangled polymer chains. (a) DNA strand bound to a bead in the entanglement. (b) Images of the fluorescent DNA strand driven by an optical tweezer. The images are taken every 0.7 s. ([26]. Thomas T. Perkins et al. 1994/Reproduced with permission from American Association for the Advancement of Science (AAAS).)

26.6.1 Moving Random Coil by Winding Through Meshwork: Gel Electrophoresis of DNA Fragments

Gels can be considered as entanglements of chain molecules in a surplus of solvent. Bonds between the chains fix the structure. The motion of a macromolecule in a gel when applying a moderate electric field (about 1 V cm^{-1}) is an important method for separating proteins and nucleic acids.

A mixture of double-stranded DNA fragments was separated by electrophoresis in a polyacrylamide gel (10 % in tris-borate buffer and 8 g NaCl per liter). As shown in Fig. 26.19a, the DNA solution was placed in a small hole in a layer of the gel and a field of $F = 5$ V cm^{-1} was applied (voltage 100 V between platinum wires at distance 20 cm). After electrophoresis for 150 min, the gel was immersed in a staining solution containing a fluorescent dye (ethidium) and then photographed under UV light. In Fig. 26.19b the displacement x is plotted versus $N_{monomer}$, the number of the base pairs (experimental data: filled circles). Can we understand this relationship?

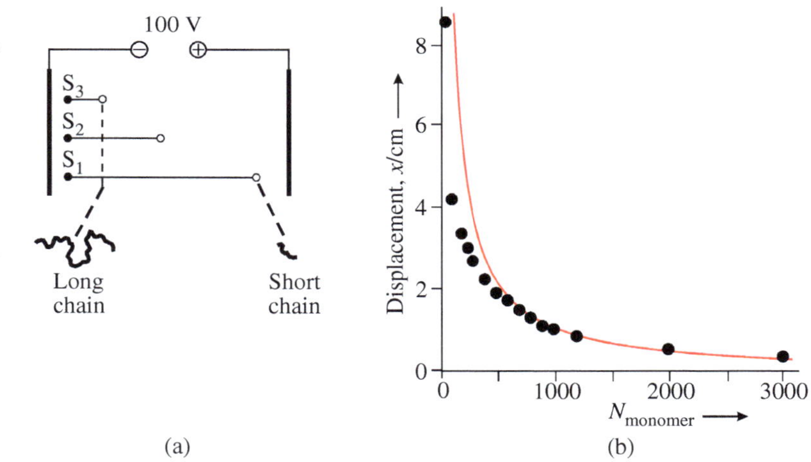

Figure 26.19 Gel electrophoresis. (a) DNA strands wind through meshwork toward anode. S_1, S_2, and S_3 are the starting points for three DNA strands with different chain lengths L. The displacements are indicated. S_1: long chain, small displacement; S_2: medium chain, medium displacement; S_3: short chain, large displacement. (b) Displacement x of the separated nucleic acids versus the number $N_{monomer}$ of monomers (defined as number of basepairs). Filled circles: experiment (courtesy Andreas Kuhn) and solid line: theory according to Equation (26.61) with $\alpha = 0.14$.

The displacement x decreases with increasing chain length because the strands have to wind through the meshwork. In Foundation 26.3, on the book's website, we show that the average speed v of the molecule through the meshwork in the direction of the applied force (x-direction) is

$$v = \frac{l}{\frac{9}{2}\pi\eta L^2} f \tag{26.56}$$

where l is the statistical chain length, L is the contour length, η is the viscosity, and f is the force acting on the strand. It is assumed that the molecule is a random coil, i.e., L is at least about $2l$, thus $N_{monomer} > 500$.

2 See [26]. As mentioned in Fig. 26.12c the force pulling the bead arises from the polarization of the bead in the electric field of the light; the bead is driven toward the point of maximum optical field strength.

In the steady state limit this average speed v can be found by balancing the force that arises from the applied field F with the hydrodynamic resistance (friction). If each monomer unit carries a charge of $-2e$ (phosphate groups in both strands of the double helix), then the total charge is

$$Q = -2e \cdot N_{\text{monomer}}$$

This charge is compensated to a large extent (to about 80%) by counter ions bound to the chain [27], and the effective viscosity in the pore will be larger than the viscosity of water in the bulk ($\eta = 1 \times 10^{-3}$ N m^{-2} s). Thus we write the effective charge Q_{eff}

$$Q_{\text{eff}} = -2e \cdot N_{\text{monomer}} \cdot \alpha \tag{26.57}$$

where α is a correction term accounting for these influences. A value of $\alpha = 0.2$ would result from an 80% compensation of the charge; however, to account for the higher viscosity in the pore, α must be smaller. A value of $\alpha = 0.14$ seems to be reasonable. Using Equation (26.56) with $f = Q_{\text{eff}} \cdot F$, we find that

$$v = \frac{l}{\frac{9}{2}\pi\eta L^2} \cdot Q_{\text{eff}} \cdot F = \frac{l}{\frac{9}{2}\pi\eta L^2} \cdot 2eN_{\text{monomer}} \cdot F\alpha \tag{26.58}$$

With $L = bN_{\text{monomer}}$ (b is the contour length per monomer), we obtain

$$\begin{aligned} v &= \frac{l}{\frac{9}{2}\pi\eta b^2 N_{\text{monomer}}} \cdot 2e \cdot F\alpha \\ &= \left(\frac{l}{\frac{9}{2}\pi\eta b^2} \cdot 2e \cdot F\alpha\right) \cdot \frac{1}{N_{\text{monomer}}} \end{aligned} \tag{26.59}$$

and the displacement x becomes

$$x = vt = \left(\frac{l}{\frac{9}{2}\pi\eta b^2} \cdot 2e \cdot Ft\right) \cdot \frac{1}{N_{\text{monomer}}} \cdot \alpha \tag{26.60}$$

With $l = 100$ nm, $b = 0.34$ nm, $\eta = 1 \times 10^{-3}$ N m^{-2} s, $e = 1.602 \times 10^{-19}$ C, $F = 5$ V cm^{-1}, and $t = 150$ min we obtain

$$x = (8822\ \text{cm}) \cdot \frac{1}{N_{\text{monomer}}} \cdot \alpha \tag{26.61}$$

This equation is plotted in Fig. 26.19b (solid line), using $\alpha = 0.14$. The theoretical curve is in good agreement with the experiment for $N_{\text{monomer}} > 500$.

The measured displacement at values of $N_{\text{monomer}} < 500$ increases continuously with decreasing N_{monomer} and it might be expected that a limit

$$x_0 = v_0 \cdot t$$

is reached; i.e., for small enough strands the motion is not impeded by the network of the gel. According to the necklace model the expected speed v_0 is, according to Equation (26.47),

$$v_0 = \frac{f}{\frac{3}{2}\pi\eta L} = \frac{2eN_{\text{monomer}} \cdot F\alpha}{\frac{3}{2}\pi\eta b N_{\text{monomer}}} = \frac{e \cdot F\alpha}{\frac{3}{4}\pi\eta b}$$

For $F = 5$ V cm^{-1}, $\eta = 1 \times 10^{-3}$ N m^{-2} s, $b = 0.34$ nm, and $\alpha = 0.14$ we obtain $v_0 = 1.40 \times 10^{-5}$ m s^{-1} and $x_0 = 13$ cm for $t = 150$ min. This is not much larger than the displacement $x = 9$ cm measured in the case $N_{\text{monomer}} = 23$.

The mobility of small DNA molecules is strongly dependent on details of the gel composition. In contrast to the present case of relatively large concentration of the gel (10%) the mobility can first increase with increasing N_{monomer} before decreasing, depending on pore size and ionic composition of the solvent [28].

26.7 Rubber Elasticity

Rubber consists of molecular chains connected by bridges made of sulfur atoms (net points are usually S atoms, see Fig. 26.20). We call L the length of the contour line of the chain between two of these bridges, or netpoints. On average such net points are at a distance $\sqrt{\overline{h^2}}$, when the system is in equilibrium.

When a force f_{total} is applied to stretch the band of rubber, the chains connecting net points are partially stretched and partially compressed. This corresponds to a decrease of the entropy. Because the molecules in the rubber have the tendency to increase their entropy by returning into the unstretched state, this results in a force counteracting f_{total}. In Problem P26.12, you show that the force to stretch a rubber band from its original length z to $z + \Delta z$ is given by

$$\boxed{f_{total} = \frac{\Delta z}{z^2} \cdot 3N_{thread}kT \quad \text{for} \quad \Delta z \ll z} \tag{26.62}$$

where N_{thread} is the number of threads of a net in the sample.

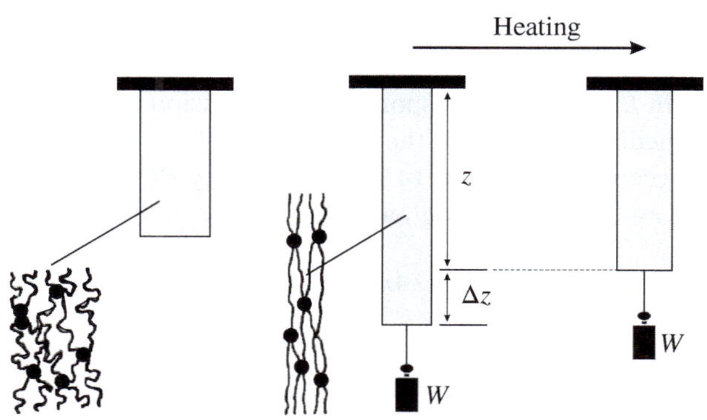

Figure 26.20 Rubber band of length z extended by weight W. Full circles: netpoints. Contraction when heating: increased thermal fluctuations.

An interesting aspect of rubber elasticity arises because the number of molecular conformations decreases in stretching; i.e., ΔS is negative. Because all conformations have similar energies, the enthalpy change ΔH upon stretching can be neglected. Hence ΔG is positive and work must be supplied to stretch the rubber band. When in contact with a temperature bath, the work supplied by stretching the rubber band is transferred as heat into the thermal bath (an isothermal process). If the stretching is performed adiabatically, the rubber band heats up; it cools down when released adiabatically.

Another aspect of this unusual behavior, which arises from the molecular network, is a rubber band's temperature dependence. A rubber sample stretched with a given weight contracts on heating:

$$\boxed{\Delta l = f_{total} \cdot \frac{z^2}{3N_{thread} \cdot k} \cdot \frac{1}{T}} \tag{26.63}$$

This behavior is in contrast to the normal case of a body extending on increasing the temperature.

Even rubber with missing sulfur bridges behaves as an elastic material when the body is subjected to a fast change in shape, giving no time for disentanglement (Fig. 26.21a). In this case, points of entangled chains act like net points. In slow processes, however, the material behaves like a viscous liquid (Fig. 26.21b).

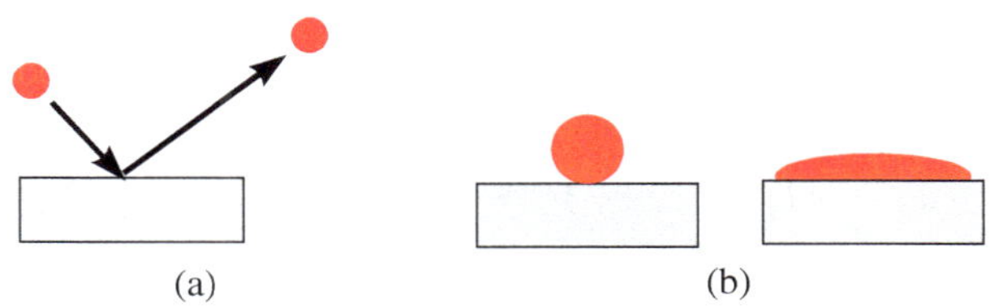

Figure 26.21 Behavior of silicone rubber with missing sulfur bridges. (a) Elastic response on a fast change of shape. (b) Liquid-like response on slow changes.

26.8 Conclusion

Understanding the behavior of macromolecules under all kinds of influences is of fundamental importance to the study of polymer materials and processes in molecular biology. We have discussed the case of chain molecules that form random coils and illustrated how the molecular shape changes, under both internal and external influences, are related to macroscopic properties. Furthermore we have illustrated how to quantify properties of the macromolecule statistically and compare them with experimentally measurable properties, such as light scattering, viscosity, and diffusion.

We have discussed fascinating experiments on single double-stranded DNA molecules (unraveling and restoring coils, pulling chains through meshworks) and shown how they can be understood using the concepts developed for a random coil.

26.9 Box 26.1. Statistical Chain Elements

We can view the conformation of a polymer molecule by considering its formation in terms of a three-dimensional random walk model, in which we start at one end of the polymer strand and decide where to place the next monomer unit in a statistically random manner. In this model, a particular polymer molecule's conformation corresponds, in a mathematical description, to a particular trajectory of a diffusing particle. In terms of diffusion theory (Chapter 14) the length of a statistical chain element l corresponds to the mean free path $\overline{\lambda}$ of a particle moving in space at random, and the number N_{stat} of statistical chain elements corresponds to the number of collisions j. Accordingly, the length of the polymer chain $L = N_{\text{stat}} \cdot l$ corresponds to the diffusion path $s = j \cdot \overline{\lambda}$ after j collisions. Thus the consideration of Chapter 14 can be immediately applied to the present case.

According to Section 14.3 the probability $\mathcal{P}(r) \cdot dr$ of finding a diffusing particle after j collisions at a distance between r and $r + dr$ from the original position is

$$\mathcal{P}(r) \cdot dr = const \cdot 4\pi r^2 \cdot \exp\left(-\frac{3r^2}{2j \cdot \lambda^2}\right) \cdot dr \quad \text{and} \quad \overline{r^2} = j \cdot \lambda^2$$

is the mean square of the displacement of the particle.

For the present case, $P(h) \cdot dh$ is the probability of finding the distance between the chain ends (end-to-end distances) between h and $h + dh$ equals

$$\mathcal{P}(h) \cdot dh = const \cdot 4\pi h^2 \cdot \exp\left(-\frac{3h^2}{2N_{\text{stat}} \cdot l^2}\right) \cdot dh \quad \text{and} \quad \overline{h^2} = N_{\text{stat}} \cdot l^2$$

is the mean square of the end-to-end distances of the polymer.

26.10 Box 26.2. Bending of a DNA Strand

We estimate the energy for bending a DNA double strand by fixing it (length l, diameter $2r$) at one end and bending it by applying a force f at the other end perpendicular to the cylinder axis, producing a shift s of the end. Modeling the DNA as an elastic material

$$f = A_{\text{Young}} \frac{3\pi r^4 s}{4l^3}$$

where A_{Young} is Young's modulus by bending, see textbooks of physics. The elastic energy is

$$E = \int_0^s f \cdot ds = A_{\text{Young}} \frac{3\pi r^4}{4l^3} \cdot \frac{s^2}{2}$$

We can write s as

$$s = l\frac{\alpha}{2}$$

where α is the bending angle. Then

$$E = A_{\text{Young}} \frac{3\pi}{32} \cdot \frac{r^4}{l} \alpha^2$$

and for $\alpha = \pi/2$ (bending by 90°) we obtain

$$E_{\alpha=\pi/2} = A_{\text{Young}} \frac{3\pi^3}{128} \cdot \frac{r^4}{l}$$

In order to estimate Young's modulus for bending a DNA strand we assume that it is the same as for bending a rubber cylinder. We measure the force to bend a rubber cylinder of length $l = 27$ cm and diameter $2r = 1$ cm by 90°; this force is $f = 0.6$ N and the elastic energy is about $E = f \cdot l/4 \approx 0.1$ J. In this way we estimate

$$A_{\text{Young}}(\text{rubber}) = \frac{128}{3\pi^3} E \frac{l}{r^4} = 5.5 \times 10^7 \text{ N m}^{-2}$$

Now consider a DNA double helix of length $l = 100$ nm and radius $r = 1.5$ nm. For the elastic energy to bend the helix by 90° we obtain

$$E_{\alpha=\pi/2} = 5.5 \times 10^7 \text{ N m}^{-2} \cdot \frac{3\pi^3}{128} \cdot \frac{(1.5 \text{ nm})^4}{100 \text{ nm}} = 2 \times 10^{-21} \text{ J}$$

This energy is on the order of the thermal energy $kT = 4 \times 10^{-21}$ J. From these considerations we conclude that the length of the statistical chain element for DNA is about 100 nm at room temperature.

26.11 Box 26.3. Force Contracting Chain Held at Chain Ends

We consider a chain in solution fixed at chain end P_1 (see Fig. 26.13). The second chain end P_2 is free to diffuse but considered to be bound to P_1 with a force f, i.e., we consider the molecule to be replaced by a spring pulling P_2 toward P_1 with the force (Hooke's law)

$$f = -k_f \cdot h \tag{26B3.1}$$

(k_f = force constant, h = elongation).

What must be k_f in order to result in distribution function (26.6)? According to Boltzmann's Law we have

$$\mathcal{P}(h) \cdot \mathrm{d}h = const \cdot h^2\, \mathrm{e}^{-E/(kT)} \cdot \mathrm{d}h \tag{26B3.2}$$

where the energy E is

$$E = -\int_0^h f \cdot \mathrm{d}h = k_f \int_0^h h \cdot \mathrm{d}h = k_f \frac{1}{2} h^2$$

so that the probability to find the chain ends at a distance between h and $h + \mathrm{d}h$ is

$$\mathcal{P}(h) \cdot \mathrm{d}h = const \cdot h^2 \cdot \exp\left(-\frac{k_f h^2}{2kT}\right) \cdot \mathrm{d}h \tag{26B3.3}$$

Correspondingly, the mean square average end-to-end distance is

$$\overline{h^2} = \frac{\int h^2 \mathcal{P}(h) \cdot \mathrm{d}h}{\int \mathcal{P}(h) \cdot \mathrm{d}h} = \frac{3}{2} \cdot \frac{2kT}{k_f} = \frac{3kT}{k_f} \tag{26B3.4}$$

From Equation (26.8) we know that

$$\overline{h^2} = L\,l$$

so that

$$k_f = \frac{3kT}{L\,l} \tag{26B3.5}$$

and the force acting between the ends of the polymer molecule is

$$f = -\frac{3kT}{L\,l} \cdot h \tag{26B3.6}$$

With this form of the force, we can write the probability distribution as

$$\begin{aligned}\mathcal{P}(h) \cdot \mathrm{d}h &= const \cdot 4\pi h^2 \exp\left(-\frac{3kT}{Ll}\frac{h^2}{2kT}\right) \cdot \mathrm{d}h \\ &= const \cdot 4\pi h^2 \exp\left(-\frac{3}{2}\frac{h^2}{Ll}\right) \cdot \mathrm{d}h\end{aligned} \tag{26B3.7}$$

in agreement with Equation (26.6).

The linear relationship between f and h given by Equation (26B3.6) holds for $h < L$. f, due to the increasing effect of thermal collisions, increases strongly toward $h = L$ and then very strongly, due to the elasticity of the stretched chain (Fig. 26.13b). The measured relation between f and h has been described by the empirical equation[3]

$$f = -\frac{kT}{2l}\left[\frac{1}{(1 - h/L)^2} - 1 + \frac{4h}{L}\right] \tag{26B3.8}$$

It reduces to Equation (26B3.6) for $h \ll L$ (see Exercise E26.18). The force f in Equation (26B3.8) becomes infinite for $h = L$, because the elasticity of the stretched chain has been neglected.

The force f can be related to the average contribution of a statistical chain element, $\overline{\Delta h}$. Defining

$$\overline{\Delta h} = \frac{h}{N_{\text{stat}}} \quad \text{or} \quad \frac{\overline{\Delta h}}{l} = \frac{h}{N_{\text{stat}}\,l} = \frac{h}{L} \tag{26B3.9}$$

3 See [29]. Note that Equation (26B3.8) is empirical.

we can substitute into Equation (26B3.8) and obtain

$$f = -\frac{kT}{2l}\left[\frac{1}{\left(1-\overline{\Delta h}/l\right)^2} - 1 + \frac{4\overline{\Delta h}}{l}\right] \tag{26B3.10}$$

26.12 Foundations

26.12.1 Foundation 26.1: Light Scattering of Macromolecules

26.12.1.1 Polarizability α of Macromolecules in the Field of an Incident Light Wave

Consider a plate capacitor in vacuum with area A and a distance d between its two plates. From elementary physics, we know that its capacitance C is given by

$$C_{\text{vacuum}} = \varepsilon_0 \frac{A}{d} = \frac{Q_{\text{vacuum}}}{Fd} \tag{26F1.1}$$

where Q_{vacuum} is the charge on the plates and F is the electric field strength. Accordingly, for a capacitor filled with a dielectric medium of relative permittivity $\hat{\varepsilon}$

$$C = \varepsilon_0 \hat{\varepsilon} \frac{A}{d} = \frac{Q}{Fd} \tag{26F1.2}$$

where Q is the charge on the plates in the dielectric with relative permittivity $\hat{\varepsilon}$. The volume polarization $\hat{P}$ of the dielectric medium (that is the dipole moment of the dielectric medium of volume $V = Ad$) is defined by

$$\hat{P} = \left(Q - Q_{\text{vacuum}}\right) d \tag{26F1.3}$$

With $Q_{\text{vacuum}} = \varepsilon_0 AF$ and $Q = \varepsilon_0 \hat{\varepsilon} AF$ we obtain

$$\hat{P} = \varepsilon_0 \left(\hat{\varepsilon} - 1\right) Ad \cdot F = \varepsilon_0 \left(\hat{\varepsilon} - 1\right) V \cdot F \tag{26F1.4}$$

For the high frequency of an incident light wave we can replace $\hat{\varepsilon}$ by $\hat{\varepsilon} = \hat{n}^2$ where $\hat{n}$ is the refractive index. Then

$$\hat{P} = \varepsilon_0 \left(\hat{n}^2 - 1\right) V \cdot F \tag{26F1.5}$$

Now we consider a solution of macromolecules. The volume polarization $\hat{P}_{\text{solution}}$ of the solution is

$$\hat{P}_{\text{solution}} = \varepsilon_0 V \left(\hat{n}^2_{\text{solution}} - 1\right) F_{0,\text{incident}} \tag{26F1.6}$$

where $F_{0,\text{incident}}$ is the amplitude of the electric field strength of the incident light wave. Accordingly, the volume polarization $\hat{P}_{\text{solvent}}$ of the pure solvent is

$$\hat{P}_{\text{solvent}} = \varepsilon_0 V \left(\hat{n}^2_{\text{solvent}} - 1\right) F_{0,\text{incident}} \tag{26F1.7}$$

The difference $\hat{P}_{\text{solution}} - \hat{P}_{\text{solvent}}$ corresponds to the contribution $\hat{P}$ of the N macromolecules in the volume V, thus

$$\hat{P} = \hat{P}_{\text{solution}} - \hat{P}_{\text{solvent}} = \varepsilon_0 V \left(\hat{n}^2_{\text{solution}} - \hat{n}^2_{\text{solvent}}\right) F_{0,\text{incident}} \tag{26F1.8}$$

On the other hand, the volume polarization $\hat{P}$ is related to the polarizability α of the macromolecules as

$$\hat{P} = N\alpha \cdot F_{0,\text{incident}} \tag{26F1.9}$$

Then for α we find that

$$\alpha = \frac{\varepsilon_0 V \left(\hat{n}^2_{\text{solution}} - \hat{n}^2_{\text{solvent}}\right)}{N} = \frac{\varepsilon_0 V \left(\hat{n}_{\text{solution}} + \hat{n}_{\text{solvent}}\right)\left(\hat{n}_{\text{solution}} - \hat{n}_{\text{solvent}}\right)}{N}$$
$$\approx \frac{\varepsilon_0 V \cdot 2\hat{n}_{\text{solution}} \left(\hat{n}_{\text{solution}} - \hat{n}_{\text{solvent}}\right)}{N}$$

26.12.1.2 Forward and Backward Light Scattering

What is the ratio of the intensities of scattered light arriving at point P_1 and P_2, respectively (diameter of molecules $<\lambda/4$)?

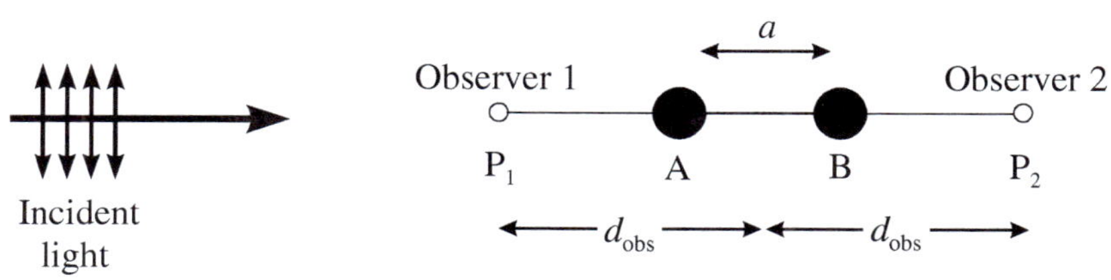

To answer this question, we superpose the light waves arising from each scattering center. The wave scattered from centers A and B arrive at P_2 with the same phase; i.e.,

$$F(s,t)|_{s=P_2} = F_0 \cos\left[\omega\left(t - \frac{s}{c_0}\right)\right]\Bigg|_{s=P_2 \text{ from A}} + F_0 \cos\left[\omega\left(t - \frac{s}{c_0}\right)\right]\Bigg|_{s=P_2 \text{ from B}}$$

$$= 2F_0 \cos\left[\omega\left(t - \frac{s}{c_0}\right)\right]\Bigg|_{s=P_2}$$

At point P_1 the wave arising from B travels an extra distance $2a$ so that it arrives with a phase shift $2a/c_0$ (c_0 = velocity of light):

$$F(s,t)|_{s=P_1} = F_0 \cos\left[\omega\left(t - \frac{s}{c_0}\right)\right]\Bigg|_{s=P_1 \text{ from A}} + F_0 \cos\left[\omega\left(t - \frac{s+2a}{c_0}\right)\right]\Bigg|_{s=P_1 \text{ from B}}$$

$$= 2F_0 \cos\left[\omega\left(t - \frac{s}{c_0}\right)\right]\Bigg|_{s=P_1} \cos\left(\frac{\omega a}{c_0}\right)$$

where $\omega/c_0 = 2\pi/\lambda$. This gives an intensity ratio for the light arriving at P_1 and P_2 as

$$\frac{I(P_1)}{I(P_2)} = \frac{F^2(P_1)}{F^2(P_2)} = \cos^2\left(\frac{2\pi a}{\lambda}\right)$$

Because of

$$\cos x = 1 - \frac{1}{2}x^2 + \cdots$$

for sufficiently small x we have

$$\cos^2 x = \left(1 - \frac{1}{2}x^2\right)^2 = 1 - x^2$$

and we can write an approximate, but simpler formula, for the intensity ratio as

$$\frac{I(P_1)}{I(P_2)} \approx 1 - \left(\frac{2\pi a}{\lambda}\right)^2$$

Note that in our model $2a = \sqrt{h^2}$ where a is the distance between points P_1 and P_2.

26.12.2 Foundation 26.2: Viscosity of Dilute Solutions of Polymers and Macroscopic Models

See the books website, www.wiley.com/go/Kuhn/physicalchemistry3e.

26.12.3 Foundation 26.3: Diffusion of a Random Coil in a Gel in the Absence of an Electric Field and Force-Induced Motion

See the books website, www.wiley.com/go/Kuhn/physicalchemistry3e.

26.13 Exercises

E26.1 (a) Show that the ratio I/I_{forward} in Equation (26.11) is one for $\varphi = 0$ (forward scattering). (b) Use Equation (26.11) for backward scattering ($\varphi = 180°$) at $\alpha < 1$ to find the ratio of backward to forward scattering. Compare to the result obtained in Equation (26.10).

E26.2 From Equation (26.11) derive an expression for the ratio I_{135}/I_{45} plotted in Fig. 26.6. What do you expect for this ratio in the cases $\sqrt{\overline{h^2}}/\lambda = 1$ and $\sqrt{\overline{h^2}}/\lambda \gg 1$?

E26.3 The following data have been obtained for ribosomes: sedimentation constant $s = 82.6 \times 10^{13}$ s, diffusion coefficient $D = 1.52 \times 10^{-7}$ cm^2 s^{-1}, partial molar volume $v_{\text{partial}} = 0.61$ cm^3 g^{-1}, and a density $\rho = 1$ g cm^{-3} at 20 °C. Calculate the molecular mass m of the ribosomes.

E26.4 Normalize the probability distribution for the end-to-end distance of a random coil, which is given by Equation (26.6).

E26.5 Evaluate the integral for the average of h^2 in Equation (26.8), by using the distribution function given in Equation (26.6).

E26.6 Using the probability distribution for a random coil, Equation (26.6), find the most probable end-to-end distance of a polymer chain. Compare this result to that we found for the mean square end-to-end distance. Provide a verbal explanation for why these two quantities are different.

E26.7 Consider the scattering data for a 0.20 g L^{-1} solution of polystyrene in toluene at 293 K, for which the excitation wavelength is 546.0 nm. Analyze these data[4] to determine the mean square end-to-end distance of the polymer chain.

$I(\varphi)/I_{\text{forward}}$	0.606	0.502	0.378	0.302	0.212	0.165	0.132
$\varphi/°$	25.8	36.9	53.0	66.4	90.0	113.6	143.1

E26.8 Measurements on dilute solutions of polyisobutylene (monomer molecular mass of 54 g mol^{-1}) in chloroform (shear viscosity η of 0.537 mPa s at the temperature of these measurements $T = 298$ K) give the diffusion coefficient data in the table. Analyze these data[5] to determine the statistical length for the polymer.

$D/10^{-11}$ m^2 s^{-1}	8.56	6.79	3.19	1.97	1.37	1.24	0.67
$\boldsymbol{M}$/kg mol^{-1}	57.1	95.0	247	610	1100	1900	3800

E26.9 Measurements on dilute solutions of polyisobutylene (monomer molecular mass of $\boldsymbol{M}_{\text{monomer}} = 54$ g mol^{-1} and $b = 250$ pm) in chloroform (at $T = 298$ K) give the Staudinger index $[\eta]$ data in the table.[6] Analyze these data to determine the mean square end-to-end distance $\overline{h^2}$ of the polymer chains and then extract the statistical length for the polymer by plotting $\overline{h^2}$ versus N_{monomer}.

$[\eta]$/cm^3 g^{-1}	29.3	38.8	65.7	107.9
$\boldsymbol{M}$/kg mol^{-1}	57.1	95.0	247	610

E26.10 Consider the Staudinger index data for different molar masses of polyisobutylene (monomer molecular mass of $\boldsymbol{M}_{\text{monomer}} = 54.088$ gmol^{-1} and $b = 250$ pm) in cyclohexane at 303 K. Use these data to determine the mean square end-to-end distance of the polymer and to determine the diameter of the random coil, as a function of the polymer's average molecular mass (see [34]).

$\boldsymbol{M}$/kg mol^{-1}	37.8	167.	333.	710.
$[\eta]$/cm^3 g^{-1}	38.8	112	181	287

E26.11 Example 26.3 analyzes diffusion data for cellulose acetate in acetone using the refined model with the translational form factor C_{transl} (Equation (26.37)). Analyze these same data with the simpler formula given by Equation (26.17) and compare your analysis to that of Example 26.3.

4 Data are adapted from [32].
5 Data are from [33].
6 Data are from [33].

E26.12 Example 26.4 analyzes sedimentation data for cellulose acetate in acetone using the refined model with the translational form factor C_{transl} (26.39). Analyze these same data with the simpler formula given by Equation (26.23) and compare your analysis to that of Example 26.4.

E26.13 The sedimentation constant s was measured for different molecular weights of polyisobutylene in cyclohexane at 293 K (see table[7]). Use these data and the facts $v_{part} = 1.091$ cm^3 g^{-1} and $\rho = 0.779$ g cm^{-3} to determine D for each of these polymers. Does the diffusion coefficient D change with $\boldsymbol{M}$ in a way you would expect?

$\boldsymbol{M}$/kg mol^{-1}	30.9	172.	672	1420
$s/10^{13}$ s	0.925	1.94	3.33	4.45

E26.14 Using your knowledge of osmotic pressure from Chapter 18 and the data in the table for polyisobutylene in cyclohexane, determine its molecular weight. The temperature is 303 K.

^{osm}P/N m^{-2}	64.8	151.0	255.0
c'/g L^{-1}	1.9	4.0	5.9

E26.15 Consider Equation (26.11), for the case of backward scattering and $C < 1$. Show that

$$\frac{I_{backward}}{I_{forward}} \simeq 1 - \frac{8\pi^2}{9}\frac{\overline{h^2}}{\lambda^2}$$

Note: This exercise is a shorter version of E26.1.

E26.16 The table provides data on small 4.3 kDa DNA double helices in a 1.5% agarose gel (taken from [36]). Show that the velocity changes linearly with the applied field.

v/cm h^{-1}	0.0090	0.014	0.017	0.023	0.035	0.052	0.080	0.137	0.180
F/V cm^{-1}	0.12	0.20	0.24	0.34	0.50	0.72	1.0	1.5	2.0

E26.17 In Box 26.2 we consider the bending of a DNA filament. Use that analysis to estimate the length of a statistical chain element for actin filaments of radius $r = 3.5$ nm.

E26.18 Simplify the empirical equation for the force f in Box 26.3 to the case that $h \ll L$.

E26.19 For incident light polarized in the z-direction, the intensity I_φ of the light scattered in the xy-plane in a direction forming an angle φ with the direction of the incident beam ($I_{\varphi=0} = I_{forward}$), is given by Equation (26.11)

$$\frac{I_\varphi}{I_{forward}} = \frac{2}{C^2}\left(e^{-C} - 1 + C\right) \quad \text{with} \quad C = \frac{8\pi^2}{3}\frac{\overline{h^2}}{\lambda^2}\sin^2\left(\frac{\varphi}{2}\right)$$

Show that this equation reduces to

$$\frac{I_{backward}}{I_{forward}} = 1 - \frac{1}{3}C = 1 - \frac{8\pi^2}{9}\frac{\overline{h^2}}{\lambda^2}$$

when the parameter C is smaller than unity.

E26.20 Using

$$\frac{I_\varphi}{I_{forward}} = \frac{2}{C^2}\left(e^{-C} - 1 + C\right) \quad \text{with} \quad C = \frac{8\pi^2}{3}\frac{\overline{h^2}}{\lambda^2}\sin^2\frac{\varphi}{2}$$

find the ratio $I_{135°}/I_{45°}$.

7 Data are from [35].

26.14 Problems

P26.1 The polymer description in this chapter was cast in terms of the mean square end-to-end distance, $\overline{h^2}$, of a polymer strand. Often, it is more convenient to use a different quantity called the radius of gyration, R_g, because it is related more directly to experimental measurables. The radius of gyration is defined as

$$R_g^2 = \frac{1}{2N^2}\sum_{j=1}^{N}\sum_{i=1}^{N}\overline{\left(\vec{r}_j - \vec{r}_i\right)^2}$$

where $\vec{r}_i$ is the vector specifying the position of statistical segment i in the chain and N is the number of statistical segments. Using this definition show that the radius of gyration for the Gaussian chain is given by

$$R_g^2 = N\, l^2/6$$

(Hint: Use the fact that each segment is statistically related to each other one.).

P26.2 Using the definition of the radius of gyration show that it may also be written as

$$R_g^2 = \frac{1}{N}\sum_{j=1}^{N}\overline{\left(\vec{r}_j - \vec{r}_{\text{center of mass}}\right)^2}$$

where $\vec{r}_{\text{cm}}$ is the vector identifying the center of mass in the laboratory frame of reference.

P26.3 Use the Stokes–Einstein relation $D = kT/(6\pi\eta r)$ for the diffusion coefficient D to prove that Equation (26.24) holds generally.

P26.4 Combine some of our developments in Section 26.3.2 to show that $s \cdot [\eta]/N_{\text{monomer}}$ should be constant for a given polymer solution. Use the data in the table[8] for different molecular weights of polyisobutylene in cyclohexane at 293 K and the facts $v_{\text{part}} = 1.091$ cm^3 g^{-1}, $\eta = 0.97$ mPa s, $\rho = 0.779$ g cm^{-3}, and $b = 250$ pm to determine this constant and evaluate the statistical length l.

$\boldsymbol{M}$/kg mol^{-1}	30.9	172	672	1420
$[\eta]$/cm^3 g^{-1}	34.2	112	287	489
$s/10^{13}$ s	0.925	1.94	3.33	4.45

P26.5 Perform Exercises E26.9 and E26.10 which are data for polyisobutylene in different solvents. How much does the characteristic size of the polymer change between these solvents? Provide a verbal explanation for any difference you observe.

P26.6 The data in the table[9] provide characteristic sizes of proteins when they are denatured in solution. Plot the radius of gyration, R_g, versus the amino acid number and compare it to the prediction of a random coil model, for which $R_g^2 = N\overline{h^2}/6$ (see Problem 26.1).

From a best fit to the random coil model extract a statistical chain length.

Protein	**GroEL**	**yPGK**	**Creatine Kinase**	**α-TS**	**Carbonic Anhydrase**	**Apomyo-globin**
N_{monomer}	549	416	380	268	260	154
R_g/nm	8.2	7.1	4.6	4.9	5.9	4.0
Protein	**ctACP**	**Protein L**	**Fyn SH3**	**Ubiquitin**	**GCN4-p2**	**drK SH3**
N_{monomer}	98	79	78	76	66	59
R_g/nm	3.05	2.6	2.6	2.5	2.4	2.2

8 Data are from [35].
9 The data are taken from [37].

Protein	Snase	Lysozyme	RNase A	pI3K SH2	pI3K SH3	mAcp
$N_{monomer}$	149	129	124	112	103	98
R_g/nm	3.7	3.6	3.3	3.0	3.1	3.0
Protein	**Protein G**	**Cyt c**	**AK-37**	**AK-32**	**AK-27**	**AK-16**
$N_{monomer}$	52	39	37	32	27	16
R_g/nm	2.3	1.8	1.7	1.5	1.3	1.0

P26.7 Obtain a rubber band and perform an experiment that will determine the heat flow, q, and the entropy change, ΔS, for the following process. In step A, stretch the rubber band six inches at constant temperature. In step B, release the rubber band so that it "snaps back" instantaneously and then allow it to warm up to the initial temperature. By performing a measurement, determine the amount of heat absorbed by the rubberband, q, in returning to its starting temperature (step B). If step A is assumed to be reversible, estimate the entropy change, which occurs in step B. From the entropy change estimate the change in the number of configurations.

P26.8 Consider the light scattering experiment whose geometry is shown in Fig. 26.6. Determine the ratio of the intensities of the scattered light arriving at points P_1 and P_2, respectively.

P26.9 In Section 26.5 (see Fig. 26.16) we showed that a protein can be considered as a compact sphere. For comparison, assume that the proteins are statistical coils and calculate the coil radius r_{coil}, using Equation (26.16) as a function of the molar mass $\boldsymbol{M}$ (use $T = 295$ K and $\eta = 10^{-3}$ Pa s). Assume the values $bl = 0.08$ nm^2 and $\boldsymbol{M}_{monomer} = 120$ g mol^{-1}.

Protein	M g mol^{-1}	D_{trans} 10^{-11} m^2 s^{-1}
Lactate dehydrogenase	145,169	5.05
Ribonuclease A	13,700	10.7
Nitrogenase MoFe	220,000	4
Citrate Synthase	97,938	5.8
Chymotrypsinogen	25,660	9.3
β-Lactoglobulin	36,730	7.82
Ovalbumin	43,500	7.96
Myoglobin	17,190	10.8
GPD	142,868	5
BPTI(q)	6158	12.9
Aldolase	156,000	4.45
Lysozyme	14,320	10.9
Catalase	230,340	4.1

P26.10 Extension of an Unraveled Coil: Consider Equation (26.52)

$$\frac{kT}{2l}\left[\frac{1}{\left(1-\overline{\Delta x}_n/l\right)^2} - 1 + \frac{4\overline{\Delta x}_n}{l}\right] = 6\pi\eta \cdot l \cdot \frac{1}{2}\sqrt{N_{stat} - n} \cdot v$$

where l is the statistical chain element, $\overline{\Delta x}_n$ is the average extension of the chain element n in the x-direction (see Fig. 26.14b), η is the viscosity, N_{stat} is the number of statistical chain elements, and v is the flow speed of the fluid. We can rewrite this equation as

$$\frac{1}{\left(1-\overline{\Delta x}_n/l\right)^2} - 1 + \frac{4\overline{\Delta x}_n}{l} = \frac{6\pi\eta \cdot l^2 \cdot v}{kT}\sqrt{N_{stat} - n}$$

a) Using the values of $L = 50\ \mu m$, $l = 100$ nm (thus $N_{stat} = L/l = 500$), $v = 14\ \mu m\ s^{-1}$, $\eta = 1 \times 10^{-3}$ N m^{-2} s, and $T = 300$ K, show that

$$\frac{1}{\left(1 - \overline{\Delta x_n}/l\right)^2} - 1 + \frac{4\overline{\Delta x_n}}{l} = 0.637 \cdot \sqrt{N_{stat} - n}$$

b) Solve this equation, either graphically or numerically, for n values of 1,100, 200,300, 400, and 500. Using the value of l given in part a) calculate $\overline{\Delta x_n}$ for each n value.

c) Using the $\overline{\Delta x_n}$ and the fact that the total extension x is obtained by summing up over all elements n, find the fractional extension x/L as a function of flow sped v.

P26.11 Time to Restore Random Coil from Unraveled Chain (see Fig. 26.15a and its discussion): After stopping the current, coiling begins at the open end where the tension along the chain is smallest. After some time a coil, including a part of the chain of contour length $L - x$, has formed. This coil grows and the displacement x of its edge from the fixed end of the chain (see Fig. 26.15a) moves with speed v.

a) Approximate the coil as a sphere of radius

$$r_{coil} = \frac{1}{2}\sqrt{(L - x)l}$$

and use Stokes law to estimate the force acting on it.

b) Thermal collisions provide a force f' between the molecular chain and the molecules of the solvent, which acts to randomize the orientation of the statistical chain element. For keeping the chain element well oriented (probability for $\varphi = 0°$ considerably larger than for $\varphi = 180°$, see Fig. 26.14b) the energy $f' \cdot l$ to turn the element must be about kT. By assuming that f' and f balance, show that the speed v is given by

$$v = \frac{kT}{3\pi\eta\sqrt{L - x} \cdot l^{3/2}}$$

c) Integrate the velocity expression to show that

$$x = L - \left(\frac{kT}{2\pi\eta}\right)^{2/3} \frac{1}{l} \cdot t^{2/3}$$

P26.12 Stretching a Chain: Here you estimate the force to stretch a piece of rubber along the z-direction. For simplicity, assume that the lines connecting the end points of the chains (net points) have the direction of the x, y, and z axes, $N_{thread}/3$ in each axis (Fig. P26.12).

Furthermore, assume that h has the average value $h_1 = \sqrt{Ll}$ before stretching; and after stretching in the z-direction, it has

$$h_2 = \sqrt{Ll} \cdot \left(1 + \frac{\Delta z}{z}\right)$$

a) Assume that the force constant for stretching is $k_f = (3kT)/(L \cdot l)$ (Equation (26B3.5) in the text); and calculate the work that is performed on the z directed chains ($\frac{1}{3}N_{thread}$ chains).

b) Assume that the volume of the rubber is unchanged by stretching along z and use it to find the change in size along x and y. Use this change in the cross sectional area to find the work performed on the x and y directed chains.

c) Combine your results from parts a) and b) to show that the total work is given by

$$w = kT\frac{3}{2}N_{thread} \cdot \left(\frac{\Delta z}{z}\right)^2$$

for $\Delta z/z \ll 1$.

d) Using the result in c), find an expression for the force.

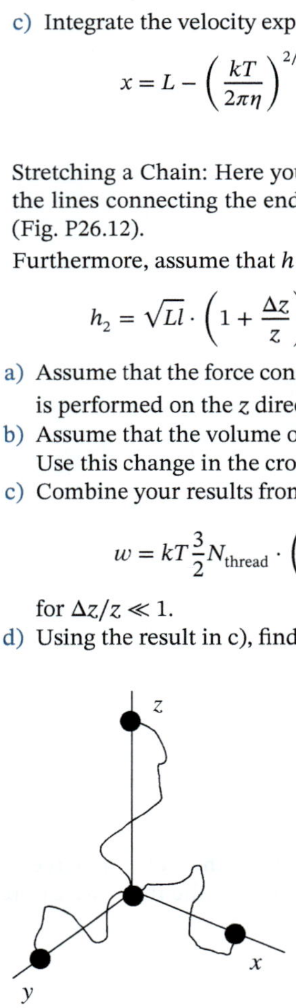

Figure P26.12 Rubber idealized by molecular chains, $N_{thread}/3$ oriented in the x direction, and $N_{thread}/3$ each in the y and z directions. Dots: net points.

References

1 J.F. Fernandez-Sanchez, R. Cannas, S. Spichiger, R. Steiger, and U.E. Spichiger-Keller, Novel nanostructured materials to develop oxygen-sensitive films for optical sensors, *Anal. Chim. Acta*, 2006, **566**, 271.

2 H. Eyring, The resultant electric moment of complex molecules, *Phys. Rev.*, 1932, **39**, 746.

3 H. Kuhn, Restricted bond rotation and shape of unbranched saturated hydrocarbon chain molecules, *J. Chem. Phys.*, 1947, **15**, 843, https://doi.org/10.1063/1.1746346.

4 R. Chiang, Intrinsic viscosity-molecular weight relationship for fractions of linear polyethylene[1], *J. Phys. Chem.*, 1965, **69**, 1645.

5 M. Yanagida, Y. Hiraoka, and I. Katsura, Dynamic behaviors of DNA molecules in solution studied by fluorescence microscopy, *Cold Spring Harb. Symp. Quant. Biol.*, 1982, **47**(2), 177.

6 S. Hoops, H. Kuhn, W. Huber, and L. Eckert, Ermittlung der äusseren Abmessungen statistisch geknäuelter Fadenmoleküle durch Ausmessung makroskopischer Modelle, *J. Polym. Sci.*, 1956, **20**, 101.

7 P. Lorrain and D. Corson, *Electromagnetic Fields and Waves* (Freeman, New York, 1970, p. 595) 2nd edition.

8 P. Debye, Molecular-weight determination by light scattering, *J. Phys. Chem.*, 1951, **51**, 18; H. Kuhn, Gestalt und Grösse gelöster Fadenmolekel aus Streulichtdepolarisationsmessungen. (Vorläufige Mitteilung), *Helv. Chim. Acta*, 1946, **29**, 432.

9 A. Einstein, Eine neue Bestimmung der Moleküldimensionen, *Ann. Phys.*, 1906, **19**, 289.

10 M.L. Mansfield, J.F. Douglas, and E.J. Garboczi, Intrinsic viscosity and the electrical polarizability of arbitrarily shaped objects, *Phys. Rev. E*, 2001, **64**, 061401; J.F. Douglas and K.F. Freed, Competition between hydrodynamic screening ("draining") and excluded volume interactions in an isolated polymer chain, *Macromolecules*, 1994, **27**, 6088–6099.

11 H. Kuhn, F. Monig, and W. Kuhn, Makroskopische Modelle statistisch geknäuelter Fadenmolekeln. Verbesserung bisheriger Ansätze über Viskosität und Strömungsdoppelbrechung von Lösungen hochmolekularer Stoffe, *Helv. Chim. A*, 1953, **36**, 731.

12 H. Kuhn, F. Moning, and W. Kuhn, Makroskopische Modelle statistisch geknäuelter Fadenmolekeln. Verbesserung bisheriger Ansätze über Viskosität und Strömungsdoppelbrechung von Lösungen hochmolekularer Stoffe, *Helv. Chim. Acta*, 1953, **36**, 731.

13 S.J. Singer, On the concentration dependence of the rates of sedimentation and diffusion of macromolecules in solution. Preliminary, *J. Chem. Phys.*, 1947, **15**, 341.

14 T.G. Fox and P.J. Flory, Second-order transition temperatures and related properties of polystyrene. I. Influence of molecular weight, *J. Appl. Phys.*, 1950, **21**, 581; J.F. Douglas and K.F. Freed, Competition between hydrodynamic screening ("draining") and excluded volume interactions in an isolated polymer chain, *Macromolecules*, 1994, **27**, 6088–6099.

15 W. Kuhn, Über die gestalt fadenförmiger moleküle in lösungen, *Kolloid Z.*, 1934, **68**, 2.

16 W. Kuhn and H. Kuhn, Die Frage nach der Aufrollung von Fadenmolekeln in strömenden Lösungen, *Helv. Chim. Acta*, 1943, **26**, 1394.

17 O. Kratky and G. Porod, Röntgenuntersuchung gelöster Fadenmoleküle, *Rec. Trav. Chim. Pays-Bas*, 1949, **68**, 1106.

18 S.B. Smith, L. Finzi, and C. Bustamante, Direct mechanical measurements of the elasticity of single DNA molecules by using magnetic beads, *Science*, 1992, **258**, 1122; D. Wirtz, Direct measurement of the transport properties of a single DNA molecule, *Phys. Rev. Lett.*, 1995, **75**, 2436.

19 F. Oesterhelt, M. Rief, and H.E. Gaub, Single molecule force spectroscopy by AFM indicates helical structure of poly (ethylene-glycol) in water, *New J. Phys.*, 1999, **1**, 6.1–6.6.

20 M. Yanagida, Y. Hiraoka, and I. Katsura, Dynamic behaviors of DNA molecules in solution studied by fluorescence microscopy, *Cold Springs Harbor Symp. Quant. Biol.*, 1982, **47** (2), 177; F. Brochard, *DNA: Observing Stems and Flowers* (Physics World, 1995, p. 24); A. Buguin and F. Brochard-Wyart, Unwinding of globular polymers under strong flows, *Macromolecules*, 1996, **29**, 4937.

21 T.T. Perkins, D.E. Smith, R.G. Larson, and S. Chu, Stretching of a single tethered polymer in a uniform flow, *Science*, 1995, **268**, 83.

22 D. Wirtz, Direct measurement of the transport properties of a single DNA molecule, *Phys. Rev. Lett.*, 1995, **75**, 2436.

23 T.T. Perkins, S.R. Quake, D.E. Smith, and S. Chu, Relaxation of a single DNA molecule observed by optical microscopy, *Science*, 1994, **264**, 822.

24 T.E. Creighton, *Proteins: Structures and Molecular Properties* (Freemen, New York, 1996) 2nd edition.

25 H. Mosimann, Über das Sedimentationsverhalten von Nitrocellulosefraktionen und die daraus ableitbare Molekelform, *Helv. Chim. Acta*, 1943, **26**, 61.

26 T.T. Perkins, D.E. Smith, and S. Chu, Direct observation of tube-like motion of a single polymer chain, *Science*, 1994, **264**, 819.

27 G.S. Manning, The molecular theory of polyelectrolyte solutions with applications to the electrostatic properties of polynucleotides, *Q. Rev. Biophys.*, 1978, **11**, 179.

28 B. Tinland, L. Meistermann, and G. Weil, Simultaneous measurements of mobility, dispersion, and orientation of DNA during steady-field gel electrophoresis coupling a fluorescence recovery after photobleaching apparatus with a fluorescence detected linear dichroism setup, *Phys. Rev.*, 2000, **E61**, 6993; A. Stellwagen and N.C. Stellwagen, Anomalously slow electrophoretic mobilities of DNA restriction fragments in polyacrylamide gels are not eliminated by increasing the gel pore size, *Biopolymers*, 1999, **30**, 309.

29 C. Bustamente, Entropy elasticity of λ-phaga DNA, *Science*, 1994, **265**, 1599.

30 H. Kuhn, W. Kuhn, and A. Silberberg, Improved relationships for diffusion and sedimentation constants and for viscosity and streaming birefringence of solutions of polymers, *J. Polym. Sci.*, 1954, **74**, 193.

31 J. Klein, Evidence for reptation in an entangled polymer melt, *Nature (London)*, 1978, **271**, 143.

32 B.H. Zimm, Apparatus and methods for measurement and interpretation of the angular variation of light scattering; preliminary results on polystyrene solutions, *J. Chem. Phys.*, 1948, **16**, 1099.

33 W. Brown and P. Zhou, Solution properties of polyisobutylene investigated by using dynamic and static light scattering and pulsed field gradient NMR, *Macromolecules*, 1991, **24**, 5151.

34 W.R. Krigbaum and P.J. Flory, Statistical mechanics of dilute polymer solutions. IV. Variation of the osmotic second coefficient with molecular weight[1a,b], *J. Am. Chem. Soc.*, 1953, **75**, 1775.

35 L. Mandelkern, W. Krigbaum, H.A. Scheraga, and P.J. Flory, Sedimentation behavior of flexible chain molecules: polyisobutylene, *J. Chem. Phys.*, 1952, **20**, 1392.

36 C. Heller, T. Duke, and J.L. Viovy, Electrophoretic mobility of DNA in gels. II. Systematic experimental study in agarose gels, *Biopolymers*, 1994, **34**, 249–259.

37 J.E. Kohn, Random-coil behavior and the dimensions of chemically unfolded proteins, *Proc. Natl. Acad. Sci. U.S.A.*, 2004, **101**, 12491.

27

Organized Molecular Assemblies

CHAPTER MENU

27.1 Introduction

We have considered simple and increasingly complex molecular systems. Now we focus on supramolecular systems where the specific molecular arrangement is crucial for the pertinent properties, in particular, the molecular arrangement at surfaces and in membranes. For these assemblies the specific arrangement (organization) of molecules gives rise to very specific properties. Organized assemblies of molecules have an enormous richness in properties, resulting from the multitude of interactions between the different kinds of molecules, and sometimes display new cooperative properties not found for individual molecules.

Many chemical and biochemical reactions occur at interfaces; these include boundaries between bulk phases (solid–gas, solid–liquid, and liquid–gas), membranes, vesicles, monolayers, and micelles. This chapter considers "soft" interfaces, which are formed by relatively weak interactions among well-defined molecular subunits. The principles of self-organizing molecular structures discussed here are basic in understanding the occurrence of complexity in natural and artificial supramolecular systems.

27.2 Liquid Surfaces and Liquid/Liquid Interfaces

A molecule in the bulk of a liquid is solvated by the neighbor molecules surrounding it. For a molecule at the liquid's surface with air, however, the neighbors on one side are missing (Fig. 27.1). Consequently, the surface localized molecules have a higher potential energy than those in the bulk. In order to minimize its energy, the liquid changes its shape to minimize its surface area, hence the number of surface localized molecules. For a given volume, a sphere has the smallest possible surface area, so that a liquid becomes spherical when other forces, which might change its shape, are absent (e.g., a drop of oil in water or a liquid in free space without gravitational forces).

Principles of Physical Chemistry, Third Edition. Hans Kuhn, David H. Waldeck, and Horst-Dieter Försterling.

Companion website: www.wiley.com/go/Kuhn/physicalchemistry3e

27.2.1 Surface Tension and Interfacial Tension

When increasing the surface area of a liquid, the number of molecules in the energy-rich interfacial region increases. For a macroscopically sized system at constant pressure and temperature, the change dG of the Gibbs energy is $dG = dGw_{rev} - PdV$. In the approximation that the density of the liquid does not change, we find that dG is equal to the differential reversible work that is needed to increase the area A of the surface by a differential amount dA is hence

$$dG = \gamma \cdot dA \tag{27.1}$$

where γ is called the *surface tension*.

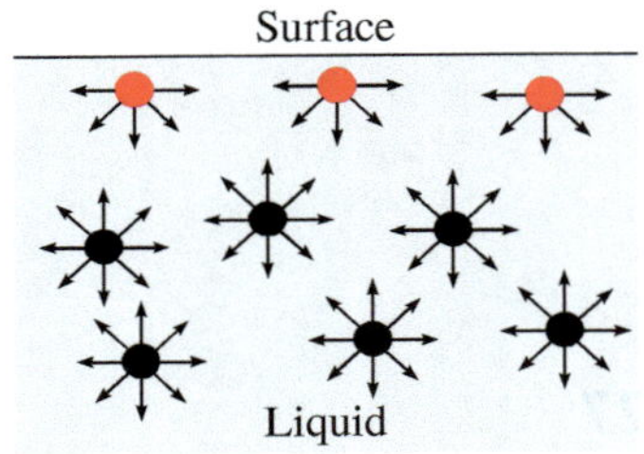

Figure 27.1 Higher energy of molecule at surface than in the bulk. The arrows represent the forces acting on a selected molecule by its neighbors. In the bulk, the forces are centrosymmetric, and the molecules can move freely. At the surface the neighbors on top are missing: a force is needed to pull a molecule beyond the surface.

The meaning of γ becomes evident if we consider the process shown in Fig. 27.2, in which we withdraw a lamella (thin film), which is supported on a rectangular frame, from a soap solution. In addition to the force to lift the movable frame and the lamella against gravity, work must be done to create the liquid surfaces on the movable frame. Writing this work as $f \cdot dh$, in the reversible limit the work is

$$f \cdot dh = \gamma \cdot dA = \gamma \cdot 2L \cdot dh \tag{27.2}$$

where L is the width of the movable frame. The factor of two arises because a surface is created on each side of the film. Thus

$$\gamma = \frac{f}{2L} \tag{27.3}$$

For most organic liquids γ is between 18 and 30 mN m^{-1}. For clean water at 20 °C, γ is as large as 72.8 mN m^{-1} because hydrogen bonds must be broken when increasing A. For molten metals or molten salts γ is large because the Coulomb attraction must be overcome when bringing an ion from the interior to the surface:

Hg at 20 °C: γ = 470 mN m^{-1}

Cu at 1130 °C: γ = 1300 mN m^{-1}

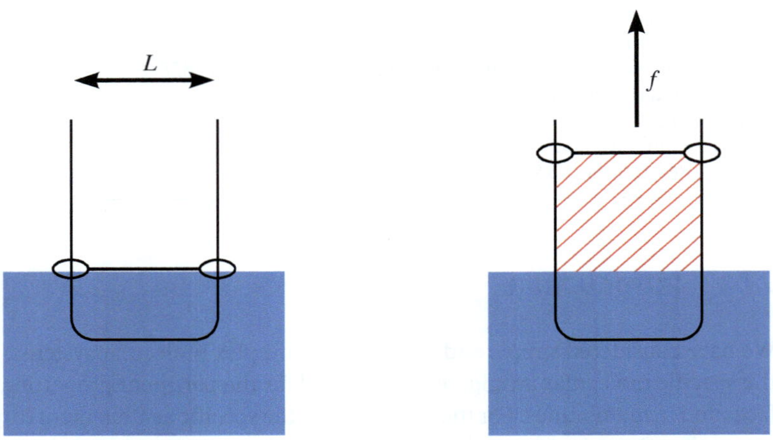

Figure 27.2 Withdrawing a lamella from a soap solution. A fixed frame made from wire is inserted into the liquid. A movable wire attached to the frame is pulled upward by applying a force $f = \gamma 2L$. We neglect the gravitational force acting on the movable wire when withdrawing the lamella from the solution.

The same relationship (Equation (27.1)) holds for the interface between two liquids

$$dG = \gamma \cdot dA \tag{27.4}$$

except that γ is called the *interfacial tension*. The surface tension of n-decane with air is 23.9 mN m^{-1}. In contrast, the interfacial tension for the n-decane/water interface is 51.2 mN m^{-1}, almost as large as the surface tension of water (Fig. 27.3). This increase occurs because hydrogen bonds in the water break when water molecules move to the decane/water interface. Because of this surface tension difference, a drop of decane on a water surface forms a lens-shaped droplet; see Fig. 27.4a.

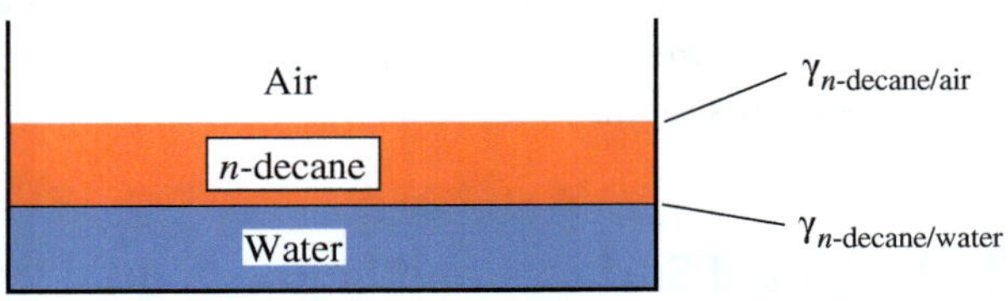

Figure 27.3 Interfacial tension. Hydrogen bonds break when water molecules move to the n-octane water interface. Therefore $\gamma_{n\text{-octane/water}}$ (51.2 mN m^{-1}) is much higher than the surface tension of n-octane (23.9 mN m^{-1}).

The shape of the lens formed between two immiscible liquids is determined by a balance of interfacial forces. For a drop of decane on water the sum of the forces acting in the direction tangential to the surface and perpendicular to the element dl on the line between the three phases is zero (Fig. 27.4a), if gravitational forces are neglected. For the horizontal components of the force f, we obtain

$$\gamma_{decane/air} \cdot \cos\alpha \cdot dl + \gamma_{decane/water} \cdot \cos\beta \cdot dl = \gamma_{water/air} \cdot dl \tag{27.5}$$

and for the vertical components we find

$$\gamma_{\text{decane/air}} \cdot \sin\alpha = \gamma_{\text{decane/water}} \cdot \sin\beta \tag{27.6}$$

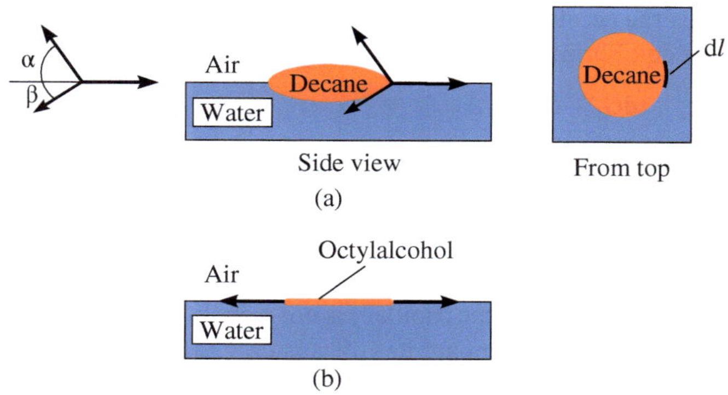

Figure 27.4 Two liquids in contact. The arrows represent the forces acting on the lens. (a) *n*-octane and water, formation of a lens, see Exercise E27.22 for the calculation of the contact angles. (b) *n*-octyl alcohol and water: spreading.

In Exercise E27.22, you show that $\alpha = 22°$ and $\beta = 10°$.

A quite different situation arises if we replace *n*-decane by an aliphatic alcohol, e.g., *n*-octyl alcohol. Because the hydrocarbon tails of the molecules at the surface face the air in both cases, the interfacial tension $\gamma_{\text{octyl alcohol/air}} = 28\ \text{mN m}^{-1}$ is only moderately larger than $\gamma_{\text{decane/air}}$. Despite this similarity for the air interface, the interfacial tension with water, $\gamma_{\text{octyl alcohol/water}} = 8.5$ mN m^{-1}, is much smaller than that for decane with water because an octyl alcohol molecule at the interface can form a hydrogen bond with a water molecule, diminishing the work to bring it to the surface. As a consequence, increasing the area of the *n*-octyl alcohol by dA decreases the area of the water/air interface, and we obtain (Fig. 27.4b)

$$\begin{aligned} \mathrm{d}G &= \gamma_{\text{water/air}} \cdot (-\mathrm{d}A) + \gamma_{\text{water/octyl alcohol}}\, \mathrm{d}A + \gamma_{\text{octyl alcohol/air}} \cdot \mathrm{d}A \\ &= \mathrm{d}A \cdot (-72.8 + 8.5 + 28)\ \text{mN m}^{-1} = -36.3\ \text{mN m}^{-1} \cdot \mathrm{d}A \end{aligned}$$

dG is negative, and octyl alcohol spreads on water (Fig. 27.4b).

27.2.1.1 Capillarity

What happens when dipping a solid bar into a liquid? A molecule at the interface is attracted both to the bar and to the bulk of the liquid. Depending on which forces predominate we can distinguish between two cases. (1) If the forces attracting the molecule to the bar dominate, then the surface of the bar (see Fig. 27.5a) is covered by a film of the liquid; i.e., the bar gets wet. An example is water on clean glass, and such a surface is called *hydrophilic.* (2) If the forces attracting the molecules of the liquid to each other dominate, the bar in Fig. 27.5b remains dry. Examples are mercury at glass and water on paraffin; the surface of paraffin is called *hydrophobic*.

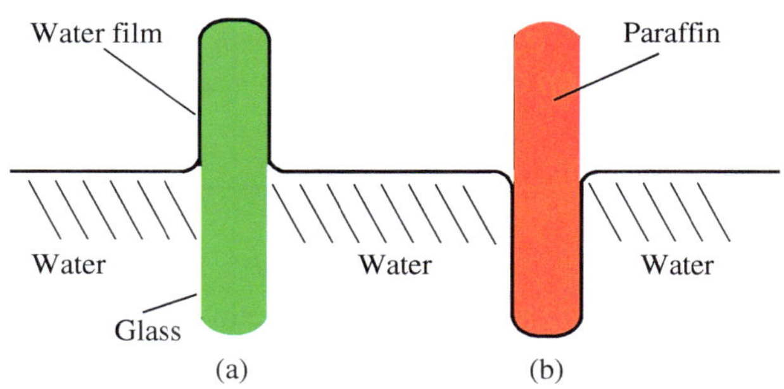

Figure 27.5 Dipping a solid (glass and paraffin) into a liquid (water). (a) Wetting the surface of glass and (b) not wetting the surface of paraffin.

When dipping a glass capillary into water, the attractive forces between the water molecules and the glass surface cause a liquid film to form on the wall of the capillary (it "wets" the surface of the glass), and correspondingly the water/air interface grows. The rise of the liquid in the interior of the capillary tube is determined by a balance of the gravitational force and the surface tension.

We can determine the height of the liquid from the equilibrium condition d$G = 0$. When the liquid water level rises by dh, the water–air interface on the inside of the capillary tube is diminished by $2\pi r$ dh, and the change in Gibbs energy is

$$\mathrm{d}G_1 = \gamma \cdot \mathrm{d}A = -\gamma \cdot 2\pi r \cdot \mathrm{d}h \tag{27.7}$$

(r is the inner radius of the capillary and $\gamma = \gamma_{\text{water/air}}$ for this particular example, see Fig. 27.6a). The liquid rises until the energy decrease by diminishing the water surface area equals the energy increase dG_2 by lifting the liquid in the capillary against gravity:

$$\mathrm{d}G_2 = \pi r^2 h\, g\, \rho \cdot \mathrm{d}h \tag{27.8}$$

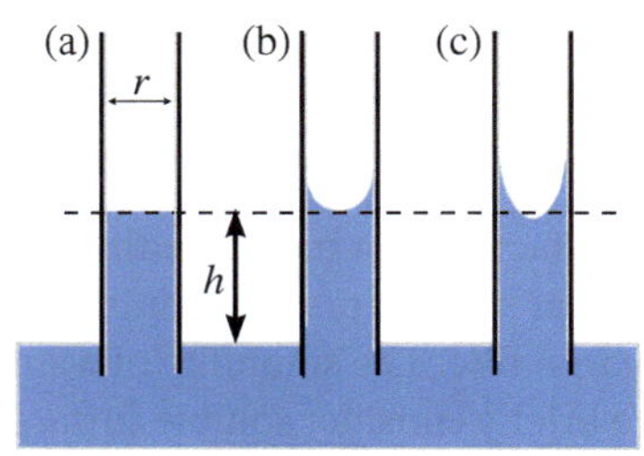

Figure 27.6 A wetting liquid rises in a capillary. (a) Meniscus approximated as being flat, (b) as being half-spherical, and (c) real shape of meniscus.

(g is the acceleration due to gravity, and ρ is the mass density of the liquid). The maximum height h_{max} is determined by the equilibrium condition that

$$dG_1 + dG_2 = 0 \tag{27.9}$$

That is, the water rises as long as the Gibbs energy change is negative and stops at the height above which the Gibbs energy change would be positive. If we solve Equation (27.9), then

$$h_{max} = \frac{2\gamma}{g\rho r} \tag{27.10}$$

The meniscus in the capillary is not planar. The reason is evident if we approximate the meniscus by that of a half sphere (Fig. 27.6b) and compare its Gibbs energy to that of a planar surface (Fig. 27.6a). From the position of the meniscus to a height r higher on the capillary, the surface area of a half-sphere is

$$A_{meniscus} = 2\pi r^2 \tag{27.11}$$

while the corresponding surface area in the case of a planar meniscus is

$$A_{meniscus} = \pi r^2 + 2\pi r \cdot r = 3\pi r^2 \tag{27.12}$$

Thus the reversible work (Gibbs energy change) to decrease the surface in proceeding from the planar condition (Fig. 27.6a) to the spherical condition (Fig. 27.6b) is

$$\Delta G = \gamma 2\pi r^2 - \gamma 3\pi r^2 = -\gamma \pi r^2 \tag{27.13}$$

The energy $\gamma \pi r^2$ is used to lift the liquid against gravity beyond h. In reality the meniscus is not a half-sphere but has a shape more like that in Fig. 27.6c.[1] In this case, G has reached its minimum.

Example 27.1 *Consider a capillary with an inner diameter of* $r = 0.5$ *mm. What is the height* h_{max} *for water, according to Equation (27.10)? Calculate* ΔG *for the formation of the meniscus, according to Equation (27.13).*

$$h_{max} = \frac{2 \cdot 72.8 \text{ mN m}^{-1}}{9.807 \text{ m s}^{-2} \cdot 1.00 \text{ g cm}^{-3} \cdot 0.5 \text{ mm}} = 2.97 \times 10^{-2} \text{m} = 2.97 \text{ cm}$$

and

$$\Delta G = -72.8 \text{ mN m}^{-1} \cdot \pi \cdot (0.5 \text{ mm})^2 = -57.2 \times 10^{-9} \text{ J}$$

For the general case, in which the liquid does not "wet" the capillary surface but forms an angle θ to the capillary wall, the height of the meniscus is given by

$$h = \frac{2\gamma \cdot \cos\theta}{g\,\rho\,r} \tag{27.14}$$

For example, for water in a glass capillary, we have $\theta = 0°$ and $\cos\theta = 1$; i.e., we obtain Equation (27.10): h is positive. In the case of mercury in a glass capillary we have $\theta = 180°$ and $\cos\theta = -1$, i.e., h becomes negative.

27.2.2 Surface Active Molecules (Surfactants)

A surfactant is a molecule that localizes at the interfacial region between two phases; e.g., the interface of air and water. Typical surfactants are amphiphilic, containing one part that is hydrophobic and a second, covalently linked part that is hydrophilic. The solubility of a surfactant molecule in the bulk phase as compared to the interface can vary substantially. We briefly consider soluble surfactants; however, most of our discussion addresses insoluble surfactants and the formation of monolayer films.

---

1 See [1] for a discussion of this issue.

27.2.2.1 Films of Soluble Surfactants

We consider a solution of a surfactant in water (e.g., Na-oleate). When increasing the concentration c of the surfactant, the *surface excess* Γ of the solute (i.e., the amount of surfactant per unit area of surface) increases and the surface tension decreases. For a dilute solution the relation

$$\Gamma = \frac{\boldsymbol{n}_{\text{surfactant}}}{A} = -\frac{c}{\boldsymbol{R}T}\frac{\mathrm{d}\gamma}{\mathrm{d}c} \tag{27.15}$$

holds (*Gibbs adsorption equation*, Box 27.1).

Γ increases with the concentration c and with the tendency of the solute to decrease the surface tension, $-\mathrm{d}\gamma/\mathrm{d}c$; furthermore, it decreases with increasing thermal motion counteracting the separation. For checking the relation, Γ can be measured by removing the surface film (i.e., stripping off a top layer of around 0.05 mm thickness) and scooping it into a sample tube with a fast moving razor blade.

This class of surfactants are very important as solubilizing agents; e.g., soap, and we discuss them more extensively in Section 27.4.

27.2.2.2 Films of Insoluble Surfactants

When a solute is weakly soluble in a liquid it may tend to segregate at the surface and form a thin film, which can be a single molecule thick. Let us consider the air–water interface and a fatty acid molecule as the surfactant, e.g., arachidic acid CH_3–$(CH_2)_{18}$–COOH,

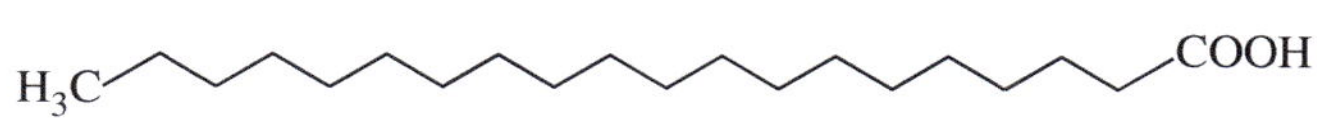

which contains a hydrophilic part (carboxyl group) and a hydrophobic part (a long hydrocarbon chain) that prevents the molecule from entering the aqueous subphase.

We can spread the fatty acid on a water surface by first dissolving it in a volatile organic solvent and then spreading the organic solution on the water surface. After evaporation of the organic solvent the fatty acid molecules are statistically distributed over the water surface. The surface density of the fatty acid molecules (number of molecules per unit area) can be controlled by a movable barrier that changes the surface area of the water (Fig. 27.7). It is possible to compress the molecules until they are densely packed forming a compact monolayer one molecule thick.

By applying a force f on the movable barrier of length L we can externally establish a surface pressure Π on the fatty acid layer. In analogy to our common three-dimensional definition of pressure, P = force/area, the surface pressure is Π = force/length. At equilibrium the forces acting on the barrier must balance so that

$$\gamma L = \gamma_s L + \Pi \cdot L$$

where the force γL pushes the barrier toward the clean surface, $\gamma_s L$ pushes the barrier toward the surfactant film, and $\Pi \cdot L$ is the externally applied force f (via the weight W in Fig. 27.7a). Rearranging this expression we find that

$$\Pi = \gamma - \gamma_s \tag{27.16}$$

i.e., the surface pressure is equal to the difference in the surface tensions on the two sides of the barrier.

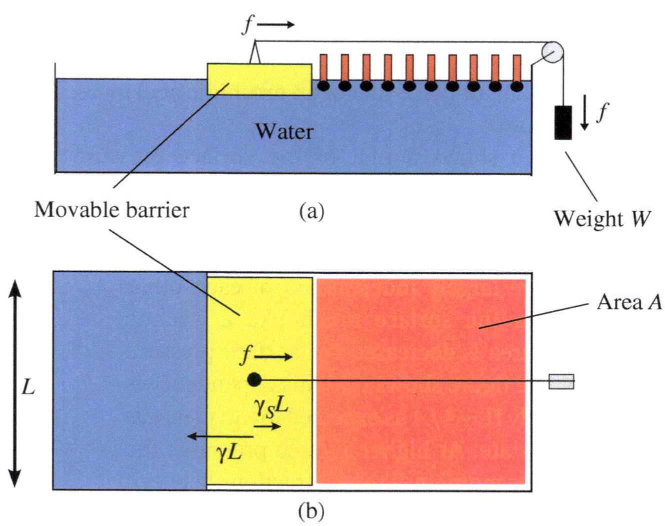

Figure 27.7 Insoluble monolayer. Fatty acid molecules are spread on a water surface and compressed by a movable barrier. (a) Side view and (b) view from top. Surface pressure $\Pi = f/L$.

When using a fatty acid as surfactant, the steep increase of Π with decreasing area occurs at an area of $A = N \times 0.2\ \text{nm}^2$, where N is the number of molecules in the layer. The value $0.2\ \text{nm}^2$ agrees well with the cross-sectional area of a fatty acid chain, which is obtained from bond lengths and van der Waals radii.

Figure 27.8 shows Π versus A data for arachidic acid and a mixture of arachidic acid and a cyanine dye (with structure shown in Fig. 27.8) in a 10:1 molar ratio. In Fig. 27.8, Π is shown as a function of the area of 10 arachidic molecules, in order to compare with the area of an additional dye molecule (Fig. 27.8b). In the experiment an amount of $\boldsymbol{n} = 1 \times 10^{-7}$ mol of

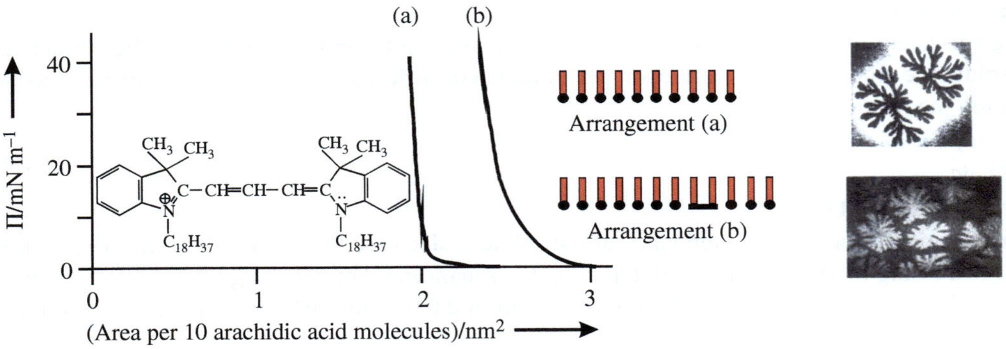

Figure 27.8 Insoluble monolayer. (a) and (b) Surface pressure Π versus area per 10 arachidic acid molecules. A compact monolayer is reached when Π is strongly increasing with decreasing area. (a) Fatty acid (cadmium salt of arachidic acid $C_{19}H_{39}$−COOH). The area per molecule is 0.2 nm^2 (cross section of the hydrocarbon chain), that is, 2 nm^2 per 10 arachidic acid molecules. (b) Fatty acid and dye in molar ratio 10:1. Additional area of 0.4 nm^2 due to the two hydrocarbon chains incorporated in the fatty acid matrix. The chromophores below the matrix are not crowded and, therefore, do not effectively contribute to the surface pressure. Panels (c) and (d) the case of phase segregation of films, in which crystalline (LC) domains and a liquid expanded phase form in a monolayer of dimyristoyl-phosphatidyl-ethanolamine on a water surface. (c) Fluorescence image: LC domains appear dark (H. Möhwald), (d) Brewster angle microscopy image: domains appear bright (D. Hönig). ((a) and (b) Adapted from [2]; (c) and (d) [3]. Bryant W. Rossiter and Roger C. Baetzold. 1993/Reproduced from JOHN WILEY & SONS, INC.)

arachidic acid is spread on the water surface, and the steep increase of Π occurs at an area of $A = 121$ cm^2. The number of fatty acid molecules in the film is $N = \boldsymbol{n} \cdot \boldsymbol{N}_A = 1 \times 10^{-7}$ mol $\cdot$ 6.03×10^{23} mol^{-1} $= 6.03 \times 10^{16}$; thus, the area per fatty acid molecule is $A/N = 121/6.03 \times 10^{16}$ cm^2 $= 2.01 \times 10^{-15}$ cm^2 $= 0.2$ nm^2.

For the surfactant solution of cyanine dye and arachidic acid, an additional area of the monolayer is observed; 2.4 nm^2 per 10 arachidic acid molecules rather than the 2.0 nm^2 per 10 arachidic molecules for the pure arachidic acid (Fig. 27.8b). If we assume that the extra area arises solely from the area displaced by the dye molecules and that the area occupied by an arachidic acid molecule is the same for the mixture as it is for the pure acid layer, then a dye molecule's area is

$$2.4 \text{ nm}^2 - 2.0 \text{ nm}^2 = 0.4 \text{ nm}^2$$

This value is about twice that of an arachidic acid molecule and is consistent with the fact that each dye molecule has two fatty acid chains.

Figure 27.9 shows a plot of the surface pressure versus the area for a Langmuir film of pentadecanoic acid, $CH_3(CH_2)_{13}COOH$. The film displays a phase behavior in two dimensions (Π versus A) that is analogous to the P/V phase diagram of a normal three-dimensional substance. At large surface areas the surfactant molecules are largely independent of each other and move in the surface much like a gas. As the surface area is decreased the surface pressure increases; the figure shows that this dependence is approximately $\Pi \sim 1/A$ as expected for an ideal gas equation of state. At higher surface pressures the area decreases more until the interactions between the adsorbates are important and a liquid-like phase begins to form; the figure shows the biphasic region. After the vapor has "condensed," further decrease of the molecular area increases the surface pressure on the liquid. At high enough surface pressure a solid-like structure with tilted alkane chains forms and a biphasic region is found. Once the liquid phase has fully converted to the "tilted" phase, further decrease of the surface area increases the surface pressure. For this particular

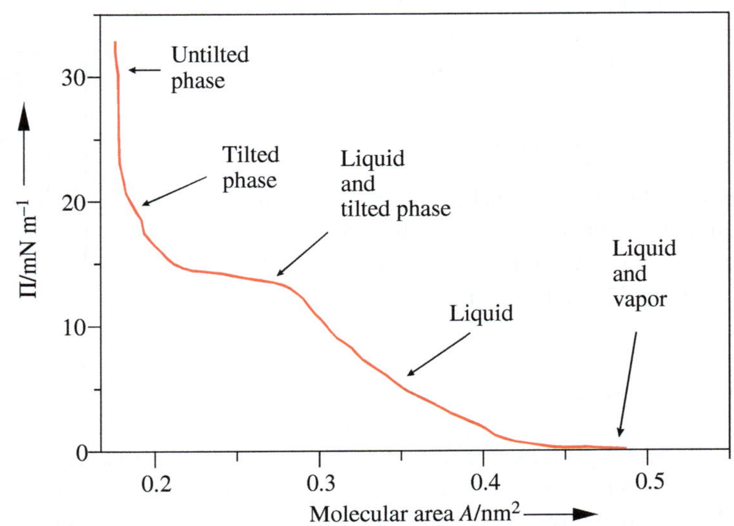

Figure 27.9 Different Langmuir–Blodgett phases. Surface pressure Π versus the molecular area A. (Data from [4].)

system, compression of the film to even higher surface pressures drives the layer into a second solid-like structure in which the alkyl chains are nearly oriented normal to the surface. When the force f reaches a critical value, sections of the monolayer begin to overturn; the film collapses and becomes irreversibly destroyed.

Phase Segregation The two-phase coexistence regions of the system can be "visualized" using fluorescence microscopy. Knobler and coworkers [5] studied the phase behavior for stearic acid molecules that were labeled with an *N*-methyl-(7-nitro-2,1,3-benzoxadiazol-4-yl) fluorescent dye molecule. When the LB film is held at pressures in the gas–liquid coexistence region fluorescence microscope images show the presence of dark and light regions in the film. The dark regions correspond to dilute fatty acid molecules (gas like) on the film and are likely to have the alkyl chain lying down (horizontal) on the surface so that the dye molecule fluorescence is quenched. The bright regions correspond to denser collections of fatty acid molecules (liquid like) in which the alkyl chains are standing up (vertical) so that the dye molecule fluorescence is apparent. Similar studies have been performed in the solid–liquid coexistence region and in different liquid–liquid coexistence regions [6].

27.3 Films on Solid Surfaces

Unless a solid is a single crystal, often its surface is not well defined, e.g., the surface of glass. When surfactant monolayers are transferred to a glass slide by the Langmuir–Blodgett method, the surface properties are controlled by the monolayer film and it becomes well-defined; i.e., the film-air interfacial properties are well defined. Alternatively, one can coat a solid substrate with a monolayer film comprising molecules that chemically bind to the surface and pack into a well-ordered array. Thus we distinguish two types of monomolecular films: *Langmuir–Blodgett (LB)* films which are formed through the application of an external force in a controlled way and transferred to a solid surface, and films which form spontaneously, called *Self-Assembled Monolayer* (*SAM*) films.

27.3.1 Langmuir–Blodgett Films (LB Films)

A monolayer of densely packed fatty acid molecules can be transferred to the surface of a solid support. For example, consider a glass slide that has been dipped in the water of a Langmuir trough before producing a film of fatty acid molecules on the surface. When taking the glass slide out of the water it passes through the monolayer film. During this transfer the slide is coated by the film because the carboxyl groups of the fatty acid molecules and associated metal ions (e.g., Cd^{2+}) bind to its surface. Such a film presents the hydrocarbon chains toward the air (Fig. 27.10, Step 1) and the surface of the coated slide is hydrophobic.

If we dip the slide from the air back through the film and into the water, it is covered by a second layer of fatty acid. The ends of the hydrocarbon chains of the fatty acids on the glass slide and those at the surfactant/air surface bind to each other (Fig. 27.10, Step 2), and the carboxyl groups are facing toward the water. When taking the slide out of the water through the film and into the air, a third monolayer of fatty acid is attached (Fig. 27.10, Step 3). This technique for producing multilayers of fatty acid (the *Langmuir–Blodgett method*) can be extended and applied to many other amphiphilic molecules (see Chapter 28). Because the process occurs through electrostatic and hydrophobic assembly of the layers, this process is quite sensitive to solution conditions (pH, ionic strength) and the surface pressure. Through careful control it is possible to assemble films containing hundreds of layers [8]. Note that layer-by-layer assembly methods [9] are similar to the more traditional LB method described here, as they rely on complementary interactions between different molecules for assembly and are often performed through sequential dipping of a substrate in different solutions.

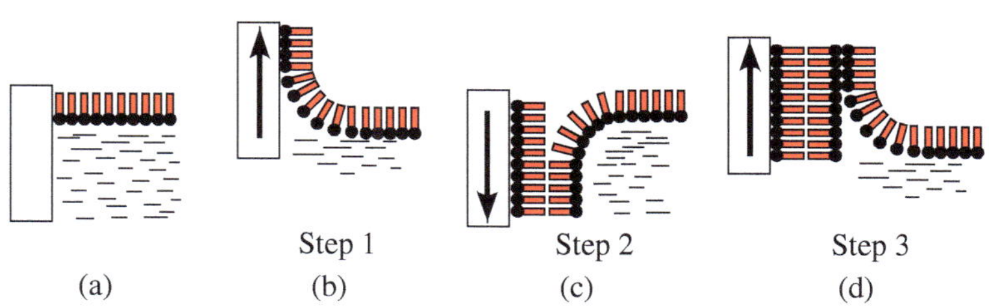

Figure 27.10 Transfer of fatty acid monolayers to a solid support (glass slide): Langmuir–Blodgett method. (a) Glass slide in contact with monolayer on a water surface. (b) Glass slide moved upwards: monolayer with hydrophilic groups attached on the glass surface. (c) Glass slide moved downwards: second monolayer with hydrophobic groups attached. (d) Glass slide moved upwards: third monolayer attached. (Adapted from [7].)

27.3.2 Self-Assembled Monolayers (SAM)

In the Langmuir–Blodgett method an external force (barrier compression) is used to compress the surfactant molecules into a monolayer film. At a high enough density the van der Waals contacts between the chains drives packing and ordering of the molecules in the film. A different strategy uses molecules with a terminal functionality that can chemically bind to the interface. This chemical force combines with the interchain forces to drive the dense packing of the molecules and formation of well-ordered monolayer films.

Typically, this process occurs spontaneously when a solid substrate is immersed into a solution of the surface active molecules comprising the film; they are called *self-assembled monolayer films*. This phenomenon was discovered by Zisman and colleagues [10] in 1946. Two common types of such films are organosilanes, which form spontaneously on oxidized and hydroxylated surfaces, and thiols or disulfides, which form spontaneously on metals.

The exposure of a glass or quartz surface to an alkyltrichlorosilane forms a monolayer film on the surface. The chlorinated silane reacts with the OH groups on the glass surface, forming Si–O bonds at the surface. Under the proper conditions a well-ordered monolayer is bound to the glass surface. The individual molecules in the monolayer are bound by covalent bonds (to be contrasted with the electrostatic binding used in fatty acid Langmuir Blodgett films), and the alkane chains pack together through van der Waals forces. Overall the surface appears hydrophobic because the nonpolar alkane chains are presented to the air.

The exposure of a Au surface to alkanethiols leads to the spontaneous formation of a monolayer thick film. The sulfur head group forms a strong Au–S bond with the substrate. On a time scale of seconds to a few minutes, the active thiol molecules form an overlayer on the surface; however, it is not compact or well ordered. Over a longer timescale of hours to days the films anneal and form compact and well-ordered layers. Figure 27.11 shows how the average thickness of an octadecanethiol monolayer film changes as a function of time. When the monolayer is first forming and contains many defects, the average thickness is much smaller than that expected from the molecular chain length. As time proceeds, however, the film becomes more uniform and compact until the average thickness asymptotes to about 2.1 nm.

Films of the SAM type, or the LB type, are not perfect and defect free. The importance of defects on the mechanics of the film is usually not significant; however, they can be quite important for electrical properties.

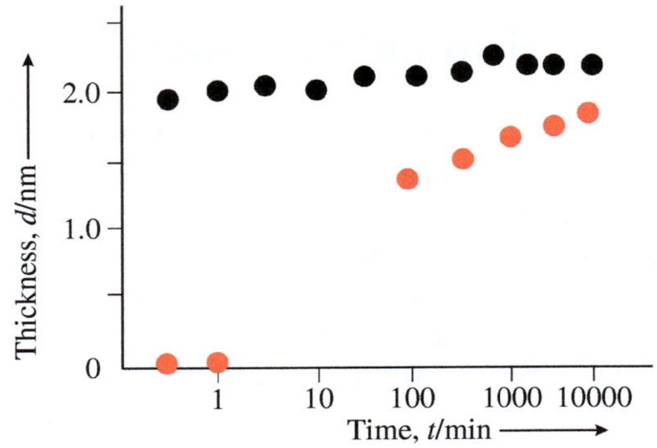

Figure 27.11 Kinetics of the adsorption of octadecanethiol from ethanol onto a gold surface [11]. The average thickness d of an octadecanethiol monolayer film, measured ellipsometrically [12], is shown as a function of time. Filled black circles: concentration of the octadecanethiol solution 1 mmol L^{-1}, filled red circles: concentration 0.001 mmol L^{-1}.

27.3.3 Contact Angle

If a drop of a liquid is placed on a flat solid surface, the liquid forms a droplet or wets the surface, depending on the difference in surface tension. The angle formed between the droplet and the surface can be used to quantify the difference in surface tension, hence the degree of hydrophobicity, or hydrophilicity, of the surface. Figure 27.12a shows the contact angle Θ that is established between a solid surface and a liquid. Θ is determined by the balance of cohesive forces measured by γ_{lg}, γ_{sl}, and γ_{sg} (energy to create a unit area of liquid–gas, solid–liquid, and solid–gas interface, respectively):

$$\gamma_{sg} = \gamma_{sl} + \gamma_{lg} \cdot \cos\Theta \tag{27.17}$$

Rearranging we find

$$\cos\Theta = \frac{\gamma_{sg} - \gamma_{sl}}{\gamma_{lg}} \tag{27.18}$$

When $\gamma_{sl} = \gamma_{lg}$ the contact angle becomes $\Theta = 90°$. An example is water on a solid surface consisting of long-chain alcohols, e.g., CH_3–$(CH_2)_{20}$–OH. When $\gamma_{sg} > \gamma_{sl}$, $\Theta < 90°$ and the liquid wets the surface; e.g., for water on a nickel surface $\Theta = 27°$. When $\gamma_{sg} < \gamma_{sl}$, $\Theta > 90°$ and the liquid beads up on the surface; e.g., for water on a paraffin surface $\Theta = 105°$.

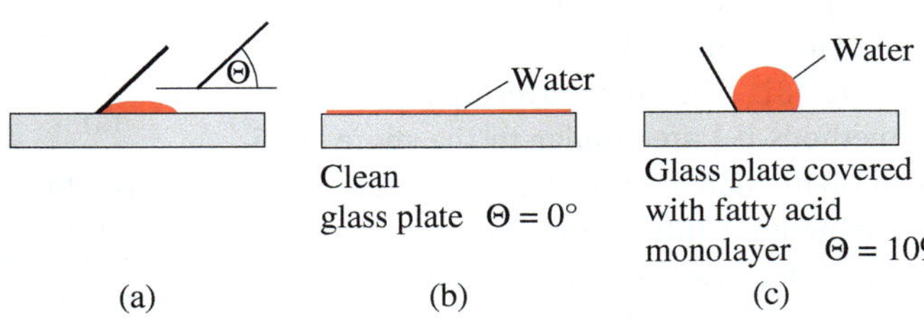

Figure 27.12 Drop of liquid on solid. (a) Definition of contact angle Θ, (b) complete wetting for water on a clean glass plate ($\Theta = 0°$), and (c) water on a fatty acid monolayer ($\Theta = 109°$).

By depositing an appropriate monolayer on top of a solid surface, almost any contact angle can be purposely achieved. For example, the contact angle for water on a clean glass surface is zero: complete wetting (Fig. 27.12b). In this case Equation (27.18) is no longer valid because the equilibrium cannot be established. The situation changes dramatically when a monolayer of a fatty acid is bound to a glass surface. In this case a contact angle of $\Theta = 109°$ or even higher can be realized (Fig. 27.12c).

Similar control over surface properties is possible using self-assembled films. The diagram in Fig. 27.13 shows how one can tune the contact angle for different monolayer films. Film A is composed of $CH_3(CH_2)_{11}SH$, B is composed of $HO(CH_2)_{11}$ SH, and A + B is a 1:1 molar ratio of A and B. In each case the thiol group reacts with the gold to make Au–S linkages and the chains pack through van der Waals forces. The contact angle changes dramatically with the nature of the end group that is presented to the air (Fig. 27.13b). In addition, the contact angle changes if the sample is immersed in octane, because the interfacial surface tension must be considered (Fig. 27.13a).

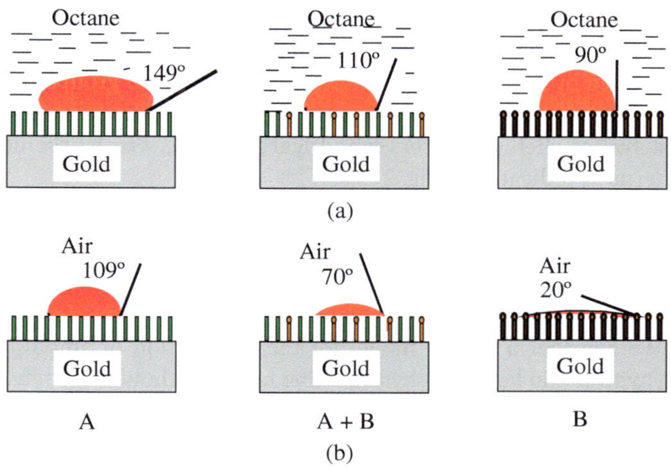

Figure 27.13 Drop of water on differently organized monolayers showing different contact angles. Left, monolayer A=$CH_3(CH_2)_{11}SH$ (CH_3 groups at surface). Right, monolayer B=$HO(CH_2)_{11}SH$ (OH groups at surface). Center, mixed monolayer of A and B in a 1:1 molar ratio [13]. (a) In octane and (b) in air.

27.4 Micelles

When a soluble surfactant dissolved in water reaches a certain concentration (e.g., 0.2 mol L^{-1} for sodium dodecyl sulfate), the molecules begin to associate (*critical micelle concentration*). The hydrophobic portion of the molecules (e.g., hydrocarbon chains) bind together and the hydrophilic headgroups (e.g., sulfate) face the water (see Fig. 27.14). Analogous structures are of core importance to understanding molecular processes in biosystems, as well as in chemistry and engineering.

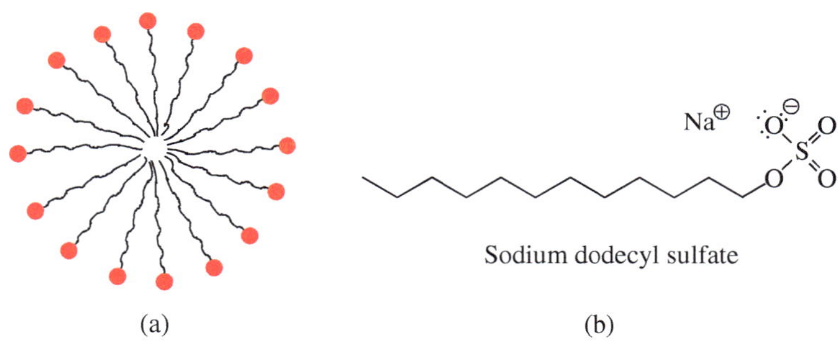

Figure 27.14 (a) Spherical micelle. (b) Structural formula of sodium dodecyl sulfate. Red filled circles represent negatively charged SO_4^- head groups.

The spontaneous aggregation of surfactant molecules into a micelle has a negative Gibbs energy change, $\Delta G < 0$, with both entropic and enthalpic contributions. The water molecules, which were originally located around the tail of the surfactant molecule, move into the bulk and form additional hydrogen bonds. This rearrangement results in a negative contribution to the enthalpy change in micelle formation. On the other hand, the charged heads become close to each other and their repulsion leads to a positive contribution to ΔH. These two effects essentially compensate, and the enthalpy of micelle formation is small. If ΔH is small, then the formation of micelles above the critical micelle concentration must be accompanied by an increase in entropy to allow ΔG to be negative. For example, $\Delta_r S = 140 \text{ J K}^{-1}\text{mol}^{-1}$ is measured at 300 K for dodecylsulfate in water.

At first the fact that $\Delta S > 0$ seems surprising because micelle formation separates two species from their mixture, causing a decrease in the entropy of mixing. The decrease in entropy from the association of surfactant molecules is counteracted by changes in the entropy of the water. Before aggregation, the water shell surrounding the surfactant molecule has an ordered ice-like structure, and this structure breaks down when the water molecules in the shell become part of bulk water, generating more water molecule configurations. Overall, this entropy production exceeds the entropy consumption in the separation process.

Micelles fulfilling the given requirements can be very differently shaped depending on the nature of the surfactant and its concentration. Spherical micelles are predominantly formed by surfactants with bulky or charged head groups. With increasing concentration however, there is a competition of the micelles for water molecules forming their solvation shell; this favors cylindrical structures which become arranged in ordered patterns as indicated in Fig. 27.15.

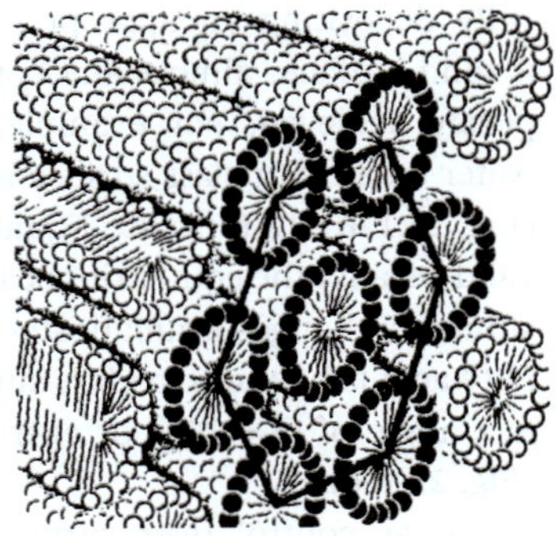

Figure 27.15 Rod-like micelles. (Courtesy: E. Sackmann).

27.4.1 Spherical Micelles. Critical Micelle Concentration

27.4.1.1 Aggregation of Amphiphilic Molecules

We consider a simple model for the aggregation of amphiphilic solute molecules (hydrophilic head and hydrophobic tail) and show that it gives rise to aggregates of a well-defined size.

At equilibrium, the chemical potentials of the different species (monomers, dimers, trimers, … n-mers) are equal; i.e.,

$$\mu_1 = \mu_2 = \mu_3 = \cdots = \mu_n = \cdots$$

in which μ_n is the chemical potential of an amphiphile in an aggregate containing n molecules. To simplify, we assume that the solution behaves ideally (replace the activity by the mole fraction (or concentration)) so we can write that

$$\mu_n^{\oplus} + \frac{\boldsymbol{RT}}{n} \ln \frac{x_n}{n} = \mu_1^{\oplus} + \boldsymbol{RT} \ln x_1 \tag{27.19}$$

where $\mu_n^{\oplus}$ is the standard chemical potential of an amphiphile in an n-mer aggregate and x_n is the mole fraction of amphiphiles contained in n-mer aggregates in solution. Rearranging Equation (27.19), we find that

$$\frac{\mu_n^{\oplus} - \mu_1^{\oplus}}{\boldsymbol{RT}} = \ln x_1 - \frac{1}{n} \ln\left(\frac{x_n}{n}\right) = \ln\left[\frac{x_1}{(x_n/n)^{1/n}}\right]$$

so that

$$\frac{(x_n/n)^{1/n}}{x_1} = \exp\left[-\frac{\mu_n^{\oplus} - \mu_1^{\oplus}}{\boldsymbol{RT}}\right]$$

and

$$x_n = n\left[x_1 \cdot \exp\left(-\frac{\mu_n^{\oplus} - \mu_1^{\oplus}}{\boldsymbol{RT}}\right)\right]^n = n\left[x_1 \cdot \exp\left(-\frac{\Delta\mu_n^{\oplus}}{\boldsymbol{RT}}\right)\right]^n \tag{27.20}$$

where $\Delta\mu_n^{\oplus} = \mu_n^{\oplus} - \mu_1^{\oplus}$ is the difference in the standard chemical potential of an amphiphilic molecule in solution and an amphiphilic molecule in the aggregate. The standard chemical potential for the n-mer aggregate is $n \cdot \mu_n^{\oplus}$.

27.4.1.2 Critical Micelle Concentration

Analysis of Equation (27.20) reveals interesting behavior. At low enough values of x_1, the quantity $y = x_1 \cdot \exp\left(-\Delta\mu_n^{\oplus}/(\boldsymbol{RT})\right)$ is much less than one ($y \ll 1$). In Equation (27.20), this means that x_n, the mole fraction of aggregates in solution is much smaller than x_1; i.e., $x_{agg} \ll x_1$, where we define

$$x_{agg} = \sum_{n=2}^{\infty} x_n$$

As the total concentration of amphiphiles increases, x_1 increases until the factor y approaches near unity. As y approaches unity, the mole fraction of aggregates x_{agg} grows much faster than x_1 (as y to the nth power). Once the concentration x_1 of free amphiphile reaches a critical value

$$x_1 = x_{\text{critical}} = \frac{1}{2}x_{\text{total}} = \frac{1}{2}\sum_{n=1}^{\infty} x_n \quad (27.21)$$

x_1 remains nearly constant as x_{agg} increases. The concentration of the aggregates (micelles) increases, and we call x_{critical} the *critical micelle concentration*. Note that x_n in Equation (27.21) is determined by Equations (27.19) and (27.20).

Experimentally, x_{critical} is the concentration of surfactant where one observes a sharp change in the physical or optical properties of the solution. Figure 27.16 shows conductivity data for a solution of sodium dodecyl sulfate in water as a function of the mole fraction of sodium dodecyl sulfate. At low mole fraction ($x_1 < 1.5 \times 10^{-4}$) the molar conductivity shows a linear increase with the surfactant concentration with a slope of 65.4 Ω^{-1} cm^2 mol^{-1}, and at high mole fraction ($x_1 > 1.5 \times 10^{-4}$) the molar conductivity still shows a linear increase with concentration but at a reduced slope of 24.8 Ω^{-1} cm^2 mol^{-1}. This change in slope results from the formation of micellar aggregates above 1.5×10^{-4} and their reduced mobility (because of their larger size).

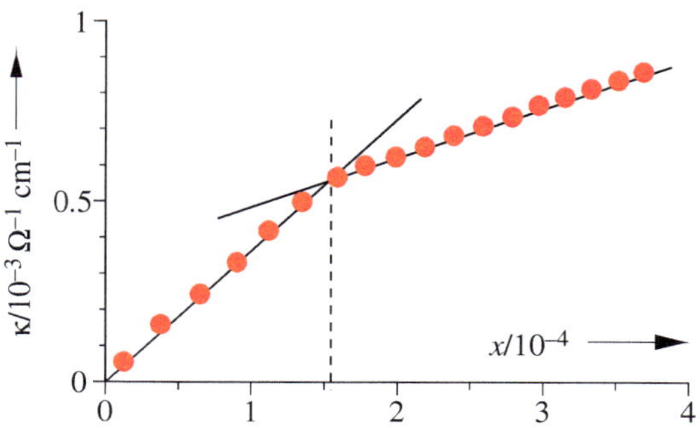

Figure 27.16 The conductivity κ of an aqueous solution of sodium dodecylsulfate at 298 K is shown as a function of the mole fraction x of the amphiphile. Note the break in the slope near $x = 1.6 \times 10^{-4}$. (Adapted from [14].)

27.4.1.3 Chemical Potential Change for Micelle Formation

To make the discussion more physical we consider sodium dodecyl sulfate as the amphiphiles and construct a model [15] for the chemical potential, partial molar Gibbs energy changes. We assume that the Gibbs energy change as a function of aggregate size has two major contributions: a favorable hydrophobic contribution ΔG_{HP} and an unfavorable head-to-head repulsive contribution ΔG_{HR}. The detailed features of these contributions will depend on the nature of the amphiphilic molecule.

Hydrophobic Gibbs Energy Change, ΔG_{HP} The hydrophobic term accounts for the decrease in chemical potential that occurs by having the nonpolar tail (alkyl chain) of the fatty acids interacting together in the core of the micelle rather than with the water molecules in solution. This chemical potential change will have a bulk contribution and a surface contribution. The bulk contribution can be obtained from the Gibbs energy of transfer between water and an alkane solvent, for the alkane molecule corresponding to the alkyl tail. These experimental data are available. Nagarajan and Ruckenstein [16] have fit such data to find an empirical relation for the free energy change as a function of temperature; at 298 K it gives

$$\Delta\mu^{\oplus}_{\text{HP,bulk}} = \left(-8.68 \text{ kJ mol}^{-1} - 3.68 \cdot n_{\text{CH}_2} \text{ kJ mol}^{-1}\right)$$

where n_{CH_2} is the number of methylene groups. For the dodecyl chain ($n_{\text{CH}_2} = 11$) this gives $\Delta\mu^{\oplus}_{\text{HP,bulk}} = -49.13$ kJ mol^{-1}. The surface contribution arises from the formation of the interface between the hydrocarbon core and the solution; however, part of the hydrophobic tails will be shielded from the solution by the hydrophilic head group. We write this contribution as

$$\Delta\mu^{\oplus}_{\text{HP,surface}} = \gamma \cdot (A - A') \cdot \boldsymbol{N}_{\text{A}}$$

where A is the average area per amphiphile on the micelle surface and A' is the area per head group of the amphiphile; i.e., the difference $(A - A')$ is the average area of the amphiphile's hydrophobic tail that is exposed to the water. We will approximate the surface tension by that found for alkane/water interfaces, e.g., $\gamma_{\text{decane/water}} = 51.2$ mN m^{-1}. Combining these two contributions, we write that

$$\begin{aligned}\Delta\mu^{\oplus}_{\text{HP}} &= \Delta\mu^{\oplus}_{\text{HP,bulk}} + \Delta\mu^{\oplus}_{\text{HP,surface}} \\ &= \Delta\mu^{\oplus}_{\text{HP,bulk}} + \gamma \cdot (A - A') \cdot \boldsymbol{N}_{\text{A}}\end{aligned}$$

Head-to-Head Repulsion Energy, ΔG_{HR} A micelle containing n surfactant molecules has an average area per amphiphile of $A = A_{\text{micelle}}/n$ and a surface area A_{micelle} given by its geometry.

In general, we must consider electrostatic repulsions and steric repulsions between the head groups and other packing factors, see Section 27.4.2. Let us assume that the repulsion between ionized head groups ΔG_{HR} is dominated by their electrostatic repulsion energy, i.e., the energy required to charge the surface layer of the micelle with a charge of $Q = n \cdot e$. If we model the micelle interface as two charged concentric spheres with a gap distance δ, we find that the electrostatic energy E_{HR} is given by

$$E_{HR} = \frac{e^2\delta}{2\varepsilon_0\hat{\varepsilon}\, A}$$

where ε_0 is the permittivity of the vacuum and $\hat{\varepsilon}$ is the relative permittivity of the solvent in the gap region (see Fig. 27.17). In Problem P27.4 you derive this result from electrostatics.

Combining these results, we write

$$\Delta\mu_n^{\oplus} = \Delta\mu_{HP,bulk}^{\oplus} + \gamma \cdot (A - A') \cdot \boldsymbol{N}_A + \frac{e^2\delta}{2\varepsilon_0\hat{\varepsilon}\, A} \cdot \boldsymbol{N}_A \qquad (27.22)$$

for the change in chemical potential of an amphiphile molecule between the monomer in solution and the n-mer aggregate.

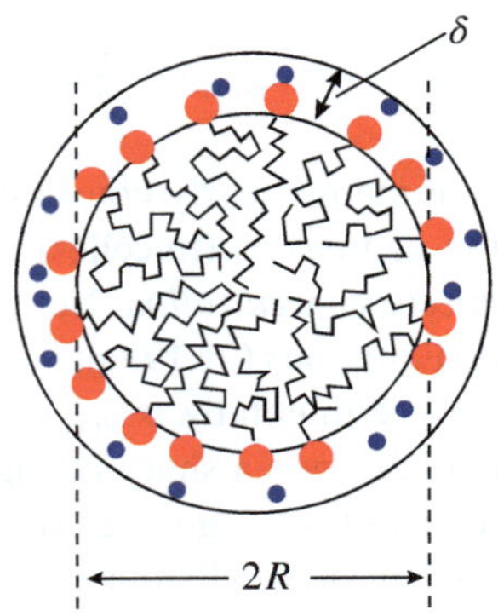

Figure 27.17 Diagram of a spherical micelle whose hydrophobic core has the radius R. We model the interfacial space charge by two concentric charge spheres with a gap of distance δ and a relative permittivity $\hat{\varepsilon}$. Red filled circles: negatively charged SO_4^- headgroups in sodium dodecyl sulfate, blue filled circles: positive charges of the Na^+ counterions.

Approximate Sodium Dodecyl Sulfate as a Spherical Micelle The aggregation number n and the surface area per amphiphile A can be related through the geometry of the micelle. The volume of the micelle core is the sum of the volumes V of the hydrophobic tails of each amphiphile, namely nV. In the case of a spherical micelle (see Fig. 27.17) with radius R of the hydrophobic core, we find that

$$\text{the core volume is } nV = \frac{4\pi}{3}R^3 \quad \text{and the core surface area is} \quad nA = 4\pi R^2$$

Thus we can write

$$A = \frac{4\pi}{n}\left(\frac{3nV}{4\pi}\right)^{2/3} = (4\pi)^{1/3} \cdot 3^{2/3} \cdot \frac{V^{2/3}}{n^{1/3}} = 4.836 \cdot \frac{V^{2/3}}{n^{1/3}}$$

and Equation (27.22) becomes[2]

$$\Delta\mu_n^{\oplus} = \Delta\mu_{HP,bulk}^{\oplus} - \gamma A' \cdot \boldsymbol{N}_A + C_1 \cdot n^{-1/3} + C_2 \cdot n^{1/3} \qquad (27.23)$$

with

$$C_1 = 4.836 \cdot V^{2/3}\gamma \cdot \boldsymbol{N}_A \quad \text{and} \quad C_2 = \frac{e^2\delta}{2\varepsilon_0\hat{\varepsilon}} \cdot \frac{\boldsymbol{N}_A}{4.836 \cdot V^{2/3}} \qquad (27.24)$$

Equation (27.23) has two competing terms that change with the aggregation number n. The destabilization of the micelle that results from exposure of the hydrophobic core to the water decreases with an increase in aggregation number n (see term $C_1 \cdot n^{-1/3}$ in Equation (27.23)), which arises because the amphiphiles assemble so that the hydrophobic core is encapsulated by the hydrophilic head group. The head groups repel each other, however, and this destabilization term increases with the aggregation number n (see term $C_2 \cdot n^{1/3}$ in Equation (27.23)). Consequently, we expect an optimal aggregation size to occur that balances these competing effects and minimizes the chemical potential. The curve in Fig. 27.18 plots the chemical potential as a function of the aggregation number n. Although the dependence is shallow in this particular case, a clear minimum at $n = 67$ exists.

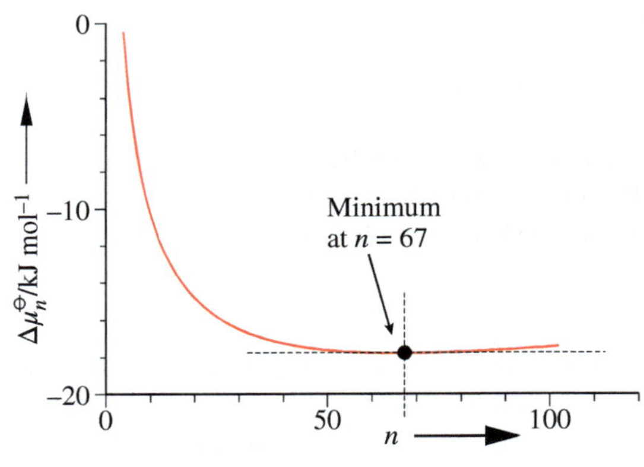

Figure 27.18 Chemical potential $\Delta\mu_n^{\oplus}$ versus the aggregation number n for the model of a spherical micelle with sodium dodecyl sulfate. The parameters used in evaluating Equation (27.23) are $\gamma = 51.2 \times 10^{-3}$ N m^{-1}, $A' = 1.7 \times 10^{-19}$ m^2, $V = 3.51 \times 10^{-28}$ m^3, $\delta = 5 \times 10^{-10}$ m, and $\hat{\varepsilon} = 40$.

In a more formal way we can find the minimum in chemical potential. We differentiate Equation (27.23) and obtain

$$\frac{d\Delta\mu_n^{\oplus}}{dn} = -\frac{1}{3}C_1 n^{-4/3} + \frac{1}{3}C_2 n^{-2/3}$$

2 Note that the model does not work for small aggregation numbers; e.g., for $n = 1$ we obtain $\Delta\mu_1^{\oplus} = 24$ kJ mol^{-1}, rather than $\Delta\mu_1^{\oplus} = 0$.

Setting this expression equal to zero and resolving for n, we obtain, using the parameters given in Fig. 23.18,

$$n_{\text{minimum}} = \left(\frac{C_2}{C_1}\right)^{3/2} = \left(\frac{74.2\ \text{kJ mol}^{-1}}{4.52\ \text{kJ mol}^{-1}}\right)^{3/2} = 67$$

in accordance with Fig. 27.18.

Distribution of Aggregate Sizes and the Critical Micelle Concentration Are most of the aggregates of size n? How sharp is the distribution of aggregate sizes? Equation (27.20) gives the distribution of aggregates in solution as a function of the monomer concentration and the change in chemical potential.

$$x_n = n\left[x_1 \cdot \exp\left(-\frac{\Delta\mu_n^{\oplus}}{RT}\right)\right]^n$$

Using Equation (27.23) for $\Delta\mu_n^{\oplus}$, this distribution is calculated for different sodium dodecyl sulfate monomer equilibrium concentrations x_1, at $T = 298$ K. For sufficiently low x_1 the amphiphiles exist practically in the monomer form (case 1 in Table 27.1, no formation of micelles). With increasing x_1 more and more n-mers are formed; see Table 27.1 and Fig. 27.19.

Table 27.1 Micelles Formed in a Model for Sodium Dodecyl Sulfate at $T = 298$ K. The Cases 2 to 4 Refer to the Three Curves in Fig. 27.19. Parameters Used in the Calculation see Fig. 27.18.

Case	x_1	n_{max}	$x_{n_{max}}$	x_{total}	% Monomers
1	0.59×10^{-3}	56	0.01×10^{-4}	0.61×10^{-3}	97
2	0.61×10^{-3}	57	0.06×10^{-4}	0.71×10^{-3}	86
3	0.62×10^{-3}	58	0.16×10^{-4}	0.88×10^{-3}	71
4	0.63×10^{-3}	58	0.40×10^{-4}	1.28×10^{-3}	49
6	0.65×10^{-3}	59	2.52×10^{-4}	4.66×10^{-3}	14
7	0.67×10^{-3}	60	15.32×10^{-4}	25.00×10^{-3}	3

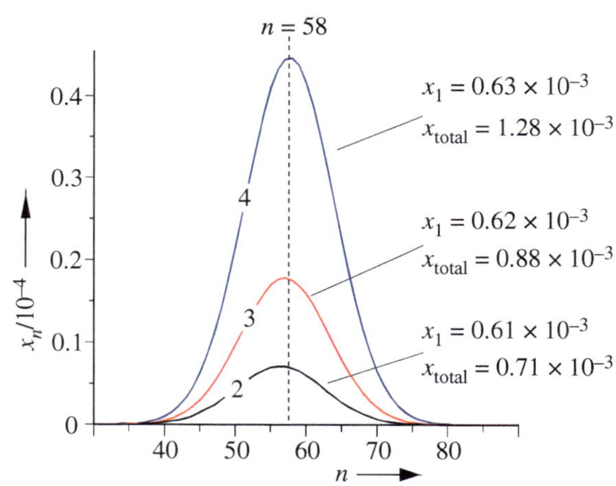

Figure 27.19 Aggregate population distribution for the model of a spherical micelle with sodium dodecyl sulfate, using the same parameters as in Fig. 27.18. The numbering of the curves is according to Table 27.1. The maxima correspond to $n = 58$.

Note the steep increase of x_{total} when x_1 approaches the value 0.63×10^{-3}, resulting in $x_{total} = 1.28 \times 10^{-3}$ (case 4). Hence, about half of the amphiphiles exist as monomers and half are aggregated with a maximum micelle size of $n_{max} = 58$. This situation is similar for the critical micelle concentration, according to Equation (27.21).

$$x_1 = x_{critical} = \frac{1}{2}x_{total} = \frac{1}{2} \times 1.28 \times 10^{-3} = 0.64 \times 10^{-3}$$

When further increasing the amphiphile concentration, nearly all amphiphiles exist as n-mers (case 6). Figure 27.19 shows that the full-width-at-half-maximum of the distribution is $\Delta n = 15$, indicating that the size of the aggregates has a variation of $\pm 25\%$.

27.4.2 Geometry of Packing

Although the thermodynamic considerations described above hold for the self assembly of amphiphiles of any shape and detailed properties, the specific model constructed for the difference in chemical potential of an amphiphile between solution and the micelle must change to reflect the detailed properties of the amphiphile. This dependence on amphiphile properties gives rise to a rich variety of shapes. Rather than develop this wide range of behaviors in detail we present some model results and refer to the literature.

Israelachvili [17] has defined three molecular parameters that are characteristic of certain amphiphiles (length l_c, area A_0, and volume V) and used these to classify the types of assemblies they form in solution. We have considered V in Section 27.4.1, l_c is the length of the amphiphilic tail (see Fig. 27.14), and A_0 is an optimal average surface area per amphiphile. Israelachvili defines an amphiphile shape factor, namely $V/(A_0 \cdot l_c)$ and uses it to rationalize the tendency for amphiphiles to form certain types of structures. In the range $V/(A_0 \cdot l_c) < 1/3$ spherical micelles are predicted, in the range $1/3 < V/(A_0 \cdot l_c) < 1/2$ cylindrical micelles, in the range $1/2 < V/(A_0 \cdot l_c) < 1$ bilayer vesicles, and in the range $V/(A_0 \cdot l_c) > 1$ planar bilayers.

Example 27.2 *As estimated in Table 27.1 the micelle formed by sodium dodecylsulfate in water has an optimal aggregation number of* $n_{max} = 58$. *For alkanes, Tanford [18] gives the critical tail length as* $l_c \lesssim \left(0.154 + 0.1265 \cdot n_{CH_2}\right)$ *nm where* n_{CH_2} *is the number of methylene groups; thus, for dodecyl with* $n_{CH_2} = 11$ *we obtain* $l_c = 1.546$ *nm, and the dodecyl tail volume is* $V = 0.351$ *nm*^{3}. *Estimate the optimal head group area and the shape factor for this micelle.*

First, we assume the micelle is spherical and determine the surface area per head group. From a hydrophobic volume of 0.351 nm^3 per alkyl chain, we estimate the core radius to be

$$R = \sqrt[3]{3Vn_{max}/4\pi} = 1.69 \text{ nm}$$

Consequently its surface area per head group will be

$$A_0 = \frac{4\pi R^2}{n_{max}} = \frac{4\pi(1.69 \text{ nm})^2}{58} = 0.62 \text{ nm}^2$$

so that

$$\frac{V}{A_0 \cdot l_c} = \frac{0.351 \text{ nm}^3}{(0.62 \text{ nm}^2)(1.546 \text{ nm})} = 0.37$$

which is a little larger than 1/3. Hence we expect that the micelle will be somewhat ellipsoidal, rather than spherical.

With increasing concentration there is a competition of the micelles for water molecules forming their solvation shell; this favors cylindrical structures which become arranged in ordered patterns as indicated in Fig. 27.15. At these high concentrations liquid crystalline phases of the assemblies are formed, see Section 27.7.

27.5 Membranes

Under appropriate conditions, surfactant molecules arrange to form thin films between two aqueous phases.

27.5.1 Liposomes

When a micellar solution of phosphatidylcholine is exposed to ultrasound the micelles are disrupted and spherical particles consisting of a bilayered membrane (called *lipid vesicles* or *liposomes*) with diameter of about 20 nm are formed (Fig. 27.20).

Typically, liposomes range in diameter from tens of nanometers to tens of micrometers. Liposomes are useful models for mimicking biomembranes (see Section 27.6 and Chapter 28).

It is of great interest to study artificial fluid state lipid bilayer networks of high geometrical complexity to create structures similar to structures in biosystems, such as the Golgi apparatus—endoplasmic reticulum network in eukaryotic cells. In Fig. 27.21 a method [19] for the construction of nanotube vesicle networks is shown. The nanotubes are between 100 and 200 nm in radius and the vesicles are 4–10 μm in diameter. Materials can be injected into specific vesicles within the network and can be further transferred to other vesicles via connecting nanotubes in a well-defined manner. The ultimate fluidic device is one that can handle single molecules and colloid particles in a distinctly planned network.

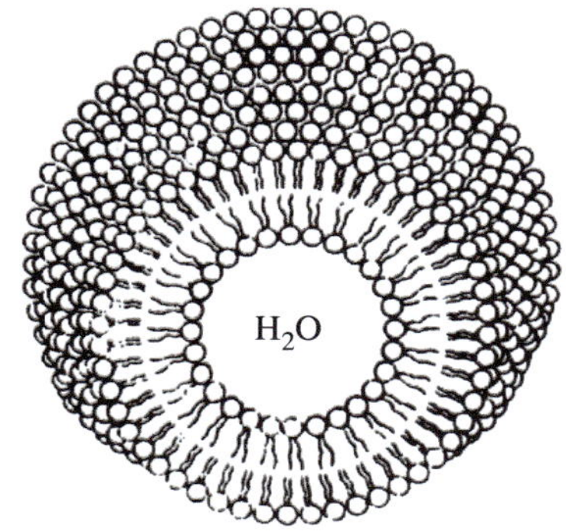

Figure 27.20 Liposome. Spherical particle: aqueous medium inside and outside. Cross-section showing double layer structure. (Courtesy: E. Sackmann).

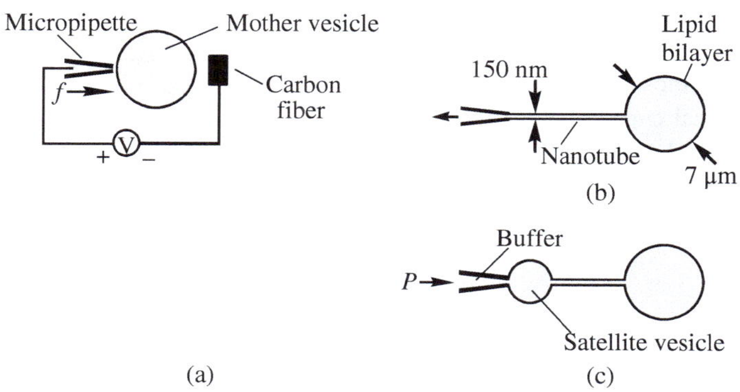

Figure 27.21 Scheme showing the beginning of a formation of lipid-bilayer nanotube-vesicle network. The polar lipid extract used consisted of a mixture of phosphatidylcholine (45.7%), phosphatidylethanolamine (27.1%), phosphatidylinositol (18.4%), phosphatidic acid (6.9%), and others (6.9%). (a) The membrane of a surface-immobilized giant vesicle is penetrated by a combination of mechanical force *f* applied from the micropipette and anodic electric pulses (voltage pulse-generator V and a carbon fiber as counter electrode). (b) A nanotube is created by pulling the micropipette away from the mother vesicle. (c) A small satellite vesicle is created at the end of the nanotube by injecting buffer solution with pressure *P* into the nanotube orifice. The additional lipid-bilayer required flows from the mother vesicle via nanotube to the growing satellite vesicle. To continue network formation vesicles can be fused and nanotube paths may cross in an overlaid fashion or may unite in a junction.

27.5.2 Soap Lamella

A soap lamella (see Fig. 27.22) can be made from a solution of sodium dodecyl sulfate in water. The lamella is stable if it is kept in air that is saturated with the solvent vapor, in order to avoid evaporation of liquid from the lamella. The lamella becomes increasingly thinner by solvent flow into the bulk of the solution and spontaneously reaches a constant thickness of almost molecular dimension. This thickness change causes a sudden disappearance of the interference colors of the lamella, well known from soap bubbles (black lamella). The hydrophobic ends of the hydrocarbon chains of the surfactant are presented to the air and the hydrophilic groups are presented to the water layer, bearing the counterions of the surfactant. In this way an equilibrium is reached, governed by the Coulomb attraction of the negatively charged surfactant monolayers with the positive counterions and the Coulomb repulsion of the two monolayers (Fig. 27.22).

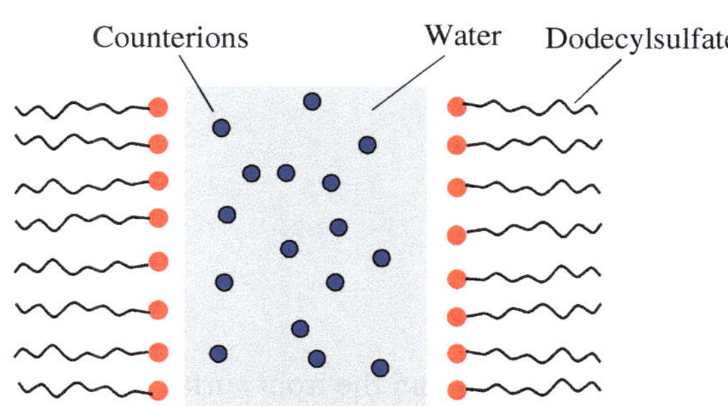

Figure 27.22 Soap lamella, equilibrium reached at 3 nm thickness. Red filled circles represent negatively charged head groups of dodecylsulfate and blue filled circles represent positively charged counterions.

27.5.3 Black Lipid Membranes

Consider a hole of about 1 mm diameter in a teflon substrate in contact with water that is covered with a tiny amount of the amphiphilic phosphatidylcholine in an organic solvent (n-decane). A film forms which becomes increasingly thinner, and finally a double layer is present. The hydrophilic groups of the surfactant face the water solutions on the left and on the right (Fig. 27.23) and the hydrocarbon chains face each other. Similar to soap lamellas, the formation of the double layer membrane can be seen by a sudden appearance of a black spot that gradually extends over the entire hole as the organic solvent is extruded. The bilayer, like the black lamella of soap, is reached at equilibration.

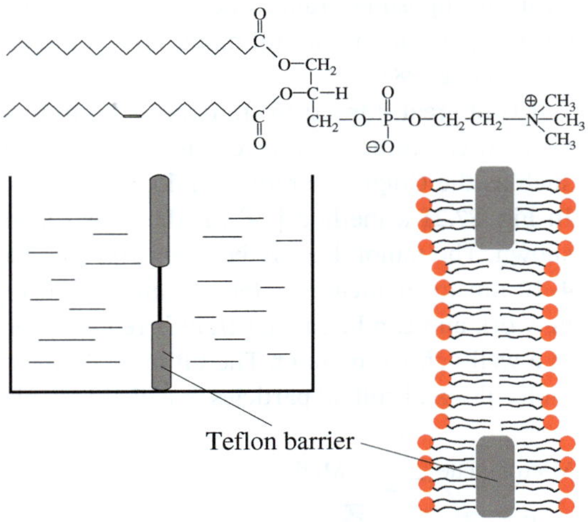

Figure 27.23 Black lipid membrane (phosphatidylcholine).

The presence of the membrane can be determined by measuring the electrical resistance, which changes from 10^8 Ω for a well-formed double layer to 1 Ω when the membrane collapses. The thickness of the structure can be deduced by measuring the electrical capacity of the film (applying an alternating voltage between the two solutions). In the approximation that the membrane forms a planar dielectric film, the capacitance C of a membrane with thickness d and area A is

$$C = \varepsilon_r \varepsilon_0 \frac{A}{d} \tag{27.25}$$

For $A = 1\ \text{cm}^2$ a capacitance of 0.5 μF and a resistance of 10^8 Ω are typical. Using $\varepsilon_r = 2.5$ we obtain a thickness $d = 4.4$ nm, which is in reasonable agreement with the assumption that the membrane consists of an ordered double layer.

Example 27.3 *Consider a thin lamella of thickness d and light incident on the lamella surface. Show that the reflected beams from the two surfaces of the film interfere and explain why the film looks dark.*

We decompose the light reflected from the film into two components: that which reflects off the front surface at point A and that which reflects off the back surface of the film at point B.

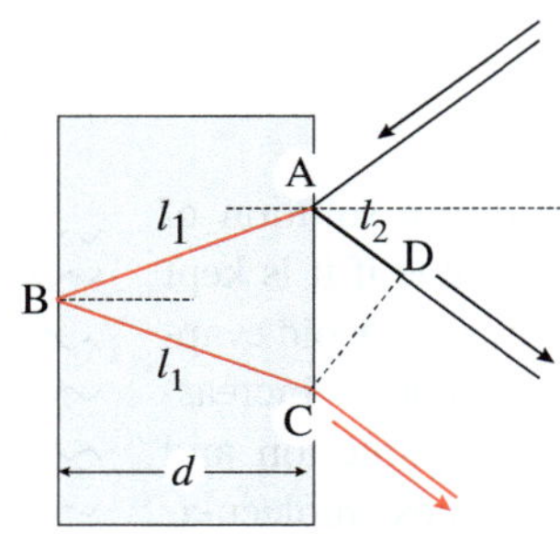

The light reflected from the front surface of the film travels a distance l from point A where it strikes the surface to your eyeball, the detector. The light traveling to the back surface travels a longer distance l'. We see that the optical difference in length $l - l'$ has three contributions. These are 1) the extra optical length from the front surface to the back surface l_1, plus 2) the extra optical length from the back surface to the front surface l_1, minus 3) the extra distance traveled by the first beam in the air l_2. Combining these terms we see that

$$l - l' = 2l_1 - l_2$$

The phase difference δ in the light waves of wavelength λ will be

$$\delta = \frac{2\pi\,(l - l')}{\lambda} + \pi = \frac{2\pi(2l_1 - l_2)}{\lambda} + \pi$$

where we include the term π to account for the phase shift upon reflection at the front surface [20]. In the limit that d is very thin, point C approaches point A so that $2\pi(2l_1 - l_2)/\lambda \ll \pi$; and the phase difference is $\delta = \pi$. Hence, both reflected waves extinguish each other, and the film looks black, or gray.

To estimate the thickness of a black membrane, let us assume that

$$2\pi(2l_1 - l_2)/\lambda \approx 0.1$$

and $\lambda = 500$ nm. Then

$$2l_1 - l_2 = 0.1\frac{\lambda}{2\pi} = 0.1\frac{500\ \text{nm}}{2\pi} = 8\ \text{nm}$$

This distance is on the order of $2d$; hence, the thickness of the "black" soap lamella should be approximately 4 nm.

27.6 Biomembranes

Biomembranes have many similarities with black lipid membranes. The resistance, capacitance, thickness, and permeability to water are similar. Essentially, biomembranes can be considered as lipid bilayers with proteins and other compounds incorporated or attached. The incorporation of proteins, their complexes, and other organic molecules form distinct domains and give rise to complex functionality (see examples in Chapter 28). Below we give examples of how molecules or ions diffuse along a membrane and through a membrane.

27.6.1 Lateral Diffusion

Small molecules can readily diffuse laterally in a membrane, indicating that the lipid matrix is liquid-like. This diffusion can be important for membrane bound reactions in biosystems. The mobility of molecules in the lipid matrix can be measured by incorporating fluorescent molecules in the membrane, then bleaching a small region on the membrane by focusing a pulsed laser beam, and measuring the recovery of the fluorescence [21]. From the time dependence of the borderline-broadening, the diffusion coefficient of the molecules can be evaluated. For example, 3,3′-dioctadecylcarbocyanine iodide was incorporated in an erythrocyte membrane, and the diffusion coefficient $D = 3 \times 10^{-9}\ \text{cm}^2\ \text{s}^{-1}$ was measured [22].

Example 27.4 *Diffusion Coefficient from borderline-broadening. In an erythrocyte membrane, the border width B increases with the square root of time. After 1 minute the border width increases to* 10 μm. *Use the Einstein–Smoluchowski relation with* $\overline{x^2} = B^2$ *to determine the diffusion coefficient.*

The Einstein–Smoluchowski relation gives

$$B^2 = 2Dt$$

and we obtain

$$D = \frac{B^2}{2t} = \frac{\left(10^{-3}\text{cm}\right)^2}{2 \times 60\ \text{s}} = 8 \times 10^{-9}\ \text{cm}^2\ \text{s}^{-1}$$

Diffusion coefficients on the order of 10^{-7} to $10^{-9}\ \text{cm}^2\ \text{s}^{-1}$ have been measured in such experiments.

27.6.2 Ion Transport Through a Membrane

The transport of ions (H^+, Na^+, K^+) and small molecules (H_2O) through membranes occurs by way of small (a few nanometer diameter) orifices, referred to as *channels*, which commonly comprise membrane spanning proteins or polypeptides. These channels may be gated by the binding of ligands or by electric fields, but can also be "open" for their characteristic time of existence. In Chapter 28, we discuss two naturally occurring membrane spanning proteins that drive proton transport across membranes. Here we discuss the case of a model system that acts as a channel for univalent ions, such as Na^+.

Gramicidin A1 is a small polypeptide (HCO-Val-Gly-Ala-Leu-Ala-Val-Val-Val-Trp-Leu-Trp-Leu-Trp-Leu-Trp-NH-CH_2 CH_2OH) that coils into a helix with its side chains projecting outward so that its core forms a small orifice (diameter ≈ 0.4 nm). The scheme in Fig. 27.24a illustrates how two gramicidin peptides dimerize to form a channel that spans a bilayer. This system has proven particularly useful for understanding basic features of ion channels because of its small size and the ability to characterize its structure. As shown in the diagram the dimer length, $l \approx 2.2$ nm, is somewhat smaller than the average bilayer thickness d_0 and causes a deformation of the bilayer.

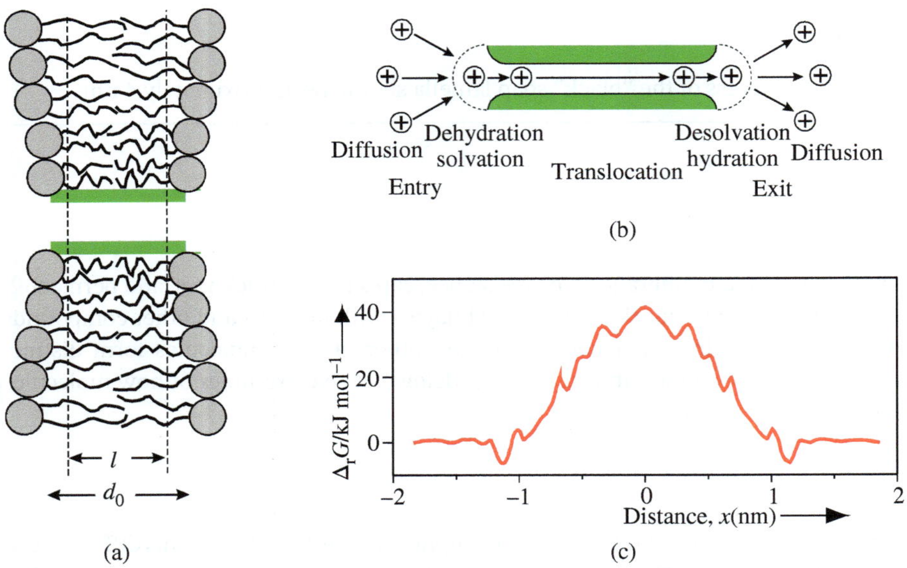

Figure 27.24 Transport of ions through a gramicidin channel. (a) Formation of a gramidicin channel. (b) Five-step mechanism of the ion transport through the channel. (c) Gibbs energy as a function of distance x from the center of the channel. Green area: surrounding of the channel. (Adapted from [23].)

The transport of ions through the channel can be decomposed into five different steps [23] (see Fig. 27.24b). First is the ion diffusion to the pore on the outside (left part in diagram) of the membrane. Second is the dehydration of the ion and its rehydration/association with the pore wall. Third is its movement through the pore to its exit point on the other side. Fourth is its dissociation from the pore and rehydration in the interior (right of diagram). Fifth is its diffusion away from the pore exit. Figure 27.24c shows the mean Gibbs energy profile obtained from molecular dynamics simulations of the transport in this case. The minima in the Gibbs energy profile near −1.1 nm and +1.1 nm correspond to binding sites for the ion at the edges of the channel, which have been observed experimentally by NMR and X-ray scattering studies. The increase in the Gibbs energy near the center of the pore arises from the higher charging energy of the ion in the nonpolar part of the bilayer.

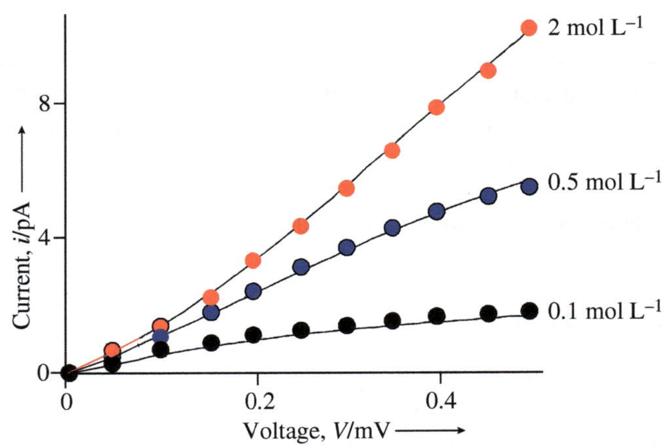

Figure 27.25 Single channel sodium current versus voltage for gramicidin pores. The data on the curves refer to the concentration of the sodium ions. (Adapted from [23].)

The ionic current through the channel can be measured using the patch clamp method described in Chapter 23. Figure 27.25 shows a plot of the single channel i/V curves for Na^+ transport through a gramicidin channel in a model bilayer as a function of sodium ion concentration. At low concentrations it is evident that the current saturates at higher bias potentials. This saturation arises because the ion translocation is limited by the diffusion of ions to the pore opening, rather than transport through the pore (see Example 27.5). As the concentration of the ions in solution becomes larger however, we see that the current becomes superlinear with the applied voltage. This latter behavior shows that the rate of transport is determined by movement of the ion in the channel and its exit, rather than the collisions with the channel opening.

Example 27.5 *At a concentration $c = 0.05$ mol L^{-1} the current in the Na^+ single channel conductance study saturates at a value of 1 pA. Assume that the rate is limited by the arrival of the sodium ions at the pore opening (diffusion constant $D = 1.4 \times 10^{-5}$ cm^2 s^{-1}) and the pore is about 2.2 nm long, calculate the effective diameter d of the pore.*

Because $i/e_0 = 1\ \text{pA}/1.6 \times 10^{-19}\ \text{C} = 6 \times 10^6\ \text{s}^{-1}$, one pA corresponds to 6×10^6 ions s^{-1} traversing the channel. That is, every $\frac{1}{6} \times 10^{-6}$ s an ion passes the channel, on average. For a 2.2 nm long channel, the average speed of the ions through the membrane is $2.2\ \text{nm}/(\frac{1}{6} \times 10^{-6}\ \text{s}) = 1.3\ \text{cm s}^{-1}$. From our development of diffusion limited reactions in Chapter 24, we write the "reaction velocity" as the successful collisions with the pore

$$1.3\ \text{cm s}^{-1} = 2 \cdot \sigma \cdot D \cdot \frac{N}{V}$$

where D is the diffusion coefficient, σ is the area of the pore, and N/V is the number density of the sodium ions. Thus with $N/V = \boldsymbol{n}\boldsymbol{N}_A/V = \boldsymbol{N}_A \cdot c$

$$\sigma = \frac{1.3\ \text{cm s}^{-1}}{2 \cdot D \cdot \boldsymbol{N}_A \cdot c} = \frac{1.3\ \text{cm s}^{-1}}{2 \cdot 1.4 \times 10^{-5}\ \text{cm}^2\ \text{s}^{-1} \cdot 6.022 \times 10^{23}\ \text{mol}^{-1} \cdot 0.05\ \text{mol L}^{-1}}$$
$$= 1.5 \times 10^{-15}\ \text{cm}^2 = 0.16\ \text{nm}^2$$

Using $\sigma = \pi d^2/4$, we find that $d = 0.45$ nm, which is in reasonable agreement with the expected value.

27.6.3 Transport of Small Protein Through a Membrane

The transport of macromolecules through membranes requires a sophisticated machinery and is not well understood. Only for some small proteins does the penetration take place without a translocation machinery; the mechanism has been elucidated by investigating a number of mutants with various patterns of charged amino acids and studying their ability to penetrate the membrane.

Figure 27.26 illustrates the proposed way of translocation of the newly synthesized coat protein of a bacteriophage through a bacterial membrane. The protein can be considered to consist of three α-helical segments 1, 2, and 3 consisting of 18, 18, and 14 amino acids, respectively (a). Segments 1 and 3 are hydrophilic, and segment 2 is strongly hydrophobic. The protein inserts into the membrane (a to b), and the region with the N-terminal is moved from the inside of the bacterial membrane to the outside (b to c).

The Gibbs energy to transfer a given segment of the protein from water into the interior of the membrane is approximately given by the sum of empirical values of the Gibbs energies of each amino acid in the segment. The contribution of a hydrophobic amino acid such as leucine is negative ($\Delta_r G^\oplus_{\text{Leu}} = -11.7$ kJ mol^{-1}), while it is strongly positive for charged amino acids such as aspartic acid ($\Delta_r G^\oplus_{\text{Asp}} = +38.5$ kJ mol^{-1}) [25].

In our example the Gibbs energy to transfer segment 1 from water into the hydrophobic interior of the membrane is

$$\Delta_r G^\oplus_1 = +21.7\ \text{kJ mol}^{-1}$$

Similarly, for segments 2 and 3 we obtain

$$\Delta_r G^\oplus_2 = -163.6\ \text{kJ mol}^{-1},$$

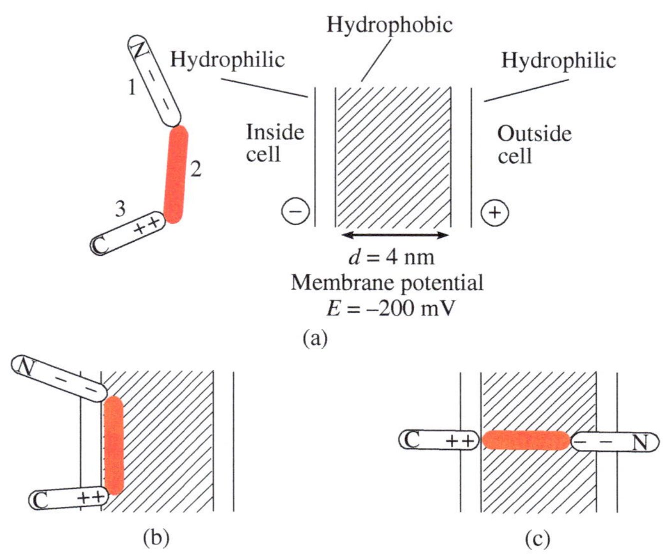

Figure 27.26 Small protein penetrating biomembrane [24]. (a) Protein and bacterial membrane. (b)–(c) Stages of penetrating the membrane. Segments 1 and 3 are hydrophilic, segment 2 is hydrophobic. The membrane potential is needed to stabilize stage c relative to stage b. (Adapted from [24].)

and

$$\Delta_r G_3^{\oplus} = +46.9 \text{ kJ mol}^{-1}$$

Thus, the transfer of segment 2 into the membrane (step a to b in Fig. 27.26) is accompanied by a strong decrease in Gibbs energy. Segment 1 can then be transferred to the outside of the membrane (step b to c) in contrast to segment 3. Stage c is stabilized due to the membrane potential $E = -0.2$ V. The membrane potential is found to be necessary for the transfer which is accompanied by a decrease of Gibbs energy (shift of negative charge $2e$ on segment 1, $\Delta_r G_{b\to c}^{\oplus}$ is on the order of $2EF = -40$ kJ mol^{-1}). In contrast, a transfer of segment 3 instead of segment 1 from inside to outside (shift of positive charge) would increase the Gibbs energy correspondingly.

27.7 Liquid Crystals

Molecules of specific shape (sufficiently long or discus-like) aggregate in a specific way that displays some long range order yet are not crystalline. A substance comprising such molecules remains in an ordered state above the melting point of its crystalline form. Although the three-dimensional lattice breaks down on heating above the melting point, the molecules are not disordered on a macroscopic scale as in an ordinary liquid, but retain a macroscopic orientational order. Only when the temperature is increased above a critical value (clearing point) does thermal motion dominate and the system liquifies. These aggregates, existing in the temperature range between the melting point and the clearing point, are called liquid crystals; see Foundation 27.1, on the book's website, for more discussion.

27.8 Conclusion

We have shown how forces acting at interfaces modify the behavior of molecules and drive self-assembly. We discussed the formation of surface films and their wetting behavior for a liquid. We considered soap lamella and black lipid membranes as model systems for an understanding of biomembranes, which provide a way to segregate and organize matter in biological organisms. In addition, we showed how basic considerations can be used to explain the formation of micelles, simple aggregates of amphiphiles. Micelles and liposomes are examples for three-dimensional organization. Understanding these simple structures represents a first step toward developing a description for the assembly of complex networks of vesicles connected by nanotubes and for describing the transport of ions and molecules (even small proteins) through a biomembrane.

27.9 Box 27.1. Surface Film; Gibbs Adsorption Equation

We derive the relation in Equation (27.15) by considering two different reversible pathways (see Figure 27B.1) for transferring a small amount $\boldsymbol{n}$ of solute from solution (a) at concentration c to solution (b) at concentration $(c + \Delta c)$. The two pathways are

I) Using an osmotic process (see Chapter 18), the Gibbs energy change for the concentration change is

$$\Delta G_I = \boldsymbol{nR}T \cdot \ln\left(\frac{c+\Delta c}{c}\right)$$
$$= \boldsymbol{nR}T \cdot \ln(1+x)$$

with $x = \Delta c/c$. In the limit where $x \ll 1$, $\ln(1+x) \approx x$, so that

$$\Delta G_I = \boldsymbol{nR}T\frac{\Delta c}{c}$$

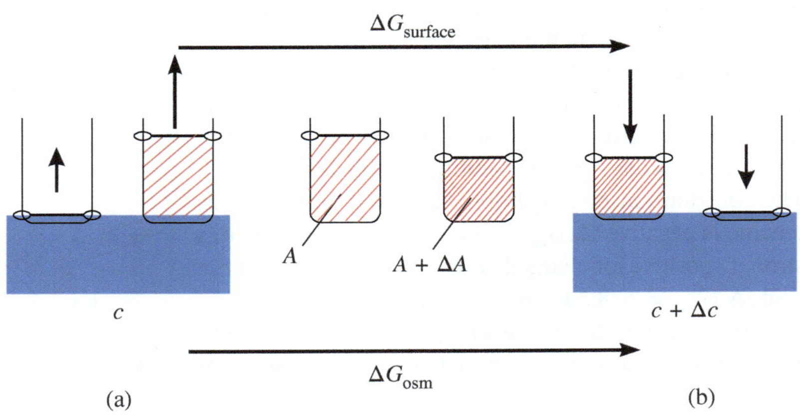

Figure 27B.1 Reversible transfer of surfactant molecules from a solution of concentration c (gray rectangle, left) to a solution of concentration $c + \Delta c$ (gray rectangle, right) as a surface film on a lamella (withdrawing lamella from solution on the left), (a). Diminishing area, transfer of lamella to solution on the right (b). Note that in this case ΔA is negative (the area A becomes smaller).

II) In the second process, we first remove the amount $\boldsymbol{n}$ by drawing a lamella (area A) from solution a and then supply this amount reversibly to solution b. The work in withdrawing the lamella is

$$\Delta G_1 = \gamma A$$

and its surface excess is $\Gamma = \boldsymbol{n}/A$. Next we diminish the area of the lamella to reach the surface excess $\Gamma + \Delta\Gamma$ (found in solution b), which requires the work

$$\Delta G_2 = \gamma \Delta A$$

Lastly, we transfer this lamella reversibly to solution b, which has the Gibbs energy change

$$\Delta G_3 = -(\gamma + \Delta\gamma) \cdot (A + \Delta A) = -\gamma A - \gamma \cdot \Delta A - \Delta\gamma \cdot A - \Delta\gamma \cdot \Delta A$$

Combining these three steps, gives the net Gibbs energy change

$$\Delta G_{II} = \Delta G_1 + \Delta G_2 + \Delta G_3 = -\Delta\gamma \cdot A - \Delta\gamma \cdot \Delta A$$

In the limit that the $\boldsymbol{n}$ is very small, ΔA is very small. In this limit we neglect the term $\Delta\gamma \cdot \Delta A$ and find

$$\boxed{\Delta G_{II} = -\Delta\gamma \cdot A}$$

By equating the work for both processes and defining $\Gamma = \boldsymbol{n}/A$, we find that

$$\boldsymbol{n}\boldsymbol{R}T\frac{\Delta c}{c} = -\Delta\gamma A = -\Delta\gamma\frac{\boldsymbol{n}}{\Gamma}$$

Taking the limit of infinitesimally small changes, we replace the differences by differentials and find

$$\Gamma = -\frac{c}{\boldsymbol{R}T} \cdot \frac{\mathrm{d}\gamma}{\mathrm{d}c}$$

27.10 Foundations

27.10.1 Foundation 27.1: Liquid Crystals

See the books website, www.wiley.com/go/Kuhn/physicalchemistry3e.

27.10.2 Foundation 27.2: Clausius–Mosotti Equation

See the books website, www.wiley.com/go/Kuhn/physicalchemistry3e.

27.11 Exercises

E27.1 In Fig. 27.2 a lamella is withdrawn from a soap solution. Given the surface tension of water $\gamma = 72.8$ mN m^{-1} calculate the work to pull the movable frame of width $L = 3$ cm to a 1 cm height, $h = 1$ cm. Compare this result to the work needed to pull the moveable frame against the gravitational force (assume the frame is made from platinum wire of thickness 0.2 mm). In addition, compare your results to the work needed to pull the liquid within the lamella against the gravitational force. Comment on the relative magnitudes.

E27.2 Calculate the contact angles α and β in Fig. 27.4 from Equations (27.5) and (27.6), using the data given in the text.

E27.3 Calculate the work to spread a water film of 1 cm^2 area on the surface of glass (take its contact angle to be 10°). Compare your result to the case of spreading water on a 1 cm^2 area of graphite (take its contact angle to be 86°) and a 1 cm^2 area of paraffin (take its contact angle to be 110°).[3] Using these results, calculate the work per molecule of water and comment on the trend through the series of surfaces.

3 The value measured for the contact angle is very sensitive to surface structure, surface impurities, and surface hydration. The values used in this exercise are taken from Table X-2 of [1].

E27.4 For low enough concentrations of surfactant c the surface pressure Π in a Langmuir trough varies linearly with the surfactant concentration, i.e.,

$$\Pi = \gamma_0 - \gamma = bc$$

where b is a constant. Show that $b = \boldsymbol{RTn}_{\text{surfactant}}/(cA)$ and the surface pressure equation has the form of a two-dimensional ideal gas law.

E27.5 In the limit that the radius of a glass capillary tube r is smaller than the height h of the liquid water inside, J. W. Rayleigh [26] found an approximate, but more exact, solution to the capillary rise, namely

$$h = \frac{2\gamma}{\rho g r} - \frac{r}{3} + \frac{0.1288\ r^2}{h} - \frac{0.1312\ r^3}{h^2} + \cdots$$

This equation is valid in the range

$$r < 0.46\ \sqrt{r\,h}$$

In this treatment, the first correction term $(r/3)$ accounts for the mass of the fluid above the meniscus and the other two terms account for the deviation of the meniscus from sphericity. Evaluate h for the case used in Example 27.1, and determine the percentage error in our approximate result.

E27.6 Calculate the rise of water in a glass capillary of different radius values and compare your findings with experiment. You should use the approximate formula $h_{\text{approx}} = 2\gamma/\rho g r$ and the more accurate formula derived by Rayleigh (see Exercise 27.5). Comment on how the error increases with the capillary radius. For water at 20° use surface tension of $\gamma = 72.8$ mN m^{-1}, a gravitational constant of g = 9.807 m s^{-2}, and a density of $\rho = 1.00$ g cm^{-3}.

E27.7 Derive Equation (27.18) by starting from Equation (27.5).

E27.8 In Example 27.2 we discussed micelles composed of sodium dodecylsulfate in water with an optimal aggregation number of $n^* = 58$. Rather than assuming a spherical shape for the micelle, assume it has an ellipsoidal shape in which the short axis diameter is twice the length l_c and the long axis has a radius b (for a prolate ellipsoid $n^* v = 4\pi b l_c^2/3$ and for an oblate ellipsoid $n^* v = 4\pi b^2 l_c/3$). Calculate the area per head group for this case.

E27.9 Consider a system like Fig. 27.6b, which shows the rise of liquid water in a glass capillary for a situation in which the water pool and the capillary are in contact with the air, but connect the capillary tube to an external pressure source P. What pressure must be applied to push the water back down to the level of that in the water pool?

E27.10 Equation (27.20) gives the distribution of aggregate sizes for an amphiphile. Use the fact that this equation also holds for the optimum aggregate size n^* and eliminate x_1 from this equation to show that

$$x_n = n \cdot \left(\frac{x_{n^*}}{n^*}\right)^{n/n^*} \cdot \exp\left(-\frac{n \cdot \left(\Delta\mu_n^{\oplus} - \Delta\mu_{n^*}^{\oplus}\right)}{\boldsymbol{RT}}\right)$$

E27.11 Explain why the critical micelle concentration of alkyl sulfates decreases with increasing alkyl chain length (an octyl chain is 130×10^{-3} mol L^{-1}, decyl chain is 33.2×10^{-3} mol L^{-1}, dodecyl chain is 8.1×10^{-3} mol L^{-1}, and a tetradecyl chain is 2.0×10^{-3} mol L^{-1}).

E27.12 Explain why the change in the critical micelle concentration per methylene for alkyl sulfates (see Exercise E27.11) is smaller than the corresponding change in the case of dialkylsulfates (e.g., dioctylmethylsulfate has a $x_{\text{critical}} = 2.35 \times 10^{-3}$ mol L^{-1} and ditetradecylmethylsulfate has a $x_{\text{critical}} = 0.08 \times 10^{-3}$ mol L^{-1}).

E27.13 Discuss the interference colors that appear when forming a black lipid membrane (see Section 27.5.3).

E27.14 A black lipid membrane has a capacitance of $C = 0.4$ μF for an area of 1 cm^2. Calculate the thickness of the membrane (see Section 27.5.3).

E27.15 Cho et al. [27] measured the diffusion constant for an aquaporin protein (assume 30 nm radius) in a red blood cell membrane to be 3.1×10^{-11} cm^2 s^{-1}. Estimate the time it takes for a mobile aquaporin to diffuse 3 μm, about the radius of the cell.

E27.16 Consider the placement of a drop of liquid A on a surface of liquid B; consider A and B to be immiscible. It is common to define a spreading coefficient Υ by the relation

$$\Upsilon_{\text{A/B}} = \gamma_{\text{B}} - \gamma_{\text{A}} - \gamma_{\text{A/B}}$$

Show that a positive spreading coefficient corresponds to a negative Gibbs energy for spreading. What is the spreading coefficient for water on benzene, given that $\gamma_{H_2O} = 72.8 \times 10^{-3}$ J m^{-2}, $\gamma_{C_6H_6} = 28.8 \times 10^{-3}$ J m^{-2}, and $\gamma_{H_2O/C_6H_6} = 35 \times 10^{-3}$ J m^{-2}.

E27.17 Use the diffusion coefficient D in Example 27.4 and the Einstein relation, $D = kT/\zeta$, to calculate the effective viscosity of a membrane $\eta_{membrane}$ (use $T = 298$ K and take the radius of the dye molecule to be $r = 0.5$ nm). Saffman and Delbrueck [28] derived an expression for the translational diffusion of a particle in a viscous planar membrane of thickness h that is surrounded by a lower viscosity medium (e.g., water which has a shear viscosity of about 0.01 N m^{-2} s). They found that the friction coefficient ζ is given by

$$\zeta = \frac{4\pi\ h\ \eta_{membrane}}{\log\left(\frac{h\ \eta_{membrane}}{r\ \eta_{solvent}} - 0.5772\right)}$$

Determine the value of $\eta_{membrane}$ for the case where h is the thickness of the membrane (~5 nm) and $\eta_{solvent}$ is the viscosity of the surrounding fluid (~0.001 N m^{-2} s). Also calculate the value of $\eta_{membrane}$ by using the formula for the translational friction of a sphere in three dimensions, i.e.,

$$\zeta = 6\pi\ r\ \eta_{membrane}$$

Compare the value you find from the three-dimensional model to that for the more appropriate two-dimensional model.

E27.18 In 1937 [8] measured optical interference fringes through Langmuir–Blodgett films of calcium stearate on glass. For light at normal incidence they showed that minima of the intensity of the reflected light occur at

$$\hat{n}N\delta = m\frac{\lambda}{4},\ m = 1,3,5\ldots$$

where $\lambda = 589.3$ nm is the light wavelength, $\hat{n} = 1.462$ is the refractive index, N is the number of monolayers, and δ is the thickness of a monolayer.

m	1	3	5	7	9	11	13
N	39	120	202	284	365	447	528

E27.19 Given that the average thickness of an octadecanethiol film is $d = 2.1$ nm, estimate the average tilt angle of the octadecane chains with respect to the surface normal of the gold surface. Assume that the thickness of a methylene group is 0.14 nm and that of the S atom is 0.2 nm.

E27.20 Calculate the work required to transfer a charged group (such as the COO^- side group of the amino acid aspartic acid) from water into a membrane.

E27.21 Consider a Na^+ ion moving through the gramicidin pore of a membrane. Compute the difference in solvation energy for a Na^+ ion in water solution (relative permittivity of 80) and that inside of a membrane (relative permittivity of 2.5). Comment on your result.

E27.22 Calculate the contact angles α and β in Fig. 27.4 (Section 27.2.1) from Equation (27.5)

$$\gamma_{decane/air} \cdot \cos\alpha \cdot dl + \gamma_{decane/water} \cdot \cos\beta \cdot dl = \gamma_{water/air} \cdot dl \tag{27.26}$$

for the horizontal components of the force f and from Equation (27.6)

$$\gamma_{decane/air} \cdot \sin\alpha = \gamma_{decane/water} \cdot \sin\beta \tag{27.27}$$

for the vertical components, using the data given in the text. Use notations $\gamma_{decane/air} = 23.9$ mN m^{-1}, $\gamma_{decane/water} = 51.2$ mN m^{-1}, and $\gamma_{water/air} = 72.8$ mN m^{-1}.

27.12 Problems

P27.1 Consider a spherical soap bubble with surface tension γ. Assuming static equilibrium, show that the pressure difference ΔP between the inside and the outside of the bubble is given by

$$\Delta P = 2\gamma/r$$

where r is the radius of the bubble.

P27.2 The table provides capacitance data for alkanethiol self-assembled monolayer films on a gold electrode. Use these data (C is the capacitance and A is the area of the capacitor) and a parallel plate capacitor model to find the monolayer film thickness as a function of n; n is the number of methylene groups in the alkanethiol $HS(CH_2)_nCH_3$. Use the relative permittivity $\hat{\varepsilon} = 2.3$ (that of polyethylene). From the bond parameters of an alkane chain, each methylene should contribute 130 pm to the length. Use this fact and the findings of your analysis to estimate the average tilt angle of the alkane chains on the gold electrode.

n	1	3	5	7	9	11	15	17
$C/A/\mu F\ cm^{-2}$	4.6	3.9	2.6	2.1	1.7	1.3	1.1	0.98

P27.3 In this model, you are asked to evaluate a simple model for the diffusion of ions through a membrane. Assume that the concentration of sodium ions inside of a liposome with an inner diameter of 100 nm is 0.1 mM and that the thickness of the bilayer is 3 nm. Estimate the number of collisions of sodium ions with the inner wall of the liposome if its diffusion coefficient is $1.3 \times 10^{-5}\ cm^2\ s^{-1}$ (see Chapter 14). Use this "collision rate" to estimate the rate of ions through the bilayer if the activation energy for the transport of an ion through a membrane is the energy that you calculated in E27.20.

P27.4 Calculate the electrostatic repulsion energy between head groups for an idealized micelle structure, like that shown in Fig. P27.4. Model the micelle interface (left) as two charged concentric spheres (right), and make the approximation that $\delta \ll R$ so that it can be treated as a parallel plate capacitor.

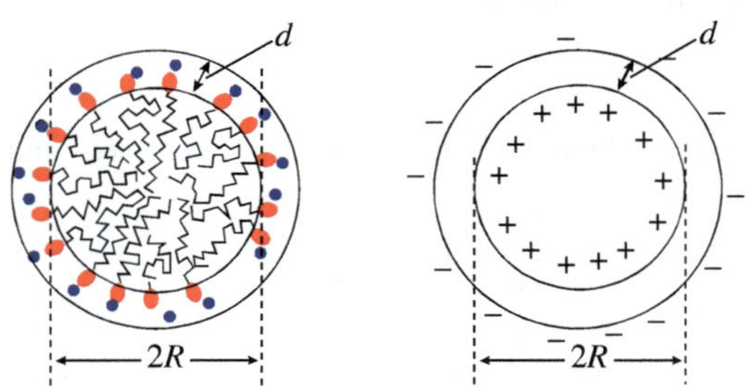

Figure P27.4 The figure shows two-dimensional cross sections for the electrostatics of two concentric spherical shells. The diagram on the left shows the molecules, and that on the right shows an idealization for the surface charges.

Part 1: Perform derivation for the electrostatic energy E in the capacitor and show that it is

$$E = \frac{Q^2}{2\hat{\varepsilon}\varepsilon_0}\frac{\delta}{4\pi R^2}$$

where Q is the charge.
Part 2: Make the correspondence with the micelle model and show that

$$E_{HR} = \frac{E}{n} = \frac{e^2}{2\hat{\varepsilon}\varepsilon_0}\frac{\delta}{A}$$

where n is the aggregation number and A is the area of the head group per molecule.

P27.5 Consider the case of a very thin (i.e., ignore thickness of plate) rectangular plate of width L that is suspended above the surface of a liquid. As it is just placed in contact with the surface of the liquid, capillary forces cause the liquid to rise. By balancing the capillary force with the gravitational force of the liquid pulled above the bulk liquid surface, show that

$$\gamma \cos\theta = weight/(2L)$$

where γ is the surface tension and θ is the contact angle. Next, assume that the surface of the liquid, which is pulled above the surface, has the shape of a cylinder (length L) and derive an expression for the contact angle θ.
Given that the density ρ of hexadecane is $0.773 \times 10^{-3}\ kg\ m^{-3}$ and its surface tension γ is $27.2\ mN\ m^{-1}$ calculate the contact angle for a plate with length $L = 3$ cm at a height of $h = 0.1$ cm.

P27.6 Consider a soap bubble. In this problem you will show that the pressure of air inside of a small soap bubble is larger than that inside of a larger soap bubble; i.e., the difference in pressure ΔP across the soap film decreases with its curvature. Consider a curved surface that has a surface tension γ and show that the pressure difference ΔP across the surface is

$$\Delta P = \gamma\left(\frac{1}{R_x} + \frac{1}{R_y}\right)$$

where R_x is the radius of curvature along the x-direction and R_y is the radius of curvature along the y-direction. Hint: Solve by finding an expression for the work of expanding the surface from the plane (x, y) outward to $(x + \delta x, y + \delta y)$, and equate this work to that for the displacement outward given by $work = force \times \delta z$, where δz is the displacement outward.

For a spherical soap bubble ($R_x = R_y$), which has two surfaces (an inner and outer surface of the soap film), and a surface tension of 0.025 N m^{-1}, calculate the pressure differential across the soap film for a radius of 5 cm and a radius of 1 cm.

P27.7 Using the result of P27.6 and elementary considerations show that the vapor pressure over a liquid with a spherically curved surface of radius a is given by

$$P = P_0 \cdot \exp\left(\frac{2V_m\gamma}{a\boldsymbol{R}T}\right)$$

where P_0 is the vapor pressure above a flat surface, V_m is the molar volume of the liquid, R is the gas constant, T is the temperature, and γ is the surface tension. Use the formula you derive to calculate the percent increase in the vapor pressure above a spherical drop of cyclohexane at 293 K which has a $\gamma = 24.95$ mN m^{-1} and a molar volume of 0.108 L mol^{-1}. Consider the case of a 0.1 mm diameter drop, a 0.2 μm diameter drop, and a 20 nm diameter drop.

P27.8 Figure 27.16 shows conductivity data for a solution of sodium dodecylsulfate in water as a function of the amount of dodecylsulfate. Some of these data are reproduced in the table. Read the paper by Jalsenjak and Tezak [14] and use the model they develop to analyze the data and obtain the aggregation number.

c_{total}/mmol L^{-1}	0	0.150	0.280	0.420	0.550	0.700	0.830	0.927
κ/mS cm^{-1}	0	0.10	0.20	0.29	0.37	0.46	0.54	0.58

c_{total}/mmol L^{-1}	1.04	1.16	1.33	1.50	1.65	1.81	1.96	2.15
κ/mS cm^{-1}	0.61	0.64	0.68	0.72	0.75	0.80	0.84	0.88

References

1 A.W. Adamson, *Physical Chemistry of Surfaces* (John Wiley & Sons, New York, 1990).

2 H. Bücher, H. Kuhn, B. Mann, D. Möbius, L.V. Szentpály, and P. Tillmann, Transfer of energy and electrons in assemblies of monolayers, *Photogr. Sci. Eng.*, 1967, **11**, 233.

3 H. Kuhn and D. Möbius, *Investigations of Surfaces and Interfaces, Part B*, eds. B.W. Rossiter and R.C. Baetzold, Physical Methods of Chemistry Series, Vol. **IXB** (John Wiley & Sons, New York, 1993) 2nd edition.

4 D. Schwartz, Langmuir-Blodgett film structure, *Surf. Sci. Rep.*, 1997, **27**, 241.

5 B. Moore, C.M. Knobler, D. Broeta, and F. Rondelez, Studies of phase transitions in Langmuir monolayers by fluorescence microscopy, *J. Chem. Soc., Faraday Trans. 2*, 1986, **82**, 1753.

6 H. Moehwald, Phospholipid and phospholipid-proetein monolayers at the air/water interface, *Annu. Rev. Phys. Chem.*, 1990, **41**, 441.

7 K.B. Blodgett, Films built by depositing successive monomolecular layers on a solid surface, *J. Am. Chem. Soc.*, 1935, **57**, 1007; I. Langmuir, Mechanical properties of monomolecular films, *J. Franklin Inst.*, 1934, **218**, 143; I. Langmuir, Pilgrim trust lecture, molecular layers, *Proc. R. Soc. London, Ser. A*, 1939, **170**, 15.

8 K.B. Blodgett and I. Langmuir, Built-up films of barium stearate and their optical properties, *Phys. Rev.*, 1937, **51**, 964.

9 Y. Li, X. Wang, and J. Sun, Layer-by-layer assembly for rapid fabrication of thick polymeric films, *Chem. Soc. Rev.*, 2012, **41**, 5998–6009.

10 W.C. Bigelow, D.L. Pickett, and W.A. Zisman, Oleophobic monolayers: I. Films adsorbed from solution in non-polar liquids, *J. Colloid Sci.*, 1946, **1**, 513.

11 C.D. Bain, E.B. Troughton, Y.T. Tao, J. Evall, G.M. Whitesides, R.G. Nuzzo, Formation of monolayer films by the spontaneous assembly of organic thiols from solution onto gold, *J. Am. Chem. Soc.*, 1989, **111**, 321.

12 H.G. Tompkins, *A User's Guide to Ellipsometry* (Dover, New York, 1993).

13 A. Ulman, *Ultrathin Organic Films* (Academic Press, San Diego, CA, 1991); S.O. Evans, P. Sanassy, and A. Ulman, Mixed alkanethiolate monolayers as substrates for studying the Langmuir film deposition process, *Thin Solid Films*, 1994, **243**, 325.

14 N. Jalsenjak and D. Tezak, A new method for simultaneous estimation of micellization parameters from conductometric data, *Chem. Eur. J.*, 2004, **10**, 5000.

15 J.N. Israelachvili, D.J. Mitchel, and B.W. Ninham, Theory of self-assembly of hydrocarbon amphiphiles into micelles and bilayers, *J. Chem. Soc., Faraday Trans. 2*, 1976, **72**, 1525.

16 R. Nagarajan and E. Ruckenstein, Theory of surfactant self-assembly: a predictive molecular thermodynamic approach, *Langmuir*, 1991, **7**, 2934.

17 J.N. Israelachvili, *Intermolecular and Surface Forces* (Elsevier, 1991) 2nd edition.

18 C. Tanford, Theory of micelle formation in aqueous solutions, *J. Phys. Chem.*, 1974, **78**, 2469.

19 M. Karlsson, K. Sott, M. Davidson, A.S. Cans, P. Linderholm, D. Chiu, and O. Orwar, Formation of geometrically complex lipid nanotube-vesicle networks of higher-order topologies, *Proc. Natl. Acad. Sci. U.S.A.*, 2002, **99**, 11573.

20 D. Halliday, R. Resnick, and J. Walker, *Fundamentals of Physics* (John Wiley & Sons, New York, 1993).

21 W. Hoppe, W. Lohmann, H. Markl, and H. Ziegler, eds., *Biophysics* (Springer-Verlag, Berlin, 1983, p. 435).

22 H.G. Kapitzka and E. Sackmann, Local measurement of lateral motion in erythrocyte membranes by photobleaching technique, *Biophys. Acta*, 1980, **595**, 56.

23 O.S. Andersen, R.E. Koeppe II, and B. Roux, Gramicidin channels, *IEEE Trans. Nanobiosci.*, 2005, **4**, 10.

24 D. Kiefer, X. Hu, R. Dalbey, and A. Kuhn, Negatively charged amino acid residues play an active role in orienting the Sec-independent Pf3 coat protein in the Escherichia coli inner membrane, *EMBO J.*, 1997, **16**, 2197.

25 D.M. Engelman, T.A. Steitz, and A. Goldman, Identifying nonpolar transbilayer helices in amino acid sequences of membrane proteins, *Annu. Rev. Biophys. Biochem.*, 1986, **15**, 321.

26 L. Rayleigh, On the theory of the capillary tube, *Proc. R. Soc. London, Ser. A*, 1916, **92**, 184–195.

27 M.R. Cho, D.W. Knowles, B.L. Smith, J.J. Moulds, P. Agre, and N. Mohandas, Membrane dynamics of the water transport protein aquaporin-1 in intact human red cells, *Biophys. J.*, 1999, **76**, 1136.

28 P.G. Saffman and M. Delbrueck, Brownian motion in biological membranes, *Proc. Natl. Acad. Sci. U.S.A.*, 1975, **72**, 3111.

28

Supramolecular Machines

CHAPTER MENU

28.1 Introduction

This chapter discusses the construction of organized systems of cooperating molecules, in which the individual molecules constitute the functional components. Such systems allow us to investigate exciting phenomena of a physicochemical nature and represent a modern subfield in physical chemistry. Organized systems have great potential as sensors and actuators, information processing devices, memories with high storage capacity, tailored catalysts for complex chemical reactions, and energy conversion systems—among others. The most intricate devices of this kind are living systems. Today's artificial supramolecular machines are very simple, but their development should pave the way to construct much more complex systems. Supramolecular engineering is a fertile research field for discovery that requires human imagination and creativity.

Modern chemistry finds itself at the birth of fascinating developments based on a new paradigm, the planned design of organized entities. Simple prototypes of supramolecular machines can be obtained by designing and synthesizing molecules that interlock like the parts of a jigsaw puzzle. They change position by diffusing and assembling in the intended way whenever opportunities arise (lock-and-key principle). The self-assembly can be guided by programmed external manipulations and by appropriate changes of the environmental conditions (programmed-environmental-change principle). Some possibilities for the assembly of identical molecules were discussed in Chapter 27. These ideas are extended here.

First attempts at molecular engineering appeared in the early 1960s when simple energy transduction devices were designed and constructed. The field was greatly accelerated by intelligently planned chemical synthesis (supramolecular chemistry) and by the development of methods to manipulate single molecules initiated by the invention of the scanning tunneling microscope (STM) in the early 1980s.[1] To illustrate this topic, we discuss some typical examples: devices to study energy transfer and electron transfer between molecules, switching devices based on photoinduced isomerization, devices for amplifying signals by inducing cooperative responses, and signal transfer through molecular wires.

28.2 Idea of a Supramolecular Machine

Tools and machines are functional devices in which different components interact to perform tasks that cannot be done by an individual component. Scissors are a simple example of a tool that is based on the cooperation of two blades (Fig. 28.1a).

1 Heinrich Rohrer and Gerd Binnig received the 1986 Nobel Prize in Physics for the invention of STM; https://www.nobelprize.org/prizes/physics/1986/summary.

Principles of Physical Chemistry, Third Edition. Hans Kuhn, David H. Waldeck, and Horst-Dieter Försterling.

Companion website: www.wiley.com/go/Kuhn/physicalchemistry3e

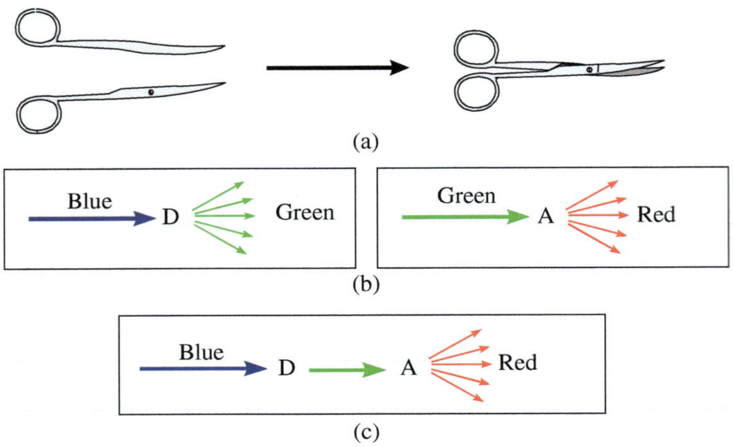

Figure 28.1 Functional units. (a) Two blades combined in a specific and unique way to give a pair of scissors. (b) Isolated dyes D and A emitting green and red light after excitation. (c) Dyes D and A combined in a specific and unique way (close proximity and mutual alignment) to give an energy transducing device.

28.2.1 A Simple Energy Transduction Device

The idea of a supramolecular machine as an entity in which single molecules comprise the cooperating functional components can be illustrated by considering two interacting dye molecules D and A that are separated by a small fixed distance. D is a donor molecule that absorbs blue light and emits green light, and A is an acceptor molecule that absorbs green light and emits red light (Fig. 28.1b). If D is excited by blue light, the excitation energy is transferred to A and A emits a quantum of red light (Fig. 28.1c). If the distance between D and A is slightly increased, however, the interaction of the two molecules becomes negligible and D itself emits a quantum of green light. The pair D and A, when close together, forms a functional unit, a simple kind of supramolecular machine. The solid parts of the supramolecular machine are the dye molecules, while the photon is the moving functional element. The entity is an energy transducing device.

Appropriate dye molecules can be switched by light (see Fig. 28.9) from an active to an inactive form and vice versa. In this case, the arrangement is a supramolecular switching element that emits red light when D and A are in the active form, green light when D is active and A inactive, and no light when D is inactive and A is active or inactive. Functional units of increasing complexity and intricacy can be constructed by appropriately arranging more and more molecules serving as functional components. The challenge is to fabricate the assembly of molecules so that their arrangement is fixed in a purposefully planned and useful manner.

28.2.2 Programmed Interlocking Molecules

Biosystems contain intricate supramolecular machines, for example, machines transforming light energy into chemical energy by electron or proton pumping, are present. They emerge by self-assembly from molecular components of complementary structure that are able to interlock in a distinct manner. This principle (Fig. 28.2) can be used to construct artificial supramolecular machines from chemically synthesized complex molecules with complementary chemical functionalities that bind to each other, assemble, and interlock (lock-and-key principle).

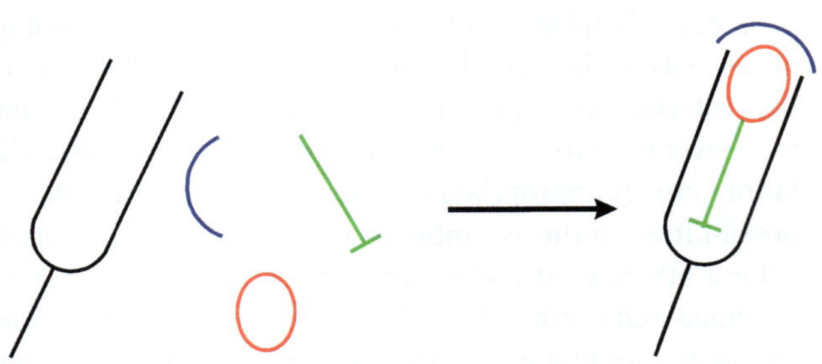

Figure 28.2 Forming complicated structures by interlocking components. Chemical synthesis of molecules that recognize each other with the aim of constructing organized molecular assemblies (supramolecular chemistry).

The attempt to construct supramolecular machines by self-assembly of molecules in solution is a modern frontier in preparative chemistry (supramolecular chemistry). This goal requires considerable synthetic and design ingenuity, however. In this section we focus on an approach demanding less preparative effort; in particular, we assist the self-assembly by exposing the molecules to programmed environmental changes (supramolecular engineering). In this way we can construct complex assemblies by using relatively simple interlocking molecules. The two basic fabrication principles we need are the lock-and-key principle and the programmed environmental change principle.

One strategy that facilitates self-assembly is the confinement of carefully designed molecules to a surface (e.g., a water surface). The molecules undergo Brownian motion on the surface and meet each other occasionally in a way that promotes mutual interlocking and binding, as planned. This process can be assisted by pushing the molecules together in order to form a compact monolayer film (Fig. 28.3). The complexity of the organized monolayer can be increased by allowing the monolayer to bind additional molecules that are present in the subphase under the water surface in a planned manner.

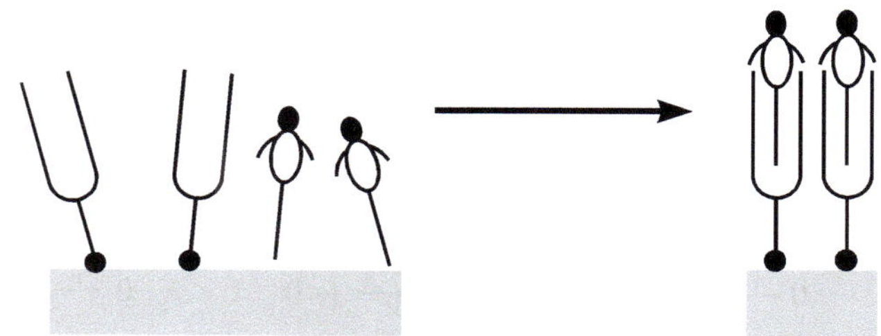

Figure 28.3 Interlocking molecules assisted by programmed environmental changes (supramolecular engineering), e.g., compression on a surface.

Such carefully designed and accordingly constructed layers can then be transferred to a solid substrate in a programmed sequence. Thus, organized supramolecular arrangements with increasing complexity can be built according to a given design plan. For example, we can fabricate multilayers of fatty acid molecules using the Langmuir–Blodgett (LB) deposition sequence (spreading solution, solvent evaporation, monolayer compression, deposition, see Chapter 27); additional techniques are crucial to obtain complex systems of planned functionality and for manipulating designed monolayer assemblies in different ways (Foundation 28.1), however. Organizing molecules at a solid surface and their subsequent chemical manipulation is another important route following the programmed environmental change principle (self-assembled monolayer films, SAM, see Chapter 27). Intelligent chemical synthesis of sophisticated interlocking compounds, combined with programmed manipulations, is a way to fabricate more and more intriguing supramolecular machines.

28.3 Manipulating Photon Motion

28.3.1 Energy Transfer Between Dye Molecules. FRET: Ruler in Nanometer Range. SNOM

28.3.1.1 Monolayer Assembly

How can we fabricate a simple supramolecular machine? A monolayer assembly for studying energy transfer from dye molecule D to dye molecule A

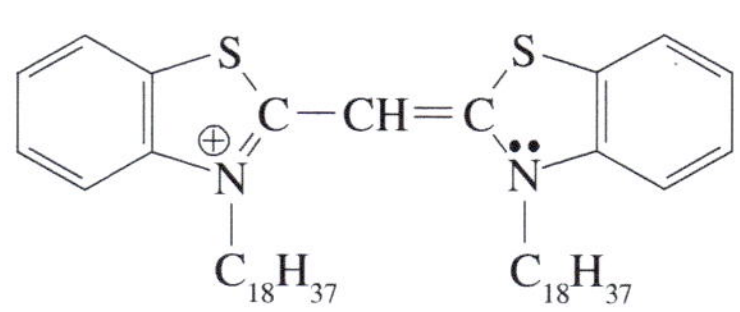

D

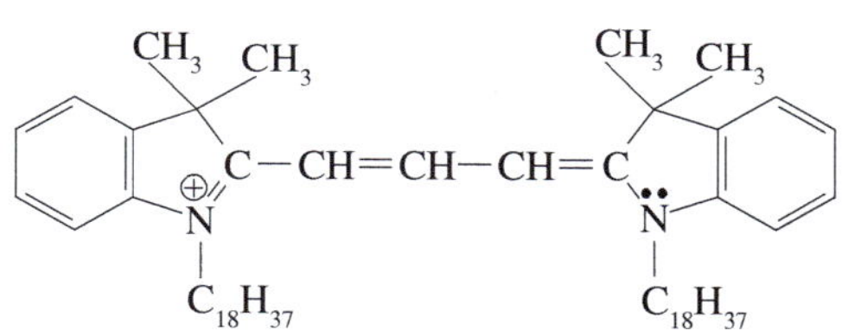

A

can be constructed using the Langmuir–Blodgett method. First we transfer a fatty acid monolayer (layer 1 in Fig. 28.4) from water to a glass slide (upward trip). The slide is covered with a mixed monolayer of D (layer 2), which absorbs blue light and emits green light, and contains the fatty acid in great excess (diluent) in a downward trip of the slide. The intended assembly is obtained by superimposing additional layers in a planned sequence—for example, two layers (3 and 4) of fatty acid (upward and downward trip) and a mixed monolayer (5) of A, which absorbs green light and emits red light (Fig. 28.4). It should be mentioned that the molecules in these layers are in a compact and well-defined arrangement. The absorption and fluorescence spectra of the individual dye molecules are shown in Fig. 28.5.

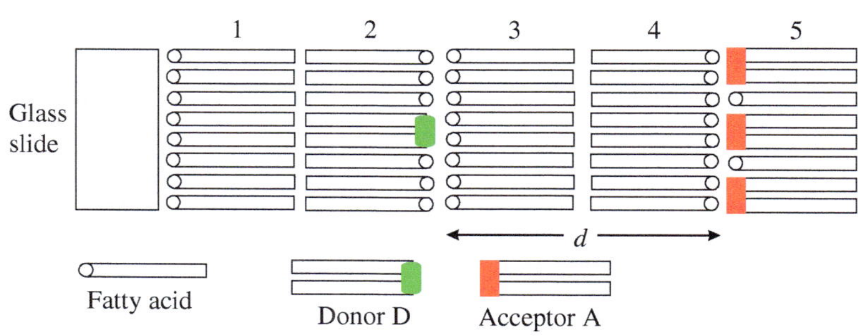

Figure 28.4 Monolayer assembly to study energy transfer from donor D to acceptor A in distance *d*.

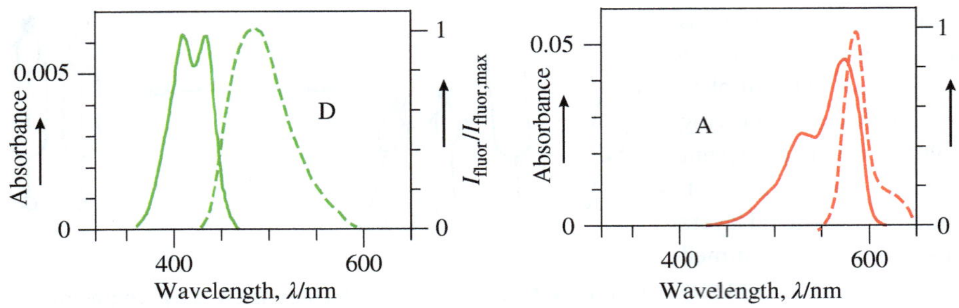

Figure 28.5 Dyes D and A: absorption spectra (solid lines) and fluorescence spectra (relative fluorescence intensity $I_{fluor}/I_{fluor,max}$, dashed lines). The spectra were measured in mixed monolayers of dye and fatty acid (molar ratio 1:20 for D, 1:1 for A) on a glass plate.

The combination of the programmed structure and the dye molecules' different electronic properties produces new behavior. Experimentally, we excite the dye molecule D with blue light. The intensities of the fluorescence of D (green) and A (red) are measured at various distances d between chromophores D and A; d is controlled by the number of fatty acid spacer layers (Fig. 28.6a). The experimental data are compared with the theoretical relations

$$\frac{I_d(\mathrm{D})}{I_\infty(\mathrm{D})} = \frac{1}{1+\left(\dfrac{d_0}{d}\right)^4} \quad \text{and} \quad \frac{I_d\,(\mathrm{A})}{I_0\,(\mathrm{A})} = \frac{1}{1+\left(\dfrac{d}{d_0}\right)^4} \tag{28.1}$$

which are derived in Foundation 28.2. In the equation on the left $I_d(\mathrm{D})$ is the intensity of the green fluorescence of D at distance d; $I_\infty(\mathrm{D})$ is the intensity in the case when the acceptor is at large distance. $I_d(\mathrm{A})$ in the equation on the right is the intensity of the red fluorescence of A at distance d, and $I_0(\mathrm{A})$ is the intensity in the case when acceptor and donor are close together. d_0 is the distance where

$$\frac{I_d(\mathrm{D})}{I_\infty(\mathrm{D})} = \frac{I_d\,(\mathrm{A})}{I_0(\,\mathrm{A})} = \frac{1}{2} \tag{28.2}$$

$I_d(\mathrm{D})/I_\infty(\mathrm{D})$ and $I_d(\mathrm{A})/I_0(\mathrm{A})$ are plotted versus distance d in Fig. 28.6b. According to Foundation 28.2 we find that the distance d_0 is given by

$$d_0 = \frac{1}{4\pi}\left(\frac{9}{4}\right)^{1/4} \frac{\lambda_D}{\hat{n}} (\phi A_\mathrm{A})^{1/4} \tag{28.3}$$

if the transition moments are statistically distributed in the layer plane; A_A is the absorbance of the layer of acceptor A for light of vacuum wavelength λ_D of the donor D's fluorescence, $\hat{n}$ is the refractive index of the medium, and ϕ is the quantum yield of fluorescence of donor D in the absence of acceptor A.

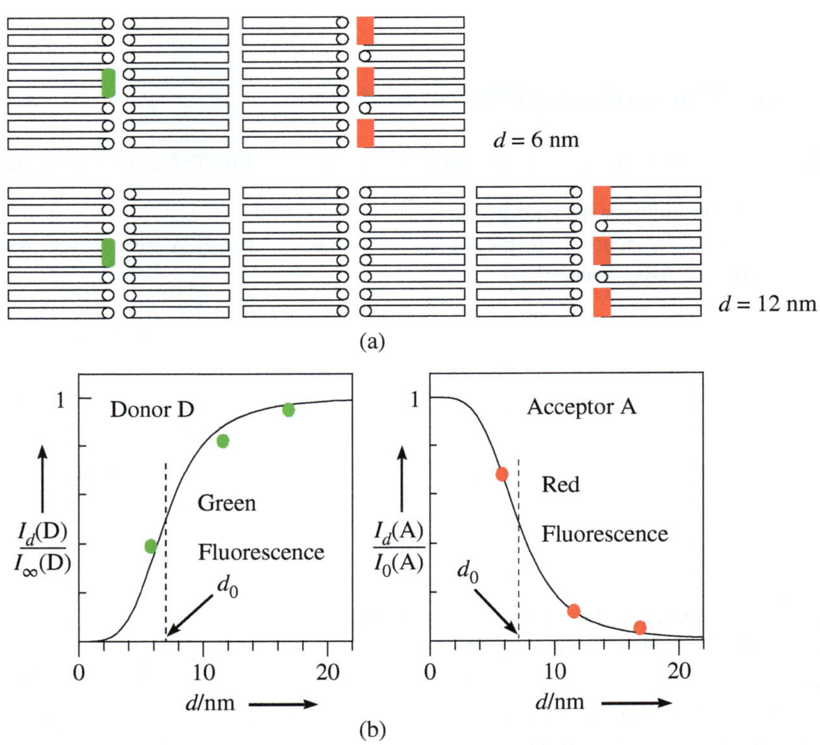

Figure 28.6 Energy transfer from donor D to acceptor A. (a) Monolayer assemblies with different numbers of fatty acid spacer layers. (b) $I_d(D)/I_\infty$ for donor fluorescence and $I_d(A)/I_0$ for acceptor fluorescence versus distance d. Dots: experiment [1]; solid lines: calculated from Equation (28.1), using $d_0 = 7$ nm.

From the value $d_0 = 7$ nm obtained in the experiment in Fig. 28.6b, a quantum yield of $\phi = 0.2$ is calculated from Equation (28.3), using $\hat{n} = 1.5$, $\lambda_D = 480$ nm, and $A_\mathrm{A} = 0.012$ (see Fig. 28.5):

$$\phi = \left(\frac{d_0\,\hat{n}\,4\pi}{\lambda_D}\right)^4 \frac{4}{9A_\mathrm{A}} = \left(\frac{7\ \mathrm{nm}\cdot 1.5\cdot 4\pi}{480\ \mathrm{nm}}\right)^4 \frac{4}{9\cdot 0.012} = 0.2$$

28.3.1.2 FRET: Near-Field Absorption

The effect of energy transfer from a donor to an acceptor as a nanometer scale ruler was pioneered while developing the techniques of manipulating monolayer assemblies [2], and it is now widely used in biophysical and biomedical research. The method is called *FRET* (*Fluorescence Resonance Energy Transfer*). This designation is somewhat misleading because the effect can be simply described as absorption of the acceptor by interacting with the electric field of the emitter, the donor molecule, in the near field [3]. The strength of the absorption is given by the square of the amplitude F_0 of the electric field at the position of the acceptor and its absorption coefficient ε_A (see Chapter 9). In usual absorption theory (the acceptor A is at a large distance r from the emitter) $F_0^2 \sim 1/r^2$ (far-field approximation); however, in the case of FRET the acceptor A is at small distance r from the emitter, and $F_0^2 \sim 1/r^6$ (near-field approximation):

$$F_0^2 = \left(\frac{1}{4\pi\varepsilon_0}\frac{2\mu_0}{r^3}\right)^2 \sim \frac{1}{r^6} \tag{28.4}$$

(see Box 13.4). μ_0 is the amplitude of the dipole moment of the emitter. In Foundation 28.2 it is shown how Equation (28.1) follows from these considerations.

In the following we focus on the essence of FRET. The mean power of energy absorbed by acceptor A is

$$\overline{\left(\frac{\mathrm{d}E}{\mathrm{d}t}\right)}_{\text{absorption of A}} = g\,\varepsilon_A F_0^2 \tag{28.5}$$

where g is a constant (derived in Foundation 28.2) and ε_A is the extinction coefficient of the absorber at the fluorescence wavelength of the donor. The mean power of energy emitted by the donor D is

$$\overline{\left(\frac{\mathrm{d}E}{\mathrm{d}t}\right)}_{\text{fluorescence of D}} = \frac{E}{\tau_0} \tag{28.6}$$

where E is the energy of the excited donor and τ_0 is the lifetime of fluorescence, calculated in Equation (10.1). At a particular distance $r = r_0$ between the donor and acceptor

$$\overline{\left(\frac{\mathrm{d}E}{\mathrm{d}t}\right)}_{\text{absorption of A}} = \overline{\left(\frac{\mathrm{d}E}{\mathrm{d}t}\right)}_{\text{fluorescence of D}} \tag{28.7}$$

In Foundation 28.2 we show that the probability $\mathcal{P}$ of energy transfer is

$$\mathcal{P} = \frac{\overline{(\mathrm{d}E/\mathrm{d}t)}_{\text{absorption of A}}}{\overline{(\mathrm{d}E/\mathrm{d}t)}_{\text{fluorescence of D}} + \overline{(\mathrm{d}E/\mathrm{d}t)}_{\text{absorption of A}}} = \frac{1}{1+\left(\dfrac{r}{r_0}\right)^6} \tag{28.8}$$

with

$$r_0 = \left(\frac{9\varepsilon_A \times 2.303 \times c_0^4}{128\pi^5 \boldsymbol{N}_\text{A}\, \nu_0^4} \cdot \frac{\kappa^2\phi}{\hat{n}^4}\right)^{1/6} \tag{28.9}$$

The factor κ measures the relative orientation of the transition moments of molecules 1 and 2 (see Foundation 28.2), the factor ϕ is the quantum yield. For the case $\phi = 1$ the energy is exclusively emitted by radiation and $\hat{n}$ is the refractive index of the medium.

Equations (28.8) and (28.9) were discovered by Theodor Förster [4] in 1948 as a result of a quantum-mechanical consideration (*Förster equation*). Here we treat the effect as usual absorption of a dye molecule A exposed to an alternating electric field of molecule D (near-field approximation). The formal identity of the classical treatment with the corresponding quantum-mechanical approach is discussed in Chapter 9.

You may ask why we have an exponent of 4 in Equation (28.1) instead of the exponent 6 in Equation (28.8). The exponent of 6 accounts for the fact that the probability of an acceptor molecule A to accept the excitation energy of a donor molecule D per unit time is proportional to r^{-6}. In the case of a monolayer the number of molecules that have a similar chance to accept the excitation energy is proportional to d^2; thus, the probability that the excitation energy is accepted by the monolayer of A per unit time is proportional to $d^{-6} \cdot d^2 = d^{-4}$.

The parameters r_0 and d_0 are related. For a statistical distribution of the transition moments of D and A in the layer plane we obtain

$$r_0 = \left(d_0^4 \cdot \frac{20}{3\pi\sigma_A}\right)^{1/6}$$

where σ_A is the number of dye molecules per unit area. In the case of Fig. 28.5 we have $\sigma_A = 0.4\ \text{nm}^{-2}$ and $d_0 = 7$ nm; thus, we obtain $r_0 = 5$ nm.

28.3.1.3 Motion of ATP Synthase

An interesting example application of FRET is its use to elucidate the motion of ATP synthase [5]. ATP synthase (Fig. 28.7a) is a membrane bound protein that uses the proton gradient across a membrane to synthesize ATP from ADP and phosphate. This enzyme undergoes a rotary motion during the course of its function.

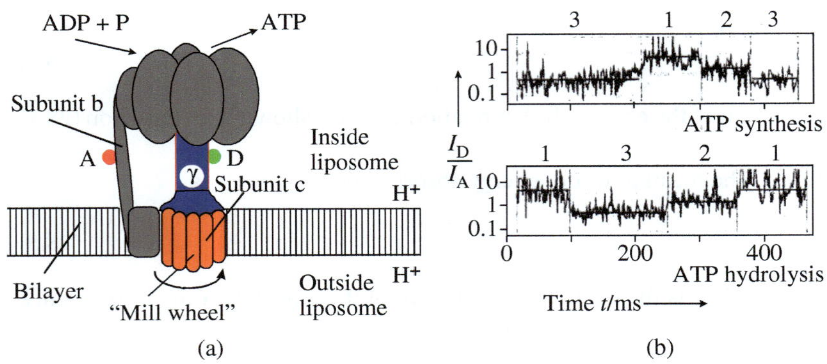

Figure 28.7 Motion of ATP synthase. (a) Fixed parts (gray) with acceptor A (red filled circle), rotating parts with subunit c (brown), subunit γ (blue), and FRET donor D (green filled circle). The H^+ gradient from outside to the inside drives the mill wheel continuously. A torque builds up on the rotating parts, fixation of ADP and P and formation of ATP occur, suddenly the rotating parts turn by 120°. (b) Monitoring the rotation by measuring the ratio I_D/I_A of the fluorescence intensities of donor D and acceptor A. Top: ATP synthesis, bottom: ATP hydrolysis. (Adapted from [5].)

The motion of the γ subunit during the enzyme operation has been demonstrated by binding an energy transfer donor (rhodamine dye molecule) to a particular cysteine, γ-T106C, of the γ subunit and a cyanine dye to the b subunit. As the γ-subunit moves relative to the b subunit, the distance between the rhodamine dye and the cyanine dye changes thereby changing the energy transfer efficiency.

By placing the fluorescent labeled enzyme complex in a liposome, the fluorescence of individual complexes can be monitored under irradiation by a laser source in a fluorescence microscope (see subsection on Scanning Near-Field Optical Microscope below). Figure 28.7b shows the photon bursts coming from the complex as a function of time.

When the donor and acceptor are far apart the energy transfer is least efficient and the emission of the photoexcited donor is strongest, and when the donor and acceptor are very close together the donor efficiently transfers its energy to the acceptor molecule whose fluorescence is most evident. The panel plots the ratio of the photon intensities from which it is apparent that three distinct levels exist. These levels correspond to three states (rotary positions) of the γ-subunit and demonstrate that the motion of the enzyme is stepwise, rather than continuous. In addition, the sequence of levels was observed to be $3 \to 1 \to 2 \to 3$, indicating directed motion. By operating the system in reverse (ATP hydrolysis) one observes that the sequence of level changes is in the opposite order, indicating an opposite sense for the rotation. The average time spent at each of the three levels corresponds well to catalytic turnover times of the enzyme.

28.3.1.4 Scanning Near-Field Optical Microscope (SNOM)

In the preceding sections we considered the energy transfer from energy donor D to energy acceptor A. D was treated as an oscillating dipole of molecular size, and A was treated as an absorbing body in the near field of the donor. The near field is the electric field of the emitter at a distance r smaller than about one-half of the wavelength (Coulomb field of the dipole). The near-field strength increases as $1/r^3$ when decreasing r. Energy transfer allows us to monitor a molecular assembly by visible light with a resolution determined by the characteristic length of energy transfer ($r_0 \approx 10$ nm), far below the wavelength of visible light.

Let us consider an optical fiber pulled to a small tip (diameter ≈ 20 nm) and illuminated by a laser beam (Fig. 28.8a, left). The tip corresponds to the fluorescent dye molecule acting as an oscillating dipole of subwavelength diameter, and the surface corresponds to the acceptor layer interacting with the near field of the emitter. This arrangement is a useful tool to address a molecular assembly with visible light (scanning near-field optical microscope, SNOM). Using this device it is possible to visualize individual dye molecules. Figure 28.8c shows the fluorescence of about 20 individual molecules when scanning an area of 4 μm^2. Bleaching after a longer illumination period does not lead to a continuous weakening of all signals, but to a sudden disappearance of individual signals in statistical sequence. This confirms that the signals belong to individual dye molecules [6].

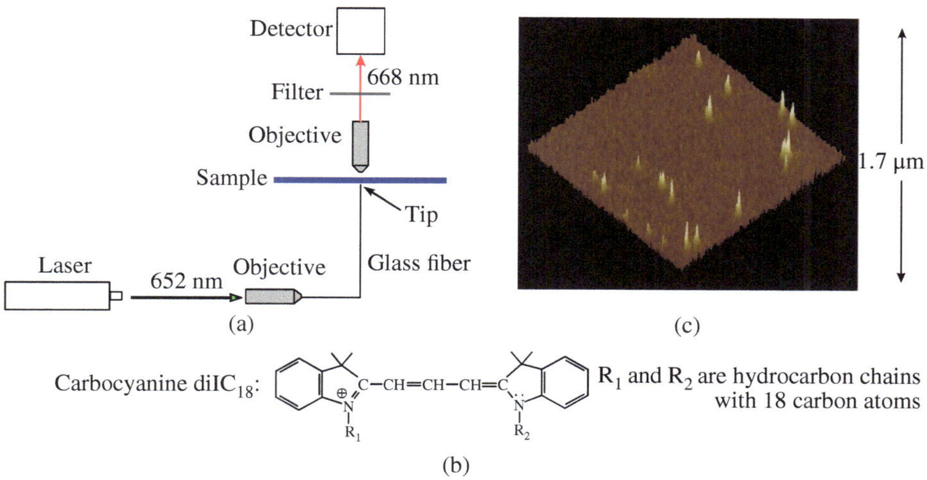

Figure 28.8 (a) Scheme of a scanning near-field optical microscope (SNOM). Light of a laser ($\lambda = 652$ nm) is coupled to the surface of the sample by using a glass fiber with a very small tip (diameter ≈ 20 nm). The outgoing light passes a filter to cut the contribution of the laser light, and the fluorescence light intensity of the sample is measured by the detector. The sample can be moved in the *x* and *y* directions; in this way the surface can be scanned. (b) The sample is a mixed monolayer containing the lipophilic carbocyanine dye diIC_{18} (absorption wavelength 652 nm, fluorescence wavelength 668 nm). The area concentration is about 20 molecules per μm^2. (c) Visualization of the fluorescence of single molecules after scanning the surface of the sample. ([7]. Eric Betzig and Robert J. Chichester. 1993/Reproduced from Elsevier).

28.3.1.5 Molecular Switch. Fluorescence of Dye A Quenched by Photoinduced Electron Transfer from B to C

Here we describe a simple supramolecular machine that uses fluorescence quenching to report on a photoinduced electron transfer event. Consider a device, shown in Fig. 28.9, consisting of an organized monolayer arrangement of the dyes A, B, and C

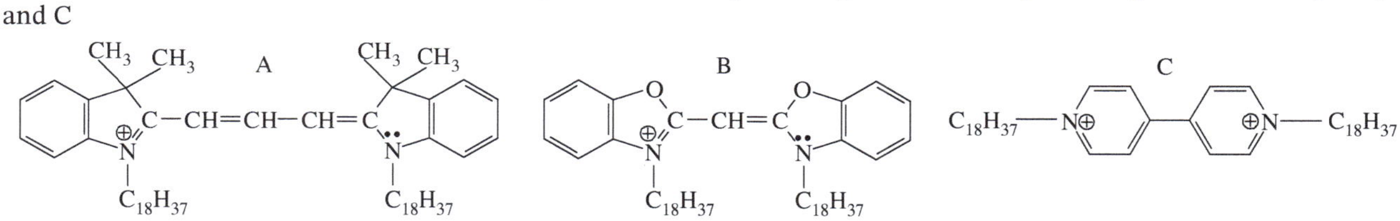

If the device is irradiated at 545 nm (absorption maximum of dye A), then fluorescence of A can be observed at 590 nm (Fig. 28.9a). If an additional irradiation at 366 nm is applied, quenching of the red fluorescence is observed (Fig. 28.9b). What is the reason for the quenching? The 366 nm light induces electron transfer from dye B to dye C (Fig. 28.9c), forming transient radicals the B^+ and C^-, which have different spectral characteristics. Whereas C absorbs in the UV, the absorbance of C^- is at 600 nm. Thus the radical C^- acts as an energy acceptor for molecule A, quenching its fluorescence at 590 nm

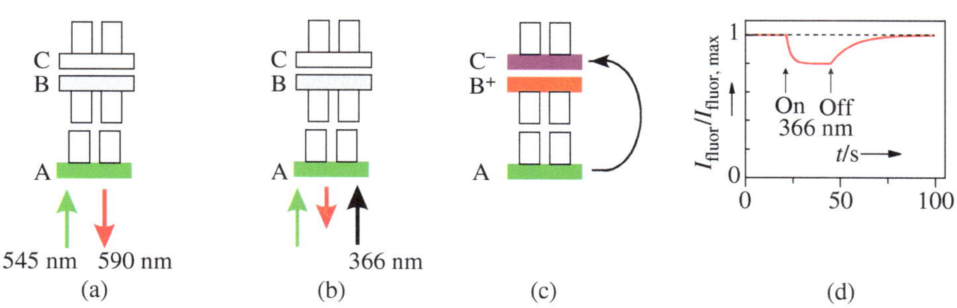

Figure 28.9 Photoinduced electron transfer. Monolayer arrangement consisting of dyes A, B, and C. (a) Irradiation of A with 545 nm results in fluorescence of A. (b) Upon additional irradiation of B at 366 nm, quenching of the red fluorescence is observed. (c) Mechanism of the fluorescence quenching: molecule C^- formed in process (b) absorbs light at 600 nm, thus acting as an energy acceptor for A. The black arrow indicates the energy transfer from A to C^-. (d) Fluorescence intensity I_{fluor} of dye A. I_{fluor} is switched off by irradiation with light of 366 nm [8].

(Fig. 28.9c). After switching off the 366-nm irradiation, the radicals C^- disappear by reaction with oxygen, and the fluorescence intensity increases to the original value (Fig. 28.9d).

This arrangement demonstrates the programmed interplay between three different molecules in a distinct geometry. It can be considered as a switch with optical read out. It is turned on and off by UV radiation absorbed by molecule B, which then triggers molecule C to change into a form that interacts with molecule A, quenching its fluorescence.

28.3.2 Coupled Chromophores and Excitons

The primary energy source of life is solar energy, which is collected by a dye molecule based antenna system consisting of donor molecules D and acceptor molecules A. The energy is funneled into the reaction center of plants and bacteria where it is converted into chemical energy. In this section we discuss the coupling of dye molecule chromophores and the propagation of electronic energy in them.

28.3.2.1 Functional Unit by Coupling Chromophores

We consider dye molecules 1 and 2 in the arrangement in Fig. 28.10a. The interaction of the two chromophores results in a change of the absorption spectrum (Fig. 28.10b,c). This change occurs for exactly the same reason as the intramolecular coupling effect discussed in Chapter 9. To understand this coupling, consider each dye molecule as an oscillator which is exposed to the oscillating electric field of the incident light and to the field generated by the other dye molecule (Fig. 28.11a). The absorption band of the dye molecule absorbing at short wavelengths is enhanced and shifted to shorter wavelengths (in-phase oscillation, Fig. 28.11b), whereas the band of the dye absorbing at long wavelengths is diminished and shifted to longer wavelengths (out-of-phase oscillation, Fig. 28.11c). The arrangement is a molecular functional unit, and the entity behaves quite differently from each part.

If molecules 1 and 2 are identical (formation of a dimer), particularly large changes of the absorption and of the fluorescence appear. The coupling strength is strongly dependent on the geometry of the molecular dimer. In the geometry of Fig. 28.12a, the absorption corresponding to the in-phase oscillation is shifted to shorter wavelengths (repulsion of the

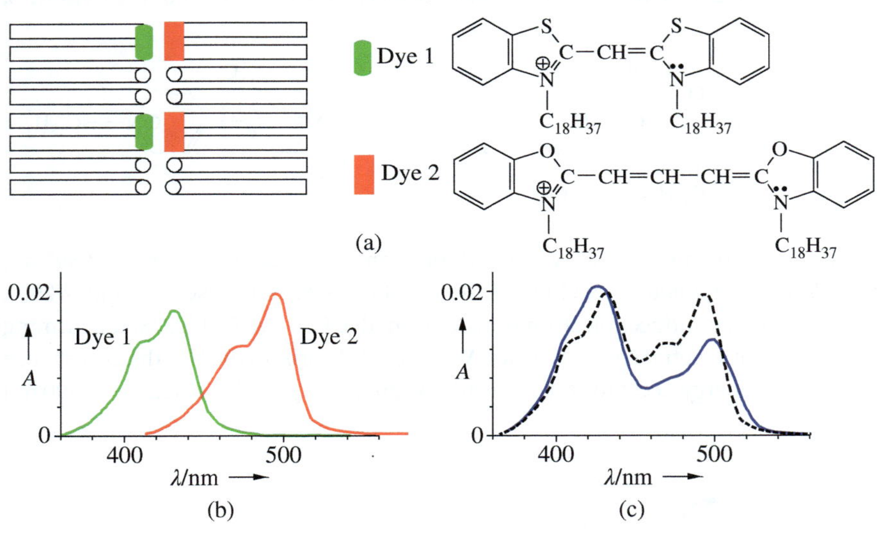

Figure 28.10 Change of absorption spectra of dyes 1 and 2 when both dyes are in a functional unit. (a) Functional unit, arrangement with distance $d = 0$. (b) Absorption spectra of isolated dyes. (c) Absorption spectra of dyes in functional unit (blue solid line); superposition of the spectra of the isolated dyes (dashed line), for comparison. (Adapted from [9].)

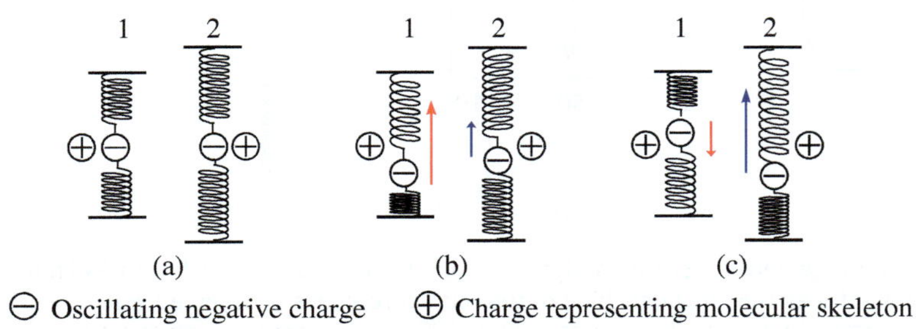

Figure 28.11 Shift of absorption spectrum in Fig. 28.10 explained by the oscillator model. (a) Two coupled oscillators at rest. (b) Oscillators vibrating in phase (strong absorption band). (c) Oscillators vibrating out of phase (weak absorption band).

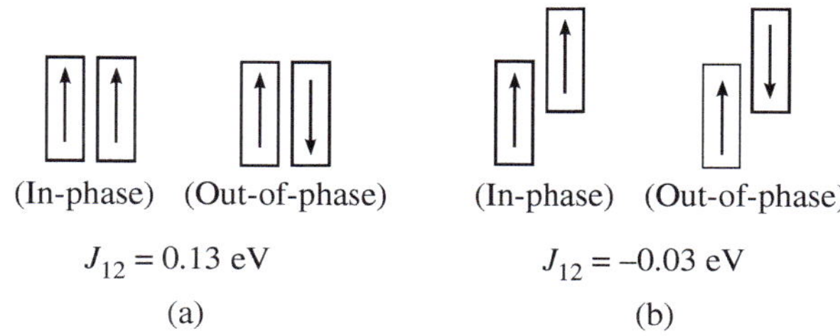

Figure 28.12 Coupling dyes in a dimer arrangement. Shift of absorption and of fluorescence bands in forming the dimer depends on geometry; interaction described by two-electron coupling integral J_{12}, see Foundation 10.3. Arrangement (a) shift of absorption and fluorescence bands to shorter wavelengths ($J_{12} > 0$). Arrangement (b) shift of absorption and fluorescence bands to longer wavelengths ($J_{12} < 0$). (Adapted from [10].)

dipoles) in contrast to the geometry in Fig. 28.12b, where the absorption corresponding to the in-phase oscillation is shifted to longer wavelengths (attraction of the dipoles). In both cases the transition moment corresponding to the out-of-phase oscillation is zero; no absorption and fluorescence bands occur (note: transitions dipoles should not be confused with permanent dipoles).

28.3.2.2 Dye Aggregate as Energy Harvesting Device

We consider an artificial energy channeling system. A 1:1 mixture of dye D

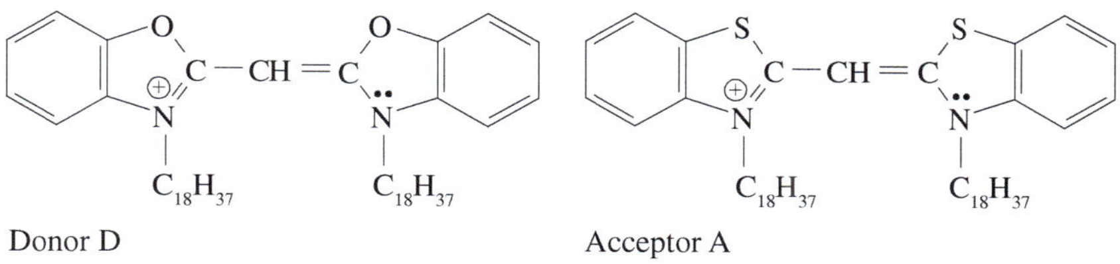

and octadecane is spread at a water surface and the layer is compressed to form a well-organized monolayer. The chromophores are tightly packed in a brickstone work arrangement, with the octadecane chains filling the holes left between the hydrocarbon substituents of the dye molecules (Fig. 28.13). A dramatic change in the absorption and fluorescence spectra occurs (Fig. 28.14) as the chromophores evolve from being independent at low surface pressure to being highly organized and packed together at high surface pressure. A very narrow and intense absorption band and an equally narrow, and nearly coincident, fluorescence band appear. The absorption band is shifted to long wavelengths relative to the absorption of the monomer in a surplus of fatty acid. These spectral changes arise from a strong coupling between the chromophores; see Sections 28.3.2.3 and 28.3.2.4 for an explanation.

Dye molecule A can be incorporated in the aggregate monolayer as an energy acceptor (Fig. 28.15). Each A molecule matches exactly into the space of a D molecule (Fig. 28.15a). A surprising phenomenon is observed even at a dilution of one molecule of A per 10,000 molecules of D ($Z = N_{donor}/N_{acceptor} = 10{,}000$). The blue fluorescence of the host is strongly

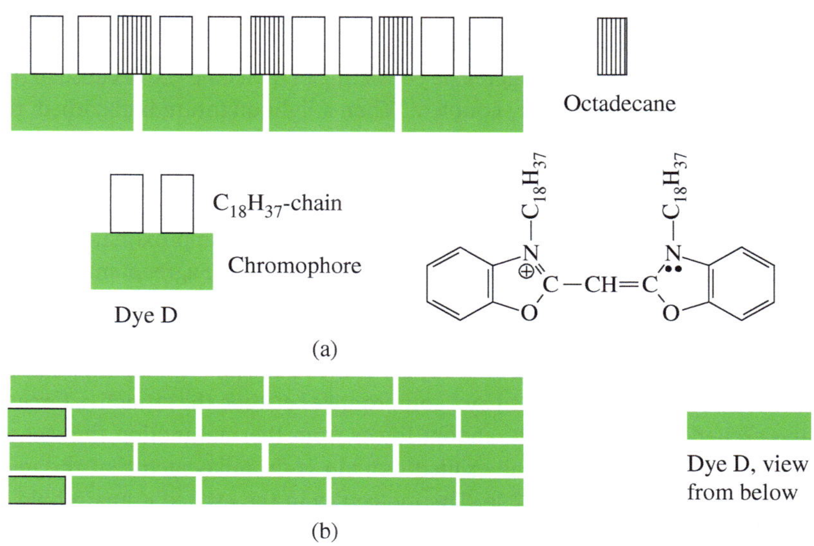

Figure 28.13 Dye D in a brickstone work arrangement. (a) Side view; the spaces between adjacent dye molecules are filled by octadecane molecules. The chromophores are perpendicular to the layer plane. (b) View from below.

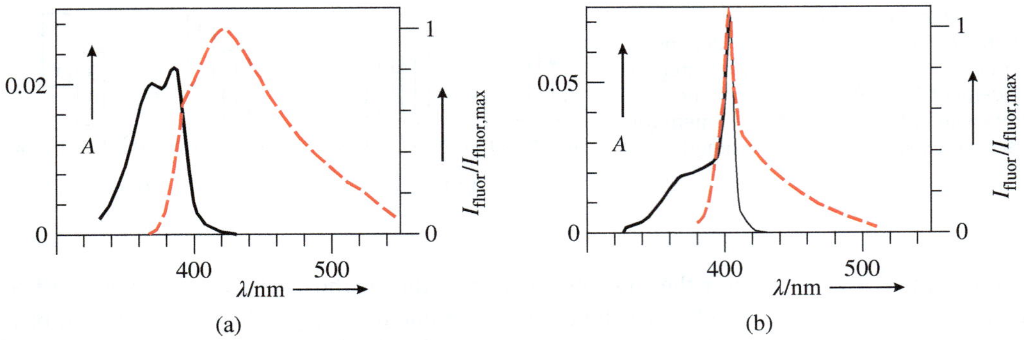

Figure 28.14 Dramatic change of absorption spectrum (solid lines) and fluorescence spectrum (dashed lines) of dye D in a monolayer when the monomers are not ordered (a), and ordered in a brickstone work arrangement (b). The spectra were measured in a mixed monolayer of the dye (with arachidic acid in case (a), and with octadecane in case (b) in a molar ratio 1:1 on a glass plate. (Adapted from [11].)

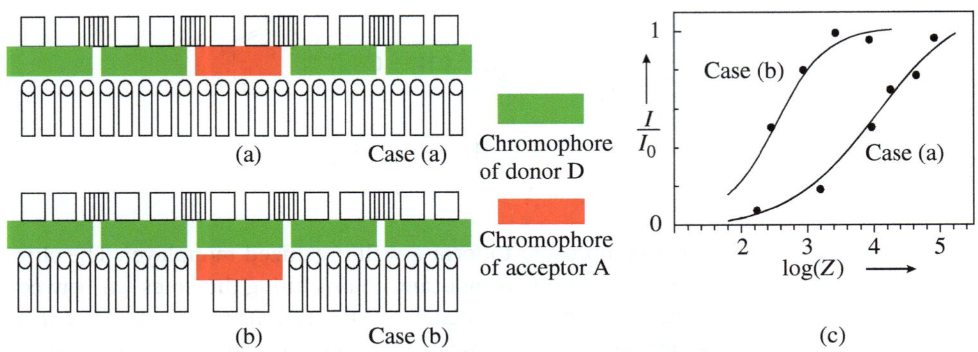

Figure 28.15 Energy transfer from exciton (donor) to acceptor. (a) and (b): acceptor molecules A incorporated in an aggregate monolayer of donor molecules D. Case (a): Acceptor A substituting one donor molecule ($Z = N_{donor}/N_{acceptor}$ = 10,000 for $I/I_0 = 0.5$). Case b: acceptor A in contact with donor molecule, but incorporated in the fatty acid layer ($Z = 300$ for $I/I_0 = 0.5$). (c) relative fluorescence intensity I/I_0 of donor versus log Z. (Adapted from [12].)

quenched (quenching by 50% for $Z = 10{,}000$ and by 85% for $Z = 1000$, see Fig. 28.15c) and the green fluorescence of the energy acceptor A appears; the energy absorbed by D moves over a large distance and is trapped by A. How can we understand this phenomenon?

28.3.2.3 Formation of Excited Domain (Exciton)

We first consider the pure aggregate consisting of molecules D. Because of the tight packing in the aggregate, the oscillators (dye molecules) are strongly coupled. When a light quantum is absorbed, the excitation is distributed over more than one molecule; i.e., it is an in-phase oscillation of the dye molecules in the domain. Nearest-neighbor molecules are coupled together (see Fig. 28.12b), and a shift to longer wavelengths occurs. Restricting the treatment to nearest-neighbor interactions (4 neighbors), the shift is $\Delta E = 8J_{12} = -0.16$ eV, where $J_{12} = -1.25 \times 10^{-19}$ J $= -0.02$ eV is the two-electron repulsion integral considered in Foundation 10.2. A more detailed approximation [13] (including all neighbors) gives $\Delta E = -0.24$ eV, which corresponds to a shift of $\Delta\lambda = 30$ nm, in reasonable agreement with the measured shift of 23 nm (from 380 to 403 nm, see Fig. 28.14).

What Determines the Size and Shape of the Excited Domain? We assume that the number N of molecules contributing to the excited domain is determined by a compromise between the attractive forces of the in-phase oscillating dipoles constituting the excited portion and the random forces from thermal motion that act to dephase and quench the cooperative dipole oscillation. The shape of the excited domain is droplike such that the region of oscillators at the border of the excited domain is as small as possible. The molecules in the excited domain hold the domain together, see Fig. 28.16 for $N = 9$. The binding energy is $-\Delta E = 0.24$ eV.

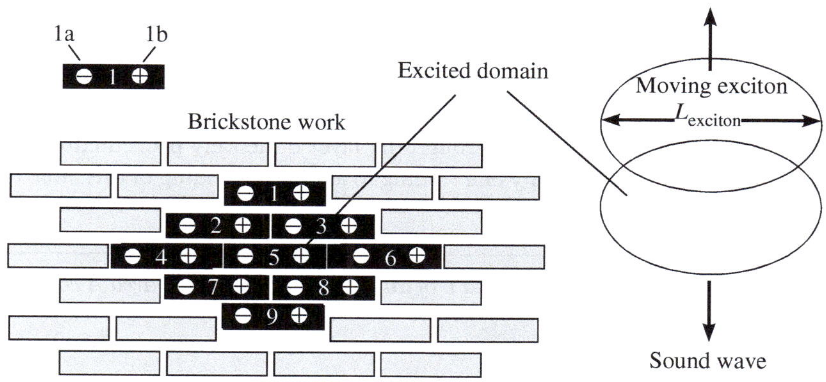

Figure 28.16 Formation of an exciton. Left: Excited domain in a brickstone work arrangement. Excited molecules replaced by coupled dipoles oscillating in phase. The charges change sign simultaneously, thus nearest-neighbor charges are always opposite; the attraction of charges 1a and 2b, 1b and 3a, 2b and 5a, etc. leads to a compression of the excited domain. Right: exciton (carried by sound wave) moving upward, and sound wave moving downward.

ΔE is the binding energy between the N coherent dipole oscillators, and the decrease in the energy of the excited state. Thus the binding energy per oscillator of the excited domain is $-\Delta E/N$. This energy, which keeps each oscillator at the border in phase with the excited domain, must be comparable to the thermal energy kT. If the binding energy becomes smaller than kT, the oscillators would be knocked out of phase. Therefore

$$-\frac{\Delta E}{N} = kT \tag{28.10}$$

or

$$N = -\frac{\Delta E}{kT} = \frac{0.24\ \text{eV}}{k} \cdot \frac{1}{T} = 3000\ \text{K} \cdot \frac{1}{T} \tag{28.11}$$

and for room temperature we obtain

$$N = \frac{3000\ \text{K}}{300\ \text{K}} = 10$$

This excited domain is called an *exciton*.

28.3.2.4 Lifetime of Fluorescence of Exciton

According to Foundation 10.1 the fluorescence lifetime of a monomer, in terms of the classical oscillator model, is

$$\tau_0 = \frac{3m\varepsilon_0 c_0^3}{2Q^2 \pi \nu_0^2 \hat{n}} \tag{28.12}$$

(m, Q, ν_0 = mass, charge, and frequency of the oscillator, ε_0 = permittivity of the vacuum, c_0 = speed of light in vacuum, $\hat{n}$ = index of refraction). Because the aggregate has N molecules oscillating in phase, the lifetime is

$$\tau_{\text{agg}} = \frac{3(Nm)\varepsilon_0 c_0^3}{2(NQ)^2 \pi \nu_0^2 \hat{n}} = \frac{1}{N}\tau_0 \tag{28.13}$$

where we replace m by Nm and Q by NQ.

Because the aggregate size depends on temperature, its fluorescence lifetime is temperature dependent. Using Equation (28.11) it follows that

$$\tau_{\text{agg}} = -\frac{kT}{\Delta E}\tau_0 = \frac{T}{3000\ \text{K}}\tau_0 \tag{28.14}$$

From Equation (28.12) we calculate $\tau_0 = 5 \times 10^{-9}$ s for $m = m_e$, $Q = e$, $\hat{n} = 1.5$, and $\nu_0 = 0.75 \times 10^{15}\ \text{s}^{-1}$. At room temperature we find

$$\tau_{\text{agg}} = \frac{300\ \text{K}}{3000\ \text{K}} \times 5 \times 10^{-9}\ \text{s} = 5 \times 10^{-10}\ \text{s}$$

As the temperature is lowered (temperature range from 90 to 300 K) the emission rate increases (called *superradiance*) because the number of oscillators in phase increases. τ_{agg} is proportional to temperature T. At much lower temperature, the lifetime becomes constant. In this case the measured lifetime is shorter than the time after absorption required to reach the fluorescent excited state (see Section 10.6, Jablonski diagram) and this step becomes rate-determining.

The predicted superradiance and the linear relationship of τ_{agg} with temperature T has been observed [14].

28.3.2.5 Speed of Exciton

How can we understand the dramatic quenching of the donor exciton by an acceptor, even at extreme distances? How is the energy package transferred to A and why is the exciton annihilated? The exciton must move with high speed through the monolayer to reach an acceptor molecule, and we assume that the exciton has the properties of a soliton (see Chapter 13) that migrates through the layer of densely packed chromophores. This assumption, supported in what follows, appears to be the only one leading to an understanding of this amazing effect.

Example 28.1 *Estimate the speed v of an exciton. Consider the area A covered by the exciton during its lifetime τ_{agg}*

$$A = L_{exciton} \cdot v \cdot \tau_{agg} \tag{28.15}$$

where $L_{exciton}$ is the characteristic width of the exciton (Fig. 28.15, $L_{exciton}$ = 5 nm at 300 K) and $v \cdot \tau_{agg}$ is the distance traveled by the exciton.

In order to reach an acceptor molecule with 50% probability, A must be equal to the average aggregate area in which just one acceptor molecule is present—namely to $N_D a$ where a is the area covered by a donor molecule and N_D is the number of donor molecules per acceptor molecule. Using $N_D a$ as the average area of the aggregate we find

$$v = \frac{A}{L_{exciton} \cdot \tau_{agg}} = \frac{N_D a}{L_{exciton} \cdot \tau_{agg}} \tag{28.16}$$

With N_D = 10,000, $a = 1500 \times 400\ \text{pm}^2$, $L_{exciton}$ = 5 nm (at 300 K) and $\tau_{agg} = 5 \times 10^{-10}$ s we obtain $v = 2.4\ \text{km s}^{-1}$. Such a high speed is required to understand the observation that only a tiny amount of acceptor molecules are required to quench the donor fluorescence.

Can we understand the fast motion of the exciton leading to these astonishing experimental findings in precisely constructed molecular assemblies? When the light quantum has been absorbed, the excited domain is compressed, because of the cohesive forces between the coherent oscillators. We have seen that the bond energy $-\Delta E$ holding the exciton together and compressing the region is 0.24 eV. This compression should generate a sound wave moving in one direction (upward in the diagram) and a corresponding sound wave moving in the other direction (downward in the diagram, conservation of momentum, see Fig. 28.16b). One wave will carry the exciton (symmetry breaking). It will hold the excited domain together. Thus this wave would be non-diffusive, the domain would constitute a soliton that moves without energy loss through the monolayer until it is annihilated by emitting a fluorescent light quantum or is trapped by exciting an acceptor molecule. For more details see Foundation 28.3.

28.3.2.6 Efficiency of Energy Transfer

In the arrangement of case (a) in Fig. 28.15a one molecule of A per N_D = 10,000 donor molecules D is sufficient for a 50% quenching of the fluorescence of the donor at room temperature. This efficiency reflects the high probability of energy transfer in this arrangement (acceptor A incorporated in the layer of D molecules).

To better appreciate the situation, consider the arrangement of case (b) in Fig. 28.15b where the D aggregate monolayer is superimposed on a monolayer of fatty acid containing traces of acceptor A, so that each A molecule is still in direct contact with D molecules at essentially equal distance from its nearest neighbors. The two cases have very different energy transfer efficiencies, because in case (a) the acceptor is exposed to a strong field of the adjacent oscillating dipoles, while in case (b) the oscillator fields partially compensate each other. The probability of energy transfer is reduced in case (b), and the effect of A as an exciton trap is considerably smaller. The fluorescence of D is quenched for one molecule of A per N_D = 300 donor molecules (Fig. 28.15c, Z = 300; see Foundation 28.3).

28.3.3 Solar Energy Harvesting in Biosystems

28.3.3.1 Purple Bacteria: Three Groups of Collector Molecules

Energy funneling in nature is less efficient than in the device of Fig. 28.15a discussed above: only about 300 dye molecules are present collecting solar energy and funneling it to a photosynthetic reaction center (instead of N_D = 10,000). However, photosynthetic antennae access a broad spectral range of sunlight, whereas the artificial system (Fig. 28.15a) absorbs light in the very narrow spectral range of the aggregate band.

Solar energy harvesting in biosystems is achieved by different types of collector molecules and different arrangements comprising a hierarchical structure. In the antenna system in purple bacteria a first group of 18 dye molecules (bacteriochlorophyll) is arranged in a ring around the reaction center (Fig. 28.17a). The chromophores are shifted with respect to each other in an arrangement that causes a strong bathochromic shift (Section 28.3.2, Fig. 28.12b). A second group of bacteriochlorophyll molecules are separated from each other in a distinct way by a protein moiety. Accordingly, this group of more isolated molecules absorb at shorter wavelengths than the first group. Furthermore, a third group, comprising carotene molecules, are incorporated in a well-organized way. They absorb at even shorter wavelengths, allowing energy transfer to the second group of bacteriochlorophyll.

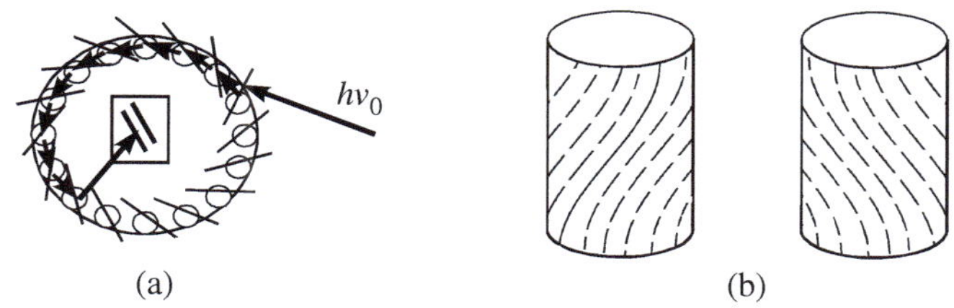

Figure 28.17 Energy funneling in nature. (a) Antenna system in purple bacterium guiding the light to the reaction center (special pair) [15]. (b) Antenna system of green bacterium. Tube of chlorophyll c molecules in a brickstone work arrangement (left-handed and right-handed species supposed to be in equal amounts).

The edge-to-edge distances between the three groups of antenna molecules are small compared to the critical distance r_0 for energy transfer (see Equations (28.8) and (28.9)). This allows efficient energy transfer from the third to the second group, and from the second to the first group absorbing at 875 nm and from there to the special pair in the reaction center (see Section 28.4.5). The special pair of bacteriochlorophyll molecules absorbs at 965 nm, as required for optimal energy transfer from the antenna molecules to the reaction center. The molecules in this pair, again, are arranged according to Fig. 28.12b and are tightly fixed to cause this large shift to 965 nm.

Both in the first group of antenna molecules and in the special pair the long wavelength absorption is reached by the very particular arrangement of bacteriochlorophyll according to Fig. 28.12b. It is remarkable that this arrangement, which allows the specific coupling of chromophores to exploit the broad spectral range of solar energy to generate chemical energy, arises through self-assembly.

28.3.3.2 Green Bacteria: Antenna Molecules in Self-Organized Brickstone Work Arrangement

The sophisticated arrangement of the chromophores in the purple bacteria antenna is achieved by fixing them in a protein matrix. In the antenna complex of green bacteria, chlorophyll c molecules are in a brickstone work arrangement forming tubes (chlorosomes, Fig. 28.17b). The tubes do not contain proteins. They result from a self-assembly process where the chlorophyll c molecules are linked by C= O···H–O···Mg bridges. Circular and linear dichroism spectra imply that left-handed and right-handed helical tubes are present in equal amounts. These tubes absorb at 750 nm, and energy is transferred via bacteriochlorophyll a (absorbing at 800 nm) to the reaction center (absorbing at 866 nm). Again we see that Nature has selected the motif of arranging the chromophores in a brickstone work, allowing coupling according to Fig. 28.12b or 28.13, to achieve the required bathochromic shifts [16].

The past 10 years have seen a resurgence of interest in exploring exciton transport in nanoscience and engineering. In 2014 researchers mimicked the natural system of green bacteria by assembling cyanine dyes into supramolecular, cylindrical nanotubes and studying their exciton properties. In addition to confirming the connection between the dye chromophore structure and the exciton transport, they showed that the nanotubes can be bundled into larger structures with little impact on the energy transport [17]. In other work, researchers are using DNA-based Watson–Crick base pairing to integrate and organize different dye-aggregates into "excitonic circuits" that direct electronic excitation energy through supramolecular constructs by rational design [18].

28.3.4 Manipulating Luminescence Lifetime by Programming Echo Radiation Field

We have seen that the lifetime of luminescence can be strongly decreased by tightly packing the molecules, so that they couple to each other, and the absorption and emission processes become cooperative. Another fascinating phenomenon is coupling a luminescent molecule to its own echo radiation field, which depends on details of the environment producing the echo.

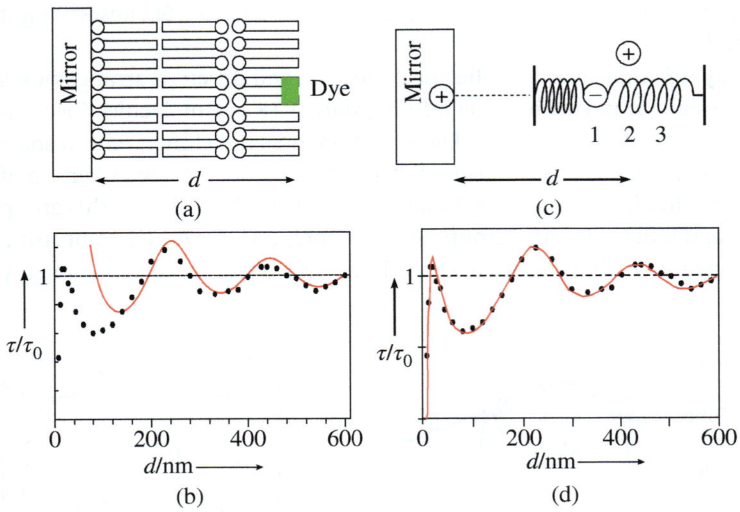

Figure 28.18 Radiation echo field. (a) Luminescent dye molecule (europium complex, $\lambda_0 = 612$ nm) at distance d in front of a mirror. (b) τ/τ_0 (τ = lifetime of the europium complex) versus distance d [3]. Filled circles: Experiment (From [19]), red solid line: Theory according to Foundation 28.4 on the book's website. (c) Illustration of the phase shift between the electron (assumed to oscillate between points 1, 2, and 3) and the induced positive charge on the mirror; the fixed positive charge in point 2 represents the molecular skeleton (as in Fig. 28.11). (d) Same experiment as in (b), but energy quenching by a mirror included in the theory. (Adapted from [20].)

Consider a luminescent molecule that is placed in front of a silver layer serving as a mirror (Fig. 28.18a). The monolayer assembly technique allows one to position the luminescent molecules at well defined distance d from the mirror. The luminescent molecules can be excited by a light flash and the number of quanta of luminescent light emitted during small time intervals after the flash can be counted. Thus the average lifetime τ of the excited state is obtained. For the europium complex

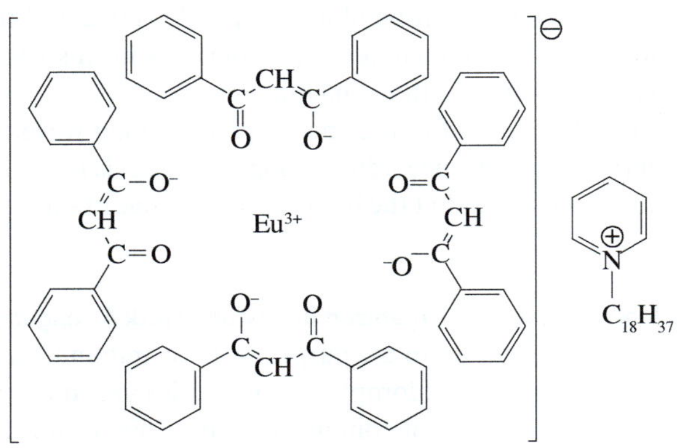

we consider the ratio τ/τ_0, where τ_0 is the average lifetime of luminescence of the sample without the mirror. τ/τ_0 is displayed in Fig. 28.18b as a function of distance d between the complex and the mirror; interestingly, τ/τ_0 depends strongly on the distance d.

How can this be the case? How can the molecule's excited state lifetime depend on objects far away? These questions can be answered by considering wave–particle duality, which is illustrated dramatically by the echo effect. According to Chapter 10 the excited molecule behaves in a way like a classical oscillator emitting light waves that are reflected at the mirror and then interact with the oscillator. This echo field interacts with the excited molecule (oscillator). Depending on the phase of the echo field relative to the phase of the oscillator an acceleration or retardation of the oscillator occurs (Foundation 28.4 on the book's website). Thus, the lifetime becomes longer or shorter depending on the distance d between oscillator and mirror.

Essentially, the light echo arrives at the oscillator with a delay time of

$$\Delta t = \frac{2d}{c} \tag{28.17}$$

(c = velocity of light in the medium), and we compare Δt with T, the period of the oscillation. First we consider the case $\Delta t = \frac{1}{4}T$ (that is for $d = \frac{1}{8}cT$), for which the oscillator is at point 1 (negative charge in Fig. 28.18c) and polarizes the metal maximally. The field of this polarized charge arrives at the oscillator with a delay, when the oscillator is just passing point 2 (Fig. 28.18c), i.e., when it has its maximum speed. In this case the oscillator is decelerated and the lifetime τ is shortened. In contrast, for $\Delta t = \frac{3}{4}T$ (that is for $d = \frac{3}{8}cT$) the oscillator is accelerated and the lifetime τ becomes longer. For $\Delta t = T$

and $\Delta t = \frac{1}{2}T$ the maximum field of the echo arrives when the oscillator speed is zero; in this case the lifetime τ remains unchanged.

The distance d corresponding to $\Delta t = \frac{1}{4}T$ is

$$d = \frac{\Delta t \cdot c}{2} = \frac{T \cdot c}{8} = \frac{\lambda}{8} = \frac{\lambda_0}{8\hat{n}} \tag{28.18}$$

where λ is the wavelength in the medium and λ_0 is the wavelength in the vacuum. With $\lambda_0 = 612$ nm and $\hat{n} = 1.5$ we obtain

$d = 51$ nm	for	$\Delta t = \frac{1}{4}T$	(τ shortened)
$d = 102$ nm	for	$\Delta t = \frac{1}{2}T$	(τ unchanged)
$d = 153$ nm	for	$\Delta t = \frac{3}{4}T$	(τ longer)
$d = 204$ nm	for	$\Delta t = T$	(τ unchanged)

This simple picture gives the maxima and minima in Fig. 28.18b at correct distances, but shifted by 70 nm. This discrepancy disappears in the refined approach discussed in Foundation 28.4 on the book's website.

28.3.4.1 Energy Transfer Quenching by a Mirror

A difference between theory and experiment still remains because silver is not an ideal (i.e., non-absorbing) reflector. This difference is of particular significance at small distances d. In addition to reflection, silver absorbs a fraction of the light energy and acts as an energy acceptor, like the acceptor dye in Fig. 28.4. This absorption acts to quench the luminescence (Fig. 28.18b, small distances: the experimental points are below the theoretical curve). A more rigorous model [21], which accounts for the energy transfer quenching by the silver, fits the data well over the entire range of distances (Fig. 28.18d).

The fluorescence lifetime has also been measured for single dye molecules near a metal surface (a carbocyanine dye in a polymer film) and also near the film air interface acting as a mirror. The above results were confirmed. The experiments illustrate very directly, at the single molecule level, how the emission can be manipulated by distinct changes of the remote neighborhood of the molecule [22].

28.4 Manipulating Electron Motion

28.4.1 Photoinduced Electron Transfer in Designed Monolayer Assemblies

28.4.1.1 Tunneling from Donor D to Acceptor A

In Fig. 28.19a an electron acceptor molecule A is positioned in the vicinity of the electron donor molecule D.

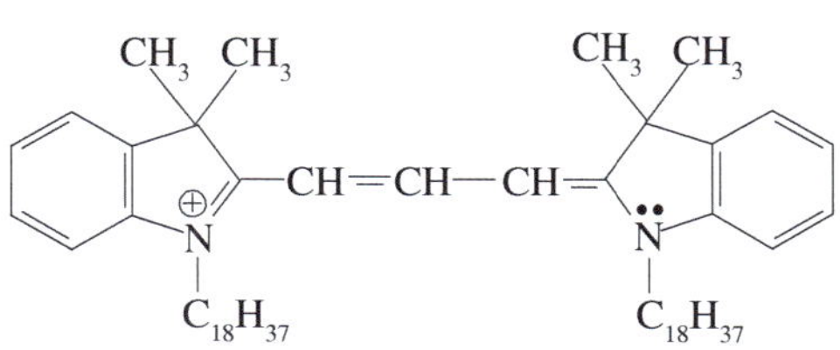

Electron donor D

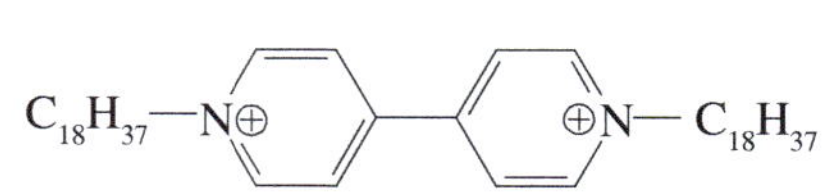

Electron acceptor A

D and A are separated by a fatty acid spacer layer with a thickness d. This thickness can be varied by changing the hydrocarbon chain length of the spacer molecule.

The donor–acceptor pair is chosen so that excitation by light at 500 nm (green light) creates an excited donor molecule D* which can undergo electron transfer to the acceptor

$$D^* + A \xrightarrow{k_{r,\text{transfer}}} D^+ + A^-$$

The electron transfer competes with the relaxation of D* and reduces the emission yield of fluorescence. The electron transfer proceeds at the expense of fluorescence, and the amount of fluorescence quenching can be used to quantify the rate

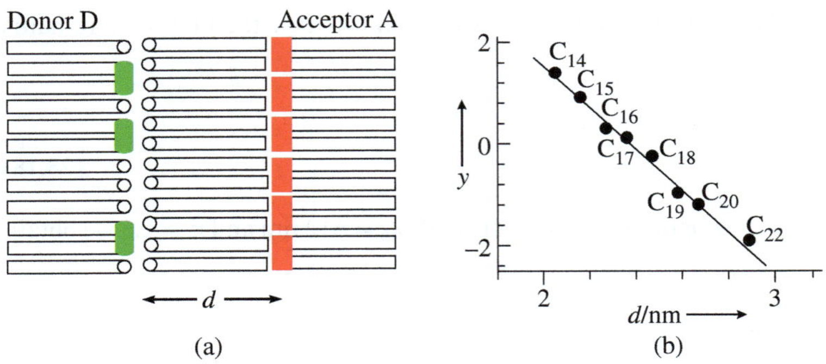

Figure 28.19 Electron transfer from excited donor D to electron acceptor A. (a) Arrangement of layer with D and layer with A separated by fatty acid molecule (14 to 22 carbon atoms). (b) $y = \ln(I_\infty/I - 1)$ for the donor fluorescence versus length d of the fatty acid chain. (Adapted from [23].)

of electron transfer. The fluorescence intensity, I, of D*, depends on $k_{r,\text{fluor}}$ (the rate constant of the decay of the fluorescence when electron transfer does not take place) and $k_{r,\text{transfer}}$, such that

$$I = \frac{k_{r,\text{fluor}}}{k_{r,\text{fluor}} + k_{r,\text{transfer}}} I_\infty \tag{28.19}$$

I_∞ is the fluorescence intensity for $d \to \infty$; i.e., no electron transfer. Equation (28.19) can be rearranged to give

$$\frac{I_\infty}{I} = 1 + \frac{1}{k_{r,\text{fluor}}} \cdot k_{r,\text{transfer}} = 1 + \tau \cdot k_{r,\text{transfer}} \tag{28.20}$$

and

$$\frac{I_\infty}{I} - 1 = \tau \cdot k_{r,\text{transfer}} \tag{28.21}$$

where τ is the fluorescence lifetime.

Electron transfer is often a tunneling process, where

$$k_{r,\text{transfer}} = a\,\mathrm{e}^{-b \cdot d} \tag{28.22}$$

Therefore we expect that the quantity

$$y = \ln\left(\frac{I_\infty}{I} - 1\right) = \ln(\tau\, a) - b \cdot d \tag{28.23}$$

depends linearly on the distance d between D and A. Figure 28.19b plots the data in this fashion and reveals an exponential dependence on distance, with a characteristic tunneling constant of $b = 4.10\ \text{nm}^{-1}$. For $d = 0$ we obtain $y = 10$; thus, we calculate the pre-exponential factor a in Equation (28.22) as $a = \mathrm{e}^{10}/\tau$. In Section 10.2.1 (see Example 10.1) we found for a thiacarbocyanine molecule with $k = 0$ that $\tau = 1.5 \times 10^{-8}$ s. Assuming that τ for the donor molecule in Fig. 28.19 is on the same order, the result for the preexponential factor is $a = \mathrm{e}^{10}/1.5 \times 10^{-8}\ \text{s} = 1.5 \times 10^{12}\ \text{s}^{-1}$. With these data we obtain

$$k_{r,\text{transfer}} = 1.5 \times 10^{12}\ \text{s}^{-1} \cdot \mathrm{e}^{-4.10 \cdot d/\text{nm}}$$

for the rate constant of electron transfer, Equation (28.22). Thus the largest value in the measured range is $k_{r,\text{transfer}} = 3 \times 10^8\ \text{s}^{-1}$ for $d = 2.1$ nm, and the smallest value is $k_{r,\text{transfer}} = 1 \times 10^7\ \text{s}^{-1}$ for $d = 2.9$ nm.

Theoretically, the tunneling probability is, according to Section 3.2.2.4, exponentially decreasing with the thickness d. Therefore, for $k_{r,\text{transfer}}$ we expect the dependence

$$k_{r,\text{transfer}} \sim \exp(-b \cdot d) \tag{28.24}$$

where

$$b = \frac{2\pi}{h}\sqrt{2m_e E_{\text{barrier}}} \tag{28.25}$$

(h = Planck constant, m_e = mass of the electron). From the measured b we calculate the barrier height E_{barrier} as.

$$\begin{aligned} E_{\text{barrier}} &= b^2 \frac{h^2}{4\pi^2 \cdot 2m_e} = (4.1 \times 10^9\ \text{m})^2 \cdot 6.104 \times 10^{-39}\ \text{J}^2\ \text{s}^2\ \text{kg}^{-1} \\ &= 1.03 \times 10^{-19}\ \text{J} = 0.64\ \text{eV} \end{aligned}$$

E_{barrier} corresponds to the difference between the ionization energy from the fluorescent excited state and the energy of the electron below the vacuum when tunneling through the dielectric medium between donor and acceptor monolayer [23].

28.4.2 Tunneling Current Through Monolayers

28.4.2.1 Tunneling Pathway in MIM Structure

The importance of covalent linkages on the tunneling pathway of an electron can also be probed by studying the thickness dependence of the tunneling current in metal-insulator-metal (MIM) contacts. Such contacts can be made by depositing a monolayer of a fatty acid (Langmuir–Blodgett technique) on a metal layer (aluminum) and covering it by a drop of mercury [21] or vapor depositing a metal layer (aluminum) on top [24]. Another possibility to realize a MIM contact is to use a SAM layer on Al covered by an Al layer [25]. If a voltage V is applied between the metal layers (area A), a current i can be measured[2] (Fig. 28.20). The current i depends on the number of carbon atoms in the fatty acid chain (determining the thickness d of the fatty acid layer), and it is quantified in terms of the conductivity κ. The conductivity is defined as the ratio of the current density, i/A, and the applied electric field, V/d; i.e.,

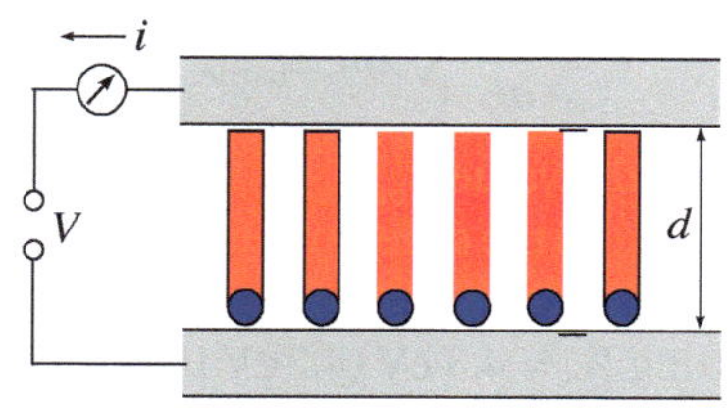

Figure 28.20 Tunneling current i through monolayer of fatty acid molecules between two parallel metal electrodes in distance d.

$$\kappa = \frac{i}{V} \cdot \frac{d}{A} = \frac{1}{R} \cdot \frac{d}{A} \tag{28.26}$$

where R is the resistance. For the insulating fatty acid films, the conductivity is observed to decrease exponentially with the thickness d, indicating electron tunneling. Thus, according to Section 28.4.1.1, we write

$$\kappa = \alpha \cdot e^{-bd} \tag{28.27}$$

and analyze the data through the relation

$$\ln\left(\kappa/\Omega^{-1}\ \text{cm}^{-1}\right) = \ln\left(\alpha/\Omega^{-1}\ \text{cm}^{-1}\right) - bd$$

Figure 28.21 plots the data in this fashion and reveals an exponential dependence on distance, with a characteristic tunneling constant of $b = 13\ \text{nm}^{-1}$. For $d = 0$ we obtain $\ln\left(\kappa/\Omega^{-1}\ \text{cm}^{-1}\right) = 2.4$; thus, we calculate the pre-exponential factor α in Equation (28.27) as $\alpha = 1.1 \times 10^3\ \Omega^{-1}\ \text{m}^{-1}$.

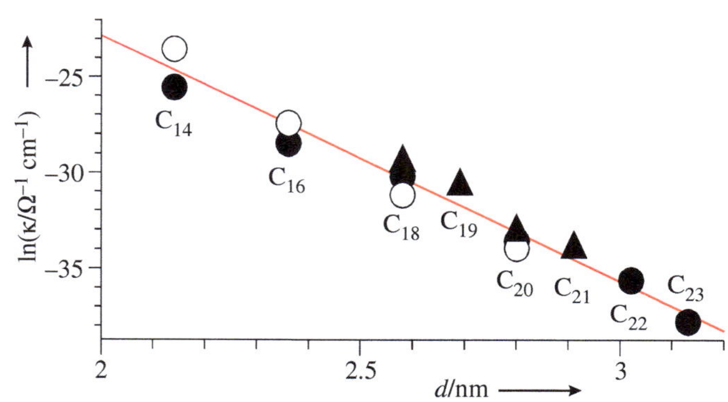

Figure 28.21 The experimental conductivity in a MIM system consisting of fatty acids (14–23 carbon atoms) enclosed between electrodes is shown; $\ln(\kappa/\Omega^{-1}\ \text{cm}^{-1})$ is plotted as a function of the distance d between the two plates. $\kappa = R^{-1}(A/d)^{-1}$ is the conductivity, where R is the resistance and A is the area of the electrodes. Open circles (from [24]) and filled triangles (from [21]): Langmuir–Blodgett technique; filled circles (from [25]): SAM technique.

Theoretically, the constant b in Equation (28.27) is connected to the tunneling barrier, E_{barrier}, similarly as in Equations (28.24) and (28.25). As derived in Foundation 28.5, we have

$$b = \frac{4\pi}{h}\sqrt{2m_e E_{\text{barrier}}}$$

and for E_{barrier} we obtain, using the experimental value $b = 13\ \text{nm}^{-1}$ in Equation (28.27),

$$E_{\text{barrier}} = b^2 \frac{h^2}{16\pi^2 \cdot 2m_e} = (13 \times 10^9\ \text{m})^2 \cdot 1.53 \times 10^{-39}\ \text{J}^2\ \text{s}^2\ \text{kg}^{-1}$$
$$= 2.57 \times 10^{-19}\ \text{J} = 1.6\ \text{eV}$$

2 Here we are using the symbol i for current to avoid confusion with the intensity I.

The constant α may be found for the case of plates of the same metal, using the free electron model approximation and assuming that a voltage V is applied [26]:

$$\alpha = \frac{e^2}{h^2} \cdot \sqrt{2m_e E_{\text{barrier}}} = 3.16 \times 10^4\ \Omega^{-1}\ \text{m}^{-1} \cdot \sqrt{\frac{E_{\text{barrier}}}{\text{eV}}} \tag{28.28}$$

Using the value $E_{\text{barrier}} = 1.6$ eV, we find that $\alpha = 3.90 \times 10^4\ \Omega^{-1}\ \text{m}^{-1}$.

E_{barrier} is the difference between the work function of the metal, W, and the energy of an electron below vacuum when tunneling through the barrier represented by the dielectric medium between the plates (2.67 eV, according to Foundation 28.5). Setting $W = 4.3$ eV (4.2 eV for aluminum, 4.5 eV for Hg) the barrier height is $E_{\text{barrier}} = (4.3 - 2.67)$ eV $= 1.63$ eV.

Another interesting way to construct a tunneling junction is binding an alkanethiol to a drop of mercury, which forms a densely packed monolayer. By bringing in contact two mercury drops covered with monolayers of alkanethiols with n_1 and n_2 CH_2 groups in the chain (e.g., $n_1 = n_2 = 10$ or $n_1 = 10, n_2 = 12$) and applying a voltage, a current is measured that decays exponentially with increasing $n_1 + n_2$, indicating tunneling [27].

28.4.2.2 Tunneling from Electrode Through Chains of Different Compositions and Lengths to Ferrocene

While MIM-based tunnel junctions provide a direct relation to electronics applications, electron tunneling through monolayer films has also been studied with electrochemical tunnel junctions. In these devices, a redox active molecule is tethered, or immobilized, on an electrode surface and the electron transfer reaction between the redox molecule and the electrode is studied. Here we discuss studies that examine how the electron tunneling probability depends on the monolayer film thickness and on its composition.

Change of Composition We compare the electron transfer rate constant for ferrocene that is tethered through chains of different compositions, which are self-assembled on gold. Alkane chains with 12 methylene groups are connected to a gold surface by a sulfur bond and to ferrocenium molecules by an ester bond:

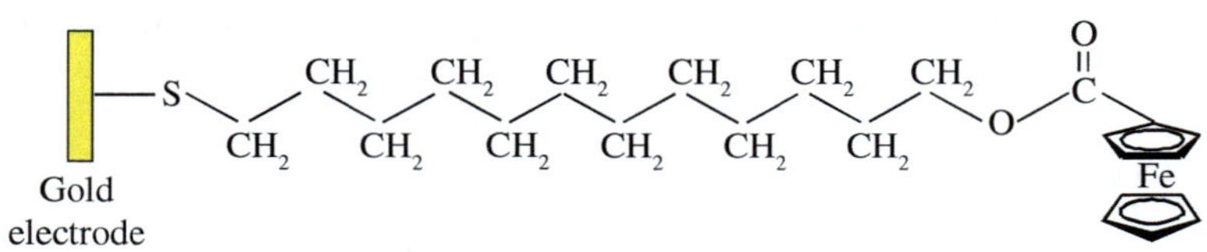

Figure 28.22a shows an arrangement with pure methylene chains, and Fig. 28.22b shows a situation in which the ferrocene tether has an ether link. Because the ferrocene tethered group is present at a concentration of only 5%, the geometry and thickness of the monolayer film is controlled by the diluent alkanethiol molecules. The rate constant observed for the ether linked ferrocene is four to five times smaller than the electron transfer through a pure methylene chain. Control studies in which the pure alkane diluent is replaced by an ether diluent (see Table 28.1) show that the change in rate constant is dominated by a change in the tunneling probability through the chains that covalently tether the ferrocene. Quantum-mechanical calculations reveal that the tunneling probability through the ether chains is smaller than through

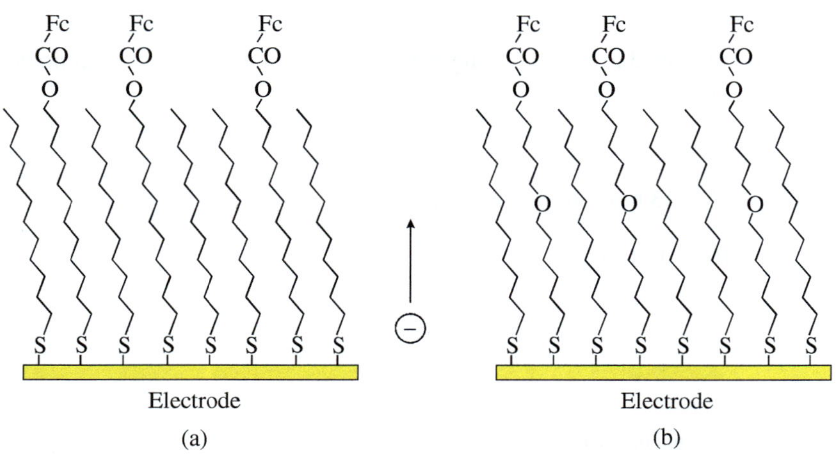

Figure 28.22 Electron tunneling from an electrode surface through an alkane chain (a) and an ether chain (b) onto ferrocenium molecules Fc.

Table 28.1 Comparison of Tunneling Rate Constant k_r for the Replacement of the Alkane Spacers by Ether Spacers.

Ferrocene Tether	Diluent	k_r/s^{-1}
$-S-(CH_2)_5-CH_2-(CH_2)_6-O-CO-Fc$	$-S-(CH_2)_6-CH_2-(CH_2)_4CH_3$	55 ± 9
$-S-(CH_2)_5-CH_2-(CH_2)_6-O-CO-Fc$	$-S-(CH_2)_6-O-(CH_2)_4CH_3$	35 ± 12
$-S-(CH_2)_5-O-(CH_2)_6-O-CO-Fc$	$-S-(CH_2)_6-CH_2-(CH_2)_4CH_3$	12 ± 3
$-S-(CH_2)_5-O-(CH_2)_6-O-CO-Fc$	$-S-(CH_2)_6-O-(CH_2)_4CH_3$	8 ± 2

the alkane chains (a higher tunneling barrier for the alkylether arising from the ether oxygen). The electron wave tunnels primarily along the covalent linkages that connect the ferrocene to the metal electrode [28].

Change of Chain Length Again we consider the reaction of a ferrocene molecule that is chemically bound to the surface of a gold electrode through an alkane chain, see Fig. 28.22a. If we quickly change the applied potential from zero to a value positive of the redox potential and measure the time-dependent current for the reaction, then the decay time τ for this transient provides the reaction rate constant $k_r = 1/\tau$ at this potential. Chidsey and coworkers [29] used this procedure to measure the electron transfer rate constant for ferrocene as a function of the alkane chain length above a gold electrode surface. These rate constants display an exponential dependence on the number of methylene units in the chain. We expect that the rate constant follows the tunneling equation

$$k_r = a\,e^{-b \cdot d} \tag{28.29}$$

or

$$\ln(k_r/s^{-1}) = \ln(a/s^{-1}) - b \cdot d \tag{28.30}$$

In Fig. 28.23 we plot $\ln(k_r/s^{-1})$ as a function of d, and the experimental points (full circles) are fit by a straight line with a slope of $b = 10.5\ nm^{-1}$, in essential agreement with the value obtained for the LB and SAM structures discussed in Section 28.4.2.1.

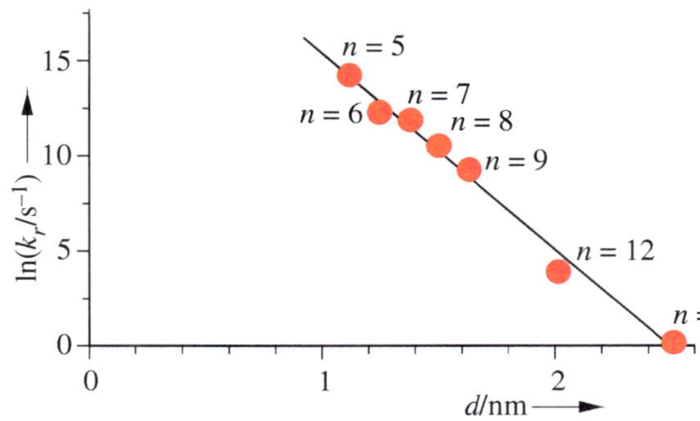

Figure 28.23 Electron transfer rate constant k_r from gold to ferrocene through alkane chains with n methylene groups (see Fig. 28.22, distance $d = (n + 4) \times 125$ pm). $\ln(k_r/s^{-1})$ is plotted as a function of d. Filled circles: experiment. Straight line: regression line with slope $-10.5\ nm^{-1}$ and intercept of 26.0.

28.4.3 Conduction Through Single Molecules

Molecular-based electronics has been a dream since the 1970's. Although it is a great challenge to measure and manipulate currents through single molecules, a number of methods have been developed to probe the conductance of single molecules [30].

28.4.3.1 Tunneling Through Octanedithiol Molecules

The STM-breakjunction method, developed by Xu and Tao [31], has proven to be a powerful method for measuring single molecule conductances at room temperature. In this method, a Au STM tip is crashed into a Au substrate, and then it is retracted from the surface. During the retraction, a thin Au wire forms between the tip and the substrates; this wire narrows as the retraction continues and eventually forms a chain of Au atoms, which is one atom thick before it breaks. By monitoring the current as a function of retraction at a fixed voltage during this process, one can determine the conductance G (inverse of resistivity). One observes a series of steps in the current; however, it is common to make a histogram plot of the conductance values. Because a cycle of crashing and retracting the tip from a surface requires only a few milliseconds, it is possible to perform many thousands of experiments and collect a large amount of data. Figure 28.24 shows a plot of the number of conductance measurements (counts) versus G/G_0, where $G_0 = 2e^2/h$ is the quantum of conductance; see Equation (13.28) in Example 13.4 (Section 13.3.3).

If molecules are present on the surface and they can bind to the Au tip, then the experimental scenario will change. After the Au wire to the substrate breaks, adsorbed molecule(s) can bridge the gap between the STM tip and the substrate, see Fig. 28.25a. In this case, the measured current does not decay directly to zero but current can be observed to flow between

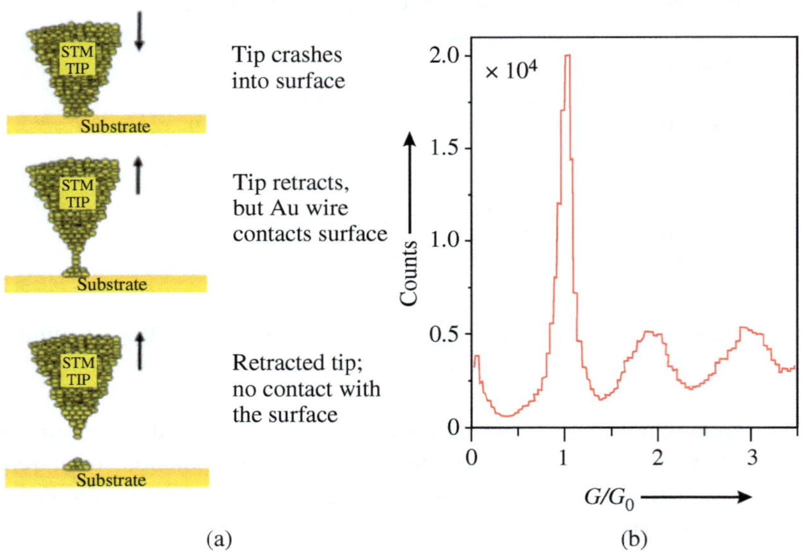

Figure 28.24 STM-breakjunction method (a) and number of conductance measurements (counts) versus G/G_0, where G_0 is the quantum of conductance (b).

the tip and the substrate by way of the molecule(s). One observes a distribution of conductance values for the molecules, which arises from differences in their binding geometry and binding strengths to the substrate and the metal tip, as well as variations in the molecule geometries. Typically, the histograms are used to obtain the positions G/G_0 of the conductance maxima, in analogy to Fig. 23.25b for a series of alkanedithiol molecules, for example

$HS\text{–}CH_2\text{–}CH_2\text{–}SH$: alkanedithiol

Figure 28.25b shows a logarithmic plot for the positions of the conductance maxima in the histograms for alkanedithiols, in units of the quantum of conductance $G_0 = 2e^2/h = 7.75 \times 10^{-5}$ S, versus the number of methylene groups in the alkyl chain (n = 5–11) from four different research groups. Although the measurements for each of the individual research groups vary in a systematic manner, a large variation in the median conductance between the different studies is evident (ranging from $1.5 \times 10^{-5} \cdot G_0$ to $25 \times 10^{-5} \cdot G_0$). This variability can arise from differences in the binding geometry and binding strength of the linker molecule to the Au surfaces, as well as instrumental differences, and underscore the need for more precise and controlled measurement platforms (see next section).

If we assume that the electron transport proceeds through a tunneling mechanism, then we can analyze these data in a manner like that we used for Fig. 28.21. Thus, the molecular conductance decays exponentially with the distance,

$$G = \alpha \cdot e^{-bd}$$

The line in Fig. 28.25, right panel, shows a linear fit to the conductance data from n = 6–11 and has a slope of −0.86 per methylene unit, which gives $b = 6.9$ nm^{-1}. This value is in reasonable agreement with the data shown in Fig. 28.23 for

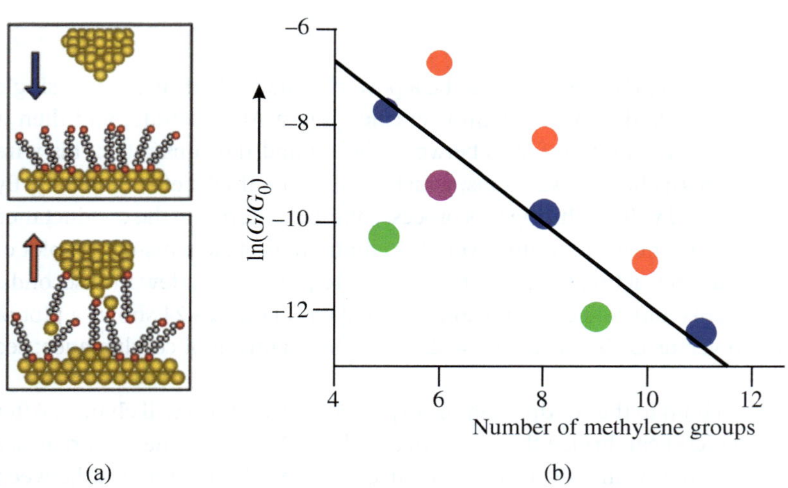

Figure 28.25 (a) Generating single Au|Alkanedithiol|Au junctions (adapted from [32]). (b) Logarithmic plot for the positions of the conductance maxima in the histogram of the measured molecular conductance versus the number of methylene groups in a molecule. The data are taken from reports of four different research groups: red filled circles [31], blue filled circles [33], green filled circles [34], and magenta filled circles [35].

alkane chains of a similar thickness, but almost a factor of two smaller than the data found for the MIM structures of much longer alkane chains.

28.4.3.2 Single Molecule Between Gold Electrodes

Current/Voltage Behavior The direct measurement of the current through a single molecule and its dependence on voltage and temperature have been achieved by Marcel Mayor and coworkers, using an experimental approach that was pioneered by Mark Reed et al. [36]. The molecule is spanned between gold electrodes binding to the thiol groups at each end [37].

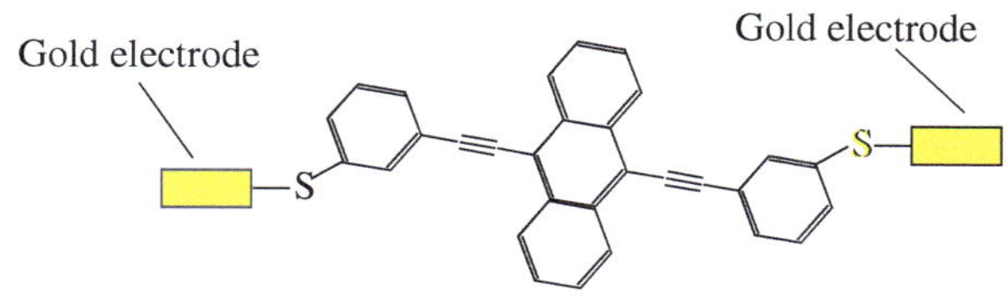

Contact and bonding of a single molecule to gold electrodes requires exact matching of the distance between the electrodes with the length of the molecule. This requirement can be achieved by controlled breaking of a lithographically fabricated junction with an area of about $50 \times 50\ \text{nm}^2$. In this way a tunable distance between the electrodes is realized. The molecules are linked with the sulfur atom to one electrode under the condition in which the electrode spacing is larger than the length of the molecule. Subsequently, the distance is decreased until the system suddenly locks into a stable behavior; i.e., the second sulfur atom reacts with the second gold electrode. d, the distance between the gold electrodes, is identical with the width of the barrier.

The current i is recorded as a function of the applied voltage V in the range of $V = -1.1$ V to $V = +1.1$ V, see Fig. 28.26a, lower curve [38]. The characteristics can be seen better in the $\mathrm{d}I/\mathrm{d}V$ plot (upper curve in the figure) than in the i/V plot. The distinct asymmetry in the current versus voltage behavior through the corresponding asymmetric compound [39] implies that the current flows through a single molecule. The i/V plot depends in a very specific way on the chemical structure. In the present case measurements at 30 K show that the current is practically zero in the potential window between −0.75 and +0.7 V and then increases suddenly. This behavior arises from the fact that the tunneling barrier is high and wide;

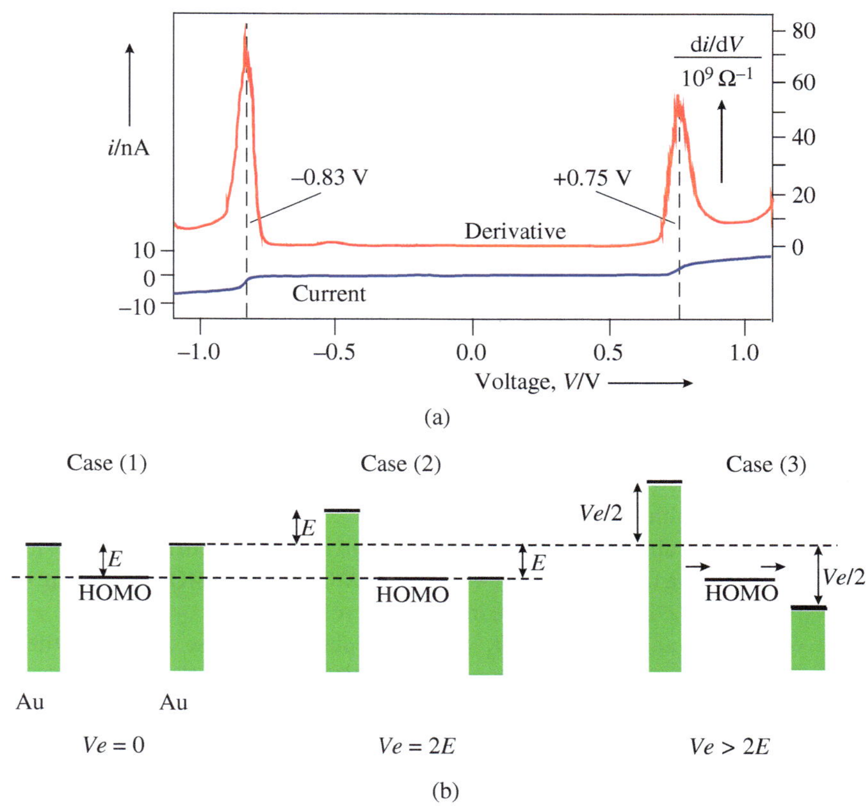

Figure 28.26 Conduction through a single molecule with a π-electron system. (a) Current i (lower curve, left-hand scale) and derivative $\mathrm{d}i/\mathrm{d}V$ (upper curve, right-hand scale) versus applied voltage V, measured at $T = 30$ K. (b) Shift of Fermi levels relative to the HOMO energy level, according to the applied voltage V. Case (1): no voltage applied; case (2): voltage $V = 2E/e$ applied; case (3): higher voltage applied (the arrows indicate flow of current). (Adapted from [38].)

hence, very little current flows, as long as the Fermi levels of the electrodes are above the HOMO of the molecule (case (1) in Fig. 28.26b). This condition holds in the range of the applied voltage between zero and $2E/e$, case (2) in Fig. 28.26b. Upon further increase of the applied potential (proceeding to case (3) in Fig. 28.26b), an electronic state of the molecule becomes accessible, e.g., an electron moves from the molecule to an empty level of the gold electrode on the right and then an electron from the gold electrode on the left replaces the electron, so that the current increases dramatically; a process referred to as resonant tunneling. According to this model the HOMO level should be $E = \frac{1}{2}Ve = \frac{1}{2}0.75$ eV = 0.38 eV below the Fermi level of gold, i.e., 4.71 + 0.38 eV = 5.09 eV below vacuum.

The current observed at bias potentials below $V = -0.83$ V and above $V = 0.75$ V appears to be proportional to the bias voltage, nearly ohmic. According to Fig. 28.26a the current is 6 nA at 1 V, implying a resistance of $R = 1\text{V}/6\text{ nA} = 1.7 \times 10^8\ \Omega = 170\ \text{M}\Omega$, corresponding to an average rate of $6\text{ nA}/e \approx 4 \times 10^{10}\ \text{s}^{-1}$ at a bias of 1 V, that is, a flow of 4×10^{10} electrons per second. Note that for a hydrocarbon chain of the same length as the considered molecule and at the same bias the flow is only 200 electrons per second. The measured value of 170 MΩ can be compared to the theoretically approximated value obtained from Equations (28.26) and (28.27)

$$R = \frac{V}{i} = \frac{1}{\kappa}\frac{d}{A} = \frac{1}{\alpha}\frac{d}{A} \cdot e^{bd}$$

using $b = 3\ \text{nm}^{-1}$ (for a π electron chain[3]), $d = 2$ nm, $\alpha = 3 \times 10^4\ \Omega^{-1}\ \text{m}^{-1}$ (see Section 28.4.2.1), and $A = 0.2\ \text{nm}^2$.

$$R = \frac{1}{3 \times 10^4\ \Omega^{-1}\ \text{m}^{-1}} \frac{2\ \text{nm}}{0.2\ \text{nm}^2} \cdot \exp(3.0 \cdot 2.0) = 3.33 \times 10^5 \cdot 403.0 = 134\ \text{M}\Omega$$

Using a similar technique dithiols with different chain lengths (6, 8, and 10 carbon atoms) were spanned between gold electrodes. However, the molecules were surrounded by an electrolyte [31]. An exponential increase of the resistance R of the single molecule with chain length was observed.

Molecular Rectifier We consider a molecule representing an A–σ–D system linked to gold electrodes:

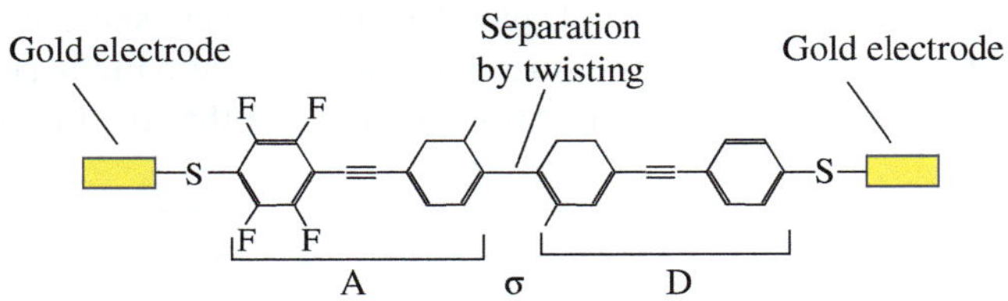

This system has been investigated using the same experimental technique as described above. It acts as a very effective molecular rectifier, as can be seen in the i/V diagram [40] in Fig. 28.27a. The current i is essentially zero between $V = -0.5$ V and $V = +0.5$ V, but then increases strongly with increasing voltage V (gold electrode on the right getting increasingly negative).

We consider the model in Fig. 28.27b. In case (1) we assume that the applied voltage is zero. The HOMO of molecule D is in its vibronic ground state (HOMO-D) below the Fermi level of gold, and the LUMO of molecule A is shown in the vibronic ground state (LUMO-A, $n = 0$); the first and second vibronically excited states are represented by LUMO-A, $n = 1$, LUMO-A, $n = 2$. We estimate the energy spacing between these vibronically excited levels to be $h\nu_0 = 0.17$ eV, which corresponds to a 1360 cm^{-1} C–C bond vibration. Although this model is highly simplified and neglects the complexity of the vibronic manifold, it appears to account for the essential features of the i/V plot.

With increasing voltage the level of HOMO-D is shifted upward relative to the Fermi level, and that of level LUMO-A, $n = 1$ is shifted downward. To simplify, we assume that the two halves of the molecule (HOMO-D and LUMO-A) move by equal amounts with the applied bias. At $V = +0.55$ V HOMO-D and LUMO-A, $n = 0$ are nearly degenerate (below Fermi level on the right, above Fermi level on the left), giving rise to a current $i_1 = 0.4$ nA, see Fig. 28.27b, case (2). When increasing the applied voltage by 0.42 V to $V = +0.97$ V, HOMO-D and LUMO-A, $n = 1$ are at an equal level, and a current of $i_2 = 1.6$ nA flows, see Fig. 28.27b, case (3). If we increase the voltage farther by 0.42 V to $V = +1.39$ V, HOMO-D and LUMO-A, $n = 2$ are at an equal level, and a current of $i_3 = 2.8$ nA is observed. These changes of the applied voltage correspond to the required shift of energy levels to reach equal heights of HOMO-D and LUMO-A, $n = 1$,2, or 3, respectively.

3 References in SK. [27]: $b = 1\ \text{nm}^{-1} - 3\ \text{nm}^{-1}$.

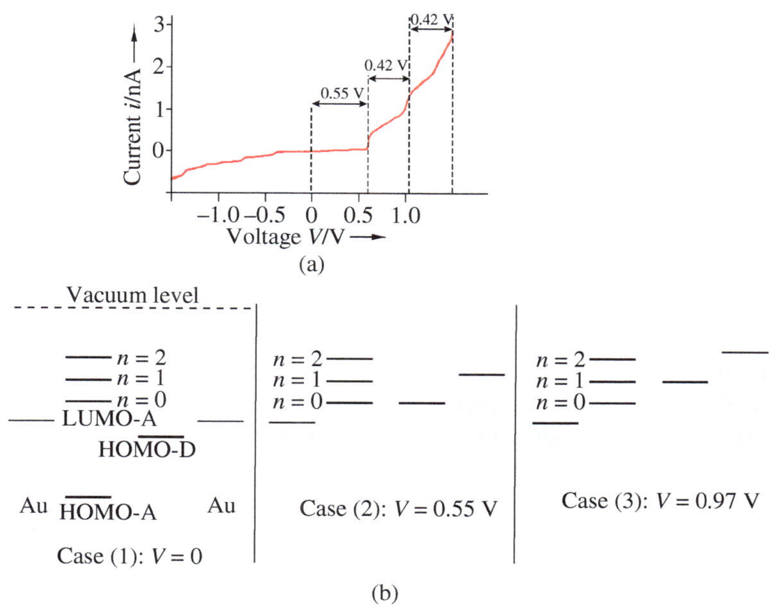

Figure 28.27 Molecular rectifier. (a) Current i versus applied voltage V, measured at $T = 30$ K. (b) Shift of Fermi levels and of HOMO-D and LUMO-A levels, according to the applied voltage V. Case (1): $V = 0$, case (2): $V = +0.55$ V, case (3): $V = +0.97$ V.

In the opposite case, when decreasing the applied voltage to $V = -0.55$ V, HOMO-D is shifted downward and LUMO-A, $n = 0$ is shifted upward. Hence the barrier between both levels increases and no current can flow (rectifier effect).

The ratios of the observed currents are

$$i_1 : i_2 : i_3 = 0.4 \text{ nA} : 1.6 \text{ nA} : 2.8 \text{ nA} = 1:4:7$$

Can we understand this finding? According to Foundation 28.5 the rate of electron transfer from D to A is proportional to $\beta^n/n!$, where β is given by the change of electron density resulting in a change in the equilibrium position of each nucleus in the π electron system. We estimate $\beta = 4$, as in the example considered in Foundation 28.5. Thus for the ratio of the currents we expect

$$i_1 : i_2 : i_3 = \frac{4^0}{0!} : \frac{4^1}{1!} : \frac{4^2}{2!} = 1:4:8$$

This is in excellent agreement with the experiment. This arrangement constitutes a fascinating model for a molecular electronic device.[4]

28.4.4 Electron Transfer in Proteins

Proteins that perform electron transfer reactions, as catalysts and/or electron shuttles, are very important in biological systems. For example, the respiration process that occurs in mitochondria proceeds through a cascade of electron transfer reactions, involving membrane bound proteins and electron transport molecules or proteins. One of these electron transfer molecules is the globular protein, cytochrome c.

28.4.4.1 Electron Tunneling in Proteins

Proteins can be modified by binding a luminescent group at exactly defined amino acids [42]. For instance, if a ruthenium complex is bound on cytochrome c to amino acids His 39 or His 72, the electron transfer rate from the excited Ru complex to the iron porphyrin complex can be measured by fluorescence quenching of the Ru complex. The rate constant for the electron transfer is similar in both cases

$$k_r = 10^6 \text{ s}^{-1} \quad \text{and} \quad k_r = 3 \times 10^5 \text{ s}^{-1}$$

even though the edge-to-edge distance between the donor and acceptor groups is different (1230 and 840 pm, respectively).

4 First steps toward molecular electronic devices (D–σ–A systems sandwiched between metal electrodes using the monolayer assembly technique) see [41].

In Foundation 28.5 we derive an expression for the electron transfer rate constant k_r (electron transfer in rigid media), namely

$$k_r = a \cdot e^{-b \cdot d} \tag{28.31}$$

with

$$a = \left(\frac{E_{\text{barrier}}}{\text{eV}}\right)^2 \cdot 3.3 \times 10^{13}\ \text{s}^{-1} \quad \text{and} \quad b = \frac{4\pi}{h}\sqrt{2m_e E_{\text{barrier}}} = \sqrt{\frac{E_{\text{barrier}}}{\text{eV}}} \cdot 10\ \text{nm}^{-1}$$

For $E_{\text{barrier}} = 2$ eV we calculate $a = 10^{14}\ \text{s}^{-1}$ and $b = 14\ \text{nm}^{-1}$; and we expect $k_r = 2 \times 10^6\ \text{s}^{-1}$ for $d = 1230$ pm and $k_r = 8 \times 10^8\ \text{s}^{-1}$ for $d = 840$ pm. This calculation is in agreement with the first case; but what is the reason for the discrepancy in the second case? From His 72 the electron has to pass an empty space of 600 pm, whereas in the derivation of Equation (28.31) it was assumed that the electron tunnels through a dielectric medium. In vacuum the wavefunction of the electron decays faster than in the dielectric medium, and the rate of electron transfer is much smaller.

The electron tunneling in proteins can be described by dissecting the protein into subunits (e.g., atoms or functional groups) and assigning them characteristic tunneling decay lengths. To determine the overall tunneling probability between two sites in the protein we must determine the tunneling probability for each possible pathway and sum over all these pathways. The probability along each pathway can be obtained by summing up the decay contributions of the subunits along a tunneling pathway. Typically, one (or a few) electron tunneling pathway makes the dominant contribution to the overall tunneling probability (i.e., the electron tunnels along the easiest path), which we call the *way of least decay*. For a uniform medium, it would be the edge-to-edge connecting line. In general, it deviates from this line because polarizable portions are preferred, and empty portions are avoided.

28.4.4.2 Wiring Proteins to Electrodes

By coating metal electrodes with organic films one can control the interaction of the protein with the surface and manipulate the electron transfer rate to the electrode. Consider the two different systems in Fig. 28.28. In case (a) cytochrome c is bound electrostatically onto a coated gold electrode (through the attraction of cytochrome c's surface lysine groups to the carboxylic acid groups of the layer), whereas in case (b) a nitrogen ligand (pyridine) displaces the native axial ligand of cytochrome c's heme (a methionine group) and coordinates directly to the heme iron. In both cases the protein is electroactive and undergoes electron transfer with the electrode.

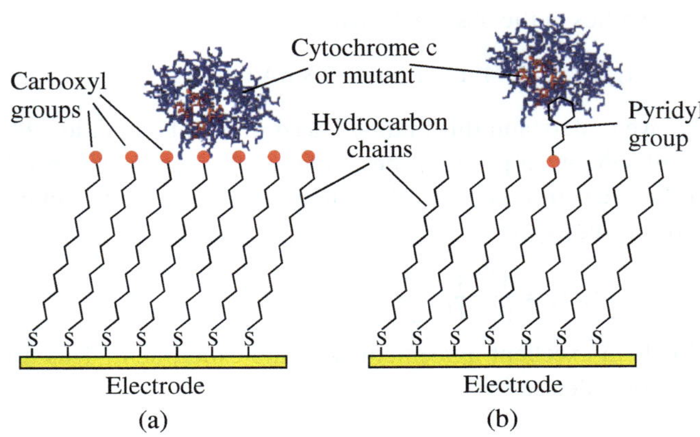

Figure 28.28 Cytochrome c attached to a monolayer of fatty acid. (a) Electrostatically fixed to carboxyl groups. (b) Chemically fixed to a pyridine group.

Mutant: 10^4 Fold Drop in the Rate Constant in Case (a), No Drop in Case (b) The difference in the tunneling pathways for the two situations is revealed by comparing the mutant form of the cytochrome c to the native form [43]. Niki and coworkers studied the electron transfer for different mutants of cytochrome c in which a lysine group is replaced by an alanine. In most cases, this modification has a small effect on the electron transfer rate constant; however, the replacement of lysine 13 with an alanine causes a 10,000 fold drop in the rate constant (see Fig. 28.29). The electron transfer rate constants are shown as a function of the length of the methylene chain in the case of the native form of cytochrome c and in the case of the mutant in the arrangements (a) and (b).

At large distances the electron transfer decays exponentially with the chain length; i.e., electron tunneling, and at shorter chain lengths the electron transfer is limited by a process other than tunneling [44]. Although the behavior with distance is similar in the four cases the system in arrangement (b) has a higher electron transfer rate at each distance than in arrangement (a). The difference in rate can be linked to the difference in the tunneling pathways for each case. In case (a) the electron must tunnel through the methylene chain to the protein interface, and through amino acid residues to reach the heme unit. In case (b), the electron tunnels through the methylene chain, into the aromatic pyridine moiety and onto the heme of the protein.

Understanding and controlling electron transfer of biomolecules remains a current frontier.

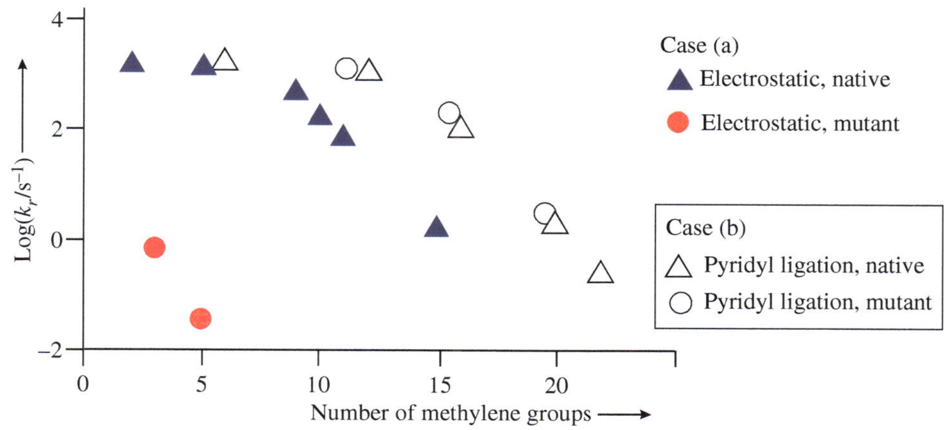

Figure 28.29 log(k_r/s^{-1}) for the electron transfer reaction according to Figure 28.28 versus the number of methylene groups. k_r is the electron transfer rate constant. Filled circles and triangles: case (a), cytochrome c electrostatically fixed. Open circles and triangles: case (b), cytochrome c with pyridyl ligation.

28.4.5 Solar Energy Conversion: The Electron Pump of Plants and Bacteria

The conversion of solar energy into chemical energy in plants is a process necessary to maintain life on earth. In photosynthesis electrons are excited by light, charges separate, and reduction and oxidation reactions take place leading to Gibbs energy-rich products (see Chapter 22).

28.4.5.1 Basic Mechanism of Electron Pumping

Let us consider the electron transfer process from a donor dye D to an acceptor dye A. As a result of the excitation of D, an electron is transferred from D to A and charge separation into D^+ and A^- occurs. In order to prevent this pair of separated charges from subsequently recombining, the distance between D^+ and A^- must be sufficiently large to inhibit back electron transfer, via electron tunneling (see Section 28.4.1). In order to transfer the electron from D* (photoexcited D) over a large distance to A fast enough to compete with deexcitation of D*, a π electron system acting as a molecular wire (W) is used to shift the electron from D* to A over a long distance. A secondary electron donor D′ must be present to restore (reduce) the oxidized dye D^+ (see Fig. 28.30). This secondary electron donor D′ should be able to reduce D^+, but not deactivate D*.

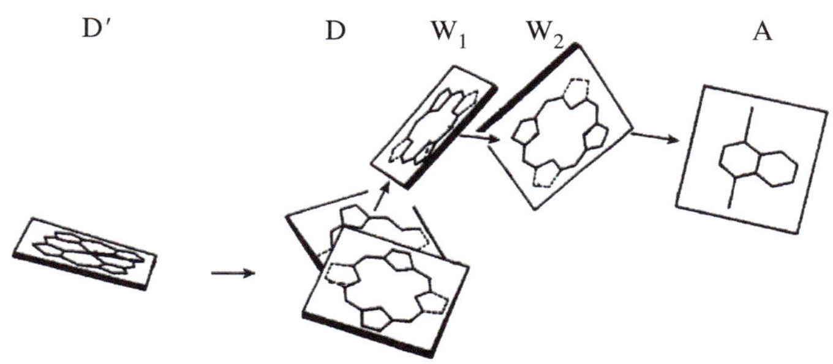

Figure 28.30 Scheme of the reaction center of purple bacterium [45]. The "special pair" of bacteriochlorophyll molecules, D, is excited (called D*) and then the electron in LUMO is transferred via molecular wires W_1 (monomer bacteriochlorophyll) and W_2 (bacteriopheophytin) to acceptor A. D is now in the oxidized form D^+ and A is in the reduced form A^-. Finally, an electron is transferred from D′ (porphyrin chromophore of cytochrome c) to D^+. D′ is now in the oxidized form D'^+ and D is in the ground state. (From [45], with permission.)

But how can the low-energy electron be transferred from D′ to D^+? A sufficiently high barrier is needed to be sure that the electron, after excitation, moves to the acceptor A. On the other hand, the barrier must allow an electron of D′ to pass the barrier later on. How can this be achieved? The high barrier must be sufficiently narrow (1 nm thickness) to allow tunneling from D′ to D^+ in due time. This mechanism [46] was proposed 12 years before Johann Deisenhofer, Robert Huber, and Hartmut Michel determined the structure of the bacterial reaction center [47].

28.4.5.2 Bacterial Electron Pump

This idealized arrangement of the essential electron transfer moieties (Fig. 28.30) is actually present in the bacterial reaction center (Fig. 28.31). The molecular wire is represented by two π electron systems W_1 (monomer bacteriochlorophyll) and W_2 (bacteriopheophytin), and the barrier between D and D′ is represented by a protein portion of 1 nm thickness, as proposed.

The initial step of electron transfer in the reaction center is particularly critical in avoiding back transfer (charge recombination). It is essential to bring the excited electron to W_2 via W_1 and, at the same time, to lose as little energy as possible. The process takes a few picoseconds, and the loss of energy is about 0.15 eV. The level of the excited state of D and the

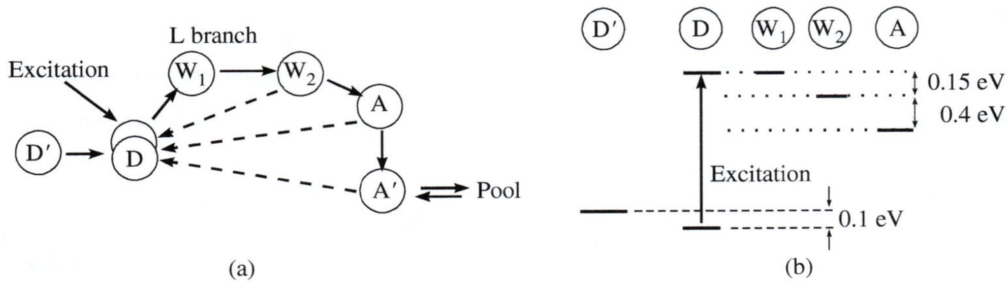

Figure 28.31 Photoinduced electron pump in reaction center of purple bacterium. (a) Photoinduced electron transfer to produce high-energy compounds (solid lines), in which the primary reaction partners are indicated in their initial state before excitation, see Table 28.2 for the primary reaction steps. The dashed lines indicate possibilities for charge recombination. (b) Balance of energy levels for optimal pumping efficiency. (Adapted from [48].)

level of W_1 coincide, allowing fast removal of the electron from D*. The next step, further charge separation by shifting the electron from W_2 to A, can be somewhat slower but is essentially completed in 230 ps.

In order to avoid losses, the precise geometrical arrangement and the matching of energy levels of each chromophore component are crucial. The probability of charge recombination by electron tunneling from W_1^- to D^+ or W_2^- to D^+ or A^- to D^+ must be kept as small as possible. Therefore, the tunneling path (dashed lines in Fig. 28.31a) is confined to the peptidic pieces to keep the tunneling probability as small as possible. The tunneling barrier is then much higher than for a path through the π electron portion. This need is met by arranging the chromophores in an arc.

The precise geometry minimizes recombination by tunneling; however, back electron transfer through the ladder of intermediates can lead to recombination. This recombination process is inhibited by the Gibbs energy gradient through the redox chain, D:W1:W2:A.

28.4.5.3 Optimum Yield. Compromise Between Minimizing Energy Loss and Minimizing Recombination Rate

W_2^- and A^- are at lower energy levels than D* and W_1^- to avoid thermally activated recombination. The lifetime of the vibronically relaxed state of W_2^- (230 ps) and the lifetime of A^- (1 ms) result from the precisely matched energy levels and matched geometry. The vibronically relaxed state of W_2^- is 0.15 eV below D*, and A^- is about (0.4 + 0.15) eV below D* (Fig. 28.31b). This allows W_2^- and A^- to exist for 230 ps and 1 ms, respectively, as required. This requirement causes an unavoidable energy loss, but optimizes the yield. Shifting W_2^- or A^- to a higher energy level leads to a higher recombination rate reducing the yield.

The requirement to keep A for 1 ms is related to the subsequent chemical steps. The electron is transferred from A to an acceptor A′ acting as a shuttle that carries electrons to a pool, i.e., to a region in the membrane where it can reach an electron acceptor. A′ is a quinone that becomes a hydroquinone after accepting two electrons; then it becomes unbound and diffuses into the pool delivering the two electrons. Afterwards it diffuses back and is trapped again in the reaction center at its original location.

The nearly optimal arrangement of the chromophores and of the matched energy levels is postulated to result from a continuous selection over the course of evolution. The electron pump in plants (photosynthesis) is closely related to the bacterial pump.

28.4.5.4 Tunneling Processes in the Bacterial Reaction Center

The optimal arrangement of the chromophores can be estimated by using a simple relationship between the electron transfer rate constant k_r and the edge-to-edge distance d between the donor and acceptor orbitals (see Foundation 28.5):

$$k_r = a \cdot e^{-bd} \tag{28.32}$$

with

$$a = \left(\frac{E_{\text{barrier}}}{\text{eV}}\right)^2 \cdot 3.3 \times 10^{13}\ \text{s}^{-1} \quad \text{and} \quad b = \frac{4\pi}{h}\sqrt{2m_e E_{\text{barrier}}} = \sqrt{\frac{E_{\text{barrier}}}{\text{eV}}} \cdot 10\ \text{nm}^{-1}$$

For $E_{\text{barrier}} = 2$ eV we calculate $a = 10^{14}\ \text{s}^{-1}$ and $b = 14\ \text{nm}^{-1}$. This relationship is based on an "effective" tunneling barrier of about 2 eV for the electron transferred from D* to W_1, W_2, A, A′ and back to D^+ from A^- and (A′). It describes the experimental data astonishingly well (Table 28.2).

Table 28.2 Calculated and Experimental [49] Electron Transfer Rate Constants for Processes in the Reaction Center of the Purple Bacterium. Calculation Based on Equation (28.32) with $a = 10^{15}$ s $^{-1}$ and $b = 14$ nm $^{-1}$; d Is the Edge-to-Edge Distance Between the Donor and Acceptor Orbitals.

Process			$\frac{d}{\text{nm}}$	$\frac{k_r}{\text{s}^{-1}}$ Calculated	$\frac{k_r}{\text{s}^{-1}}$ Experiment
$D^* + W_1$	→	$W_1^- + D^+$	0.55	4×10^{11}	3×10^{11}
$W_1^- + W_2$	→	$W_1 + W_2^-$	0.48	1×10^{12}	2×10^{12}
$W_2^- + A$	→	$W_2 + A^-$	0.96	1×10^{9}	1×10^{9}
$A^- + A'$	→	$A + (A')^-$	1.35	1×10^{7}	4×10^{7}
$A^- + D^+$	→	$A + D$	2.24	2×10^{1}	6×10^{1}
$(A')^- + D^+$	→	$A' + D$	2.34	6×10^{0}	2×10^{0}
$D'^- + D^+$	→	$(D')^{+-} + D$	1.8	2×10^{6}	2.6×10^{6}

To connect with Equation (28.32), consider the excited electron at location a in Fig. 28.32. This location corresponds to the antinode of the electron wavefunction at the edge of the chromophore. The probability of finding the electron at antinode 1 is $\mathcal{P} = c_a^2 = 0.1$ in the case of bacteriochlorophyll,

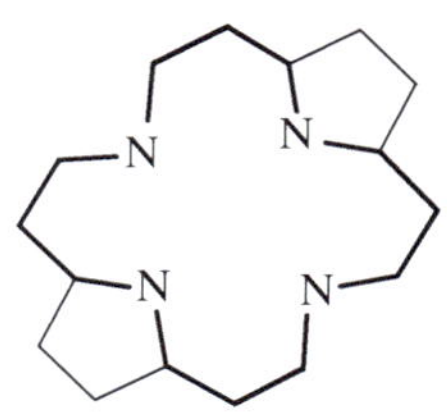

Bacteriochlorophyll

because the wave of the electron in the excited state or in the reduced state has 10 essentially equal antinodes—similar to the inner ring of phthalocyanine with its 18 π-electrons (see Fig. 9.18, sites 5a and 5b). The electron behaves as if it would hit the wall with a frequency

$$\nu = \frac{v}{L} \tag{28.33}$$

where L is the extension of the p orbital and the velocity v is given by the de Broglie relation.

$$v = \frac{h}{m_e \Lambda} \quad \text{with} \quad \Lambda = L \tag{28.34}$$

The frequency ν is

$$\nu = \frac{1}{L}\frac{h}{m_e \Lambda} = \frac{h}{m_e L^2} \tag{28.35}$$

According to Section 3.2.2.3 we obtain

$$\mathcal{P} = e^{-bd} \quad \text{with} \quad b = \frac{4\pi\sqrt{2m_e}}{h}\sqrt{E_{\text{barrier}}} \tag{28.36}$$

where $\mathcal{P}$ is the probability of the electron to tunnel through a barrier in one hit, and where E_{barrier} and d are the height and the width of this barrier. Hence, the tunneling rate constant k_r becomes

$$k_r = \nu \cdot c_a^2 \cdot \mathcal{P} = \frac{h}{m_e L^2} \cdot c_a^2 \cdot e^{-2bd} \tag{28.37}$$

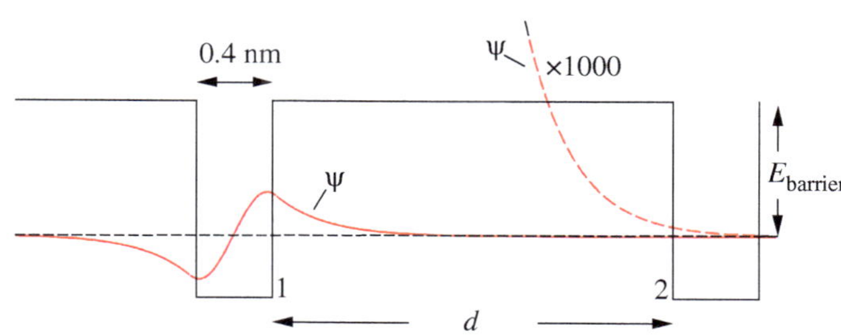

Figure 28.32 Tunneling of electron from location 1 (antinode of electron wavefunction in donor) to location 2 (acceptor) in distance $d = 2$ nm through a rectangular potential barrier of height $E_{\text{barrier}} = 1.3 \times 10^{-19}$ J = 0.7 eV. Red solid and dashed lines: wavefunction ψ of tunneling electron. We use $E_{\text{barrier}} = 0.7$ eV, because for 2 eV the initial exponential decrease of the wavefunction is too steep for illustrating. The value of ψ at location 1 corresponds to the probability $\mathcal{P} = 0.1$ of finding the electron at this location.

With $L = 0.4$ nm, $c_a^2 = 0.1$, and $E_{barrier} = 2$ eV (Foundation 28.5) we obtain

$$k_r = 0.5 \times 10^{15} \cdot \exp(-14\ \text{nm}^{-1} \cdot d) \quad \text{s}^{-1}$$

28.4.6 Electron Transfer in Soft Medium

28.4.6.1 Marcus Equation

The analysis in the preceding section has assumed that D, W_1, W_2, and A are in a thermalized, but rigid, medium; however, this assumption is not always fulfilled, e.g., D′ is in a soft environment. This idea is supported by the fact that D′ is not fixed; instead, it can diffuse between its location in the reaction center (where it is oxidized) and a pool (where the reduced form is restored by accepting an electron).

Actually, the surroundings of D′ are soft. How can we obtain k_r in this case? For both rigid and soft media the electron transfer occurs during energetic match of the electronic energy levels; i.e., the transition state. Although the tunneling probability at the transition state may be the same for both media, the probability of reaching it is quite different. In the soft medium this probability depends on the Gibbs energy of activation, $\Delta G^{\ddagger}$. As shown in Foundation 28.6

$$\Delta G^{\ddagger} = \frac{(\Delta G^{\ominus} + \lambda)^2}{4\lambda} \tag{28.38}$$

where $\Delta G^{\ominus}$ is the change of the Gibbs energy of reaction, and λ is the reorganization energy. The rate equation

$$k_r = \frac{2\pi}{\hbar}\varepsilon^2_{\text{electronic}} \cdot \left[\frac{\beta^n}{n!} e^{-\beta} \cdot \frac{1}{E_{\text{fluct}}}\right] \tag{28.39}$$

derived in Foundation 28.5 for the rigid medium can be simplified in the present case by setting $n = 0$. The π electron, before and after its transition from D′ to D^+, is spread over a large cloud; thus, the lattice vibration can be neglected and we can write

$$k_r = \frac{2\pi}{\hbar}\varepsilon^2_{\text{electronic}} \cdot \frac{1}{E_{\text{fluct}}}$$

As shown in Foundation 28.6, in the soft medium $1/E_{\text{fluct}}$ is replaced by

$$\frac{1}{\sqrt{4\pi kT\lambda}} \cdot \exp\left(-\frac{\Delta G^{\ddagger}}{kT}\right) \tag{28.40}$$

Using Equation (28.38), we obtain the classical *Marcus Equation*

$$k_r = \frac{2\pi}{\hbar}\varepsilon^2_{\text{electronic}} \cdot \frac{1}{\sqrt{4\pi kT\lambda}} \cdot \exp\left(\frac{-(\Delta G^{\ominus} + \lambda)^2}{kT \cdot 4\lambda}\right) \quad \text{(soft medium, no lattice vibration)} \tag{28.41}$$

Rudolph Marcus was awarded the Nobel Prize in Chemistry (1992) for the development of this model.

28.4.6.2 Electron Tunneling from D′ to D+

$\Delta G^{\ominus}$ for the electron transfer from D′ to D^+ (Fig. 28.31) can be varied by changing the environment of D (e.g., using mutants with changed amino acids in the surroundings of D). In fact, the measured rate constants k_r increase much more strongly with increasing $-\Delta G^{\ominus}$ than they would according to Foundation 28.5 on the book's website. This indicates that the assumption of a rigid medium of D′ is inadequate in this case. If we treat λ and $\varepsilon_{\text{electronic}}$ in Equation (28.41) as adjustable parameters; the measured dependence of k_r on $\Delta G^{\ominus}$ is well described by $\lambda = 0.5$ eV and $\varepsilon_{\text{electronic}} = 2 \times 10^{-5}$ eV (Fig. 28.33).

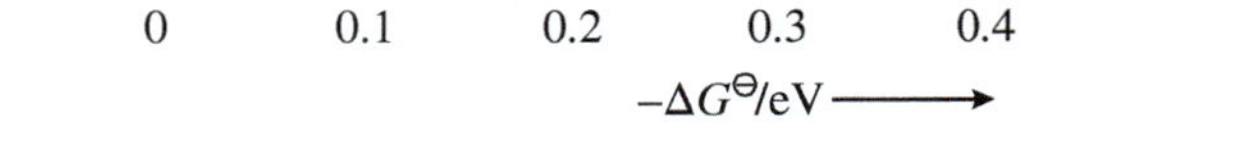

Figure 28.33 Electron transfer from D′ to D^+ for various mutants differing in $-\Delta G^{\ominus}$. Rate constant k_r versus $-\Delta G^{\ominus}$. Soft medium. Full circles: experiment. Solid line: Marcus equation (28.41), with $\varepsilon_{\text{electronic}} = 21 \times 10^{-5}$ eV, $\lambda = 0.5$ eV.

28.4.7 Inverted Region of Electron Transfer Reactions

In considering the soft medium, Marcus made the interesting prediction that the reaction rate first increases and then decreases with decreasing $\Delta G^{\ominus}$, according to Equation (28.41). The activation barrier is given by

$$\Delta G^{\ddagger} = \frac{\left(\Delta G^{\ominus} + \lambda\right)^2}{4\lambda} \tag{28.42}$$

From this formula the rate constant should be a maximum when $\Delta G^{\ominus} = -\lambda$, and it should decrease as the reaction Gibbs energy moves to either side of this equality. The decrease in rate as $\Delta G^{\ominus} > -\lambda$ is typical and corresponds to a less exoergic reaction. Hence it is called the "normal region," and many examples are known for the reaction under these conditions. In addition, Equation (28.41) predicts the unusual circumstance that the reaction becomes slower as $\Delta G^{\ominus}$ becomes more negative, for $\Delta G^{\ominus} < -\lambda$. This prediction is opposite to that commonly found in reactions and is called the "inverted region." In step W_2^- to A (rigid medium, see Equation (28.39)), discussed above qualitatively, the same behavior of the reaction rates is predicted (see Foundation 28.5). It is caused by the fact that k_r is proportional to $\beta^n/n! \cdot e^{-\beta}$, according to Equation (28.39).

In generalizing Equation (28.41) to include the vibration of the molecular lattice of donor and acceptor we can replace the term $1/E_{\text{fluct}}$ in Equation (28.39) by a corresponding factor in the Marcus equation:

$$k_r = \frac{2\pi}{\hbar}\varepsilon_{\text{electronic}}^2 \cdot \left[\frac{\beta^n}{n!}\, e^{-\beta}\right] \cdot \frac{1}{\sqrt{4\pi kT\lambda}} \cdot \exp\left(\frac{-\left(\Delta G^{\ominus} + \lambda + nh\nu_0\right)^2}{kT \cdot 4\lambda}\right) \tag{28.43}$$

(soft medium, lattice vibration included)

where n is the integer for which $\Delta G^{\ominus} + \lambda + nh\nu_0$ is next to zero.

In 1988 experimental evidence for the inverted region was reported by Closs and coworkers [50]. In their study, they covalently attached an electron donor (4-biphenyl) and different electron acceptor groups onto a rigid scaffold (dissolved in methyltetrahydrofuran) so that they could systematically change the electron acceptor group without changing important features of the reaction such as the distance d between donor and acceptor (Figure 28.34 and Table 28.3).

They irradiated the sample with an electron beam to generate the anion of the donor moiety, and they measured the electron transfer to the acceptor. Their data show that the rate constant has a maximum value at an intermediated reaction Gibbs energy (Fig. 28.34, full circles). Figure 28.34 (triangles) shows the result for the calculation according to Equation (28.43), using the parameters $\varepsilon_{\text{electronic}} = 7.68 \times 10^{-4}$ eV, $\beta = 2.42$, $\lambda = 0.75$ eV, and $h\nu_0 = 0.186$ eV. The solid line in Fig. 28.34 is the result of a more rigorous calculation [51], using the equation

$$k_r = \frac{2\pi}{\hbar}\varepsilon_{\text{electronic}}^2 \cdot \frac{1}{\sqrt{4\pi kT\lambda}} \sum_{n=0}^{\infty}\left[\frac{\beta^n}{n!}\, e^{-\beta} \cdot \exp\left(\frac{-\left(\Delta G^{\ominus} + \lambda + nh\nu_0\right)^2}{kT \cdot 4\lambda}\right)\right] \tag{28.44}$$

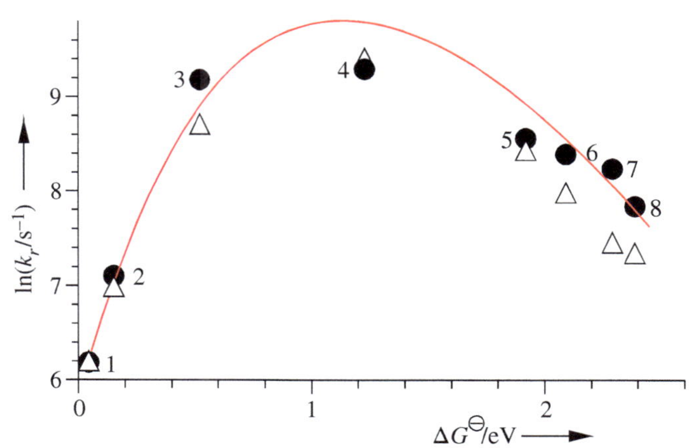

Figure 28.34 Inverted region of electron transfer reactions. $\log(k_r/s^{-1})$ as a function of the Gibbs energy of reaction, $\Delta G^{\ominus}$ for the systems 1–8 in Table 28.3. Filled circles: experiment [50]. Triangles: calculated from Equation (28.43). Solid line: calculated from Equation (28.44).

Table 28.3 Inverted Region. $-\Delta G^{\ominus}$ and $\log(k_r/s^{-1})$ for Some Electron Acceptor Groups.

Acceptor	**1**	**2**	**3**	**4**	**5**	**6**	**7**	**8**
$-\Delta G^{\ominus}$/eV	0.05	0.16	0.52	1.23	1.93	2.10	2.29	2.40
$k_r/10^9\ s^{-1}$	0.0015	0.013	1.5	2	0.38	0.25	0.17	0.07
$\log(k_r/s^{-1})$	6.18	7.11	9.18	9.30	8.58	8.40	8.23	7.85

1 2 3 4 5 6 7 8

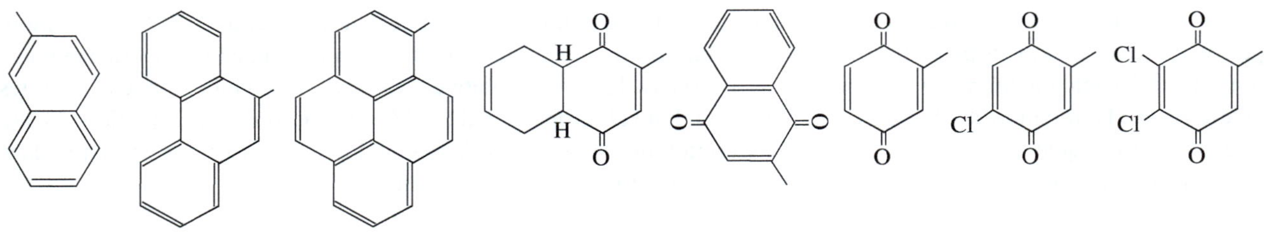

28.4.8 Artificial Photoinduced Electron Pumping

An electron pump based on the mechanism discussed in Section 28.4.5 can be realized by appropriately assembling monolayers. Another approach links the functional components D′, D, W, and A together in a "supermolecule." Figure 28.35a illustrates this strategy. Optical excitation of the chromophore D causes charge transfer from D via W to A. Subsequently charges transfer from D′ to D^+. In this system an intermediate energy transfer step takes place (from b to c in Table 28.4). The charges remain separated as long as 55 μs in state f. Thus the basic requirement for an electron pump can be achieved: a sufficiently long-lived charge separated state to allow chemical oxidation and reduction processes.

An intermediate energy transfer step is avoided in the arrangement in Fig. 28.35b, see Table 28.5. The charges remain separated for 280 ns. The quantum yield of formation of state d is 80%; the residual processes are de-excitation (b → a) and early recombination (c → a).

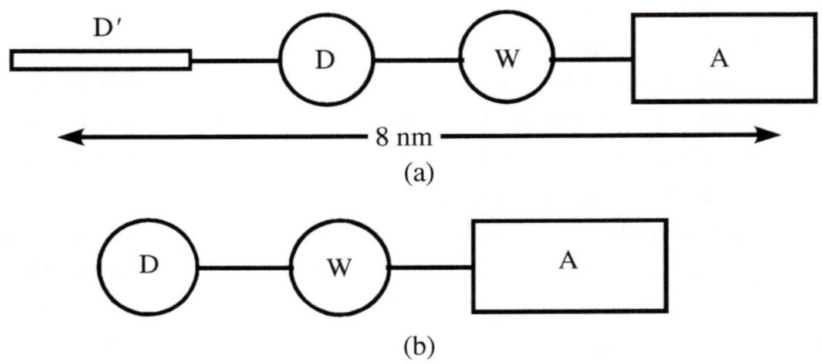

Figure 28.35 Artificial photosynthetic reaction center. (a) Donor D = zinc-porphyrin (see Section 9.2.7), wire W = free base porphyrin, acceptor A = naphthoquinone-benzoquinone derivative, donor D′ = β-carotene [52]. (b) Donor D = zinc-porphyrin, wire W = free base porphyrin, acceptor A = pyromellitimid [53].

Table 28.4 Photoinduced Electron Transfer from D′ to A in the Artificial Reaction Center Displayed in Fig. 28.35a.

State	**Functional Components**				**Transfer Step**	k_r/s^{-1}
a	D′	D	W	A		
b	D′	D*	W	A	Excitation of D	
c	D′	D	W*	A	Energy transfer to W	$k_r = 2 \times 10^{10}$
d	D′	D	W^+	A^-	Electron transfer from W* to A	$k_r = 7 \times 10^8$
e	D′	D^+	W	A^-	Electron transfer from D to W^+	
f	$(D')^+$	D	W	A^-	Electron transfer from D′ to D^+	

Table 28.5 Photoinduced Electron Transfer from D to A in the Artificial Reaction Center Displayed in Fig. 28.35b.

State				Process Leading to this State	k_r/s^{-1}
a	D	W	A		
b	D*	W	A	Excitation of D	
c	D^+	W^-	A	Electron transfer to W	$>5\times10^{10}$
d	D^+	W	A^-	Electron transfer from W^- to A	4×10^{9}

28.5 Manipulating Nuclear Motion

28.5.1 Light-Induced Change of Monolayer Properties

The creation of polymeric and liquid crystal materials that respond to external stimuli is an active area of research relevant to fundamental studies of biomechanics as well as many commercial applications. in order to actuate mechanical and electrical responses. The *cis*-to-*trans* isomerization of small chromophores provides a useful way to transduce light energy into a mechanical response of a material [54].

28.5.1.1 Periodic Change of Monolayer Area

Azobenzene can be converted from the *cis* form to the *trans* form and vice versa by illuminating with 365 and 436 nm, see Fig. 28.3a. This optically induced isomerization can be used for switching. If a *trans*-azobenzene derivative is bound to a polymer (esterification of the R_2 group with polyvinyl alcohol) and spread on a water surface at a constant surface pressure (8 mN m^{-1}), then *cis–trans* isomerization causes a change in the film's area when the surface pressure is held constant. Because the isomerization reaction can be driven by optical excitation, the film expands to three times its area when illuminated with 436-nm light and contracts again to the original area when illuminated with 365-nm light (Fig. 28.36b).

If such a monolayer is kept at constant area, illumination with light of 436 nm (365 nm) causes the surface pressure and surface potential to increase (decrease). This change in surface pressure arises because the *cis* form is more bulky, and the surface potential changes because *trans*-azobenzene has no dipole moment, while the dipole moment of *cis*-azobenzene is 3 Debye.

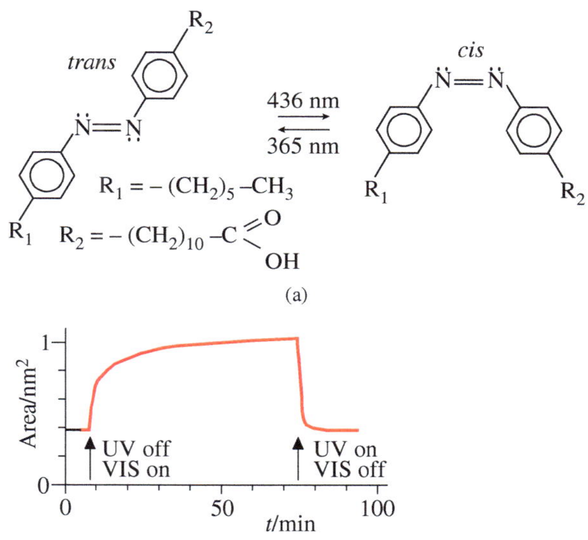

Figure 28.36 Panel (a) shows the photoisomerization between the *trans* and *cis* forms of azobenzene. Panel (b) shows the light-induced expansion and contraction of monolayers of azobenzene induced by illumination with VIS light (436 nm) and UV light (366 nm). (Adapted from [55].)

28.5.1.2 Monolayer Commanding Liquid Crystal Orientation

The diagram in Fig. 28.37 illustrates a monolayer of *trans*-azobenzene (with the same substituents R_1 and R_2 as in the previous example) on a solid surface (polyethyleneimine) that is covered by a nematic liquid crystal. The molecules in the liquid crystal are aligned perpendicular to the solid surface. When illuminated with 436-nm light, the *trans–cis* isomerization of azobenzene induces an alignment parallel to the surface. The reverse effect is

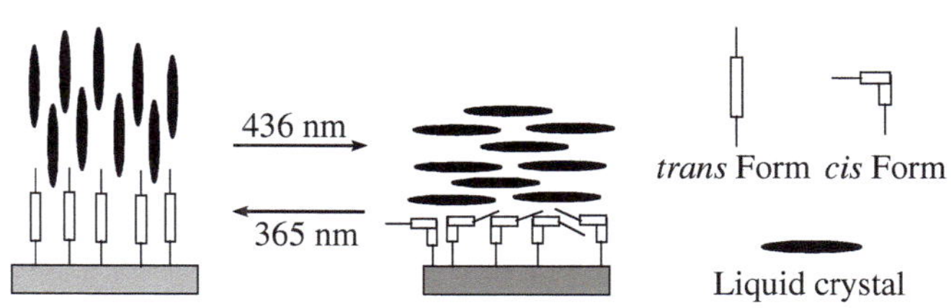

Figure 28.37 Command surface effect. Azo dyes are switched by light from the *trans* into the *cis* form and back; as a consequence, a liquid crystal changes its orientation. (Adapted from [56].)

induced by 365 nm light. The photoinduced *trans–cis* and *cis–trans* isomerization is greatly amplified by the changing alignment of the liquid crystal.

28.5.1.3 The Visual System and *Cis–Trans* Isomerization

The initial reaction in the vision process is the photoinduced *cis–trans* isomerization of a retinal chromophore that is coupled to a chemical amplification process [57]. Human rod cells contain retinal which, in the dark, is in the 11-*cis* form attached to a protein (opsin) by a protonated Schiff base linkage to a lysine side chain. The complex is called *rhodopsin*.

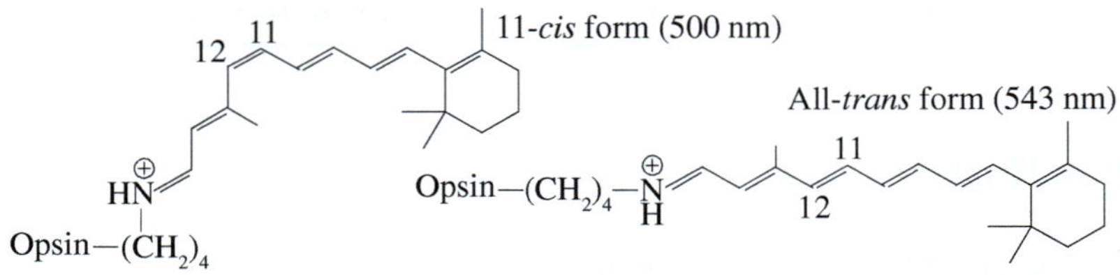

The absorption peak shifts from 500 to 543 nm within 1 ps. The change in shape causes a conformational change of the surrounding protein, resulting in a shift of the absorption peak from 543 to 497 nm within nanoseconds, along with a further shift to 480 nm within microseconds. These conformational changes in the protein environment activate a protein (transducin) within milliseconds. About 1000 molecules of activated transducin per second act to block a huge number of specific membrane channels for Na^+; this change in the membrane conductivity constitutes the event of sensory transduction and ultimately leads to an actively propagated nerve impulse. Thus a single light quantum absorbed in a human rod cell can trigger a neural impulse in a tremendously efficient chemical amplification process.

Photoactivated modulation of ion transport is an important motif in microbes and other biological organisms [58].

28.5.2 Mechanical Switching Devices

Supramolecular chemistry offers the fascinating possibility of designing machines based on photoinduced electron transfer to trigger distinct shifts of molecular parts. Bistability of a molecule is an essential property for realizing molecular information storage. Figure 28.38 shows a molecule (called *rotaxane*) synthesized by threading a strand with electron acceptors A_1 and A_2 through a ring with an electron donor D and then placing bulky groups B to the strand ends. The structure forms by self-assembly, under the appropriate conditions, in a mixture containing three ingredients: the rings, the strands, and the plugs that prevent the strands from becoming unthreaded. A_1 is the stronger acceptor; thus, normally the ring is bonded to A_1 (Fig. 28.38a). By irradiation with light, photosensitive molecules in the solution transfer an electron to A_1, thus reducing A_1 and weakening the A_1–D bond. Consequently, the system can switch to the A_2 station. By switching light on and off, the ring moves from A_1 to A_2 and vice versa (Fig. 28.38b–d).

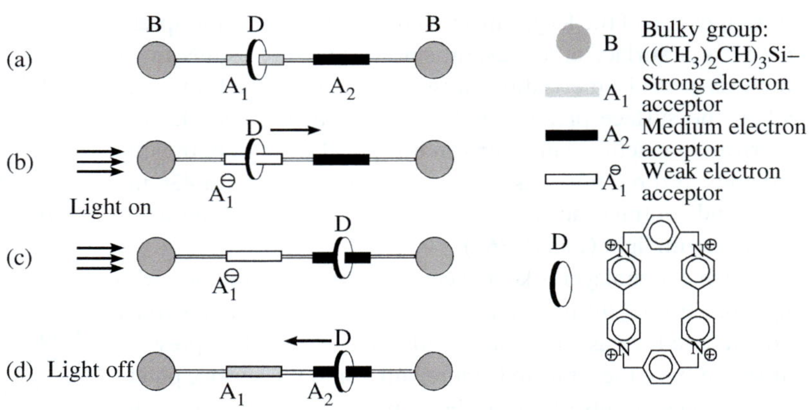

Figure 28.38 Supramolecular device (*rotaxane*) for photoinduced movement of ring D (electron donor) on a strand with electron acceptors A_1 and A_2. B is a bulky group preventing unthreading of the strand. (a) Because A_1 is a strong acceptor and A_2 is a medium acceptor only, the ring is bonded to A_1. (b) Upon irradiation with light, photosensitive molecules in the solution (not shown in the figure) reduce A_1 to the weak electron acceptor $A_1^{\ominus}$. (c) Consequently, the ring D moves to the stronger electron acceptor A_2. (d) By switching the light off, the D–A_1 bond is reformed. (Adapted from [59].)

Figure 28.39 shows a catenane, consisting of two interlocked rings. The rings are connected by Cu^+ ions which form tetrahedral bonds. If Cu^+ is oxidized to Cu^{2+} (step a → b), then the complex becomes unstable because Cu^{2+} forms a solvation shell, and the link between the two rings opens up. The compound undergoes a complete reorganization process, and the system stabilizes again forming the stable 5-coordinate Cu^{2+} complex (step b → c). Upon reduction a conformational change takes place which regenerates the starting complex (steps c → d and d → a).

The actuation of mechanical displacement by chemical stimulus is an important new direction in chemistry. While motors and pumps are well known on the macroscopic, and even submillimeter, length scales, the creation of artificial mechanical devices on the submicron length scale is only beginning to be explored. Progress in the elucidation and understanding of naturally occurring biomotors (see Section 28.5.5) underscores the richness of phenomena and behaviors yet to be discovered [61].

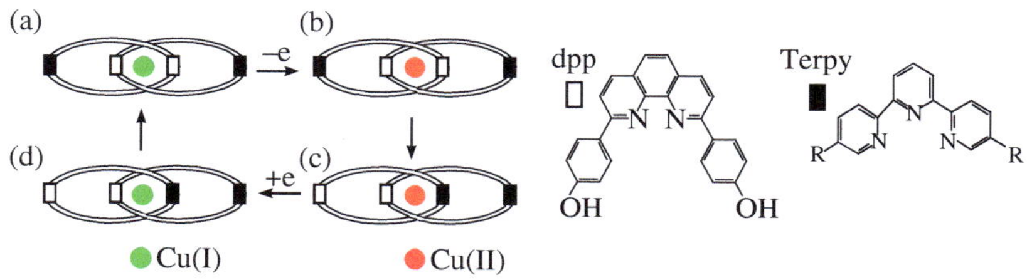

Figure 28.39 Catenane ring rotating upon oxidation and reduction process ($Cu^+ \rightarrow Cu^{2+} \rightarrow Cu^+$). dpp: 2,9-diphenyl-1,10-phenanthroline; terpy: 2,2′,6′,2″ -terpyridine. The coordination number of Cu(I) is 4, and of Cu(II) is 5. Then Cu(II) prefers a five-fold coordinated complex with a dpp and a terpy unit, whereas Cu(I) prefers a four-fold coordinated complex with two dpp units. (Adapted from [60].)

28.5.3 Solar Energy Conversion in Halobacteria

When illuminating a halobacterium, protons are transferred from the inside of the cell to the outside by a membrane-bound device. A membrane potential is created and a proton back-transfer drives the formation of energy-rich compounds (especially ATP). It has been shown that mitochondrial ATP-synthase is fastened to the membrane and its axis rotates in a proton gradient to produce ATP in a highly cooperative process. In a surplus of ATP the process can be reversed and the rotation of the axis can be visualized by binding an actin filament to the axis and observing its fluorescence in a light microscope [62].

This system is the energy source for a motor that drives the rotation of the flagella which propels the bacterial motion, see Fig. 28.40. The flagella are fastened to the membrane through a molecular bearing. The rotational motion of the flagella can be visualized by fastening it on a flat surface. When fastened, the cell body rotates and the rotation can be observed in a light microscope.

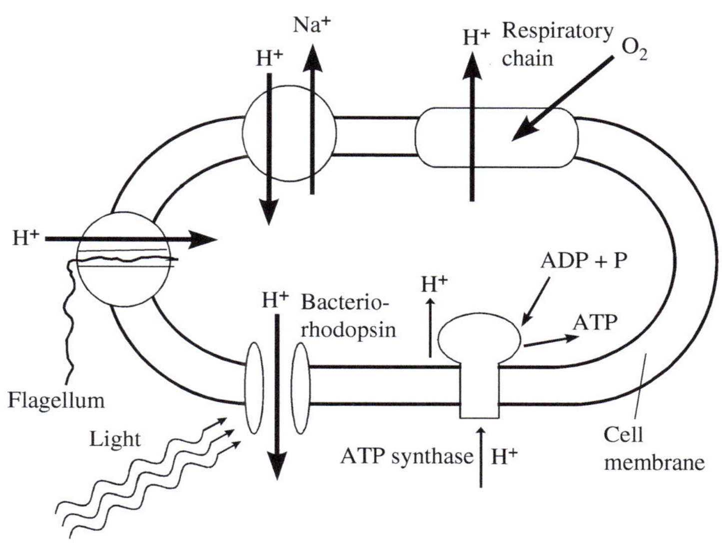

Figure 28.40 Halobacterial cell: scheme of the cell. Transport of H^+ by a light-driven proton pump (bacteriorhodopsin, see Section 28.5.3.1) produces an H^+ gradient between inside and outside the cell. The gradient ΔpH and the membrane potential $\Delta\varphi$ drive the ADP → ATP reaction: $-\Delta_r G = \boldsymbol{RT} \cdot 2.303 \cdot \Delta\text{pH} - \boldsymbol{F} \cdot \Delta\varphi$.

28.5.3.1 Photoinduced Proton Pump

Trans *to* Cis *Isomerization and Proton Translocation* Retinal in halobacteria is attached to a protein by a protonated Schiff base linkage to the side chain of the amino acid lysine.

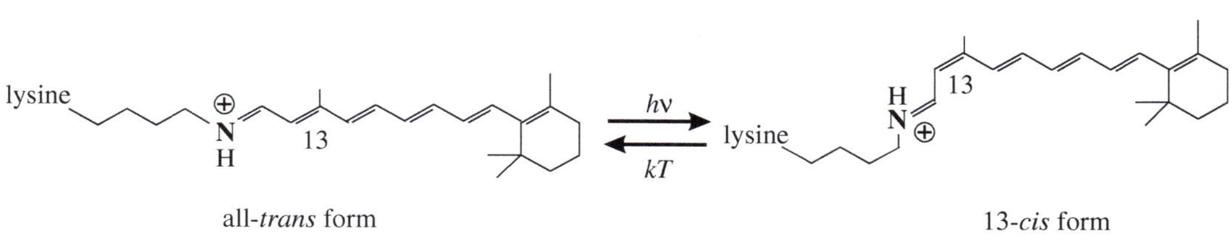

This complex is called *bacteriorhodopsin*. Bacteriorhodopsin is incorporated in the cell membrane (Fig. 28.40). Light isomerizes the all-*trans* chromophore into the 13-*cis* form. Later in the cycle the all-*trans* form is reestablished by thermal isomerization [63].

The chromophore spans two parts of a channel that connects the inside of the cell (through the *cytoplasmic channel*) to the outside (through the *extracellular channel*). It is assumed that the *trans* isomer can exchange protons only with the outside part; the *cis* isomer, correspondingly, is connected only to the inside part (Fig. 28.41). In each cycle a proton is pumped from inside the cell to the outside. Under bright light about 50 protons are pumped per second by each bacteriorhodopsin molecule.

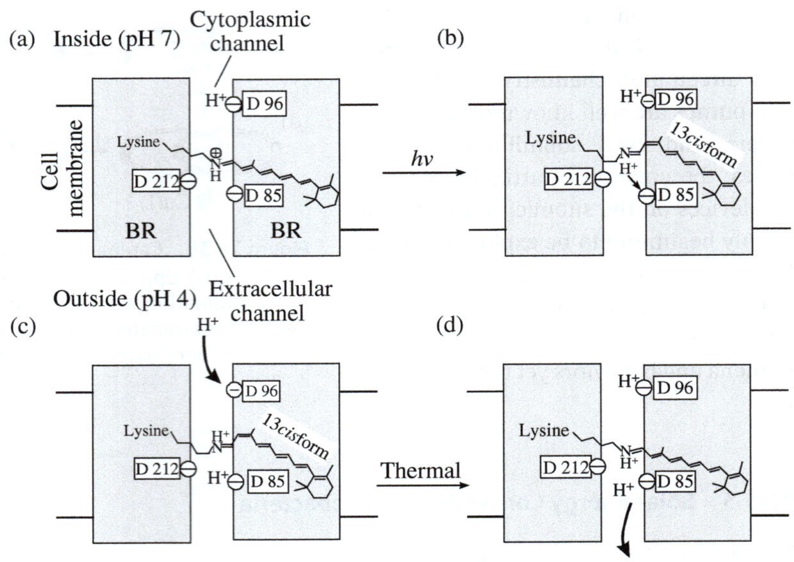

Figure 28.41 Proton pump in halobacterial cell. H^+ ions are pumped from inside to outside by means of light induced *trans* → *cis* and thermal *cis* → *trans* isomerization of retinal bound to protein (called *bacteriorhodopsin*). Acidification from pH 7 to pH 4 is achieved in an unbuffered saline suspension of cells.

A photoinduced transfer of a proton from the cytoplasmic channel (pH > 10) to the extracellular channel (pH < 3.5) takes place. How can this extreme pK change of the Schiff base happen? We are here in a situation typical for research in progress: a final answer cannot be given, the present experimental facts allow various explanations stimulating further experimental work. In the usually considered model the strong pK change of the Schiff base is related to a conformational change (in proceeding from the spectroscopically defined stage L550 to stage M412) in a protein which occurs at a distinct stage in the cycle beginning with the photoinduced *trans*- to *cis*-isomerization [64].

An aspartic acid in the extracellular channel (D85, pK = 3.5) acts as a proton acceptor of the protonated Schiff base of the excited chromophore (Fig. 28.41, step a to b). Therefore, the Schiff base in the excited chromophore must have a pK_a lower than 3.5. On the other hand, an aspartic acid in the cytoplasmic channel (D96, $pK_a = 10$) acts as a proton donor for the Schiff base of the chromophore in the ground state which must then have a pK_a higher than 10 (actually, it is 12). The Schiff base accepts the proton which is then translocated from the cytoplasmic into the extracellular channel due to the thermal -to-*trans* reisomerization of the chromophore (Fig. 28.41, step c to d). The aspartate form of D96 accepts a proton from the inside of the cell and the protonated D85 releases the proton to the outside. Thus, effectively, a proton is pumped from the inside (pH 7) to the outside (pH 4).

Schiff Base. Photoinduced Change from pK > 10 to pK < 3.5 How can this extreme pK change of the Schiff base (from pK > 10 to pK < 3.5) be achieved by exciting the chromophore? Experiments show that the chromophore conformation and protonation state strongly affect the excited state dynamics which drives the proton translocation process; however, detailed features of the mechanism are still debated [65]. Here we describe a less well-known model [66], in which we consider the photoinduced pKa of the chromophore's N atom. In Box 28.1 we describe how the excited chromophore's π electron cloud becomes highly polarizable, and it shifts away from the N atom because of the repulsion of the charged amino acids in the extracellular channel (Fig. 28.42, step 1 → 2). Consequently, the N atom becomes more acidic and a H^+ is released into the extracellular channel.

A transient absorption peak at 430 nm appears [67] in less than 150 fs, and it is considered to represent the deprotonated excited *trans* form T430 (step 2 → 3). From infrared transient absorption spectroscopy it can be concluded [68] that the *trans*- to *cis*-isomerization takes place with a characteristic time of 500 fs. The form J625 appears on this timescale and must be a ground-state species. In the present view, step 3 → 4 (*trans*- to *cis*-isomerization of the unprotonated excited form) takes place in 500 fs and the deexcitation and protonation in a time short compared to 500 fs. In this picture the Schiff base accepts a proton from a water molecule of the hydrogen bounded network present in the cytoplasmic channel [69] leaving OH^- behind that reacts later with a H^+ in the network.

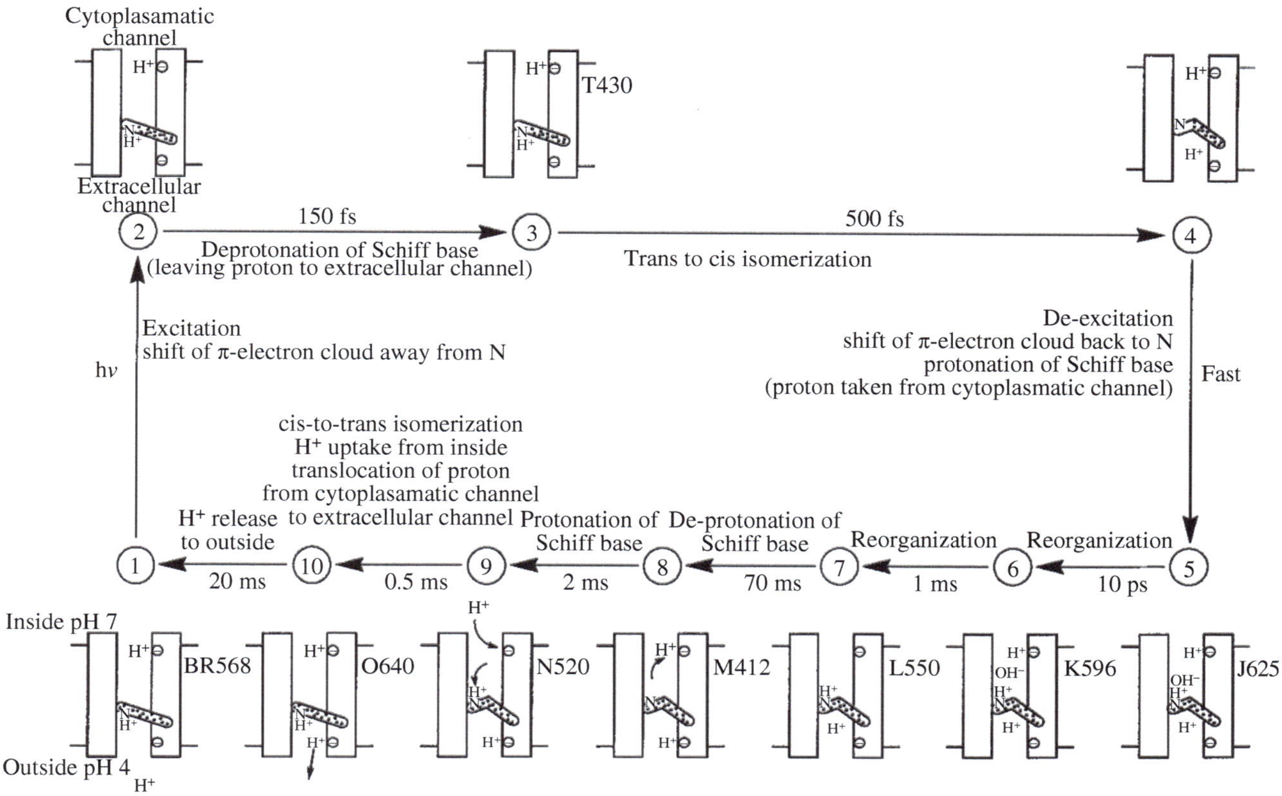

Figure 28.42 Photoinduced proton pump. Overall mechanism. π electron cloud of chromophore indicated by dots. 1 → 2: Excitation of bacteriorhodopsin BR568 by a light pulse. Shift of π electron cloud away from N atom (causing a pK change of protonated Schiff base from 12 to 3). 2 → 3: Deprotonation of Schiff base, formation of transient T430. 3 → 4: *Trans*-to-*cis* isomerization. 4 → 5: Deexcitation, shift of π electron cloud back to Schiff base (pK change from 3 to 12). Protonation of Schiff base. 4 → 5 → 7: Reorganizations. 7 → 8: Deprotonation of Schiff base. 8 → 9: Protonation of Schiff base. 9 → 10: *cis*-to *trans*-isomerization, translocation of H^+ from cytoplasmic to extracellular channel, uptake of H^+ from inside. 10 → 1: Release of H^+ to outside.

28.5.4 Translocation of Proton from Cytoplasmic to Extracellular Channel

In the steps considered above the form J625 is obtained. Several optically observable reorganization steps (5 → 6 → 7 → 8 → 9, Fig. 28.42) lead to the protonated *cis* isomer N520. To complete the cycle the proton must move from the cytoplasmic to the extracellular channel. This process occurs in the thermally activated re-isomerization of the *cis* form into the *trans* form (Fig. 28.42, step 9 → 10). Uptake of a proton from the cytoplasm (step 9 → 10), release of a proton to the outside, and reorganization of O640 forming the original dark state BR568 (step 10 → 1).

Mutant with Chromophore Blocked in *Trans*-Form

A good argument for the basic assumption in the presented model (deprotonation of Schiff base in the initial step) is the photoinduced transient absorption of a mutant in which the *all trans* form of the chromophore is blocked. A sample was excited by a flash at 580 nm (close to the absorption maximum of BR586) and probed at 480 nm (close to the absorption maximum of T430 and to the absorption maximum of the deprotonated *trans*-form in the ground state, 412 nm) [70]. In the wild type the absorption increases on a 150 fs timeframe: step 1 → 2 → 3, (release of proton into extracellular channel) and decreases in 500 fs forming J625. The mutant shows the same increase in absorption as the wild type (release of proton into extracellular channel), but it decays with a time constant of 16 ps to its original value, showing a slow recombination of the deprotonated *trans* form with a H^+ after its deexcitation where it is accepted into the hydrogen bonding network of water molecules (Fig. 28.43).

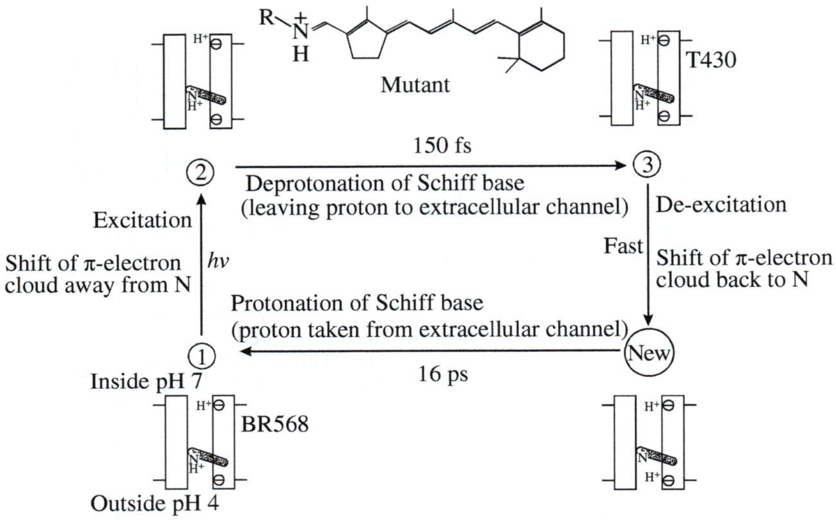

Figure 28.43 Mutant with chromophore blocked in *trans* form: photoinduced release of H^+ into extracellular channel as in wild type. Recombination slow.

Driving Force of Proton Pump

The driving force of the proton pump can be found by calculating the reversible work to transfer protons from the outside of the cell (pH_{out}) to the inside (pH_{in}). The change of the molar Gibbs energy of reaction for transporting protons between the two regions will have two contributions: the movement of charge against the membrane potential and the movement against the pH difference. Hence, we can write

$$-\Delta_r G = -\boldsymbol{F} \cdot \Delta\varphi + 2.303 \cdot \boldsymbol{R}T \cdot \Delta\text{pH}, \tag{28.45}$$

where $\Delta\text{pH} = \text{pH}_{\text{in}} - \text{pH}_{\text{out}}$ and the membrane potential is $\Delta\varphi = \varphi_{\text{in}} - \varphi_{\text{out}} = 280$ mV. For this contribution a value of $\boldsymbol{F} \cdot 280\ \text{mV} = 27\ \text{kJ mol}^{-1}$ (corresponding to 10 kT per molecule) was experimentally determined [71]. The contribution of the pH change ($\Delta\text{pH} = 3.5 - 10 = -6.5$) is $2.303 \cdot \boldsymbol{R}T \cdot (-6.5) = -16\ \text{kJ mol}^{-1}$; thus, $-\Delta_r G = (-27 - 16)\ \text{kJ mol}^{-1} = -43\ \text{kJ mol}^{-1}$.

In summary, the essence of the pumping mechanism is the photoinduced π electron shift arising from the increase in polarizability of the electron in the excited state. This shift causes the proton to separate from the chromophore and to remain behind in the extracellular channel. In the thermal *cis*-to-trans isomerization step a new proton that has been linked to the chromophore, is shifted from the cytoplasmic channel into the extracellular channel.

Halobacteria have a chloride ion pump and sensory receptors incorporated in their membrane. Their operation is based on a mechanism that is closely related to that of the proton pump, Foundation 28.7 on the book's website.

28.5.5 Biomotors

Molecular motors are of basic importance in biology. Investigating how they work and how artificial molecular motors can be obtained is a big challenge. We consider three examples: (i) a motor used in building a bacteriophage, one of the simplest organisms; (ii) a motor called kinesin, used for cargo transport from one end of a cell to the other (e.g., vesicles, about 200 nm diameter), and (iii) a motor called ATP synthase, which generates the energy rich molecule ATP that provides energy to drive chemical reactions in biosystems. Both kinesin and ATP synthase are motors present in all biological cells.

28.5.5.1 Motor Portal to Bacteriophage Capsid Packaging Newly Synthesized DNA

Bacteriophage $\varphi 29$ is formed by packaging newly replicated DNA ($L = 6.6$ µm) into a capsid (protein based shell) at nearly crystalline density (Fig. 28.44). To achieve this task entropic, electrostatic, and bending energies of DNA must be overcome. A portal enzyme, which hydrolyzes ATP, acts as a force generating motor [72]. This motor must work against loads of up to about 50 pN; a value deduced from pulling experiments on a DNA strand with an optical trap. The DNA is packaged into the capsid when applying a force below that limit, and the capsid is emptied when applying larger forces. A limit of 50 pN makes this portal enzyme one of the strongest molecular motors. When filling the capsid the work done by the motor against an applied force of 50 pN is 50 pN ×6.6 µm = 3.3×10^{-16} J. In addition, the motor must compress the DNA in the capsid (entropic, electrostatic, and bending energies); however, the accumulated energy can be later used for the ejection of the DNA from the capsid into the interior of the bacterium.

Is the energy for packaging, in addition to the work done by pulling the strand against the applied force, available? It is observed that each base pair, to be incorporated, requires one ATP molecule that is hydrolyzed. According to Equation (18.50), we can write

$$\Delta_r G = \Delta_r G^{\ominus} + \boldsymbol{R}T \ln Q \quad \text{with} \quad Q = \frac{\hat{c}_{\text{ADP}} \cdot \hat{c}_P}{\hat{c}_{\text{ATP}}} \tag{28.46}$$

where $\Delta_r G^{\ominus} = 30.5$ kJ mol^{-1}. Estimating the concentrations of ADP, P, and ATP to be 0.1, 5, and 1 millimol per Liter, respectively, we find $\Delta_r G = 50.0$ kJ mol^{-1} or 20 kT per base pair.

Taking the length of a base pair b to be 0.34 nm, a 6.6 μm DNA strand has $L/b = 6.6$ μm $/0.34$ nm $= 2 \times 10^4$ base pairs. Hence, the energy supply is $2 \times 10^4 \times 20\ kT = 40 \times 10^4 \times 1.38 \times 10^{-23}$ J K^{-1} $\times$ 300 K $= 17 \times 10^{-16}$ J. About four times more energy is used for packaging the DNA into the capsid than the 3.3×10^{-16} J required to move the chain end against an applied force of 50 pN. Even though the working mechanism of this remarkable motor is not yet known, some general energy considerations can be applied to its function.

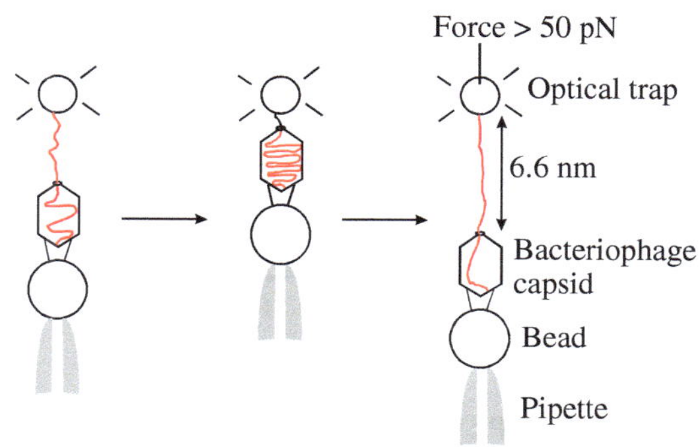

Figure 28.44 Bacteriophage φ29. Packaging DNA in capsid against internal force up to 57 pN.

28.5.5.2 Molecular Motor Kinesin Walking Along a Microtubule

The cytoskeleton of eukaryotic cells comprises three primary types of filaments: actin filaments, intermediate filaments, and microtubules. Microtubules are 25 nm diameter tubes made of the protein tubulin obtained by interlocking of α- and β-subunits (Fig. 28.45). Microtubules are polar; one end is fast growing (+ end), the other end is slow growing (− end). Here we consider the motion of a motor protein, kinesin, along microtubules.

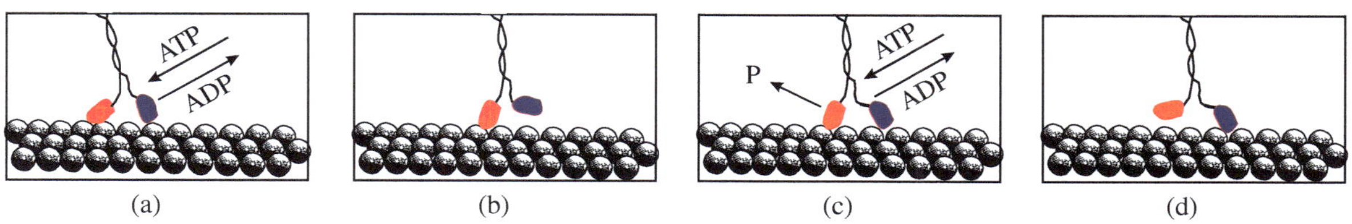

Figure 28.45 Kinesin walking along microtubule.

The protein kinesin moves from the − end to the + end of the microtubule; it is an ATP driven motor. The motor domain of kinesin consists of two "head" units that are connected by "neck" linkers to a coil that extends to a cargo (a vesicle). The heads interact with the α- and β-subunits. The heads cooperate in a manner that leads the kinesin to "walk" along the microtubule (see Fig. 28.45). Consider the beginning to be each of the motor heads carrying an ADP molecule, and the heads loosely bound to the microtubule. In one of the heads the bound ADP is replaced by an ATP molecule, which leads to a conformational change that causes stronger bonding of the head to the microtubule. The bound head pulls the other ADP carrying motor head forward, so it can bind to the next microtubule binding site. This results [73] in the movement of the cargo by about 8 nm [74].

The second head now binds to the microtubule by losing its ADP which is replaced by another ATP molecule. Meanwhile the first head hydrolyzes its ATP and is then snapped forward. The coordinated hydrolysis of the two motor heads is the key to understand the walking, which proceeds with a speed of about one micrometer per second.

Here we compare the work that must be performed for this motion to the Gibbs energy released by ATP hydrolysis. To keep the kinesin at rest requires a force of $f = 5$ pN. The work done per step is $f \times s = 5$ pN $\times$ 8 nm $= 0.4 \times 10^{-19}$ J ($= 10\ kT$ at room temperature). The Gibbs energy that is released by hydrolyzing one ATP molecule under physiological conditions is 20 kT at room temperature (see above). Thus the yield in converting chemical energy into mechanical work is about 50%. For comparison, typical combustion engines operate at 25% to 30% efficiencies.

Using biomotors as a functional component in an engineered environment is a great challenge in the fast developing field of nanotechnology [75]. Figure 28.46 shows an arrangement where kinesin is immobilized on a surface of glass by being incorporated in a monolayer of casein. Microtubules, with an aqueous solution of ATP, are moved by the kinesin molecules. In this experiment, the microtubules are linked to a double-stranded DNA coil so that the moving microtubule, driven by the kinesin motors, uncoils the DNA. This is observed by labeling the DNA with a fluorescent dye [76].

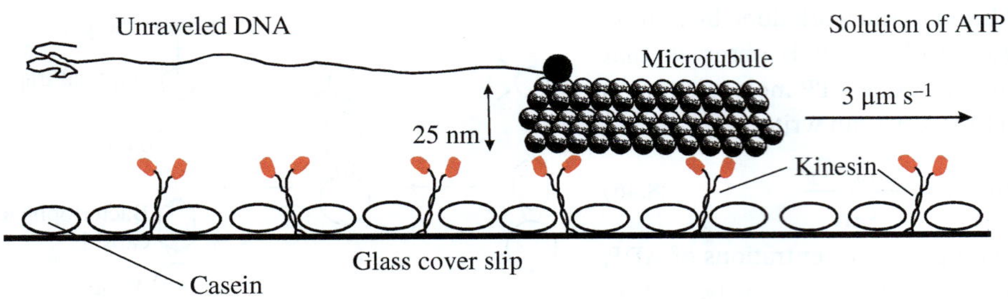

Figure 28.46 Surface coated with motor kinesin. ATP fueled transport of microtubule moving with speed 3 μm s^{-1}. DNA (labeled with a fluorescent dye) bound at its chain end to a microtubule. The coil is stretched due to the motion of the microtubule.

28.5.5.3 ATP Synthase

We have considered ATP synthase in Section 28.3.1.3, as an example where FRET is used as a supramolecular measurement tool. Here we consider ATP synthase as a membrane-bound machine driven by the electrochemical potential difference E given by the membrane potential difference and the difference in pH between the solutions on the positive and the negative sides of the membrane. Depending on the organism the electric or entropic contribution is predominant. Depending on the cell's requirement ATP synthase produces ATP (consuming electrochemical energy) or hydrolyzes ATP (delivering electrochemical energy). In the following we restrict ourselves to the synthesis of ATP, the universal "fuel" of biological systems.

ATP synthase consists of two coupled motors F_0 and F_1. F_0 is located in the membrane and driven by the motion of a proton through the membrane from the electrostatically positive to the negative side. The motor F_1 is located in the solution on the negative side attached to the membrane via F_0, and in bacteria sticking out into the cytoplasm. It is driven by a shaft (the γ-subunit) that is connected to F_0. Mechanical energy builds up by forming a torque in the portion between the "mill wheel" and the solid portion of F_1. The mechanical energy is transferred to F_1 and transformed into chemical energy: F_1 synthesizes ATP from ADP [77].

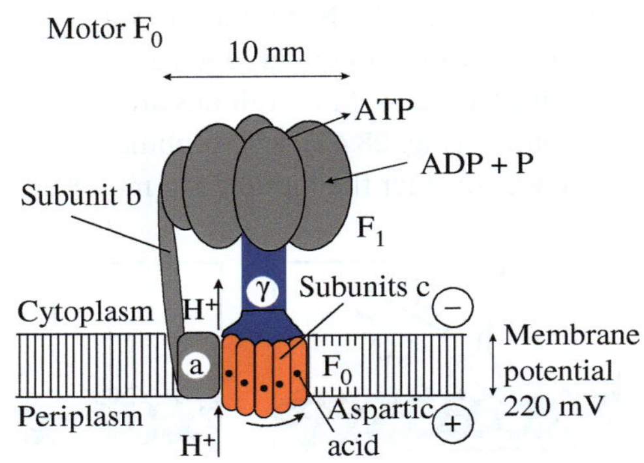

Figure 28.47 ATP synthase. Stator (subunits a, b, F_1: gray), rotor (subunits c: brown, subunit γ: blue). Protons flow from periplasm to cytoplasm, driving rotation of mill wheel. Energy source is the membrane potential and the ion-gradient. Torque transferred to ATP producing device by subunit γ.

Motor F_0 The "mill wheel" in Fig. 28.47 is composed of 10–14 (depending on the organism) copies of subunit c that each contain an essential aspartic or glutamic acid. The acid groups are protonated with the exception of the residue in the interface between subunit a and subunit c that is unprotonated. The transfer of protons from the positive side of the membrane to the negative side causes a rotation of the "mill wheel" that is counterclockwise (seen from the negative side of the membrane).

Motor F_1 A molecule of ATP is made from ADP and P (Fig. 28.48), and this is continuously repeated on the timescale of milliseconds. The formation of ATP under physiological conditions requires $\Delta G = 50$ kJ mol^{-1} $= 20\ kT$ at $T = 298$ K, see above. About 25 kT must be available, which is sufficiently more than $\Delta G = 20\ kT$, to drive the chemical reaction taking place with the given speed. Three ATP molecules are synthesized in a full turn. Thus $3 \cdot 25\ kT$ of mechanical energy must be transferred to F_1. We consider the special case that 12 polypeptides (subunits c) are present that contain an aspartic acid. Thus, to produce 3 ATP molecules, 12 protons are shifted from the solution on the positive side of the membrane to the solution on the negative side. An electrochemical potential difference

$$E = \frac{3 \cdot 25\ kT}{12\ e} = 161\ \text{mV}$$

is minimally required to allow production of ATP.

On what mechanism is the torque-producing motor based? A model must be in accord with the following facts:

1) In a surplus of ATP the process is reversed, ADP and P are formed and protons are pumped from the cytoplasm to the periplasm [78].
2) The rotation of the axis can be visualized by binding an actin filament to the axis and observing its fluorescence in a light microscope [62].
3) The synthesis of ATP from ADP and P can be shown to take place in turning the axis by applying an external force instead of using the energy produced by the ion-transfer from the periplasm to the cytoplasm [79].

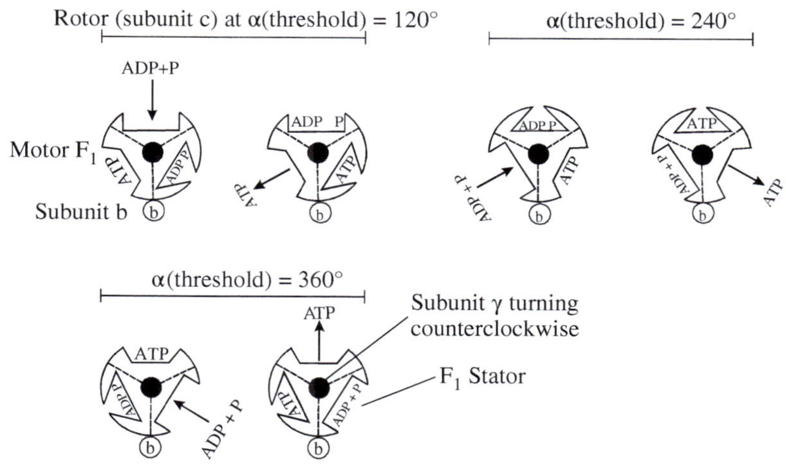

Figure 28.48 ATP producing device F_1, seen from top of Fig. 28.48. Formation of ATP from ADP and P in three steps driven by torque in subunit γ produced by rotation of subunit c (motor F_0, Fig. 28.48). Three ATP molecules are formed in one full turn of subunit c.

28.5.5.4 Mechanism of F_0 Motor

The mechanism of this process is still in discussion. We consider the model proposed by Wolfgang Junge et al. (two-channel model [77]). Location A in Fig. 28.49 is connected by a channel to the solution on the negative side of the membrane. In turning the motor counterclockwise a torque builds up in 4 steps as shown in Fig. 28.49. Then, reaching a turn of 120°, an ATP molecule is released from the F_1 motor and in any further turn of 120° another ATP molecule appears. We have estimated that the mechanical energy to be transferred to F_1 in a 120° turn is 25 kT (see above). The distance for a rotation by 30° is

$$d = \frac{2\pi r}{12} \tag{28.47}$$

Thus for the radius $r = 5$ nm we obtain $d = 2.6$ nm and the average force $\overline{f}_{\text{torque}}$ acting on the aspartate between subunits a and c is

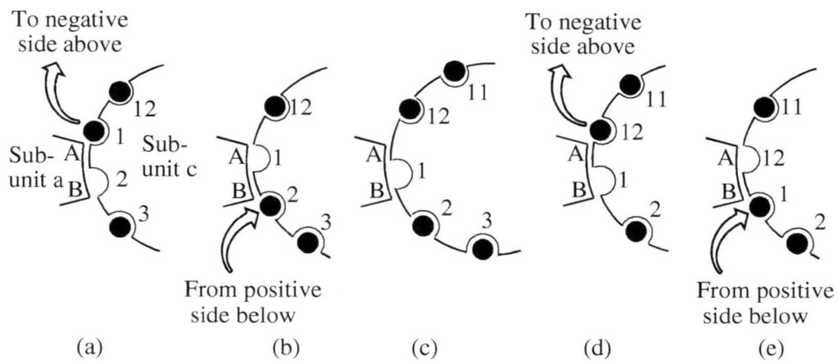

Figure 28.49 Torque producing motor F_0 based on proton flow through membrane. Model showing the proposed mechanism schematically. Subunit a (stator), subunit c (rotor with 10 sites of aspartic acid). Cross-section in the middle of membrane seen from top of Fig. 28.48. Periplasm below paper plane, cytoplasm above paper plane. c rotates counterclockwise during ATP synthesis. Channel to cytoplasm at point A (electrostatic negative), channel to periplasm at point B (electrostatic positive). The electrostatic field between A and B is not sufficient, and a mechanism including Brownian motion is considered to overcome this puzzling situation. The radius of the rotating wheel is $r = 5$ nm.

$$\overline{f}_{\text{torque}} = -\frac{25\ kT}{4\ d} = -10\ \text{pN}$$

where $4d$ corresponds to a rotation by 120°. We estimate the force $\overline{f}_{\text{torque}}$ by the torque in the last of the 4 steps forming an ATP molecule, to be 40% larger than the average because the torque is largest in these steps. That is, the torque is $\overline{f}_{\text{torque}} = -14$ pN.

We focus on the step between Fig. 28.49b,d. Besides the force by the torque tending to turn the rotor clockwise, the force caused by the membrane potential difference acts on the negative charge of the aspartate anion 1 counterclockwise. In our model we estimate the membrane potential as $E_{\text{membrane}} = 130$ mV; thus, the force is

$$f_{\text{field}} = \frac{E_{\text{membrane}} \cdot e}{d} = \frac{130\ \text{mV} \cdot 1.602 \times 10^{-19}\ \text{C}}{2.6\ \text{nm}} = 8\ \text{pN}$$

The total force f acting on anion 1

$$f = f_{\text{torque}} + f_{\text{field}} = -14\ \text{pN} + 8\ \text{pN} = -6\ \text{pN}$$

acts clockwise. How can the rotor then turn counterclockwise against force f?

Modeling Brownian Motor Brownian motion is responsible, leading to a transition from the situation in Fig. 28.49b to the situation in Fig. 28.49d. We consider anion 1 as a "particle" moving stochastically between point A in Fig. 28.49 ($x = 0$) and point B ($x = x_1$). This process is considered in Box 28.2, where it is shown that anion 1, on average, reaches a small interval at $x = 0$ many times ($n_0 = 3751$ times) before reaching an equal interval at $x = x_1$, where an ATP molecule is induced to be released. The stochastic process requires $n_1 = 10{,}372$ steps, on average.

The probability $\mathcal{P}_0$ that a system at $x = 0$ accepts a proton from the solution on the negative side of the membrane and releases a proton into the solution on the positive side (inducing a clockwise turn by 30°) is much smaller than the probability $\mathcal{P}_1$ of the process in Fig. 28.49d,e (inducing a counterclockwise turn by 30°). The ratio of these probabilities is

$$\frac{\mathcal{P}_0}{\mathcal{P}_1} = \exp\left(-\frac{e\,E_{\text{electrochemical}}}{kT}\right)$$

where $E_{\text{electrochemical}}$ is the electrochemical potential difference

$$E_{\text{electrochemical}} = E_{\text{membrane}} + 2.303\frac{\boldsymbol{RT}}{\boldsymbol{F}}\left(\text{pH}_{\text{negativeside}} - \text{pH}_{\text{positiveside}}\right)$$

As an example, we set $\text{pH}_{\text{negativeside}} = 5$. Then with $E_{\text{membrane}} = 130$ mV we obtain $E = 130\text{ mV} + 2 \cdot 59.5\text{ mV} = 249$ mV and find

$$\frac{\mathcal{P}_0}{\mathcal{P}_1} = \exp\left(-\frac{249\text{ mV} \cdot 1.602 \times 10^{-19}\text{ C}}{4.14 \times 10^{-21}\text{ J}}\right) = \text{e}^{-9.6} = 6.5 \times 10^{-5} \tag{28.48}$$

This is a most important fact: a backward motion is essentially prohibited, the behavior is that of a ratchet. Thus the ratio of the occurrence of a clockwise or a counterclockwise turn by 30°, respectively, is

$$Ratio = \frac{n_0 \cdot \mathcal{P}_0}{\mathcal{P}_1} = 3751 \cdot 6.5 \times 10^{-5} = 0.24$$

Now we calculate the probabilities $n_0 \cdot \mathcal{P}_0$ and $\mathcal{P}_1$ for a clockwise and counterclockwise rotation by 30°. Because of

$$n_0 \cdot \mathcal{P}_0 + \mathcal{P}_1 = 1 \qquad \text{and} \qquad Ratio = \frac{n_0 \cdot \mathcal{P}_0}{\mathcal{P}_1}$$

we obtain

$$\mathcal{P}_1 = \frac{1}{1 + Ratio} = \frac{1}{1 + 0.24} = 0.81 \quad \text{and} \quad n_0 \cdot \mathcal{P}_0 = 0.19$$

This means that 19% of the turns will be clockwise and 81% will be counterclockwise, and we can practically neglect the clockwise turns.

Time of Rotation To estimate the speed of rotation u we assume that the rotor is surrounded by a liquid with the viscosity of a lipid bilayer ($\eta \approx 0.1$ kg m^{-1} s^{-1}).

Thinking in terms of hydrodynamics (Section 26.3), f_{viscous} is proportional to η, proportional to the area $2\pi rl$, and proportional to the velocity gradient, which is roughly given by u/r as you can see by rotating a cylinder with diameter r and length l (a pencil or a pill-box) in honey

$$f_{\text{viscous}} = \eta \cdot 2\pi rl \cdot \frac{u}{r} = \eta \cdot 2\pi l \cdot u$$

Then

$$u = \frac{f_{\text{viscous}}}{2\pi\eta \cdot l}$$

The time Δt for each step c is given by the Einstein–Smoluchowski relation

$$\overline{(\Delta x)^2} = 2D\Delta t$$

where Δx at time Δt is the distance of the "particle" from its position in the beginning, and D is the diffusion coefficient. D is given by the relation

$$D = \frac{kT}{f/u} = \frac{kT}{2\pi\eta l}$$

and we set (in analogy to the treatment in Box 14.4),

$$a^2 = \left(\overline{\Delta x}\right)^2 = \frac{1}{2}\overline{(\Delta x)^2}$$

Thus

$$\Delta t = \frac{\overline{(\Delta x)^2}}{2D} = \frac{2a^2}{2D} = 2a^2\frac{\pi\eta l}{kT}$$

With $a = 0.26$ nm, $\eta = 0.1$ kg m^{-1} s^{-1}, and $l = 6$ nm we obtain $\Delta t = 6 \times 10^{-8}$ s. Thus the rotation by 30° requires

$$\Delta t \cdot n_1 = 6 \times 10^{-8}\ \text{s} \cdot 7066 = 4 \times 10^{-3}\ \text{s} = 0.4\ \text{ms}$$

The full counterclockwise rotation (synthesis of three ATP molecules) requires 3×0.4 ms = 1.2 ms since the previous steps in the ATP formation require less time. This is on the order of the experimental finding.

28.5.6 Conclusion

We have discussed a novel subfield of physical chemistry, the design and construction of entities with planned functionalities where interlocking single molecules are the functional components. This allows one to investigate a great variety of new and exciting phenomena. For example, we have considered an aggregate of exactly interlocking dye molecules. When exciting with light an exciton is formed that moves with high speed through the aggregate and is annihilated by meeting and exciting a molecule of an acceptor present in extremely high dilution. This process takes place only if the acceptor is incorporated in the aggregate in a very specific manner. Investigating such processes is based on building distinctly planned arrangements of different kinds of molecules. We discussed a number of different arrangements to manipulate photon motion, electron motion, and nuclear motion.

28.6 Box 28.1. Polarization of Retinal in the Coulomb Field of Amino Acids D85 and D212

The chromophore has alternating single and double bonds and the electron in the HOMO is largely confined to the double bonds, whereas the electron in the LUMO is more delocalized. Thus the chromophore's excited state is much more polarizable than in its ground state. When the *trans*-isomer (protonated Schiff base) is excited, the π-electron cloud is polarized in the electrostatic field of the negatively charged side groups of two amino acids 1 and 2 located close to the Schiff base in the extracellular channel (D85 and D212, see Fig. 28.41).

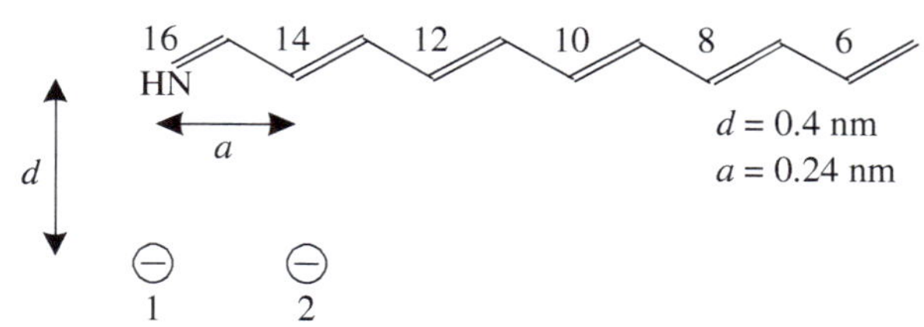

The Coulomb field acting on the π electron cloud is approximately 4×10^9 V m^{-1}, and it is restricted to the section of the chromophore chain between atoms 8 to 14 (see Exercise E28.3). The π electron density distribution of the chromophore in this electrostatic field can be estimated according to the methods described in Chapter 8. In the calculation it is found that the π electron charge at the N atom decreases by $0.2e$ for the excited chromophore. Accordingly, the bonding of the proton to the Schiff base nitrogen is weakened by 0.5 eV when the *trans*-isomer is excited.

We can quantify the equilibrium

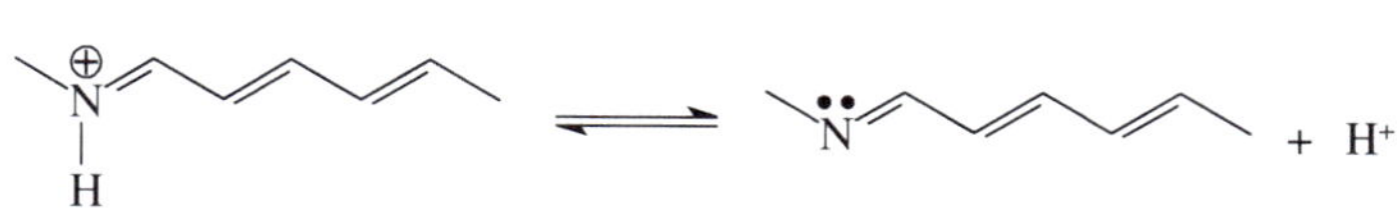

by the equilibrium constant. K_1 is the equilibrium constant before excitation, and K_2 is the equilibrium constant after excitation:

$$\Delta_r G_1^{\ominus} = -\boldsymbol{R}T \cdot \ln K_1 \quad \text{and} \quad \Delta_r G_2^{\ominus} = -\boldsymbol{R}T \cdot \ln K_2$$

The change in pK is

$$\Delta pK = -\frac{1}{2.303} \cdot (\ln K_2 - \ln K_1) = -\frac{1}{2.303} \frac{\Delta_r G_2^{\ominus} - \Delta_r G_1^{\ominus}}{\boldsymbol{R}T}$$

With $\Delta_r G_2^{\ominus} - \Delta_r G_1^{\ominus} = \boldsymbol{N}_A \times 0.5$ eV $= 48$ kJ mol^{-1} we obtain $\Delta pK = -8.3$, slightly larger than the difference between the experimentally determined pK values of D85 and D96 (3.5 and 10).

28.7 Box 28.2. Modeling Brownian Motion

The problem is treating the diffusion of a "particle" driven toward smaller values of x by a force f. Our approach is in the spirit of the considerations in Chapter 14, Section 14.3. We divide the range from $x = 0$ to $x = x_1$ in 10 equal intervals of length a. In the beginning the "particle" is in the first interval, between $x = 0$ and $x = a$. Then it moves stochastically from interval to interval with probabilities $\mathcal{P}_+$ and $\mathcal{P}_-$ in the positive and in the negative direction, respectively. $\mathcal{P}_+$ and $\mathcal{P}_-$ are given by the Boltzmann law as shown below. In the course of the process the "particle" reaches the first interval (between $x = 0$ and $x = a$) many times before reaching the last interval. We assume that each time it is reflected, it leaves the range with a very small probability, according to Equation (28.48).

We solve the given stochastic problem using a computer program. We ask for the number n_1 of steps until the interval between $x_1 - a$ and x_1 is reached, on average. We also ask for the number n_0 of reflections at $x = 0$ during that time.

28.7.1 Calculation of Probabilities $\mathcal{P}_+$ and $\mathcal{P}_-$

According to the Boltzmann law the probability $\mathcal{P}(x) \cdot dx$ to find the "particle" in the interval between x and $x + dx$ in equilibrium condition is given by

$$\mathcal{P}(x) \cdot dx = const\ \exp(-E(x)/(kT)) \cdot dx$$

with

$$E(x) = -f \cdot \frac{x}{x_1} \cdot \frac{2\pi r}{12}$$

where f is the force, r is the radius of the rotating wheel in Fig. 28.49, and $2\pi r/12$ is the distance calculated in Equation (28.47) for a rotation by 30°. With the data $f = -6$ pN and $r = 5$ nm (see text in Fig. 28.49) we obtain

$$E(x) = 6\ \text{pN} \cdot \frac{x}{x_1} \cdot \frac{2\pi \cdot 5\ \text{nm}}{12} = \frac{x}{x_1} \cdot 1.57 \times 10^{-20}\ \text{J}$$

and

$$\frac{E(x)}{kT} = \frac{x}{x_1} \cdot 3.79 \qquad \text{for} \qquad T = 300\ \text{K}$$

Thus

$$\mathcal{P}(x+a) = const \cdot \exp\left(-\frac{E(x+a)}{kT}\right) = const\ \exp\left(-\frac{x}{x_1} \cdot 3.79 - \frac{a}{x_1} \cdot 3.79\right)$$

$$= \mathcal{P}(x)\ \exp\left(-\frac{a}{x_1} \cdot 3.79\right) = \mathcal{P}(x) \cdot \exp(-0.379)$$

where we have used the fact that $a/x_1 = 0.1$, because the range between $x = 0$ and $x = x_1$ is divided into 10 equal intervals.

In order to maintain this probability distribution in equilibrium P_+ and P_- must be given by the relations

$$\mathcal{P}_+ = \frac{\mathcal{P}(x+a)}{\mathcal{P}(x+a) + \mathcal{P}(x-a)} \quad \text{and} \quad \mathcal{P}_- = 1 - \mathcal{P}_+$$

Thus we obtain

$$\mathcal{P}_+ = \frac{e^{-0.379}}{e^{-0.379} + e^{0.379}} = 0.319 \quad \mathcal{P}_- = 1 - 0.319 = 0.681$$

In this way we obtain $n_1 = 10{,}372$ and $n_0 = 3751$, see Problem P28.11. This means it needs a very great number of diffusion steps for the particle to move to the final position $x = x_1$, and about one-third of them are reflections at the position $x = 0$.

28.8 Foundations

28.8.1 Foundation 28.1: Manipulation of Monolayer Assemblies

In Fig. 28F.1 different manipulation processes with monolayers are shown. The precision of the procedures can be checked by using energy transfer as a molecular ruler.

Figure 28F.1 Manipulation of monolayer assemblies. (Adapted from [80].)

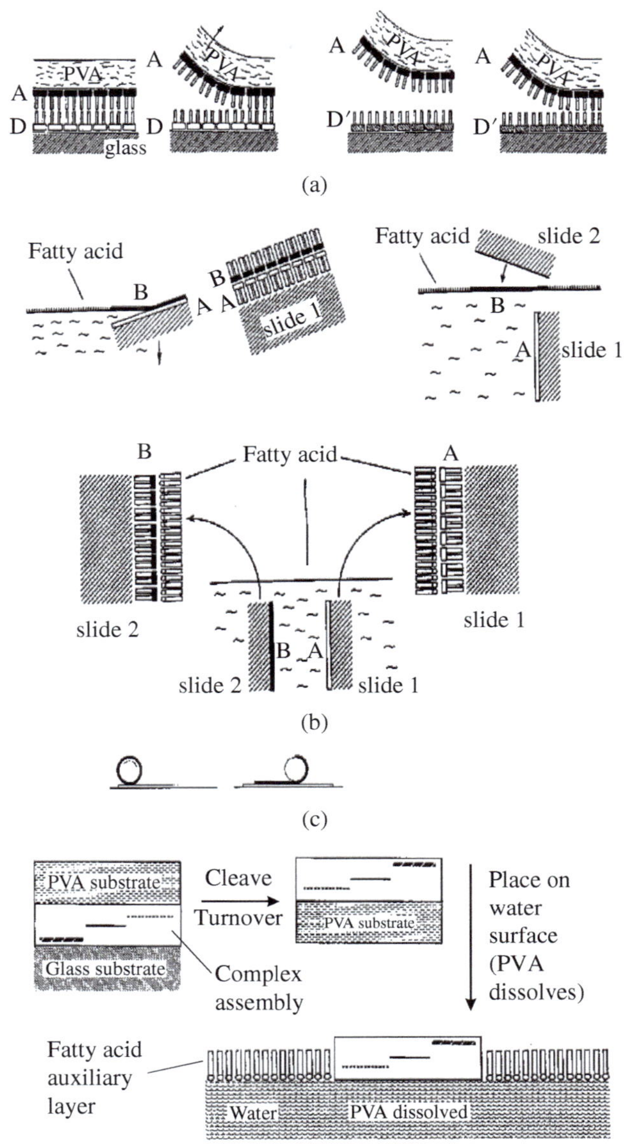

28.8.1.1 Cleavage at Hydrophobic Interface (Fig. 28F.1a)

Left: Monolayer assembly of D and A on a glass plate is covered with a polyvinyl alcohol (PVA) film. Cleavage at the hydrophobic interface between distinct monolayers is induced by lifting the PVA film. The PVA film serves as a transfer vehicle.

Before cleavage: when shining light on the assembly, energy is absorbed by D and transferred to A, resulting in a fluorescence of A. After cleavage: shining the light on the glass slide results in a fluorescence of D; shining the light on the PVA film results in a fluorescence of A. This shows molecular precision in cleavage.

Right: Monolayer A carried by the PVA film is allowed to contact a different dye molecule D′ on a glass plate.

28.8.1.2 Cleavage at Hydrophilic Interface (Fig. 28F.1b)

We consider an assembly of two monolayers A and B on glass slide 1. The assembly is immersed in a water trough, where the water surface is covered with a fatty acid monolayer. A cleavage at the hydrophilic interface between distinct monolayers A and B occurs, allowing monolayer B to float on the water surface.

Subsequently, slide 2, with a hydrophobic surface, is allowed to contact the floating monolayer, B. Both slides (slide 1 with A and slide 2 with B) are then coated with a layer of fatty acid by producing a fatty acid layer on the surface of the water and lifting the slides upward.

28.8.1.3 Dry Transfer from Glass to PVA Film (Fig. 28F.1c)

Dry transfer of a monolayer assembly from a glass tube to a PVA film covered with a monolayer. The transfer is implemented by rolling the tube on the PVA film.

28.8.1.4 Transfer of Organized Structure (Fig. 28F.1d)

The multilayer assembly made by sequentially depositing monolayers on a glass slide is overlaid with a PVA thin film. The assembly and the PVA film are stripped off the glass slide, turned upside down, and placed on the water surface. The PVA film, which serves as a vehicle for this transfer, is then dissolved, leaving the complex assembly on the water surface. A fatty acid monolayer surrounding the assembly aids further manipulation, shown here for deposition on a AgBr crystal. The AgBr crystal, while being lifted upward, is then coated with the monolayer assembly. This method allows the dye to be positioned at various distinct distances from the surface of the AgBr crystal.

28.8.2 Foundation 28.2: Energy Transfer

28.8.2.1 One Donor and One Acceptor in Distance *r*

We consider an excited dye molecule (energy donor D) and an energy acceptor molecule (A) at a distance r. To simplify, we assume that the transition moments of the fluorescence of D and of the absorption of A are in the direction of r. We begin with the classical approach, which is formally identical to the quantum-mechanical treatment as outlined in Foundation 9.1 for related cases. We replace molecule D by a classical oscillator: frequency ν_0, charge Q, mass m, amplitude ξ_0, and energy

$$E = \frac{1}{2} k_f \, \xi_0^2 \tag{28F2.1}$$

The electric field acting on acceptor A is the Coulomb field of the oscillator used to model D (field of a dipole with $\mu = Q\,\xi_0$, see Equation (13.43):

$$F = F_0 \cos 2\pi\nu_0 t \quad \text{and} \quad F_0 = \frac{1}{4\pi\varepsilon_0}\frac{2Q\,\xi_0}{r^3} \tag{28F2.2}$$

where ε_0 is the permittivity of vacuum. It is assumed that r is small compared to the wavelength $\lambda = c_0/\nu_0$, but large compared to the dimension of D (then the point dipole approximation is sufficiently accurate). The mean power absorbed by the acceptor A is proportional to ε_A (molar extinction coefficient of A at the wavelength of the donor fluorescence) and proportional to the square of the field F_0 of the donor at the acceptor (Foundation 9.1).[5]

$$\overline{\left(\frac{dE}{dt}\right)}_{\text{absorption of A}} = g\varepsilon_A F_0^2 \quad \text{with} \quad g = \frac{3}{2}\frac{2.303 \times c_0\varepsilon_0}{\boldsymbol{N}_\text{A}} \tag{28F2.3}$$

The power emitted by the oscillator D (Hertz antenna equation) is, according to Foundation 10.1,

$$-\overline{\left(\frac{dE}{dt}\right)}_{\text{fluorescence of D}} = \frac{1}{4\pi\varepsilon_0}\frac{2}{3}\cdot\frac{Q^2}{m}\cdot\frac{4\pi^2\nu_0^2}{c_0^3}E = \frac{1}{\tau_0}E \quad \text{with} \quad \tau_0 = \frac{3}{2}\frac{m}{Q^2}\frac{\varepsilon_0 c_0^3}{\pi\nu_0^2} \tag{28F2.4}$$

where

$$E = \frac{1}{2}k_f\xi_0^2 \quad \text{and} \quad \frac{k_f}{m} = 4\pi^2\nu_0^2 \tag{28F2.5}$$

At a certain distance $r = r_0$ the emission is quenched to half of the value without acceptor A; namely

$$\overline{\left(\frac{dE}{dt}\right)}_{\text{absorption of A}} = -\overline{\left(\frac{dE}{dt}\right)}_{\text{fluorescence of D}} \tag{28F2.6}$$

5 The factor 3 in the present case takes into account the fact that the transition moment of the acceptor molecule A is parallel to field F. In contrast, the molar decadic absorption coefficient ε_A of A as defined in Foundation 9.1 refers to a statistical spatial distribution of the transition moments of the absorbing molecules (or to an isotropic oscillator).

at $r = r_0$. Hence

$$g\varepsilon_A F_0^2 = g\varepsilon_A \left(\frac{2Q\xi_0}{4\pi\varepsilon_0 r_0^3} \right)^2 = g\varepsilon_A \frac{4Q^2\xi_0^2}{16\pi^2\varepsilon_0^2 r_0^6} = \frac{1}{\tau_0} E \tag{28F2.7}$$

Therefore we obtain

$$r_0 = \left(\frac{9\varepsilon_A \times 2.303 \times c_0^4}{128\pi^5 \mathbf{N}_A v_0^4} \cdot 4 \right)^{1/6} \tag{28F2.8}$$

This relation holds for quantum yields of fluorescence in the absence of the acceptor, $\phi = 1$ (energy emitted exclusively by radiation), for refractive index of the medium $\hat{n} = 1$ (vacuum), and for orientation of the transition moments parallel to r. In the general case [81] the factor 4 in the brackets has to be replaced by $\kappa^2\phi/\hat{n}^4$, where κ is a measure of the relative orientation of the transition moments of molecules 1 and 2: $\kappa = 2\cos\Theta_1 \cos\Theta_2 - \sin\Theta_1 \sin\Theta_2 \cos\xi$.

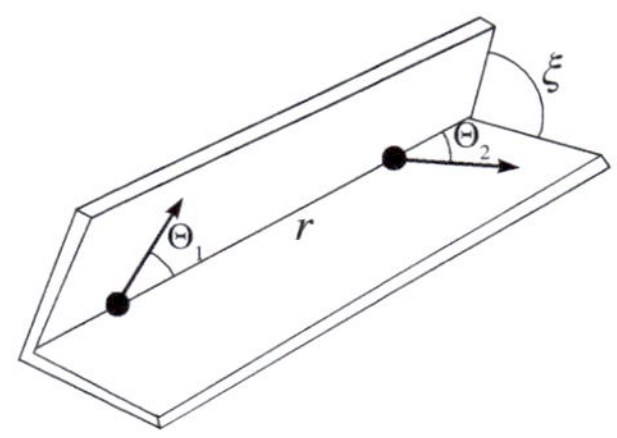

In terms of quantum theory, the probability of energy transfer is

$$\mathcal{P} = \frac{\overline{(\mathrm{d}E/\mathrm{d}t)}_{\text{absorption of A}}}{\overline{(\mathrm{d}E/\mathrm{d}t)}_{\text{fluorescence of D}} + \overline{(\mathrm{d}E/\mathrm{d}t)}_{\text{absorption of A}}} = \frac{1}{1+C} \tag{28F2.9}$$

with

$$C = \frac{\overline{(\mathrm{d}E/\mathrm{d}t)}_{\text{absorption of A}}}{\overline{(\mathrm{d}E/\mathrm{d}t)}_{\text{fluorescence of D}}} = \left(\frac{r}{r_0} \right)^6 \tag{28F2.10}$$

Then we obtain

$$\mathcal{P} = \frac{1}{1 + (r/r_0)^6} \tag{28F2.11}$$

Apparently, r_0 is the distance where $\mathcal{P} = \frac{1}{2}$. Correspondingly, the probability of emission of light is

$$1 - \mathcal{P} = \frac{1}{1 + (r_0/r)^6} \tag{28F2.12}$$

(*Förster Equation*).

In the actual experiment we measure the intensity ratio $I_r(\mathrm{A})/I_0(\mathrm{A})$ of the fluorescence of the acceptor molecules A, where $I_r(\mathrm{A})$ is the intensity at distance r and $I_0(\mathrm{A})$ is the intensity at distance $r = 0$. This ratio equals the probability $\mathcal{P}$. On the other hand, $I_r(\mathrm{D})/I_\infty(\mathrm{D})$ is the intensity ratio of the fluorescence of the donor molecules, where $I_\infty(\mathrm{D})$ is the intensity at distance $r = \infty$. Thus we obtain

$$\frac{I_r(\mathrm{D})}{I_\infty(\mathrm{D})} = \frac{1}{1 + \left(\frac{r_0}{r}\right)^6} \quad \text{and} \quad \frac{I_r(\mathrm{A})}{I_0(\mathrm{A})} = \frac{1}{1 + \left(\frac{r}{r_0}\right)^6} \tag{28F2.13}$$

28.8.2.2 Molecules in a Planar Layer

If the donor molecule D is spherically surrounded by a layer of acceptor molecules A, then $\mathrm{d}E/\mathrm{d}t$ is proportional to the number of molecules A in the layer; this number increases with r^2 changing the $1/r^6$ dependence into a $1/r^4$ dependence. There is no essential change in the result if the acceptor molecules are in a plane in distance d from the donor [3]:

$$\mathcal{P} = \frac{1}{1 + (d/d_0)^4} \quad \text{where} \quad d_0 = \frac{1}{4\pi} \left(\frac{9}{4} \right)^{1/4} \frac{\lambda_D}{\hat{n}} (\phi A_\mathrm{A})^{1/4} \tag{28F2.14}$$

if the transition moments are statistically distributed in the layer plane. A_A is the absorbance of the layer of acceptor A for light of vacuum wavelength λ_D of fluorescence of donor D, $\hat{n}$ is the refractive index of the medium, and ϕ is the quantum yield of fluorescence of donor D in the absence of acceptor A.

The probability of fluorescence of the donor D is $(1 - \mathcal{P})$; thus,

$$\frac{I_d(\mathrm{D})}{I_\infty(\mathrm{D})} = 1 - \mathcal{P} = \frac{1}{1 + \left(d_0/d\right)^4} \tag{28F2.15}$$

where $I_d(\mathrm{D})$ is the fluorescence intensity of the donor at distance d (Fig. 28.6). This is the result in Equation (28.1). Correspondingly, for the fluorescence of the acceptor A we obtain

$$\frac{I_d\ (\mathrm{A})}{I_0\ (\mathrm{A})} = \mathcal{P} = \frac{1}{1 + \left(d/d_0\right)^4} \tag{28F2.16}$$

28.8.2.3 Relation Between r_0 and d_0

From Equations (28F2.8) and (28F2.14) we obtain

$$r_0^6 = \frac{9\varepsilon_A \times 2.303 \times c_0^4}{128\pi^5 \mathbf{N}_\mathrm{A} \nu_0^4} \cdot \frac{\kappa^2 \phi}{\hat{n}^4}$$

and

$$d_0^4 = \frac{1}{4^4\pi^4} \frac{9}{4} \frac{\lambda_D^4}{\hat{n}^4} \phi A_\mathrm{A}$$

Thus we find

$$\frac{r_0^6}{d_0^4} = \frac{9\varepsilon_A \times 2.303 \times c_0^4 \cdot \kappa^2 \phi \cdot 4^4 \pi^4 \cdot 4\hat{n}^4}{128\pi^5 \mathbf{N}_\mathrm{A} \nu_0^4 \hat{n}^4 \cdot 9\lambda_D^4 \cdot \phi A_\mathrm{A}}$$
$$= \frac{\varepsilon_A \times 2.303 \times c_0^4 \cdot \kappa^2 \cdot 4^4 \pi^4 \cdot 4}{128\pi^5 \mathbf{N}_\mathrm{A} \nu_0^4 \cdot \lambda_D^4 \cdot A_\mathrm{A}}$$
$$= \frac{8\varepsilon_A \times 2.303 \times c_0^4 \cdot \kappa^2}{\pi \mathbf{N}_\mathrm{A} \nu_0^4 \cdot \left(c_0^4/\nu_0^4\right) \cdot A_\mathrm{A}} = \frac{8\varepsilon_A \times 2.303 \times \kappa^2}{\pi \mathbf{N}_\mathrm{A} \cdot A_\mathrm{A}}$$

Using

$$A_A = 1 - T_A, \quad T_A = \frac{I}{I_0}$$

where A is the absorption and T is the transmission, and

$$\text{absorbance} = -\log T = \varepsilon_A c l$$

(c = concentration, l = optical path length), we find for small enough absorption that

$$\log T = \log\left(1 - A_A\right) = -\frac{1}{2.303} A_A$$

Thus

$$\varepsilon_A c l = \frac{1}{2.303} A_A \qquad \text{or} \qquad \varepsilon_A = \frac{1}{2.303 \cdot cl} A_A$$

and with

$$cl = \frac{\boldsymbol{n}}{V} l = \frac{N}{\mathbf{N}_\mathrm{A} V} l = \frac{N}{\mathbf{N}_\mathrm{A} V/l} = \frac{N}{\mathbf{N}_\mathrm{A} \cdot \text{area}} = \frac{\sigma_A}{\mathbf{N}_\mathrm{A}}$$

(σ_A is the number of acceptor molecules per unit area) we obtain

$$\frac{r_0^6}{d_0^4} = \frac{8 \times 2.303 \times \kappa^2}{\pi \mathbf{N}_\mathrm{A} \cdot A_\mathrm{A}} \cdot \frac{1}{2.303 \cdot cl} A_A$$
$$= \frac{8\kappa^2}{\pi \mathbf{N}_\mathrm{A}} \cdot \frac{1}{cl} = \frac{8\kappa^2}{\pi \mathbf{N}_\mathrm{A}} \cdot \frac{\mathbf{N}_\mathrm{A}}{\sigma_A} = \frac{8\kappa^2}{\pi \sigma_A}$$

In the case that we can set $\kappa = 1$, we find

$$\frac{r_0^6}{d_0^4} = \frac{8}{\pi \sigma_A} \quad \text{or} \quad r_0 = \left(d_0^4 \frac{8}{\pi \sigma_A}\right)^{1/6}$$

28.8.2.4 Comparison of Absorbed Power in Far Field and Near Field

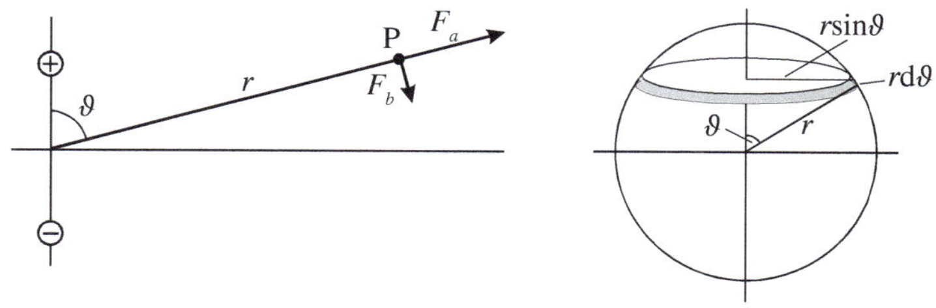

(a) (b)

According to Equation (28F2.3) the absorbed power is

$$\overline{\left(\frac{\mathrm{d}E}{\mathrm{d}t}\right)}_{\text{absorption of A}} = g\varepsilon_A F_0^2 \quad \text{with} \quad g = \frac{3}{2}\frac{2.303 \times c_0\varepsilon_0}{\boldsymbol{N}_\mathrm{A}}$$

According to Foundation 10.1 the amplitude of the field strength of an oscillating dipole perpendicular to the direction of the dipole ($\vartheta = 90°$) is

$$F_0 = F_{b,0} = \frac{Q\xi_0}{4\pi\varepsilon_0} \cdot \left\{ -\frac{\omega^2}{c_0^2 r} - \frac{\omega}{c_0 r^2} + \frac{1}{r^3} \right\}$$

From this relation we obtain for the far field

$$F_0 = F_{b,0} = \frac{Q\xi_0}{4\pi\varepsilon_0} \cdot \frac{\omega^2}{c_0^2 r} \qquad \text{(far field)}$$

and for the near field

$$F_0 = F_{b,0} = \frac{Q\xi_0}{4\pi\varepsilon_0} \cdot \frac{1}{r^3} \qquad \text{(near field)}$$

for the near field. Thus the ratio of the absorbed powers is

$$\frac{\overline{\left(\frac{\mathrm{d}E}{\mathrm{d}t}\right)}_{\text{absorption of A, far field}}}{\overline{\left(\frac{\mathrm{d}E}{\mathrm{d}t}\right)}_{\text{absorption of A, near field}}} = \frac{\left(\frac{Q\xi_0}{4\pi\varepsilon_0} \cdot \frac{\omega^2}{c_0^2 r}\right)^2}{\left(\frac{Q\xi_0}{4\pi\varepsilon_0} \cdot \frac{1}{r^3}\right)^2} = \left(\frac{\omega}{c_0} r\right)^4 = \left(\frac{2\pi}{\lambda} r\right)^4$$

In the case that $r > \lambda$ the far-field absorption is greater than the near-field contribution. This is the usual case in spectroscopy in measuring the extinction coefficient ε. In the opposite case that $r < \lambda$ the near-field absorption is greater than the far-field contribution leading to a greater value of the extinction coefficient. For example, for a wavelength of $\lambda = 500$ nm and a distance $r = 2$ nm (as used in the monolayer experiments) we have ε(near field) $= 4 \times 10^7$ ε(far field). Because of this fact we have a high absorbance of a monolayer of absorber molecules in spite of its extremely small thickness.

28.8.3 Foundation 28.3: Energy Transfer from Exciton to Acceptor

See the book's website,www.wiley.com/go/Kuhn/physicalchemistry3e.

28.8.4 Foundation 28.4: Radiation Echo Field

See the book's website,www.wiley.com/go/Kuhn/physicalchemistry3e.

28.8.5 Foundation 28.5: Electron Transfer Between π-Electron Systems

In Foundation 6.3 (oscillation of electron between two protons) we have seen that an electron, first at proton a, oscillates between protons a and b located at a distance d. The probability $\mathcal{P}_b$ of finding the electron at proton b is

$$\mathcal{P}_b = \sin^2\left(\frac{\varepsilon_2 - \varepsilon_1}{2\hbar} t\right) = \sin^2\left(\frac{\varepsilon}{\hbar} t\right) \quad \text{with} \quad \varepsilon = \frac{1}{2}(\varepsilon_2 - \varepsilon_1) \tag{28F5.1}$$

and ε_1 and ε_2 are the energies of the system in the ground state and the first excited state, respectively.

28.8.5.1 General Case

We generalize this consideration for an electron oscillating between two different molecules a and b. According to Foundation 6.3,

$$2\varepsilon = \varepsilon_2 - \varepsilon_1 = 2(H_{aa}S_{ab} - H_{ab})$$

In that case, for reasons of symmetry we have

$$H_{ab} = H_{ba} \quad \text{and} \quad H_{aa} = H_{bb}$$

In our present case, however,

$$H_{ab} \neq H_{ba} \quad \text{and} \quad H_{aa} \neq H_{bb}$$

In this more general case we must analyze

$$2\varepsilon = H_{aa}S_{ab} - H_{ab} + H_{bb}S_{ab} - H_{ba}$$

where

$$H_{ab} = \int \psi_a \left[-\frac{h^2}{8\pi^2 m} \left(\frac{\partial^2}{\partial x^2} + \frac{\partial^2}{\partial y^2} + \frac{\partial^2}{\partial z^2} \right) + V_a \right] \psi_b \cdot \mathrm{d}\tau$$
$$= \int \psi_a \mathcal{H}_b \psi_b \cdot \mathrm{d}\tau + \int \psi_a V_a \psi_b \cdot \mathrm{d}\tau = E_b S_{ab} + \int \psi_a V_a \psi_b \cdot \mathrm{d}\tau$$

and correspondingly,

$$H_{ba} = E_a S_{ab} + \int \psi_a V_b \psi_b \cdot \mathrm{d}\tau, \quad H_{aa} = E_a, \quad H_{bb} = E_b$$

Then it follows that

$$2\varepsilon = E_a S_{ab} - E_b S_{ab} - \int \psi_a V_a \psi_b \cdot \mathrm{d}\tau + E_b S_{ab} - E_a S_{ab} - \int \psi_a V_b \psi_b \cdot \mathrm{d}\tau$$
$$= -\int \psi_a \left(V_a + V_b \right) \psi_b \cdot \mathrm{d}\tau$$

and for ε we obtain

$$\varepsilon = -\int \psi_a \left(\frac{V_a + V_b}{2} \right) \psi_b \cdot \mathrm{d}x\,\mathrm{d}y\,\mathrm{d}z \quad (28F5.2)$$

ψ_a and ψ_b are the wavefunctions and V_a and V_b are the potential energies of the electron in the field of molecules a and b, respectively; ε is called the *transfer matrix element*. As in Foundation 6.3, ε depends exponentially on the distance d between the two molecules:

$$\varepsilon \sim \mathrm{e}^{-\alpha d} \quad \text{with} \quad \alpha = \frac{2\pi}{h}\sqrt{2m_e E_{\text{barrier}}} \quad (28F5.3)$$

where E_{barrier} is the energy barrier to be surmounted by the electron.

In the derivation of the transfer probability $\mathcal{P}_b$ in Equation (28F5.1), the energies E_a and E_b of the electron in molecules a and b have been assumed to be equal. In the general case

$$\Delta E = E_b - E_a \neq 0$$

the result of the corresponding consideration is

$$\mathcal{P}_b = \frac{1}{1 + \left(\frac{\Delta E}{2\varepsilon} \right)^2} \cdot \sin^2 \left[\sqrt{1 + \left(\frac{\Delta E}{2\varepsilon} \right)^2} \frac{\varepsilon}{\hbar} t \right] \quad (28F5.4)$$

This means that $\mathcal{P}_b$ decreases strongly with increasing ΔE. For $\Delta E = 0$ a maximum of $\mathcal{P}_b$ is achieved. The situation is similar to two coupled penduli when the first one is excited at $t = 0$. The excitation is transferred back and forth between both penduli only in the case of resonance (when both penduli are identical).

28.8.5.2 Rate of Electron Transfer

As an example, we consider the electron transfer from bacteriopheophytin to a quinone, like that in the reaction center of the photosynthetic apparatus:

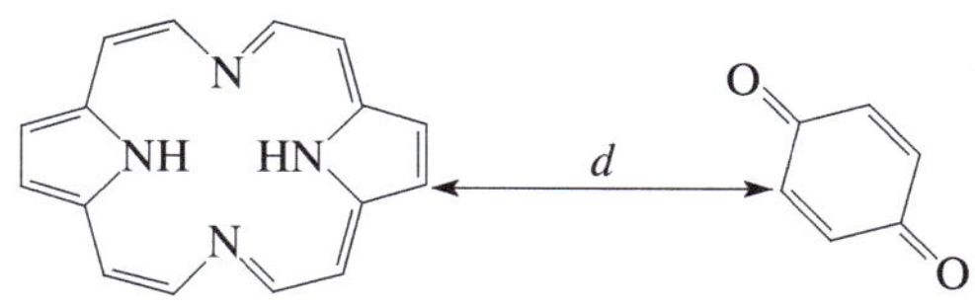

The electron transfer is accompanied by a decrease in Gibbs energy G ($\Delta G^{\ominus} = -0.47$ eV or $\Delta_r G^{\ominus} = -45$ kJ mol^{-1}). As mentioned above, the maximum electron transfer rate is achieved when the energy of the electron is the same in both systems (resonance). This condition can be realized by a transition from the original donor ground state (wavefunction ψ_a, energy E_a) to a vibronically excited state of the acceptor at the same energy (wavefunction ψ_b, energy $E_b = E_a$). E_a and E_b are fluctuating in a fluctuating medium (protein); thus, ΔE is fluctuating as well. The energy match required for electron transfer is only occasionally reached for a short characteristic time t_c (time between fluctuations; t_c typically ranges from picoseconds to femtoseconds).

We consider the limit where the characteristic time t_c is short compared with the oscillation period of the electron. From Equation (28F5.4) at $\Delta E = E_b - E_a = 0$, it follows that:

$$\mathcal{P}_b = \sin^2\left(\frac{\varepsilon}{\hbar}t\right) \tag{28F5.5}$$

so that the oscillation period T is

$$\frac{\varepsilon}{\hbar} = \frac{\pi}{T} \quad \text{or} \quad T = \frac{\pi\hbar}{\varepsilon}$$

A typical value for ε (see below) is $\varepsilon = 1 \times 10^{-4}$ eV $= 2 \times 10^{-23}$ J, so that $T = 10$ ps. In such a case, we can simplify the expression for $\mathcal{P}_b$ for small enough time t:

$$\mathcal{P}_b = \sin^2\left(\frac{\varepsilon}{\hbar}t\right) \approx \left(\frac{\varepsilon}{\hbar}t\right)^2 \tag{28F5.6}$$

The probability $\mathcal{P}_{\Delta E}$ of an energy match can be estimated by the ratio between the uncertainty in energy, given by the Heisenberg uncertainty relation (Section 3.6)

$$\Delta E = \frac{h}{t_c} \tag{28F5.7}$$

and the range of fluctuations, E_{fluct},

$$\mathcal{P}_{\Delta E} = \frac{\Delta E}{E_{\text{fluct}}} = \frac{h/t_c}{E_{\text{fluct}}} \tag{28F5.8}$$

Then the rate constant k_r for electron transfer is

$$k_r = \begin{array}{l}\text{number}\\ \text{of attempts}\\ \text{per unit time}\end{array} \times \begin{array}{l}\text{probability } \mathcal{P}_{\Delta E}\\ \text{of energy}\\ \text{match}\end{array} \times \begin{array}{l}\text{probability}\\ \text{of electron transfer}\\ \text{during match}\end{array}$$

or

$$k_r = \frac{1}{t_c} \times \frac{h}{t_c\, E_{\text{fluct}}} \times \left(\frac{\varepsilon}{\hbar}t_c\right)^2 = \frac{2\pi}{\hbar}\varepsilon^2\frac{1}{E_{\text{fluct}}} \tag{28F5.9}$$

For $E_{\text{fluct}} = 0.1$ eV $= 0.16 \times 10^{-19}$ J, and $\varepsilon = 1 \times 10^{-4}$ eV $= 1.602 \times 10^{-23}$ J, the rate constant is $k_r = 9.5 \times 10^8$ s^{-1}.

28.8.5.3 Evaluation of ε

As in Section 11.5, we treat the wavefunctions of molecules a and b as products of the electronic part ψ_{el} (depending on the electron coordinates x, y, z) and the vibronic part ψ_{vib} (depending on the nuclear coordinates ξ):

$$\psi_a = \psi_{a,\text{el}} \cdot \psi_{a,\text{vib}}, \quad \psi_b = \psi_{b,\text{el}} \cdot \psi_{b,\text{vib}}$$

(Born–Oppenheimer approximation). To simplify, we consider one nuclear coordinate. Then for ε we obtain

$$\varepsilon = -\int \psi_{a,\text{el}} \cdot \psi_{a,\text{vib}} \left(\frac{V_a + V_b}{2}\right) \psi_{b,\text{el}} \cdot \psi_{b,\text{vib}} \cdot \, dx\, dy\, dz \cdot \, d\xi$$

We apply the same procedure as for the calculation of transition moments in Section 11.5: we use the wavefunctions of the harmonic oscillator (Fig. 11.12) and evaluate the overlap integral of the vibrational wavefunctions. Thus we obtain

$$\varepsilon = \varepsilon_{\text{electronic}} \cdot \left(\frac{\beta^n}{n!} \, \mathrm{e}^{-\beta} \right)^{1/2} \tag{28F5.10}$$

with

$$\varepsilon_{\text{electronic}} = \int \psi_{a,\text{el}} \left(\frac{V_a + V_b}{2} \right) \psi_{b,\text{el}} \cdot \, \mathrm{d}x \, \mathrm{d}y \, \mathrm{d}z \quad \text{and} \quad \beta = \frac{E^*}{h\nu_0} \tag{28F5.11}$$

n is the quantum state of vibration in molecule b reached after electron transfer; E^* is the vibration energy of all vibrating nuclei.

Example 28.2 *Contribution of Vibration to* ε.

During the electron transfer from a bacteriopheophytin to a quinone, the quinone is reduced to a semiquinone:

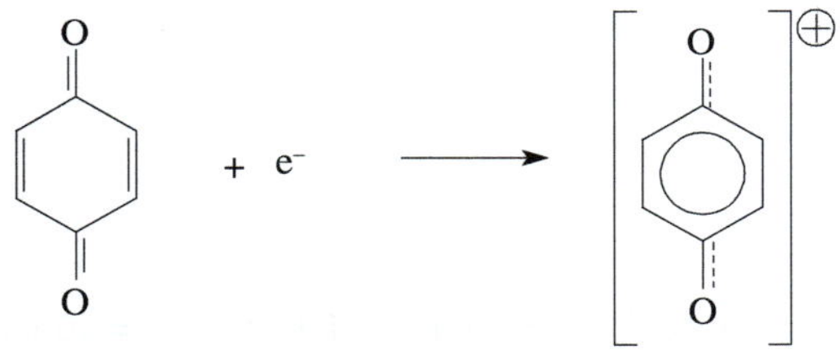

Thus alternating bond lengths are transformed into equal bond lengths giving rise to bond vibrations indicated by arrows:

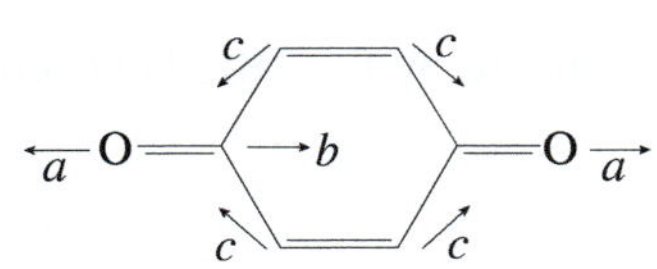

$a = 6$ pm
$b = 2$ pm
$c = 4$ pm

The bond lengths in bacteriopheophytin, because of the large extension of the π system, remain nearly unchanged. We approximate all vibrations in the quinone by one effective frequency ν_0 (C–C valence frequency, $\nu_0 = 4 \times 10^{13}\ \text{s}^{-1}$) with elongations a, b, and c. We express E^* by the force constants k_f and the elongations ξ_i of all vibrating nuclei (assuming the same mass $m = m_C$)

$$E^* = \sum \tfrac{1}{2} k_f \, \xi_i^2 = 2\pi^2 \nu_0^2 \, m \sum \xi_i^2$$

(k_f expressed by the frequency ν_0). Then

$$\beta = \frac{2\pi^2 \nu_0 \, m}{h} \sum \xi_i^2$$

With

$$\sum \xi_i^2 = 2\xi_O^2 + 4\xi_{C,1}^2 + 2\xi_{C,2}^2 = 2a + 4c + 2b = 144\ \text{pm}^2$$

we obtain $\beta = 3.4$; including the donor in the calculation results in $\beta = 4$. The quantum number $n = 3$ is estimated as the next integer to $-\Delta G^{\ominus}/(h\nu_0) = 2.8$ (for $\Delta G^{\ominus} = -0.47$ eV). Then we obtain

$$\varepsilon = \varepsilon_{\text{electronic}} \times \left(\frac{\beta^n}{n!} \, \mathrm{e}^{-\beta} \right)^{1/2} = \varepsilon_{\text{electronic}} \times \left(\frac{4^3}{6} \, \mathrm{e}^{-4} \right)^{1/2}$$
$$= \varepsilon_{\text{electronic}} \times 0.44$$

28.8.5.4 Evaluation of $\varepsilon_{\text{electronic}}$

In evaluating the expression

$$\varepsilon_{\text{electronic}} = \int \psi_{a,\text{el}} \left(\frac{V_a + V_b}{2} \right) \psi_{b,\text{el}} \cdot \mathrm{d}x\, \mathrm{d}y\, \mathrm{d}z \tag{28F5.12}$$

we restrict to the contribution of the shaded antinodes, where the edges of donor and acceptor are closest together, to $\psi_{a,\text{el}}$ and $\psi_{b,\text{el}}$:

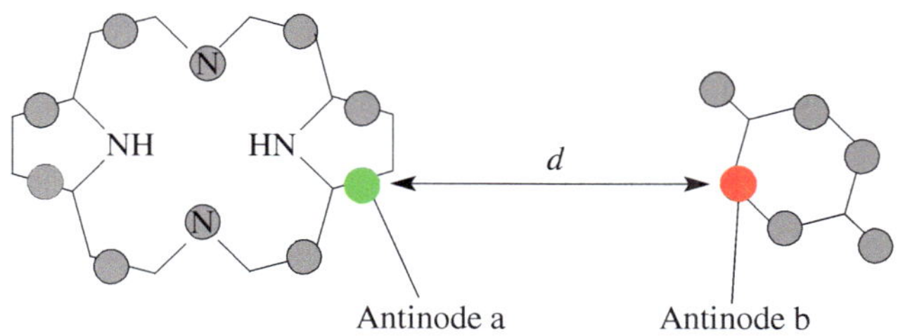

We describe the overlap of the wavefunctions by the coefficients c_a and c_b, where c_a^2 and c_b^2 are the probabilities of finding the electron in antinode a and b, respectively (before and after transfer).

Furthermore, the overlap depends on the geometry, which can be described by a geometrical factor κ. Then using the relation derived in Foundation 6.3 for an electron tunneling from proton a to proton b, we obtain

$$\varepsilon_{\text{electronic}} = E_{\text{barrier}} \cdot c_a c_b \kappa \cdot \mathrm{e}^{-\alpha d} \quad \text{with} \quad \alpha = \frac{2\pi}{h} \sqrt{2 m_e E_{\text{barrier}}} \tag{28F5.13}$$

Example 28.3 *Electron transfer from bacteriopheophytin to quinone.*

In the considered case of electron transfer from bacteriopheophytin to quinone $c_a = 0.303$ and $c_b = 0.28$ (calculated from the π electron densities), $\kappa = 0.5$ (from the geometry available from X-ray experiments) and $d = 0.96$ nm. We assume that E_{barrier} is the energy to transfer the electron from the donor into a dielectric medium. This energy is less than the energy E_{ion} required to transfer the electron into the vacuum.

In treating π electrons (see Chapters 8 and 9) we have considered the polarization of the electron cloud by assuming that the C atom next to the π electron is unshielded. This means that a spherical hole is formed near the π electron (Fermi hole with radius $r = 0.154$ nm, corresponding to the bond distance in an alkane). Then, according to a consideration analogous to that given in Section 23.4.2 (Born model) the energy to bring the electron from the vacuum into a medium with the relative permittivity $\hat{\varepsilon}$ is

$$\begin{aligned}\Delta E &= -\frac{e^2}{8\pi\varepsilon_0 r}\left(1 - \frac{1}{\hat{\varepsilon}}\right) = -7.49 \times 10^{-19}\ \text{J} \cdot \left(1 - \frac{1}{\hat{\varepsilon}}\right) \\ &= -4.68\ \text{eV} \cdot \left(1 - \frac{1}{\hat{\varepsilon}}\right)\end{aligned}$$

With $\hat{\varepsilon} = 2.34$ for hydrocarbon[6], we find $\Delta E = -2.67$ eV. Thus the upper level of the barrier is about 2.67 eV below the vacuum level.

Note that the value $\Delta E = -2.67$ eV follows from Fermi's ingenious idea (to account for the influence of a polarizable medium on the wavefunction of an electron singled out); it should not be confused with the bonding energy of an electron to a hydrocarbon chain.

This approximation is appropriate for the rigid portion of the protein matrix between the chromophores. E_{ion} can be evaluated from the reduction potential of the bacteriopheophytin: $E_{\text{Ion}} = 3.9$ eV. Then we obtain

$$E_{\text{barrier}} = 3.9\ \text{eV} - 4.0\ \text{eV}\left(1 - \frac{1}{2.25}\right) = (3.9 - 2.2)\ \text{eV} \approx 2\ \text{eV}$$

6 We neglect the contribution of permanent dipoles to the polarization of the medium. In this case for the relative permittivity we find $\hat{\varepsilon} = n^2$, where n is the index of refraction. For hydrocarbons, $n = 1.53$, thus $\hat{\varepsilon} = 2.34$.

and $\alpha = 7\ \text{nm}^{-1}$. Thus the result for $\varepsilon_{\text{electronic}}$ is

$$\varepsilon_{\text{electronic}} = 2\ \text{eV} \times 0.303 \times 0.28 \times 0.5 \cdot \text{e}^{-7\ \text{nm}^{-1} \cdot d} = 0.085\ \text{eV} \cdot \text{e}^{-7\ \text{nm}^{-1} \cdot d}$$

28.8.5.5 Evaluation of Rate Constant k_r

In summary, we obtain for the rate constant k_r of electron transfer

$$k_r = \frac{2\pi}{\hbar}\varepsilon^2 \cdot \frac{1}{E_{\text{fluct}}} = \frac{2\pi}{\hbar}\varepsilon^2_{\text{electronic}} \cdot \frac{\beta^n}{n!}\ \text{e}^{-\beta} \cdot \frac{1}{E_{\text{fluct}}}$$
$$= \frac{2\pi}{\hbar} \cdot \left[E^2_{\text{barrier}} \cdot (c_a c_b \kappa)^2 \cdot \text{e}^{-2\alpha d}\right] \cdot \frac{\beta^n}{n!}\ \text{e}^{-\beta} \cdot \frac{1}{E_{\text{fluct}}} \tag{28F5.14}$$

or

$$k_r = a \cdot \text{e}^{-b \cdot d} \tag{28F5.15}$$

with

$$a = \frac{2\pi}{\hbar} \cdot \left[E^2_{\text{barrier}} \cdot (c_a c_b \kappa)^2\right] \cdot \frac{\beta^n}{n!}\ \text{e}^{-\beta} \cdot \frac{1}{E_{\text{fluct}}} \tag{28F5.16}$$

and

$$b = \frac{4\pi}{h}\sqrt{2m_e E_{\text{barrier}}} = \sqrt{\frac{E_{\text{barrier}}}{\text{eV}}} \cdot 10\ \text{nm}^{-1} \tag{28F5.17}$$

Example 28.4 *Calculate rate constants.*

In the present case, with $E_{\text{barrier}} = 2$ eV, $c_a c_b \kappa = 0.042$, $\beta = 4$, $n = 3$, and $E_{\text{fluct}} = 0.1$ eV we obtain

$$a = \frac{2\pi}{\hbar}\left[E^2_{\text{barrier}} \cdot 0.042^2\right] \cdot \frac{64}{6} \cdot \text{e}^{-4} \cdot \frac{1}{0.1\ \text{eV}}$$
$$= 2.06 \times 10^{32}\ \text{J}^{-1}\ \text{s}^{-1}\ \text{eV}^{-1} \cdot E^2_{\text{barrier}}$$
$$= 3.30 \times 10^{13}\ \text{s}^{-1} \cdot \left(\frac{E_{\text{barrier}}}{\text{eV}}\right)^2 = 1.3 \times 10^{14}\ \text{s}^{-1}$$

and

$$b = \frac{4\pi}{h}\sqrt{2m_e E_{\text{barrier}}} = 1.4 \times 10^{10}\ \text{m}^{-1} = 14\ \text{nm}^{-1}$$

Then for $d = 0.96$ nm we obtain

$$k_r = 1.3 \times 10^{14}\ \text{s}^{-1} \cdot \text{e}^{-14 \cdot 0.96} = 2 \times 10^8\ \text{s}^{-1}$$

which is about a factor of ten lower than the experimental value ($k_r = 4 \times 10^9\ \text{s}^{-1}$).

The natural quinone in the reaction center can be replaced by many other quinones. What is the consequence for the corresponding rate constant k_r? By changing the quinone we change $\Delta G^\ominus$ for the electron transfer. Such a change also changes n, the next integer to $\Delta G^\ominus / h\nu_0$, and n affects k_r through the term $\beta^n/n!$. With $\beta = 4$ and 5, respectively, and $h\nu_0 = 0.17$ eV we obtain the following values:

$\frac{-\Delta_r G^\ominus}{\text{kJ mol}^{-1}}$	$\frac{-\Delta G^\ominus}{h\nu_0}$	n	$\frac{\beta^n}{n!}$ For $\beta = 4$	$\frac{\beta^n}{n!}$ For $\beta = 5$
0.0	0.0	0	1	1
10	0.6	1	4	5
30	1.8	2	8	21
40	2.4	3	11	26
60	3.5	4	11	26
80	4.7	5	5	22

If we consider $\varepsilon_{electronic}$, β and E_{fluct} to be unchanged, we expect that k_r first increases with increasing $(-\Delta G^{\ominus})$ and, after a maximum at $n = 3$, slowly decreases with increasing $(-\Delta G^{\ominus})$. Such a behavior is observed in experiment (Fig. 28F.2).

$$\log k_r = const + \log \frac{\beta^n}{n!}$$

The experimental points are widely spread, essentially because $\varepsilon_{electronic}$ depends on structural details at the quinones and d is not strictly constant.

From this consideration we conclude that the dependence of k_r on $\Delta G^{\ominus}$ is not dramatic. Thus the electron tunneling rate in rigid σ electron media can be well described by

$$k_r = 10^{14}\ \mathrm{s}^{-1} \cdot \exp\left(-14\ \mathrm{nm}^{-1} \cdot d\right)$$

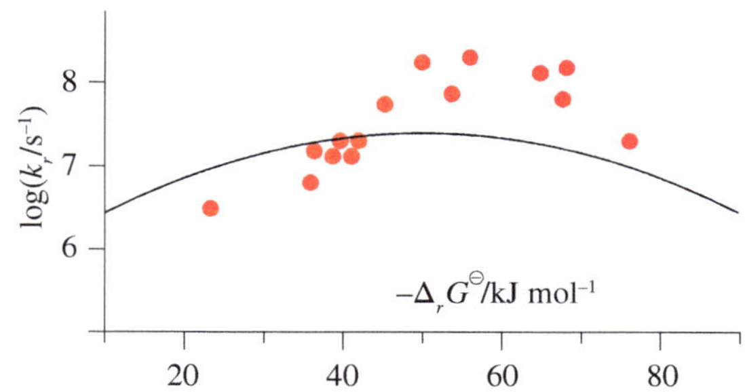

Figure 28F.2 Electron transfer from W_2^- to A for various artificial acceptors A differing in their reduction potential. Rate constant k_r versus $-\Delta_r G^{\ominus}$. Rigid medium: moderate dependence on $-\Delta_r G^{\ominus}$. Individual properties of acceptors prevail. Dots: experiment (same reference as in Table 28.2). Solid line: theory.

28.8.6 Foundation 28.6: Marcus Equation

To understand the Marcus equation we consider, as an example, the electron transfer from Cu^+ to Fe^{3+}, where the ions are fixed at a given distance (see Fig. 28F.3). The counter ions (OH^-, solvent water) are indicated. The electron transfer (step 2 to 3) can happen when the solvent ion has reached point P (activated state, situation 2). Then reorganization of the solvent takes place (step 3 to 4). The process is approximated by describing the potential energy change during nuclear displacement, before and after electron transfer, by harmonic oscillators (force constant k_f). We identify $-\Delta_r G^{\ominus}$ with the difference of the minima of the potential energy curves.

Before electron transfer:

$$E_1 = a + \frac{k_f}{2} x^2 \tag{28F6.1}$$

After electron transfer:

$$E_2 = \frac{k_f}{2}(x - d)^2 \tag{28F6.2}$$

Along the abscissa, the position x_P of point P is determined by the condition $E_1 = E_2$:

$$a + \frac{k_f}{2} x_P^2 = \frac{k_f}{2}\left(x_P - d\right)^2 = \frac{k_f}{2}\left(x_P^2 - 2x_P d + d^2\right)$$

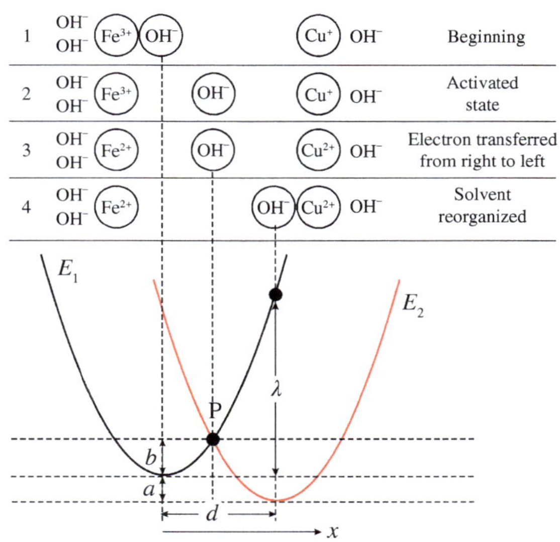

Figure 28F.3 Example for electron transfer in a soft medium: $Fe^{3+} + Cu^+ \rightarrow Fe^{2+} + Cu^{2+}$. Relation between the activation energy, $\Delta_r G^{\ddagger} = b$, the change of Gibbs energy of reaction, $\Delta_r G^{\ominus} = a$, and the reorganization energy, λ.

Thus

$$a = \frac{k_f}{2}\left(-2x_P d + d^2\right) \tag{28F6.3}$$

The reorganization energy λ, according to Fig. 28F.3, is given by

$$\lambda = \frac{k_f}{2} d^2 \tag{28F6.4}$$

Thus

$$a - \lambda = -2x_P d \frac{k_f}{2} \tag{28F6.5}$$

and for x_P we obtain

$$x_P = \frac{-a + \lambda}{k_f d} \tag{28F6.6}$$

Accordingly, the result for b is

$$b = \frac{k_f}{2}x_P^2 = \frac{k_f}{2}\left(\frac{-a+\lambda}{k_f d}\right)^2 = \frac{1}{2k_f d^2}(-a+\lambda)^2 = \frac{(-a+\lambda)^2}{4\lambda} \tag{28F6.7}$$

$-a$ corresponds to the change of Gibbs energy of the reaction, $\Delta G^{\ominus}$, and b corresponds to the activation energy, $\Delta G^{\ddagger}$:

$$-a = \Delta G^{\ominus}, \quad b = \Delta G^{\ddagger} \tag{28F6.8}$$

Thus we obtain

$$\Delta G^{\ddagger} = \frac{\left(\Delta G^{\ominus} + \lambda\right)^2}{4\lambda} \tag{28F6.9}$$

In analogy to Equation (28F5.9) in Foundation 28.5 the rate k_r is given by

$$k_r = \begin{matrix}\text{number}\\ \text{of attempts}\\ \text{per unit time}\end{matrix} \times \begin{matrix}\text{probability } \mathcal{P}_{\Delta E}\\ \text{of energy}\\ \text{match}\end{matrix} \times \begin{matrix}\text{probability}\\ \text{of electron transfer}\\ \text{during match}\end{matrix}$$

$$k_r = \frac{1}{t_c} \times \mathcal{P}_{\Delta E} \times \left(\frac{\varepsilon}{\hbar}t_c\right)^2 \tag{28F6.10}$$

Here $\Delta\mathcal{P}$ is the probability of the electron to be in an energy interval h/t_c at

$$x = \frac{-a+\lambda}{k_f\, d}$$

To estimate the uncertainty in energy, ΔE, we take the product of the force $\left[k_f x + k_f(d-x)\right]$ and the uncertainty in position, Δx, to find

$$\Delta E = k_f\, d \cdot \Delta x = \frac{h}{t_c}$$

or

$$\Delta x = \frac{h}{t_c k_f\, d}$$

We approximate the probability $\mathcal{P}_{\Delta E}$ of being within Δx of the crossing point P by a Boltzmann distribution in the classical limit.

$$\begin{aligned}\mathcal{P}_{\Delta E} &= \left(\frac{k_f}{2\pi kT}\right)^{1/2} \cdot \exp\left(-\frac{k_f x^2}{2kT}\right) \cdot \Delta x \\ &= \left(\frac{k_f}{2\pi kT}\right)^{1/2} \cdot \exp\left(-\frac{\Delta G^{\ddagger}}{kT}\right) \cdot \frac{h/t_c}{k_f d}\end{aligned} \tag{28F6.11}$$

With $\lambda = k_f d^2/2$ or $d^2 = 2\lambda/k_f$ we can write

$$k_f d = k_f \sqrt{\frac{2\lambda}{k_f}} = \sqrt{2\lambda\, k_f} \tag{28F6.12}$$

Thus for the probability we obtain

$$\begin{aligned}\mathcal{P}_{\Delta E} &= \sqrt{\frac{k_f}{2\pi kT}} \cdot \exp\left(-\frac{\Delta G^{\ddagger}}{kT}\right) \cdot \frac{h/t_c}{\sqrt{2\lambda k_f}} \\ &= \sqrt{\frac{1}{4\pi kT\lambda}} \cdot \exp\left(-\frac{\Delta G^{\ddagger}}{kT}\right) \cdot \frac{h}{t_c}\end{aligned} \tag{28F6.13}$$

and

$$k_r = \frac{1}{t_c} \times \mathcal{P}_{\Delta E} \times \left(\frac{\varepsilon}{\hbar}t_c\right)^2 \tag{28F6.14}$$

$$= \frac{1}{t_c} \times \sqrt{\frac{1}{4\pi kT\lambda}} \cdot \exp\left(-\frac{\Delta G^\ddagger}{kT}\right) \cdot \frac{h}{t_c} \times \left(\frac{\varepsilon}{\hbar} t_c\right)^2$$
$$= \frac{2\pi\varepsilon^2}{\hbar} \sqrt{\frac{1}{4\pi kT\lambda}} \cdot \exp\left(-\frac{\Delta G^\ddagger}{kT}\right)$$

In the present case (no lattice vibration) we have $\varepsilon = \varepsilon_{electronic}$. Then, using Equation (28F6.9), we obtain

$$k_r = \frac{2\pi}{\hbar} \varepsilon^2_{electronic} \cdot \frac{1}{\sqrt{4\pi kT\lambda}} \cdot \exp\left(-\frac{(\Delta G^\ominus + \lambda)^2}{4kT\lambda}\right) \tag{28F6.15}$$

This result is the Marcus equation.

Example 28.5 *A particularly simple case of the Marcus equation is the ferric/ferrous exchange reaction, $Fe^{3+} + Fe^{2+} \rightarrow Fe^{2+} + Fe^{3+}$. Here we have $\Delta G^\ominus = 0$ and the rate constant expression (28F6.15) simplifies to*

$$k_r = A \cdot \exp\left(\frac{-\lambda}{4kT}\right) \tag{28F6.16}$$

By mixing ferric ions of the radioisotope Fe^{55} (half-life of 2.95 years) with natural ferrous ions and analyzing the concentration of the Fe^{55} after a known amount of time, the exchange kinetics can be followed. Measured rate constants in 0.547 M $HClO_4$ as a function of temperature are given in the table. Assume that the preexponential factor is independent of temperature and use these data to determine the reorganization energy and frequency factor for the reaction.

k_r/s^{-1}	1.35	2.73	5.34	9.97
T/K	273.2	280.4	287.8	294.8

Figure 28F.4 plots these data, as $\ln(k_r/s^{-1})$ versus $1/T$, according to Equation (28F6.16). A best fit yields a slope of -7.50×10^3 K and an intercept of 27.8. From the slope, we find

$$\lambda = -4k \cdot slope = -4 \cdot 1.38 \times 10^{-23}\ \text{J K}^{-1} \cdot (-7500\ \text{K})$$
$$= 4.14 \times 10^{-19}\ \text{J} = 2.6\ \text{eV}$$

and from the intercept we find $A = 1.2 \times 10^{12}\ s^{-1}$. This result is in good agreement with a more refined analysis [82].

Marcus first derived a mathematical expression for the solvent reorganization energy in the approximation that intermolecular rearrangements can be neglected and the electrostatic polarization of the solvent controls the reorganization energy [83]. Using a dielectric continuum model to treat the solvent and approximating the molecules by spheres of radius r, he found that

$$\lambda = \frac{e^2}{4\pi\varepsilon_0}\left(\frac{1}{\hat{\varepsilon}_\infty} - \frac{1}{\hat{\varepsilon}_s}\right)\left(\frac{1}{r} - \frac{1}{d}\right)$$

where d is the distance between the centers of the two reactants, $\hat{\varepsilon}_s$ is the relative dielectric constant at zero frequency, and $\hat{\varepsilon}_\infty$ is the high frequency dielectric constant, often approximated by the refractive index squared ($\hat{\varepsilon}_\infty \simeq \hat{n}^2$). If we assume that the two spherical cavities representing the reactants are touching, then $2r = d$ and

$$\lambda = \frac{e^2}{4\pi\varepsilon_0 r}\left(\frac{1}{\hat{\varepsilon}_\infty} - \frac{1}{\hat{\varepsilon}_s}\right)$$
$$= \frac{e^2}{2\pi\varepsilon_0 d}\left(\frac{1}{\hat{\varepsilon}_\infty} - \frac{1}{\hat{\varepsilon}_s}\right)$$

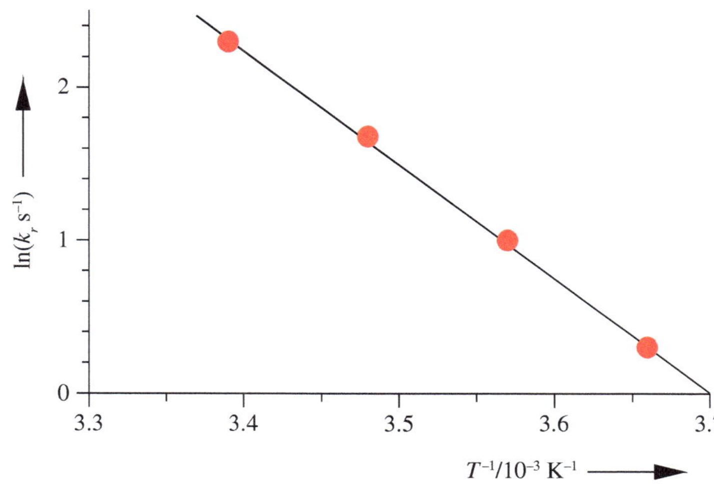

Figure 28F.4 Electron transfer reaction $Fe^{3+} + Fe^{2+} \rightarrow Fe^{2+} + Fe^{3+}$. $\ln(k_r/s^{-1})$ as a function of $1/T$.

Note that a comparison with Equation (28F6.9) shows that the force constant for the solvent potentials (parabolas) in Fig. 28F.3 is

$$k_f = \frac{e^2}{\pi \varepsilon_0 d^3} \left(\frac{1}{\hat{\varepsilon}_\infty} - \frac{1}{\hat{\varepsilon}_s} \right)$$

Let us use the continuum model to calculate the reorganization energy within the picture of Fig. 28F.3. If we take the ionic radius of Fe^{2+} to be 82 pm, that of Fe^{3+} to be 67 pm, and the diameter of a hydroxide ion to be 110 pm, then we find $d = 259$ pm for the electron transfer distance of two iron ions, and using $\hat{\varepsilon}_\infty = 3.79$[7] and $\hat{\varepsilon}_s = 78.4$, we find

$$\lambda = \frac{(1.602 \times 10^{-19})^2}{2\pi\,(8.854 \times 10^{-12})\,(259 \times 10^{-12})} \left(\frac{1}{3.79} - \frac{1}{78.4} \right) \text{J}$$
$$= 4.47 \times 10^{-19}\ \text{J} = 2.8\ \text{eV}$$

This value is in reasonable agreement with that found from the analysis in Fig. F28.4.

28.8.7 Foundation 28.7: Chloride Ion Pump and Sensory Receptor of Halobacteria

See the book's website,www.wiley.com/go/Kuhn/physicalchemistry3e.

28.9 Exercises

E28.1 Using Equation (28.3) for the distance d_0 in a donor–acceptor system, calculate the quantum yield ϕ from the measured $d_0 = 7$ nm. The wavelength of the light emitted by the donor is $\lambda = 480$ nm, the absorbance of the layer is $A_A = 0.012$, and the index of refraction is $\hat{n} = 1.5$.

E28.2 Consider the two molecules fluorescein (emitter) and tetramethylrhodamine (absorber) and calculate the characteristic Förster distance, r_0, using Equation (28.9). Assume that the quantum yield of fluorescein is 0.95, that the extinction coefficient of the tetramethylrhodamine is 72,000 $\text{mol}^{-1}\ \text{L cm}^{-1}$, and the wavelength at the absorbance maximum is $\lambda = 550$ nm.

E28.3 Consider the electric field that acts on a chromophore from nearby charges, e.g., charged amino acids. Assume that the negatively charged amino acids D212 and D85 (denoted as 1 and 2 in Fig. E28.3) are at distance $d = 0.4$ nm from the N atom and from C atom 14, respectively. What is their field strength F at the chromophore?

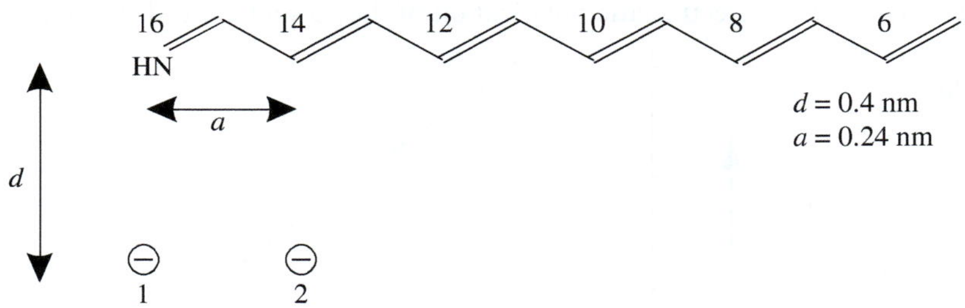

Figure E28.3 Chromophore in the electric field of two negative charges.

E28.4 The circular dichroism of chlorosomes changes sign within the absorption band at 740 nm (panel c of Fig. E28.4), similar to the example in Section 9.3. However, the linear dichroism of stretched samples is constant within the absorption band [84]—in contrast to the situation in Section 9.3 where the coupled oscillations are polarized perpendicular to each other. Try to rationalize this surprising effect and the values

$$\left(\frac{\Delta\varepsilon}{\varepsilon}\right)_{750\ \text{nm}} = -10^{-2}, \qquad \left(\frac{\Delta\varepsilon}{\varepsilon}\right)_{720\ \text{nm}} = +10^{-2}$$

taken [84] from the figure. Assume that the exciton has the same size as the exciton in the brickstone work aggregate in Section 28.3 (distributed over about 10 molecules at 300 K). For simplicity consider the exciton as confined to molecules 1 and 2 (panel a of Fig. E28.4; that is, replace the excited portion by two coupled oscillators (panel b).

7 This value is at 1 GHz frequency and 298K. Taken from CRC Handbook of Chemistry and Physics.

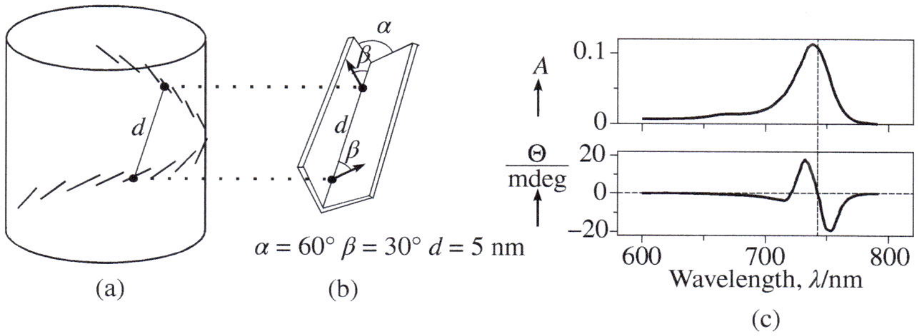

Figure E28.4 Model for the circular dichroism of chlorosomes.

E28.5 The data in the table[8] show the temperature dependence for the exciton of a dye molecule (pseudocyanine bromide) in an ethylene glycol/water glass as a function of temperature; for which the lifetime is $\tau_0 = 3.7 \times 10^{-9}$ s. Analyze this temperature dependence using Equation (28.14) and determine the binding energy of the exciton.

$k_r/10^9\ \mathrm{s}^{-1}$	0.90	1.5	2.2	2.9	4.1	3.8	5.4
T/K	222	200	174	143	133	114	114

$k_r/10^9\ \mathrm{s}^{-1}$	6.4	9.0	11.6	14.8	14.7	14.7	15.0
T/K	105	80	62	45	32	19	10

E28.6 In Section 28.3.2 we calculate the speed of an exciton as $v = 2.4$ km s^{-1} and its length as $l = 5$ nm. How long does it take for the exciton to move a distance of 10 times its own length? Compare this time to the period of a 1000 cm^{-1} vibration; a 100 cm^{-1} vibration; and a 10 cm^{-1} intermolecular motion.

E28.7 Consider a monolayer thick tunneling barrier that contains a certain fraction f of defects. Consider the electrical current through the defects to have the same magnitude as the tunneling current through a 0.5 nm thick tunneling barrier. Consider the true reciprocal decay length of the current to be $b = 14$ nm^{-1}. Calculate the current as a function of film thickness (from 0 to 10 nm) for the cases of no defects, 0.1% defects, and 1% defects.

E28.8 An important technical issue in electron tunneling experiments is the possibility of tunneling through defect sites. Consider a tunneling experiment for a SAM (decanethiol, C10) coated Au electrode (1 mm^2 area) in contact with the $Fe(CN)_6^{3-}/Fe(CN)_6^{4-}$ redox couple and an observed tunneling current of 100 nA at an applied bias potential of 0.25 V versus Ag/AgCl. Assume that some number of defects exist in the SAM film and that a defect has a current that is 10 times higher than that which occurs by tunneling through a chain. What percentage of defects can exist in the film before they contribute 1% to the current? Use this result to calculate the current per defect site. If you assume that the defects have the same tunneling efficiency independent of the carbon chain length of the SAM, what fraction of defects are needed to contribute 1% to the current through an octadecanethiol (C18) film?

E28.9 Estimate the height of a tunneling barrier formed by a dielectric medium, by determining how far it lies below the vacuum level. In particular, find the polarization energy for transferring an electron from vacuum into a dielectric medium of relative permittivity $\varepsilon_r = 2.5$. Estimate this polarization energy by modeling the electron as a charged sphere of radius a) $r = 0.2$ nm and b) $r = 0.5$ nm.

E28.10 In Exercise E28.9, you estimate the polarization energy for an electron that is localized to a certain size in a dielectric medium. By making the charged sphere smaller you can increase the polarization stabilization; however, this decrease in size increases the electron's kinetic energy. Use the de Broglie relation to estimate the kinetic energy of the electron for the two cases in Exercise E28.9. Is it possible to find an optimal size for the electron?

E28.11 Consider the case of a monolayer film in which the hydrocarbon chains of length l are perpendicular to the layer plane ($d = l$) and $2r = 0.475$ nm is the distance between the axes of adjacent chains (Fig. E28.11a). What is expected to happen to the current through the film when the hydrocarbon chains are inclined and the distance becomes $d' = l \cos \varphi$, where φ is the tilt angle (panel b)?

8 These data are read from the graph in Figure 5 of [14].

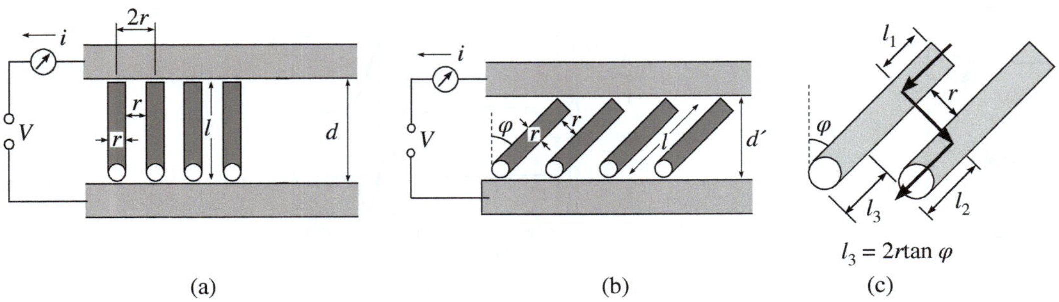

Figure E28.11 Monolayer film with hydrocarbon chains perpendicular (a) and inclined (b) to the layer plane. A close up view of the geometry for two chains is shown in (c).

E28.12 Tivanski et al. [86] measured the conductivity through conjugated molecules with two different types of linkers. Use their results to estimate the difference in the resistance of the two types of thiol contacts.

E28.13 The pseudorotaxane system shown in Figure E28.13 forms a bound complex of the TTF-polyether dumbbell with the cationic cyclophane when the TTF is neutral. Upon oxidation of the TTF unit, the system dissociates with a rate constant of 0.2 s^{-1} at 298 K in CH_3CN ($\eta = 0.37$ mPa s). Use the Smoluchowski limit of the Kramers model (see Foundation 25.2) and this rate constant to estimate the barrier to dissociation.

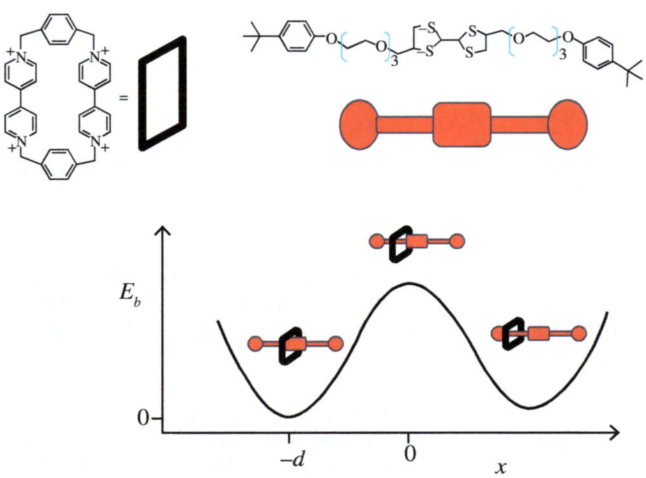

Figure E28.13 Pseudorotaxane system.

E28.14 For the catenane ring rotation in Fig. 28.39, the switching time for (b) to (c) is 14 hours at 293 K in CH_3CN and from (d) to (a) is a few seconds. Rationalize why these times are so different. If the catenane changes its orientation by diffusion, with a rotational diffusion coefficient given approximately by $D_{rotn} = kT/6V\eta$ (where V is the hydrodynamic volume and η is the viscosity of the solvent), estimate the time for it to switch. This expression for the diffusion coefficient is found by using the Einstein relation $D_{rotn} = kT/\zeta_{rotn}$ and using a hydrodynamic (Stokes model) for the rotational friction coefficient [87]. Compare your result to the observed rates in CH_3CN at 293 K ($\eta = 0.38$ mPa s). What is likely to limit the rate?

E28.15 Consider a kinesin protein that is transporting a 200 nm diameter vesicle through a cell at 1 μm s^{-1}. Estimate a viscosity for the medium and compare the frictional force to that provided by the kinesin, ca. 5 pN. At what viscosity would you expect these values to be equal?

28.10 Problems

P28.1 Straight et al. [88] synthesized and studied a molecule that performs the AND logic operation. Describe how their system works. Propose an alternative molecular logic element that uses chemical binding events for inputs.

P28.2 The review article by Balzani et al. [89] contains descriptions of a number of different supramolecular devices. Investigate one of these in detail, explain its operation principles, and delineate its specifications; i.e., performance characteristics. Lastly, propose one way to improve the machine.

P28.3 In Section 28.3, we discussed the effect of a mirror on an Eu luminophore. Consider the situation in which the Eu is placed at the center of two opposing perfect mirrors, i.e., in an optical cavity, and describe how the radiative rate of the Eu* is expected to change as the distance between the mirrors changes. Consider the mirror distances to vary over a range of 10 nm to 10 μm and only consider cavity modes along the axis which is normal to the mirror surfaces.

P28.4 Catalytic Nanomotors: Wang et al. reported the autonomous movement of bimetallic nanorods (e.g., Pt: Au dyad) in 5 wt.% aqueous peroxide solutions, see [90]. Read this paper and explain the mechanism by which these rods move preferentially toward their one end (e.g., Pt end of Pt:Au).

P28.5 Bhosale et al. [91] reported an artificial light driven proton pump, which is based on a π-stacked assembly of chromophores that are inserted into the lipid bilayer of a vesicle. Read and analyze this paper. Provide a brief (2 paragraphs or less) description of its operation.

P28.6 Artificial Bacterial Photosynthesis: Steinberg-Yfrach et al. [92] report an artificial supramolecular assembly that mimics bacterial photosynthesis. Read and describe the findings of this paper. Be sure to report which aspects of the system are artificial and which are natural.

P28.7 Switching Device. He et al. [93] reported a light driven electrical switch. They investigated a molecule that exists in an open form (a) and a closed form (b):

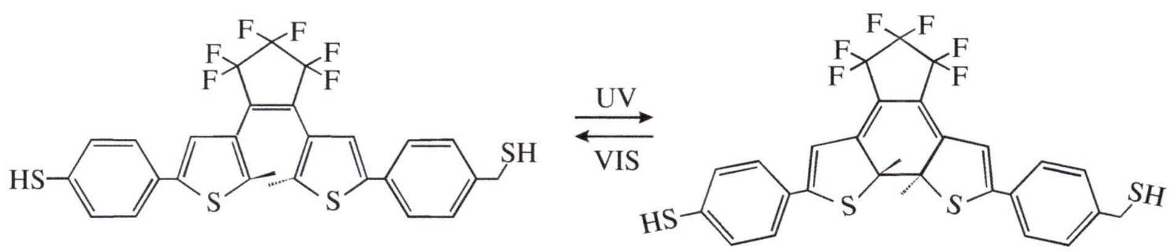

(a) (b)

By binding the thiol groups at both ends to small gold particles they were able to measure the change in the molecule's resistance. Read and analyze this paper. Provide a short (two paragraphs or less) explanation of their claim and the observations which support it.

P28.8 Sequencing Individual DNA Molecules: W. J. Greenleaf and S. M. Block report a method for measuring the sequence of oligonucleotides in [94]. Provide a short (two paragraphs or less) explanation of their claim and the observations which support it.

P28.9 van Delden et al. [95] have reported a molecular motor which is driven by a combination of photochemical and thermal isomerization processes. Read and analyze this paper. Provide a short (less than or equal to two paragraphs) explanation of the motor's operation principle and their observations which support it.

P28.10 In the paper [96], Paddon-Row and coworkers demonstrated the tunneling nature of an electron transfer reaction by studying how the reaction rate changes as a function of the distance between a naphthalene (electron donor) and an acceptor (electron acceptor). However, the reaction Gibbs energy and the activation energy for the reaction changes with the donor to acceptor distance. Evaluate this effect for the system studied by Paddon-Row and assess whether it can be neglected.

System	k_{CS}/s^{-1}	R/pm
6σ	3.3×10^{11}	680
8σ	3.0×10^{10}	940
10σ	2.4×10^{9}	1150
12σ	1.6×10^{8}	1350

P28.11 Here you consider the Brownian motion problem in Box 28.3, in which the Brownian "particle" is driven toward smaller values of x by a force f. In the spirit of Section 14.3, the computer program "P28.11.exe" divides the range from $x = 0$ to $x = x_1$ into 10 equal intervals of length a. The "particle" starts between $x = 0$ and $x = a$; and it moves stochastically from interval to interval with a probability of $\mathcal{P}_+$ in the positive direction and a probability of $\mathcal{P}_-$ in the negative direction. In the course of the process the "particle" reaches the first interval (between $x = 0$ and $x = a$) many times before reaching the last interval between $x = x_1 - a$

to $x = x_1$. We assume that each time it is reflected at this position according to Equation (28.48); hence, it leaves this range with a very small probability. In the computer program, we ask for the number n_1 of steps until the interval between $x_1 - a$ and x_1 is reached, on average. We also ask for the average number n_0 of reflections at $x = 0$ during that process.

References

1 D. Möbius, Abstandsgesetz des Energieübergangs aus Messung der sensibilisierten Fluoreszenz, *Z. Naturforsch.*, 1969, **24a**, 251.

2 H. Bücher, K.H. Drexhage, M. Fleck, H. Kuhn, D. Möbius, F.P. Schäfer, J. Sondermann, W. Sperling, P. Tillmann, and J. Wiegand, Controlled transfer of excitation energy through thin layers, *Mol. Cryst.*, 1967, **2**, 199.

3 H. Kuhn, Classical aspects of energy transfer in molecular systems, *J. Chem. Phys.*, 1970, **53**, 101.

4 Th. Förster, Zwischenmolekulare Energiewanderung und Fluoreszenz, *Ann. Phys.*, 1948, **2**, 55.

5 M. Dietz, B. Zimmerman, M. Borsch, M. Konig, E. Schweinberger, S. Steigmiller, R. Reter, S. Felekyan, V. Kudryavtsev, C.A.M. Seidel, and P. Graber, Proton-powered subunit rotation in single membrane-bound F0F1-ATP synthase, *Nat. Struct. Mol. Biol.*, 2004, **11**, 135.

6 E. Betzig and R.J. Chichester, Single molecules observed by near-field scanning optical microscopy, *Science*, 1993, **262**, 1422; W.E. Moerner and T. Basché, Optical spectroscopy of single impurity molecules in solids, *Angew. Chem. Int. Ed. Engl.*, 1993, **32**, 457; T. Basché, Detection and imaging of single molecules by optical near-field microscopy, *Angew. Chem. Int. Ed. Engl.*, 1994, **33**, 1723.

7 O.L. Mooren, E.S. Erickson, N.E. Dickenson, and R.C. Dunn, Extending near-field scanning optical microscopy for biological studies, *JALA: Technol. Rev.*, 2006, **11**, 268.

8 D. Möbius, Designed monolayer assemblies, *Ber. Bunsenges. Phys. Chem.*, 1978, **82**, 848.

9 H. Kuhn and D. Möbius, in: *Investigations of Surfaces and Interfaces, Part B*, eds. B.W. Rossiter and R.C. Baetzold, Physical Methods of Chemistry Series, Vol. **IX B**, (Wiley, 1993, pp. 375–541) 2nd Edition (Fig. 6.38, p. 415).

10 V. Czickely, H.D. Försterling, and H. Kuhn, Light absorption and structure of aggregates of dye molecules, *Chem. Phys. Lett.*, 1970, **6**, 11.

11 H. Bücher and H. Kuhn, Scheibe aggregate formation of cyanine dyes in monolayers, *Chem. Phys. Lett.*, 1970, **6**, 183.

12 D. Möbius and H. Kuhn, Energy transfer in monolayers with cyanine dye Sheibe aggregates, *J. Appl. Phys.*, 1988, **64**, 5138.

13 H. Kuhn and C. Kuhn, "Chromophore Coupling Effects", in: *J-Aggregates*, ed. T. Kobayashi (World Scientific, Singapore, New Jersey, London, Hong Kong, 1996, pp. 1–40).

14 S. de Boer and D.A. Wiersma, Dephasing-induced damping of superradiant emission in J-aggregates, *Chem. Phys. Lett.*, 1990, **165**, 45; J.R. Durrant, J. Knoesler, and D.A. Wiersma, Local energetic disorder in molecular aggregates probed by the one-exciton to two-exciton transition, *Chem. Phys. Lett.*, 1994, **222**, 450.

15 W. Kühlbrandt, Many wheels make light work, *Nature*, 1995, **374**, 497; S. Karrasch, P. Bullough, and R. Ghosh, The 8.5 A projection map of the light-harvesting complex I from Rhodospirillum rubrum reveals a ring composed of 16 subunits, *EMBO J.*, 1995, **14**, 631.

16 A.R. Holzwarth and K. Schaffner, On the structure of bacteriochlorophyll molecular aggregates in the chlorosomes of green bacteria. A molecular modelling study, *Photosynth. Res.*, 1994, **41**, 225.

17 D.M. Eisele, D.H. Arias, X. Fu, E.A. Bloemsma, C.P. Steiner, R.A. Jensen, P. Rebentrost, H. Eisele, A. Tokmakoff, S. Lloyd, K.A. Nelson, D. Nicastro, J. Knoester, and M.G. Bawendi, Robust excitons inhabit soft supramolecular nanotubes, *Proc. Natl. Acad. Sci. U.S.A.*, 2014, **111**, E3367–E3375.

18 E. Boulais, N.P.D. Sawaya, R. Veneziano, A. Andrewoni, J.L. Banal, T. Kondo, S. Mandal, S. Lin, G.S. Schlau-Cohen, N.W. Woodbury, H. Yan, A. Aspuru-Guzik, and M. Bathe, *Nat. Mater.*, Programmed coherent coupling in a synthetic DNA-based excitonic circuit, 2017, **17**, 159–168.

19 K.H. Drexhage, Influence of a dielectric interface on fluorescence decay time, *J. Lumin.*, 1970, **1–2**, 693.

20 K.H. Tews, Zur Variation von Lumineszens-Lebensdauern, *Ann. Phys.*, 1973, **29**, 97; D.H. Waldeck, A.P. Alivisatos, and C.B. Harris, Nonradiative damping of molecular electronic excited states by metal surfaces, *Surf. Sci.*, 1985, **158**, 103.

21 B. Mann and H. Kuhn, Tunneling through fatty acid salt monolayers, *J. Appl. Phys.*, 1971, **42**, 4398.

22 X.S. Xie and R.C. Dunn, Probing single molecule dynamics, *Science*, 1994, **265**, 361; W.P. Ambrose, P.M. Goodwin, J.C. Martin, and R.A. Keller, Alterations of single molecule fluorescence lifetimes in near-field optical microscopy, *Science*, 1994, **265**, 364.

23 H. Kuhn and D. Möbius, in: *Investigations of Surfaces and Interfaces, Part B*, eds. B.W. Rossitter and R.C. Baetzold, Physical methods of Chemistry Series, Vol. **IX B** (Wiley, New York, 1993, p. 465) 2nd Edition; H. Kuhn, Organized monolayer assemblies and their role in constructing supramolecular devices and in modelling evolution of early life, *IEEE Eng. Med. Biol. Mag.*, 1994, **13**, 33–44, DOI: 10.1109/51.265773.

24 E.E. Polymeropoulos, Electron tunneling through fatty-acid monolayers, *J. Appl. Phys.*, 1977, **48**, 2404.

25 E.E. Polymeropoulos and J. Sagiv, Electrical conduction through adsorbed monolayers, *J. Chem. Phys.*, 1978, **69**, 1836.

26 A. Sommerfeld and H. Bethe, in: *Handbuch der Physik*, Vol. **24** (Springer-Verlag, Berlin, 1933, p. 450; J.G. Simmons, Electric tunnel effect between dissimilar electrodes separated by a thin insulating film, *J. Appl. Phys.*, 1963, **34**, 2581.

27 K. Slowinski, H.K.Y. Fong, and M. Madja, Mercury–mercury tunneling junctions. 1. Electron tunneling across symmetric and asymmetric alkanethiolate bilayers, *J. Am. Chem. Soc.*, 1999, **121**, 7257.

28 A.M. Napper, H. Liu, and D.H. Waldeck, The nature of electronic coupling between ferrocene and gold through alkanethiolate monolayers on electrodes: the importance of chain composition, interchain coupling, and quantum interference, *J. Phys. Chem.*, 2001, **B105**, 7699–7707.

29 J.F. Smalley, S.W. Feldberg, C.E.D. Chidsey, M.R. Linford, M.D. Newton, and Y.P. Liu, The kinetics of electron transfer through ferrocene-terminated alkanethiol monolayers on gold, *J. Phys. Chem.*, 1995, **99**, 13141; C.E.D. Chidsey, Free energy and temperature dependence of electron transfer at the metal-electrolyte interface, *Science*, 1991, **351**, 919.

30 F. Evers, R. Korytar, S. Tewari, and J.M. van Ruitenbeek, Advances and challenges in single-molecule electron transport, *Rev. Mod. Phys.*, 2020, **92**, 035001 H.B. Akkerman and B. de Boer, Electrical conduction through single molecules and self-assembled monolayers, *J. Phys. Condens. Matter*, 2008, **20**, 013001.

31 B. Xu and N.J. Tao, Measurement of single-molecule resistance by repeated formation of molecular junctions, *Science*, 2003, **301**, 1221.

32 C. Li, I. Pobelov, T. Wandlowski, A. Bagrets, A. Arnold, and F. Evers, Charge transport in single Au— alkanedithiol—Au junctions: coordination geometries and conformational degrees of freedom, *J. Am. Chem. Soc.*, 2008, **130**, 318–326, DOI: 10.1021/ja0762386

33 E. Wierzbinski, X. Yin, K. Werling, and D.H. Waldeck, The effect of oxygen heteroatoms on the single molecule conductance of saturated chains, *J. Phys. Chem. B*, 2013, **117**, 4431.

34 W. Haiss, R.J. Nichols, H. van Zalinge, S.J. Higgins, D. Bethell and D.J. Schiffrin, Measurement of single molecule conductivity using the spontaneous formation of molecular wires, *Phys. Chem. Chem. Phys.*, 2004, **6**, 4330.

35 M. Suzuki, S. Fujii, and M. Fujihira, Measurements of currents through single molecules of alkanedithiols by repeated formation of break junction in scanning tunneling microscopy under ultrahigh vacuum, *Jpn. J. Appl. Phys.*, 2006, **45**, 2041.

36 M.A. Reed, C. Zhou, C.J. Muller, T.P. Burgin, and J.M. Tour, Conductance of a molecular junction, *Science*, 1997, **278**, 252.

37 J. Reichert, R. Ochs, D. Beckmann, H.B. Weber, M. Mayor, and H.v. Löhneysen, Driving current through single organic molecules, *Phys. Rev. Lett.*, 2002, **88**, 176804.

38 M. Mayor, H.B. Weber, J. Reichert, M. Elbing, C. von Hänisch, D. Beckmann, and M. Fischer, Electric current through a molecular rod—relevance of the position of the anchor groups, *Angew. Chem. Int. Ed.*, 2003, **42**, 5834.

39 J. Reichert, H.B. Weber, and M. Mayor, Low-temperature conductance measurements on single molecules, *Appl. Phys. Lett.*, 2003, **82**, 4137.

40 M. Elbing, R. Ochs, M. Koentopp, M. Fischer, C. von Hänisch, F. Weigand, F. Evers, H.B. Weber, M. Mayor, *Proc. Natl. Acad. Sci. U.S.A.*, 2005, **102**, 8815.

41 H. Kuhn, Energieübertragung in monomolekularen Schichten, *Naturwissenschaften*, 1967, **54**, 429.

42 D.R. Casimiro, J.H. Richards, J.R. Winkler, and H.B. Gray, Electron transfer in ruthenium-modified cytochromes c. σ-tunneling pathways through aromatic residues, *J. Phys. Chem.*, 1993, **97**, 13073; D.S. Wuttke, M.J. Bjerrum, J.R. Winkler, and H.B. Gray, Electron-tunneling pathways in cytochrome c, *Science*, 1992, **256**, 1007; D.N. Beratan, J.N. Betts, J.N. Onuchic, Protein electron transfer rates set by the bridging secondary and tertiary structure, *Science*, 1991, **252**, 1285.

43 J.J. Wei, H. Liu, K. Niki, E. Margoliash, and D.H. Waldeck, Probing electron tunneling pathways: electrochemical study of rat heart cytochrome *c* and its mutant on pyridine-terminated SAMs, *J. Phys. Chem.*, 2004, **108B**, 16912.

44 D.E. Khoshtariya, J. Wei, H. Liu, H. Yue, and D.H. Waldeck, Charge-transfer mechanism for cytochrome c adsorbed on nanometer thick films. Distinguishing frictional control from conformational gating, *J. Am. Chem. Soc.*, 2003, **125** (25), 7704–7714; D.V. Matyushov, Dynamical effects in protein electrochemistry, *J. Phys. Chem. B*, 2019, **123**, 7290–7301.

45 H. Kuhn, Electron transfer mechanism in the reaction center of photosynthetic bacteria, *Phys. Rev.*, 1986, **A34**, 3409.

46 H. Kuhn, Electron tunneling effects in monolayer assemblies, *Chem. Phys. Lipids*, 1972, **8**, 401.

47 J. Deisenhofer, O. Epp, K. Miki, R. Huber, and H. Michel, X-ray structure analysis of a membrane protein complex. Electron density map at 3 A resolution and a model of the chromophores of the photosynthetic reaction center from Rhodopseudomonas viridis, *J. Mol. Biol.*, 1984, **180**, 385.

48 H. Kuhn, Organized monolayer assemblies/spl minus/their role in constructing supramolecular devices and in modeling evolution of early life, *IEEE Eng. Med. Biol.*, 1994, **13**, 33.

49 C.C. Moser, J.M. Keske, K. Warncke, R.S. Farid, and P.L. Dutton, Nature of biological electron transfer, *Nature*, 1992, **355**, 796; W. Zinth, private communication.

50 J.R. Miller, L.T. Calcaterra, and G.L. Closs, Intramolecular long-distance electron transfer in radical anions. The effects of free energy and solvent on the reaction rates, *J. Am. Chem. Soc.*, 1984, **106**, 3047; G.L. Closs and J.R. Miller, Intramolecular long-distance electron transfer in organic molecules, *Science, New Ser.*, 1988, **240**, 440.

51 N.R. Kestner, J. Jortner, and J. Logan, Thermal electron transfer reactions in polar solvents, *J. Phys. Chem.*, 1974, **78**, 2148; S.F. Fisher and R.P. Van Duyne, On the theory of electron transfer reactions. The naphthalene-./TCNQ system, *Chem. Phys.*, 1977, **26**, 9.

52 D. Gust, T.A. Moore, and A.L. Moore, Photosynthesis mimics as molecular electronic devices, *IEEE Eng. Med. Biol.*, 1994, **13**, 58; D. Gust, T.A. Moore, and A.L. Moore, Molecular mimicry of photosynthetic energy and electron transfer, *Acc. Chem. Res.*, 1993, **26**, 198.

53 A. Osuka, S. Nakajima, T. Okada, S. Taniguchi, K. Nozaki, T. Ohno, I. Yamazaki, Y. Nishimura, and N. Mataga, A sequential electron-transfer relay in diporphyrin–porphyrin–pyromellitimide triads analogous to that in the photosynthetic reaction center, *Angew. Chem. Int. Ed. Engl.*, 1996, **35**, 92.

54 T. Ikeda, J. Mamiya, and Y. Yu, Photomechanics of liquid-crystalline elastomers and other polymers, *Angew. Chem.*, 2007, **46**, 506.

55 T. Seki and T. Tamaki, Photomechanical effect in monolayers of azobenzene side chain polymers, *Chem. Lett.*, 1993, **22**, 1739.

56 K. Ichimura, Y. Suzuki, T. Seki, A. Hosoki, and K. Aoki, Reversible change in alignment mode of nematic liquid crystals regulated photochemically by command surfaces modified with an azobenzene monolayer, *Langmuir*, 1988, **4**, 1214; T. Seki, M. Yokoi, R.I. Fukuda, H. Hiramatsu, and T. Tamaki, Photoregulation of nematic liquid crystal alignment by LB films of azobenzene/polyethylenimine derivatives, *Mol. Cryst. Liq. Cryst.*, 1995, **267**, 429.

57 K.P. Hofman and T.D. Lamb, Rhodopsin, light-sensor of vision, *Prog. Retin. Eye Res.*, 2023, **93**, 101116.

58 H. Kandori, Ion-pumping microbial rhodopsins, *Front. Mol. Biosci.*, 2015, **2**, 52.

59 R. Ballardini, V. Balzani, M.T. Gandolfi, L. Prodi, M. Venturi, D. Philp, H.G. Ricketts, and J.F. Stoddart, Eine photochemisch betriebene molekulare Maschine, *Angew. Chem.*, 1993, **105**, 1362; D. Philp and J.F. Stoddart, Self-Assembly in natural and unnatural systems, *Angew. Chem.*, 1996, **108**, 1243; P.R. Ashton, P.T. Glink, J.F. Stoddart, P.A. Tasker, A.I. White, and D.J. Williams, Self-assembling [2]-and [3] rotaxanes from secondary dialkylammonium salts and crown ethers, *Chem. Eur. J.*, 1996, **2**, 729; V. Balzani, A. Credi, G. Mattersteig, O.A. Matthews, F.M. Raymo, J.F. Stoddart, M. Venturi, A.J.P. White, and D.J. Williams, Switching of pseudorotaxanes and catenanes incorporating a tetrathiafulvalene unit by redox and chemical inputs, *J. Org. Chem.*, 2000, **65**, 1924.

60 A. Livoreil, C.O. Diedrich-Buchecker, and J.P. Sauvage, Electrochemically triggered swinging of a [2]-catenate, *J. Am. Chem. Soc.*, 1994, **116**, 9399.

61 Y. Feng, M. Ovalle, J.S.W. Seale, C.K. Lee, D.J. Kim, R.D. Astumian, and J.F. Stoddart, Molecular pumps and motors, *J. Am. Chem. Soc.*, 2021, **143**, 5569.

62 J.E. Walker, ATP-Synthese durch Rotations-Katalyse (Nobel-Vortrag), *Angew. Chem.*, 1998, **110**, 2438; M. Yoshida, R. Yasuda, M. Yoshida, and K. Kinosita Jr., Direct observation of the rotation of F_1-ATPase, *Nature*, 1997, **386**, 299; W. Junge, H. Lill, and S. Enelbrecht, ATP synthase: an electrochemical ransducer with rotatory mechanics, *Tibbs*, 1997, **22**, 420.

63 D. Oesterhelt and W. Stoeckenius, Functions of a new photoreceptor membrane, *Proc. Natl. Acad. Sci. U.S.A.*, 1973, **70**, 2853; U. Haupts, J. Tittor, and D. Oesterhelt, General concept for ion translocation by halobacterial retinal proteins: the isomerization/switch/transfer (IST) model, *Biochemistry*, 1997, **36**, 2; J.K. Lanyi, Proton translocation mechanism and energetics in the light-driven pump bacteriorhodopsin, *Biochem. Biophys. Acta*, 1993, **1183**, 241; D. Oesterhelt, Structure and function of halorhodopsin, *Isr. J. Chem.*, 1995, **35**, 475.

64 D. Kaur, U. Khaniya, Y. Zhang, and M.R. Gunner, Protein motifs for proton transfers that build the transmembrane proton gradient, *Front. Chem.*, 2021, **9**, 660954.

65 O.P. Ernst, D.T. Lodowski, M. Elstner, P. Hegemann, L.S. Brown, and H. Kandori, Microbial and animal rhodopsins: structures, functions, and molecular mechanisms, *Chem. Rev.*, 2014, **114**, 126–163.
66 H. Kuhn and C. Kuhn, On the mechanism of proton pumping, *Chem. Phys. Lett.*, 1996, **253**, 61.
67 J. Dober, W. Zinth, W. Kaiser, and D. Oesterhelt, Excited-state reaction dynamics of bacteriorhodopsin studied by femtosecond spectroscopy, *Chem. Phys. Lett.*, 1988, **144**, 215.
68 J. Herbst, K. Heyne, and R. Diller, *Science*, Femtosecond infrared spectroscopy of bacteriorhodopsin chromophore isomerization, 2002, **217**, 822.
69 E. Pebay-Peyroula, G. Rummel, J.P. Rosenbusch, and E.M. Landau, X-ray structure of bacteriorhodopsin at 2.5 angstroms from microcrystals grown in lipidic cubic phases, *Science*, 1997, **277**, 1676; J. LeCoutre, J. Tittor, D. Oesterhelt, and K. Gerwert, Experimental evidence for hydrogen-bonded network proton transfer in bacteriorhodopsin shown by Fourier-transform infrared spectroscopy using azide as catalyst, *Proc. Natl. Acad. Sci. U.S.A.*, 1995, **92**, 4962.
70 A. Aharoni, I. Weiner, M. Ottolenghi, and M. Sheves, Bacteriorhodpsin experiences light-induced conformational alterations in nonisomerizable C_{13}=C_{14} pigments: a study with EPR, *J. Biol. Chem.*, 2000, **275**, 21010.
71 H. Michel and D. Oesterhelt, Light-induced changes of the pH gradient and the membrane potential in *H. halobium*, *FEBS Lett.*, 1976, **65**, 175.
72 D.E. Smith, S.J. Tans, S.B. Smith, S. Grimes, D.L. Anderson, and C. Bustamante, The bacteriophage ϕ29 portal motor can package DNA against a large internal force, *Nature*, 2001, **413**, 748; C. Bustamante and S. Yan, The development of single molecule force spectroscopy: from polymer biophysics to molecular machines, *Q. Rev. Biophys.*, 2022, **55**, e91.
73 C.L. Asbury, A.N. Fahr, and S.M. Block, Kinesin moves by an asymmetric hand-over-hand mechanism, *Science*, 2003, **302**, 2130.
74 J.O. Wolff, L. Scheiderer, T. Engelhardt, J. Engelhardt, J. Matthias, and S.W. Hell, MINFLUX dissects the unimpeded walking of kinesin-1, *Science*, 2023, **379**, 1004–1010.
75 A. Goel and V. Vogel, Harnessing biological motors to engineer systems for nanoscale transport and assembly, *Nat. Nanotechnol.*, 2008, **3**, 465–475; F. Haque, J. Li, H.-C. Wu, X.-J. Liang, and P. Guo, Solid-state and biological nanopore for real-time sensing of single chemical and sequencing of DNA, *Nano Today*, 2013, **8**, 56–74.
76 S. Diez, C. Reuther, C. Dinu, R. Seidel, M. Mertig, W. Pompe, and H. Howard, Stretching and transporting DNA molecules using motor proteins, *Nano Lett.*, 2003, **3**, 1251.
77 P.D. Boyer, The ATP synthase—a splendid molecular machine, *Annu. Rev. Biochem.*, 1997, **66**, 717; P.D. Boyer, *Nobel Lecture*, December 8, 1997; W. Junge, H. Lill, and S. Engelbrecht, ATP synthase: an electrochemical transducer with rotatory mechanics, *Trends Biochem. Sci.*, 1997, **22**, 420; W. Junge and N. Nelson, Nature's rotary electromotors, *Science*, 2005, **308**, 642; P. Dimroth, C. von Ballmoos, and T. Meier, Catalytic and mechanical cycles in F-ATP synthases, *EMBO Rep.*, 2006, **7**, 276; T. Meier, P. Polzer, K. Diederichs, W. Welte, and P. Dimroth, Structure of the rotor ring of F-Type Na^+-ATPase from *Ilyobacter tartaricus*, *Science*, 2005, **308**, 659; T. Elston, H. Wang, and G. Oster, Energy transduction in ATP synthase, *Nature*, 1998, **391**, 510.
78 G. Oster and H. Wang, Reverse engineering a protein: the mechanochemistry of ATP synthase, *Biochim. Biophys. Acta*, 2000, **1458**, 482.
79 H. Itoh, A. Takahashi, K. Adachi, H. Noji, R. Yasuda, M. Yoshida, and K. Kinosita, Mechanically driven ATP synthesis by F_1-ATPase, *Nature*, 2004, **427**, 465.
80 O. Inacker, H. Kuhn, D. Möbius, and G. Debuch, Manipulation in molecular dimensions, *Z. Phys. Chem. Neue Folge*, 1976, **101**, 337; D. Möbius, Molecular cooperation in monolayer organizates, *Acc. Chem. Res.*, 1981, **14**, 63; H. Kuhn, Information, electron and energy transfer in surface layers, *Pure Appl. Chem.*, 1981, **53**, 2105.
81 Th. Förster, Intermolecular energy migration and fluorescence, *Ann. Phys.*, 1948, **2**, 55.
82 B.S. Brunschwig, J. Logan, M.D. Newton, and N. Sutin, A semiclassical treatment of electron-exchange reactions. Application to the hexaaquoiron (II)-hexaaquoiron (III) system, *J. Am. Chem. Soc.*, 1980, **102**, 5798.
83 R.A. Marcus, On the theory of oxidation-reduction reactions involving electron transfer. I, *J. Chem. Phys.*, 1956, **24**, 966.
84 F. van Mouvik, K. Griebenow, B. van Haeringen, A.R. Holzwarth, and R. van Grondelle, in: *Current Research in Photosythesis*, Vol. **11**, eds. M. Baltscheffsky (Kluver Academic Publishers, 1990, p. 141).
85 H. Tamiaki, M. Amakawa, Y. Shimono, R. Tanikaga, A.R. Holzwarth, and K. Schaffner, Synthetic zinc and magnesium chlorin aggregates as models for supramolecular antenna complexes in chlorosomes of green photosynthetic bacteria, *Photochem. Photobiol.*, 1996, **63**, 92.
86 A.V. Tivanski, Y. He, E. Borguet, H. Liu, G.C. Walker, and D.H. Waldeck, Conjugated thiol linker for enhanced electrical conduction of gold–molecule contacts, *J. Phys. Chem. B*, 2005, **109**, 5398.
87 P. Debye, *Polar Molecules* (Dover, New York, 1929).

88 S.D. Straight, J. Andréasson, G. Kodis, S. Bandyopadhyay, R.H. Mitchell, T.A. Moore, A.L. Moore, and D. Gust, Molecular AND and INHIBIT gates based on control of porphyrin fluorescence by photochromes, *J. Am. Chem. Soc.*, 2005, **127**, 9403.

89 V. Balzani, A. Credi, F.M. Raymo, and J.F. Stoddart, Artificial molecular machines, *Angew. Chem.*, 2000, **39**, *3348*.

90 Y. Wang, R.M. Hernandez, D.J. Bartlett, J.M. Bingham, T.R. Kline, A. Sen, and T.E. Mallouk, Bipolar electrochemical mechanism for the propulsion of catalytic nanomotors in hydrogen peroxide solutions, *Langmuir*, 2006, **22**, 10451.

91 S. Bhosale, A.L. Sisson, P. Talukdar, A. Fürstenberg, N. Banerji, E. Vauthey, G. Bollot, J. Mareda, C. Röger, F. Würthner, N. Sakai, and S. Matile, Photoproduction of proton gradients with π-stacked fluorophore scaffolds in lipid bilayers, *Science*, 2006, **313**, 84.

92 G. Steinberg-Yfrach, J.-L. Rigaud, E.N. Durantini, A.L. Moore, D. Gust, and T.A. Moore, Light-driven production of ATP catalyzed by F0F1-ATP synthase in an artificial photosynthetic membrane, *Nature*, 1998, **392**, 478.

93 J. He, F. Chen, P.A. Liddell, J. Andreasson, S.D. Straight, D. Gust, T.A. Moore, A.L. Moore, J. Li, O.F. Sankey, and S.M. Lindsay, Switching of a photochromic molecule on gold electrodes: single-molecule measurements, *Nanotechnology*, 2005, **16**, 695.

94 W.J. Greenleaf and S.M. Block, Single-molecule, motion-based DNA sequencing using RNA polymerase, *Science*, 2006, **313**, 801.

95 R.A. van Delden, M.K.J. ter Wiel, M.M. Pollard, J. Vicario, N. Koumura, and B.L. Feringa, Unidirectional molecular motor on a gold surface, *Nature*, 2005, **437**, 1337.

96 H. Oevering, M.N. Paddon-Row, M. Heppener, A.M. Oliver, E. Cotsaris, J.W. Verhoeven, and N.S. Hush, Long-range photoinduced through-bond electron transfer and radiative recombination via rigid nonconjugated bridges: distance and solvent dependence, *J. Am. Chem. Soc.*, 1987, **109**, 3258.

29

Origin of Life. Matter Carrying Information

CHAPTER MENU

29.1 Introduction

Life is the most fascinating form of matter and it is useful to see how physicochemical ideas can contribute to our understanding of this form of matter in all its diversity. A fundamental problem is to identify the origin of living forms and the process leading to the first manifestation of life. The origin of life has many different facets, but the question, "What is the basic mechanism that leads to the emergence of systems that carry information?," is definitely a fundamental question in physical chemistry. One might call the emergence of systems that carry information the "tiny bang" facilitating the vast diversity of forms of life's matter.

In the preceding chapters we developed the scientific world view of physical chemists and their struggle to understand the behavior of molecular and material systems. What are the unique features of living systems in physicochemical terms, how was it possible that life emerged, and what was the mechanism? Living organisms are certainly the most intricate supramolecular entities we know of, and insights from the study and fabrication of supramolecular devices (Chapter 28) should help elucidate the mechanism leading to the origin of life. The underlying design principles in fabricating supramolecular devices are the lock-and-key concept and the programmed-environmental-change concept; and we use these two concepts as guiding principles to understand how life can arise. The "program," driving the origin of life, is considered to be given by chance at some particular location. A huge variety of conditions existed on the early earth; however, the particular situation in this particular location led to the emergence of life.

Developing a mechanism for the origin of life is similar to the type of modeling used in physical chemistry, i.e., the search for extremely simplified descriptions that capture the essence of the phenomenon. For the origin of life, one searches for the basic mechanism behind the process; and for this, it is useful to search for the basic mechanisms behind supramolecular chemistry and the principles for developing artificial self-organizing complex systems. Later in the chapter, we discuss how the ideas on the mechanism of the origin of life may be helpful in tracing the limits of physicochemical thinking.

29.1.1 Definition of Life in the Present Context

Physical objects that carry information come into being. They behave as if they have a purpose, an intention, and an aim. This phenomenon begins with the first object that evolves in a particular environment by multiplication, variation, and selection. Let us call physical objects with this fundamentally new property (not present in any ancestral form in a prebiotic universe) *living*.

Principles of Physical Chemistry, Third Edition. Hans Kuhn, David H. Waldeck, and Horst-Dieter Försterling.

Companion website: www.wiley.com/go/Kuhn/physicalchemistry3e

29.2 Investigation of Complex Systems

Rigorous predictions on the basis of quantum mechanics encounter insurmountable problems, even in very simple cases. In order to proceed toward understanding the appearance of life on the basis of physical chemistry, simplifications have to be made. The search for ingenious models is a central task in physical chemistry. These models must be as simple as possible but sufficiently detailed for a reasonable description of molecular systems.

29.2.1 Increasing Simplification with Increasing Stages of Complexity

The treatment of increasingly complex systems requires further and further simplifications in order to not lose the overview. We used this strategy already when we proceeded from describing the properties of simple molecules, which must be described by quantum mechanical considerations, to molecular assemblies, where statistical and deterministic thinking schemes were employed. For example, describing molecules by their chemical structural formula plays an important role in understanding assemblies; however, this extremely useful tool becomes too complicated in describing the network of biochemical metabolism. The description again becomes too complicated when attempting to grasp general aspects about the origin of life. All irrelevant details must be abstracted in order to condense the problem to its essential features.

29.3 Can Life Emerge by Physicochemical Processes?

In the course of investigating biological systems, no behavior has been observed which contradicts the laws of physical chemistry. This observation supports the hypothesis that all material processes in living matter are of a physicochemical nature.

29.3.1 Bioevolution as a Process of Learning How to Survive as a Species

Living individuals produce other individuals that are essentially copies, but by being somewhat different from the original some copies can be more fit to survive in the given environment than is the original. In this way the species (the sum of the individuals) adapts to the environment and learns how to survive better. We use the term "learning" in the sense of finding out what makes individuals fit, not in the sense of memorizing.

29.3.1.1 Multiplication, Variation, Selection

According to the results of molecular biology, every living organism carries a prescription to build copies. The message is carried in the form of a very specific sequence of four types of building blocks (monomers) arranged to form a molecular chain. According to that message the organism is built by using a very particular information translation mechanism and a replication mechanism for the message to be copied (Fig. 29.1).

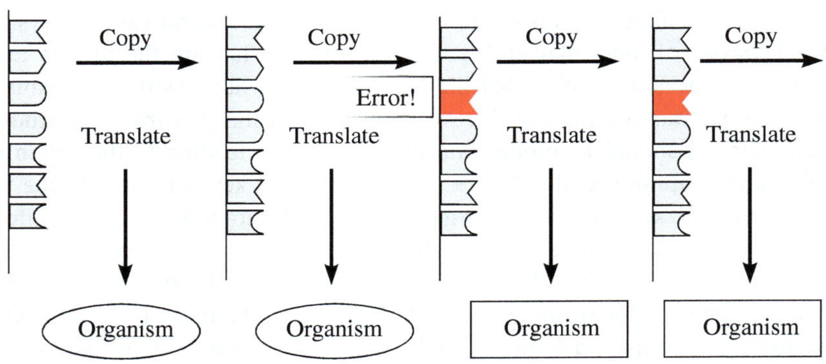

Figure 29.1 Bioevolution as a process of learning. The genetic message is copied and translated. The translation product is the organism which interacts with the environment. In the multiplication process a mismatch can lead to a copying error. The message is changed, and a correspondingly changed translation product is formed.

Occasionally an error, arising from a mismatch during the copying process, can occur which does not affect the power of this copying process – that is, a wrong monomer is introduced at some point on the chain. The biological copying process is extremely exact, and such events are extremely rare. When this kind of error does occur in copying the genetic message, however, the individual built by translating the changed message will become somewhat different. In most cases this change is detrimental because the organism is a complex functional structure and small changes can have catastrophic consequences.

Consequently, the changed individual is often unable to survive. This yields an important error filter: Individuals with changed genetic message are eliminated and do not further disturb the species. Rarely, a copying error results in improved survival chances of the changed individual. Thus, individuals will survive which are better adjusted to the environment. This adaptation and adjustment proceeds through generations. The resulting biological evolution process is a process of learning that requires many individuals and many generations; the ultimate goal of evolution is to acquire the message to survive in an increasingly complex environment. Further and further adaptation of the species to the environment ensues.

29.3.2 Model Case for the Learning Mechanism

29.3.2.1 Drawing Structures Adapting to a Tub

To simplify and focus on the essence of the learning mechanism (finding out what is fit), we consider a game involving drawn structures that adapt to the shape of a tub, which symbolizes the environment. The drawn structures are obtained by following an algorithm.

- First, we proceed from left to right at a set of equally spaced stations, and at each station we flip a coin to determine whether the next step should be up ("heads") or down ("tails"). This process generates a zigzag path; see Fig. 29.2. We proceed to generate a large number of structures (Fig. 29.2 only shows four).

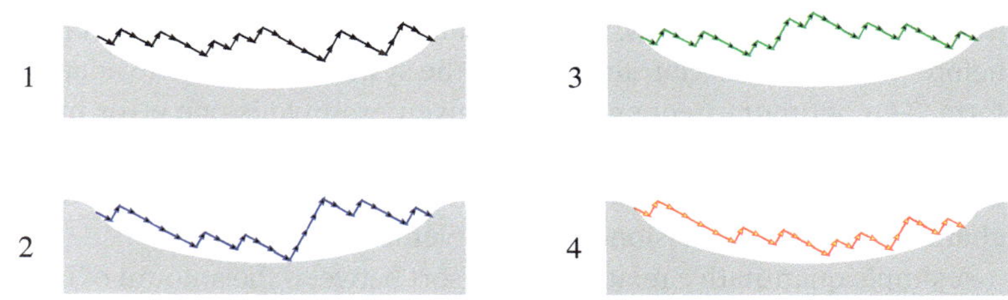

Figure 29.2 Various structures 1–4 randomly formed. Structure 4 is best fitted to the given form of a tub; thus, its survival chances are increased in our game. The a priori probability for structure 4 (red solid line) to survive is 50% higher than that for structure 3 (green solid line).

- Second, half of the structures obtained in the first step are eliminated. The structures are eliminated by chance; however, the probability of elimination is statistically weighted so that structures which are better adapted to the tub's shape have a higher chance of survival. In particular, the structures having a smaller area between themselves and the tub's rim have a greater chance to survive. Therefore structure 4 in Fig. 29.2 has a better survival chance.
- Structures with a section below the rim of the tub cannot survive.
- In the next stage of the game, all of the surviving structures are duplicated; however, a copying error in the duplication process is assumed to occur with a small probability – for example, a step up instead of a step down or vice versa (Fig. 29.3) so that the altered chain is shifted up (in panel a) or down (in panel b). Or the chain occasionally becomes longer by a link (panel c) or shorter.
- Doubling and selection of structures continues for many cycles, in this loop.

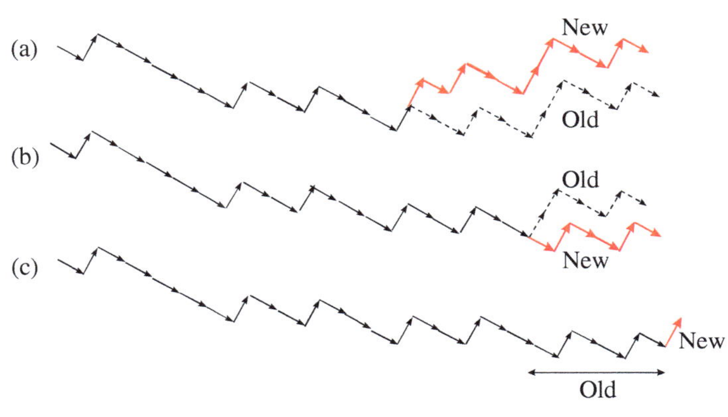

Figure 29.3 Various types of copying errors, which occur with low probability (e.g., one error in 30 copying steps), are shown for three cases: a) chain shifted up, b) chain shifted down, and c) chain elongated by a segment.

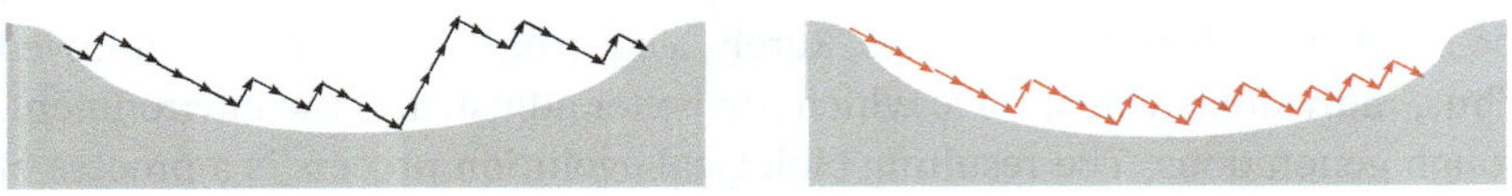

Figure 29.4 Adaptation of the structure to the given form of a tub after many generations.

According to the game instruction, drawn structures which are close to the given tub are selected for in the game, because they disappear less frequently than others (Fig. 29.4). If the shape of the tub is changed (i.e., environmental conditions change) and the process of multiplication and selection continued, then after a sufficient number of generations we will find mainly structures which are closely adapted to the new form of the tub (Fig. 29.5).

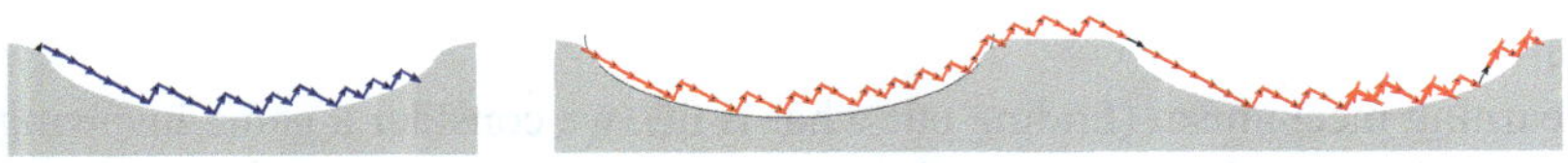

Figure 29.5 Adaptation to a changing form of the tub.

This example illustrates a mechanism in which structures are created by chance, and by a process of multiplication, variation, and selection the structures adapt to the environment. The system behaves as if it would have an aim, to adjust to the environment.

29.3.3 Optimum Error Frequency

From the examples in Figs. 29.2–29.5, we infer that the frequency of errors in an adapting (learning) system should be neither too high nor too low, in order to allow adaptation. When errors occur too often the system forgets what it has learned before. When errors occur too seldom, the system is not able to adapt. It does not adjust fast enough to changes in the tub form. The optimum frequency in the given example is one error in about every 30th multiplication, which corresponds to a 97% probability that no error will occur in each step. Thus enough error-free copies are present in each cycle, so that the information is preserved in the population, but also enough errors occur to ensure adaptation. Accordingly, in longer chains the copying precision per chain element has to be better.

A simple quantitative relation of this sort between the amount of information and the optimal error frequency should be valid for any learning system. Therefore it can be applied to the process of evolution for living organisms as well. Evolution is a process of learning; the system evolves so that its ability to survive in an increasingly complex environment improves over the course of many generations. For living organisms, the amount of transferred genetic information is very large; therefore, errors in one step are extremely rare. No fundamental reasons oppose the hypothesis that sophisticated organisms were formed from first learning systems, i.e., first systems that carry information. On the one hand, increasingly complex forms developed (higher forms of life), and on the other hand forms which are robust because of their relative simplicity were formed (viruses and bacteria).

29.4 Modeling the Emergence of the Genetic Apparatus

29.4.1 Basic Questions and their Refinement

The crucial question concerning the origin of life is whether learning systems can emerge from inanimate matter (Section 29.3.1). The problem of how life originated has a chemical aspect, to understand how the building blocks were formed on the early planet and how their replication started. Also it has a biological aspect: how the evolutionary explosion of life took place after the first simple living cells were present. The origin of life also has a physicochemical and a molecular engineering aspect (Fig. 29.6), and we will focus our discussion on this aspect. We aim to understand how the biomachinery developed from the given building blocks – that is,

- how it was possible for evolution to start and proceed;
- by what basic mechanism was the genetic apparatus and the genetic code formed,
- how must the particular molecular building blocks be constructed, independent of their specific chemical composition;
- how must the environment be designed to promote the process.

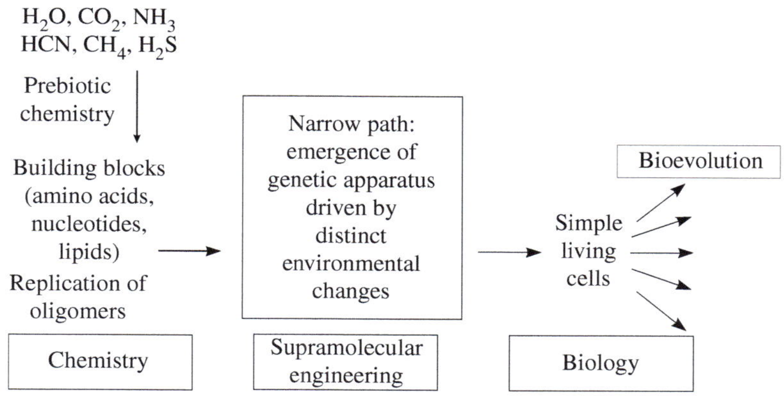

Figure 29.6 Prebiotic chemistry, supramolecular engineering (emergence of genetic apparatus), bioevolution.

Findings in prebiotic chemistry give good reasons to assume that the building blocks of early life – the amino acids: glycine, alanine, valine, aspartic acid, and glutamic acid; the nucleic bases: uracil, adenine, cytosine, guanine, and thymine; and ribose –

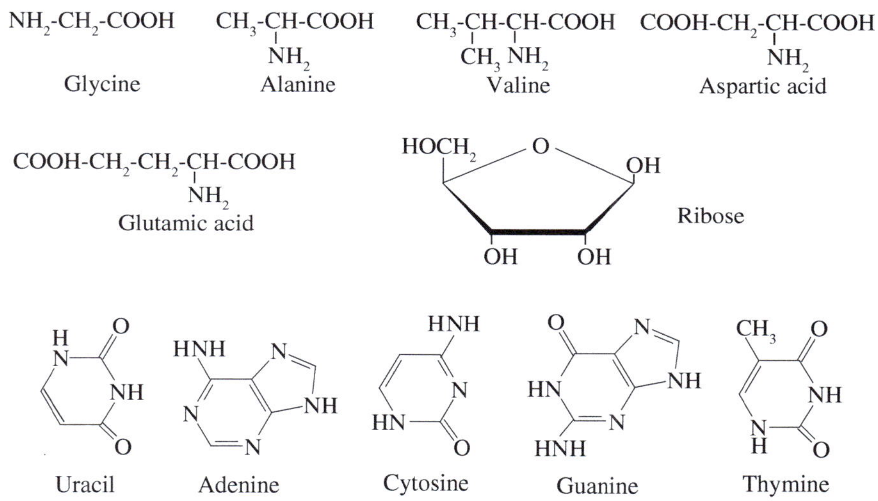

were present on a prebiotic earth. Furthermore it is reasonable to assume that short strands of nucleic acids or similar compounds were present at particular locations and were able to replicate. This assumption is not unreasonable in view of the recent progress in attempts to replicate short nucleic acid strands without an enzyme [1]. The basic mechanisms of bioevolution are known and the phenomenon is well established. In contrast, there is no information about the stages between the prebiotic phase and the phase where simple living cells were present, possessing the incredibly complex and intricate genetic apparatus (Box 29.1). Although no intermediates, which indicate how such systems might have evolved, have been identified, the benefits of cellular structures can guide our thinking about plausible mechanisms [2].

The task of elucidating how life emerged is quite different from the familiar tasks in physics, chemistry, and biology where the pertinent questions are approached by solving many special, well-defined problems, each being checked by comparison with facts. The difficult situation in analyzing the origin of life is similar to analyzing the evolution of the universe. Here also the evolutionary product is known – our universe – but we have no direct evidence on how it emerged.

29.4.1.1 Evolution of the Universe and Evolution of Life: The Big Bang and the Tiny Bang

Steven Weinberg, in his famous book *The First Three Minutes* [3] has emphasized the importance of theories on the origin of the universe: "Our mistake is not that we take our theories too seriously, but that we do not take them seriously enough" and he provides strong evidence for that statement. In attempting to understand the evolution of the universe, a long sequence of events has been proposed, namely, a logical chain of hypothetical processes based on our present knowledge. This long chain leads to a most astonishing result at the end, a possible explanation of basic facts in astrophysics which cannot be rationalized otherwise, such as the cosmic background radiation and the relation between the redshift in spectra and the distance of stars. Correspondingly, an attempt to model the basic mechanisms in the origin of life should be taken as an earnest task in physical chemistry.

Now, let us try to approach the problem of the origin of life in the same manner, to begin with some prebiotically reasonable conditions and reasonable properties of the building blocks and to see what kind of processes could occur, according to what is expected from our knowledge of physics and chemistry. Can we rationalize the emergence of life's intricate machinery (i.e., the genetic apparatus) by inventing a long sequence of logically related physicochemical processes? The goal is not to find the detailed historic path but to show that a reasonable path can actually be given (by specifying the fundamental changes in structure and functionality required to finally reach a life-like genetic apparatus). We want to understand why and by what mechanism(s) these changes should occur, and by what environmental influences they are driven by chance and necessity.

The kind of theoretical modeling required here is familiar in physical chemistry, to start with reasonable simplifying assumptions and to investigate carefully the logical consequences. Thus, modeling basic mechanisms in the evolution of the genetic machinery is a useful exercise in physical chemistry and should help elucidate the important principles and stimulate future attempts in supramolecular engineering.

The crucial goal is to find appropriate conditions leading to a tiny bang: the evolution of a life-like system at the end of a long sequence of consistent steps, with each step being simple. Einstein's advice to scientists – a theory should be as simple as possible but not simpler – must always be kept in mind, however. Searching for a theory on the origin of life is searching for a sequence of physicochemical steps that lead to a life-like genetic apparatus. Particularly relevant is the search for conditions that drive a continuous increase in complexity. Equally relevant is the search for why and how a genetic apparatus emerged that realizes Francis Crick's Central Dogma [4] : the roles of DNA, RNA, and proteins.

29.4.1.2 Paradigm of Present Attempt to Understand the Origin of Life

The given considerations lead to the paradigm: First, attempt to find the logical pattern of a process that leads to a bio-like genetic apparatus. Try to find out, by systematic thinking, how that ingenious and complex machinery could have evolved, by chance and necessity, in a sequence of many distinct steps. Then try to give a sequence of hypothetical steps that are based on this logical pattern, on the existing chemistry and physics, and in accord with prebiotic conditions and biochemical requirements. Develop such a hypothesis to reach an increasingly reliable, realistic, detailed, and experimentally supported description of the process. According to that paradigm, life's emergence is guided by a logical construct, not through blind trial and error.

29.4.2 General Conditions for Life to Come into Being: Periodicity in Time, Compartmentalization, and Structural Diversity

We are looking for what is required, just from logical considerations, to stimulate a process leading to bio-like forms. These requirements should hold anywhere in the universe at appropriate locations; the chemistry could be similar to that on the early earth, or quite different.

- A living organism is a complex aggregate of interlocking molecules acting as a functional unit. Hence, we must look for the basic conditions that are required to stimulate a process leading to aggregates of interlocking molecules of increasing complexity and intricacy.
- A first condition is the presence of a molecule that is self-reproducing in the given environment, whereby variations (copies with a defect) can still replicate, e.g., an oligomer R (Fig. 29.7). This requires energy-rich building blocks that can assemble to form oligomers R under particular conditions; and it requires a particular spatial structure and a distinct periodicity in time which imposes a cycle of repeated self-reproduction (multiplication) of oligomers R, variation, and selection. The extremely specific environmental conditions driving these first and later steps in the evolutionary process are assumed to be present in a very special location somewhere on the prebiotic earth, and anywhere else in the universe where corresponding conditions hold.
- Further conditions to form aggregates of increasing functional intricacy must be given as follows:
  (i) *Periodicity*. Oligomers R must be induced to multiply and subsequently to aggregate in a very specific way. To be exposed to a selection process, to dissociate, and then to be ready again to multiply, requires distinct environmental conditions. Basically different situations must follow each other, a most specific periodic change of conditions is needed to enable that particular process. For example, it may be driven by the day–night periodicity.

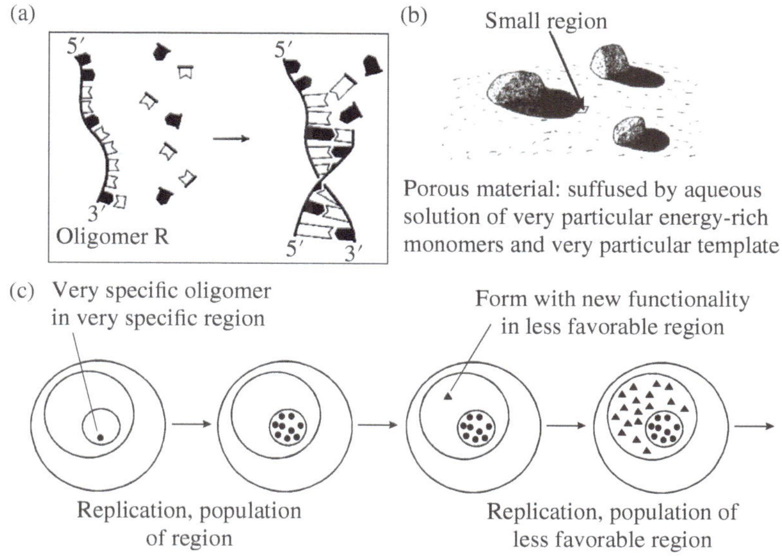

Figure 29.7 (a) Modeling emergence of first replicating strand. (b) Very particular location on the prebiotic planet. Arrow: small region, very particular cyclic change of temperature and many other special conditions given by chance just here. (c) Evolution of increasingly complex self-reproducing forms by populating increasingly unfavorable regions.

(ii) *Compartmentalization*. The single oligomers occurring in the multiplication phase should not be able to escape easily; otherwise, no aggregates would be formed. The region must be compartmentalized (e.g., porous rock), keeping the oligomers together.

(iii) *Microdiversity*. Microdiversity of the environment is the fundamental driving force for a process that leads to the emergence of entities that are more and more functionally sophisticated. Neighborhood regions with slightly different (less favorable) structural properties cannot be populated in the beginning but by casually occurring slightly improved forms. More and more complex forms evolve by colonizing further and further regions. In this way, one avoids difficulties in rationalizing the evolution of increasingly complex forms. Microdiversity serves as an evolutionary gradient.

The continuous extension of the populated area in the highly diversified world, beginning with oligomers R emerging at that very special location, is the basic mechanism leading to the origin of life. A continuous drive to increasing complexity is given by the necessity of increasing sophistication required to populate decreasingly favorable regions.

The given conditions, in the present view, are fundamental requirements for the emergence of life in the universe. Thus they must be fulfilled in attempts to model the emergence of a genetic apparatus as a sequence of physicochemical steps that are, per se, in agreement with thermodynamics [5].

29.4.3 Modeling a Continuous Sequence of Physicochemical Processes Leading to a Genetic Apparatus

Having set the basic conditions in Section 29.4.2, we can only indicate important breakthrough events. A description of a pathway would be a much too involved enterprise. We first focus on engineering aspects and then reflect on the chemical realization of the proposed construction principles.

29.4.3.1 Strategy: Search for Sequences of Small Steps

To illustrate the probability considerations in the search for a meaningful model we consider the probability for the emergence of a replicating strand. We expect (see Example 29.1) that a short strand, later serving as a template for replication, will appear among a reasonable number of spontaneously formed strands. In contrast, a longer replicable strand will not appear (the probability of its appearance is almost zero) even among a huge number of spontaneously formed strands. Long strands can be easily obtained by linking small replicable strands, however, once such strands have accumulated.

Example 29.1 *Consider a solution of monomers that can interlink to form strands and assume that a strand which is able to replicate is formed only when the monomers are connected in a distinct manner. In addition, assume that the probability for a monomer to attach to a strand is* 1/100. *Calculate the probability of finding a strand made of* 60 *monomers (a* 60*-mer) directly and compare this probability to that for first forming six* 10*-mers which then attach to form a* 60*-mer.*

The probability of directly forming a 60-mer that is capable of replicating is $(1/100)^{60} = 10^{-120}$. Out of 10^{120} strands, only one is a correct sample. This probability is so small that if we could fill the the whole universe with such strands (assuming each strand would require 10^{-20} cm^3), it would still be extremely improbable to have a correct sample among them.

The probability of obtaining a correct strand of 10 monomers under the same conditions is $(1/100)^{10} = 10^{-20}$. In one millimole of substance (6×10^{20} strands) we would find, on average, six correct strands. Hence we should find at least one correct strand with a probability very close to 1. Now let the special strand with 10 monomers multiply by replication. Some strands fuse under plausible conditions, forming strands of 60 monomers that are able to replicate. In this way the result, which would have been reached only with minute probability spontaneously, is obtained with a probability close to 1.

Example 29.1 shows that one is able to understand the formation of a complex system in many small evolutionary steps under suitable plausible conditions, whereby each step occurs with a probability close to 1. With large steps there is only a minute probability that the same system can be formed. This illustrates the importance of viewing small steps in proposing reasonable thought experiments. Hence, we attempt to model the origin of life with short strands and consider many small steps. This indicates the strategy: Search for a gapless sequence of small prebiotically reasonable physicochemical model steps and see what the thought experiments predict when setting appropriate initial conditions and boundary conditions. Can the clearly defined goal be reached, a genetic apparatus?

Recall that we are searching for basic principles. The actual historical processes in the origin of life may have been much more complicated.

29.4.3.2 Beginning: Replication of short Strands that are Homochiral. Symmetry Breaking

In order to be replicable in the given environment, a spontaneously formed strand must allow a precise template-assisted interlocking to form a daughter strand. It is reasonable to assume that this precision is reached when the monomers forming the template strand are chiral and have the same chirality, and this must also be the case for the monomers forming the daughter strand. Monomers of incorrect chirality are rejected in the build-up of the daughter strand (lock-and-key principle).

The origins of biological homochirality in prebiotic chemistry remain an open question. To be concrete, let us assume that left-handed and right-handed monomers were present in equal amounts under prebiotic conditions so that it was equally probable that the strand which initiates an evolutionary process is left-handed or right-handed (frozen accident). In this view the fact that all life on earth is homochiral would be a consequence of a single molecular process. Then the emergence of life on earth is a breaking of symmetry to be compared with the breaking of symmetry in the evolution of the universe (e.g., the upheaval, by an accident, of a system containing neutrinos and neutrons into electrons and protons, while a transition into positrons and antiprotons would be a priori equally probable).

29.4.3.3 Sketching the Path

A priori, we cannot predict the detailed chemical mechanism, rather we can only sketch the path. First we focus on the logical frame of the process—that is, the search for the properties of the components and the conditions needed to drive self-assembly. The goal is to find conditions that lead to the simplest possible path of obvious conceptual steps that suggest themselves.

The basic problem occurring throughout the evolutionary process is to store and read information, the recipe to construct the machinery developed so far—that is, how to copy the genetic message with the required accuracy. The systems of increasing complexity and sophistication that emerge in the course of evolution require increasing accuracy in attaching each monomer at the template, in order to reach the optimum copying-error frequency as discussed in Section 29.3.2. Again and again a new mechanism will be needed that allows a better suppression of copying errors than the presently available mechanism. The occurrence of an improved mechanism will lead to an evolutionary boom that is followed by a stagnation until a new and better solution of the error suppression problem appears.

We elucidate this evolutionary search procedure by starting with the simplest assumptions and investigating what environmental conditions might generate further evolutionary steps; the underlying engineering principles are the lock-and-key concept and the programmed environmental change concept. The analysis in Foundation 29.1 tells us that two kinds of monomers are needed to avoid an early evolutionary barrier. It is assumed that they must be complementary to each other and form strands that interlock in an antiparallel fashion. In this case we ask: Can an aggregate, serving as translation device, be constructed? A model is considered in Foundation 29.1, while still restricting our discussion to engineering principles, rather than specific chemical aspects. We show that the creation of a translation apparatus appears as a by-product in the evolution of an aggregate that emerges because of its power to act as an error filter.

29.4.3.4 Basic Evolutionary Principle and Paradigm for Increasing Complexity

In the present theoretical approach, we assume that a molecular aggregate acts as a distinct first machine to initiate the consecutive and sustained evolution of machines of increasing complexity. The step initiating this process is the formation of an aggregate of precisely interlocking molecules, accompanied by the disappearance of molecules that do not match this condition. This step is of fundamental importance; a main difficulty in the search for reasonable model pathways is the huge variety of harmful side reactions which lead to a dead end. Besides the highly specific aggregate-filtering process these side reactions are assumed to play no appreciable role. We neglect them, i.e., we apply Occams' razor (preferring the simplest of competing theories to the more complex ones).

Our main task in the search for a possible pathway, which leads to the emergence of a genetic apparatus, is to construct environmental situations that lead to the formation of increasingly sophisticated aggregates. The formation of aggregates of precisely interlocking molecules appears to be Nature's trick to overcome side reactions. The first aggregate is the first supramolecular system that acts as a functional unit constituting an individual, i.e., an entity exposed to selection.

The scheme in Fig. 29.8 is thought to provide the fundamental changes in the system structure and their logical connections, which promote colonization of more and more different regions. This growth requires an increasing complexity, ultimately connected with the increasing populated area. The thought experiment considered in Foundation 29.1 emphasizes this basic structure of the evolutionary process and is supported by computer simulation [6]. The computer simulation is helpful in overcoming the fundamental difficulty of identifying a feasible mechanism for how the genetic machinery could have emerged, but the simplicity of the simulation, in contrast to the prebiotic reality, should be kept in mind.

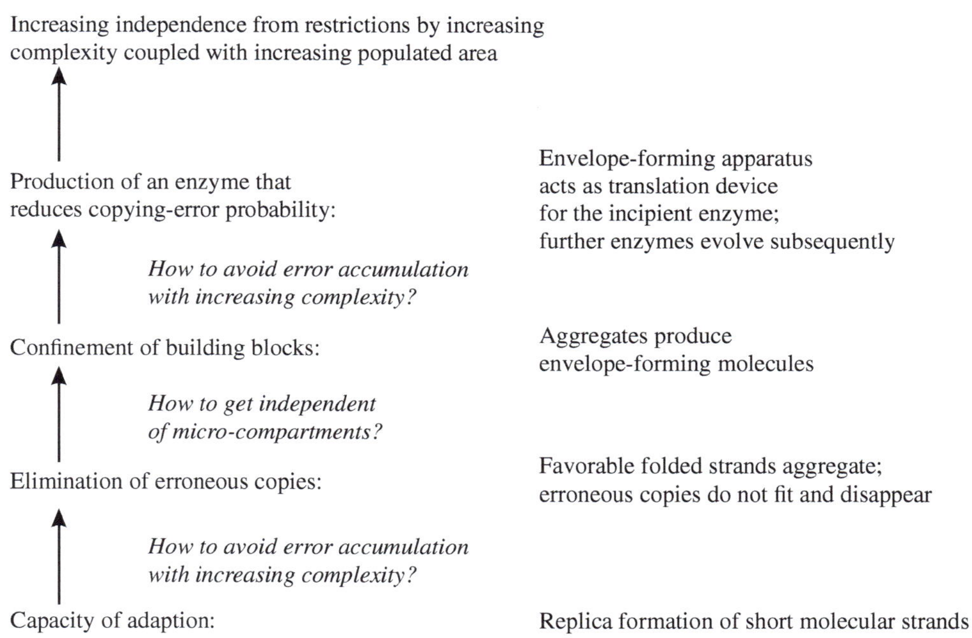

Figure 29.8 Modeling evolution of early life: logical requirements, their realization in the thought experiment, and barriers to overcome.

29.4.3.5 Chemical Features

Special chemical features are considered in Foundation 29.2. It is shown that the genetic code can be rationalized by introducing amino acids in the approximate order of their availability. This reconstruction of the modern code supports the model approach considered. It remains a most challenging problem to understand how the conditions resulting from these engineering considerations can be met with prebiotic conditions.

The basic structure of the genetic apparatus according to the present hypothesis had to evolve in three unavoidable steps:

(1) RNA replication.
(2) RNA replication → protein translation: a system unable to develop into systems of increasing complexity and sophistication.
(3) DNA replication, DNA → RNA translation, RNA → protein translation: a system able to form more and more complex systems by integrating more and more subsystems. At this point the confined evolutionary path opens up and a divergent evolution in many different directions takes place: the explosive evolution of life.

Although we use the terms RNA, DNA, and protein, the proposed process is not tied to a specific terrestrial realization; rather, we use these terms in the generalized sense of their attributes for information storage and physicochemical function. Important features do not depend on specific monomers, although, of course, quite definite requirements must be met by suitable monomers, as discussed in Foundation 29.1. Basic is the fact that a huge variety of situations existed on the early planet, and at many locations a large number of spontaneously formed oligomers were present. At some locations, the condition to replicate was fulfilled, which initiated the emergence of life.

29.4.3.6 Later Evolutionary Steps: Emergence of an Eye with Lens

In Foundation 29.3, we describe a more sophisticated computer simulation that accounts for the evolutionary steps in the emergence of an eye with a lens.

29.5 General Aspects of Life's Emergence and Evolution

29.5.1 Information and Knowledge

According to Section 29.4 the basic mechanism leading to lifelike processes is found in the occurrence of entities, which we call "individuals" (the first simply replicating strands), that interact with the "environment." An individual either survives the "selection phase" or disappears; only the surviving individuals have a chance to produce copies in the following "multiplication phase." The process is repeated many times, and new forms of individuals arise because of replication errors. Forms which survive and evolve are "fit." In this way individuals of increasing intricacy are created. The genetic information of an individual (the message to produce copies in the given environment) and the individual's knowledge (how its physicochemical properties enhance its survival and probability to replicate in a given environment) increase accordingly. Genetic information and knowledge describe the quantity and quality of the genetic message, respectively. The appearance of matter that carries information and accumulates knowledge constitutes the fundamental change from inanimate to animate in the present context.

29.5.2 Processing Information, Genesis of Information and Knowledge, and the Maxwell Demon

29.5.2.1 Computer and Biosystem. Genetic Information I and Knowledge K

A computer, a device for processing information, can be described, in principle, as a system of switches; each switch is represented by a particle in a modulated potential, changing periodically, thus driving the switch through a switching phase, a storage phase, and a reset phase (Fig. 29.9). In the switching phase the switch is set by the directing field of other switches that are in the storage phase. Accordingly, in its storage phase the switch directs other switches that are in the switching phase. In the reset phase the switch is made ready for its use in the next period. Each switch, in the storage phase, is said to carry one bit of information. Rolf Landauer [7] identified this principle for the digital computer.

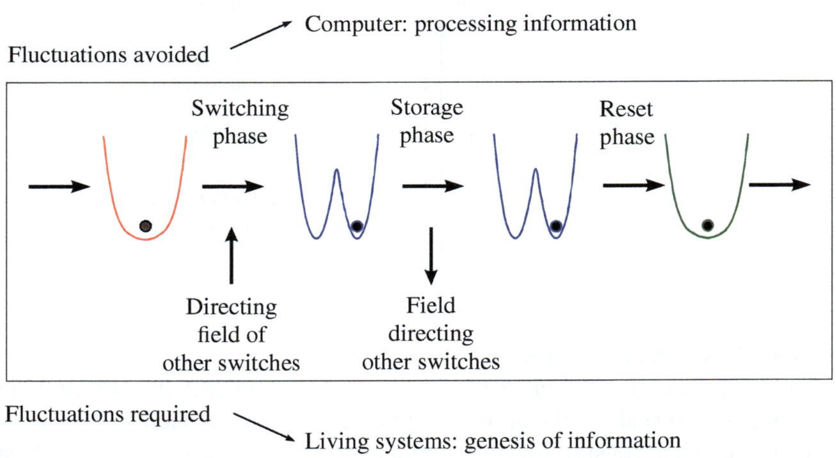

Figure 29.9 Switching processes in a computer and in a biosystem.

It is fascinating to see that this mechanism is equally basic to biosystems where each switch is represented by a complementary pair of nucleotides. The replication process represents the switching phase. The switch is set by the directing field of the template strand, i.e., the field of the hydrogen bond forming groups. This field determines which of the two complementary nucleotides is incorporated in the growing daughter strand. In the storage phase the considered nucleotide, now incorporated in a template strand, directs further processing. The decomposition of the strand into monomers represents the reset phase [8].

A fundamental difference exists between the computer and the living system. In the computer the probability of a copying error (particle jumping from the one into the other minimum by thermal noise) is so small that the probability of an error during the complete computation is still very small. In the living system the probability of errors (incorporation of a nucleotide that is not complementary to the nucleotide in the template) is small but finite; it cannot be too small or too large. The error probability must be small enough to preserve the accumulated information, but large enough to produce variations by occasional copying errors.

In the computer, information is handled in a deterministic process. In the biosystem information is created by variation, and the quality of the created information, the knowledge K, grows stepwise in the course of the evolution. Information is measured according to Claude E. Shannon [9] in number of bits, and knowledge is also measured in number of bits. Genetic information I is the Shannon information transferred from one generation to the next. Knowledge K is the total genetic information to be thrown away during the process leading to a given evolutionary stage. It increases in low steps when small improvements in functionality occur, and in high steps when revolutionary changes in the machinery occur that require many trials by chance.

Example 29.2 *Calculate the knowledge K to create a first replicable 10-mer strand.*

In Example 29.1 we considered an environment allowing a continuous de-novo formation of oligomers consisting of 10 monomers. We found that 10^{18} strands, on average, must appear until a strand shows up by chance that has the property to be replicable. Assuming that 10% of such strands actually survive the reproduction and selection process, $10^{19} = 2^{63}$ strands have to be discarded. Thus, the knowledge K when such a particular strand appears, increases suddenly from zero to 63 bits.

29.5.2.2 Entropy Production in Information Processing and in Information and Knowledge Creation

What is the minimum energy dissipation of information processing in biosystems?

Information Processing and Information Creation The replication process, in principle, can be performed reversibly. Nevertheless, entropy is ultimately produced in the reset process, the degradation of the strands into monomers. The information represented by the distinct sequence of the monomers in the strand is destroyed. Thus, creating information, in contrast to processing information, is necessarily irreversible.

We consider, for simplicity, a strand with only two kinds of monomers, say, G and C. Then the minimum entropy produced is the entropy of mixing (Section 17.5)

$$\Delta S = k(N_G + N_C) \cdot \ln 2$$

(energy dissipation per bit $kT \ln 2$). In principle, a reversible degradation would be possible by binding the strand to a complementary strand and reversing the replication process. One can imagine a computer, based on this idea, that works in reversible steps throughout the computation; it stores the result and processes backward in reversible steps (reversible computing) [10].

Entropy S and Knowledge K A formal similarity exists between the change in entropy in heating a body and knowledge. The entropy S is zero at $T = 0$ and increases with temperature T; the knowledge K is zero before the Darwinian process begins (before the first replicate appears and multiplies) and it increases with time t. The entropy S increases suddenly at phase changes; the knowledge K increases suddenly at changes of structure and functionality. The minimum entropy increase when degrading a strand into monomers is related to K because the number of representations (number of possible structures of the oligomer) is 2^K. Thus $\Omega_1 = 1$ (in the case of our simplifying assumptions that only one out of 2^K de novo formed structures is replicable and apt to survive) and $\Omega_2 = 2^K$. Therefore, $\Delta S = \ln \Omega_2 = k \ln 2^K = k K \ln 2$ (minimum energy dissipation $Q_{\text{rev}} = T\Delta S = kT K \ln 2$).

29.5.2.3 Minimum Energy Dissipation in a Measurement and the Maxwell Demon

In any measurement the measuring instrument is exposed to some directing field produced by the object to be investigated. The measuring process, in principle, can be represented by a switch that is set by this directing field in the switching phase.

The bit of information is processed in the storage phase, and in the reset phase the measuring instrument is made ready for the next measurement. Again, the measuring process can be reversible, but entropy ($k \cdot \ln 2$ per bit, when the two possible results are a priori equally probable) is ultimately produced in the reset phase [11].

This consideration is important in understanding the famous Maxwell Demon Paradox. In the beginning, the molecules in a gas are equally distributed between two equal volumes separated by a wall with a door. When a molecule approaches from the left, the demon quickly opens the door and quickly closes it when the molecule has passed (Fig. 29.10a). In this way more and more gas molecules accumulate in the volume on the right, in seeming disagreement with the principle of entropy increase.

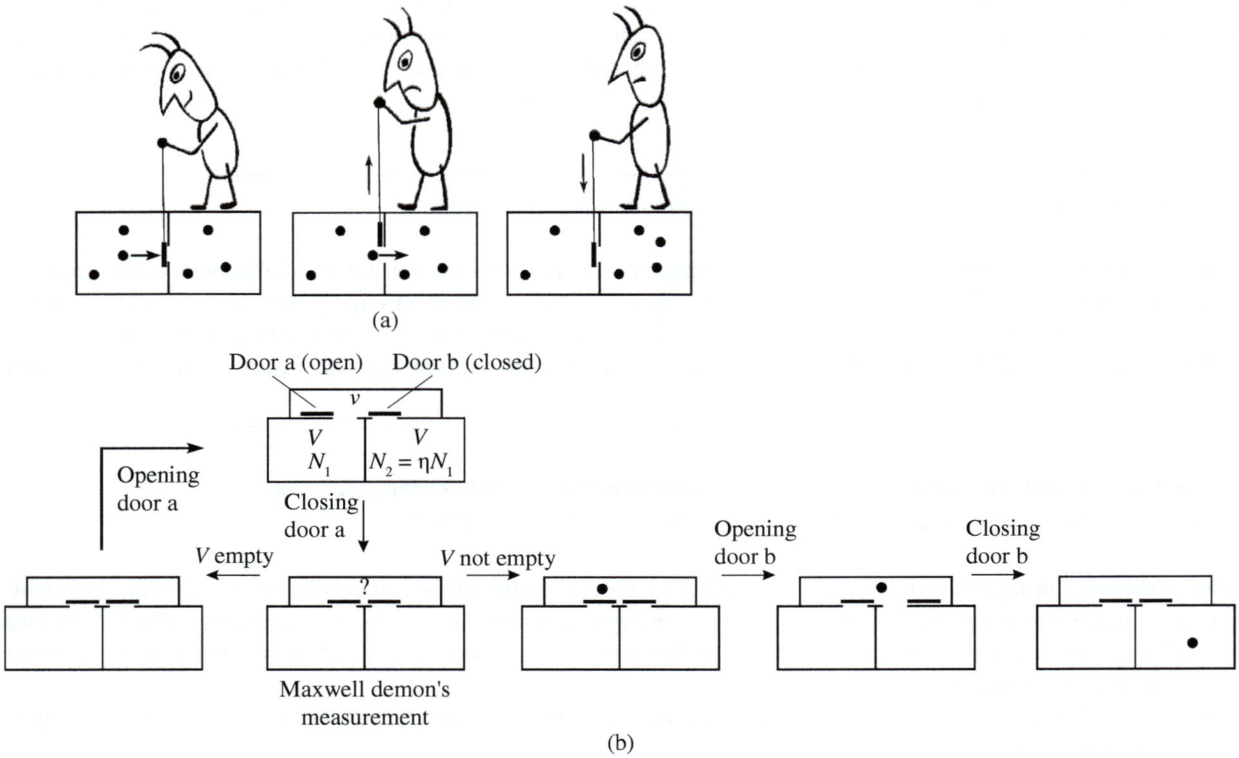

Figure 29.10 Maxwell Demon paradox. Panel a shows the Demons gating operation, and panel b illustrates the logic of the Demon's measurement and operations.

The demon must perform a measurement. Even if this measurement can be performed without entropy production, entropy is produced in the reset phase, when getting the measuring device ready for the next measurement. It can be shown that the minimum entropy produced in this way is larger than the entropy decrease by accumulating the gas molecules on the right. No entropy decrease takes place in the isolated system when the demon is included as part of the system.

Figure 29.10b illustrates the Maxwell demon "paradox." We consider ideal gases in two containers of volume V separated from each other by a wall. Both containers are in contact with a temperature bath of temperature T. N_1 particles are in the left container, and $N_2 = \eta N_1$ particles are in the right container, where N_1 is a large number and $\eta \geq 1$. Additionally, a small compartment of volume v is present which is connected to the two volumes by doors a and b. In the beginning door a is open and door b is closed. v is chosen such that it is much smaller than V/N_1 or V/N_2, the volume per molecule in the left and right container; then volume v is usually empty.

The Maxwell Demon (Fig. 29.10a) is replaced by a device that closes door a and then performs a measurement deciding whether volume v is empty or not (Fig. 29.10b). If volume v is empty, door a is opened again. The process (opening and closing door a and performing a measurement) is repeated until v is found to be not empty. When v is occupied, door b is opened and later closed again. Then in most cases the particle has entered the container on the right because of $v \ll V/N_2$. Door a is then opened and the cycle is repeated. This means the demon is replaced by a device that is testing continuously volume v for being empty or not.

The decrease of the entropy of the system, when shifting a particle from the left to the right (corresponding to an isothermal compression of the ideal gas) is

$$-\Delta S_{\text{system}} = k \ln \left(\frac{N_2}{N_1} \right) = k \cdot \ln \eta$$

We call n the number of measurements that must be taken, on average, until υ is found to be not empty. The entropy produced in the surroundings by n times resetting the measuring instrument is $nk \cdot \ln 2$. In the ideal gas approximation no heat is produced when the particle moves from υ to the right container, so that it does not contribute to the entropy and we find that

$$\Delta S_{\text{surroundings}} = nk \cdot \ln 2$$

For the ratio $r = \left(-\Delta S_{\text{system}}\right) / \Delta S_{\text{surroundings}}$ we find

$$r = \frac{k \cdot \ln \eta}{nk \cdot \ln 2} = \frac{1}{n} \cdot \frac{\ln \eta}{\ln 2}$$

and $r = 1/n$ for $\eta = 2$. In Foundation 29.4 we show that r cannot be smaller than $r = 1$. This means that the entropy of the isolated system (consisting of volumes, Maxwell demon, and temperature bath) cannot decrease.

Although our analysis is restricted to the connection between classical information (bits) and thermodynamics, theoretical investigations of the information and thermodynamics for quantum systems (e.g., a particle in a box) show that the generation of entropy through "erasure" or through the action of a quantum measurement ("collapse of wavefunction") manifests there also [12].

29.5.3 Limits of Physicochemical Ways of Thinking

In order to address this question, we consider a simplified physical model rather than the real complex system.

29.5.3.1 Device That Develops an Internal Model of the World

Following our earlier development we consider a model like that shown in the diagram, in which a "survival of the fittest" algorithm selects a system of most fit structures that comprise the internal model of the world.

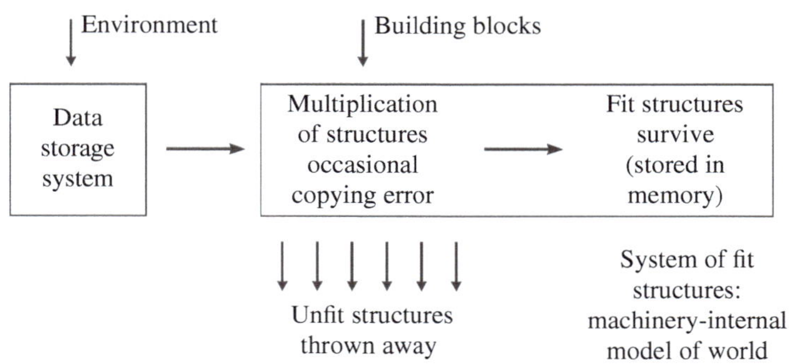

Figure 29.11 Model of a learning system.

We consider an apparatus made of an observation-device (which gives data that may be described by points in a plane), a device for storing these data (memory), and a device which processes the data in a manner analogous to the biological evolution process (Figs. 29.11 and 29.12). The apparatus produces structures according to an algorithm and selects those structures which happen to correspond to a multitude of stored data. Through multiplication of the structure with occasional errors in the copying process and selection, the adaptation becomes increasingly better. This process is illustrated in Fig. 29.12a, in which the circles are stored data and the curves are trial functions. When an adapted function has been obtained in this manner the observation data are deleted in the memory (Fig. 29.12c) so the memory (of limited size) is again ready to store further observation data.

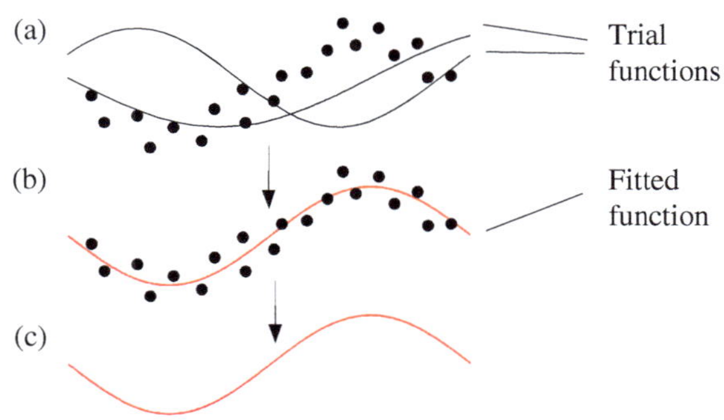

Figure 29.12 Fitting of trial functions to stored observation data. (a) Observed data (dots) and arbitrary trial functions. (b) Fitted trial function. (c) Discharge of the memory by deleting observation data, but storing the fitted trial function.

Creating the Laws of Nature The thought experiments, just described, illustrate the way in which the laws of nature are found (created). The observation-devices, our measuring instruments, give experimental data that are recorded.[1] Structural

1 Note that the basic step in the measuring process is a yes-or-no decision, similar to the survival and multiplication or disappearance in the evolution process: The initiating stimulus is either amplified and recorded or there is no response.

schemes covering more and more recorded data gradually develop, i.e., theories of increasing generality are created by a human, in a way corresponding to the way new forms emerge in bioevolution.

29.5.3.2 Development of Human Internal Model of an External Physical World

Let us assume that the functioning of our brain is based on producing and selecting schemes accommodated to sensory data. Possible reactions are tried within our internal model of an external physical world before they are released as impulses for action. The internal representation of an external physical world cannot be more than the availability of these structural schemes. Our brain does not seem to supply us with any other information about the external physical world. Thus the claim that our brain works similar to an extremely sophisticated form of the considered device cannot be readily contradicted.

How to Cope with Situations of Great Uncertainty Uncertainty is experienced in many forms, [13] and being in a situation of physical darkness is one of them. There are two types of photoreceptor cells in the retina of our eyes: (i) cones, which are more located in the fovea a central area of the retina, where we place the focus of our attention and see the world we already know, and (ii) rods, which are located more in the periphery of the retina, where objects of the world come into sight in surprise and that subsequently drives us to refocus our attention there, to the surprise and to the unknown. Rods function mainly in dim light and provide black-and-white vague vision. This is a situation of great uncertainty our brain must cope with. Furthermore in order to survive in an uncertain and changing environment cognition and intelligent behavior are fundamental. Our only access to knowledge about the world is through our senses, which provide information that is usually corrupted by random fluctuations originating from all possible sources of variation along the path: from the object itself in the world out there all the way to our cognition – technically termed "noise with a spectral density" [14]. Thus our perception may provide ambiguous information about the possible states of the environment.

How then can our nervous system exactly operate in situations of uncertainty? The Bayesian approaches to brain function assume that the nervous system maintains internal probabilistic models (prior) that are updated (posterior) by neural processing of sensory information (likelihood) using methods approximating those of Bayesian inference (see Fig. 29.13). Bayesian inference is a well-known technique in statistics that describes, in mathematical terms, the scientific method:

- Propose a hypothesis from the given theory by deduction.
- Extract data from experiment.
- Use the proposed hypothesis and data to create an updated hypothesis and a revised theory by induction.

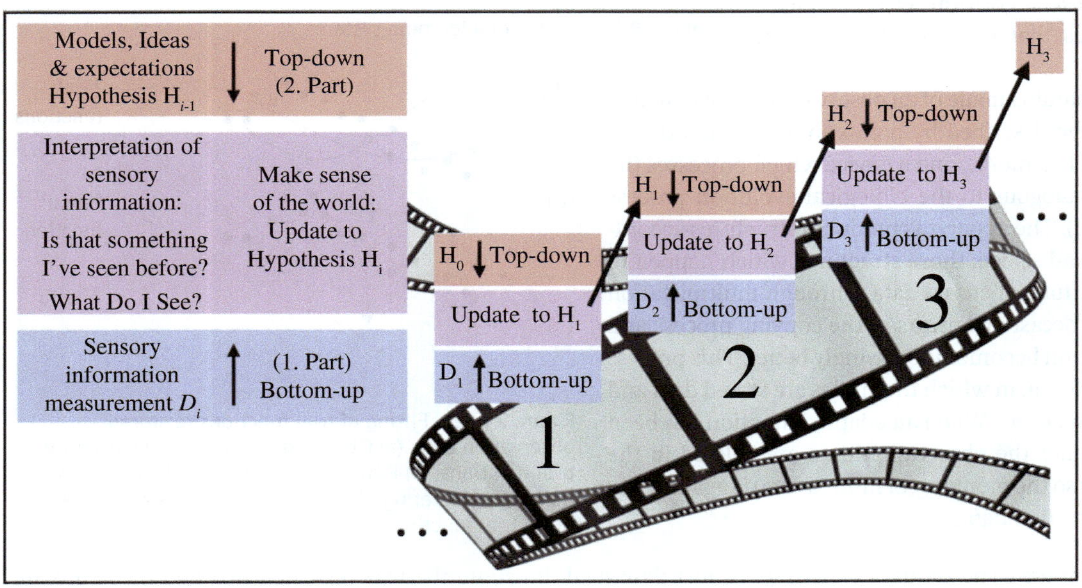

Figure 29.13 The Bayes brain paradigm. A motion-picture scheme illustrates the theorem of Bayes applied recursively. Note the bipartite nature of the theorem of Bayes. The recursive Bayesian inference defined as the posterior of the previous step equated to the prior of the next step in the sequence of consecutive hypotheses. The rapid succession of hypotheses gives the illusion of natural movement.

Bayesian Brain Paradigm The Bayesian brain paradigm [15] — conjectures that the brain's functional unit is a Bayesian device.

Our mind tends to perceive visual impressions as objects in space. This feature is illustrated by ambiguous images, where the impression allows two different interpretations. The moment that our brain flips from one interpretation to the other interpretation is not controlled by our will. Only by integrating an ambiguous figure into a particular context does the appropriate interpretation persist and the ambiguity resolves itself. By way of example, consider the 12 lines in Fig. 29.14, which is known as the "Necker cube" illusion.

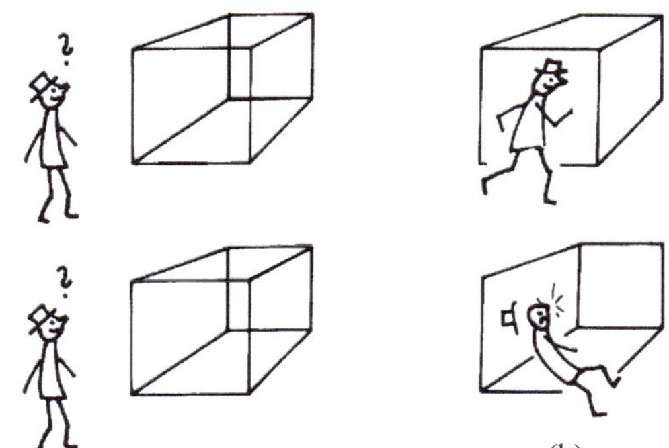

(a) (b)

Figure 29.14 (a) The stickman considers whether the object is in the way of his trajectory. (b) The result of the stickman's trajectory is shown for the two cases.

In panel a of Fig. 29.14, the 12 lines indicate an object (the so called "Necker cube") that we perceive either as a big block at some distance on our left side or a thick beam at the height of our head in front of us. The stickman represents our observing consciousness. Disturbed by the flip between two different scenes we then ask ourselves: "what is reality?" In the attempt to resolve the question the stickman (now acting and testing) makes the assumption "I can get past it" and starts walking. In the top panel, the stickman has a carefree experience (his assumption was correct) and casually passes the block. In the bottom panel, the stickman has a painful experience (his assumption was not correct) and he bumps into the beam. See Foundation 29.5 for a more formal definition of conditional probabilities and the Necker cube.

The ordering of impressions in daily life is limited, as described in Chapter 1. We cannot describe quantum effects by assuming the existence of distinct objects, but we still assume that processes take place in space and time. When including other experimental facts the existence of a universal time and space must be rejected also, and replaced by describing phenomena in the framework of the theory of relativity. In this theory, space and time are restricted to describe local phenomena and motions of material particles with speeds well below the speed of light.

29.5.3.3 Limitations in Physical Theory

The axioms of geometry are conventions using words like "straight line," which are useful in describing visual impressions. They are not statements about the nature of physical space. The general laws of physics are free inventions of the human mind which are selected on the basis of their utility. We cannot say whether these inventions are true or false, but merely whether they are useful schemes in describing facts. We have illustrated this in introducing the Schrödinger Equation.

Similar to the human mind, a sophisticated version of the model device considered above should create useful schemes that describe a multitude of facts, all based on trial, random changes, and selection, in a multitude of repetitions. A remarkable aspect is that the device considered can be described within the framework of classical physics (quantum effects played no role in the process we described for knowledge accumulation). On the other hand, the accumulating knowledge (the evolving scheme ordering the input signals) depends on the received data and can be, for example, the scheme of quantum mechanics.

One of the grand challenges in science is the search for a detailed knowledge of the processes in the brain [16], see Fig. 29.15 as a very crude sketch for what the brain does.

Let us assume for a moment that we possess detailed knowledge and know the correlation of every facet of our consciousness (the feeling of happiness or sadness, the feeling of having a free will to act, seeing colors, listening to music) with the corresponding state of the cortical network in all its physicochemical detail. Then we would know all the correlations of our immediate feelings and impressions with processes in the brain, but we would not see behind the mystery of this correlation. The situation is not different in physics, where we know the general laws but do not understand why they are the way they are. Any question beyond the scheme to describe Nature cannot be answered and is therefore meaningless within the framework of the scientific theory. In Chapters 1–3 we considered this issue. It is important to see this limitation in each physical theory.

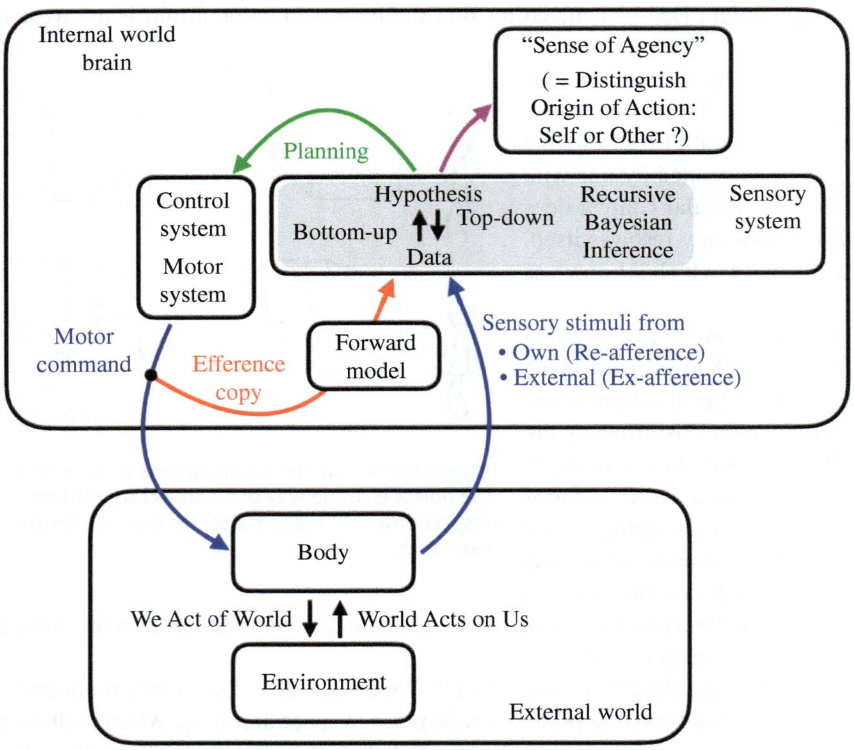

Figure 29.15 The brain and the external world. In the simplest model, the brain consists of a sensory system, a control system, and a motor system. The "external world" is the body, where the brain resides, and the environment. The sensory and motor pathways by which we perceive and act on the world are shown in blue and the process of making plans in green. Note the bipartite nature of the sensory system provided by the recursive Bayesian inference: bottom-up and top-down. The efference copy (in red) of the motor command is an essential pathway – conjointly with a forward model and the sensory system (final path in violet) – by which we distinguish through the "sense of agency" in situations when the world acts on us.

29.6 Conclusion

The genetic apparatus and the genetic code are essentially identical in all living forms. Understanding how the genetic code evolved is essential for understanding the origin of life. We have attempted to model a sequence of processes that should help to enlighten that process. The systems evolve, in many cycles of multiplication, variation, and selection. We have asked how life can originate. According to the given model, a short strand appears that replicates in a particular environment. Occasionally, by mismatch in the replication process, copies appear that have a changed sequence of monomers (variations) and a selection process takes place. The sequence is continuously repeated.

Specific environmental conditions are required to initiate and drive the evolutionary process for knowledge accumulation. Microdiversity is crucial to drive evolution toward always increasing complexity and intricacy. These particular conditions are assumed to have been present at some very specific locations on the prebiotic earth. Through processes like those described, a new form of matter emerges, matter carrying information that continuously increases in an evolutionary process. The information carriers are physical objects that act as if they would have a purpose, an intention, an aim (to survive and to evolve to always increasing intricacy). Living systems are carriers of information, namely the recipe to form copies in the given environment.

29.7 Box 29.1. Genetic Apparatus

The information, which is coded in the DNA strand as a sequence of four kinds of monomers (Fig. 29.1), is multiplied by replication based on complementary pairing, and it is translated into proteins – that is, linear sequences of 20 possible kinds of amino acids. This process is accomplished by first transcribing the message from the DNA to the *messenger ribonucleic acid* (mRNA) (Fig. 29B.1). The translation into proteins takes place by means of adapter or transfer ribonucleic acid (tRNA) molecules, typically 70 to 100 bases long. For every amino acid a_1, a_2, a_3, …, at least one specific tRNA molecule, to which the amino acid can become attached, exists and carries a specific *anticodon* nucleotide triplet. The anti-codon segment of the tRNA couples by complementary base pairing to a mRNA strand that contains the message for the protein in question,

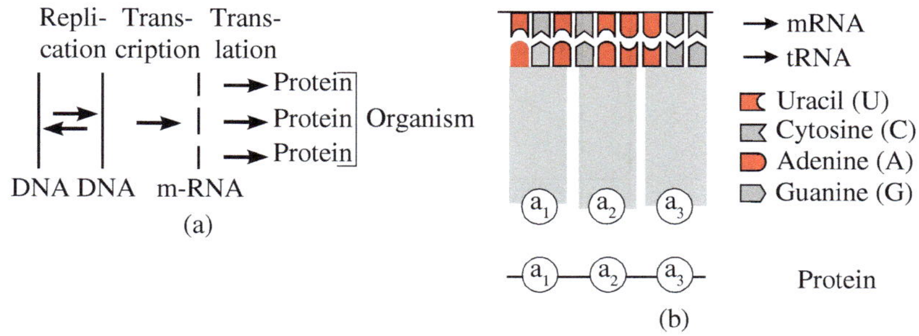

Figure 29B.1 Simplified version of present-day understanding of protein synthesis. (a) Polypeptide bonds are formed from amino acids. The proteins form the organism and take part in the mechanism of replication, transcription, and translation. (b) The synthesis is guided by a messenger nucleic acid strand, to which adapter molecules carrying the amino acids a_1, a_2, and a_3 are attached.

mRNA. Attachment of the anticodon triplet of the tRNA molecule to a base triplet on a messenger nucleic acid strand can only occur if corresponding base pairs are *complementary*. The four bases G (guanine), C (cytosine), A (adenine), and U (uracil) are pairwise complementary; bases G and C can readily form three hydrogen bonds with each other, while bases A and U can readily form two hydrogen bonds (see Chapter 13, Section 13.5). Put more simply, if there is a C on mRNA strand, there must be a G in the corresponding place on the tRNA molecule, and so on.

For example, the first codon triplet UCU on the messenger strand (read in the 5′-3′-direction) must correspond to an adapter molecule with the anticodon triplet AGA (read in the 3′5′-direction), which carries, for example, the amino acids a_1 (Fig. 29B.1). Amino acids are therefore linked into protein chains with an amino acid sequence that is determined by the sequence in the messenger. These proteins then form the functional structure of the organism. Proteins constitute essential components in all parts of the genetic apparatus.

29.8 Foundations

29.8.1 Foundation 29.1: Search for Logical Conditions Driving the Emergence of a Genetic Translation Device

29.8.1.1 A Thought Experiment

The chart shown here illustrates a sequence of possible evolutionary steps to proceed from a pool of monomers (nucleobases) to a replicating strand, which can be organized by an assembler molecule for reading of the information. At first we consider simple monomers of one kind with the property of forming strands that can replicate under appropriate conditions by lateral bonding of monomers to a template strand and by linking adjacent monomers, but have no evolutionary power. Later we ascribe an additional property to the monomers and to the environment, and then investigate how far the evolutionary process can proceed. After that we add another property, and so on.

	Properties of monomers required for evolution Lock and key concept	**Features of environment required for evolution** Programmed environmental chance concept	**Result of evolution process**
(a)	• One kind of R-monomer Strand formation: strong head to tail bonding Copying: weak lateral bonding 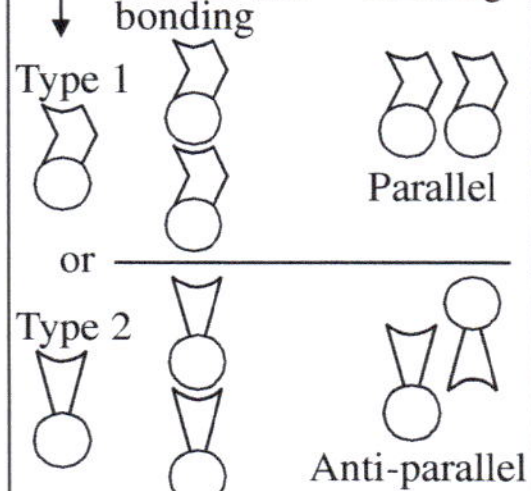 Type 1 … Parallel or Type 2 … Anti-parallel	• Space: Milieu appropriate for copying	• Few monomers bind head to tail: a short strand forms and copies by lateral bonding • Strands multiply by copying • Length of strands limited by occasional mismatch (preventing links between monomers)

	Properties of monomers required for evolution Lock and key concept	**Features of environment required for evolution** Programmed environmental chance concept	**Result of evolution process**
(b)	Additionally to (a): • Two kinds of R-monomer (R_2 complementary to R_1) 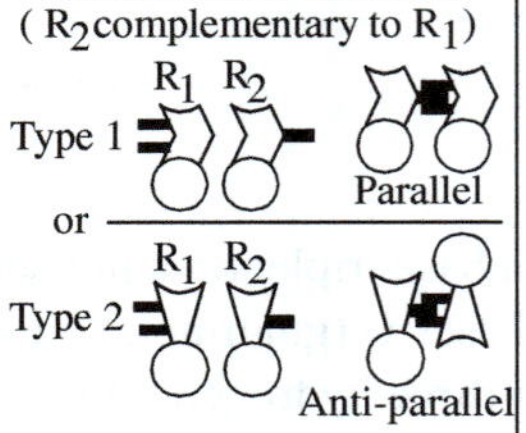 	Additionally to (a): • Time: periodic (e.g. day/night) Driving force for Darwinian process 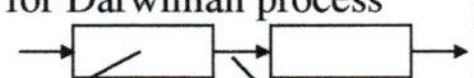 • Space: compartmentalized (e.g. porous rock) microheterogeneity 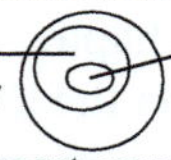 	• Replication • Folding structures 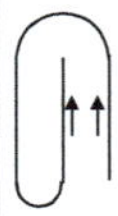 Folding structures with monomers of type 1 (parallel case) hairpin not possible  Hairpin (most compact folding structure) possible only with monomers of type 2 (anti-parallel case)
(c)	Additionally to (b): • Hairpins formed of monomers of type 2 • Monomers of type 2 allow lateral binding of hairpins	Additionally to (b): • Time:  • Space:  	• Aggregate of hairpins 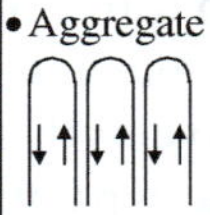 Aggregate of hairpins preserves know how to make hairpins: erroneous copies rejected (error filter)
(d)	Additionally to (c): • Monomers allow binding of hairpins to assembler strand	Additionally to (c): • Region of even larger-pores compartment populated by hairpins assembler devices (HA-devices)	• HA-device 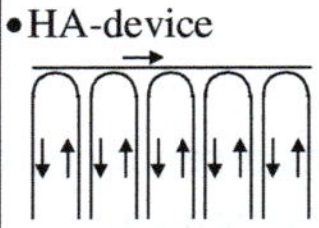 HA-device (most compact aggregate) hairpins and assembler with reading frame
(e)	Additionally to (d): • Monomers of a new variety (a-monomers a_1 and a_2) bind to open end of hairpins and subsequently interlink forming a-oligomers of unspecific sequence • a-oligomers act as impediments (agglutinate)	Additionally to (d): • Less restricted regions (populated by HAE-devices forming agglutinate coating)	• HAE-device 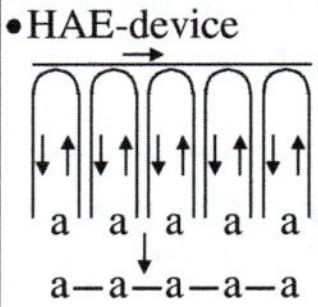 HAE-device catalyzes formation of a-oligomers (agglutinate E_0) • Potential of becoming a translation device as a by-product
(f)	Additionally to (e): • Enzyme E_R (a special a-oligomer with distinct sequence) intervenes in replication process and lowers the probability of mismatch in replication	Additionally to (e): • Less restricted regions (populated by HAE-devices forming enzymes)	• Translation device and first enzyme E_R preserved • Breakthrough of a translation apparatus for further enzymes E_1, E_2, ...

We consider each of the evolutionary steps in this supramolecular engineering problem to provide a scenario for how a simple information transfer apparatus might emerge.

Step (a): A Short Template Marks the Beginning We consider a solution with two kinds of monomers capable of assembling covalent links into short polymer strands (Fig. 29F1.1a and b).

The two kinds of monomers are complementary. The first short polymer strand that is, by chance, structured in a most particular manner, serves as a template for the synthesis of further strands. Once the template strand has formed, monomers are attached because of their lock-and-key properties (e.g., hydrogen bonding, Fig. 29F1.1c), and form a double strand (Fig. 29F1.1d). We suppose that an unusual event occurs to initiate formation of the first template strand; for instance, some environmental change occurs during the drying and redissolving of a solution so that the first short template strand forms.

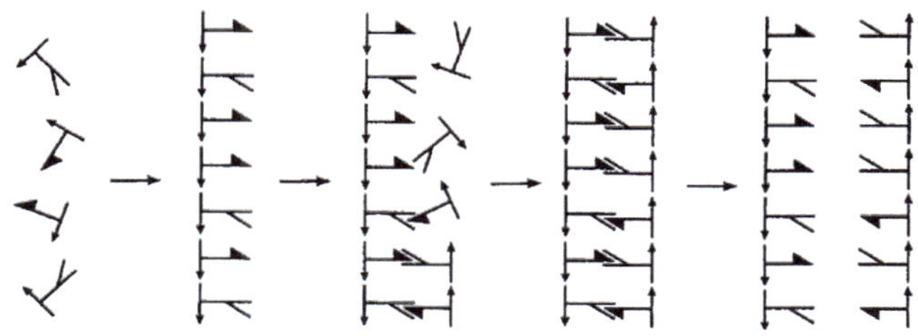

Figure 29F1.1 Monomers capable of assembling covalent links into short polymer strands. (a) Monomers. (b) Monomers assembled to a strand by chance. (c) and (d) Strand as matrix for the assemblage of a double strand at low temperature. (e) Splitting of the double strand into single strands at high temperature (replication).

Intricate Cycles of Temperature Drive Replication We postulate that distinct cyclical temperature changes occur in some location and drive strand replication. These may be given by day-and-night cycle, particular shadow-casting cycles and pores in rocky material (see Fig. 29.7 in Section 29.4.4). A double strand which is formed at low temperature can be split into two single strands at high temperature (the weak hydrogen bonds break, but the strong covalent bonds persist); replication occurs when lowering the temperature (Fig. 29F1.1e). The arrangement of the monomers in the copy is complementary to that in the template strand (with regard to their lock-and-key properties).

Strand Evolution Requires Rare Replication Errors The arrangement of the monomers in the first template strand is determined by chance, so all of the copies have the same or the complementary accidental structure. During the replication process, however, errors can occur. Most of these errors are detrimental; by chance and very rarely, however, such an error imparts an advantage upon the replicate strand, improving its adaptation to environmental conditions (e.g., to the porous structure in rocky material). Eventually these descendant strands replace their less efficient ancestor strands.

Step (b): Strand Lengthening—Colonization of Larger Porous Regions We consider short strands which are well adapted to the fine pore structure in rocky material. By chance and under very special circumstances (e.g., partial drying and redissolution), two short strands can combine to make a longer one. What is important at this point is the presence of a neighboring region, suffused by monomers, with pores that offer almost – but not quite – adequate confinement for the existing short strands. The longer strands are then better adapted to the large pore size region, because they are constrained from being lost by diffusion; the populated area extends by the colonization of pores with larger openings by longer strands. However, replication errors set an upper limit for this process as well. It cannot be avoided that for longer strands the replication error probability is much larger than for the shorter ones, and we need mechanisms to eliminate error copies.

Strand Folding: Hairpins—Most Resistant Structure A long strand is exposed to chemical attack from the environment. How can the reactive groups along the strand be protected? A very effective possibility for the strand is to fold back onto itself to form weak bonds between complementary groups. In Fig. 29F1.2a, as an example, four adjacent groups are complementary. The most favorable conformation is a complete fitting of all groups along the strand (Fig. 29F1.2b) forming a perfect hairpin. The hairpin conformation requires an antiparallel monomer orientation. With monomers oriented parallel, the formation of hairpins is impossible (Fig. 29F1.2c). A most significant aspect is strand chirality discussed in Foundation 29.3. Hairpins have helical twist stabilizing the structure. This stability is important for further evolution. Hairpins with opposite twist do not interlock properly.

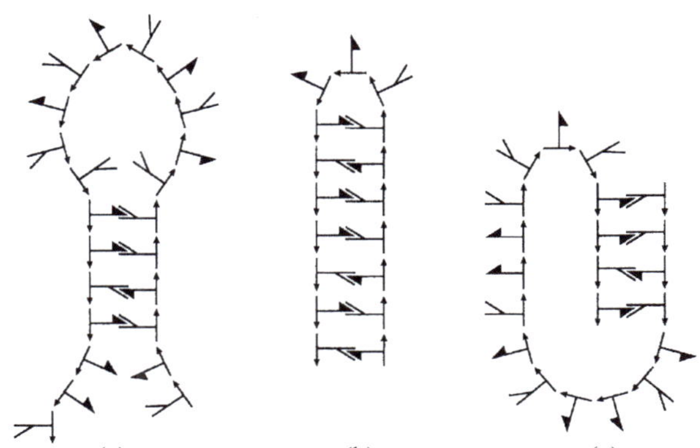

Figure 29F1.2 Protection of a strand by folding. (a) Partial hairpin, (b) perfect hairpin: folding structure with largest pairing domain in the case of anti-parallel weak lateral bonding between monomers, (c) no hairpins possible in the case of parallel weak lateral bonding.

Step (c): Aggregation of Hairpins: An Error Filter A strand folded into a hairpin is already well protected against attacks from outside; however, this protection can be improved by aggregation (Fig. 29F1.3, picket-fence-like structure). The aggregates are much larger than single hairpins; hence the aggregates are able to colonize the larger pores, whereas single hairpins cannot. Another advantage of aggregation is that erroneous copies of hairpins, which do not fit into the aggregate, will be eliminated. On the other hand, errors that do not affect incorporation into the aggregate are not lethal; this means that the further evolution of the strands can occur. Thus, aggregates of hairpins constitute a supramolecular functional unit that preserves the knowledge of how to make hairpins.

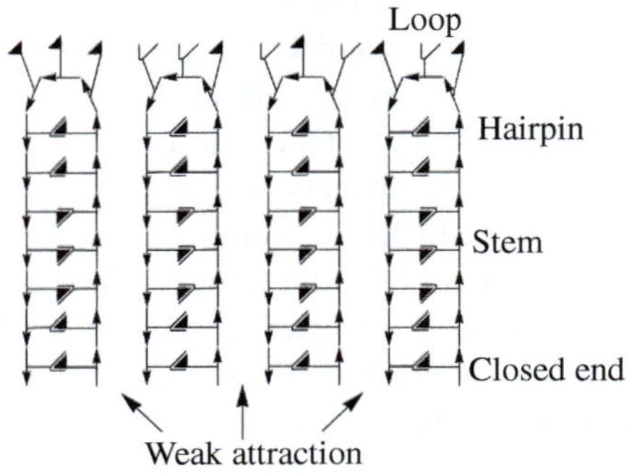

Figure 29F1.3 Aggregation of hairpins (picket-fence like structure).

Step (d): Assembler Strand Guides Hairpin Aggregation The evolution toward picket-fence-like hairpin aggregates is a very important step to stabilize the strands. Nevertheless, the aggregation can still be improved. The weakly bonding groups in the hairpin loop are fixed to complementary groups on an open (i.e., unfolded) strand (Fig. 29F1.4). Such a strand (assembler strand or A-strand) greatly speeds the formation of interlocking picket-fence-like aggregates: The three-dimensional diffusion of the hairpins is converted into a one-dimensional diffusion along the A-strand.

Open strands always exist, because of errors in the replication of hairpins. Usually they are lost by diffusion, but if, by chance, one of them has the appropriate sequence for serving as an A-strand, then it becomes part of the hairpin-assembler (HA) device. Once in existence, such a strand-aided assemblage of hairpins is greatly favored by selection.

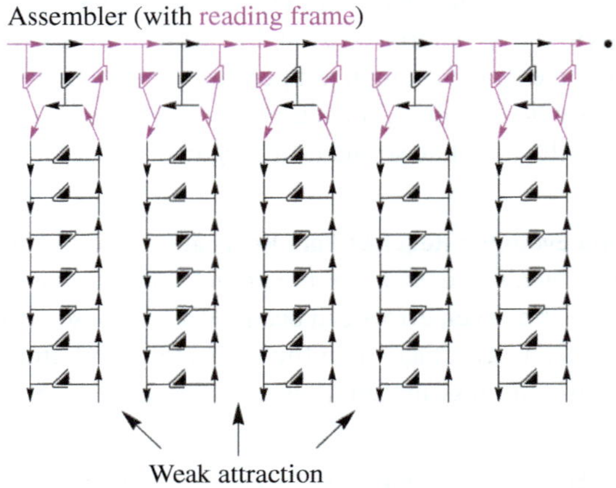

Figure 29F1.4 Hairpins fixed to complementary groups (loop of hairpin) on an unfolded strand (assembler). At the outset these loops are triplets of unspecific sequence leaving the avenue for gaps between the hairpins. The emerging reading frame (loop of specific sequence, the R-monomers are colored in magenta) assures gapless aggregation of hairpins (HA-device).

Step (e): A New Variety (a-Monomers) Appears: a-Oligomers Formed by HA Device Enable Colonization of Larger Porous Regions The replicate of a hairpin can be different in its open end without significantly affecting its stability and formation rate. We can imagine that replication errors lead to an unmatched pair of monomers in the open end of the hairpin (Fig. 29F1.5). Then the open end is capable of binding a new variety of monomers (a-monomers). We assume that these new monomers line up along the hairpin ends and that covalent bond formation occurs between neighbors (formation of a-oligomers). These oligomers are assumed to be capable of agglutination; they assist the restraining function

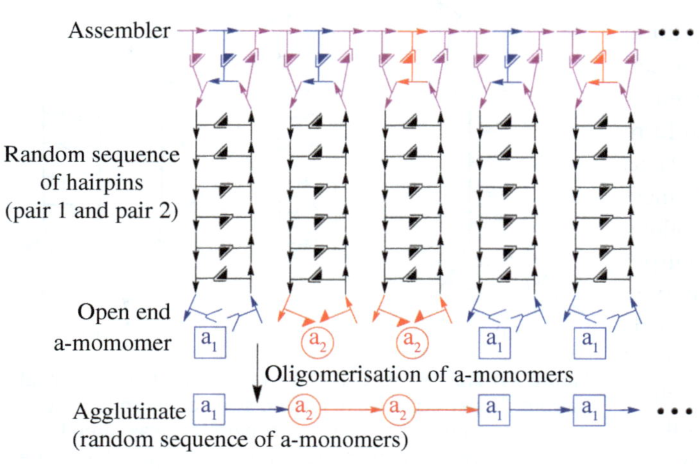

Figure 29F1.5 Formation of a-oligomers (agglutinate) by the HAE-device, where a-monomers are bound to the open ends of hairpins (a_1-monomer in blue and a_2-monomer in red). No mapping between the a-monomers and the R-monomers in the middle of the hairpin's loop by coexistence of hairpins of pair 1 and pair 2.

of pores by narrowing pore openings and making larger pores accessible to colonization, which provides an evolutionary gradient. We assume that variety a_1 binds to the end of the (+)-strand and that variety a_2 binds to the end of the (−)-strand (Fig. 29F1.6).

Assembler-strands carry the pattern, which provides the prescription for the a-oligomers. In this way a translation apparatus has come into existence as a by-product of an error filter. However, this device is of no immediate benefit. Any special sequence of monomers in the Assembler-strand that leads to an oligomer with an especially beneficial sequence of a_1 and a_2 is not preserved, because of copying errors.

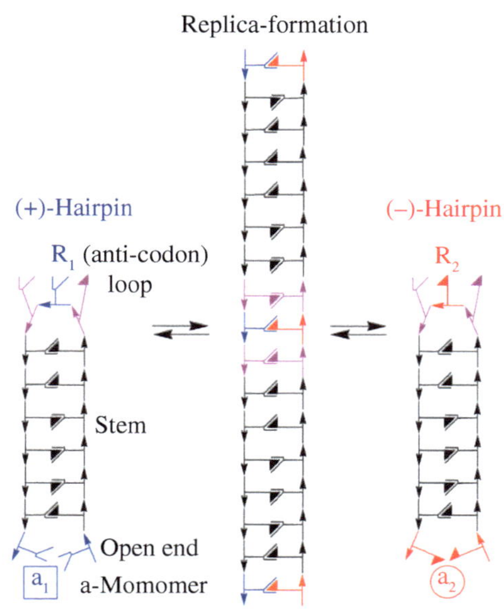

Figure 29F1.6 The replicate of a hairpin forms a hairpin again. The (+)-hairpin and (−)-hairpin are complementary to each other and form a pair of hairpins with unequivocal correspondence between the monomer in the middle of the loop and the open end of the hairpin (colored in blue and red, case of pair 1).

Step (f): Breakthrough of Translational Device A special a-oligomer improves the replication by intercalation. It acts as a replicase; it is called enzyme E_R and the device producing E_R is called an HAE$_R$-device. In this way the prescription for the a-oligomers can be preserved because the replication error probability can be held below a certain critical value (Fig. 29F1.7). The required firm relationship between the monomer sequence of the assembler strand and the sequence of the a-oligomer requires the linkage of three adjacent monomers in the hairpin loop to three complementary monomers along the assembler strand, in addition to a weak lateral attraction between neighboring hairpins. The monomers of an assembler strand are arrayed into adjoining triplets, and to each triplet is attached a complementary triplet in the loop of a hairpin.

A triplet provides the smallest number of complementary R-monomers for a firm and stable aggregate. Triplets offer sufficient coding possibilities also, and unless it is really required for coding needs, an attachment by more than three monomer pairs would represent an evolutionary burden rather than an advantage.

Figure 29F1.7 Formation of an a-oligomer with unique sequence (first enzyme) by the HAE$_R$-device. The (+)-hairpins and (−)-hairpins translate the recipe (sequence of R-monomers of the assembler) into an a-oligomer of corresponding sequence (enzyme E_R reducing the error probability in replication of R-oligomers). The replica of the (+)-assembler (called (−)-assembler) does not produce E_R but an a-oligomer E_0 (agglutinate) that is not enzymatic.

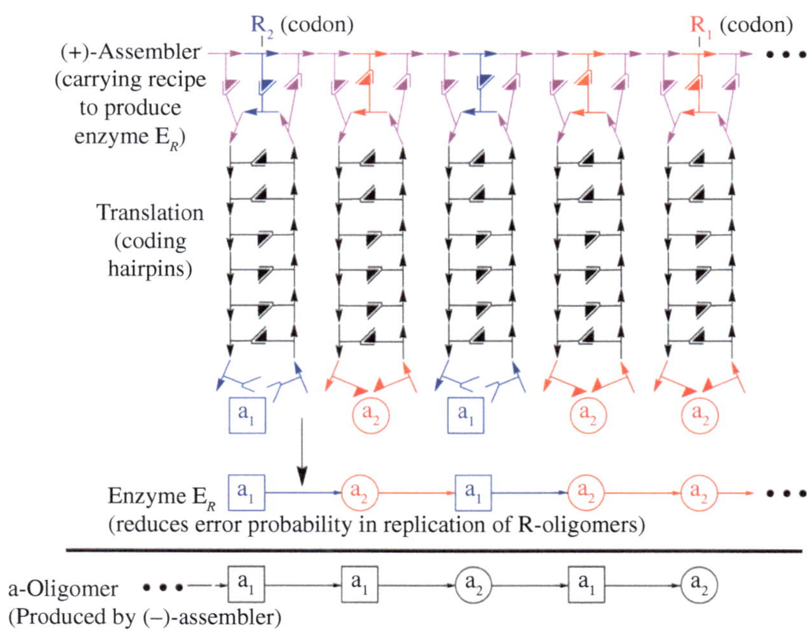

The stability of HA-devices, of course, depends on quite definite requirements to be met by suitable monomers. Other HAE devices can develop by taking advantage of enzyme E_R and produce useful enzymes E_i (Fig. 29F1.8). An entity of cooperating HAE-devices of increasing complexity gradually evolves.

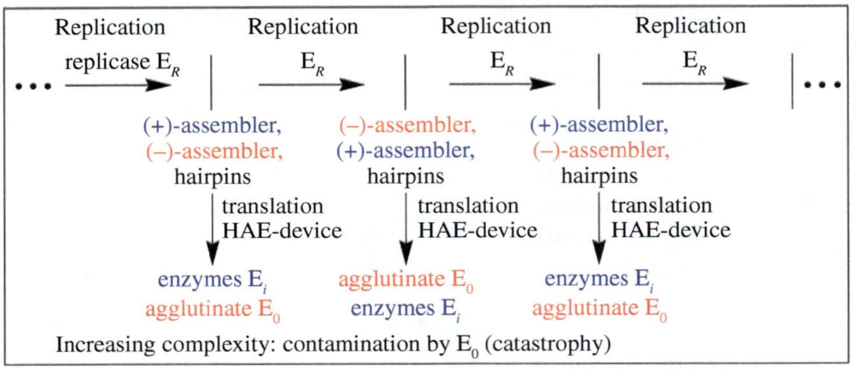

Figure 29F1.8 Cooperative HAE-devices. A mixture of (+)-assemblers, (−)-assemblers, and hairpins is formed by replication. The strands interlock specifically building HAE-devices that produce useful a-oligomers E_R (reducing the error probability in replication of R-oligomers) and E_i (other enzymes improving the functionality of the cooperative). Oligomers E_0 (products from all the (−) assemblers) that may first be beneficial as agglutinate, accumulate in the more complex cooperatives and contaminate the system.

29.8.1.2 Breakthrough of an Integrative Translation Device

The cooperation of an increasing number of HAE-devices, each providing a beneficial oligomer E_R or E_i, is only possible by enclosure in an envelope and by overcoming a dead end that is reached at a certain level of complexity, with the pollution by (−) assembler strands producing nonsensical oligomers E_0. By continuing the search for further small modeling steps in the way shown above for the earlier steps, it is found that the barrier is overcome by a basic transformation in the genetic machinery. This breakthrough is achieved by the appearance of a third monomer variety [17] D that is a modified version of the original monomers of the first variety. Fig. 29F1.9 shows how to avoid pollution by accumulating E_0. Cooperative entities of increasing complexity evolve. But soon another barrier is reached, the loss of information carriers.

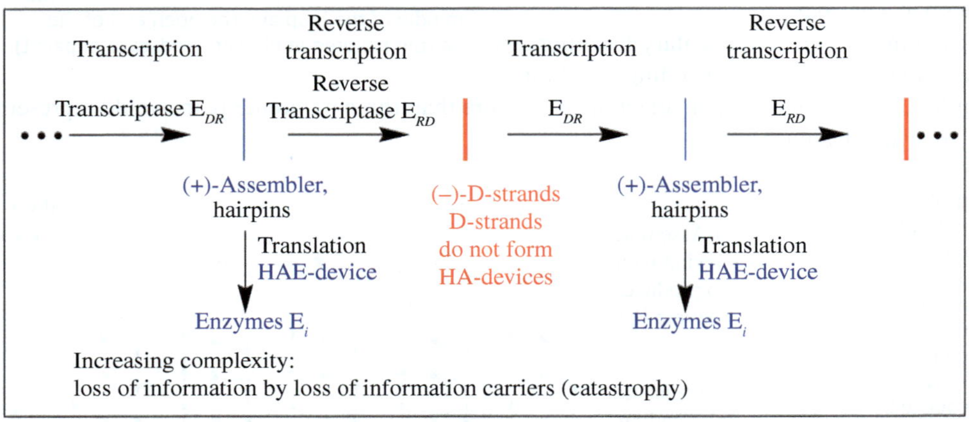

Figure 29F1.9 Breakthrough avoiding pollution by a-oligomers E_0. HAE_R-device evolves into devices that produce enzymes E_{DR} and E_{RD} (differentiation). E_{RD} catalyzes formation of a D-strand replicate (using D-monomers) at an R-strand template and E_{DR} catalyses formation of an R-strand replicate at a D-strand template. D-strands do not form HA-devices. A cooperative entity evolves where (+)-assemblers form (−)-D-strands and (−)-D-strands form (+)-assemblers. No oligomer E_0 is produced.

D-Superstrand Emerges in Theoretical Modeling and Serves as a Template for Many R-Strands This fundamental change leads to a system in which the genetic message for constructing all HA-devices of the cooperative device is carried by a strand made of D-monomers. This D-strand replicates, and in this way the genetic message of the cooperative device (given by the monomer sequence in the D-strand) is multiplied. The detailed model path leading to the machinery cannot be given here. The machinery has basically the same organizational structure as today's genetic apparatus of the biosystems (Box 29.1). It is remarkable that theoretical reasoning leads to that result.

We just give a theoretical model showing how the genetic message of all HAE-devices of the cooperative device can be stored in a D-superstrand (Figs. 29F1.10 and 29F1.11). We model the emergence of a replicase E_D and the gradual disappearance of E_{RD} (Fig. 29F1.12). In this way the thought experiment leads to a machinery that allows a continuous evolutionary process. For details see [18].

Figure 29F1.10 Formation of a (–)-D-superstrand using alternating (+)-R-strands (information carriers with end caps y and z) and (+)-R-bridges (with end caps z and y). The (+)-R-bridge adheres to the constitutive (–)-D-superstrand with end cap y′ (sequence complementary to y) and the (+)-R-strand adheres to the constitutive (–)-D-superstrand end cap z′ (sequence complementary to z).

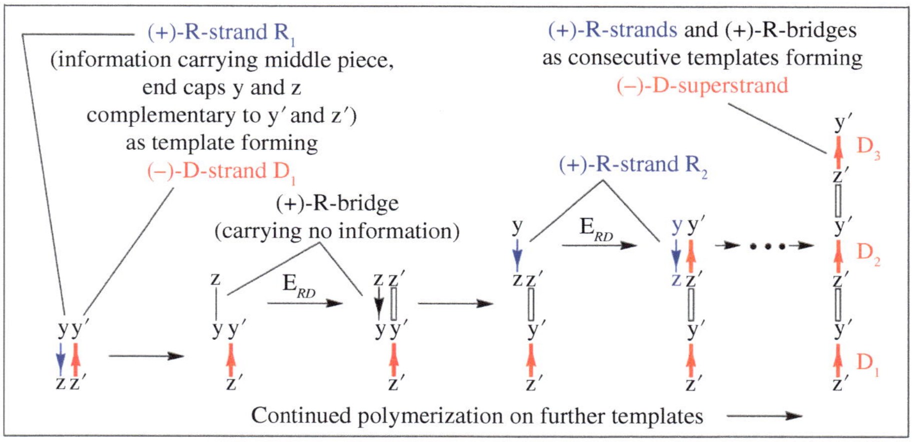

Figure 29F1.11 D-superstrands. Machinery like that of Fig. 29F1.9, but information carriers (–)-D-strands are linked to D-superstrand. Complexity of cooperative is no longer limited by loss of information carriers.

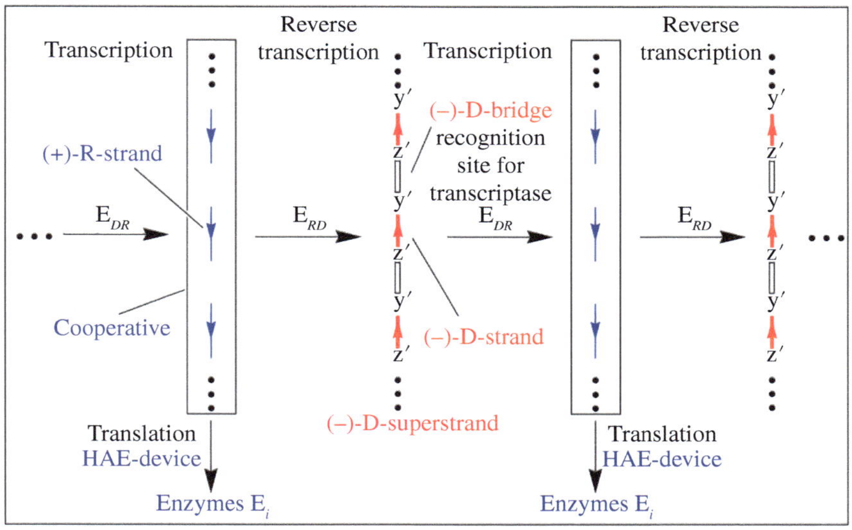

29.8.1.3 Computer Simulation of Thought Experiment

The simulation [18] is carried out as follows:

Storing fitness parameters. Many devices are taken into account as possibilities that may or may not be realized at some evolutionary level. Each individual device (defined by the particular sequences of the strands involved) is represented at the end of the construction phase by values of parameters that describe its survival chance during the selection phase and its multiplication fitness (proliferation rate and replication precision) during the consecutive multiplication phase. The fitness parameters (obtained during the construction phase) are stored in a dictionary relating fitness to devices (during the selection phase) and strands (during the multiplication phase), respectively.

Applying an evolutionary search program. The following processes are simulated on a computer. Isolated strands are exposed to a periodically changing environment. Each cycle consists of three phases: (i) Multiplication phase: The strands are driven to replicate with occasional variation by copying mismatch. (ii) Construction phase: An aggregate of strands is formed by many random choices and random trials to interlock the strands. At the end of such an aggregation process, the final aggregate constitutes a device with its functionality and fitness values assigned to the device accordingly. This scheme is maintained for all remaining strands until the end of the construction phase. (iii) Selection phase: A random selection of devices biased by their fitness parameters takes place. The surviving aggregates dissociate into isolated strands. This completes the cycle and the next cycle begins with the subsequent multiplication phase. Increasingly complex aggregates constituting devices[2] of increasing survival capability and evolve over the course of very many cycles.

2 Note that in this context the term "aggregate" focuses on its construction and the term "device" focuses on its function.

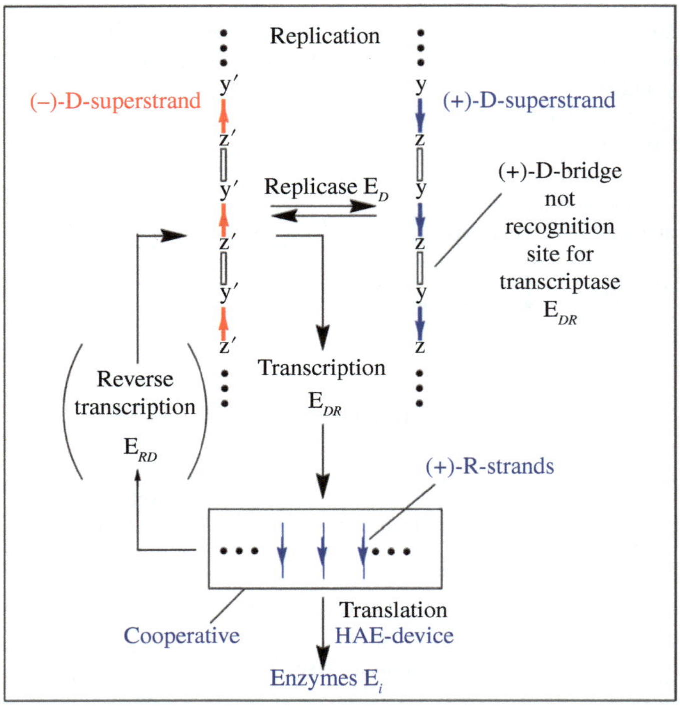

Figure 29F1.12 Appearance of replicase E_D. A fundamental change of the machinery. Reverse transcriptase is not required and no longer produced. D-superstrand carries information to reproduce cooperative. Transcription from (−)-D-superstrand to (+)-R-strands and translation by (+)-R-strands. Unlimited increase in complexity.

Repeated random sampling (Monte Carlo method). The result of a computer simulation singled out from the pool of repeated random sampling is shown in Fig. 29F1.13. Starting with a population of open strands (with sequences of monomers of the first kind R1 and R2) hairpins emerge (see Fig. 29F.13a), after some number of generations. The hairpins, together with open strands, build up incomplete HA-devices and some isolated complete HA-devices build up by chance. Complete HA-devices take over the population (see Fig. 29F.13a) only when a reading frame emerges (R_1R_2, see Fig. 29F1.13b). Introducing monomers of the second kind a_1 and a_2 from generation number 1301 on. After some number of generations, agglutinate producing HAE-devices build up (see Fig. 29F1.13c), Agglutinate is a polymer with a random sequence of monomers of the second kind. Isolated enzymes (polymers of specific sequence of monomers of the second kind) are formed by chance. Enzyme producing HAE-devices take over the population (see Fig. 29F1.13c) only when there is the recipe on the assembler strands and when a translation code emerges (code R_1a_1 and R_2a_2, see Fig. 29F1.13d).

Tracing the family tree back to the universal common ancestor. Figure 29F1.14 presents an analysis of the computer simulation shown in Fig. 29F1.13. Within generation number 6500 four strands numbered 7 and 20 (assembler strands) and numbered 78 and 188 (hairpins) are singled out from the population of strands (assembler strand 7 and hairpin 78 belong to a device producing enzyme E, assembler strand 20 and hairpin 188 belong to a device producing enzyme M, see Fig. 29F1.14). Their family trees are traced back asking generation by generation: to which device each of their ancestors belong. Note examples in which the ancestor is not the fittest of its population (bottleneck).

It is fascinating to observe the computer program automatically developing, in the course of several thousand cycles, a genetic apparatus and a genetic 2-letter code. This process, however, occurs only in a small fraction of runs. With other random seeds the evolution becomes stuck. This supports the idea that many starts of an evolutionary process were needed until the break-through leading to the explosion of life occurred.

29.8.2 Foundation 29.2: Attempts to Model the Origin of Life

It is often assumed that life's origin, in essence, is a self-organizing process in a homogeneous phase and in a steady state far from thermodynamic equilibrium. By contrast, the theoretical approach discussed here emphasizes the importance of the particularity of the environment driving a complex chemical process that is, per se, in agreement with thermodynamics.

Can all the requirements of the model discussed in Foundation 29.1 be fulfilled by the molecules comprising the genetic apparatus in biosystems? Can such considerations contribute to the understanding of the origin of life on that basis?

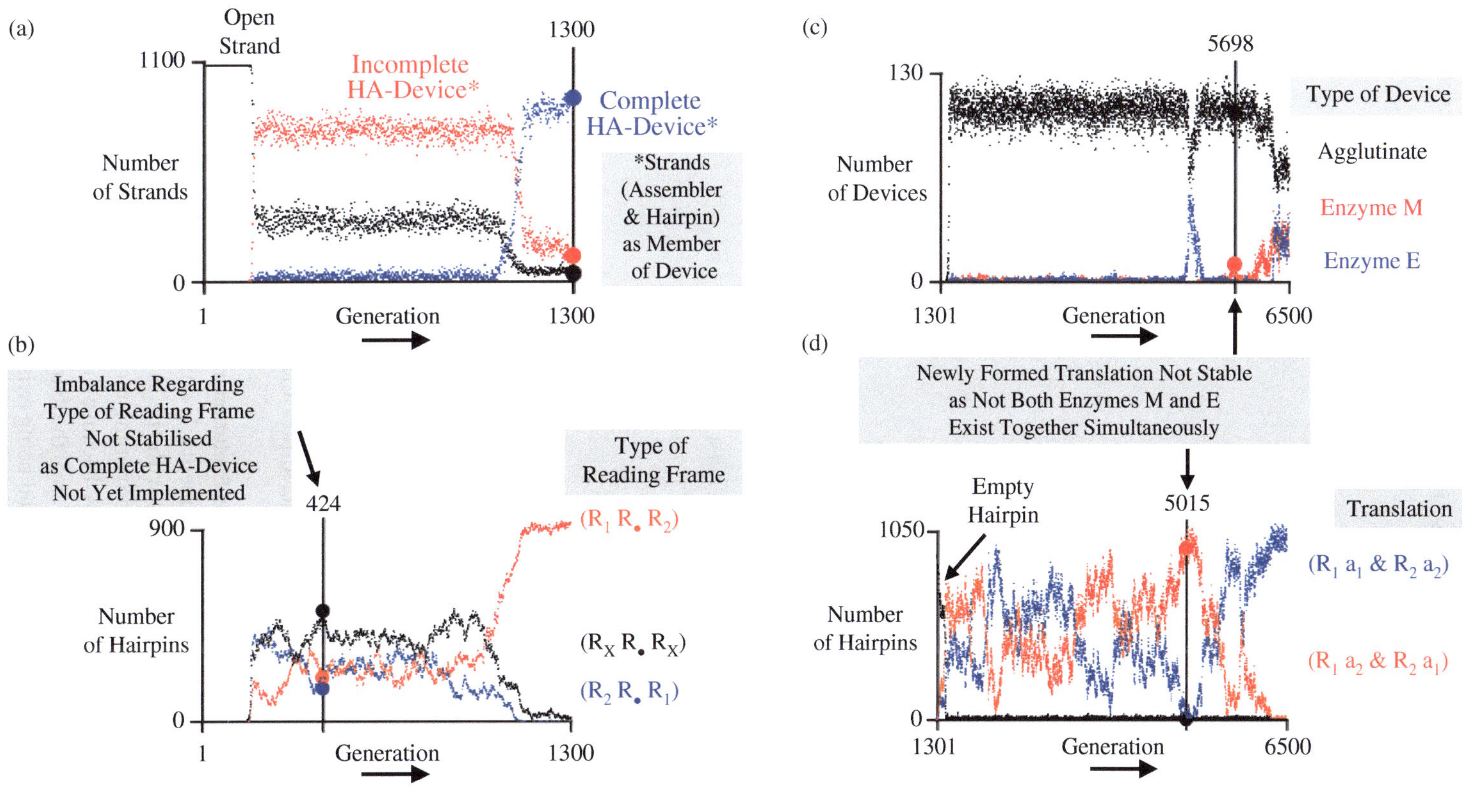

Figure 29F1.13 Starting with a population of open strands (with sequences of monomers of the first kind R_1 and R_2) the computer simulation shows (panel a) the emergence of complete HA-devices that take over the population only (panel b) when a reading frame emerges. Introducing monomers of the second kind, a_1 and a_2, the computer simulation shows (panel c) the emergence of enzyme producing HAE-devices that take over the population only when there is a recipe on the assembler strands and (d) when a translation code emerges. Further description is given in the text.

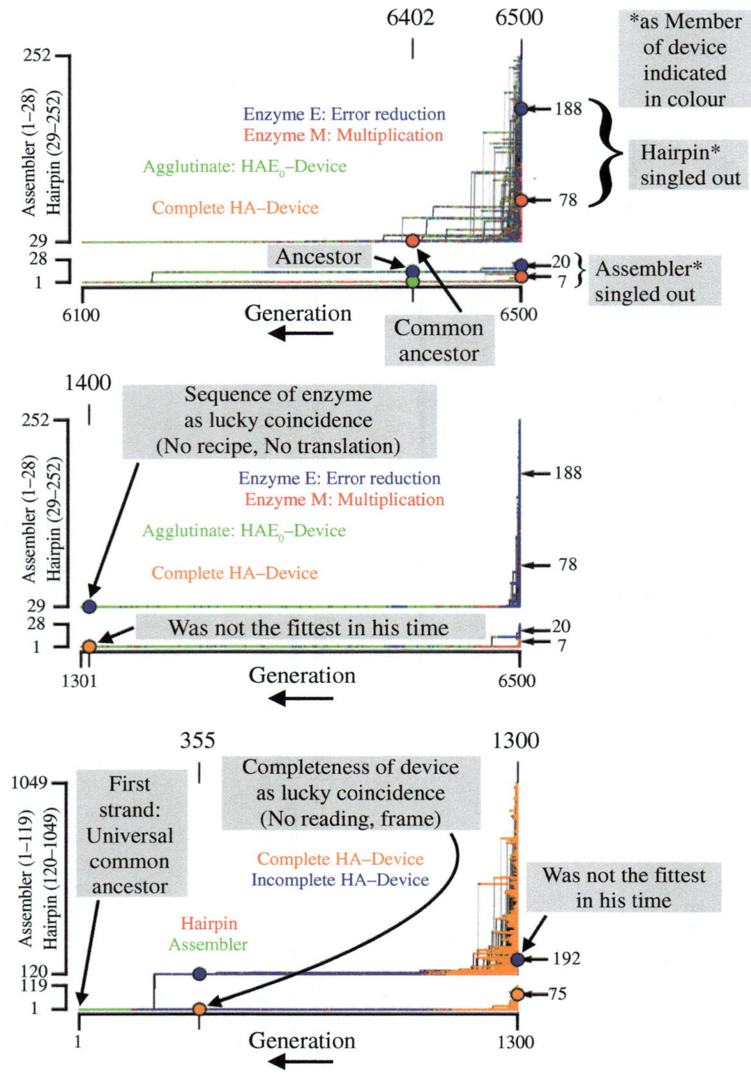

Figure 29F1.14 The family tree of four singled out strands is traced back asking generation by generation: to which device each of their ancestors belong, respectively. Note examples in which the ancestor is not the fittest of its population (bottleneck). Further description in the text.

29.8.2.1 The Earliest Phase

Among candidates for the monomers of the first variety are the nucleotides guanine (G), cytosine (C), adenine (A), and uracil (U). We favor G and C for the beginning, because they basepair with three rather than two hydrogen bonds. Polymer strands have the essential features required for HA-devices (forming hairpins, lateral attraction of hairpins by the intercalation of Ca^{2+} and Mg^{2+} ions). For the monomers a_1 and a_2 we choose the most prevalent amino acids alanine and glycine. The earliest "code" in our model involves just these two nucleotides and two amino acids.

29.8.2.2 Breaking Symmetry: Chirality of Nucleotides and Its Evolutionary Benefit

Nucleotides are chiral because ribose, one of their components, is chiral, and nucleotides with D-ribose have opposite chirality from those containing the L-isomer. Nucleotides with the same chirality, either D or L, are more appropriate for precise template-assisted replication than are strands containing monomers of mixed chirality. Monomers of incorrect chirality are rejected in this buildup of new daughter strands, because they do not precisely interlock at the growth position created by the template strand and the already existing portion of the daughter strand (*lock-and-key principle*). The precision in binding the appropriate monomer is reached by the strand being twisted.

If D-ribose and L-ribose were present in equal amounts on prebiotic earth, and this would therefore be true for nucleotides as well. In this view the fact that present-day RNA contains the D-isomer rather than the L-isomer of ribose is commonly considered the result of a "frozen accident" [17]; recently, however workers have conjectured that it may arise from the enantiospecific interaction of chiral molecules with magnetized surfaces [19]. Thus the chirality of the amino acids entering the genetic machinery (L-amino acids) is a consequence of the chirality of nucleotides (lock-and-key principle). The process could have started at many locations on the early earth resulting, occasionally, in systems with L-ribose. They occasionally interacted with D-systems; thereby one chiral form became stronger at the expense of the other one and finally the D-forms survived the struggle.

The view of symmetry breaking by a frozen accident is supported by the finding [20] that pyranosyl-RNA (which is a stronger and more selective pairing system than the natural furanosyl-RNA) is obtained in pure L or pure D form by oligomerization of G-C-C-GcP tetramers (where cP means cyclophosphate). These tetramers consist of D and L nucleotides linked at random (mixed D and L tetramers besides pure D and pure L tetramers). Alone pure D (or pure L) tetramers oligomerize by self-assembly:

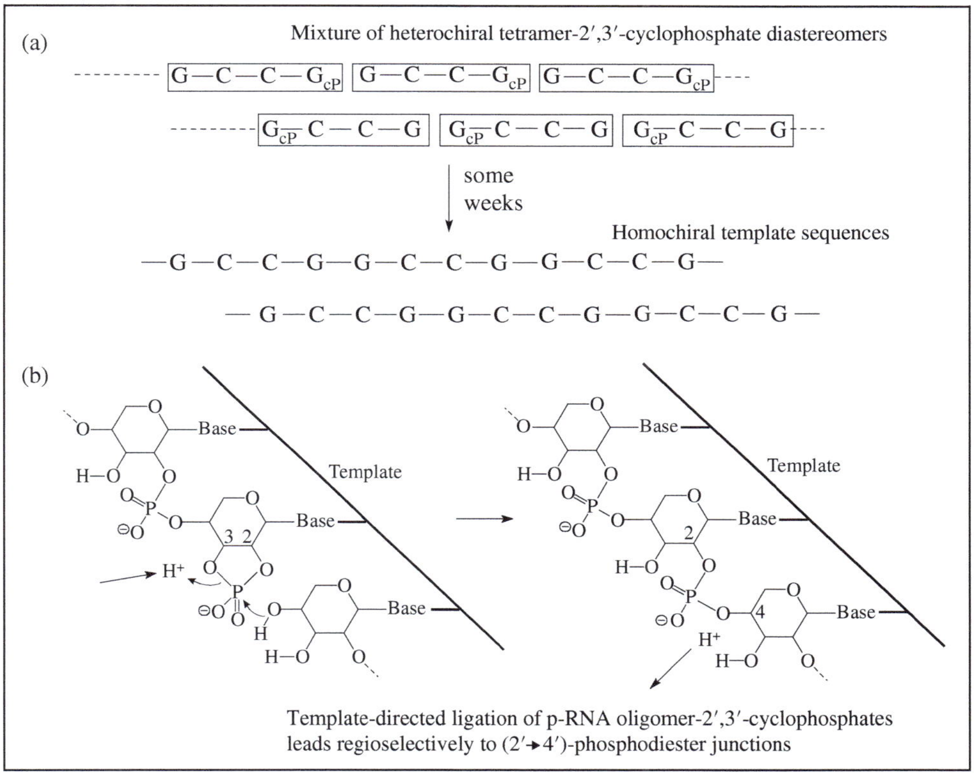

Each tetramer carries cP that opens up by linking the phosphate group to the 4′ end of the adjacent tetramer. This reaction occurs under mild conditions and only in the presence of the template.

Replication and exponential amplification of DNA analogues has been achieved, [1] emphasizing the idea that cycles of particular environmental changes are the basic driving force in the origin of life (see Box 29.2). Templates have been exposed to a periodically changing environment, allowing, in succession, complementary binding of oligomers at template strands, chemical ligation, and liberation of new strands from templates. Original templates and new strands, after liberation, are immobilized to avoid formation of stable duplexes. Cyclic repetition allows an exponential increase in the amount of strands.

29.8.2.3 The Earliest HA-Device

The fact that complementary hairpins differ only in the middle of their loops is of crucial importance for the earliest HA-device considered as a thought experiment (Foundation 29.1). The earliest HA-device contained just the near-identical (+) and (−) versions of the "same" hairpin differing at the very end of their legs. They must be slightly different at their leg ends because (+) and (−) hairpins have to carry different amino acids; the complementarity requirement for monomers at the leg ends must therefore be relaxed. The stability of HA-devices, built from such near-identical hairpin neighbors, is, of course, favored by their close fit.

For reasons detailed below, we make the assumption that hairpins with C in the loop were invested with glycine and that "G-hairpins" were invested with alanine. Therefore, looking at the middle nucleotide of a triplet along an A-strand,

G would "code" for glycine and C would code for alanine. There is, however, another requirement to be met, because there is a need for recognizing the location of different code letters on the A-strand, a need for a "reading frame."

The Reading Frame Requirement In the early stages of evolution, there is the demand for a way of telling where a triplet along the A-strand begins and ends; that is, there has to be a reading frame. Specifically, our model requires that in all triplets along the A-strand, the first nucleotide must be G and the third must be C, assigning any coding function to the middle nucleotide. The reading frame has therefore the form 5′-GNC-3′, where N first stands for G or C and later for A or U as well. For a corresponding hairpin loop, the complementary sequence is 3′-CN′ G-5′ or 5′-GN′ C-3′, N′ being the nucleotide complementary to N. However, as the loops of complementary (+) and (−) hairpins differ in the middle nucleotide only, they are both of the same type, namely, 3′-CN′ G-5′. This implies that the same GNC reading frame applies to both (+) and (−) hairpins.

Table 29F2.1 Hypothetical Stages in the Development of a Genetic Code.

Amino acids	Stage 1	Stage 2	Stage 3	Stage 4		Stage 5
Decreasing Availability ↓	**GNC**	**PuNN**	**Third Position Opens for Coding**	**First Position Opens for Py**		**Modern Genetic Code**
Gly	GGC	GGN				GGN
Ala	GCC	GCN				GCN
Asp	GAC	GAN	GAPy			GAPy
Glu	GAC	GAN	GAPu			GAPu
Val	GUC	GUN				GUN
Stop	UPuPu	UPuN				UGA, UAPu
Leu		AUN				
Ile		AUN				AUPy, AUA
Ser		AGN	AGPy			AGPy
Thr		ACN				ACN
Lys		AAN	AAPu			AAPu
Arg			From Ser AGPu			AGPu
Gln			From Lys AAPy			
Asn			From Lys AAPy			AAPy
Pro				CCN		CCN
				Arg CGN		CGN
				Gln CAN		CAPu
				Leu CUN		CUN
				Leu UUN		UUPu
				Ser UCN		UCN
Phe					From Leu	UUPy
Tyr					From Stop	UAPy
His					From Gln	CAPy
Cys					From Stop	UGPy
Met					From Ile	AUG
Trp					From Stop	UGG
Sec					From Stop	UGA

G: guanine, A: adenine, C: cytosine, U: uracil.
Pu stands for purine G or A. Py stands for pyrimidine C or U.

29.8.2.4 Stage-by-Stage Evolution of the Code

The translation device becomes increasingly sophisticated by incorporation of increasingly complex translation products stabilizing the attachment of hairpins to the assembler strand. Strength in base pairing should be no longer beneficial. An original pyranosyl RNA based system should be replaced by the more flexible and fast furanosyl RNA system. The reading frame gradually decreases in importance. Triplet position 1 is also pressed into coding service, and finally all three positions are used for this.

Each step of relaxation of constraints on the code by improvement of the translation machinery confers a selective advantage, increasing the number of amino acids that can be coded. The modern code can be constructed if we assume that (1) the order in which amino acids are introduced into the code proceeds in the order of their availability, (2) that the code, first restricted to GNC, changes stagewise (see Table 29F2.1) ; and (3) A in position 2 is associated with the most polar amino acid to be introduced, G, C, U are associated with amino acids in the order of decreasing polarity. This is a generalization of the initial codes GAC for Glu, Asp; GGC for Gly; GCC for Ala; GUC for Val. Table 29F.1 lists the amino acid in decreasing order of availability in column 1. The modern code can be obtained in five stages:

- *Stage 1.* Code GNC (G=glycine, N = G,C,A,U, C=alanine) for the five most abundant amino acids Gly, Ala, Val, and Glu; Asp (not distinguished at this stage). Oligomerization is stopped by the best suppressing-binding triplet UPuPu considered as a stop codon.
- *Stage 2.* In stage 2, the code, first restricted to GNC (N = G,C,A,U), changes over to GNN (first position sufficient for docking); code PuNN. Purine G or A in first position of triplet allows coding for the 10 most abundant amino acids (Asp; Glu and Leu; Ile, not distinguished at this stage). The third position is no longer used as a reading frame and the code changes to PuNN (Pu = G,A; a purine is sufficient for docking).
- *Stage 3.* In this stage, the third position is used for coding and the first position is opened for pyrimidines Py = C, U. The third position opens for coding: distinction between Asp and Glu; AGPu and AAPy open for Arg and Gln; Asn (not distinguished at this stage).
- *Stage 4.* First position opens for Py: Code for Pro and additional codes for particularly important amino acids. UPuN is reserved for Stop at this stage.
- *Stage 5.* Emergence of modern code: UUPy, UAPy, CAPy, UGPy open for special amino acids; finally AUG, UGG, UGA open for latest amino acids.

The modern code results from these postulates [21] when taking into consideration that important amino acids are provided with an additional codon at the stage when position 1 is opened for C and U (Table 29F2.1). Later, part of the codons for Leu, Gln, and Stop are used, instead, for special amino acids Phe, Tyr, His, and Cys (see Table 29F2.2). Finally, one of the

Table 29F2.2 The 21 Amino Acids. Single Letter Codes[3], Abbreviations and Full Names.

The 21 Amino Acids					
Code	**Abbreviation**	**Full Name**	**Code**	**Abbreviation**	**Full Name**
A	Ala	Alanine	N	Asn	Asparagine
C	Cys	Cysteine	P	Pro	Proline
D	Asp	Aspartic acid	Q	Gln	Glutamine
E	Glu	Glutamic acid	R	Arg	Arginine
F	Phe	Phenylalanine	S	Ser	Serine
G	Gly	Glycine	T	Thr	Threonine
H	His	Histidine	V	Val	Valine
I	Ile	Isoleucine	W	Trp	Tryptophane
K	Lys	Lysine	Y	Tyr	Tyrosine
L	Leu	Leucine		Sec	Selenocysteine
M	Met	Methionine			

3 Do not confuse the letter codes A, C, and G for the amino acids alanine, cysteine, and glycine with the corresponding letter codes for the nucleic acids adenine, cytosine, and guanine.

codons for Ile and two of the codons for Stop are used for the latest amino acids Met, Trp, and Sec (Sec is amino acid number 21, selenocysteine). The insight thus afforded strongly suggests that the ancestral self-reproducing system resembled the HA-device.

It is, however, important to recognize that a basic change in the structure of the genetic machinery is required before the evolution of the genetic code can proceed, the breakthrough of an integrative translation device. In the special case of bioevolution the third monomer variety, D, is deoxyribonucleic acid. This leads to a device with the basic features of the genetic apparatus of biosystems (Box 29.1). The considerations in Box 29.1 and Foundation 29.1 indicate that the emergence of such a device should not be restricted to the special case of bioevolution based on nucleotides and amino acids.

29.9 Exercises

E29.1 In Foundation 29.4 we consider the entropy change

$$\Delta S = k \cdot \left(\mathcal{P} \cdot \ln \frac{1}{\mathcal{P}} + \mathcal{Q} \cdot \ln \frac{1}{\mathcal{Q}} \right)$$

where $Q = 1 - \mathcal{P}$ and $\mathcal{P}$ is a probability with a value between 0 and 1. Show that ΔS has a maximum for $\mathcal{P} = 1/2$.

29.10 Problems

P29.1 Aggregation of Folded Strands. A requirement for selection is that after the growth phase the folded strands must reach each other and precisely interlock in a time span short compared to the decay phase. If we estimate a time $t = 10$ s required for aggregating two molecules, use Fig. P29.1 and the Einstein–Smoluchowski equation to estimate the volume V of a compartment that would include both molecules to allow the process in the given time. Suppose that the aggregation of the folded strands A and B requires that the points a and b approach each other more closely than a critical distance d of about 0.1 nm. Assume that strand A remains stationary and that b is at the center of strand B; then point b participates only in the translational diffusion of B. The strands must also have the correct mutual orientation; perform your analysis for the case of $\alpha = 1$ and $\alpha = 0.1$, where α is the steric factor, i.e., the probability (when the points a and b have reached the critical distance d) that the strands A and B are oriented correctly to allow bonding.

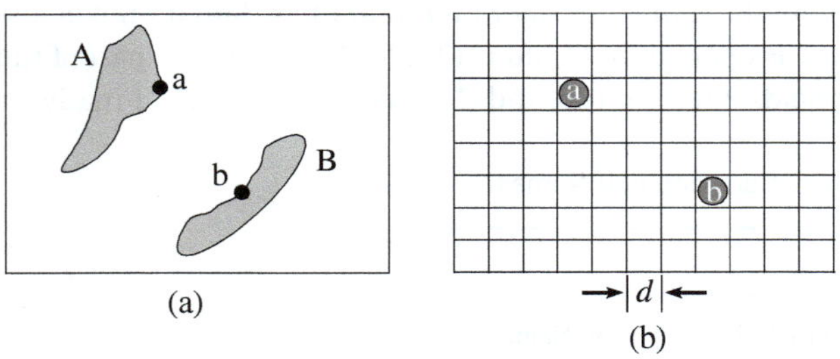

Figure P29.1 Diffusion of two folded strands to form an aggregate. Panel (a) shows the aggregates, and panel (b) shows the strands on a grid which identifies the critical distance d for the strands' aggregation.

P29.2 Replication Error Rate of a Bacterium and a Human. Estimate the replication error probability per nucleotide base in a bacterium consisting of 10^3 proteins (three nucleotides code for a single amino acid). What is different in a human with about 40×10^3 proteins? Assume the protein has about 300 amino acids and estimate the replication probability per base to be 10^{-6}.

P29.3 Time Needed to Evolve a Bacterium. Evolution up to the period in which genetic machinery appeared must have occurred rapidly, because of the large error frequency that dominated these early stages. This time is small compared to the time required for the instruction of the about 1000 proteins of a bacterium. Estimate the time needed to evolve a bacterium.

Assume that one protein after another is inserted into the form existing at the time. This is achieved by lengthening of the DNA strand by $N = 10^3$ nucleotides for each new protein (about 300 amino acids). To simplify, assume that each protein is instructed in some hundred steps of optimization. Between each of these steps we must wait until a new random sequence in the non-instructed section of the DNA strand is reached, i.e., $1000 \cdot \frac{1}{\mathcal{P}}$ generations. Then each base in that section had a chance to mutate. This implies $1/\mathcal{P} = 10^6$ generations per step, if $\mathcal{P} = 10^{-6}$ is the replication error probability per base estimated for an early bacterium (see Problem 29.2).

P29.4 Maximum Genetic Information Carried by DNA. Assume that the replication error probability per base in a human is about $\mathcal{P} = 10^{-10}$ (see Problem 29.2). Also assume that the loss of information, caused by thermal processes $\mathcal{P}'$, during time t between two subsequent cell divisions is not larger than the loss by replication $\mathcal{P}$. Assume that the $\mathcal{P}'$ error probability provides an upper bound on the information storage and estimate the maximum genetic information that can be stored in a human by DNA.

References

1 A. Luther, R. Brandsch, and G. Von Kiedrowski, Surface-promoted replication and exponential amplification of DNA analogues, *Nature*, 1998, **396**, 245; L.E. Orgel, Prebiotic chemistry and the origin of the RNA world, *Crit. Rev. Biochem. Mol. Biol.*, 2004, **39**, 99; M.P. Robertson and G.F. Joyce, The origins of the RNA world, *Cold Spring Harb. Perspect. Biol.*, 2012, **4**, a003608.

2 S. Mann, Systems of creation: the emergence of life from nonliving matter, *Acc. Chem. Res.*, 2012, **45**, 2131–2141; J.P. Schrum, T.F. Zhu, and J.W. Szostak, The origins of cellular life, *Cold Spring Harb. Perspect. Biol.*, 2010, **2**, a002212.

3 S. Weinberg, *The First Three Minutes* (Basic Books, New York, 1977).

4 F. Crick, Central dogma of molecular biology, *Nature*, 1970, **227**, 561.

5 H. Kuhn, Origin of life - Symmetry breaking in the universe: emergence of homochirality, *Curr. Opin. Colloid Interface Sci.*, 2008, **13**, 3.

6 C. Kuhn, Computer modeling origin of simple genetic apparatus, *Proc. Natl. Acad. Sci. U.S.A.*, 2001, **98**, 8620.

7 R. Landauer, Irreversibility and heat generation in the computing process, *IBM J. Res. Dev.*, 1961, **5**, 183.

8 H. Kuhn, Origin of life and physics: diversified microstructure-inducement to form information carrying and knowledge-accumulating systems, *IBM J. Res. Dev.*, 1988, **32**, 37.

9 C.E. Shannon, *The Mathematical Theory of Communication* (University of Illinois Press, Urbana, IL, 1949).

10 C.H. Bennett, Logical reversibility in computing, *IBM J. Res. Dev.*, 1973, **17**, 525; R. Landauer, *Nature*, 1988, **335**, 779.

11 C.H. Bennett, Maxwell Demon, Szilard engine, and the second law, *Sci. Am.*, 1987, **255**, 107.

12 A. Aydin, A. Sisman, and R. Kosloff, Landauer's Principle in a quantum Szilard engine without Maxwell's Demon, *Entropy*, 2020, **22**, 294.

13 T. O'Hagan, Dicing with the unknown, *Significance*, 2004, **1**, 132–133.

14 K.P. Körding and D.M. Wolpert, Bayesian decision theory in sensorimotor control learning, *TRENDS Cognit. Sci.*, 2006, **10**, 319–326

15 G. Westheimer, Was Von Helmholtz a Bayesian? *Perception*, 2008, **39**, 642–650.

16 K. Weigmann, Our sense of self, *EMBO Rep.*, 2013, **14**, 765–768.

17 H. Kuhn and J. Waser, in: *Lock- and Key Principle*, ed. J.P. Behr (Wiley, Chichester, 1994, pp. 247–306).

18 C. Kuhn, A computer-glimpse of the origin of life, *J. Biol. Phys.*, 2005, **31**, 571–585.

19 S.F. Ozturk and D.D. Sasselov, On the origins of life's homochirality: inducing enantiomeric excess with spin-polarized electrons, *Proc. Natl. Acad. Sci. U.S.A.*, 2022, **119**, e2204765119 B.P. Bloom, A.R. Waldeck, and D.H. Waldeck, Homochirality and chiral-induced spin selectivity: a new spin on the origin of life, *Proc. Natl. Acad. Sci. U.S.A.*, 2022, **119**, e2210505119.

20 M. Bolli, R. Micura, and A. Eschenmoser, Pyranosyl-RNA: chiroselective self-assembly of base sequences by ligative oligomerization of tetra nucleotide-2′,3′-cyclophosphates (with a commentary concerning the origin of biomolecular homochirality), *Chem. Biol.*, 1997, **4**, 309.

21 H. Kuhn and J. Waser, On the origin of the genetic code, *FEBS Lett.*, 1994, **352**, 259.

Appendices

CHAPTER MENU

A Data Tables

A.1 Inorganic Compounds

Enthalpy of formation $\Delta_f H^{\ominus}_{298}$, entropy $S^{\ominus}_{m,298}$, Gibbs energy of formation $\Delta_f G^{\ominus}_{298}$, and heat capacity $C^{\ominus}_{P,298}$ at 298.15 K. Solids (s) and liquids (l) under P =1 bar; gases (g) under P = 1 bar in ideal state; substances dissolved in water (aq) under c = 1 mol L^{-1} in ideal state. The data are corrected for an assumed ideal gas under P = 1 bar and an assumed ideally diluted solution under c = 1 mol L^{-1}.

Substance	State	M g mol^{-1}	$\Delta_f H^{\ominus}_{298}$ kJ mol^{-1}	$S^{\ominus}_{m,298}$ J K^{-1}mol^{-1}	$\Delta_f G^{\ominus}_{298}$ kJ mol^{-1}	$C^{\ominus}_{P,m,298}$ J K^{-1}mol^{-1}
Ag silver						
Ag	s	107.88	0	42.7	0	25.5
Ag^+	aq	107.88	105.9	42.70	77.1	

Principles of Physical Chemistry, Third Edition. Hans Kuhn, David H. Waldeck, and Horst-Dieter Försterling.

Companion website: www.wiley.com/go/Kuhn/physicalchemistry3e

Substance	State	M g mol^{-1}	$\Delta_f H^{\ominus}_{298}$ kJ mol^{-1}	$S^{\ominus}_{m,298}$ J K^{-1}mol^{-1}	$\Delta_f G^{\ominus}_{298}$ kJ mol^{-1}	$C^{\ominus}_{P,m,298}$ J K^{-1}mol^{-1}
AgF	s	126.88	−202.9	84.0	−184.9	
AgCl	s	143.34	−126.1	96.1	−108.7	50.8
AgBr	s	187.80	−99.2	107.1	−95.5	52.4
AgI	s	234.79	−62.2	114.2	−68.2	54.4
$AgNO_3$	s	169.89	−120.7	140.9	−29.7	93.1
Ag_2CO_3	s	275.77	−506.0	167.3	−436.9	112.1
AgCN	s	133.90	146.1	83.7	163.9	
$Ag(CN)_2^-$	aq	150.92	269.9		301.46	
Al aluminum						
Al	s	26.98	0	28.3	0	24.3
Al^{3+}	aq	26.98	−524.7	−313.4	−481.2	
Al_2O_3 (corundum)	s	101.96	−1675.0	50.9	−1581.5	79.0
AlF_3	s	83.98	−1488.0	66.5	−1408.1	75.1
$AlCl_3$	s	133.35	−705.6	167.0	−647.0	89.1
$AlBr_3$	s	266.73	−526.2	184.0	−502.1	102.5
$Al_2(SO_4)_3$	s	342.16	−3434.0	239.2	−3091.5	259.3
Ar argon						
Ar	g	39.94	0	154.8	0	20.8
As arsenic						
As	s	74.91	0	35.1	0	24.6
AsH_3	g	77.93	75	222.5	67.1	38.5
As_2O_3	s	197.82	−656.8	107.1	−575.7	95.7
As_2O_5	s	229.82	−918.0	105.4	−775.3	117.5
AsO_4^{3-}	aq	129.90	−870.3	−144.8	−636.0	
$HAsO_4^{2-}$	aq	130.91	−898.7	3.8	−707.1	
$H_2AsO_4^-$	aq	131.92	−904.6	117.2	−748.5	
Au gold						
Au	s	197.0	0	47.5	0	25.4
Au^{3+}	aq	197.0			137.0	
$AuCl_4^-$	aq	338.83	−325.5		−235.1	
$AuBr_4^-$	aq	516.67	−190.4		−159.4	
B boron						
B	s	10.82	0	5.9		11.1
B	g	10.82	407	153.3	3623.0	20.8
BH_3	g	13.85	75.2	188	79.0	
B_2H_6	g	27.69	28.2	232.9	79.0	56.4
H_3BO_3	s	61.84	−1089	88.7	−962.8	81.3
Ba barium						
Ba	s	137.36	0	64.9	0	26.4
$BaCl_2$	s	208.28	−859.8	125.0	−810.9	75.3

(Continued)

Substance	State	M g mol^{-1}	$\Delta_f H^\ominus_{298}$ kJ mol^{-1}	$S^\ominus_{m,298}$ J K^{-1}mol^{-1}	$\Delta_f G^\ominus_{298}$ kJ mol^{-1}	$C^\ominus_{P,m,298}$ J K^{-1}mol^{-1}
$BaSO_4$	s	233.43	−1465.0	131.8	−1352.0	101.8
$BaCO_3$	s	197.37	−1202	112.1	−1123.5	85.4
Be beryllium						
Be	s	9.01	0	9.6	0	14.4
Be	g	9.01	320	136.2	282.3	20.9
BeF_2	s	47.01	−1011.7	57.3	−965.5	
$BeCl_2$	s	79.91	−456.9	90.0	−414.4	
Bi bismuth						
Bi	s	209.00	0	56.8	0	25.5
Bi	g	209.00	207.9	186.9	169.4	20.9
Bi_2O_3	s	466.00	−578.0	151.2	−497.4	113.5
$BiCl_3$	s	315.37	−379.1	191.6	−319.5	
Br bromine						
Br	g	79.92	96.4	174.9	80.8	20.8
Br_2	l	159.84	0	152.3	0	75.7
Br_2	g	159.84	0	245.4	0	36.1
Br^-	aq	79.92	−120.9	80.71	102.9	
HBr	g	80.92	−36.2	198.4	−49.8	29.1
BrCl	g	115.38	14.7	239.9	−0.9	35.0
BrO_3^-	aq	127.91	−40.2	162.8	45.6	
C carbon						
C (graphite)	s	12.01	0	5.7	0	8.5
C (diamond)	s	12.01	1.9	2.4	2.9	6.1
C	g	12.01	714.99	157.9	669.58	20.8
CO	g	28.01	−110.5	197.4	−137.0	29.2
CO_2	g	44.01	−393.5	213.6	−394.3	37.1
CO_2	s	44.01		129.6		
CO_3^-	aq	60.01	−676.3	−53.1	−528.1	
HCO_3^-	aq	61.02	−691.1	95.0	−587.1	
CN^-	aq	26.02	151.0	118.0	165.7	
CF_4	g	88.01	−933.20	261.31	−888.54	
CCl_4	l	153.84	−139.3	214.43	−68.6	
CCl_4	g	153.84	−95.98	309.70	−53.67	
CS_2	l	76.14	87.9	151.04	63.6	
CS_2	g	76.14	117.07	237.79	66.91	
COS	g	60.07	−138.41	231.47	−165.64	
HCN	l	27.03	105.5	112.9	121.3	71.0
HCN	g	27.03	135.14	201.72	124.71	
$(CN)_2$	g	52.04	−308.9	242.3	−320.5	56.9

Substance	State	M g mol^{-1}	$\Delta_f H^{\ominus}_{298}$ kJ mol^{-1}	$S^{\ominus}_{m,298}$ J K^{-1}mol^{-1}	$\Delta_f G^{\ominus}_{298}$ kJ mol^{-1}	$C^{\ominus}_{P,m,298}$ J K^{-1}mol^{-1}
Ca calcium						
Ca	s	40.08	0	41.6	0	26.3
Ca^{2+}	aq	40.08	−543.0	−55.2	−553.0	
CaO	s	56.08	−635.1	39.7	−603.8	42.8
CaH_2	s	42.10	−192.5	42.0	−153.6	
$Ca(OH)_2$	s	74.10	−986.2	83.4	−898.2	87.5
CaF_2	s	78.08	−1214.0	68.9	−1161.3	67.0
$CaCl_2$	s	110.99	−785.8	113.8	−740.5	72.6
CaS	s	72.15	−478.3	56.5	−472.9	47.4
$CaSO_4 \cdot 2H_2O$	s	172.18	−1762.0	184.0	−1564.5	178.6
$Ca(NO_3)_2$	s	164.10	−936.9	193.2	−741.1	149.2
CaC_2	s	64.10	−62.7	70.3	−67.8	62.3
$CaCO_3$ (calcite)	s	100.09	−1206.0	92.9	−1127.9	81.9
$CaCrO_4$	s	156.09	−1379.0	133.9	−1277.4	
Cd cadmium						
Cd	s	112.41	0	51.8	0	26.0
Cd^{2+}	aq	112.41	−72.4	−61.09	−77.7	
CdO	s	128.41	−256.1	54.8	−226.3	43.4
$Cd(OH)_2$	s	146.43	−557.4	95.4	−470.2	
$CdCl_2$	s	183.32	−389.0	115.3	−341.5	76.6
CdS	s	144.48	−144.3	71.0	−140.5	55.2
$CdSO_4$	s	208.48	−925.9	123.1	−815.0	99.6
$CdCO_3$	s	172.42	−747.0	105.4	−669.9	
Cl chlorine						
Cl	g	35.46	121.1	165.1	105.1	21.8
Cl_2	g	70.92	0	223.0	0	33.8
Cl^-	aq	70.92	−167.5	55.10	−131.2	
HCl	g	36.47	−92.3	186.9	−95.3	29.1
HCl	aq	36.47	−167.5	55.10	−131.2	
ClO_2^-	aq	86.91	104.60	257.1	385.90	
ClO_2	g	86.91	104.6	251.3	123.9	41.8
ClO_3^-	aq	83.46	−98.32	163.2	−2.59	
ClO_4^-	aq	99.46	131.42	182.0	−10.75	
ClF	g	54.46	−56.2	217.8	−57.6	32.1
ClF_3	g	92.46	−163.1	284.7	−124.0	65.1
Co cobalt						
Co	s	58.94		30.0		25.0
CoO	s	74.94	−238.9	53.0	−215.1	55.3
Co^{2+}	aq	58.94	−67.4	−155.2	−51.0	
$CoCl_2$	s	129.85	−325.4	106.6	−281.5	78.6

(Continued)

Substance	State	M g mol^{-1}	$\Delta_f H^\ominus_{298}$ kJ mol^{-1}	$S^\ominus_{m,298}$ J K^{-1}mol^{-1}	$\Delta_f G^\ominus_{298}$ kJ mol^{-1}	$C^\ominus_{P,m,298}$ J K^{-1}mol^{-1}
Cr chromium						
Cr	s	52.01	0	23.8	0	23.6
Cr_2O_3	s	152.02	−1141.0	81.1	−1058.8	118.7
Cr_2O_7	s	216.02	−1444	295.0	−1303.2	
K_2CrO_4	s	194.21	−1383	200.0	−1274.0	146.0
$K_2Cr_2O_7$	s	294.22	−2033	291.2	−1852.4	219.7
Cs cesium						
Cs	s	132.91	0	84.3	0	31.4
Cs	g	132.91	78.8	175.4	51.7	20.8
Cs^+	aq	132.91	−247.7	133.1	−292.02	−10.5
CsF	s	151.91	−530.8	82.8	−530.4	50.7
CsCl	s	168.37	−432.9	100.0	−437.6	52.6
CsBr	s	212.83	−394.4	121.0	−405.3	51.8
CsI	s	259.82	−336.7	130.0	350.3	51.9
Cu copper						
Cu	s	63.54	0	33.3	0	24.5
Cu^{2+}	aq	63.54	64.4	−98.7	65.0	
Cu^{1+}	aq	63.54	51.9	−26.4	50.2	
Cu_2O	s	143.08	−170.7	94.3	−148.3	63.6
CuCl	s	99.00	−134.7	91.6	−118.8	
CuBr	s	223.37	−105.0	96.1	−101.0	54.7
CuI	s	190.45	−68.2	96.6	−69.7	54.0
Cu_2S	s	159.15	−83.4	120.9	−90.1	76.3
CuO	s	79.54	−165.3	42.6	−136.9	42.3
$CuSO_4$	s	159.61	−771.1	113.3	−662.8	102.0
$CuSO_4 \cdot 5H_2O$	s	249.69	−2277.0	305.3	−1879.9	281.1
$[Cu(NH_3)_4]^{2-}$	aq	131.67	−334.4	806.7	−256.1	
$[Cu(CN)_4]^{3-}$	aq	167.62			566.5	
F fluorine						
F	g	19.00	79.1	158.6	62.0	22.8
F_2	g	38.00	0	202.7	0	31.3
F^-	aq	19.00	−329.1	−9.6	−276.5	
HF	g	20.01	−268.5	173.7	−270.6	29.1
F_2O	g	54.00	31.8	246.8	49.2	43.3
Fe iron						
Fe	s	55.85	0	27.2	0	25.2
Fe^{2+}	aq	55.85	−87.9	−113.4	−84.94	
Fe^{3+}	aq	55.85	−47.7	−293.3	−10.5	
FeO	s	71.85	−272.0	60.7	−251.4	48.1
Fe_2O_3	s	159.70	−822.25	87.4	−739.9	103.7
Fe_3O_4	s	231.55	−1116.0	146.4	−1012.9	143.4

Substance	State	M g mol^{-1}	$\Delta_f H^{\ominus}_{298}$ kJ mol^{-1}	$S^{\ominus}_{m,298}$ J K^{-1}mol^{-1}	$\Delta_f G^{\ominus}_{298}$ kJ mol^{-1}	$C^{\ominus}_{P,m,298}$ J K^{-1}mol^{-1}
$FeCl_2$	s	126.76	−340.9	120.1	−301.9	76.3
FeS, *α*	s	87.92	−96.2	60.3	−96.0	50.5
FeS_2, pyrite	s	119.98	−177.8	53.1	−166.5	61.9
Ga gallium						
Ga	s	69.72	0	41.1	0	26.1
Ga	g	69.72	276.1	169.0	238.0	25.3
Ga_2O_3	s	187.44	−1815.0	84.6	−1723.0	91.8
GaN	s	83.73	−104.2	46.0	−77.2	
$GaCl_3$	s	176.09	−524.7	169.5	−463.3	
Ge germanium						
Ge	s	72.60	0	31.1	0	23.4
Ge	g	72.60	328.1	167.7	287.4	30.7
GeO_2	s	104.60	−540.6	55.3	−486.3	52.1
H hydrogen						
H	g	1.01	217.9	114.6	203.3	20.8
H_2	g	2.02	0	130.6	0	28.8
H^+	aq	1.01	0	0	0	
OH	g	17.01	42.1	183.6	37.4	29.9
OH^-	aq	17.01	−229.95	−10.54	−157.32	
H_2O	l	18.02	−285.84	69.94	−237.19	75.2
H_2O	g	18.02	−241.83	188.72	−228.60	33.6
D	g	2.01	221.68	123.24	206.51	
HD	g	3.02	0.16	143.7	−1.63	29.2
D_2O	l	20.03	−294.6	72.4	−244.2	84.4
D_2O	g	20.03	−249.21	198.23	−234.58	34.3
HDO	l	19.02	−290.3	79.3	−242.2	78.9
HDO	g	19.02	−245.7	199.4	−255.8	33.8
He helium						
He	g	4.00	0	126.0	0	20.8
Hg mercury						
Hg	l	200.61	0	76.1	0	27.8
Hg	g	200.61	60.8	174.9	31.4	20.8
Hg_2^{2+}	aq	401.22			153.6	
Hg^{2+}	aq	200.61			174.6	
Hg_2Cl_2	s	472.13	−264.8	195.7	−211.3	101.6
Hg_2Br_2	s	561.05	−206.7	212.9	−179.7	
Hg_2I_2	s	655.04	−120.9	239.3	−112.1	105.8
HgO, yellow	s	216.61	−90.2	73.2	−58.7	
HgO, red	s	216.61	−90.8	70.2	−52.5	44.0
HgS, red	s	232.68	−58.6	76.6	−48.1	50.0
HgS, black	s	232.68	−53.6	84.1	−46.5	
Hg_2SO_4	s	497.29	−741.7	201.0	−624.3	132.0

(Continued)

Substance	State	M g mol^{-1}	$\Delta_f H^\ominus_{298}$ kJ mol^{-1}	$S^\ominus_{m,298}$ J K^{-1}mol^{-1}	$\Delta_f G^\ominus_{298}$ kJ mol^{-1}	$C^\ominus_{P,m,298}$ J K^{-1}mol^{-1}
I iodine						
I	g	126.91	106.6	180.7	70.4	20.8
I_2	s	253.82	0	116.1	0	54.4
I_2	g	253.82	62.4	260.7	19.1	36.9
I^-	aq	126.91	−55.9	109.4	51.7	
I_3^-	aq	380.73	−51.9	173.6	−51.5	
ICl	g	162.37	17.51	247.5	−5.72	35.6
IBr	g	206.83	40.88	258.7	3.71	36.5
ICl_3	s	233.28	−88.5	172.0	−22.6	
HI	g	127.92	25.9	206.3	1.2	29.2
In indium						
In	s			58.1		26.7
K potassium						
K	s	39.10	0	64.4	0	29.5
K	g	39.10	90.4	160.2	61.9	20.8
K^+	aq	39.10	−251.2	102.5	−282.0	
KH	g	40.11	125.5	197.9	105.1	30.8
KF	s	58.10	−562.4	66.5	−532.5	49.0
KCl	s	74.56	−435.7	82.7	−408.0	51.5
$KClO_3$	s	122.56	−391.1	142.9	−290.4	100.2
$KClO_4$	s	138.56	−433.3	151.0	−303.7	110.2
KBr	s	119.02	−392.0	96.4	−378.8	53.6
$KBrO_3$	s	167.02	−332.2	149.1	−242.9	104.9
KI	s	166.01	−327.5	104.6	−321.8	55.1
KIO_3	s	214.01	−508.4	151.4	−436.9	106.4
K_2SO_4	s	174.27	−1433.0	175.7	−1314.8	129.9
KNO_3	s	101.11	−492.5	132.9	−392.5	96.3
$KMnO_4$	s	158.03	−813.4	171.7	−713.58	
Kr krypton						
Kr	g	83.80	0	164.0	0	20.8
Li lithium						
Li	s	6.94	0	28.0	0	23.6
Li^+	aq	6.94	−278.5	14.2	−293.8	
LiH	s	7.95	−89.3	24.7	−68.8	34.7
LiOH	s	23.95	−487.8	42.8	−441.9	49.6
LiF	s	25.94	−611.9	35.7	−583.5	41.9
LiI	s	133.86	−270.1	85.7	−295.6	
$LiCl \cdot H_2O$	s	60.42	−712.3	103.7	−632.6	97.9
Li_2CO_3	s	73.89	−1215.0	90.3	−1131.2	97.4

Substance	State	M g mol^{-1}	$\Delta_f H^\ominus_{298}$ kJ mol^{-1}	$S^\ominus_{m,298}$ J K^{-1}mol^{-1}	$\Delta_f G^\ominus_{298}$ kJ mol^{-1}	$C^\ominus_{P,m,298}$ J K^{-1}mol^{-1}
Mg magnesium						
Mg	s	24.32	0	32.6	0	24.8
MgO	s	40.32	−601.2	26.8	−568.6	37.8
MgF_2	s	62.32	−1102	57.2	−1048.4	61.6
$MgCl_2$	s	95.23	−641.1	89.5	−591.5	71.3
$MgSO_4$	s	120.39	−1278	91.6	−1163.1	96.5
$MgCO_3$	s	84.33	−1105	65.7	−1021.2	75.5
Mn manganese						
Mn	s	54.94	0	31.8	0	26.3
Mn^{2+}	aq	54.94	−462.0	−118.0	−223.4	
MnO	s	70.94	−384.8	59.7	−362.4	44.1
MnO_2	s	86.94	−519.7	53.1	−464.9	54.0
$MnCl_2$	s	125.85	−466.9	117.1	−425.5	72.9
$MnSO_4$	s	150.99	−1063.0	112.1	−955.1	100.0
$MnCO_3$	s	114.95	−894.7	85.7	−816.8	81.5
$KMnO_4$	s	158.03	−813.4	171.7	−713.4	119.2
MnO_4^{2-}	aq	118.93	−518.4	190.0	−425.1	
N nitrogen						
N	g	14.01	470.6	153.1	453.3	20.8
N_2	g	28.02	0	191.5	0	29.1
NO	g	30.01	90.4	210.6	86.6	29.8
NO_2	g	46.01	33.3	239.8	51.5	37.1
N_2O_4	g	92.02	9.4	304.3	97.9	79.0
NO_3^-	aq	62.01	−206.56	146.4	−110.50	
$N_2O_2^{2-}$	aq	60.02	−10.84	27.61	138.1	
NH_3	g	17.04	−45.9	192.5	−19.5	35.5
NH_4^+	aq	18.05	−132.8	112.8	−79.50	
N_2H_4	l	32.05	50.4	121.2	149.2	98.8
HNO_3	l	63.02	−173.0	156.1	−79.8	109.8
$HNO_3 \cdot H_2O$	l	81.04	−472.6	216.9	−327.8	182.5
NH_4Cl	s	53.51	−314.6	94.6	−203.1	84.1
$(NH_4)_2SO_4$	s	132.15	−1399.6	220.3	−899.9	187.5
Na sodium						
Na	s	22.99	0	51.4	0	28.2
Na	g	22.99	108.6	153.6	78.1	20.8
Na^+	aq	22.99	−239.7	60.2	−261.9	
NaF	s	41.99	−570.3	51.3	−539.7	46.8
NaCl	s	58.44	−410.9	72.4	−383.5	49.7
NaOH	s	40.0	−426.6	64.2	−380.3	59.5
$NaOH \cdot H_2O$	s	58.02	−732.7	84.5	−623.4	59.5
Na_2SO_3	s	126.05	−1090.0	146.0	−1001.3	120.1
Na_2SO_4	s	142.05	−1384.0	149.4	−1265.3	127.7

(Continued)

Substance	State	M g mol^{-1}	$\Delta_f H^{\ominus}_{298}$ kJ mol^{-1}	$S^{\ominus}_{m,298}$ J K^{-1}mol^{-1}	$\Delta_f G^{\ominus}_{298}$ kJ mol^{-1}	$C^{\ominus}_{P,m,298}$ J K^{-1}mol^{-1}
$Na_2SO_4 \cdot 10H_2O$	s	322.21	−4323	585.6	−3644	575.7
$NaNO_3$	s	85.00	−466.5	116.3	−365.7	93.1
Na_2CO_3	s	105.99	−1129.0	136.0	−1045.1	110.5
$NaHCO_3$	s	84.01	−947.7	102.1	−849.3	87.7
Na_2SiO_3	s	122.07	−1518.0	113.8	−1423.2	111.8
$NaBH_4$	s	37.85	−183.3	101.5	−118.5	86.9
Ne neon						
Ne	g	20.18	0	146.2	0	20.8
Ni nickel						
Ni	s	58.71	0	29.9	0	26.1
NiO	s	74.71	−239.7	38.0	−211.5	44.3
Ni^{2+}	aq	58.71	64.0	−159.4	−46.4	
$NiCl_2$	s	129.62	−315.8	97.4	−269.3	71.7
$Ni(CO)_4$	g	170.75	−163.6	399.2	−144.5	136.0
O oxygen						
O	g	16.00	249.2	160.9	229.9	21.9
O_2	g	32.00	0	205.0	0	29.4
O_3	g	48.00	143.2	238.8	163.6	39.2
OH	g	17.02	42.1	183.6	37.4	29.9
OH^-	aq	17.02	−229.9	−10.5	−157.3	
H_2O	l	18.02	−285.9	69.9	−236.1	75.2
H_2O	g	18.02	−241.8	188.7	−228.5	33.6
P phosphorus						
P, white	s	30.98	0	44.4	0	23.2
P, red	s	30.98	−18.4			24.7
P	g	30.98	314.6	163.1	279.2	20.8
PH_3	g	34.00	9.3	210.2	18.2	37.1
H_3PO_4	s	98.00	−1281	110.5	−1119.6	106.1
PCl_5	s	208.26	−463.2			
Pb lead						
Pb	s	207.21	0	64.9	0	26.8
Pb^{2+}	aq	207.21	1.63	21.34	−24.31	
PbO, yellow	s	223.21	−217.8	67.4	−187.9	45.8
PbO, red	s	223.21	−219.2	65.3	−188.8	45.8
PbO_2	s	239.21	−276.6	76.4	−218.9	64.4
Pb_3O_4	s	685.63	−734.5	211.2	−617.1	147.0
$PbCl_2$	s	278.12	−359.1	136.4	−313.8	77.8
$PbBr_2$	s	367.04	−277.0	161.4	−260.2	80.1
PbI_2	s	461.03	−175.1	176.9	−173.8	81.2
PbS	s	239.28	−94.3	91.2	−92.6	49.5
$PbSO_4$	s	303.28	−918.1	147.2	−810.3	104.3
$PbCO_3$	s	267.22	−700.0	131.0	−626.3	87.4
$PbSiO_3$	s	283.30	−1082.8	113.0	−1000.0	90.0

Substance	State	M g mol^{-1}	$\Delta_f H^{\ominus}_{298}$ kJ mol^{-1}	$S^{\ominus}_{m,298}$ J K^{-1}mol^{-1}	$\Delta_f G^{\ominus}_{298}$ kJ mol^{-1}	$C^{\ominus}_{P,m,298}$ J K^{-1}mol^{-1}
Pd palladium						
Pd	s		0	37.8	0	26.2
Pt platinum						
Pt	s	195.09	0	41.8	0	26.6
K_2PtCl_6	s		−1259	333.8	−1107.7	205.4
Re rhenium						
Re	s	186.22		37.2		25.7
Rb rubidium						
Rb	s	85.48	0	76.2	0	30.9
Rb	g	85.48	85.8	169.9	57.9	20.8
RbF	s	104.48	−549.1	75.3	−547.9	50.7
RbCl	s	120.94	−430.4	88.7	−400.8	51.2
RbBr	s	165.40	−389.1	108.2	−377.6	53.0
RbI	s	212.39	−328.3	118.0	−325.4	52.3
Rn radon						
Rn	g	222.00	0	176.2	0	20.8
Ru ruthenium						
Ru	s	101.1	0	28.5	0	24.3
S sulfur						
S (rhombic)	s	32.07	0	31.9	0	22.6
S (monocline)	s	32.07	0.30	32.6	0.10	23.6
S	g	32.07	222.1	167.8	176.9	23.7
S_2	g	64.14	129.1	228.1	70.6	32.5
S^{2-}	aq	32.07	41.8	22.2	83.7	
SO	g	48.07	80.8	222.2	54.5	30.3
SO_2	g	64.07	−296.9	248.1	−300.1	39.9
SO_3	g	80.07	−395.2	256.0	−370.2	50.6
SO_3^{2-}	aq	80.07	−624.3	43.5	−497.1	
SO_4^{2-}	aq	96.07	−907.5	17.2	−741.99	
$S_2O_3^{2-}$	aq	112.14	−644.3	121.3	−532.2	
H_2S	g	34.08	−20.7	205.5	−33.0	34.2
HS^-	aq	33.08	−17.66	61.1	12.59	
HSO_3^-	aq	81.07	−627.98	132.4	−527.31	
HSO_4^-	aq	97.07	−885.75	126.9	−752.87	
SF_6	g	146.07	−1096	290.4	−1183.0	96.1
Sb antimony						
Sb	s	121.76	0	45.7	0	25.4
Sb_2O_5	s	323.52	−880.0	125.1	−737.3	117.7
Se selenium						
Se	s	78.96		42.4		25.4

(Continued)

Substance	State	M g mol^{-1}	$\Delta_f H^\ominus_{298}$ kJ mol^{-1}	$S^\ominus_{m,298}$ J K^{-1}mol^{-1}	$\Delta_f G^\ominus_{298}$ kJ mol^{-1}	$C^\ominus_{P,m,298}$ J K^{-1}mol^{-1}
Si silicon						
Si	s	28.09	0	18.7	0	19.8
SiH_4	g	32.12	−61.0	204.5	−39.5	42.8
SiO_2, quartz	s	60.09	−859.3	42.1	−804.7	44.4
$SiCl_4$	l	169.93	−577.4	239.7	−509.9	145.3
Sn tin						
Sn	s	118.70	0	51.4	0	26.4
Sn	g	118.70	301.0	168.3	266.2	21.3
SnO	s	134.70	−286.0	56.7	−246.2	44.3
SnO_2	s	150.70	−580.8	52.3	−519.7	52.6
$SnCl_4$	l	260.53	−544.9	258.5	−473.5	165.2
Sr strontium						
Sr	s	87.63	0	54.4	0	25.1
SrO	s	103.63	−589.1	54.4	−175.6	45.0
$SrCl_2$	s	158.54	−879	117.0	−261.9	79.7
$SrSO_4$	s	183.70	−1444	121.7	−1332.1	107.8
$SrCO_3$	s	147.64	−1218	97.0	−1136.9	81.4
Te tellurium						
Te	s	127.61		49.7		25.6
Te	g	127.61	199.0	182.6	159.4	20.8
Ti titanium						
Ti	s		0	30.7	0	25.0
TiO	s		−518.4	34.8	−488.7	40.0
TiO_2	s		−943.9	50.2	−888.0	55.1
Tl thallium						
Tl	s	204.39	0	64.2	0	26.4
Tl	g	204.39	181.3	180.9	146.5	20.8
Tl_2O	s	239.9	−178.0	99.5	−138.7	
TlCl	s	424.8	−204.9	111.2	−185.7	50.9
Xe xenon						
Xe	g	131.30	0	169.6	0	20.8
Zn zinc						
Zn	s	65.38	0	41.6	0	25.5
Zn^{2+}	aq	65.38	−152.4	−106.5	−147.3	
ZnO	s	81.38	−349.0	43.5	−318.6	40.3
$ZnCl_2$	s	136.29	−415.7	108.3	−368.9	76.5
$ZnBr_2$	s	225.21	−326.9	137.4	−309.9	35.9
ZnI_2	s	319.20	−209.0	158.9	−209.2	
ZnS	s	97.45	−201.0	57.7	−196.1	46.0

Substance	State	M g mol^{-1}	$\Delta_f H^\ominus_{298}$ kJ mol^{-1}	$S^\ominus_{m,298}$ J K^{-1}mol^{-1}	$\Delta_f G^\ominus_{298}$ kJ mol^{-1}	$C^\ominus_{P,m,298}$ J K^{-1}mol^{-1}
$ZnSO_4$	s	161.45	−978.2	124.6	−873.1	117.0
$ZnSO_4 \cdot 7H_2O$	s	287.56	−3075.0	388.7	−2560.2	380.3
$ZnCO_3$	s	125.39	−810.7	82.4	−729.2	80.17
$[Zn(CN)_4]^{2-}$	aq	169.46	343.1		446.9	
Zr zirconium						
Zr	s	91.22	0	38.9	0	25.2
ZrO_2	s	123.22	−1094.0	50.32	−1035.8	55.8
$ZrCl_4$	s	233.05	−982.0	186.1	−881.8	119.9

References:

1 Landolt-Börnstein, Zahlenwerte und Funktionen, 2. Bd., 4. Teil, Springer, Berlin, 1961.
2 F.D. Rossini, D.D. Wagman, W.H. Evans, S. Levine, I. Jaffe, *Natl. Bur. Stand. Circ.* 1952, **500**.
3 D.D. Wagman, W.H. Evans, V.B. Parker, I. Halow, S.M. Bailey, R.H. Schumm, Selected Values of Chemical Thermodynamic Properties, Technical Notes 270-3 and 270-4 (Natl.Bur.Stand., Washington 1968, 1969).
4 I. Barin, O.Knacke, *Thermochemical Properties of Inorganic Substances* (Springer, Berlin, 1973).
5 F.D. Rossini, K.S. Pitzer, R.L. Arnett, R.M. Braun, G.C. Pimentel, *Selected Values of Physical and Thermodynamic Properties of Hydrocarbons and Related Compounds* (Carnegie Press, Pittsburgh, 1953).
6 D.R. Stull, H. Prophet, *Natl. Stand. Ref. Data Ser., Natl. Bur. Stand.* 1971, **37**.
7 H.D. Brow, *Biochemical Microcalorimetry* (Academic Press, New York, 1969).
8 R. Chang, *Physical Chemistry with Applications to Biological Systems* (Macmillan, London, 1978).
9 R.C. Weast, *Handbook of Physics and Chemistry* (CRC Press, Cleveland, 1998).
10 G. Kortüm, H. Lachmann, Einführung in die chemische Thermodynamik, Verlage Chemie, Weinheim, und Vandenhoeck & Ruprecht, Göttingen 1981.
11 NIST Chemwebbook. https://webbook.nist.gov/chemistry/.

A.2 Organic Compounds

Enthalpy of formation $\Delta_f H^\ominus_{298}$, entropy $S^\ominus_{298}$, Gibbs energy of formation $\Delta_f G^\ominus_{298}$, and heat capacity $C^\ominus_{P,298}$ at 298.15 K. Solids (s) and liquids (l) under $P = 1$ bar; gases (g) under $P = 1$ bar in ideal state; substances dissolved in water (aq) under $c = 1$ mol L^{-1} in ideal state. The data are corrected for an assumed ideal gas under $P = 1$ bar and an assumed ideally diluted solution under $c = 1$ mol L^{-1}.

Substance	State	$\Delta_f H^\ominus_{298}$ kJ mol^{-1}	$S^\ominus_{298}$ J K^{-1}mol^{-1}	$\Delta_f G^\ominus_{298}$ kJ mol^{-1}	$C^\ominus_{P,298}$ J K^{-1}mol^{-1}
acetaldehyde	l	−166.4	265.7	−133.7	
acetate	aq	−485.6		−376.9	
acetic acid	aq	−485.3		−404.1	
acetic acid	l	−487.0	159.8	−392.5	123.4
alanine	aq	−554.9		−371.3	
alanine	s	−563.7	119	−372.0	115.2
benzene	g	82.9	269.2	129.7	81.7
benzene	l	49.0	173.2		136.1
n-butane	g	−124.7	310.0	−15.6	98.8
iso-butane	g	−131.6	294.6	−17.9	96.8
butadiene-(1,3)	g	111.9	278.7	152.4	79.5
butanol-(1)	l	−2670	252	−172.5	179
cyclohexane	g	−123.1	298.2	31.7	106.3

(Continued)

Substance	State	$\Delta_f H^\ominus_{298}$ kJ mol^{-1}	$S^\ominus_{298}$ J K^{-1}mol^{-1}	$\Delta_f G^\ominus_{298}$ kJ mol^{-1}	$C^\ominus_{P,298}$ J K^{-1}mol^{-1}
cyclohexane	l	−156.2	204.1	26.8	156.5
cyclohexene	g	−7.1	310.7	105.1	105.0
cyclohexene	l	−37.8	216.2		149
cyclopentane	g	−77.2	292.9	38.7	82.9
cyclopropane	g	53.2	237.9	104.1	55.9
ethane	g	−84.7	225.5	−32.9	52.7
ethanol	aq	−278.7		−182.0	
ethanol	l	−227.6	161.0	−174.8	112.0
ethene	g	52.3	219.5	68.1	43.6
ethyne	g	226.7	200.8	209.8	43.9
formic acid	l	−409.2	129.0	−346.0	99.0
formiate	aq	−410.0		−334.7	
glucose (α-D)	aq	−1263.1	264.0	−914.5	
glucose (α-D)	s	−1274.5	212.1	−910.6	218.9
glycine	aq	−523.1		−379.9	
glycine	s	−537.3	103.5	−379.9	95
glycyl-alanine	aq			−733.9	
n-hexane	g	−167.2	386.8	0.3	146.7
n-hexane	l	−198.8	296.0	−4.3	195.0
hydroquinone	s	−363.0			142.0
lactate ion	aq			−516.72	
lactic acid	aq			−538.77	
methane	g	−74.9	186.2	−40.6	35.8
methanol	aq	−245.9		−175.2	
methanol	l	−238.7	126.8	−166.3	81.6
octane	g	−208.4	463.7	17.4	194.9
octane	l	−250.0	361.2	6.4	254.1
pentane	g	−146.4	348.4	−8.1	122.6
pentane	l	−173.1	262.7	−9.2	171.5
phenol	s	−162.8	142.0	−47.5	134.7
polyethylene	s	−53.7	45.0	14.1	51.0
propane	g	−103.9	269.9	−23.4	73.5
propanol-(1)	l	−2017	192.9		148.6
propene	g	20.4	226.9	74.6	63.9
quinone	s	−186.8			132.0
sucrose	s	−2221	360	−1542	425

A.3 Redox Reactions and Standard Potentials

Change of Gibbs energy $\Delta_r G^\ominus_{298}$ and $\Delta_r G^{\ominus'}_{298}$ (at pH7) and standard potentials $E^\ominus_{298}$ at 298.15 K for reactions in aqueous solution, according to the scheme: reactant + $\frac{1}{2}H_2$ → product + H^+ [1].

Reaction			$\Delta_r G^{\ominus}_{298}$ kJ mol^{-1}	$E^{\ominus}_{298}$ mV	$\Delta_r G^{\ominus\prime}_{298}$ kJ mol^{-1}
$Ag^+ + \frac{1}{2}H_2$	→	$Ag + H^+$	−77.2	+800	117.2
$\frac{1}{3}Al^{3+} + \frac{1}{2}H_2$	→	$\frac{1}{3}Al + H^+$	+164.6	−1670	−124.6
$\frac{1}{3}Au^{3+} + \frac{1}{2}H_2$	→	$\frac{1}{3}Au + H^+$	−137	+1500	177.0
$\frac{1}{3}[AuCl_4]^- + \frac{1}{2}H_2$	→	$\frac{1}{3}Au + \frac{4}{3}Cl^- + H^+$	−95.5	+994	135.9
$\frac{1}{2}Br_2 + \frac{1}{2}H_2$	→	$Br^- + H^+$	−104.9	1065	144.9
$\frac{1}{2}Ca^{2+} + \frac{1}{2}H_2$	→	$\frac{1}{2}Ca + H^+$	276.9	−2870	−236.9
$\frac{1}{2}Cl_2 + \frac{1}{2}H_2$	→	$Cl^- + H^+$	−131.0	1358	171.0
$Cu^+ + \frac{1}{2}H_2$	→	$Cu + H^+$	−50.4	522	−90.4
$\frac{1}{2}Cu^{2+} + \frac{1}{2}H_2$	→	$\frac{1}{2}Cu + H^+$	−32.8	345	−72.8
$\frac{1}{2}Cd^{2+} + \frac{1}{2}H_2$	→	$\frac{1}{2}Cd + H^+$	38.9	−403	−1.1
$\frac{1}{2}F_2 + \frac{1}{2}H_2$	→	$F^- + H^+$	−276.9	2870	−316.9
$\frac{1}{2}Fe^{2+} + \frac{1}{2}H_2$	→	$\frac{1}{2}Fe + H^+$	42.5	−441	2.5
$\frac{1}{3}Fe^{3+} + \frac{1}{2}H_2$	→	$\frac{1}{3}Fe + H^+$	3.5	−36	−36.5
$Fe^{3+} + \frac{1}{2}H_2$	→	$Fe^{2+} + H^+$	−74.3	770	−114.3
$\frac{1}{2}Hg_2^{2+} + \frac{1}{2}H_2$	→	$Hg + H^+$	−76.8	789	−116.8
$\frac{1}{2}Hg^{2+} + \frac{1}{2}H_2$	→	$\frac{1}{2}Hg + H^+$	−87.3	905	−127.3
$\frac{1}{2}I_2 + \frac{1}{2}H_2$	→	$I^- + H^+$	−51.6	535	−91.6
$K^+ + \frac{1}{2}H_2$	→	$K + H^+$	282.1	−2924	242.1
$Li^+ + \frac{1}{2}H_2$	→	$Li + H^+$	293.8	−3024	253.88
$Na^+ + \frac{1}{2}H_2$	→	$Na + H^+$	261.6	−2714	221.6
$\frac{1}{2}Pb^{2+} + \frac{1}{2}H_2$	→	$\frac{1}{2}Pb + H^+$	12.2	−126	−27.8
$\frac{1}{2}Zn^{2+} + \frac{1}{2}H_2$	→	$\frac{1}{2}Zn + H^+$	73.6	−762	33.6
$\frac{1}{2}[Zn(CN)_4]^{2-} + \frac{1}{2}H_2$	→	$\frac{1}{2}Zn + 2CN^- + H^+$	108	−1120	68.0
$\frac{1}{2}$ quinone $+ \frac{1}{2}H_2$	→	$\frac{1}{2}$ hydroquinone	−67.5	700	−67.5
$\frac{1}{2}$[methyleneblue]$^+ + H_2$	→	$\frac{1}{2}$ leucomethylene blue $+ H^+$	−41.0	425	−41.0
$\frac{1}{2}NAD^+ + H_2$	→	$\frac{1}{2}NADH + H^+$	21.9	−113	−18.1
Cytb(Fe^{3+}) $+ \frac{1}{2}H_2$	→	Cytb(Fe^{2+}) $+ H^+$	−3.9	40	−43.9
Cytc$_1$(Fe^{3+}) $+ \frac{1}{2}H_2$	→	Cytc$_1$(Fe^{2+}) $+ H^+$	−21.2	220	−61.2

A.4 Biochemical Reactions

$\Delta G^{\ominus\prime}$ for Hydrolysis [2] and Oxidation [3] Reactions. Change of Gibbs energy $\Delta G^{\ominus\prime}$ at pH= 7 and at $T = 298.15$ K.

Reaction	*Hydrolysis Reactions*	$\Delta G^{\ominus\prime}$/kJ mol^{-1}
1	ATP → ADP + P	−30.5
2	glucose-6-phosphate → glucose + P	−13.8
3	acetylphosphate → acetic acid + P	−43.9
4	glycyl-alanine → glycine + alanine	−17.3
	Oxidation Reactions	
5	$NADH + \frac{1}{2}O_2 + H^+ \rightarrow NAD^+ + H_2O$	−220
6	$NADPH + \frac{1}{2}O_2 + H^+ \rightarrow NADP^+ + H_2O$	−220
7	3-phosphoglycerol-aldehyd + $\frac{1}{2}O_2$ → 3-phosphoglyceric acid	−263

A.5 Thermophysical Properties of Water

A.5.1 Fixed Point/Phase Transition Properties

	T_c/°C	P_c/MPa	ρ/(g cm^{-3})	
Critical Point	373.99	22.064	0.322	
	T/°C	P/Pa	ρ_l/(g cm^{-3})	ρ_s/g (cm^{-3})
Triple Point	0.01	611.73	0.99978	1.1055
Melting point	0.0		0.9998	0.9168
Properties at 1 bar	$\Delta_{fus}H$/kJ mol^{-1}	T_{mp}/°C	$\Delta_{vap}H$/(kJ mol^{-1})	T_{bp}/°C
	6.01	0.00	40.657	100.00

These data are taken from the CRC Handbook of Chemistry and Physics, 75th edition, 1994.

A.5.2 Density of Water at 1 bar as a Function of Temperature

Temperature in °C	T/K	ρ_l/g cm^{-3}	ρ_s/g cm^{-3}
−80	193.15		
−60	213.15		
−40	233.15		
−20	253.15		
0	273.15	0.99984	0.9168
10	283.15	0.99970	
20	293.15	0.99821	
30	303.15	0.99565	
40	313.15	0.99222	
50	323.15	0.98803	
60	333.15	0.98320	
70	343.15	0.97778	
80	353.15	0.97182	

Temperature in °C	T/K	ρ_l/g cm^{-3}	ρ_s/g cm^{-3}
90	363.15	0.96535	
100	373.15	0.95840	

These data are taken from the CRC Handbook of Chemistry and Physics, 75th edition, 1994.

A.5.3 Vapor Pressure of Water

Temperature in °C	T/K	Vapor Pressure in mmHg (Torr)	Vapor Pressure in mbar	P_{vap}/Pa
−80	193.15	0.00041	0.00055	0.055
−60	213.15	0.008101	0.0108	1.08
−40	233.15	0.09631	0.1284	12.84
−20	253.15	0.77451	1.0326	103.26
0	173.15	4.5840	6.1115	611.15
20	293.15	17.546	23.393	2392.3
25	298.15	23.776	31.699	3169.9
40	313.15	55.391	73.849	7384.9
60	333.15	149.61	199.46	1.9946×10^4
80	353.15	355.63	474.14	4.7414×10^4
100	373.15	760.00	1013.3	1.013×10^5

The data are taken from: https://www.internetchemie.info/chemie-lexikon/daten/w/wasser-dampfdruck.php. ©

A.6 Figure A.1 Shows Extinction Coefficients of Carbocyanine Dyes

See data files Cy0_1.dat, Cy1_1.dat, Cy2_1.dat. Cy3_1.dat in the Appendix directory.

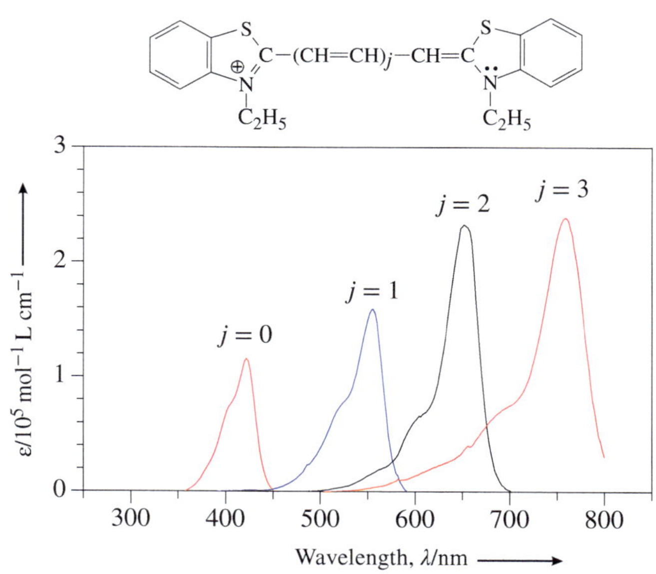

Figure A.1 Shift of the absorption maximum of symmetric cyanine dye cations with increasing chain length (i.e., with increasing j). ε is the decadic absorption coefficient and λ is the wavelength.

B Mathematical Tables

B.1 Summations

Note that Below we give a list of summations that are used in the text and solutions. A more complete list can be found in Gradshteyn and Ryzhik [4].

$$\sum_{n=1}^{n_{\max}} \sum_{j=1}^{n_{\max}} \cos\left(\frac{n\pi x}{L}\right) \cdot \cos\left(\frac{j\pi x}{L}\right) = \frac{1}{4} \frac{\sin^2\left(\pi\left(n_{\max}+1\right)x/L\right)}{\sin^2\left(\pi x/L\right)}$$

$$\sum_{k=0}^{(n_{\max}-1)/2} (2k+1)^2 = n_{\max}(n_{\max}+1)(n_{\max}+2)/6$$

$$\sum_{k=0}^{\infty} x^k = \frac{1}{1-x} \qquad \text{for } |x| < 1$$

B.2 Indefinite Integrals

Note that each of these integrals has an undetermined integration constant that is not explicitly given here.

$$\int \frac{1}{a-x} \cdot \mathrm{d}x = -\ln(a-x)$$

$$\int \frac{1}{(a-x)(b-x)} \cdot \mathrm{d}x = \frac{1}{b-a} \cdot \ln\frac{b-x}{a-x}$$

$$\int \frac{1}{ax-x^2} \cdot \mathrm{d}x = -\frac{1}{a} \cdot \ln\left(1-\frac{a}{x}\right)$$

$$\int \frac{1}{ax^2+bx} \cdot \mathrm{d}x = \frac{1}{b} \cdot \ln\frac{ax}{ax+b} = -\frac{1}{b} \cdot \ln\left(1+\frac{b}{ax}\right)$$

$$\int \frac{x}{\left(ax^2-b\right)^2} \cdot \mathrm{d}x = -\frac{1}{2a} \cdot \frac{1}{ab^2-b}$$

$$\int \sqrt{1-x} \cdot \mathrm{d}x = -\frac{2}{3}\sqrt{(1-x)^3}$$

$$\int \sqrt{a^2-x^2} \cdot \mathrm{d}x = \frac{1}{2}a^2 \cdot \left(\arcsin\frac{x}{a} + \frac{x}{a}\sqrt{1-\frac{x^2}{a^2}}\right)$$

$$\int \frac{1}{\sqrt{a-x}} \cdot \mathrm{d}x = -2\sqrt{a-x}$$

$$\int \frac{x}{\sqrt{1-x^2}} \cdot \mathrm{d}x = -\sqrt{1-x^2}$$

$$\int x\cos x \cdot \mathrm{d}x = x\sin x + \cos x$$

$$\int x^2\cos x \cdot \mathrm{d}x = \left(x^2-2\right)\sin x + 2x\cos x$$

$$\int x\cos ax \cdot \mathrm{d}x = \frac{x\sin ax}{a} + \frac{\cos ax}{a^2}$$

$$\int x^2\cos ax \cdot \mathrm{d}x = \left(\frac{x^2}{a} - \frac{2}{a^3}\right)\sin ax + \frac{2x\cos ax}{a^2}$$

$$\int x^n\cos x \cdot \mathrm{d}x = x^n\sin x - n\int x^{n-1}\sin x \cdot \mathrm{d}x$$

$$\int \sin x \cos x \cdot \mathrm{d}x = \tfrac{1}{2}\sin^2 x$$

$$\int \sin nx \cos kx \cdot \mathrm{d}x = \frac{1}{2} \cdot \left[\frac{-\cos(n+k)x}{n+k} - \frac{\cos(n-k)x}{n-k}\right] \quad \text{for} \quad k \neq n$$

$$\int \sin nx \sin kx \cdot \mathrm{d}x = \frac{1}{2} \cdot \left[\frac{-\sin(n+k)x}{n+k} + \frac{\sin(n-k)x}{n-k}\right] \quad \text{for} \quad k \neq n$$

$$\int \sin^2 x \cdot \mathrm{d}x = \tfrac{1}{2}\,(x - \cos x \sin x) = \tfrac{1}{2}\left(x - \tfrac{1}{2}\sin 2x\right)$$

$$\int (\sin ax)^3 \mathrm{d}x = -\frac{1}{3a}\cos(ax)\left[(\sin(ax))^2 + 2\right]$$

$$\int \cos^2 x \cdot \mathrm{d}x = \tfrac{1}{2}\,(x + \cos x \sin x) = \tfrac{1}{2}\left(x + \tfrac{1}{2}\sin 2x\right)$$

$$\int \cos^2 x \cdot \sin x \cdot \mathrm{d}x = -\frac{1}{3}\cos^3 x$$

$$\int \cos^3 x \cdot \sin x \cdot \mathrm{d}x = -\frac{1}{4}\cos^4 x$$

$$\int \cos x \cdot \sin^2 x \cdot \mathrm{d}x = \frac{1}{3}\sin^3 x$$

$$\int x \sin x \,\mathrm{d}x = \sin x - x\cos x$$

$$\int x \cos x \,\mathrm{d}x = \cos x + x \sin x$$

$$\int x^2 \sin x \,\mathrm{d}x = 2x\sin x - \left(x^2 - 2\right)\cos x$$

$$\int x^2 \cos x \,\mathrm{d}x = 2x\cos x + \left(x^2 - 2\right)\sin x$$

$$\int x \sin^2 x \,\mathrm{d}x = \frac{x^2}{4} - \frac{x\sin(2x)}{4} - \frac{\cos(2x)}{8}$$

$$\int x^2 \sin^2 x \,\mathrm{d}x = \frac{x^3}{6} - \frac{1}{4}\left[-\frac{\sin(2x)}{2} + x\cos(2x) + x^2\sin(2x)\right]$$

$$\int x^2 \cos^2 x \,\mathrm{d}x = \frac{x^3}{6} + \frac{1}{4}\left[-\frac{\sin(2x)}{2} + x\cos(2x) + x^2\sin(2x)\right]$$

$$\int \sin x \sin 2x \cdot \mathrm{d}x = -\frac{2}{3}\sin x\cos 2x + \frac{1}{3}\cos x \sin 2x$$

$$\int \sin 2x \cos x \cdot \mathrm{d}x = \tfrac{1}{3}\sin^3 x$$

$$\int \sin(i-j)x \cdot x \cdot \sin(i+j)x \cdot \mathrm{d}x = \frac{1}{2(i-j)^2}\left[x(i-j)\cdot\sin(i-j)x + \cos(i-j)x\right]$$
$$- \frac{1}{2(i+j)^2}\left[x(i+j)\cdot\sin(i+j)x + \cos(i+j)x\right]$$

$$\int \arcsin x \cdot \mathrm{d}x = x\arcsin x + \sqrt{1-x^2}$$

$$\int x \cdot \mathrm{e}^{-x} \cdot \mathrm{d}x = -(x+1)\cdot \mathrm{e}^{-x}$$

$$\int x^2 \cdot e^{-x} \cdot dx = -(x^2 + 2x + 2) \cdot e^{-x}$$

$$\int x^3 \cdot e^{-x} \cdot dx = -(x^3 + 3x^2 + 6x + 6) \cdot e^{-x}$$

$$\int x^4 \cdot e^{-x} \cdot dx = -\left(x^4 + 4x^3 + 12x^2 + 24x + 24\right) \cdot e^{-x}$$

$$\int x^5 \cdot e^{-x} \cdot dx = -\left(x^5 + 5x^4 + 20x^3 + 60x^2 + 120x + 120\right) \cdot e^{-x}$$

$$\int x^n \cdot e^{-x} \cdot dx = -x^n \cdot e^{-x} + n \int x^{n-1} \cdot e^{-x} \cdot dx$$

$$\int x \cdot e^{-x^2} \cdot dx = -\frac{1}{2} e^{-x^2}$$

$$\int \ln x \cdot dx = x \cdot (\ln x - 1)$$

$$\int x \cdot \ln x \cdot dx = \tfrac{1}{2} x^2 \left(\ln x - \tfrac{1}{2}\right)$$

$$\int \frac{\ln x}{x} \cdot dx = \frac{1}{2} (\ln x)^2$$

B.3 Definite Integrals

$$\int_0^\infty x^n \cdot e^{-x} \cdot dx = n!$$

$$\int_0^\infty \frac{x^n}{e^x - 1} \cdot dx = n! \cdot \sum_{k=1}^{\infty} \frac{1}{k^{n+1}}$$

$$\int_0^\infty \frac{x}{e^x - 1} \cdot dx = 1 \cdot \left(1 + \frac{1}{4} + \frac{1}{9} + \cdots\right) = 1.6400$$

$$\int_0^\infty \frac{x^2}{e^x - 1} \cdot dx = 2 \cdot \left(1 + \frac{1}{8} + \frac{1}{27} + \cdots\right) = 2 \cdot 1.2020 = 2.4040$$

$$\int_0^\infty \frac{x^3}{e^x - 1} \cdot dx = 6 \cdot \left(1 + \frac{1}{16} + \frac{1}{81} + \cdots\right) = 6 \cdot 1.0823 = 6 \cdot \frac{\pi^4}{90} = 6.4939$$

$$\int_0^\infty \frac{x^4}{e^x - 1} \cdot dx = 24 \cdot \left(1 + \frac{1}{32} + \frac{1}{243} + \cdots\right) = 24 \cdot 1.0369 = 24.886$$

$$\int_0^\infty e^{-x^2} \cdot dx = \frac{1}{2} \sqrt{\pi}$$

$$\int_0^\infty x \cdot e^{-x^2} \cdot dx = \frac{1}{2}$$

$$\int_0^\infty x^2 \cdot e^{-x^2} \cdot dx = \frac{1}{2} \int_0^\infty e^{-x^2} \cdot dx = \frac{1}{4} \sqrt{\pi}$$

$$\int_0^\infty x^3 \cdot e^{-x^2} \cdot dx = \frac{1}{2}$$

$$\int_0^\infty x^4 \cdot e^{-x^2} \cdot dx = \frac{3}{2} \int_0^\infty x^2 \cdot e^{-x^2} \cdot dx = \frac{3}{8} \sqrt{\pi}$$

$$\int_0^{\infty} x^{2n} \cdot e^{-x^2} \cdot dx = \frac{1 \times 2 \times 3 ..... \times (2n-1)}{2^{n+1}} \sqrt{\pi}$$

$$\int_0^{\infty} \cos bx \cdot e^{-(ax)^2} \cdot dx = \frac{\sqrt{\pi}}{2a} e^{-b^2/(4a^2)}$$

$$\int_0^{\infty} \sqrt{x} \cdot e^{-x} \cdot dx = \frac{1}{2}\sqrt{\pi}$$

$$\int_0^{\infty} x^{3/2} \cdot e^{-x} \cdot dx = \frac{3}{4}\sqrt{\pi}$$

$$\int_0^{\infty} x^{-1/2} \cdot e^{-x} \cdot dx = \sqrt{\pi}$$

B.4 Trigonometric Functions

$$\sin\alpha = \cos\left(\frac{\pi}{2} - \alpha\right) = \sqrt{1 - \cos^2\alpha}$$

$$\sin(\alpha + \beta) = \sin\alpha\cos\beta + \cos\alpha\sin\beta \quad \text{and} \quad \sin(\alpha - \beta) = \sin\alpha\cos\beta - \cos\alpha\sin\beta$$

$$\cos(\alpha + \beta) = \cos\alpha\cos\beta - \sin\alpha\sin\beta \quad \text{and} \quad \cos(\alpha - \beta) = \cos\alpha\cos\beta + \sin\alpha\sin\beta$$

$$\sin(2\alpha) = 2\sin\alpha\cos\alpha \quad \text{and} \quad \cos(2\alpha) = \cos^2\alpha - \sin^2\alpha$$

$$\sin(3\alpha) = 3\sin\alpha - 4\sin^3\alpha \qquad \text{and} \qquad \cos(2\alpha) = 4\cos^3\alpha - 3\cos\alpha$$

$$\sin(4\alpha) = 8\cos^3\alpha - 4\cos\alpha\sin\alpha \qquad \text{and} \qquad \cos(4\alpha) = 8\cos^4\alpha - 8\cos^2\alpha + 1$$

$$\sin\alpha + \sin\beta = 2\sin\left(\frac{\alpha+\beta}{2}\right)\cos\left(\frac{\alpha-\beta}{2}\right) \quad \text{and} \quad \sin\alpha - \sin\beta = 2\cos\frac{\alpha+\beta}{2}\sin\frac{\alpha-\beta}{2}$$

$$\cos\alpha + \cos\beta = 2\cos\left(\frac{\alpha+\beta}{2}\right)\cos\left(\frac{\alpha-\beta}{2}\right) \quad \text{and} \quad \cos\alpha - \cos\beta = -2\sin\frac{\alpha+\beta}{2}\sin\frac{\alpha-\beta}{2}$$

$$\sin\alpha\sin\beta = -\frac{1}{2}\cos(\alpha+\beta) + \frac{1}{2}\cos(\alpha-\beta)$$

$$\cos\alpha\cos\beta = \frac{1}{2}\cos(\alpha+\beta) + \frac{1}{2}\cos(\alpha-\beta)$$

$$\sin\alpha\cos\beta = -\frac{1}{2}\sin(\alpha+\beta) + \frac{1}{2}\sin(\alpha-\beta)$$

$$\sin(3\alpha)\sin(\alpha) = \frac{1}{2}[-\cos(4\alpha) + \cos(2\alpha)]$$

$$\sin\alpha + \cos\beta = 2\sin\left(\frac{\pi}{4} + \frac{\alpha-\beta}{2}\right)\cos\left(-\frac{\pi}{4} + \frac{\alpha+\beta}{2}\right)$$

$$\sin\alpha - \cos\beta = 2\cos\left(\frac{\pi}{4} + \frac{\alpha-\beta}{2}\right)\sin\left(-\frac{\pi}{4} + \frac{\alpha+\beta}{2}\right)$$

$$\sin^2\alpha = \frac{1}{2}[1 - \cos(2\alpha)]\cos^2\alpha = \frac{1}{2}[1 + \cos(2\alpha)]$$

$$\sin^3\alpha = \frac{1}{4}[3\sin\alpha - \sin(3\alpha)]\cos^3\alpha = \frac{1}{4}[\cos(3\alpha) + 3\cos\alpha]$$

$$\sin^4\alpha = \frac{1}{8}[\cos(4\alpha) - 4\cos(2\alpha) + 3] \quad \text{and} \quad \cos^4\alpha = \frac{1}{8}[\cos(4\alpha) + 4\cos(2\alpha) + 3]$$

$$\sinh(x) = \frac{e^x - e^{-x}}{2} \qquad \text{and} \qquad \cosh(x) = \frac{e^x + e^{-x}}{2}$$

$$\tanh(x) = \frac{e^x - e^{-x}}{e^x + e^{-x}} \qquad \text{and} \qquad \coth(x) = \frac{e^x + e^{-x}}{e^x - e^{-x}}$$

B.4.1 Some Fundamental Relations for a Plane Triangle

Figure B.1 shows a plane triangle and defines its angles and lengths.

Law of Cosines

$$a = \sqrt{b^2 + c^2 - 2 \cdot c \cdot b \cdot \cos\alpha}$$

Law of Sines

$$\frac{\sin\alpha}{\alpha} = \frac{\sin\beta}{\beta} = \frac{\sin\gamma}{\gamma}$$

Law of Tangents

$$\frac{a-b}{a+b} = \frac{\tan\left((\alpha-\beta)/2\right)}{\tan\left((\alpha+\beta)/2\right)}$$

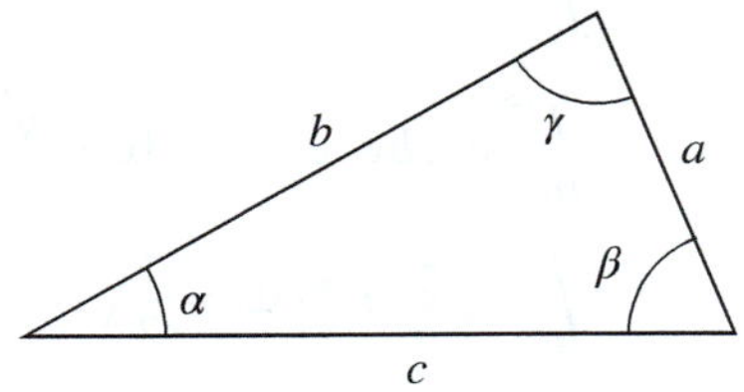

Figure B.1 Arbitrary triangle with side lengths a, b, and c and angles α, β, and γ.

B.4.2 Calculation of Tetrahedral Angle θ

Let us consider a CH_4 molecule as indicated in Fig. B.2. The tetrahedral angle is the angle enclosed by the two red lines of length d (C–H bond length) in the triangle H2–C–H3. In this triangle we have the relation

$$b^2 = 2d^2 - 2d^2 \cdot \cos\theta = 2d^2\left(1-\cos\theta\right) \tag{B.1}$$

For the triangle with the two blue sides we find with $\beta = 120°$

$$b^2 = 2r^2 - 2r^2 \cdot \cos\beta = 2r^2\left(1-\cos\beta\right) = 2r^2\frac{3}{2} = 3r^2 \tag{B.2}$$

For the rectangular triangle with the side lengths r, b, and $(c+d)$ we find

$$b^2 = r^2 + (c+d)^2 \qquad \text{with} \qquad c^2 = d^2 - r^2 \tag{B.3}$$

Thus we have from Equations (B.1) and (B.3)

$$2d^2\left(1-\cos\theta\right) = r^2 + (c+d)^2 = r^2 + c^2 + 2cd + d^2$$

$$2d^2 - 2d^2\cos\theta = d^2 + 2cd + d^2$$

$$\cos\theta = \frac{2cd}{2d^2} = \frac{c}{d} = \frac{\sqrt{d^2-r^2}}{d}$$

With Equations (B.2) and (B.1) we find

$$r^2 = \frac{1}{3}b^2 = \frac{2}{3}d^2\left(1-\cos\theta\right)$$

to obtain

$$\cos\theta = \frac{\sqrt{d^2 - \frac{2}{3}d^2\left(1-\cos\theta\right)}}{d} = \frac{d\sqrt{1-\frac{2}{3}\left(1-\cos\theta\right)}}{d} = \sqrt{1-\frac{2}{3}\left(1-\cos\theta\right)}$$

Thus

$$\cos^2\theta = \frac{1}{3} + \frac{2}{3}\cos\theta$$

and

$$\cos^2\theta - \frac{2}{3}\cos\theta = \frac{1}{3}$$

We solve this equation for $\cos\theta$

$$\cos\theta = \frac{1}{3} \pm \sqrt{\frac{1}{3}+\frac{1}{9}} = \frac{1}{3} \pm \frac{2}{3}$$

leading to

$$\cos\theta = 1 \qquad \text{or} \qquad \cos\theta = -\frac{1}{3}$$

The first solution gives $\theta = 0$, which is a trivial solution. The second solution gives $\theta = 109.471°$, which is the tetrahedral angle.

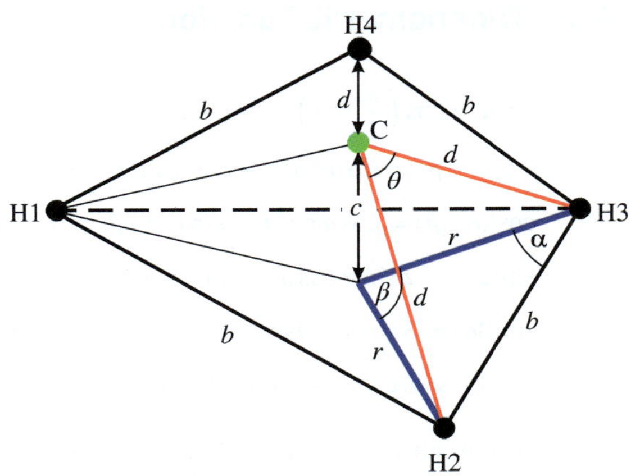

Figure B.2 Sketch of a CH_4 molecule. θ is the tetrahedral angle. The green filled circle indicates the position of the C atom; the black filled circles indicate the positions of the four H atoms; d is the C–H bond length; b is the distance between two H atoms; c is the distance between the C atom and the plane defined by the H atoms 1, 2, and 3. The three H atoms 1, 2, and 3 form an equilateral triangle with angles $\alpha = 30°$ and $\beta = 180° - 2 \cdot 30° = 120°$.

B.5 Power Series

$$y(a+x) = y(a) + \frac{1}{1!}\left(\frac{dy}{dx}\right)_{x=a} \cdot x + \frac{1}{2!}\left(\frac{d^2y}{dx^2}\right)_{x=a} \cdot x^2 + \frac{1}{3!}\left(\frac{d^3y}{dx^3}\right)_{x=a} \cdot x^3 + \cdots\cdots$$

$$y(x) = y(0) + \frac{1}{1!}\left(\frac{dy}{dx}\right)_{x=0} \cdot x + \frac{1}{2!}\left(\frac{d^2y}{dx^2}\right)_{x=0} \cdot x^2 + \frac{1}{3!}\left(\frac{d^3y}{dx^3}\right)_{x=0} \cdot x^3 + \cdots\cdots$$

$$e^x = 1 + x + \frac{1}{2!}x^2 + \frac{1}{3!}x^3 + \cdots$$

$$\ln(1+x) = x - \frac{1}{2}x^2 + \frac{1}{3}x^3 - \frac{1}{4}x^4 + \cdots \quad \text{for} \quad -1 < x \le 1$$

$$\ln(1-x) = -x - \frac{1}{2}x^2 - \frac{1}{3}x^3 - \frac{1}{4}x^4 + \cdots \quad \text{for} \quad -1 \le x < 1$$

$$\ln\frac{1+x}{1-x} = 2x + \frac{2}{3}x^3 + \frac{2}{5}x^5 + \cdots \quad \text{for} \quad |x| < 1$$

$$\ln\frac{x+1}{x-1} = 2\frac{1}{x} + \frac{2}{3}\frac{1}{x^3} + \frac{2}{5}\frac{1}{x^5} + \cdots \quad \text{for} \quad |x| > 1$$

$$\sin x = x - \frac{1}{3!}x^3 + \frac{1}{5!}x^5 + \cdots$$

$$\cos x = 1 - \frac{1}{2!}x^2 + \frac{1}{4!}x^4 + \cdots$$

$$(1+x)^n = 1 + \frac{n}{1!}x + \frac{n(n-1)}{2!}x^2 + \frac{n(n-1)(n-2)}{3!}x^3 + \cdots \text{for} \qquad |x| \le 1$$

$$\frac{1}{1+x} = (1+x)^{-1} = 1 - x + x^2 - x^3 + \cdots \text{for} \qquad |x| \le 1$$

$$\sqrt{1+x} = (1+x)^{1/2} = 1 + \frac{1}{2}x - \frac{1}{8}x^2 + \frac{1}{16}x^3 - \cdots \text{for} \qquad |x| \le 1$$

$$\frac{1}{\sqrt{1+x}} = (1+x)^{-1/2} = 1 - \frac{1}{2}x + \frac{3}{8}x^2 + \frac{5}{16}x^3 - \cdots \text{for} \qquad |x| \le 1$$

$$\frac{1}{(1+x)^2} = 1 - 2x + 3x^2 - 4x^3 \ldots\ldots.$$

Geometric series

$$\sum_{n=0}^{\infty} x^n = x^0 + x^1 + x^2 + \cdots = \frac{1}{1-x} \ldots\ldots \text{ for} \qquad |x| < 1$$

$$\sinh(x) = x + \frac{x^3}{3!} + \frac{x^5}{5!} + \frac{x^7}{7!} + \cdots$$

$$\cosh(x) = x + \frac{x^2}{2!} + \frac{x^4}{4!} + \frac{x^6}{6!} + \cdots$$

$$\tanh(x) = x - \frac{1}{3}x^3 + \frac{2}{15}x^5 - \frac{17}{315}x^7 + \frac{62}{2835}x^9 - \cdots + \cdots$$

$$\coth(x) = \frac{1}{x} + \frac{1}{3}x - \frac{1}{45}x^3 + \frac{2}{945}x^5 - \frac{1}{4725}x^7 + \cdots - \cdots$$

B.6 Miscellaneous Relations

$$\exp(i \cdot x) = \cos(x) + i \cdot \sin(x) \quad \text{and} \quad \exp(-i \cdot x) = \cos(x) - i \cdot \sin(x)$$

$$\ln(xy) = \ln x + \ln y \qquad \text{and} \qquad \ln\left(x^y\right) = y \ln x$$

B.7 Coordinate Transformations

B.7.1 Polar Coordinates

$x = r\cos\varphi, \qquad y = r\sin\varphi$

area element: $r\,\mathrm{d}r\,\mathrm{d}\varphi$

B.7.2 Spherical Coordinates

Figure B.3 shows the definition of the spherical coordinates on a cartesian axis system.

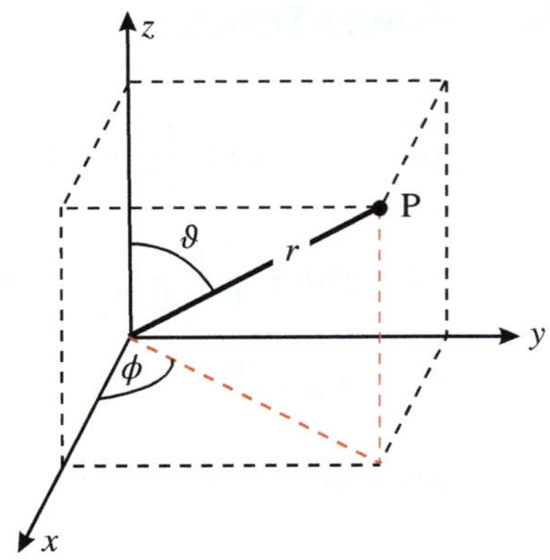

Figure B.3 Conversion of Cartesian coordinates into spherical coordinates: $z = r\cos\vartheta$, $y = r\sin\vartheta\sin\varphi$, and $x = r\sin\vartheta\cos\varphi$, volume element: $r^2\sin\vartheta\,\mathrm{d}r\,\mathrm{d}\varphi\,\mathrm{d}\vartheta$

C Mathematical Miscellany

C.1 Differentiation, Differential Equations, Partial and Total Differentials

C.1.1 Differentiation Rules

$$y = a\cdot f(x) \qquad \frac{\mathrm{d}y}{\mathrm{d}x} = a\cdot\frac{\mathrm{d}f(x)}{\mathrm{d}x}$$

$$y = a + f(x) \qquad \frac{\mathrm{d}y}{\mathrm{d}x} = \frac{\mathrm{d}f(x)}{\mathrm{d}x}$$

$$y = f(x) + g(x) \qquad \frac{\mathrm{d}y}{\mathrm{d}x} = \frac{\mathrm{d}f(x)}{\mathrm{d}x} + \frac{\mathrm{d}g(x)}{\mathrm{d}x}$$

$$y = f(x)\cdot g(x) \qquad \frac{\mathrm{d}y}{\mathrm{d}x} = \frac{\mathrm{d}f(x)}{\mathrm{d}x}\cdot g(x) + \frac{\mathrm{d}g(x)}{\mathrm{d}x}\cdot f(x)$$

$$y = \frac{f(x)}{g(x)} \qquad \frac{\mathrm{d}y}{\mathrm{d}x} = \frac{\frac{\mathrm{d}f(x)}{\mathrm{d}x}\cdot g(x) - \frac{\mathrm{d}g(x)}{\mathrm{d}x}\cdot f(x)}{g(x)^2}$$

$$y = g[f(x)] \qquad \frac{\mathrm{d}y}{\mathrm{d}x} = \frac{\mathrm{d}g}{\mathrm{d}f}\cdot\frac{\mathrm{d}f}{\mathrm{d}x}$$

C.1.2 Calculation of the Extrema of a Function

As an example, we consider the function

$$y = -1 - x + x^2 + x^3$$

displayed in Figure C.1. At an extremum (maximum or minimum) the slope of the curve is zero. Because the slope is identical with the first derivative, we calculate

$$\frac{\mathrm{d}y}{\mathrm{d}x} = -1 + 2x + 3x^2$$

and set the expression equal to zero:

$$0 = -1 + 2x_1 + 3x_1^2 \qquad \text{or} \qquad x_1^2 + \frac{2}{3}x_1 = \frac{1}{3}$$

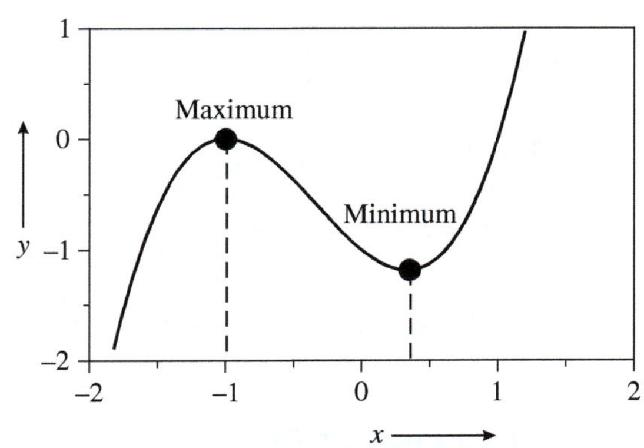

Figure C.1 Sketch of a curve with a maximum and a minimum.

where the subscript indicates that this is an equation for a special value of x. This equation is solved for x_1

$$x_1 = -\frac{1}{3} \pm \sqrt{\frac{1}{3} + \frac{1}{9}} = -\frac{1}{3} \pm \sqrt{\frac{4}{9}} = -\frac{1}{3} \pm \frac{2}{3}$$

and we obtain the two solutions $x_{1a} = \frac{1}{3}$ and $x_{1b} = -1$.

At a maximum, the second derivative of the function is negative, and at a minimum it is positive. Thus we have to calculate the second derivative of y.

$$\frac{\mathrm{d}^2 y}{\mathrm{d}x^2} = 2 + 6x$$

and to insert our two solutions:

$$\left(\frac{\mathrm{d}^2 y}{\mathrm{d}x^2}\right)_{1a} = 2 + 6 \cdot \frac{1}{3} = +4, \qquad \left(\frac{\mathrm{d}^2 y}{\mathrm{d}x^2}\right)_{1b} = 2 - 6 \cdot 1 = -4$$

Hence we conclude that the solution x_{1a} corresponds to a minimum, and x_{1b} corresponds to a maximum.

C.1.3 Partial and Total Differentials

Consider some function u of r independent (state) variables; e.g.,

$$u = F(x_1, x_2, \ldots, x_r)$$

where the x_i are independent variables and u is a dependent variable. Let x_1 increase from x_1 to $x_1 + \Delta x_1$ with the rest held constant. Then the change in u is

$$\Delta u = F(x_1 + \Delta x_1, x_2, \ldots x_r) - F(x_1, x_2, \ldots x_r)$$

The infinitesimal change is given by the partial derivative. Geometrically the partial derivative corresponds to the slope of a plot of u versus x_1 at selected values of $x_2, x_3, \ldots, x_r$. For functions that are continuous and differentiable, we can write

$$\left(\frac{\partial u}{\partial x_1}\right)_{x_2, x_3, \ldots x_r} = \lim_{\Delta x_1 \to 0} \left[\frac{F(x_1 + \Delta x_1, x_2, \ldots, x_r) - F(x_1, x_2, \ldots, x_r)}{\Delta x_1}\right]$$

The total differential of u is written as

$$\mathrm{d}u = \left(\frac{\partial u}{\partial x_1}\right)_{x_2, x_3, \ldots x_r} \mathrm{d}x_1 + \left(\frac{\partial u}{\partial x_2}\right)_{x_1, x_3, \ldots x_r} \mathrm{d}x_2 + \cdots$$

Example C.1 *For $P = P(V, T)$ the total differential is*

$$dP = \left(\frac{\partial P}{\partial V}\right)_T dV + \left(\frac{\partial P}{\partial T}\right)_V dT$$

C.1.4 Properties of Exact (or Total) Differentials:

Consider $u = f(x, y)$, then

$$\mathrm{d}u = \left(\frac{\partial u}{\partial x}\right)_y \mathrm{d}x + \left(\frac{\partial u}{\partial y}\right)_x \mathrm{d}y = M\ \mathrm{d}x + N\ \mathrm{d}y \quad \text{and} \quad \left(\frac{\partial M}{\partial y}\right)_x = \left(\frac{\partial N}{\partial x}\right)_y$$

which is called the *Euler criterion* and is a sufficient condition that $\mathrm{d}u$ be exact.

C.1.5 Some Useful Relations for Partial Derivatives

Here we note three useful relationships. They are

$$1: \quad \left(\frac{\partial x}{\partial y}\right)_z \left(\frac{\partial z}{\partial x}\right)_y \left(\frac{\partial y}{\partial z}\right)_x = -1$$

$$2: \quad \left(\frac{\partial x}{\partial y}\right)_z = \frac{1}{\left(\frac{\partial y}{\partial x}\right)_z}$$

and the result for a change of variables

3. $$\left(\frac{\partial f}{\partial x}\right)_g = \left(\frac{\partial f}{\partial x}\right)_y + \left(\frac{\partial f}{\partial y}\right)_x \left(\frac{\partial y}{\partial x}\right)_g$$

To show relation 3, we start with a function $f(x, y)$. If we change x by Δx and y by Δy for the change Δf we obtain

$$\Delta f = \left(\frac{\Delta f}{\Delta x}\right)_y \Delta x + \left(\frac{\Delta f}{\Delta y}\right)_x \Delta y$$

Then, keeping a variable g constant, we find that

$$\left(\frac{\Delta f}{\Delta x}\right)_g = \left(\frac{\Delta f}{\Delta x}\right)_y + \left(\frac{\Delta f}{\Delta y}\right)_x \left(\frac{\Delta y}{\Delta x}\right)_g$$

and

$$\lim_{\Delta x \to 0}\left(\frac{\Delta f}{\Delta x}\right)_g = \left(\frac{\partial f}{\partial x}\right)_g \qquad \lim_{\Delta x \to 0}\left(\frac{\Delta f}{\Delta x}\right)_y = \left(\frac{\partial f}{\partial x}\right)_y \qquad \lim_{\Delta x \to 0}\left(\frac{\Delta f}{\Delta y}\right)_x \frac{\Delta y}{\Delta x} = \left(\frac{\partial f}{\partial y}\right)_x \left(\frac{\partial y}{\partial x}\right)_g$$

Thus we obtain

3. $$\left(\frac{\partial f}{\partial x}\right)_g = \left(\frac{\partial f}{\partial x}\right)_y + \left(\frac{\partial f}{\partial y}\right)_x \left(\frac{\partial y}{\partial x}\right)_g$$

Example: for the enthalpy $H(T, V)$ we have

$$\mathrm{d}H = \left(\frac{\partial H}{\partial T}\right)_V \mathrm{d}T + \left(\frac{\partial H}{\partial V}\right)_T \mathrm{d}V$$

Thus

$$\left(\frac{\partial H}{\partial T}\right)_P = \left(\frac{\partial H}{\partial T}\right)_V + \left(\frac{\partial H}{\partial V}\right)_T \left(\frac{\partial V}{\partial T}\right)_P$$

C.2 Differential Equations

C.2.1 Analytical Solutions

We start with a first-order homogeneous differential equation with the constant coefficient a, as it might appear in chemical kinetics.

$$\frac{\mathrm{d}y}{\mathrm{d}x} + ay = 0$$

We rearrange

$$\frac{\mathrm{d}y}{y} = -a\,\mathrm{d}x \tag{C.1}$$

and integrate on both sides

$$\int \frac{\mathrm{d}y}{y} = -a \int \mathrm{d}x$$

leading to the solution

$$\ln y = -ax + C \tag{C.2}$$

C is the integration constant and must be determined using the condition that $y = y_0$ for $x = 0$ (initial condition). Thus we have

$$\ln y_0 = C$$

and

$$\ln \frac{y}{y_0} = -ax$$

Solving for y, we can write that

$$y = y_0\,\mathrm{e}^{-ax}$$

This expression for y is called the solution of the homogeneous first-order differential equation.

An inhomogeneous first-order differential equation has the structure

$$\frac{\mathrm{d}y}{\mathrm{d}x} + ay = b \tag{C.3}$$

where a and b are both constants. We can solve this equation by first solving it in the limit where $b = 0$ and then constructing a more general solution. For the case where $b = 0$, the differential equation is homogeneous (called the corresponding homogeneous equation)

$$\frac{\mathrm{d}z}{\mathrm{d}x} + az = 0 \tag{C.4}$$

and it has the solution

$$z = z_0\,\mathrm{e}^{-ax}$$

In the next step we make the ansatz that

$$y = z \cdot f$$

where f is a function of x. Applying the product rule, we find

$$\frac{\mathrm{d}y}{\mathrm{d}x} = \frac{\mathrm{d}}{\mathrm{d}x}(z \cdot f) = f \cdot \frac{\mathrm{d}z}{\mathrm{d}x} + z \cdot \frac{\mathrm{d}f}{\mathrm{d}x}$$

If we insert this expression in the original differential Equation (C.3), then we find

$$f \cdot \frac{\mathrm{d}z}{\mathrm{d}x} + z \cdot \frac{\mathrm{d}f}{\mathrm{d}x} + a \cdot z \cdot f = b$$

$$f \cdot \left(\frac{\mathrm{d}z}{\mathrm{d}x} + az\right) + z \cdot \frac{\mathrm{d}f}{\mathrm{d}x} = b$$

With Equation (C.4) it follows that

$$\frac{\mathrm{d}f}{\mathrm{d}x} = \frac{b}{z} = \frac{b}{z_0} \cdot \mathrm{e}^{ax}$$

This equation can be solved directly by integration, and we find that

$$f = \frac{b}{a \cdot z_0} \cdot \mathrm{e}^{ax} + C_1$$

Using this result for f and our earlier result for z, we can now construct the expression for y and find

$$y = zf = z_0 \cdot \mathrm{e}^{-ax}\left(\frac{b}{az_0}\mathrm{e}^{ax} + C_1\right)$$

$$= \frac{b}{a} + C_1 \cdot z_0 \cdot \mathrm{e}^{-ax} = \frac{b}{a} + C_2 \cdot \mathrm{e}^{-ax}$$

With the initial condition $y = y_0$ for $x = 0$, we fix the integration constant C_2 as $C_2 = y_0 - b/a$ and the final result is

$$y = \frac{b}{a} + \left(y_0 - \frac{b}{a}\right) \cdot \mathrm{e}^{-ax}$$

C.2.2 Numerical Solutions

Euler–Cauchy Method

Most differential equations do not have an analytical solution. In many of these cases we can integrate numerically. In order to understand this method we consider the differential equation

$$\frac{\mathrm{d}y}{\mathrm{d}x} = -2 \cdot x \cdot y \text{ with the initial condition } y = 1 \text{ for } x = 0 \tag{C.5}$$

Note that this equation has the analytical solution $y = \exp(-x^2)$, and we will compare our numerical result with the exact analytical one as a check of our method.

For the numerical solution we replace the differentials by differences and write

$$\frac{dy}{dx} = \frac{\Delta y}{\Delta x}$$
$$\Delta y = \frac{dy}{dx} \cdot \Delta x = -2 \cdot x \cdot y \cdot \Delta x$$

We start the calculation with the initial value $y = 1$ for $x_0 = 0$. We obtain the next values $y_1, y_2, \ldots$ as

$$y_1 = y_0 + \Delta y = y_0 - 2x_0\, y_0\, \Delta x$$
$$y_2 = y_1 + \Delta y = y_1 - 2x_1\, y_1\, \Delta x$$
$$y_3 = y_2 + \Delta y = y_2 - 2x_2\, y_2\, \Delta x$$

where $x_0 = 0, x_1 = \Delta x, x_2 = 2\Delta x, x_3 = 3\Delta x$. With $y_0 = 1$ and $\Delta x = 0.1$ we obtain the following data

	Numerical				Exact	
i	x_i	y_i	Δy	y_{i+1}	y_i	y_{i+1}
0	0	1	0	1.00	1.0000	0.9900
1	0.1	1.00	−0.02	0.98	0.9900	0.9608
2	0.2	0.98	−0.04	0.94	0.9608	0.9139
3	0.3	0.94	−0.06	0.88	0.9139	0.8521
4	0.4	0.88	−0.07	0.81	0.8521	0.7788
5	0.5	0.81	−0.08	0.73	0.7788	0.6977

Using the value $\Delta x = 0.1$ we find that the result for $x = 0.5$ differs by 5% from the exact one. A reduction of Δx improves the result considerably: $\Delta x = 0.01$ reduces the error to 0.4%. In this way we are able to solve first-order differential equations.

In order to solve a differential equation of higher order we transform it into first-order differential equations. As an example we consider the equation

$$\frac{d^2y}{dx^2} = -y$$

with the initial conditions

$$y = 0 \quad \text{and} \quad \frac{dy}{dx} = 1 \quad \text{for}\, x = 0$$

and the exact solution $y = \sin(x)$.

With the substitution

$$z = \frac{dy}{dx}$$

we obtain a system of two coupled first-order differential equations.

$$\frac{dy}{dx} = z$$
$$\frac{dz}{dx} = -y$$

We start the calculation with $y_0 = 0$ and $z_0 = 1$ for $x_0 = 0$. Thus

$$\Delta y = \frac{dy}{dx} \cdot \Delta x = z \cdot \Delta x$$
$$\Delta z = \frac{dz}{dx} \cdot \Delta x = -y \cdot \Delta x$$

and

$$y_1 = y_0 + \Delta y = y_0 + z_0 \cdot \Delta x$$
$$z_1 = z_0 + \Delta z = z_0 - y_0 \cdot \Delta x$$

With $\Delta x = 0.1$ we obtain the following data:

Numerical								Exact			
i	x_i	y_i	z_i	Δy	Δz	y_{i+1}	z_{i+1}	y_i	y_{i+1}	z_i	z_{i+1}
0	0	0	1	0.1	0	0.10	1	0	0.0998	1	0.9950
1	0.1	0.10	1.0	0.1	−0.01	0.20	0.99	0.0998	0.1987	0.9950	0.9891
2	0.2	0.20	0.99	0.098	−0.02	0.30	0.0.97	0.1987	0.2955	0.9891	0.9553
3	0.3	0.30	0.97	0.096	−0.03	0.40	0.0.94	0.2955	0.3894	0.9553	0.9211
4	0.4	0.40	0.94	0.093	−0.04	0.49	0.90	0.3894	0.4794	0.9211	0.8776
5	0.5	0.49	0.90	0.089	−0.05	0.58	0.85	0.4794	0.5646	0.8776	0.8253

The numerical result at $x = 0.5$ differs by 2% for y and 3% for z from the exact solution.

The Euler–Cauchy method can be significantly refined by including more intermediate steps in the calculation of Δy and Δz.

Runge–Kutta Method

Mostly used and implemented in computer programs is the Runge–Kutta method. In the Euler–Cauchy method we calculated y_1 from y_0 in the case of the differential Equation (C.5) by

$$y_1 = y_0 + (\mathrm{d}y/\mathrm{d}x)_{x_0}\Delta x = y_0 - 2x_0y_0\Delta x$$

In the Runge–Kutta method we use different slopes at $x = x_0$, $x = x_1$, and $x = x_2$, where $x_1 = x_0 + \Delta x/2$ and $x_2 = x_0 + \Delta x$. Then we have

$$\left(\frac{\mathrm{d}y}{\mathrm{d}x}\right)_1 = -2x_0y_0 \text{ and } y_1 = y_0 - 2x_0y_0\Delta x/2.0$$

$$\left(\frac{\mathrm{d}y}{\mathrm{d}x}\right)_2 = -2x_1y_1 \text{ and } y_2 = y_0 - 2x_1y_1\Delta x/2.0$$

$$\left(\frac{\mathrm{d}y}{\mathrm{d}x}\right)_3 = -2x_1y_2 \text{ and } y_3 = y_0 - 2x_1y_2\Delta x$$

$$\left(\frac{\mathrm{d}y}{\mathrm{d}x}\right)_4 = -2x_2y_3 \text{ and } y_4 = y_0 - 2x_1y_3\Delta x$$

and the final value y_1 for $x_1 = x_0 + \Delta x$ is calculated as

$$y_1 = y_0 + \frac{1}{6}\left[\left(\frac{\mathrm{d}y}{\mathrm{d}x}\right)_1 + 2\left(\frac{\mathrm{d}y}{\mathrm{d}x}\right)_2 + 2\left(\frac{\mathrm{d}y}{\mathrm{d}x}\right)_3 + \left(\frac{\mathrm{d}y}{\mathrm{d}x}\right)_4\right]\Delta x$$

With the initial values $x_0 = 0$ and $y_0 = 1$ we obtain

Numerical		Exact
x_i	y_i	y_i
0	1.000000	1.000000
0.1	0.990049	0.990049
0.2	0.960789	0.960789
0.3	0.913931	0.913931
0.4	0.852139	0.852143
0.5	0.778782	0.778800
1.0	0.367488	0.367879
1.5	0.104589	0.105399
2.0	0.017763	0.018315
2.5	0.001765	0.001930

The numerical result agrees with the exact one almost completely up to $x = 0.5$, within 0.1% at $x = 1.0$ and within 1% at $x = 1.5$, whereas the error in the Euler–Cauchy method was already 5% at $x = 0.5$ with the same value $\Delta x = 0.1$. This demonstrates the great power of the Runge–Kutta method.

Gear Method

The Euler–Cauchy and the Runge–Kutta methods use a fixed value of Δx. In many situations this is not a disadvantage. However, there are critical situations which cannot be handled in this way. For example, in reaction kinetics we often have a system of coupled reactions with a wide range of rate constants. Such systems are called "stiff." In order to follow the fast reactions we have to use a very small value of Δx. If we use this small value throughout the whole range of reaction time, then an extremely high number of integration steps and an extremely large computing time is needed. Therefore it is much better to use an integration method where the integration steps are modified throughout the calculation: small steps if the concentrations in the system are changing fast and large steps if the concentration changes are small. One of the best methods with this property is the Gear method [5] (predictor–corrector method).

As an example, consider a system of two coupled autocatalytic reactions with three compounds A, B, and C, which is a strongly nonlinear problem.

$$\text{Reaction 1:} \quad \mathrm{A} \xrightarrow{k_1} 2\,\mathrm{A} \qquad k_1 = 1\ \mathrm{s^{-1}} \tag{C.6}$$

$$\text{Reaction 2:} \quad \mathrm{A} + \mathrm{B} \xrightarrow{k_2} 2\mathrm{B} \qquad k_2 = 0.01\ \mathrm{s^{-1}}$$

$$\text{Reaction 3:} \quad \mathrm{B} \xrightarrow{k_3} \mathrm{C} \qquad k_3 = 1\ \mathrm{s^{-1}}$$

The differential equations are

$$\frac{\mathrm{d}N_\mathrm{A}}{\mathrm{d}t} = k_1 \cdot N_\mathrm{A} - k_2 \cdot N_\mathrm{A} \cdot N_\mathrm{B} \tag{C.7}$$

$$\frac{\mathrm{d}N_\mathrm{B}}{\mathrm{d}t} = k_2 \cdot N_\mathrm{A} \cdot N_\mathrm{B} - k_3 \cdot N_\mathrm{B}$$

$$\frac{\mathrm{d}N_\mathrm{C}}{\mathrm{d}t} = k_3 \cdot N_\mathrm{B}$$

where N_A, N_B, and N_C are the numbers of the reactants. Figure C.2 shows the time dependence that is found for the populations of B and C. The reaction starts with an autocatalytic formation of reactant A. As soon as the number N_A has reached a critical value, reactant B reacts autocatalytically as well. As soon as the number N_B has reached a critical value, the decay of A by reaction with B is faster than its formation, and eventually N_A becomes very small. As a consequence the formation of B by reaction with A slows down and N_B becomes very small as well because of its reaction to form the final product C. After some time, the population N_B is so small that the reactant A increases autocatalytically and the whole process is repeated. During the calculation using the Gear method it turns out that 187 integration steps are taken in the time interval from 1 to 4 s (fast change of N), but only 40 steps are taken in the time interval from 5 to 9 s (slow change of N).

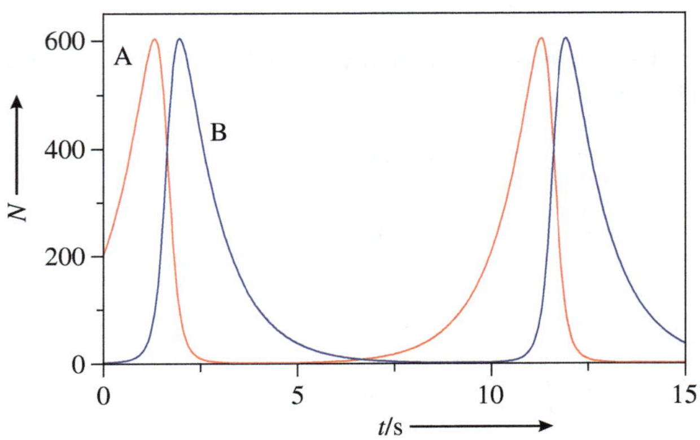

Figure C.2 Time dependence of the numbers N of the reactants A (red solid line) and B (blue solid line) in the nonlinear reaction system 6. The initial concentrations are: $N_{A,0} = 200$, $N_{B,0} = 2$, $N_{C,0} = 0.0$; and the rate constants are $k_1 = 1.0\ \mathrm{s^{-1}}$, $k_2 = 0.01\ \mathrm{s^{-1}}$, and $k_3 = 1\ \mathrm{s^{-1}}$. The differential equations (C.7) were solved by using the Gear method.

Boundary Values Problems

To illustrate the solution of boundary value problems, let us consider the example of a standing wave in a one-dimensional box with box length $L = 1000$ pm, for which we know the exact analytical solution. As shown in Chapter 3 of the text the wavefunctions and energies of an electron in such a box are found by solving the one-dimensional Schrödinger equation

$$-\frac{h^2}{8\pi^2 m_e}\frac{\mathrm{d}^2\psi}{\mathrm{d}x^2} = E\psi \tag{C.8}$$

The analytical result for the wave function is

$$\psi = A\sin(kx) + B\cos(kx)$$

which has three unknown constants: A, B, and k. By applying the boundary condition at $x = 0$

$$\psi = 0 \quad \text{for} \quad x = 0$$

we find that $B = 0$; and applying the boundary condition at $x = L$

$$\psi = 0 \quad \text{for} \quad x = L$$

we find that

$$A \sin(kL) = 0$$

which gives a nontrivial solution for particular values of k,

$$kL = n\pi \qquad \text{or} \qquad k = \frac{n\pi}{L} \qquad \text{where} \qquad n = 1, 2, 3, \ldots$$

The constant A is determined by applying the normalization condition: $A = \sqrt{2/L} = 0.0447\ \text{pm}^{-1/2}$. Thus the final analytical solution for ψ and for the energy E in the lowest quantum state ($n = 1$) is

$$-\psi = A \sin\left(\frac{\pi}{L}x\right) \qquad \text{and } E = \frac{h^2}{8m_e L^2} = 0.603 \times 10^{-19}\ \text{J}. \tag{C.9}$$

How can we solve the Schrödinger Equation (C.8) by numerical integration? As an example we use $L = 2000$ pm and start the solution with the boundary condition $\psi = 0$ at $x = 0$. Note that the Schrödinger equation is a second-order differential equation and we need an additional condition for $d\psi/dx$. Arbitrarily we choose $d\psi/dx = 0.1$ for $x = 0$. This choice does not affect the energy of the system, it only determines the amplitude of the standing wave, as the constant A does in Equation (C.9).

At this stage of the analysis, the energy E of the system is an unknown parameter. Therefore we start with a guess for E and use it to calculate the wavefunction ψ. We compare the value of ψ to the boundary condition $\psi = 0$ at $x = L$, and make a new guess for E to improve the agreement between the calculated value of ψ at $x = L$, and the requirement that $\psi = 0$ at $x = L$. By proceeding iteratively, we can improve the agreement with the boundary condition and make our solution converge arbitrarily close to the exact result. Figure C.3 shows the outcome of this calculation. The guess 1) $E = 0.8 \times 10^{-19}$ J results in too small of a wavelength (i.e., $\psi \to 0$ at $x < L$), whereas the guess 5) $E = 0.4 \times 10^{-19}$ J results in too large of a wavelength (i.e., ψ does not go to zero until a value of $x > L$). Thus we know that the correct energy must be between these two values. Therefore we try with a value of $E = 0.7 \times 10^{-19}$ J, case 2), the wavelength is too short but closer to the correct value, and for a value of $E = 0.5 \times 10^{-19}$ J, case 4) it is again too long but closer to the right value. By repeating the procedure we can refine the result, case 3) is for $E = 0.603 \times 10^{-19}$ J, which gives the correct behavior at the right boundary.

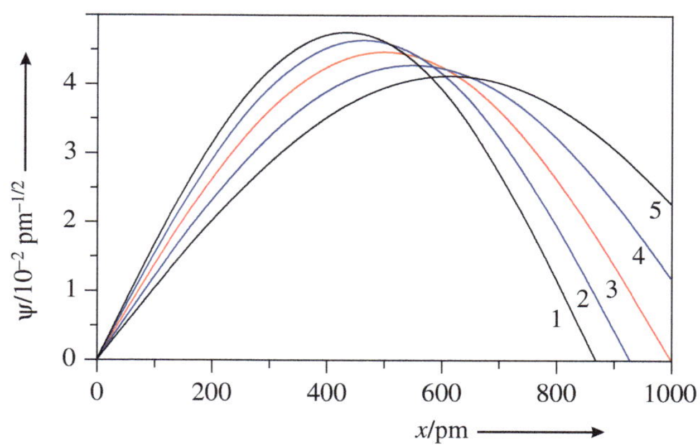

Figure C.3 Numerical solution of differential Equation (C.8) with $L = 1000$ pm. Normalized wavefunction ψ versus x. Initial value $\psi = 0$ for $x = 0$, boundary condition $\psi = 0$ for $x = L$. Values for the parameter E: 1) $E = 0.8 \times 10^{-19}$ J, 2) $E = 0.7 \times 10^{-19}$ J, 3) $E = 0.603 \times 10^{-19}$ J, 4) $E = 0.5 \times 10^{-19}$ J, 5) $E = 0.4 \times 10^{-19}$ J.

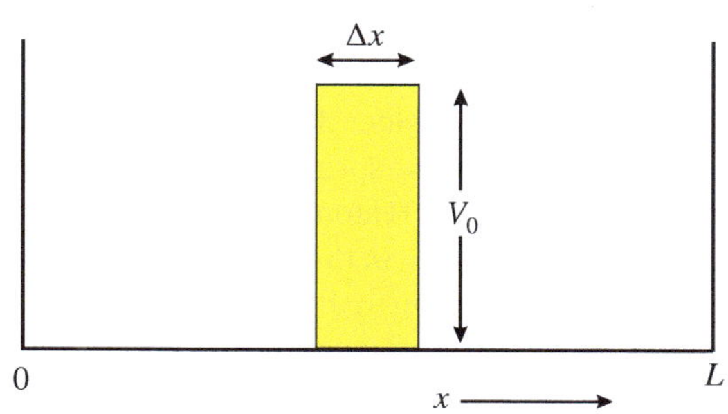

Figure C.4 One-dimensional box with walls of infinite height and a barrier with height V_0 and width Δx. The box length is L.

How does the wavefunction and the energy change if we introduce a rectangular barrier of height V_0 and width Δx in the center of the box (Fig. C.4)? We find the wavefunctions and energies of an electron in this box by solving the one-dimensional Schrödinger equation

$$-\frac{h^2}{8\pi^2 m_e}\frac{d^2\psi}{dx^2} + V\psi = E\psi \tag{C.10}$$

where $V = 0$ outside the barrier and $V = V_0$ inside the barrier:

$$V = 0 \qquad \text{for} \quad 0 < x < \frac{L - \Delta x}{2} \quad \text{and} \quad L > x > \frac{L + \Delta x}{2} \tag{C.11}$$
$$V = V_0 \qquad \text{for} \quad \frac{L + \Delta x}{2} > x > \frac{L - \Delta x}{2}$$

We imagine that an electron is confined inside the box. For a box length of $L = 1000$ pm the energy of the electron in its ground state in the absence of the barrier ($V_0 = 0$) is

$$E_0 = \frac{h^2}{8m_eL^2} = 0.603 \times 10^{-19}\ \text{J}$$

and its corresponding de Broglie wavelength is $\Lambda = 2L = 2000$ pm. We solve the Schrödinger equation for $V_0 = 8 \times 10^{-19}$ J and $\Delta x = 125$ pm using the Gear method with the initial conditions $\psi = 0$ and $d\psi/dx = 0.1$ for $x = 0$ and the boundary condition $\psi = 0$ for $x = L$. To make an informed initial guess for the energy, we consider that the box splits into two smaller boxes with box lengths of 437.5 pm for $V_0 = \infty$, resulting in an energy of $E_0 \cdot (1000/437.5)^2 = 3.15 \times 10^{-19}$ J. Thus, we guess that the energy must be greater than the energy 0.603×10^{-19} J for the case of no barrier but smaller than the value 3.15×10^{-19} J for an infinitely high barrier.

Therefore it seems reasonable to start the calculation with a first guess $E = 2.0 \times 10^{-19}$ J and to refine the result (Fig. C.5) by requiring to satisfy the boundary condition. The guess 1) $E = 2.0 \times 10^{-19}$ J results in too small of a wavelength, whereas the guess 5) $E = 1.7 \times 10^{-19}$ J results in too large of a wavelength. Thus we know that the correct energy must be between these two cases. Therefore, with 2) we try with a larger value of $E = 1.85 \times 10^{-19}$ J and in 4) with a somewhat smaller value of $E = 1.75 \times 10^{-19}$ J. This looks already better, but the procedure has to be refined until 3) $E = 1.7975 \times 10^{-19}$ J gives the correct behavior at the right boundary. Note that also the correct symmetry of the wavefunction is reached with this last value.

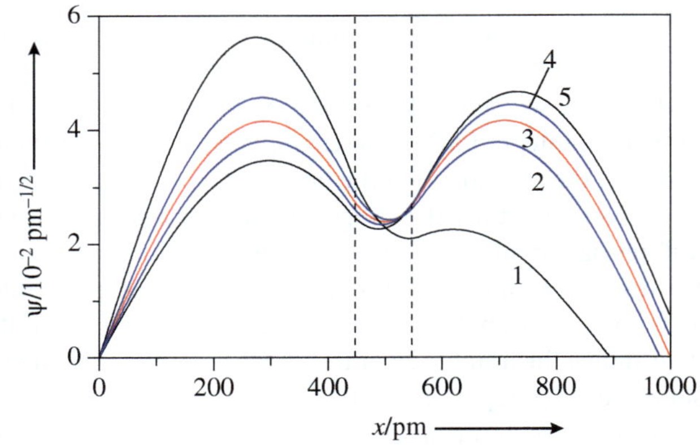

Figure C.5 Numerical solution of differential Equation (C.10) with $V_0 = 8.0 \times 10^{-19}$ J and $L = 1000$ pm. Normalized wavefunction ψ versus x. Initial value $\psi = 0$ for $x = 0$, boundary condition $\psi = 0$ for $x = L$. Values for the parameter E: 1) $E = 2.0 \times 10^{-19}$ J, 2) $E = 1.85 \times 10^{-19}$ J, 3) $E = 1.7975 \times 10^{-19}$ J, 4) $E = 1.75 \times 10^{-19}$ J, 5) $E = 1.7 \times 10^{-19}$ J. The dashed lines indicate the position of the barrier.

Up to this stage we have considered the case in which the boundary on the right-hand side has a finite value. What happens if there is no such boundary? An example is the Schrödinger equation for the harmonic oscillator.

$$-\frac{h^2}{8\pi^2 m_H}\frac{d^2\psi}{dx^2} + V\psi = E\psi \tag{C.12}$$

with

$$V = \frac{1}{2}k_f x^2 \tag{C.13}$$

To be specific, we consider the case where $k_f = 258\ \text{Nm}^{-1}$ is the force constant of the spring and $m_H = 1.673 \times 10^{-27}$ kg is the mass of a vibrating H atom. The potential energy function is symmetric about $x = 0$; thus, we expect the same symmetry for the wavefunction in the lowest energy state. This leads to the initial condition $\psi = 1.0$ and $d\psi/dx = 0$ for $x = 0$. For $x \to \infty$ the wavefunction must vanish; thus, the boundary condition is $\psi = 0$ for $x \to \infty$. Fig. C.6 shows the result of the calculation. The guess 1) $E = 3.0 \times 10^{-20}$ J results in too small of a wavelength, and the guess 5) $E = 2.9 \times 10^{-20}$ J results in too large of a wavelength. Thus we know that the correct energy must lie between these two cases. Therefore we try with a larger value 2) $E = 2.94 \times 10^{-20}$ J and a smaller value 4) $E = 2.92 \times 10^{-20}$ J. This looks already better, but the procedure has to be refined until 3) $E = 2.9285 \times 10^{-19}$ J

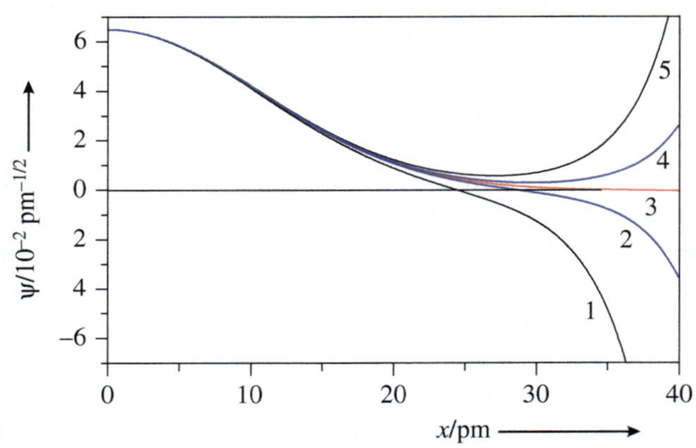

Figure C.6 Numerical solution of differential Equation (C.12) with $V = \frac{1}{2}k_f x^2 = 258\ \text{N m}^{-1} \cdot x^2$. Normalized wavefunction ψ versus elongation x. Initial value $\psi = 0.10$ for $x = 0$, boundary condition $\psi = 0$ for $x \to \infty$. Values for the parameter E: 1) $E = 3.0 \times 10^{-20}$ J (black), 2) $E = 2.94 \times 10^{-20}$ J (blue), 3) $E = 2.9285 \times 10^{-20}$ J (red), 4) $E = 2.92 \times 10^{-20}$ J (blue), 5) $E = 2.90 \times 10^{-20}$ J (black).

gives the correct behavior at the right boundary. This value is identical with the result of the analytical solution

$$E = \frac{1}{2}h\nu_0 = \frac{1}{2}h\sqrt{\frac{k_f}{m_H}} = 2.9283 \times 10^{-19}\ \text{J}$$

C.2.3 Differential Equation of the Vibrating String

We approximate the string by point masses (mass Δm) connected by a very thin wire of length L (Fig. C.7). The ends of the wire are fixed in such a way that a constant force f_0 is acting on the wire in the x -direction. If a mass point at position x is deflected by an elongation $A(x, t)$ perpendicular to the x-direction, a restoring force f_A acts on the mass point. We search for the elongation $A(x, t)$ at time t, after an initial elongation at time $t = 0$. For the calculation of $A(x, t)$ we consider the elongations at three adjacent mass points, namely, A_{n-1}, A_n, and A_{n+1} at the mass points $(x - \Delta x)$, x, and $(x + \Delta x)$, see Fig. C.8

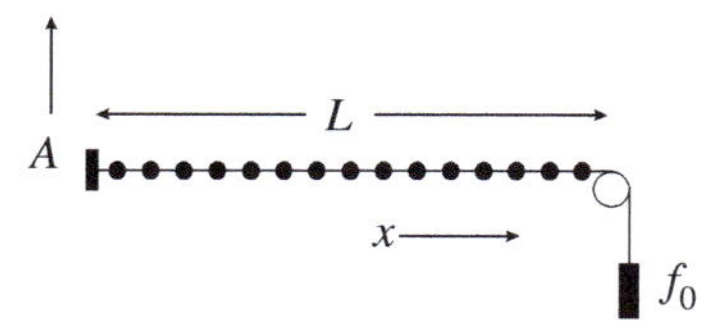

Figure C.7 A string is approximated by point masses (mass Δm) connected by a very thin wire of length L. A constant force f_0 (realized by a weight) acts on the wire in the x-direction. $A(x, t)$ is the elongation of a point mass perpendicular to the x-direction at the position x and at time t.

For sufficiently small elongations the force f_A acting on the mass point n in the direction of A is

$$f_A = -f_0 \cdot \sin\alpha_1 + f_0 \cdot \sin\alpha_2 \approx -f_0 \cdot \tan\alpha_1 + f_0 \cdot \tan\alpha_2$$

where $\tan\alpha_1$ and $\tan\alpha_2$ are the slopes of the string

$$\tan\alpha_1 = \frac{A_n - A_{n-1}}{\Delta x} \quad \text{and} \quad \tan\alpha_2 = \frac{A_{n+1} - A_n}{\Delta x}$$

$$f_A = f_0\left[-\frac{A_n - A_{n-1}}{\Delta x} + \frac{A_{n+1} - A_n}{\Delta x}\right] = \frac{f_0}{\Delta x}\left[A_{n+1} - 2A_n + A_{n-1}\right]$$

The restoring force f_A in the direction of $-A$ at the location x leads to an acceleration of mass point n, and we describe it using Newton's law

$$\Delta m\frac{\partial^2 A}{\partial t^2} = f_A$$

or

$$\frac{\partial^2 A}{\partial t^2} = \frac{f_0}{\Delta m \cdot \Delta x}\left[A_{n+1} - 2A_n + A_{n-1}\right]$$

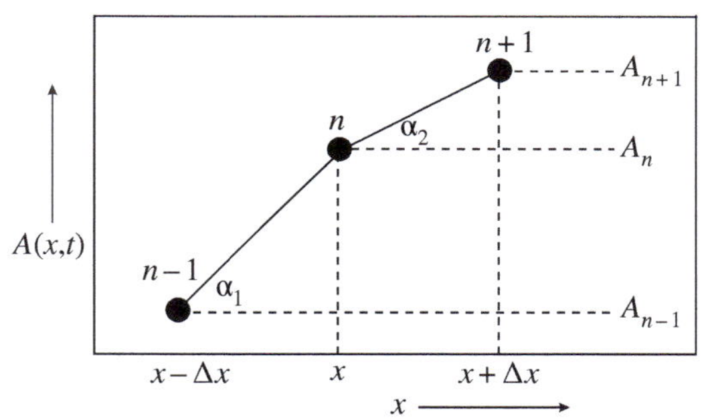

Figure C.8 Elongation $A(x, t)$ perpendicular to the x-direction at a fixed time t for three adjacent mass points at the locations x, $x - \Delta x$, and $x + \Delta x$.

If we express the elongations by a Taylor series expansion:

$$A_{n+1} = A_{x+\Delta x} = A_x + \frac{\partial A}{\partial x}\Delta x + \frac{1}{2}\frac{\partial^2 A}{\partial x^2}\Delta x^2$$

$$A_{n-1} = A_{x-\Delta x} = A_x - \frac{\partial A}{\partial x}\Delta x + \frac{1}{2}\frac{\partial^2 A}{\partial x^2}\Delta x^2$$

we obtain the result

$$A_{n+1} - 2A_n + A_{n-1} = \frac{\partial^2 A}{\partial x^2}\Delta x^2$$

If we substitute this result into the expression for the acceleration, we find that

$$\frac{\partial^2 A}{\partial t^2} = \text{factor}\frac{\partial^2 A}{\partial x^2} \quad \text{with} \quad \text{factor} = f_0\frac{\Delta x}{\Delta m} \tag{C.14}$$

This is a partial differential equation for the elongation $A(x, t)$ of the string. In the following we will solve this equation by the method of **separation of variables**.

We start by assuming that the function $A(t, x)$ can be represented by a product of functions $B(t)$ and $C(x)$, each depending on t or on x

$$A(t, x) = B(t) \cdot C(x)$$

and introduce this product into Equation (C.14). The partial derivatives of this product function are

$$\frac{\partial^2 A}{\partial t^2} = \frac{d^2 B(t)}{dt^2} \cdot C(x) \quad \text{and} \quad \frac{\partial^2 A}{\partial x^2} = \frac{d^2 C}{dx^2} \cdot B(t)$$

Thus we can write

$$\frac{d^2B(t)}{dt^2} \cdot C(x) = \text{factor} \cdot \frac{d^2C}{dx^2} \cdot B(t)$$

and it rearranges to give

$$\frac{d^2B(t)/dt^2}{B(t)} = \text{factor} \cdot \frac{d^2C/dx^2}{C(x)} = const$$

We recognize that the left-hand side of this equation depends on t only and the right-hand side depends on x only. Note that if we change the variable t on the left-hand side, this does not affect the right-hand side and vice versa. This means that both sides must equal a common constant, *const*. Thus we obtain two independent ordinary differential equations.

$$\frac{d^2B(t)}{dt^2} = const \cdot B(t) \qquad \text{and} \qquad \frac{d^2C(x)}{dx^2} = \frac{const}{\text{factor}} \cdot C(x) \tag{C.15}$$

From our daily experience with a vibrating string we expect that the string's elongation B changes periodically with time. Therefore we try as a solution of the differential equation for B

$$B = B_0 \cdot \sin(\omega t)$$

where ω is the angular frequency. Insertion of this trial function in the differential equation for B gives

$$-B_0 \cdot \omega^2 \cdot \sin(\omega t) = const \cdot B_0 \cdot \sin(\omega t)$$

and

$$const = -\omega^2$$

This implies that *const* must be a negative quantity.

Now we turn to the differential equation for C using the trial function

$$C = C_0 \cdot \sin(\pi \cdot \frac{nx}{L})$$

where $n > 0$ is an integer and L is the length of the string (see Fig. C.7). This means that the sine function is zero for $x = 0$ and for $x = L$ for arbitrary values of n. Insertion of this trial function in the differential equation gives

$$-C_0 \cdot \pi^2 \frac{n^2}{L^2} \cdot \sin\left(\pi \cdot \frac{nx}{L}\right) = \frac{const}{\text{factor}} \cdot C_0 \cdot \sin\left(\pi \cdot \frac{nx}{L}\right)$$

and

$$const = -\text{factor} \cdot \pi^2 \frac{n^2}{L^2} = -f_0 \frac{\Delta x}{\Delta m} \cdot \pi^2 \frac{n^2}{L^2} = -\omega^2 = -4\pi^2 \upsilon^2$$

Thus for $A(t, x)$ we obtain the equation for a standing wave

$$A(t, x) = B_0 \cdot C_0 \cdot \sin\left(\pi \cdot \frac{nx}{L}\right) \cdot \sin(\omega t)$$

with amplitude $B_0 \cdot C_0 \cdot \sin(\pi \cdot \frac{nx}{L})$ and wavelength $\Lambda = 2L/n$. For the product $\upsilon\Lambda$ we obtain

$$\nu\Lambda = \sqrt{f_0 \frac{\Delta x}{\Delta m}} = v$$

where v denotes the wave's velocity. The velocity increases with increasing force f_0 on the string and it decreases with increasing mass density $\Delta m/\Delta x$ of the string.

C.3 Stirling's Formula: $\ln N! = N \cdot \ln N - N$

Calculating $N!$ in the case when N is a large number is a difficult problem. However, in most cases it is appropriate to evaluate $N!$ by using a simple approximation (Stirling's formula).

Beginning with

$$N! = 1 \cdot 2 \cdot 3 \cdot \cdots \cdot N$$

we can take the logarithm and obtain

$$\ln N! = \ln 1 + \ln 2 + \cdots + \ln N = \sum_{i=1}^{N} \ln i$$

If we approximate the summation by an integral, we find that

$$\ln N! = \int_0^N \ln i \cdot \mathrm{d}i = |i \ln i - i|_0^N = N \ln N - N$$

By exponentiating the result we obtain

$$N! = \mathrm{e}^{N \ln N - N} = \left(\mathrm{e}^{\ln N}\right)^N \cdot \mathrm{e}^{-N} = N^N \cdot \mathrm{e}^{-N}$$

These are two equivalent forms of Stirling's formula.

A more sophisticated form of Stirling's formula is

$$\ln N! = N \cdot \ln N - N + \tfrac{1}{2} \cdot \ln(2\pi N), \qquad N! = \sqrt{2\pi N} \cdot N^N \cdot \mathrm{e}^{-N}$$

Example C.2 *Calculate N! for different values of N*

N	$N!$	$\ln N!$	$N^N \cdot e^{-N}$	$\sqrt{2\pi N} \cdot N^N \cdot e^{-N}$	$N \ln N - N$	$N \ln N - N + \frac{1}{2} \cdot \ln(2\pi N)$
10	3.36×10^6	15.1	4.54×10^5	3.60×10^6	13.0	15.1
60	8.32×10^{81}	189	4.45×10^{80}	8.64×10^{81}	186	189

An even more sophisticated formula for $N!$ and $\ln N!$ is

$$N! = \left(\frac{N}{\mathrm{e}}\right)^N \sqrt{2\pi n} \cdot \left(1 + \frac{1}{12N} + \frac{1}{288N^2} + \cdots\right)$$

$$\ln(N!) = \left(N + \frac{1}{2}\right) \ln N - N + \ln \sqrt{2\pi}$$

C.4 Operators

What is an operator? An operator is a rule by which, given a function, we can generate a corresponding function. Consider an operator $\mathcal{A}$ that acts on a function $f(x)$ to create a new function $g(x)$. We would write

$$\mathcal{A}f(x) = g(x)$$

By way of example we could define the operator $\mathcal{A}$ as $2\times$, so that for a function $f(x) = 3x$ we find

$$\mathcal{A}\left[f(x)\right] = 2 \times [3x] = 6x;$$

that is $g(x) = 6x$. In general we will use the Calligraphic font to indicate an operator. A number of textbooks that develop properties of operators and their use in Physical Chemistry are available, e.g., Mortimer [6] or Barrante [7]. Below we outline some essential aspects of operators and operator algebra.

Consider the two operators $\mathcal{A}$ and $\mathcal{B}$, and consider the function $f(x)$. To begin, we define a sum of operators by

$$(\mathcal{A} + \mathcal{B})f(x) = \mathcal{A}f(x) + \mathcal{B}f(x)$$

and we define a product of operators as

$$\mathcal{A}\mathcal{B}f(x) = \mathcal{A}(\mathcal{B}f(x))$$

It is important to understand that an operator needs to have something to operate on; i.e., a function. When performing manipulations on operators, it is strongly encouraged that you use a function (for the general case use $f(x)$). This procedure will dramatically decrease the number of errors that can occur.

Example: Let $\mathcal{D} \equiv \frac{\mathrm{d}}{\mathrm{d}x}$ and $\mathcal{T} \equiv x\cdot$ and use them to operate on the function $f(x) = (x^2 + 3\mathrm{e}^x)$.

$$\begin{aligned}(\mathcal{D}+\mathcal{T})(x^2+3\mathrm{e}^x) &= 2x + 3\mathrm{e}^x + x^3 + 3x\mathrm{e}^x \\ &= x\left(2+x^2\right) + 3\mathrm{e}^x\,(x+1)\end{aligned}$$

and

$$\begin{aligned}(\mathcal{D}\mathcal{T})(x^2+3\mathrm{e}^x) &= \mathcal{D}(x^3+3x\mathrm{e}^x) \\ &= 3x^2 + 3\mathrm{e}^x + 3x\mathrm{e}^x \\ &= 3x^2 + 3\mathrm{e}^x\,(1+x)\end{aligned}$$

We note that

$$\begin{aligned}(\mathcal{T}\mathcal{D})(x^2+3\mathrm{e}^x) &= \mathcal{T}(2x+3\mathrm{e}^x) \\ &= 2x^2 + 3x\mathrm{e}^x \\ &= 2x^2 + 3x\mathrm{e}^x\end{aligned}$$

and the order of the two operations on this function gives different results. We say that the operators do not commute.

C.4.1 Properties of Operators and an Operator Algebra

- We define the unit operator $\mathcal{I}$ to be $1\cdot$, so that $\mathcal{I}f(x) = 1\cdot f(x) = f(x)$. This operator corresponds to an identity operator for the multiplication of operators.
- Define the null operator $\mathcal{N}$ to be $0\cdot$, so that $\mathcal{N}f(x) = 0\cdot f(x) = 0$. This operator corresponds to an identity operator for the addition of operators.
- The associative laws of addition and multiplication will hold for operators. In short, we have that

  $$\mathcal{A}(\mathcal{B}\mathcal{N}) = (\mathcal{A}\mathcal{B})\mathcal{N} \qquad \text{and} \qquad (\mathcal{A}+\mathcal{B})+\mathcal{N} = \mathcal{A}+(\mathcal{B}+\mathcal{N})$$

- Addition of operators is commutative; namely

  $$(\mathcal{A}+\mathcal{B})f(x) = (\mathcal{B}+\mathcal{A})f(x)$$

- Multiplication of operators is NOT necessarily commutative; i.e.,

  $$\mathcal{A}\mathcal{B} \neq \mathcal{B}\mathcal{A}$$

  We will define a commutator as $[\mathcal{A},\mathcal{B}]f(x) = \mathcal{A}\mathcal{B} - \mathcal{B}\mathcal{A}$; so that if $[\mathcal{A},\mathcal{B}] = 0$ then $\mathcal{A}$ and $\mathcal{B}$ commute, and if $[\mathcal{A},\mathcal{B}] \neq 0$, then $\mathcal{A}$ and $\mathcal{B}$ do not commute.

- The square of an operator is defined as

  $$\mathcal{A}^2 = \mathcal{A}\mathcal{A}$$

- An operator is linear if

  $$\mathcal{A}\,(f(x)+g(x)) = \mathcal{A}f(x) + \mathcal{A}g(x) \quad \text{and} \quad \mathcal{A}\,(N\cdot f(x)) = N\cdot(\mathcal{A}f(x))$$

  where N is a constant. We note that the operators we define in quantum mechanics are linear, and this property allows us to perform superpositions (i.e., sums) of wavefunctions without changing the mathematical structure of the result of an operation.

C.4.2 Hermitian Operators

An operator is Hermitian if

$$\int_{-\infty}^{\infty} \psi_i^* \mathcal{A}\psi_j \,\mathrm{d}\vec{r} = \int_{-\infty}^{\infty} \psi_i \left(\mathcal{A}\psi_j\right)^* \mathrm{d}\vec{r}$$

where ψ_i and ψ_j are wavefunctions and ψ_i^* and ψ_j^* are their complex conjugates. This condition is important because it means that the value returned by the integral is a real number. Integrals of this sort are used in quantum mechanics to calculate physical properties of a system, such as the energy.

Although the text is careful to keep the notation simple and explicit, you are likely to run into a variety of notations in your external reading. Of particular importance is "Dirac" notation, namely

$$\int_{-\infty}^{\infty} \psi_i^* \mathcal{A} \psi_j \mathrm{d}\vec{r} \equiv \langle \psi_i | \mathcal{A} | \psi_j \rangle \equiv \langle i | \mathcal{A} | j \rangle$$

Hence the previous requirement (Hermitian property) becomes

$$\mathcal{A}_{ij} = \langle i | \mathcal{A} | j \rangle = \langle \psi_i | \mathcal{A} | \psi_j \rangle = \langle \psi_j | \mathcal{A} | \psi_i \rangle = \langle j | \mathcal{A} | i \rangle = \mathcal{A}_{ji}^*$$

In this equation the quantity is given; e.g., A_{ij}, is called a matrix element and it is the value of the integral. Often times, it is a number. The fact that the operator is Hermitian means that $A_{ij} = A_{ji}^*$ and if $j = i$ then we have proved that the matrix element A_{ij} is a real number.

Example: Show that the Hermitian property of a 2×2 matrix operator$\mathcal{A}$ implies that the matrix is equal to its complex conjugate transpose, for a wavefunction Ψ that is a superposition of two basis functions (or eigenfunctions) such that $\Psi = c_1\psi_1 + c_2\psi_2$ where the ψ_i are the basis set functions and the c_i are the coefficients.

Solution: Our approach is to evaluate $\int_{-\infty}^{\infty} \Psi^* \mathcal{A} \Psi \mathrm{d}\vec{r}$ and $\int_{-\infty}^{\infty} \Psi(\mathcal{A}\Psi)^* \mathrm{d}\vec{r}$; and then equate them. First, we find

$$\begin{aligned} \int \Psi^* \mathcal{A} \Psi \, \mathrm{d}\vec{r} &= \int \begin{bmatrix} \psi_1^* & \psi_2^* \end{bmatrix} \begin{bmatrix} \mathcal{A}_{11} & \mathcal{A}_{12} \\ \mathcal{A}_{21} & \mathcal{A}_{22} \end{bmatrix} \begin{bmatrix} \psi_1 \\ \psi_2 \end{bmatrix} \mathrm{d}\tau \\ &= \int \begin{bmatrix} \psi_1^* & \psi_2^* \end{bmatrix} \begin{bmatrix} \mathcal{A}_{11}\psi_1 + \mathcal{A}_{12}\psi_2 \\ \mathcal{A}_{21}\psi_1 + \mathcal{A}_{22}\psi_2 \end{bmatrix} \mathrm{d}\tau \\ &= \int \left(\psi_1^* \mathcal{A}_{11} \psi_1 + \psi_1^* \mathcal{A}_{12} \psi_2 + \psi_2^* \mathcal{A}_{21} \psi_1 + \psi_2^* \mathcal{A}_{22} \psi_2 \right) \mathrm{d}\tau \end{aligned}$$

Next we find $\int_{-\infty}^{\infty} \Psi(\mathcal{A}\Psi)^* \mathrm{d}\vec{r}$

$$\begin{aligned} \int_{-\infty}^{\infty} \Psi(\mathcal{A}\Psi)^* \mathrm{d}\vec{r} &= \int \begin{bmatrix} \psi_1 & \psi_2 \end{bmatrix} \left(\begin{bmatrix} \mathcal{A}_{11} & \mathcal{A}_{12} \\ \mathcal{A}_{21} & \mathcal{A}_{22} \end{bmatrix} \begin{bmatrix} \psi_1 \\ \psi_2 \end{bmatrix} \right)^* \mathrm{d}\tau \\ &= \int \begin{bmatrix} \psi_1 & \psi_2 \end{bmatrix} \left(\begin{bmatrix} \mathcal{A}_{11}\psi_1 + \mathcal{A}_{12}\psi_2 \\ \mathcal{A}_{21}\psi_1 + \mathcal{A}_{22}\psi_2 \end{bmatrix} \right)^* \mathrm{d}\tau \\ &= \int \begin{bmatrix} \psi_1 & \psi_2 \end{bmatrix} \left(\begin{bmatrix} \mathcal{A}_{11}^*\psi_1^* + \mathcal{A}_{12}^*\psi_2^* \\ \mathcal{A}_{21}^*\psi_1^* + \mathcal{A}_{22}^*\psi_2^* \end{bmatrix} \right) \mathrm{d}\tau \\ &= \int \left[\psi_1 \mathcal{A}_{11}^* \psi_1^* + \psi_1 \mathcal{A}_{12}^* \psi_2^* + \psi_2 \mathcal{A}_{21}^* \psi_1^* + \psi_2 \mathcal{A}_{22}^* \psi_2^* \right] \mathrm{d}\tau \end{aligned}$$

Comparison of the two results shows us that

$$\mathcal{A}_{11} = \mathcal{A}_{11}^*; \qquad \mathcal{A}_{12} = \mathcal{A}_{21}^*; \qquad \mathcal{A}_{21} = \mathcal{A}_{12}^*; \qquad \mathcal{A}_{22} = \mathcal{A}_{22}^*$$

Thus we see that a hermitian matrix is equal to its complex conjugate transpose (also called the adjoint of a matrix). In our abbreviated notation, we can write

$$\mathcal{A} = \mathcal{A}^* \quad \text{is} \quad \begin{bmatrix} \mathcal{A}_{11} & \mathcal{A}_{12} \\ \mathcal{A}_{21} & \mathcal{A}_{22} \end{bmatrix} = \begin{bmatrix} \mathcal{A}_{11}^* & \mathcal{A}_{21}^* \\ \mathcal{A}_{12}^* & \mathcal{A}_{22}^* \end{bmatrix}$$

We can extend this example to the case of the product of two matrix operators. In this case, we find that the complex conjugate transpose of the matrix swaps the order of the operators; namely $(\mathcal{AB})^T \to \mathcal{BA}$. Using our abbreviated matrix notation

$$\begin{aligned} \left(\begin{bmatrix} \mathcal{A}_{11} & \mathcal{A}_{12} \\ \mathcal{A}_{21} & \mathcal{A}_{22} \end{bmatrix} \begin{bmatrix} \mathcal{B}_{11} & \mathcal{B}_{12} \\ \mathcal{B}_{21} & \mathcal{B}_{22} \end{bmatrix} \right)^* &= \left(\begin{bmatrix} \mathcal{A}_{11}\mathcal{B}_{11} + \mathcal{A}_{12}\mathcal{B}_{21} & \mathcal{A}_{11}\mathcal{B}_{12} + \mathcal{A}_{12}\mathcal{B}_{22} \\ \mathcal{A}_{21}\mathcal{B}_{11} + \mathcal{A}_{22}\mathcal{B}_{21} & \mathcal{A}_{21}\mathcal{B}_{12} + \mathcal{A}_{22}\mathcal{B}_{22} \end{bmatrix} \right)^* \\ &= \left(\begin{bmatrix} \mathcal{A}_{11}^*\mathcal{B}_{11}^* + \mathcal{A}_{12}^*\mathcal{B}_{21}^* & \mathcal{A}_{21}^*\mathcal{B}_{11}^* + \mathcal{A}_{22}^*\mathcal{B}_{21}^* \\ \mathcal{A}_{11}^*\mathcal{B}_{12}^* + \mathcal{A}_{12}^*\mathcal{B}_{22}^* & \mathcal{A}_{21}^*\mathcal{B}_{12}^* + \mathcal{A}_{22}^*\mathcal{B}_{22}^* \end{bmatrix} \right) \end{aligned}$$

$$= \left(\begin{bmatrix} \mathcal{A}_{11}\mathcal{B}_{11} + \mathcal{A}_{21}\mathcal{B}_{12} & \mathcal{A}_{12}\mathcal{B}_{11} + \mathcal{A}_{22}\mathcal{B}_{12} \\ \mathcal{A}_{11}\mathcal{B}_{21} + \mathcal{A}_{21}\mathcal{B}_{22} & \mathcal{A}_{12}\mathcal{B}_{21} + \mathcal{A}_{22}\mathcal{B}_{22} \end{bmatrix} \right)$$

$$= \begin{bmatrix} \mathcal{B}_{11} & \mathcal{B}_{12} \\ \mathcal{B}_{21} & \mathcal{B}_{22} \end{bmatrix} \begin{bmatrix} \mathcal{A}_{11} & \mathcal{A}_{12} \\ \mathcal{A}_{21} & \mathcal{A}_{22} \end{bmatrix}$$

C.4.3 Eigenvalue Equations

Some of the equations encountered in physical chemistry are called eigenvalue equations; e.g., the Schrödinger equation is an eigenvalue equation. If an operator $\mathcal{A}$ acts on a function $f(x)$ and returns a constant a times the same function $f(x)$, we say that $f(x)$ is an eigenfunction of the operator $\mathcal{A}$ and that the constant a is the eigenvalue. We call the equation

$$\mathcal{A} f(x) = a f(x)$$

an eigenvalue equation. This mathematical structure is especially useful in the quantum mechanical description of atoms and molecules (see Chapters 1 through 12), but it can also be useful in describing coupled rate processes and coupled equilibria.

C.4.4 Laplacian Operator. Conversion in Spherical Coordinates

In Cartesian coordinates (see Fig. C.9), the Laplacian operator is

$$\nabla^2 = \frac{\partial^2}{\partial x^2} + \frac{\partial^2}{\partial y^2} + \frac{\partial^2}{\partial z^2}$$

We transform this operator into spherical coordinates using the relations

$$x = r \cdot \sin\vartheta \cdot \cos\varphi; \quad y = r \cdot \sin\vartheta \cdot \sin\varphi; \quad z = r \cdot \cos\vartheta \tag{C.16}$$

and

$$r = \sqrt{x^2 + y^2 + z^2}\,; \quad \cos\vartheta = \frac{z}{\sqrt{x^2 + y^2 + z^2}}; \tag{C.17}$$

$$\sin\vartheta = \frac{\sqrt{x^2 + y^2}}{\sqrt{x^2 + y^2 + z^2}}; \quad \cos\varphi = \frac{x}{\sqrt{x^2 + y^2}}; \quad \sin\varphi = \frac{y}{\sqrt{x^2 + y^2}}$$

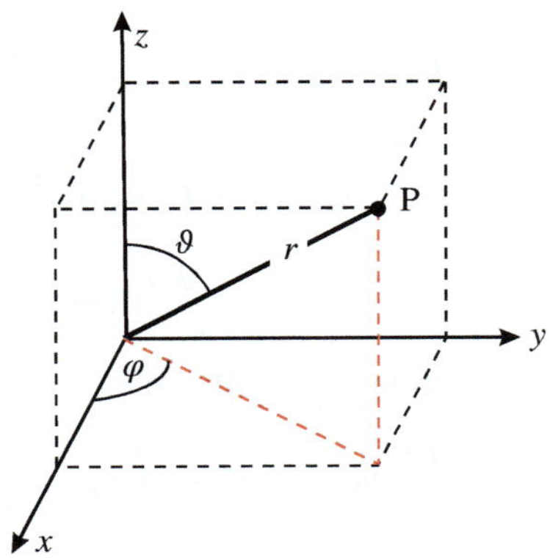

Figure C.9 Coordinates in the Cartesian and the Spherical coordinate system.

Next we calculate

$$\frac{\partial r}{\partial x} = \frac{2x}{2\sqrt{x^2 + y^2 + z^2}} = \frac{x}{r} = \sin\vartheta \cdot \cos\varphi \tag{C.18}$$

$$\frac{\partial r}{\partial y} = \frac{2y}{2\sqrt{x^2 + y^2 + z^2}} = \frac{y}{r} = \sin\vartheta \cdot \sin\varphi$$

$$\frac{\partial r}{\partial z} = \frac{2z}{2\sqrt{x^2 + y^2 + z^2}} = \frac{z}{r} = \cos\vartheta$$

For the differential quotient $\partial\vartheta/\partial x$ with Equations (C.16) and (C.17) we obtain

$$\frac{\partial \cos\vartheta}{\partial x} = \frac{\partial \cos\vartheta}{\partial\vartheta}\frac{\partial\vartheta}{\partial x} = -\sin\vartheta\frac{\partial\vartheta}{\partial x} = \frac{2\ zx}{-2\left(x^2 + y^2 + z^2\right)^{3/2}} = -\frac{zx}{r^3} \tag{C.19}$$

$$\frac{\partial\vartheta}{\partial x} = \frac{zx}{r^3}\frac{1}{\sin\vartheta} = \frac{r\cos\vartheta \cdot r \cdot \sin\vartheta \cdot \cos\varphi}{r^3}\frac{1}{\sin\vartheta} = \frac{\cos\vartheta \cdot \cos\varphi}{r}$$

In analogy, for $\partial\vartheta/\partial y$ and $\partial\vartheta/\partial z$ we obtain

$$\frac{\partial\vartheta}{\partial y} = \frac{\cos\vartheta \cdot \sin\varphi}{r} \quad \text{and} \quad \frac{\partial\vartheta}{\partial z} = -\frac{\sin\vartheta}{r} \tag{C.20}$$

The result for the differential quotient $\partial\varphi/\partial x$ is

$$\frac{\partial \cos\varphi}{\partial x} = \frac{\partial \cos\varphi}{\partial\varphi}\frac{\partial\varphi}{\partial x} = -\sin\varphi\frac{\partial\varphi}{\partial x} = \frac{1}{\sqrt{x^2 + y^2}} - \frac{2\ x^2}{2\sqrt{\left(x^2 + y^2\right)^3}} \tag{C.21}$$

$$\frac{\partial\varphi}{\partial x} = -\frac{1}{\sin\varphi}\left[\frac{1}{\sqrt{x^2+y^2}} - \frac{2x^2}{2\sqrt{(x^2+y^2)^3}}\right]$$
$$= -\frac{1}{\sin\varphi}\left[\frac{1}{r\sin\vartheta} - \left(\frac{x}{\sqrt{(x^2+y^2)}}\right)^2\frac{1}{\sqrt{x^2+y^2}}\right]$$
$$= -\frac{1}{\sin\varphi}\left[\frac{1}{r\sin\vartheta} - \cos^2\varphi\frac{1}{r\sin\vartheta}\right] = -\frac{1}{\sin\varphi}\frac{1-\cos^2\varphi}{r\sin\vartheta}$$
$$= -\frac{1}{\sin\varphi}\frac{\sin^2\varphi}{r\sin\vartheta} = -\frac{\sin\varphi}{r\sin\vartheta}$$

In analogy, for $\partial\varphi/\partial y$ and $\partial\varphi/\partial z$ we obtain

$$\frac{\partial\varphi}{\partial y} = \frac{\cos\varphi}{r\sin\vartheta} \quad \text{and} \quad \frac{\partial\varphi}{\partial z} = 0 \quad (\varphi \text{ does not depend on } z) \tag{C.22}$$

In the following step we calculate the first derivatives using Equations (C.18), (C.19), (C.20), and (C.21):

$$\frac{\partial}{\partial x} = \left[\frac{\partial}{\partial r}\cdot\frac{\partial r}{\partial x} + \frac{\partial}{\partial\vartheta}\cdot\frac{\partial\vartheta}{\partial x} + \frac{\partial}{\partial\varphi}\cdot\frac{\partial\varphi}{\partial x}\right]$$
$$= \left[\sin\vartheta\cdot\cos\varphi\cdot\frac{\partial}{\partial r} + \frac{\cos\vartheta\cdot\cos\varphi}{r}\cdot\frac{\partial}{\partial\vartheta} - \frac{\sin\varphi}{r\sin\vartheta}\cdot\frac{\partial}{\partial\varphi}\right]$$
$$\frac{\partial}{\partial y} = \left[\sin\vartheta\cdot\sin\varphi\cdot\frac{\partial}{\partial r} + \frac{\cos\vartheta\cdot\sin\varphi}{r}\cdot\frac{\partial}{\partial\vartheta} - \frac{\cos\varphi}{r\sin\vartheta}\cdot\frac{\partial}{\partial\varphi}\right]$$
$$\frac{\partial}{\partial z} = \left[\cos\vartheta\cdot\frac{\partial}{\partial r} - \frac{\sin\vartheta}{r}\cdot\frac{\partial}{\partial\vartheta}\right]$$

Next we calculate the second derivatives, again using Equations (C.18), (C.19), (C.20), and (C.21). We start with $\partial^2/\partial z^2$:

$$\frac{\partial^2}{\partial z^2} = \frac{\partial}{\partial z}\left[\frac{\partial}{\partial z}\right] = \frac{\partial}{\partial r}\cdot\cos\vartheta - \frac{\partial}{\partial\vartheta}\cdot\frac{\sin\vartheta}{r}\cdot\left[\frac{\partial}{\partial r}\cdot\cos\vartheta\cdot-\frac{\partial}{\partial\vartheta}\cdot\frac{\sin\vartheta}{r}\cdot\right]$$
$$= \cos\vartheta\cdot\frac{\partial}{\partial r}\left[\frac{\partial}{\partial r}\cdot\cos\vartheta - \frac{\partial}{\partial\vartheta}\cdot\frac{\sin\vartheta}{r}\right] - \frac{\sin\vartheta}{r}\cdot\frac{\partial}{\partial\vartheta}\left[\frac{\partial}{\partial r}\cdot\cos\vartheta - \frac{\partial}{\partial\vartheta}\cdot\frac{\sin\vartheta}{r}\right]$$
$$= \cos\vartheta\cdot\left(\frac{\partial^2}{\partial r^2}\cdot\cos\vartheta + \frac{\partial}{\partial r}\cdot\frac{\partial\cos\vartheta}{\partial r} - \frac{\partial^2}{\partial r\partial\vartheta}\cdot\frac{\sin\vartheta}{r} - \frac{\partial}{\partial\vartheta}\frac{\partial}{\partial r}\frac{\sin\vartheta}{r}\right)$$
$$-\frac{\sin\vartheta}{r}\cdot\left(\frac{\partial^2}{\partial\vartheta\partial r}\cdot\cos\vartheta + \frac{\partial}{\partial r}\frac{\partial}{\partial\vartheta}\cdot\cos\vartheta - \frac{\partial^2}{\partial\vartheta^2}\cdot\frac{\sin\vartheta}{r} - \frac{\partial}{\partial\vartheta}\cdot\frac{\partial}{\partial\vartheta}\frac{\sin\vartheta}{r}\right)$$
$$= \cos\vartheta\cdot\left(\frac{\partial^2}{\partial r^2}\cdot\cos\vartheta - \frac{\partial^2}{\partial r\partial\vartheta}\cdot\frac{\sin\vartheta}{r} + \frac{\partial}{\partial\vartheta}\cdot\frac{\sin\vartheta}{r^2}\right)$$
$$-\frac{\sin\vartheta}{r}\cdot\left(\frac{\partial^2}{\partial\vartheta\partial r}\cdot\cos\vartheta - \frac{\partial}{\partial r}\sin\vartheta - \frac{\partial^2}{\partial\vartheta^2}\cdot\frac{\sin\vartheta}{r} - \frac{\partial}{\partial\vartheta}\cdot\frac{\cos\vartheta}{r}\right)$$
$$= \cos^2\vartheta\frac{\partial^2}{\partial r^2} - \frac{\cos\vartheta\sin\vartheta}{r}\frac{\partial^2}{\partial r\partial\vartheta} + \frac{\cos\vartheta\sin\vartheta}{r^2}\frac{\partial}{\partial\vartheta}$$
$$-\frac{\sin\vartheta\cos\vartheta}{r}\frac{\partial^2}{\partial\vartheta\partial r} + \frac{\sin^2\vartheta}{r}\frac{\partial}{\partial r} + \frac{\sin^2\vartheta}{r^2}\frac{\partial^2}{\partial\vartheta^2} + \frac{\sin\vartheta\cos\vartheta}{r^2}\frac{\partial}{\partial\vartheta}$$
$$= \cos^2\vartheta\frac{\partial^2}{\partial r^2} - \frac{2\cos\vartheta\sin\vartheta}{r}\frac{\partial^2}{\partial r\partial\vartheta} + \frac{2\cos\vartheta\sin\vartheta}{r^2}\frac{\partial}{\partial\vartheta} + \frac{\sin^2\vartheta}{r^2}\frac{\partial^2}{\partial\vartheta^2} + \frac{\sin^2\vartheta}{r}\frac{\partial}{\partial r}$$

The derivative $\partial^2/\partial x^2$ follows in a corresponding way.[1]

$$\frac{\partial^2}{\partial x^2} = \frac{\partial}{\partial x}\left[\frac{\partial}{\partial x}\right]$$
$$= \sin^2\vartheta\cos^2\varphi\frac{\partial^2}{\partial r^2} + \frac{\cos^2\vartheta\cos^2\varphi}{r^2}\frac{\partial^2}{\partial\vartheta^2} + \frac{\sin^2\varphi}{r^2\sin^2\vartheta}\frac{\partial^2}{\partial\varphi^2}$$

1 https://qudev.phys.ethz.ch/static/content/science/BuchPhysikIV/PhysikIVap6.html

$$
\begin{aligned}
&+\left[\frac{\cos^2\vartheta\cos^2\varphi}{r}+\frac{\sin^2\varphi}{r}\right]\frac{\partial}{\partial r}+\left[\frac{\cos\vartheta\sin^2\varphi}{r^2\sin\vartheta}-\frac{2\sin\vartheta\cos\vartheta\cos^2\varphi}{r^2}\right]\frac{\partial}{\partial\vartheta}\\
&+\left[\frac{\cos\varphi\sin\varphi}{r^2}+\frac{\cos^2\vartheta\cos\varphi\sin\varphi}{r^2\sin^2\vartheta}+\frac{\cos\varphi\sin\varphi}{r^2\sin^2\vartheta}\right]\frac{\partial}{\partial\varphi}\\
&+\frac{2\sin\vartheta\cos\vartheta\cos^2\varphi}{r}\frac{\partial^2}{\partial r\partial\vartheta}-\frac{2\sin\varphi\cos\varphi}{r}\frac{\partial^2}{\partial r\partial\varphi}\\
&-\frac{2\cos\vartheta\cos\varphi\sin\varphi}{r^2\sin\vartheta}\frac{\partial^2}{\partial\vartheta\partial\varphi}
\end{aligned}
$$

The derivative for $\partial^2/\partial y^2$ follows from that for $\partial^2/\partial x^2$ essentially by exchanging the $\sin\varphi$ terms by $\cos\varphi$ terms and vice versa:

$$
\begin{aligned}
\frac{\partial^2}{\partial y^2}&=\frac{\partial}{\partial y}\left[\frac{\partial}{\partial y}\right]\\
&=\sin^2\vartheta\sin^2\varphi\frac{\partial^2}{\partial r^2}+\frac{\cos^2\vartheta\sin^2\varphi}{r^2}\frac{\partial^2}{\partial\vartheta^2}+\frac{\cos^2\varphi}{r^2\sin^2\vartheta}\frac{\partial^2}{\partial\varphi^2}\\
&+\left[\frac{\cos^2\vartheta\sin^2\varphi}{r}+\frac{\cos^2\varphi}{r}\right]\frac{\partial}{\partial r}+\left[\frac{\cos\vartheta\cos^2\varphi}{r^2\sin\vartheta}-\frac{2\sin\vartheta\cos\vartheta\sin^2\varphi}{r^2}\right]\frac{\partial}{\partial\vartheta}\\
&-\left[\frac{\cos\varphi\sin\varphi}{r^2}+\frac{\cos^2\vartheta\cos\varphi\sin\varphi}{r^2\sin^2\vartheta}+\frac{\cos\varphi\sin\varphi}{r^2\sin^2\vartheta}\right]\frac{\partial}{\partial\varphi}\\
&+\frac{2\sin\vartheta\cos\vartheta\sin^2\varphi}{r}\frac{\partial^2}{\partial r\partial\vartheta}+\frac{2\sin\varphi\cos\varphi}{r}\frac{\partial^2}{\partial r\partial\varphi}\\
&+\frac{2\cos\vartheta\cos\varphi\sin\varphi}{r^2\sin\vartheta}\frac{\partial^2}{\partial\vartheta\partial\varphi}
\end{aligned}
$$

Finally we add the resulting 23 terms together. Note that the terms with $\partial/\partial\varphi$ and the mixed terms $\partial^2/\partial r\partial\vartheta$, $\partial^2/\partial r\partial\varphi$, $\partial^2/\partial\vartheta\partial\varphi$ add up to zero. For the sum of the remaining 13 terms we obtain:

$$
\begin{aligned}
&\frac{\partial^2}{\partial x^2}+\frac{\partial^2}{\partial y^2}+\frac{\partial^2}{\partial z^2}\\
&=\cos^2\vartheta\frac{\partial^2}{\partial r^2}+\frac{2\cos\vartheta\sin\vartheta}{r^2}\frac{\partial}{\partial\vartheta}+\frac{\sin^2\vartheta}{r^2}\frac{\partial^2}{\partial\vartheta^2}+\frac{\sin^2\vartheta}{r}\frac{\partial}{\partial r}\\
&+\sin^2\vartheta\cos^2\varphi\frac{\partial^2}{\partial r^2}+\frac{\cos^2\vartheta\cos^2\varphi}{r^2}\frac{\partial^2}{\partial\vartheta^2}+\frac{\sin^2\varphi}{r^2\sin^2\vartheta}\frac{\partial^2}{\partial\varphi^2}\\
&+\left[\frac{\cos^2\vartheta\cos^2\varphi}{r}+\frac{\sin^2\varphi}{r}\right]\frac{\partial}{\partial r}+\left[\frac{\cos\vartheta\sin^2\varphi}{r^2\sin\vartheta}-\frac{2\sin\vartheta\cos\vartheta\cos^2\varphi}{r^2}\right]\frac{\partial}{\partial\vartheta}\\
&+\sin^2\vartheta\sin^2\varphi\frac{\partial^2}{\partial r^2}+\frac{\cos^2\vartheta\sin^2\varphi}{r^2}\frac{\partial^2}{\partial\vartheta^2}+\frac{\cos^2\varphi}{r^2\sin^2\vartheta}\frac{\partial^2}{\partial\varphi^2}\\
&+\left[\frac{\cos^2\vartheta\sin^2\varphi}{r}+\frac{\cos^2\varphi}{r}\right]\frac{\partial}{\partial r}+\left[\frac{\cos\vartheta\cos^2\varphi}{r^2\sin\vartheta}-\frac{2\sin\vartheta\cos\vartheta\sin^2\varphi}{r^2}\right]\frac{\partial}{\partial\vartheta}
\end{aligned}
$$

We collect all terms containing $\partial^2/\partial r^2$

$$
\begin{aligned}
&\cos^2\vartheta\frac{\partial^2}{\partial r^2}+\sin^2\vartheta\cos^2\varphi\frac{\partial^2}{\partial r^2}+\sin^2\vartheta\sin^2\varphi\frac{\partial^2}{\partial r^2}\\
&=\frac{\partial^2}{\partial r^2}\left(\cos^2\varphi+\sin^2\varphi\right)\sin^2\vartheta+\cos^2\vartheta\frac{\partial^2}{\partial r^2}\\
&=\frac{\partial^2}{\partial r^2}
\end{aligned}
$$

We collect all terms containing $\partial/\partial r$

$$
\begin{aligned}
&\frac{\sin^2\vartheta}{r}\frac{\partial}{\partial r}++\left[\frac{\cos^2\vartheta\cos^2\varphi}{r}+\frac{\sin^2\varphi}{r}\right]\frac{\partial}{\partial r}\\
&+\left[\frac{\cos^2\vartheta\sin^2\varphi}{r}+\frac{\cos^2\varphi}{r}\right]\frac{\partial}{\partial r}\\
&=\frac{\partial}{\partial r}\left[\frac{\sin^2\vartheta}{r}+\frac{\cos^2\vartheta\cos^2\varphi}{r}+\frac{\sin^2\varphi}{r}+\frac{\cos^2\vartheta\sin^2\varphi}{r}+\frac{\cos^2\varphi}{r}\right]
\end{aligned}
$$

$$= \frac{\partial}{\partial r}\left[\frac{1}{r} + \frac{\cos^2\vartheta}{r} + \frac{\sin^2\vartheta}{r}\right] = \frac{\partial}{\partial r}\left[\frac{1}{r} + \frac{1}{r}\right] = \frac{2}{r}\frac{\partial}{\partial r}$$

We collect all terms containing $\partial^2/\partial\vartheta^2$

$$\frac{\sin^2\vartheta}{r^2}\frac{\partial^2}{\partial\vartheta^2} + \frac{\cos^2\vartheta\cos^2\varphi}{r^2}\frac{\partial^2}{\partial\vartheta^2} + \frac{\cos^2\vartheta\sin^2\varphi}{r^2}\frac{\partial^2}{\partial\vartheta^2}$$
$$= \frac{\partial^2}{\partial\vartheta^2}\left[\frac{\sin^2\vartheta}{r^2} + \frac{\cos^2\vartheta\cos^2\varphi}{r^2} + \frac{\cos^2\vartheta\sin^2\varphi}{r^2}\right]$$
$$= \frac{\partial^2}{\partial\vartheta^2}\left[\frac{\cos^2\vartheta}{r^2} + \frac{\sin^2\vartheta}{r^2}\right] = \frac{\partial^2}{\partial\vartheta^2}\left[\frac{1}{r^2}\right] = \frac{1}{r^2}\frac{\partial^2}{\partial\vartheta^2}$$

We collect all terms containing $\partial/\partial\vartheta$

$$\frac{2\cos\vartheta\sin\vartheta}{r^2}\frac{\partial}{\partial\vartheta} + \left[\frac{\cos\vartheta\sin^2\varphi}{r^2\sin\vartheta} - \frac{2\sin\vartheta\cos\vartheta\cos^2\varphi}{r^2}\right]\frac{\partial}{\partial\vartheta}$$
$$+ \left[\frac{\cos\vartheta\cos^2\varphi}{r^2\sin\vartheta} - \frac{2\sin\vartheta\cos\vartheta\sin^2\varphi}{r^2}\right]\frac{\partial}{\partial\vartheta}$$
$$= \frac{\partial}{\partial\vartheta}\left[\frac{2\cos\vartheta\sin\vartheta}{r^2} + \frac{\cos\vartheta\sin^2\varphi}{r^2\sin\vartheta} - \frac{2\sin\vartheta\cos\vartheta\cos^2\varphi}{r^2} + \frac{\cos\vartheta\cos^2\varphi}{r^2\sin\vartheta} - \frac{2\sin\vartheta\cos\vartheta\sin^2\varphi}{r^2}\right]$$
$$= \frac{\partial}{\partial\vartheta}\left[\frac{\cos\vartheta}{r^2\sin\vartheta} - \frac{2\sin\vartheta\cos\vartheta}{r^2} + \frac{2\cos\vartheta\sin\vartheta}{r^2}\right] = \frac{\cos\vartheta}{r^2\sin\vartheta}\frac{\partial}{\partial\vartheta}$$

We collect all terms containing $\partial^2/\partial\varphi^2$

$$\frac{\sin^2\varphi}{r^2\sin^2\vartheta}\frac{\partial^2}{\partial\varphi^2} + \frac{\cos^2\varphi}{r^2\sin^2\vartheta}\frac{\partial^2}{\partial\varphi^2}$$
$$= \frac{\partial^2}{\partial\varphi^2}\left[\frac{\sin^2\varphi}{r^2\sin^2\vartheta} + \frac{\cos^2\varphi}{r^2\sin^2\vartheta}\right] = \frac{1}{r^2\sin^2\vartheta}\frac{\partial^2}{\partial\varphi^2}$$

Finally we sum up:

$$\nabla^2 = \frac{\partial^2}{\partial x^2} + \frac{\partial^2}{\partial y^2} + \frac{\partial^2}{\partial z^2}$$
$$= \frac{\partial^2}{\partial r^2} + \frac{2}{r}\frac{\partial}{\partial r} + \frac{1}{r^2}\frac{\partial^2}{\partial\vartheta^2} + \frac{\cos\vartheta}{r^2\sin\vartheta}\frac{\partial}{\partial\vartheta} + \frac{1}{r^2\sin^2\vartheta}\frac{\partial^2}{\partial\varphi^2}$$

C.5 Vector Algebra

We consider two vectors $\vec{A}$ and $\vec{B}$ in the x,y-plane and define their vector product $\overrightarrow{AB}$ and their dot product $|AB|$ as

$$\overrightarrow{AB} = \vec{A}\times\vec{B}; \qquad |AB| = \left|\vec{A}\right|\cdot\left|\vec{B}\right|\cdot\cos(\alpha) \tag{C.23}$$

where α is the angle between the vectors $\vec{A}$ and $\vec{B}$.

Furthermore we define that the direction of the vector $\overrightarrow{AB}$ is perpendicular to the xy-plane; this is the z-direction. Now there are two possibilities for the direction of the vector $\overrightarrow{AB}$, namely, in the positive or in the negative z-direction. Thus we need an additional definition, which is demonstrated in Figure C.10. If the shortest way we can move vector $\vec{A}$ to vector $\vec{B}$ is a clockwise movement, vector $\overrightarrow{AB}$ is oriented to the negative z-direction; if the shortest way we can move vector $\vec{A}$ to vector $\vec{B}$ is an anticlockwise movement, vector $\overrightarrow{BA}$ is oriented to the positive z-direction.

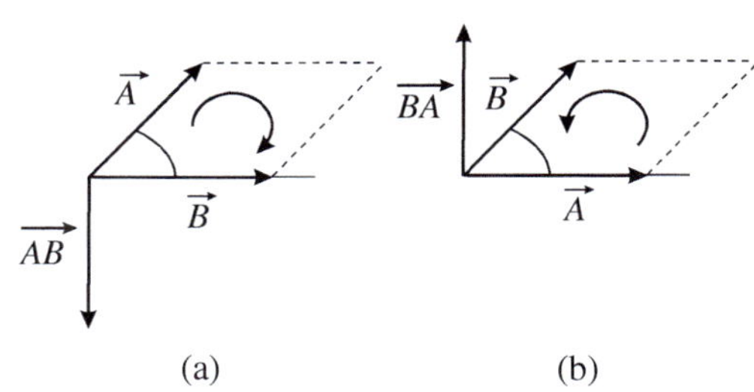

Figure C.10 Vector product of two vectors: $\overrightarrow{AB} = \vec{A}\times\vec{B}$ (a) and of $\overrightarrow{BA} = \vec{B}\times\vec{A}$ (b).

This means that an exchange of two vectors $\vec{A}$ and $\vec{B}$ results in a change of the direction of vector $\vec{P}$:

$$\vec{A}\times\vec{B} = -\vec{B}\times\vec{A} \tag{C.24}$$

If the vectors $\vec{A}$ and $\vec{B}$ are parallel (that is, $\sin(\vec{A},\vec{B}) = 0$), it follows that $\overrightarrow{AB} = 0$. In the special case $\vec{A} = \vec{B}$ we find that

$$\vec{A}\times\vec{A} = 0 \tag{C.25}$$

Furthermore we define unit vectors $\vec{e_x}$ and $\vec{e_y}$. They have, according to the preceding definitions, the following properties:

$$\vec{e_x} \times \vec{e_x} = \vec{e_y} \times \vec{e_y} = 0 \qquad \vec{e_x} \times \vec{e_y} = \vec{e_z} \tag{C.26}$$

Vectors can be represented by their components in the x-, y-, and z -directions. Using the relations in Equation (C.26) we obtain:

$$\begin{aligned} \overrightarrow{AB} = \vec{A} \times \vec{B} &= \left(A_x\vec{e_x} + A_y\vec{e_y}\right) \times \left(B_x\vec{e_x} + B_y\vec{e_y}\right) \\ &= A_xB_x\vec{e_x} \times \vec{e_x} + A_xB_y\vec{e_x} \times \vec{e_y} + A_yB_x\vec{e_y} \times \vec{e_x} + A_yB_y\vec{e_y} \times \vec{e_y} \\ &= A_xB_y\vec{e_x} \times \vec{e_y} + A_yB_x\,\vec{e_y} \times \vec{e_x} \\ &= A_xB_y\vec{e_x} \times \vec{e_y} - A_yB_x\,\vec{e_x} \times \vec{e_y} = (A_xB_y - A_yB_x)\,\vec{e_z} \end{aligned}$$

Thus we arrive at the result

$$\overrightarrow{AB} = \vec{A} \times \vec{B} = (A_xB_y - A_yB_x)\,\vec{e_z}$$

that is, the components of vector $\overrightarrow{AB}$ are A_xB_y and $-A_yB_x$.

In the text we have the task to represent the vector for the angular momentum,

$$\vec{L} = \vec{r} \times \vec{p}$$

by its components x, y, p_x, and p_y, see Equation (3.67). Following our result for $\vec{A} \times \vec{B}$ we can write that

$$\vec{L_z} = \vec{r} \times \vec{p} = (xp_y - yp_x)\,\vec{e_z}$$

This is Equation (3.68) in the text.

C.6 Probability and Statistics

C.6.1 Discrete Probability Distributions

For a variable (or physical property) that can take on discrete values A_i, one can define the probability (or likelihood) that a particular value is realized. If we sample the values of A some large number N times, then we can plot a histogram of the number of realizations (or frequency of occurrence) n_i of a given A_i as a function of A_i The probability is proportional to the number of realizations; however, it is commonly normalized so that the probability of observing any of the possible values is unity. The normalization condition is $N = \sum_i n_i$, and we can define the probability $\mathcal{P}_i$ to observe the value x_i by

$$\mathcal{P}_i = n_i/N$$

The collection of all such probabilities of the variable comprise its probability distribution.

Example: By way of example if we roll two six-sided dice (which has the values 1, 2, 3, 4, 5, and 6 uniquely on each side) then a range of values between 2 and 12 are possible (Fig. C.11).

Value, A_i	2	3	4	5	6	7	8	9	10	11	12
# of realizations	1	2	3	4	5	6	5	4	3	2	1
probability, $\mathcal{P}_i$	1/36	1/18	1/12	1/9	5/36	1/6	5/36	1/9	1/12	1/18	1/36

Figure C.11 The figure plots the probability for observing a particular sum (given on the abscissa) if two dice are rolled.

Examples of Discrete Probability Distributions

Binomial Distribution The binomial distribution describes the distribution of probabilities for the outcomes of a measurement which has two possibilities (e.g., the flipping of a coin to give "heads" or "tails"). For a set of N measurements (i.e., N coin flips) the probability distribution describing the likelihood of N_1 "heads" and N_2 "tails" (with $N_1 + N_2 = N$) takes the form

$$\mathcal{P}(N_1, N_2; N) = \frac{N!}{N_1!\,N_2!} = \frac{N!}{N_1!\,(N - N_1)!}$$

Figure Distributions shows the probability P versus N_1 (red full circles).

Poisson Distribution The Poisson distribution describes the distribution of events that occur at a constant average rate and are independent of the time between the events; e.g., the spontaneous decay of a radionuclide. A Poisson distribution law for a discrete property N_i has the form

$$\mathcal{P}(N_i) = \frac{\overline{N}^{N_i}}{N_i!} \exp\left(-\overline{N}\right)$$

where $\overline{N}$ is the average value of all N_i; in this special case knowledge of the average value determines the distribution. As an example, we consider the following set of $n_{\text{data}} = 20$ data points:

i	1	2	3	4	5	6	7	8	9	10	11	12	13	14	15	16	17	18	19	20
N_i	1	2	3	4	5	6	7	8	9	10	11	12	13	14	15	16	17	18	19	20

We calculate the average $\overline{N}$ as

$$\overline{N} = \frac{\sum N_i}{n_{\text{data}}} = \frac{210}{20} = 10.5$$

Figure C.12 shows this distribution (blue full circles).

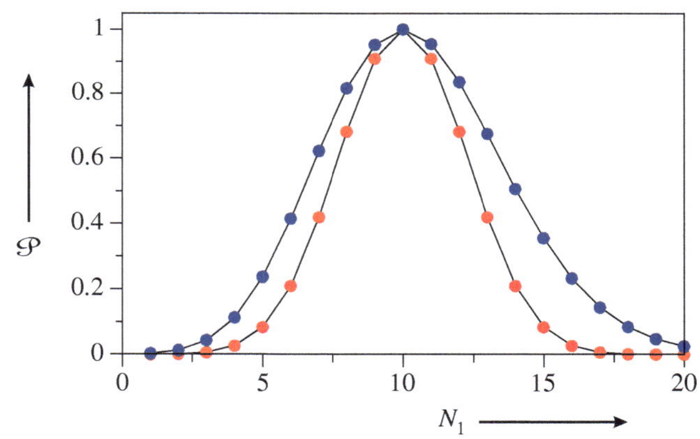

Figure C.12 Plot of the probability P versus N_1 for $N = 20$. Red filled circles: Binomial distribution, blue filled circles: Poisson distribution.

Multinomial Distribution For measurements with more than two possible outcomes, for example i discrete values of a property $(x_1, x_2, ..., x_i, ..)$, the probability distribution takes the form

$$\mathcal{P}_j(N_1, N_2, ...N_i; N) = \frac{\omega_j}{\sum \omega_j}$$

with

$$\omega_j = \frac{N!}{N_1!\,N_2! \cdot\cdot\cdot\cdot\cdot N_i!}$$

where the N_i are the number of times the property x_i manifests and ω_j is the number of realizations for type j.

As an example we consider a situation with three different occupation numbers N_1, N_2, and N_3 with the restrictions that $N_1 + N_2 + N_3 = N$ and $N_1E_1 + N_2E_2 + N_3E_3 = U$, where E_1, E_2, and E_3 are the energies of the three energy states and U is the internal energy. Using $N = 100$, $E_1 = 0$, $E_2 = a$, and $E_3 = 2a$ and $U = 40a$ we have a total of 21 different population types:

j	1	2	3	4	5	6	7	8	9	10
N_3	20	19	18	17	16	15	14	13	12	11
N_2	0	2	4	6	8	10	12	14	16	18
N_1	80	79	78	77	76	75	74	73	72	71
$\omega_j/10^{32}$	0	0	0	0	0	0	0.1	0.4	1.5	4.3
$\mathcal{P}_j$	0	0	0	0	0	0	0.001	0.006	0.022	0.063

j	11	12	13	14	15	16	17	18	19	20	21
N_3	10	9	8	7	6	5	4	3	2	1	0
N_2	20	22	24	26	28	30	32	34	36	38	40
N_1	70	69	68	67	66	65	64	63	62	61	60
$\omega_j/10^{32}$	8.8	13.4	15.0	12.6	7.8	3.6	1.2	0.3	0	0	0
P_j	0.128	0.194	0.217	0.183	0.113	0.0522	0.0174	0.004	0	0	0

For the sum over all types j we find

$$\sum \omega_j = 69.0 \times 10^{32}$$

and the result for $\mathcal{P}_j$ is

$$\mathcal{P}_j = \frac{\omega_j}{69.0 \times 10^{32}}$$

This distribution is plotted in Fig. C.13.

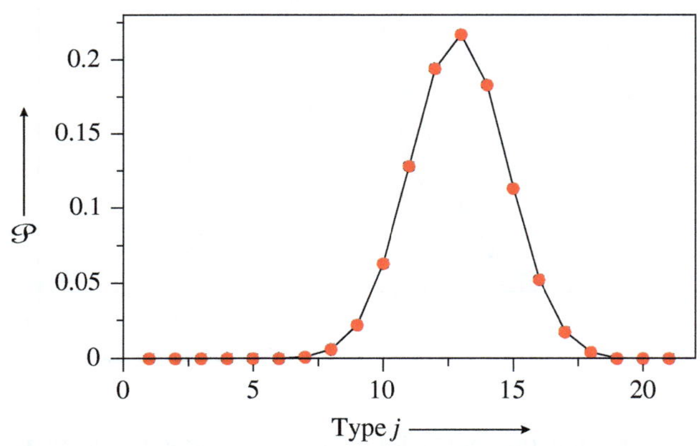

Figure C.13 Plot of the probability $\mathcal{P}$ versus type j for three occupation numbers N_1, N_2, and N_3, using $N = 100$ and $U = 40a$.

C.6.2 Metrics for Probability Distributions

A number of useful quantities (parameters) can be defined that provide a measure of the probability distribution. Some quantities used to provide a measure of probability distributions are the mean, the mode, the median, the variance, and the standard deviation.

Mean, or Average, Value

The average value of a quantity A is

$$\overline{A} = \frac{1}{N}\sum_{i=1}^{i=N} A_i$$

where A_i is the measured value and N is the total number of measurements. Alternatively, we can use the definition of the probability and write

$$\overline{A} = \sum_{i=1}^{i=k} A_i \cdot \mathcal{P}_i$$

where the sum is performed over the set of possible values of A, $\{A_1, A_2, ...A_k\}$. For the case of the dice, the average value is 7.

Mode

The mode is the value of A_i that is the most likely, i.e., the highest $\mathcal{P}_i$.

Median

The median is the value of A_i that divides a distribution into two equal halves.

Variance

The variance of a distribution provides a measure of its width, or spread, and is defined as

$$\sigma_A^2 = \frac{\sum_{i=1}^{i=N} \left(A_i - \overline{A}\right)^2}{N}$$

for the sum over measurements, or

$$\sigma_A^2 = \sum_{i=1}^{i=k} \left(A_i - \overline{A}\right)^2 \cdot \mathcal{P}_i$$

for the sum over the set of possible values of A, $\{A_1, A_2, \ldots A_k\}$.

Standard Deviation

The standard deviation of a quantity A is the square root of the variance, so that

$$\sigma_A = \sqrt{\frac{\sum_{i=1}^{i=N} \left(A_i - \overline{A}\right)^2}{N}}$$

for the sum over measurements, or

$$\sigma_A = \sqrt{\sum_{i=1}^{i=k} \left(A_i - \overline{A}\right)^2 \cdot p_i}$$

for the sum over the set of possible values of A, $\{A_1, A_2, \ldots A_k\}$.

The values given above assume that the number of measurements N is very large. For finite sample sizes the standard deviation is defined as

$$\sigma_A = \sqrt{\frac{\sum_{i=1}^{i=N} \left(A_i - \overline{A}\right)^2}{N - 1}}$$

where A_i is the measured value i, N is the total number of measurements, and $\overline{A}$ is the mean value of A. Note that for $N = 1$ we have $A_1 = \overline{A}$ and the expression for σ_A becomes indefinite. This means we need at least two measurements to extract a value for the standard deviation.

We can rewrite this expression for the standard deviation as

$$\sigma_A = \sqrt{\frac{\sum_{i=1}^{i=N} \left(A_i^2 - 2A_i\overline{A} + \overline{A}^2\right)}{N - 1}}$$

$$= \sqrt{\frac{\sum_{i=1}^{i=N} \left(A_i^2\right)}{N - 1} - 2\frac{\sum_{i=1}^{i=N} \left(A_i\overline{A}\right)}{N - 1} + \frac{\sum_{i=1}^{i=N} \left(\overline{A}^2\right)}{N - 1}}$$

$$= \sqrt{\overline{A^2} - 2\overline{A}\frac{\sum_{i=1}^{i=N}(A_i)}{N-1} + \overline{A}^2} \qquad \text{for} \qquad N \gg 1$$

$$= \sqrt{\overline{A^2} - 2\overline{A}\cdot\overline{A} + \overline{A}^2} = \sqrt{\overline{A^2} - \overline{A}^2}$$

C.6.3 Continuous Probability Distributions

If a variable (or physical property) can take on a continuous range of values, then the probability distribution is continuous. In this case the summations in the definitions given above for discrete distributions are replaced by integrals. For example, the average value of $\overline{A}$ for a probability distribution $\mathcal{P}(A)$ is

$$\overline{A} = \int A \cdot \mathcal{P}(A)\, \mathrm{d}A$$

Correspondingly the variance is

$$\sigma_A^2 = \int \left(A - \overline{A}\right)^2 \cdot \mathcal{P}(A)\, \mathrm{d}A$$

Of particular importance in physical chemistry are the Gaussian distribution and the Lorentzian distribution. For a variable x the Gaussian probability distribution has the form

$$\mathcal{P}(x) = \frac{1}{\sigma\sqrt{2\pi}} \exp\left(-\frac{\left(x - \overline{x}\right)^2}{2\sigma^2}\right)$$

where σ^2 is its variance and $\overline{x}$ is its mean value. Its shape is symmetric about the mean and the entire distribution's shape is determined by σ^2 and $\overline{x}$. For a variable x the Lorentzian probability distribution has the form

$$\mathcal{P}(x) = \frac{1}{\pi} \frac{\gamma/2}{\left(x - \overline{x}\right)^2 + (\gamma/2)^2}$$

where γ is its half-width and $\overline{x}$ is its mean value.

C.6.4 Standard Deviation σ as a Measure of Uncertainty

We can compare the half-width of a Gaussian function to the standard deviation σ, the square of which is defined as

$$\sigma^2 = \int_{-\infty}^{+\infty} \left(x - \overline{x}\right)^2 \cdot g(x)\, \mathrm{d}x$$

where $\overline{x}$ is the mean value of x. and $g(x)$ is a normalized Gaussian distribution function. For the case in which the Gaussian function is centered at its maximum, $\overline{x} = 0$, we can write that

$$\sigma^2 = \int_{-\infty}^{+\infty} x^2 \cdot const \cdot \mathrm{e}^{-(bx)^2}\, \mathrm{d}x = const \int_{-\infty}^{+\infty} x^2 \cdot \mathrm{e}^{-(bx)^2}\, \mathrm{d}x$$

The constant is evaluated by the condition

$$const \int_{-\infty}^{+\infty} \mathrm{e}^{-(bx)^2} \mathrm{d}x = 1$$

leading to

$$const = \frac{1}{\int_{-\infty}^{+\infty} \mathrm{e}^{-(bx)^2}\, \mathrm{d}x}$$

Using the substitution

$$b^2x^2 = u^2, \qquad x^2 = \frac{1}{b^2}u^2, \qquad x = \frac{1}{b}u$$

we find that

$$\int_{-\infty}^{+\infty} \mathrm{e}^{-(bx)^2}\, \mathrm{d}x = \frac{1}{b}\int_{-\infty}^{+\infty} \mathrm{e}^{-u^2}\, \mathrm{d}u = \frac{1}{b}\sqrt{\pi}$$

and

$$const = \frac{b}{\sqrt{\pi}}$$

Accordingly, we find

$$\int_{-\infty}^{+\infty} x^2 e^{-(bx)^2}\, dx = \frac{1}{b^3}\int_{-\infty}^{+\infty} u^2 e^{-u^2}\, du = \frac{1}{b^3}\frac{1}{2}\sqrt{\pi}$$

Thus we obtain

$$\sigma^2 = const \int_{-\infty}^{+\infty} x^2 \cdot e^{(-bx)^2}\, dx$$

$$= \frac{b}{\sqrt{\pi}} \cdot \frac{1}{b^3}\frac{1}{2}\sqrt{\pi} = \frac{1}{2b^2}$$

and the standard deviation is

$$\sigma = \sqrt{\frac{1}{2b^2}} = \frac{1}{\sqrt{2}}\frac{1}{b}$$

Thus the result for the half-width is

$$\Delta x(\sigma) = 2\sigma = \sqrt{2}\frac{1}{b} = 1.414 \cdot \frac{1}{b}$$

Compared to Δx(half-width)

$$\Delta x(\text{half-width}) = 2\sqrt{\ln 2}\frac{1}{b} = 1.386 \cdot \frac{1}{b}$$

this is only a small difference.

C.6.5 Error Function

The integral

$$\int_0^x \exp(-t^2)\, dt$$

has not been solved in a closed form, yet it is very important in the physical sciences and statistics. Note that the integrand is a Gaussian function and the integral represents the area under the curve from the lower limit of zero to a value x. Because of its importance it has been evaluated numerically and is tabulated in a form called the error function, erf(x), where

$$\text{erf}(x) = \frac{2}{\sqrt{\pi}}\int_0^x \exp(-t^2)\, dt$$

This form is convenient because

$$\lim_{x\to\infty} \text{erf}(x) = \lim_{x\to\infty} \frac{2}{\sqrt{\pi}}\int_0^x \exp(-t^2)\, dt = 1$$

see Section B.3 on definite integrals. One sometimes finds use of the complementary error function, erfc(x), which is defined by

$$\text{erfc}(x) = 1 - \text{erf}(x)$$

Some useful integrals involving the error function are

$$\int_0^\infty e^{-at}\text{erf}\left(\sqrt{bt}\right) dt = \frac{1}{a}\sqrt{\frac{b}{a+b}} \qquad \text{where } a, b > 0$$

$$\int_0^\infty e^{-at}\text{erf}\,(bt)\, dt = \frac{1}{a}\exp\left(\frac{a^2}{4b^2}\right)\text{erfc}\left(\frac{a}{2b}\right) \qquad \text{where } a, b > 0$$

and

$$\int_0^\infty e^{-at}\text{erfc}\left(\sqrt{\frac{b}{t}}\right) dt = \frac{1}{a}\exp\left(-2\sqrt{ab}\right) \qquad \text{where } a, b > 0$$

For more information on the mathematical properties of the error function see M. Abramowitz and I. A. Stegun, Handbook of Mathematical Functions (NBS, 1970).

C.6.6 Permutations and Combinations

In general for N electrons and G orbitals the number of equivalent orbital arrangements is given by $G!/[(N!(G-N)!]$.

Permutations of distinguishable objects refer to the different ways of arranging them. For example, the permutation of the three integers 1, 2, and 3 in a sequence is

$$(1,2,3); \quad (1,3,2); \quad (2,1,3); \quad (2,3,1); \quad (3,2,1); \quad (3,1,2)$$

In general the number of permutations that exist for n distinguishable objects is $n!$. We see that this holds for our example of three integers; i.e., $3! = 6$. More generally, one can ask about arranging objects in groups rather than one object at a time; and in such a case the number of permutations P for n objects taken in groups of k objects is given by

$$P = \frac{n!}{(n-k)!}$$

In contrast to permutations, a combination of objects does not distinguish the order of the objects. This case is important for indistinguishable particles (e.g., gas molecules and electrons) for which we cannot discern an order. The number of combinations C of n indistinguishable objects, which are taken in groups of k, is

$$C(n,k) = \frac{n!}{k!\,(n-k)!}$$

This result can be understood by considering that the number of combinations $C(n,k)$ here is the same as finding the number of ways for dividing n objects into two groups of k and $(n-k)$. Even more generally, the number of ways to divide n indistinguishable objects into multiple groups is

$$\frac{n!}{n_1!\,n_2!\,..n_k!}$$

where n_i is the number of objects in group i and k is the total number of groups.

C.7 Determinants

Determinants are used extensively in quantum mechanics calculations and they are used in the text with regard to the Hückel model for π-electron systems. Determinants are useful for solving systems of linear equations, i.e.,

$$\begin{array}{ccccccc}
a_{11}x_1 & + & a_{12}x_2 & \ldots + & a_{1n}x_n & = & b_1 \\
a_{21}x_1 & + & a_{22}x_2 & \ldots + & a_{2n}x_n & = & b_2 \\
\cdot & & \cdot & & \cdot & & \cdot \\
\cdot & & \cdot & & \cdot & & \cdot \\
\cdot & & \cdot & & \cdot & & \cdot \\
a_{n1}x_1 & + & a_{n2}x_2 & \ldots + & a_{nn}x_n & = & b_n
\end{array}$$

by use of Cramers rule; namely

$$x_k = \frac{\begin{vmatrix} a_{11} & a_{12} & \cdots & a_{1,k-1} & b_1 & a_{1,k+1} & \cdots & a_{1n} \\ a_{21} & a_{22} & \cdots & a_{2,k-1} & b_2 & a_{2,k+1} & \cdots & a_{2n} \\ \cdot & \cdot & & & & \cdot & & \cdot \\ \cdot & \cdot & & & & \cdot & & \cdot \\ \cdot & \cdot & & & & \cdot & & \cdot \\ a_{n1} & a_{n2} & \cdots & a_{n,k-1} & b_k & a_{n,k+1} & \cdots & a_{nn} \end{vmatrix}}{\begin{vmatrix} a_{11} & a_{12} & a_{13} & \cdots & a_{1n} \\ \cdot & & & & \cdot \\ \cdot & & & & \cdot \\ \cdot & & & & \cdot \\ a_{n1} & a_{n2} & a_{n3} & \cdots & a_{nn} \end{vmatrix}}$$

If any constant b_i is not equal to zero then the system of equations is inhomogeneous and if all of the b_i constants are zero then the system is homogeneous. If all of the b coefficients are zero (a system of n linear homogeneous equations) the determinant must be equal to zero. Let

$$\det\left(a_{ij}\right) = \begin{vmatrix} a_{11} & a_{12} & a_{13} & \cdots & a_{1n} \\ \cdot & & & & \cdot \\ \cdot & & & & \cdot \\ \cdot & & & & \cdot \\ a_{n1} & a_{n2} & a_{n3} & \cdots & a_{nn} \end{vmatrix}$$

and

$$\begin{vmatrix} a_{11} & a_{12} \\ a_{21} & a_{22} \end{vmatrix} = a_{11}a_{22} - a_{12}a_{21}$$

Then we can write that

$$\det\left(a_{ij}\right) = \sum_{l=1}^{n} a_{kl}N_{lk}$$

where $k = 1$ or 2 or … or n. $N_{lk} = (-1)^{l+k}M_{lk}$ is the cofactor of a_{ij} and M_{lk} is the minor of a_{ij} (i.e., the determinant which is left after crossing out the ith row and the jth column).

C.7.1 Some Properties of Determinants

- If every element of a row (or column) is zero, the value of the determinant is zero.
- Interchanging any two rows (or columns) multiplies the value of the determinant by −1.
- If any two rows (or columns) of a determinant are identical, the determinant is zero.
- Multiplication of any one row (or column) by a constant k multiplies the value of the determinant by k.
- Addition to each element of one row, of the same constant multiple of the corresponding element of another row leaves the value of the determinant unchanged (same for columns).
- Interchange of all corresponding columns and rows leaves the value of the determinant unchanged (i.e., an even number of times).

C.8 Complex Numbers

A complex number z is an ordered pair of real numbers (x, y) and is typically written in either Cartesian form

$$z = x + \mathrm{i}y \quad \text{where} \quad \mathrm{i}^2 = -1$$

or in polar form

$$z = r\exp(\mathrm{i}\theta) = r(\cos\theta + \mathrm{i}\sin\theta) \quad \text{where} \quad \mathrm{i}^2 = -1$$

We also define the operation $Re(z)$ (the real part of z) by

$$Re(z) = x = r\cos\theta$$

and $Im(z)$ (the imaginary part of z) by

$$Im(z) = y = r\sin\theta$$

The modulus, or magnitude, of z is given by

$$|z| = \sqrt{x^2 + y^2} = r$$

The relationship between these two forms can be observed by considering Figure C.14.

We define the complex conjugate of a complex number z^* by

$$z^* = x - \mathrm{i}y = r\exp(-\mathrm{i}\theta)$$

Hence we see that the product of a complex number with its complex conjugate is always real, namely

$$\begin{aligned} z \cdot z^* &= (x + \mathrm{i}y)(x - \mathrm{i}y) \\ &= x^2 + \mathrm{i}yx - \mathrm{i}yx - \mathrm{i}^2y^2 \\ &= x^2 - \mathrm{i}^2y^2 = x^2 + y^2 \end{aligned}$$

and

$$\begin{aligned} z \cdot z^* &= r\exp(\mathrm{i}\theta)\left[r\exp(-\mathrm{i}\theta)\right] \\ &= r^2\exp(\mathrm{i}\theta - \mathrm{i}\theta) = r^2 \end{aligned}$$

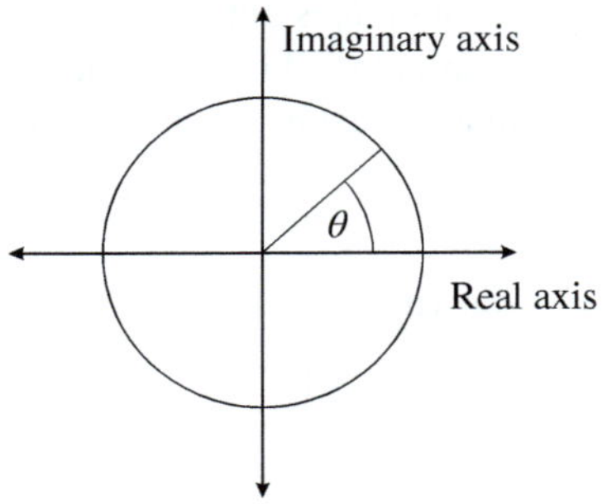

Figure C.14 Real and imaginary axes.

References

1 D. Dobos, *Electrochemical Data* (Elsevier, Amsterdam, 1975); I.G. Morris, *Physikalische Chemie für Biologen* (Verlag Chemie, Weinheim, 1976).

2 Reaction 1: A.L. Lehninger, *Biochemie* (Verlag Chemie, Weinheim, 1979, p. 324); Reaction 2: A.L. Lehninger, *Biochemie* (Verlag Chemie, Weinheim, 1979, p. 323); Reaction 3: I.G. Morris, *Physikalische Chemie für Biologen* (Verlag Chemie, Weinheim, 1976, p. 222); Reaction 4: I. Klotz, *Energy Changes in Biochemical Reactions* (Academic, New York, 1967, p. 22).

3 Reactions 5,6: R.C. Bohinski, *Modern Concepts in Biochemistry* (Verlag Chemie, Weinheim, 1979, p. 458); Reaction 7: A.L. Lehninger, *Biochemie* (Verlag Chemie, Weinheim, 1979, p. 350).

4 I.S. Gradshteyn and I.M. Ryzhik, *Tables of Integrals, Series, and Products* (Academic Press, New York, 1980).

5 C.W. Gear, *Numerical Initial Value Problems in Ordinary Differential Equations* (Prentic Hall, Englewood Cliffs, NJ, 1971); Y.X. Wang and J.M. Wen, Gear method for solving differential equations of gear systems, *J. Phys.: Conf. Ser.*, 2006, **48**, 026.

6 R.G. Mortimer, *Mathematics for Physical Chemistry* (MacMillan, New York, 1981).

7 J.R. Barrante, *Applied Mathematics for Physical Chemistry* (Prentice-Hall, New York, 2004).

Index

Principles of Physical Chemistry, Third Edition. Hans Kuhn, David H. Waldeck, and Horst-Dieter Försterling.

Companion website: www.wiley.com/go/Kuhn/physicalchemistry3e

C

d

e

f

i

j

m

n

o

Greek Alphabet

Alpha	A	α	Eta	H	η	Nu	N	ν	Tau	T	τ
Beta	B	β	Theta	Θ	θ	Xi	Ξ	ξ	Upsilon	Υ	υ
Gamma	Γ	γ	Iota	I	ι	Omnicron	O	o	Phi	Φ	ϕ
Delta	Δ	δ	Kappa	K	κ	Pi	Π	π	Chi	X	χ
Epsilon	E	ε	Lambda	Λ	λ	Rho	P	ρ	Psi	Ψ	ψ
Zeta	Z	ς	Mu	M	μ	Sigma	Σ	σ	Omega	Ω	ω

Mathematical Relations

$$\pi = 3.14159265359 \qquad \mathrm{e} = \lim_{n\to\infty}\left(1+\frac{1}{n}\right)^n = 2.71828182846$$

$$\ln x = \ln(10)\cdot\log x = 2.302585\cdot\log x \qquad \ln(\mathrm{e}) = 1$$

$$N! = 1\cdot 2\cdot 3\ldots\cdot N \qquad \text{Stirling's formula: } \ln N! = N\ln N - N$$

$$\ln(xy) = \ln x + \ln y \qquad \ln\left(\frac{x}{y}\right) = \ln x - \ln y \qquad \ln x^y = y\ln x$$

$$\mathrm{e}^x\,\mathrm{e}^y = \mathrm{e}^{(x+y)} \qquad \frac{\mathrm{e}^x}{\mathrm{e}^y} = \mathrm{e}^{(x-y)} \qquad (\mathrm{e}^x)^y = \mathrm{e}^{xy} \qquad \mathrm{e}^{\pm ix} = \cos x \pm i\sin x$$

$$\tan\alpha = \frac{\sin\alpha}{\cos\alpha} \qquad \cot\alpha = \frac{\cos\alpha}{\sin\alpha} \qquad \sin^2\alpha + \cos^2\alpha = 1$$

$$\sin(\alpha\pm\beta) = \sin\alpha\cos\beta \pm \cos\alpha\sin\beta \qquad \cos(\alpha\pm\beta) = \cos\alpha\cos\beta \mp \sin\alpha\sin\beta$$

$$\mathrm{d}(f+g) = \mathrm{d}f + \mathrm{d}g \qquad \mathrm{d}(fg) = g\mathrm{d}f + f\mathrm{d}g \qquad \mathrm{d}\left(\frac{f}{g}\right) = \frac{g\mathrm{d}f - f\mathrm{d}g}{g^2}$$

$$\frac{\mathrm{d}}{\mathrm{d}x}(x^n) = nx^{n-1} \qquad \frac{\mathrm{d}}{\mathrm{d}x}(\mathrm{e}^{ax}) = a\mathrm{e}^{ax} \qquad \frac{\mathrm{d}}{\mathrm{d}x}(\ln x) = \frac{1}{x}$$

$$\frac{\mathrm{d}}{\mathrm{d}x}\sin(ax) = a\cos(ax) \qquad \frac{\mathrm{d}}{\mathrm{d}x}\cos(ax) = -a\sin(ax)$$

$$\int x^n\mathrm{d}x = \frac{x^{n+1}}{n+1} + const \qquad \int\frac{1}{x}\mathrm{d}x = \ln x + const$$

$$\int\sin(ax)\,\mathrm{d}x = -\frac{1}{a}\cos(ax) + const \qquad \int\cos(ax)\,\mathrm{d}x = \frac{1}{a}\sin(ax) + const$$

$$\int\sin^2(ax)\,\mathrm{d}x = \frac{1}{2}x - \frac{a\sin(2ax)}{4} + const \qquad \int_0^\infty x^n\,\mathrm{e}^{-ax}\mathrm{d}x = \frac{n!}{a^{n+1}}$$

$$\text{Taylor series:} \quad f(x) = \sum_{n=0}^{\infty}\frac{1}{n!}\left(\frac{\mathrm{d}^n f}{\mathrm{d}x^n}\right)_{x=x_0}(x-x_0)^n$$

$$\mathrm{e}^x = 1 + x + \frac{1}{2}x^2 + \ldots \qquad \ln(1+x) = x - \frac{1}{2}x^2 + \frac{1}{3}x^3 - \ldots$$

$$\frac{1}{1+x} = 1 - x + x^2 - \ldots \qquad \sqrt{1+x} = 1 + \frac{1}{2}x - \frac{1}{8}x^2 + \ldots$$